9781421406336
W0028370

Frogs of the United States and Canada

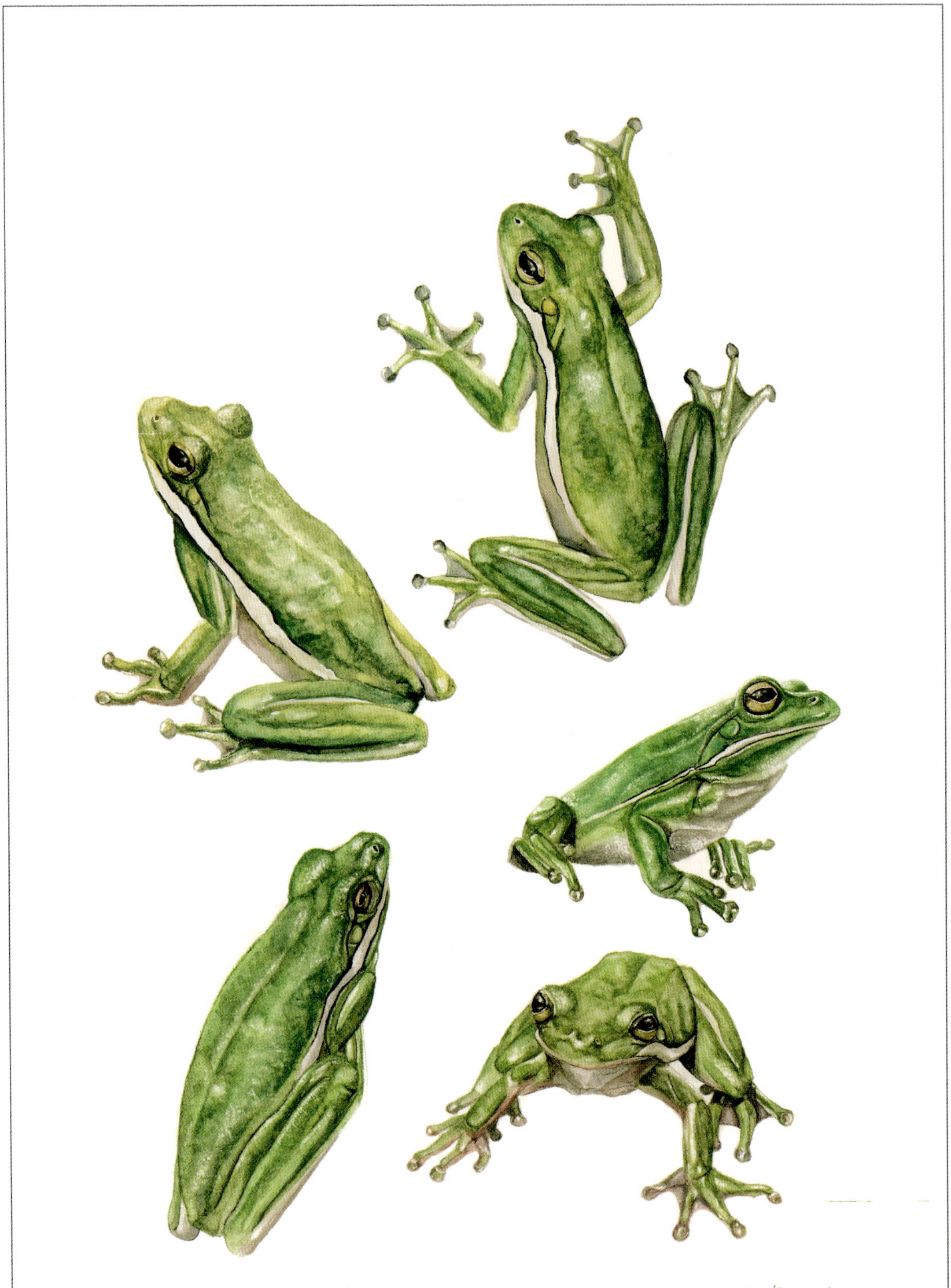

Frogs of the United States and Canada

VOLUME 1

C. Kenneth Dodd Jr.

The Johns Hopkins University Press
Baltimore

This book was brought to publication with the generous support of the U.S. Geological Survey's Amphibian Research and Monitoring Initiative, the Center for Biological Diversity, and the Herpetologists' League.

© 2013 The Johns Hopkins University Press
All rights reserved. Published 2013
Printed in China on acid-free paper
9 8 7 6 5 4 3 2 1

The Johns Hopkins University Press
2715 North Charles Street
Baltimore, Maryland 21218-4363
www.press.jhu.edu

Library of Congress Cataloging-in-Publication Data

Dodd, C. Kenneth.
  Frogs of the United States and Canada / C. Kenneth Dodd Jr.
    p. cm.
  Includes bibliographical references and indexes.
  ISBN 978-1-4214-0633-6 (hdbk. : alk. paper) — ISBN 1-4214-0633-0 (hdbk. : alk. paper)
  1. Frogs—United States. 2. Frogs—Canada. I. Title.
  QL668.E2D57 2013
  597.8′9—dc23   2012017648

A catalog record for this book is available from the British Library.

Frontispiece: "Rain Response" by Audrey K. Owens. Drawn from photos by Bill Hilton, Jr. (www.hiltonpond.org), Jane M. Rohling, and May Lattanzio.

Special discounts are available for bulk purchases of this book. For more information, please contact Special Sales at 410-516-6936 or specialsales@press.jhu.edu.

The Johns Hopkins University Press uses environmentally friendly book materials, including recycled text paper that is composed of at least 30 percent post-consumer waste, whenever possible.

To the memory of Albert Hazen Wright and Anna Allen Wright,
pioneers in the study of North American frogs

and to Marian, for her love of nature

Little Frog Came a-hopping!

Little tree-frog . . .
   . . . on the door . . .
      . . . in the house . . .
         . . . on my arm!
Sticky-cold
   Clinging to warm-n-dry.
      Go, frog! Go!
         Back to the garden, frog!
         The house and I are no place for you!
But, here he comes again ~
   . . . on the wall . . .
      . . . on the door . . .
         . . . on my head . . .
            . . . to the porch tiles . . .
               . . . to the garden! (Sigh!)
Sing, frog!
   Sing, if you can.
      Be well
         And do good
            Tending the garden.

                                                Marian Lovene Griffey

Contents

Preface ix
Introduction xi
List of Abbreviations xxix

SPECIES ACCOUNTS

Family Ascaphidae

Ascaphus montanus 1
Ascaphus truei 7

Family Bufonidae

Anaxyrus americanus 17
Anaxyrus baxteri 43
Anaxyrus boreas 47
Anaxyrus californicus 65
Anaxyrus canorus 70
Anaxyrus cognatus 78
Anaxyrus debilis 88
Anaxyrus exsul 92
Anaxyrus fowleri 96
Anaxyrus hemiophrys 113
Anaxyrus houstonensis 120
Anaxyrus microscaphus 127
Anaxyrus nelsoni 132
Anaxyrus punctatus 136
Anaxyrus quercicus 144
Anaxyrus retiformis 149
Anaxyrus speciosus 152
Anaxyrus terrestris 155
Anaxyrus woodhousii 166
Ollotis alvaria 177
Ollotis nebulifer 180
Rhinella marina 186

Family Craugastoridae

Craugastor augusti 192

Family Eleutherodactylidae

Eleutherodactylus cystignathoides 197
Eleutherodactylus guttilatus 199
Eleutherodactylus marnockii 201

Family Hylidae

Acris blanchardi 205
Acris crepitans 219
Acris gryllus 226
Hyla andersonii 235
Hyla arenicolor 239
Hyla avivoca 245
Hyla chrysoscelis 250
Hyla cinerea 262
Hyla femoralis 274
Hyla gratiosa 280
Hyla squirella 288
Hyla versicolor 294
Hyla wrightorum 309
Pseudacris brachyphona 313
Pseudacris brimleyi 319
Pseudacris cadaverina 322
Pseudacris clarkii 328
Pseudacris crucifer 331
Pseudacris feriarum 348
Pseudacris fouquettei 357
Pseudacris illinoensis 363
Pseudacris kalmi 367
Pseudacris maculata 371
Pseudacris nigrita 385
Pseudacris ocularis 391
Pseudacris ornata 395
Pseudacris regilla 400
Pseudacris streckeri 416
Pseudacris triseriata 421
Smilisca fodiens 428
Smilisca baudinii 431

Family Leptodactylidae

Leptodactylus fragilis 436

Family Microhylidae

Gastrophryne carolinensis 439
Gastrophryne olivacea 448
Hypopachus variolosus 455

Family Rhinophrynidae

Rhinophrynus dorsalis 458

Preface

There are two reasons I became a biologist. First, nature fascinates me. I have always been astonished at the diversity of life and how the sum total of its parts, from basic chemistry through physiology and genetics all the way to immense ecosystems, still cannot explain the essence and "why" of life. Through herpetology, I have tried to make sense of how even a small portion of nature works, and I have never understood people who have no interest in what makes them and our world what they are. There is nothing more fascinating than the organization and evolution of life. The second reason for becoming a biologist was the dread of working in an office building. I wanted to be outdoors, not anchored to a desk; if there were canyons and forests and wild animals "out there," why be inside? I was not always successful at avoiding the tedium of paperwork and administration—but then I was able to take to the woods, creeks, deserts, and mountains. Many people are drawn to the beauty of nature, but I was drawn also to its silence. I am never happier than when I am in some wild beautiful quiet place.

I cannot say when I saw my first frog, but I must have been very young. Growing up in mostly rural northern Virginia in the late 1950s and 1960s provided a wealth of habitats to explore. I remember the singing American Toads and catching leopard frogs along the creek near my house in places that no longer exist. That started me on a long journey that has taken me to six continents, fifty states, three Canadian provinces, and many Caribbean islands. At every turn I found frogs, and each place left its own special memories: listening to Bird-voiced Treefrogs in a Mississippi swamp, trying to photograph a large ranid in Kenya and suddenly realizing I was laying on an ant mound (do the dance!), searching unsuccessfully for gastric-brooding frogs in Australia's tropical rainforest, finding rare frogs in the Seychelles under a starry sky with fruit bats squawking in the trees, finding *Scutiger* tadpoles at 5,000 m in Tibet and foam-nesting *Rhacophorus* in Taiwan, and seeing my first dart-poison frogs in Costa Rica. I still get a thrill seeing a Barking Treefrog or hearing the soft chirping of *Eleutherodactylus planirostris* around our front porch in Florida. Frogs are truly gentle beings and, as the New Zealanders say, "fact: the survival of the Earth depends on frogs."

Throughout my career, I have been fortunate to work with many creative, enthusiastic, and knowledgeable friends, colleagues, and students. Although they may not have contributed directly to this book, it could not have been written without their guidance and friendship at least at some point over the past 30 years. I specifically thank Butch Brodie, my major advisor at Clemson University so many decades ago. There is nothing to stimulate interest and excitement better than someone who is interested and excited about their work. Ronn Altig secured my first teaching position, Hobart Smith provided encouragement when I needed it, Jim Williams got me a job in conservation, Ernie Liner talked books and made great food, Bob Shoop and Carol Ruckdeschel offered perspective and wine, and Dick Franz got me started again in research after an eight-year administrative stint. Thanks to you all.

I thank the following persons, in particular, for providing help with literature and information used in this volume: Kraig Adler, Ronn Altig, Kim Babbitt, Jamie Barichivich, Breck Bartholomew, Aaron Bauer, James Bettaso, Jeff Briggler, Robin Jung Brown, Charles Bursey, Bruce Bury, Brian Butterfield, Christine Campbell, Celia Chen, Michael Conlon, Steve Corn, Christopher Distel, Nathan Engbrecht, Edward Ervin, Gary Fellers,

Don Forester, Tony Gamble, Dana Ghioca, James Gibbs, Harry Greene, Patrick Gregory, Jackson Gross, Gavin Hanke, John Himes, Steve Johnson, Larry Jones, Tom Jones, William Karasov, Vicky Kjoss, Kenney Krysko, Michael Lannoo, James Lazell, Emily Lemmon, Lawrence Licht, Ernie Liner, Lauren Livo, Mickey Long, Clark Mahrdt, Chris McAllister, Malcolm McCallum, Jonathan Micancin, Joseph Mitchell, John Moriarty, Erin Muths, Max Nickerson, Deb Patla, Thomas Pauley, Greg Pauly, Chris Pearl, David Pilliod, Andrew Price, Louis Porras, Michael Redmer, Phil Rosen, Mark Roth, Ray Saumure, David Seburn, Lynnette Sievert, Hobart Smith, Lora Smith, Scott Smyers, Mike Sredl, Dennis Suhre, Dean Thompson, Stan Trauth, Susan Walls, Kent Wells, and Richard Yahner.

I thank Monica McGarrity for preparing the maps, Camilla Pizano and Breck Bartholomew for the illustrations, and Audrey Owen for the beautiful frontispiece. I appreciate the photographs sent by many colleagues, even though I was unable to use them all. Specific thanks go to: Ronn Altig, Jennifer Anderson-Cruz, Adam Backlin, Jamie Barichivich, Sean Barry, Breck Bartholomew, James Beck, Dawne Becker, Steve Bennett, David Bishop, Steve Brady, Koen Breedveld, Christopher Brown, Jennifer Brown, Travis Brown, Jason Butler, Robert Byrnes, Jan Caldwell, Alessandro Catenazzi, Adam Clause, Steve Corn, John Cossel, Alan Cressler, Eric Dallalio, Nina D'Amore, Ray Davis, Chris Dellith, David Dennis, Dana Drake, Andrew Durso, Anna Farmer, Gary Fellers, Mike Forstner, Justin Garwood, Carl Gerhardt, Jim Godwin, Earl Gonsolin, Tom Gorman, Mike Graziano, Matt Greene, Kerry Griffis-Kyle, Kyle Gustafson, Robert W. Hansen, James Harding, Chris Harrison, Marc Hayes, Valentine Hemingway, Aubrey Heupel, Jody Hibma, Brian Hubbs, John Jensen, Steve Johnson, Mike Jones, James Juvik, Kris Kendell, Ceal Klingler, Roland Knapp, Fred Kraus, Brian Kubicki, Jeff LeClere, Twan Leenders, Kirk Lohman, Patrice Lynch, Bryce Maxwell, Jonathan Mays, Brome McCreary, Melanie McFarland, Maija Meneks, Gabe Miller, Joe Mitchell, Martin Morton, Gary Nafis, Robert Newman, Justin Oguni, Charles Painter, Cindy Paszkowski, David Patrick, Seth Patterson, Joe Pechmann, Ryan Peck, Maria Pereyra, Jeff Petersen, Todd Pierson, David Pilliod, Brian Pittman, Jesse Poulos, Jim Rorabough, Francis Rose, Kevin Rose, Mark Roth, Paddy Ryan, Rob Schell, Sara Schuster, Cecil Schwalbe, Betsy Scott, Richard Seigel, Brad Shaffer, Nathan Shepard, Brent Sigafus, Bill Stagnaro, David Steen, Cameron Stevens, Dirk Stevenson, Jim Stuart, Dennis Suhre, Cynthia Tate, Robert A. Thomas, Stan Trauth, John Tupy, Michael van Hattem, Sara Viernum, Laurie Vitt, Kenny Wray, and Bob Zappalorti. Lee Brumbaugh of the Nevada Historical Society, Kraig Adler, and Mark Jennings were helpful in obtaining photos of *Lithobates fisheri* habitat in the Las Vegas Valley; Stephanie Munson of Cornell University Press granted permission to reproduce the Wrights's photographs of *Lithobates fisheri*.

Ellin Beltz permitted me to use her information on amphibian etymologies and updated generic names. Other colleagues provided advice and technical assistance, particularly Kraig Adler, Jamie Barichivich, Sherry Bostick, Priya Nanjappa, Ken Sulak, Susan Walls, and Jim Williams. This book could never have been completed without the outstanding services provided by the University of Florida library system, particularly in finding theses, dissertations, and sometimes very obscure papers. Vincent Burke of the Johns Hopkins University Press got me into this project in the first place, and Helen Myers ensured the readability and accuracy of the manuscript—I can't thank her enough. To all of you, I offer my sincere thanks.

Finally, I thank Marian Griffey, Dick Daniell, Antoinette Marie, Benjamin Silas, Christopher Robin, Etta Mae, Fallicity Sue, Frederick Hercules, Guinevere Faye, Gwyneth Grace, Herman Bartholomew, MacMurphy O'Leary, Sir Reginald Michael, and Mortisha Marie for their special support and encouragement.

Introduction

I have always liked frogs... I like the looks of frogs, and their outlook, especially the way they get together in wet places on warm nights and sing about sex.

Archie Carr, *The Windward Road*

As this book is completed (April 2013), there are >6,800 species of frogs known worldwide with new species being described at a rapid pace. North of the United States–Mexico border, however, only 100 species are known (1.7% of the world total), and 10 of them barely enter the United States. All of Hawaii's frogs are introduced species, and only 27 species occur in Canada, with many only just entering the country. Only the genera *Acris*, *Ascaphus*, *Pseudacris*, and *Spea* may be considered endemic to our region, although *Pseudacris* and *Spea* also cross the border into northern Mexico. The regions of temperate North America with the greatest species richness of frogs are the south Atlantic Coastal Plain, south Texas, and the Pacific Northwest. A surprising diversity of frogs is found in the arid Southwest. Although it might appear that the diversity of North American frogs is reasonably well known, one new species from the Southeast will be described, perhaps by the time this book is published (see *Lithobates catesbeianus* account) and a number of subspecies or isolated populations are likely to be reevaluated as full species, especially in the *Anaxyrus boreas* complex.

No one can mistake a frog for any other vertebrate. All frogs are tetrapods and have the same basic body plan (Figs. 1, 2), with short heads, large bulging eyes (reduced in some fossorial forms), little or no neck, and a short compact body. The front legs have four toes on each foot; the rear legs have five toes on each foot and are often webbed to some extent except in terrestrial

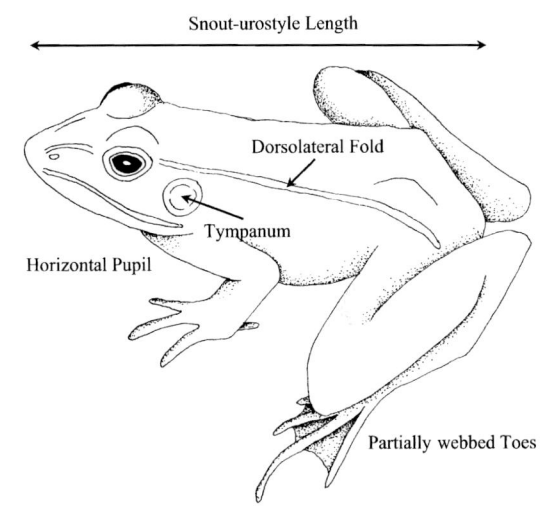

Fig. 1. Basic body plan of an aquatic frog. Illustration by Camila Pizano

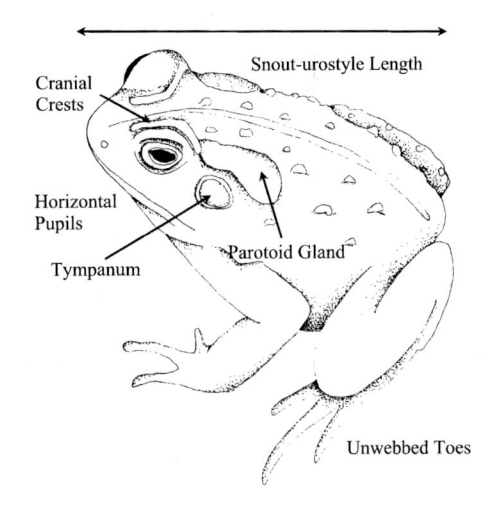

Fig. 2. Basic body plan of a toad. Illustration by Camila Pizano

forms. Frogs, of course, lack tails as adults. Their short powerful bodies with specialized hind legs are made for hopping, jumping (sometimes in great leaps), running, climbing, swimming, or burrowing. Toads (*Anaxyrus, Ollotis, Rhinella*), for example, usually walk or hop, whereas the so-called true frogs (*Lithobates, Rana*) and chirping frogs (*Eleutherodactylus*) are expert leapers; *Scaphiopus* and *Spea* are burrowers with specialized digging tubercles (Fig. 3); *Xenopus* is a swimmer; and *Hyla, Smilisca,* and *Osteopilus* are expert climbers.

The toes of frogs normally are long and thin, except within the chorus and treefrog family (Hylidae). In this group, the ends of the toes are expanded slightly in cricket and chorus frogs (*Acris* and *Pseudacris*) and greatly in treefrogs (*Hyla, Osteopilus, Smilsca*). The treefrogs use their expanded toe tips to climb trees and smooth surfaces; the expanded surface area of the toe pad exerts hydric friction toward the surface on which the frog walks, enabling it to hold on and climb. The largely ground-dwelling chorus frogs do not need such expanded toe pads. Aquatic species usually have a membranous web between their hind toes that facilitates swimming, but *Xenopus* has webs between the front toes as well. A comparison of the hind feet of bufonids, hylids, and ranids is provided in Fig. 4.

Frogs do not have external ears, although they are very attuned to sound. The opening to the inner ear is covered by a thin, membranous sheath of skin, the tympanum, located behind the eyes. The tympanums of American Bullfrogs, for instance, are rounded and large; they are conspicuous on either side of the head. The tympanums of some other species are less easily seen or may not be present (*Ascaphus*). Many frogs have "warts," bumps or ridges on the back (the dorsum), dorsolaterally, or on the upper portions of the limbs. These bumps and ridges usually contain mucous or granular glands and are important for moisture retention and defense. The parotoid glands of toads are kidney bean–shaped structures located on the head behind the eyes, and their shape and size are important, in conjunction with the configuration of the cranial crests, for the identification of the different species of toads (see Fig. 2).

Male frogs of many species can be distinguished from females by the presence of vocal pouches and a darkened coloration on their throats, at least during the breeding seasons. Males also develop enlarged and roughened thumbs during the reproductive season, which are useful while amplexing females (that is, when the male clasps the female during courtship). Female frogs are often much larger than males, and usually do not produce the "warning croak" when picked up. Eggs may be visible through the ventral body wall. Outside the breeding season, differentiating the sexes may be difficult. Juveniles usually resemble miniature adults.

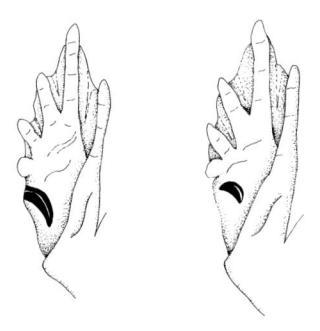

Fig. 3. Configuration of the spade on the left hind foot in the genus *Scaphiopus* (*left*) and *Spea* (*right*). In *Scaphiopus* the tubercle is cycle-shaped, whereas in *Spea* it is wedge-shaped. Illustration by Camila Pizano

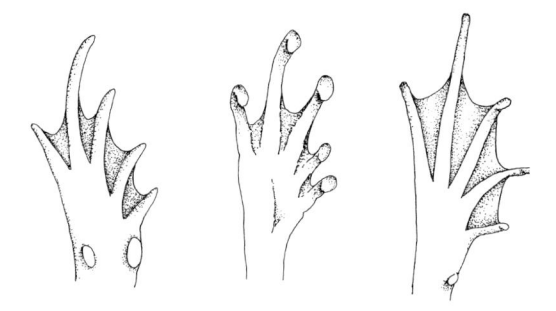

Fig. 4. A comparison of the hind feet of bufonids (*left*), hylids (*center*), and ranids (*right*). Note the differences in the extent of webbing and the presence of toepads in hylids. Illustration by Camila Pizano

Anuran Evolution

The first vertebrates to leave the water did so during the Devonian Period, some 350 to 370 million years ago. These animals were transitional between the lobe-finned fishes and the true Amphibia. These amphibian ancestors moved around on land, based on the fossilized trackways that they left in soft muds. Their bony fossils clearly demonstrate that they were not adapted solely for living in the water, although they may have had fish-like tails. Some of the transitional forms, such as the giant flat-headed *Ichthyostega*, had impressive rows of teeth and a vertebral structure designed for flexibility and mobility outside water. The first true amphibians, the Labyrinthodonts, appeared in the Carboniferous Period and survived until the Early Cretaceous Period, a span of about 230 million years. The Labyrinthodonts gave rise to the modern Amphibia and to the reptiles, and were present throughout the age of the dinosaurs. Another group of primitive amphibians, the Lepospondyls, became extinct in the Early Permian Period, and left no modern descendants.

Modern amphibians evolved at least by the mid-Mesozoic Era, and Holman (2003) has discussed the importance of paedomorphosis and locomotion in the evolution of the first frogs. Frog-like amphibians, *Triadobatrachus* (from Madagascar) and *Czatkobatrachus* (from Poland; Evans and Borsuk-Białynicka, 2009), are known from the Early Triassic Period (about 225–245 million years ago), but the earliest known true frog (*Prosalirus bitis*) was found in Lower Jurassic Kayenta Formation deposits in Arizona (Shubin and Jenkins, 1995). These deposits date from 190 million years ago. Thus, both primitive modern amphibians were present throughout the Mesozoic Era, and the first known true frog likely evolved in North America. The basic body plan of modern frogs was set nearly 200 million years ago, although the earliest frog-like amphibians likely walked more than they hopped or jumped. When dinosaurs ruled the world, frogs called from the swamps.

Life History

North of the Mexican border, frogs occur from sea level to the high Rocky Mountains, and from the south Florida Keys to the Arctic Ocean. They occur in the tropical lowlands, grassland prairies, deserts, and in alpine-tundra habitats. Although thought of as entirely freshwater in nature, a few frogs and tadpoles have been found associated with saline habitats near oceans and in isolated desert wetlands. Some species have restricted habitat requirements, whereas others occur contiguously from the arid Plains (*Pseudacris maculata*) or humid Southeastern forests (*Lithobates sylvaticus*) to the high tundra.

In the "typical" amphibian life cycle of most U.S. and Canadian frogs, adults move to water to breed and lay eggs, then emigrate back to terrestrial or other aquatic habitats to forage and overwinter. The eggs develop into larvae that remain in water for a period of time, then the larvae metamorphose, and the tiny juveniles disperse. The only entirely aquatic species in North America is the introduced African Clawed Frog, *Xenopus laevis*. All members of the mostly tropical genus *Eleutherodactylus* deposit eggs terrestrially. The larval period occurs entirely within the egg, and hatchlings resemble miniature adults. Only a few other frogs depart from the typical pattern. In the nonindigenous *Dendrobates auratus* of Hawaii, eggs are oviposited in moist situations and fertilized by the male who then guards them until they hatch. He then carries the tadpoles on his back to an appropriate developmental site, such as a tree hole or bromeliad cup. In *Leptodactylus fragilis*, eggs are oviposited into a foam nest and tadpoles are released as water eventually fills the shallow nest depression.

All frogs in Canada and the United States have external fertilization except for *Ascaphus* (internal fertilization via an intromittant organ) and *Eleutherodactylus* (through cloacal apposition). The process of holding onto a female during reproduction is termed amplexus. The male frog grabs the female dorsally either under the armpits (axial or pectoral amplexus; Fig. 5) or just in front of her hind legs (inguinal or pelvic amplexus) and holds on as strongly as he can. The location where the male grabs the female is species-specific. As eggs are extruded from the female's vent, the male releases clouds of sperm over them. Male frogs often amplex the wrong species, other males, or inanimate objects. If another male is amplexed, he gives a warning croak and/or vibration to let the courting male know that he has erred in his mate choice. Otherwise, a male amplexing an inappropriate object often has a long and frustrating reproductive season. Spent females also may give a warning vibration, but many amplexing males seem to be able to determine whether a female is gravid by her girth and perhaps the firmness of her body as she carries eggs.

Nearly all North American anurans lay eggs

Fig. 5. Axillary (pectoral) amplexus in aquatic frogs. Illustration by Camila Pizano

in water. A few (*Dendrobates auratus*, *Eleutherodactylus* sp.) deposit their eggs in moist terrestrial locations. The eggs are coated by one or more jelly-like capsules surrounded by a firm membrane (Fig. 6). The capsules and membrane help protect the eggs from infection and desiccation while allowing for the exchange of oxygen, carbon dioxide, and nitrogenous wastes (mostly ammonia). Eggs may be deposited individually, in small clusters, in strings, in thin sheaths that float on the surface of water, or in large compact jelly masses (Figs. 7–9). In addition, eggs may or may not be attached to aquatic vegetation or to the wetland bottom. Depending on the species, the number of eggs deposited can be a few to as many as tens of thousands. All these characteristics are important in helping to identify species in the field, and are mentioned in more detail in the species accounts.

In the early stages of development, most anuran eggs have a dark animal pole and a light cream to yellowish vegetal pole, although in a few species (*Eleutherodactylus* sp.) the eggs are entirely white. The cells of the animal pole will eventually overgrow the vegetal pole and form the developing embryo, whereas the yolk forms from cells of the vegetal pole. The nutrient-rich yolk supplies all the energy needed to complete development prior to hatching. Hatching occurs as partially developed larvae (tadpoles) in aquatic situations, and as miniature adults in terrestrial nests (*Eleutherodactylus* sp.). All larvae have gills for respiration; in larval frogs, the gills are hidden shortly after hatching within an internal chamber covered by an operculum on the side of the head.

As the larva grows, it eventually reaches the stage where it metamorphoses into an adult. Metamorphosis involves a complex set of morphological and physiological changes that radically alter the body plan. It is also the time when the animal is most likely to be affected by chemicals or pollutants in the environment that interfere with the intricate developmental changes. Anurans remain as larvae from a few days (e.g., Eastern Spadefoot) to 3–5 yrs (e.g., Coastal and Rocky Mountain Tailed Frogs). Most larvae transform within a few months to a year, however. If breeding occurs in spring or early summer, transformation usually occurs in late summer to autumn. If breeding occurs in the fall, the larvae overwinter and metamorphose the following spring or summer. There is a great deal of variation in the larval period among species, however. Some of the variation undoubtedly is derived from hereditary factors, although the availability of high-quality food and the duration of the hydroperiod also influence the length of the larval period. Unlike salamanders, there are no paedomorphic frogs.

Tadpoles are mostly detritivores in that they eat a wide variety of benthic organic matter. They graze on algae or aquatic vegetation using a specialized mouth structure that rasps small bits of vegetation into the mouth. This material is funneled into the gut for digestion. Some tadpoles are filter feeders and feed in the water column or at the surface, specializing on phytoplankton and zooplankton, but others are carnivorous or even cannibalistic. Adult frogs are all carnivorous, usually eating a wide array of insects, spiders, and other invertebrate animals. A few species, such as American Bullfrogs, consume large prey such as small mice or snakes, whereas others such as the ant-loving Eastern Narrow-mouthed Frog are specialist feeders.

Larvae are largely free-swimming and do not form social aggregations. Spadefoots, some toads (*Anaxyrus* spp.), and the River Frog (*Lithobates heckscheri*) are exceptions, however, in that their tadpoles form large schools that move and feed together throughout a pond or lake. Schooling helps to churn up the bottom and expose more food for consumption, provides a measure of protection, and maintains a favorable thermal environment. Spadefoot schools in particular are initially comprised of kin, that is, tadpoles that hatched from eggs deposited by the same parents. They apparently keep in contact via pheromones, chemicals that aid them in sibling recognition. Tadpoles also are known to respond to alarm chemicals within their environment. If another tadpole is attacked and injured, the chemical released by the injured tadpole alerts the rest of the tadpoles to danger in the area.

Larval anurans have a smooth and moist skin that readily allows water to enter and waste products to be expelled. During the terrestrial phase of the life cycle, however, frogs must remain in moist situations because their skins are only semi-impermeable to water loss. In dry conditions, moisture is lost to the surrounding air from the animal's body, and frogs will desiccate if too much water is lost. Anuran skins are usually moist to the touch because of mucous secreted to keep them damp. The skins of some species, such as toads, may appear rather dry, however. A few

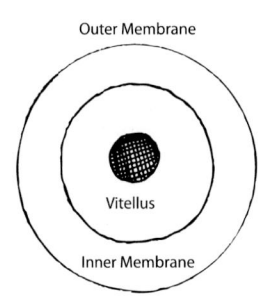

Fig. 6. Diagram of a frog egg. The vitellus is surrounded by two jelly envelopes enclosed within encapsulating membranes. Illustration by Camila Pizano

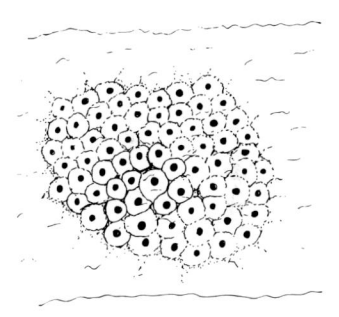

Fig. 8. Surface film of frog eggs. A number of ranids and microhylids deposit their eggs in single-layer surface films. Illustration by Camila Pizano

Fig. 7. A frog egg mass. Egg masses of this type are typical of many ranids. Illustration by Camila Pizano

Fig. 9. Toad egg strings. Note that in some species the eggs are partitioned into separate chambers (*left*), whereas in others the eggs are continuous, forming one or more rows without partitioning (*right*). Illustration by Camila Pizano

species are able to spread a lipid-based secretion across the body to impede water loss.

All North American frogs possess sac-like lungs, but gas exchange takes place across the skin of larvae and adults. Some areas of the skin are particularly well vascularized (e.g., the pelvic patch of certain frogs), and gases and water readily diffuse across the thin skin membranes. Frog skin also is remarkable in terms of defense. Many species possess either toxic (*Dendrobates auratus*, *Rhinella marina*) or noxious (e.g., *Anaxyrus* sp., *Scaphiopus* sp., *Lithobates palustris*) secretions that help them to avoid predators and parasites. Noxious and toxic skin secretions are produced from granular glands in the skin of frogs. Such glands may be concentrated, such as in the "warts" of toads, or they may be more diffusely distributed over the dorsal part of the body. Toads (Fig. 2) and spadefoots (Fig. 10) have kidney bean–shaped glands behind the head that contain glands producing noxious secretions. These parotoid glands have even been observed to squirt secretions in the direction of an attacking predator, especially as the predator bites down on the frog. Other anurans (many *Lithobates*, *Ascaphus* sp.) have antimicrobial peptides in their skins that help ward off disease and fungi. Chemical communication is not as important in terrestrial frogs as it appears to be for salamanders, at least as adults. Larvae and perhaps aquatic adults seem to be keenly aware of chemical cues in water.

Most male frogs in the United States and Canada gather at breeding sites from winter to late summer, and send out chirps, peeps, and trills depending on species. Only *Ascaphus* is truly silent. The calls serve two main functions: to entice the females to the pond and to inform other males that the caller has occupied a particular portion of the breeding site. Thus what sounds to a human listener like a single call conveys multiple messages to conspecifics. For example, a female may assess male quality by the sound of his call, how long it is, how loud it is, or how deep it is. Inasmuch as some frogs are territorial at breeding sites, the calls also serve to alert other males that the caller is ready to defend his calling space. Females often prefer the largest males when presented with a choice. However, "satellite" males (males that do not call but sit near a calling male) sometimes intercept females on their way to breed, and thus avoid the competitive chorus.

Frogs generally return to the same ponds or wetlands to breed from one year to the next, although there are exceptions. In some frogs, both sexes may use the same ponds each breeding season. In other species the males are site-specific, but the females are not. In this way, a female can choose the best-fit male among all the males that she can hear, regardless of where he is located. Male pond-breeding frogs usually arrive at the ponds first and establish their calling sites and territories, and such males may overwinter closer to the ponds than females, presumably so they can get to the breeding sites early. Males also stay at ponds longer than females, who frequently stay only long enough to mate and deposit their eggs. In explosive breeders (that is, frogs with short reproductive periods in which nearly all adults breed at the same time), males and females arrive nearly simultaneously.

At least some species of frogs breed in about any type of wetland. Frogs that breed in small temporary or semipermanent ponds do so because these habitats lack fish and may have fewer invertebrate predators, especially early in the season. In areas lacking aquatic vertebrate predators, larvae are often conspicuous; in habitats where predators are abundant, larvae are secretive and cryptic. Many of the ponds and streams in North America are temporary or intermittent, especially in the Great Plains and West. Even in the East small breeding pools often disappear by summer. As the year progresses, these habitats dry up making them unsuitable for breeding. Drought and low-rainfall years can also pose problems, and in

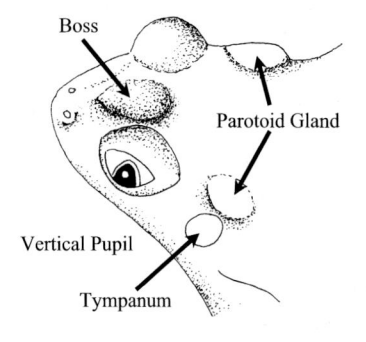

Fig. 10. Head morphology of a Spadefoot showing the position of the boss, parotoid gland, tympanum, and vertical eye pupil. Illustration by Camila Pizano

some years frogs may not be able to breed. Thus, whereas frogs breeding in temporary wetlands may reduce predation risk to their larvae, there is a trade-off as to whether the wetlands will fill and allow sufficient time for the larvae to develop and metamorphose.

Many frogs travel short distances from their breeding sites, whereas others travel hundreds of meters or even several kilometers from where they passed the larval stage. Terrestrial habitats may be as close as streamsides, making certain frogs (e.g., *Ascaphus*) seem semiaquatic. Other frogs (e.g., the Western Toad, *Anaxyrus boreas*) travel extensively before establishing a nonbreeding terrestrial home range. Here, they spend their time in surface, subsurface, cliff face, or arboreal habitats before returning to the area where, presumably, they hatched. As a rule, the shorter lived a frog is, the sooner it begins to breed. Still, not every adult frog breeds every year, and little is known about reproductive periodicity through their life span. I have observed a female Florida Eastern Narrow-mouthed Frog skip a breeding season, even though the radioactive-tagged animal was near a pond, and the life span of this species is only about 4 yrs. Females, in particular, seem to skip breeding in years of harsh environmental conditions and poor food availability, and instead keep their energy reserves to stay alive.

Anurans are especially active at night, although those species with noxious secretions (especially toads) and a few others (e.g., *Acris*) are often active by day. Aquatic frogs (e.g., *Lithobates*, *Rana*) are active diurnally especially on cloudy and rainy days, but they are usually in water and tend to be very alert. Frogs congregate around all kinds of wetlands, emerge from under rocks and logs, and forage arboreally or through the terrestrial leaf litter or along stream edges. Frogs often can be spotted with a flashlight sitting on buildings, at the mouths of burrows and hiding places, and in water or along the shoreline waiting for unsuspecting invertebrates. At night, larvae also become more conspicuous as they leave daytime refugia for open and shallow water. Nocturnal activity presumably makes some frogs less prone to predation, especially by visually oriented predators. Only on rainy nights can the abundance of anurans can be appreciated as they call and forage.

Frog Conservation

Frogs are now at greater peril worldwide than at any time in recent geologic history (Stuart et al., 2004; Sodhi et al., 2008; Wake and Vredenburg, 2008). Indeed, the Earth's biota may already be well into the sixth mass extinction event since life began on this planet (Barnosky et al., 2011). It is beyond the scope of this introduction to discuss in depth the many threats to anuran populations in Canada and the United States and the ways to mitigate and manage these threats. Much more detail can be found in Stebbins and Cohen (1995), Kingsbury and Gibson (2002), Semlitsch (2003), Lannoo (2005), Bailey et al. (2006), Mitchell et al. (2006), Pilliod and Wind (2008), Dodd (2009), and in the references in the species accounts.

Threats to anuran populations derive from local, regional, and global sources. Wetland and associated terrestrial habitats are being lost and fragmented (or shredded, as some biologists have termed it) at alarming rates because of expanding human populations and economic and political considerations fostering rapid and often poorly regulated development (Hamer and McDonnell, 2008). In the United States alone, 185,400 ha of wetlands were destroyed per year from the mid-1950s to the 1970s, with another 117,400 ha lost per year from the mid-1970s to the 1980s; another 155,200 ha were lost from 1986 to 1997 (Dodd and Smith, 2003). These numbers are frightening when considering the potential loss of aquatic diversity and individual frogs. Habitat changes affect amphibians over a long time period, so it is necessary to consider both historic and current landscape effects on frog populations in order to conserve them (Piha et al., 2007).

Anurans with their unprotected eggs, aquatic larval development, permeable skins, complex endocrinological and morphologic changes associated with metamorphosis, diverse life histories, and biphasic life cycles requiring both terrestrial and aquatic habitats are being saturated by a host of lethal and sublethal toxic substances (Quaranta et al., 2009). Emerging infectious diseases such as amphibian chytrid fungus (*Batrachochytrium dendrobatidis*), iridoviruses, and novel alveolate pathogens threaten worldwide impacts (e.g., Fisher and Garner, 2007; Vredenburg et al., 2010b). Indeed, amphibian chytrid is now so widespread that it can be considered endemic across North America (Lannoo et al., 2011). Threats such as global and regional climate change affect both temperature and precipitation patterns and further imperil frogs, especially those with limited distributions and dispersal capabilities (Corn, 2005). Fully one-third of all amphibians are now considered threatened worldwide (Stuart et al., 2004), and 168 amphibians have become extinct within the last two decades. Clearly, these are treacherous times for many frogs, as many of the species accounts will indicate.

Anuran conservation requires an integrated landscape approach to management, rather than a species-oriented focus, except under dire circumstances. The reason for this is simple: frogs do not live in a biotic or physical vacuum in nature. They frequently move from wetlands to terrestrial sites, and movements may cover hundreds of meters. This necessitates protection of breeding sites, movement corridors, and surrounding terrestrial habitats used for foraging and overwintering. Simply "protecting" breeding sites will not conserve anuran populations. It once was thought that if frogs and habitats could be protected against take or destruction, then frog populations would survive. Threats from disease, toxic chemicals, and climate change clearly have demonstrated the inadequacy of such an approach. The genetic consequences of habitat fragmentation also are much better understood, so that connectivity among even protected sites must be maintained despite serious habitat fragmentation (Cushman, 2006; Becker et al., 2007). Habitats cannot be surrounded by a fence with the assumption they will maintain their ecological integrity.

Frog management options range from the rather simple and inexpensive to the very complex and expensive. When planning, the overriding consideration should be "first, do no harm" to the species, its community, or its habitat. High technology-based approaches may work no better than simple and inexpensive approaches, and care should be exercised to maximize the benefits from the human resources and funds available. The primary objectives of management should always focus on the species or community of concern, and not on peripheral or extended objectives, such as positive publicity.

All state and provincial governments have statutes aimed at protecting endangered and threatened species. The effectiveness of regulations varies considerably among jurisdictions, and rarely is habitat considered on an equal basis with individual protection. For listed species, wildlife and conservation agencies frequently attempt to develop conservation and recovery plans, but these plans are vastly underfunded. When compared to highly publicized species such as manatees, frogs receive a pitiable amount of financial support for research and conservation. Unfortunately, many individuals and organizations are opposing environmental land buying and protection under the guise of economic recovery, masking, such as in Florida, a deep-seated opposition to impediments to wealth accumulation. Such myopic foresight will have devastating consequences for frogs and their habitats.

As of May 2011, the following species are protected under provisions of the U.S. Endangered Species Act of 1973, as amended: Wyoming Toad (*Anaxyrus baxteri*), Arroyo Toad (*A. californicus*), Houston Toad (*A. houstonensis*), Chiricahua Leopard Frog (*Rana chiricahuensis*), California Red-legged Frog (*R. draytonii*), Mountain Yellow-legged Frog (*R. muscosa*), and Dusky Gopher Frog (*Lithobates sevosus*). The following are considered candidate species: Yosemite Toad (*Anaxyrus canorus*), Arizona Treefrog (*Hyla wrightorum*), Relict Leopard Frog (*Lithobates onca*), Columbia Spotted Frog (*Rana luteiventris*), Oregon Spotted Frog (*R. pretiosa*), and Sierra Nevada Yellow-legged Frog (*R. sierrae*). A few (Amargosa Toad, *Anaxyrus nelsoni*; Illinois Chorus Frog, *Pseudacris illinoensis*) seem inexcusably absent from the candidate list. Blanchard's Cricket Frog (*Acris blanchardi*) is the only frog listed by the Scientific Committee on the Status of Endangered Wildlife in Canada (COSEWIC).

Frogs have had a very long association with humans, whether through mythology, spirituality, art, literature, or scientific investigation. Their effigies occur on totem poles in Alaska, pueblo pottery (Fig. 11), and on pipes ceremonially smoked by native peoples in the Southeast (Fig. 12). Images of frogs from North America first appeared in Catesby's magnificent *The Natural History of Carolina, Florida, and the Bahamas*, published in parts from 1729 to 1742. They

Fig. 11. Zuni bowl featuring tadpole motif. Pitt Rivers Museum, Oxford. Photo: C.K. Dodd Jr.

Fig. 12. Frog pipe, Hopewell Culture, central United States. British Museum, London. Photo: C.K. Dodd Jr.

continue to provide us with immense inspiration and enjoyment with their calls signaling an end to winter and the rebirth of dormant nature. They eat massive quantities of injurious insects; they are subjects of research in development, regeneration, tissue repair, and genetics; they are prey for countless other animals within their community. Despite the presence of sometimes noxious or toxic secretions, frogs and toads are gentle creatures that can be handled safely.

The rise of human population coupled with agricultural and industrial development has been termed the Anthropocene. Although lasting only a few thousand years, direct and indirect human activity during the Anthropocene has dramatically threatened the environment of frogs and indeed much of nature. At the same time, human action is necessary to prevent the continued loss of frogs and biodiversity in general; if frogs are to survive, humans must take effective measures to ensure their future. There are many ways to do this, from active participation in research to joining organizations trying to conserve what remains of Earth's biotic diversity. Concerned individuals may become "citizen scientists," learning the methods of research and cooperating with state, province, and national conservation organizations and government agencies in monitoring species' status. People need to be aware of how land-use proposals impact the function and diversity of nature, and to oppose those that are detrimental to natural systems. Individuals must learn to think critically, keep abreast of scientific developments, develop a passion for the value of nature unto itself, and to not be afraid to oppose ignorance and ideologically based hyperbole, such as that surrounding climate change and the impact of pesticides on sexual development and health. In short, be involved.

Etymology

Etymology refers to the derivation, or meaning, of a name. Species names are always italicized or underlined. They consist of both a genus and specific epithet, with the genus, but not the specific epithet, capitalized (e.g., *Acris gryllus*). Species names are derived from Latin or Greek roots and in accordance with Classical grammar. When they are used to honor a person or place, specific epithets must conform in gender and number to both the generic name and the name of the person or place they honor. Although there are a number of frivolous names in the scientific literature, most authors attempt to convey some important characteristic of the animal, its location or habitat, or the people important to its recognition and discovery.

The etymologies of specific epithets are provided in the species accounts. In order to avoid duplication, generic etymologies are provided below, including the names proposed by Frost et al. (2006a); these are derived from Beltz (2007; personal communication) and other sources.

Especially with older names, authors often did not provide an etymology with the description, so the exact meaning of the name is sometimes unclear.

Acris. From the Greek *akris*, meaning 'locust.' The name refers to the call of cricket frogs, which sounds like insect calls.

Anaxyrus. Possibly from the Greek *anax*, meaning 'leader' or 'chief,' and *urus*, meaning 'tail.' What Tschudi was referring to when applying this name is unclear. There are several other possible meanings, including the Greek *an*, meaning 'without,' the Greek *axyr*, meaning 'being cut' or 'an anchor,' and the Greek *ax*, meaning 'axel.'

Ascaphus. From the Greek *a*, meaning 'without,' and *skaphis*, meaning 'spade.' Literally, without a spade, denoting the absence of a metatarsal spade.

Craugastor. From the Greek *krauros*, meaning 'hard' or 'brittle,' and *gaster*, meaning 'of the belly.'

Dendrobates. From the Greek *dendro*, meaning 'leaf,' and *bates*, meaning 'one who treads or climbs'; literally, leaf-climber.

Eleutherodactylus. From the Greek *eleutheros*, meaning 'free' or 'unbound,' and *dactylos*, referring to 'finger' or 'toe.' These frogs lack a web between their digits.

Gastrophryne. From the Greek *gastros*, meaning 'belly,' and *phryne*, meaning 'toad.' The name may refer to the fat bellies of these frogs.

Glandirana. The name literally means 'glandular frog' and is in reference to the numerous glands found on both tadpoles and postmetamorphs.

Hyla. Greek for Hylas, the companion of Hercules. Hylas was one of the mythological Argonauts who sailed with Jason looking for the Golden Fleece. Exactly what Laurenti had in mind when he coined the name is unclear.

Hypopachus. From the Greek *hypo*, meaning 'under,' 'beneath,' or 'lesser,' and *pachos*, meaning 'thickness.'

Leptodactylus. From the Greek *leptos*, meaning 'fine,' 'slender,' or 'thin,' and *daktylos*, meaning 'fingers' or 'digits.'

Lithobates. From the Greek *lithos*, meaning 'rock,' and *bates*, meaning 'one who treads or climbs'; literally, rock climber.

Ollotis. From the Greek *olla*, meaning a 'pot' or 'jar,' and *itis*, meaning 'like' or 'pertaining to.'

Osteopilus. From the Greek *osteon*, meaning 'bone,' and *pileos*, meaning 'cap.' The name refers to the co-ossified skin on the top of the head.

Pseudacris. From the Greek *pseudes*, meaning 'false' or 'deceptive,' and *akris*, meaning 'locust.' Hence, 'false *Acris*,' that is, a false cricket frog.

Rana. Latin for 'frog.'

Rhinella. From the Greek *rhinos*, meaning 'nose,' and *ella*, meaning 'diminutive'; literally, little nose.

Rhinophrynus. From the Greek *rhinos*, meaning 'nose,' and *phrynos*, meaning 'toad.'

Scaphiopus. From the Greek *skaphis*, meaning 'shovel' or 'spade,' and *pous*, meaning 'foot.' The reference is to the spade of each hind foot, which is used in digging.

Smilisca. From the Greek *smiliskos*, meaning 'little knife.' The name refers to the sharp pointed frontoparietal processes.

Spea. From the Greek *speos*, meaning 'cave' or 'cavern.'

Xenopus. From the Greek *xenos*, meaning 'foreign' or 'stranger,' and *pous*, meaning 'foot.'

About the Book

This volume is not a field guide. There are no keys to the genera or species of anurans, although important distinguishing characteristics are provided in each species account. Many regional, state, and provincial field guides provide keys to adults and tadpoles, and additional information regarding identification can be found on numerous Internet sites. General references to many other aspects of the biology of frogs are included in the list of books, Internet sites, and atlases at the end of this section.

My objective in this book is to synthesize the literature on all frogs of North America north of the Mexican border through May 2011. In that regard, I have concentrated on what is known about frogs within our region, even when the range extends farther south as it does for many tropical species that reach their northern limits just north of the Mexico–United States border. The life history of tropical species may be different at the northern periphery of their range than it is hundreds or thousands of kilometers southward. Summary references to the biology of peripheral and nonnative species are provided in the species accounts.

The following synopsis outlines the basis for the information provided, in order to delineate the book's scope and limitations.

Nomenclature. I follow Frost et al. (2006a, 2007) when using generic names, updating scientific names as necessary based on revisions published since 2006. I realize, of course, that there is considerable controversy in the use of certain amphibian generic names, particularly *Anaxyrus* and *Lithobates*; I will leave future systematists to argue those battles. English common names follow Crother (2008); French common names follow Desroches and Rodrigue (2004); and Hawaiian common names follow McKeown (1996). I include the scientific names in the popular field guides of Conant and Collins (1998) and Stebbins (2003) inasmuch as these guides are widely in use. I also provide a brief list of species synonyms in the scientific literature. Full synonymies are published in the *Catalogue of American Amphibians and Reptiles* put out by the Society for the Study of Amphibians and Reptiles and online through AmphibiaWeb or the American Museum of Natural History (see below). Some accounts contain additional nomenclatural notes, particularly when there is confusion in the literature concerning species identification.

Etymology. As noted above, most etymologies follow Betlz (2007). In a few cases, I have added additional information or amended Beltz's accounts for clarification.

Identification. The identification of most species of North American frogs is relatively straightforward, and the verbal descriptions and photographs of postmetamorphs should prove sufficient for identification. However, many species have multiple color morphs and dorsal patterns that change among and even within individuals. Tadpoles also are highly variable in coloration and pattern depending upon growth stage, population, and region. Lighting and humidity influence the appearance of both adults and tadpoles of many species. In a few species, juveniles vary greatly from adults. Any description is likely to be insufficient to identify some individuals. Regional field guides may have multiple photographs of some variants, but in this work it is impossible to include more than a hint of the color and pattern variation within frog populations. I have, however, pointed out some of the most common variants within the accounts.

The identification of tadpoles is truly an art as much as a science. Many tadpoles look alike especially when small, and even good diagnostic characters may only be apparent among the larger size classes. Researchers often rely on the features of the oral apparatus to recognize species (Fig. 13). In this book, however, I have not provided much information on labial teeth, tooth rows, and the position and number of oral papillae. Although important, these characters of tadpole identification will be covered in much more detail in a forthcoming book on larval biology (*Handbook of Larval Amphibians of the United States and Canada* by R. Altig and R.W. McDiarmid, Cornell University Press, Ithaca, NY). Nomenclature for tadpole external morphology follows

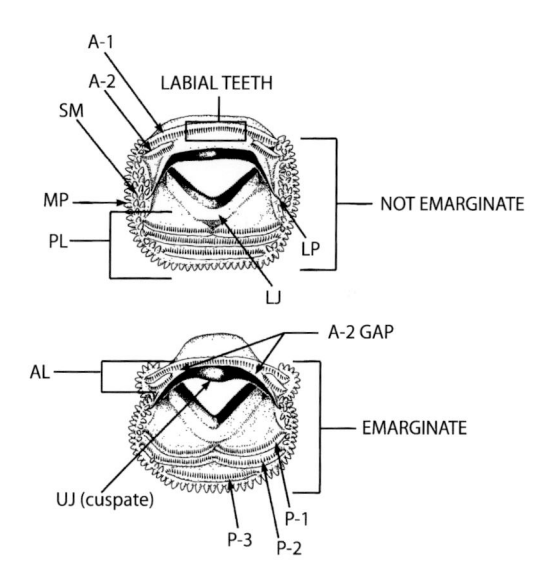

Fig. 13. Oral disk (mouthparts) of a tadpole. AL = anterior (upper) labium; A-1 and A-2 = first and second anterior (upper) tooth rows; A-2 GAP = medial gap in second anterior tooth row; LJ = lower jaw sheath; LP = lateral process of upper jaw sheath; MP = marginal papilla; PL = posterior (lower) labium; P-1, P-2, P-3 = first, second, and third posterior (lower) tooth rows; SM = submarginal papilla; UJ = upper jaw sheath. Terminology is that of Altig and McDiarmid (1999:35). Reprinted with permission of the University of Chicago Press

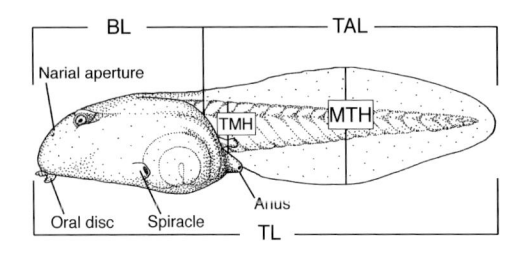

Fig. 14. Body morphology of a tadpole. TL = total length; BL = body length; TAL = tail length; TMH = the maximum height of the tail musculature; MTH = maximum height of the tail, including both tail fins and tail musculature; IND = distance between the narial apertures (internarial distance); IOD = distance between the eyes (the interorbital distance); TMW = maximum width of tail. Terminology is that of Altig and McDiarmid (1999:26). Reprinted with permission of the University of Chicago Press

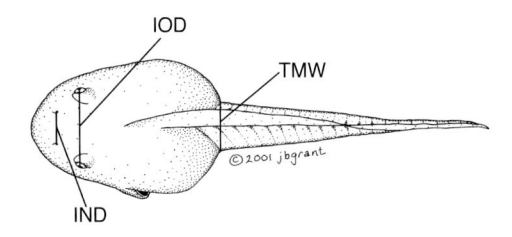

Altig and McDiarmid (1999:35) (Fig. 14). Field experience and knowledge of natural history are the best teachers when identifying larvae. A combination of verbal description, known range, breeding site, and season of observation will help in identifying tadpoles.

Distribution. Species range is based on the latest field guides and primary systematic literature available as this work is completed. Readers should keep in mind that all range accounts and maps are approximate. Although maps may appear contiguous, frog populations are not evenly distributed within an area, especially since many species are found only in specific habitats that are themselves patchily distributed. Even in "ideal" habitats, colonization and extinction may change distribution patterns through time; frogs may be present at a site one year but absent the next. More detailed information on distribution is contained in many regional field guides or online atlases, especially those that are georeferenced with dot distribution maps. I do not mention the field guides by Conant and Collins (1998) and Stebbins (2003) as important references on distribution within each account; their inclusion should be considered automatic and unnecessary to repeat.

Fossil Record. Most information on fossil frogs is based on Holman (2003). More detailed information and additional references are in Holman's book.

Systematics and Geographic Variation. Information on systematics comes from the latest primary peer-reviewed literature. Important geographic phenotypic variation is included, whether regionally based or as unusual morphs such as the different genetically based color patterns of *Lithobates pipiens*. Information on hybridization is included within this section.

Adult Habitat. The adult habitat refers to the macroenvironment of the species, specifically the physiography and vegetative components of its ecosystem.

Terrestrial and Aquatic Ecology. This section focuses on the microenvironmental components of a species' life history. Information on daily and seasonal activity, movement patterns, juvenile dispersal, overwintering, physiological mechanisms employed to cope with harsh environmental conditions, orientation, and sensory perception is included within this heading. The title for this section may change depending upon the life history of the species under discussion (e.g., some species are entirely terrestrial, hence the section labeled Terrestrial Ecology).

Calling Activity and Mate Selection. Seasonal and daily calling patterns, descriptions and specifications of calls, environmental factors affecting calling, male calling behavior, female response behavior, courtship, and intraspecific aggression with regard to calling are covered in this section.

Breeding Sites. The physical and biotic components of breeding sites are covered, including the types of habitats used and physical and biotic characteristics of those habitats.

Reproduction. This section includes information of reproductive activity, such as breeding season, environmental factors stimulating the onset of breeding, oviposition, clutch size, total fecundity, and time to and size at hatching. I include information from throughout the range of the species to illustrate the effects that latitude, altitude, and changes in weather patterns have on reproductive behavior. Despite this, little information is available on the variability of reproductive traits, and what is known is often based on small sample sizes from one or at best a few locations. This caveat applies to virtually all data on anuran life history from North America and makes it difficult to assess the importance of differences among literature reports. There is an urgent need for more extensive information over long time periods from multiple locations.

Larval Ecology. All aspects of larval ecology are included within this section, such as the duration of the larval period, growth, factors influencing larval ecology and behavior, and the influence of predators on larval activity. I also include data on sizes of tadpoles and recent metamorphs. Some aspects of larval ecology may be included under the sections labeled Predation and Defense or Community Ecology, depending upon the nature of the interaction.

Diet. Information on the diet of tadpoles, recent metamorphs, and adults is presented in this section. Most amphibians consume a wide array of invertebrates, particularly beetles. Even when information is not reported in the literature, invertebrates probably are the main prey.

Predation and Defense. All amphibians are preyed upon opportunistically by a wide array of vertebrates (snakes, birds, mammals). Much information in the literature, however, reflects opportunistic sightings rather than detailed analyses. Tadpoles face countless aquatic predators, such as turtles, snakes, and predaceous aquatic invertebrates. To avoid predation, larval and adult anurans have evolved a variety of defensive morphological, color-related, behavioral, and biochemical responses. These are discussed in this section.

Population Biology. Information on growth and demography is included here, with emphasis on population structure, age and size at maturity, sex ratios, recruitment, the effects of immigration and emigration, population genetics, survivorship, and longevity.

Community Ecology. This section focuses on interactions between anuran species within the aquatic and terrestrial habitat with specific emphasis on competition and habitat partitioning. It does not contain a list of other frogs within the area or same habitat type, nor does it attempt to list all potential biotic interactions. Indeed, so little is known about the community ecology of most anurans in North America that this section is often omitted from species accounts.

Diseases, Parasites, and Malformations. As the heading implies, information is provided on anuran diseases, endo- and ectoparasites, and malformations. Malformations include those induced by unknown developmental causes as well as those thought to be induced by toxic substances or parasites. Malformations due to injuries are not reported. I have attempted to update parasite nomenclature, but it is likely I have inadvertently used synonyms on occasion.

Susceptibility to Potential Stressors. Many amphibians are affected by substances (e.g., toxic chemicals) or environmental variables (e.g., acidity, conductivity) arising from both anthropogenic and natural sources. These stressors can have both lethal and sublethal effects that differ among species and life stage. Some stressors act paradoxically, that is, small amounts may have more serious effects than large amounts. In this section, I report on potential stressors, as not all studies of stressors on anurans have been demonstrated to have effects on wild populations. The section includes subheadings on metals, chemicals (pesticides, industrial chemicals, drugs), nitrates and nitrites (e.g., fertilizers), pH (acidity), conductivity, alkalinity, UV radiation, and others as appropriate. I have made no attempt to standardize measurements; the measurement units are those reported in the publication cited.

Status and Conservation. This final section covers all aspects of the status and trends of species as understood based on published literature. I acknowledge that there is a great deal of information that is unpublished or available only in reports, files, and non-peer-reviewed Internet sites. In addition, the legal and regulatory status of species is constantly being revised and updated by federal, state, and provincial wildlife agencies. Each governmental organization maintains an Internet site with up-to-date information on the protected status of species within their jurisdiction. Readers are advised to check on regulations prior to searching for frogs.

Bibliography. The bibliography includes references from the 1700s to mid-2011. All references have been examined for accuracy of citation and content. Many authors apparently have not examined original copies of the papers they referenced since I found many incorrect citations of titles, journals, books, dates, and page numbers as I read the more than 4,500 references included in the bibliography. As it is, I made no attempt to cite every paper, note, thesis, or dissertation ever published on North American anurans; as Ernie Liner once told me "You can't get them all!" I have tried to cite all pertinent papers, however, and of course I take full responsibility for oversights. Readers are urged not to take my word for the brief summaries provided in the species accounts, but to verify information in which they are interested. Many obscure publications are available on the Internet, but by no means all. Just because information is not readily available does not mean it should be ignored or that it is not important or insightful.

Measurements, Precision, and Generalizations

Measurements of body size, weight, distance, and area have been converted to metric units throughout the book as appropriate. Values for toxicological effects are largely given in LC_{50} values, and these generally are based on exposures for 24–96 hrs. Amphibian biologists have begun to recognize that short-term LC_{50}s are grossly inadequate to assess the effects of toxic substances on anurans. Toxic substances may have both lethal and sublethal effects (on activity, feeding behavior, survivorship when faced by predators, body mass, larval period, postmetamorphic fitness) that cannot be measured using traditional short-term toxicological protocols. Toxics may affect eggs, hatchlings, larvae, and postmetamorphs differently and in paradoxical ways (lower concentrations may actually have more serious consequences than higher doses; eggs may be far less prone to toxic effects than late-stage larvae). Results may vary between even closely related species, and the effects or lack of effects on one species cannot be extrapolated to another species. Clearly, toxicological studies need to be more rigorous.

While combing through literally thousands of research papers, it became evident that authors frequently were imprecise in the terminology they used, a situation that has sometimes led to apparent contradictions in the literature. In order to clarify and standardize information, I have attempted to interpret what the authors meant when confusion or contradiction seemed apparent. Below, I provide several examples of such confusion. As these examples demonstrate, precise meaning is important.

• *Calling versus oviposition.* Authors frequently confuse calling with breeding (oviposition). Many species call early and continue to call well after oviposition has ended. In fact, many species are opportunists with short breeding bouts over an extended period of months. Just because a species was heard calling does not mean oviposition occurred.

• *Breeding season.* Some authors assume breeding occurs continuously if records are available in spring and autumn ("I heard calls in March and October, therefore the breeding season extends March to October"). In fact, some species have a biphasic breeding cycle (spring and autumn), and the strength of the biphasic cycle may vary with latitude. For example, the breeding season may be biphasic in southern Arizona but monophasic during midsummer in northern Colorado. Researchers who work on single populations often fail to recognize and appreciate this distinction.

• *Clutch size versus fecundity.* Authors often confuse clutch size with fecundity when reporting egg counts. Clutches may be oviposited minutes, days, weeks, or even months apart. Total fecundity (total number of eggs deposited during a breeding season) may actually be a compilation of many bouts of egg deposition. Thus, some species appear to have widely varying "clutch sizes" when the authors actually were reporting on different phenomena. Uncertainty about what constitutes a clutch occasionally influences conclusions about whether a species oviposits multiple clutches.

• *Time to sexual maturity and larval duration.* Authors often use different methods of reporting these time-based traits. For example, one author might say "the time to maturity is two years" whereas another might say "the time to maturity is three years," yet both authors apparently mean the same thing. For example, a frog metamorphosing in August 2011 might not reach maturity until June 2013, which is a period of nearly two years, although the frog will be in its third year of life. Authors often fail to make this distinction; some might report two years whereas others would report three years for the larval duration. Time confusion frequently leads to apparent contradictions in the literature.

In preparing this volume, I endeavored to determine where data originated, especially when recounted in general field guides. I quickly discovered that the basis for many reported life history traits rested on little empirical data; once a "fact" got into the literature it tended to be repeated. Unfortunately, sample sizes were often very small and data originated from a single population or study. In one species, for example, the clutch size was repeatedly said to be 600–800 eggs in book after book. I soon discovered that the actual number was based on counts of one ovary from two individuals, then doubled to get counts of 684 and 760 (thus, 600–800). Such imprecision makes it difficult to understand life history evolution or build predictive models. Researchers need to understand the basis and limitations of data used to test hypotheses, or their meta-analyses will not be of much value.

Although much statistical analysis and theory is based on variance, many researchers still do not appreciate its importance and are quick to make generalizations. The validity of models and data interpretation depends on understanding variance. Life history traits of North American frogs are poorly known. Season, year, sex, climate, physiography, latitude, and elevation all contribute to the local population dynamics, status, and trends of North American frogs. Just because a species has a larval duration of 2–3 yrs in Ohio, for example, does not mean it does so in Louisiana or Québec (think of American Bullfrogs).

All too often, biologists consciously or unconsciously embrace the concept of ecological typology whereby if a species does something in one location, it must do the same thing elsewhere. Such narrow thinking needs to be countered by comprehensive long-term studies in a variety of locations and under varying environmental conditions. Amphibian biologists need to understand whether observed variance is based on sampling errors and biases, resource-based variation, genetics, or landscape differences. Unfortunately, such work is often tedious and consigned to "natural history" that does not immediately lead to generalizations suitable for publication in prestigious journals. Researchers also should not be hasty to suggest inaccuracy in the work of others when results do not coincide.

As the reader will soon learn when using this work, an initial impression of a great amount of information quickly fades to a realization about how so little is known about the basic natural history of North American frogs. Natural-history study has lost much of its following as a result of the impact of genetics, population modeling, meta-analysis, and a shift in focus to short-term experimental studies with high potential for research funding. Regardless, natural-history data form the basis upon which hypotheses are conceived and conservation planned. I hope that this volume restimulates interest in the natural history of anurans before many of these species disappear from our land- and soundscapes.

FOR FURTHER INFORMATION

No single reference can cover all information available on frogs. Additional information can be obtained through the following sources.

Books

Bailey, M.A., J.N. Holmes, K.A. Buhlmann, and J.C. Mitchell. 2006. Habitat Management Guidelines for Amphibians and Reptiles of the Southeastern United States. Partners in Amphibian and Reptile Conservation, Technical Publication HMG-2, Montgomery.

Conant, R., and J.T. Collins. 1998. A Field Guide to the Reptiles and Amphibians. Eastern and Central North America. 3rd exp. ed., Houghton Mifflin, Boston.

Dodd, C.K., Jr. 2009. Amphibian Ecology and Conservation. A Handbook of Techniques. Oxford University Press, Oxford.

Dorcas, M., and W. Gibbons. 2011. Frogs. The Animal Answer Guide. Johns Hopkins University Press, Baltimore.

Duellman, W.E. 1986. Biology of Amphibians. McGraw Hill, New York.

Elliott, L., C. Gerhardt, and C. Davidson. 2009. The Frogs and Toads of North America. Houghton Mifflin Harcourt, Boston. [contains a CD with frog breeding calls]

Hillman, S.S., P.C. Withers, R.C. Drewes, and S.D. Hillyard. 2009. Ecological and Environmental Physiology of Amphibians. Oxford University Press, Oxford.

Kingsbury, B., and J. Gibson. 2002. Habitat Management Guidelines for Amphibians and Reptiles of the Midwest. Partners in Amphibian and Reptile Conservation, Technical Publication HMG-1, Montgomery.

Lannoo, M.J. 2005. Amphibian Declines. The Conservation Status of United States Species. University of California Press, Berkeley.

McDiarmid, R.W., and R.G. Altig. 1999. Tadpoles. The Biology of Anuran Larvae. University of Chicago Press, Chicago.

Mitchell, J.C., A.R. Breisch, and K.A. Buhlmann. 2006. Habitat Management Guidelines for Amphibians and Reptiles of the Northeastern United States. Partners in Amphibian and Reptile Conservation, Technical Publication HMG-3, Montgomery.

Pilliod, D.S., and E. Wind. 2008. Habitat Management Guidelines for Amphibians and Reptiles of the Northwestern United States and Western Canada. Partners in Amphibian and Reptile Conservation, Technical Publication HMG-4, Montgomery.

Semlitsch, R.D. 2003. Amphibian Conservation. Smithsonian Books, Washington, DC.

Society for the Study of Amphibians and Reptiles. 1963–2012. Catalogue of American Amphibians and Reptiles. [individual species accounts; see website below]

Stebbins, R.C. 1962. Amphibians of Western North America. 2nd printing. University of California Press, Berkeley.

Stebbins, R.C. 2003. A Field Guide to Western Amphibians and Reptiles. 3rd ed. Houghton Mifflin, Boston.

Stebbins, R.C., and N.W. Cohen. 1995. A Natural History of Amphibians. Princeton University Press, Princeton.

Wells, K.D. 2007. The Ecology & Behavior of Amphibians. University of Chicago Press, Chicago.

Wright, A.H., and A.A. Wright. 1949. Handbook of Frogs and Toads. 3rd ed. Comstock Publishing, Ithaca, New York. [reprinted 1995]

Internet Sites

These websites contain a wealth of information, and some have sound files so that listeners can hear breeding calls.

Altig, R., R.W. McDiarmid, K.A. Nichols, and P.C. Ustach. 2011. Tadpoles of the United States and Canada. A Tutorial and Key.
http://www.pwrc.usgs.gov/tadpole/

Amphibian Research and Monitoring Initiative (ARMI). U.S. Geological Survey.
http://armi.usgs.gov/

Amphibian Species of Canada. University of Guelph, Ontario.
http://www.carcnet.ca/

AmphibiaWeb. Museum of Vertebrate Zoology, University of California–Berkeley.
http://amphibiaweb.org/

California/Nevada Amphibian Population Task Force. c/o Dr. David Bradford.
http://www.canvamphibs.com/

Canadian Amphibian and Reptile Conservation Network.
http://www.carcnet.ca/

Center for North American Herpetology.
http://www.cnah.org/

Frost, D.R. 2011. Amphibian Species of the World. An Online Reference. Version 5.5 (31 January 2011). American Museum of Natural History, New York.
http://research.amnh.org/vz/herpetology/amphibia/

North American Amphibian Monitoring Program (NAAMP).
http://www.pwrc.usgs.gov/naamp/

Partners in Amphibian and Reptile Conservation (PARC).
http://www.parcplace.org/

Save the Frogs! c/o Dr. Kerry Kriger, Santa Cruz, CA.
http://www.savethefrogs.com/

Herpetological Atlases

Herpetological atlases are being developed in many states and provinces to record and monitor the distribution of amphibians within their region. Different organizations may take the lead on atlas projects, including local herpetological societies, museums, universities, or state and provincial wildlife agencies. The URLs of these projects frequently change, so when looking for a state or provincial atlas it is easiest to conduct a quick Internet search using "[State or Province] Herp Atlas" to find the latest web pages. At the time of writing (March 2011), some atlas websites include:

Alabama: http://www.ag.auburn.edu/~guyercr/ahap/
Colorado: http://ndis.nrel.colostate.edu/herpatlas/coherpatlas/
Kansas: http://webcat.fhsu.edu/ksfauna/herps/
Manitoba: http://www.naturenorth.com/Herps/Manitoba_Herps_Atlas.html
Massachusetts: http://www.massherpatlas.org/data_collection/sighting.php
Missouri: http://atlas.moherp.org/
New York: http://www.dec.ny.gov/animals/7140.html
Ontario: http://nhic.mnr.gov.on.ca/herps/ohs_site_map.html
Québec: http://redpath-museum.mcgill.ca/Qbp/herps/herps.html
Tennessee: http://apbrwww5.apsu.edu/amatlas/
Washington: http://www1.dnr.wa.gov/nhp/refdesk/herp/index.html

Wisconsin: http://www4.uwm.edu/fieldstation/herpetology/atlas.html

Sound Recordings

Many Internet sites contain links to anuran advertisement calls. Of particular importance is the Macaulay Library of Natural Sounds at the Cornell Laboratory of Ornithology (http://macaulaylibrary.org/browse/scientific/10521002). There are also a variety of compact disks available. The list below certainly does not contain all such disks, but it will give the reader an idea of the recordings currently accessible.

The Frogs and Toads of North America. Lang Elliott, Carl Gerhardt, and Carlos Davidson. 2009. Houghton Mifflin Harcourt.

Frog Talk. 1990. NorthSound/NorthWood Press, Inc.

Frog and Toad Calls of the Rocky Mountains. Carlos Davidson, Library of Natural Sounds, Cornell Laboratory of Ornithology.

Frog and Toad Calls of the Pacific Coast. Carlos Davidson, Library of Natural Sounds, Cornell Laboratory of Ornithology.

West Central Florida's Frogs and Toads. 2001. Hillsborough River Greenways Task Force.

Calls of the Wild, Vocalizations of Georgia's Frogs. Georgia Department of Natural Resources.

Field Recordings of Maine Frogs and Toads. University of Maine Press.

Michigan Frog Calls. Stan Tekela, Adventure Publications.

Minnesota Frog Calls. Stan Tekela, Adventure Publications.

Mississippi Frog Songs. Mississippi Museum of Natural Science.

Calls of New Jersey Frogs and Toads. New Jersey Division of Fish and Wildlife.

The Frogs and Toads of North Carolina. Michael Dorcas, Davidson University.

Vocal Calls of Ohio Frogs and Toads. 2002. Ohio Biological Survey.

Frog and Toad Calls of South Dakota. South Dakota Department of Game, Fish and Parks.

Wisconsin Frogs. 2001. Wisconsin Audubon, Inc.

Professional Herpetological Societies in North America

American Society of Ichthyologists and Herpetologists (ASIH)
http://www.asih.org/

Canadian Association of Herpetologists (CAH/ACH), c/o Dr. Jacqueline D. Litzgus, Department of Biology, Laurentian University, 935 Ramsey Lake Road, Sudbury, Ontario P3E 2C6 Canada; email: JLitzgus@laurentian.ca

Herpetologists' League (HL)
http://www.herpetologistsleague.org/en/index.php

Society for the Study of Amphibians and Reptiles (SSAR)
http://www.ssarherps.org/

Abbreviations

AT	air temperature	mya	million years ago
BP	before present	NALMA	North American Land Mammal Age
ca.	approximately	nDNA	nuclear DNA
cm	centimeter	ng	nanogram (one billionth of a gram)
cps	cycles per seconds	nm	nanometer
CTmax	maximum critical temperature	PCBs	polychlorinated biphenyls
CTmin	minimum critical temperature	PCi g	picogram (one trillionth of a gram)
DOC	dissolved oxygen content	PCR	polymerase chain reaction
DOR	dead on road	pH	a measure of hydrogen ion concentration and hence, acidity
F_1	first filial generation		
g	grams	PO_2	partial pressure of oxygen
ha	hectares (= 2.2 acres)	ppb	parts per billion
hr, hrs	hour, hours	ppm	parts per million
Hz	hertz (a measure of sound frequency)	ppt	parts per thousand
J	joules (a measure of light intensity)	sec	second, seconds
kHz	kilohertz (a measure of sound frequency)	SUL	snout-urostyle length
		T_b	body temperature
km	kilometers (0.6 miles)	THz	terahertz, a measure of extremely high frequency light waves
LC_{50}	lethal concentration at which 50% of the test animals die		
		TL	total length
m	meter, meters	TL_{50}	lethal concentration at which 50% of the test animals die
m^2	square meters		
mg/L	milligrams per liter	µg/L	micrograms per liter
mhos	a unit of measurement equal to 1 ampere per volt (a measure of conductivity)	UVB	ultraviolet light in the beta (10–400 nm) wavelengths; these short, high frequency wavelengths are invisible to humans
ml/L	milliliters per liter	yr, yrs	year, years
mm	millimeter, millimeters	♀	female
ms	milliseconds	♂	male
mtDNA	mitochondrial DNA	‰	parts per thousand
my	million years		

Frogs of the United States and Canada

Family Ascaphidae

Ascaphus montanus
Mittleman and Myers, 1949
Rocky Mountain Tailed Frog

ETYMOLOGY

montanus: from the Latin *montanus* meaning 'belonging to a mountain.'

NOMENCLATURE

Synonyms: *Ascaphus truei montanus*

Nearly all information pertaining to this species has been published under the name *Ascaphus truei*.

IDENTIFICATION

Adults. Dorsal coloration is light tan, reddish brown, medium to dark brown, or deep blackish brown with a tinge of green. Light spots may be present dorsally, but some individuals are uniformly unpatterned. Undersides are grayish white, yellowish white, or pinkish. Faint black bands may be evident from the shoulder through the eye and nostril to the upper jaw in some populations, whereas others have a light area that extends across the head, anterior to the eyes. A light mid-dorsal line may or may not be present. There is no tympanum. The eyes are large, and the pupils are oval to elliptical and vertical. Black spots may be present on the sides, back, and upper portions of the limbs, giving the species a rugose appearance. Male secondary sex characteristics (spines within the cloaca, horny black nuptial pads on front limbs and chest, rim of warts on the chin, engorged tail) are evident only during certain times of the year (Metter, 1964a; Black and Black, 1968). Minor variation in hind foot width, the extent of webbing between toes, head width, and internasal distance may be observed among populations (Metter, 1964a). Although they cannot be observed directly, *Ascaphus* have short ribs, unlike all our other frogs.

The most striking feature of the tailed frogs is the possession of the "tail," really a copulatory organ that is an extension of the cloaca and is found only in males. The tail is controlled by two unique muscles, the pyriformis and the caudalipuboischiotibiales, which allow it to be turned forward for insertion into the female's cloaca. Sperm is transferred along the surface of the tail allowing internal fertilization in the fast flowing water currents of turbulent northwestern streams. Because of the tail, determination of sex is possible immediately after transformation, although maturity may not be attained for an additional 4–5 yrs.

Females are larger than males. In Oregon, males were 22.1–36.2 mm SUL (mean 30.3 mm) and females 22.3–45.1 mm SUL (mean 30.9 mm) from mainstream habitats, whereas males were 22–34.2 mm SUL (mean 26.3 mm) and females 22.2–36.2 mm SUL (mean 26.3 mm) from small tributary streams (Landreth and Ferguson, 1967). In Montana, sexually mature males were 40–52 mm SUL and females 44–56 mm SUL (Daugherty and Sheldon, 1982a).

Larvae. Larvae are easily identified by their large sucker-like mouth. At hatching, larvae are dark but with a large amount of yolk. Normal larval coloration is slate gray to black or occasionally reddish brown. The dorsum and sides of the tail are patterned variably with light and dark markings. Tadpoles usually have a small cream or white spot at the terminus of the tail, but occasionally the spot may be large or even absent. Franz (1971) noted larvae from Mineral County, Montana, that had reduced pigmentation on the body and tail, giving a very mottled appearance; such tadpoles also had large orange tail spots.

Venters are white and flattened. The spiracle (external opening for the exit of respiratory water) is ventromedial rather than on the left side as in other North American anurans. Tadpoles are fully pigmented by 18 mm TL, but the yolk is not fully absorbed until 20–21 mm TL. Tadpoles reach 64 mm TL (Franz, 1971). Larvae are illustrated by Franz (1971), and Metter (1964a) has photographs of larvae at various sizes.

Eggs. Eggs are creamy white to yellowish, unpigmented, and measure 4 mm in diameter. They are oviposited in rosary-like strings. Most eggs have a large (1.2–1.8 mm) translucent spot, a product of light refraction caused by formation of a cavity beneath the surface of the egg (Franz, 1970). Eggs have two gelatinous capsules, with a third capsule forming the egg string. Franz (1970) provided an illustration of the eggs.

DISTRIBUTION

Ascaphus montanus occurs in the northern Rocky Mountains of extreme southeastern British Columbia, western Montana, northern Idaho, northeastern Oregon, and southeastern Washington (Rocky, Blue, Wallowa, and Seven Devils mountains). Important distributional references include Linsdale (1933b), Slater (1941b, 1955), Metter (1960), Black and Black (1968), Black (1970), Franz and Lee (1970), Franz (1971), Nussbaum et al. (1983), McAllister (1995), Marnell (1997), Nielson et al. (2001), Maxwell et al. (2003), Werner et al. (2004), Jones et al. (2005), and Matsuda et al. (2006).

FOSSIL RECORD

No fossils are known.

SYSTEMATICS AND GEOGRAPHIC VARIATION

The genus *Ascaphus* comprises the monotypic basal lineage of all anurans and retains many primitive morphological characters (Ford and Cannatella, 1993). The diploid chromosome number is 46, and *Ascaphus* has the largest number of chromosomes of all diploid anurans (Green et al., 1980). Mittleman and Myers (1949) found small, but what they considered noteworthy, morphological variation among *Ascaphus* populations in the coastal Pacific Northwest, Rocky Mountains, and northern California. They divided the species into subspecies based on these differences (*truei, montanus, californicus*). *Ascaphus truei* and *A. montanus* are now considered distinct species, but recognition of *A. t. californicus* is not justified based on molecular data. Using a randomly amplified polymorphic DNA (RAPD) survey, Ritland et al. (2000) noted the uniqueness of inland populations but did not at that time recommend species recognition.

Ascaphus montanus is distinctive, based on analyses of mitochondrial DNA (Nielson et al., 2001), skin peptides (Conlon et al., 2007b), allozymes (Daugherty, 1979), and morphology (Mittleman and Myers, 1949; Pauken and Metter, 1971; but see Metter, 1967). This species has been isolated from its nearest relative, *A. truei*, since the late Miocene to early Pliocene, probably in response to the Cascade Mountain orogeny. As such, suggestions of a Pleistocene connection between these species are not tenable (Metter and Pauken, 1969; Pauken and Metter, 1971; Nielson et al., 2001). Additional genetic structuring among populations reflects isolation in the late Pliocene to early Pleistocene. Closely related haplotypes among *A. montanus* populations reflect expansion following glaciation.

Not surprisingly, Metter (1967) found considerable variation among populations of "*A. truei*" from throughout its range. Although he did not think so at the time, two species were involved in his analysis. Still, there appears to be a great deal of morphological and color pattern variation within the two species.

Distribution of *Ascaphus montanus*

Eggs of *Ascaphus montanus*. Photo: Kirk Lohman

ADULT HABITAT

The Mountain Tailed Frog is found in high-gradient headwater streams in mesic to moist coniferous forest, and prefers old-growth and mature habitats. In the Palouse region of Idaho, *A. montanus* may be found in slower-moving streams with a sandier substrate (Metter, 1964a). As with *A. truei*, the species requires cold-flowing water free of siltation. Streams should be rocky bottomed with cobbles and boulders and little aquatic vegetation; streams with sandy, gravel, or organic debris as substrates are not inhabited except as noted above (Franz and Lee, 1970). Streams with rocks that are flat (slabs) are preferred over streams with small irregular smooth-edged or round rocks. *Ascaphus montanus* is found in pools less often than *A. truei* (Karraker et al., 2006). The area of the basin is important, as Rocky Mountain Tailed Frogs are most often found in basins 0.3–100 km^2 in area, with maximum numbers in basins to 35 km^2 (Dupuis and Friele, 2006). Basins of this size result in ideal fluvial conditions for tailed frogs and their larvae, such as runoff, sedimentation, physiography, and temperature.

Ascaphus montanus prefers streams with a canopy overstory that provides protection against high temperatures. Frequent riparian trees include dogwood (*Cornus stolonifera*), alder (*Alnus incana*), western red cedar (*Thuja plicata*), hemlock (*Tsuga heterophylla*), fir (*Abies grandis*), mountain pine (*Pinus monticola*), and larch (*Larix occidentalis*). Vegetation may vary among sites (Metter, 1964a). If a dense canopy is not present, the species will be absent. Shaded streams have abundant diatoms on which larvae feed. Franz and Lee (1970) provided a list of aquatic insects inhabiting *Ascaphus montanus* streams.

Even in favorable habitat, Rocky Mountain Tailed Frogs may not occur in every stream. In the Flathead River Basin of British Columbia, for example, they were found in only 39% of the streams sampled; in the Yahk region, they occurred in 80% of the streams surveyed (Dupuis and Friele, 2006). In Glacier National Park, *A. montanus* is found from 1,045 to 2,140 m (Marnell, 1997), whereas in the Flathead River Drainage it is found at elevations of 914–1,829 m (Franz and Lee, 1970). The highest elevation in Montana is 2,557 m (Werner et al., 2004). The species does not occur < 914 m.

TERRESTRIAL AND AQUATIC ECOLOGY

The Rocky Mountain Tailed Frog is a mostly aquatic species of the cool moist forests of the northern Rocky Mountains. Adults may leave the water and wander terrestrially far from streams, although most individuals stay in proximity (within 12 m) to streamside habitats, where they feed along creek banks and on the nearby forest floor. Some terrestrial activity occurs both night and day from May to early October, at least in western Montana (Daugherty and Sheldon, 1982b; Werner et al., 2004). Overwintering occurs in the sand bottom or under large rocks. During spring flooding, frogs move to quiet areas, away from the strong current (Black and Black, 1968). Aquatic activity may begin earlier in the season, even when streams are snowbound. Most terrestrial activity occurs within the first several hours after dark.

There are sometimes contradictory statements in the literature concerning movement patterns of *Ascaphus*, with some authors reporting little movement, some reporting movement up- or downstream but not both, and others reporting movements in both directions. As pointed out by Adams and Frissell (2001), differences in movement patterns likely reflect different thermal regimes among locations. For example, Daugherty and Sheldon (1982b) noted that *A. montanus* juveniles usually remained along stream margins after metamorphosis, with older individuals (5–8 yrs) gradually extending their range of activity. By maturity, frogs began to move less, with most activity confined to about 20 m along a stream. Males and females exhibited the same level of philopatry both within and between seasons (Daugherty and Sheldon, 1982b). The greatest distance moved was 360 m over a 12.5-month period, and long distance movements up- or downstream occurred only prior to maturity.

Daugherty and Sheldon (1982b) measured the neighborhood size of *A. montanus* as ranging from 20 to 350 m. In another area of Montana, *A. montanus* appears to make seasonal migrations downstream of up to 3 km in response to changing thermal conditions (Adams and Frissell, 2001).

Displaced Rocky Mountain Tailed Frogs are capable of returning to their release site using visual rather than olfactory or geotactic cues (Landreth and Ferguson, 1967). Under natural conditions, rheotactic cues also are used. Tailed frogs also can use celestial (sun) cues via a Y-axis orientation mechanism to locate a home shore. These results suggests that sun-compass orientation appeared early in anuran evolution. Movements occur between the main stream channel and tributary streams; rheotactic cues are used in stream movements, whereas the celestial cues are used to reorient during terrestrial activity.

Desiccation is a potential problem for this species since it tends to lose water even when on a moist substrate, and in experimental conditions tailed frogs prefer water over moist substrate (Claussen, 1973a). Tailed frogs can only lose up to ca. 39% of their body water, and they rehydrate more slowly than other species. The species does not like warm temperatures, and its CTmax is only 27.6–29.6°C (Claussen, 1973b). However, Adams and Frissell (2001) found *A. montanus* in water as high as 21°C. These authors noted that streams had complex thermal profiles,

Tadpole of *Ascaphus montanus*. Photo: Kenny Wray

Adult male *Ascaphus montanus*. Note the "tail." Photo: Kirk Lohman

especially where cold springs entered the channel, and that *A. montanus* appeared to be able to use behavioral thermoregulation to migrate seasonally and thus avoid warm stream temperatures. Overwintering by adults and larvae occurs under large rocks in stream beds (Bull and Carter, 1996).

Rocky Mountain Tailed Frogs are monotonically photonegative in their phototactic response to white light, suggesting that they prefer to avoid daylight (Jaeger and Hailman, 1973). They show a U-shaped spectral response and as such cannot be demonstrated to use color vision in phototaxis (Hailman and Jaeger, 1974, 1978). Hailman and Jaeger (1978) term the U-shaped spectral response the "green antimode response," because its peak is at 589 THz.

CALLING ACTIVITY AND MATE SELECTION

This species has no advertisement call or tympanum. Amplex has been observed 2 hrs after dark in September in Glacier National Park (Marnell, 1997). Mating behavior is similar to that reported for *A. truei* (Metter, 1964b).

BREEDING SITES

Oviposition usually occurs in shallow tributaries of main stream channels. Like *A. truei*, the species deposits its eggs under large rocks within the stream, where they will be sheltered from the current.

REPRODUCTION

Most references state that breeding occurs in September and October, with oviposition the following summer in June to August (Metter, 1964a, 1964b; Daugherty and Sheldon, 1982a). Werner et al. (2004), however, mentioned spring breeding, and Matsuda et al. (2006) suggested that mating occurs year-round. Karraker et al. (2006) reported oviposition from mid-June to mid-July. As such, the female Rocky Mountain Tailed Frog is capable of storing sperm for a considerable time, up to 2 yrs (Metter, 1964b). Eggs are deposited in double strands in nests on the underside of boulders and rocks within the stream channel. The outer membranes of the two strands are firmly attached to one another. Egg masses are globular and are held suspended in water-filled depressions with some current, which keeps the eggs oxygenated; they are placed in nests under very large rocks unlikely to be disturbed by the current.

The clutch size of *A. montanus* may be larger than *A. truei* (mean 66.6 vs. 41.9) (Karraker et al., 2006). In Karraker et al.'s (2006) summary, clutch sizes were 7–200, but the larger clutches likely represent multiple clutches and the smaller values incomplete clutches; most clutch sizes were 40–80 eggs. Clutch size in 2 Montana clutches was 64 and 86 eggs with a larval period of >46 days (Franz, 1970). Metter (1964a) reported clutch sizes of 50–85 (mean 68), whereas Werner et al. (2004) give the clutch size as 30–90 eggs. In dissected females, total egg complement varies from 388 to 1,360 eggs (Karraker et al., 2006). Within-egg development requires six to eight weeks. At hatching in August to September (ca. 30 days), larvae are 12.5–15 mm TL (Noble and Putnam, 1931; Metter, 1964a; Franz, 1970). Metter (1964a) suggested that eggs are deposited every other year.

LARVAL ECOLOGY

Larvae remain in the nest cavity until the yolk is fully absorbed. They are filter feeders, eating diatoms almost exclusively, which they scrape off rocks, as well as pollen in midsummer. They attach themselves to rocks with a highly suctorial oral apparatus that enables them to feed while moving on rocks, despite strong currents. Larvae even move above the water's surface at night on wet rocks within the spray zone of waterfalls, splashing water by using their suctorial mouthparts.

Larvae can only be found in cold water (<16°C) and will move easily and feed in water 1–2°C (Franz and Lee, 1970; Franz, 1971). Overwintering occurs in ice- and snow-covered streams. Larvae also tend to be found in streams with low alkalinity. Marnell (1997) suggested that the larval period might be 3–5 yrs at the higher elevations of Glacier National Park. In other areas, larvae have legs by 2 yrs of age, but transformation does not occur until 3 yrs (Metter, 1964a; Daugherty and Sheldon, 1982a) or perhaps longer. The process of transformation takes 60 days and is usually completed by September. However, some transformed individuals may have a short tail remnant until the following spring.

DIET

Ascaphus montanus eats a wide variety of invertebrates, the composition of which changes seasonally with the abundance of prey. Specific items include snails, spiders, ticks and mites, centipedes, collembolans, adult and larval stoneflies, adult and larval caddisflies, other types of flies, adult and larval beetles, ants, adult and larval lepidopterans, grasshoppers, true bugs, and other miscellaneous species (Metter, 1964a). Feeding occurs both under water and terrestrially.

PREDATION AND DEFENSE

When disturbed in the water, Rocky Mountain Tailed Frogs are sluggish and assume a posture whereby the limbs are folded into the body. This immobile posture allows them to ride currents downstream with a minimal chance of detection (Metter, 1964a). At night, *A. montanus* is agile terrestrially and able to escape by jumping into the water. Larvae and adults are eaten by the salamander *Dicamptodon tenebrosus* (Metter, 1963), garter snakes (*Thamnophis* spp.) (Metter, 1964a; Marnell, 1997), and trout (*in* Metter, 1964a).

POPULATION BIOLOGY

The Rocky Mountain Tailed Frog has an extended larval period, with reproductive maturity not attained until 7–8 yrs. Females are first observed with eggs at 44–48 mm SUL; 7 yr males are 40–46 mm SUL. Longevity may be as long as 14 yrs (3 as larvae, 11 as postmetamorph) under natural conditions (Daugherty and Sheldon, 1982a). Sex ratios may be even or female-biased, but few data are available. In Metter's (1964a) study, one population had a 1:1 sex ratio, whereas the other had 1 male per 1.7 females.

Genetic diversity appears to be high among populations of *A. montanus*, regardless of

Habitat of *Ascaphus montanus*. Photo: Steve Corn

habitat disturbance, at least in north-central Idaho (Spear and Storfer, 2010). Genetic differentiation actually increased in timber-harvested regions, and overall genetic diversity has not as yet been reduced. In this regard, gene flow is still maintained along riparian corridors, unlike with *A. truei* (Spear and Storfer, 2008). The extent of gene flow in *A. montanus* depends on the type of habitat disturbance, with suggestions that gene flow continues overland in fire-impacted areas, whereas it is obligately confined to riparian areas in harvest-fragmented habitats. Prior history is thus important in explaining population genetic responses to recent stressors.

Flash floods may cause temporary loss or reduction in the number of tadpoles or juveniles present along a stream (Metter, 1968). Severe flooding may cause physical injuries to adults. It appears that such flooding may affect population structure locally for a time, but probably has no long-term effects as populations recover as the riparian habitats recover. Densities among populations vary considerably, and it is possible that previous disturbances may influence perceived differences in adult density.

DISEASES, PARASITES, AND MALFORMATIONS
The ciliate protozoan *Protoopalina* is found in tadpoles (Metcalf, 1928), with abundance decreasing with age. Unidentified trematode metacercariae were reported from under the skin of larvae and adults (Metter, 1964a), and the trematode *Bunoderella metteri* later was described from this species (Schell, 1964; Anderson et al., 1965). Other trematodes include *Euryhelmis pacifica* and *E. squamula* (Schell, 1964). Waitz (1961) reported no endoparasites from *A. montanus*.

SUSCEPTIBILITY TO POTENTIAL STRESSORS
pH. Streams with a pH >7.7 are not inhabited by *A. montanus* larvae (Franz and Lee, 1970).

STATUS AND CONSERVATION
The main threat to this species is clearcut logging and others forms of habitat alteration affecting headwater streams. Wildfires may alter stream characteristics and thus affect Rocky Mountain Tailed Frogs. For example, tadpole abundance in streams affected by wildfires was half of that in unaffected streams in northwestern Montana 2 yrs after the fires (Hossack et al., 2006b). Although the reason for the decrease was not ascertained, increased temperature and ammonium concentrations were thought to be involved. Despite the decrease, the fires were not expected to have long-term consequences on the status of affected *A. montanus* populations, and indeed, Spear and Storfer (2010) suggested that fire was not a hindrance to gene flow overland across previously burned habitat. In contrast, timber harvesting had a much more dramatic effect, with gene flow confined to riparian corridors. As a result, Spear and Storfer (2010) recommended the maintenance of riparian corridors in areas affected by timber harvest.

This species is considered common within its range (Werner et al., 2004; Jones et al., 2005). However, it is a secretive species. Recent sampling innovations have allowed detection of this species using environmental DNA, a technique that may revolutionize the monitoring of rare species in aquatic habitats (Goldberg et al., 2011).

Ascaphus truei Stejneger, 1899
Coastal Tailed Frog

ETYMOLOGY
truei: a patronym honoring Frederick W. True (1858–1914), who was head curator at the Department of Biology, U.S. National Museum. True later became Assistant Secretary of the Smithsonian Institution.

NOMENCLATURE
Synonyms: *Ascaphus truei californicus*

IDENTIFICATION
Adults. Dorsal color variation is variable, with various shades of brown, olive brown, rose to brick red, pink, gray, and cinnamon to almost black. Patterns may be obscured in very light or dark individuals. The head is flat, slightly broader than long, and has a dark almost triangular

marking. The eyes are large and the pupils are oval to elliptical and vertical. There is no tympanum. There is a dark glandular ridge or rows of glands on the side; these are dark, tipped with golden yellow. Warts are present on the sides and legs and are cinnamon to buff. Venters are whitish with a slight clouding of mustard yellow, especially across the chest. Females are more brightly colored than males. Breeding males have very enlarged forearms and inner palmar tubercle, a white horny patch of the forearm where the palmar tubercle touches it; several areas on the forelimbs are covered in black spines. Although they cannot be observed directly, *Ascaphus* have ten vertebrae and short ribs, unlike all our other frogs. Van Denburgh (1912) provides additional details on morphology.

The most striking feature of the tailed frogs is the possession of the "tail," really a copulatory organ that is an extension of the cloaca and is found only in males. The tail is controlled by two unique muscles, the pyriformis and the caudalipuboischiotibiales, which allow it to be turned forward for insertion into the female's cloaca. Sperm is transferred along the surface of the blood-engorged tail allowing internal fertilization in the fast flowing water currents of turbulent northwestern streams. Because of the tail, determination of sex is possible immediately after transformation, although maturity may not be attained for many additional yrs.

Females are slightly larger than males. In Washington, females were 33–50 mm SUL (mean 42.1 mm), whereas males were 29–40 mm SUL (mean 35.4 mm) (Gaige, 1920). In northern California, adult males average 36.7 mm SUL, whereas females average 44.4 mm SUL (Burkholder and Diller, 2007). Tails of males are 3–9 mm (mean 5.4 mm) in length, 4 mm in width, and 3.5 mm in thickness at the base (Van Denburgh, 1912).

Larvae. Larval *A. truei* are readily identified by their large sucker-like oral disks. At hatching, they are translucent white and retain a considerable amount of yolk. Pigmentation takes several weeks to develop. Mature larvae are round, black to dark brown, with a speckled black pattern. No iridescence is present. Tails are long (twice the length of the body) and either dark or spotted with a creamy white coloration. Tail fins are not conspicuous and not translucent. Tail tips may be white with a dark band just behind it. The light tail tip coloration may even be rose or flame, making them conspicuous. Mouths are very large and round. The spiracle (external opening for the exit of respiratory water) is ventromedial rather than on the left side, as in other North American anurans. The highly specialized tadpole's feeding adaptations are discussed by Altig and Brodie (1972). Maximum size is 56–60 mm TL (Nussbaum et al., 1983; Bury and Adams, 1999). Hatchlings and tadpoles at various stages of development are illustrated by Brown (1990).

Eggs. The eggs are white and unpigmented at deposition. They average 6.1 mm in diameter (range 4.9 to 7.1 mm) (Adams, 1993), although Gaige (1920) reported a mean vitellus diameter of 5 mm and an overall diameter of 8 mm. Wernz and Storm (1969) gave a mean diameter of 3–4 mm and noted the vitellus was surrounded by 3 membranes, the outer of which formed a sticky tough tube; these authors described the development of the eggs through hatching. Noble and Putnam (1931) reported freshly deposited eggs had a vitellus of 4 mm and a total diameter of 4.5 mm. Because development takes a long time, the diameter of the egg likely varies with age. Eggs are oviposited in rosary-like strings.

DISTRIBUTION

Ascaphus truei occurs from coastal British Columbia south to northern California and west of the Cascade Mountains. Important distributional references include Slater (1955), Metter (1960), Bury (1968), Nussbaum et al. (1983), Leonard et al. (1993), Jennings and Hayes (1994b), McAllister (1995), Nielson et al. (2001), Jones et al. (2005), and Matsuda et al. (2006).

FOSSIL RECORD

No fossils are reported, although there is a left ilia known from the Wyoming Cretaceous that might belong to the same family as *Ascaphus* (Holman, 2003).

SYSTEMATICS AND GEOGRAPHIC VARIATION

The genus *Ascaphus* comprises the monotypic basal lineage of all anurans and retains many primitive morphological characters (Ford and Cannatella, 1993). The diploid chromosome number is 46, and *Ascaphus* has the largest number of chromosomes of all diploid anurans (Green et al., 1980). *Ascaphus truei*, described

Distribution of *Ascaphus truei*

Mittleman and Myers (1949) found small, but what they considered noteworthy, morphological variation among populations in the coastal Pacific Northwest, Rocky Mountains, and northern California. They divided the species into subspecies based on these differences (*truei, montanus, californicus*). Not surprisingly, Metter (1967) also found considerable variation among populations of "*A. truei*" from throughout its range. Although he did not think so at the time, two species were involved in his analysis. Still, there appears to be a great deal of morphological and color pattern variation within the two species. *Ascaphus truei* and *A. montanus* are considered distinct species, but recognition of *A. t. californicus* is not justified based on molecular data.

ADULT HABITAT

The Coastal Tailed Frog is found in cold high-gradient headwater streams, without filamentous algae, in moderate to wet coniferous forest (Gaige, 1920; Noble and Putnam, 1931; Bury, 1968; Metter and Pauken, 1969; Brown, 1975c; Bury et al., 1991a; Wahbe and Bunnell, 2003). It is well adapted to such habitats, with reduced lungs to minimize buoyancy. The species prefers humid old-growth and mature habitats, but can tolerate young forest as long as protective canopy cover insures cold flowing water, free of siltation (Bury and Corn, 1988b; Welsh, 1990; Aubry and Hall, 1991; Bury et al., 1991a, 1991b; Bosakowski, 1999; Matsuda and Richardson, 2005). Stream substrates should consist of boulders and cobble, but *A. truei* is also found in pools more often than *A. montanus* (Karraker et al., 2006). Higher elevations with a southern exposure are preferred in some areas, as are wet taluses (Aubry and Hall, 1991); in other areas, a northern exposure is preferred (Adams and Bury, 2002). A maritime wet environment allows *A. truei* to descend to lower elevations near the coast than in inland regions.

Even in favorable habitat, Coastal Tailed Frogs may not occur in every stream. In Olympic National Park, for example, they were found in only 60% of the streams sampled (94 of 168) using belt transects by Bury et al. (2001). Nests were only found in five streams. Also in Washington, Hayes et al. (2006c) found Coastal Tailed Frogs in only 19% of streams sampled (25 of 131), with most occurrences in third- and fourth-order

by Stejneger (1899) based on a single specimen, is distinctive from inland populations of *Ascaphus*, based on analyses of mitochondrial DNA (Nielson et al., 2001), skin peptides (Conlon et al., 2007b), allozymes (Daugherty, 1979), life history (Karraker et al., 2006), and morphology (Mittleman and Myers, 1949; Pauken and Metter, 1971; but see Metter, 1967). This species has been isolated from its nearest relative, *A. montanus*, since the late Miocene to early Pliocene, probably in response to the Cascade Mountain orogeny. As such, suggestions of a Pleistocene connection between these species are not tenable (Metter and Pauken, 1969; Pauken and Metter, 1971; Nielson et al., 2001). Additional genetic structuring among populations reflects isolation in the late Pliocene to early Pleistocene as well as the impacts of habitat fragmentation due to timber harvest (Wahbe et al., 2005). For example, Ritland et al. (2000) noted significant genetic diversity between mid/north-coastal and south-coastal populations in British Columbia, using a randomly amplified polymorphic DNA (RAPD) survey.

streams. In first- and second-order streams in managed Redwood forest in California, Diller and Wallace (1999) found *A. truei* larvae in 75% of the streams examined (54 of 72). The species occurs from near sea level to > 1,981 m in elevation (Bury, 1968; Bosakowski, 1999).

TERRESTRIAL AND AQUATIC ECOLOGY

The Coastal Tailed Frog is a species of the cool, moist, mature, and old-growth forests of the Pacific Northwest. Adults frequently leave the water and forage terrestrially, especially along a stream bank. Most terrestrial activity occurs in the autumn (September to October), with little summer activity (Wahbe et al., 2004). Some juveniles and adults even wander overland (Bury and Corn, 1988a), with juveniles being found in clearcuts far more often than adults, who tend to remain in old-growth forest (Wahbe et al., 2004; Matsuda and Richardson, 2005). Most individuals stay in proximity to streamside habitats, where they feed along creek banks and on the nearby forest floor to within 100 m of the stream. In the water, Coastal Tailed Frogs are found in shallow water (mean 4.1 cm in depth) under large rocks (mean cover area of 1,011 cm^2) (Bury et al., 1991b), where they crawl rather than swim across the stream bottom. The thermal and water relations of this species are likely similar to *A. montanus* (Claussen, 1973a, 1973b), indicating a preference for cold mountain streams.

Ascaphus truei prefers specific habitat conditions, although they may be found at lower densities at less favorable sites. For example, in Washington Coastal Tailed Frogs are associated with higher elevations, moderate stream gradients, cobble substrates, substrates covered by 10% coarse woody debris, consolidated substrates, canopy covers >50%, and northern aspects (Bosakowski, 1999; Adams and Bury, 2002). The adjacent forest is often composed of Douglas fir (*Pseudotsuga menziesii*), red cedar (*Thurja plicata*), firs (*Abies amabilis*, *A. procera*), and western hemlock (*Tsuga heterophylla*) with mixed hardwoods such as red alder (*Alnus rubra*).

Ascaphus truei are found at different abundances in various stream segments and stream orders. For example, older life stages are most often found in higher elevation headwater streams than younger life stages. Higher-order streams tend to have more evidence of reproduction than low-order streams, and many first-order streams may be too small to maintain *Ascaphus* populations. *Ascaphus* are also vulnerable to fish predation in lower-order streams, and can be maintained only where there is sufficient flow for at least 600 m above where fish distribution stops (Hayes et al., 2006c). Higher-order stream basins usually have more of such areas than lower-order stream basins.

Large-scale seasonal movements also are evident in some areas, as both larvae and adults move upstream in late summer from positions occupied early in the summer. For example, Hayes et al. (2006c) found that larvae moved a median distance of 733 m upstream by late summer, and adults 406 m upstream, from portions of the stream they occupied earlier in the year. Hayes et al. (2006c) speculated that adults moved downstream before breeding activity and returned upstream immediately afterward. In contrast, Burkholder and Diller (2007) found that most individuals were site philopatric, with little upstream or downstream movement. Movements were usually 0–30 m within the stream channel, with adult females making longer movements than males and immatures of both sexes. Still, occasional adult movements of up to 110 m were noted both up- and downstream.

Coastal Tailed Frogs are likely to be monotonically photonegative in their phototactic response to white light, as is *A. montanus*, suggesting that they prefer to avoid daylight (Jaeger and Hailman, 1973). They likely show a U-shaped spectral response and do not likely use color vision in phototaxis (Hailman and Jaeger, 1974).

CALLING ACTIVITY AND MATE SELECTION

Ascaphus truei does not have a mating call as it would be useless over the roar of the cascades in which it lives. It is not known how the male and female locate each other, although they may move to a favored nest location where they are likely to come into contact. Visual cues are not used, but waterborne chemical cues may assist in locating mates or drawing them to a nest location (Asay et al., 2005).

A male touches a female and holds onto her leg as he slowly works his way into position to grasp her thigh in inguinal amplexus. Axial and midbody attempts at amplexus are unsuccessful. During this time, the female is quiet with her legs extended; her thighs form a channel directed toward her cloaca. The male clasps his hands around the female's body and then turns his "tail" inward and arches his back to insert the tip of the "tail" into the female's cloaca. This occurs about 1.5 hrs following initial amplexus. During this time, the female keeps her nictitating membrane tightly closed as if in a trance and remains immobile. Amplexus lasts from 70 hrs to 7 days under laboratory conditions (Wernz, 1969; Brown, 1975c). After mating, the female dislodges the male and moves rapidly away, although the male may attempt to remain clasped. The mating sequence is described by Slater (1931), Noble and Putnam (1931), and Wernz (1969).

BREEDING SITES

Oviposition usually occurs in shallow tributaries of mainstream channels. In Washington, females move into the tributaries in July for a late July oviposition. Mainstream channel waters are usually 4–10°C, with tributaries slightly warmer. However, eggs may be oviposited in both the main stream and tributaries. Eggs are placed under large flat rocks or boulders.

REPRODUCTION

Mating occurs from spring to fall, with oviposition occurring in the summer after the spring snowmelt and before the winter rains. For example, Gaige (1920) found females with eggs and males in breeding condition from late June to early September, although Wernz (1969) found gravid females in early April. Burkholder and Diller (2007) suggested most breeding in northern California occurs in spring with egg deposition from July to August. In western Washington, Palmeri-Miles et al. (2010) found eggs at a communal site in late July, and Bury et al. (2001) found nests with eggs from late July to mid-August. In the North Cascades, oviposition occurs from mid- to late July (Brown, 1990). Noble and Putnam (1931) reported clasping pairs in June to July, but Nussbaum et al. (1983) thought most reproduction takes place in the autumn. Karraker et al. (2006) reported oviposition from early June to late August.

Unlike all other North American frogs, frogs of the genus *Ascaphus* have internal fertilization. Eggs are oviposited in nests on the underside of stream boulders and cobbles, where they readily stick to the surface because of their sticky outer membrane (Gaige, 1920; Noble and Putnam, 1931; Brown, 1975c; Adams, 1993; Karraker and Beyersdorf, 1997; Bury et al., 2001; Palmeri-Miles et al., 2010). Substrates under nest boulders may be sand, silt, or gravel. Under laboratory conditions, oviposition takes from 15 to 25 hrs with a fertilization rate of >90%. Nussbaum et al.

Adult female *Ascaphus truei* extruding eggs. Photo: Amber Palmeri-Miles

(1983) reported that copulation requires 24 to 30 hrs. During oviposition, females aggregate with other females; Brown (1975c) reported from 5 to 20 females under a single rock with developing embryos in early stages of cleavage. Nests may contain more than 1 clutch (Bury et al., 2001; Palmeri-Miles et al., 2010); these latter authors reported 4 females and 183 eggs beneath a single boulder. Bury et al. (2001) noted nests with 96 and 182 eggs.

The clutch size of *A. truei* may be smaller than *A. montanus* (mean 41.9 vs. 66.6; Karraker et al., 2006), but females of both species may deposit eggs only every other year in some populations (Metter, 1967; Burkholder and Diller, 2007). In contrast, Bury et al. (2001) suggested populations in the Olympic Peninsula deposited fewer eggs but on a yearly cycle. Clutch size is reported from 37 to 82 eggs in the Washington Cascades (Brown, 1975c). Adams (1993) reported a single clutch of 27 eggs in Oregon and P.S. Corn (*in* Adams, 1993) found a clutch of 38 eggs. Karraker and Beyersdorf (1997) reported a single clutch of 28 eggs in California. In the Olympic Mountains, clutch size averaged 37 eggs (range 28 to 47) (Noble and Putnam, 1931) and 48 eggs (range 40 to 55) (Bury et al., 2001), whereas Gaige (1920) observed clutches of 35 and 49 eggs. Palmeri-Miles et al. (2010) later reported clutch sizes of 70, 68, 47, and 24 eggs, also on the Olympic Peninsula. In Karraker et al.'s (2006) summary, clutch sizes were 15–182, but the larger clutches (>89) likely represent multiple clutches; most clutch sizes were <60 eggs. Nussbaum et al. (1983) gave the clutch size as 33–98, but these figures may include *A. montanus* clutches. In dissected females, total egg complement varies from 44 to 1,298 eggs (Karraker et al., 2006). The density of egg clutches is not high. Bury et al. (2001) reported a density of 1 nest per 286 m of stream searched (1 nest per 34 streams searched) in Olympic National Park.

Eggs will not develop normally at temperatures >19°C, and only about 50% develop normally at temperatures <7.6°C (Brown, 1975c). Embryonic development requires ca. six weeks under field conditions. Increasing temperatures causes an increase in developmental rate. At hatching, larvae are a mean of 11.5 mm TL (range 10.5 to 12.1 mm) (Adams, 1993) to 13.5 mm TL (Noble and Putnam, 1931).

LARVAL ECOLOGY

Hatching occurs in late summer (Brown, 1990; Adams, 1993; Palmeri-Miles et al., 2010). At hatching, embryos still have a substantial quantity of yolk and are incapable of swimming. Some yolk even may remain the following May from tadpoles hatching the previous September (Brown, 1990). By 38 days after hatching in water of 11°C, the rasping, suctorial tadpole mouth is functional. In Washington and Oregon, larvae are found in fast-flowing headwater streams with a mean depth of ca. 4.9 cm. In California, larvae are associated with riffles and step runs (a sequence of runs separated by riffles over a cobble and boulder substrate), stream flows of > 1,500 cm^3/sec, low siltation, high gradients, and low water temperature (Welsh and Ollivier, 1998; Diller and Wallace, 1999), and have even been found in waterfalls (Gaige, 1920). They use large rocks for cover, with a mean area of 415 cm^2, and they are found in deeper water but under smaller rocks than adults (Bury et al., 1991b). Hawkins et al. (1988) noted a preference for substrate cobble 10–30 cm in diameter, with highest larval densities in open regions just below headwater forests. In British Columbia, larval densities decrease with increasing levels of fine sediment, rubble, detritus, and wood; larval densities increase with bank width (Dupuis and Steventon, 1999).

Cool waters are necessary for larval development. During the first year of life, laboratory observations suggest larvae prefer water <10°C, whereas during the second year they select water 10–22°C (de Vlaming and Bury, 1970). In the field, most stream temperatures average <14°C. Not surprisingly, overwintering occurs in streams covered by ice and snow, especially at higher elevations. According to Brown (1990), larvae cluster (three to five individuals) beneath large rocks and are soon covered by gravel and sand, where they remain not feeding until the following spring. Larvae acclimated at 5°C have a mean CTmax of 29.6°C.

Tadpoles hide under rocks and in crevices during the day and emerge to feed from 20:00 to 01:00 hrs at night (Feminella and Hawkins, 1994). They are rasping feeders, eating diatoms almost exclusively, which they scrape off rocks. They attach themselves to the rocks with a highly suctorial oral apparatus that enables them to feed

Tadpole of *Ascaphus truei*.
Photo: Brome McCreary

despite strong currents. In addition, the body profile reduces resistance, and the body is flexible to move with the current. In strong water currents, tadpoles face upstream. They are capable of swimming short distances upstream into the current and, when dislodged, they quickly reattach themselves to a rock. In old-growth habitats, larvae move a mean distance of 1.1 m/day; movements are much curtailed in clearcuts, averaging only 0.15 m/day (Wahbe and Bunnell, 2001).

Growth of the body averages 0.15 mm/day and of the tail 0.12 mm/day over the first 2 months after hatching under laboratory conditions (Brown, 1975c). There appears to be considerable variation in the duration of the larval period, with larvae requiring 1–4 yrs to reach metamorphosis, depending on stream conditions and location (Brown, 1990; Bury and Adams, 1999). Indeed, larvae may be on a 2 yr cycle, then switch to a 1 yr cycle (or vice versa), within the same stream. Streams in close proximity may contain larvae on different larval schedules, at least in the coastal areas of northern California (Wallace and Diller, 1998). At higher elevations and latitudes along the coast, most larvae are on a 2 yr schedule, whereas at low elevations and latitudes most larvae are on a 1 yr schedule. Longer larval periods (to 4 yrs) are found in the colder inland portions of the species' range, such as in the North Cascades (Metter, 1967; Brown, 1990). Larval transformation requires 60 days to complete, at which time newly metamorphosed frogs are 23–28 mm SUL (Nussbaum et al., 1983). In California, metamorphs are observed from late June to early August (Wallace and Diller, 1998).

DIET
Larvae feed largely on diatoms. Little is known of the diet of postmetamorphs, but it includes beetles and spiders (Gaige, 1920; Fitch, 1936; Nussbaum et al., 1983). The species probably has a diet and feeding behavior similar to *A. montanus*.

PREDATION AND DEFENSE
Coastal Tailed Frogs are cryptic in coloration and difficult to observe on stream bottoms. When a rock is upturned, they tend to float passively downstream and thus escape detection (Gaige, 1920). They also jump into the water and swim to the bottom, where they blend in with the substrate (Van Denburgh, 1912). On land, however, they are clumsy and make little effort to escape if removed directly from water. Terrestrially active individuals are quite capable of hopping to safety over logs and debris or jumping into water. This species is eaten by garter snakes (*Thamnophis sirtalis*) (Schonberger, 1945).

Coastal Tailed Frogs have dermal peptides that may aid in deterring predators and protecting the frog against microbial invaders (Conlon

Adult male *Ascaphus truei*.
Photo: David Dennis

et al., 2004, 2005a). These results suggest that antipredator and antimicrobial secretions evolved early in anuran phylogeny.

POPULATION BIOLOGY

Densities of Coastal Tailed Frogs can be rather high in stream habitats. For example, Bury et al. (1991b) recorded densities of 0.65–0.86 (mean 0.76) frogs per m^2 at 23 survey locations in the Oregon Coastal Range, 0.10–0.38 individuals/m^2 (mean 0.25) at 18 locations in the Oregon Cascades, and 0.35–4.32 individuals/m^2 (mean 1.72) at 18 locations in the Washington Cascades. Overall, the mean density was 0.90 frogs/m^2 at 59 survey sites, with a biomass of 1.56 g/m^2. In southern Washington, one stream had from 1.1 to 9.5 individuals/m^2 (mean 4.5 individuals/m^2); indeed, one 10 m stretch had 109 Coastal Tailed Frogs, most of which were larvae (Bury, 1988). In riparian habitats, Gomez and Anthony (1996) captured a mean of 5 to 25 tailed frogs per transect in western Oregon. Despite relatively high densities, most populations are somewhat isolated from one another, and dispersal among populations is rare.

Growth occurs year-round in postmetamorphic *A. truei*, although growth rates decrease during the winter months (Burkholder and Diller, 2007). In the summer, growth averages 1.47 mm/month, whereas it slows to 0.89 mm/month in winter. Maturity is delayed for several years in this species, with females reaching maturity 3 yrs after metamorphosis (at ca. 42 mm SUL) and males 2 yrs after metamorphosis (at ca. 34 mm SUL) in California (Burkholder and Diller, 2007). It likely has somewhat similar age and demographic characteristics as *A. montanus*, although the tadpole stage is variable and not as long in coastal populations.

COMMUNITY ECOLOGY

Although Coastal Tailed Frogs inhabit streams with large adult and larval salamanders (*Dicamptodon, Rhyacotriton*), there is no correlation between adult frog density and salamander density (Adams and Bury, 2002). Tadpoles are able to detect chemical cues from *Dicamptodon*, brook trout (*Salvelinus fontinalis*), and cutthroat trout (*Salmo clarki*), and reduce their activity significantly (Feminella and Hawkins, 1994). They are unable to detect cues from the predatory shorthead sculpin (*Cottus confusus*) and are readily eaten by this species. Feminella and Hawkins (1994) suggested that the absence of *Ascaphus* from low-gradient streams might be the result of their inability to detect the sculpin.

Under controlled conditions, larval *A. truei* appear to have little effect on stream periphyton biomass, although they may be food-limited in headwater streams in nature (Kiffney and Richardson, 2001). Larvae may reduce the abundance of small insects, such as mayflies, when nutrients are limited by out-competing them or interfering with larval insect grazing. When nutrients are abundant, however, tadpole grazing has no effect on the aquatic insect community.

DISEASES, PARASITES, AND MALFORMATIONS
No information is available.

SUSCEPTIBILITY TO POTENTIAL STRESSORS
No information is available.

STATUS AND CONSERVATION
Coastal Tailed Frogs are negatively impacted when the canopy cover is removed around stream habitats. This often occurs in connection with clearcutting, and density and biomass of tailed frogs and their larvae are greater in mature uncut forests than in those affected by timber harvesting (Bury and Corn, 1988b; Corn and Bury, 1989; Welsh, 1990; Gomez and Anthony, 1996; Dupuis and Steventon, 1999; Wahbe and Bunnell, 2003; Matsuda and Richardson, 2005). Such habitat disruption, even if the stream per se is unaffected, increases temperatures, decreases humidity, increases the potential for algal blooms, which can disrupt larval feeding ecology, and leads to increases in siltation, which negatively affects this species (Noble and Putnam, 1931; Bury, 1983; Bury and Corn, 1988a; Walls et al., 1992; Jennings and Hayes, 1994; Welsh and Ollivier, 1998; Dupuis and Steventon, 1999).

Movement patterns vary with the stage of forest succession and with life stage, with postmetamorphic frogs staying much closer to streams (< 25 m) in clearcuts than in old-growth forest (to > 100 m) (Wahbe et al., 2004). Larvae also move much farther in streams flowing through old-growth habitats than streams flowing through clearcuts, where logjams may create considerable barriers to aquatic movement (Wahbe and Bunnell, 2001). Long distance overland movements are more likely in contiguous mature forest, leading Wahbe et al. (2004) to suggest that watersheds should be managed for the conservation of *Ascaphus* rather than streams. This recommendation is congruent with findings of Dupuis and Friele (2006) on the importance of individual basins to maintaining populations of *Ascaphus*. Differences in adult movement patterns between old-growth and mature forest are reflected in the differences in the genetic diversity of populations in the two forest types, where gene flow is restricted among surviving clearcut populations. Gene flow is associated with low solar radiation and closed forest (Wahbe et al., 2005; Spear and Storfer, 2008) and in fragmented forests there is less gene flow.

Although certain forms of timber harvest are detrimental to *A. truei*, the effects depend on elevation, geologic formation, stream size, and the percent cover of sand, bounders, runs, and riffles (Diller and Wallace, 1999; Wahbe and Bunnell, 2003). Tailed frogs affected by clearcutting may survive in the cold and wet northern portions of their range, especially in headwater streams or in uncut refugia downstream from logging occurring upstream, but may be extirpated in the warmer and dryer southern regions. Coastal populations may be less susceptible to clearcutting than inland populations because of the maritime conditions (Bury, 1968). Undoubtedly this is due to dry conditions, which restrict overland movement among forest remnants (Spear and Storfer, 2008). Likewise, tailed frogs in portions of streams at lower elevations, impacted by harvesting, appear

Habitat of *Ascaphus truei*. Photo: C.K. Dodd Jr.

more vulnerable than those at higher elevations, due to a combination of decreased gradient, higher temperatures, and increased siltation. Unlike *A. montanus* (Spear and Storfer, 2010), *A. truei* does not maintain genetic connectivity along riparian corridors in clearcut forests.

The effects of timber harvest last a long time, with significantly fewer Coastal Tailed Frogs in habitats affected by harvest even 37–60 yrs postharvest (Corn and Bury, 1989; Ashton, 2002; Ashton et al., 2006). Gomez and Anthony (1996) suggested maintaining a riparian buffer zone of 75–100 m on both sides of headwater streams to protect the amphibian fauna, and that upslope and old-growth forests should be managed for this and other amphibian species. The long-term effects of harvesting are mirrored in the genetic structure of fragmented populations. For example, populations in old-growth habitats are more genetically diverse than in cutover habitats (Wahbe et al., 2005). Populations in cutover areas also may go through genetic bottlenecks. In clearcut areas, there is no correlation between genetic relatedness and physical distance. Homogeneity probably results from greater dispersal of frogs in clearcuts, trying to find better habitat. In old-growth forest, genetic relatedness is inversely correlated with physical distance.

Ascaphus truei survived the catastrophic eruption of Mount St. Helens in 1980 (Hawkins et al., 1988; Crisafulli et al., 2005). Surveys in 10 streams within the blowdown zone conducted 5–7 yrs following the eruption revealed survival, although mortality had been widespread and severe shortly after the blast. Mean densities often were high, even 4–5 yrs after the eruption, as were tadpole densities; tadpole densities were correlated with the amount of forest cover (Hawkins et al., 1988). In areas where little canopy remained, *Ascaphus* were scarce or absent, largely as a result of increased water temperature. By 1998, however, they were present in nearly every stream sampled (Crisafulli et al., 2005). Coastal Tailed Frogs probably survived in refugia protected from scour and abrasion, such as cascades and water chutes, and in areas where some canopy remained. Populations likely will recover in time as the canopy forest regenerates.

The species is considered "Threatened" or of "Special Concern" in California (Jennings and Hayes, 1994).

Family Bufonidae

Anaxyrus americanus
(Holbrook, 1836)
American Toad
Crapaud d'Amérique

ETYMOLOGY
americanus: Latinized name for America, meaning of or belonging to America.

NOMENCLATURE
Conant and Collins (1991): *Bufo americanus*
　Synonyms: *Bufo americanus copei, Bufo copei, Bufo lentiginosus, Bufo lentiginosus americanus, Bufo terrestris americanus, Bufo terrestris charlesmithi, Chilophryne americana, Incilius americanus*

IDENTIFICATION
Adults. This is a medium to large toad (normally ca. 50–90 mm SUL; maximum 155 mm SUL, Harding, 1997); females may be much larger than males. The dorsum is gray, olive brown, or a rich brown to nearly brick red, and it is well covered with knobs and bumps; the elliptical parotoids and cranial crests are prominent, with postorbital branches of the cranial crests extending posteriorly in front of the parotoid gland. There is usually one knob or wart per dark spot. Some individuals have a cream to yellow line extending from the head down the middle of the back. The bellies are white with some mottling around the sides, although there is considerable geographic variation in the extent of ventral spotting or mottling (Blair, 1943a). The most extensive ventral mottling is found on northern toads. Limbs are short and stocky, in keeping with its hopping and walking mode of travel, and the hind feet are moderately webbed between the digits. Males have a dark subgular vocal sac, and they have nuptial pads on the thumb during the breeding season. The iris of the eye is bronze or gold. Juveniles are colored like adults, but their skins are much smoother, the cranial crests are not as prominent, and the tympanum is not evident. Albino American Toads have been collected in Minnesota (Oldfield and Moriarty, 1994) and Virginia (*in* Dyrkacz, 1981).

Sexual size dimorphism is present, with males being smaller than females (Wilbur et al., 1978; Minton et al., 1982). In Indiana, northern animals are larger (males 65–80 mm SUL; females 75–87 mm SUL) than southern animals (males 50–65 mm SUL; females 60–75 mm SUL) (Minton et al., 1982). In central Indiana, males averaged 62.8 mm and females 72.7 mm (Howard, 1988a). Other records include: males averaging 67 mm SUL (range 59 to 78 mm) and females 75 mm SUL (64–89 mm) in Ontario (Licht, 1976); males 60–83 mm SUL in Maine (Sullivan, 1992a); males 51–72 mm SUL (mean 61 mm) and females 68–85 mm SUL (mean 75 mm) in Connecticut (Klemens, 1993); males 58–81 mm SUL (mean 68 mm) and females 63–93 mm SUL (mean 84 mm) on Prince Edward Island (Cook, 1967); males 49–76 mm SUL and females 57–105 mm SUL in Nova Scotia (Gilhen, 1984). Blair (1941a) provided mean adult SULs for 18 populations; they ranged from 54.5 to 72.8 mm SUL.

Larvae. The tadpole's dorsum is uniformly dark brown to black without a light mark behind each eye; very fine silvery or gold spots may overlay the ground color; the throat is largely pigmented; the snout is sloping in lateral view; and the eyes are small. The belly may have more extensive aggregations of gold or copper spots than the dorsum or laterally. Tails are short and bicolored. Tail musculature is usually bicolored and is greater in height than either the upper or lower tail fin.

The fins themselves are "cloudy transparent" in coloration. For mouthparts, the ratio of the first posterior tooth row to the third posterior tooth row is two or less. The upper jaw is not noticeably angulate, and the submarginal papillae at the tips of the third posterior tooth rows are reduced or absent. Descriptions of tadpoles are in Wright (1929) and Altig (1970). Descriptions and illustrations of larval mouthparts are in Hinckley (1882) and Dodd (2004).

Eggs. Individual eggs are separated by a thin connecting jelly strand, thus forming a tubular rosary bead-like string. The eggs are dark brown to black above and white to cream below. They are surrounded by two jelly envelopes, which are not scalloped. The outermost envelope is 2.86–4 mm in New York; the inner envelope is 1.6–2.2 mm (Livezey and Wright, 1947). There may be some geographic variation in egg size and number. Dundee and Rossman (1989) recorded egg diameters as 2.9–3.1 mm in Louisiana for the outermost envelope, and 2.1 mm for the inner envelope. According to Livezey and Wright (1947), there are 15–17 eggs per each 2.54 cm of string in New York, but there are only 7.5 per 2.54 cm in Louisiana (Dundee and Rossman, 1989). Egg strings can be very long, to > 60 m in length. Hatching occurs in two to seven days, depending on water temperature. An illustration of the spacing of eggs is in Miller (1909a).

DISTRIBUTION

The range of *Anaxyrus americanus* extends from southern Labrador across Québec to Hudson and James Bays, through most of Ontario to southeastern Manitoba, and south, paralleling the western border of Minnesota and Iowa. The species barely enters North and South Dakota, and perhaps Nebraska (Lynch, 1985, noted previous identification errors involving *A. americanus* in Nebraska [Hudson, 1942] and suggested that its status and possible introgression with *A. woodhousii* needed clarification). The range includes eastern Kansas and east-central Oklahoma, and from northeast Texas east to northern Louisiana (Louisiana specimen disputed by Dundee and Rossman, 1989, but confirmed by Himes and Bryan, 1998). American Toads are found throughout central and northwest Mississippi south to the Florida Parishes of Louisiana. Boyd and Vickers (1963) restricted this species to Tishomingo County in northeast Mississippi, but Lazell and Mann (1991) confirmed its presence along the loess bluffs east of the Mississippi River at least as far north as Regantown.

Distribution of *Anaxyrus americanus*

The species appears absent from northwestern Alabama, but is found in northeastern Alabama south through the Coosa Valley. The southern limit of the range generally follows the southern boundary of the upper Piedmont and Blue Ridge Province from east-central Alabama through southeastern Virginia, with isolated populations in coastal North Carolina and the Delmarva Peninsula. It ranges from sea level to nearly 1,800 m in the Southern Appalachians (Dodd, 2004).

American Toads are found on islands in western Lake Erie (Langlois, 1964; King et al., 1997; Hecnar et al., 2002), the Apostle Islands of Lake Superior (Bowen and Beever, 2010), Walpole Island in Lake St. Clair (Woodliffe, 1989), northern Michigan including Isle Royale (Ruthven, 1908; Harding, 1997), James Bay (Hodge, 1976), the mouth of the St. Lawrence River (Fortin et al., 2004a), Prince Edward Island (Cook, 1967), Muskeget Island off the coast of Massachusetts (J.A. Allen, 1868), Martha's Vineyard (Lazell, 1976), and Long Island, New York (Overton, 1914). The species has been introduced into Newfoundland (Maunder, 1983, 1997), where it has been recorded at 52 localities (Campbell et al., 2004).

Important distributional references include: Alabama (Mount, 1975; Redmond and Mount, 1975), Arkansas (Black and Dellinger, 1938;

Trauth et al., 2004), Canada (Bleakney, 1954, 1958a), Connecticut (Klemens, 1993), Georgia (Williamson and Moulis, 1994; Jensen et al., 2008), Great Lakes region (Harding, 1997), Illinois (Schmidt and Necker, 1935; Smith, 1961; Phillips et al., 1999), Indiana (Lynch, 1964a; Minton et al., 1982; Minton, 2001; Brodman, 2003), Iowa (Christiansen and Bailey, 1991), Kansas (Collins, 1993; Collins et al., 2010), Labrador (Maunder, 1983; Markle and Green, 2009), Louisiana (Dundee and Rossman, 1989), Maine (Hunter et al., 1999), Minnesota (Oldfield and Moriarty, 1994), Missouri (Johnson, 2000; Daniel and Edmond, 2006), Nebraska (Hudson, 1942; Lynch, 1985; Ballinger et al., 2010; Fogell, 2010), New Brunswick (Gorham, 1964; McAlpine, 1997a), New England (Hoopes, 1930), New Hampshire (Oliver and Bailey, 1939; Taylor, 1993), New York (Gibbs et al., 2007), North Carolina (Dorcas et al., 2007), North Dakota (Wheeler and Wheeler, 1966), Nova Scotia (Bleakney, 1952; Richmond, 1952; Gilhen, 1984), Nunavut (Hodge, 1976), Ohio (Walker, 1946; Allen, 1963; Pfingsten, 1998; Davis and Menze, 2000), Ontario (Ashton et al., 1973; Schueler, 1973; Johnson, 1989; MacCulloch, 2002), Pennsylvania (Hulse et al., 2001), Prince Edward Island (Cook, 1967), Québec (Harper, 1956; Ashton et al., 1973; Denman and Denman, 1985; Bider and Matte, 1996; Desroches and Rodrigue, 2004; Galois and Ouellet, 2005), South Carolina (Chamberlain, 1939; Montanucci, 2006), South Dakota (Fischer, 1998; Kiesow, 2006), Tennessee (Redmond and Scott, 1996), Texas (Whiting and Price, 1994; Dixon, 2000), Vermont (Andrews, 2001), Virginia (Tobey, 1985; Mitchell and Reay, 1999; White and White, 2007), and Wisconsin (Suzuki, 1951; Vogt, 1981).

FOSSIL RECORD

American Toads are recorded from Pleistocene deposits throughout the central and eastern portions of North America. Irvingtonian fossils occur from Maryland, Nebraska, Pennsylvania, and West Virginia. Rancholabrean fossils are much more numerous, from Alabama, Arkansas, Georgia, Michigan, Ohio, Ontario, Pennsylvania, Tennessee, Virginia, West Virginia, and Wisconsin. Most of these records come from cave deposits. A complete list is found in Holman (2003), along with a discussion of the osteological differences that separate *americanus* from *fowleri* and *terrestris*. Holman (2003) considered reports of *Americanus* group of toads from the Miocene of Kansas (Sanchiz, 1998) to represent an undescribed taxon. Tihen (1962a) and Holman (2003) note other distinguishing osteological characteristics among *A. fowleri*, *A. americanus*, and *A. terrestris*.

SYSTEMATICS AND GEOGRAPHIC VARIATION

Anaxyrus americanus is a member of the North American *Americanus* clade within the Nearctic clade of toads, and is a sister taxon of *A. woodhousii*. It is closely related to *A. houstonensis*, more distantly so to *A. fowleri*, *A. terrestris*, *A. hemiophrys*, and *A. baxteri* (Pauly et al., 2004). Goebel (2005) also includes *A. woodhousii*, *A. microscaphus*, and *A. californicus* within the *Americanus* clade. Although Blair (1963b) and others hypothesized that *A. americanus* and *A. terrestris* were sister groups, mtDNA analyses confirm that *A. americanus* is more closely related to *A. woodhousii* than *A. terrestris* (Masta et al., 2002; Pauly et al., 2004).

There is widespread genetic compatibility within the *americanus* group. Females hybridize and produce viable offspring with other members of the clade, including male *A. fowleri*, *A. hemiophrys*, *A. houstonensis*, *A. microscaphus*, *A. terrestris*, and *A. woodhousii* in artificial crosses (A.P. Blair, 1941a; Moore, 1955; W.F. Blair, 1961a, 1963a, 1972; Cook, 1983). However, there is both a geographic and sex component to the effectiveness of hybrid crosses, with some geographic populations (e.g., Louisiana ♀ *A. fowleri* × North Carolina ♂ *A. americanus*) being less successful than others (North Carolina ♂ *A. fowleri* × North Carolina ♀ *A. americanus*) (Volpe, 1955a).

Male hybrids are fertile with female *A. americanus* and *A. fowleri* in backcrosses. Male *A. americanus* also hybridize and may produce fertile offspring with female *A. woodhousii* (A.P. Blair, 1946; W.F. Blair, 1963a) and *A. terrestris* (W.F. Blair, 1959). Female *A. americanus* produce at least some metamorphs in crosses with male *A. punctatus* (Blair, 1959), *A. boreas* (Moore, 1955), *A. cognatus* (Blair, 1959), and *Ollotis nebulifer* (Volpe, 1959a), although malformations of larvae are common. Crosses between male *Anaxyrus americanus* and female *A. boreas*,

A. compactilis, A. debilis, A. punctatus, A. speciosus, and *Ollotis nebulifer* are not successful (Blair, 1959, 1961a, 1963b). Blair (1972) reviewed hybridization crosses in this species.

Hybridization in nature is widespread between *Anaxyrus americanus, A. fowleri, A. terrestris,* and *A. woodhousii* (Myers, 1927; Blair, 1941a, 1942, 1947a; Neill, 1949a; Volpe, 1952, 1959a, 1959b; Cory and Manion, 1955; Brown, 1964, 1969; Ideker, 1968; Zweifel, 1968a; Jones, 1973; Norton and Harvey, 1975; Wilbur et al., 1978; Weatherby, 1982; Green, 1981a, 1984, 1989, 1996; Green and Pustowka, 1997; Minton, 2001; Masta et al., 2002; Green and Parent, 2003). Hybridization occurs in contact zones with *A. hemiophrys* in eastern Manitoba (Cook, 1983; Green, 1983; Green and Pustowka, 1997; Roy, 2009) and in eastern South Dakota (Henrich, 1968; Ideker, 1968).

Hybrids are intermediate in both morphology and call characteristics with the parental species (Blair, 1946; Cory and Manion, 1955; Green, 1981a; Cook, 1983; Green and Pustowka, 1997). Green and Parent (2003) noted that hybridization occurs in a mosaic, that is, at some locations, but not others, and at some times, but not others. In the case of hybridization between *A. americanus* and *A. hemiophrys* in southeastern Manitoba, the < 20 km wide hybrid zone shifted westward 9.6 km over a period of 11 yrs, but there was no change in width (Green and Pustowka, 1997).

American Toads show considerable phenotypic variation in color pattern, spotting, extent of ventral pigmentation, and size. This has led to the description of three subspecies: *americanus, charlesmithi,* and *copei.* Some authors have suggested that the *copei* phenotype is so distinctive that it warrants recognition as a full species (Gaige, 1932; reviewed by Sanders, 1987), but specific recognition is not warranted based on molecular analyses (Guttman, 1975). Molecular analyses only weakly support recognition of *charlesmithi* (Masta et al., 2002).

Toads from far northern Ontario are boldly mottled in various shades of brick red, white, yellow, and black, have a light mid-dorsal stripe, and are stocky. Some individuals may have a pinkish or rose color in the axillary region. This description conforms to the phenotype of toads described as *Bufo copei* (*B. americanus copei*) (Schueler, 1973) and is typical of far northern populations.

Bleakney (1952) also reported this phenotype throughout most of Nova Scotia, with the exception of the Annapolis Valley. Here, toads appeared more typical of the *americanus* phenotype occurring farther south. In Maine, the *copei* phenotype is more common in the north than in the south. Ashton et al. (1973) noted that 75% of the toads collected in northern Ontario and Québec were typical *copei* in pattern, 20% were smoky gray with a strong mid-dorsal white stripe, and the remaining 5% were solid light reddish brown.

The "Dwarf American Toad" (*charlesmithi*) of the central and south-central United States appears much like its more typical northern relative, except that it is much smaller (ca. 50–60 mm SUL), more reddish brown, and has few or no black spots dorsally; the venter is cream colored, and there may be a small number of gray spots. Calls also vary geographically, with the pulse rates of calls from *A. americanus* from northern areas higher than those of conspecifics from Oklahoma and Arkansas, even after adjusting for temperature differences (Zweifel, 1968a). The "Dwarf American Toad" in Missouri also has a slightly higher pitch.

Harding (1997) noted that giant American Toads (to 15.5 cm) reside on islands in northern Lake Michigan; the local name *alani* has been used for these toads (Long, 1998), but the name has no taxonomic standing.

ADULT HABITAT

The American Toad is found in a variety of habitats, and habitat preferences may reflect the dominant habitat types within its landscape. The species seems to prefer open deciduous forests and grasslands as opposed to dense forests. For example, in New Hampshire *A. americanus* is commonly found in red maple (*Acer rubrum*), northern hardwood, and balsam fir (*Aibes balsamea*) forest types in both upland and streamside settings, although it is found significantly less often in the balsam fir (DeGraaf and Rudis, 1990). Occupancy is often quite high in some areas and is related to wetland proximity, but it is difficult to predict which habitat variables are most important to the species (Roloff et al., 2011).

In the upper Midwest, American Toads are associated with grassland habitats rather than forested habitats, and these reflect the predominant habitats within the prairie states of Illinois, Indiana, Iowa, and Wisconsin (Smith, 1961; Knutson et al., 2000; Minton, 2001). American Toads are often found in urban, agricultural, and even mined sites (Smith, 1961; Lacki et al., 1992; Kolozsvary and Swihart, 1999; Knutson et al., 2004; Anderson and Arruda, 2006). Several authors have suggested that American Toads are generalists in their habitat preference (Bonin et al., 1997a; Hecnar and M'Closkey, 1997a); that they do not exhibit habitat-based nestedness patterns supports this hypothesis (Hecnar and M'Closkey, 1997a).

American Toads are frequently found in disturbed habitats, for example, in both open- and closed-canopy conifer plantations, in clearcuts and wildlife clearings (Pais et al., 1988), and in agricultural landscapes (Knutson et al., 2004). Abundance within conifer plantations in New Brunswick actually was greater than in natural forest (Waldick et al., 1999), although most adults and juveniles were found in pitfall traps facing natural forest. Toads are also commonly observed in old fields (Clawson and Baskett, 1982), and in Indiana were reported as ubiquitous in farmland habitats (Kolozsvary and Swihart, 1999).

TERRESTRIAL ECOLOGY

Assessments of habitat associations of American Toads have provided mixed results. In Ontario, American Toad abundance was not correlated with the amount of forest cover within 2,000 m of a breeding pond (Eigenbrod et al., 2008), and in Maine, occupancy decreased with the extent of forest within 300 m of breeding ponds (Guerry and Hunter, 2002). However, in Minnesota, occupancy at breeding ponds was positively correlated with forest cover at 500 m and 2,500 m, suggesting the importance of forests as corridors during terrestrial movement and as foraging and overwintering sites (Lehtinen et al., 1999). Around the Great Lakes, their presence is positively correlated with herbaceous uplands within 3,000 m of breeding sites; within 100 m, presence was positively associated with the extent of grass and sedge cover, and negatively associated with the amount of forest cover (Price et al., 2004). These latter authors found no significant correlations with any of the other habitat variables they measured within 500 m of toad breeding sites. Differing results may reflect differences in methodology as well as differences in the landscapes investigated surrounding breeding ponds.

American Toads spend most of their time in upland habitats, well away from breeding sites. Outside the breeding period, they are routinely found at distances > 400 m from wetlands. In a Maryland study, female toads had home ranges from 434 to 1,305 m^2 in the surrounding uplands (mean 717 m^2), and used up to 10 well-defined activity centers of from 30 to 100 m^2 each. When moving between these centers, females sometimes traveled several hundred meters in a 24 hr period (Forester et al., 2006). Rainfall is highly correlated with daily activity and localized movements (Ewert, 1969; Forester et al., 2006). Warm-season activity is devoted to feeding, and as the season progresses, to the selection of overwintering sites. During hot, dry periods, toads seek shelter under leaf litter and woody debris, or burrow into the soil.

The microenvironmental correlates of terrestrial habitat use are not well understood, but probably include friable soils in which to dig, abundant coarse woody debris for shelter, and appropriate thermal and hydric conditions. Optimal adult temperature preferences tend to be 21–34°C, with juvenile preferences ranging between 19 and 31°C (most 24–27°C) (Tracy et al., 1993). This species seems tolerant of dry conditions, although in the laboratory they choose the moister substrate in paired-choice trials (Tracy et al., 1993), and seek a balance between optimal thermal and hydric conditions (Bundy and Tracy, 1977). American Toads can rehydrate moisture directly from the soil, regaining 94–99% of their original body weight at 1–1.5 atmospheres within 12 hrs (Walker and Whitford, 1970). When stressed, desiccated juveniles will flatten themselves on a moist substrate. However, they do not necessarily choose terrestrial habitats in nature with higher soil moisture content. Instead, they choose soils that are less acidic when a mosaic of soil pHs are available (Wyman, 1988). Thus, the microhabitat of toads reflects a balance among a number of environmental variables.

American Toads move long distances: 6.4 km (Hamilton, 1934), 4 km (Maynard, 1934), 1 km (Maunder, 1983), 594 m (Oldham, 1966), 548 m (Blair, 1943b), 348 m (Ewert, 1969), and 235 m (Dole, 1972a). Smith and Green (2005) found no evidence of a classical metapopulation structure in

American Toads, and concluded that metapopulations were unlikely since there was probably too much dispersal occurring among habitat patches separated by 10 km or less. Extensive interpond movements both within and between seasons supports this hypothesis.

Terrestrial activity occurs whenever environmental conditions are favorable, but American Toads are commonly found during the dry and warm parts of the activity season. In the early part of the season, activity is correlated with temperature, but as the season progresses and mean temperatures rise, temperature plays less of a role in initiating activity. Both temperature and rainfall help stimulate activity, although their effects vary by latitude; there may be a 24 hr lag between stimulus and activity. Fitch (1956a) reported activity in Kansas between 17.5 and 32.3°C, with a preferred range of 26.5 to 31°C; these values are similar to those (25–30°C) reported from Minnesota (Tester et al., 1965). Brattstrom (1963) recorded an 11.2°C body temperature in a toad during the summer, but specifics of time and place were not provided.

American Toads are primarily nocturnal, although individuals may be active on cloudy days or during rainfall. Peak activity occurs shortly after nightfall, depending on location, with activity steadily declining or with a secondary activity peak shortly before until shortly after dawn (Higginbotham, 1939). For example, activity peaks from 19:00 to 23:00 hrs in southern Québec. Activity then peaks later (20:00–03:00 hrs) as latitude increases, due to the increasing day length (Bider and Morrison, 1981); toads also are active daily earlier in July than in August. In southern Québec, adult American Toads are mostly nocturnal during the warm season, with only 2.6% being active diurnally. In contrast, juveniles are more diurnal (18.6%) (Fitzgerald and Bider, 1974c). Diurnal activity may be more thermally favorable for small toads, and it may reduce predation by large toads or competition for prey or cover.

Anaxyrus americanus is active throughout the warmer parts of the year, and thus the extent of activity is greater in the South than in the North. Activity begins before the migration to breeding ponds, as weather permits, and initial localized movements need not be directed toward the breeding site (Ewert, 1969). For example, in South Dakota toads emerge from dormancy in April, although breeding does not commence until May or June (Fischer, 1998). In New England, activity occurs from March to October (Klemens, 1993). In the Mid-Atlantic, activity extends from early February to late October in Maryland (Smithberger and Swarth, 1993) and to late October in Virginia (Buhlmann and Mitchell, 1997). In the Midwest, activity occurs from early January to late November in Ohio (Wilcox, 1891) and from mid-April to mid-October in Minnesota (Ewert, 1969). American toads have been observed in mid- to late November in New York (Hamilton, 1934) and Illinois, but this latter activity was associated with severe flooding (Tucker et al., 1995).

Farther north, activity extends from snowmelt throughout the growing season (Bider and Morrison, 1981; Fischer, 1998; Waldick et al., 1999). American Toads are active until late October to early November in New Brunswick, where they are found under stones or rotten wood debris (Gorham, 1964). In other areas of Canada, activity extends from late April to late September, depending on latitude (Fitzgerald and Bider, 1974a; Bider and Morrison, 1981). For example, the activity period is five to six weeks shorter at James Bay than in Lac Carre, Québec. Activity usually ends at least a month before first snowfall.

American Toads usually overwinter in terrestrial habitats, where they dig burrows protecting them from frost penetration. They also overwinter in ant mounds (Carpenter, 1953a). *Anaxyrus americanus* is neither freeze tolerant nor does it have a cryoprotectant (Holzwart and Hall, 1984); hence, it escapes freezing temperatures by digging below the frost line. Allen (1868) reported finding American Toads 30 cm below the surface in sandy soils in Massachusetts during the first week of September. He also found "great numbers" of toads under loose stones in an unfrozen spring in February. In the upper Midwest, American Toads bury from 12 to 58 cm into mostly sandy soils, a depth that takes them below the frost line (Ewert, 1969). Ewert (1969) noted that toads sometime enter dormancy in the dry summer (to 28 cm below the surface), and remain buried in the same location throughout the winter without resurfacing. He further reported a mortality rate of 32% in the 28 toads he tracked through the winter.

American Toads are regularly found at the entrances and in the passages of limestone caves

(Franz, 1967; Garton et al., 1993; Dodd et al., 2001). They also were found dormant in July down a well in Massachusetts, buried in mud as far as 60 to 90 cm below the surface (Allen, 1868). There is one recorded instance of an American Toad found 1 m above ground on top of a vertical tree stump (Cochran, 2008).

American Toads are photopositive in their phototactic response, suggesting they use both sunlight and moonlight when feeding and going about their daily activities (Pearse, 1910; Jaeger and Hailman, 1973). They are sensitive to light in the greenish-blue spectrum ("blue-mode response," but at 625 THz), which apparently helps them orient toward areas of increasing illumination, such as the open horizon above lakes and ponds (Hailman and Jaeger, 1974, 1978). American Toads likely have true color vision.

MIGRATION TO BREEDING SITES

American Toads migrate to breeding ponds, mostly from forested habitats within several hundred meters of the pond, although Maynard (1934) observed large numbers moving to a breeding pond aquatically via a stream. Waldick et al. (1999) suggested that most toads in a mix of plantation-natural forest habitats in New Brunswick migrated from natural forest within 400 m of breeding ponds, and Oldham (1966) recorded movements from nearly 600 m (mean 110 m) away. Migrations take place in a series of short, directed movements toward a breeding pond. Some toads migrate to a wetland entirely within a single night, whereas others take days to reach the breeding site. Toads do not necessarily move in a straight line or over the shortest possible distance, and they may be able to take advantage of roads as corridors (Ewert, 1969). Ewert provides a number of case histories on movements to ponds, highlighting the individual variation associated with migratory movements.

American Toads in at least some locations are able to use geotaxis during both immigration (positive) and dispersal (negative) to overwintering sites, an advantage in areas of hilly terrain (Fitzgerald and Bider, 1974b). Fitzgerald and Bider (1974a) found that toads moved downhill toward breeding sites in May, but uphill toward terrestrial forest overwintering sites in mid-September. In between, terrestrial movements were random during the warm season. Oldham (1966), however, could not demonstrate geotaxis as influencing orientation toward breeding ponds, suggesting geographic and possibly habitat-related differences in this ability.

Large male American Toads tend to arrive earlier at breeding ponds than smaller males, and they remain at breeding ponds longer than smaller males (Gatz, 1981a). Males may appear initially (Licht, 1976), but Gatz (1981a) noted there was no pattern regarding the arrivals of the sexes at his study site, as males and females initially arrived in about the same proportion; both males and females arrive in greater numbers as the season continues. Choruses may be large or small; at some sites large numbers of toads will chorus, whereas at others only a few males will call. In Oklahoma, Blair (1943b) recorded a mean of 19.3 toads (16.6 males, 2.6 females) per site at 15 breeding sites. The number of toads at a pond can also change annually. Ewert (1969) recorded 6 males and 4 females at a Minnesota breeding site one year, whereas 96 males and 12 females were at the same site the following year.

BREEDING SITES

American Toads use seasonal temporary ponds and permanent wetlands (bogs, fens, marshes), river backwaters, and flooded meadows as breeding sites as long as they are not too acidic; they can even use very small pools, such as rain pools, road ruts, sinkhole ponds, or shallow borrow pits (Blair, 1943b; Adam and Lacki, 1993; Klemens, 1993; Mitchell and Buhlmann, 1999). In South Dakota, for example, 50% of breeding sites were seasonal or semipermanent wetlands (Fischer, 1998). Toads readily use man-made impoundments, reservoirs, stormwater retention ponds, and beaver ponds (Blair, 1943b; Jones, 1973; Klemens, 1993; Metts et al., 2001; Snodgrass et al., 2008; Simon et al., 2009; Brand and Snodgrass, 2010). They also breed in small streams and backwater pools adjacent to streams. In Kentucky, Holomuzki (1995) found them breeding in second-order streams and adjacent runs where males chose sites with few predaceous fishes, and eggs were deposited along the edge of the stream in shallow water, often in small recesses.

A number of variables influence breeding site choice. For example, toads may prefer breeding sites with high dissolved oxygen content (Campbell et al., 2004), whereas Hecnar and M'Closkey

Egg strings of *Anaxyrus americanus*. Photo: John Jensen

Tadpole of *Anaxyrus americanus*. Photo: David Dennis

(1996b) found American Toads preferred sites with high nutrient contents and low levels of water hardness. Although a wide variety of different-sized wetlands are used for breeding, there is a positive correlation between the size of the wetland (and area containing cattail) and the amount of peatland habitat and toad occupancy, at least in New Brunswick (Stevens et al., 2002). However, an experimental study suggested that pond size is inversely correlated with larval survivorship, growth rates, and mass at metamorphosis; increasing pond depth leads to lower survivorship but greater mass at metamorphosis (Pearman, 1993). Whether this study reflects pond choice in nature remains to be determined. Toads may or may not breed in wetlands with predatory and nonpredatory fish. For example, they tend to avoid ponds with predatory fish in New Hampshire (Babbitt et al., 2003), but not in Ontario (Hecnar, 1997; Hecnar and M'Closkey, 1997b). How they identify the presence of fish is unknown.

American Toads do not breed uniformly in wetlands throughout an area, as many seemingly good breeding ponds are unoccupied. Occupancy and colonization of breeding sites are influenced both by current-year conditions (fill date) and the historical average hydroperiod (Church, 2008). American Toads were found in only 6 of 61 wetlands during a 2 yr survey in New Hampshire (Herrmann et al., 2005), 52 of 90 potential sites in Newfoundland (Campbell et al., 2004), and 83 of 116 ponds in Maine (Guerry and Hunter, 2002). In Québec, they were heard about on about 43% of transects along call-survey routes (Lepage et al., 1997), whereas in Ontario, occupancy was >45% for 180 ponds surveyed over 3 yrs (Hecnar and M'Closkey, 1998). In New Hampshire, ponds with breeding toads were equally divided between temporary and permanent wetlands, and most toads were in wetlands that had 0–40% forest cover within 100 m (Herrmann et al., 2005). Also in New Hampshire, larvae were associated with wetlands containing high invertebrate numbers, species richness, and long hydroperiods (Babbitt et al., 2003). Brodman (2009) also noted considerable variation in annual occupancy over a 14 yr study in Indiana.

The sensory cues used by American Toads to locate breeding ponds have not been conclusively demonstrated. Toads are able to locate breeding ponds whether or not an existing chorus is present, and olfaction does not seem to be involved in determining the location of the pond. Furthermore, topography, vision, or humidity do not seem to play key roles in locating breeding ponds (Oldham, 1966). In sensory experiments, American Toads discriminated between bog and fen water (perhaps keying in on the water's acidity), and there was an orientation bias toward the home breeding area's odor (Karns, 1983). In contrast, toads did not orient toward mud, water, or vegetation from their "home" pond in another set of experiments (Tester et al., 1965). The discrepancy between these results may reflect differences in methodology. American Toads are able to return to their original capture area when displaced up to 235 m, and neither blinding nor ablation of the olfactory tracts had an effect on homing ability (Dole, 1972a). Toads sometimes took a nondirect route, but orientation was possible in both clear and overcast conditions.

There is a considerable amount of turnover at a breeding pond, with continuous immigration and emigration of adults during the breeding season.

Oldham (1966) recaptured only 29% of the males and 8% of the females at his Ontario breeding site over a 4 yr period, with a mean duration of 2.9 nights per stay; many toads (34%) remained 1 night or less. In Maine, males remained from 1 to 15 nights at breeding sites (Sullivan, 1992a). The duration of time spent at the breeding ponds does not necessarily enhance the probability of mating, but instead affords more mating opportunities for more days within a breeding season (Gatz, 1981a). Unsuccessful males are smaller than males that breed successfully.

Males have a tendency to remain within one area of the breeding pond during a single night rather than move from location to location; they often change locations from one night to the next, however (Gatz, 1981a). There is no assortive mating with respect to genotypes, even though males tend to remain within a particular area (Christein et al., 1979). Indeed, breeding success may be correlated with the activity of the male as he moves around a pond from one night to the next, as this could increase the probability of encountering a female. The areas of most chorusing also can change within and between seasons. Not surprisingly, males tend to remain longer at a pond than females (e.g., 7 days vs. 4.8 for breeding females and 6.2 for nonbreeding females in Ohio, Christein and Taylor, 1978; to 23 days for Maryland males, Forester and Thompson, 1998) since the probability of reproductive success is much less in males than females.

Most American Toads return to the same breeding pond annually rather than choose different breeding ponds from one year to the next. However, there are important exceptions, and at some ponds >50% of the males visit different breeding sites both within and between seasons, traveling as far as 230 m in a straight line between wetlands over a period of a few days (Ewert, 1969). Female movement between breeding sites is very rare within a season, as most females are amplexed rapidly at a breeding site and deposit all of their eggs. However, females may visit different breeding sites between years (Ewert, 1969); one female moved 762 m between breeding sites in successive years.

When displaced, toads usually orient toward the breeding site from which they were removed instead of moving to the nearest breeding site (Oldham, 1966). In a study of toad genetics, there were significant differences in mitochondrial DNA haplotypes between 5 breeding sites located at least 500 m apart. Significant differences between haplotypes were not recorded from one year to the next within a breeding pond, however, confirming site philopatry (Waldman et al., 1992). These results seem at odds with the movements recorded by Ewert (1969), and suggest the possibility of regional or population differences in genetic structure and within- and between-season movement patterns.

CALLING ACTIVITY AND MATE SELECTION

Male American Toads attract mates both by calling and by active search, which they alternate at wetland breeding sites (Fairchild, 1984; Wells and Taigen, 1984; Sullivan, 1992a). Males vary considerably in the amount of time they spend calling, searching for females, and in clasping attempts. Aerobic capacities of American Toads are higher than those of other anurans (Taigen and Pough, 1981; Taigen et al., 1982), and allow this species to vigorously pursue mates during the short reproductive period. However, aerobic capacity is not correlated with the time spent in any behavior, although individual males differ significantly in their aerobic capacity. Neither is aerobic capacity correlated with male body size. Thus, these physiological variables do not account for differences in the levels of male activity (Wells and Taigen, 1984).

Calling occurs along the shoreline of breeding sites or in shallow water, and certain locations within a breeding site may be favored over other locations. For example, 85% of an Ohio chorus called from an area constituting only 30% of the available shoreline (Fairchild, 1984). Toads may switch calling sites, that is, go from shoreline to water and back. Calls primarily serve as a species identifier and indicate the location of a breeding chorus. The call may also function to stimulate the final stages of oogenesis and thus bring the sexes together during the optimum time for reproduction (Christein and Taylor, 1978). The call is a high-pitched trill lasting 8–30 sec. Calling activity is short, usually occurring over a period of only a few days (Gatz, 1981a; Sullivan, 1992a; Howard and Young, 1998). Howard and Young (1998) classified 50.5% of the males at an Indiana breeding site as callers, and from 33–64% of the males present were chorusing at any one time over 7 breeding seasons. In Ohio, calling activity could not be correlated with successful amplexus, as many noncalling males were also successful (Gatz, 1981a).

Call characteristics vary among individuals and populations, and may or may not be correlated with morphology and environmental variables. For example, dominant call frequency (1,400–1,900 cps) decreases with body length (Sullivan, 1992a) and mass, but only weakly with age. Water temperatures from 10 to 23°C have no effect on dominant call frequency (Howard and Young, 1998). However, pulse rates (normally

Calling male *Anaxyrus americanus*. Photo: David Dennis

18–59 pulses/sec; Zweifel, 1968a) are weakly correlated with body length and mass, but not age. Pulse rates are, however, strongly correlated with water temperatures near a calling male, and with body temperature (Zweifel, 1968a). In discrimination tests, females did not discriminate between high and low call frequencies, but instead preferred high calling effort (duration × call rate) (Sullivan, 1992a). The dominant frequency of the release call is also correlated negatively with male mass and body size (Sullivan, 1992a). Regardless, males make the same release call whether they are amplexed by conspecifics or by artificial means. The same pulse rates and dominant frequencies are used for release calls regardless of the species of the amplexing intruder (Leary, 1999).

Call duration varies annually (but not within a season; Sullivan, 1992a), and there is no correlation with male body mass, length, condition, or age. Call duration is weakly correlated with water temperature and negatively correlated with pulse rate, but it is unrelated to dominant call frequency (Howard and Young, 1998). Call durations lasted a mean of 7.8 sec (range 1.1 to 18.6 sec), with 1.6 calls per min in one Indiana population (Howard and Palmer, 1995), whereas in the Northeast, call duration was from 4 to 11 sec (Zweifel, 1968a). Mean and maximum call rates varied between 0 and 3.2 calls per min, but rates were not related to age, body length, or mass (Howard and Young, 1998). These authors noted that the social context of calling affected characteristics of the call, such as a reduction in dominant frequency when nearby males' calls overlapped. In such circumstances, females tend to choose the male with the lowest dominant frequency. However, call duration was not affected by the presence of a nearby calling male, and females do not have preferences for certain dominant frequencies, regardless of their age or past breeding experience (Howard and Palmer, 1995). They do, however, tend to prefer the leading caller when mixed choruses or paired males are present.

Male American Toads have individual differences in mating calls, which are evident at the population level, such that females may be able to discriminate relatives from nonrelatives at a breeding pond. Only 2 of 86 pairings were by toads genetically confirmed as relatives in ponds in Massachusetts, a ratio far below what would be expected if mating was completely random (Waldman et al., 1992). After adjusting for temperature and size, pulse duration, interpulse interval, pulse rise time, and call duration varied among breeding ponds (Waldman et al., 1992). Individuals that were genetically similar had similar call characteristics (except for dominant frequency), with genetically dissimilar individuals having much more varied call characteristics. These variables help ensure that despite breeding site philopatry, matings between relatives likely rarely take place.

In areas where closely related members of the *americanus* group of toads come into contact and hybridize, call characteristics are intermediate between the parental species (Zweifel, 1968a; Cook, 1983). For example, *A. woodhousii* × *A. americanus* hybrids in Oklahoma call at 57 pulses/sec (vs. 29–33 pulses/sec for *A. americanus* and no distinct pulses for *A. woodhousii*) and 5.3 sec in duration (vs. 7.5–14.5 sec for *A. americanus* and 0.9–2.6 sec for *A. woodhousii*) at 20.5°C (Blair, 1956a). In Ontario at 16°C, *A. americanus* calls at a mean of 33 pulses/sec for a mean of 6.5 sec (or 216 pulses per call) compared with *A. fowleri* (90 pulses/sec, 1.8 sec in duration, 157 pulses per call) and hybrids (49 pulses/sec, 4.2 sec call duration, 203 pulses/call) (Green, 1982). The intermediate nature of the calls of hybrids is likely due to the intermediate morphology of the laryngeal cartilages that affect call pulse (Green, 1982).

Females move through a group of chorusing and nonchorusing males as they migrate to a breeding site. If not amplexed, they move toward a calling male and initiate clasping by touching him. Sullivan (1992a) noted that females always choose calls with higher rates or longer duration in discrimination tests. Licht (1976) proposed that size-assortive mating occurs in *A. americanus*, but there is no evidence of this (Wilbur et al., 1978; Gatz, 1981a; Kruse, 1981a; Howard and Young, 1998). Instead, females generally select large males over smaller males, particularly while they are chorusing, indicating a degree of sexual selection by the female. It is not entirely clear why she does so, however, since neither male size nor arm length (for holding on during amplexus) is correlated with egg clutch size or mating success. Whereas there are heritable differences in the larval offspring among different mated pairs in terms of survival to metamorphosis, mass at metamorphosis, and timing of metamorphosis, these fitness variables are not correlated with size or age of the male. Thus, there appears to be no support for the hypothesis that adult size and age are correlated with larval genetic superiority or mating success (Kruse, 1981a; Kalb and Zug, 1990; Howard et al., 1994).

Either sex may initiate reproduction at the breeding site, including noncalling males. In Indiana, females initiated 43% of the 68 pairings observed by Howard and Young (1998), with males initiating 57%; an earlier study noted females initiated reproduction in 31% of the encounters observed (Howard, 1988a). Prior breeding experience had no effect on the size of the males that females chose, and females did not choose larger males with successive pairings. Females may mate with the same male more than once (22% of observed pairings) and as many as 3 times (Howard and Young, 1998). In Maryland, as many as 70% of the females approaching a pond are intercepted by terrestrial males, and these are often smaller than calling males. Thus, small males may have mating opportunities, even though their calls are not as attractive to females (Forester and Thompson, 1998).

Many males are not successful. In Ohio, Gatz (1981a) estimated an average of 0.2 matings per male per breeding season. Mating success may vary depending on the number of toads calling. Fairchild (1984) suggested that in large populations, calling males and noncalling males have an equal probability of finding a mate. In small populations (<20 calling males), the callers are more likely to mate successfully. At his Ohio breeding site, calling males were outnumbered by noncalling males, and never more than 27% were calling at any point; of 294 toads, only 8.1% were calling on the night of the largest breeding aggregation (Fairchild, 1984). He did not report the success rate of noncalling males.

REPRODUCTION

American Toads are an explosive breeder, with most reproductive activity occurring on only a small number of nights (for example, 7–14 in Ontario, Licht, 1976; mean 3.7 nights, also in Ontario, Oldham, 1966; <2 nights in Indiana, Howard, 1988a; 4 nights between 12 May and 16 June in Maine, Sullivan, 1992a; 3 nights between 24 April and 9 May in Ohio, Fairchild, 1984). Calling thus may occur over an extended period of weeks within the breeding season, but actual mating takes place on only a few nights.

The breeding season occurs from late winter to midsummer, depending on latitude. Breeding has been recorded from January to April (Alabama: Mount, 1975), February and March (North Carolina: Murphy, 1963), February to May (Tennessee: Dodd, 2004), February to June (Louisiana: Dundee and Rossman, 1989), March and April (Missouri: Johnson, 2000; Virginia: Kalb and Zug, 1990), March to May (Arkansas: Trauth et al., 1990), April (Ohio: Gatz, 1981a; Lovich and Jaworski, 1988), April and May (Illinois: Smith, 1947; Kansas: Fitch, 1958; New Hampshire: Taylor, 1993; New York: Wright, 1914; Raney and Lachner, 1947; Ohio: Langlois, 1964; Christein and Taylor, 1978; Ontario: Licht, 1976; Bishop et al., 1997; Pennsylvania: Hulse et al., 2001), April to June (Ohio: Walker, 1946; Rhode Island: Anonymous, 1918; Wisconsin: Vogt, 1981), April to July (Maine: Hunter et al., 1999; Massachusetts: Miller, 1909a; New Brunswick: Gorham, 1964; southern Ontario: Piersol, 1913; Prince Edward Island: Cook, 1967; Québec: Lepage et al., 1997), early May (Massachusetts: Allen, 1868), late May (Nova Scotia: Bleakney, 1952), and May and June (Maine: Sullivan, 1992a; Minnesota: Ewert, 1969; northern Ontario: Schueler, 1973; South Dakota: Fischer, 1998).

Amplexus is either axillary or supra-axillary, depending on the size of the female (Aronson, 1944). During amplexus, the male may make an "amplexus call," a series of clicks made as the male presses his chin to the female's parotoid region. Price and Meyer (1979) speculated that this sound may help induce oviposition. Once amplexus occurs, mating pairs tend to stay together until oviposition, approximately for 12–24 hrs (*in* Wells, 1977a). Male displacement during amplexus is uncommon (for example, 2 of 221 opportunities; Howard et al., 1994), but may happen when a larger male displaces a smaller male.

Both the female and the primary amplexing male discourage additional suitors by vigorously kicking them with their back legs. However, some smaller toads may learn an alternative strategy to surreptitiously fertilize a clutch rather than by direct confrontation. They do this by backing up to a mated pair so that the second male's cloaca is immediately next to that of the mating male. Presumably as eggs are released by the female, both males then would be in a position to fertilize them (Kaminsky, 1997). How prevalent this behavior is in mating choruses is unknown.

As with other toads, males have a warning sound (termed a "chirp" by Aronson, 1944) and associated vibrations that alert arduous males as to the sex of a male erroneously courted or amplexed (Aronson, 1944; Blair, 1947b; Cory and Manion, 1955; Oldham, 1966; Licht, 1976; Gatz, 1981a). After one male attempts unsuccessful amplexus with another male, the roles may be reversed and the sequence of behaviors repeated. Afterward, the males swim away from one another and begin calling at an extended distance from the site of the encounter. Precocious amplexus also has been recorded in juvenile American Toads (Huheey, 1965a).

At breeding ponds, males greatly outnumber females at any particular time (Wright, 1914; Blair, 1943b; Ewert, 1969; Dirig, 1978; Sullivan, 1992a). For example, Raney and Lachner (1947) recorded 451 males to 33 females at a breeding pond in upstate New York over a 2-week period; in Oklahoma Blair (1943b) recorded 767 males and 40 females based on observations at 50 ponds from 22 March to 26 May; Dirig (1978) recorded 579 males and 21 females at a Catskill breeding pond; Oldham (1966) noted that only 15% (of 5,937 over 4 yrs) of the toads at a breeding site in Ontario were female; Christein and Taylor (1978) reported a sex ratio of 7 males per female at an Ohio breeding pond; Murphy (1963) reported 2.1 males per female at a North Carolina pond. As noted above, the majority of males may not mate successfully in any one breeding season (Blair, 1943b; Howard, 1988a), although a few males mate more than once (*in* Wells, 1977a; Howard, 1988a). Gravid females at breeding ponds usually find a mate successfully.

Mating may occur diurnally or, more commonly, at night. Eggs are deposited on the bottom in shallow fresh water (10–30 cm in depth), or entwined in branches, leaves, aquatic plants, twigs, and other bottom debris around the margins of wetlands. However, *A. americanus* also may oviposit eggs in somewhat brackish water in coastal areas (Schueler, 1973; Kiviat and Stapleton, 1983; Klemens, 1993). Fertilization approaches 100%, but not all eggs are viable, and males are capable of fertilizing several clutches in a breeding season (Kruse and Mounce, 1982). Williams (1969) recorded hatching rates of 57–62% in Minnesota. Eggs usually hatch synchronously over a period of a few days, depending on water temperature (Wright, 1914).

The oviposition sequence and the behavior of the courting pair have been described in detail (Wright, 1914; Aronson, 1944; Licht, 1976), and consist of amplexus, preovulatory behavior (restlessness and abdominal muscular contractions lasting from a few minutes to several hours), and oviposition (back arching, extrusion of eggs with simultaneous fertilization, movement along the wetland bottom as egg strings are extruded). Aronson (1944) described one complete sequence lasting 3 hrs and 30 min; Howard (1988a) noted successful matings lasting 3.6–27.1 hrs. If the egg strings break during deposition, amplexus ends and the mating sequence is then either reinitiated with the same or a different male. A recently spent female will arch her back in a very pronounced manner to indicate she has finished spawning, although it may take several tries before the male releases her and moves elsewhere.

Eggs are deposited in long strings (ca. 5 mm in width), one string from each oviduct. Estimates of the number vary from 1,700 to 20,000 per female, and 4,000 to 12,000 per string (Miller, 1909a; Anonymous, 1918; Treat, 1948; Livezey and Wright, 1947; Wright and Wright, 1949). Specific records include: mean 4,701 (range 1,840 to 13,982) in Arkansas (Trauth et al., 1990); 3,341 in Louisiana (Dundee and Rossman, 1989); 2,200–6,500 in Michigan (Brockelman, 1968); 1,767–9,765 in Indiana (Howard, 1988a); 3,929–15,835 in Massachusetts (Miller, 1909a). Clutch size is strongly correlated with female body size, but not with the size of the male (Howard, 1988a).

American Toads hybridize and produce viable offspring with a number of other toads that may be spatially sympatric. Hybridization normally is avoided because of temporal differences in the breeding season (American Toads usually breed earlier than congeners) and differences in breeding site preferences. Differences in calls, adult size, and behavior are often insufficient to maintain reproductive isolation with sympatric *Anaxyrus* in mixed choruses.

LARVAL ECOLOGY

American Toad larvae are usually found in rather shallow water (but note a depth record of 8 m; Richmond et al., 1999), where they are observed in large numbers due to the large number of eggs oviposited by the female. Larvae have been used to study many aspects of pond ecology, particularly the ways in which density, competition, conspecific interaction, and predation influence vital components of growth, size, survivorship, and the duration of larval periods. As with many other species producing large numbers of tadpoles, the time to metamorphosis, individual growth rate, and larval mortality of American Toads are related to the initial density of the tadpoles present in intraspecific trials. Thus, an increase in density retards growth in experimental pens and increases the time to transformation (Brockelman, 1969; Wilbur and Fauth, 1990). Size at metamorphosis is inversely related to density, presumably as tadpoles at lower densities have reduced competition for food resources. Food addition, then, results in a density-dependent increase in growth rate, as it partially counters the effects of competition. Adding predators increases mortality rates (Wilbur and Fauth, 1990), and thus negates the density effects (Brockelman, 1969). Tadpoles may even use visual cues to determine the density of conspecifics (Rot-Nikcevic et al., 2006).

American Toad tadpoles are often observed forming dense aggregations in shallow water near the shore, where they cover between 20 and 90% of the substrate (Dodd, 1979). The function of the aggregations may be related to feeding efficiency (larvae are benthic filter feeders), defense, enhancing the thermal environment, or even as a way to prolong water retention in shallow soft-bottomed pools (Black, 1975). The propensity to aggregate follows a daily cycle, whereby tadpoles, scattered in deeper water at night, come together to form dense aggregations as light intensity increases. These aggregations move toward the warmer and shallower parts of the pond as the day progresses. Beiswenger (1977) found that in Michigan, tadpole aggregations prefer shallow warm water (26–29°C) that is 1–3°C higher in temperature than the rest of the pond. Individuals then disperse back to deeper water with decreasing light intensity late in the day. When days are heavily overcast, tadpoles do not form dense aggregations, and activity is severely curtailed.

Tadpole aggregations initially form as a result of both visual and thigmotactic cues between individuals immediately after hatching. Visual cues and sensory input from the lateral line system then help maintain the schools, an association that is strengthened through kin-based olfactory input (Dawson, 1982). There is no evidence, however, that vision plays a role in kin recognition per se. Thus, group cohesion is visual, whereas group formation is based on sibling recognition. Habitat differences do not influence schooling association.

Larval aggregations may be stationary, where they feed upon dead tadpoles or carrion (necrophagous aggregations), periphyton or other debris (feeding aggregations), or gather immediately proceeding metamorphosis (metamorphic aggregations). Stationary aggregations may deepen or enlarge shallow pools, forming what Black (1975) describes as "tadpole nests." In this regard, Heinen (1993) suggested that aggregations functioned to retard desiccation, since desiccating tadpoles in dense aggregations had more chances to survive and minimize weight loss than single tadpoles. Tadpoles also form moving aggregations, where they swim parallel to one another, moving from one location to the next within a pond and do not feed (termed streams), or in slow, dense clusters stirring up food particles (schools) (Beiswenger, 1975).

American Toad tadpoles tend to associate with siblings rather than nonsiblings in both field and laboratory settings, regardless of whether they are raised in mixed assemblages or in social isolation (Waldman and Adler, 1979; Waldman, 1981, 1982a, 1986a, 1991; Dawson, 1982). In experimental trials, tadpoles displayed no differences in positive association with siblings, whether the siblings were familiar (that is, they were raised together) or unfamiliar (that is, they were raised in separate tanks) (Waldman, 1985a). Larvae appear to recognize siblings based on "shared recognition traits" rather than on prior social experience (Waldman, 1986a).

The positive aggregative association depends, however, on the period of development. When both siblings and nonsiblings are raised in close contact during early development, the positive association toward siblings tends to disappear. It appears that tadpoles are better able to discriminate maternal half-sibs than paternal half-sibs, especially if there is limited contact during the early stages of larval development. Waldman (1981) suggested that the mother might impart some means of chemical recognition to the tadpoles to facilitate kin recognition and schooling behavior. Indeed, visual cues do not seem to assist in recognition, and the cues used do appear to involve olfaction (Dawson, 1982; Waldman, 1985b). Very large schools form and subdivide into smaller schools (Waldman, 1991). The large schools may not consist entirely of siblings, however, although the subdivision of such schools might reflect kinship recognition and social pairing (Beiswenger, 1975; Waldman, 1982a).

Indeed, the level of interaction occurring within dense aggregations may have effects on growth and time to metamorphosis, regardless of food levels and density per se. Breden and Kelly (1982) reported that increased conspecific interaction increases the variation in the time to metamorphosis, thus decreasing a tendency to have synchronous metamorphosis. They also determined that increased intraspecific interactions result is an increase in body mass at metamorphosis.

Even having huge numbers of tadpoles is no guarantee than any will survive to metamorphosis. Seale estimated that 200,000 *A. americanus* eggs hatched in her Missouri study pond in each of 2 years, yet no larvae survived, presumably because of predation by salamander (*Ambystoma tigrinum*) larvae. Williams (1969) found a mean survivorship of only 24.5% in field cages in Minnesota.

Metamorphosis occurs in 39 days in North Carolina (Wilbur, 1977b), 37–52 days in Michigan (Brockelman, 1968), 41–66 days in New York (Wright, 1914), 49–63 days in Rhode Island (Anonymous, 1918), and 42–70 days in Indiana (Minton, 2001). In Arkansas, recently metamorphosed toads have been found as late as 26 July (Trauth et al., 1990). In upstate New York, metamorphosis occurs from June to August (Wright, 1914), and from mid-July to late August in Nova Scotia (Bleakney, 1952).

Larvae are tolerant of high temperatures (40–43°C) at least for a short time, and survival is 100% at 36°C for 24 hrs. Their preferred temperatures appear to be in the range of 14 to 26°C (Williams, 1969); optimal temperatures for locomotion are >17°C (Tracy et al., 1993). Forming dense aggregations in shallow water helps ensure that the black heat-absorbing larvae will have a proper thermal environment for development, even during the cool early spring or at northern latitudes.

DISPERSAL

American Toads are tiny at metamorphosis and are incapable of rapid dispersal. Immediately after metamorphosis, however, juvenile American Toads rapidly increase their endurance and aerobic activity; increases in hematocrit, hemoglobin concentration, and heart mass also help to improve their activity capabilities prior to dispersal (Pough and Kamel, 1984). Anaerobic scope, however, is independent of body size (Taigen and Pough, 1981). These latter authors suggested that basking by juvenile toads prior to dispersal hastens physiological development and increases their capacity for aerobic metabolism. This in turn gives them the capacity to disperse.

Most juvenile dispersal takes place over about a two-week period, depending on location and weather conditions, and is usually completed within six weeks of the initiation of dispersal (Murphy, 1963). Emigration is not correlated with the site of egg deposition, that is, juveniles do not disperse near where they hatched in the pond (Christein et al., 1979). Emigrating juveniles show a preference for forest habitats, where they are less exposed to the dangers of dehydration, at least when experimental ponds where development is completed are located at a forest edge (Rothermel and Semlitsch, 2002; Walston and Mullin, 2008). However, Rothermel (2004) could not demonstrate specific orientated movements using the same experimental design from ponds located in pastures; recapture rates were low, likely due to heat-related mortality. She suggested that celestial cues during the entire larval period were not essential for postmetamorphic orientation toward forested habitats. Further, the study emphasized the need for connectivity between breeding sites and forested habitats, as survival was inversely related to the distance from the nearest forest. It is imperative for tiny juveniles to reach cover as soon as possible, and distances > 50 m from a forest edge may be insurmountable under adverse weather conditions (heat, direct sun, exposure to desiccating wind) during emigration. Open areas thus may form barriers to dispersal by juveniles, even though they do not do so for adults.

Adult dispersal occurs diurnally, especially in areas where nighttime temperatures drop below freezing, or nocturnally. Toads may use riparian corridors to move among habitats, although adult toads do not prefer to travel through inundated swamps and wetlands (Burbrink et al., 1998). Rainfall is not required to initiate adult dispersal. In a study of 16 female *A. americanus* tracked by radio-telemetry, Maryland toads dispersed from 250 to 1000 m from the breeding pond, and most traveled > 400 m. Dispersal was nonrandom and linear, with toads interrupting travel with sedentary periods (Forester et al., 2006). Most long-distance dispersal took place within 10 days of leaving the breeding pond, with one toad moving 610 m in a single day.

Movements usually begin within a week after chorusing ends. In Minnesota, Ewert (1969) recorded dispersal taking place over 109 days one year and for >73 days the next. One toad took 22 days to move 701 m and was still moving away from its breeding site when last observed. Another toad traveled 1,005 m to its overwintering site, and another took 68 days to reach its final overwintering location. Movements of > 200 m per 24 hr period were common in Ewert's population; most movements were ca. 26 m per 24 hr period, however, as toads moved to overwintering sites. Summer movements were much the same as in Forester et al.'s (2006) Maryland population.

DIET

Larvae begin feeding at Gosner (1960) stage 25. Tadpoles eat filamentous algae, blue-green algae, periphyton, diatoms, soft tissues from vascular plants, eggs of crustaceans, carrion (including other tadpoles), and fecal material (Munz, 1920). Larvae also filter detritus suspended in the water column (Test and McCann, 1976). Feeding ceases about Gosner stage 42. In experimental trials, American Toad tadpoles chose foods higher in calcium than in controls, which suggests a degree of selectivity in tadpole food choice (Botch et al., 2007).

Juveniles eat a wide variety of small invertebrates (Hamilton, 1930). For example, in a study in Québec, the diet of recently metamorphosed American Toads consisted primarily of springtails (Collembola), mites (Acarina), parasitoid wasps and ants (proctotrupoid and formicid Hymenoptera), and beetles (staphylinids and carabids) (Leclair and Vallières, 1981). Other prey included spiders, larval lepidoptera, larval and adult flies, and various other insects (Hemiptera, Homoptera, Neuroptera, Trichoptera, Thysanoptera).

American Toads are both active and sit-and-wait foragers. Adults eat a great variety and quantity of invertebrates, including ants, beetles, grasshoppers, isopods, millipedes, spiders, gypsy moths, tent caterpillars, diptera larvae, snails, slugs, butterflies, moths, wasps, and virtually any small animal that can be captured with its protractible tongue (Kirkland, 1897, 1904; Garman, 1901; Dickerson, 1906; Miller, 1909a, 1909b; Hine, 1911; Surface, 1913; Smith and Bragg, 1949; Gilhen, 1984; Klemens, 1993). Their large appetite makes them valuable in controlling agricultural pests. Kirkland (1897; repeated by Dickerson, 1906; and Patch, 1918), reported that a single toad could eat 9,936 injurious insects and 368 beneficial insects in a single season, making its economic value $19.88 in *1897* dollars! Miller (1909a) disputed that claim somewhat, recalculating the yearly value of a toad at $5.00.

PREDATION AND DEFENSE

Chief among the antipredator defenses of American Toads are the granular glands that secrete poison when the gland is mechanically stimulated, as when the toad faces a severe threat from a large predator. During normal handling, a toad does not secrete poison. The granular glands are concentrated dorsally as somewhat elliptical or oblong parotoids behind the eyes and in the warty protuberances of the skin. Scatterings of granular gland occur elsewhere, although they are not prominent. Juveniles also have these glands, and, when handled by invertebrate predators, juveniles become immobile and are often released as the predator is exposed to the skin secretion (Brodie et al., 1978). Details of glandular structure of the American Toad are in Muhse (1909).

American Toads, both larvae (Kats et al., 1988) and postmetamorphs, are toxic or noxious to many potential predators. Interestingly, recently hatched larvae and tadpoles approaching metamorphic climax are unpalatable to both vertebrate and invertebrate predators, but intermediate-stage larvae are not (Brodie et al., 1978; Brodie and

Formanowicz, 1987). The toxicity serves to discourage predation, rather than to deliver a lethal toxin to the predator. Toxin is also deposited in the eggs via the female's blood. They may be eaten by certain salamander larvae, but they are often spit out after initial contact, whereupon the larva, rather than the jelly capsule, is consumed.

Larvae are also unpalatable to many potential predators, but the degree of unpalatability may depend upon the predator (Voris and Bacon, 1966; Walters, 1975; Kruse and Stone, 1984; Holomuzki, 1995; Smith et al., 1999) as well as the larval stage of development. Paradoxically, larvae may be more successful in certain ponds with fish predators than in ponds where fish are removed. Fish depress the abundance of other tadpole predators, and when predaceous fish are removed, increases in the level of invertebrate predation result in decreased larval American Toad survival (Walston and Mullin, 2007b).

American Toad larvae approaching metamorphosis have an immobility response, whose duration is inversely correlated with stage of development (Dodd and Cupp, 1978). When disturbed, they draw their limbs into the body and refrain from any movement. However, just prior to metamorphosis, the immobility response is very short, as all attention is directed toward metamorphosis. Recent metamorphs, however, again adopt an immobile response that allows concealment from certain predators.

The mosquito *Culex territans* feeds on American Toads and even can use the toad's call to locate its next blood meal (Bartlett-Healy et al., 2008). Other predators include Chipping Sparrows (*Spizella passerina*) on recent metamorphs (Gorham, 1964), and Eastern Hognose Snake (*Heterodon platirhinos*), Eastern Garter Snake (*Thamnophis sirtalis*), Northern Watersnake (*Nerodia sipedon*), Blotched Watersnake (*N. erythrogaster*), Queensnake (*Regina septemvittata*), Snapping Turtles (*Chelydra serpentina*), hawks, grackles, geese, chickens, guinea fowl, owls, herons and waterfowl, skunks (*Mephitis mephitis*), and raccoons (*Procyon lotor*) on adults (Miller, 1909b; Burt, 1935; Oldham, 1966; Schaaf and Garton, 1970; Forester et al., 2006). *Heterodon* in particular is a toad specialist, containing specialized dentition for manipulating and deflating toads. Several incidents regarding mass predation on breeding toads have been reported, with either raccoons or striped skunks (*Mephitis mephitis*) implicated as responsible (Groves, 1980).

Predators of larvae include leeches (*Desserobdella picta*), odonate larvae (*Anax, Libellula, Leucorrhinia, Sympetrum, Eurythemis, Plathemis*), backswimmers (notonectids), predacious diving beetles (*Belostoma, Dytiscus*), waterscorpions (*Ranatra*), other frog larvae (*Lithobates*), salamanders (*Ambystoma opacum, Notophthalmus viridescens*), and Least Sandpipers (Brockelman, 1969; Walters, 1975; Stangel, 1983). American Toads, which breed four to ten weeks after Wood Frogs in North Carolina, avoid breeding ponds inhabited by Wood Frog larvae (Petranka et al., 1994). Not surprisingly, Wood Frog tadpoles totally consumed eggs and hatchlings of the toad quickly in a series of natural and mesocosm experiments, and none survived (Petranka et al., 1994).

Larval American Toads respond to the potentially debilitating cercariae of *Echinostoma* trematodes by alternating periods of swimming fast with extremely rapid twisting, turning, and tumbling movements (Taylor et al., 2004). These movements have been described as "massive, violent and multiplanar" (Taylor et al., 2004). Such rapid movements last 4–8 sec. Presumably these rapid and vigorous movements help to dislodge cercariae crawling on the skins of tadpoles and lodging in places where they could cause limb malformations. Taylor et al. (2004) observed cercariae attempting to enter the spiracle of toad larvae, where buccal pumping rates are not as high as they are in other species (Wassersug and Hoff, 1979). That toad larvae are distasteful probably allows them to move more vigorously (and thus conspicuously) in their attempts to dislodge these parasites, more so than a less palatable species might do.

In the presence of potential predators, there are conflicting reports on the behavior of American Toad larvae. American Toad larvae do not modify their activity levels or swimming behavior in experimental trials where olfactory cues of fish, odonate larvae, or combinations thereof were present but not in direct physical proximity (Richardson, 2001; Smith et al., 2009). Likewise, toad larvae did not reduce the amount of time swimming in the presence of nonindigenous mosquito fish (*Gambusia affinis*) or native bluegill (*Lepomis macrochirus*), nor did they spend more time in vegetation than exposed on gravel substrates (Smith et al., 2008; Smith and Awan, 2009).

In contrast, Holomuzki (1995) reported reduced larval activity in the presence of fish, and Relyea (2001b) found that toad larvae reduced their level of activity in the presence of both fish (*Umbra*) and dragonfly larvae (*Anax*), but not in the presence of newts (*Notophthalmus*). In addition, morphological changes took place in response to the predator's presence. Tadpoles developed shallower and longer tails in the presence of *Umbra*, and shallower tails with *Anax*, than they did with newts and predaceous beetles (*Dytiscus*) (Relyea, 2001b). Presumably, this phenotypic plasticity in response to predator presence has some adaptive component, although most of these predators tend to avoid *Anaxyrus* tadpoles. Likewise, American Toad larvae decrease activity levels in the presence of *Anax* predators, as well as when food availability is increased. Activity levels of larvae exposed to various types of predators do not change with predator density or satiation in laboratory trials, or with larval body size, despite differences in noxiousness (Anholt et al., 1996). In contrast, activity levels increase in the presence of water beetles (*Dysticus*) (Smith and Awan, 2009). Body size further affects the activity of larvae in the presence of predators; small tadpoles are more active and use open water more often than large tadpoles.

Chemoreception plays an important role in larval defense. Larvae are able to recognize some predators innately using chemoreception. American Toad larvae, both wild-caught and laboratory-reared, decrease activity and increase aggregation in the presence of water-borne cues emanating from fish (*Lepomis macrochirus*) and odonate (*Anax*) predators. This response does not require previous experience with the predators, as both naïve and experienced larvae exhibit the same behavior. In contrast, no such behaviors are observed in response to chemical cues from the newt *Notophthalmus viridescens* (Gallie et al., 2001). In addition, larvae are able to recognize cues released from injured conspecifics, and will avoid an area where injured tadpoles occur (Petranka, 1989).

In trials featuring beetles (*Dytiscus*), fish (*Umbra*), newts (*Notophthalmus*), and dragonfly larvae (*Anax*), larval toads had low levels of predation. Toads were easily captured by *Umbra* and *Anax*, but were immediately rejected; however, the most common response by these predators was to ignore the tadpoles. Newts also found toad larvae unpalatable, whereas *Dysticus* readily captured and ate toad tadpoles in about half their attempts (Relyea, 2001a). The beetles also took a long time to consume the tadpoles they ate (mean 17.2 sec).

Adults make use of the following behaviors when confronted by a predator: fleeing rapidly, remaining immobile, taking a crouching stance, tucking the chin downward toward the pectoral region, body inflation, digging, hiding, and backing away (Marchisin and Anderson, 1978; Hayes, 1989; Heinen, 1994, 1995). Body postures and inflation presumably make the toad appear larger or more difficult to handle. Postures may be oriented toward an approaching predator, perhaps to direct the dorsum containing the parotoid glands and skin secretions to the face (and eyes) of the predator. By remaining immobile, both adult and recently metamorphosed juveniles may be able to avoid detection, especially when the potential predator is visually oriented. American Toads also exhibit an ability to blend into surrounding litter by slowly changing their background coloration.

POPULATION BIOLOGY

Newly metamorphosed American Toads are 10 mm and weigh approximately 0.05 g in New York (Miller, 1909a; Hamilton, 1934; Raney and Lachner, 1947; Pough and Kamel, 1984). They are 12–15 mm in Indiana (Minton, 2001), but only 6–8 mm in Louisiana (Dundee and Rossman, 1989). By first overwintering after metamorphosis, they are approximately 20 mm SUL, and grow to 60 mm SUL in their first full season. They add approximately 18 mm during the second full season, with negligible growth thereafter (Hamilton, 1934). Of course, these growth rates will vary geographically, depending on the extent of seasonal activity, but the pattern of initial rapid growth followed by decreasing growth to sexual maturity should be consistent among populations.

Toads tend to grow slowly, at least in upstate New York. Small males (63–72 mm SUL) grew at an annual rate of 4.8 mm/yr, 73–76-mm males grew 4.0 mm/yr, 77–81-mm males grew 2.9 mm/yr, and 82–94-mm males grew 0.1 mm/yr (Raney and Lachner, 1947). These authors sug-

Adult *Anaxyrus americanus charlesmithi*. Photo: David Dennis

gested that male toads in this region reached their maximum length at three or more years, but they did not have comparable data for females.

Most American Toads in breeding choruses are 2–4 yrs in age, with a few individuals reaching 5 yrs (Acker et al., 1986) or older. In Indiana, males first breed when they are 2 yrs of age, and females at 3 yrs. Males in an active chorus ranged from 2 to 6 yrs of age, and were 49–70 mm SUL (mass 14–48 g) (Howard and Young, 1998). Indiana females varied between 3 and 6 yrs in age, with older females generally larger than younger females. Body length is positively correlated with both age and mass. The growth rates reported by Hamilton (1934) for upstate New York suggest a similar age at first breeding. In Virginia, skeletochronology revealed most of the breeding population consisted of males at 3–4 yrs and females at 4–5 yrs (Kalb and Zug, 1990); age was not correlated with size in this southern population.

COMMUNITY ECOLOGY

Both predators and interspecific competitors may have significant effects on larval size, the length of the larval period, and survivorship, depending on the densities of the various members of the community (Wilbur and Fauth, 1990). In experimental trials, the growth rates of American Toads are negatively affected by the density of other larvae (*Lithobates palustris*, *Lithobates sphenocephalus*, *Scaphiopus holbrookii*), including conspecifics (Wilbur, 1977b; Alford and Wilbur, 1985; Alford, 1989a, 1989b; but see Wilbur, 1987) or even the density of snails (Holomuzki and Hemphill, 1996). In turn, the density of American Toad tadpoles has a negative effect on the density of *Hyla chrysoscelis* tadpoles (Alford, 1989a) and on snail reproduction (Holomuzki and Hemphill, 1996), but no effect on *Lithobates palustris*, regardless of when tadpoles enter the pond (Alford, 1989b). *Lithobates sphenocephalus*, in particular, significantly affects both the abundance and biomass of *Anaxyrus* tadpoles, but the effects vary depending on which species bred first in the pond (Alford and Wilbur, 1985); when this species is not present or eliminated from experimental mesocosms by predation, the mass and survivorship of *Anaxyrus* increases. By day 51 of development, however, American Toad tadpoles are of a sufficient size that even the presence of the newt *Notophthalmus viridescens* has no effect upon them. For tadpoles surviving to this stage, metamorphosis occurs at greater mass than if newts had not been present to reduce initial densities of tadpoles. Even when *Hyla chrysoscelis* tadpoles were reduced in number, the biomass of American Toads increased, suggesting a degree of competition between these species (Alford, 1989a). As might be expected, the effects of competition are more pronounced at higher tadpole densities.

Wilbur (1987) achieved similar results to the later study by Alford (1989a) regarding interspecific competition, although he also superimposed an experimental situation in which mesocosms were drawn down simulating pond desiccation. At low densities, most American toad tadpoles completed metamorphosis, but at high densities, competition between the toads and *Lithobates sphenocephalus* larvae prevented successful metamorphosis. The addition of newts mediated these results. Although fewer (28%) American Toads survived at low densities in the presence of *Notophthalmus viridescens* as water was drawn down, 44% actually survived at higher densities as newt predation removed toad tadpoles during the desiccation process (Wilbur, 1987). These complimentary studies illustrate how density, competition, community composition, predation, and environmental stochasticity interact in larval survival.

Other taxa that compete with American Toad tadpoles in similar habitats, such as snails (*Physella integra*) in small streamside pools, tend not to co-occur, even within spatially proximate pools, if tadpoles are present (Holomuzki and Hemphill, 1996). In this case, both compete for benthic algae as food, and their presence or absence affects the species composition of algae within the shallow pools. By avoiding each other, the snails and tadpoles eliminate the potential for competition.

Larval toads tend to metamorphose at smaller sizes in the presence of a predator, such as dragonfly larvae (*Anax*) or newts (*Notophthalmus*), as well as when food is scarce, when compared to situations where predators are absent and food is adequate. In experimental trials, the presence of dragonfly larvae did not have an effect on the larval period, but food scarcity significantly increased the duration of the larval period. The presence of potential predators does affect larval behavior, as larvae tend to decrease activity and segregate themselves away from contact with the predator. As a result, larvae in the presence of a predator will have a decreased growth rate. The metamorphic response of tadpoles is mediated through behavioral effects on growth, which in turn affects size at metamorphosis (Brockelman, 1969; Wilbur, 1987; Skelly and Werner, 1990). Larvae also may metamorphose earlier in the presence of an aquatic predator (Wilbur and Fauth, 1990), suggesting a trade-off between future fitness (i.e., larger size at metamorphosis) and current survivorship.

Habitat complexity may affect various life history aspects of larval development. In contrast to *Hyla versicolor*, for example, complex habitat structure increased time to metamorphosis, decreased mass at metamorphosis, and increased survivorship among treatments of *Anaxyrus americanus* tadpoles, an effect that was compounded by the addition of *Lithobates pipiens* tadpoles to experimental enclosures (Purrenhage and Boone, 2009). With *Lithobates*, mass at metamorphosis decreased at high densities compared with low competitor densities, regardless of the amount of time to metamorphosis. Survivorship also increased in the presence of *Lithobates* competitors in complex habitat structures, especially at low competitor densities.

Wetland vegetation may have profound effects on tadpole development, and recent experiments suggest that nonindigenous species may alter survival, developmental rates, and diet of American Toad tadpoles (Maerz et al., 2005b; Brown et al., 2006). In both laboratory and field experiments, tadpoles developed slower in the presence of both water extracts and vegetative debris composed of Eurasian purple loosestrife (*Lythrum salicaria*) compared with native cattails (*Typha latifolia*). Survival and developmental rates were more variable in the presence of loosestrife than cattails, although survival was not affected in mesocosms and field enclosures. Different plant species result in different algal communities, and thus food quality and quantity available to tadpoles; this in turn affects performance (C.J. Brown et al., 2006). Purple loosestrife did this by reducing the nutrients available for algae. Maerz et al. (2005b) and C.J. Brown et al. (2006) further suggested that *Lythrum* caused toxicity due to its high tannin concentrations, perhaps due to damage of the gills. In another example of inhibitory effects by plants, larval American Toads failed to develop through metamorphosis when exposed to a species of the green alga *Spirogyra*. Exposure to this alga not only inhibits development in natural populations, but extracts from the alga, built up through time, can be acutely toxic to developing larvae (Wylie et al., 2009).

In addition to larval effects, recently metamorphosed American Toads may experience competition from other recently metamorphosed species. Recently metamorphosed Green Frogs (*Lithobates clamitans*) do not eat recently metamorphosed *Anaxyrus americanus*, but when reared together American Toads have a smaller body mass and lower survivorship when in proximity to Green Frogs than when they are reared separately; activity levels also are reduced. Sams and Boone

(2010) suggested that interspecific competition between recently metamorphosed anurans may occur, but that in nature such effects are ameliorated by spatial segregation in terrestrial habitats.

DISEASES, PARASITES, AND MALFORMATIONS

The only disease reported from *A. americanus* is a case of xanthoma cancer (Counts and Taylor, 1977). However, American Toads are parasitized by a wide variety of protozoans and protozoan-like organisms, including include *Hepatozoon climate* (Kim et al., 1998), *Opalina obtrigonoidea* (Odlaug, 1954; Delvinquier and Desser, 1996), *Myxidium* (McAllister and Trauth, 1995), and *Toxoplasma* (Stone and Manwell, 1969). As noted by Green (2005), blood parasitism is common, especially involving *Trypanosoma fallisi*.

Nematodes include *Gyrinicola batrachiensis* on tadpoles (Adamson, 1981a, 1981b, 1981c) and *Cosmocercoides* sp., *C. dukae*, *C. variabilis*, *Oswaldocruzia leidyi*, *O. pipiens*, *Physaloptera ranae*, *Rhabdias* sp., *R. americanus*, *R. bufonis*, and *R. ranae* on adults (Baker, 1977, 1978a, 1978b, 1979a; Ashton and Rabalais, 1978; Vanderburgh and Anderson, 1987a, 1987b; Dyer, 1991; Joy and Bunten, 1997; Yoder and Coggins, 2007). Additional helminths include the trematodes *Allassostomoides* sp., *Clinostomum marginatum*, *Echinostoma* sp., *Fibricola texensis*, *Glypthelmins quieta*, *Gorgodera bilobata*, *Gorgoderina attenuata*, *G. bilobata*, *G. translucida*, *Haematoloechus similiplexus*, *Megalodiscus temperatus*, *Mesocoelium monas*, *Ostiolum medioplexus* (Bouchard, 1951; Ulmer, 1970; Brooks, 1975; Coggins and Sajdak, 1982; Dyer, 1991; Cross and Hranitz, 1999; Yoder and Coggins, 2007; McAllister et al., 2008) and the cestodes *Cylindrotaenia americana*, *Ophiotaenia saphena*, *Mesocestoides* sp. and others (Ulmer and James, 1976a; Dyer, 1991; Yoder and Coggins, 2007; see reference list in Green, 2005).

Pathogens include viruses, bacteria, and fungi. Water molds (*Saprolegnia*) form on eggs, especially unfertilized eggs (Miller, 1909b), and tadpoles (Bragg, 1962a). The fungi *Basidiobolus ranarum* and *Dermosporidium penneri* have been recorded from American Toads in the Midwest (Nickerson and Hutchison, 1971; Jay and Pohley, 1981), and dermosporidiosis is suspected in toads from Virginia (Green et al., 2002). In addition, amphibian chytrid fungus (*Batrachochytrium dendrobatidis*) has been found on *Anaxyrus americanus* (14.6%) from Maine, where it has been recorded from toe-webbing and the skin of the pelvis (Longcore et al., 2007). *Batrachochytrium dendrobatidis* has been reported in toads from Québec (4.3%), spanning a period between 1960 and 2001 (Ouellet et al., 2005a), and from Minnesota (Martinez Rodriguez et al., 2009). The deadly pathogen iridovirus has been found in *Anaxyrus americanus* (Wolf et al., 1968), and the bacterium *Aeromonas hydrophila* has been reported from a wild population (Dusi, 1949).

Biflagellated algae similar to *Chlorogonium* have been found along the lateral margins of American Toad tadpoles in Missouri (Drake et al., 2007) and Arkansas (Tumlison and Trauth, 2006). The alga may function to increase the amount of oxygen available to tadpoles, especially during thermal stress. In turn, the facultative symbiosis benefits the alga by making CO_2 available to it from the tadpole's waste products.

Juvenile American Toads may be parasitized by the green blowflies *Lucilla elongata* and *L. silvarum*, but infections do not appear to be common (Anderson and Bennett, 1963; Bleakney, 1963; Briggs, 1975; Bolek and Coggins, 2002). Mortality is high, however, in infected individuals. Other ectoparasites include the leech *Desserobdella picta* (Briggler et al., 2001; Bolek and Janovy, 2005).

American Toads living in areas where there are high concentrations of pesticides often have a considerable prevalence of hind limb deformities. In Québec, Ouellet et al. (1997) found, from three locations adjacent to agricultural fields, that 17% of 252 metamorphosing American Toads had missing or partly missing hind limb (ectromelia) or missing toes (ectrodactyly); at one location, the percentage was 69%. However, malformations have also been found on national wildlife refuges, especially affecting the hind limbs and feet (Converse et al., 2000). In Minnesota, malformations were found in 6.7% of American Toads examined (Hoppe, 2005). Toad larvae exposed to "cocktails" of pesticides also may have delayed metamorphosis, experience low survivorship, or be unable to breed at severely impacted sites.

SUSCEPTIBILITY TO POTENTIAL STRESSORS

Metals. Aluminum toxicity is inversely correlated with pH in American toads (Freda et al., 1991).

However, total aluminum (31–1,155 µg/L) and labile aluminum (1–1,073 µg/L) had no effect on transplanted *Anaxyrus americanus* embryos or tadpoles at 16 ponds in Ontario (Freda and McDonald, 1993). In laboratory trials, embryos were most stressed by aluminum at pH 4.2. Cadmium at environmentally relevant doses has no effect on adult survival, percentage of body mass loss during winter dormancy, or locomotor performance (James et al., 2004). However, toads fed mealworms with 4.7 µg/g dry weight had only a 56% survivorship compared with 100% for controls. Aqueous cadmium decreased larval survivorship under experimental conditions, but had no effect on larval body mass at metamorphosis. Instead, the time it took to reach metamorphosis was increased (James et al., 2005). These authors hypothesized that cadmium has indirect effects on larvae by decreasing the abundance of detritus and periphyton. Doses of cadmium > 18 µg/L appear to be very toxic. In Ontario, the presence of American Toads is negatively correlated with nickel concentrations (Glooschenko et al., 1992). American Toad larvae do not avoid detrimental concentrations of lead under laboratory conditions (Steele et al., 1991).

Other elements. Boron reduces hatching success in *A. americanus*. At low concentrations (50 mg L^{-1}), 46.7% of eggs hatched, whereas at high concentrations (100 mg L^{-1}), only 11.3% hatched in experimental trials in Pennsylvania. In control trials, 82% of eggs hatched (Laposata and Dunson, 1998). Since wastewater has been measured at 169 mg L^{-1}, such concentrations could adversely affect toads breeding at sites where wastewater effluent is disposed.

pH. Even short-term exposure of American Toad larvae to pH's of 3.0–3.5 cause 100% mortality (Leftwich and Lilly, 1992). In a study of 16 ponds in Ontario, survivorship of larval American Toads was significantly reduced (5–28%) in 4 of 7 ponds with a pH <4.78 and 100% at 3 ponds with a pH of 4.15–4.81. Three ponds with a pH of 4.56–4.76 had high survivorship (>80%), however (Freda and McDonald, 1993). In laboratory trials, there is a significant increase in embryo mortality at a pH of 4.2 and 0 µg/L of aluminum (Freda and McDonald, 1993). Freda (1986) and Freda et al. (1991) reviewed the literature on the effects of pH on American Toad embryos and suggested that lethal pH's were 3.8–4.2, with critical pH's at 4–4.2. Karns (1983) reported 100% egg fertilization in bog water with a pH of 4.2, although these embryos died of developmental abnormalities. Based on caged, transplant field experiments, Dale et al. (1985) recorded 100% mortality in *A. americanus* eggs at a pH of 4.1, and 10–58% mortality at a pH of 6.3. In laboratory trials, American Toad larvae increase their activity in the presence of acidified water and were able to avoid water with lethal and sublethally low pH (Freda and Taylor, 1992).

DOC. DOCs of 4–29 mg/L have no effect on larval survivorship of American Toad tadpoles (Freda and McDonald, 1993). Larvae can survive at low oxygen levels (< 1 mg/L), at least for short periods. As oxygen decreases, they float, lose their righting ability, gulp air, and gas bubbles may be visible in their gut (Williams, 1969).

Nitrates. In experimental trials in Pennsylvania, low, medium, and high concentrations of nitrate (10, 25, and 40 mg L^{-1}) had no effect on hatching success of *A. americanus* eggs (Laposata and Dunson, 1998). Likewise, un-ionized ammonia at 0–2 mg NH$_3$/L for 3–5 days had no effect on hatching success (Jofré and Karasov, 1999), and Earl and Whiteman (2009) found no effects on larval survivorship in either steady-state concentrations (1, 2.5, 5 mg/L) or in pulsed exposure (early, middle, or late pulses of 5 mg/L). In exposure to median lethal concentrations of ammonium nitrate fertilizer, the LC$_{50}$ for *A. americanus* tadpoles was 13.6 mg/L at one Ontario location, but 39.3 mg/L at a second location, suggesting possible variation in resistance to pesticide poisoning (Hecnar, 1995). Mean larval body mass decreased at both locations during the 96 hr exposure period.

Chemicals. Pesticides may have direct, indirect, interactive, or no effects on larval American Toads, depending on the pesticide used. Further, these effects may be shaped by the presence of predators. The insecticides Sevin (carbaryl, a neurotoxin), malathion, and 2,4-D have little or no effect on American Toad tadpole survival when used at the manufacturer's recommended concentrations (Relyea, 2005b; Bulen and Distel, 2011). However, Boone and James (2003) reported reduced survival of larvae by 20% at carbaryl exposures of 5 mg/L. Carbaryl significantly lengthens the larval period, thus making tadpoles vulnerable when ponds dry down. Similar results were reported even at exposures of

2 mg/L carbaryl (Distel and Boone, 2009, 2010). Mass at metamorphosis decreased when tadpoles were kept at a constant water level, but actually increased when water levels simulated a drying pond. Mass and survivorship also increased when toads exposed to carbaryl were raised at high densities (Distel and Boone, 2009, 2011). There was no latent effect of carbaryl on overwintering survival or mass at spring emergence of toadlets exposed to carbaryl as larvae (Distel and Boone, 2010). Toad survival increased when Northern Leopard Frog (*Lithobates pipiens*) larvae were present, but there was no indication that carbaryl affected heterospecific competition (Distel and Boone, 2011). Multiple species within a pond may ameliorate the effects of the pesticide on any one species, and conspecific density appears more important than heterospecific density in the way carbaryl affects American Toad larvae.

In mesocosms, the insecticide malathion (at a concentration of 0.315 mL/m^2) by itself caused only a small reduction in survival of American Toad larvae (the $LC_{50\ (16\ day)}$ for malathion is 5.9 mg/L; Relyea, 2004b), but by adding the herbicide Roundup™ (1.3 mg AI/L), toad survivorship dropped dramatically by 71% (Relyea et al., 2005). Smith et al. (2011) obtained essentially the same result, but noted that larval duration was increased and that larvae were smaller at metamorphosis than controls; adding nitrate to malathion-exposed tadpoles had no effect on the larvae. If newts (*Notophthalmus*) were present, few toads survived in any pesticide treatments. If beetles (*Dytiscus*) were used as a predator, larval toad survival increased dramatically, as expected, as the insecticide killed the predator; adding Roundup™ then had no further effect on the survival of toad larvae. The pesticides also had no effect on the growth of larval toads (Relyea et al., 2005). However, Roundup™ by itself is extremely toxic to toad tadpoles at ecologically relevant concentrations (Relyea, 2005a, 2005b). Evidence suggests that it is not the glyphosate herbicide component of Roundup™ or Vision™ that is toxic, but the POEA (polyethoxylated talloamine) surfactant (Edginton et al., 2004; Howe et al., 2004).

The pesticide fenitrothion has no appreciable effect on American Toads at 4–8 ppm (Berrill et al., 1997). The organochlorine endosulfan affects American Toads by causing paralysis of larvae exposed to concentrations of 0.03–0.4 mg/L for a period of 96 hrs. Postexposure mortality is high after 2-week-old larvae are exposed, even at low concentrations (0.04–0.05 mg/L). American toad larvae at developmental stages just prior to metamorphosis did not survive exposure to endosulfan (Berrill et al., 1998). In another study, the rate of malformations, chiefly of the eyes, was higher in metamorphosing toads exposed to endosulfan, although there was no acute toxicity to larvae exposed to 2.3–2.5 mg active ingredient/L or less (Harris et al., 2000).

When fed to adult American Toads in food or water, the insecticide methoxychlor yielded no signs of organ pathology, or effects on organ weights, feeding, behavior, or survival, but most experiments were run only one to six days, and sample sizes were small (Hall and Swineford, 1979). Levels of the now banned insecticides aldrin and dieldrin were found in young toads at levels of 2.31–8.3 ppb (mean 4.6 ppb), and 1.28 and 1.54 ppb in 2 adults (Korschgen, 1970). Residues of DDE and dieldrin have been found in adult American Toads from Iowa (Punzo et al., 1979).

The herbicide atrazine can be absorbed directly from moist soil by postmetamorphic American Toads through the pelvic patch; postmetamorphs do not avoid atrazine-contaminated soils (Storrs Méndez et al., 2009). The majority of the herbicide is concentrated in the intestines and gallbladder. Atrazine has deleterious effects on the survival of larval American Toads. Exposure of tadpoles to atrazine causes edema as a result of the herbicide's effects on renal function (Howe et al., 1998). Allran and Karasov (2001) found no effect of atrazine at doses 2.59–20 mg/L on hatchability of toads or 96 hr posthatching survival, although the proportion of deformed larvae increased dramatically with increasing atrazine concentration (to 35% at 20 mg/L). At low concentrations (3 ppb), however, Storrs and Kiesecker (2004) reported survivorship was nil after 6 days of exposure for larvae exposed early in development, whereas at 30 ppb and 100 ppb, at least a very small percentage of tadpoles survived a 30-day trial. For American Toads exposed to atrazine later in development, the concentration of the pesticide had no significantly different effects on survivorship, which continued to be low.

In contrast to the results above, larvae at Gos-

Anaxyrus americanus breeding habitat. Great Smoky Mountains National Park, Tennessee. Photo: C.K. Dodd Jr.

ner stage 40 were more sensitive to atrazine (LC$_{50}$ 26.5 vs. 10.7 mg/L) and alachlor (LC$_{50}$ 3.9 vs. 3.3 ml/L) than at stage 29, suggesting that the onset of metamorphosis might increase sensitivity to the herbicides (Howe et al., 1998); significant chemical synergy increases toxicity to American Toads when these chemicals are mixed. Atrazine at 200 µg/L also significantly reduces tadpole mass at metamorphosis (Boone and James, 2003). Birge et al. (1980) determined an LC$_{50}$ of 48 ppm for American Toads 4 days after hatching. Although counterintuitive, the primary deleterious effects of atrazine are inversely proportional to the dose and may depend on the developmental stage at which the tadpole is exposed. Such a pattern is characteristic of the way endocrine disruptors affect amphibian larval survivorship.

Finally, the fungicide mancozeb (one ethylene-bisdithiocarbamate) is very toxic to toads at 0.08 mg/L or greater, and there is a high incidence of skeletal deformities (lateral flexure and tail kinking) and neurological problems in those toads that hatch (Harris et al., 2000). This is a common fungicide used in agriculture.

UV light. UVA has no effect on eggs or larvae of *Anaxyrus americanus* (Grant and Licht, 1995). Ecologically relevant doses of UVB had no effect on embryo hatching (Crump et al., 1999), the developmental period, duration of metamorphosis, or mass at metamorphosis, and some larvae and recently metamorphosed American Toad juveniles can survive even high doses of UVB (Grant and Licht, 1995; Licht, 2003). However, exposure for 2 min 3 times per week of unshielded UVB (0.113 J/cm^2) significantly reduced larval survivorship and killed most recently metamorphosed juveniles. Older metamorphs survive better than younger animals. Crump et al. (1999) and Licht (2003) concluded that there is no evidence that ambient levels of UVB light impacts *A. americanus* in nature. In this species, the jelly envelopes surrounding the eggs help to reduce the adverse effects of ultraviolet light.

STATUS AND CONSERVATION

The status of American Toads seems relatively secure in most areas throughout its range (Weller and Green, 1997; Brodman and Kilmurry, 1998; Mierzwa, 1998; Mossman et al., 1998; Green, 2005). For example, Gibbs et al. (2005) surveyed 300 sites in western upstate New York, and found that this species had increased region-wide compared with surveys >20 years previously. Toads were found generally at higher elevations in the western and southern portions of the survey area and were associated with open habitats (pastures), mixed deciduous forests, nondeveloped areas, and low levels of acid and base deposition; toads tended to be absent from evergreen forest. In Illinois, no trends were discernable based on call counts between 1986 and 1989, and American Toads were heard on >85% of call-survey routes (Florey and Mullin, 2005). However, Minton (2001) noted that they were not as abundant as they were 40 yrs ago in Indiana farmland and urban settings. In the Northeast, populations were considered declining in 4 states (Delaware, Massachusetts, Pennsylvania, West Virginia) based on 7 yrs of data using occupancy modeling (Weir et al., 2009).

Like most amphibians, the primary threat to American Toads comes from habitat loss or alteration. As early as 1909, concern was being expressed about the effects of agriculture and wetland loss on the distribution of this species (Miller, 1909a). The presence of American Toads is negatively correlated with the extent of human-based development as well as the extent of forested land within 3000 m of breeding sites around the Great Lakes (Price et al., 2004). In Iowa, a survey of collections made prior to 1950 compared with collections made after 1950 suggested a 29% decline in toads, possibly due to continuing loss of habitat (Christiansen, 1998). In contrast, there appeared to be no change in status of this species between past and recent surveys in Iowa (Christiansen, 1981; Lannoo et al., 1994); the differing results may be due to survey methodology rather than changes in occupancy or abundance.

Habitat fragmentation may be more of a threat to juvenile dispersal than to adults, as juveniles tend to avoid open habitats and prefer deciduous leaf litter substrates to those of conifers or bare soil (Rothermel and Semlitsch, 2002; Smith and Schulte, 2008). The effects of fragmentation, such as occurs in agriculturally dominated landscapes, may be ameliorated by providing shrubby or wooded riparian strips allowing for dispersal (Maisonneuve and Rioux, 2001). Ponds in agricultural areas also can be enhanced as breeding sites by limiting nitrogen influx, limiting access by cows, and by not stocking fish (Knutson et al., 2004).

While American Toads are affected by some forms of habitat disturbance, the response is not always negative. When habitats are opened up, such as when peat bogs are mined, toad abundance may actually increase around wetland margins. This might result from a greater tolerance of American Toads to dry conditions, or reflect an actual preference for the habitat conditions found in mined peat bogs (Mazerolle, 2003). The presence of cattle in breeding ponds had little effect on postmetamorphic American Toads, and Burton et al. (2009) suggested that toads may benefit from controlled grazing. In the Southern Appalachians, American Toad abundance did not differ among control forest sites and sites where there were gaps created by wind disturbance or salvage logging (Greenberg, 2001a). After 1–2 yrs postharvest, American Toad abundance actually increased on timber-harvested sites in Maine, although toads were still more abundant in 11 and 23 m buffer strips along headwater streams than in the clearcuts per se (Perkins and Hunter, 2006). In Missouri, however, most American Toads were captured leaving an area that had been clearcut during 2 yrs following the clearcut (Semlitsch et al., 2008). In response to fire, toads burrow into the ground, where they may suffer from superficial burns (Pilliod et al., 2003). They do survive in burned areas (Floyd et al., 2002), and their terrestrial abundance later may actually increase as habitats are opened up (Kirkland et al., 1996).

The ability to detect amphibian species at breeding sites is not constant from year to year. Skelly et al. (2003) used data from American Toads, in part, to suggest that resurvey results vary depending on the duration and spatial scale of the detection effort. American Toads had only a 0.3 probability of presence (naïve estimate), leading these authors to suggest that a population may shift breeding sites through time, despite the tendency to be site philopatric in breeding site choice from one year to the next. Further, the importance of multiple site visits to estimate occupancy throughout a breeding season is emphasized by the results of MacKenzie et al. (2002). They computed detection probabilities for *A. americanus* at 29 Maryland sites visited from 2 to 66 times (mean 9.6) throughout a breeding season, and determined an occupancy estimate of 0.49, 44% higher than the proportion of sites where toads were actually observed.

Hecnar and M'Closkey (1996a) noted short-term increases in occupancy over a 2 yr period, which may suggest shifting breeding sites, but it is difficult to make trend statements based on short-term data. That American Toads may shift breeding sites if fish become established, and that they rapidly colonize newly created ponds, lends credence to the "shifting breeding site through time" hypothesis. The likelihood of changing breeding sites temporally, spatial scale, duration of surveys, and the number of visits per site all need to be incorporated into the design of monitoring programs for this species.

Head pattern of *Anaxyrus americanus*. Illustration by Camila Pizano

Roads and other high-volume transportation corridors do have significant effects on American Toads, as they do on many other frog species. For example, 433 *A. americanus* were killed on a 3.6 km stretch of highway in Ontario over a 4 yr period (Ashley and Robinson, 1996), and 111 *A. americanus* were counted dead on 4 Indiana survey routes over a 1 yr period (Glista et al., 2008). As might be expected, most mortality is seasonal, associated with spring immigration and juvenile emigration. American Toads may not use existing culverts to cross some roads (Patrick et al., 2010), so more experimental work is necessary to facilitate traverse across roads and highways.

The spatial effects of highway mortality are most pronounced within 500 m of a breeding site. Near Ottawa, Ontario, for example, American Toad abundance was significantly negatively correlated with traffic density, at least within 2,000 m of breeding ponds (Eigenbrod et al., 2008). Since American Toads easily disperse as far as 1,000 m from wetlands (Forester et al., 2006), the negative effects of roads with high traffic volume extend considerable distances across a landscape. The area of immediate adverse impact (the "road-zone" effect) extends 200–300 m from the pavement (Eigenbrod et al., 2009). On the plus side of transportation corridors, American Toads will use terrestrial habitats and wetlands located in power-line rights-of-way as breeding sites if such corridors go through deciduous forest (Yahner et al., 2001; Fortin et al., 2004).

A most unusual source of mortality of American Toads was reported by Miller (1909b). He noted that sewer traps caught large numbers of toads, and that sewer cleaners removed "piles of toads" each fall and spring. He went on to estimate that 24,000 toads were caught in the sewers of Worcester, Massachusetts, alone, and that mortality could approach 50,000 toads per year. It is likely that such human-made traps have caught toads in virtually every municipality throughout the species' range, even to the present day.

As *A. americanus* often breeds in fishless ponds, the introduction of fish may lead to the disappearance of breeding toads (Sexton and Phillips, 1986). Nonindigenous anurans also may impact American Toads. Howard (1988a) noted 10 males at an Indiana breeding site one year, where there had been 200 males and 60 females 2 years earlier; the decline coincided with the introduction of American Bullfrogs (*Lithobates catesbeianus*) into the pond. In captivity at least, the Cane Toad (*Rhinella marina*) out-competes American Toads (Boice and Boice, 1970), but since these species do not occur together, this result may be more pertinent to the status of Southern Toads (*Anaxyrus terrestris*) in Florida than to American Toads in nature.

American Toads readily and rapidly colonize newly constructed or restored breeding ponds (Briggler, 1998; Kline, 1998; Mierzwa, 2000; Stevens et al., 2002; Touré and Middendorf, 2002; Weyrauch and Amon, 2002; Brodman et al., 2006; Barry et al., 2008; Shulse et al., 2010). Water in retention ponds, however, may have chemicals resulting in sublethal effects, such as reduced size at metamorphosis (Snodgrass et al., 2008). If golf courses can be managed so as to reduce effects from American Bullfrogs and fish, even these highly artificial wetlands can serve as breeding habitat for American Toads (Mifsud and Mifsud, 2008). However, if bullfrog larvae are allowed to overwinter, then survivorship to metamorphosis is reduced. One way to reduce effects is to manage these wetlands by creating variable hydroperiods that do not allow bullfrog overwintering or the presence of predatory fish (Boone et al., 2008).

Anaxyrus baxteri (Porter, 1968)
Wyoming Toad

ETYMOLOGY
baxteri: a patronym honoring George T. Baxter (1919–2005), a University of Wyoming professor who discovered the species in 1946 and first called attention to its near extinction while working on his Master's degree.

NOMENCLATURE
Stebbins (2003): *Bufo baxteri*
  Synonyms: *Bufo hemiophrys baxteri*

IDENTIFICATION
Adults. The Wyoming Toad is a small (to 68 mm SUL), dark brown, gray, or greenish toad with dark blotches and an indistinct, narrow, mid-dorsal stripe. The elongate boss between the eyes is prominent (narrow and high), and the parotoid gland is more distinct, elevated, smoother, and frequently separated from the indistinct postorbital crests than in the closely related Canadian Toad. Tubercles on the rear feet are well developed for digging. Males have a dark throat patch. Males are slightly smaller than females, ranging between 46 and 57 mm SUL as opposed to 47 and 68 mm SUL (Withers, 1992). Smith et al. (1998) gave an adult size range of 47.0–59.5 mm SUL (mean 52.7 mm). Sanders (1987) and Smith et al. (1998) discussed additional aspects of morphology and cranial osteology.
  Larvae. The larvae are small and black, reaching 25–27 mm TL just prior to metamorphosis.
  Eggs. The eggs are small (2–3 mm in diameter) and black. They are deposited in long gelatinous strings, one from each oviduct. There are about 10–12 eggs per cm of string (Withers, 1992).

DISTRIBUTION
This species is endemic to the Laramie Basin of Wyoming and is currently known only from the vicinity of Mortenson Lake. Historically it occurred at Porter Lake and other nearby localities, but no longer does so. The range of this species is shown by Baxter and Stone (1985).

FOSSIL RECORD
There are no fossils reported for this species.

SYSTEMATICS AND GEOGRAPHIC VARIATION
Anaxyrus baxteri is a member of the *Americanus* clade of North American bufonids, a group that includes *A. americanus, A. fowleri, A. hemiophrys, A. houstonensis, A. terrestris,* and *A. woodhousii.* It is closely related to *A. hemiophrys* and somewhat more distantly to *A. fowleri, A. americanus,* and *A. houstonensis* (Pauly et al., 2004). This species was originally described as a subspecies of the Canadian Toad, based on differences in morphology, parotoid venoms, and advertisement calls (Porter, 1968). Relying on a variety of morphological and call characteristics, Smith et al. (1998) elevated it to specific status. High percentages of metamorphs were produced in laboratory crosses between *A. baxteri* and *A. hemiophrys* (Porter, 1968).

ADULT HABITAT
The Wyoming Toad is known historically only from short-grass communities near ponds, seepage lakes, and in adjacent floodplains of the Laramie Basin. The species was associated with wetlands in wind-erosion basins that contained vegetated sand dunes used for burrowing. At Mortenson Lake, it is found in a mixed-sedge (*Eleocharis palustris-Scirpus americanus*) shoreline community. The species is associated with short (< 90 cm), thin-stemmed vegetation near water, with a moderate canopy cover and a moist substrate. Withers (1992) described the vegetation in detail.

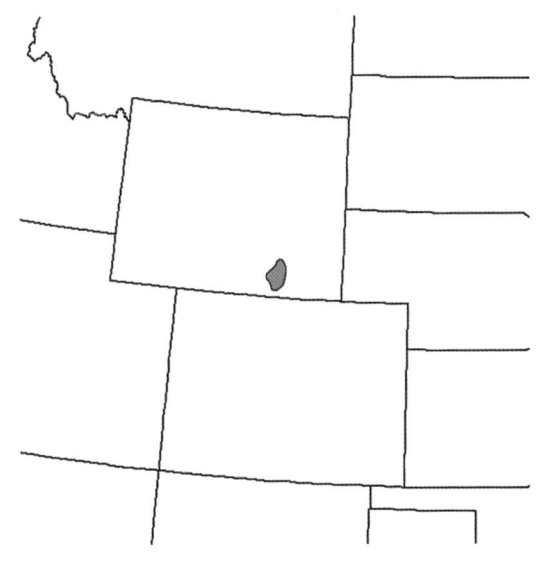

Distribution of *Anaxyrus baxteri*

TERRESTRIAL ECOLOGY

Terrestrial activity occurs in close proximity to Mortenson Lake, where toads are found on saturated substrates. Adults occur in vegetation with a greater percentage of canopy cover than juveniles, and they tend to wander farther from the lake shore (mean 1.32 m) than juveniles (mean 1.04 m). The repatriated population of adults uses denser vegetation with more canopy cover than did the wild adult population (Withers, 1992; Parker, 2000; Parker and Anderson, 2003). This species does not exhibit long-distance movements, with most activity confined to an area 30 m × 500 m along the south to east lake shore. Daily movements > 10 m from shore are rare, and the distance from calling site to overwintering site may only be 30 m. After breeding, the toads often disperse from the calling sites southward to a nearby ditch, the berms of which provide overwintering sites.

Emergence occurs in May, when daytime temperatures reach 22°C. Activity occurs throughout the warm season, at least to October, with preferred substrate temperatures reaching >20°C. Juveniles prefer slightly warmer substrates than adults. Wyoming Toads are diurnal and tend to become dormant at night. During the warmest part of the day, they may dig themselves into shallow depressions. They also spend a considerable amount of time down mammal burrows (Parker, 2000). Some toads go back and forth between the burrows and foraging areas, whereas others may remain in burrows for many consecutive days. Toads likely use ground squirrel or pocket gopher burrows for overwintering in addition to the berm near a ditch used by calling toads (Withers, 1992). Parker (2000) excavated one burrow and found a toad 36 cm below the surface. Dispersal by recent metamorphs appears rather limited, and does not occur even to the north shore of Mortenson Lake.

CALLING ACTIVITY AND MATE SELECTION

Calls are heard from late May through June, with a calling season lasting from 9 to 36 days (Withers, 1992). Calls appear to be initiated by rising temperatures and usually occur after the last spring frost, although a sudden cold spell can interrupt the breeding season. Males arrive before females at the breeding pond. Males call while sitting in shallow water (4–11 cm) or floating in leaf litter at the water's surface. Calling occurs in the vicinity of emergent vegetation at water temperatures of 18–22°C. Peak calling air temperatures are 21–27°C, but calls have been heard at water temperatures as low as 10°C (Withers, 1992). At Mortenson Lake over a 4 yr period, males called from 1 to 6 calling centers with each center containing 2–15 males; occasional lone males also were heard calling (Withers, 1992). These calling centers may shift during the breeding season and annually.

The call of the Wyoming Toad is a buzzing trill characterized by its longer duration (4–12 sec at 15°C), slower repetition rate (37–48 pulses/sec; mean 41.7), and a dominant frequency of 1,450–1,700 cps when compared with the closely related *A. hemiophrys* (Porter, 1968). There may be a 7–9 sec interval between calls. The calls can be heard by a human observer 200 m away. Calling occurs both diurnally and at night, but is strictly temperature-dependent. When grasped, male Wyoming Toads have a release vibration and warning call.

BREEDING SITES

Breeding occurs only in the shallow littoral waters, mostly on the eastern and southern sides of Mortenson Lake. Calling also occurs at a nearby ditch, but breeding has not been documented there.

REPRODUCTION

Oviposition occurs from mid-May through early June, depending upon environmental conditions. Eggs are deposited in long strings in shallow water (3.5–6.3 cm) among low vegetation. The strings may be placed in a helical clump or deposited linearly through the vegetation. The pH at Mortenson Lake is 6.6–8.6, and conductivity should be < 1,700 μmhos. Eggs hatch in 4–6 days when the larvae are 5–7 mm TL. Clutch size is 1,000–6,000 eggs per female (Withers, 1992), and hatching rates appear to be high except for one recorded by Withers (1992) as having a 10% fertilization rate. Rarely, no eggs will be fertilized.

LARVAL ECOLOGY

The jet-black larvae become active about one day after hatching. They initially remain near the gelatinous strings and slowly disperse into the surrounding water. They appear to prefer warm (25–30°C) shallow water by day, and frequently

Adult *Anaxyrus baxteri*.
Photo: Steve Corn

form dense aggregations. At night, they move to deeper waters as temperatures decrease. Wyoming Toad tadpoles are detritivores and grazers, feeding on algae and diatoms (Withers, 1992). Growth is rapid, so that the small larvae reach 25–27 mm TL after ca. 30 days. Metamorphosis occurs in August.

DIET
Wyoming Toads feed on ants, beetles, and other small arthropods (Baxter and Stone, 1985). Recent metamorphs feed on small black flies (Withers, 1992).

PREDATION AND DEFENSE
Like all toads, this species has granular glands concentrated in the parotoids and warts on the dorsal surface of the skin. Still, predation by mammals, particularly mustelids (skunks, weasels, badgers), may be substantial (Parker et al., 2000; Parker and Anderson, 2003), and Withers (1992) reported a number of toads with wounds. Gulls and herons also may attack toads. When disturbed on land, the toad readily takes to water as a means to escape (Parker, 2000). Larvae are apparently eaten by trout in Mortenson Lake, but they may be regurgitated due to their noxious secretions.

POPULATION BIOLOGY
Recent metamorphs are 11–15 cm. Wyoming Toads grow rapidly following metamorphosis in mid-July, and attain ca. 30 mm SUL by late August (Withers, 1992). No growth occurs later in the year until the next activity season, even though toads may be observed well into October. Growth is rapid during the first full summer following metamorphosis, averaging 10 mm/month and peaking in mid-August. By mid- to late July, toads are 45–55 mm SUL and are of adult size. Wyoming Toads likely breed for the first time in the second year following metamorphosis. In the third summer, toads may grow an additional 10 mm. Longevity may be as much as 8 yrs in captivity (*in* Odum and Corn, 2005).

DISEASES, PARASITES, AND MALFORMATIONS
The virulent amphibian pathogen *Batrachochytrium dendrobatidis* (misidentified as *Basidiobolus ranarum*) was reported from Wyoming Toads beginning in 1989, although it likely was present well prior to this date (Taylor, 1998; Taylor et al., 1999c, 1999d). Mortality has occurred both in the wild and captive populations, and this pathogen is thought to be a leading cause of the near extinction of this species. In addition to the chytrid pathogen, Taylor et al. (1999c) isolated 32

species of bacteria from captive and wild-caught toads, and Taylor et al. (1999d) reported mortality from mycotic dermatitis caused by *Mucor* sp. Withers (1992) reported mortality resulting from bacterial redleg (*Aeromonas* sp.), but this might have been a secondary infection. Small toads appeared to be more susceptible to this disease than large toads, and the infection seemed to be triggered by decreasing temperatures. Unfertile eggs are quickly covered with a fungus, presumably *Saprolegnia* sp. Unidentified helminth endoparasites have been reported from the captive population, but none have been reported from wild toads (Odom and Corn, 2005). Withers (1992) noted two toads with missing right eyes.

SUSCEPTIBILITY TO POTENTIAL STRESSORS
Chemicals. Although pesticides have been implicated in the decline and disappearance of the Wyoming Toad (see Status and Conservation), there are no empirical data on pesticide loads or the effects of pesticides on embryos, larvae, or adults.

STATUS AND CONSERVATION
This species is critically endangered, and there may be no self-sustaining wild populations. Through the 1970s, this species was considered common within the Laramie Basin, but populations quickly declined thereafter (Baxter et al., 1982; Lewis et al., 1985; Beiswenger, 1986). Toads were extremely rare in surveys from 1976 to 1978, and a few were observed as late as 1983 to 1984 (Lewis et al., 1985; Beiswenger, 1986). By the mid-1980s, the species was thought to be extinct in the wild, but a small population was discovered at Mortenson Lake in 1987. The land on which the population was found was purchased by the Nature Conservancy and is now Mortenson Lake National Wildlife Refuge. However, no reproduction occurred after 1991, and remaining wild toads were brought into a captive breeding program in 1993.

Wyoming Toads are now dispersed among seven approved AAZPA rearing facilities, plus the Saratoga National Fish Hatchery and the Wyoming Game and Fish Sybille Wildlife Research Unit. Efforts to establish a population at Lake George in 1992–1993 proved unsuccessful, despite the release of 3,963 larvae or metamorphosed toadlets and 56 juveniles (Odom and Corn, 2005). Between 1995 and 1998, more than 9,500 postmetamorphs were released at Mortenson Lake, and a few wild egg masses were observed at least through 1999. No reproduction was observed in 2000, and dead toads were found at the release site. The "wild" population is not self-sustaining and relies on a continual influx of captive-bred animals. Parker and Anderson (2003) noted that only 20 adult toads have been seen each year since repatriation began (but see Odum and Corn, 2005).

There are a number of hypotheses regarding the decline of this species, chiefly the effects of land alteration and the lethal effects of pesticide (fenthion) spraying. The timing of anuran disap-

Breeding habitat of *Anaxyrus baxteri*. Mortensen Lake, Wyoming. Photo: Steve Corn

pearances in the Laramie Basin coincided with the initiation of fenthion spraying for mosquitoes. Not only were Wyoming Toads affected, but Northern Leopard Frogs as well (Baxter et al., 1982). However, there is no direct evidence that this pesticide caused the population declines. Later, the amphibian chytrid fungus was identified from Wyoming Toads (see Diseases) and is now thought to have played a decisive role in the near extinction of this species (Odum and Corn, 2005).

The species was federally listed as "Endangered" in 1984 after the toad was thought extinct in the wild, although listing was sought much earlier (CKD, memorandum, Office of Endangered Species, 1982). A Recovery Plan was approved in 1991, but a Recovery Team was not appointed until 2001, 17 years after the species was listed. The population has been monitored continuously, and some management recommendations have been implemented. Limited sight-based count surveys were used to judge the results of the captive release program, but these efforts provided data only for the years 1990–1992 and were considered of limited value. Cattle were allowed to graze in the area in an attempt to control vegetation as per the recommendations of Withers (1992), but cattle may put dormant toads at risk and trample dense vegetation (Parker, 2000; Parker and Anderson, 2003). Withers (1992) recommended a number of additional management actions, including relocating a boat ramp that bisected the Wyoming Toad's habitat and phasing out fishing within the lake.

Despite successes in captive propagation, the reintroduction program has met with repeated setbacks and unsatisfying results. In general, recovery efforts have been hampered by bureaucratic infighting, lack of rigorous habitat management, reliance on non-peer-reviewed interpretations of data, often questionable restrictions on research, and poor communication (Dreitz, 2006). Despite >26 years of ineffective management, the Wyoming Toad continues to survive, at least in captivity.

Anaxyrus boreas (Baird and Girard, 1852)
Western Toad
Boreal Toad (*A. b. boreas*); California Toad (*A. b. halophilus*)

ETYMOLOGY
boreas: from the Greek *boreas* meaning 'north wind' or 'northern'; *halophilus*, from the Greek *halos* meaning 'sea' or 'salt,' and *philos* meaning 'having an affinity for.'

NOMENCLATURE
Stebbins (2003): *Bufo boreas*
Synonyms: *Bufo boreas halophilus, Bufo canagicus, Bufo canagicus halophilus, Bufo columbiensis, Bufo columbiensis halophilus, Bufo halophila, Bufo halophilus, Bufo lamentor, Bufo lentiginosus pictus, Bufo nestor, Bufo pictus, Bufo politus, Rana canagica*

IDENTIFICATION
Adults. The Western Toad has a wide range of dorsal coloration, from light gray to greenish to a dull black; most are brownish gray. This medium-sized toad has numerous warts (granular glands) covering its dorsum, which are light brown. The parotoid glands are oval and smooth, and they and the warts may be tinged in red. The Western Toad is distinguished from other toads within its range by its complete lack of supraorbital and postorbital cranial crests. Adults have a white vertebral stripe down the middle of the dorsum, and the venter is dull white with various amounts of spotting. There may be a light area under the eye and a conspicuous black blotch between the thighs ventrally. The limbs are barred or blotched with melanin. Tubercles and toes may be tipped with orange. Males have an enlarged nuptial pad on the thumb during the breeding season, but not a dark throat as do many other *Anaxyrus*; there is no vocal pouch (but see Awbrey, 1972).

Anaxyrus boreas halophilus is less blotched (reduced dorsal melanin) than *Anaxyrus boreas boreas*, has a wider head and larger eyes, smaller feet, and a weaker development of the margins along the dorsal stripe. In addition, there is considerable regional variation in the extent of dorsal and ventral melanin and in the relative width of the dorsal stripe (Karlstrom, 1962). Some of this morphological variation may stem from the different evolutionary histories of the various toad populations currently recognized under the name *A. boreas*.

Juveniles are patterned like the adults, but may have red warts dorsally, and they lack the white mid-dorsal stripe. They also have bright yellow or orange flecks on the bottoms of their feet and body. Burger and Bragg (1947) provided a detailed description of recently transformed toadlets.

In general, Western Toads from higher elevations (mountains) are smaller than conspecifics from low elevations (along the coastal Pacific Northwest). Karlstrom (1962) noted that there was considerable variation among populations in terms of mean and maximum sizes. The largest male he found was 112 mm SUL and the largest female 120 mm SUL. Mean sizes (in mm SUL) for males and females are as follows: Alaska (59.5, 68.7); British Columbia (84.5, 86.2); Oregon (78.3, 93.1); Butte and Tehama counties, California (85.9, 87.7); Contra Costa and Alameda counties, California (84.3, 93.3); Alpine, Placer, and Plumas counties, California (80.5, 95.6); Mariposa County, California (81.3, 90.6); Mariposa, Madera, and Merced counties, California (94.3, 99.1); Fresno and Tulare counties, California (76.1, 86.2); Kern County, California (75.2, 83.1); Colorado (68.5, 80.2); Utah (83.4, 94.1); Mono and Mineral counties, California (71.5, 81.6); Inyo County, California (67.3, 81.4); Los Angeles and San Bernardino counties, California (78.4, 85.7); San Bernardino County, California (73.1, 75.7); and Baja California, Mexico (70.4, 77.4) (Karlstrom, 1962). Additional size records are available for southern Utah (males: mean 86.8 mm SUL, range 75 to 98 mm; females: mean 96.3 mm SUL, range 81 to 111 mm) (Robinson et al., 1998), western Montana (adults 55–125 mm SUL; means 89 to 95 mm) (Adams et al., 2005), Idaho (mean SULs of males 69.9 mm and females 78.4 mm) (Bartelt et al., 2004), and Oregon (depending on location, males a mean of 83 to 95 mm SUL, maximum 125 mm; females a mean of 93 to 119 mm SUL, maximum 130 mm) (Bull, 2006). Some individuals may reach 145 mm SUL (Jones et al., 2005).

Larvae. The larvae are usually jet black, and they have no iridescence on the body. However, they sometimes are lighter in coloration than the deep black color of other *Anaxyrus*, and they have a very transparent tail. The tail musculature is not bicolored in lateral view, and the tail fin appears clouded. In Colorado, some larvae become olive brown just prior to metamorphosis and have pigmented tail fins. Livo (1999) suggested that this coloration was an environmentally induced phenotype in response to predation by garter snakes (*Thamnophis*). In profile, the snout is flattened, and the tail tip is pointed and deepest at one-fourth of its length. When viewed from above, the body is broadest between the snout and the spiracle. The largest *A. b. halophilus* larvae is 56 mm TL (Storer, 1925), although larvae in Colorado are 34–37 mm TL just prior to metamorphosis (Burger and Bragg, 1947). Albino larvae were reported by Hensley (1959) from Washington. A description of the larvae is in Burger and Bragg (1947).

Eggs. The Western Toad oviposits 3,000–8,000 eggs per clutch in long, mostly double, strings. Jelly partitions do not separate individual eggs from one another within the continuous string tube as they do in *Anaxyrus canorus*. In *A. boreas halophilus*, the diameter of the outer envelope is 4.9–5.3 mm; the inner envelope is 3.5–3.8 mm; the diameter of the ovum is 1.65–1.75 mm. There are 13–52 eggs per 2.54 cm of string (Karlstrom, 1962); Savage and Schuierer (1961) counted 120 eggs per 10 cm for 2 strings. Werner et al. (2004) noted that the egg strings from a single female can be 20 m in total length. Egg strings oviposited near algal mats may incorporate algae into the gelatinous string, giving a green hue to the eggs. Additional details on the eggs of this species are in Karlstrom (1962). Illustrations of eggs are in Savage and Schuierer (1961).

DISTRIBUTION

Western Toads are among the farthest northward-distributed amphibians in North America and occur from near sea level along the West Coast to 3,655 m in California and Colorado (elevation records in Stejneger, 1893; Karlstrom, 1962; Campbell, 1970d; Salt, 1979; Livo and Yeakley, 1997; Keinath and Bennett, 2000; Muths et al., 2008). In the Rockies, they occur at elevations > 2,100 m. *Anaxyrus b. boreas* is found from southern Alaska and the Yukon Territory south to northern California and west-central Nevada. The range includes much of British Columbia, west-central Alberta, western Montana and

Distribution of *Anaxyrus boreas*

Wyoming, and northern Utah. Isolated populations are scattered in Nevada and southwestern Utah; Thompson et al. (2004) reported populations in 12 geographic areas of Utah. Historically, the species occurred from south-central Wyoming south through the Rocky Mountains to northern New Mexico. Many of these populations have disappeared, and the species likely no longer occurs throughout much of its former range in the central and southern Rockies. *Anaxyrus b. halophilus* occurs from northern California and western Nevada south into Baja California. Scattered populations are found in isolated areas of the Mojave Desert.

Western Toads are found on islands, including Vancouver Island, the Queen Charlotte Islands, the Alexander Archipelago, and islands in Prince William Sound and Puget Sound (Bainbridge, Blakely, Camano, Cypress, Fidalgo, Harstine, Lopez, Orcas, San Juan, Shaw, Whidbey). Island locations are in Slevin (1928), Myers (1930a), Swarth (1936), Brown and Slater (1939), and Slater (1941a).

Important distributional references include: Multiple states or provinces (Slevin, 1928; Logier and Toner, 1961; Jones et al., 2005), Alaska (Van Denburgh, 1898; Hock, 1957; Hodge, 1976), Alberta (Salt, 1979; Eaton et al., 1999; Russell and Bauer, 2000), British Columbia (Logier, 1932; Cowan, 1939; Carl, 1943; Carl and Hardy, 1943; Herreid, 1963; Cook, 1977; Matsuda et al., 2006), California (Storer, 1925; Karlstrom, 1958, 1962; Lemm, 2006), Colorado (Maslin, 1959; Smith et al., 1965; Hammerson, 1999), Idaho (Slater, 1941b; Tanner, 1941), Mexico (López et al., 2009), Montana (Black, 1970, 1971; Franz, 1971; Marnell, 1997; Maxwell et al., 2003; Werner et al., 2004), Nevada (Ruthven and Gaige, 1915; Linsdale, 1940; Hovingh, 1997), New Mexico (Campbell and Degenhardt, 1971; Stuart and Painter, 1994; Degenhardt et al., 1996), Oregon (Gordon, 1939; Pearl et al., 2009a), Utah (Van Denburgh and Slevin, 1915; Tanner, 1931; Ross et al., 1995; Hovingh, 1997; Robinson et al., 1998; Thompson et al., 2004), Washington (Slater, 1955; Metter, 1960; McAllister, 1995), Wyoming (Baxter and Stone, 1985; Koch and Peterson, 1995), and the Yukon (Cook, 1977).

FOSSIL RECORD

The Western Toad is known from Pleistocene (Rancholabrean) deposits in California (at Rancho La Brea and Shasta counties; Brattstrom, 1953, 1958), Colorado, and Nevada (Holman, 2003). Holman (2003) notes that a combination of characters is useful in identifying fossils of this toad, such as a lack of cranial crests on the frontoparietal, a very narrow distil one-third of the humerus, and a moderately curved ilium; the anterior margin of the ventral acetabular expansion has a hemispherical curve with the shaft's posterior, the apices are pointed rather than curved in the dorsal and ventral acetabular expansions, and the dorsal prominence is low with two or three small tubercles. Holman (2003) also contains illustrations of some of these features.

SYSTEMATICS AND GEOGRAPHIC VARIATION

The Western Toad is a member of the Nearctic clade of New World toads. Two subspecies are generally recognized (*A. b. boreas, A. b. halophilus*), although systematic evaluation of toads referred to *A. boreas* is ongoing. It is likely that additional species may be described based on the results of molecular studies, and it seems certain that current taxonomy does not reflect the levels of genetic divergence within Western Toads. Many Western Toad populations in the Southwest are found in isolated remnants of desert springs and riparian areas or located in wetter mountainous regions, where they have differentiated as intervening habitat has become dryer and unsuitable for toad survival.

Anaxyrus boreas has long been recognized as being related evolutionarily to a morphologically similar group of toads, including *A. canorus, A. exsul,* and *A. nelsoni*, together constituting the *Boreas* clade. Precise relationships have been evaluated differently depending on the type of data and analyses used (W.F. Blair, 1959, 1964; Tihen, 1962a; Graybeal, 1993, 1997; Shaffer et al., 2000; Stephens, 2001; Pauly et al., 2004; Goebel, 2005). At times, even *Ollotis alvaria* has been considered a member of the group (W.F. Blair, 1959, 1963b, 1964), although it is not considered so at present. It seems, however, that a consensus is growing that toads previously known as *Anaxurys boreas* and *A. canorus* are paraphyletic. Thus, the origin of the toads used in systematic research is important for understanding phylogeny, and this may account for differing hypotheses of relationships.

Goebel (2005) and Goebel et al. (2009) have identified three major lineages in the *A. boreas* group of toads, including a northwestern (NW), an eastern (E), and a southwestern (SW). Each of these major lineages contains minor lineages. The toad currently recognized as *A. b. boreas* largely falls into the NW and E clades, with most *A. b. halophilus* allocated to the SW clade. However, it is clear that *A. canorus* is paraphyletic and related to both the NW and SW groups of *A. boreas* (also see Stephens, 2001), that *A. nelsoni* is related to *A. b. halophilus*, and that *A. exsul* is quite unique but also distantly related to *A. b. halophilus* and *A. nelsoni*. The Western Toads from Darwin Canyon in Inyo County, California, may represent a fully supported lineage as differentiated from NW *A. boreas* as are *A. nelsoni* and SW *A. boreas*. These lineages reflect differentiation among isolated populations of a much more contiguous Western Toad distribution that extended throughout western North America prior to Pleistocene climate changes.

Hybridization is not uncommonly reported in the literature, where different species or lineages come into contact. Natural hybrids between *A. boreas* and *A. canorus* are known from Mono County, California (Morton and Sokolski, 1978). A suspected hybrid was reported from Alpine County, California (Mullally and Powell, 1958), although the identification has been questioned (Karlstrom, 1962). Natural hybrids have been reported between *A. boreas* and *A. punctatus* in Darwin Canyon, California (Feder, 1979). Artificial hybrids have been produced between *A. boreas* and *A. americanus, A. speciosus* (as *A. compactilis*), *A. punctatus*, and *Ollotis valliceps*; at least some development occurs in crosses with *Anaxyrus debilis* and *Ollotis canaliferus* (A.P. Blair, 1955; W.F. Blair, 1959, 1963b) and even South American and European bufonids (W.F. Blair, 1964).

Intrapopulation genetic structure has been studied in a few Western Toad populations. Gene frequencies among adults tend to be consistent from one year to the next within a population, but they may be different from the larval cohorts inhabiting the same population (Samollow, 1980). This result might reflect selection acting for or against genotypes, especially since 90–95% of the larvae will never reach sexual maturity. Gene frequencies among cohorts also vary among individual genetic loci.

In British Columbia and Alberta, female Western Toads are dimorphic in color, with some yellowish and others reddish. Sympatric males are only yellowish in color. According to Schueler (1982b), the proportion of reddish females increases as one moves toward the coast. On Vancouver Island and in the lower Fraser Valley, however, this dimorphism is not apparent. The reddish-yellowish coloration may reflect differences in habitat use during the short activity season (Bartelt et al., 2004), with reddish females inhabiting the forest floor, where reddish conifer needles are predominant, and males remaining mostly in the vicinity of breeding sites, characterized by yellowish muddy bottoms (Schueler, 1982b).

A *canorus*-like phenotype has been reported for the higher elevations of Glacier National Park (Black, 1971). These toads are smaller than typical *Anaxyrus boreas*, have large dorsal warts set in blotches, and have a vertebral stripe that is broken rather than continuous. Toads from the southern Rocky Mountains, however, do not have any dichromatic color or pattern variation between the sexes. Western Toads from Fish Lake Valley, Esmeralda County, Nevada, have a very spotted throat (Linsdale, 1940).

ADULT HABITAT

This species occurs at higher elevations in the southern Rockies and other high mountains of the West in spruce-fir, mountain fir, aspen, corkbark fir, Englemann spruce, bog birch, willow, and lodgepole pine forests interspersed with wet or moist open meadows. They are found from the mountain foothills to subalpine meadows. In coastal areas, they have been found among sand dunes, shrubs, and moist Sitka spruce climax forest (Pimentel, 1955). Throughout much of their range, Western Toads prefer open habitats (>50% open canopy), minimal ground cover, and areas with a more southern exposure (Bull, 2006). However, Browne et al. (2009) found that abundance was associated with closed deciduous and mixed-boreal forests or tall (but not short) shrubs, at least at the 50–100 m scale in Alberta. They also prefer sites with high densities of refugia nearby. In a study of female movement patterns, for example, Bartelt et al. (2004) found that females spent far more time in open-canopied habitats than in canopies where cover was >50%. Females also preferred forest edges, avoided clearcuts, and tended to seek refuge under shrubs providing protection from dehydration.

Historically, a mosaic of open forests and meadows existed throughout the range of this species, as overgrown areas were reopened due to periodic wildfires or other disturbances. After a fire, ground cover decreases, although coarse woody debris may actually increase due to tree fall. After an area is burned, *A. boreas* quickly colonizes the burned or partially burned areas (Guscio et al., 2007), even as unburned habitats go unoccupied. Such colonization may occur even when the closest breeding ponds are at a considerable distance. The terrestrial wandering tendencies of some Western Toads may facilitate recolonization of burned or disturbed areas and allow them to respond quickly to disturbance events, as long as cover sites are available to protect them against predators and dehydration. Hossack and Corn (2007b) and Rochester et al. (2010) noted, however, that populations of Western Toads declined after initial colonization of burned sites and 2 yrs postburn, respectively. Thus, it remains to be determined whether toad colonization is a short-term response to fire disturbance or whether there will be long-term benefits to the population.

Adult *A. boreas* have been found in saline Pyramid Lake, Nevada, where they take refuge from the heat and dry conditions in an area otherwise nearly devoid of shade or vegetation. They also take refuge in pockets of tufa washed by wave action. Brues (1932) stated, "due to the conditions prevailing here they appear to have become as highly aquatic as frogs." Tolerance of somewhat saline conditions allows them to inhabit an otherwise uninhabitable environment.

TERRESTRIAL ECOLOGY

Western Toads walk rather than hop. When not breeding, adults are found commonly in shallow depressions called forms. The animals dig the form in a concealed location into moist soil that allows them to maintain a favorable thermal microenvironment as well as absorb moisture directly from the soil. In addition, toads in forms are difficult to locate, presumably lessening predation chances and enabling them to be good sit-and-wait ambush predators. Campbell (1976) found that females were more likely to dig forms than males, and that as many as 25% of the females and 13% of the males could be found in forms.

Anaxyrus boreas normally is active from spring through November, depending on weather conditions and elevation, but activity begins in January along the coast (e.g., Nussbaum et al., 1983; Lemm, 2006). Activity may be somewhat bimodal, especially at higher elevations, with breeding and feeding occurring from spring until the dry conditions of late summer inhibit daily activity. With cooler temperatures in the autumn, toads reemerge from shelters and feed until weather conditions turn cold. In Yosemite Valley, for example, California Toads become active from mid-April to early May, and at Big Bear Lake, California, activity occurs from March to October (Mullally, 1952). After breeding, they depart the breeding sites and assume a more terrestrial existence, although many remain in the vicinity (within 100–300 m) of streams and other moist and sheltered locations. Most retire for the season in September and early October, although warm autumns and the lack of storms may prolong the activity season into November. Year-round activity has been reported at a warm spring (15°C) in northwestern Utah (Thompson, 2004) and in lowland parts of its range in California (Storer, 1925).

Activity may occur at any time of the day, with peaks at midday and after dark (Sullivan et al., 2008). They frequently move short distances, and will orient themselves to bask in the warm sun. Adults at high elevations have a somewhat biphasic daily activity period in early spring. On cold spring mornings, Western Toads will sit at the entrances to cover sites and warm themselves by basking (Karlstrom, 1962). They remain active until the warmest part of the day (12:00–14:00), when they return to cover under loose soil, in mammal burrows, or under surface debris. Toads will reemerge in the late afternoon and remain active until the dry mountain air cools. This pattern is repeated in the autumn as daily temperatures become cooler. Activity occurs at temperatures as low as 3°C as toads seek shelter for the night, and toads routinely are active at temperatures <12°C in the autumn (Mullally, 1952); the upper range of normal terrestrial daytime activity is 30°C (Brattstrom, 1963). Smits (1984) measured the body temperature of free-ranging toads and found that most activity occurred over a temperature range of from 10 to 25°C. Sullivan et al. (2008) noted that daily movements occurred more often at lower temperatures than at warmer temperatures.

Adult Western Toads become nocturnal as the season progresses throughout much of their range, spending their daylight hours under coarse woody debris, in rodent burrows, under rocks, or in other sites offering shelter. However, individuals at high latitudes and elevations are frequently diurnal (Mullally, 1958). In Alaska and at higher latitudes in Alberta, for instance, they are mostly diurnal, but will occasionally be active on cool nights. Even at high elevations, toads may venture away from the vicinity of wetlands more at night than they do in the daytime, as long as conditions are mild. Toads are also more likely to be active diurnally on rainy or cloudy days than they might be on sunny days. Western Toads lose water rapidly (Thorson, 1955) and are not as adept at rehydration as are more desert-adapted toads (Fair, 1970), so it is not surprising that they are more active at night than by day, especially as the warm and dry summer progresses. Bartelt and Peterson (2005) have developed a physical model that simulates thermal and evaporative properties of Western Toads and helps to account for daily and seasonal activity patterns.

The extent of movement among individuals is quite variable. *Anaxyrus boreas* often makes extensive movements, with records of toads traveling 1 to 13 km over the course of the summer or between captures (Pimentel, 1955; Bartelt, 2000; Muths, 2003; Bartelt et al., 2004; Thompson, 2004; Bull, 2006; Schmetterling and Young, 2008). Others, however, tend to remain within a more limited area along watercourses for days at a time, and then move only short distances (Carpenter, 1954a; Campbell, 1976). Sullivan et al. (2008) recorded short-distance movements of 40 m back and forth between terrestrial and aquatic microsites on a daily basis. Males tend to be more sedentary than females and generally do not travel long distances from the breeding sites (usually no more than ca. 1 km). Female home ranges are sometimes 4 times as large as male home ranges (246,000 m^2 versus 58,300 m^2; Muths, 2003).

Reports of long-distance movements are frequent. For example, Thompson (2004) recorded a male that moved 5 km during a drought, between isolated springs from summer to fall, and Bull (2006) recorded a male moving 3.87 km. Females have a tendency to wander far from the breeding sites through open forested habitats over the course of the summer, movements which often include extremely inhospitable terrain of cliffs and mountain passes. The farthest distance a female has been tracked is 13 km in Montana (Schmetterling and Young, 2008). Other movement records include: in Oregon, males traveled a mean of 1 km, whereas females traveled a mean of 2.54 km (Bull, 2006); in Idaho, males moved a mean of 581 m, whereas females moved a mean of 1,107 m (Bartelt et al., 2004); in Colorado, males traveled a mean of 218 m (mean range 131 to 462 m) and females traveled a mean of 721 m (mean range 393 to 905 m) (Muths, 2003); in Montana, radio-tagged toads moved a median distance of 2.9 km (Schmetterling and Young, 2008). Toads may cross rugged territory and even cross from one drainage basin to another. Movements frequently occur in conjunction with precipitation.

As they depart from a breeding site, adults may average > 200 m/day (Bull, 2006) and travel > 400 m in a single day (Bartelt et al., 2004); Schmetterling and Young (2008) recorded median movements of 152–162 m/day, although most movements are < 50 m. Not all movements are terrestrial, as Adams et al. (2005) observed adults and juveniles travel hundreds of meters (median 294 to 353 m) by floating or swimming downstream in small streams in Montana. One toad floated 1.5 km downstream over a period of 6 days. Western Toads frequently use stream and riverbanks as riparian dispersal corridors, allowing them to remain close to water; movements are usually downstream. Adults take from 16 to 83 days to reach their summer terrestrial habitat after departing from the breeding site (Bull, 2006).

After metamorphosis, young toads form extremely dense aggregations on bare ground along the shoreline, often in two or more layers deep, in the direct sunlight (Patch, 1922; Burger and Bragg, 1947; Turner, 1952; Lillywhite and Wassersug, 1974; Arnold and Wassersug, 1978; Karlstrom, 1986; Leonard et al., 1993; Koch and Peterson, 1995; Marnell, 1997; Corkran and Thoms, 2006). Toads within the aggregations are slightly warmer than adjacent water and ambient air temperatures, and the aggregations have been attributed to basking (Black and Black, 1969). In Yellowstone National Park, however, Turner (1952, 1955) noted one such aggregation that included large numbers of recent metamorphs that died of dehydration (or heat?) in an aggregation along the shoreline, even though the toadlets were within cm of water.

Metamorph mass aggregations ("flood puddles") tend to disperse rather quickly, giving rise to reports of mass movements across the landscape by juvenile toads. The ground may appear literally to be moving or alive with toadlets. Mullally (1958) stated that juvenile toads were primarily diurnal in their dispersion, with virtually all activity ceasing by dark. Movements occur especially in late afternoon, after the peak of the daily maximum temperatures. During the warmest part of the day, the newly metamorphosed toads huddle in shade under protective vegetation. As they grow to 20–35 mm SUL, juveniles take up residence near wetlands and adjacent to the breeding site.

Recent metamorphs may disperse great distances. In northeastern Oregon, Bull (2009) recorded juveniles dispersing as far as 2,720 m from the natal pond, a journey that took 8 weeks. On average, toads traveled 84 m/day and used moist drainage corridors for dispersal; she recorded one toadlet traveling 610 m in 7 days and a juvenile moving 840 m upstream in 40 days. Juveniles routinely traveled more than 1,070 m. During this time, they were preyed upon by birds, and many died of desiccation. In addition, toadlets were killed by vehicles and trampled by cattle. Adults may use riparian areas for foraging (e.g., Okonkwo, 2011), although they tend to forage well into forested habitats away from water.

Winter retreats are usually in mammal burrows, such as those constructed by ground squirrels, or in small chambers under rocks or debris. For example, Bull (2006) found that Western Toads used rodent burrows (38%) or went under large rocks (27%), under logs and roots (19%), and under banks adjacent to streams and rivers (15%) as overwintering sites. In California, Western Toads were found in about half of the Golden-mantled Ground Squirrel (*Spermophilus lateralis*) burrows, excavated at distances 10–60 cm from the entrance (Mullally, 1952). However, Western Toads are adept at digging using the keratinized tubercles on their back feet; they are able to dig backward into the ground much like spadefoots.

Western Toads can move considerable distances to reach a suitable overwintering retreat. Adults take a rather direct route, whereas juveniles wander to a greater extent. Bull (2006) found that winter dormancy sites were 180–6,230 m (mean 1,968 m) from the breeding sites, whereas Campbell (1970c) recorded movements between 900 and 2,975 m from summer habitat to winter dormancy sites. In Alberta, dormancy sites were < 320 m from foraging areas (Browne and Paszkowski, 2010). Such retreats are usually insulated by deep snowpack and, with groundwater seepage, keep the temperature above freezing. Juveniles arrive at overwintering sites prior to adults, and adult males arrive earlier than females. However, arrival date is correlated with SUL in this sexually size-dimorphic species, suggesting that sex differences do not really exist; it is likely that larger and more experienced toads forage longer than small individuals (Browne and Paszkowski, 2010). Arrival date is correlated with temperature and day length. Adults and juveniles enter dormancy at the same time. Emergence from winter retreats is correlated with rising temperatures, a loss of snow cover, a daily minimum temperature above freezing, and a rising water table as a result of snowmelt or spring rains.

Western Toads are photopositive in their phototactic response, suggesting they use both sunlight and moonlight when feeding and going about their daily activities (Jaeger and Hailman, 1973). They are sensitive to light in the blue spectrum ("blue-mode response"), which apparently helps them orient toward areas of increasing illumination, such as the open horizon above lakes and ponds (Hailman and Jaeger, 1974). Western Toads likely have true color vision.

CALLING ACTIVITY AND MATE SELECTION

Male Western Toads are often assumed not to produce an advertisement call. However, a number of reports of calling are available. For example, Awbrey (1972) reported on a calling male *A. b. halophilus* from San Diego County, California, complete with a vocal pouch. Likewise, Long (2010) in Alberta presented evidence of males producing a series of rapid, extended trills characteristic of other toad advertisement calls. Other reports are in Black and Brunson (1971) and Cook (1983). Male Western Toads are reported to make a "series of low mellow tremulous notes. In chorus the notes may be compared to the voicings of a brood of young domestic goslings" (Storer, 1925:177). Another analogy is to that of a "soft chirp," which they utter in a chorus singly or as a series of chirps. The voice of the Western Toad in such aggregations is weak and does not carry more than a few dozen meters (Karlstrom, 1962).

Western Toads chirp when handled or amplexed by another male, suggesting that the sound functions to discourage amplexus. They also have an encounter call made in response to the presence of another male. They may occasionally chirp from daytime retreats or in response to an insect crawling on them (Storer, 1925; Sullivan et al., 2008). These calls do not appear to be advertisement calls. Hammerson (1999) hypothesized that the soft chirping sounds may function in the formation of male breeding aggregations.

The Western Toad is a spring breeder, but the dates of initiation of the breeding season vary by latitude, elevation, and local environmental conditions. Literature reports of breeding dates include January to July (Pacific Northwest: Nussbaum et al., 1983; Jones et al., 2005), late March to early May (Utah: Thompson, 2004), March to June (Montana: Black, 1970), April and May (Alberta: Salt, 1979; California: Yosemite Valley, Martin, 1940), throughout the month of May (Montana: Glacier National Park in Montana, Marnell, 1997), May and June (Colorado: Holland et al., 2006; Alberta: Long, 2010), May to July (Alaska: Juneau, Hodge, 1976), and June and July (British Columbia: Logier, 1932). The breeding season within a region lasts longer than it does at an individual site. For example, the regional breeding season lasted six weeks in northwestern Utah, but only about four weeks at any one site (Thompson, 2004). Males arrive at the breeding sites prior to females by about a week or two. Most males select calling sites in shallow water along the shoreline, but a few males may ride on vegetation and call from deeper portions of the breeding site.

Breeding populations normally are not large, consisting of 12–15 breeding pairs in Yosemite Valley (Karlstrom, 1962), although other breeding aggregations may be considerably larger. Males space themselves at about 30 cm from one another and sit in the water facing the shoreline, with only their eyes and front legs above water (Black and Brunson, 1971). This positioning allows them good advantage to detect arriving females. Males usually outnumber females at a breeding site, as females depart the breeding site shortly after depositing their egg clutch. Males also usually breed annually, whereas females likely skip breeding seasons in order to yolk clutches of eggs (Carey, 1976). Variation in breeding frequency (see below) may skew perceptions about sex ratios within a population based solely on the numbers of adults at a breeding site.

There is some indication that males prefer larger females than smaller females, perhaps since large body size may be an indication that the female is gravid, although the results are not uniform. Olson et al. (1986) noted that sexual size preferences varied among sampling sites, and even among days sampled. What appeared to be a preference for large females on one day was not evident on the next, even at a single breeding site. At other times and at other sites, preferences seemed to be random. Size-assortive mating (large males mating with large females; smaller males with smaller females) was also observed at some sites and on some days, but not always. According to Olson et al. (1986), these results indicate that anuran mating patterns are neither species-specific nor population-specific attributes. Under experimental conditions, males do not discriminate whether a female is gravid or not as long as they are of equal size (Marco et al., 1998). Since females are generally larger than males, moving toward a large animal may facilitate finding a female and avoid amplexus by other males. When a male is amplexed, he emits a warning chirp and is usually released within 3 sec. However, males are extremely tenacious when amplexing females and will vigorously repel mating attempts by another male rather than just chirping.

A pair in amplexus attracts other males, and often a swirl or ball of toads will surround an amplexing pair. Such swarms can be fatal to females, as so many males will attempt to mate as to drown the female. A swirl of amorous males will set off a chorus of chirping as males try to ward off other amplexing males, and this chirping seems to attract more males (Black and Brunson, 1971). An amplexed pair may hide in the vegetation in order to avoid other males.

BREEDING SITES

The Western Toad breeds in a variety of aquatic habitats, from ephemeral to permanent in hydroperiod. They often use shallow water meadow pools and overflow pools along rivers and streams that result from high water spring runoff. Other breeding sites include high alpine glacial lakes, kettle ponds, oxbow ponds, muskeg ponds, small temporary ponds, ponds created by beavers or geologic disturbances, slow-moving channels along rivers and streams, and human-created cattle tanks, ditches, ponds, and reservoirs. The toad prefers breeding sites with a mud or silt bottom, and may be found in pools with almost no vegetation. In addition, toads select sites with open canopies with little or no shading, which might decrease temperatures during critical larval developmental periods. In Alberta, Western Toads frequently use borrow-pits for breeding, but these habitats may serve as population sinks, as recruitment is low, and because they often dry prior to larval metamorphosis; beaver ponds are especially favored (Stevens et al., 2006b).

Historically, the breeding sites used by Western Toads, particularly high-elevation lakes, did not contain native species of fish. However, game fishes (eastern brook trout, rainbow trout, cutthroat trout, kokanee) have been stocked in many of these lakes and streams by agencies promoting recreational fisheries. Western Toads probably prefer habitats free of predaceous fishes, but they are also found in larger fish-stocked lakes (Bull and Marx, 2002; Thompson et al., 2004).

REPRODUCTION

Breeding begins in the spring, with the date of first egg deposition varying by latitude, elevation, and in certain circumstances water level (e.g., Metter, 1961). At higher elevations, deposition may take place as late as July (Livo, 1998) or early August (Degenhardt et al., 1996; 3,240 m, Fetkavich and Livo, 1998). Degenhardt et al. (1996) even reported a gravid female on 4 September. In northwestern Utah, breeding occurred from late March to early May, with breeding occurring earlier in ponds fed by warm springs than in ponds of normal water temperature (Thompson, 2004). Metter (1961) recorded an instance where an *A. boreas* population delayed breeding until spring flood water levels had dissipated. Oviposition usually occurs at night, although some

Egg strings and two pairs of amplexing *Anaxyrus boreas*.
Photo: Brome McCreary

females will oviposit diurnally in the early spring (Mullally, 1958).

Eggs are oviposited into shallow water (ca. 3–30 cm in depth), and cover an area of roughly one square meter or greater. Oviposition may take several hours, as a female will deposit several cm of eggs, swim to the surface for a breath, and then resume egg laying; this cycle repeats itself until the full egg complement is deposited at the breeding site. Total clutch size ranges to 16,500 tadpoles per clutch (Storer, 1925; Karlstrom, 1962), but Samollow (1980) reported a mean clutch size of 12,000 in Oregon. In Colorado, clutch size ranged between 3,239 and 8,663 (mean 5,213) (Carey, 1976). Thus, there may be a degree of geographic variation in clutch size that either reflects different environmental conditions or evolutionary history. Clutches may or may not be deposited in proximity to one another. Thompson (2004) reported finding from 28 to 45 egg strings annually at breeding sites in northwestern Utah, whereas Holland et al. (2006) found from 1 to 13 strings (mean 6) at sites in Colorado. Eggs swell to double their oviposition size in about 5 min as the surrounding jelly absorbs water.

Most males return to a breeding site from one year to the next (Corn et al., 1997; Bull and Carey, 2008). Some males will frequent a breeding site up to 4 consecutive years, but there are few males 3 yrs or older at a breeding pond. However, this is not always the case. Breeding site occupancy may vary among sites and years. Muths et al. (2006) reported temporary emigration rates of 10–29% at one site, and from 3 to 95% at another site. However, males usually return the following year after being away from a site for a season. Either the males are skipping breeding seasons or they are moving from one breeding site to another, although that might not always be an option due to the spatial scale of the breeding ponds. According to Muths et al. (2006), the length of the activity period during the previous season and the depth of the snowpack prior to the breeding season may influence whether or not a male will skip a breeding season. These authors also suggested that a disease outbreak at one site may have triggered emigration, as many animals would have been too ill to breed had they remained at the breeding site. Numbers also may vary annually due to environmental conditions such as drought (Thompson, 2004).

Female Western Toads are capable of breeding in two to three consecutive years (Bull and Carey, 2008). However, some females also will skip a breeding season (Carey, 1976), with at least a few skipping two to three breeding seasons. Females that breed in consecutive years are smaller than females that skip breeding seasons, which suggests a greater breeding frequency among younger females. As females get larger and older, they may skip breeding seasons. Campbell (1976) suggested that reproduction was successful only about half of the time.

There are a number of cases of interspecific attempts at mating, either with a male *A. boreas* attempting to mate with another species (Brown, 1977) or another species attempting to mate with a female *A. boreas* (Brodie, 1968). These two cases involved different species of ranid frogs that share breeding sites with Western Toads. Interspecific amplexus also has been observed between *A. boreas* and *A. hemiophrys* (Eaton et al., 1999).

LARVAL ECOLOGY

Larval Western Toads are found in the shallow warm waters of the breeding sites, where they receive maximal thermal insolation. Larval access to shallow warm-water pockets is critical at high elevations, where water remains cold and ambient air temperatures, even in the summer, can be cool. The warm-water environment provides sufficient heat for growth and development, critical factors that are directly correlated with temperature. Indeed larval growth rates are inversely correlated with elevation (and thus temperature) (Salt, 1979). Larvae can tolerate rather high temperatures for a cold-adapted species. Although unusual situations, larvae have been found in 34°C water where they were associated with a hot spring (Ellis and Henderson, 1915), and they have been found in slightly saline waters (Brues, 1932).

Western Toad larvae form dense aggregations. Larvae within these aggregations tend to swim together and may be seen in huge numbers (Samollow, 1980; Koch and Peterson, 1995). Nussbaum et al. (1983) recorded a single aggregation extending for 300 m along a shallow-water shoreline! At times, small pockets of the aggregation may break away from the main group and swim together for a while (a few minutes to hours) before rejoining the main aggregation. The aggregations may serve a variety of functions, from predator satiation to providing a thermal or feeding advantage, as the many larvae stir up detritus. In Alberta, Salt (1979) reported densities of up to 550 m^2, especially at elevations of 970–1,220 m.

Aggregations are based initially, at least in part, on kin recognition. Larvae reared in isolation tend to associate with full-siblings over paternal half-siblings, and with maternal half-siblings over nonsiblings. However, they do not distinguish between maternal half-siblings and full-siblings, or between paternal half-siblings and nonsiblings (Blaustein et al., 1990; Blaustein and Waldman, 1992). In contrast, larvae in mixed-rearing experimental choice tests (i.e., larvae raised both with kin and without kin), or with only nonkin, associate randomly with both siblings and nonsiblings (O'Hara and Blaustein, 1982). Kin recognition is not based on experience (age) or ontogeny (developmental stage). The formation of the aggregations initially may involve kin recognition, but the continuance and size of the aggregations is likely based on species recognition. As larvae hatch they are most likely exposed to kin, but as the aggregations grow and consolidate, species preference tends to outweigh sibling preference. They literally "go with the flow."

A major source of larval mortality is the desiccation of breeding ponds prior to metamorphosis (Martin, 1940; Karlstrom, 1962). If ponds and pools dry quickly, literally thousands of tadpoles will die, significantly affecting annual recruitment. Winterkill also kills larvae, especially if the larvae do not hatch until July or August. Late season larvae do not have the time to develop to metamorphosis, and even if they did, the metamorphs could not gain enough energy before snow falls to survive the long winter.

Larvae complete metamorphosis in >60 days, depending on thermal conditions; metamorphosis occurs at higher elevations from mid-July to mid-September. There are literature references to overwintering by *A. boreas* larvae at high eleva-

Tadpole of *Anaxyrus boreas*. Photo: Chris Brown

tions. Campbell (1970a, 1976) suggested larval overwintering based on finding two different size classes of tadpoles at high elevations. Given the range of oviposition dates reported by Fetkavich and Livo (1998), it is possible that the smaller larvae were oviposited late in the season and may not have survived the winter to metamorphose the following summer. Overwintering by larvae of this species has yet to be demonstrated conclusively. New metamorphs measure 10–20 mm SUL (Blair, 1951; Hammerson, 1999).

As might be expected from an animal that deposits so many eggs, literally thousands of metamorphs can be produced in a single season. In northeastern Oregon, Bull (2009) recorded 750,000 toadlets dispersing from a single locality after an exceptionally good year of moisture. Production varies among ponds and years, however. At the same pond, only 50,000 toadlets were produced 2 yrs later, whereas another pond produced 90,000 toadlets one year, and only 4,000 2 yrs later. Desiccation can claim a high number of larvae; 20,000 larvae died when a reservoir was drained at one locality. Desiccation of breeding sites prior to metamorphosis is likely a main cause of mortality among Western Toad populations.

POPULATION BIOLOGY

Male Western Toads probably reach sexual maturity by 4 yrs of age and females by 6 yrs, with longevity at >8 yrs (Carey, 1976). Adult populations of *A. boreas* at any one site may be small. At 5 sites in Utah, a mark-recapture study estimated that there was a mean of 40–243 adults per site, and that the population size was stable, at least over a short-term period (Thompson, 2004). Sex ratios reported in the literature are male-biased (Campbell, 1976; Salt, 1979; Samollow, 1980; Olson et al., 1986; Olson, 1988; Thompson, 2004), which may at least in part be accounted for by differences in breeding frequency and the amount of time adults remain at the breeding ponds (males 1–4 weeks, females 1–2 days; Bull, 2006). In Utah, for example, the sex ratio varied between 1.2–4 males per female depending on site and year sampled (Thompson, 2004), whereas in Colorado it was 2.3–4.6 males per female (5 yr average 2.7:1) (Campbell, 1976).

DIET AND FEEDING

Tadpoles are herbivorous, eating a variety of blue-green and green algae, including *Oscillatoria*, *Mougeotia*, and *Desidium* (Franz, 1971). Juvenile and recently metamorphosed Western Toads feed on small arthropods with lengths < 5 mm. In the study by Bull and Hayes (2009), more than 20 Families and 10 Orders were represented in the diet. True bugs, ants, and spiders were particularly well represented in the diet of metamorphs, whereas juveniles ate collembolans, beetles, and ants. Prey is selected nonrandomly, with certain prey being found more in stomach contents than in community abundance. Metamorphs selected aphids, ants, lepidopteran larvae, beetles, and spiders more than they were present in the environment, but ate flies in less proportion than availability might suggest. Juveniles selected ants, beetles, flies, and spiders, but focused less on collembolans. A few snails were also consumed. Most toads had eaten recently (99%), and the mean number of prey items was 15.2 per toad with an average total mass of 4.7 g of prey per toad (Bull and Hayes, 2009).

Adult toads eat a wide variety of mostly ground-dwelling or low-flying taxa, including ants, beetles, mosquitoes, flies, gastropod snails, lepidopteran larvae, grasshoppers, spiders, sow bugs, and even crayfish (Ruthven and Gaige, 1915; Tanner, 1931; Schonberger, 1945; Moore and Strickland, 1955; Turner, 1955; Campbell, 1970b; Miller, 1978; Bull, 2006). Males and females have similar diets and prefer prey < 15 mm in length. Hayes and Hayes (2004) recorded *A. boreas* feeding nocturnally on a column of carpenter ants (*Camponotus modoc*); toads sat alongside the column and ate more than 60 ants each over a 5 hr period. Toads fed at 3–18 min intervals, and appeared to choose ants that strayed from the immediate column. Feeding lasted until shortly before sunrise. The scats resulting from such feeding bouts are recognizable, and have been used to document presence during visual encounter surveys. Because of the amount of insects this species eats, Storer (1914) considered this species an economic asset to California.

Western Toads apparently can learn the odor of prey to which they have had previous exposure. In experimental trials, they approached an odor-filled filter paper, lowered their head, and extruded the tongue toward filter papers with odors of prey that they had previously eaten or of lipid extracts from such prey (Shinn and Dole, 1979a, 1979b; Dole et al., 1981). In contrast, they ignored filter papers with odors of unfamiliar prey.

PREDATION AND DEFENSE
The eggs of this species are unpalatable (Licht, 1969a), but the larvae are palatable to predaceous insects (*Lethocerus americanus*, *Dysticus*) (Peterson and Blaustein, 1992). Eggs may be eaten by salamander larvae (*Ambystoma gracile*, *Taricha granulosa*), but in much less proportion to their abundance, and they are often spit out after being ingested (Peterson and Blaustein, 1991). Juvenile and adult Western Toads, like other toads, have noxious or toxic skin secretions that may deter predators. The toad also has a distinct musky odor. If a toad is attacked, it will inflate its body, making swallowing by a predator difficult (Vestal, 1941), and/or void its bladder. To what extent this is successful in thwarting a predator such as a snake is unknown. Toads are cryptic in coloration and are often difficult to detect on a matching substrate.

Larval Western Toads tend to form aggregations in response to the presence of chemicals either emanating from potential garter snake predators or from conspecifics attacked by garter snakes (*Thamnophis*). Presumably these aggregations minimize the potential for any one tadpole to be eaten by the snake. Larvae also tend to metamorphose and emerge from the breeding site in synchrony in the presence of a predator. Since garter snakes tend to aggregate at breeding sites awaiting the metamorphs, synchronous emergence may reduce the probability of predation on individuals, as snakes can only eat one toad at a time (DeVito et al., 1998). When snakes are present, young toads also emerge from the water and depart faster than they do when snakes are not present.

Larval Western Toads are preyed upon by larval and adult predaceous beetles (*Dytiscus, Agabus, Rhantus, Graphoderus*), notonectids, garter snakes (*Thamnophis elegans, T. sirtalis*), and possibly larval Tiger Salamanders (*Ambystoma mavortium*) (Kiesecker et al., 1996; Livo, 1999). Robins and mallards are also known to eat tadpoles in desiccating pools. Adult and juvenile Western Toads are preyed upon by Tiger Salamanders, Oregon Spotted Frogs (*Rana pretiosa*), garter snakes (*Thamnophis*) of several

Adult *Anaxyrus boreas*.
Photo: Chris Brown

species, birds (Common Raven, Common Crow, Magpie, Spotted Sandpiper, Grey Jay, Steller's Jay, Red-tailed Hawk, Loggerhead Shrike, Northern Shrike, Pygmy Owl), mammals (red foxes, domestic dogs, badgers, raccoons, coyote, short-tailed weasels, mink, marten) (Karlstrom, 1954, 1962; Cunningham, 1955a; Long, 1964; Arnold and Wassersug, 1978; Salt, 1979; Beiswenger, 1981; Olson, 1989; Jennings et al., 1992; Corn, 1993; Koch and Peterson, 1995; Hammerson, 1999; Jones et al., 1999; Keinath and Bennett, 2000; Pearl and Hayes, 2002; Bull, 2006). Large *Anaxyrus boreas* will cannibalize smaller toads (Cunningham, 1955a; Mullally, 1958). Predators often eat the entrails of the toad and leave the toxic or noxious skin.

COMMUNITY ECOLOGY

Larval Western Toads produce an alarm substance when attacked that alerts conspecifics to the threat of the predator. When the substance is detected, larvae increase their activity and move away from the vicinity of the predatory attack (Hews and Blaustein, 1985; Hews, 1988). Larvae do not react when heterospecifics are attacked. In experimental trials, dragonfly larvae (*Anax*) are less successful in capturing Western Toad larvae when exposed to the alarm substance than in its absence. The chemoreceptive response makes the toad larvae more likely to avoid predation. Chemoreception of predators does not end at the larval stage. Juvenile Western Toads avoid the vicinity of garter snakes (*Thamnophis sirtalis*) previously fed other juvenile *Anaxyrus boreas*, but not snakes fed larvae or other types of prey (Belden et al., 2000).

Larval Western Toads also may change their phenotype in response to the presence of predator cues within the water. Larvae may become olive brown immediately prior to metamorphosis due to the proximity of garter snakes. Tadpoles exposed to feces from garter snakes fed *A. boreas* tadpoles grow slower and reach maturity later than controls. These tadpoles had different morphologies than controls (wider in proportion to length), and swam more rapidly than controls (Livo, 1999). In contrast, larvae fed in the presence of predator alarm cues emanating from injured conspecifics attacked by aquatic predators metamorphosed quicker than controls (Chivers et al., 1999).

Such phenotypic plasticity is associated with an increased ability to avoid predation when in the presence of different types of predators.

The presence of fish may or may not influence survivorship of *A. boreas* larvae. In Alberta, for example, larval recruitment is not affected following winters in which severe fish kills occur. Reduced fish abundance may even decrease the abundance of developing toad tadpoles, perhaps because of complex trophic interactions among different types of fish and insect abundance (Eaton et al., 2005a). Experiments have not demonstrated that small-bodied fish can have significant predatory effects on toad larvae in boreal habitats, perhaps because the toad tadpoles are distasteful and not an important part of fish diets under normal circumstances.

The presence of a voracious predator, such as ravens, at a breeding site may alter behavior and habitat use by breeding Western Toads. Toads in deeper waters suffered no predation in comparison to those breeding in shallow water (Olson, 1989). Males in amplexus tend to release their mates sooner than they might otherwise when ravens are present. Large males were more likely to do this than small males, and males tended to clasp smaller females than larger females. The readiness to unclasp a female might have survival value if attacked by a raven, but it also reduces the male's mating success. Releasing a smaller female may be easier than a large female, and at the same time less of a reproductive cost than the release of a larger female, presumably with more eggs. From a toad's perspective, the presence of ravens tends to place survival ahead of optimal reproductive outcome (Olson, 1989).

The decline of *Anaxyrus boreas* and other amphibians at high elevations throughout the West could have serious consequences for those species dependent upon amphibians for food. For example, recent metamorphs comprise a significant portion of the seasonal diet of garter snakes (*Thamnophis*) (Arnold and Wassersug, 1978). There is some evidence of decline in garter snake (*Thamnophis elegans*) populations associated with high-elevation anuran communities, and this decline could be tied in to the decline of anuran prey. However, studies indicate that *Thamnophis* are more likely to prey on *Pseudacris regilla* than Western Toads, so a decline in members of the

Western Toad group may not have serious impacts on this snake, at least in the Sierras (Jennings et al., 1992).

DISEASES, PARASITES, AND MALFORMATIONS

The amphibian chytrid fungus (*Batrachochytrium dendrobatidis*) is thought to be a main cause in the decline of the Western Toad throughout much of the West, particularly in the Rocky Mountains. It has been found in many populations in the Rockies, and is particularly associated with cooler northern populations than with populations in the south (Muths et al., 2008). In northern populations, the fungus was present at lower elevations, whereas in southern populations, it occurred at higher elevations. Such a distribution reflects the life history of the fungus, a pathogen of cooler temperatures. Records of chytrid infecting *Anaxyrus boreas* are available for British Columbia (Deguise and Richardson, 2009), California (Green et al., 2002), Colorado (Green et al., 2002; Green and Muths, 2005; Young et al., 2007; Hasken et al., 2009), Oregon (Bull, 2006; Adams et al., 2010), Utah (Thompson et al., 2004), and Wyoming (Green et al., 2002; Young et al., 2007; Murphy et al., 2009). Larvae experimentally infected with amphibian chytrid fungus show no evidence of behavioral fever or altered thermoregulation, and uninfected larvae aggregate freely with infected larvae (Han et al., 2008).

Water molds (Oomycetes, *Saprolegnia*) are known to infect the eggs and skin of *Anaxyrus boreas* (Green and Muths, 2005) and can lead to significant mortality (Blaustein et al., 1994a; Kiesecker and Blaustein, 1997b; Kiesecker et al., 2001b). Other fungi isolated from this species include *Aspergillus, Basidiobolus, Cladosporium, Fusarium, Mucor, Penicillium,* and *Rhizopus* (Green and Muths, 2005). The bacterium *Aeromonas hydrophila* was isolated from a Colorado *Anaxyrus boreas* (Green and Muths, 2005).

Unidentified trematodes in the urinary bladder have been reported from Western Toads from Colorado (Green and Muths, 2005), and dermal metacercariae from larvae in California (Green et al., 2002). Other trematodes include *Glypthelmins* sp., *G. shastai, Gorgoderina* sp., *Haematoloechus kernensis, Megalodiscus microphagus,* and *Ribeiroia* sp. (Ingles, 1936; Koller and Gaudin, 1977; Goldberg et al., 1999a).

Cestodes include *Cylindrotaenia americana* and *Distoichometra bufonis*, and nematodes include *Aplectana itzocanensis, Cosmocercoides variabilis, Falcaustra inglisi, F. pretiosa, Ozwaldocruzia pipiens, Physaloptera* sp., *Rhabdias* sp., and *R. americanus* (Ingles, 1936; Walton, 1941; Koller and Gaudin, 1977; Goldberg et al., 1999a). Adults may occasionally be parasitized by blowfly larvae (*Lucilla elongata, L. silvarum*) (James and Maslin, 1947; Eaton et al., 2008) and leeches (Koch and Peterson, 1995). Pea clams (*Pisidium*) have been found attached to the toes of *Anaxyrus boreas* in Colorado, suggesting a means of dispersal for the clam (Hammerson, 1999).

Western Toads with supernumerary legs were first reported by Washburn (1899) and Crosswhite and Wyman (1920). Western Toads with missing limbs, mostly affecting the hind limbs, have also been reported from California (Johnson et al., 2001b); however, malformation frequencies were low (<2.8%). Metacercariae of the trematode *Ribeiroia ondatrae* can induce malformations experimentally in *Anaxyrus boreas*, especially of the limbs (Johnson et al., 2001a).

SUSCEPTIBILITY TO POTENTIAL STRESSORS

Metals. Lethal levels of metals to Western Toad larvae are available for iron (> 20 mg/L), zinc (39 mg/L), and copper (3.7 mg/L). The actual lethal concentrations are likely lower than these values for zinc and copper, as a range of concentrations was not tested (Porter and Hakanson, 1976).

pH. Western Toads are sensitive to pH, and all larvae die at a pH <4.0. The LC_{50} for pH is 4.5 (Corn et al., 1989). In Colorado, acid mine drainage is much lower than this, hence *A. boreas* is absent from drainages affected by acid mine runoff (Porter and Hakanson, 1976).

Chemicals. Larval mortality as a result of coal oil poured in ditches and meadow ponds was reported from Yosemite Valley (Karlstrom, 1962). The insecticides Parathion, Bayer 37289, Guthion, Methyl parathion, G-30494, G-30493, and Ronnel were reported as "safe" for larval *A. boreas* at high doses (Mulla, 1962). Mulla (1962) recommended that the extremely toxic pesticides toxaphene, DDT, and thiodan be used to control "infestations" of juvenile and adult Western Toads on golf courses and residential areas of California. Apparently they were quite successful.

Thermal tolerance in juvenile Western Toads can be lowered by exposure to sublethal concentrations of organophosphorus insecticides, such as Abate™ (60 ppb), fenthion (60 ppb), chlorpyrifos-methyl (30 and 60 ppb), chlorpyrifos-ethyl (30 and 60 ppb), and methylparathion (25 and 50 ppb), and the growth regulator Altosid™ (100 and 200 ppb), even at one-half the recommended field concentrations. Fenthion had the greatest effect and Altosid™ the least (Johnson and Prine, 1976). Exposure to these chemicals also lowered the activity levels of the toads. The herbicide Roundup® is toxic to larvae, with an $LC_{50\ (24\ hrs)}$ of 2.66 mg/L (King and Wagner, 2010). The LC_{50} decreases at 15 days to 1.95 mg/L.

Nitrate and nitrite. The median lethal concentration of nitrite for larvae is 1.75 mg $N-NO_2$/L at 15 days, 5.4 at 7 days, and >7.0 at 4 days (Marco et al., 1999). *Anaxyrus boreas* larvae are not particularly sensitive to the effects of nitrates. Juveniles may be sensitive to urea-based fertilizers. In laboratory experiments, juvenile *A. boreas* avoided paper towels saturated with urea but not soils dosed with urea. However, juvenile mortality increased significantly over a five-day period on urea-dosed soils when compared with controls. In addition, juveniles exposed to urea had a lowered feeding rate than controls (Hatch et al., 2001). Thus, juveniles may suffer adverse effects because they apparently cannot detect urea-based fertilizers under natural conditions.

UV light. Experiments on the effects of ambient levels of UVB light on embryos have had mixed results. In Colorado, Corn (1998) found no effects of up to 100% ambient UV radiation on hatching success of Western Toads under natural conditions. In contrast, Blaustein et al. (1994b) reported mortality and reduced hatching success of Western Toad embryos in Oregon under conditions of natural sunlight. *Anaxyrus boreas* has decreased levels of the enzyme photolyase, which is known to be involved in photo-damaged DNA repair, in comparison with some other anurans. Blaustein et al. (1994b) speculated that decreased levels of this enzyme might be responsible for decreased hatching success. Corn (1998) suggested that differences between these studies might stem from differences in experimental design, the presence of water mold (*Saprolegnia*), or genetic variation among populations. In any case, there is no evidence to suggest that enhanced levels of UVB light by themselves are responsible for the region-wide disappearance of this species in montane regions of the West (Hossack et al., 2006a).

Under experimental conditions, juvenile Western Toads exposed to full spectrum lighting with ambient levels of UVB for 10 hr days have greater mortality than do controls where UV light is excluded. Changes in behavior occur very rapidly, with mortality occurring after four days. After 7 days, mortality in the UVB treatment was 70% compared with 5% in the control (Blaustein et al., 2005).

Kiesecker et al. (2001b) hypothesized that lowered water tables could cause increased exposure to UV light, and thus subject developing embryos to an increased possibility of attack by water molds (*Saprolegnia*). They noted that embryos developing in shallow water were more prone to infection by water molds than embryos living in deeper waters, and that shallow water depths were positively associated with UVB light penetration. Larval *A. boreas* do not avoid areas of high UV light, either in the field or under experimental conditions, however. Instead, they tend to choose warmer water regardless of UV light levels (Bancroft et al., 2008).

These results suggest that long-term drying weather patterns, water depth, temperature, and susceptibility to the pathogen *Saprolegnia* may be interrelated with UVB light exposure, and that Western Toad declines result from complex interactions among these and other factors rather than from a single factor alone. In addition, the spectral characteristics of water in amphibian breeding ponds indicate that UVB effects would be mediated and not likely reach levels to affect embryos (Palen et al., 2002). UVB likely plays a minimum role in amphibian declines throughout the West (but see Blaustein et al., 2004).

STATUS AND CONSERVATION

Western Toads have declined or disappeared throughout much of their former range, particularly in the Rocky Mountains and eastern portion of their historic distribution (Corn et al., 1989; Corn, 1994, 2003; Koch and Peterson, 1995; Ross et al., 1995; Corn et al., 1997; Livo and Yeakley, 1997; Hammerson, 1999; Davis and Gregory, 2003; Corn et al., 2005). These declines probably began in the early 1970s and continued

Habitat of *Anaxyrus boreas*. The dark line in the water is schooling tadpoles. Photo: Brome McCreary

through the early 1980s. Literature accounts suggest Western Toads were very common in areas where they are now extirpated or quite rare (e.g., Ellis and Henderson, 1915, in Colorado). Reports indicate that 80% of the historically known populations in northern Colorado and southern Wyoming have disappeared, that 11 populations have become extinct in central Colorado, that the species is absent from around 50% of historically known sites in the northern Great Basin, and that Western Toads have been extirpated from New Mexico (Corn et al., 1989; Carey, 1993; Stuart and Painter, 1994; Degenhardt et al., 1996; Keinath and Bennett, 2000; Muths et al., 2003; Wente et al., 2005). Occupancy rates are low in Alberta (13% of 120 sites sampled; Stevens et al., 2006b), but such northern populations may be naturally small and sparsely distributed (Hannon et al., 2002; Stevens et al., 2006b).

Literature reports indicate that the Western Toad was fairly common in northern Utah, with disjunct populations in the south; unfortunately, many reports lacked empirical data or specimens (Ross et al., 1995; Thompson et al., 2004). Western Toads were also common in Yellowstone National Park (Turner, 1955) and in the Central Valley of California (Fisher and Shaffer, 1996), where they are now much rarer. Reports from western Washington, the Oregon Cascades, Vancouver Island, Montana, and western Wyoming indicated severe declines in the 1980s (Blaustein et al., 1994a; Koch and Peterson, 1995; Richter and Azous, 1995; Adams et al., 1998; Davis and Gregory, 2003), although some populations had recovered between the 1980s and the 2000s (Olson, 2001). In California, the species has declined in the high Sierras (Drost and Fellers, 1996) and from the Central Valley (Fisher and Shaffer, 1996). In 1925, Storer reported the California Toad as being extremely abundant in both natural and agricultural areas of the Valley, areas where populations are now decimated.

Not all populations are experiencing declines, however. In the Cascade Range of the Pacific Northwest, Pearl et al. (2009a) found *A. boreas* at 76% of historically reported sites. Using occupancy models, they estimated that Western Toads had an occupancy rate of 85% at historic sites and concluded that this species had not experienced broad population declines in the Oregon Cascades. Detection probabilities varied annually depending upon the presence of snowpack and introduced fishes. Weller and Green (1997) found no evidence of declines in Alberta and British Columbia.

Causes of declines in Western Toad populations are complex and may result from the interaction of many factors (Carey et al., 2005). Disease, especially amphibian chytridiomycosis and the water mold *Saprolegnia*, UV light, climate change (especially drought), habitat alteration and human disturbance, roads, trampling by cattle, and the effects of introduced fishes and American Bullfrogs all may affect toad populations (Storer, 1925; Turner, 1955; Banta and Morafka, 1966; Koch and Peterson, 1995; Maxwell and Hokit, 1999; Blaustein and Belden, 2003; Corn, 1994, 2003, 2007; McMenamin et al., 2008; Bull, 2009; Hayes et al., 2010). For example, Carey (1993) hypothesized that an environmental stress caused suppression of the immune system, which made Western Toads more susceptible to the bacterium *Aeromonas hydrophila*. *Aeromonas* is not now thought to be the primary pathogen, as previous mortality events were likely the result of amphibian chytridiomycosis. However, the idea that immunosuppression is involved with the spread of this pathogenic fungus cannot be ruled out. In another example, Hayes et al. (2010) reported on the decline of a Western Toad population that likely was caused by habitat alteration, but not the elimination of breeding adults. Recruitment was essentially blocked by changes in water flow and the silting-in of breeding sites, although adults persisted in the area for years.

Although acidification has been suggested as a possible factor influencing population decline, there is no empirical support for the hypothesis (Corn and Vertucci, 1992; Vertucci and Corn, 1996). Likewise, the effects of UVB radiation experiments suggest mixed results, but enhanced UV radiation by itself does not appear likely as a reason for massive region-wide declines (Corn, 1998; Hossack et al., 2006a). Sublethal effects, however, could be important.

It now seems likely that amphibian chytrid fungus, perhaps in concert with increasing aridity, is responsible for the decline of Western Toads in many regions, particularly in the Rocky Mountains (Muths et al., 2003; Green and Muths, 2005; Bull, 2009; Pilliod et al., 2010). Models of weather and pathogen infection provide little support for hypotheses relating declines to environmental conditions (e.g., cold during breeding seasons), but do support amphibian chytrid infection as the cause of decline (Scherer et al., 2005). Infection spreads through contact with zoospores, and direct physical contact between infected toads is not required for transmission (Carey et al., 2006). The body size of toads, duration of exposure, and the dosage of infectious zoospores all influence mortality. The observation that *Anaxyrus boreas* in the central Rockies first disappeared from higher elevations, conditions ideal for amphibian chytrid, lends support for the chytrid hypothesis of decline (Livo and Yeakley, 1997).

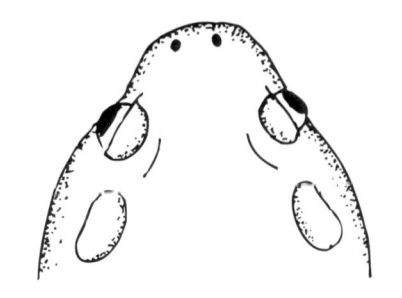

Head pattern of *Anaxyrus boreas*. Illustration by Camila Pizano

Chytrid may persist in populations for many years, and it has significant effects even without killing the entire population. Pilliod et al. (2010) found that amphibian chytrid reduced annual survivorship by 31–42% in infected toads. In the infected Rocky Mountain populations, declines were on the order of 5–7% per yr over a period of 6 yrs, whereas uninfected populations were stable. Thus, rapid declines are not generally observed in chytrid-infected toad populations. The long-term consequences of low-level chronic infection are instead slow declines and possible increased susceptibility to other sublethal stressors in the environment.

Western Toads will breed in newly constructed or restored ponds, often not long after they have been modified. The dimensions of the newly created ponds do not seem to be important, as the number of larvae produced is not proportional to the size of the new pond. At the ponds studied by Pearl and Bowerman (2006), immigrating toads would have had to travel between 0.1 and 4.8 km across substantially arid habitats from the nearest natural pond to colonize the newly constructed ponds. This indicates a significant amount of wandering away from breeding sites and nonphilopatric tendencies, even under dry conditions. Western Toads also have been observed to recolonize ponds rapidly after forest fires or volcanic eruptions (Karlstrom, 1986; Crisafulli et al., 2005; Pearl and Bowerman, 2006; Guscio et al., 2007; Hossack and Corn, 2007a, 2007b). As such, pond creation or restoration might prove a potential mitigation measure to help conserve this species.

In contrast, Monello and Wright (1999) suggested that *A. boreas* was not particularly successful at using artificial ponds in northern Idaho.

Repatriation of boreal toads to several locations in Rocky Mountain National Park has not proved successful. In a program using eggs, recent metamorphs, and adults, researchers were unable to reestablish populations of Western Toads (Muths et al., 2001). The reasons for this are unclear, but may have been due to extreme weather conditions during the winters following release.

There can be little doubt that the conservation of Western Toads requires an extensive amount of habitat, particularly since this species uses summer foraging and overwintering dormancy sites that may be considerable distances from the breeding sites. Management requires attention to the breeding sites, migration routes, summer habitats, and overwintering sites. Inasmuch as Bartelt (1998) noted extensive mortality of recent metamorphs after a herd of sheep passed through a breeding pond as froglets were beginning to disperse, management will require regulation of stock access to important breeding sites. Protecting narrow corridors along riparian strips does not seem to be an effective strategy for this species because of its patterns of movement and terrestrial habitat use (Hannon et al., 2002). Fortunately, the toad is a breeding site generalist with a preference for open habitats. If amphibian chytridiomycosis can be arrested, its wetland breeding sites maintained, and American Bullfrogs and other exotics kept from its habitats, the toad can probably survive most human-related disturbances. However, it may not survive the prolonged and erratic drought associated with global climate change. Conservation and management plans for this species are available for Wyoming (Keinath and Bennett, 2000).

COMMENT
Nearly all life history data on Western Toads refer to inland populations at high elevations. There is little information available on the coastal populations from Alaska to southern California.

Anaxyrus californicus (Camp, 1915)
Arroyo Toad

ETYMOLOGY
californicus: pertaining to California.

NOMENCLATURE
Stebbins (2003): *Bufo californicus*
   Synonyms: *Bufo cognatus californicus, Bufo compactilis californicus, Bufo microscaphus californicus, Bufo woodhousii californicus*

IDENTIFICATION
Adults. The Arroyo Toad has a light olive green, gray, or tannish brown ground color with light patches immediately posterior to the parotoid glands. Black-colored small spots are present irregularly throughout the dorsum. No vertebral stripe is present. The head is short and thick, and the nasal region is slightly elevated into a bony protuberance. The cranial crests are more or less united across the medial region and slightly divergent; no boss is present, but there is a light V-shaped pattern between the eyes. The small bicolored parotoids are oval and very broad, and there are numerous white tubercles below and just posterior to the tympanum. Tubercles are also present dorsally, and these are encircled by small black rings at their base and may be tipped in red. The sides are mottled in light and dark. The rear legs are comparatively short, and there is a small pointed tubercle on the inner edge of the hind foot. Venters are creamy white, unspotted, and not granular; they are never mottled with pigmentation. The vocal sac is round when inflated. Juveniles are ashy white with salmon-colored or olive sides.

Males are smaller than females. Myers (1930b) noted than an exceptionally large female measured 58 mm SUL, and Slevin (1928) examined 5 toads that were 42–55 mm SUL. Sweet and Sullivan (2005) reported calling males from 51 to 67 mm SUL and breeding females from 66 to 78 mm SUL.

Larvae. Upon hatching (to 12 mm TL), larvae are black as in most other toads, but they become progressively lighter after 3 weeks of age. Tan crossbars then appear on the tail base. By 18–20 mm TL, the tadpoles are tan with dark crossbars on the tail base. There is extensive pale gold stippling dorsally, which may produce a mottled appearance. Tail fins are nearly transparent, with only minor black markings giving an appearance of having a lateral stripe on the tail. Venters are white with a pinkish iridescence. At this time they are highly cryptic and blend in well with their sandy substrate. Viewed dorsally, the bars on the tail musculature are visible on mature larvae. Larvae reach 34–40 mm TL.

Eggs. The eggs are black dorsally and gray ventrally. The eggs are oviposited as tangled strings in a gelatinous casing. The strings contain one to two rows of eggs. One jelly envelope surrounds the vitellus. This envelope is 3.3–4.2 mm in diameter and the vitellus averages 1.7 mm in diameter (Sweet and Sullivan, 2005). According to Livezey and Wright (1947) there are 42 eggs per 25 mm of string. Lemm (2006) provides a figure of the eggs.

Distribution of *Anaxyrus californicus*

DISTRIBUTION

The Arroyo Toad is historically known only from the Coastal Ranges of southern California (Salinas River drainage southward) to Baja California Norte, Mexico. Many populations have been extirpated (76% of its historic range according to Jennings and Hayes, 1994b), and the current distribution consists of scattered, isolated populations. There are six currently known populations: Mojave River, Little Rock Creek, Whitewater River, San Felipe Creek, Vallecito Creek, Pinto Canyon. Important distributional references include Slevin (1928), Patten and Myers (1992), Jennings and Hayes (1994b), USFWS (1999), Grismer (2002), Sweet and Sullivan (2005), Lemm (2006), and López et al. (2009).

FOSSIL RECORD

No fossils are known.

SYSTEMATICS AND GEOGRAPHIC VARIATION

Anaxyrus californicus is a member of the North American *Americanus* clade within the Nearctic clade of toads (Goebel, 2005). It is most closely related to *A. microscaphus* and only more distantly to other *Americanus* clade species (Pauly et al., 2004). It was originally described as a subspecies of *A. cognatus* (Camp, 1915) and elevated to species status by Myers (1930b).

Under laboratory conditions, this species produces successful metamorphs in crosses with *A. microscaphus, A. speciosus,* and *A. woodhousii* (Moore, 1955).

ADULT HABITAT

As its common name suggests, this is a species closely associated with intermittent or perennial streams in arroyos and canyons of the coastal mountains of southern California. It prefers a semiarid to moist environment, where it lives along small rocky canyon streams in the Upper Sonoran Life Zone. It requires exposed overflow pools adjacent to flowing streams of low current velocity. Scattered trees (California sycamore, Fremont's cottonwood, coast live oak, willows) occur along its riparian habitat, but the species does not prefer thick riparian vegetation (Cunningham, 1955c; Jennings and Hayes, 1994b). Cunningham (1955c) described the best habitat as that containing fine sandy and gravelly beaches and washes extending for up to 100 m from the stream, interspersed with scattered mulefat (*Baccaris viminea*), sedges, grasses, and annuals near the stream. Griffin (1999) noted a preference for sand substrates for burrowing, but no distinct overall substrate preference. Toads are more common where there are large boulders. The species occurs from near sea level to 2,440 m (Welsh, 1988).

TERRESTRIAL ECOLOGY

Activity is largely confined to the immediate vicinity of streams along rocky and sandy washes. Adult movement patterns are poorly known, but Sweet (*in* Jennings and Hayes, 1994b) reported movements of > 0.8–1 km along streams. They rarely disperse away from the stream margins farther than adjacent upland terraces in narrow canyon habitats, but can travel > 1 km within a stream channel over a 42-day period (Griffin, 1999). In coastal localities, however, they may move more extensively overland (to 1.2 km) into upland grasslands and sage scrub (*in* Sweet and Sullivan, 2005). Griffin (1999) also noted upstream dispersal prior to the breeding season by females.

Except during the breeding season, activity is largely nocturnal from January to early August, and body temperatures (<21°C) reflect the cooler nights (Cunningham, 1955c). Most adult activity ceases by August, but subadults may be active as late as October and even later. Burrows may be > 100 m distant from summer activity areas (Griffin, 1999). During the day and in winter, adults seek refuge in root channels, rodent burrows, along stream terraces, or in moist areas associated with the underground passage of water; burrows may be horizontal or vertical (Griffin, 1999). Many individuals can occupy a single refugium. Cunningham (1955c) reported individuals buried in sand up to 45.7 cm below the surface. Thunderstorms may bring them rapidly to the surface. Locomotion is by hopping rather than walking as in many toads.

After metamorphosis, froglets remain near water for a week, then move to dryer sand bars for about three to eight weeks; there, they hide in vegetation and under surface debris. Juveniles and newly metamorphosed toadlets are diurnal but will take refuge in existing burrows during the hottest part of the day. Their body temperatures can reach 26–37°C. Small juveniles often sit in the direct sun, even on extremely hot substrates, without apparent distress; evaporative cooling may assist in keeping body temperatures lower than the surrounding environmental temperatures. At about 30 mm SUL, dispersal begins to areas dominated by willows; it is at this size that they become capable of burrowing themselves. They will burrow 10–18 cm below the surface and remain inactive there for 6–8 months (Griffin, 1999; Sweet and Sullivan, 2005). By August, juveniles assume a nocturnal activity pattern.

CALLING ACTIVITY AND MATE SELECTION

Males emerge from overwintering sites prior to females. Calling may begin in late March and extend to late July. The call of *Anaxyrus californicus* has been described as a "sweet trill" (Myers, 1930b) or as a clear, prolonged musical trill (Stebbins, 1962). The call has a dominant frequency of ca. 1.42 kHz, a mean pulse rate of 41.5 pulses/sec, and a mean call duration of 6.62 (range 2 to 14) sec (Stebbins, 1962; Sullivan, 1992b). Males also have a warning call that is somewhat like the musical trill of the advertisement call. Males position themselves along pools and call from exposed locations, where they display a high level of site fidelity. From one to three males may call from a pool on any one night (S. Sweet, *in* Jennings and Hayes, 1994b). Calling ceases when temperatures are <13–14°C.

BREEDING SITES

Breeding occurs in quiet pools along slowly flowing intermittent streams in rocky canyons. They avoid deep and swift water, tree-canopy cover, and steeply incised banks. Substrates are usually sand or gravel. Arroyo Toads only rarely breed in pools isolated from flowing water.

REPRODUCTION

Females forage for several weeks prior to breeding in order to yolk their egg clutches. Breeding commences after the winter and spring rains and

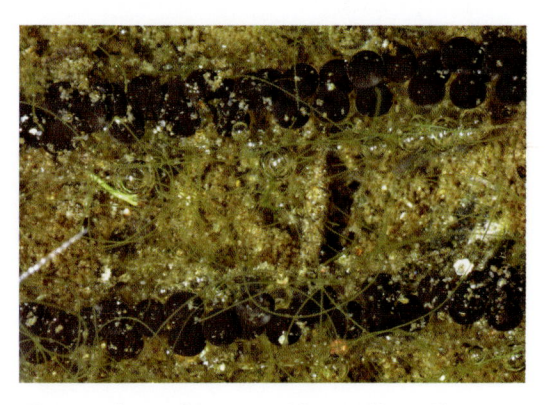

Close-up of eggs of Anaxyrus californicus. Photo: Chris Brown

appears to be triggered by increasing temperature. Most reproduction occurs from early March to late June, depending on weather and water flow conditions (Myers, 1930b; Jennings and Hayes, 1994b; Griffin, 1999), but gravid females have been observed as late as 20 July and calling to 29 July, indicating an extended breeding season (Cunningham, 1955c; Griffin, 1999). Such an extended season allows adults to take advantage of precipitation events in a region where precipitation can vary considerably from one year to the next.

Eggs are deposited in two long strings (3–10.7 m) on the bottom of shallow (mean 9 cm, range 3.1 to 31.8 cm) pools on top of leaves, mud, and subsurface debris; the mean clutch size is 4,714 eggs (range 2,013 to 10,368) (Sweet and Sullivan, 2005). Eggs are deposited on silt in water with no current, and egg strings are not twisted around branches or debris. A female only oviposits one clutch per breeding season. Hatching occurs in four to six days.

LARVAL ECOLOGY
Newly hatched larvae cannot swim for 15–18 days after hatching. Indeed, larvae do not move much during the entire larval period. Feeding occurs by sifting the substrate for organic detritus and algae, bacteria, protozoans, and fungi; they do not eat macroscopic algae or aquatic vegetation (Sweet and Sullivan, 2005). They prefer open habitats on a stream bottom of sand or gravel. The duration of the larval period is 65–85 days, and metamorphosis requires 4 days to complete. Newly metamorphosed froglets are 9–15 mm SUL, rarely to 22 mm SUL (Cunningham, 1955c; Sweet and Sullivan, 2005).

DIET
Cunningham (1955c) reported instances of adults eating juvenile conspecifics, and Sweet (*in* Jennings and Hayes, 1994b) reports the diet of juveniles to consist mostly of ants. At about 20–25 mm SUL, the diet switches mostly to small beetles. Lemm (2006) also includes caterpillars, moths, Jerusalem crickets, and snails as part of the diet.

PREDATION AND DEFENSE
Both the tadpoles and juveniles are highly cryptic, making it difficult to observe them on sandy substrates. Eggs and larvae may be distasteful.

Tadpole of *Anaxyrus californicus*. Photo: Chris Brown

Large toads also have noxious skin secretions, and will readily hide in vegetation or, if calling, dive beneath the water's surface to escape. Natural predators of eggs and larvae include fish (*Gambusia*, *Lepomis cyanellus*, *Cottus asper*), frogs (*Lithobates catesbeianus*), garter snakes (*Thamnophis hammondii*, *T. sirtalis*), birds, aquatic insects (*Abedus indentatus*), and crayfish (*Procambarus clarkii*) (Griffin, 1999; Sweet and Sullivan, 2005). Killdeer (*Charadrius vociferous*) take a large toll on recent metamorphs and juveniles. American Bullfrogs and garter snakes attempt to eat adult toads, but even some toads that escape subsequently die from their injuries.

POPULATION BIOLOGY
Most populations are small (30–100 adults) making this species vulnerable to extinction. Sweet and Sullivan (2005) reported low abundance (12 adults/ha) along second- to fourth-order streams in mountains and foothills, but from 10–100 adults/ha along coastal streams in suitable habitat. There is a considerable amount of variation among localities in adult densities.

Males can grow to sexual maturity in 1–2 yrs (mostly 2), depending on favorable weather conditions, but females require 2 yrs to achieve sexual maturity (i.e., they breed during their third activity season; Sweet and Sullivan, 2005). Longevity in wild populations may approach 5 yrs. Sweet (*in* Jennings and Hayes, 1994b) reported that breeding and recruitment frequently failed because of instability in stream flows, such as when eggs or larvae are washed away by flash floods or during periods of prolonged drought.

COMMUNITY ECOLOGY
American Bullfrogs may target calling males, leading to highly skewed sex ratios and localized

extirpation (Sweet and Sullivan, 2005). These authors noted that bullfrogs even eat amplexed pairs and are a major threat to Arroyo Toad populations.

DISEASES, PARASITES, AND MALFORMATIONS
No diseases are known. Larvae are often infected with unidentified encysted trematode metacercariae and occasionally by an unidentified cestode in the body wall musculature (Sweet and Sullivan, 2005).

SUSCEPTIBILITY TO POTENTIAL STRESSORS
No information is available.

STATUS AND CONSERVATION
This species is critically "Endangered" and is listed as such under the U.S. Endangered Species Act of 1973 and California state law. A Recovery Plan has been developed for this species (USFWS, 1999). Jennings and Hayes (1994b) estimated that Arroyo Toads had lost 76% of their historically known populations, and that populations in the northern, eastern, and central parts of its range had disappeared. After federal protection, however, some populations, especially those on federal lands, have recovered, whereas others have continued to decline, particularly on private, urbanizing lands (Sweet and Sullivan, 2005).

Adult *Anaxyrus californicus*. Photo: Brian Pittman

Habitat of *Anaxyrus californicus*. Photo: Chris Brown

Threats include habitat destruction, particularly from suction placer/dredge mining, sedimentation, human-based manipulation of hydrological regimes, urban development, human recreation (mountain biking), cattle grazing, the introduction of American Bullfrogs, exotic fish and crayfish, road construction and maintenance, drought, gravel mining, military maneuvers, trampling, and fire (Jennings and Hayes, 1994b; Griffin, 1999; Sweet and Sullivan, 2005). Ervin et al. (2006) recorded mortality due to off-road vehicles. Jennings and Hayes (1994b), USFWS (1999), and Sweet and Sullivan (2005) have suggested a number of ways to conserve this species, especially protecting its habitat from continued destruction and restoring altered habitat. Griffin (1999) advised protecting considerable linear distances in stream canyon habits in order to ensure survival. This species would appear particularly vulnerable to changes in precipitation patterns.

Anaxyrus canorus (Camp, 1916)
Yosemite Toad

ETYMOLOGY
canorus: from the Latin *canorus*, meaning 'tuneful.' The name refers to the trilling mating call.

NOMENCLATURE
Stebbins (2003): *Bufo canorus*
  Synonyms: *Bufo canorus*

IDENTIFICATION
Adults. Males and females are strikingly dichromatic, with the females being more colorful. Males gradually become more olive greenish as they grow, and they have a reduction in dorsal and ventral spotting. They have fewer and smaller warts, and the parotoid glands are reduced in size when compared with females; the body has small, scattered black spots, which are rimmed in white. Dorsal spotting is accentuated in females, but the venter becomes whiter with a more immaculate tint. Females also have prominent black bars or spots on the legs. Both sexes tend to lose the vertebral white stripe that is prominent in the young animals, with males losing it faster than females. Karlstrom (1962) describes the ontogeny of color change in great detail. Juvenile males and females are similar in coloration, having black spots or blotches on a brownish, greenish-brown, or grayish ground color.

Males generally are slightly smaller than females, and there is some size variation among populations. Specific size records include a mean of 63.4 mm SUL for females (range 56 to 71 mm SUL) and 59.3 mm SUL (range 53 to 68 mm) for males from Kaiser Pass, Fresno County. In contrast, females are a mean of 52.1 mm SUL (range 48 to 56 mm) and males 49.3 mm SUL (range 45 to 53 mm) from Tioga Pass in Yosemite National Park. Also from Yosemite, adult males had a mean SUL of 60.6–73.2 mm, depending on population, whereas females were 61.7–75.7 mm SUL (Kagarise Sherman, 1980). Storer (1925) recorded males 50–64 mm SUL and females 57–74 mm SUL from throughout Yosemite National Park. The largest toads were estimated to be four years old.

Larvae. The larvae are very dark, with slightly transparent tails that may be marked by large branched melanophores. In profile, the snout is rounded and the tail tip is broadly rounded and deepest at its midpoint (Karlstrom and Livezey, 1955). The largest larvae reported are 35 mm TL.

Eggs. The eggs of *Anaxyrus canorus* are much larger than many other species of *Anaxyrus*. They are deposited in strings that may be single- or double-stranded, or they may be clustered in a network four- or five-eggs deep (Karlstrom and Livezey, 1955). Jelly partitions separate individual eggs from one another within the continuous string tube. The eggs are brownish black to deep black dorsally, and gray or tannish gray ventrally. The diameter of the outer envelope is 3.7–4.6 mm (mean 4.1 mm); the inner envelope is 3.4–4.1 mm (mean 3.8 mm); the diameter of the ovum is 1.7–2.7 mm (mean 2.1 mm). There are 6–8 eggs per 2.54 cm of string (Karlstrom, 1962); Savage and Schuierer (1961) reported 28 eggs per 10 cm of string. Total clutch size is 1,500–2,000 eggs per female. Additional details on the eggs of this species are in Karlstrom (1962). Illustrations of eggs are in Karlstrom and Livezey (1955) and Savage and Schuierer (1961).

Distribution of *Anaxyrus canorus*. Dark gray indicates extant populations; light gray indicates extirpated populations.

DISTRIBUTION

The Yosemite Toad is found at high elevations in the mountains of the central Sierra Nevada of California. It occurs at elevations of 1,950–3,445 m, although most records are from between 2,590 and 3,050 m. Important distributional references include Slevin (1928), Karlstrom (1962), and Jennings and Hayes (1994b).

FOSSIL RECORD

There are no specific fossils recorded for this species. See the *Anaxyrus boreas* account for a listing of fossils attributed to the *A. boreas* species group.

SYSTEMATICS AND GEOGRAPHIC VARIATION

The Yosemite Toad is a member of the *Boreas* group of toads (with *A. boreas, A. exsul,* and *A. nelsoni*), which are all members of the Nearctic clade of New World toads. Current species nomenclature does not represent evolutionary history, as molecular data suggest that *A. canorus* is paraphyletic (Graybeal, 1993, 1997; Shaffer et al., 2000; Stephens, 2001; Pauly et al., 2004; Goebel et al., 2009). Several studies have offered differing hypotheses of evolutionary relationships with other members of the *A. boreas* group. Some authors, for example, have suggested that *A. canorus* and *A. exsul* are sister taxa (Graybeal, 1993, 1997; Shaffer et al., 2000) within the SW clade of *A. boreas*. Others (e.g., Stephens, 2001) have recognized that even the *A. canorus* related to SW *Anaxyrus boreas halophilus* represented a paraphyletic assemblage.

There are two distinct lineages of *A. canorus*, one centered on Kings Canyon National Park (south) and the other on Yosemite National Park (north). Shaffer et al. (2000) reported considerable genetic substructuring at Yosemite, focused on major drainages and breeding ponds. At Kings Canyon, however, different genetic assemblages were associated with breeding sites, but not aligned with major drainages. The highly different genetic structuring of these toads suggests different modes of colonization and emphasizes the need for a breeding-site specific approach to management.

As currently recognized, the northern evolutionary lineage of "*A. canorus*" is genetically related to *A. b. boreas* from the central part of the range of the Western Toad, whereas the southern lineage is more closely allied with the SW group of *A. b. halophilus* and *A. nelsoni* (Goebel et al., 2009). According to Goebel et al. (2009), "*A. canorus*" either represents multiple taxa or is derived from multiple divergent mtDNA lineages. As a result of continued investigation, the taxonomy of the Yosemite Toad is likely to soon change. Like other isolated western toad populations, the genetically distinct populations of *A. canorus*, at least in part, are the products of isolation resulting from Pleistocene climate change.

Hybridization is not uncommonly reported in the literature where Western Toads come into contact with other members of the *A. boreas* group. Natural hybrids between *A. boreas* and *A. canorus* are known from Mono County, California (Morton and Sokolski, 1978). A suspected hybrid was reported from Alpine County, California (Mullally and Powell, 1958), although the identification has been questioned (Karlstrom, 1962).

ADULT HABITAT

The habitat of the Yosemite Toad consists of high mountain meadows surrounded by dense coniferous forests of lodgepole and whitebark pines in Canadian, Hudsonian, and Lower Alpine life zones. It is most often found in wet meadow bottoms and meadow areas near talus slopes (Morton and Pereyra, 2010). The grassy meadows contain cold snow-fed streams and breeding pools, some of which are temporary and dry during the summer whereas others are permanent. Low willow bushes grow in the wet areas.

TERRESTRIAL ECOLOGY

Adults are active early in the season, initially in the forest and near the forest meadow ecotone where conditions are moist. As snow melts in the alpine meadows, they wander farther into these habitats as the forests become dryer and the meadows become warmer and moist from melting snows. Adults are active at a wide range of temperatures, from −1 to 38°C, but they prefer body temperatures >ca. 13°C and avoid ambient temperatures >28–30°C. They are able to survive a mild freeze, presumably by converting body water to ice crystals. According to Karlstrom (1962), they show no particular thermal preferences, and are likely to be active anytime during normal ambient temperature conditions for this region and season (2–30°C).

Yosemite Toads are diurnal. At night during cold temperatures, they shelter under logs and other coarse woody debris or in rodent (*Marmota flaviventris, Microtus montanus, Spermophilus beldingi, Thomomys monticola*) burrows several cm below the surface. Overwintering also occurs in such sites, as well as in cracks and crevices in rocks beneath the soil or among the roots of vegetation. When the body temperature reaches 8–10°C, the toads become active. Toads are commonly seen sitting at the entrance to their shelters, where they may bask for 30–60 minutes before leaving, especially at lower temperatures (Mullally and Cunningham, 1956a). Basking serves to warm the animal as it begins its daily activity, and toads may achieve body temperatures of 18°C, despite the morning chill. Karlstrom (1962) also noted crepuscular and nocturnal activity, especially during the breeding season.

Males tend to remain in the vicinity of the breeding pools, whereas females disperse more widely after their eggs are deposited. For example, Morton and Pereyra (2010) noted that females tended to move to the vicinity of a talus slope 800 m from the nearest breeding pond, but that males did not. Mullally (1953) noted that many Yosemite Toads stay near watercourses, where they feed as they move several meters up and down the shoreline. Movements away from water can be extensive. Activity continues throughout the summer as late as September to early October. Winter retreats include rodent burrows and may be located in forested areas not far from the breeding sites.

Amplexing adults and egg strings of *Anaxyrus canorus*. Photo: Earl Gonsolin

As with *A. boreas*, Yosemite Toads likely are photopositive and sensitive to light in the blue spectrum ("blue-mode response"). They probably have true color vision.

CALLING ACTIVITY AND MATE SELECTION

Toads emerge from dormancy in late spring to early summer, depending on weather conditions and when the pond ice melts. Kagarise Sherman (1980) reported first emergences from 29 April to 4 June, depending upon year. They do not dig out from under snow but must wait for it to melt from the entrance to their winter dormancy site. Emergence likely occurs later at higher elevations as snow and ice melts, allowing access to the breeding ponds. However, even during the breeding season, temperatures may fall below freezing leaving a thin crust of ice on the breeding ponds. There are four to eight days between emergence and the arrival at breeding sites and the first deposition of eggs (Karlstrom, 1962; Kagarise Sherman, 1980). Calling may begin sporadically at temperatures of 1–2°C, with full choruses trilling at temperatures of 11–12°C and especially >14°C. Chorusing peaks at midday and in the afternoon, but Karlstrom (1962) recorded calling even at near-freezing temperatures at 21:50 hrs. Calling usually ends by mid-June, but in years of later emergence, it continues to mid-July (Kagarise Sherman, 1980).

Unlike the closely related *A. boreas*, the male's voice is a long melodious trill. The trill consists of a series of 26–51 (mean 38) evenly spaced notes lasting from 0.8 to 3.8 sec (mean 2.6). There are 6 note harmonics, with a frequency range to 8,000 cps and a dominant frequency of 1,550 cps at 24–25°C (Karlstrom, 1962). There is the potential for sexual selection based on size, as larger males have a voice with lower tones, and larger males do seem to be more successful at mating than smaller males.

Males may call in close proximity to one another or space themselves at considerable distances along a wetland margin. At least during the day, they tend to remain at the same calling station, although they may move about searching for females. Indeed, silent males tend to search near calling males and thus attempt to intercept a female approaching the caller. Such satellite male behavior is not uncommon in the Anura. There is some support for the hypothesis that in low-density situations, males tend to call more than search. As the density of calling males increases to a certain level, some males may alter their tactics by ceasing to call and trying to intercept a female moving toward a calling male (Kagarise Sherman and Morton, 1984).

Calling occurs from the shallow water along the margins of the pond or stream. Male–male aggression (fighting, wrestling, making a low-volume clucking sound) occurs should a nearby male approach a calling male. Male territoriality seems directed toward the protection of adequate space in which to incept a female rather than selection of a favorable oviposition site (Kagarise Sherman and Morton, 1984). Aggressive bouts may only last 20 sec, but aggressive encounters may last 5 min. Large males tend to win more fights than smaller males, which usually give up and return from whence they came.

The calling male rears up in a nearly vertical position, and the vocal sac is prominent as he trills. Chorusing by one male tends to trigger additional males to begin calling. Calls last for several seconds, and intervals between calls can last several minutes. Calls have good carrying capacity and may be heard > 30 m away. According to Karlstrom (1962), chorusing may be more or less continuous among toads occupying a large meadow. Calling males are easily silenced if they are approached at night or as a result of loud noises, such as traffic. During the day, however, they are less easily disturbed.

Males recognize females by their girth (full of eggs) and by their silence if contacted. The dichromatic coloration does not appear to play a part in sex recognition. Amplexus is pectoral. Males will attempt to amplex any moving object in their vicinity; if another male is contacted, he will emit his "warning vibration" and utter a release cluck. This alerts the amplexing male to the sex of the animal it has clasped. Males also discourage amplexus by other males through pushing and kicking. Despite these efforts, balls of toads have been observed during peak mating periods as excited males vie for females and desperately grasp any object or multiple objects within reach (Kagarise Sherman and Morton, 1984).

Females may have some choice of males. A female may approach a calling male directly and actually touch him, and initiate amplexus. Females are known to move among calling males as

Egg strings of *Anaxyrus canorus*. Photo: Marty Morton and Maria Pereyra

if seeking a specific caller. Female choice seems to be effective in about 50% of the observed cases, with the remainder subject to random male interception. The males do not seem to be particularly selective in their choice of females.

BREEDING SITES

Optimal breeding sites include alpine lakes and pools of clear and cold shallow water surrounded by meadows. Yosemite Toads also breed in shallow slow-moving runoff streams. They initially prefer breeding sites near the forest, as forested habitat provides cover during the still cold spring nights experienced immediately after emergence. As the temperatures increase, they move farther into the meadow and occupy the borders of wetlands and streams as the forested areas become drier. The presence or absence of nonnative trout does not appear to have an effect on whether *A. canorus* occupies a breeding site (Knapp, 2005).

REPRODUCTION

Breeding normally occurs from early May to mid-June in the high Sierras, as soon as snow and ice melt at the nearly alpine breeding pools. However, Kagarise Sherman and Morton (1984) noted that breeding may not commence until as late as 20 June following winters with exceptional snowfall, and Wiggins (1943) and Mullally (1953) recorded gravid females on 11 August and 18 August, respectively. Even spring storms can interrupt a breeding season for weeks after it has started. Yosemite Toads routinely move to breeding sites when snow still covers much of the ground, as it does for eight to nine months of the year, often tiptoeing across snow patches (Kagarise Sherman and Morton, 1984). As might be expected, the date of breeding is dependent upon elevation and local environmental conditions, with breeding commencing later at higher elevations.

Egg deposition occurs in shallow water (7–8 cm) along a wetland or stream margin. Preferred sites have a silt bottom that allows the mated pair to swim about, leaving egg strings anchored among sedges. Eggs are deposited over an area of 1–2 m^2, and a single egg clutch produces 1,500–2,000 larvae (Karlstrom, 1962). Hatching occurs at 5–12 days after oviposition, as the sun speeds up development of the eggs despite the cold water temperatures. A female only produces one clutch per year and may skip years between reproductive bouts in order to obtain sufficient energy to provide yolk for a clutch (Kagarise Sherman, 1980; Kagarise Sherman and Morton, 1984).

Most oviposition occurs during the afternoon and lasts only long enough for females to extrude their eggs and males to fertilize them. Amplexed pairs may be seen moving about a shallow wetland (perhaps for a few meters), seeking an oviposition site, and several mated pairs may deposit their eggs in close proximity to one another. After amplexus, the adults seek shelter for the night.

There appear to be many more males than females in attendance at any one time at a chorus. For example, Karlstrom (1962) stated that there were ten males per female at the chorus he studied at Tioga Pass, California. However, Kagarise Sherman and Morton (1984) noted that the sex ratio may change through time at a site (6.3 males per female to 1 male per 2 females at their site over a period of 6 yrs), and that the degree of polygyny varies accordingly. Females presumably come into a breeding site, are immediately amplexed, deposit their eggs, and leave. Males, however, likely remain chorusing at the breeding site for 7–17, days in order to intercept as many females as possible.

Male success appears to be tied to when he arrives and the duration of his stay; thus, males arriving earlier and staying longer have a better chance of successful reproduction than do later arrivals (Kagarise Sherman and Morton, 1984). Males do not remain throughout the breeding period, as the short activity season necessitates maximum foraging in order to survive the long winter. Males may mate with more than one female (to three), or not be successful at all. Male mating success was zero in 58.7–88.3% of the time in Kagarise Sherman's (1980) 4 yr study, whereas 10.5–32% of males mated with 1 female, 1.2–9.3% mated with 2 females, and only 0.5% mated with 3 females, and that in only 1 year. Females, however, mate with only a single male annually, and some may skip breeding seasons. Because of the variance in mating success, the operational sex ratio is actually 10–38 males per female (Kagarise Sherman, 1980).

LARVAL ECOLOGY

Eggs are deposited in shallow water, which provides a positive thermal environment for embryonic development. Karlstrom (1962) considered the primary high elevation reproductive adaptation of *A. canorus* to be the placement of eggs where they could be warmed through shallow-water heat insolation. The sun's rays warm the eggs and accelerate development despite the overall cold water and air temperatures early in the season. Larvae tend to remain in the shallow warm waters during the day, but retreat to deeper water after dark (Mullally, 1953). Presumably the deeper water holds heat longer than the shallow waters along the shoreline.

Significant larval mortality occurs if water temperatures reach 31°C. The longer larvae remain within the breeding pond, the greater the temperatures to which they are able to acclimate. Thus, tadpoles easily survive waters at 32°C during the late summer stages of development. The larval CTmax is 36–38°C (Karlstrom, 1962). Still, ambient temperatures rarely get this high, especially in the generally cool high elevations inhabited by the Yosemite Toad.

The large amount of yolk in *A. canorus* eggs allows the larvae to grow to 9–10 mm body length (26 mm TL) prior to feeding. Thus, larvae are able to attain one-third of their final body size even before feeding begins. Such parental investment ensures the likelihood of successful metamorphosis despite the extreme high-elevation mountain conditions. Metamorphosis occurs in from 40 to 50 days in the laboratory (e.g., Jennings and Hayes, 1994b) and 56–110 days in

Tadpoles of *Anaxyrus canorus*. Photo: Marty Morton and Maria Pereyra

nature, depending upon thermal conditions. New metamorphs are 8.4–13.2 mm SUL.

Aside from insect predators, the greatest threat to larval Yosemite Toads is desiccation, as ponds and wetlands dry during the summer (Jennings and Hayes, 1994b). Ponds are more likely to dry when the snowpack is shallow and no rain falls during the spring. If the previous winter's snowfall was heavy, ponds are likely to be deeper and hold water longer to allow metamorphosis. Exceptionally dry springs and summers also cause ponds to draw down quickly.

Larval *A. canorus* do not exhibit any changes in behavior when exposed to chemical stimuli from the nonnative brook trout *Salvelinus fontinalis* (Grasso et al., 2010).

POPULATION BIOLOGY

After metamorphosis, toadlets must grow quickly in order to obtain sufficient body reserves to last their first long winter. Sexual maturity is reached in about 3–5 yrs in males and 4–6 yrs in females (Kagarise Sherman, 1980; Kagarise Sherman and Morton, 1984; note that Karlstrom, 1962, incorrectly speculated that maturity is reached in 2–3 yrs). Sex ratios are unknown, but Morton and Pereyra (2010) recorded 237 females and 225 males over a 7 yr period at Yosemite National Park. Morton and Pereyra (2010) further suggested that the cold weather and short activity period results in an irregular breeding cycle for this species, since females must acquire energy for egg production and to survive hibernation in a rather short period of time. Yosemite Toads may have long lives; Kagarise Sherman and Morton (1984) record maximum female longevity as at least 15 yrs, and males as 12 yrs.

In addition to avian predators, Kagarise Sherman (1980) reported mortality from multiple amplexus, exposure, unidentified illness or pathogens (possibly *Aeromonas*), and from trampling by large mammals. She also found dead toads with no apparent injury or cause of death.

DIET

Yosemite Toads are ambush predators, eating a variety of invertebrates including ants, bees, wasps, beetles, millipedes, flies, mosquitoes, lepidopteran larvae, dragonfly larvae, and spiders (Mullally, 1953; Wood, 1977). Males tend to reduce feeding activities during the breeding season (Wood, 1977).

PREDATION AND DEFENSE

Eggs, larvae, and postmetamorphic *Anaxyrus canorus* are unpalatable to some predators, such as the nonnative brook trout (*Salvelinus fontinalis*) (Grasso et al., 2010). Larvae are preyed upon by dragonfly naiads and predaceous diving beetles. Birds eat larvae that are stranded by lowering water levels. Adult and subadult *Rana sierrae* and *R. muscosa* also feed on larval Yosemite Toads (Mullally, 1953; Pope and Matthews, 2002). Adults are preyed upon by California gulls (*Larus californicus*) and Clark's nutcrackers (*Nucifraga columbiana*). Adults are particularly vulnerable as they cross snowfields.

Adults have a noxious secretion produced by parotoid and granular glands on the dorsum.

DISEASES, PARASITES, AND MALFORMATIONS

Kagarise Sherman (1980) reported mortality from unspecified illness or pathogens, and speculated that red-leg (the bacterium *Aeromonas*) might be responsible. Yosemite Toads are parasitized by the cestode *Cylindrotaenia americana* and the nematode *Cosmocercoides variabilis* (Walton, 1941).

SUSCEPTIBILITY TO POTENTIAL STRESSORS

pH. Survival of embryos and hatchling Yosemite Toads is affected by low pH. In experimental trials, Bradford et al. (1992) estimated the $LC_{50\,(7\,day)}$ for embryos as 4.7 and for hatchlings as 4.3.

Female adult *Anaxyrus canorus*. Photo: Ceal Klingler

Habitat of *Anaxyrus canorus*. Photo: Gary Fellers

Survival was normal at a pH of 5.0. However, a pH of 5.0 causes earlier hatching than normal.

Metals. Exposure of embryos and larvae of aluminum at 39–80 µg/L had no effect on survivorship (Bradford et al., 1992). However, sublethal effects included a reduced body size in tadpoles and an earlier hatching by embryos.

UVB radiation. UVB radiation does not affect hatching success or developmental rates of embryos (Vredenburg et al., 2010a).

STATUS AND CONSERVATION

The Yosemite Toad has declined dramatically (by 50%) over the last 30 years, despite protection of habitat within large national parks (Martin, 1991; Kagarise Sherman and Morton, 1993; Stebbins and Cohen, 1995; Drost and Fellers, 1996; Corn, 2003; Morton and Pereyra, 2010). Even in surviving populations, abundance has decreased substantially. In times of drought, egg deposition sites may not be available, or larvae may not have sufficient time to complete metamorphosis (Myers, 1930a; Karlstrom, 1962). At such times, substantial mortality may occur and considerably set back recruitment. Another potential killing agent is winterkill, especially in years of low snowpack, which serves to insulate the subsurface retreat sites. Drought and winterkill over several seasons could selectively target early arriving males or their offspring and thus alter the breeding sex ratio in subsequent years (Kagarise Sherman and Morton, 1984).

The potential for climate change (drought, alteration of rainfall patterns) leading to reductions or even extinction of this species is a real possibility. Given the extensive genetic structuring among populations based on breeding sites and drainages (Shaffer et al., 2000), this species must be conserved on a population-by-population basis. Every population becomes important, and the survival of the species will not be ensured unless each population receives specific attention. Although acidification has been suggested as a possible factor influencing population decline, there is no empirical support for the hypothesis (Bradford et al., 1994a). Grasso et al. (2010) suggested that removal of nonnative trout from Yosemite Toad habitats would have no effect on this species because it is unpalatable.

The Yosemite Toad is protected by the National Park Service and the State of California as "Endangered." It is a candidate for protection under the federal Endangered Species Act.

Anaxyrus cognatus (Say, 1823)
Great Plains Toad

ETYMOLOGY
cognatus: from the Latin word *cognatus*, meaning 'related by birth.'

NOMENCLATURE
Conant and Collins (1998) and Stebbins (2003): *Bufo cognatus*

Synonyms: *Bufo dipternus, Bufo lentiginosus cognatus, Chilophryne cognata, Incilius cognatus*

The species *Anaxyrus californicus* was originally described as *Bufo cognatus californicus* (Camp, 1915).

IDENTIFICATION
Adults. Great Plains Toads are medium to large. The ground color is olive to pale gray brown, and there are large, well-defined blotches of olive, dark green, or dusky coloration over the dorsum. These blotches are bordered with a light tan, brown, or green ring. Darker individuals have an obvious mid-dorsal light stripe that may not be apparent in light individuals. The entire dorsum is covered with small warty bumps, and toads may have a faint white mid-dorsal stripe running down the back. The prominent cranial crests and parotoids are large and together form an L-shape which comes together to form a V-shaped osseous boss (a bump) between or behind the nares; the boss is not well developed until the toad is ca. 37 mm SUL. Parotoids are large and oval-shaped and are contacted by the postorbital crests.
The bellies are white, usually with no blotches or spots, but rarely spotted. Juveniles, but not adults, have small brick-red tubercles. The rear feet have prominent metatarsal tubercles, which aid in digging. Males have a dark loose-skinned throat, which forms an apron in appearance, and cornified thumbs. When the vocal sac is extended, it is sausage-shaped and juts up over the snout. Melanistic juveniles were reported by Bragg (1958a).

Although occasional individuals are > 100 mm SUL, most Great Plains Toads are smaller. Males are on average slightly smaller than females. In Oklahoma, calling males were 72–90 mm SUL (mean 80 mm) (Bragg, 1940c); Bragg (1950e) later reported Oklahoma males at 62–103 mm SUL (mean 79.4 mm) and females at 49–112 mm SUL (mean 85.8 mm). In Krupa's (1994) study, the mean size of males was 69.5–79.8 mm SUL (range 56 to 98 mm) and females 71.8–85 mm SUL (range 60 to 115 mm). Nebraska adults averaged 65.1 mm SUL (range 52 to 78 mm) (Ballinger et al., 2010). In Arizona, males were 68.8 mm in mean SUL, whereas females were 72.8 mm SUL (Sullivan and Fernandez, 1999). Wright and Wright (1949) reported males as 47–95 mm SUL and females as 60–99 mm SUL. Smith (1934) recorded a 114-mm SUL female in Kansas. SUL is directly correlated with body mass.

Larvae. At hatching, the tadpoles are black with the ventral coloration lighter than the dorsal coloration. As the tadpoles grow larger, they become grayer and mottled brown with silver chromatophores, which may be present throughout the dorsum. Venters are gray, and there are golden chromatophores scattered throughout the belly. Viscera may be seen through the belly early in development, but become obscure as the tadpole grows. Eyes are large, with a golden iris. Tail fins are highly arched and contain scattered dendritic pigment patterns on the dorsal fins. Ventral tail fins are clear. Bragg (1936) provided detailed illustrations and a description of the larvae during development. They usually reach 29–30 mm TL prior to transformation.

Eggs. The small eggs (about 1.2 mm in diameter) are black dorsally and lighter ventrally, although the distribution of the pigment makes the eggs appear dark all over. They are enclosed in a gelatinous capsule within an extended string of tough elastic gelatin, and each egg is separated from its neighbor by a partition. Rarely are there two rows of eggs within a single string (Bragg, 1936). Egg strings may be single or double. Two scalloped envelopes surround the vitellus, the outer envelope averaging 1.7 to 2.1 mm and the inner envelope averaging 1.6 mm (Bragg, 1937b; Livezey and Wright, 1947).

DISTRIBUTION
As its common name implies, this is a grassland and semiarid species that occurs throughout the Great Plains from southern Alberta, Saskatchewan, and southwestern Manitoba through north and west Texas, south to Aguascalientes and San Luis Potosi, Mexico. It occurs in the semiarid West through intermediate elevations along the

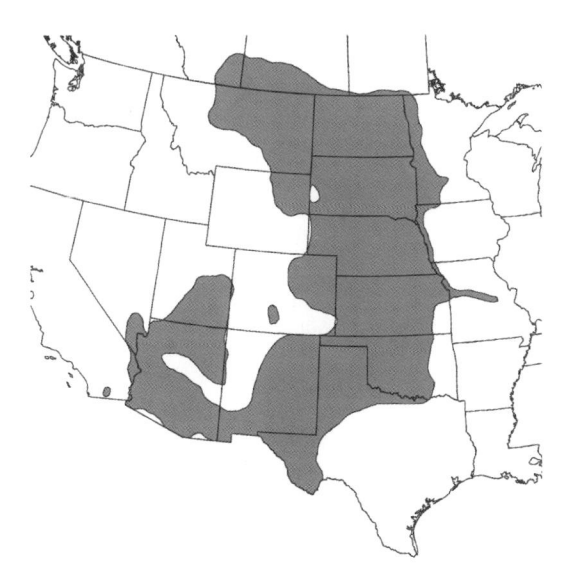

Distribution of *Anaxyrus cognatus*

Wasatch Front, southern and southeastern Utah, and northern Arizona, the Sonoran Desert in southern Arizona, the Mojave Desert, and Imperial Valley of southeastern California, and much of New Mexico (excluding the southern Rockies), especially along the Rio Grande. The species is absent from the Rocky Mountains, the Black Hills, and from the high mountains of central Arizona and west and central New Mexico. Isolated populations occur in southern Colorado and north of the Great Salt Lake in Utah. The range extends to its farthest point east along the Missouri River into central Missouri.

Important distributional references include: Alberta (Russell and Bauer, 2000; Pearson, 2009), California (Vitt and Ohmart, 1978), Colorado (Hammerson, 1999), Kansas (Smith, 1934; Collins, 1993; Collins et al., 2010), Mexico (López et al., 2009; Lemos-Espinal and Smith, 2007b), Montana (Black, 1970, 1971), Minnesota (Oldfield and Moriarty, 1994), Missouri (Johnson, 2000; Daniel and Edmond, 2006), Nebraska (Lynch, 1985; Ballinger et al., 2010; Fogell, 2010), Nevada (Linsdale, 1940), New Mexico (Degenhardt et al., 1996), North Dakota (Wheeler and Wheeler, 1966; Hoberg and Gause, 1992), Oklahoma (Sievert and Sievert, 2006), Saskatchewan (Cook, 1960), South Dakota (Fischer, 1998; Ballinger et al., 2000; Kiesow, 2006), Texas (Dixon, 2000), Utah (Tanner, 1931), and Wyoming (Baxter and Stone, 1985).

FOSSIL RECORD

This species is known from many fossil deposits from the Miocene (Kansas, Nebraska), Pliocene (Kansas, Nebraska, Texas), and Pleistocene (Colorado, Kansas, Sonora, Texas) (Rogers et al., 1985; Rogers, 1987; Holman, 2003). In the fossil literature, bones are often referred to this species as *Bufo* cf. *B. cognatus*. Tihen (1962b) and Holman (2003) gave identifying characters, such as its distinctive frontoparietals, but noted that the ilia may be confused with those of other species.

SYSTEMATICS AND GEOGRAPHIC VARIATION

The Great Plains Toad is a member of the *Cognatus* clade of North American bufonids. This clade also includes *A. compactilis* of Mexico and *A. speciosus*, but Masta et al. (2002) have shown that the clade is paraphyletic. The *Cognatus* clade is a sister group to the larger *Americanus* clade of toads. Rogers (1972, 1973) provided information on morphologic and genetic variation in the species.

A single hybrid *A. cognatus* × *A. hemiophrys* was found in Minnesota (Brown and Ewert, 1971), and a single *A. cognatus* × *A. woodhousii* hybrid was reported from Arizona (Gergus et al., 1999). Hybridization has been achieved in laboratory crosses between *A. cognatus* and *A. americanus*, *A. boreas*, *A. debilis*, *A. punctatus*, *A. terrestris*, *A. woodhousii*, *Ollotis alvaria*, and *O. nebulifer* (Moore, 1955; Blair, 1959, 1972). Artificial crosses may produce few larvae, however, with *Anaxyrus americanus*, *A. quercicus*, or *A. punctatus* (Blair, 1972). Chihuahuan Desert populations of *A. cognatus* exhibit little between-population genetic variation, and most variation occurs within populations (Jungels et al., 2010). It appears that *A. cognatus* uses river corridors for dispersal, thus reducing genetic differentiation among distant populations.

ADULT HABITAT

Anaxyrus cognatus is a species mostly of the dry short-grass or mixed-grass prairies of the Great Plains and the arid and semiarid deserts and grasslands of the American Southwest (Bragg, 1940c, 1950b; Bragg and Smith, 1943). Both tall grass and short grass prairies are occupied, as are mesquite (*Prosopis* sp.) grasslands, riparian areas in deserts, desert scrub, creosote bush (*Larrea tridentata*) desert, sagebrush (*Artemisia tridentata*) plains, pine forests, and mesquite woodlands. As one proceeds westward in the short grass prairies, the species becomes more restricted to riparian habitats. In the eastern part of its range it favors prairie uplands, where it is associated with loess hills and heavy floodplain soils (Timken and Dunlap, 1965). In the Great Plains, it is especially associated with sand plains and sandhill habitats that are necessary so that the toads can dig below the frost line. This species is frequently encountered in cultivated and agricultural fields. Great Plains Toads are found to 1,900 m in New Mexico (Degenhardt et al., 1996).

Within this extensive region, the species is subject to extreme variation in temperature, from very hot summers to very cold winters. Precipitation is often scant and spottily distributed and can vary greatly from year to year. This species occupies about 65% of potential breeding sites in west Texas, regardless of whether the adjacent land is in crops or open range (Anderson et al., 1999). Occupancy of a site is positively correlated with the presence of adjacent wetlands, but landscape structure per se does not have much influence on this species (Gray et al., 2004a). This species occasionally may be found in or near cave entrances (Black, 1973a; Collins, 1993).

TERRESTRIAL ECOLOGY

Activity occurs throughout the warm season. In Alberta, this extends from April through September (Pearson, 2009), whereas in Oklahoma it extends from March to September. Most activity occurs at night or during overcast weather, but they may be active diurnally after heavy rain. During the day, they remain dormant in shallow (< 5 cm) depressions or burrows (Bragg, 1937b; Ewert, 1969); a toad may spend up to five to six days without moving from one depression to another (Ewert, 1969). Juveniles shelter in cracks in drying mud or under debris. Foraging is especially prevalent during favorable weather in the spring and early summer along streams and ponds. Daily movements are usually < 3 m, although movements > 30 m are not uncommon, especially in early summer. Indeed, Ewert (1969) recorded 10 daily movements > 131 m, with the longest being 808 m. Movements occur only sporadically during the late summer hot and dry season, when environmental conditions are not optimal. Then, toads go into a period of dormancy in protected refugia or emerge only at night, and many do not emerge at all, even during periods of favorable rainfall (Bragg, 1945a). They burrow backward into the soil and rest alert below the surface, or they may occupy cracks in mud. A light rain may bring them back to the surface, and they can readily rehydrate from soil moisture alone (Walker and Whitford, 1970).

Long-distance movements frequently occur between breeding ponds, foraging areas, and overwintering sites, usually at night. Great Plains Toads can travel from summer foraging areas to winter dormancy sites over a considerable distance. For example, Ewert (1969) noted a mean movement distance of 229 m (range 100 to 1,036 m) between release points at foraging areas and winter dormancy sites. During long distance movements, toads may use man-made landmarks, such as roads or the borders of agricultural fields, as dispersal corridors.

Juveniles can disperse over considerable distances after metamorphosis. For example, Ewert (1969) found toadlets > 900 m from the nearest breeding pond. Juvenile dispersal occurs during mass movement events, when tens of thousands of small toads (30–52 mm SUL) may be observed moving unidirectionally, often north (Breckenridge, 1944; Bragg and Brooks, 1958). During dispersal, toadlets often stop to feed for a short time, but movements are steady and occur even during the heat of the day. These mass movements do not appear correlated with humidity, moonlight, or the opportunistic presence of smooth roads offering few obstacles to travel (Smith and Bragg, 1949).

Dense cover is preferred as a microhabitat in choice tests, although the species does not avoid open habitats (Tester et al., 1965). Foraging mode may change in relation to vegetative cover, with juvenile toads in dense vegetation adopting a sit-and-wait foraging strategy, while those in open vegetation actively pursuing prey (Flowers, 1994). *Anaxyrus cognatus* does not appear to regulate its body temperature to any extent by selecting a particular density of vegetation, as its body temperature is very similar to ambient temperature in experimental trials. It also does not seem to actively avoid high temperatures (Tester et al., 1965). Indeed, this species prefers high temperatures (31°C), even more so than some tropical toads (Sievert, 1991).

Postmetamorphic juveniles often aggregate among conspecifics in terrestrial situations; aggregations consist of four to seven toadlets. This tendency occurs during daylight hours but not at night, and therefore suggests a role for vision in the formation of juvenile aggregations. In addition, toadlets aggregate in areas previously occupied by conspecifics, suggesting a role for chemoreception in their spatial distribution. The aggregations are not formed in response to optimal feeding areas or to certain preferred habitat structure, and perhaps they serve an antipredator function (Graves et al., 1993).

Anaxyrus cognatus is often found in very cold environments, but it is not tolerant of freezing temperatures and possesses no cryoprotectants (Swanson et al., 1996). During freezing, liver glucose levels may be elevated but not enough to serve as a cryoprotectant. Swanson et al. (1996) suggested that Great Plains Toads might be able to supercool within protected burrows, but empirical support has not been forthcoming. Most activity occurs at temperatures >15°C (Hammerson, 1999), with mortality at 41–42°C (Paulson and Hutchison, 1987).

Great Plains Toads take refuge from both hot and cold temperature extremes by burrowing deep within sandy soils. When first emerging after winter, toads remain in the winter burrow about a day before exiting, presumably resting, and depart from the site of the winter burrow within six days. Summer retreats may be up to 55 cm below the surface during extended periods of hot and dry weather (Ewert, 1969). The summer burrow may be envisioned as an inverted question mark with the toad situated in the upper part of the short end (C.E. Burt *in* Tihen, 1937).

Toads enter winter burrows from late summer to fall (e.g., August to September in Minnesota; Ewert, 1969), usually at night. Summer burrows may be converted to winter dormancy burrows. In Minnesota, overwintering sites were located in raised areas along roads or in soil banks in the prairie; such areas frequently had a minimum of snow cover compared with *A. americanus* sites in the same general area. Winter retreats extend 74–104 cm below the surface to ensure the toad is below the frost line (Graves and Krupa, 2005), but toads will move both horizontally and vertically within the burrow depending on conditions. A rising ground water level in spring may kill *A. cognatus*. *Anaxyrus cognatus* is frequently found in mammal burrows, such as those of pocket gophers (*Geomys bursarias, Thomomys talpoides*), old badgers (*Taxidea taxus*), and prairie dogs (*Cynomys ludovicianus*) (Ewert, 1969; Lomolino and Smith, 2003).

CALLING ACTIVITY AND MATE SELECTION

After emergence, males move toward the breeding sites. Ewert (1969) noted that toads traveled > 805 m to reach a particular breeding site, and that a male might bypass an apparently suitable site with other calling males in order to reach a

Large fat adult *Anaxyrus cognatus*. Photo: Jonathan Mays

breeding pool. The toads appear to know where they are going, as movements are not random. Adults may alter their direction once movements have begun.

Calling is closely tied to precipitation. Males arrive at breeding sites prior to females, both seasonally and daily. Calling can begin from late afternoon to early evening, but most calling begins ca. 30 min after dark. Choruses function both to stimulate females and to orient males to a breeding site. A few males begin to call, then more and more males arrive until many toads are calling in full chorus. Breeding aggregations can be quite large, and males often return to the same breeding site from one year to the next. No breeding may take place during years of drought.

If one pool has many chorusing males and a nearby pool has few males, the latter will move to the larger chorus. By ca. 21:30, females arrive at the breeding site, and most breeding occurs within 1–3 hrs after sunset (Krupa, 1994). Calling starts to taper off after 1:00. Choruses have been heard at air temperatures of 8°C, but chorusing is initiated only at temperatures >12°C (Bragg, 1940c), with most favorable temperatures between 16.5–21°C (Ballinger et al., 2010). The mean body size of breeding adults increases at a pool as the season progresses, indicating that small toads arrive at breeding sites prior to larger toads (Krupa, 1994).

The advertisement call of *A. cognatus* is an ear-splitting trill that can be heard at least 2 km away. Standing next to a chorus of calling males is deafening. In Minnesota, the call has a dominant frequency of 2,250–2,725 cps, a duration of 5.0–42.7 sec, and a pulse rate of 16–18 pulses/sec (Blair, 1957). In Texas, the call has a dominant frequency of 1,775–2,075 cps, a duration of 10.4–53.8 sec, and a pulse rate of 13–15 pulses/sec (Blair, 1956b). The dominant frequency of the male's call is negatively influenced by body length, but is unrelated to water temperature (Krupa, 1990); pulse rate (positively) and call duration (weakly negative), however, are influenced by temperature but not body length. The dominant frequency of the call may vary among individual males, even over the course of a single night.

Males call from many types of sites, although females prefer larger breeding sites and will bypass males calling from small or unsuitable locations. Even if a male from a small breeding site amplexes her, she will attempt escape or move to a larger site; eggs are rarely deposited in small inappropriate locations. Calling occurs from around the borders of pools and from within the water of shallow (< 45 cm) wetlands. While calling, males may hold onto vegetation for stability, where they are positioned at right angles to the water's surface with only the head projecting (Linsdale, 1940). In other instances, calling occurs from very shallow water in a normal sitting position, where the body is held at about a 45° angle.

In large choruses, satellite males are frequently near calling males. This is especially true at high calling densities (toads < 0.5 m from one another), when as many as 95% of calling males have from 1 to 5 satellite males nearby (Sullivan, 1982b). In Oklahoma, Krupa (1989) noted that 0–57% of males exhibited satellite behavior. When toads are calling in low densities (mean distances ca. 2 m between calling males), all males call, and satellite behavior does not occur. In contrast, Krupa (1989) found that density accounted for only 17% of the variation in satellite behavior, and that 74% of males switched back and forth between calling and satellite status. Calling males may try to amplex satellite males but quickly let go when the warning vibration (5–9.5 vibrations/sec; Blair, 1947b) is given. If a calling male stops calling or is successful, a satellite male may move to his position and begin calling; other satellite males then move to his vicinity.

Despite the abundance of satellite males, females generally avoid them and move only toward calling males. If approached by a satellite male, females quickly move away (Krupa, 1989), but if amplexed by a satellite male, they are likely to be unresponsive. Only rarely (ca. 8%; Krupa, 1989) is a satellite male successful, and this usually occurs only after a female has contacted a calling male (Sullivan, 1982b). Females must actually touch a calling male to get him to initiate amplexus, but satellite males are nonspecific in their choice of mates and will amplexus any female that moves within their vicinity. Not surprisingly, interspecific amplexus is not uncommon. For example, Bragg (1939) observed amplexus between *A. cognatus* and *A. woodhousii*.

Amplexus is axillary. During amplexus, large males arch their backs, whereas small males hold the female more posteriorly than the large males and do not arch the back (Krupa, 1988). Males hold their legs in such a position as to form a basket. As eggs are extruded, he uses his legs to gather them within the basket, where they are held and fertilized; this process takes about 3 min. Krupa (1988) provided an illustration of the basket and noted that the male's position and behavior was similar to that described for *A. americanus* by Miller (1909a, 1909b).

Female Great Plains Toads tend to prefer males with long-call durations, and satellite males also associate with long-call males (Krupa, 1989). There is a weak negative correlation across nights in the levels of the hormone corticosterone and call duration, but not pulse rate. Males with low corticosterone have long calls that attract both females and potential interlopers. Callers without satellite males have higher average corticosterone levels than males with satellites. Androgen levels are similar, however, among calling and satellite males (Leary et al., 2006). Since males may have up to five satellites nearby (Sullivan, 1982; Krupa, 1989), there appears to be a fitness tradeoff between attracting females and potential rivals.

Although *A. cognatus* is an explosive breeder, the actual amount of time breeding occurs at a single location is short and dependent on rainfall. For example, Sullivan (1989) noted a mean chorus duration of 2.6 days in the Sonoran Desert of Arizona, with an average chorus size of 25 males. He later recorded chorusing on only 17 nights over a period of 6 breeding seasons at one Arizona location (Sullivan and Fernandez, 1999). In Oklahoma, Krupa (1994) also found considerable annual variation in the length of the breeding season, from 6 to 14 days. Afterwards, occasional choruses would form, but they would last only one to two days. Chorus size is variable, with small congress means of <10–38, depending upon location (Ewert, 1969; Sullivan and Fernandez, 1999). However, Brown and Pierce (1967) reported several desert choruses of 200 to 500 males, as well as choruses of ca. 75, 20, and <15 individuals. Chorus size appeared to influence male spacing patterns, that is, random spacing in large choruses but more even spacing at 3–4 m apart in smaller choruses.

BREEDING SITES

Breeding primarily occurs in temporary clear shallow pools, although the species also uses permanent and semipermanent wetlands. These can be small to large pools, and may contain considerable submerged vegetation such as pondweeds (*Potamogeton* sp.). Deeper wetlands in open playas are preferred sites. *Anaxyrus cognatus* also has been reported to use buffalo wallows (Bragg, 1937b, 1940c; Bragg and Smith, 1943), pools in intermittent streams, and occasionally pools in floodplains (Bragg, 1941c). The species is not associated with silty or muddy habitats (Bragg, 1937b, 1940c; Bragg and Smith, 1942, 1943). They may use man-made wetlands, such as irrigation ditches, waterfowl management ponds, and dikes in shallow drainages (Anderson et al., 1999). They prefer areas surrounded by natural vegetation, but are also found in agricultural areas. Wetland breeding sites do not appear to be influenced by vegetation, but by temperature. There are reports of the species using saline habitats in Alberta (*in* Pearson, 2009).

REPRODUCTION

Breeding takes place during the early to midsummer, the timing of which depends on latitude and environmental conditions. In Alberta the breeding season occurs from early to mid-June and lasts about ten days once initiated (Bennett, 2003), but occurs from May to July in Colorado (Hammerson, 1999) and Minnesota (Ewert, 1969). In Montana, breeding occurs from April through August (Black, 1970), whereas in Arizona most breeding occurs in June and July (e.g., Brown and Pierce, 1967). In Oklahoma, by contrast, breeding begins in March and continues to August, although most breeding occurs in April and May (Bragg, 1950a; Krupa, 1994). There is a single New Mexico record for gravid females found in early September, with tadpoles metamorphosing by October (Lewis, 1950).

Breeding is stimulated by precipitation, especially warm, heavy rainfall, and most oviposition occurs on the first two nights following the rain (Krupa, 1994). In the Sonoran Desert, for example, the violent thunderstorms of the summer monsoons stimulate chorusing, as breeding pools suddenly fill, and breeding occurs only after > 25 mm of rain (Sullivan and Fernandez, 1999). In Alberta, however, precipitation is not as necessary to bring forth chorusing males as in the southern portions of the range (*in* Pearson, 2009). This discrepancy may be a false perception, and it is likely that the effect of rainfall on breeding is dependent on local or regional weather patterns. In wet years, precipitation has little effect on stimulating breeding activities, since breeding sites are likely to contain water, but in dry years rainfall becomes a necessity.

Females select the oviposition site, and oviposition begins at dawn and continues through midmorning, regardless of when amplexus occurs (Krupa, 1994). The same sites may be used by different females throughout the breeding season and from one year to the next, but the characteristics that make individual sites attractive are unknown. Amplexus lasts many hours, with Krupa (1994) reporting a mean of 800 minutes (range 448 to 1,032 min) for oviposition. Eggs are deposited in long gelatinous strings, one from each oviduct, a few eggs at a time. Females move from one location to the next as eggs are deposited (Bragg, 1937b). They are deposited communally in larger pools along the substrate and over, around, and through submerged vegetation and debris.

Clutch size varies between 1,300 and 45,000 eggs per clutch (Bragg, 1937b, 1940c; Krupa, 1994; Graves and Krupa, 2005), but Krupa (1986a) noted that some females oviposit multiple clutches in a single year. The exceptionally high numbers of eggs counted from a few females, therefore, may reflect multiple rather than single clutches. The number of eggs is directly proportional to female body size, but at an exponential rather than a linear rate. Females < 80 mm SUL usually oviposit <10,000 eggs, with larger females ovipositing the much larger clutches. Natural fertilization rates are ca. 89% (Krupa, 1988).

Eggs hatch in two to seven days (Graves and Krupa, 2005), and hatching occurs very quickly. Hatching success is high. Embryos are about 3 mm TL (range 1.7 to 3.5 mm) at hatching (Bragg, 1940c). According to Bragg (1940c), freezing weather delays development but does not kill the embryos, and eggs and tadpoles develop normally in pools that may reach 37°C. Ballinger and McKinney (1966) reported normal development between 14 and 34.5°C, with absolute lethal limits slightly higher. The ovarian cycle is described by Clark and Bragg (1950).

LARVAL ECOLOGY

The larval period of the Great Plains Toad is 17–49 days, depending upon temperature and developmental conditions (Bragg and Smith, 1943; Hahn, 1968; Krupa, 1994). As might be expected, larvae hatching in June have a much faster rate of development than larvae that hatched early in the season when water temperatures are cooler. With so many tadpoles potentially occupying a single wetland, density exerts a major influence on survivorship. High densities result in lower survivorship, due to competition, predation, and even susceptibility to desiccation as ponds start to dry. This is because high densities increase the larval period due to decreased food resources. If pools dry quickly, tadpoles at high densities have little opportunity to reach metamorphosis.

Metamorphosis occurs rapidly. Tadpoles (at 25 mm TL) tend to reach the metamorphic stage synchronously, with subsequent mass emergence (Krupa, 1994). Bragg (1937a) noted that the largest tadpoles were 29 mm TL at metamorphosis, but that most were ca. 21 mm TL. Newly metamorphosed toadlets average 11.2 mm TL (Bragg, 1940c). The greatest threat to developing larvae throughout much of this species' range is inadequate hydroperiod within which development can be completed. Krupa (1994) found that metamorphosis successfully occurred only 9 times at 26 pools after 18 instances of rainfall.

DIET

Tadpoles are suspension feeders as well as scavengers and will graze on algae and other organic and inorganic detritus (Bragg, 1940c; Graves and Krupa, 2005). Adults are sit-and-wait predators. Postmetamorphs (both males and females) are opportunistic and eat a wide variety of invertebrates (especially beetles) and even small vertebrates. Insects comprise nearly all of the diet, depending on season, although they do not appear to favor earthworms, or at least some species of earthworms (Hartman, 1906; Tanner, 1931; Little and Keller, 1937; Bragg, 1940a; Smith and Bragg, 1949). Specific prey items include spiders, mites, beetles, flies, true bugs, moths, cut-worms, termites, and ants. Great Plains Toads may consume many prey items in a short period. In California, for example, Dimmitt and Ruibal (1980a) reported that 13 toads consumed 958 termites, 134 beetles, 1,024 ants, and a variety of other prey over an 8-night period. Great Plains Toads have a conversion efficiency of 18–26%, depending on temperature, and a single toad can eat 12% of its body weight in a single feeding (Dimmitt and Ruibal, 1980a).

Prey composition may change annually depending upon availability (Flowers, 1994), and the types of prey consumed may be dependent on the habitat, with newly metamorphosed toads in natural habitats eating a more diverse array of invertebrates than those from areas surrounded by cultivation (Smith et al., 2004). Juveniles eat much the same prey as adults, except that small ants, collembolans, and mites make up a much greater proportion of the diet, the exact composition of which can change annually (Flowers, 1994; Flowers and Graves, 1995). Smaller invertebrates are preferred by juveniles. Feeding occurs by day and night, with prey selection based on availability. Because this species occurs in the vicinity of agricultural fields and has a voracious appetite, Smith and Bragg (1949) considered it to be of potential economic importance.

PREDATION AND DEFENSE

Great Plains Toads are cryptic and possess noxious skin secretions, especially on the parotoids and dorsal warts. Toads generally remain under cover by day and often do not move much while awaiting prey. When approached, adults may crouch down against the substrate and remain motionless for some time before attempting escape (Bragg, 1945a). In choice tests, juvenile *A. cognatus* will avoid the odors of one of their primary predators, the garter snake *Thamnophis radix* (Flowers, 1994; Flowers and Graves, 1997). The ability to use chemoreception to detect predators could be an important survival mechanism in this and other toad species. Eaton (1935) described what appears to be a defensive posture, as the toad "was puffed up with air, legs straight out behind and snout against the ground" when found.

Predators of postmetamorphs include snakes (*Heterodon nasicus, Thamnophis* sp.), birds (crows), and mammals (badgers) (Bragg, 1940a, 1940c; Jense and Linder, 1970). There is a report of *A. cognatus* being eaten by Sonoran Desert Toads (*Ollotis alvaria*) (Gates, 1957). Tadpoles are eaten by invertebrates, particularly dragonfly naiads, water beetles (*Hydrophilus triangularis*), larval giant water bugs, birds (crows), and larvae of spadefoots (*Spea bombifrons, S. hammondii*).

POPULATION BIOLOGY

Recruitment varies considerably at any one breeding site from one year to the next. During dry years or after periods of drought, breeding may be postponed or even skipped altogether until heavy precipitation fills breeding pools (Bragg, 1950e; Sullivan and Fernandez, 1999). For example, Bragg (1950e) reported a range of breeding success over the period 1935 to 1947, with no or very poor reproduction in 8 yrs and moderate to very successful production of metamorphs in 5 yrs. Dependency on temporary pools for breeding thus could make this species vulnerable to changes in precipitation patterns due to climate change or simply periodic drought (see Status and Conservation).

Growth is rapid in recently metamorphosed toadlets, but there may be variation in growth rates depending on when in the season tadpoles metamorphose. They can increase in size from 11 mm TL to 20 mm TL in one week (Bragg, 1940c). By 4 months, they can be 50 mm TL, although some individuals grow at slower rates, and late season metamorphs may only reach 30 mm prior to their first winter dormancy. Most growth occurs during the second year, and males grow at slightly slower rates than females (Krupa, 1994).

Toads metamorphosing early in the season start to exhibit secondary sex characteristics, such as the male warning vibration and the dark subgular pouch, by the end of their first growing season (Bragg, 1950e). Although emerging males may have some secondary sex characteristics the spring following metamorphosis, these animals are not present at breeding ponds. Instead, breeding by males does not begin until the second year following metamorphosis (i.e., the third season of life), with females not breeding until 3 yrs following metamorphosis (the fourth season of life) (Bragg, 1950e). It is likely that age at sexual maturity is variable among populations, but more specific data are needed.

Longevity may be > 10 yrs in Alberta (Pearson, 2009), but Sullivan and Fernandez (1999) only counted from 1 to 6 lines of arrested growth (LAG) in an Arizona population. In Colorado, Rogers and Harvey (1994) counted only two LAGs and noted that fossil *Anaxyrus cognatus* had a similar pattern of LAGs. These data suggest considerable variation among populations in

Breeding habitat of *Anaxyrus cognatus*. Photo: Kerry Griffis-Kyle

demographic characters. Size and age are not directly correlated, although older toads are usually larger than younger toads.

Great Plains Toads in Alberta may be able to survive long periods of unfavorable conditions in the cold North, only to emerge years later to breed. Breckenridge (1944) observed that Great Plains Toads were absent from the western prairies during the great droughts of the early 1930s, yet appeared in large numbers after plentiful rains in 1937. In contrast, other researchers have reported much shorter apparent life spans (Rogers and Harvey, 1994; Sullivan and Fernandez, 1999). In these latter populations, longevity does not appear to be correlated with an extended reproductive lifespan in response to variable desert or otherwise harsh conditions. Clearly, some toads must be able to survive the long periodic droughts of the Great Plains. More population data are needed for this species, especially in the North and in drought-prone areas.

COMMUNITY ECOLOGY

This species is allopatric or comes into contact with a number of other closely related toads within the *A. americanus* complex. Hybridization may occur along contact zones between species, but introgression may not occur as extensively as hybridization experiments might suggest. Closely related toads often have different breeding phenology or habitat preferences that may minimize hybridization potential. Differences in habitat preference can separate the species, even though in close proximity. Woodhouse's Toad tends to be replaced by *A. cognatus* in mixed-grass prairies (Bragg, 1940a, 1940b), and mixed breeding assemblages only rarely occur between *A. cognatus* and *A. woodhousii* (Bragg, 1940a; Bragg and Smith, 1942; *in* Collins, 1993).

There may be some differences in diet between closely related toads in areas where ranges overlap. For example, *A. cognatus* eats fewer small ants, harvester ants, sowbugs, and spiders than *A. woodhousii*, but more lepidopteran larvae, weevils, and centipedes (Smith and Bragg, 1949). Whether such dietary differences reflect habitat differences, preferences, or availability is unknown.

In the Great Plains, there are a number of spadefoot species (*Spea*, *Scaphiopus*) that breed in playas and temporary wetlands that potentially may be used by *Anaxyrus cognatus* for breeding. However, *A. cognatus* tends to avoid wetlands containing these species, as spadefoot larvae are competitive, aggressive, and carnivorous. It is not surprising then that wetland occupancy by *A. cognatus* is negatively correlated with the presence of spadefoots (Gray et al., 2004a).

DISEASES, PARASITES, AND MALFORMATIONS

A variety of pathogens have been identified from Great Plains Toads, including the bacteria *Mycobacterium marinum* (Shively et al., 1981). Ewert (1969) reported winter mortality associated with bacterial red-leg (*Aeromonas*) infection. During the winter, cold inhibits the immune response making the toads susceptible to the pathogen.

The intestinal protozoan *Opalina* sp. occurs in this species (Trowbridge and Hefley, 1934). Trematodes were not identified in Great Plains Toads from Iowa (Ulmer, 1970), Oklahoma (Bragg, 1940c), or New Mexico (Goldberg et al., 1995). However, *Clinostomum attenuatum* was found in *Anaxyrus cognatus* from Texas (Miller et al., 2004), where it was twice as prevalent in grasslands as in agricultural areas (Gray et al., 2007c). The shorter hydroperiods in agricultural areas may disrupt the life cycle of the trematode parasite. Cestodes include *Cylindrotaenia americana*, *Ophiotaenia magna*, and *Distoichometra bufonis* (Trowbridge and Hefley, 1934; *in* Bragg, 1940c; Goldberg and Bursey, 1991b; Goldberg et al., 1995). Nematodes include *Aplectana incerta*, *A. itzocanensis*, *Oswaldocruzia pipiens*, *Physaloptera* sp., *Rhabdias americanus*, and unidentified species (Little and Keller, 1937; Goldberg and Bursey, 1991b; Goldberg et al., 1995).

SUSCEPTIBILITY TO POTENTIAL STRESSORS

Chemicals. Survivorship of juvenile Great Plains Toads is reduced when they are exposed to the glyphosate-based herbicides Roundup Concentrate® and Roundup Ready-to-Use Plus® (Dinehart et al., 2009). Survivorship was not affected in exposure to Roundup WeatherMAX® or the glufosinate-based herbicide Ignite 280SL®. Experimental concentrations mimicked exposure to direct spraying to simulate a worst-case scenario. Significant mortality occurred within 48 hrs of exposure.

STATUS AND CONSERVATION

Threats to *Anaxyrus cognatus* include drought and climatic changes, modification of wetlands and groundwater hydrology, habitat alteration and destruction, habitat fragmentation, the use of pesticides, highway mortality, oil and gas exploration-related development, and pathogenic disease (Pearson, 2009). In Alberta, for example, drought in the Great Plains in the 1970s and 1980s is thought to have reduced populations by 50% (*in* Pearson, 2009), although toads may appear in large numbers when rains finally arrive after a drought (Bennett, 2003). Smith and Bragg (1949) reported that *A. cognatus* used roadways as dispersal corridors, which certainly leads to mass mortality. Great Plains Toads also have been trapped in gas well caissons, but the extent of the problem is unknown (*in* Pearson, 2009); they are able to get out of irrigation pits (Anderson et al., 1999).

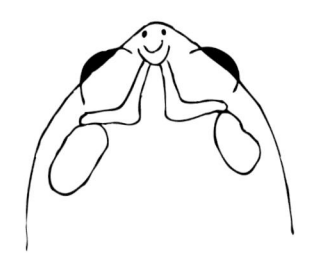

Head pattern of *Anaxyrus cognatus*. Illustration by Camila Pizano

Populations in some areas have declined or disappeared, whereas in other areas Great Plains Toads may be simply rare (e.g., Hossack et al., 2005). This species is protected in Alberta, where an estimated 2,100–10,000 individuals occur in the southeastern portion of the province (Pearson, 2009). Populations in Manitoba and Saskatchewan are rare, and their status is unknown (Weller and Green, 1997). There may be evidence of some range expansion to the east in Iowa (Lannoo et al., 1994; Hemesath, 1998). However, Christiansen (1998) reported an overall decline in the number of historic populations, despite an earlier suggestion that populations were stable (Christiansen, 1981). In Kansas, surveys in the Flint Hills located no specimens although the species was reported there historically (Busby and Parmelee, 1996). In southern Nevada, historical records are available, but Bradford et al. (2005a) could find no extant populations. In general, this species is not considered to be declining, but more data are needed, especially in the North and at the periphery of the species' range.

Anaxyrus debilis (Girard, 1854)
Green Toad

ETYMOLOGY
debilis: from the Latin *debilis* meaning 'lame,' 'crippled,' or 'frail.' The name is probably in reference to the slender body and limbs.

NOMENCLATURE
Conant and Collins (1998) and Stebbins (2003): *Bufo debilis*

Synonyms: *Bufo debilis insidior, Bufo insidior*

The nomenclatural history of this species is discussed by Sanders and Smith (1951), Bogert (1962), and Degenhardt et al. (1996).

IDENTIFICATION
Adults. Adults are small with a bright to very pale green or yellowish-green ground coloration containing variable patterns of dark flecks or even round yellow spots. The combination of green, black, and yellow gives many individuals a reticulated appearance. The body and head are flattened, and the head is broader than long. Viewed dorsally, the head is wedge-shaped and sharply truncate at the tip. Cranial crests are reduced and surmounted by a discontinuous series of black-tipped warts. Parotoid glands are large and elongated, and sit obliquely on the shoulders. The tympanum is small. There are many dark dorsal warts that give the impression of black flecking or spots. Venters are light with some dark pectoral spots. Males have a dusky or black throat, whereas in females the throat is white or yellow. Females are also yellowish in comparison to the more green males. Collins et al. (2010) provide photographs of the variation in pattern.

Females are larger than males. The maximum length is 54 mm SUL (Sanders and Smith, 1951). In Arizona, Sullivan (1984) recorded a mean male size of 42.2 mm SUL (range 37 to 46 mm) and a mean female size of 50.3 mm SUL, whereas in Oklahoma and Texas, the mean size was 34.5 mm SUL (range 30 to 42 mm) for males and 39.4 mm (range 28 to 48 mm) for females (Savage, 1954). In Colorado, males averaged 42 mm SUL and females 50 mm SUL (Hammerson, 1999). In Kansas Smith (1934) recorded adults as 24.6–34.5 mm SUL.

Larvae. Tadpoles are light in coloration and stippled with black, with golden patches on the dorsum and tail musculature. They are rounded with the eyes situated dorsally, and appear paler than most other toad tadpoles. Tail fins are moderately developed; they are clear initially but develop dark spots or lines as they grow larger. The tail musculature is unicolored, and larger tadpoles have melanophores on the ventral tail fin, giving a stippled appearance. The belly is dark laterally with golden flecking. There is no coloration on the throat and midbelly. Zweifel (1970b) noted that tadpoles are relatively transparent, even when full grown. Larvae reach a maximum size of ca. 25.1 mm TL (Zweifel, 1970b). Zweifel (1970b) provides illustrations of larvae at various stages of development.

Eggs. The eggs are light cream to yellow, with a dark band located one-third the distance from the top of the animal pole (Ferguson and Lowe, 1969; Zweifel, 1970b). Scattered melanin occurs over the entire egg. There are two gelatinous membranes surrounding the vitellus, which measures ca. 1.15 mm in diameter. Eggs are deposited singly or in short strands. The outer envelope of the eggs have an agglutinating property that causes them to clump together, giving the appearance of a small egg mass rather than single eggs or a small string.

DISTRIBUTION
The Green Toad occurs from western Kansas (Smoky Hill River drainage) and southeastern Colorado south through Texas to the Gulf Coast, and westward to southeastern Arizona to Zacatecas, Mexico. Isolated populations are in southeastern Colorado and in the Arbuckle uplift of Oklahoma. Important distributional references include: Arizona (Brennan and Holycross, 2006), Colorado (Hammerson, 1999), Kansas (Smith, 1934; Collins et al., 2010), Mexico (Lemos-Espinal and Smith, 2007b; López et al., 2009), New Mexico (Degenhardt et al., 1996), Oklahoma (Sievert and Sievert, 2006), and Texas (Dixon, 2000).

FOSSIL RECORD
No fossils have been identified.

SYSTEMATICS AND GEOGRAPHIC VARIATION
Anaxyrus debilis is most closely related to *A. retiformis* within the Nearctic clade of North

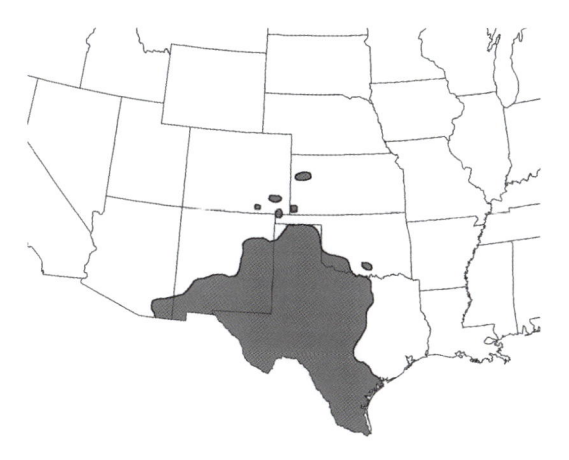

Distribution of *Anaxyrus debilis*

American bufonids (Pauly et al., 2004), and together with *A. kelloggi* in Mexico, these species form the *Debilis* clade. The *Debilis* clade is most closely related to *A. punctatus*. Two subspecies, *Bufo debilis insidior* and *B. d. debilis* have been described (Savage, 1954), but subspecific recognition is not warranted. As Crother (2008) noted, the nominal subspecies may reflect nothing more than clinal variation. Savage (1954) noted substantial "intergradation" between the two subspecies.

Hybridization between *Anaxyrus debilis* and *A. punctatus*, *A. retiformis*, or *A. kelloggi* can produce larvae that complete metamorphosis under laboratory conditions (Ferguson and Lowe, 1969), but the success of the crosses may depend on which species is male and which is female (Blair, 1961a). Some larvae may be produced in crosses between ♀ *Ollotis canaliferus* and ♂ *Anaxyrus debilis*, but few survive to metamorphosis (Blair, 1961a). Blair (1959) provides a photo of a postmetamorph hybrid from a ♂ *A. debilis* crossed with a ♀ *A. terrestris*, although few larvae survived to metamorphosis. In any case, hybridization experiments frequently base their conclusions on small sample sizes.

ADULT HABITAT

Anaxyrus debilis is associated with clay loam soils with high water-holding capacity in short-grass and mixed-grass prairie and open plains and in subhumid valleys. It is a species of the Chihuahuan Desert Scrub and semiarid grasslands. In the Big Bend region, most records are from 700 to 900 m, although a few records are available at lower elevations along the Rio Grande (Dayton and Fitzgerald, 2006). Dayton and Fitzgerald (2006) developed a habitat suitability model that correctly identified 89% of the locations of breeding sites for this species within Big Bend National Park; there, it is a species of the short-grass prairie and creosote vegetative communities (Dayton et al., 2004). In the Sierra Vieja Range of southwestern Texas, Jameson and Flury (1949) found it associated with the tabosa-gramma plant association in sandy habitats. In New Mexico, it is an inhabitant of desert grasslands and mesquite-creosote bush vegetative associations (Degenhardt et al., 1996). Sievert and Sievert (2006) report it to occur in gypsum caves. In Arizona and most of New Mexico, it occurs from 792 to 1,500 m; there is a report of 1,830 m from New Mexico (Ellis and Henderson, 1913) and 1,829 m in Mexico (Bogert, 1962). Hammerson (1999) reports most Colorado populations at 1,220–1,525 m.

TERRESTRIAL ECOLOGY

Activity is almost entirely nocturnal, and the species is rarely observed during dry weather. In Kansas, most activity occurs from mid-May to late August (Collins et al., 2010), whereas Hammerson (1999) reports activity from May to September in Colorado. Cruesere and Whitford (1976) found that adults preferred dense grasses near breeding pools during the nonbreeding season, but will venture into sparse grass habitats within 300 m of the pool. During the dry season and winter, the Green Toad wedges into crevices, buries into the soil, or occupies the burrows of rodents or other animals; Green Toads are occasionally found within the burrows of the prairie dog *Cynomys ludovicianus* (Lomolino and Smith, 2003) and nests of the ant *Novomessor cockerelli* (Cruesere and Whitford, 1976). The weak limbs of this species make digging its own burrows difficult. Hammerson (1999) found individuals active at 11°C.

After metamorphosis, young toads remain within the vicinity of breeding pools, and activity occurs both day and night (Smith and Bragg,

Tadpole of *Anaxyrus debilis*. Photo: Ronn Altig

1949). In the Big Bend region, Minton (1958) recorded activity as early as 18 March, with Cruesere and Whitford (1976) observing active adults from May to September. Juveniles are active longer in the season than adults, when they find shelter in vegetative cover and the cracks in mud formed by drying playas and pond bottoms.

Green Toads are photopositive in their phototactic response, suggesting they use both sunlight and moonlight when feeding and going about their daily activities (Jaeger and Hailman, 1973). They are sensitive to light in the blue spectrum ("blue-mode response") (Hailman and Jaeger, 1974). Green Toads likely have true color vision.

CALLING ACTIVITY AND MATE SELECTION

Breeding is opportunistic and associated with plains or desert rainfall (Bogert, 1962; Creusere and Whitford, 1976). Most chorusing occurs by night, but Degenhardt et al. (1996) and Hammerson (1999) reported chorusing by day after warm summer rains. Choruses form rapidly and are often small. In the Sonoran Desert, for example, the mean chorus size was 11, and the chorus only lasted a mean of 2.6 days (Sullivan, 1984, 1985a, 1989). Males call from wet, muddy ground within 2 m of the shoreline or from shallow water in temporary rain pools; they prefer sparse or dense vegetation from which to call (Cruesere and Whitford, 1976). Despite the small chorus size, males tend to aggregate 0.5–3 m from one another, even in large pools, and may call, even in close proximity to another calling male (Sullivan, 1984).

Calling occurs at temperatures of 19–22°C (Zweifel, 1968b). The call is a high frequency buzz with a dominant frequency of 2,795–3,600 cps, a duration of 2.2–10.4 sec, and a mean of 84–139 trills/sec (Blair, 1956b; Bogert, 1962; Ferguson and Lowe, 1969; Sullivan et al., 2000). Males call at air temperatures of 16–21°C (Ferguson and Lowe, 1969). There are no correlations between dominant frequency, pulse rate, or call duration with the size of the male, nor is there a relationship between the size of the male and mating success, that is, larger males are not more successful at mating than are smaller males (Sullivan, 1984).

A female approaches a calling male and initiates amplexus by touching or jumping on him. Males continue to vocalize until touched, but amplexus is rapid afterward. There is no evidence that males employ a satellite strategy to intercept incoming females.

BREEDING SITES

Breeding occurs primarily in temporary pools in alluvial or playa floodplains, but Green Toads also have been observed in small streams and cattle tanks. Breeding adults prefer pools with grass clumps, and permanent sites should be free of fish. Hammerson (1999) reported breeding in muddy pools with rocky and muddy bottoms.

REPRODUCTION

Breeding occurs in spring and summer, depending on rainfall. For example, Bragg and Smith (1942) reported calling in April to August in Oklahoma, although eggs were not observed. Wright and Wright (1938) reported breeding from March to June in Texas, and Bogert (1962) noted breeding as late as August in New Mexico. According to Degenhardt et al. (1996) and Collins et al. (2010), most breeding occurs in mid-June to early July. Eggs are deposited singly or in small strings attached to grass and other submerged vegetation. The eggs tend to clump together. Collins et al. (2010) reported clutch sizes of up to 1,610 in a single female from Kansas. *Anaxyrus debilis* requires temperatures of 18.8–33.8°C for normal egg and larval development (Zweifel, 1968b).

Adult *Anaxyrus debilis*. Photo: Aubrey Heupel

Habitat of *Anaxyrus debilis*.
Photo: Jan Caldwell

Hatchlings are pale tan and are 3.1–3.4 mm TL (Zweifel, 1970b).

LARVAL ECOLOGY
Metamorphosis occurs in as little as 7 to 20 days (Strecker, 1926; Collins et al., 2010) and is temperature-dependent. Toadlets are 14.8–15.9 mm SUL upon emergence (Degenhardt et al., 1996).

DIET
This species likely eats a variety of small invertebrates, but only anecdotal information is available. Little and Keller (1937) reported a specimen had eaten insects, and Hammerson (1999) noted that nearly all DOR specimens he examined only had ants in their stomachs; one beetle also was observed.

PREDATION AND DEFENSE
Larvae and young are eaten by Plains Garter Snakes (*Thamnophis radix*) (Collins et al., 2010). Stuart (1995) noted predation on an adult by an American Bullfrog.

POPULATION BIOLOGY
Nothing is reported on the population biology of this species. Taylor (1929) reported mortality from hailstones. Colorado populations are small, with surveys reporting about 25 adults at any one time (*in* Hammerson, 1999).

DISEASES, PARASITES, AND MALFORMATIONS
The Green Toad is parasitized by the cestode *Distoichometra bufonis* and the nematodes *Aplectana incerta, A. itzocanensis, Cosmocercoides variabilis, Physaloptera* sp., and *Rhabdias americanus* (McAllister et al., 1989; Goldberg et al., 1995). In addition to helminths, the protozoan *Nyctotherus cordiformis* and the myxozoan *Myxidium serotinum* also parasitize *Anaxyrus debilis* (McAllister et al., 1989).

SUSCEPTIBILITY TO POTENTIAL STRESSORS
No information is available.

STATUS AND CONSERVATION
According to Collins et al. (2010), this species was not discovered in surveys in Morton County in the 1980s, where it had been present before the drought that created the dust storms of the 1930s. They mention an attempt to repatriate this species into Morton County in 1992, but the success of this effort was not known. This species is protected by state law in Kansas. Hammerson (1999) reported Colorado populations as secure, with no known existing threats.

Anaxyrus exsul (Myers, 1942)
Black Toad

ETYMOLOGY
exsul: from the Latin *exsul*, meaning 'exile' or 'refugee.' The name refers to the species' isolated distribution in Deep Springs Valley.

NOMENCLATURE
Stebbins (2003): *Bufo exsul*
    Synonyms: *Bufo boreas exsul*

IDENTIFICATION
Adults. Black Toads are indeed black. They are extremely mottled, with the black pigment coalescing to form an almost continuous black dorsum. This black coloration is superimposed upon a lighter cream color giving the toad a marbled appearance. Schuierer (1963) suggested that the increased melanism was an adaptation to high levels of solar insolation in the arid Deep Springs Valley. The venter is white.

Anaxyrus exsul is a small toad, but there is some variation among populations. The toads at Antelope Springs are much larger than those in the southeastern part of Deep Springs Valley. Specific size records from the southeastern population include: a mean of 51.1 mm SUL for males (range 42 to 60 mm) and 51.4 mm SUL for females (range 47 to 71 mm) (Karlstrom, 1962); males with a mean of 50 mm SUL and females with a mean of 52 mm SUL (Schuierer, 1963); males with a mean of 60.4 to 64.2 mm SUL and females 62.0–64.4 mm SUL over a 2 yr period (Kagarise Sherman, 1980); and a mean of 51.6 mm SUL for males and 50.9 mm SUL for females (Murphy et al., 2003). In a short visit, Schuierer and Anderson (1990) recorded a mean of 48.6 mm SUL for males (range 45 to 52 mm). Murphy et al. (2003) suggested that some of these size differences resulted from including smaller adults captured outside the breeding season with breeding adults.

At nearby Antelope Springs, adults averaged 74.7 mm SUL (range 64 to 82 mm) (Murphy et al., 2003). In earlier studies, Antelope Springs's males averaged 69 mm SUL, whereas females averaged 72 mm SUL (Schuierer, 1963); Kagarise Sherman (1980) reported males averaging 80.7 mm SUL (range 71 to 98 mm) and females 78.0 mm SUL (range 61 to 90 mm).

Larvae. There are no published larval descriptions. Presumably, Black Toad larvae resemble the larvae of *A. boreas* in being generally small and black.

Eggs. The eggs of *A. exsul* are deposited in single or double strings within a continuous gelatinous envelope. The normally double-stranded strings are loosely attached to vegetation such as sedges in water 16–25 cm in depth (Savage and Schuierer, 1961). There are two envelopes surrounding the eggs, with each egg separate from its neighbor. The linear arrangement of the eggs is such to give a zigzag appearance, much like the eggs of *A. boreas*. Partitions between eggs are incomplete. The outer envelope is 4.0–6.9 mm in diameter, with the inner envelope 2.0–3.3 mm in diameter. The ovum is brownish black to very black, and is 1.2–1.4 mm in diameter. The vegetal pole is cream colored to gray. There are roughly 60 eggs per 10 cm of string length. Illustrations of eggs are in Livezey (1960) and Savage and Schuierer (1961).

DISTRIBUTION
The Black Toad is known only from Antelope Springs and marshes and freshwater springs (Corral Springs, Buckhorn Springs, Bog Mound Springs), feeding Deep Springs Lake at the north and southeast ends of Deep Springs Valley, Inyo County, California. A fifth population was reported by Simandle (2006) in Birch Creek near Antelope Springs. Deep Springs Lake is not inhabited, as it is the ephemeral and alkaline lakebed terminus of the Deep Springs Basin. The valley is entirely enclosed between the White and Inyo mountains. The 2 presumed natural populations are about 9 km apart, with Bog Mound Springs being located closest to Antelope Springs; these springs may occasionally be connected by watercourses during periods of high rainfall. Most studies of this species have centered on Corral Springs in the southeast. The elevation of Deep Springs Valley is approximately 1,525 m.

Black Toads have been introduced to a flowing well in Saline Valley in Death Valley National Park (Murphy et al., 2003). According to Schuierer (1963), Black Toads also may have been translocated by Deep Springs College students to Antelope Springs and to Cottonwood Springs in the Owens Valley. However, it is known that there is some natural population movement between

Distribution of *Anaxyrus exsul*

Antelope Springs and the Deep Springs Lake marsh complex (Simandle, 2006). An additional nonblack population has also been discovered (D. Tracy, personal communication), although its location has not been disclosed as of May 2011. Important distributional references include Schuierer (1962, 1963).

FOSSIL RECORD
There are no specific fossils recorded for this species. See the *A. boreas* account for a listing of fossils attributed to the *A. boreas* species group.

SYSTEMATICS AND GEOGRAPHIC VARIATION
The Black Toad has long been recognized as a member of the *Boreas* group of toads (with *A. boreas*, *A. canorus,* and *A. nelsoni*). Some authors have suggested that *A. canorus* and *A. exsul* are sister taxa (Graybeal 1993, 1997; Shaffer et al., 2000) within the SW clade of *A. boreas*, whereas others (Pramuk et al., 2007) place it as the sister species of *A. boreas* in general. New molecular analyses now establish it as the sister species of the SW group of Western Toads (mostly *A. boreas halophilus*), and most related to *A. nelsoni*, a southern group of toads currently included within *A. canorus*, and the Darwin Canyon population of *A. boreas halophilus* (Goebel et al., 2009). The species is the product of genetic differentiation in isolation in an extremely remote desert valley during the Pleistocene (Myers, 1942; Schuierer, 1963; Murphy et al., 2003).

ADULT HABITAT
Almost all activity takes place along the slow-flowing watercourses and marshes surrounding the few remnant springs in Deep Springs Valley. Toads do not venture far from these wetlands. The habitat is open, with little sheltering vegetation, and is dominated by Great Basin sagebrush (*Artemisia*), saltbush (*Atriplex*), and rabbit brush (*Chrysothamnus*). The Antelope Springs location is more heavily shaded by dense stands of *Salix* than the open Deep Springs marsh complex. Watercourses are lined with rushes (*Juncus*), bulrushes (*Scirpus*), sedges (*Carex*), and other aquatic plants. The toads prefer unobstructed access to water. Photographs of Antelope Springs in the early 1960s are in Schuierer (1963) and of the Deep Springs marsh complex in the early 1960s in Schuierer (1962) and 1970s in Busack and Bury (1975).

TERRESTRIAL ECOLOGY
The Black Toad is a mostly diurnal species, although in May and June they may be active after dark. They remain in close proximity to wetlands, and readily dive into the water when disturbed. The toads are active beginning in early March until early summer, when temperatures begin to climb. They also are active in the fall, at least until November. Activity begins about one to two weeks before breeding actually commences. During periods of excessive heat or winter cold, toads retreat to burrows, which may be used communally. Between 20 and 30 toads have been found occupying a single burrow, where the temperature was 12°C (Schuierer, 1962; Busack and Bury, 1975).

Most activity occurs in the morning and early evening at temperatures 17–22°C (Schuierer, 1962). Adults can tolerate temperatures >39°C for very short periods (Straw, 1958), although it is unlikely any toad would be active willingly at such high temperatures. Black Toads may lose up to 35% of their body weight in water and recover. If a toad loses much more than this, death occurs rapidly. The rate of dehydration is linear under controlled conditions, indicating no special physiological adaptation to prevent or slow down water loss. Black toads do not venture more than a meter or so from a water source, which likely helps prevent dehydration in an environment where ambient humidity is 10%.

As with *Anaxyrus boreas*, Black Toads likely are photopositive and sensitive to light in the blue spectrum ("blue mode response"). They probably have true color vision.

CALLING ACTIVITY AND MATE SELECTION

Toads move to the breeding site in early spring (March), and there is no chorus. Schuierer (1962) noted that the soft "chirping" call made by some males is a warning call, not an advertisement call. These soft sounds (termed "clucks" by Kagarise Sherman, 1980) are made by amplexed males, while an amplexed pair is moving around and when an amplexed pair is attacked by another male. Kagarise Sherman (1980) suggested they also might attract females to areas of male activity, thus acting like advertisement calls. The clucks consist of 3 to 15 short pulses of energy emitted over a range of frequencies.

Males and females arrive nearly simultaneously, with males arriving about three days prior to females (Kagarise Sherman, 1980). Breeding adults presumably find each other by active search in breeding aggregations that form in shallow waters of the marsh. Females move rather inconspicuously around the wetland, and may avoid certain males by diving, hopping away, or becoming motionless. Females choose a male by hopping directly toward him, but they also are able to dislodge or scrape off males if they so choose, at least in some instances. Kagarise Sherman (1980) suggested that indirect male selection is the most common way mate choice occurs in *A. exsul*.

BREEDING SITES

Breeding occurs in the freshwater marshes and sloughs associated with Deep Springs Lake, which is fed by ten springs flowing into the lake, in the slow-moving shallow watercourses emanating from these springs, and in former irrigation ditches that crisscross the marsh. A smaller population breeds in the shallow watercourses associated with three springs at Antelope Springs.

REPRODUCTION

Breeding occurs in the early spring over a period of several weeks, particularly from mid-March to late April. However, Schuierer (1962) noted an instance when breeding did not begin until 30 April. Water temperature at Deep Springs remains at a fairly constant 20–21°C, but air temperatures may be considerably cooler. Savage and Schuierer (1961) reported finding breeding toads at air temperatures of 10°C; however, most reports indicate ambient temperatures of 19–21°C during breeding. Breeding is diurnal, as night temperatures may be below freezing during the breeding season.

Clutch size ranges from ca. 400 to 2,600 eggs per female, with cutch size positively correlated with female SUL (Kagarise Sherman, 1980). Eggs are oviposited in strings placed in water 10–50 cm in depth. They cover an area of 30 by 35 cm, although some eggs may extend 45 cm from the main string mass. Strings are entwined around submerged vegetation in the upper section of the water column and near the shoreline. Eggs hatch in four to nine days (Kagarise Sherman, 1980). Most breeding occurs over a period of 15 to 17 days, even though the season extends nearly 40 days. Indeed, 73–78% of females oviposited during this abbreviated time period, with 25–45% ovipositing on the 2 peak days of breeding (Kagarise Sherman, 1980).

Males remain at breeding sites from a mean of 14 to 17 days. Very few males are actually successful when mating. Over a 2 yr period, 70.3% males were unsuccessful, 23.8% mated with one female, 5.3% mated with 2 females, 0.5% mated with 3 females, and 0.2% had 4 mates. Combined, between 19.3–38.5% of males mated at least once (Kagarise Sherman, 1980). Not all

Amplexing adult *Anaxyrus exsul*. Photo: Robert Hansen

males are present annually. In Kagarise Sherman's (1980) study, only 23% of breeding males were at Corral Springs in consecutive years. Males that visited breeding sites in consecutive years were more successful than males that were observed only once.

LARVAL ECOLOGY

Larvae tolerate temperatures to 29°C with little problem. From 30–35°C they become quite active, but by 39°C periods of rest are interspersed with violent movements. Mortality occurs when larvae are exposed to temperatures of 41–42°C (Straw, 1958). Metamorphosis occurs in 65–100 days and is usually completed by early to late June. Recent metamorphs are 14–19 mm SUL (Schuierer, 1962).

DIET

Adults and juveniles are largely insectivorous, feeding on ants, fly larvae, leafhoppers, and beetles (Livezey, 1962; Schuierer, 1962; Busack and Bury, 1975). Fly larvae, aphids, and ants, in particular, form a large part of the juvenile diet, inasmuch as these are small prey. Other prey include springtails, spiders, mites, fairy shrimp, and snails.

PREDATION AND DEFENSE

Predators on adults include black-billed magpies (*Pica pica*) and ravens (*Corvus corax*). Kagarise Sherman (1980) did not record any predation on larvae, although a number of potential snake, bird, and mammal predators are known from Deep Springs Valley. Black Toads readily dive into the shallow water streams associated with its open marshy habitat. In this, they behave somewhat more like frogs than toads. The black coloration may help in concealment among the dark watercourses or aid in screening excessive ultraviolet radiation in its exposed environment.

POPULATION BIOLOGY

Black Toads grow quickly, feeding on the abundant invertebrates associated with its desert oasis. Young toadlets remain within the vicinity of the breeding site. By their first winter dormancy, they reach 22–33 mm SUL, and by the following summer they are 44 mm SUL. These lengths suggest that maturity is reached in about 2 yrs following oviposition (Kagarise Sherman, 1980).

Early accounts probably underestimated the number of Black Toads in Deep Springs Valley. Myers (1942) suggested there were only 700 toads, whereas Schuierer (1962) estimated the population size at 1,200–3,600 between 1954 and 1961. A population census in 1971 suggested there were 4,000 Black Toads in Deep Springs Valley (Schuierer, 1972; Busack and Bury, 1975). Censuses at Corral Springs (1977: 7,897–9,744 [Kagarise Sherman, 1980]; 8,419 [Murphy et al., 2003]) suggest that the population is doing well. No major changes to the habitat appear to have occurred over a period of 48 yrs (Schuierer and Anderson, 1990; Murphy et al., 2003). The effective population size at Antelope Springs is ca. 7,700 toads and at the Buckhorn-Corral Springs complex ca. 10,000 toads (Simandle, 2006).

The sex ratio of Black Toads at Corral Springs in 1999 was slightly female-biased, with females making up 62.7% of the population (Murphy et al., 2003). However, Kagarise Sherman (1980) recorded sex ratios of 2 males per female over a 2 yr period among the breeding population at Corral Springs during the late 1970s.

Although most Black Toads remain close to water in their home springs, genetic analysis has revealed that three populations have significant genetic differentiation from one another: Antelope Springs, Bog Mound Springs, and the Buckhorn-Corral Springs complex (Simandle, 2006). As such these sites represent discreet breeding populations. However, there appears to be some bidirectional gene flow, with low levels of migration, and much of it from Antelope Springs to Bog Mound Springs. Presumably such migration occurs during rare flood events. Simandle (2006) did not find any evidence of genetic bottlenecks over the past four generations.

In addition to predation, Kagarise Sherman (1980) recorded mortality from drowning or injury during amplexus, desiccation in a habitat 100 m from water, larval trampling by cattle, and from unspecified disease or parasites.

DISEASES, PARASITES, AND MALFORMATIONS

An unidentified nematode was found in the intestine of a Black Toad, and leeches are reported on them occasionally (Schuierer, 1962).

SUSCEPTIBILITY TO POTENTIAL STRESSORS

No information is available.

Habitat of *Anaxyrus exsul*.
Photo: Dawne Becker

STATUS AND CONSERVATION

The precarious status and need for conservation management of *Anaxyrus exsul* has long been recognized (Bury et al., 1980). In the past, the Black Toad was threatened by cattle grazing and the potential for diverting its water sources for irrigation, thus leaving tadpoles to die of desiccation (Schuierer, 1972; Bury et al., 1980; Kagarise Sherman, 1980). The population at Antelope Springs even was thought to have been extirpated, but this fortunately is not the case. Most of the Black Toad habitat is owned by Deep Springs College, which has developed a conservation program to remove potential threats, especially through controlled cattle grazing. However, the effectiveness of some of these practices (e.g., cattle exclosures leading to an increase in vegetation density and dense stands of sedges and cattails) needs to be evaluated. It is possible that the dense vegetation that has grown up within the enclosures actually inhibits use by Black Toads. As with other anurans isolated in the Desert Southwest, there is concern about the possible introduction of the American Bullfrog into Black Toad habitat, especially since it could introduce amphibian chytrid fungus. The species is protected as "Threatened" by the State of California.

Anaxyrus fowleri (Hinckley, 1882)
Fowler's Toad

ETYMOLOGY

fowleri: a patronym honoring Samuel Page Fowler (1800–1888), a naturalist from Massachusetts who was one of the founders of the Essex County Natural History Society, later the Essex Institute. Information on Samuel Fowler and how his name came to be attached to Fowler's Toad is in Lazell (1968).

NOMENCLATURE

Conant and Collins (1998): *Bufo fowleri*
  Synonyms: *Bufo compactilis fowleri, Bufo hobarti, Bufo lentiginosus fowleri, Bufo velatus, Bufo woodhousei fowleri, Bufo woodhousii fowleri, Bufo woodhousii velatus*

  There is some confusion in the early literature on the identification of *Anaxyrus fowleri* as distinct from *A. americanus* and *A. woodhousii*. For example, Burt (1935) considered *A. fowleri* and *A. woodhousii* to be identical and therefore included information on both species under discussions of *A. woodhousii*. For this reason, biologists should consult the primary literature to ensure that citations are appropriate to the location and question of interest.

IDENTIFICATION

Adults. Anaxyrus fowleri normally is a medium-sized grayish to light brown toad, often confused with the slightly larger American Toad in areas of sympatry. They are found in a range of earth-tone colors, however, from reddish brown to greenish gray to dark gray to brown. Indeed, colors tend to darken with cold weather, as they often do in frogs and toads. Cranial crests are not as conspicuous as they are in *A. americanus* and *A. terrestris*, and the long oblong to oval parotoid glands are not large or conspicuous. These parotoid glands are low and rarely descend onto the neck. Cranial crests lack the posterior knob found on *A. americanus* and *A. terrestris*, and the postorbital crests directly touch the parotoid glands. Fowler's Toads have more than two (generally three to five) warts per spot. Like *A. americanus*, there may be a tan, white, or yellowish vertebral stripe running down the back, but this stripe is narrow and of uniform width in *A. fowleri*; in *A. americanus*, the stripe is broad and irregular. The skin is dry to the touch. Venters vary considerably, from unspotted through singular spots in the pectoral region to spots present throughout the anterior third of the venter. The incidence of spotting varies among populations and between the sexes. The hind feet are webbed, and they have enlarged inner and outer metatarsal tubercles used for digging. During the breeding season, males have swollen and darkened (cornified) patches on the thumbs that enable them to grasp females during amplexus. Males also have darkened throat pouches. Albinos or partial albinos have been reported from Louisiana, Ohio, North Carolina, and Texas (Hensley, 1959; Vance and Talpin, 1977; Palmer and Braswell, 1980; Dyrkacz, 1981). Additional morphological details are discussed by Netting (1929).

In general, females are slightly larger than males. However, there is no apparent sexual size dimorphism in some populations (Hranitz et al., 1993). Adults normally are 45–82 mm SUL. The largest recorded Fowler's Toad was 92 mm SUL and found in Massachusetts (Lazell, 1968), although Conant and Collins (1998) gave a figure of 95 mm SUL but without locality data. Other reports include means of 58.6 mm SUL for males and 59.9 mm SUL for females (Netting, 1929); means of 57.4 mm SUL for males (range 50 to 67 mm) and 63 mm SUL for females (range 48 to 76 mm) in Connecticut (Klemens, 1993); a mean of 57 mm SUL for males and 64 mm SUL for females (Brown, 1970); a mean of 55.7 mm for males and 71.5 mm for females in Pennsylvania (Hulse et al., 2001); males 53–70 mm SUL and females 61–83 mm SUL in Ohio (Walker, 1946); males 42–65 mm SUL (mean 54 mm) and females 55–68 mm SUL (mean 60 mm) in Indiana (Minton, 2001); 70 mm SUL for males and 80 mm SUL for females in Virginia (Mitchell, 1986); a mean of 44 mm SUL (range 26 to 70 mm SUL) in Maryland (Smithberger and Swarth, 1993); a mean of 38 to 57 mm SUL for males and 37–60 mm SUL for females in eastern Virginia (Hranitz et al., 1993); male means of 50.9 to 56.6 mm SUL over a 7 yr period in Ontario, with a maximum of 63 mm SUL (Green, 1997). Blair (1941a) provided means for 25 locations throughout the range; they were 51–61 mm SUL, but sample sizes often were small and sexes were combined. Hranitz et al. (1989, 1993) and Brill (1993) found that mainland toads were larger than toads on adjacent islands.

Larvae. Tadpoles are small and uniformly colored dark brown to black. However, there may be some mottling and metallic or reddish flecking on the dorsum. Bodies are ovoid. Tail fins are about equal in size dorsally and ventrally from the tail musculature. Tails are short and not distinctly bicolored. The snout is rounded in lateral view, and the eyes are large. Descriptions of the tadpole are in Siekmann (1949), Brown (1956), and Altig (1970, as *A. woodhousii* in part). Nichols (1937) provided an extensive account of the oral features of larval *A. fowleri*. Wright (1929) also described larvae of "*Bufo* from Raleigh, N.C." that were likely this species. The tadpoles of *Anaxyrus fowleri* may be impossible to distinguish from those of *A. americanus*.

Eggs. The eggs of *A. fowleri* are black dorsally and tan to yellow ventrally (Livezey and Wright, 1947). Eggs are deposited in continuous, gelatinous strings with a singular tubular membrane and no intercellular partition between adjacent eggs. Strings may be singular or paired, with 17 to 25 eggs per 25 cm in length; Brown (1956) gives a figure of 30 eggs per 40 cm in length. Only one jelly envelope < 5 mm in diameter (2.6–4.6 mm; mean 3.5 mm) surrounds the egg (Livezey and Wright, 1947; Brown, 1956). The vitellus is 1–1.5 mm in diameter. The clutch size is from 4,000 to 16,000 eggs per female.

DISTRIBUTION

Fowler's Toad occurs from southern New Hampshire and eastern Vermont south through southeastern New York, with an extension northward along the Hudson River to the Albany Pine Bush (Stewart and Rossi, 1981). There is also a disjunct population along the eastern shore of Lake Erie. The range extends westward across the southern two-thirds of Pennsylvania, around Lake Erie, and west through southern and western Michigan. The range skirts the southern end of Lake Michigan and crosses the southern two-thirds of Illinois to southeastern Iowa. It then continues southward across much of eastern and southern Missouri and Arkansas. Allen (1932) considered this to be a rare species in Mississippi. *Anaxyrus fowleri* occurs along the northern Gulf Coast eastward to the Florida Panhandle, thence northeastward across the Piedmont of Georgia and South Carolina north of the Fall Line. The southern-most range on the Atlantic Coastal Plain includes nearly all of North Carolina.

The distribution of Fowler's Toad in Texas and Louisiana is not clear. Sanders (1986, 1987) reviewed available information and concluded that *A. fowleri* was confined to the eastern Florida Parishes and did not occur in Texas. Conant and Collins (1998), however, suggested that *A. fowleri* occurred into eastern Texas and Oklahoma, albeit with a great deal of introgression with *A. woodhousii* to the west. Sanders (1986, 1987) and Dixon (2000) used the name *A. velatus* for those toads in east Texas, attributed by Conant and Collins (1998) as hybrids between *A. fowleri* and *A. woodhousii*. However, Pauly et al. (2004) found that the *velatus*-like toads were more related to *A. americanus* and *A. houstonensis*

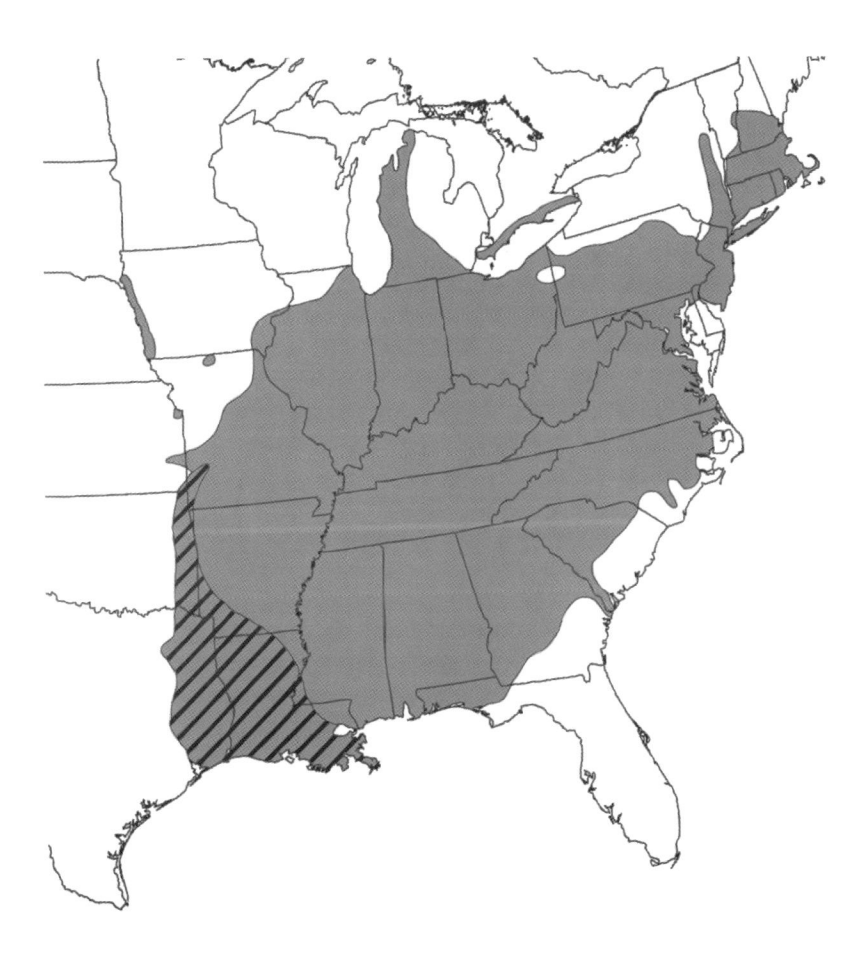

Distribution of *Anaxyrus fowleri*. The hatched area indicates an extensive area of hybridization with *A. woodhousii*. It may not be possible to assign specific status to toads in this area.

than to *A. fowleri*. None of the names proposed by Sanders (1987) for various *A. fowleri* populations are currently recognized (Crother, 2008). It may not be possible to determine accurately the identity of a specimen in Louisiana, eastern Oklahoma, east Texas, or southwestern Arkansas, or to determine whether it is a hybrid (see Vogel, 2007; Vogel and Johnson, 2008). Additional information on this complex assemblage of related *Americanus* group of toads is in Volpe (1959b), Meacham (1962), Dundee and Rossman (1989), and Trauth et al. (2004).

There are a number of reports of *Anaxyrus fowleri* being found in the Coastal Plain near Charleston, South Carolina (Schmidt, 1924; Chamberlain, 1939), but the identification of these toads has been disputed. I also have seen toads that appear to be *A. fowleri* breeding at Kingfisher Pond on Savannah National Wildlife Refuge just north of the Savannah River. Netting and Goin (1945) reported individuals from White Springs, Hamilton County, Florida, but these animals represent either a deliberate introduction or an unexplained disjunct population.

Fowler's Toads were or are found on islands, including Naushon, Pasque, Penikese, Muskeget, Nashawena, Cuttyhunk, Tuckernuck, Nantucket, and Martha's Vineyard in Massachusetts (Lazell, 1976); Pelee, Gibralter, and Kelleys islands in Lake Erie (Logier and Toner, 1961; Langlois, 1964; King et al., 1997); Long Island and Staten Island, New York (Miller and Chapin, 1910; Overton, 1914); Kent Island, Maryland (Grogan and Bystrak, 1973b); Assateague Island, Maryland and Virginia (Lee, 1972, 1973b; Brill, 1993; Mitchell and Anderson, 1994); Chincoteague, Wallops, Hog, Paramore, and Smith islands, Virginia (Mitchell, 2002; Mitchell and Anderson, 1994), and Hatteras, Bodie, Ocracoke, and Roanoke islands, North Carolina (Engels, 1942; Brothers, 1965; Gaul and Mitchell, 2007).

Important distributional references include: Alabama (Brown, 1956; Mount, 1975), Arkansas (Black and Dellinger, 1938; Trauth et al., 1990), Connecticut (Klemens, 1993), Delaware (White and White, 2007), Florida (Netting and Goin, 1945), Georgia (Williamson and Moulis, 1994; Jensen et al., 2008), Illinois (Schmidt and Necker, 1935; Smith, 1961; Phillips et al., 1999), Indiana (Minton, 2001; Brodman, 2003), Iowa (Hemesath, 1998), Kansas (Collins et al., 2010), Kentucky (Barbour, 1971), Louisiana (Dundee and Rossman, 1989), Maryland (Harris, 1975), Massachusetts (Hoopes, 1930; Lazell, 1976), Michigan (Harding, 1997), Missouri (Johnson, 2000; Daniel and Edmond, 2006), New Hampshire (Oliver and Bailey, 1939; Taylor, 1993), New Jersey (Schwartz and Golden, 2002), New York (Gibbs et al., 2007), North Carolina (Dorcas et al., 2007), Ohio (Walker, 1946; Pfingsten, 1998; Davis and Menze, 2000), Ontario (Green, 1989; MacCulloch, 2002), Rhode Island (Hoopes, 1930), South Carolina (Chamberlain, 1939), Tennessee (Redmond and Scott, 1996), Vermont (Andrews, 2001), Virginia (Tobey, 1985; Mitchell and Reay, 1999), and West Virginia (Green and Pauley, 1987).

FOSSIL RECORD

There are many fossil localities for this species. It has been found in Pleistocene Irvingtonian localities in Arkansas, Maryland, and West Virginia. Most localities are from Pleistocene Rancholabrean sites in Florida, Georgia, Missouri, Pennsylvania, Tennessee, and Virginia (Holman, 2003). Most fossils come from limestone cave deposits. The most common identifying bone is the ilium; *A. fowleri* has a much higher and better-developed ilial process than *A. terrestris*. However, the ilia of *A. fowleri* and *A. americanus* can be difficult to identify with certainty. The base of the ilial process is narrower in *A. fowleri* than *A. americanus*, but it takes a series of comparative material of equal sizes to differentiate this characteristic. Tihen (1962a, 1962b) and Holman (2003) note other distinguishing osteological characteristics among *A. fowleri*, *A. americanus*, and *A. terrestris*.

SYSTEMATICS AND GEOGRAPHIC VARIATION

Anaxyrus fowleri is a member of the *Americanus* group of toads, which includes *A. americanus*, *A. houstonensis*, *A. microscaphus*, *A. terrestris*, and *A. woodhousii*. This group in turn is a member of the Nearctic clade of North American bufonids (Pauly et al., 2004). Based on molecular analysis, the species is most closely related to *A. terrestris*, a relationship that is different from earlier phylogenetic hypotheses solely based on morphology and call similarity (e.g., Blair, 1963b). Fowler's Toad is paraphyletic, with three somewhat geographically distinct clades (north, central, south), suggesting allopatric differentiation during the Pleistocene. Some divergences were evident > 2 mya (Masta et al., 2002).

Hybridization among bufonids is widely reported in North America, and the introgression of genes among *A. americanus*, *A. woodhousii*, and *A. fowleri* is complex (Masta et al., 2002; Vogel, 2007). Caution is advised in accepting some reports of hybridization, however. Several authors have examined toads morphologically and concluded the variation observed must be due to hybridization without actually demonstrating it (e.g., Allard, 1908; Miller and Chapin, 1910; Hubbs, 1918; Myers, 1927; Pickens, 1927a, 1927b; Chamberlain, 1939; Blair, 1947a). In fact, introgression has occurred in a number of areas where there is no phenotypic means to discern the hybrids (Masta et al., 2002; Vogel, 2007; Vogel and Johnson, 2008). Discerning natural variation from variation due to hybridization is not easy when examining or hearing a toad in the field, or even a series of toads from different localities. In other words, not all unusually appearing toads are hybrids, nor are morphologically identifiable toads lacking in genetic evidence of past introgression.

Natural and laboratory-reared hybrids have been documented between *A. fowleri* and *A. americanus* (Blair, 1941a, 1942; Walker, 1946; Volpe, 1952, 1955a; Cory and Manion, 1955; Brown, 1964; Zweifel, 1968a; Jones, 1973; Green, 1981a, 1981c, 1984, 1989; Smithberger and Swarth, 1993), *A. terrestris* (Blair, 1941a, 1942; Moore, 1955; Gosner and Black, 1956; Volpe, 1958, 1959b; Brown, 1969), *A. woodhousii* (Blair, 1942; Meacham, 1958, 1962) and *Ollotis nebulifer* (Volpe, 1960). *Ollotis nebulifer* (formerly *Bufo valliceps*) males can produce viable hybrids with *Anaxyrus fowleri* females, but all are male and sterile; the reciprocal cross is not viable (Volpe, 1960). However, Vogel (2007) and Vogel and Johnson (2008) reported natural hybrids between these species and noted that hybrids were frequently unable to be distinguished using phenotypic morphological characters alone. They suggested that hybridization has contributed to the decline of *A. fowleri* on the Gulf Coast.

Hybrids may be intermediate in morphology (e.g., Blair, 1941a), call characteristics (Green, 1981a), temperature tolerance and developmental rate (Volpe, 1952), blood proteins, and characteristics of the egg strings (Cory and Manion, 1955). They may have characteristics of both species.

In southeastern Louisiana and adjacent Mississippi, for example, Volpe (1959b) demonstrated extensive hybridization between *A. terrestris* and *A. fowleri*. In some populations, "hybrid swarms" (individuals with an extensive diversity of character combinations) were observed, whereas in other populations repeated backcrossing of hybrids with parental species was apparent.

In contrast, Blair (1941a) reported interspecific mating between *A. fowleri* and *A. americanus* in 9.4% of observed matings, but 30 years later Jones (1973) found no evidence of ongoing hybridization at the same location. Instead, premating isolating mechanisms appeared effective at preventing interspecific matings, although both species were smaller than in the earlier study and showed substantial morphological convergence. This conclusion was challenged by Loftus-Hills (1975), who noted the difficulty in discriminating hybrids using morphological characteristics and pointed out that no temporal overlap in breeding between these species occurred during the course of Jones's study. As such, hybrids may have been missed or been rare, when at other times they would be expected to be more common. W.F. Blair (1974) hypothesized that human-caused habitat disturbances may have led to hybridization at A.P. Blair's (1941a) study site, and that after 30 yrs, selection had operated to reestablish isolating mechanisms giving the results reported by Jones (1973).

The effectiveness of premating isolating mechanisms and differences in breeding phenology and habitat preferences prevent *A. americanus* and *A. fowleri* from introgression in areas of sympatry. Meacham (1962) arrived at the same conclusion for populations of *A. fowleri* and *A. woodhousii* in contact in east Texas; he postulated that ecological segregation was the primary factor separating the species. These species freely hybridize in the laboratory and produce fertile offspring capable of backcrosses with the parental species. Volpe (1955a) noted that postmating isolating mechanisms were more poorly effective at preventing hybridization in sympatric populations of *A. fowleri* and *A. americanus* than between allopatric populations. Climate change that alters the breeding phenology of *Americanus*-group toads could result in increased hybridization.

Laboratory crosses between *Ollotis nebulifer* ♀♀ and *Anaxyrus fowleri* ♂♂ are inviable, but crosses between *Ollotis nebulifer* ♂♂ and *Anaxyrus fowleri* ♀♀ are viable, proceed through metamorphosis, and are intermediate between the parents in pattern and morphology (Volpe, 1956b). Interspecific matings in nature between these species have been reported by Orton (1951), Liner (1954), and Volpe (1956b). Crosses between ♀ *A. fowleri* and ♂ *A. hemiophrys* produced some metamorphs, but few survived very long (Blair, 1959). A few studies have reared hybrid metamorphs through adulthood. For example, some hybrids are fertile and capable of backcrossing with parental species, such as the progeny of ♀ *A. woodhousii* × ♂ *A. fowleri* (Meacham, 1958) and *A. americanus* × *A. fowleri* hybrids with *A. americanus* (Blair, 1942).

Most hybrids are F_1 progeny based on molecular and acoustic analyses (Brown, 1969; Green, 1981a, 1981c, 1984), and a few have been found to be fertile (Green, 1981c, 1984). Hybridization may be prevalent and ongoing at some locations, depending on extent of geographic overlap and breeding phenology, but is rare and affects only a few individuals at other locations. At Long Point, Ontario, the extent of hybridization varies annually, depending on weather conditions (Green, 1996). Even when weather allows an overlap in breeding phenology with *A. americanus*, the frequency of hybridization varies among nearby populations. The chance of hybridization also depends on the frequency of parental species at a breeding site, and is thus density-dependent to a certain extent.

As noted in Identification, spot patterns on the venter vary among populations and between the sexes. Blair (1943a) recorded ventral pattern variation among Fowler's Toad males from 23 populations. Seven populations had no ventral pattern, 11 had spots only in the pectoral region, and 5 had spots on the anterior third of the venter. Variation occurs even among populations in close proximity. In the Midwest, ventral spot patterns increase from south to north (Smith, 1961; Minton, 2001). Such morphological variation affects many characteristics, from coloration to various morphometric measurements involving size and shape of parotoids to limb proportions.

ADULT HABITAT

Adults are often found living in open habitats with friable soils for digging, particularly those with sand or gravel or a high sand content. Such habitats include well-drained woodlands, dry scrub, lake shores, and terraces, Midwest oak savannas, sand and beach dunes, coastlines, and river banks and terraces (Logier, 1931; Oliver and Bailey, 1939; Stille, 1952; Cory and Manion, 1955; Lazell, 1968; Zweifel, 1968a; Jones, 1973; Green, 1989; Klemens, 1993; Brodman and Kilmurry, 1998; Johnson, 2000; Brodman et al., 2002). Fowler's Toads frequently are found in riparian habitats (Rudolph and Dickson, 1990; Burbrink et al., 1998; Pauley et al., 2000; Metts et al., 2001) and are often associated with human-modified areas such as fields and pastures. Throughout a landscape, populations of Fowler's Toads are most common in areas with wetlands of varying hydroperiods occurring in clusters than they are in areas with isolated wetlands (Brodman, 2009). They occur from near sea level through 1,295 m in northern Georgia (Harper, 1935a) and 1,494 m in Tennessee (Stevenson, 1959), but they are not generally considered a toad of higher elevations. At higher latitudes, they reach elevations of 115 m in New Hampshire (Oliver and Bailey, 1939) and 351 m in Connecticut (Klemens, 1993), although most Connecticut populations were below 152 m.

TERRESTRIAL ECOLOGY

Fowler's Toads may be active at any time of the year in the southern portion of its range, but the extent of annual activity varies by latitude. Northern activity patterns are much more restricted. In many northern regions, they are active only for about five to six months (April or May to September or October) (Walker, 1946; Clarke, 1974a; Green, 1989; Klemens, 1993; Hulse et al., 2001; Minton, 2001). In Maryland, activity occurs from April until mid-September on the mainland (Smithberger and Swarth, 1993), but from April to October on Assateague Island (Mitchell and Anderson, 1994). However, they may be observed even in northern regions if a significant disturbance, such as a flood, stirs them from winter retreats (Tucker et al., 1995). Immature toads are seen about two weeks after adults have emerged and begun to chorus. Juveniles are also active later in the season than adults.

Movement patterns of adult Fowler's Toads involve migrations to breeding sites, movement to summer terrestrial feeding areas, diel movements between daytime refugia and nighttime feeding areas, and movements to overwintering sites. *Anaxyrus fowleri* is capable of extensive movements during each of these events, often over a very short time. Toads may travel considerable distances between breeding ponds and summer feeding areas, and they do not appear to use topographically based migratory corridors between breeding and terrestrial sites. In Michigan, they were found to travel more than 1,609 m from the nearest breeding pond to their summer terrestrial locations near the Lake Michigan beach (Stille, 1952). In Connecticut, Clarke (1974a) recorded movements of 312 m between a breeding pond and terrestrial refugia. Juveniles also dispersed about the same distances after metamorphosis and took up feeding home ranges similar to those of adults.

Fowler's Toads are familiar with landmarks within their home range, such as retreat sites, shelter, and optimal feeding sites. When disturbed, they will move directly to places offering cover. If displaced from a home chorus, many individuals are capable of returning to the release site, especially if displacement is < 137 m. At greater distances, returns are far fewer, although Ferguson (1963) recaptured one toad displaced 1,207 m. Some displaced toads moved in the wrong direction or stopped at intermediate ponds. One displaced toad even traveled 1,227 m to a different pond. If displaced from summer terrestrial habitats, return percentages are much higher even at a distance of 335 m from the summer habitat (Ferguson, 1963). Movements are generally direct, and toads do not appear to wander as they move between habitats.

Fowler's Toads are primarily nocturnal (Higginbotham, 1939; Clarke, 1974a), although individuals may be active on cloudy days or during rainfall. Indeed, it may be encountered nearly at any time during the day. Body temperature closely mirrors substrate or water temperature (Brattstrom, 1963; Hadfield, 1966). Inactive toads are warmer than active toads, as the active toads may have just emerged from retreat sites and actively seek warmer temperatures. Once optimal sites are found, they tend to stay put. Peak activity occurs shortly after nightfall, with activity steadily declining or with a secondary activity peak shortly before or after dawn. In Ontario, for example, they are primarily nocturnal and active at ambient air temperatures of 14–25°C (Green, 1989). Rainfall does not seem to stimulate activity (Clarke, 1974a).

Individual Fowler's Toads may not be active on successive nights. At high temperatures, toads seek shelter and become dormant. This reaction is mirrored by their sensitivity to light. At favorable activity temperatures, they tend to be photopositive, but at higher or lower temperatures they are photonegative (Martof, 1962b). They are adept at feeding at night, and may gather under light sources, where they consume nocturnal invertebrates drawn to the light (Brimley, 1944; Ferguson, 1960). Most adult toads visit the same streetlight night after night, with very little movement among adjacent streetlights; juveniles, however, are more transient. Most activity under lights occurs on rainy or overcast nights, with the least activity under a full moon with clear conditions.

When not breeding, Fowler's Toads occupy small to moderate terrestrial home ranges. They are usually found within a linear distance of from 21 to 32 m between recaptures, which translates to 103–1,245 m^2 (Clarke, 1974a). Nighttime sojourns may involve both rehydration and feeding, and the destinations may not be in the vicinity of daytime retreats. Some individuals, however, travel a considerable distance between daytime retreats and moist habitats, where they replenish body water by sitting on a wet substrate (Claussen, 1974). They are adept at absorbing water through a highly vascularized area near the cloaca; Fowler's Toads do not drink water.

Fowler's Toads are commonly found along lake shores (Necker, 1939; Stille, 1952; Breden, 1988; Green, 1989), where they tend to remain within a well-defined home range from one night to the next. Along the shores of Lake Michigan, however, toads apparently move to the beach shortly after dark from terrestrial sites 61–213 m away, and invariably return to the same area to feed or replenish body water from the moist sand. Females precede males to the beach. Stille (1952) found Fowler's Toads usually within about 8 m of where they were previously captured at night along the shoreline, and in no case was a toad further than 152 m away from the site of its previous capture. Toads seemed to need to replenish water about every five days, depending on environmental conditions.

When toads are active, they chose the warmest substrates on which to sit while they await potential prey. When inactive, they seek retreats in mammal burrows, holes in the leaf litter, substrate, or old tree root channels, under logs and other types of surface debris and vegetation, or simply dig themselves into the substrate (e.g., Logier, 1931). Overwintering occurs terrestrially by burrowing into the substrate below the frost line, and there are reports of *A. fowleri* found 1.8 m below the surface (Latham, 1968b). Cagle (1942) reported them dug into clay 15–55 cm below the soil surface, and Latham (1968b) found them from 8 to 20 cm below the surface in sandy-loam soils. Bossert et al. (2003) found them using nest chambers of the Northern Diamondback Terrapin (*Malaclemys terrapin*) as refugia and perhaps to overwinter. Burrowing occurs using the rear legs to scoop soil out of the way, much like *Scaphiopus*. Retreats may be occupied by many toads at one time. Allard (1908) reported finding 15–20 Fowler's Toads under a single stone doorstep. Fowler's Toads are occasionally found at the entrance of caves (Blanchard, 1925; Dodd et al., 2001).

Fowler's Toads are monotonically photopositive in their phototactic response, suggesting they use both sunlight and moonlight when feeding and going about their daily activities (Pearse, 1910; Jaeger and Hailman, 1973). They are sensitive to light in the blue spectrum ("blue-mode response"), which apparently helps them orient toward areas of increasing illumination, such as the open horizon above lakes and ponds (Hailman and Jaeger, 1974). Fowler's Toads likely have true color vision.

CALLING ACTIVITY AND MATE SELECTION

Calling begins shortly after males emerge from dormancy. Toads orient toward and move to breeding ponds using a combination of celestial (lunar, stellar; Ferguson, 1966a; Ferguson and Landreth, 1966; Landreth and Ferguson, 1968) and olfactory cues (Grubb, 1973). Indeed, toads are able to discriminate the odors from their home ponds over adjacent ponds. However, they may be diverted to other breeding ponds by a nearby chorus. At night during migrations, sun compass Y-axis orientation is of little value to the adults, although laboratory trials indicate an ability to use the sun for directional orientation. Auditory and olfactory orientation would be particularly useful supplements when celestial cues are unavailable due to cloud cover, although Ferguson and Landreth (1966) noted that most migrations occurred on clear nights following rainfall. Toads apparently can use landmarks in their environment to move around their home ranges and presumably back to a breeding pond.

Males have a loud explosive advertisement call which functions to attract females to a breeding site. Calls have the following characteristics: in Ontario a mean number of 90 pulses/sec, a mean call duration of 1.8 sec, and a mean number of 157 pulses/sec at 16°C (Green, 1981a). In New Jersey, the mean pulse rate is 126 pulses/sec, the mean call duration is 1.76 sec, and the mean dominant frequency is 1.93 kHz, after adjusting for temperature and size (Sullivan et al., 1996a). In Texas, the dominant frequency ranges between 1,775 and 2,100 cps with a mean duration from 1.0 to 3.2 sec (Blair, 1956b). Thus, the advertisement calls of male *Anaxyrus fowleri* have a shorter duration and higher dominant frequency than related species within the *A. woodhousii* complex. The call (a "bleat," Green, 1989; an unmusical "waaaah," Bartlett and Bartlett, 1999) is influenced by temperature, with pulse rates increasing and the duration of the call decreasing with increasing temperature (Zweifel, 1968a).

Males call from scattered clumps of grasses, reeds, and sedges along the shoreline. They also use other forms of cover at breeding sites, such as rocks and other shoreline debris, or they may call from floating mats of debris or algae (*Cladophora*). Calling also has been reported from shallow pools and sedge islands in an otherwise marshy habitat (Logier, 1931; Green, 1989).

Choruses form in late spring to early summer, the exact timing of which varies annually by environmental conditions. Most breeding occurs at the peak of intense chorusing, with virtually no amplexing pairs observed during the latter part of the calling season. In Ontario, choruses at Long Point were observed initially as early as the beginning of May and as late as the end of May (Green 1981c, 1989), whereas in Maryland chorusing begins in late April (Grogan and Bystrak, 1973a). Temperature plays an important role in the start of chorusing. Choruses will not commence at temperatures less than 10–14°C (Breden, 1988; Green, 1989), and at least in Indiana, large choruses do not form until temperatures reach

16°C. Calls may occur late in the season; Hoopes (1930) reported hearing calls in Massachusetts in late August.

Chorus size varies among populations from a few individuals to breeding populations of hundreds of calling males (300–500 in Ontario; Green, 1989). In Mississippi, Ferguson and Landreth (1966) reported 3,000 toads breeding at a 0.14 ha pond. Of 934 collected, only 10 were females. Also in Mississippi, Ferguson (1960) estimated 1,500 calling males at a single pond in early March. The size of the chorus is probably correlated with the extent of available breeding habitat, but it is clear that great numbers of toads can attempt to use even a small amount of habitat. According to Ferguson and Landreth (1966), choruses are large and widely spaced early in the breeding season, but smaller and more closely spaced as the season progresses.

Males spend only a few days at the breeding site. In New Jersey, the median tenure of chorusing males was only 3 days, with a range of 1 to 15 days (Given, 2002). Males tend to vocalize 11–14 sec per minute while at a breeding site. Some males chorus with minimal effort, and these males tend to gain body mass while at the breeding pond. Continuously calling males tend to lose body mass, however, suggesting that prolonged calling is energetically expensive. Growth rates are not correlated with calling effort. Also, the amount of time a male spends at the breeding site is not correlated with calling effort, body size, or growth rate. However, chorus tenure is positively correlated with mating success (Given, 2002). Why some males remain at breeding ponds for variable amounts of time is unknown.

The dominant frequency of the advertisement call is correlated with body size (Sullivan et al., 1996a), and this is important since females tend to choose large males as preferred mates. Females cue in to the dominant frequency to gain an indication of male size. Males should prefer to call at cool temperatures or at cold microhabitats if given the choice, as the temperature would influence the dominant frequency of the call. This is indeed the case, with large males selecting cooler calling sites than smaller males (Fairchild, 1981). Behavioral thermoregulation allows males to appear larger than they are by altering their call characteristics to influence female perception of large size.

Females prefer males that have a consistently intense advertisement call rather than those with modulated advertisement calls. However, their ability to discriminate between the two types of calls decreases with increasing proximity to a calling male (Given, 1996). Other call properties, such as call duration, pulse rate, and dominant frequency, do not vary among males within a population or from nearby populations. The ability to discriminate males may be enhanced in choruses where comparisons are possible, such as when males are widely spaced. In dense choruses, females have more difficulty in discriminating particular males based on the advertisement call.

Mate selection and oviposition sequence is essentially the same as in *A. americanus*. Amplexus is axillary or supra-axillary, depending on the size of the female. Females are attracted by the calls of the males and move toward them, although they may be amplexed at any time by a satellite male or indeed any male that intercepts them on the way to a breeding site. The oviposition sequence and the behavior of the courting pair were described by Aronson (1944) and consist of amplexus, preovulatory behavior (restlessness and abdominal muscular contractions lasting from a few minutes to several hours), and oviposition (back arching, extrusion of eggs with simultaneous fertilization, movement along the wetland bottom as egg strings are extruded). Aronson (1944) described one complete sequence lasting 5 hrs and 10 min. If the egg strings break during deposition, amplexus ends, and the mating sequence is reinitiated with the same or a different male. A recently spent female will arch her back in a pronounced manner to indicate she has finished spawning, although it may take several tries before the male releases her and moves elsewhere.

Anaxyrus fowleri has a release call that serves to alert both conspecific and heterospecific males that the amplexed animal is a male (Aronson, 1944; Blair, 1947b). Indeed, the release calls may be convergent in call structure among toad species so as to broaden their applicability, even as mate advertisement calls diverge in structure to allow for species isolation (Leary, 2001). The convergence centers on the periodicity of release vocalizations, which are not significantly different. Character displacement in the periodicity of the advertisement calls of *A. fowleri*, however, is pronounced both in sympatry and allopatry with *A. terrestris*. Advertisement call periodicity is much higher in *A. terrestris* (mean 16.8 to 18.3 ms) than in *A. fowleri* (mean 7.3 ms).

BREEDING SITES

Breeding site preferences are wide-ranging, from small temporary puddles to the shores of large lakes. Permanent and semipermanent wetlands are preferred (Tupper and Cook, 2008). Breeding occurs frequently in the shallower portions of large water bodies such as lakes and beaver ponds (Metts et al., 2001), as well as man-made impoundments such as reservoirs (Jones, 1973), retention ponds, golf course ponds, goldfish ponds (Neill, 1950a; Clarke, 1974a, 1974b, 1974c, 1977; Scott et al., 2008), and possibly wetlands in the vicinity of mine-drainage treatment ponds (Lacki et al., 1992). In Ontario, for example, breeding occurs in marshy shallow pools adjacent to sand dunes, savannas, or open forest, and in bays and inlets along Lake Erie (Green, 1989). They also breed in small human-created pools such as road ruts (Adam and Lacki, 1993) and stock ponds (Cagle, 1942). In addition to standing water, *A. fowleri* deposits eggs in the shallow waters of streams where there is little or no current, such as quiet pools, backwaters, and along the stream margin (Allard, 1908; Blair, 1942; Dodd, 2004). The often sympatric toad *A. americanus* does not breed in streams.

Breeding sites may or may not contain fish as many species of fishes avoid the eggs and larvae of this species. Breeding has even been reported in slightly saline pools found on barrier islands, such as Assateague Island, Maryland (Lee, 1972). Wetlands used for breeding typically have shallow shores, decreased amounts of emergent vegetation, an open-canopy cover, contain few organic acids, have a higher pH (4.4–7.9), and are warmer than randomly available sites (Tupper and Cook, 2008).

Most toads return to their natal pond when they begin to reproduce, with lesser numbers opting for breeding sites close to the natal pond (Breden, 1987). A few toads, however, move considerable distances from the pond where they metamorphosed as tadpoles. Toads also do not remain at a breeding site throughout their lifetime. Breden (1987) suggested that there was a 49% per-generation migration rate among nearby breeding ponds. The same breeding site may or may not be used in consecutive years, as about 17% of the population of toads will opt to breed in adjacent sites from one year to the next (Breden, 1987). Ferguson (1963) reported that 14 of 135 marked toads were found later at a different breeding site located 389 m away.

REPRODUCTION

Breeding occurs in the spring and summer, the start of which depends on latitude and environmental conditions. Specific breeding dates include March to May (Arkansas: Trauth et al., 1990; Maryland: Grogan and Bystrak, 1973a), March to July (Alabama: Brown, 1956; Louisiana: Dundee and Rossman, 1989), late March to August (Mississippi: Ferguson, 1960; Ferguson and Landreth, 1966), April to June (Georgia: Allard, 1908; Illinois: Smith, 1961; Missouri: Johnson, 2000), April to July (Indiana: Brodman and Kilmurry, 1998; Minton, 2001; Ohio: Walker, 1946), April to August (Georgia: Jensen et al., 2008; Illinois: Cagle, 1942; New Hampshire: Oliver and Bailey, 1939), early May to June (Indiana: Breden, 1988; Ontario: Green, 1989, 1997; New Hampshire: Taylor, 1993; Pennsylvania: Hulse et al., 2001; Virginia: Lee, 1973a), early May to July (Virginia: Mitchell, 1986), and May to August (Connecticut: Babbitt, 1937). Unusually warm weather may cause *A. fowleri* to call much earlier than normal.

The testicular and ovarian cycles of *A. fowleri* were described by Smith (1975). Spermatogenesis is greatest in summer, although the male spermatogenic cycle is somewhat continuous. Ovarian size is smallest immediately after egg deposition, and gradually increases throughout the autumn and winter through the next oviposition the following year. The size and mass changes of the ovary are mirrored somewhat by the size and mass of the female's fat bodies, which are greatest in summer. Fat bodies are lowest in mass in spring after the period of overwintering; they may gain ten times their spring mass by summer (Smith, 1975).

The eggs are deposited in long, tangled strings around vegetation and debris in shallow water. During oviposition, the amplexed pair moves through the shallow water entwining eggs around vegetation and debris. Oviposition may last several hours. Clutch size is variable and may be from 1,000 to 16,000 eggs per female (Wright and Wright, 1949; Clarke, 1974a; Green, 1989;

Mitchell and Anderson, 1994; Hulse et al., 2001). Specific records include 5,600 eggs for an Ontario female (Green, 1989); 6,050 from an Illinois female (Cagle, 1942); a mean of 8,175 eggs in Arkansas (range 3,067 to 15,618) (Trauth et al., 1990). Clutch size is positively correlated with female body size (Clarke, 1974a). Eggs hatch in as little as two to five days, depending on temperature.

There are far more males than females at breeding sites at any one time, as females remain only long enough to deposit their eggs (Aronson, 1944; Zweifel, 1968a; Jones, 1973; Breden, 1988). However, Breden (1988) suggested that overall sex ratios do not differ from 1:1. In contrast, Hranitz et al. (1993) found a sex ratio of 1:1 one year and 1:2 the next. The amount of time a male toad remains at a breeding site varies annually. In Ontario, for example, males spent from 9 to 25 nights at a breeding pond over a 7 yr period (Green, 1997). In addition, the number of males breeding varies annually, with Green (1997) recording from a low of only 12 breeding males in one year to between 229 and 294 males in 3 other years. In Indiana, Breden (1988) estimated mean male breeding populations of from 35 to 360 individuals at 7 study sites, and 157 females for the entire study area. He further estimated that these 157 females produced 819,697 eggs resulting in 65,081 (range 37,196 to 104,553) metamorphs.

LARVAL ECOLOGY

Larvae are benthic feeders and apparently can discriminate and choose foods of higher quality over those of lower quality (Taylor at al., 1995). Fowler's Toad larvae develop rapidly, but not as rapidly as the early breeding American Toad (Moore, 1939). They have a high temperature tolerance (to 43°C), depending on developmental stage and acclimation temperature (Cupp, 1980). As might be expected, larvae with low acclimation temperatures (10°) have a lower CTmax than those acclimated at higher temperatures. At metamorphosis, however, the CTmax drops off quickly, and recently transformed toadlets are less tolerant of extreme temperatures. Brill (1993) measured natural pond temperatures during development in Virginia from 11–30°C.

Anaxyrus fowleri larvae form dense aggregations in shallow water (Wright and Wright, 1949; Breden et al., 1982). Larvae tend to aggregate by size, and most tadpoles of a similar size maintain a specific orientation. However, the size-assortive aggregations are not exclusive, with much intermingling occurring among tadpoles of different size classes. Thus, an aggregation of tadpoles observed in the field may represent a compilation of aggregates rather than a single large panmictic school. As with other anuran schools, aggregates of *A. fowleri* larvae may function in trophic, defensive, and/or physiological capacities.

Larvae are capable of assessing preferred water depth, and they can use celestial cues to orient toward such locations. Larvae as young as five days have this ability (McKeown, 1968). During warmer parts of the day, they move to the shallows to take advantage of thermal insolation. As waters cool at night, they move to deeper waters that retain heat longer.

Metamorphosis occurs from mid-June to mid-August, depending upon location and when the eggs were deposited. Larvae take about 21 to 60 days to complete development (Wright and Wright, 1949; Mitchell and Anderson, 1994).

Egg strings of *Anaxyrus fowleri*. Photo: Stan Trauth

Tadpole of *Anaxyrus fowleri*. Photo: Stan Trauth

Tadpoles reach a maximum length of about 21 mm, and recent metamorphs are usually 10–15 mm SUL (e.g., Trauth et al., 1990; Brill, 1993; Mitchell and Anderson, 1994; Minton, 2001). Ferguson (1960), however, reported that metamorphs in Mississippi only measured 7.2 mm SUL, and Labanick and Schlueter (1976) gave a mean of 10.7 mm SUL for 6 recent metamorphs in Indiana. Recent metamorphs may rest at the edge of water before dispersing with the remnants of the tail in the water. This presumably aids them in remaining hydrated at such a small size.

DISPERSAL

Juvenile toads tend to disperse far more than adult toads throughout the surrounding environment. This dispersal results in toads often breeding in distant sites from their natal ponds. For example, Breden (1987) found that 27% of metamorphs did not return to their natal pond to breed, and that one juvenile dispersed 2 km (Breden, 1988). Once toads reach a subadult size, they tend to remain within a defined area. Along the shores of Lake Michigan, juveniles disperse from the breeding sites a median of 174 m to moist areas along the lake shore before settling on overwintering sites. These juveniles generally stayed within 29 m of their first lake shore capture location and remained at the lake shore until late September, well after adults had retreated to terrestrial refugia for the winter (Breden, 1988).

Celestial orientation is important upon initial dispersal, as toadlets move perpendicularly away from their natal pond shore (Ferguson, 1963, 1966a; Ferguson and Landreth, 1966). Recent metamorphs are primarily diurnal in their movements, and a Y-axis vector allows them to disperse at right angles to their transformation shoreline using the position of the sun. The young toads also are capable of lunar and stellar orientation, but as movements do not occur at night for toads < 38–45 mm SUL, nocturnal celestial cues are probably not important during dispersal (Ferguson and Landreth, 1966). Instead, they are probably more important for adults returning to breeding ponds during reproduction.

DIET

Adult Fowler's Toads are sit-and-wait predators, although juveniles are more active predators (Clarke, 1974a). These toads are highly insectivorous, but eat a wide variety of invertebrates in proportion to their availability. Virtually any animal that can be captured will be consumed, although beetles and ants appear to be favorites. Klimstra and Myers (1965) found 11 insect Orders and 67 Families represented in 497 stomachs. Specific dietary items include monarch butterflies, lepidopteran larvae, crickets, centipedes, beetles, ants, true bugs, harvestmen, sowbugs, and spiders (Cagle, 1942; Stille, 1952; Bush and Menhinick, 1962; Klimstra and Myers, 1965; Latham, 1968a; Clarke, 1974b; Green, 1989). Mollusks and earthworms are minor components of the diet. Juveniles eat much the same prey as adults, especially ants, but at smaller sizes. They also consume mites, flies, collembolans (springtails), and aphids (Clarke, 1974b; Timm and McGarigal, 2010a).

Individual Fowler's Toads may consume a substantial number of insects. In a mostly agricultural area in Arkansas, R.L. Brown (1974) found a mean number of 17 prey in toad stomachs at one location, and 52 at another. Metcalf (1921) reported a single toad eating 177 May beetles (*Lachnosterna ephilida*) over a 5-day period. Toads use their hands to cram food into the mouth.

PREDATION AND DEFENSE

Anaxyrus fowleri is a gray to brown squat frog that blends in well with its substrate. When not active, the toads bury themselves beneath the substrate or hide under litter and surface debris. Crypsis and an ambush predator lifestyle make them difficult to locate unless they move. Juvenile Fowler's toads also have an immobile response to the approach of a predator, and even the shadow of an over-flying bird may cause them to cease activity and assume an immobile posture (Dodd,

1977). When forced to flee, adults tend to move toward shapes, which suggest a wide perimeter and the greatest likelihood of escape (Martof, 1962b).

Like all toads, *A. fowleri* has secretions that are noxious or toxic to potential predators. These secretions are produced by granular glands located on the dorsal surface of the body and appendages, particularly in the parotoid glands and the warts of the skin. Toxic secretions are effective at thwarting predation attempts, at least for some predators (Tucker and Sullivan, 1975), and the biochemical properties of the protein toxins are different among toad species (Pollard et al., 1973; Mahan and Biggers, 1977). Toads also will inflate the body by inhaling air, lower the head, and raise the hindquarters toward the predator, head butt, or tilt the body in the direction of an intruder. These behaviors place the skin secretions in direct proximity to an intruder, and may assist in directing the toxic substances into the predator's mouth and eyes. Unfortunately for the toads, several species of snakes are immune to the poisons of the secretions. Some predators also learn to attack toads through their ventral surfaces, which lack poison glands. Clarke (1974c) suggested that the large numbers of toads with scars indicated that secretions are effective in diverting predation attempts.

Larvae are distasteful to many predators, and even a slightly noxious tadpole may steer a predator toward more palatable prey. Largemouth bass (*Micropterus salmoides*) readily spit larvae out and rapidly learn to avoid them with repeated exposure (Kruse and Stone, 1984). The level of predator hunger may mediate the effectiveness of the skin secretions as a deterrent, however, as experiments have shown that the bass, when hungry, are more likely to attempt to consume toad tadpoles than when they have eaten previously. Larvae also reduce their activity levels or even become immobile when in the presence of predators, even predators that are known to find them distasteful such as the sunfish *Enneacanthus obesus* (Lawler, 1989).

The primary predators of Fowler's Toads are Eastern Hognose Snakes (*Heterodon platirhinos*) and garter snakes (*Thamnophis sirtalis*). *Heterodon* specializes on toads. They have specialized dentition to hold onto toads during swallowing, and they are immune to the toxic secretions. As much as 75% of the diet of *Heterodon* consists of toads, often this species. Other predators include snapping turtles (*Chelydra serpentina*), watersnakes (*Nerodia*), Black Racers (*Coluber constrictor*), birds (shrikes, bitterns, owls, herons), and mammals (raccoons, skunks). Small water snakes (*Nerodia sipedon*), in particular, may consume many recent metamorphs (Cagle, 1942).

POPULATION BIOLOGY

Fowler's Toad metamorphs grow rapidly during the first month following transformation (Labanick and Schleuter, 1976; Claussen and

Adult *Anaxyrus fowleri*.
Photo: David Dennis

Layne, 1983), with size increases up to 6.6 fold in Connecticut after 1 yr of feeding (Clarke, 1974c). In Ohio, Labanick and Schleuter (1976) recorded growth rates of 0.36 mm SUL/day during the initial 2 months after metamorphosis. Clarke (1974c) observed maximum incremental growth rates of 0.1–0.18 mm/day at toadlet sizes of 10–22.4 mm SUL. After 22.5 mm SUL, growth rates were reduced considerably. Females grow faster than males and reach larger sizes (Clarke, 1974c). The content of the diet influences growth rates in the laboratory (Claussen and Layne, 1983), but the variety of prey consumed in nature probably compensates for the variation in prey quality.

Few metamorphs survive the first winter. Of the 65,081 metamorphs Breden (1988) estimated to have left his breeding pond, only 1,440 were thought to have survived the first winter based on a mark-recapture assessment. In continuing his analyses, Breden (1988) predicted there were 1,814 adults in his population, and that 762 new recruits were added to the breeding population each year. Clearly, only a tiny fraction of tadpoles survive to the adult stage, and population turnover is considerable. The probability of surviving as a 2 yr old adult was only 9.3×10^{-4}! These data suggest that toad populations might be vulnerable to disturbances to breeding sites, whether the disturbances are human-created or the result of environmental variability (e.g., drought, climate change).

The age at first reproduction varies by latitude, with some males attaining secondary sexual characteristics 1 yr following metamorphosis in Indiana (Breden, 1987). A few females also may attain sexual maturity after 1 yr (second autumn after metamorphosis), although later in the year than males (Clarke, 1974c). Thus, many toads do not commence breeding until the second spring following metamorphosis. In northern latitudes, reproduction may not begin until a year later than it does at more southern latitudes. In Ontario, for example, males first breed at 50 mm SUL as 2 yr olds; Ontario females are larger and older than males at first reproduction (Green, 1989). In Pennsylvania, males attain sexual maturity at 48–50 mm SUL and females at 57–60 mm SUL (Hulse et al., 2001), whereas in Indiana Breden (1988) estimated sexual maturity at 59 mm SUL (sex not specified).

Longevity appears potentially to be >3 yrs following metamorphosis (Clarke, 1977), and Green (1997) recorded a single 4 yr old. However, most animals in the Ontario population were 2 to 3 yrs old. Juveniles have a very high mortality rate immediately after metamorphosis, with subsequent annual survivorship of about 30% (Clarke, 1977). Clarke (1977) suggested that males may have lower survivorship than females based on recapture rates. Breden (1987) tracked 17,000 larvae using electrophoretic techniques to determine their fate after release into an Indiana pond. He subsequently recovered 8,539 recently metamorphosed toadlets, but only 37 of these as breeding adults. Clearly, only a very few tadpoles ever reach the adult stage. Decreased numbers of toads may be especially noticeable after an unusually harsh winter.

Toe-clipping as a marking technique has been widely debated among amphibian biologists as to whether there is an effect on subsequent survival and recapture probability. Clarke (1972) was among the first to demonstrate that recapture probability of Fowler's Toads declined with an increasing number of toes clipped (up to eight). The effect was slightly more pronounced on smaller individuals. Other researchers observed that researchers rarely cut more than two or three toes, and that Clarke (1972) assumed that failure to recapture toads was entirely due to toe-clipping, which is not likely the case.

Toad populations were once thought to follow a classic metapopulation structure, with source and sink populations scattered throughout a local or regional landscape. It has now been demonstrated that there is considerable movement by toads across the land, with the level of philopatry not nearly as strong as previously believed. Smith and Green (2005) noted that the assumptions of metapopulation structure were not often met when Fowler's Toads were studied using long-term mark-recapture and genetic data. In part, this stems from the tendency of individuals to move long distances. The maximum distances reported for movement range between several hundred meters (Blair, 1943b; Ferguson, 1960; Clarke, 1974a) to more than 1 km (Stille, 1952; Breden, 1987); Smith and Green (2005) recorded a single adult female moving 34 km! Gene flow resulting from juvenile dispersal may actually prevent genetic divergence among localized toad populations (Hranitz, 1993).

Population processes in addition to dispersal, such as overwintering success and migration, may be more important in structuring toad populations than what goes on at the breeding pond. Although Smith and Green (2005) noted that Ontario Fowler's Toads might have high extinction rates (31 extinction events; 0.25) and colonization rates (38 colonization events; 0.28), there appeared to be no metapopulation effects on toad population dynamics. Extinction events tend to be deterministic rather than stochastic, as breeding ponds change through time. The tendency of juveniles to disperse considerable distances, with many breeding at distant sites, also argues against a metapopulation structure (Breden, 1987). Populations simply are not isolated to any extent.

COMMUNITY ECOLOGY

Anaxyrus fowleri often breeds in ponds or wetlands where other species of amphibians either have bred or are likely to breed later in the season. Whether adults arrive before or after another species has bred, and the length of time between arrivals, determines what effects toad larvae have on other species. For example, if *A. fowleri* breeds first, experiments have demonstrated that their presence prolongs the developmental period, reduces body mass, and delays growth of Spring Peeper larvae (Lawler and Morin, 1993). However, the presence of Spring Peeper larvae in a pool has no effect on toad larvae. As long as the Spring Peepers precede the toad larvae, the presence of the toad larvae also has no effects on subsequent Spring Peeper larval development and growth. Breeding phenology thus determines the nature of the interactions between species in a larval community.

Anaxyrus americanus and *A. fowleri* often use the same habitats for breeding, despite differences in habitat preferences (forest vs. open; uplands vs. lowlands) and temporal breeding seasons (early spring vs. early summer). A particularly cold spring could delay American Toad breeding and result in temporal overlap with Fowler's Toad. Although call differences help to maintain isolation when mixed species assemblages are present, amplexus occurs, frequently leading to hybridization. Satellite males in particular will intercept females of another species, and male toads are notorious for their inability to discriminate other anuran species (or even inanimate objects) from conspecifics.

In the south-central United States, Fowler's Toads are declining because of hybridization with *Ollotis nebulifer* as it expands its range northward. Recent evidence has suggested that human disruption of breeding sites has facilitated this hybridization (Vogel, 2007). In addition, larval *Anaxyrus fowleri* are competitively disadvantaged when reared with *Ollotis nebulifer* larvae, which causes them to metamorphose at smaller size and weights than normal under drying conditions (Vogel, 2007). Survivorship also is less than normal when in competition with *O. nebulifer*. The duration of the larval period was unaffected, however. Vogel (2007) and Vogel and Pechmann (2010) have suggested that the superior competitive advantage of *O. nebulifer* larvae results in ecological displacement of *Anaxyrus fowleri* in human-disturbed habitats.

In areas of sympatry such as the Florida Parishes of Louisiana, there may be some degree of habitat partitioning between closely related toad species, with *A. fowleri* preferring the bottomlands and *A. terrestris* the uplands. *Anaxyrus fowleri* and *A. terrestris* may occasionally be found breeding in the same pond (e.g., Gosner and Black, 1956), however. Experiments have shown that competition occurs between the larvae of these species at high and medium densities, but not at low densities (Wilbur et al., 1983). Competition occurs because of the similarities between the species in feeding ecology and habitat preference. High densities also affect the size differential between the larvae of the two species and the length of the larval period (Wilbur et al., 1983). Thus, the amount of habitat available and breeding phenology becomes important to larval ecology if the two species occupy the same breeding pond. The presence of a predator, such as the newt *Notophthalmus*, can actually minimize the effects of competition among anuran larvae by eating many small larvae and thus reducing larval densities. In addition, the presence of other amphibians may preclude breeding by *Anaxyrus fowleri*. For example, there is a negative correlation between the presence of the salamander *Ambystoma maculatum* and *Anaxyrus fowleri* reproduction, although sample sizes were small (Briggler, 1998).

Breeding habitat of *Anaxyrus fowleri*. *Gastrophryne carolinensis* also bred in this pond. Great Smoky Mountains National Park, Tennessee. Photo: C.K. Dodd Jr.

As with any frog breeding in a pond community, larvae are likely to encounter competitors and predators among both heterospecific larvae and aquatic insects. In experimental trials, the presence of insects and *A. fowleri* larvae decrease the mean individual body mass of *Hyla andersonii* at metamorphosis. The presence on insect larvae also can significantly reduce the mean body mass of *Anaxyrus fowleri* larvae (Morin et al., 1988). The results suggest that competition with invertebrates can play an important role in larval ecology. In another experiment, the newt *Notophthalmus* completely eliminated larval *Anaxyrus fowleri*, creating an assemblage dominated by *Pseudacris* (Kurzava and Morin, 1998). Thus the patchy distributions of predators and prey can result in rather different anuran communities across a landscape.

DISEASES, PARASITES, AND MALFORMATIONS

The fungus *Basidobolus ranarum* was reported from *Anaxyrus fowleri* in Arkansas and Missouri (Nickerson and Hutchison, 1971). Fowler's Toads are parasitized by the protozoans *Hexamita (Octomitus) intestinalis*, *Nyctotherus cordiformis*, *Opalina* sp., *O. obtrigonoidea*, *O. triangulata*, *O. virguloidea*, *Trichomonas augusta*, and *Trypanosoma rotatorium* (Brandt, 1936; Campbell, 1968). Trematodes include *Brachycoelium hositale*, *Gorgodera amplicava*, *Gorgoderina attenuata*, *G. simplex*, *G. translucida*, *Haematoloechus* sp., and *Megalodiscus temperatus* (Brandt, 1936; Campbell, 1968). Cestodes include *Cylindrotaenia americana* and *Distoichometra bufonis*, and nematodes include *Agamonema* sp., *Cosmocercoides dukae*, *Foleyella americana*, *Oswaldocruzia pipiens*, *Oxysomatium ranae*, *Physaloptera* sp., *Rhabdias* sp., *R. bufonis*, *R. ranae*, and *Spinitectus gracilis* (Brandt, 1936; Campbell, 1968; Jilek and Wolff, 1978; Dyer, 1991). They are parasitized by the acanthocephalan *Centrorhynchus* sp. and the mite *Hannemania penetrans* (Brandt, 1936; Campbell, 1968). A mean of 4.8 mites was found per individual *Anaxyrus fowleri* from North Carolina.

Malformations do not appear to be common in this species. A single juvenile with an extra malformed limb was reported from Virginia (Mitchell and Burwell, 2004).

SUSCEPTIBILITY TO POTENTIAL STRESSORS

Fowler's Toad tadpoles reared in stream water resulting from an urban setting in Columbus, Georgia, were smaller in size and metamorphosed faster than those reared from water in a forested setting (Barrett et al., 2010). The reason for this was unknown, but the results suggested that the urban stream water contained some form of environmental stressor that caused tadpoles to accelerate development.

Metals. Mercury affects both embryos and larvae of *A. fowleri*. The $LC_{50\ (7\ day)}$ for embryos is 0.065 mg/L and for larvae is 0.025 mg/L (*in* Sparling, 2003).

Other elements. The metalloid boron is toxic to Fowler's Toad embryos at an $LC_{50\ (8\ day)}$ concentration of 123 mg/L (*in* Sparling, 2003). This element is commonly used in pesticides.

pH. The lethal pH of *A. fowleri* is 4.0, with a critical pH of 4.2 (Freda and Dunson, 1985, 1986; Freda et al., 1991). Under natural conditions, pH values can range to 8.2 (Brill, 1993).

Salinity. In natural ponds, salinity values can range to 2 ppt without any effect on larval growth and development. In laboratory trials, tadpoles raised at 6 ppt had higher growth rates and mortality than tadpoles raised at 0–3 ppt. However, survivors were the same size as those reared at the lower salinity levels (Brill, 1993). Brill (1993) concluded that the small size of adults on Assateague Island compared to mainland populations was not due to environmental stress during development.

Chemicals. A number of industrial chemicals may impact Fowler's Toad. Carbon tetrachloride causes reduced survivorship of larvae after 4 days at > 4.98 mg/L (LC_{50} 2.83 mg/L); chloroform at > 40 mg/L (LC_{50} 35.14 mg/L); trisodium nitrilotriacetic acid (NTA) at > 206 mg/L (LC_{50} 175.5 mg/L); phenol 1 at > 0.53 mg/L (LC_{50} 2.45 mg/L) (Birge et al., 1980). Some of these chemicals had little effect on hatching, however. The $LC_{50\ (4\ day)}$ for methylene chloride is > 32 mg/L.

The herbicide paraquat has an $LC_{50\ (96\ hr)}$ of 15 mg/L affecting Fowler's Toad tadpoles (Mayer and Ellersieck, 1986). Residues of the insecticide fenvalerate were found in *A. fowleri* at concentrations of 0.02 ppm in Arkansas, although the sample size was low (Bennett et al., 1983). The insecticide had been applied five days previous to sampling. Sanders (1970) provided TL_{50} values for DDT administered to *A. fowleri* larvae from 24 to 96 hrs after exposure at ages of from 1 to 7 weeks. These values decreased with age and time after exposure. For example, the TL_{50} for 1-week-old larvae after 24 hrs of exposure was 5.3 mg/L. However, the TL_{50} for 7-week-old larvae 96 hrs after exposure was only 0.03 mg/L. Exposure of larvae to DDT in small doses clearly was lethal to this species.

Sanders (1970) provided additional TL_{50} values for a host of pesticides, including trifluralin, endrin, toxaphene, guthion, TDE, methoxychlor, heptachlor, dieldrin, DEF, malathion, aldrin, hydrothol 191, benzene hexachloride, lindane, silvex 2-(2,4,5-T), molinate, and paraquat, for 4–5-week-old larvae at 24–96 hrs after exposure. Lindane was the least toxic pesticide, whereas endrin was the most toxic. In another comparative study, Ferguson and Gilbert (1968) found that aldrin and dieldrin were less toxic to adult Fowler's Toads after 36 hrs of exposure than were DDT and toxaphene in treated cotton fields. Toads residing in areas previously treated by the pesticides showed up to a 200-fold increase in resistance compared with toads from untreated sites.

STATUS AND CONSERVATION

Roads have helped in monitoring Fowler's Toads populations, but they also have detrimental impacts, especially if located near breeding sites or on migratory pathways. Fowler's Toads are easily monitored by listening for their calls on prescribed survey routes. Their loud call is easily detected at a considerable distance, and 5 min call listening protocols have proven nearly as effective as 10 min protocols in allowing researchers to detect this species (Burton et al., 2006). On the negative side, toads are frequent victims of highway-related mortality, and perhaps tens of thousands are killed annually (Ferguson, 1960; Campbell, 1969; Sutherland et al., 2010).

Throughout much of its range, *A. fowleri* is a very common species. However, there is no doubt that populations of Fowler's Toads have been extirpated or declined at many locations. This is not surprising, since much of its range occurs in the heavily urbanized eastern United States and Canada. For example, Lazell (1976) notes its disappearance from Penikese, Muskeget, Cuttyhunk, Nantucket, and Tuckernuck islands in Massachusetts because of pesticide spraying from the 1940s to the 1960s. Davis and Menze (2000) suggested populations might be declining in Ohio. In contrast, populations may be increasing in Illinois (Florey and Mullin, 2005), and they are considered common in Indiana (Brodman and Kilmurry, 1998), Georgia (Jensen et al., 2008), and Arkansas (Trauth et al., 2004). Klemens (1993) considered them secure in eastern New England but vulnerable in western New England due to human activity.

Fowler's Toad is considered rare in Ontario because of its restricted distribution (Green, 1989). It has a discontinuous distribution along the north shore of Lake Erie, and several populations have disappeared since the 1950s. Populations at

Point Pelee Park disappeared ostensibly because of human disturbance, and populations on Pelee Island disappeared due to intensive agriculture associated with pesticide use. Other populations are stable, however, and the population at Long Point appears to be large but with occasional fluctuations (Logier, 1931; Evans and Roecker, 1951; Green, 1989, 1992). The conservation status of this species in Canada and its potential critical habitat has been summarized by Green (1989) and Weller and Green (1997).

Like many other toads, Fowler's Toad appears resilient to certain types of habitat disturbances, such as agriculture, silviculture, and suburbanization (Ferguson, 1960; Foster et al., 2004; Jensen et al., 2008), as long as breeding sites are maintained and suitable overwintering sites are available. The presence of cattle in breeding ponds had little effect on postmetamorphic Fowler's Toads, and Burton et al. (2009) suggested that toads may benefit from controlled grazing. This may result from the toad's preferences for open habitats, in contrast to the American Toad, which prefers forested woodlands. For example, Fox et al. (2004) found few Fowler's Toads compared to American Toads in diversely managed forested watersheds in Arkansas, regardless of type of management used. Still, clearcutting may be detrimental to the species, at least over a short-term period (McLeod, 1995; McLeod and Gates,

Head pattern of *Anaxyrus fowleri*. Illustration by Camila Pizano

1998). Fowler's Toads also appear to tolerate prescribed burning inasmuch as this opens up the habitat, making it more favorable from the toad's perspective (McLeod, 1995; McLeod and Gates, 1998; Floyd et al., 2002).

Fowler's Toads rapidly colonize newly constructed or restored wetlands (Briggler, 1998; Mierzwa, 1998; Merovich and Howard, 2000; Brodman et al., 2006; Palis, 2007), and have been found in reclaimed surface mine sites 19 to 29 yrs after mining had ceased (Myers and Klimstra, 1963). They have been successfully repatriated to several locations owned by the National Park Service in the New York City metropolitan area, using mostly larvae for release (Cook, 2008). The only unsuccessful release site was impacted by saltwater overwash.

Anaxyrus hemiophrys (Cope, 1886)
Canadian Toad

ETYMOLOGY
hemiophrys: from the Greek words *hemi* meaning 'half' and *ophrys* meaning 'brow' or 'eyebrow.' The name might refer to the eye shape.

NOMENCLATURE
Conant and Collins (1998) and Stebbins (2003): *Bufo hemiophrys*
  Synonyms: *Bufo woodhousii hemiophrys*

IDENTIFICATION
Adults. Canadian Toads are brown to brownish green to gray, but occasional individuals may be reddish to reddish brown (Cook, 1964b). The cranial crests fuse to form a conspicuous boss (bump) between the eyes, and postorbital crests are weak or absent. A vertebral stripe is present down the midline of the dorsum. Parotoids are wide and long and are not greatly elevated. The species has numerous (10–14) dark spots dorsally, with 1–2 warts per spot. Warts are darker than the ground color, although they are lighter than the dark patch that surrounds the warts. Venters are dirty white or slightly yellowish with variable dark flecking or spotting. Males have a swollen thumb and are darker than females all year-round; during the breeding season they also have a dark throat. Underhill (1962) provided additional information on morphology.

Males are generally smaller than females. In

contrast, Underhill (1962) found no statistical difference in the size of South Dakota adults, with a male mean of 51.5 mm SUL and a female mean of 52.6 mm SUL. Other specific records include: 58–68 mm SUL for males and 56–80 mm SUL for females (Wright and Wright, 1949); in Minnesota, the largest male measured by Breckenridge and Tester (1961) was 60 mm SUL, whereas the largest female was 68 mm SUL; Cook (1964b) reported a maximum size of 85 mm SUL for males and 91 mm SUL for females.

Larvae. The tadpole is darkly pigmented dorsally. The tail musculature is bicolored, the tail fins are unpigmented, and the ventral tail fin is narrow. The throat area is clear, and the venter is light, extending to the gut. TL is ca. 30 mm.

Eggs. The eggs have not been described. Presumably they are similar to those of the closely related *Anaxyrus americanus*.

Distribution of *Anaxyrus hemiophrys*

DISTRIBUTION

The range of the Canadian Toad extends from the extreme southern portion of the Northwest Territories south throughout eastern Alberta to just south of the border with Montana. The range then extends eastward through western North Dakota, then south across eastern North Dakota to northeastern South Dakota and western Minnesota. From there it occurs northwest across southwestern Manitoba and much of Saskatchewan. The species barely enters the extreme southwestern corner of Ontario. Harper (1963) corrected some previously erroneous records from the northern portion of the range.

Important distributional references include: Alberta (Russell and Bauer, 2000), Manitoba (Harper, 1963), Minnesota (Oldfield and Moriarty, 1994), Montana (Black, 1970, 1971), North Dakota (Wheeler and Wheeler, 1966; Hoberg and Gause, 1992), Northwest Territories (Preble, 1908; Harper, 1931; Timoney, 1996; Fournier, 1997, undated), and South Dakota (Peterson, 1974; Fischer, 1998; Ballinger et al., 2000; Kiesow, 2006).

FOSSIL RECORD

Fossils of the Canadian Toad have been reported from Pleistocene Irvingtonian deposits from Kansas and Texas, and from Rancholabrean deposits in Alberta (Tihen, 1962b; Holman, 2003). Bones are often referred to this species as *Bufo* cf. *B. hemiophrys*.

SYSTEMATICS AND GEOGRAPHIC VARIATION

Anaxyrus hemiophrys is a member of the *Americanus* clade of North American bufonids, a group that includes *A. americanus*, *A. baxteri*, *A. fowleri*, *A. houstonensis*, *A. terrestris*, and *A. woodhousii*. It is closely related to *A. baxteri* and somewhat more distantly to *A. fowleri*, *A. americanus*, and *A. houstonensis* (Pauly et al., 2004). Suggestions that *A. hemiophrys* is more closely related to *A. microscaphus* than the *Americanus* clade (Blair, 1963b) are not supported.

The Canadian Toad hybridizes in nature with a number of other toads in narrowly defined contact zones. These include *A. americanus* (Cook, 1983; Henrich, 1968; Green, 1983, 1996; Green and Pustowka, 1997), *A. boreas* (Cook, 1983), *A. cognatus* (Brown and Ewert, 1971), and *A. woodhousii* (Meacham, 1962). Interspecific amplexus has been reported under natural conditions (Cook, 1983; Eaton et al., 1999). Hybridization in the laboratory has been reported with *A. americanus*, *A. houstonensis*, *A. microscaphus* (small % reach metamorphosis), *A. terrestris* and *A. woodhousii*, but not with *A. punctatus* (Moore, 1955; Blair, 1961a, 1963a, 1972). Even hybrids backcrossed with the parental species frequently produce larvae that survive through metamorphosis.

The rusty color phase of *A. hemiophrys* has been described in detail and mapped by Cook (1964b). The frequency of toads with rusty coloration varies among populations, but most populations with rusty toads are found in west-central Manitoba and adjacent east-central Saskatchewan. In populations containing rusty toads, between 6–35% of the individuals observed are reddish. The rusty color phase appears absent from Alberta and much of southern Saskatchewan and southwestern Manitoba. Whatever its origin, the rusty coloration is not associated with substrate color.

ADULT HABITAT

This species is usually found in open sites close to water (e.g., 80% within 8 m; Breckenridge and Tester, 1961), but some limited dispersal occurs into adjacent mixed upland forests. *Anaxyrus hemiophrys* is usually found in damp open areas adjacent to more permanent water bodies such as lakes and rivers. Dense cover is preferred as microhabitat in experimental choice tests, although the species does not avoid open substrates (Tester et al., 1965). Preferred habitats include grass (e.g., *Andropogon* sp.), meadows and willow (*Salix* sp.), bogs near water and in the surrounding poplar (*Populus tremuloides*) and willow forests and aspen parklands of the Transition and Canadian (Boreal) life zones (Stebbins, 1954; Breckenridge and Tester, 1961; Roberts and Lewin, 1979; Cook, 1983). In other areas, it is found associated with prairie ponds and lakes (Underhill, 1961; Henrich, 1968; Williams, 1969). Canadian Toads avoid the Jack Pine (*Pinus banksiana*) forests of Alberta (Roberts and Lewin, 1979), as they do not like dry substrates. The open habitats preferred by this species experience more rapid temperature fluctuations than closed-in forest habitats.

TERRESTRIAL ECOLOGY

This is definitely a cold-adapted species. The activity season may be short, especially at the northern portion of the species' range. In the Northwest Territories, for example, toads exit their hibernacula around the first week of May and return the first week of September, making the activity season 3.5 months (Kuyt, 1991). Emergence in Manitoba occurs in May, although full chorusing does not occur until June (Tamsitt, 1962). In Minnesota, emergence occurs from late April to mid-June over a five- to six-week period, with the peak of emergence in mid-May (Tester and Breckenridge, 1964a; Kelleher and Tester, 1969). First emergence dates can vary by as much as 20 days, depending on environmental conditions, and emergence only occurs when all traces of soil frost have disappeared (Kelleher and Tester, 1969). Emergence may be triggered by a period of warming weather (>15–21°C) or precipitation. Adults emerge prior to juveniles, but juveniles are active later in the season than adults.

As with many amphibians, Canadian Toads do not occupy every wetland that appears suitable. In Alberta, for example, they were found at 9 of 25 sites sampled, bred at only 7 sites, and were common at only 3 sites (Roberts and Lewin, 1979). Canadian Toads tend to be found near water and appear less purely terrestrial than many other bufonids. However, there is some disagreement in the literature as to the extent of summer nonbreeding activity in relation to water. For example, Breckenridge and Tester (1961) reported most activity within close proximity to the margins of prairie ponds and lakes in Minnesota, whereas others (Cook, 1983; Constible et al., 2010) have noted that in boreal forest populations this does not appear to be the case.

Toads wander extensively in upland forest, and combined short-term movements can translate into long-distance directional movements that vary in timing and tortuosity of the routes taken. In one study, toads moved from 4.7 to 116.6 m/day and traveled up to 4.7 km with a maximum displacement of 1,836 m (Constible et al., 2010). In a study of toads in a prairie region, the maximum distance between captures was 341 m (Breckenridge and Tester, 1961). Movements are not stimulated by rainfall, and movements among individuals do not vary synchronously; they occur both by day and night. Toads frequently stay in one place for several days, then suddenly move to a new location.

Differences in nonbreeding habitat use and activity may reflect variation in the habitats occupied by the toads and perhaps other environmental differences. Daily movements are generally < 50 m, but occasional movements of ca. 100 m can occur in habitats adjacent to boreal forest (Constible et al., 2010); one toad moved an astounding 480 m in 4.6 hrs! In a prairie habitat, most toads were captured relatively close to water, although some toads moved at least 61 m (Breckenridge and Tester, 1961), and one toad moved 229 m.

Movements are not confined to aquatic areas, nor are movements simply going from one moist area to another; a variety of habitats are used during movement, and dry areas may be crossed. One form of daily movement is the migration between retreat sites or nearby marshes and lakeside beach habitats, usually located within < 123 m from the diurnal activity area (Tamsitt, 1962). Movements occur at night, much in the way *A. fowleri* has been described moving to and from beaches along Lake Michigan (Stille, 1952).

After metamorphosis, juveniles disperse quickly into upland forests in order to feed and find overwintering sites. In Alberta, for example, juveniles may be abundant around breeding ponds in early September, but by mid-September they are difficult to locate (Roberts and Lewin, 1979). In contrast, many metamorphs in Minnesota remained in the vicinity of the natal pond, whereas others dispersed more widely (Breckenridge and Tester, 1961).

In general, adults are more common on beaches surrounding lakes early in the season during daylight hours, but as the season progresses adults disappear from beaches and juveniles become more common (Tamsitt, 1962). At night, both adults and juveniles may be encountered in numbers foraging on the beaches along lakeshores, providing the temperatures are appropriate. On cold sunny days, toads bury themselves into soft sand with only the eyes protruding. After the breeding season, toads do not enter the water.

Postmetamorphs use terrestrial sites for overwintering, and bury themselves well into the soil. They enter dormancy in late August to early September (Breckenridge and Tester, 1961). During the winter, they continually dig vertically into the soil to position themselves below the frost line. Tester and Breckenridge (1964b) reported that the minimum temperature a toad was exposed to be ca. 2.7°C. Constible et al. (2010) found toads at 42 cm and 50 cm below the surface in well-packed substrate in sandy soils. In this study, the overwintering locations were 654–1,386 m from the breeding site. In Minnesota, Tester and Breckenridge (1964b) reported toads 1–1.25 m below the surface; toads in this latter population overwintered within 35 m of the breeding pond.

In the Great Plains, overwintering often occurs in mima mounds, that is, in low (30–60 cm high), flattened circular to oval (3–30 m in diameter) domelike mounds composed of loose, sandy loam or loamy sand. Many toads will use the same mound. A communal overwintering site in an open, south-facing, sparsely vegetated sandy hillside in the Northwest Territories contained >600 tunnels where Canadian Toads spent the winter (Kuyt, 1991). This site measured 112 m by 12–15 m in one area with a smaller area about 30 m away (Timoney, 1996). Timoney (1996) counted 1,144 burrows in the main area and another 25 in the smaller area. Toads bury themselves backward down the tunnels using their rear feet to dig. Both emigration from and return to the tunnels occurred en masse within a few days, with male and female activity occurring simultaneously in the extreme North. In Minnesota, males emerge two to four days prior to females (Kelleher and Tester, 1969).

Canadian Toads usually return to the same overwintering sites from one year to the next (rates of 88–95%) (Kelleher and Tester, 1969). Indeed, toads often return to the same portion of a mound indicating well-developed orientation abilities. Occasionally, however, toads will switch mima mounds and may remain at the second mound over a period of years. A very few toads might use additional mounds in subsequent years. Most mound switching occurred after the first year of dormancy, but even adults occasionally switch mounds.

CALLING ACTIVITY AND MATE SELECTION

Males call while sitting in shallow water. The call of the Canadian Toad is a trill characterized by its short duration (1.3–5 sec), rapid repetition rate (82–90 pulses/sec), and a dominant frequency of 1,400–1,725 cps (Blair, 1957). Breeding aggregations of *A. hemiophrys* are often small, consisting only of one or a few individuals at any one time. Apparently toads disperse over a large area to breeding sites, although a large number of toads may use the same communal overwintering site. Males move to the breeding ponds shortly after emergence and begin calling. Females are attracted to the choruses and move toward the calling males within two to four days. When a female approaches a male, he will amplex her. The approach and movement triggers breeding attempts, that is, the female may orient toward a specific male, but the male is nonselective. After oviposition, males remain within the breeding chorus, whereas spent females leave the immediate area and move to the margins of the pond. After the breeding season ends, males also move to the margins of the wetlands. In prairie regions, both sexes remain near water, whereas in boreal forests they begin a more terrestrial foraging activity pattern.

BREEDING SITES

Breeding occurs at a variety of sites, including ponds, man-made borrow pits and ditches, slow-flowing water in streams and rivers, and along the margins of lakes (Roberts and Lewin, 1979; Cook, 1983). In Minnesota, Williams (1969) observed breeding in wetlands with a mean pH of 8.7 and an alkalinity of 231 m/L $CaCO_3$. Breeding sites are located in open habitats, such as prairies and aspen parklands, and not in dense forest.

REPRODUCTION

Breeding occurs from spring into summer depending upon latitude, and takes about two months once initiated. Specific dates include early May in the Northwest Territories (Kuyt, 1991) and May to July in Alberta and Manitoba (Tamsitt, 1962; Roberts and Lewin, 1979; Cook, 1983). The onset of breeding is very dependent on temperature, as frosts may occur into June in northern areas. Calls have been heard into mid-September near Lake Manitoba (Tamsitt, 1962). Eggs are deposited in long strings that may or may not be associated with vegetation. Clutch sizes vary, with reports of 3,354–5,842 in Alberta (Roberts and Lewin, 1979). Laboratory observations suggest that fertilization rates are high, and that hatching success ranges from 68 to 70% (Williams, 1969).

LARVAL ECOLOGY

Under field cage experimental conditions, survivorship varies considerably among treatments and years, but is less than 30% and not infrequently 0% (Williams, 1969). However, these results may have been due to poor experimental protocols rather than reflect actual low survivorship in nature. Optimal larval temperature preference is 18–23°C (mean 20.5°C), and larvae can tolerate temperatures >38°C for only 30 min or less (Williams, 1969). When tadpoles encounter high temperatures, they immediately move toward cooler portions of the wetland.

The length of the larval period for free-ranging tadpoles has not been determined. When the posterior limbs were present (18 June), tadpoles in Manitoba were 22–27 mm TL (mean 25 mm) (Tamsitt, 1962); metamorphosis occurred by early July. Newly metamorphosed Canadian Toads averaged 12.4 mm SUL (range 11.9 to 13.2 mm) in Alberta (Roberts and Lewin, 1979), 11.3 mm SUL in Manitoba (Tamsitt, 1962), and 9–15 mm SUL in Minnesota (Breckenridge and Tester, 1961).

DIET

The diet of adult *A. hemiophrys* consists mostly of arthropods, almost all of them insects. Beetles and ants are particularly favored; other prey includes grasshoppers, locusts, mayfly and damselfly

Adult *Anaxyrus hemiophrys*.
Photo: Jeffrey LeClere

nymphs, cicadas, true bugs, various types of flies, ichneumonids, spiders, and rarely snails (Moore and Strickland, 1954). However, there is one report of a large Canadian Toad attempting to eat a young Red-winged Blackbird (*Agelaius phoeniceus*) (Cook and Cook, 1981). Juveniles consume smaller prey, such as beetles, midges and other small flies, springtails, mites, and ants. Flies are very important in the juvenile diet. The types of prey reflect the terrestrial habits of the toad.

PREDATION AND DEFENSE

Like other toads, Canadian Toads have noxious skin secretions concentrated in the granular glands of the warts and parotoids. The toads are well camouflaged, especially when buried in the sands with only the eyes showing. When approached, they may leave their subsurface retreats and move rapidly into the surrounding vegetation. They readily take to water and run quickly through vegetation to make their escape. However, two cases of death-feigning have been reported (Nero, 1967; McNicholl, 1972), and toads often remain immobile hunkered toward the ground when approached.

Juveniles are eaten by a wide range of predators, including garter snakes (*Thamnophis* sp.), birds (Red-tailed Hawk), and mammals (raccoons, badgers) (Breckenridge and Tester, 1961; Tamsitt, 1962; Tester and Breckenridge, 1964a). There are many reports of which predators might eat Canadian Toads, but few actual observations.

POPULATION BIOLOGY

There is a certain degree of regional variation in life history characteristics of this species. The largest Canadian Toads are generally found in the south, whereas populations with the smallest toads are located in three populations in the middle of the latitudinal range (Eaton et al., 2005b). These latter populations also contained the oldest individuals. Despite regional variation, size and age are positively correlated within a population. Populations may be male- or female-biased. For example, toads caught emerging from overwintering sites at a drift fence in Minnesota had a sex ratio of 1.13:1 (n = 540), whereas hand-caught toads had a ratio of 4.1:1 (n = 117). Hand catching toads could yield a biased sex ratio because of male visibility and activity, and Testor and Breckenridge (1964a) suggested that females actually left the pond margins and were therefore less likely to be captured. In other data presented by these authors, the overall sex ratio was 1:1 in one year and 1.3:1 a second year. Additional observations suggested that males made up 33–38% of the population, but that sex ratios varied considerably among sites and years (Kelleher and Tester, 1969).

Growth occurs rapidly, but growth rates also vary among populations (Eaton et al., 2005b). Toadlets metamorphosing at 11–12 mm SUL reach 19.3–28 mm SUL in Alberta (Roberts and Lewin, 1979) and 22.3–28.4 in Manitoba (Tamsitt, 1962) by August, and a mean of 31 mm by September in Minnesota (Breckenridge and Tester, 1961). Even small individuals can overwinter successfully, however, as Roberts and Lewin (1979) found 22 mm SUL juveniles in June that must have metamorphosed the previous year. In Minnesota, males reach maturity by 38–45 mm SUL, as indicated by the presence of motile sperm; females are mature by 43–45 mm SUL, as indicated by the presence of pigmented oocytes (Tester and Breckenridge, 1964a).

Thus, it is possible that males could reach a minimum breeding size midway to late in the first breeding season following their metamorphosis. Female Canadian Toads reach maturity by the second summer after metamorphosis (Eaton et al., 2005b). The age at maturity may vary, however, depending on when tadpoles metamorphose. If early in the season, a long growing season allows them to reach maturity by the end of the first summer; those metamorphosing late in the season would be delayed by the necessity of overwintering and resuming growth the following spring. In this case, maturity may not be reached until after the breeding season, so that the first opportunity for breeding may occur well after maturity is reached. At one population in Alberta, however, females did not reach sexual maturity until 4 yrs, although sample size was very small, and in another population both males and females did not reach maturity until 3 yrs. These data suggest the potential for considerable variation among populations in life history traits, perhaps depending on environmental conditions. Maximum longevity is 7–12 yrs, but again there is much variation among populations (Eaton et al., 2005b). Simply put, toads in some populations live much longer than in other populations.

Breeding habitat of *Anaxyrus hemiophrys*. Photo: Cindy Paszkowski

The density of Canadian Toads can be rather high. Roberts and Lewin (1979) recorded 12/1,000 m² in Alberta, with maximum densities found within 50 m of water. Densities become lower as the toads disperse into mixed upland forests. Other reports suggest that populations are not as large as those of other toad species, with adults being more widely dispersed. In Minnesota, the ratio of juveniles to adults was 1.24–14.25:1, depending upon year sampled (Tester and Breckenridge, 1964a; Kelleher and Tester, 1969).

Mortality of the small size classes is high, especially due to water loss, temperature extremes, and predation. Mortality of overwintering toads occurs when frost is deep, especially as a result of lower snowfall amounts than normal. In such conditions, juveniles might not be as able as adults to burrow beneath the deepening frost level. Over one winter sampled, however, survival rates were about equal for males, females, and juveniles. Kelleher and Tester (1969) estimated annual survivorship at 24–44%, depending upon year and location.

COMMUNITY ECOLOGY

The presence of fish may or may not influence survivorship of *Anaxyrus hemiophrys* larvae. In Alberta, for example, larval recruitment is not affected following winters in which severe fish kills occur. Reduced fish abundance may even decrease the abundance of developing toad tadpoles, perhaps because of complex trophic interactions among different types of fish and insect abundance (Eaton et al., 2005a). Experiments have not demonstrated that small-bodied fish can have significant predatory effects on toad larvae in boreal habitats, perhaps because the toad tadpoles are distasteful and not an important part of fish diets under normal circumstances.

The range of this species may be allopatric or sympatric to that of other toads. In Minnesota, for example, *A. hemiophrys* is allopatric with *A. americanus*. Whereas *A. hemiophrys* is found in prairie habitats, *A. americanus* is confined to forested habitats. Neither species breeds in ponds in the ecotone between these habitats. Thus, there appears to be habitat segregation between them in contact zones in the Itasca and Waubun region (Williams, 1969). Not surprisingly, larval prairie-dwelling *A. hemiophrys* have a higher thermal tolerance than the forest-dwelling *A. americanus*, since they are more likely to be exposed to sunlight in the open prairie habitat.

In other regions, ecotones may form an important area for hybridization between closely related species. In southeastern Manitoba, for example, a 3 km wide zone of hybridization occurs between the prairie species *A. hemiophrys* and the eastern forest species *A. americanus* (Green, 1983). Unlike the area in Minnesota, hybridization occurs extensively within this narrow band, where most toads are intermediate between the parental species. The hybrid zone probably formed about 8,000 years ago, after the retreat of glacial Lake Agassiz, which brought these species into secondary contact. Although there is a degree of stability in the hybrid zone, it appears to be shifting westward.

DISEASES, PARASITES, AND MALFORMATIONS
The amphibian pathogen *Batrachochytrium dendrobatidis* (misidentified as *Basidiobolus ranarum*) was reported from Canadian Toads originally captured in North Dakota (Taylor et al., 1999b). Canadian Toads have been used as a surrogate species for understanding the effects of the chytrid pathogen on *Anaxyrus baxteri*, a critically endangered species. For example, transmission of the fungus from infected to healthy toads was confirmed via clinical experiments using this species (Taylor et al., 1999e). Myiasis from green blowfly (*Lucilla silvarum*) larvae has been reported from Canadian Toads in Alberta (Eaton et al., 2008). A single Canadian Toad with an unspecified malformation was reported by Converse et al. (2000).

SUSCEPTIBILITY TO POTENTIAL STRESSORS
No information is available.

STATUS AND CONSERVATION
There are few data on the effects of habitat alteration on this species. Canadian Toads breed in open areas, including recent clearcuts. One study on the effects of clearcutting around ponds could not demonstrate that the width of a buffer strip was important for conserving Canadian Toads. The authors (Hannon et al., 2002) therefore suggested that in some habitat types, a buffer zone around wetlands may be small and yet be adequate to protect this species. *Anaxyrus hemiophrys* will colonize restored wetlands, although they must be

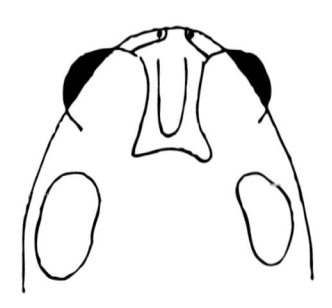

Head pattern of *Anaxyrus hemiophrys*. Illustration by Camila Pizano

within the relatively short dispersal distance of the species (Lehtinen and Galatowitsch, 2001).

Anaxyrus hemiophrys is uncommon and not protected in the Northwest Territories, but most populations occur in Wood Buffalo National Park (Fournier, 1997). Timoney (1996) recommended population monitoring and complete protection of the northernmost overwintering site from human disturbance. Primary concerns involved road modification, vehicular traffic, and vegetation intrusion. There is no evidence of declines in Saskatchewan (Didiuk, 1997), but declines have been reported in central and southern Alberta and in parts of Manitoba, perhaps due to drought and the destruction of wetlands (Weller and Green, 1997; Russell and Bauer, 2000). These latter authors noted declines in Elk Island National Park, Alberta, from 1971 to the mid-1980s, suggesting that habitat alteration alone is not responsible for the species' decline. Drought, however, decreases the amount of snowpack and thus may expose dormant toads to excessive cold temperatures normally mediated by snow's insulation capacity.

Anaxyrus houstonensis Sanders, 1953
Houston Toad

ETYMOLOGY
houstonensis: named after Houston, Texas.

NOMENCLATURE
Conant and Collins (1998): *Bufo houstonensis*
  Synonyms: *Bufo americanus houstonensis*

IDENTIFICATION
Adults. The Houston Toad is a small brown, gray, or purplish gray toad with a somewhat herringbone pattern of dark brown or greenish blotches separated by light areas between them. The postorbital crest is large. Parotoids are elongated and about twice as long as broad. There are from 1 to 5 warts per spot, but rarely >3. Warts are smooth and rounded. A mid-dorsal light stripe may be present. Venters have some midventral

spotting anteriorly but not posteriorly, and are otherwise pale. The dorsal surface of the femur is dark striped. Males have dark throats during the breeding season. Sanders (1953) described the osteology in detail.

Males are slightly smaller than females. Males were ca. 52–64 mm SUL (mean 57 mm) in one Bastrop County study and were 50–66 mm SUL (mean 57 mm) in the Houston area (Brown, 1971b). In another study in Bastrop County, males averaged 61 mm SUL (range 53 to 77 mm) and females 66 mm SUL (range 54 to 84 mm) (Jacobson, 1989). The smallest female from Bastrop County producing eggs was 59.5 mm SUL, and the largest was 81 mm SUL; males were 53–61.8 mm SUL (Quinn and Mengden, 1984). A fourth study in Bastrop found that males averaged 57.1 mm SUL and females 63 mm SUL (Hillis et al., 1984).

Larvae. The body of the tadpole is dark without contrasting markings. Venters are evenly pigmented. The throat is largely pigmented, and there are no gaps in the dark pigment on the dorsal part of the tail musculature. The light ventral part of the tail musculature is narrow. Tail fins, however, are mostly unpigmented. The snout is sloping in lateral view. TL is < 30 mm.

Eggs. Eggs are dark brown dorsally and light brown ventrally. The vitellus is about 1.4–1.9 mm in diameter, and eggs are oviposited in long strings that are separated from one another by compartments. The outer jelly envelope is somewhat scalloped and is 3.2–3.5 mm in diameter (Sanders, 1953). Sanders (1953) provided a photograph of an egg string.

DISTRIBUTION

Houston Toads are known historically only from a small area of southeastern Texas, including Austin, Bastrop, Burleson, Colorado, Fort Bend, Harris, Lavaca, Lee, Leon, Liberty, Milam, and Robertson counties (Sanders, 1953; Brown, 1971b; Shepard and Brown, 2005). The largest remaining populations are in and near Bastrop and Buescher State Parks. Toad populations in Burleson, Fort Bend, Harris, and Liberty counties are likely extirpated. Important distributional references include Hillis et al. (1984), Dixon (2000), and Shepard and Brown (2005).

FOSSIL RECORD

No fossils have been identified.

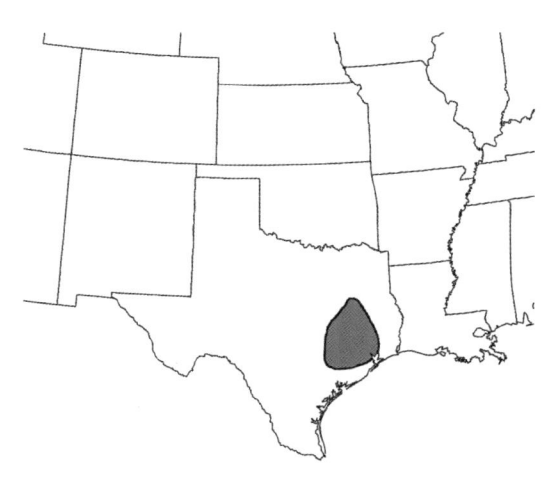

Distribution of *Anaxyrus houstonensis*

SYSTEMATICS AND GEOGRAPHIC VARIATION

Anaxyrus houstonensis is a member of the North American *Americanus* clade within the Nearctic clade of toads. It is closely related to *A. americanus*, more distantly so to *A. fowleri*, *A. hemiophrys*, and *A. baxteri* (Pauly et al., 2004). According to Sanders and Cross (1964), the diploid number of chromosomes is 21, unlike other *Anaxyrus* that have 22 diploid chromosomes.

Under laboratory conditions, crosses between *A. houstonensis* and *A. americanus* or *A. terrestris* produce viable metamorphs (Blair, 1959, 1963a). A successful cross also has been reported between ♀ *A. houstonensis* and ♂ *Ollotis nebulifer* (Kennedy, 1962). Male hybrids from a cross between ♀ *Anaxyrus terrestris* and ♂ *A. houstonensis* backcrossed with a ♀ *A. terrestris* also successfully produced metamorphs (Blair, 1963a). F_1 hybrids from *A. houstonensis* × *Ollotis nebulifer* crosses are sterile, whereas F_1 hybrids from *Anaxyrus houstonensis* × *A. woodhousii* crosses are fertile (Brown, 1971b).

Natural hybridization has been reported with *A. woodhousii* and *Ollotis nebulifer* (Brown, 1971b; Hillis et al., 1984). The call of *Anaxyrus houstonensis* × *A. woodhousii* hybrids is intermediate in pulse rate, dominant frequency, and duration from the parental species, although it is very difficult to tell them apart morphologically. Hybrids between *A. houstonensis* and *Ollotis nebulifer* are intermediate in morphology between the parentals, but the calls are abnormal (Brown, 1971b). Hillis et al. (1984) found that <1% of the population consisted of hybrids.

ADULT HABITAT

The Houston Toad is a relict species found in the southeastern Coastal Plain of Texas. It frequents deep, Carrizo sandy substrates characterized by the presence of Loblolly Pine (*Pinus taeda*) in the Post Oak Savanna region of Texas.

TERRESTRIAL ECOLOGY

Houston Toads may make movements of considerable distances, even across adjacent drainages. Price (2003) recorded a number of sizeable movements by adults covering distances of 0.95–1.85 km. Immediately after metamorphosis, toadlets may return to water briefly before dispersing. Larvae metamorphose simultaneously in large numbers and may disperse rapidly away from the breeding pond. Dispersal occurs both day and night. After 48 hrs, toadlets disperse about 3.2 m from water (range 0.7 to 5.13 m) and seek refuge buried under grass or sedge tussocks (Swannack et al., 2006). By the 13th to the 19th day, they can travel at least 8 m, and by day 30 they can be up to 35 m away (Greuter, 2004). In contrast to Swannack et al.'s (2006) observations, however, Greuter (2004) noted that juveniles often stayed in the vicinity of water for up to three weeks prior to beginning to disperse. They ultimately are capable of dispersing as far as 100 m from the breeding site, and may use physical corridors, such as gulleys and ravines, to facilitate dispersal (Hillis et al., 1984; Thomas and Allen, 1997). Juveniles prefer areas with moisture and shade; are found on both clay and sandy substrates; and occupy both pine and mixed oak-juniper vegetation communities (Greuter, 2004).

CALLING ACTIVITY AND MATE SELECTION

Males call beginning in January, although actual breeding may not occur until much later. Kennedy (1962) recorded calling from 22 February to 26 June in the Houston area, but Hillis et al. (1984) never observed gravid females before early May in the Bastrop population, despite calls heard from March to April. Calling is initiated in the absence of rain, when nighttime temperatures do not drop below 14°C the previous night and humidity is high (Hillis et al., 1984; Jacobson, 1989); Kennedy (1962), however, stated that calling was initiated by heavy rain and warm temperatures.

Houston Toads are somewhat of an explosive breeder, with most males appearing over a short number of nights (three to five) early in the season. Jacobson (1989) found that males were at a breeding pond for only 24 nights over a 4-month period. Some amplectant males stayed as long as 10 nights (median 5 nights), but nonamplectant males only stayed for a median of 3.5 nights (Jacobson, 1989). About one-third of the males are at breeding ponds only once during a season, with others showing up multiple times depending upon rainfall (Kennedy, 1962; Hillis et al., 1984; Price, 2003).

Males begin calling just before sunset, often initially from terrestrial burrows from 1 to 40 m from the breeding site (Hillis et al., 1984). The burrows are not constructed by the toads, but consist of rodent burrows, old root burrows, or spaces under downed logs. These retreats are frequently located in gulleys leading to the breeding pond. After a night of calling, the toad will retreat back to a burrow until the next night. Movement to breeding sites occurs just about sunset and continues until after midnight. Females arrive several hours after sunset and are quickly amplexed. By 02:00, females cease moving to ponds.

Males mate from one to three times per season, but only a very few mate more than once. In Jacobson's (1989) study, 56% of the males were never observed in amplexus. Males that are successful in amplexus and fertilization may not be different in size than unsuccessful males in some years (Jacobson, 1989), although in other years successful males are larger than nonsuccessful males (Hillis et al., 1984). These authors suggested that male experience or dominance were more important in mating success than female preference for large males. A few females also were observed in amplexus with more than one male during the breeding season.

Males space themselves randomly around breeding pools and call from the shoreline, shallow water, or from on top of debris well above the water's surface. The call of *A. houstonensis* is a long musical trill. The dominant frequency in one study was 2,300 cps with a mean duration of 7.3 sec (range 3.8 to 11.2) and a mean of 32 trills/sec at 19.5°C (Blair, 1956b). Brown (1971b) recorded calls with a dominant frequency of 1,915–2,089 cps with a duration of 9.5–16.6 sec and 18.7–32.3 trills/sec at 14.5–22.1°C.

Amplexus may last 24 hrs, and Jacobson (1989) recorded one pair that stayed together 34 hrs.

Amplexus can take place for up to 6 hrs prior to oviposition (Hillis et al., 1984). Once a female begins oviposition, she does not stop until the entire clutch is laid, even if this means continuing well into daylight. Males do not release females for as long as 0.5 hrs after oviposition has ceased. Males grasp any other toad within their range in their attempts at amplexus, including already amplectant pairs. Males are rarely (only about 8% of the time) successful in attempts to displace other males. Females also have been observed displacing unwanted males. Interspecific amplexus has been observed between Houston Toads and *Ollotis nebulifer* (Brown, 1971b).

Houston Toads do not appear to use the same breeding ponds from one year to the next. The lack of breeding-site fidelity allows for genetic exchange within the population.

BREEDING SITES

The Houston Toad breeds in temporary rain pools and ditches located within its forest habitat. Pools must persist at least 30 days to allow for larval development.

REPRODUCTION

Houston Toads breed in the spring and early summer, with records extending from February to June (Kennedy, 1962; Brown, 1971b; Hillis et al., 1984; Price, 2003). Breeding aggregations often form well before actual oviposition takes place. Despite the long season, oviposition may occur on only a very few nights and may be delayed until late in the season if spring rains do not arrive. Jacobson (1989) found females at a breeding site on only 15 nights in 4 months, and oviposition only occurred on 7 nights. Nearly all females were present on only one or two consecutive nights, and virtually all females visited a pond only once during the breeding season. Nearly all eggs are oviposited during the second and third night of the main chorusing. Oviposition occurs from 19:00 until noon the following day, with most occurring after 03:00.

Clutch sizes in captivity ranged from 513 to 6,199 (Quinn and Mengden, 1984). Kennedy (1962) reported a single clutch size of 728 from a hybrid cross with *O. nebulifer*. Greuter (2004) obtained estimates of a mean of 1,279–2,098 eggs, depending upon which counting technique was used, with 2 exact counts of 2,807 and 4,211. It has been suggested that females will oviposit more than one clutch per season, but it has not been demonstrated. Hatching occurs in two to eight days and is temperature dependent (Hillis et al., 1984; Quinn and Mengden, 1984). The warmer the temperature, the sooner the eggs hatch. At hatching, larvae are 6.1–6.7 mm TL.

LARVAL ECOLOGY

Larvae consume algae and pine pollen and have been observed eating the jelly envelopes of previously hatched eggs. During the day, tadpoles may form large aggregations, but these aggregations tend to disperse at night. Tadpoles tend to remain on the substrate by day, but at night they forage

Egg strings of *Anaxyrus houstonensis*. Photo: Michael Forstner

Tadpole of *Anaxyrus houstonensis* undergoing metamorphosis. Photo: Jim Godwin

on detritus and algae attached to aquatic vegetation along the shoreline or at the pond surface. Under cultivation, the duration of the larval period ranges between 15 and 100 days (Quinn and Mengden, 1984), whereas the larval period lasts about 60 to 65 days in a wild population (Hillis et al., 1984). Larvae reach a maximum size of 20–22 mm. Newly metamorphosed toadlets are 7–12.5 mm SUL.

DIET

Houston Toads feed on many invertebrates, including beetles, flies, lacewings, and small moths (Bragg, 1960b). They appear to exhibit some degree of selectivity by avoiding certain types of scarab beetles. Juveniles are both sit-and-wait and active foragers, particularly feeding on small ants (Thomas and Allen, 1997). Juveniles use shallow burrows as ambush sites, which they create by digging backward into the sand.

PREDATION AND DEFENSE

Tadpoles are eaten by the snakes *Nerodia erythrogaster* and *Thamnophis proximus*; newly metamorphosed toadlets are killed by nonindigenous fire ants (*Solenopsis invicta*) (Freed and Neitman, 1988; Greuter, 2004). McHenry et al. (2010) reported an American Bullfrog consuming an adult Houston Toad.

POPULATION BIOLOGY

In a study at breeding ponds in Bastrop County, Texas, the sex ratio of *Anaxyrus houstonensis* was male-biased (11 males per female), but in traps set away from the ponds, the sex ratio was quite different (2.8 males per female). The probability of capturing females in traps was 3.5 times greater than when capturing them at ponds, whereas the disparity was only 0.28 times for males after adjusting for sampling effort. A male-biased sex ratio at ponds is not surprising, since males likely remain at ponds longer than females, who deposit their eggs and leave soon after oviposition. Still, the overall sex ratio was highly male-biased and changed from one year to the next. This led Swannack and Forstner (2007) to suggest that populations of Houston Toads are male-biased because of differences in the age at first reproduction between males and females. Quinn and Mengden (1984) suggested females reach maturity in 14–15 months and probably first breed the second spring following transformation; males likely breed the first spring following metamorphosis.

As might be expected, larval and early stage metamorph survival is quite low. Greuter (2004) found a larval survivorship of only 4.73%. From metamorphosis to 13 weeks postmetamorphosis, the population estimate declined by 15% further. In arrays around natural and artificial ponds, Greuter (2004) captured 15 to 332 metamorphosed juveniles; numbers captured varied by year and location. At hatching, juveniles are < 0.1 g but grow to 0.8–2.0 g by the 11th week. Growth is rapid, with an average growth rate of 0.019 g/day over the first few months of life (Greuter and Forstner, 2003). Juvenile survivorship is vital to population persistence, but the probability of juvenile survival to the adult stage is only between 0.75 and 1.5% (Swanack et al., 2009). Male annual survivorship is only about 15%.

Adults likely breed over a period of several years. For example, Price (2003) marked 220 males in 1995, of which 62 were recaptured in 1996, 18 in 1997, and 6 in 1998. Of 103 marked females, 14 were recaptured in 1996 and 4 in

Adult *Anaxyrus houstonensis*. Photo: Robert A. Thomas

1997. No marked animals were observed after these dates. Because of differences in the age at maturity, both sexes probably have a maximum longevity of only 4 yrs., although most males breed only 1 to 2 times (Hillis et al., 1984). The breeding population, therefore, turns over rather rapidly making the species vulnerable to stochastic changes in environmental conditions, such as drought during the breeding season.

Populations may be large at some breeding sites. Using an open population Jolly-Seber model, Duarte et al. (2011) estimated there were between 201 and 307 males at the Griffith League Ranch breeding site in 2010. Using a predetermined sex ratio of 5:1, the total adult population was estimated at 241–368.

COMMUNITY ECOLOGY

Two other species of toads, *Anaxyrus woodhousii* and *Ollotis nebulifer*, are frequently found at the same breeding sites as *Anaxyrus houstonensis*. These species do not currently segregate by habitat or calling site, and their breeding seasons overlap to some extent. Occasional hybrids may be produced, but the main isolating mechanisms are their call differences and differences in adult size (Brown, 1971b). Brown (1971b) speculated that extensive habitat alteration allowed these species to come into contact more than they would otherwise, and thus facilitated occasional hybridization (see Status and Conservation). The presence of larval *A. speciosus* has no effect on the growth and survival of larval *A. houstonensis* under laboratory conditions (Licht, 1967).

DISEASES, PARASITES, AND MALFORMATIONS

Houston Toads are parasitized by the trematode *Brachycoelium storeriae* and the nematodes *Cosmocercoides dukae*, *Ozwaldocruzia pipiens*, *Physaloptera ranae*, and *Rhabdias ranae* (Thomas et al., 1984). The virulent fungal pathogen *Batrachochytrium dendrobatidis* also has been found in this species at a 17% rate of infection in 2006 (Gaertner et al., 2010).

SUSCEPTIBILITY TO POTENTIAL STRESSORS

No information is available.

STATUS AND CONSERVATION

Houston Toads are critically endangered by habitat destruction and alteration, particularly the cutting of pine forests, the modification of its habitat during the development of state parks, and the outright destruction of populations in the rapidly urbanizing Houston metropolitan area (Brown, 1975; Bury et al., 1980; Hillis et al., 1984; Shepard and Brown, 2005). Habitat alteration may have facilitated hybridization between Houston Toads and other toad species since *Anaxyrus houstonensis* moved from its natural temporary rain pools to the more permanent constructed stock ponds that are frequented by the other species (Brown, 1971b). Six disjunct metapopulations are currently known from seven

Breeding habitat of *Anaxyrus houstonensis*.
Photo: Gary Nefis

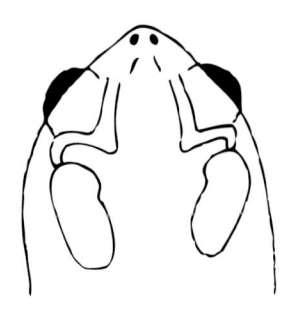

Head pattern of *Anaxyrus houstonensis*. Illustration by Camila Pizano

counties. The species has been extirpated from Houston (Harris and Fort Bend counties), where the species was discovered, and from the majority of its historic range. Populations in Burleson and Liberty counties also have been extirpated; the only remaining sizeable population is in Bastrop County.

The species is protected as "Endangered" under the Endangered Species Act of 1973, and Critical Habitat has been determined to include populations in Bastrop and Burleson counties, Texas. The species is protected by state law. A Recovery Plan was developed by the U.S. Fish and Wildlife Service in 1984, and two separate Recovery Teams were formed, but the recommendations of the Recovery Plan have gone largely ignored. The political maneuvers concerning the designation of Critical Habitat are discussed by Brown and Mesrobian (2005).

As demonstrated by Swanack et al. (2009), the survival of the juvenile cohort of Houston Toads is critical to population persistence. Research on methods to enhance survivorship should be supported. Habitat Suitability Models (HSI) perform well in predicting the occurrence of Houston Toads in high and medium suitability categories (using soil type, canopy cover, and distance to water as variables) and in determining absence in very low and none categories. Buzo (2008) suggested that HSI models could be used to identify habitats currently unoccupied that might serve as locations for repatriation as part of a conservation program.

An extensive captive propagation program was initiated by the Houston Zoo beginning in 1978, but the reintroduction program never had much success (reviewed by Dodd and Seigel, 1991). Monitoring is critical to assessing the status and effectiveness of conservation efforts, and call surveys have been used to monitor wild populations. Jackson et al. (2006) found that a minimum of 11 surveys per site were necessary to be 95% confident that conclusions about the absence of the species were valid. In addition, the variability in not detecting Houston Toads when actually present decreased dramatically from 6 to 16 surveys. Detection probabilities, as expected, varied considerably annually. Simply put, the more call surveys are conducted, the greater reliability of the data.

Even in "protected" areas, breeding populations have declined, despite some annual fluctuation in the number of breeding adults (Price, 2003; Brown and Mesrobian, 2005). In Bastrop State Park, the numbers have varied from 58 to 437 from 1990 to 2002, but most of the lower numbers have been observed during the latter part of the sampling period; the decline in the number of breeding females is particularly troublesome (8–11 in 2000–2002). A Population Viability Assessment (PVA) was developed to assist in Houston Toad recovery (http://www.cbsg.org/cbsg/workshopreports/23/ houston_toad_phva_(1994).pdf); the PVA reviewed threats and provided management guidelines, including the possibility of constructing breeding ponds and providing for movement corridors between existing populations. Key recommendations have been ignored, however. The development of ecopassages due to toad mortality on State Highway 21, for example, were nixed in favor of poorly designed and ineffective fences designed to channel toads to existing culverts.

Much of the problems that have faced the Houston Toad have resulted from attempts to convert its scant remaining habitat into golf courses, housing developments, and to extract its pine trees in a region where biodiversity conservation often is not highly valued by public officials. As Brown and Mesrobian (2005) have reviewed, the management of the Houston Toad has had more to do with political influence than science. In short, federal and state agencies have fallen far short of their legal and ethical mandate in conserving the critically "Endangered" Houston Toad. Because of his efforts to conserve Houston Toads despite many obstacles, this account is dedicated to the memory of Andy Price.

Anaxyrus microscaphus
(Cope, 1867 "1866")
Arizona Toad

ETYMOLOGY
microscaphus: from the Greek *mikros* meaning 'small' and *skaphis* meaning 'shovel.' The name refers to the tarsal spades on the hind feet.

NOMENCLATURE
Stebbins (2003): *Bufo microscaphus*

Synonyms: *Bufo columbiensis* [in part], *Bufo compactilis* [in part], *Bufo lentiginosus microscaphus*, *Bufo woodhousii microscaphus*

Information on this species is sometimes listed under *Bufo compactilis* (e.g., Slevin, 1928; Tanner, 1931; Linsdale, 1940). A.P. Blair (1955) reviewed the nomenclatural history. It is best to check on distribution when using the older literature.

IDENTIFICATION
This species has sometimes been confused with *Anaxyrus woodhousii*. Associating a specific name with the phenotype of a collected toad has been especially difficult in the area where the states of Arizona, Nevada, and Utah intersect (in the region of the Virgin River; see Linsdale, 1940, for example) because of the widespread hybridization between *A. microscaphus* and *A. woodhousii* in this area.

Adults. Adults are squat with a ground color described as pinkish drab, gray, greenish gray, or light brown. Dull irregular spots are on the dorsum, and the back is covered with light-tipped tubercles. There is no light mid-dorsal stripe. Cranial crests are absent or poorly defined. Parotoid glands are elongate and sometimes smooth. Venters are light, and a tarsal fold is present. Breeding males have dark throats. A.P. Blair (1955) provided measurements on a variety of morphological characters.

Males are smaller than females. In Arizona, adults are 34–86 mm SUL (mean 61.4 mm) (Goldberg et al., 1996b), whereas Tanner (1931) measured 5 adults at 43–56 mm SUL. Schwaner et al. (1997) reported males 52–69 mm SUL and females 56–96 mm SUL. Wright and Wright (1949) gave the range as 52–78 mm SUL for males and 54–91 mm SUL for females, although it is possible that specimens of *A. speciosus* were included in the measurements. A.P. Blair (1955) recorded mean male SULs of 52.8–61.9 mm from 11 populations; the largest male was 69.6 mm SUL and the largest female was 83.4 mm SUL. Males begin to show secondary sexual characteristics at 46 mm SUL.

Larvae. The dorsal part of the body is only lightly pigmented and drab or light grayish olive, and the tail musculature is light with dark blotches. Small brassy flecks may or may not be present. Bellies are pale cinnamon pink. Tail fins are clear. TL is < 56 mm.

Eggs. Eggs are brown above and yellow below, and deposited in long strings. The jelly tube is long and narrow. The vitellus is about 1.4 mm in diameter.

DISTRIBUTION
The Arizona Toad is found in extreme southwestern Utah and adjacent Arizona, along the lower Virgin River and its tributaries. A more extensive population is found on the Mogollon Plateau from west-central Arizona across the state to southwest-central New Mexico (Catron, Grant, Sierra, Socorro counties). The species does not enter the lowland deserts. Although there are historical records for the Arizona Toad from the Vegas Valley in Nevada (Linsdale, 1940; Wright and Wright, 1949), the toad has not been collected there for more than five decades

Distribution of *Anaxyrus microscaphus*

Eggs of *Anaxyrus microscaphus*. Photo: Breck Bartholomew

(Bradford et al., 2005a). The species is reported from Fort Mohave, Arizona, southern Nevada (Slevin, 1928), and California, but these isolated populations have either been extirpated or the identifications are in error. Important distributional references include: Arizona (Stebbins, 1962; Sullivan, 1993; Brennan and Holycross, 2006), Nevada (Linsdale, 1940), and New Mexico (Findley, 1964; Degenhardt et al., 1996).

FOSSIL RECORD
No fossils have been identified.

SYSTEMATICS AND GEOGRAPHIC VARIATION
Anaxyrus microscaphus is a member of the North American *Americanus* clade within the Nearctic clade of toads (Goebel, 2005). It is most closely related to *A. californicus* and only more distantly to other *Americanus* clade species (Pauly et al., 2004).

This species hybridizes naturally with *A. woodhousii* (A.P. Blair, 1955; Sullivan, 1986a, 1995; Sullivan and Lamb, 1988; Goldberg et al., 1996b; Lamb et al., 2000; Malmos et al., 2001; Schwaner and Sullivan, 2009). The extent of hybridization may change through time, and contact zones may be dynamic areas of gene exchange. For example, *A. woodhousii* hybridizes with *A. microscaphus* along the Beaver Dam Wash where it intersects the Virgin River in southwestern Utah. Based on collections from 1949 to 1953, hybrid individuals possessed a mostly *A. microscaphus*-like phenotype. By 2001, however, hybrids within the contact zone exhibited mostly *A. woodhousii*-like phenotypes (Schwaner and Sullivan, 2009). Molecular data confirmed a "hybrid swarm" at the contact zone, and demonstrated considerable *A. woodhousii* introgression with populations of *A. microscaphus* that extended a considerable distance up Beaver Dam Wash. The toad population at the confluence of the rivers consists of the parental species, F_1 hybrids, and individuals resulting from reciprocal backcrossing.

Despite differences in the calls of the parental species, the calls may not be different enough to prevent interspecific mating (Malmos et al., 2001). Many hybrid toads have an intermediate phenotype between the parental species. In contrast to morphological intermediacy, the calls of hybrid *A. woodhousii* × *A. microscaphus* are highly repeatable in call duration and pulse rate (65–80 pulses/sec), much more so than their parental species (Malmos et al., 2001). This suggests that hybrids do not call less than the parentals nor are their calls more variable. In Arizona, 25% of a chorus was hybrid *A. woodhousii* × *A. microscaphus* with the remainder all *A. microscaphus* (Malmos et al., 2001). In northwestern Arizona, mtDNA analysis has shown that directional introgression is occurring between ♂ *A. microscaphus* and ♀ *A. woodhousii*.

In laboratory settings, crosses between ♂

A. *punctatus* and ♀ *A. microscaphus* produce some metamorphs, although the reciprocal cross is not viable (A.P. Blair, 1955). Crosses between *A. microscaphus* and *A. americanus* or *A. terrestris* can produce some viable metamorphs (W.F. Blair, 1959, 1963a), but crosses between *A. microscaphus* and *A. woodhousii* are not viable according to some authors but not others (see Moore, 1955). This species produces successful metamorphs in crosses with *A. californicus*, *A. boreas*, and *A. speciosus* (Moore, 1955). The sex of the parents is important in determining whether hybrid crosses will be successful. Some hybridization experiments also appear to rely on very small sample sizes.

ADULT HABITAT

This is a species of the Ponderosa Pine forest and riparian woodland habitats. It is usually found in the vicinity of slowly running shallow water or permanent ponds and impoundments, and may be found along tributaries of larger streams in rocky canyons. The preferred habitat is unaltered riparian areas with cottonwoods and sycamores, and it is found in grasslands, pinyon-juniper, and ponderosa pine communities. Occasionally it is found in near agricultural fields, irrigation ditches, and in meandering broad river bottoms (Degenhardt et al., 1996). A.P. Blair (1955) recorded it at an elevation of 1,768 m in Utah, and Degenhardt et al. (1996) found it at 1,900–2,700 m in New Mexico.

TERRESTRIAL ECOLOGY

Arizona Toads have been observed from February to September (Schwaner and Sullivan, 2005). Arizona Toads remain near flowing water and riparian areas, and most do not move far into adjacent desert scrub or dry forests. They are active at night, feeding along stream shorelines. After metamorphosis, juveniles have been observed to migrate 50–200 m to nearby irrigated fields; adults also returned to these fields following the breeding season (Schwaner et al., 1997). Overwintering habits are unknown but may be similar to *A. californicus*.

CALLING ACTIVITY AND MATE SELECTION

Breeding often occurs shortly after desert rainfall, although Sullivan (1992b) reported calling in the absence of previous rainfall. The initiation of calling may be stimulated by a warming trend early in the season, and males have been observed calling at air temperatures of 8–18°C (Sullivan, 1992b). Choruses form rapidly after warm rains and are often small, with males spread out at about 1 m intervals, calling from along the shoreline; they call with their rear legs in the water and their front legs on shore. In the Sonoran Desert, for example, the chorus size ranged from 2 to 25 males (Sullivan, 1989, 1992b). The call is a loud and shrill trill, similar to that of *A. americanus* in the East. The trill lasts 3.9–10 sec, and has a pulse rate of 43.6–60 pulses/sec. The dominant

Tadpoles of *Anaxyrus microscaphus* scavenging a fish. Photo: Breck Bartholomew

Adult *Anaxyrus microscaphus*. Photo: Dennis Suhre

Habitat of *Anaxyrus microscaphus*. Photo Breck Bartholomew

frequency has a mean of 1.38 to 1.48 kHz (Sullivan, 1992b). The pulse rate and call duration, but not frequency, are positively correlated with temperature (Sullivan, 1992b). However, the dominant frequency is correlated with the male's SUL. Calling occurs almost exclusively at night (dusk to midnight) and rarely during rainfall.

Males will try to amplex any frog within their vicinity. A.P. Blair (1955) recorded male *A. microscaphus* amplexing female *A. punctatus* and *Spea intermontana*. A lack of male discrimination undoubtedly contributes to hybridization, especially in contact zones where habitats have been modified by human activity. When a male *Anaxyrus microscaphus* is amplexed by a conspecific, it will give a release call of ca. 70 pulses/sec. Males also continue to call for some time after females have ceased to be receptive.

BREEDING SITES
Breeding occurs in slow-flowing streams, shallow pools or ponds, and in artificial cattle tanks and reservoirs. According to A.P. Blair (1955), marshes and sloughs along rivers are used sparingly or not at all.

REPRODUCTION
Most breeding occurs from early spring after snowmelt to summer from February to July (Sullivan, 1992b; Degenhardt et al., 1996; Schwaner et al., 1997), although late summer rain can stimulate calling. Amplexed pairs have been observed under ice (*in* Degenhardt et al., 1996). Breeding is explosive, but rainfall is not necessary for breeding to commence (A.P. Blair, 1955). The

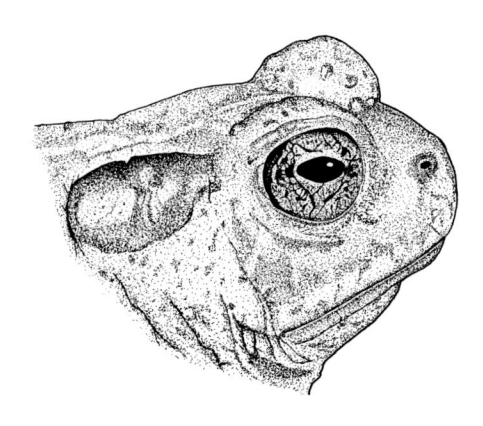

Lateral head view of *Anaxyrus microscaphus*. Illustration by Breck Bartholomew

Head pattern of *Anaxyrus microscaphus*. Illustration by Camila Pizano

eggs are deposited in long strings in shallow water along the substrate, with about 14–20 eggs per 30 mm of string. A.P. Blair (1955) recorded clutch sizes of 3,153–4,279 in 5 clutches from Utah. A great many eggs may be oviposited during the breeding season at a single site. For example, 227 egg strings were observed in a 2 km section of stream over a period of 16 days in southern Utah; the mean clutch size was 4,500 eggs (Schwaner et al., 1997). Hatching occurs in two to six days depending on water temperature.

LARVAL ECOLOGY

After hatching, larvae remain immobile for several days before beginning to feed at 3–4 mm TL (Schwaner et al., 1997). Larval development takes 40–60 days after hatching, depending upon temperature. Newly metamorphosed toadlets are 14.2–16.5 mm SUL (Degenhardt et al., 1996; Schwaner et al., 1997).

DIET

The food habits of *A. microscaphus* are largely unstudied, but likely include a wide range of invertebrates. Tanner (1931) recorded beetles, a stink bug, ants, bees, moth larvae, snails, and a sand cricket from five individuals in Utah. Larvae may consume algae and small unicellular organisms within the substrate.

PREDATION AND DEFENSE

Garter snakes (*Thamnophis* sp.) are important predators of both juvenile and adult Arizona Toads, and killdeers (*Charadrius vociferous*) take larvae and metamorphs. Juveniles can detect the odor of garter snakes in laboratory trials and tend to avoid areas where snake odors are present (Flowers and Graves, 1997). Adults are also killed by raccoons (Schwaner and Sullivan, 2005). Like all toads, the species is cryptically colored and possesses noxious skin secretions in its warts and parotoid glands.

POPULATION BIOLOGY

After metamorphosis growth averages 0.19 mm/day (Schwaner et al., 1997). In southwestern Utah, Schwaner et al. (1997) estimated there were three generations of toads within the population at any one time. Most males were in their second year of life, whereas females were in their third year of life. A relatively short life span in a hostile environment could make local populations vulnerable to habitat alteration and climate change.

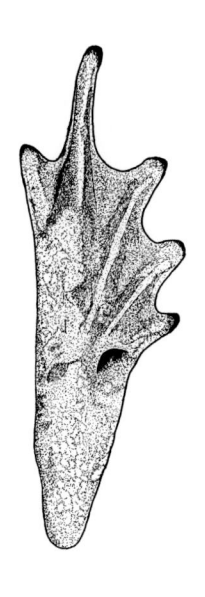

Hind foot of *Anaxyrus microscaphus*. Illustration by Breck Bartholomew

DISEASES, PARASITES, AND MALFORMATIONS

Parasites include the trematodes *Glypthelmins quieta*, *Haematoloechus coloradensis*, and an unidentified Plagiorchiinae; the cestodes *Cylindrotaenia americana*, *Cysticercus* sp., *Distoichometra bufonis*, and *D. kozloffi*; and the nematodes *Aplectana incerta*, *A. itzocanensis*, *Cosmocercoides dukae*, an unidentified Ascarididae sp., *Physaloptera* sp., *Physocephalus* sp., and *Rhabdias americanus* (Goldberg et al., 1996b). The species is parasitized by the mite *Hannemania hegeneri* and by the protozoans *Chilomastix* sp., *Hexamita intestinalis*, *Karatomorpha swazyi*, *Nyctotherus cordiformis*, *Opalina* sp., *Trichomonas* sp., *Tritrichomonas batrachorum*, and *Zelleriella* sp. (Parry and Grundman, 1965).

SUSCEPTIBILITY TO POTENTIAL STRESSORS

No information is available.

STATUS AND CONSERVATION

This species has disappeared from riparian areas affected by urbanization, land alteration, and impoundments (Sullivan, 1993). Dams are particularly a threat to this species, as is water diversion during the breeding season. Throughout much of their range, Arizona Toads are threatened by widespread urbanization, agriculture, and the influence of hybridization with *Anaxyrus woodhousii* in areas of human disturbance (Schwaner and Sullivan, 2005). Unless riparian areas are protected, this species will continue to lose habitat necessary for its survival.

Anaxyrus nelsoni (Stejneger, 1893)
Amargosa Toad

ETYMOLOGY

nelsoni: a patronym honoring Edward William Nelson. Stejneger (1893) named the species for Nelson "for his valuable zoographical work both in the extreme south and the extreme north of our country." Nelson served in a variety of positions with the U.S. Biological Survey, and was a member of the Death Valley Expedition that secured the first specimens of this species.

NOMENCLATURE

Stebbins (2003): *Bufo nelsoni*
  Synonyms: *Bufo boreas nelsoni*, *Bufo halophilus nelsoni*

IDENTIFICATION

Adults. The Amargosa Toad is a small to medium-sized toad. The dorsal coloration is brown to olive buff with a yellow to buff stripe down the midback. The skin is smooth, and the warts (granular glands) are small and weakly developed. Specific size records include a mean of 56.5 mm SUL for males (range 50 to 64 mm) and 70.6 mm SUL for females (range 56 to 83 mm) (Karlstrom, 1962), and a mean adult size of 75.9 mm SUL (Altig and Dodd, 1987).

Larvae. The tadpole is dark to light, with the venter lighter than the dorsum. The tail fins are moderately pigmented, and the tail musculature is unicolored. However, there does not appear to be a good description of this tadpole available. Altig (1970) included it with *Anaxyrus boreas* and *A. exsul* in his descriptive key.

Eggs. The eggs of *A. nelsoni* are deposited in single or (normally) double strings. There are two envelopes surrounding the eggs. The inner envelopes of adjacent eggs are usually in contact

Distribution of *Anaxyrus nelsoni*

with one another forming partitions. The outer envelope is 4.3–5 mm in diameter, with the inner envelope 1.4–2.7 mm in diameter. The ovum is blackish above and somewhat lighter at the vegetal pole and is 1.3–1.8 mm in diameter. There are roughly 60 eggs in single file per 10 cm of string length. Illustrations of eggs are in Savage and Schuierer (1961).

DISTRIBUTION

The Amargosa Toad is known only from the Oasis Valley, Nye County, Nevada. Toads inhabit a number of springs and associated wetlands in the region, and historically have been found along the Amargosa River. This river may dry both spatially and temporally, depending upon environmental conditions. The toad is extinct at least at two historically known locations at the northern limit of its distribution. Important distributional references include Jones (2004) and Simandle (2006).

Previous authors (Stejneger, 1893; Linsdale, 1940; Wright and Wright, 1949) identified toads from locations in Hot Creek Valley, Nye County, Nevada, the Pahranagat Valley, Lincoln County, Nevada, and Owens Valley, Inyo County, California, as belonging to *A. nelsoni*. These reports have not been verified or may refer to populations of other toad species. Altig and Dodd (1987) found no toads in Hot Creek Valley or Pahranagat Valley during brief surveys in 1981.

FOSSIL RECORD

There are no specific fossils recorded for this species. See the *A. boreas* account for a listing of fossils attributed to the *A. boreas* species group.

SYSTEMATICS AND GEOGRAPHIC VARIATION

The Amargosa Toad is a member of the *Boreas* group of toads (with *Anaxyrus boreas*, *A. canorus*, and *A. exsul*). Molecular data place it nested in a clade that includes the SW group of the California Toad (mostly *A. boreas halophilus*). It is genetically closely related to *A. exsul*, a population of *A. b. halophilus* from Inyo County, California, and a southern group of toads currently included within the name *A. canorus* (Goebel et al., 2009). The species has a relictual distribution as a result of genetic differentiation in isolation during Pleistocene climate change.

Egg strings of *Anaxyrus nelsoni*. Photo: Ronn Altig

ADULT HABITAT

The Oasis Valley is a part of the Mojave Desert, and conditions are harsh. The vegetation is Mojave Desert scrub dominated by creosote bush in upland habitats adjacent to the Amargosa River and between the spring sites. Amargosa Toads inhabit riparian portions of the Amargosa River and the springs, ponds, shallow streams, irrigation ditches, and wet meadows associated with the various known populations. The river dries as the season progresses, but some deeper pools hold water year-round. The river channel is bordered by sedges (Cyperaceae) and tule (*Schoenoplectus acutus*), and sections may be heavily vegetated by mesquite, cottonwood, tamarisk, bamboo, cattail, and young willow trees. Jones (2004) provides a description of many sites inhabited by this species.

TERRESTRIAL ECOLOGY

Anaxyrus nelsoni is active from the early spring breeding season into at least November. After breeding, adults stay in proximity to the Amargosa River and adjacent springs and pools. As nighttime temperatures rise, adults become increasingly nocturnal, and by mid-April, diurnal activity ceases. Most toads remain close to the watercourses, although they do not necessarily use riparian habitats for dispersal and take refuge in burrows or under rocks and debris. Overwintering apparently occurs in the muddy bottom of springs (Altig and Dodd, 1987) and in rodent burrows.

The home range of Amargosa Toads ranges between 800 and 16,000 m^2 (mean 6,000 m^2), with no differences between males and females or

Tadpoles of *Anaxyrus nelsoni*. Photo: C.K. Dodd Jr.

by size (Jones, 2004). Toads tend to move after rainfall, either as short-term feeding bouts or during movements to and from water sources. Most activity occurs between May and September, but even then movements are generally 15 m or less (Jones, 2004). They are found especially closer to water (usually within 1 m) from February to April during the breeding season and rarely more than ca. 25 m from water during the rest of the year. The farthest an Amargosa Toad has been found to move was 400 m (Jones, 2004). Females travel farther from water than males during the non-breeding season. Migration between populations occurs infrequently, and then is a result of juvenile dispersal rather than adults moving between breeding populations. Jones (2004) provided detailed maps of the movements of 20 Amargosa Toads at two locations.

As with *A. boreas*, Amargosa Toads likely are photopositive and sensitive to light in the blue spectrum ("blue-mode response"). They probably have true color vision.

CALLING ACTIVITY AND MATE SELECTION
There are no data available for this species. Presumably mate selection is similar to that among other members of the *A. boreas* group of toads.

BREEDING SITES
Breeding occurs in shallow waters associated with the ephemeral Amargosa River and in a series of springs, spring outflows, and associated wetlands in the Oasis Valley. Jones (2004) provided detailed individual site descriptions. Most of the habitat is open, with little canopy cover. Toads prefer areas without much emergent vegetation or algae, and tend to avoid areas with concentrations of American Bullfrogs and crayfish (Jones, 2004).

REPRODUCTION
Breeding occurs over a period from February to April and is diurnal, as night temperatures may approach freezing during the early part of the breeding season. Most breeding occurs over a period of about four weeks from late February to late March, with the majority of the remainder through late April. Jones (2004) reported clutches found as late as mid-July after a rainstorm; these clutches did not survive.

Clutch size ranges between 4,200 and 6,200 eggs (mean 4,987) (Altig and Dodd, 1987). Eggs are oviposited in shallow water pools (to 23 cm; mean 6.5 cm), usually with no or only minimum flow. For the most part, eggs are deposited without selection for canopy cover (usually 15% or less within the habitat) or oviposition substrate, but females prefer open areas rather than sites with emergent vegetation (Jones, 2004). Water temperatures within the shallow river vary, but are usually ca. 15–19°C, with air temperatures around 20°C.

LARVAL ECOLOGY
Larvae have been observed as early as 17 February and as late as the middle of July (Jones, 2004). The earliest date was at a warm spring

location, and the tadpoles observed late in the season died of desiccation. Placing eggs in shallow ephemeral pools probably exposes them and larvae to the potential for desiccation as the season progresses. No other information is available on larval ecology.

DIET
As with other *A. boreas*-group toads, the diet is probably highly insectivorous and composed of ants, beetles, and invertebrate larvae. There are no published studies of diet.

PREDATION AND DEFENSE
Amargosa Toads are known to have been eaten by introduced American Bullfrogs (*Lithobates catesbeianus*) (Jones et al., 2003). The secretions of the toad were not successful in repelling attacks, and the bullfrogs suffered no apparent adverse reactions to the secretions. Amargosa Toads probably are eaten by snakes, birds, and mammals.

POPULATION BIOLOGY
The Amargosa Toad is one of the least-known North American amphibians in terms of its

Adult *Anaxyrus nelsoni*.
Photo: C. K. Dodd Jr.

Habitat of *Anaxyrus nelsoni*.
Photo: Ronn Altig

basic life history. Genetic analysis suggests that several of the springs where this species is found form discrete populations, showing significant genetic differentiation (Simandle, 2006). These include the Crystal Springs, River site, Brothel, and Parker locations. A fifth population includes Amargosa Toads at Goss Springs, the Oleo Road site (Mullen), and the Torrance site. Some movement occurs among toads inhabiting these latter three sites, and between them and Crystal Springs. Simandle (2006) found evidence of recent genetic bottlenecks at the River site through the town of Beatty and at Crystal Springs. Effective population sizes vary between 600 toads at the Parker site to 1,300 toads at the River site.

DISEASES, PARASITES, AND MALFORMATIONS
No helminths were observed in two adults from north of Beatty (Altig and Dodd, 1987).

SUSCEPTIBILITY TO POTENTIAL STRESSORS
No data are available on the effects of toxic conditions or substances on this species.

STATUS AND CONSERVATION
The precarious status of this species has been known for a long time (Bury et al., 1980). Historically, there are reports of a great many Amargosa Toads breeding along the Amargosa River near Springdale, Nevada (Savage, 1959; Savage and Schuierer, 1961). These authors stated there were "literally thousands of this extremely rare toad feeding and moving about the river bottom among the sedges and tules." It appears this species suffered a serious population decline between Savage's (1959) observations and Altig and Dodd's (1987) surveys in 1981. Huge breeding numbers no longer occur, and the entire population is estimated at 2,000 individuals. Populations at Indian Springs (impacted by a municipal water well and feral burros) and Springdale are no longer extant (Simandle, 2006). However, Amargosa Toads were found at Indian Springs at least as late as 1981 (Altig and Dodd, 1987). Amargosa Toads have also undergone serious declines at extant populations such as Crystal Springs.

Threats to the Amargosa Toad revolve primarily around its need for water. Virtually all known sites have been modified by human activity for livestock grazing and irrigation, and the Amargosa River is subject to water removal (ground water pumping) and flash flooding (flood control projects in Beatty). Trampling of larvae by cattle is known to occur, and road-widening was perceived as a threat in the early 1980s (Altig and Dodd, 1987). It is not known if roadwork subsequently impacted the toad. Improper management, such as fencing that prevents animals from light grazing and thus allow choking vegetation to invade watercourses and spring sites, also may have deleterious effects on the Amargosa Toad. Some of these fences have now been removed as have feral burros that impacted the sites. Other potential threats include introduced American Bullfrogs (*Lithobates catesbeianus*), crayfish (*Procambarus clarkii*), and habitat damage by off-road vehicles.

Some of the habitat occupied by the Amargosa Toad is on land managed by the USDI Bureau of Land Management and The Nature Conservancy.

Anaxyrus punctatus
Baird and Girard, 1852
Red-spotted Toad

ETYMOLOGY
punctatus: from the Latin *punctatus* meaning 'spotted.'

NOMENCLATURE
Conant and Collins (1998) and Stebbins (2003): *Bufo punctatus*
   Synonyms: *Bufo beldingi*

IDENTIFICATION
Adults. Adults are pale gray to reddish brown with numerous dull gray or black dorsal warts topped by red spots. A few individuals may be slightly greenish. The reddish or orange color of the tip of the warts on the back is more pronounced in juveniles than in adults. The base of the warts is dusky. The head has a generally flattened appearance. Cranial crests are lacking or only barely developed. The parotoid glands are subtriangular and small, not larger than the eye. Venters are white to cream colored. Males have a dark subgular vocal sac. Juveniles are colored like

the adults but have ventral spotting and yellow on the undersides of the feet.

Males are slightly smaller than females. In southern California, Johnson et al. (1948) recorded males from 40 to 51.9 mm SUL and females to 60.3 mm SUL. In Arizona, males from a single population had a mean SUL of 52 mm whereas females were 59.2 mm (Sullivan and Fernandez, 1999). In another Arizona study, Sullivan (1984) recorded males 47–55 mm SUL (mean 52.7 mm) and a female mean of 57.5 mm SUL. Tanner (1931) measured 5 adults at 40–61 mm SUL. Conant and Collins (1998) give the maximum size as 76.2 mm SUL.

Larvae. Tadpoles of the Red-spotted Toad are small and black with a bronze-flecked venter; larger tadpoles may be faintly mottled. Viewed dorsally, they have an oval body. Eyes are placed well up on the head. The tail musculature is dark and does not have light dorsal saddles, but the tail fin is translucent with evenly spaced spots. Dorsal and ventral tail fins and the tail fin musculature are of equal height. Eyes are dorsally located. Throats are unpigmented. Venters are dark with much gold flecking. Larval are < 40 mm TL and usually 30–32 mm TL. Luepschen (1981) observed albino larvae in Colorado.

Eggs. The eggs are heavily pigmented dorsally, with the dark pigment extending much farther toward the vegetal pole than in many other species. The vegetal pole is white. There are two gelatinous membranes surrounding the vitellus according to Ferguson and Lowe (1969), but only one according to Livezey and Wright (1947) and Hammerson (1999). The vitellus is 1–1.3 mm in diameter, and the envelope 3.2–3.6 mm in diameter (Livezey and Wright, 1947).

DISTRIBUTION

The Red-spotted Toad occurs from southern Nevada (Virgin and Dead mountains and along the Virgin River) eastward to south-central and southwestern Kansas, southwestern and southeastern Colorado, western Oklahoma, and the more arid portions of Texas. Populations occur in the Mojave Desert of southern California well southward into Mexico. An isolated population is in the Arbuckle uplift of Oklahoma. Black and Dellinger (1938) reported this species from Arkansas, but the record is erroneous (Trauth et al., 2004). Important distributional references include: Arizona (Brennan and Holycross, 2006), California (Storer, 1925; Lemm, 2006), Colorado (Ellis and Henderson, 1915; Hammerson,

Distribution of *Anaxyrus punctatus*

1999), Kansas (Smith, 1934; Collins et al., 2010), Mexico (Grismer, 2002; López et al., 2009; Lemos-Espinal and Smith, 2007b), Nevada (Linsdale, 1940; Bradford et al., 2003), New Mexico (Van Denburgh, 1924; Degenhardt et al., 1996), Oklahoma (Sievert and Sievert, 2006), and Texas (Dixon, 2000).

FOSSIL RECORD

Fossils have been reported from Pleistocene deposits in Arizona, California, New Mexico, and Nevada (Holman, 2003). Tihen (1962b) and Holman (2003) provide osteological characters used to identify this species in fossil deposits.

SYSTEMATICS AND GEOGRAPHIC VARIATION

Anaxyrus punctatus is a member of the Nearctic clade of North American bufonids (Pauly et al., 2004). It appears to be only distantly related to other North American toads, particularly with *A. debilis*, *A. retiformis*, and *A. kelloggi*. Ferguson and Lowe (1969) suggested these constituted the *Punctatus* group of toads based on a suite of morphological and call characters.

Natural hybridization between *A. punctatus* and *A. woodhousii* occurs in Colorado (McCoy et al., 1967; Malmos et al., 1995; Hammerson, 1999) and with *A. boreas* in California (Feder, 1979). In laboratory settings, crosses between ♂ *A. punctatus* and ♀ *A. microscaphus* produce some metamorphs, although the reciprocal cross is not viable (A.P. Blair, 1955, but see Moore, 1955). Hybridization with ♀ *A. terrestris* also produces metamorphs, but hybridization between ♀ *A. punctatus* and ♂ *A. americanus*, *A. cognatus*, *A. terrestris*, *A. speciosus*, *A. hemiophrys*, *A. debilis*, or *Ollotis caniliferus* is not successful (W.F. Blair, 1961a). Under laboratory conditions, this species will hybridize and produce larvae that complete metamorphosis with *Anaxyrus punctatus*, *A. debilis*, and *A. kelloggi* (Ferguson and Lowe, 1969), but the success of the crosses may depend on which species is male and which is female (Blair, 1961a). In any case, hybridization experiments frequently base their conclusions on often very small sample sizes, resulting in contradictory results.

ADULT HABITAT

Adults occupy a wide variety of mostly dry rocky habitats throughout their range. For example, Bragg and Smith (1943) recorded it from oak-hickory forest, oak-hickory savanna, short-grass plains, and in the ecotone between the short-grass and mixed-grass prairie. East of Phoenix, I have observed them commonly along desert streams in very rocky canyons. McCoy et al. (1967) recorded the habitat as rocky desert canyons and dry mesas containing plunge pools and seepage areas along canyon bottoms. They occur from the depths of Death Valley (Stejneger, 1893) and the Grand Canyon (Miller et al., 1982; personal observation) to > 2,150 m in elevation in New Mexico. In California they occur from near sea level to 1,980 m (Lemm, 2006).

Information on habitat preferences is available from a variety of locations throughout its range. In Big Bend, they are a species of the Rio Grande floodplain, desert flats, and desert foothills but not prairies and the high Rocky Mountains (Minton, 1958). Dayton and Fitzgerald (2006) developed a Habitat Suitability Model using soil characteristics, slope, elevation, and proximity to drainage channels that correctly identified 59% of the locations of breeding sites for this species within Big Bend National Park. As such, *A. punctatus*'s preferred habitat included lower elevations, < 50 m from a drainage channel, flat to moderate slopes, and cobble or gravel soils. Compared with some other park anurans, the relatively low accuracy of categorization was due to the wide range of habitats occupied by the toad. *Anaxyrus punctatus* is associated with creosote and mixed scrub vegetation within the park; vegetation communities include creosote, lechuguilla, and to a lesser extent mesquite thicket, mixed scrub, and sotol (Dayton et al., 2004). In Colorado, it is present in juniper and shrub habitats (Hammerson, 1999). In Kansas, they occur in rocky areas of dry prairies and canyons (Collins et al., 2010). In southern Nevada, habitat patch occupancy is associated with the size of the habitat patch, lower elevations, latitude (to the south), rocky terrain, periodic scouring by water, and the presence of ephemeral water (Bradford et al., 2003).

TERRESTRIAL ECOLOGY

Adults are frequently associated with rocky canyons along streams in well-drained rocky soils. In isolated desert wetlands, they congregate around water sources, and most toads do not move far from them, preferring instead to move up- or

Eggs of *Anaxyrus punctatus*. Notice how the eggs accumulate silt thus becoming difficult to locate. Photo: Breck Bartholomew

Tadpole of *Anaxyrus punctatus*. Photo: Chris Brown

downstream. For example, McClanahan et al. (1994) tracked toads to rock crevices up to 100 m from a water source, whereas Weintraub (1974) found a mean distance between male recaptures of 17.4 m and between female recaptures of 9.3 m. Most toads remain within a segment of stream rather than moving between segments. Toads move back and forth from burrows located on flats above a stream, although activity does not occur every night.

Occasional movements may be substantial, however. Movements of up to 366 m over a 5-day period were recorded among desert wetlands in a spatially limited environment in Death Valley; mean movements were < 91 m, however (Turner, 1959c). Parker (1973) observed a maximum distance traveled of 91 m in a small Arizona canyon, but Tevis (1966) recorded a maximum movement of 823 m by a female in a California canyon. Some males migrated from a drying breeding site 610 m upstream to more favorable pools, whereas others sought shelter underground or in rocky crevices; some females moved great distances from breeding sites, whereas others did not move at all. Males had a linear home range of 183 m and females 457 m during the breeding season (Tevis, 1966). One toad moved 98 m in a single day. In California, Weintraub (1974) recorded a movement of 457 m by one toad. Thus it appears that there is a great deal of individual variation in movement patterns, although females tend to move greater distances than males. Toads displaced up to 900 m upstream or downstream from their home stream segment are able to return (Weintraub, 1974), indicating familiarity with their environment. However, toads displaced into the desert flats home with much less success.

Toads are mostly active at night (30–60 min after sunset), taking refuge during the day under rocks, surface debris, rocky crevices, rodent burrows, or in burrows that they dig themselves. Such burrows are horizontal and shallow, only about

0.5 m in depth (Weintraub, 1974). Toads may return repeatedly to the same burrow. Activity occurs throughout the warmer part of the year, and toads may appear in large numbers following warm summer rains. In California and Colorado, for example, activity occurs from April/May to October (Weintraub, 1974; Hammerson, 1999), and Lazaroff et al. (2006) reported activity in the Sonoran Desert as late as November. During the dry season and winter, they also take refuge under stones, vegetation, boards, trash or woody debris, or they may burrow into the ground. Multiple toads may occupy the same site (Weintraub, 1974; McClanahan et al., 1994). Emergence from winter retreats occurs at evening temperatures of ca. 18°C, although breeding does not begin until the temperatures are several degrees higher (Tevis, 1966).

As befitting a desert anuran, *A. punctatus* is able to absorb water rapidly from a moist substrate and thus rehydrate quickly in response to periodic rainfall (Fair, 1970). It does this by having a highly vascular patch of skin ("seat patch") ventrally that facilitates water resorbtion and storage in the bladder, even when the toad is just sitting on a moist substrate (Brekke et al., 1991). Red-spotted Toads can store up to 40% of their body weight as dilute urine (McClanahan et al., 1994). Despite disappearing during long periods during hot dry weather, they may become active immediately after late summer flash floods in order to rehydrate.

Red-spotted Toads cannot tolerate temperatures >35°C and can lose up to a further 40% of their body water even when the bladder is empty (McClanahan et al., 1994). Their lowest temperature for activity is 14–18°C (Zweifel, 1968b). Most activity occurs at night when body temperatures are less than they are even in some secretive hiding places (Moore and Moore, 1980; Hammerson, 1999). However, the toad can remain somewhat cool by hiding in crevices during unfavorable conditions and by using evaporative cooling during periods of travel above ground. In the winter, habitat selection allows toads to remain comfortable. For example, McClanahan et al. (1994) found the body temperature of a toad in a crevice at 25°C, whereas outside air temperatures were 4–12°C.

Red-spotted Toads are photopositive in their phototactic response, suggesting they use both sunlight and moonlight when feeding and going about their daily activities (Jaeger and Hailman, 1973). They are sensitive to light in the blue spectrum ("blue-mode response") (Hailman and Jaeger, 1974). Red-spotted Toads likely have true color vision.

CALLING ACTIVITY AND MATE SELECTION

Breeding is opportunistic and associated with rainfall and will continue several weeks after heavy precipitation. However, activity may occur for several weeks prior to breeding. For example,

Adult *Anaxyrus punctatus*.
Photo: Aubrey Heupel

toads have been observed in Death Valley in February, even though calling did not commence until late March (Turner, 1959c). Temperature likely plays an important role, with toads normally breeding only at temperatures >21–24°C. Gehlbach (1965), however, recorded chorusing at a water temperature of only 12°C. Calling begins at dusk, that is, after 18:00 hrs, and continues to ca. 02:00 hrs or later. Calling does not occur at temperatures >28°C.

Males call from at or near the edge of the water, with amplexus occurring in water or occasionally on land. Some males also call diurnally from rock crevices or holes along the bank near the breeding pool (Brown and Pierce, 1965; Hammerson, 1999). Other males may call from shallow water, periodically submerging between calling bouts (Sullivan, 1984). Ferguson and Lowe (1969) recorded them calling from exposed boulders and rocks at the water's edge. Choruses form rapidly and are often small (<10). In the Sonoran Desert, the mean chorus size was only three toads but the chorus lasted a mean of 20 days (Sullivan, 1984, 1985a, 1989). Sullivan and Fernandez (1999) recorded males at 2 Arizona breeding sites for only 20 nights over a 6 yr period; from 2–65 toads were present. At Cow Creek in Death Valley, Turner (1959c) estimated the breeding population as consisting of only 37–40 adults. No breeding may take place during years of drought.

Males tend to occupy the same 1–2-m long calling sites on successive nights and are separated from other males by 1.5–4 m (Sullivan, 1984). The advertisement call is a series of short musical or melodious trills with a dominant frequency of 2,262–2,700 cps, a duration of 2–9.7 sec, and 46–59 trills/sec (Blair, 1956b; Ferguson and Lowe, 1969; Sullivan et al., 2000). Trills are interrupted by periods of silence, with some males alternating trills while others may call simultaneously. There are no correlations between dominant frequency, pulse rate, or call duration with the size of the male, nor is there a relationship between the size of the male and mating success, that is, larger males are not more successful at mating than are smaller males (Sullivan, 1984). The dominant frequency (+), pulse rate (+), call rate (+), and call duration (−) are all significantly correlated with temperature.

A female approaches a calling male and initiates amplexus by touching or jumping on him. Males

Habitat of *Anaxyrus punctatus*. Photo: Chris Brown

continue to vocalize until touched, but amplexus is rapid afterward. Amplexus is axillary. Aggression is not uncommon, with calling males approaching one another and engaging in wrestling bouts lasting as long as six minutes before returning to previous calling sites. Males frequently try and amplex other males, who respond by giving a loud warning trill and struggling to fight off the amplexing male. There is no evidence that males employ a satellite strategy to intercept incoming females.

BREEDING SITES

Breeding occurs in a wide variety of both temporary and permanent rain-formed ponds and pools. These include permanent springs, bedrock pools, pools in canyons and on low-gradient flood plains, along intermittent desert streams in canyon bottoms, plunge pools in sandstone, and in tanks along the edges of steeply sloped streambeds (Turner and Wauer, 1963; Douglas, 1966; McCoy et al., 1967; Dayton and Fitzgerald, 2006). The pools can be very small and shallow, but contain some algae or other organic matter. For example, Bragg and Smith (1943) reported eggs within a rocky pool measuring only 30 cm × 30 cm and 51 mm in depth. Parker (1973) observed eggs in water 25–51 mm in depth. McCoy et al. (1967) also noted occasional breeding along desert rivers, and Tevis (1966) recorded breeding in shallow, slow-flowing, sandy-bottomed canyon streams. In streams, egg deposition occurs along the stream margin, coves, or in quiet backwaters.

REPRODUCTION

Breeding begins in spring and continues throughout the summer as weather permits. It has been recorded as early as mid-March in Arizona (Parker, 1973), April to June in California (Johnson et al., 1948; Tevis, 1966; Weintraub, 1974), early August in Colorado (Douglas, 1966), and from early April to September in Kansas (Collins et al., 2010). In the Sonoran Desert, most breeding is tied to the summer monsoon season. Oviposition occurs in shallow quiet water 25–75 mm in depth. Eggs are deposited singly or in short strings that may give the appearance of a loose film or cluster along the substrate bottom. Tevis (1966) recorded from 30 to 5,000 (mean 1,500) eggs per mass. These eggs can stand a wide range of temperatures, from 14 to 34°C (Ballinger and McKinney, 1966; Zweifel, 1968b), but eggs will not hatch below 21.9°C. Hatching occurs in two to five days, depending upon temperature.

LARVAL ECOLOGY

Larvae develop rapidly in small pools that are often exposed to direct sunlight. The temperature of the pools can range from 12 to 30°C (Johnson et al., 1948; Tevis, 1966; Parker, 1973). The duration of the larval period is temperature dependent, ranging from ca. 60 days in larvae hatching in early spring (Tevis, 1966) to 35 days for those hatching in summer (Parker, 1973). During early development, tadpoles form small aggregations in shallow water, which may help them absorb solar radiation and speed up development. As the developmental period progresses, they move to the deeper water that is most likely to remain the longest as temporary pools desiccate. Toward the end of development, aggregations break up as individual larvae disperse across the substrate to maximize feeding efforts (Tevis, 1966). As metamorphosis approaches, aggregations again form at the edges of breeding pools, where larvae pile up facing the shoreline and engage in constant swimming behavior. Presumably such behavior maximizes temperature and facilitates the biochemical processes resulting in the profound morphological changes associated with transformation. Metamorphosis occurs from April to May in Arizona and New Mexico and to as late as October in Colorado (Douglas, 1966; Degenhardt et al., 1996). Larvae metamorphose at 13–18 mm SUL, but recent metamorphs lacking tails are only ca. 7 mm SUL.

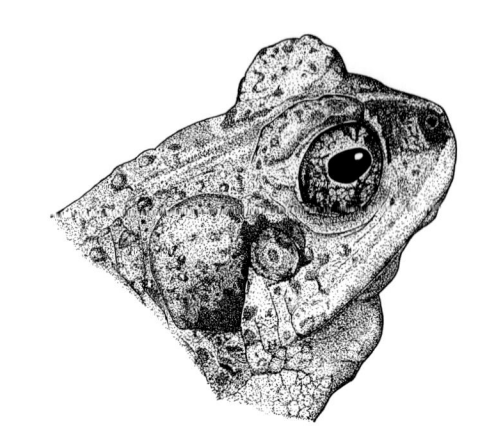

Lateral head view of *Anaxyrus punctatus*. Illustration by Breck Bartholomew

DIET

The diet includes a variety of invertebrates, including beetles, true bugs, bees, and ants (Tanner, 1931; Smith, 1934; Little and Keller, 1937). Metamorphs are especially fond of ants.

PREDATION AND DEFENSE

When disturbed, adults may jump into the water and dive to the bottom, where they remain immobile. This untoad-like behavior is more reminiscent of ranid escape tactics. Hammerson (1999) provides a photograph of an *A. punctatus* in a head-down defensive posture. This allows the toad to direct its noxious or poisonous secretions directly toward an approaching threat. At the same time, it emits a series of body vibrations similar to its release call. Predation reports are rare, with a juvenile *Ollotis alvaria* eating a young *Anaxyrus punctatus* (Lazaroff et al., 2006), a

Sonoran Mud Turtle *Kinosternon sonoriensis* eating an adult (Ligon and Stone, 2003), and the Black-necked Garter Snake *Thamnophis cyrtopsis* eating tadpoles (Stone and Ligon, 2011). Hammerson (1999) noted a predation event that was consistent with a raccoon attack.

POPULATION BIOLOGY

As noted above, breeding populations appear to be small. At Cow Creek in Death Valley, Turner (1959c) estimated there were 8 toads per 0.45 ha. Turner (1959c) also noted that *Anaxyrus punctatus* appeared to be distributed as "isolated colonies" within its range, certainly around desert wetlands, and Parker (1973) considered the population he studied in an Arizona canyon as isolated and small. Bradford et al. (2003) also mapped out the distribution in southern Nevada, which suggested a series of isolated populations rather than a traditional metapopulation structure across the landscape. The median distance of the extent of water was only 200 m in an area of 72 m², and the nearest-neighbor median distance was 1.8 km (range 0.4–22 km) straight-line distance and 6.8 km (range 0.5 to 64.9 km) by stream channel. Populations may have genetic interchange within a mountain range, but populations between mountain ranges are likely isolated from one another. Whether populations are isolated in more contiguous urban habitats is unknown, but Parker (1973) noted that the species was abundant around artificial irrigation ditches in the Phoenix area. If the general rule is for natural populations to be isolated within a mountain range, this species could be particularly vulnerable to habitat alteration and climate change, especially changes in precipitation patterns.

Head pattern of *Anaxyrus punctatus*. Illustration by Camila Pizano

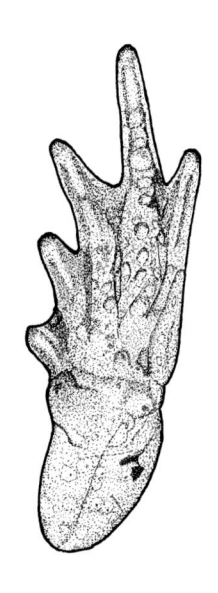

Hind foot of *Anaxyrus punctatus*. Illustration by Breck Bartholomew

Sexual maturity is reached at 2 yrs (third yr of life), with most growth occurring during the second season of life. Sullivan and Fernandez (1999) reported a mean of 2–2.3 lines of arrested growth in adults, suggesting that they have a potential longevity of at least 5 yrs.

COMMUNITY ECOLOGY

Decreases in body mass occur in larval *A. punctatus* when reared with *Scaphiopus couchii* larvae (Dayton and Fitzgerald, 2001). Data suggest that *S. couchii* are able to out-compete *Anaxyrus punctatus* larvae and may exclude them from shallow water desert pools in some areas, such as Big Bend National Park.

DISEASES, PARASITES, AND MALFORMATIONS

Parasites include the cestode *Distoichometra bufonis* and the nematodes *Aplectana itzocanensis* and *Oswaldocruzia pipiens* (Goldberg and Bursey, 1991a). It is parasitized by the mite *Hannemania hegeneri* (Parry and Grundman, 1965).

SUSCEPTIBILITY TO POTENTIAL STRESSORS

No information is available.

STATUS AND CONSERVATION

The Red-spotted Toad is a common species throughout its range, but is protected by state law in Kansas because of its limited distribution. There does not appear to be any widespread decline (Hammerson, 1999; Bradford et al., 2005a), although populations likely have been lost to urbanization, riparian modification, and dam building. Turner (1959c) noted highway mortality in Death Valley. In steep canyons, eggs and tadpoles may be washed away during sudden flashfloods, so breeding may not result in local recruitment. Local populations also may be impacted by introduced fish and crayfish, which eat eggs and larvae.

Anaxyrus quercicus (Holbrook, 1840)
Oak Toad

ETYMOLOGY
quercicus: from the Latin *quercus* meaning 'oak.' The name refers to the species' association with oak woodlands.

NOMENCLATURE
Conant and Collins (1998): *Bufo quercicus*
 Synonyms: *Bufo dialophus, Chilophryne dialopha, Incilius dialophus*

IDENTIFICATION
Adults. This is the smallest North American toad. The ground color is gray, brown, or black, with two to five dark paired, unconnected blotches on either side of the mid-dorsal stripe. The mid-dorsal stripe is white, cream, light red, or orange. Fine red to orange tubercles may be present dorsally, and the granular warts are evenly distributed. Cranial crests are low, and the parotoid glands are large for such a small toad. The bellies are white with mottled dark pigment ventrolaterally; a few individuals may be more heavily pigmented. Lighter coloration may extend in a lateral band from the parotoid to the groin, and the groin may be tinged with yellow. Males have dark oval vocal sacs.

Females are slightly larger than males. In north Florida, males averaged 27.8 mm SUL (range 23 to 32 mm) whereas females averaged 30.3 mm SUL (range 23 to 35 mm) (Dodd, 1994). In Miami, males averaged 26.5 mm SUL whereas females were 27.9 mm SUL; on Big Pine Key, both males and females averaged 24.2 mm SUL (Duellman and Schwartz, 1958). Wright (1932) recorded males from 19 to 30 mm SUL and females from 20.5 to 32 mm SUL in the Okefenokee. In eastern North Carolina, males were 2.2–2.7 mm smaller than females (Wilbur et al., 1978); males averaged 28 mm SUL whereas females averaged 30.2 mm SUL.

Larvae. The tadpole is dark and blotched in coloration. Dark pigmentation covers the upper portion of the tail musculature, but the ventral portion is unpigmented. In preservative, six to seven dark saddles on the upper tail musculature may be observed (Wright and Wright, 1949), but this pattern may not be as evident in live tadpoles (Volpe and Dobie, 1959). Dorsal tail fins are lightly spotted, but become increasingly pigmented as development proceeds. Larvae reach a maximum length of 19–28 mm TL. The mouthparts are described in Volpe and Dobie (1959).

Eggs. Oak Toads oviposit small eggs that are black or brown dorsally and white to yellow ventrally. The vitellus ranges from 0.8 to 1.13 mm in diameter (mean 1.09 mm), and there is a single jelly envelope measuring 1.2 to 1.4 mm (mean 1.25 mm) in diameter surrounding it (Wright, 1932; Volpe and Dobie, 1959). A thin jelly membrane of loose consistency unites adjacent eggs into a strand. This jelly membrane is drawn and pinched off between egg strands.

DISTRIBUTION
The Oak Toad occurs on the Atlantic and Gulf Coast plains from southeastern Virginia to the Florida Parishes of eastern Louisiana. The species' range extends north in central Alabama in the Coosa Valley and includes the entire Florida peninsula through the Florida Keys (Boca Chica, Big Pine Key; the status of these populations is unknown), and Marco Island. Oak Toads also are reported from Sapelo Island, Georgia (Martof, 1963), but reports of it on Cumberland and St. Catherines islands (Laerm et al., 2000) are in error (Shoop and Ruckdeschel, 2003). Important distributional references include: Alabama (Mount, 1975; Redmond and Mount, 1975), Florida (Duellman and Schwartz, 1958; Lazell, 1989), Georgia (Williamson and Moulis, 1994;

Distribution of *Anaxyrus quercicus*

Jensen et al., 2008), North Carolina (Dorcas et al., 2007), and Virginia (Tobey, 1985; Mitchell and Reay, 1999).

FOSSIL RECORD

Fossil *Anaxyrus quercicus* are reported from a Pleistocene Rancholabrean site in north-central Florida (Holman, 2003). Tihen (1962b) and Holman (2003) discuss its unique osteological characters. *Anaxyrus pliocompactilis* and *A. tiheni*, both from the Miocene, may be its closest fossil relatives.

SYSTEMATICS AND GEOGRAPHIC VARIATION

Anaxyrus quercicus is a member of the Nearctic clade of North American bufonids (Pauly et al., 2004). It is probably more closely related to *A. punctatus* than to other southeastern toads.

ADULT HABITAT

The Oak Toad is a species of the dry uplands of the Coastal Plain. It inhabits a variety of communities such as xeric sandhills, mesic mixed pine-deciduous hardwood forest, pine flatwoods, rockland pine woods, Florida scrub, maritime forest, pocosins, and rarely in xeric hammock (e.g., Wright, 1932; Carr, 1940a; Duellman and Schwartz, 1958; Dodd, 1992; Enge and Wood, 1998, 2001; Branch and Hokit, 2000; Meshaka et al., 2000; Means and Franz, 2005; Dorcas et al., 2007). Soils should be sandy-loam and friable, and there should be extensive areas of low vegetation, grasses, debris, and downed logs, which serve as cover and provide moist microenvironments. The species prefers habitats that are relatively open and is rare in dense forested habitats. Habitats should contain shallow temporary wetlands for breeding that are free of fish, although fish may rarely be present in ponds (e.g., Eason and Fauth, 2001).

TERRESTRIAL ECOLOGY

Adults are active mostly during the humid, warm summer months of May to July, with activity decreasing significantly during the dry late summer months. Seasonal activity increases toward the subtropical south but largely ceases during temperate cool and cold months in the north. Dodd (1994) observed at least some activity throughout the year in north Florida, and Branch and Hokit (2000) recorded toads from February to November in scrub habitat in central Florida. Juveniles are active throughout the cool autumn and winter months in Florida, especially after a year of high recruitment. Overwintering occurs in downed logs, under tree bark on fallen trees, under surface debris, or in small burrows made by the toads.

Oak Toads enter breeding ponds from upland communities, and many will immigrate to and emigrate from ponds nonrandomly. Males and females occupy the same terrestrial habitats. Juveniles tend to disperse toward uplands rather than to other wetlands, but they may use small temporary ponds as stopping points along their dispersal routes (Dodd, 1994). Oak Toads may disperse far from the nearest pond after breeding. For example, Dodd (1996) recorded them at a mean of 574 m (range 404 to 914 m) from the nearest wetland in dry Florida sandhills.

Most activity occurs by day, except during breeding migrations. These are triggered by warm rains in spring and early summer, and heavy thunderstorms can stimulate activity by large numbers of Oak Toads. During the nonbreeding season, Oak Toads tend to remain within a relatively small area and may rarely be observed despite their numbers. The farthest distance Hamilton (1955) recaptured a marked toad was 27 m from the original point of capture, although most toads were < 3 m from the first capture point.

Oak Toads are photopositive in their phototactic response, suggesting they use both sunlight and moonlight when feeding and going about their daily activities (Jaeger and Hailman, 1973). They are sensitive to light in the blue spectrum ("blue-mode response"), which apparently helps them orient toward areas of increasing illumination, such as the open horizon above lakes and ponds (Hailman and Jaeger, 1974). Oak Toads likely have true color vision.

CALLING ACTIVITY AND MATE SELECTION

Calling normally occurs from the spring throughout the warm summer months (April to September), although Carr (1940b) heard a chorus on 26 February in north Florida. Calling occurs till late October in south Florida (Duellman and Schwartz, 1958), and Jensen et al. (2008) record calling in November in Georgia. The high-pitched advertisement call sounds like baby chicks continuously peeping. The call duration is short (0.12–0.19 sec), and the call frequency is high (4,300–5,200 cps) (Blair, 1956b). Calling occurs from both exposed and concealed locations along the shoreline or in shallow water.

Larger males are more successful, on average, in amplexing a female and fertilizing eggs than smaller males. However, there is no evidence of size assortative mating, that is, large males do not necessarily mate with large females (Wilbur et al., 1978). Amplexus is axillary. A reproductive bout consists of the male drawing his legs up and fertilizing several egg bars in two to three exertions as they are extruded, then relaxing his legs for one to two minutes before repeating the process. The amplexed pair then moves to another site close by and goes through another reproductive bout. This behavior continues until the entire egg compliment is oviposited.

BREEDING SITES

Breeding occurs in shallow temporary ponds, rain pools, flooded fields in pastures, borrow pits, and ditches that are free of fish (Dodd, 1992; Gunzburger et al., 2010). In south Florida, Babbitt et al. (2006) found that presence at breeding sites was positively correlated with the proximity of nearby woodlands (i.e., within < 200 m, but most < 100 m), when the breeding sites were located

Tadpole of *Anaxyrus quercicus*. Photo: Steve Johnson

in open habitats. They do not breed in wetlands surrounded by improved pasture. Preferred breeding sites should have low pH, low conductivity, be small, and lack predaceous fish. At their study site, 17% of the potential breeding sites were occupied by larvae. In South Carolina, 58% of the fishless ponds studied by Eason and Fauth (2001) contained *A. quercicus*.

REPRODUCTION

Although calling can occur as early as late winter and into autumn, breeding mostly occurs during the warm summer months (June to September), even in south Florida (Babbitt and Tanner, 2000). Egg strands are deposited in shallow water (2.5–15 cm) in short (2–7 mm) bars of 3–6 eggs, which tend to remain together on the substrate or become attached to branches, grass blades, or other pond debris (Volpe and Dobie, 1959). These bars may remain separate or form small clusters. The total clutch size is 500 to >700 eggs, but a female deposits only 14–30 eggs per reproductive bout. Wright (1932) notes clutches of 610 and 766 eggs, but these were derived of counts from one ovary from two toads and then doubled. Female body size is not correlated with clutch size, nor is mean egg diameter (Wilbur et al., 1978). Hatching occurs within 24–80 hrs.

LARVAL ECOLOGY

The larval period is 4–8 weeks, with Wright (1932) recording metamorphosis between 33 and 44 days. Nothing is known of the larval ecology. Larval density at a series of south Florida wetlands was 0.042 tadpoles/m^2 of habitat (Babbitt et al., 2006). The primary cause of egg and larval mortality is desiccation as shallow pools dry prior to metamorphosis. Newly metamorphosed froglets are 7–8.9 mm SUL (Wright, 1932; Volpe and Dobie, 1959).

DIET

Oak Toads are sit-and-wait predators. The diet consists of small invertebrates, mostly ants of many species (Hamilton, 1955; Duellman and Schwartz, 1958; Punzo, 1995). Other prey includes spiders (Crosby and Bishop, 1925), termites, beetles, aphids, lepidopteran larvae, true bugs, millipedes, centipedes, and even mollusks. Juveniles feed primarily on collembolans, ants, spiders, and mites. Soft bodied prey are digested in 60–72 hrs whereas heavily chitinized prey take 96 hrs at 25°C (Punzo, 1995). Feeding is initiated by prey movement. Stomach contents may increase the overall body weight by one-third.

PREDATION AND DEFENSE

Anaxyrus quercicus likely has poisonous skin secretions as do other toads. When faced by a predator, they may take flight, remain motionless, crouch, walk, hide, or inflate the body (Marchisin and Anderson, 1978). The effectiveness of these behaviors is unknown. Larvae of this species are generally unpalatable to fish, although a few native fish will eat them in small amounts (Baber, 2001; Baber and Babbitt, 2003). Predators of adults are hognose snakes (*Heterodon platirhinos*, *H. simus*), ribbon snakes (*Thamnophis sauritus*), and water snakes (*Nerodia*). Wright (1932) noted, "the gopher frog has also been given as inordinately fond of oak toads."

POPULATION BIOLOGY

The Oak Toad likely has a rather short life span. For example, Dodd (1994) captured only a few Oak Toads the year following marking and never recaptured individuals 2 yrs after marking despite marking 400 adults over a 5 yr period. During a severe drought, the population size-class structure shifted, reflecting juvenile recruitment, adult growth, and eventually lack of recruitment or activity during a severe drought year. Both SULs and weights decreased as the drought progressed, and successful reproduction occurred in only 1 yr from 1986 to 1990. These results suggest that populations are sensitive to environmental perturbations of even short duration, which could make them vulnerable to changes in climatic patterns.

Oak Toads occupy even small temporary ponds, but often not in great numbers. For example, Dodd (1994) recorded 274 adults and 177 juveniles over a 5+ yr study, although most captures occurred in only 2 yrs. The overall sex ratio was 1.7 males per female. In southeastern Georgia, Cash (1994) recorded only 37 *A. quercicus* over a 2 yr period at a temporary sandhills pond, and in South Carolina, Gibbons and Bennett (1974) captured 57 *A. quercicus*, also over a 2 yr period at a single pond. Also in South Carolina, Russell et al. (2002a) recorded 2–52 *A. quercicus* (total of 87) at 5 ponds over a 2 yr period. In contrast,

Adult *Anaxyrus quercicus*, Putnam County, Florida. Photo: C.K. Dodd Jr.

Breeding habitat of *Anaxyrus quercicus*. Other species breeding at this site included *Acris gryllus*, *Anaxyrus terrestris*, *Gastrophryne carolinensis*, *Hyla femoralis*, *Hyla squirella*, *Lithobates capito*, *Lithobates sphenocephalus*, *Pseudacris ocularis*, and *Scaphiopus holbrookii*. Photo: C.K. Dodd Jr.

Greenberg and Tanner (2005b) captured 2,452 Oak Toads at 8 ponds in Florida sandhills from 1994 to 2001; from 68 to 999 *A. quercicus* were captured per pond.

DISEASES, PARASITES, AND MALFORMATIONS

Oak Toads are parasitized by the nematodes *Ascarops*, *Cosmocercoides* sp., *C. dukae*, *C. variabilis*, Filariidae gen. unident., *Ozwaldocruzia pipiens*, and *Physaloptera* (Walton, 1938; Hamilton, 1955; Goldberg and Bursey, 1996). The acanthocephalan *Polymorphus* is also known from this species.

SUSCEPTIBILITY TO POTENTIAL STRESSORS

Chemicals. Oak Toad embryos are sensitive to trisodium nitrilotriacetic acid (NTA) at concentrations > 203 mg/L. NTA prevents hatching altogether at > 400 mg/L (Birge et al. 1980).

STATUS AND CONSERVATION

Historically, *Anaxyrus quercicus* was considered an extremely abundant species, especially in Florida. Hamilton (1955) described it as extraordinarily abundant, and Carr (1940a) stated that it was possible to drive across the central Florida peninsula and never be out of the sound of breeding choruses. With extensive landscape alteration and the destruction of unique communities and breeding ponds, those days are gone.

The Oak Toad is particularly susceptible to habitat loss, as it depends on small temporary pools in which to breed. Such habitats are easily overlooked or ignored, and most are unprotected. Likewise, the species cannot tolerate fish, so the introduction of fish into formerly fish-free wetlands is detrimental to this species. Because of its small size, the species also may be sensitive to the presence of imported fire ants (*Solenopsis victa*), which have invaded its habitat. Oak Toads do not tolerate urbanization well, and abundance may be reduced in silvicultural and agricultural areas (Delis et al., 1996; Surdick, 2005). For example, Oak Toads were common in northwest Gainesville, Florida, in the early 1980s, but have virtually disappeared as small wetlands were destroyed. Finally, alteration of precipitation patterns leading to extended periods of drought make this essentially annual species vulnerable to regional climate change.

Oak Toads may be found in cypress flatwoods ponds after ditching (Vickers et al., 1985), although whether reproduction takes place has not been documented. They also colonize areas previously clearcut 1.5–4 yrs earlier (Enge and Marion, 1986; Russell et al., 2002b). It should be noted, however, that there was no replication in some of these studies and that conclusions about short-term tolerance to clearcutting were based on presence, not reproduction or demographic analysis. The presence of *Anaxyrus quercicus* is indicative of low-intensity land use (Surdick, 2005).

The Oak Toad inhabits fire-maintained ecosystems throughout its range, and prescribed fire regimes seem to enhance the habitat quality for this species (Langford et al., 2007). Native grasses increase following prescribed fire, and clumps of grasses provide ideal microhabitat for Oak Toads (Baxley and Qualls, 2009). The species readily recolonizes wetlands that experience saltwater overwash during tropical storms (Gunzburger et al., 2010), but do not seem to take to newly constructed ponds (Pechmann et al., 2001). The species is considered rare and a species of "Special Concern" in Virginia (Pague and Mitchell, 1987; Mitchell and Reay, 1999). In North Carolina, Dorcas et al. (2007) noted that they have undergone dramatic declines in recent years and attributed declines to habitat loss, acidification of breeding ponds, and the potential impacts of disease and imported red fire ants. Enge (2005a) reported 2,131 *A. quercicus* were harvested for the pet trade from 1990 to 1994 in Florida.

Anaxyrus retiformis (Sanders and Smith, 1951) Sonoran Green Toad

ETYMOLOGY
retiformis: from the Latin *retis* meaning 'of a net' and *forma* meaning 'shape.' The name is in reference to the reticulated dorsal pattern of black lines.

NOMENCLATURE
Stebbins (2003): *Bufo retiformis*
　　Synonyms: *Bufo debilis retiformis*

IDENTIFICATION
Adults. This is an attractive small yellow-green toad with a dark reticulated pattern covering the dorsum and upper portions of the limbs. The dorsum is not spotted as in *Anaxyrus debilis*. The head is flat. Parotoid glands are large in relation to the size of the toad and are patterned like the dorsum. Venters are unspotted. Males are smaller than females. Savage (1954) reported a mean of 43.8 mm SUL (range 40 to 47 mm) for males and 47 mm (range 45 to 49 mm) for females but sample sizes were small. Sullivan et al. (2000) reported means of 48.1 to 52.1 for calling males at 3 populations. Maximum size is 64 mm SUL.

Larvae. At hatching, larvae are yellow with scattered dark pigment. Larger tadpoles are pale yellow and stippled with black and golden patches on the dorsum and tail musculature. They are rounded in appearance with the eyes situated dorsally, and are much paler than most other toad tadpoles. The dorsal tail musculature has more abundant melanophores than the ventral musculature, giving a somewhat striped appearance when viewed laterally. Dorsal tail fins may have scattered melanophores, but ventral tail fins are clear. The lateral side of the body may be darkly pigmented behind the eyes, and the belly has golden flecking. There is no coloration on the throat and midbelly. Zweifel (1970b) noted that tadpoles are relatively transparent even when full grown. Zweifel (1970b) provides illustrations of larvae at various stages of development.

Eggs. The eggs are light cream to yellow with a dark band located one-third the distance from the top of the animal pole (Ferguson and Lowe, 1969; Zweifel, 1970b). Scattered melanin occurs over the entire egg. There are two gelatinous membranes surrounding the vitellus, which is ca. 1.15 mm in diameter. Eggs are deposited singly or in short strands. The outer envelope of the eggs have an agglutinating property, which causes them to clump together, giving the appearance of a small egg mass rather than single eggs or a small string.

Distribution of *Anaxyrus retiformis*

Larval *Anaxyrus retiformis*. Photo: Cecil Schwalbe

Adult male *Anaxyrus retiformis*. Photo: Rob Schell

DISTRIBUTION

Anaxyrus retiformis is only known from south-central Arizona and west-central Sonora, Mexico. Important distributional references include Nickerson and Mays (1968), Sullivan et al. (1996b), Blomquist (2005), Brennan and Holycross (2006), and López et al. (2009).

FOSSIL RECORD

No fossils have been identified.

SYSTEMATICS AND GEOGRAPHIC VARIATION

The Sonoran Green Toad was described originally as a subspecies of *A. debilis* (Sanders and Smith, 1951). It is indeed most closely related to *A. debilis* within the Nearctic clade of North American bufonids (Pauly et al., 2004), and together with *A. kelloggi* these species form the *Debilis* clade within the *Punctatus* group.

Under laboratory conditions, this species will hybridize and produce larvae that complete metamorphosis with *Anaxyrus punctatus*, *A. debilis*, and *A. kelloggi* (Ferguson and Lowe, 1969). Hybridization in nature with *A. punctatus* also has been reported (Bowker and Sullivan, 1991; Sullivan et al., 1996b). Calls of hybrids between *A. retiformis* and *A. punctatus* are aberrant and similar to *A. punctatus* (Sullivan et al., 2000).

Adult female *Anaxyrus retiformis*. Photo: David Dennis

ADULT HABITAT

The Sonoran Green Toad is a species of the desert flats, valleys, and gently sloping bajadas of the lower Sonoran Desert. It occurs in Colorado River Desert Scrub, semidesert grassland, and in Arizona Upland Desert Scrub.

TERRESTRIAL ECOLOGY

Sonoran Green Toads are rarely observed and nocturnal in behavior. They remain hidden during the day, presumably in rodent burrows, root channels, and in underground crevices. Very little is known of the species' life history.

Sonoran Green Toads are photopositive in their phototactic response, suggesting they use both sunlight and moonlight when feeding and going about their daily activities (Jaeger and Hailman, 1973). They are sensitive to light in the blue spectrum ("blue-mode response") (Hailman and Jaeger, 1974). Sonoran Green Toads likely have true color vision.

CALLING ACTIVITY AND MATE SELECTION

Breeding is opportunistic and associated with desert rainfall. Choruses form rapidly and are often small. In the Sonoran Desert, for example, the chorus size ranged from 3 to 50, and the chorus only lasted a mean of 2 days (Sullivan, 1989). Sullivan et al. (1996b) later reported sizeable breeding aggregations consisting of >200 adults.

The advertisement call is a short insect-like buzz with a dominant frequency of 2,714–3,376 cps, a duration of 2.0–4.3 sec, and 193–238 trills/sec (Ferguson and Lowe, 1969; Sullivan et al., 1996b, 2000). Savage (1954) described the call as "a rising crescendo of a single drawn-out note, not unlike the buzzer of an electric alarm clock, with a slight trill giving the effect of a vibrating police whistle." The call duration is not correlated with temperature, but the pulse rate and dominant frequency are correlated with temperature (Sullivan et al., 2000). Males call at air temperatures of 26–30°C. The call duration is not correlated with SUL, but the pulse rate and dominant frequency are negatively correlated with SUL (Sullivan et al., 2000).

Nocturnal calling occurs on land 1–5 m from the water's edge, and amplexus occurs on dry, damp, or wet substrates. Ferguson and Lowe (1969) recorded amplexus up to 18 m from water. Males call from under vegetation such as shrubs and grasses and will try and amplex any female moving in their vicinity. Females then carry the smaller male to the breeding site. Satellite males are common at breeding sites, with up to three satellites for a single calling male (Sullivan, 1996b).

BREEDING SITES

Breeding occurs in temporary pools and ditches that form after summer thunderstorms. These include roadside pools in desert washes and cattle tanks.

REPRODUCTION
Most breeding occurs in July to August, associated with the summer monsoon rains. The clutch size of this species is unknown. Hatching occurs in two to three days after oviposition.

LARVAL ECOLOGY
Laval diets are unknown. The duration of the larval period also is unknown but probably short; Savage (1954) noted metamorphs 13 days after previously observing breeding in the area. At hatching, larvae are 3.1–3.4 mm TL (Zweifel, 1970b).

DIET
Nothing specific is known of the diet of *A. retiformis*. It presumably eats small invertebrates such as ants and beetles.

PREDATION AND DEFENSE
The Sonoran Green Toad presumably has noxious or toxic skin secretions as do other bufonids. Nothing is known concerning defensive behaviors or predators.

POPULATION BIOLOGY
No information is available.

DISEASES, PARASITES, AND MALFORMATIONS
Sonoran Green Toads are parasitized by the cestode *Distoichometra bufonis* and the nematodes *Aplectana incerta*, *A. itzocanensis*, *Oswaldocruzia pipiens*, *Physaloptera* sp., and *Rhabdias americanus* (Goldberg et al., 1996a).

SUSCEPTIBILITY TO POTENTIAL STRESSORS
No information is available.

STATUS AND CONSERVATION
Anaxyrus retiformis presently inhabits most of the same area as reported from historical collections. Sullivan et al. (1996b) suggested that hybridization with *A. punctatus* might be facilitated by the construction of cattle tanks, which draw these normally habitat-separated species to the same breeding sites. No information is available on status and population trends (Blomquist, 2005), although Bury et al. (1980) noted the need for a status survey 30 yrs ago, because so little was known concerning this species.

Anaxyrus speciosus Girard, 1854
Texas Toad

ETYMOLOGY
speciosus: from the Latin *speciosus* meaning 'showy' or 'beautiful.' Toads, indeed, are beautiful creatures.

NOMENCLATURE
Conant and Collins (1998) and Stebbins (2003): *Bufo speciosus*

Synonyms: *Bufo compactilis speciosus*, *Bufo lentiginosus speciosus*, *Bufo pliocompactilis*, *Bufo spectabilis*

IDENTIFICATION
Adults. The ground coloration of *Anaxyrus speciosus* is olive, gray, or gray brown dorsally. There is a well-defined pattern of dusky spots or somewhat elongated blotches dorsally. Cranial crests are low and poorly defined; warts are numerous and low-to-rounded without spines. The parotoid glands are oval and less than twice as long as wide. No mid-dorsal stripe is present. Venters almost always lack spots, but Sievert and Sievert (2006) reported a black spot on the chest. This toad has two distinctive black spades on each hind foot that are used for digging; the inner spade is sickle-shaped. Females are larger than males. Adults are 65–100 mm SUL.

Larvae. The dorsal part of the tadpole is lightly pigmented, and the tail musculature is light with dark lateral blotches that tend to form a stripe, giving a bicolored appearance. The body is an elongate oval. Tail fins are low with the greatest height midway down the tail. Eyes are situated dorsally. The maximum size is ca. 30 mm TL.

Eggs. The eggs are brown or dark gray dorsally and yellow ventrally. The eggs are oviposited as fine tightly coiled strings in a gelatinous casing. One slightly scalloped jelly envelope surrounds the vitellus. This envelope is 1.8–2.4 mm in diameter, with the vitellus 1.2–1.6 mm in diameter. According to Livezey and Wright (1947) there are 11–17 eggs per 25 mm of string.

DISTRIBUTION

The Texas Toad occurs from southwestern Oklahoma south through Texas to the Gulf Coast. It occurs north into New Mexico along the Pecos River drainage, and southward in Chihuahua and northeastern Mexico. This species is found on Mustang Island, Texas (Moore, 1976). Important distributional references include Smith and Sanders (1952), Degenhardt et al. (1996), Dixon (2000), Sievert and Sievert (2006), Lemos-Espinal and Smith (2007b), and López et al. (2009).

FOSSIL RECORD

Fossils of this species have been reported from Miocene deposits in Nebraska, Pliocene deposits in Kansas, and Pleistocene deposits in Texas (Tihen, 1962b; Holman, 2003). According to Holman (1969, 2003), some of the fossil records now referred to this species were originally identified as belonging to *A. cognatus*. These species are closely related, and the slight osteological differences between them make specific identification tentative.

SYSTEMATICS AND GEOGRAPHIC VARIATION

Anaxyrus speciosus is most closely related to *A. cognatus* within the Nearctic clade of North American bufonids (Pauly et al., 2004). These species form the *Cognatus* clade, together with *Anaxyrus compactilis* of Mexico, but this clade is considered paraphyletic by Pauly et al. (2004). Cope (1889) and Smith (1947) regarded *A. speciosus* as a subspecies of *A. compactilis*, but evidence of their specific distinctiveness is based on calls (Bogert, 1960), morphology (Rogers, 1972), and protein polymorphism (Rogers, 1973).

Laboratory crosses between *A. speciosus* and *A. californicus, A. cognatus, A. woodhousii, A. microscaphus,* or *Ollotis nebulifer* can produce successful metamorphs (A.P. Blair, 1955; W.F. Blair, 1959), but not always (Moore, 1955). The color and pattern of the hybrids is intermediate between those of the parents, but developmental abnormalities are common among larvae that do not survive. Crosses between *Anaxyrus terrestris* or *A. punctatus* and *A. speciosus* are not successful (Moore, 1955; Blair, 1959). The sex of the parents is important in determining whether hybrid crosses will be successful. Some hybridization experiments also appear to rely on small sample sizes.

ADULT HABITAT

The Texas Toad is a species of the short-grass plains, where it prefers sandy soils. In the Sierra Vieja Range of southwestern Texas, Jameson and Flury (1949) found it associated with the salt cedar-mesquite and catclaw-creosote bush plant associations in low sandy washes and along the banks and floodplain of the Rio Grande. In Big Bend National Park, it is a species of mesquite scrub vegetation; vegetation communities include lechuguilla, creosote, mesquite thickets, and mixed scrub (Dayton et al., 2004). In New Mexico, it occurs from 900 to 1,300 m (Degenhardt et al., 1996).

TERRESTRIAL ECOLOGY

Most activity occurs at night when air temperatures are >17°C. Activity can occur any time of the year, at least in south Texas, temperature and precipitation permitting (Moore, 1976). For example, Minton (1958) recorded juvenile activity on 26 February in the Big Bend region. Rainfall is necessary to stimulate emergence in the spring. After metamorphosis, young toads remain within the vicinity of breeding pools for up to several months, as long as breeding pools retain water.

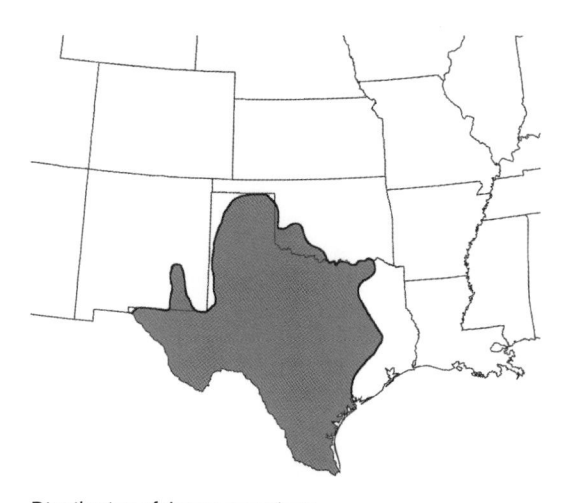

Distribution of *Anaxyrus speciosus*

Adult *Anaxyrus speciosus*. Photo: Robert Hansen

CALLING ACTIVITY AND MATE SELECTION

Breeding choruses are often very large, with males sometimes moving about continuously calling and searching for females (Bragg, 1940b) or calling from prominent positions around the shoreline (Moore, 1976). Males precede females to the breeding sites and will amplex any toad within their vicinity. Sometimes this behavior results in a "seething mass" of struggling toads with up to five males piled on top of one another (Bragg, 1950b). According to Bragg (1945a), females will bypass males calling from small pools in favor of those calling from nearby larger pools. The small pools undoubtedly have a much shorter hydroperiod for tadpole development than the larger pools, but how the female makes this discrimination is unknown.

Calling occurs both night and day but is most intense at night. The call is a very short, loud, explosive trill with a dominant frequency of 2,400–2,800 cps, a duration of 0.4–0.6 sec, and a mean of 39–57 trills/sec at 23–27°C (Blair, 1956b). As in *A. cognatus*, amplexed male *A. speciosus* have a slow series of warning vibrations to alert a conspecific as to the sex of grasped animal (Blair, 1947b). Wiest (1982) reported calling at air temperatures of 18.5–23.1°C.

Females will move toward calling males via triangulation and initiate amplexus by touching him (Axtell, 1958). At this time, the male stops calling and grabs the female. Satellite males in the vicinity will also attempt to amplex the female, but the largest male usually wins the struggle. If grasped by a small male, a female will attempt to dislodge him, although not always successfully.

BREEDING SITES

Breeding occurs in both clear and muddy shallow temporary ponds and pools in open fields or near streams, irrigation ditches, as well as in cattle tanks and even buffalo wallows. Ponds should be open-canopied and may be free of vegetation or contain a considerable amount of aquatic vegetation (Bragg and Smith, 1942).

REPRODUCTION

Breeding occurs from the spring throughout the summer and is stimulated by heavy rainfall. For example, breeding has been observed in early April in Texas and as late as August in Oklahoma and September in Texas (Bragg, 1940b; Bragg and Smith, 1942; Moore, 1976; Wiest, 1982); most reproduction probably occurs from May to July. The eggs are deposited in long strings in shallow pools. No data are available on clutch size, but there are 14–20 eggs per 30 mm of string (Wright and Wright, 1949). Degenhardt et al. (1996) reported additional data on reproduction of *A. speciosus*, but these observations were based on Bragg's (1955) reports on "*B. compactilis*" from southwestern Utah, which is actually *A. microscaphus*.

LARVAL ECOLOGY

Nothing is known of the larval ecology of this species except that larvae eat algae scraped from rocks. Metamorphosis occurs in as quickly as 18 days (Moore, 1976).

DIET

The adult diet consists almost entirely of arthropods, particularly ants, lepidopteran adults and larvae, and many types of beetles; other prey include spiders, millipedes, centipedes, true bugs, weevils, flies, crickets, and homopterans (e.g., cicadas, aphids) (Smith and Bragg, 1949). Specific dietary items change somewhat between spring and summer, presumably in response to availability rather than selection. Juveniles eat mostly ants, beetles, and flies (Smith and Bragg, 1949). Texas Toads are reported to feed commonly under streetlights on insects drawn to the light.

PREDATION AND DEFENSE

Tadpoles of *Spea hammondii* avoid attacking live or dead larvae of *Anaxyrus speciosus* (Bragg,

1960a). Like many other toads, the Texas Toad is cryptically colored. When approached by a predator, it may crouch low against the substrate and only slowly seek cover under vegetation (Bragg, 1945a). No information is available on predators, but they likely include snakes, birds, and carnivorous mammals.

POPULATION BIOLOGY
Growth is rapid during the first 3 months following metamorphosis, averaging 10 mm per month in Texas (Moore, 1976). Growth rates then decrease to almost zero during the cold winter months. It seems likely that sexual maturity is reached by the second spring following metamorphosis. The large disparity in size between males and females suggests that females reach maturity later than males, assuming growth rates are equal.

COMMUNITY ECOLOGY
Anaxyrus speciosus and *A. cognatus* are sympatric in the ecotone between short-grass and mixed-grass prairies, with *A. speciosus* replacing *A. cognatus* to the west (Bragg and Smith, 1943). The presence of *A. speciosus* larvae tends to inhibit growth and survival of larvae of *Gastrophryne olivacea* and Gray Treefrogs (*Hyla versicolor*/*H. chrysoscelis* complex) but not *Anaxyrus houstonensis* under laboratory conditions. The presence of *A. woodhousii* larvae, however, had no effect on *A. speciosus* (Licht, 1967). Decreases in body mass occur in larval *A. speciosus* when reared with *Scaphiopus couchii* larvae (Dayton and Fitzgerald, 2001). Data suggest that *S. couchii* are able to out-compete *Anaxyrus speciosus* larvae and may exclude them from shallow-water desert pools in some areas, such as Big Bend National Park.

DISEASES, PARASITES, AND MALFORMATIONS
The myxozoan *Myxidium serotinum* parasitizes *Anaxyrus speciosus* (McAllister and Trauth, 1995) as do two species of cestodes and three species of nematodes (Kuntz, 1941).

SUSCEPTIBILITY TO POTENTIAL STRESSORS
No information is available.

STATUS AND CONSERVATION
The Texas Toad appears to be widespread and abundant throughout its range, and no large-scale population declines have been reported (Dayton and Painter, 2005). Some populations in the agricultural regions of south Texas may be declining due to pesticide use (Dixon, 2000).

Anaxyrus terrestris (Bonnaterre, 1789) Southern Toad

ETYMOLOGY
terrestris: from the Latin *terrestris* meaning 'pertaining to the Earth.'

NOMENCLATURE
Conant and Collins (1998): *Bufo terrestris*
  Synonyms: *Bufo clamorus, Buffo clamosus, Bufo erythronotus, Bufo lentiginosa, Bufo lentiginosus, Bufo lentiginosus pachycephalus, Bufo musicus, Bufo rufus, Chilophryne lentiginosa, Incilius lentiginosus, Rana lentiginosa, Rana musica, Rana terrestris, Telmatobius lentiginosus*

IDENTIFICATION
Adults. This is a medium-sized toad, with coloration ranging from reddish to olive green to gray to nearly black. Red toads in the South are usually this species. The head is relatively broad with a sharp snout; females have broader heads than males. Like the American Toad, there is usually one wart per spot, and the cranial crests are well developed and quite obvious. The cranial crests have a large knob at the posterior ends of the parallel interorbital crests that increases in prominence with age, and this feature is usually more prominent in females. The posterior portion of the crest has a relatively long extension called the preparotoid ridge, which extends at right angles and touches the parotoid gland. Parotoids are of various shapes, from oval to elongate to kidney-shaped. The white mid-dorsal line is not

Distribution of *Anaxyrus terrestris*

well developed. The Southern Toad is usually larger and darker than the sympatric Fowler's Toad. Dorsal warts are prominent. Venters may be spotted in the pectoral region or unmarked. Limbs are relatively short, with the dorsal surface of the rear legs barred. The rear toes are webbed, except for the end of the fourth toe, which extends beyond the webbing. Males normally have a dark subgular vocal sac, but this is not always darkly colored. An albino was reported from Duval County, Florida (Dyrkacz, 1981).

Males are smaller than females. In central Florida, for example, males averaged 58.5 mm SUL (range 48 to 65 mm), whereas females averaged 73.8 mm SUL (Bancroft et al., 1983). In north-central Florida, males were 43–70 mm SUL (mean 53 mm) whereas females were 43–79 mm SUL (mean 58 mm) (Dodd, 1994). Brown (1956) recorded an adult mean of 48 mm SUL (range 56.3 to 75.4 mm) in Alabama.

Larvae. The larvae are small, ovoid in appearance, and black. Small purplish dots may be present dorsally. Venters are black with somewhat purplish spots scattered freely and not forming a continuous mass. The upper tail fin is somewhat spotted, but the lower tail fin is unspotted and clear to yellowish; tail fins are about equal in depth but wider than the tail musculature. The lower edge of the tail musculature is light cream or light yellow. The tail is short with a rounded tip. The eyes are positioned dorsally and they are close together. According to Dundee and Rossman (1989), there is a light oblique mark behind each eye. The body of the tadpole is 5–10 mm in length; total lengths approach 24–28 mm. Descriptions of larvae are in Wright (1929) and Siekmann (1949).

Eggs. The eggs of *A. terrestris* are black dorsally and grayish white ventrally (Livezey and Wright, 1947). Eggs are deposited in continuous gelatinous strings with a singular tubular membrane and no intercellular partition between adjacent eggs. Strings may be singular or paired, with 8–12 eggs per 25 cm in length. There are 2 jelly envelopes surrounding the egg (outer envelope 2.6–4.6 mm; inner envelope 2.2–3.4 mm in diameter) (Brown, 1956), although upon deposition the inner envelope may be difficult to see. The vitellus is 1.0–1.4 mm in diameter and may be elliptical. The normal clutch size is from 2,500 to 3,000 eggs per female.

DISTRIBUTION

The Southern Toad occurs on the Atlantic and Gulf Coast Coastal Plain from southeastern Virginia through eastern Mississippi, and southward through the Florida Parishes of Louisiana. An isolated population occurs in the upper Piedmont on the border of Georgia and South Carolina (e.g., Chamberlain, 1939). Most records of Southern Toads, however, are below the Fall Line. Populations of Southern Toads are found in the Florida Keys (Big Pine, Cudjoe, Sugarloaf: Duellman and Schwartz, 1958; Lazell, 1989) and on various Atlantic and Gulf Coast barrier islands. These include Bodie and Smith islands, North Carolina (Lewis, 1946; Gaul and Mitchell, 2007), Kiawah Island, South Carolina (Gibbons and Coker, 1978), Sapelo, Little Cumberland, and Cumberland islands, Georgia (Martof, 1963; Gibbons and Coker, 1978; Shoop and Ruckdeschel, 2006), St. George and St. Vincent islands, Florida (Irwin et al., 2001), and Cat Island, Mississippi (M.J. Allen, 1932).

Important distributional references include: Alabama (Brown, 1956; Mount, 1975), Florida (Lazell, 1989; Bartlett and Bartlett, 1999), Georgia (Williamson and Moulis, 1994; Jensen et al., 2008), Louisiana (Dundee and Rossman, 1989), North Carolina (Dorcas et al., 2007), and Virginia (Tobey, 1985; Mitchell and Reay, 1999).

FOSSIL RECORD

Fossil Southern Toads are known from Pleistocene Irvingtonian sites in Florida and Pleistocene Rancholabrean sites in Florida, Georgia, and Tennessee. The species is particularly common in Florida fossil deposits (Holman, 2003). Large knobs on the frontoparietal bone distinguish this species from other fossil *Anaxyrus* within its range. The dorsal process of the ilium is lower than that of *A. americanus*. Tihen (1962a, 1962b) and Holman (2003) note other distinguishing osteological characteristics among *A. fowleri*, *A. americanus*, and *A. terrestris*. Meylan (2005) named a new species of late Pliocene toad as *Bufo defensor* (*Anaxyrus defensor*), which appears to be related to *Anaxyrus terrestris*. *Anaxyrus defensor* has the largest supraorbital crest on the frontoparietals of all New World bufonids.

SYSTEMATICS AND GEOGRAPHIC VARIATION

Anaxyrus terrestris is a member of the *Americanus* group of toads, which also includes *A. americanus*, *A. houstonensis*, *A. microscaphus*, *A. fowleri*, and *A. woodhousii*. This group in turn is a member of the Nearctic clade of North American bufonids (Pauly et al., 2004). Based on molecular analysis, the species is most closely related to the southern clade of *A. fowleri* (Masta et al., 2002), a relationship that is different from earlier phylogenetic hypotheses based on morphology and call similarity (e.g., Blair, 1963b). Divergence probably took place within the last 2 my, during the Pleistocene.

Natural and experimental hybridization has been reported with *A. fowleri* (Gosner and Black, 1956; Volpe, 1959b; Brown, 1969) and *A. americanus* (Wilbur et al., 1978). In southeastern Louisiana and adjacent Mississippi, Volpe (1959b) demonstrated extensive geographical areas of hybridization between *A. terrestris* and *A. fowleri*. In some populations, "hybrid swarms" (individuals with an extensive diversity of character combinations) of toads were observed, whereas in other populations repeated backcrossing of hybrids with parental species was apparent. Thus the extent of hybridization and its effects on phenotype vary a great deal in some areas, making identification difficult.

Laboratory experiments have demonstrated hybridization capability between ♀ *A. terrestris* and ♂ *A. hemiophrys*, *A. houstonensis*, *A. punctatus*, *A. microscaphus*, *A. woodhousii*, and *Ollotis nebulifer* (Blair, 1959, 1961a, 1963a). Crosses between ♀ *Anaxyrus punctatus* or *A. woodhousii* and ♂ *A. terrestris* are generally infertile, and those between ♀ *Incilius coccifer* and ♂ *Anaxyrus terrestris* may produce larvae but few if any metamorphs (Blair, 1959, 1961a). A cross between a ♀ *A. terrestris* and a ♂ *A. speciosus* (as *A. compactilis*) was unsuccessful (Blair, 1959). Some hybrids are fertile and capable of backcrossing with parental species, such as the F_1 progeny of ♀ *A. terrestris* and ♂ *A. hemiophrys* or *A. woodhousii* (Blair, 1963a).

ADULT HABITAT

Southern Toads occur in a wide variety of terrestrial and semiaquatic habitats, including seashore scrub, seashore dunes, pine flatwoods, scrub, sandhills, xeric hammock, dome swamps, coastal hydric hammock, inland hydric hammock, basin swamps, depression marshes, forested wetlands, basin marshes, steephead ravines, pine flatwoods, wet prairies, upland hardwood forests, and mixed-upland forests (Carr, 1940a; Harima, 1969; Buhlmann et al., 1993; Enge et al., 1996; Enge, 1998a, 1998b; Enge and Wood, 1998; Smith et al., 2006; Baxley and Qualls, 2009). They occur in ruderal and disturbed areas, such as sand pine plantations, other types of silviculture, agricultural areas, and urban and suburban habitats (Carr, 1940a; Neill, 1950a; Surdick, 2005). According to Dundee and Rossman (1989), Southern Toads occupy dryer and higher sites when they are sympatric with *Anaxyrus fowleri*.

TERRESTRIAL ECOLOGY

Southern Toads are active year-round, weather permitting, especially in the southern portion of the species' range. Rainfall is not necessary to stimulate terrestrial activity, and in extreme drought toads may remain in terrestrial refugia rather than chance movements to what may be a dry breeding site (Dodd, 1994). During cold or dry weather, they take refuge in mammal burrows and holes, under surface debris and leaf litter, in rock piles, or buried into the soil. They also seek shelter during the hottest part of the day in leaf litter or burrowed into the substrate. Sometimes they are found with just the eyes and top of the head protruding above the substrate surface. Most activity occurs at night, with breeding migrations occurring almost exclusively at night (Todd and Winne, 2006). However, they are not immune to diurnal activity and are sometimes found hopping around during the middle of the day.

Southern Toads are familiar with their environment, and some individuals are able to return to a release point, even after being displaced as far as 1,609 m from the point of collection (Bogert, 1947). The direction of displacement does not appear to influence homing ability, with an estimated 50% of the toads eventually returning from a 640 m displacement over a period of several months. However, Bogert (1947) reported one individual traveling 790 m in a 24 hr period during favorable rainy conditions. Southern Toads are frequently found far from the nearest breeding ponds. In north-central Florida, for example, toads were captured at a mean distance of 515 m from water (range 46 to 914 m) (Dodd, 1996).

Southern Toads are monotonically photopositive in their phototactic response, suggesting they use both sunlight and moonlight when feeding and going about their daily activities (Jaeger and Hailman, 1973). They are sensitive to light in the blue spectrum ("blue-mode response"), which apparently helps them orient toward areas of increasing illumination, such as the open horizon above lakes and ponds (Hailman and Jaeger, 1974). Southern Toads likely have true color vision.

CALLING ACTIVITY AND MATE SELECTION

Male *Anaxyrus terrestris* have a loud and melodic ("musical," Dundee and Rossman, 1989) trill; a chorus can be deafening. Call characteristics in Alabama include a mean pulse rate of 51.8–66 pulses/sec, a call duration of 6.9–8.8 sec, and a dominant frequency of between 1,878 and 2,042 cps (Brown, 1969). In Florida, call characteristics include a pulse rate of 68–78 pulses/sec, a call duration of 1.5–8.3 sec, and a dominant frequency of between 2,000 and 2,300 cps (Blair, 1956b). Whereas pulse rates are lower and call duration is much longer than sympatric *A. fowleri*, the dominant frequencies are about the same. The advertisement call serves to attract females, and call differences help maintain species isolation.

Anaxyrus terrestris has a release call that serves to alert both conspecific and heterospecific males that the amplexed animal is a male. Indeed, the release calls may be convergent in call structure so as to broaden their applicability, even as mate advertisement calls diverge in call structure to allow for species isolation (Leary, 2001). The convergence centers on the periodicity of release vocalizations, which are not significantly different among some toad species. However, character displacement in the periodicity of the advertise-

Egg strings of *Anaxyrus terrestris*. Photo: Steve Johnson

ment calls of *A. fowleri* is pronounced both in sympatry and allopatry with *A. terrestris*. Advertisement periodicity is much higher in *A. terrestris* (mean 16.8 to 18.3 ms) than in *A. fowleri* (mean 7.3 ms).

Males call from a variety of open locations, usually along the shoreline of a breeding site. They may face either the wetland or the shoreline, depending upon the positioning of aquatic vegetation. However, Wright (1932) recorded them calling perched on cypress logs, cypress knees, and stumps, resting on aquatic plant stems, and sitting in shallow water along a pond margin. They do not prefer sites with obstructing vegetation, such as thick cattail. Males generally do not float while calling. Rainfall is not necessary to stimulate calling, but particularly large choruses call loudly during and after heavy rains. As the season progresses, rainfall may be more of a requirement to stimulate calling. Calling occurs by night or day. Wright (1932) recorded calling at temperatures of from 14 to 31°C, and Dundee and Rossman (1989) noted that temperatures had to be >18°C in Louisiana for calling to occur.

Mate selection and oviposition sequence is essentially the same as in *A. americanus*. Amplexus is axillary or supra-axillary, depending on the size of the female. Females are attracted by the calls of the males and move toward them, although they may be amplexed at any time by a satellite male or indeed any male that intercepts them on the way to a breeding site. The oviposition sequence and the behavior of the courting pair have been described by Aronson (1944) and consist of amplexus, preovulatory behavior (restlessness and abdominal muscular contractions lasting from a few minutes to several hours), and oviposition (back arching, extrusion of eggs with simultaneous fertilization, movement along the wetland bottom as egg strings are extruded). If the egg strings break during deposition, amplexus ends and the mating sequence is either reinitiated with the same or a different male. A recently spent female will arch her back in a very pronounced manner to indicate she has finished spawning, although it may take several tries before the male releases her and moves elsewhere.

Mating male Southern Toads may try to amplex virtually any species in their vicinity, from Southern Leopard Frogs to large hylids. Occasionally, males may be able to displace other males already amplexing a female, although the extent to which this occurs in nature and its significance remains unquantified (Lamb, 1984a).

BREEDING SITES

Southern Toads breed in a variety of wetlands, from small, isolated, temporary pools and ponds to the shorelines of large permanent lakes. They breed in old field ponds, stream overflow basins, marshes, cypress savannas, and cypress/gum ponds, and even in the quiet waters of shallow coastal plain streams (Wright, 1932; Liner et al., 2008). They readily breed in man-made habitats, such as suburban retention ponds, roadside ditches, farm ponds, road ruts, borrow pits, and golf course ponds (Scott et al., 2008). Isolated ponds may be particularly important to reproduction and perhaps as areas of refuge and feeding. In South Carolina, for example, Russell et al. (2002a) captured nearly 2,000 *A. terrestris* at 5 wetlands < 1.06 ha over a 2 yr period, and Cash (1994) captured 154 and 242 *A. terrestris* at a temporary pond in southeastern Georgia over a 2 yr period. Other reports of numbers of toads at a pond include 114–817 at Sun Bay on the Savannah River Site in South Carolina over a 4 yr period (Pechmann et al., 2001).

REPRODUCTION

Breeding most frequently occurs from spring to summer, the timing of which depends on latitude and environmental conditions. In Florida and Mississippi, for example, calls are heard beginning in March (M.J. Allen, 1932; Carr, 1940a, 1940b), and in North Carolina, Brimley (1944) recorded breeding from March through June. In Alabama, Brown (1956) gave the breeding period as March to September. Dundee and Rossman (1989) recorded calling males in Louisiana from February through August. In central Florida, choruses were heard from February to September (Bancroft et al., 1983). However, Semlitsch et al. (1996) indicated that in South Carolina, breeding can occur in all months of the year. It seems likely that in the southern portion of its range, breeding by Southern Toads can occur opportunistically at times outside the "normal" peak season of reproduction.

Eggs are deposited in shallow water in long strings coiled around grass, vegetation, and debris, or they may be deposited loosely on the

Tadpole of *Anaxyrus terrestris*. Photo: Jamie Barichivich

bottom of shallow pools. When deep pools are used, the eggs are attached to debris at the surface. Wright (1932) reported a single complement of ca. 2,888 eggs and suggested that clutch sizes ranged from 2,500 to 3,000 eggs (Wright and Wright, 1949). Eggs hatch in two to four days. The biggest threat to the eggs may be desiccation as shallow ponds dry.

LARVAL ECOLOGY

The larval period of Southern Toads is rather short (38–55 days; Wright, 1932) and is dependent on water temperature, larval rearing conditions, density of conspecifics, and the presence of predators during development. In field and experimental trials, larvae reared in high-density conditions had a longer duration of larval development than larvae reared in lower density conditions (Nicholson, 1980; Wilbur et al., 1983). Density dependence also affects survivorship and the size at metamorphosis, with decreased densities resulting in higher survivorship and larger-sized larvae at metamorphosis (Morin, 1983; Wilbur et al., 1983). Indeed, there is a negative correlation between survivorship and mass at metamorphosis. Individual cohorts reached metamorphosis in as little as 17 days or as long as 38 days during experimental trials in Florida, when snout-vent lengths were between 7.4 and 10.3 mm (Nicholson, 1980). In South Carolina, Daszak et al. (2005) give a minimum hydroperiod of 56 days necessary to complete metamorphosis in the field.

Other factors influencing the duration of the larval period are food resources and the presence of predators. When food resources are abundant, development proceeds quickly and larvae reach larger sizes than they do when food is scarce. When food is scarce, the larvae grow slower, thus extending development time, and they metamorphose at smaller sizes. Gape-limited predators, such as the newt *Notophthalmus viridescens*, also exert a strong influence on *Anaxyrus* survival, especially during the initial stages of development (Wilbur et al., 1983). As tadpoles grow, however, they eventually reach a stage where they are much less vulnerable to predation. Not surprisingly, as predator density increases in experimental trials, survivorship of toad larvae decreases (Morin, 1983).

As might be expected, the timing of reproduction and the timing of newts coming to occupy a pond have important implications for larval survivorship. If newts arrive first, survivorship will be decreased. However, if toads breed before the arrival of newts, allowing tadpoles to grow, the predator will have much less of an effect on larval survival. Predation has no effect, however, on the size of larvae at metamorphosis. This is because survivors metamorphose at the same size as larvae reared without predators, as long as food does not become limiting. Whereas size (mass) at metamorphose does not change, the number metamorphosing is much less when predators are around.

Larval Southern Toads may be affected by a variety of physical and biotic conditions (food resources, oxygenation, siltation), which influence growth rates, size at metamorphosis, and other measures of larval success. The location in a pond where eggs are oviposited thus becomes a potentially important component of larval ecology. In a series of field experiments, Travis and Trexler (1986) demonstrated that the location where eggs developed (various levels of water circulation, siltation, food resources, oxygenation) affected larval growth and influenced whether density had an effect on larval development. Increased larval density generally inhibited growth rates in poor habitats, but tadpoles reared in good habitats actually had higher growth rates with increasing density. Increased densities of tadpoles increased the time to metamorphosis in harsh environments, but had no effect on the larval period in good habitats. Further, the initial size of hatchlings also affected metamorphic success, with small hatchlings never achieving similar growth rates with large hatchlings, regardless of conditions. Thus, the effect of larval density on metamorphic success is variable, depending on the initial size of the larva when exposed to varying physical and biotic components of the location where development takes place.

New metamorphs may vary substantially in size. In South Carolina, for example, metamorphs varied between 9.3 and 24.0 mm SUL (Pechmann et al., 2001), whereas in the Okefenokee Swamp they were 6.5–10.5 mm SUL, with a mode of 8 mm (Wright, 1932).

DISPERSAL

Since metamorphosis occurs synchronously, large numbers of juveniles disperse at the same time. Vast numbers congregate along the shoreline, and waves upon waves of metamorphs begin dispersing across the ground surface. Dispersal is largely diurnal, as opposed to movement of adults to and from breeding sites (Todd and Winne, 2006). However, many metamorphs will continue to disperse after dark. Rain is a poor predictor of juvenile movements, although metamorphs are more likely to move during daytime rainy weather than when it rains at night.

Dispersal may or may not be random, depending on the types of surrounding habitat, sex, and life stage. Dodd (1994) found that most Southern Toads dispersed generally randomly from a depression marsh of variable hydroperiod toward surrounding xeric hammock and sandhill habitats in north-central Florida. Immigration and emigration patterns appeared similar between males and females. However, juveniles entered and left the pond in slightly different directions. Emigrating males and females also exited via different pathways, with males tending to move toward the direction of distant hammocks. Some hammock habitat surrounded the pond in all directions, however, so it is possible that back-and-forth movements between the pond basin and surroundings habitats obscured orientation patterns. Since no reproduction occurred at the pond, the dispersal of metamorphs could not be ascertained.

DIET

Anaxyrus terrestris is a sit-and-wait ambush predator, and feeding occurs opportunistically. Movement is necessary to elicit feeding behavior, although toads will feed if their perceptions of motion are artificially induced (Kaess and Kaess, 1960). They even have been reported to scavenge on the carcasses of juvenile *Lithobates heckscheri*, although in all probability the toad was drawn to the carcass by the motion of ants (Beane and Pusser, 2005). Southern Toads eat a wide variety of invertebrates, particularly beetles, ants, true bugs, grasshoppers, katydids, and worms (Krakauer, 1968; Punzo, 1992a; Meshaka and Mayer, 2005; Meshaka and Powell, 2010), with Shaw (1802) indicating they are particularly attracted to luminous prey. They are able to discriminate between palatable and unpalatable prey, and can learn to use brightness (whether of prey or substrate background) as a cue toward

Amplexing *Anaxyrus terrestris*. Photo: Matt Greene

palatability, at least under laboratory conditions. Learning occurs slowly, however, and odor does not appear to influence prey choice (Mikulka et al., 1980).

PREDATION AND DEFENSE

Larval and adult Southern Toads possess a noxious skin secretion that assists in deterring predation. In larvae, this secretion is effective to the extent that larval activity does not change appreciably in the presence of predators (dragonfly naiads, newts, and sunfish) (Richardson, 2001). In adults, the skin secretion is produced in granular glands located primarily in the parotoid glands and warts located on the dorsal surface of the back and appendages.

In response to a terrestrial predator, *Anaxyrus terrestris* will flee or become immobile. They may go into a crouching posture with the chin tucked downward and into the body, or they may inflate the body with air and rear up on their hind legs (Marchisin and Anderson, 1978). Presumably these postures help orient the body's secretions toward the predator and make the toad more difficult to attack or handle.

The most important predator of Southern Toads is probably the Eastern Hognose Snake (*Heterodon platirhinos*), a toad specialist. Other predators include racers (*Coluber constrictor*), Eastern Indigo Snakes (*Drymarchon couperi*), garter snakes (*Thamnophis* sp.), water snakes (*Nerodia*), and even a Black-crowned Night Heron (Jones et al., 2010). Turtles also consume larvae and adults if they can catch one.

POPULATION BIOLOGY

The age and size at metamorphosis may have important implications for recent metamorph Southern Toads. In a series of experimental trials, sprint speed (movement bursts measured over short distances), metabolic rate, and the physiological endurance of metamorphs were positively correlated with body mass at metamorphosis, but negatively correlated with age at metamorphosis (Beck and Congdon, 2000). This means that decreases in the length of the larval period and increases in mass at metamorphosis may help a metamorph's ability to escape a predator, but that there may be an adverse energy cost on maintenance. As might be expected, the body mass and duration of the larval period are affected by food and temperature; as these factors increase, body mass increases and the duration of the larval period decreases. Metabolic rates of metamorphs decrease when larvae experience high food levels and high temperatures, but are higher compared with larvae raised at low temperatures but high food levels. Not only can age and size of larvae affect metamorph fitness, but also the larval environment that impacts its physiology (Beck and Congdon, 2000).

Experiments by Beck and Congdon (1999) have shown that the survivorship and growth rates of postmetamorphic individuals are not related to initial body size or to metabolic rate. However, the size at metamorphosis is positively correlated with survivorship to first census after a period of two weeks (that is, over a short-term period). The age and size at metamorphosis is not related to survivorship from metamorphosis to a second census after two months, however, and the size at metamorphosis is only marginally correlated with survivorship to two months. In addition, age and size at metamorphosis are not significantly correlated with total growth. Taken together, these results suggest that an initial advantage of large size at metamorphosis on postmetamorphic survival and growth is lost within a relatively short period. Although age and size at metamorphosis affect metabolism, they are not related to fitness through their effects on the subsequent survival and growth of the toadlets (Beck and Congdon, 1999).

Wright (1932) divided toads from Florida into various size classes and speculated that Southern Toads might take up to 8 yrs of age to reach the largest sizes, with maturity at ca. 3 yrs. Such a classification suggests slow growth rates for Southern Toads, but the classifications based on size used by Wright (1932) have no empirical basis.

Populations of Southern Toads can be large. For example, Gibbons and Bennett (1974) recorded >2,000 Southern Toads at 2 ponds over a 2 yr period, and Dodd (1992) captured 331 different toads over a 5+ yr period in a very small depression marsh in central Florida. Mass metamorphosis can result in hundreds of thousands of recent metamorphs moving across a landscape simultaneously (referred to in Florida as a "jubilee"). However, this does not necessarily mean that females produce large numbers of juveniles every year or at every breeding site. For example, Semlitsch et al. (1996) reported 816 females captured over a 16 yr period at a pond in South Carolina, but only 693 metamorphs. In Florida, Greenberg and Tanner (2005b) captured 1,321 adult Southern Toads at 8 ponds over a 7 yr period, and 1,359 metamorphs. However in both studies, most of the metamorphs resulted from a single breeding year. There is no correlation between the number of adults using a breeding site and the number of juvenile recruits to the population.

COMMUNITY ECOLOGY

Anaxyrus fowleri and *A. terrestris* may occasionally be found breeding in the same pond (e.g., Gosner and Black, 1956). Experiments have shown that competition occurs between the larvae of these species at high and medium densities, but not at low densities (Wilbur et al., 1983). Competition occurs because of the similarities between the species in feeding ecology and habitat preference. High densities also affected the size differential between the larvae of the two species and the length of the larval period (Wilbur et al., 1983). Thus, the amount of habitat available and the timing of breeding become important to larval ecology if the two species occupy the same breeding pond. The presence of a predator, such as *Notophthalmus*, can actually minimize the effects of competition among anuran larvae by eating many small larvae and thus reducing larval densities.

Larval *Anaxyrus terrestris* are intense competitors with larval *Pseudacris crucifer* and *Hyla gratiosa* (Morin, 1983). However, *Anaxyrus terrestris* larvae are sensitive to the density of larvae of other anurans, not just toad larvae. This sensitivity is manifested in slower growth rates, depending on density, although most individuals will survive and metamorphose. Plasticity in the size at metamorphosis also helps mediate conditions within the breeding site. If competition becomes too intense, larvae can metamorphose at a smaller size than they might normally.

Southern Toads also may be impacted by nonindigenous species such as the Cuban Treefrog (*Osteopilus septentrionalis*). The presence of larval *O. septentrionalis* causes reduced growth rates and delayed metamorphosis in *Anaxyrus terrestris*, at least under experimental trials (Smith, 2005a). In addition, the body mass of *A. terrestris* larvae at metamorphosis is decreased in the presence of Cuban Treefrog larvae. These species are syntopic in peninsular Florida, often using the same temporary small breeding sites.

DISEASES, PARASITES, AND MALFORMATIONS

Amphibian chytridiomycosis was not found on *A. terrestris* from specimens collected in South Carolina between 1940 and 1970 (Daszak et al., 2005), nor was it known from elsewhere in the Southeast through 2008 (Rothermel et al., 2008). Rizkalla (2010) reported the first positive tests for Florida from specimens collected in 2009. The fungus *Basidiobolus* sp., however, was found in Southern Toads from Florida (Okafor et al., 1984). The cestode *Distoichometra bufonis* was described from this species (Dickey, 1921), and the species is parasitized by the trematodes *Brachycoelium hospitale* and *Megalodiscus temperatus* (Manter, 1938). Biting midges of the genus *Corethrella* feed upon *Anaxyrus terrestris* and may be attracted by the frog's advertisement call (McKeever and French, 1991).

SUSCEPTIBILITY TO POTENTIAL STRESSORS

Nitrate. Nitrates in water can affect the duration of the larval period. High levels of nitrates (30 mg/L NO_3-N, or even fluctuations between 0–30 mg/L NO_3-N) act as a stressor, speeding up developmental rates of Southern Toads and causing larvae to metamorphose at smaller sizes than they would if not subject to nitrates (Edwards et al., 2006). However, other chemicals in natural water also may act to decrease growth rates and thyroxine concentrations, and nitrates may help modify a tadpole's response to environmental stressors. Thus, the type of water (e.g., spring water vs. reverse osmosis filtered water) in which laboratory studies are conducted is important when assessing the effects of nitrates on larval development.

Metals. Coal ash settling basins may have high concentrations of toxic chemicals, particularly arsenic, cadmium, chromium, copper, selenium, and other substances. In a series of transplant experiments, Rowe et al. (2001) demonstrated that Southern Toad larvae raised in such waters have a high rate of mortality compared to reference sites. This result involved both embryos and larvae, and stemmed from the toxic effects of the chemicals and the depauperate algal food sources at the coal ash site. If embryos survived to hatching (about 34%), then they had nothing to eat and were increasingly exposed to the toxic mix. Rowe et al. (2001) suggested that such basins could be population sinks for those species attempting to use them for breeding.

Toxic elements at coal ash settling basins also concentrate in adult toads frequenting these sites. Increased levels of selenium (17.4 ppm), arsenic (1.58 ppm), and vanadium (1.24 ppm) are observed in Southern Toads collected from coal ash basins, and transplant experiments suggested that these chemicals can be concentrated in significant amounts after only 7–12 weeks of exposure (Hopkins et al., 1998). These results indicate that Southern Toads bioaccumulate toxic substances and that bioaccumulation can occur rather rapidly.

Elevated levels of toxic substances in coal ash settling basins can lead to increased levels of corticosterone B in resident toads, whether they are calling or not. Testosterone levels are also higher in all months than at control sites, suggesting altered androgen production, utilization, and clearance. Males call into August at ash basins but not at control sites, suggesting that the toxic substances affect hormone levels, which in turn extends calling behavior to later in the season. Elevated levels of corticosterone B are evident in as little as ten days when toads are transplanted from control sites to coal ash settling basins (Hopkins et al., 1997).

Heat. Southern Toads oviposit eggs in thermally heated reservoirs associated with nuclear power plants. Most such eggs die because of thermal loading, and thus these "cooling" reservoirs become population sinks. However, some eggs and tadpoles may survive in shallow areas receiving cooling inflows from surrounding springs, seeps, or creeks. In these areas, temperatures may be sufficiently cool to allow development. However, resulting larvae tend to grow faster than conspecifics in unaffected surrounding areas, and the larvae transform at a smaller size than normal (Nelson, 1974).

STATUS AND CONSERVATION

Undoubtedly, many populations of Southern Toads have been lost as wetlands and adjacent uplands have been destroyed or modified over the last 500 yrs. This species appears to be somewhat resilient, however, and is still frequently observed in suburban and agricultural regions throughout the Southeast (Delis, 1993). Highway-related mortality must be very high, as these toads are frequently found on paved roads on warm rainy nights (M.J. Allen, 1932; Sutherland et al., 2010). Even in suburban settings, mortality can be high where roads are located next to retention ponds, and toads are drawn to nearby streetlights where they congregate to feed. Southern Toads readily use culverts, when available, to traverse under highways. The placement of tunnels is crucial to their effectiveness in toad conservation as toads may use topographical features such as lake basin rims to facilitate movement (Dodd et al., 2004).

There are a variety of other factors that likely

Anaxyrus terrestris breeding habitat, McIntosh County, Georgia. Photo: C.K. Dodd Jr.

influence Southern Toad populations, including climate change. There is growing evidence that changes in rainfall patterns in the Southeast may have affected reproduction. Despite regularly stable numbers of immigrating females, for example, some populations produced none (Dodd, 1994, over a 5 yr period) or few metamorphs as hydroperiods decreased or ponds did not fill (Semlitsch et al., 1996; Daszak et al., 2005; Greenberg and Tanner, 2005b). More data are needed from long-term studies to differentiate what level of reproduction is "normal" (i.e., in terms of frequency and recruitment) for a toad population to maintain itself through time.

Silvicultural activities impact much of the habitat occupied by Southern Toads on the southeastern coastal plains. Long-term effects in a variety of habitat types from various silvicultural treatments are few as long as suitable nearby habitat is available; *A. terrestris* may be one of the most dominant amphibians within its environment, even in human-disturbed habitats (Dodd et al., 2007). Results of studies that measured short-term effects on the numbers of toads captured within 1.5–2 yrs of cutting also have suggested little impact on the species at wetlands surrounded by various levels of treatment (Clawson et al., 1997; Russell et al., 2002b). The latter authors used counts rather than abundance to assess treatment effects, and the treatments were conducted in such a manner as to cause minimal disturbances to wetlands and to minimize surface disturbances even in cutover treatments. Toads are frequently found in open habitats created during forestry operations, such as in canopy gaps and along skidder trails (Cromer et al., 2002). Such areas may contain high insect abundance.

Likewise, a study that followed the effects of clearcutting in a Florida pine flatwoods also could not detect short-term effects in the numbers of Southern Toads using the site (Enge and Marion, 1986). Southern Toads may even be found more often in previously ditched cypress ponds than around cypress ponds holding water (Vickers et al., 1985). In contrast, Hanlin et al. (2000) resurveyed an area studied by Bennett et al. (1980) and found many fewer toads than in the previous study despite a more extended sampling period (3,312 vs. 700). Habitat changes associated with pond restoration may have influenced subsequent use by toads. As with previous assessments, relative abundance was assumed based solely on count data. Simply counting toads, however, provides an inexact picture of the short-term effects of clearcutting. Southern Toads appear to tolerate prescribed burning and are usually found soon after burning (Floyd et al., 2002; Moseley et al., 2003).

Studies that provide data on habitat use and demographics of toad populations give a somewhat different picture of the short-term effects of silviculture, and Todd and Rothermel (2006) have argued for the importance of examining metrics other than abundance in determining site impacts on amphibians resulting from silviculture. For example, Southern Toads will use breeding ponds adjacent to or in clearcut habitats. Although juveniles are readily produced from such ponds, they have lower survivorship than conspecifics reared in forested habitats (Todd and Rothermel, 2006). In addition, growth rates are less in clearcuts than in forested areas and, as a result, juveniles found in clearcuts are smaller than those in adjacent forests. Todd and Rothermel (2006) noted the importance of examining metrics other than abundance in determining site impacts on amphibians resulting from silviculture.

Southern Toads also use habitat differently in clearcuts than they do in forested habitats. They move farther in forests than in adjacent clearcuts, but movement patterns are rather similar regardless of sex, body size, or environmental cues (humidity, wind, air temperature, precipitation, soil moisture, sky conditions) (Graeter, 2005; Graeter et al., 2008). Movements occur in straighter paths in forested habitats than in clearcuts, where toads wander in no particular direction. Southern Toads

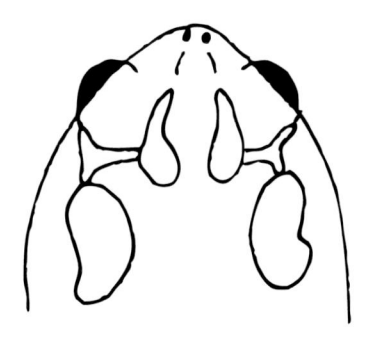

Head pattern of *Anaxyrus terrestris*. Illustration by Camila Pizano

may treat large clearcuts as filters to extensive movements, but they do not avoid them entirely and readily cross the forest-clearcut edge. In an experimental choice test, toads moved between 3.5 and 324 m (mean 46 m) over a 5-night period when released into a managed forest in South Carolina. Toads traversed small clearcuts easily within a 24 hr period (Graeter, 2005; Graeter et al., 2008). In large clearcuts, however, the increased searching behavior by Southern Toads could expose them to predation or desiccation, and discourage them from traversing the area.

Anaxyrus terrestris readily colonizes restored or newly created mitigation wetlands (Pechmann et al., 2001). However, not all such ponds are successful in producing juveniles. Of the four ponds studied by Pechmann et al. (2001), only one pond produced substantial numbers of metamorphs, and then in only three of eight years during which the pond was studied. Populations even can recover quickly in some coastal areas that receive hurricane overwash with resulting temporary increases in salinity (Gunzburger et al., 2010).

A total of 16,005 *A. terrestris* were exported from Florida between 1990 and 1994 for the pet trade (Enge, 2005a).

Anaxyrus woodhousii (Girard, 1854) Woodhouse's Toad Rocky Mountain Toad (*A. w. woodhousii*); Southwestern Woodhouse's Toad (*A. w. australis*)

ETYMOLOGY

woodhousii: a patronym honoring the first collector of this toad, S.W. Woodhouse (1821–1904). Woodhouse was the naturalist and Assistant Army Surgeon who accompanied Captain L. Sitgreaves on his expedition to the Zuni and Colorado rivers in 1851, during which the toad was collected (Woodhouse, 1854). Woodhouse had previously participated in western exploring expeditions with Sitgreaves in 1849–1850.

NOMENCLATURE

Conant and Collins (1998) and Stebbins (2003): *Bufo woodhousii*

Synonyms: *Bufo aduncus, Bufo antecessor, Bufo compactilis woodhousii, Bufo dorsalis, Bufo frontosus, Bufo lentiginosus frontosus, Bufo lentiginosus woodhousei, Bufo planiorum, Bufo velatus, Bufo woodhousei, Bufo woodhousei australis, Bufo woodhousei bexarensis, Bufo woodhousii velatus, Incilius woodhousei*

This species was described as *Bufo dorsalis*, but this name had been used previously and was thus unavailable for the toad. The name *woodhousii* is attributed to Girard (1854).

IDENTIFICATION

Adults. This is a medium to large dry-skinned, brown to tan to olive toad (although some animals may be greenish to gray), with small dark brown to black irregular spots containing one wart per spot. Light brown warts are observed within the darker blotches, and these blotches tend to increase in number with the size of the toad (Keeton and Carpenter, 1955). Parotoids are oblong, elongated, narrow, and elevated and touch the cranial crests. A light mid-dorsal stripe is present. Venters are mostly white to yellowish and unmarked, except that some animals may have a spot or spots in the pectoral region or on the anterior third of the venter (Blair, 1943a). Males have conspicuous dark throats during the breeding season, but these lighten considerably during the remainder of the year. They also develop nuptial pads.

Adult males are usually slightly smaller than females. In Arizona, adults were measured at 49–91 mm SUL (mean 74.5 mm) (Goldberg et al., 1996b) in one study, with males in another population having a mean of 83 to 88.1 mm SUL and females a mean of 93.2 to 104.2 mm SUL, depending on year (Sullivan, 1987). Blair (1941a) provided mean SULs for adults from 9 populations; they ranged from 62 to 81 mm. Other specific size measurements include: females with a mean of 83.3 mm SUL from Texas (Meacham, 1962); a female mean of 88.8 mm SUL in Oklahoma (Bragg and Sanders, 1951); female means of 97.3 mm SUL (range 64 to 113 mm) and male means of 73.5 mm SUL (58–69 mm) in Nebraska (Ballinger et al., 2010); and male means of 55.2 to

64.4 mm SUL in southwestern Utah and adjacent Nevada (A.P. Blair, 1955). Conant and Collins (1998) reported a maximum size of 127 mm SUL.

Larvae. Tadpoles are small and dark brown to gray or slate, often with light mottling or gold flecking. The dorsal musculature of the tail is lighter than the body, and the ventral musculature is immaculate. The tail fins have a few scattered dark flecks, which are more numerous on the dorsal tail fin than on the ventral tail fin. The tail tip is rounded. Venters are lighter than the dorsum. At hatching, the larvae are tiny (2.5–3 mm) but reach a maximum size of 23 mm TL (Youngstrom and Smith, 1936). Stuart (1991) reported amelanistic tadpoles from New Mexico.

Eggs. Eggs are deposited in two long strings within a continuous gelatinous casing and no partitions between the eggs. Within the strings, eggs are normally in single rows, although they occasionally may be "crowded" together (Smith, 1934). The vitellus is 1–1.5 mm in diameter and is black dorsally and tan to yellow ventrally. There is a single gelatinous capsule surrounding the vitellus, which is 2.6–4.6 mm in diameter (mean 3.5 mm) (Livezey and Wright, 1947).

DISTRIBUTION

Woodhouse's Toad is a species of the Great Plains and semiarid west; *A. woodhousii* is absent from the Rocky Mountains. It occurs from North Dakota and Montana southward to the Gulf Coast in eastern Texas, and its range covers much of northern and central Texas, Utah and western Colorado, northern Arizona and eastern Wyoming, Colorado, and New Mexico. In Missouri, Woodhouse's Toad is found in the western part of the state, but the range extends along the Missouri River east nearly to the Mississippi River. Populations extend along the middle Colorado River into Nevada and northwestern Arizona. Isolated populations occur in Washington and Idaho (Snake and Columbia rivers), along the lower Colorado River in California, Arizona, and Mexico, west Texas, and along the lower Rio Grande. The species overlaps broadly with *Anaxyrus fowleri* in east Texas, southwestern Arkansas, and Louisiana; hybridization may be extensive, and toads from this region may actually include genes from *A. fowleri* and *A. americanus* as well as *A. woodhousii* (Pauly et al., 2004). It overlaps with *A. speciosus* throughout much of

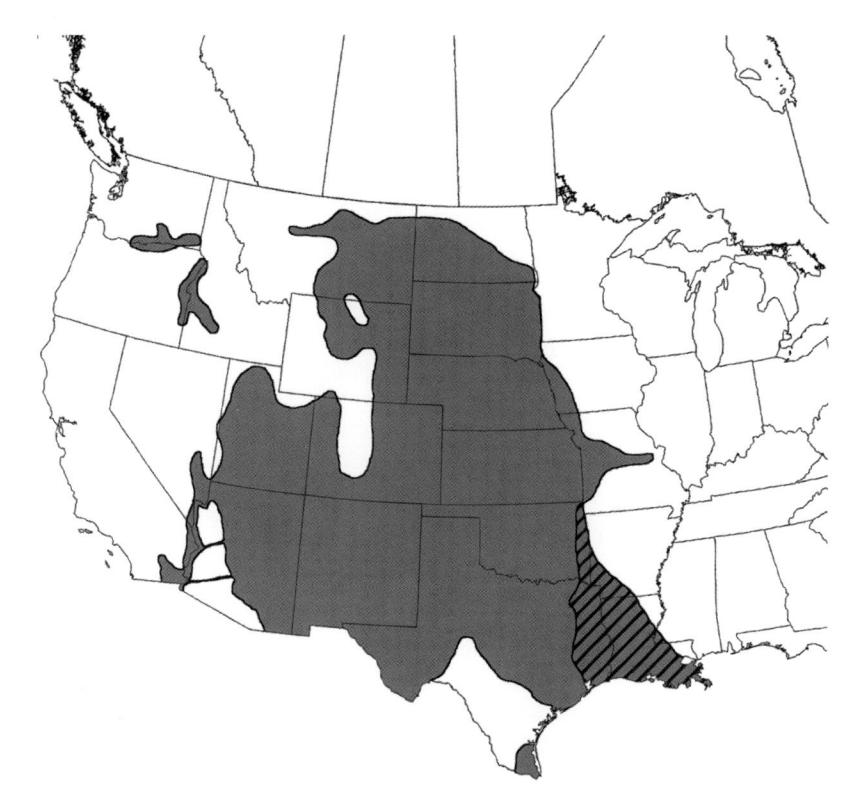

Distribution of *Anaxyrus woodhousii*. The hatched area indicates an extensive area of hybridization with *Anaxyrus fowleri*. It may not be possible to assign specific status to toads in this area.

New Mexico and southeastern Arizona, where hybridization also may be extensive.

Important distributional references include: Arizona (Eaton, 1935; Vitt and Ohmart, 1978), California (Vitt and Ohmart, 1978), Colorado (Hammerson, 1999), Idaho (Slater, 1941b; Nussbaum et al., 1983), Kansas (Smith, 1934; Collins, 1993; Collins et al., 2010), Missouri (Johnson, 2000; Daniel and Edmond, 2006), Mexico (López et al., 2009), Montana (Black, 1970, 1971), Nebraska (Lynch, 1985; Ballinger et al., 2010; Fogell, 2010), Nevada (Linsdale, 1940; Bradford et al., 2005a), New Mexico (Van Denburgh, 1924; Degenhardt et al., 1996), North Dakota (Wheeler and Wheeler, 1966; Hoberg and Gause, 1992), Oklahoma (Sievert and Sievert, 2006), Oregon (Nussbaum et al., 1983; Leonard et al., 1993; Jones et al., 2005), South Dakota (Peterson, 1974; Fischer, 1998; Ballinger et al., 2000; Kiesow, 2006), Texas (as a mixture of species and subspecies; Dixon, 2000), Utah (Tanner, 1931), Washington (Slater, 1939b, 1955; Metter, 1960; Nussbaum et al., 1983; Leonard et al., 1993; Jones et al., 2005; McAllister, 1995), and Wyoming (Baxter and Stone, 1985).

FOSSIL RECORD

Fossil *A. woodhousii* have been reported from Miocene deposits in Arizona, Pliocene deposits in Arizona, Kansas, and Texas, late Pliocene–early Miocene deposits in Kansas, and Pleistocene deposits in Arizona, Colorado, Kansas, Nevada, New Mexico, South Dakota, and Texas (Tihen, 1962b; Holman, 2003). The species may be referred to as *Bufo* cf. *B. woodhousii* in the fossil literature. A very large (100–160 mm SUL) fossil subspecies, *Bufo woodhousii bexarensis*, was described from the late Pleistocene of Texas (Mecham, 1958).

SYSTEMATICS AND GEOGRAPHIC VARIATION

Anaxyrus woodhousii is a member of the *Americanus* clade of North American bufonids (Blair, 1963b), a group that includes *A. americanus*, *A. baxteri*, *A. fowleri*, *A. hemiophrys*, *A. houstonensis*, and *A. terrestris*. It is closely related to *A. americanus* and somewhat more distantly to *A. fowleri* and *A. terrestris* (Masta et al., 2002). Allopatric divergence among these lineages probably occurred during the Pleistocene. *Anaxyrus woodhousii* evolved in the Great Plains and expanded into the southwest during the mid- to late Pleistocene. Masta et al. (2003) reviewed the evolutionary history of *A. woodhousii* and its expansion to its present geographic distribution.

There are two distinct clades with *A. woodhousii*, one in the southwestern United States, originally described as *Bufo woodhousii australis* (Shannon and Lowe, 1955), and the other comprising the remainder of the range (Masta et al., 2003). *Anaxyrus w. australis* has distinct narrow interorbital ridges separated by a frontal trough, whereas in *A. w. woodhousii* the interorbital ridges are shallow and in contact, and there is no frontal trough. The Southwestern Woodhouse's Toad occurs from central Colorado through New Mexico and Arizona and along the Rio Grande of southwestern Texas. Most authors (e.g., Brennan and Holycross, 2006) currently treat it as a subspecies of *A. woodhousii*, although it is likely that *A. w. australis* will be considered a full species as more data become available.

The subspecies *A. w. velatus* was described by Bragg and Sanders (1951) and elevated to specific status by Sanders (1986). The taxon was said to be smaller, less spotted, and darker than *A. fowleri*, to possess a different call, and to have a pectoral velum (a darkened area ventrally between the front limbs). This phenotype was confirmed by Masta et al. (2002) as representing a hybrid population between *A. woodhousii* and *A. fowleri*. It is not currently recognized, despite the phenotypic-based arguments in Dixon (2000). Although Sanders (1987) described two additional species (as *Bufo antecessor* and *B. planiorum*) based on morphological data, they are clearly referable to *A. woodhousii,* and their recognition is not supported by molecular data.

Natural hybridization with other *Anaxyrus* species occurs throughout the range of *A. woodhousii* and has been reported with *A. americanus* (Blair, 1956a; Ideker, 1968; Brown, 1970), *A. cognatus* (Gergus et al., 1999), *A. fowleri* (Meacham, 1962), *A. houstonensis* (Brown, 1971b; Hillis et al., 1984), *A. microscaphus* (A.P. Blair, 1955; Sullivan, 1986a, 1995; Sullivan and Lamb, 1988; Goldberg et al., 1996b; Lamb et al., 2000; Schwaner and Sullivan, 2009), *A. punctatus* (McCoy et al., 1967; Malmos et al., 1995; Hammerson, 1999), *Ollotis alvaria* (Gergus et al., 1999), and *O. nebulifer* (Thornton, 1955; Brown, 1971a; Sanders, 1986). Although hybridization occurs "naturally," many hybrid populations are found in areas disturbed by human activity, indicating a breakdown in ecological isolating mechanisms that normally would keep the various species apart (Sullivan, 1986a). Hybridization may also be facilitated by the scramble competition mating strategy used by some toads.

The advertisement calls of hybrids in nature tend to be intermediate in characteristics between the parental species (Blair, 1956a, 1956b; Ideker, 1968), although in some cases the calls may not be different enough to prevent interspecific mating (Malmos et al., 2001). Some suspected hybrids, however, have atypical calls that are little more than clicks, whereas others have very extended trills (Ideker, 1968). In contrast to intermediacy, the calls of hybrid *Anaxyrus woodhousii* × *A. microscaphus* are highly repeatable in call duration and pulse rate (65–80 pulses/sec), much more so than their parental species (Malmos et al., 2001). This suggests that hybrids do not call less than the parentals nor are their calls more variable. In a Texas fish hatchery, 6% of the calling males consisted of *A. woodhousii* × *Ollotis nebulifer* hybrids (Brown, 1971a) and in Arizona, 25% of a chorus was hybrid *Anaxyrus woodhousii* × *A. microscaphus*, with the remainder all *A. microscaphus* (Malmos et al., 2001).

The extent of hybridization may change through time, and contact zones may be dynamic areas of gene exchange. For example, *A. woodhousii* hybridizes with *A. microscaphus* along the Beaver Dam Wash, where it intersects the Virgin River in southwestern Utah and adjacent Arizona. Based on collections from 1949 to 1953, hybrid individuals possessed a mostly *A. microscaphus*-like phenotype. By 2001, however, hybrids within the contact zone exhibited mostly *A. woodhousii*-like phenotypes (Schwaner and Sullivan, 2009). Sullivan and Lamb (1988) indicated that *A. woodhousii* appeared to be replacing *A. microscaphus* in this area. Molecular data confirmed a "hybrid swarm" at the contact zone, and demonstrated considerable *A. woodhousii* introgression with populations of *A. microscaphus* that extended a considerable distance up Beaver Dam Wash. The toad population at the confluence of the rivers consists of the parental species, F_1 hybrids, and individuals resulting from reciprocal backcrossing. Note that these results are in contrast to earlier suggestions of temporal stability in this region (Sullivan, 1995), although this study was based solely on morphological and acoustic characters.

Hybrids between this species have been produced in numerous laboratory experiments with the following species (reviewed by Blair, 1972): *A. americanus* (A.P. Blair, 1941a, 1946; W.F. Blair, 1963a); *A. boreas* (Moore, 1955; Blair, 1959); *A. californicus* (Moore, 1955); *A. cognatus* (Blair, 1959); *A. fowleri* (Blair, 1941a; Meacham, 1962); *A. hemiophrys*, *A. houstonensis*, *A. punctatus* (Moore, 1955); *A. microscaphus*, *A. speciosus* (Moore, 1955); *A. terrestris* (Blair, 1961a, 1963a); *Ollotis nebulifer* (Thornton, 1955; Blair, 1956a, 1959). However, some crosses work only with one parent (e.g., ♂ *O. nebulifer* and ♀ *Anaxyrus woodhousii*), and even then resulting males and the offspring of the reciprocal cross are sterile (Thornton, 1955). In other cases (e.g., with *A. fowleri*), hybridization produces fully fertile offspring that are capable of backcrossing with other hybrids or the parental species and still produce offspring (Blair, 1941a; Meacham, 1962; Blair, 1963a). Some crosses, however, are not successful, such as those with other members of the *Ollotis nebulifer* group of toads (Blair, 1966).

ADULT HABITAT

Woodhouse's Toad occupies a wide variety of mostly mesic habitats from the central and southwestern Great Plains to arid habitats in the American Southwest. Habitat types include plains grasslands, mountain grasslands, pinyon-juniper woodland, riparian woodland, ponderosa pine forest, spruce-fir forest, open fields, and suburban environments (Bragg, 1950b; Aitchison and Tomko, 1974; Ballinger at al., 2010). This species is often the most abundant toad in sandy habitats, although not restricted to them, and is partial to sandy floodplains and bottomlands with deep friable soils (Bragg and Smith, 1942, 1943; Timken and Dunlap, 1965; Hammerson, 1999). It is found in the depths of the Grand Canyon (Miller et al., 1982; personal observation) and occurs at elevations to 2,133 m in Montana (Black, 1970), 2,440 m in Colorado (Hammerson, 1999), and 900–2,400 m in New Mexico (Degenhardt et al., 1996).

TERRESTRIAL ECOLOGY

Activity occurs from March or April through September or October, depending upon latitude (e.g., Collins, 1993; Engeman and Engeman, 1996). This species is largely nocturnal in its foraging habits throughout much of the warm season. Diurnal activity may occur on cloudy days and in wet weather, and small toads tend to be more diurnal than large adults. Woodhouse's Toads frequently congregate under streetlights or other bright lights to which insects are drawn. Dispersal of recent metamorphs occurs en masse, with diurnal activity occurring even during very hot weather; some dispersal may occur at night. The young toadlets seek moist locations as they move away from a breeding pond and avoid the driest locations. As the toads grow, they become less diurnal. Adults are active only at night except during cloudy or rainy weather. This species is capable of considerable movement. They have been found up to 1.9 km straight-line distance from original point of capture in a mark release study in Oklahoma (King, 1960), and Thornton (1960) noted a movement of 643 m from the site where a toad was originally marked.

Activity occurs over a range of temperatures, with literature reports of from 15 to 33.7°C (Scott and Carpenter, 1956; Fitch, 1956a; Stebbins, 1962; Brattstrom, 1963; Hammerson, 1999). However, Hammerson (1999) noted activity as low as 8°C. Although this species is often found in very cold environments, it is not tolerant of freezing temperatures and possesses no cryoprotectants (Swanson et al., 1996). They are sometimes found in mammal burrows, such as those of prairie dogs (*Cynomys ludovicianus*) (Taylor, 1929; Kretzer and Cully, 2001; Lomolino and Smith, 2003), or they dig shallow burrows where they can escape midday heat.

Woodhouse's Toads are photopositive in their phototactic response, suggesting they use both sunlight and moonlight when feeding and going about their daily activities (Jaeger and Hailman, 1971, 1973). They are sensitive to light in the blue spectrum ("blue-mode response"), which apparently helps them orient toward areas of increasing illumination, such as the open area above a pond (Hailman and Jaeger, 1974). Woodhouse's Toads likely have true color vision.

CALLING ACTIVITY AND MATE SELECTION

Anaxyrus woodhousii is a spring to summer breeder, with most breeding occurring between February and May in Texas and Oklahoma (Force, 1930; Bragg, 1940a; Thornton, 1960; Meacham, 1962) and April to August in Montana (Black, 1970). Even in Texas, however, calls may be heard as late as July, and Bragg (1940a) reported chorusing in Oklahoma in early August. The advertisement call is a loud burst, but not a melodic trill as in related members of the *Americanus* group of toads. Calls are short, ranging from 0.9 to 2.6 sec, and cover a wide range of frequencies (dominant frequency 1,500–1,874 cps) (Blair, 1956b). Bragg (1940a) also described a call he termed a "whew," that is, a high-pitched, long-duration call with a rising inflection. Calling may be initiated by rainfall, especially after long dry periods, but rainfall is not necessary for calling or breeding to occur (Thornton, 1960). Calling occurs at temperatures >14.5°C (Meacham, 1962), and Thornton (1960) and Sullivan (1982a) noted optimal temperatures for calling of 17–30°C. Most chorusing occurs in the early evening.

Males call from shallow water near the shoreline, although they may call farther from shore if the water is not deep. Calling rarely occurs on land. In some areas, chorusing males tend to be clumped in groups around a breeding site (Sullivan, 1982a), and males in these large choruses call more often than males in small choruses (Woodward, 1984). Sullivan (1985a) characterized this as a lek mating system, a system, which in this case, is not shaped by female preference for large choruses (Sullivan, 1986b). In other areas, males call from dispersed locations and do not tend to clump near one another (Bragg, 1940c). Calling sites are defended, and males act aggressively toward intruders (Sullivan, 1982a). At an intruder's approach, there may be a series of back-and-forth calls between the rivals, followed by amplexus of the intruder by the resident male and extended wrestling. Calling males invariably win such matches. While they are calling, males are very alert and will dive into the water if approached. It is not unusual to hear only a few males calling at once, but choruses in the hundreds have been reported.

Males arrive earlier in the evening at breeding ponds than females. They are nonspecific in their choice of mates, and will amplexus any female that moves within their vicinity. Not surprisingly, interspecific amplexus is not uncommon. For example, Brown (1971a) noted that as many as 8.8% of amplexed pairs at a Texas fish hatchery consisted of mixed *A. woodhousii* × *Ollotis nebulifer* individuals (also see Thornton, 1960), and Bragg (1939) observed amplexus between *Anaxyrus cognatus* and *A. woodhousii*. Males have a characteristic release call, which is uttered in conjunction with body vibrations when the toad is amplexed or handled (84–104 vibrations/sec; Blair, 1947b). The pulse rate during the release call is positively correlated with temperature, but the duration of the release vibrations and the frequency of the release call is inversely proportional to temperature (Brown and Littlejohn, 1972). A male can continue the release calls and vibrations for a considerable time. Small females may utter a short "peep" when surprised, but the call is not repeated (Bragg, 1940a).

Despite the extended breeding season, covering several weeks, individual males and females may spend only a few days at the breeding pond. For example, males spent from 1.9 to 2.4 nights at a breeding site in New Mexico, whereas females spent from 1.0 to 1.4 nights at the pond (Woodward, 1984); chorus tenure was positively associated with male mating success. Similar residency was observed in Arizona (Sullivan, 1982a), but the length of stay at a chorus was not an important determinant of mating success (Sullivan, 1985a). Chorusing does not occur every night at all sites; Sullivan (1982a) noted that choruses formed from 6 to 36 nights (mean 23) at 5 breeding choruses in Arizona. In 2 other populations, choruses lasted from 23 to 45 days (Sullivan, 1986b). He later (1989) noted a mean chorus duration of 19.8 days, with an average chorus size of only 5–6 males at another desert location. Some males and even females move among nearby choruses rather than remain at a single site throughout the breeding season. Males may revisit a chorus several times during the breeding season (Woodward, 1982b). Breeding population size varies annually. For example, Thornton (1960) marked 9–87 toads (6–64 males, 3–23 females) over a 3 yr period at a site in Texas.

Although larger males may have greater reproductive success than small males (Woodward, 1982a, but see Sullivan, 1983), the sequence of arrival is not dependent on the size of the males (i.e., large males do not arrive earlier in the season than small males), and the operational sex ratio (1 male per 0.26–0.34 females) remains about the same at a breeding site throughout the season (Woodward, 1984). In contrast, Sullivan (1987) found that larger males tended to arrive earlier in the season than smaller males. The size of the male is not associated with chorus tenure, so that large males do not spend more time in a chorus than small males (Woodward, 1982b).

Male call rates are positively correlated with mating success (Sullivan, 1982a, 1983, 1987), and the call rate is not limited by aerobic capacity

Egg strings of *Anaxyrus woodhousii*. Photo: Dana Drake

(Sullivan and Walsberg, 1985). This suggests that the male's call rate is not an accurate indicator of his phenotypic vigor. Males lower their call rates and avoid acoustic overlap with all stimuli centered at 1.4 kHz (Sullivan, 1985b; Sullivan and Leek, 1986). This reduces the extent of overlap with nearby calling males, and could lead to new chorus formation as the density of calling males increases. In this regard, male call rates are inversely proportional with chorus density (Sullivan, 1985b), but the intensity of sexual selection increases with chorus size (Sullivan, 1986b). Calling male size may or may not be correlated with mating success. In some years and at least some locations, the correlation is positive, whereas in other years it is not (Sullivan, 1983, 1987). In addition, there is no evidence of size-assortive mating (i.e., large females mating with large males) (Woodward, 1982a; Sullivan, 1983). In acoustic choice tests, females do not prefer the calls of large males over those of small males.

Most males do not successfully breed within a season, whereas nearly all females do. Woodward (1984) calculated a male's chance of finding a receptive female each night at 14%, whereas the female nightly success rate was 71–95% and 100% through the season. In Woodward's (1984) study, the number of calling males varied between 1 and 120 per night. Males amplex from 0 to 3 females during a breeding season (Sullivan, 1982a; Woodward, 1982b), and mating success can vary from one location to another (Sullivan, 1986b).

Relatively few males (5–10%) and females (5–15%) captured one year at a pond return to the same pond to breed in subsequent years (Sullivan, 1987). Sullivan (1987) attributed the low recapture rate to high population turnover rather than toads breeding at other sites.

BREEDING SITES

Breeding sites include a wide variety of habitats, including swamps, small ponds, river backwaters and sloughs, flowing desert streams, and shallow waters in the littoral zone of small lakes (Bragg, 1941b; Bragg and Smith, 1942). Both temporary and permanent sites are used (Woodward, 1987a). Woodhouse's Toad readily breeds in man-made habitats, such as constructed ponds or lakes, stock tanks, irrigation ditches, borrow pits, garden ponds, gravel pits, and flooded fields. This species appears to prefer muddy or silty water, but does not breed in buffalo wallows.

REPRODUCTION

Breeding occurs over an extended period during the warm season, even if most breeding occurs in late spring to early summer. For example, most breeding occurs in May and June in Nebraska (*in* Ballinger et al., 2010) and from April to June in Colorado (Hammerson, 1999). In Oklahoma, breeding extends from March to September (Bragg, 1950a). Despite the extended breeding season, successful breeding actually occurs on only a few nights (Woodward, 1982b). Rainfall is not necessary to stimulate breeding activity, although reproduction frequently occurs after periods of rain.

The female chooses the oviposition site away from the main chorus of calling males. Eggs are deposited in two long strings, one from each ovary, along the substrate and among vegetation near the shoreline. Although breeding may occur throughout a pond, eggs are often oviposited by different females in close proximity to one another. Clutch sizes range as high as 25,644 (Smith, 1934), with Woodward (1987a) giving a mean clutch size of 10,469 in New Mexico, Collins (1993) providing a mean clutch size of 8,500 for 2 Kansas clutches, and Krupa (1995) recording a clutch of 28,500 in Oklahoma. The body weight of the female is positively correlated with clutch weight, egg weight, and the number of eggs per clutch (Woodward, 1987a). Hatching occurs ca. three days following oviposition (Youngstrom and Smith, 1936). The ovarian cycle is described by Clark and Bragg (1950).

LARVAL ECOLOGY

The larvae become free swimming when they attain a length of 6.5–7.0 mm after 5 to 6 days (Youngstrom and Smith, 1936). Developmental rate varies with temperature, food resources, and conspecific density and size. The length of the larval period is a minimum of 34 days and can take as many as 70 days, although Hammerson (1999) gives the larval period as 4 to 7 weeks. The front limbs appear after the maximum size has been attained. Metamorphosis begins when the tadpole is about 30 mm TL, and recent metamorphs average 10–15 mm SUL (Bragg, 1940a; Degenhardt et al.,

Larval *Anaxyrus woodhousii*.
Photo: John Cossel

1996; Hammerson, 1999). In South Dakota, Malaret (1978) measured 3 recent metamorphs at 15, 15.9, and 17.9 mm SUL, but these likely had already grown several mm.

Larvae isolated from conspecifics have the same growth rates as larvae raised in pairs. However, large larvae tend to inhibit the growth rates of smaller conspecifics and may negatively impact survivorship. Woodward (1987b) suggested that the larvae of clutches oviposited late in a season would, therefore, be at a disadvantage when compared with larvae hatching from earlier clutches. Regardless, hatchlings tend to metamorphose at similar sizes.

DIET

Larvae eat algae and organic detritus and may consume animal matter. Postmetamorphic *Anaxyrus woodhousii* eat a wide variety of prey, nearly all of it insects (Tanner, 1931; Smith, 1934; Smith and Bragg, 1949; Stebbins, 1962; Flowers, 1994; Flowers and Graves, 1995; Hammerson, 1999). Prey composition may change from one year to the next depending on availability. Specific items include ants, many types of beetles, lepidopterans, cicadas, crickets, grasshoppers, weevils, cutworms, spiders, phalangids, sowbugs, true bugs, millipedes, and centipedes. Juveniles eat many types of insects, particularly beetles and mites (Flowers, 1994). Gehlbach and Collette (1959) suggested that Woodhouse's Toad might feed underwater or at the water's surface since prey included dytiscid, hydrophilid, and stratiomyid larvae. A large female in captivity was reported to eat an unidentified hatchling turtle (King, 1960).

Males may have fewer items in the stomach per individual than females, and juveniles eat much the same prey as adults, except that small ants, collembolans, and mites make up a much greater proportion of the diet (Flowers and Graves, 1995). Of course, small invertebrates are preferred. Regional differences may occur in the diets of this species. For example, certain ground beetles (*Agonum* sp.) make up a high percentage of the diet of *Anaxyrus woodhousii* from eastern Oklahoma, but only a tiny percentage from western populations. Just the reverse is seen in the prevalence of small ants in the diet (Smith and Bragg, 1949). Such differences likely reflect differences in prey availability.

PREDATION AND DEFENSE

Woodhouse's Toads are cryptic and possess noxious skin secretions, especially on the parotoids and dorsal warts. These secretions are effective at deterring predators, as evident by the unsuccessful predation attempt on a Woodhouse's Toad by an American Bullfrog (Brown, 1974). When approached, adults may crouch down against the substrate and remain motionless for some time before attempting escape (Bragg, 1945a).

Predators of postmetamorphs include frogs (*Lithobates catesbeianus*), snakes (*Heterodon nasicus, Nerodia sipedon, Pantherophis* sp., *Pituophis catenifer, Thamnophis cyrtopsis, T. marcianus, T. sirtalis*), birds (hawks), and mammals (skunks, raccoons) (Bragg, 1940a; Gehlbach and Collette, 1959; Woodward and Mitchell, 1990; *in* Hammerson, 1999; *in* Johnson, 2000). Woodward and Mitchell (1990) noted that males were more likely to be eaten than females, and that mammalian predators tended to disembowel toads and leave the skin with the noxious granular glands intact. Larvae are eaten by a wide variety of aquatic invertebrates and by the larvae of spadefoots (*Spea, Scaphiopus*).

POPULATION BIOLOGY

Individual growth is rapid in *Anaxyrus woodhousii*, with Oklahoma toads reaching 43–65 mm SUL (mean 53.4 mm) by October prior to their first winter dormancy (Bragg, 1940a). The estimated growth rate in Bragg's (1940a) study was 0.3 mm/day. In general, toadlets double their SUL within the first month after metamorphosis. First reproduction likely occurs the first (males) or second (females) year after metamorphosis (Thornton, 1960).

As with many frogs, the sex ratio at a breeding pond is usually male-biased. Bragg (1940a), however, suggests that the overall sex ratio actually may be female-biased, based on captures at foraging locations. Because some females retain eggs late in the breeding season and female sex ratios at breeding ponds are very different from sex ratios in foraging areas, it has been suggested that some females may skip a breeding season, even when environmental conditions are favorable (Bragg, 1940a; Bragg and Smith, 1943). Longevity in the wild is unknown, but Engeman and Engeman (1996) observed a toad trapped in a window well in Denver living outdoors for 19 yrs.

COMMUNITY ECOLOGY

This species is allopatric or comes into contact with a number of other closely related toads within the *Anaxyrus americanus* complex. As noted above, hybridization may occur along contact zones between species, but introgression may not occur as extensively as hybridization experiments might suggest. For one thing, females are able to discriminate advertisement calls. These calls are often different among species and serve as a premating isolating mechanism. Advertisement calls also help females discriminate conspecifics from hybrids. For example, females respond

Adult *Anaxyrus woodhousii*. Photo: C.K. Dodd Jr.

positively to the calls of their own males but not to the calls of hybrids (Awbrey, 1965).

Closely related toads also often have different breeding phenology or habitat preferences that may minimize hybridization potential (Blair, 1942). Differences in habitat preference can separate the species, even though in very close proximity, and mixed breeding assemblages generally do not occur (Bragg, 1940a; Bragg and Smith, 1942; but see Collins, 1993). In Texas, for example, *A. woodhousii* contacts *A. fowleri* along the western margin of the eastern deciduous forest, and hybridization occurs along the 16–30-km-wide contact zone (Meacham, 1962). Fowler's Toad prefers sandy soils and mesic forest habitats, whereas Woodhouse's Toad prefers blackland prairie soils in more open habitats, although it is more versatile in its choice of habitats. Breeding site differences also are evident, with *A. fowleri* preferring creeks and *A. woodhousii* preferring prairie ponds (Blair, 1941a). Under pre-European colonization conditions, habitats presented a dendritic pattern along the contact zone, or they changed abruptly at the prairie-forest interface. The demarcation between species was also abrupt. As humans have modified the habitats in this region, the species are more likely to come into contact forming broad rather than narrow contact zones. Since *A. fowleri* tends to out compete *A. woodhousii*, the nature of the interaction between species along the contact zone changes with the nature of the habitat, and the extent of introgression varies from one location to the next.

There also may be differences in diet between certain toads in areas where they come into contact. For example, *A. cognatus* eats fewer small ants, harvester ants, sow bugs, and spiders than *A. woodhousii*, but more lepidopteran larvae, weevils, and centipedes (Smith and Bragg, 1949). Whether such dietary differences reflect habitat differences, preferences, or availability is unknown. In general, *A. cognatus* tends to replace *A. woodhousii* in the mixed-grass prairies (Bragg, 1940a, 1940b), so a combination of factors may influence perceived dietary differences.

In some anurans, the presence of a large number of heterospecific larvae may inhibit their growth. In *A. woodhousii*, however, the density of heterospecifics does not appear to inhibit growth, and *A. woodhousii* larvae do not contain the growth-inhibiting cell observed in many other frog species (Licht, 1967). In contrast, the presence of large numbers of *A. woodhousii* larvae may inhibit the growth of the larvae of some species (*A. houstonensis*, *Gastrophryne olivacea*, *Lithobates* sp.) but not others (*Anaxyrus debilis*, *A. speciosus*, *Hyla chrysoscelis*, *Scaphiopus couchii*). Density is inversely proportional to growth among conspecific *Anaxyrus woodhousii*.

DISEASES, PARASITES, AND MALFORMATIONS

The amphibian pathogen *Batrachochytrium dendrobatidis* was not found at sites sampled in Colorado and Oregon (Green and Muths, 2005; Adams et al., 2010). Endoparasites include the trematodes *Glypthelmins quieta*, *Gorgoderina amplicava*, *Haematoloechus* sp., *Haematoloechus coloradensis*, *H. complexus*, *H. medioplexus*, *H. parviplexus*, and *Megalodiscus temperatus*, the cestodes *Cylindrotaenia americana* and *Distoichometra bufonis*, the cestode *Distoichometra bufonids*, and an unidentified metacestode, and the nematodes *Aplectana incerta*, *A. itzocanensis*, *Cosmocercoides variabilis*, *Ozwaldocruzia pipiens*, *Physaloptera* sp., *Rhabdias americanus*, and *Rhabdias* sp. (Trowbridge and Hefley, 1934; Parry and Grundman, 1965; Brooks, 1976; McAllister et al., 1989; Goldberg et al., 1996b; Green and Muths, 2005). Waitz (1961) found no endoparasites in two animals. Green and Muths (2005) mentioned additional unidentified trematodes, nematodes, and cestodes. The myxozoan parasite *Myxidium serotinum* has been reported from *Anaxyrus woodhousii* as have the protozoans *Hexamita intestinalis*, *Karatomorpha swazyi*, *Nyctotherus cordiformis*, *Opalina* sp., *Trichomonas* sp., and *Zelleriella* sp. (Parry and Grundman, 1965; McAllister et al., 1989). Ectoparasites of larvae include the leech *Desserobdella picta* (Bolek and Janovy, 2005). A toad with an extra leg was reported from Oklahoma (King, 1960), and toads with missing eyes were observed in Colorado (*in* Hammerson, 1999).

Breeding habitat of *Anaxyrus woodhousii*. Photo: David Pilliod

SUSCEPTIBILITY TO POTENTIAL STRESSORS

Chemicals. Under normal conditions (no insecticide), tadpole body mass is inversely proportional to density. The addition of the carbamate insecticide carbaryl (3.5 or 7.0 mg/L) actually increased the survivorship of *Anaxyrus woodhousii* larvae in one set of experimental treatments, proportionally more so at high tadpole densities than at low densities (Boone and Semlitsch, 2002; Boone et al., 2004b). Body mass also was greater at metamorphosis in carbaryl-treated ponds than in those not treated with the insecticide. In contrast, an earlier study had suggested that carbaryl decreased survivorship of tadpoles to metamorphosis, an effect that was more pronounced at high tadpole densities than at low densities (Boone and Semlitsch, 2001). Even at low densities, survivorship greatly decreased at 7.0 mg/L carbaryl, although survivorship at 3.5 mg/L actually increased as in the later study. Increased density also reduced mass at metamorphosis. As zooplankton are killed, there is a greater amount of algae available to the tadpoles, and they actually grow larger, even at high densities. In essence, the pesticide reduces or eliminates competition for food resources from zooplankton, allowing greater tadpole growth and survival. The insecticide itself appeared to have no immediate adverse effects on the larvae. Boone and Semlitsch (2002) urged caution when interpreting the results, noting that effects on a community from an unnatural change in community structure could be subtle and occur long after the larval period.

The insecticide malathion causes morbidity and mortality in adult *A. woodhousii*, even at sublethal doses (0.011, 0.0011 mg/g of toad), when administered in conjunction with amphibian red-leg (*Aeromonas hydrophila*) disease; no effects were seen in toads challenged by the bacterium alone (Taylor, 1998; Taylor et al., 1999a). The insecticide appears to cause toads to be more susceptible to naturally occurring disease pathogens than they would be when the pesticide was not present. Malathion at doses > 0.110 mg/g toad is directly lethal when applied to the ventral skin. Malathion causes a decrease in brain cholinesterase activity levels.

STATUS AND CONSERVATION

Threats to Woodhouse's Toad include habitat destruction, alteration, and fragmentation, mortality from roads and other transportation corridors, and pesticides. Declines have been noted in urban areas as early as 1944 (Bragg, 1952). Bragg (1940a) noted that recent metamorphs were used as fish bait in Oklahoma, and that thousands of *Anaxyrus woodhousii* were killed on roads every year. Road kill of immense proportions has occurred at least for 90 years, especially considering the extensive range of this species. The species generally avoids pine plantations, preferring instead the mixed-hardwood riparian areas that bisect them (Rudolph and Dickson, 1990). When present in clearcuts or intensely managed silvicultural areas, numbers are low (Fox et al., 2004).

There does not appear to be any indication of declines throughout substantial portions of this species' range (Christiansen, 1981; Busby and Parmelee, 1996; Hammerson, 1999; Hossack et al., 2005). However, surveys in Big Bend National Park between 1998 and 2004 failed to detect *A. woodhousii*, despite the availability of historical records (Dayton et al., 2007). The species has been monitored in a variety of ways. Corn et al. (2000) noted that automated recording devices and observer-based call surveys did a poor job at estimating abundance of this species, perhaps because the devices were located at some distance from the chorus being monitored. A mark-recapture study yielded a population estimate of 213 toads, considerably larger than what had been estimated based on the call surveys alone. This species will occasionally colonize man-made ponds during habitat restoration efforts (Briggler, 1998).

Head pattern of *Anaxyrus woodhousii*. Illustration by Camila Pizano

Ollotis alvaria (Girard, 1859)
Sonoran Desert Toad

ETYMOLOGY
alvaria: from the Latin *alvus* meaning the 'womb' or 'belly' and *-arius* meaning 'belonging to.'

NOMENCLATURE
Stebbins (2003): *Bufo alvarius*
  Frost et al. (2006a): *Cranopsis alvaria*
  Frost et al. (2006b): *Ollotis alvaria*
  Synonyms: *Phrynoidis alvarius*

IDENTIFICATION
Adults. This is a large toad, attaining a maximum length of ca. 190 mm SUL. The ground color is olive green to dark brown, and the skin is smooth and leathery. Small, low and rounded tubercles are scattered on the dorsum. The parotoid gland is large, twice as long as wide, and kidney-shaped. Distinct cranial crests curve above the eye. The hind legs are covered with large warts, and there is a conspicuous light-colored wart at the angle of the jaw. Venters are white. Juveniles have small orange-tipped tubercles.

Males are smaller than females. In Arizona, males from 2 populations had a mean SUL of 127.1 and 129.7 mm (Sullivan and Fernandez, 1999). Sullivan and Malmos (1994) observed calling males with a mean of 141.6 mm SUL, whereas swimming males were a mean of 134.1 mm SUL. Brennan and Holycross (2006) note a maximum size of 191 mm SUL.

Larvae. The body of the tadpole is lightly pigmented and brassy in life and somewhat flattened. There is no pigmentation on the center of the belly. The tail musculature is evenly pigmented with large spots or blotches, but the tail fins are largely clear, although there may be a few small spots on the dorsal fin. Tail tips are rounded. Jaws are coarsely serrate. TL is < 56 mm.

Eggs. The eggs are black or dark brown dorsally and light tan ventrally. They are oviposited as tightly coiled strings in a long gelatinous casing. A nonscalloped jelly envelope surrounds the vitellus. This envelope is 2.12–2.25 mm in diameter, with the vitellus 1.4–1.7 mm in diameter. According to Livezey and Wright (1947) there are 12–18 eggs per 25 mm of string. Illustrations of eggs are in Savage and Schuierer (1961).

DISTRIBUTION
The Sonoran Desert Toad is known historically from southeastern California (including the southern Imperial Valley), southern Arizona, and extreme southwestern New Mexico (Peloncillo Mountains, Hidalgo County) south through Baja California Norte, Sonora, extreme northwestern Chihuahua, and northern Sinaloa, Mexico. Stebbins (2003) records isolated populations in southern Nevada, northern Arizona, and on islands in the Gulf of California. The species may have extended from the Colorado River at the Mexican border northward to southern Nevada near Fort Mohave, but these populations are no longer extant (Jennings and Hayes, 1994b). Important distributional references include Storer (1925), Cole (1962), Jennings and Hayes (1994b), Degenhardt et al. (1996), Brennan and Holycross (2006), and López et al. (2009).

FOSSIL RECORD
Ollotis alvaria is referred to fossils from Pliocene deposits in Arizona and is known from Pleistocene deposits in Sonora, Mexico (Tihen, 1962b; Holman, 2003). Fossils are similar to those of *Rhinella marina* but differ in details of the frontoparietal, ilia, and scapula.

SYSTEMATICS AND GEOGRAPHIC VARIATION
The Sonoran Desert Toad is a member of the Middle American clade of toads, and is most closely related to *Ollotis occidentalis* and *O. taca-*

Distribution of *Ollotis alvaria*. Dark gray indicates extant populations; light gray indicates extirpated populations.

nensis (Pauly et al., 2004). Fouquette (1968) reviewed and discussed the confusion surrounding the identity of the type specimen.

Natural hybridization has been reported with *Anaxyrus woodhousii* in central Arizona (Gergus et al., 1999). The ability of hybridize with this species has been maintained despite 6 my of independent evolution.

ADULT HABITAT

The Sonoran Desert Toad is a species of the arid southwestern deserts, where it is normally found in the vicinity of water (streams and permanent springs). It prefers broad flat expanses of desert scrub habitat dominated by creosote bush (*Larrea tridentata*) and mesquite (*Prosopis* sp.). It is also found in rocky riparian habitats with cottonwood and sycamore trees and around muddy man-made impoundments with abundant aquatic vegetation. Habitats include mesquite bosque and paloverde-saguaro desert scrub. In New Mexico, it is found from 1,250 to 1,387 m (Degenhardt et al., 1996).

TERRESTRIAL ECOLOGY

This species is nocturnal and frequents underground retreats, such as rodent burrows, to escape the desert heat and dryness by day and during cold weather. Lazaroff et al. (2006) noted diurnal activity during the breeding season. Activity occurs throughout the warm season as long as retreats are nearby; precipitation is not necessary for activity. Brennan and Holycross (2006) note that Sonoran Desert Toads are often found far from water, and Slevin (1928) noted individuals 1.6 km from the nearest water in the Sonoran Desert. Bogert (*in* Stebbins, 1962) mentioned observations of Sonoran Desert Toads 3.3–5 km from water. *Ollotis alvaria* somehow has found its way even to very remote tinajas such as those in the Cabeza Prieta Range of southern Arizona, which is surrounded by seemingly totally inhospitable desert (Childs, 2000).

Sonoran Desert Toads are photopositive in their phototactic response, suggesting they use both sunlight and moonlight when feeding and going about their daily activities (Jaeger and Hailman, 1973). They are sensitive to light in the blue spectrum ("blue-mode response") (Hailman and Jaeger, 1974). Sonoran Desert Toads likely have true color vision.

CALLING ACTIVITY AND MATE SELECTION

Despite speculation to the contrary, *O. alvaria* has an advertisement call, which has been described as a clucking sound made by contented chickens. Thorber (repeated by Storer, 1925) even noted how loud choruses were at Sabino Canyon in Arizona, although he may have been confusing *O. alvaria* calls with those of other species. Calls are very short in duration (0.6–0.8 sec) with a fundamental frequency of ca. 1,083–1,096 cps; the pulse rate is 61 pulses/sec, and persistent callers have a call rate of around 10–15 calls per minute (Sullivan and Malmos, 1994). The dominant frequency and call duration are not correlated with SUL or temperature; pulse rate is correlated with temperature but not SUL (Sullivan and Malmos, 1994).

Blair and Pettus (1954) noted that *O. alvaria* occasionally made a much softer call that was difficult to detect over the noise of other toad choruses. Because of the softness of the latter call and the prevalence of the "clucks" produced by the species in breeding aggregations, Blair and Pettus (1954) suggested that calls did not have a role in mate attraction. According to some authors the vocal pouches are more or less vestigial, but McAlister (1961) found a large larynx and well-developed vocal cords. The calls definitely function in an advertisement capacity (Sullivan and Malmos, 1994).

Breeding is associated with desert rainfall, although Degenhardt et al. (1996) noted instances of breeding in the absence of precipitation. Choruses form rapidly, are often small, and males call both by night and day. About 25–50% of the males at a chorus will produce an advertisement call, but only about 19% will call consistently (Sullivan and Malmos, 1994). In the Sonoran Desert, for example, the chorus size ranged from 3 to 75 and the chorus only lasted a mean of 2 days (Sullivan, 1989). In a later study, Sullivan and Fernandez (1999) noted that chorus size ranged from 3 to 40 males at 2 Arizona breeding sites; males were at these sites for only 12 nights over a 6 yr period. No breeding may take place during years of drought.

Many males call from shallow water or from along the shoreline. Others, however, continuously swim around the pool actively searching for females. The larger males tend to call, whereas smaller males tend to adopt an active pursuit behavior. Despite the proximity of swimming and calling males, satellite behavior has not been observed. Mate choice is largely by scramble competition, that is, males will clasp any other toad as they swim around a breeding site. Males will attempt to clasp any frogs within their reach. Clasped males produce the warning "cluck," a slow series (mean 5/sec) of warning vibrations to alert a conspecific as to the sex of the grasped animal (Blair, 1947b). Sullivan and Malmos (1994) reported the release call to have a pulse rate of 216.5 pulses/sec, 10.6 pulse groups/sec, and a frequency of 0.994 kHz. Blair and Pettus (1954) reported finding clasped pairs both in the water and as far away as 15 m from the shoreline. Males may attempt to clasp other species, such as introduced American Bullfrogs (Grogan and Grogan, 2011).

BREEDING SITES

Breeding occurs in rain-formed temporary desert pools, pools in intermittent stream channels, water pockets, springs, low-lying depressions, in clear streams (to 45 cm in depth), and in stock tanks. Dikes used to stem floods also may have artificial pools formed behind them that can be used for breeding by this species.

REPRODUCTION

Most oviposition occurs in summer (May to July) after the first heavy rains (> 25 mm) of the summer monsoon, although the breeding season may be extended. For example, the presence of tadpoles in October suggests that breeding can occur through September (Degenhardt et al., 1996). Chorusing may be delayed till the second or third night after excessive rainfall events. Eggs are oviposited in long (to > 1 m in length) jelly-coated strings in temporary pools or shallow streams in water 30–45 cm in depth. Clutch size is 7,500–8,000 eggs (Livezey and Wright, 1947).

LARVAL ECOLOGY

The larval period is short (<30 days), but otherwise, data on the larval ecology of this species are lacking. A major threat to larvae is flashfloods that sweep them away from their breeding pools out into desert flats. Tadpoles metamorphose at about 15 mm SUL.

Tadpole of *Ollotis alvaria*. Photo: Robert Byrnes

DIET

The diet of this large toad consists of a wide variety of invertebrates, particularly insects. Prey include spiders, a wide variety of beetles, many species of ants, wasps, velvet ants, seed bugs, assassin bugs, stink bugs, termites, lepidopteran larvae and adults, grasshoppers, locusts, spiders, scorpions, solpugids, centipedes, millipedes, other amphibians (*A. cognatus, Scaphiopus couchii*), small lizards (*Holbrookia* sp.), and mice (Slevin, 1928; Gates, 1957; Cole, 1962; Stebbins, 1962). Beetles make up the vast majority of the diet, and King (1932) noted that these toads may have an important value to agriculture because of the large number of fruit beetles they consume.

PREDATION AND DEFENSE

This large toad contains a secretion that is poisonous to some vertebrates, such as small dogs (Musgrave, 1930; Stebbins, 1962). Threats are responded to by puffing up the body, orienting the body toward the threat, and by making a hissing noise (Hanson and Vial, 1956). The toad is capable of ejecting the secretion 0.3–3.7 m from the parotoid gland when pressure is exerted on the glands (Hanson and Vial, 1956). If the secretion is ingested by a small animal, such as when biting a toad, the result can be temporary paralysis or even death. The secretion likely is also noxious in order to deter an attack. The toads themselves are gentle and not aggressive. When handling this species, care should be taken to avoid contact with the eyes and mouth, and hands should be washed immediately. Despite the poisonous secretion, *Ollotis alvaria* are eaten by raccoons. The predators learn to avoid the dorsal secretions by turning the toad on its back and eviscerating the belly (Wright, 1966).

Adult *Ollotis alvaria*. Photo: Rob Schell

POPULATION BIOLOGY

Sexual maturity is reached at 2 yrs, with most growth occurring during the second season of life. Sullivan and Fernandez (1999) reported a mean of 3.69 lines of arrested growth in males suggesting that they have a potential longevity of at least 6 yrs of age.

DISEASES, PARASITES, AND MALFORMATIONS

The Sonoran Desert toad is parasitized by the cestode *Nematotaenia dispar* and the nematodes *Aplectana itzocanensis*, *Oswaldocruzia pipiens*, *Physaloptera* sp., *Physocephalus* sp., and *Rhabdias americanus* (Goldberg and Bursey, 1991b).

SUSCEPTIBILITY TO POTENTIAL STRESSORS

No information is available.

STATUS AND CONSERVATION

Although this species historically is known from specimens in the southern Imperial Valley and along the Colorado River in California and Nevada, no sightings have occurred since 1955, and Jennings and Hayes (1994b) could find no populations during their survey. The species has been observed on the opposite side of the Colorado River in Arizona, however. Extirpation likely occurred as a result of habitat modification and the possible effects of extensive pesticide application in the Imperial Valley. Introduced crayfish may impact eggs and larvae at breeding sites. Jennings and Hayes (1994b) also noted that specimens have been taken from the wild for the hallucinogenic properties of the skin secretions. This species is protected as "Endangered" in New Mexico and California.

Ollotis nebulifer (Girard, 1854)
Gulf Coast Toad

ETYMOLOGY

nebulifer: from the Latin *vallis* meaning 'hollow' and *ceps* referring to 'head.' The name refers to the depression between the cranial crests.

NOMENCLATURE

Conant and Collins (1998): *Bufo valliceps*
  Frost et al. (2006a): *Cranopsis nebulifer*
  Frost et al. (2006b): *Ollotis nebulifer*
  Synonyms: *Bufo granulosus*, *Bufo valliceps* [in part], *Chilophryne nebulifera*, *Cranopsis valliceps*, *Incilius nebulifer*

Much of the available information on this

species in the United States has been published under the name *Bufo valliceps*. Based on an analysis of mtDNA, Mulcahy and Mendelson (2000) recognized *valliceps* as a morphologically variable tropical species and *nebulifer* as a morphologically uniform temperate species. All members of the *valliceps* complex in the United States are referred to *Ollotis nebulifer*.

IDENTIFICATION

Adults. Ollotis nebulifer is a striking medium-sized toad with prominent cranial crests and triangular parotoid glands. The dorsal coloration can be almost black with patches that are orange yellow to yellow brown with white to yellow spots. There is a broad conspicuous light-colored band that extends from the area of the parotoid glands toward the groin, and a light-colored stripe runs down the middle of the back. The dorsum and venter are covered with closely set tubercles. The cranial crests are sharply defined and are parallel to a deep depression down the center of the head. The cranial crests have divergent ends posteriorly which form a Y-shape. Males have a yellowish-green throat and a large round subgular vocal sac.

Males are smaller than females. Blair (1941a) reported an adult mean SUL of 62.5 mm in Texas. Also in Texas, Brown (1971a) noted males 75–118 mm SUL at a fish hatchery and 57–98 mm in the surrounding area; corresponding values for females were 72–132 mm SUL and 63–118 mm SUL. In south Texas, Salinas (2009) recorded a mean male SUL of 67 mm and a mean female SUL of 71.5 mm. Females also weigh more than males.

Larvae. Larvae are dark in coloration. The tail musculature is evenly pigmented, and pigmentation extends as an elaborate reticulated pattern on the dorsal and to a lesser extent on the ventral tail fins. The pattern on the musculature may resemble 3–10 light/dark dorsal saddles because of the presence of xanthophores among the darker pigment. Embryo and larval developmental stages and tadpole mouthparts are described by Limbaugh and Volpe (1957).

Eggs. Eggs are brown to black dorsally and white or cream ventrally. They are deposited in a double row within a jelly tube, and there are no partitions between the eggs. There are two gelatinous envelopes around the vitellus. The outer envelope is indistinct and is 2.7–3.2 mm in diameter. The inner envelope is distinct and is ca. 2.6 mm and close to the outer envelope. The vitellus is ca. 1.2 mm in diameter (Livezey and Wright, 1947; Grubb, 1972). Each 25.4 mm of string contains 20–30 eggs.

DISTRIBUTION

The Gulf Coast Toad is found from southern Mississippi west through much of southeastern Texas and up the Rio Grande and into Mexico. Dayton et al. (2007) do not list it from Big Bend National Park, although Dixon (2000) indicates records as far as the El Paso region. Isolated populations occur in south-central Mississippi, northeastern Louisiana, and adjacent Arkansas (single known location in Union County). Important distributional references include: Arkansas (Smith and Langebartel, 1949; Trauth et al., 2004), Louisiana (Dundee and Rossman, 1989), Mexico (López et al., 2009; Lemos-Espinal and Smith, 2007b), and Texas (Brown, 1950; Dixon, 2000).

FOSSIL RECORD

Ollotis nebulifer has been recorded from Pliocene and Pleistocene sites in Texas (Holman, 2003).

SYSTEMATICS AND GEOGRAPHIC VARIATION

Ollotis nebulifer is a member of the Middle American clade of bufonids, whose closest relatives are four Mexican species that together form the *O. valliceps* group (Pauly, 2004). The divergence between *O. nebulifer* and *O. valliceps* occurred on the Mexican Neovolcanic Plateau

Distribution of *Ollotis nebulifer*. The extent of the range along the western Rio Grande is uncertain.

between 4.2 and 7.6 mya (Mulcahy and Mendelson, 2000). There appear to be two lineages within the species, one in the North and the other in Mexico.

Hybridization in nature has been reported between *O. nebulifer* and *Anaxyrus woodhousii* (Thornton, 1955; Brown, 1971a; Sanders, 1986), although such hybridization may be facilitated by environmental disturbances caused by human activity. Likewise, the species hybridizes with *A. fowleri* in Louisiana, and again environmental disturbances facilitate hybridization (Vogel, 2007). Hybrids have been produced in laboratory experiments between *Ollotis nebulifer* and the following: *Anaxyrus americanus* and *A. terrestris* (A.P. Blair, 1941a; Volpe, 1959a; W.F. Blair, 1961a), *A. fowleri* (Moore, 1955; Volpe, 1956b), and *A. woodhousii* (Thornton, 1955; Blair, 1956a). However, some crosses work only with one parent (e.g., ♂ *Ollotis nebulifer* and ♀ *Anaxyrus woodhousii*, *A. americanus*, *A. terrestris*, or ♀ *A. fowleri*), and even then resulting males and the offspring of the reciprocal cross are sterile (Thornton, 1955; Volpe, 1956b, 1959a). Crosses between *Ollotis nebulifer* and *Anaxyrus hemiophrys*, *A. debilis*, *A. speciosus*, or *Ollotis alvaria* are not very successful (Blair, 1959). In addition, the species appears genetically compatible with some Mexican species, but not others, in laboratory crosses (Blair, 1961a, 1966; Ballinger and McKinney, 1966).

ADULT HABITAT

Throughout its range this species is found in a diversity of climates, from tropical and subtropical humid to very xeric. Rainfall patterns are often variable and are a controlling factor in breeding activity. Gulf Coast Toads are found in many different habitats, but prefer open areas and wet hardwood forests. The species tolerates agriculture and other disturbed habitats well. It does not seem to prefer pine forests.

TERRESTRIAL ECOLOGY

Activity occurs nearly throughout the year, especially in south Texas, where Salinas (2009) found them active from February through December. Activity is positively associated with rainfall and temperature. High barometric pressures tend to decrease toad activity (Salinas, 2009).

Gulf Coast Toads generally remain within a

Tadpole of *Ollotis nebulifer*. Photo: Ronn Altig

limited area during the nonbreeding season (ca. 225 m^2), although they occasionally move well beyond the home area for a period of time before returning (Awbrey, 1963). They sometimes move considerable distances over inhospitable terrain, with Blair (1953) and Thornton (1960) noting movements by males 805 m and 743 m, respectively, away from where the toads were marked. They are capable of returning to the original capture site when displaced as far away as 222–237 m (Blair, 1953; Awbrey, 1963). *Ollotis nebulifer* appear to use a combination of visual and olfactory cues in their orientation, and can use either if the other cue is disabled. However, they cannot home when both visual and olfactory cues are unavailable (Grubb, 1970). Dispersal away from breeding sites may or may not occur immediately after metamorphosis, depending on environmental conditions. During drought toads may remain near water or move only to nearby moist or watered areas. In Texas, Blair (1953) recorded dispersal by newly metamorphosed toads up to 72 m away; dispersal occurred rapidly during rainfall. Overwintering occurs under logs and surface debris in hardwood forests.

One unusual facet of the life history of *O. nebulifer* is that it climbs trees. In Texas, Neill and Grubb (1971) observed adults and juveniles in live oak (*Q. virginiana*) trees from 2–10 m above the ground. Many toads can occupy the same tree, where they are found in holes and cavities. Climbing may be facilitated by the gentle slopes of some of the larger limbs of this tree. Climbing is accomplished by pulling with the front limbs, holding on in bark crevices, and pushing with the rear legs against the rough bark surface. Gulf Coast Toads are also adept at climbing rock walls. Toads may return repeatedly to the same cavity.

Gulf Coast Toads are photopositive in their phototactic response, suggesting they use both sunlight and moonlight when feeding and going about their daily activities (Jaeger and Hailman, 1973). They are sensitive to light in the blue spectrum ("blue-mode response"), which apparently helps them orient toward areas of increasing illumination, such as the open horizon above lakes and ponds (Hailman and Jaeger, 1974). Gulf Coast Toads likely have true color vision.

CALLING ACTIVITY AND MATE SELECTION

Males arrive at breeding pools prior to females and remain at the site for an extended period. Males almost invariably call from the shoreline, with few calling from water. In Texas, calls have been heard from March to late September (e.g., Wiest, 1982), although the actual breeding period is much shorter. Calling most often occurs when nighttime temperatures are >17°C, with calls heard at evening temperatures as high as 35°C (Thornton, 1960; Blair, 1961b). Wiest (1982) noted calling at 18.5–28.3°C over a 44-day period in Texas. Males may return to breed at a pond several times during a season when conditions permit. In addition, some males will call early in the season whereas others call later in the season. Thus, not all males are present at a breeding pool, even under favorable conditions.

The call of *O. nebulifer* is a sharp trill of moderate length. Trill rates vary from 30–43/sec at a dominant frequency of 1,250–1,700 cps; the trill lasts from 1.9–5 sec (Blair, 1956b) and is repeated at 1–4 sec intervals. The sound has been likened to a wooden rattle. It is similar to the call of *Anaxyrus americanus* but more guttural and less musical.

BREEDING SITES

Gulf Coast Toads breed in intermittent streams, temporary ponds, shallow pools, and virtually any available water body. Breeding also occurs in artificial pools, gravel pits, irrigation ditches, and stock tanks (Blair, 1960a; Hubbs and Martin, 1967). Breeding has been recorded in brackish water (Burger et al., 1949; *in* Neill, 1958).

REPRODUCTION

The peak breeding season in Texas is April to September, with toads first appearing at breeding ponds in February to April and the last reproduction occurring in September; variation in the timing of breeding is dependent on temperature, with toads in the south active earlier and later in the season than toads in the north (Blair, 1960a; Thornton, 1960; Salinas, 2009). In Louisiana, calls are heard from April to September, although there are reports of breeding in March (Dundee and Rossman, 1989). Breeding is initiated by rainfall in the spring that fills the breeding ponds, and reproduction is sporadic and dependent on precipitation throughout the season. For example, Blair (1960a, 1961b) recorded only one rain event over a 2-day period that resulted in breeding in 1956, but in the wet year of 1957 the season extended 125 days. Although breeding may occur as late as August, the onset of dry weather may preclude successful metamorphosis. Females are capable of ovipositing more than one clutch per breeding season (Blair, 1960a). Spent females also return to breeding sites after periods of rainfall.

Eggs are deposited in shallow water in long gelatinous strings about a meter in length. These strings can either rest on the substrate or be wound around vegetation or other submerged debris. Greuter (2004) reported exact counts of 1,887–5,614 eggs. Hatching occurs in 1.5–3 days. Embryos are capable of developing between 18–35°C, with an optimal temperature of 20–30°C (Volpe, 1957c; Hubbs et al., 1963; Ballinger and McKinney, 1966). Abnormalities increase at temperatures >31.5°C. Rates of development are similar among populations from various parts of its range in Texas and Louisiana.

LARVAL ECOLOGY

The larval period is 20–30 days under laboratory conditions, with tadpoles reaching 25 mm prior to metamorphosis (Volpe, 1957c). Salinas (2009) mentions a larval period of 2–10 weeks. No other information is available on larval ecology.

DIET

There are no published studies on the diet of this species. Presumably it eats a wide variety of invertebrates.

PREDATION AND DEFENSE

Gulf Coast Toads have noxious and poisonous skin secretions that are effective in deterring predators. For example, Brown (1974) reported an unsuccessful predation attempt by an American Bullfrog on an adult *Ollotis nebulifer,* and Salinas (2009) noted an opossum carrying a toad. The

Adult Ollotis nebulifer. Photo: David Dennis

Breeding habitat of Ollotis nebulifer. Photo: Mike Forstner

eggs are toxic to some species when injected (Licht, 1968) but not when eaten by mosquitofish (*Gambusia*) (Grubb, 1972). Garter snakes (*Thamnophis*) may be immune to the toxins in the egg. Western Ribbonsnakes (*T. proximus*) have been observed eating young toads (Wright and Wright, 1949).

POPULATION BIOLOGY

Growth occurs rapidly in young toads (0.55 mm/day), slows during the winter (0.11 mm/day), then accelerates in spring (0.2 mm/day) until the animals reach maturity. After maturity, growth slows to 0.04 mm/day). Males attain breeding size the summer following metamorphosis, at about 10 months of age, at 61–78 mm SUL (Blair, 1953; Thornton, 1960). Age of first reproduction in females is not known. Greuter (2004) reported a survivorship of <1.4% of larvae that emerged as juveniles in natural enclosures.

Breeding population size may vary substantially from one year to the next, depending on precipitation. For example, Thornton (1960) marked 27–85 toads (20–52 males, 7–33 females) over a 3 yr period at a site in Texas. Likewise, sex ratios vary, although they are male biased at breeding sites. For example, Blair (1960a) recorded sex ratios of 1.94 males per female over 3 yrs, but 4:1 one yr and 13.3:1 another yr over a 5 yr study. In south Texas, Salinas (2009) recorded a male skewed sex ratio of 1.96:1 over a 1 yr survey. Skewed ratios result from different amounts of time spent at breeding ponds, with females staying only until egg deposition, and different climatic conditions during the breeding seasons. Thus, females may be missed during sampling while males are more likely to be encountered by researchers. In addition, a drought-shortened breeding season may result in only a portion of the adults visiting the breeding site, and some toads may skip a breeding season. In Blair's report (1960a), the entire breeding population was estimated at 97–208 adults but was still suspected of being male biased, whereas Salinas (2009) estimated her population at 2,935 individuals. Survivorship is likely greater in females, based on mark-recapture data. Gulf Coast Toads are relatively long-lived, with males surviving as long as 8 yrs and females 5 yrs (Blair, 1960a).

COMMUNITY ECOLOGY

In a series of comparisons where larvae of one species were reared in water crowded by the larvae of another species, the prior presence of *Lithobates pipiens, Scaphiopus couchii,* or *Hyla chrysoscelis* larvae appeared to have an inhibitory effect on the subsequent growth of *Ollotis nebulifer* larvae. In contrast, the presence of larval *O. nebulifer* did not inhibit the subsequent growth of larval *Anaxyrus woodhousii* (Licht, 1967). The inhibition results from the effects of an inhibitory substance found in an alga-like cell in the feces of tadpoles.

In the south-central United States, Fowler's Toads are declining because of hybridization with *Ollotis nebulifer* as it expands its range northward. Recent evidence has suggested that human disruption of breeding sites has facilitated this hybridization (Vogel, 2007). In addition, larval *Anaxyrus fowleri* are competitively disadvantaged when reared with *Ollotis nebulifer* larvae, which causes them to metamorphose at smaller size and weights than normal under drying conditions (Vogel, 2007). Survivorship also is less than normal when in competition with *O. nebulifer*. The duration of the larval period was unaffected, however. Vogel (2007) and Vogel and Pechmann (2010) have suggested that the superior competitive advantage of *O. nebulifer* larvae also results in ecological displacement of *Anaxyrus fowleri* in human-disturbed habitats.

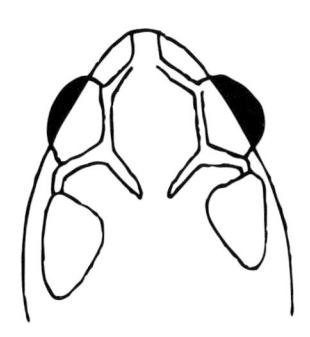

Head pattern of *Ollotis nebulifer*. Illustration by Camila Pizano

Males have a warning vibration to deter intraspecific amplexus, but this is not entirely effective at deterring other bufonids (Blair, 1947b). Sexual interference may arise when interspecific amplexus occurs. At a fish hatchery in Texas, for example, 7.9–8.8% of amplectant pairs consisted of *Ollotis nebulifer* and *Anaxyrus woodhousii* (Thornton, 1955; Brown, 1971a). This results in hybridization with as many as 6% of calling males being hybrids. Interspecific amplexus between *Ollotis nebulifer* and *Anaxyrus fowleri* also has been observed (Orton, 1951; Liner, 1954; Volpe, 1957d). Tadpoles resulting from hybridizing parents may exhibit heterosis (hybrid advantage) since they tend to reach larger sizes and metamorphose sooner than the parental species (Thornton, 1955; Volpe, 1960). Hybrid adults also may be larger than adults of one of the parental species (Brown, 1971a).

DISEASES, PARASITES, AND MALFORMATIONS

The myxozoan parasite *Myxidium serotinum* has been reported from *Ollotis nebulifer* as have the protozoan *Opalina* sp., the apicomplexan *Eimeria*, and an *Adelina*-like coccidian, the cestode *Mesocestoides* sp., and the nematode *Cosmocercoides variabilis* (McAllister et al., 1989). The virulent fungal pathogen *Batrachochytrium dendrobatidis* also has been found in this species at a 83% rate of infection in south-central Texas in 2006 (Gaertner et al., 2010).

SUSCEPTIBILITY TO POTENTIAL STRESSORS

No information is available.

STATUS AND CONSERVATION

Although this species likely has lost populations due to habitat destruction and alteration, it appears to be common throughout its range. Still, there are no long-term data showing status and trends. In urban situations, Gulf Coast Toads may be unable to breed in managed ponds, especially when water levels are drawn down or chlorinated during the breeding season, killing eggs and tadpoles (Salinas, 2009). In addition, toads may be killed on roads, walkways, and by lawn maintenance equipment. Recent evidence has suggested that human disruption of breeding sites has facilitated hybridization, and even may be contributing to the decline of other toad species as *Ollotis nebulifer* extends its range (Vogel, 2007).

Rhinella marina (Linnaeus, 1758)
Cane Toad
Poloka (Hawaiian)

Information in this account pertains primarily to Florida, Hawaii, and Texas. There is an extensive literature on the biology of this species, particularly on its impacts to native fauna in Australia. Readers should consult Zug and Zug (1979) and Lever (2001) for more extensive reviews of Cane Toad biology.

ETYMOLOGY
marina: from the Latin *marinus* meaning 'pertaining to the sea.' Linnaeus described the Cane Toad from a drawing in Seba (1734), who mistakenly believed the species lived on land and in the sea.

NOMENCLATURE
Pramuk et al. (2007): *Chaunus* moved to *Rhinella*
  Frost et al. (2006a): *Chaunus marinus*
  Synonyms: *Bufo gigas*, *Bufo horridus*, *Bufo horribilis*. Complete synonymy in Amphibian Species of the World 5.1, an online reference.

IDENTIFICATION
Adults. This is a large, heavy-bodied toad, and full-grown adults cannot be confused with any other species. The ground color usually is a uniform light brown in males, although some individuals may appear yellowish or dark brown. Females and juveniles have cream-colored and dark patches dorsally in a somewhat interlaced reticulated pattern. Ovate to triangular parotoids are large and distinct, and large warts are present dorsally and on the tops of the legs. Male warts tend to be spiny whereas female warts are smooth. The top of the head is smooth with ridges present from the nose to the back of the head. The maximum body width is nearly three-fourths of the body length. Venters are mottled in adults, but in the Tampa region the venters of newly metamorphosed *R. marina* are black.

Males are smaller than females. Adults are usually 100–180 mm SUL, but the largest Cane Toad on record is 230 mm SUL from Suriname. Rossi (1981) indicated the largest *R. marina* from Florida was 168 mm SUL. Meshaka et al. (2004) reported males to 150 mm SUL and females to 175 mm SUL from Highlands County, Florida. In 5 south Florida locations, males averaged 99.3 to 113.2 mm SUL (range 75 to 135 mm) and females averaged 107.9 to 140.2 mm SUL (range 89 to 165 mm) (Meshaka et al., 2004). According to these authors, the mean adult body size is now smaller than it was in the 1980s. Krysko and Sheehy (2005) reported a single male 171 mm SUL from Key West. Krakauer (1968) reported a mean adult size of 108 mm SUL. In Taylor and Wright's (1932) series from south Texas, toads were 66–168 mm SUL.

Larvae. Tadpoles are small, round, dark brown to black, and usually 10–25 mm TL. Throats are pigmented. Viewed laterally, the body appears flattened. There is a distinctive pale cream stripe on the lower edge of the tail musculature, but there is no evidence of saddles on the dorsal part of the tail musculature. Tail fins are a uniform translucent gray. Eyes are dorsal in position.

Eggs. The eggs are small (1.7–2 mm in diameter) and black. They are oviposited in very long gelatinous strings.

DISTRIBUTION
Rhinella marina occurs naturally from south Texas throughout Central America into the Amazon Basin of northern South America. Cane Toads are native in the United States only in a few

Distribution of *Rhinella marina*. The population in Florida is introduced.

Texas counties bordering the lower Rio Grande, although Brown (1950) noted deliberate releases of *R. marina* from Monterey, Mexico, into Hogg County, Texas. Introductions throughout tropical areas of the world were conducted, ostensibly to control sugarcane beetles and cutworms. Although Cane Toads ate the beetles, they were ineffective at controlling this pest because they are generalists that do not specialize on sugarcane beetles or cutworms. Despite repeated failure as pest control agents, introductions have continued resulting in the pest-control agent becoming the pest.

Cane Toads were introduced from Puerto Rico and other areas unsuccessfully several times in Florida beginning in 1936, with intentional releases from the pet trade supplementing agriculture-related releases. The present population resulted possibly from a 1955 accidental release from an animal dealer but was supplemented by intentional releases in 1963 and 1964 (Riemer, 1958; King and Krakauer, 1966; Krakauer, 1970a; Lever, 2001). These latter releases included toads from Suriname and Colombia. Riemer (1958) also mentions unsuccessful releases of the giant toads *Rhinella arenarum* and *R. paracnemis*.

Cane Toads are currently reported from about ten south Florida counties, and have extended their range northward to Pasco County along the Gulf Coast. Calling was reported from Orlando in Orange County in 1979, but this population may not be extant (Rossi, 1981). Johnson and McGarrity (2010) reported an isolated population from Bay County along Florida's northern Gulf Coast, but this population likely was extirpated by cold weather. Range expansion may continue along coastal areas, but northward expansion likely will be curtailed by periodic cold weather. Introductions into Louisiana were not successful (Lever, 2001).

Cane Toads were introduced into Hawaii from Puerto Rico in March and April 1932; the site of introduction was the Manoa Valley and Waipio on the island of Oahu. The introduction was immediately successful, and young toads were observed by August. The population exploded by the thousands. Subsequent introductions took place in 1933 to Hawaii, Kauai, Maui, and Molokai. The history and success of introductions is reviewed by Pemberton (1933, 1934), Oliver and Shaw (1953), and Lever (2001). The species does not occur on the other Hawaiian islands. In addition to islands in Hawaii, Cane Toads are found on islands in the Florida Keys (Stock Island and Key West; Krakauer, 1970a; Lazell, 1989).

In addition to its native range, Cane Toads occur in Australia, Bermuda, many Caribbean islands (including Puerto Rico), the Chagos Archipelago, Guam, Japan, the Philippines, Papua New Guinea, throughout the Pacific islands, and Taiwan (Zug and Zug, 1979; Easteal, 1981; Lever, 2001; Kraus, 2009). Important distributional references include: Florida (Krakauer, 1970a), Hawaii (McKeown, 1996), and Texas (Taylor and Wright, 1932; Dixon, 2000).

FOSSIL RECORD

Miocene fossils attributed to *R. marina* are reported from Kansas (Holman, 2003).

SYSTEMATICS AND GEOGRAPHIC VARIATION

The Cane Toad is a member of the South American toad radiation and is most closely related to *Rhinella jimi*, *R. poeppigii*, *R. cerradensis*, *R. veredas*, and *R. schneideri* (Pauly et al., 2004; Pramuk et al., 2007; Maciel et al., 2010; Vallinoto et al., 2010). This group diverged from the rest of the *Rhinella marina* clade during the late Miocene, about 10.5 mya. Fossil evidence suggests it may have had a more extensive range in North America during the Miocene.

Laboratory crosses between *R. marina* and *Anaxyrus woodhousii* or *A. fowleri* are not successful (Blair, 1959).

ADULT HABITAT

In south Texas, Cane Toads occur along the Rio Grande Valley. Introduced populations of Cane Toads in Florida are found mostly in moist open urban and agricultural environments as opposed to natural habitats (Krakauer, 1968; Meshaka et al., 2004; Surdick, 2005). They do not occur in wet prairies, dry sandy areas, or in natural habitats within the Everglades (Meshaka et al., 2000). Rossi (1981) recorded them in disturbed riparian habitats in the Tampa area, suburban yards, agricultural areas, areas dominated by exotic plants, brushlands dominated by castor bean (*Ricinus communis*), willows, and strangler fig (*Ficus aurea*) and pine flatwoods. In Hawaii, Cane Toads are found in the lowlands throughout the islands on which they have been introduced. They occur from sea level to 300–600 m on Maui and Oahu, depending on location (*in* Lever, 2001).

TERRESTRIAL ECOLOGY

Rhinella marina usually does not stray far from water, and activity is stimulated by heavy rainfall. Most adult and subadult activity occurs at night between March and November in Florida (Krakauer, 1968; Rossi, 1981; Meshaka et al., 2004), although activity may be extended in the most southern portions of the peninsula and Keys. For example, Krakauer (1968) observed them active in winter sitting along canals with full stomachs. Metamorphs tend to be diurnal. In Hawaii, activity is both diurnal and nocturnal, but most are active after sunset. Toads hide under surface debris such as rocks, logs, pine needles, and in canal banks. Dispersal occurs along roads and highways in Australia; toads avoid heavily vegetated habitats (Brown et al., 2006). It is not known how dispersal occurs in Florida or Hawaii, but it seems reasonable that Cane Toads use transportation corridors to facilitate movements. Rossi (1981) also mentions deliberate and inadvertent transport by people.

Throughout much of its range, the ambient temperatures experienced by *R. marina* are rather stable. The species can tolerate a wide range of temperatures. For example, the CTmax for Florida adults is 40–40.8°C and the CTmin is 5–10°C, depending upon acclimation temperature (Krakauer, 1968, 1970b). Under experimental conditions, Cane Toads select temperatures from 21–28°C, with neither light nor season influencing mean temperature selection (Sievert, 1991). Still, the species may experience cold spells at the northern limit of its range, and this may prohibit expansion northward. Krakauer (1970b) recorded 6 of 10 adults died at 4.2°C over a 96 hr period. Interestingly, cold weather does not seem to result in immunosuppression in this species as it does in other ectotherms (Carey et al., 1996).

Cane Toads are photopositive in their phototactic response, suggesting they use both sunlight and moonlight when feeding and going about their daily activities (Jaeger and Hailman, 1973). They are sensitive to light in the blue spectrum (probable "blue-mode response"), which apparently helps them orient toward areas of increasing illumination, such as the open horizon above lakes and ponds (Hailman and Jaeger, 1974). Cane Toads likely have true color vision.

CALLING ACTIVITY AND MATE SELECTION

Males have what has been described as a deep, slow, booming melodic trill. Calling has been reported from December through September in Florida, depending on location (Krakauer, 1968; Rossi, 1981; Meshaka et al., 2004). Calling has been heard at ambient temperatures as low as 15°C (Meshaka et al., 2004). It is likely that some reproduction occurs year-round in the extreme south, as gravid females are found in all months of the year. In Hawaii, calling occurs year-round. There appears to be no information on the Texas population.

Amplexed male *R. marina* have a slow series (mean 6.9/sec) of warning vibrations to alert a conspecific as to the sex of grasped animal (Blair, 1947b). Male *Anaxyrus terrestris* have been observed in amplexus with female *Rhinella marina* in Florida (*in* Lever, 2001).

BREEDING SITES

Breeding occurs in a wide variety of habitats, from small shallow pools to open wetlands. It has been recorded in ponds, puddles, rock pits, and man-made canals. Krakauer (1970a) and

Adult *Rhinella marina*, Hillsborough County, Florida. Photo: C.K. Dodd Jr.

La Rivers (1948) mentioned breeding in slightly brackish water and lily ponds, and successful development in salinities of 10–15% dilute sea water was noted by Ely (1944).

REPRODUCTION

Most oviposition occurs from March to September in Florida and year-round in Hawaii. No information is available for Texas. Krakauer (1968) found Florida females with mature eggs year-round, but suggested that cold weather might limit oviposition. Breeding may be stimulated by hurricanes that result in high amounts of rainfall over a short period (Meshaka, 1993), but rainfall is not necessary for reproduction. Eggs are deposited in very long strings containing from 5,000 to 32,000 eggs (Krakauer, 1968, 1970a). These strings are attached loosely to vegetation and aquatic debris. Hatching occurs in 1.5–4 days.

LARVAL ECOLOGY

Larvae are herbivores consuming algae and other plant detritus. Larval Cane Toads form large aggregations (Mares, 1972; photo in McKeown, 1996). They swim actively both day and night, and schools can cover a large distance. Tadpoles can tolerate rather high water temperatures, with the CTmax from Florida larvae being 41.6–42.5°C (Krakauer, 1970b). Larvae acclimated at 7°C can survive normally after 72 hrs exposure, with half being killed after 96 hrs. These results suggest that Cane Toads from northern populations can tolerate low and even freezing temperatures better than their tropical conspecifics. The duration of the larval period is about 45–60 days in south Florida (Krakauer, 1970a; Meshaka et al., 2004) and 30 days in Hawaii (Oliver and Shaw, 1953). New metamorphs are 9.75–13 mm SUL (Meshaka, 1993; Meshaka et al., 2004).

DIET

Cane Toads eat just about any invertebrate or vertebrate they can catch and stuff into their mouth; they have a prodigious appetite. They even readily consume stationary items such as pet food left out in a dish, garbage, fruits, and vegetables (Alexander, 1965). Apparently they can use olfactory as well as visual cues to orient toward and identify food (Rossi, 1981, 1983), and the forelimbs are used to stuff large prey into the mouth. Toads are reported to see prey up to a meter away (Pemberton, 1934).

Specific items in the diet of Florida toads include snails, isopods, land crabs, ants, cockroaches, lepidopterans, dragonflies, true bugs, beetles, earwigs, spiders, flies, juvenile *R. marina*, other frogs (*Anaxyrus quercicus, A. terrestris, Hyla squirella*), and small snakes (*Diadophis punctatus, Thamnophis sauritus, Ramphotyphlops braminus*) in Florida (Krakauer, 1968; Rossi,

1981; Meshaka et al., 2004; Meshaka and Powell, 2010). In Hawaii, the diet includes grasshoppers, armyworms (*Spodoptera mauritia*), cutworms, borers, weevils, beetles (including large quantities of the rose beetles *Adoretus sinicus* and *Pantomorus godmani*), centipedes, millipedes, spiders, true bugs, earthworms, scorpions, wasps, moths, bees, ants, caterpillars, slugs, snails, cockroaches (including the exotic pest *Pycnoscelus surinamensis*), sow bugs, and geckoes (Pemberton, 1934; Illingworth, 1941; Oliver and Shaw, 1953; Lever, 2001). Rossi (1981) also found mammal hair, bird bones, and another unidentified bufonid and observed them scavenging dead fish and eating dog feces. Plant material is frequently ingested. In a laboratory setting, they consumed every vertebrate offered, including various frogs and small snakes. Although Cane Toads consume large quantities of insect pests, their effectiveness at controlling pests is questionable.

PREDATION AND DEFENSE

These giant toads likely have few major predators as adults, particularly because of the toxic and noxious skin secretions that are produced in copious amounts from the parotoids and other granular glands. When threatened, they inflate their body and orient it in such a manner as to present the attacker with the toxin-containing parotoid glands, and they may even jump toward the attacker. However, a number of predators on *Rhinella marina* have been reported from Florida, including birds (American Crow, Bluejay, Mockingbird, Red-shouldered Hawk) and snakes (*Drymarchon couperi*, *Heterodon platirhinos*, *Nerodia fasciata*, *Thamnophis sauritus*, *T. sirtalis*) (Rossi, 1981; Meshaka, 1994). Birds and some mammals learn to roll the toads onto their backs and eviscerate them, thus avoiding the poisonous secretions. In Hawaii, mongooses (*Herpestes*) may occasionally eat Cane Toads (La Rivers, 1948; Oliver and Shaw, 1953), but they are not a major portion of the diet, and mongooses are not an effective control agent.

Tadpoles are noxious or poisonous to some predators but are eaten by Bullhead Catfish (*Ameiurus natalis*). Under laboratory conditions, fertilized eggs of *Rhinella marina* are toxic to gastropods (*Lymnaea* sp.), goldfish (*Carassius auratus*), and larval frogs (*Anaxyrus terrestris*, *Hyla cinerea*, *Osteopilus septentrionalis*, *Lithobates sphenocephalus*, *Scaphiopus holbrookii*) (Punzo and Lindstrom, 2001). However, many aquatic invertebrates, including crayfish and predaceous water bugs, and larval *Gastrophryne carolinensis* eat *Rhinella marina* eggs with no ill effects.

POPULATION BIOLOGY

Growth is rapid, with toads reaching 60–80 mm SUL in Hawaii within 3 months of hatching. Individuals reach sexual maturity in 1 yr in Hawaii (Pemberton, 1934). Longevity in wild populations is unknown, but Pemberton (1949) kept a captive for 16 yrs in Hawaii.

COMMUNITY ECOLOGY

The presence of *R. marina* larvae did not affect the growth, development, or survivorship of larval *Anaxyrus terrestris* or *Hyla cinerea* under experimental conditions (Smith, 2005a). Larval *Osteopilus septentrionalis* consume larval *Rhinella marina*, and eating the toxic toad larvae has no effect on the subsequent survival of the Cuban Treefrog larvae, although tadpoles had symptoms of intoxication for 24 hrs following ingestion (Smith, 2005b). The diets of adult *Anaxyrus terrestris* and *Rhinella marina* overlap considerably (0.86) suggesting the potential for competition between these species should resources become limiting (Meshaka and Powell, 2010).

DISEASES, PARASITES, AND MALFORMATIONS

The fungus *Basidiobolus* sp. has been found on Cane Toads from Tampa, Florida (Okafor et al., 1984). The exotic tick *Amblyomma rotundatum* may have been introduced with *Rhinella marina* into south Florida and is now established there (Oliver et al., 1993). Other reported tick parasites include *A. americana* and *Dermacentor* sp. (Rossi, 1981). Eggs of the hookworm parasite *Aclyostoma caninum*, the threadworm *Strongiloides* sp., the coccidian *Isopora* sp., and large numbers of unidentified protozoans have been reported from toad feces; these likely resulted from the ingestion of dog and cat feces (Rossi, 1981). In other parts of its range, *Rhinella marina* is parasitized by protozoans, helminthes, and ticks, and a variety of pathogens affect this species, including viruses, bacteria, and fungi (Zug and Zug, 1979; Lever, 2001). Neoplasias (cancers) also have been reported.

In south Florida, increased gonadal abnormalities and in the frequency of intersex, Cane Toads are correlated with agricultural activities. Gonadal abnormalities in turn are associated with impaired gonadal function, especially regarding the effects of testosterone. Secondary sex characteristics were either feminized (increased skin mottling) or demasculinized (reduced forearm width and nuptial pad number) in intersex toads. Male toads had hormone concentrations and secondary sex traits intermediate between intersexes and toads from nonagricultural areas (McCoy et al., 2008). Taken together, these results suggest that toads from agricultural areas have reduced fitness and reproductive success.

SUSCEPTIBILITY TO POTENTIAL STRESSORS

Chemicals. Cane Toads in Hawaii were poisoned by strychnine after ingesting flowers of the tree *Strychnos nuxvomica* (Arnold, 1968).

STATUS AND EFFECTS ON NATIVE FAUNA

The status of the native Texas population is not reported in the literature. Throughout the world, the nonaggressive *Rhinella marina* has had detrimental effects on native faunas wherever they have been introduced (e.g., Burnett, 1997; Lever, 2001; Griffiths and McKay, 2007; Letnic et al., 2008), despite glowing support for their introduction to control pests particularly among organic farmers (Pfeiffer, 1949). Little is known of their effects on the fauna of Florida and Hawaii. In Australia where most research on invasive toads has occurred, Cane Toads have caused high rates of mortality of native fauna and drastically altered community composition and structure. Adverse effects are largely due to the toxic nature of the toad's skin secretion coupled with the tendency of naïve predators to attack novel prey. This may not be as much of a problem in Florida, where other bufonids occur to which predators have been exposed. Effects on the endemic Hawaiian bird and insect fauna are not known.

Cane Toads may have negative impacts on native frogs and toads. Because of their large size and appetite, they consume almost any small invertebrate or vertebrate within range. In laboratory feeding trials, they aggressively out-compete domestic toads by striking them into submission with their tongue (Boice and Boice, 1970). Cane Toads have lateral vision and can observe an approaching competitor in order to slap it with their tongue and discourage further approach toward a prey item (Robins et al., 1998). Whether feeding competition occurs in nature or was an artifact of captivity is not known. In addition to adults, larval Cane Toads are toxic to many naïve predators, including other tadpoles (e.g., Crossland et al., 2008). Indeed, predators exposed to Cane Toad larvae even learn to avoid native palatable prey after an encounter with the *R. marina* larvae (Nelson et al., 2010).

Domestic animals are routinely poisoned by the toad's secretions resulting from the effect of cardiac glycosides, especially cats and dogs that refuse to let go of the toad (Otani et al., 1969). This can sometimes result in the death of the pet. As long as humans wash their hands after handling a Cane Toad, there is minimal danger of poisoning. Children, however, should be advised to leave Cane Toads alone. After handling *R. marina*, a person should avoid contact with their eyes or mucous membranes. Domestic animals should be treated by a veterinarian.

It seems highly unlikely that Cane Toads in Florida and Hawaii will be eradicated. As early as 1956, a Dade County official proposed a bounty on the toad because of its threat to pets, but the toad is still present and even expanding its range.

Family Craugastoridae

Craugastor augusti
(Dugès *in* Brocchi, 1879)
Barking Frog
Western Barking Frog
  (*C. a. cactorum*); Balcones Barking
Frog (*C. a. latrans*)

ETYMOLOGY

augusti: patronym honoring Auguste Henri Andre Dumeril (1812–1870), Professor of Herpetology and Ichthyology at the Museum National d'Histoire Naturelle, Paris; *cactorum*: from the Greek word *kaktos* meaning 'prickly plant' and the Latin suffix -*orum* meaning 'of the.' Literally, a frog of the cactus; *latrans*: from the Latin word meaning 'barking' and referring to the frog's call.

NOMENCLATURE

Stebbins (2003): *Eleutherodatylus augusti*
  Synonyms: *Eleutherodactylus augusti fuscofemora, Eleutherodactylus augusti, Eleutherodactylus augusti cactorum, Eleutherodactylus augusti latrans, Eleutherodactylus bolivari, Eleutherodactylus cactorum, Eleutherodactylus fuscofemora, Eleutherodactylus latrans, Hylactophryne augusti, Hylodes augusti, Hylodes latrans, Lithodytes latrans*

IDENTIFICATION

Adults and juveniles. The ground coloration is brownish gray, tan, reddish brown, or olive gray. There is regional variation in ground color, with *C. a. cactorum* from Arizona primarily brown and *C. a. latrans* from Texas and New Mexico from sulphur yellow to beige (Goldberg et al., 2004b). A dorsolateral fold is present, and the granular skin lacks warts or tubercles. A distinguishing fold of skin is present across the back of the head. Heads are broad. Large brown or black spots may be present with pale greenish or light centers. The outer and inner surfaces of the limbs are bright yellowish green; this color may extend along the sides of the body. Indistinct bars are present on the upper surfaces of the limbs (illustrated by Zweifel, 1956b). A black line is present on the underside of the lip. Venters are pale gray or pinkish and dotted in white. Prominent tubercles are present on the underside of the feet below the joints. Toes are not webbed. During the calling season, male *C. a. cactorum* have dark throats with dark tympanums, whereas female throats are white with pink tympanums; male throats later change to mottled gray. Sexually dimorphic throat coloration is not present in *C. a. latrans* (Goldberg et al., 2004b). Adults are 64–75 mm SUL. Although based on a small sample size, Goldberg (2002) noted males were 71.4–74.4 mm SUL and females were 77.2–82.8 mm SUL in an Arizona population. Schwalbe and Alberti (1998) mentioned an 85 mm SUL female. The largest Texas male was 77.2 mm SUL and the largest female was 94 mm SUL (Zweifel, 1956b).

Juveniles are patterned differently from adults. They have a dark head and shoulder region, a light band across the middle of the back, and a dark posterior third of the body (Zweifel, 1956b; Brennan and Holycross, 2006). Dark spots may be visible through the dark dorsal pattern. The band disappears in the adults, although it may be very faintly visible in some individuals. Throats are white and bellies are translucent. Juveniles in New Mexico have been reported to be strikingly green and ivory or white (Koster, 1946; *in* Wright and Wright, 1949), or only with the dorsum of the legs yellow tinged in green (*in* Degenhardt, 1996).

Larvae. There is no free-swimming larva. The

larval period is passed within the egg, and froglets hatch as miniature adults.

Eggs. Eggs are deposited terrestrially. They are large, with a diameter of 6–7.5 mm.

DISTRIBUTION

The Barking Frog occurs from southern Arizona (Santa Rita, Patagonia, Quinlan, Pajarito, Huachuca mountains, possibly Sierra Ancha), New Mexico (Pecos Valley, Otero County), and west-central Texas south to Oaxaca, Mexico. Important distributional references include Koster (1946), Bezy et al. (1966), Degenhardt et al. (1996), Dixon (2000), Murray and Painter (2003), and Brennan and Holycross (2006).

FOSSIL RECORD

Fossil Barking Frogs have been reported from Pleistocene deposits in New Mexico, Texas, and Sonora in Mexico (Holman, 1969, 2003). Holman (2003) provided a description of the bones useful in identification of fossils.

SYSTEMATICS AND GEOGRAPHIC VARIATION

Craugastor augusti is a member of the *Craugastor* (*Hylactophryne*) *augusti* species series that mostly inhabits Mexico (Hedges et al., 2008). It has been associated phylogenetically with the Middle American clade of eleutherodactylines (genus *Craugastor*) (Heinicke et al., 2007); Hedges et al. (2008) elevated this clade to family status and placed *C. augusti* within the subgenus *Hylactophryne*. The subspecies *C. a. latrans* has been referred to populations in Texas and New Mexico and *C. a. cactorum* for Arizona populations (Zweifel, 1956b). *Craugastor a. latrans* is distinguished from *C. a. cactorum* by differences in relative tympanum size, body size, coloration, mtDNA, and skin toxicity (Zweifel, 1956b; Goldberg, 2002; Goldberg et al., 2004b). Goldberg et al. (2004b) suggested that these differences could represent sufficient justification to recognize *latrans* and *cactorum* as separate species, but did not formally designate them as such. In the United States, three different clades were revealed by mtDNA analysis within *C. augusti* (Goldberg et al., 2004b). Additional molecular data are needed from throughout the species' range.

Juvenile *Craugastor augusti* have a very different pattern from adults. Photo: Jay Withgott

ADULT HABITAT

The Barking Frog is associated almost exclusively with limestone caves, cliffs, crevices, and rock outcroppings in extreme southern Arizona, New Mexico, and Texas (Strecker, 1910b). The frogs rarely move between even adjacent rock outcroppings, and population size on any particular outcrop is small (two to ten in Goldberg's, 2002 study). *Craugastor augusti* is also found associated with rodent burrows in creosote bush flats near limestone and gypsum outcrops in New Mexico (Koster, 1946; Degenhardt et al., 1996) and along bluffs bordering streams in the prairies of west Texas (Strecker, 1910b). In Arizona, the habitat is Oak Woodland–Grassland ecotones characterized by junipers, manzanita, oaks, sumac, yucca, agave, sotol, and prickly pear cactus (photographs in Bezy et al., 1966). In Arizona, they are found at least from 1,397 to 1,890 m in elevation (Rorabaugh, 2004).

TERRESTRIAL ECOLOGY

Activity is nocturnal, with frogs leaving protected underground retreats only during the summer

Distribution of *Craugastor augusti*

Adult *Craugastor augusti*.
Photo: James Stuart

rainy season. Daylight hours (05:00–17:00) are spent in rock crevices, usually of limestone but also in crevices of granite, caves, and even mine tailings. Goldberg (2002) found most frogs either on or within 30 m of limestone outcrops, with home ranges of 112.5–2,327 m² depending upon estimator used. Nocturnal movements were usually short, averaging only 4.6 m (range 2.6 to 7.7 m). Occupied crevices are not oriented in any particular direction, but west-facing crevices may provide suboptimal conditions. In Arizona, they return to their subterranean refugia from August to October depending on weather conditions (Goldberg, 2002). Body temperatures range from 15 to 28°C.

Barking Frogs are photonegative in their phototactic response, suggesting they are highly nocturnal (Jaeger and Hailman, 1973).

CALLING ACTIVITY AND MATE SELECTION

The call is similar to the barking of a dog ("a resounding bark"; Wright and Wright, 1938) or the caw of a raven at a distance, but more of a "whurr" at closer range (Schwalbe and Alberti, 1998). Advertisement calls from Arizona *Craugastor augusti cactorum* are longer in duration (mean 573 ms vs. 343–376 ms), higher in frequency (mean 1,137 Hz vs. 919–995 Hz), and have longer pulses (mean 9.26 ms vs. 4.28–5.6 ms) than those of *Craugastor augusti latrans* from Texas (fundamental frequency of 200–300 cps, a duration of 0.24–0.28 sec, and a call rate of 40 calls/min) and New Mexico (Fouquette, 1960; Goldberg et al., 2004b). Calls can be heard by a human observer 1.6 km distant in Texas (*C. a. latrans*) and 600 m in Arizona (*C. a. cactorum*).

Calling occurs only from a two to four week period in Arizona, with most males calling on the first few nights following the initiation of summer rains (Goldberg, 2002). Calling also is associated with high humidity, no wind, lower temperatures, and high hourly rainfall. In a 2 yr study, Goldberg (2002) heard frogs on only 8 and 16 nights. In contrast, Jameson (1954) reported hearing calls from February to August in Texas. Calls are very loud, but frogs will stop calling if an observer approaches within 14–22 m. Calling occurs from within crevices, from under rocks, or from small chambers. They also have been observed perched on boulders and calling from late afternoon (17:00) into the evening (Rorabaugh, 2004). Males move up to 50 m from their retreats to call.

Calls are often made in tandem whereby one frog initiates calling followed by other nearby males in sequence. This sequential chorusing extends to all chorusing males, with one particularly loud frog taking the role of the leader. If the lead male does not call, a second loud frog will assume the leadership role (Jameson, 1954). Males sometime vie back and forth in an apparent calling contest to determine which will take the lead; weaker-voiced males within the chorus then follow. Females in some populations also have a startle vocalization (central Texas) when grasped, whereas others do not (New Mexico, west Texas, Arizona) (Jameson, 1954; Goldberg et al., 2004b).

OVIPOSITION SITES
Eggs are deposited terrestrially deep in moist cracks, crevices, and fissures or under rocks in the moist earth. Males attend to the clutches to protect them from predators and desiccation, even urinating on the clutch to keep it moist. Strecker (1910b) noted a pair in amplexus near a stream and incorrectly inferred that eggs were deposited in water; he suggested that eggs were deposited in wet leaves along the stream margins. Other early investigators also have incorrectly assumed aquatic breeding (*in* Jameson, 1950).

REPRODUCTION
Egg clutches are unpigmented and not deposited in a frothy mass. Clutch size is 6–80 eggs (Livezey and Wright, 1947; Schwalbe and Alberti, 1998). Jameson (1950) reported a single clutch of 67 eggs in Texas, and a single clutch of 60 eggs was reported from a New Mexico female (*in* Degenhardt et al., 1996). Hatching from the eggs as miniature adults is estimated to require 25–35 days (Jameson, 1950). Schwalbe and Alberti (1998) reported a single hatchling of 21 mm SUL, and a 16 mm SUL specimen was observed in New Mexico (*in* Degenhardt et al., 1996). Strecker (1910b) reported eggs being deposited in February in aquatic habitats. He further provided information on tadpoles, suggesting he was confusing this species with some other species. Information in this paper should be used with caution. Livezey and Wright (1947) gave the breeding season as February to May, but males have been heard calling later in the season (Jameson, 1954).

DIET
Prey in the scat of Barking Frogs includes centipedes, field crickets, scorpions, leafhoppers, mites, grasshoppers, spiders, ant lions, and katydids (Goldberg, 2002). Other prey include kissing bugs, cave crickets, and land snails (*Bulimulus, Succinea*) (Olson, 1959; Schwalbe and Alberti, 1998). Strecker (1910b) reported beetles, ants, and spiders in the diet of Texas frogs.

PREDATION AND DEFENSE
Barking Frogs have been described as "prodigious" jumpers when disturbed (Schwalbe and Alberti, 1998). They noted leaps of 70 cm between boulders, even when carrying a radio transmitter. Arizona *C. a. cactorum* does not possess toxic skin secretions, but *C. a. latrans* from Texas and New Mexico does (Goldberg et al., 2004b). They also puff up their body to enhance contact with the skin secretions and make the frog difficult to handle by a predator. Jameson (1954) noted that females will deliver a "blaring screech" when handled, which presumably serves to startle a would-be predator.

Rocky habitat favored by *Craugastor augusti*. Photo: Cecil Schwalbe

POPULATION BIOLOGY

Using a variety of techniques, Goldberg (2002) and Goldberg and Schwalbe (2004) estimated a survival probability of 66% between years and a capture probability of 13% for a population of *C. augusti* in the Huachuca Mountains of Arizona. Survival rate was estimated at 93%. They further estimated density at 2 populations as one frog per 736 m^2 at 1 location and one per 3,507 m^2 at another location, noting that it took 7.6 hrs to find a frog at the first and 5.1 hrs at the second. Longevity is unknown but likely longer than most frogs.

DISEASES, PARASITES, AND MALFORMATIONS

No information is available.

SUSCEPTIBILITY TO POTENTIAL STRESSORS

No information is available.

STATUS AND CONSERVATION

Sampling and monitoring this species in difficult terrain is not an easy task considering its limited calling and period of activity. Distance sampling, visual encounter surveys, and call surveys are not useful, and mark-recapture is labor intensive and does not give sufficient statistical power for monitoring (Goldberg, 2002; Goldberg and Schwalbe, 2004). Using an occupancy approach focusing on rock outcrops might be the best method to follow status and trends. Fortunately, the species is abundant throughout much of its extensive range in Mexico.

Family Eleutherodactylidae

Eleutherodactylus cystignathoides (Cope, 1878 "1877")
Rio Grande Chirping Frog

ETYMOLOGY
cystignathoides: According to Lynch (1970), the name is derived from *Cystignathus*, an old generic name for several leptodactylid frogs. *campi*: a patronym honoring R.D. Camp who presented the type specimen to the U.S. National Museum.

NOMENCLATURE
Frost et al. (2006a): *Syrrhophus cystignathoides*
    Synonyms: *Eleutherodactylus campi, Eleutherodactylus cystignathoides campi, Phyllobates cystignathoides, Syrrhophus campi*

IDENTIFICATION
Adults. This is a small, rather nondescript frog with a brownish gray to brownish green ground color. The body is elongate and flattened, and the snout is pointed. A dark bar is present from the eye to the end of the nose, but there are no interorbital bars. Dark spots are present dorsally on a finely granular skin. The rear part of the body has irregular flecking. Vertical bars are present on the legs. Toes on the digits are only slightly expanded. Venters are smooth and translucent, and the ventral vein appears as a dark line down the middle of the belly. Adult males are 16.3–23.5 mm SUL, whereas females are 16–25.8 mm SUL (Lynch, 1970). Eleven individuals captured in Houston ranged from 7 to 23.7 mm SUL (Quinn, 1979). Adults are 15–25 mm SUL.
    Larvae. There is no free-swimming larva. The larval period is passed within the egg, and froglets hatch as miniature adults. Hayes-Odum (1990) describes development within the egg.
    Eggs. The eggs are small and unpigmented, measuring 3–3.5–5.0 mm in diameter (Livezey and Wright, 1947; Hayes-Odum, 1990).

DISTRIBUTION
The Rio Grande Chirping Frog is found naturally from the Rio Grande embayment south to Nuevo Leon, Tamaulipas, San Luis Potosi, and Veracruz in Mexico. Scattered populations are reported from a variety of locations in east Texas (Corpus Christi, Dallas, Houston, Kingsville, Tyler, San Antonio, La Grange) (Quinn, 1979) into Louisiana (Hardy, 2004); these were probably introduced through the potted plant trade. An important distributional reference is Dixon (2000).

FOSSIL RECORD
Fossil Rio Grande Chirping Frogs are reported from Pleistocene deposits in Tamaulipas, Mexico

Distribution of *Eleutherodactylus cystignathoides*

Adult *Eleutherodactylus cystignathoides*. Photo: Kenny Wray

(Holman, 2003). Holman (2003), however, noted that osteological characters useful in identification of fossils of this species are not available.

SYSTEMATICS AND GEOGRAPHIC VARIATION
Rio Grande Chirping Frogs were described as *Syrrhophus campi* by Stejneger (1915) and later assigned as a subspecies of *S. cystignathoides* (Martin, 1958). This relationship needs to be reexamined using molecular analysis. In his 1970 review of the genus *Syrrhophus*, Lynch placed this species in the *leprus* species group. Hedges (1989) synonymized the genus *Syrrhophus* with *Eleutherodactylus*, and Heinicke et al. (2007) noted the Caribbean clade of eleutherodactyline frogs, included frogs formerly considered *Syrrhophus*, and that the proper generic name was *Eleutherodactylus*. Hedges et al. (2008) placed *E. cystignathoides* within the *Eleutherodactylus (Syrrhophus) leprus* species series. Some authors, however, continue to recognize *Syrrhophus* as a genus rather than as a subgenus.

ADULT HABITAT
The Rio Grande Chirping Frog is found along sandy river alluvium rather than in rock outcrops as are the other Texas chirping frogs. Wright and Wright (1938) described the habitat as "moist earth under boards, brick, or stone piles, under walks, or any cover of yard, field, grass, or brush."

TERRESTRIAL ECOLOGY
Nothing has been reported on the terrestrial ecology of this species.

CALLING ACTIVITY AND MATE SELECTION
In Louisiana, Hardy (2004) reported calling in April and from August to November. Calling occurred mostly at night, but calling also occurred in the morning from 07:30 to 09:15 hrs. The minimum air temperature at which calling takes place is ca. 20°C (Hayes-Odum, 1990). Calling occurs from perches as high as 22 cm above the substrate early in the evening (Hayes-Odum, 1990). As the night progresses, most calling occurs at lower perch heights and from beneath vegetation, where males call from crevices in leaf litter and mulch. The crevices appear to penetrate into the soil, and frogs retreat deeper into the burrows and tunnels should they be disturbed or after calling ceases at dawn. Crevices in sidewalks, rock walls, and buildings also are used. Calling occurs both in rainy and dry weather, but calling is more intensive on rainy nights. Females move toward calling males and may initiate amplexus by lightly touching the male. An amplexed pair may move around until an appropriate oviposition site is found.

OVIPOSITION SITES
Oviposition occurs at or just below the substrate surface in moist soil or under surface debris.

Hayes-Odum (1990) found eggs in a shallow flowerbed, and in captivity eggs were deposited at or just below the soil surface.

REPRODUCTION
According to Wright and Wright (1938) and Livezey and Wright (1947), the clutch size is 6–13 with a breeding season from April to May. It may be that this species deposits multiple clutches over a longer breeding season than present information suggests. Eggs are deposited in a clump, which the female then spreads around and covers using her rear legs (Hayes-Odum, 1990). Development takes 14–16 days at 27–33°C. Hatchlings are dark brown and ca. 6 mm SUL (Hayes-Odum, 1990).

DIET
The diet includes small beetles and spiders (Hardy, 2004), but a comprehensive assessment of diet has not been undertaken.

PREDATION AND DEFENSE
This frog is very adept at running or scurrying under vegetation and into rock crevices to avoid capture.

POPULATION BIOLOGY
No information is available.

DISEASES, PARASITES, AND MALFORMATIONS
Unidentified nematodes were reported by Hardy (2004).

SUSCEPTIBILITY TO POTENTIAL STRESSORS
No information is available.

STATUS AND CONSERVATION
Population status and trends have not been investigated.

Eleutherodactylus guttilatus (Cope, 1879)
Spotted Chirping Frog

ETYMOLOGY
guttilatus: from the Latin *guttula* meaning 'spotted' or 'flecking.'

NOMENCLATURE
Frost et al. (2006a): *Syrrhophus guttlilatus*
   Synonyms: *Eleutherodactylus petrophilus, Eleutherodactylus smithi, Malachylodes guttilatus, Syrrhophus gaigeae, Syrrhophus petrophilus, Syrrhophus smithi*
   Information on *Eleutherodactylus guttilatus* is frequently included with *E. marnockii*, as the species were once considered conspecific. Collecting locations should be verified when using the literature.

IDENTIFICATION
Adults. This is a small frog with a wide head and light brown to black flecking and vermiculation on the dorsal pattern. The ground color is cream, gray, or brown. The eyes are prominent, and an interorbital bar is present. The dorsal skin is smooth. The thighs are usually not banded, and the digital pads are slightly expanded. Adult males are 20.6–29 mm SUL, whereas females are 25.7–31 mm SUL (Lynch, 1970). Gaige (1931) reported specimens 24–32 mm SUL, and Milstead et al. (1950) gave 20.9–26.5 mm SUL for 4 adults.

   Larvae. There is no free-swimming larva. The larval period is passed within the egg and froglets hatch as miniature adults.

Distribution of *Eleutherodactylus guttilatus*

Adult *Eleutherodactylus guttilatus*. Photo: Robert Hansen

Eggs. The eggs are small and unpigmented, measuring ca. 4 mm in diameter (Gaige, 1931). However, Gaige may have been referring to *E. marnockii* since she considered populations in central Texas to be conspecific with those in the Chisos Mountains.

DISTRIBUTION
Spotted Chirping Frogs are known only from the Big Bend region of Texas and several scattered localities in Sierra Madre Oriental in Mexico. Dixon (2000) also mentioned calls heard from the east side of the Davis Mountains. Important distributional references include Dixon (2000) and Dayton et al. (2007).

FOSSIL RECORD
No fossils are known.

SYSTEMATICS AND GEOGRAPHIC VARIATION
In his 1970 review of the genus *Syrrhophus*, Lynch placed this species in the *marnockii* species group. Hedges (1989) synonymized the genus *Syrrhophus* with *Eleutherodactylus*, and Heinicke et al. (2007) noted the Caribbean clade of eleutherodactyline frogs, included frogs formerly considered *Syrrhophus*, and that the proper generic name was *Eleutherodactylus*. Hedges et al. (2008) placed *E. guttilatus* within the *Eleutherodactylus (Syrrhophus) marnockii* species group. Some authors, however, continue to recognize *Syrrhophus* as a genus rather than as a subgenus. Schmidt and Smith (1944) described *Syrrhophus gaigae* from the Chisos Mountains, but the name *S. guttilatus* (Cope, 1879) has nomenclatural priority.

ADULT HABITAT
This species is found under rocks, on rock bluffs and among rock slides, and along stream beds in canyons (Gaige, 1931; Jameson and Flury, 1949).

TERRESTRIAL ECOLOGY
Activity appears to be stimulated by rainfall.

CALLING ACTIVITY AND MATE SELECTION
The call is described as a metallic note repeated at irregular intervals (Gaige, 1931), but Dayton et al. (2007) noted there are actually 2 distinct calls: a short, sharp chirp and a 1–2 sec trill. Calling occurs in the evening after rain at temperatures of 15.6–26.7°C (Dayton et al., 2007).

OVIPOSITION SITES
Oviposition sites are unknown but are presumed to be in moist rock crevices or in cavities beneath the soil under rocks.

Habitat of *Eleutherodactylus guttilatus*, Big Bend National Park. Photo: C.K. Dodd Jr.

REPRODUCTION
Breeding occurs from April to July (Dayton et al., 2007). Individual clutches contain from 6 to 20 eggs. A female with five eggs was observed on 20 July (Gaige, 1931).

DIET
Few data are available. Gaige (1931) observed ants, beetles, termites, and an isopod in the gut of a single female.

PREDATION AND DEFENSE
Like other chirping frogs, *E. guttilatus* is agile and rapidly leaps to safety at the approach of an observer.

POPULATION BIOLOGY
No information is available.

DISEASES, PARASITES, AND MALFORMATIONS
The mite *Hannemania hylae* is an ectoparasite of this species in Texas (Gaige, 1931; Lynch, 1970; Jung et al., 2001).

SUSCEPTIBILITY TO POTENTIAL STRESSORS
No information is available.

STATUS AND CONSERVATION
No information is available. Dayton et al. (2007) suggested that the species might be more widely distributed than is currently recognized.

Eleutherodactylus marnockii (Cope, 1878)
Cliff Chirping Frog

ETYMOLOGY
marnockii: a patronym in honor of G.W. Marnock of Helotes, Texas, who collected the original specimen.

NOMENCLATURE
Frost et al. (2006a): *Syrrhophus marnockii*
  Synonyms: *Eleutherodactylus marnockii, Hylodes marnockii, Syrrhophus marnochii*

IDENTIFICATION
Adults. Adults are small stout-bodied and flattened, measuring 19 to 38 mm SUL. The head is noticeably large. The ground color is greenish brown with dark flecks throughout the dorsum. Mohr (1948) noted that individuals appeared white in caves but turned a normal color in light. An interorbital bar is present. The dorsal skin is smooth to weakly pustular, but the venter is smooth and white. The thighs are banded and the digital pads are expanded. Adult males are 18.4–28.9 mm SUL, whereas females are 20.4–35.4 mm SUL (Lynch, 1970). Milstead et al. (1950) reported adults from the Edwards Plateau at 18–32.7 mm SUL and the Stockton Plateau at 28.8–37.7 mm SUL.

Larvae. There is no free-swimming larva. The larval period is passed within the egg, and froglets hatch as miniature adults.

Eggs. The eggs are small and unpigmented, measuring ca. 4 mm in diameter (Livezey and Wright, 1947).

DISTRIBUTION
This species is endemic to Texas and is known only from the Edwards Plateau and the edge of the Stockton Plateau. Calls also have been reported from Upton County. Important distributional references include Milstead et al. (1950) and Dixon (2000).

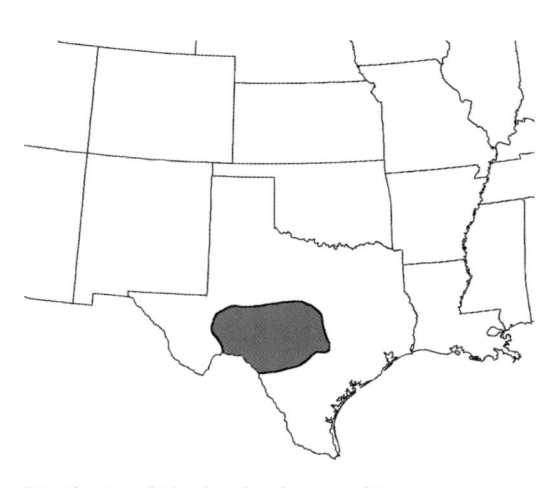

Distribution of *Eleutherodactylus marnockii*

Adult *Eleutherodactylus marnockii*. Photo: Dirk Stevenson

FOSSIL RECORD

Cliff Chirping Frogs are reported from Pleistocene deposits in a variety of locations in Texas (Holman, 1969, 2003). Holman (2003) provided a description of the ilium useful in identifying fossils of this species.

SYSTEMATICS AND GEOGRAPHIC VARIATION

In his 1970 review of the genus *Syrrhophus*, Lynch placed this species in the *marnockii* species group. Hedges (1989) synonymized the genus *Syrrhophus* with *Eleutherodactylus*, and Heinicke et al. (2007) noted the Caribbean clade of eleutherodactyline frogs, included frogs formerly considered *Syrrhophus*, and that the proper generic name was *Eleutherodactylus*. Hedges et al. (2008) placed *Eleutherodactylus marnockii* within the *Eleutherodactylus (Syrrhophus) marnockii* species group. Some authors, however, continue to recognize *Syrrhophus* as a genus rather than as a subgenus.

ADULT HABITAT

This is a species of limestone caves and crevices throughout the Edwards and parts of the Stockton plateaus. These areas are heavily dissected by canyons and stream channels, surrounded by blackland prairie. The topography consists of escarpments, terraced hills, bluffs, talus slopes, rolling grasslands, and rock outcrops. Jameson (1955) discusses topography and vegetation in connection with *E. marnockii* occupancy in the different geologic formations, and Milstead et al. (1950) noted its presence in the cedar-oak association on the Stockton Plateau. Rock crevices in the vicinity of creeks appear to be preferred habitat. The species is common in caves (Nicholson, 1932; Mohr, 1948; Jameson, 1955).

Cliff Chirping Frogs are terrestrial, inhabiting rock outcrops and the substrate adjacent to rocky cliffs. They also have been found to 2.4 m off the ground in trees and *Smilax* vines (Jameson, 1955) and along stream banks. Cliff faces may be isolated from one another by grasslands, cedar breaks, and deep soils.

TERRESTRIAL ECOLOGY

Activity is primarily nocturnal, beginning shortly after dusk, with peak activity periods from April to early June and from September to October corresponding to moderate temperatures. Cliff Chirping Frogs are more active when rainfall is above normal than during droughts, and spring drought may delay spring activity into early summer. Some activity, however, occurs year-round, with frogs becoming active in winter as temperatures rise following a period of cold weather. Frog activity also is associated with high humidity, low wind, low light intensity, and high substrate moisture (Jameson, 1955). Frogs may be observed moving sluggishly in crevices at temperatures of 4.4–10°C. At higher temperatures, they move around much more often. Jameson (1955) observed frogs from 1.1 to 33.3°C and at a humidity of only 35%.

Other than juvenile dispersal, movements probably are limited, with most frogs remaining in the immediate vicinity of one location their entire life. Indeed, breeding season home ranges averaged only 0.034–0.07 ha at 12 Texas sites, with no differences between the sexes (Jameson, 1955). The largest home ranges occurred on open slopes. In Jameson's study, marked juveniles dispersed 112.5–300 m (mean 211 m), with most animals moving parallel to the bluff. A few frogs established immediate residency. Dispersal likely depends on topography, available habitat, weather conditions, and time of year. Jameson (1955) further showed that dispersal readily occurs into areas of low population density and away from areas of high population density.

CALLING ACTIVITY AND MATE SELECTION

Calling occurs from protected sites, such as cracks, fissures, caves, and from under houses. Most calling occurs at night, but calls occasionally are heard during the day. Even when in amplexus, a male may continue calling. The usual call is likened to a cricket chirping and can be heard by human observers up to 30 m distant. Calls are often made in tandem among individuals (trios or quartets), whereby one frog initiates calling followed by other nearby males in sequence. If the lead male does not call, the others also do not call (Jameson, 1954, 1955). This sequential calling pattern is heard throughout the year and not just during the breeding season. It is possible that the sequence represents a territorial system or a means of nearest-neighbor recognition. Jameson (1955) noted that the advertisement call is much more rapid, clear, and sharp and is only made during the breeding season. This latter call is pronounced when males are already in amplexus or when there is a female close by.

Males approach a female and scratches her back, neck, and legs with his hind legs. If the female does not move away, he will clasp her in axillary amplexus. The male continues to stimulate the female with his hind legs. She will dig a trench with her forelegs and deposits the eggs as the clasped pair move along the trench. Both the male and female cover the eggs with soil after they are deposited. Eight to twenty eggs are deposited at a time (Jameson, 1955).

OVIPOSITION SITES

Despite intensive searches, Jameson (1955) was unable to locate egg clutches in nature. Presumably they are deposited deep in moist soil in protected crevices and fissures underground.

REPRODUCTION

Gravid females have been reported from February to December, suggesting an almost year-round breeding season. Most breeding occurs during

Habitat of *Eleutherodactylus marnockii*. Photo: Dirk Stevenson

the spring and fall rains, based on the presence of tiny juveniles; juveniles have been observed from February to November (Jameson, 1955). Eggs are deposited shortly after mating. Jameson (1955) found no evidence of nest guarding. Individual clutches contain from 6 to 20 eggs. Multiple clutches may be oviposited, suggesting a total reproductive output of 60 or more eggs per season.

DIET
Cliff Chirping Frogs eat a variety of invertebrates, including ants, beetles, crickets, termites, and spiders (Jameson, 1955). They likely consume any small animal they can.

PREDATION AND DEFENSE
As with other chirping frogs, this species is adept at running into cracks and crevices in limestone in order to avoid capture. Natural predators include snakes (*Agkistrodon contortrix*, *Crotalus atrox*, *Thamnophis eques*) and wolf spiders (Jameson, 1955).

POPULATION BIOLOGY
Females reach sexual maturity at 19–22 mm SUL, whereas males reach maturity at 18–22 mm SUL (Jameson, 1955). Sex ratios are difficult to determine under natural circumstances. Of 245 frogs marked by Jameson (1955), 48 were juveniles, 66 were females (with eggs), and 141 were classified as "probable males." For another study, Jameson had collectors search for frogs and found 57 males and 53 females.

Density varies among populations, with Jameson's 12 study sites containing densities of 1.2–8.9 frogs per 0.45 ha. Density estimates also vary annually and within a season. Populations appear greater in the spring than in the fall. Jameson (1955) estimated there were from 3 to 31 frogs per "average home range" at 12 study locations. Population turnover is considerable between seasons and years, but sample sizes (years and locations observed) are very small. Jameson (1955) suggested a complete population turnover every 2.5–3 yrs.

DISEASES, PARASITES, AND MALFORMATIONS
The mite *Hannemania hylae* is an ectoparasite of this species in Texas (Lynch, 1970).

SUSCEPTIBILITY TO POTENTIAL STRESSORS
No information is available.

STATUS AND CONSERVATION
This species appears to be reasonably common throughout its range, but there are no studies on its status and population trends.

Family Hylidae

Acris blanchardi Harper, 1947
Blanchard's Cricket Frog

ETYMOLOGY
blanchardi: a patronym honoring Frank N. Blanchard (1888–1937), a herpetologist at the University of Michigan.

NOMENCLATURE
Synonyms: *Acris crepitans blanchardi, Acris crepitans paludicola, Acris gryllus blanchardi, Acris gryllus paludicola, Hyla ocularis blanchardi*

There is considerable confusion in the primary literature about whether the species *A. crepitans* or *A. blanchardi* is being discussed. Until recently (Gamble et al., 2008), these taxa were considered subspecies of a single wide-ranging species, and many authors did not make a distinction between them. Overlap may not be extensive, however, although some distributional details remain to be determined.

IDENTIFICATION
Adults. This is a slender small frog (maximum 35 mm SUL) with a blunt snout and at least one black line on the posterior portion of the thigh. The upper black line on the back of the thigh is usually irregular; the lower area of the thigh consists of black stippling suggestive of a second line. Background colors are brown to gray. The dorsal stripes of Blanchard's Cricket Frogs are variable in coloration, with colors of gray to brownish red to green vertebral stripes. Down the back, the stripe is often bordered on either side by black spots. These colors may be vivid to pale, and the band itself can be continuous or interrupted. A few rare individuals have both red and green dorsal coloration. The coloration is not related to sex and can be variable within the population. Even frogs with a nongreen vertebral stripe may have occasional green spots, and frogs rarely have a red-green stripe (Pyburn, 1961a).

The dorsal surface of the skin has a profusion of tiny warts of irregular size. Small black spots may be present, which are rimmed in white. A dark triangle is present between the eyes, although in some specimens the triangle may appear pale. Upper jaws are black to brown, and from the angle of the eye to the jaw a white line is usually present. Bellies are white, and some individuals have a very pale yellow throat. The undersides of limbs are white. The tips of the fingers have slightly enlarged disks and the fingers lack webbing. The feet are well webbed. Males have a yellow to dark subgular vocal pouch. There are no morphological characters that definitively separate *A. crepitans* from *A. blanchardi*.

Females are slightly larger than males. In Wisconsin, adults range from 20 to 30 mm SUL (mode 25–26 mm) (Dernehl, 1902). In South Dakota and Nebraska, males have a mean SUL of 22.6 mm (range 15.5 to 26.7 mm), whereas the female mean is 24.5 mm SUL (range 16.3 to 31.5 mm SUL) (McCallum and Trauth, 2004). In Indiana, males are 20.3–25.5 mm SUL (mean 22.6 mm), whereas females are 20–30 mm SUL (mean 24.5 mm) (Minton, 2001). Nevo (1973b) recorded the following mean male and female SUL lengths in mm: west Texas (25.5, 28.7), central Texas (23.6, 26.4), west Gulf Coast Plain (22.5, 24.0), northern interior lowlands (22.5, 24.3) and northern Great Plains (25.8, 27.2).

Larvae. Larvae are olive to brown, and are speckled with small black markings. The mottled upper and lower tail fins are narrow, and the tail musculature may have a dark bordering line at the junction between the upper tail fin and the tail musculature. The tail usually has a conspicuous black tip, as in other cricket frog tadpoles.

It is also much longer than the body and has a pointed tail tip. The throat is gray in the center, and the body is mottled toward the side and belly; the belly itself is white to pale yellow. The total length may reach 40–50 mm. Cricket frogs have the distinction of having the largest tadpole in relation to metamorph size of all North American anurans. A description of the tadpole contrasting it with larval *A. gryllus* is in Orton (1947).

Eggs. The eggs are small and dark brown to black dorsally and somewhat tan ventrally. In Ohio, the mean diameter of the vitellus is 0.85–1.0 mm (Dickson, 2002) and 1.13 mm in Texas (range 1.06 to 1.17 mm). There are two gelatinous envelopes surrounding the egg, although the inner envelope may not be visible without the aid of a microscope. The inner envelope is 2.38–2.95 mm in Ohio (Dickson, 2002) and 2.34–2.74 mm (mean 2.6 mm) in Texas; the outer envelope is 2.98–3.7 mm (mean 3.34 mm). The total number of eggs ranges between ca. 150 to >400, but the sample sizes of actual counts are small. The eggs were described by Livezey (1950).

Distribution of *Acris blanchardi*. An isolated population once existed in extreme southeastern Arizona but is no longer extant. The hatched area in Mississippi indicates uncertainty in the identity of *Acris* in this region.

DISTRIBUTION

Blanchard's Cricket Frog occurs from central Ohio southward into western West Virginia and northern Kentucky. Morse (1904) reported *A. blanchardi* from several locations in Ohio where Walker (1946) could not relocate them. The species has not been observed in West Virginia since 1948 and is apparently extirpated within the state (Dickson, 2002). The range once included a small part of southwestern Ontario (now extirpated). Populations occur in southern Michigan and southwestern Wisconsin and once extended up the Mississippi River to include Chisago County in eastern Minnesota. The latter population has been extirpated, but there is a more recent record for Hennepin County (Moriarty et al., 1998). Populations in southeastern and southwestern Minnesota may have been extirpated, except perhaps in Winona County. The range includes southeastern South Dakota and most of Iowa, south through much of central Nebraska, most of Kansas and Oklahoma, and into the eastern three-quarters of Texas and the adjacent Rio Bravo drainage in Mexico. There were populations along the Republican and Platte rivers in northern Colorado, but these disappeared in the 1970s, and the last cricket frog was observed in Colorado in 1979 (Hammerson, 1999; Hammerson and Livo, 1999). The range extends into New Mexico along the Pecos River, and *A. blanchardi* may have once reached as far west as Cochise County, Arizona (Frost, 1983); this population at the springs and cienegas of San Bernardino no longer exists. To the east, *A. blanchardi* is found throughout Louisiana (except the Florida Parishes) and may cross into Mississippi in the counties bordering the Mississippi River in the Delta Region. There is no evidence, however, of it crossing the Mississippi and Ohio rivers between Memphis and about the vicinity of Louisville, Kentucky. Many *A. blanchardi* populations have disappeared since the 1970s, and the current range no longer includes much of the area previously occupied, especially in the West and upper Midwest.

Blanchard's Cricket Frogs have been recorded on North Bass, Middle Bass, Kelley's, and Pelee islands in Lake Erie (Walker, 1946; Logier and Toner, 1961; Langlois, 1964; Weller and Green, 1997; Davis and Menze, 2000; Hecnar et al., 2002).

Important distributional references include: Arizona (Frost, 1983), Arkansas (Burt, 1935; Trauth et al., 2004), Colorado (Hammerson, 1999; Hammerson and Livo, 1999), Great Lakes region (Harding, 1997), Illinois (Schmidt and Necker, 1935; Smith, 1961; Phillips et al., 1999), Indiana

(Minton, 2001), Iowa (Hemesath, 1998), Kansas (Smith, 1934; Collins, 1993), Louisiana (Dundee and Rossman, 1989), Michigan (Ruthven, 1912), Minnesota (Oldfield and Moriarty, 1994), Missouri (Johnson, 2000; Daniel and Edmond, 2006), Nebraska (Hudson, 1942; Lynch, 1985; McCallum and Trauth, 2004; Ballinger et al., 2010; Fogell, 2010), New Mexico (Van Denburgh, 1924; Degenhardt et al., 1996), Ohio (Walker, 1946; Pfingsten, 1998; Davis and Menze, 2000), Oklahoma (Burt, 1935; Sievert and Sievert, 2006), South Dakota (Ballinger et al., 2000; McCallum and Trauth, 2004; Naugle et al., 2005; Kiesow, 2006), Texas (Hardy, 1995; Dixon, 2000, as both *A. crepitans* and *A. blanchardi*), West Virginia (Green and Pauley, 1987), and Wisconsin (Suzuki, 1951; Vogt, 1981; Mossman et al., 1998). Dixon (2000) and Trauth et al. (2004) include a separate map for *A. crepitans*, but only *A. blanchardi* occurs in Texas and Arkansas, respectively (Gamble et al., 2008). The map in Dundee and Rossman (1989) includes both *A. crepitans* and *A. blanchardi*.

FOSSIL RECORD

Pliocene fossils attributed to *A. blanchardi* (as *A. crepitans*) are known from Texas, as are Pleistocene Irvingtonian fossils. Pleistocene Rancholabrean fossils are more common, having been described from many localities in Kansas and Texas (Holman, 2003).

SYSTEMATICS AND GEOGRAPHIC VARIATION

Acris blanchardi has long been considered a subspecies of *A. crepitans* based on morphological data (Harper, 1947). However, McCallum (2003) and McCallum and Trauth (2006) reported on the results of an extensive morphological study (including analyses of webbing, spot patterns, stripe patterns, presence of warts, morphological measurements) and concluded that there were no meaningful differences between these taxa; they recommended that *A. c. blanchardi* not be recognized as a valid taxon. Using both mtDNA and nDNA, however, Gamble et al. (2008) concluded that there were enough differences in the molecular data to warrant recognition of *A. blanchardi* as a full species. There are also regional differences in physiological variables between the species (e.g., Salthe and Nevo, 1969).

The Coastal Cricket Frog, *A. gryllus paludicola*, was described by Burger et al. (1949), based on specimens found from near Sabine Pass, Jefferson County, Texas. The subspecies was described based on its relatively smooth skin, a lack of or reduced presence of anal warts, ill-defined postfemoral stripes, and a rose-pink coloration of the male's vocal sac. Based upon a reanalysis of seven newly found specimens from the type locality, Rose et al. (2006) concluded that recognition of the subspecies was warranted based on mtDNA comparisons, but that it was allied with *A. crepitans* instead of *A. gryllus*. Gamble et al. (2008), however, did not recognize the subspecies, since it was nested within the *A. blanchardi* clade.

Ward et al. (1987) examined 16 enzyme loci for populations in north-central (n = 14) and south-central Texas (n = 7). Ten of the loci were monomorphic. They concluded that differentiation among remaining loci was low between northern and southern populations. However, there was a considerable amount of allelic heterozygosity, indicating high levels of differentiation at the population level. Genetic similarity was not related to geographic proximity, and gene flow between even adjacent populations was minimal. Ward et al. (1987) hypothesized that inbreeding and/or stochastic events related to environmental factors were responsible for the maintenance of the genetic variation.

In Kansas, Gorman and Gaines (1987) examined 17 loci from 16 populations occurring across the state. Some populations were found along the margins of ponds, whereas others occurred along streams. Allelic frequencies varied considerably among populations, and differences were evident in the frequencies of alleles between adults and juveniles. These differences were maintained across years, and suggested strong selection. In general, genetic variation declined toward the western portion of the state. It appears that temperature and moisture exert significant influence upon allele frequencies. In addition, there were strong differences in allele frequencies between cricket frog populations living at ponds or along streams. There were no differences in allele frequencies among different color morph stripes (Gorman, 1986). Gorman and Gaines (1987) hypothesized that selection operated on physiological characteristics of Blanchard's Cricket Frogs rather than through predator-mediated selection.

Whereas Nevo and Capranica (1985) differ-

Adult female *Acris blanchardi* with egg packets. Photo: Sara Viernum

entiated three different call patterns in *Acris*, roughly corresponding to forests (*A. crepitans*), grasslands (*A. blanchardi*), and pinelands (*A. gryllus*), it is clear that these patterns do not hold in forested areas in the southern part of the range of *A. blanchardi*. In Louisiana and Texas, for example, *A. blanchardi* has a "forest" call similar to *A. crepitans,* where populations occur in forested habitats. Similarities in calls relate more to habitat characteristics (open versus closed habitats) than to phylogeny.

Dorsal coloration can include various degrees of color intensity of the grayish, red, and green morphs. Vogt (1981) even has a photograph of a very light colored yellowish-brown individual. In Wisconsin, for example, Dernehl (1902) recorded 8 frogs with very little red coloration, 3 with medium shades of red, and 12 with intense shades of red. Corresponding values for the green morph were six, one, and five. Thirty-five of the 100 frogs he examined had at least some shades of red or green. Nongreen morphs predominate. For example in Texas, 83% of 521 recent metamorphs were nongreen and 17% were green (Pyburn, 1961b), whereas in Indiana Isaacs (1971) recorded 273 gray-striped frogs, 134 red-striped frogs, and only 110 green-striped frogs. Pyburn (1961b) reported on an additional collection in Texas in which 28% were green and 72% were nongreen. In Kansas, western populations are composed almost entirely of gray-striped frogs, with the proportions of green- and red-striped frogs increasing toward the east (Gorman, 1986).

Pattern polymorphism is common in cricket frogs. Some authors have suggested that there are different selection pressures on frogs on seasonally changing backgrounds or habitat. Burkett (1984) noted that the proportion of green striped frogs seemed highest when water levels were high and vegetation was abundant. Green frogs also were scarcer in open situations than in wooded habitats. As with *Pseudacris maculata*, the frequencies of these morphs may change seasonally and annually within a population (Nevo, 1973a), and may reflect different selection pressures depending on seasonal changes in background coloration. However, color changes do not vary seasonally in all populations (Isaacs, 1971; Gray, 1983; Gorman, 1986), and there may be different escape behaviors associated with different stripe colors. For example, Wendelken (1968) noted that red-striped *A. blanchardi* tended to jump more often toward water than land (but see Gray, 1978). Thus, pattern polymorphism in cricket frogs could be maintained via substrate matching, predator avoidance, and different escape behaviors (Milstead et al., 1974).

Not all authors agree. In rather simple experiments, Gray (1978) could find no differences among color morphs in predation susceptibility, substrate, or water preferences, or in distribution along the shoreline. He found no differences in color morphs in desiccation rates, resistance to prolonged stress, thermal preference, or CTmax among cricket frogs in Illinois (Gray, 1977). Finally, he found no differences in movement patterns, dispersal, growth, survivorship, or seasonal frequencies of stripe pattern at various study sites (Gray, 1983, 1984). However, there were differences geographically and annually in color-morph frequencies among locations. Isaacs (1971) also found no seasonal differences in the frequencies of color morphs over a three-month study at a single site in Indiana. Gray (1983, 1984) suggested simple chance might account for variation in color-morph frequencies in Blanchard's Cricket Frog, as effective population size was small in his Illinois populations.

Two basic patterns of dorsal stripe coloration in *A. blanchardi*, from a genetic point of view, are green and nongreen (red or gray). The red stripe appears dominant to gray, but its genetic relationship to green stripes is unclear. The green coloration is determined by either a heterozygous or homozygous condition at a single gene, and the green stripe results from simple dominance (Pyburn, 1961a). Thus, crossing a pair of heterozygous green-stripe frogs should produce a roughly 3:1 ratio of green to nongreen F_1 offspring, and this prediction is easily verified in laboratory crosses (Pyburn, 1961b).

A.P. Blair (*in* Moore, 1955) attempted to hybridize female "*A. gryllus crepitans*" (presumably *A. blanchardi*) with male *Pseudacris streckeri* and *P. clarki*. The attempt was unsuccessful.

ADULT HABITAT

Acris blanchardi is a frog of open habitats throughout much of its western range. In the Midwest and in east Texas and adjacent Louisiana, Blanchard's Cricket Frogs are found around permanent ponds, lakes, marshes, and small streams surrounded by both deciduous and piney woods habitats with substantial amounts of edge habitat (Knutson et al., 2000). They are found less often around impoundments, swales, and in riverine situations, at least in the South (Lichtenberg et al., 2006). They can also be found associated with agricultural crops (Brown, 1974; Anderson and Arruda, 2006), gravel pits (Schmidt and Necker, 1935), or formerly strip-mined areas (Myers and Klimstra, 1963; Anderson and Arruda, 2006). In the arid High Plains, they are more confined to river bottoms than wetlands directly imbedded in the grasslands (see, for example, the distribution map in Lynch, 1985). They are rarely found far from water, and they do not frequent temporary wetlands unless they are in proximity to permanent wetlands. These frogs prefer moist and muddy substrates in close proximity to shelter sites, such as rocks (Smith et al., 2003). When along streams, they prefer pool to riffle habitats and use both sunny and shaded microhabitats close to water. Blanchard's Cricket Frogs have been reported at elevations as high as 1,275 m in New Mexico (Degenhardt et al., 1996).

AQUATIC AND TERRESTRIAL ECOLOGY

Activity normally occurs throughout the warmer months of the year, although activity extends year-round in many areas if environmental conditions permit (e.g., Pyburn, 1958). This species is often active at cold temperatures when "normal" frogs are dormant, but they also are active at temperatures approaching 38°C (Clarke, 1958). Activity normally occurs well into the autumn. For example, *A. blanchardi* in Ohio were observed to be active from February to late March until early September (Wilcox, 1891) or mid-November (Walker, 1946; Brenner, 1969). In Kansas and New Mexico, activity normally occurs into October or November (Smith, 1934; Heinrich and Kaufman, 1985; Collins, 1993; Degenhardt et al., 1996). In southern Illinois, cricket frogs have been observed in January (Gray, 1983). Other reports of winter activity are available for Kansas (Linsdale, 1927; Busby et al., 2005), Illinois (Rossman, 1960; Smith, 1961; Tucker et al., 1995), Indiana (Blatchley, 1892), and Ohio (Dickson, 2002). In Texas, activity tends to decline greatly in December and January, but the frogs never really disappear as long as temperatures permit.

Overwintering occurs terrestrially in crayfish burrows and cracks in the mud banks along ponds (Gray, 1971; Irwin et al., 1999; Swanson and Burdick, 2010), under surface debris (Garman, 1892; Walker, 1946, 1963; Pope, 1964; Bayless, 1966), among tree roots (Walker, 1946), under rocks near springs (Blair, 1951), or in gravel slopes along and above streams (McCallum and Trauth, 2003b). These frogs are capable of surviving for short periods in oxygenated water, but not in hypoxic conditions, and they even may survive short periods (<6 hrs) at temperatures as low as –1.5 to –2.5C° (Swanson and Burdick, 2010). They do not possess any special physiological properties to ward off freezing temperatures, that is, they are not freeze tolerant, and they are not able to supercool (Irwin et al., 1999; Irwin, 2005; Swanson and Burdick, 2010). Frogs become dormant when air temperatures reach 4°C and water temperatures are between 3–6°C. Frogs emerge from dormancy when air temperatures reach 10°C (Brenner, 1969). The amount of body fat and access to favorable overwintering sites are critical factors in determining whether frogs will survive the winter dormancy.

Blanchard's Cricket Frogs usually stay close to water, especially since they do not appear to be tolerant of desiccation (Ralin and Rogers, 1972). Lateral movements along the shoreline are not uncommon. Most frogs are found within 8–15 m or so of point of original capture, although some frogs may travel > 90 m away (Pyburn, 1958; Burkett, 1984). Movements of these distances can occur over a period of a few days to months. Individuals seem to remain within an area for an extended period, with long-distance movements back and forth between distant locations occurring periodically. Movements occur during periods of wet weather, and individuals move between adjacent ponds, pools, or other moist habitats. They even have been observed 60 m from the nearest water during rainfall. The greatest distance Pyburn (1958) recorded movement was 167 m, whereas Burkett (1984) recorded a single frog moving 213 m and Fitch (1958) recorded movements up to 183 m. These frogs will readily return to near the point of capture when displaced over distances > 30 m.

Despite most observers noting the adherence of this species to moist areas near water, there are discrepancies in the literature. For example, Collins and Collins (2006) stated that "Northern Cricket Frogs evidently wander great distances from water during both dry and wet weather, and many apparently die during these wanderings, keeping populations at an optimal level." It seems doubtful, however, that selection would operate so forcefully for the group as opposed to the individual. Fitch (1958) noted that most movements occurred along pond margins and moist ravines, and that frogs could move considerable distances under wet conditions. Many frogs likely die during dispersal, but not to keep populations "optimal."

The body length is positively correlated with body mass, and the larger body size of *A. blanchardi* compared with *A. crepitans* and *A. gryllus* may be associated with the tendency of large frogs to lose water more slowly than small frogs (Nevo, 1973b). This is important, since much of the range of *A. blanchardi* is in the semi-arid grasslands of the Plains and upper Midwest. The CTmax for this species is 38.8–39.5°C, and the thermal preference is 27–28°C (Gray, 1977).

Blanchard's Cricket Frogs likely are photopositive in their phototactic response as is the phylogenetically related *A. crepitans* (Jaeger and Hailman, 1973), suggesting they use both sunlight and moonlight when feeding and going about their daily activities. They probably are sensitive to light in the blue spectrum ("blue-mode response") (Hailman and Jaeger, 1974), which apparently helps them orient toward areas of increasing illumination, such as the open horizon above lakes and ponds. Blanchard's Cricket Frogs also likely have true color vision.

CALLING ACTIVITY AND MATE SELECTION

Males produce an advertisement call that both attracts females and provides information about his position and calling status to rival males. Males tend to call from the same location on consecutive nights and if displaced will return to the original calling location (Perrill and Shepherd, 1989).

Calling occurs both by day and night. In Arkansas, calling occurred at temperatures between 13 and 24°C (Briggler, 1998), whereas in Texas calls were heard at temperatures from 6 to 34°C. Wiest (1982), also in Texas, noted calling at 8.7–28.3°C over an 85-day period. Diurnal calling occurs at the lowest temperatures (<12°C) (Blair, 1961b) and immediately after frogs have emerged from dormancy sites (Dickson, 2002). Temperature preferences at which breeding occurs may be more narrow. In a savanna region in Texas, calls were heard from March to September (85 nights), but breeding only occurred from April to June (Wiest, 1982). When calling, air temperatures ranged between 8.7 and 28.3°C; breeding, however, occurred on only 5 nights at temperatures of between 18.5 and 23.2°C (Wiest, 1982). As the season progresses, diurnal calls gives way to nocturnal calls. Nocturnal calling continues until 02:00–03:00 in the morning (Perrill and Shepherd, 1989). Call rates tend to increase with air temperature, but not with relative humidity or barometric pressure (Jackson, 1952).

Males may begin calling weeks prior to actual reproduction. For example, Pyburn (1958) recorded calling behavior in early February, but egg deposition did not begin until mid-March. Calls tend to grade between attractant and aggressive, and males do not have a specific aggressive call (Wagner, 1989a). The call sounds like a loud click, with the clicks being produced in series or bouts; the distinctive call gives the species its common name, sounding like a chorus of crickets calling from around a pond or lake. Clicks may be continuous or divided into two or more pulse groups. Each group contains from 1 to 58 calls ranging from 0.5 to 11.5 sec.

In Texas, calls range from 22 to 168 calls/min with a dominant frequency of 2.8–3.9 kHz (Wagner, 1989a). Texas females preferred calls of 3.5 kHz. In contrast, Capranica et al. (1973) recorded a mean dominant frequency of 2.9 kHz in South Dakota; they also provided a sound spectrogram of the call's frequency and amplitude. The variation in Wagner's study may result from collecting frequency data from frogs living in different habitat types (open vs. closed). Increasing call complexity without increasing its overall amplitude actually decreases its attractiveness to females, unlike in many other species of anurans (Witte et al., 2001). The auditory nervous system of the Northern Cricket Frog is closely matched to the spectral energy of frogs within a localized population (Capranica et al., 1973).

Certain morphological characteristics of cricket frogs do appear correlated with call variation. The spectral and temporal characteristics of the call are associated with the size of the larynx, which in turn is correlated with body size (McClelland et al., 1996). Ryan et al. (1995) concluded that there was no evidence that morphologically symmetrical frogs were more successful than less symmetrical frogs, whether or not variables measured were directly related to call production.

There is considerable variation in the call characteristics of *A. blanchardi* (Nevo and Capranica, 1985), leading some researchers to assume that there are two intergrading species, *A. crepitans* and *A. blanchardi*, perhaps separated by habitat preferences (grassland vs. forest), particularly in east Texas and adjacent Louisiana. According to Gamble et al. (2008) and T. Gamble (personal communication, 2009), however, based on molecular analysis, *A. crepitans* does not occur in this region. Instead, females from populations even rather close to one another can have different preferences in call frequency and in the fine tuning of the basilar papilla necessary for call detection (Ryan and Wilczynski, 1988). Females show a preference for the calls of their own population.

Populations from Texas tend to show clinal variation in call characteristics. Frogs from open habitats tend to produce calls with lower frequency, longer duration, and slower call rates than frogs from more closed habitat populations in the east (Ryan and Wilczynski, 1991). Habitat differences certainly explain some of the variation in call structure in that call characteristics appear to have evolved differently in different habitats in order to enhance call transmission (Ryan et al., 1990; Ryan and Wilczynski, 1991; Witte et al., 2005). Females are "tuned" below the frequency of their local population, and tuning is correlated with body size. Small females prefer males with high frequencies, and large females prefer males with lower frequencies. According to Ryan et al. (1992), this variation in female preference contributes to the variation heard in call frequency within populations.

Males are spread out along the shoreline, calling in a density of 0.1–0.9 males per meter (Wagner, 1989a) and rarely within 50 cm of one another (Perrill and Shepherd, 1989). Males also call while floating in shallow water or sitting on floating vegetation such as lily pads. They tend to defend their "acoustic" space rather than a specific resource territory. If one male violates the acoustic space of another (within ca. 30–40 cm), the frogs will fight in a wrestling match, stop calling, or simply tolerate the intruder's presence (Burmeister et al., 1999b). In high-calling densities, males are more likely to tolerate an intruder than in more widely spaced choruses. Males also are more likely to abandon calling early in the season, or tolerate an intruder, than they are later in the breeding season (Burmeister et al., 1999b). Instead, they are more likely to attack as the season progresses, presumably as opportunities for successful mating decrease.

Males are able to determine distance from rivals by changes in sound pressure levels, with the pressure becoming greater the closer a rival approaches (Wagner, 1989c). Calls become increasingly aggressive as the distance between males decreases. In agonistic encounters, rival male cricket frogs increase the number of calls per bout, increase call-bout duration, and lower the call rate within the bout. Individual calls become longer with a lower pulse rate. Dominant frequencies also may be lowered (Perrill and Shepherd, 1989; Wagner, 1989b, 1992), and callers that lower the dominant frequency of the call are more effective at repelling intruders than callers that do not lower the frequency (Wagner, 1992).

Aggressive calls tend to be more aggressive when a male is presented with an aggressive call; this indicates that males perceive temporal characteristics of the call. The response also varies with the dominant frequency of the resident's call. If a resident receives an attractant call rather than an aggressive call, the response call is similar. Burmeister et al. (1999a) thus suggested that the resident's experience with different types of calls influences the behavior of male cricket frogs.

Aggressive intent may be signaled in a number of ways by *A. blanchardi*. For example, dominant frequency is correlated with the male's SUL (Wagner, 1989a). Although the call may provide information about size of an intruder or resident through changes in dominant frequency, such information does not appear to influence a male's decision to fight, flee, or tolerate another male. Changes in call frequency, however, can be used by males to assess the agonistic intent of another male (Wagner, 1992). These changes in the temporal calling patterns may help facilitate mutual assessments of size and, therefore, the outcome of a wrestling match. Although changes in temporal calling patterns do not predict the outcome of a fight, they do predict whether a male will tolerate another male (Burmeister et al., 2002). In a fight, the larger male usually wins (Wagner, 1989c). Thus, this complex interchange of information between males in effect helps them avoid needless contests. If large males usually win fights, and size can be roughly assessed through temporal characteristics of the call, why waste energy and time fighting?

There may also be a bit of subterfuge in the interactions between male Blanchard's Cricket Frogs. Small males tend to try and lower the dominant frequency of their call. In this way they appear larger than the really are and thus give a false impression of their wrestling ability (Wagner, 1989b). Losers in wrestling matches may become satellite males located nearby (within 15 cm) calling males (Perrill and Mager, 1988). Losers may also adopt other mating strategies. Thus both the call structure and the size (or perceptions of size) are important in male competition and the evolution of mating behavior.

The advertisement call may have another way in which it functions to attract females. Although it has been assumed that the escalation of calling by males functions only in a male–male context by signaling aggressive intent, the escalation in calling might also increase a male's attractiveness to females. Indeed, females tend to prefer the calls of interacting males. Thus, males may use changes in their temporal calling characteristics to increase their attractiveness to females when in the presence of other males (Kime et al., 2004).

Only a portion of the male breeding population may call from a breeding site on any particular night. For example, Perrill and Magier (1988) recorded between 4 and 27 (mean 12) calling males and about one satellite male every 2 nights at a breeding site in Indiana over a period of 5 weeks. The following year at a different site, they recorded from 5 to 44 calling males (mean 18) and nearly 1 satellite male per night over a 2-week period. The total number of calling males per year was 282 and 421, respectively. Satellite and calling males may change roles, and there is no difference in size between satellites and callers. Satellite males are occasionally successful in intercepting incoming females (Perrill and Magier, 1988).

Amplexus normally occurs in shallow water along the shoreline, although amplexing pairs have been observed away from water (Pyburn, 1958). The female controls the movement of the pair, going from the shoreline area to a location where eggs will be deposited. Curiously, amplexed females tend to respond to phonotaxis choice tests more strongly than unamplexed females. However, unamplexed females do have the same overall preferences for male callers as amplexed females (Witte et al., 2000).

BREEDING SITES

Breeding occurs in mostly open permanent ponds, lakes, bogs, marshes, and along slow-moving streams. They are frequently observed along the level shoreline, or may sit on floating mats of vegetation; they do not frequent ponds with steep-sided banks. Ponds with a good supply of algae and emergent vegetation are preferred. In some ponds, they will venture far from the shoreline as they move from lily pad to lily pad or through other thick emergent vegetation, but they do not breed in deep water. They normally do not breed successfully in temporary wetlands such as bison wallows, although they may be found in these unique prairie habitats (Bragg, 1943a, 1950b; Gerlanc, 1999; Gerlanc and Kaufman, 2003).

REPRODUCTION

The breeding season varies by latitude, which is not surprising given the extensive range of this species. Most reproduction appears to occur during a peak breeding season from late spring to early summer. Calling does not indicate successful reproduction, as calls are heard both before and after the main breeding season. In some areas of Texas, for example, calling extends from early January to October with a peak from April to July (Livezey, 1950; Pyburn, 1958; Blair, 1961b). In Oklahoma, Bragg (1950a) recorded breeding dates from March to August. In Illinois, most breeding occurs in April to May (Gray, 1983), whereas in Missouri, Johnson (2000) gives the breeding dates as April to July; Seale (1980) recorded egg deposition in Missouri from April to June. Calls in Ohio are heard from April to October, although gravid females were found only from mid-April to mid-June (Walker, 1946; Dickson, 2002). Some authors appear to equate calling with breeding, and these clearly are not synonymous in Blanchard's Cricket Frogs.

Tadpole of *Acris blanchardi*. The tadpole of *Acris crepitans* is identical with *Acris blanchardi* in appearance. Photo: Stan Trauth

Gular pouch formation in males may occur as early as October in Illinois (Gray, 1983). McCallum (2003) and McCallum et al. (2011) noted that some male Blanchard's Cricket Frogs are capable of reaching sexual maturity by late summer. Fisher and Richards (1950) described the ovarian cycle for Blanchard's Cricket Frogs in Oklahoma. In Arkansas, ovarian development occurs in spring and summer, with oviposition in May and June; all evidence of vitellogenic ova have disappeared by mid-July (McCallum, 2003; McCallum and Trauth, 2004; McCallum et al., 2011). As such, the breeding season in this latter region extends only from late May to possibly early July, the same as in Minnesota (Oldfield and Moriarty, 1994), Kansas (Burkett, 1984), and the Great Lakes region (Harding, 1997). In contrast, males begin to show gular pouch coloration in May in South Dakota and Nebraska, and females are ready to deposit eggs in June.

Most authors report a single egg-laying event. However, Burkett (1969; repeated by Busby et al., 2005) suggested that eggs may be deposited twice each breeding season, or that partial clutches may be deposited multiple times. In possible verification, Trauth et al. (1990) reported on two females in captivity that laid smaller clutches after previously depositing a clutch. Either they deposited two clutches or they deposited partial clutches on two occasions.

It is unlikely that reproduction takes place in autumn despite the continuance of calling outside the main breeding season. Most records indicate cessation of breeding activity by June or July, depending upon location. In any event, the amount of body fat and the size of the gonads are inversely correlated, which suggests a lack of autumnal egg deposition. Even if it occurred, it is unlikely metamorphs would survive the winter.

Eggs are deposited singly into small loose clusters containing two to seven eggs each (Livezey, 1950). The irregular egg masses are held together by the gelatinous envelopes. Debris also may adhere to the egg mass, making it difficult to observe. The eggs may adhere to vegetation via the gelatinous envelopes, or they may rest quietly on the substrate. Eggs hatch in three to four days, and the tiny larvae are inconspicuous. Eggs deposited in water <17.4°C will not hatch; they will hatch up to 33.7°C (Ballinger and McKinney, 1966).

Actual counts of clutch size are few. Smith (1934) reported a single female ovipositing 248 eggs; Burkett (1969) recorded a mean clutch size of 323 from 3 females; Gorman (in Collins, 1993) reported a mean of 170 eggs from 6 females; Trauth et al. (1990) reported a mean clutch size of 266 (range 174 to 431) from 9 females; Trauth et al. (1990) also reported a clutch size of from 231–298 (mean 258) from 3 "*A. c. crepitans*" in Arkansas. In Trauth et al.'s (1990) study, 2 female "*A. c. blanchardi*" produced additional clutches of 120 and 134 eggs after depositing an initial clutch.

LARVAL ECOLOGY

The larval period is 5–10 weeks in Kansas (Burkett, 1969) and up to 61 days in Ohio (Dickson, 2002). Total larval length reaches a mean of 35–40 mm (Minton, 2001; Dickson, 2002). Metamorphosis occurs at ca. 7.0 mm tibial length (range 5.3 to 8.5 mm) in Texas (Bayless, 1969a), or about 13.5–15 mm TL for a froglet in Ohio (Walker, 1946; Dickson, 2002).

DISPERSAL

In Illinois, Gray (1983) reported that some frogs marked as they left their breeding ponds dispersed to other ponds 0.8–1.3 km away.

DIET

Prey consists of small invertebrates, particularly collembolans, beetles, flies and fly larvae, ants, true bugs, lepidopteran larvae, aphids, mites, spiders, grasshoppers, small crustaceans, miscellaneous other invertebrates, and even a small crayfish (Garman, 1892; Hartman, 1906; Jameson, 1947; Gehlbach and Collette, 1959; Johnson and Christiansen, 1976; Labanick, 1976; Burkett, 1984). Prey appears to be consumed in proportion to their availability, that is, there does not appear to be any specific prey selection. Prey consumption tends to be greatest in female frogs, large frogs, reproducing frogs, frogs with small fat bodies, and frogs captured in June and July (Johnson and Christiansen, 1976). It takes about 8 hrs for food to pass through the digestive tract of *A. blanchardi*.

PREDATION AND DEFENSE

Adult cricket frogs are expert jumpers that are quick to change direction when approached. They hop along the shoreline and veer toward water or the bank, invariably turning rapidly back toward the shore, regardless of whether on land or water. Cricket frogs in the water also may submerge into the substrate or hide among thick vegetation (Stebbins, 1951; Pyburn, 1958; Stebbins and Cohen, 1995; Dickson, 2002). On land, their coloration is quite cryptic as long as they are sitting still. Death-feigning has been reported for the species when roughly handled, at least in captivity (McCallum, 1999b). Frogs assumed a tight posture with the limbs folded in toward the body. When released, they remained in this position for a few seconds before hopping away.

Larvae are fully palatable, but many have black tail tips, which serve as disruptive coloration, that is, it alters the apparent shape of the larva. The tail tip may also function to distract a potential predator away from the head and body. Caldwell (1982) demonstrated that larvae with black tipped tails were found most often in ponds, whereas those without this coloration were found more frequently in lakes and creeks. Whether larvae in a population have black tail tips may be associated with the presence of certain predators. In ponds, dragonfly larvae are the primary predator, and the black tail tip draws their attention. In creeks and lakes, fish are the primary potential predators, and uniform body coloration may help in concealment (Caldwell, 1982). The polymorphism is present even in nearby populations and is predator-mediated but habitat specific.

Larvae are preyed upon by a variety of invertebrate predators, particularly dragonfly naiads. Recent metamorphs may be killed by fishing spiders (*Dolomedes sexpunctatus*) as they leave the breeding site. Predators of adults include fish

(*Micropterus salmoides*), American Bullfrogs (*Lithobates catesbeianus*), snakes (*Nerodia, Thamnophis*), birds (great blue herons, green herons, American Bitterns), raccoons (*Procyon lotor*), and opossums (*Didelphis virginianus*) (Fitch, 1958; *in* Langlois, 1964; Burkett, 1984). Perrill and Magier (1988) found that Banded Watersnakes (*N. fasciata*) targeted calling males more than noncalling males.

The eggs of *Acris blanchardi* contain no toxins or substances that would deter predators. However, the large egg capsule may confer mechanical protection against predation by some predators (Grubb, 1972). The eggs are consumed by small fishes, depending on their level of satiation.

POPULATION BIOLOGY

After metamorphosis, froglets grow rapidly (Burkett, 1984) and at relatively uniform rates (Dickson, 2002). At a rate of 0.1 mm/day, they reach adult size at a tibial length of 11.0 mm. Growth rates are linear for males but curvilinear for females (McCallum, 2003; McCallum and Trauth, 2004; McCallum et al., 2011). Adult size can be reached as quickly as 40–60 days following metamorphosis (Bayless, 1969a; Burkett, 1984) in contrast to Brenner's (1969) suggestion that they must overwinter first. In the North, at least, growth does not occur in winter. Blanchard's Cricket Frogs attain their greatest body mass and size between May and July, after which mean body sizes in a population tend to decline with an infusion of juveniles (McCallum and Trauth, 2004). Blanchard's Cricket Frog is essentially an annual species with near complete population turnover every year (Pyburn, 1958; McCallum and Trauth, 2004). Gray (1983), however, noted that a few frogs survived a second winter.

Sex ratios may be skewed toward males around a site, or more likely the skewed ratios reported in the literature are biased by sampling. In one example, Pyburn (1958) determined the sex of 283 frogs captured around a small pond-ravine complex in Texas during the breeding season and found a ratio of two males per female. At another location, however, the sex ratio was equal, based samples taken at different times of day. In Ohio, Dickson (2002) captured 98 males, 56 females, and 81 juveniles (a sex ratio of 1.75:1). In Illinois, Gray (1984) estimated the sex ratio to be 2.7:1 at his study sites. Perceptions of sex ratios may change with life stage. Burkett (1984) noted that the sex ratio of newly metamorphosed froglets was one male per four females, but after three months it was male dominated. Such a skewed sex ratio may be real or it may reflect seasonal differences in behavior and thus sampling bias.

A population of Blanchard's Cricket Frogs around a pond or lake may contain many individuals, especially after metamorphosis. For example, Pyburn (1958) estimated there were 1,173 frogs around a 34 m × 24 m pond in Texas based on a 2-sample Lincoln Index and 1,053 using the Hayne Index. In Kansas, Burkett (1984) estimated

Adult *Acris blanchardi*.
Photo: Brian Hubbs

a "minimum living population" of 1,260 at one site; the number of frogs counted varied dramatically through the season, however, with peak numbers in the autumn as juveniles entered the population. The fewest frogs generally were observed in June and July. Also in Kansas, Anderson and Arruda (2006) using visual encounter survey techniques estimated there were a mean of 127, 91, and 187 frogs per ha in mined, agricultural, and natural habitats, respectively. Estimates based on aural survey techniques were 130, 251, and 546 frogs per ha in the same three habitat types. Finally, Gray (1983, 1984) suggested that the effective breeding population might be rather small. He observed a maximum of 54–213 frogs in most of his Illinois populations; however, one population was estimated at 900 frogs.

COMMUNITY ECOLOGY

There are probably many factors that influence whether *A. blanchardi* co-occurs with fish, including the trophic niche of the fish. Predaceous fishes (*Lepomis, Micropterus, Ictalurus*) can eliminate *Acris blanchardi* from ponds that previously did not contain these species (Sexton and Phillips, 1986). However, not all fish introductions are detrimental. For example, the introduction of *Notemigonus crysoleucas* and *Pimephales promelas* may have no impact on *Acris blanchardi* presence, and *A. blanchardi* can disappear even from fishless ponds (Sexton and Phillips, 1986).

DISEASES, PARASITES, AND MALFORMATIONS

Amphibian chytrid fungus (*Batrachochytrium dendrobatidis*) has been detected in seven specimens of *A. blanchardi* from Arkansas, Louisiana, and Texas (Rothermel et al., 2008; Saenz et al., 2010). Although other diseases have not been identified, McCallum and Trauth (2007) demonstrated that activation of the immune response, even to a sublethal pathogen or toxin, may reduce male fertility and thus adversely affect reproduction in this species. Even without a morbidity or mortality event, pathogens thought harmless might impact anuran populations in ways that are not obvious.

Parasites include unidentified mites from specimens in South Dakota and Nebraska (McCallum and Trauth, 2004) and metacercariae from the fluke *Zeugorchis megacystis* (Burkett, 1984). In one Kansas population, 97.5% had this species in their abdominal cavity, where the number of metacercariae was inversely proportional to the frog's body size. Other reported parasites include ciliate protozoans (*Opalina obtrigonoidea, Opalina* sp.) (Trowbridge and Hefley, 1934; Odlaug, 1954) and the myxidia *Myxidium serotinum* (McAllister and Trauth, 1995) and *M. melleni* (Jirků et al., 2006; McAllister et al., 2008). Helminths include the trematodes *Glypthelmins quieta*, unidentified kidney trematode metacercariae, and the cestodes *Brachycoelium salamandrae, Cylindrotaenia americana*, and *Ophiotaenia magna* (Trowbridge and Hefley, 1934; Najarian, 1955; Ulmer, 1970; Brooks, 1976; Ulmer and James, 1976a; Dyer, 1991; McCallum, 2003). Nematodes include *Physaloptera ranae, Rhabdias ranae*, and *Rhabdias* sp. (Trowbridge and Hefley, 1934; Ashton and Rabalais, 1978).

Malformations have been reported not uncommonly in *Acris blanchardi*. These include malformations of the jaws, fused or missing toes, missing whole or parts of limbs, extra digits or limbs, unusual eyes, abnormal foot development, reduced size of femur, unresorbed tail, and a greatly enlarged testis (Smith and Powell, 1983; Gray, 2000a, 2000b; Dickson, 2002; McCallum and Trauth, 2003a, 2004; Lannoo, 2008). High frequencies of malformations in museum samples were evident in the late 1950s, but the sample size was small. From 1957 to 1979, the frequency was 3.33%; the frequency increased after that reaching a high in Arkansas and Missouri museum records from 2000 when it was 8.48%. Frogs from the Ozark region had the highest number of malformations. In contrast, field collections in the 1970s and again in 1998 showed low frequencies (0.39%) of malformations in Illinois (Gray, 2000a). He further reported (Gray, 2000b) only 7 malformations in 10,000 frogs examined. No malformations were observed in 153 *A. blanchardi* in surveys in the north-central United States (Converse et al., 2000).

Intersexuality and abnormal testicular development have been recorded from natural populations of this species (Reeder et al., 1998, 2005; McCallum, 2003).

SUSCEPTIBILITY TO POTENTIAL STRESSORS

Chemicals. The use of chemical pesticides, particularly DDT and PCBs, may have contributed significantly to population declines in Blanchard's

Cricket Frogs (see Status and Population Trends). Reeder et al. (1998) found that 2.6% of the *A. blanchardi* collected over 3 yrs in Illinois were intersexes. The herbicide atrazine was correlated with intersexuality, but the exact mechanism of interaction was uncertain. In addition, the presence of PCBs and PCDF (polychlorinated dibenzofuran) was correlated significantly with sex-ratio reversal. The sex ratio of juveniles from contaminated ponds was skewed toward males, whereas that from natural ponds was skewed toward females. These contaminants adversely affect the gonads of *A. blanchardi* probably by mimicking the activity of estrogenic endocrine compounds during development.

The chemical perchlorate is known to disrupt thyroid activity in Blanchard's Cricket Frogs in east Texas. Perchlorate is an oxidizer used in solid-fuel rockets, and water where it was used or processed is often contaminated with it. Cricket frogs at such sites show greater than normal levels of follicle cell hypertrophy, especially at concentrations > 10 μg/L (Theodorakis et al., 2006).

STATUS AND CONSERVATION

At one time, *A. blanchardi* was one of the most ubiquitous and abundant anurans within its range (e.g., Wright and Wright, 1949); this is sadly no longer the case. Blanchard's Cricket Frog has experienced significant alarming population declines, especially in the upper Midwest agricultural regions. Declines or complete disappearances have been documented in Arizona (Frost, 1983), Colorado (Hammerson, 1999; Hammerson and Livo, 1999), northern Illinois (Mierzwa, 1998), northern Indiana (Brodman and Kilmurry, 1998; Minton, 2001; Brodman et al., 2002), Iowa (Lannoo et al., 1994; Hemesath, 1998; Christiansen, 1998), Michigan (Lehtinen, 2002), Minnesota (Oldfield and Moriarty, 1994), Nebraska (McLeod, 2005), Ohio (*in* Lehtinen, 2002; Lehtinen and Skinner, 2006), Ontario (Oldham, 1992; Weller and Green, 1997), South Dakota (Naugle et al., 2005; Kiesow, 2006), West Virginia (Dickson, 2002) and Wisconsin (Vogt, 1981; Jung, 1993; Casper, 1998; Hay, 1998; Mossman et al., 1998). Most of the declines appear to have been triggered in the 1970s and perhaps early 1980s.

Populations appear to be doing much better in the southern portion of its range. In southern Illinois, for example, Wilson (2000) found them still to be abundant along the Kaskaskia River, and Davis et al. (1998) found them still common in southern Ohio. In addition, Brodman (2009) found small numbers of *A. blanchardi* in Jasper County, Indiana, from 1999 to 2007 in an area where surveys from 1994 to 1998 failed to locate them. In a review of status between 1970 and 1980 in Iowa, Christiansen (1981) considered the status unchanged; by the late 1990s, however, they had declined by 26% from historically reported localities (Christiansen, 1998). In the Kansas Flint Hills, Viets (1993) and Busby and Parmalee (1996) found *A. blanchardi* to be still abundant and the status to have changed little over the previous 60–70 yrs.

There are a number of factors, perhaps acting separately or in concert that may be responsible for the disappearance of this species throughout the northern portions of its range. These have been reviewed by Greenwell et al. (1996) and Gray and Brown (2005), and include biocides (especially those affecting gonad development), toxic substances, changes in pH, the effects of introduced or subsidized predators, UVB radiation, parasites affecting kidneys and limb development, climate change (drought and cold), changes to habitats (siltation, habitat alteration, agricultural practices, development, dredging, outright habitat destruction), habitat fragmentation, disease, plant succession, and human indifference. Many of these factors likely affect populations simultaneously at different locations. Coupled with life history constraints (e.g., short life span and high population turnover), these factors combine to make it impossible for this species to survive throughout large portions of its former range. Beasley et al. (2005) performed a risk assessment, evaluating many of these factors and concluded that phytotoxic herbicides could disrupt pond ecology and lead to decreased recruitment, especially when coupled with all the other contaminants that affect breeding ponds in the agricultural Midwest.

The pH has been suggested as an important factor affecting the distribution of this species as well as influencing its decline. However, distributional data from Alabama (Mecham, 1964) and Ohio (Lehtinen and Skinner, 2006) offer no support for the acidity effects hypothesis. In Wisconsin, Jung (1993) found no real differences in environmental

Habitat of *Acris blanchardi*. Photo: Jennifer Anderson-Cruz

conditions at sites with cricket frogs compared with sites without them, except that ponds where frogs were no longer present were located in proximity to agricultural fields.

As Reeder et al. (2005) have shown, declines in Blanchard's Cricket Frog populations correspond to the period of heaviest use of DDT and PCBs (1946–1959) and are associated with increased intersexuality. Despite the prohibitions on DDT and the decrease in use of PCBs, populations of this frog have continued to decline. Reeder et al. (2005) hypothesized that declines in *A. blanchardi* were initiated through the effects of these chemicals acting as endocrine disrupters. Declines have been most pronounced, and the incidence of intersexes greatest, in the heavily industrialized, urbanized, and intensively farmed regions of northern Illinois, compared with the more diverse and less chemically pulverized areas of southern Illinois. In addition, the vast quantity of chemicals used in agricultural areas could affect the phytoplankton in pond communities, and this would impact both potential predators and food sources of cricket frogs. The presence of estrogenic compounds also is highly correlated with trematode parasitism, which has implications for the development of malformations (Beasley et al., 2005). The geographical trends in declining populations are consistent with the endocrine-disruption hypothesis.

Still, the effects of climate change may have drastic consequences for a frog with a relatively short life span (McCallum, 2010). Lannoo (1998) has pointed out that conditions of drought superimposed on a drastically anthropogenically altered habitat could be responsible for disappearances of Blanchard's Cricket Frog. As permanent ponds may be filled with fish and as temporary ponds are lost to periodic drought or habitat destruction, the short-lived cricket frogs are unable to reproduce successfully. These effects occur concurrently with increasing levels of fertilizer and pesticide pollution. Whereas all populations may not be affected simultaneously, they will tend to "wink-out" through time. Managing wetlands to include fishless permanent wetlands could help to reduce the potential for extirpation.

Another hypothesis for the decline in Blanchard's Cricket Frogs in the upper Midwest is related to changes in precipitation coupled with severe cold. A combination of severe winter weather coupled with drought (and lack of snow cover) could lead to direct contact with killing cold and desiccation (Irwin, 2005). Frogs killed in winter retreats would never be observed, and this observation is consistent with declines in this species, that is, numbers of dead or dying frogs are not observed. Irwin (2005) has noted that there is a correlation between climate changes related to the severity of drought and cold and the disappearance of this species.

Finally, habitat destruction or alteration by itself has undoubtedly resulted in the disappearance of this species throughout much of its range. As prairies were turned into agricultural fields, temporary wetlands were lost. Indeed, wetlands throughout the prairies and upper Midwest were destroyed by the tens of thousands, and with them their resident anuran communities. In the South, silvicultural activities may have had mixed effects. Urbanization is another potentially detrimental factor, with *A. blanchardi* occurring more frequently in rural areas than in the vicinity of urban development (e.g., Foster et al., 2004). Natural catastrophic disturbances such as hurricanes also result in decreased abundance, particularly in forested habitats (Schriever et al., 2009), at least over a short-term period.

Not all habitat changes are detrimental to Blanchard's Cricket Frog, however. In Arkansas, well-managed forestry has actually opened up habitats increasing Blanchard's Cricket Frog abundance, at least over a short-term period (Fox et al., 2004). In the latter case, watersheds were managed to protect wetlands, and ponds had been created to assist in fire management. Keeping fish out of such ponds would be one way to enhance their value to the amphibian community. As a habitat mitigation measure, the species benefits from having natural riparian buffers that traverse pine plantations or agricultural areas (Burbrink et al., 1998), and the wider the buffer the more beneficial it is to *A. blanchardi* (Rudolph and Dickson, 1990). In addition, Blanchard's Cricket Frogs may use restored or constructed woodland ponds in numbers greater than the other cricket frog species (Briggler, 1998; Palis, 2007; Shulse et al., 2010). For example, Palis (2007) recorded 7,807 *A. blanchardi* over a 4 yr period at 3 constructed wetlands in southern Illinois. In Missouri, Shulse et al. (2010) found them in 88% of the constructed ponds they sampled.

As with many frogs, populations of *A. blanchardi* have been monitored using aural surveys along predesignated road routes. In Michigan, this species was heard at only 2.8% (43 of 1739) to 2.9% (42 of 1851) of the sites surveyed in 1996 and 1997 (Sargent, 2000). In Illinois, surveys from 1986 to 1989 suggested an increasing population (Florey and Mullin, 2005). Lotz and Allen (2007) noted that observers frequently confused the call of this species with crickets, thus lending a source of observer bias to call surveys using volunteers.

Acris crepitans Baird, 1854
Northern Cricket Frog

ETYMOLOGY
crepitans: from the Latin *crepitans* meaning 'clattering,' referring to the male's advertisement call.

NOMENCLATURE
Synonyms: *Acris crepitans blanchardi, Acris crepitans paludicola, Acris gryllus crepitans, Acris gryllus blanchardi, Acris gryllus paludicola, Hyla ocularis blanchardi, Hylodes gryllus, Rana pumila*

Much of the literature referring to "*A. crepitans*" actually pertains to *A. blanchardi*. Some of the earliest literature also includes information on *A. gryllus* when discussing the Northern Cricket Frog. Very little information is available on the life history and ecology of *A. crepitans* as delimited by Gamble et al. (2008). Readers should consult locations and the original literature in order to be certain which species is being addressed. In areas such as in western Mississippi, northern Kentucky, or Louisiana, it may not be possible to determine which species an author was discussing.

IDENTIFICATION
Adults. This is a slender small frog with a blunt snout and at least one black line on the posterior portion of the thigh. However, the head and body are proportionally wider than in *A. gryllus*. The ground color is gray, tan, light brown, or olive, with colors of yellow to orange also being reported (Bayne, 2004), and the dorsum is rugose. The species has a dark triangle between the eyes pointing caudally, followed sometimes by a median stripe of gray, red, brownish red, rust, or green, and a Y-figure down the back. The upper parts of the limbs retain the ground color, and they may have dark bands; the undersides are white. The upper black line on the back of the thigh is usually very irregular; the lower area of the thigh consists of black stippling suggestive of a second line. The webbing of the hind feet of Northern Cricket Frogs is extensive, and the fourth toe may not extend beyond the third or fifth toe phalanges. Anal warts may or may not be present, and the belly is white or cream. In some individuals there may be a very light yellow coloration around the groin and the base of the forelegs. Males have a yellowish vocal sac during the breeding season. During the nonbreeding

season, this sac can be identified by a series of folds on the throat.

In the related and sometimes sympatric *A. gryllus*, the body is slender with a sharp snout, the upper black line on the thigh is sharply defined, the lower black stripe ranges from a vague stippling to a well-defined regular line, and the limbs are proportionally longer. Unfortunately, the striping pattern is often not sufficient to distinguish between cricket frog species. Neill (1950b) discusses morphological differences between these species. There are no morphological characters that definitively separate *A. crepitans* from *A. blanchardi*.

Adults are 16–35 mm SUL, with females being slightly larger than males. Nevo (1973b) recorded the following mean male and female lengths in mm SUL: south Atlantic and eastern Gulf Coastal Plains (22.6, 25.5), Mid-Atlantic Coastal Plain (23.3, 26), and northeastern uplands (22.5, 24.3). Feliciano (2000) gave a mean adult size of 16.4 mm SUL in New York and 22.6 mm SUL in the Florida Panhandle. Adults are 18–25.6 mm SUL in Alabama (mean 22.2 mm) (Brown, 1956), a mean of 21.6 mm SUL in North Carolina (Micancin and Mette, 2009), and a mean of 21.4–21.9 mm SUL in West Virginia (Bayne, 2004). In Virginia, Mitchell (1986) recorded a mean male SUL of 23 mm, whereas the mean female size is 26 mm SUL.

Larvae. The tadpoles are medium-sized light to dark gray or greenish, and have long tails with low tail fins. Bellies are lighter than the dorsal surface. There is a light stripe bordered by dark stripes along the side of the snout. Throats are light with a dark band extending across the chest. Many of these tadpoles are easily distinguished by their black tail tips, although some tadpoles lack the "black flag." Tails are mottled or reticulate in pattern. There may be dark spots on the tail musculature and the dorsal tail fin. The dark spots on the dorsal tail musculature may result in a saddle-like appearance. Large tadpoles range between 30 and 46 mm TL in length, but TL is usually about 25 mm. The spiracular tube is shorter in *A. crepitans* than in *A. gryllus*. Cricket frogs have the distinction of having the largest tadpole in relation to metamorph size of all North American anurans.

Eggs. According to Livezey and Wright (1947), the eggs of "*Acris gryllus crepitans*" are identical to those of "*Acris gryllus gryllus*." However, it is clear that he included *A. blanchardi* within his concept of *A. crepitans* as well. A female deposits 200–400 eggs per season.

DISTRIBUTION

The Northern Cricket Frog is a mostly upland species that occurs from southern New York (lower Hudson River valley) southward through Georgia, Alabama, and Mississippi. In the northern part of its range, it occurs in the Coastal Plain, Piedmont, and into the Ridge and Valley Province south into Virginia. It is absent from southwestern Virginia and adjacent North Carolina, but it appears to be more common in the upper Coastal Plain of North Carolina than previously appreciated (Micancin and Mette, 2009). In South Carolina and elsewhere, the species may occur along river valleys extending into the Blue Ridge Mountains (Montanucci, 2006). The range also extends well into the Coastal Plain along the Savannah and Altamaha rivers and in the middle and lower Coastal Plain of South Carolina (Harrison et al., 1979). The western limit of its range is probably the Mississippi River, whereas the Midwestern northern limit is the Ohio River; the identity of cricket frogs bordering these rivers remains to be determined, especially in the Mississippi Delta where *A. blanchardi* is known to occur. According to Boyd (1964), it does not penetrate very far into the loess hills bordering the Delta. Much of

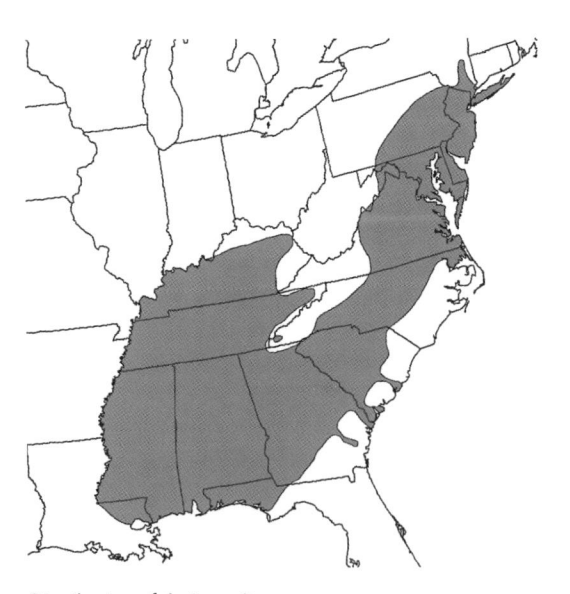

Distribution of *Acris crepitans*

its range is in the Piedmont above the Fall Line along the south Atlantic Coastal Plain, although the range extends well onto the Gulf Coastal Plain including the Florida Panhandle, southernmost counties of Alabama and Mississippi, and possibly the Florida Parishes of Louisiana. The range includes most of Tennessee and Kentucky, although the identity of populations in northern Kentucky needs verification.

The species no longer occurs on Long and Staten islands, New York. In addition, an isolated population once occurred in Allegheny County, Pennsylvania, 195 km west of its known range within the rest of the state; this population apparently no longer exists (Hulse et al., 2001). Hulse et al. (2001) attribute this population to deliberate introduction rather than being a natural population. Records for St. Simons Island, Georgia (Williamson and Moulis, 1994), are questionable. The species does occur on Kent Island, Maryland (Grogan and Bystrak, 1973b).

Important distributional references include: Alabama (Brown, 1956; Mount, 1975), Delmarva Peninsula (White and White, 2007), Georgia (Williamson and Moulis, 1994; Jensen et al., 2008), Louisiana (Dundee and Rossman, 1989), Maryland (Harris, 1975), Mississippi (Boyd, 1964), New Jersey (Schwartz and Golden, 2002), New York (Gibbs et al., 2007), North Carolina (Dorcas et al., 2007; Micancin and Mette, 2009), Pennsylvania (Hulse et al., 2001), South Carolina (Harrison et al., 1979), Virginia (Tobey, 1985; Mitchell and Reay, 1999), and West Virginia (Green and Pauley, 1987; Bayne, 2004). Maps in Williamson and Moulis (1994) present a confused distribution pattern between *A. crepitans* and *A. gryllus* in Georgia. It seems likely that many of the locations are based on misidentified specimens. The map in Dundee and Rossman (1989) includes both *A. crepitans* and *A. blanchardi*.

FOSSIL RECORD

Holman (2003) lists a number of Pliocene and Pleistocene fossils locations for *Acris crepitans*, but the locations of the deposits suggest that these belong to *A. blanchardi* or an ancestor thereof.

SYSTEMATICS AND GEOGRAPHIC VARIATION

At one time, all cricket frogs in North America were thought to be part of a single wide-ranging species. Comparisons of blood and liver proteins (Dessauer and Nevo, 1969), other physiological and molecular data (Salthe and Nevo, 1969), chromosomes (Bushnell et al., 1939), call structure, distributional patterns (Neill, 1954), morphology, and ecology have conclusively demonstrated that *A. crepitans* is a distinct species.

There are two major lineages within *A. crepitans*, an eastern clade that extends across northern Alabama and Georgia through New York, and a western clade that includes populations in Kentucky and Tennessee (Gamble et al., 2008). As with *A. gryllus*, the boundary between clades has not been determined. Gamble et al. (2008) suggested that it might prove to be in the same area as that of *A. gryllus*, that is, in the Mobile Basin and Tombigbee River system of the Gulf Coastal Plain. Morphological variation between these clades may not be evident. However, there appears to be some morphological variation within the species, giving rise, for example, to Jordan et al.'s (1968b) statement that frogs in central Tennessee had "characteristics" of both *A. crepitans* and *A. blanchardi*.

Laboratory experiments indicate a high degree of genetic compatibility between *A. crepitans* and *A. gryllus*, regardless of which species is male or female (Mecham, 1964). However, few larvae could be reared through metamorphosis. None released into the environment were subsequently recovered. Backcrossed hybrids had low fertilization rates and very poor development, and the hybrids appeared stunted. Hybridization does not appear to occur in nature, perhaps because of differences in mating calls and limited geographic overlap. However, there is at least one observation of interspecific aggression between males of these species at a breeding pond (Micancin and Mette, 2010).

There is disagreement in the literature whether cricket frogs have metachrosis, that is, the ability to change dorsal coloration. Authors have thought metachrosis can occur on the entire dorsum (Abbott, 1882), the ground color, but not the vertebral stripe (Cope, 1889; Dernehl, 1902; Dickerson, 1906), the green but not the red vertebral stripe (Pyburn, 1961a; *A. blanchardi*), and both the red and green vertebral stripe (Gray, 1972; *A. blanchardi*). Pyburn (1961a) in particular noted metachrosis based on recaptures of marked frogs. In any case, there is considerable variation among populations as to the frequencies

Adult *Acris crepitans*. Photo: Todd Pierson

of the grayish frogs without any vertebral stripes and those with red or green stripes (Nevo, 1973a). In New York, for example, most frogs have a vivid green dorsal stripe, whereas in Florida frogs are more subdued in coloration, with few exhibiting bright dorsal stripes (Feliciano, 2000).

As with many anurans, Northern Cricket Frogs tend change color shading to match their background coloration. In the direct sunlight, they tend to be pale and to minimize the vivid coloration if red or green. After dark or in the shade, individuals tend to darken almost to the point of obscuring the dorsal pattern, but the red and green colors become more intense (Abbott, 1882).

ADULT HABITAT

Northern Cricket Frogs occupy a variety of both permanent and seasonal ponds, lakes, marshes, and other wetlands in open-canopied forest and meadows. The surrounding habitats may include deciduous, pine, or mixed pine-deciduous forests, mixed hardwood forest, bottomland swamps, cedar-swamps, open meadows, and even agricultural or urbanized spaces (Harima, 1969). In the North, inhabited marshes and meadows usually have rather dense vegetation, but further south a greater variety of more open-water habitats is occupied. They also may inhabit the margins of slow-moving streams and ditches. They prefer muddy substrates in close proximity to shelter.

AQUATIC AND TERRESTRIAL ECOLOGY

Northern Cricket Frogs spend most of their adult lives near the shallow breeding ponds and normally do not venture far from water. They are commonly observed along the shoreline and on floating vegetation mats and water lilies. Prior to or after breeding, they are found in terrestrial moist sites, particularly along streams, in the vicinity of springs, and in lowland swamp forest. They may move short distances back and forth between nearby habitats. In the lowland swamps, overwintering frogs tend to occupy the muddy bottoms. They also may occupy moist ravines or man-made drainage ditches (Enge et al., 1996). Northern populations are most active from early spring through autumn, with some individuals being active as late as 27 December in New Jersey (Abbott, 1882). Activity occurs year-round as long as environmental conditions permit, even in the middle of winter.

When at the breeding site, Northern Cricket Frogs move very little from one part of the pond to another. In general, they are found within 5 m of previous capture; Bayless (1966) found the mean movement to be 2.1 m (range 0.6 to 6.5 m). Greater movements do occur, as one *A. crepitans* in Bayless's (1966) study moved 29.5 m in consecutive days. Another frog moved ca. 43 m in ca. 37 days. According to Gibbs et al. (2007), dispersal occurs to overwintering sites within 100 m of the breeding site. The greatest distance Bayless (1966) recorded movement was 350 m over a period of ca. 50 days between captures.

Overwintering or periodic dormancy occurs in cracks and crevices in mud fissures and under surface debris in muddy areas. Overwintering may occur in adjacent swamps or in areas in close proximity to the breeding ponds. Frogs are active at the breeding sites in early spring, having emerged in response to temperature, rainfall, and sexual development. Movements to breeding sites are probably precipitated by rainfall. Emergence may precede calling by as much as six to eight weeks (Gibbs et al., 2007).

Northern Cricket Frogs are able to use Y-axis celestial orientation to determine dispersal and migratory directions to and away from a perpendicular shoreline or within a pond (Ferguson, 1966a; Ferguson et al., 1967). This ability requires a learned shore position, a view of a celestial cue (sun, moon, and even stars), and an internal clock phased to local time. Displaced frogs are able to return to the shoreline where captured or to the point of release. Celestial orientation may not be the only form of orientation, and alternate methods (e.g., use of familiar landmarks, olfaction) may work better in different habitats or under different environmental conditions.

Northern Cricket Frogs are photopositive in their phototactic response, suggesting they use both sunlight and moonlight when feeding and going about their daily activities (Jaeger and Hailman, 1973). They are sensitive to light in the blue spectrum ("blue-mode response"), which apparently helps them orient toward areas of increasing illumination, such as the open horizon above lakes and ponds (Hailman and Jaeger, 1974). Northern Cricket Frogs likely have true color vision.

CALLING ACTIVITY AND MATE SELECTION

Calling begins a few weeks before actual egg deposition. Most calling occurs from spring through early summer. However, calls may be heard in the autumn and even in midwinter (Abbott, 1882). Populations of calling males may be fairly large. In West Virginia, for example, Bayne (2004) estimated there were 200 calling males at each of two different sites in June. Estimates in May were between 100 at one site and 150 at the second. Only about 20–25 males were calling in August. Most calling occurred at air temperatures of 15°C and when soil and water temperatures were ≥20°C (Bayne, 2004). However, calling occurred at air temperatures as low as 12°C.

As with other cricket frogs, the advertisement call consists of a series of clicks that reminds a listener of crickets calling. These clicks have a dominant frequency of about 3.55 kHz. Other characteristics of the call include: a mean click duration of 159 ms (range 92 to 241 ms); a mean of 5.5 pulses (range 3.9 to 7.1); a mean interpulse interval of 16 ms (range 11 to 21 ms) (Micancin and Mette, 2009). Capranica et al. (1973) provided a sound spectrogram of the call's frequency and amplitude. Much of the information relating to call characteristics, female preference, and variation in male call characteristics described for *A. blanchardi* likely applies to this species (see the *A. blanchardi* account). The auditory nervous system of the Northern Cricket Frog is closely matched to the spectral energy of frogs within a localized population (Capranica et al., 1973).

Mate selection is probably similar to that of *A. blanchardi*. Amplexus is axillary.

BREEDING SITES

Breeding sites include pond and lake margins, shallow open or grassy pools, and even man-made wetlands such as farm or retention ponds. They commonly are found in beaver ponds (Metts et al., 2001). Breeding sites should be open-canopied. As long as seasonal wetlands are available, they will even breed on golf courses (Scott et al., 2008).

REPRODUCTION

Breeding dates vary by location and latitude, but Northern Cricket Frogs normally breed throughout the spring to early summer. In Alabama, for example, calling begins in late March, with first egg deposition occurring by mid-April. The peak breeding season is from April to June, but calls are occasionally heard until August (Mecham, 1964). In Louisiana, most reproduction occurs in March and April, although calling may continue until September. Bayless (1966) found no clasping pairs after 8 April in Louisiana, and considered this species possibly to be an explosive breeder whose breeding activity was triggered by heavy rainfall. In Virginia, the breeding season extends from late April to July (Mitchell, 1986). In West Virginia, reproduction occurs mostly in May and June, but amplexing pairs have been observed in July (Bayne, 2004). In New York, reproduction occurs between mid-May and mid-July (Gibbs et al., 2007).

Eggs are deposited singly in shallow water in small masses attached to blades of grass and other vegetation or free on the substrate. Each irregular cluster may contain from 2 to 20 eggs. Females may produce multiple egg masses per season. In West Virginia, Bayne (2004) counted between 50–400 eggs per total clutch. Hatching occurred in eight to nine days following deposition. Hatchling size is 4–4.25 mm TL (Gosner and Black, 1957b).

LARVAL ECOLOGY

Larval *A. crepitans* feed along the substrate of the breeding site, but as they approach metamorphosis, they swim near the surface close to the shoreline. The larval period lasts between 40 and 50 days in Louisiana (Bayless, 1966), about 60 days in New Jersey (Black and Gosner, 1957b), and from 5–14 weeks in West Virginia (Bayne, 2004). Transformation takes two days. Recent metamorphs were a mean of 12.9 mm SUL in New York (Feliciano, 2000). Wright and Wright (1949) give transformation size as 9–15 mm SUL.

DISPERSAL

Recent metamorphs disperse from shallow wetlands toward damp locations in terrestrial habitats, and may be found near springs or along small streams flowing through woodlands. They also disperse toward other ponds or lakes nearby, and can use streams and rivulets as dispersal corridors. In the South, *Acris* will move to swamp bottomland forest areas, as these areas often retain moisture during the summer and autumn. Dispersal normally occurs in rainy weather.

DIET

Northern Cricket Frogs eat a variety of small invertebrate prey. Specific items include collembolans, ants, flies and fly larvae, spiders, mites, beetles, true bugs, aphids, termites, and miscellaneous larval and adult insects (Bayless, 1969b). In areas of sympatry, the food items of *A. crepitans* and *A. gryllus* overlap, showing no prey specialization. Feeding occurs year-round when weather conditions permit.

PREDATION AND DEFENSE

Northern Cricket Frogs are eaten by a wide variety of predators, including fishes, snakes (*Heterodon platirhinos*), turtles, and birds (Abbott, 1882). Abbott (1882) suggested that snakes congregated in wet meadows during the breeding season to feed on the abundance of cricket frogs. Larvae are eaten by Rusty Crayfish, Bluegill Sunfish, and Grass Carp under experimental conditions (Ade et al., 2010).

Like many frogs, Northern Cricket Frogs stop calling upon the approach of an intruder, only to resume their calls when the intruder walks away. Their cryptic coloration can make them difficult to locate in shallow grassy pools. If disturbed, they are capable of making long and very quick leaps away from the source of disturbance (mean 52.8 cm: Blem et al., 1978; to 89 cm: Green and Pauley, 1987; 1 m high and a distance of 1.2 m: Gibbs et al., 2007), and they are able to change the course of their escape very rapidly. They generally escape toward water, where they hide on the bottom or among dense vegetation. According to Brown (1956), however, they are not as easily disturbed as *A. gryllus*. Interestingly, the characteristics important in hind-limb function are less variable than those traits not associated specifically with locomotion (Salthe and Crump, 1977). This suggests that selection has been more intense on the locomotion traits highlighting their importance to survival.

Adults also may assume a death-feigning posture where they float motionless in the water, venter up (Dickerson, 1906; Feliciano, 2000). Marchisin and Anderson (1978) also recorded crouching, hiding, climbing, and remaining motionless when approached by a snake predator under experimental conditions.

POPULATION BIOLOGY

The Northern Cricket Frog is an abundant species and often the most noticeable species within an area. Abbott (1882) describes their numbers in New Jersey as "incalculable." In New York, Feliciano (2000) estimated a population of 628 (range 343 to 1,681) frogs along a 150 m transect using mark-recapture methods, whereas in the Florida Panhandle she estimated 259 (range 217 to 313) frogs along two 150 m transects. Also in New York, Gibbs et al. (2007) stated that a chorus of 50 calling males represented a large population.

As with other cricket frogs, maturity is achieved within a few months of transformation. Populations reach their greatest abundance in early summer and then decline rapidly, and Abbott (1882) went so far as to suggest that adults died shortly after breeding, due to an excessive expenditure of energy during the breeding season. For whatever reason, the species does appear to experience nearly complete population turnover annually.

COMMUNITY ECOLOGY

In many areas of the South, populations of *A. gryllus* and *A. crepitans* overlap geographically (Viosca, 1944; Mecham, 1964; Bayless, 1966, 1969b; Micancin and Mette, 2009). However, it appears these species have different habitat preferences where they occur together. For example, *Acris crepitans* is found in bottomland swamps or water close to the shore of lakes, whereas *A. gryllus* occurs more in uplands or along drainage ditches where the species overlap (Norton

Breeding habitat of *Acris crepitans*, Arkansas County, Arkansas. Photo: Kevin Rose

and Harvey, 1975; Dundee and Rossman, 1989). *Acris gryllus* tends to avoid shrub-dominated pond margins whereas *A. crepitans* prefers such areas; although *A. gryllus* will breed in ponds, it prefers open shorelines. *Acris gryllus* also prefers to breed in shallow, grassy meadow pools whereas *A. crepitans* does not. Neither species breeds in lowland swamps, but both species may be found simultaneously year-round in such habitats especially during the nonbreeding season. According to Bayless (1969b), *A. gryllus* tends to occur along the swamp margins, with *A. crepitans* occurring in the muddy bottoms. These differences may help segregate the species, at least during the breeding season, although other authors (e.g., Mecham, 1964) could find no consistent differences in ecological separation between the species.

Despite possible different preferences, however, both species may occur at a pond. Bayless (1966) found that abundance was about equal between species in the winter, but that during the warm months of the year, *A. crepitans* far exceeded *A. gryllus* in abundance. Newly metamorphosed young tend to swell abundance in midsummer, but these frogs disperse soon thereafter. In addition, *A. crepitans* tends to occupy the more shaded (or less exposed portions of the pond), whereas *A. gryllus* occupies the more exposed sections (Bayless, 1966).

In addition to habitat differences, there are differences in the advertisement calls of the species. The calls sound like a series of clicks, and the dominant frequencies of the calls overlap. However, there are differences in the temporal structure and amplitude of the clicks (Micancin and Mette, 2009). Females are able to recognize the advertisement calls of conspecific males (Micancin, 2008).

DISEASES, PARASITES, AND MALFORMATIONS
No information is available.

SUSCEPTIBILITY TO POTENTIAL STRESSORS
Metals. In Northern Cricket Frog tadpoles, concentrations of aluminum, iron, magnesium, and manganese may be quite concentrated (Sparling and Lowe, 1996). Body concentrations of barium, beryllium, iron, magnesium, manganese, nickel, lead, and strontium increase with concentrations in the soil; body concentrations of beryllium and strontium are actually decreased with acidification. Sparling and Lowe (1996) suggested that the tadpoles could concentrate metals in their environment to the extent that the larvae were toxic to their predators. Sparling (2003) attributes toxic levels of mercury in *A. crepitans* to "Birge et al. (1978)." The reference date is cited incorrectly (should be 1979), and there is no mention of *A. crepitans* in this paper.

pH. Survival of Northern Cricket Frogs is reduced in acidified water, even at a pH of 5–5.2, which is well above lethality (Sparling et al., 1995). Acidification affects the concentration of other elements, such as magnesium and aluminum, which in turn affect amphibian larvae. However, the substrate may play a role in neutralizing or ameliorating the effects of acidity on survival. For example, larvae have slightly higher rates of survival on clay-lined substrates than they do on substrates composed of loamy soils. The lethal pH for Northern Cricket Frog larvae is 4.1 (Gosner and Black, 1957a; Sparling et al., 1995), and the critical pH is between 4.2 and 4.6 (Gosner and Black, 1957a). In Louisiana, however, Bayless (1966) routinely found *A. crepitans* in water with a pH of 4–5. In West Virginia, the mean pH in breeding ponds was 7.4 (Bayne, 2004).

Chemicals. Agricultural pesticides are lethal to Northern Cricket Frogs at a wide range of dosages. In Mississippi, the 36 hr total LC_{100} to adult *A. crepitans* for various pesticides was: endrin

(0.04–0.06 mg/L), dieldrin (0.2–0.85 mg/L), aldrin (0.2–0.75 mg/L), toxaphene (0.5–5.4 mg/L), DDT (0.62–50.0 mg/L) (Ferguson and Gilbert, 1968). These authors suggested that the wide variance might be due to resistance built up through time to repeated exposures to pesticides. Note that some of the data reported for *A. crepitans* by Ferguson and Gilbert (1968) might be referable to *A. blanchardi*, since collections included frogs from the Mississippi Delta. Larval Northern Cricket Frog survivorship is significantly reduced at concentrations of 9 mg/L of the insecticide imidacloprid (Ade et al., 2010).

Measurements of the observed effects of toxic pesticides on larvae can be very different depending upon the time survivorship is followed. For example, the mean $LC_{50\ (24\ hr)}$ for larvae exposed to toxaphene is > 1000 ppb and for endrin is 23 ppb. However, the mean $LC_{50\ (96\ hr)}$ for toxaphene is only 76 ppb and for endrin is 10 ppb (Hall and Swineford, 1981). These results were for continuous flow toxicity tests, and survivorship was determined after eight days.

STATUS AND CONSERVATION

Acris crepitans often is an abundant frog, and there is no report of region-wide declines. Populations at the northern extent of its range are vulnerable, however, and some populations have been extirpated. Northern Cricket Frogs disappeared from Long and Staten islands, New York, by the 1930s and 1970s, respectively (Gibbs et al., 2007). On Long Island, Overton (1914) characterized them as "extremely abundant and noisy in several marshes in the vicinity of Jamaica." Only a few populations remain in New York on private lands. Other populations are threatened by agriculture and urbanization. In the western Piedmont of North Carolina, Rice et al. (2001) were only able to document *A. crepitans* at one site, although up to eight sites had been documented historically prior to 1970.

In terms of mitigation, Northern Cricket Frogs may use constructed ponds, but their abundance is not great at restored ponds. Merovich and Howard (2000) recorded only 10 *A. crepitans* using ponds > 4–5 yrs old over a 2 yr monitoring period in Maryland. No use was recorded in constructed ponds < 4 yrs old.

Northern Cricket Frogs have been monitored using aural surveys along predetermined highway survey routes. Five-minute listening stops detect most choruses, but in mixed choruses with other anurans the call of this species, which does not carry well, may be drowned out and overlooked (Burton et al., 2006).

Acris gryllus (LeConte, 1825)
Southern Cricket Frog

ETYMOLOGY

gryllus: from the Latin *gryllus* meaning 'cricket' or 'grasshopper.'

NOMENCLATURE

Synonyms: *Acris acheta, Acris gryllus achetae, Acris gryllus dorsalis, Hylodes gryllus, Rana dorsalis, Rana gryllus*

Wright (1932) provided a comprehensive review of the nomenclatorial history of this species, including its inclusion under the name *Acris crepitans*.

IDENTIFICATION

Adults. This is a slender small frog with a sharp snout and two black lines on the posterior portion of the thigh. The ground color is gray, brown, black, or olive, and there is usually a colorful vertebral stripe of bright gray, yellow, red, or green running down the middle of the back. The sides are dark and may have blotches. A dark triangle between the eyes points rearward, but a few individuals lack this distinctive characteristic. The head region appears long and narrow. The dorsum has varying degrees of rugosity. Limbs are long and slender, and the tips of the third and fifth toe do not reach the end of the fourth toe on the hind foot. The limbs are dark above with black bands, but white underneath. The upper black line on the back of the thigh is usually narrow and sharply defined, whereas the lower black stripe ranges from a vague stippling to a well-defined regular line. Anal warts are usually smaller and less distinct than in *A. crepitans*. Males have dark throats during the breeding season. *Acris crepitans* is more squat, with a rounded snout, and the upper thigh line is very irregular; the lower area of the thigh consists of black stippling suggestive of a second line. Still, the thigh striping is not always reliable in separating these species. Neill (1950b) discusses morphological differences between *A. gryllus* and *A. crepitans*.

Females are slightly larger than males. In North Carolina males are > 19.0 mm SUL and females are 20–24.5 mm SUL (Alexander, 1966). Nevo (1973b) recorded the following mean male and female lengths in mm SUL: western Gulf Coastal Plain (22.8, 24.3), eastern Gulf Coastal Plain (22.2, 23.4), southern Florida (22.1, 22.5). Males range between 15 and 29 mm SUL and females 16–33 mm SUL in Georgia (Jensen et al., 2008). Adults are 16–22 mm SUL in Alabama (mean 19.7 mm) (Brown, 1956) and a mean of 21.1 mm SUL from another North Carolina study (Micancin and Mette, 2009).

Larvae. The tadpoles are medium-sized, light to dark gray, and have long tails with low tail fins. The body is slightly depressed, the nostrils are large, and the tail fins do not have bold markings although they are finely flecked. Throats are dark. Many larvae have black tail tips, but a few do not, even within the same pond. The free part of the spiracle's tube is longer in *A. gryllus* than in *A. crepitans*. Note that the larval description for *A. gryllus* in Wright (1929) is a composite that includes *A. crepitans* and *A. blanchardi*. Cricket frogs have the distinction of having the largest tadpole in relation to metamorph size of all North American anurans. A description of the tadpole contrasting it with larval *A. blanchardi* is in Orton (1947).

Eggs. The eggs are small and deep brown to buffy olive above and cream to white below. The vitellus is 0.9–1.43 mm in diameter (mean 1.16 in Alabama; Brown, 1956). There is a single, firm gelatinous coating around the egg measuring 2.4 to 3.6 mm in diameter and no inner envelope according to Wright and Wright (1949). However, Brown (1956) noted three gelatinous membranes surrounding eggs deposited in captivity. The second envelope was firm but not gelatinous. Brown (1956) recorded the mean diameter of the inner envelope as 1.71 mm (range 1.62 to 1.81), the second envelope as 1.96 mm (range 1.9 to 2), and the third envelope as 2.68 mm (range 2.57 to 2.95).

Distribution of *Acris gryllus*. It is likely that *A. gryllus* crosses the border into east-central Alabama.

DISTRIBUTION

Southern Cricket Frogs occur on the Coastal Plain from southeastern Virginia through the Florida Parishes of Louisiana. In Georgia and Alabama, the range extends northward into the Ridge and Valley Province, for example, the Coosa Valley and Cumberland Plateau. Populations may extend into the Piedmont along river floodplains. In the vicinity of the Fall Line, the range of this species overlaps somewhat with *A. crepitans* but perhaps not as much as statements in the older literature indicate. The Mississippi River largely forms the western boundary of this species, but its presence in the Delta Region of Mississippi has been questioned (Boyd, 1964). The range extends northward to southwestern Tennessee, then eastward again south of the Fall Line from Alabama back though southeastern Virginia. Southern Cricket Frogs may be found on larger islands, such as Roanoke and Bodie islands, North Carolina (Braswell, 1988; Gaul and Mitchell, 2007), and St. Simons and Cumberland islands, Georgia (Williamson and Moulis, 1994; Shoop and Ruckdeschel, 2006).

Important distributional references include: Alabama (Brown, 1956; Mount, 1975), Georgia (Williamson and Moulis, 1994; Jensen et al., 2008), Louisiana (Viosca, 1944; Dundee and Rossman, 1989), Mississippi (Boyd, 1964), North Carolina (Dorcas et al., 2007; Micancin and Mette, 2009), and Virginia (Tobey, 1985; Mitchell and Reay, 1999). Note that the maps in Williamson and Moulis (1994) present a confused distribution pattern between *A. crepitans* and *A. gryllus* in Georgia. Many of the dot locations likely are based on misidentified specimens (see Jensen et al., 2008).

FOSSIL RECORD

Pleistocene Rancholabrean fossils of *A. gryllus* have been reported from north-central Florida (Holman, 2003). A Miocene *Acris* was also described from northern Florida as *A. barbouri*, based on differences in the shape of the ilium (Holman, 1967).

SYSTEMATICS AND GEOGRAPHIC VARIATION

At one time, all cricket frogs in North America were thought to be part of a single wide-ranging species. Comparisons of blood and liver proteins (Dessauer and Nevo, 1969), other molecular data (Gamble et al., 2008), call structure (Nevo and Capranica, 1985), chromosomes (Bushnell et al., 1939), distributional patterns (Neill, 1954), ecology, and morphology have conclusively demonstrated that *A. gryllus* is a distinct species. There are two major clades within *A. gryllus*, an eastern clade occurring from the Florida Panhandle to southeastern Virginia and a western clade found in Mississippi and adjacent Tennessee (Gamble et al., 2008). The boundary between these clades is undetermined, but is likely in the Mobile Basin and Tombigbee River system. The distinction between these clades does not follow previous subspecific nomenclature (*A. g. gryllus* throughout most of the range and *A. g. dorsalis* in Florida). Despite some morphological differences (Neill, 1950b), the recognition of *A. g. dorsalis* is not supported by molecular data.

Laboratory experiments indicate a high degree of genetic compatibility between *A. crepitans* and *A. gryllus*, regardless of which species is male or female (Mecham, 1964). However, few larvae could be reared through metamorphosis, and none released into the environment were subsequently recovered by Mecham (1964). Backcrossed hybrids had low fertilization rates and poor development, and the hybrids appeared stunted. Hybridization does not appear to occur in nature, perhaps because of differences in mating calls and limited geographic overlap.

However, there is at least one observation of interspecific aggression between males of these species at a breeding pond (Micancin and Mette, 2010).

Pattern polymorphism is common in cricket frogs and may reflect different selection pressures on different seasonally changing backgrounds (see discussion in *Acris crepitans* account). In Southern Cricket Frogs, Nevo (1973a) found that between 24 and 100% of frogs were gray, 22.6–62.5% had a red stripe, and from 9 to 30% had a green stripe at the 9 populations sampled.

ADULT HABITAT

Acris gryllus is a wide-ranging species occupying many different habitats throughout the southeastern United States. Primarily a lowland species, it prefers well-drained habitats in the vicinity of permanent or temporary ponds, wet prairies, seepage bogs, sloughs, canals, sawgrass prairies, bayheads, lakes, swamps, and grassy meadows. Examples of surrounding habitats include longleaf pine-turkey oak sandhills, mixed pine-oak woodlands, hardwood forest, mesic and xeric hammocks, pine flatwoods, subtropical tree islands, agricultural fields, and silvicultural operations (Anderson et al., 1952; Harima, 1969; Dodd, 1992; Enge, 1998a, 2002; Enge and Wood, 1998; Meshaka et al., 2000; Baber et al., 2005; Greenberg and Tanner, 2005b; Means and Franz, 2005; Surdick, 2005; Dodd and Barichivich, 2007). The species ventures into the uplands in many areas, marking the division between the lowland Coastal Plain and the more upland Piedmont provinces. Southern Cricket Frogs are abundant along the shorelines of ponds and wetlands, especially where there is an abundance of low-growing grasses such as *Eleocharis* (Bancroft et al., 1983). It readily is found on mats of floating vegetation.

AQUATIC AND TERRESTRIAL ECOLOGY

The first field notes on this little terrestrial frog were recorded by Bartram (1791:278) as he traveled in the southeastern United States, although he may have mixed in observations of other species as he characterized its habits. For example, *Acris gryllus* does not climb on vegetation or trees, although Bartram certainly got the sound of the call correct ("a feeble note sounding like crickets"). Throughout the year, they congregate around the margins of ponds and lakes and in grassy wetlands in savannas in considerable

Larval *Acris gryllus*. Some larvae have the black tail flag (*left*), whereas others do not (*right*). Photo: C.K. Dodd Jr.

numbers. During the nonbreeding season, they are frequently found in nearby lowland cypress swamps where they occupy the margins of the swamp. They normally do not occur very far from water and are among the most aquatic of anurans with the exception of the ranids (*Lithobates, Rana*). They frequently occur on emergent and floating vegetation and along wet trails in prairie, swampy habitats, and along streams in steephead ravines (Wright, 1932; Enge, 1998b). Indeed, Wright (1932) characterized them as being found them "in almost every type of plant habitat which has any moisture at all." When in wetlands, they do not always remain close to shore. For example, Beck (1948) frequently found them on floating vegetation 91 m from shore.

Southern Cricket Frogs are active year-round, weather permitting. They tolerate colder weather better than most other southern frogs. In laboratory experiments, the mean CTmin is ca. 3°C and the mean CTmax is ca. 38°C (John-Alder et al., 1988). Southern Cricket Frogs are active at every temperature in between these values.

When at the breeding site or sitting along a pond shoreline, Southern Cricket Frogs move very little from one part of the pond or wetland to another. In general, they are philopatric with a small home range, preferring to remain within 5 m of the site of previous occurrence. Bayless (1966) found the mean movement distance to be 1.3 m (range 0.5 to 3 m). Some frogs do make extensive movements over the course of an activity season, however. For example, one frog in Louisiana moved ca. 33 m in ca. 35 days, and the greatest distance Bayless (1966) recorded movement was ca. 65 m over a period of 180 days between captures. In Florida, Telford (1952) encountered them often at distances of 91 m or more from the nearest water, and Carr (1940a) recorded them 274 m from water. Also in Florida, Dodd (1996) recorded *A. gryllus* from 255 to 492 m (mean 383 m) from the nearest water in sandhills habitat. In North Carolina, Micancin (2010) found a single female 562 m from the nearest breeding chorus. The extent to which *A. gryllus* uses terrestrial habitats needs further investigation.

Like Northern Cricket Frogs, Southern Cricket Frogs are able to use Y-axis celestial orientation to determine dispersal and migratory directions to and away from a perpendicular shoreline or within a pond. This ability requires a learned shore position, a view of a celestial cue (sun, moon, and stars), and an internal clock phased to local time. Orientation is most effective by day when the sun is visible and by night when the moon is visible (Ferguson et al., 1965; Ferguson, 1966a). If only the stars are visible, frogs show a bipolar orientation. Displaced frogs are able to return to the shoreline where captured or to the point of release, using the position of the sun for orientation. Ferguson (1963) reported that displaced frogs easily returned 91 m to the point of capture, and that many frogs returned as far as 223 m although sometimes stopping at intervening ponds along the way. Celestial orientation can be mediated through the eyes, but extraoptic light receptors in the brain allow proper orientation and setting of biological clocks even in blind *A. gryllus* (Taylor and Ferguson, 1970).

Southern Cricket Frogs are photopositive in their phototactic response, suggesting they use both sunlight and moonlight when feeding and going about their daily activities (Jaeger and Hailman, 1973). They are sensitive to light in the blue spectrum ("blue-mode response"), which apparently helps them orient toward areas of increasing illumination, such as the open horizon above lakes and ponds (Hailman and Jaeger, 1974). Southern Cricket Frogs likely have true color vision.

CALLING ACTIVITY AND MATE SELECTION

Male Southern Cricket Frogs call both by day and night, but the majority of calls are made between 19:30 and 02:00 (Mohr and Dorcas, 1999). Calling extends throughout the year in Southern Cricket Frogs, but nearly all successful reproduction takes place from late spring throughout the summer. For example, calling begins in April and extends until August in North Carolina (Alexander, 1966) and Mississippi (Smith, 1975), and from February to October in Georgia (Jensen et al., 2008). In Florida, calls may be heard year-round depending on location and weather conditions, although Bancroft et al. (1983) only reported calls from February to September in central Florida. In most locations, calling likely precedes actual reproduction by several weeks as it does in other cricket frogs. The reason for extensive calling late in the season is unknown.

As with other cricket frogs, the advertisement call sounds like a series of clicks or "ticks," which are composed of pulses repeated at intervals of 0.2 sec (Blair, 1958a). The dominant frequency of the call is a mean of 3.55 kHz. Some characteristics of the call include: a click rate of 1.2–6.21 (mean 2) per entire call, a mean click duration of 34–36 ms, a mean number of pulses as 10.2–10.9, a call duration of 12–28 sec (mean 16.5 sec), a total number of clicks as 13–40 (mean 27.8), and an average click rate of 1.4–3.8. Interpulse intervals average 1.7 to 2.1 sec. The number of clicks increases with the duration of the call. Nevo and Capranica (1985) and Micancin and Mette (2009) provided extensive information on regional variation in both gross and fine structure of the call of *A. gryllus*.

Males call from along the shoreline of a pond or lake and, like *A. blanchardi*, tend to defend an acoustic calling space. In Georgia, calling males occupied a territory of only 0.56 m² (range 0.028 to 1.362 m²) (Forester and Daniel, 1986). The size of the territory is not correlated with the number of nights a male is present or with mating success. Males remain within a particular area from one night to the next, and movements are very small. Forester and Daniel (1986) recorded mean movements of only 52 cm (range 0 to 205 cm).

Females move toward a calling male. She will turn her head from side to side in the direction of the caller as if this assists her in orientation, and she will perform a series of circling hops if she has difficulty in locating his position. The female does not initiate contact. In time, the male usually observes the female and moves toward and amplexes her. Forester and Daniel (1986) described such a courtship sequence that included 8 female circles over a 5 min period. A male may successfully amplex more than one female (Forester and Daniel, 1986).

BREEDING SITES

Southern Cricket Frogs breed in a wide variety of temporary to permanent wetlands, including small, grassy meadow wetlands, depression marshes, cypress savannas, cypress-gum ponds, woodland ponds, and the quiet shallows of very large lakes (e.g., Liner et al., 2008). They tend to prefer shorelines with minimum or no vegetation when breeding along pond or lake margins. Fish may or may not be present. *Acris gryllus* also breeds in human-created wetlands, such as those that form under power line rights-of-way, in roadside ditches, and along trails through wet areas. As long as seasonal wetlands are available, they will breed on golf courses (Scott et al., 2008). Breeding sites often have extensive mats of floating vegetation or lily pads on which frogs sit and call. The vegetation also provides concealment from predators and from direct solar insolation.

A generalist approach to breeding-site choice may allow this species to adapt to extreme environmental changes across a landscape. For example, Babbitt and Tanner (2000) studied breeding at a site in south Florida that experienced a drought one year followed by flooding the next. The flooding allowed fish to invade formerly fishless wetlands. *Acris gryllus* was able to use temporary isolated ponds during the first year's drought as well as the extensive wetlands resulting from the flooding during the second year. Reproductive output was reduced during the drought as some ponds dried, but it was successful both years despite the extremes in precipitation.

REPRODUCTION

Southern Cricket Frogs normally breed throughout the spring to early summer, with the exact timing depending on environmental conditions. Calling occurs throughout the warmer period of the year and extends well into the autumn or even winter, but the extent of successful reproduction after the primary calling period is unknown. Carr

(1940a) stated that they bred every month of the year in Florida. In Louisiana, however, most reproduction occurs in March and April, although calling may continue until September. Bayless (1966) found no clasping pairs after 8 April and considered this species possibly to be an explosive breeder whose breeding activity was triggered by heavy rainfall. In Georgia, calling occurs from February to October, but most breeding occurs from April to June (Jensen et al., 2008).

As noted above, it is unlikely that midsummer to autumn calling results in amplexus and reproduction except perhaps well south in the Florida peninsula. Testis mass and male fat body mass are lowest in winter and highest in spring, and testis mass decreases in summer and increases in autumn. Ovarian mass is correlated with both body mass and snout-urostyle length. In contrast, male fat body mass was correlated with body mass but not snout-urostyle length. Female fat body mass reaches its maximum in winter and is used to yolk developing eggs in the coming spring. These factors, together with the timing of spermatogenesis and ovarian cycles, do not lend themselves to successful reproduction after the spring to early summer breeding season throughout much of the species' range (Smith, 1975).

In central and south Florida, however, some reproduction likely occurs year-round, even if greatly decreased during the winter. Tadpoles of *A. gryllus* are present year-round at Loxahatchee National Wildlife Refuge (Baber et al., 2005) and at the MacArthur Agro-Ecology Research Center (Babbitt and Tanner, 2000). Tadpole densities ranged between 0.06 and 0.43 m^2, but varied annually and monthly. The location at which breeding grades from solely in spring and summer to year-round in Florida has not been determined.

Eggs are deposited singly and sometimes in small groups of three to four. The eggs may stick together in a loose mass of seven to ten eggs per mass. These masses cling to vegetation via the gelatinous envelopes, or eggs may rest freely upon the substrate in shallow water. Actual counts of egg clutches are few in the literature. Females oviposited from 99 to 156 eggs per clutch in North Carolina (Alexander, 1966), and Wright (1932) reported a single clutch count of 241 eggs. Many authors state that clutch size can reach 250, perhaps referring to Wright's count. Hatching occurs quickly, within four days of deposition.

LARVAL ECOLOGY

The larval period lasts from 32 to 94 days (Wright, 1932; Bayless, 1966; Jensen et al., 2008). Wright (1932) recorded a mean of 67 days for the larval period, but stated that 50–90 days seemed the "probable" period. Metamorphs are 9–15 mm SUL (Wright, 1932; Alexander, 1966; Jensen et al., 2008).

Adult *Acris gryllus*, green phase. Photo: Alan Cressler

DIET

The diet of Southern Cricket Frogs includes small invertebrates, especially ants. In a Florida study, 70% of the stomachs contained ants and they made up 68% of the prey ingested. Other prey included mosquitoes, leaf bugs, spiders, chironomid flies, and beetles (Franz, 1972). In the latter study, frogs captured between 22:00 and 07:00 had no food in their stomachs, suggesting solely diurnal feeding. In Louisiana, specific items include collembolans, ants, flies and fly larvae, spiders, mites, beetles, true bugs, aphids, termites, and miscellaneous larval and adult insects (Beck, 1948; Bayless, 1969b). Carr (1940a) recorded them feeding on emerging midges, mayflies, and chironomids. He further noted that they sometimes attacked prey too large to be eaten. In areas of sympatry, the food items of *A. crepitans* and *A. gryllus*, overlap showing no prey specialization. Feeding occurs year-round when weather conditions permit.

PREDATION AND DEFENSE

Larval *A. gryllus* appear to be palatable, and even in the presence of chemicals from potential fish predators do not spend more time in refugia than they do when fish are absent (Kats et al., 1988). They may, however, reduce their activity in the presence of dragonfly naiads, fish, or Eastern Red-spotted Newts (Richardson, 2001). As with *A. blanchardi*, many have black tail tips, which serve as disruptive coloration or help to distract a potential predator away from the head and body. Caldwell (1982) demonstrated that larvae with black tipped tails were found most often in ponds, whereas those without this coloration were found more frequently in lakes and creeks. In ponds, dragonfly larvae are the primary predator. The black tail tip draws their attention, and larvae in wetlands containing *Anax* have much higher frequencies of black-tipped tails than in their absence. In creeks and lakes, the uniform body coloration may help in concealment from predaceous fishes (Caldwell, 1982). This polymorphism is present, even in nearby populations, and is thus predator-mediated but habitat specific.

The Southern Cricket Frog jumps readily at the approach of an intruder and makes quick, rapid leaps. The first leap is usually the longest (mean 72.7 cm: Blem et al., 1978). Jumping distances are correlated with temperature, with jumps of 20 cm at 5°C to nearly 1 m at 35°C (John-Alder et al., 1988). A frog will jump toward the water and swim rapidly among vegetation, circling back toward shore. If disturbed again, it may dive and bury itself among vegetation or mud, but it often pops back up rather rapidly. During jumping, the frog is able to quickly change directions making pursuit difficult. Dodd (1991) noted that they were capable of easily crossing a drift fence by simply jumping on it and walking the remaining way to the top, at least in laboratory experiments; there appeared to be less tendency to trespass the fence in the field.

Predators undoubtedly include a variety of invertebrates, ranid frogs, snakes (*Thamnophis* sp.), birds (particularly wading birds, Florida grackles), and mammals (pigs). Few specific observations are available, most notably for garter and ribbon snakes (*T. sirtalis, T. sauritus*) (Wright, 1932). Goin (1943) reported a large *Dolomedes* spider eating an adult cricket frog.

POPULATION BIOLOGY

Southern Cricket Frogs are among the most abundant frogs in their environment, where a short walk along a lake or pond shore results in dozens of these tiny frogs hopping in every direction. Wright (1932) called it the most abundant frog in the Okefenokee Swamp, and it is certainly one of the most abundant frogs around ponds and lakes in north-central Florida. Distribution, however, is somewhat nonrandom in the vicinity of more contiguous shallow grassy pools and associated terrestrial habitat.

Robust population estimates or even empirical counts are few in the literature. In a Louisiana population, Turner (1960a) found that individuals were spaced at intervals ranging from 1.7 to 1.9 m from their nearest neighbor in and around a grassy wetland, and that densities changed with time. He estimated that there were 0.066 frogs per m^2 for a total population of about 90 in the 1,350 m^2 wetland in December. By April, the density increased to 0.085 frogs per m^2 with a population estimate of about 140. In Florida, Greenberg and Tanner (2005b) captured 888 adults and 297 juveniles around 8 isolated temporary ponds surrounded by dry sandhills over a 7 yr period. Also in Florida, Dodd (1992) captured 255 *A. gryllus* over a 5+ yr study at a temporary pond surrounded by xeric hammock and sandhills. Gibbons and Bennett (1974) captured 340 *A. gryllus* around 2 ponds in 2 yrs; most frogs, however, were captured at one of the ponds in only 1 year. Also in South Carolina, Russell et al. (2002a) documented 1,298 *A. gryllus* using 5 small isolated wetlands over a 2 yr period; tadpoles also were observed in the wetlands.

Both *A. crepitans* and *A. blanchardi* are essentially annual species with very few individuals surviving beyond the first breeding season. Some authors also state that *A. gryllus* reaches maturity quickly and is ready to breed in the spring following metamorphosis (e.g., Jensen et al., 2008). However, this may not be the case with *A. gryllus*, as longevity was >3 years in the North Carolina population studied by Alexander (1966). He found that between 50 and 70% of females failed to reach sexual maturity one year after metamorphosis and consequently first bred the second year. Thus, the normal life-span may be greater in Southern Cricket Frogs than in the other two species of *Acris*.

As with many frogs, the number of males is greater at breeding ponds than females, but overall population sex ratios remain poorly documented. In Georgia, Forester and Daniel (1986) reported an operational sex ratio of 5.6 males per 0.3 females, but acknowledged that this represented an underestimation of the number of females in the population.

COMMUNITY ECOLOGY

In many areas of the South, populations of *A. gryllus* and *A. crepitans* overlap geographically (Viosca, 1944; Neill, 1954; Mecham, 1964; Bayless, 1969b; Micancin and Mette, 2009). However, it appears these species have different habitat preferences where they occur together. For example, *Acris crepitans* is found in bottomland swamps or water close to the shore of lakes, whereas *A. gryllus* occurs more in uplands or along drainage ditches, where the species overlap (Norton and Harvey, 1975; Dundee and Rossman, 1989). *Acris gryllus* tends to avoid shrub-dominated pond margins whereas *A. crepitans* prefers such areas; although *A. gryllus* will breed in ponds, it prefers open shorelines. *Acris gryllus* also prefers to breed in shallow grassy meadow pools whereas *A. crepitans* does not. Neither species breeds in lowland swamps, but both species may be found simultaneously year-round in such habitats, especially during the nonbreeding season. According to Bayless (1969b), *A. gryllus* tends to occur along the swamp margins, with *A. crepitans* occurring in the muddy bottoms. These differences may help segregate the species, at least during the breeding season, although other authors (e.g., Mecham, 1964) could find no consistent differences in ecological separation between the species.

Despite different preferences for microhabitats, however, both species may occur at a pond. Bayless (1966) found that abundance was about equal between species in the winter, but that during the warm months of the year, *A. crepitans* far exceeded *A. gryllus* in abundance. Newly metamorphosed young tend to swell in abundance in midsummer, but these frogs disperse soon thereafter. In addition, *A. crepitans* tends to occupy the more shaded (or less exposed) portions of the pond, whereas *A. gryllus* occupies the more exposed sections (Bayless, 1966). This suggests the potential for observer bias when assessing abundance through visual encounter surveys.

In addition to habitat differences, there are differences in the advertisement calls of the species. The calls sound like a series of clicks, and the dominant frequencies of the calls overlap. However, there are differences in the temporal structure and amplitude of the clicks (Micancin and Mette, 2009). Females are able to recognize the advertisement calls of conspecific males (Micancin, 2008).

DISEASES, PARASITES, AND MALFORMATIONS

The fungal pathogen *Basidiobolus* sp. has been reported from Southern Cricket Frogs from Florida (Okafor et al., 1984). As of 2008, the amphibian chytrid fungus had not been detected in *A. gryllus* in the Southeast (Rothermel et al., 2008). However, Rizkalla (2010) reported the first records from central Florida.

SUSCEPTIBILITY TO POTENTIAL STRESSORS

pH. In Louisiana, Bayless (1966) routinely found *A. crepitans* in water with a pH of 4–5. In the related *A. crepitans*, the toxic effects of pH are evident at pH values <4.6.

Chemicals. Agricultural pesticides are lethal to Southern Cricket Frogs at relatively small dosages. In Mississippi, the 36 hr total lethal concentration to adult *A. gryllus* for various pesticides

Habitat of *Acris gryllus*.
Photo: C.K. Dodd Jr.

was: endrin (0.02–0.045 mg/L), dieldrin (0.3–0.4 mg/L), aldrin (0.2–0.3 mg/L) (Ferguson and Gilbert, 1968).

STATUS AND CONSERVATION

There is no indication that Southern Cricket Frogs have experienced regional population declines except in the northern Coastal Plain of North Carolina. There, the species has disappeared from the three most northern river basins (Roanoke, Tar, Neuse), where it had been abundant in the 1960s (Micancin and Mette, 2009). The reason for the decline is unknown. Undoubtedly many populations have been lost to habitat destruction and alteration as wetlands have been destroyed or altered throughout the South. Populations may recover quickly in coastal areas that receive hurricane overwash, with resulting temporary increases in salinity (Gunzburger et al., 2010). They can survive and perhaps even thrive in some types of habitat alteration associated with silviculture, but it is not clear that occupancy equals viability, since reproduction is often not documented and sample sizes often small (e.g., Vickers et al., 1985; Enge and Marion, 1986; Russell et al., 2002b). Some studies examining the effects of clearcutting on Southern Cricket Frogs also have failed to separate treatment effects from other environmental change, or they have been conducted over a short time following perturbation (Russell et al., 2002b). The numbers captured pre- and postcut are sometimes rather similar, although relatively few frogs may have been captured during both sampling periods (O'Neill, 1995).

In contrast, Russell et al. (2002a) noted the importance of small isolated wetlands in commercially managed forests, where *A. gryllus* could be abundant and reproduce successfully. Harvesting in a managed forest may actually increase the numbers of cricket frogs captured during surveys (Clawson et al., 1997). Harvesting likely opens up habitats and causes ground water to rise, resulting in many shallow potential breeding sites at least for a short time following harvest.

In Florida, Surdick (2005) found decreasing abundance from natural reference sites (85%) through lands affected by silviculture (73%), agriculture (41%), and urbanization (15%). These results suggest that habitat occupancy is inversely proportional to the extent of land-use intensity. Delis et al. (1996) did not find any appreciable effects of urbanization on *A. gryllus* in an area of the Tampa Bay region, although the species was uncommon at both the developed site and an undeveloped reference site. In southwest Florida, however, there appeared to be a slight decrease in populations based on road surveys conducted over 5 yrs in an urbanizing area (Pieterson et al., 2006).

In terms of wetland mitigation, a few Southern Cricket Frogs may subsequently be found around restored or created wetlands. However, in the one study to date the numbers were small (e.g., Pechmann et al., 2001). Some forms of habitat management, such as prescribed burning (Schurbon and Fauth, 2003; Langford et al., 2007), appear to have no detrimental impacts on the species and may be necessary to keep habitats open. Wellfield pumping may adversely affect reproduction even when calling indices suggest no differences in occupancy between affected and unaffected wetlands (Guzy et al., 2006).

Individuals are certainly killed on roads and highways, but their small size in relation to automobile size makes counting DOR individuals impossible. Cricket frogs use culverts under highways (Dodd et al., 2004; Smith et al., 2005), and thus the construction of these corridors probably benefits this species to some extent. According to Enge (2005a), a total of 4,225 *A. gryllus* were collected for the pet trade from Florida between 1990 and 1994.

Hyla andersonii Baird, 1854
Pine Barrens Treefrog

ETYMOLOGY
andersonii: named for Anderson, South Carolina, where the first specimen was supposedly found. The species has never been observed in this part of South Carolina, however. It is likely that the first specimen was shipped from Anderson, but that it was collected in the Carolina Sandhills (Neill, 1947; Brown, 1980).

NOMENCLATURE
Synonyms: None

IDENTIFICATION
Adults. *Hyla andersonii* is certainly one of North America's most attractive frogs. The dorsum and upper surfaces of the legs are green. The general coloration can fade to olive under varying light conditions; when cold, they become much darker. There is a prominent chocolate-brown to plum-colored band that passes laterally from the nostril through the eye to the hind limb. This band is bordered by a white to yellow stripe. The venter is white, with the chin of males light purplish gray. There are bright yellow-orange spots in the axilla, on the inside surfaces of the hind legs, and along the lower sides of the body. These spots are not visible unless the frog moves. Males have a distinctive subgular vocal pouch.

Females are larger than males. The species ranges from 33.6 to 40.9 mm SUL (mean 37.2 mm) in Florida, 32.3–39.3 mm SUL (mean 36 mm) in the Carolinas, and 31–38.3 mm SUL (mean 35.5 mm) in New Jersey (Means and Longden, 1976). Noble and Noble (1923) reported New Jersey males 35–38 mm SUL and females 38–44 mm SUL.

Larvae. Pine Barrens Treefrog tadpoles are brown to dark olive with scattered dark spots. Bellies are greenish yellow or golden. There is no light stripe anterior to the eye. Both the tail fins and musculature have dark blotches and spots posteriorly. Along the upper part of the tail musculature, the dark spots and blotches coalesce to form a dark stripe with irregular borders; this line continues about halfway down the tail. Tails are short and narrow and may be somewhat yellowish. There is no narrow filament at the end of the tail as there is in *H. femoralis*. Newly hatched larvae are 4.5–4.8 mm TL and are pale yellow finely stippled with brown (Gosner and Black, 1957b). Larvae reach 35–43 mm TL prior to transformation.

Eggs. The eggs are dark brown dorsally and creamy white ventrally; the brown coloration becomes lighter as the egg develops. There are two jelly capsules surrounding the egg. The vitellus is 1.2–1.4 mm in diameter; the inner capsule is 1.9–2 mm in diameter; and the outer capsule is 3.5–4 mm in diameter (Noble and Noble, 1923).

DISTRIBUTION
The Pine Barrens Treefrog is found in three disjunct populations: southern New Jersey, the Carolina Sandhills of North and South Carolina, and the Florida Panhandle and adjacent Alabama. A record exists from Richmond County, Georgia (Neill, 1947), but attempts to relocate the population have not been successful. Important distributional references include: Alabama (Moler, 1981), Florida (Christman, 1970; Means and Longden, 1976; Moler, 1981), New Jersey (Anderson et al., 1978; Schwartz and Golden, 2002), North

Distribution of *Hyla andersonii*

Carolina (Bullard, 1965; Dorcas et al., 2007), and South Carolina (Tardell et al., 1981; Cely and Sorrow, 1986).

FOSSIL RECORD
No fossils are known.

SYSTEMATICS AND GEOGRAPHIC VARIATION
Hyla andersonii is most closely related phylogenetically to *H. avivoca* (Hedges, 1986), although Blair (1958c) had suggested an affinity with *H. gratiosa* and *H. cinerea* based on call structure. Northern populations have a white stripe bordering the brown lateral band, but in Florida populations the stripe is lemon yellow. Means and Longden (1976) also noted slight morphometric and other coloration differences between Florida populations and populations further north. Florida populations also tend to have a high-pitched advertisement call.

Laboratory crosses between *H. andersonii* and *H. cinerea* may produce larvae, but only a portion may complete metamorphosis; metamorphs resemble both parents (Gerhardt, 1974b; Means and Longden, 1976). Anderson and Moler (1986) reported single hybrids between *H. andersonii* and *H. femoralis* and *H. cinerea* from Florida.

ADULT HABITAT
This is a species of pocosins, acid bogs, shrub bogs, and seepage areas in otherwise sandy habitats. In Florida and South Carolina, the predominant vegetation consists of titi (*Cliftonia monophylla, Cyrilla racemiflora*), *Ilex coriacea*, sweetbay (*Magnolia virginiana*), sweet-pepper-bush *(Clethra alnifolia),* and sparkleberry (*Vaccinium arboretum*) (Means and Longden, 1976; Moler, 1981; Tardell et al., 1981; Cely and Sorrow, 1986; Enge, 2002). These seeps are botanically rich, especially when undisturbed by human activity, and they frequently contain *Juncus, Drosera, Lycopodium, Sphagnum,* and pitcher plants (*Sarracenia*). Adjacent habitats consist of sandhills (longleaf pine- [*P. palustris*] wiregrass- [*Aristida*] turkey oak [*Q. laevis*]) or mixed pine- (*Pinus*) deciduous forest (*Acer, Liriodendron, Nyssa, Quercus*). In New Jersey, the species occurs in the Pine Barrens, an extensive dry, sandy area dominated by pitch pines (McCormick, 1970). In North Carolina, Gosner and Black (1956) noted longleaf pine, blackjack oak (*Q. marilandica*), red maple, azalea (*Rhododendron*), blueberry (*Vaccinium*), and alder (*Alnus*). In South Carolina, the species has been found in power-line and gas-line rights-of-way and in recent clearcuts (Tardell et al., 1981).

TERRESTRIAL ECOLOGY
Activity may begin in March in Florida or early May in New Jersey, well before calling begins. Individuals are active at least until September in New Jersey and North Carolina (Davis, 1904; Bullard, 1965). Pine Barrens Treefrogs have an extraepidermal layer of mucous and lipids that help retard water loss (Barbeau and Lillywhite, 2005). The mean CTmax temperature is 40.2°C (John-Alder et al., 1988).

CALLING ACTIVITY AND MATE SELECTION
Calling occurs from March to September in Florida and is usually associated with high humidity and precipitation (Means and Longden, 1976; Moler, 1981). South Carolina males call from April to September (Cely and Sorrow, 1986). In New Jersey, calling begins in May and continues to August (Davis, 1907; Noble and Noble, 1923; Wright, 1932). The call has been described in various ways as "keck-keck," "peep-peep," "quack-ack," "whang," "aquack-aquack-aquack," or "quănk," which is given in 10–20 resonant nasal notes that increase in volume (*in* Noble and Noble, 1923). The fundamental call frequency of Florida males is 80–100 Hz higher (485 Hz) than

Tadpole of *Hyla andersonii*. Photo: John Bunnell

in New Jersey males. Dominant frequencies are in harmonics 3–6 at 1,100–2,440 Hz in Florida and North Carolina, whereas energy is more evenly distributed in New Jersey males to 8,000 Hz (Gosner and Black, 1967; Gerhardt, 1974b; Means and Longden, 1976). The call duration averages 0.12 sec with a call repetition rate of 0.41 sec (Gerhardt, 1974b). Gerhardt (1974b) provides an oscillogram of the call. The call can be heard > 1 km away.

Males call from vegetation within and surrounding acid seeps and bogs. Most perch heights are located 1–1.5 m off the ground, rarely above 2 m (Bullard, 1965; Means and Longden, 1976). Males leave their arboreal calling perches well after dark and move to the margins of slow-moving streams and seepage bogs. There, they begin calling from the substrate near an oviposition site. They frequently call in duets, with the second male calling immediately after the first. The call has a ventriloquist quality that makes it difficult for a human observer to locate the male. Calling begins shortly after sunset with most calling occurring 2–3 hrs after sunset. Occasional males may be heard around daybreak.

A female moves toward a calling male and makes contact by striking him one or more times. This usually occurs on the ground, although initial contact may be made on a calling perch since amplexed pairs have been observed in trees. Amplexus is axillary, with the male riding on her back with his toes resting on the top of her thighs. The female initiates oviposition by bowing her back (head and posterior up, abdomen down) and thrusting her legs laterally. This places her cloaca in close proximity to that of the male's cloaca. The male moves his legs to her sides in a stroking manner, whereupon she straightens her back and four to nine unattached single eggs are extruded suddenly and fertilized (Noble and Noble, 1923; Means and Longden, 1976). Noble and Noble (1923) illustrate a mating pair.

Males appear to be evenly spaced about 5–10 m apart. Calling *H. andersonii* males are territorial and have an aggressive encounter call if an intruder approaches too closely. The encounter call is similar to the advertisement call but has an increased repetition rate made by shortening the interval between calls, much as *H. squirella* does (Fellers, 1979a).

BREEDING SITES
Breeding occurs in acidic hillside seepage bogs and shrub bogs, although the plant composition varies among the disjunct populations. Water should be clear, shallow, and slow moving; stagnant waters are avoided. Breeding sites in North Carolina are mostly temporary to semipermanent ditches, streams, and pools.

REPRODUCTION
Breeding occurs in shallow water, with eggs lying free on the substrate and unconnected to one another. Clutch size is 800–1,000 in New Jersey (Noble and Noble, 1923), and Means and Longden (1976) reported a female ovipositing 206 eggs in Florida. Hatching occurs within three days.

LARVAL ECOLOGY
Larvae prefer shallow water where they largely remain immobile during daylight hours. They tend to reduce their activity patterns when exposed to predators (*Notophthalmus viridescens*, *Enneacanthus obesus*, *Pantala*) in experimental trials and assume a more benthic position in the water column than they otherwise would (Lawler, 1989). The larval period is estimated to be 49–75 days (Wright, 1932; Gosner and Black, 1957b). Newly metamorphosed froglets are 11–15 mm SUL.

DIET
The adult diet consists of beetles, grasshoppers, ants, flies, and larval insects (Noble and Noble, 1923).

PREDATION AND DEFENSE
When handled, *H. andersonii* emits an odor of "raw peas," which could have an antipredator function (Noble and Noble, 1923). The green coloration and stripe of postmetamorphs suggest crypsis with surrounding vegetation. The stripe on the tail of the tadpole may function in countershading, which aids in crypsis.

POPULATION BIOLOGY

Growth of postmetamorphs is rapid, and juveniles can reach breeding size by the first summer following metamorphosis, especially if hatched from eggs oviposited early in the season. It seems likely, however, that juveniles hatched from eggs oviposited late in the season might not actually breed until the spring of their second year. Larvae of different sizes can be found throughout the summer, indicating a prolonged breeding season.

COMMUNITY ECOLOGY

Green Treefrogs and Pine Barrens Treefrogs have very similar calls, call from similar locations and microhabitats, and may occasionally be found at the same pond. Call rates of *Hyla cinerea* may be as fast as *H. andersonii*, but call durations and repetition rates are shorter in *H. andersonii*; lower peak frequencies are higher in *H. andersonii* compared with *H. cinerea*, but upper peak frequencies are lower in *H. andersonii*. Female *H. cinerea* are not attracted to calls of male *H. andersonii*. However, some female *H. andersonii* are attracted to the calls of male *H. cinerea*. Differences in female call discrimination capability could lead to interspecific hybridization if both species called from the same location (Gerhardt, 1974b).

The Pine Barrens Treefrog is found in acidic habitats. However, low pH has no effect on the competitive ability of larval *H. andersonii* when raised experimentally with larval *Lithobates sphenocephalus* or *Hyla versicolor*. Competition from *Lithobates sphenocephalus* causes a decrease in survival and body mass of larval *Hyla andersonii*, whereas competition from larval *H. versicolor* decreases body mass but not survival (Pehek, 1995).

As with any frog breeding in a pond community, larvae are likely to encounter competitors and predators among heterospecific larvae and aquatic insects. In experimental trials, the presence of insects and toad (*Anaxyrus fowleri*) larvae decreased the mean individual body mass of *Hyla andersonii* at metamorphosis and the cumulative biomass of larvae transforming from a pond; they did not affect survivorship or the duration of the larval period, however (Morin et al., 1988). In another experiment, the sunfish *Enneacanthus* completely eliminated larval *H. andersonii* (Kurzava and Morin, 1998).

The timing of reproduction also may influence the outcome of competition between *H. andersonii* larvae, aquatic insects, and heterospecific anurans. Morin et al. (1990) found that larvae from early reproductive bouts survived better, grew and developed more rapidly, and metamorphosed at a larger size than those from a later cohort. Competition, however, was actually greater in the earlier cohort because of the reduced availability of periphyton as food. The advantage of variation in breeding phenology may depend on local environmental conditions and the extent to which females oviposit multiple times within a season.

Amplexing *Hyla andersonii*.
Photo: John Bunnell

Breeding habitat of *Hyla andersonii*. Photo: John Bunnell

DISEASES, PARASITES, AND MALFORMATIONS
No information is available.

SUSCEPTIBILITY TO POTENTIAL STRESSORS
pH. Low pH (3.9) has no effect on survival, body mass, duration of the larval period, or growth of larval *H. andersonii* (Pehek, 1995). A pH >3.8 is easily tolerated by *H. andersonii*, with the critical pH from 3.6 to 3.8 (Gosner and Black, 1957a). Successful hatching can occur at a pH of 3.7 (Freda and Dunson, 1986). The estimated lethal pH is just over 3.4. Still, larval *H. andersonii* grow significantly slower at a low pH (40% slower at 3.75 and 25% at 4.0) than they do at higher pH (Freda and Dunson, 1986). The pH of most natural ponds is >4.0.

STATUS AND CONSERVATION
The precarious status and need for conservation management of *H. andersonii* has long been recognized (Bury et al., 1980). Pine Barrens Treefrog habitat may be threatened by the encroachment of woody, canopy vegetation, especially in areas where fires have been excluded. In managed habitats, this species may persist even in some areas of commercial forestry and along energy corridors where shrubby vegetation is maintained. The use of herbicides to control vegetation, however, is not advisable until the lethal and sublethal effects on eggs and larval development are understood. Pine Barrens Treefrogs are protected by state laws throughout their range.

Hyla arenicolor Cope, 1866
Canyon Treefrog

ETYMOLOGY
arenicolor: from the Latin *arena* meaning 'sand' and *color* meaning 'color.' The name refers to the frog's ground color.

NOMENCLATURE
Synonyms: *Hyla affinis*, *Hyla coper*, *Hyla copii*, *Hyliola digueti*

In early literature (e.g., Storer, 1925; Slevin, 1928; Wright and Wright, 1949; Stebbins, 1951), accounts of *H. arenicolor* may include information on *Pseudacris cadaverina* intermingled with it. This has often resulted in incorrect range delineation and confused descriptions, such as statements that *Hyla arenicolor* is sometimes greenish.

IDENTIFICATION
Adults. The ground color is light gray to dark brown, depending upon substrate and tempera-

ture, with a rugose dorsal surface. Individuals may exhibit a lichen-like pattern that allows them to blend in well with their background and, like many other treefrogs, they readily change their color and pattern to match their substrate. A dark-edged white spot is present under the eye, and there is a fold of skin above the typanum. The skin is granular with many small tubercles. The toe tips are expanded, allowing them to move on slick canyon walls. The fingers are not webbed, but the toes are partially webbed. Venters are creamy white, with bright yellow or orange on the undersides of the hind legs; venters are closely granular. The maximum size is 57 mm SUL (Conant and Collins, 1998). Wright and Wright (1949) give the male SUL as 29–53 mm and the female SUL as 30–54 mm.

Larvae. Newly hatched larvae are yellow brown dorsally and yellow ventrally. Mature tadpoles are dark brown to golden brown, somewhat globular in shape, with abdomens that appear black; the area anterior to the abdomen is unpigmented. When viewed dorsally, the tail musculature does not have regularly spaced transverse light bars or saddles. Dorsal tail fins have slight reticulations, but ventral fins are unpigmented; the extent of tail fin pigmentation increases with development, especially as the rear legs appear. In addition, there appears to be regional variation in tadpole pigmentation, with Texas specimens having tail fins that are much more heavily blotched than more western populations. There also appears to be regional variation in tadpole size, with tadpoles from the Chiricahuas smaller (to 38.3 mm TL) than from along the Virgin River (to 47.3 mm TL) (Zweifel, 1961). Minton (1958) mentions tadpoles 35–40 mm TL and Wright (1929) noted tadpoles to 50 mm TL, both from Texas populations. Albino larvae and a recently metamorphosed juvenile were reported from Texas (Van Devender, 1969). Zweifel (1961) provided figures of larvae at various stages of development and diagrams of the mouthparts.

Eggs. The vitellus is 1.8–2.4 mm in diameter (mean 2.07 mm) with a single envelope 3.9–5 mm in diameter (mean 4.4 mm) (Livezey and Wright, 1947).

DISTRIBUTION

Canyon Treefrogs are found from southern Utah southward into Mexico as far as northern Oaxaca. Isolated populations occur in southeastern Colorado, eastern New Mexico, and in west Texas (Chisos and Davis mountains). It occurs in many deep canyons, including the Grand Canyon (Eaton, 1935; personal observation). Although it occurs on the Arizona side of the Colorado River, there are no records from California. Important distributional references include Van Denburgh (1924), Degenhardt et al. (1996), Hammerson (1999), Dixon (2000), and Brennan and Holycross (2006).

Distribution of *Hyla arenicolor*

FOSSIL RECORD

Pleistocene fossils of *H. arenicolor* are known from New Mexico and Sonora, Mexico (Holman, 2003).

SYSTEMATICS AND GEOGRAPHIC VARIATION

Barber (1999) identified three divergent lineages within *H. arenicolor,* based on analyses of mtDNA. These areas corresponded to populations in Texas, New Mexico, eastern and central Arizona, southern Utah, and the Chiricahua Mountains (clade 1), sky islands of the Sonoran Desert and adjacent Mexican Highlands (clade 2), and in the Grand Canyon (clade 3). Clade 2 has two subgroups, one in the Pinaleño, Ricon, and Catalina mountains and Arivipa Creek and the other in the Santa Rita, Huachuca, and Harshaw mountains. Clade 3 was highly divergent from clades 1 and 2. Indeed, the level of divergence among the three clades suggested they might represent different species. In contrast, Klymus et al. (2010) reexamined the phylogeography of *H. arenicolor* using nuclear DNA and call analyses. They concluded that nuclear data and variation in call patterns were more congruent with a two-clade scenario. These authors found evidence of past mtDNA introgression with *H. wrightorum* in Barber's (1999) Grand Canyon populations, even though *H. wrightorum* no longer occurs in the area.

Under laboratory conditions, crosses of *H. arenicolor* with *Smilisca baudinii* or *Pseudacris cadaverina*, *P. clarkii*, *P. regilla*, *Hyla cinerea*, or *H. versicolor* are generally unsuccessful, and no metamorphs are produced; crosses with *H. femoralis*, *H. squirella*, or *H. chrysoscelis* will produce some metamorphs, but backcrosses are unsuccessful (Pierce, 1975).

ADULT HABITAT

As their name implies, Canyon Treefrogs occupy rocky canyons and arroyos with permanent to semipermanent water in the Desert Southwest. Despite the name "treefrog," they really are a rock wall frog and rarely are found on trees in riparian habitats, although Lemos Espinal and Smith (2007a) note them on trees and shrubs on rainy days. During the day, they seek refuge in rocky crevices in canyon walls, but can be found in numbers in shallow water and ditches along a creek bed. Minton (1958) mentions them in "slide rock" at an unspecified "considerable" distance from a stream; Swann (2005) notes they are found in talus on Arizona's "Sky Islands"; and Degenhardt et al. (1996) observed them in talus 1 km from the nearest water in New Mexico's Animas Mountains. Inhabited canyons often have

Eggs of *Hyla arenicolor*.
Photo: Breck Bartholomew

Tadpole of *Hyla arenicolor*.
Photo: James Rorabough

Calling adult male *Hyla arenicolor*. Photo: Dennis Suhre

cottonwood trees and other plants frequently associated with desert riparian areas. Adjacent habitats include pinyon-juniper woodland, pine-oak associations, and semiarid mesquite grasslands (Aitchison and Tomko, 1974). Aitchison and Tomko (1974) observed them at elevations of 1,676–1,798 m near Flagstaff, and Degenhardt et al. (1996) reported them from 1,200 to 2,500 m in New Mexico. It occurs from near sea level to 3,048 m (Stebbins, 2003).

TERRESTRIAL ECOLOGY

Hyla arenicolor usually are found perched on boulders and cliffs above pools of water. Activity occurs from February to November depending upon location and weather conditions (Gates, 1957; Swann, 2005; Lazaroff et al., 2006). Canyon Treefrogs are largely nocturnal, but may become active a few hours before dusk as canyon shading deepens. They also bask in the direct sun at temperatures of up to 35°C (Wylie, 1981). For example, Jameson and Flury (1949) recorded them sitting in the direct sun on boulders and solid rock stream banks. Wylie (1981) suggested sitting on exposed rock faces allows them to avoid certain predators, such as garter snakes (*Thamnophis*). During very dry weather, they may aggregate in moist situations. Swann (2005) mentions up to 100 frogs on rock surfaces over shrinking water pools, where they moved back and forth in order to soak in the remaining water. They are capable of water uptake through the ventral skin surface. During the summer rains, Canyon Treefrogs can forage away from streams for a considerable distance. Swann (2005) mentions finding them up to 100 m from water, tucked into crevices, and Lazaroff et al. (2006) noted reports of them in a cave far from water.

Canyon Treefrogs are photopositive in their phototactic response, suggesting they use both sunlight and moonlight when feeding and going about their daily activities (Jaeger and Hailman, 1973). They are sensitive to light in the blue spectrum ("blue-mode response"), which apparently helps them orient toward areas of increasing illumination (Hailman and Jaeger, 1974). Canyon Treefrogs likely have true color vision.

CALLING ACTIVITY AND MATE SELECTION

Calling begins in March at temperatures as low as 10°C (Gates, 1957) and extends into August (Livezey and Wright, 1947; Lazaroff et al., 2006; Dayton et al., 2007). Gehlbach (1965), however, suggested two calling periods, which coincide with local precipitation patterns. Zweifel (1968b) found males calling usually at 21–25.2°C in the Chiricahua Mountains. Calling occurs in or immediately adjacent to a stream, such as under flat ledges located along a stream margin or from smooth rock walls at and above the water line

Breeding habitat of *Hyla arenicolor*. Photo: Breck Bartholomew

(Degenhardt et al., 1996). They are heard diurnally calling from crevices or from spaces between rocks. Most calls are heard from just before dark into the night.

The call of the Canyon Treefrog has been described as "that of a slightly hoarse lamb: Ba-a-a" (Eaton, 1935). Intercall intervals in one Arizona study were 0.20–0.28 sec, calls lasted 0.13–0.14 sec, the pulse repetition rate was 60–75 pulses/sec, and the dominant frequency was 0.50 kHz; in Utah, the intercall interval was ca. 0.59 sec, calls lasted 0.33–0.36 sec, and the pulse repetition rate was 60–240 pulses/sec (Pierce and Ralin, 1972). In another study, the number of pulses was 15–17.9, the duration ranged between 37–39.4 ms, the call period lasted 15.15–16.3 sec, and the high frequency peak was 2,093–2,487 Hz (all data are means; Klymus et al., 2010). Some of the variation in means represent different phylogenetic histories and perhaps the influence of past introgression with *H. wrightorum* (Klymus et al., 2010). Males also possess an encounter call made in response to other males calling nearby or when another male attempts amplexus. Pierce and Ralin (1972) liken this call to a series of high-pitched "erps" or "yips." This suggests that males estab-

Crevice habitat of *Hyla arenicolor*. Photo: Breck Bartholomew

lish a calling territory that they will defend against intruders. Pierce and Ralin (1972) described both calls in detail.

Females approach calling males in response to the advertisement call. A male can detect an approaching female and move to amplex her as she approaches (Brown and Pierce, 1965).

BREEDING SITES

Oviposition occurs in shallow water in potholes and pools in intermittent canyon streams. Preferred breeding sites lack fish and crayfish. Eggs are attached singly to weeds and brush or to rock surfaces on the bottom, although they may clump together (Campbell, 1934). Occasional eggs may break off from the substrate and float to the surface.

REPRODUCTION

Eggs are oviposited attached to leaves and other debris on or near the bottom of shallow streams and potholes. Eggs will occasionally float on the surface or clump together, giving rise to reports of surface film egg clutches. Clutch size is several hundred eggs. Hatching occurs in ca. three days.

LARVAL ECOLOGY

Larval development occurs normally at temperatures >15.5°C, with the minimum developmental temperature ca. 13°C; Zweifel (1968b) suggested that 31°C was the maximum temperature for normal larval development. The duration of the larval period is 30–74 days, depending on water temperature (Zweifel, 1961; Wylie, 1981; Dayton et al., 2007), but some tadpoles may overwinter (*in* Swann, 2005). Newly transformed froglets are ca. 15 mm TL (Zweifel, 1961).

DIET

The diet includes a variety of insects, including beetles, ants, true bugs, worms, and caddisflies (*in* Degenhardt et al., 1996).

PREDATION AND DEFENSE

Canyon treefrogs are wary and readily jump toward crevices to escape. Their ground color and lichen-like patterns aid them in blending in with canyon habitats. In addition, their ability to change color rapidly allows them to blend into substrates under changing conditions. Color changes quickly from light to dark, more so than from dark to light. Not surprisingly, Canyon Treefrogs choose darker substrates when given a choice, at least at lower temperatures (Swann, 2005). At higher temperatures, they will choose a light substrate. Canyon Treefrogs also have skin secretions that cause irritation of the eyes and mucous membranes.

Tadpoles and adults are eaten by garter snakes (*Thamnophis cyrtopsis, T. eques*) and giant water bugs (Jameson and Flury, 1949; Wylie, 1981; Swann, 2005).

POPULATION BIOLOGY

No information is available.

COMMUNITY ECOLOGY

Hyla wrightorum and *H. arenicolor* frequently are found at the same breeding sites in the mountains of central Arizona and northern Mexico. Differences in calls, breeding season (although they do overlap), and response to rainfall normally are sufficient to prevent hybridization, although Klymus et al. (2010) found evidence of past introgression between them in at least one region.

DISEASES, PARASITES, AND MALFORMATIONS

The fungal pathogen *Batrachochytrium dendrobatidis* was reported from two Arizona Canyon Treefrogs (Bradley et al., 2002). Canyon Treefrogs are parasitized by the cestodes *Cylindrotaenia americana* and *Distoichometra kozloffi*; the nematodes *Cosmocercella haberi, Cosmocercoides*

Lateral head view of *Hyla arenicolor*. Illustration by Breck Bartholomew

dukae, and *Physaloptera* sp.; and larval forms of the trematode subfamily Plagiorchiinae (Parry and Grundman, 1965; Goldberg et al., 1996c). The mite *Hannemania hylae* is an ectoparasite of this species in Texas (Jung et al., 2001), and *H. hegeneri* is found on *Hyla arenicolor* from Utah (Parry and Grundman, 1965). The species also is infected by an undetermined species of trypanosome and the protozoans *Zelleriella* sp., *Nyctotherus cordiformis*, *Karatomorpha swazyi*, and *Hexamita intestinalis* (Parry and Grundman, 1965).

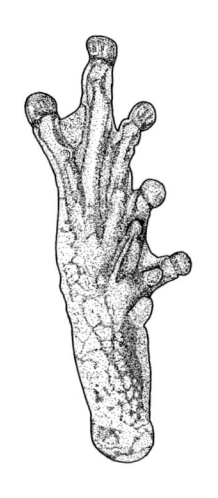

Hind foot of *Hyla arenicolor*. Illustration by Breck Bartholomew

SUSCEPTIBILITY TO POTENTIAL STRESSORS
No information is available.

STATUS AND CONSERVATION
The Canyon Treefrog does not seem to have experienced population declines and is still rather abundant in habitats throughout the Desert Southwest. Individual populations may be adversely affected by introduced fish and crayfish, altered stream flow patterns, sand deposition, and human disturbance (Lazaroff et al., 2006). Drought may also impact local populations.

Hyla avivoca Viosca, 1928
Bird-voiced Treefrog

ETYMOLOGY
avivoca: from the Latin *avis* meaning 'bird' and *voca* meaning 'call.' The name refers to the male's advertisement call.

NOMENCLATURE
Synonyms: *Hyla avivoca ogechiensis*, *Hyla phaeocrypta ogechiensis*, *Hyla versicolor phaeocrypta*

Cope (1889) originally assigned the name *H. phaeocrypta* to this frog, but it appears he was describing a small *H. versicolor*. Thus, Viosca's (1928) name *H. avivoca* is the proper name for the Bird-voiced Treefrog.

IDENTIFICATION
Adults. The Bird-voiced Treefrog is very similar in appearance to members of the Gray Treefrog complex, but it is smaller and has an entirely unique voice. The dorsum is gray, greenish, or brown with a variable lichen-like pattern, and there is a greenish-white spot directly under the eye. If green, the coloration is entirely dorsal and may be bordered by an intermittent dorsolateral thin stripe. The concealed portions of the thigh and sides are greenish, and the dark markings on the thigh sometimes form a vermiculate pattern. The body is slender with a blunt snout, and the eyes are protuberant. The dorsal skin surface is smooth, but the ventral skin may be granular. Venters are dull white with throats peppered by small dark spots. Dark bars are present on both fore and hind limbs. Toe pads are conspicuous, and hind toes are webbed. The groin and posterior portion of the thighs have a light green-spotted flash coloration. Mittleman (1945) described differences between *H. avivoca* and members of the Gray Treefrog complex.

Maximum size is 52.5 mm SUL (Neill, 1948), and males are smaller than females. Mittleman (1945) gave a male size range of 29 to 40 mm SUL (mean 33.9 mm). In Oklahoma, Secor (1988) recorded calling males at 29.7–40.5 mm SUL (mean 35.3 mm), and in Arkansas males were 28.7–37.5 mm SUL (mean 35 mm) (Trauth and Robinette, 1990). In Tennessee, Parker (1951) recorded males 32–42.8 mm SUL (mean 35.5 mm) and females to 52.5 mm SUL. Hellman (1953) recorded a single female 50 mm SUL in Florida.

Alabama frogs were 29.9–37.7 mm SUL (mean 33.7 mm) (Brown, 1956).

Larvae. The tadpole is distinctive. The body and tail are jet black with very light flecks of gold, and there is a yellow preorbital stripe extending from the eye to the tip of the snout. A pale, short postorbital stripe also is present. Larvae have three to seven copper-red saddles on the dorsal tail musculature. Bellies are black. The tail is long with a pointed tip, and the tail fins are grayish with black mottling. The iris has four gold spots. Tadpoles reach a maximum of 31–35 mm TL. Tadpoles are described by Siekmann (1949), Parker (1951), Hellman (1953), and Volpe et al. (1961). Mouthparts are figured in Volpe et al. (1961).

Eggs. There are two envelopes surrounding the egg, which is dark brown dorsally and cream to white ventrally. The vitellus is 1.1–1.4 mm in diameter; the inner envelope is 1.6–2.2 mm in diameter; and the outer envelope is 4–5.5 mm in diameter (Parker, 1951; Hellman, 1953; Volpe et al., 1961). In Tennessee, Parker (1951) recorded the vitellus as 0.8–0.9 mm in diameter, although the other measurements were similar. This suggests the possibility of geographic variation in size. Volpe et al. (1961) suggested that the jelly film surrounding the outer envelope could be considered a third envelope; it is 0.22 mm in thickness.

DISTRIBUTION

The Bird-voiced Treefrog occurs from the South Carolina Coastal Plain across Georgia to the Florida Panhandle, then west to the Florida Parishes of Louisiana. Populations occur west of the Mississippi River in Louisiana, southeastern Oklahoma (Little River drainage), and southwestern and central Arkansas. The range includes the Black Belt of Alabama, most of Mississippi, and northward along the Mississippi River to southern Indiana (possibly extirpated) and Illinois. Isolated populations also occur in Tennessee (lower Cumberland River near Ashland City), north-central Georgia, and in the Coosa Valley of Alabama.

Important distributional references include: Arkansas (Davis and Hollenback, 1978; Trauth and Robinette, 1990; Trauth, 1992; Fulmer and Tumlison, 2004; Trauth et al., 2004), Georgia (Harper, 1933; Williamson and Moulis, 1994; Jensen et al.,

Distribution of *Hyla avivoca*. This species may occur in adjacent regions of Texas and northwestern Louisiana.

2008), Illinois (Phillips et al., 1999; Redmer et al., 1999a), Indiana (Minton, 2001), Louisiana (Davis and Hollenback, 1978; Dundee and Rossman, 1989), Oklahoma (Blair and Lindsay, 1961; Krupa, 1986b; Sievert and Sievert, 2006), and Tennessee (Redmond and Scott, 1996).

FOSSIL RECORD
No fossils are known.

SYSTEMATICS AND GEOGRAPHIC VARIATION

Hyla avivoca was described by Viosca (1928), who noted that the species had been confused previously with members of the Gray Treefrog complex. It is most closely related to *H. andersonii* and slightly more distantly to the Gray Treefrog complex (Hedges, 1986). Contrary to suggestions based on call structure, there is no *H. versicolor* group (containing *H. avivoca*) as proposed by Blair (1958c). There are two clades within *H. avivoca*, a western clade and an eastern clade, which was described as *Hyla phaeocrypta ogechiensis* (Neill, 1948; Hedges, 1986). The primary difference among subspecies involves postfemoral coloration (Neill, 1954). The validity of this taxon needs clarification.

In the laboratory, *Hyla avivoca* hybridizes successfully with *H. cinerea* and *H. chrysoscelis* (Mecham, 1965; Fortman and Altig, 1973). Fortman and Altig (1973) described the hydrid tadpoles. Crosses with *H. femoralis*, *H. gratiosa*, and *H. squirella* are generally unsuccessful, but success may depend on the sex of the parents (Mecham, 1965). Natural hybridization with *H. chrysoscelis* was reported by Mecham (1960b) in Alabama. The diploid chromosome number is 24, and Bushnell et al. (1939) described the chromosomes.

ADULT HABITAT

This is a species of the cypress- (*Taxodium*) tupelo-gum headwater swamps, swampy floodplains, bottomland and slope forest, and swampy streams, lakes, rivers, and ponds of the Southeast. It frequents buttonbush (*Cephalanthus*), reed thickets, and other woody plants in riparian habitats. These habitats are often flooded in winter but hold only a little water in summer. The canopy is usually dense and the understory community diverse. Habitat descriptions are in Carr (1940a), Enge et al. (1996), and Redmer et al. (1999a).

TERRESTRIAL ECOLOGY

During the nonbreeding season, Bird-voiced Treefrogs are found in lowland forests and swamps adjacent to summer habitats, where they are observed hiding in tree crevices, in stumps, and occasionally sitting on palmetto fronds. Juveniles are found in late summer to autumn perched in low shrubs (< 2 m) near breeding habitats. Both adults and juveniles are occasionally encountered in terrestrial situations outside the breeding season and likely overwinter under leaf litter or buried into the soil in protected locations. Bird-voiced Treefrogs engage in a complex series of movements whereby the limbs are used to wipe the head and body with an extraepidermal layer of mucous and lipids, which help retard water loss (Barbeau and Lillywhite, 2005). Body wiping is important in allowing frogs to remain exposed in arboreal perches.

Bird-voiced Treefrogs are photopositive in their phototactic response, suggesting they use both sunlight and moonlight when feeding and going

Tadpole of *Hyla avivoca*. Photo: Stan Trauth

Adult *Hyla avivoca*, gray phase. Photo: Aubrey Heupel

Adult *Hyla avivoca*, green phase. Photo: Kenny Wray

about their daily activities (Jaeger and Hailman, 1973). They are sensitive to light in the blue spectrum ("blue-mode response"), which apparently helps them orient toward areas of increasing illumination, such as the open area above a pond (Hailman and Jaeger, 1974). Bird-voiced Treefrogs likely have true color vision.

CALLING ACTIVITY AND MATE SELECTION

Males begin calling from arboreal habitats sometimes a few weeks prior to moving to breeding ponds. Advertisement calling occurs from April to September throughout much of its range (Parker, 1937; Carr, 1940a; Wright and Wright, 1949; Mount, 1975; Dundee and Rossman, 1989), April to July in Oklahoma (Krupa, 1986b), and from May to mid-July in Illinois (Redmer et al., 1999a). Dundee and Rossman (1989) noted calling in March in Florida. The call is very distinctive, consisting of a ringing bird-like whistle. The prolonged whistle may be preceded by a slower series of notes. The call has a dominant frequency of 4,500 cps and a duration of ca. 2 sec (Blair, 1958a). Notes are repeated at a rate of 9.7/sec.

Calling begins just before dusk and only occurs at temperatures >18.4°C. Males call from woody vegetation, trees, and downed logs at perch heights of 0.1–7 m, mostly above the water (from 2 m over land to 15 m into the water) (Secor, 1988; Redmer et al., 1999a); most males are 1–2 mm above the water. A few males call from the shoreline, whereas others are found to > 3 m above water. Perch sites range from narrow vines to the trunks of large trees, but most males call from low branches and vines.

Females approach a calling male and initiate amplexus by touching him. She will make short, frantic movements toward him after each call until she is close to him (Trauth and Robinette, 1990). As soon as he is touched, the male turns and amplexes the female in an axillary position. After initiating amplexus, the male continues to make a raspy call for a while. The amplexed pair then move head-first down from the perch site to the water. At the water's surface, the female turns around so that she is one-third submerged and the male is half submerged in the water. Oviposition begins immediately. The female arches her head and back and extends her rear legs in the water. Eggs are extruded in packets in 1–4 min intervals (Redmer, 1998b). Oviposition takes > 1.5 hrs.

Males appear to be territorial and will engage another male that encroaches too closely to a calling perch. Secor (1988) recorded intercalling male mean distances of 3.9–7.8 m; however, occasional calling males may be much closer. Once a calling male detects an intruder (from as far away as 45 cm), he switches from the advertisement call to a short trilling chirp (Altig, 1972b). If the intruder does not depart, the calling male will approach and challenge him. A wrestling match ensues, with attempted amplectant behavior and constant chirping by the resident male. Resident

Breeding habitat of *Hyla avivoca*. Photo: Stan Trauth

males usually win wrestling matches, which end with the intruder's departure. Encounters last up to 15 min.

BREEDING SITES
Breeding occurs in riparian habitats, river bottomlands, and around ponds in cypress-tupelo swamps. Water below calling sites is shallow, usually 0.1–0.7 m (mean 0.3 m) (Redmer et al., 1999a).

REPRODUCTION
Eggs are oviposited in packets of from 3 to 15 eggs (Parker, 1951; Redmer, 1998b). Hellman (1953) estimated total clutch sizes of 567 and 720 eggs from 2 dissected specimens, and Redmer (1998b) estimated a clutch size of 150–180 from a single female observed in a 40 min period. Trauth and Robinette (1990) found a mean clutch size of 838 from 3 females and 315 from another female. It appears that females are capable of ovipositing several times during a breeding season. Eggs are oviposited in small clumps that quickly break apart and sink to the bottom of the substrate or adhere to nearby vegetation. Volpe et al. (1961) described hatching and early larval development in detail. Hatching occurs at ca. 40 hrs after deposition.

LARVAL ECOLOGY
Larvae reared in the laboratory at 28.5–35.5°C required only 29 days to complete development. Recently transformed froglets measure 9.4–13.2 mm SUL (Volpe et al., 1961).

DIET
The diet consists of small invertebrates, particularly adult and larval beetles and lepidopterans. Other prey includes ants, treehoppers, leafhoppers, bark lice, assassin bugs, mites, and spiders (Jamieson et al., 1993; Redmer et al., 1999b). Dietary items reflect an arboreal existence.

PREDATION AND DEFENSE
The dorsal coloration is cryptic and makes the frogs difficult to locate. Predators have not been reported, but likely include snakes. Larvae are undoubtedly eaten by aquatic invertebrates.

POPULATION BIOLOGY
No information is available.

DISEASES, PARASITES, AND MALFORMATIONS
Biting midges of the genus *Corethrella* feed upon *Hyla avivoca* and may be attracted by the frog's advertisement call (McKeever, 1977; McKeever and French, 1991).

SUSCEPTIBILITY TO POTENTIAL STRESSORS
No information is available.

STATUS AND CONSERVATION
The protection of riparian corridors is absolutely essential for this species, as it does not disperse

widely away from bottomland swamps and riparian habitats (Burbrink et al., 1998). Disturbances may or may not have significant impacts on populations of Bird-voiced Treefrogs. Following Hurricanes Ivan and Katrina in Louisiana, Bird-voiced Treefrog numbers actually increased substantially in swamp habitats (Schriever et al., 2009). Florey and Mullin (2005) could detect no trends in *H. avivoca* populations in Illinois during surveys from 1986 to 1989. Bird-voiced Treefrogs are protected in Illinois, Kentucky, Oklahoma, and South Carolina.

Hyla chrysoscelis Cope, 1880
Cope's Gray Treefrog

ETYMOLOGY

chrysoscelis: from the Greek *chrysos* meaning 'gold' and *kelis* meaning 'spot' or 'stain.' The name refers to the golden spots on the back of the thigh of this gray treefrog. The name Cope in the common name refers to famed herpetologist and paleontologist Edward Drinker Cope who described this species in 1880.

NOMENCLATURE

Synonyms: *Hyla femoralis chrysoscelis, Hyla versicolor chrysoscelis, Hyla versicolor sandersi*

There are many publications where the identity of the species of "Gray Treefrog" is uncertain between *H. versicolor* and *H. chrysoscelis*. For example, Fellers (1975) discusses intermale behavior and aggression in *H. versicolor*, but in a later paper (Fellers, 1979a) the characteristics described seem more associated with *H. chrysoscelis* (for example, in intermale calling distances). For this reason, this information is included in the *H. chrysoscelis* account. In other situations, it may not be possible to identify the species involved, especially when ranges overlap (e.g., Bowers et al., 1998). Many characteristics (morphology, calling season, behavior, some life history traits, and desiccation rates) may be applicable to both species. It is best to check the context of the original citation for confirmation.

IDENTIFICATION

This species is morphologically identical with *H. versicolor*. It can be differentiated by the following characteristics: chromosome number (*H. versicolor* is tetraploid [n = 48] whereas *H. chrysoscelis* is diploid [n = 24]); call rate (*H. versicolor* has a slow trill [17–35 notes/sec], whereas *H. chrysoscelis* is fast trilling [34–69 notes/sec]); cytology (cells of *H. versicolor* are larger with more nucleoli than those of *H. chrysoscelis*) (Cash and Bogart, 1978). There are no morphological characters related to body proportions or coloration that accurately separate these species (Matson, 1990).

Adults. Cope's Gray Treefrog is a small to medium-sized frog with a distinctive lichen-like dorsal pattern. The lichen coloration is composed of various gray to buff patches; sometimes the patches may be distinctively greenish. There is a conspicuous light patch underneath the eye, and the eyes are prominent. The toes are tipped by conspicuous toe pads; the rear toes are partially webbed, but there is only a slight trace of webbing between the fingers. The concealed portions of the inner thigh, shanks, groin, and axilla are bright orange and unspotted. The undersides of the body are unpigmented, but the throat of the male is dark during the breeding season. A blue (axanthic) *H. chrysoscelis* has been reported from Minnesota (Oldfield and Moriarty, 1994). Juveniles are a yellowish to olive tan to gray dorsally, frequently with a dark band between the eyes. The belly is white to cream, with the dark-colored intestinal vein obvious. Males are smaller than females, and the male's vocal sac appears as a transverse fold when not inflated. Adults reach maximum size quickly and show little growth between breeding seasons (Ritke et al., 1990). Sexual maturity is reached in two years (Jensen et al., 2008). Size reports vary for Virginia, with Mitchell (1986) reporting males averaging 42 mm SUL and females 48 mm SUL and Hoffman (1946) reporting them to reach sexual maturity at 34–40 mm SUL (mean 37.2 mm). In Wisconsin, males are 27–42 mm SUL (mean 35.2 mm SUL) (Jaslow and Vogt, 1977). In Nebraska, males are 31–47 mm SUL (mean 38 mm SUL) and females 40–50 mm SUL (mean 44 mm SUL) (Lynch, 1985), whereas in Tennessee males are 39–53 mm SUL (mean 45.9 mm SUL, mass 7.6 g) and females 45–62 mm SUL (mean 52.4 mm SUL, mass 11.2 g) (Ritke et al., 1990).

Larvae. Hatching occurs at a total body length of approximately 4.1–4.7 mm in Florida, and by 19 days the total length increases to 11.5 mm (Hellman, 1953). Jensen et al. (2008) report hatching at 6–7 mm TL in Georgia. The general body coloration may be several shades of brown, with numerous small black and gold flecks scattered across body and tail. Body coloration is not affected by the color of the water in which development occurs (Akers, 1997). Tail fins are basically colorless in terms of the background, but they are heavily mottled with black pigment. There is a pair of preorbital stripes, but they are not distinctly outlined and only become apparent after two weeks. The venter is immaculate anteriorly with some gold flecking, and the intestines are readily visible through the body wall. As the tadpole grows, the venter posteriorly becomes more cream colored with heavy intrusion of gold flecks. The iris is bright gold. Total tadpole lengths may reach 64–65 mm just before metamorphosis (Jensen et al., 2008), but the largest larva measured by Wright (1929) in the Okefenokee Swamp was 46.6 mm TL. Tadpole descriptions are in Hellman (1953) and Altig (1970).

There is a unique color morph of *H. chrysoscelis* tadpoles that appears only when predators are relatively inactive at a breeding pond. Such tadpoles develop bright red tail fins with dark margins. When predators, such as dragonfly larvae, are absent, the bright red coloration does not develop.

Eggs. Most descriptions of the eggs of Gray Treefrogs do not differentiate between *H. chrysoscelis* and *H. versicolor*. There appears to be some variation in egg size, number of eggs per packet, and total clutch size, but whether these variables are species-specific or reflect intraspecific variation is unknown. The eggs of *H. chrysoscelis* likely are similar to those described for *H. versicolor*. Eggs are deposited initially in small floating packets of <20 to 90 eggs each (Ritke et al., 1990), which may submerge to about 70 mm through time, or they may be attached to small twigs or stems near the surface. Hatching occurs quickly (3 days in Tennessee; Ritke et al., 1990). The vitellus averages 1.23–1.38 mm (mean 1.26 mm). The mean diameter of the egg capsule is 3.5 mm (range 3.1 to 5.0 mm) in Texas (Grubb, 1972) and 5.86 mm (range 5.4 to 6.6 mm) in Alabama (Brown, 1956). Total clutch size is 1,500–4,800 eggs, depending on location (see Reproduction).

DISTRIBUTION

Determination of the distribution of this species is compounded by confusion with *H. versicolor*. Many publications make no distinction between the two (e.g., Bragg, 1943a; Walker, 1946; Suzuki, 1951; Smith, 1961; Hoberg and Gause, 1992), and later distributional publications often refer to the "Gray Treefrog complex" rather than differentiate between the two morphologically identical forms (e.g., Tobey, 1985; Dundee and Rossman, 1989; Lehtinen et al., 1999; Phillips et al., 1999; Davis and Menze, 2000; Dixon, 2000; Minton, 2001; Lichtenberg et al., 2006; Sievert and Sievert, 2006). In some places, Gray Treefrogs are not sympatric, whereas in others they overlap to a greater or lesser extent. In determining distribution, I followed Holloway et al. (2006) as much as possible, correcting locations based on the primary literature. To be absolutely certain about identification, it is best to listen carefully to the calls (slow trills vs. fast trills), obtain a karyotype (diploid vs. tetraploid), or examine a blood smear to determine cell size.

Hyla chrysoscelis is found from Cape May, New Jersey, the Delmarva Peninsula, and the Potomac River valley southward through the Piedmont and Coastal Plain of Virginia, and throughout most of the Southeast. The species extends as

Distribution of *Hyla chrysoscelis*

far south as Marion County, Florida (Means and Simberloff, 1987). Populations of Gray Treefrogs consisting only of *H. chrysoscelis* are found from southern Ohio, Indiana, Illinois, and Missouri through much of Arkansas, Louisiana, and a small portion of east Texas. An additional band of *H. chrysoscelis* populations occurs from southwest Iowa, and northwestern Missouri to central Oklahoma and Texas. An isolated population of only *H. chrysoscelis* may occur in north-central Illinois (Guderyahn et al., 2004), and it is likely that only *H. chrysoscelis* occurs in the prairie regions of northeastern and southeastern South Dakota (Ballinger et al., 2000, but see Fischer, 1998). The species is sympatric with *H. versicolor* in a broad area from Maryland and south-central Virginia to southeastern West Virginia, and westward in an ark all the way from eastern Kansas to east Texas. It occurs throughout much of West Virginia and southeastern Ohio (lower Scioto River) in the unglaciated Allegheny Plateau (Little et al., 1989). Additional sympatric populations occur widely in southeastern Manitoba, North Dakota, Wisconsin, Minnesota, and Iowa. Isolated populations occur in south-central and northwestern Michigan where they are sympatric with *H. versicolor* (but see Bogart and Jaslow, 1979). In Wisconsin, *H. chrysoscelis* is more prevalent in the southeastern prairie regions of the state than *H. versicolor* (Casper, 1998). Cope's Gray Treefrogs are found only rarely on islands, for example, Bodie Island in North Carolina (Braswell, 1988).

Important distributional references include: Alabama (Brown, 1956; Mount, 1975), Arkansas (Trauth et al., 2004), Georgia (Williamson and Moulis, 1994; Jensen et al., 2008), Illinois (Brown and Brown, 1972b; Guderyahn et al., 2004), Iowa (Hemesath, 1998; Oberfoell and Christiansen, 2001), Kansas (Hillis et al., 1987; Collins, 1993; Busby and Parmelee, 1996; Collins et al., 2010), Kentucky (Barbour, 1971; Burkett, 1989), Manitoba (Preston, 1982), Maryland (Noble and Hassler, 1936; Harris, 1975), Michigan (Bogart and Jaslow, 1979), Minnesota (Oldfield and Moriarty, 1994), Missouri (Johnson, 2000; Daniel and Edmond, 2006), Nebraska (Lynch, 1985; Ballinger et al., 2010; Fogell, 2010), New Jersey (Zweifel, 1970a; Anderson et al., 1978), North Carolina (Dorcas et al., 2007; Gaul and Mitchell, 2007), Ohio (Walker, 1946; Pfingsten, 1998), South Dakota (Dunlap, 1963; Fischer, 1998; Ballinger et al., 2000; Kiesow, 2006), Tennessee (Burkett, 1989), Texas (Flury, 1951; Smith and Sanders, 1952; Hardy, 1995), Virginia (Hoffman, 1946, 1996; Mitchell and Reay, 1999), West Virginia (Little, 1983; Little and Pauley, 1986; Green and Pauley, 1987), and Wisconsin (Jaslow and Vogt, 1977; Vogt, 1981; Casper, 1996; Mossman et al., 1998).

FOSSIL RECORD
There is no way to differentiate members of the *Hyla versicolor–Hyla chrysoscelis* complex osteologically, so there is no way to differentiate their fossils. Fossil members of the complex are found in Pleistocene (Irvingtonian) deposits from Nebraska and West Virginia. The complex also is found in Pleistocene (Rancholabrean) sites from Georgia, Kansas, Texas, and Pennsylvania (Holman, 2003).

SYSTEMATICS AND GEOGRAPHIC VARIATION
Previous suggestions that polyploidy in Gray Treefrogs was produced via a single speciation event among consistent progenitors are not supported based on a suite of analyses of molecular and call advertisement data. *Hyla versicolor* arose though polyploidy via several speciation events from *H. chrysoscelis*-like diploid ancestors and two other now extinct lineages of tree frogs (Holloway et al., 2006). Geographic variation in albumins (Maxson et al., 1977; Maxson and Maxson, 1978; Ralin, 1978), morphology and coloration (Smith and Brown, 1947), call characteristics (Gerhardt, 1974a; Ralin, 1968, 1977; Bogart and Jaslow, 1979) and allele frequencies (Ralin and Selander, 1979; Romano et al., 1987) undoubtedly stem from the different evolutionary histories of the *H. chrysoscelis*-like progenitor species. Members of the *H. versicolor/H. chrysoscelis* complex are most closely related to *H. andersonii* and *H. avivoca* (Hedges, 1986).

Natural hybrids between *H. chrysoscelis* (presumably) and *H. femoralis* are suspected from Mississippi and Texas (Pyburn, 1960) and occur with *H. avivoca* in Georgia (Jensen et al., 2008) and Alabama (Mecham, 1960b). Laboratory crosses between *H. versicolor* and *H. chrysoscelis* produce high rates of mortality (Johnson, 1959, 1963; Ralin, 1976). Crosses with *Acris crepitans, Gastrophryne olivacea,* and *Anaxyrus compactilis* do not survive (Pyburn and Kennedy, 1960). Crosses with *Pseudacris crucifer*, *P. triseriata*, *P. streckeri*, and *P. nigrita* result in development to a larval stage, but no further (Moore, 1955). Metamorphosis may occur in crosses between *Hyla chrysoscelis* and *Smilisca baudinii*, *Pseudacris ornata*, *P. clarki*, or *Hyla arenicolor* (Pyburn and Kennedy, 1960; Littlejohn, 1961a; Pierce, 1975). Male *H. chrysoscelis* produce at least some larvae in crosses with female *H. avivoca*, *H. femoralis*, *H. cinerea*, and *H. gratiosa*, and female *H. chrysoscelis* may or may not produce some larvae with male *H. femoralis*, *H. cinerea*, *H. squirella*, and *H. gratiosa* (Moore, 1955; Pyburn, 1960; Pyburn and Kennedy, 1960; Littlejohn, 1961a; Mecham, 1965). No larvae are produced in crosses between male *H. chrysoscelis* and female *H. squirella*. In most of these crosses, high levels of abnormal development and morphology were observed.

In the southwestern portion of its range in Texas, the rear of the thigh is light to orange colored with fine white flecks in Cope's Gray Treefrog. There are no dark markings on the rear of the thigh except at the border. The skin is smooth, and the fingers generally are not webbed. Smith and Brown (1947) described such frogs as *H. versicolor sandersi*. Flury (1951) documented much interpopulation variation in color pattern and morphometrics within the *H. versicolor* complex in Texas and demonstrated that this taxon was invalid. There was no evidence of variation, even within the Balconian and Texan biotic provinces most likely to contain pure populations of *H. chrysoscelis*, suggesting that different taxa might be involved.

ADULT HABITAT

Cope's Gray Treefrogs tend to prefer deciduous hardwood forests (oak, hickory, maple) over pine forests (Bennett et al., 1980; Enge, 1998a; McLeod and Gates, 1998). They sometimes are found in specialized forested habitats, such as steephead ravines in the Gulf Coastal Plain (Enge, 1998b), bottomland and slope forests (Enge et al., 1996), hardwood swamps, mesophytic hammock, titi, bogs, cypress savanna, and cypress/gum ponds (Liner et al., 2008), and in dome swamps and inland hydric hammocks (Enge and Wood, 1998). Although they may occupy small, isolated temporary ponds for reproduction (Russell et al., 2002a), Cope's Gray Treefrog does not remain at ponds during the nonbreeding season.

Hyla chrysoscelis is associated with prairie and oak savanna habitats in the upper Midwest and western part of its range, whereas the tetraploid *H. versicolor* is more often found in forested habitats (Vogt, 1981; Oldfield and Moriarty, 1994; Ballinger et al., 2000; Knutson et al., 2000). Vogt (1981) also noted an association with dry and dry-to-mesic northern hardwoods, but they do not penetrate heavily forested areas. In the south, this distinction is not apparent, and different habitat preferences may reflect the different evolutionary lineages within the Gray Treefrog complex. In addition, the once-extensive longleaf pine (*Pinus palustris*) savanna of the Southeastern Coastal Plain no longer exists, blurring possible habitat differences. In south-central Louisiana, Gray Treefrogs (presumably *H. chrysoscelis*) are found associated with woody litter and plant cover, but not with herbaceous plant cover (Lichtenberg et al., 2006).

TERRESTRIAL ECOLOGY

Cope's Gray Treefrog is generally considered a lowland species, but individuals have been seen or heard as high as 1,508 m in the Great Smoky Mountains (Dodd, 2004). Away from the breeding season, Cope's Gray Treefrogs are found in the deciduous forest canopy, where they conceal themselves under bark or occupy cracks, crevices, tree holes, woodpecker holes, and other types of cavities. These frogs engage in a complex series of movements whereby the limbs are used to wipe the head and body with an extraepidermal layer of mucous and lipids, which help retard water loss (Barbeau and Lillywhite, 2005). Body wiping probably is important in allowing frogs to inhabit the forest canopy. *Hyla chrysoscelis* generally remains close to the breeding pond, and they spend considerable time moving back and forth from breeding ponds to terrestrial/arboreal sites away from the pond, primarily to feed. The juxtaposition between forest cover and breeding pond is thus very important in population dynamics. In Kansas, however, Ritke et al. (1990) recorded dispersal as far as 500 m from a breeding pond.

According to Fitch (1956a), the minimum voluntary temperature for this species in Kansas is 19.7°C, with summer body temperatures of 25.7–33.7°C. He also recorded individuals sitting directly in the sun. However, others have recorded calling at 15°C (see below). During the winter and under laboratory conditions, Cope's Gray Treefrogs survive freezing temperatures in terrestrial habitats under leaf litter buried in loose soil (Layne and Romano, 1985; Schmid, 1986; Costanzo et al., 1992; Burkholder, 1998). Glucose is used as a primary cryoprotectant rather than glycerol as in northern *H. versicolor*. A considerable portion of the body water may be frozen, and Cope's Gray Treefrogs can survive 24–40 hrs at body temperatures of −2.5 to −2.9°C.

Cope's Gray Treefrogs are photopositive in their phototactic response, suggesting they use both sunlight and moonlight when feeding and going about their daily activities (Jaeger and Hailman, 1973). They are sensitive to light in the blue spectrum ("blue-mode response"), which apparently helps them orient toward areas of increasing illumination, such as the open area above a pond (Hailman and Jaeger, 1974). Cope's Gray Treefrogs likely have true color vision.

Longevity in nature is unknown, but captive members of the Gray Treefrog complex have been reported to live seven years, ten months (Snider and Bowler, 1992).

CALLING ACTIVITY AND MATE SELECTION

Cope's Gray Treefrogs have an extended calling season that continues throughout the late spring and summer months. Calls normally are heard from April through August in many areas (Smith, 1934; Parker, 1937; Cagle, 1942; Godwin and Roble, 1983; Ritke et al., 1990; Burkett, 1991; Dodd, 2004), but most reproduction occurs early in the season. In the South, calling may begin as early as February (Carr, 1940b) and March (Brown, 1956; Littlejohn, 1958; Dundee and Rossman, 1989; Trauth et al., 2004; Jensen et al., 2008). During the day, males may call from high in the trees, with adults descending to the ground to the breeding site after dark.

Calling usually begins at dusk or during periods of diurnal rainfall, with some males initially calling as far away as 25 m from a pond (Godwin and Roble, 1983). Ambient temperatures during calling are usually >15°C. Arboreal calling occurs from trees or shrubby vegetation bordering the breeding sites, where males call from perches located 0.5–6 m (usually 1–2 m) above the water level; perch diameters range from 1 to 30 cm (Godwin and Roble, 1983; Ritke et al., 1990). Perches tend to be horizontal or located at a 45° angle, but Cope's Gray Treefrogs also can call from vertical tree trunks. In prairie regions, Cope's Gray Treefrogs call from the ground next to breeding pools, where they have been observed calling from bare ground, grass, and floating algae. Males call from the same general area throughout the breeding season, although not necessarily from the same perch or site each night. In areas of sympatry, *H. versicolor* calls from higher locations than *H. chrysoscelis* (Johnson, 1966).

As with Gray Treefrogs, dominant males are territorial and aggressively defend calling sites by kicking, shoving, head-butting, or jumping on the interloper. Dominant males space themselves at intervals of ca. 75 cm (Fellers, 1979a). Few (<2%) dominant males have a satellite male nearby, although many Cope's Gray Treefrog choruses (ca. 21%) have at least one satellite male near a calling male (Roble, 1985b). Satellite or subordinate males may occupy favored perches after a successful male has departed for the breeding pond. Other males, however, may remain near the departed male's perch and begin calling from their current position (Ritke and Semlitsch, 1991). Calling usually begins within 4 min of the dominant male's departure. However, if a dominant male returns, the subordinate male stops calling (Fellers, 1975).

Calling males are aggressive toward interlopers if the intruder ventures too close to the dominant male. Most aggressive bouts are of short duration (<1 min) and occurred when one male approached too closely to a calling male. In rare instances, a satellite male may dislodge an amplexing male, and the presence of a silent satellite male near a calling male is not uncommon at many ponds (Godwin and Roble, 1983; Ritke and Semlitsch, 1991; but see Roble, 1985b).

The literature suggests that there are slight differences in mating call characteristics (duration, repetition rate, frequency) among populations of Cope's Gray Treefrog from different portions of its range. These differences may, in part, be a result of the different phylogenetic history of the progenitor species to the Gray Treefrog complex. In addition, call characteristics vary with temperature, so comparisons of call characteristics must take this environmental parameter into consideration. Another variable to take into consideration

Tadpole of *Hyla chrysoscelis*. Photo: Dirk Stevenson

is the size of the breeding chorus and other environmental noises. Under noisy conditions, for example, *H. chrysoscelis* males change their call's rate and duration, but not the amplitude (Love and Bee, 2010).

Calls of *H. chrysoscelis* have mean pulse rates between 44.6 and 59.3 pulses/sec at 24°C, depending on location (Zweifel, 1970a; Jaslow and Vogt, 1977; Gerhardt, 1974a, 1978a; Ralin, 1977; Bogart and Jaslow, 1979), and they may be as high as 75 pulses/sec (Pierce and Ralin, 1972; temperature not recorded). The pulse rate was 51 pulses/sec at 23°C in Illinois (Brown and Brown, 1972b). In Michigan, the mean call duration was 0.46 sec (range 0.34 to 0.59 sec) at a dominant frequency of 2.52 (range 2.17 to 3.0) kHz (Bogart and Jaslow, 1979), but in Mississippi it was 0.84 sec at 2.44 kHz (Blair, 1958a). Inter-call intervals in Texas were 0.22–0.30 sec, calls lasted 0.14–0.25 sec, and dominant frequencies ranged between 1.88 and 2.11 kHz (Pierce and Ralin, 1972).

Female *H. chrysoscelis* are most responsive to frequencies of 5 and 6 kHz when presented with the opportunity to choose from a variety of call frequencies; no response was made at <2.5 and >12 kHz. They prefer average-length calls to short-duration calls, and they prefer longer-duration calls to average-duration calls (Bee, 2008a). Female call preference is directional, linear, and limited by environmental noise. Indeed, female discrimination of male calls is complex (Bee and Schwartz, 2009), especially in a noisy environment. Signal recognition has been likened to discriminating conversation at a cocktail party and, like humans, female Gray Treefrogs orient better toward call signals when the signals are spatially separated from masking background noises (Bee, 2007a, 2008b; Bee and Riemersma, 2008).

Female Cope's Gray Treefrogs are not attracted to the calls of male Gray Treefrogs. In choice experiments, female *H. chrysoscelis* readily distinguish the calls of conspecific males from male *H. versicolor*, and will move toward those calls even when in the immediate vicinity of male Gray Treefrogs (Littlejohn and Fouquette, 1960). In *H. chrysoscelis*, the faster pulse rate (50 pulses/sec) is an important cue for female mate selectivity, and the pulse rise time does not play an important role as it does for *H. versicolor* (Bush et al., 2002).

Females choose males by a combination of their mating call characteristics and perch site, and move directly toward the male's location (e.g., Littlejohn, 1958). A female initiates contact and subsequent amplexus by nudging the male or even crawling onto his back, moves with him on her back to the breeding pond, and chooses an oviposition site. Amplexus occurs at the male's calling location, although the female moves a short distance after amplexus. The couple may remain in this location 3–5 hrs before moving to the pond.

Females tend to choose males that have a lower fundamental call frequency, at least in choice situations, which may give her an indication of the male's size and potential fitness. In natural settings, male body size is negatively correlated with fundamental frequency, which suggests females may choose larger males with which to mate. There is not much difference in male fundamental call frequencies, however. Morris and Yoon (1989) hypothesized that frequency is probably not a good predictor of male mating success.

Whether females choose larger males with which to mate is unclear. Morris (1989) reported that females tended to initiate mating with larger males and that mated males were usually larger than unmated males. This result is in contrast with the findings of Ritke and Semlitsch (1991), who found no evidence that male aggression influenced how females chose males with which to mate, and that females did not appear to choose larger males for amplexus. Size also does not appear to influence whether a male is dominant or subordinate (Roble, 1985b).

In addition to mating and encounter calls, Cope's Gray Treefrogs are often heard calling high from the tree canopy outside the breeding season. Such calls are frequently heard in conjunction with rainfall or an approaching storm front, and the term "rain call" has been applied to this vocalization. Rain calls have been reported in September in Arkansas (Trauth et al., 1990) and throughout the summer in North Carolina (Murphy, 1963), but such calls can be heard at many times of the year depending on weather conditions. Their function is unknown.

BREEDING SITES

Hyla chrysoscelis breeds in small permanent ponds, temporary woodland pools, overflow pools in floodplains, pools in riparian corridors, in open pools in flooded fields, and in flooded ditches. Larvae have been found in road ruts, roadside ditches, and artificial ponds created by human activity. However, not all such sites are acceptable. For example, calling may occur around strip-mine ponds, but successful reproduction may not occur there (Turner and Fowler, 1981). Adults call from along lakes and swales, but do not breed in rivers and impoundments, at least in south-central Louisiana (Lichtenberg et al., 2006). Metts et al. (2001) did not find them around beaver ponds, although beaver ponds and impoundments are reported by Jensen et al. (2008) to be used in Georgia. They readily use stormwater retention ponds and other urban artificial wetlands (Brand and Snodgrass, 2010). In some wetlands, the presence of fish likely precluded selection as oviposition sites.

Males tend to avoid calling from wetlands in which conspecifics have already bred and in which tadpoles are present, and they avoid ponds containing certain predators, such as the sunfish *Enneacanthus chaetodon*. However, the presence of other predators, such as salamander larvae (*Ambystoma maculatum*) or adults (*Notophthalmus viridescens*), ranid tadpoles, or even dragonfly larvae (*Tramea*), does not appear to influence male calling site choice (Resetarits and Wilbur, 1991a). Although females avoid oviposition pools where sunfish and conspecific tadpoles are present, they also avoid pools containing larval *Ambystoma maculatum* (Resetarits and Wilbur, 1989). Resetarits and Wilbur (1991a) suggested that the sexes respond independently to environmental variables, and that neither sex controls the behavior of the other in terms of mate choice or oviposition site.

Hyla chrysoscelis tends to exhibit site fidelity in breeding ponds, with frogs returning to the same pond over multiple years. Most Cope's Gray Treefrogs remain around a single breeding pond within a season, but a few individuals move among breeding ponds (Roble, 1985a; Ritke et al., 1991). Over a multiyear study, a few males moved 0.1–0.5 km among adjacent breeding sites in western Tennessee, and a very few females moved from 0.1 to 0.63 km among ponds. Most interpond movement involved adjacent ponds connected by dense or uniform types of vegetation. Brodman (2009) noted considerable variation in annual occupancy over a 14 yr study in Indiana.

REPRODUCTION

Eggs are yolked late in the season in preparation for deposition the following breeding season. Egg oviposition occurs throughout the extended breeding season. Eggs of females of the *H. versicolor*/*H. chrysoscelis* complex have been found as early as 12 March in Arkansas (Trauth et al., 2004) and as late as early September in Louisiana (Dundee and Rossman, 1989). The normal reproduction period probably extends from March to August throughout much of the species' range, particularly in the South. Other aspects of reproduction are probably similar to *H. versicolor*.

After amplexus, the mated pair moves to the

breeding site during the early morning hours, arriving as late as 09:15. When ready to oviposit, the female arches her back and thrusts her legs rearward, during which time her head is slightly below water. By arching her back, she is able to raise her cloaca above the water surface and into close apposition with the male. Egg packets are released and fertilized, after which the female's cloaca submerges and the eggs float away. The female uses her front legs for stabilization and to move to different locations. After deposition, eggs swell to maximum size in about 5 min. The female's entire compliment of eggs can be oviposited in 5–10 min.

The mean clutch size varies among reports in the literature: ca. 1,500 in North Carolina (Resetarits and Wilbur, 1991a); 2,060 (range 628 to 4,208) in Tennessee (Ritke et al., 1990); 2,600 in Virginia (Mitchell, 1986); and 3,401 (range 1,086 to 4,797) in Arkansas (Trauth et al., 1990). Female body size is positively correlated with clutch size, with each increase of 1.0 mm allowing an additional 130 eggs per clutch (Ritke et al., 1990). Egg mass accounts for a considerable amount (45%) of the postgravid female's body mass. A very few females may deposit two clutches per season, in which case the second clutch is smaller than the first.

Reproduction occurs throughout the warm summer season, depending on latitude. Individual males call from one to seven (usually one or two) nights per season, and mate from zero to four times, although most males only mate once per season. Many males are unsuccessful throughout the breeding season. There is no correlation between mating success or the number of matings and the body size of males (Godwin and Roble, 1983). Mating success also is not related to when a male starts calling. Both early starters and late arrivals appear to have the same probability of mating successfully. However, males that call five or more nights per season have a better chance of successfully mating than those that call on only a few nights.

Females mate one to three times per season, although most females only mate once. A female does not remain at the breeding pond except on the night she mates. As soon as oviposition is completed, she will return to a terrestrial cover site away from the pond. As a result, sex ratios recorded at breeding ponds are highly skewed toward males at any one point, although they may approach 1:1 through an entire season (Murphy, 1963). Chorus size will often be small. At a breeding pond in Kansas for example, chorus size ranged from 0 to 27 (median 7) males calling per night (Godwin and Roble, 1983). However, Murphy (1963) marked 118 adults over a single season at a small pond in North Carolina. The longer a male calls, the more likely he is to mate successfully.

LARVAL ECOLOGY

The larval period lasts for approximately 30 to 60 days, depending on temperature and environmental conditions. In Florida, transformation occurs at a tadpole length of 32–33 mm, although tadpoles may grow as large as 40 mm TL. Reports of tadpoles 64–65 mm TL are available for Georgia, suggesting a considerable amount of variation in the minimum size at which a tadpole can transform. This plasticity likely varies with the hydroperiod and food resources available at the breeding site. The smallest metamorphs retain a small length of tail bud. Metamorphs are 12.3–15.0 mm SUL in Florida (Hellman, 1953), 15–16 mm in Arkansas (Trauth et al., 1990), 15.8 mm SUL (range 12 to 19 mm SUL) in Tennessee (Ritke et al., 1990), 13–20 mm SUL in the Great Smokies (Dodd, 2004), and as small as 13 mm in southeastern Georgia (Wright, 1932).

DISPERSAL

Dispersal by recently metamorphed *H. chrysoscelis* probably is similar to that of *H. versicolor*. However, Fitch (1958) reported that dispersal did not occur rapidly in northeastern Kansas, but that juveniles remained in the vicinity of the natal pond, living on broad-leaved plants. Froglets were still in the vicinity of the pond when cold weather began, suggesting that they overwintered near the pond. Occasional juveniles were found > 200 m from the pond late in the year (*in* Roble, 1979).

DIET

Hyla chrysoscelis eats a wide variety of insects, particularly beetles (Coleoptera), click beetles (Elateridae), and ants (Formicidae, *Pogonomyrmex*). After an analysis of diet in Texas, Ralin (1968) concluded that *Hyla chrysoscelis* foraged more arboreally than *H. versicolor*.

Like many nektonic tadpoles, those of *H. chrys-*

oscelis consume algae and detritus rasped from plant material and the substrate in breeding ponds. They also consume organic molecules and microorganisms (bacteria, fungi, small metazoans) attached to clay particles, and growth is accelerated in such turbid habitats when compared to growth of tadpoles in clear water habitats (Akers, 1997; Akers et al., 2008). Indeed, feeding activities that help stir up substrate materials into suspension may assist in nutrient uptake and allow tadpoles to exploit otherwise poor developmental habitats.

PREDATION AND DEFENSE

Eggs have been reported to be eaten by Marbled Salamanders (*Ambystoma opacum*) (Burkett, 1991), and they are consumed by the fish *Gambusia affinis* in experimental trials (Grubb, 1972). Larval *Hyla chrysoscelis* are palatable (Kats et al., 1988) and are readily eaten by a wide variety of predators, including salamander larvae (*Ambystoma*) (Cortwright and Nelson, 1990; Burkett, 1991); there appear to be no larval chemical defenses. Adults are undoubtedly preyed upon by many predators, including birds, garter snakes (*Thamnophis*), water snakes (*Nerodia*), American Bullfrogs, and meso-mammals. There is a scarcity of empirical information, however.

Cope's Gray Treefrog is an extremely cryptic species sitting on a tree trunk or among shrubby vegetation. When disturbed, they may jump away, remain immobile, crouch toward the predator, inflate the body, or climb away (Marchisin and Anderson, 1978). As the frog jumps, the sudden display of bright coloration on the inside of the thighs may startle a would-be predator long enough for the frog to escape. Adults have a noxious skin secretion that may help deter predators. Persons handling *H. chrysoscelis* should avoid contact with the eyes or mucous membranes until they have had the chance to wash off the secretion.

COMMUNITY ECOLOGY

Anuran tadpoles are important in trophic interactions within breeding ponds, both as consumers and as prey. Many factors will affect survivorship, such as the presence of predators, food resources, density, and the phenology of pond colonization by predators and competitors. For example, tadpoles are particularly prone to predation by dragonfly larvae and by fish (e.g., *Lepomis macrochirus*). In experimental situations, larger Cope's Gray Treefrog tadpoles have a greater chance of survival than smaller tadpoles (Alford, 1985), and tadpoles that are injured (on the tail fin) have reduced chances of completing development to metamorphosis, regardless of size (Semlitsch and Gibbons, 1988; Semlitsch, 1990). The presence of salamanders (*Notophthalmus*) significantly reduces the chances of survivor-

Adult *Hyla chrysoscelis*.
Photo: Dirk Stevenson

ship and decreases the time to metamorphosis in experimental mixed-species assemblages, but interestingly increases body mass and growth rate (Morin, 1983). Presumably the survivors are able to grow faster to a larger size as conspecifics and heterospecifics are eliminated. In mixed-species assemblages, newts can also increase the chances of Cope's Gray Treefrog survival by eliminating competitors.

Unlike some other frogs, there is no effect on tadpole survivorship when siblings are injured, that is, there do not appear to be any alarm substances or allelochemicals released by injured larvae that might alert siblings to the presence of a nearby predator. Larger fish are more successful at capturing tadpoles than smaller conspecifics, and they consume both large numbers and biomass of tadpoles. As might be expected, large Cope's Gray Treefrogs are better able to avoid fish than are small tadpoles (Semlitsch and Gibbons, 1988).

Like many anurans, larval *Hyla chrysoscelis* have effects on both conspecifics and heterospecifics within breeding ponds, and these effects vary by species and rearing conditions, such as density. In experimental trials with mixed tadpole assemblages, larval *H. chrysoscelis* seemed to have positive effects on the growth of *Lithobates pipiens* complex larvae and negative effects on *Anaxyrus woodhousii* larvae when crowded (Licht, 1967). As expected, crowding by conspecifics also decreases growth rates.

Competition with both siblings and other species (e.g., *Anaxyrus americanus*, *Lithobates clamitans*, *L. sphenocephalus*, *Scaphiopus holbrooki*) reduces survivorship, growth, body mass, and the length of the larval period (Steinwascher, 1981; Alford, 1986, 1989a; Wilbur, 1987); the extent of competition is especially apparent when mixed-species assemblages occur at high densities and food is limited (Wilbur, 1987). The method of competition occurs through effects on a pond's trophic structure (food availability), and this effect can persist even after *Anaxyrus americanus*, for example, has completed metamorphosis. In experimental trials, even the form of food (particulate vs. solid) affected growth rates, with tadpoles eating particulate food growing faster than those eating rabbit pellets (Steinwascher, 1981).

The priority in which different species breed at a site thus can have consequences for survivorship, body mass, and the length of the larval period of Cope's Gray Treefrogs (Wilbur and Alford, 1985). If competition is intense, then it can delay metamorphosis long enough so that a decreasing hydroperiod could result in mass mortality (Wilbur, 1987). The best ponds for breeding are those that are used immediately after filling or which have not been used previously by other species. In this way, competition is reduced and larvae should be able to complete their developmental period before seasonal temporary ponds completely dry.

As noted in the larval description, tadpoles of *Hyla chrysoscelis* growing in relatively predator free ponds develop a bright red tail coloration, which is highlighted by its bordering black margin. In situations where predators are abundant, the bright tail coloration does not develop or is a more muted yellow to orange red. In addition, the shape of the tail fin changes when predators are in close proximity; predators induce a relatively deep tail fin as opposed to a more streamlined tail fin in larvae reared without predators. The presence of crayfish predators can even result in alteration of the shape of a tadpole, causing them to have deeper bodies than in the absence of crayfish (Akers, 1997).

The presence of predators under experimental conditions also alters tadpole behavior, with tadpoles in the presence of predators swimming and feeding less often than tadpoles reared in predator-free situations (McCollum and Van Buskirk, 1996). Dragonfly larvae tend to reduce tadpole survivorship and retard size at metamorphosis. Taken together, the larval phenotypic plasticity allows tadpoles to be less available or conspicuous when the threat of predation is high, and to swim better in the presence of the predator (McCollum and Leimberger, 1997).

Such inducible phenotypes seem to be associated with animals that live in highly variable environments, such as the small pools and wetlands favored by Cope's Gray Treefrog for breeding. The phenotypic changes are induced by noncontact cues as *Hyla* tadpoles preyed upon by dragonflies. Such changes may be in response to substances released as a dragonfly larva digests a tadpole (McCollum and Leimberger, 1997). Damage to a tail fin per se does not induce the predator-avoidance tadpole phenotype.

DISEASES, PARASITES, AND MALFORMATIONS

The intestinal tracts of *H. chrysoscelis* naturally contain the bacterium *Staphylococcus sciuri*. This bacterium is not detrimental to the frog and may actually assist in resistance to cadmium, a toxic substance (AbuBakr and Crupper, 2010).

The fungus *Basidiobolus ranarum* has been found in treefrogs of the *Hyla versicolor/H. chrysoscelis* complex in Arkansas and Missouri (Nickerson and Hutchison, 1971). The amphibian chytrid fungus *Batrachochytrium dendrobatidis* has been found in one individual of the *Hyla versicolor/H. chrysoscelis* complex in Louisiana (Rothermel et al., 2008). In controlled situations, larvae orally inoculated or exposed to water containing Frog Virus 3 or an FV3-like isolate develop disease symptoms, and death occurs in three to seven days. Tadpoles exhibited moderate to severe edema, erythema, hemorrhaging, and had pale livers and kidneys (Hoverman et al., 2010).

Under experimental conditions, *H. chrysoscelis* larvae infected by the amphibian chytrid fungus have longer developmental periods than noninfected larvae, and they were smaller at metamorphosis than control larvae. Body mass decreased significantly, suggesting that food intake was impaired by the loss of the keratinized tadpole mouthparts. In addition, tadpoles that metamorphosed carried the disease into the metamorph stage, which presumably would further decrease the chances of surviving to adulthood (Parris and Beaudoin, 2004).

The leech *Desserobdella picta* has been observed on adults from Wisconsin (Bolek and Janovy, 2005). Biting midges of the genus *Corethrella* feed upon *Hyla chrysoscelis* and may be attracted by the frog's advertisement call (McKeever and French, 1991). Endoparasites include the trematodes *Allassostomoides chelydrae*, *Glypthelmins pennsylvanicus*, *Haematoloechus coloradensis*, and *Polystoma nearcticum* (Campbell, 1968; Brooks, 1976; Bolek and Coggins, 1998) and the nematodes *Cosmocercoides haberi*, *C. dukae*, *Foleyella americana*, *Oxysomatium variabilis*, and *Physaloptera ranae* (Campbell, 1968). Parasitic protozoans include *Balantidium* sp., *Hexamita intestinalis*, *Nyctotherus cordiformis*, *Opalina hylaxena*, *O. triangulata*, *Tritrichomonas augusta*, and *Trympanosoma rotatorium* (Campbell, 1968).

SUSCEPTIBILITY TO POTENTIAL STRESSORS

Cope's Gray Treefrogs are susceptible to a variety of toxic agents, as are most anurans. Westerman et al. (2003a) classified them as very sensitive to metals and organic compounds. In some situations, individual chemicals may not have detrimental effects at field concentrations. However, chemical mixtures or environmental stressors may produce lethal effects when the individual components do not. In addition, the timing of application, concentration, and frog life stage may all affect survival. For example, atrazine has little effect on hatching, whether in ambient or UV-filtered light, when applied in mid-June. However, hatching success is markedly decreased if the application occurs in mid-July (Britson and Threlkeld, 2000). In another example, mortality is less at low concentrations of chlorpyrifos, but increases at higher concentrations. Both low and high concentrations of atrazine result in less mortality, whereas intermediate concentrations increase mortality when compared to other concentrations (Britson and Threlkeld, 2000). To complicate results further, competition affects the survivorship of *H. chrysoscelis* larvae when subjected to mixtures of stressors such as methyl mercury and atrazine.

Metals. Mercury is toxic to Cope's Gray Treefrog embryos at a 7-day LC_{50} of 0.002 mg/L (*in* Sparling, 2003). Mercuric chloride is known to be teratogenic (Birge et al., 1983).

Chemicals. Combinations of the insecticides phenyl saliginen cyclic phosphate, leptophos-oxon, tri-o-tolyl phosphate, and paraoxon produce spinal malformations in *H. chrysoscelis* (Fulton and Chambers, 1985).

Nitrates. *H. chrysoscelis* exposed to pulses of nitrates (5.0 mg/L) showed decreased levels of developmental stability when the pulses occurred during the middle and late stages of development. Pulses of nitrogen during the early stages of development, however, had no effect on developmental stability, nor did steady-state concentrations of 1–5 mg/L (Earl and Whiteman, 2009).

Phosphate. Phosphate at concentrations up to 200 mg/L PO_4-P for 15 days had no effect on the survival, growth, or development of Cope's Gray Treefrogs (Earl and Whiteman, 2010). Phosphate increased the pH levels in test waters.

UV light. *Hyla chrysoscelis* does not appear to be sensitive to ambient levels of UVB light.

Breeding habitat of *Hyla chrysoscelis*. Great Smoky Mountains National Park, Tennessee. Photo: C.K. Dodd Jr.

Larval deformities occur at slightly higher levels in ambient light than in light shielded from UVB radiation, but percentages were not significantly different (Starnes et al., 2000).

STATUS AND CONSERVATION
Hyla chrysoscelis does not seem to be declining throughout its range (e.g., Brodman and Kilmurry, 1998; Davis et al., 1998; Jensen et al., 2008). It is infrequently taken in many habitat-based studies using traps or drift fences, and this sometimes leads to its being underrepresented in surveys. As such, lack of apparent abundance may be an artifact of sampling procedure rather than a reflection of true abundance. The status of the species at a landscape scale is probably secure, although individual populations are threatened by habitat destruction and alteration. Based on call surveys over a 4 yr period (1986–1989), populations of Cope's Gray Treefrog were considered to be increasing in Illinois (Florey and Mullin, 2005), but Christiansen (1981) considered them declining in Iowa because of the loss of prairie marshes, and Mossman et al. (1998) reported a decline in Wisconsin over a 10 yr period.

Habitat loss has taken its toll on members of the Gray Treefrog complex, and the species are susceptible to adverse effects from urbanization (Lehtinen et al., 1999) and by even selective tree cutting, at least over a short-term period after harvest (McLeod, 1995; Clawson et al., 1997). On intensively managed plots, Cope's Gray Treefrogs may survive in numbers, providing a variety of habitats are maintained, particularly in riparian situations (Rudolph and Dickson, 1990; Fox et al., 2004; Muenz et al., 2006). Even in heavily impacted areas, riparian areas that are periodically flooded may provide sufficient habitat to maintain populations of this species (Burbrink et al., 1998). Abundance of members of the Gray Treefrog complex is reduced in agricultural versus natural habitats, but Gray Treefrogs still use ponds located in agricultural landscapes quite frequently (Anderson and Arruda, 2006).

There are many variables associated with urbanization, not just habitat loss or alteration. Cope's Gray Treefrogs tend to have reduced foraging success under conditions of artificial lighting. Detection of prey, orientation toward prey, and the number of prey captured are all adversely affected by artificial light or by the sudden appearance of light (Buchanan, 1993). These findings suggest that lights associated with human habitation or highways may be detrimental to Cope's Gray Treefrog. In addition, road noise masks the perception of acoustic signals by female *H. chrysoscelis* and could thus disrupt the "chorus space" during the breeding season (Bee and Swanson, 2007).

There are fewer egg masses deposited in ponds receiving runoff from pastureland and oil brine pits than in reference ponds, suggesting that *H. chrysoscelis* actively avoids these and other contaminated sites (Westerman et al., 2003b). The average number of eggs per mass and percentage of survival also decreases in contaminated sites when compared to unaffected areas. Cope's Gray Treefrogs also tend to avoid ponds containing predaceous fish. When fish are introduced, such as the green sunfish (*Lepomis cyanellus*), members of the Gray Treefrog complex often disappear (Sexton and Phillips, 1986).

Mitigation measures, such as the creation or restoration of breeding sites, have mixed results for this species (Kline, 1998; Merovich and Howard, 2000; Mierzwa, 2000; Lehtinen and Galatowitsch, 2001; Pechmann et al., 2001; Weyrauch and Amon, 2002; Foster et al., 2004; Brodman et al., 2006; Barry et al., 2008; Shulse et al., 2010). Poor colonization may be an artifact of a short period during which monitoring was conducted, sampling protocol (e.g., using drift fences), or procedural problems (e.g., small numbers released in restored sites) associated with the restoration-recolonization project. When large wetlands are involved (e.g., > 8 ha in Palis, 2007) and followed over a multiyear period, successful colonization is more likely to be documented.

Call surveys used to monitor this species can be successful if either 5 or 10 min listening stops are used to record presence (Sargent, 2000; Burton et al., 2006). However, calls of this species have been misidentified for other species during call surveys (Lotz and Allen, 2007).

A total of 7,268 Cope's Gray Treefrogs were collected commercially in Florida from 1990 to 1994 (Enge, 2005a).

Hyla cinerea (Schneider, 1799)
Green Treefrog

ETYMOLOGY
cinerea: Latin for 'ash-colored,' referring to the color of the frog in preservative.

NOMENCLATURE
Synonyms: *Calamita cinereus, Calamita lateralis, Hyla blochii, Hyla carolinensis, Hyla carolinensis semifasciata, Hyla cinerea evittata, Hyla cinerea semifasciata, Hyla evittata, Hyla holmani, Hyla lateralis, Hyla semifasciata, Rana bilineata, Rana lateralis*

The nomenclature of this species was reviewed by Rhoads (1895).

IDENTIFICATION
Adults. *Hyla cinerea* normally is a moderately large bright green slender smooth frog with distinct dorsolateral white stripes and a somewhat pointed snout. The dorsal body color may vary, however, from olive or brownish (especially in the South) to gray, silvery green, or yellowish, depending upon temperature, light conditions, and stress level. The normally white stripes may be infused with yellow, and occasional individuals lack the white dorsolateral stripe altogether. Bellies are white or cream colored. In general, frogs are light on light backgrounds and dark on dark backgrounds, changes that are brought about by interacting melanophores, iridophores, and xanthophores (Nielsen and Dyck, 1978). The stripes do not blend into the venter as they do in *H. squirella*. The head and dorsal body surface may have small yellow spots that are bordered in black. Eyes are large and protuberant. All toes have obvious toe pads; the front fingers are very slightly webbed, whereas the hind toes are fully webbed. Males have a subgular vocal pouch. A photograph of a blue (axanthic) Green Treefrog was published in Florida Naturalist (Anonymous, 2007), but albinos are unknown. The diploid chromosome number is 24 (Bushnell et al., 1939).

Females tend to be only slightly larger than males based on both SUL and tibiofibula length (Gunzburger, 2006). Adults are < 66 mm SUL. Literature records on size include: males between 37 and 59 mm SUL in north-central Florida (Goin, 1958); males 37–51 mm SUL (mean 45.7 mm SUL) in central Florida (Bancroft et al., 1983); males at the Savannah River Site in South Carolina 44–60 mm SUL and females 43–63 mm SUL (McAlpine, 1993); males 47–63 mm SUL (mean 53.1 mm SUL) and females 40–59 mm SUL (mean 48.9 mm) in Illinois (Garton and Brandon, 1975); Florida adults 37–66 mm SUL (mean 45 mm SUL) and weight a mean of 5.3 g (Zacharow et al., 2003). Aresco (1996) reported the following adult size variation: 25–49 mm SUL in the Coastal Plain of South Carolina, 44–56 mm SUL in northern Alabama, 43–51 mm SUL in the Piedmont of Alabama, 43–56 mm SUL in the Coastal Plain of Alabama, and 28–53 mm SUL in the Florida Panhandle.

Larvae. Very small larvae of *H. cinerea* are light yellowish brown. As they grow, they become various shades of green and have a yellow preorbital stripe extending from the eye to the tip of the snout. Small tadpoles develop a gold midbody crossband of yellow iridophores, which disappears with age. Both the preorbital stripe and crossbands are apparent in hybrids with *H. gratiosa*, *H. squirella*, *H. femoralis*, *H. avivoca*, and *H. chrysoscelis* and are dominant traits in larvae (Fortman and Altig, 1973); these characters are not apparent in preservative. The tail musculature and fins are mottled or reticulated with dark pigment and do not have a clear area next to the musculature. Bellies are light yellow or buff colored. The jaws are narrow to medium in width, and the eyes are lateral as with other treefrog larvae. Tadpoles are described by Wright (1929), Orton (1947), Brown (1956), and Altig (1970, 1972a).

Eggs. The eggs are black or brown above and white to cream underneath. Eggs are deposited in small packets of 10–50 at or near the water's surface; egg packets usually are attached to floating vegetation. Total clutch size varies between 288 and 2,903 eggs, and there appears to be evidence of regional variation in total clutch size. The inner envelope is 2–3.4 mm (mean 2.5 mm) in diameter, with a vitellus 0.8–1.6 mm; the outer envelope diameter is 3.2–5.0 mm, and is poorly defined. Eggs hatch within two to three days. A description of the eggs is in Livezey and Wright (1947) and Brown (1956).

DISTRIBUTION

Hyla cinerea is found from the Delmarva Peninsula (White and White, 2007) southward on the Atlantic Coastal Plain, throughout the lowlands of the southeastern United States. The range extends northward up river valleys from the Deep South, such as the Coosa Valley of Alabama and the Mississippi River floodplain to southern Illinois, extreme western Kentucky (Reelfoot Lake), and southeastern Missouri. The species occurs throughout much of eastern and southern Arkansas, eastern (along the Arkansas River Valley) and southeastern Oklahoma, and the forested lowlands of east Texas to the Rio Grande.

Distribution of *Hyla cinerea*

According to Redmer et al. (1999a), the range has expanded northward over the last 35 years. A similar conclusion was reached by Platt et al. (1999) concerning range extensions into the Piedmont along the Savannah River, Jensen et al. (2008) for expansions into the Georgia Piedmont, and Friebele and Zambo (2004) for areas on the eastern shore of Maryland. Scattered populations may extend above the Fall Line elsewhere in the South, such as in Alabama. The species may have been translocated to the Brownsville, San Benito, and Harlingen areas (Lever, 2003) and introduced into Camden and Johnson counties, Missouri (Johnson, 2000; Daniel and Edmond, 2006). Another introduced population occurs at the Rio Grande Visitor Center in Big Bend National Park, Texas (Leavitt and Fitzgerald, 2009). In Florida, the species is found southward throughout the Florida Keys. Lever (2003) mentions failed translocations to Puerto Rico and England. Other introductions include Germany, Kansas, and Massachusetts (Kraus, 2009).

The Green Treefrog is found on islands, including Kent Island, Maryland (Grogan and Bystrak, 1973b), Assateague and Chincoteague in Maryland and Virginia (Lee, 1972; Conant et al., 1990; Mitchell and Anderson, 1994), and the barrier islands of North Carolina (Bodie, Cape Hatteras, Roanoke, Smith), South Carolina (Kiawah), Georgia (Blackbeard, Cumberland, Little Cumberland, Ossabaw, Sapelo, St. Catherines, Wassaw), Florida (St. George, St. Vincent, Cape St. George, Seahorse Key), and Alabama (Dauphin) (Brimley, 1926; Jackson and Jackson, 1970; Blaney, 1971; Iverson, 1973; Gibbons and Coker, 1978; Laerm et al., 2000; Irwin et al., 2001; Shoop and Ruckdeschel, 2006; Gaul and Mitchell, 2007). They are found on a number of islands in the Florida Keys (Lazell, 1989).

Important distributional references include: Alabama (Mount, 1975; Redmond and Mount, 1975), Arkansas (Black and Dellinger, 1938; Trauth et al., 2004), Florida (Lazell, 1989), Georgia (Williamson and Moulis, 1994; Jensen et al., 2008), Illinois (Phillips et al., 1999; Redmer et al., 1999a), Louisiana (Dundee and Rossman, 1989), Maryland (Harris, 1975), Missouri (Johnson, 2000), North Carolina (Dorcas et al., 2007), Oklahoma (Sievert and Sievert, 2006), Tennessee (Redmond and Scott, 1996), Texas (Hardy, 1995; Dixon, 2000), and Virginia (Tobey, 1985; Mitchell and Reay, 1999).

FOSSIL RECORD

Fossils of the Green Treefrog are known from the Pliocene (Meylan, 2005) and Pleistocene Irvingtonian and Rancholabrean from Florida (references in Holman, 2003). Holman (2003) illustrates several bones useful in identifying fossil *H. cinerea*.

SYSTEMATICS AND GEOGRAPHIC VARIATION

Hyla cinerea is most closely related phylogenetically to *H. gratiosa* and *H. squirella* (Maxson and Wilson, 1975; Hedges, 1986), although Blair (1958c) had suggested an affinity with *H. andersonii* based on call structure. Frogs from the Potomac River basin were said to lack the dorsolateral body stripe and stripes on the legs, and were described as a new species, *H. evittata* (Miller, 1899). This taxon is not currently recognized. As noted below, there is considerable variation in the extent of the dorsolateral stripe in many areas, from no stripe to a stripe running the entire length of the body. Leg stripes also may be absent in a few individuals.

There is regional variation among populations of *H. cinerea* in tibiofibula length, SUL, percent of white stripe on body, and where the point at which the stripe begins anteriorly. The length of the white dorsolateral stripe is variable among populations, with some stripes only reaching midbody in about 10.5% of the frogs (Aresco, 1996); in other populations, the stripe extends down the entire length of the body in all frogs, or the stripe may be broken into sections. The stripe begins at the snout tip in 25–89.5% of the frogs, depending on population; otherwise, it begins beneath the eye.

Natural hybrids are found between *H. cinerea* and *H. gratiosa*, and backcrosses with the parental species produce fertile offspring (Mecham, 1960a, 1965; Gerhardt et al., 1980; Schlefer et al., 1986). Hybridization in nature is not excessively common, since females prefer the calls of conspecific males (Oldham and Gerhardt, 1975), and there is a degree of ecological segregation based on differences in calling location and the permanency of the wetland used for breeding (Lamb and Avise, 1986). Hybrids show a great deal of morphological variability, from completely intermediate between parental species to nearly completely resembling one of the parental species. Schlefer et al. (1986) and Lamb and Avise (1987) noted that 38 to >40% of known hybrids could be misclassified as pure parental species based on phenotype alone, had not genetic analyses proved otherwise. However, there is no evidence that morphological asymmetry is greater in hybrids than in the parental species, suggesting that hybrids are not subject to decreased levels of developmental stability (Lamb et al., 1990). Hybrids may exhibit spectral variability in call characteristics, thus resembling either of the parental species, although the ranges of variation tend to overlap with *H. cinerea* (Gerhardt et al., 1980). Most introgressive hybridization occurs between male *H. cinerea* and female *H. gratiosa*.

Hybridization experiments with other species (*H. arenicolor*, *H. versicolor* complex, *Pseudacris clarkii*) have not proven successful (e.g., Littlejohn, 1961a; Pierce, 1975). However, Gerhardt (1974b) reported a single laboratory cross between a female *Hyla cinerea* and a male *H. andersonii* that resulted in normally developing larvae, and Anderson and Moler (1986) reported a single hybrid between *H. andersonii* and *H. cinerea* from Florida.

ADULT HABITAT

When not at or near wetlands during the breeding season, *Hyla cinerea* lives in forested areas adjacent to the freshwater breeding sites. They are frequently observed in upland hardwood forest, hardwood hammock, bottomland forest, old fields, ecotones between forested and open areas, broad-leaved marshes, sawgrass marsh, sloughs, canals, wetland forest, woody shrub habitats along stream channels, and in ravines (Enge et al., 1996; Enge, 1998a; Enge and Wood, 1998; Meshaka et al., 2000; Donnelly et al., 2001), and in riparian areas (Rudolph and Dickson, 1990; Burbrink et al., 1998; Muenz et al., 2006). For much of the year, adult habitat consists of the margins of the wetlands in which they breed. They move to mostly permanent lakes and ponds in spring (March to April), and return to terrestrial feeding areas and refugia in the autumn, usually from September to November. Green Treefrogs are also associated with habitats that may be saline, such as mangrove forest (Carr, 1940a; Meshaka et al., 2000), coastal marshes (Allen, 1932; Burger et al., 1949; Werler and McCallion, 1951; White and White, 2007), brackish areas of the Florida Keys (Peterson et al., 1952), rivers with saltwater tidal fluxes (Dunn, 1937), salt marshes (Neill, 1958), and in splash pools and marshes near Chesapeake Bay (Noble and Hassler, 1936; Cooper, 1953; Hardy, 1953) and Mobile Bay (Neill, 1958).

TERRESTRIAL ECOLOGY

Hyla cinerea becomes active in late spring in the north and remains active until October (e.g., Rossman, 1960). As one proceeds south, the activity season is extended, and in the southern Coastal Plain and Florida, frogs are active year-round. Green Treefrogs are primarily active at night, at dusk, and on cloudy or rainy days with high humidity. It is during these times when they are most likely to be feeding or moving to and from calling perches. During the day, they remain hidden in refuges or clinging to vegetation, where they are exposed directly to the elements. Freed (1980b) termed this basking behavior, and showed that growth rates are increased by basking. Under laboratory conditions, frogs spend considerable time basking, and may raise their body temperatures >6°C above nonbasking frogs.

Individual frogs are often difficult to observe because of their green coloration, lack of movement, and the tucked-in position of the limbs. Indeed, Hobson et al. (1967) suggested that Green Treefrogs may be sleeping. Once one is seen, an observer may develop a search image and see dozens more attached to vegetation, especially cattail *Typha* stems, branches, or palm fronds. In extreme southern Florida, small specimens are frequently found arboreally in bromeliads (Neill, 1951).

Green Treefrogs remain in the vicinity of the calling site throughout the breeding season. Spring and summer retreat and feeding sites are usually located in close proximity to the calling site. Daily movement from calling perches to the diurnal retreat site occurs between 01:00 and 03:00, and may involve traveling to and from cattail stems only a few meters away. Green Treefrogs are frequently associated with human habitations, and they readily take shelter near buildings or in pipes, gutters, electric or cable boxes, or other suitable places associated with anthropogenic structures (Goin, 1958). They may return to the same refuges after spending the evening feeding, but they also may choose other sites or move around from location to location. Occasionally frogs may be active at temperatures approaching 4°C, but most activity occurs at temperatures >13°C. The CTmin is 3.6–4.6°C and the CTmax is 36.6°C (John-Alder et al., 1988).

In the southern portion of its range, *H. cinerea* never becomes dormant for long periods. During cold or dry weather, it shelters under coarse woody debris or boards, under bark, in sawdust piles, treehole crevices, or in other protected places (Fontenot, 2011). The extent to which Green Treefrogs move away from breeding sites is unknown, although Dodd (1996) recorded *H. cinerea* from 457 to 914 m (mean 545 m) from the nearest possible breeding site in north Florida sandhills. In northern Florida, there appears to be a peak in movement to winter retreats in November, but even in winter frogs have been recorded moving as far as 238 m between retreat sites (Zacharow et al., 2003). In extreme south Florida, Green Treefrogs tend to occupy cypress and marsh habitats during the dry season rather than prairie habitats, but in the wet season such habitat differences are not apparent (Waddle, 2006). Capture probabilities are inversely correlated with marsh depth, suggesting that Green Treefrogs disperse during the wet season (summer) and are more concentrated in the dry season (autumn and winter).

Hyla cinerea is not particularly adept at absorbing water from moist soil, which probably inhibits

its activity during drought or dry weather (Walker and Whitford, 1970). However, they are able to slow down evaporative water loss better than most nonarboreal species (Wygoda, 1984). To do this, Green Treefrogs engage in a complex series of movements whereby the limbs are used to wipe the head and body with an extraepidermal layer of mucous and lipids, which help retard water loss (Barbeau and Lillywhite, 2005). Body wiping is important in allowing frogs to remain exposed in arboreal perches.

Green Treefrogs are photopositive in their phototactic response, suggesting they use both sunlight and moonlight when feeding and going about their daily activities (Jaeger and Hailman, 1973). They are sensitive to light in the blue spectrum ("blue-mode response"), which apparently helps them orient toward areas of increasing illumination, such as the open area above a pond (Hailman and Jaeger, 1974). Green Treefrogs likely have true color vision.

CALLING ACTIVITY AND MATE SELECTION

In the South, occasional calls are heard beginning in late February (Carr, 1940b), although the breeding season does not begin in earnest until later in April; choruses have been heard as late as early September (Bancroft et al., 1983; Dundee and Rossman, 1989; Gunzburger, 2006) or October (Einem and Ober, 1956) in Florida and Louisiana. In the northern portion of the range, such as in Illinois and North Carolina, calling begins in May and extends to August (Garton and Brandon, 1975; Redmer et al., 1999a; Gaul and Mitchell, 2007). Environmental conditions such as temperature appear to control the onset of breeding.

Arboreal or ground-level calling occurs from shrubby vegetation surrounding a lake or pond or from emergent vegetation (to about 30 cm from the water's surface) around a lake's shoreline and shallows. In Florida, Bancroft et al. (1983) recorded calling from cattail, pickerel weed (*Pontederia*), and water hyacinth (*Eichhornia*). Green Treefrogs rarely call from the water-shoreline interface or from water as does the closely related Barking Treefrog. They may occasionally call from such locations, however, if the shrubby area around the breeding site is removed or altered. In these cases, the habitat change and subsequent shift in calling site may cause them to come into contact and breed with *H. gratiosa*, resulting in hybridization (Lamb, 1987).

A chorus of Green Treefrogs usually begins with one or a few males initiating a calling bout. More and more males join into a chorus, until virtually all males are calling and the pond or lake reverberates in a deafening cacophony of noise. A bout may continue for several minutes, followed by a sudden cessation of calling. A number of minutes then pass in silence until the chorus bout repeats itself. The cycle of slow buildup to a full chorus, followed by silence, continues until the early morning hours. On a large lake, calling bouts among Green Treefrogs give the impression of waves of noise as the cycle repeats itself among resident populations. Carr (1940a) noted that choruses of Green Treefrogs may continue unbroken for 10 to 12 km of river front.

Calling begins in earnest at dusk and continues to after midnight. In South Carolina in June and July, most calling occurred between 22:00 and 01:30 (Mohr and Dorcas, 1999). Calls are also heard on overcast or rainy days. Male Green Treefrogs produce two types of calls during the breeding season, a mating call and a "pulsed" call. The mating call duration varies between 0.10 and 0.29 pulses/sec, with repetition rates of 0.27–1.1 sec (Blair, 1958a; Oldham and Gerhardt, 1975). The high frequency part of the mating call centers around 3.0 to 3.8 kHz, and the low frequency peak ranges from 0.7 to 1.250 kHz (Gerhardt, 1982). In playback experiments, females prefer calls with low frequencies of 0.8–1.0 kHz and high frequencies of 2.4–3.6 kHz (Gerhardt, 1987), preferences that are independent of female body size. Female preference allows for species-specific discrimination, so that although *H. cinerea* and *H. gratiosa* are often found calling from the same site, mating between the two does not normally occur unless the habitat has been altered or satellite males intercept a heterospecific female.

The pulsed call is an agonistic or encounter call made by a dominant male toward a subordinate or nearby male. Males may occasionally produce a call that is intermediate between these two. The calling male will turn to face the direction of the intruder. The encounter call is then produced when the approaching male comes with ca. 170 cm of the primary calling male (Fellers, 1979a). Females are able to discriminate among the calls and are attracted primarily to the male's breeding call (Oldham and Gerhardt, 1975). Indeed, her call discriminatory ability is graded in such a manner as to facilitate identification among calling males, rather than responding to categorical cues that would only identify a male as being a male (Gerhardt, 1978b).

Calls are projected omnidirectionally, thus making it possible to attract females no matter where they are located in relation to the calling male. Females, however, locate calling males by using lateral head movements, and by elevating and lowering the head as they move toward a calling male. These head movements help them to locate arboreally calling males, perhaps by using a sound pressure gradient system similar to that used by insects to localize sounds (Gerhardt and Rheinlaender, 1982). This is important, since large choruses (>200 calling males to literally thousands around large lakes) likely mask the acoustic signals of all but the closest males. Females thus choose from among the three to five males closest to her, essentially allowing her to select the best that is available close at hand (the "best of N"; Gerhardt and Klump, 1988). As might be expected under such circumstances, there is no evidence for size-assortive mating (Gerhardt, 1982, 1987).

Satellite males frequently accompany calling male H. cinerea (to 18% at a Georgia study site, Perrill et al., 1978). A single calling male may have one to three satellite males, usually within 50 cm of the caller. Calling males attempt to thwart satellite males through encounter calls to warn the interloper away, but they may also resort to wrestling, head-butting, or chasing the intruder away. If a calling male leaves or is removed from a pair-wise confrontation, the satellite male will assume a mating call (Perrill et al., 1982). Calling males also may become satellites in the presence of another nearby calling male. Satellite males are occasionally successful in intercepting females approaching a calling male, a strategy known as sexual parasitism. The propensity of satellite males to intercept females has led some authors to suggest that the directional introgression seen in hybrid populations of H. cinerea and H. gratiosa results from satellite male H. cinerea intercepting H. gratiosa females on their way to the water/water-shoreline calling sites of conspecific males (Lamb and Avise, 1986).

As with other hylid treefrogs, calling is an energetically expensive activity that is almost entirely aerobic and may exceed the level of energy required for locomotion (Prestwich et al., 1989). Calling requires seven times the metabolic activity of a Green Treefrog at rest. This is not surprising, since Green Treefrogs can utter 3,500 calls/hr (at 27°C) during peak calling performance.

Axillary amplexus is initiated when a female approaches and touches a calling male or crawls over his back. Males stop calling after amplexus, but there is one report of a male calling while still in amplexus under natural conditions (McCallum and Trauth, 2009). At her approach, he turns and faces the female, and his call rate and pitch increase. After she touches him, he immediately amplexes her. At this point ovulation begins as denoted by muscular contractions along the female's flank. Egg deposition occurs within 4–5 hrs.

BREEDING SITES

Green Treefrogs breed around a wide range of mostly permanent wetlands, from ponds and swamps to the margins of large lakes. Abundance is positively associated with wetland size and depth when considered in a landscape of seasonally flooded ponds, and the most used breeding sites are those located within 50–100 m of forested habitat (Babbitt et al., 2006). In this latter study, pasture ponds were deeper, larger, of higher pH (6.0 vs. 5.3), and of greater conductivity (120 µs vs. 40 µs) than wetlands located within woodlands. *Hyla cinerea* readily uses anthropogenic sites, such as retention ponds, impoundments, golf course ponds, and drainage ditches (Lichtenberg et al., 2006; Dodd and Barichivich, 2007; Scott et al., 2008). Reproduction may occur in small isolated wetlands, such as freshwater marshes, cypress savannas, or cypress/gum ponds (Babbitt and Tanner, 2000; Eason and Fauth, 2001; Russell et al., 2002a; Surdick, 2005; Liner et al., 2008).

Green Treefrogs prefer wetlands with substantial amounts of emergent vegetation, including water hyacinth (Kilby, 1936; Goin, 1943), bordering shrubs, and other types of low vegetation around the shoreline, from which they call. Sites with dense mats of floating and subsurface vegetation are preferred, as these sites provide protection for eggs and larvae. For example, they are abundant in the wet prairies of the Okefenokee Swamp (Wright, 1932).

As with many species, many potential breeding sites are not occupied annually for reproduction. For example, Babbitt et al. (2006) recorded breeding by *H. cinerea* at 45% of the 78 seasonally flooded wetlands surveyed on an agriculturally modified tract of land in southern Florida. The presence or absence of fish does not appear to influence the choice of breeding sites.

REPRODUCTION

Reproduction takes place from early spring throughout the summer, when temperatures are >20°C. Eggs have a minimum temperature tolerance of 20°C and a maximum tolerance of 34–39°C (Ballinger and McKinney, 1966). Mated pairs are found in March in Alabama and Florida, although most breeding occurs between April and August throughout much of the species' range (Wright, 1932; Carr, 1940a; Moulton, 1954; Mecham, 1960a; Mount, 1975; Gunzburger, 2006). Breeding is nocturnal, and rainfall is not required to trigger it. However, the greatest amount of breeding occurs during warm rain events (e.g., Richmond and Goin, 1938). Diurnal breeding activity is also triggered during humid and overcast conditions.

After the female initiates amplexus, the mated pair remains stationary for several hours before entering the water. Ovulation occurs 4–5 hrs after the female touches the male, leading most oviposition to occur during the early morning hours (02:00–04:00). Indeed, Garton and Brandon (1975) suggested auditory stimuli are responsible for initiating ovulation, whether or not a female enters into amplexus. Ovulation takes 15–35 minutes. Females depress their back and extend their limbs downward. This causes the cloaca to rise and to be near to that of the male's cloaca. The male then fertilizes the eggs as they are extruded. Eggs (10–50 at a time) are sent backward from the female's cloaca under the water's surface, where they adhere to floating vegetation. The eggs must remain near the surface, as the warm waters of the pond and lake are often oxygen depleted even a short distance below the water's surface. Hatching occurs in two to three days. Garton and Brandon (1975) provided a detailed description of mating behavior and oviposition.

The degree of heterozygosity in *H. cinerea* is important to reproductive success, but it varies between the sexes (McAlpine, 1993). In wild-caught females under laboratory conditions, clutch size and the number of offspring surviving through metamorphosis was correlated positively with heterozygosity, and the percentage of offspring that hatched from eggs showed a similar trend, although it was not statistically significant. In contrast, body size was not correlated with heterozygosity in either males or females, and the number of eggs that hatched was not correlated with the number of heterozygous loci in males. The total reproductive success of females was correlated with heterozygosity, but not with fitness traits such as body size (McAlpine, 1993). McAlpine and Smith (1995) concluded that multilocus heterozygosity was not a good indicator of either survival or mating success in Green Treefrogs.

In Florida, clutch size ranges from 359 to 2,658 eggs (mean 1,214) (Gunzburger, 2006). In Illinois, the mean number of eggs per clutch is 700 (Garton and Brandon, 1975), whereas in coastal Georgia the mean is 790 eggs per clutch (Perrill and Daniel, 1983), and in Arkansas it is 2,152 (range 1,348 to 3,946) (Trauth et al., 1990). In South Carolina, the mean is 1,472 eggs, and hatching success is high (mean 87%, range 33 to 100%) (McAlpine, 1993). Some of the variation in the literature reports of clutch size may result from small sample sizes and the confusion of singular clutch counts with the number of ova produced per season. Perrill and Daniel (1983) noted that female *H. cinerea* may deposit from 1 to 3 clutches per season, with a mean of 19.3 days between clutches. Single clutches ranged from 275 to 1,160 eggs (mean 275 eggs). Female size is positively correlated with clutch size and weakly correlated with the size of the amplexing male.

LARVAL ECOLOGY

Larvae hatch into shallow areas with dense submergent or floating vegetation, such as hyacinth mats in Florida. The dense vegetation provides both cover and food for the growing larvae. This is important since the late hatching time means that the larvae will be exposed to predators that may have been in the wetland for a time, and be of a size to easily consume the vulnerable larvae. Babbitt et al. (2006) found a larval density of 0.078/m^2, a value lower than most other summer season breeders.

The duration of the larval period (normally five to nine weeks) and the size of the larvae as they

Tadpole of *Hyla cinerea*. Photo: Dirk Stevenson

Tadpole of *Hyla cinerea* with black-spotted tail fins. Photo: Joe Mitchell

approach metamorphosis are dependent on food levels and the temperature at which development occurs, among other factors. Under experimental conditions, larvae raised at high temperatures (30°C) develop more quickly (24.4 days) but are smaller at metamorphosis (0.90 g) than larvae raised at lower temperatures (25°C, 1.18 g, 35.7 days). Likewise, *H. cinerea* larvae raised under high food availability are much larger (1.22 g) and complete metamorphosis sooner (29.4 days) than larvae raised under more restricted food intake (0.67 g, 36.8 days) (Blouin, 1992a). Size at metamorphosis is correlated positively with the length of the larval period and larval growth rates, and there appears to be a genetic component between the size at metamorphosis and larval growth rates (Blouin, 1992b). According to Wright (1929), the largest tadpoles grow to 40 mm TL just prior to metamorphosis. Newly transformed *H. cinerea* are 11.5–17 mm SUL (Wright, 1932).

Hyla cinerea larvae occasionally are vulnerable to pond or lake drying, even though they prefer permanent water bodies for reproduction. Whether a pond dries slowly or rather fast does not influence the duration of the larval period or the body size at metamorphosis per se under experimental conditions. This is not surprising given that this species normally does not face desiccating conditions and that selection may favor short larval periods for a species cohabiting with fish predators. However, larval density increases as water evaporates, and the increase in density may result in extended larval periods or in metamorphosis at a smaller size than normal (Leips et al., 2000). Under experimental conditions, the larval period lasts from 35 to 58 days. In Illinois, Garton and Brandon (1975) report a larval period of 4–6 weeks, whereas in Georgia the larval period is 55–63 days (Wright, 1932).

Green Treefrog larvae have been observed in water that is saline, to 8.3‰ (Diener, 1965). Normally lethal salinity levels are 8–12‰ (Schreiver, 2007). The developmental rate of tadpoles is inversely correlated to the salinity of the water. At 2‰, the larval period takes a mean of 25 days, whereas at 6‰, the mean is 30.8 days. Tadpoles that survive 8‰ take a mean of 34 days to complete development. In addition, tadpoles developing at higher salinities have smaller body masses at metamorphosis (Schreiver, 2007). These data suggest that salinity may have sublethal effects even among surviving larvae. Under normal circumstances, body mass and TL are positively correlated with the length of the larval period.

DISPERSAL

Recently metamorphosed *H. cinerea* tend to remain in the immediate vicinity of the natal breeding site during their first winter period, although they have been recorded as far as 90 m from the breeding site into surrounding forest (*in* Jensen et al., 2008). The metamorphs grow quite rapidly, and some individuals approach adult size before they enter their first winter dormancy (Garton and Brandon, 1975). Wright's (1932) interpretation of the relationship between size-classes and age is probably in error.

DIET

Feeding occurs primarily during dusk and at night, as the frog is far more active nocturnally than during the day. During cloudy or rainy days of high humidity, the frogs will be diurnally active and more likely to feed later in the afternoon. Feeding occurs throughout the reproductive season, as mated pairs both contain food items (Kilby, 1945).

The diet consists of prey indicative of the shrubs and herbs on which the species is found. Insect prey primarily includes flies, moths, crickets, dragonflies, butterfly adults and larva, ants, bees and wasps, and beetles. Green Treefrogs are primarily insectivorous, although they readily consume mollusks (snails), spiders, mites, collembolans, millipedes, and phalangids (daddy-longlegs). Inorganic debris and epidermis are found frequently in the digestive tract (Haber, 1926). Cannibalism also has been reported (Höbel, 2011). In frogs obtained from the Okefenokee Swamp in Georgia, Haber (1926) reported a high percentage (92%) of frogs with food in their digestive tracts, including recently eaten insects (68%), orthopterans (24%), coleopterans (24%), lepidopterans (21%), hymenopterans (15%), hemipterans (13%), spiders (24%), and epidermis (35%). Many digestive tracts contained nematodes (41%) and parasitic protozoans (7%).

In a Florida study, Green Treefrogs particularly ate larvae of lepidoptera (*Spodoptera*) and beetles (*Chauliognathus*), stink bugs (*Euschistus*), crawling flea beetles (*Disonycha*), spiders (*Clubiona*), and ants (*Crematogaster*); prey of 35 invertebrate families were included in the diet (Freed, 1982). The diet of the introduced population at Big Bend National Park consists of beetles, cockroaches, grasshoppers, crickets, ants, spiders, and even scorpions (Leavitt and Fitzgerald, 2009). Additional information on diet is contained in Haber (1926), Kilby (1945), and Brown (1974). Prey is selected based on size, activity, and frequency of occurrence (Freed, 1980a, 1982). Green Treefrogs have prey preferences, such as choosing flies (*Musca*) over mosquitoes; the more a prey item is likely to be active, the more likely it is to be eaten. Juveniles and adults have basically the same diet, but one attuned to the size of the frog.

PREDATION AND DEFENSE

Larval predators include dragonfly naiads (*Tramea, Anax*), dytiscid larvae, notonectids, belostomatids (giant water bugs), crayfish, fish (*Centrarchus, Micropterus, Umbra, Fundulus*), newts (*Notophthalmus*), and musk and mud turtles (*Sternotherus, Kinosternon*). Adult Green Treefrogs likely are eaten by a wide variety of vertebrates. Specific reports include killdeers (Schardien and Jackson, 1982), rat snakes (*Elaphe obsoleta*) (Neill, 1951), black racers (*Coluber constrictor*), ribbon snakes (*Thamnophis sauritus*), water snakes (*Nerodia sipedon*), snapping turtles (*Chelydra serpentina*), herons, raccoons, and foxes (Wright, 1932; Mitchell and Anderson, 1994).

The best antipredator defense of postmetamorphic Green Treefrogs is their ability to blend in with vegetation and avoid diurnal movement over long periods. When disturbed, they may hop or jump away (a 4 g frog may hop 73 cm; John-Alder et al., 1988), but their inner thighs do not contain the flash colors possessed by other hylids that might startle would-be predators. Contact with snakes may elicit body inflation (Marchisin and Anderson, 1978), and attacked individuals will use their limbs in an attempt to prevent ingestion (Höbel, 2011). There is no indication that their skin contains toxins or noxious substances that might deter predators.

POPULATION BIOLOGY

Sexual maturity is reached in one year (Garton and Brandon, 1975). Although the Green Treefrog is considered to be extremely abundant, there are no published data on population size. Literally thousands of Green Treefrogs can be heard in some areas calling on warm summer nights. In Louisiana, Pham et al. (2007) estimated there were 143 *Hyla cinerea* in a small urban office complex prior to metamorphosis. After metamorphosis, the population estimate was 446.

COMMUNITY ECOLOGY

Green Treefrogs breed later in the season than many other species, and they prefer permanent bodies of water over temporary ponds. This may be because temporary ponds tend to build up a diverse array of predators as the season progresses, especially since temporary ponds lack fishes, which might otherwise keep many invertebrate predators in check. As such, variable predation pressure might be very important in determining both larval occupancy and reproductive success. Gunzburger and Travis (2005) showed that larval *H. cinerea* had low but constant survival rates when exposed to many types of predators and suggested that predation rate increases linearly with predator density. Not surprisingly, smaller tadpoles are more vulnerable to predators than are medium to large tadpoles (Gunzburger and Travis, 2004), and tadpoles tend to decrease activity when exposed to potential predators (*Anax, Lepomis, Notophthalmus*) under experimental conditions (Richardson, 2001).

Green Treefrogs and Pine Barren Treefrogs have very similar calls, call from similar locations and microhabitats, and may occasionally be found at the same pond. Call rates of *H. cinerea* may be as fast as *H. andersonii*, but call durations and repetition rates are shorter in *H. andersonii*; lower peak frequencies are higher in *H. andersonii* compared with *H. cinerea*, but upper peak frequencies are lower in *H. andersonii*. Female *H. cinerea* are not attracted to calls of male *H. andersonii*. However, some female *H. andersonii* are attracted

Adult *Hyla cinerea*. Photo: Alan Cressler

Habitat of *Hyla cinerea*, McIntosh County, Georgia. Photo: C.K. Dodd Jr.

to the calls of male *H. cinerea*. Differences in female call discrimination capability could lead to interspecific hybridization if both species called from the same location (Gerhardt, 1974b).

The presence of *Rhinella marina* larvae did not affect the growth, development, or survivorship of larval *Hyla cinerea* under experimental conditions (Smith, 2005a).

DISEASES, PARASITES, AND MALFORMATIONS
The fungus *Basidiobolus ranarum* has been reported from *Hyla cinerea* in 6 of 14 animals examined in Missouri/Arkansas (Nickerson and Hutchison, 1971). Amphibian chytrid fungus is as yet unreported from this species (Rothermel et al., 2008). An unidentified *Myxidium* (a myxozoan parasite of the gallbladder) has been observed in tadpoles from the southeastern United States (Green and Dodd, 2007), and unidentified nematodes and protozoans (*Opalina*) have been reported from the gut (Haber, 1926). The trematode *Gyrodactylus* sp. has been observed on larvae (Green and Dodd, 2007) and the nematode *Cosmocercella haberi* on adults (McAllister et al., 2008). Haber (1926) also reported ectoparasitic mites. Biting midges of the genus *Corethrella* feed upon *Hyla cinerea* and may be attracted by the frog's advertisement call (McKeever, 1977; McKeever and French, 1991). Höbel and Slocum (2010) reported a Green Treefrog with malformed vocal and dorsal sacs.

SUSCEPTIBILITY TO POTENTIAL STRESSORS
Metals. Aluminum at sublethal concentrations slows the growth rate of *H. cinerea* larvae, and the effect is more pronounced at a low pH (4.5) than at a higher pH (5.5). The $LC_{50\ (96\ hr)}$ is 277 µg/L at a pH of 4.5. Larval mortality increases with increasing aluminum concentration at low pHs, but not at higher pHs. High aluminum concentrations result in a reduced body size, which in turn makes larvae swim at a slower speed and increases the likelihood of predation (Jung and Jagoe, 1995). However, it is not just the smaller body size that causes a reduction in swimming speed; swim speed reduction is a direct result of the aluminum itself.

Other contaminants. Green Treefrogs in a contaminated nuclear power facility outflow swamp had radiocesium levels at a mean of 204.2 pCi g^{-1} dry weight (Dapson and Kaplan, 1975). The biological half-life of radiocesium is 30.1 days at 20–30°C.

pH. Green Treefrogs are found in Carolina bays at pHs as low as 4.3 and in farm ponds with pHs at 7.9 (Jung and Jagoe, 1995). The lower limit of pH tolerance is 4.2, and Green Treefrogs are found commonly in ponds with a pH >4.5 (Eason and Fauth, 2001).

Salinity. The mean LC_{50} value for early stage embryos is 5.2‰, for late-stage embryos 8.0‰, and for larvae 9.2‰. Salinity is lethal at 8–12‰. Salinity levels affect body mass and larval period (see Larval Ecology) and may have sublethal effects for surviving larvae (Schreiver, 2007).

Chemicals. Bennett et al. (1983) reported no significant amounts of the pyrethroid insecticide fenvalerate in *H. cinerea* living in the vicinity of a cotton field sprayed five days previously. However, they only examined one frog.

Petroleum crankcase oil had no effect on Green Treefrog hatching success, growth, or successful metamorphosis at concentrations from 0–55 mg/L (Mahaney, 1994). At 100 mg/L, tadpole growth was slowed considerably, although not at the lower concentrations, and no successful metamorphosis occurred. However, oil concentrations decreased through time as experimental mesocosms received rainfall, and oil was chemically and biologically degraded; thus, the adverse effects of intermediate concentrations of oil were masked by dilution. High oil concentrations also reduced the amount of floating algae, but not the overall algal standing crop. As a result, Mahaney (1994) suggested that Green Treefrog larvae would not metamorphose successfully at breeding sites contaminated by > 50 mg/L of oil, since growth would be so delayed that winter low temperatures would inhibit thyroxin output and prevent complete metamorphosis.

STATUS AND CONSERVATION

Hyla cinerea is a widespread and abundant species. Populations appear to be expanding in Illinois, the Piedmont of the South, and along the eastern seaboard (Platt et al., 1999; Rice et al., 2001; Friebele and Zambo, 2004; Florey and Mullin, 2005). They are common in rural areas and are able to persist in some abundance in disturbed (urbanized, silvicultural, agricultural) settings (Brown, 1974; Wilson and Porras, 1983; Delis et al., 1996; Surdick, 2005). For example, Pieterson et al. (2006) recorded a substantial increase in call detection over a 5 yr period in southwest Florida, despite intensive urbanization between 2000 and 2004. However, roads can cause significant mortality (Smith et al., 2005).

Changes in vegetation structure surrounding breeding ponds may lead to a greater potential for hybridization between *H. cinerea* and *H. gratiosa*. In particular, a surrounding border of arboreal calling sites is necessary for Green Treefrog reproduction so that they do not come into contact with Barking Treefrogs; clearing vegetation from the shoreline and borders of ponds and lakes is not conducive to Green Treefrog reproduction. In the South, these frogs readily survive prescribed burns and recolonize burned wetlands, as long as retreat sites are available (Schurbon and Fauth, 2003; Langford et al., 2007).

Because of their propensity for using cavities as retreats, the use of ground or tree-placed PVC pipes has proven effective for monitoring this species and examining its movement patterns (Boughton et al., 2000; Zacharow et al., 2003; Waddle, 2006; Campbell et al., 2010). The lack of observations or captures of few of these frogs during extended surveys may reflect sampling biases (e.g., use of inappropriate techniques) rather than scarcity. Researchers in monitoring studies show that survivorship decreases by as much as 11% when 3 or 4 toes are clipped during mark-recapture studies, although toe-clipping does not influence capture probability (Waddle, 2006). In terms of wetland mitigation measures, Green Treefrogs readily colonize newly created ponds and wetlands (Merovich and Howard, 2000; Palis, 2007; Shulse et al., 2010).

Disturbances may or may not have significant impacts on populations of Green Treefrogs. Following Hurricanes Ivan and Katrina in Louisiana, Green Treefrog numbers plummeted in swamp, marsh, and levee habitats, presumably due to saltwater intrusion (Schriever et al., 2009). In contrast, populations can recover quickly in some coastal areas that receive hurricane overwash, with resulting temporary increases in salinity (Gunzburger et al., 2010). In noncoastal habitats, canopy gaps may increase insect prey abundance, and Green Treefrogs may be more abundant in open-canopied situations than they are in closed-canopy forest habitats (Horn et al., 2005). Presumably the frogs are in the open-canopy gaps because of better feeding opportunities.

A total of 31,265 *H. cinerea* were exported from Florida between 1990 and 1994 for the pet trade (Enge, 2005a).

Hyla femoralis Bosc in Daudin, 1800
Pine Woods Treefrog

ETYMOLOGY
femoralis: from the Latin *femuralis* meaning 'pertaining to the hindleg.' The name likely refers to the bold markings on the back of the thighs.

NOMENCLATURE
Synonyms: *Auletris femoralis*, *Calamita femoralis*

IDENTIFICATION
Adults. Pine Woods Treefrogs are mostly gray, tan, or reddish brown with distinct bark-like dark dorsal markings, with broad heads and short, rounded snouts. Some individuals adopt a greenish coloration when on a green background. Frogs are often lighter with more uniform markings at night than during the day, but there is considerable variation among individuals. There is a dark mark between the eyes that is variable in shape. A thin black line may be present on the upper jaw. Venters are white and unspotted. The rear of the thighs is marked by bright yellow-orange spots. Toe pads are apparent but are not as conspicuous as in many other *Hyla*. Males have a dark vocal sac located more posteriorly than other *Hyla*; the sac forms a transverse fold pectorally when collapsed. According to Brown (1956), females are more reddish brown and males are more often gray than vice versa.

Males are smaller than females (Delis, 2001); central Florida males are 21–36 mm SUL (mean 27.1 mm) and females are 22–38.5 mm SUL (mean 31.1 mm). In north Florida, frogs are 19–40 mm SUL (mean 32.1 mm) (O'Neill, 1995). South Florida males average 30.7 mm SUL and females 34.2 mm SUL (Duellman and Schwartz, 1958). For Alabama frogs, Brown (1956) gave a size range of 28.2–40.5 mm SUL (mean 33.5).

Larvae. Very small larvae are brownish yellow with a distinct lateral stripe and sharply bicolored tail musculature. As the tadpole grows, the dorsolateral body stripes are lost, and the tail stripe becomes more distinctive. The body assumes a dark olive to black coloration with a light (often bright yellow) venter. A pale postorbital stripe is present, but there is no interorbital bar. The tail musculature is distinctly striped, and the tail fins are flecked or blotched although a clear area remains near the tail musculature. There is a well-developed clear flagellum at the tip of the tail. Larvae exposed to certain predators develop orange-red coloration on the tail fins and have deeper bodies and shorter tails than larvae not exposed to predators. Tadpoles reach about 22–38 mm TL (Brown, 1956; Jensen et al., 2008). Siekmann (1949) described the tadpole and Altig (1972a) described early development.

Eggs. The eggs are brown dorsally and yellow ventrally. The vitellus is 0.8–1.2 mm in diameter, the inner envelope is 1.4–2 mm in diameter, and the outer envelope is 4–8 mm in diameter (Livezey and Wright, 1947; Delis, 2001). This outer envelope is loose, indistinct, and sticky. Eggs are deposited in a circular or elliptical mass 1.5–10 cm × 10–18 cm, which floats on the surface. Altig (1972a) noted that egg masses covered ca. 45 cm^2.

DISTRIBUTION
Hyla femoralis occurs on the Atlantic Coastal Plain from southeastern Virginia throughout much of Florida (to Broward and Collier counties), then west to the Florida Parishes of Louisiana. An isolated population occurs in the Coosa Valley of central Alabama. There is a single, questionable record from Calvert County, Maryland (Fowler and Orton, 1947; Cooper, 1953). The species occurs on Roanoke Island, North Carolina (Gaul and Mitchell, 2007), Harris Neck, St. Catherine's,

Distribution of *Hyla femoralis*

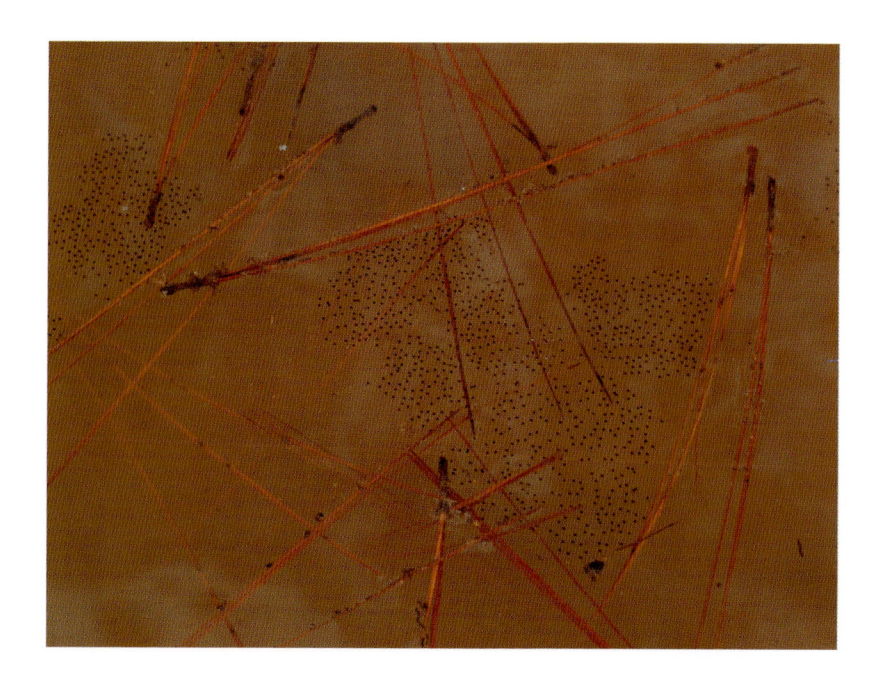

Eggs of *Hyla femoralis*.
Photo: Ronn Altig

Sapelo, and Cumberland islands, Georgia (Martof, 1963; Laerm et al., 2000; Shoop and Ruckdeschel, 2006; Dodd and Barichivich, 2007), and Marco and St. George islands, Florida (Duellman and Schwartz, 1958; Irwin et al., 2001).

Important distributional references include: Alabama (Mount, 1975; Redmond and Mount, 1975), Florida (Means and Simberloff, 1987), Georgia (Williamson and Moulis, 1994; Jensen et al., 2008), Louisiana (Dundee and Rossman, 1989), North Carolina (Dorcas et al., 2007), and Virginia (Tobey, 1985; Mitchell and Reay, 1999).

FOSSIL RECORD

Pleistocene fossils of *H. femoralis* are reported from northern Florida (Holman, 2003). Holman (2003) provided characters of the ilium that separate this species from other hylids.

SYSTEMATICS AND GEOGRAPHIC VARIATION

Hyla femoralis is most closely related to *H. squirella*, *H. gratiosa*, and *H. cinerea* (Hedges, 1986). In the laboratory, *H. femoralis* hybridizes successfully with *H. arenicolor*, *H. cinerea*, *H. avivoca*, *H. gratiosa*, *H. squirella*, and *H. chrysoscelis* (Mecham, 1965; Fortman and Altig, 1973; Pierce, 1975), although success depends on the sex of the parents (Mecham, 1965). Fortman and Altig (1973) described the hybrid tadpoles. Jensen et al. (2008) reported natural hybrids with *H. chrysoscelis*, and Anderson and Moler (1986) reported a single hybrid between *H. andersonii* and *H. femoralis* from Florida.

ADULT HABITAT

As its name implies, the Pine Woods Treefrog is closely associated with the great pine forests that once covered the Southeastern Coastal Plain. Today it is found in remnants of these habitats, as well as in mixed pine deciduous hardwood forests including mesic and xeric hammocks, bottomland and slope forest, creek swamps, dome swamps, cypress bays, pine flatwoods, and sphagnum bogs (Carr, 1940a; Enge et al., 1996; Enge and Wood, 1998, 2000, 2001; Smith et al., 2006). The unifying theme in habitat is the presence of pine trees, although it may be found on live oaks and magnolias (Harper, 1932). Occasional individuals are found among palmetto fronds. The species does not venture into the open wetlands favored by other southeastern hylids, although it may occur in shrubby wetlands surrounding wet prairies. For example, Babbitt et al. (2005) found no *H. femoralis* in wetlands > 150 m from the nearest woodlands.

TERRESTRIAL ECOLOGY

Pine Woods Treefrogs are active year-round, weather permitting (O'Neill, 1995). This species is largely arboreal on pine and hardwood trees, although individuals may be found on the ground as they make their way to and from breeding sites. Movement patterns and distance traveled are largely unknown, but in Florida sandhills, Dodd (1996) found *H. femoralis* 42–815 m (mean 317 m) from the nearest water body. They frequent tree canopies, but little is known of their activity in high branches. Pine Woods Treefrogs take refuge in treeholes, cracks in trees, titi knot holes, and under bark (Harper, 1932). They also are found in the structure of old dilapidated wood buildings, such as cabins, and in pine stumps. Harper (1932) speculated that only one frog occupied each tree, but this has not been confirmed. They sit with their limbs folded into the body to minimize moisture loss.

Pine Woods Treefrogs engage in a complex series of movements whereby the limbs are used to wipe the head and body with an extraepidermal layer of mucous and lipids, which help retard water loss (Barbeau and Lillywhite, 2005). Body wiping is important in allowing frogs to remain exposed in arboreal perches. As a result, evaporative water loss is significantly lower in *H. femoralis* than in many terrestrial anurans (Wygoda, 1984).

Pine woods Treefrogs are photopositive in their phototactic response, suggesting they use both sunlight and moonlight when feeding and going about their daily activities (Jaeger and Hailman, 1973). They are sensitive to light in the blue spectrum ("blue-mode response"), which apparently helps them orient toward areas of increasing illumination, such as the open area above a pond (Hailman and Jaeger, 1974). Pine Woods Treefrogs likely have true color vision.

CALLING ACTIVITY AND MATE SELECTION

Calling normally occurs from April to September when air temperatures are >21°C, although advertisement calls have been heard as early as February in north and central Florida (Carr, 1940b; O'Neill, 1995; Delis, 2001). In south Florida, breeding occurs from June to October (Duellman and Schwartz, 1958; Babbitt and Tanner, 2000), whereas in Georgia calls are heard from March to September (Jensen et al., 2008). There are two distinctive calls. One is made from the trees (to ca. 9 m in height) and occurs throughout the warm activity season, whereas the other is a more typical male advertisement call. Canopy calls occur both day and night, especially before, during, and after summer rain showers. At such times, the pine woods can erupt in choruses of *H. femoralis*. Males call from tree trunks at 1–4 m above the ground surface, located in or around breeding pools; they also may call from tufts of grass near the shore. They wedge the rear part of their body into the tree bark and extend the front legs fully while calling. This allows expansion of the large vocal sac, which can be nearly as large as the male's body.

At breeding ponds, males produce an advertisement call from logs and floating debris, or from the pond margin on the ground. The advertisement call can be deafening, with Harper (1932) noting that one may find "the incessant din quite oppressive to the auditory organs." Harper (1932) further noted that the "kek-kek-kek-kek" call can be repeated as long as 11 min without pausing. At the end of a series of "keks," a male may even increase the rapidity of the call. The call has a dominant frequency of 4,800 cps, a duration of ca. 2.35 sec, and a repetition rate of 12.2 notes/sec (Blair, 1958a). Large choruses (>50 calling males) are not uncommon (Delis, 2001). Most advertisement calling occurs at night.

BREEDING SITES

Breeding occurs in temporary ponds, woodland pools, and isolated freshwater marshes scattered throughout the pine woods and uplands. Sites include cypress ponds, roadside ditches, rain pools in fields, branch and creek swamps, Carolina Bays, depression marshes, cypress savannas, and cypress-gum ponds (Harper, 1932; Eason and Fauth, 2001; Greenberg and Tanner, 2005b; Liner et al., 2008). In south Florida, breeding habitats are associated positively with the presence of nearby woodlands (within 20 m) and negatively with low pH and the presence of fish (Babbitt et al., 2006). Not all breeding sites may be used in any one year. For example, Babbitt and Tanner (2000) found *H. femoralis* tadpoles in 5 of 12 potential breeding sites in south Florida over a 21-month sampling period. According to W. Brode (*in* Neill, 1958), they breed in brackish water around Bay St. Louis, Mississippi.

REPRODUCTION

Although calling can be prolonged, most breeding occurs in early summer, especially after heavy rainfall and moderate temperatures. Eggs are deposited in a small surface film containing 17–68 or more eggs per mass (Wright, 1932; Altig, 1972a) or just below the water's surface attached to vegetation or debris. A female deposits several masses, however, so that the total fecundity was estimated at 500–800 eggs per female; this estimate was based on counts of single ovaries of only two females and extrapolated (Wright, 1932). However, Delis (2001) found clutch sizes of 205–1,948 (mean 924) in Florida. Clutch size is not correlated with SUL, but it is positively correlated with clutch mass. Fertility is about 73%, with 97% of clutches showing at least some development (Delis, 2001). Eggs cannot survive dehydration, but they can withstand heat shock as high as 42°C; survivorship decreases at 8°C (Delis, 2001). Hatching occurs in three days.

LARVAL ECOLOGY

As with other hylids, the conditions faced by Pine Woods Treefrog larvae in breeding ponds have varied effects on larval life history traits. For example, larval *H. femoralis* have decreased survival and size at metamorphosis and an increased larval period when densities are high. They also have low survivorship and a decreased size at metamorphosis at low pH (4.3); there are no interactive effects between pH and density, however (Warner et al., 1991). In the presence of dragonfly (*Anax*) nymphs, *H. femoralis* larvae decrease activity levels in laboratory trials, but in the presence of newts (*Notophthalmus*) or predatory fish (*Lepomis*), activity actually increases (Richardson, 2001). Larvae are likely to encounter *Anax* in their breeding ponds, but not the latter predators.

Phenotypic plasticity is common in *Hyla femoralis* tadpoles and involves body shape, tail length, and larval coloration, all of which have different survival value depending on the environment in which the tadpole develops. Larval appearance can change in response to the presence of chemicals from potential predators. For example, larvae exposed to chemical cues from dragonfly (*Anax*) nymphs are deeper bodied with shorter tails than those not exposed to nymphs. Exposed larvae also develop an orange-red color pattern on the tail with a more pronounced black tail outline (LaFiandra and Babbitt, 2004). Larval phenotypes do not change in response to conspecific densities, but the density of predators has a significant effect on phenotypic plasticity. When larval densities are low in the presence of a predator (a high degree of threat to the tadpole), tadpole morphology and coloration are much more likely to change to one fostering higher survivorship than when larval densities are high (a comparatively low degree of threat) (McCoy, 2007). Tadpoles develop a similar phenotype to alarm cues from attacked conspecifics that develop in response to *Anax*, and the response to predator and alarm cues can be additive. Presumably there is an advantage to the induced phenotype (perhaps by allowing faster swimming), although such larvae experience slower growth and development than "normal" phenotypes. The induced phenotype will develop, even when food resources are low, suggesting a high survival value.

The larval period is estimated at 40–70 days (Wright, 1932). The maximum tadpole size is 33–36 mm TL, and newly metamorphosed froglets are 11.5–14 mm SUL based on a sample size of 4 (Wright, 1932).

Tadpole of *Hyla femoralis*.
Photo: Aubrey Heupel

Adult *Hyla femoralis*. Photo: Steve Johnson

DIET AND FEEDING
Allen (1932) noted that Pine Woods Treefrogs congregated around drops of pine sap and fed on insects drawn to the sap, and Carr (1940a) reported them congregating around dung piles and feeding on insects. Food records are scarce. Carr (1940a) noted feeding on moths and a crane fly. Duellman and Schwartz (1958) noted grasshoppers, a cricket, beetles, caddisflies, ants, wasps, and a spider in 20 *H. femoralis*. Presumably they consume a variety of insects.

PREDATION AND DEFENSE
The Pine Woods Treefrog is well camouflaged, with its bark-like dorsal pattern and tucked-in resting position. It has an ability to change color from brown to black to green to match the color of its resting spot. Crypsis and remaining immobile appear to be its main protection against visually oriented predators. *Hyla femoralis* also jump, crouch, inflate the body, and climb away from potential predators (Marchisin and Anderson, 1978). The stripe on the tail of the tadpole may function in counter-shading which aids in crypsis, and the induced red tail coloration may be aposematic, mimetic, or disruptive. Predators include a variety of snakes (*Coluber constrictor, Elaphe obsoleta, Thamnophis sauritus, T. sirtalis*) (Wright, 1932). Tadpoles are eaten by aquatic invertebrates and by some native fish (Baber and Babbitt, 2003).

POPULATION BIOLOGY
Delis (2001) found that female *Hyla femoralis* devoted a mean 29% of their body mass to reproduction, with one female devoting 51%. Pine Woods Treefrogs reach sexual maturity within three months after metamorphosis, with first breeding the following season. Populations can be large, but precise estimates are not available. From 1994 to 2001, Greenberg and Tanner (2005b) recorded 488 adults and 258 juveniles at 8 breeding ponds in central Florida.

COMMUNITY ECOLOGY
Interactions in larval hylid communities are complex, with larval *H. femoralis* often found in ponds containing other hylid larvae. In experimental mesocosms with *H. gratiosa* larvae, *H. femoralis* are smaller at metamorphosis than in mesocosms lacking this species, but survivorship and the duration of the larval period are not affected by interspecific competition. In turn, larval *H. femoralis* in mesocosms with a high pH decrease the survivorship of *H. gratiosa* larvae and cause an increase in larval duration

for that species (Warner et al., 1993). Under low pH conditions, interspecific competition at high larval densities actually resulted in an increase in survivorship and size at metamorphosis of *H. femoralis*. The size at metamorphosis is thus the developmental trait most strongly affected by competition with *H. gratiosa*. As a result, *H. gratiosa* has a more negative competitive effect on *H. femoralis* than vice versa (Travis, 1980; Wilbur, 1982), even though *H. gratiosa* seems to have a positive priority effect on subsequent development by *H. femoralis* (Warner et al., 1991). Differences between the results of Warner et al. (1991) and others (Travis, 1980; Wilbur, 1982) may be due to differences in experimental procedure.

DISEASES, PARASITES, AND MALFORMATIONS
Biting midges of the genus *Corethrella* feed upon *Hyla femoralis* and may be attracted by the frog's advertisement call (McKeever and French, 1991). As of 2007, amphibian chytrid had not been reported from this species (Rothermel et al., 2008).

SUSCEPTIBILITY TO POTENTIAL STRESSORS
pH. *Hyla femoralis* is less affected by variation in pH than in other hylids (Warner et al., 1991). Eggs can withstand a pH of 5.0, but survivorship is decreased < pH 7.0 (Delis, 2001).

UV light. Eggs and embryos exposed to UV light stress show decreased survivorship over ambient levels of UV light (Delis, 2001).

STATUS AND CONSERVATION
The loss of temporary breeding ponds undoubtedly has affected *H. femoralis* populations, but in general the species is still widespread and common. Any factors affecting breeding sites and associated uplands would impact this species. In that regard, Pine Woods Treefrogs tend to decline in urban and agricultural areas where temporary breeding ponds are not protected (Delis et al., 1996; Surdick, 2005). For example, Pieterson et al. (2006) noted declines from 2000 to 2004, based on call-monitoring surveys in heavily urbanized southwest Florida.

Vickers et al. (1985) implied that ditching cypress ponds had little effect on this species, but the authors only measured adult occupancy over a short period, not reproductive success. Surdick (2005) found them to occur frequently in silvicultural areas, but numbers can be quite reduced on clearcuts, even when uncut habitat is in close proximity (O'Neill, 1995). Initial studies of the effects of well draw-down zones in west-central Florida have noted little effect on tadpoles or call survey indices (Guzy et al., 2006). Some disturbances may not have significant impacts on Pine Woods Treefrogs. Following Hurricane Dennis in Florida, populations recovered quickly in some coastal areas that received hurricane overwash, despite temporary increases in salinity (Gunzburger et al., 2010). Pine Woods Treefrogs also readily populate forest habitats managed by prescribed fire (Langford et al., 2007).

Breeding habitat of *Hyla femoralis*, McIntosh County, Georgia. Photo: C.K. Dodd Jr.

Small numbers of *H. femoralis* may be captured in drift fences around breeding ponds (Enge and Marion, 1986; Dodd, 1992; Cash, 1994; Pechmann et al., 2001; Russell et al., 2002a; Baxley and Qualls, 2009), presumably because they can easily cross a fence. This species is best sampled using PVC pipes placed either in the ground or on nearby trees (O'Neill, 1995; Means and Franz, 2005).

A total of 55 *H. femoralis* were exported from Florida from 1990 to 1994 for the pet trade (Enge, 2005a).

Hyla gratiosa LeConte, 1857 "1856"
Barking Treefrog

ETYMOLOGY

gratiosa: from the Latin *gratiosa* meaning 'favored' or 'beloved.' The meaning in reference to the Barking Treefrog is unknown.

NOMENCLATURE

Synonyms: *Epedaphus gratiosus*

IDENTIFICATION

Adults. The Barking Treefrog is a robust treefrog with a background color of bright green, gray, brown, or purplish gray. The dorsum is evenly covered by dark spots, and there is a light stripe along the side of the body. This species can change color according to environmental conditions, and the spots may be very faded in some animals. In the bright green phase, a few scattered yellow spots may be apparent. Darker phases have dark spots and small gold dots dorsally. The head is short and broad with a rounded snout. The tympanum is relatively large. Toe pads are well developed, and webbing between the toes on the hind legs is extensive. Some webbing may be evident between the fingers on the front limbs. The skin is smooth to coarsely granular. A transverse fold is evident in the pectoral region. Upper forearms, femurs, and tibia have dark bars of varying widths. Venters are immaculate. Males have dark throats, although some females also have pigment on the throat.

Males are slightly larger than females (Delis, 2001); central Florida males are 53–67 mm SUL (mean 59.8 mm) and females are 52–67 mm SUL (mean 59.2 mm). South Florida males are 54.4–70.3 mm SUL (mean 61.5 mm) (Duellman and Schwartz, 1958). Alabama frogs are 52.8–62.3 mm SUL (mean 57.5 mm) (Brown, 1956).

Larvae. There are two color patterns of *Hyla gratiosa* larvae. Less than 29 mm TL, larvae are translucent or white, except for the black eyes and gut, and they have a very well-defined black saddle midway down the tail musculature. They have high, clear or very lightly pigmented tail fins. The overall effect is to make the saddle appear independent of the rest of the body. Small tadpoles also change color from day to night, with those collected at night having black opaque fins with a metallic luster that becomes translucent by day. For larvae > 30 mm TL, the saddle disappears, and the larvae develop a translucent coloration with a dark lateral stripe. The body is wide and deep. There is a yellow preorbital stripe extending from the eye to the tip of the snout. A pale postorbital stripe is present. Tail fins are clear or slightly speckled as the tadpole becomes larger. Tadpoles reach a TL of 50 mm in the field, although Dundee and Rossman (1989) reported laboratory-reared larvae of 71 mm TL. Altig (1972a) describes the tadpole.

Eggs. Eggs are greenish brown dorsally and yellowish ventrally. They are deposited in a continuous sheet and have a single gelatinous envelope (Altig, 1972a). Wright and Wright (1949; often repeated by others, e.g., Mount, 1975; Jensen et al., 2008) stated that eggs are deposited singly. The vitellus is 1.2–1.8 mm in diameter (Livezey and Wright, 1947; Travis, 1983; Delis, 2001), and the outer envelope is 2.3–5 mm in diameter (Livezey and Wright, 1947).

DISTRIBUTION

Hyla gratiosa occurs from the Coastal Plain of North Carolina throughout much of Florida (to Broward and Collier counties), then westward to the Florida Parishes of Louisiana. The range includes most of central Alabama northward to the Tennessee River, then east and west along this river in southern Tennessee. Isolated populations occur on the extreme southern tip of New Jersey (Black and Gosner, 1958; now extirpated,

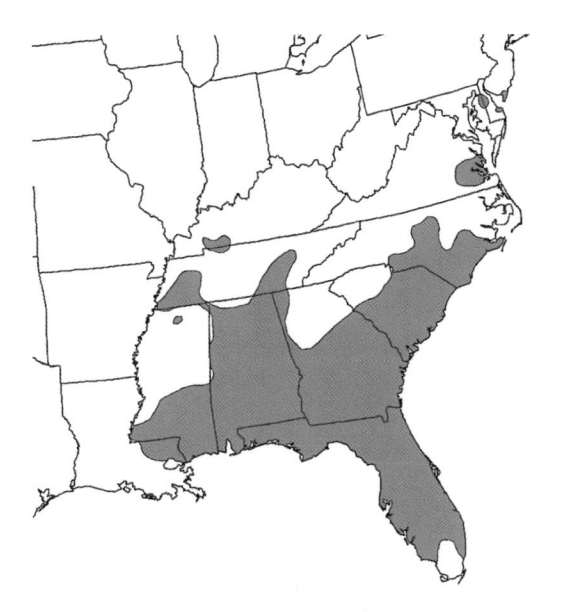

Distribution of *Hyla gratiosa*. The population in extreme southern New Jersey has been extirpated.

R. Zappalorti, personal communication), east of the Chesapeake Bay (White, 1988; White and White, 2007), southeastern Virginia, and western Kentucky and adjacent Tennessee, along the lower Cumberland River. The species occurs on Harris Neck, Ossabaw, Little Cumberland, and Cumberland islands, Georgia (Gibbons and Coker, 1978; Williamson and Moulis, 1994; Shoop and Ruckdeschel, 2006; Dodd and Barichivich, 2007).

Important distributional references include: Alabama (Mount, 1975), Delmarva (White, 1988; White and White, 2007), Florida (Means and Simberloff, 1987), Georgia (Williamson and Moulis, 1994; Jensen et al., 2008), Kentucky (Monroe and Taylor, 1972; Monroe and Giannini, 1977), Louisiana (Dundee and Rossman, 1989), North Carolina (Dorcas et al., 2007), Tennessee (Scott and Harker, 1968; Redmond and Scott, 1996), and Virginia (Tobey, 1985; Mitchell and Reay, 1999).

FOSSIL RECORD

Pleistocene fossils of *H. gratiosa* are reported from Alabama and Florida (Holman, 2003).

SYSTEMATICS AND GEOGRAPHIC VARIATION

Hyla gratiosa is most closely related phylogenetically to *H. cinerea* and *H. squirella* (Blair, 1958c; Maxson and Wilson, 1975; Hedges, 1986). Natural hybrids are found between *H. cinerea* and *H. gratiosa*, and backcrosses with the parental species produce fertile offspring (Mecham, 1960a, 1965; Fortman and Altig, 1973; Mount, 1975; Gerhardt et al., 1980; Schlefer et al., 1986; Lamb, 1987; Burkett, 1991). Hybridization in nature is not excessively common, as females prefer the calls of conspecific males (Oldham and Gerhardt, 1975), and there is a degree of ecological segregation based on differences in calling location and the permanency of the wetland used for breeding (Lamb and Avise, 1986; Lamb, 1987). Hybrids show a great deal of morphological variability, from completely intermediate between parental species to nearly completely resembling one of the parental species. Schlefer et al. (1986) and Lamb and Avise (1987) noted that 38 to >40% of known hybrids could be misclassified as pure parental species based on phenotype alone, had not genetic analyses proved otherwise. However, there is no evidence that morphological asymmetry is greater in hybrids than in the parental species, suggesting that hybrids are not subject to decreased levels of developmental stability (Lamb et al., 1990). Hybrids may exhibit spectral variability in call characteristics, thus resembling either of the parental species, although the ranges of variation tend to overlap with *H. cinerea* (Gerhardt et al., 1980). Most introgressive hybridization occurs between male *H. cinerea* and female *H. gratiosa*.

In the laboratory, *H. gratiosa* hybridizes successfully with *H. chrysoscelis* (Moore, 1955; Mecham, 1965; Fortman and Altig, 1973). Fortman and Altig (1973) described the hybrid tadpoles. Crosses with *H. avivoca* are generally unsuccessful (Mecham, 1965). Crosses with ♂ *H. femoralis* are unsuccessful, but the reciprocal cross is highly successful. Crosses with ♀ *H. squirella* are unsuccessful, but the reciprocal cross is highly successful (Mecham, 1965).

Specimens from south Florida do not have the prominent white line along the upper lip posterior to the tympanum that is seen in northern populations. There are a number of other minor color variations between south Florida frogs and their counterparts in the Carolinas and Louisiana (Duellman and Schwartz, 1958).

ADULT HABITAT

Hyla gratiosa is found in a variety of Coastal Plain and inland habitats, including pine flatwoods, longleaf pine sandhills, xeric hammock, mixed pine deciduous forest, and basin and depression marshes (Carr, 1940a; Enge and Wood, 1998, 2000, 2001). Habitats should contain temporary wetlands for breeding.

TERRESTRIAL ECOLOGY

During the day, adults are found normally in the surrounding forest, often in the canopy, and move to the margins of ponds and wetlands at dusk or after sunset. Males arrive before females. Individuals are frequently observed clinging vertically to vegetation surrounding a pond or wetland during the day. The limbs are tucked into the body to minimize moisture loss, and the eyes are closed tightly. Barking Treefrogs in such a position are often quite exposed to sunlight and the elements. After finishing calling, males either return to the forest canopy or remain at the breeding site, presumably to forage. All movements to and from breeding ponds occur at night (Todd and Winne, 2006).

After the breeding season, Barking Treefrogs seek shelter in dense forested habitats within close proximity to wetland breeding sites. They take up residence in trees at heights > 2.5 m above the ground, but ca. one-third of the frogs are found buried underground in shallow pits or burrows covered by soil or pine needles. With cold weather, individuals are found in cavities under downed logs or surface debris (e.g., Franz, 2005), buried under the soil surface, or in treeholes or similar protected sites. Carr (1940a) even found four individuals buried in an Indian mound > 1 m below the surface.

Females do not disperse away from wetlands as far as males, tending to move 30% shorter distances. The furthest distance Delis (2001) tracked a male was 330 m (37.6–329.8 m, mean 207.1 m) from its point of release; the farthest distance a female moved was 289 m (8.7–288.7 m, mean 121.6 m). Movements may be considerable over a 24 hr period, with one male moving 287.5 m. Movement is not correlated with rainfall. Frogs assume a dormant posture whereby the limbs are tucked tightly into the body and the eyes are closed. Juvenile dispersal occurs mostly at night (Todd and Winne, 2006).

Barking Treefrogs engage in a complex series of movements whereby the limbs are used to wipe the head and body with an extraepidermal layer of mucous and lipids, which help retard water loss (Barbeau and Lillywhite, 2005). Body wiping is important in allowing frogs to remain exposed in arboreal perches. As a result, evaporative water loss is significantly lower in *H. gratiosa* than in many terrestrial anurans (Wygoda, 1984).

Barking Treefrogs are photopositive in their phototactic response, suggesting they use both sunlight and moonlight when feeding and going about their daily activities (Jaeger and Hailman, 1973). They are sensitive to light in the blue spectrum ("blue-mode response"), which apparently helps them orient toward areas of increasing illumination, such as the open area above a pond (Hailman and Jaeger, 1974). Barking Treefrogs likely have true color vision.

CALLING ACTIVITY AND MATE SELECTION

Chorusing by *H. gratiosa* begins at dusk and continues through daybreak, peaking from 21:00 to 24:00 hrs (Mohr and Dorcas, 1999); calling does not occur during daylight. The call of *H. gratiosa* is a hollow-sounding "bonk-bonk-bonk" and has been compared to a dog barking. The initial part of the call is 150 cps, followed by a series of harmonics up to 8,000 cps (Blair, 1958a). Calling is primarily an aerobic activity and is energetically costly for a breeding male. Calls are produced at the rate of 3,600/hr with a mean duration of 0.16–0.18 sec at dominant frequencies of 450–1,900 Hz (Blair, 1958a; Prestwich et al., 1989). Calls are repeated at intervals of 0.8 sec. The trunk musculature associated with calling makes up 10% of the total body mass. Calls are produced in an omnidirectional pattern and have a long carrying capacity (to 2.5 km; Allen, 1932).

The calling season of *H. gratiosa* extends throughout the warm season from March to September, depending upon location. In south Florida, calling extends into October (Babbitt and Tanner, 2000). Calls also have been heard in October in Louisiana (Dundee and Rossman, 1989). It is likely that this species calls both well before and after the period when most reproduction occurs. Males call while floating in the water rather than from the surrounding vegetation or shoreline. Calls have been heard at temperatures as low as 17.8°C.

An individual male usually calls from 1 to 4 hrs per night and spends a median of 2–3 nights (nearly all <6–8; maximum 46) at the breeding pond within a season (Murphy, 1994a). Mortality is high while chorusing, with as many as 20% of calling males succumbing to predation. The number of matings in which a male participates (up to 13; Murphy, 1994a) and the probability of mating are correlated with chorus tenure. However, body size is not correlated with the amount of time at a pond. Males can increase their chorus tenure by 2.5 nights by moving to another pond and calling, but Murphy (1994b) found that only 16% of calling males frequented more than one breeding pond. The number of Barking Treefrogs in a chorus is often small (<20), and choruses of 10–49 frogs are uncommon, at least in central Florida (Delis, 2001).

Body condition plays an important role in male mating success, with those males in good condition remaining longer at a site than those that lose weight rapidly. On any given night, a male that spends an intermediate amount of time at a pond had a longer chorus tenure than those that spent either short or long amounts of time calling. The energy cost of calling is high and is the primary determinant of chorus tenure (Murphy, 1994b); those males that maximize their energy expenditure through time tend to have higher mating

Early stage tadpole of *Hyla gratiosa*. Photo: Gabe Miller

Late stage tadpole of *Hyla gratiosa*. Photo: Gabe Miller

success. Thus, a male that is able to maintain his body weight is in better condition to remain at a chorus. Regardless of mating success on any one night, males frequently return the next night or for several nights thereafter.

Females approach a calling male in order to initiate amplexus. She nudges him directly, after which he stops calling and amplexes her. Satellite males may be present, but they generally do not intercept females or try to disrupt amplexus. Amplexus lasts for up to 5 hrs. As a result, a male will mate only once per night.

BREEDING SITES

Breeding sites are temporary to semipermanent ponds. Barking Treefrogs are frequently found in Carolina Bays, depression marshes, cypress savannas, and cypress-gum ponds (Liner et al., 2008). Breeding sites may or may not contain grasses and emergent vegetation around the margins, or they may have cypress and black gum trees throughout. Water depth can be > 1 m. In south Florida, breeding habitats are associated positively with the presence of nearby woodlands (within 20 m) and pond depth and negatively with the presence of fish (Babbitt et al., 2006). Not all breeding sites may be used in any one year. For example, Babbitt and Tanner (2000) found *H. gratiosa* tadpoles in 8 of 12 potential breeding sites in south Florida over a 21-month sampling period. According to W. Brode (*in* Neill, 1958), they breed in brackish water around Bay St. Louis, Mississippi.

REPRODUCTION

The breeding season is prolonged, beginning in spring, with tadpoles of different sizes found within a pond all summer, hydroperiod permitting. In Alabama and central Florida, for example, the breeding season is April to July, but its actual length varies annually (Mount, 1975; Delis, 2001). Eggs are deposited in a continuous sheet covering ca. 500 cm^2. Other reports have eggs being deposited singly on the substrate or in small clumps (R. Altig, personal communication). It might be that eggs are initially oviposited in a film or in small clumps that quickly break apart and submerge. Livezey and Wright (1947) give a clutch size of 2,084, although Delis (2001) found clutch sizes of 557–4,034 (mean 1,990) in Florida. Clutch size is not correlated with SUL, but it is positively correlated with clutch mass.

Fertility is about 65% in the field, with 77% of clutches showing at least some development (Delis, 2001). Eggs can survive dehydration and heat shock as high as 42°C, but survivorship decreases significantly at 8°C (Delis, 2001). Hatchlings are 5.9–6.2 mm TL (Travis, 1983).

LARVAL ECOLOGY

Larvae prefer midwater habitats during the day. Barking Treefrog larvae grow quite rapidly through the early larval stages in order to reach a size that reduces predation risk. As in most anurans, environmental conditions affect larval growth, size at metamorphosis, and the duration of the larval period. For *H. gratiosa*, survival and size at metamorphosis decrease with increasing larval density, and the duration of the larval period increases (Morin, 1983; Warner et al., 1991, 1993). Likewise, high densities of heterospecifics (*Scaphiopus holbrookii*, *Lithobates sphenocephalus*, *Anaxyrus terrestris*, *Hyla chrysoscelis*) also have adverse effects on larval growth of *H. gratiosa* (Morin, 1983). High pH generally has no effect on size at metamorphosis, but low pH decreases survival and size at metamorphosis (Warner et al., 1991). Food affects development in several ways. A low food level decreases larval growth and mean size at metamorphosis and increases the duration of the larval period (Blouin, 1992a). This is important because the length of the larval period is correlated with size at metamorphosis, an important fitness trait, but only at low food levels (Travis, 1984). At normal food levels, low temperatures during development increase the duration of the larval period which in turn increases the size at metamorphosis compared to larvae that metamorphose faster at high temperatures (Blouin, 1992a).

Larval growth rates also can vary among kin groups in the latter stages of development, although there does not appear to be any effect early in development. This suggests that larvae can respond to decreasing hydroperiod during development, but that some tadpoles will be more effective than others in their response, indicating phenotypic variation in fitness potential depending upon hydroperiod. Higher growth rates lead to shorter larval periods and smaller body size at metamorphosis, but this allows some larvae to metamorphose before pond desiccation (Travis, 1980, 1983). Interestingly, experimental ma-

Adult Hyla gratiosa, spotted phase. Photo: Alan Cressler

Adult H. gratiosa, green phase. Photo: Dirk Stevenson

nipulation of hydroperiod has no effect on larval period and body size by itself, but it may result in increased density, which in turn affects these traits (Leips et al., 2000). As expected, the temporary pond-breeding *H. gratiosa* is more plastic in its response to varying environmental conditions than the permanent pond-breeding *H. cinerea*.

Travis (1983) recorded annual variation in reproductive traits among *H. gratiosa* families, such as initial body size of larvae, which in turn was correlated with female body size. Growth rates also varied considerably but were not correlated with initial body size. In addition, there is no correlation between mean egg size and the average growth rate of a female's offspring. Early growth rates of larvae are inversely correlated with the duration of the larval period, and larvae that have the longest larval periods naturally grow to larger sizes at metamorphosis. Phenotypic plasticity in reproductive traits allows *H. gratiosa* to respond to environmental conditions, which vary considerably from one year to the next.

Another factor affecting tadpole ecology is the presence of predators. In experimental field enclosures, predators such as dragonfly (*Tramea*) nymphs and salamanders (*Ambystoma opacum*, *Notophthalmus viridescens*) reduce survival by as much as 24% (Travis et al., 1985b) and increase the duration of the larval period (Morin, 1983). However, the effect varies by tadpole density, with low densities resulting in negligible effects on survival. Indeed, growth rates of *H. gratiosa* larvae can actually increase as predators remove competitive larvae (Morin, 1983). In the presence of dragonfly (*Anax*) nymphs, newts (*Notophthalmus*), or predatory fish (*Lepomis*), *Hyla gratiosa* larvae decrease activity levels in laboratory trials. Larvae may encounter these predators in their breeding ponds.

Multiple predators result in an additive effect, but in general predator effects are mediated by tadpole size; the larger the tadpole, the more likely it is to survive (Caldwell et al., 1980; Travis et al., 1985b). Caldwell et al. (1980) also noted that larval *H. gratiosa* tended to prefer midwater habitats during the day and have a very disruptive color pattern when young. Habitat differences and cryptic pattern thus help larvae avoid even low predation pressure. Survivorship is thus related to conspecific density, growth rates, and the types and numbers of predators within a pond community. These factors naturally change from pond to pond as well as annually.

There are conflicting estimates of the duration of the larval period. Wright (1932) estimated it at 41–65 days, but Dundee and Rossman (1989) reported laboratory-reared larvae required 131 days to metamorphose. Newly transformed froglets are 18–23 mm SUL (Wright, 1932).

DIET
The diet likely consists of a variety of invertebrates, but no specific data are available.

PREDATION AND DEFENSE
Tadpoles are cryptically or disruptively colored, especially during the vulnerable small stages. These small larvae "hang" with their heads held high in the midwater column, where they exhibit an immobility response when disturbed that does not draw the attention of a predator (Caldwell et al., 1980). The black saddle on a translucent background also helps to obscure the outlines of the tadpole. Postmetamorphs are cryptically colored.

Predators likely include wading birds, water snakes (*Nerodia* sp.), and turtles (*Deirochelys reticularia*). Larvae are probably eaten by a variety of aquatic invertebrates and vertebrates.

POPULATION BIOLOGY
Delis (2001) found that female *Hyla gratiosa* devoted a mean 21% of their body mass to reproduction, with one female devoting 32%. Barking Treefrogs probably breed for the first time the second summer after metamorphosis (Delis, 2001).

Breeding habitat of *Hyla gratiosa*, Putnam County, Florida. Photo: C.K. Dodd Jr.

COMMUNITY ECOLOGY

Hyla gratiosa often breeds in ponds used by other hylid frogs. The interaction between species under varying environmental conditions can affect life history attributes associated with larval development. For example, adding larval *H. femoralis* to an experimental mesocosm containing *H. gratiosa* at high pH decreases the survivorship of *H. gratiosa* and lengthens the duration of the larval period (Warner et al., 1993). However, interspecific competition with *H. femoralis* larvae only causes a decrease in the size at metamorphosis when *H. gratiosa* is at higher densities. *H. gratiosa* tadpoles also have a significant negative effect on the rate of metamorphosis of *H. femoralis* tadpoles (Wilbur, 1982). The larval period is thus the most strongly affected developmental trait affected by competition with *H. femoralis*. As a result, *H. gratiosa* has a more negative competitive effect on *H. femoralis* than vice versa, perhaps due to the larger body size of the Barking Treefrog larvae. Faster growing *H. gratiosa* grow to larger sizes later in development and thus out-compete *H. femoralis* for limited food resources (Travis, 1980).

H. cinerea and *H. gratiosa* often use the same ponds for breeding. Differences in calling position and in the advertisement calls of males are usually sufficient to prevent widespread hybridization between these closely related species. The potential for hybridization with *H. cinerea* occurs as *H. gratiosa* females move through the surrounding perimeter of calling *H. cinerea* males (Lamb, 1987). Intercepted females may successfully produce hybrid offspring with ardent Green Treefrog males. Hybrids, in turn, select calling positions intermediate (e.g., along the shoreline) between the parental species.

DISEASES, PARASITES, AND MALFORMATIONS

Biting midges of the genus *Corethrella* feed upon *Hyla gratiosa* and may be attracted by the frog's advertisement call (McKeever, 1977; McKeever and French, 1991). An unidentified microsporidian (*Glugea* sp. ?) has been reported in larvae from Florida and a myxozoan parasite (*Myxidium* sp.) has been reported from Georgia (Green and Dodd, 2007). As of 2007, amphibian chytrid had not been reported from this species (Rothermel et al., 2008).

SUSCEPTIBILITY TO POTENTIAL STRESSORS

pH. Eggs cannot withstand a pH of <5.0 (Delis, 2001).

UV light. Eggs and embryos exposed to UV light stress show decreased survivorship over ambient levels of UV light (Delis, 2001).

STATUS AND CONSERVATION

The loss of temporary breeding ponds undoubtedly has affected *H. gratiosa* populations, but in general the species is still widespread and common. Any factors affecting breeding sites and associated uplands would impact this species. In that regard, Barking Treefrogs tend to decline in urban, agricultural, and silvicultural areas where temporary breeding ponds are not protected (Delis et al., 1996; Surdick, 2005), but they may do reasonably well in certain managed forests (Bennett et al., 1980; Hanlin et al., 2000). Initial studies of the effects of well draw-down zones in west-central Florida noted little effect on tadpoles or call survey indices (Guzy et al., 2006). Pieterson et al. (2006), however, noted declines from 2000 to 2004, based on call monitoring surveys in heavily urbanized southwest Florida. Some disturbances may not have significant impacts on Barking Treefrogs. In Florida following Hurricane Dennis, populations recovered quickly in some coastal areas that received hurricane overwash, despite temporary increases in salinity (Gunzburger et al., 2010). The species is considered rare and protected in Delaware, Maryland, and Virginia (Pague and Mitchell, 1987; White and White, 2007).

Mecham (1960a) noted that habitat modification probably led to increased contact between *H. cinerea* and *H. gratiosa* at a well-studied hybridization site near Auburn, Alabama. By decreasing surrounding vegetation and constructing permanent ponds in an area where they did not naturally occur, an expanding population of *H. cinerea* had an increased chance of coming into contact with the resident population of *H. gratiosa*.

Small numbers of *H. gratiosa* may be captured in drift fences around breeding ponds (Gibbons and Bennett, 1974; Enge and Marion, 1986; Cash, 1994; Pechmann et al., 2001; Greenberg and Tanner, 2005b; Baxley and Qualls, 2009), presumably because they can easily cross a fence.

This species is better sampled using PVC pipes placed either in the ground or on nearby trees.

A total of 12,242 *H. gratiosa* were exported from Florida from 1990 to 1994 for the pet trade (Enge, 2005a).

Hyla squirella Bosc, 1800
Squirrel Treefrog

ETYMOLOGY
squirella: from the English word for squirrel and the Latin *ella* meaning 'diminutive' or 'little.' The name is in reference to the chattering squirrel-like call of the male.

NOMENCLATURE
Synonyms: *Auletris squirella*, *Calamita squirella*, *Dendrohyas squirella*, *Hyla delitescens*, *Hyla flavigula*, *Hyla goini*

IDENTIFICATION
Adults. *Hyla squirella* is a mostly small smooth-skinned green treefrog that lacks the conspicuous dorsolateral white body stripe of *H. cinerea*. It is also a more slender frog, with proportionally longer limbs. Squirrel Treefrogs may be blotched or uniform in color, with a green, gray, brown, or light tan background color. Spots or yellow flecks may be present. The dorsal coloration changes with temperature, activity, and stress, often rapidly. Dark blotched patterns may cause confusion with *H. femoralis*, but *H. squirella* lacks the bright orange spots seen on the backsides of the thighs of the Pinewoods Treefrog. There is a white line on the upper lip and under the eye, which may extend across the shoulder. The venter is a translucent white, and there may be yellow coloration in the groin. Toe pads are distinct, and the toes have some webbing at their base. Males have a deep yellow throat in the breeding season; female throats are lighter.

Females are usually larger than males, with a maximum SUL of 45 mm (Jensen et al., 2008). Alabama frogs are 24.7–35 mm SUL (mean 29.2 mm) (Brown, 1956), whereas Florida adults are 24–35 mm SUL (mean 28.7 mm) (Zacharow et al., 2003). In north Florida in 1982, Brugger (1984) recorded females 28–36.6 mm SUL, whereas males were 28.6–34 mm SUL; in 1983, females were 28–40.1 mm SUL, whereas males were 21–42 mm SUL. Mean SULs of males were 33 mm (Charleston, South Carolina), 31.6 mm (Marion County, Florida), 33.9 mm (Miami, Florida), and 26.6 (Collier County, Florida); corresponding mean SULs of females were 33.4 mm (Charleston, South Carolina), 33.4 mm (Marion County, Florida), 31.5 mm (Miami, Florida), and 27.8 mm (Collier County, Florida) (Duellman and Schwartz, 1958). In south Florida, males are 24.2–43 mm SUL (mean 32.6 mm) and females 23.3–35.1 mm SUL (mean 29 mm) (Duellman and Schwartz, 1958).

Larvae. Tadpoles are drab green to brown, medium sized, and deep bodied, reaching a maximum of 32–38 mm TL. A pale postorbital stripe is present. Tails are not distinctly striped. The dorsal tail fin is somewhat clear; it is lower in height than the tail musculature and is marked with dark spots or blotches. The tail musculature is light to olive and irregularly pigmented. Tail tips are acuminate. The throat and belly are yellowish with a black center. A tail flagellum is not well developed. Siekmann (1949) described the tadpole, and Orton (1947) and Altig (1972a) described early development.

Eggs. Eggs are small and brown dorsally and white to cream ventrally. They are deposited singly or in small clumps and have two gelatinous envelopes. The vitellus is 0.8–1 mm in diameter; the inner envelope is 1.2–1.6 mm in diameter; and the outer envelope is 1.4–2 mm in diameter (Livezey and Wright, 1947).

DISTRIBUTION
The range of *Hyla squirella* extends from southeastern Virginia throughout peninsular Florida, and west on the Coastal Plain to south of Corpus Christi Bay in Texas. The range extends northward in eastern Texas to extreme southeastern Oklahoma and through the Coosa Valley of central Alabama. Isolated populations occur in northern Virginia, the Piedmont of North and South Carolina (Brown, 1992), and Oktibbeha

County, Mississippi. This species has been introduced to the Bahamas (Kraus, 2009). Reports of this species from Vermont (Thompson, 1842), New York, and Massachusetts (Allen, 1868) undoubtedly refer to newly metamorphosed or green *H. versicolor*.

Hyla squirella is frequently found on islands, including: Dauphin Island, Alabama (Jackson and Jackson, 1970); throughout the Florida Keys, Egmont Key, St. George, and St. Vincent islands, Florida (Duellman and Schwartz, 1958; Lazell, 1989; Franz et al., 1992; Irwin et al., 2001); Harris Neck, Tybee, Little Tybee, Wassaw, Ossabaw, St. Catherine's, Blackbeard, Sapelo, Little Cumberland, and Cumberland islands, Georgia (Martof, 1963; Gibbons and Coker, 1978; Williamson and Moulis, 1994; Laerm et al., 2000; Shoop and Ruckdeschel, 2006; Dodd and Barichivich, 2007); Horn and Cat islands, Mississippi (Allen, 1932); Shackleford Banks, and Smith, Hatteras, Bodie, and Roanoke islands, North Carolina (Lewis, 1946; Engels, 1952; Gaul and Mitchell, 2007); and Kiawah Island, South Carolina (Gibbons and Coker, 1978).

Important distributional references include: Alabama (Mount, 1975; Redmond and Mount, 1975), Florida (Duellman and Schwartz, 1958), Georgia (Williamson and Moulis, 1994; Jensen et al., 2008), Louisiana (Dundee and Rossman, 1989), North Carolina (Dorcas et al., 2007), Texas (Hardy, 1995; Dixon, 2000), and Virginia (Tobey, 1985; Mitchell and Reay, 1999).

Distribution of *Hyla squirella*

FOSSIL RECORD
Pleistocene fossils of *H. squirella* are reported from northern Florida (Holman, 2003).

SYSTEMATICS AND GEOGRAPHIC VARIATION
Hyla squirella is most closely related phylogenetically to *H. gratiosa* and *H. cinerea* (Maxson and Wilson, 1975; Hedges, 1986), not *Pseudacris regilla* as suggested by Blair (1958c). In the laboratory, *Hyla squirella* hybridizes successfully with *H. arenicolor*, *H. cinerea*, *H. femoralis*, and *H. chrysoscelis* (Blair, 1958c; Fortman and Altig, 1973; Pierce, 1975), although success depends on the sex of the parents (Mecham, 1965). Fortman and Altig (1973) described the hydrid tadpoles. Unsuccessful crosses occur with *Pseudacris regilla* and *Hyla wrightorum*; crosses between ♀ *H. squirella* and ♂ *H. chrysoscelis* from Texas also were unsuccessful (Littlejohn, 1961a). Crosses with *H. avivoca* are unsuccessful (Mecham, 1965).

ADULT HABITAT
Hyla squirella is found in a variety of habitats, particularly in longleaf pine sandhills, mesic and xeric hammocks, coastal and inland hydric hammock, deciduous forest, dome and basin swamps, upland mixed forest, ecotones between forest and open areas, sloughs, wet prairies, south Florida rocklands, sawgrass marsh, and pine flatwoods (Carr, 1940a; Brown, 1956; Harima, 1969; Enge, 1998a; Enge and Wood, 1998, 2000, 2001; Meshaka et al., 2000; Smith et al., 2006). They also are found in the vicinity of lakes and impoundments, even if they do not breed there (Lichtenberg et al., 2006). Duellman and Schwartz (1958) even suggested they might occupy mangrove habitats in extreme south Florida. Habitats should contain small temporary wetlands for breeding.

TERRESTRIAL ECOLOGY
Squirrel Treefrogs are active year-round in the southern portion of their range, with activity in the north confined to the warmer seasons. Both adults and juveniles are active at night, much more so than *H. cinerea*. Foraging occurs within areas nearby diurnal retreat sites, to which the frogs return by day. Squirrel Treefrogs may use the same diurnal retreat for many days, returning each night after foraging. They may select new sites over an activity season, and even return to

Eggs of *Hyla squirella*, Alachua County, Florida. Photo: C.K. Dodd Jr.

previously occupied sites (e.g., Neill, 1957c). During daylight, Squirrel Treefrogs tuck their limbs into the body and close their eyes tightly, often while positioned on exposed locations such as palm fronds. They will not move unless disturbed, leading Hobson et al. (1967) to suggest they sleep during the day. Squirrel Treefrogs engage in a complex series of movements whereby the limbs are used to wipe the head and body with an extraepidermal layer of mucous and lipids, which help retard water loss (Barbeau and Lillywhite, 2005). Body wiping is important in allowing frogs to remain exposed in arboreal perches.

Adults move to breeding ponds in spring (mid-April in Florida) and remain in the vicinity of wetlands throughout the summer; they return to terrestrial habitats by autumn (e.g., late September to November in Florida; Goin, 1958; Boughton et al., 2000). During the nonbreeding season, Squirrel Treefrogs move well into the forest surrounding wetlands and may be found at considerable distances from water. For example, Dodd (1996) found *H. squirella* 446–914 m (mean 594 m) from the nearest water body in Florida sandhills. They occur more frequently on deciduous trees than pines, and are usually found at heights > 2 m off the ground (Boughton et al., 2000). Still, Beck (1948) recorded Squirrel Treefrogs from the middle of a lake and from water near shore in February.

Movements in autumn appear to be triggered by a combination of dry weather, a sudden drop in temperature, and rainfall associated with an approaching weather front. Frogs are active and forage throughout the winter, weather permitting, and are found in tree cavities, under bark and leaves, in palmetto grooves, in bromeliads (Neill, 1951), and in other secluded locations at night and during cold weather or the dry season (Zacharow et al., 2003; Campbell et al., 2010). They are frequently associated with buildings and sheds, and sometimes come out at night to feed on insects drawn to lights. Even during the winter, Zacharow et al. (2003) recorded movements as far as 238 m between retreat sites. The same retreat may be used repeatedly by a frog for a long time. Goin (1958) recorded activity at temperatures as low as 7°C, although most activity occurs at temperatures >15°C. When temperatures drop below 15°C, frogs move to retreats. The mean CTmax is 36.5°C and the CTmin is 2.9°C (John-Alder et al., 1988).

Dispersal by juveniles occurs immediately after metamorphosis, when large numbers may be found moving away from breeding ponds into mesic hammocks. They seek retreats in much the same sites as adults, especially in the tree canopy. Juveniles may be active at lower temperatures than adults, and juveniles have been observed in retreats at temperatures of 1.1–3.3°C.

Tadpole of *Hyla squirella*. Photo: Steve Bennett

Squirrel Treefrogs are photopositive in their phototactic response, suggesting they use both sunlight and moonlight when feeding and going about their daily activities (Jaeger and Hailman, 1973). They are sensitive to light in the blue spectrum ("blue-mode response"), which apparently helps them orient toward areas of increasing illumination, such as the open area above a pond (Hailman and Jaeger, 1974). Squirrel Treefrogs likely have true color vision.

CALLING ACTIVITY AND MATE SELECTION

Males usually arrive at breeding sites about sunset, and remain 2–4 hrs. Females usually arrive 2–3 hrs after sunset (Brugger, 1984). Choruses last 3–6 hrs. Calling occurs within 1–2 m of a breeding site, and males may change calling position several times during a night. On very warm nights choruses may not commence until 01:00 and last until 04:30. On such occasions, females arrive 1–2 hrs after males begin chorusing. Most males call 1–2 hrs per night. When chorus densities are high, calls may be loosely synchronous among neighboring males.

Calls may be heard at any time during the warm season, particularly from late spring to August (Duellman and Schwartz, 1958; Mount, 1975; Babbitt and Tanner, 2000; Jensen et al., 2008). However, advertisement calls have been heard as early as February in north Florida (Carr, 1940b) and from March to December in Louisiana (Dundee and Rossman, 1989). Air temperatures must be >19.5°C; warm rains stimulate extensive breeding activity. In northern Florida, Brugger (1984) found choruses from March to September, with most frogs calling from roadside ditches from May to September rather than from semipermanent ponds that were frequented earlier in the season.

Males have a "rain call" that is made from the tree canopy throughout the warm months, especially during high humidity and precipitation; the species is sometimes referred to as the "rain frog." The function of the rain call is not known. They also have a typical male advertisement call made from the margins of breeding ponds and pools. Calling is primarily an aerobic activity and is energetically costly for a calling male. Calls are produced at the rate of 6,000/hr with a mean duration of 0.2 sec at dominant frequencies of 1,200 and 2,950–3,457 Hz (Blair, 1958a; Prestwich et al., 1989). The fundamental frequency is ca. 133 cps, and there is a large series of harmonics ranging up to 7,500 cps (Blair, 1958a). As with other hylids, call characteristics vary with temperature and are primarily aerobic (Brugger, 1984). The trunk musculature associated with calling makes up 10% of the total body mass. Calls are produced in an omnidirectional pattern.

Calling *H. squirella* males are territorial and

Adult *Hyla squirella*, green phase. Photo: Aubrey Heupel

have an aggressive encounter call if an intruder approaches too closely. The minimum distance between calling males is ca. 110 cm, and males are usually evenly spaced. The encounter call is similar to the advertisement call but has an increased repetition rate, made by shortening the interval between calls (Fellers, 1979a). If an intruder continues to approach, a wrestling match ensues. Satellite males may be nearby (usually within 50 cm; Brugger, 1984), but females bypass these males as they approach a calling male. The female initiates amplexus by touching the calling male (Brugger, 1984). Amplexus is axillary. Males and females may remain in amplexus on the bottom of a pond for more than 2 hrs prior to egg deposition, which takes 35–65 minutes to complete (Brugger, 1984). The SULs of amplexed pairs are not correlated, suggesting random mating rather than selection based on size.

BREEDING SITES

Breeding occurs in small temporary pools, ditches, and shallow grassy wetlands in fields, and is associated with summer precipitation. They are frequently found in Carolina Bays, depression marshes, cypress savannas, pools in pastures, and cypress-gum ponds (Babbitt et al., 2005; Liner et al., 2008). They are capable of breeding in very small pools, and almost any available small water body may contain larvae. Duellman and Schwartz (1958) even reported calling from a prairie adjacent to a marine canal. They do not normally breed in pools containing fish, but when they do, larval densities are much lower than in pools without fish. In south Florida, breeding habitats are associated negatively with the presence of nearby woodlands (preferred 21–100 m) and positively with conductivity (Babbitt et al., 2006). Although fairly ubiquitous, not all breeding sites may be used in any one year. For example, Babbitt and Tanner (2000) found *H. squirella* tadpoles in 11 of 12 potential breeding sites in south Florida over a 21-month sampling period. Frog populations calling from semipermanent ponds are larger than those calling from roadside ditches (Brugger, 1984). Brugger (1984) also found that about 25% of adults returned to the same breeding site in successive years, but that interpond movements were quite frequent.

REPRODUCTION

Breeding occurs throughout the warm spring and summer months, and males may amplex more than one female during the breeding season (Brugger, 1984). Eggs are deposited singly and are scattered on the substrate or pasted onto grass blades (Brugger, 1984). Other reports have eggs being deposited in a film or in small clumps (Wright, 1932). It might be that eggs are initially oviposited in a film or in small clumps that quickly break apart and submerge. Eggs may cling to one another. Wright (1932) reported clutches of 942 and 972 eggs, whereas Brugger (1984) recorded a mean of 1,059 eggs for 5 clutches observed in nature. In additional laboratory experiments, Brugger (1984) recorded a mean of 900 eggs per clutch (range 361–2,003). Female body size may be correlated with clutch size. Hatching occurs in 36–48 hrs.

LARVAL ECOLOGY

Larvae of *H. squirella* are often the most abundant tadpoles in small temporary pools. They are active foragers, even in the presence of invertebrate predators. Survival, however, is greater in habitats with complex vegetative structure than in less complex habitats (Babbitt and Tanner, 1997). Tadpoles become more difficult to find, and predators have reduced foraging success when cover is available. In the presence of dragonfly (*Anax, Pachydiplax, Tramea*) nymphs, *Hyla squirella* larvae decrease activity levels in laboratory trials, but in the presence of newts (*Notophthalmus*) or predatory fish (*Lepomis*) activity actually increases (Richardson, 2001; McCoy and Bowker, 2008). Larvae are likely to encounter dragonfly nymphs in their breeding ponds, but not the latter predators.

Environmental conditions affect larval growth, size at metamorphosis, and the duration of the larval period. For *Hyla squirella*, low temperatures during development increase the duration of the larval period, but have no effect on size at metamorphosis (Blouin, 1992a). In general, however, *H. squirella* larvae are as active, have slower growth rates, and metamorphose at smaller sizes than other temporary pond-developing hylids (Richardson, 2001).

Larvae, not infrequently, are found in coastal areas affected by salt spray or occasional overwash. According to W. Brode (*in* Neill, 1958),

Color variation of *Hyla squirella*. Photo: Steve Johnson

they breed in brackish water around Bay St. Louis, Mississippi. Webb (1965) observed tadpoles in a shallow pool with a salinity of 4.7 ppt in coastal North Carolina. The larval period is unknown, but possibly 40–50 days or more; newly transformed froglets are 11–13 mm TL (Wright, 1932).

DIET
H. squirella likely consume a wide variety of mostly arboreal insects. Duellman and Schwartz (1958) found beetles, small crayfish, a spider, an ant, and a cricket in a few south Florida Squirrel Treefrogs. Both males and females appear to feed during the breeding season (Brugger, 1984).

PREDATION AND DEFENSE
This species is cryptically colored and blends in well with its background. The ability to change color allows it to blend in with a variety of background substrates. Predators likely include a wide range of vertebrates. Larvae are readily eaten by some native fish (Baber and Babbitt, 2003).

POPULATION BIOLOGY
Breeding choruses can be small or large, depending on the breeding site used. Brugger (1984) found as many as 220 chorusing males in a single semipermanent pond, whereas roadside ditches contained 7–48 chorusing males. Brugger (1984) also found that operational sex ratios around a pond changed nightly, with more males arriving several days after a heavy rain. Immediately after a rain, the sex ratio was 2.6 males per female, whereas several days later it was 33 males per female.

DISEASES, PARASITES, AND MALFORMATIONS
Biting midges of the genus *Corethrella* feed upon *Hyla squirella* and may be attracted by the frog's advertisement call (McKeever and French, 1991). An unidentified myxozoan parasite (*Myxidium* sp.) has been reported from Georgia (Green and Dodd, 2007). As of 2007, amphibian chytrid had not been reported from this species (Rothermel et al., 2008).

SUSCEPTIBILITY TO POTENTIAL STRESSORS
No information is available.

STATUS AND CONSERVATION
This species is widespread, common, and even may be expanding its range into the Piedmont and elsewhere (e.g., Pague and Mitchell, 1987). It tolerates urban, suburban, and agricultural areas fairly well (Delis et al., 1996; Hanlin et al., 2000; Surdick, 2005; Muenz et al., 2006), and is frequently seen at night perched on buildings and lighted windows, feeding on insects. Still, Pieterson et al. (2006) noted declines from 2000 to 2004 based on call monitoring surveys in heavily urbanized southwest Florida, and occupancy is reduced in silvicultural areas (Surdick, 2005). Initial studies of the effects of well draw-down zones in west-central Florida noted little effect on tadpoles or call-survey indices (Guzy et al., 2006).

Breeding habitat of *Hyla squirella*. Charlton County, Georgia. Photo: C.K. Dodd Jr.

Disturbances may have significant impacts on populations of Squirrel Treefrogs. Following Hurricane Dennis in Florida, Squirrel Treefrogs had not recolonized wetlands inundated with hurricane overwash 3 yrs following the storm (Gunzburger et al., 2010). However, Squirrel Treefrogs readily populate forest habitats managed by prescribed fire (Langford et al., 2007). Roads may impact Squirrel Treefrogs (Dodd et al., 2004), but the small size of this species makes it difficult to assess mortality, as carcasses are quickly obliterated or scavenged.

Although numbers may change annually, small numbers of *H. squirella* may be captured in drift fences around breeding ponds (Gibbons and Bennett, 1974; Bennett et al., 1980; Enge and Marion, 1986; Dodd, 1992; Russell et al., 2002a; Greenberg and Tanner, 2005b; Baxley and Qualls, 2009), presumably because they can easily cross a fence. This species is better sampled using PVC pipes placed either in the ground or on nearby trees (Zacharow et al., 2003; Means and Franz, 2005; Campbell et al., 2010). Waddle (2006) found little effect of toe clipping on either survival or capture probabilities of Squirrel Treefrogs.

A total of 5,362 *H. squirella* were exported from Florida between 1990 and 1994 for the pet trade (Enge, 2005a).

Hyla versicolor LeConte, 1825
Gray Treefrog
Rainette versicolore

ETYMOLOGY
versicolor: From the Latin word *versi* meaning 'changing' or 'variable,' and the Latin word *color* meaning 'color.' The name refers to the ability of individuals to change color.

NOMENCLATURE
Synonyms: *Calamita verrucosus, Dendrohyas versicolor, Dryophytes versicolor, Hyla phaeocrypta, Hyla richardii, Hyla verrucosa, Hyla versicolor phaeocrypta*

There are many publications where the identity of the species of "Gray Treefrog" is uncertain between *H. versicolor* and *H. chrysoscelis* (see *H. chrysoscelis* account for example; Roble, 1979). Therefore, it may not be possible to identify the species involved in literature discussions of Gray Treefrog biology, especially when ranges appear to overlap. Many characteristics (morphology, calling season, behavior, some life history traits) may be applicable to both species. It is best to check the context of the original citation for confirmation.

IDENTIFICATION

This species is morphologically identical with *H. chrysoscelis*. It can be differentiated by the following characteristics: chromosome number (*H. versicolor* is tetraploid [n = 48] whereas *H. chrysoscelis* is diploid [n = 24]); call rate (*H. versicolor* has a slow trill [17–35 notes/sec], whereas *H. chrysoscelis* is fast trilling [34–69 notes/sec], depending on temperature); cytology (cells of *H. versicolor* are larger with more nucleoli than those of *H. chrysoscelis*) (Cash and Bogart, 1978). There are no morphological characters related to body proportions or coloration that accurately separate these species (Matson, 1990).

Adults. The Gray Treefrog is a small- to medium-sized frog with a distinctive lichen-like dorsal pattern; they appear to reach a larger size than their sister taxon *H. chrysoscelis*. The ground color is brownish to gray to greenish, with a distinct lichen-like pattern. The lichen coloration is composed of various gray to buff patches; sometimes the patches may be distinctively greenish. There is a conspicuous light patch or spot underneath the eye, and there may be a dark crescent dorsally behind the eyes. All toes are tipped by conspicuous toe pads; a mucous layer is produced by the toe pad surface cells, which causes the pads to become sticky and allows the frog to climb vertically (Green, 1981b). The rear toes are partially webbed, but there is only slight webbing between the fingers. Dorsally, the rear legs have a dark banding pattern. The inner thigh is bright yellow orange to orange and unspotted. Bellies are unpigmented, but the throat of a calling male is black.

Males generally are smaller than females. Sizes also vary geographically and perhaps from one pond to another. In Maine, adult males range in size from 50 to 54 mm SUL (Sullivan and Hinshaw, 1992) with females to 60 mm SUL (Hunter et al., 1999); in Virginia, adults are 44–51 mm SUL (mean 47.7 mm) (Hoffman, 1946); in Connecticut, males are 36–51 mm SUL (mean 46.3 mm SUL) and females 43–60 mm SUL (mean 49.9 mm SUL) (Klemens, 1993); and in Wisconsin, adults are 40–46 mm SUL (mean 42.6 mm SUL) (Jaslow and Vogt, 1977). Mean sizes of males in Ohio ranged from 43.2 to 46 mm SUL, and females 48.9–52.6 mm SUL, depending upon pond sampled (Gatz, 1981b).

Gray Treefrogs change from dark to light (and vice versa) depending on temperature and light conditions. At night, melanophores in the skin contract producing a light colored frog. As temperatures decrease or as ambient light increases, the melanophores expand producing a darker-colored frog (Edgren, 1954). Thus, a dark frog may change to a very light lichen-colored pattern rather quickly (Babbitt, 1937).

Metamorphs are usually green, gradually changing into the adult shades of gray and brown. The bright light spot under the eye is conspicuous, and should aid in the identification of recent metamorphs. In coloration, they readily match the substrates surrounding a breeding site, such as grass, tree bark, or soil substrate. The skin may be completely smooth or granulated. As the metamorphs age, there is less of a tendency to be green. Confusion over the color pattern may have led Allen (1868) to record this species as *H. squirella* in Massachusetts, a place far from the distribution of the Squirrel Treefrog.

Larvae. The description of the tadpole of *H. versicolor* is essentially the same as *H. chrysoscelis*; Altig (1970) does not differentiate the species. Gills are completely absorbed by six days after hatching. The general body coloration may be several shades of brown to olive green, with numerous small black and gold flecks scattered across the body and tail. The tail musculature does not have dorsal saddles. Tail fins may be colorless or blotched. There is a pair of preorbital stripes, but they are not distinctly outlined and only become apparent two weeks after hatching. The venter is immaculate anteriorly with some gold flecking, and the intestines may or may not be readily visible through the body wall. As the tadpole grows, the venter posteriorly becomes more cream colored with heavy intrusion of gold flecks. The iris is bright gold or bronze. Maximum size is 42 mm TL. Descriptions of larvae and or their mouthparts are in Hinckley (1880, 1881), Wright (1914, 1929), Babbitt (1937), Walker (1946), Gosner and Black (1957b), and Altig (1970).

There is a unique color morph of *H. versicolor* tadpoles that appears only when predators are relatively inactive at a breeding pond. Such tadpoles develop bright red tail fins with dark margins. When predators, such as dragonfly larvae, are absent, the bright red coloration does not develop. Bright tail fins are seen only on the oldest and largest larvae, usually within about two weeks of metamorphosis.

Eggs. Most descriptions of the eggs of Gray Treefrogs do not differentiate between *H. chrysoscelis* and *H. versicolor*. There appears to be some variation in egg size and total clutch size, but whether these variables are species-specific or reflect intraspecific variation is unknown. When deposited, eggs are drab in coloration and roughly 1.1–2 mm (mean 1.7 mm) in diameter (4–8 mm with the envelope, mean 5.2 mm). After a few hours the vegetal pole becomes white to cream or yellow, the extent of which increases with development. There is only a single jelly capsule surrounding the egg (illustration in Tyler, 1994). Eggs may be oviposited singly or in small packets of 4–40 eggs along the surface of submerged vegetation (depth 0.2–0.4 cm) or at the edge of the breeding site. Packets may be separated by 15–30 cm (Wright, 1914; Livezey and Wright, 1947). The total egg complement per female is approximately 1,000–2,600 eggs. Hatching occurs from two to five days after deposition. Egg descriptions are in Wright (1914), Livezey and Wright (1947), and Tyler (1994).

DISTRIBUTION

Determination of the distribution of this species is compounded by confusion with *H. chrysoscelis*. Many publications make no distinction between the two (e.g., Bragg, 1943a; Suzuki, 1951; Smith, 1961; Tobey, 1985; Dundee and Rossman, 1989; Hoberg and Gause, 1992; Phillips et al., 1999; Davis and Menze, 2000; Dixon, 2000; Hulse et al., 2001; Minton, 2001; Sievert and Sievert, 2006), and later distributional publications often refer to the "Gray Treefrog complex" rather than differentiate between the two morphologically identical forms. In some places, Gray Treefrogs are not sympatric, whereas in others they overlap to a greater or lesser extent. In determining distribution, I generally followed the range map in Holloway et al. (2006). To be absolutely certain about identification, it is best to carefully listen to the calls (slow trills vs. fast trills), obtain a karyotype (diploid vs. tetraploid), or examine cell size (see above).

The Gray Treefrog is found in Canada from western New Brunswick and southern Québec and Ontario west to southern and western Manitoba and eastern Saskatchewan. Pure populations of *H. versicolor* occur southward to southern New Jersey and Pennsylvania, west-central Virginia, eastern West Virginia, central Ohio, Indiana, and Illinois westward to northeast Missouri, and in scattered areas in eastern Wisconsin north to northern Minnesota and southern Manitoba. Isolated *H. versicolor* populations occur in western Wisconsin, northeastern Iowa, South Dakota (Kiesow, 2006), northwestern Arkansas, and east-central Oklahoma. The species overlaps with *H. chrysoscelis* in a broad area from Maryland and south-central Virginia to southern Ohio, and westward in an arc all the way from eastern Kansas to east Texas. *Hyla versicolor* also is found in Breckenridge, Hardin, and Meade counties, Kentucky (Burkett, 1989) and in Warren and Caswell counties, North Carolina (Dorcas et al., 2007). Additional mixed-species populations occur widely in eastern North Dakota, Wisconsin, Minnesota, and Iowa. In Wisconsin, *H. chrysoscelis* is more prevalent in the southeastern prairie regions of the state than *H. versicolor* (Vogt, 1981; Casper, 1998).

The distribution pattern suggests that the polyploid *H. versicolor* may be adapted to colder and dryer conditions (more northern, and at higher elevations) than *H. chrysoscelis*, and adaptation to harsher conditions may have played a role in the evolution and establishment of polyploidy in this species complex (Otto et al., 2007a). *Hyla versicolor* may be able to tolerate more variation in climatic conditions, leaving most of the range

Distribution of *Hyla versicolor*

of the progenitor species in more southerly areas of lower elevation, where climate extremes are not as common. The model used by Otto et al. (2007a), however, did not include northwestern populations of *H. chrysoscelis*; hence the applicability of the climate model origin may be limited to the northeastern tetraploid lineage of Holloway et al. (2006).

Gray Treefrogs occur on the island of Ile Perrot, Québec (McCoy and Durden, 1965), Long Island, New York (Overton, 1914), the Elizabeth Islands of Massachusetts (Lazell, 1974), Walpole Island in Lake St. Clair (Woodliffe, 1989), islands at the Lake Ontario end of the St. Lawrence River, the Apostle Islands in Lake Michigan, and islands in Georgian Bay of Lake Huron (Hecnar et al., 2002). Hecnar et al. (2002) found them on ca. 23% of 107 islands surveyed. In New Hampshire, they occur from 75 to 195 m in elevation (Oliver and Bailey, 1939).

Important distributional references include: Arkansas (Black and Dellinger, 1938; Trauth et al., 2004), Connecticut (Klemens, 1993), Delaware (Zweifel, 1970a; Otto et al., 2007a), eastern Canada (Bleakney, 1958a; Logier and Toner, 1961), Illinois (Brown and Brown, 1972b), Iowa (Hemesath, 1998; Oberfoell and Christiansen, 2001), Kansas (Hillis et al., 1987; Collins, 1993; Collins et al., 2010), Kentucky (Burkett, 1989), Maine (Hunter et al., 1999), Manitoba (Preston, 1982; Taylor, 2009), Maryland (Noble and Hassler, 1936; Harris, 1975; Otto et al., 2007a), Massachusetts (Lazell, 1974), Michigan (Bogart and Jaslow, 1979), Minnesota (Oldfield and Moriarty, 1994), Missouri (Johnson, 2000; Daniel and Edmond, 2006), New Brunswick (Cox, 1898, 1899; Bleakney, 1954; McAlpine, 1997a), New Hampshire (Oliver and Bailey, 1939; Taylor, 1993), New Jersey (Schwartz and Golden, 2002), New York (Gibbs et al., 2007), North Carolina (Dorcas et al., 2007), North Dakota (Wheeler and Wheeler, 1966), Ohio (Walker, 1946; Little et al., 1989; Pfingsten, 1998), Ontario (Johnson, 1989; MacCulloch, 2002), Pennsylvania (Little et al., 1989), Québec (Bider and Matte, 1996; Desroches and Rodrigue, 2004), Saskatchewan (Taylor, 2009), South Dakota (Kiesow, 2006), Texas (Flury, 1951; Hardy, 1995), Vermont (Andrews, 2001), Virginia (Hoffman, 1946; Little, 1983; Mitchell and Reay, 1999; Otto et al., 2007a), West Virginia (Little, 1983; Green and Pauley, 1987; Little et al., 1989), and Wisconsin (Jaslow and Vogt, 1977; Vogt, 1981; Casper, 1996; Mossman et al., 1998).

FOSSIL RECORD

There is no way to differentiate members of the *Hyla versicolor/Hyla chrysoscelis* complex osteologically, so there is no way to differentiate their fossils. Fossil members of the complex are found in Pleistocene (Irvingtonian) deposits from Nebraska and West Virginia. The complex also is found in Pleistocene (Rancholabrean) sites from Georgia, Kansas, Texas, and Pennsylvania (Holman, 2003).

SYSTEMATICS AND GEOGRAPHIC VARIATION

Previous suggestions that polyploidy in Gray Treefrogs was produced via a single speciation event among consistent progenitors are not supported based on a suite of analyses of molecular and call advertisement data. *H. versicolor* arose though polyploidy via several speciation events from *H. chrysoscelis*-like diploid ancestors and two other now extinct lineages of tree frogs. According to Holloway et al. (2006), these three groups then freely interbred to produce the single species *H. versicolor*. There are four distinct haplotype lineages, with *H. avivoca* also involved in the evolution of the tetraploid *H. versicolor*, but not in all four lineages. Today, however, gene flow is restricted between *H. avivoca* and the Gray treefrog complex. Despite the different origins of the lineages of tetraploid *H. versicolor*, the lineages are not reproductively isolated from one another. According to Keller and Gerhardt (2001), the reduction in pulse rate in *H. versicolor* is the direct consequence of ploidy level. Geographic variation in albumins (Maxson et al., 1977; Maxson and Maxson, 1978; Ralin, 1978), call characteristics (Gerhardt, 1974a; Ralin, 1968, 1977; Bogart and Jaslow, 1979), and allele frequencies (Ralin and Selander, 1979; Romano et al., 1987) undoubtedly stem from the different evolutionary histories of the progenitor species, as well as comparisons uncorrected for temperature (Johnson, 1966). Members of the *H. versicolor/H. chrysoscelis* complex are most closely related to *H. andersoni* and *H. avivoca* (Hedges, 1986).

Laboratory crosses between *H. versicolor* and *H. chrysoscelis* produce high rates of mortality (Johnson, 1959, 1963; Ralin, 1976). Crosses between female *H. versicolor* and male *H. squirella*,

Pseudacris cadaverina, and *P. fouquettei* hatch and may produce a tiny percentage of metamorphs. Crosses between female *Hyla versicolor* and male *H. cinerea, Pseudacris streckeri, P. clarkii,* and *P. regilla* do not reach metamorphosis; crosses between *Hyla arenicolor* or female *H. cinerea* and male *H. versicolor* are likewise unsuccessful (Pierce, 1975; Ralin, 1976).

Flury (1951) documented much interpopulation variation in color pattern and morphometrics within the *H. versicolor* complex in Texas. However, there was no evidence of variation, even within different biotic provinces, suggesting that different taxa might be involved.

ADULT HABITAT

Preferred habitats include southern mesic hardwoods, southern and northern lowland forest, boreal forest, northern mesic, and dry-mesic hardwoods (Vogt, 1981). In the upper Midwest, *H. versicolor* is considered a forest species, whereas *H. chrysoscelis* is more associated with prairies, grassland, and oak savanna habitats (Vogt, 1981; Oldfield and Moriarty, 1994; Knutson et al., 2000). Price et al. (2004) suggested that this species preferred "highly irregular cover class patches dispersed throughout large areas of forest."

As Grays Treefrogs normally do not venture far from breeding ponds, habitat variables influencing distribution apply both to breeding and nonbreeding habitats. In much of their range, they prefer sites with forest cover nearby (> 40 m from a breeding pond; Herrmann et al., 2005) but not necessarily directly surrounding a pond. Along the Great Lakes, for example, Gray Treefrogs are positively associated with the type of wetland and the relative area of grass and sedge cover, but negatively associated with the extent of forest cover within 100 m of breeding ponds; however, occupancy is positively associated with the amount of forest cover within 500 m of the pond. Within 3000 m, there is a negative correlation between the amount of urban habitat and recreational grassland.

TERRESTRIAL ECOLOGY

Gray treefrogs grow quickly, and show little growth after attaining adult sizes. Sexual maturity is reached in 2 yrs. According to Babbitt (1937), metamorphs grow to > 25 mm SUL before undergoing their first overwintering season. However, Wright (1932) recorded them at 20–30 mm SUL at 1 yr of age (mean 25 mm), 30–41 mm SUL at 2 yrs of age (mean 35 mm), and 41–51 mm SUL (mean 45 mm) thereafter. Differences could reflect differences in growth rates. During the first season following metamorphosis, Gray Treefrogs probably spend most of the time near the forest floor (Roble, 1979). As they get older, they ascend into the deciduous forest canopy where they spend most of the time in cracks, crevices, treeholes, woodpecker holes, and other types of cavities. They generally remain within ca. 200 m of the breeding pond, with most frogs living in close proximity to the pond; the maximum recorded distance a frog moved from a pond was 330 m in Missouri (Johnson, 2005). Females occupy areas farther out from the pond than males, and thus move over greater distances. Gray Treefrogs spend considerable time moving back and forth from breeding ponds to terrestrial/arboreal sites away from the pond, primarily to feed. The juxtaposition between forest cover and breeding pond is thus very important in population persistence.

In Missouri, Johnson (2005) found that there was significant variation in population genetic structuring, such that populations > 30 km apart were identifiable. Under 3 km, however, significant structuring was not evident. Johnson (2005) hypothesized that Gray Treefrogs were structured as classic metapopulations at regional distances, but that populations inhabiting a localized series of ponds formed a "patchy" metapopulation structure whereby some cohesiveness was established through interpond movement. Thus, populations are structured differently, depending upon regional scale of distance and adjacent habitats.

Gray Treefrogs become active in spring, with the timing depending on weather conditions. In Arkansas, individuals have been observed in early March (Trauth et al., 1990), and it seems likely they become active earlier when temperatures permit. The time they cease activity varies by latitude. In Massachusetts, New York, Ohio, and Connecticut, for example, they are active until October (Hinckley, 1880; Wright, 1914; Walker, 1946; Klemens, 1993). *Hyla versicolor* is active at temperatures between 15 and 34°C in New Jersey, with a CTmax of 38.7°C (John-Alder et al., 1988). However, they call at temperatures as low as 8°C in Ontario, and it is probable that they are active at temperatures lower than 15°C in other parts of their range.

Adults are able to survive freezing temperatures (5–7 days at –6°C) through the use of glycerol as a cryoprotectant (Storey and Storey, 1986), although subadults use both glycerol and glucose to survive freezing temperatures (Schmid, 1982; Storey and Storey, 1985). Costanzo et al. (1992) noted, however, that glucose was the sole cryoprotectant in Indiana populations of *H. versicolor*, so there may be some regional variation in the physiological means by which this species survives freezing temperatures. In addition, Gray Treefrogs mobilize glycerol differently from year to year (Layne and Stapleton, 2009). The body temperature stabilizes at 0°C with the heart continuing to beat at five beats per minute when the frog is frozen; the body temperature rises slowly in recovery, with an increase in heart rate as the body thaws (Layne and Lee, 1995). As much as 35–41.5% of the frog's body water may be contained as ice at such temperatures (Schmid, 1982; Storey and Storey, 1985), although it may reach 50% depending on temperature, freeze conditions, and season (Layne and Lee, 1989).

Gray Treefrogs are photopositive in their phototactic response, suggesting they use both sunlight and moonlight when feeding and going about their daily activities (Jaeger and Hailman, 1973). They are sensitive to light in the blue spectrum ("blue-mode response"), which helps them orient toward areas of increasing illumination, such as the open area above a pond (Hailman and Jaeger, 1974). Gray Treefrogs likely have true color vision.

CALLING ACTIVITY AND MATE SELECTION

Calling in Gray Treefrogs begins in spring and may extend throughout the summer months; most reproduction occurs during the early part of the breeding season. The timing of calling varies by latitude and environmental conditions. For example, calling begins in late April in Rhode Island (Anonymous, 1918) and in early to late May in Massachusetts and Ontario (Allen, 1868; Hinckley, 1880; Bertram and Berrill, 1997), and extends to June in Maine and Québec (Sullivan and Hinshaw, 1992; Lepage et al., 1997), July in Ontario and Wisconsin (Piersol, 1913; Bishop et al., 1997; Mossman et al., 1998), and to early August in New Hampshire (Oliver and Bailey, 1939). In Texas, Gray Treefrogs call from March to August (Wiest, 1982). Some literature records may confuse breeding calls with rain calls (see below).

Males call from ground-level or low shrubby vegetation surrounding a pond at ambient temperatures ranging between 13 and 26°C in Missouri (Gerhardt, 1978a), although some males call from emergent vegetation within a pond. In Ontario, males did not call at ambient temperatures <8°C (Bertram and Berrill, 1997). Wiest (1982) noted calling at 13–28.1°C over a 51-day period in Texas. In Maine, males did not make use of a particular calling location from one night to the next (Sullivan and Hinshaw, 1992). Choruses tend to be of short duration (12–23 days in Maine, Sullivan and Hinshaw, 1992; 13–37 days in Ontario, Bertram and Berrill, 1997), and only a few males (<10) may call in any particular chorus. In Indiana, Brodman and Kilmurry (1998) recorded choruses of from 0–8 calling males per night over a period of 30 days. Chorus size may be larger; Bertram and Berrill (1997) recorded a resident population of 30–77 calling males, with a mean of 10–22 males (1–42) calling in nightly choruses over a 3 yr period. When females were in the pond breeding, chorus size ranged from 5 to 40 calling males. Male size is not correlated with the extent of chorus activity.

Calling male *H. versicolor* are territorial (minimum distance between calling males about 70 cm), and the mating call is used not only to attract females but to establish calling male residency and spacing. Optimal calling perches generally have less vegetation surrounding them than perches of lesser quality, presumably allowing better acoustic transmission and space for amplexus (Fellers, 1979b). Such perches are occupied early in the season, and they generally remain occupied throughout the season. Calling perches are defended by dominant males, and both dominant and subordinate males have an encounter call (2.8 dB) when the subordinate approaches near the dominant male (call and encounter described by Overton, 1914). Most calling sequences featuring encounter calls last < 2 min, and may result in the subordinate male departing the area. Males will fight (wrestle, shove, kick, jump upon, headbutt) other males should a subordinate encroach upon an occupied territory; fights normally last

30–90 sec. If a male loses a fight, it will cease calling. Amplexus may last up to 4 hrs or more, a good portion of which is on or near the calling perch (Fellers, 1979a). When the dominant male is in the pond fertilizing eggs, a satellite male may occupy the dominant male's calling perch and resume calling. The mating sequence has been described in detail by Fellers (1979a).

There appears to be slight differences in call characteristics (duration, repetition rate, frequency) among populations of Gray Treefrogs from different portions of its range. These differences may, in part, be a result of the different phylogenetic history of the progenitor species to the Gray Treefrog complex. In addition, call characteristics vary with temperature, so comparisons of call characteristics must take this environmental parameter into consideration. Indeed, female Gray Treefrogs will choose males calling from a similar temperature as that experienced by her, rather than conspecific calls recorded at a different temperature. Thus, male calls and female preferences tend to be coupled by temperature (Gerhardt, 1978a). In addition, male calling rates, the number of pulses per call, and pulse effort vary individually, may change from one night to the next, and likely decrease in intensity as the breeding season progresses (Runkle et al., 1994). Males also decrease the number of hours spent in energetically expensive calling later in the season compared with earlier in the season.

Calls of *H. versicolor* have mean pulse rates between 24.2 and 29.1 pulses/sec at 24°C, depending on location (Zweifel, 1970a; Jaslow and Vogt, 1977; Ralin, 1977; Gerhardt, 1978a; Bogart and Jaslow, 1979). In Maine, pulse rates varied between 20 and 21 pulses/sec at 20°C (Sullivan and Hinshaw, 1992), whereas in Illinois they were 14.3–23.5 pulses/sec at temperatures of 16–22.2°C (Brown and Brown, 1972b). In Michigan, the mean call duration was 0.57 sec (range 0.34 to 0.73 sec) at a dominant frequency of 2.29 (range 2.13 to 2.55) kHz (Bogart and Jaslow, 1979), whereas in Maine it was 0.63–0.81 sec at a dominant frequency of 1.94–2.05 kHz (Sullivan and Hinshaw, 1992). Intercall intervals in Texas are 0.24–0.38 sec, calls last 0.14–0.24 sec, and dominant frequencies range between 2.10 and 2.39 kHz (Pierce and Ralin, 1972). In Maryland, calls range from 94 to 108 dB (mean 103 dB) (Fellers, 1979a). The call of the Gray Treefrog is among the loudest calls ever measured for North American frogs (Gerhardt, 1975).

Females may not be in attendance during every night of calling. In Maine, Sullivan and Hinshaw (1992) found them on 33–65% of the nights sampled over a several year study. Presence of females tends to be correlated with chorus size and density. Female Gray Treefrogs initiate mating by slowly approaching a calling male and touching him. She will then turn at a 90° angle, whereupon the male clasps her. Females are most responsive to frequencies of 2.0 kHz when presented with the opportunity to respond to a variety of call frequencies; response rates are close to 0 at > 5.0 kHz. As a result, female Gray Treefrogs are not attracted to the calls of male Cope's Gray Treefrogs. In *H. versicolor*, both the pulse rise time (especially when > 9 ms) and the slower pulse rate (20 pulses/sec) are important cues for female mate selectivity (Bush et al., 2002).

Females appear to discriminate among conspecific males, preferring males with higher pulse rates and longer calls (Klump and Gerhardt, 1987). Such males presumably would be fitter than males with low pulse rates or short calls, since the former are energetically more costly to produce. Female preference is thus important to both prevent hybridization and to select the best males for breeding. However, evidence that females choose larger males for mating is mixed (Fellers, 1979b; Gatz, 1981b).

Calling usually begins shortly before dusk, and continues for two to four hours; calls may occur as late as 03:00 in northern populations. Mating calls are uttered at 3–7 sec intervals. In Connecticut, males calling early in the evening have call rates of 600/hr, whereas this rises to 1,400 calls per hour by 21:00 hrs (Taigen and Wells, 1985). In one night, a male may make 1,200–1,300 calls per hour over a 2–4 hr period (Wells and Taigen, 1986). As with Cope's Gray Treefrogs, *H. versicolor* males do not remain at a pond throughout the breeding season, but appear to prefer short stays interspersed by movements to terrestrial feeding sites. This is not surprising, as calling is a very energetically expensive activity in Gray Treefrogs (Taigen and Wells, 1985), and males have a significant loss of body mass after only a few nights of calling (Fellers, 1976). Thus, males probably need to feed before resuming a calling bout.

Call activity is influenced by chorus size. In dense choruses, males call about twice as long as single males, but the call rate decreases somewhat proportionately (Wells and Taigen, 1986). This results in a trade-off such that calling effort and energetic costs become independent of chorus density. However, male–male competition in choruses produces changes in energetic costs, inasmuch as males tend to respond in kind (long call responses to long calls, for example) to the call characteristics of adjacent males. Thus, the social interactions of calling males produce effects on the ability of males to call and on the duration of calling bouts. Wells and Taigen (1986) suggested that calling time may be limited in males by the rate at which energy reserves are used up, but Schwartz and Rahmeyer (2006) found no evidence of this. Instead, they suggested that pulse rate rather than duration might better reflect limitations of the cardiovascular and respiratory systems.

Calling and aggressive behavior are influenced significantly by neuro-peptides, particularly arginine vasotocin (AVT) (Semsar et al., 1998; Klomberg and Marler, 2000). AVT appears to be important in calling and male–male competition by increasing call duration and the number of pulses in a call. It does not influence dominant frequency, call rate, or pulse effort. Subordinate males under the influence of AVT call and behave like dominant males. In many animals, AVT interacts with testosterone to influence behavior, and in treefrogs it may serve to mediate individual behavioral encounters between males.

In addition to the breeding and male–male encounter calls, Gray Treefrogs often call outside the breeding season from high in the tree canopy. As such calls frequently are heard prior to or during rainfall or on cloudy days, they have been termed "rain calls." The function of these calls is unknown. They have been heard as late as mid-September in Massachusetts (Allen, 1868). As in other anurans, Gray Treefrogs have a release call that alerts an amplexing male when he attempts to mate with another male (Pierce and Ralin, 1972).

BREEDING SITES

Gray Treefrogs breed in a variety of small wetlands, such as red maple and shrub swamps and woodland pools. Breeding ponds usually are located adjacent to woodlands, and may have low height shrubs, trees, or other vegetation surrounding them for use as calling perches. Wright (1914) noted that males frequently called from ponds full of lily pads, *Potamogeton*, and floating leaves of large surface area, which served as resting places or calling perches. They also may use anthropogenic disturbed habitats, such as ditches, pasture ponds, quarries, and sand pit ponds.

Hydroperiods of breeding ponds tend to be of intermediate (>4 months but not permanent) to long (permanent wetlands) duration (Herrmann et al., 2005). Not surprisingly, therefore, wetlands used for breeding by Gray Treefrogs usually contain a high invertebrate richness and abundance (Babbitt et al., 2003). Their presence tends to be negatively associated with the presence of predaceous fish, although they may be found inhabiting ponds containing fish (Babbitt et al., 2003; Baber et al., 2004), particularly nonpredatory fish (Hecnar, 1997; Hecnar and M'Closkey, 1997b). In experimental trials, predaceous fishes easily eliminated tadpoles of *H. versicolor*, and fish abundance was directly correlated with predation rate (Kurzava and Morin, 1998; Smith et al., 1999). When fish such as the green sunfish (*Lepomis cyanellus*) are introduced to a breeding site, Gray Treefrogs tend to disappear (Sexton and Phillips, 1986). In Ontario, Hecnar and M'Closkey (1996b) found Gray Treefrogs preferred sites with low conductivity.

As with most anurans, not all available breeding sites are used in any one year, and occupancy can change from one year to the next (Hecnar and M'Closkey, 1996a). In New Hampshire, for example, 41% of the potential breeding ponds surveyed by Herrmann et al. (2005) were occupied by *H. versicolor*, whereas <20% of wetlands were occupied in Ontario (Hecnar, 1997; Hecnar and M'Closkey, 1998). Over a 5 yr period during which 32 ponds were sampled, Skelly et al. (2003) recorded a 60% probability of detecting Gray Treefrogs at a pond in a single year on the E.S. George Reserve in Michigan. The authors used such data to demonstrate that the duration of a resurvey influences whether declines are detected. Short-term resurveys tend to overestimate the magnitude and potential for declines. Brodman (2009) also noted considerable variation in annual occupancy over a 14 yr study in Indiana.

In some cases, only a very small number of Gray Treefrogs may use a site. Over a 4 yr period,

Hocking et al. (2008) found a mean of 4–7 Gray Treefrog adults and 2–7 metamorphs per year at 7 ponds in Missouri. These sites represented constructed rather than natural ponds, as no natural ponds were available for reproduction. At other sites, choruses may be large enough to be deafening to the ear.

REPRODUCTION

The oviposition sequence has not been described, but it is undoubtedly similar to that described for *H. chrysoscelis*, on which the following is based. When ready to oviposit, the female arches her back and thrusts her legs rearward, during which time her head is slightly below water. By arching her back, she is able to raise her cloaca above the water surface and into close apposition with the male. The position of the male causes fertilization to occur at the water's surface. Egg packets are released and fertilized, after which the female's cloaca submerges and the eggs float away. The time between successive fertilizations varies between 10 sec and 2 min. The female uses her front legs for stabilization and to move to different locations. After deposition, eggs swell to maximum size in about 5 min. The female's entire compliment of eggs can be oviposited in 5–10 min, although Wright (1914) noted egg-laying could take up to 1 hr, and mated pairs often remained in amplexus several hrs after oviposition.

Egg deposition takes place throughout the calling season (generally late April to early August, depending on latitude), but most reproduction occurs early in the season. In Missouri, for example, eggs were deposited between 23 May and 6 August (Johnson and Semlitsch, 2003), and Brodman and Kilmurry (1998) noted intense breeding between May and July. Eggs are oviposited within two weeks of initial calling. Clutch sizes reported in the literature include 1,018 in Missouri (*in* Johnson and Semlitsch, 2003), a mean of 2,070 (range 1,288 to 2,604) in Arkansas (Trauth et al., 1990), and an estimate of 1,800–2,000 in New York (Wright, 1914). Hatching occurs in 2–3 weeks in the Northeast at 4.6–4.8 mm TL (Gosner and Black, 1957b).

Female *H. versicolor* arrive at breeding ponds shortly after males. Females may visit a breeding site multiple times within a season, but they only stay at the breeding site the night that they deposit eggs (Godwin and Roble, 1983; Sullivan and Hinshaw, 1992). They then return to terrestrial habitats, usually within 200 m of the pond, where they remain until the next visit to the breeding pond. As such, there is considerable movement by females back and forth between breeding and terrestrial sites within the breeding season.

Males spend a greater amount of time at the breeding sites than females (hence, operational sex ratios are male-biased), but they too go back and forth between the breeding site and terrestrial refugia (Fellers, 1979a); Ritke and Semlitsch, 1991). It is not known if they move among other close-by breeding sites. Only a small percentage (25.7% in Fellers, 1979b) of males successfully attract a female. Fellers (1979b) could not demonstrate size-assortive mating, and successful males did not call on more nights than unsuccessful males. Males mate from 0 to 4 times per breeding season.

LARVAL ECOLOGY

Under laboratory conditions, larvae take up to 60 days to reach metamorphosis, which occurs over a six-day period (metamorphic climax). Under natural conditions, the larval period lasts 40–60 days (Gosner and Black, 1957b). Recently hatched larvae weigh ca. 23 mg (Relyea, 2001b). At metamorphosis, mature larvae weigh a mean of 0.39 g (Grant and Licht, 1995). Recent metamorphs are 13.6–20 mm SUL (mean 16 mm) at Ithaca, New York (Wright, 1914).

Larvae may be found in a variety of temporary to permanent small ponds and woodland pools. Under experimental conditions, the extent of canopy cover appears to have little effect on either percentage of larvae surviving through metamorphosis and virtually no effect on the time to metamorphosis. Food resources, whether grass or leaf litter, do have an effect on survivorship. When leaf litter is used as a food resource, tadpoles have a greater survivorship than they do with grass as a food source. However, the time to metamorphosis is not affected by whether the substrate is grass or leaf litter (Williams et al., 2008). Body mass is not different among tadpoles raised in various levels of canopy cover, but body mass is higher in Gray Treefrog larvae raised on grass as opposed to leaf litter. These results suggest a complex interaction between canopy cover and substrate on various life history parameters of Gray Treefrog tadpoles.

In some tadpoles, habitat complexity may affect

Tadpole of *Hyla versicolor*.
Photo: David Dennis

various life history aspects of larval development. In experimental trials, however, habitat structure had no effect on time to metamorphosis, mass at metamorphosis, or survivorship among treatments of *H. versicolor* tadpoles (Purrenhage and Boone, 2009). As competition increases (measured by an increase in tadpole density), time to metamorphosis increases, and mass at metamorphosis decreases; no effects on survivorship have been observed. Larvae also tend to be more active at low densities compared with high densities.

DISPERSAL

In Arkansas, metamorphs are 17–20 mm when they transform (Trauth et al., 1990), and in Wisconsin, 14–30 mm SUL when recaptured to within one month after metamorphosis (Roble, 1979). Froglets begin their dispersal usually within a week of completing metamorphosis (Bragg, 1943a; Roble, 1979), and most dispersal occurs at night. Frogs usually disperse to < 80 m (mean 13.7 m, range 0 to 115.2 m *in* Roble, 1979) from the breeding pond. In this latter study, froglets dispersed at a rate of 1.58 m per day. Juvenile *H. versicolor* tend to disperse within the confines of forested habitats, and to avoid old fields and habitats that form a forest-old field matrix (Johnson, 2005). They are found on vegetation 20–50 cm above the ground surface, although occasional individuals may be found at greater heights, and even in trees. Large numbers are sometimes seen diurnally resting on vegetation. In Wisconsin, most froglets were observed on sedges (*Carex*), false nettle (*Boehmeria*), reed canary grass (*Phalaris*), and swamp white oak (*Quercus bicolor*).

DIET

Hyla versicolor eats a wide variety of insects, particularly beetles (Coleoptera: Elateridae, Curculionidae, Scarabaeidae, Carabidae) and ants (Formicidae, *Pogonomyrmex*). Other prey items include crickets, moths, roaches, true bugs, spiders, daddy-longlegs, and other invertebrates in much less frequency. Plant material may be ingested incidentally. Johnson (2005) found diets consisting of 31.6% ants and 30% beetles; males and females had similar diets. After an analysis of diet in Texas, Ralin (1968) concluded that *Hyla versicolor* foraged more on or near the ground surface than *H. chrysoscelis*. Recently metamorphosed Gray Treefrogs begin feeding on insects (flies, ants, beetles, damsel bugs, lace bugs, plant lice, bees, leafhoppers) and other tiny invertebrates (soil mites) immediately after leaving the water (Munz, 1920). Their stomachs frequently also contain pieces of shed epidermis. Gray Treefrogs consume their shed skin almost simultaneously with molting. Surface (1913) and Sweetman (1944) provide additional dietary information.

PREDATION AND DEFENSE

Larval Gray Treefrogs are fully palatable, and do not appear to have any chemical defenses; they are eaten by a wide variety of both vertebrate and invertebrate predators. Predators of juveniles include spiders (*Argiope* sp.) (Groves and Groves, 1978). Adults are undoubtedly preyed upon by many predators, including birds, garter snakes (*Thamnophis*), water snakes (*Nerodia*), American Bullfrogs, and meso-mammals. However, there is a scarcity of empirical information. The mosquito

Culex territans feeds on Gray Treefrogs and even can use the frog's call to locate its next blood meal (Bartlett-Healy et al., 2008).

The Gray Treefrog is an extremely cryptic species sitting on a tree trunk or among shrubby vegetation. When handled, they may assume a death-feigning posture with the limbs tucked closely to the body (Banta and Carl, 1967:317) that they remain in for up to 15 sec; whether they do this in proximity to a predator under natural conditions is unknown. Exposed juveniles also assume a crouched posture with limbs held tight to the body; their greenish coloration makes them difficult to see on vegetation (Roble, 1979). Gray Treefrogs may jump away when disturbed. The mean initial distance that individuals can jump is approximately 56 cm (John-Alder et al., 1988), but the distances jumped depend upon ambient temperature. The sudden display of the bright coloration on the inside of the thighs as the frog jumps may serve to startle a would-be predator long enough for the frog to escape. Adults also have a noxious skin secretion that may help deter predators. Persons handling *H. versicolor* should avoid contact with the eyes or mucous membranes until they have had the chance to wash off the secretion.

POPULATION BIOLOGY

It would appear that most male Gray Treefrogs have the opportunity to breed only once in their lifetime. In Ontario, 83% of calling males were first-year breeders, and only 11–21% (3 of 27; 16 of 77) visited breeding sites a second year. A very few (6 of 77, 7.8%) were found a third year after first breeding (Bertram and Berrill, 1997). Chorus size varied considerably by year, and it appeared temperature had a major effect on the number of nights males called. Longevity in nature is unknown, but captive members of the Gray Treefrog complex have been reported to live 7 yrs, 10 months (Snider and Bowler, 1992).

COMMUNITY ECOLOGY

Hyla versicolor tadpoles may occasionally develop in ponds containing other larvae. However, oviposition site likely is influenced by the presence of predators and conspecifics, as it is in *H. chrysoscelis*. Seale (1980) demonstrated a great degree of overlap in larval diet with *Acris crepitans*, *Lithobates pipiens*, and *L. catesbeianus*, at least in terms of particle size. This suggests the potential for competition, selectivity in breeding sites, and the avoidance of sites containing preexisting tadpole populations. Indeed, Seale (1980) rarely reported breeding by *Hyla versicolor* in the ponds she studied over a several-year period. Competition from larval *H. versicolor* also decreases body mass but not survival of larval *H. andersonii* (Pehek, 1995).

Gray Treefrogs tend to alter their preferences for habitats within a pond depending on whether or not predators are in the vicinity. For example, *H. versicolor* tadpoles prefer complex habitats structurally when predators are absent and tend to group into such areas. In experimental

Adult *Hyla versicolor*, amplexing gray adults. Photo: David Dennis

trials, adding a predator (water beetle *Dyticus*) causes the tadpoles to disperse more evenly around a chamber, especially since the predator also preferred the structurally complex habitats (Formanowicz and Bobka, 1989). Presumably, dispersion lessens the risk of capture for any one individual tadpole.

The presence of certain predators, such as dragonfly larvae (*Anax, Pantala*), newts (*Notophthalmus*), and fish (mudminnow, *Umbra*; sunfish, *Enneacanthus*), tends to result in decreased levels of activity by Gray Treefrog tadpoles, at least under experimental conditions (Lawler, 1989; Richardson, 2001; Relyea, 2001b; Smith et al., 2009). However, this response is predator dependent, and not all experimenters have found similar results (e.g., Smith et al., 2009, regarding responses to odonate larvae). In the presence of water beetles (*Belostoma*) or salamander larvae (*Ambystoma*), for example, Gray Treefrog larvae did not reduce their activity levels (Relyea, 2001b). When cages of all four predators were inserted individually into experimental chambers, *Hyla versicolor* larvae avoided the vicinity of the predators. Newly hatched tiny larvae, however, tend to remain immobile whether or not predators are in their vicinity.

In experimental trials, Gray Treefrog tadpoles are able to discriminate chemical cues from conspecific and heterospecific individuals attacked by dragonfly larvae, and to reduce their activity accordingly. Schoeppner and Relyea (2009) suggested that the chemical cues emitted by amphibians following an attack may be quite similar to one another, and different from the cues emitted by other taxa.

Morphological changes also occur in the presence of different predators, such as the development of longer tails, wider muscles, and shallower bodies (to *Umbra*); longer and deeper tails, wider muscles, and shallower and shorter bodies (to *Anax*); longer tails, wider muscles, and shallower bodies (to *Ambystoma*); and shallower and longer tails, wider muscles, and wider bodies (to *Belostoma*) (Relyea, 2001b). Morphological plasticity presumably allows tadpoles to adapt to different predator contexts, and to better avoid predation. This is important, since some predators (*Ambystoma, Anax*) are better predators of *Hyla versicolor* larvae than others (*Umbra, Belostoma*) (Relyea, 2001a).

Certain invasive plant compounds have been demonstrated to adversely affect the development of native species of anuran larvae (see *Anaxyrus americanus* account). However, larvae of Gray Treefrogs were not affected in any way by extracts of either Broad-leaved Cattail (*Typha latifolia*), a native species, or by the nonindigenous Purple Loosestrife (*Lythrum salicaria*) (Maerz et al., 2005b).

DISEASES, PARASITES, AND MALFORMATIONS

Frog Virus 3-like infections have been reported in natural populations of Gray Treefrog larvae from Ontario (Duffus et al., 2008). The fungus *Basidiobolus ranarum* has been found in treefrogs of the *Hyla versicolor/H. chrysoscelis* complex in Arkansas and Missouri (Nickerson and Hutchison, 1971). Amphibian chytrid fungus (*Batrachochytrium dendrobatidis*) was not observed in samples from the northeastern United States (Longcore et al., 2007). However, amphibian chytrid has been found in one individual *Hyla versicolor* in Québec (Ouellet et al., 2005a).

Trypanosome protozoans (*Trypanosoma* sp., *T. andersoni*, and *T. grylli*) have been found in *Hyla versicolor* (Reilly and Woo, 1982; Woo and Bogart, 1984; Shannon, 1988). Enteric protozoans include *Tritrichomonas augusta, Nyctotherus cordiformis*, and *Urophagus* sp. from Minnesota (Anderson and Buttrey, 1962). Trematode metacercariae (unidentified, *Alaria* sp., *Echinostoma* sp.) have been reported from *Hyla versicolor* in Iowa and Ontario (Ulmer, 1970; Dyer, 1991; Koprivnikar et al., 2006). Gray Treefrogs are parasitized by the nematodes *Gyrinicola batrachiensis* and *Oswaldocruzia pipiens* (Baker, 1977; Adamson, 1981c) and a fluke infecting the urinary tract (Olsen, 1962). Other parasites found in Missouri include the trematodes *Clinostomum marginatum, Haematoloechus* sp., and *Polystomum nearcticum*; the nematodes *Dorylaimus* sp., *Cosmocercella* sp., and *Physaloptera* sp.; and unidentified proteocephalid cestodes and polymorphid acanthocephalians (Shannon, 1988; Hausfater et al., 1990). The level of trematode infection in tadpoles has been correlated with agricultural activity, but not with forest cover, pond size, road density, nitrate level, or the presence of atrazine (Koprivnikar et al., 2006). In an extensive analysis of the effects of parasitism on mate choice, Hausfater et al. (1990) could find

Adult *Hyla versicolor*, green phase. Photo: David Dennis

no effects of helminth parasitism on male calling activity or mating success.

Malformed Gray Treefrogs have been reported from federal lands in the upper Midwest and Northeast. Of 174 frogs examined in the Midwest, 9 had malformations; in the Northeast, 3 of 55 frogs examined had malformations (Converse et al., 2000). Malformations included missing or partial hind limbs, malformed hind limbs, missing eye, missing or partial front limbs, and malformed front limbs. In Minnesota, only a single *Hyla versicolor* was reported with a malformation by Hoppe (2005).

SUSCEPTIBILITY TO POTENTIAL STRESSORS

Metals. Gray Treefrog tadpoles accumulate moderate to high concentrations of metals, including aluminum, iron, magnesium, and manganese, depending on soil substrates in ponds (Sparling and Lowe, 1996). Tadpoles concentrate metals, which could be concentrated as tadpoles are ingested by predators.

Other elements. The contaminants selenium (Se) and vanadium (V) have no effect on larval Gray Treefrogs at concentrations of 7.5–32.7 μg/g dry weight Se and 132.1–485.7 μg/g dry weight V. Growth, survivorship, metabolic rate, and lipid content were unaffected by exposure to these elements (Rowe et al., 2011).

pH. The critical pH level is 3.5 to 4.3 in larval Gray Treefrogs and 3.8 in embryos (Gosner and Black, 1957a; Grant and Licht, 1993). There is some indication of regional variation in pH tolerance, with larvae from Ontario more tolerant than in more southern areas. Experimental acidification of mesocosms to pHs between 5.0 and 5.2 made them less desirable as breeding sites for Gray Treefrogs, but this preference was affected by soil type (loam soils preferred over clay) (Sparling et al., 1995). Low pH (3.9) has no effect on survival, body mass, duration of the larval period, or growth of larval *H. versicolor* from New Jersey (Pehek, 1995).

Chemicals. Westerman et al. (2003a) classify *H. versicolor* as very sensitive to metals and organic compounds. Survivorship and growth of Gray Treefrog larvae when exposed to toxic substances varies by chemical, concentration, the presence of competitors and predators, and the availability of food resources. For example, Relyea (2009) exposed Gray Treefrog larvae to complex mixtures of low concentrations of ten pesticides in the presence of Northern Leopard Frog larvae under varying conditions of potential food resources (zooplankton, phytoplankton, algae). The chemical mixture had no effects on tadpole survivorship, and actually increased their body mass when reared with atrazine, a mix of insecticides, and a mixture of all pesticides. By eliminating predators and competitors and secondarily enhancing the availability of food resources, the pesticides actually caused the tadpoles to grow larger.

Combinations of pesticides may be far more toxic than concentrations of individual pesticides by themselves. Using high concentrations of a mixture of atrazine, metolachlor (both herbicides), and chlorpyrifos (an insecticide) killed 100% of *H. versicolor* larvae; mixtures of these pesticides at lower doses caused higher mortality than the pesticides alone. Low concentrations of the herbicide-insecticide combinations produced lethargy or reduced growth, suggesting the potential for sublethal effects even in the absence of mortality. Such results may be obtained both in laboratory and in naturally colonized experimental wetlands (Mazanti, 1999; Sparling, 2003).

The pesticide carbaryl apparently has no adverse effects on tadpole survivorship at 0.955 mL/m^2 (Relyea, 2005b). However, survivorship is known to be affected both by concentration and the length of exposure. For example, the $LC_{50\ (2\ day)}$ is 12.9 mg/L (Bridges, 1997) and for 4-day exposures, it is 2.5 mg/L (Zaga et al., 1998). These short-term LC_{50s} give the impression that lessened toxicity might be expected of much smaller doses. However, larvae exposed to carbaryl for 10–16 days at 3–4% of the $LC_{50\ (4\ day)}$ die off at rates of 10–60%. Mortality rates increase even higher (to 60–98%) in the presence of predator cues (larval salamanders, *Ambystoma maculatum*) (Relyea and Mills, 2001). It would seem, therefore, that the standard four-day trial period for many toxicity tests is far too brief to assess the effects of pesticides on amphibians.

The ability to tolerate carbaryl varies among individual Gray Treefrogs, even within a population, and this variation in tolerance appears to have a genetic basis (Semlitsch et al., 2000). In contrast, those tadpoles that are tolerant of carbaryl have a decreased probability of surviving when the chemical is absent during development in the field. Thus, there appears to be a trade-off in fitness between those larvae that are tolerant of the pesticide and those that are not. Carbaryl at 1–2.5 mg/L significantly reduced activity and the amount of time larvae spent feeding and swimming, suggesting that sublethal concentrations might have adverse impacts on growth and time to metamorphosis (Bridges, 1997, 1999a; Relyea and Edwards, 2010).

Paradoxically, carbaryl tends to reduce survivorship of Gray Treefrog larvae, but survivors have increased mass at metamorphosis. Larval density also affects the response to carbaryl, with proportionately more larvae surviving from low-density ponds in experimental trials, an effect that is further complicated by the addition of a predator (Boone and Semlitsch, 2001). More Gray Treefrog larvae might actually survive in high densities than in low densities in the presence of predators, as long as the predators' response and feeding are slowed down by high concentrations of carbaryl.

Carbaryl appears to interact with ultraviolet radiation to increase mortality in Gray Treefrog tadpoles. Under control conditions of simulated solar radiation, carbaryl at 2.51 mg/L caused 5% mortality. However, when combined with a low irradiance (4μW/cm^2), mortality increased to 100% (Little et al., 2000). Carbaryl by itself resulted in 10–21% mortality at 1.24 to 2.51 mg/L after 48 hrs of exposure in the absence of UV light. Low levels of UV light alone resulted in 10–22% mortality, whereas high levels of irradiance (65μW/cm^2) resulted in 78% mortality.

Application of the insecticide malathion at ecologically relevant concentrations (0.315 mg/L) may actually benefit Gray Treefrog tadpoles by killing predaceous insects, at least under experimental conditions (Relyea et al., 2005), although activity is reduced at concentrations of 0.1–1 mg/L (Relyea and Edwards, 2010). At even higher concentrations (10–20 mg/L), however, malathion also decreases tadpole survivorship, even in the absence of predators (Relyea, 2004b). At concentrations greater than the LC_{50} ($LC_{50\ (16\ day)}$ 2.0–4.1 mg/L), the presence of predators decreases survivorship. Since malathion normally occurs at much lower concentrations than these under field conditions, it is doubtful that it has any deleterious effects on Gray Treefrog development.

The lampricide TFM has a $LC_{50\ (96\ hr)}$ of 1.98 mg/L to Gray Treefrog larvae (Chandler and Marking, 1975).

The herbicide Roundup™ causes significant tadpole mortality (at 1.3 mg active ingredient/L or 15.3 mL/m^2), regardless of the presence of predators or the abundance of algae, a food resource (Relyea, 2005a, 2005b; Relyea et al., 2005). In experimental trials, Roundup™ decreased larval survivorship from 75 to 2% after 20 days; even in terrestrial trials, Roundup™ decreased survivorship from 100 to 18% after just 24 hrs at concentrations of 6.5 mL per 10 L plastic container (Relyea, 2005a). The herbicide 2,4-D has no effect on Gray Treefrog larvae (Relyea, 2005b).

Nitrate. Concentrations of nitrate to 20 mg/L had no effect on Gray Treefrog larval survivorship, final body mass, or activity levels over a 15-day study period (Vaala et al., 2004).

UV light. Ecologically relevant levels of UVB light have no effect on the developmental period, duration of metamorphic climax, or body mass at metamorphosis of *H. versicolor* (Grant and Licht, 1995; Crump et al., 1999). Artificially high doses of UVB kills >60% of embryos exposed for > 15 min (Licht, 2003) and 80% of 2–4-week-old larvae after 5–7 days of exposure (Crump et al., 1999). The jelly envelope absorbs about 80% of the incoming UV radiation (280–320 nm). UV light may interact with toxic substances to increase their lethal effects.

STATUS AND CONSERVATION

Gray Treefrogs are widely distributed throughout their range, and most populations appear stable, with no evidence of decline at a landscape scale (Weller and Green, 1997; Mossman et al., 1998; Casper, 1998; Brodman et al., 2002; Brodman, 2003). A few reports suggest localized declines or that the species is uncommon (Mierzwa, 1998). However, habitat loss has taken its toll on individual populations of the Gray Treefrog complex, and both species are susceptible to adverse effects from urbanization (Babbitt, 1937; Klemens, 1993; Lehtinen et al., 1999; Knutson et al., 2000), prolonged drought, and roads. For example, in Ontario Gray Treefrog abundance is inversely related to the volume of traffic on roads within 200 m of breeding ponds (Eigenbrod et al., 2008, 2009). Other road mortality reports are in Ashley and Robinson (1996); numbers reported killed are often small, probably reflecting almost immediate obliteration rather than a small amount of mortality.

Gray Treefrogs may persist in the vicinity of many types of adjacent land uses, even in fragmented habitats, as long as sufficient forested habitat remains around breeding sites. Gray Treefrogs also may survive, providing riparian buffer zones are maintained (Rudolph and Dickson, 1990). Abundance of members of the Gray Treefrog complex is reduced in agricultural vs. natural habitats, but Gray Treefrogs still use ponds located in agricultural landscapes quite frequently (Knutson et al., 2004; Anderson and Arruda, 2006). For example, Kolozsvary and Swihart (1999) reported Gray Treefrogs to be ubiquitous in 30 forest patches of various sizes and degrees of isolation in the agricultural midwestern United States.

Gray Treefrogs naturally occupy restored wetlands (Kline, 1998; Lehtinen and Galatowitsch, 2001; Foster et al., 2004; Brodman et al., 2006; Shulse et al., 2010), such as those created at former mine sites (Myers and Klimstra, 1963; Lacki et al., 1992). In addition, they may be candidates for inclusion in large-scale restoration programs; Gray Treefrogs have been successfully translocated to areas within the Gateway National Recreation Area, mostly using larvae for release (Cook, 2008). Based on an analysis of movement patterns, protecting a 60 m area around the breeding site may allow Gray Treefrog populations to persist when surrounding habitats are disturbed (Johnson and Semlitsch, 2003; Johnson, 2005).

Gray Treefrogs also occupy retention ponds in urban areas, although the proportion of impervious land surface is negatively correlated with

Breeding habitat of *Hyla versicolor*. Photo: Jennifer Anderson-Cruz

occupancy (Simon et al., 2009). However, salts tend to accumulate in such ponds and make them less than optimal as breeding sites inasmuch as survival is negatively correlated with conductivity (Brand et al., 2010). Exposure to retention-pond sediments reduces embryo survival but does not affect larval survival. Instead, surviving larvae tend to develop faster and reach larger size at metamorphosis (Brand et al., 2010).

Gray Treefrogs also occupy retention ponds in urban areas, although the proportion of impervious land surface is negatively correlated with occupancy (Simon et al., 2009). However, salts tend to accumulate in such ponds and make them less than optimal as breeding sites, as survival is negatively correlated with conductivity (Brand et al., 2010). Exposure to retention-pond sediments reduces embryo survival, but does not affect larval survival. Instead, surviving larvae tend to develop faster and reach larger size at metamorphosis (Brand et al., 2010).

Gray Treefrogs tend to deposit more eggs in open-canopied breeding sites than those located in forested or selectively cut habitats (Hocking and Semlitsch, 2007). Areas near a forest-open area ecotone appear favored, even when the breeding site is located 50 m from the forest edge. Thus, this species tends to do well in areas that were previously clearcut, as long as source habitats are located nearby. Isolation diminishes the potential of wetlands to serve as good breeding sites in clearcuts. In silvicultural operations, it is best to leave a mosaic of habitats in order to benefit Gray Treefrog populations, particularly wetlands located in open-canopied sites near the ecotone of forest and clearcut. Gray Treefrogs also are found in power-line rights-of-way, which provide a broad extent of ecotonal habitats (Fortin et al., 2004b).

Gray Treefrogs commonly are monitored using call surveys (Bishop et al., 1997; Bonin et al., 1997a; Lepage et al., 1997; Mossman et al., 1998; de Solla et al., 2005). As the number of sampling nights increases, the more likely it is that Gray Treefrogs will be detected. De Solla et al. (2005) determined that 10 sampling nights were needed in order to achieve detection probabilities >80% in Ontario, and Lepage et al. (1997) reported that it would take 334 listening stations to estimate presence at ±5% accuracy and 1,716 stations to estimate abundance at ±10% accuracy. Based on call surveys over a 4 yr period (1986–1989), populations of the Gray Treefrog were considered stable in Illinois (Florey and Mullin, 2005); in upstate New York, they were considered to be increasing (Gibbs et al., 2005).

Hyla wrightorum Taylor, 1938
Arizona Treefrog

ETYMOLOGY
wrightorum: This species is named in honor of Albert Hazen Wright and Anna Allen Wright in recognition of their pioneering work on frogs.

NOMENCLATURE
Stebbins (2003): *Hyla eximia*
Synonyms: *Hyla eximia* [in part], *Hyla eximia wrightorum*, *Hyla regilla wrightorum*

IDENTIFICATION
Adults. The Arizona Treefrog is bright green with a dark line extending from the snout through the eye onto the side of the body. The posterior half of the lower jaw is darkly pigmented, and the posterior side of the femur is lightly and evenly pigmented. The underside of the thigh and groin are orange or gold with a greenish tinge. Throats of males are dull greenish and tan, whereas female throats are white. At transformation, metamorphs are brown but change to a green coloration within a day or two after the tail is resorbed. Males averaged 41.3 mm SUL (range 37 to 47 mm) in one Arizona study (Sullivan, 1986c) and 37.3 mm SUL (31–45 mm) in another (Renaud, 1977). Chapel (1939) noted adults 25–48 mm SUL.

Larvae. Newly hatched larvae are yellow brown dorsally and yellow ventrally. Dusky pigmentation develops early on the body and tail fin. Mature larvae are brown and have a deep, globose belly, wide head, blunt snout, and lateral eyes; venters become dusky. The dorsal tail fin is low, and there are dark reticulations on both the posterior dorsal and ventral fins. Maximum length is 38 mm

TL. Note that descriptions of the larvae in early publications (e.g., Wright and Wright, 1949) may include information pertaining to *H. arenicolor*. Zweifel (1961) provided figures of larvae at various stages of development and diagrams of the mouthparts.

Eggs. The eggs have not been described but presumably are similar to those of *H. arenicolor*.

DISTRIBUTION

Hyla wrightorum occurs in the mountains of central Arizona and western New Mexico. Isolated populations occur in extreme southeastern Arizona (Huachuca Mountains and Canelo Hills) and in the Sierra Madre Occidental of northern Mexico. A record from 37.7 km north-northeast of Yuma likely represents an introduction (Vitt and Ohmart, 1978), and a record from Santa Fe, New Mexico (Van Denburgh, 1924; Wright and Wright, 1949), represented a point of shipment but not a collection locality. Important distributional references include Van Denburgh (1924), Degenhardt et al. (1996), Gergus et al. (2004), and Brennan and Holycross (2006).

FOSSIL RECORD

No fossils are known.

SYSTEMATICS AND GEOGRAPHIC VARIATION

The Arizona Treefrog was described by Taylor (1938) as distinct from the Mexican species *H. eximia* to which it previously had been referred. It was later considered a subspecies of *Pseudacris regilla* (Jameson et al., 1966; Jameson and Richmond, 1971), but this arrangement is not supported by allozyme and mtDNA data and analyses of advertisement calls (Case et al., 1975; Gergus et al., 2004). There are no fixed allozymes among *Hyla wrightorum* populations, and the amount of genetic variation among populations is small. *H. wrightorum* is a member of the *H. eximia* species group, a primarily tropical assemblage from Mexico.

Under laboratory conditions, crosses between members of the Gray Treefrog complex (*H. chrysoscelis/versicolor*) or *H. squirella* and *H. wrightorum* are largely unsuccessful, although some eggs produce larvae, which then fail to develop. Many structural abnormalities are evident (Littlejohn, 1961a). Klymus et al. (2010) found evidence of past mtDNA introgression between *H. wrightorum* and *H. arenicolor* in the region of the Grand Canyon.

ADULT HABITAT

Arizona Treefrogs are found in oak-Ponderosa pine and spruce-Douglas fir forests above 1,520 m (Chapel, 1939; Aitchison and Tomko, 1974). Other dominant trees include juniper, pinon, Mexican white pine, and white fir. They occur along small streams, in wet meadows and cienegas, and in roadside ditches. Williams and Chrapliwy (1958) heard them at 2,438 m in Coconino County, Arizona. Aitchison and Tomko (1974) observed them at elevations of 2,133–2,560 m near Flagstaff, and Degenhardt et al. (1996) recorded them at 2,000–2,750 m in New Mexico.

TERRESTRIAL ECOLOGY

Arizona Treefrogs may be found in the surrounding forest before and after the breeding season (Chapel, 1939), but little is known of their movements. They are found both on the forest floor and in trees. Chapel (1939) noted one individual 23 m off the ground.

Arizona Treefrogs are likely photopositive in their phototactic response as is *H. eximia*, suggesting they use both sunlight and moonlight when feeding and going about their daily activities (Jaeger and Hailman, 1973). They are likely sensitive to light in the blue spectrum ("blue-mode response") (Hailman and Jaeger, 1974). Arizona Treefrogs likely have true color vision.

Distribution of *Hyla wrightorum*

Adult *Hyla wrightorum*, color variation. Photo: Cecil Schwalbe

CALLING ACTIVITY AND MATE SELECTION

The call of *H. wrightorum* is rapidly pulsed and similar but of lower dominant frequency than that of *H. squirella* from the eastern United States. It has been liked to a low-pitched metallic trill. The advertisement call has a dominant frequency of 1,900–2,200 cps at air temperatures of 6–16°C; there are 100–120 pulses/sec with a mean duration of 0.15–0.17 sec (Blair, 1960b). In other studies (Renaud, 1977; Sullivan, 1986c), dominant frequencies were 1,600–2,300 cps, pulse rates were 77.5–156 pulses/sec, and call durations were 0.12–0.24 sec. Dominant frequency is negatively correlated with male SUL (Sullivan, 1986c). Gergus et al. (2004) noted that temperature exerted an important influence on call variables, and Sullivan (1986c) noted that Renaud's (1977) inability to correlate pulse rate with body temperature was inconsistent with most studies on temperature effects. Whereas they found similar pulse rates to Blair (1960b), the dominant frequency dropped to near 1,600 cps.

Choruses are not prolonged, but form after each rainfall event and last two to four nights. Chorusing occurs all night, but is greatest before midnight. Although the size of the male often influences female choice in anurans, this does not appear to be the case in *H. wrightorum*. There are no significant differences in the SULs of mated and unmated males, nor is there a correlation between the size of males and females in amplexus (Sullivan, 1986c).

BREEDING SITES

Breeding sites mostly consist of large shallow grassy rainwater pools formed after summer rains, but they also breed in stock tanks. Chapel (1939) noted that they congregated prior to breeding by lakes, ponds, streams, wells, and in any area likely to hold water once rainfall begins.

REPRODUCTION

Reproduction commences with the first summer rains from June to August, the exact date depending on weather conditions. Eggs are attached in small loose clusters to vegetation in shallow water. Eggs may be swept away after sudden rainstorms. There is no information available on clutch size or other aspects of reproduction.

LARVAL ECOLOGY

At hatching, larvae are 4.9–5.2 mm TL. They develop in shallow pools and have been found in streams, perhaps washed in from nearby pools after summer storms. Mortality occurs when water evaporates. Larvae forage in vegetation in the warmest part of the pools, and Chapel (1939) noted that they aggregate to feed in great numbers around cow manure. The duration of the larval period is unknown. Newly transformed froglets are 10–13 mm SUL.

DIET

Very few data are available on diet. Chapel (1939) noted beetles, a spider, an earthworm, a fly, and

Breeding habitat of *Hyla wrightorum*. Photo: Cecil Schwalbe

"ips" (bark beetles, an insect pest) from a small number of individuals.

PREDATION AND DEFENSE
This species is protected to some extent by its camouflage coloration. Predation of *H. wrightorum* has been recorded by American Bullfrogs, and they are considered a threat to this species (Jones and Timmons, 2010). The species also has a noxious skin secretion that causes burning to eyes and likely the mucous membranes (*in* Degenhardt et al., 1996). Postmetamorphs are probably eaten by garter snakes and the larvae by garter snakes and predaceous aquatic insects.

POPULATION BIOLOGY
No information is available.

COMMUNITY ECOLOGY
Hyla wrightorum and *H. arenicolor* frequently are found at the same breeding sites in the mountains of central Arizona. Differences in calls, breeding seasons (although they do overlap), habitats, and responses to rainfall normally are sufficient to prevent hybridization, although Klymus et al. (2010) found evidence of past introgression between them in at least one region.

Under experimental conditions, larval salamanders (*Ambystoma tigrinum*) severely reduce survivorship of larval *H. wrightorum*, but had no effect on mass at metamorphosis, length of the larval period, or growth rates (Sredl and Collins, 1992). These species occasionally occur together in permanent ponds, and the larval periods overlap to the extent that large salamander larvae could cause significant predation on the small frog larvae.

DISEASES, PARASITES, AND MALFORMATIONS
Arizona Treefrogs are parasitized by the cestode *Cylindrotaenia americana* and the nematodes *Cosmocercella haberi* and *Physaloptera* sp. (Goldberg et al., 1996c).

SUSCEPTIBILITY TO POTENTIAL STRESSORS
No information is available.

STATUS AND CONSERVATION
Arizona Treefrogs in the Huachuca Mountains are in danger of extirpation and have been identified as needing urgent conservation management. Status and trends of other populations have not been reported.

Pseudacris brachyphona (Cope, 1889)
Mountain Chorus Frog

ETYMOLOGY
brachyphona: from the Greek word *brachys*, meaning 'short,' and *phōnē*, meaning 'voice.' The name refers to the short trilling call of this species.

NOMENCLATURE
Synonyms: *Chorophilus feriarum brachyphonus*

IDENTIFICATION
Adults. This is a small frog, with adults averaging approximately 30 mm SUL. Ground coloration ranges through several shades of tan to brown. Adults have a generally broad head, more rounded snout, and longer hind legs than sympatric *Pseudacris*. Toes have distinct disks with minimal webbing. There is a conspicuous triangle between the eyes when viewed dorsally (apex pointing posteriorly), and there is a broad stripe extending posteriorly from the tympanum that arches inward toward the midline of the dorsum. These stripes may connect, giving a cruciform appearance similar to that of *P. crucifer*. Some individuals have a spot above the vent. In addition, a few individuals have an irregular spotting pattern without the two dorsolateral stripes. The venter is cream colored and usually unmarked, although some individuals have a few dark speckles. The underside of the limbs has a yellowish coloration, and is most pronounced in breeding individuals. The male's vocal sac is dusky to nearly jet black. Newly transformed froglets have the adult pattern.

This species is sometimes confused with *P. crucifer*, especially in those individuals where the dorsolateral stripes fuse to form a cruciform pattern. However, the skin of *P. brachyphona* is rougher than that of *P. crucifer*, the toe disks are smaller, it frequently has a yellowish coloration on the hind limbs, and the call is distinctive between these species (Netting, 1933).

There is a degree of sexual dimorphism, with males generally olive brown and females lighter and more reddish brown. Females are also larger than males. In a large sample from Kentucky, the mean was 29.8 mm SUL, with a range from 25 to 32.2 mm (Barbour, 1957); on Big Black Mountain, males averaged 30 mm SUL (range 27 to 33 mm) and females 34.8 mm SUL (range 31 to 37 mm) (Barbour, 1953). Males ranged from 24 to 32 mm SUL (mean 27 mm) in North Carolina (Schwartz, 1955), and averaged 24.6 mm SUL in Pennsylvania; females in this latter population averaged 30.3 mm SUL (Hulse et al., 2001). In Ohio, males measured 24.5 to 30 mm SUL and females 26–34 mm SUL (Walker, 1946).

Larvae. The larvae are small and deep bodied, and reach a length of about 25–30 mm SUL prior to metamorphosis (19 mm when rear legs become evident). The eyes are located dorsolaterally, and the spiracle is not obvious, and it is located on the left side of the body. The body is generally black to dark brassy brown dorsally, with the venter dark brown with numerous bronze iridescent specks. The tail fin is low and has scattered pigment dots, and there is a small amount of pigment on the lower edge of the tail musculature. There are two rows of labial teeth above the mouth (termed anterior) and three rows below the mouth (termed posterior). Additional descriptions of the tadpole are in Green (1938b) and Altig (1970).

Eggs. The vegetal pole of the eggs of *P. brachyphona* is creamy white to buff, whereas the dorsal animal pole is brown. Eggs have a single gelatinous envelope, and range from 6 to 8.5 mm (mean 7 mm) in total diameter. The vitellus is 1.6 mm. The envelopes of adjacent eggs adhere to one another providing a degree of cohesion to the loosely formed mass. Descriptions of eggs are in Green (1938b).

DISTRIBUTION
The Mountain Chorus Frog is found from southwestern Pennsylvania (as far north as Jefferson County), through the Allegheny Mountains and Cumberland Plateau of West Virginia, southeast Ohio, eastern Kentucky (and extending west along the Green River and Rolling Fork drainages to Edmonson and Jefferson counties), east-central Tennessee (Cumberland Mountains, Blue Ridge Mountains), and southwest Virginia. The range includes most of northern and central Alabama (north of the Fall Line, including the Talladega Uplands), extreme northeast Mississippi, and into Georgia. Isolated populations are found in the Southern Appalachians (Iron Mountain, Virginia, and Cherokee County, North Carolina: Floyd and Kilpatrick, 2002), Ohio (Carroll and Jefferson counties), and the Panhandle of West

Distribution of *Pseudacris brachyphona*

specific status was not recognized until Walker's (1932) examination of additional specimens from Ohio. Although no subspecies have been described, there is a distinct difference in mitochondrial 12S and 16S rRNA genes between northern and southern populations (Lemmon et al., 2007b). The northern clade includes populations from Tennessee, Kentucky, and West Virginia northward, whereas the southern clade includes populations in North Carolina, Georgia, Alabama, and presumably Mississippi.

Based on a cladistic analysis of morphology and karyology, Cocroft (1994) suggested that *P. brachyphona* was more closely related to *P. feriarum* and *P. kalmi* than to other *Pseudacris*, but this arrangement is not supported by molecular data. Instead, *P. brachyphona* is most closely related to *P. brimleyi* (Moriarty and Cannatella, 2004). These latter two species diverged from a common ancestor during the Pliocene (4.6 mya), perhaps as a consequence of interspecific competition with a *Nigrita* clade ancestor (Lemmon et al., 2007a). Together, they are the sister clade of the *Nigrita* clade (*nigrita, feriarum, triseriata, kalmi, fouquettei, clarkii, maculata*), which together form the Trilling Chorus Frog clade. This large clade shares a similar albumin phylogeny (Maxson and Wilson, 1975), and all members possess a cuboidal intercalary cartilage (Paukstis and Brown, 1987).

Natural hybridization occurs with *P. feriarum* in northeast Mississippi and south-central Kentucky, and with *P. triseriata* from central Kentucky (Lemmon et al., 2007b). Tadpoles can be produced in laboratory crosses between *P. brachyphona* and *P. triseriata*, and between ♀ *P. crucifer* and ♂ *P. brachyphona*. Crosses were not viable between ♂ *P. crucifer* and ♀ *P. brachyphona* (Green, 1952). In contrast, Mecham (1965) reported almost the exact opposite results to Green (1952) involving *P. brachyphona* and *P. crucifer*. Laboratory experiments also produced viable tadpoles in crosses with *P. nigrita*, *P. feriarum*, *P. ornata*, and *P. brimleyi* (Mecham, 1965). In most cases, metamorphosis was completed successfully, and frogs grew normally, appearing intermediate in phenotype between the parental species. Lemmon et al. (2007b) found molecular evidence of crosses in nature between *P. brachyphona* and *P. triseriata* and *P. feriarum*.

There may be regional or population variation

Virginia (Berkeley County). Only a single population remained in Maryland as of 1999 of nine historically known populations (Forester et al., 2003). The species occurs from 365 m to 1,220 m in elevation (Green, 1938b; Barbour, 1953; Hoffman, 1981). Suggestions by Neill (1954) that the species might occur in the Florida Panhandle have not been verified.

Important distributional references include: Range-wide (Lemmon et al., 2007b), Alabama (Mount, 1975; Redmond and Mount, 1975; Graham, 2010), Georgia (Martof and Humphries, 1955; Williamson and Moulis, 1994; Jensen et al., 2008; Graham, 2010), Kentucky (Barbour, 1957, 1971), Maryland (Harris, 1975), Mid-Atlantic (Beane et al., 2010), Mississippi (Ferguson, 1961), North Carolina (Schwartz, 1955; Dorcas et al., 2007), Ohio (Walker, 1946; Pfingsten, 1998; Davis and Menze, 2000), Pennsylvania (Hulse et al., 2001), Tennessee (Gentry, 1955; Barbour, 1956; Redmond and Scott, 1996; Wilmhoff et al., 1999), Virginia (Hoffman, 1955, 1981; Tobey, 1985; Mitchell and Reay, 1999), and West Virginia (Green and Pauley, 1987).

FOSSIL RECORD

No fossils have been referred to this species.

SYSTEMATICS AND GEOGRAPHIC VARIATION

This species was described by Cope (1889) as *Chorophilus feriarum brachyphonus*, but its

in the extent of the X-pattern on the dorsum. About 50% of the specimens had fused stripes into either an X- or a U-pattern in northern Georgia (Martof and Humphries, 1955). In Ohio, however, Walker (1932) reported that only about 20% of the frogs had fused dorsolateral stripes. In West Virginia, 55% of 1,251 individuals examined had a crescent pattern, 35% had a cruciform (fused) pattern, and 10% had no dorsal markings (Green, 1969). There is no correlation between dorsal markings and sex or size.

ADULT HABITAT

Mountain Chorus Frogs are associated almost entirely with deciduous woodlands, and individuals are not usually found away from woodland habitats. The amount of tree cover appears important to this species, regardless of whether it is in upland or valley habitats. When areas are cleared, this species is replaced by *P. feriarum*. Hulse et al. (2001) noted that Mountain Chorus Frogs are sometimes seen moving after summer rains, and that they are found under surface debris in the leaf litter.

TERRESTRIAL ECOLOGY

Walker (1932) records this species as occurring in "deep woods" during the summer. Upon emergence, they may spend several weeks in transit to breeding sites, often pausing at small wetlands along the way (Green, 1952). After breeding, they return to terrestrial sites farther away from the early-season breeding sites, but may pause along the way and breed later in the summer at these intermediate wetlands. Green (1952) recorded marked individuals traveling 610 m between terrestrial overwintering sites and breeding ponds, and 1220 m between breeding ponds in successive years. This is a terrestrial species, with no evidence of climbing.

Small adult males grow more rapidly than larger frogs, and growth essentially ceases in the largest individuals. A 23-mm frog might be expected to increase from 3 to 9 mm, whereas a 30-mm male might grow 0–4 mm in a year (Green, 1964). The oldest animals even appear to shrink by 1–2 mm as they reach the largest size classes.

Like most frogs, Mountain Chorus Frogs have a blue-mode phototactic response, indicating they have true color vision (Hailman and Jaeger, 1974). They are monotonically photopositive, meaning that given a choice, individuals will seek out optimal illumination (Jaeger and Hailman, 1973). Presumably, these characteristics assist in terrestrial movements and in locating prey.

CALLING ACTIVITY AND MATE SELECTION

Mountain Chorus Frogs emerge from dormancy in late winter (December to March), depending on location and temperature. Males arrive at the breeding sites two to eight days before the females, and immediately begin calling (Green, 1938b; Barbour and Walters, 1941). The call of *P. brachyphona* has been compared with that of *P. triseriata*, but at a more rapid rate and at a higher pitch. Barbour (1971) describes it as a rasping "reke-rake." The dominant frequency is 1,050–3,000 cps (mean 2,290 cps) (Thompson and Martof, 1957; Forester et al., 2003). The mean duration of the call is approximately 220–460 milliseconds (range 400 to 508), which is much shorter than *Pseudacris* of the Nigrita clade (mean >600 milliseconds) (Thompson and Martof, 1957; Forester et al., 2003). Notes are repeated 50–70 times a minute, and they may be sustained over several minutes. From 24–26 notes (or pulses) comprise each call. Breaks last 15–20 sec between calls. Call length and the number of calls per minute are negatively correlated with temperature, but the number of pulses per sec is positively correlated with temperature (Forester et al., 2003). The midpoint dominant frequency of the call is negatively correlated with the male's body mass.

Most calling occurs between March and July (Wright and Wright, 1949). However, they have been heard calling in early December in Alabama (Mount, 1975), January in West Virginia

Eggs of *Pseudacris brachyphona*. Photo: J. Michael Butler

Tadpole of *Pseudacris brachyphona*. Photo: David Dennis

(Green and Pauley, 1987), and late February in the Hiawassee River floodplain of southwestern North Carolina (Schwartz, 1955). At this latter time, the air temperature was 15°C, and the water temperature was 10°C. Calling begins when water temperatures reach 7.5°C and corresponding air temperatures are at least 5°C (Barbour and Walters, 1941). In Maryland, calling took place over a 26-day period in 1996 at temperatures of 5.4–18°C (Forester et al., 2003). Mean chorus attendance by males was 13 days.

Calling occurs both diurnally and at night. Males often call from hidden locations under dead leaves and debris along the shoreline of ditches and shallow temporary wetlands, from short grass clumps, or while floating on algal mats or other vegetation. Individuals may call from water, where only the head may be visible, or from land. Males often float in the water with their back legs outstretched while calling, as is common in other species of *Pseudacris*, and Barbour and Walters (1941) reported that males were distributed throughout a wetland rather than being confined to areas near the shore. Barbour (1953) recorded a solitary male calling from the forest floor, 183 m from the nearest potential breeding site.

While calling, males tend to face away from the pond if calling in water, or toward the pool if they are calling from along the shore. The diffuse nature of the call makes them difficult to locate when they are concealed by vegetation. According to Green (1938b), however, males make no attempt to conceal their calling sites, and they space themselves in such a manner that there may be 3 or 4 males every 60 cm.

The male makes no attempt at amplexus until actually contacted by the female. Females approach and back into calling males between the male's forelegs (Green, 1952). The male then clasps the female, and the female swims to deeper water for oviposition. Amplexus is axillary, and the female usually rests with legs outstretched throughout amplexus. Oviposition is initiated as the female flexes her back. Amplexed males struggle vigorously to dislodge an inappropriate suitor.

BREEDING SITES

Both quiet and slow-moving temporary water may be used for breeding; permanent water is avoided. Breeding occurs in shallow, flooded pools, grassy pastures, the vicinity of small springs, and in roadside ditches. Mountain Chorus Frogs also have been observed calling from slow flowing waters adjacent to a culvert (Schwartz, 1955), road ruts (Barbour, 1971; Adam and Lacki, 1993; Barry et al., 2008), and from swampy areas, along small woodland streams, and along farm ponds (Barbour, 1957). Martof and Humphries (1955) recorded it breeding in open fields, although most accounts indicate that it prefers wooded areas. Males sometimes move between nearby breeding sites within a breeding season, but males also commonly breed in the same location from one year to the next (Green, 1952).

REPRODUCTION

Breeding occurs from late winter to midsummer, depending on latitude and elevation, although

Barbour (1953) recorded egg-bearing females as late as 14 August. Breeding commences shortly after the initiation of calling. In North Carolina, choruses ranged from approximately 30–35 calling males (Schwartz, 1955), and Green (1938b) noted that 15–20 males may call from a shallow pool only 1.5 m in diameter. Males remain at breeding sites during the entire breeding season, whereas females only visit sites on a single night to mate and oviposit their eggs. As a result, reports of sex ratios are highly male-biased (6 males per female in West Virginia, Green, 1938b; 4.4 males per female in eastern Kentucky, Barbour, 1953). The extended breeding-site residency also suggests that males mate more than once during a breeding season (Green, 1952).

Eggs are deposited in small soft gelatinous masses of 3–50 eggs, with most masses containing about 14–34 eggs (Green, 1938b; Barbour and Walters, 1941; Forester et al., 2003). The eggs are attached to vegetation, pool debris, or submerged grasses. Masses do not float, and eggs are often found on the pool substrate (Brown, 1956; Mitchell and Pauley, 2005). Females deposit from several hundred to more than 1,000 eggs during a breeding season. Forester et al. (2003) counted clutches of 90, 108, and 118 eggs. In West Virginia, Green (1938b) recorded total egg counts of 318, 383, 406, and 1,479. In Kentucky, the total number of eggs was 983–1,202 (mean 1,092) (Barbour and Walters, 1941). Oviposition occurs over a period of several hours. Eggs hatch in 3–5 days at laboratory temperatures (18–22°C), but at 7–10 days in the field (2–13°C) (Barbour and Walters, 1941). The newly emerged larvae measure 4.5–5 mm in total length.

LARVAL ECOLOGY

Larval development occurs over a period of about 45 to 60 days (Green, 1938b; Barbour and Walters, 1941; Walker, 1946; Green, 1952). Newly metamorphosed frogs are 8 mm SUL, and frogs 11–13 mm were found in Ohio from mid-June to mid-August.

POPULATION BIOLOGY

Sexual maturity occurs at 22 mm SUL for males and 28 mm SUL for females. Based on a sample of 1,189 marked individuals followed over a 6 yr period, the population of Mountain Chorus Frogs studied by Green (1952) in West Virginia was composed of 71% 1 yr old frogs, 20% 2 yr old frogs, 6.8% 3 yr old frogs, and 1.8% 4 yr old frogs. Green (1952) recorded 2 Mountain Chorus Frogs 5 years after they were initially marked, but suggested that the mean life span was only 1.4 yrs after metamorphosis. These results imply that maturity is reached the first breeding season after metamorphosis, but this has not been verified.

DIET

Prey consists of small ground-dwelling invertebrates. Green (1952) reported that beetles (45%), spiders (25%), and true bugs (Hemiptera) (13%) formed the bulk of the prey items found in 42 individuals. Other prey included ants, leaf hoppers, flies, centipedes, earthworms, and larval lepidoptera (butterflies and moths).

PREDATION AND DEFENSE

According to E.C. Lemmon (*in* Jensen et al., 2008), predators of larvae and adults include dragonfly larvae, aquatic beetles, fishing spiders, fish, salamander larvae (*Ambystoma talpoideum*), Eastern Red-spotted Newts (*Notophthalmus viridescens*), Eastern Garter Snakes (*Thamnophis sirtalis*), water snakes (*Nerodia* sp.), and birds. Predation on adults by American Bullfrogs (*Lithobates catesbeianus*) was recorded by Barbour (1957).

When disturbed while calling, males cease calling or dive to the bottom of the breeding pool and attempt to conceal themselves under debris and grass. In terrestrial situations, they are reported to jump vigorously, resembling juvenile Wood Frogs.

Adult *Pseudacris brachyphona*. Photo: J. Michael Butler

Breeding habitat of *Pseudacris brachyphona*, Powell County, Kentucky. Photo: J. Michael Butler

COMMUNITY ECOLOGY

As noted above, there is a sharp difference in microhabitat preferences between *Pseudacris brachyphona* and the often sympatric *P. feriarum*. The Mountain Chorus Frog prefers wooded areas, whereas the Upland Chorus Frog is found only in open habitats. Although these frogs are often in close proximity to one another, they only rarely breed in the same wetlands or ditches (Wilson, 1945; Walker, 1946; Martof and Humphries, 1955; Barbour, 1957). Even then, they may breed at different times, with the Mountain Chorus Frog later than the Upland Chorus Frog (Walker, 1932). The Spring Peeper, Wood Frog, and American Toad are often found in association with the Mountain Chorus Frog.

DISEASES, PARASITES, AND MALFORMATIONS

No information is available.

SUSCEPTIBILITY TO POTENTIAL STRESSORS

No information is available.

STATUS AND CONSERVATION

Green (1938b) considered this species "fairly common throughout the central and southern part of West Virginia," and Barbour and Walters (1941) described it as "one of the most abundant spring frogs to be found in northeastern Kentucky." However, there is circumstantial evidence that the species has declined since the 1930s, primarily due to the loss of habitat (e.g., Mitchell and Reay, 1999; Davis and Menze, 2000). McClure (1996) found none of Green's dissertation sites had populations of *P. brachyphona* in the early 1990s and that nearby populations were small, scattered, threatened by habitat loss, and skewed toward larger individuals. Still, Weir et al. (2009) found no significant trends in occupancy for West Virginia populations followed over a 7 yr period. Hulse et al. (2001) stated that "all of the reports in the state [of Pennsylvania] are historical. No specimens have been reported in the past 20 yrs or so." Only a single population in Savage River State Forest remained in western Maryland as of 1999 (Forester et al., 2003). Mitchell and Pauley (2005) stated that the Mountain Chorus Frog no longer occurred in North Carolina where Schwartz (1955) found them, but this species was rediscovered in the state in 2001 in Cherokee County (Floyd and Kilpatrick, 2002). A range-wide assessment of the status of this species clearly needs to be conducted. The species is listed as a "Species of Concern" in Georgia and North Carolina.

Pseudacris brimleyi
Brandt and Walker, 1933
Brimley's Chorus Frog

ETYMOLOGY
brimleyi: a patronym honoring Clement S. Brimley (1863–1946), North Carolina zoologist who wrote extensively on southeastern amphibians and reptiles (Cooper, 1979).

NOMENCLATURE
Synonyms: *Hyla brimleyi*

IDENTIFICATION
Adults. This is a long-legged chorus frog with three dark stripes on the dorsum, bordered by a dark dorsolateral line running from the snout through the eye to the groin. The ground color is yellowish to reddish shades of brown; some individuals are dark brown. The lateral two dorsal stripes are often well defined, but the median stripe may be lighter; in other frogs all the dorsal stripes are light. The skin is smooth. A light line is present on the upper jaw and extends to the shoulder. There is no dark triangle between the eyes. The tympanum is distinct and smaller than the eye. There are longitudinal rather than transverse markings on the legs, and there is a dark line along the outer edge of the tibia. There is no webbing on the front digits, and the rear digits are only weakly webbed. The toe disks are only slightly enlarged. The venter is distinctly yellow with small dark spots on the chest; the amount of spotting is variable. Males have a dark vocal sac. Females are larger than males. Wright and Wright (1949) give a male range of 24–28 mm SUL and a female range of 27–35 mm SUL.

Larvae. The ground color of larval *P. brimleyi* is dark with a greenish tinge, often with scattered gold, yellow, or brassy flecks over the body and dorsal tail musculature. Bodies are deep with a broadly rounded head. The chin and throat are darkly pigmented, with flecks or blotches. The tail musculature is sharply bicolored or striped, with a dorsal light stripe that extends forward through the eye to near the snout. The dark tail muscle pigmentation and striped pattern may be red orange. Tail fins are sparsely speckled. Venters are heavily spotted. Tadpoles reach about 30 mm TL.

Eggs. The eggs are dark brown to black dorsally and white ventrally. There is a single jelly envelope surrounding the egg, which averages 7.5 mm in diameter (range 6.75 to 8.64 mm); the mean vitellus diameter is 1.45 mm (range 1.3 to 1.71 mm) (Gosner and Black, 1958). Eggs are deposited in loose clumps.

DISTRIBUTION
Brimley's Chorus Frog occurs from eastern Virginia south to eastern Georgia on the Atlantic Coastal Plain. The record for northern Georgia (Brandt and Walker, 1933) is in error. The species has been recorded from Roanoke Island, North Carolina (Gaul and Mitchell, 2007). Important distributional references include: Georgia (Jensen et al., 2008), North Carolina (Dorcas et al., 2007), and Virginia (Tobey, 1985; Mitchell and Reay, 1999).

FOSSIL RECORD
No fossils are known.

SYSTEMATICS AND GEOGRAPHIC VARIATION
A number of workers have suggested hypotheses concerning the evolutionary relationship

Distribution of *Pseudacris brimleyi*

Tadpole of *Pseudacris brimleyi*. Photo: Steve Bennett

between *P. brimleyi* and other trilling chorus frogs (Hedges, 1986; Cocroft, 1994; Da Silva, 1997), although the results often were not congruent. It is clearly allied with other *Pseudacris* based on its albumins (Maxson and Wilson, 1975). *Pseudacris brimleyi* is a member of the Trilling Chorus Frog clade and is most closely related to *P. brachyphona* of the Appalachian Mountains (Moriarty and Cannatella, 2004). These species diverged during the Miocene about 4.6 mya, thus the Appalachian orogeny had no impact on speciation since it occurred much earlier (Lemmon et al., 2007a). Instead, competition between an ancestral *P. brachyphona-brimleyi* species and an expanding ancestral *P. feriarum-kalmi-triseriata* species may have bisected the range leading to *P. brachyphona* in the mountains and *P. brimleyi* on the coast. The *P. brachyphona–P. brimleyi* clade is sister group to the *Nigrita* clade.

Successful hybridization in the laboratory has been reported with *P. feriarum, P. ornata, P. nigrita,* and *P. brachyphona* (Mecham, 1965).

ADULT HABITAT
Brimley's Chorus Frog is a species of low swampy woodlands along the Atlantic Coastal Plain.

TERRESTRIAL ECOLOGY
Little is known of the terrestrial ecology of this species. Presumably they disperse to fields and swampy woodlands surrounding breeding sites. They forage terrestrially, hiding under woody debris, leaf litter, and downed logs by day and emerge at night to feed.

Brimley's Chorus Frogs are photopositive in their phototactic response, suggesting they use both sunlight and moonlight when feeding and going about their daily activities (Jaeger and Hailman, 1973). They are sensitive to light in the blue spectrum ("blue-mode response"), which apparently helps them orient toward areas of increasing illumination (Hailman and Jaeger, 1974). Brimley's Chorus Frogs likely have true color vision.

CALLING ACTIVITY AND MATE SELECTION
Males call from the base of grass clumps and adjacent shallow water. The call is described as a short, rasping trill of 15–22 pulses ("kr-r-r-a-k") lasting about 0.25 sec (Jensen et al., 2008); this is uttered in a series with short intervals in between. Males compete for calling positions and will drive off intruders. Calling may precede egg deposition by one to two weeks (Mitchell, 1986). Amplexus is supra-axillary.

BREEDING SITES
Breeding occurs in heavily wooded shallow (15–20 cm) grassy temporary pools in wet forest and floodplains. Gosner and Black (1958) recorded breeding in a flooded field adjacent to a flooded pine woodland, at the edge of red maple swamps, and in scrub thickets and roadside ditches. Breeding sites may be located at some distance from the nearest woodlands.

REPRODUCTION
Breeding occurs in late winter to spring at air temperatures >4.5°C. Calling, amplexus, and egg deposition have been observed from February to April in North Carolina (Brandt and Walker, 1933; Gaul and Mitchell, 2007) and Virginia (Werler and McCallion, 1951; Mitchell, 1986). Gosner and Black (1958) counted egg clutches of 264 and 290 in 2 North Carolina females. The

Adult *Pseudacris brimleyi*. Photo: Todd Pierson

Breeding habitat of *Pseudacris brimleyi*, Hampton County, South Carolina. Photo: C.K. Dodd Jr.

eggs are deposited in multiple clumps on vegetation and woody debris just below the water's surface. Eggs hatch in 4.5–5.5 days at 18–20°C, although Jensen et al. (2008) gave a figure of 1–2 weeks.

LARVAL ECOLOGY
The larval period is about 35–60 days (Gosner and Black, 1958). Newly metamorphosed froglets are 9–11 mm SUL.

DIET
Nothing appears to be published on the feeding habits of *P. brimleyi*. They likely feed on small invertebrates, especially insects.

PREDATION AND DEFENSE
The ground coloration makes this species cryptic and hard to see around breeding ponds and in woodland terrestrial habitats. Adults are preyed upon by Eastern Garter Snakes (*Thamnophis sirtalis*), watersnakes (*Nerodia* sp.), and other vertebrates. Larvae are eaten by mole salamanders (*Ambystoma* sp.), newts (*Notophthalmus viridescens*), fishing spiders, dragonfly larvae, predaceous beetles, fish, and birds (Jensen et al., 2008).

POPULATION BIOLOGY
Sexual maturity is probably attained by the first spring after metamorphosis, but nothing else is known about the species' demography and population characteristics.

COMMUNITY ECOLOGY
The Savannah River Site (SRS) is one of the best-studied locations in the United States in terms of its amphibian community. For > 50 yrs, no records of *P. brimleyi* were made at this site despite intensive long-term sampling. In 2007, however, Luhring (2008) recorded Brimley's Chorus Frog from the southern part of SRS. Either the frogs were absent for decades and had recolonized the site, or they had just been overlooked because of their small, scattered populations. Luhring (2008) suggested that the rediscovery of *P. brimleyi* illustrated "hidden biodiversity" that can be overlooked during even intensive faunal surveys.

DISEASES, PARASITES, AND MALFORMATIONS
Parasites include the protozoans *Nyctotherus cordiformis*, *Octomitus intestinalis*, *Opalina chorophili*, *O. hylaxena*, *O. oblanceolata*, *O. obtrigonoidea*, *O. pickeringii*, *O. virguloidea*, *Trichomonas augusta*, *Trypanosoma rotatorium*, and an unidentified flagellate (Brandt, 1936). Other parasites include the trematodes *Brachycoelium hospitale* and *Diplodiscus temperatus*; the nematodes *Agamascaris odontocephala*, *Agamonema* sp., *Cosmocercoides dukae*, *Ozwaldocruzia pipiens*, *Physaloptera* sp., *Rhabdias* sp., and *R. ranae*; and the acanthocephalan *Centrorhynchus* sp. (Brandt, 1936).

SUSCEPTIBILITY TO POTENTIAL STRESSORS
No information is available.

STATUS AND CONSERVATION

Brimley's Chorus Frog appears to have declined on the southeastern Coastal Plain, especially in areas of urban expansion. The species is treated as of "Special Concern" in Georgia, where there are few recent records. In North Carolina, it is considered "as a species in need of monitoring." An assessment of the species' status is overdue.

Pseudacris cadaverina (Cope, 1866)
California Treefrog

ETYMOLOGY

cadaverina: from the Latin *cadaver* meaning 'corpse,' and *-ina* meaning 'having the appearance of.' The name refers to the pale coloration of the species.

NOMENCLATURE

Synonyms: *Hyla affinis* [in part], *Hyla arenicolor* [in part], *Hyla cadaverina*, *Hyla californiae*, *Hyla nebulosa* [in part]

The name *H. cadaverina* was assigned to the species by Cope (1866) as a replacement name for *H. nebulosa* Hallowell, 1859, since the name *H. nebulosa* was already occupied by a South American frog. Gorman (1960) incorrectly stated that Cope (1866) used *H. cadaverina* as a replacement name for *H. affinis* and cited a publication title by Cope that is nonexistent. Gorman (1960) then described this species as *Hyla californiae* and noted confusion in the identification of specimens recorded as *Hyla arenicolor* from California in 1875. In the early literature (e.g., Storer, 1925; Slevin, 1928; Stebbins, 1951), accounts of *H. arenicolor* may include information on *H. cadaverina* intermingled with it.

IDENTIFICATION

Adults. This is a small to medium-sized (to about 51 mm SUL) treefrog. The gray to sandy dorsum is slightly blotched or spotted, and the skin appears somewhat warty (more evenly spaced on males than females); there are no parallel dorsal stripes. This coloration blends well with the rocks on which it sits, gray being common in granite areas with sandy or light brown colors on sandstone backgrounds. This species does not have a black or brown lateral stripe through the eye. The toes have larger discs on the ends of the digits, and the webbing, while reduced from that of aquatic species, is more extensive than that of the Pacific Treefrog. Venters are white with the undersides of the legs, groin, and lower abdomen a lemon yellow. Males have a dusky or yellowish throat and are smaller than females. The mean male size ranged from 27.9 to 36.4 mm SUL at 26 populations; females averaged 34.5 to 42.4 mm SUL at 10 populations (Ball and Jameson, 1970). Frogs from xeric habitats are larger than those from mesic habitats (Ball and Jameson, 1970). These authors provide additional morphological measurements.

Larvae. Pigmentation is from light brown to dark brown. Golden flecks may be evident dorsally and laterally, and these may increase during ontogeny. Larval *Pseudacris cadaverina* are more elongate than larval *P. regilla*, and the tail fins are not as high. The greatest tail fin depth occurs well behind the anus, down about half or more of the length of the tail. The body tapers slightly from a point immediately behind the eyes toward the posterior part of the body when viewed dorsally, and it is more flattened than the body of *P. regilla* larvae. The dorsal tail fin tapers to a point at its tip, and the ventral tail fin is somewhat parallel to the body when viewed laterally. The eyes are small and entirely dorsally oriented when viewed from above. The dorsal part of the body is pigmented, but the venter and ventral tail fin are much less so, especially as the tadpole grows. When viewed dorsally, the tail musculature has regularly spaced transverse light bars or chevrons, and these become more pronounced with size. The dorsal tail fin is mottled. The tail musculature is heavily pigmented, much more so than the dorsal tail fin. The intestines are visible through the ventral body wall as the tadpole grows, but may not be visible during early growth stages. Gaudin (1964, 1965) provided descriptions of larvae and larval ontogeny and illustrations of larvae and mouthparts.

Eggs. The eggs of *P. cadaverina* are bicolored, dark dorsally and white ventrally. Eggs are deposited singly, and each egg has a single gelatinous envelope. The egg is 1.8–2.4 mm in diameter (mean 2.07 mm), and the gelatinous envelope is 3.9–5 mm in diameter (mean 4.4 mm) (Storer, 1925; Gaudin, 1965). Although the eggs are deposited singly, they tend to stick together.

DISTRIBUTION

The California Treefrog occurs from the lower Salinas River canyon and associated tributaries in central California to about 29°N latitude in Baja California, Mexico. The eastern limits of its range are the western fringes of the Mojave Desert northwest of the Salton Sea. Important distributional references include Storer (1925), Gorman (1960), and Lemm (2006).

FOSSIL RECORD

There are no fossils known for this species.

SYSTEMATICS AND GEOGRAPHIC VARIATION

Although described as *Hyla cadaverina*, this species was transferred to the genus *Pseudacris* by Hedges (1986), based on an analysis of allozyme phylogeny. Hedges (1986) further recognized the distinctiveness of a western clade of *Pseudacris* that included *P. regilla* and *P. cadaverina*. Cocroft (1994) suggested that a generic reallocation was unnecessary based on morphological data, and recommended that *P. cadaverina* remain in *Hyla*. Da Silva (1997) used additional morphological information to return *cadaverina* to the genus *Pseudacris*.

The most recent assessment of *Pseudacris* phylogeny is the study by Moriarty and Cannatella (2004) that examined 12S and 16S mtDNA from 38 populations of chorus frogs from throughout North America. They concluded that the California Treefrog was a member of the West Coast Chorus Frog clade, along with *P. regilla*. The recognition of a monophyletic West Coast clade, thus, is supported by immunological (Maxson and Wilson, 1975), allozyme (Hedges, 1986), morphological (Cocroft, 1994; Da Silva, 1997), and mtDNA (Moriarty and Cannatella, 2004) data.

The current distributional pattern of *Pseudacris cadaverina* was established during the Pleistocene. The center of speciation was the eastern Transverse Ranges of southern California. There are three major haplotype groupings (northern, central, southern), with a major break within the Transverse Range, as populations were fragmented by mountains and watersheds. Limited amounts of gene flow are evident, with some desert populations having been isolated over a considerable period (Phillipsen and Metcalf, 2009).

Hybridization is rare in nature, except perhaps in one general area. Brattstrom and Warren (1955) reported a possible *P. regilla* × *P. cadaverina* hybrid from San Diego Co., California. Gorman (1960) noted that hybrids seemed "especially plentiful" in central Los Angeles Co. Phillipsen and Metcalf (2009) did not mention any evidence of hybridization. Laboratory crosses with *Hyla arenicolor* were not successful (Pierce, 1975).

ADULT HABITAT

Pseudacris cadaverina prefers shady, rocky areas with granite boulders adjacent to swift-moving permanent streams and spring-fed warm water ponds. It occupies rocks within the swift current, or it forages among rocks and rock crevices immediately adjacent to shallow water. The banks bordering streams are often steep. This species is seldom found very far from water, although Cunningham (1955b) reported an adult 46 m from a stream down a rodent burrow and a juvenile 69 m from water. It occurs from near sea level to 2,290 m (Lemm, 2006).

Distribution of *Pseudacris cadaverina*

Embryos of *Pseudacris cadaverina*. Photo: Chris Brown

Tadpole of *Pseudacris cadaverina*. Photo: Chris Brown

TERRESTRIAL ECOLOGY

Pseudacris cadaverina is active both by day and night, although there is conflicting information in the literature about whether it is primarily diurnal or nocturnal (Cunningham, 1955b; Lemm, 2006). Activity occurs from mid-February through early October, and activity is influenced by temperature and moisture. At some locations, individuals may be active year-round to some extent. Populations near the California coast are more likely to be active for a longer period than are populations in the cooler high deserts. Individuals are tolerant of desiccation, and can lose up to 35% of their body water before succumbing to moisture loss (Cunningham, 1955b). They can also store up to 25% of their body weight in the form of dilute urine (McClanahan et al., 1994).

California Treefrogs do not venture far from water, often sitting on boulders in the sun but within a quick jump of the water. While sunning, adult body temperatures may be considerably higher than ambient air and water temperatures.

For example, Brattstrom (1963) recorded body temperatures of frogs sitting in the direct sun of 20.8–26.2°C when the water temperature was 21°C and the air temperature 19.8°C. California Treefrogs can maintain their body temperature at <30°C through evaporative cooling, even when air temperatures are 40°C (McClanahan et al., 1994).

California Treefrogs are likely to be photopositive in their phototactic response, suggesting they use both sunlight and moonlight when feeding and going about their daily activities. They have not been tested, however. They probably are sensitive to light in the blue spectrum ("blue-mode response"), which would help them orient toward areas of increasing illumination. California Treefrogs likely have true color vision.

CALLING ACTIVITY AND MATE SELECTION

The call of the California Treefrog has been described as a duck-like quack (Lemm, 2006), although the call also has been characterized as "soft and monotonous" (Gorman, 1960). Males call from rocks or the shoreline immediately adjacent to breeding sites while facing the stream. Only very rarely do they call from water. Calls are uttered for a short period, after which males are silent for a spell before calling again. They are territorial and will defend their calling position from other males; the call serves to warn other males of a caller's territory. If an intruder approaches too closely, a resident male will approach an intruder (Ervin, 2005), and a vigorous wrestling match may ensue.

The characteristics of the mating call of *P. cadaverina* have been described by a number of authors (Snyder and Jameson, 1965; Ball and Jameson, 1966; Littlejohn, 1971), sometimes drawing different conclusions from one another as to the distinctiveness of the call of this species from *P. regilla*. Littlejohn (1971) demonstrated, however, that there are differences in call duration, pulse repetition rate, and dominant frequency, especially where these species are sympatric. Indeed, Straughan (1975) suggested the pulse repetition rate was the only variable necessary for call discrimination. In *P. cadaverina*, the call duration averages 148.4 ms, the pulse repetition rate averages 142.4 pulses/sec, and the dominant frequency ranges between 1,950 and 2,200 Hz at 16°C (Littlejohn, 1971). Gorman (1960) recorded two major dominant frequencies, one at 1,200 Hz and the other at 3,600 Hz; the full range of frequencies was recorded as 800–4,000 Hz. He also noted a great amount of variation among individuals.

BREEDING SITES
Breeding occurs in quiet pools associated with swift-moving permanent streams in rocky canyon country. Quiet pools and stream reaches without fish are preferred.

REPRODUCTION
The initiation of the breeding season and its duration is somewhat flexible and dependent on the late winter to early spring rainfall. Breeding occurs a few days after rainstorms, when very turbulent waters have ceased being a danger to the frogs. When deep pools have formed and sedimentation has ceased, reproduction commences. Calls begin in mid-March in the western-facing canyons, and females with eggs are found from mid-April to mid-August. Cunningham (1955b) found amplexing pairs from mid-April to late June. In the eastern canyons facing the deserts, breeding occurs later in the year and perhaps only for a very short time following rainfall.

Adult *Pseudacris cadaverina*, green phase. Photo: Brian Hubbs

Eggs are deposited singly and are attached to vegetation or debris at or near the surface of the water, although they may form loose clusters. Eggs also may be found resting on the bottom of quiet pools within rocky streambeds. The jelly coat of the eggs is very sticky so that they readily attach to objects. Eggs may be distributed downstream by the current of the stream or by the amplectant pair moving from one location to another. Hatching occurs in 5–6 days, and hatchlings are about 8.0 mm TL (Storer, 1925).

LARVAL ECOLOGY

Development requires about 75 days, and recent metamorphs are 18.0–20.3 mm SUL (Storer, 1925). Little is known about larval ecology. They appear to prefer warmer water along the margins of streams or pools, and retreat quickly into deeper water or among algal mats and submerged vegetation when threatened.

The main threats to embryonic and larval California Treefrogs are excessive heat or desiccation of their breeding pools. Larvae have been found in water temperatures ranging from 14 to 32°C, and they tend to occupy the warmer parts of the breeding sites (Cunningham, 1955b). Cunningham (1955b) further noted that they would lie on floating vegetation in such a manner that their backs were out of water and exposed to the direct sunlight. Water at breeding sites can reach much higher temperatures than 32°C, especially in the direct desert sun, at which time larvae need to find shade or deeper water. When disturbed, the larvae flee to deep water and attempt to bury themselves vertically in the substrate.

DIET

The diet consists of small invertebrates, including spiders, grasshoppers, ants, terrestrial beetles, moths, isopods, neuropterans, true bugs, and even diving beetles (Cunningham, 1955b).

PREDATION AND DEFENSE

Pseudacris cadaverina is a cryptic species that blends in well on a rocky substrate, especially if they remain immobile. They generally remain

Color variation in *Pseudacris cadaverina*. Photo: Chris Brown

close to water. When disturbed, California Treefrogs jump to the bottom of the water where they submerge themselves in the substrate. However, they do not remain submerged for long and climb back up to boulders or rocks within a few minutes. Jumping into a swift current may carry them downstream. They will void their bladder when handled and assume an immobile response with the limbs held tightly into the body and the eyes closed. The body is curved in such a manner as to assume a balled-up posture. Juveniles will skitter across the water for a short ways before diving toward the substrate. Like the adults, they do not remain submerged for long, and quickly return to the surface (Cunningham, 1955b).

The primary predator is the Two-striped Garter Snake *Thamnophis hammondii*. Fish (Rainbow Trout, Largemouth Bass, Green Sunfish) also have been reported as predators (Ervin et al., 2000; Hovey and Ervin, 2005). Other snakes, birds, and mammals likely prey on postmetamorphs, and predaceous invertebrates on the aquatic larvae.

COMMUNITY ECOLOGY

California and Pacific Treefrogs occupy the same general regions and stream drainages, but tend to partition the habitat according to specific microenvironmental preferences (e.g., along slow-moving vs. fast-moving waters). California Treefrogs are also much less abundant than sympatric Pacific Treefrogs. However, as drought conditions cause habitats to dry in summer, individuals of these two species may be brought into close contact. Despite this, there is no evidence of extensive hybridization between them, except possibly in some areas of Los Angeles County (Gorman, 1960), even though they may simultaneously use the same waters for reproduction. Despite assertions in Ball and Jameson (1966), differences in call characteristics provide a premating isolating mechanism to prevent introgression between these closely related taxa.

DISEASES, PARASITES, AND MALFORMATIONS

California Treefrogs are parasitized by the trematodes *Alaria* sp., *Fibricola* sp., *Gorgoderina* sp., *Langeronia burseyi*, *Ribeiroia* sp.; the cestode *Distoichometra bufonis*; and the nematodes *Physaloptera* sp. and *Rhabdias ranae* (Dailey and Goldberg, 2000; Goldberg and Bursey, 2001a,

Habitat of *Pseudacris cadaverina*. Photo: Chris Brown

2002a; Ervin, 2005). Two protozoans (*Opalina* sp., *Balantidium* sp.) have been found in the intestines (Ervin, 2005). Chiggers (*Hannemania hylae*) embed themselves under the skin (Welbourn and Loomis, 1975; Ervin, 2005). A single *P. cadaverina* with extra legs was reported by Ervin (2005).

SUSCEPTIBILITY TO POTENTIAL STRESSORS

UV light. Embryos exposed to direct ambient levels of UVB radiation experience high levels of mortality, a situation that can be reversed by filtering UVB wavelengths. However, UVB radiation has no effect on the duration of development on surviving embryos (Anzalone et al., 1998).

Ozone. Four-hour exposures of the California Treefrog to ozone concentrations of 0.2–0.8 ppm causes an alteration in respiration, including depressed lung ventilation (measured by buccal pumping and flank movements) and reduced oxygen consumption. Frogs also assume body postures that reduced the exposed body surface to the atmosphere (Mautz and Dohm, 2004). Ozone irritates the wet tissue membranes such as the eyes and respiratory epithelia.

STATUS AND CONSERVATION

There are no studies on the status and trends of this species. At present, however, there does not appear to be cause for concern about California Treefrog populations. Presumably, individuals hide deep in canyon crevices during the periodic wildfires that sweep through its habitat.

Pseudacris clarkii (Baird, 1854)
Spotted Chorus Frog

ETYMOLOGY
clarkii: a patronym honoring Lt. John Henry Clark (*b* 1830), a zoologist with the U.S.–Mexican Boundary Survey. Clark made extensive collections, including many new species.

NOMENCLATURE
Synonyms: *Chorophilus triseriatus clarkii, Helocaetes clarkii, Heloecetes clarkii, Holocoetes clarkii, Hyla clarkii, Pseudacris nigrita clarkii, Pseudacris triseriata clarkii*

Information on *P. clarkii* may be listed in older publications as simply *P. nigrita* or *P. triseriata* (see Lord and Davis, 1956, for discussion).

IDENTIFICATION
Adults. Pseudacris clarkii is an attractive frog with green dorsal markings surrounded by thin black or dark gray borders on a light gray or greenish background. The green markings may appear as rows of spots and even coalesce to give the appearance of stripes or bars; the borders of the stripes are quite irregular. Some frogs may have a dark or green triangle between the eyes. There is a distinct dark or greenish band beginning at the snout and continuing through the eye toward the groin. A light line is present on the upper jaw. The limbs have dark green bands or spots. Venters are white. Males have dark brown throats. Both sexes can have spots or stripes, despite Wright and Wright's (1933) speculation concerning sexual dimorphism in this character. The mean adult size in Texas was 29.1 mm SUL (Lord and Davis, 1956), but Collins et al. (2010) give a normal adult size of 19–28 mm SUL in Kansas. Smith (1934) measured 4 individuals from 29.7–31 mm SUL in Kansas. Wright and Wright (1949) reported males at 20–29 mm SUL and females at 25–31 mm SUL.

Larvae. The ground color of a mature tadpole is grayish olive. The tail musculature is darkly pigmented above the lateral line, and the tail fins are splotched with pigment, the dorsal fin more so than the ventral fin. The snout is rounded in lateral view. Maximum size is 30–31 mm TL. Bragg (1943b) and Eaton and Imagawa (1948) illustrated larval mouthparts, embryos, and larvae, and described developmental stages.

Eggs. Eggs are pale brown, brownish gray, or dark gray dorsally and light gray, white, or ivory yellow ventrally. The eggs have a mean diameter of 2.3 mm (range 2.1 to 2.7) (Grubb, 1972), although Bragg (1943b) gave the mean diameter as 1.04 mm (range 0.99 to 1.3 mm), and Eaton and Imagawa (1948) gave a mean of 1.28 mm (range 1 to 1.66 mm); it seems likely that these latter measurements did not include one or more of the jelly envelopes, whereas Grubb's observations did. Wright and Wright (1949) reported the vitellus as 0.65–0.9 mm in diameter, with two surrounding jelly envelopes. The outer envelope is loose and 2.2–2.4 mm in diameter, whereas the inner envelope is 1.4–1.8 mm in diameter (Bragg, 1943b).

DISTRIBUTION
Spotted Chorus Frogs occur from south-central Kansas south to the Texas Gulf Coast and slightly into adjacent Mexico. They occur throughout much of the Texas Panhandle and east of the Balcones Escarpment almost to Arkansas. An isolated population was reported from central Montana (Black, 1970) but was based on misidentified *P. maculata* (see Corn, 1980b; Maxwell et al., 2003). Important distributional references include: Kansas (Collins et al., 2010), Oklahoma (Sievert and Sievert, 2006), and Texas (Burt, 1936; Brown, 1950; Dixon, 2000).

FOSSIL RECORD
Fossils of *P. clarkii* are known from Pleistocene locations in Nebraska and Texas (Holman, 1969,

2003). Differences in the scapula, radio-ulna, sacral condyles, and ilium separate this species from other *Pseudacris*.

SYSTEMATICS AND GEOGRAPHIC VARIATION
Pseudacris clarkii is a member of the Trilling Chorus Frog clade and is most closely related to *P. maculata* of central North America (Moriarty and Cannatella, 2004) rather than *P. nigrita* as indicated by Hedges (1986), Cocroft (1994), and Da Silva (1997). This group of frogs has the lowest amounts of genetic variation among members of the genus *Pseudacris* (Lemmon et al., 2007a). Lemmon et al. (2007a) attributed this to aridification of the Great Plains. It is clearly allied with other *Pseudacris* based on its albumins (Maxson and Wilson, 1975).

Burt (1936) suggested intergradation with *P. fouquettei* (as *P. triseriata*) in southern Kansas and eastern Texas, but this is disputed by Smith (1934) and Bragg (1943a). Laboratory hybridization between *P. streckeri* and *P. clarkii* produces tadpoles that complete metamorphosis successfully (Mecham, 1957). Crosses between *P. clarkii* and *Hyla versicolor* complex, *Pseudacris fouquettei* (possibly listed as *P. triseriata* by Moore, 1955), *P. ornata* or *P. feriarum* (listed as *P. nigrita* by Mecham, 1957) may produce larvae and metamorphs of uncertain viability. Crosses with *Acris blanchardi, Hyla cinerea, H. arenicolor,* or *Pseudacris regilla* are not successful (Moore, 1955; Littlejohn, 1961a; Pierce, 1975).

ADULT HABITAT
The Spotted Chorus Frog is a species of the short grass and mixed prairies and prairie islands in wooded savanna. Along the eastern border of its range, it may be found along woodland borders and islands of prairie in the oak-hickory savanna, but this is not a woodland or floodplain species (but see Bragg, 1941c).

TERRESTRIAL ECOLOGY
Most activity occurs during the spring and summer after rains. In Kansas, activity occurs from March to September (Collins et al., 2010). Spotted Chorus Frogs have been found under rocks and debris near breeding ponds and may use tunnels and burrows of other animals during the nonbreeding season. They migrate from terrestrial sites to breeding ponds, perhaps using olfactory cues. Grubb (1973) noted that reproductively active male *P. clarkii* were capable of detecting odors in water from ponds in which they had previously bred as opposed to water from "foreign" ponds or distilled water. This ability persisted in trials weeks after breeding and suggests the possibility that breeding adults might return to the ponds from which they metamorphosed, or at least return to ponds at which they bred previously. *Pseudacris clarkii* is occasionally found around the entrances of caves (Black, 1973a).

Most nonbreeding activity occurs at night, when frogs forage in pastures and fields (Bragg, 1943b). Activity tends to cease in the hot, dry late summer. Spotted Chorus Frogs are photopositive in their phototactic response, suggesting they can use both sunlight and moonlight when feeding and going about their daily activities (Jaeger and Hailman, 1973). They are sensitive to light in the blue spectrum ("blue-mode response"), which apparently helps them orient toward areas of increasing illumination (Hailman and Jaeger, 1974). Spotted Chorus Frogs likely have true color vision.

Distribution of *Pseudacris clarkii*

CALLING ACTIVITY AND MATE SELECTION
Pseudacris clarkii is an opportunistic breeder that can form breeding choruses just about any time except during midsummer (Bragg, 1943b). Most calling occurs in the winter and spring. For example, Blair (1961b) recorded calling in Texas from February to May, but noted a few instances of calling from late August to early October depending upon year. Wiest (1982) heard calls

Tadpole of *Pseudacris clarkii*. Photo: Laurie Vitt

from January to June in Texas. Lindsay (1958) also recorded amplexus in October. Bragg (1950a) reported calling in Oklahoma from March to September. Further to the north in Kansas, breeding occurs from March to August with a peak in April to June (Collins et al., 2010). Choruses form rapidly with peak calling occurring soon after heavy rainfall, and chorusing continues both day and night. Choruses are heard at air temperatures of 8–27°C (Blair, 1961b). Wiest (1982) noted calling at 3–23.2°C over a 60-day period in Texas.

Males call from the center of dense grass clumps, which makes them difficult to locate. Calls are made rapidly and in succession. The dominant frequency is 2,850 cps with a trill rate of 75/sec and a duration of 0.21 sec (Michaud, 1962). In choice tests, female *P. clarkii* favor the calls of conspecific males over those of *P. fouquettei* (Michaud, 1962). Amplexus is axillary.

BREEDING SITES

Breeding occurs in open shallow clear temporary ponds and pools of the arid Great Plains, such as those that form in meadows, on playas, or in pastures and roadside ditches (Bragg, 1943b). Such pools have grassy vegetation onto which frogs hold while calling. The species also uses buffalo wallows and pools in agricultural regions, as long as tadpoles have enough time to complete metamorphosis. Other wetland types include mesquite ponds, shallow water lily ponds, and occasionally ponds in floodplains (Bragg, 1941c). Ponds should have relatively tall (36 cm) and dense (213 stems/m^2) vegetative cover. *Pseudacris clarkii* prefers breeding sites with extensive vegetative cover (52%), low aluminum concentrations, a more neutral pH (7.2 vs. 7.7), and deeper water (mean 22.5 cm), at least in the playas of west Texas (Anderson et al., 1999); adjacent land use, oxygen, conductivity, temperature, nitrate, and phosphate did not affect occupancy. *Pseudacris clarkii* does not occur in ponds with fish, as its eggs are readily palatable (Grubb, 1972), but it is occasionally found at permanent ponds.

REPRODUCTION

Egg masses (6–37 eggs/mass) are deposited on upright vegetation, such as dead weed stems, sedges, and grasses, just below the water's surface. Bragg (1943b) provided a total clutch size of 916 eggs from a single female, whereas Wright and Wright (1949) recorded a clutch of 154 eggs (14–37 eggs/mass) from a single female. Blair (1961b) suggested that females may produce more than one set of eggs per year because of the prolonged and opportunistic breeding season. Eggs hatch in two to ten days. At hatching the gray larvae are 3.8–4.7 mm TL (Bragg, 1943b).

LARVAL ECOLOGY

Nothing is known about larval ecology. After metamorphosis, young may remain in the vicinity of the breeding pool for three to four days before gradually dispersing (Bragg, 1943b). Dispersal by recent metamorphs is diurnal.

DIET

No published information is available, but Spotted Chorus Frogs likely opportunistically consume a variety of small invertebrates.

PREDATION AND DEFENSE

Upon disturbance, males cease calling and dive underwater, where they hide in the substrate and among submerged vegetation. The call has a ventriloquist effect that makes it difficult to locate a calling male. The eggs of this species are readily palatable to mosquitofish (*Gambusia*) (Grubb, 1972). Postmetamorphs likely are eaten by a wide variety of vertebrate predators.

POPULATION BIOLOGY

No information is available.

Adult *Pseudacris clarkii*.
Photo: C.K. Dodd Jr.

COMMUNITY ECOLOGY

Strecker's and Spotted Chorus Frogs often are found at the same breeding sites, as their habitat preferences and breeding seasons overlap considerably. The main isolating mechanisms between them appear to involve call discrimination and female preference. In experimental trials, female *P. clarkii* are able to discriminate and choose the calls of male conspecifics over those of *P. streckeri* males (Littlejohn, 1961b). Lord and Davis (1956) noted that *P. fouquettei* (as *Pseudacris nigrita triseriata*) nearly always bred at the same pools as *P. clarkii*; they never observed interspecific amplexus, however.

DISEASES, PARASITES, AND MALFORMATIONS

The myxozoan parasite *Myxidium serotinum* is known from *P. clarkii*, as are the protozoans *Opalina* sp., *Hexamita intestinalis*, *Tritrichomonas augusta*, and *Nyctotherus cordiformis*, the cestode *Cylindrotaenia americana*, the nematode *Cosmocercoides variabilis*, and the mite *Hannemania* sp. (McAllister, 1991).

SUSCEPTIBILITY TO POTENTIAL STRESSORS

No information is available.

STATUS AND CONSERVATION

Spotted Chorus Frog populations are relatively tolerant of agriculture and rangeland as long as vegetative cover is maintained and hydroperiods of breeding ponds are sufficient for larval development (Anderson et al., 1999). Nothing is known concerning status and population trends, although populations undoubtedly have been lost especially in rapidly urbanizing areas (e.g., Bragg, 1952).

Pseudacris crucifer (Wied-Neuwied, 1838)
Spring Peeper
Rainette crucifère

ETYMOLOGY

crucifer: from the Latin *cruces*, meaning 'cross,' and *-ifer*, meaning 'bearer.' The name refers to the X on the dorsum of this small frog.

NOMENCLATURE

Synonyms: *Acris pickeringii*, *Hyla crucifer*, *Hyla crucifer bartramiana*, *Hyla crucifera*, *Hyla pickeringii*, *Hyliola pickeringii*, *Hylodes pickeringii*, *Parapseudacris crucifer*

IDENTIFICATION

Adults. Adult Spring Peepers are light tan to dark brown with a distinctive dark-colored X on the back. In a few individuals, the arms of the X may approach one another but not actually come into contact. The frogs may appear darker during the day than at night. A dark V-shaped line connects the eyes. The belly is light and normally unmarked, although a few animals may have a slightly spotted venter. The legs are banded above but light underneath, and the underside of the rear legs is light to lemon yellow on the femur. The toe tips are only faintly expanded, and there is no webbing between the toes. Males have a black vocal pouch that is evident throughout the year, although it is more pronounced in the breeding season. Males also have darkly pigmented testes. Juveniles are colored like adults, but the pattern may not be as evident, and the ground color is usually light tan.

Adult males are slightly smaller than adult females. In Florida, males are 23–30 mm SUL (mean 27.7 mm) and females 29–34 mm SUL (mean 31.3 mm) (Owen, 1996); in New York, males are 18–30 mm (mean 24.4 mm) and females 23–33 mm SUL (mean 28.3 mm) (Oplinger, 1963); and in Ohio, a male mean was 28.6 mm SUL and a female mean 30.1 mm SUL (Gatz, 1981b). Other size data include males > 23.0 mm SUL and females 26.2–29 mm SUL in North Carolina (Alexander, 1966); males 21–34 mm SUL (mean 27.9 mm) in Maine (Sullivan and Hinshaw, 1990); males 25–33 mm SUL (mean 29.9 mm) in Connecticut (Flores, 1978); males 21–30 mm SUL (mean 24.6 mm) and females 23–35 mm SUL (mean 28 mm) in Connecticut (Klemens, 1993); males 25–32 mm SUL (mean 27.8 mm) in Nova Scotia (Bleakney, 1952); males 20–33 mm and females 27–37 mm in Nova Scotia (Gilhen, 1984); males 27–35 mm (mean 31.4 mm) and females 34–36 mm (mean 35 mm) on Prince Edward Island (Cook, 1967). Body length and mass are highly correlated for both males and females (Owen, 1996).

Larvae. Tadpoles of this species are small and deep bodied with a medium-sized tail. The tail musculature is mottled, but the fins are clear or with blotches. Gosner and Black (1957b) noted two distinct tail morphologies: one narrow and thin and the other with a very wide tail fin. There are no dots on the grayish to light brown body. When viewed from above, the snout is square-like. The larval mouthparts have two rows of marginal papillae, and the second anterior tooth row is longer than the first anterior tooth row. Larvae in the south are 8–10% larger than northern larvae (Gosner and Rossman, 1960). Mean length at hatching is 4.46 mm in Florida and 4.21 mm in New Jersey (range 4 to 4.25 mm; Gosner and Black, 1957b). At transformation, larvae are 33–39 mm in length (Harper, 1939a; Wright and Wright, 1949; Gosner and Black, 1957a; Gosner and Rossman, 1960). Larvae from northern and southern populations have the same coloration, however. Detailed descriptions of tadpoles are in Wright (1929) and Gosner and Black (1957b). Descriptions and illustrations of larval mouthparts are in Hinckley (1882) and Dodd (2004).

Eggs. The eggs have a mean diameter of 1.5 mm (1.4–2 mm) in the North and 2.56 mm (2.45–2.8 mm) in the South (Gosner and Rossman, 1960) and a vitellus of 0.9–1.13 mm. They are deposited singly or in small bunches attached to submerged vegetation near the bottom. Females oviposit both large and small eggs in a single clutch. As many as 1,500 eggs may be oviposited by a single female during the breeding season (Wright, 1914; Livezey and Wright, 1947; Gilhen, 1984), usually in water 0.25–0.5 m in depth. The eggs are black or brown dorsally and white or cream ventrally, and have two gelatinous envelopes. Most authors report that eggs hatch in from 5 to 14 days, depending on temperature, although MacCulloch (2002) states that hatching requires up to 3 weeks in Ontario.

DISTRIBUTION

Spring Peepers are widely distributed in eastern North America. They occur from the Canadian Maritimes across southern Québec, around James Bay, and westward across Ontario to southeastern Manitoba. There is a single report from far northwestern Ontario near Sachigo Lake (Weller and Green, 1997). Observations from Labrador are questionable (Maunder, 1983). The western extent of the range includes eastern Minnesota and Iowa, most of Missouri (but not the northwest tip), to eastern Oklahoma and Texas. The species barely enters Kansas. *Pseudacris crucifer* occurs throughout the eastern and southeastern United States, and southward on the Florida peninsula to Orange and Sumter counties (Stevenson and Crowe, 1992; Owen, 1996). Elevations range from sea level to 1,650 m in the Great Smoky Mountains (Mathews and Echternacht, 1984).

Spring Peepers are found on islands, including Ile d'Orleans in Québec (Fortin et al., 2004a),

Distribution of *Pseudacris crucifer*

Prince Edward Island (Cook, 1967), Pelee Island in Lake Erie (Langlois, 1964; Hecnar et al., 2002), the Apostle Islands in Lake Superior (Hecnar et al., 2002; Bowen and Beever, 2010), Isle Royale (Ruthven, 1912), the Georgian Bay islands in Lake Huron (Hecnar et al., 2002), Walpole Island in Lake St. Clair (Woodliffe, 1989), Martha's Vineyard and Nantucket (Lazell, 1976), Long Island, New York (Overton, 1914), Kent Island, Maryland (Grogan and Bystrak, 1973b), and the Pleistocene barrier islands of Georgia (Shoop and Ruckdeschel, 2006). Reports of introductions to Cuba are apparently not accurate (Estrada and Ruibal, 1999).

Important distributional references include: Alabama (Mount, 1975), Arkansas (Trauth et al., 2004), Canada (Bleakney, 1954, 1958a), Connecticut (Klemens, 1993), Georgia (Williamson and Moulis, 1994; Jensen et al., 2008), Illinois (Smith, 1961; Mierzwa, 1998), Indiana (Brodman and Kilmurry, 1998; Minton, 2001; Brodman, 2003), Iowa (Vandewalle et al., 1996), Kansas (Rundquist, 1978; Collins, 1993; Collins et al., 2010), Louisiana (Dundee and Rossman, 1989), Maine (Hunter et al., 1999), Massachusetts (Lazell, 1976), Michigan (Ruthven, 1912), Minnesota (Oldfield and Moriarty, 1994), Missouri (Johnson, 2000; Daniel and Edmond, 2006), New Brunswick (Gorham, 1970), New Hampshire (Oliver and Bailey, 1939; Taylor, 1993), New York (Gibbs et al., 2007), North Carolina (Dorcas et al., 2007), Nova Scotia (Gilhen, 1984), Ohio (Walker, 1946; Pfingsten, 1998; Davis and Menze, 2000), Ontario (Schueler, 1973; MacCulloch, 2002); Pennsylvania (Hulse et al., 2001), Prince Edward Island (Cook, 1967), Québec (McCoy and Durden, 1965; Bider and Matte, 1996; Desroches and Rodrigue, 2004), Tennessee (Redmond and Scott, 1996; Butterfield et al., 2009), Texas (Smith and Sanders, 1952; Dixon, 2000), Vermont (Andrews, 2001), Virginia (Tobey, 1985; Mitchell and Reay, 1999), West Virginia (Green and Pauley, 1987), Wisconsin (Suzuki, 1951; Vogt, 1981; Mossman et al., 1998).

FOSSIL RECORD

This species is known from Pleistocene (Irvingtonian and Rancholabrean) fossil deposits in Georgia, Maryland, Pennsylvania, Tennessee, Virginia, and West Virginia (Holman, 2003). Most of the fossil sites represent cave deposits.

SYSTEMATICS AND GEOGRAPHIC VARIATION

Two subspecies have generally been recognized: the Northern Spring Peeper (*Pseudacris crucifer crucifer*) and the Southern Spring Peeper (*Pseudacris crucifer bartramiana*). Populations of the Spring Peeper in southern Georgia and Florida differ phenotypically from northern populations by the presence of ventral spotting, possession of dark spots instead of a stripe on the upper jaw, having a richer coloration, and by broader dorsal stripes (Harper, 1939a). However, there is no molecular evidence to support the recognition of these subspecies as distinct evolutionary lineages (Austin et al., 2002; Moriarty and Cannatella, 2004). Gilhen (1984) also noted that populations in the Canadian Maritimes have a more distorted X-mark dorsally than populations farther south, and that the X may be fragmented or connected to other dark markings.

Pseudacris probably evolved from a *Hyla*-like ancestor in the early Tertiary (Hedges, 1986). Moriarty and Cannatella (2004) suggested that there are four evolutionary lineages in *Pseudacris*. *Pseudacris crucifer* is more closely related to *Pseudacris ocularis* in the *Crucifer* clade than it is to the Trilling Chorus Frog clade (most eastern *Pseudacris*), the Fat Chorus Frog clade (e.g., *P. streckeri*), or the West Coast clade (e.g., *P. regilla*). Within *P. crucifer*, there are three major phylogenetic clades (a complex and diverse eastern, a southwestern, and a western) reflecting

Eggs of *Pseudacris crucifer*.
Photo: Stan Trauth

the complex effects Pleistocene climatic fluctuation had on distribution, particularly as frogs dispersed from refugia in the Ozark Central Highlands and from areas within the Southern Appalachians (Austin et al., 2002; Austin et al., 2004a). These authors noted that diversification probably originated in the Pliocene, but that phylogenetic diversity was amplified by isolation in refugia through the Pleistocene glacial cycles. Some of the highest levels of haplotype diversity occur in northern populations, where rapidly expanding lineages have come into contact and introgression has occurred (Austin et al., 2002; Austin et al., 2004a).

Spring peepers usually have been assigned to either the genera *Pseudacris* or *Hyla* over the last several decades. Hardy and Burroughs (1986) compared literature data on osteology, external morphology, internal anatomy, biochemistry, and life history and concluded that Spring Peepers were intermediate between *Pseudacris* and *Hyla*. They proposed a new genus, *Parapseudacris*, to accommodate this species. This arrangement has not received subsequent support. Hedges (1986) first suggested that Spring Peepers should be placed in *Pseudacris* based on allozyme data, but this transfer was disputed by Cocroft (1994), who analyzed a series of morphological characters. After adding additional morphological characters to the Cocroft (1994) dataset, Da Silva (1997) concluded that Spring Peepers belonged in *Pseudacris*, an alignment that also coincides with chromosomal banding patterns (Wiley, 1982). The expanded morphological data and the more recent phylogenetic analyses of Moriarty and Cannatella (2004) confirm the placement of Spring Peepers within *Pseudacris*.

Spring Peepers can be hybridized artificially in the laboratory with a number of other *Pseudacris*. Crosses of male *P. feriarum*, *P. brachyphona*, *P. kalmi*, *P. triseriata*, or *P. streckeri* with female *P. crucifer* produce a few tadpoles, but most are not viable, and developmental abnormalities are common (Blair, 1941b; Gosner, 1956; Mecham, 1957, 1965). Likewise, hybridization with *P. nigrita* does not result in larvae that metamorphose (Moore, 1955). Crosses between male *P. crucifer* and female *P. ornata*, *P. feriarum*, *P. kalmi*, or *P. brachyphona* also may produce a few tadpoles. Very few tadpoles are produced in crosses between male *P. crucifer* and female *P. streckeri*, and a high proportion of developmental abnormalities prevent successful metamorphosis (Mecham, 1957). Crosses between male *P. crucifer* and female *P. triseriata* result in about 10% of the eggs producing larvae, but survival to metamorphosis is extremely rare, again because of severe developmental abnormalities (Blair, 1941b). Mecham (1965) noted that sexual development in hybrids is abnormal, and that hybrids that survive do not appear to be fertile. Gosner (1956) found no evidence of hybridization between *P. crucifer* and *P. kalmi* in a large sample from a natural population.

ADULT HABITAT

The Spring Peeper is a frog of eastern forested habitats (Knutson et al., 2000; Price et al., 2004), and the extent of forests at various scales (to 3,000 m) is a good predictor of the presence of peepers at individual breeding sites. Likewise, the extent of open habitats is inversely correlated with Spring Peeper occupancy. They are found in a wide variety of terrestrial communities, including xeric hammock, upland hardwood forest, mixed hardwood and pine forest, mesophytic forest, Great Lakes pine barrens, and forested ravines (Carr, 1940a; Marshall and Buell, 1955; DeGraaf and Rudis, 1990; Enge et al., 1996; Enge, 1998a, 1998b; Enge and Wood, 1998; Varhegyi et al., 1998; Evrard and Hoffman, 2000). Peepers are present, but tend to be less common in tamarack (*Larix laricina*) and coniferous forest than in hardwood or mixed-hardwood forest. In otherwise developed land, they occur in riparian forest along streams (Burbrink et al., 1998). Occupancy is often quite high in south central Michigan and is related to wetland proximity, but it is difficult to predict which habitat variables are most important to the species (Roloff et al., 2011). At the southern end of its range, *P. crucifer* prefers mesic hardwood hammocks with some degree of topographic relief; the lack of topographic relief south of the Orlando area, coupled with their cool-weather breeding cycle, may limit their distribution on the peninsula. Likewise, their close association with forested habitats limits their distribution into the prairie and more arid regions to the west.

TERRESTRIAL ECOLOGY

Away from breeding wetlands, Spring Peepers are not commonly observed. They are found in the terrestrial leaf litter and among surface debris, where they forage for small invertebrates. McAlister (1963) noted an arboreal aggregation of Spring Peepers in September, when numerous frogs were feeding on small arthropods in bushes along a road. Carr (1940a) also noted arboreal retreat sites under bark and in knot holes, and they are occasionally observed in pitcher plants (*Sarracenia purpurea*) (Russell, 2008). They appear to establish small home ranges (diameter of 1.7–5.4 m) around downed logs, woody debris, stumps, and surface vegetation (Delzell, 1958). Home ranges overlap with one another, but away from the breeding site individuals do not interact. Overwintering sites may be as far away as 300 m from the summer home range.

Activity occurs throughout the warm season, even in Florida where I have occasionally observed them hopping among litter in a mesic hammock in August. As noted below, Spring Peepers are heard calling during periodic warm weather in autumn and early winter, even in the far North. This suggests that they routinely remain active late in the season, depending upon weather conditions. The CTmax of *P. crucifer* is 34.8–36°C (John-Alder et al., 1988), so this species must find refuge during very hot days in summer. Individual Spring Peepers have occasionally been found at the entrances to and passages within caves (Franz, 1967; Black, 1973).

Most activity away from breeding sites is devoted to feeding and growth during the warm activity season. In nature, feeding occurs in a bimodal pattern, that is, in early morning and late afternoon. Oplinger (1967) noted that peepers often did not feed after 30 min of concentrated feeding, even though prey was still available. Presumably, these individuals were temporarily satiated. Most individuals do not have food in their stomachs first thing in the morning, suggesting that feeding does not occur at night. By midmorning, most individuals have fed. The fact that Spring Peepers have color vision (Hailman and Jaeger, 1974) also suggests that most feeding activity occurs during daylight hours.

The CTmax is approximately 36°C in wild-caught New York specimens, so this cold-breeding species can tolerate warm temperatures. In Minnesota, Brattstrom (1963) recorded body temperatures 3.5–6°C warmer than ambient air and soil temperatures and suggested that the frogs were purposely absorbing solar radiation through their exposed position in the vegetation. Since peepers feed during the early morning hours, increasing temperature through basking could facilitate digestion. Spring Peepers appear to prefer higher temperatures when in the vicinity of conspecifics than when alone, although the significance of this preference is not known (Gatten and Hill, 1984).

The presence of Spring Peepers is positively correlated with forested habitats within 100–1,000 m of breeding sites (Herrmann et al., 2005; Eigenbrod et al., 2008), depending on location and method used to identify habitat variables, suggesting that they spend much of their non-breeding activity in terrestrial habitats in rather close proximity to wetlands. Owen (1996) also hypothesized that in Florida they spent most of their time in terrestrial habitats adjacent to the forested wetlands in which they bred. Density seems to increase with the percentage of forest cover, with greatest densities at >80% forest cover. On a localized scale (< 10 km), Spring Peeper populations do not likely conform to a classical metapopulation structure (Smith and Green, 2005).

Spring Peepers overwinter in terrestrial refugia, although deep winter retreats have not been described. On a cold November day in Michigan, Blanchard (1933) found 114 peepers under leaf litter in shallow depressions, clumped together with a variety of other amphibians. Peepers are capable of surviving freezing temperatures, when as much as 35% of their body water may be frozen as ice. Under laboratory conditions, *P. crucifer* survived five days frozen at –6°C, although movements or other vital signs were not observed until two to four days after thawing (Schmid, 1982). Storey and Storey (1986) reported survival after two weeks at temperatures below 0°C. Glucose is used as a cryoprotectant, and it is rapidly produced for at least 1 hr when the frog is exposed to freezing temperatures. Organs, including the heart, then rapidly accumulate the glucose produced by the liver (Churchill and Storey, 1996). The ability to supercool in terrestrial refugia allows Spring Peepers to extend their range far into northern latitudes. In addition, Spring Peepers have reduced cutaneous evaporative water loss when compared to nonarboreal anurans (Wygoda, 1984), and they have greater desiccation tolerances at low temperatures and a shortened daylight cycle than at warmer temperatures with longer day lengths (Farrell, 1971). These physiological adaptations probably facilitate both arboreal and terrestrial cool-season activity.

MIGRATION AND DISPERSAL

Spring Peepers are often the first frogs of the year to breed in eastern North America, and their calls are a sure sign of the arrival of spring. Migration to breeding sites occurs at night, with most movements occurring around midnight (Pechmann and Semlitsch, 1986; Todd and Winne, 2006).

Exactly how Spring Peepers locate the breeding site is unknown. Olfaction does not appear to be involved (Martof, 1962a), although the experimental design Martof used to study the influence of olfaction on orientation might not have been appropriate for this arboreal hylid; his results could have been an artifact of design rather than a lack of response to odors of the home pond. However, Spring Peepers are able to respond positively to different light intensity (Jaeger and Hailman, 1973; Hailman and Jaeger, 1974), and that may help with nocturnal orientation both to and from breeding sites. Spring Peepers tend to choose the lightest sites in terms of the amount of light that strikes a wetland, and growth rates of peepers are actually higher in lighter wetland breeding sites than they are in more darkened habitats (Halverson et al., 2003). Positive phototaxis may allow the animal to find the lightest breeding sites and to locate wetlands in otherwise forested habitats.

Migration to breeding ponds extends over periods of weeks depending on the weather conditions; they are not an explosive breeder. For example, immigration occurred on 27 consecutive days in Michigan (Cummins, 1920). During the course of a breeding season, males and females move back and forth between breeding sites and terrestrial locations near the breeding site; both immigration and emigration are strongly correlated with precipitation and high humidity (Cummins, 1920; Murphy, 1963; Pechmann and Semlitsch, 1986).

Adult dispersal is usually completed within about a month after the last adults arrive at the breeding site. In North Carolina, for example, most breeding occurs between February and April, but adults are captured dispersing from wetlands into early June (Murphy, 1963). Both adult and juvenile dispersal occurs at night, with most activity occurring from around midnight to dawn (Pechmann and Semlitsch, 1986; Todd and Winne, 2006). In Missouri, metamorph dispersal is completed from mid-June to mid-July (Hocking et al., 2008).

CALLING ACTIVITY AND MATE SELECTION

Throughout their range, Spring Peepers are a winter to early spring breeder, depending on rainfall, latitude, and elevation. Calling begins prior to actual mating, and may extend 7–21 days before amplexing pairs are observed (Sullivan and

Hinshaw, 1990). In the South, calling is greatest from late November to February, with occasional individuals calling into April (Gerhardt, 1973). Cool warm-season temperatures in the North allow it to breed much later and into the summer season. Specific breeding periods include: late October to April (Florida: Carr, 1940b; Owen, 1996), November to March (Georgia: Harper, 1939a), November to April (South Carolina: Semlitsch et al., 1996), November to May (Louisiana: Dundee and Rossman, 1989), December to May (Alabama: Mount, 1975), January to April (Arkansas: Trauth et al., 1990), January to May (Georgia: Martof, 1955), February to April (North Carolina: Murphy, 1963; Alexander, 1966; Gaul and Mitchell, 2007), February to July (Tennessee: Dodd, 2004), March to May (Connecticut: Flores, 1978; Klemens, 1993; Indiana: Minton, 2001; New York: Wright, 1914), March to June (Illinois: Smith, 1961; Massachusetts: Hinckley, 1884; Pennsylvania: Hulse et al., 2001; Wisconsin: Vogt, 1981), March to July (Québec: Desroches and Rodrigue, 2004), April (Kansas: Rundquist, 1978), April to May (Illinois: P.W. Smith, 1947; southern Ontario: Piersol, 1913), April to June (Manitoba: Preston, 1982; New Hampshire: Oliver and Bailey, 1939; Nova Scotia: Gilhen, 1984; Prince Edward Island: Cook, 1967; Wisconsin: Mossman et al., 1998), April to July (New Brunswick: Gorham, 1964; Nova Scotia: Bleakney, 1952); June to July (northern Ontario: Moore, 1965; Schueler, 1973). Peepers continue to call into early July at higher elevations in North Carolina, such as around Highlands, and elsewhere in the Southern Appalachians. There is a single report of successful egg deposition in late September in northern Georgia, after a sudden drop in temperature (Martof, 1960).

Pseudacris crucifer is often extremely abundant at breeding ponds, and indeed may be the most commonly observed species (Murphy, 1963). From 1979 to 1994, for example, 2,499 female *P. crucifer* produced 9,245 metamorphs at a 1 ha Carolina Bay (Semlitsch et al., 1996). However, only 17 and 51 peepers were recorded at 2 small South Carolina ponds during a 2 yr period (Gibbons and Bennett, 1974). Large numbers of both males and females may be captured around breeding ponds, with most immigration occurring at the start of the reproductive season.

Breeding can occur in rather cold water. In Florida, breeding pairs have been found in water at 12°C, and activity occurs in water of 8°C in New Jersey (*in* John-Alder et al., 1989). The muscles of Spring Peepers are uniquely adapted for activity at cooler temperatures. John-Alder et al. (1989) noted that *P. crucifer* attained maximum swimming velocity at 17°C, and at 13°C Spring Peepers can attain 80% of their maximum jumping distance (a mean of 64 cm) (John-Alder et al., 1988). Even at 0.7°C, peepers can jump 40% of their maximum distance. The muscles in the legs of Spring Peepers retain a large force-generating capacity at low temperatures, and this facilitates migration and reproduction at some of the coldest activity temperatures experienced by North American frogs. The thermal adaptations of the muscles are similar between northern and southern populations, and they likely account for the differences in latitudinal breeding phenology. Cool temperature for optimal muscular activity is thus another variable limiting the distribution of Spring Peepers to north of central Florida.

Females usually initiate amplexus by approaching a calling male and nudging him. In a few cases, the male will grasp a female in close proximity without being touched (Fellers, 1979a). Amplexus is axillary (male clasping female just behind the forelimbs). The pair may remain in amplexus for several hours prior to oviposition, which begins with lateral contractions in the female's abdominal area. The female arches her back and extends her rear legs, bringing her and the male's cloacae into close proximity (Flores, 1978). Fertilization occurs as the amplexed pair floats in the water or as the female grasps vegetation, with the male fertilizing each egg as it is extruded by positioning his cloaca over the point of emergence. Eggs are extruded singly or in small masses of 2–30 eggs (Flores, 1978). Oviposition may last several hours, and pairs may remain in amplexus for several hours after oviposition has been completed.

Fertilization and hatchling size varies among amplexing pairs. In Florida, between 62–100% of the clutches examined by Travis et al. (1987) were fertilized. Most of the variation in fertilization rate was related to the female, that is, the eggs of some females had high fertilization rates, whereas the eggs of other females were consistently lower. Embryo mortality occurred most frequently at Gosner stages 11–16 (gastrulation to neurulation). Hatchling sizes ranged between 0.3 and 0.43 mg body mass, with most of the variation associated with maternal effects. These authors noted that maternal effects accounted for only 26% of the variance, however, as dominance genetic effects accounted for 70% of the variance (Travis et al., 1987). Differences in egg viability and hatchling size do not have a high level of heritability.

The advertisement call of the Spring Peeper is a high-pitched "peep, peep, peep" repeated over and over. Frogs may call in duets or, frequently, in triplets; the first frog calls, followed by the second in 40–70 ms, and then the third. After the third frog calls, the cycle is repeated. A chorus is thus a series of peepers calling mostly in sets of two's or three's, although quartets and pentets have been recorded. The Spring Peeper also has a trilling call, which is uttered in order to induce calling by duet or triplet members. Goin (1948) noted that the initial call by the leading male is usually in a different octave than the calls of succeeding males.

In addition to breeding calls, solitary peepers are sometimes heard calling in single notes from the tree canopy during the autumn. I have heard calls from high up in live oak trees (*Quercus virginiana*) in late October and November in Alachua County, Florida, and from the tree canopy in Cades Cove in the Great Smoky Mountains in September (Dodd, 2004). Calls outside the breeding season or away from breeding ponds have been reported from Connecticut, Maine, Massachusetts, New Brunswick, and New Hampshire sporadically from August to December (Allen, 1868; Oliver and Bailey, 1939; Gorham, 1964, 1970; Klemens, 1993; Hunter et al., 1999) and in midwinter in Michigan (Ruthven, 1912). Hinckley (1884) reported calling all months of the year except July and January. Such calling occurs mostly during unusually humid, warm, and cloudy days, or just prior to precipitation. The purpose of the tree canopy and other late-season arboreal calls is unknown.

The majority of calling occurs in the late afternoon and at night, usually between 16:00 and 22:00 hrs (Owen, 1996; Steelman and Dorcas, 2010), although sporadic calls or even chorusing may occur throughout the daylight hours, depending on weather conditions (Wright, 1914). The first choruses are heard from 1 to >20 days after the first individual calls are heard. Chorusing occurs at a rather constant level for several months thereafter (Owen, 1996; Todd et al., 2003) until the end of the breeding season, when it tapers off with warmer weather. The optimum temperature for calling is 10–20°C, but calls can be heard from <5 to >30°C (Steelman and Dorcas, 2010).

Calling sequences are highly variable in terms of the number of calls per sequence; Gerhardt (1973) recorded a male whose sequence consisted of 66 consecutive calls. Call duration, dominant frequency, and the number of calls per minute are similar throughout the species' geographic range, after adjusting for the effects of temperature. For example, calls in South Carolina lasted a mean of 0.18 sec, with a mean number of 49 calls/min (range 30 to 70). The midpoint dominant frequency at 15°C was 2,600 Hz (range 2,200 to 2,900). These values are close to those reported for *P. crucifer* in Maine (mean 2,809 cps, 0.21 sec duration, call rate 53.9 calls/min; Sullivan and Hinshaw, 1990), Maryland (2,625–3,549 cps, mean 3,061 cps; Forester and Czarnowsky, 1985), Missouri (2,800–3,360 cps, 0.09–0.25 sec duration; Doherty and Gerhardt, 1984), New York (2,588–3,212 cps, mean 2,895 cps; Wilczynski et al., 1984), Texas (2,400–2,600 cps, mean 2,467 cps, 0.14 sec duration; Blair, 1958a), and North Carolina (3,200 cps, 0.13 sec duration; Martof, 1961).

The dominant call frequency and call rate are positively correlated with temperature, but the duration of the call is negatively correlated with temperature (Sullivan and Hinshaw, 1990). Dominant frequency is negatively correlated with male body size (Lykens and Forester, 1987). Call characteristics are specific to individuals, that is, a male with certain call characteristics one night will have the same characteristics on subsequent nights. Females have definite call preferences (calls of 0.15–0.30 sec duration and 2,875 cps), but the female's auditory perceptions are only roughly attuned to the temporal and spectral properties of the male's call (Doherty and Gerhardt, 1984). In another series of experiments, females had no real preferences in the frequency of the male's call if no background noise was evident, but in situations mimicking natural calling background noise, they chose callers with high frequencies (Schwartz and Gerhardt, 1998). This makes the male's call audible to females in a variety of environmental conditions, particularly a variety of temperatures and levels of background noise. However, males can barely detect their own calls (Wilczynski et al., 1984).

In terms of hearing, the extent of the male's call is roughly 1.1–3.8 m, whereas females have an auditory space of 2.2–11.2 m. Calling from trees or brush, even at only 50 cm, increases the male's auditory space to 1.8–11.6 m and that of females from 4.6 to 69.4 m (Brenowitz et al., 1984). Thus, males space themselves generally < 2 m from one another (see below), which allows them to keep track of other males in the immediate vicinity but not throughout a chorus. Females can hear chorusing males at some distance from the breeding pond, which allows them to focus on a particular male's call and orient toward him.

In Florida, preferred calling sites are on plant leaves with a dry surface that have a steep angle (ca. 41%) and are located at least 20 cm above the ground surface (Owen, 1996). Peepers throughout their range call from the lower branches of vegetation surrounding a pond, from along the water's edge, from grass tussocks, and on floating mats of vegetation (Allen, 1963). Most males call at temperatures <20°C (Gerhardt, 1973; Owen, 1996). Males are capable of calling in very cold weather, at just below (–0.6°C, Allen, 1868) or just above freezing (0.4°C). Even in Florida, calling occurs at air temperatures as low as 1.7°C (Carr, 1940a). Calling males are most often associated with soil temperatures >9.5°C (Owen, 1996) and air temperatures between 4–22.8°C (Alexander, 1966; Brown and Brown, 1977; Flores, 1978; Gerhardt et al., 1989; Owen, 1996). Most calling occurs at water temperatures of a minimum of 5–7°C. Since male body temperature is more closely correlated with water temperature than to air temperature (Brown and Brown, 1977), water temperature is probably of greater significance than air temperature to reproductive timing and success.

Males often call in close proximity to one another. In Missouri, Gerhardt et al. (1989) recorded distances of 21–124 cm (mean ca. 50 cm) between calling males. The amplitude of the calls allowed males to hear their rivals and thus space themselves to reduce agonistic encounters. Resulting densities thus were 1–4.5 males/m^2 at the Missouri breeding ponds. Peepers spaced themselves at mean distances of 24–113 cm in Maryland, depending on density (Fellers, 1979a), 40–140 cm (mean 110 cm) in Québec, and a mean of 210 cm in New York (Brenowitz et al., 1984). At low densities, males tend to space themselves evenly around a site, whereas at high densities, spacing becomes more random. The minimum distance between calling males is about 10 cm. Call rates do not increase in response to another male in close proximity (Sullivan and Hinshaw, 1990), but the length of the mating call does increase (Fellers, 1979a).

Still, males often engage in aggressive encounters if a rival comes too close, and males have an "encounter call" (termed a trill) to warn intruders. Rosen and Lemon (1974) noted trilling when adjacent males were within 80 to 100 cm of one another, but that they could be much closer; either or both males may trill. Males thus call aggressively in response to the calls of nearby rivals, whereby the call duration and the number and proportion of aggressive calls increase with changes in the calling intensity of rivals (Schwartz, 1989). In playback experiments, Marshall et al. (2003) found a positive correlation between the amplitude of advertisement calls of a male's nearest neighbor and the amplitude at which a male first produces an aggressive call. If a neighbor repeatedly calls above the first male's aggressive threshold, he responds by rapidly decreasing aggressive signaling, and the threshold of his aggressive call increases. Through time, however, some degree of habituation occurs.

Males may be energetically constrained during calling (Schwartz, 1989; Marshall et al., 2003), and aggressive calls may be given if the male's calling level is already at its upper limit; thus, males might be able to assess the status and density of opponents (Schwartz, 1989). Trill encounters may lead to brief fights between males. Males grapple with one another for < 20 sec, until one of the males departs, and then peeping begins anew. Since males tend to position themselves in the same location from one night to the next, the trilling also may serve a calling site territorial function.

The size of calling males is not normally distributed. For example, a population in Maine had a bimodal size-class structure, suggesting the possibility of two or more year-classes calling in any one breeding season (Sullivan and Hinshaw, 1990). This seems likely since the larger animals may comprise more than one age class (Lykens and Forester, 1987). Males that mated successfully in Maine or in Maryland, however, were not larger on average than those that did not mate (Forester and Czarnowsky, 1985; Sullivan and Hinshaw, 1990). However, in Ohio (Gatz, 1981b) and Maryland (Fellers, 1979a, as corrected by Gatz, 1981b), slightly larger males (mean 29.1 mm SUL) were more often successful in amplexing females than smaller males (mean 28.1 mm SUL). There is no indication of size-assortive mating; this result is not unexpected, as there may be no correlation between female size and fecundity.

Noncalling males are sometimes found in close proximity to calling males. Forester and Lykens (1986) noted that satellite males made up <14% of the males at a chorus, and that they were smaller (about 6%) than calling males. However, Lance and Wells (1993) could not find any difference between satellite and calling males in body length, body mass, trunk mass, leg mass, or ventricle mass. Satellite males are sometimes successful in securing females, which makes their sexual parasitism advantageous, especially since the calls of small males are not as attractive to females as are the calls of larger males (Forester and Lykens, 1986). When in close proximity, the satellite males assume a posture close to the ground, whereas the calling male is oriented at about a 40–45° angle. Presumably, this allows them to track the close approach of a female.

BREEDING SITES

Spring Peepers are found at many types of breeding sites. They occur around lakes, artificial impoundments, bogs, marshes, retention ponds, sinkhole ponds, swales, basin and seepage swamps, beaver ponds, cypress/gum wetlands, lowland riverine forests, roadside ditches, and even in golf course ponds (Crump, 1981a; Dale et al., 1985; Enge and Wood, 1998; Pauley et al., 2000; Metts et al., 2001; Liner et al., 2008; Mifsud and Mifsud, 2008; Scott et al., 2008; Brand and Snodgrass, 2010). They occur in ponds with open water as well as in ponds completely covered to some extent in surface vegetation (Goin, 1943). Smithberger and Swarth (1993) noted that a freshwater tidal stand of common reeds (*Phragmites australis*) supported a dense concentration of Spring Peepers. Some breeding sites may be very small (Allen, 1963; Russell et al., 2002a). Hecnar and M'Closkey (1996b) found Spring Peepers in Ontario preferred sites with low conductivity and low levels of turbidity.

Although found in wetlands with a variety of hydroperiods, Spring Peepers prefer breeding sites with hydroperiods >4 months that are not permanent (Babbitt et al., 2003). Indeed, the best predictor of occupancy and colonization is the current-year trends in hydrology, rather than the hydrological history of a potential breeding site (Church, 2008). Both temporary and permanent wetlands are used, however. For example, in a New Hampshire survey there was a 50:50 split in breeding sites with regard to permanency (Herrmann et al., 2005). In agricultural landscapes, Spring Peepers have a tendency to be associated with temporary wetlands (Kolozsvary and Swihart, 1999). Not surprisingly, medium-length hydroperiod ponds, although temporary, are also associated with high invertebrate abundance. Spring Peepers breed at sites that may contain predatory and nonpredatory fish (Sexton and Phillips, 1986; Babbitt et al., 2003), but they are found more commonly at sites without predatory fish (Hecnar, 1997; Hecnar and M'Closkey, 1997b).

Pseudacris crucifer prefers open-canopied wetlands, at least where forest cover has not completely closed off the sky. They tolerate some canopy closure, but in central Florida the canopy over forested wetlands must be at least 5 m in height for occupancy (Owen, 1996). Further, Spring Peepers tend to disappear from breeding sites when the canopy closes over during succession (Skelly, 2001). This suggests that the mere presence of peepers at breeding sites with some canopy cover does not mean the populations will persist through time or are breeding in preferred habitat. This species grows much faster in open-canopied situations than under closed-forest canopies, although survivorship (68% under field experimental conditions) was not affected by canopy cover (Skelly et al., 2002). In contrast, Williams et al. (2008) reported decreased survivorship in shaded ponds, but not in the time it took to reach metamorphosis. Open-canopied wetlands are much more productive and diverse in terms of periphyton (diatoms and algae), and generally have higher dissolved oxygen concentrations than closed-canopied ponds. Whether the pond contains a grass or leaf litter substrate makes no difference in survivorship, although the time to metamorphosis is a little faster on grassy substrates (Williams et al., 2008).

Pseudacris crucifer may occupy many of the regionally available breeding sites from one year to the next. For example, during a survey of breeding choruses in the Mississippi River Alluvial Valley in Louisiana, they were heard chorusing at 13 of 31 sites in one year and 23 of 28 sites the next year (Lichtenberg et al., 2006). In an extensive survey near the southern limit of the species' range, Owen (1996) found Spring Peepers at 167 of 332 sampled sites, whereas in Ontario they occurred at 30–40% of 180 wetlands surveyed (Hecnar, 1997; Hecnar and M'Closkey, 1998) and at 64% of 61 wetlands in New Hampshire (Herrmann et al., 2005). Based on repeated site visits, MacKenzie et al. (2002) used models to estimate potential occupancy at 85% for Spring Peepers surveyed at 32 Maryland wetlands, which was close to what was actually observed (83%). In Michigan, Skelly et al. (2003) estimated the probability of presence at 79% in any one pond during a single year over a 5 yr study period. Brodman (2009) also noted considerable variation in annual occupancy over a 14 yr study in Indiana.

Spring Peepers may be able to use certain landscape features or topography as they move toward and away from breeding ponds. For example, the movements of peepers at a North Carolina pond suggested that they were using old furrows from pine plantations as migratory pathways to the breeding site (Murphy, 1963).

REPRODUCTION

Males are sexually mature by approximately 22 mm SUL. Spermatozoa are found in the testes in the autumn, before the frogs enter hibernation, and are retained in seminiferous tubules during the winter (Rugh, 1941). Eggs are present in females at 23 mm SUL, and they are also evident during the autumn prior to the next breeding season. The mean clutch size was 847 in Arkansas (range 505 to 1,201). Clutch sizes between 200 and 1,050 are reported for New York (Oplinger, 1966), clutches of 850 and 900 in Connecticut (Flores, 1978), a single cutch of 702 in Kansas (Lorraine, 1984), from 659 to 1,285 in North Carolina (Alexander, 1966), and from 396 to 1,581 in Nova Scotia (Gilhen, 1984). There is no correlation between clutch size and female body size in Arkansas (Trauth et al., 1990), but there

Tadpole of *Pseudacris crucifer*. Photo: Stan Trauth

does appear to be a positive correlation in New York (Oplinger, 1966). Ovarian mass was a mean of 0.57 g (0.22–1.52 g) for Arkansas frogs. The females use the energy stored in fat bodies during the summer to provide energy to their eggs. Ovulation precedes amplexus.

The literature reports considerable variation in egg size. However, there also is substantial variation in egg size within a single clutch of eggs deposited by a female. For example, the diameter of eggs in a Florida study was 1.08–1.48 mm, where volumetrically the largest eggs could be 157% larger than the smallest egg from the same female (Crump, 1981b). Larger eggs result in larger hatchlings, but larger eggs do not take longer than smaller eggs to develop to the hatchling stage. In addition, hatchlings from larger eggs do not have a greater survivorship to metamorphosis than their smaller siblings; individuals from larger eggs do not necessarily metamorphose at a larger size than their small-egged siblings; and individuals from larger eggs do not require less time to reach metamorphosis (Crump, 1981b). Crump (1981b) hypothesized that under conditions of scarcity, large eggs might have a fitness value higher than smaller eggs, but during "normal" conditions, there were no advantages to having larger eggs.

LARVAL ECOLOGY

Larval *P. crucifer* are mostly benthic detritivores, and relatively inactive in breeding ponds. They do not form social aggregations, and they display no sibling recognition (Fishwild et al., 1990). As such, they act largely independently of the actions of other conspecific larvae. However, they may be found in close proximity to one another, possibly as the result of the synchronous hatching of eggs from a single pair that deposited eggs within a circumscribed area. Brattstrom (1962) reported that the temperature within such aggregations of about 75 individuals each was higher than in the surrounding water. Densities in wetlands may vary considerably, with breeding sites in North Carolina containing from 12 to 464 larvae per cubic meter of pond water (Morin, 1983). Conspecific density affects the timing of metamorphosis and the size at metamorphosis, with higher densities increasing developmental time and decreasing the size at metamorphosis (Crump, 1981a).

As with the larvae of other frog species, the amount of food available affects both age and size at metamorphosis; increasing food causes tadpoles to metamorphose earlier and at larger sizes than tadpoles with more restricted amounts of food. However, changes in growth rates are subject to the timing of food increases, where increasing food early in development increases the growth rate, but increasing food later in development (at Gosner stages 35–37) has no effect on growth rate. Thus, growth rates are more plastic early in development than later in development, at least for this species (Hensley, 1993). Growth rates are also affected by the presence of predators. In experimental situations, larval growth rates decrease in the presence of dragonfly (*Anax*) larvae, but appear to have no effect on survivorship (Van Buskirk, 2000).

The amount of food available to larvae during development not only affects larval growth rates but postmetamorphic growth rates as well. In experimental trials, increased food during the larval period increased postmetamorphic growth efficiency and led to increased mass (Van Allen et al., 2010). A 10% increase in food resulted in a 10% decrease in the larval period, a 9% increase in mass conversion efficiency, and a 12% increase in juvenile mass 50 days postmetamorphosis. Van Allen et al. (2010) term this a "carry-over effect." As a result, the resources available during the larval period can have important fitness consequences well after the tadpole has metamorphosed.

Even without the effects of food and predators, growth rates vary among the offspring produced by different parents, such that some "families" of tadpoles grow at only 70% the rate of others. Dominant genetic variance accounts for most of the phenotypic variance; this same pattern is repeated for the length of the larval period. Thus, females tend to influence growth rates and the length of the larval period of their offspring. However, the differences in growth rates do not have a high level of heritability. In contrast, the size at metamorphosis is more associated with paternal influences, where the male's contribution accounts for between 13 and 22% of the size variance. The size at metamorphosis was also the only trait among those examined by Travis et al. (1987) that had a high heritable component (51–87%) and a low dominance genetic variance (2%).

Metamorphosis normally occurs from 45 to 65 days after hatching. MacCulloch (2002) states the larval period may extend to 3 months in Ontario, and in New York it extends to 90–100 days (Wright, 1914). Thus, metamorphs are still leaving ponds in late September at the northern limit of the range (Bleakney, 1952). Newly metamorphosed juveniles are 9.4–11 mm in Florida, 10.5–17 mm (mean 13.5 mm) in South Carolina (Pechmann et al., 2001), 9–13.1 mm in North Carolina (Alexander, 1966), 10–11 mm in Maryland (Lykens and Forester, 1987), 9–14 mm (mean 11 mm) in New York (Wright, 1914) and New Jersey (Gosner and Black, 1957b), a mean of 10.9 mm in New Jersey (Gosner and Rossman, 1960), and 13.5–14.5 mm in Nova Scotia (Gilhen, 1984).

The production of juveniles varies considerably from one year to the next, and by location. For example, breeding populations of 324, 495, 288, and 57 adults produced 50, 166, 75, and 377 juveniles, respectively, at a single South Carolina pond over a 4 yr period (Pechmann et al., 2001). In Missouri, very small numbers of metamorphs (9–22) were produced over a 4 yr period in 2 forested breeding ponds (Hocking et al., 2008).

DIET

Larval Spring Peepers eat diatoms, filamentous algae, blue-green algae, protozoans, and eggs of crustaceans (Munz, 1920). Adults eat a great variety of small invertebrates, including aphids, ants, mosquitoes, larval lepidopterans, spiders, phalangids, beetles, ticks, springtails, mites, terrestrial gastropods, fungus gnats, and crane flies (Surface, 1913; Marshall and Buell, 1955; McAlister, 1963; Oplinger, 1967; Gilhen, 1984). Recently metamorphosed peepers eat small prey such as ants, larval lepidopterans, spiders, beetles, springtails, mites, terrestrial gastropods, and lacebugs (Oplinger, 1967); only a very few stomachs (<9%) were empty. Adults also consume their shed skins (see Johnson, 1989, for a description of the process); skin sheds occur most frequently by volume early in the season. Very little if any feeding occurs during the breeding season, but from May to September, only 13–38% of stomachs were empty in a study in upstate New York (Oplinger, 1967).

PREDATION AND DEFENSE
Larval Spring Peepers are palatable (Kats et al., 1988) and do not have any chemical defenses, so their only chance of survival is to blend in with their environment and control the timing of their activity. In experimental tanks involving controls (no predators) and predators (the newt *Notophthalmus viridescens* and the predatory fish *Enneacanthus obsesus*), larval *Pseudacris crucifer* were relatively inactive compared with the larvae of other species. With predators, there was a slightly greater tendency to remain immobile or inactive than in control situations, and larval Spring Peepers generally had higher survival rates than the larvae of *Hyla* and *Anaxyrus* (Lawler, 1989). Similar results were obtained by Richardson (2001) using dragonfly larvae (*Anax*) and sunfish (*Lepomis* spp.) as predators. The timing of reproduction (very early in the breeding season) and long developmental period of *Pseudacris crucifer* allows them to be less active than later arrivals and thus less likely to draw the attention of potential predators.

Adults use a variety of behaviors to escape predators, including flight (jumping), remaining motionless, crouching down to the ground, body inflation, walking away, hiding, and climbing away from the predator (Marchisin and Anderson, 1978). They are excellent jumpers. Perhaps the best defense of Spring Peepers is their ability to change color, either lighter or darker, to match the substrate. These little tan frogs with various stripes are quite difficult to see within the leaf litter, and crypsis undoubtedly helps to account for the paucity of observations during the nonbreeding season. The speed of color change is temperature-dependent, with more rapid changes occurring at warmer temperatures than at 4–10°C (Kats and Van Dragt, 1986).

Predators include a wide variety of vertebrates and invertebrates, including brown trout (Cochran and Cochran, 2003), snakes (*Clonophis kirtlandi*, Tucker, 1977; *Thamnophis butleri*, Test, 1958; *Thamnophis sauritus*, in Klemens, 1993), birds (Barred Owls, Woodward and Mitchell, 1990), mammals (*Spilogale putorius*, Huheey and Stupka, 1967), ground beetles (G.R. Smith, 2002), giant water bugs (*Lethocerus americanus*) and dytiscid diving beetles (Hinshaw and Sullivan, 1990). The latter predators killed Spring Peepers as they were calling at a breeding site. Undoubtedly, birds and other predators take large numbers, especially the juveniles. The mosquito *Culex territans* feeds on Spring Peepers and even uses the frog's call to locate its next blood meal (Bartlett-Healy et al., 2008).

Adult *Pseudacris crucifer*, amplexus with extruded egg. Photo: David Dennis

POPULATION BIOLOGY

Sex ratios appear slightly male biased at breeding sites. For example, at a North Carolina pond Murphy (1963) recorded 834 males and 583 females for a 1.43:1 sex ratio one year, and 1.2 males per female the next year. However, at a New York breeding site, Oplinger (1966) found 416 males and only 45 females. Based on internal examination of immature peepers in autumn, the sex ratio was 288 males to 266 females, or nearly 1:1. This discrepancy could suggest differential mortality between the sexes. More likely, however, the difference was an artifact of biased sampling procedures; Oplinger (1966) suggested that males were more easily captured than females at the breeding site.

Peepers are capable of reproducing the first (Delzell, 1958) or second (Lykens and Forester, 1987) summer after metamorphosis. Alexander (1966) noted that in North Carolina, between 36–67% of female Spring Peepers do not reach sexual maturity their first year. The maximum longevity of Spring Peepers is 4 yrs. Males at a Maryland breeding site were 22–28 mm SUL at 2 yrs, 25–30 mm at 3 yrs, and 27–33 mm at 4 yrs, based on an analysis of growth rings in the long bones of the hind leg. Females were 23–30 mm SUL at 2 yrs, 25–35 mm at 3 yrs, and 28–33 mm at 4 yrs (Lykens and Forester, 1987). In the general population, females were older than males, and they were larger at 3 and 4 yrs of age than similarly aged males. As might be expected, population age- and size-class structure can vary with location. At one Maryland site, for example, Lykens and Forester (1987) noted that 87% of the population was 2 yrs old, whereas at another population, older males were much more common. Peepers in their first year of life are 14–15.5 mm as they enter their first winter dormancy.

COMMUNITY ECOLOGY

A number of *Pseudacris* species breed in the same general area and at the same time of year, but the species may have different habitat preferences. For example, *P. crucifer* prefers forested wetlands, whereas sympatric *P. ocularis* and *P. nigrita* occupy ephemeral marshes (Owen, 1996). In Ontario, *P. triseriata* prefers ponds with short hydroperiods, whereas *P. crucifer* tends to occur in ponds with longer hydroperiods, since it tolerates predators better. Similarly, *P. feriarum* prefers shallow temporary rain pools, whereas the slower developing *P. crucifer* prefers larger bodies of water (Burkett, 1991). Thus, the separation of these species is the result of differences in habitat preference rather than competition (Hecnar and M'Closkey, 1997a).

Call differences also separate these early breeding species. For example, *P. crucifer* and *P. ornata* also often breed at the same wetlands within the southeastern Coastal Plain. Their calls are very similar, and the males of the two species sometimes call in close proximity to one another. However, female *P. crucifer* are able to discriminate conspecific calls, and they prefer them over the calls of *P. ornata* (Gerhardt, 1973). Gerhardt's results were different from those obtained by Martof (1961), although the differences are likely due to experimental design rather than regional differences in female call discrimination ability. Female *P. crucifer* did choose the calls of conspecific males over calls of *P. feriarum*, however (Martof, 1961).

The timing of breeding may have important consequences on Spring Peeper larval fitness. When Spring Peepers precede *Anaxyrus woodhousii* at breeding ponds, larvae are unaffected by the later appearance of the toads. However, when *Anaxyrus* precedes *Pseudacris crucifer*, Spring Peepers have a more extended larval period, and their growth rate and mass at metamorphosis is reduced. This competitive effect is most pronounced when toads arrive approximately seven days in advance of the Spring Peepers, and the competitive effect seems to decrease if Spring Peepers arrive approximately 14 days after the toads. Lawler and Morin (1993) hypothesized that the more active toads were able to grow quickly to the point where they had no appreciable effect on the less active Spring Peeper larvae, and thus reduced competition. By arriving prior to toads at breeding ponds, Spring Peeper larvae were able to grow to the point where the subsequent appearance of toad larvae had no effect. The subtle interplay of reproductive timing, therefore, has effects on the level of interspecific competition (Lawler and Morin, 1993).

There is some disagreement in the literature concerning the competitive ability of Spring Peeper larvae, with laboratory or mesocosm studies suggesting inferior competitive ability due to small body size, decreased activity levels, or poor food gathering abilities (Seale and Beckvar, 1980;

Adult *Pseudacris crucifer*.
Photo: Aubrey Heupel

Morin and Johnson, 1988; Werner, 1991; Skelly, 1995a). A number of authors, for example, have noted that Spring Peepers and Green Frogs (*Lithobates clamitans*) do not co-occur in the same breeding sites (Collins and Wilbur, 1979; Morin, 1983; Dale et al., 1985) and inferred that Green Frogs may out-compete Spring Peeper larvae. In field simulations, however, Green Frog larvae have no effect on survivorship, larval period, or growth rates, and only a slight effect on size at metamorphosis (Skelly, 1995b). Alternatively, competition with *Anaxyrus*, *Hyla*, *Lithobates sphenocephalus*, and *Scaphiopus* reduced survivorship to zero even without predators (Morin, 1983). In a further experiment, the sunfish *Enneacanthus* completely eliminated larval *Pseudacris crucifer* (Kurzava and Morin, 1998).

Other mesocosm studies have suggested that the competitively inferior Spring Peeper larvae actually have greater body mass, survivorship, and abundance in the presence of predators (e.g., *Notophthalmus viridescens*) than they do when confined with nonpredatory tadpoles of other species (Morin, 1983). It may be that newts selectively remove the more active larvae of other species, leaving more food for the relatively inactive peeper larvae that do not attract the attention of the visually oriented predator. Still, that *Pseudacris crucifer* tadpoles require higher critical and threshold concentrations of algae for effective suspension feeding compared with the larvae of other genera (*Anaxyrus*, *Lithobathes*) suggests the potential for exploitative competition for food resources (Seale and Beckvar, 1980).

Thus, the bottom line is that the survivorship, growth, size at metamorphosis, and duration of the larval period of Spring Peepers likely depends on which species are present, the density of conspecifics and other larvae, predation pressure (Morin, 1986), and in the extent of the temporal overlap within natural ponds. Competition is a matter of context.

DISEASES, PARASITES, AND MALFORMATIONS

Water molds (*Leptolegnia* sp.) are pathogenic to Spring Peeper eggs (Ruthig, 2009). An *Ichthyophonus*-like fungal infection has been found on a single Spring Peeper from Québec (Mikaelian et al., 2000) and from larvae in Maine (Gahl, 2007). The virulent amphibian chytrid fungus *Batrachochytrium dendrobatidis* was not detected on Spring Peepers in the Northeast (Longcore et al., 2007), but it is now known from Maine (Gahl, 2007) and a single Texas specimen (Saenz et al., 2010). Ranavirus is known from Spring Peepers in Maine (Green et al., 2002; Gahl, 2007) and Frog Virus 3-like infections have been reported in natural populations of Spring Peeper larvae from Ontario (Duffus et al., 2008). An alveolate pathogen affecting larvae was reported from Maine (Gahl, 2007). Malformations also have not been found in Spring Peepers from the north-central and northeastern United States (Converse et al., 2000).

Parasitic protozoans include *Nyctotherus cordiformis, Octomitus intestinalis, Opalina* sp., *Opalina chorophili, O. hylaxena, O. obtrigonoidea, O. pickeringii, O. virguloidea, Trichomonas augusta,* and *Trypanosoma rotatorium* (Brandt, 1936). Spring Peepers are parasitized by the trematodes *Alaria mustelae, Brachycoelium hospitale, Diplodiscus temperatus, Fibricola texensis, Glypthelmins* sp., *Glypthelmins pennsylvanicus, G. quieta, Haematoloechus* sp., *H. varioplexus, Megalodiscus temperatus,* and larval flukes (Brandt, 1936; Odlaug, 1954; Najarian, 1955; Cheng, 1961; Catalano and White, 1977; Ashton and Rabalais, 1978; Coggins and Sajdak, 1982; Muzzall and Peebles, 1991; Joy and Dowell, 1994; Yoder and Coggins, 2007; McAllister et al., 2008), the nematodes *Agamascaris odontocephala, Agamonema, Cosmocercoides* sp., *C. dukae, Oswaldocruzia pipiens, Physaloptera, Rhabdias* sp., *Rhabdias ranae,* and *Spiroxys* sp. (Brandt, 1936; Odlaug, 1954; Muzzall and Peebles, 1991; Yoder and Coggins, 2007), and an undescribed nematode (Joy et al., 1996). Other helminths include acanthocephalans (*Centrorhynchus* sp.), the nematode *Mesocestoides* sp., and proteocephalid cestode cysts (Brandt, 1936; Yoder and Coggins, 2007). Biting midges of the genus *Corethrella* feed upon *Pseudacris crucifer* and may be attracted by the frog's advertisement call (McKeever and French, 1991).

A single malformed *P. crucifer* was reported from Minnesota by Hoppe (2005).

SUSCEPTIBILITY TO POTENTIAL STRESSORS

Metals. Spring Peepers are considered very sensitive to heavy metals, including aluminum, arsenic, barium, beryllium, iron, gallium, mercury, magnesium, lead, tin, titanium, and zirconium (Westerman et al., 2003a). Copper concentrations at 5.5 µg/L cause significant reduction in larval survivorship; survivorship is further decreased when this level of exposure is coupled with high levels of UVB light (Baud and Beck, 2005). The presence of Spring Peepers is negatively correlated with atmospheric deposition of cadmium, nickel, other correlated metals, and sulphate (Glooschenko et al., 1992).

pH. Spring Peeper larvae develop best in water with a pH >4.2 (Gosner and Black, 1957a). In the field, they tend to choose ponds with lower levels of acidity. For example, they are found most often in ponds with a mean pH of 5.1 in Nova Scotia (Dale et al., 1985) and a pH >6.3 in Florida (Owen, 1996). In addition, tadpole activity levels decrease at a pH <4.5 (Freda and Taylor, 1992). They are very sensitive to high levels of acidity at breeding ponds. A pH of 3.8–4.2 is lethal to *P. crucifer* embryos, and hatching success decreases at a pH of 4.3 or lower (Karns, 1983; Freda, 1986). The tadpole LC_{50} is a pH of 3.74. Although tadpoles do not avoid water at a pH of 4.5, they do so at a pH of 4 (Freda and Taylor, 1992). Still, Dale et al. (1985) recorded calling from sites with a pH as low as 3.9 in Nova Scotia, although that does not mean that successful reproduction occurred at those sites. The presence of Spring Peepers increases in ponds with a high buffering status; buffering status is inversely correlated with acidity (Glooschenko et al., 1992).

Chemicals. In experimental laboratory trials, two insecticides (carbaryl, malathion) and two herbicides (glyphosate [Roundup™], 2,4-D) had no effects on Spring Peeper larval survivorship at the manufacturers recommended dosage (Relyea, 2005b). Still, pesticide residues may persist over a long period. In Ontario, dieldrin, DDT, DDE, and DDD persisted in animals at mean concentrations of 199.8, 160.6, 1,001, and 26.5 µg/kg wet weight, respectively, 26 yrs after they were last used (Russell et al., 1995). The combined concentrations of DDT and their metabolites exceed human health recommendations.

The herbicide atrazine has its greatest effects at low dosages (3 ppb), regardless of whether larvae are in early or late stages of development for exposures over a 30-day period. Survivorship is significantly decreased at low doses when compared to high doses (30, 100 ppb). Low and medium doses have their greatest effect early in development; medium doses in late development have no effects on survivorship. At medium and low doses, survivorship goes to 0 at about 29 days (Storrs and Kiesecker, 2004). These results suggest that atrazine at low doses can have significant negative effects on the larval development of Spring Peepers in nature.

Spring Peepers are extremely sensitive to chloroform. Complete mortality occurs at 7.34 mg/L, with mortality and teratogenesis occurring in concentrations of as little at 0.0087 mg/L. Larval survival was 88% at 0.0087 mg/L, but increased to 46% at 0.69 mg/L (Birge at al., 1980).

UV light. Eggs of *P. crucifer* exposed to artificially high does of UVB radiation have a 40% mortality rate after 60 min of exposure. The jelly capsule surrounding the egg absorbs about 80% of the UVB light and helps protect the embryo (Licht, 2003). Larval survivorship decreases in the presence of both low (19–21 mW/cm^2) and high (28–30 mW/cm^2) levels of UVB light under laboratory conditions (20 min/day for 7 days), especially when exposed to copper (Baud and Beck, 2005).

STATUS AND CONSERVATION

Pseudacris crucifer is considered a relatively secure species throughout much of its range, except at the margins of its distribution. In New York, for example, surveys >20 years apart revealed increasing and stable populations (absence from 38 sites, disappearance from 50 sites, appearance at 100 sites, persistence at 111 sites). Populations persisted in areas of higher elevation, on less acidic soils, and in less developed and agricultural areas. Increases in mixed forested habitats were correlated with population appearances (Gibbs et al., 2005). The species is also reported as increasing in abundance or stable in various other portions of its range (Minton et al., 1982; McAlpine, 1997a; Weller and Green, 1997; Rice at al., 2001; Brodman et al., 2002).

However, Christiansen and Burken (1978) and Christiansen (1981) considered it uncommon in Iowa due to the loss of suitable breeding sites (but see Hemesath, 1998), and it has experienced a "small but significant" decline in Wisconsin (Mossman et al., 1998). Populations also tend to disappear from heavily urbanized areas. In the Northeast, populations were considered declining in 6 states (Delaware, Massachusetts, New Hampshire, Pennsylvania, Vermont, West Virginia) based on 7 yrs of data using occupancy modeling (Weir et al., 2009). It is considered threatened in Kansas at the periphery of its range and protected in New Jersey.

Spring Peepers are a commonly monitored species during amphibian call surveys (e.g., Burton et al., 2006). Detection probabilities are usually high because of their high-pitched whistle, with probabilities of >80% after only 3 sampling nights during Ontario road transect surveys (de Solla et al., 2005), 79–86% in Québec (Bonin et al., 1997a; Lepage et al., 1997), and 76% in Wisconsin (Mossman et al., 1998). During call surveys in Illinois over a 4 yr period (1986–1989), populations appeared to be increasing (Florey and Mullin, 2005). Interestingly, Spring Peeper occupancy of breeding ponds does not seem to be affected by nearby road traffic density (Eigenbrod et al., 2008). That does not mean, however, that significant mortality does not occur on roads. Over a 3-day period in March, Carpenter and Delzell (1951) recorded 1,080 Spring Peepers from a road approximately 10 m from a breeding pond. They noted that 75% of the amphibians they observed (10 species) were found dead. Mortality of these small frogs is probably quite high on roads, but their squashed delicate bodies probably do not remain long and are thus vastly under-counted in road-effect studies. The area of immediate adverse impact (the "road-zone" effect) extends 200–300 m from the pavement (Eigenbrod et al., 2009).

Spring peepers prefer hardwood sites to cut-over hardwoods, pines, and burned sites (McLeod, 1995; McLeod and Gates, 1998). They tend to avoid recently mined sites, and their numbers may be severely reduced in agricultural areas, especially if forested habitats are not in close proximity (Knutson et al., 2004; Anderson and Arruda, 2006). However, through time as succession occurs, old mine sites may be recolonized; Myers and Klimstra (1963) recorded *P. crucifer* as common on sites where coal mining had ceased 19–29 yrs previously. Peepers disappear from clearcut sites, but may return if suitable breeding sites remain nearby (Clawson et al., 1997). However, it is difficult to evaluate Clawson et al.'s (1997) results and gauge the extent of the effects of clearcutting since numbers were not provided, and the study was conducted for only 6-months postharvest; presence is not equivalent to having no effects or minimal effects.

In another study in New Brunswick, conversion of natural forests to conifer plantations had detrimental impacts on *P. crucifer*. Breeding sites in converted areas tended to have short hydroperiods, which resulted in poor recruitment (Wal-

Breeding habitat of *Pseudacris crucifer*, Noxubee County, Mississippi. Photo: C.K. Dodd Jr.

rights-of-way, as long as there are forest patches immediately nearby (Fortin et al., 2004b).

There is evidence that Spring Peepers are adversely affected by cattle using their breeding ponds. Larvae from such sites are smaller than comparable sites where cattle do not have access. Cattle-access ponds have high turbidity, which may inhibit tadpole development and survival (Schmutzer et al., 2008).

Repatriation efforts were carried out for *P. crucifer* at the Gateway National Recreation Area, New York, between 1980 and 1990. A total of 68 adults and 12,700 larvae was used in the repatriation. According to Cook (1989, 2008), this repatriation has been successful at three of four reintroduction sites. Spring Peepers normally colonize restored and newly created ponds (Briggler, 1998; Nyberg and Lerner, 2000; Pechmann et al., 2001; Stevens et al., 2002; Touré and Middendorf, 2002; Brodman et al., 2006; Palis, 2007; Shulse et al., 2010), but not in all circumstances, at least over the time monitored (Kline, 1998; Lehtinen and Galatowitsch, 2001). Numbers at recent mine treatment and reclamation sites may be low, although these sites may harbor calling males (Lacki et al., 1992; Anderson and Arruda, 2006). In South Carolina, a series of three 3 created ponds produced 1,500 juveniles over a 7 yr period (Pechmann et al., 2001).

While it is difficult to imagine commercial trade in Spring Peepers, Enge (2005a) reported that 2,361 *P. crucifer* were collected for the pet trade from 1990 to 1994 in Florida.

dick et al., 1999). Spring Peepers are somewhat resistant to habitat fragmentation, occurring in a wide variety of forest fragmentation gradients (Gibbs, 1998b; Kolozsvary and Swihart, 1999). They may breed in wetlands within power-line

Pseudacris feriarum (Baird, 1854)
Upland Chorus Frog

ETYMOLOGY

feriarum: Beltz indicates that the name is derived from the Latin *feriarum* meaning 'holidays' or 'leisure.' However, I suggest Baird was more likely referring to the Latin *ferrum*, meaning 'of iron,' 'sword.' The name would thus refer to the parallel iron-colored stripes down the back of the frog.

NOMENCLATURE

Synonyms: *Helocaetes feriarum, Chorophilus feriarum, Chorophilus nigritus feriarum, Hyla feriarum, Pseudacris nigrita triseriata* (in part), *Pseudacris triseriata feriarum*

There is considerable taxonomic confusion in the literature on chorus frogs. In the past authors may have assigned a name to a chorus frog population under study that has no resemblance to the current nomenclature based on molecular analysis. For example, Smith and Smith (1952) compared various morphological characteristics of

"*Pseudacris nigrita feriarum*" with "*P. n. triseriata*." Upon looking at their collecting localities, it is clear that what we now know as *P. feriarum, P. fouquettei, P. triseriata,* and *P. kalmi* were all included under their name *P. n. feriarum,* and that *P. fouquettei, P. triseriata,* and *P. maculata* were included under their name *P. n. triseriata.* In another example, Platz (1988a, 1989) analyzed calling characteristics of several species of striped chorus frogs from throughout their range, using the then current subspecific designations. He noted significant differences in many call characteristics, and suggested a revision of the subspecies. Based on genetic evidence (Lemmon et al., 2007a), Platz's species now are known to have included *P. maculata, P. fouquettei, P. feriarum* (possibly in western Kentucky), and *P. kalmi.* Always check the location where specimens originated to verify correct taxonomic allocation.

The name *P. triseriata* has often been used for this species in the published literature, and a number of authors have discussed its confusing nomenclatural history (e.g., Brown, 1956). In the following sections, I have based accounts only on studies with precise locations in order to ensure proper taxonomic allocation.

IDENTIFICATION

Adults. This is a small ground-dwelling frog that is tan, medium to dark brown, or gray with three dark parallel bands down its back. Postmetamorphs have a dark triangle between the eyes, but this trait is often lacking in Coastal Plain populations from South Carolina. The upper lip has a white line, there may be a dark band extending through the eye and continuing to just above the forelimb, and the belly is white to cream in contrast to the darker dorsal color; the belly may have small brown flecks. There may be some dark coloration on the chest. The toes have almost imperceptible expanded pads, and webbing between toes is virtually lacking. Some specimens may completely lack dorsal markings. Males during the breeding season have a brown to yellowish coloration on the throat where the vocal sac is located.

Males are slightly smaller than females, with South Carolina Coastal Plain populations having the largest frogs. Adult males are 23–31 mm SUL, whereas in North Carolina females are 25–33 mm SUL (Alexander, 1966). Mitchell (1986) gives maximum sizes of 31 mm SUL for males and 32 mm SUL for females in Virginia, and Brown (1956) recorded a size range of 22 to 31 mm SUL (mean 28 mm) in Alabama. In West Virginia, males reach 24.9 mm SUL, whereas females reach 28.7 mm SUL (Sias, 2006). Pennsylvania males average 24.1 mm SUL (range 20 to 27 mm) and females average 25.1 mm (23–27 mm) (Schwartz, 1957). South Carolina Coastal Plain *P. feriarum* males are 25–35 mm SUL and females are 33–39 mm SUL; males from the Piedmont are smaller (21–29 mm SUL), as are females (22–31 mm) (Schwartz, 1957). An albinistic adult was reported in Virginia (Ackroyd and Hoffman, 1946; Hensley, 1959).

Larvae. The tadpoles are small and olive, bark brown, or nearly black, and they have gold flecking uniformly across the dorsum; they have a whitish or silver to bronze coloration on the belly. The tail is medium in length. Tail fins may have small dark flecks, and the dorsal tail fin is slightly higher than the ventral tail fin. The snout is rounded when viewed from the side. Tadpoles reach 34–36 mm TL prior to metamorphosis. The mouthparts are figured in Dodd (2004).

Eggs. The eggs are white to yellow on the vegetal pole, with darkly pigmented brown hemispheres. They are contained in a loose irregular jelly mass about 25 mm in diameter, attached to submerged stationary objects such as leaves, twigs, and branches. Egg masses are usually located within 25 cm of the water's surface. Egg counts in the literature vary considerably. Alexander (1966) recorded from 4 to 159 eggs per mass, whereas Mitchell (1986) gave a figure of 600 as maximum clutch size. Brown (1956) counted 10 to 40 eggs per mass. A full breeding season compliment of eggs will comprise many clusters, with a possible output of about 1,500 eggs during a breeding season. Eggs contain only a singular gelatinous envelope 5–7.8 mm in diameter; the vitellus is 0.9–1.3 mm (Livezey and Wright, 1947; Brown, 1956).

DISTRIBUTION

Upland Chorus Frogs are found in the Piedmont and in the Ridge and Valley Province and foothills from central Pennsylvania southward in an arc to northwest Mississippi, western Tennessee, and

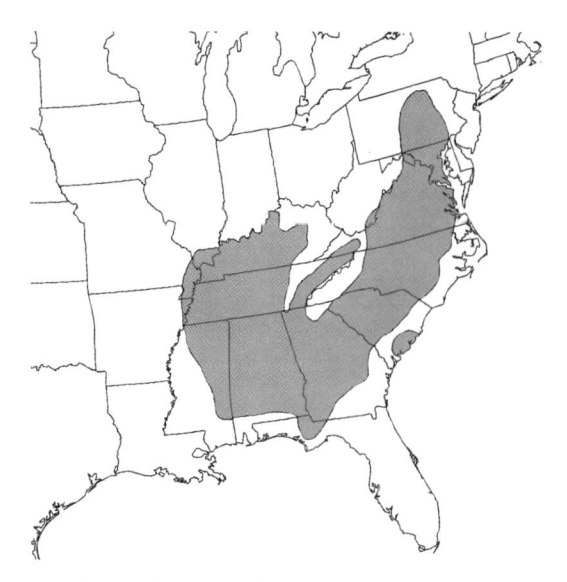

Distribution of *Pseudacris feriarum*

central Kentucky. Although generally found in the Piedmont, populations in Georgia follow the Savannah and Altamaha rivers into the Atlantic Coastal Plain in the East, and down the Chattahoochee and Flint river drainages into the Gulf Coastal Plain. In North Carolina, they follow the Cape Fear River southeast toward the Coastal Plain. They barely enter southern Illinois, but not in the area indicated by Cagle (1942). They occur west of the Mississippi River in extreme northeastern Arkansas and the Missouri boot heel. A record from Lafayette County, Arkansas (Black and Dellinger, 1938), is not valid. These frogs occur in east Tennessee in the Tennessee River Valley and Smoky Mountain foothills, but appear to be sparse or absent from the Cumberland Plateau and Allegheny Mountains. They follow the Apalachicola and Yellow rivers south into the Florida Panhandle, thus verifying the observations by Carr (1940a) and Neill (1949b). A disjunct population of *P. feriarum* occurs in Berkeley, Dorchester, and Charleston counties., South Carolina (Schwartz, 1957; Lemmon et al., 2007b).

Because of recent taxonomic changes, the best reference on distribution is the systematic study by Lemmon et al. (2007b). Other useful references include: Alabama (Mount, 1975), Florida (Bartlett and Bartlett, 1999), Georgia (Williamson and Moulis, 1994; Jensen et al., 2008), Illinois (Phillips et al., 1999), Maryland (Harris, 1975), Mid-Atlantic (Beane et al., 2010), Missouri (Johnson, 2000), North Carolina (Dorcas et al., 2007), Pennsylvania (Hulse et al., 2001), South Carolina (Schwartz, 1957), Tennessee (Redmond and Scott, 1996), Virginia (Tobey, 1985; Mitchell and Reay, 1999), and West Virginia (Sias, 2006).

FOSSIL RECORD

No fossils have been specifically attributed to this species or to any member of the striped trilling frog species complex within the known range of *P. feriarum*.

SYSTEMATICS AND GEOGRAPHIC VARIATION

Pseudacris feriarum is a member of the Trilling Chorus Frog clade of the genus *Pseudacris*, a group that includes *P. brachyphona, brimleyi, clarkii, fouquettei, kalmi, maculata, nigrita,* and *triseriata* (Moriarty and Cannatella, 2004). These species are distinguished by their advertisement call structures, color patterns, genetics, the presence of a cuboidal intercalary cartilage (Paukstis and Brown, 1987), and similar albumin phylogeny (Maxson and Wilson, 1975). Upland Chorus Frogs are the sister taxon of Western Chorus Frogs. These two species plus *P. kalmi* are the sister taxa of *P. nigrita* and *P. fouquettei*. In addition, there are two geographically separate lineages within *P. feriarum*, an inland and a coastal lineage roughly separated by the Altamaha River in Georgia (Lemmon et al., 2007b).

Natural hybridization occurs with *P. brachyphona* in northeast Mississippi and south-central Kentucky (Lemmon et al., 2007b). Fouquette (1975) found evidence of call-based character displacement and no evidence of a hybrid zone at a contact zone between *P. nigrita* and *P. feriarum* along the Apalachicola River, despite low levels of hybridization (Crenshaw and Blair, 1959). This result is not surprising since these are nonsister taxa (Lemmon et al., 2007b). In contrast, Gartside (1980) found evidence of high levels of hybridization between *P. nigrita* and *P. fouquettei* along the Pearl River. Since these are closely related taxa, the level of hybridization at the contact zone between them is more easily explained than it would be without the phylogenetic interpretation of Lemmon et al. (2007b).

In the laboratory, *P. feriarum* readily hybridizes with *P. nigrita, P. ornata, P. brimleyi,* and *P. brachyphona,* and it doesn't appear to make much difference whether the *P. feriarum* is male or female (Mecham, 1965). However, crosses with *P. crucifer* are much less successful, but more so when the *P. feriarum* is female. This species can be hybridized with *P. clarkii*, and some larvae will metamorphose. Crosses with *Acris crepitans* are unsuccessful, usually dying prior to hatching (Littlejohn, 1961a).

ADULT HABITAT

During the nonbreeding season, Upland Chorus Frogs are found in nearby swamp forest, moist forests, or river bottomland forest. They tend not to disperse very far from breeding sites when these sites are in proximity to nonbreeding forested habitats. However, they will cross an extensive amount of open area to reach breeding sites in grassy marshes and temporary wetlands surrounded by grasslands. They inhabit the leaf litter, and shelter under surface debris, tree bark on the ground, or in small animal burrows.

TERRESTRIAL ECOLOGY

Upland Chorus Frogs spend most of the year in swamp or moist forests, where they forage in the leaf litter. They are usually well hidden in thick vegetation and are rarely encountered outside the breeding season, leading authors to speak of their "disappearance." Despite this, they are likely active throughout the warmer parts of the year and may be active year-round in the southern portion of its range. For example, Dodd (2004) found occasional individuals in a drying wetland from July to September in Tennessee, and Alexander (1966) found food in stomachs all year in North Carolina. Individuals forage on moist shaded forest floors or semifossorially, and they hide under surface debris. The species is capable of extensive terrestrial movements, and Ferguson (1963) found that displaced Upland Chorus Frogs were capable of correct orientation from the release site, moving 422 m to the site of original capture.

Upland Chorus Frogs use sun-compass (i.e., celestial or Y-axis) orientation to direct their movements to breeding sites, as well as auditory cues (Ferguson, 1963, 1966a, 1966b). Indeed, auditory cues may be more important in finding breeding ponds than celestial cues, since most movement occurs at night or on overcast days. Migrating frogs are capable of locating a breeding site by moving toward calling conspecifics. As they move toward the site, they will occasionally utter a call. However, if a chorus is interrupted, they will call more frequently, as if waiting for a response. Martof (1962a) noted their ability to discriminate odors from different habitats, suggesting a third potential orientation mechanism, but this has not been studied since his observations. Ferguson's (1966b) experiments on celestial orientation eliminated odor as a potential cue.

Like *Pseudacris nigrita*, Upland Chorus Frogs

Eggs of *Pseudacris feriarum*.
Photo: Dana Drake

probably have a blue-mode phototactic response, indicating they have true color vision (Hailman and Jaeger, 1974).

CALLING ACTIVITY AND MATE SELECTION

The call of the male Upland Chorus Frog sounds like running a finger along the tines of a pocket comb. General characteristics of the call include a mean duration of 0.8 sec (range 0.5 to 1.25), a mean interval of 1.2 sec between calls, notes averaging 17 in number (range 15 to 23), and a dominant frequency of 2,800 cps (range 1,100 to 3,100) (Thompson and Martof, 1957; Jensen et al., 2008). Notes tend to be closer together at the beginning of the call compared with the end of the call. Martof and Thompson (1964) produced a series of "artificial" calls by altering call interval, intensity, uniformity and number of notes, duration of call, and frequency. They found that the number and spacing of notes and patterns of spacing and frequency were not necessary by themselves to induce proper choices by Upland Chorus Frog females. Calls needed to be at least 0.5 sec in duration with short intervening pauses for correct discrimination, but no one characteristic of the call was effective in eliciting discriminatory behavior.

Mecham (1961) has discussed differences between the call of this species and *P. nigrita*. On the Coastal Plain in areas of overlap between the species, the call is fast and shorter (43 pulses/sec, duration of 0.62 sec) than in areas of nonoverlap (22 pulses/sec, duration of 0.75 sec). Coupled with changes in the call of *P. nigrita* in the zone of overlap, these differences may help maintain premating isolating mechanisms.

Females exert strong selection in terms of mate choice, preferring conspecific males over heterospecific males, especially in areas of contact with other members of the Trilling Chorus Frog clade. Signal diversification occurs in these areas of sympatry, especially in terms of pulse rate and the number of pulses, with *P. feriarum* males increasing their calling effort when compared to areas where sympatry does not occur. In areas of sympatry, females have evolved a greater within species discriminatory ability as a consequence of between-species discrimination. Such a directional selection is metabolically costly to the males because of the increased energy expenditure needed to increase the number and rate of pulses. As a result, males may have to reduce the amount of time spent calling or adopt other behavior that will increase their fitness, despite the cost. Lemmon (2009) discusses these options in detail.

Males that are amplexed by another male give off a harsh and penetrating warning call that causes conspecific males to let go. Males will struggle to escape the grasp of a conspecific, often appearing to wrestle the intruder. When males approach one another, they become motionless at about 30 cm apart, appearing to watch one another. After no longer than about 6 min, one of the males will turn away from the other male. Male–male interactions have been described by Martof (1958) under laboratory conditions.

Calling occurs from late winter to early spring, and Upland Chorus Frogs are among the first frogs heard calling each year. Occasional calls may be heard into the summer or autumn, well past the breeding season. For example, Burt (1933) recorded an individual calling in June, Murphy (1963) heard calls in late autumn, and Mount (1975) noted calling in summer during cool rains. The timing of calling is variable and dependent on environmental conditions. Peak breeding occurs in December and January in South Carolina (Schwartz, 1957). Scattered calls were heard by Alexander (1966) from October to May in North Carolina, but full breeding choruses occurred usually from January to March, and rarely in December. In Mississippi, calls are heard from late November through mid-March (Landreth and Ferguson, 1966), whereas in Tennessee calling occurs from January to April (Burkett, 1991). Snow and ice may remain on the ground and in ponds when breeding begins.

Chorusing usually occurs for several days prior to the onset of breeding. Calling occurs throughout the day, with peaks around 12:30, 18:00, and 24:00 hrs; the least calling occurs around 08:00 hrs (Todd et al., 2003; Steelman and Dorcas, 2010). The intensity of calling does not occur synchronously even between closely situated breeding sites; calling intensity may be high at one location yet greatly decreased at the next, and calling may extend longer in the season at one location than another.

Temperature has a direct influence on calling behavior. Calling is initiated at ca. 10°C air temperature and at 6°C water temperature, but it is infrequent at these cold temperatures. Strong choruses are heard when water temperatures reach 6.5°C, even when air temperatures remain as cold as 5°C. As temperatures warm to 7–10°C, chorusing occurs both day and night and continues as long as temperatures remain ca. 10–25°C. Calls are heard from ca. 5 to <25°C with an optimum of 10°C, at least in North Carolina (Steelman and Dorcas, 2010).

Males call from the water or shoreline, using their front legs to balance and hold onto pond detritus. They do not climb vegetation. The rear legs float out behind the body, or the frog may sit on them in very shallow water. The head is thrust out of the water during calling, exposing the balloon-like subgular vocal sac, which is single. The vocal sac is partially deflated between calls or when the frog is disturbed. A female deliberately orients directly toward a calling male and eventually moves directly in front of him and actually contacts him. The male stops calling and immediately amplexes the female. Amplexus occurs dorsal to the forelimbs (i.e., it is axillary). Ovulation occurs immediately after amplexus. Should amplexus be disrupted, the female may re-pair with the same male to continue egg laying (Alexander, 1966) rather than choose a different male. Alexander (1966) describes the courtship sequence.

Males tend to congregate within particular areas of the breeding pool, rather than be randomly dispersed. During warmer weather, they will leave the water and call from the shoreline, but as temperatures drop they move back to the water and continue calling until the temperature approaches freezing.

BREEDING SITES

Breeding sites are usually small temporary pools and puddles in open grassy fields, although some frogs will breed at small woodland pools. Other reported breeding sites include cypress/gum temporary ponds in sandhill habitats, marshes, cypress savanna (Liner et al., 2008), roadside ditches, cedar glades, cow footprints, and wheel ruts. They do not breed in permanent water. Although breeding does not occur until late winter or spring, adults begin to congregate in the vicinity of breeding sites in the fall and remain there throughout the winter. Thus, there is no breeding migration to ponds in the spring as there is with most other anuran species.

Frogs breeding in open situations usually commence breeding earlier than those breeding in woodland pools. Large breeding congregations use the same pools and wetlands from one year to the next, with new sites more likely to be used by fewer individuals later in the year. Frogs tend to choose sites that have warmer water in which larvae can develop rapidly, hence the preference for open-canopied sites.

REPRODUCTION

Breeding occurs from winter through spring depending upon environmental conditions and characteristics of the breeding site. Specific breeding months include December to March (Alabama: Brown, 1956), January to March (Mississippi: Landreth and Ferguson, 1966), January to April (North Carolina: Murphy, 1963), February to April (North Carolina: Alexander, 1966), to late April in Tennessee (Jordan et al., 1968b), and March to April (West Virginia: Sias, 2006). Frogs may call for some time before actual egg deposition. For example, Landreth and Ferguson (1966) noted calling from 20 to 25 November in Mississippi, although actual mated pairs were not first observed until between 23 January and 4 February. At their study sites, large choruses were not observed after 15 to 23 March, and the breeding season lasted from 63 to 71 days. Although the Upland Chorus Frog calls over an extended time, most reproduction at any particular pool occurs on only a few nights.

Reproduction occurs at temperatures greater than initial calling temperatures, which helps explain the delay between the initiation of calling and breeding. According to Alexander (1966), the optimum range for breeding is 10–15°C, with reproduction occurring to 22°C. In Mississippi, there were between 20 and 33 days during the 70-or-so-day breeding season when temperatures were below freezing. Rainfall appears to be important in stimulating breeding, especially since this species uses in temporary pools, puddles, and shallow grassy wetlands. Eggs are deposited in elongate clusters attached to sticks, leaves, or other subsurface debris.

Males begin breeding at 8–10 months after metamorphosis, whereas females often delay breeding, and may not breed until the second year (Alexander, 1966). In the autumn, females have developing ova, but the oviducts are not distended. The number of eggs oviposited per female

Tadpole of *Pseudacris feriarum*. Photo: Gabe Miller

ranges from 163 to 534, with larger females producing more eggs than smaller females. Several clutches may be deposited during a single breeding season. Females in their first breeding season have clutches of 179–275 eggs, whereas females breeding in their second season have clutches of 282–463 eggs (Alexander, 1966). Most eggs are successfully fertilized during amplexus. In Alexander's (1966) study, only 20 of 2,798 eggs were not fertilized. Egg deposition lasts over a period of several days. Eggs hatch in 8–13 days in West Virginia (Sias, 2006).

Martof (1958) described the oviposition sequence. The female straddles a stem, branch, or other support, and clasps it with her hands. Prior to egg extrusion, she arches her dorsum backward, which brings her cloaca nearly into contact with the supporting structure and the vent of the male. As the eggs are extruded, the male fertilizes them. These eggs form a loose mass, each containing 8 to 35 eggs. In Martof's (1958) observations, the mean number of egg masses per female was 6.8. His females deposited a total of 78–157 eggs (mean 111), a process that took about 2 hrs.

There appears to be only a single reference on sex ratios at breeding ponds. In a pond in North Carolina, Murphy (1963) found 74 males and 51 females, for a sex ratio of 1.45:1. Jensen et al. (2008) noted that females remain at a pond only one night before they return to terrestrial habitats, whereas males remain around the pools for several weeks.

LARVAL ECOLOGY

Larvae at a site tend to result from eggs deposited only on a few nights. As a result, they represent an even-aged cohort as they proceed through larval growth and development (Murphy, 1963; Alexander, 1966). Temperature has an important effect on larval activity. At 1–6°C, larvae are inactive, whereas they are most active at water temperatures of 11–25°C. In ponds directly exposed to the sun, larval mortality begins at about 38°C (with 10–40% mortality); by 40°C, complete mortality occurs (Alexander, 1966). Upland Chorus Frog larvae are able to acclimate to various temperatures, and their ability to tolerate high temperatures is directly related to acclimation temperature. Thus, the CTmax for acclimation temperatures <20°C is 38°C, whereas larvae acclimated at 30°C can withstand temperatures to 41°C (Cupp, 1980). Adults have lower CTmax than larvae, and the CTmax of larvae decreases as they approach metamorphosis. Considering the type of breeding ponds selected, it is important for larvae to complete metamorphosis before spring temperatures heat the shallow pools.

Upland Chorus Frog larvae do not form aggregations, as they are uniformly distributed throughout a breeding pool. They tend to swim rapidly, and upon contacting an object within their path, they suddenly stop swimming and sink motionlessly to the bottom of the pool. Larvae feed within the bottom mud and plant detritus, and they will feed on dead animal material. They feed at a sharp angle to the substrate, with rapid jerks of the tail. The jaws are used to rasp detritus and other potential food items, which is then picked up by a stream of mucous and carried into the esophagus. This type of feeding is termed mucous cord microphagy, and it is common among detritus-feeding larvae.

Larval *P. feriarum* are detritivores. Although pine and oak pollen is copious during the spring developmental period, Britson and Kissell (1996) found that a diet that included pollen was inadequate nutritionally to allow for proper development. Tadpoles fed pollen at any time during the larval period were smaller than nonpollen-eating tadpoles, with a higher percentage of individuals that failed to initiate metamorphosis. Thus, *P. feriarum* larvae need to avoid ingesting pollen despite its availability. The presence of pollen and the possible scarcity of phytoplankton early in the year may have led to a detritus-based feeding mode rather than to suspension feeding.

Larvae develop rapidly, with the tadpole stage lasting 7–14 weeks. Specific literature records include 36 days in West Virginia (Sias, 2006), 50 days in Virginia (Mitchell, 1986), 48–80 days in Georgia (Jensen et al., 2008), and 40–95 (generally about 70) days in Alabama (Brown, 1956). There is considerable variation in growth rates and in the length of larval period among the progeny of different males and females, suggesting an important parental component to these traits. Maternal effects seem stronger than paternal effects, and Travis (1981) suggested that nongenetic maternal effects were the primary source of phenotypic variability. The size at metamorphosis also has a genetic component that is independent of both growth rate and the length of the larval period (Travis, 1981). Metamorphosis is complete by ca. 9.5 mm SUL (range 7.8 to 12 mm) in North Carolina (Alexander, 1966) or 9.8 mm SUL (range 8.8 to 12 mm) in Alabama (Brown, 1956). However, Sias (2006) recorded larvae hatching at 6–9 mm TL in West Virginia. Alexander (1966) provides illustrations of developmental changes through metamorphosis. Froglets begin to feed, even when there is about 2 mm of tail yet to be absorbed.

One of the primary sources of egg and larval mortality is desiccation, as shallow wetlands dry. Frogs that breed early have adopted a bet-hedging strategy whereby the first arrivals may get a headstart on subsequent arrivals and avoid certain poikilotherm predators, such as snakes, and competitors, such as other anuran larvae. However, if rains do not continue and wetlands do not have sufficient hydroperiods, then larvae will die. Frogs breeding later in the season might be able to choose deeper and more stable waters, but ones that have already been colonized by potential predators and competitors. They also may face

Adult *Pseudacris feriarum*.
Photo: Aubrey Heupel

suboptimal thermal conditions as waters warm beyond tolerance. As pools dry down, oxygen levels are depleted as crowding increases. In such circumstances, tadpoles are able to gulp air at the surface of the water. The other major source of tadpole mortality is cold weather.

DISPERSAL
Alexander (1966) studied the dispersal of Upland Chorus Frogs in North Carolina. Dispersal began when the metamorphs reached about 20 mm SUL and took several weeks to complete. Almost all the metamorphs moved to a swamp forest, with only minor dispersal toward a mesic woodland. Frogs remained terrestrial rather than arboreal and tended to hide in very thick vegetation. At least in Alexander's (1966) population, the sex ratio of dispersing metamorphs was slightly female biased. Murphy (1963) recorded juveniles over a 51-day period at his study site before they completely left the vicinity of the pond. Most movements occur during or immediately after rainfall. Dispersal occurs to about 200 m from the breeding site (Jensen et al., 2008).

FEEDING AND DIET
Within thick vegetative cover, metamorphs remain stationary for long periods as sit-and-wait predators. When prey is sighted, they swiftly hop directly toward it. Fifty percent of the diet of metamorphs includes mites (Acarina) and springtails (Collembola). Adults are also sit-and-wait predators, and consume everything they can within about 1 m from their vantage point. Their diet includes various small arthropods, spiders, gastropods, caterpillars, dipteran larvae, beetles, and ants. Prey varies seasonally, as different types of prey become available. Alexander (1966) provides a complete summary. Feeding apparently occurs year-round when temperatures permit. Almost all frogs have some food in their stomachs, but the percentage decreases in winter. Males tend not to feed during the breeding season.

PREDATION AND DEFENSE
Upland Chorus Frog larvae approaching metamorphosis have an immobility response whose duration is inversely correlated with stage of development (Dodd and Cupp, 1978). When disturbed, they draw their limbs into the body and refrain from any movement (see Fig. 1 in Dodd and Cupp, 1978). However, just prior to metamorphosis, the immobility response decreases slightly in duration, as all energy is directed toward metamorphosis.

Larval predators include invertebrates (predaceous diving beetles, dragonfly larvae), larvae of the Marbled Salamander (*Ambystoma opacum*), newts (*Notophthalmus viridescens*), water snakes (*Nerodia*), and garter snakes (*Thamnophis*). Mosquitofish (*Gambusia holbrooki*) readily consume larvae under experimental conditions and likely do so in nature (Stanback, 2010). Recently metamorphosed froglets and small adults are killed by spiders (*Dolomedes triton, Lycosa*) (Mitchell, 1990), snakes, and birds (Jensen et al., 2008).

COMMUNITY ECOLOGY
According to Jensen et al. (2008), *Pseudacris feriarum* breeds in temporary pools, flooded fields, ditches, and other shallow-water habitats, where there is an open canopy throughout much of its range. Where populations are sympatric with *P. nigrita*, they switch to forested gum or cypress swamp habitats. This difference suggests a level of habitat partitioning, but the nature of the interaction is unknown. Jensen et al. (2008) noted that Coastal Plain populations had higher call pulse rates (to 35 pulses/sec) than interior populations, in contrast to Southern Chorus Frogs at 5–11 pulses/sec. The difference in call pulse rates could facilitate species-specific female discrimination where populations of these chorus frogs are sympatric. However, the peak breeding seasons of these species do not overlap, as *P. feriarum* usually breeds earlier than *P. nigrita* (Schwartz, 1957).

DISEASES, PARASITES, AND MALFORMATIONS
Larvae are susceptible to the epizoic protozoan *Vorticella* (Alexander, 1966). Mitchell and Georgel (2005) report an anophthalmic *P. feriarum* from southeastern Virginia. No other information is available.

SUSCEPTIBILITY TO POTENTIAL TOXICS
pH. Upland Chorus Frogs have been recorded breeding in a reclaimed mine pond with a pH of 4.0, which is probably near the lethal limit for this species, although the majority of such ponds used had a pH >6.0 (Turner and Fowler, 1981).

UV light. In the Southern Appalachians, con-

Breeding habitat of *Pseudacris feriarum*, northern Virginia. Photo: C.K. Dodd Jr.

trolled experiments suggested that Upland Chorus Frogs experience only a very slight risk from UV light. Malformation rates were slightly higher in unscreened rearing tanks compared with tanks where UV light was excluded. Survivorship to hatching, however, was not affected by field levels of UV light (Starnes et al., 2000).

STATUS AND CONSERVATION
Pseudacris feriarum populations appear to be stable, and there is no indication of extensive population declines. Undoubtedly populations have been lost to wetland habitat destruction and alteration, and highway mortality may affect populations located adjacent to transportation corridors. Populations also may disappear as vegetation succession occurs from old fields to forest (Dodd, 2004). Still, the few studies that have been undertaken that include data on this species have not reported regional declines (e.g., Rice et al., 2001), with the possible exception of an area in northern Virginia (Pollio and Kilpatrick, 2002) and in southeastern West Virginia (Sias, 2006). This species may be fairly adaptable to shifting habitat availability, as long as some areas are present for cover and contain shallow wetlands for breeding. For example, Upland Chorus Frogs even have been observed to breed successfully in reclaimed surface strip mines, as long as the pH is >6.0 (Turner and Fowler, 1981), and in restored or mitigation wetlands (Palis, 2007).

Pseudacris fouquettei
Lemmon, Lemmon, Collins, and Cannatella, 2008
Cajun Chorus Frog

ETYMOLOGY
fouquettei: The specific epithet *fouquettei* is a patronym honoring Martin J. Fouquette of Arizona State University, who conducted extensive studies on chorus frogs in the 1960s and 1970s.

NOMENCLATURE
This morphologically cryptic and newly described species has been referred to in the literature as *Pseudacris triseriata*, *P. nigrita*, *P. feriarum*, or even *P. clarkii*, depending upon the date of publication and location within its range. For example, Platz (1988a, 1989) analyzed calling characteristics of several species of striped chorus frogs from throughout their range, using the then current subspecific designations. He noted significant differences in many call characteristics, and suggested a revision of the subspecies. Based

on genetic evidence (Lemmon et al., 2007a), Platz's species now are known to have included *P. maculata*, *P. fouquettei* (his *P. feriarum*), *P. feriarum*, and *P. kalmi*. Among the authors that have discussed the confusing nomenclature of frogs now referred to as *P. fouquettei* is Bragg (1943a), who early on recognized the distinctness of *P. fouquettei* (as *P. triseriata*) from *P. clarkii*. Always check the location where specimens originated to verify correct taxonomic allocation.

IDENTIFICATION

Adults. The Cajun Chorus Frog is a small slender tan to medium-brown member of the Trilling Chorus Frog clade. Other than by genetic data (Lemmon et al., 2008) and distribution, it can be identified by its subacuminate snout, a pattern of three medium- to dark-brown longitudinal dorsal stripes or rows of spots on a pale tan to gray background, and the presence of a white iridescent labial stripe. The head is slightly narrower than the body and the tympanum is distinct. The arms are long and robust, with the legs slender and of moderate length. The toe tips are only slightly wider than the digits. Dorsally, the skin is weakly granular, but ventrally it is noticeably granular. The belly is cream colored, with scattered dark flecks, and the throat is yellowish brown. Males (maximum 30 mm SUL) are slightly larger than females (maximum 27 mm SUL), with adults ranging from 22 to 30 mm SUL. Lemmon et al. (2008) provided a series of photographs showing variation in color and stripe pattern in comparison with some of the other trilling chorus frogs. Males have a yellowish-orange or dark gray coloration in the area of the subgular vocal sac.

Larvae. Like other members of the Trilling Chorus Frog clade, the tadpoles are small and dark brown or gray to black. Hatching occurs at 6 to 6.5 mm TL. The snout is rounded in lateral view, and there is a light line posterior to the eye. Tail fins may be pigmented via fine mottling. The dorsal tail musculature is darkly pigmented, whereas the ventral portion of the tail musculature is not pigmented. The tail is about twice as long as the body. The maximum tadpole size is 36–39 mm TL (Siekmann, 1949). Siekmann (1949) describes larvae in greater detail, including tooth formulas and variation in oral morphology.

Eggs. Eggs are gray brown to deep brown, with white to cream on the vegetal pole. Bragg (1948) noted that eggs in turbid waters were brown, whereas eggs in clear water were much darker. Siekmann (1949) noted light tan-colored eggs in Louisiana. Eggs are deposited in a loose irregular oblong cluster of about 25 mm in diameter, and a full breeding season compliment of eggs will comprise many clusters of 8–300 eggs each. Females produce approximately 1,500 eggs during a breeding season, perhaps during 2 separate periods of amplexus. Eggs contain only a singular gelatinous envelope measuring 3 to 6.1 mm in diameter (mean 4.6 mm); the vitellus is 1.2–1.5 mm (Livezey, 1952). Egg measurements in Smith (1934) and Livezey and Wright (1947) refer to other members of the Trilling Chorus Frog clade or confuse species.

DISTRIBUTION

The Cajun Chorus Frog is found from central Oklahoma near the border with Kansas southward through central Texas. It occurs throughout the states of Arkansas and Louisiana, and in southwest Mississippi. It does not occur farther east than the Pascagoula–Leaf River system in southern Mississippi. The species barely enters southern Missouri. Important distributional references include Lemmon et al. (2008) as well as: Arkansas (Trauth et al., 2004), Louisiana (Dundee and Rossman, 1989), Oklahoma (Sievert and Sievert, 2006), and Texas (Smith and Sanders, 1952; Hardy, 1995; Dixon, 2000).

FOSSIL RECORD

No fossils have been specifically attributed to this species. However, fossil *P. triseriata* have been described from Rancholabrean sites in Oklahoma (Holman, 2003). The major lineages within the genus *Pseudacris* were established prior to the age of the fossil deposits. Therefore, these fossils are within the range of *P. fouquettei* as currently understood (Lemmon et al., 2007a).

SYSTEMATICS AND GEOGRAPHIC VARIATION

Pseudacris fouquettei is a member of the morphologically conservative Trilling Chorus Frog clade of the genus *Pseudacris*, a group that includes *P. brachyphona*, *P. brimleyi*, *P. clarkii*, *P. feriarum*, *P. kalmi*, *P. maculata*, *P. nigrita*, and *P. triseriata* (Moriarty and Cannatella, 2004). These species are distinguished by their advertisement call structures, color patterns, the presence of a cuboidal

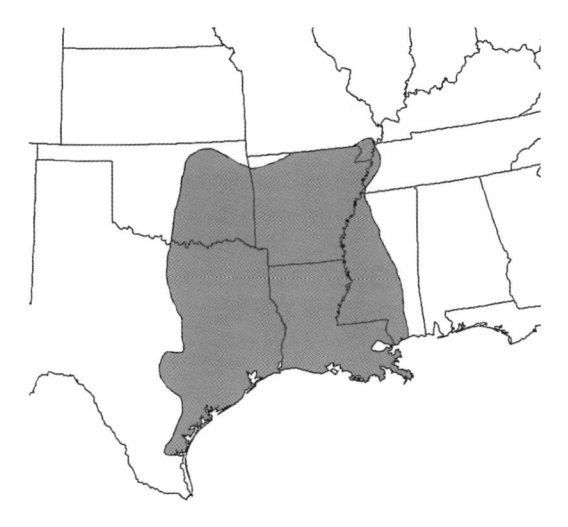

Distribution of *Pseudacris fouquettei*

intercalary cartilage (Paukstis and Brown, 1987), and genetics. Cajun Chorus Frogs are the sister taxon of Southern Chorus Frogs. Together, these two species are the sister taxa of a group that includes *P. triseriata*, *P. kalmi*, and *P. feriarum*.

Natural hybridization occurs with *P. nigrita* in southeast Mississippi along the Pearl River floodplain (Gartside, 1980; Lemmon et al., 2007b). Smith and Smith's (1952) intergrades between *P. n. feriarum* and *P. n. triseriata* are, in part, based on contact zones between *P. fouquettei* and *P. maculata*. Burt (1936) recorded intergradation between "*P. triseriata*" and *P. clarkii* in southern Kansas on the Oklahoma border and perhaps in east Texas; these instances would involve *P. fouquettei* and *P. maculata* in the former, but it is unclear whether intergrades actually occur in the latter case. Under laboratory conditions, *P. fouquettei* readily hybridizes with *P. clarkii* (Lindsay, 1958, but see Lord and Davis, 1956).

There is a considerable amount of variation in the patterning of stripes and spots on the dorsum of this species. Lemmon et al. (2008) noted strong three-stripe patterns, a three-stripe pattern with dark spots bounding the stripes, a broken three-stripe pattern, and even patterns with no stripes at all, except for markings on the legs; various numbers of transverse bars (2–15) are present on the legs. Bragg (1948) noted that there was equal representation between striped and spotted dorsal patterns in eastern Oklahoma. Although most frogs have a cream-colored venter, some have gray pigment.

ADULT HABITAT

Pseudacris fouquettei generally is a frog of the moist forests, woodlands, and savannas within its range, as opposed to *P. clarkii*, which occurs primarily in prairies and grasslands. However, Bragg (1943a) found *P. fouquettei* in tall-grass prairie and mixed-grass prairie in Oklahoma. The species requires open-canopied breeding sites in proximity to woodlands.

TERRESTRIAL ECOLOGY

Little is known of the terrestrial ecology of this species. It is likely a surface-dwelling frog that forages in the moist leaf litter during the nonbreeding season, although a semifossorial existence cannot be excluded. For example, Black (1973a) found a Cajun Chorus Frog in a limestone crack near a cave entrance. These frogs are rarely encountered outside the breeding season. Like *P. nigrita*, Cajun Chorus Frogs probably have a blue-mode phototactic response, indicating they have true color vision (Hailman and Jaeger, 1974).

CALLING ACTIVITY AND MATE SELECTION

Males precede females to the breeding sites by about a day or so, and there is usually a slight time lag from when males are first heard calling until the first eggs are deposited. Males call from the water within 5 to 20 cm of the shoreline and are usually spaced around a breeding site, although a few may call from the bank. They generally face toward the pond or wetland while calling from very shallow water. Males sit with about three-quarters of their body out of water, with the rear legs folded and floating at right angles to the body. The front limbs are used to maintain balance on the substrate or vegetation. This posture essentially thrusts the male upward as he emits his call and is maintained throughout the calling bout over a period of hours. If the frog relaxes his posture, he slips down into the water with his back legs underneath his body. This altered posture allows him to jump should a predator approach, or to approach a rival. Wiest (1982, as *P. triseriata*) noted calling at −2.5 to 22°C over an 84-day period in Texas.

The Cajun Chorus Frog has a trilling call somewhat similar to running a finger over the tines of a comb. The species has a slower mean call rate (0.34 calls/sec, range 0.14 to 0.43 calls/sec) than allopatric *P. feriarum* and *P. maculata*, a higher mean call-duty cycle (0.36, range 0.26 to 0.44) than *P. nigrita*, a long mean call length (1,115.4 ms, range 867 to 1,554 ms), and a mean pulse number (13.1, range 9.9 to 15.7) intermediate between *P. maculata*, *P. feriarum*, and *P. nigrita*. The mean dominant frequency is 3,138 Hz (range 2,846 to 3,900 Hz) with a duration of 0.74–0.85 sec (Michaud, 1962, 1964; Lemmon et al., 2008).

As for frogs in general, temperature affects various parameters of Cajun Chorus Frog calls. For example, the duration of the call and the interval between calls are inversely proportional to temperature. The trill rate increases with temperature, but the number of trills remains constant regardless of temperature. Even in areas of sympatry with other members of the Trilling Chorus Frog clade, the call of *P. fouquettei* is distinctive and does not deviate from its basic characteristics, giving any evidence of character displacement (Michaud, 1964). In choice tests, female *P. fouquettei* favor the calls of conspecific males over those of *P. clarkii* (Michaud, 1962). Males stop calling after amplexus, but there is one report of a male calling under laboratory conditions while still in amplexus (McCallum and Trauth, 2009).

BREEDING SITES

Cajun Chorus Frogs breed in shallow temporary pools and wetlands, similar to those of other members of the Trilling Chorus Frog clade. They may be found in various habitat types, from forested areas to open fields and grassy swales, and even in shallow roadside ditches or cultivated fields. Livezey (1952) also recorded them breeding in deep semipermanent ponds. Breeding sites are characterized by having an open canopy that allows sunlight to warm the shallow water, especially since ambient temperatures during the breeding season often are quite cold, especially at night. The wetland substrate may be covered with grasses or vegetation, or rotting detritus may be present. Most breeding ponds are small, although Livezey (1952) measured one that was 10 m by 25 m.

REPRODUCTION

Breeding normally occurs from December through May, depending on location and environmental conditions. Livezey (1952) suggested that breeding activity begins as early as November in east Texas, and Dundee and Rossman (1989) reported

Tadpole of *Pseudacris fouquettei*. Photo: Stan Trauth

calling in late October in Louisiana. Most reproduction occurs early in the breeding season rather than throughout the winter and spring, and it may occur on only a few days. Initially and as the season progresses, occasional egg masses are found, but they are usually few. Breeding patterns are variable and weather dependent, and the breeding phenology one year may be quite different from the following year.

Breeding frogs prefer ambient temperatures between 4 and 24°C, and rainfall usually triggers calling and mating. As noted by Livezey (1952), reproduction often occurs as temperatures rise and rain falls following a previous sudden drop in temperature. Liner (1954) noted an instance in Louisiana where snow and ice were on the ground as Cajun Chorus Frogs were breeding on 7 February, and Livezey (1952) observed intense breeding as temperatures rose following an 18 cm snowfall. Most egg deposition occurs before minimum temperatures are about 10°C.

Amplexus occurs in water. Females will approach the vicinity of a calling male, at which point he swims toward her and grasps her in an axillary position. Females control the movement of the mated pair. She may remain in the vicinity of the male's calling site, or she may move elsewhere within the wetland. Females deposit their eggs in shallow temporary wetlands and pools, and the exact oviposition site is chosen by the female (Livezey, 1952). Prior to deposition, the female will grasp the twig with her front limbs, with the hind limbs free and somewhat off to the side. Egg deposition is preceded by abdominal contractions, which alternate from side to side. Abdominal contractions in conjunction with the female arching her back eject the eggs slightly upward toward the male's cloacal opening. Eggs are emitted in short strings, with a mass being composed of eggs emitted over at least three abdominal contraction sequences.

The eggs are attached to slanting debris such as twigs, sticks, or branches, and usually not to vegetation such as sedges or grasses. As the eggs are extruded, the female climbs upward along the stem to allow for more eggs to be deposited. As eggs are extruded, the male squeezes the female and hunches his body toward hers, assuming a concave position, moving his vent toward hers. As eggs are extruded, they are fertilized. More than one egg mass may be attached to a branch. The egg mass is placed 2.5–10 cm below the water's surface. Egg masses may be clumped in distribution, or they may be scattered throughout the wetland depending on water depth and the spacing of available branches. Livezey (1952) provides a comprehensive description of mating behavior and egg deposition.

Pseudacris fouquettei females may deposit up to nine egg masses during a single night on a variety of attachment substrates. This process takes up to about 3 hrs (Livezey, 1952); a considerable time (12–60 min) may pass between successive ovipositions. Indeed, the number of egg masses produced by a female may be dependent on the number of attachment sites. When few attachment sites are available, there will be fewer and larger egg masses; when attachment sites are plentiful, egg masses are smaller and more scattered (Bragg, 1948; Livezey, 1952). Egg masses range from a few to dozens, depending on the size of the breeding site. Livezey (1952) found from 3 to 58 egg masses per site, even in wetlands as small as < 2 m^2.

Adult *Pseudacris fouquettei*, amplexus. Photo: James Beck

In Arkansas, clutch size averaged 1,002 (range 445 to 1,380) in February and 857 (range 555 to 1,218) in March (Trauth et al., 1990). A single female in April had a clutch of 507 eggs. In Texas, ovarian counts ranged from 178 to 1,321, but counts of eggs per egg mass were 7–176 (mean 42.8) (Livezey, 1952); in Louisiana, Siekmann (1949) noted only 8–15 eggs per mass. In Oklahoma, Bragg (1948) counted clutches of 620–1,081 eggs from females ovipositing in the laboratory. Trauth et al. (1990) suggested that *P. fouquettei* might deposit two clutches per season, although they could not rule out the oviposition of partial clutches at various times during the season. In fact, it is important to differentiate the number of eggs per egg mass, the number of eggs deposited during a single amplexus sequence, and the number of eggs deposited during the entire breeding season; these differences sometimes are not clear in the literature. Perceptions of fecundity may be quite different, as the counts above demonstrate. Eggs hatch in 4–11 days (Bragg, 1948; Livezey, 1952). The hatching process is described by Bragg (1948).

LARVAL ECOLOGY

The medium-brown to dark-brown larvae hatch at 4–6 mm TL. The larval period lasts from 48 to >80 days, depending upon temperature. Larvae can reach a maximum of 40–43 mm TL prior to hatching (Livezey, 1952; Dundee and Rossman, 1989). Juveniles emerge from the water with a significant amount of tail remaining; this disappears after three or four days. They remain within close proximity to the breeding site until the tail is resorbed. At this time they are about 12.4 mm SUL; Siekmann (1949) reported a range of from 11 to 13.5 mm SUL. After metamorphosis, the young froglets disperse immediately into surrounding habitats, usually within about 5–10 m of the wetland. They remain in proximity to the wetland for about four weeks (Livezey, 1952), where they are active by day and night.

DIET

Little or no food is consumed by juveniles until the tail has been completely reabsorbed. Livezey (1952) recorded copepods, springtails, beetles, and water mites from juveniles, and beetles, fly larvae, lepidopteran larvae, and true bugs from the stomachs of adults. Adults consume shed skin.

Breeding habitat of *Pseudacris fouquettei*. Photo: Chris Brown

It seems likely that Cajun Chorus Frogs eat any small invertebrate they can capture.

PREDATION AND DEFENSE

Observations of predation on this species are nearly nonexistent. Dundee and Rossman (1989) reported that a garter snake (*Thamnophis proximus*) regurgitated a Cajun Chorus Frog. The species probably relies on its cryptic coloration and jumping ability, especially in cold weather, to avoid predators. A calling male also is very difficult to locate, due to the ventriloquist-like nature of the call.

COMMUNITY ECOLOGY

Cajun Chorus Frogs occur in the same pools with *P. clarkii* in central Pontotoc County, Oklahoma, in a narrow band where these species come into contact. According to Bragg (1943a), the transition between these species is rather abrupt with little geographic overlap. A similar area of contact once occurred just east of Dallas, Texas, but this area has been highly disturbed in recent years. The exact nature of interspecific interaction is unknown, but differences in female call preference and male mating calls probably serve as effective isolating mechanisms.

DISEASES, PARASITES, AND MALFORMATIONS

Amphibian chytrid fungus (*Batrachochytrium dendrobatidis*) has been found in five specimens from Texas (Saenz et al., 2010).

SUSCEPTIBILITY TO POTENTIAL TOXICS

pH. Briggler (1998) reports calling from a pond with a pH of 6.0. It is likely that this species has an acid tolerance similar to *Pseudacris kalmi*, that is, about 4.0.

STATUS AND CONSERVATION

The Cajun Chorus Frog tolerates silviculture relatively well, as long as forest cover is retained during the nonbreeding part of the year. This species prefers breeding in open-canopied temporary wetlands of a type which form in clear- and selectively-cut areas. As such, they are frequently found in clearcuts, sometimes even far from the nearest forest cover. In such areas, they prefer flooded, brushy areas with sparse overstory (Fox et al., 2004). As trees are cut, the water table tends to rise providing numerous flooded pools, especially for a few seasons after the cut. Intensive forest management may actually benefit this species by providing open breeding sites in a mosaic of habitats.

Pseudacris fouquettei rarely has been observed calling from permanent constructed ponds in forested habitats (Briggler, 1998). It is unlikely that such ponds would be a significant asset to maintenance of Cajun Chorus Frog populations unless hydroperiod could be varied and fish prohibited.

Pseudacris illinoensis Smith, 1951
Illinois Chorus Frog

ETYMOLOGY
illinoensis: referring to the State of Illinois.

NOMENCLATURE
Synonyms: *Hyla streckeri illinoensis, Pseudacris streckeri illinoensis, Pseudacris streckeri illinoisensis, Pseudacris triseriata illinoensis*

IDENTIFICATION
Adults. The Illinois Chorus Frog is a stout frog with a variable ground color of gray, tan, or brown with dark brown or black marks on the dorsum. There is a distinct V-shaped pattern between the eyes, a dark stripe from the snout to the shoulder, and a dark spot below each eye. Most individuals have a dark, inverted Y-shaped mark on one or both shoulders. The forearms are enlarged and used for burrowing, and the front digits lack a terminal disk. There is no yellow groin coloration as there is in *P. streckeri*. Venters are white. There is virtually no webbing on the hind feet.

Males in Arkansas had a mean SUL of 36.9 mm, whereas Illinois males averaged 35 mm SUL (Trauth et al., 2007). In another Arkansas study, males averaged 38 mm SUL (range 35 to 41) whereas females averaged 39 mm SUL (range 37 to 40) (Butterfield, 1988). Tucker (1995) reported an adult size of 32–43 mm SUL in Illinois. The maximum size is 48 mm SUL.

Larvae. The brownish-gray tadpoles are heavy bodied with a high tail fin containing many small, scattered markings. The total length is 38 mm. Johnson (2000) provided a figure of the tadpole and Smith (1951) an illustration of tadpole mouthparts.

Eggs. The eggs are brownish gray dorsally and white ventrally. There is a single jelly envelope surrounding the egg. Eggs are 1.9–2.6 mm (mean 2.3 mm) in diameter (Butterfield, 1988).

DISTRIBUTION
The Illinois Chorus Frog is known from three scattered populations in west-central Illinois, southwest Illinois, and the junction of southern Illinois, southeastern Missouri, and northeastern Arkansas (Clay County). Important distributional references include: Arkansas (S.E. Trauth at al., 2004; J.B. Trauth et al., 2006), Illinois (Smith, 1961; Holman et al., 1964; Brown and Brown, 1973; Axtell and Haskell, 1977; Brown and Rose, 1988; Brandon and Ballard, 1998; Phillips et al., 1999), and Missouri (Smith, 1955; Johnson, 2000; Daniel and Edmond, 2006).

FOSSIL RECORD
No fossils are known.

SYSTEMATICS AND GEOGRAPHIC VARIATION
Pseudacris illinoensis was described as a subspecies of *P. streckeri* by Smith (1951). Differences between the subspecies were said to include a lack of bright coloration in the groin in *P. illinoensis*, a more uniform distribution of pigment, a reduction in the dark lateral stripe, a more general pallid color, and a smaller 5th row of labial teeth in the larvae (Smith, 1951). Except for the lack of groin coloration, variation in these other characters make them less useful in separating the species. Trauth et al. (2007) examined a number of morphological characters and identified differences among *P. illinoensis* and *P. streckeri* populations, but none strong enough to support status as a separate species. Based on mtDNA, however, Moriarty and Cannatella (2004) concluded that specific status was warranted.

The species is a member of the Fat Chorus Frog clade and is most closely related to *P. streckeri* of the south-central United States (Hedges, 1986;

Distribution of *Pseudacris illinoensis*

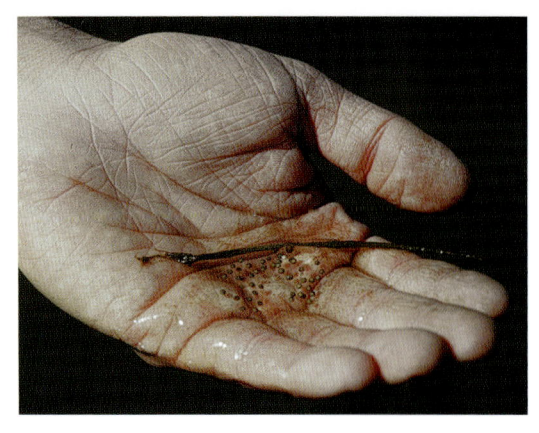

Eggs of *Pseudacris illinoensis*. Photo: Stan Trauth

Moriarty and Cannatella, 2004). The species is likely a remnant of ancestral *P. streckeri*, that extended into Illinois along river floodplains with the expansion of the Prairie Peninsula.

ADULT HABITAT

The Illinois Chorus Frog is found on river floodplains that contain sands or sandy soils deposited either by water or wind. It is a remnant of sand prairie habitat that once extended into Illinois but has largely been eliminated by agriculture. In areas where the sand prairies no longer exist, *P. illinoensis* may survive in agricultural fields. It does not occur on substrates that inhibit burrowing, such as sod (Brown et al., 1972). Many localities are associated with old glacial meltwater and river terraces that were deposited during the Pleistocene (Brown and Rose, 1988).

TERRESTRIAL ECOLOGY

This species is largely fossorial, digging into deep sands where it spends most of the year, except when breeding. Burrows are made in open sandy habitats, and frogs remain 15–20 cm below the surface (Axtell and Haskell, 1977). The burrows are dug using synchronized movements of the forelimbs rather than the hind limbs; as a consequence, the forelimbs are stout and well-muscled (Brown et al., 1972). It takes a frog from 85 to 142 sec to complete a burrow. During the winter, Illinois Chorus Frogs must burrow beneath the frost line (> 25 cm) because they are intolerant of freezing (Packard et al., 1998). Feeding likely occurs underground within burrows, an ability facilitated by the frog using its front limbs to catch and hold prey (Brown, 1978; McCallum and Trauth, 2001b). Traces of the burrows may be present on the surface of the substrate.

Dispersing individuals tend to occupy old-field habitats. Tucker (1998) found marked individuals a mean of 0.52 km (range 0 to 0.9 km), with females tending to disperse slightly but not significantly farther than males. Breeding individuals may not return to the same breeding site from one year to the next. Movement is by short toad-like hops, quite unlike the jumps of other hylid frogs.

Illinois Chorus Frogs are photopositive in their phototactic response, suggesting they use both sunlight and moonlight when feeding and going about their daily activities (Jaeger and Hailman, 1973). They are sensitive to light in the blue spectrum ("blue-mode response"), but with a slight departure from monotonicity at low frequencies. The blue-mode response would help them orient toward areas of increasing illumination (Hailman and Jaeger, 1974). Illinois Chorus Frogs likely have true color vision.

CALLING ACTIVITY AND MATE SELECTION

The Illinois Chorus Frog is one of the earliest breeding frogs in its range, and breeding can commence with snow still on the ground. The call is a short high-pitched bird-like whistle that is repeated rapidly. Calls are heard from March to April (33 days) in Illinois (Brown and Rose, 1988) and from February to April in Arkansas (Butterfield, 1988). Within a region, calling may occur on various nights so that calls may be heard in one area but not in an adjacent area. However, males call virtually every night in large choruses. Calling may be stimulated by heavy rainfall. Calling occurs at night early and later in the season (productive at 18:30–00:50 at air temperatures of 6–15°C) (Brandon and Ballard, 1998), but during peak breeding in early April, males call during the day. Males likely remain at or near a breeding site throughout the spring, with females returning to their burrows soon after oviposition.

BREEDING SITES

Most of the natural habitat of this species has been modified by agriculture for a long time. Breeding occurs in remnant shallow pools (< 40 cm), especially flooded depressions in fields. They usually are associated with small marshy areas, but breeding can occur in lentic water, drainage and roadside ditches, borrow pits, agricultural

Tadpole of *Pseudacris illinoensis*. Photo: Stan Trauth

fields, and pastures. Occasionally frogs are found in permanent ponds. The species avoids flowing waters such as creeks and small streams, as well as floodplain lakes.

Males call while floating in shallow water or while sitting on grassy vegetation; they are capable of hanging onto grasses while calling, using their strong front forelimbs. Amplexus occurs around the female's midbody, and egg deposition occurs as the amplexing pair floats in the water. Females may grasp vegetation for stability. As eggs are extruded, they are fertilized.

REPRODUCTION

Eggs are deposited in small compact masses of 8–79 eggs at a time (mean 41), with a mean total clutch size of 469 eggs (range 148 to 1,012) in northeastern Arkansas (Butterfield, 1988; Butterfield et al., 1989). Brandon and Ballard (1998) counted 15 to 100 eggs per mass (mean 56) in Illinois. Others have estimated clutch size at 200–700 eggs (Smith, 1951 [2 females depositing 900 eggs in captivity], 1961; Johnson, 1977; *in* Butterfield, 1988) or from 400–1,000 eggs (Johnson, 2000). Egg masses are deposited in shallow water attached to twigs, branches, or submerged vegetation. Gravid females have been observed in November, leading Axtell and Haskell (1977) to question whether the species might breed in autumn. Hatching occurs in about a week (Johnson, 2000).

LARVAL ECOLOGY

Larvae are detritivores, although cannibalism has been observed under captive conditions (McCallum and Trauth, 2001a). Growth rates may vary individually even when larvae are reared under the same conditions with similar diets. McCallum and Trauth (2001a) suggested the potential for a cannibalistic morph but did not describe differences in larval morphology, except for size. The developmental period (egg deposition to metamorphosis) is estimated at 35–50 days (*in* Brown and Rose, 1988). Newly metamorphosed froglets usually contain a tail stub, but are 12–25 mm TL after the stub has been resorbed (Smith, 1951; Tucker, 1995). Tucker (1995) captured a total of 727 froglets departing a single breeding pond in Illinois.

DIET

In an Illinois study, Tucker (1997) found the diet to be dominated by larval dingy cut-worms (*Feltia ducens*), a lepidopteran, both in numbers and mass. Other prey included spiders, beetles, adult and larval flies, ants, an unidentified wasp or bee, and hemipterans in the family Gerridae.

PREDATION AND DEFENSE

The secretive nature of this species may be its best defense. Nothing is reported on predators, but a variety of vertebrates likely eat this frog.

POPULATION BIOLOGY

Initial growth is rapid. Immediately after metamorphosis, froglets grow at the rate of 1.18 mm/day (Tucker, 1995). Within a year, males grow 15.6 mm to 78% of their maximum adult size and females 18.1 mm to 90% of their maximum adult size (Tucker, 2000). After the first year, however, growth slows considerably, and adults only add an additional 2–3.3 mm depending on sex. Survivorship from metamorphosis to sexual maturity is only about 2.8%, and from 1 yr to 2 yrs of age only 0.8% (Tucker, 1998).

Breeding occurs during the first spring after metamorphosis, and maximum longevity is 2–5 yrs (Tucker, 1995, 2000), although Butterfield

Adult *Pseudacris illinoensis*. Photo: Stan Trauth

(1988) estimated age of first reproduction as 2 yrs with a maximum of 2–3 yrs longevity. As such, most *Pseudacris illinoensis* breed only one to two times during their lifetime. A short life span with limited reproductive opportunity suggests that this species is particularly vulnerable to perturbations, such as changes in precipitation patterns that might influence reproduction. Tucker (2000) estimated annual adult survivorship at 26%. This would mean that only 7 frogs are estimated to survive from a cohort of 1,000 that metamorphosed 5 yrs earlier. However, only about 2.8% survive the first year after metamorphosis (Tucker, 1998).

DISEASES, PARASITES, AND MALFORMATIONS
The myxozoan parasite *Myxidium serotinum* is known from *P. illinoensis* (McAllister and Trauth, 1995). The coccidian protozoan *Isospora delicatus* occurs in *P. illinoensis* (Upton and McAllister, 1988). McCallum et al. (2001) reported on an Illinois Chorus Frog that was unable to deflate its vocal sac and thus floated at the water's surface.

SUSCEPTIBILITY TO POTENTIAL STRESSORS
No information is available.

STATUS AND CONSERVATION
The Illinois Chorus Frog is declining throughout its range, and the need for conservation research and management has long been recognized (Bury et al., 1980). The species requires small flooded depressions in fields, and its propensity to use agricultural fields makes it particularly vulnerable to tillage and pesticide application (Brown and Rose, 1988; Johnson, 2000). Other effects include soil compaction, adverse effects from nitrogenous fertilizers and farm animal waste, road mortality, off-road vehicle use, sand mining, and the increasing severity and alteration of flood patterns and wetland hydrology. If the frogs breed before tillage and have enough time for tadpoles to complete metamorphosis, some populations may persist in the areas of intensive agriculture that are ubiquitous throughout its range. However, agricultural extension agents often advocate precision land-leveling to increase agricultural yields, and this "best management practice" has the potential to destroy the wetlands needed by *Pseudacris illinoensis* (Trauth et al., 2006).

Habitat loss and destruction are the primary threats, but lethal and sublethal effects of agricultural chemicals likely have taken a considerable toll on isolated populations. Coupled with all of the above is the effect of a long-term drought that considerably impacted small temporary wetlands from 1987 to 2004. In Arkansas alone, the range of this species decreased 61% between 1992 and 2004; choruses were once quite large, but by 2004 averaged <8 calling males per location (Trauth et al., 2006), and in southern Illinois Tucker (1998) and Brandon and Ballard (1998) noted

Breeding habitat of *Pseudacris illinoensis*. Photo: Stan Trauth

populations were once much more widespread than they are at present. Florey and Mullin (2005) could not establish trends in isolated *P. illinoensis* populations in Illinois, but their study only encompassed 2 yrs. The future of the Illinois Chorus Frog is precarious. The species is considered as "Threatened," "Rare," or of "Special Concern" in all the states within its range, but obstacles in protection of its habitat are illustrated by attempts to drain Sand Lake, Illinois, a pristine temporary wetland breeding site for Illinois Chorus Frogs that is prone to flooding nearby residents of Havana (Brown and Cima, 1998).

Pseudacris kalmi Harper, 1955
New Jersey Chorus Frog

ETYMOLOGY

kalmi: a patronym in honor of Peter Kalm (1716–1779), author of *Travels in North America* (1753–1761) and an early student of New Jersey frogs.

NOMENCLATURE

Synonyms: *Hyla kalmi, Pseudacris feriarum kalmi, Pseudacris nigrita kalmi, Pseudacris triseriata corporalis, Pseudacris triseriata kalmi*

Some of the literature on this species may refer to this species as *P. triseriata*. Always check the location of specimens to verify the correct taxonomic allocation.

IDENTIFICATION

Adults. This is a robust chorus frog somewhat resembling the Western Chorus Frog, *P. maculata.* It has clearly defined longitudinal dark olive to gray stripes on a smooth-skinned light brown to grayish dorsum, with ventral areolae (nonpigmented areas) that do not extend very far along the femur. The dorsal stripes are more ragged and broader than in allopatric *P. feriarum,* and they are usually continuous rather than broken into spots. There is a distinct triangular blotch between the eyes (interorbital spot) that extends onto the eyelids. New Jersey Chorus Frogs have an ivory-yellow stripe from the tip of the snout extending along the upper lip (labial stripe) to the arm insertion, and there is a dark lateral stripe from the snout to the groin. The limbs are longer than in related striped chorus frogs, and they are spotted or barred with blotches of dark olive gray. There is a usually a distinct dusky stripe along the outer edge of the tibia that is absent in *P. feriarum*. The ventral cream-colored areolae are conspicuous and extend from the posterior portion of the throat to the femur. The anterior of the throat is dusky in males during the breeding season (but not females), and a dusky speckling may extend across the midabdomen. Adult males average 28 to 32 mm SUL, with females averaging 29 to 36 mm SUL.

Larvae. The tadpoles are dark brown, with bronze flecking on the body, a bronze iridescent belly, and translucent tail fins with very small black spots. The tail musculature is bicolored with dark above and light below. The lighter ventral portion may have small spots (Gosner and Black, 1957b).

Eggs. The eggs are brown to black dorsally and white on the vegetal pole. The mean vitellus diameter is 1.32 mm (range 1.29 to 1.34 mm), with a mean total diameter (vitellus plus single egg capsule) of 3.10 mm (Gosner and Rossman, 1959). Time to hatching is presumably similar to the eggs of the other striped trilling chorus frogs.

DISTRIBUTION

The New Jersey Chorus Frog is found only in New Jersey and on the Delmarva Peninsula of Maryland, Delaware, and Virginia. A population from Bucks County, Pennsylvania, also has been reported (Hulse et al., 2001). The original description (Harper, 1955) listed the distribution as including Staten Island, New York, but the status of this population is extirpated (not mentioned by Gibbs et al., 2007). Lemmon et al. (2007b) found evidence of hybridization with *P. nigrita* in southeastern Virginia, but the animals appeared morphologically similar to the Southern Chorus Frog, and no pure population of *P. kalmi* is found west of the Chesapeake Bay. The species occurs on Kent Island, Maryland (Grogan and Bystrak, 1973b), and Assateague and Chincoteague Islands in Maryland and Virginia (Lee, 1973b; Mitchell

Distribution of *Pseudacris kalmi*. The status of the population in extreme southeastern Pennsylvania is uncertain.

and Anderson, 1994). Important distributional references include: Delmarva Peninsula (White and White, 2007), Maryland (Harris, 1975), New Jersey (Schwartz and Golden, 2002), Pennsylvania (Hulse et al., 2001), and Virginia (Tobey, 1985; Mitchell and Reay, 1999).

FOSSIL RECORD
No fossils have been specifically attributed to this species or to any member of the *P. triseriata* species complex within the known range of *P. kalmi*.

SYSTEMATICS AND GEOGRAPHIC VARIATION
The New Jersey Chorus Frog was described as a subspecies of *P. nigrita* by Harper (1955). Based on an examination of 33 loci, Hedges (1986) recommended that *P. kalmi* be recognized at the subspecific level as distinct from *P. feriarum*, thus *P. f. kalmi*. Current understanding of the systematics of *P. kalmi* place it as a distinct species within the Trilling Chorus Frog clade of the genus *Pseudacris*, a group that includes *P. brachyphona*, *P. brimleyi*, *P. clarkii*, *P. feriarum*, *P. fouquettei*, *P. maculata*, *P. nigrita*, and *P. triseriata* (Moriarty and Cannatella, 2004). These species are distinguished by their advertisement call structures, color patterns, the presence of a cuboidal intercalary cartilage (Paukstis and Brown, 1987), and genetics. The New Jersey Chorus Frog is the sister taxon of the Western and Upland Chorus Frogs.

In turn, these three species are the sister taxa of *P. nigrita* and *P. fouquettei*. P.W. Smith (1957) proposed that *P. kalmi* evolved as a relictual population of *P. triseriata* that had expanded east from the Great Lakes region to the East Coast. The molecular analysis of Lemmon et al. (2007b) does not support this scenario. Natural hybridization with *P. nigrita* was reported by Lemmon et al. (2007b).

ADULT HABITAT
New Jersey Chorus Frogs prefer open habitats for breeding, although they likely spend the nonbreeding season on the floor of moist or wet forested areas. Habitats should include a mix of forests and grassy floodplains where shallow breeding pools form in the winter and early spring. They inhabit cutover hardwood forests, presumably because these disturbed areas contain open-canopied (>50%) breeding sites in spring (McLeod, 1995; McLeod and Gates, 1998).

TERRESTRIAL ECOLOGY
Little is known about the terrestrial ecology of this species, as the frogs are rarely encountered outside the breeding season. They presumably lead a terrestrial or semifossorial existence in moist woodlands, similar to that of *P. feriarum*. Activity can occur over a wide range of temperatures, despite the species' preference for breeding when ambient temperatures are often quite low. The CTmin for *P. kalmi* is 0.25°C, whereas the CTmax is 35.8°C (John-Alder et al., 1988). Like

Embryos of *Pseudacris kalmi*. Photo: John Bunnell

Adult *Pseudacris kalmi*.
Photo: John Bunnell

P. nigrita, New Jersey Chorus Frogs probably have a blue-mode phototactic response, indicating they have true color vision (Hailman and Jaeger, 1974).

CALLING ACTIVITY AND MATE SELECTION

The call of the New Jersey Chorus Frog has been likened to a "ratchety trill that rises in pitch, sounding similar to but much louder than the sound produced by scraping a fingernail over a comb from the coarse to the fine teeth" (White and White, 2007:88). Males call both diurnally and nocturnally. Mate selection probably is similar to that of *P. feriarum*. Calling chorus frogs are extremely difficult for a human observer to locate. In addition, the frogs stop calling upon the approach of an observer. These characteristics presumably also make them difficult for a predator using auditory cues.

BREEDING SITES

Like other striped chorus frogs, this species breeds in shallow temporary pools that form in open-canopied habitats in winter and spring. Pools in grassy meadows are preferred, but breeding may occur in ephemeral woodland pools open to the sky. Cope (1889) noted that they prefer open temporary water bodies surrounded by dense stands of vines and scrub oaks.

REPRODUCTION

Breeding normally begins in February, reaches a peak in March, with occasional calls continuing into early May. Lee (1973a), however, recorded calling as early as 6 December. Most breeding takes place over a rather short time, despite the extended calling season. Males and females arrive at the breeding pools at about the same time, with males calling both by day and night as temperatures permit. Males call while floating on the water's surface or as they hold onto submerged vegetation.

Egg deposition appears similar to that of other striped chorus frogs. An amplexed female will grasp a grass stem or branch with both hands, and bring her body upward in a longitudinal axis in relation to the male. Once in position, she floats free. Oviposition occurs when the female arches her body concavely, thus bringing her vent upward against the male. The male then thrusts downward and backward, hunching his body, and slides his venter along the female's dorsum. This thrust is more prolonged and drawn out than the similar pelvic thrust of *P. crucifer*, and it may enable the male to fertilize several eggs rather than a single egg as in Spring Peepers. About a dozen or so eggs at a time are fertilized as they are extruded, accompanied by spasms of the female's abdomen.

Breeding habitat of *Pseudacris kalmi*. Photo: Joe Mitchell

Females deposit their eggs in flimsy masses attached to submerged vegetation or debris such as stems and branches. As she extrudes her eggs, the female pulls herself along the supporting stem and positions the eggs. This process is then repeated at the same support, or the pair may move elsewhere within the wetland. The amplexed pair may take several minutes for rest between oviposition bouts before resuming egg deposition. Gosner and Rossman (1959) described and illustrated the sequence of egg deposition in detail.

Each egg mass contains between 8 and 143 eggs (mean 40). Gosner and Rossman (1959) recorded a mean of 215 eggs (range 110 to 575) in an ovarian egg count, with additional clutches in the laboratory of 223 and 317 eggs. Hatching occurs at about 4.7 mm (Gosner and Black, 1957b).

LARVAL ECOLOGY
There is no information available in the literature on the larval ecology of this species. Larval ecology likely is similar to that of other phylogenetically related striped chorus frogs. Larvae reach a maximum size of 30 mm and recently transformed froglets are 8–12 mm SUL; the larval period is 40–60 days (Gosner and Black, 1957b).

DIET
Like most frogs, New Jersey Chorus Frogs probably eat a variety of invertebrates. There have been no studies of its diet.

PREDATION AND DEFENSE
New Jersey Chorus Frogs are adept jumpers. Even at 0.6°C, they are able to jump with about 40% effectiveness in terms of distance. Perhaps this is not surprising since the species has a low CTmin, is active at cold temperatures, and jumps better at cooler temperatures than most other species of hylids (John-Alder et al., 1988). Chorus frogs may jump toward deeper waters at breeding ponds, since few poikilotherm predators are active at its breeding sites early in the year. Predators of larvae and adults include crayfish, water spiders, water snakes (*Nerodia sipedon*), birds, raccoons, and foxes (Mitchell and Anderson, 1994).

DISEASES, PARASITES, AND MALFORMATIONS
No information is available.

SUSCEPTIBILITY TO POTENTIAL STRESSORS
pH. The lethal pH for *P. kalmi* is 3.8, with a critical pH of 4.1 (Gosner and Black, 1957a).

STATUS AND CONSERVATION
The New Jersey Chorus Frog is abundant on the Delmarva Peninsula. Mitchell and Anderson (1994) suggested that the population observed by Lee (1973b) on Assateague Island might be extirpated. However, there have been no published studies on the species' status and trends.

Pseudacris maculata (Agassiz, 1850)
Boreal Chorus Frog
Rainette faux-grillon boréale

ETYMOLOGY
maculata: from the Latin meaning 'spotted.'

NOMENCLATURE
Synonyms: *Chorophilus septentrionalis, Chorophilus nigritus septentrionalis, Hyla canadensis, Hyla septentrionalis, Hylodes maculatus, Pseudacris nigrita septentrionalis, Pseudacris nigrita maculata, Pseudacris septentrionalis, Pseudacris triseriata maculata*

Considerable taxonomic confusion exists in the literature on striped chorus frogs about which species is being discussed, at least as the systematics are currently understood. Much literature includes *P. maculata* within *P. triseriata* (e.g., Harding, 1997), and discussions proceed as if these were the same species. In the past, authors may have assigned a name to a chorus frog population that has no resemblance to the current nomenclature based on molecular analysis. For example, Smith and Smith (1952) compared various morphological characteristics of "*Pseudacris nigrita feriarum*" with "*P. n. triseriata*." Upon looking at their collecting localities, it is clear that what we now know as *P. fouquettei, P. triseriata*, and *P. maculata* were included under their name *P. n. triseriata*. In another example, Platz (1988a, 1989) analyzed calling characteristics of several species of striped chorus frogs from throughout their range, using the then current subspecific designations. He noted significant differences in many call characteristics, and suggested a revision of the subspecies. Based on genetic evidence (Lemmon et al., 2007a), Platz's species now are known to have included *P. maculata* (populations from Arizona and the Great Plains), *P. fouquettei* (his *P. feriarum*), *P. feriarum* (possibly in western Kentucky), and *P. kalmi*.

Further discussions of the nomenclatural history of this species are in Smith (1956) and Cook (1964a). It is essential for authors using the literature on the striped chorus frog complex to have a clear understanding of the origin of the frogs in question in order to make a proper taxonomic allocation.

IDENTIFICATION
Adults. The Boreal Chorus Frog is a small slender frog characterized by three relatively broad parallel dark (black, brown, or gray) dorsal stripes on a lighter background color that ranges from gray to tan to brown; some individuals may have a green or reddish background color. Some frogs may have the pattern broken up into three rows of spots. There is considerable variation in dorsal patterning, with individuals containing only parallel stripes, stripes and spots (most commonly), and only spots (Ellis and Henderson, 1915). The dorsal and ventral skin is granular in appearance. This species also exhibits color dichromatism whereby frogs are dark by day and at cold temperatures and much lighter at night or at warm temperatures.

The head is pointed when viewed from above, and there is a dark stripe on the side of the body extending from the snout, through the eye, to the groin. A white labial stripe is present. A triangular dark spot is usually present dorsally between the eyes. The tops of the legs are patterned with dark transverse stripes or spots, and toe discs have only the barest expansions at the tip. The legs are shorter than some of the other striped chorus frogs, such as *P. triseriata*. There is usually no webbing between the toes, although some individuals have very slight webbing between the toes on the rear feet. Venters are white and may have some dark gray speckling on the throat or chest.

Males exhibit secondary sexual characteristics. Males throughout much of the species' range have a dark or yellowish throat coloration during the breeding season. However, breeding males from northern Manitoba have plain unpigmented throats (Harper, 1963). Male throats are wrinkled with transverse folds, as the subgular vocal pouch expands considerably during calling. Males also have a pad on the inside of the thumb during the breeding season, which aids in amplexus.

Females are slightly larger than males, and there is variation in adult size depending upon elevation. Individuals from high elevation populations are larger than individuals from nearby lower elevation populations, and there may be a gradual change in size with elevation (Pettus and Spencer, 1964). For example, females from the Colorado mountains average 31.4 mm SUL (range 26.4 to 38.4 mm), whereas those from the prairie average 24.1 mm SUL (range 18.8 to 28.9 mm) (Pettus and Angleton, 1967). Populations between these two are intermediate in mean size. Other published data on size include: Isle Royale, Michigan, where males average 27 mm SUL (range 22 to 31 mm) and females average 30 mm SUL (range 25 to 35 mm) (Smith, 1987); in Alberta, adults range between 19.3 and 28.2 mm SUL (Roberts and Lewin, 1979); and in southeastern South Dakota adults range between 20.7 and 33.1 mm SUL (mean 28.5 mm) (Nickum, 1961). The largest specimens have been recorded in Missouri (38 mm SUL; *in* Johnson, 2000), New Mexico (39 mm SUL; Degenhardt et al., 1996), and in Colorado (52 mm SUL; Hammerson, 1999).

Albino (amelanic) adults made up 7–12% of a breeding chorus of this species at certain sites in Colorado (Corn, 1986). Albinos had very faint stripes and a reddish ground color.

Larvae. The tadpole is dark olive, brown, or gray, uniformly stippled with bronze, with a transparent body. The intestinal coil is visible beneath the belly skin. The tail fins have finely scattered pigment areas, and the tail tip is obtuse and rounded. The dorsal musculature of the tail is heavily pigmented, but the ventral musculature is less so. Tadpoles of the Boreal Chorus Frog may reach a length of 30–43 mm TL prior to metamorphosis (Claflin, 1962). A description of the larvae is in Youngstrom and Smith (1936). Albino tadpoles were reported by Corn (1986).

Eggs. Eggs are a creamy tan to completely black, with solid black eggs most common at higher elevations. The vegetal pole is cream in color. Eggs from high-elevation females are larger in diameter (1.04–1.32 mm, mean 1.15 mm) than are eggs from low-elevation females (0.76–0.85 mm, mean 0.80 mm) (Pettus and Angleton, 1967). The outer envelope is about 3.0 mm in diameter. Smith (1934) recorded a second envelope of 2.1 mm in diameter, but no other authors have reported a second jelly envelope surrounding the egg. Like other striped chorus frogs, the eggs are oviposited in loose jelly masses containing 15 to 190 eggs per mass, and a female deposits a number of jelly masses per breeding event. Albino eggs were reported by Corn (1986) in Colorado and Cahn (1926) in Illinois.

DISTRIBUTION

Pseudacris maculata is the most widely distributed member of the Trilling Chorus Frog clade. There are two divisions within its range. An eastern group occurs from the St. Lawrence River Valley west to Lake Huron; populations barely enter extreme upstate New York (Gibbs et al., 2007) and Vermont. Lakes Ontario and Erie form much of the southern limit of this population. Boreal Chorus Frogs were also introduced to Newfoundland (Maunder, 1983), but apparently no longer occur there (Maunder, 1997). The western group occurs from extreme northwestern Québec (South Shore from Ile Perrot to Boucherville) and western Ontario perhaps to Great Bear Lake in the Northwest Territory, south through Alberta and much of Montana, Idaho, Utah, Colorado, and northwestern and western New Mexico. The species enters British Columbia only near Dawson Creek and Fort Nelson, and it scarcely enters Michigan's Upper Peninsula. Boreal Chorus Frogs occur westward along the Snake River in western Idaho, throughout the Sevier River drainage and marshes of southwestern Utah, and in the highlands of east-central Arizona at least to the Flagstaff vicinity (Aitchison and Tomko, 1974). The southern limit of the range of the Boreal Chorus Frog is about the Kansas–Oklahoma border, east along the Missouri–Arkansas border, then northeastward across most of Illinois and a sliver of northwestern Indiana to the southern end of Lake Michigan. Many older publications treat *P. maculata* and *P. triseriata* as discrete taxa (usually as subspecies), and indicate that both taxa occur throughout the Midwest (e.g., Vogt, 1981; Collins, 1993; Oldfield and Moriarty, 1994; Casper, 1996; Kline, 1998). This distinction is not borne out by molecular analysis, as *P. maculata* is the only striped chorus frog within the region.

The Boreal Chorus Frog occurs on some islands, such as Akimiski Island and Shipsand Island in James Bay (Cook, 1964a; Schueler, 1973), islands in Lake Athabaska, Manitoba–Alberta (Harper, 1931), Isle Royale and the Apostle Archipelago in Lake Superior (Smith, 1983; Hecnar et al., 2002), the Georgian Bay islands in Lake Huron (Hecnar et al., 2002), and islands in the mouth of the St. Lawrence River where it enters Lake Ontario (Hecnar et al., 2002).

Important distributional references include: Continent-wide (Smith, 1956), Alberta (Preble, 1908; Harper, 1931; Russell and Bauer, 2000), Arizona (Brennan and Holycross, 2006), British Columbia (Cowan, 1939; Nussbaum et al., 1983;

Distribution of *Pseudacris maculata*

Weller and Green, 1997; Matsuda et al., 2006), Canada (Logier and Toner, 1961), Colorado (Maslin, 1959; Smith et al., 1965; Hammerson, 1999), Idaho (Slater, 1941b; Tanner, 1941; Nussbaum et al., 1983), Illinois (Schmidt and Necker, 1935; Necker, 1939; Smith, 1961; Phillips et al., 1999), Indiana (Schmidt and Necker, 1935; Necker, 1939), Iowa (Hemesath, 1998), Kansas (Smith, 1934; Heinrich and Kaufman, 1985; Collins et al., 2010), Manitoba (Smith, 1953; Harper, 1963; Preston, 1982), Minnesota (Oldfield and Moriarty, 1994), Missouri (Johnson, 2000), Montana (Black, 1970; Maxwell et al., 2003; Werner et al., 2004), Nebraska (Lynch, 1985; Fogell, 2010; Ballinger et al., 2010), New Mexico (Degenhardt et al., 1996), New York (Gibbs, 1957; Gibbs et al., 2007), North Dakota (Wheeler and Wheeler, 1966), Northwest Territories (Hodge, 1976; Fournier, 1997, undated), Ontario (Cook, 1964a; Schueler, 1973; MacCulloch, 2002 [in part]), Québec (Bleakney, 1958a, 1959; Bider and Matte, 1996; Daigle, 1997; Fortin et al., 2003; Desroches and Rodrigue, 2004 [in part]; Ouellet et al., 2009), Saskatchewan (Cook, 1966), South Dakota (Peterson, 1974; Ballinger et al., 2000; Kiesow, 2006), Utah (Tanner, 1931), Vermont (Andrews, 2001), Wisconsin (Suzuki, 1951; Vogt, 1981), and Wyoming (Baxter and Stone, 1985; Koch and Peterson, 1995). In some publications, authors have combined *P. maculata* with *P. triseriata* in their range maps and species accounts (listed as *in part* above); these data must be used with caution.

FOSSIL RECORD

Fossils attributed to *P. triseriata* are known from Irvingtonian deposits in Colorado and Kansas and from numerous Rancholabrean deposits in Kansas and New Mexico (Holman, 2003). The major lineages within the genus *Pseudacris* were established prior to the age of the fossil deposits. Therefore, these fossils are within the range of *P. maculata* as currently understood (Lemmon et al., 2007a).

SYSTEMATICS AND GEOGRAPHIC VARIATION

Pseudacris maculata is a member of the morphologically conservative Trilling Chorus Frog clade of the genus *Pseudacris*, a group that includes *P. brachyphona, P. brimleyi, P. clarkii, P. feriarum, P. fouquettei, P. kalmi, P. nigrita,* and *P. triseriata* (Moriarty and Cannatella, 2004). These species are distinguished by their advertisement call structures (Platz, 1988a, 1989), dorsal stripes and color patterns, the presence of a cuboidal intercalary cartilage (Paukstis and Brown, 1987), and genetics. The *Nigrita* clade within the trilling chorus frogs includes all the species above except *P. brimleyi* and *P. brachyphona*. In turn, this clade has three subdivisions, with *P. triseriata* and *P. maculata* west of the Mississippi River forming a western group. The lineage leading to *Pseudacris maculata* was among the earliest to split off from the ancestral trilling chorus frogs, a split that occurred about 9.7 mya (Lemmon et al., 2007a).

Call characteristics, such as pulse rate and call duration, are very different between northern and southern populations of *P. maculata* from the Great Plains. In northern *P. maculata*, the calls are long in duration and low in pulse rate. Southern *P. maculata* have a much shorter call, with pulse rates twice those of northern *P. maculata* (Platz, 1989). Based on call characteristics and morphology (various body ratio measurements), Platz (1989) recognized the distinctiveness of these populations, referring them to *P. maculata* (northern) and *P. triseriata* (southern) species. However, this separation is not in accordance with the genetic data (Lemmon et al., 2007a). Smith and Smith's (1952) intergrades between *Pseudacris nigrita feriarum* and *P. n. triseriata* are, in part, based on contact zones between *P. triseriata* and *P. maculata*.

In Larimer County, Colorado, there are three color morphs of this species in certain montane populations: a brown dorsum with green or brown spots, a reddish dorsum with green or brown spots, and a green dorsum with green or brown spots. Matthews and Pettus (1966) proposed that there were three loci responsible for color polymorphism, a red locus, a green locus, and a green-spot locus. Red is dominant to brown, green is recessive to brown, and green-spot is dominant to brown-spot. The interaction of the dominant-recessive alleles at the three loci results in the different color and spot patterns seen in these montane populations. The most common background color of adults is red (49–61% of adults over a 3 yr period), followed by brown (25–33%), green (11–14%), and red-green (2–4%). In terms of spot pattern, brown-spotted individuals are most common (71–78%) followed by green-spotted frogs (22–29%) (Matthews, 1971). Matthews (1971) provided additional color and spot frequencies for yearlings and juveniles.

Multiple color morphs (green, tan, gray brown, rusty pink, reddish) also have been reported from northern Ontario (Schueler, 1973) and Yellowstone National Park (Koch and Peterson, 1995). Green is the predominant color morph in these forested regions. In Minnesota, the brown color morph is predominant in the in the warm prairie region, whereas greens and reds are more common in the forested and cooler parts of the state (Hoppe, 1981). Selection has favored a background color that matches the primary habitat type within which these frogs reside, and the adult color polymorphism may be maintained by differential predation pressure.

Hybridization with other *Pseudacris* species has not been reliably reported. Nickum (1961) considered populations in the Midwest (from northern Wisconsin to much of Nebraska) to form a zone of intergradation between what were then considered *P. nigrita maculata* and *P. n. triseriata* (see map in Conant and Collins, 1998). Despite some variation in external phenotype, all the populations studied by Nickum (1961) are now considered to represent *P. maculata*.

ADULT HABITAT

Boreal Chorus Frogs occur from rather low elevations along the St. Lawrence River to the alpine meadows of the high Rockies, and from the humid Midwest to the high-latitude boreal climates of the southern Northwest Territories. Habitat preferences vary by latitude. The species prefers areas with grassy meadows, swamps, and small shallow ponds for breeding throughout much of its range. On the Great Plains, especially, it is an upland grassland prairie species. The species also is associated with periodically inundated riparian forests, where it occupies temporary overflow ponds and woodland pools (Burbrink et al., 1998; Wilson, 2000). Along the shoreline of the Great Lakes, *Pseudacris* is associated with a high perimeter-to-area ratio of emergent wetlands and negatively with edge density of water (a measure of the total edge length of a cover type divided by the total landscape area) (Price et al., 2004). Essentially, this translates to small wetlands with abundant short grassy vegetation.

Pseudacris maculata is associated with a variety of forest types, from deciduous to mixed boreal-coniferous forests, especially in the North. In one example, Marshall and Buell (1955) found this species most often in spruce (*Picea*) and fir- (*Aibes*) ash (*Fraxinus*) forests surrounding a bog in Minnesota. In Alberta, willows and poplars are commonly associated with *P. maculata* habitat (Roberts and Lewin, 1979), and abundance of this species is related to closed deciduous forests, mixed forest, and even urban habitats (Browne et al., 2009). Associations are often scale-dependent, that is, the importance of different habitat associations changes depending upon the distance from a breeding site.

During the nonbreeding season, frogs prefer moist forested areas, old fields, or upland grasslands, where they forage in leaf litter. They remain in riparian areas to some extent in the extreme North (Okonkwo, 2011), although in fewer numbers than some other boreal species. *Pseudacris maculata* may be found in proximity to forest patches, and most individuals do not travel far from the breeding sites (Roberts and Lewin, 1979). The habitat should contain a complex, heterogeneous environment of grasses, sedges, high shrubs, and herbs, and a low or open-forest canopy cover (Constible et al., 2001; Ouellet et al., 2009).

AQUATIC AND TERRESTRIAL ECOLOGY

Activity occurs throughout the warmer months following the breeding season, when most Boreal Chorus Frogs forage in leaf litter adjacent to the wetland breeding sites. Vogt (1981) also recorded foraging off the ground in short shrubs and grasses. The maximum dispersal distance is reported as > 600 m (Spencer, 1964a), but most frogs remain within 20–50 m of their breeding site. Movements may favor a particular direction, and a maximum distance between recaptures has been recorded as ca. 580 m for a movement rate > 61 m/day (Spencer, 1964c). During the nonbreeding season, Boreal Chorus Frogs take refuge in rodent burrows, such as those of the pocket gopher *Thomomys* or prairie dog *Cynomys*, under surface debris (rocks, human-created refuse), in thick grass clumps, and within the vegetation of wetlands. Population estimates are unknown, but in Northern Alberta, Roberts and Lewin (1979) estimated there were 2.3 *P. maculata* per 1000 m^2.

The end of the activity season is determined by the arrival of cold weather, which is latitude- and elevation-dependent. When weather permits, some frogs may be active at any time of the year. Specific records for autumnal activity include late September in Wisconsin (where individuals were observed in a streambed; Edgren, 1944), Kansas (where activity was observed in an agricultural field on a sunny day; Busby et al., 2005), and northern Alberta (Harper, 1931). Activity occurs through September at high elevations in Colorado (Spencer, 1964a) and October in Yellowstone (Koch and Peterson, 1995), but extends into November on the prairies (Hammerson, 1999). Tucker et al. (1995) observed this species active in Illinois in late November after being displaced by floodwaters. In Kansas, Taggart (*in* Collins, 1993) found an adult active on 23 December. According to Harper (1931), Boreal Chorus Frogs were easily observed and captured after the breeding season along the shores of wetlands and in marshes in northern Alberta, although most reports indicate they are uncommonly observed after the breeding season.

The Boreal Chorus Frog occurs at much higher elevations (to 3,720 m; Spencer, 1971) and latitudes (65° N) than nearly all other North American anurans. As such, it must have a method for surviving cold temperatures for long periods. Overwintering occurs in animal burrows, ant mounds, and root channels below decaying stumps. Moving underground below the frost line may be an option in parts of the species' range, but the presence of permafrost may limit vertical movement. Boreal Chorus Frogs survive freezing by producing a cryoprotectant in the form of glucose, which is made available via liver glycogen. The cryoprotectant allows extracellular water to freeze while preserving intracellular function. After about 2 hrs at freezing temperatures, glucose levels in the liver increase substantially (Edwards et al., 2000) as a result of a generalized stress response, the heart continues to beat, and the glucose is circulated throughout the body. The core body organs contain the most glucose and are most protected. The use of glucose in conjunction with changes in plasma osmolality allow Boreal Chorus Frogs to survive temperatures of –3 to –2°C for a considerable period (Hunka, 1974; MacArthur and Dandy, 1982), although survivorship is dependent on the level of supercooling prior to freezing (Swanson et al., 1996).

Reserves of glycogen are limited, however, so repeated cycles of freezing and thawing could limit the ability of the frog to produce the cryoprotectant. However, Jenkins (2000) found no significant differences in glycogen levels in frogs exposed to two or three freeze-thaw cycles, although her sample size was small. Survivorship may be low in some particularly cold winters, especially if liver glycogen levels are low. Laboratory studies have demonstrated that glycogen levels vary among frogs acclimated to cold and that survivorship is low when glycogen levels are low (Edwards et al., 2000; Jenkins, 2000). To what extent these levels vary in nature and by size class is undetermined.

Although the Boreal Chorus Frog inhabits an environment that may be rather cool during the peak activity season, the CTmax for specimens acclimated at 5°C is 37.1°C, and at 20°C, it is 38.7°C (Miller and Packard, 1977). As elevation increases, the CTmax decreases so that frogs from higher elevations have a lower CTmax than frogs from low elevations, regardless of acclimation temperature. Clarke (1958) recorded a wide range of temperatures (4.4–32°C) at which this species is active.

Populations can be rather large and cover a substantial area at a landscape scale, especially across extensive areas of the contiguous western plains. Spencer (1964a), however, considered populations to be discreet and to display a certain degree of genetic isolation from one another. It seems likely that the structure of populations may vary with habitat, with more isolated populations in the mountains and more contiguous populations in the rolling grasslands, where there are few barriers to dispersal. Whether Boreal Chorus Frogs form metapopulations across the grassland prairies and adjacent highlands is unknown. Spencer (1964a, 1964c) found that most frogs remained near wetlands, but that they might move 183–244 m over cutover upland habitat to reach adjacent wetland sites.

CALLING ACTIVITY AND MATE SELECTION

The phenology of calling by Boreal Chorus Frogs depends on environmental conditions, elevation, and latitude. Calling begins in late March in Arizona (Bezy et al., 2004) and occurs from April to mid-June in Colorado (Vertucci and Corn, 1996), when patchy ice still covers the breeding site, but most calling occurs between May and June (Corn and Muths, 2002). Indeed, ice is not uncommon on breeding sites at the start of the breeding season, and Boreal Chorus Frogs have been reported calling during snowfall when air temperatures were –1°C to 0.7°C (Degenhardt et al., 1996; Bezy et al., 2004). In southern Ontario, calling occurs from early to mid-April (Piersol, 1913). In northwestern Ontario, calling begins in late April and continues through June (Cook, 1964a), whereas calls are heard May to July in northern Manitoba (Harper, 1963) and into July in North Dakota (Bowers et al., 1998) and South Dakota (Peterson, 1974). Calls have been recorded at the southern end of James Bay, Québec, from late May to mid-June (Ouellet et al., 2009). Occasional calls have even been reported in August and mid-September in northern Alberta (Harper, 1931), although breeding had long ceased by this time.

The mean duration of the call of *P. maculata* from Churchill, Manitoba, is 578 ms (range 572 to 584 ms). The call has 18 notes and a dominant frequency of 3,069 cps (range 2,750 to 3,300) (Thompson and Martof, 1957). Platz (1989) found the number of pulses to vary from 14.3 to 17.4 in a transect from the northern to southern Great Plains; dominant frequencies varied clinally, from 2.91 (south) to 3.58 (north) kHz. Call duration (0.59–1.16 sec) and pulse rate (12.7–23.7 pulses per sec) also varied geographically (Platz, 1989). As might be expected, the call of this species is most similar to that of phylogenetically related *P. triseriata*. In mixed-species choice tests, *P. feriarum*, which is unlikely to encounter *P. maculata* in nature, easily preferred conspecifics to the calls of this species (Martof, 1961).

Males arrive at a breeding site for a week or more before females are initially observed. The calling behavior of males is like that of other striped chorus frogs in that males call from grass tussocks and are well hidden from human observers. When conditions prohibit calling from grass tussocks or from emergent vegetation, *P. maculata* will call from the shoreline, open water, cracks within mud, and even away from water as far as 4 m (Shepard and Kuhns, 2000). Calling occurs day or night, and nighttime calling extends all night, although there is a peak in calling activity from dusk till about 10:00 in North Dakota (Bowers et al., 1998) and from dusk to 02:00 in Wyoming (Koch and Peterson, 1995).

The presence of satellite males is not uncommon in this species. Many choruses have satellite associations involving one to two calling males, and choruses rarely have circumstances where ten calling males are shadowed by satellite males. Even when choruses are very small, some males will choose to form satellite associations rather than call. In Kansas, Roble (1985b) found that 60.5% of *P. maculata* choruses had at least one satellite male association, and if ten calling males were present, fully 77.4% of the choruses had satellites. Satellite males are commonly larger than the calling male, and generally sit 0.5–25 cm from the caller. Satellite males may remain near a calling male for as long as 2 hrs (Roble, 1985b). Calling males may even switch between calling and satellite status. Satellite males attempt to intercept females moving toward the calling male, and they are occasionally successful.

BREEDING SITES

Boreal Chorus Frogs breed in small shallow (< 35 cm) grassy ponds, prairie potholes, swamps, thicket swamps, marshes, playa lakes, glacial kettle ponds, bog ponds, ponds fed by snowmelt, beaver ponds, flooded fields, and other small wetlands (e.g., Okonkwo, 2011). In the extreme North, they occupy sedge-bordered ponds in the open forest–tundra transition zone (Smith, 1953; Ouellet et al., 2009). They occur over a range of wetland hydroperiods, from temporary to permanent (Kolozsvary and Swihart, 1999). They also have been recorded breeding in man-made wetlands, such as roadside ditches, stock ponds, wheel ruts and borrow pits, ponds in agricultural settings (Knutson et al., 2004), and even in bison wallows (Gerlanc, 1999; Gerlanc and Kaufman, 2003). Breeding sites almost always do not contain fish, although there are rare reports of Boreal Chorus Frogs being observed at sites containing fish (Hecnar, 1997; Hecnar and M'Closkey, 1997b). If fish are added to a former breeding site, Boreal Chorus Frogs disappear (Sexton and Phillips, 1986). Like other frogs, *P. maculata* does not occupy all likely breeding sites during any one year. For example, Corn (2007) suggested that Boreal Chorus Frogs occupied 32–43% of all potential breeding sites within the Greater Yellowstone Ecosystem of Wyoming and Montana.

REPRODUCTION

Breeding occurs from spring throughout the early summer, depending upon elevation and latitude. Ice may still cover a portion (<50%) of the breeding site when breeding begins, but excessive ice will delay the start of breeding. The total length of the breeding season is about six weeks, with females present throughout the period. Frogs arriving early in the season appear to remain longer than frogs arriving late in the season. For example, Corn (1980a) reported that frogs arriving between 21 and 24 April in Colorado stayed a mean of 21.8 days at a breeding site, whereas those arriving between 10 and 17 May only stayed a mean of 6.3 days. Corn (1980a) also noted that frogs with the green spot color morph stayed slightly longer at the breeding site than those with the brown spot color morph.

The phenology of breeding depends on elevation and latitude, as well as environmental conditions. In New Mexico, for example, breeding occurs from February through August, depending upon elevation (Degenhardt et al., 1996). Most breeding occurs within a relatively narrow time frame of less than six weeks, despite a somewhat extended breeding period that can include four or more months. Most peak breeding begins in March and extends through April, but there are exceptions. In Wisconsin, calling begins in March in the south and April in the north; the breeding season then extends to May and June, depending on latitude (Vogt, 1981). In Kansas, breeding occurs from February to May (Smith, 1934), whereas in North Dakota it occurs from April to June (Wheeler and Wheeler, 1966). In northern Ontario, breeding occurs from May to early July, and calls have been heard as late as 23 July (Schueler, 1973). In northern Alberta, mating occurs from mid-May through the end of June (Harper, 1931). The initiation of the breeding season also may be different from one year to the next, with as much as a month starting difference between years, depending upon environmental conditions. In addition, Western Chorus Frogs may breed earlier at some sites than others. On

Tadpole of *Pseudacris maculata*. Photo: William Leonard

the Konza Prairie, for example, Gerlanc (1999) recorded breeding at small intermittent streams four to six weeks earlier than at nearby bison wallows.

Breeding populations may consist of only a few individuals to several hundred adults. In Colorado, Tordoff and Pettus (1977) reported 3 populations with estimates of 37–83 adults, 49–262 adults, and 471–580 adults over a 5 yr period. Corn et al. (1997) estimated 2 populations in Rocky Mountain National Park as containing 136 and 161 adults, respectively. In Montana, chorus size ranges between 10 and 600 (Werner et al., 2004). The number of breeding adults may fluctuate considerably from one year to the next.

Egg masses are placed on leaning stems, grass blades, and branches in the shallow water 7.5–20 cm below the surface; the individual masses are 10–20 cm wide by 30–40 cm in length (Vogt, 1981). A single female may deposit more than one egg mass, and amplexus may last 8–10 hours (Werner et al., 2004). Eggs hatch in 3–5 days in Kansas, and the larvae are 4.5–5 mm TL at hatching (Youngstrom and Smith, 1936). Vogt (1981) gave a hatching size of 7–9 mm TL in Wisconsin, whereas Pettus and Angleton (1967) record hatching at 4.8–6.5 mm TL in Colorado. Heinrich (1985) recorded hatching in 5–8 days at 17–25°C. In the cold waters of Yellowstone, Turner (1955) recorded hatching in "about two weeks," and Vogt (1981) reported hatching in 6–18 days. Werner et al. (2004) recorded a range of 5 to 27 days until hatching.

In determining clutch size, it is important to differentiate between the number of eggs per egg mass, the number deposited during a single period of amplexus, and total fecundity during the breeding season. The number of eggs per egg mass varies from 15 to 190 in Utah (mean 66) (Pack, 1920) and 20–100 in Wisconsin (Vogt, 1981); in Colorado, the total egg compliment varies from 137 to 793, with lower-elevation females depositing slightly more eggs than high-elevation females (Pettus and Angleton, 1967). Even after adjusting for egg size and elevation, larger females have larger egg clutches than small females. Smith (1934), however, stated that clutches contained between 110 and 300 eggs per mass in Kansas, with a modal number of 140. He further noted that a single female oviposited 1,459 eggs in one night in the laboratory. Collins (1993) included an observation of a female depositing 1,169 eggs. On Konza Prairie, Kansas, females deposited total egg compliments of 223 and 294 eggs in masses containing from 5 to 157 eggs per mass (Heinrich, 1985).

LARVAL ECOLOGY

Larvae are herbivorous, consuming algae; they tend to swim at or near the surface of the water. Crowding (or density) can increase the length of the larval period, reduce growth and developmental rates, decrease the size at metamorphosis, and decrease larval survivorship as well as the proportion of survivors that actually attain metamorphosis. In small containers at high densities, pure sibling cohorts grew faster with a higher proportion attaining metamorphosis than in mixed-relationship cohorts (Smith, 1990). These results suggest that kin recognition may play a role in larval development, especially in small pools where tadpoles are likely to be at high densities.

Because Boreal Chorus Frogs breed in small temporary pools, and a single pool may contain larvae of various sizes (Claflin, 1962), one pool may contain progeny from the mating of a single male and female. As such, all the tadpoles in some pools may be full siblings, and even larger pools may contain the progeny of only a few adults. Smith (1990) calculated that the actual coefficients of relationship among larvae in breeding pools on Isle Royale, Michigan, and found values from 0.1 to 0.5 (mean 0.35). Despite this, experiments suggested that there were no kin effects that affected larval period or size at metamorphosis.

Larval Boreal Chorus Frogs are sensitive to temperature. The CTmax of larvae at lower elevations is higher than the CTmax from conspecifics living at high elevations. At low elevations, the CTmax ranges between 38.6 and 39.1°C depending on larval stage, whereas at high elevations, the CTmax is between 37.5 and 38.4°C (Hoppe, 1978). These results suggest that larvae approaching metamorphosis at low elevations, where temperatures could be quite warm, exhibit a thermal tolerance that allows them to complete metamorphosis despite the increasing temperatures of summer.

Larvae reach a TL of 25–30 mm just prior to transformation (Vogt, 1981). In Kansas, recent metamorphs are 7.5 mm SUL (Youngstrom and Smith, 1936) 9–15 mm SUL (mean 11.5 mm) on Isle Royale, Michigan (Smith, 1987), 12.6–14.8 mm SUL (mean 13.7 mm) in northern Alberta (Roberts and Lewin, 1979), 13–14 mm SUL in northwestern New Mexico (Gehlbach, 1965), and ca. 15 mm SUL just south of Hudson Bay (Smith, 1953). Body size tends to decrease with the date of metamorphosis, so that metamorphs from long-developing larvae are smaller than those that grow rapidly and metamorphose sooner. According to Turner (1955), metamorphosis occurs in 40 to 50 days in Yellowstone, whereas Breckenridge (1944) recorded metamorphosis in 8 to 10 weeks in Minnesota, and Werner et al. (2004) recorded metamorphosis after 6 to 10 weeks in Montana.

DIET

The diet of the Boreal Chorus Frog consists primarily of small invertebrates, mostly insects. Beetles and flies (gnats, chironomids) form the bulk of the diet, but many types of invertebrates are consumed, including crickets, grasshoppers, true bugs, ichneumonids, and ants. Noninsects include springtails, spiders, mites, and centipedes. Information on food is contained in Cragin (1881), Hartman (1906), Tanner (1931), Moore and Strickland (1954), and Christian (1976). Large insects are avoided, and prey consists of both semiaquatic and terrestrial species.

PREDATION AND DEFENSE

The Boreal Chorus Frog is a cryptic species that blends in well with its grassy surroundings and substrate. When approached, males stop calling and remain motionless, making them difficult to locate. In addition, the call has ventriloquist-like qualities, making it extremely difficult to pinpoint the location of calling males. Males remain silent until an intruder has moved away.

Birds may be a significant predator of Boreal Chorus Frogs. Gray jays (*Perisoreus canadensis*) have been shown to readily eat *Pseudacris maculata*, and these visually oriented predators are adept at locating chorus frogs, particularly if the frogs' background color does not match the substrate (Tordoff, 1980). Differential predation pressure could thus account for some of the differences in color polymorphism among postmetamorphic frogs.

Other predators of adults and juveniles include other species of birds (ravens, great blue heron, black-crowned night heron, cranes, pied-billed grebe, American robin), the Oregon Spotted Frog (*Rana luteiventris*), Northern Leopard Frog (*Lithobates pipiens*), garter snakes (*Thamnophis cyrtopsis, T. radix, T. sauritus, T. sirtalis*), water snakes (*Nerodia*), and possibly Eastern Foxsnakes (*Pantherophis gloydi*) (Coues and Yarrow, 1878; Turner, 1955; Vogt, 1981; Larsen and Gregory, 1988; Koch and Peterson, 1995; Livo et al., 1998; Hammerson, 1999). Larvae are eaten by Barred Tiger Salamanders (*Ambystoma mavortium*), garter snakes, bass (*Micropterus* sp.), and predacious invertebrates, such as diving beetles (*Dytiscus*) and insect larvae.

POPULATION BIOLOGY

Like other striped chorus frogs, many Boreal Chorus Frogs likely breed the first spring following their metamorphosis. Tordoff and Pettus

(1977) estimated that two of every three frogs at a breeding site were first-year arrivals, and that only a relatively small percentage bred in multiple years. Even in northern Manitoba, young-of-the-year reach about 19 mm SUL by September (Harper, 1963), suggesting that they could attain a minimum breeding size by July the following year, assuming they feed for a period of weeks during the spring and early summer. Maturity may not occur until the second summer in some animals, especially in the extreme north or due to the early onset of cold weather (Roberts and Lewin, 1979). Werner et al. (2004) noted that one Boreal Chorus Frog lived six years during an unreferenced study, but that normal longevity was about three years.

There are many factors that influence the population biology of Boreal Chorus Frogs. For example, small size at metamorphosis and an increase in the duration of the larval period both act to delay sexual maturity. Individuals that are large as larvae mature fast and retain a size-based advantage, even a year after metamorphosis. Early maturing male frogs, for example, are likely to mate the first spring following metamorphosis. Metamorphosing quickly essentially has the same effect, that is, it allows some males a longer time to grow and thus breed the first spring following metamorphosis. Smith (1987) estimated that about 63% of males matured the first spring, whereas 37% matured the second spring. Male survivorship from metamorphosis to maturity was estimated at 19%.

Frogs that metamorphose too late either cannot survive the first winter or are too small to breed the first spring following metamorphosis. However, the initial male size advantage disappears with age so that survivorship for frogs that are recaptured 2 yrs after metamorphosis is not related to size or duration of the larval period (Smith, 1987). Most females may actually delay initial breeding 1 yr longer than most males. A few females apparently are able to reach maturity the first spring after metamorphosis, but they are the smallest breeders at the pond, and their success rate is unknown. All such females result from larvae that grow quickly to a large size and metamorphose quickly. After the second spring following metamorphosis, females are all about the same size at a breeding site (Smith, 1987).

The effect of various color patterns on the life history and evolution of Boreal Chorus Frogs has been studied intensively in Colorado. Most life history traits show no variation between green and brown color morphs. However, larvae from green females metamorphose more rapidly than those from brown females. Green juveniles are slightly larger at metamorphosis than brown juveniles and grow slightly faster during the month following metamorphosis, resulting in a significant difference in size among chorus frogs of the same age (Hoppe and Pettus, 1984). Why these differences occur is unknown.

Despite the possibility of selection favoring one color under some circumstances and another color under different circumstances, mating appears to be random among color morphs, and there is no evidence of interaction between color polymorphism and sex (Matthews, 1971). Matthews (1971) could find no patterns in changes related to the frequencies of the different background colors, but there is an ontogenetic change in frequency in spot patterns. The highest frequency of green-spot phenotypes occurs in adults, and the lowest in recent metamorphs. In other words, green spotting appears to be selectively superior to the brown-spot phenotype after metamorphosis. However, brown-spot phenotypes had the highest survivorship during the larval stage. Thus, the color-spot pattern polymorphism could be maintained by differential selection between aquatic and terrestrial life stages.

Selection undoubtedly is responsible for maintaining the frequencies of the various combinations of background coloration and spot patterns seen at the Larimer County sites (Tordoff and Pettus, 1977). The polyphenism patterns appear to remain stable through time (Tordoff and Pettus, 1977), and Tordoff et al. (1976) demonstrated that there were significant differences among seven populations in the frequencies of the alleles controlling the color polymorphisms. In laboratory trials, Tordoff (1980) demonstrated that frogs that mismatched their background suffered higher predation rates from an avian predator than did those frogs that matched their substrate coloration. Thus, predation may play a role in the maintenance of the different color morphs, although it may not be the only contributing factor.

COMMUNITY ECOLOGY

As with other anurans, Boreal Chorus Frog larvae are affected by both predators and competitors, although there is not much published on the role of competition in the larval life history. In one example, aquatic snails may compete with Boreal Chorus Frog tadpoles for limited resources when snail abundance is high, resulting in a reduction in larval growth rates and a decrease in survivorship (Woodward and Mitchell, 1992, *in* Koch and Peterson, 1995). Empirical confirmation is needed, however.

Larval Boreal Chorus Frogs exhibit a certain degree of phenotypic plasticity, depending upon whether or not predators are within their environment. In the presence of dragonfly naiads (*Anax*), for example, larvae with a shallow and narrow body, a deep tail fin, and wide muscle have better survivorship than larvae without these traits. However, when *Anax* are not present, larvae with narrow tail musculature grow faster than those with deep tail fins, and larvae with shallow tail fins and a deep body grow faster than those with the opposite traits of shallow bodies with deep tail fins. Presumably, the changes in the tail structure allow the tadpole to swim faster and escape predation when predators are around; when no predators are present, the tadpole puts its resources into growth rather than an escape-based tail morphology. Tadpoles are able to adjust their tail shape, but not their body shape, to the presence of predators. Thus, some important components of fitness are phenotypically plastic, whereas others are developmentally rigid (Van Buskirk et al., 1997).

The pH of a wetland breeding site may have sublethal effects on the success rate of predatory encounters between Boreal Chorus Frog larvae and would-be predators, such as larval Tiger Salamanders (*Ambystoma mavortium*). Low pH tends to make the predator less successful in predation attempts on the frog larvae, while having no direct impact on the larvae themselves (Kiesecker, 1996).

Predators may play an important role in determining the spatial breeding patterns of Boreal Chorus Frogs. In areas where there is a range of available breeding sites in terms of hydroperiod, *Pseudacris maculata* will often choose those of intermediate-length hydroperiod. This allows them to avoid desiccation in shallow pools that tend to dry quickly, and to survive predation from invertebrates (dragonflies, *Anax*) or vertebrates (salamanders, *Ambystoma laterale*), which is more intense in the ponds with a longer hydroperiod. Even intermediate-duration hydroperiods may allow for some aquatic predators to become established, so the sequence of colonization becomes important. In addition, intraspecific competition in the remaining pools leads to variation in growth rates and larval success as larval density varies. These factors lead to a mosaic of occupancy patterns, as demonstrated, for example, along the shores of Lake Superior where small, temporary pools along the shoreline grade into permanent pools as one proceeds inland (Smith, 1983).

Adult *Pseudacris maculata*, brown phase. Photo: James Stuart

Still, the density of salamander predators and tadpoles in mixed-species experimental trials has no effect on larval mass at metamorphosis, length of the larval period, or growth rates of *Pseudacris maculata* larvae. It would appear that microhabitat differences and breeding phenology (priority effects) have more effects on Boreal Chorus Frog larval development than the interaction between larval density and single-species predator density per se (Sredl and Collins, 1991). High levels of predator densities significantly decrease larval survivorship, as expected, especially in field enclosures.

In general, Boreal Chorus Frogs do not show positive nestedness distributional patterns with other anurans, primarily because of their almost total reliance on shallow-water temporary breeding sites (Hecnar and M'Closkey, 1997a). The other frogs within its range use both temporary and permanent wetlands, but even when temporary wetlands are used as breeding sites, they are usually larger than the sites chosen by *P. maculata*.

DISEASES, PARASITES, AND MALFORMATIONS

Egg masses are affected by mold (saprolegniasis), unidentified pyriform protozoa, and red larval insects (*Ablabesmyia* sp). Tadpoles are subject to helminth parasitism by nematodes (*Gyrinicola batrachiensis*) (Adamson, 1981c) and renal metacercariae consistent with *Echinostoma* sp.; unidentified encysted metacercariae have been found in the body cavity (Green and Muths, 2005). Green and Muths (2005) also found the liver in a single specimen to be infected by intracellular protozoa.

The amphibian chytrid fungus *Batrachochytrium dendrobatidis* was first reported in *Pseudacris maculata* in Colorado by Rittmann et al. (2003). In Canada, 54 of 143 *P.* "*triseriata*" (most likely *P. maculata*) examined from 1993 to 2001 were identified as being infected by amphibian chytrid (Ouellet et al., 2005a). In Colorado and Wyoming, 6 of 14 *P. maculata* examined were infected (Muths et al., 2008), and 37 of 82 adults from Rocky Mountain National Park had mild to severe amphibian chytridiomycosis (Green and Muths, 2005). No declines have been attributed to the disease, however. Bacteria isolated from *P. maculata* include *Aeromonas encheleia*, *A. hydrophila*, *Bacillus* sp., *Enterococcus* sp., *Hafnia alvei*, *Pseudomonas* sp., and *Sphingomonas paucimobilis* (Green and Muths, 2005).

An unidentified beetle was found in the dorsal lymphatic sac overlying the urostyle in a single Boreal Chorus Frog from Colorado (Green and Muths, 2005), and blowflies (*Lucilla silvarum*) have been reported from *Pseudacris maculata* in Alberta (Roberts, 1998; Eaton et al., 2008). Many enteric protozoans (*Giardia agilis*, *Hexamita intestinalis*, *Monocercomonas* sp., *Monocercomonoides* sp., *Nyctotherus cordiformis*, *Octomastix* sp., *Octomitus* sp., *Opalina* sp., *Phacus pleuronectes*, *P. torta*, *Tritrichomonas augusta*, *T. batrachorum*, *Urophagus* sp.) have been reported from Boreal Chorus Frogs in Minnesota (Anderson and Buttrey, 1962). The Boreal Chorus Frog is parasitized by the trematodes *Apharyngostrigea pipientis*, *Falcaustra catesbeianae*, *Fibricola cratera*, *Glypthelmins pennsylvanicus*, *G. quieta*, and *Megalodiscus temperatus*, and the nematodes *Aplectana* sp., *Ozwaldocruzia* sp., *O. leidyi*, and *Rhabdias ranae* (Ubelaker et al., 1967; Ulmer, 1970; Brooks, 1976; Dyer, 1991; Goldberg et al., 1996c, 2002; Bolek and Coggins, 1998; Bolek and Janovy, 2008). Unidentified trematodes and nematodes were reported by Green and Muths (2005). No cestodes were found by Ulmer and James (1976a) after examining eight Boreal Chorus Frogs in Iowa, but Goldberg et al. (2002) reported *Cylindrotaenia americana* from this species in Alberta. The myxozoan species *Myxidium melleni* was described from *Pseudacris maculata* in Nebraska (Jirků et al., 2006). In Colorado, Spencer (1964b) found small clams (*Pisidium*) attached to the toes of Boreal Chorus Frogs, suggesting that the frogs might serve as vectors in distributing the clams.

SUSCEPTIBILITY TO POTENTIAL STRESSORS

pH. In Colorado, the lethal pH for *P. maculata* is 4.8, and the lowest pH tolerated with no mortality is 5.2 (Corn et al., 1989; Corn and Vertucci, 1992). In Minnesota, the lethal pH was reported as 4.2 (Karns, 1983). However, Kiesecker (1996) raised Boreal Chorus Frog larvae at a pH of 4.5, 5.5, 6.0, and 7.0 and reported no effects on larval growth or developmental time. The pH of breeding ponds in Colorado varied between 6.0 and 7.2 during a study of the effects of acidity on the status of this species (Vertucci and Corn, 1996).

Calling male *Pseudacris maculata*, green phase. Photo: Steve Corn

Chemicals. The tolerance levels (amount of chemical it takes to kill 50% of the animals in the specified time period) of larval *P. maculata* to a wide range of pesticides after 24, 48, and 96 hrs exposure were provided by Sanders (1970). The following are means in mg/L after 96 hrs exposure: carbophenothion (0.028), dieldrin (0.03), endrin (0.09), methoxychlor (0.19), malathion (0.33), TDE (0.09), DDT (0.21), parathion (0.30), toxaphene 0.10), piperonyl butoxide (0.10), naled (0.50), lindane (1.4), 6-chloro-2-picolinic acid (2.0), Silvex (4.0), and paraquat (21–36). For paraquat, the $LC_{50\ (96\ hrs)}$ has also been reported as 28 mg/L (Mayer and Ellersieck, 1986).

UV light. Although UVB radiation has increased in the high mountains of Colorado, Corn and Muths (2002) showed that embryos of *P. maculata* are not necessarily exposed to detrimental levels of UVB in nature. Breeding by Western Chorus Frogs occurs earlier in dry years than in years with normal or greater snowpack. As such, levels of UVB, even in shallow water, are less than they would be later in the season. The timing of reproduction thus tends to offset potential UVB effects in dry seasons. In addition, the spectral characteristics of water in amphibian breeding ponds indicate that UVB effects would be mediated and not likely reach levels to affect embryos (Palen et al., 2002). UVB likely plays a minimum role in amphibian declines throughout the West (but see Blaustein et al., 2004).

STATUS AND CONSERVATION

Boreal Chorus Frogs appear to be secure throughout the vast majority of their extensive North American range. Populations of a number of anuran species have disappeared in the Rocky Mountains over the last decades. However, there is no evidence that episodic acidification is responsible for declines of the Boreal Chorus Frog (Vertucci and Corn, 1996). McMenamin et al. (2008) stated that significant declines occurred in Yellowstone National Park due to wetland loss and desiccation, but this assertion has been both criticized (Patla et al., 2009) and defended (McMenamin et al., 2009). Contention centers on how well frogs can survive a long-term drought and what the impacts will be in the future. There is little doubt that the frog was common in the past. Turner (1955:17) noted that they "breed in vast numbers in the swamp between the bridge and Yellowstone Lake, as well as in the marshy

Prairie breeding habitat of *Pseudacris maculata*, also used by *Lithobates sylvaticus*. Photo: Robert Newman

mouth of Pelican Creek and countless other sloughs and rain pools in the area."

Some populations of *P. maculata* may be in decline, but supporting data often are not rigorous. Exceptions are Daigle's (1997) study in Québec and Seburn et al.'s (2008) survey in eastern Ontario. In comparison with historical data, *P. maculata* declined significantly in a large area southeast of Montréal from 1959 to 1993. In Ontario, occupancy declined from 58% in 1990 to only 12% in 2007. Populations in Illinois decreased over a 4 yr period in the late 1980s, based on aural road surveys (Florey and Mullin, 2005); Christiansen (1981) considered them declining in Iowa; declines also have been noted in the eastern forested regions of Wisconsin (Mossman et al., 1998). At other locations, populations appear stable with no evidence of significant declines (Busby and Parmelee, 1996; Corn et al., 1997; Weller and Green, 1997; Brodman and Kilmurry, 1998; Casper, 1998; Hemesath, 1998; Mierzwa, 1998; Mossman et al., 1998; Hammerson, 1999; Hossack et al., 2005; Brodman, 2009).

Like the other striped chorus frogs, the Boreal Chorus Frog appears to tolerate considerable habitat disturbance, including silviculture, forest burns, and even suburbia, as long as shallow breeding sites are available and a suitably complex environment of grasses, high shrubs and herbs, and a low canopy cover remains (Constible et al., 2001). They will colonize restored or mitigation wetlands (Kline, 1998; Mierzwa, 2000; Lehtinen and Galatowitsch, 2001; Brodman et al., 2006; Shulse et al., 2010).

Road-based or pond-based call surveys have been used to monitor Boreal Chorus Frog populations at a number of locations (e.g., Bishop et al., 1997; Lepage et al., 1997; Bowers et al., 1998; Mossman et al., 1998; Corn et al., 2000). The species is easily detected if surveys are conducted early in the season. De Solla et al. (2005) found that only three sampling nights were needed to detect *Pseudacris* (both *P. triseriata* and *P. maculata*) with 80% probability in southern Ontario. Near 100% detection was achieved after ten sampling nights. Even in pond-based surveys, detection probabilities may be high if surveys are conducted at the right time of year. In the Greater Yellowstone Ecosystem, for example, detection probabilities were >88% with adjusted occupancy rates at 32–43% (Corn et al., 2005).

Pseudacris nigrita (LeConte, 1825)
Southern Chorus Frog

ETYMOLOGY
nigrita: from Latin, meaning 'blackened.'

NOMENCLATURE
Synonyms: *Acris nigrita, Chorophilus nigrita, Chorophilus nigritus, Chorophilus verrucosa, Cystignathus nigritus, Hyla nigrita, Hyla nigrita floridensis, Hyla verrucosa, Rana nigrita*

There is considerable taxonomic confusion in the literature on chorus frogs. In the past authors may have assigned a name to a chorus frog population under study that has no resemblance to the current nomenclature based on molecular analysis. For example, Smith and Smith (1952) compared various morphological characteristics of "*Pseudacris nigrita feriarum*" with "*P. n. triseriata*." Upon looking at their collecting localities, it is clear that what we now know as *P. feriarum, P. fouquettei, P. triseriata,* and *P. kalmi* were all included under their name *P. n. feriarum*, and that *P. fouquettei, P. triseriata,* and *P. maculata* were included under their name *P. n. triseriata*. Indeed, they may not have examined any *P. nigrita* as now defined. This species mixing makes the paper uninterpretable. Always check the location where specimens originated to verify correct taxonomic allocation.

IDENTIFICATION
Adults. This species is a member of a morphologically conservative group of chorus frogs characterized by a slender olive, gray, brown, or black body with three rows of rounded dark spots that may fuse to form longitudinal stripes. The mid-dorsal stripe extends almost to the tip of the snout. *Pseudacris nigrita* has a long and narrow head and sharply pointed snout. There is a dark lateral band running from the snout to midway between the arm and the groin. Like other small striped chorus frogs, there is a white to cream-colored labial stripe on the upper jaw; this line may be continuous or broken. The skin on the dorsum is somewhat granular with numerous small bumps. The venter is immaculate. Although similar to *P. feriarum*, *P. nigrita* does not have a dark triangular mark between the eyes, its body is less stout than *P. feriarum*, and its snout is more pointed than in the Upland Chorus Frog. Males have a dark subgular vocal sac.

Males are slightly smaller than females. In Alabama, adults are 23–28 mm SUL (mean 24.8 mm) (Brown, 1956). South Carolina females range between 25 and 33 mm SUL (mean 29.5 mm), whereas males range between 26 and 32.5 mm SUL (mean 28.3 mm) (Caldwell, 1987). In central Florida, males averaged 27.7 mm SUL (range 25 to 30.1 mm SUL) (Owen, 1996), and in south Florida males measured 25 to 29 mm SUL and females measured 27 to 30 mm SUL (Duellman and Schwartz, 1958).

Larvae. In larval *P. nigrita*, the dorsum is uniform black to dark brown, and dark spots are not apparent. Both tail fins are lower than the tail musculature, and they are blotched or mottled with small dark flecks. The venter is pigmented with a golden or brassy coloration, and the snout is rounded. A description of the larva is in Altig (1970).

Eggs. The eggs are brown above and white, cream, or yellow underneath. They are contained in a loose irregular jelly mass about 25–30 mm in diameter attached to submerged stationary objects. Livezey and Wright (1947) recorded a clutch size of 160 eggs, but this likely did not represent a full season's breeding effort. Jensen et al. (2008) provided a figure of 500–1,500 eggs deposited annually in masses containing a few dozen to more than 100 eggs per mass. Eggs contain only a singular gelatinous envelope measuring a mean of 2.6–4.5 mm in diameter (usually about 2.6 mm); the vitellus is 0.5–1 mm (Livezey and Wright, 1947).

DISTRIBUTION
The Southern Chorus Frog is found on the Southeastern Coastal Plain from North Carolina to the Florida peninsula, and west to the Pascagoula River in southern Mississippi. A disjunct population occurs in southeastern Virginia. Because of recent taxonomic changes, the best reference on distribution is the systematic study by Lemmon et al. (2007b). Other useful references include: Alabama (Mount, 1975), Florida (Bartlett and Bartlett, 1999), Georgia (Williamson and Moulis, 1994; Jensen et al., 2008), North Carolina (Dorcas et al., 2007), South Carolina (Schwartz,

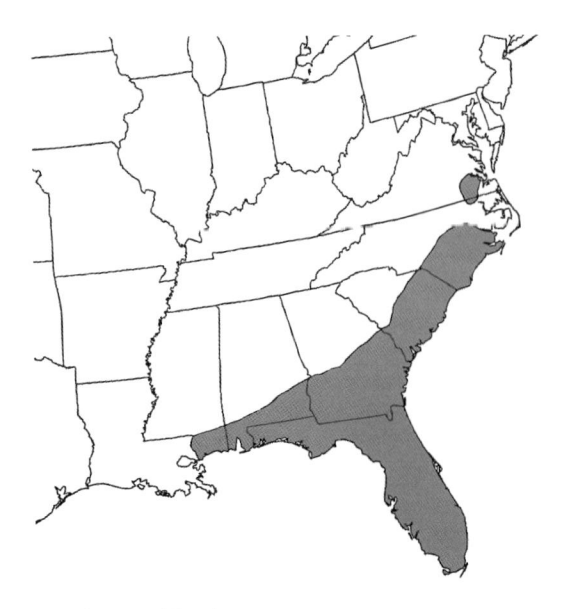

Distribution of *Pseudacris nigrita*

1957), and Virginia (Hobson and Moriarty, 2003). According to Gibbons and Coker (1978), Williamson and Moulis (1994), and Laerm et al. (2000), Southern Chorus Frogs occur on Sapelo and Cumberland Islands, Georgia. However, these records are incorrect (Shoop and Ruckdeschel, 2003).

FOSSIL RECORD

Fossils of the Southern Chorus Frog are known from Rancholabrean deposits in Marion County, Florida. The ilium of this species has a wider acetabular expansion than either *P. ornata* or *P. triseriata*, and this characteristic was used to identify the Florida fossils to species. Illustrations of the ilium, sacrum, and humerus are in Holman (2003). Wilson (1975) is the only researcher to conclusively identify fossils of this species.

SYSTEMATICS AND GEOGRAPHIC VARIATION

The evolution of the main lineages within the genus *Pseudacris* occurred during the Tertiary (2.6–34 mya), prior to the onset of Pleistocene glaciation. The inundation of the Mississippi Embayment was one of the driving geological events influencing subsequent divergence, and there is no evidence that mountain uplift or Quaternary climate change had an impact on speciation. Rivers, however, have played major roles in influencing speciation within *Pseudacris* by reducing gene flow, and this is evident in the current ranges of these species. Periodic sea level rise and fall had demographic effects on the Coastal Plain species as ranges expanded and contracted, and gave rise to the levels of genetic variation seen today (Lemmon et al., 2007a).

Pseudacris nigrita is a member of the Trilling Chorus Frog clade of the genus *Pseudacris*, a group that includes *P. brachyphona*, *P. brimleyi*, *P. clarkii*, *P. feriarum*, *P. fouquettei*, *P. kalmi*, *P. maculata*, and *P. triseriata* (Moriarty and Cannatella, 2004). These species are distinguished by their advertisement call structures, color patterns, genetics, the presence of a cuboidal intercalary cartilage (Paukstis and Brown, 1987), and similar albumin phylogeny (Maxson and Wilson, 1975). The Southern Chorus Frog is the sister taxon of the Cajun Chorus Frog. Together, these two species are the sister taxa of a group that includes *P. triseriata*, *P. kalmi*, and *P. feriarum*. Two subspecies of Southern Chorus Frogs have been described, *P. n. nigrita* and *P. n. verrucosa* (Brady and Harper, 1935). However, Moriarty and Cannatella (2004) found no justification for recognizing subspecies.

Natural hybridization occurs with *P. fouquettei* in the Pearl River drainage of southeast Mississippi and Louisiana and with *P. kalmi* in southeastern Virginia (Lemmon et al., 2007b). Fouquette (1975) found evidence of call-based character displacement and no evidence of a hybrid zone at a contact zone between *P. nigrita* and *P. feriarum* along the Apalachicola River, despite low levels of hybridization (Crenshaw and Blair, 1959). This result is not surprising since these are nonsister taxa (Lemmon et al., 2007b). In contrast, Gartside (1980) found evidence of high levels of hybridization between *P. nigrita* and *P. fouquettei* along the Pearl River. Since these are closely related taxa, the level of hybridization at the contact zone between them is more easily explained than it would be without the phylogenetic interpretation of Lemmon et al. (2007b).

In the laboratory, *P. nigrita* readily hybridizes with *P. feriarum*, *P. ornata*, *P. brimleyi*, and *P. brachyphona*, and it doesn't appear to make much difference whether the *P. feriarum* is male or female (Mecham, 1965). Crosses between *P. nigrita* and *Hyla versicolor* survive to the larval stage (Moore, 1955). This species can be hybridized with *Pseudacris clarkii*, and some larvae will metamorphose. Crosses with *Acris crepitans* are unsuccessful, usually dying prior to hatching (Littlejohn, 1961a).

ADULT HABITAT

The Southern Chorus Frog is an upland species of sandy and loamy soils (Jensen et al., 2008). Unlike many of its morphologically similar sister taxa, it tends to avoid floodplains and habitats with wet, mucky soils. It is found in pinelands and in mixed pine-hardwoods and in many disturbed habitats. In Florida, this species has been observed in pine flatwoods, pine rockland, sand pine plantation, xeric hammock, dome swamp, and inland hydric hammock habitats (Carr, 1940a; Owen, 1996; Enge and Wood, 1998; Meshaka et al., 2000).

TERRESTRIAL ECOLOGY

Southern Chorus Frogs are terrestrial during much of the year, and they forage at night in the leaf litter and under surface debris. Brown (1956) recorded an instance of specimens found beneath the bark of a large upright stump, but this species likely does not venture into the trees. There is a possibility that this species spends much of the year in fossorial conditions. Carr (1940a) noted the tendency of captive individuals to dig into the substrate, and he found several buried underground in natural settings. They are active from the early winter breeding season throughout the warmer months of the year. For example, Wright (1932) recorded them in August and September in southeast Georgia. It seems likely that this little frog remains active year-round throughout much of its range.

Like most frogs, Southern Chorus Frogs have a blue-mode phototactic response, indicating they have true color vision (Hailman and Jaeger, 1974). They are monotonically photopositive, meaning that given a choice individuals will seek out optimal illumination (Jaeger and Hailman, 1973). Presumably, these characteristics assist in terrestrial movements and in locating prey during both day and night.

CALLING ACTIVITY AND MATE SELECTION

Calling occurs during the breeding season, and choruses reach their peak from late December through early April. However, calls may occur in autumn and even in the summer following heavy rainfall (Einem and Ober, 1956; Mount, 1975; Owen, 1996). Indeed, breeding in south Florida may occur at any time of the year, depending upon rainfall (Duellman and Schwartz, 1958; Bartlett and Bartlett, 1999). However, it is unknown if breeding actually occurs in summer north of Florida.

Males call from clumps of grass while they are half submerged in shallow water. Despite their loud call (likened to "ik-ik-ik-ik-ik" by Jensen et al., 2008), they are difficult to locate, because they blend so well into their background and because of the ventriloquist-like nature of the call. Notes are trilled in rapid succession. Calling characteristics of *Pseudacris nigrita* at 27°C include a mean dominant frequency of 3,325 cps, a mean duration of 0.39 sec, and a mean repetition rate of 28.6 calls per sec (Blair, 1958a). The mean calling interval is 0.9 sec (Blair, 1958a).

Mecham (1961) has discussed differences between the call of this species and *P. feriarum*. On the Coastal Plain in areas of geographic overlap between the species, the call is very slow (11 pulses/sec, duration of 0.82 sec). However, south of the overlap zone the call is much faster. Coupled with changes in the call of *P. feriarum* in the zone of overlap, these differences may help maintain premating isolating mechanisms. Mate selection and calling behavior have not been described for this species. They are probably similar to the descriptions presented in the *P. feriarum* and *P. fouquettei* accounts.

BREEDING SITES

Breeding normally occurs during the winter and early spring in shallow ephemeral wetlands. *Pseudacris nigrita* has been recorded breeding in open-canopied cypress/gum temporary ponds in sandhill habitats, sinkhole habitats, limestone sinks, roadside ditches, vehicle tracks, marshes, wet grasslands, cypress savanna and ponds, Carolina Bays, semi-improved pasture, and in rainwater pools (Cash, 1994; Jensen et al., 2008; Liner et al., 2008). Owen (1996) noted that preferred breeding sites in central Florida were shallow marshes of about 1 ha. However, some of the breeding habitats can be quite small. Brown (1956) mentioned occasional breeding in permanent ponds, but they avoid ponds with predatory fish. The species is not uniformly distributed across a landscape, but it does occupy a large number of sites. For example, *P. nigrita* occurred in 109 of 332 potential breeding localities surveyed in central Florida over a 2 yr period (Owen, 1996).

REPRODUCTION

The Southern Chorus Frog is among the earliest breeders on the Southeastern Coastal Plain. Breeding in northern Florida and southern Alabama, for example, may occur as early as October, although the normal breeding season is from late November to April throughout much of its range, depending upon environmental conditions (Allen 1932; Carr, 1940b; Brown, 1956; Mount, 1975; Caldwell, 1987; Semlitsch et al., 1996; Babbitt and Tanner, 2000; Jensen et al., 2008). In South Carolina, breeding occurs to mid-May (Schwartz, 1957).

An interesting shift seems to occur from a winter to summer breeding phenology as one proceeds south on the Florida peninsula. In the south, most breeding occurs in summer, reaching its peak in June and July in south Florida (Duellman and Schwartz, 1958), whereas in the north, most breeding occurs in winter. Owen (1996) noted that most breeding occurred in central Florida during the cooler months of the year (December to February), but that chorus frogs were likely to be at breeding ponds during any time of the year. He suggested that frogs in this area were transitional in breeding phenology between winter and warm weather. The shift may be correlated with seasonal rainfall patterns, as winter is the normal dry season in south Florida and summer the wet season. Potential breeding sites are usually dry in winter in south Florida.

Migration to breeding ponds may occur during any time of the day, but the vast majority of movement occurs after dark (Pechmann and Semlitsch, 1986). Movements during the day are associated with overcast and rainy conditions. Most females are gravid when they enter a wetland for the first time, but not all females will deposit their eggs when breeding sites are scarce, such as during a drought. Caldwell (1987) suggested that females that did not deposit eggs at a particular site might postpone breeding to the following year, even though *P. nigrita* is normally an annual species.

Males and females enter and leave the pond at about the same time and spend the same amount of time at the breeding site in any one year, in contrast with *P. feriarum*. However, the time spent at a site may vary considerably. In South Carolina over a 3 yr period, adults spent from 1 to 61 days at a site, but the means were quite different (13–17 in yr 1, 4–5 in yr 2, 23–25 in yr 3). The overall and operational sex ratio of adults normally is 1:1 at a breeding site (Caldwell, 1987).

Breeding populations can be rather large. Cash (1994) captured 759 *P. nigrita* in 544 days at a pond in southeast Georgia, and Pechmann and Semlitsch (1986) captured 66 individuals in 6 days at a pond in South Carolina. However, only 318 females were captured over a 16 yr study, also at a site in South Carolina, and these females produced only 237 metamorphs (Semlitsch et al., 1996). Most females were captured during the first 8 yrs. Successful reproduction occurred in only 3 yrs, and a single year accounted for the vast majority of metamorphs. In a related example, Gibbons and Bennett (1974) found 37 *P. nigrita* over 2 yrs at a permanent pond, but 73 frogs at a nearby temporary pond. Captures were more evenly distributed in the former, but the second year of sampling accounted for 69 captures at the temporary pond. Caldwell (1987) recorded from 62 to 129 breeding adults per year over a 4 yr period at a Carolina Bay in South Carolina. It seems likely that such annual fluctuations in numbers of breeding adults and metamorphs are normal for populations of *P. nigrita*.

LARVAL ECOLOGY

Eggs hatch in <7 days, with metamorphosis occurring in 6 to 8 weeks (Jensen et al., 2008), although Caldwell (1987) put the larval period of *P. nigrita* as approximately 4 weeks. Tadpoles reach a maximum of 35 mm TL prior to metamorphosis, and new metamorphs are 12–16 mm TL (Jensen et al., 2008).

Tadpole of *Pseudacris nigrita*. Photo: Steve Bennett

Adult *Pseudacris nigrita*.
Photo: Matt Greene

DISPERSAL

Movement away from breeding ponds occurs primarily at night (Pechmann and Semlitsch, 1986). Nocturnal movement patterns may expose the frogs to less predation by visually oriented predators such as birds, and allow them to take advantage of better environmental conditions, particularly the increase in humidity that occurs after dark. These frogs are capable of moving at least 200 m from a breeding site, and they may move farther (Palis and Aresco, 2007). Although the sample size was small, these authors found females moved the farthest distances.

POPULATION BIOLOGY

Sexual maturity is reached within eight to ten months after metamorphosis, such that breeding commences the first winter following the larval period. Males have a lower survival rate at breeding sites than females, and very few adults (<3.2%) take part in more than a single breeding season. In Caldwell's (1987) 3 yr study, male survival at the breeding site ranged between 16 and 34%, whereas the female survival rate was between 27 and 55%. Caldwell (1987) could not explain the reasons for the differential survival, especially since adults spent the same amount of time at the breeding sites and were exposed to similar environmental conditions.

The number of metamorphs produced at a pond is quite variable from one year to the next, with annual recruitment unlikely at many locations. In South Carolina, Caldwell (1987) reported between 180 and 7,345 metamorphs were produced over a 4 yr period, although no larvae reached metamorphosis in one year of the study. Metamorphs average from 11.5 to 16 mm SUL.

DIET

Postmetamorphs of this species undoubtedly feed on a wide array of small invertebrates, although its feeding habits have not been described in detail. Carr (1940a) noted adults feeding on grasshopper nymphs, and Duellman and Schwartz (1958) recorded ants and small beetles in the stomachs of ten frogs.

PREDATION AND DEFENSE

Larvae are probably eaten by a variety of predaceous invertebrates (spiders, water bugs, diving beetles, dragonfly naiads), although its winter-spring cold-weather breeding habits may reduce aquatic predation at least outside of south Florida. Adults and metamorphs are undoubtedly eaten by water snakes, garter snakes, small turtles, and birds, but specific records are not available. Like the other striped chorus frogs, this species is cryptic in its grassy breeding wetlands, and it is good at concealing itself. In addition, the ventriloquist-like nature of its call makes it very difficult to

Breeding habitat of *Pseudacris nigrita*, Columbia County, Florida. Photo: C.K. Dodd Jr.

locate a calling male, even when an observer thinks they are looking directly at the frog.

COMMUNITY ECOLOGY

Jensen et al. (2008) noted that Coastal Plain populations of *P. feriarum* had higher call pulse rates (to 35 pulses/sec) than interior populations, in contrast to Southern Chorus Frogs whose call pulse rate is much slower (5–11 pulses/sec). The difference in pulse rates could represent an example of character displacement, whereby populations of these morphologically similar sympatric chorus frogs maintain their genetic identity through female call discrimination. However, the peak breeding seasons of these species do not overlap, as *P. nigrita* usually breeds later than *P. feriarum* (Schwartz, 1957).

DISEASES, PARASITES, AND MALFORMATIONS

The trematode *Glypthelmins pennsylvanicus* has been reported from *Pseudacris nigrita* (Sullivan and Byrd, 1970).

SUSCEPTIBILITY TO POTENTIAL STRESSORS

Salinity. Populations can recover quickly in coastal areas that receive hurricane overwash with resulting temporary increases in salinity (Gunzburger et al., 2010).

STATUS AND CONSERVATION

The Southern Chorus Frog is found in rural areas, protected areas within more developed areas, and even in some residential areas in west-central Florida (Delis, 1993; Delis et al., 1996). They occasionally may be trapped in flatwoods that had been clearcut (Enge and Marion, 1986). They do not appear to breed in restored or created ponds, although small numbers of adults may rarely be observed at such sites (Pechmann et al., 2001). These frogs use recently burned habitats, as prescribed burning helps keep habitats open and reduces dense stands of vegetation (Langford et al., 2007).

Caldwell (1987) hypothesized that rapid development followed by extensive juvenile dispersal allows this annual species to take advantage of the many presumed breeding sites in close proximity to the natal site. Such a breeding strategy might make this species especially sensitive to climate change that involves decreases in precipitation or changes in rainfall patterns during the winter to spring breeding season.

Pseudacris ocularis
(Bosc and Daudin, 1801)
Little Grass Frog

ETYMOLOGY
ocularis: from the Latin word *ocularis* meaning 'eye.' The name refers to the dark stripe that runs through the eye.

NOMENCLATURE
Synonyms: *Acris ocularis, Auletris ocularis, Calamita ocularis, Chorophilus angulatus, Chorophilus ocularis, Cystignathus ocularis, Hyla ocularis, Hyla oculata, Hylodes ocularis, Limnaeodus ocularis*

This little frog was sometimes referred to as the "savanna cricket" by early explorers. Harper (1939b) reviewed the nomenclatural history.

IDENTIFICATION
Adults. This is the smallest frog in the United States and Canada. Adults are only 10–20 mm TL (Harper, 1939b). The ground color is a uniform gray, greenish, brown, bronze, or reddish with a thick dark line running from the eye to shoulder and sometimes farther to the groin. Wright (1932) suggests that the lateral stripe is more prominent and prevalent in females than males. The skin is smooth, and the snout is pointed. Venters are yellowish to white. Terminal disks are better developed than they are in terrestrial chorus frogs, and the tibia is long in proportion to the frog's SUL. Hind feet are only slightly webbed. There is a black line along the lower portion of the tibia.

Males are slightly smaller than females, and they have a dark throat during the breeding season. In central Florida, males averaged 13.8 mm SUL whereas females were 15.3 mm SUL (Owen, 1996). Body mass was correlated with SUL. In the Okefenokee Swamp, Wright (1932) recorded males as 11.5–13.5 mm SUL and females as 12–17.5 mm SUL.

Larvae. Hatchlings are initially light brown and cream colored. The mature tadpole generally has a light to medium brown or drab and olive-green ground color broken by a number of stripes. Scattered black spots are distributed throughout the dorsum. The dorsal part of the tail musculature has a dark stripe in lateral view bordered by a ventral light stripe. A light stripe extends forward to the eye from the tail musculature; this stripe is paralleled by a darker stripe underneath it. The throat and chest are mottled. The dorsal part of the tail musculature is sometimes banded. Tail fins are clear or contain a few blotched melanophores. The dorsal tail fin is higher than the ventral fin. Tails are long and the tips are sharply acuminate. Tadpoles grow to about 23 to 25 mm TL. Gosner and Rossman (1960) provided figures of tadpoles at different stages of development.

Eggs. The eggs are brown dorsally and cream colored ventrally and deposited singly or in small clusters totaling 100–200 per female. Eggs are left on the substrate or attached to vegetation in small clusters of ca. 25 eggs. There is a single jelly envelope surrounding the vitellus. The vitellus has a mean diameter of 0.95 mm (range 0.88 to 1.03 mm), whereas the envelope's mean diameter is 1.57 mm (range 1.5 to 1.68 mm) (Gosner and Rossman, 1960). Wright (1932) gave 0.6–0.8 mm for the diameter of the vitellus (mean 0.75 mm) and 1.2–1.8 mm in diameter for the envelope (mean 1.36 mm). The outer envelope is sticky and collects debris.

DISTRIBUTION
Pseudacris ocularis occurs on the Atlantic Coastal Plain from southeastern Virginia southward throughout peninsular Florida; others are incor-

Distribution of *Pseudacris ocularis*

rect in listing Lake County as the southernmost county (Telford, 1952; Means and Simberloff, 1987). They extend west midway through the Florida Panhandle barely entering Alabama (four locations in Houston County; Jensen, 2004). Records from Mississippi and Texas (Harper, 1939b) are in error. The species has been recorded from Roanoke and Smith islands, North Carolina (*in* Gibbons and Coker, 1978; Gaul and Mitchell, 2007), and Sapelo and Cumberland islands, Georgia (Martof, 1963; Shoop and Ruckdeschel, 2006). Important distributional references include: Range-wide (Harper, 1939b), Alabama (Mount, 1975; Jensen, 2004), Florida (Johnson and McGarrity, 2010), Georgia (Jensen et al., 2008), North Carolina (Dorcas et al., 2007), and Virginia (Tobey, 1985; Mitchell and Reay, 1999).

FOSSIL RECORD
No fossils are known.

SYSTEMATICS AND GEOGRAPHIC VARIATION
Pseudacris ocularis is a member of the Trilling Chorus Frog clade and is most closely related to *P. crucifer* and the *Nigrita* clade of *Pseudacris* in eastern North America, although published phylogenies do not agree about its exact placement (Hedges, 1986; Cocroft, 1994; Da Silva, 1997; Moriarty and Cannatella, 2004).

ADULT HABITAT
Little Grass Frogs are found in a variety of habitats, including low wet pine flatwoods, sand pine plantations, xeric hammock, dome swamps, river hammocks, creek swamps, wet prairies, cypress savannas, cypress- (*Taxodium*) gum (*Nyssa*) ponds and bays, sphagnum bogs, drainage and roadside ditches, and depression and basin marshes (Wright, 1932; Harper, 1939b; Carr, 1940a; Owen, 1996; Enge and Wood, 1998; Smith et al., 2006; Dodd and Barichivich, 2007; Liner at al., 2008). They prefer moist grassy areas around temporary and semipermanent pools and ponds, such as those around cypress dome ponds. Surrounding habitats include sandhills, flatwoods, and mixed pine-deciduous forest.

AQUATIC AND TERRESTRIAL ECOLOGY
This frog normally does not stray far from its breeding sites and resides year-round in shallow grassy wetlands such as cypress domes and other temporary ponds. However, Dodd (1996) reported a single individual 434 m from the nearest water in a sandhills habitat in north Florida. *Pseudacris ocularis* is more arboreal than most other *Pseudacris*, being frequently found in grass, woody shrubs, and small trees at heights of 30 cm or more. It sits on stems and branches diagonally and can rotate its head and neck on a longitudinal access in such a manner as to scan surrounding vegetation. This ability may allow it to leap quickly among vertical grass stems and other vegetation (Harper, 1939b). The species is active year-round, weather permitting. Harper (1939b) observed an individual under a log in a dried cypress pond, and Telford (1952) found a frog under *Sphagnum*, but nothing else is known

Tadpole of *Pseudacris ocularis*. Photo: Kenny Wray

Adult *Pseudacris ocularis*.
Photo: Aubrey Heupel

concerning its terrestrial habits during drought or cold weather.

Little Grass Frogs are photopositive in their phototactic response, suggesting they use both sunlight and moonlight when feeding and going about their daily activities (Jaeger and Hailman, 1973). They are sensitive to light in the blue spectrum ("blue-mode response"), which apparently helps them orient toward areas of increasing illumination (Hailman and Jaeger, 1974). Little Grass Frogs likely have true color vision.

CALLING ACTIVITY AND MATE SELECTION

Males call from elevated clumps of wet grass or from as high as 0.9–1.2 m in shrubs and woody vegetation surrounding a breeding pond or from debris near the surface of the water, such as on sticks, pine needles, grasses, and sedges. Calling and amplexus occur both day and night. The call of *P. ocularis* is a faint, shrill insect-like "ts-r-e-ek" or "s'lick," and calls often blend into the background noise when many species are calling. Harper (1939b) likened it to tinkling beads. The dominant frequency is ca. 7,125 cps (and possibly higher) with a duration of 0.16–0.18 sec and a repetition rate of 50 calls/sec (Blair, 1958a). The high frequency call is complex in structure, and the high pitch may be difficult to discern among some human observers. Calls consist of two parts. The first part of the call rises from 6,700 to 7,125 cps and lasts ca. 0.04 sec; after a very short lapse, 4–5 notes are repeated rapidly at 50 notes/sec with only a slight increase in frequency. When calling, the male's transparent vocal sac expands to the size of a blueberry. Calling occurs at temperatures >10°C, but especially at 21–26°C. Amplexus is axillary, and more than one male may attempt amplexus with a single female (Wright, 1932).

BREEDING SITES

Breeding occurs in mostly in shallow grassy temporary pools and ponds and in flooded fields and wet pine flatwoods. The Little Grass Frog frequently is found in savannas, river swamps, roadside pools, power-line rights-of-way, bogs, and previously cut areas in pine plantations. Only occasionally will it be found in permanent ponds and then only in the shallow grasses around the margin. Breeding sites should be free of fish. In central Florida, Owen (1996) found that 69% of the sites studied had no canopy cover.

REPRODUCTION

Breeding occurs throughout the year, at least in Georgia (Harper, 1939b) and Florida (Carr, 1940a; Owen, 1996; Babbitt and Tanner, 2000). In northern areas, most breeding occurs in winter and spring, particularly from March to April. Eggs are deposited in shallow grassy pools and hatch in 1.75–4 days depending on temperature. Hatchlings are 3.5 mm TL (range 3.32 to 3.6 mm) (Gosner and Rossman, 1960).

LARVAL ECOLOGY

According to Kehr (1997) the mean larval duration from hatching to metamorphosis is only 7.3 days, making this one of the fastest developing species in North America. In contrast, Wright (1932) estimated the larval period as 45–70 days (although he never actually raised tadpoles), and Gosner and Rossman (1960) reported a larval period of 42 days under laboratory conditions. The overall survival rate was 10.3% in a natural pond, with a newly hatched tadpole having a life expectancy of 3.1 days. Survivorship was constant regardless of the tadpole's developmental stage (Kehr, 1997). During the day, tadpoles select the margins of ponds where the water temperature is higher rather than cooler deeper waters. Recently metamorphosed froglets are 7–9 mm SUL (Wright, 1932; Gosner and Rossman, 1960).

DIET
The diet consists of very small soil and leaf litter arthropods, with about 50% consisting of collembolans (springtails). Other prey include ants, parasitic wasps, plant hoppers, beetles, roaches, and walking sticks (Marshall and Camp, 1995). Juveniles eat a wider array of prey, especially mites, than adult males. Tadpoles graze on algae and aquatic vegetation.

PREDATION AND DEFENSE
Cryptic coloration and the ventriloquist nature of the advertisement call make this tiny species difficult to locate. The small size of both larvae and adults makes Little Grass Frogs vulnerable to predators, but their ability to rapidly leap away from danger undoubtedly helps adults escape from predators. It does not leap into the water, but among shoreline vegetation or emergent vegetation within the wetland. It may be that the short larval period, if verified, may allow larvae to escape a predator-rich environment as rapidly as possible. The lateral stripes of the tadpole probably assist in camouflaging it among short grassy stems and pond debris. Many species likely consume adults and larvae, but specific records are not available.

POPULATION BIOLOGY
Nothing is reported in the literature on the abundance and population size of Little Grass Frogs. They can appear quite common during choruses, and Harper (1939b) noted after a heavy rainfall that they could be heard "every rod of the way" across lumber railways in south Georgia. Population size likely varies considerably from one pond and year to the next. For example, Cash (1994) recorded 81 *P. ocularis* one year, but only 8 the next, when hydroperiods were short during 544 days of sampling in southeastern Georgia; Greenberg and Tanner (2005b) recorded only 179 adults and 65 juveniles at 8 ponds in central Florida over a 7 yr period; Dodd (1992) captured 172 *P. ocularis* over a 5+ yr study at a north Florida temporary pond.

Sexual maturity is likely reached the spring following metamorphosis, but nothing is known concerning its demography or longevity.

Breeding habitat of *Pseudacris ocularis*, Putnam County, Florida. Photo: C.K. Dodd Jr.

DISEASES, PARASITES, AND MALFORMATIONS
Pseudacris ocularis is parasitized by the nematode *Falcaustra catesbeiana* (Walton, 1938).

SUSCEPTIBILITY TO POTENTIAL STRESSORS
pH. In one study in Florida, the pH of a natural pond was 6.5–6.7 (Kehr, 1997), but the tolerance limits are not known.

STATUS AND CONSERVATION
Populations are threatened by habitat loss and urbanization, particularly at the northern extent of its range (Pague and Mitchell, 1987). They tolerate some habitat disturbance, such as in former clearcuts and areas modified for improved and semi-improved pastures, especially when canopies are opened (Babbitt and Tanner, 2000; Surdick, 2005). In such areas, they can occupy nearly all available shallow wetlands (Babbitt and Tanner, 2000; Eason and Fauth, 2001). They are found in ditched cypress ponds (Vickers et al., 1985), but reproduction and demographic effects of ditching have not been addressed. They readily recolonize wetlands affected by hurricane overwash (Gunzburger et al., 2010). However, they do not tolerate urbanization and agricultural habitats, where abundance is severely reduced (Delis et al., 1996; Surdick, 2005). Jensen et al. (2008) considered this species common in Georgia.

Pseudacris ornata (Holbrook, 1836)
Ornate Chorus Frog

ETYMOLOGY
ornata: from the Latin word *ornata* meaning 'ornate' or 'decorated.' The name refers to the dorsal coloration of the frog.

NOMENCLATURE
Synonyms: *Chorophilus copii, Chorophilus occidentalis, Chorophilus ornatus, Cystignathus ornatus, Hyla copii, Hyla occidentalis, Hyla ornata, Hyla weberi, Litoria occidentalis, Pseudacris copii, Pseudacris occidentalis, Rana ornata*

IDENTIFICATION
Adults. Ornate Chorus Frogs derive their common name from the attractive dorsal coloration of green, gray, pinkish, red, or chestnut brown providing an ornate appearance. A few individuals may have more than one color. Color patterns are variable, with dark pigment dorsally; this pigment tends to form two dorsolateral longitudinal bars from the neck region to the thighs, although the dark bars may be greatly reduced in some individuals. Bars may be complete or broken. These bars may be bordered by pale cream to yellow halos. A narrow black line is present from the tip of the snout through the eye to the shoulder. The species is more slender, has a more pointed snout, and has longer arms and legs than the somewhat similar *P. streckeri* and *P. illinoensis*. Dark crossbars are present on the forelimbs, femurs, and tibia, but these do not connect ventrally. Venters are cream colored, with the concealed portion of the groin and hind limbs bright yellow. Adults exhibit a definite metachrosis, that is, they quickly change color in relation to their background. On light backgrounds, they become pale and on dark backgrounds, the ground color becomes much darker (Carr, 1940a). Males have dark throats in the breeding season.

Males are only slightly smaller than females. Males from Georgia and Florida average 29.4 mm SUL, whereas females average 31 mm SUL (Blair and Littlejohn, 1960). Jensen et al. (2008) give the range for males as 25–39 mm SUL and for females 28–40 mm SUL. In central Florida, males averaged 33.8 mm SUL (Owen, 1996).

Larvae. Tadpoles are reddish brown with a slender body and a long, acuminate tail. Throats are not pigmented. The tail fin is high with scattered black dots and inserts just behind the eye; the dark tail musculature is bisected by a light gold or brassy stripe, which extends onto the dorsum of the body. Tadpoles reach a maximum length of 23–43 mm TL (Siekmann, 1949; Dundee and Rossman, 1989).

Eggs. The eggs are brown dorsally and cream to white ventrally. There is a single jelly envelope surrounding the egg. The vitellus is 0.95 mm in diameter (range 0.9 to 1 mm), and the envelope is 3.6–4.2 mm in diameter (Livezey and Wright, 1947). The envelope is loose, elastic, and sticky, and debris readily adheres to it.

Distribution of *Pseudacris ornata*

DISTRIBUTION

Ornate Chorus Frogs occur on the Atlantic Coastal Plain from North Carolina through northern Florida (Lake County), and west to the Florida Parishes of Louisiana (Washington and St. Tammany). Populations extend into central Alabama, but not in the Black Belt region. Important distributional references include: Alabama (Mount, 1975), Georgia (Jensen et al., 2008), Louisiana (Dundee and Rossman, 1989), and North Carolina (Dorcas et al., 2007).

FOSSIL RECORD

P. ornata is known from Pleistocene Rancholabrean fossils from Florida and Georgia (Holman, 2003). Identification is made based on characteristics of the ilium.

SYSTEMATICS AND GEOGRAPHIC VARIATION

Pseudacris ornata is a member of the Fat Chorus Frog clade and is most closely related to *P. streckeri* and *P. illinoensis* of the central United States (Hedges, 1986; Cocroft, 1994; Da Silva, 1997; Moriarty and Cannatella, 2004). It is clearly allied with other *Pseudacris*, based on its albumins (Maxson and Wilson, 1975). Within the species, there are three major evolutionary clades: a northern (North Carolina), central (Coastal Plain from central South Carolina to central Georgia), and southern (northern Florida westward) clade. Although corresponding to three different geographic regions, isolation is based on distance across the range rather than geographic barriers (Degner et al., 2010). Historically this species was widespread throughout the southeastern United States, and a combination of genetic drift and migration formed the basis of genetic divergence.

Hybridization with *P. streckeri* can occur in laboratory experiments, although backcrosses are unsuccessful, as hybrids are sterile (Mecham, 1957, 1959). *Pseudacris ornata* hybridizes and produces successful offspring with *P. crucifer, P. brimleyi, P. feriarum, P. brachyphona, P. nigrita,* and *P. brimleyi* in laboratory trials (Mecham, 1965). Many crosses with *P. crucifer*, however, are not successful. Artificial crosses with Gray Treefrogs (*Hyla versicolor* complex) and *Pseudacris clarkii* may occasionally produce metamorphs, but these are probably not viable; crosses with *Acris blanchardi* also are not successful (Littlejohn, 1961a).

A single color morph may dominate one population, whereas other populations contain a mixture of color morphs in various ratios. The basis for the color polymorphism has not been determined.

ADULT HABITAT

Ornate Chorus Frogs are a species of the pinewoods and sandhills of the southeastern Coastal Plain. Carr (1940a) gives the habitat associations as pine flatwoods, mixed hammock, and ruderal. They are associated with a variety of wetland habitats, including marshes, cypress savannas, and cypress (*Taxodium*) and gum (*Nyssa*) ponds (Liner et al., 2008).

TERRESTRIAL ECOLOGY

The Ornate Chorus Frog is fossorial, spending most of its time buried in loose sandy soil. Burrows can be located in open areas or within the moist root systems of grasses and other nonwoody vegetation. Burrowing is accomplished using synchronized movements of the forelimbs; sand so moved may then be distributed behind the frog through movements of the hind limbs (Brown and Means, 1984). Hind limbs also may occasionally be used while burrowing. Frogs excavate shallow depressions, shallow (4–5 cm) sloping pits, deeper (10 cm) tunnels, or cavities beneath the surface with no external openings. In deep tunnels, entrances remain open, and the frog occupies the deepest portion. Feeding likely

Tadpole of *Pseudacris ornata*. Photo: Kenny Wray

occurs at the surface or while underground, and *Pseudacris ornata* may be able to communicate via a short (<0.2 sec) chirp-like subterranean call. About four of these calls may be repeated in succession followed by long intervals without calling. Except in the breeding season, this species is hardly ever observed by day, although they might come to the surface at night to feed.

Pseudacris ornata is found in a variety of habitats including longleaf pine sandhills, basin swamps, and depression marshes (Enge and Wood, 1998). They may occupy burrows in pine barrens, old fields, and savannas at considerable distances from the nearest breeding pond and move over many substrate types to find suitable sand for burrowing. Brown and Means (1984) recorded dispersing juveniles and adults moving hundreds of meters away from breeding ponds, with the greatest distances at 425–480 m. Movements and dispersal occur at night (Todd and Winne, 2006).

Ornate Chorus Frogs are photopositive in their phototactic response, suggesting they use both sunlight and moonlight when feeding and going about their daily activities (Jaeger and Hailman, 1973). They are sensitive to light in the blue spectrum ("blue-mode response"), which apparently helps them orient toward areas of increasing illumination (Hailman and Jaeger, 1974). Ornate Chorus Frogs likely have true color vision.

CALLING ACTIVITY AND MATE SELECTION
Calling is initiated by winter rains, with breeding occurring from November to March (Allen, 1932; Harper, 1937; Carr, 1940a; Mount, 1975). Calls are similar to those of *P. crucifer* but not as musical, with more of a "kik-kik-kik-kik" sound. The dominant frequency is 2,290–2,830 cps, call duration is 0.03–0.06 sec, and the interval between calls is 0.34–0.4 sec (Blair, 1958a; Blair and Littlejohn, 1960; Gerhardt, 1973). Calling normally occurs at air temperatures of 3–27°C, and the number of calls per minute (85–140) increases with temperature; water temperatures, however, may be slightly higher (Harper, 1937; Gerhardt, 1973). Harper (1937) recorded a single individual calling at −3°C. Gerhardt (1973) noted that the longest calling sequence consisted of 91 calls.

Adults migrate to breeding ponds normally only after dark (Pechmann and Semlitsch, 1986; Todd and Winne, 2006). Calling occurs from shallow water, grass clumps along the shoreline, and on surface debris within the wetland. Males call while floating or sitting upright at an angle of 50–60°, balanced on their front fingertips or as they hold onto grass clumps. Even when not calling, the vocal sac remains inflated. Most calling occurs after dusk, although males will call diurnally on cloudy days and during rain showers. Choruses can be heard at a distance of 640 m.

BREEDING SITES
Breeding occurs in shallow temporary pools, cypress ponds, marshy ponds, sinkhole ponds, rain-flooded meadows, Carolina Bays, borrow pits, and wet pine barrens. Wetlands should contain grassy or other nonwoody vegetation for support

Color variation among adult *Pseudacris ornata*. Photo: Justin Oguni

and concealment. Branch and creek swamps are avoided. This species does not occupy wetlands with fish present (e.g., Eason and Fauth, 2001).

REPRODUCTION

The Ornate Chorus Frog is a winter breeding species. In South Carolina and Louisiana, breeding occurs from November to April (Pechmann and Semlitsch, 1986; Dundee and Rossman, 1989; Semlitsch et al., 1996), whereas Jensen et al. (2008) give the breeding season as November to March with peak activity in January and February. In Georgia, eggs are deposited in masses of usually 20–40 eggs/mass (range 21 to 106; mean 40.5) (Seyle and Trauth, 1982) and 6–100 in Louisiana (Dundee and Rossman, 1989). In Alabama, Brown (1956) counted a single mass of 25 eggs. These masses are attached to submerged vegetation (grasses and sedges) in shallow water open to full sunlight. In Georgia, gravid females contained a mean of 443 yolked ovarian follicles (range 214 to 848); the number of follicles was weakly correlated with female SUL (Seyle and Trauth, 1982). Hatching occurs in about one week.

LARVAL ECOLOGY

The larval period is 90–120 days (Dundee and Rossman, 1989; Jensen et al., 2008). Although successful metamorphosis is dependent upon sufficient hydroperiod, an unusually long hydroperiod early in the season is not conducive to larval recruitment. This may be due to excessive predation on larvae by established predators. Likewise, wetlands with short hydroperiods fail to produce metamorphs. It appears that optimal hydroperiods for larval development are intermediate in duration between the extremes (Semlitsch et al., 1996). The eggs of this species could be vulnerable to salamander (*Ambystoma*) predation, but early breeding allows development before salamanders become established. Newly metamorphosed froglets are 11–17 mm SUL.

The temperature at which larvae develop has important effects on growth and development. In laboratory studies, larval mass at metamorphosis and time to metamorphosis decrease with increasing temperature. Survivorship is highest at temperatures of 20–25°C when compared to lower or higher temperatures, and larvae raised at these temperatures have fewer deformities (Harkey and Semlitsch, 1988). Interestingly, the proportion of brown color morphs increases with increasing developmental temperature, but temperature has no effect on the proportion of the gray color morphs. Thus, the fitness and phenotype of the metamorph can be substantially affected by depositing eggs early or late in the season, when larvae are subject to differing thermal regimes.

DIET
Carr (1940a) noted that the diet of newly emerged froglets consisted largely of orthopterous nymphs. Captives eat a wide variety of insects. Tadpoles are grazers on algae and other aquatic vegetation.

PREDATION AND DEFENSE
Ornate Chorus Frogs are difficult to locate when calling, and they quickly stop calling at the approach of an observer. The voice has a ventriloquist effect, making it appear that the male is closer than he actually is. It is likely that the color pattern serves as a disruptive or camouflage function. Predators of larvae include mole salamanders (*Ambystoma*) and many types of predaceous invertebrates. Adults are eaten by the Southern Hognose Snake (*Heterodon simus*) and undoubtedly by other vertebrates.

POPULATION BIOLOGY
Breeding populations of Ornate Chorus Frogs can be large or small. Pechmann and Semlitsch (1986), for example, recorded 78 adults immigrating and 377 emigrating from a 1 ha wetland in South Carolina during six two-day sampling periods from 1982 to 1984. This study was extended to a total of 16 yrs, with 1,592 adults and 19,182 metamorphs recorded; in the last 7 yrs, few metamorphs were captured (Semlitsch et al., 1996). In Georgia, Cash (1994) recorded 62 *Pseudacris ornata* one year and 116 the next.

The number of females that breed at a pond can vary substantially from one year to the next, as can annual recruitment. In South Carolina, relatively large numbers (100 to >300) of females entered a Carolina Bay in 7 of 12 yrs, but substantial recruitment occurred in only 4 yrs, and 3 yrs had no recruitment at all (Pechmann et al., 1991). Indeed, two-thirds of the recruitment over an 8 yr period occurred in only 2 yrs, with as many as 7,345 juveniles produced in a single year (Pechmann et al., 1989). Recruitment failure was largely due to the effects of periodic drought. The size of the breeding population is correlated with the extent of hydroperiod. However, recruitment is correlated with the size of the female breeding population and not directly with hydroperiod (Pechmann et al., 1989).

COMMUNITY ECOLOGY
The Ornate Chorus Frog and the Spring Peeper breed at the same time and frequently use the same breeding site. The calls of these species are superficially similar, and males call as close to

Habitat of *Pseudacris ornata*, Wakulla County, Florida. Photo: C.K. Dodd Jr.

one another as 1 m (Gerhardt, 1973). However, gravid *P. crucifer* will choose male conspecifics over calling *P. ornata* (Gerhardt, 1973). Call discrimination appears to be an important isolating mechanism between these species.

DISEASES, PARASITES, AND MALFORMATIONS
No information is available.

SUSCEPTIBILITY TO POTENTIAL STRESSORS
No information is available.

STATUS AND CONSERVATION
Protection of existing breeding sites is absolutely essential for the preservation of this species. This is especially true since Ornate Chorus Frogs do not appear to colonize ponds constructed to mitigate existing wetland loss (Pechmann et al., 2001). *P. ornata* appears to be declining in many areas, particularly in peninsular Florida and in North Carolina. For example, it used to be abundant in the pinelands around Gainesville but is no longer. Dundee and Rossman (1989) noted that known populations were being eliminated in Louisiana because of urbanization.

The reasons for declines likely include habitat loss and changes in winter precipitation patterns. Ornate Chorus Frogs can persist on previously cut timberlands (e.g., Means and Means, 2005), but persistence is likely tied to the extent of subsurface habitat disturbance during felling and site preparation. Long-term climate change has the potential to seriously affect the distribution of this species. In other areas, such as St. Marks National Wildlife Refuge and the Savannah River Site, populations are still reasonably abundant. They do not readily recolonize wetlands affected by hurricane overwash (Gunzburger et al., 2010). A total of 901 *P. ornata* were collected in Florida for the pet trade from 1990 to 1994 (Enge, 2005a).

Pseudacris regilla (Baird and Girard, 1852) Pacific Treefrog

ETYMOLOGY
regilla: from Latin, meaning 'regal' or 'splendid.'

NOMENCLATURE
Synonyms: *Hyla curta, Hyla nebulosa* [in part], *Hyla regilla, Hyla regilla cascadae, Hyla regilla curta, Hyla regilla deserticola, Hyla regilla hypochondriaca, Hyla regilla laticeps, Hyla regilla pacifica, Hyla regilla palouse, Hyla regilla regilla, Hyla regilla sierrae, Hyla scapularis, Hyla scapularis hypochondriaca, Pseudacris hypochondriaca, Pseudacris pacifica, Pseudacris sierrae*

The nomenclatural history of this species was reviewed by Test (1898). The taxa *P. hypochondriaca* and *P. sierrae* may represent valid species (see Systematics and Geographic Variation).

IDENTIFICATION
Adults. Pacific Treefrogs are small- to medium-sized frogs, with maximum SULs of approximately 51 mm. The ground color of Pacific Treefrogs may be quite variable, from whitish to gray, various shades of tan to brown to black, green, and red. Within these general color schemes, coloration ranges from dark green to grass green, yellowish tan to dark brown, or russet to bright red. Some frogs may even be brown or reddish dorsally, but green laterally. Pattern and color frequencies are different among populations (Test, 1898). Blue color morphs are reported from southern Oregon (that became reddish silver as they grew) (Altig and Brodie, 1968; Nussbaum et al., 1983), and adult albinos and amelanistic frogs are known from Oregon (Jameson and Myers, 1949), Nevada (Wright and Wright, 1949), and California (Brattstrom and Warren, 1955; Resnick and Jameson, 1963; Kay, 1969).

Pacific Treefrogs have a conspicuous dark-colored stripe extending through the eye to the shoulder, with dark triangular, V- or Y-shaped marks on the head. Black markings may or may not be present dorsally on the granular skin, and they are variable in size, shape, and color. Spots may be present in both the lighter and darker color phases, but lighter color phases may not have any spotting patterns. Spotting patterns may change according to background; in a uniform background, spots tend to disappear. In a variegated background, however, spots tend to become more prominent (Brattstrom and Warren, 1955).

Dark bars or blotches occur on the dorsal side of the legs. Venters are unspotted and whitish to pale yellow or cream. Pacific Treefrogs have obvious toe pads on the ends of their digits, and the hind toes are webbed. Like other *Pseudacris*, they possess an intercalary cartilage that provides flexibility to the digits. Sexually mature males have a dark, dusky, or yellowish throat patch.

Pacific Chorus Frogs are polyphenic, that is, they can change shades of color and spot pattern in response to substrate conditions, light, and temperature (Marimon, 1923). A light colored frog placed in a light chamber will immediately begin to lighten, and a light frog at night will darken if placed in a darkened chamber (Marimon, 1923; Brattstrom and Warren, 1955). Temperature causes brown morphs to darken at cooler temperatures and lighten at high temperatures. Green color morphs, in contrast, are darker at both cool and high temperatures but lighten at intermediate temperatures. Intensive illumination causes both color morphs to become yellowish. However, the basic coloration does not change, that is, a green frog will not change its background color to brown under any circumstances.

Females are larger than males. Specific measurements recorded in the literature include males 26.5–32.5 mm SUL (mean 29.2 mm) and females 28.5–36.5 mm SUL (mean 33.6 mm) in Nevada (Banta, 1961); males 29–39 mm SUL (mean 33.4 mm) in California (Perrill, 1984); in Idaho, males range between 28 and 43 mm SUL, whereas females range between 37 and 46 mm SUL (Schaub and Larsen, 1978). Test (1898) provided mean adult SULs (24.5–38.1 mm) for many populations throughout the range of the species. Among-year measurements suggest that the mean SUL at a population may be different from one year to the next.

Larvae. The larvae of Pacific Treefrogs are rather uniformly pigmented dark brown to black, and there is no definite pattern. Some golden flecks occur on the body and tail musculature. The bodies of Pacific treefrog larvae are somewhat ovoid when viewed dorsally. Tail fins are high in relation to the tail musculature, and the tail tip is pointed. The fins are deepest at a point about one-third down the tail. The tail musculature is more pigmented than the tail fins, and the dorsal fin is slightly more mottled than the ventral fin. The eyes are large and laterally placed when viewed from above. Snouts are somewhat pointed when viewed laterally. The intestines are not visible through the ventral body wall. Maximum TL is 53 mm (Logier, 1932), although most are well below this size. Gaudin (1965) provided a description of larval ontogeny and an illustration of a larva.

Jameson and Myers (1957) raised albino larvae through metamorphosis from unpigmented eggs; the metamorphs did not grow or survive through the winter. The larvae appeared sensitive to ambient light, and many died during transformation.

Eggs. Individual eggs are bicolored, olive brown dorsally and yellow to cream ventrally. Eggs are deposited in masses, with each egg surrounded by two gelatinous envelopes. The ovum is 1.2–1.4 mm (mean 1.3 mm); the inner gelatinous envelope is 1.9–2.7 mm (mean 2 mm); and the outer gelatinous envelope is 4.7–6.7 mm (Livezey and Wright, 1947). When viewed individually, the gelatinous envelope appears cuboidal (Gaudin, 1965) rather than round as depicted in Livezey and Wright (1947). An illustration of the egg capsule is in Gaudin (1965).

DISTRIBUTION

Pacific Treefrogs occur from southwestern and central British Columbia (to Mt. Scriven, Bute Inlet, Quesnel, and McBride) south through the tip of Baja California Sur. The range is generally west of the Continental Divide in British Columbia and Montana, although a few populations are known from east of the Continental Divide in Montana. For example, they are very rare at Glacier National Park (Marnell, 1997). They are found from Idaho's Panhandle south throughout much of western, central, and northeastern Nevada. There are scattered populations in the Mojave Desert of southeastern California and southern Nevada (Bradford et al., 2005a), especially along the Colorado River. They have been reported in extreme southwestern Utah along the Muddy and Virgin rivers, at Snowville on the northern border with Idaho, and in the Raft River Mountains of northwestern Utah (Tanner, 1931; Bradford et al., 2005a). Pacific Treefrogs occur at isolated desert springs (Turner and Wauer, 1963; Bradford et al., 2005a) and were heard at Needles, California, in 1954 (Banta, 1961). Vitt and Ohmart (1978) did not locate any specimens along the Colorado River. The Pacific Treefrog occurs on some islands, including the Channel Islands (e.g., Santa Cruz, Santa Rosa, Santa Catalina) (Van Denburgh, 1905; Van Denburgh and Slevin, 1914; Schoenherr et al., 1999), Brooks Island in San Francisco Bay (Schoenherr et al., 1999), Vancouver and adjacent islands (e.g., Gabriola, Newcastle), several islands off the west coast of Baja California (e.g., Cerros), and islands in Puget Sound (Anderson, Bainbridge, Blakely, Camano, Cypress, Decatur, Fidalgo, Fox, Harstine, Indian, Ketron, Lopez, Marrowstone, Maury, McNeil, Orcas, Raft, Saddlebag, San Juan, Shaw, Stretch, Sucia, Vashon, Waldron, Whidbey) (Brown and Slater, 1939; Slater, 1941a). They were introduced into the Queen Charlotte Islands, British Columbia (Graham and Moresby islands) (Reimchen, 1990), and there is an introduced population at Ketchikan, Alaska (Waters et al., 1998). Individuals may turn up well outside their native range, especially in conjunction with horticultural transport (Livo et al., 1998; Brennan and Holycross, 2006).

Important distributional references include Storer (1925), Slevin (1928), Stebbins (1951), and Nussbaum et al. (1983) as well as: Alaska (Waters et al., 1998), Arizona (Brennan and Holycross, 2006), British Columbia (Logier, 1932; Reimchen, 1990; Matsuda et al., 2006), California (Lemm, 2006), Idaho (Slater, 1941b; Tanner, 1941), Mexico (Grismer, 2002), Montana (Black, 1970; Maxwell et al., 2003; Werner et al., 2004), Nevada (Ruthven and Gaige, 1915; Linsdale, 1940; Banta, 1961), Utah (Tanner, 1931), and Washington (Slater, 1955; Metter, 1960; McAllister, 1995; Adams et al., 1998; Adams et al., 1999).

FOSSIL RECORD

Pseudacris regilla is known from a Pleistocene (Rancholabrean) site in Orange County, California (Holman, 2003).

SYSTEMATICS AND GEOGRAPHIC VARIATION

Blair (1958c) placed *Hyla squirella* as the closest relative of *Pseudacris regilla*, based on call structure. The first major range-wide review of the *P. regilla* complex was undertaken by Jameson et al. (1966), based on a discriminant function analysis of ten morphological characters. They identified ten populations of Pacific Treefrogs and suggested they be considered subspecies of *Pseudacris* (then *Hyla*) *regilla*: *cascadae*, *curta*, *deserticola*, *hypochondriaca*, *pacifica*, *palouse*, *regilla*, and *sierrae*. They further placed *Hyla wrightorum* and *H. lafrentzi* in synonymy with *H. regilla*, and designated them as subspecies of *H. regilla*; neither classification is currently accepted. The arrangement suggested by Jameson et al. (1966) is not supported by molecular data, although this early study did identify differences between northern and southern types and between inland and coastal populations.

Although described as *Hyla regilla*, this species was transferred to the genus *Pseudacris* by Hedges (1986), based on an analysis of allozyme phylogeny. Hedges (1986) further recognized the distinctiveness of a western clade of *Pseudacris* that included *P. regilla* and *P. cadaverina*. Cocroft (1994), based on a morphological analysis, suggested that placing the western clade within *Pseudacris* was unnecessary and recommended that *P. regilla* remain in *Hyla*. Da Silva (1997) used additional morphological information to return *regilla* to the genus *Pseudacris*.

The most recent assessment of *Pseudacris* phylogeny is the study by Moriarty and Canna-

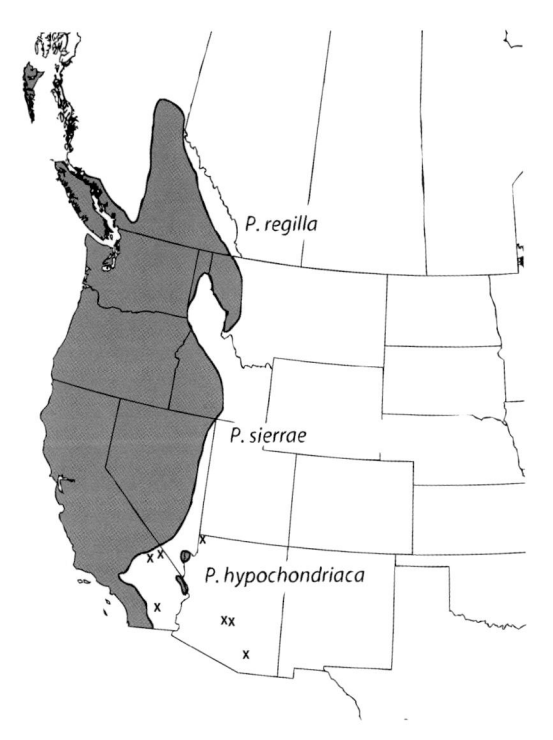

Distribution of *Pseudacris regilla*. The distribution of the three genetic species in the *P. regilla* complex is ill-defined and likely complex.

tella (2004), which examined 12S and 16S mtDNA from 38 populations of chorus frogs from throughout North America. They concluded that *Pseudacris* represented a valid taxon within the family Hylidae and concurred that the Pacific Treefrog was a member of the West Coast Chorus Frog clade, along with *P. cadaverina*. The recognition of a monophyletic West Coast clade is supported by immunological data (Maxson and Wilson, 1975), allozyme data (Hedges, 1986), morphological data (Cocroft, 1994; Da Silva, 1997), and mtDNA data (Moriarty and Cannatella, 2004).

There is a certain degree of genetic differentiation within the taxon currently recognized as *Pseudacris regilla*. Populations are differentiated within Baja California, in particular, where Pacific Chorus Frogs from Baja California Sur are genetically different from those farther north, based on a separation that occurred 0.9–1 mya. There is additional partitioning between populations in the Pacific Northwest, central California, and southern California, based on an analysis using 609 base pairs of the mitochondrial cytochrome *b* gene. Recuero et al. (2006a) suggested recognizing the three northern groups as distinct species: *P. hypochondriaca* (south), *P. regilla* (central), and *P. pacifica* (Pacific Northwest). Recuero et al. (2006b) later corrected the taxonomy to *P. hypochondriaca* (south), *P. sierrae* (central) (Jameson et al., 1966), and *P. regilla* (Pacific Northwest). Recognition of these taxa may be warranted, but more data are needed. Recuero et al.'s (2006a) data were based on a single mitochondrial gene and a relatively small dataset. The geographic sampling failed to cover areas along the California coast with common biogeographic barriers, and the relatively poor geographic sampling coupled with the occurrence of widespread haplotypes across the reported species boundaries suggest that recognizing distinct species is premature (G. Pauly, personal communication).

An indication of additional partitioning is evident in the Pacific Northwest, where coastal and inland populations show substantial variation. Based on an analysis of 725 bp, also of the mitochondrial cytochrome *b* gene, Ripplinger and Wagner (2004) noted that inland populations had different allele frequencies than populations farther west. This type of variation occurs in other Pacific Northwest amphibians and may result from vicariance events during the High Cascade orogeny of the Pliocene, 2.5–3.3 mya. The inland group appears to have much lower genetic diversity than the coastal group, emphasizing its isolation.

There is a considerable amount of color polymorphism among populations of the Pacific Treefrog (Test, 1898). Most Pacific Treefrogs are tan to various shades of gray, brown, or black, and there is much variation in shading and spot patterns. However, some populations have substantial percentages of green and red color morphs. For example, 90% of the frogs at one northwestern Nevada site were various shades of green with black flecks and eye stripes (Weitzel and Panik, 1993), and in the Willamette Valley of Oregon, 81–97% of the frogs have some green coloration (Nussbaum et al., 1983). In contrast, red colors have not been observed in Montana (Werner et al., 2004). Red colors result from a homozygous condition of an autosomal recessive gene, whereas green coloration is determined by genes at two loci, each of which must have at least one dominant gene. The brown color morph is controlled by at least one dominant gene, but the genetics of nongreen or red coloration is likely more complex than the relatively simplicity of red-green dominants and recessives (Resnick and Jameson, 1963).

Color frequencies may be different among populations, sexes, seasons, and years. The frequency of the red and green color morphs certainly varies among some populations (Resnick and Jameson, 1963; Schaub and Larsen, 1978). These frequencies may be stable from one year to the next at any one population, or they may vary. In Idaho, for example, green color morphs made up 40% of the population one year, but 55% the next (Schaub and Larsen, 1978). Shifts in frequencies of color morphs may be sex-related, with female color morphs more stable than those of males. In contrast to stable population frequencies of color morphs, there appear to be differences in the frequencies of color morphs during the activity season at a location, such that green colors are more prevalent in the spring, whereas red colors are more prevalent in the summer and autumn (Jameson and Pequegnat, 1971; Schaub and Larsen, 1978).

In addition to color polymorphism, populations of *P. regilla* differ in morphology, hemoglobin content, metabolism, oxygen consumption, developmental rate, embryonic tolerance to heat, and seasonal respiration rate, depending upon habitat type and elevation (Jameson et al., 1970, 1973; Brown, 1975a). Such variation allows physiological acclimation to a variety of different habitat types and environmental conditions. Such variation, particularly involving metabolic rate and morphology, may have a genetic basis and not be a result of physiological acclimation. Correlations between morphology and various weather conditions (Jameson et al., 1973) also need further investigation in light of more recent genetic analysis.

Hybridization is rare in nature. Brattstrom and Warren (1955) reported a possible *P. regilla* × *P. cadaverina* hybrid from San Diego County, California. Gorman (1960), however, reported extensive hybridization with *P. cadaverina* in Los Angeles County, California. Laboratory attempts at hybridization between *P. regilla* and *P. clarkii*, *Hyla squirella*, or *H. arenicolor* were unsuccessful (Littlejohn, 1961a; Pierce, 1975).

ADULT HABITAT

The Pacific Treefrog inhabits terrestrial areas adjacent to ponds and slow-moving streams, where it forages in marshy vegetation and along wetland margins. It is one of the most common western frogs and occupies a wide variety of habitats. Preferred habitats include low-density forests, chaparral, woodlands, desert oases, and even areas surrounded to some extent by agriculture; nonpreferred habitats include sagebrush, grasslands, dense forests, and areas with intensive human development (Goldberg and Waits, 2009).

Pacific Treefrogs usually are found within the vicinity of wetlands, but they can travel considerable distances. In rocky habitats at nonbreeding times of the year, it occupies moist riparian foliage off the ground, or it is found in rock crevices in canyon walls along a stream corridor (Morafka and Banta, 1976). Temporary and semipermanent ponds are preferred habitat, as they do not contain fish predators. In old-growth forest stands, *Pseudacris regilla* is often associated with ponds, pools, and even lakes, although it occasionally breeds in slow-moving streams (Walls et al., 1992). The species is found from near sea level to at least 2,560 m in Nevada (Linsdale, 1940) and 3,800 m in California (Matthews et al., 2001). However, it tends to be more common at lower elevations.

TERRESTRIAL ECOLOGY

Activity may occur throughout the year depending on environmental conditions (e.g., Morafka and Banta, 1976). Frogs are most likely to be seen during the extended breeding season, but it is not uncommon to encounter them foraging during other times of the year (Weitzel and Panik, 1993). During the nonbreeding season, the frogs disperse up to 1.7 km from the breeding site and hide under woody debris or rocks, in rock crevices or animal burrows, or under other forms of surface debris, as long as there is some moisture present. Other overwintering sites or sites used during unfavorable moisture conditions include debris piles, mammal burrows, arboreal nests of the vole *Arborimus*, crevices, cavities in logs, and under human dwellings (Forsman and Swingle, 2007; McCreary and Pearl, 2010). Pacific Treefrogs often are found communally in cavities and crevices and may use these sites as they disperse through the forest. Brattstrom and Warren (1955) even found 40 Pacific Treefrogs hiding within a horse skull near a stream. Recent metamorphs have been recorded dispersing at least 270 m from a breeding site (Jameson, 1956).

Pacific Treefrogs are familiar with the general features of their habitat. Most frogs remain in the vicinity of their breeding site and are usually recaptured there. However, frogs may move between adjacent ponds, and Schaub and Larsen (1978) recaptured frogs as far away as 400 m from the site of original capture. Movements between ponds often occurred rapidly, such as 230 m in 48 hrs. In a series of experiments, Jameson (1957) displaced adults up to 305 m from their capture site to different ponds and in different directions. While not all returned to the exact same location, a substantial percentage moved back to the vicinity of the point of capture. These displaced frogs often had to cross inhospitable territory to return to the home pond. In a remarkable feat of movement, Crisafulli et al. (2005) reported *P. regilla* to colonize habitats within the erupted crater of Mount St. Helens in Washington, even though the nearest known breeding ponds were 10 km away.

Pacific Treefrogs are active at a variety of temperatures and preferentially seek favorable thermal microhabitats. Brattstrom and Warren (1955) recorded activity in water of 10–12°C, even when air temperatures were 5–10°C. Most frogs regulate their temperature behaviorally, and rarely do body temperatures exceed 24°C. The preferred temperature is about 15°C (Brattstrom, 1963), and the CTmax is 34.8–35.2°C (Claussen, 1973b), although this depends upon acclimation temperature (see Jameson et al., 1973). If cold weather affects a site, they will move to deeper water and a more thermally stable environment. If a site becomes too warm (>20°C), they will move away from the pond to find cooler vegetation or to areas where the water is cooler, such as streams fed by springs or seeps, or hide in cool cracks, crevices, or animal burrows. *P. regilla* lose water slowly and can tolerate ca. 51% loss of body water; rehydration is rapid (Claussen, 1973a).

Much activity of *P. regilla* during the nonbreeding season occurs nocturnally, as nighttime has more favorable temperatures and humidity than daytime conditions. However, recent metamorphs disperse diurnally, and their body temperatures may reach 34°C. The metamorphs control their body temperature by moving to favorable microclimates and avoiding direct solar insolation when too warm, or by moving into sunny areas during cool weather. The CTmax is 36–37.7°C for adults, whereas the CTmin is −1 to −2°C (Cunningham and Mullally, 1956). Pacific Treefrogs are capable of using glucose as a cryoprotectant to survive bouts of freezing weather (Croes and Thomas, 2000). Freezing, especially in the fall, causes an increase in plasma glucose production from glycogen stored in the liver. As the winter season progresses, glycogen stores are depleted, thus reducing the effectiveness of responding to subfreezing temperatures.

Pacific Treefrogs appear to be photopositive in their phototactic response, suggesting they use both sunlight and moonlight when feeding and going about their daily activities (Jaeger and Hailman, 1973). They are sensitive to light in the blue spectrum ("blue-mode response"), which apparently helps them orient toward areas of increasing illumination, such as the open area above a pond (Hailman and Jaeger, 1974). Pacific Treefrogs likely have true color vision.

CALLING ACTIVITY AND MATE SELECTION

Males enter breeding sites as many as several weeks prior to females and may call in substantial choruses. At the beginning of the season, most calling occurs during the day. However, as the season progresses and temperatures rise, most chorusing occurs after dark. Then, males call in full chorus for up to 2 hrs; a second bout of strong chorusing may occur for 2 hrs after sunrise. Chorusing normally occurs at water temperatures from 8 to 19°C, despite reports suggesting calling does not occur below 9.8°C (Brattstrom and Warren, 1955). In fact, calls have been heard at water temperatures as low as 3.8°C (Cunningham and Mullally, 1956; Snyder and Jameson, 1965) and air temperatures of 0.5°C (Schaub and Larsen, 1978). Calling males even have been reported to sing from under ice sheaths or during falling snow. Water temperatures >20°C inhibit chorusing (Brattstrom and Warren, 1955).

Calling often occurs in bouts, that is, a chorus will call for several minutes, be silent for several minutes, then start again. Calls may be heard at any time of the day, but peak chorusing occurs several hours after dark. At least some males are highly territorial and will return to a favored calling site if displaced (Perrill, 1984). Certain call characteristics, such as dominant frequencies, call repetition rates, and call duration vary significantly among populations (Snyder and Jameson, 1965); call characteristics in the south tend to be more variable than those in the north. These differences appear to correlate well with the different phylogenies of the various populations, perhaps lending credence to taxonomic recognition.

The characteristics of the mating call of *P. regilla* have been described by a number of authors (Snyder and Jameson, 1965; Ball and Jameson, 1966; Littlejohn, 1971; Allan, 1973; Straughan, 1975; Rose and Brenowitz, 2002), sometimes drawing different conclusions from one another as to the distinctiveness of the call of this species from that of *P. cadaverina*. Littlejohn (1971) demonstrated, however, that there are differences in call duration, pulse repetition rate, and dominant frequency between these species, especially where they are sympatric. Indeed, Straughan (1975) suggested the pulse repetition rate was the only variable necessary for call discrimination. In *P. regilla*, the call duration averages 237.7 ms, the pulse repetition rate averages 66.2 pulses/sec, and the dominant frequency ranges between 2,350 and 2,650 Hz at 16°C (Littlejohn, 1971).

Male Pacific Treefrogs have both a mate attractant call and an encounter call, that is, a call made upon the approach of a conspecific male. The advertisement call is of two types. In stable choruses in which the number and position of calling males is constant, they produce a diphasic advertisement call. The diphasic call consists of 9–13 pulses followed after 50 ms by 3–4 pulses. The monophasic advertisement call is a longer series of 20–25 pulses uttered at the approach of a female. Since females prefer a monophasic call to a diphasic call (Straughan, 1975), and in contrast to the conclusions of Allan (1973), males switch from a diphasic advertisement call to a monophasic call at the approach of a female.

A second type of call is the encounter call. This is an aggressive call and serves to notify an intruder that he is encroaching upon a resident male (Awbrey, 1978; Whitney, 1980). Both the resident and the intruder may use the encounter call, the resident to warn the intruder and the intruder to notify the resident of his intent to approach. If the intruder does not retreat, a fight (head-butting, wrestling) between males may ensue. One function of the encounter call is thus to maintain spacing, presumably preventing calling males from being too close to one another and thus reducing the chance of attracting a female (Whitney, 1980).

A resident male will issue his encounter call when the intruder's encounter call exceeds a certain amplitude. Presumably the level of call amplitude provides a resident with distance information about the intruder's position within his proximity. This helps to maintain calling male spacing patterns in stable choruses by notifying the intruder not to come closer. However, if chorus size increases, spacing patterns may change. A resident male may then respond differently to an intruder's advertisement call by tolerating a higher amplitude than he would normally tolerate. The aggressive threshold for advertisement calls thus varies with the amplitude of the neighbor's call (Rose and Brenowitz, 1991). Indeed, he may tolerate an amplitude four times that he normally would, a response known as an accommodation. An aggressive response threshold (that is, the amplitude at which the encounter call is made) thus will differ between calling males in a dense population and those in a sparsely distributed chorus (Rose and Brenowitz, 2002). Such accommodation may occur rapidly. Plasticity in aggressive thresholds appears to affect responses to encounter calls much more strongly than responses to advertisement calls (Rose and Brenowitz, 1997).

Pacific Treefrogs cannot use call duration to differentiate advertisement calls from encounter calls; if males did so, the advertisement call would sound similar to the encounter call, especially since both types of calls are uttered at the same rate of about 100 cps. Females also prefer the advertisement call to a male's encounter call (Brenowitz and Rose, 1999). Therefore, the plasticity in male accommodation (that is, changing the threshold for aggression) allows males to maximize their time in producing advertisement calls and minimize time spent in aggressive encounters, especially in response to changes in male spacing patterns at a chorus.

Males tend to space themselves at regular intervals around a breeding site. In British Columbia, Whitney (1980) recorded regular spacing by calling males about 50 cm apart, but males may call from 20 to 30 cm of one another when the density of calling males at a breeding site is greater. If additional males are added to a well-established chorus, fewer males call than in a control situation (Whitney and Krebs, 1975b). This suggests that there may be a maximum density of chorusing males at a breeding site, at least for established choruses. If other males encroach within 50 cm, resident males issue the encounter call.

Females approach a calling male, almost to the point of touching him, before he ceases calling and attempts amplexus. Females tend to prefer males that initiate calling bouts (Whitney and Krebs, 1975a) and that are large (Benard, 2007). Certain males not only initiate calling bouts, but they tend to call longest within a calling bout; they also call at a faster rate, are louder, and they are more likely to call outside the bout when other males are not calling. These cues provide a receptive female with extensive information about the fitness of the calling males. The frog that can expend the most energy on calling is likely the most successful male. However, the risk of predation is substantial during the mating season, and smaller males have a better chance of surviving attacks by giant water bugs than larger males (Benard, 2007). Thus, there is a trade-off between female choice and predation risk with regard to male body size. Male body condition is not correlated with either mating success or predation risk (Benard, 2007).

Males appear to adopt one of three strategies at a breeding site: become a calling male in a fixed location within the chorus; become a satellite male; become an opportunist and switch back and forth between calling and satellite status. Up to 17% of males at a chorus may be satellite males, that is, they do not call but position themselves near a calling male. In this way, they may be able to intercept a female moving toward the calling male, and satellites are occasionally successful (Perrill, 1984).

Satellite males may become calling males, and these males may switch back and forth from calling to satellite from one night to the next (Perrill, 1984). These opportunistic males also may be successful in obtaining mates. Calling males may see satellite males as females and attempt to amplex them (Whitney, 1981). If a male is clasped by another male, he will give a trilling call that serves to identify his sex and causes the clasping male to release him (Brattstrom and Warren, 1955; Allan, 1973).

Male Pacific Treefrogs sometimes call during off-season rains or at other times during the nonbreeding season. Calls may be uttered from seasonal retreat sites when the frogs are nowhere near their normal breeding sites. Brattstrom and Warren (1955) observed males in January calling from small holes. The frogs would sing a few notes, then shimmy back down the hole to safety, only to emerge and call a few minutes later. The holes were about 50 cm in depth, and were sometimes occupied by a Common Side-blotched Lizard (*Uta stansburiana*). The purpose of these calls, if any, is unknown.

BREEDING SITES

Pacific Chorus Frogs breed in slow-moving streams (<5 cm/sec) and in small quiet ponds. Streams may be permanent, intermittent, or seasonally flooded, as treefrogs will occupy isolated pools and oxbows as a stream dries during the summer. The eggs are deposited in pools and in backwater sections of these streams. They also breed in shallow-water ponds with emergent vegetation, especially those with high percentages of silt in near-shore habitat (Munger et al., 1998; Matthews et al., 2001; Goldberg and Waits, 2009). Ponds may be temporary, such as overflow ponds adjacent to streams, or permanent as long as fish are not present. This species prefers breeding sites with an open canopy that are somewhat isolated from other breeding sites. They have been recorded from snowmelt alpine pools to tidal pools near the coast. In times of poor snowfall or excessive drought, snowmelt pools and small streams may not fill, causing reproductive failure. *Pseudacris regilla* also breeds in man-made wetlands, such as roadside ditches, irrigation canals, and stormwater retention ponds (Ostergaard et al., 2008). Breeding occurs in the shallow-water portions of these pools where vegetation cover is substantial.

As with most frog species, breeding does not take place in every potential wetland in every year. In Idaho, for example, Goldberg and Waits (2009) estimated that 28–32% of the wetlands they surveyed were used as breeding sites. In the Puget Sound region, Richter and Azous (1995) found *P. regilla* in 12 of 19 (63%) wetlands surveyed.

REPRODUCTION

Breeding occurs from winter through late summer, the initiation of which depends upon latitude, elevation, and other environmental conditions. In some areas near the coast, Pacific Treefrogs call as early as November, whereas in other areas, breeding continues until mid-August or later (Myers, 1930a; Jameson, 1957; Nussbaum et al., 1983; Jones et al., 2005). The breeding season may be quite extended over a period of months at any one site, but the majority of breeding may occur in a span of only a few weeks. In southern California, for example, breeding begins in January and continues until May, although most breeding occurs from February to April. In western Washington, breeding occurs from February to late May, with most reproduction taking place in March and April (Brown, 1975a). Other records include late February to early May (Nevada: Weitzel and Panik, 1993), February to July (British Columbia: Carl, 1943), early March to mid-June (Washington: Karlstrom, 1986), early April to mid-June (Idaho: Linder and Fichter, 1977; Schaub and Larsen, 1978), and May to July (Montana: Black, 1970).

Although amplexus may occur at 8°C (Cunningham and Mullally, 1956), actual egg deposition only occurs when water temperatures are >12°C and <15°C (Brattstrom and Warren, 1955). Pacific Treefrogs go about their breeding activities as long as water temperatures are above 8–12°C, despite considerably colder air temperatures, even below freezing. They are able to swim slowly in water at 0°C, although at 2°C they are helpless on land (Cunningham and Mullally, 1956). Tolerance to cold temperatures allows activity despite the sudden onset of cold weather during the winter–spring breeding season.

The amount of time an individual remains at a breeding site is quite variable. Nussbaum et al. (1983) stated that males remain 1–90 days at a breeding site (mean 33), whereas females stay a much shorter period (1–27 days, mean 9.6). Females enter the breeding sites when they are ready to deposit their eggs. In the morning, almost all frogs are male; as the day continues, however, females increasingly enter the pond in time for the main period of reproduction, which occurs after sundown.

Males will amplex females from 4 to perhaps 10 hrs prior to oviposition. Amplexus is axillary. The male holds tightly onto the female with his rear legs folded into a sitting position. Prior to deposition, the male extends his rear legs and brings his cloaca into position with that of the female. Eggs are fertilized as they are extruded in a cloud of sperm, after which the male refolds his limbs in the sitting position. As the eggs are extruded, the female brings them into contact with the substrate to which they are attached. Eggs are positioned by the female using her hind limbs, and any sticky eggs around her cloaca are removed using a slow extension of her hind feet. Intervals between deposition of eggs range from 2 to 10 min, during which time the female may be very active. The entire oviposition process lasts 8–40 hrs. The oviposition sequence has been described by Smith (1940).

Oviposition has been recorded from December to September (Jones et al., 2005). Eggs are deposited in soft masses attached to any kind of vegetative debris, either submerged or at the surface. Clutch size is variable, but as many as 1,250 eggs may be deposited by a single female during a single amplectant bout; a mean of 21.5 eggs was recorded for the 58 clusters examined (Storer, 1925). Smith (1940) stated there were 500–750 eggs deposited over the course of a single amplectant bout, with a mean of 16 (range 5 to 60) eggs per clutch (mass). Werner et al. (2004) reported masses of from 18 to 120 eggs; Jones et al. (2005) recorded 10–80 eggs per cluster; Nussbaum et al. (1983) reported 9–70 eggs per mass (mean 25). As oviposition nears completion, clutch size (eggs/mass) decreases. Time-to-hatching records are quite variable, from two to three days in the laboratory to two to three weeks (Storer, 1925; Brown, 1975a; Jones et al., 2005), three to five weeks (Nussbaum et al., 1983), or one to five weeks (Lemm, 2006) under natural conditions. Hatchlings are 6–8 mm TL (Nussbaum et al., 1983).

Pseudacris regilla in oviposition. Note the brown and reddish color patterns and eggs. Photo: Justin Garwood

Embryonic Pacific Treefrogs may experience very different water temperatures during development, depending upon latitude and elevation. According to Schechtman and Olson (1941), eggs can survive at temperatures of −7°C for 2 hours, and up to 38°C depending on location and stage of development. Survival does not equate with normal development, however. For example, in Washington, normal development occurs at ambient water temperatures of 6–20°C, and normal embryonic temperature tolerance is 8–27°C. In warmer California, embryonic temperature tolerance is slightly greater, from 8 to 29°C. The lowest temperature permitting normal development is 6°C. Embryos developing in southern California at temperatures of 29–31.5°C experience an inhibition of developmental rate at these higher temperatures, although they may be somewhat more heat resistant than larvae from cooler environments. For example, water temperatures in Washington >28°C result in substantial embryonic mortality (mean 85%, range 70 to 97%), whereas embryos developing at 29°C in California experience normal development with hatching successes of 80–100% (Brown, 1975a).

LARVAL ECOLOGY

Larvae are herbivorous, feeding particularly on filamentous green algae. In studies of growth rates, Kupferberg et al. (1994) found that larvae grew fastest on a diet of *Cladophora* (a filamentous green alga that has diatom epiphytes). Larvae grew at lower rates on filamentous green algae that do not have diatom epiphytes (e.g., *Mougeotia*), and slowest on flocculent detritus and ambient seston. Larvae fed other algae (*Zygnema*, *Oedogonium*) had between 40–53% survivorship compared with 85% for larvae fed *Cladophora* with epiphytes. Since body mass and length of the larval period are inversely correlated, a high-quality diet reduces the amount of time spent as a tadpole. Kupferberg et al. (1994) suggested that alteration of streams by human activities could affect the resulting algal communities, which in turn affects the quality of the food available to anuran larvae.

Larvae prefer the deeper water (to 1.2 m) of ponds and pond outflows to shallow water. Larvae can tolerate water to 34°C without acclimation (Brown, 1969), and they are normally exposed to a wide range of temperatures daily and seasonally (Cunningham and Mullally, 1956). In one extreme, larvae have been recorded in the desert in somewhat saline water at 33.4°C (Brues, 1932). Acclimation can raise the lethal temperature to 36°C. However, compared with desert and tropical species, larval *Pseudacris regilla* are not particularly heat resistant or tolerant.

Aggregations of Pacific Treefrog larvae have been reported to form intermittently, such as Brattstrom and Warren's (1955) observations of an aggregation of 150–180 tadpoles swimming in a circle about 61 cm in diameter. Most tadpoles faced in the same direction in water 7.5 cm in depth. Larvae do not show a preference for kin in forming an aggregation (O'Hara and Blaustein, 1988). Such aggregations may serve thermal, feeding, or defensive functions.

The larval period is 2–3 months (Jones et al., 2005), and tadpoles may reach 45–55 mm TL (Nussbaum et al., 1983). At metamorphosis, young Pacific Treefrogs are 11.2–16.5 mm SUL (Stebbins, 1951; Jameson, 1956; Gaudin, 1965).

DIET

Larval Pacific Treefrogs are omnivorous, feeding on filamentous algae and associated diatom epiphytes and bacteria suspended in the water. Adult Pacific Treefrogs eat a variety of small invertebrates, including beetles (larvae and adults), midges, tabanid flies, crane flies, true bugs, leafhoppers, ants, parasitic wasps, ichneumons, spiders, mites, pillbugs, isopods, snails, dragonfly naiads, and adult and larval flies (Needham, 1924; Tanner, 1931; Brattstrom and Warren, 1955; Johnson and Bury, 1965). Feeding occurs above water, and these small treefrogs may climb 1.2 m into bushes and branches as they forage. They also make use of floating vegetation as platforms for ambush, and will sit motionless in the shade of vegetation until an insect lands nearby.

PREDATION AND DEFENSE

The eggs and larvae of Pacific Treefrogs are palatable and do not appear to have any chemical defenses (Licht, 1969a). They are preyed upon by backswimmers, predaceous water bugs, and salamander larvae (*Ambystoma gracile*, *Taricha granulosa*) (Peterson and Blaustein, 1991). It is not surprising, therefore, that this species generally avoids breeding in permanent ponds, especially those with fish. However, larvae tend to

Tadpoles of *Pseudacris regilla*. Photo: Breck Bartholomew

form aggregations in response to the presence of chemicals either emanating from potential garter snake predators or from conspecifics attacked by garter snakes (DeVito et al., 1999). Presumably these aggregations minimize the potential for any one tadpole to be eaten by the snake. Once metamorphosed, however, juveniles do not form aggregations as do some toads, regardless of whether a predator is present.

Larval *Pseudacris regilla* tend to avoid areas where conspecifics and even other anuran larvae have been attacked by predators, such as predaceous diving beetles (*Dytiscus*). They presumably are able to detect larval damage-release chemicals in the water. Adams and Claeson (1998) showed that larval Pacific Treefrogs avoided traps where damaged larvae were present and that it made no difference whether the injured larvae were conspecifics or *Rana*.

Adult and juvenile Pacific Treefrogs are cryptic on many types of substrates, especially blending with green and tan backgrounds in spring and with autumnal reds and russets in the summer and fall. Color-morph frequencies change with seasonal changes in background color, suggesting selection toward seasonal substrate matching. The dorsal patterns of lines and spots further aid in crypsis. When disturbed by movement, Pacific Treefrogs will cease calling or jump away from a potential predator. Dill (1977) suggested a slightly leftward bias in jumping preference, with frogs generally jumping within a 70° arc from the frog's initial bearing. Defensive behavior also may involve remaining motionless so as not to draw the attention of a predator. An adult was reported to become immobile in response to being poked while sitting on a cactus (Banta, 1974), and Brattstrom and Warren (1955) noted that if an adult was tossed into the water and landed on its back, it would remain motionless, with its limbs folded into the body, and float.

A great many predators probably prey on Pacific Treefrogs, including invertebrates, snakes, birds, and mammals. Specific references include giant water bugs (Benard, 2007), American Bullfrogs (*Lithobates catesbeianus*), Belted Kingfishers (*Megaceryle alcyon*), herons, egrets, Mallards (*Anas platyrhynchos*), garter snakes (*Thamnophis elegans, T. sirtalis*), and raccoons, skunks, opossums, river otters, domestic house cats (*Felis sylvestris*), and dogs (*Canis familiaris*) (Storer, 1925; Arnold and Wassersug, 1978; Nussbaum et al., 1983; Jennings et al., 1992; Weitzel and Panik, 1993; Werner et al., 2004; Rombough and Bradley, 2010). In addition to native predators, introduced mosquitofish (*Gambusia affinis*) readily consume Pacific Treefrog larvae and may have an effect on stream-dwelling populations (Goodsell and Kats, 1999). Larvae also are eaten by predaceous aquatic insects and other anurans (*Rana muscosa*) (Pope, 1999).

POPULATION BIOLOGY

Adult population size varies from a few dozen to several hundreds individuals at a breeding site. For example, Weitzel and Panik (1993) estimated an adult annual breeding population of 60 at a small pond in northwestern Nevada, and Schaub and Larsen (1978) estimated the size of the male breeding population at study sites in Idaho as 360 one year and 160 the next. Jameson (1957) estimated there were 484 adult males at a pond in Oregon in April. At a pond in California, Jameson and Pequegnat (1971) estimated a population size of about 322 frogs in the first year (range 245 to 435 as the season progressed) and 366 during the second year (313–435 as the season progressed) of a 2 yr study.

Most *Pseudacris regilla* mate only once in their lifetime, but a small percentage return to a breeding site in more than one year. Jameson and Pequegnat (1971) estimated an annual mortality rate of 90%, making this species essentially an annual species with a short life span (three years or less, Jameson, 1956). A total of 38 of 373 (10.2%) marked frogs were recaptured one year later at Jameson's (1957) Oregon site, and Schaub and Larsen (1978) recaptured 13.8% of adults marked one year during the following year. Jones et al. (2005) reported that most Pacific Treefrogs do not reach maturity until their second year (presumably the first spring after metamorphosis).

Males probably remain for days to weeks at a breeding site, although not perhaps throughout the entire breeding season. In Idaho, Schaub and Larson (1978) recorded male stays from an average of 10 to 13 days, depending upon year surveyed. *Pseudacris regilla* males may mate with multiple females, and Perrill (1984) recorded successful amplexus by males on consecutive nights. Even after mating with several females, males will continue calling throughout the peak breeding season. Females may remain only as long as it takes to oviposit her complement of eggs. The size of the male does not appear to influence mating success, and there is no size-assortive mating between males and females. This might be expected of a species that normally has only one chance at reproduction and for which size differences are not very great between males and females.

For those frogs that do breed more than once, subsequent breeding make take place at a different pond than the pond originally selected (Storer, 1925). Since the availability of breeding sites varies annually, a lack of site philopatry allows the frogs to breed at whatever sites are available. It seems probable that long-term persistence at any one location is precarious, as environmental perturbations could easily eliminate populations, at least temporarily. A cycle of persistence for a number of years followed by temporary extirpation appears to have occurred at an isolated population in northwestern Nevada, where successful reproduction occurred during 12 of 15 years the population was studied (Weitzel and Panik, 1993). In the other three years, the population was wiped out by severe flooding, only to be rapidly recolonized. Indeed, the cycle of successful reproduction-extirpation appears to have been repeated frequently over a period of 84 years due to flooding, desiccation, and extreme fluctuations in water temperatures.

The red-green color frequencies of the metamorphs do not match the color frequencies of either the parental population or the population that breeds the following year. These results suggest that polymorphism is maintained by differential selection among color types and that this selection pressure varies seasonally and by location (Jameson and Pequegnat, 1971), perhaps in much the way color morph frequencies vary in Boreal Chorus Frogs. Jameson and Pequegnat (1971) further speculated that the plasticity in color-morph frequencies among and within populations allowed adaptation to rapidly changing environmental conditions.

COMMUNITY ECOLOGY

Pacific and California Treefrogs occupy the same general regions and stream drainages, but tend to partition the habitat according to specific microenvironmental preferences (e.g., along slow-moving vs. fast-moving waters). However, as drought conditions cause habitats to dry in summer, individuals of these two species may be brought into close contact. Despite this, there is little evidence of hybridization between them, even though they may simultaneously use the same waters for reproduction. Despite assertions in Ball and Jameson (1966), differences in call characteristics provide a premating isolating mechanism to prevent introgression between these closely related taxa.

The presence of Pacific Treefrogs appears to be important to the distribution of garter snakes, particularly *Thamnophis elegans*. Indeed the presence of *Pseudacris regilla* is a major indicator of the presence of this snake (Knapp, 2005). Anurans are a major portion of the diet of *T. elegans*, and snakes appear to congregate along streams and ponds when metamorph Pacific Treefrogs are beginning dispersal (Arnold and Wassersug, 1978). The distribution of and declines in garter snake populations may be correlated with the presence or absence of *P. regilla* populations.

Garter snakes also may have a secondary effect on the activity and behavior of larval Pacific Treefrogs. When given a choice, larvae feed preferentially on the green alga *Cladophora*, which may contain diatom epiphytes, rather than lower nutritional quality filamentous algae such as *Mougeotia*, which does not contain epiphytes. However, larvae switch to *Mougeotia* in the presence of garter snakes. This is because *Mougeotia* forms a floating cloud-like structure with numerous hiding locations, whereas *Cladophora* forms a much more exposed floating mat. If larvae are forced to remain in the *Mougeotia*, their growth rate declines and the larval period may increase because of the decrease in available nutritional resources (Kupferberg, 1997b).

Larval life history characteristics can also be affected is by the presence of other anuran larvae.

In a California river, for example, the presence of American Bullfrog larvae decreased the size at metamorphosis of Pacific Treefrog larvae by 16% (Kupferberg, 1997a, 1997b). Kupferberg (1997a) attributed the effect to competition for algae between these species. The decreased size at metamorphosis in turn could have adverse impacts on this species, as fitness may be a function of size at metamorphosis. Thus, even though tadpole survivorship was not affected, larvae of the nonindigenous American Bullfrog could have long-term indirect detrimental impacts on populations of Pacific Treefrogs.

Despite the results from Kupferberg et al. (1994) and Kupferberg (1997b), Adams (2000) and Pearl et al. (2005b) could not demonstrate any direct effects of American Bullfrogs on Pacific Treefrog survivorship or presence in the Pacific Northwest. They noted that Pacific Treefrogs were virtually absent from permanent ponds, but attributed their absence to the presence of predaceous exotic sunfish in the permanent habitats. Adams (2000) suggested that gradients in habitat characteristics may be more important in assessing the effects of nonindigenous species on distributional patterns than broad-scale characterizations. Still, indirect effects may play a critical role by altering food availability, access, and behavior, as noted by Kupferberg (1997b).

Many species of anuran larvae are able to detect the presence of potential predators within their environment. Pacific Treefrog larvae, for example, can detect chemicals emanating from the native Redside Shiner (*Richardsonius balteatus*) and reduce their activity accordingly. They are also able to detect chemicals from injured hetero- and conspecifics and avoid them, presumably reducing their chance of a predatory encounter (Adams and Claeson, 1998). However, they do not respond to chemicals from nonnative fish (*Lepomis*) or crayfish (*Procambarus*) (Pearl et al., 2003). This inability to detect nonindigenous predators exposes them to predation and may help to limit their distribution to temporary wetlands. It also helps explain why the introduction of nonindigenous species has a detrimental effect on the treefrogs.

The presence of grazing *Pseudacris regilla* tadpoles may have effects on the periphyton community within a wetland, at least for the duration of the developmental period. In field experiments, Kupferberg (1997c) demonstrated that larval Pacific Treefrogs could cause as much as an 18% decrease in area-specific primary production due to grazing on periphyton. Grazing had no effect on periphyton biomass-specific productivity. Community effects were based on consumption, rather than through the recycling of nutrients.

DISEASES, PARASITES, AND MALFORMATIONS

A few morbidity events have been recorded for the Pacific Treefrog. Green et al. (2002) recorded one event in Oregon involving about 5% of the observed larvae of Pacific Treefrogs and Western Toads that had hypopigmentation and gigantism. Another morbidity event involved dermal metacercariae and possibly amphibian chytrid fungus affecting four larval species. In neither report is it possible to discern whether these abnormalities or pathogens affected all the species listed from the particular event. The amphibian fungal pathogen *Batrachochytrium dendrobatidis* was found in museum specimens collected from California from the 1970s to the 2000s (Padgett-Flohr and Hopkins, 2009). Larvae experimentally infected with amphibian chytrid fungus show no evidence of behavioral fever or altered thermoregulation (Han et al., 2008).

Parasites include the trematodes *Alaria* sp., *A. mustelae*, *Clinosternum* sp., *Distoichometra bufonis*, and *Megalodiscus microphagus* (Goldberg and Bursey, 2001b, 2002a; Zamparo and Brooks, 2005); the cestode *Distoichometra bufonis* (Koller and Gaudin, 1977); the nematodes *Cosmocercoides variabilis*, *Oswaldocruzia pipiens*, *Physaloptera* sp., *Rhabdias* sp., and *R. ranae* (Koller and Gaudin, 1977; Goldberg and Bursey, 2001b, 2002a); and the acathocephalid *Centrorhynchus californicus* (Millzner, 1924).

Limb malformations have been reported commonly in Pacific Treefrogs. Both field and laboratory studies have linked these malformations with interference in development by the digenetic trematode *Ribeiroia ondatrae* (Sessions and Ruth, 1990; Johnson et al., 1999, 2002; Green et al. 2002; Blaustein and Johnson, 2003; Johnson and Sutherland, 2003; Lannoo, 2008). Malformations include missing limbs (ectromely) and digits (ectrodactyly), extra digits (polydactyly) and limbs (polymely), femoral projections, bony triangles (taumely), missing eyes (anophthalmy), abnormal jaws (mandibular hypoplasia), un-

Adult *Pseudacris regilla*.
Photo: Aubrey Heupel

usual skin webbings, and others. Hind limbs are affected far more often than forelimbs. In heavy experimental infestations, nearly three abnormalities per abnormal frog were recorded (Johnson et al., 1999). Other reports of malformations are found in Hebard and Brunson (1963), Miller (1968), Reynolds and Stephens (1984), and Lannoo (2008).

SUSCEPTIBILITY TO POTENTIAL STRESSORS

Metals. Mercury has been found in *Pseudacris regilla* from California at concentrations of 0.063–0.22 μg/g wet weight (mean 0.12) (Hothem et al., 2010).

pH. Acidification does not appear to have affected the distribution of *P. regilla* in the Sierra Nevada Mountains of California (Bradford et al., 1994a).

Nitrates, nitrites, and ammonia. Prolonged exposure to ammonium and nitrate compounds from agricultural runoff could have detrimental effects on *P. regilla* prior to metamorphosis. Mortality of *P. regilla* larvae is complete at concentrations of 15,000 mg/L N after a 10-day exposure to urea. Little mortality occurs at concentrations of 6,000 mg/L N or less (Schuytema and Nebeker, 1999a). The $LC_{50\,(10\,day)}$ was estimated at 7,105 to 9,921 mg/L N for Pacific Treefrog larvae. For embryos, the $LC_{50\,(10\,day)}$ ranged between 25 and 32.4 mg/L NH_4-N, whereas the $LC_{50\,(10\,day)}$ for sodium nitrate was 578 mg/L NO_3-N (Schuytema and Nebeker, 1999b). Ammonium sulfate, another component of fertilizers, also has detrimental effects on larval growth rates and body weight, but not so much on survival; larvae are more sensitive to ammonium sulfate during the early stages of larval development than at later stages (Nebeker and Schuytema, 2000). The median lethal concentration of nitrite for larvae is 1.23 mg $N-NO_2$/L at 15 days, 3.6 at 7 days and >5.5 at 4 days (Marco et al., 1999). These results suggest that there may be sublethal effects on Pacific Treefrog larvae from ammonium fertilizers if exposure occurs throughout the developmental period. *Pseudacris regilla* larvae are not particularly sensitive to the effects of nitrates alone, however.

Chemicals. Pesticides pose a serious threat to Pacific Treefrog populations. Cholinesterase activity in tadpoles is depressed in populations exposed to organophosphorus pesticides (e.g., malathion, chlorpyrifos, diazinon), as these pesticides bind with cholinesterase to inhibit neural functioning. In the Sierras of California, east of areas with heavy pesticide use as well as in areas downwind of the agricultural San Joaquin Valley, cholinesterase of Pacific Treefrogs is particularly depressed. In such areas, >50% of the population had organophosphorus pesticide residues, with concentrations as high as 190 ppb (Sparling et al., 2001).

Chlorpyrifos (to 17.4 ppb in larvae) and chlorothalonil (to 47.7 ppb in larvae and 4.5 ppb in eggs) have been found in *P. regilla* from natural populations in California (Datta et al., 1998). Larvae experimentally exposed to field concentrations of chlorpyrifos, methyl parathion, temephos, fenthion, and malathion experienced a decreased thermal tolerance compared with controls. Malathion was least toxic, whereas chlorpyrifos and methyl parathion were the most active toxicants (Johnson, 1980). Methyl parathion and malathion also resulted in depressed activity in comparison with controls. Endosulfan causes slight decreases in larval survival, larval deformities (kinked tails, loss of pigmentation), and changes in the behavior of survivors at environmentally relevant doses (Westman et al., 2010); it does not affect egg development, however. At concentrations > 50 ng/L, azinphosmethyl had decreased survivorship, but no changes in behavior were evident (Westman et al., 2010).

Heavy concentrations of noncholinesterase-inhibiting pesticides have been found in the agricultural areas of California. These include endosulfan (in as many as 86% of the frogs sampled at some populations), 4,4′-dichlorodiphenyl-dichloroethylene, 4,4′-DDT, and 2,4′-DDT (DDTs in as many as 40% of the frogs sampled), and α- and γ-hexachlorocyclohexane in their tissues (Sparling et al., 2001). Organochlorine pesticide residues (PCBs, DDE, DDT, toxaphene) have been recorded in eggs, tadpoles, and adults from natural populations in California (Datta et al., 1998; Sparling et al., 2001; Angermann et al., 2002). Concentrations ranged as high as 258 ppb DDE and 229 ppb PCBs on the campus of the University of California at Davis (Datta et al., 1998). Toxicant levels are much higher at low elevations in the Sierras than at high elevations, but the presence of these toxicants at high elevations at long distances from their point of origin suggests significant transport on wind and air currents.

In a series of transplant experiments, Cowman (2005) examined pesticide uptake at three high elevation national parks in California. She found that DDE was present in 97% of her samples from Yosemite, 84% from Sequoia, and 15% from Lassen Volcanic national parks. Total endosulfans were detected in 3% of the Sequoia frogs, 9% of the Lassen Volcanic frogs, and 24% of the Yosemite frogs. Organophosphorus pesticides were not detected. Contaminants were accumulated directly as the tadpoles fed. Cowman (2005) further demonstrated chromosomal breakage in juvenile Pacific Treefrogs, mostly at Yosemite and Sequoia, which are closer to areas of high agricultural contamination than is Lassen Volcanic. Sublethal effects of pesticide exposure included shorter SULs, longer developmental periods, and lower survivorship at metamorphosis, especially at Sequoia National Park, than at the control site. Aerially transported pesticides likely have had severe impacts on frog populations in the Sierra Nevada Mountains.

Other contaminants include the organophosphorus pesticides Guthion (lethal to tadpoles at 8.7–9.7 mg/L) and Guthion 2S (lethal to tadpoles at 1.4–1.5 mg/L) (Schuytema et al., 1995). Guthion affects total length, hind limb length, and mass at 3.6 mg/L. Guthion 2S is significantly more toxic than Guthion and has greater adverse effects on larval growth at lower concentrations than Guthion (Nebeker et al., 1998). The herbicide Roundup® is toxic to larvae, with an $LC_{50\ (24\ hrs)}$ of 0.43 mg/L (King and Wagner, 2010). The LC_{50} decreases at 15 days to 1.30 mg/L.

UV light. Eggs and larvae exposed to ambient UVB radiation, as well as enhanced UVB radiation 15 and 30% above the ambient level, showed no differences from controls in terms of hatching success. Enhanced levels of UVB radiation reduced larval survivorship by 18.4%, but controls and ambient levels of UVB radiation had no effect on larval survivorship (Ovaska et al., 1997). Other studies have found similar results regarding development and hatching success (Blaustein et al., 1994b; Anzalone et al., 1998; Vredenburg et al., 2010a) and together suggest that *P. regilla* is relatively tolerant of UVB radiation. However, UVB may have effects on larvae when in combination with other sublethal stressors, such as nitrate. For example, larval mass is reduced when these stressors are combined in field experiments at low elevations; at high elevations, the combination reduces larval survivorship (Hatch and Blaustein, 2003). Neither of the stressors had an effect on larvae by themselves.

Larval *P. regilla* do not avoid areas of high UV light either in the field or under experimental conditions. Instead, they tend to choose warmer water regardless of UV light levels (Bancroft et al.,

2008). Eggs of Pacific Treefrogs have a relatively high content of photolyase, an enzyme known to be important in repairing UVB radiation damage to DNA (Blaustein et al., 1994b).

STATUS AND CONSERVATION

Like all species of North American Anura, Pacific Treefrogs have disappeared from wetlands lost to development, agriculture, drainage, or habitat modification. In urban areas, remnant treefrog populations are particularly vulnerable, especially in semidesert and desert regions. For example, treefrogs have virtually disappeared from the San Francisco Bay region, except in remnant parks and preserves (Banta and Morafka, 1966). Roads also may take a heavy toll (e.g., Watson et al., 2003). Still, the species is not vulnerable on a landscape scale, particularly away from human population centers.

Many species and populations of anurans have disappeared from the high mountains of California during the past few decades. However, these declines do not appear to have affected the presence of *P. regilla* at historically identified sites, although there may have been a decrease in abundance at many of the sites (Drost and Fellers, 1996). Exotic trouts (*Oncorhynchus*, *Salvelinus*, *Salmo*) and other nonnative fish species have a negative impact on the presence of *Pseudacris regilla* in the high mountains of California and Oregon and in the Willamette Valley of Oregon (Matthews et al., 2001; Bull and Marx, 2002; Knapp, 2005; Pearl et al., 2005b; Welsh et al., 2006). Pacific Treefrogs are 2.4–3 times as likely to be found in waters without trout as opposed to those with trout (Matthews et al., 2001; Welsh et al., 2006). Wetland and riparian modification has undoubtedly affected many populations. For example, Banta (1961) noted that populations of Pacific Treefrogs had disappeared along the Colorado River as a result of dam building and reservoir filling.

Pacific Treefrogs may be among the first amphibians to recolonize habitats affected by major catastrophic disturbances. After the 1980 eruption of Mt. St. Helens, Pacific Treefrogs were breeding successfully within the blast area by 1985. By 2000, they were present at more than 100 ponds. They currently are the most common amphibian throughout the area (Karlstrom, 1986; Crisafulli et al., 2005). In a similar vein, they readily occupy open-canopied wetlands in commercial silvicultural forest and in early succession stages of habitats previously clearcut or in a sapling stage (Raphael, 1988; Bosakowski, 1999). As forests mature into old growth, abundance decreases with canopy closure. *Pseudacris regilla* also will readily colonize artificial ponds (Monello and Wright, 1999).

This species is not considered to be of conservation concern in Canada (Weller and Green, 1997) or in any U.S. state.

Habitat of *Pseudacris regilla*. Photo: Chris Brown

COMMENT

The call of the Pacific Treefrog is one of the most recognized frog calls in the world. This species' call is invariably used as background in Hollywood movies and television programs, regardless of where the movie or program is supposed to take place. Listen for its repetitive "kreek-ek" call, especially in older movies whenever a frog call is used to enhance the "naturalness" of a scene.

Pseudacris streckeri Wright and Wright, 1933
Strecker's Chorus Frog

ETYMOLOGY

streckeri: a patronym honoring John K. Strecker (1875–1933), a Texas naturalist and curator of the Baylor University Museum (now Strecker Museum), who wrote one of the earliest accounts of the amphibians and reptiles of Texas (1915).

NOMENCLATURE

Synonyms: *Hyla streckeri*, *Pseudacris occidentalis* [in part]

IDENTIFICATION

Adults. This is a short squat burrowing frog with powerful front limbs that are used for digging. The background color is usually gray, grayish brown, or olive brown with dark brown to black markings on the dorsum. These dark markings are sometimes more distinct laterally with the mid-dorsal mark much less so, and in some cases the mid-dorsum may appear to be unmarked. The dark markings may be connected to give the appearance of stripes, they may be broken into bars, or they may present no indication of a longitudinal pattern. Some individuals may be bright green. A dark triangle spot is present between the eyes. A dark line extends from the snout to the shoulder and may extend posteriorly to the groin. The groin and posterior surfaces of the hind legs may have a greenish-yellow tint. The undersurfaces of the limbs are pinkish purple. Venters are white and unmarked. The skin is smooth dorsally and granular ventrally. Male vocal sacs are greenish yellow and dark during the breeding season.

Males are only slightly smaller than females. Males from Texas and Oklahoma averaged 31.9 mm SUL, whereas females averaged 33.9 mm SUL (Blair and Littlejohn, 1960). Males in Arkansas had a mean SUL of 34.5 mm; Oklahoma males averaged 34.9 mm SUL; and Texas males averaged 32.6 mm SUL (Trauth et al., 2007). In another Texas study, Burt (1936) recorded adults as 31–40 mm SUL. Wright and Wright (1949) give the normal male range as 25 to 41 mm SUL and the female range as 32 to 46 mm SUL.

Larvae. The ground color of mature larvae is brownish gray. The snout is sloping in lateral view. The tail musculature is brownish gray above the lateral line but light below it. The tail fin is clear and unpigmented, and the tail fin is only lightly pigmented with dark flecks. Throats are not pigmented, and venters are chalky white. Tails are long with attenuated tips. Larvae reach about 32–35 mm TL (Bragg, 1942).

Eggs. The eggs are brownish gray dorsally and white ventrally. There is a single jelly envelope surrounding the egg. The vitellus is 1.5 mm in diameter (range 1.2 to 1.8 mm) and the envelope is 3–5 mm in diameter (Livezey and Wright, 1947). The envelope is loose, elastic, and sticky, and debris readily adheres to it.

DISTRIBUTION

Pseudacris streckeri is known from south-central Kansas (between Chikaskia and Medicine Lodge Rivers) to the Gulf Coast of Texas. The range extends eastward into central Arkansas (along the Arkansas River) and northwest Louisiana (Caddo and De Soto parishes) and westward to the dry shortgrass prairies of central Texas at the Balcones Escarpment. Isolated populations occur in south Texas and in western Oklahoma. A number of early distribution records for this species were based on misidentification (e.g., in Arkansas: Trauth et al., 2004). Important distributional references include: Arkansas (Trauth et al., 2004), Kansas (Gray and Stegall, 1986; Collins et al., 2010), Louisiana (Morizot and Douglas, 1970; Dundee and Rossman, 1989), Oklahoma (Sievert

Distribution of *Pseudacris streckeri*

and Sievert, 2006), and Texas (Brown, 1950; Smith and Sanders, 1952; Hardy, 1995; Dixon, 2000).

FOSSIL RECORD
Strecker's Chorus Frog is known from Pleistocene Rancholabrean fossil deposits in Texas (Holman, 1969, 2003).

SYSTEMATICS AND GEOGRAPHIC VARIATION
Pseudacris streckeri is a member of the Fat Chorus Frog clade and is most closely related to *P. illinoensis* and *P. ornata* (Hedges, 1986; Cocroft, 1994; Da Silva, 1997; Moriarty and Cannatella, 2004). Trauth et al. (2007) examined a number of morphological characters and identified minor differences among *P. streckeri* populations. The species is clearly allied with other *Pseudacris*, based on its albumins (Maxson and Wilson, 1975).

Hybridization with *P. fouquettei* occurs in nature and in the laboratory, and successful F_1 backcrosses with ♀ *P. fouquettei* produce offspring very similar in appearance to that species (Lindsay, 1958). Successful crosses are made with *P. ornata* and *P. clarkii*, although backcrosses are unsuccessful as hybrids are sterile (Moore, 1955; Mecham, 1957, 1959). *Pseudacris streckeri* also hybridizes successfully with *P. brachyphona* (Mecham, 1965; in seeming contrast to Moore, 1955). Moore (1955) reports successful hybridization and development into adults with members of the *Hyla versicolor* complex. *Pseudacris streckeri* does not hybridize successfully in laboratory trials with *P. crucifer* or with *Acris blanchardi* (*in* Moore, 1955; Mecham, 1957).

ADULT HABITAT
Strecker's Chorus Frog is a species of prairies and open fields. It prefers sandy habitats such as occur along river terraces. It also occurs in cultivated fields and pastures.

TERRESTRIAL ECOLOGY
Activity occurs throughout most of the year to some extent, weather permitting. During the nonbreeding season, Strecker's Chorus Frogs have been found under rocks and rotten woody debris. They migrate from terrestrial sites to breeding ponds perhaps using olfactory cues. Grubb (1973) noted that both male and female *Pseudacris streckeri* often (but not always) were capable of detecting odors in water from ponds in which they had previously bred as opposed to water from "foreign" ponds or nearby creeks. This ability declined in trials after several days.

Strecker's Chorus Frogs likely are photopositive in their phototactic response as are Illinois Chorus Frogs, suggesting they use both sunlight and moonlight when feeding and going about their daily activities (Jaeger and Hailman, 1973). They likely are sensitive to light in the blue spectrum ("blue-mode response"), but with a slight departure from monotonicity at low frequencies. The blue-mode response would help them orient toward areas of increasing illumination (Hailman and Jaeger, 1974). Strecker's Chorus Frogs likely have true color vision.

CALLING ACTIVITY AND MATE SELECTION
Pseudacris streckeri is primarily a winter and spring breeding species, depending on latitude. Calls are heard from the autumn to spring, but calls can begin weeks in advance, prior to actual breeding (Wiest, 1982). For example, in Texas calls begin from October to January and continue until March and April (Burt, 1936; Blair and Littlejohn, 1960; Blair, 1961b; Wiest, 1982). In Louisiana, calls occur from January to March

Tadpole of *Pseudacris streckeri*. Photo: Stan Trauth

Adult *Pseudacris streckeri*, gray phase. Photo: David Dennis

Adult *Pseudacris streckeri*, green phase. Photo: David Dennis

Adult *Pseudacris streckeri*, reddish phase. Photo: C.K. Dodd Jr.

(Morizot and Douglas, 1970). In Kansas, breeding begins in February and continues until May with a peak in March and April (Gray and Stegall, 1986; Collins et al., 2010), whereas in nearby Oklahoma, it occurs from January to May (Bragg, 1942, 1950a). In Arkansas, calling occurs from February to April (Trauth et al., 1990).

Calling occurs both by day, especially early in the season, and night and is initiated by heavy winter rains. Calls usually cease at temperatures <4.4°C and do not begin until daytime temperatures are >15°C. However, Blair (1961b) recorded calling males when air temperatures were 0.5–3°C; Bragg (1950b) mentions males calling in a sleet storm at 0°C; and Wiest (1982) noted calling at 1.8–21.8°C over an 89-day period in Texas. Amplexus has been observed at air temperatures of 15–22°C.

The call of *P. streckeri* has been likened to "a rhythmic series of sharp pipping notes" (Burt, 1936) or as "clear and bell-like" (Bragg, 1942). A few short hoarse chirps may precede the main advertisement call. The dominant frequency is 2,225–3,900 cps and does not change with temperature (Blair and Littlejohn, 1960; Michaud, 1964). Call duration (0.04–0.06 sec) and the interval between calls (0.21–0.35 sec) are inversely proportional to temperature, and the trill rate (10–20.6 trills/sec) increases with temperature; the number of trills, however, remains constant (Blair and Littlejohn, 1960; Michaud, 1964). Males call while hanging onto vegetation, from vegetation above the water, and while sitting on the bank (Bragg, 1942).

Calling begins slowly at the beginning of the season, with a few individuals initiating it and choruses growing larger as the season progresses. Initial chorusing occurs in the late afternoon. With increasing rainfall, chorusing occurs both day and night for days on end, as long as environmental conditions permit. One male will initiate calling with others rapidly joining him. Chorusing will continue for some time until all frogs stop. After 10 or 15 minutes, the cycle is repeated. Bragg (1942) noted that this synchronized chorusing occurs more often by day than night. After dark, calling is nearly uninterrupted. Males remain at or near the breeding pool throughout the season. Females presumably depart once oviposition is completed.

BREEDING SITES

Breeding occurs in virtually any shallow temporary grassy pool, pond, or ditch. Pools are frequently located adjacent to pastures, fields, or agricultural tillage. They should be clear and unpolluted, not contain fish, and the hydroperiod should be sufficient to support a considerable amount of vascular water plants. In Oklahoma, Bragg and Smith (1942) characterized breeding habitat as including clear water sloughs on the floodplains of large rivers, but also noted breeding in muddy pools and cattle tanks (Bragg, 1942). The species does not breed in flowing water. Blair (1961b) noted that *P. streckeri* will breed in ponds that once contained fish but will no longer do so if fish are restocked. This species must therefore be able to detect fish and move elsewhere when fish are present, even if a wetland was used previously.

REPRODUCTION

Breeding is opportunistic. Eggs are deposited normally in small masses of 20–50 eggs per mass (Bragg, 1942), although the range is 2 to 250 eggs. Masses are oviposited below the water's surface, attached to vegetation; these masses may hang lobately in the water. Total counts of eggs per female are few, with the range often repeated as 400–700 (Livezey and Wright, 1947). Trauth et al. (1990) recorded a mean clutch size of 440 eggs (range 372 to 503) from 3 females in Arkansas. Normal development occurs at water temperatures of 8–27°C (Hubbs et al., 1963). Time to hatching is directly related to temperature, with eggs at 8°C requiring 28 days and at >19°C requiring 3–5 days to hatch. Hatchlings are 4.7–6.2 mm TL (mean 5.7 mm) (Bragg, 1942).

LARVAL ECOLOGY

Very little is known of the species' larval ecology. Feeding on algae begins about two days after hatching. Larval growth may be affected by the presence of other species of tadpoles. For example, tadpoles reared in water crowded by *Lithobates pipiens* weigh less than controls, which presumably could have effects on the duration of the larval period and size at metamorphosis (Licht, 1967). Larvae tend to hide in aquatic vegetation. Transforming froglets are 16–25 mm SUL (Trauth et al., 1990, 2004).

Breeding habitat of Pseudacris streckeri. Photo: Mike Forstner

DIET
No published information is available on diet. The species is probably opportunistic and eats a wide variety of small invertebrates.

PREDATION AND DEFENSE
Calling frogs immediately stop calling at the approach of an observer during daylight or on moonlit nights. The call also has a ventriloquist effect making it difficult to locate a calling male. Presumably these behaviors also help avert predation. Frogs also readily dive under water and hide in submerged vegetation. No information is available on predators.

POPULATION BIOLOGY
Occasional dry years may result in complete reproductive failure, but nothing is known of the species' population biology and demography.

COMMUNITY ECOLOGY
This species broadly overlaps with *Pseudacris fouquettei* in a north–south strip about 80–160 km wide along the eastern margin of its range. These species have strongly differentiated calls, however, and there is no evidence of call reinforcement in areas of sympatry (Michaud, 1964). Strecker's and Spotted Chorus Frogs often are found at the same breeding sites, as their habitat preferences and breeding seasons overlap considerably. The main isolating mechanisms between them also involve call discrimination and female preference. Female *P. streckeri* prefer the calls of male conspecifics rather than other *Pseudacris* whose calls are very similar except in dominant frequency (Littlejohn and Michaud, 1959).

DISEASES, PARASITES, AND MALFORMATIONS
Bragg (1962a) noted the fungus *Saprolegnia* on dead tadpoles in Oklahoma, but it is not clear whether the fungus caused the death or grew postmortem. The myxozoan parasite *Myxidium serotinum* (McAllister, 1987) and the coccidian protozoans *Eimeria flexuosa, E. streckeri,* and *Isospora delicatus* occur in *Pseudacris streckeri* (Upton and McAllister, 1988). Other parasites include the protozoans *Opalina* sp. and *Nyctotherus cordiformis,* the cestode *Mesocestoides* sp., and the nematode *Ozwaldocruzia* sp. (McAllister, 1987).

SUSCEPTIBILITY TO POTENTIAL STRESSORS
No information is available.

STATUS AND CONSERVATION
This species is considered reasonably abundant throughout much of its range, although habitat loss undoubtedly has destroyed many populations, particularly in highly disturbed agricultural and urban areas. Dixon (2000) noted a general decline in Texas, where it had disappeared from the post-oak savannas of Brazos County by 1976.

Preference for shallow temporary breeding sites makes this species vulnerable to "best management practices," which call for precision tillage. Changes in precipitation patterns and alteration of hydroperiod also make local populations vulnerable to drought. Coupled with a short life span, Strecker's Chorus Frogs may be subject to considerable population fluctuation and associated demographic effects on species' persistence. In Kansas it is protected by state law.

Pseudacris triseriata (Wied-Neuwied, 1838)
Western Chorus Frog
Rainette faux-grillon de l'ouest

ETYMOLOGY

triseriata: from the Latin *tri* meaning 'three' and *seriata* meaning 'lines.' The name refers to the three dark stripes on the dorsum.

NOMENCLATURE

Synonyms: *Chorophilus triseriatus, Chorophilus nigritus triseriatus, Helocaetes triseriatus, Hyla triseriata, Pseudacris nigrita triseriata*

There is considerable taxonomic confusion in the literature on striped chorus frogs. In the past authors may have assigned a name to a striped chorus frog population under study that has no resemblance to the current nomenclature based on molecular analysis. For example, Smith and Smith (1952) compared various morphological characteristics of "*Pseudacris nigrita feriarum*" with "*P. n. triseriata*." Upon looking at their collecting localities, it is clear that what we now know as *P. feriarum, P. fouquettei, P. triseriata,* and *P. kalmi* were all included under their name *P. n. feriarum*, and that *P. fouquettei, P. triseriata,* and *P. maculata* were included under their name *P. n. triseriata*. This species mixing makes the paper uninterpretable. Other authors have combined different species under the name "chorus frog" (e.g., Brodman, 2003). It is essential for authors using the literature on the striped chorus frog complex to have a clear understanding of the origin of the frogs in question in order to make a proper taxonomic allocation.

IDENTIFICATION

Adults. Like other striped chorus frogs, the Western Chorus Frog is a small slender frog characterized by having three broad parallel dark dorsal stripes on a lighter ground color that ranges from gray to tan, greenish brown, or olive. Occasional specimens may be russet or reddish. Some frogs may have the pattern broken up into spots or even may be unpatterned. In Indiana, for example, Whitaker (1971) reported that nearly 80% of the frogs examined had the typical striped pattern, whereas only 5% were spotted or unpatterned; the rest were intermediate between striped and spotted. The head is pointed when viewed from above, and there is a dark stripe on the side of the body extending from the snout, through the eye, to the groin. A white labial stripe is present. A triangular dark spot is present dorsally between the eyes. There is virtually no webbing between the toes, and the toes have only the slightest expansions at the tip. The tops of the legs are patterned with dark stripes or spots. Venters are cream colored and may have some dark speckling.

Sexual dimorphism is apparent. During the breeding season, males have a noticeable subgular vocal pouch that is greenish yellow to dark olive. Males also have a nuptial pad on the inside of the thumb that is used to hold onto the female during axillary amplexus. Females are slightly larger than males, although few specific measurements are available with adequate sample sizes. For example, Indiana adults were measured at a mean of between 26.6 and 28.1 mm SUL (Whitaker, 1971). In Ohio, males were 22.5–30 mm SUL, whereas females were 23.5–32 mm SUL (Walker, 1946).

Larvae. Larvae hatch at 4.8–6.1 mm TL (mean 5.7 mm) (Whitaker, 1971). Initially, they are yellow brown with tiny black spots, and they have tiny external gills; the gills are resorbed after three days. As larvae grow, the dorsum becomes darker (gray, black, or brown) and speckled with golden flecks, and the venter becomes more infused with a bronze color. The iris is golden. The venter is also dark with golden flecks, but the intestines are visible through the abdominal wall. Toward the end of development, the belly becomes quite brassy in color. The tail is long, and its tip is acute, tail fins are clear with dark flecks, and tail flecks are not concentrated along the edge of the fin. Metamorphosis can begin when larvae reach about 20 mm TL, but Hay (1889) recorded metamorphosis beginning at 27 mm TL, of which 16 mm was tail. The developmental process was described by Hay (1889). Larvae are described in greater detail by Wright (1929).

Eggs. Eggs are dark brown to black dorsally and white ventrally. The diameter is about 1.0 mm. The egg is surrounded by a gelatinous envelope 4–7 mm in diameter. As in other striped chorus frogs, eggs are deposited in loose gelatinous masses of 25–75 eggs per mass (Walker, 1946), which are attached to emergent vegetation, sticks, and branches in shallow water.

DISTRIBUTION

Pseudacris triseriata occurs from western New York south of Lake Ontario, west across extreme southern Ontario (Essex and adjacent counties), the entire state of Michigan, south throughout much of Indiana to extreme southern Illinois, western Kentucky, and scarcely into west-central Tennessee. It occurs throughout Ohio but does not cross the Ohio River throughout the eastern portion of the river drainage. The Western Chorus Frog occurs in Pennsylvania only along the western quarter of the state in the Allegheny Plateau; the range includes only the most northern tip of West Virginia. The Western Chorus Frog is known to occur on Kelleys Island in southwestern Lake Erie (Hirschfeld and Collins, 1961; Langlois, 1964; King et al., 1997; Hecnar et al., 2002) and on Walpole Island in Lake St. Clair (Woodliffe, 1989).

Many older publications treat *P. maculata* and *P. triseriata* as discrete taxa (usually as subspecies), and indicate that both taxa occur throughout the Midwest (e.g., Vogt, 1981; Collins, 1993; Oldfield and Moriarty, 1994; Casper, 1996; Kline, 1998). This distinction is not borne out by molecular analysis, as *P. maculata* is the only striped chorus frog within the region.

Important distributional references include: Illinois (Smith, 1961), Indiana (Minton, 2001 [in part]; Brodman, 2003), New York (Gibbs et al., 2007), Ohio (Walker, 1946; Pfingsten, 1998; Davis and Menze, 2000), Ontario (Cook, 1964a; MacCullouch, 2002 [in part]), Pennsylvania (Hulse et al., 2001) and Québec (Desroches and Rodrigue, 2004 [in part]). In some publications, authors have combined *P. maculata* with *P. triseriata* in their range maps and species descriptions (listed as *in part* above); these data should be used with caution.

FOSSIL RECORD

Fossils attributed to *P. triseriata* are known from Irvingtonian deposits in Maryland and from Rancholabrean deposits in Ohio (Holman, 2003). The major lineages within the genus *Pseudacris* were established prior to the age of the fossil deposits. Therefore, these fossils are within the range of *P. triseriata* as currently understood (Lemmon et al., 2007a).

SYSTEMATICS AND GEOGRAPHIC VARIATION

Pseudacris triseriata is a member of the morphologically conservative Trilling Chorus Frog clade of the genus *Pseudacris*, a group that includes *P. brachyphona, P. brimleyi, P. clarkii, P. feriarum, P. fouquettei, P. kalmi, P. maculata,* and *P. nigrita* (Moriarty and Cannatella, 2004). These species are distinguished by their advertisement call structures, color patterns, the presence of a cuboidal intercalary cartilage (Paukstis and Brown, 1987), and genetics. The *Nigrita* clade within the trilling chorus frogs includes all the species above except *P. brimleyi* and *P. brachyphona*. In turn, the *Nigrita* clade has three subdivisions, with *P. triseriata* and *P. maculata* west of the Mississippi

Distribution of *Pseudacris triseriata*

River forming one group, and *P. triseriata* east of the Mississippi aligning with *P. feriarum* (in part), and *P. kalmi* in forming a northeastern group.

Natural hybrids between *P. triseriata* and *P. brachyphona* occur in central Kentucky (Lemmon et al., 2007b). Smith and Smith's (1952) intergrades between *P. n. feriarum* and *P. n. triseriata* are, in part, based on contact zones between *P. triseriata* and *P. maculata*. In the laboratory, hybrids between male *P. crucifer* and female *P. triseriata* survive to the larval stage, but the reverse cross (male *P. triseriata* and female *P. crucifer*) is largely unsuccessful (Blair, 1941b). Male *P. triseriata* sperm are generally unsuccessful at penetrating the eggs of *P. crucifer*.

ADULT HABITAT

Adult habitat for *P. triseriata* consists of moist deciduous forests and woodlands interspersed with grassy meadows and open areas containing shallow wetlands. Woodland pools may be used by the species, but these should be relatively open canopied during the winter and early spring breeding period. Along the shoreline of the Great Lakes, chorus frogs are associated with a high perimeter-to-area ratio of emergent wetlands, and negatively with edge density of water (a measure of the total edge length of a cover type divided by the total landscape area) (Price et al., 2004). Such an association refers to small shallow grassy wetlands.

TERRESTRIAL ECOLOGY

During the nonbreeding season, Western Chorus Frogs are terrestrial, where they inhabit moist woodlands, usually within close proximity to water. They forage within the leaf litter, mostly at night, usually within 100 m of the likely breeding site; Kramer (1983) found that a few frogs traveled as far as 183 m from a wetland where he had tagged them. Like other striped chorus frogs, they are rarely encountered outside the breeding season, suggesting a rather sedentary and possibly semifossorial existence hidden under coarse woody debris or rocks. Using a Cobalt[60] tagging system, Kramer (1974) determined home ranges of from 641 to 6,024 m². Home ranges always included a wetland breeding pool.

Activity occurs throughout the warmer parts of the year, from the early spring breeding season until the cool weather of autumn. In Indiana and Ohio, for example, activity occurs through October to early November (Walker, 1946; Whitaker, 1971). Terrestrial overwintering near the breeding ponds occurs in underground crevices, animal burrows, or cavities below the frost line, although *P. triseriata* is tolerant of freezing to −1.2 to −1.6°C (Packard et al., 1998). Other overwintering sites include ant mounds and crayfish holes (Carpenter, 1953a) and under surface debris (Blatchely, 1899). Kramer (1973) reported most activity as occurring between dusk and dawn during the nonbreeding season.

CALLING ACTIVITY AND MATE SELECTION

Males call from late winter through the spring, the timing of which is dependent on environmental conditions. Indeed, Walker (1946) reported hearing occasional calling males during all winter months. Males hide themselves in grass clumps or within masses of floating vegetation and only rarely call from open water. They also call from the shoreline and occasionally while floating in the water next to surface mats of vegetation. Calling occurs both by day and night, temperature permitting. Daytime calls on clear days are especially prevalent early in the season, but as the season progresses and temperatures increase, calling occurs increasingly at night and on cloudy days. In Indiana, Whitaker (1971) recorded calling at temperatures <4°C, although most calling occurred at temperatures >10°C; calling can begin when ice still covers part of the breeding site.

The calls are analogous to running one's fingers along the tines of a comb producing 8 to 12 clicks at a time. Call rates are temperature-dependent. The mean duration of the call of *P. triseriata* from Rochester, New York, is 600 ms (range 488 to 728 ms). The call has a mean of 19 notes (range 14 to 19) and a dominant frequency of 2,582 cps (range 200 to 4,000) (Thompson and Martof, 1957). As might be expected due to shared phylogeny, the call of this species is most similar to that of *P. maculata*. Interestingly, *P. feriarum* did not prefer conspecifics to the calls of *P. triseriata* in mixed-species choice tests (Martof, 1961). In cases of range contact, such lack of discrimination could contribute to the potential for hybridization.

If calling is interrupted by a period of cold weather, Western Chorus Frogs do not return to terrestrial habitats. Instead, they remain at breeding sites, underwater, holding onto vertical vegetation. As temperatures drop, they retreat further beneath the surface. When warm weather returns, they climb back up the stems to resume calling. Calls are occasionally heard in the late summer or autumn, especially on overcast days or after a heavy rain. Breeding does not occur at this time, however.

BREEDING SITES

Breeding occurs in both permanent and temporary shallow-water wetlands, including woodland pools, grassy marshes, ditches, and small swamps. They have even been reported to use shallow wetlands on golf courses (Mifsud and Mifsud, 2008). Western Chorus Frogs prefer fishless breeding sites with an open canopy and emergent vegetation. In a mixed hydroperiod situation, Hay (1889) reported that numerous egg masses were found in a deep tree stump hole adjacent to the edge of a temporary pond. Occupancy is not uniform at wetlands throughout a region. For example, Hecnar and M'Closkey (1996a) reported about 20% occupancy of 122 ponds surveyed in southwestern Ontario. Also in Ontario, Hecnar (1997) reported <20% occupancy in 174 ponds surveyed over a 2 yr period. In Michigan, the probability of detecting *P. triseriata* at least once at any single pond of 32 ponds surveyed over a 5 yr period was 40% (Skelly et al., 2003).

REPRODUCTION

Pseudacris triseriata is a late-winter-to-spring breeding frog, and it is one of the earliest breeding frogs within its range. Specific breeding dates include late February to late April (Indiana: Whitaker, 1971), late February to early May (Illinois: Smith, 1961; Indiana: Minton, 2001), March and April (Michigan: Cummins, 1920), and March to May (Illinois: Cagle, 1942; Ohio: Walker, 1946). Breeding occurs in shallow grassy wetlands and often occurs during rainfall. Most eggs are deposited over a short period (for example, six days in Indiana; Whitaker, 1971), despite the extended breeding season (e.g., 35 days in Michigan; Cummins, 1920). Males and females tend to enter the breeding sites simultaneously. However, females usually leave the wetland a day or so after oviposition, whereas males remain much longer, perhaps throughout the breeding season.

Like other striped chorus frogs, eggs are deposited in loose masses of from a dozen or so to several hundred eggs per mass. In Indiana, Whitaker (1971) recorded 12–245 eggs per mass (mean 80) after examining 96 egg masses; the total egg compliment from 16 females was between 373 and 983 eggs (mean 642). In Ohio, Walker (1946) stated that there were usually 25–75 eggs per mass. Egg masses are from 10 mm wide to 110 mm in length. Eggs are placed in water 15–45 cm in depth, but the eggs themselves are 2.5–20 cm (usually 5–10 cm) below the water's surface. The egg masses are placed on slanting or horizontal stems, branches, or leaf petioles.

As with other frogs, water temperature affects the developmental rate and the time to hatching. Thus, at warm temperatures, eggs may hatch in 8–13 days, whereas in cold temperatures, hatching takes considerably longer, from 15 to 27 days (Whitaker, 1971).

LARVAL ECOLOGY

The duration of the larval period is also dependent on water temperature. In Indiana, the larval period ranged from 50 to 55 days after hatching in Whitaker's (1971) study, with metamorphosis occurring by June. Walker (1946) recorded a developmental period of from 60 to 82 days based on data from Hay (1892). Larvae prefer shallow water (11–20 cm in depth) with sparse emergent vegetation. Tadpoles tend to avoid both extremely shallow water and deep water. All the larvae from within a clutch tend to metamorphose within about five days of one another. At this time, in Indiana, they are about 20 mm TL, and metamorphs are from 9.4 to 12.7 mm SUL (mean 10.4 mm) (Whitaker, 1971). In Ohio, Walker (1946) recorded total larval lengths of 25–30

Eggs of *Pseudacris triseriata*. Photo: Dana Drake

Tadpole of *Pseudacris triseriata*. Photo: T. Travis Brown

mm immediately prior to metamorphosis, with metamorphs about 11 mm TL.

DISPERSAL

After the breeding season, adults disperse gradually into the surrounding moist forest leaf litter. Most individuals remain relatively close to the breeding site, but some move a considerable distance away. Kramer (1973) found that frogs moved about 3.5 m per day (range 0.25–11.2 m/day), but that dispersal could be much faster; one frog moved 42 m in a single day. Males moved at a slower rate (3.2 m/day) than females (4.1 m/day). Frogs could make extensive movements between successive captures, from 8 m/day to > 30 m/day. The frogs sometimes moved into terrestrial habitats, only to return to the original breeding pool. Other Western Chorus Frogs moved between adjacent breeding pools, with one adult moving a total distance of 206 m from its original breeding pool to another. It is not known whether these frogs bred in the secondary pools or whether they were using them as dispersal corridors or foraging sites. While they are dispersing, Western Chorus Frogs hide in the leaf litter in a variety of locations in addition to sapling roots, under logs or bark, in hollow logs, in animal burrows, or in cracks in the substrate (Kramer, 1973).

POPULATION BIOLOGY

Recently metamorphosed Western Chorus Frogs grow rapidly, reaching 22 mm SUL by the beginning of their first winter season. They probably reach sexual maturity by the first spring after metamorphosis (Whitaker, 1971). Growth rates average 3–4 mm per month.

DIET

The larval diet consists of filamentous and non-filamentous algae. During metamorphosis, the froglets do not feed. However, they soon begin feeding on very small invertebrates such as springtails, ants, fly larvae, mites, and small insects; metamorphs also consume shed skin. Adults feed on a variety of small invertebrates, including lepidopteran larvae, spiders, midges, and snouted beetles. Other prey include slugs, ants, crickets, snails, other species of beetles, sowbugs, and a host of available invertebrates. Feeding occurs while at the breeding site, unlike some other species of frogs, and about 65% of the adults have food within their digestive tract. Whitaker (1971) provided a comprehensive list of prey for both adults and juveniles.

PREDATION AND DEFENSE

Larvae are palatable, and they do not show any avoidance response to chemical cues emanating from fish (Kats et al., 1988). This makes them particularly vulnerable to fish and likely helps explain their preference for breeding in temporary wetlands.

Garter snakes (*Thamnophis*) are likely major predators of *Pseudacris triseriata* (Whitaker, 1971), especially as these frogs undergo metamorphosis. Larvae are capable of sustained swimming, which might help them elude predators, but as they approach metamorphosis the development of the limbs alters body form and makes sustained swimming more difficult. The rear legs act as a drag on the body, whereas the eruption of the front limbs disrupts the body contour and makes controlled swimming erratic. Likewise, postmetamorphic Western Chorus Frogs are better jumpers than froglets in the early stages of metamorphosis. Thus, metamorphosis is a particularly vulnerable stage in the transition from larval to adult stage (Wassersug and Sperry, 1977). Other predators include water snakes (*Nerodia sipedon*) (Whitaker, 1971). In the laboratory, mudminnows (*Umbra*) readily consume both eggs and newly hatched larvae (Whitaker, 1971). Eggs and larvae are probably eaten by a host of aquatic invertebrates, although timing reproduction for the cold late winter and early spring may help reduce both predation and competition.

COMMUNITY ECOLOGY

Both *Pseudacris crucifer* and *P. triseriata* call from the same ponds. However, these species may have a certain degree of temporal partitioning, whereby *P. triseriata* calls by day and night and *P. crucifer* calls mostly at night. The Western Chorus Frog also prefers temporary ponds, whereas the Spring Peeper prefers more permanent ponds. Western Chorus Frogs have greater growth and survivorship than Spring Peepers in temporary ponds, whereas the opposite is true in permanent ponds. Skelly (1995a) found no evidence for competition between the two species, with pond hydroperiod limiting the success of *P. crucifer* in temporary ponds and predation limiting the success of *P. triseriata* in permanent ponds. Further, Whitaker (1971) noted that the two species tended to use different parts of a wetland when they occurred together, although there was still broad spatial overlap in calling position. In southern Ontario, Hecnar and M'Closkey (1996b) suggested differences in habitat use between these species (possibly including both *P. maculata* and *P. triseriata* under the name *P. triseriata*) were more related to the presence or absence of fish than to any aspect of water chemistry.

DISEASES, PARASITES, AND MALFORMATIONS

Frog Virus 3-like infections have been reported in natural populations of Western Chorus Frog larvae from Ontario (Duffus et al., 2008). Parasites include the ciliate protozoan *Opalina obtrigonoidea*, the trematodes *Brachycoelium salamandrae* and *Glypthelmins quieta*, and the nematodes *Aplectana* sp., *Cosmocercoides dukae*, *C. variabilis*, *Oswaldocruzia leidyi*, and *O. pipiens* (Odlaug, 1954; Whitaker, 1971; Baker, 1977; Ashton and Rabalais, 1978). Males and females are parasitized at about the same rate, with 7.2 trematode individuals per male and 3.9 individuals per female (Whitaker, 1971). There are usually one or two nematodes per frog. No malformations were observed in *Pseudacris triseriata* during a survey of frogs from National Wildlife Refuges in Pennsylvania and western New York (Converse et al., 2000).

SUSCEPTIBILITY TO POTENTIAL STRESSORS

Nitrates. Western Chorus Frog larvae exposed to chronic levels of ammonium nitrite at 10 mg/L for 100 days had lower survivorship and fewer metamorphs than controls or larvae exposed to lower concentrations. However, time to metamorphosis was not affected. Acute exposure resulted in reduced activity, weight loss, and malformations. Acute exposure also resulted in high rates of mor-

Adult *Pseudacris triseriata*.
Photo: T. Travis Brown

Breeding habitat of *Pseudacris triseriata*. Photo: T. Travis Brown

tality at concentrations > 10 mg/L, especially at > 25 mg/L (Hecnar, 1995). These concentrations are exceeded commonly in agricultural settings.

STATUS AND CONSERVATION

Western Chorus Frog populations appear to be doing well over much of their range (Weller and Green, 1997; Gibbs et al., 2005). Minton et al. (1982) felt they were declining in Indiana, although Minton (2001) later felt they were recovering from declines that occurred from 1975 to 1985. Undoubtedly populations have been lost due to habitat destruction and wetland alteration. For example, call surveys conducted in northern and western New York from 1973 to 1980 and again in 2001 to 2002 reported overall stable populations (disappearance from 76 sites, new appearance at 63 sites), with most declines occurring in the North (Gibbs et al., 2005). Individual populations were likely affected by roads, transportation corridors, acid deposition, and succession from open habitats to more forested habitats (Gibbs et al., 2005). At the same time, there did not appear to be regional declines. Still, it must be kept in mind that resurvey efforts are sensitive to the type of historical data and the length of the resurvey period when attempting to estimate regional declines and changes in the distribution of this species (Skelly et al., 2003).

Fortunately, the species does not appear to be affected by the serious amphibian chytrid fungus pathogen. Considering that amphibian chytrid has been found in nearby populations of *P. maculata*, it is likely that *P. triseriata* also may be colonized by this pathogen. In terms of other variables impacting populations, road mortality likely affects some populations. However, Ashley and Robinson (1996) found only 12 dead *P. triseriata* on Long Point Causeway, Ontario, over a 4 yr period. Either mortality was low, the bodies of these small frogs did not remain on highways long, or surveys were not conducted during their migration period. The area of immediate adverse impacts from roads (the "road-zone" effect) extends 1,100–2,400 m from the pavement (Eigenbrod et al., 2009). Roads may alter the choice of breeding sites, even when direct mortality is not a threat.

Pseudacris triseriata breeds in a variety of man-made wetlands, including ditches, pastures, road ruts in fields, depressions in agricultural and silvicultural areas, mitigation wetlands, wetlands in suburban areas, and in reclaimed mine sites (Walker, 1946; Myers and Klimstra, 1963; Lacki et al., 1992; Foster et al., 2004; Gibbs et al, 2005). Western Chorus Frogs appear to tolerate disturbances rather well as long as moist forest terrestrial habitat is nearby. The early breeding season may limit contact with some human activities, allowing reproduction and metamorphosis to occur prior to the onset of summer habitat disturbance. The species may be found in urban environments, but it is not as common as it would be in rural habitats (Foster et al., 2004).

As in other *Pseudacris*, this species may be monitored by using aural surveys. Varhegyi et al. (1998) noted, however, that call reception by automated frog call recorders decreased drastically at distances > 20 m from a chorus, depending on the location of the recorder in relation to the chorus. Aural surveys conducted where frogs are distant from a listening site likely underestimate occupancy.

Smilisca fodiens (Boulenger, 1882)
Lowland Burrowing Treefrog

ETYMOLOGY
fodiens: from the Latin *fodio* meaning 'to dig' or 'dig up,' in reference to the spade-like inner metatarsal tubercles, which are an adaptation for digging.

NOMENCLATURE
Stebbins (2003): *Pternohyla fodiens*
  Synonyms: *Hyla fodiens, Pternohyla fodiens*
Most of what is known of the biology of *Smilisca fodiens* is based on observations from much further south in Mexico. Duellman (2001) and Sredl (2005b) review the general biology of this species from throughout its range.

IDENTIFICATION
Adults. *Smilisca fodiens* is a large yellow-brown frog with reddish-brown blotches surrounded by cream or yellow coloration. Snouts are conspicuously rounded, and there is a distinctive skin fold along the back of the head. The skin on the top of the head is noticeably hard and firmly attached to the skull (termed cranial co-ossification), identifying it as a member of the tropical casque-headed treefrogs. Juveniles are pale green with scattered brown flecks or spots dorsally and a dark stripe between the eyes and the nostrils; dorsal patterns are less prominent than they are in adults. Venters are creamy white. Males have a dark patch on either side of the throat that is absent on females. In Arizona, adults averaged 55.1 mm SUL (range 45 to 70 mm) (Goldberg et al., 1999b). The maximum size of males in Mexico is 62.6 mm SUL, whereas females reach 63.7 mm SUL (Duellman, 2001).

Larvae. Tadpoles are dull tan with olive-brown mottling and dusty white venters. The body of the tadpole is as wide as it is deep and has a bluntly rounded snout. The eyes are small and laterally placed. The caudal musculature is slender, and the dorsal tail fin does not extend to the body. Both are lightly pigmented. Larvae reach a maximum size of 45–50 mm TL. Descriptions and illustrations of the larvae and mouthparts are in Duellman (2001).

Eggs. The eggs have not been described.

DISTRIBUTION
The Lowland Burrowing Frog occurs from Pinal County, Arizona, southward into Jalisco and Michoacan, Mexico. Important distributional references include Chrapliwy and Williams (1957), Sullivan et al. (1996b), Enderson and Bezy (2000), Duellman (2001), and Brennan and Holycross (2006).

FOSSIL RECORD
Pleistocene fossils of *S. fodiens* are known from Sonora, Mexico (Holman, 2003).

Distribution of *Smilisca fodiens*

SYSTEMATICS AND GEOGRAPHIC VARIATION

Originally described within the genus *Pternohyla* (Boulenger, 1882), it was relegated to the genus *Smilsca* by Faivovich et al. (2005). It is a member of the Tribe Hylinae within the Middle American/Holarctic clade of New World hylids.

ADULT HABITAT

In Arizona, *Smilsca fodiens* is associated with mesquite-lined desert washes. Surrounding vegetation is desert scrub or semidesert grassland. Farther south, it is a species of arid tropical scrub forests. It occurs from sea level in Mexico to at least 2,300 m.

TERRESTRIAL ECOLOGY

This species is largely fossorial in the winter and dry months, emerging only after heavy summer monsoon thunderstorms to breed and forage. In that regard, the head (with the skin attached to the skull) can be used to plug the burrow and thus retard water loss and protect the species from terrestrial predators. Whether the frog digs its own burrows is unknown but seems likely, based on hind foot morphology. The frog can also shed its skin in layers and form a cocoon in order to create a protective barrier, insulating it from water loss (Ruibal and Hillman, 1981). Brennan and Holycross (2006) stated that it is adept at climbing mesquite trees but is mostly terrestrial. After metamorphosis, juveniles disperse by day and night.

Lowland Burrowing Treefrogs are photopositive in their phototactic response, suggesting they use both sunlight and moonlight when feeding and going about their activities (Jaeger and Hailman, 1973). They are sensitive to light in the blue spectrum ("blue-mode response") (Hailman and Jaeger, 1974). Lowland Burrowing Treefrogs likely have true color vision.

CALLING ACTIVITY AND MATE SELECTION

Smilsca fodiens is an explosive breeder, with males calling in large numbers following summer thunderstorms. Males call from in or immediately adjacent to pools, from mud flats, in sparse vegetation, or even from under rocks or grass clumps (Duellman, 2001). The advertisement call is described as a loud raspy "wonk-wonk-wonk" repeated at a high rate. It is uttered at 81–115 notes/min with a duration of 0.21–0.28 sec and a pulse rate of 118–125 pulses/sec; the dominant frequency is 2,200–2,278 cps (Duellman, 2001). Males also have a territorial call which sounds like the advertisement call of *Pseudacris maculata*,

Tadpole of *Smilisca fodiens*. Photo: Ronn Altig

Adult *Smilisca fodiens*. Photo: Rob Schell

that is, a finger running over comb teeth (Sullivan et al., 1996b).

Calling occurs within only a short period of time after rainfall ends. For example, Sullivan et al. (1996b) noted that they never heard calling males more than 36 hrs after rain had ceased. Males have an encounter call made in response to other males calling within 60–90 cm; this call also is made in response to recordings of advertisement calls from other *S. fodiens* males (*in* Wells, 1977a). These results suggest that males establish a calling territory that they will defend against intruders.

BREEDING SITES
Breeding occurs in rain-formed temporary pools in desert washes, along roads, and in cattle tanks (Sullivan et al., 1996b; Sredl, 2005b). Vegetation in the vicinity of breeding pools may be dense.

REPRODUCTION
Breeding occurs opportunistically from June to September depending upon rainfall. Eggs are oviposited in jelly envelopes scattered on the pool bottom. Clutch size and other aspects of reproduction are unknown. Larvae are ca. 10 mm TL upon hatching.

Adult male (*left*) and female (*right*) *Smilisca fodiens*. Note the difference in tympanum coloration. Photo: Cecil Schwalbe

Calling male *Smilisca fodiens*. Photo: Brent Sigafus

Breeding habitat of *Smilisca fodiens*. Photo: Cecil Schwalbe

LARVAL ECOLOGY
Larvae may feed at the surface of the water, but the diet is unknown. They tend to float within the water column tail down. When disturbed, they swim away or float to the bottom of the pool. The duration of the larval period and other aspects of larval ecology have not been reported. Metamorphosis occurs at 18–24 mm SUL.

DIET
Specific data are not available for specimens from Arizona, but it likely eats a wide variety of invertebrates.

PREDATION AND DEFENSE
The hard head is used to seal burrows and thus protect the species from both predators and desiccation. The frog also has an unken reflex whereby it flexes its head downward and elevates its limbs causing it to rest directly on the belly. Nothing is known of its predators.

POPULATION BIOLOGY
Juveniles lack the cranial co-ossification, which suggests they have a different terrestrial life history from the adults. Nothing is known about the demography of adults.

DISEASES, PARASITES, AND MALFORMATIONS
Smilisca fodiens is parasitized by the cestode *Distoichometra bufonis* and the nematodes *Aplectana itzocanensis*, *Cosmocercella haberi*, *Rhabdias americanus*, *Physaloptera* sp., and *Skrjabinoptera* sp. (Goldberg et al., 1999b).

SUSCEPTIBILITY TO POTENTIAL STRESSORS
No information is available.

STATUS AND CONSERVATION
No information is available.

Smilisca baudinii
(Duméril and Bibron, 1841)
Mexican Treefrog

ETYMOLOGY
baudinii: a patronym honoring Monseur Baudin, a French commander in Mexico who donated the type specimen to the Museum National d'Histoire Naturelle in Paris.

NOMENCLATURE
Synonyms: *Hyla baudinii*, *Hyla beltrani*, *Hyla manisorum*, *Hyla muricolor*, *Hyla pansosana*, *Hyla vanvlietii*, *Hyla vociferans*, *Smilisca daudinii*, *Smilisca daulinia*

IDENTIFICATION

Adults and juveniles. This is an attractive pale green to tan treefrog with dark, irregular dorsal spots and a network of black-on-yellow lateral coloration. Heads are wide and flat, and snouts are rounded and short. There is a dark marking back from the eye that becomes a black vertical shoulder bar; a distinctive light yellow or green spot is present below the eye. A black interocular bar and a distinct creamy-white anal stripe is usually present. The throat is greenish or yellow except in breeding males, when it is gray. Limbs have three to four dark transverse bands. The rear of the thigh has greenish-yellow to purplish-russet reticulations. Toes have moderately large disks, and are three-fourths webbed. The vocal sac is paired and subgular. Venters are white. Males are 47.3–75.9 mm SUL (mean 58.7) from throughout its range; females reach a maximum size of 90 mm SUL from Sinaloa (Duellman, 2001). Juveniles are dull olive green dorsally and white ventrally. There are faint brown transverse bars on the limbs. A distinctive white suborbital spot is present.

Larvae. Tadpoles are dark brown with a pale crescent mark posteriorly and a body that is wider than deep. Snouts are rounded. Eyes are widely separated and located dorsolaterally. The caudal musculature is pale tan with a dark brown longitudinal streak. Prominent tail fin markings are absent, as are any indications of dorsal saddles; small brown flecks may be present, however. Tails are long in mature larvae, and tail fins are deepest at one-third down the length of the tail. Venters are transparent, with brown flecks anteriorly below the eye. Larvae reach 35–37.5 mm TL. Tadpoles are described in detail by Duellman (2001).

Eggs. Eggs have a diameter of ca. 1.3 mm and are encased in a single membrane of 1.5 mm in diameter (Duellman, 2001).

DISTRIBUTION

Smilisca baudinii occur from south Texas through Costa Rica. This species occurs only along the lower Rio Grande Valley in the United States, with additional reports from two nearby counties to the north (Bexar, Rufugio), which probably represent introductions. Important distributional references include Dixon (2000) and Duellman (2001).

Distribution of *Smilisca baudinii*. The two northern populations are likely introduced.

FOSSIL RECORD

No fossils are known.

SYSTEMATICS AND GEOGRAPHIC VARIATION

Smilisca baudinii is a member of a tropical group of hylid frogs, the Middle American/Holarctic clade (Faivovich et al., 2005). *Smilsca* reaches its northernmost distribution in Texas and southern Arizona. Duellman (2001) provided characteristics defining the genus, which includes at least seven species. There is substantial morphological and color variation among *S. baudinii* populations over its considerable range, particularly with regard to maximum size and larval coloration. These have been discussed by Duellman (2001).

ADULT HABITAT

The habitat in Texas is described by Wright and Wright (1938) as "on houses, in meadows, overflow lands, on trees of forests, in palm groves." Throughout its range, it is a species of xeric and subhumid habitats that have a prolonged dry season.

TERRESTRIAL ECOLOGY

Smilisca baudinii occupies tree cavities and refugia under debris during prolonged dry periods. It does not tolerate freezing temperatures. Mexican Treefrogs are photopositive in their phototactic response, suggesting they use both sunlight and moonlight when feeding and going about their daily activities (Jaeger and Hailman, 1973). They are sensitive to light in the blue spectrum ("blue-mode response"), which apparently helps them orient toward areas of increasing illumination (Hailman and Jaeger, 1974). Mexican Treefrogs likely have true color vision.

CALLING ACTIVITY AND MATE SELECTION
The advertisement call is described as a series of short explosive notes of "wonk-wonk-wonk." Two to 15 notes comprise a call group, with call groups spaced from 15 sec to several minutes apart. The duration of the notes is 0.09–0.13 sec with a fundamental frequency of 135–190 cps. The lowest major frequency is 175–495 cps, whereas the highest major frequency is 2,400–2,725 cps (Duellman, 2001). Males usually call from the ground near water but also occasionally from bushes and trees. Males call in duets, with each chorus made up of several pairs of calling males. Successive choruses are initiated by the same duet. Males have an encounter call that is directed toward other males that come within 30 to 50 cm of the primary calling male (*in* Wells, 1977a). Males will engage encroaching intruders aggressively.

BREEDING SITES
Breeding occurs in overflow pools, wet grassy meadows, resacas, or streams. Duellman (2001) noted that it breeds in nearly any body of water.

REPRODUCTION
Most breeding records in Texas are from June to July, but the breeding season may be longer than currently understood. In Mexico it extends from June to October. Eggs are oviposited in a surface

Eggs of *Smilisca baudinii*.
Photo: Cecil Schwalbe

Tadpole of *Smilisca baudinii*.
Photo: Seth Patterson

Adult *Smilisca baudinii*, bronze phase. Photo: Cecil Schwalbe

Amplexing *Smilisca baudinii*, green phase. Photo: Cecil Schwalbe

Calling male *Smilisca baudinii*, patterned. Photo: Seth Patterson

Breeding habitat of *Smilisca baudinii* in south Texas. Photo: Mike Forstner

film, with each deposition having several hundred eggs. Total reproductive output was reported as 2,620–3,320, but the sample size was very small (Duellman, 2001). Hatchlings are 5.1–5.4 mm TL.

LARVAL ECOLOGY
Nothing is known of the larval ecology. Newly metamorphosed froglets are 12–15.5 mm SUL.

DIET
This species probably eats a wide variety of invertebrates.

PREDATION AND DEFENSE
Both males and females have a high-pitched distress call. Predators have not been reported, but likely include snakes, birds, and mammals.

POPULATION BIOLOGY
Throughout most of its range, this is an extremely abundant species with sometimes thousands of frogs calling at a single site (Duellman, 2001). Nothing has been reported on its demography or life history.

DISEASES, PARASITES, AND MALFORMATIONS
No information is available.

SUSCEPTIBILITY TO POTENTIAL STRESSORS
No information is available.

STATUS AND CONSERVATION
This species is considered "Threatened" by Texas. There is virtually nothing known of the species' biology in Texas.

Family Leptodactylidae

Leptodactylus fragilis (Brocchi, 1877)
Mexican White-lipped Frog

ETYMOLOGY
fragilis: from the Latin *frag* meaning 'brittle.' The meaning of the name in relation to the frog is unknown.

NOMENCLATURE
Synonyms: *Cystignathus caliginosus, Cystignathus fragilis, Leptodactylus albilebris, Leptodactylus caliginosus, Leptodactylus gracilis, Leptodactylus labialis* [in part], *Leptodactylus melanotus, Leptodactylus mystaceus*

There is some discussion in the literature as to which specific epithet, *fragilis* or *labialis*, has priority for this species. I follow the nomenclature recommended by Heyer (2002), Heyer et al. (2006), and Crother (2008) in using *fragilis*. Heyer (2002) and Heyer et al. (2006) discussed the nomenclatural history in detail and provided a synonymy.

IDENTIFICATION
Adults. This is a very streamlined frog with a pointed snout. The ground coloration is olive yellow, olive, brown, or reddish brown, with dark dorsal irregular spots. These spots may or may not be circled in white. There is a distinctive white or cream stripe on the upper jaw and a dark stripe from the nostril to the tympanum. Dorsolateral and lateral folds are present. There is no webbing on the long fingers and toes, and there are no terminal disks on the digits. There is a distinct disk on the belly. Sexual dimorphism is not present. Males are 27–43 mm SUL and females are 30–44 mm SUL (Heyer et al., 2006).

Larvae. At hatching, tadpoles are nearly invisible except for their yolk sac. After several hours, they begin to attain a light brown coloration. Mature tadpoles are dark brown with an elongate snout. Nostrils are nearer the eyes than the snout. Tails are long, ending in an obtuse point. The tail musculature is cream and heavily pigmented with brown. Venters are pale brown. The tail fins do not extend onto the body, are translucent, and heavily marked with brown, especially on the dorsal fin. Mulaik (1937) described and figured the larval mouthparts. Heyer et al. (2006) described the larva in detail.

Eggs. Eggs are light yellow and lack gelatinous envelopes. The vitellus is about 1.5 mm in diameter (Mulaik, 1937).

DISTRIBUTION
This species occurs from extreme southern Texas in the Rio Grande Valley (Cameron, Hidalgo, and Starr counties) to central Colombia and northern Venezuela. Important distributional references include Dixon (2000) and Heyer et al. (2006).

Distribution of *Leptodactylus fragilis*

Adult *Leptodactylus fragilis*, northern El Salvador. Photo: Twan Leenders

FOSSIL RECORD
Fossil Mexican White-lipped Frogs are reported from Pleistocene deposits in Tamaulipas, Mexico (Holman, 2003; Heyer et al., 2006).

SYSTEMATICS AND GEOGRAPHIC VARIATION
Leptodactylus fragilis is a member of the *fuscus* group of the genus *Leptodactylus*, a diverse assemblage of >23 species that inhabit the Neotropics (Heyer, 1978).

ADULT HABITAT
The habitat of this species in Texas was described as "moist meadows, irrigated cane fields, drains and gutters in towns, beneath stones, logs, in sandy banks and fields; near streams and marshy places" (Wright and Wright, 1938).

TERRESTRIAL ECOLOGY
Nothing is known concerning the life history of *L. fragilis* at the northern end of its range. Mexican White-lipped frogs are unclassifiable with regard to spectral response, but the response is similar to the "blue-mode response" but shifted to green (Hailman and Jaeger, 1974). Mexican White-lipped Frogs may have color vision.

CALLING ACTIVITY AND MATE SELECTION
Calling occurs from concealed positions at the base of grass hummocks. Males construct shallow cavities that will be used for egg deposition and call from these cavities. Mulaik (1937) described the call as resembling the "plunk-plunk" from a drop of water falling into a cave pool, although Heyer et al. (2006) described it as "a rising harsh whistle." Calls are given at the rate of 120–150 calls/min with a call duration of 0.16–0.2 sec. The call begins about 600–750 Hz and has a fast rise in intensity during the first third of the call, a weak increase in the next third, and another weak increase followed by a sharp drop in intensity at the end. The call rises to 1,000–1,200 Hz. The dominant fundamental frequency is 740–780 Hz in Texas (Fouquette, 1960; Heyer et al., 2006). Males call in the late afternoon and night during the rainy season.

Males will also call away from the nest depression. The female responds to the male's call and moves toward him. As she moves next to him, he leads her to the shallow depression. During this movement, the female makes what appears to be a reciprocal call, consisting of a rapid series of short notes or trill that is barely audible (Bernal and Ron, 2004).

BREEDING SITES
Mulaik (1937) found a clutch in a shallow depression 4 cm in diameter and 3 cm deep at the base of a grass hummock 30 cm from the nearest water. Male frogs of the genus *Leptodactylus* typically excavate similar shallow cavities located near ponds or pools for egg deposition.

REPRODUCTION
Eggs are deposited terrestrially in a frothy mass in small excavations near water at the base of vegetation (Livezey and Wright, 1947). Males construct these shallow depressions next to pools into which females deposit their eggs. The female whips up the foam mass from her body secretions, and the foam will protect the tadpoles even during dry weather. However, the normal sequence is to have the depression flooded, which releases the tadpoles into water. The clutch size is 80 eggs or greater, with Mulaik (1937) reporting one clutch of 86 eggs. Hatching occurs within 40 hrs of deposition, when larvae are 6.6 mm TL. After 24 hrs, they are ca. 8.1 mm TL.

LARVAL ECOLOGY
Larvae are free-living benthic feeders living in nonflowing water. The larval period is ca. 30–35 days (Mulaik, 1937). Transforming juveniles average 16.1 mm SUL.

DIET
Nothing is reported on the diet of this species in Texas. Presumably it eats a wide variety of small invertebrates.

PREDATION AND DEFENSE
Leptodactylus fragilis is a superb jumper, giving rise to a common name, rocket frog. Nothing is known concerning its predators, but they likely include snakes, birds, and mammals.

POPULATION BIOLOGY
No information is available.

DISEASES, PARASITES, AND MALFORMATIONS
No information is available.

SUSCEPTIBILITY TO POTENTIAL STRESSORS
No information is available.

STATUS AND CONSERVATION
There is some question whether this species still occurs in Texas, but its extirpation has not been confirmed, and it may still be present within the state (A.G. Gluesenkamp, personal communication). Dixon (2000) stated that the use of organophosphates could have caused its extirpation. This species is very common from Mexico to South America.

Family Microhylidae

Gastrophryne carolinensis (Holbrook, 1836)
Eastern Narrow-mouthed Frog

ETYMOLOGY

carolinensis: Latinized noun referring to Carolina. The species was described from specimens found near Charleston, South Carolina.

NOMENCLATURE

Synonyms: *Engystoma carolinense, Engystoma rugosum, Gastrophryne rugosum, Microhyla areolata, Microhyla carolinensis, Stenocephalus carolinensis*

Although commonly called a toad, this species is in the family Microhylidae, not the family Bufonidae, the true toads. Hecht and Matalas (1946) and Nelson (1972) reviewed the nomenclatural history of *Gastrophryne*.

IDENTIFICATION

Adults. The Eastern Narrow-mouthed Frog is a small smooth-skinned rotund pointy-snouted frog. The dorsum is brownish red to blue black and may be highly mottled with coppery or silvery pigmentation. A reddish (or chestnut) color may give the appearance of dorsolateral bands; the reddish pigmentation also may be present on the front legs. The head is flattened and small. There is a transverse dermal groove behind the eyes that may be covered by a fold of skin. This fold becomes more pronounced when the frog is irritated. The belly is medium to dark gray; brown splotches or light mottling may be present. There is no tympanum nor is there webbing between the toes. The rear feet have small spades that are used for digging into friable soils.

Sexual dimorphism is apparent. Males have a dark throat patch that never really disappears, although it may intensify in coloration during the breeding season. Most males also develop conical chin tubercles (1–28) along the ventral surface of the lower jaw (Mittleman, 1950) in at least some populations, and the area of the subgular throat patch appears wrinkled. These latter two characteristics are evident only during the breeding season. A few males may not have a dark throat or chin tubercles, however, and some breeding females may have tubercles in the perianal region (Nelson, 1972). Adults are 22–39 mm SUL. Albino adults from Louisiana were reported by Gordon (1955).

Males are generally smaller than females in SUL and body mass. There are literature records for males and females from Texas (means 27.4 mm SUL, 29.5 mm SUL) and Florida (means 23.4 mm SUL, 25.9 mm SUL) (Blair, 1955a). Loftus-Hills and Littlejohn (1992) recorded males with a mean SUL of 28.6 mm (range 29.9 to 32 mm), also from Texas, and Meshaka and Woolfenden (1999) recorded a mean of 25.7 mm SUL for males (range 19.8 to 29.3 mm) and 26.4 mm SUL for females (range 20 to 33 mm) from south Florida. Also in south Florida, Duellman and Schwartz (1958) recorded a mean of 26.2 mm SUL for males (range 18.8 to 30.5 mm) and 28.3 mm SUL (range 22.4 to 32.5 mm) for females. In north-central Florida, males ranged between 22 and 33 mm SUL and females between 21 and 35 mm SUL (Dodd, 1995). In Missouri, the largest males normally are 32.2 mm SUL, whereas the largest females normally are 34.3 mm SUL (Anderson, 1954). However, there is one record of an individual 39 mm SUL (*in* Johnson, 2000). In Arkansas, the mean size for males was 27.6 mm SUL and for females 29.6 mm SUL; a few individuals of both sexes reached 36.5 mm SUL (Trauth et al., 1999). In Alabama, adults were 22.9–33.6 mm

SUL (Brown, 1956). There is considerable overlap among age classes in terms of SUL.

Larvae. Tadpoles are small and jet black, with lateral distinct white to pink stripes on either side of the posterior portion of the body, extending onto the tail musculature. Bellies are deep and light in coloration. Viewed from the side, the head comes to a fairly sharp point. The jaws do not have keratinized teeth, and no oral disk is present. Tails are 1–1.5 times the length of the body and have a pointed tip. The maximum TL is 38 mm although normally the TL is 15–30 mm. Albino tadpoles were reported from Louisiana (Anderson, 1951). Descriptions of the tadpole are in Wright (1929, 1932), Orton (1946), and Siekmann (1949).

Eggs. Individual eggs are small and dark brown or black dorsally and white, yellowish, or brownish ventrally. The mean diameter of the envelope is 3.35 mm (range 2.8 to 4.0 mm); the vitellus is 0.7–1.3 mm (Wright, 1932; Livezey and Wright, 1947; Trauth et al., 1999). The eggs are deposited in a small surface film with a mosaic-like structure; the egg mass is round or squarish, with 10–150 eggs deposited in each mass. The egg jelly is truncate (i.e., flattened or appearing to be cut off). If disturbed, the surface film breaks up into individual eggs. Total annual clutch size varies considerably (see Reproduction), and females likely deposit partial clutches rather than a full complement with each oviposition.

Distribution of *Gastrophryne carolinensis*

DISTRIBUTION

The Eastern Narrow-mouthed Frog occurs from the Delmarva Peninsula (White and White, 2007) and southeastern Virginia throughout much of the Piedmont and Southeastern and Gulf Coastal Plains to the southern tip of Florida and the Florida Keys. In the mid-South, populations are found in Tennessee, Kentucky, and southwestern Virginia along tributaries of the Tennessee River. The species barely enters Illinois. The range extends across the Mississippi River through southern Missouri and into eastern Oklahoma and Texas. A population also occurs in extreme southeastern Kansas. Eastern Narrow-mouthed Frogs have been introduced into the Bahamas (Grand Bahama, New Providence) and to Grand Cayman Island (Lever, 2003).

The species naturally occurs on islands, including Cape Hatteras, Shackleford Banks, Smith, Roanoke, and Bodie islands in North Carolina (Lewis, 1946; Engels, 1952; Brothers, 1965; Gibbons and Coker, 1978; Gaul and Mitchell, 2007), Kiawah Island in South Carolina (Gibbons and Coker, 1978), Blackbeard, Ossabaw, St. Catherines, Tybee, Sapelo, Little Cumberland, and Cumberland islands in Georgia (Martof, 1963; Laerm et al., 2000; Shoop and Ruckdeschel, 2006), Big Pine Key, Cudjoe Key, Egmont Key, Key Largo, Key West, Lignumvitae Key, Little Torch, Matecumbe Key, Stock Island, Sugarloaf Key, Summerland Key, St. George Island, and St. Vincent Island in Florida (Duellman and Schwartz, 1958; Lazell, 1989; Franz et al., 1992; Irwin et al., 2001), and Dauphin Island, Alabama (Jackson and Jackson, 1970).

Important distributional references include: Range-wide (Nelson, 1972), Alabama (Brown, 1956; Mount, 1975), Arkansas (Black and Dellinger, 1938; Trauth et al., 2004), Georgia (Williamson and Moulis, 1994; Jensen et al., 2008), Illinois (Smith, 1961; Phillips et al., 1999), Louisiana (Dundee and Rossman, 1989), Kansas (Collins, 1993; Collins et al., 2010), Kentucky (Barbour, 1971), Maryland (Cooper, 1953; Harris, 1975), Missouri (Johnson, 2000; Daniel and Edmond, 2006), North Carolina (Dorcas et al., 2007), Oklahoma (Sievert and Sievert, 2006), Tennessee (Redmond and Scott, 1996), Texas (Hardy, 1995; Dixon, 2000), and Virginia (Fowler and Hoffman, 1951; Tobey, 1985; Mitchell and Reay, 1999).

FOSSIL RECORD

Miocene fossils referred to this species are known from Hemingfordian deposits in Florida (Auffenberg, 1956). Pliocene fossils are reported from Florida (Meylan, 2005), and Pleistocene Rancholabrean fossils are known from Florida and Georgia (Holman, 2003). The ilium of this species is identifiable by its well-developed triangular dorsal prominence.

SYSTEMATICS AND GEOGRAPHIC VARIATION

Frogs of the genus *Gastrophryne* are derived phylogenetically from the South American clade of microhylid frogs (Meijden et al., 2007). There are five species within this genus, but only two enter the United States. *Gastrophryne* is most closely related to the genus *Hypopachus* (Nelson, 1972).

Natural hybridization with *G. olivacea* has been reported in a narrow contact area in eastern Texas and Oklahoma (Hecht and Matalas, 1946), although the larger *G. carolinensis* males are reluctant to clasp the much smaller *G. olivacea* females. Strecker's (1909) description of *Microhyla areolata* appears to represent these hybrid individuals. Pattern phenotypes intermediate between *Gastrophryne olivacea* and *G. carolinensis* are known from a number of locations even where the species are not sympatric. Therefore, pattern variation is not based on hybridization alone (contrary to Hecht and Matalas, 1946), and care must be taken when identifying "hybrids" solely on morphological grounds (Nelson, 1972). Most pattern variation relates to the extent of mottling on the belly, sides, and throat. Laboratory crosses with *G. olivacea* may result in tadpoles (Blair, 1950). The diploid chromosome number is 22.

There appear to be differences in the size of adult *G. carolinensis* in varying parts of its range. Such variation becomes important when examining demographic traits, such as size and age at maturity. For example, Blair (1955a) provided measurement data for males from nine locations in Florida, Oklahoma, and Texas. He noted that there was a geographic trend in size, with smaller individuals in the East and larger individuals in the West. In addition, there appears to be variation in size by habitats. Nelson (1972) showed that males and females from lowland habitats were smaller than males from upland habitats. This size variation was consistent throughout the range of the species. Variation in size is noted in age at sexual maturity among populations (see Population Biology). The basis for the variation could be genetic or result from differences in environmental conditions.

In addition to size, morphological variation among populations is evident in the presence or absence of tubercles in males and in dorsal pattern. Individuals in some populations of *G. carolinensis* lack the chin tubercles (e.g., in Illinois; Smith, 1961) whereas other populations have only a small percentage of individuals with them (e.g., 10% in Alabama; Mount, 1975). Florida Keys populations of *G. carolinensis* also have dorsal patterns that differ from mainland populations. Most Keys frogs have two prominent light tan dorsolateral stripes bordered by a distinct dark margin on a tan background. Others have a *G. olivacea*-like color pattern. Frogs with a typical *G. carolinensis*-like pattern are found in the Keys, but they make up less than 25% of the population (Duellman and Schwartz, 1958). Similar pattern variation is evident in other *G. carolinensis* populations, but not with the same high frequencies of atypical patterns as in the Keys.

ADULT HABITAT

Gastrophryne carolinensis inhabits a wide range of environments, from flooded swamps, steephead ravines, and open pastures and fields to upland dry pine, mixed deciduous, or hardwood forests (Carr, 1940a; Anderson et al., 1952; Harima, 1969; Bennett et al., 1980; Enge et al., 1996; Enge, 1998a, 1998b; Enge and Wood, 1998; Baxley and Qualls, 2009). Different habitat types are occupied in different parts of the range, depending on availability. In Missouri, for example, Anderson (1954) identified four major habitats: cypress-gum swamps, bottomland hardwoods, live-oak ridges, and pine-oak woodlands. In uplands, it is frequently associated with sandy friable soils.

Gastrophryne carolinensis inhabits coastal marshes, caves and sinkhole depressions, springs, and river bluffs, which serve as refugia from extreme weather conditions. They may be found along small streams (Metts et al., 2001), and there are reports of this species in close proximity to beaches (Viosca, 1923) or brackish habitats (Noble and Hassler, 1936; Werler and McCallion, 1951; Hardy, 1953; Neill, 1958). In the West,

G. carolinensis tends to occupy lowland riverine forests extending its distribution into the prairie region.

The Eastern Narrow-mouthed Frog requires constant moisture, as frogs will quickly desiccate when exposed to very dry conditions or when left in traps without a moisture source. They generally favor habitats with dense leaf letter and surface debris covering organic soils. If retreats with high humidity (animal burrows, crevices, root channels) are available, however, they will occupy dry sandy areas. They often are found under surface debris, rocks, boards, logs, decaying vegetation (such as mats of water hyacinth), and in decaying stumps and rotten wood that remains moist or saturated. I have found them in palm stumps in mixed palm-hardwood hammocks on dry offshore islands in west-central Florida. They even occur in Florida muskrat (*Neofiber alleni*) houses in flooded wet prairies and in Florida woodrat (*Neotoma floridana*) middens.

TERRESTRIAL AND AQUATIC ECOLOGY

These are mostly semifossorial or fossorial frogs, although Anderson (1954) found an individual 2.4 m off the ground in a dead stump. The Eastern Narrow-mouthed Frog is capable of making short burrows under the surface of the ground, or they may occupy burrows of other animals (e.g., crayfish, mammals). Individuals will sit with only the top portion of the head visible at the surface, waiting for prey, normally ants, to pass by. *Gastrophryne* occurs in just about every terrestrial situation available during the nonbreeding season, from short-grass fields to dense meadows, and from cypress-gum swamps to upland forests and stands of palmetto flatwoods. They occur in anthropogenic habitats such as golf courses, lawns, gardens, urban lots, and even in agricultural fields.

By day, the frogs remain concealed in retreats. Most activity occurs at dusk or during the early evening, and surface activity is especially common during and immediately after precipitation. After about 22:00 hrs, activity is reduced and ceases by about 24:00 hrs. Activity occurs year-round in Florida (Dodd, 1995), with usually only brief periods of inactivity due to cold weather. Seasonal activity occurs from May through September or October at more northern locations (Gentry, 1955; Dodd, 1995).

Eastern Narrow-mouthed Frogs are photopositive in their phototactic response, suggesting they use both sunlight and moonlight when feeding and going about their daily activities (Jaeger and Hailman, 1973). They are sensitive to light in the blue spectrum ("blue-mode response") (Hailman and Jaeger, 1974). Eastern Narrow-mouthed Frogs likely have true color vision.

CALLING ACTIVITY AND MATE SELECTION

Males call throughout the breeding season when temperatures exceed 19.5°C, and water and air temperatures can approach 36°C and 33°C, respectively, when calling (Dundee and Rossman, 1989). In south Florida, calling may extend from March to October, whereas in North Carolina it occurs from June to August (Gaul and Mitchell, 2007). Records are available for Virginia for July and August (Mitchell, 1986), eastern Tennessee from June to September (Dodd, 2004), Arkansas from April to August (Trauth et al., 2004), and Oklahoma from May to July (Bragg, 1950a).

Both males and females move back and forth between retreat sites and breeding sites on a daily basis. Immigration and emigration patterns at breeding ponds may change annually and by sex and life stage (Dodd and Cade, 1998). Movements are nonrandom, depending on the type and availability of surrounding habitats. *Gastrophryne carolinensis* does not appear to use movement corridors between nearby terrestrial refugia and aquatic breeding sites, although juveniles use temporary wetlands as stopping points as they disperse from other wetlands. Terrestrial habitats may be occupied at considerable distances from the nearest wetland. In north-central Florida, for example, Dodd (1996) captured *G. carolinensis* from 42–914 m (mean 420 m) from the nearest possible breeding site while trapping for snakes in dry sandhills habitats.

Most terrestrial movements occur at dusk. Males arrive at breeding ponds by about 10 to 20 min prior to females and immediately begin calling from very shallow positions in the water. Calling occurs in bouts, with periods of silence in between bouts. Choruses usually consist of two or more males calling in tandem, although not all frogs call in each bout, and different combinations and numbers of frogs may call in successive choruses. Certain individuals may call far more often than others. Single males may call outside the bouts, but these individuals call less often than those in larger choruses. Anderson (1954) provided examples of male chorus calling patterns.

Males use vegetation, branches, detritus, bark, logs, or nearly any other object in the water for cover. The vegetation serves to conceal the male as well as offer him support and balance while calling. When calling, males raise their body in a vertical or near-vertical position and use their forelimbs for stability as they call. The substrate at the calling site may consist of mud, clay, sand, or vegetative debris. Indeed Carr (1940a) noted a unique calling behavior in this species whereby males buried their bodies in soft sand in very shallow water and called with only their snouts protruding.

The call has been compared to the bleating of a sheep, but some authors have likened it to a buzz with up to 40 consecutive harmonics; it is difficult to describe verbally. The mean call duration ranges from 0.04 to 2.4 sec at a mean frequency of 2.4–3.9 kHz, depending on temperature (Blair, 1955b; Nelson, 1972, 1973), and Anderson (1954) noted a duration of about 1.4 sec at 25°C in Missouri. Frequency tends to increase with temperature. The harmonic interval is 160–250 Hz. Nelson (1972) provided data on call characteristics from throughout the species' range. When a full chorus sings, the sound can be quite intense and has been described as a "din."

During the evening, males tend to remain in the same place and do not change their calling station even over many hours. After mating, they may leave their station. Calling males tolerate one another, even in close proximity, and calling site territoriality is not evident. Anderson (1954) recorded nearest males calling within 25 mm of one another to 6.1 m away from one another.

Females approach males and swim with their bodies floating and their rear legs widely extended. At the approach of the female, the male increases his calling rate. The cues for triggering amplexus are unknown, but males often remain in position calling even as females approach in very close proximity. Soft grunts and croaks are often heard just prior to amplexus. A male will approach a female from behind and swim into an amplexing position (either semipectoral or axillary), inching his way slowly up the female's back. Males are able to remain clasped to females by using their arms and by the adherence of special dermal secretions; essentially, they become glued to one another. Glands on the male's venter in the region of the sternum secrete an adhesive substance (Conaway and Metter, 1967), and the "breeding secretion" allows rotund males to remain in amplexus despite their short little arms. Amplexus may be necessary to trigger ovulation, and oviposition occurs 1.5–2 hrs after the initiation of amplexus.

Males may try to amplex other males or unreceptive females, resulting in inguinal amplexus. In successful amplexus with a receptive female, the male firmly holds the female with his thumbs, and his palms are turned out and upward. The male's back will ride just above water, the thighs are flexed, and the feet are held laterally and anteriorly. The female's body is held at an angle of 45° from the water's surface, her thighs are extended, and her feet also are held laterally and slightly anteriorly. An amplexed male and female will initially swim together, but the male soon stops and holds his legs tightly anterior. The female will swim in short, jerky motions with the male firmly clasping her. Shortly before oviposition, the female will stop moving and raise her hind limbs and rear part of her body toward the water's surface. This results in a backward flexed posture with the rear limbs at right angles and even with the water's surface. She moves backward a little, then extends her legs rigidly in a ventrolateral position. At this time, the cloaca will be above the water surface. The male then slides forward bringing his cloaca into close proximity with that of the female's cloaca. The eggs are fertilized as they burst from the female's cloaca in groups of about 30 eggs per ejaculation. After oviposition is completed, the male may remain clasped for a short time period before releasing the female. Anderson (1954) described mating behavior in detail.

BREEDING SITES

The species breeds in a variety of wetlands, including woodland pools, marshes, sloughs, seasonally flooded forest, depression wetlands, sinkhole ponds, cypress savannas, seasonally inundated wetlands in pastures, small to large permanent and temporary ponds, large semipermanent lakes, cypress/gum ponds, beaver ponds, and a variety of anthropogenic wetlands (Cash, 1994; Dodd, 1995; Babbitt and Tanner, 2000; Metts et al., 2001; Surdick, 2005; Babbitt et al., 2006;

Lichtenberg et al., 2006; Liner et al., 2008). They are frequently found in borrow pits and roadside ditches, and they can even breed in wheel ruts. Wetlands may contain deep water, but breeding only takes place in relatively shallow water (< 30 cm) covered by vegetation such as water hyacinth (*Eichhornia*) or alligator weed (*Alternanthera*). Mortality may be substantial if temporary breeding sites dry earlier than normal.

Nearby habitats may be important to the selection of breeding sites. For example, *Gastrophryne carolinensis* larvae are abundant in pasture wetlands located from 21 to 500 m from the nearest woodlands and in wetlands with relatively high pH and conductivity (Babbitt et al., 2006). Abundance is much less at distances closer than 20 m and greater than 500 m from woodlands. These results imply that optimal breeding sites may be associated with relatively open habitats, although the species is by no means restricted to such habitats. Predatory fish may be present, but wetlands without fish are preferred.

REPRODUCTION

Breeding occurs from spring through autumn, depending upon location. In Louisiana and Florida, it normally occurs from late April or May to mid-September (Carr, 1940a; Anderson, 1954; Bancroft et al., 1983; Meshaka and Woolfenden, 1999), although Meshaka and Woolfenden (1999) reported a gravid female in October from southern Florida. In Georgia, breeding occurs from April to October, although there are records for March and November on the Coastal Plain (Jensen et al., 2008). There may be temporal variation in seasonal breeding patterns between localities and years, depending upon precipitation and other environmental conditions. For example, Babbitt and Tanner (2000) only found tadpoles from June to August at a south Florida site, despite observations from other researchers of a more extended breeding season at this latitude.

The literature on reproduction sometimes presents differing accounts of diel reproductive behavior and the effects of environmental conditions on breeding. Different reports of the effects of precipitation, and the timing of calling (or breeding), may reflect differences in when the observations occurred (early in the season vs. later) or terminology (calling vs. breeding). Choruses usually are initiated by rain (Wright, 1932; Barbour, 1941; Anderson, 1954; Meshaka and Woolfenden, 1999; personal observation), and most calling occurs at night, especially early in the season. As the season progresses, rainfall becomes less important in stimulating at least some reproductive behavior by individuals, and calling becomes more likely during daylight hours, regardless of precipitation, especially at the height of the breeding season. Thus, Bragg's (1950c) suggestion that precipitation was not associated with rainfall in Oklahoma is technically accurate, since his observations were made in midsummer rather than early in the season. Amplexus and oviposition usually occur at night.

Spermatogenesis and oogenesis have been de-

Eggs of *Gastrophryne carolinensis*. Photo: C.K. Dodd Jr.

scribed in detail by Anderson (1954) and Trauth et al. (1999). Spermatozoa are produced in fall and winter, reaching their peak abundance in February. Secondary spermatocytes and spermatids appear in March and are soon ready to replace expended spermatozoa. By the end of the breeding season (e.g., August in Arkansas), however, most spermatozoa have disappeared from the male reproductive tracts, although even the smallest male (> 21 mm) has some remaining spermatozoa year-round.

Anderson (1954) described three stages of oocyte development, the last of which results in pigmented oocytes. Oocytes grow little in the fall and early winter, and pigmented oocytes first appear in January. Oocyte growth occurs steadily under normal conditions, but it can be delayed due to cold weather. Growth then accelerates to make up for the delayed growth. A single female may contain oocytes destined for two or three breeding seasons, and thus eggs may be at very different sizes and stages of development.

Egg counts vary considerably. In Missouri, Anderson (1954) counted 152–1,089 eggs per female (overall mean 510); the means were 378 for 2 yr old females (23–24 mm SUL), 466 for 3 yr old females (24.5–26.9 mm SUL), and 681 for 4 yr old females (> 27 mm). In Arkansas, counts ranged between 186–1,459 (mean 673); neither clutch size nor ovum size increased with female body size (Trauth et al., 1999). Wright and Wright (1949) reported clutches of up to 850 eggs, and Deckert (1914) noted that 100–150 eggs were oviposited at a time. The wide variation in reports of clutch size could be due to oviposition of partial clutches or reflect regional and individual variation. Early development was described by Ryder (1891).

LARVAL ECOLOGY

Hatching occurs in from 1.5–3 days after oviposition (Ryder, 1891; Wright, 1932). Larvae grow rapidly to achieve metamorphosis prior to pond drying; the larval period is 23–67 days (Wright, 1932). Recently transformed metamorphs are 10 mm SUL in Missouri (Anderson, 1954), 11 mm SUL in Florida (Dodd, 1995), and 7–12 mm SUL in Georgia (Wright, 1932). Tadpoles are filter feeders (on zoo- and phytoplankton) that do not possess rasping mouthparts.

Tadpoles of *G. carolinensis* are capable of acclimating to various thermal regimes, even to

Tadpole of *Gastrophryne carolinensis*. Photo: Stan Trauth

temperatures of 43–44°C (Cupp, 1980). Newly hatched larvae are capable of tolerating higher temperatures than later-stage larvae. As they approach metamorphosis, larvae can only tolerate temperatures to 39–40°C.

DIET AND FEEDING

The Eastern Narrow-mouthed Frog is an ant specialist. Individuals will often sit next to or even within an ant hill (Wood, 1948) and gorge themselves on ants leaving the mound. Other small invertebrates also may be consumed, such as mites, collembolans, isopods, spiders, snails, beetles, and termites (Anderson, 1954; Brown, 1974). In Arkansas, Brown (1974) found a mean of 35 to 45 prey per gut in *G. carolinensis*, depending upon sampling location.

PREDATION AND DEFENSE

This species is more of a walker than a hopper, and terrestrial escape attempts are awkward and cover only short distances. However, they can be surprisingly agile in short bursts and can quickly burrow into leaf litter or loose soil to escape. Individuals often will remain immobile, crouch, or try to walk away when faced by a predator (Marchisin and Anderson, 1978). When floating in the water, a disturbed animal or amplexed pair will dive to escape. *Gastrophryne* have a noxious skin secretion that makes them less prone to predation than other small frogs (Garton and Mushinsky, 1979). The secretion also protects them from attack by the ants on which they feed. It causes a burning or irritating sensation to delicate tissues such as the eyes and mucous membranes of the mouth and throat, and entangles attacking ants. Anderson (1954) noted that other frogs (*Hyla cinerea*, *H. squirella*) placed in close proximity with Eastern Narrow-mouthed Frogs could die from prolonged exposure. Secretions are produced by granular glands located throughout the skin; these

Adult *Gastrophryne carolinensis*. Photo: Todd Pierson

glands first appear late in development, coinciding with the eruption of the forelimbs.

Anderson (1954) reported that *Gastrophryne carolinensis* makes a faint clicking or chirp when roughly handled. Whether these sounds function in a defensive or alarm context is unknown, although it seems unlikely that such a relatively soft sound would deter a predator.

There are a few reports of predation by snakes (*Thamnophis* sp., *Agkistrodon contortrix*) and unidentified mammals (Wright, 1932; Anderson, 1942). Aquatic invertebrates are likely predators on larvae, but older larvae are distasteful or noxious to many predators. However, larvae are readily eaten by some native fish (Baber and Babbitt, 2003).

POPULATION BIOLOGY

Sexual maturity is reached at 21–26.7 mm SUL, depending upon location, in the second spring (first spring following metamorphosis), although some females may not deposit eggs until their third spring (second spring following metamorphosis). Secondary sex characteristics in males (black chins, mature spermatozoa, development of gonads) occur by 21–22.9 mm SUL in Missouri (Anderson, 1954) and 21.5 mm SUL in Arkansas (Trauth et al., 1999); pigmented eggs are first observed in females by 23 mm SUL in Missouri and 26.7 mm SUL in Arkansas. Based on recapture data in north-central Florida, 2 yr and 3 yr old *G. carolinensis* are a minimum of 27 mm SUL; a few *G. carolinensis* have been found 4 yrs after having been marked as adults and were > 32 mm SUL (Dodd, 1995).

Sex ratios appear to be male-biased both at breeding ponds and in many terrestrial samples (Anderson et al., 1952; Anderson, 1954; Dodd, 1995; Trauth et al., 1999; but see Meshaka and Woolfenden, 1999). For example, Trauth et al. (1999) found a sex ratio of 1.52 males per female in Arkansas, and Dodd (1995) found 1.3 males per female over a 5 yr period of continuous sampling (1,154 males, 929 females). Perceptions of skewed sex ratios may result from biased sampling, differential movement between the sexes, or differences in the amounts of time each sex spends at a breeding site. However, males may indeed outnumber females as a result of differences in survivorship or fertilization rates.

Whereas most females frequent breeding ponds every season, reproduction is not successful every year. Indeed, metamorphs are produced only when sufficient hydroperiod is maintained to complete larval development, and there may be many years when few or no metamorphs are produced at any particular breeding site or when no breeding occurs at all (Brown, 1956; Dodd, 1995; Semlitsch et al., 1996; Daszak et al., 2005). As a result, there is likely considerable fluctuation in recruitment and survivorship among *G. carolinensis* populations, making them vulnerable to drought and climate change.

There have been no mark-recapture attempts to document population sizes of *G. carolinensis* at any breeding site. Populations often appear large, but there may be considerable variation annually and among sites (e.g., Gibbons and Bennett, 1974; Greenberg and Tanner, 2005b). Dodd (1992) recorded 4,573 different *G. carolinensis* using a very small sandhill depression marsh in north-central Florida over a 5 yr period, despite a severe drought which limited reproduction to only 1 yr. Greenberg and Tanner (2005b) captured 3,117 *G. carolinensis* adults but only 168 juveniles during a 7 yr study in central Florida. In South Carolina, Semlitsch et al. (1996) recorded 3,072 adults and 2,930 metamorphs at a temporary pond over a 16 yr period, but most captures occurred in only 3 yrs for adults and 1 yr for metamorphs. These studies have shown that individuals can use even small shallow temporary ponds for breeding and as refugia in otherwise dry habitats, despite stochastic environmental conditions. Eastern Narrow-mouthed Frogs can be ubiquitous within a landscape when wetlands are abundant. In south Florida for example, Babbitt et al. (2006) found them at 79% of the sites they sampled.

Terrestrial abundance in forests also can be high. For example, Bennett et al. (1980) captured 4,195 *G. carolinensis* during two summers trapping in three forest types, and Dodd et al. (2007) found Eastern Narrow-mouth Toads to be the dominant amphibian on the forest floor of a variety of mesic, mixed palm-pine-deciduous forest associations on the northern Gulf Coast.

COMMUNITY ECOLOGY

The ranges of *G. carolinensis* and *G. olivacea* overlap in eastern Texas and Oklahoma. Although hybridization may occur in areas of contact, differences in call structure and body size help to minimize gene exchange. In zones of contact with *G. olivacea*, males and females are both larger than they are in areas where the two species are not sympatric (Blair, 1955a). The structure of the advertisement calls is also different between the species, with *G. carolinensis* having calls of lower frequencies and shorter duration than the calls of *G. olivacea* (Blair, 1955b; Awbrey, 1965; Nelson, 1972; Loftus-Hills and Littlejohn, 1992). Hybrids have call characteristics intermediate between the parental species. In choice tests, females prefer the calls of males of their own species to those of *G. olivacea* (Awbrey, 1965).

The tadpoles of *G. carolinensis* are palatable, but the behavior of the tadpole aids in predator avoidance. For example, mosquitofish (*Gambusia holbrooki*) readily forage effectively for tadpoles in structurally complex environments. However, tadpoles of Eastern Narrow-mouthed Frogs have greater survivorship in structurally complex habitats than in habitats with fewer places to hide (Baber and Babbitt, 2004). This is because the larvae are much less active than the tadpoles of other species and thus tend not to draw attention to themselves. With *Gastrophryne carolinensis* larvae remaining immobile much of the time, fish tend to focus on species more readily detectable.

DISEASES, PARASITES, AND MALFORMATIONS

Larval Eastern Narrow-mouthed Frogs inoculated orally with Frog Virus 3 or an FV3-like isolate developed disease symptoms, but exposure to water inoculated with the viruses did not result in infection (Hoverman et al., 2010). Infected frogs were 3.8 times as likely to die during development as uninfected larvae, and mild to moderate edema was the only sign of infection. The fungus *Basidiobolus ranarum* was reported from *G. carolinensis* in Arkansas and Missouri (Nickerson and Hutchison, 1971).

SUSCEPTIBILITY TO POTENTIAL STRESSORS

Metals. Selenium, mercury, silver, zinc, chromium, lead, cadmium, copper, germanium, cobalt, nickel, aluminum, and arsenic are toxic to embryos of *G. carolinensis* (Birge, 1978; Birge et al., 1979; Birge et al., 1983). The $LC_{50 \, (7 \, day)}$ for mercury and silver is < 0.01 ppb and for arsenic is 0.04 ppb. The LC_{ls} for mercury and silver are 0.1 and 0.6 ppb, respectively, and for lead is 3.2 ppb. Selenium is particularly toxic. Since these heavy metals are by-products of coal extraction, amphibians were projected to be good indicators of sensitivity to toxic waste effluents (Birge, 1978).

Chemicals. The pesticide phenyl saliginen cyclic phosphate (PSCP) causes edema and spinal deformities of *G. carolinensis* embryos at 500 ppb (Fulton and Chambers, 1985).

STATUS AND CONSERVATION

Gastrophryne carolinensis appears to be widespread and abundant, and there are no identified population declines. Undoubtedly populations are being lost to urbanization and agriculture, and many frogs likely are killed on highways. They often tolerate agriculture, urbanization, and silviculture rather well as long as breeding sites and terrestrial refugia are available (e.g., Surdick, 2005). However, population size may be severely reduced in rapidly urbanizing areas (Delis, 1993; Delis et al., 1996), and they may be uncommon in some agricultural settings and absent from former mine sites (Anderson and Arruda, 2006). Individuals are frequently encountered in managed forestry operations (Bennett et al., 1980; Vickers et al., 1985; Enge and Marion, 1986; Clawson et al., 1997; Hanlin et al., 2000; Russell et al., 2002b; Fox et al., 2004), although in some studies sample sizes are small. Eastern Narrow-mouthed Frogs are frequently found in open habitats created during forestry operations, such as in canopy gaps and along skidder trails (Cromer et al., 2002). Natural catastrophes such as hurricanes may affect population status for a while (Schriever et al., 2009), and it is unknown how long it will take for populations to recover. From 1990 to 1994, 197 *G. carolinensis* entered the pet trade from Florida (Enge, 2005a).

In terms of management, *G. carolinensis* will colonize created wetlands and successfully produce metamorphs (Briggler, 1998; Pechmann et al., 2001; but see Shulse et al., 2010). If fish are present, this species will not colonize restored wetlands. Protection of small, isolated wetlands during silvicultural operations helps maintain populations of this species (Russell et al., 2002a) as does the provision of wide (> 50 m) riparian zones (Rudolph and Dickson, 1990). The species also appears to tolerate prescribed burns and is frequently found in fire-maintained habitats (e.g., Moseley et al., 2003). Because of its peripheral range, the species is protected in Kansas.

Gastrophryne olivacea (Hallowell, 1857 "1856")
Western Narrow-mouthed Frog

ETYMOLOGY
olivacea: from Latin, referring to the olive-green dorsal coloration.

NOMENCLATURE
Synonyms: *Engystoma areolata, Engystoma texense, Engystoma olivaceum, Gastrophryne areolata, Gastrophryne carolinensis mazatlanensis, Gastrophryne texana, Gastrophryne texense, Gastrophryne texensis, Microhyla areolata, Microhyla carolinensis mazatlanensis, Microhyla carolinensis olivacea, Microhyla mazatlanensis, Microhyla olivacea, Microhyla olivacea mazatlanensis*

Gastrophryne olivacea has had a confusing nomenclatorial history, especially in Texas where several species were described (Strecker, 1909) that were subsequently referred to this species. Hecht and Matalas (1946) and Nelson (1972) reviewed the nomenclatural history of *Gastrophryne*. Although commonly called a toad, this species is in the family Microhylidae and thus is not a true toad.

IDENTIFICATION
Adults. This is a round smooth-skinned pointy-snouted frog with a dorsal uniform pale tan to olive to greenish-gray coloration that is paler in females than males. There are no prominent dorsal markings. Temperature and moisture affect the darkening of the skin, with frogs in low moisture and high temperature situations lighter than well-hydrated frogs at lower temperatures. There is a transverse dermal groove behind the eyes that may be covered by a fold of skin. This fold becomes more pronounced when the frog is irritated. Black spots are present dorsally and on the upper surface of the hind legs. Venters are white with slightly gray to brown pigmentation in the gular region of some specimens. No webbing is present between the digits and no tympanum is evident.

Gastrophryne olivacea is sexually dimorphic. Males have a black throat patch that never really disappears, although it fades during the nonbreeding season. The color of the throat even may intensify during the breeding season. Males develop conical chin tubercles along the ventral surface of the lower jaw, and the subgular throat patch appears wrinkled (when calling, the throat sac is much larger than the frog's head). These latter two characteristics are evident only during the breeding season. Males also develop a nuptial adhesive gland covering the thorax and inner surface of the arms (Fitch, 1956b) that secretes an adhesive substance that helps the male stay attached to the female during amplexus. Some breeding females may have prominent tubercles in the perianal region (Taylor, 1940).

Males are generally smaller than females. There are literature records for males and females from Texas (means 24.1 mm SUL; 27.4 mm SUL respectively) (Blair, 1955a). Blair (1955a) included measurement data for males from 18 locations in 8 Texas counties. He noted that there appeared to be a geographic trend in size, with smaller individuals in the east and larger individuals in the west. Loftus-Hills and Littlejohn (1992) recorded males with a mean SUL of 26.3 mm (range 23 to 30.8 mm), also from Texas. The largest recorded *G. olivacea* measured 42 mm SUL (Conant and Collins, 1998).

Larvae. Tadpoles are grayish brown or grayish tan dorsally and may or may not have an indistinct stripe on the side of the tail. The venter is light in coloration but may be mottled with gray. Viewed from the side, the head comes to a fairly sharp point. The jaws do not have keratinized teeth, and no oral disk is present. A single spiracle is present that opens midventrally, and the tail tip is dark. Aspects of the morphology of the tadpole were described by Nelson and Cuellar (1968).

Eggs. Individual eggs are small and brown or black dorsally and white ventrally. The diameter of the outer envelope is 2.8–3 mm, but there may be a second envelope near the vitellus. The vitellus is 0.8–0.9 mm (Livezey and Wright, 1947). Eggs are oviposited in a surface film whose jelly may be truncate (i.e., flattened or appearing to be cut off) or nontruncate. Total clutches normally contain from 500 to >2,100 eggs, but partial clutches are usually deposited in a single jelly film.

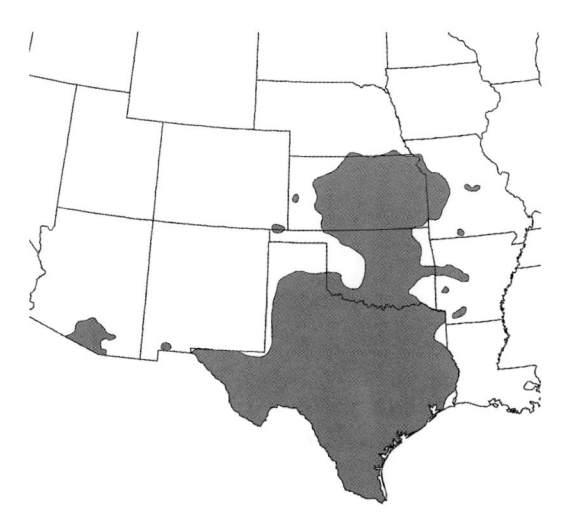

Distribution of *Gastrophryne olivacea*

DISTRIBUTION

The Western Narrow-mouthed Frog occurs from western Missouri and Kansas south throughout much of Oklahoma and eastern and central Texas, then as far south as Tamaulipas and San Luis Potosí, Mexico. It occurs eastward along the Missouri River into central Missouri and barely enters extreme western Arkansas. To the north, it occurs in a very small area of southern-central Nebraska. Populations occur westward through the Oklahoma Panhandle to southeastern Colorado. Isolated populations occur in the Guadalupe Mountains of west Texas and in southern New Mexico. An additional disjunct population occurs from Nayarit, Mexico, into southern Arizona.

Important distributional references include: Range-wide (Nelson, 1972), Arizona (Sullivan et al., 1996b; Brennan and Holycross, 2006), Arkansas (Trauth et al., 2004), Colorado (Hammerson, 1999), Kansas (Smith, 1934; Collins, 1993; Collins et al., 2010), Mexico (Lemos-Espinal and Smith, 2007b), Missouri (Johnson, 2000; Daniel and Edmond, 2006), Nebraska (Lynch, 1985; Ballinger et al., 2010; Fogell, 2010; Geluso and Wright, 2010), New Mexico (Degenhardt et al., 1996), Oklahoma (Sievert and Sievert, 2006), and Texas (Dixon, 2000).

FOSSIL RECORD

Pleistocene Rancholabrean fossils are known from Texas and Sonora, Mexico (Holman, 2003). The transverse dorsal prominence of the ilium is lower and less triangular than in *G. carolinensis*, making discrimination possible.

SYSTEMATICS AND GEOGRAPHIC VARIATION

Frogs of the genus *Gastrophryne* are derived phylogenetically from the South American clade of microhylid frogs (Meijden et al., 2007). There are five species within this genus, but only two enter the United States. The genus *Gastrophryne* is most closely related to frogs of the genus *Hypopachus* (Nelson, 1972). Two species described from Texas (*areolata, texensis*) were synonymized with *Gastrophryne olivacea* and may represent hybrids, variations in color patterns, or misidentified specimens (see discussions in Smith, 1933; Burt, 1937).

Laboratory-based hybridization between *G. olivacea* and *Chiasmocleis panamensis* is not very successful (Littlejohn, 1959) and with *Hypopachus* may produce a very few metamorphs although hatching success is low (Wilks and Laughlin, 1962). Natural hybridization with *Gastrophryne olivacea* has been reported in a narrow contact area in eastern Texas and Oklahoma (Hecht and Matalas, 1946), although the larger *G. carolinensis* males are reluctant to clasp the much smaller *G. olivacea* females. Strecker's (1909) description of *Microhyla areolata* appears to represent these hybrid individuals. Pattern phenotypes intermediate between *Gastrophryne olivacea* and *G. carolinensis* are known from a number of locations, even where the species are not sympatric. Therefore, pattern variation does not result from hybridization alone (contrary to Hecht and Matalas, 1946), and care must be taken when identifying "hybrids" solely on morphological grounds (Nelson, 1972). Most pattern variation relates to the extent of mottling on the belly, sides, and throat. Laboratory crosses with *G. carolinensis* may produce tadpoles (Blair, 1950).

In addition to sexual size dimorphism, Nelson (1972) showed that males from lowland habitats were smaller than males from upland habitats. This size variation was consistent throughout the range of the species. He did not have sufficient data to determine whether this trend applied to females, however.

ADULT HABITAT

Gastrophryne olivacea is a frog of the Great Plains and grassland deserts (such as Big Bend National Park; Dayton and Fitzgerald, 2006), where it occupies a wide variety of habitats and is not particularly restricted to any special type of habitat. It prefers open wooded habitats with abundant cover, such as provided by the flat limestone slabs, rock outcrops, and bluffs of eastern Kansas (Fitch, 1956b; Heinrich and Kaufman, 1985). In other areas, the species occupies mesquite flats, prairie grasslands, and overgrazed desert scrub dominated by mesquite, creosote bush, and desert grasses, sometimes with little surface cover (Smith, 1950; Degenhardt et al., 1996). It is not generally a frog of extensive river floodplains, although it may occupy sloughs along streamcourses. Adjacent habitat provides terrestrial cover and foraging sites, and the frogs even may occupy cultivated fields and other agricultural habitats if ants are in abundance.

The species occupies underground cracks and crevices, including burrows of mammals such as pocket gophers (*Geomys bursarius*), prairie dogs (*Cynomys ludovicianus*), and voles (*Microtus ochrogaster*), or lizards (*Crotaphytus collaris*) (McAllister and Tabor, 1985; Lomolino and Smith, 2003). Considering that this species often is found in limestone areas, it is not surprising that individuals are found in caves (Black, 1973a) or near the entrance to caves. The species occupies habitats below 1,342 m in New Mexico (Degenhardt et al., 1996) and 1,525 m in Colorado (Hammerson, 1999) and is considered a frog of the lowlands.

TERRESTRIAL ECOLOGY

The Western Narrow-mouthed Frog may be more widespread and abundant than it appears, due to its fossorial or semifossorial habits. *Gastrophryne olivacea* is active in warm weather, especially under conditions of high humidity or during or immediately after warm rains. Most activity occurs nocturnally when frogs leave their shelters to forage, but diurnal activity may occur when night temperatures are too low for normal activity or during cloud cover and precipitation. Preferred temperatures range between 24 and 31°C. Surface activity occurs at temperatures >20°C, although activity at as low as 16°C has been observed (Fitch, 1956b). The upper limit for activity is 37.6°C. This species is thus one of the most sensitive anurans in the Great Plains to cold but one of the least sensitive to heat.

Activity occurs throughout the warm season. For example, *G. olivacea* have been recorded from April to October in Kansas, with rare captures in December (Fitch, 1956b). Activity also may be curtailed by drought, causing them to cease surface movements several months earlier than normal.

Fitch (1956b) characterized four movement patterns within *G. olivacea*: routine short daily movements; home range shifts occurring over long or short periods either gradually or abruptly; movements to and from breeding ponds; dispersal by recent metamorphs. It appears these small frogs are capable of considerable movement. For example, Freiburg (1951) noted a female that moved 229 m straight-line distance over a period of 55 days between recaptures; a male moved 102 m in 3 days. Movements > 30 m appear quite common, and females and males move to the same extent and distances, although Fitch (1956b) suggested males were more vagile than females. This species is often found far from the nearest source of water, and Fitch (1956b) recorded the greatest distance moved as 610 m from water.

During cold or otherwise inhospitable weather, this species takes refuge in loose moist soil under large flat rocks, under the bark of fallen trees, in rock or mud crevices, in and under moist logs, and under surface debris (Freiburg, 1951). They will often occupy the burrows of other animals, such as invertebrates, lizards, and mammals. Strecker (1909) found them overwintering with two copperheads (*Agkistrodon contortrix*), the lizard *Scincella lateralis*, and the frogs *Lithobates sphenocephalus* (?), *Hyla cinerea*, *H. squirella*, and *Ollotis nebulifer*. Many other vertebrates and invertebrates share their terrestrial refugia (Freiburg, 1951).

Like most frogs that are active both day and night, Western Narrow-mouthed Frogs are generally photopositive in their phototactic response, suggesting they use both sunlight and moonlight when feeding and going about their daily activities (Jaeger and Hailman, 1973). They are particularly sensitive to light in the blue spectrum ("blue-mode response"), which apparently helps them orient toward areas of increasing illumination, such as the open horizon above lakes and ponds (Hailman and Jaeger, 1974). Western Narrow-mouthed Toads likely have true color vision.

CALLING ACTIVITY AND MATE SELECTION

Calls are heard throughout the warmest season of the year (generally May to August), but they occasionally are heard as early as March in Texas (Blair, 1961b) and April in Oklahoma (Bragg, 1943a, 1950a). In Kansas, calls have been recorded from June to September (Freiburg, 1951), and females with eggs have been found in July and August (Smith, 1934). In New Mexico, calling occurs from June to August (Degenhardt et al., 1996). Thus, calling may precede and follow the primary dates of egg deposition (Wiest, 1982). Calling occurs most often from dusk until about midnight, after which it tapers off appreciably. Occasional calls may be heard during daylight, but most individuals withdraw to nearby retreat sites during daylight hours.

Precipitation is necessary to initiate large choruses, especially at the start of the breeding season. In Kansas, such storms usually occur from May to mid-June (Fitch, 1956b), although frogs may appear as early as mid-April. Large choruses are sometimes heard after a considerable time with no chorusing; these later choruses are triggered by heavy rainfall, especially after substantial periods without precipitation. This suggests an opportunistic breeding strategy to take advantage of optimum breeding conditions. Wiest (1982) noted calling at 11.3–27.2°C over a 34-day period in Texas.

Calls have been described as "a high shrill buzz of some two to three seconds" (Smith, 1934) or as insect-like and difficult to describe (Fitch, 1956b).

Calls are uttered at a mean rate of 7 per minute. They do not have much carrying capacity, and it is difficult to hear the calls at any distance away from a breeding site. The mean call duration ranges from 0.9–3.7 sec at a mean frequency of 2.6–5.0 kHz, depending on temperature (Bragg, 1950c; Blair, 1955b; Nelson, 1972, 1973; Loftus-Hills and Littlejohn, 1992). The harmonic interval is 155–280 Hz. Call frequency tends to increase with temperature. There may be geographic differences in call characteristics. For example, the frequency is less in Arizona populations and the call duration shorter when compared with Texas populations. Nelson (1972) provided data on call characteristics from throughout the species' range.

Males call while floating in shallow water among the vegetation or hidden among dense grass clumps. In Kansas, they have also been heard calling from under protective rock ledges adjacent to breeding sites. The calling posture is similar to that described for *Gastrophryne carolinensis*, with males almost vertical as they call. There does not appear to be any calling territoriality, as many males may call in close proximity to one another (Bragg, 1943a; Freiburg, 1951).

Amplexus may occur shortly after calling begins. Females approach calling males and likely initiate breeding. Amplexus is axillary. Males are able to remain clasped to females using their arms and by the adherence of dermal secretions produced from glands located on the male's venter; essentially, they become glued to one another (Fitch, 1956b). After oviposition, it takes up to 15 min for the secretions to break down and allow the pair to separate.

A short (0.5 sec) nasal-pitched buzz has been described from *G. olivacea* occupying terrestrial

Eggs of *Gastrophryne olivacea*. Photo: Dana Drake

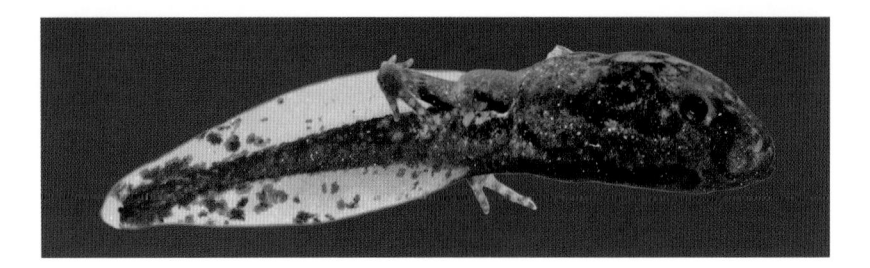

Tadpole of *Gastrophryne olivacea*. Photo: Laurie Vitt

refugia in Texas. Individuals making this call were not associated with a breeding chorus, and Dayton (2000) interpreted the calls as agonistic territorial calls. Such calls could establish calling hierarchies prior to the start of breeding choruses, or they may have nothing to do with reproductive behavior.

BREEDING SITES

Breeding occurs in a wide association of wetlands, including tinajas, temporary pools, large permanent ponds, inundated alluvial floodplains, roadside ditches, and even in man-made habitats such as stock tanks and irrigation pits (Smith, 1934; Sullivan et al., 1996b; Anderson et al., 1999; Dayton and Fitzgerald, 2006). Campbell (1934) noted calling from pools at the base of trees and suggested that the tree roots offered diurnal retreats. They frequently call from bison wallows, but these sites may not hold water long enough for successful metamorphosis (Gerlanc, 1999). In the desert, breeding occurs in pools located in dense stands of mesquite shrubs in river washes. In the Great Plains, this species breeds in wetlands with a pH of 6.9–7.5, low aluminum concentration, low dissolved oxygen content (0.7 ppm), and with dense vegetation forming a considerable amount of cover across a wetland (Anderson et al., 1999).

REPRODUCTION

Breeding occurs over an extended season and is largely dependent upon precipitation. Some females breed during the first rains, whereas others stagger breeding throughout the summer as do many xeric-adapted anurans. Thus, each precipitation event results in some breeding. Eggs are attached to vegetation in shallow water at the water's surface in an irregular oblong film. Clutch sizes are quite variable, with reports of from several hundred to more than 2,000 eggs per female. For example, clutch size ranged from 59 to 2,174 eggs (mean 1,166) in 28 Texas females (Henderson, 1961). Except for a single female with 59 eggs, all other clutches contained >519 eggs. In Kansas, clutch size ranged from 532 to 1,217 with most clutches >800 eggs during the breeding season; 50 and 75 tiny immature eggs were found in spent ovaries in August, presumably after the breeding season (Freiburg, 1951). Wright and Wright (1949) give a clutch size of 645. Some of the literature-based variation in clutch size may reflect counts of partial clutches.

Cold temperatures inhibit hatching, and eggs must be at temperatures >17–19°C in order to hatch (Hubbs and Armstrong, 1961; Ballinger and McKinney, 1966). Prolonged drought and decreased rainfall are probably the biggest threats to successful reproduction as they eliminate breeding ponds and/or decrease the duration of the hydroperiod, leaving eggs or larvae to perish. *Gastrophryne olivacea* may skip breeding during periods of drought.

LARVAL ECOLOGY

Hatching occurs in two to three days following egg deposition. Tadpoles are filter feeders on phyto- or zooplankton, and their mouthparts are incapable of rasping algae or other plant surface detritus. The length of the larval period is dependent on temperature and resource availability, and there are reports of 24 days (Fitch, 1956b) and 30–50 days to complete development (Wright and Wright, 1949). Newly metamorphosed froglets range from 10 to 12 mm SUL (Wright, 1932; Wright and Wright, 1949). Froglets with a small vestige of a tail stump just prior to metamorphosis may be 14.5–16 mm SUL as they aggregate around the margins of a pond (Fitch, 1956b).

DIET AND FEEDING

The Western Narrow-mouthed Frog is an ant specialist (termed myrmecophagy), especially foraging on *Crematogaster*, a subterranean species (Smith, 1934; Tanner, 1950; Freiburg, 1951; Fitch, 1956b). Like *Gastrophryne carolinensis*, individuals will often sit near an ant-hill or colony and feed on ants leaving the mound or forage near it (Tanner, 1950; Carpenter, 1954b). Freiburg (1951) found from 7 to 71 (mean 32.5) ant heads in the digestive tracts of 15 adults in Kansas. Other prey include small beetles.

PREDATION AND DEFENSE

The primary defenses of this species involve its cryptic coloration, secretive behavior, and noxious secretions. *Gastrophryne olivacea* tends to walk, run, or move in short hops, so saltation cannot be used to escape predators. They are capable of short, elusive bursts of movement followed by complete immobility. The Western Narrow-mouthed Toad has noxious skin secretions that make them less prone to predation than other small frogs and protects them from attack by the ants on which they feed. It causes a burning or irritating sensation to delicate tissues such as the eyes and mucous membranes of the mouth and throat, and entangles attacking ants. If a frog is handled, copious secretions are exuded making it slippery. As in *G. carolinensis*, secretions are produced by granular glands located throughout the skin, which first appear late in development coinciding with the eruption of the forelimbs.

Frogs at breeding ponds are often very difficult to locate as they call hidden in vegetation. At the approach of a predator (or observer), they stop calling and do not resume until the predator has moved several meters away. They may remain immobile or dive under water and hide in dense bottom vegetation. Tadpoles also remain immobile in the water column or hang motionless at the water's surface.

Known or suspected predators include other frogs (*Lithobates blairi*, *L. catesbeianus*), snakes (*Agkistrodon contortrix*, *Thamnophis cyrtopsis*, *T. proximus*), and possibly shrews (*Blarina brevicauda*) (Anderson, 1942; Freiburg, 1951; Fouquette, 1954; Fitch, 1956b; Fitch, 1960; *in* Collins, 1993). Eggs are fully palatable to fish (Grubb, 1972).

POPULATION BIOLOGY

In Kansas, sexual maturity is reached by 25 mm SUL (Freiburg, 1951). This size may be reached late in the first season following metamorphosis in frogs that metamorphose early in the season. Indeed, *Gastrophryne olivacea* can reach 19–28 mm SUL before the first overwintering period, due to rapid growth (Freiburg, 1951; Fitch, 1956b). At least some *G. olivacea* probably breed initially about one year following metamorphosis, whereas others (e.g., those that metamorphose late in the breeding season) may require a second growing season to reach maturity. Whether temporal differences in age at first reproduction reflect sexual differences in the timing of maturity (e.g.,

Breeding adult *Gastrophryne olivacea*. Photo: Dennis Suhre

Breeding habitat of *Gastrophryne olivacea*. Photo: Mike Forstner

females reaching maturity later than males), individual variation in growth rates (Fitch, 1956b), or differences in the timing of metamorphosis remains to be clarified. The bulk of a population consists of 3 yr olds. *Gastrophryne olivacea* in their 4th year are 30–38 mm SUL (males to 38 mm and females to 42 mm), and some frogs may live even longer. Frogs have been held as captives for six years.

Like many observations on sex ratios in frogs, males appear to be more abundant than females, but this perception could be biased by sampling and differences in activity patterns between the sexes. In Kansas, Freiburg (1951) found sex ratios of from 1:1–4:1 males per female depending upon month sampled.

COMMUNITY ECOLOGY

The ranges of *G. carolinensis* and *G. olivacea* overlap in eastern Texas and Oklahoma. Although hybridization may occur in areas of contact, differences in call structure and body size help to minimize gene exchange. In zones of contact with *G. carolinensis*, males and females are both smaller than they are in areas where the two species are not sympatric (Blair, 1955a). The structure of the advertisement calls is also different between the species, with *G. olivacea* having calls of higher frequencies and longer duration than the calls of *G. carolinensis* (Blair, 1955b; Awbrey, 1965; Nelson, 1972; Loftus-Hills and Littlejohn, 1992). Hybrids have call characteristics intermediate between the parental species. In choice tests, females prefer the calls of males of their own species to those of *G. carolinensis* (Awbrey, 1965).

Decreases in body mass occur in larval *G. olivacea* when reared with *Scaphiopus couchii* larvae (Dayton and Fitzgerald, 2001). Data suggest that *S. couchii* are able to out-compete *Gastrophryne olivacea* larvae and may exclude them from shallow water desert pools in some areas, such as Big Bend National Park.

Western Narrow-mouthed Frogs form an unusual association with tarantula spiders (*Dugesiella hentzi*) (Blair, 1936; Yeary, 1979; Hunt, 1980; Dundee, 1999). Tarantulas and *Gastrophryne* occupy the same burrows. Despite their small size, the narrow-mouthed toads are not attacked by the spider. If a frog predator attacks, the toads will huddle under the spider for protection. Likewise, narrow-mouthed toads will not eat baby tarantulas, no matter how small they are. In return for protection, the toads eat ants and termites that might otherwise attack the spider's eggs and hatchlings; they also find shelter in a favorable environment. This unlikely cohabitation is an example of mutualistic symbiosis. Many

Gastrophryne may occupy the burrow of a single spider.

DISEASES, PARASITES, AND MALFORMATIONS
No evidence of ectoparasites has been reported. Freiburg (1951) recorded unidentified nematodes in the gastrointestinal tract. The nematodes *Aplectana incerta* and *A. itzocanensis* were found in *Gastrophryne olivacea* from Arizona (Goldberg et al., 1998b). The coccidian protozoan *Isospora fragosum* also occurs in *Gastrophryne olivacea* (Upton and McAllister, 1988).

SUSCEPTIBILITY TO POTENTIAL STRESSORS
No information is available.

STATUS AND CONSERVATION
The Western Narrow-mouthed Frog appears to be widespread and generally abundant, though secretive. As with all amphibians, populations may have been lost due to habitat destruction and alteration. Because of its peripheral range, the species is considered "Endangered" in New Mexico.

Hypopachus variolosus (Cope, 1866)
Sheep Frog

ETYMOLOGY
variolosus: from the Latin *variola* meaning 'variegated' or 'with small spots' in reference to the dorsal pattern.

NOMENCLATURE
Synonyms: *Engystoma inguinalis, Engystoma variolosum, Hypopachus alboventer, Hypopachus alboventer reticulatus, Hypopachus caprimimus, Hypopachus championi, Hypopachus cuneus, Hypopachus cuneus nigroreticulatus, Hypopachus globulosus, Hypopachus inguinalis, Hypopachus maculatus, Hypopachus ovis, Hypopachus oxyrhinus, Hypopachus oxyrrhinus ovis, Hypopachus oxyrrhinus taylori, Hypopachus oxyrhynchus, Hypopachus oxyrrhinus, Hypopachus reticulatus, Hypopachus seebachi, Hypopachus variolosus inguinalis, Systoma variolosum*

IDENTIFICATION
Adults. Sheep Frogs are globular in shape with no neck and pointed snouts. There is a small fold of skin just behind the eyes, and there is no visible tympanum. The ground color is greenish brown to olive, usually with an orange or yellow middorsal stripe. The ground color shades to brown laterally, and there may be dark reticulations on the side. An oblique white band is present from the eye to the shoulder. The skin is smooth and exceptionally thick, but there are scattered white-tipped tubercles on the dorsal and lateral surfaces and on the forelegs and margins of the lower jaw. These tubercles may serve an adhesive function. The venter is mottled gray or yellowish with a midventral white line; dark reticulations may or may not be present. There are well-developed palmar tubercles at the base of the first and fifth toes of the hind foot. Throats are black in males. Adults from Texas are 27–42.5 mm SUL (mean 35.8 mm) (Nelson, 1974). Wright and Wright (1949) reported males as 25–37.5 mm SUL and females as 29–41 mm SUL. Lemos Espinal and Smith (2007a) mention specimens to 47 mm SUL from Guerrero.

Larvae. Tadpoles are medium-sized and darkly pigmented dorsally but immaculate ventrally on the body. This pigmentation difference is sharply delineated laterally. The tail musculature is darkly pigmented. The dorsal tail fin is highest halfway down the tail and evenly but darkly pigmented or with dark spots running together. The ventral tail fin has scattered spots but is not as darkly or evenly pigmented as the dorsal tail fin. The tail ends in a sharp tip. Jaws are without keratinized structures, and the oral disk and labial teeth are absent. The edge of the labial flaps are scalloped or with distinct papillae. Larvae are 27–30 mm TL. Tadpoles are illustrated by Wright (1929).

Eggs. The eggs are black dorsally and white ventrally with a single jelly envelope 1.5–2 mm in diameter; the vitellus alone is 1 mm in diameter (Mulaik and Sollberger, 1938). Eggs are deposited in a surface film.

Distribution of *Hypopachus variolosus*

DISTRIBUTION
The Sheep Frog occurs in the United States only in extreme south Texas (15 counties) along the Gulf Coast. The native range of the species extends southward to Costa Rica, but the Texas population appears disjunct from those farther south in Mexico. Important distributional references include Dixon (2000) and Judd and Irwin (2005).

FOSSIL RECORD
No fossils are known.

SYSTEMATICS AND GEOGRAPHIC VARIATION
Originally described as *Hypopachus cuneus* by Cope (1889), the Texas *Hypopachus* was synonymized with *H. variolosus* by Nelson (1974). It is closely related to the genus *Gastrophryne*. Laboratory crosses between *Hypopachus variolosus* and *Gastrophryne olivacea* produce a small percentage of larvae and metamorphs (Wilks and Laughlin, 1962).

ADULT HABITAT
Hypopachus variolosus is a frog of humid open woodlands, savannas, mature coastal brushlands, and pasturelands with short grass cover. Throughout much of its range, it is a frog of thorn scrub and savannas.

TERRESTRIAL ECOLOGY
Sheep Frogs are fossorial and rarely observed and then only during heavy rains from April to October (Mulaik and Sollberger, 1938; Judd and Irwin, 2005). These authors reported them in rodent burrows, pack rat middens, and the hollows under trees. They burrow backward and move deeper as moisture decreases. After metamorphosis, froglets migrate to upland burrows, often stopping along the way in cow dung, surface detritus, and litter (*in* Wright and Wright, 1933). Dispersal occurs in high humidity and during rainfall, even diurnally.

CALLING ACTIVITY AND MATE SELECTION
Males call while floating in the water or from the shoreline of ponds within 24 hrs after heavy rainfall during the warm season; Garrett and Barker (1987) give the breeding season as March to September. The call of *Hypopachus variolosus* is described as a bleat, similar to a sheep, lower in tone than that of *Gastrophryne olivacea* (Mulaik and Sollberger, 1938). Calls last 0.8–6 sec (mean range 1 to 4.5 sec) and are repeated at >15 sec intervals. The dominant frequency is 1,500–3,900 Hz (mean 2,620) with a harmonic interval of 100–220 Hz (mean 180) (Nelson, 1973, 1974). Calling occurs from 18–30°C.

BREEDING SITES
Breeding occurs in shallow temporary to permanent pools. They also breed in roadside ditches, railroad right-of-way ditches, and pothole basins (Judd and Irwin, 2005).

REPRODUCTION
Reproduction is stimulated by heavy rainfall, but can occur in response to the sudden irrigation of agricultural fields (Judd and Irwin, 2005). Oviposition occurs within 24 hrs of rainfall. The eggs are oviposited in a surface film that is loosely held together. The film has a somewhat truncate or flattened appearance similar to the surface films of *Gastrophryne*. Mulaik and Sollberger (1938) reported a single clutch of 700 eggs, whereas Wilks and Laughlin (1962) obtained 452 eggs from a single female. Eggs hatch in ca. 12 hrs after deposition (Mulaik and Sollberger, 1938).

LARVAL ECOLOGY
The larval period lasts about 30 days (Mulaik and Sollberger, 1938). Newly metamorphosed froglets are 11–16 mm SUL.

Adult *Hypopachus variolosus*, south Texas. Photo: Seth Patterson

DIET
The diet of adults consists of ants, termites, and small flies (Mulaik and Sollberger, 1938; Garrett and Barker, 1987). Larvae are likely generalist feeders of organic and inorganic matter.

PREDATION AND DEFENSE
Predators include ribbon snakes (*Thamnophis sauritus*) (Wright and Wright, 1949).

POPULATION BIOLOGY
No information is available.

DISEASES, PARASITES, AND MALFORMATIONS
The mite *Caeculus hypopachus* was described from this species (Mulaik, 1945).

SUSCEPTIBILITY TO POTENTIAL STRESSORS
No information is available.

STATUS AND CONSERVATION
Sheep frogs are considered "Threatened" by the State of Texas, but are abundant in several counties in south Texas (Judd and Irwin, 2005). No information is available on status or population trends.

Family Rhinophrynidae

Rhinophrynus dorsalis
Duméril and Bibron, 1841
Mexican Burrowing Toad

ETYMOLOGY
dorsalis: from the Latin *dorsalis*, possibly referring to the dorsal pattern.

NOMENCLATURE
Synonyms: *Rhinophryne rostratus*, *Rhinophrynus rostratus*

IDENTIFICATION
Adults. Rhinophrynus dorsalis is a rotund frog with very short but powerful limbs, a pustulose and loose skin, a short little head, tiny eyes with vertical pupils, and no tympanum. The ground color is dark brown to black and appears somewhat translucent. There is a distinctive light yellow to red-orange mid-dorsal line down the back with similarly colored small spots on both sides. Venters are lighter than the dorsum but still dark in coloration. The hind legs are partially enclosed in loose body skin. Large spade-like tubercles are present on the rear feet. The snout is protuberant. They reach about 50–90 mm SUL at maturity, with females larger than males (*in* Fouquette, 2005).

Larvae. The head is broad and depressed, with small eyes. The mouth is unique in that it is a wide slit bordered by 11 short barbels, giving it the appearance of whiskers. There are no keratinized structures associated with the jaws, and the oral disk and labial teeth are absent. The spiracles are paired and laterally positioned unlike in other native North American frogs. The tail fins and musculature are well-developed and taper to a narrow pointed tip. Larvae reach to 39.5 mm TL. Orton (1943) describes the tadpole and illustrates the unique oral barbels surrounding the mouth.

Eggs. The eggs have not been described.

DISTRIBUTION
The Mexican Burrowing Toad occurs from south Texas to northwestern Costa Rica. In the United States, *R. dorsalis* is known only from Starr and Zapata counties, Texas. Important distributional references include James (1966) and Dixon (2000).

FOSSIL RECORD
Pleistocene fossils of *R. dorsalis* are known from a cave in Tamaulipas, Mexico (Holman, 2003).

SYSTEMATICS AND GEOGRAPHIC VARIATION
Rhinophrynus is a monotypic genus allied within a highly derived, primitive group of frogs, the aglossal pipids of the suborder Pipoidea. The diploid chromosome number is 22 (Bogart and Nelson, 1976).

Distribution of *Rhinophrynus dorsalis*

Tadpole of *Rhinophrynus dorsalis*. Photo: Seth Patterson

Adult *Rhinophrynus dorsalis*, southern Texas. Photo: Seth Patterson

ADULT HABITAT

Garrett and Barker (1987) noted that Mexican Burrowing Toads prefer areas with loose soil for digging. In Texas it is found in agricultural areas and gardens in the Matamoran District of the Tamaulipan Biotic Province. The area in Texas consists of rolling hills of sand and gravel over thin soils. These hills are interspersed by deep arroyos fed from shallow ravines and washes. Vegetation is arid to semiarid trees and shrubs. James (1966) recorded breeding in arroyos surrounded by very thick tangles of thorny vegetation consisting of cacti, acacias, and retama (*Parkinsonia aculeata*).

TERRESTRIAL ECOLOGY

As its common name implies, *Rhinophrynus dorsalis* is almost entirely fossorial and rarely comes to the surface. They can remain for long periods underground. Surface activity is opportunistic and occurs only in response to rainfall. They dig backward into the soil with digging spades on the rear feet. They also twist their body extensively back and forth and inflate their bodies to facilitate penetration into the soil. Body inflation also allows them to maintain a cavity space as the soil settles around them. They also can effect some forward motion using their widely spaced, powerful spatulate forelimbs and tuberculate hands (Trueb and Gans, 1983). Nothing is known concerning migration or movement patterns.

CALLING ACTIVITY AND MATE SELECTION

The call of *R. dorsalis* is a loud but low-pitched guttural moan (James, 1966; Garrett and Barker, 1987). Calling occurs from within burrows, with males emerging after heavy rains to form large breeding choruses. Males then call from the

Adult *Rhinophrynus dorsalis* peering from burrow. Photo: Seth Patterson

surface of the water or on soil or among short vegetation bordering flooded sites. When calling, James (1966) likened males to inflated balloons floating on the water. James (1966) mentioned very large and loud breeding choruses.

BREEDING SITES
Breeding takes place in shallow temporary pools and flooded areas formed after heavy rainfall. These ponds may contain extensive vegetation. James (1966) also mentioned breeding in stock tanks and drainage ditches.

REPRODUCTION
Eggs are deposited in small clumps, which separate and float to the surface. No information is available on clutch size or any other aspect of reproduction.

LARVAL ECOLOGY
The duration of the larval period is at least two months (*in* Fouquette, 2005). In other parts of its range, tadpoles form large aggregations that are maintained by visual or olfactory cues.

DIET
Rhinophrynus dorsalis is an ant and termite specialist. The species has a number of unique morphological specializations for feeding on these species, including an epidermal armor on the snout, ornately folded buccal and esophageal lining folds, an ability to "double-close" the lips, and a specialized tongue apparatus for handling small prey in subterranean burrows (Trueb and Gans, 1983). The spade-like tubercles on the hind feet are used to dig into termite mounds. Larvae feed on phytoplankton.

PREDATION AND DEFENSE
When threatened, the Mexican Burrowing Frog inflates its body thus obscuring the head and limbs. There is no information available on predators.

POPULATION BIOLOGY
No information is available.

DISEASES, PARASITES, AND MALFORMATIONS
Parasitic opalinid protozoans have been reported from this species in Mexico.

SUSCEPTIBILITY TO POTENTIAL STRESSORS
No information is available.

STATUS AND CONSERVATION
Dixon (2000) stated that there were no records of this species since 1984 except for a chorus heard in 1998 after a tropical rainstorm. Texas protects this species as "Threatened," but there is nothing known concerning its status and trends (Fouquette, 2005).

Frogs of the United States and Canada

Frogs of the United States and Canada

VOLUME 2

C. Kenneth Dodd Jr.

The Johns Hopkins University Press
Baltimore

This book was brought to publication with the generous support of the U.S. Geological Survey's Amphibian Research and Monitoring Initiative, the Center for Biological Diversity, and the Herpetologists' League.

© 2013 The Johns Hopkins University Press
All rights reserved. Published 2013
Printed in China on acid-free paper
9 8 7 6 5 4 3 2 1

The Johns Hopkins University Press
2715 North Charles Street
Baltimore, Maryland 21218-4363
www.press.jhu.edu

Library of Congress Cataloging-in-Publication Data

Dodd, C. Kenneth.
　Frogs of the United States and Canada / C. Kenneth Dodd Jr.
　　p. cm.
　Includes bibliographical references and indexes.
　ISBN 978-1-4214-0633-6 (hdbk. : alk. paper) — ISBN 1-4214-0633-0 (hdbk. : alk. paper)
　1. Frogs—United States. 2. Frogs—Canada. I. Title.
　QL668.E2D57 2013
　597.8′9—dc23　　　2012017648

A catalog record for this book is available from the British Library.

Special discounts are available for bulk purchases of this book. For more information, please contact Special Sales at 410-516-6936 or specialsales@press.jhu.edu.

The Johns Hopkins University Press uses environmentally friendly book materials, including recycled text paper that is composed of at least 30 percent post-consumer waste, whenever possible.

Contents

List of Abbreviations   vii

SPECIES ACCOUNTS

Family Ranidae

Lithobates areolatus   461
Lithobates berlandieri   466
Lithobates blairi   472
Lithobates capito   479
Lithobates catesbeianus   486
Lithobates chiricahuensis   515
Lithobates clamitans   522
Lithobates fisheri   547
Lithobates grylio   551
Lithobates heckscheri   556
Lithobates okaloosae   561
Lithobates onca   565
Lithobates palustris   568
Lithobates pipiens   578
Lithobates septentrionalis   608
Lithobates sevosus   617
Lithobates sphenocephalus   621
Lithobates sylvaticus   637
Lithobates tarahumarae   669
Lithobates virgatipes   674
Lithobates yavapaiensis   681
Rana aurora   687
Rana boylii   697
Rana cascadae   707
Rana draytonii   715
Rana luteiventris   723
Rana muscosa   733
Rana pretiosa   739
Rana sierrae   747

Family Scaphiopodidae

Scaphiopus couchii   753
Scaphiopus holbrookii   761
Scaphiopus hurterii   772
Spea bombifrons   777
Spea hammondii   786
Spea intermontana   791
Spea multiplicata   798

ESTABLISHED NONNATIVE SPECIES

Dendrobates auratus   809
Eleutherodactylus coqui   812
Eleutherodactylus planirostris   815
Glandirana rugosa   819
Osteopilus septentrionalis   822
Xenopus laevis   828

Glossary   833
Bibliography   837
Index of Scientific and Common Names   975
Index of Potential Stressors   981

Frogs of the United States and Canada

Family Ranidae

Lithobates areolatus
(Baird and Girard, 1852)
Crawfish Frog
Northern Crawfish Frog
 (*L. a. circulosus*); Southern
 Crawfish Frog (*L. a. areolatus*)

ETYMOLOGY
areolatus: from the Latin *aesopus* meaning 'dwarf.' The name refers to the hunchback appearance of the frog.

NOMENCLATURE
Conant and Collins (1998): *Rana areolata*
 Dubois (2006): *Lithobates* (*Lithobates*) *areolatus*
 Synonyms: *Rana areolata, Rana circulosa*

IDENTIFICATION
Adults. Lithobates areolatus is a large stocky frog with a large head and prominent eyes. The ground color is light gray to brown, with the dorsum covered by large round dark spots that are surrounded by light halos. The spots are interspersed by dark reticulations. Dorsolateral folds are present, which may or may not be prominent. The skin is smooth or rugose. The rear limbs are banded with dark markings bordered by light stripes. Venters are off-white. During the breeding season, males have enlarged thumbs and paired vocal sacs, which are evident behind and below the tympanum.

Females are generally larger than males. Records include males 64–117 mm SUL (mean 93 mm) and females 75–118 mm SUL (mean 98 mm) in Illinois (Smith et al., 1948) and adults 57–110 mm SUL, also in Illinois (Smith, 1961). In another Illinois population, males averaged 82.8 mm SUL (range 71 to 90 mm) and females 89.6 mm SUL (range 79 to 102 mm) (Redmer, 2000). Length is positively correlated with body weight and age. Collins et al. (2010) report a Kansas specimen of 122 mm SUL.

Larvae. The tadpole of *L. areolatus* is large (to 65 mm TL; Smith et al., 1948) and deep-bodied, and dark brown to various shades of green (light, dark, olive) dorsally. Viewed dorsally it is ovoid and chunky with a rounded blunt snout. Upper and lower tail fins may or may not be heavily marked with small diffuse spots, and the dorsal tail fin is attenuated and elongate. In general, the tail fin is equally pigmented above and below the tail musculature. The lower jaws are wide, and the mouth's beak is broadly marginated. Throats are unpigmented or uniformly pigmented. The gut may or may not be visible through a white venter. Bragg (1953) offers a number of additional identification characters. The larvae are similar to those of leopard frogs (*L. blairi, L. pipiens, L. sphenocephalus*) and may be impossible to distinguish.

Eggs. Eggs are medium in size and dark colored. Two gelatinous envelopes are present; the outer envelope is 3.3–5 mm in diameter and the inner envelope is 1–3.3 mm in diameter; the vitellus is 1–3.8 mm in diameter (Smith, 1934; Livezey and Wright, 1947; Smith et al., 1948; Bragg, 1953). Egg measurements are smaller in Illinois than in Kansas, indicating the potential for regional variation, although differences in the condition of the specimens may account for the disparity. Eggs are deposited in an ovoid firm jelly mass and attached to vegetation below the water's surface. These masses may be invaded by algae, giving them a green appearance.

Distribution of *Lithobates areolatus*

SYSTEMATICS AND GEOGRAPHIC VARIATION

Crawfish Frogs are in the *Nenirana* clade of North American ranid frogs, a group that includes *L. palustris*, *L. capito,* and *L. sevosus* (Hillis and Wilcox, 2005). They are only distantly related to the *Scurrilirana* clade. Two subspecies have been identified, *L. areolatus areolatus* and *L. areolatus circulosus*. *L. a. areolatus* is said to be larger with a more rounded snout and a greater dorsal rugosity than *L. a. circulosus*. Other differences involve body proportions and the prominence of the dorsolateral folds. Bragg (1953) found some of these characters useful in separating subspecies, but noted a large degree of overlap. Calls of these subspecies are very similar.

Lithobates areolatus does not appear to hybridize with other *Lithobates* in nature. Under laboratory conditions, viable hybrids have been produced between ♂ *L. sphenocephalus* and ♀ *L. areolatus*. However, crosses between ♂ *L. areolatus* and ♀ *L. sphenocephalus* are inviable or only produce a small number of larvae (Cuellar, 1971). Other laboratory studies have produced a small to large number of larvae in crosses between *L. areolatus* and *L. blairi*, *L. sphenocephalus*, *L. palustris*, *L. pipiens*, and a number of Mexican *Lithobates* (Moore, 1949a; McAlister, 1961; Mecham, 1969; Cuellar, 1971; Hillis, 1988). Crosses between *L. berlandieri*, *L. catesbeianus*, *L. clamitans*, or *L. sylvaticus* and *L. areolatus* are unsuccessful (Moore, 1949a, 1955; Mecham, 1969; Cuellar, 1971).

DISTRIBUTION

The Northern Crawfish Frog occurs from southern Iowa (isolated populations in western and northwestern Iowa; Hemesath, 1998), Illinois and western Indiana, south to central Mississippi and southeastern Arkansas. A disjunct population occurs in southeastern Indiana. In the west, it occurs in eastern Kansas and Oklahoma, where it intergrades with the Southern Crawfish Frog. The Southern Crawfish Frog occurs from eastern Oklahoma and southwestern Arkansas south to the Texas Gulf Coast. This species is often associated with river floodplains, such as along the Arkansas and Missouri Rivers. Distribution is often spotty.

Important distributional references include: Arkansas (Bacon and Anderson, 1976; Trauth et al., 2004), Illinois (Smith, 1961; Phillips et al., 1999), Indiana (Swanson, 1939; Minton, 2001; Brodman, 2003; Engbrecht and Lannoo, 2010), Iowa (Bailey, 1943; Christiansen, 1998; Hemesath, 1998), Kansas (Collins et al., 2010), Louisiana (Dundee and Rossman, 1989), Missouri (Johnson, 2000; Daniel and Edmond, 2006), Oklahoma (Sievert and Sievert, 2006), Tennessee (Redmond and Scott, 1996), and Texas (Dixon, 2000).

FOSSIL RECORD

Fossils of *Lithobates areolatus* are reported from Miocene (Kansas) and Pliocene (Texas) deposits (Holman, 2003).

ADULT HABITAT

Southern Crawfish Frogs are found in open damp areas, wooded valleys, oak-hickory woodlands, floodplains, and open and brushy fields (Bragg, 1953; Clawson and Baskett, 1982). The Northern Crawfish Frog is a species of tallgrass prairie and grasslands in some areas (Johnson, 2000), although Engbrecht and Lannoo (2010) noted that many of these areas were forested prior to European settlement. Thompson (1915) recorded the species in Illinois in an area of rolling hills and agriculture with few streams and no natural ponds or lakes. In eastern Kansas, Crawfish Frogs are found in gentle terrain with deep (0.9–1.5 m) clay soils, whereas in western Tennessee it was considered common in flat sandy or semiswampy areas (Gentry, 1955). Areas with shallow soils and intensive mechanized agricultural activity are not favored.

TERRESTRIAL ECOLOGY

During the nonbreeding season, the Crawfish Frog is largely fossorial, although frogs may be active throughout the day and night at burrow entrances and even leave the burrow for long periods of time (Hoffman et al., 2010). They prefer the burrows of crayfish (particularly *Cambarus diogenes*, which is often not associated with wetlands), but have been found in mammal burrows, under logs, and in tunnels in road cuts. Occupied crayfish burrows may extend 90 cm below the surface (Thompson, 1915; Bailey, 1943). The sides of the burrows tend to become slick as the frog moves up and down, and the bottom of the burrow will be full of frog feces. They likely will utilize any burrow available, including tree root channels. More than one frog may occupy a burrow, and burrow occupancy can be extensive; Bailey (1943) reported all 12 burrows observed within a 15 m radius were occupied by one or more frogs.

Juveniles in particular are quick to use preexisting burrows, which helps to minimize moisture loss; they are capable, however, of digging their own burrows using only their hind limbs (Parris, 1998). In Illinois, individuals have been found in agricultural fields 15 cm below the surface not long after the breeding season (Smith et al., 1948). Crawfish Frogs forage at the entrance of their burrows as ambush predators. They frequently leave the burrow and forage in terrestrial litter. For example, Hoffman et al. (2010) reported a crawfish frog in Indiana that spent 87% of its time (237 hrs) outside the burrow and only 13% (36 hrs) in its burrow over a period of just under 13 days. Another frog spent as much as 91% of its time (162 of 179 hrs) within its burrow. From mid- to late summer, periods of inactivity tended to be shorter than periods of activity outside burrows. As the season progresses, activity periods become shorter. Hoffman et al. (2010) observed one frog outside its burrow as late as 13 December.

Emergence in spring is related to increasing temperatures and rainfall. Several days with temperatures ≥15°C followed by moderate rainfall are enough for frogs to emerge and move to breeding sites. Cool temperatures early in the spring may inhibit activity, even if breeding is underway. Activity occurs from March to September in Kansas (Collins et al., 2010) and from March to October in Missouri (Johnson, 2000).

CALLING ACTIVITY AND MATE SELECTION

Rainfall is necessary to stimulate movement to breeding ponds, and calling occurs nocturnally from January to July, depending on weather and latitude. The call is a deep guttural snore, but Bragg (1953) also noted a short call, which he likened to a dog barking. Calls are made from shallow water while sitting on the bottom, although some will call from the bank. In deeper water, males call while floating on the surface. Bragg and Smith (1942) noted that calling occurred while the male was "sprawled out" in the water. The call produces a distinct vibration as the male's paired vocal sacs beat the water's surface. Chorusing is most intense during and shortly after nightfall and during peak chorusing can last all night. Occasional males will call during cloudy or rainy days. Males will call at temperatures as low as 2–8°C, but temperatures ≥13°C are necessary to initiate calling (Busby and Brecheisen, 1997).

Males arrive 5–14 days prior to females and remain at the breeding site throughout the breeding season, which lasts 22–55 days (Smith et al., 1948; Bacon and Anderson, 1976; Busby and Brecheisen, 1997; Trauth et al., 2004). Peak chorusing only lasts three to four consecutive nights and usually occurs about a week after chorus initiation. Choruses often are small and consist of <10 males, although occasionally much larger choruses of up to several hundred males are heard. Large choruses are associated with native grassland habitats. Females depart immediately after oviposition. Males will continue to call for some time after most breeding has ceased. Interestingly, Cagle (1942) mentions "calling from their holes in the fields" following the main breeding season.

Males clasp females shortly after the latter arrive. Amplexus is axillary.

BREEDING SITES

Crawfish Frogs normally breed in temporary fishless ponds that contain water for at least five months to allow larval development. Ponds usually have grassy margins. Both natural (prairie sloughs and depressions) and artificial ponds are used, such as stock ponds and old buffalo wallows (Thompson, 1915; Busby and Brecheisen, 1997). They have been reported from managed ponds where fish have been stocked, but the stocking occurred after the larvae had reached a size sufficient to minimize predation (Palis, 2009). Breeding sites also include large ponds with depths up to 2 m, roadside pools, long shallow ditches, overflow pools from clear streams, deep muddy pools, golf course ponds, and flooded fields (Bragg and Smith, 1942; Bragg, 1953). Bailey (1943) noted they preferred ponds in stream valleys. Vegetation may be scant within the pond, but algae, pondweed, and other vegetation should be present in at least a portion of it to serve as food and shelter for tadpoles. Bragg (1953) noted that considerable vegetation fringed breeding sites in Oklahoma.

REPRODUCTION

Oviposition occurs in shallow water that is often very cold during peak breeding in late February to early April. However, breeding records extend from January to July, depending on location (Smith, 1934; Bragg and Smith, 1942; Cagle, 1942; Bragg, 1953; Dundee and Rossman, 1989; Trauth et al., 1990, 2004; Johnson, 2000; Minton, 2001; Collins et al., 2010). Air and water temperatures must be >8°C for breeding to commence, with most activity occurring at >12°C. Diminished calling and breeding may occur at ≤6°C (Smith et al., 1948; Bacon and Anderson, 1976).

Eggs are oviposited in a plinth 120–210 mm in diameter and ca. 25 mm thick. The mass is attached under water, and water depths are usually 150–200 mm (Busby and Brecheisen, 1997). Egg masses may be communally located (Bragg, 1953), with Busby and Brecheisen (1997) reporting 22 masses in a 1 m² area. Clutch size ranges between 2,000 and 7,000 eggs per egg mass. Specific counts include 3,192–6,807 in Illinois (Smith et al., 1948) to >7,000 in Indiana (Wright and Myers, 1927), 3,208–6,807 (mean 4,868) in Illinois (Redmer, 2000), single clutches of 2,233 in Arkansas (Trauth et al., 1990), and 3,801 in Oklahoma (Bragg, 1953). Clutch size is strongly positively correlated with female SUL but only weakly with age; clutch size is negatively cor-

Tadpole of *Lithobates areolatus*. Photo: Laurie Vitt

related with ovum size (Redmer, 2000). Hatching occurs in three to four days, although Johnson (2000) indicated it took seven to ten days.

LARVAL ECOLOGY

Larvae grow rapidly, with Bragg (1953) reporting growth rates of 1.06 mm/day during the first 15 days. Growth rates gradually decrease to 0.9–0.6 mm/day. Bragg (1953) estimated that the growth rate for 59 days following the initiation of feeding (4 days after hatching) was 0.76 mm/day. The length of the larval period was 63 days after hatching in Bragg's (1953) study. Recent metamorphs are 22–30 mm SUL (Wright and Myers, 1927; Cagle, 1942; Smith, 1961). Bailey (1943) reported finding large larvae in early spring in Iowa, suggesting that some larvae overwinter and transform the following summer. In a within-pond field enclosure experiment, Williams et al. (2012) noted that larvae reared at low densities were larger at metamorphosis and survived better than larvae reared at high densities. When released terrestrially, large juveniles resulting from large tadpoles had higher survivorship than small juveniles resulting from small tadpoles.

DIET

The diet includes beetles, spiders, crickets, ants, millipedes, centipedes, and small crayfish (Thompson, 1915; Smith, 1934; Smith et al., 1948). The type of beetles eaten indicates nocturnal feeding. Any animal that can fit into the mouth will likely be consumed.

PREDATION AND DEFENSE

When disturbed, Crawfish Frogs quickly retreat down their burrows. The hind limbs in particular are used as wedges to prevent extraction from the burrow. They are extremely wary in breeding ponds and readily become quiet and submerge at the approach of an intruder. Thompson (1915) included a photo of a *Lithobates areolatus* with the body arched and inflated in what might be a defensive posture. Altig (1972b) also provided a photograph and noted that the posture was assumed in response to snake and small mammal predators and often accompanied by a loud scream. This species has a noxious odor that may serve an antipredator function. Crawfish Frogs also have antimicrobial peptides in their skin, which may assist in protecting the frog against microorganisms (Ali et al., 2002). Larvae and postmetamorphs are likely eaten by a variety of invertebrate and vertebrate predators, but no information is available. Adults are eaten by hognose snakes (*Heterodon platirhinos*) (Engbrecht and Heemeyer, 2010) and raccoons (Heemeyer et al., 2010).

POPULATION BIOLOGY

Not surprisingly, sex ratios at breeding ponds are highly skewed toward males. For example, Smith et al. (1948) recorded a sex ratio of 6.08:1 in an Illinois pond. At some ponds, populations appear to have been quite large at one time. For example, Cagle (1942) noted a pond in southern Illinois that contained 500 breeding adults that deposited 179 egg masses; the pool only measured 36 m × 91 m. In 3 other pools, there were 115, 125, and 75 egg masses. Cagle (1942) also reported large numbers of adults being collected (289 from several small ponds) and removing the 179 egg masses. Such large-scale collecting could have adversely affected the local population. Smith (1961) noted breeding aggregations contained as many as several hundred animals, and Barbour (1971) noted that breeding populations were large in western Kentucky.

Males mature earlier than females, with a mean male age of 3.53 yrs (range 2 to 5 yrs) and a female mean of 3.83 yrs (range 3 to 5 yrs) in Illinois (Redmer, 2000).

COMMUNITY ECOLOGY

Lithobates areolatus may breed in ponds occupied by other ranid species. In experimental ponds, interspecific competition with *L. blairi* and *L. sphenocephalus* resulted in an increased larval period and a decreased body mass of metamorphic *L. areolatus* (Parris and Semlitsch, 1998). Competition has a density and species-specific component. For example, survivorship of *L. blairi* larvae is decreased in the presence of *L. areolatus* larvae at high density. In contrast, the presence of *L. areolatus* seems to facilitate growth of *L. sphenocephalus* larvae. Complex interactions such as these are important components of larval amphibian communities.

DISEASES, PARASITES, AND MALFORMATIONS

The fungal pathogen *Batrachochytrium dendrobatidis* (Bd) has been found on *Lithobates areolatus* from Indiana. A total of 53% of adults tested over a 2 yr period tested positive for Bd, with more adults exiting the pond testing positive than adults entering the pond. Mortality occurred especially when frogs contained >10,000 zoospores. Infection rates are near zero at the end of

Adult *Lithobates areolatus circulosus*. Photo: David Dennis

summer, but increase to >25% following overwintering in crayfish burrows; rates then double again following breeding, when mortality occurs (Kinney et al., 2011).

SUSCEPTIBILITY TO POTENTIAL STRESSORS
Chemicals. Death occurs in about 17 hrs at 30 mg/L of carbaryl, a broad-spectrum insecticide, and significantly reduces larval activity levels at 2.5 mg/L (Bridges and Semlitsch, 2000). This is among the most sensitive species to carbaryl among ranids so far tested.

STATUS AND CONSERVATION
The Crawfish Frog was once widely distributed and reasonably common, but through the years many populations have been lost due to habitat loss and degradation. For example, many records in Indiana are >50 yrs old, and there are no current records for many counties (Minton, 2001; Brodman, 2003; Engbrecht and Lannoo, 2010). Christiansen (1998) reported no recent records at 9 historic locations in Iowa. Minton (2001) indicated populations began declining about 1970 for no obvious reason. In contrast, Florey and Mullin (2005) noted increasing detection during road-call surveys in Illinois from 1986 to 1989; Phillips et al. (1999) later indicated many populations were no longer extant in much of the state. This species is considered "Endangered" in Indiana (Engbrecht and Lannoo, 2010) and "Threatened" in Iowa (Christiansen, 1981).

Threats to Crawfish Frogs include pond draining, road mortality, habitat loss, and fragmentation. They may be evicted from burrows during plowing and mowing, and this appears to have occurred commonly in the past (Hurter, 1911; Thompson, 1915). Even in 1913, the species was reported as becoming increasingly rare in Illinois due to agricultural activities (Thompson, 1915). Minton (2001) also noted that the species was used for food in Indiana and Illinois, where it was easily collected. Although present in some agricultural settings, the species is more abundant in natural habitats and absent from mined habitats (Anderson and Arruda, 2006). Populations in Indiana were destroyed by mining (Minton et al., 1982). Bragg (1953) noted that the lights and noise from a nearby highway did not seem to disturb the frogs.

In order to conserve this species, breeding and upland habitats must be protected, especially in the prevention of lowered water tables due to ground water pumping. Crawfish Frogs will occasionally occupy newly created mitigation wetlands as long as a source population is nearby (Palis, 2007), and Johnson (2000) noted that the Missouri Department of Conservation was in the process of constructing fishless ponds in managed prairies for this species.

Lithobates berlandieri (Baird, 1859)
Rio Grande Leopard Frog

ETYMOLOGY
berlandieri: a patronym honoring Jean Louis Berlandier (1805–1851). Berlandier was a French naturalist who worked for the Mexican Government surveying eastern Texas in 1827–1828. His extensive collections were the first made in Texas.

NOMENCLATURE
Conant and Collins (1998): *Rana berlandieri*
  Dubois (2006): *Lithobates (Lithobates) berlandieri*
  Synonyms: *Rana austricola, Rana halecina, Rana pipiens berlandieri, Rana virescens*

References to this species in the literature are often incorrect. For example, Smith and Sanders (1952) discussed the distribution of "*Rana pipiens berlandieri*" in Texas, but included specimens from what is now known to be *Lithobates blairi*. The nomenclatural history of this species is discussed by Hillis (1988). Readers should verify the locations of specimens in order to ensure correct species identification. Degenhardt et al. (1996) noted that the date of the description of this species is often incorrectly listed as 1854.

IDENTIFICATION
Adults. This light brown, olive-green, tan, or grayish-colored leopard frog has well-developed dorsolateral folds that are discontinuous and curve medially toward the rear of the frog. A

supralabial stripe is present, but it is indistinct anterior to the eye. Light borders may or may not surround the dark dorsal spots, but they are usually faint when present. No spots usually are present on the nose in front of the eyes, although a few individuals may have a single spot or faint mottling. A tympanic light spot is usually absent. Throats and the anterior portions of the chest are mottled, especially in older frogs, and the level of mottling appears to increase after dark (Sanders and Smith, 1971). Venters are cream colored. The posterior part of the thigh has sharply contrasting dark reticulations. Males have paired external vocal sacs and prominent vestigial oviducts. Hind toes are webbed. Males are generally smaller than females, with a mean body length of 64.4 mm SUL and 73.5 mm SUL for females (*in* Degenhardt et al., 1996). The maximum size is 114 mm SUL (Brennan and Holycross, 2006).

Larvae. The tadpole is long and slim and has a variable color pattern. The overall color is a dark grayish black in small tadpoles that becomes olive with a yellowish cast as the tadpoles grow. The lateral line system is generally obscure, especially on the head. Spots on the side of the body are gray. The tail is moderately deep with a narrow pale gray tail muscle. The belly and throat are white. The tail pattern consists of discrete dark olive and pale spots, or it may have a strikingly dark olive reticulated pattern enclosing pale spots. The iris is gold and contains black flecks. The tadpole was described by Hillis (1982) and Scott and Jennings (1985).

Eggs. Eggs are deposited in a firm jellied mass that is about 70–90 mm across (*in* Degenhardt et al., 1996). The eggs likely are bicolored, white on the bottom and dark on top, although Dayton et al. (2007) describe them simply as black. The egg capsule is 3.2–5.1 mm in diameter (mean 3.9 mm) (Grubb, 1972). Clutch size has not been reported.

DISTRIBUTION

Lithobates berlandieri occurs naturally in southwest Texas and southeastern New Mexico (lower Pecos drainage of Eddy County) into the adjacent Mexican states of Coahuila and Chihuahua and well south to Veracruz and Oaxaca. Although "*L. berlandieri*" has been identified as far south as Nicaragua, it is likely that frogs in southern Mexico and Central America represent diverse

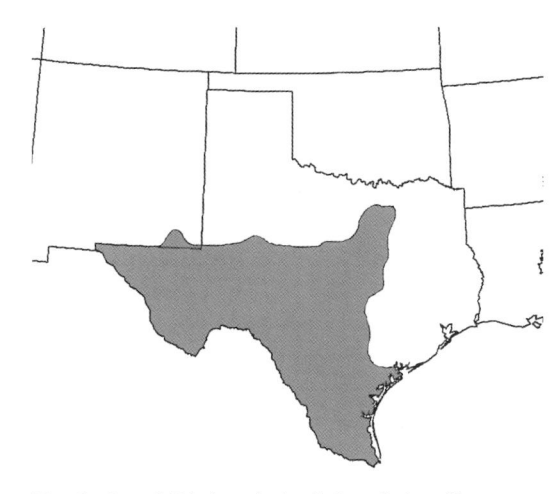

Distribution of *Lithobates berlandieri*, exclusive of introduced populations

taxa (reviewed by Rorabaugh, 2005b). The species has been introduced into the southwestern United States across southern Arizona (Gila, Salt, and Agua Fria drainages) and along the southern Colorado River of Arizona and California (Clarkson and Rorabaugh, 1989; Platz et al., 1990). It has also spread into Utah and Mexico (Lemos-Espinal and Smith, 2007b; Kraus, 2009). Important distributional references include Platz et al. (1990), Degenhardt et al. (1996), Dixon (2000), and Brennan and Holycross (2006).

FOSSIL RECORD

Pleistocene fossils referred to as "*Rana pipiens*" by Holman (1969) from Llano and Bexar counties, Texas, may be referable to *Lithobates berlandieri*. Similar fossils reported for Denton County, Texas, are in the contact zone between *L. blairi* and *L. berlandieri* (Dixon, 2000) and may be referable to either species.

SYSTEMATICS AND GEOGRAPHIC VARIATION

Lithobates berlandieri is a member of the *Novirana* clade of North American ranid frogs, and it is classified within the *Pantherana*, a group that includes the leopard and gopher frogs and comprises the *Rana pipiens* complex. Mating calls within the *Pantherana* are highly complex in structure, and include elements described as chuckles, grunts, and snores (Hillis and Wilcox, 2005). Mating calls are only produced during breeding choruses, whereas other elements of the call are produced at other times of the year in the *Pantherana*. Rio Grande Leopard Frogs are more closely related phylogenetically to *Lithobates sphenocephalus*, *L. blairi*, *L. yavapainensis*, *L. onca*, and several Mexican species (the *Scurrilirana*, frogs that produce a chuckle-like call) than they are to *Lithobates pipiens* and *L. chiricahuensis*.

Lithobates berlandieri hybridizes in nature with other members of the Leopard Frog complex, such as *L. sphenocephalus* and *L. blairi* in central and west Texas (McAlister, 1962; Littlejohn and Oldham, 1968; Sage and Selander, 1979; Platz, 1981; Kocher and Sage, 1986). Contact hybrid zones may be rather narrow in extent (8 km in Texas) and stable through time, although hybridization can occur over a wider area (36–75 km) as demonstrated by biochemical analysis (Sage and Selander, 1979). Backcrossing into the parental populations is not extensive. Hybridization rates are low; for example, Platz (1981) recorded 5.8% hybrids at a contact zone between *L. blairi* and *L. berlandieri* in Texas. He further noted a significant change in the ratio of *L. berlandieri* to *L. blairi* through time (2 to 1 in 1969 and 19 to 1 in 1975), which suggested a rather dynamic interaction between the species. Differences in premating isolating mechanisms likely make hybrids unsuccessful in spreading into habitats occupied by the parental species.

In laboratory crosses, ♂ *L. berlandieri* can produce successful larvae when crossed with ♀ *L. palustris* but not with ♀ *L. montezumae* (Mecham, 1969). Larvae of the ♂ *L. berlandieri* × ♀ *L. palustris* and ♂ *L. berlandieri* × ♀ *L. blairi* crosses developed macrocephaly and several tail fin abnormalities, but these disappeared with further development. Larvae resulting from crosses between *L. berlandieri* and *L. forrei*, *L. magnaocularis*, *L. spectabilis*, or *L. sphenocephalus* exhibited mild to severe hybrid inferiority (references in Hillis, 1988). Crosses between *L. berlandieri* and *L. areolatus*, *L. chiricahuensis*, *L. megapoda*, or *L. pipiens* were unsuccessful (Mecham, 1969; Cuellar, 1971; references in Hillis, 1988).

ADULT HABITAT

This species is associated with a wide variety of mostly clear-water aquatic habitats, including ponds, springs, streams, rivers, permanent pools in intermittent streams, temporary pools with extended hydroperiod, and stock tanks. Habitats should contain nearby retreat sites, such as permanent water, root systems, rock cracks, or burrows. Along rivers, this species is most often found sitting on mud banks and rarely on rocks, sand, or in the water (Jung et al., 2002). On the Rio Grande, they were observed most often in open habitats or near seepwillow (*Baccharis salicifolia*) and giant reeds, but much less often among willows, tamarisk, or mesquite. In the Chisos Mountains of Texas, Minton (1958) found it at elevations < 1,219 m, whereas Dayton et al. (2007) gave an elevation < 1,676 m; in New Mexico, it is found from 900 to 1,450 m (Degenhardt et al., 1996).

AQUATIC AND TERRESTRIAL ECOLOGY

The species is primarily nocturnal in activity (Degenhardt et al. 1996), but it also is active by day. Activity may occur year-round. No empirical data have been published on its movement patterns or any other aspect of its ecology during the nonbreeding season. Dayton et al. (2007) noted that the species does not burrow, but that it often appears at temporary ponds that fill after a prolonged drought. This suggests they are able to remain dormant in nearby protected habitats, perhaps for considerable periods of time. Dayton et al. (2007) further suggested that *L. berlandieri* is capable of long-distance dispersal during rainy periods, which thus may at least partially account for their sudden appearance in isolated temporary ponds. In that regard, Rorabaugh (2005b) found them 1.6 km from the nearest water source in Arizona.

CALLING ACTIVITY AND MATE SELECTION

There are two types of calls emitted by *L. berlandieri*, an advertisement trill and a chuckle call that is made in response to other frogs, both con- and heterospecific (Gambs and Littlejohn, 1979). The advertisement call of *L. berlandieri* is a trill with the following properties: call duration 0.47–0.83 sec (mean 0.64); 26–31.3 pulses/sec (mean 28.2); pulse duration 16–22 msec (mean 19.1) (Littlejohn and Oldham, 1968). It serves to identify the calling male and to serve notice that a calling territory is occupied. The chuckle call is a territorial call given in response to the direct approach of an intruder. Although the chuckle call may be given when another species (e.g., *Ollotis valliceps*) calls, it may be because of the similarities in the advertisement calls of these species rather than a response to a different species per se.

In Texas, calls may be heard at most any time of the year, and amplexus has been observed from February through December (Minton, 1958; Blair, 1961b; Hillis, 1981). There is considerable annual variation in reproductive activity, depending upon environmental conditions. Calling may be stimulated by heavy rain or warm temperatures, or a combination of both. A few males will begin calling early in the season, with chorus numbers growing steadily as the season progresses. The first males call in the afternoon, but as temperatures permit, calling will continue after dark. As the season progresses, calling occurs mostly at night through the hottest part of the year, when chorusing peaks after midnight. Rainfall stimulates chorusing, especially after a dry period late in the summer. Calling occurs over a wide range of temperatures (5–34°C; median 19.4°C), with an optimum of ca. 11°C (Blair, 1961b). However, Blair (1961b) recorded calling as low as 1.7°C.

BREEDING SITES

Breeding occurs in both ponds and streams, but there may be regional differences depending upon habitat availability and the presence of other species. This habitat separation may be most evident in areas of sympatry with other members of the Leopard Frog complex. In Texas, for example, this species is considered a stream breeder (Hillis, 1981, 1982), where it is sympatric with *Lithobates sphenocephalus*, a pond breeder. In New Mexico, springs are an important breeding site (Scott and Jennings, 1985); springs and spring runs provide year-round water at a constant temperature and are normally free from nonindigenous predaceous fish. Scott and Jennings (1985) noted that Rio Grande Leopard Frogs, like other leopard frogs, breed in nearly any type of available aquatic habitat to some extent as long as it is free from vertebrate predators.

REPRODUCTION

Data on oviposition are scarce throughout much of this species' range in the United States. Calls may be heard from late winter into the autumn. In Texas, oviposition occurs in the winter to spring (February to May) and the fall (late August to early October) (Blair, 1961b); equal reproductive effort occurs between the breeding periods. In New Mexico, however, Degenhardt et al. (1996) noted that calls have been heard from March to August, but that eggs had been found only from April through early July. Scott and Jennings (1985) also reported no reproductive activity after August. Minton (1958) suggested that breeding may occur year-round in the Big Bend region, and Dayton et al.'s (2007) reports of breeding in January, March, May to October, and December would seem to corroborate an extended breeding season. Still, Blair (1961b) considered *L. berlandieri* to be primarily a fall and winter–spring breeder in Texas, and the data provided by Hillis (1981) suggested a bimodal breeding pattern except where the species is sympatric with other species of leopard frogs. Then, breeding occurred primarily from August to December.

Eggs are transferred to the oviducts ca. 24 hrs prior to amplexus. Globular egg masses are attached to emergent vegetation in shallow (9–15 cm) water. When eggs are deposited in streams, they are placed in lentic water. Oviposition occurs at temperatures of 11–31°C (median 21°C). Apparently there are no records of clutch size, time to hatching, or any other information related to reproduction.

LARVAL ECOLOGY

Larvae are morphologically stream-adapted, featuring a streamlined body with well-developed tail musculature that allows them to forage in benthic habitats along a stream bottom (Hillis, 1982). Whether tadpoles from pond habitats have a similar morphology, or whether there is any

Tadpole of *Lithobates berlandieri*. Photo: Jim Rorabough

Adult Lithobates berlandieri.
Photo: Chris Brown

habitat-based phenotypic plasticity, is unknown. The tadpole diet consists of benthic diatoms and much inorganic and organic matter, including bacteria, small metazoans, protozoans, and carrion.

The larval period is from four to nine months (Dayton et al., 2007). Larvae hatching from eggs deposited late in the season may not transform until the following winter. For example, in Big Bend, Minton (1958) noted large tadpoles in mid-February, with large numbers of recent metamorphs by late April. In the Rosas Mountains, tadpoles were nearing metamorphosis by early April. Such larvae could have resulted from eggs deposited during the autumn or winter, lending credence to Minton's (1958) and Dayton et al.'s (2007) suggestions of a year-round breeding season.

The type of breeding ponds may affect larval development. Dayton et al. (2007) noted that larvae developing in cool permanent ponds took longer to reach metamorphosis and were larger than tadpoles developing under warm conditions with shorter hydroperiods.

DIET

There are no reports on the diet of *L. berlandieri*. However, McCoid (2005) observed frogs feeding on insects in a saline area along the Texas coast, where they were exposed to salinities of 39‰; they were drawn to the area by insects feeding on rotting vegetation (*Halodule* sp.). Postmetamorphs likely consume a wide range of invertebrates in proportion to their availability.

PREDATION AND DEFENSE

Like other members of the *Lithobates pipiens* complex, *L. berlandieri* readily jump into the water when threatened by a predator and dive into the substrate for concealment. Their coloration undoubtedly aids in crypsis from vision-oriented predators. There is no evidence of noxious or toxic chemicals in the skin, and postmetamorphs are highly palatable. However, postmetamorphs have peptides in their skin secretions that have antimicrobial properties that may aid in preventing infection (Goraya et al., 2000). These peptides inhibit the growth of the bacteria *Escherichia coli*, *Staphylococcus aureus*, and *Candida albicans*.

The eggs of *Lithobates berlandieri* are palatable, but their large size, rigid egg capsules, and adherence to one another may inhibit predation by some small predators, such as mosquitofish (*Gambusia* sp.) (Grubb, 1972). After several attempts to open a capsule, these voracious little fish tend to give up. The aquatic beetles *Cybister fimbriolatus* and *Hydrophilus triangularis* have been observed to prey on tadpoles in field and laboratory settings (Ideker, 1979). Predators of postmetamorphs include garter snakes (*Thamnophis eques, T. marcianus*), grackles (*Quiscalus mexicanus*), and likely many other invertebrates and vertebrates (Jameson and Flury, 1949; Sanders and Smith, 1971; Ideker, 1976; Rorabaugh, 2005b).

POPULATION BIOLOGY

No information is available on demography or population biology. Rio Grande Leopard Frogs can be quite common along rivers and streams. In canoe-based visual counts along the Rio Grande, Jung et al. (2002) recorded 2,779 *Lithobates berlandieri* over a 2 yr period in 4 river sections totaling nearly 25 km. Counts increased with air temperature and were inversely proportional to river gauge level.

COMMUNITY ECOLOGY

Lithobates berlandieri sometimes overlaps in geographic range with the leopard frogs *L. blairi* and *L. sphenocephalus*. Interbreeding occurs occasionally but is rare because of differences in habitat preference, call structure, and reproductive patterns (Hillis, 1981). In Texas, for example, *L. berlandieri* is primarily a stream breeder, whereas *L. blairi* breeds in standing water (Hillis, 1981). In addition, calling patterns vary when species are in sympatry. Then, *L. berlandieri* switches from a bimodal pattern to breeding from August to December, whereas *L. sphenocephalus* breeds from January to March and *L. blairi* from April to June. In areas of sympatry with *L. berlandieri*, *L. sphenocephalus* also tends to forgo its brief autumn breeding period that occurs in areas of allopatry in Texas.

DISEASES, PARASITES, AND MALFORMATIONS

Frogs from a morbidity and mortality event in Arizona exhibited signs of the bacteria *Aeromonas* (red-leg disease), but histological examination revealed the presence of the amphibian chytrid fungus *Batrachochytrium dendrobatidis* (*in* Rorabaugh, 2005b).

Tetrathyridia of the cestode *Mesocestoides* sp. have been found in *Lithobates berlandieri* from Texas (McAllister and Conn, 1990). Unidentified leeches were reported on *L. berlandieri* from the Sierra Vieja Mountains of Texas (Jameson and Flury, 1949). An undescribed mite (*Hannemania* sp.) was reported to be common on Rio Grande Leopard Frogs in Big Bend, Texas; mites were found more often on males than females (Jung et al., 2001). In the lower Rio Grande, Sanders and Smith (1971) noted that about 1% of *Lithobates berlandieri* had papilla-like structures which they termed "skin tags." These were thought to have originated from wounds.

SUSCEPTIBILITY TO POTENTIAL STRESSORS

Chemicals. Rio Grande Leopard Frog tadpoles fed paraquat-treated plants suffered significant mortality beginning 7 days after exposure, and only 19.4% survived to day 15 in studies by Bauer Dial and Dial (1995). Tadpoles ingested the herbicide through contaminated plants. Tadpoles that survived often had malformations of the tail (flexed and shortened), exhibited abnormal swimming behavior, and were smaller than controls. Different species of plants concentrated paraquat differently. The concentrations varied from 72.6 mg/L to 1,011 mg/L depending upon plant species, but the effects on tadpoles did not vary by plant species ingested.

STATUS AND CONSERVATION

Information on the status of this species is lacking. It is not protected within the United States by any state or federal law. The Government of Mexico considers this species as a "Species of Special Protection." The impact of this introduced species on native fauna of the Gila and Colorado river systems is unknown. Platz et al. (1990) suspected the introduction to have taken place 14–16 yrs prior to their surveys. The suspected introduction likely occurred via fish-transplanting operations from stock originating from National Fish Hatcheries. The likelihood that tadpoles have been widely dispersed to other regions of the country via fish-stocking programs is currently under appreciated.

Lithobates blairi
(Mecham, Littlejohn, Oldham, Brown, and Brown, 1973)
Plains Leopard Frog

ETYMOLOGY
blairi: a patronym in honor of W. Frank Blair (1912–1984), a University of Texas herpetologist who pioneered the study of the systematics and evolutionary biology of frogs. Blair first suggested that there were cryptic species of leopard frogs in the southwestern United States.

NOMENCLATURE
Conant and Collins (1998): *Rana blairi*
  Dubois (2006): *Lithobates* (*Lithobates*) *blairi*
  Synonyms: *Rana pipiens brachycephala*
  References to this species in the literature are often incorrect. For example, Smith and Sanders (1952) discussed the distribution of "*Rana pipiens berlandieri*" in Texas, but included specimens from what is now known to be *Lithobates blairi*; specimens now referred to as *L. blairi* were discussed under the name "*Rana pipiens brachycephala*" in this publication. Burt (1935) likely combined three species (*Lithobates blairi, L. pipiens, L. sphenocephalus*) in his discussion of leopard frogs from the Midwest, and Bragg (1950d) and Bragg and Dowell (1954) called this species *L. berlandieri*. Many early state guides (e.g., Hurter, 1911; H.M. Smith, 1934; Hudson, 1942; P.W. Smith, 1961) include information on this species within accounts of *L. pipiens*. Other papers with confusion as to taxonomy include Moore (1944) and McAlister (1961). This species often is simply referred to in the literature as the "southern plains form" of the *L. pipiens* complex. The complex nomenclatural history of this species is discussed by Hillis (1988). Readers should verify the locations of specimens in order to ensure correct species identification.

IDENTIFICATION
Adults. The ground color of *L. blairi* is tan to pale brown to dull olive, with large round or slightly wider than long dark brown spots dorsally. There is usually a single dark spot on the snout. Spots may or may not have light halos, and the margins of the spots may be somewhat indistinct. Dorsolateral folds are light in color and tend to break up and turn inward posteriorly. There is a usually a white or ivory-colored spot in the center of the tympanum, as well as a white line on the upper lip. Venters are cream colored. The ventral portions of the lateral sides of the body, groin, and hind limbs are generally light yellow. Throats are often mottled. Bars on the femurs and tibia tend to be narrow and complete; the posterior of the thigh has dark but indistinct reticulations on a light background. The areas around the cloaca and the undersides of the thighs have many tubercles. Adult males do not have a vestigial oviduct. Males are smaller than females and have enlarged thumbs. Adults normally reach 50–95 mm SUL, with Gillis (1975) reporting a mean adult size of 71.8 mm SUL in Colorado. In New Mexico, males averaged 64.4 mm SUL, whereas females averaged 75.5 mm SUL (*in* Degenhardt et al., 1996). In Indiana, males were 47.0–61.7 mm SUL and females 50.5–59.3 mm SUL, but the sample size was small (Minton, 2001). The maximum reported size is 114 mm SUL (Creel, 1963).

Larvae. The tadpole of *L. blairi* is olive with numerous obscure dark dots, but the overall impression is of a gray-brown tadpole. It is the palest of all leopard frog tadpoles. The belly and throat are white with a hint of pinkish bronze coloration ventrolaterally. The tail musculature is pale olive with large bronze blotches; tail fins are clear olive with pale spots. The entire tail may have a smudged appearance. The iris is black with medium-gold flecks. Korky (1978) and Scott and Jennings (1985) described the tadpole in more detail. An albino tadpole was reported from Oklahoma (Hensley, 1959).

Eggs. The eggs are black or brownish above and white below and surrounded by two to three jelly envelopes. The outer envelope is 2.5–3.5 mm in diameter, whereas the inner envelope is 1.5–2 mm; the vitellus is 1–1.5 mm in diameter (Smith, 1934). The eggs are deposited in a jelly mass 5–15 cm in diameter. The dark colored eggs located close together may cause the egg mass to appear black. According to Hammerson (1999), there are several hundred eggs per mass, but Phillips et al. (1999) give the number as 3,000–7,000. Hatching occurs in about five days (Smith, 1934).

DISTRIBUTION

Lithobates blairi is a prairie grassland species whose range extends from western Indiana across Illinois (the Prairie Peninsula), southern Iowa through southeastern and southwestern South Dakota (three localities), eastern and central Nebraska to eastern Colorado and New Mexico, the panhandle and central Texas, western and central Oklahoma, Kansas, and northern Missouri. The range narrowly extends southward along the Mississippi River to just below its confluence with the Ohio River (Brown et al., 1993) into Mississippi County, Arkansas (Trauth et al., 1992). Isolated populations occur in southeastern Illinois, southern Indiana (Tipton County), northern (Rio Arriba County, possibly extirpated) and southern (Sierra County) New Mexico, and southeastern Arizona (Sulphur Springs Valley). It is apparently extending its range northward in Iowa (Christiansen, 2001). *Lithobates blairi* may have been introduced in areas outside its natural range, such as in Weld County, Colorado (Livo et al., 1998; Hammerson, 1999) and near Flagstaff, Arizona (Clarkson and Rorabaugh, 1989).

Important distributional references include: central Great Plains (Dunlap and Kruse, 1976), Arizona (Clarkson and Rorabaugh, 1989; Brennan and Holycross, 2006), Arkansas (Trauth et al., 1992), Colorado (Post and Pettus, 1966; Hammerson, 1999), Illinois (Phillips et al., 1999), Indiana (Minton, 2001), Kansas (Smith, 1934; Collins, 1993; Collins et al., 2010), Missouri (Johnson, 2000; Daniel and Edmond, 2006), Nebraska (Lynch, 1978, 1985; Ballinger et al., 2010; Fogell, 2010), New Mexico (Degenhardt et al., 1996), Oklahoma (Sievert and Sievert, 2006), South Dakota (Ballinger et al., 2000; Ernst, 2001; Kiesow, 2006), and Texas (Dixon, 2000).

FOSSIL RECORD

Pleistocene (Irvingtonian) fossils referred to *L. blairi* are known from Nebraska (Holman, 2003). Identification of the *L. pipiens* complex fossils as *L. blairi* was made on the basis of zoogeographic considerations. Pleistocene fossils of "*Rana pipiens*" reported for Lubbock, Foard, and Hardeman counties, Texas (Holman, 1969), are currently within the range of *Lithobates blairi*. Similar fossils reported for Denton County, Texas, are in the contact zone between *L. blairi* and *L. berlandieri* (Dixon, 2000) and may be referable to either species.

SYSTEMATICS AND GEOGRAPHIC VARIATION

The systematics of leopard frogs, including *L. blairi*, has been especially confusing. Species concepts have varied through the years (Moore, 1975; Hillis, 1988), and many authors have tried to make sense of the Leopard Frog complex with varying degrees of success. Morphological characters proved difficult to use (e.g., Moore, 1944), especially without precise geographic information. Fortunately a combination of morphological, serum protein, developmental, acoustic, molecular, and life history information, coupled with precise geographic data, now allow the species to be identified and their relationships to be reasonably well understood (Littlejohn and Oldham, 1968; Brown and Brown, 1972a; Mecham et al., 1973; Gillis, 1975; Kruse and Dunlap, 1976; Axtell, 1976; Hillis et al., 1983).

Lithobates blairi is a member of the *Novirana* clade of North American ranid frogs, and it is classified within the *Pantherana*, a group that includes the leopard and gopher frogs and comprises the *Rana pipiens* complex. Mating calls within the *Pantherana* are highly complex in structure, and include elements described as chuckles, grunts, and snores (Hillis and Wilcox, 2005). Mating calls are only produced during breeding choruses, whereas other elements of the

Distribution of *Lithobates blairi*

call are produced at other times of the year in the *Pantherana*. Plains Leopard Frogs are more closely related phylogenetically to *Lithobates sphenocephalus, L. berlandieri, L. yavapainensis, L. onca,* and several Mexican species (the *Scurrilirana,* frogs that produce a chuckle-like call) than they are to *Lithobates pipiens* and *L. chiricahuensis.*

Hybridization with some other members of the Leopard Frog complex occurs in nature, and these hybrids may be identified by intermediate phenotypes. Hybridization rates often are low. For example, in Texas Platz (1981) recorded 5.8% hybrids at a contact zone between *L. blairi* and *L. berlandieri.* He further noted a significant change in the ratio of *L. berlandieri* to *L. blairi* through time (2 to 1 in 1969 and 19 to 1 in 1975), which suggested a rather dynamic interaction between the species. *Lithobates pipiens* × *L. blairi* hybrids occur at a rate of 1–16% in overlapping Nebraska populations (Lynch, 1978) and at 1–4.4% in Iowa and South Dakota populations (Dunlap and Kruse, 1976). In eastern Colorado, Gillis (1975) found much more evidence of hybridization, approaching 60–70% in some ponds, using morphological criteria to determine hybridization. In a follow-up study, Cousineau and Rogers (1991) noted the continual presence of hybrids in one pond studied by Gillis (1975), but that *L. pipiens* had disappeared from the area. Apparently, the hybrids were breeding with *L. blairi* or were self-sustaining. Throughout much of the area where contact occurs between these species in the north, there has been asymmetrical genetic swamping of *L. pipiens* nuclear haplotypes by *L. blairi* haplotypes (Di Candia and Routman, 2007). Although hybrids are often fertile, they do not backcross successfully with the parental species. Zones of hybridization appear stable in some contact zones.

In contrast, hybridization is rare or absent altogether in many areas where different leopard frog species come into contact or when ranges are allopatric. Such is the case in southeastern Arizona between *L. blairi* and *L. chiricahuensis* (Frost and Bagnara, 1977a), with *L. pipiens* in central Colorado (Post and Pettus, 1966, 1967), and with *L. sphenocephalus* in Illinois and Missouri (Axtell, 1976). Hybridization is prevented or minimized throughout much of the species' range by a combination of both pre- and postmating isolating mechanisms, including differences in reproductive phenology, habitat preferences, call structure, and developmental constraints.

In laboratory crosses, ♂ *L. blairi* can produce larvae when crossed with ♀ *L. palustris* or ♀ *L. montezumae* (Mecham, 1969). Larvae from crosses between female *L. blairi* and ♂ *L. megapoda, L. montezumae,* or *L. sphenocephalus* appeared normal, as are crosses between *L. blairi* and *L. pipiens* (Gillis, 1975). However, larvae from crosses of ♀ *L. blairi* with ♂ *L. berlandieri* developed macrocephaly and several tail fin abnormalities, but these disappeared with further development. Crosses with *L. chiricahuensis* may produce some hybrids successfully or result in complete developmental failure (Frost and Bagnara, 1977a; Frost and Platz, 1983). In crosses with *L. areolatus,* hybrids are successfully produced whether *L. blairi* is the male or female parent (Cuellar, 1971).

ADULT HABITAT

The Plains Leopard Frog is a species of the semi-arid Great Plains, southwestern grasslands, and Madrean evergreen woodlands. It favors areas with loess soils rather than areas with sandy soils, and it frequently occurs along river drainages, especially to the north and west. Such habitats have very turbid ponds and streams that are favored by this species. It occurs in nearly all temporary and permanent wetlands (ponds, pools in rocky canyons, river marshes), and is frequently observed along streams and rivers. It frequents man-made habitats, such as irrigation ditches and cattle tanks. This species is associated with a variety of nonaquatic habitats, including prairie grasslands, river bottom forest, and upland woodlands. Surrounding land uses may include agriculture, such as cropland or rangeland, although abundance is not as high as when surrounded by natural habitats. In New Mexico, it is found from 1,000 to 2,250 m (Degenhardt et al., 2006), whereas in Colorado most populations are < 1,525 m but extend to 1,830 m (Hammerson, 1999).

AQUATIC AND TERRESTRIAL ECOLOGY

The Plains Leopard Frog is a species of temporary and semipermanent ponds and wetlands in grassland and semiarid habitats of the Great Plains. It is not infrequently found away from water, and may be observed in both upland forest and

Adult *Lithobates blairi*.
Photo: Jody Hibma

old-field habitats (Clawson and Baskett, 1982; Heinrich and Kaufman, 1985). Dispersal likely is facilitated by water drainages or irrigation ditches and may occur rapidly over long distances in waves during and immediately after rainfall (Fitch, 1958). Blair (1961b) noted that some frogs tended to remain near a wetland all year-round, whereas others moved to a breeding site only for reproduction. He noted one female that moved 30 m from its nonbreeding stock tank to another stock tank to breed, then returned to her original location.

Viets (1993) noted that Plains Leopard Frogs wandered "great distances" from water into the surrounding grasslands, where they occupy low-lying moist areas in green meadows. In eastern Colorado, Gillis (1975) found marked *L. blairi* from 3 to 8 km from the point of original capture; *L. blairi* × *L. pipiens* hybrids were found from 3 to 14 km from point of capture. When at an aquatic site, they tend to sit at the water's edge, where they have both good foraging and an escape route to water should a predator approach.

Recent metamorphs are frequently encountered in the fall along streams and rivers. They are capable of burrowing or of using the burrows of other animals to escape dry conditions in their semiarid environment (Parris, 1998). Burrowing is accomplished by using only the back feet and allows them to conserve water efficiently. As might be expected, digging requires a substantially longer period of time than occupying an existing burrow. This species is prone to water loss and can die if it loses too much water, so burrowing into moist soil could assist in survival, even if this species is relatively inefficient in constructing burrows compared to some other ranid species (Parris, 1998). Although prone to water loss, juvenile *L. blairi* lose water at a slower rate than *L. pipiens* (Gillis, 1975, 1979), perhaps accounting for its dispersal in more arid habitats than the Northern Leopard Frog.

Activity occurs throughout the warmer parts of the year, although Fitch (1956a) found a Kansas individual active in mid-December when the temperature had reached −6.6°C the previous night; Plains Leopard Frogs later were active at the water's edge in late December when the air temperature was 14°C. In Kansas, Plains Leopard Frogs normally are observed from early February or March to mid-November (Smith, 1934; Clarke, 1958). Emergence occurs as much as several weeks prior to the start of the breeding season and can vary considerably from one year to the next. The species can tolerate high temperatures, as the CTmax is 36–39°C (mean 37.3°C) (Gillis, 1975). Fitch (1956a) recorded them active up to 33.6°C. However, Plains Leopard Frogs may take refuge in mammal burrows, such as those of prairie dogs (*Cynomys ludovicianus*)

(Lomolino and Smith, 2003), during hot and dry weather. They also are common around the entrances of limestone and gypsum caves (Black, 1973a). Overwintering occurs in the mud and debris in the bottom of ponds and streams (Johnson, 2000). Mortality may occur if frogs emerge from winter dormancy and are unable to return to protected sites during a sudden subsequent freeze (Fitch, 1956a; Heinrich and Kaufman, 1985).

CALLING ACTIVITY AND MATE SELECTION

Males call while floating at the surface of the water. Calls may be heard from spring through autumn, but oviposition may not occur in mid- to late summer. Calls are a series of clucks with a low pulse rate and a dominant frequency of 1.2 kHz (Frost and Platz, 1983). The advertisement call has the following properties: call duration 0.48–0.89 sec (mean 0.66); 4.6–6.8 pulses/sec (mean 5.6); pulse duration 23–35 msec (mean 27) (Littlejohn and Oldham, 1968; Brown and Brown, 1972a; Dunlap and Kruse, 1976). The trill call usually consists of 3 notes (range 1–4), with the first the longest and lasting 0.5–1.0 sec. Frost and Platz (1983) provided an audiospectrogram of the call. Calling has been recorded at temperatures as low as 4.4°C (Clarke, 1958). In call-back experiments, neither female *L. pipiens* nor *L. pipiens* × *L. blairi* hybrids respond positively to calls from male *L. blairi*, suggesting that the advertisement call is at least a partially effective premating isolating mechanism in areas of species contact (Kruse, 1981b).

BREEDING SITES

Breeding occurs in a variety of mostly permanent prairie and semidesert wetlands, such as the playa wetlands of west Texas and the prairie potholes of the Great Plains. It breeds in nearly all available aquatic habitats, from ponds and streams to man-made irrigation and roadside ditches. The species prefers warm standing water that is often turbid with large amounts of silt, but it also is found in highly turbid and silt-laden streams (Lynch, 1978; Hillis, 1981).

REPRODUCTION

Lithobates blairi is a prolonged, opportunistic breeder that times reproduction in accordance with the availability of precipitation. Reported breeding dates in the literature vary, and there is some confusion about whether the species is a continuous or bimodal breeder. Most breeding occurs in the spring to summer in the North, but reproduction likely peaks bimodally to the South. Fall season breeding dates may be indicative of an extended, opportunistic warm season breeding period, or they might reflect a bimodal breeding season. It is likely that apparent discrepancies in various author's accounts stem from differences in annual or regional precipitation patterns, rather than reflect actual geographic differences in breeding seasons. If rainfall is heavy at any point from spring to fall, it is likely that some Plains Leopard Frogs will breed. As a result, eggs and large tadpoles may be found together at a site, reflecting different breeding bouts at the same pond.

Reports of breeding in the literature include the following: February to October (Pace, 1974); February to September in Oklahoma (Bragg, 1950a); March and April in Illinois (Phillips et al., 1999); March to June in South Dakota (Kiesow, 2006); March in southern Nebraska, but in northern Nebraska it does not begin until late April to May (Lynch, 1985); peak breeding from March to May in Kansas (Smith, 1934); north of the Arkansas River in eastern Colorado, breeding may begin in late April (Post, 1972), but most breeding occurs in July and August in conjunction with summer precipitation (Gillis, 1975); south of the Arkansas River in Colorado, most breeding occurs from May to July (Post, 1972; Hammerson, 1999); in Texas and Oklahoma, the primary breeding season is March to June with no fall breeding (Hillis, 1981); oviposition has been observed from March to August in Arizona, but there may be two breeding peaks, one from March to June and a second from August to October (Frost and Platz, 1983); in New Mexico, Scott and Jennings (1985) suggested that breeding likely commenced in February, and they indicated that there was no evidence of breeding after August (also see Degenhardt et al., 1996); reproduction may occur as late as mid-September in Nebraska (Lynch, 1985) or even in early October in Oklahoma (Bragg and Dowell, 1954), when egg masses or newly hatched larvae have been found. Possible fall breeding also has been reported from Kansas (*in* Collins, 1993).

The firm egg masses are reported to contain between "several hundred" and 3,000–7,000 eggs per mass (e.g., Smith, 1934; Hammerson, 1999;

Phillips et al., 1999), but there are no empirical data giving actual egg counts. Egg masses are deposited in shallow water and are attached to vegetation.

LARVAL ECOLOGY
Tadpoles of *L. blairi* are more slender and less robust than larval *L. pipiens* with which they are sometimes sympatric (Korky, 1978). The subterminal oral disk suggests this species is a benthic feeder ingesting bacteria, small metazoans, protozoans, or even carrion. Differences in tadpole morphology among leopard frog species are likely correlated with differences in feeding behavior and habitat preference (i.e., silty and turbid waters vs. clear waters). Larval development requires 50–60 days (Lynch, 1985), and recent metamorphs are 27–30 mm SUL (Degenhardt et al., 1996; Johnson, 2000). Larvae are reported to overwinter in Colorado (Gillis, 1975) and Oklahoma, as tadpoles were found in January and February (Bragg and Dowell, 1954) or early in the spring. Overwintering tadpoles result from late-season breeding events.

As with many species of anurans, a low larval density results in an increase in survivorship of *L. blairi*, at least under in experimental conditions. High densities tend to increase the larval period and decrease the body mass at metamorphosis (Boone and Semlitsch, 2002). Constant hydroperiods also result in larger sizes at metamorphosis than in ponds with decreasing water levels.

DIET
Invertebrates probably make up the major portion of the diet of postmetamorphic individuals. Prey items include various types of flies, beetles, worms, snails, grasshoppers, and crickets (Hartman, 1906; Black, 1973a; Hammerson, 1982). Plains Leopard Frogs may be common around bat guano piles in caves, where they feed on the abundant invertebrate fauna (Black, 1973a). Creel (1963) reported a large *L. blairi* feeding upon a bat (*Pipistrellus subflavus*).

PREDATION AND DEFENSE
Like all leopard frogs, this species is cryptic and readily jumps into nearby water when disturbed. As it jumps or when seized, it emits a loud distress call that can easily startle a would-be predator. Degenhardt et al. (1996) noted a tendency to jump away from water rather than toward it when approached by a predator. *Lithobates blairi* possesses cytolytic antimicrobial peptides in their skin secretions that may assist in protecting the species from bacterial infection (Conlon et al., 2009).

Reports of predators are few. Predation has been reported by the garter snakes *Thamnophis cyrtopsis* (Hammerson, 1999) and *T. sirtalis* (*in* Johnson, 2000). Ehrlich (1979) observed predation by *Lithobates catesbeianus* tadpoles on eggs and newly hatched larvae of *L. blairi*. Raccoons, opossums, and striped skunks also prey on this species (Shirer and Fitch, 1970).

POPULATION BIOLOGY
The abundance of *L. blairi* at ponds varies considerably among ponds and years. For example, Gillis (1975), using capture-mark-recapture techniques, estimated there were 31–57 frogs at one pond, 185–451 frogs at a second pond, and 71–401 frogs at a third pond in Colorado over a 4 yr period. However, his estimates included *L. blairi*, *L. pipiens*, and their hybrids.

COMMUNITY ECOLOGY
Lithobates blairi sometimes overlaps in geographic range with the leopard frogs *L. pipiens*, *L. berlandieri*, *L. sphenocephalus*, and *L. chiricahuensis*. Interbreeding occurs occasionally but is rare because of differences in habitat preference, call structure, and reproductive patterns. For example, in Nebraska Lynch (1978) noted that *L. blairi* occurred in more turbid waters than *L. pipiens* when the species were sympatric; in South Dakota the species also is found in silt-laden habitats (Ernst, 2001). In Texas, *L. blairi* is most associated with warm, turbid pools, whereas the sympatric *L. sphenocephalus* is associated with clear and cool water (Hillis, 1981). In Arizona, *L. blairi* is often found in temporary and semipermanent wetlands, whereas *L. chiricahuensis* is found in permanent streams and stock ponds (Frost and Bagnara, 1977). In addition, the species may come in contact only during certain portions of the year. In Nebraska, *L. blairi* and *L. pipiens* are often syntopic in the late summer and fall, but not during the breeding season.

Calling patterns may vary when species of the leopard frog complex are in sympatry. Then, *L. berlandieri* switches from a bimodal pattern to breeding from August to December, whereas

L. sphenocephalus breeds from January to March, and *L. blairi* breeds from April to June. The switch in breeding phenology by *L. berlandieri* minimizes reproductive overlap and allows coexistence even though both *L. blairi* and *L. sphenocephalus* have the same phenology, regardless of whether species are allopatric or sympatric (Hillis, 1981).

In experimental ponds, interspecific competition with *L. blairi* and *L. sphenocephalus* resulted in an increased larval period and a decreased body mass of metamorphic *L. areolatus* (Parris and Semlitsch, 1998). Competition has a density and species-specific component. For example, survivorship of *L. blairi* larvae is decreased in the presence of *L. areolatus* larvae at high density.

DISEASES, PARASITES, AND MALFORMATIONS

Plains Leopard Frogs are parasitized by the trematodes *Allassostomoides chelydrae*, *Cephalogonimus americanus*, *Glypthelmins quieta*, *Gorgoderina amplicava*, *G. attenuata*, *G. simplex*, *G. translucida*, *Haematoloechus coloradensis*, *H. complexus*, *H. longiplexus*, *H. medioplexus*, *H. parviplexus*, and *Megalodiscus temperatus*; the cestodes *Ophiotaenia filaroides*, *O. magna*, and *Mesocestoides* sp.; and the nematodes *Cosmocercoides variabilis*, *Falcaustra catesbeianae*, *Oswaldocruzia pipiens*, *Physaloptera* sp., *Rhabdias ranae*, and *Spinitectus gracilis* (Brooks, 1976; Brooks and Welch, 1976; Goldberg et al., 2000; Bolek and Janovy, 2008). The leech *Desserobdella picta* has been recorded on larval *Lithobates blairi* from Nebraska (Bolek and Janovy, 2005). The fungus *Saprolegnia* was reported from larvae (as "*L. berlandieri*," but likely *L. blairi*) in Oklahoma (Bragg, 1962a).

SUSCEPTIBILITY TO POTENTIAL STRESSORS

Chemicals. Esfenvalerate, an insecticide, has an $LC_{50\ (96\ hr)}$ of 7.79 µg/L and causes larval convulsions in concentrations as low as 3.6µg/L. Even at 1.3µg/L, activity is decreased (Materna et al., 1995). It should be noted, however, that Materna et al. (1995) did not differentiate between *L. blairi*, *L. pipiens*, and *L. sphenocephalus* in presenting results. Temperature magnifies the effect of the pesticide, and additional effects on frog populations may result from the pesticide's lethal effects on fish and invertebrates.

Death occurs in about 18 hrs at 30 mg/L of carbaryl, a broad-spectrum insecticide, and significantly reduces larval activity levels at 2.5 mg/L (Bridges and Semlitsch, 2000). At low densities, the addition of carbaryl might actually increase overall survivorship, and it certainly appears to enhance survivorship at high tadpole densities (Boone and Semlitsch, 2002). Carbaryl tended to decrease the larval period, but overall had no significant effect on *L. blairi*.

Breeding habitat of *Lithobates blairi*. Photo: Jody Hibma

UV light. Embryos exposed to high levels of UVB light (84% transmittance) have the same hatching success as those exposed to low levels (58% transmittance). Hatchlings from high exposure levels were smaller than those from low exposure levels. Tadpoles exposed to high levels of UVB also had slower growth rates and development than those exposed to low UVB levels, although survivorship was not affected. These results suggested that UVB could have sublethal effects on growing larvae and thus on the complex life cycle of the species (Smith et al., 2000).

STATUS AND CONSERVATION

Some populations of *L. blairi* have declined or disappeared, particularly in peripheral populations. For example, Plains Leopard Frogs have declined in Cochise County, Arizona, where Clarkson and Rorabaugh (1989) found it at only 2 of 13 historic locations and at 1 new site. Brennan and Holycross (2006) reported it from only two remaining sites in Arizona. Populations of *L. blairi* had disappeared from some areas of eastern Colorado by the late 1970s, possibly due to bullfrog introductions (Hammerson, 1982). The species has disappeared from some areas of the Prairie Peninsula in Indiana, where it has not been observed since the 1930s (Brodman et al., 2002). In Illinois, most of its prairie habitat was destroyed by agriculture, although the species remains widespread but not abundant (Phillips et al., 1999). In Iowa, Christiansen (1981) suggested its status was indeterminate but later indicated that its range was actually expanding northward (Christiansen, 2001). Throughout much of its range, however, this species is still considered common or abundant (e.g., Busby and Parmelee, 1996; Hammerson, 1999) or even increasing (Florey and Mullin, 2005).

Reasons for declines of some Plains Leopard Frog populations include competition or predation by American Bullfrogs and exotic fishes, drought, habitat loss and alteration, groundwater pumping, and agricultural development (Hammerson, 1982; Livo, 1984; Clarkson and Rorabaugh, 1989; Cousineau and Rogers, 1991; Livo et al., 1998). These effects may be more severe in some locations than others, such as in the isolated population in Sulphur Springs Valley, Arizona. Populations in the Prairie Peninsula of Indiana and Illinois undoubtedly were lost to agriculture. Empirical data are lacking for determining the cause of decline in most areas. One possible management option to benefit this species is to construct artificial ponds that leopard frogs readily colonize (Shulse et al., 2010).

The species is listed as a "Species of Special Concern" in Indiana.

Lithobates capito (LeConte, 1855)
Gopher Frog

ETYMOLOGY
capito: from the Latin *capito* meaning 'one that has a large head' and referring to the large head of the frog.

NOMENCLATURE
Conant and Collins (1998): *Rana capito*
  Dubois (2006): *Lithobates* (*Lithobates*) *capito*
  Synonyms: *Rana areolata aesopus*, *Rana areolata capito*, *Rana capito stertens*
  Information on *L. capito* is sometimes contained in accounts of frogs identified as *R. c. sevosa* (e.g., Bailey, 1991).

IDENTIFICATION
Adults. Lithobates capito is a medium-sized stocky frog with a large head and short hind limbs. The background coloration is off-white to gray to brown, and the dorsum, sides, and legs have dark spots of varying shades, shapes, and sizes. The skin is smooth to rugose, and the varying warts and folds give the frog a wrinkled appearance. Prominent dorsolateral folds are present; these may be gray, orange yellow, or dark. The venter is white, cream, or yellow and mottled or spotted with dusky pigment. A yellowish color may be present in the groin region. Breeding males have enlarged thumbs. Recent metamorphs are not mottled on the belly.

Males are generally smaller than females. In Florida, for example, males were 61–93 mm SUL (mean 80.2 mm) whereas females were 78–112 mm SUL (mean 93.5 mm) (Palis, 1998). Females have a much larger body mass than males, but eggs can account for one-third of the body weight. Adults in South Carolina were 59–91 mm SUL (mean 76.1 mm) (Semlitsch et al., 1995); Alabama females were 70–94 mm SUL and males 61–87 mm SUL (Bailey, 1990).

Larvae. The larvae are yellowish to olive green with scattered large dark blotches or spots throughout the body and tail, including the tail fin. Some individuals lack fine spotting, and tail fins are sometimes clear (Gregoire, 2005). Throats are unpigmented, and the translucent snout does not have a light line extending away from each corner of the jaw. The gut may or may not be visible through a cream to yellowish venter. Maximum size is 90 mm TL prior to transformation (Gregoire, 2005). The larvae are very similar to those of *L. sphenocephalus* and may be impossible to distinguish.

Eggs. Eggs are relatively large and colored gray to gray black. Two gelatinous envelopes are present; the outer envelope is 4.4–6.0 mm in diameter, with the inner envelope 3.1–4.4 mm in diameter. The vitellus is 1.8–2.4 mm in diameter (Wright, 1932). Phillips (1995) gives a mean egg diameter of 6.5 mm. Palis (1997) gives a range of 2.9–3.8 mm for embryos from western Florida. Eggs are deposited in an ovoid to spherical jelly mass 8–12 cm in diameter and 2.5–7 cm in depth (Bailey, 1990; Phillips, 1995).

DISTRIBUTION

Lithobates capito occurs on the Atlantic Coastal Plain from southern North Carolina throughout much of the Florida peninsula to Broward County (Means and Simberloff, 1987) and west to Mobile Bay. Disjunct populations are reported in central Alabama (Shelby County) and Tennessee (Coffee County). Populations in North and South Carolina are widely scattered. Important distributional references include: Wright (1932), Mount (1975), Redmond and Mount (1975), Palis (1997), Bailey and Means (2004), Dorcas et al. (2007), and Jensen et al. (2008).

FOSSIL RECORD

Fossils of *L. capito* are reported from Pliocene and Pleistocene deposits in Florida (Holman, 2003; Meylan, 2005).

Distribution of *Lithobates capito*

SYSTEMATICS AND GEOGRAPHIC VARIATION

Gopher Frogs are in the *Nenirana* clade of North American ranid frogs, a group that includes *Lithobates palustris*, *L. areolatus*, and *L. sevosus* (Hillis and Wilcox, 2005). They are only distantly related to the *Scurrilirana* clade. The species has had a complex nomenclature reflecting differing interpretations of phenotypic characters that show much variation. At various times it has been considered a full species or as a subspecies of *Lithobates areolatus* (Neill, 1957b). For example, Harper (1935b) reviewed the nomenclatural history of this species and synonymized *Rana areolata aesopus* with *R. capito*. Schwartz and Harrison (1956) described *Rana capito stertens* from North and South Carolina based on differences in coloration and pattern from Florida *Lithobates capito*. There does not appear to be any phylogenetic or biogeographic justification for recognizing subspecies within this taxon, although there are two distinct clades within *L. capito* (Young and Crother, 2001).

Laboratory crosses of *L. capito* with *L. clamitans* were unsuccessful (Moore, 1949a, 1955).

ADULT HABITAT

The best habitat for Gopher Frogs is fire-maintained pine savanna sandhills or flatwoods dominated by longleaf pine (*Pinus palustris*) and wiregrass (*Aristida beyrichiana*), interspersed with temporary, open-canopied shallow wetlands for breeding (Roznik and Johnson, 2009a). Surrounding habitats should contain stump holes or small mammal or Gopher Tortoise (*Gopherus polyphemus*) burrows for shelter. The species also occurs in savannas invaded by hardwoods, but in less abundance.

TERRESTRIAL ECOLOGY

Gopher Frogs are primarily fossorial, spending most of their time in burrows of Gopher Tortoises (Franz, 1986), crayfish, and mammals (*Geomys*, *Podomys* sp.). They likely use any available tunnel, especially in areas not occupied by Gopher Tortoises in the northern part of the range. I even found an adult *Lithobates capito* under a sunken pitfall bucket with no obvious external connection. Underground retreats offer shelter from cold, drought, and periodic fires (Roznik and Johnson, 2007), and they are inhabited by a wide variety of potential prey. These frogs are frequently observed sitting at the mouth of the burrow, but they may forage away from the burrow's entrance at night or on rainy days. Activity at burrow entrances may occur throughout the year, weather permitting, especially in Florida (Branch and Hokit, 2000).

Lithobates capito do not necessarily remain at a single burrow throughout the year; some frogs move among burrows and other terrestrial refugia. Blihovde (2006) found that they used 1–4 shelters in central Florida, with total terrestrial movements covering 0–35 m except for a single male that moved 286 m. Other frogs showed strong site fidelity toward specific pocket gopher and Gopher Tortoise burrows and remained at a single burrow for 14 months. Of the burrows examined, 31.8% contained Gopher Frogs at least some time during the study. Upland home ranges extended from 0 to 222 m^2 with a mean of 45 m^2. In another area of Florida, Gopher Frogs occupied about 10% of Gopher Tortoise burrows examined, with as many as 5 frogs in a single burrow (Kent et al., 1997).

Gopher Frogs may occur at considerable distances from breeding ponds (e.g., > 2 km; Franz et al., 1988). Phillips (1995) also recorded 2 frogs in Georgia moving distances of 95–102 m from their breeding ponds. They make use of Gopher Tortoise burrows and other retreats for temporary shelter as they migrate to and from breeding ponds (Bailey, 1989). Underground burrows are also crucial for dispersing juveniles, with survivorship greatly increased in areas with high numbers of burrows (Roznik and Johnson, 2009b). Precipitation over a period of days likely facilitates such movements.

Juvenile dispersal occurs during the hot summer months, and may not be correlated with rainfall during the humid nights (Greenberg, 2001b); some juveniles may be active even in fall and winter, however. Juveniles emigrate from breeding ponds nonrandomly toward fire-maintained

Eggs of *Lithobates capito*.
Photo: Steve Johnson

longleaf pine habitats rather than to habitats that are fire-suppressed. Fire-maintained habitats have a more open-canopy forest (ca. 18%; Phillips, 1995), fewer hardwood trees, smaller amounts of leaf litter, and larger amounts of wiregrass than fire-suppressed habitats. They also contain more Gopher Tortoise and small mammal burrows than fire-suppressed habitats, thus offering migratory Gopher Frogs shelter and permanent retreats. In Florida, dispersing juveniles moved to 691 m (mean 173 m) from the natal pond and sometimes used dirt roads as movement corridors (Roznik and Johnson, 2009a, 2009b). The mean distance between successive moves was 60.4 m. Long movements sometimes occur in a short period, and juveniles will use other ponds in the vicinity as stopovers during dispersal.

Gopher Frogs are sensitive to light in the blue spectrum ("blue-mode response"), which apparently helps them orient toward areas of increasing illumination, such as the open horizon above lakes and ponds (Hailman and Jaeger, 1974). They likely have true color vision.

CALLING ACTIVITY AND MATE SELECTION

Males and females arrive simultaneously at breeding ponds, but males arrive in greater numbers early in the breeding season. Although movements may occur over an extended time, most frogs move on only a few nights. Movements to breeding sites are correlated with rainfall, and most frogs arrive 1–5 hrs after dark or around sunrise (Bailey, 1990). Males remained longer at a breeding site in Alabama than females (males, mean 25 days; females, mean 9 days), although males spent up to 59 days and females 37 days at the pond (Bailey, 1990, 1991). Nearly half the males remained at the pond ≥30 days, but only 6% of the females did so. In Florida, mean residency for males was 14.6 days (range 1 to 78 days) and for females 9.5 days (range 1 to 95 days) (Palis, 1998). Movements occur back and forth between the breeding site and terrestrial habitats throughout the breeding season depending on rainfall, and these frogs usually emigrate in the same general direction as they arrived.

Calling occurs from thick grass tussocks and sedges in shallow water, pondside debris, and the base of stumps, gum, and cypress trees. Males call at night, with peaks after dark and before dawn, and calls may occasionally be heard in the late afternoon. The call of *L. capito* is a deep snore lasting about 2 sec. It has good carrying capacity, and may be heard 0.4 km distant (Wright and Wright, 1949). Gopher Frogs also call from underwater, in which case the call may not be audible to human observers > 10 m away (Jensen et al., 1995).

BREEDING SITES

Breeding occurs in temporary to semipermanent, mostly fishless ponds (but see Phillips, 1995) dominated by short herbaceous vegetation (maidencane, panic grass, bluestem, yellow-eye, pipewort, various rushes). Short woody vegetation often surrounds the pond margins, but there should be an open canopy throughout much of the pond. Depression wetlands range in size from 0.10 to 33.5 ha (Bailey, 1990; Dodd, 1992; Cash, 1994; Palis, 1997; Greenberg, 2001b; Greenberg and Tanner, 2005b). Other sites include sinkhole ponds, cypress ponds, cypress/gum ponds, pine savanna wetlands, Carolina Bays, ditches, and borrow pits. Ponds are usually shallow (< 1.5 m) with a pH <6.0 (Bailey, 1990; Palis, 1997).

REPRODUCTION

Calls may be heard at any time of the year, although most reproduction takes place during the winter and early spring (Carr, 1940b; Bailey, 1990, 1991; Semlitsch et al., 1995; Palis, 1998). Breeding may be more opportunistic than literature records suggest, with the season longer in peninsular Florida than in the north. For example, Palis (1998) recorded successful reproduction in October, February, and April, when water temperature was 15–26°C; other records are available during the summer and autumn, especially after tropical storms (Wright, 1932; *in* Palis, 1998; Jensen et al., 2008). Reproductive effort (number of breeding females, number of egg masses) is positively correlated with the amount of precipitation just prior to (December to January) and during (February to March) the main part of the breeding season (Palis, 1998; Jensen et al., 2003). During years of drought, no breeding will occur.

Eggs are oviposited in a plinth 8–15 cm in diameter and ca. 2.5–7 cm thick, and occasionally larger (Wright, 1932). The mass is attached to vegetation (grass, sedges, pickerel weed, rushes) about 0.5 m from shore. Egg masses are placed from 0 to 20 cm below the surface in water 17–78 cm deep (Bailey, 1990; Phillips, 1995; Palis,

1998). Wright (1932) counted half of a single female's egg compliment and estimated there were 5,008 eggs in the clutch, whereas Palis (1998) counted 540–4,825 (mean 2,210) eggs per mass. In Georgia, clutch size was 988, 1,281, and 1,463 for 3 clutches (Phillips, 1995), and in Alabama, Bailey (1990) counted a single clutch of 1,709 eggs. Clutches may be deposited in a clumped distribution, although not necessarily communal (Phillips, 1995). Only one clutch is deposited per year, and hatching success is generally high (ca. 93.5%) (Phillips, 1995). Hatching occurs in four to seven days. Hatchlings are 12–13 mm TL, and have a deep tail fin (Palis, 1997).

LARVAL ECOLOGY

The larval period lasts from 3 to >6 months. In South Carolina, for example, Semlitsch et al. (1995) estimated it at 87–113 days, whereas Wright (1932) estimated it as 85–106 days. Phillips (1995) recorded 175–225 days before most transformation was completed in Georgia, although some larvae still had not transformed after 290 days in the laboratory. Jensen et al. (2008) give a larval period of 145–169 days. Newly transformed *L. capito* are 26–43 mm SUL (Wright, 1932; Phillips, 1995; Semlitsch et al., 1995; Palis, 1998; Jensen et al., 2008). They may begin dispersing prior to resorption of the tail bud.

In experimental trials, larval *L. capito* are readily consumed by predaceous fish and may be injured even by fish too small to consume them. In the presence of such species, *L. capito* reduces its activity, which could lead to decreased size or delayed metamorphosis among survivors. Although all fish do not have a negative impact on Gopher Frogs, susceptibility to larval predation likely accounts for their preference of fishless ponds for breeding (Gregoire and Gunzburger, 2008). In laboratory trials, as tadpoles become larger, predation rates by some insects decrease either because of predator satiation or because the tadpoles have become too large for the predators to handle (Cronin and Travis, 1986).

DIET

The diet includes just about any animal that can be crammed into the mouth. This includes earthworms, beetles, hemipterans, grasshoppers, other species of frogs (*Anaxyrus*), and even birds (Dickerson, 1906; Carr, 1940a).

PREDATION AND DEFENSE

Lithobates capito has a peculiar odor when handled that may function in predator deterrence. They are wary when approached and rapidly retreat down burrows to safety. Wright (1932) noted a number of accounts whereby Gopher Frogs rapidly changed background color, presumably to better match light and substrate conditions; this would undoubtedly aid in crypsis. Adults also assume a defensive stance similar to *L. areolatus*, and may become immobile when handled. Larvae are vulnerable to a variety of invertebrate predators, especially when small. As they grow large, however, predators have a more difficult time handling them, and predation rates decrease in proportion to increasing body size (Travis et al., 1985a).

Eggs are attacked by caddisfly larvae and newts (*Notophthalmus viridescens*) (Bailey, 1990). *Lithobates capito* normally breeds in fishless ponds. However, mosquitofish (*Gambusia*) were introduced historically into some Gopher Frog ponds, and these fish proved difficult to eradicate. Although this fish does not eat the eggs of *L. capito*, they nip at embryos, causing an inability to swim and, eventually, death (Braid et al., 1994; Gregoire and Gunzburger, 2008). Predators of postmetamorphs include snakes (*Coluber constrictor, Nerodia fasciata, Thamnophis sirtalis*), turtles (*Apalone ferox, Kinosternon subrubrum*), birds (owls), and mammals (raccoons) (Jensen et al., 2008; Roznik and Johnson, 2009b). Larvae are likely consumed by a wide array of invertebrate and vertebrate predators. However, Phillips (1995) found no differences in the swimming behavior between controls and larvae exposed to predator chemical cues.

Tadpole of *Lithobates capito*. Photo: Todd Pierson

POPULATION BIOLOGY

The sex ratio within a population will appear male-biased on any one night at a breeding pond, although it does not differ from 1:1 based on combined captures during a breeding season (Palis, 1998). However, in southern Alabama, Bailey (1990, 1991) reported capturing 93 males and 176 females using a drift fence at a breeding site (1:1.9 ratio) during a single breeding season. Mortality is high for dispersing juveniles, especially during the first 12–30 days following metamorphosis (Roznik and Johnson, 2009b). Survivorship varies among breeding ponds, and Roznik and Johnson (2009b) suggested that the presence of underground refuges along dispersal routes significantly impacts juvenile survival. Indeed, all surviving juveniles in their study located a burrow within eight days of leaving a pond.

Juvenile recruitment tends to be much higher in open longleaf pine savanna than in similar areas invaded by extensive hardwoods (Greenberg, 2001b). Presumably there are more retreat sites in areas with high numbers of Gopher Tortoises, burned-out tree holes, and root cavities. As expected, juvenile recruitment varies annually and among breeding sites, where it may or may not be associated with the number of adults during the previous breeding season (Greenberg, 2001b; Greenberg and Tanner, 2005b). During years of poor recruitment, recent metamorphs tend to be smaller and weigh less than in good years.

Breeding populations of *Lithobates capito* may be extremely small. At the Savannah River Site, only 7 ponds are known to contain Gopher Frogs despite intensive sampling for 41 yrs, and no more than 10 adults have ever been collected at a single site (Semlitsch et al., 1995). Gibbons and Bennett (1974) recorded 62 *L. capito* at one of these ponds in 1969–1970, but this figure undoubtedly included juveniles. Despite small numbers, there has been no evidence of declines at the known sites. In other examples, Dodd (1992) captured 44 *L. capito* at a small pond in Florida in a single year, and Cash (1994) recorded 154 *L. capito* in 544 days of trapping at a 2.5 ha pond in southeastern Georgia. At 8 small ponds in central Florida, Greenberg and Tanner (2005b) captured 123 adults and 1,338 juveniles from 1994 to 2001. Abundance can be high in optimum terrestrial upland sandhill habitats (Branch and Hokit, 2000; Enge and Wood, 2001).

DISEASES, PARASITES, AND MALFORMATIONS

The tick *Ornithodoros turicata* is reported on *Lithobates capito* from Florida (*in* Young and Goff, 1939). This species is parasitized by the

Adult *Lithobates capito*.
Photo: Aubrey Heupel

Breeding habitat of *Lithobates capito*, Putnam County, Florida. Photo: C.K. Dodd Jr.

trematodes *Brachycoelium hospitale* and *Megalodiscus temperatus* (Manter, 1938).

SUSCEPTIBILITY TO POTENTIAL STRESSORS
pH. Phillips (1995) measured pHs of 4.6–5.7 (mean 5.0) at a southeast Georgia site, but no data are available on sublethal or lethal pH levels.

STATUS AND CONSERVATION
Populations of Gopher Frogs have declined throughout the species range due to habitat loss and degradation, particularly from agriculture, silviculture, and urbanization. As of the mid-1990s, populations were known from 7 sites in Alabama, 32 historic sites in North Carolina, 2 areas in South Carolina, and only 5 areas in Georgia (Bailey, 1991; Semlitsch et al., 1995; Palis, 1997). In North Carolina, 5 sites had been destroyed, 10 were inactive, 6 represented road collections where the breeding pond was unknown, and 11 were active (Braswell, 1993). Most populations occur in northern and central Florida and the species is doing well on protected lands. However, it has disappeared from many areas as wetlands and uplands are developed and Gopher Tortoises decline (e.g., Delis et al., 1996). The species is protected in Alabama, Florida, and North Carolina.

Primary threats include habitat loss, degradation, and fragmentation. In Florida, the species is usually not found in pine plantations and agricultural areas, and only rarely in urbanized settings (Means and Means, 2005; Surdick, 2005). Other threats include fire suppression, site preparation, road construction and mortality, the introduction of fishes, and mud-bogging by off-road vehicles (Bailey, 1991; Palis, 1997; Roznik and Johnson, 2009a, 2009b). Populations also may have declined as Gopher Tortoise populations declined. Management options include protecting known breeding sites and adjacent uplands, instituting prescribed fires to control invasive vegetation, stopping fish introductions and removing them from breeding sites when present, and prohibiting wanton destruction of wetlands by off-road vehicles. Buffer zones may have to extend a considerable distance around breeding sites, and local populations of Gopher Tortoises need to be managed to ensure their viability.

Lithobates catesbeianus (Shaw, 1802)
American Bullfrog
Ouaouaron
Poloka lana (Hawaiian)

ETYMOLOGY
catesbeianus: named for Mark Catesby (1679/83–1749), an English naturalist who described the fauna and flora of the Southeastern Coastal Plain and the Bahamas in his book, *Natural History of Carolina, Florida, & the Bahama Islands*, issued in parts from 1729 to 1747.

NOMENCLATURE
Conant and Collins (1998) and Stebbins (2003): *Rana catesbeiana*

Dubois (2006): *Lithobates (Aquarana) catesbeianus*

Synonyms: *Rana conspersa, Rana mugicus, Rana mugiens, Rana scapularis, Rana taurina*

IDENTIFICATION
Adults. American Bullfrogs are large light olive, dark green, or brown, heavy-bodied frogs that have a large tympanum and lack a dorsolateral fold. They may be distinctly mottled to nearly uniformly colored without a pattern dorsally. Snouts are often bright green, and there is usually a pronounced sacral hump. Distinct black bars are present on the dorsal surface of the thighs and lower legs. The bellies are cream colored to white, although some individuals possess dusky reticulations or mottling that cover various amounts of the belly. A few individuals have coarse dark mottling or oscillations covering the entire belly. Individuals with such variable ventral patterns can occur in the same location (e.g., Rhoads, 1895). The tympanic fold is well developed. The eyes are large, and the rear feet are highly webbed, although the front feet are not. The longest toe (fourth) of the hind foot lacks webbing at its most distil end, and the thigh is patterned, but not striped (unlike Pig Frogs); there is no white spotting on the lips (unlike River Frogs). Adults normally are 85–200 mm SUL. Lutterschmidt et al. (1996) recorded a female from Oklahoma at 204 mm SUL weighing 909 g.

Adults are sexually size-dimorphic (Howard, 1981), and there are additional distinguishing characteristics between males and females during the breeding season. Males are generally smaller than females. They have larger forearms than females, and the thumbs become enlarged during the breeding season. These changes assist a male in holding on to a female during amplexus. Adult males also have a larger tympanum than females, which probably functions to assist him in projecting his voice rather than for receiving sounds (Purgue, 1997). In females, the eye and the tympanum are about the same size, whereas in males the tympanum is larger than the eye. This difference becomes apparent as frogs reach sexual maturity (Schroeder, 1966). The male's throat color changes from a pale cream to a deep yellow during the breeding season.

Specific size records include: males 91–152 mm SUL in Michigan (Collins, 1975); males to 156 mm SUL and females to 172 mm SUL in Michigan (Howard, 1981); males to 151 mm SUL and females to 162 mm SUL in Québec (Bruneau and Magnin, 1980b); male mean of 151 mm SUL and female mean of 140 mm SUL in New Jersey (Ryan, 1980); males 95–145 mm SUL (mean 121 mm) and females 102–137 mm SUL (mean 125 mm) in Indiana (Minton, 2001); males to 170 mm SUL and females to 181 mm SUL in Missouri (Schroeder and Baskett, 1968); males to 171 mm SUL and females to 184 mm SUL in Louisiana (George, 1940); adults to 180 mm SUL along the lower Colorado River (Clarkson and DeVos, 1986); males 48–61 mm SUL and females 60–82 mm SUL in California (Twedt, 1993). From 1986 to 1993 in Ontario, female mean sizes ranged between 93 and 116 mm SUL, with maximum SUL between 118 and 157 mm SUL (Shirose and Brooks, 1997); these values suggest considerable annual variation in mean and maximum body size. Body masses of the largest individuals in Québec are 252 g for males and 335 g for females (Bruneau and Magnin, 1980b).

Juveniles are pigmented much like adults, normally with an olive green color dorsally with many small black spots. These spots may be retained as adults (as brown instead of black), although most juveniles lose the spots as they grow. Like the adults, the dorsum may range from deep olive to light green.

Albino and yellow-colored bullfrog adults and larvae have been reported from throughout the species' range, including Alabama (Mount, 1975),

Arizona (Echternacht, 1964; Cecil Schwalbe photo, this volume), Arkansas (Trauth et al., 2004), California (Peters and Ruth, 1962; Lemm, 2006), central Pennsylvania (Phillips and Boone, 1976), Virginia (Mitchell and McGranaghan, 2005), Louisiana, Maryland (Harris, 1968), New York, Ohio, Oklahoma, Ontario, Oregon, and Washington (reviews in Hensley, 1959; Dyrkacz, 1981). Axanthic (blue) bullfrogs are also known (Berns and Uhler, 1966; Bechtel, 1995; photo, this volume).

Larvae. Bullfrog larvae are the second largest anuran larvae in North America; only River Frog tadpoles are larger. However, maximum size varies considerably through the species' range and may depend on the duration of the larval period. There appear to be two major phenotypes of American Bullfrog larvae. Tadpoles throughout much of the species' range are large, especially after their first season, and grayish green to olive in coloration. There are small, widely spaced black flecks throughout the body and tail, which helps distinguish them from other ranid tadpoles of similar size. More flecks are found on the dorsal fin than the ventral fin. The belly is off-white or cream to straw colored in pigmentation, and the viscera are indistinct. The anus is dextral. Tails are long and pointed, with a well-developed fin. Eyes are dorsolateral. The total length (body + tail) may reach 162 mm, although most larvae undergo metamorphosis at ca. 100 mm TL. The largest larva ever recorded was 197 mm TL, and was found in an Arizona stock pond (Schooley et al., 2008). This tadpole had been impacted developmentally, likely by the addition of copper sulfate to the water. Small larvae (< 25 mm TL) have transverse gold bands on the snout and body and are black.

A second phenotype, found in Florida and along the southern sections of the Atlantic Coastal Plain, is much more olive to dark green dorsally, with bright yellow pigmentation ventrally. A somewhat similar phenotype has been reported from the Pacific Northwest, perhaps suggesting a Coastal Plain origin for this population (Jones et al., 2005). Larvae also develop a certain degree of phenotypic plasticity in response to the presence of predators. In the presence of mudminnows (*Umbra*) and *Ambystoma* salamander larvae, the tail fins of bullfrog tadpoles are longer than they are in the absence of these predators (Relyea, 2001b). Presumably, longer tail fins assist in escaping from these tadpole predators. Descriptions and illustrations of larval mouthparts are in Hinckley (1882), Wright (1914), Walker (1946), and Dodd (2004). Descriptions of larvae are in Wright (1914, 1929) and Altig (1972a).

Eggs. Eggs are laid in a large surface film in the form of a disk, normally with 10,000 to 12,000 eggs per disk. In Louisiana, the film can be from 0.45 m × 0.45 m to 0.9 m × 0.6 m (George, 1940). The film is placed over submergent vegetation or among sticks and branches. After about 20 min, the surface film sinks and settles over the substrate. It may remain relatively flat (termed 2D), or it can snag on vegetation and branches where it clumps (termed 3D). The vitellus is 1.2–1.7 mm, and the outer membrane 6.0–10.4 mm in diameter. There is only one gelatinous envelope, and it is indistinct, often merging with the envelopes of adjacent eggs. More than one clutch may be oviposited per season, especially if the female is > 130 mm SUL, accounting for some of the literature records of exceptionally large egg counts in gravid females. Eggs hatch in 2–5 days; the larvae have a 3–4 mm body length and a total length of 7 mm.

DISTRIBUTION

American Bullfrogs occur widely in eastern North America and have established populations throughout the lower 48 United States, Hawaii, and western Canadian provinces as a result of human activities. They occur from sea level to ca. 1,900 m, but prefer lower elevations. The highest elevation recorded is 2,745 m at Hot Springs Creek in Colorado, a warm water stream in an area otherwise too cold for reproduction (Hammerson, 1999). Bury and Whelan (1984) summarized the distribution of American Bullfrogs in detail.

The natural distribution encompasses Nova Scotia, west across extreme southern Quèbec and Ontario (including the Laurentians), the Upper Peninsula of Michigan, much of Wisconsin, and southeast to include southern Iowa and eastern Nebraska, Kansas, Oklahoma, and Texas. The northern Great Plains (most of Minnesota, north of central South Dakota) are too cold for this species, and much of Maine is not inhabited. Populations in mountainous regions are widely scattered and limited to lower elevations. The dis-

Distribution of *Lithobates catesbeianus*

tribution to the North is limited by cold temperatures, whereas the range to the West historically was limited by the availability of aquatic habitat (especially along river floodplains), as the land becomes more arid. The natural western limits of the species are unknown, but it has been widely introduced (see below). The southernmost locality on the Florida peninsula is in Charlotte County (Means and Simberloff, 1987).

American Bullfrogs may occur on islands, such as those in western Lake Erie (Pelee, South Bass, Middle Bass) (Langlois, 1964; King et al., 1997), the Apostle Archipelago (Vogt, 1981), Georgian Bay islands, islands in the St. Lawrence (Hecnar et al., 2002), Walpole Island in Lake St. Clair (Woodliffe, 1989), Long Island, New York (Overton, 1914), Kent Island, Maryland (Grogan and Bystrak, 1973b), Assateague Island, Virginia (Conant et al., 1990; Mitchell and Anderson, 1994), Roanoke, Hatteras, and Bodie islands, North Carolina (Braswell, 1988; Gaul and Mitchell, 2007), Ossabaw Island, Georgia (Williamson and Moulis, 1994), and St. Vincent Island, Florida (Blaney, 1971). Some populations (e.g., Assateague, Ossabaw) may be introduced or augmented from mainland source ponds. They are not found on the barrier islands of South Carolina (Gibbons and Coker, 1978).

Important distributional references include: Alabama (Mount, 1975), Arizona (Clarkson and DeVos, 1986; Brennan and Holycross, 2006), Arkansas (Black and Dellinger, 1938; Trauth et al., 2004), British Columbia (Weller and Green, 1997; Orchard, 1999; Matsuda et al., 2006), California (Bury and Luckenbach, 1976; Lemm, 2006), Canada (Bleakney, 1958; Logier and Toner, 1961), Colorado (Maslin, 1959; Smith et al., 1965; Hammerson, 1999), Connecticut (Klemens, 1993), Georgia (Williamson and Moulis, 1994; Jensen et al., 2008), Illinois (Smith, 1961), Indiana (Minton, 2001; Brodman, 2003), Kansas (Smith, 1934; Collins, 1993; Collins et al., 2010), Louisiana (Dundee and Rossman, 1989), Maine (Hunter et al., 1999), Maryland (Harris, 1969), Michigan (Ruthven, 1912), Minnesota (Oldfield and Moriarty, 1994), Missouri (Johnson, 2000; Daniel and Edmond, 2006), Montana (Black, 1970; Werner et al., 2004), Nebraska (Lynch, 1985; Ballinger et al., 2010; Fogell, 2010), Nevada (Linsdale, 1940), New Brunswick

(McAlpine, 1997a), New Hampshire (Oliver and Bailey, 1939; Taylor, 1993), New Mexico (Degenhardt et al., 1996), New York (Gibbs et al., 2007), North Carolina (Dorcas et al., 2007), Nova Scotia (Gilhen, 1984), Ohio (Walker, 1946; Pfingsten, 1998; Davis and Menze, 2000), Oklahoma (Sievert and Sievert, 2006), Ontario (MacCulloch, 2002), Oregon (Ferguson et al., 1958), Pacific Northwest (Nussbaum et al., 1983; Leonard et al., 1993; Jones et al., 2005), Pennsylvania (Hulse et al., 2001), Québec (Bider and Matte, 1996; Desroches and Rodrigue, 2004), South Dakota (Malaret, 1978; Fischer, 1998; Ballinger et al., 2000; Kiesow, 2006), Tennessee (Redmond and Scott, 1996), Texas (Hardy, 1995; Dixon, 2000), Utah (Tanner, 1931), Vermont (Andrews, 2001), Virginia (Tobey, 1985; Mitchell and Reay, 1999), Washington (Metter, 1960; McAllister, 1995), West Virginia (Green and Pauley, 1987), Wisconsin (Suzuki, 1951; Vogt, 1981; Mossman et al., 1998), and Wyoming (Baxter and Stone, 1985; Koch and Peterson, 1995).

In addition, the American Bullfrog has been widely introduced throughout the world (Lever, 2003; Kraus, 2009), including central and western North America (e.g., Slater, 1941b; Vitt and Ohmart, 1978; Clarkson and DeVos, 1986; Livo et al., 1998; Adams et al., 1999; Christiansen, 2001; Krupa, 2002; Rosen and Schwalbe, 2002; Suhre, 2010; Funk et al., 2011), Hawaii (McKeown, 1996), Mexico (Conant, 1977; Luja and Rodríguez-Estrella, 2010), South America (Giovanelli et al., 2008; Barrasso et al., 2009; Sanabria and Quiroga, 2010), Oceania, the Caribbean, Asia (e.g., Li and Xie, 2004; Wu et al., 2004; Minowa et al., 2008; Liu et al., 2010), and Europe (e.g., Albertini and Lanza, 1987; Stumpel, 1992). They were imported to Idaho in 1890 (Tanner, 1941) and established in Oregon by the early 1920s, possibly from stock originating in Idaho and Louisiana (Lampman, 1946), and they were established at Carlsbad Caverns National Park, New Mexico, by 1959 (Krupa, 2002). They have even been introduced onto islands in Puget Sound (Anderson, Whidby) (Brown and Slater, 1939; Slater, 1941a). Information on the history, distribution, and impacts of bullfrog introductions in the United States and Canada are available for Arizona (Vitt and Ohmart, 1978), British Columbia (Orchard, 1999), California (Storer, 1922; Banta and Morafka, 1966; Bury and Luckenbach, 1976; Vitt and Ohmart, 1978; Jennings and Hayes, 1985, 1994a), Colorado (Livo et al., 1998; Hammerson, 1999), Hawaii (Oliver and Shaw, 1953; McKeown, 1996), Idaho (Slater, 1941b; Tanner, 1941), Iowa (Hemesath, 1998), Montana (Black, 1970), Nevada (Jennings and Hayes, 1994a), Oregon (Lampman, 1946; Funk et al., 2011), Utah (Hovingh, 1993a), and Washington (Slater, 1939b, 1955; Adams et al., 1998). Introduced bullfrogs originated from a wide variety of locations in the East (e.g., Funk et al., 2011), and even from Hawaii (to California).

FOSSIL RECORD

The earliest fossils of the American Bullfrog are known from Miocene deposits in Indiana and Nebraska, and fossil bullfrogs have been described from Pliocene deposits in Kansas and Nebraska and late Pliocene deposits in Florida (Meylan, 2005). Pleistocene deposits containing this species are widespread. They include Irvingtonian specimens from Colorado, Florida, Kansas, Nebraska, South Dakota, and West Virginia; Rancholabrean specimens occur from Alabama, Arkansas, Florida, Georgia, Indiana, Missouri, Texas, and Virginia. Most of these localities represent cave deposits. Holman (2003) and Meylan (2005) provide a comprehensive list of locations and associated references, as well as osteological criteria to separate American Bullfrog fossils from those of other ranids. Specimens referred to *L. catesbeianus* from Miocene and Pliocene deposits actually may represent the ancestral lineage to the *L. catesbeianus* species group (see below).

SYSTEMATICS AND GEOGRAPHIC VARIATION

Ranoid frogs are composed of two major lineages in North America, an eastern (*Novirana*) and a western (*Amerana*) clade (Hillis and Wilcox, 2005). Eastern North American ranoids evolved from a western European lineage in the Eocene, whereas the ancestor of the *catesbeianus* group diverged from the *pipiens* and *tarahumara* groups in the Oligocene (Case, 1978). The American Bullfrog species group (*Aquarana* clade) consists of seven closely related species (*L. catesbeianus, L. clamitans, L. okaloosae, L. septentrionalis, L. virgatipes, L. heckscheri, L. grylio*) that evolved rapidly from a common ancestor during the late Miocene to early Pliocene (Austin et al., 2003a). A previous suggestion that this group is paraphyletic (Pytel, 1986) is not supported. The group originated on the Southeastern Coastal Plain, with dispersal and vicariance accounting for subsequent speciation. Speciation within the group began ca. 12 mya, with the split between *L. catesbeianus* and *L. clamitans* occurring approximately 9 mya.

Based on an analysis of 925 base pairs of the cytochrome *b* gene, there are 2 major lineages within *L. catesbeianus*, an eastern and a western clade (Austin et al., 2004a; Austin and Zamudio, 2008). The western clade originated via range expansion from an eastern clade ancestral lineage on the Southeastern Coastal Plain, followed by isolation by the Mississippi River drainage during the Pleistocene. The lower Gulf Coastal Plain west of the Mississippi served as a refuge for the western clade during the Pleistocene. Refuge areas for the eastern clade have not been identified. The distribution of American Bullfrog cytochrome *b* haplotypes is mapped by Austin et al. (2004a). These authors noted the potential of human-mediated dispersal to account for certain anomalies in the genetic structure of American Bullfrog populations, although artificial gene flow was not considered as a sufficient explanation for large-scale phylogenetic structure.

Hybrids between *L. catesbeianus* ♂ and *L. clamitans* ♀ have been reported in laboratory crosses, with larvae reared through metamorphosis. In nature, this does not occur because the *L. clamitans* jelly capsule surrounding the egg normally blocks the entry of sperm from other ranid species (Elinson, 1975a). However, this barrier can be circumvented by high concentrations of *L. catesbeianus* sperm (Elinson, 1975b).

There appears to be regional variation in tadpole coloration. However, it seems likely that the brightly colored tadpoles of southern Georgia and Florida may represent a cryptic species.

ADULT HABITAT

American Bullfrogs are a highly aquatic species and are associated with wetlands containing a high invertebrate abundance (Babbitt et al., 2003). They are found in a variety of habitats over the course of a season, from large lakes and marshes to small backyard ornamental ponds, along streams and creeks (Enge et al., 1996; Burbrink et al., 1998; Enge, 1998b; Metts et al., 2001), beaver ponds (Metts et al., 2001), and even vernal pools (e.g., Corser, 2008). In Florida and south Georgia, for example, they occur in dome swamps, cypress and gum marshes, cypress savanna, coastal and inland hydric hammocks, basin swamps, and depression marshes (Carr, 1940a; Enge and Wood, 1998; Liner et al., 2008), although adults do not favor hyacinth covered ponds and streams (Goin, 1943). During the nonbreeding season, they reside along the shorelines of wetlands of varying sizes, either in close proximity to breeding sites or at considerable distances; movements between sites may or may not be extensive. Movements during the nonbreeding season tend to be around wetlands, along the courses of small streams and creeks, or from wetland to nearby wetland across terrestrial habitats. In the desert Southwest, however, movements between cattle tanks often occur over long distances in extremely inhospitable dry countryside.

In eastern North America, American Bullfrogs are associated with forested habitats (e.g., Enge, 1998a), particularly where forested habitats comprise >80% of the area within 1 km of the home pond (Herrmann et al., 2005). In the Great Plains along the western extent of its natural range, bullfrogs probably followed forested riparian habitats along river floodplains into the prairies. Today in the West, however, bullfrogs are widely introduced into many types of habitats, from California rivers to isolated cattle tanks in Arizona deserts to prairie potholes throughout the upper Great Plains. They are found at densities of 9/km along the lower reaches of the Colorado River, primarily in areas with >50% of the riverbank in vegetation (Clarkson and DeVos, 1986).

American Bullfrogs have occasionally been found in wells or caves (Lee, 1969a; Black, 1973a; Mount, 1975; Garton et al., 1993; Rimer and Briggler, 2010).

ECOLOGY DURING NONBREEDING SEASON

Adults and juveniles presumably occupy the same general types of habitats, but there may be some degree of habitat partitioning among the life stages. For example, Goin (1943) reported that adult bullfrogs avoided the hyacinth-covered ponds and streams of Florida, but that juveniles were common in this habitat in spring.

American Bullfrogs are active usually from spring through the autumn in the North, depending on weather conditions and latitude. They tend to emerge from overwintering sites when water temperatures reach 14°C and air temperatures reach 20–24°C (Willis et al., 1956; Treanor and Nicola, 1972; Wiese, 1985). In the southern parts of their range, they are active all year, both by day and night (e.g., George, 1940), or they become active sometimes during the midwinter depending on temperature. Literature reports of first activity (summarized by Bury and Whelan, 1984; also see Willis et al., 1956; Smithberger and Swarth, 1993) include January along the Gulf Coast and in California; February in Kansas, Florida, Missouri, and Texas; March in Colorado, Connecticut, Texas, Missouri, Illinois, and Ohio; April in Indiana, Maryland, and New York; and May in Colorado (Wiese, 1985). Some of these reports probably represent sampling biases, as bullfrogs are active year-round in the South, even as far north as eastern North Carolina (Gaul and Mitchell, 2007).

In the central and northern parts of their range, American Bullfrogs are active until the cold weather of autumn, for example, mid- to late October in Illinois (Durham and Bennett, 1963), Ohio (Walker, 1946), Colorado (Wiese, 1985), and Connecticut (Klemens, 1993), and late October to early November in Missouri (Willis et al., 1956). Gorham (1964) found juveniles under stones along a lake in October and around cold springs in November in New Brunswick. Adults usually begin winter dormancy well before the onset of freezing weather, whereas juveniles may be active until the first freezing temperatures of winter. Still, a few individuals may be observed in northern locations almost any time there is an unusual warm spell in winter. For example, Nussbaum et al. (1983) reported a massive ball of 186 semitorpid bullfrogs in an Oregon pond in February. Winterkill can be a major source of mortality.

Lithobates catesbeianus are most active in the spring, and terrestrial activity is associated with night and rainfall (George, 1940; Raney, 1940; Gibbons and Bennett, 1974). This may be because adult bullfrogs are unable to tolerate the loss of much body water (Thorson and Svihla, 1943; Thorson, 1955), so they must remain close to water or move when humidity is high. Unless exposed directly to sunlight, larvae and adults tend to reflect the temperature of the water in which they are located (Brattstrom, 1963). They spend more time on land on warm rainy nights than on cold rainy nights (Currie and Bellis, 1969). On cool or windy nights, they tend to float lower in the water than they do on warm and still nights, perhaps to avoid unfavorable temperatures (Wiese, 1985).

Basking has been reported in American Bullfrogs, and they tolerate high thermal temperatures (26–33°C, mean 30°C, CTmax = 38.2°C in California; Lillywhite, 1970). Postmetamorphs often sit exposed in sunlight near ponds with little indication of heat stress. Juveniles and adults have similar preferred body temperatures (roughly 30°C), although acclimation history affects preferences, such that frogs acclimated at lower temperatures prefer cooler environmental temperatures (Lillywhite, 1971a).

Temperature regulation is by behavioral means and is mediated by the hypothalamus. Juveniles and adults move back and forth between the pond shoreline and water in order to regulate their body temperature, although juveniles also change postures more than adults to achieve thermal preferences. The heads of bullfrogs are particularly sensitive to heat. As they warm, the anterior hypothalamus directs mucous glands in the skin to increase mucous secretion, which modulates evaporative water loss as the temperature rises (Lillywhite, 1971b). The circulatory system also plays a vital role in maintaining skin hydration, perhaps more so than mucous secretions during basking (Lillywhite, 1975).

Movements within and between seasons are sometimes extensive. An adult was recorded moving 1,219 m within a period of 19 days in New York, and another 380 m in 8 days (Ingram and Raney, 1943). In this same study, an individual was recaptured 1,600 m from where it was marked the previous year, with many other individuals moving > 350 m from one season to the next. In Missouri, Willis et al. (1956) found that most bullfrogs remained within a wetland but that overland movement could occur to 1.25 km. In addition, many individuals also remain close (within 30 m) to where originally marked, both within and between seasons (Raney, 1940; Ingram and Raney, 1943; Wiese, 1985). Other frogs stay in one general location for a season or more, then change locations, or they move back and forth between a series of locations (Durham and Bennett, 1963). Raney (1940) noted that bullfrogs inhabiting small wetlands and streams tended to remain at these locations during the course of the summer, although he also noted occasional long-distance exceptions. In arid conditions, adults and juveniles may not leave the shoreline for more than a few meters, except when they are dispersing, due to a lack of suitable terrestrial habitats with favorably humid microenvironments.

Movements may occur from one side of a large pond to the other (Willis et al., 1956). In a postbreeding study in Ontario, Currie and Bellis (1969) recorded a mean activity radius of 2.6 m (0.61–11.3 m) over a ca. 2-month observation period, with males having a slightly larger activity radius than females. When ponds are at a low density of bullfrogs, individual activity radii are larger than they are when bullfrog densities are high.

Displaced bullfrogs home readily (McAtee, 1921; Schroeder and Baskett, 1965), moving distances of 205 m in 48 h (Ingram and Raney, 1943), although some animals move to other locations (Durham and Bennett, 1963). Based on a small number of recaptures, Raney (1940) initially thought the tendency to home was not well developed, but these results probably reflected biases resulting from a short study season and a small sample size.

In the Deep South, bullfrogs do not become dormant in the winter. In the North, overwintering occurs in water (Smith, 1934; Willis et al., 1956; Treanor and Nicola, 1972; Collins, 1993; Stinner et al., 1994; Minton, 2001; Chance, 2002), although there is a report of dormancy under leaf litter on land throughout the winter (Bohnsack, 1952). Bullfrogs bury into the mud, and their cavities have been described as pits or cave-like holes (*in* Bury and Whelan, 1984). They tend to overwinter in shallow areas of a pond and do not bury into the substrate; although such areas are colder than deeper waters, they have a higher dissolved oxygen content (Chance, 2002). Bullfrogs tend not to remain in one location but move around considerably under water, even under ice, despite the cold conditions (Friet, 1993; Stinner et al., 1994; Chance, 2002). Bullfrogs also will leave drying pools or stream channels and move to dormancy sites in springs, wells, and moist crevices in the ground, or to other wetlands if available. Juveniles may excavate single-occupancy pits, which serve a dual function of concealment and retarding desiccation (Thrall, 1971).

During the activity season, American Bullfrogs do not exhibit any characteristics of sleep behavior, and appear alert at all times despite intervals of rest (Hobson, 1966). Like most frogs that are active both day and night, American Bullfrogs are photopositive in their phototactic response, suggesting they use both sunlight and moonlight when feeding and going about their daily activities (Jaeger and Hailman, 1973). They are particularly sensitive to light in the blue spectrum ("blue-mode response"), which apparently helps them orient toward areas of increasing illumination, such as the open horizon above lakes and ponds (Hailman and Jaeger, 1974). Bullfrogs likely have true color vision.

CALLING ACTIVITY AND MATE SELECTION

Male American Bullfrogs are highly territorial (Wiewandt, 1969; Emlen, 1977), and both male calling behavior and territorial defense may be influenced by hormone levels. Levels of androgen and luteinizing hormone generally increase at the start of the breeding season, and corticosterone peaks as the most intense period of chorusing begins. However, hormone levels fluctuate throughout the breeding season, even in nonterritorial males. Mendonça et al. (1985) suggested that stresses associated with territorial behavior and aggressive encounters might result in an inhibitory effect on androgen production. Thus, the effects of hormones on reproductive activity and behavior in American Bullfrogs are not as clear as they are in some other animals.

There are a number of distinct types of calls in the American Bullfrog: mating (or advertisement), territorial (sex specific), territorial (produced by both sexes), release, warning, and distress (Capranica, 1968). As their names imply, each has a specific function, with their own call characteristics and sounds. Mating calls are those most familiar to people, that is, the deep sonorous snore (vaarrhummm) of the male's voice vibrating across a water body at dusk. Mating calls are composed of 3–15 rising and falling notes in rapid succession, during which the male's paired vocal sacs are inflated. Individual notes last 0.6–1.5 sec, with intervals between notes of 0.5–1.0 sec. Calls have a periodicity of 90–110/sec; low frequencies of 200–300 cps rise to high frequencies of

1,400–1,500 cps. Capranica (1965, 1968) and Bee and Gerhardt (2001a) provide comprehensive summaries of the mating call characteristics.

The male uses the advertisement call to attract females to mate, although it can be induced in laboratory settings at other times of the year. There is an inverse relationship between the size of the male and the resonance of his mating call. Thus, the call transmits information to the female of his potential as a mate (assuming larger males have greater fitness) as well as his location. In addition, male bullfrogs are able to discriminate the calls of familiar and unfamiliar males (Bee and Gerhardt, 2002), and thus become familiar with the location and characteristics of adjacent (and perhaps rival) males. Bee and Gerhardt (2001b) suggested such neighbor recognition could lead to reduced aggression among adjacent males, as males become habituated to a neighbor's call. Bee (2001) could provide only partial support for this hypothesis. However, American Bullfrogs do exhibit some degree of stimulus-specific habituation in regard to encounter calls, aggressive movements, and tendency to approach a rival (Bee, 2003).

Territorial calls are used by males and females to establish the bounds of their territories to members of their own sex. The male's territorial call challenges nearby or rival males and asserts ownership of the territory, but calls are not used to assess the fighting ability of rivals as they are in other species (Bee, 2002). If the territorial call is unheeded by an approaching frog, an attack may follow (Wiewandt, 1969; Ryan, 1980; photo of wrestling in Howard, 1988b). As might be expected, most aggressive encounters occur early in the breeding season as territories are being established, and they are usually won by the older and larger male (Howard, 1978b). A female's territorial call serves the same function. In both instances, however, the opposite sex does not respond by calling in response to the other's territorial call, although they may move away from the area. A third call, termed the male/female territorial call by Capranica (1968), is a more generalized call issued by the winner of a contest or at the distant approach of a conspecific.

Territorial males occur at regular intervals (every 2–6 m) around the breeding pond, and readily defend their calling station. Wiewandt (1969) noted that the most responsive bullfrogs in his study defended from 9 to 25 m of shoreline. According to Emlen (1976), centrally located territories are occupied by larger (older) males, with smaller males peripheral. Howard (1978b), however, disputed a size-based distribution pattern to chorus structure, and noted that it is often difficult to define the spatial configuration of a chorus.

Calling occurs during a discrete breeding season beginning in late spring and continuing through midsummer, depending on latitude. Bullfrogs are able to call both entirely on land and while partially submerged. Partial submergence allows the calls to resonate throughout the breeding pond (Boatwright-Horowitz et al., 1999), and allows males to transmit information about their location and physical attributes farther than if calling occurred solely through the atmosphere. Males arrive first at the breeding ponds, and larger males usually arrive before smaller males. Calling begins when spring air temperatures are >21°C and water temperatures are >20°C (Fitch, 1956a; Oseen and Wassersug, 2002). Bullfrogs prefer calling when wind speeds are low (Oseen and Wassersug, 2002).

Choruses draw bullfrogs of both sexes toward a breeding pond (Emlen, 1976). At first, calling is sporadic and begins in the late afternoon, although calls may occur at any time during the day. As the season progresses, however, choruses (defined by Howard, 1978b, as a group of acoustically interacting males) become much more intense, with peak calling after midnight (e.g., >12:00 in Michigan; 02:50–08:00 in Québec; 01:30–06:30 in South Carolina). Chorusing occurs all night, tapering off at dawn (Oseen and Wassersug, 2002; Mohr and Dorcas, 1999).

Chorusing occurs steadily throughout the breeding season, but may be inhibited by low temperatures and high winds along the water's surface. A chorus begins with a few males calling, followed by a steadily increasing number of males calling until a veritable roar emanates from the breeding site. Males begin to drop out of the chorus after a short while, and calling decreases to only a few males. The chorus may be silent for several minutes before the cycle starts again.

Peak calling occurs over a period of three to five nights within an individual chorus, but choruses may form and dissolve at different locations around a breeding pond during the course of the breeding season. Emlen (1976) provides a diagram of chorus location on ten different evenings, illustrating the fluctuating position of American Bullfrog choruses around a single pond. Chorus shifts may occur in response to leech population buildup, changes in vegetation structure, or thermal changes as the season progresses. Not all bullfrog populations have shifting chorus locations, however. The positions of chorus locations also can change annually. Males sometimes move between chorusing aggregations. In contrast to Emlen's (1976) assertion, the structure of bullfrog choruses in not analogous to lek formations in birds, inasmuch as chorus structure is not easily defined spatially, nor is it size-based (Howard, 1978b).

Females initiate mating by approaching and physically contacting the male. They may not accept the first male contacted, but instead approach several males before choosing to mate, a process that may take several hours (Emlen, 1976). Sexual selection is weakly size related, with large females selecting large males, although smaller females are more generalist in their mate choice (Howard, 1988b). As might be expected, larger males tend to mate more often than smaller males, presumably in part because of their higher-quality territories and physical superiority. Small males, however, may become satellite males, trying to intercept females attracted to the high-quality territories of large males. Satellites remain in a crouched or "low" position within ca. 1 m of a territorial male, do not call, and avoid encounters with large males; their mating success is very low. A third group of opportunistic males may call, but they do not establish territories and readily flee when challenged by territorial males. These males tend to move around a great deal, avoiding aggressive encounters; they usually have intermediate mating success between territorial and satellite males.

The number of nights a male will engage in territorial behavior and calling varies, and shifts by males among choruses do not affect the amount of time spent chorusing. Dominant males spend about 50% of their time chorusing (that is, on 26 nights of a 46-night breeding season in Ontario, with 15 nights in dominant tenure; Judge and Brooks, 2001), whereas many males spend much less time in doing so. Males in better initial body condition tended to have longer dominant tenures at choruses and lost their body condition slowly, but had poorer body condition at the end of the season than males with shorter periods of dominant tenure. Thus, there appears to be an energy constraint to chorusing that affects the extent a male will remain in dominant tenure during the breeding season.

Release calls are issued by males or unreceptive females in response to amplexus attempts by other males or would-be suitors. The sound is usually made after a short struggle, and the amplexed animal is quickly released. Release calls last 0.5–1.0 sec, and are repeated every 1.5–2.0 sec. Calls are repeated at 60–85/sec at about 500 cps (Capranica, 1968).

BREEDING SITES

Breeding sites are large, permanent bodies of water, such as lakes, oxbows, natural and artificial ponds and reservoirs, and quiet waters of rivers, streams, and canals; bullfrogs also may occupy semipermanent wetlands with long hydroperiods (Eason and Fauth, 2001; Babbitt et al., 2003). In a typical example, most American Bullfrogs found in New Hampshire occurred in permanent ponds (95%), with only 5% found in temporary wetlands (Herrmann et al., 2005). In addition, they are frequently found in association with agricultural and urban settings, but much less so with silviculture (Surdick, 2005). Brodman (2009) noted considerable variation in annual occupancy of breeding sites over a 14 yr study in Indiana.

In the South, American Bullfrogs are associated with cypress, tupelo, buttonbush, willows, and grasses, although it is difficult to identify preferred habitats by vegetation alone (George, 1940). They readily breed in human-created wetlands, including reclaimed surface mines (Myers and Klimstra, 1963; Turner and Fowler, 1981; Lacki et al., 1992; Ultsch et al., 1999; Anderson and Arruda, 2006), farm and stock ponds (Galois and Ouellet, 2005; Anderson and Arruda, 2006), quarries, golf course ponds (Boone et al., 2008; Scott et al., 2008; Mifsud and Mifsud, 2009), artificial urban wetlands (Neill, 1950a; Delis et al., 1996; Surdick, 2005), fish hatcheries, power-line wetlands (Fortin et al., 2004b), constructed or restored wetlands (Briggler, 1998; Merovich and Howard, 2000; Pechmann et al., 2001; Foster et al., 2004; Brodman et al., 2006; Henning and Schirato, 2006; Palis, 2007; Shulse et al., 2010), and retention ponds (Foster et al., 2004; Surdick, 2005; Ostergaard et al., 2008).

American Bullfrogs oviposit in shallow water (15–60 cm), preferably with substantial, uniformly distributed emergent and submerged vegetation. They tend to choose warmer sites early in the breeding season, but cooler sites as the water temperature increases as the season progresses (Howard, 1978b). Breeding sites are usually open-canopied and relatively devoid of tree cover.

REPRODUCTION

Breeding begins in early to late spring and occurs through the summer, depending on latitude. However, adults will become active much earlier than when calls are first heard. There also appears to be some confusion in the literature as to the breeding season, as calling is often equated with breeding; hence, some "breeding" dates in the literature may not be accurate. For example, bullfrogs in east-central Illinois first appear at breeding sites in late March to mid-April, but eggs are not observed until late May (Durham and Bennett, 1963). In Québec, calling begins in May, but eggs are not observed until mid-June (Bruneau and Magnin, 1980b). In Louisiana, calling occurs in February, but egg deposition occurs in mid-April (George, 1940), and in Indiana, calling occurs from April to early July, although peak calling occurs from early May to mid-June (Brodman and Kilmurry, 1998).

Literature records of calling dates include: February to August (Louisiana: George, 1940), March to October (Florida: Carr, 1940a; Texas: Blair, 1961b), April to June (New Jersey: Ryan, 1980), April to July (Indiana: Brodman and Kilmurry, 1998; Virginia: Mitchell, 1986), May to July (Michigan: Emlen, 1976; North Carolina: Gaul and Mitchell, 2007; Ohio: Walker, 1946; Québec: Bruneau and Magnin, 1980b); May to August (Oregon: Nussbaum et al., 1983), June and July (Minnesota: Oldfield and Moriarty, 1994; New York: Raney, 1940; Nova Scotia: Gilhen, 1984; Ohio: Varhegyi et al., 1998; Ontario: Piersol, 1913; Judge and Brooks, 2001; Québec: Lepage et al., 1997; South Dakota: Fischer, 1998; Wisconsin: Vogt, 1981); July to September (Colorado: Wiese, 1985). Parker (1937) also reported hearing calls of American Bullfrogs as late as the first week of September after a period of rain in western Tennessee. Egg deposition occurs in April to May in Alabama (Brown, 1956), May to June in Tennessee (Gentry, 1955), June to July in Kentucky (Viparina and Just, 1975), April to July in Arizona (Dowe, 1979), and May to August in Oregon (Brown, 1972); in Oregon, however, most eggs are oviposited from late June to mid-July.

Egg mass of *Lithobates catesbeianus*. Photo: Dana Drake

The normal breeding season (i.e., with egg deposition) extends over a period of one to two months throughout most of the range of the American Bullfrog, despite differences in when it is initiated. However, there may be two peaks of egg laying within the breeding season, the first in spring to early summer and the second about three weeks later as large females get ready to deposit a second clutch (Bruggers and Jackson, 1974; Emlen, 1977; Dowe, 1979); Emlen noted 5 of 73 bullfrogs followed during a breeding season in Michigan produced a 2nd clutch.

George (1940) suggested that choruses only lasted a few days after egg deposition in Louisiana, although occasional calls could be heard until August. Females ovulate over a very short period of time (one night), but ovulation within the population is asynchronous and occurs throughout the breeding season (Emlen, 1976). Thus, males may stay throughout the breeding period at the breeding pond, whereas females stay just long enough to mate and move back and forth to the main breeding pond. The sex ratio around a breeding pond is highly male-skewed at any one time during the breeding season, since females are receptive for only a very brief time and do not remain in close proximity to the highly territorial males. However, the sex ratio may be 1:1 for frogs sampled throughout the breeding season at multiple sites (Schroeder, 1966; Schroeder and Baskett, 1968; Suhre, 2010). Still, Wiese (1985) found a sex ratio of 1.3 females per male in a Colorado population.

Amplexus is axillo-pectoral, occurs in water, and lasts from 17 to 155 min (mean 49 min) in Michigan (Howard, 1978b). During oviposition, the female extends her legs backward, downward, and laterally; her body arches concavely, her head is under water, and her cloaca is arched upward (illustration in Aronson, 1943a). The male appears to stimulate the female using his hind limbs to stroke her body, at which time the female's abdominal muscles begin to contract and the first eggs are extruded. The male fertilizes the eggs, then uses his rear legs to spread them out in a surface film. The duration of amplexus (ca. 2 hrs in Aronson's observations) is positively correlated with female body size, but not with male body size.

Males mate with more than one female during a breeding season, although Howard (1988b) found that 48% of the males in his population were unsuccessful at mating. Females, however, all mated at least once, and some twice, over a 3 yr period. Mating success is low in the first year of maturity (11% for males, 10% for females), but increases from 55 to 100% for 2 to 5 yr old males. Females >1 yr in age had a 100% chance of mating successfully (Howard, 1988b).

Eggs are normally deposited within the territory of the fertilizing male (Ryan, 1980). The number of eggs produced varies with female size, such that larger and older females produce far more eggs than younger and smaller females (Collins, 1975; Howard, 1978a; McAuliffe, 1978; Bruneau and Magnin, 1980b). For example, Howard (1988b) recorded Michigan females as producing means of 2,007 eggs in yr 1, 3,372 in yr 2, 7,228 in yr 3, 10,238 in yr 4, and 11,147 in yr 5; the maximum number was ca. 20,000 eggs. Female body weight is correlated with clutch weight and the number of eggs per clutch, but not with female SUL (Woodward, 1987a).

In Québec, the number of eggs ranged from 3,826 to 23,540 (Bruneau and Magnin, 1980b), whereas a 128 mm SUL female deposited 16,640 eggs in Nebraska, and a 179 mm SUL female possessed 47,480 eggs (McAuliffe, 1978). George (1940) estimated clutch sizes at 8,000–15,000 (mean 12,000) eggs in Louisiana; Trauth et al. (1990) recorded 12,756–43,073 (mean 22,944, n = 7) eggs per female in Arkansas; Ryan (1980) recorded a mean of 7,360 eggs/mass in New Jersey; between 11,585 and 21,510 eggs were counted in Nova Scotia (Gilhen, 1984); Brown (1972) counted 16,491 eggs in a single Oregon clutch; and Woodward (1987a) recorded a mean of 11,126 eggs per clutch in New Mexico. Numbers >20,000 or so probably represent multiple clutches. Willis et al. (1956) provide detailed descriptions of the eggs, oviducts, and ovaries of bullfrogs through the reproductive cycle.

Males generally are able to fertilize more eggs as they grow larger and older, from 2,732 in yr 1 to 19,346 in yr 5 (Howard, 1988b). Large females also produce larger eggs (1.58 mm in diameter) than smaller females in their first clutch of the year, but not in the second clutch (1.48 mm in diameter). Female size and total fecundity are not correlated, although such a relationship exists if first and second clutches are distinguished (Howard, 1978a). The second clutch of small females, however, also had smaller eggs than their first clutch of the season. Eggs fertilized by larger males also may have higher hatching success in some years than eggs fertilized by small males.

American Bullfrogs have a rather narrow range of temperatures in which normal embryo development takes place. Eggs do not hatch or develop in water <15°C (Moore, 1942a; Viparina and Just, 1975). Eggs are not deposited until the water temperature reaches ca. 22°C in Michigan (Howard, 1978a). The maximum water temperature normally should not exceed 32°C for successful development. Abnormalities are seen in developing embryos at temperatures >32°C, with death at 33–36°C (Moore, 1942a; but see Dowe, 1979, where eggs developed at 33°C). The egg mass temperature may be slightly lower than the surrounding water temperature, thus providing a measure of protection from high temperatures (Ryan, 1978).

LARVAL ECOLOGY

Larvae are considered primarily herbivorous, grazing on algae and plant material. Larvae were considered detritus feeders by Thrall (1972), whereas others have reported them grazing on algae and bacteria as well as detritus (Brown, 1972; Kupferberg, 1997a), or acting as suspension feeders on phytoplankton (Seale, 1980).

Tadpole of *Lithobates catesbeianus*. Photo: David Dennis

Thrall (1972) found no evidence of morphological or physiological specializations for herbivory. In contrast, Pryor (2003, 2008) described the gut as containing bacteria (*Edwardsiella tarda*, *Clostridium* sp.), ciliates, and nematodes, particularly the nematode *Gyrinicola batrachiensis*. Under laboratory conditions, the presence of the nematode aided in fermentation, except for cellulose, and Pryor (2003) considered it a mutualistic symbiont of bullfrog tadpoles. Munz (1920) recorded diatoms, mud, a number of species of algae, and epidermis in tadpole guts.

In Oregon, American Bullfrogs tend to feed on algae in proportion to availability within the pond. However, there may be among-habitat differences in the percentage of species consumed, a diet that changes during the season. Larvae may eat species in proportion to their occurrence in one pond, but not in adjacent ponds. In other situations, some algae (e.g., *Spirogyra*, *Mougeota*) may be found in much greater proportions in guts than in pond water, suggesting preferential selection (Brown, 1972). Types of algae in the diet of bullfrogs include members of the Oedogoniales, Bacillariophycea, Desmidacea, and Zygnematales.

Larval American Bullfrogs are opportunistically carnivorous, feeding on the dead bodies of other animals, including larval conspecifics. They sometimes consume the eggs and hatchlings of other *Rana* species (Ehrlich, 1979), despite the inability of Funkhouser (1976) to induce them to take animal material. Disagreements as to the categorization of diet of bullfrog larvae may reflect regional differences among populations of bullfrogs or in food availability. Under laboratory conditions, larvae are coprophagous, a situation that enhances growth rates (Steinwascher, 1978a; Pryor, 2003).

In the North, larvae grow from around 19–26 mm body length in midsummer to 32 mm by autumn of the 1st year (Willis et al., 1956; Bruggers and Jackson, 1974), to 40–52 mm by the 2nd year, and 53–56 mm after the 3rd winter (in Québec: Bruneau and Magnin, 1980a). Larval growth rates are dependent on temperature, oxygen levels, larval density, food resource amount and availability, and possibly extent of sedimentation (George, 1940; Raney and Ingram, 1941; Licht, 1967; Brown, 1972; Bruggers and Jackson, 1974; Corse and Metter, 1980). These latter authors found that supplemental feeding greatly increased growth rates and size at metamorphosis and decreased the length of the larval period. Larvae do not grow in the winter, but they will continue to feed at much reduced levels. For example, in Oregon nearly all growth occurs from May to November (Brown, 1972).

Larger larvae tend to choose water temperatures of 24–30°C (Brattstrom, 1962, 1963; Lucas and Reynolds, 1967), but preferences and developmental rates change with previous thermal history and stage of development (Hutchison and Hill, 1978; Crawshaw et al., 1992). In Michigan, however, small bullfrog tadpoles tended to prefer more open habitats in medium-depth water, whereas larger tadpoles prefer deeper water

Tadpole of Florida-Georgia *Lithobates catesbeianus.* Photo: Ronn Altig

and show no preferences regarding the extent of cover, at least under laboratory conditions (Smith, 1999). These results seem counter to other observations obtained from field observations of temperature and habitat use. The length of photoperiod does not affect temperature preferences.

As they approach metamorphosis, premetamorphic larvae move to the warmest part of the pond prior to leaving. During the summer, tadpoles may be found in both deep and shallow areas of ponds and lakes, a habitat choice that may result from different predation pressures in shallow water habitats, that is, if predation pressure is high, larvae may move to deeper waters. Inasmuch as deep water may be anoxic or severely hypoxic, bullfrog larvae switch from facultative air breathing to obligate air breathing in deep-water lakes and ponds (Ultsch et al., 1999). However, Nie et al. (1999) suggested that diurnal and annual shifts in habitat use between deep and shallow water is driven by temperature selection, rather than by predation pressure. In any case, larger larvae tend to choose more structured habitats than smaller larvae, but there is a great deal of individual variation in habitat use and activity (Smith, 1999; Smith and Doupnik, 2005).

American Bullfrog larvae often aggregate in large numbers, but these aggregations do not appear to have a social context as they do in some other ranids (Wassersug, 1973). Bullfrog larvae are more active in groups when in proximity to predators than are solitary tadpoles (Smith and Awan, 2009); this suggests that presence in larval groups reduces individual predation threat They are able to use plane-polarized light in spatial orientation (Auburn and Taylor, 1979) as well as sun-compass orientation using extraocular photoreceptors (Justis and Taylor, 1976), both of which help in learning the physiography of the pond and in setting biological clocks. In this way, they are able to ascertain the location of shorelines, an important consideration for escape to deeper waters and foraging and thermoregulation in shallower water. Larval bullfrogs are photopositive toward the color green, which may help them associate with green plants, which offer a source of both food and cover (Jaeger and Hailman, 1976).

Generalizations about the length of the larval period of this wide-ranging species are misleading, as American Bullfrogs occur from the subtropics to the northern hardwood forest. Much more research is available on northern bullfrogs than southern bullfrogs, and statements on generalized life history characteristics are often based on northern populations. In addition, literature records are sometimes unclear in their use of the term "year." Thus, "transformation occurs in two years" could mean after two calendar years (i.e., in the third summer of life) or in the second season (i.e., in the second summer of life).

American Bullfrogs overwinter one or two years prior to metamorphosis (i.e., metamorphosis in their second or third summer) throughout much of their range, but may require three overwintering periods in Canada (Bruneau and Magnin, 1980a) and New Hampshire (Oliver and Bailey, 1939). In some areas, however, larvae do not overwinter. In Louisiana, metamorphosis occurs 4 months after hatching (George, 1940); in Arizona, transformation occurs after 3 months (Dowe, 1979); Seale (1980) suggested transformation at <3 months in Missouri; Cagle (1942) noted that some larvae complete metamorphosis within the first season in Illinois; and metamorphosis occurs in <6 months in Hawaii (McKeown, 1996). In Tennessee, Gentry (1955) stated that about 50% of the tadpoles transform the first summer, with the remaining overwintering, whereas in central Kentucky, 95% of larvae metamorphose in 12–14 months and 5% complete metamorphosis in 3–4 months (Viparina and Just, 1975). In California, bullfrogs transformed after six to seven months as a reservoir dried-down, thus suggesting a decreasing hydroperiod might speed up metamorphosis (Cohen and Howard, 1958).

Since the duration of the larval period is dependent on environmental conditions, food resources, and the timing of reproduction (i.e., late season reproduction most likely results in a prolonged larval period), transformation can occur over a range of sizes, depending on the status of the habitat and the resources available during development. For example, George (1940) reported metamorphs of 32–45 mm SUL, depending on whether the tadpoles metamorphosed from a drying pond or were completing metamorphosis without the stress of decreasing water levels; Walker (1946) recorded newly transformed frogs at 29–53 mm SUL; Dowe (1979) found newly transformed Arizona bullfrogs to be 41–46 mm SUL and weigh 6.3–9.4 g; and Wiese (1985) found that Colorado metamorphs were 35–54 mm SUL (range 30–60 mm SUL). Seale (1980) noted wet mass body weights of 7.2 g for froglets metamorphosing after 3 months, and 24.6 g for froglets that had overwintered and transformed the following June.

The location within a pond where larval American Bullfrogs overwinter may be influenced by vegetative cover, dissolved oxygen, and temperature (Hargis et al., 2008). In northern California, nonindigenous populations of American Bullfrog larvae overwinter in quiet pools along streams. Densities of large tadpoles that overwintered the previous year ranged from 1 to 52 larvae/m^2 (mean 18; n = 17 quadrats) in mid-June to ca.1.4 larvae/m^2 in late July, as transformation occurs (Kupferberg, 1997a). In contrast, the density of small first-year tadpoles was 0–310 larvae/m^2 (mean 91; n = 20 quadrats) in mid-June, decreasing to a mean of 42 larvae/m^2 (range 4 to 120, n = 5 quadrats) in late July (Kupferberg, 1997a).

Bullfrog larvae reach densities of 0.9–13.2/m^2 under natural conditions, depending upon pond location and time of year. Cecil and Just (1979) estimated larval biomass at 11–103 g/m^2 at their study ponds in Kentucky. Survivorship varies among locations, and Cecil and Just (1979) estimated it as 11.8–17.6% at 3 Kentucky ponds over a 2 yr period. In experimental ponds, density was inversely correlated with survivorship, but not body mass at metamorphosis or time to metamorphosis. However, larval body mass was less and developmental stage slowed down when larvae were raised at high densities than when raised at low densities (Provenzano and Boone, 2009). Biomass tends to decrease during fall and winter and increase rapidly with the warm weather of spring. Biomass decreases are associated with larvae leaving the pond, and with mortality from predation or disease through the latter part of the year. Transforming American Bullfrogs congregate near the shoreline and emerge usually at night. They begin feeding immediately.

DISPERSAL

Both subadult and adult American Bullfrogs use celestial cues, both day and night, in orienting toward or away from a breeding pond (Ferguson et al., 1968). This ability disappears during complete cloud-cover. Displaced adults use a Y-axis orientation to locate home ponds, whereas dispersing subadults use the same mechanism to move away from breeding sites, particularly at night. Orientation ability changes seasonally, and it is more evident in spring and fall than during the summer breeding season (Ferguson, 1966a). After the breeding season, resident adults may disperse 90–180 m to other wetlands, and they can move back and forth between large and small wetlands during the remainder of the activity season. As winter approaches, the frogs tend to become more sedentary, and they orient back toward larger water bodies, again using a Y-axis.

Dispersal from breeding ponds is extensive, and American Bullfrogs are known to disperse long distances, even across extremely inhospitable territory. Based on an analysis of seven polymorphic DNA microsatellite loci, Austin et al. (2003b) suggested that dispersal across habitats was strongly female-biased, that is, females are primarily responsible for dispersal across landscapes. The identity of distinct genetic populations often is not possible, even over distances of tens of kilometers (Austin et al., 2004b). Colonizing American Bullfrogs do not appear to be more genetically bottlenecked than native populations. They have low levels of polymorphic loci, with a wide variation in allele frequencies (Wiese, 1990). These characteristics suggest that bullfrogs do not experience severe genetic consequences due to small population size, which probably assists them in colonizing new habitats and establishing populations.

The tendency of American Bullfrogs to disperse far and rapidly has allowed them to extend their range throughout eastern North America, as well as from areas where they have been introduced. American Bullfrogs do not have a metapopulation structure with a series of source and sink populations. Instead, populations tend to be connected by extensive gene flow and have little genetic structuring, especially in the northern portions of the range.

Reports of small numbers of American Bullfrogs captured in upland habitats (e.g., Bennett et al., 1980; McLeod and Gates, 1998; Greenberg and Tanner, 2005b) probably represent dispersing individuals rather than animals that live directly in forested habitats for extended periods. Even in the deserts of the American Southwest, dispersal routinely occurs across large distances, with a maximum of 12.8 km reported (Suhre, 2010). Suhre (2010) recorded dispersal of marked individuals from 1 to 10.2 km from the largest stock tank monitored, although most individuals moved < 4 km. Dispersal often occurred directly through much unfavorable habitat, and was correlated with the number of stock tanks in the area. Dispersal occurred at night during the monsoonal summer rainfall.

DIET

Feeding in American Bullfrogs occurs both by day and night (Carpenter and Morrison, 1973). They eat a wide variety of vertebrate and invertebrate prey, and basically any creature that can fit into the mouth will be consumed. Prey are consumed in proportion to the size of the frog, that is, the larger the frog, the more variety and greater sizes of prey likely to be eaten. Studies of the diet of American Bullfrogs are numerous and have been conducted in a variety of habitats: farm ponds (Korschgen and Moyle, 1955; McCoy, 1969), fish hatcheries (Dyche, 1914; Lewis, 1962; Corse and Metter, 1980), stock ponds (Schwalbe and Rosen, 1988), ponds (Brooks, 1964; Bruggers, 1973; Twedt, 1993), rivers (McKamie and Heidt, 1974; Clarkson and DeVos, 1986), streams (Korschgan and Baskett, 1963; McKamie and Heidt, 1974), springs (Krupa, 2002), impoundments (Cohen and Howard, 1958; Korschgen and Baskett, 1963), and strip-mine pits (McKamie and Heidt, 1974). Bury and Whelan (1984) provide a comprehensive summary.

American Bullfrogs begin feeding at metamorphosis (and before the tail is reabsorbed), eating small insects, snails (*Physa*), springtails, honeybees, and a variety of small prey (Munz, 1920; George, 1940). Prey of small bullfrogs includes ants (in considerable numbers), other insects, spiders, and snails (Munz, 1920; Frost, 1935; McAuliffe, 1978; Bruneau and Magnin, 1980b). As the frogs become larger, the size and diversity of prey increases. For example, newly metamorphosed bullfrogs in Nebraska eat small invertebrate prey, but by the time they are 1–2 yrs old, crayfish make up a larger portion of the diet. Older and larger bullfrogs then consume a much greater proportion of large invertebrates, especially crayfish, as well as a variety of vertebrates (McAuliffe, 1978). Juvenile bullfrogs eat a larger percentage of aquatic prey than sympatric Northern Green Frogs, reflecting a difference in

Adult *Lithobates catesbeianus*. Photo: David Dennis

Adult *"Lithobates catesbeianus,"* undescribed species from Florida and Georgia. Photo: Joe Mitchell

habitat preference between these species (Werner et al., 1995).

The bulk of the diet is composed of insects, although the prey of adult American Bullfrogs includes a great variety of items (summarized by Bury and Whelan, 1984): worms (including leeches, *Lumbricus*), spiders (*Lycosa*), scorpions (*Hadrurus*), insects (terrestrial and aquatic beetles, dragonflies, larval lepidoptera, damselflies, caddisflies, sawflies, mayflies, grasshoppers, cockroaches, syrphid flies, honey bees, bumble bees, wasps, true bugs, mosquitoes, water boatmen, leaf hoppers, moths, fireflies, walking sticks, midges, backswimmers, water striders, horseflies, crickets, mole crickets), snails (*Physa, Planorbella, Succinea*), clams (*Corbicula*), millipedes, isopods, crayfish (*Cambarus, Orconectes immunis, Procambarus lucifugus*), fish (*Fundulus chrysotis, Gila purpurea, Ictalurus, Lepomis* sp., *L. macrochirus, Micropterus salmoides, Notemigonus crysoleucas, Notropis* sp., *Pimephales promelas, Poeciliopsis occidentalis*), salamanders (*Ambystoma, A. talpoideum, Desmognathus auriculatus, Gyrinophilus palleucus, Notophthalmus viridescens, Taricha granulosa*), frogs (*Acris blanchardi, A. crepitans, Anaxyrus americanus, A. houstonensis, Hyla cinerea, H. gratiosa, H. versicolor, H. wrightorum, Lithobates berlandieri, L. catesbeianus, L. clamitans, L. pipiens, Pseudacris crucifer, P. regilla, Rana draytonii, Scaphiopus holbrooki*), anuran tadpoles (*Lithobates catesbeianus, L. clamitans, Rana draytonii*), snakes (*Crotalus atrox, Lampropeltis getula, Leptotyphlops dulcis, Micrurus fulvius, Nerodia* sp., *Thamnophis* sp., *T. eques, T. sauritus*), lizards (*Plestiodon fasciatus, Sceloporus*), juvenile turtles (*Apalone spinifera, Chrysemys picta, Chelydra serpentina, Sternotherus odoratus, Terrapene carolina, T. ornata, Trachemys scripta*), a young alligator (*Alligator mississippiensis*) (Haber, 1926), mice (*Mus, Peromyscus*), cotton rat (*Sigmodon*), voles (*Microtus ochrogaster*), moles (*Parascalops breweri*), bats (*Lasiurus borealis, Myotis austroriparius*), shrews (*Blarina brevicauda, Sorex* sp.), muskrat (*Ondatra zibethica*), young mink (*Mustela vison*), and birds (*Agelaius phoeniceus, A. tricolor, Ampelis cedrorum, Anas rubripes, Pipilo fuscus*, warblers, swallow, sparrows, grackles, House Finch, Red-winged Blackbirds). Some of the more important references on diet are: Allen, 1868; Needham, 1905; Dickerson, 1906; Surface, 1913; Dyche, 1914; Frost, 1924, 1935; Force, 1925; Heller, 1927; Carr, 1940; George, 1940; Baker, 1942; Cagle, 1942; Pope, 1947; Minton, 1949; Hewitt, 1950; Howard, 1950; Penn, 1950; Korschgen and

Moyle, 1955; Holman, 1957; Cohen and Howard, 1958; W.M. Lewis, 1962; Korschgen and Baskett, 1963; Brooks, 1964; Fulk and Whitaker, 1969; Lee, 1969a; Taylor and Michael, 1971; Bruggers, 1973; Carpenter and Morrison, 1973; McKamie and Heidt, 1974; Pine, 1975; McAuliffe, 1978; Gilhen, 1984; Wiese, 1985; Clarkson and DeVos, 1986; Hinshaw and Sullivan, 1990; Ferner et al., 1992; Stuart and Painter, 1993; Twedt, 1993; Beringer and Johnson, 1995; Cook, 1997b; Krupa, 2002; Brown and Brown, 2009; Jones and Timmons, 2010; McHenry et al., 2010. Diet and dietary overlap have also been studied in feral populations outside of North America (Perez, 1951; Hirai, 2004; Wu et al., 2005; Díaz de Pascual and Guerrero, 2008; Filho et al., 2008; Da Silva et al., 2009, 2010).

Whereas American Bullfrogs consume a wide variety of prey, including the occasional toad (*Anaxyrus*) (Munz, 1920; Korschgen and Moyle, 1955; Huheey, 1965b; McCoy, 1969; Green and Pauley, 1987 [an amplexed pair]; Stuart, 1995), they may not consume these poisonous animals in proportion to their abundance at breeding ponds, nor are bullfrogs always successful. Brown (1974) and Tucker and Sullivan (1975) recount instances where toad poisons (bufogenins) were successful in repelling attacks by bullfrogs. The success of such an attack may be size dependent, with the larger toads more successful in surviving a predation attempt than metamorphs or small individuals.

Vegetation, pebbles, and surface detritus are found commonly in stomachs, presumably as they are ingested incidentally to predation attempts (Krupa, 2002). Not all bullfrogs have prey in their stomachs. Percentages of empty stomachs include 7.3% in Illinois (R.J. Lewis, 1962), 6.6% in Virginia (Brooks, 1964), and 6.1% in New Mexico (Krupa, 2002). Diets may change with seasonal availability within a site, and may be slightly different among sites (in species composition, proportion, volume) even within a season, again presumably in response to availability (Brooks, 1964; McKamie and Heidt, 1974). The American Bullfrog exhibits no dietary specialization. Occasionally gluttony catches up with an individual; Stevenson (2007) found an American Bullfrog that died while attempting to eat a Brown Bullhead catfish.

PREDATION AND DEFENSE

Eggs are not eaten by some predators (e.g., raccoons, the newt *Notophthalmus viridescens,* and larvae of the salamander *Ambystoma opacum*; Nelson, 1980), but are readily consumed by others (larvae of the salamander *Ambystoma gracile*; the fish *Salmo clarkii* and *Gasterosteus aculeatus*; Licht, 1969a). The thick jelly capsule provides some protection from the leech *Macrobdella decora* (Howard, 1978a). Howard (1978a) described the feeding behavior of leeches on eggs, where as much as 43% of the clutch may be consumed. The configuration of the egg mass, that is, whether it is flat (2D) or tends to clump on vegetation (3D), also affects the ability of leeches to attack eggs. Thus, the female's choice of territorial male, and thus the substrate configuration near the oviposition site, in turn has implications for hatching success. If leech densities build up during the breeding season at particular oviposition sites, adults may shift the location of choruses (Howard, 1978a, 1978b).

Chemoreception is very important to bullfrog larvae for defense and predator deterrence. American Bullfrog larvae may emit a chemical substance that deters predaceous fishes from attacking them. In experimental trials, for example, larval bullfrogs are attacked much less often than other similarly-sized species, and some fishes (*Ictalurus melas*) do not attack larvae (Kruse and Francis, 1977). However, American Bullfrog larvae do not respond to chemical cues from some nonindigenous species that may prey on smaller larvae, such as the mosquitofish *Gambusia affinis* (Smith et al., 2008).

American Bullfrog larvae can detect some predator odors in the water in experimental trials, resulting in decreased activity and growth rates (Peacor and Werner, 2000). Small larvae grow slowly in the presence of dragonfly (*Anax*) larvae and are less active than they are in the absence of the predator. With bluegill sunfish (*Lepomis*) or a combination of predators, growth and activity are intermediate between the *Anax* and predator-free trials. However, size matters. Large larvae neither modulate growth or activity in the presence of *Anax* or *Lepomis*, presumably because they are too large to be eaten (Eklöv, 2000). The success rate of predators, handling time, and palatability of larvae to predators varies considerably among predators, and is, in part, a function of the size of the tadpole (Relyea, 2001a).

Predators comprise a wide variety of invertebrates and vertebrates on both larvae and postmetamorphic frogs, as bullfrogs do not have noxious or poisonous secretions to deter predators (Formanowicz and Brodie, 1979). Larvae, however, are unpalatable to some predators such as sunfish (*Lepomis*) (Kats et al., 1988; Szuroczki and Richardson, 2011). The mosquito *Culex territans* feeds on American Bullfrogs in the lab but cannot use the frog's call to locate its next blood meal (Bartlett-Healy et al., 2008). Other predators include invertebrates (predaceous diving beetles, dragonfly larvae, spiders [*Dolomedes* sp.]), fish (*Esox, Micropterus*), other frogs, snakes (*Nerodia sipedon, N. fasciata, N. erythrogaster, Thamnophis hammondii, T. sauritus, Coluber constrictor, Agkistrodon contortrix*), turtles, alligators, birds (bald eagles, large waterbirds, wading birds, herons, crows, kingfishers, owls, hawks, osprey), and mammals (raccoons, *Procyon*; mink, *Mustela*; otters, *Lutra*; opossums, *Didelphis*; coyotes, *Canis*) (Howard, 1978a; reviewed by Bury and Whelan, 1984; Applegate, 1990; Rogers, 1996; Ervin et al., 2003; Rombough, 2010). Interestingly, adult, juvenile, and larval bullfrogs appear resistant to the venom of copperheads (*Agkistrodon contortrix*), a resistance that increases ontogenetically. They are somewhat less resistant to the venom of cottonmouths (*A. piscivorus*) (Heatwole et al., 1999).

Like many other ranids, postmetamorphic American Bullfrogs emit a sudden cry ("scream of distress," Dickerson, 1906; a mournful "yeowal" Vogt, 1981) as they leap into the water when disturbed or are attacked by predators (Formanowicz and Brodie, 1979). The cry is emitted with the mouth closed. This warning cry presumably disorients a potential predator, distracting it from the location of the frog as it leaps to safety. However, the distress cry of young bullfrogs and other ranids may attract the attention of adults seeking prey (Smith, 1977). The distress cry is usually a brief (100 ms) and loud (1,500 cps) call (Capranica, 1968), and has been likened to a human scream. However, Hoff and Moss (1974) stated that the scream can last up to 9 sec. If injured, bullfrogs have been reported to make a distress call, which could alert conspecifics to the presence of a predator. This would make sense if nearby frogs were kin, that is, that they were related to nearby individuals. In addition to screams, a disturbed bullfrog makes a great amount of splashing and noise as it leaps and skitters across the water surface to dive into deep water safety. According to Ferguson et al. (1968), they tend to use the same pathways of escape.

POPULATION BIOLOGY

Growth is rapid in juvenile bullfrogs (Durham and Bennett, 1963) and exceeds the rate of larval growth (Werner, 1986). For example, mean size and body mass in Québec are: 1 yr (59 mm SUL,

Albino adult *Lithobates catesbeianus*. Photo: Cecil Schwalbe

Axanthic (blue) *Lithobates catesbeianus*. Photo: Stan Trauth

18 g), 2 yr (81 mm, 40 g), 3 yr (108 mm, 101 g for females, 96 g for males), 4 yr (125 mm), 5 yr (137 mm, 215 g for females, 194 g for males), and 6 yr (143 mm) (Bruneau and Magnin, 1980b). In Arizona, Suhre (2010) estimated juvenile growth rates at 0.168 mm/day. Growth rates are equal between males and females, but males reach sexual maturity at smaller sizes than females (95–112 vs. 108–118 mm SUL) and suffer higher predation rates at the larger size classes, as they defend territories (Howard, 1981). Howard (1981) noted that size at sexual maturity varied annually. Survivorship between breeding seasons was equal between the sexes in Howard's (1981) Michigan study.

Most male American Bullfrogs reach sexual maturity 1 yr after metamorphosis, whereas females require 1 to 2 yrs to reach sexual maturity. In Québec, however, sexual maturity is not reached until 3 yrs, with some frogs maturing at 2 yrs, but others at 4 yrs (Bruneau and Magnin, 1980b). In Ontario, maturity is not reached until 5 yrs postmetamorphosis at 113 mm SUL for females and 3 yrs postmetamorphosis and 91 mm SUL for males (Shirose et al., 1993). George (1940), however, stated that Louisiana bullfrogs reached sexual maturity 20 months after metamorphosis. He noted that females 114–126 mm SUL had eggs in the autumn that would be deposited the next year when females were > 127 mm SUL. These data are similar to those reported by Willis et al. (1956) in Missouri, where spent females are found at 123–125 mm SUL. In Colorado, females mature at 110–115 mm SUL (Wiese, 1985). Taken together, these data suggest that males reach sexual maturity faster than females throughout the range, and that northern animals reach maturity at smaller sizes than southern animals.

The normal life span is 5 to 8 yrs after metamorphosis in Michigan (Howard, 1978b; Howard, 1988b) and to 9 yrs in Québec (Bruneau and Magnin, 1980b). In Howard's (1988b) study, males had a 42–52% probability of surviving to the next year during their first 4 yrs of life, but this probability dropped to 23% between the fourth and fifth years. Female survivorship was 42–69%, but again there was a decrease to 26% between the fourth and fifth year. Shirose et al. (1993) reported an inverse relationship between sex ratio and age, indicating a decreased survivorship of large males compared with females. This decrease undoubtedly results from the necessity of males to maintain visible and defended territories, and thus a potential for prolonged exposure to predators (also see Howard, 1981).

The numbers of animals may fluctuate substantially at a breeding pond (Durham and Bennett, 1963), but good population estimates are lacking. In some instances, it is impossible to determine whether authors are talking about juveniles or adults when recording captures at ponds. For example, Gibbons and Bennett (1974) recorded 105 bullfrogs at a pond one year, and 31 the next; in a second pond, the numbers were 16 and 18, respectively. Presumably, these were dispersing juveniles rather than breeding adults, but the life stage was not indicated in the text. Small numbers of American Bullfrogs have been recorded in a number of other studies without data on life stage, including some purporting to assess the effects of habitat disturbance on amphibian communities (Russell et al., 2002a).

Programs aimed at reducing populations of introduced American Bullfrogs have focused primarily on removing larvae and reducing the number of adults. Govindarajulu et al. (2005) demonstrated, however, that partial removal of larvae actually increased the survivorship of remaining larvae and that removal of large adults allowed juveniles to survive and move into the unoccupied territories. Neither strategy would be successful in regulating populations. Instead, these authors recommended focusing eradication measures on newly metamorphosed juveniles, in order to eliminate future generations.

COMMUNITY ECOLOGY

The presence of larval American Bullfrogs may have diverse effects on other vertebrates. In wetlands in Massachusetts, for example, fish species richness appears inversely proportional with the abundance of *Lithobates catesbeianus* larvae. In laboratory experiments, the presence of water containing extracts from bullfrogs inhibited reproduction in the fish *Poecilia reticulata*, although this is a nonindigenous species in Massachusetts (Boyd, 1975). Since boiling and carbon filtration reduced the effectiveness of the tadpole extract, a large heat-labile molecule likely is responsible for the inhibitory effect. Whether American Bullfrog larvae inhibit native fish reproduction is unknown.

Sunfish (*Lepomis*) do not appear to eat American Bullfrog tadpoles, and bullfrogs are commonly found in ponds and lakes with predaceous fishes (e.g., Sexton and Phillips, 1986; Hecnar, 1997; Hecnar and M'Closkey, 1997b; Babbitt et al., 2003). Indeed, bullfrogs tend to thrive in ponds with *Lepomis*, and larval growth rates actually increase when these predators are present. Adding *Lepomis* to ponds where they formerly were not present also increases bullfrog larval survivorship. Presumably the fish eat the invertebrates (particularly dragonfly larvae) that prey upon the anuran larvae, freeing the larvae from a significant predation pressure (Werner and McPeek, 1994; Smith et al., 1999). The presence of tadpoles of other species had no interactive effects on the bullfrog larvae.

The presence of predators in both pond and mesocosm settings may or may not have significant influences on tadpole growth and activity, but it depends upon which predators are present and their abundance, and a host of other factors (e.g., vegetation, food resources, spatial arrangement, the presence and density of other species of tadpoles, and size of the larvae) (Licht, 1967; Woodward, 1983; Werner, 1991, 1994; Werner and McPeek, 1994; Relyea and Werner, 1999; Anholt et al., 2000; Van Buskirk, 2000; Relyea, 2001a, 2001b; Richardson, 2001; Smith and Awan, 2009). The presence of dragonfly larvae (*Anax*) in particular tend to cause bullfrog larvae to decrease their activity, which in turn affects growth rates. Other potential predators, such as mudminnows (*Umbra*), cause moderate changes in activity and growth, whereas activity and growth are not affected by bluegills (*Lepomis*) (Relyea and Werner, 1999; Smith and Awan, 2009). However, American Bullfrogs suffer high predation rates from larval Tiger Salamanders (*Ambystoma tigrinum*); salamanders have about a 75% capture success rate and quickly (1–2 min) consume larvae (Relyea, 2001a).

Sometimes the effects of predators may be counterintuitive at first. For example, bullfrog larvae actually grow faster in the presence of caged dragonfly larvae than in the absence of the predator, presumably in order to reach a large size where predation attempts are unsuccessful (Werner, 1991; Anholt et al., 2000). Food resources also come into play, as larval bullfrogs grow faster in the presence of increasing food resources, regardless of the presence of predators. Indeed, high food levels and high predator densities tend to act additively, such that larval bullfrogs reach larger sizes faster on high food levels as predators remove competing species of *Lithobates* (Anholt et al., 2000).

American Bullfrog larvae may be negatively impacted by predators, even if they are not killed outright. Brown (1972) found that tadpoles whose bodies or tails had been injured had low proportions of food within their guts in comparison to uninjured tadpoles. This suggests that the injured tadpoles were either not feeding or not feeding effectively, which could affect their ability to transform in a timely manner. In addition, injuries likely affect the ability to escape future attacks.

Since bullfrog larvae are found in the same ponds as other *Lithobates*, it is not surprising that the presence and density of these species may affect growth in bullfrogs. Larval bullfrogs grow slower and have lower survivorship in mesocosm experiments with high densities of small Northern Green Frogs than they do without this species or with this species at low densities (Peacor and Werner, 2000), presumably as a result of competition for limited resources. Bullfrog larvae also impact other anuran larvae within a breeding pond. For example, American Toad and Southern Leopard Frog larvae have greater survivorship after overwintering bullfrog larvae are removed, suggesting competition for space or resources (Boone et al., 2008). Southern Leopard Frog larvae have a smaller body mass at metamorphosis when overwintering with American Bullfrogs, and they are subject to bullfrog larval predation (Boone et al., 2004a). These effects are particularly evident in semipermanent ponds, when hydroperiods are shortened or when the density of bullfrog larvae is high.

Likewise, the presence of adults may have diverse effects on habitat use by other sympatric amphibians. Adult Northern Green Frogs and Mink Frogs are more abundant and evenly distributed in habitats where American Bullfrogs are absent than when they occur within the same lake as bullfrogs (Courtois et al., 1995). In mixed assemblages of ranids in New Brunswick, Northern Green Frogs and Northern Leopard Frogs select areas of cooler water, denser vegetation, and areas that are closer to the shoreline when bullfrogs are present (McAlpine and Dilworth, 1989). The presence of bullfrogs causes microhabitat segregation among these species, as they try to avoid contact with the large predaceous bullfrog. Habitat partitioning also has been reported in California between *Rana draytonii* and introduced *Lithobates catesbeianus* (Twedt, 1993; Cook, 1997b). Whether this is the result of interactions between the species or reflects differences in breeding phenology and pond habitat characteristics is unknown.

The presence of disease also may affect interspecific interactions at a breeding pond. In laboratory and field experiments, American Bullfrog larvae exposed to dead bacteria selected warmer temperatures than controls, at least until the bacteria cleared their bodies. This increase in thermal preference in response to an infection is termed behavioral fever and presumably helps the tadpole's immune system in fighting the infection. However, tadpoles exhibiting behavioral fever were less likely than controls to seek refuge in vegetation, thus leading to increased predation by newts (*Taricha granulosa*) (Lefcort and Eiger, 1993). As such, there may be a subtle interplay between physiological requirements, infection, and predation risk in a larval amphibian community.

In the northern part of its range, the American Bullfrog is the dominant species of frog in its community. In the South, however, bullfrogs are much less common on the Southeastern Coastal Plain, where their range overlaps with the highly aquatic Pig Frog (*Lithobates grylio*). Pig Frogs are much more common than bullfrogs (e.g., Carr, 1940a), and when the species co-occur (such as at Kingfisher Lake on Savannah National Wildlife Refuge in South Carolina), bullfrog numbers are usually few; Pig Frogs remain in water, whereas American Bullfrogs are found along the shoreline. In many areas, the species do not occur in the same locations, with Pig Frogs seeming to exclude American Bullfrogs (Viosca, 1923; George, 1940). The nature of the interaction is unknown.

Introductions of American Bullfrogs have often been identified as a major reason for the decline in native species, particularly ranid frogs (Dumas, 1966; Moyle, 1973; Hammerson, 1982, 1999; Schwalbe and Rosen, 1988; Corn, 1994; Panik and Barrett, 1994; Pearl et al., 2004) and garter snakes (*Thamnophis eques*) (Schwalbe and Rosen, 1988). Presumably, bullfrogs are predators of the other species, and they are a source of interference or resource competition. Direct empirical evidence is often lacking, however (Hayes and Jennings, 1986). For example, there was no evidence to suggest that bullfrogs were responsible for changes in the distribution of *Rana aurora* in the Puget Sound region (Adams, 1999), and the presence of bullfrog larvae had no effect on larval *Pseudacris regilla* and decreased survivorship of *Rana aurora* larvae only at low densities (Adams, 2000). In contrast, Northern Green Frogs increased by a factor of four in an Ontario park after bullfrogs became extinct, suggesting either predation or competition by bullfrogs had been keeping populations of *Lithobates clamitans* in check (Hecnar and M'Closkey, 1997c).

There is evidence that native frogs alter their behavior after they have been exposed to the presence of chemical cues from adult bullfrogs. Red-legged Frog tadpoles from ponds where they are sympatric with American Bullfrogs are less active in experimental trials when exposed to water-borne bullfrog chemical cues than tadpoles that have never been exposed to bullfrogs. In addition, bullfrog predation on Red-legged Frogs is lower when experimental animals are tested from areas where they are sympatric as opposed to areas where they are allopatric (Kiesecker and Blaustein, 1997a). These results might explain why bullfrogs do not appear to have much impact on populations when contact has occurred for a long time. However, the introduction of the predator to Red-legged Frog populations never previously associated with bullfrogs could have short-term negative consequences for the native species, at least until survivors "learn" to avoid bullfrogs. In addition, the distribution of resources, that is, whether they are clumped or evenly distributed, could affect the ways bullfrogs interact with native species (Kiesecker et al., 2001a). Decreased activity may have no adverse consequences if resources are widely distributed; if resources are clumped, however, bullfrogs tend to out-compete *Rana aurora* larvae. In that regard, American Bullfrogs seem to have no effect on Red-legged Frog survivorship and metamorph body mass, although the presence of bullfrogs increased time to metamorphosis, suggesting the potential for resource competition (Kiesecker and Blaustein, 1998; Lawler et al., 1999). Adult bullfrogs, however, can still exert considerable predation pressure on juvenile *Rana aurora*.

Male *Rana pretiosa* and *R. aurora* have frequently been observed to amplex female or juvenile *Lithobates catesbeianus* in areas where bullfrogs have been introduced. Pearl et al. (2005a) suggested that such interference competition is probably widespread, and that interspecific amplexus may interfere with the reproductive activity and success of native ranids. Other studies have suggested that *Rana pretiosa* are more vulnerable to bullfrog predation than *R. aurora*, since Spotted Frogs are more aquatic than Red-legged Frogs, and they have shorter jump distances as a means of escape (Pearl et al., 2004).

Adverse interactions between native and nonindigenous populations of American Bullfrogs may not result solely from competition after meta-

Habitat of *Lithobates catesbeianus*, eastern North America. Photo: John Bunnell

Habitat of *Lithobates catesbeianus*, desert Southwest. Photo: C.K. Dodd Jr.

morphosis. In a California river system, Northern Yellow-legged Frogs were significantly less abundant where American Bullfrogs had been introduced compared with locations where bullfrogs were absent (Kupferberg, 1997a). Mesocosm experiments demonstrated that larval bullfrogs reduced the survivorship of native *R. boylii* by 48%, but not of *Pseudacris regilla*. However, the mass at metamorphosis of both species declined by 24% and 16%, respectively. Further measurements revealed that bullfrogs significantly reduced the amount of algae available as food for native tadpoles (Kupferberg, 1997b). Resource competition, rather than interference competition or chemically induced changes in behavior, thus accounts, in part, for the decreased abundance of some native frogs in northern California when bullfrogs are present.

River and floodplain habitat modification throughout western North America undoubtedly have aided the spread of American Bullfrogs. As rivers are dammed and water flow is decreased, habitats become more favorable for bullfrogs. This is especially true where backwaters and retention ponds are located adjacent to river channels. Some such habitats have even been created during restoration projects, such as those relating to mine restoration (Fuller et al., 2011). Returning rivers to a natural hydrologic regime, without the creation of artificial backwaters, could reduce the spread of bullfrogs and make areas where they currently reside unfavorable. Native ranids would then be likely to recolonize former habitats.

One thing is certain. In the southwestern deserts and elsewhere, introduced bullfrogs can reach extremely high population densities. In Arizona, Schwalbe and Rosen (1988) estimated 718 bullfrogs per ha (or 215.5 kg/ha) in a small pond. At San Bernadino National Wildlife Refuge, Schwalbe and colleagues removed 552 adults and juveniles during 4 sampling periods in 1986 and 1987, an estimated 55–80% of the frogs on the refuge at that time. At 6 stock tanks and along Arivaca Creek on the refuge, individual locality estimates ranged from 19 to 7,710 over a 4 yr period (Suhre, 2010). These values translated to an unheard of 271–20,365 bullfrogs per ha! Efforts to eradicate bullfrogs from this area have not been successful as of this date.

DISEASES, PARASITES, AND MALFORMATIONS

Lithobates catesbeianus is subject to a wide variety of diseases and parasites, both in natural populations and in culture. These include rickettsial-like cytoplasmic viruses of erythrocytes, which primarily affect young frogs but not adults (Desser and Barta, 1984a; Gruia-Gray et al., 1989; Gruia-Gray and Desser, 1992; Faeh et al., 1998), tadpole edema virus (Faeh et al., 1998), ranaviruses (Frog Virus 3, RCV-Z) (Majji et al., 2006; Gahl, 2007; Gray et al., 2007a; Miller et al., 2007; Gray et al., 2009; Mazzoni et al., 2009; Une et al., 2009), unspecified iridovirus (Green et al., 2002), an *Ichythophonus*-like fungal infection affecting both tadpoles and juveniles (Mikaelian et al., 2000; Green et al.,

2002; Gahl, 2007), water molds (*Saprolegnia* sp. on eggs) (Ruthig, 2009), the fungus *Basidiobolus ranarum* (Nickerson and Hutchison, 1971), a *Dermosporidium*-like fungus (Green et al., 2002), the bacterium *Escherichia coli* (Gray et al., 2007b), parasitic coccidian protozoans (*Eimeria*) (Chen and Desser, 1989), the erythrocytic parasite *Lankesterella* (Brandt, 1936; Desser et al., 1989), the myxozoan *Sphaerospora ohlmacheri* found in larval kidney tubules (Desser et al., 1986), a virulent alveolate pathogen affecting larvae (Gahl, 2007), anchor worms (a crustacean) (Green et al., 2002), and leeches (*Desserobdella picta*) (Barta and Desser, 1984; Bolek and Janovy, 2005). Other parasitic protozoans include *Cytamoeba bactifera, Entamoeba ranarum, Haematogregarina* sp., *Leptotheca ohlmacheri, Nyctotherus cordiformis, Hexamita (Octomitus) intestinalis, Opalina virguloidea, Tritrichomonas augusta,* and *Trypanosoma rotatorium* (Brandt, 1936; Campbell, 1968). In live American Bullfrogs on sale for the U.S. domestic food market, the incidence of ranavirus infection was 8.5% in 3 major cities (New York, Los Angeles, San Francisco) (Schloegel et al., 2009).

American Bullfrogs are infected by the amphibian chytrid fungus (*Batrachochytrium dendrobatidis*) in natural populations, international commercial frog farms, fish hatcheries (Mazzoni et al., 2003; Daszak et al., 2005; Ouellet et al., 2005a; Gahl, 2007; Green and Dodd, 2007; Longcore et al., 2007; Peterson et al., 2007; Rothermel et al., 2008; Rizkalla, 2009, 2010; Adams et al., 2010; Rimer and Briggler, 2010; Saenz et al., 2010; Schloegel et al., 2010; Tupper et al., 2011) and even in introduced populations (Adams et al., 2008; Padgett-Flohr and Hopkins, 2009). In the northeastern United States, Longcore et al. (2007) found incidence rates <10%, depending on the tissues examined (n = 141). Daszak et al. (2005) found 2 of 13 bullfrogs collected from 1940 to 2001 to be infected. In South Carolina, *B. dendrobatidis* was found in 64% of the bullfrog tadpoles examined from 1 pond, although tadpoles from 2 other ponds were not infected (Peterson et al., 2007). In northern California and Oregon, Adams et al. (2010) reported 10.5% of postmetamorphs and 17.3% of larvae tested positive for chytrid. Finally, the fungus was found in museum specimens collected from California as early as 1961, with a combined incidence rate of 10.5% (18 of 171 specimens examined) from the 1960s to the 2000s (Padgett-Flohr and Hopkins, 2009).

The apparent resistance (Daszak et al., 2004) to the often lethal effects of this emerging infectious disease has caused concern, especially since bullfrogs have proven disease vectors to frog populations in many parts of the world (Garner et al., 2006; Bai et al., 2010; Schloegel et al., 2010). Introduced bullfrogs in western Canada and the United States, Brazil, China, England, France, Italy, Uruguay, and Venezuela have all tested positive for *B. dendrobatidis*. In live American Bullfrogs on sale for the U.S. domestic food market, the incidence of chytrid infection was 62% in 3 major cities (New York, Los Angeles, San Francisco) (Schloegel et al., 2009).

Mortality events have been recorded not infrequently in larval bullfrogs. In Oregon, Brown (1972) recorded more than 100 larvae that were grossly swollen with nearly spherical bodies, and that were covered with patches of fungus. Up to 50% of the body weight resulted from edema fluid. Analyses ruled out tadpole edema virus and bacteria as causes. Brown (1972) suggested infection by the water mold *Saprolegnia* as a possible causative agent, perhaps facilitated by low oxygen levels.

Adult American Bullfrogs are parasitized by a variety of helminths, including the trematode species *Alaria* sp., *A. arisaemoides, A. mustelae, Allassostomoides chelydrae, A. parvus, Apharyngostrigea pipientis, Auridistomum chelydrae, Brachycoelium louisianae, Cephalogonimus americanus, C. brevicirrus, Clinostomum* sp., *C. attenuatum, C. marginatum, Cystagora tetracystis, Diplodiscus temperatus, D. intermedius, Echinostoma trivolvis, Glypthelmins linguatula, G. proxima, G. quieta, G. subtropica, Gorgodera amplicava, G. bilobata, G. circava, G. cygnoides, G. minima, G. vivata, Gorgoderina attenuata, G. bilobata, G. simplex, Gyrodactylus* sp., *G. catesbeiana, G. jennyae, Haematoloechus* sp., *H. breviplexus, H. buttensis, H. complexus, H. floedae, H. longiplexus, H. medioplexus, H. parviplexus, H. variegates, H. varioplexus, Halipegus* sp., *H. amherstensis, H. eccentricus, H. occidualis, H. ovocaudatus, Levinseniella ophidea, Loxogenes arcanum, L. provitellaria, Loxogenoides bicolor, L. loborchis, Megalodiscus* sp., *M. ferrissianus, M. intermedius, M. micro-*

phagus, M. temperatus, Megalogonia octaluri, Mesostephanus kentuckiensis, Pleurogenoides sp., *Proterometra albacauda, Pseudosonsinotrema catesbeianae, Strigea elegans,* and *Teloporia aspidonectes* (Trowbridge and Hefley, 1934; Brandt, 1936; Manter, 1938; Odlaug, 1954; Najarian, 1955; Waitz, 1961; Campbell, 1968; Ulmer, 1970; Brooks, 1976; Ashton and Rabalais, 1978; Dyer, 1991; Andrews et al., 1992; Bursey and DeWolf, 1998; McAlpine and Burt, 1998; Goldberg and Bursey, 2002a; Bolek and Janovy, 2008; Marcogliese et al., 2009; Paetow et al., 2009; King et al., 2010); the cestodes *Bothriocephalus* sp., *Cylindrotaenia americana, Distoichometra kozloffi, Ophiotaenia filaroides, O. gracilis, O. magna, O. saphena,* and an undescribed proteocephalan (Brandt, 1936; Campbell, 1968; Andrews et al., 1992; Crawshaw, 1997; McAlpine and Burt, 1998; Goldberg and Bursey, 2002a); the nematodes *Abbreviata* sp., *Agamascaris odontocephala, Agamonema* sp., *Contracaecurn* sp., *Cosmocercoides dukae, C. variabilis, Dioctophyma renale, Dujardiniascaris* sp., *Eustrongylides* sp., *E. wenrichi, Falcaustra* sp., *F. catesbeianae, F. inglisi, Filaria quadrituberculata, Foleyellides* sp., *F. americana, F. flexicauda, F. ranae, Hedruris* sp., *H. pendula, H. siredonis, Microfilaria* sp., *Oxysomatium longicaudata, Oxysomatium* sp., *O. americana, O. longicaudata, Ozwaldocruzia* sp., *O. pipiens, O. variabilis, Physaloptera* sp., *P. ranae, Rhabdias* sp., *R. ranae, Spinitectus* sp., *S. gracilis, Spironoura catesbeianae, Spiroxys* sp., *S. contortus, Strongyloides* sp., and *Strongyluris ranae* (Brandt, 1936; Campbell, 1968; Ashton and Rabalais, 1978; Dyer, 1991; Andrews et al., 1992; Bursey and DeWolf, 1998; Goldberg et al., 1998a; McAlpine and Burt, 1998; Goldberg and Bursey, 2002a; McAllister et al., 2008; King et al., 2010); and the acanthocephalans *Centrorhynchus* sp., *Fessisentis friedi,* and *Neoechinorhynchus rutili* (Brandt, 1936; Campbell, 1968; McAlpine, 1996; McAlpine and Burt, 1998).

Larval American Bullfrogs serve as hosts for the tadpole-specific nematode *Gyrinicola batrachiensis*, a worm that apparently aids in fermentation (Adamson, 1981c; Bursey and DeWolf, 1998; Pryor, 2003). Leeches (*Macrobdella ditetra*), mites (*Hannemania penetrans*), and blowfly larvae (*Lucilla silvarum*) also parasitize American Bullfrogs (Brandt, 1936; Hall, 1948). The introduced Asian copepod *Lernaea cyprinacea* may have been introduced to *Rana boylii* via *Lithobates catesbeianus* (Hayes et al., 2010).

An anomalous *L. catesbeianus* with expanded digits giving the appearance of toe pads was reported from Alabama by Smith and List (1951). Ectromelia (absence of all or part of a limb) has been recorded in Québec (Ouellet et al., 1997), and Meteyer (2000) has a photograph of a bullfrog with a mass of multiple limbs. Lannoo (2008) provides extensive documentation and photographs of many types of malformations affecting American Bullfrog larvae and postmetamorphs. Other reports include malformations observed in the Midwest and northeastern sections of the United States (Converse et al., 2000; Green et al., 2002), Georgia (Houck and Henderson, 1953), Illinois (Lopez and Maxson, 1990), and Colorado (Hammerson, 1999).

Malformations may become particularly evident in metamorphosing frogs. In such cases, an association has been found between the presence of limb abnormalities and trematode metacercariae (Christiansen and Feltman, 2000). Bullfrogs with supernumerary or missing limbs were found only in Iowa flood pools with abundant metacercariae, whereas pools without the parasitic stage produced normal frogs. These metacercariae are accumulated throughout larval development. They interfere with limb/organ bud development by dividing it or by blocking it altogether from chemically influencing further development. The stage at which interference occurs determines the type and severity of malformation. Ocular malformations are also associated with Frog Virus 3 infection (Burton et al., 2008), which is in turn more prevalent in cattle-access ponds than in similar ponds without cattle access (Gray et al., 2007a).

In addition to those resulting from biological sources, malformations involving larvae have been found (axial, spinal curvature, excessive fluid accumulation in lymph spaces) (Hedeen, 1976), including some in areas contaminated by a coal-burning power plant (Hopkins et al., 2000). In this latter study, malformed tadpoles comprised 18–37% of the sample from polluted sites and 0–4% at reference sites. This suggests that toxic substances may contribute to certain types of malformations during development.

SUSCEPTIBILITY TO POTENTIAL STRESSORS

Metals. Larval American Bullfrogs are very sensitive to many metals; Westerman et al. (2003a) provide LC_{50} values for 14 metals. Larval American Bullfrogs living in wetlands adjacent to highways may contain concentrations of lead (0.07–270 mg kg^{-1} dry weight). Lead concentrations are positively correlated with traffic volume and are about 20–25% of the amount found in sediments (Birdsall et al., 1986). These amounts are much less than those found in effluent from lead mines and smelters, but their effects on larvae are undetermined. Long-term exposure of cadmium at 0.4–0.8 mg/L increases B-lymphocytes in the liver and mesonephros of larval livers (Zettergren et al., 1991).

Metals associated with coal ash settling basins near coal-fired electric generating plants include arsenic, cadmium, chromium, copper, vanadium, strontium, and selenium, as well as a variety of other metals and pollutants. These metals are accumulated by bullfrog larvae and may increase or decrease depending on metal and time of the year (Snodgrass et al., 2003; Unrine et al., 2007). American Bullfrog larvae found in such basins have reduced numbers of labial teeth and malformations of the labial papillae when compared to reference sites. These oral deformities make the larvae less effective at food gathering than nonaffected larvae, resulting in slower growth rates and limiting the types of food resources that can be consumed (Rowe et al., 1996). Axial malformations are associated with American Bullfrog larvae from coal ash polluted sites, presumably due to the effects of metal contamination. Tadpoles in such polluted sites also have 40–97% higher metabolic rates than in reference sites, although Rowe et al. (1998) found no correlation between metabolic rate and survival.

Tadpoles inhabiting areas subject to coal ash pollution also have slower response times to external stimuli, poorer swimming ability, and decreased abilities to escape from predators than larvae from reference sites (Raimondo et al., 1998). These authors speculated that selenium, a common contaminant in coal ash polluted sites, inhibited tail function by being incorporated into proteins in place of the normal sulfur molecule. This would result in changing the functional protein structure, making muscle activity much less effective in responding to predators.

Mercury has been found in American Bullfrog tadpoles in natural ponds in Maine at a mean of 19.1 ng/g; methyl mercury comprised 14.7–38.1% of the total (Bank et al., 2007). Mercury also has been found in *L. catesbeianus* larvae and juveniles from California at concentrations of 0.016–0.6 µg/g wet weight (Hothem et al., 2010). Values > 0.3 µg/g wet weight exceed EPA standards for mercury concentrations in fish. The highest concentrations are found in the liver, but substantial amounts are also found in leg muscle.

pH. Bullfrogs normally do not live in acidic waters at a pH of <4.0, although Freda and Dunson (1984) reported they could survive chronic exposure at pH 4.0 at least for 7 days. Indeed, early stage bullfrogs do not appear tolerant of some bog waters, even when pH is >4.0, suggesting a source of toxicity other than pH alone (Saber and Dunson, 1978). The embryo lethal and critical pH levels are 3.5–3.8 and 4–4.5, respectively; for larvae, they are 3.5–3.8 and 4.0, respectively (Grant and Licht, 1993). In Nova Scotia, larvae occur in waters with a pH range from 4.5 to 9, although adults were found at a pH of 4.1 (Dale et al., 1985). Sodium efflux is inversely correlated with pH, that is, the lower the pH, the more sodium is pumped from the body (Freda and Dunson, 1984).

Chemicals. The organophosphate insecticide parathion is toxic to *L. catesbeianus* larvae at concentrations > 10 mg/L, but not less, during 96 hr exposure trials and followed for 18 days after exposure. Mortality was high (>70%) at concentrations > 30 mg/L (Hall, 1990). The insecticide fenitrothion kills or leaves bullfrog larvae paralyzed at exposures of 4.0 and 8.0 ppm; larvae are unresponsive to touch at exposures as little as 0.5 ppm, and they do not recover (Berrill et al., 1994; Berrill et al., 1997). Azinphos-methyl (Guthion™) kills bullfrog tadpoles at 1.8 kg/ha in field applications (Mulla, 1962), but Meyer (1965) found no effect at 1.0 mg AI/L. TFM (a lampricide) has an $LC_{50\ (96\ hrs)}$ of 3.55 mg/L (Chandler and Marking, 1975). The now banned DDT caused both mortality and delayed tail regeneration in American Bullfrog larvae (Weis, 1975). Other insecticides known to kill bullfrog larvae are endrin, heptachlor, dieldrin, aldrin, toxaphene, thiodan, Bayer 38920, trithion, and GC-3582 (Mulla, 1962, 1963). Fenthion at 5 mg/L for an exposure of 96 hrs had no effect on bullfrog larvae, but mallard ducklings that fed on the tadpoles died (Hall and Kolbe, 1980).

Some insecticides also have sublethal effects beyond lethal toxicity. For example, the $LC_{50\,(16\,day)}$ for malathion is 1.5 mg/L, which suggests that it is not lethal to American Bullfrogs under field conditions (Relyea, 2004b). However, concentrations of 0.1–1 mg/L can reduce activity by 19–27% (Relyea and Edwards, 2010). Likewise, carbaryl at the same concentrations also significantly reduces activity, likely making larvae more susceptible to predation (Relyea and Edwards, 2010). Interestingly, parasite species richness and diversity is lower in American Bullfrogs from wetlands where parasites have been sprayed than in insecticide-free wetlands (King et al., 2010).

The herbicide acetochlor appears to affect thyroid hormone gene expression, and therefore possibly brain function, in American Bullfrog larvae. However, no effects on metamorphosis or on escape behavior could be detected (Helbing et al., 2006). The herbicide hexazinone has no lethal effects on American Bullfrog embryos or tadpoles, although they may not respond to stimuli (Berrill et al., 1994, 1997). However, the herbicide triclorpyr kills newly hatched tadpoles at 2.4 and 4.8 ppm (Berrill et al., 1994, 1997). Atrazine causes decreased hatching and survivorship (LC_{50} = 0.41 mg/L; Birge et al., 1980) and is lethal to larvae at 200 mg/L (Boschulte, 1993). Atrazine also has indirect effects on larvae at much less dosage, particularly on larval biomass when other grazers are present (DeNoyelles et al., 1989). The American Bullfrog appears resistant to the herbicide paraquat, perhaps due to stress-induced increases in antioxidant enzyme activity (Jones et al., 2010).

American Bullfrog embryos and larvae are sensitive to toxicity from: carbon tetrachloride (LC_{50} = 0.90 mg/L), methylene chloride (LC_{50} = 17.78 mg/L), nitrilotriacetic acid (NTA) (LC_{50} = 113.4 mg/L), phenols (LC_{50} = 0.23 mg/L) (Birge et al., 1980); other chemical geometric mean LC_{50} values include: acridine (1.24 mg/L), CCl_4 (1.26 mg/L), methylene chloride (22.9 mg/L), NTA (63.8 mg/L), B-naphtal (18.1 mg/L) (Westerman et al., 2003a); as well as 2-chloroethanol, hexachloroethane, pentachlorophenol, permethrin, and 2,2,2-trichloroethanol (Thurston et al., 1985).

American Bullfrog tadpoles experience a suite of adverse behavioral and morphological changes when exposed to crude oil (McGrath and Alexander, 1979). They will float on the water's surface, regardless of oil concentration, and they swell, causing bulges on the lateral body surfaces. Tadpoles become lethargic and are unable to dive. Larvae exposed to high oil concentrations rapidly swim at the water's surface with the head above water, while vigorously fanning their tail. After a short period of time, the tadpoles collapse back under the water's surface. The eyes become bloodshot indicating hemorrhage, and the forepart of the body takes on an unusual heart-shape when viewed dorsally, due to grossly inflated lungs. Upon dissection, oil is common in the digestive tract, and the liver appears bright yellow ("fatty liver"), due to the presence of oil droplets. Later-stage tadpoles are more sensitive to oil than early stage larvae. Taken together, these results suggest that crude oil in the environment is detrimental to American Bullfrog larvae.

UV light. American Bullfrog eggs have high levels of hatching success under field conditions of incident light, incident light with UVB light blocked, and even when UVB light is artificially enhanced (Crump et al., 1999). Significant mortality occurs after 10 min exposure at high levels of UVB (> 936 mJ/cm^2) (Licht, 2003). At 300 nm, 40% of UVB is absorbed by the jelly capsule surrounding the egg.

COMMERCIAL USE

American Bullfrogs are one of the most heavily exploited amphibians in the world and have been so for more than a century. The primary purpose of commercial use has been for frog's legs for human consumption. They have also been used extensively in science teaching labs and for medical research (Culley, 1973). Although frog "gigging" continues to supply frog legs for local consumption, much of today's commercial trade results from vast "frog farms," both in the United States (Arkansas, Louisiana, Florida) and in numerous other countries.

Attempts at frog farming began in the United States and Canada before 1900, and perhaps earlier than 1888 in Ontario (Meehan and Andrews, 1908; Priddy and Culley, 1971). For example, Anonymous (1899) reported on a small frog farm in New York; doubtless many others existed throughout the country. Chamberlain (1897) noted that a frog farm in the Trent River Basin, Ontario, had been in operation for "about 20 years." However, frog farms have often operated more like a vast network of persons collecting frogs and transporting them to a central location than true closed-cycle operations.

Since the late 1800s, American Bullfrogs have been recognized by private, state, and federal agencies as one of the primary species to fulfill the demand for frog's legs (Meehan, 1906, 1908a, 1908b; Dyche, 1914; Louisiana Department of Conservation, 1931; Viosca, 1931; Stoutamire, 1932; Hannaca, 1933; AFCC, 1936; Baker, 1942; Brashears and Brashears, 1950; Broel, 1950; Florida Department of Agriculture, 1952; Brown, 1953; USDI, 1965; also see Storer, 1933; for an opposing view, see Schmidt, 1946). These publications supply a wealth of information on the life history of *L. catesbeianus* in addition to their marketing.

State and federal publications also provide evidence of the massive dispersal of frogs throughout the United States and elsewhere. The earliest trade statistics were supplied by Chamberlain (1897), who reported an annual catch of about one million frogs worth $50,000. Anonymous (1892) noted that shipments of 30,000–40,000 at a time were sent to Vienna, Austria, and that New Yorkers consumed 60,000 pounds/yr at 30¢ per pound. In Texas, annual consumption was estimated at 300,000 pounds (Baker, 1942). Storer (1933) noted that that there was already a plentiful supply of American Bullfrogs in California, and that attempts to start frog farms in that state began in 1898, based on a supply of imported bullfrogs "from somewhere in the eastern States." American Bullfrogs were known from a frog farm in Contra Costa County, California, in 1896, however, they received imported stock from Maryland and Florida (Heard, 1904; Jennings and Hayes, 1985).

More than 5,000 live American Bullfrogs were shipped to Japan in the 1920s to create frog farms, although Japanese frog populations were subsequently seriously depleted by DDT. In 1945, 300,000 pounds of frog's legs, worth >$100,000, were exported to Cuba, even though *L. catesbeianus* had been released in Cuba in 1915 and had spread throughout the island (Martinez, 1948). Broel (1950) lists American Bullfrogs as being shipped to China, France, Germany, Hungary, Italy, the Philippines, and South America either as food or as stock for frog farms.

The value of the frog industry has been substantial, even before the advent of commercial farms (see, for example, discussion in Dundee and Rossman, 1989). In 1908, for example, 113,636 kg of frog's legs (worth $42,000) were reported taken in the entire country. In 1928, 325,245 kg of bullfrogs (worth $107,331) were harvested from Louisiana, and these figures increased to 2.75 million pounds worth $650,000 in 1936 (George, 1940). Some of the figures, however, include other ranid species, particularly *Lithobates pipiens*.

By the early 1970s, 9 million ranid frogs were harvested annually (360 tons), nearly all from wild-caught stock. Major suppliers were in Wisconsin, Minnesota, and Vermont. Commercial take in the Midwest, in particular, led to regional declines and to calls to enact conservation measures (Gibbs et al., 1971). Concern for wild populations had been expressed much earlier, however. Chamberlain (1897) noted that bullfrog populations near areas of market and transport were decimated by unregulated harvest. Meehan and Andrews (1908) even stated that Northern Leopard Frogs might be the best species to use in a frog farm, due to "the natural supply [of bullfrogs] being apparently doomed to exhaustion." The primary frog markets were in New York, Chicago, St. Louis, San Francisco, Boston, Philadelphia, Washington, and New Orleans.

Pennsylvania established the first major state-led effort at commercial frog farming in 1899, but many problems were encountered, and the attempt was not successful (Meehan and Andrews, 1908). By the 1930s, interest in commercial frog farming appeared to gain strength, in part due to the state of Louisiana's interest in aquaculture, which had been initiated around 1917 (Viosca, 1931, 1934). Louisiana State University began extensive research into bullfrog culture in the late 1960s (Priddy and Culley, 1971). Successful frog culture was carried out in Arkansas and Louisiana by the early 1970s. Today, American Bullfrogs are highly sought after for commercial farms, and they have been introduced throughout the world, often with known or suspected deleterious effects on native fauna (Orchard, 1999; Mazzoni et al., 2003; Giovanelli et al., 2008; Wang and Li, 2009). Commercial bullfrog farming is still promoted by the UN/FAO (http://www.fao.org/fishery/culturedspecies/Rana_catesbeiana).

STATUS AND CONSERVATION

Despite a long history of exploitation, American Bullfrogs remain a common and widely distributed species. Most surveys indicate populations that are stable or increasing, even when surveys of particular areas were conducted decades apart (e.g., Christiansen, 1981; Busby and Parmelee, 1996; Weller and Green, 1997; Mierzwa, 1998; Mossman et al., 1998; Brodman et al., 2002; Florey and Mullin, 2005). On the eastern seaboard, for example, populations were considered increasing in 4 states (Delaware, New Jersey, Virginia, West Virginia) based on 7 yrs of data using occupancy modeling (Weir et al., 2009). Throughout many areas, introductions from a variety of sources (but principally by state wildlife and fisheries agencies) have resulted in expanding populations (Lannoo et al., 1994; Lannoo, 1996; Christiansen, 1998).

Still, habitat loss to urbanization, agriculture, silviculture, transportation corridors, wetlands drainage, and other causes is the primary threat to most bullfrog populations. For example, American Bullfrogs disappeared around Baton Rouge in the 1930s as ponds and wetlands were drained (George, 1940); wetland modification leading to premature or unusual pond drying contributed to population declines. High levels of toxic agricultural pesticides are associated with smaller sizes, body masses, tibia length, size of tympanum, and younger ages than conspecifics found in nonagricultural areas (Spear et al., 2009). In addition, agricultural chemicals can alter physiological activity and compromise immune function, and these in turn can be mediated through parasite activity (Marcogliese et al., 2009). These data suggest that a soup of agricultural chemicals causes decreased longevity and growth rates in impacted populations, even if populations are not outright eliminated. The presence of cattle in breeding ponds also reduces the abundance of postmetamorphic American Bullfrogs (Burton et al., 2009).

In Ontario, evidence suggests that some populations are declining. Hecnar (1997) found American Bullfrogs in only 2.9% (5 of 174) ponds in southern Ontario and noted declines or extirpations in areas where they were formerly abundant. Shirose and Brooks (1997) also noted unstable population structures in Algonquian Provincial Park, suggesting increased mortality prior to 1985 and from 1987 to 1991. Declines have been attributed to the effects of otters, overharvest, shoreline modification, and the use of pesticides (Shirose and Brooks, 1997; Weller and Green, 1997). Declines elsewhere have been reported for Wisconsin (Casper, 1998), and the species is listed as of "Special Concern" in Minnesota (Oldfield and Moriarty, 1994).

Bullfrogs may be found in forested wetlands and then disappear following clearcutting. According to some studies, initial declines are followed by eventual recolonization (e.g., Clawson et al., 1997). However, it is often difficult to assess impacts of clearcutting, as follow-up studies

Extreme abundance of *Lithobates catesbeianus* at a desert stock tank. Photo: Dennis Suhre

usually are of short duration, and they often use presence rather than abundance or demographic data to evaluate the effects of habitat disturbance. This is a poor metric to use, since it tends to ignore the functional context of a species, opting instead for mere occurrence. In studies on the effects of silviculture on bullfrogs, it is sometimes impossible to know whether 1 frog is involved, or 10,000. However, American Bullfrogs are occasionally found in open habitats created during forestry operations, such as in canopy gaps and along skidder trails (Cromer et al., 2002). They will colonize artificial ponds, but Monello and Wright (1999) found that they were not particularly successful at doing so in northern Idaho.

Road traffic undoubtedly takes a heavy toll on populations, and little is understood of its long-term effects. In Indiana, bullfrogs were the most common species found road-killed over a 17-month study, with 1,671 recorded (Glista et al., 2008), whereas 1,345 bullfrogs were killed on 3.6 km of the Long Point Causeway in Ontario over a 4 yr period (Ashley and Robinson, 1996). In the latter study, mortality was seasonally bimodal, as frogs immigrated in spring and emigrated in the autumn. As with other amphibians, bullfrogs are frequently surveyed using road-based call surveys (Burton et al., 2006).

In addition, American Bullfrog populations undoubtedly were decimated or extirpated by the unregulated harvest for frog's legs throughout vast areas of North America. Schroeder and Baskett (1968), for example, noted an absence of large individuals in areas heavily harvested in comparison with areas not harvested. It is clear that many authors were concerned about the effects of the trade, although no data are available on specific populations. Most states and provinces regulate the take of bullfrogs today, although there are still few data on the population effects of harvest. Despite a variety of potential threats, however, the American Bullfrog remains one of the most widespread and common of North American amphibians, and has become a considerable nuisance throughout the western states and provinces.

Bullfrog populations may be transient, depending upon local conditions. Hecnar and M'Closkey (1996a) noted a small regional decline among ponds surveyed in Ontario over a 3 yr period. Rather than indicate concern, such changes may be indicative of normal population turnover, especially near the northern limit of the species' range.

The U.S. Fish and Wildlife Service developed a Habitat Suitability Index (HSI) model for this species for use in habitat evaluation procedures on federal lands (Graves and Anderson, 1987). Such models incorporate literature reports of habitat requirements, in this case employing 11 variables, to determine relationships between habitat variables, model components ("life requisites"), and HSI values that can be used to ascertain whether a habitat supports American Bullfrogs.

American Bullfrogs usually are sampled through visual encounter surveys of larvae or adults, or by detecting them using automated frog call recorders. In France, however, researchers have developed the ability to identify American Bullfrog presence by detecting its DNA in water samples. Environmental DNA is amplified using PCR techniques, and tests reveal it is highly reliable at identifying wetlands where the species is present (Ficetola et al., 2009).

Lithobates chiricahuensis (Platz and Mecham, 1979)
Chiricahua Leopard Frog

ETYMOLOGY
chiricahuensis: the name is derived from the Apache word Chiricahua, in reference to the Chiricahua Mountains, where the holotype was collected and in recognition of the Chiricahua Indians who inhabited the region.

NOMENCLATURE
Stebbins (2003): *Rana chiricahuensis*
Dubois (2006): *Lithobates (Lithobates) chiricahuensis*
Synonyms: *Rana chiricahuensis*, *Rana subaquavocalis*

Leopard frogs have long been recognized for their phenotypic variation. Different phenotypes have been considered to reflect polytypic variation of a wide-ranging species (*Lithobates pipiens*), the result of clinal variation, or as members of a wide-ranging multispecies complex (the Leopard Frog complex) (Ruibal, 1957; Moore, 1975; Hillis, 1988). Some assessments (e.g., Moore, 1944; Ruibal, 1957) mixed specimens from a variety of locations, resulting in considerable taxonomic confusion. In addition, these species often may be sympatric (e.g., Frost and Bagnara, 1977a; Frost and Platz, 1983), further complicating identity. Much of the scientific literature uses the name *Rana pipiens* for frogs now recognized as *Lithobates chiricahuensis*. Readers should verify locations when using older literature.

IDENTIFICATION

Adults. This is a rather stocky leopard frog. The ground color is light to dark olive or brown with numerous small black spots. These spots usually lack a light halo, and they are present anterior to the eye. There is an incomplete white stripe on the upper lip, which is diffuse in front of the eye. Dorsolateral folds are present; these folds are interrupted and tend to align medially toward the frog's posterior. Blunt tubercles are present between the dorsolateral folds. The skin is rough. A light stripe on the upper lip is either faint or absent. Throats are mottled gray (unlike other Southwestern leopard frogs), and this gray color may extend onto the chest. Yellow pigmentation occurs in the groin region and ventrally on the thighs, and often extends onto the rear part of the venter. Venters are dull and melanistic; gray mottling may be present. The posterior concealed portion of the thigh is darkly pigmented except for scattered small light spots, each containing a tubercle. Rear toes are broadly webbed. Males usually have small external vocal sacs (but see Platz et al., 1997, who characterized the vocal sacs of the Huachuca Mountain population > 80 mm as "well-developed"), and the folds of the vocal sacs may be darkened. Additional information on morphology and coloration is provided by Ruibal (1957), Mecham (1968), Platz and Platz (1973), and Platz (1976).

Females are generally larger than males, growing to 125 mm SUL in the Huachuca Mountains; males grow to slightly more than 100 mm SUL (Platz et al., 1997). There, the mean male length was 83 mm SUL, whereas the female mean was 105 mm SUL (Platz, 1993). In New Mexico, Fritts et al. (*in* Degenhardt et al., 1996) gave a mean male size of 64.3 mm SUL and a mean female size of 76.9 mm SUL. Brennan and Holycross (2006) gave the maximum size as 135 mm SUL.

Larvae. Larval *L. chiricahuensis* are darker than other leopard frog tadpoles in the Southwest. The dorsal coloration is a dusky olive gray, and the dorsum contains faint black spots. The lateral sides of the tadpole are olive with large dark spots, and there are large bronze splotches toward the venter. The venter itself is grayish white with a pinkish-bronze sheen. Tails tend to be olive gray with large but dull olive spots. The iris is bronze. The maximum size is ca. 80 mm TL, and the tail is about 1.5 times the body length. A detailed description of the larvae and mouthparts is in Scott and Jennings (1985), and Jennings and Scott (1993) illustrated color and morphological differences between stream and pond larvae (see Larval Ecology).

Eggs. There are no descriptions of the eggs of this species. Eggs masses are spherical and are attached to vegetation in shallow water. Clutch sizes range from 300 to 1,485 (*in* Sredl and Jennings, 2005).

DISTRIBUTION

Lithobates chiricahuensis is found in a small part of southwest New Mexico (mostly the Gila and San Francisco river drainages), with disjunct populations occurring in southwest Arizona and adjacent Hidalgo County in New Mexico in the

Distribution of *Lithobates chiricahuensis*. Dark gray indicates extant populations; light gray indicates extirpated populations.

Animas, Peloncillo, Huachuca, Dragoon, Pajarito, and Chiricahua mountains. This species has disappeared from much of its historic range in the U.S. In Mexico, Chiricahua Leopard Frogs are found in the Sierra Madre Occidental of Chihuahua, northern Durango, and eastern Sonora. Important distributional references include Campbell (1934), Frost and Bagnara (1977a), Clarkson and Rorabaugh (1989), Degenhardt et al. (1996), Sredl and Jennings (2005), Brennan and Holycross (2006), Lemos Espinal and Smith (2007a), and Hekkala et al. (2011).

FOSSIL RECORD

No fossils are known. Holman (2003) noted Miocene (Hemphillian) fossils from Navajo County in northern Arizona that belonged to the *Lithobates pipiens* complex.

SYSTEMATICS AND GEOGRAPHIC VARIATION

Lithobates chiricahuaensis is a member of the *Novirana* clade of North American ranid frogs. It is an associate of the mostly Mesoamerican *Lithobates montezumae* group (or *Lacusirana*) (Hillis and Wilcox, 2005). Its closest relative in the United States is the Northern Leopard Frog (*L. pipiens*).

Variation in leopard frog phenotypes from the American Southwest has been recognized for some time, with *L. chiricahuensis* often referred to as the "southern type" or form (Mecham, 1968; Mecham et al., 1973; Frost and Bagnara, 1976, 1977). The species can be separated from other leopard frogs by a combination of morphological, biochemical, auditory, and genetic characteristics (Mecham, 1968; Platz and Platz, 1973; Platz and Mecham, 1979; Frost and Platz, 1983). Frost and Bagnara (1976) presented a table comparing various phenotypes among leopard frog populations. In their description, Platz and Mecham (1979) noted regional variation in the presence of vestigial oviducts, with northern males tending to lack them and southern males having rudimentary oviducts. Populations of frogs in the Huachucas, Dragoon, Pajarito, and Chiricahua Mountains were thought to form a southern clade within *L. chiricahuensis*, with populations in the White Mountains and along the Mogollon Rim forming a northern clade. These northern clade forms have now been identified as conspecific with *L. fisheri*, a species known previously only from the Vegas Valley and thought to be extinct (Hekkala et al., 2011).

In 1988, Platz (1993) described a leopard frog from the Huachuca Mountains in southern Arizona as *Rana subaquavocalis* (Ramsey Canyon Leopard Frog). One of the most distinctive characteristics of the species was its underwater advertisement call (hence its name), a character unique among leopard frogs. The range included only Ramsey Canyon and two other locations. Both individual and population-level genetic heterozygosity was low in the Ramsey Canyon and Barachas Ranch populations, and these populations disappeared by 1996 (Platz and Grudzien, 2003). The results of further molecular analyses have demonstrated that *Lithobates subaquavocalis* and *L. chiricahuensis* are conspecific (Goldberg et al., 2004a).

Natural hybrids have been identified between *L. chiricahuensis* and *L. pipiens* in Arizona (Mecham, 1968; Platz and Platz, 1973; Platz, 1976). No evidence of hybridization between *L. chiricahuensis* and *L. blairi* has been demonstrated in wild populations (Frost and Bagnara, 1977). In laboratory crosses, *L. chiricahuensis* produces hybrids with *L. blairi*, *L. berlandieri*, *L. magnaocularis*, and *L. pipiens*; the percentage of embryos developing and levels of abnormalities are variable (Mecham, 1968; Purcell, 1968; Frost and Bagnara, 1977; Frost and Platz, 1983). Surviving hybrids have very low sperm counts.

ADULT HABITAT

Lithobates chiricahuensis is found in semidesert grassland, Madrean evergreen woodland, pinyon-juniper conifer forest, and montane conifer forest habitats. Streams and associated plunge pools in rocky canyons constitute the preferred habitat for this species. It also inhabits grassy streams and slow-moving creeks, rocky pools, springs, pools alongside streams, beaver ponds, and even stock tanks and ditches.

AQUATIC AND TERRESTRIAL ECOLOGY

This is a highly aquatic species that rarely ventures far from water. At night, however, individuals may venture from water's edge to forage. Chiricahua Frogs are sometimes observed floating on algal mats or other floating vegetation. Frogs are generally inactive from November through February, but frogs at geothermal sites may be active year-round. During the warm season, activity follows temperature, with frogs being most active in the morning, prior to when temperatures increase. As the season progresses and water temperatures warm, nocturnal activity becomes more prevalent. Activity also is associated with calm winds. Refugia from cold weather and drought have not been described, but Chiricahua Leopard Frogs likely seek refuge in rock cracks and crevices, under tree roots, and in undercut stream banks.

Long-distance movements take place along watercourses. Sredl and Jennings (2005) offer the following information on home range. The home range size varies between the dry and wet seasons. For males, the mean is 161 m^2 in the dry season and 375.7 m^2 in the wet season. The largest male home range recorded was 23,390 m^2 for an individual that used a stream corridor 10 m wide × 2,339 m in length. Another male was recorded to move 3.5 km. The largest female home range recorded was 9,500 m^2 (10 m wide × 950 m in length). Males tend to have greater home ranges than females.

CALLING ACTIVITY AND MATE SELECTION

Advertisement calls consist of a long snore-like trill. Calls consist of a single note lasting 1–2 sec with a dominant frequency of 0.9 kHz (Frost and Platz, 1983). They are characterized by a high pulse repetition rate (17–39 pulses/sec) and a high number of pulses (19–68/call). Pulse durations are short (6–28 msec) and rise with time (0.4–6.6 msec) (Platz and Mecham, 1979; Platz, 1993), producing a brief audible rise in pitch. Frost and Platz (1983) and Platz (1993) provide sonograms comparing this species' advertisement call with those of other Southwest leopard frogs.

Most *L. chiricahuensis* call above water from the shorelines of streams and associated water pockets, tanks, or pools. In the Huachuca Mountains, however, Chiricahua Leopard Frogs call from 1.0 to 1.3 m underwater. This call is inaudible in air. Whether other populations of this species, particularly those of the southern clade, have an underwater call has not been published, although the possibility is mentioned by Norman J. Scott (*in* Degenhardt et al., 1996). Aggressive behavior between males has been recorded during the breeding season (Sredl and Jennings, 2005), but apparently does not occur at other times.

BREEDING SITES

Breeding occurs in a wide variety of slow-moving or lentic waters, including streams, rivers, pools in intermittent streams, beaver ponds, marshy wetlands, and springs. They also use man-made water sources, such as stock tanks, irrigation sloughs, wells, and backyard ponds. Thermal springs may be a particularly important breeding type for this species, since such springs may allow year-round breeding and activity and are free from predaceous nonindigenous fish.

REPRODUCTION

Eggs may be deposited throughout the warm season, with records in New Mexico from April and September (Scott and Jennings, 1985) and from February to September in Arizona (Brennan and Holycross, 2006). Indeed, elevation plays an important role in the timing of reproduction. At low elevations (< 1,800 m), breeding occurs in the spring through late summer (especially in June), whereas at higher elevations (> 1,800 m) it occurs mostly from June to August (Frost and Platz, 1983). According to Sredl and Jennings (2005), breeding may occur year-round in warm-water geothermal springs. In Arizona, the northern clade of frogs breeds later than the southern clade frogs. An extended breeding season probably allows Chiricahua Leopard Frogs to take advantage of favorable environmental conditions for larval development in an arid land.

Eggs are attached to vegetation and are deposited with 5 cm of the water's surface. Under experimental conditions, embryos can develop at water temperatures from 12 to 31.5°C (Zweifel, 1968b). Egg mass temperatures in nature have been measured from 12.6 to 29.5°C (Sredl and Jennings (2005). Hatching occurs within 14 days of deposition in the Huachuca Mountains populations of this species, but at geothermal springs hatching may occur within 8 days (Sredl and Jennings, 2005).

LARVAL ECOLOGY

Tadpoles have been found in New Mexico from February through November, although most probably metamorphose by September. Metamor-

Egg mass of *Lithobates chiricahuensis*. Photo: Brent Sigafus

Tadpole of *Lithobates chiricahuensis*. Photo: Cecil Schwalbe

phosis occurs at 35–40 mm SUL. The bimodal breeding season and differences in thermal conditions probably account for the high proportion of the year when tadpoles may be found. According to information in Sredl and Jennings (2005), the larval period lasts three to nine months. The difference in the length of the larval period may reflect thermal conditions during development. In warm-water springs, continuous growth occurs toward metamorphosis, whereas in cold-water habitats development proceeds slower and larvae may overwinter. Tadpoles have been observed under ice in water 5°C (Sredl and Jennings, 2005).

Tadpoles of this species may be found in a variety of habitats, from streams to ponds. The habitat in turn influences larval coloration and morphology. Larvae from streams have more contrasting and blotched patterns on the tail. They have thicker dorsal tail fins and larger tail muscles than larvae taken from ponds. Even within streams, the level of tail thickening can vary, perhaps in response to stream flow. These differences are illustrated by Jennings and Scott (1993).

DIET
Larvae are herbivorous. The diet of postmetamorphic Chiricahua Leopard Frogs has not been examined. Like other leopard frogs, it probably feeds on a variety of invertebrates and even small vertebrates in relation to their availability.

PREDATION AND DEFENSE

Nothing has been reported on the predators of this species. It is likely eaten by a wide variety of snakes, birds, and mammals. Larvae may be eaten by aquatic invertebrates. Upon the approach of a potential predator, postmetamorphs jump into the water. In low temperatures and reflectance, the ventral skin coloration tends to darken, which may make this species less obvious to potential predators.

POPULATION BIOLOGY

Chiricahua Leopard Frog populations consist of a series of small subpopulations (<10) living along a drainage system or series of interconnected drainage systems in close proximity (within a few km) to one another. Individuals may move among these subpopulations along drainage corridors during favorable weather. This population structure is especially obvious in the Madrean (an evergreen forest and woodland community) mountain zones, and may allow the mountain populations to persist despite periodic extinction of individual populations (Sredl and Howland, 1995). In non-Madrean habitats, populations are more isolated with little possibility of individuals moving among them. This isolation makes them vulnerable to catastrophic declines or stochastic events.

Chiricahua Leopard Frogs may have a rather long life span. In the Huachuca Mountains, adults may live 10 yrs after metamorphosis. Males (> 68 mm SUL) were identifiable at 2 yrs of age, whereas females (> 72 mm SUL) were at least 3 yrs of age (Platz et al., 1997). The smallest mature males observed by Jennings (*in* Sredl and Jennings, 2005) in New Mexico were 53.5 mm and 56.2 mm SUL. However, Platz et al. (1997) suggested that sexual maturity was not reached until 6 yrs postmetamorphosis. There is considerable variation in the correlation between age and body size, suggesting that size is not directly correlated with age. However, large frogs are mostly older than smaller frogs. Although males can reach 10 yrs, most older frogs tend to be female. Growth rates are slow and vary among populations, as does population size-class structure (Platz et al., 1997).

DISEASES, PARASITES, AND MALFORMATIONS

The virulent pathogen *Batrachochytrium dendrobatidis* has been found in Chiricahua Leopard Frogs in Arizona (Milius, 1998), especially from the San Bernardino National Wildlife Refuge in Arizona (Bradley et al., 2002). Sredl and Howland (1995) reported the bacterial pathogen *Aeromonas* (red-leg) caused 50–80% mortality of frogs at a site in southern Arizona; whether red-leg was the primary or secondary cause of mortality is unknown. Cold weather may exacerbate the effects of disease, and Sredl and Jennings (2005) used the term "post-metamorphic death syndrome" to indicate high levels of mortality associated with overwintering. This syndrome likely affected frogs with weakened immune systems and was probably due to *Batrachochytrium dendrobatidis*.

Adult *Lithobates chiricahuensis*. Photo: Brent Sigafus

Habitat of *Lithobates chiricahuensis*. Photo: James Rorabough

Chiricahua Leopard Frogs are parasitized by the trematodes *Cephalogonimus brevicirrus, Clinostomum* sp., *Glypthelmins quieta, Gorgoderina attenuata, Haematoloechus complexus,* and *Megalodiscus temperatus.* It is parasitized by the nematode *Physaloptera* sp. (Goldberg et al., 1998a).

SUSCEPTIBILITY TO POTENTIAL STRESSORS
No information is available.

STATUS AND CONSERVATION
Like many western ranids, the Chiricahua Leopard Frog has disappeared from much of its historic range. In surveys from 1983 to 1987, Clarkson and Rorabaugh (1989) found the species at only 2 of 36 historic locations in Arizona from sites where it was present in the 1960s and 1970s. Sredl and Howland (1995) reported substantial population mortality at several southern Arizona localities in the early 1990s. Sredl (1998) later noted that since 1990, *Lithobates chiricahuensis* had disappeared from 82% of its reported historical localities; often only a few frogs remained at occupied sites. In the Galiuro Mountains in Coronado National Forest, only 2 of 21 historically reported sites still contained *L. chiricahuensis* as of 2003 (Jones and Sredl, 2005). Scott (in Degenhardt et al., 1996) could not locate *L. chiricahuensis* at eight historic populations in Hidalgo County, New Mexico, in the early 1990s. Chiricahua Leopard Frogs clearly are in serious trouble.

The likely causes of decline include a variety of factors, especially habitat alteration and outright destruction as springs, riparian environments, and rivers were modified and water reallocated for human use. For example, Platz and Grudzien (2003) noted that extreme modification of the San Pedro River likely contributed to the loss of populations described as *Rana subaquavocalis*. Aquatic modification results in increased sediment loads, high concentrations of hydrogen sulfide, high water temperatures, lowered water tables, and decreased levels of dissolved oxygen. Ranid declines in Arizona also are correlated with the presence of nonindigenous crayfish, American Bullfrogs (*Lithobates catesbeianus*), a variety of predaceous fishes, and amphibian pathogens, including but perhaps not limited to *Batrachochytrium dendrobatidis* (Rosen et al., 1995; Sredl and Howland, 1995; Witte et al., 2008). Declines are thus the product of multiple causes acting independently or cumulatively on a population-by-population basis.

In an extensive simulation analysis, perturbations involving disease, drought (in terms of four levels of base flow), and catastrophic floods (in terms of four levels of peak flow) were modeled for their effects on population persistence of

Lithobates chiricahuensis. The results demonstrated that amphibian chytrid disease was most likely the cause of the serious declines observed within the range of this species, but that these other variables could act synergistically to hasten declines (Boykin, 2006; Boykin and McDaniel, 2008). Although models predictive abilities are limited, they provide a theoretical framework with which to examine ongoing declines.

Because of the perceived uniqueness of the Ramsey Canyon population of this species, a captive breeding program was started through the Nature Conservancy and Phoenix Zoo (Demlong, 1997). At that time, the "species" was known only from a few man-made impoundments and may have had the smallest range of any known ranid. Less than 50 adults were observed in 1995. Captive reared metamorphs and larvae were released in October 1995, and despite unusually dry conditions some frogs survived. More were released in 1996.

Much information on the ecology and life history of *L. chiricahuensis* is contained in unpublished reports. Sredl and Jennings (2005) summarize some of this information. The species is protected in Arizona, and is considered "Threatened" in Mexico. It is listed as "Threatened" under the U.S. Endangered Species Act.

Lithobates clamitans
(Latreille *in* Sonnini and Latreille, 1801)
Green Frog
Northern Green Frog
(*L. c. melanota*); Bronze Frog
(*L. c. clamitans*)
Grenouille verte

ETYMOLOGY

clamitans: from the Latin *clamito* meaning 'to call loudly,' or *clamator* meaning 'bawler' or 'shouter.' Both names are in reference to the vocal sounds produced by this frog. *melanota*: from the Greek *melaina* meaning 'black' and *nota* meaning 'mark.'

NOMENCLATURE

Conant and Collins (1998) and Stebbins (2003): *Rana clamitans*
   Dubois (2006): *Lithobates* (*Aquarana*) *clamitans*
   Synonyms: *Rana clamata, Rana clamator, Rana flaviviridis, Rana fontinalis, Rana horiconensis, Rana melanota, Rana nigrescens, Rana nigricans, Ranaria melanota*

IDENTIFICATION

Adults. There are two distinct color phases of *Lithobates clamitans*, the northern green phenotype (Northern Green Frog) and the southern bronze phenotype (Bronze Frog). Calling *L. clamitans* in the South a "Green Frog" is definitely a misnomer, and the common name Green Frog is not used there. In the North, Green Frogs are medium-sized green to olive green to brownish frogs with a somewhat rugose dorsal skin surface and a dorsolateral fold that extends from the rear of the eye halfway down the body. Heads are usually bright green. The dorsum, legs, and sides may be flecked with black spots. The tops of the thighs and legs do not possess black bars. On the side of the body, there is a dark, wavy, indistinct band that gives way ventrally to a mottled pattern. No light line is present on the upper lip. The center of the tympanum is boldly marked by a white or yellow spot. The belly is light (grayish white to enamel white) but slightly mottled around its margins, or it can have rather dusky vermiculation. Smith (1961) noted that within Illinois between 33.5 and 84.5% of the frogs he examined had unmarked bellies and that the percentage varied by region. The throat may be bright yellow (in males) to light gray or white; the yellow coloration can extend onto the male's belly during the breeding season (Fleming, 1976).

In the South, Bronze Frogs are more slender and smaller than their northern counterparts, and they are more reddish brown to bronze dorsally. The upper jaw, however, is bright green, providing an obvious contrast with the rest of the dorsal coloration. In contrast with Northern Green Frogs, there is no dorsal spotting between the dorsolateral folds, and the dorsal skin surface is not as rugose. The venter is usually silvery white, but it may have a considerable amount of pigmentation (dusky vermiculation). However, Mecham (1954) noted that the belly pigmentation from southern *L. clamitans* at the western extent of its range more closely resembles the venter of Northern Green Frogs. Throats are light yellow rather than boldly colored. For subadults, the belly may be dark to black on some individuals (Wright, 1932). Juveniles may have dorsal spotting initially, but these spots are lost as the frogs grow. Juveniles, however, are more spotted ventrally than Northern Green Frogs.

In addition to these regional phenotypic differences, some northern populations of *L. clamitans* are extremely dark olive to black, giving rise to the common name, Black Frog. Very dark colored frogs occur around the Great Lakes, Québec, Ontario, Vermont, New Hampshire, and in the Adirondack Mountains of New York (Mecham, 1954). In such cases, the dorsal spots are obscured, the contrast between the black dorsum and the white venter is quite distinct, and a yellow streak may be obvious on the side of the head. The vermiculate dorsal pattern might appear to resemble that of *L. septentrionalis* and thus lead to confusion in distinguishing these species. This darkening of the dorsum is not uniform within a region, and normally colored Northern Green Frogs are found within the same region as the black phenotype.

In areas where the green and bronze phenotypes intergrade, individuals have intermediate color patterns and markings. For example, Green Frogs from southern Illinois have a heavily pigmented venter, often forming a black mottled pattern on the belly, pectoral region, and throat. Occasional individuals, however, have green jaws and yellow throats, and frogs from adjacent counties often have unmarked bellies (Rossman, 1960). In both phenotypes, however, the feet are broadly webbed for swimming; the distil two joints of the fourth toes of the rear feet are free of webbing.

In addition to throat coloration, there is sexual dimorphism in the size of the eye in relation to the tympanum. In males, the eye is smaller than the tympanum, but in females, the eye and tympanum are about the same size. The thumbs of males also are enlarged at the base for use in amplexus, and the forearms are hypertrophied, again to enable the male to hold onto the female during amplexus.

Albino, partial albino, and yellowish adults are not unusual, and have been observed mostly from the northern portion of its range (Deckert, 1915; Fowler, 1918; Wright, 1932; Hensley, 1959; Oldfield and Moriarty, 1994). Pinder (2010) also reported an unusual yellowish-orange and olive individual from New York. Berns (1966) and Berns and Narayan (1970) reported on two blue (axanthic) *L. clamitans*. In these frogs, the blue coloration resulted from the nearly complete absence or large reduction of the yellow pigment pteridines and carotenoids, respectively. Other blue *L. clamitans* are known from Massachusetts (Lazell, 1976), Delaware (Arndt, 1977), and Virginia (Berns and Uhler, 1966). Berns (1966) noted considerable variation in the extent and shade of blue coloration on individual frogs. Photos of blue Green Frogs are in Harding and Holman (1992) and Desroches and Rodrigue (2004).

There is a regional difference in size between the Northern Green Frog and the Bronze Frog, with frogs in the South being smaller. Green Frogs are reported to reach 108 mm (Gilhen, 1984). Other reports of sizes include: 58–70 mm SUL (mean 64 mm) in New Jersey vs. 52–61 mm SUL (mean 57 mm) in Florida (Mecham, 1954); these figures likely include subadults. Louisiana females >10 months in age average 66.4 mm SUL, with gravid females larger (mean 77.4 mm SUL, range 60.6 to 87.9 mm) than nongravid females (Meshaka et al., 2009a). In Michigan, males are 60–103 mm and females 66–105 mm (Martof, 1956a), whereas males reach 90 mm and females 98 mm in New York (Ryan, 1953). In Connecticut, males are 52–84 mm SUL (mean 68.8) and females 52–94 mm SUL (mean 64.8 mm) (Klemens, 1993). Smithberger and Swarth (1993) measured Green Frogs at 30–70 mm SUL (mean 58 mm) and 2.4–59 g (mean 21.4 g), but these figures obviously included subadults. On Prince Edward Island, males averaged 84.1 mm SUL (range 68 to 100 mm) and females 84.4 mm SUL (range 70 to 96 mm) (Cook, 1967), whereas in Nova Scotia males were 70–98 mm and females 67–108 mm (Gilhen, 1984). Finally, Shirose and Brooks (1997) provided mean SULs of females in Algonquin Park, Ontario, that ranged from 63 to 80 mm SUL, depending on year sampled, over a 6 yr period.

Mecham (1954) noted that Northern Green Frogs were routinely > 85 mm SUL, but that

Bronze Frogs rarely exceeded 75 mm SUL. Males and females are about the same maximum size in some areas (Martof, 1956a; Smith, 1961; Fleming, 1976), although females have been reported to be larger than males in others (Ryan, 1953; Jenssen and Klimstra, 1966; Meshaka et al., 2009a). The reverse is also sometimes true. Differences in mean sizes are small, however, and not much more than about 3 mm. Thus, this species cannot be generalized to be sexually size-dimorphic.

Larvae. The tadpoles are large (80–100 mm TL; Logier, 1952) but not deep bodied, and are olive green with small to large dark markings. Throats are usually white and bellies are a deep cream color with no iridescence (but see Systematics and Geographic Variation). The tail is green and mottled with brown, and without pinkish to buff spotting. The dorsal fin terminates posterior to the spiracle, which is located on the left side of the body. There are no stripes on either the dorsal fin or on the tail musculature. The eyes are dorsally located, and the oral disc is emarginate and narrowly pigmented. The anus is dextral rather than located medially. Green Frog larvae may be in the same developmental stages, but still be very different in size, depending upon the pond conditions in which they develop (Rogers, 1999). Descriptions of larvae are in Wright (1914, 1929, 1932), Altig (1970), and Priestley et al. (2010). Yellowish-white tadpoles were reported from Tennessee (Dyrkacz, 1981), an unusually colored specimen from Michigan (Bowen and Beever, 2010), and albinos from Québec (Saumure, 1993).

Eggs. The black-and-white eggs are laid in a large (for example, 12 cm × 17 cm or 15 cm × 21 cm) irregular surface film; each film (maximum diameter 30 cm) contains from 1,000 to >5,000 eggs amid emergent or floating vegetation. The surface film may float free, or it may be attached to aquatic vegetation. Surface films sometimes break apart, giving the impression of smaller clutch sizes, or films oviposited by separate females may merge, extending the extent of the film and increasing perceptions of the numbers of eggs (Wright, 1914). Although somewhat similar to the surface film egg masses of American Bullfrogs, there are many fewer eggs in Green Frog masses. Surface films allow for better oxygenation at the warm water temperatures in which this species breeds. The eggs are deposited in relatively shallow water (< 50 cm), where hatching occurs in a few days.

Eggs have two gelatinous envelopes. The vitellus is 1.2–1.8 mm, and the eggs are black dorsally and white ventrally. The inner capsule is 2.8–4.0 mm and the outer capsule 5–6 mm. The outer capsule is indistinct and may merge into the surface film. Hatching occurs in three to five days. Descriptions of eggs are in Livezey and Wright (1947).

The presence of the species-specific jelly capsule surrounding the egg of the Green Frog prevents fertilization by the sperm of other ranid frogs. However, if the jelly of another ranid (*L. pipiens*, for example) is artificially transposed with the jelly of a Green Frog under laboratory conditions, fertilization may then occur (Elinson, 1974, 1975a). Thus, the jelly capsule plays an important role in preventing hybridization with most other ranid frogs. This protective device is not foolproof, however, as native *L. clamitans* jelly will not prevent fertilization by *L. catesbeianus* sperm (Elinson, 1975b).

DISTRIBUTION

In the East, Green Frogs occur from the Canadian Maritimes and St. Lawrence River Valley south throughout all of North Carolina, except for the extreme southeastern Coastal Plain in that state. They occur throughout much of southern Québec and Ontario to extreme southeastern Manitoba, then south through eastern Minnesota, Iowa, and Missouri. Green Frogs are found throughout much of southern Missouri southwestward into eastern Oklahoma and northern Arkansas. Their southern limit appears to be the Tennessee River and associated drainages from Mississippi through northern Georgia. Further east, they occur in the Piedmont and uplands of South Carolina across to southeastern North Carolina.

Green Frogs appear to be scarce from much of central Illinois, where many records are from prior to 1980 (Smith, 1961; Phillips et al., 1999), and the species is associated with river drainages; it is also present in a small area of adjacent southwest Indiana. Intergradation with the Bronze Frog phenotype occurs extensively in central Alabama and Georgia.

In the South, Green Frogs are replaced by the Bronze Frog. Bronze Frogs occur from eastern Texas, southeastern Oklahoma, and southern

Distribution of *Lithobates clamitans*

Arkansas north to the Missouri boot along both sides of the Mississippi River. They occur south of the Tennessee River drainage throughout Mississippi, then eastward on the Coastal Plain through southeastern North Carolina. In Florida, they are found from Polk County northward (Means and Simberloff, 1987).

The Green Frog was deliberately introduced to the island of Newfoundland, and people have played an important role in its dispersal around the island (Maunder, 1983). Likewise it was introduced to the French islands of Saint Pierre and Langlade off the southern coast of Newfoundland. In the West, Green Frogs have been successfully introduced into British Columbia on southern Vancouver Island and along the lower Fraser River valley from Hope to Richmond (Matsuda et al., 2006), West Ogden, Utah (Behle and Erwin, 1962), and at Toad Lake, Whatcom County, Washington (Slater, 1955). It has also been introduced into Arizona, Iowa, Montana, the Bahamas, and The Netherlands (Kraus, 2009).

Green Frogs occur commonly on islands, including islands in western Lake Erie (Pelee, Kelleys, Middle, Middle Bass, North Bass), at the mouth of the St. Lawrence River (Île d'Orléans), in Georgian Bay, Isle Royale, the Apostle Islands in western Lake Superior, Cape Breton Island, and Prince Edward Island (Ruthven, 1908, 1912; Richmond, 1952; Cook, 1967; King et al., 1997; Hecnar et al., 2002; Fortin et al., 2004a; Bowen and Beever, 2010). They are found on Long Island, New York (Overton, 1914), and still occur on Nantucket, Martha's Vineyard, Nashawena, and Naushon off Cape Cod (Lazell, 1976). In the South, they occur on New Kent Island, Maryland (Grogan and Bystrak, 1973b), Assateague Island, Virginia (Conant et al., 1990; Mitchell and Anderson, 1994), and Bodie Island, North Carolina (Braswell, 1988).

Important distributional references include: Alabama (Mount, 1975), Arkansas (Black and Dellinger, 1938; Trauth et al., 2004), British Columbia (Matsuda et al., 2006), Canada (Bleakney, 1958a; Logier and Toner, 1961; Weller and Green, 1997), Chicago metropolitan area (Schmidt and Necker, 1935; Necker, 1939), Connecticut (Klemens, 1993), Florida (Ashton and Ashton, 1988), Georgia (Williamson and Moulis, 1994; Jensen et al., 2008), Great Lakes

Region (Harding, 1997), Illinois (Smith, 1961; Phillips et al., 1999), Indiana (Minton, 2001; Brodman, 2003), Iowa (Hemesath, 1998), Kansas (Smith, 1932; Collins, 1993; Collins et al., 2010), Louisiana (Dundee and Rossman, 1989; Hardy, 1995), Manitoba (Preston, 1982), Massachusetts (Lazell, 1976), Michigan (Ruthven, 1912), Minnesota (Fleming, 1976; Oldfield and Moriarty, 1994), Missouri (Johnson, 2000; Daniel and Edmond, 2006), New Brunswick (McAlpine, 1997a), New England (DeGraaf and Rudis, 1983), Newfoundland (Maunder, 1997; Campbell et al., 2004), New Hampshire (Oliver and Bailey, 1939; Taylor, 1993), New York (Gibbs et al., 2007), North Carolina (Dorcas et al., 2007), Nova Scotia (Richmond, 1952; Gilhen, 1984), Ohio (Walker, 1946; Pfingsten, 1998; Davis and Menze, 2000), Oklahoma (Sievert and Sievert, 2006), Ontario (Johnson, 1989), Pennsylvania (Hulse et al., 2001), Prince Edward Island (Cook, 1967), Québec (Bider and Matte, 1996), Tennessee (Redmond and Scott, 1996), Texas (Raun and Gehlbach, 1972; Hardy, 1995; Dixon, 2000), Vermont (Andrews, 2001), Virginia (Tobey, 1985; Mitchell and Reay, 1999), Washington (Slater, 1939b, 1955), West Virginia (Green and Pauley, 1987), and Wisconsin (Suzuki, 1951; Vogt, 1981 Casper, 1996; Mossman et al., 1998).

FOSSIL RECORD

Green Frog fossils are found from the Miocene Barstovian (11–15 my BP) of Nebraska, the Pleistocene Irvingtonian (1.9 my BP–150,000 BP) of Maryland, Nebraska, and West Virginia, and the Pleistocene Rancholabrean (150,000 BP–10,000 BP) of Georgia, Indiana, Michigan, Ohio, Pennsylvania, Virginia, and West Virginia (Holman, 2003). Holman (2003) also noted osteological differences between *L. clamitans* and *L. catesbeianus*, as these species have sometimes been confused in the literature on fossils.

SYSTEMATICS AND GEOGRAPHIC VARIATION

Lithobates clamitans is a member of the *Aquarana* clade of ranid frogs within the New World *Novirana* clade (Hillis and Wilcox, 2005). It is most closely related to *Lithobates okaloosae*, which is probably a Pleistocene derivative of *L. clamitans* (Austin et al., 2003a). The *Aquarana* clade includes *L. catesbeianus, L. grylio, L. heckscheri, L. okaloosae, L. septentrionalis,* and *L. virgatipes* and, with the exception of *L. okaloosae*, probably was derived during the Pliocene. These data suggest that reports of *L. clamitans* from the Miocene may refer to an ancestral member of the *L. catesbeianus* group rather than *L. clamitans sensu stricto*.

The different coloration, sizes, and habits among *L. clamitans* populations in the North and South have led to different nomenclatures based on the green/bronze phenotypic differences. Holbrook (1842), for example, described *Rana clamitans* from the South, and *R. fontinalis* from the North. A complex nomenclatorial history (Rhoads, 1895; Mecham, 1954; Stewart, 1984) led eventually to the recognition of two subspecies, *Lithobates clamitans clamitans* (Bronze Frog) and *L. c. melanota* (Northern Green Frog). Although phenotypically distinctive, molecular data do not support the recognition of these subspecies within *L. clamitans* (Austin et al., 2003). Indeed, there are three main evolutionary lineages, Coastal Plain-Appalachian, Louisiana, and Northern (Austin and Zamudio, 2008), reflecting Pliocene isolation followed by dispersal in the Pleistocene.

Tadpoles are described with throats that are usually white and bellies that are a deep cream color with no iridescence. Priestley et al. (2010), however, picture tadpoles from within the range of the Bog Frog (*L. okaloosae*) that are quite similar to that species. Both have dark chins and bellies, but *L. clamitans* has fewer iridescent spots that are more bronze and less sharply defined than *L. okaloosae*. The ground color also is more dusky than black. In addition, the tails of *L. clamitans* are unspotted whereas those of *L. okaloosae* have diffuse dark spots.

Crosses between ♀ *L. clamitans* and ♂ *L. septentrionalis* are not viable, and larvae do not develop. However, crosses between ♀ *L. clamitans* and ♂ *L. catesbeianus* are fully viable, and the resulting tadpoles undergo normal metamorphosis (Elinson, 1975a). This result tends to confirm the close phylogenetic relationship between *L. clamitans* and *L. catesbeianus*. Natural hybrids between *L. clamitans* and *L. okaloosae* were noted by Moler (1992).

ADULT HABITAT

Green Frogs are very common frogs found in a variety of aquatic habitats, including large lakes and ponds, small ponds, springs and seeps, woodland pools and streams, steephead ravines, peatlands, and in streamside riparian zones (Stockwell and Hunter, 1989; DeGraaf and Rudis, 1990; Rudolph and Dickson, 1990; Klemens, 1993; Burbrink et al., 1998; Pauley et al., 2000; Metts et al., 2001; Babbitt et al., 2003; Lichtenberg et al., 2006). For instance, they were found in 104 of 117 sites surveyed in Ontario, indicating a rather ubiquitous presence throughout the region (Hecnar and M'Closkey, 2005), and other reports also indicate high levels of regional occupancy (Hecnar, 1997; Hecnar and M'Closkey, 1998). Although the probability of any one pond being occupied is generally high, there is a considerable amount of turnover at a site from one season to the next, so that a pond may be inhabited by a population one year but not the next (Hecnar and M'Closkey, 1996a). In a 5 yr survey of Michigan ponds, there was an approximately 45% probability of finding Green Frogs at the 32 sites in any one year (Skelly et al., 2003).

Likewise, Bronze Frogs have been recorded in a variety of habitats, including mesic hammocks, mixed hardwood forests, fluctuating flatwoods ponds, bayheads, cypress and titi swamps, cypress and gum swamps, cutoff pools of river swamps, steephead ravines, freshwater ponds, small isolated wetlands, seepage bogs, and rivers and small streams (in wet and mesophytic hammocks) in the South (Wright, 1932; Carr, 1940a; Harima, 1969; Gibbons and Bennett, 1974; Enge et al., 1996; Enge, 1998a, 1998b, 2002, 2005b; Russell et al., 2002a). The wetlands are generally the same as those chosen for breeding sites, and fish may or may not be present.

Green Frogs generally prefer habitats that are relatively unaffected by urbanization and agriculture, although they are sometimes not uncommon at such locations. They have been found in wetland and terrestrially associated silvicultural areas, as well as in agricultural and urban locations (Bennett et al., 1980; Pais et al., 1988; Russell et al., 2002b; Surdick, 2005), although the mere presence of individuals at disturbed sites does not indicate that they will survive through time or that populations at disturbed and undisturbed sites are functionally equivalent.

Certain habitat tendencies are evident. For example, Green Frogs are negatively associated with development and urban/recreational grassland with 3,000 m from wetlands, but positively associated with the extent of woody wetlands, submergent vegetation, forested uplands, herbaceous wetlands, and other variables reflecting undisturbed habitat, especially within 1,000 m and 3,000 m, in the western Great Lakes (Price et al., 2004). In Wisconsin but not Iowa, Green Frogs also are positively associated with lake variables (area of wetland-emergent vegetation, area of nonpermanent wetlands, diversity of habitat patches), and negatively with urban variables (urban lands, extent of edges between agricultural habitats and wetlands) (Knutson et al., 2000). This discrepancy results from differences in the availability of habitats, with agricultural lands predominant in Iowa and forests and lakes predominant in Wisconsin. Finally, in Ontario, Green Frogs are associated with forest cover within 300 m (Eigenbrod et al., 2008) and, in Wisconsin, with a variety of habitats (Mossman et al., 1998).

Green Frogs are occasionally observed in the entrances to caves or in small recesses in limestone outcroppings (Duellman, 1951; Franz, 1967; Black, 1973a; McDaniel and Gardner, 1977; Trauth and McAllister, 1983; Garton et al., 1993: Buhlmann, 2001), and even in abandoned mines (McAllister et al., 1995b; Pauley and Pauley, 2007). They also can be found in human-dominated landscapes, including impoundments, storm-water catchment basins, golf course ponds, and in beaver ponds (Metts et al., 2001; Birchfield, 2002; Birchfield and Deters, 2005; Mifsud and Mifsud, 2008; Scott et al., 2008). They are not often found in the thickly vegetated water hyacinth communities of the Deep South (Goin, 1943).

AQUATIC AND TERRESTRIAL ECOLOGY

Adults primarily stay near water throughout the year (Martof, 1953a, called them "bank" frogs for their tendency to reside on the banks of ponds and streams), preferring wetlands with long hydroperiods and high invertebrate abundance (Babbitt et al., 2003). They tend to remain within a particular area, occasionally moving up or down a small stream or brook (Breder, 1927), or changing positions along a shoreline (Martof, 1953a). In Florida, Gorman et al. (2009) found that individuals moved a mean of 3.3 m along a shoreline. Both adults and subadults have home ranges, but the adult and subadult microhabitats are not the same. Subadults prefer shallow areas with thick vegetation, whereas adults move into deeper waters, unless American Bullfrogs are present in the deep water. If bullfrogs are close to shore, however, Green Frogs will move away from the shoreline to avoid the predator (Birchfield, 2002). Green Frogs are most often found in areas with shrubby shoreline vegetation, which provides them with protective cover, although they will call from shorelines with open understory (Hennig and Remsburg, 2009).

Breeding and nonbreeding habitats may not be identical, although they both center on aquatic habitats. Oldham (1967) recorded Ontario Green Frogs moving up to 457 m from terrestrial locations to breeding sites, suggesting that they inhabit terrestrial areas or small streams and wetlands, sometimes at considerable distances from breeding sites. When displaced, most Green Frogs will return directly to familiar territory, usually within a few days, depending on intervening terrain and distance displaced. They tend to avoid forested areas when homing if given a choice, instead preferring the path of least resistance, such as short-grass habitats; they do not follow topographic features (Birchfield, 2002; Birchfield and Deters, 2005). The homing ability is dependent on the distance frogs are displaced. Frogs displaced 550 m were able to orient toward a homeward path, but frogs displaced more than 3.2 km were unable or unlikely to do so (Oldham, 1967). This olfactory-based homing tendency suggests they are familiar with the environment in the general vicinity of their home range.

Green Frogs are often abundant within both aquatic and adjacent terrestrial habitats during the nonbreeding portions of the year, and their abundance increases in the absence of American Bullfrogs (Courtois et al., 1995), a major predator. The number of Green Frogs around a wetland is likely dependent on the size of the wetland and the availability of male territories in addition to predator proximity. In Michigan, Martof (1956b) estimated there were about 500 frogs around one pond (6.1 ha) in Michigan, and 100 around a second pond (2.7 ha). Both populations remained stable from May to mid-July, when the population at the first pond decreased to 300 as metamorphs departed, whereas the second increased to 200–300, reflecting the immigration of breeding adults. Thus, the number of frogs at the site is a balance between residents that do not move, juveniles, and a seasonal influx of adults into the breeding site from other wetlands or terrestrial habitats. A high rate of turnover (via mortality and movement) of both adults and juveniles at a site makes estimating the size or definition of a population difficult.

Lithobates clamitans is seasonally active as long as temperatures permit. In the South, this means that Bronze Frogs regularly are active year-round (Meshaka et al., 2009a, 2009b). As latitude increases northward, the activity season becomes more restricted. In Ohio and Maryland, for example, Green Frogs usually are active from March through late October (Walker, 1946; Smithberger and Swarth, 1993) and possibly later, depending on environmental conditions; in Minnesota, activity occurs from late April to early October (Fleming, 1976); in Michigan, New York, and New England, the activity season spans approximately 220 days from March to November (Martof, 1953a, 1956b; Klemens, 1993; Gibbs et al., 2007). Minton (2001) noted a Green Frog active in December in central Indiana. Juveniles are usually the first to appear in the spring and the last to be seen before winter sets in.

Green Frogs often forage in terrestrial habitats, particularly during periods of rainfall, when they can be observed in abundance in upland forested habitats and even in old fields (Allen, 1868; Martof, 1953a; Clawson and Baskett, 1982; Buhlmann et al., 1993). In Florida, they are most active at night during the warmer parts of the year, but they are also commonly observed sitting quietly around and in small bodies of water during the day. Like Martof (1953a), I have observed Bronze Frogs foraging up to 20 m away from their resident wetland during and immediately after precipitation; there they learned to congregate around a compost pile offering a variety of invertebrate prey.

Martof (1956b) observed in Michigan that juveniles were more active by day and adults by night. Body temperatures tend to be slightly higher than ambient air temperatures (Brattstrom, 1963), suggesting thermoregulation. They have been recorded as active between 10 and 29°C (*in* Brattstrom, 1963), but this range seems to be an artifact of sampling rather than an accurate evaluation of temperature-activity limits. The CTmax is slightly >36°C. Martof (1953a) occasionally found frogs at temperatures below 15°C, but such frogs were inactive or rarely encountered.

Green Frogs are photopositive in their phototactic response, suggesting they use both sunlight and moonlight when feeding and going about their daily activities (Pearse, 1910; Jaeger and Hailman, 1973). They are sensitive to light in the blue spectrum ("blue-mode response"), which apparently helps them orient toward areas of increasing illumination, such as the open area above a pond (Hailman and Jaeger, 1974). Green Frogs likely have true color vision.

Adult and juvenile Green Frogs often remain near water during the postbreeding season. However, they forage extensively in terrestrial habitats where food resources are greater than in proximity to water. During these forays, they gain body mass. In New York, Lamoureux et al. (2002) recorded from 0 to 7 forays per frog from mid-August until late October. Movements between pond and terrestrial foraging areas occurred with precipitation and in less than 24 hrs. Frogs moved between 10 and 99 m (mean 36 m), and forays lasted from 7 to 408 hrs (mean 88 hrs). Foraging sites had dense terrestrial vegetation with thick layers of leaf litter; frogs were usually observed well hidden, with only their noses visible.

Green Frogs move to protected sites in the fall and enter winter dormancy from mid-September to mid-November in the North. Such sites may not be in the immediate proximity of the summer activity centers. For example, frogs in New York moved 100–560 m to reach overwintering sites in distant streams, seeps, and a beaver pond (Lamoureux and Madison, 1999), and from 550 to 900 m in Missouri (Birchfield, 2002). The length of dormancy changes with latitude and environmental conditions, but can last more than five to six months; overwintering in Michigan occurs from late October to early November until late March to early April (Martof, 1953a, 1956b). Frogs become active when daily maximum temperatures reach 15°C for several days, and mean daily temperatures are >4°C. Rainfall helps stimulate an end to winter dormancy.

Overwintering normally occurs in aquatic habitats in mud and bottom debris (Dickerson, 1906; Walker, 1946; Wright and Wright, 1949; Logier, 1952; Lamoureux and Madison, 1999; Birchfield, 2002). Streams seem to be preferred, as the flowing water remains unfrozen and provides oxygenation during the long winter months. Males and females choose the same overwintering sites and remain there until dormancy ends. Gorham (1964) noted in New Brunswick that *L. clamitans* are observed around springs in November. Green Frogs also overwinter in terrestrial situations in soil pockets below the leaf litter. In such a situation, Bohnsack (1951) noted air temperatures of 0–2.5°C surrounding an overwintering Green Frog in Michigan, and that ice crystals were occasionally found in the soil near the frog. Winterkill may occur in such sites, however; Lannoo et al. (1998) recorded a mass mortality event of subadult Green Frogs found adjacent to a wetland in Iowa, and suggested that winterkill resulting from a poorly chosen terrestrial site was the most likely explanation.

CALLING ACTIVITY AND MATE SELECTION

Calling occurs over an extended period in spring and summer. Male Green Frogs are territorial, with larger males occupying larger and better quality habitats than smaller males. During the course of a breeding season, males often occupy more than one territory (usually two to five, mean 3.7; Wells, 1977b), and they may spend from one to as long as seven weeks within a particular territory. Males move back and forth between defended territories. Large males also occupy their territories longer throughout the breeding season than small males. Occasionally, a group of males will maintain a spatial relationship to one another (Martof, 1953b), but with the shifting of territories, most spatial relationships change constantly, especially during the breeding season (Shepard, 2004). This may explain the seemingly random nonterritorial distribution pattern reported by Shepard (2002), especially since his study was conducted for only a short-term period.

Territories (4–6 m in diameter) are defended during the breeding season, and males aggressively challenge intruders, who sometimes remain as satellite males in the anticipation of a territory becoming available (Wells, 1978; Shepard, 2004). In such encounters, males use patrolling, splashing displays, vocalizations ("growling": Jenssen and Preston, 1968), chases, attacks, and wrestling to establish dominance over their territory (Brode, 1959; Schroeder, 1968; Wells, 1978). Aggressive encounters often involve the display of the yellow throat patch directed toward the intruder, and vocalizations occur continuously during wrestling and shoving matches. Most encounters are short-lived, but some can last a considerable time (45 min). Larger males are normally successful in aggressive encounters.

Territory quality is determined by the extent of vegetation in the shallow water of the potential oviposition site, and females choose males based on the quality of the habitat and male size. In turn, males with high-quality territories mate more often (to five times) than those with lower-quality territories, although some males are not successful despite the quality of their territory. Wells (1977b) suggested that the intense male–male competition for mates evolved as a result of the prolonged breeding season, and described the territorial behavior of males in detail.

Calling sites are located in shoreline vegetation where ample cover is provided. Males call from vegetation near the shoreline or while floating on the water's surface, and they may swim between nearby vegetation mats between calls. Males usually are spaced about 2–3 m from one another, but they may call from much closer proximity (Martof, 1953b). If calling from the shoreline, the male often creates a small pool to sit in by rotating the hind limbs back and forth. Wells (1977b) noted that pool construction occurred when pond water levels were low or decreasing.

Wells (1978) delimited four types of calls, in addition to a release call. These are: Type I (spontaneous calls given day or night, one single note); Type II (a high intensity advertisement call of three to four notes delivered in rapid succession, usually in response to a disturbance in the territory); Type III (similar to Type I but more explosive, directed toward an opponent in an agonistic encounter); and Type IV (a long low-frequency call often given by the winner of a wrestling bout, perhaps as a warning). The release call (Type V) may be given both by males and females. The dominant frequency of the call of male Green Frogs is 416–544 Hz (Given, 1990), and the dominant frequencies of the advertisement calls are inversely correlated with the caller's snout-vent length (Ramer et al., 1983).

Males can assess the dominant frequency of the calls of conspecific males, and are thereby able to assess the size of a rival. If the rival is a large male, they will lower the dominant frequency of their call. However, if the rival is a small male, they do not change their own dominant frequencies (Bee et al., 1999). Ramer et al. (1983) found that small males increased the rate of agonistic vocalizations directed toward other small males, whereas large males decreased their rate of baseline calling. Large males directed agonistic calls to other large males, but responded little if at all to small males. In these ways, males know the sizes of males in adjacent territories, and this helps mediate or direct territorial aggression. The calls used in territorial aggression are Type III calls, which have lower dominant frequencies and a longer note duration than the somewhat similar Type I advertisement calls (Bee and Perrill, 1996). Males do not respond to heterospecific calls, despite the close proximity of calling males (Given, 1990).

Calling occurs both diurnally and nocturnally, but the greatest intensity of calling occurs from shortly after midnight until dawn (Mohr and Dorcas, 1999). Northern Green Frogs generally call in the spring to midsummer, whereas Bronze Frogs will extend calling into late summer or even early autumn. In New Brunswick, for example, calling occurs from late April to mid-June (Gorham, 1964), and in Michigan calling occurs from May through mid-August. In Louisiana, calling occurs from March to September (Dundee and Rossman, 1989; Meshaka et al., 2009a), although calls have been heard there as early as January (Meshaka et al., 2009b).

Males often begin calling from pools and isolated portions of streams early in the season, but they will eventually leave these temporary calling locations after two to four weeks and move to larger lakes and ponds, where calling begins in earnest. Males frequently move during dry weather, but females appear to require both warm weather and precipitation for movement and breeding activity.

Calling occurs at temperatures from 11 to 29°C, with most activity at ca. 22–24°C (Meshaka et al., 2009b). Environmental conditions influence the timing of calling, but the way they impact call activity varies seasonally. At the beginning of the season in Nova Scotia (Oseen and Wassersug, 2002), for example, males generally wait for warm water temperatures (>22°C) between sunset and sunrise before beginning to call in earnest, and calling is associated with high relative humidity and high barometric pressure. As the season progresses, water temperature becomes less important as summer temperatures climb, and frogs call at lower barometric pressures associated with rainfall. Humidity also becomes unimportant later in the season, as frogs attempt to reproduce before cool weather sets in. Precipitation is important in initiating spring movements to breeding ponds, and later in the season in initiating calling when water levels are decreasing; it is not important in triggering calling activity in the spring. Calling is not affected by windy or cloudy conditions.

In Michigan, males tend to congregate in the ponds or other suitable sites during the entire breeding season, whereas females are found more along small streams when they are not at the breeding sites (Martof, 1956b). Martof (1956b) suggested that females rarely spent more than a week at the main breeding sites, quickly returning to the nonbreeding habitats after ovipositing their eggs. Thus, there appears to be a degree of adult habitat partitioning, which allows spent or nonreproductive females to avoid contact with reproductive males.

At the breeding site, females in a head-down posture approach the territory of a calling male, and they may visit several territories prior to selecting a mate. The cues with which females choose males may vary, but Schulte-Hostedde and Schank (2009) have shown that males with larger forearms and intermediate shades of yellow on the throat are in better body condition than other males. In addition, the size of the tympanum may provide the female with cues for mate selection. These multiple cues likely help the female refine her choice based on the condition of a calling male.

The male Green Frog approaches the female from behind while holding her hind limbs, and then guides her body underneath his forelimbs as he floats in water. Amplexus is axillo-pectoral, that is, halfway between axillary and pectoral. The male holds on firmly to the female by pressing his thumbs into the female's lateral sides. Spawning takes place in the male's territory as the pair floats just beneath the water surface. Just prior to spawning, the female lowers her head below the water's surface, arches her back concavely, and brings her cloaca to just above the water surface. The male then slides backward and rotates his thighs upward. This brings the cloacas of the pair into near (3 mm) contact. Eggs are extruded at 30 to 50 at a time, and the male uses his rear feet to direct the eggs to his cloaca, where they are fertilized. The male then uses his toes and feet to push the eggs away from the amplexed pair. The pair then repeats the deposition/fertilization process until all the eggs are oviposited. Oviposition lasts 10–25 min. Males release the females as soon as she abandons the oviposition posture. Males and spent females have a warning croak indicating nonreceptivity to a suitor male (call Type V), but the female does not regain her voice until about 24 hrs after oviposition. In the meantime, she may be clasped by another male. The mating sequence and postures of Green Frogs are described and illustrated in detail by Aronson (1943a). Amplexus by males with females of other ranids (*L. pipiens, L. palustris, L. sylvaticus*) has been reported (Wright, 1914).

BREEDING SITES

Green Frogs are found in a wide variety of breeding sites, usually with long hydroperiods. They prefer lakes, ponds, and slow moving permanent streams, although they can be found in relatively small woodland pools. They use artificial breeding sites, such as retention ponds (Brand and Snodgrass, 2010), and canals or ditches with slow-moving water. In experimental mesocosms, increased depth and area led to lower survivorship but greater growth of the survivors when densities were kept constant (Pearman, 1993), but observations on Green Frogs in the field do not suggest that wetlands are selected on the basis of size for breeding, much less that small ponds produce metamorphs of higher fitness.

In Newfoundland, the presence of Green Frogs is negatively correlated with dissolved oxygen content (preferred DOC 5–10 mg/L), but positively correlated with pond permanence and the presence of human residences nearby (Campbell et al., 2004). The positive association between Green Frogs and human occupation, however, likely reflects human-mediated dispersal in this area. In Ontario, Hecnar and M'Closkey (1996b) could find no appreciable differences in pond water chemistry between occupied and unoccupied sites.

Brodman (2009) noted considerable variation in annual occupancy over a 14 yr study in Indiana.

REPRODUCTION

The breeding period of Green Frogs is prolonged throughout the spring and summer at many locations (but see Berven et al., 1979), rather than occurring over a short period of days to weeks as in many other anurans. Literature records of breeding dates include the following: March to July (northern Louisiana: Meshaka et al., 2009a), March to September, with a peak of April and May (southern Louisiana: Meshaka et al., 2009b), April and May (Rhode Island: Anonymous, 1918), April to August (North Carolina: Gaul and Mitchell, 2007), May and June (Indiana: Brodman and Kilmurry, 1998), May to July (Maryland: Lee, 1973a; New Hampshire: Oliver and Bailey, 1939; Ohio: Walker, 1946), May to August (Michigan: Martof, 1956a, 1956b; Minnesota: Fleming, 1976; Oldfield and Moriarty, 1994; New Brunswick: Gorham, 1970; New England: Klemens, 1993; New York: Wright, 1914; Gibbs et al., 2007; Ontario: Piersol, 1913), May to September (Maryland and Virginia: Berven et al., 1979), June and July (Illinois: Cagle, 1942; Nova Scotia: Gilhen, 1984; Ohio: Varhegyi et al., 1998; Virginia: Berven et al., 1979; West Virginia: Rogers, 1999), June to August (New York: Wells, 1977b; Tennessee: Gentry, 1955).

It is likely that the actual breeding season (that is, when eggs are deposited) extends beyond or is less than some of the monthly records found in the literature, depending on weather conditions. Calling usually precedes mating by up to a month or so, which in turn affects perceptions of the timing of the breeding season. For example, calling begins in April in lowland sites in the Shenandoah Valley, Virginia, although breeding does not actually occur until mid-May (Berven et al., 1979). Many literature reports do not make the distinction between when calls are heard and when breeding actually takes place.

In addition to a latitudinal effect on breeding dates in such a widespread species, elevation also affects the amount of time available for reproduction. In lowland areas of Maryland and Virginia, for example, breeding is prolonged from mid-May to September. In the higher elevations of the Alleghany Mountains in Virginia, the time of egg deposition is much shorter, extending only between July and August (Berven et al., 1979).

The timing of the ovarian cycle likely varies according to region. In Louisiana, gravid females are observed early in the season and increase rapidly in spring and summer. Some females yolking eggs may be found at any time of the year, but advanced-stage gravid females are usually not found in any frequency in the autumn. It is at this time they begin acquiring lipids needed for the

Eggs of *Lithobates clamitans*. Photo: Dana Drake

eggs to be deposited the following breeding season, a process that will continue right up to egg deposition (Meshaka et al., 2009b). Fat reserves are usually depleted by July or August.

Clutch size may vary with phenotype (and hence latitude) and size of the female, but data are limited. Trauth et al. (1990) recorded clutch sizes of 4,924 and 5,730 in two Bronze Frogs, and 2,851 from a single Green Frog; Martof (1956b) recorded 3 clutches of 3,800, 4,100, and 4,300 eggs in Michigan; clutch size ranged from 1,401 to 5,289 in 10 Green Frog clutches from Nova Scotia (Gilhen, 1984); Wright (1932) recorded a single clutch of 1,451 eggs from a Bronze Frog in the Okefenokee Swamp; Meshaka et al. (2009a) estimated a mean of 2,550 eggs (range 1,600 to 4,200) in Louisiana. Small females may deposit as few as 1,000 eggs, perhaps reflecting regional variation in female size (Pope, 1964). Green Frogs often have two sets of eggs developing within an ovary, one set mature and the other in a much less advanced state of development. Martof (1956b) therefore suggested that maturation of eggs requires one full year prior to oviposition. Green Frogs normally oviposit one clutch per season, but double-clutching has been reported (Wells, 1976).

Eggs kept in cold water (<10°C) do not reach the gastrulation stage of development, but eggs kept at temperatures between 12.2 and 33.4°C develop normally. The upper limit of temperature tolerance for eggs is 35°C (Moore, 1939). Developmental rates are slower than most other ranids, including *L. sylvaticus, L. pipiens,* and *L. palustris,* and are directly correlated with water temperature. For example, eggs developing at 15°C require 238–287 hrs to reach Pollister stage 20 (see Pollister and Moore, 1937), whereas they require only 58–65 hrs to reach this stage at 25°C (Moore, 1939). Eggs hatch in three to six days, and at an earlier morphological stage than many other ranids (Moore, 1940).

LARVAL ECOLOGY

Green Frog larvae are generally among the most active tadpoles, and they often swim faster than the larvae of other ranid species, such as Wood Frogs and American Bullfrogs. Small larvae tend to reduce the amount of activity in the presence of predators (*Anax* dragonfly nymphs), more so than larger Green Frog larvae (see Community Ecology). Food supplementation does not affect swimming speed, but activity tends to decrease with the availability of food in this generally permanent water species (Anholt et al., 2000). Larvae do not maintain discrete spacing distances from one another under laboratory conditions, either to conspecifics or larvae of other species (Smith and Jennings, 2004). This suggests that intraspecific competition plays no role in tadpole dispersal patterns within aquatic habitats.

The length of the larval period varies according to environmental conditions and elevation, with many reports from northern populations indicating an extended time spent as larvae. In contrast, Meshaka et al. (2009b) suggested only a two-month larval period in southern Louisiana. When conditions permit, tadpoles may be found year-round (Liner, 1954), and a single pond may have different sized larvae at various developmental stages, depending on when eggs were deposited (Berven et al., 1979; Rogers, 1999; Meshaka et al., 2009b). Under stressful conditions, for larvae developing in warm water at low elevation, or when eggs are deposited in early spring, metamorphosis occurs within the same season as egg deposition.

Transformation in the laboratory (Ting, 1951) and in the field can occur within the first (Martof, 1952, 1956a; Richmond, 1964; Berven et al., 1979) to the second or even third summer. Tadpoles developing in the first summer often overwinter and metamorphose during the second summer. In Newfoundland, for example, metamorphosis occurs in July and August of the second summer (Bleakney, 1952). Some authors suggest an even more extended larval period; Anonymous (1918) stated that metamorphosis takes 2–3 yrs in Rhode Island, as did Babbitt (1937) in Connecticut. Berven et al. (1979) noted that Virginia larvae metamorphose two full years after egg deposition (that is, in the third summer season), as did Rogers (1999) in West Virginia.

Elevation also has important effects on developmental time, growth, and size at metamorphosis. In the Mid-Atlantic region, low-elevation larvae may complete metamorphosis within the same season if hatched early (before mid-July), whereas larvae hatching later in the season (August and September) will not transform until May or June of the following season, depending upon the hatching date the previous year. In montane ponds, overwintering a second season might be necessary; Berven et al. (1979) found that 21% of the Green Frog larvae hatched late in one year could not complete metamorphosis by September of the second summer, and hence overwintered a second time. The total extent of the larval developmental time period stretches from 300 to 670 days at high elevations, as opposed to the approximately 90 to 300 days for lowland Green Frogs. Presumably such extended larval periods also occur at more northern latitudes, thus accounting for some of the extended larval periods mentioned in the literature (Anonymous, 1918; Babbitt, 1937).

Green Frog larvae feed at higher rates at temperatures >23°C than they do at 20°C or less (Warkentin, 1992a), and thus feed more actively during the middle of the day than at night or in the morning, when temperatures are cooler. It is not surprising, then, that larvae exhibit a daily rhythm in their tolerance to high temperatures, a rhythm that mirrors daily water temperatures. Larvae acclimated to specific temperatures have a higher CTmax than larvae that are not acclimated, and this allows them to adjust rapidly to changing temperatures within a pond (Willhite and Cupp, 1982). Unlike temperature, illumination per se has no effect on larval feeding, a result that is not unexpected for a suspension feeder (Warkentin, 1992a).

Microhabitat choice plays an important role in larval feeding ecology, and likely on subsequent growth and the time to metamorphosis. During the day, tadpoles tend to concentrate in vegetated areas to feed, but at night some tadpoles, particularly larger tadpoles, move to deeper waters where feeding efficiency is not as great. Feeding rates are not different between vegetated and more open water areas during the daylight, and small tadpoles do not move to open areas at night (Warkentin, 1992b). These results suggest that feeding efficiency is not the only consideration in the choice of microhabitat selection. Perhaps larvae use the vegetated areas during the day because of increased cover, and with larger larvae moving to deeper waters at night to escape shoreline predators.

Variation in growth rates is genetically based, such that larvae are adapted to metamorphose as rapidly as possible given ambient temperature constraints (Berven et al., 1979). Larvae grow slower but longer at high elevations, and presumably in northern latitudes. Larval size is also affected by length of development. In montane ponds, larvae are consistently larger at any developmental stage than are low-elevation larvae, and thus juveniles resulting from high-elevation ponds are larger at metamorphosis than those from lower-elevation ponds. Berven et al. (1979) noted that size at metamorphosis and mean pond temperature were significantly negatively correlated under field conditions.

Stress can help speed up metamorphosis. Martof (1956a) noted that Green Frogs transformed within the first season (within 70–85 days of oviposition) in Michigan when the developmental pond dried up. Presumably, such metamorphs would be smaller than those that develop normally, and this may account for some of the extremes in variation observed in the size of metamorphs. For example, I found very small (15 mm SUL) metamorphic Green Frogs in shallow drying woodland pools in the Great Smoky Mountains National Park, yet large (35 mm SUL) metamorphs emigrating from a permanent pond about 20 km away. The large metamorphs likely originated from the previous year's reproductive cohort, whereas the small metamorphs were escaping from a currently desiccating pool (Dodd, 2004).

As a result of variation in larval period, the age of metamorphs and age at maturity will vary with the conditions under which larvae developed, even within a geographic region. Large spring-transforming metamorphs will be at least nine months older than small fall-transforming metamorphs, and they will probably reach sexual maturity faster after metamorphosis than small metamorphs. In addition, the larger juveniles should have greater survivorship than the smaller juveniles, especially as winter approaches. Still, transforming at a smaller size and delaying maturity may be less of a risk than remaining as larvae in a drying pond.

In situations where larvae develop in the absence of environmental stress, the size at metamorphosis is the same for those individuals completing development within a single season as those individuals that overwinter (Martof, 1956a). Size at metamorphosis may vary with latitude and elevation in accordance with develop-

Tadpole of *Lithobates clamitans*. Photo: David Dennis

Tadpole of *Lithobates clamitans* approaching metamorphosis. Photo: Joe Mitchell

mental pond temperatures, and the fact that tadpoles and metamorphs can be of greatly different sizes is often confusing to naturalists. Literature records of size at metamorphosis include 24–36 mm (mostly 26–30 mm) in West Virginia (Rogers, 1999), 28.4–36.3 mm (mean 32.6 mm) in Michigan (Martof, 1956a), 28–38 mm (mean 32 mm) in New York (Wright, 1914; also see Ryan, 1953), 28–39 mm (mean 33.2 mm) in Nova Scotia (Gilhen, 1984), 19–32.4 mm (mean 27.3 mm) in northern Louisiana (Meshaka et al., 2009a), and 19.6–47.0 mm SUL (mean 28.3 mm) in southern Louisiana (Meshaka et al., 2009b). In laboratory experiments, montane Green Frogs from Virginia transformed at means of 19.8 to 36.1 mm, depending on constant temperature raised, whereas lowland Green Frogs transformed at means of 22.2 to 46.3 mm (Berven et al., 1979). As noted above, frogs occasionally metamorphose at smaller sizes (Walker, 1946; Martof, 1956a; Dodd, 2004).

The percentage of individuals successfully completing metamorphosis varies considerably from year to year, depending upon pond and weather conditions. In Ontario, Shirose and Brooks (1997) recorded between 32.5 and 85.3% of larvae transforming successfully over a 6 yr period. In Missouri, 2 breeding ponds produced 3,206 metamorphs from 86 females (Hocking et al., 2008). Temporary breeding sites may occasionally dry before metamorphosis is completed, leading to mass mortality (Tinkle, 1959) and, although occasionally used, such temporary sites are not preferred for reproduction.

DISPERSAL

Dispersal occurs rapidly after metamorphosis, and Schroeder (1976) found that nearly all frogs dispersed within 27 days after the first frogs transformed. However, Martof (1956b) suggested that nearly all frogs dispersed from a Michigan pond within a week of transformation. Dispersal from a West Virginia site occurred primarily from late June to early August (Rogers, 1999), and from Missouri ponds as late as mid- to late October (Hocking et al., 2008). Unlike American Toads and Wood Frogs, Green Frogs show no special changes in activity metabolism or aerobic capacity immediately prior to dispersing from the natal pond (Pough and Kamel, 1984).

Juveniles do not take any particular route, but disperse in all directions from the breeding pond (Schroeder, 1976). Dispersal pathways take advantage of topographical features, however, such as stream drainages and other wetlands interspersed throughout the landscape. Most of the dispersal at Martof's (1953a) study site occurred upstream away from the natal site. A few frogs may not disperse at all, but remain in close proximity to the pond in which they developed. Most frogs dispersed 183–448 m from Schroeder's (1976) Virginia pond, but several frogs moved up to 4.8 km from the breeding pond. Other long-distance movements are 457 m (Oldham, 1967), 600 m (Martof, 1953a), and 560 m (Lamoureux and Madison, 1999). Displaced frogs do not have a tendency to home toward their original capture site if moved long distances, but they often move to streams and creeks where overwintering will occur.

In Louisiana, juveniles are found in terrestrial situations throughout the late winter and early spring (Liner, 1954). Liner (1954) stated, "sometimes the woods seem to be covered with them." At a 325-m^2 pond in Virginia, Schroeder (1976) estimated there were 512 newly transformed Green Frogs dispersing at about the same time. In Michigan, Martof (1956b) estimated 10,500 newly transformed Green Frogs within a 6,090 m^2 area. In contrast, 106 females produced only 2,214 metamorphs over a 16 yr period at a 1 ha Carolina Bay in South Carolina (Semlitsch et al., 1996); substantial numbers of metamorphs were produced in only 2 of the 16 years. This latter site did not fill with water in a number of years, and the numbers highlight the disparity in reproductive success of *L. clamitans* at temporary vs. permanent breeding sites.

Green Frogs disperse randomly, and take up residence at ponds, streams, and wetlands they encounter along the way. This means that frogs originating from one particular pond may be found at virtually any other pond within an area and ensures colonization of a wide variety of breeding sites and a fairly panmictic population. Because of this extensive dispersal, there is no reason to suggest Green Frog populations are structured as metapopulations either locally or regionally (Smith and Green, 2005). Instead,

Adult *Lithobates clamitans.*
Photo: David Dennis

colonization and extinction estimates tend to be constant through time, except that wetlands with longer hydroperiods tend to have lower extinction probabilities than wetlands with short hydroperiods (Mattfeldt et al., 2009). Mortality of juveniles is quite high during the dispersal phase.

DIET

Green Frogs feed opportunistically year-round when weather conditions permit. The greatest amount of feeding occurs in the spring, and adult females consume more food during the breeding season than adult males. Juveniles tend to forage more in the winter than the adults. Aquatic organisms make up about one-third of the diet of both adults and juveniles. The most important prey are lepidoptera (moths, including caterpillars), beetles, and snails and slugs, but spiders and flies also are eaten year-round, depending on availability. Prey include both ground and water beetles, lymnaeid snails, crane flies, millipedes, katydids, grasshoppers, dragonflies and dragonfly nymphs, wasps, ants, and wolf spiders. Green Frogs are visually oriented predators that have been observed to stalk crayfish from a distance of 12 m (Hamilton, 1948).

In winter, amphipods become the most important prey, at least in Illinois (Jenssen and Klimstra, 1966). These latter authors also found Southern Cricket Frogs (*Acris gryllus*) and a small amount of plant material, perhaps inadvertently ingested, eaten by some frogs. Gilhen (1984) recorded Spring Peepers (*Pseudacris crucifer*) in the diet, and a Green Frog adult eating a conspecific tadpole. DeGraaf and Nein (2010) recorded a hatchling spotted turtle (*Clemmys guttata*) eaten by an adult Green Frog, and Vergeer (1948) reported a Green Frog eating a jumping mouse (*Zapus hudsonius*). While the general prey items are similar in different areas of North America for both adults and juveniles (Surface, 1913; Hamilton, 1948; Whitaker, 1961; Jenssen and Klimstra, 1966; Stewart and Sandison, 1972; Gilhen, 1984; Werner et al., 1995; Forstner et al., 1998; Rogers, 1999), the frequency and volume of the prey taken change with availability and local invertebrate faunas. Regardless of frog size, frogs select the largest prey available that they can ingest.

Tadpoles feed on a variety of diatoms and filamentous algae, with blue-green algae, protozoa, and microcrustaceans occasionally consumed (Munz, 1920; Jenssen, 1967). These items are grazed or picked up in muddy sediments from the pond or lake bottom. Larvae also have been observed eating the eggs of Wood Frogs (Jennette, 2010). Feeding occurs throughout the overwintering period. As a tadpole metamorphoses, epidermis is found within the gut, but feeding does not commence until the forelegs are well developed and the body measures greater than approximately 32 mm. It is then possible to find a wide variety of small invertebrates within a juvenile's stomach, especially terrestrial crustaceans, land hemiptera (true bugs), spiders, flies, and beetles (Munz, 1920); it appears that juveniles ingest just about any small animal they can. Rogers (1999) found that insects comprised 93% of the 341 prey found in 65 metamorph stomachs; nearly all prey were of terrestrial origin.

PREDATION AND DEFENSE

The best defensive attribute of Green Frogs is their protective coloration, which allows them to blend in with both terrestrial and emergent aquatic vegetation. Simply by sitting quietly they avoid detection. The most common active defense used by Green Frogs at the approach of a predator is to jump into the nearest body of water and give a short high-pitched cry that tends to startle or disorient the would-be attacker. As a result, the predator tends to lose track of the frog's entry into water, and the frog quickly swims under water and buries into submerged debris and leaf litter. Alternatively, a frightened frog will swim sharply to one side and reemerge a few meters away, where it will sit motionless blending in with pond vegetation (Bragg, 1945a).

Juvenile Green Frogs will also crouch and cease to move upon the approach of snakes, which makes them less likely to be detected and attacked (Marchisin and Anderson, 1978; Heinen and Hammond, 1997). They also retain the tendency to be immobile and crouch for a considerable time after the threat has left, thus making them less likely to be detected should the predator backtrack. When contacted, they will usually rapidly hop two to five times during the escape attempt. Body inflation also has been reported in both adults and juveniles; this allows the frogs to appear larger and perhaps more formidable than they really are (Marchisin and Anderson, 1978). Based on predator trials, it does not appear that

Lithobates clamitans is distasteful or noxious to vertebrate predators (Formanowicz and Brodie, 1979). Likewise, the eggs of *L. clamitans* are highly palatable to fish (Licht, 1969a).

The mosquito *Culex territans* feeds on Green Frogs and even can use the frog's call to locate its next blood meal (Bartlett-Healy et al., 2008). Other predators of adults include frogs (*Lithobates catesbeianus*), turtles (*Chelydra serpentina, Emydoidea blandingii*), snakes (*Agkistrodon piscivorus, Nerodia sipedon, Thamnophis sirtalis, T. sauritus*), crows (*Corvus*), herons (Ardeidae), and hawks (Wright, 1932; Richmond, 1952; Martof, 1956b; Werner et al., 1995; Birchfield, 2002). Mammals such as mink, otters, and raccoons undoubtedly take a considerable toll (e.g., Lamoureux and Madison, 1999). Large *Lithobates clamitans* also will consume small *L. clamitans*. Larvae are eaten by predaceous diving beetles (*Dytiscus*), water beetles (Hydrophylidae), whirly gigs (Gyrinidae), adults and nymphs of giant water bugs (Belostomatidae), water scorpions (Nepidae), backswimmers (Notonectidae), dragonfly naiads (Aeschnidae, Libellulidae), and turtles (*Glyptemys insculpta*; Ernst, 2001). Catfish (*Ictalurus*) and bass (*Micropterus, Ambloplites rupestris*) prey upon both tadpoles and small frogs, although Green Frogs are often found in ponds and lakes with fish predators (Hecnar, 1997; Hecnar and M'Closkey, 1997b; Babbitt et al., 2003). The larvae may be somewhat unpalatable to certain predators, such as some species of sunfish (*Lepomis*) (Kats et al., 1988; Szuroczki and Richardson, 2011), although larvae are eaten by Rusty Crayfish, Bluegill Sunfish, and Grass Carp under experimental conditions (Ade et al., 2010). Eggs are eaten by four species of ostracods (Gray et al., 2010).

POPULATION BIOLOGY

Juvenile Green Frogs grow rapidly following metamorphosis, regardless of whether it occurs early or late in the season. In Michigan, the highest growth rates occur in midsummer, when juveniles grow from 0.17 to 0.29 mm per day. Growth rates in spring and autumn are much less, from 0.03 to 0.09 mm per day (Martof, 1956a). Some adult frogs appear to stop growing before others, so that two frogs of the same age may be of very different sizes. As might be expected, growth rates are fastest for frogs in their first summer after metamorphosis (mean 33.6 mm/year), declining to 17.8 mm/year in the 60–70 mm adult size class. Growth still occurs at sizes greater than 70 mm SUL, but it decreases to only 2.1 mm/year in frogs greater than 90 mm SUL (all data for Michigan population; Martof, 1956a). Martof (1956a) concluded that it would take a Michigan Green Frog 4–5 yrs following metamorphosis to reach its maximum size (103 mm in males, 105 mm in females). Growth also occurs rapidly in New York Green Frogs. Examples include 40–78 mm in 13 months, 44–74 mm in 12 months, 40–70 mm in 5 months, and 40–57 mm in 3 months (Ryan, 1953). Growth patterns appear similar, despite the geographic differences.

Most males in Michigan reach sexual maturity prior to 70 mm SUL (perhaps as small as 58.6 mm); the smallest female with eggs measured 65.7 mm (Martof, 1956a). In New York, the smallest calling male was 63 mm, whereas the smallest female with eggs was 71 mm (Wells, 1977b). Another estimate is 52 mm for males and 58 mm for females (Wright and Wright, 1949). Since females require a full year for the maturation of eggs, at least in Michigan (Martof, 1956b), females do not oviposit the first year of their maturity. Otherwise, in most populations maturation occurs the first year after transformation.

In the Deep South, Bronze Frogs appear to have a rather different life cycle. Meshaka et al. (2009a) recorded mature females at 56.7 mm 7 months after metamorphosis in northern Louisiana, and suggested sexual maturity was reached after only 4 months at a SUL of 45.2 mm. In southern Louisiana, Meshaka et al. (2009b) estimated maturity in males occurs after only 3 months at 39.9 mm SUL and that they attain their mean adult body size (56.8 mm SUL) after only another 3 months. The smallest females in southern Louisiana reached maturity at 4 months at 43.1 mm SUL and attain their mean adult body size (59.8 mm SUL) after only 2 or 3 more months.

COMMUNITY ECOLOGY

When calling, *Lithobates virgatipes* and *L. clamitans* form mixed-species aggregations along a shoreline. There is a slight degree of habitat partitioning when calling, with *L. clamitans* calling from vegetation at the shoreline and *L. virgatipes* calling from locations 0.1–0.5 m into the wetland

(Given, 1990). Male *L. clamitans* are capable of displacing male *L. virgatipes* from calling sites. When choruses are mixed, *L. virgatipes* tend to associate with conspecifics rather than the larger and heavier *L. clamitans*; *L. clamitans* shows no such preference. Florida *L. clamitans* and the closely related *L. okaloosae* also tend to be clumped in distribution, showing a positive degree of interaction. Both species selected similar calling positions and appeared to exclude neither conspecifics nor heterospecifics; interfrog distances ranged from a mean of 6.5 to 9.2 m (Gorman et al., 2009).

Adult Green Frogs, Northern Leopard Frogs (*L. pipiens*), Mink Frogs (*L. septentrionalis*), and American Bullfrogs (*L. catesbeianus*) are often found occupying the same general nonbreeding habitat, that is, along the shoreline of larger lakes and ponds. These four species tend to select slightly different microhabitats, presumably because of competition, overlap in prey size and composition, and the threat of predation by the large bullfrogs on the smaller ranids, particularly subadults (Courtois et al., 1995). For example, Green Frog diets in New York overlapped by 63% with Mink Frogs and 24% with American Bullfrogs (Stewart and Sandison, 1972).

Green Frogs likely compete with Northern Leopard Frogs and to a lesser extent with American Bullfrogs for prey. In turn, Green Frogs are preyed upon by the larger ranid. For example, Northern Green Frogs increased fourfold in an Ontario park after bullfrogs became extinct, suggesting either predation or competition by bullfrogs had been keeping populations of *L. clamitans* in check (Hecnar and M'Closkey, 1997c). In aquatic situations where they overlap spatially, Green Frogs occupy less dense vegetation than Northern Leopard Frogs, and they occupy microhabitats closer to shore, of lower water temperature, and in areas with a higher vegetation canopy than American Bullfrogs. American Bullfrogs breed in the more open water, whereas Green Frogs breed closer to the shoreline. Green Frogs also select areas close to the shoreline allowing rapid escape, and in less vegetated areas than Northern Leopard Frogs (McAlpine and Dilworth, 1989), when on land. Mink Frogs, however, are usually far off in deeper waters associated with floating lily pads. Thus, the four species partition open water and lake shorelines allowing for coexistence despite overlap in diet and habitat occupancy.

Since adults of these species occupy the same lakes or ponds, their larvae often overlap spatially within aquatic habitats. Both Green Frog and bullfrog tadpoles may overwinter and, as a result, various size classes of both species often are present at the same time. In experimental trials, competition is strong between the larvae of these species, except that neither small nor large Green Frog tadpoles have any effect on small American Bullfrog tadpoles (Werner, 1994). In general, American Bullfrogs are better competitors because of their greater activity levels; they have significant effects on resources as large individuals, as well as significant impacts as smaller larvae because of their great overall biomass.

Green Frog larvae have the ability to respond morphologically to predators, at least in experimental trials. When raised with mudminnows (*Umbra*), they have shorter and narrower tail muscles and wider bodies than they do in the absence of these fish; with dragonfly larvae (*Anax*), they also had shorter tails and wider bodies; with salamanders (*Ambystoma*), they had shorter and shallower tails and wider and longer bodies; and with predacious water bugs (*Belostoma*), they had shorter and shallower tails and deeper and wider bodies (Relyea, 2001b). These phenotypic responses to different predators suggest that tadpoles respond to perceived threats by altering locomotor and/or vulnerability traits (e.g., by altering the predator's ability to handle the prey or the extent of tail fin exposed to attack). Whether these responses occur in nature, and whether they actually confer advantages in survivorship in mixed-species assemblages, has yet to be demonstrated. In general, Green Frog tadpoles do not respond to overall predation risk, but to predator-specific situations (Relyea, 2001a).

Larval behavior also changes in the presence of aquatic predators (*Anax*). In laboratory trials, Green Frog larvae greatly reduced their activity and shifted the area occupied away from the caged predator (Werner, 1991; Relyea and Werner, 1999; Peacor and Werner, 2000; Fraker, 2010). This response reduced the growth rate of the tadpole and resulted in a much smaller animal than the American Bullfrog larvae, whose size and growth rate were comparable in the absence of the predator (Werner, 1991).

The presence of the caged predator *Anax* also reduced the activity of small Green Frog larvae but not large larvae, at low densities but not at high densities (Peacor and Werner, 2000). Large larvae presumably could avoid or reduce predation chances by size alone, regardless of density. Small larvae, however, had better chances of surviving when they were in a dense group of conspecifics, and thus did not change their activity levels. Small larvae at low densities would be particularly vulnerable to predation, and thus reduced their activity when predator odors were detected. In response to bluegill sunfish (*Lepomis macrochirus*), no reduction in activity or spatial avoidance occurred in experimental trials, and only moderate changes in activity and avoidance occurred in the presence of mudminnows (*Umbra*) (Relyea and Werner, 1999). Taken together, these results suggested that predator avoidance and activity responses in Green Frog larvae are predator-, stage-, and density-specific.

In contrast, Smith et al. (2010) noted that larvae responded similarly (by reduced activity) whether predator cues (from odonate larvae, mosquitofish [*Gambusia*]) were present singly or in mixed assemblages. Based on sibship experiments, these authors suggested that variation in response to predators might be genetically based. In particular, variation in larval reaction occurred to nonnative *Gambusia* and suggested an evolutionary response to an invasive species. As regards many such experiments, the context of the predation experiment becomes important in interpreting the results.

In addition to larval responses to predators, there is some evidence that eggs respond to chemical cues of predators. In laboratory experiments, eggs exposed independently to water from an egg predator (crayfish, *Procambarus*) and a tadpole predator (*Anax* naiad) had a decreased time to hatching compared with eggs reared in predator-free water. These results suggest that the eggs respond to a generalized predator cue, rather than to a life-stage specific predator cue (Anderson and Brown, 2009). Hatching success also decreases in the presence of crayfish, even though there is no direct contact.

DISEASES, PARASITES, AND MALFORMATIONS

The bacterial disease "red-leg" (*Aeromonas hydrophila*) has been reported to kill Green Frogs in Michigan (Martof, 1956b) and Québec (Bonin et al., 1997b). Frog Virus 3 (a *Ranavirus*) has been found in Green Frogs, particularly in ponds with access by cattle (Gray et al., 2007a; Schmutzer et al., 2008). Ranavirus has been recorded from *Lithobates clamitans* in Maine, Minnesota, and New York (Green et al., 2002; Gahl, 2007; Brunner et al., 2011). They are infected frequently by amphibian chytrid fungus (*Batrachochytrium dendrobatidis*). In Maine, for example, 17.2% of the frogs examined (34 of 197) were infected in the toe webbing, 8.3% (3 of 36) in the skin of the tibia, and 12.8% (22 of 172) on the skin of the pelvis (Longcore et al., 2007; see also Gahl, 2007), whereas in Québec prevalence was 20% for 718 individuals collected from 1900 to 1999 (Ouellet et al., 2005a). The earliest recorded case was from a Green Frog collected in 1961 at Saint-Pierre-de-Wakefield. Amphibian chytrid also occurs on Green Frogs from the upper Midwest, New Hampshire, Massachusetts, and Georgia (Timpe et al., 2008; Sadinski et al., 2010; Tupper et al., 2011). Fungal *Ichthyophonus*-like spores were found on 17 of 83 Green Frogs in Québec; these spores caused mild to severe myositis (Mikaelian et al., 2000). The significance and prevalence of this disease and its long-term effects are unknown. It has also been found on Green Frogs from Vermont (Green et al., 2002), Maine (Gahl, 2007), and New Jersey (Monsen-Collar et al., 2010). Finally, a *Dermocystidium*-like fungus is reported from Green Frogs in Minnesota (Green et al., 2002).

Green Frogs are parasitized by >13 species of nematodes, 36 species of trematodes (of which 12 affect larvae), 3 cestodes, a copepod on tadpoles, mites, 3 species of flies, and 32 species of protozoans (Walton, 1947; McAlpine, 1997b; McAlpine and Burt, 1998). Intraerythrocytic parasites include *Aegyptianella ranarum*, *Hepatazoon* sp., *Lankestterella minima*, *Trypanosoma rotatorium*, and *T. ranarum* (Bonin et al., 1997b), and viruses (Desser and Barta, 1984a; Gruia-Gray et al., 1989). A pathogenic protozoan (*Brugerolleia algonquinensis*) was described from Green Frogs in Canada (Desser and Jones, 1985; Desser et al., 1993); the ciliate protozoan *Nyctotherus cordiformis* was reported from Ohio and Virginia Green Frogs (Odlaug, 1954; Campbell, 1968); other protozoans include *Entamoeba ranarum*, *Haematogregarina* sp., *Hexamita intestinalis*, and

Tritrichomonas augusta (Campbell, 1968). The myxozoan *Myxidium serotinum* also parasitizes *Lithobates clamitans* (McAllister et al., 2008). A virulent alveolate pathogen affecting larval Green Frogs was reported from Maine (Gahl, 2007).

Trematode cysts are common in the kidneys (Martin and Conn, 1990), and Gilhen (1984) noted the presence of unidentified nematodes in the stomach. Species include *Alaria arisaemoides, Apharyngostrigea pipientis, Cephalogonimus americanus, C. vesicaudus, Clinostomum attenuatum, Echinostoma trivolvis, Glypthelmins quieta, Gorgodera amplicava, Gorgoderina attenuata, G. simplex, G. subtropica, G. tanneri, G. translucida, Gyrodactylus* sp. (on larvae), *Haematoloechus breviplexus, H. longiplexus, H. parviplexus, H. similiplexus, H. varioplexus, Haematoloechus* sp., *Halipegus eccentricus, H. occidualis, Hichinastomia trivolvis, Loxogenes arcanum, Loxogenoides bicolor, Megalodiscus temperatus, Pneumobitis parviplexus,* and *Pneumonoeces parviplexus* (Irwin, 1929; Bouchard, 1951; Odlaug, 1954; Najarian, 1955; Campbell, 1968; Goater et al., 1990; Dyer, 1991; Wetzel and Esch, 1996; McAlpine, 1997b; Bursey and DeWolf, 1998; McAlpine and Burt, 1998; Zelmer et al., 1999; Green and Dodd, 2007). The cestode *Ophiotaenia saphena* is common in Green Frogs (Sutherland, 2005), as are *Cylindrotaenia americana, Mesocestoides* sp., *Ophiotaenia gracilis, O. saphenus,* and *Proteocephalus saphena* (Williams and Taft, 1980; Dyer, 1991; McAlpine, 1997b; Bursey and DeWolf, 1998; McAlpine and Burt, 1998). Nematodes include *Cosmocercoides dukae, C. variabilis, Falcaustra inglisi, Foleyella americana, Gyrinicola batrachiensis, Oxysomatium variabilis, Ozwaldocruzia pipiens, O. waltoni, Physaloptera* sp., *P. ranae, Rhabdias ranae,* and *Spiroxys contortus* (Odlaug, 1954; Campbell, 1968; Adamson, 1981c; Dyer, 1991; McAlpine, 1997b; Bursey and DeWolf, 1998; McAlpine and Burt, 1998). Acanthocephalan larvae are reported from the mesenteries of the body cavity (Odlaug, 1954), and Campbell (1968) recorded the species *Centrorhynchus* sp. and *Centrorhynchus wardae* from Virginia. The mite *Hannemania penetrans* is known from this species in Virginia (Campbell, 1968). Martof (1956b) observed that 80% of the male Green Frogs at Michigan ponds were infested by leeches, especially on the webs of the hind feet, and leeches (*Desserobdella picta*) are found commonly on Canadian Green Frogs (Barta and Desser, 1984).

Green Frogs with malformations, including hind feet of abnormal shape or proportion, missing feet or partial limbs (ectromelia), missing front legs, missing digits (ectrodactyly), missing eyes, jaw deformities, and even a missing tympanum, have been found in many locations (Martof, 1956b; Bonin et al., 1997b; Ouellet et al., 1997; Converse et al., 2000; Helgen et al., 2000; Hoppe, 2005; Lannoo, 2008). A photograph of an *Lithobates clamitans* with multiple forelimbs is in Meteyer (2000). In Minnesota, 80 of 219 frogs examined had some type of malformation (Hoppe, 2005). Most malformations affected the hind limbs at Martof's study site in Michigan, and 23 of 428 (5.3%) frogs had missing body parts or exhibited abnormal development. Most injuries, however, affected digits or limbs and appeared to represent unsuccessful predation attempts. The Rock Bass (*Ambloplites rupestris*) probably was responsible for many of the predation attempts (Martof, 1956b).

In Québec, malformations and abnormal DNA profiles were associated with agricultural activity, and agricultural contaminants (especially carbofuran) were suspected as the cause (Bonin et al., 1997b; Lowcock et al., 1997; Ouellet et al., 1997). The levels and types of abnormalities were consistent with acute or cumulative pesticide toxicity. Exposure to coal ash contaminated waste water also results in larval malformations (Snodgrass et al., 2004). In addition to malformations, Green Frogs infected with ranavirus show higher levels of fluctuating asymmetry than uninfected frogs suggesting that diseases can have sublethal effects on development (St.-Amour et al., 2010). Frogs infected by amphibian chytrid were not asymmetrical.

In addition to external malformations, developmental anomalies are found in Green Frogs from urban and residential areas in higher frequencies than in natural populations. Such malformations involve the development of intersexes in males, with males from the Connecticut River Valley having gonads containing testicular oocytes. Up to 21% of male Green Frogs from suburban landscapes had abnormal testicular development; frogs from nearby agricultural areas had low levels of abnormalities (Skelly et al., 2011). It seems reasonable to hypothesize that gonadal abnormalities are associated with the high levels of pesticides used in residential areas.

SUSCEPTIBILITY TO POTENTIAL STRESSORS

Metals. In a study of lead in tadpoles and their habitats near highways, lead concentrations ranged between 0.90–240 mg kg^{-1}. Lead concentrations in tadpoles were positively correlated with the amount of lead in sediments and mean daily traffic volume. It is not known whether these concentrations adversely affect Green Frog larvae, although similar concentrations have been associated with adverse physiological and reproductive effects in other vertebrates (Birdsall et al., 1986). In addition, cadmium (0.1–0.19 ppm wet weight), copper (0.93–1.2 ppm), lead (1.4–1.5 ppm), mercury (0.04–0.1 ppm), zinc (3.7–6 ppm), magnesium (14–29 ppm), and manganese (1.1 ppm) have been reported from Green Frog tadpoles (Hall and Mulhern, 1984). Mercury has been found in Green Frog tadpoles in natural ponds in Maine, with a mean of 25.1 ng/g; methyl mercury comprised 7.6–40% of the total (Bank et al., 2007). In a series of mine-restored and seminatural areas in Ohio, metal concentrations (means in ppm) ranged from 41 to 113 for manganese, 289–423 for iron, 59–90 for aluminum, and 91–112 for zinc (Lacki et al., 1992). The presence of Green Frogs in Ontario is negatively correlated with aluminum concentrations (Glooschenko et al., 1992).

Under laboratory conditions, Green Frog larvae accumulate heavy metals (arsenic, cadmium, iron, selenium, strontium, and vanadium) when raised in water contaminated with coal combustion wastes. Exposure to polluted water decreased growth and developmental rates, reduced survivorship by 26%, and decreased metamorphic success by 45%. Furthermore, the duration of the larval period increased, and the size at metamorphosis decreased by 10% (Snodgrass et al., 2004). These results suggest severe effects on amphibian larval development in water polluted by coal ash power plants.

pH. Higher pH values are positively associated with Green Frog abundance (Mazerolle, 2003). The lethal pH for Green Frog embryos is 3.7–3.8, with a critical pH of 4.1 (Gosner and Black, 1957a; Freda and Dunson, 1986; Freda et al., 1991). Larval Green Frogs are generally unable to acclimate to low pH (≤5.0). Tadpoles are unable to initially regulate ionic (Na^+, Cl^-, Ca^{2+}, K^+) losses at low pH (Freda and Dunson, 1984), but they slowly recover after 7 hrs; juveniles do not, however, and sodium uptake is reduced by 60%. Chronic ion losses occur primarily in the gills of developing larvae. Thus, chronic acid exposure does not allow larvae to adapt to low pH (McDonald et al., 1984).

There may be some regional variation in pH tolerance, as adults have been found in Nova Scotia in waters of pH 3.5, and larvae at pH 3.9 (Dale et al., 1985); most frogs were found in waters with higher pH content (3.88–7.26, mean 4.95). Freda and Dunson (1984) stated that Green Frog larvae can live for "several weeks" in water at pH 3.5. In experimental trials, Green Frog larvae are able to detect and move away from water that is very acidic (Freda and Taylor, 1992; Vatnick et al., 1999). Green Frogs do not appear to be affected by soil type in experimental situations, although the chemical properties of soils may affect acidity within breeding ponds (Sparling et al., 1995). The presence of Northern Green Frogs increases in ponds with a high buffering status; buffering status is inversely correlated with acidity (Glooschenko et al., 1992).

Alkalinity. The presence of Northern Green Frogs is positively correlated with alkalinity concentrations in Ontario (Glooschenko et al., 1992).

Ammonia and Nitrates. As might be expected, larvae metamorphosing from wetlands with high nitrate content have higher concentrations of nitrates than larvae metamorphosing from wetlands with low nitrate content (Jefferson and Russell, 2008). Green Frog larvae are sensitive to concentrations of > 30 mg/L ammonium nitrate in acute exposure trials; the $LC_{50\ (96\ hr)}$ is 32.4 mg/L. Tadpoles exposed to higher concentrations of nitrate generally become sluggish, and they may forgo feeding. They do not appear to be sensitive to chronic exposure at exposures < 10 mg/L, and do not experience weight loss after 100 days when compared to controls (Hecnar, 1995).

When exposed to un-ionized NH_3 in excess of 0.6 mg/L, Green Frog larvae (at 20 days) had reduced hatching and survivorship, the presence of deformities increased (body curl, asymmetry, curled spine, short tails, deformed tails, abnormal tail fins), and growth and development were slow when compared to controls in both acute- and chronic-exposure laboratory trials (Jofré and

Karasov, 1999; Rosenshield et al., 1999; Jofré et al., 2000). Mortality at higher concentrations of NH_3 occurred after 20 days of chronic exposure. In combination with PCB 126, NH_3 significantly reduced hatching success and survivorship, but growth was not affected (Jofré et al., 2000).

NaCl. Lithobates clamitans appears to be tolerant of road deicing salts, with only a small reduction in survivorship at even high levels of conductivity (3000 µS). Still, Karraker (2007) found that 15% of larvae were malformed (mostly abdominal edema for hatchlings, dorsal, and lateral tail bends) at the highest conductivity levels; malformed embryos and larvae do not survive. Deicing salts also had no effect on the survival of eggs affected by water molds (Karraker and Ruthig, 2009). In contrast, both NaCl and $MgCl_2$ both negatively affected Green Frog larval survival in laboratory studies using deicer concentrations that might be expected in the field (Dougherty and Smith, 2006); the sample size (five per treatment) in this study was small.

Chemicals. The herbicide atrazine detrimentally affects Green Frog larval survival at 3 ppb for 30 days, but not at higher concentrations (30, 100 ppb); this is especially true for exposure of late-stage larvae (Storrs and Kiesecker, 2004). Although Hecker et al. (2003, *in* Hayes, 2004) reported no effects of atrazine on *L. clamitans*, there were many problems with this study, including contamination of the "control" by atrazine, extremely high mortality rates (80%) in the controls suggesting poor husbandry, and poor evaluation of the results (Hayes, 2004).

The $LC_{50\ (16\ day)}$ for malathion is 3.7 mg/L (Relyea, 2004b). As such, it should not have deleterious effects on *R. clamitans* since concentrations normally are much lower under field conditions. At concentrations of 1 mg/L malathion, larval activity decreases. Activity was not affected by the presence of caged water bugs or dragonfly naiads without pesticides, and malathion had no additional effect on the already reduced activity levels when larvae were in the presence of the newt *Notophthalmus viridescens* (Relyea and Edwards, 2010).

The glyphosate herbicide Vision™ (also marketed as Roundup™) induces malformations (abnormal face, eye, and gut development, tail damage) in larval Green Frogs under laboratory conditions, but it appears that the teratogenic effects result from an interaction between pH and the POEA (polyethoxylated tallowamine) surfactant of the herbicide mixture (Howe et al., 2004); toxicity effects become more severe at higher pHs (Edington et al., 2004). Hatching is also delayed at pHs of 4.5–5.0, about 24 hrs, when exposed to Vision™. Tadpoles exposed to the glyphosate-surfactant mixture are smaller than controls, indicating possible effects on growth. In field circumstances, Wojtaszek et al. (2004) and Thompson et al. (2004) could not demonstrate any long-term effects of Vision™ on Green Frogs when used under product label guidelines, although they noted mortality and abnormal avoidance responses at high concentrations, which would not be expected under normal environmental conditions. Response varied by site, particularly with regard to pH and the amount of suspended material in the water.

The carbamate insecticide carbaryl is lethal to Green Frog larvae at high concentrations (> 7.2 mg/L), and this response is temperature dependent. At 30 mg/L, carbaryl kills all larvae in about 19 hrs and significantly reduces larval activity at 2.5 mg/L (Bridges and Semlitsch, 2000). As temperatures increase, survival decreases. Larvae can tolerate higher concentrations at lower temperatures (Boone and Bridges, 1999). Carbaryl at concentrations of 0.1–1 mg/L reduces activity levels 6–10%, including when larvae are in the presence of predators (Relyea and Edwards, 2010). Paradoxically, sublethal amounts (3.5 mg/L) lead to increased survivorship and precocious metamorphosis; these outcomes may result from direct effects on hormones, or from indirect effects on the tadpoles' food supply (Boone et al., 2001). The presence of carbaryl at sublethal concentrations extends the larval period, but does not increase size at metamorphosis (Boone and Semlitsch, 2002). Thus, the metamorphs are smaller than they would be in the absence of the pesticide. Metamorphs from low-density ponds and larvae from high-density ponds were not affected by the presence of a predator, the carbaryl, or an interaction among them in pond-based experimental trials (Boone and Semlitsch, 2001).

The insecticide diazinon is toxic to *Lithobates clamitans* at an LC_{50} of 2.8–5.0 µg/L in commercial preparations. Other pesticides (< 0.01 ml/L Basudin™500EC, technical grade diazinon, Dithane™DG) adversely affected mortality, malformations, and growth rates, whereas Imidan™50WP, Guthion™50WP, and Nova™40W produced similar effects at 5–10 mg/L (Harris et al., 1998b). While some tadpoles metamorphosed in caged field-experiments when exposed to these pesticides, it is likely that chemical mixtures have sublethal effects that could add stressors to the species' larval ecology and development. Bromoxynil is toxic to larvae at 0.01 ppm, triallate and trifluralin at 8 ppm, and triclopyr ester at 4.8 ppm; lower concentrations of these pesticides often had little or nonlasting effects (Berrill et al., 1997).

Potentially sublethal effects of permethrin and fenvalerate (pyrethroid insecticides), affecting size, growth, and the prevalence of bent backs affecting swimming behavior, were observed at 1.0–2.0 mg/L for a 96 hr exposure (Berrill et al., 1993). Even exposure to lesser concentrations resulted in weak or paralyzed tadpoles; recovery was possible, but dependent on chemical concentration and temperature (Berrill et al., 1997). Low concentrations of fenitrothion had no effect on *L. clamitans* larvae, but high concentrations (> 4 ppm) resulted in malformations, paralysis, and mortality (Berrill et al., 1997).

In addition to these pesticides, Green Frog larvae are sensitive to temephos (Abate™) ($LC_{50\ (96\ hr)}$ at 4.24 µg/L; Sparling et al., 1997), triclopyr (100% mortality at 2.4 mg/L; Berrill et al., 1994), fenitrothion (generally unresponsive > 4.0 ppm; Berrill et al., 1994), and endosulfan (30–80% mortality 12 days postexposure; Berrill et al., 1998). They are not sensitive to hexazinone (Berrill et al., 1994). Larval Green Frog survivorship is not significantly reduced at concentrations of 9 mg/L of the insecticide imidacloprid (Ade et al., 2010).

Combinations of pesticides often have greater effects than might be expected based on single-chemical exposures. In experimental wetlands, a combination of 0.2 mg/L atrazine, 0.25 mg/L metolachlor, and 0.1 mg/L chlorpyrifos given in low or high herbicide/low insecticide combinations led to the extirpation of Green Frog tadpoles (Mazanti, 1999).

PCB 126 (3,3′, 4,4′, 5-pentachlorobyphenyl) does not cause embryo mortality at concentrations of 0–50 µg/L, and larval survivorship was reduced at 50 mg/L. Edema was common in tadpoles at this concentration, and growth was slowed. As PCB 126 concentrations increase, fewer larvae successfully complete metamorphosis (Jofré et al., 2000). In Michigan, Green Frogs had total PCB concentrations ranging from 11 to 568 ppb in adults to 200 to 826 ppb in larvae; there was a nonsignificant trend between decreased density and increased levels of PCB in sediments (Glennemeier and Begnoche, 2002).

A number of chemicals show increased toxicity during Green Frog metamorphosis, particularly those with high hydrophobicity values (DDT, PCBs, polybrominated diphenyl ethers, dioxins, furans) (Leney et al., 2006). This is especially true as tails are resorbed. As tadpoles undergo metamorphosis, lipid content is decreased, which in turn leads to increased blood and organ concentrations of toxicants. This increase will lead to heightened levels of toxicity.

UV light. Green Frog eggs and larvae experience substantially increased mortality when exposed to artificially high levels of UVB for periods of 15–60 minutes. Most larvae died, but even those that survived the trials never metamorphosed. Deformities included thick, pigmented corneas, increased melanin on the dorsum, and tail curvature (Grant and Licht, 1995). Posthatching larvae are much more sensitive than embryos because an embryo's jelly capsule protects it from adverse effects of UVB (Licht, 2003). Adverse effects of natural UVB on Green Frog embryos and larvae have not been demonstrated (Crump et al., 1999), although intense direct sunlight in shallow water, where tadpoles have no shade or cover, can reduce survivorship in experimental trials (Tietge et al., 2001). Green Frog eggs and larvae are not susceptible to UVA light (Grant and Licht, 1995).

STATUS AND CONSERVATION

Green Frogs are often one of the most abundant frogs throughout their extensive range. In certain areas, populations seem to have increased in abundance or extent. For example, the species was considered rare in northwest Indiana in the 1930s, but surveys in the 1990s in the same areas describe it as relatively widespread (Brodman, 2003). Records of Green Frogs in two North Carolina counties were greatly increased as a result of concentrated sampling (Rice et al., 2001). Past sampling biases undoubtedly influence perceptions of status, and it is likely that Green and Bronze Frogs are much more widely distributed than museum or literature records would suggest. In most areas, populations appear stable (Hecnar, 1997; McAlpine, 1997a; Shirose and Brooks, 1997; Weller and Green, 1997; Brodman and Kilmurry, 1998; Mierzwa, 1998; Mossman et al., 1998; Florey and Mullin, 2005). Cortwright (1998) suggested there was a slight decline in tadpole production at his Indiana sites from 1984 to 1994, but variances were large as might be expected. In Iowa, however, populations are reported as declining (Christiansen, 1981), and they are considered peripheral in Kansas (Platt, 1973; Collins, 1993).

Green Frogs can tolerate habitat disturbance and alteration better than many other frogs. They persist in mined bogs in Nova Scotia, although abundance is much higher in adjacent unmined sites (Mazerolle, 2003), and they tolerate terrestrial tree harvesting, including selective cutting and controlled burning, as long as suitable wetland sites are located nearby or interspersed within the landscape (McLeod, 1995; McLeod and Gates, 1998; Floyd et al., 2002; Russell et al., 2002b; Fox et al., 2004; Semlitsch et al., 2008). Green Frogs are frequently found in open habitats created during forestry operations, such as in canopy gaps and along skidder trails (Cromer et al., 2002). There is a tendency to move away from clearcuts (Semlitsch et al., 2008), but colonization of disturbed habitats is rapid (Russell et al., 2002b). They are also present on clearcut forested wetlands (Clawson et al., 1997), although it is difficult to assess long-term status and trends on such plots. Studies assessing silvicultural practices are usually conducted only for a relatively short time, and they often use metrics of presence or simple counts, rather than rigorous estimates of abundance or population demography, to assess treatment effects. The results of short-term studies must be used with caution.

Disturbances may have significant impacts on populations of Green Frogs. Following Hurricanes Ivan and Katrina in Louisiana, Green Frog numbers plummeted in swamp habitats, presumably due to saltwater intrusion (Schriever et al., 2009).

In contrast, it is not the amount of development per se that affects Green Frog abundance in developed areas, but the quality of available habitat. In Wisconsin, for example, lakeshore development decreased the quality of available habitat, which in turn decreased Green Frog abundance (Woodford and Meyer, 2003). Still, Palorski (2008) pointed out that even in this type of habitat, factors that affect Green Frog presence are complex, and it is not just the development alone that affects frog populations. Thus, conservation plans need to target crucial habitats and protect them within a development-based landscape, regardless of the extent of development.

Green Frogs may be detrimentally affected by agricultural activities, in addition to direct habitat loss. Larval and postmetamorphic abundances are less in ponds where cattle have access than in cattle-free ponds, and the presence of Frog Virus 3 is associated with ponds with cattle access (Gray et al., 2007a; Schmutzer et al., 2008; Burton et al., 2009). Many types and combinations of pesticides may be used in a region, and these, if not directly lethal, often have suboptimal effects (see Susceptibility to Potential Stressors). For example, in an Ontario study of apple orchards, pesticide use had mixed effects on larval development, depending upon concentration, stage of development, and type of pesticide used (Harris et al., 1998b). However, there was no evidence of a reduction in genetic variation due to the orchard environment, perhaps since animals were not site philopatric but extensively exchanging individuals with populations in nonorchard surroundings. Elevated levels of organochlorines indicated that frogs were accumulating pesticide loads at three of four orchard sites, and that suboptimal conditions were present at some of the orchard sites. Still, in an indication of how difficult it is to assess the effects of contaminants on frog populations, elevated levels of organochlorines and individual size differences of frogs among populations could not be conclusively demonstrated to have resulted from pesticide use at the orchards under study (Harris et al., 1998a).

The effects of nonindigenous species on Green Frogs have only recently been recognized. In field experiments, Green Frogs lost body mass in areas invaded by Japanese knotweed (*Fallopia japonica*) when compared to noninvaded areas, and Maerz et al. (2005a) suggested that this loss was due to decreased arthropod abundance in the invaded area.

Roads undoubtedly take a heavy toll on Green Frogs, especially during the breeding migration and juvenile dispersal. Roads per se are not barriers to Green Frog movements, as they read-

Unusually colored adult *Lithobates clamitans*, Bruce Peninsula, Ontario. Photo: C.K. Dodd Jr.

ily will attempt to cross both major and minor roadways (deMaynadier and Hunter, 2000). Few empirical data are available on mortality, however. Ashley and Robinson (1996) only recorded 57 dead *Lithobates clamitans* on a 3.6-km stretch of Ontario highway over a 4 yr period, and Glista et al. (2008) counted 172 road-killed Green Frogs on 4 survey routes over a 17-month period in Indiana. There is a weak negatively correlated effect of traffic density on Green Frog relative abundance within 300 m of ponds, and abundance tends to increase linearly the farther one moves away from a heavily traveled highway. Thus, roads may have an adverse effect (the "road-zone" effect) at considerable distance from the pavement (Eigenbrod et al., 2008, 2009).

Methods to mitigate Green Frog mortality on roadways could include the erection of drift fence barriers or by providing tunnels or culverts to facilitate crossing under the highway. Green Frogs prefer tunnels > 0.5 m in diameter that possess a soil or gravel-lined substrate with ambient light penetration. They prefer shorter tunnels over long tunnels. Fences 0.6–0.9 m in height also prevented trespass (Woltz et al., 2008).

Roads, however, also are used extensively to monitor Green Frog populations based on call surveys over prescribed routes at regular intervals (Bishop et al., 1997; Bonin et al., 1997a; Lepage et al., 1997; Mossman et al., 1998; Sargent, 2000; Nelson and Graves, 2004). The probability of hearing Green Frogs during an aural survey is usually good, assuming individuals are calling that night. In Ontario, de Solla et al. (2005) estimated that one could detect Green Frogs using aural surveys on only 3 sampling nights, and still be 80% certain of detecting the species' presence within the area of interest. Still, there is some degree of annual variation in the frequency with which Green Frogs are heard along standardized calling survey routes (Bishop et al., 1997). Green Frog abundance is correlated with the number of calls per minute, and abundance is positively correlated with calling index; these results suggest that calling index values are useful measures of abundance during road survey transects (Nelson and Graves, 2004).

Green Frogs are able to use wetlands found under power-line rights-of-way, as long as forested habitat abuts the transmission corridor (Fortin et al., 2004b). They may colonize created or restored wetlands, as long as there are source populations nearby (Lacki et al., 1992; Briggler, 1998; Stevens et al., 2002; Touré and Middendorf, 2002; Weyrauch and Amon, 2002; Brodman et al., 2006). In contrast, Lehtinen and Galatowitsch (2001) did not find *L. clamitans* in ponds restored up to 20 months previously, although the restored ponds appeared to be in highly disturbed areas, including wetlands that had not held water for 50 yrs. Palis (2007) also recorded only small numbers of Green Frogs colonizing three restored wetlands in southern Illinois, as did Pechmann et al. (2001) at four ponds in South Carolina. Merovich and Howard (2000) found that it took Green Frogs at least 4–5 yrs to colonize small constructed ponds, and even then frogs were not abundant. Taken together, these results suggest mixed effects of wetland restoration on this species. It seems likely that Green Frogs are more likely to become established at larger and more permanent created wetlands than at small ponds with varying hydroperiods.

Success or failure of Green Frogs to colonize small restored wetlands also undoubtedly depends on location within the landscape in relation to source wetlands and nearby breeding populations, as well as on the viability of populations that had been displaced. In some cases, it may be necessary to repatriate adults or larvae to establish a population. Repatriation has proven unsuccessful at the Gateway National Recreation Area in New York (Cook, 2008), despite initial indications of success (Cook, 1989).

Green Frog larvae may be moved throughout a landscape through purposeful or inadvertent anthropogenic mediation in connection with sport fish stocking. The Ohio Division of Conservation stocked Green Frog tadpoles obtained in fish hatcheries throughout the state (Walker, 1946), and this undoubtedly has occurred elsewhere. In terms of economic importance, Green Frogs have been a minor source of commercial frog's legs, and they have been harvested by biological supply houses for use in laboratories. A total of 32 *L. clamitans* were reported commercially collected for the pet trade in Florida from 1990 to 1994 (Enge, 2005a).

Lithobates fisheri (Stejneger, 1893) Vegas Valley Leopard Frog

ETYMOLOGY

fisheri: a patronym honoring Dr. A.K. Fisher (1856–1948), "in recognition of his share in the herpetological success of the Death Valley Expedition." Fisher was a medical doctor and ornithologist on the expedition, and likely captured many herpetological specimens.

NOMENCLATURE

Stebbins (2003): *Rana onca*
  Dubois (2006): *Lithobates (Lithobates) fisheri*
  Synonyms: *Lithobates chiricahuensis* (in part), *Rana onca fisheri*, *Rana pipiens fisheri*

IDENTIFICATION

Adults. The dorsal coloration is grayish to olive green with numerous small distinct dark greenish olive spots (in three to four rows) surrounded by lighter rings. Anteriorly, males tend to lose their dorsal spots and assume a bright green coloration, and females are more spotted than males. Dark olive spots extend along the sides of the body, which are grayish olive. The tympanum is large (greater than the distance between the nostrils and eyes) with no black patch. A dorsolateral fold, bordered by light stripes, is present but not strongly developed; it only extends about halfway down the back, making this and *L. onca* distinct from other Southwestern leopard frogs. The hind legs are relatively short, and the hind toes are two-thirds webbed. Along the front and rear of the hind limbs are many reticulations of deep olive to pale gray. The throat is light green and spotted, at least in smaller individuals, and the venter pinkish cinnamon. Ventral coloration of the hind limbs is honey yellow to buff. No external vocal sac is apparent. Males have an enlarged thumb. Males range between 44 and 64 mm SUL and females 46–74 mm SUL (Wright and Wright, 1949), suggesting sexual size dimorphism. Slevin (1928) recorded 6 specimens from 45 to 60 mm SUL (as *L. onca*). Additional specimens in collections are described by Wright and Wright (1949). The only known photographs of living specimens from the Vegas Valley are in Tanner (1931), Wright and Wright (1949), and Jennings and Hayes (1994a).

Larvae. Wright (1929) and Wright and Wright (1949) described the colorful tadpole as buffy olive to "dull citrine" with pale greenish-yellow clusters over it and a greenish-yellow tail that was heavily mottled. The venter is semitransparent, pure white to a faint cinnamon color. The tail is elongate with a rounded tip, "oil yellow" in color, and the TL reaches ca. 85 mm. Tadpoles have small black spots or mottling on the tail. Wright (1929) figured the mouthparts and has a lateral black-and-white photograph of a live tadpole.

Eggs. No information is available, but presumably they are similar to other members of the leopard frog complex.

DISTRIBUTION

Until recent genetic analysis of museum collections (Hekkala et al., 2011), this species was known only from the Vegas Valley in southern Nevada. Collections were made west of the then small town (Wright and Wright, 1949), about 1.6 km north of the town (Slevin, 1928), and in Tule Springs, 25.7 km north of Las Vegas (A. Vanderhorst collections; Jennings, 2005). Stejneger's and Slevin's collections were made at Las Vegas Ranch. This population is extinct. Specimens are in the collections of the Smithsonian Institution (USNM), the California Academy of Science, and the Museum of Vertebrate Zoology at the University of California–Berkeley. Hekkala et al. (2011) extended the range to include Arizona's Mogollon Rim, barely entering extreme western New Mexico. Many of these populations have been extirpated.

Distribution of *Lithobates fisheri*. The X indicates the disjunct and now extirpated population in the Vegas Valley.

FOSSIL RECORD

No fossils are known. Holman (2003) noted Miocene (Hemphillian) fossils from Navajo County in northern Arizona that belonged to the *Lithobates pipiens* complex. The relationship between these fossils and *L. fisheri* is unknown.

SYSTEMATICS AND GEOGRAPHIC VARIATION

Lithobates fisheri was described by Stejneger (1893) from the Vegas Valley of southern Nevada. In the description, he noted that other leopard frogs were found ("tolerably common") along Beaverdam Creek, near its junction with the Virgin River, but he could not determine whether they were *L. fisheri*. This taxon has been variously considered as a species, a subspecies, or as conspecific with *L. onca*. Hekkala et al. (2011) extracted DNA from preserved frogs collected from the Vegas Valley and determined that *L. fisheri* was conspecific with populations of leopard frogs from the Mogollon Rim heretofore considered *L. chiricahuensis*. Additional molecular research is under way to elucidate the relationships of the Mogollon Rim frogs with *L. chiricahuensis* and *L. fisheri*. *L. fisheri* is a member of the *Novirana* clade of North American ranid frogs.

ADULT HABITAT

According to Linsdale (1940), this species inhabited artesian springs and open sedge-bordered short streams in the vicinity of Las Vegas, at an elevation of 610 m. Wright and Wright (1949) recount a collecting trip to the Vegas Valley in 1925. *Lithobates fisheri* was found in a small spring outlet about 1.0–1.2 m across, covered in algae, and bordered by sedges. Other *L. fisheri* were found along small spring holes in areas that looked "marly and alkali" and in a number of springheads or well holes. Most locations appeared to be rather small, based on available descriptions. The largest populations were located at three large springs at the headwaters of Las Vegas Creek, which had a flow of nine million m^3 annually. Prior to 1938, riparian vegetation along Las Vegas Creek and surrounding springs included cottonwoods, willows, bulrushes, sedges, and cattails. The creek itself was reported to be 1–6 m in width with a sand, gravel, and mud bottom. Jennings and Hayes (1994a) provided historical photos of Las Vegas Creek, Kiel's Spring run, and another small spring in Las Vegas Valley, all taken in 1903. Along the Mogollon Rim, the species is found in Madrean evergreen woodland, pinyon-juniper conifer forest, and montane conifer forest habitats. Streams and associated plunge pools in rocky canyons constitute the preferred habitat for this species.

AQUATIC AND TERRESTRIAL ECOLOGY
Little is known concerning the ecology of this species. Vegas Valley frogs were said to be most active in spring and early summer, with capture dates from January to mid-August (Stebbins, 1962). It likely does not venture far from water, and activity may be curtailed from November to February depending on weather.

CALLING ACTIVITY AND MATE SELECTION
No information is available, but the behavior of *L. fisheri* may be similar to that of *L. chiricahuensis*. When handled, *L. fisheri* in Las Vegas made a "semicroak" or a "very low croak" (Wright and Wright, 1949).

BREEDING SITES
Presumably the species bred in the marshes, meadows, springs, and short streams that once were present in the Vegas Valley. These areas may have resembled the marsh and spring systems in Ash Meadows to the west. Photographs of spring and creek habitats are in Jennings and Hayes (1994a), but even by the time the photos were made, the habitats had already been highly modified. Breeding habitats along the Mogollon Rim includes stock tanks, ponds, and along slow-moving streams and pools.

REPRODUCTION
Lithobates fisheri probably breeds in spring (February to April), but no other information is available.

LARVAL ECOLOGY
Linsdale (*in* Wright and Wright, 1949) recorded transformation sizes of 28–30.5 mm. Slevin (*in*

Tadpole of *Lithobates fisheri*. Photo: A.H. Wright collection, Kroch Library, Cornell University

Plate 96. *Rana (Lithobates) fisheri*. 1 male, 2–4 females. From Handbook of Frogs and Toads by A.H. and A.A. Wright, 3rd ed., Cornell University Press, Ithaca, NY. Reprinted with permission.

Las Vegas River. Historic habitat of *Lithobates fisheri*, ca. 1903. Courtesy of Mark Jennings

Wright and Wright, 1949) found transforming froglets of 30–35 mm, some with remaining tail nubs, on 1 May 1913. Tadpoles also were observed on 20 August 1925 (Wright and Wright, 1949). If the species bred in late March and April, transformation likely occurred in summer. However, the observation of small frogs (ca. 30 mm SUL) in April and early May (reviewed by Wright and Wright, 1949) suggests a long larval period (perhaps even overwintering as larvae), little growth by metamorphs from summer transformation until the following spring (unlikely), or a rather extended breeding and larval period. Jennings (2005) suggested that metamorphosis could have occurred throughout much of the year.

DIET
Presumably, this species eats a variety of invertebrates found in the vicinity of springs and marshes.

PREDATION AND DEFENSE
Wright and Wright (1949) noted that *L. fisheri* jumped into the water when approached, but rested on the bottom and made no attempt to bury into the substrate. The species was easily captured.

POPULATION AND COMMUNITY ECOLOGY
No information is available.

DISEASES, PARASITES, AND MALFORMATIONS
No information is available, but parasites are probably similar to those of *L. chiricahuensis*.

SUSCEPTIBILITY TO POTENTIAL STRESSORS
No information is available.

STATUS AND CONSERVATION
This species is presumed locally extinct in the Vegas Valley, due to the destruction of its habitat during the expansion of Las Vegas in the 1940s. The water table was lowered due to groundwater pumping, and springs and wells were capped or destroyed. J.R. Slevin in 1913, C.L. Camp in 1923, A.H. Wright in 1925, J. Linsdale in 1934 and 1936, and Stanford biologists in 1934 and 1938 were able to locate springs and waterholes with *L. fisheri* and make collections (Wright and Wright, 1949). A. Vanderhorst also collected 13 frogs from Tule Springs in January 1942 (Jennings, 2005). On a return trip in May 1942, A.H. Wright had difficulty locating sites and only heard a few frogs splash into the water. He noted severe habitat modification, as well as the presence of

Spring in the Las Vegas Valley, ca. 1903. Historic habitat of *Lithobates fisheri*. Courtesy of Mark Jennings

American Bullfrogs (*L. catesbeianus*) and crayfish. By this time, 13 km long Las Vegas Creek had been heavily polluted. He expressed grave concern for the survival of this species and, unfortunately, his concern proved prophetic. If it was *L. fisheri* that jumped into the water during his 1942 collecting trip, it was the last recorded observation of this species in the Vegas Valley. Robert Stebbins attempted to find this species in 1949 without success and noted the loss of virtually all riparian vegetation by 1938. Populations along the Mogollon Rim have apparently declined dramatically over the last decades (see *L. chiricahuensis* account), but there is still some question as to whether they are conspecific with *L. fisheri*.

Lithobates grylio (Stejneger, 1901) Pig Frog

ETYMOLOGY
grylio: from the Greek *grylos* meaning 'pig.' The name refers to the species' call.

NOMENCLATURE
Conant and Collins (1998): *Rana grylio*
  Dubois (2006): *Lithobates (Aquarana) grylio*
  Synonyms: *Rana grylio*
  In the older literature this species is referred to as the Southern Bullfrog or as "Joe Browns."

IDENTIFICATION
Adults. This is a large olive to dark green frog with few black spots, at least in the East. Juveniles have a light tan dorsolateral band. This band is absent in eastern adults, but it tends to be reduced and persist in western adults. The Pig Frog is very similar to *Lithobates catesbeianus* and often has been confused with that species. Snouts tend to be more pointed and narrower than in *L. catesbeianus*, but this character is not always reliable. Throats are cream to yellowish, and bellies and the undersides of legs are white or yellowish white with black mottling toward the groin. The toes of *L. grylio* are longer than those of *L. catesbeianus*, but the fourth toe is the longest. When the fourth and fifth toes are adpressed, the tip of the fifth toe should reach beyond the base of the next-to-the-last joint of the fourth toe. If it doesn't, the species is *L. catesbeianus* (Dundee, 1974). Males often have a bright yellow throat and a tympanum larger than the eye.

Males are smaller than females. In the Everglades, the largest female captured was 157 mm SUL (431 g) whereas the largest male was 131 mm SUL (228 g) (Ligas, 1960). In Ugarte's (2004) Everglades study, the maximum female size was 130 mm SUL. Differences probably reflect harvest pressure. Duellman and Schwartz (1958) record males 97–117 mm SUL (mean 106 mm) and females 98–136 mm SUL (mean 115 mm). Jensen et al. (2008) give an adult range of 83 to 162 mm SUL, but Mount (1975) gives a maximum size of 165 mm SUL.

Larvae. The mature tadpoles are olive green with dark spots throughout the body and tail. About 9.5 days after hatching, the initially black larvae develop prominent transverse brassy bands dorsally. One is located anterior to the eyes, whereas the other is posterior to the eyes; the bands disappear by 35 mm TL. A gold spot may be visible at the base of the tail. Snout stripes are poorly developed if present at all. Tadpoles are rounded in a dorsal view, have a more pointed snout, and are less depressed than other large ranids (Altig, 1972a). Guts are partially visible through a heavily pigmented venter. The throat is light colored. Maximum size is 100 mm TL.

Eggs. The eggs are black above and white below and oviposited in a surface film about 30 cm × 30 cm or larger. The vitellus is 1.4–1.8 mm in diameter; the inner envelope 2.8–4 mm in diameter; the outer envelop 3.8–7 mm in diameter (Wright, 1932; Livezey and Wright, 1947).

DISTRIBUTION
Lithobates grylio occurs on the Atlantic Coastal Plain from southern South Carolina throughout the Florida peninsula and west to southeastern Texas. It has also been introduced into the Bahamas (Neill, 1964), China (in the early 1980s), and Puerto Rico (Kraus, 2009). Pig Frogs have been recorded on Ossabaw, Sapelo, and Cumberland islands, Georgia (Martof, 1963; Williamson and Moulis, 1994; Shoop and Ruckdeschel, 2006). Important distributional references in-

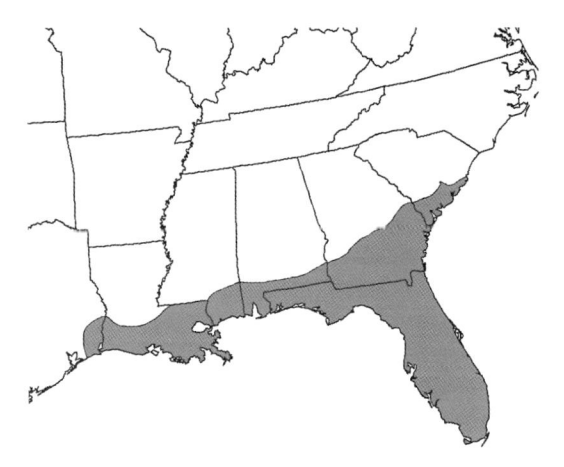

Distribution of *Lithobates grylio*

clude: Range-wide (Wright, 1932), Alabama (Mount, 1975), Georgia (Williamson and Moulis, 1994; Jensen et al., 2008), Louisiana (Dundee and Rossman, 1989), South Carolina (Beane et al., 2010), and Texas (Livezey and Johnson, 1948; Dixon, 2000).

FOSSIL RECORD

Pleistocene fossils of this species are known from deposits in Florida (Holman, 2003). The species is distinguished from *L. catesbeianus* and *L. heckscheri* by differences in the maxilla and ilium.

SYSTEMATICS AND GEOGRAPHIC VARIATION

Lithobates grylio is a member of the American Bullfrog species group (*Aquarana* clade) that consists of seven closely related species (*L. catesbeianus, L. clamitans, L. grylio, L. heckscheri, L. okaloosae, L. septentrionalis, L. virgatipes*) that evolved rapidly from a common ancestor during the late Miocene to early Pliocene (Austin et al., 2003a; Hillis and Wilcox, 2005). It is closely related to *L. heckscheri*.

Duellman and Schwartz (1958) noted that south Florida Pig Frogs were less brown in coloration and had more ventral pigmentation than Pig Frogs from north Florida and Mississippi. However, there is considerable variation in these characters even within the same population.

ADULT HABITAT

This is a species of large water bodies (ponds, lakes, wet prairies, freshwater marshes, Carolina Bays) throughout the southern Coastal Plain and peninsular Florida. Other habitats include sloughs, canals, sinkhole ponds, permanent ditches, and sawgrass marshes (Meshaka et al., 2000). Preferred habitats contain extensive amounts of emergent and floating vegetation in open-canopied wetlands of varying depths. Wetlands frequently have different vegetation zones, such as an emergent herb zone, a grass-herb zone, and a cypress-hardwood zone. Vegetation provides cover and support for foraging and basking frogs. Surrounding habitats consist of bayheads, mesic hammock, mixed deciduous forest, and upland longleaf pine sandhills. They may be present in flatwoods ponds but numbers are low (Vickers et al., 1985). Pig Frogs frequently occur with many species of fish. This species is the most common big frog of large wetlands such as Okefenokee Swamp and the Everglades.

AQUATIC AND TERRESTRIAL ECOLOGY

Lithobates grylio is a highly aquatic species that rarely leaves the water. Adults can lose no more than ca. 27% of their body water before succumbing (Thorson and Svihla, 1943). However, individual Pig Frogs, mostly juveniles, have been found dispersing through sandhills habitats at some distance from the nearest large wetlands (Dodd, 1992; Branch and Hokit, 2000; Enge and Wood, 2001; Langford et al., 2007). These frogs may occupy seasonally temporary wetlands, presumably to feed rather than to breed (Babbitt and Tanner, 2000; Greenberg and Tanner, 2005b). During the nonbreeding season, male and female Pig Frogs inhabit waters along the fringes of the pond and lake margins, such as in the cypress and hardwood zones that fringe many coastal plain lakes. As the breeding season commences, males move out into the grass-herb zones 40–50 m from shoreline; females remain in the shallow waters. The only time females move away from the shore is to mate with a male calling from the more open emergent vegetation. After oviposition, they return to the pond margins. Pig Frogs are active year-round, weather permitting, especially in Florida. They will move around within large aquatic systems. For example, Ligas (1960) recorded movement of a tagged frog 411 m in 52 days between recaptures.

Pig Frogs are generally photopositive in their phototactic response, suggesting they use both sunlight and moonlight when feeding and going about their daily activities (Jaeger and Hailman,

1973). They are sensitive to light in the blue spectrum ("blue-mode response"), which apparently helps them orient toward areas of increasing illumination, such as the open area above a pond or lake (Hailman and Jaeger, 1974). Pig Frogs likely have true color vision.

CALLING ACTIVITY AND MATE SELECTION

Calling may occur at any time during the warm season. In Georgia, this will extend from late March to early September (Wright, 1932; Lamb, 1980, 1984b), but in central Florida calls are heard from early March to November (Bancroft et al., 1983). In south Florida, calling will occur all year (Duellman and Schwartz, 1958; Ligas, 1960), although breeding likely takes place only from April to July. Calls in Louisiana are heard from February to August (Dundee and Rossman, 1989). The majority of chorusing occurs at night, but calls can be heard at most any time of the day.

The call is a guttural series of grunts reminiscent of a pig grunting, hence its common name. Males also produce a loud single grunt, which may serve a territorial function. Unreceptive females produce a grunt that is similar to a male's call. If a male approaches a calling male too closely, agonistic behavior consisting of the assumption of an agonistic posture, chasing, and wrestling ensues. Although unstudied, the vocal repertoire of this species and its territorial behavior are probably as complex as in *L. catesbeianus*. Calling occurs while floating in the water or while perched on floating vegetation, such as lily pads or lotus. Males tend to form loose choruses in thick vegetation, which provides concealment. Males tend to remain at one location. Chorusing occurs at temperatures >21°C.

BREEDING SITES

Breeding occurs in a variety of wetlands, from small to large ponds and lakes to large wet prairies. Pig Frogs favor quiet open waters with an abundance of emergent vegetation (see Adult Habitat). In the Everglades, Pig Frogs are more abundant in sloughs and sawgrass prairies than they are in large wet prairies (Baber et al., 2005).

REPRODUCTION

In southwest Georgia, mature spermatozoa are present in male testes year round, whereas mature ova are present in females from April to July (Lamb, 1980, 1984b); thus, the breeding season is likely about four months in the north. Oogenesis begins in August. Even in south Florida, most breeding takes place from March to September with a peak in June (Ligas, 1960), although breeding there can take place year-round (Babbitt and Tanner, 2000; Ugarte, 2004; Baber et al., 2005). Clutch size ranges from 6,000 to 34,000, with clutch size proportional to female body size (Wright, 1932; Livezey and Wright, 1947; Ligas, 1960). The mean clutch size in Ugarte's (2004) Everglades study was 9,149 eggs. Egg diameter is not correlated with female SUL. Hatching occurs in two to four days.

LARVAL ECOLOGY

Larvae begin feeding 6.5 days after hatching (Ligas, 1960) and feed extensively on unicellular algae. Tadpoles spend most of their time in

Tadpole of *Lithobates grylio*. Photo: David Dennis

Adult *Lithobates grylio*. Photo: Alan Cressler

shallow water and are abundant among stands of *Juncus* (Lamb, 1980). The larval period is 3–15 months and varies considerably among individuals and with latitude. In southwest Georgia, for example, larvae overwinter and transform after 10–15 months (Lamb, 1980); in the Everglades, transformation occurs as soon as 3 months (Ligas, 1960). Newly metamorphosed froglets are 30–49 mm SUL (Wright, 1932; Ligas, 1960).

DIET

The diet consists of many types of invertebrates, principally beetles, larval and adult dragonflies, and crayfish (*Procambarus*) (Carr, 1940a; Duellman and Schwartz, 1958; Ligas, 1960; Lamb, 1980, 1984b; Ugarte, 2004). Other items include grass shrimp, snails, millipedes, centipedes, spiders, crickets, true bugs, wasps, grasshoppers, cicadas, alligator fleas, larval and adult lepidopterans, grasshoppers, leeches, and flies. Pig Frogs eat vertebrates including lizards (*Eumeces laticeps*), snakes (*Nerodia fasciata*), frogs (*Acris gryllus*, *Hyla cinerea*, *Lithobates grylio*, *L. sphenocephalus*), salamanders (*Eurycea quadridigitata*), and fish (mosquito fish, least killifish, sailfin molly, marsh killifish) (Duellman and Schwartz, 1958; Lamb, 1980, 1984b; Ugarte, 2004). They likely consume any animal they can cram into the large mouth, much as American Bullfrogs do, but in the Everglades crayfish comprised 75% by volume of the prey consumed (Ligas, 1960).

During the nonbreeding season, males and females have the same diet, consisting mostly of aquatic species. Fish assume more importance during droughts and the dry season; males tend to eat more crayfish in the wet season, whereas females eat more frogs, at least in the Everglades (Ugarte, 2004). However, there also are sex-related differences in the relative types and abundance of prey eaten during the breeding season, reflecting the differences in the habitats of males and females at this time (Lamb, 1980). Males eat infrequently during the breeding season, and most frogs lack fat bodies during the wet summer season.

PREDATION AND DEFENSE

The skin secretions of *Lithobates grylio* contain antimicrobial peptides, which may assist in defense against microorganisms (Kim et al., 2000). Pig Frogs are wary and difficult to approach, but Carr (1940a) indicated this might be the result of hunting pressure. Their coloration makes them difficult to see in emergent vegetation. Allen (1932) noted that Pig Frogs give off a musty odor when handled and that its "slime" is bitter to the taste. Still, many vertebrates likely eat Pig Frogs, especially alligators, snakes (*Nerodia fasciata*, *N. erythrogaster*), wading birds, and mammals (raccoons) (Wright, 1932). The tadpoles appear to be fully palatable and are eaten by snakes and predaceous invertebrates.

POPULATION BIOLOGY

Females must be 94–96 mm SUL to begin breeding in the Everglades, with males reaching maturity at 70–75 mm SUL (Ligas, 1960; Ugarte, 2004). Ovaries may have developing eggs and oocytes at 74 and 86 mm SUL, respectively. In the Everglades, adult sizes are reached in 1 yr following metamorphosis in males and after 1.5 yrs in females. This suggests that males are capable of breeding the first summer after metamorphosis, but that females first breed the second summer following transformation. In the north, age at maturity is probably later because of the more extended larval period and winter cold season, although Mount (1975) recorded a calling male in Alabama of only 52 or 59 mm SUL. Maximum longevity is 5 yrs based on growth curves (Ugarte, 2004).

In north-central Florida, female Pig Frogs have higher rates of survivorship than either adult males or juveniles (Wood et al., 1998). Survivorship rates do not change monthly, although low capture probabilities vary among sampling periods. Survivorship varies among sampling sites, with frogs at some locations having much higher rates of monthly survival than at other locations. Abundance also varies considerably among sampling periods based on mark-recapture studies. At one pond in Gainesville, Wood et al. (1998) estimated there were as many as 715 juveniles, 219 males, and 499 females, but at a smaller pond the maximum estimates were 39 juveniles, 16 males, and 19 females.

COMMUNITY ECOLOGY

The Pig Frog is the common large ranid throughout much of its range. *Lithobates catesbeianus* often occurs within the area, but it is not nearly abundant and often inhabits smaller ponds. Occasionally the species are found together, however, such as at Kingfisher Pond on Savannah National Wildlife Refuge in South Carolina. The interaction between the Florida Bullfrog and the Pig Frog deserves study.

DISEASES, PARASITES, AND MALFORMATIONS

This species is parasitized by the trematodes *Clinostomum marginatum*, *Gyrodactylus* sp., and *Haematoloechus longiplexus* (Manter, 1938; Paetow et al., 2009). *Ranavirus* has been found in Pig Frogs introduced into Chinese aquaculture (Zhang et al., 2001). As of 2007, amphibian chytrid fungus had not been detected on Pig Frogs (Rothermel et al., 2008).

SUSCEPTIBILITY TO POTENTIAL STRESSORS

Metals. Mercury is common in Pig Frogs from south Florida (Ugarte et al., 2005). Ugarte (2004) found a mean of 121.4 ng/g total mercury (maximum 2.3 mg/kg wet mass), with liver levels 2–5 times the levels in leg muscle. The amount of mercury is not correlated with SUL. High levels are recorded in both protected and unprotected sites, and Ugarte et al. (2005) recommended no harvesting in areas with high mercury content.

STATUS AND CONSERVATION

This species is hunted heavily for food in Florida and Georgia. Ligas (1960) noted that literally "hundreds of thousands" were taken by commercial hunters in the Everglades throughout the year. Frogs were spotlighted and gigged, dip-netted, or captured by hand, much as they are today. Catches were as high as 34–45 kg/night in 1959 (Ligas, 1960). The effect of long-term unregulated harvest is seen in the maximum size decrease from the 1950s to the 2000s. Harvest selectively removes the largest frogs (Ugarte, 2004). In north Florida, Beck (1948) indicated that Pig Frog populations in unprotected areas had been decimated by frog hunters.

Habitat of *Lithobates grylio*, Jasper County, South Carolina. Photo: C.K. Dodd Jr.

Ligas (1960) recommended a series of habitat and administrative measures to ensure and monitor the harvest, including the implementation of a licensing program and closure to harvest during the breeding season, at least in the Everglades. Although a commercial license is now required, there are still no season, size, or quantity limits on any large ranid species in Florida, and no recreational license is required. Ugarte (2004) noted that both harvest and periodic water draw-downs affected Pig Frog demography. Frogs become less common after draw-downs, and survivorship decreases, with highest catches in areas with the longest hydroperiod. This is particularly evident among the juvenile population, where recruitment is low following draw-down.

The species is relatively tolerant of urbanization as long as wetlands are maintained (Delis et al., 1996). Since it does not venture far from the shoreline, development of uplands may not affect the species as long as pollution and urban runoff is controlled. Still, populations may not be as extensive or abundant in urban, silvicultural, or agricultural settings as they are in native ecosystems (Surdick, 2005). Because the diet consists largely of crayfish, and crayfish populations may be reduced in disturbed habitats, diet rather than surrounding habitat use per se may be responsible for reduced abundance in some areas. The species quickly recovered from saltwater overwash after Hurricane Dennis hit the Gulf Coast of Florida (Gunzburger et al., 2010), and they were present following hurricanes Ivan and Katrina in Louisiana, although abundance estimates are not available (Schriever et al., 2009). Viosca (1923) noted they could tolerate moderate salinity.

Roads likely take a considerable toll on Pig Frogs, especially when they cross expansive marshes and wet prairies; kills were especially prevalent from April to August (Smith and Dodd, 2003). Culverts and underpasses help reduce mortality considerably (Dodd et al., 2004). From 1990 to 1994, 613 Pig Frogs were collected in Florida for the pet trade (Enge, 2005a).

Lithobates heckscheri (Wright, 1924)
River Frog

ETYMOLOGY

heckscheri: named for the Heckscher Foundation for the Advancement of Research, established at Cornell University by August Heckscher. The Heckscher Foundation supported A.H. Wright's work in the Okefenokee Swamp.

NOMENCLATURE

Conant and Collins (1998): *Rana heckscheri*
   Dubois (2006): *Lithobates (Aquarana) heckscheri*
   Synonyms: *Rana heckscheri*

IDENTIFICATION

Adults. This is a large brown to grayish olive frog with evenly spaced tubercles occurring throughout its dorsum, which gives it a textured appearance. On the sides, the color blends to a more olive or even cinnamon-brown color. The dorsum is mottled in black. White spots may be present on the dark lower jaw. Dorsolateral folds are absent. Venters are spotted white to light gray to dark and are more heavily mottled on the throat and between the front legs. A pale crescent-shaped line may girdle the groin. There are narrow dark bars on the upper sides of the legs. The ultimate phalange of the fourth hind toe is free of webbing. Males are smaller than females but have wider tympanums. Adults are 83–155 mm SUL.

Larvae. The tadpoles are initially small and black with gold to white transverse bands on the snout and body. These bands disappear when the tadpole is about half grown. Tails are transparent. Mature tadpoles are dark greenish olive to olive with fine pale greenish-yellow spots or flecks over the dorsum. Venters are pigmented and the gut is not visible. The lateral line system is easily visible. Tails are elongate, and the top fin is not as deep as the musculature. The tails have a prominent black band that extends along the upper tail musculature about two-thirds of its length. The entire tail is rimmed with black pigment. The black band and tail border present a striking contrast to the nearly clear tail fins. Tadpole lengths reach

160 mm TL (Mount, 1975). Older tadpoles and metamorphs have brick-red eyes; the iris eventually turns golden in adults. Altig (1972a) described tadpole development. Albino tadpoles were observed in Georgia (Wright, 1924).

Eggs. The eggs have not been described in detail, but presumably they are similar to *L. grylio*. The egg mass is probably oviposited in a sheet-like surface film as in other members of the *Aquarana* clade (Wright and Wright, 1949), although Allen (1938) stated that wild-caught females deposited eggs in a "mass" in captivity. The vitellus is 1.5–2 mm in diameter (Wright, 1932).

DISTRIBUTION

The species historically occurred from the Lumber and Cape Fear river systems in North Carolina (not seen since 1975; Beane, 1998) south along the Atlantic Coastal Plain below the Fall Line, throughout the Florida peninsula, and westward to the Wolf River in southern Mississippi. Means and Simberloff (1987) gave the southernmost location as Marion County, Florida, but Punzo (1991b, 1992b) used specimens from the Hillsborough River in the Tampa Bay area in his experiments. The identity of these specimens is questionable. It has also been introduced into China (Kraus, 2009). Important distributional references include: Alabama (Mount, 1975), Georgia (Williamson and Moulis, 1994; Jensen et al., 2008), Mississippi (Allen, 1932), North Carolina

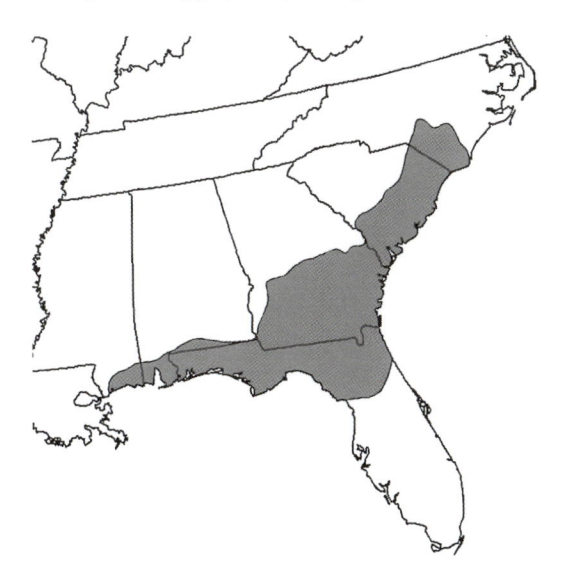

Distribution of *Lithobates heckscheri*

Schooling tadpoles of *Lithobates heckscheri*. Photo: Dirk Steveson

(Simmons and Hardy, 1959; Beane, 1998; Dorcas et al., 2007), and South Carolina (Leiden et al., 1999; Beane et al., 2010).

FOSSIL RECORD
No fossils are known.

SYSTEMATICS AND GEOGRAPHIC VARIATION
Lithobates heckscheri is a member of the American Bullfrog species group (*Aquarana* clade), consisting of seven closely related species (*L. catesbeianus, L. clamitans, L. grylio, L. heckscheri, L. okaloosae, L. septentrionalis, L. virgatipes*) that evolved rapidly from a common ancestor during the late Miocene to early Pliocene (Austin et al., 2003a; Hillis and Wilcox, 2005). It is closely related to *L. grylio*. Laboratory crosses of *L. heckscheri* with *L. clamitans* were unsuccessful (Moore, 1949a, 1955).

ADULT HABITAT
The River Frog is a species of flowing streams, rivers, swampy backwaters, fluvial swamps, and medium- to large-sized ponds and lakes; it also

Recent metamorph of *Lithobates heckscheri*. Photo: C.K. Dodd Jr.

occurs in small streams flowing between larger water bodies, oxbows, sinkhole ponds, and permanently flooded borrow pits. Shaded banks with good cover are preferred. Surrounding forest often consists of mesic hammock and riparian cypress and hardwoods (Harima, 1969).

AQUATIC AND TERRESTRIAL ECOLOGY
River Frogs are active at temperatures of 18–35°C, with an optimum of 25°C (Hansen, 1957). They frequent banks and shorelines with high moisture content, and often are found in sphagnum. They usually sit within 23 cm of water and face the open water. Frogs will change position along the shore, but usually stay within a particular area. For example, Hansen (1957) recorded movements of about 10 m between recaptures, with no differences between males and females; juveniles tended to move farther than adults. Activity areas averaged 17 m × 1.5 m. Allen (1938) found a single juvenile dispersing 183 m from water a month after transformation; few juveniles remained around the larval lake. Adults also may be observed sitting on logs or debris in the water.

River Frogs are photopositive in their phototactic response, suggesting they use both sunlight and moonlight when feeding and going about their daily activities (Jaeger and Hailman, 1973). They are sensitive to light in the blue spectrum ("blue-mode response"), which apparently helps them orient toward areas of increasing illumination, such as the open area above a river or pond (Hailman and Jaeger, 1974). River Frogs likely have true color vision.

CALLING ACTIVITY AND MATE SELECTION
Calling occurs from April to August. The call is a loud, distinctive snore or snort. Males call from both the shoreline and from shallow water.

BREEDING SITES
Breeding occurs along the shoreline of rivers, swampy backwaters, and ponds (see Adult Habitat).

REPRODUCTION
Breeding occurs from spring through late summer. Reports of clutch size include about 5,000 eggs per mass (Allen, 1938) and 14,000 eggs estimated from a single female (Wright, 1932). Hatching occurs in 10–15 days. According to Allen (1938), about 10% of the eggs fail to hatch.

LARVAL ECOLOGY
Lithobates heckscheri tadpoles form large schools consisting of hundreds of individuals. For example, I captured 1,132 tadpoles in 8 trap-nights at a pond on the Savannah National Wildlife Refuge; 3 traps accounted for >300 tadpoles each on a single night. These schools move parallel to the shoreline and remain in shallow water during the day. At night, tadpoles move into the deeper water, although the tadpoles are photonegative in laboratory experiments (Altig and Christensen,

1981). Allen (1938) noted that tadpole schools tended to remain within a defined area of about 46 m in linear distance. Tadpoles respond positively to the presence of food and food odors by increasing feeding activity, even when food is abundant. Tadpoles in schools exhibit a degree of social facilitation whereby swimming speeds are increased and avoidance of unpleasant stimuli is enhanced in the presence of groups of conspecifics (Punzo, 1991b, 1992b). They can be heard making a smacking sound as dozens of tadpoles simultaneously gulp air at the surface as a school moves through shallow water.

The larval period extends throughout the winter (Wright and Wright, 1949). For example, in Florida eggs hatching on 8 June resulted in metamorphs by 10 April the following year (Allen, 1938). Allen (1938) estimated that as many as 20% died during transformation.

DIET

Lithobates heckscheri eats a variety of invertebrates, particularly crayfish and insects. Specific items include millipedes, centipedes, spiders, beetles, true bugs, tabanid flies, cockroaches, and caddisflies (Lamb, 1980). It seems likely that a River Frog would eat any prey item that it could fit into its mouth.

PREDATION AND DEFENSE

Adult skin secretions of *L. heckscheri* contain antimicrobial peptides, which may assist in

Recent metamorph of *Lithobates heckscheri*.
Photo: Dirk Stevenson

Adult *Lithobates heckscheri*.
Photo: C.K. Dodd Jr.

Habitat of *Lithobates heckscheri*, Jasper County, South Carolina. *Acris gryllus*, *Anaxyrus fowleri*, *Anaxyrus terrestris*, *Lithobates catesbeianus*, *Lithobates clamitans*, *Lithobates grylio*, *Lithobates sphenocephalus*, and *Hyla cinerea* also bred in this pond. Photo: C.K. Dodd Jr.

defense against microorganisms (Conlon et al., 2007a). However, the larval skin does not appear to contain toxic or noxious properties (Altig and Christensen, 1981). This species is easily approached by human observers and may go limp when handled.

Predators of tadpoles and recent metamorphs include water snakes (*Nerodia fasciata*), turtles (*Sternotherus minor*), and Boat-tailed Grackles (Wright, 1932; Allen, 1938). Tadpoles respond to predator odors and the odors of injured tadpoles by swimming wildly and erratically away from the source of the odor (Altig and Christensen, 1981). They also attempt to dive and hide.

POPULATION BIOLOGY

Juvenile growth rates apparently are rapid. For example, Hansen (1957) recorded growth in juveniles of 11 mm in 30 days, 6 mm in 21 days, and 4 mm in 27 days. Juveniles continue to grow through October, and growth slows appreciably in winter. The age at maturity is unknown but presumably similar to other large Southeastern ranids.

DISEASES, PARASITES, AND MALFORMATIONS

Larval River Frogs are parasitized by the copepods *Argulus americanus* (Goin and Ogren, 1956) and *A. diversus* (Clark, 2001). The trematode *Gorgodera amplicava* has been reported from *Lithobates heckscheri* (Parker, 1941).

SUSCEPTIBILITY TO POTENTIAL STRESSORS

Metals. Mercuric chloride inhibits fertilization at > 1 mg/L, with total prevention at 5 mg/L (Punzo, 1993a). The $LC_{50\ (3\ hr)}$ is 1.43 mg/L for eggs and the $LC_{50\ (96\ hr)}$ for tadpoles is 0.68 mg/L. Mercury also impairs the development of oocytes in adult *L. heckscheri* (Punzo, 1993b). Developmental abnormalities (kinked and curved tails) are common in *L. heckscheri* exposed to mercuric chloride concentrations > 1 mg/L (Punzo, 1993a).

STATUS AND CONSERVATION

There are no published studies on the status of this species in much of its range. As a large ranid, it presumably has been hunted for food in the rural South. Undoubtedly populations have been lost as habitats were destroyed and river floodplains altered. According to Jensen et al. (2008), the species is relatively common in Georgia. Aresco (2004) noted only 6 breeding localities in Alabama and that there have been no records for 25 yrs. Beane (1998) could find no extant populations in North Carolina despite extensive surveys. Between 1990 and 1994, 32 River Frogs were collected in Florida for the pet trade (Enge, 2005a).

Lithobates okaloosae (Moler, 1985)
Florida Bog Frog

ETYMOLOGY
okaloosae: named for Okaloosa County, Florida, where the species was first discovered.

NOMENCLATURE
Conant and Collins (1998): *Rana okaloosae*
  Dubois (2006): *Lithobates (Aquarana) okaloosae*
  Synonyms: *Rana okaloosae*

IDENTIFICATION
Adults. The Florida Bog Frog is a small yellowish-greenish to yellowish-brown frog that is superficially similar to *Lithobates clamitans*. Light-colored dorsolateral folds are present, but they do not extend to the groin. The dorsum is uniformly colored and covered with small white tubercles, but the white venter is smooth with dark vermiculation. The tympanum is flat in both sexes. Throats may be infused with yellow in both sexes, and there are light spots on the lower jaw. The toe webbing is distinctive, with three phalanges of the fourth toe and two phalanges of all other toes free of webbing. In sympatric ranids, toe webbing is more extensive. Males have internal vocal sacs, a slightly larger tympanum, and swollen thumbs.

Males are only slightly smaller than females. In the type series, males were 34.8–45.8 mm SUL (mean 40.6 mm) and females 38.2–48.8 mm SUL (mean 44.6 mm) (Moler, 1985). Bishop (2005) recorded a mean male size of 40.2 mm SUL (range 34.4 to 56.9 mm) and female size of 41.5 mm SUL (range 33.5 to 48.8 mm). Gorman (2009) recorded *L. okaloosae* with a mean of 39.7 mm SUL (range 26.6 to 49.8 mm). Body mass increases with SUL.

Larvae. The tadpole is olive brown with numerous dusky spots on the tail; tadpoles of *L. clamitans* lack these spots. The black venter is marked by well-defined white spots, and the coiled intestines are not visible. The white ventral spotting is distinctive and separates them from sympatric *L. clamitans* larvae, whose spots are more gold on a deep brown background. Tadpoles reach a maximum length of 56 mm TL (Moler, 1985). Moler (1985) illustrates the tadpole and its mouthparts, and Priestley et al. (2010) illustrate differences between sympatric *L. okaloosae* and *L. clamitans* larvae.

Eggs. The eggs of this species have not been described. Presumably they are similar to those of *L. clamitans*. Eggs are oviposited in a single-layer surface film.

DISTRIBUTION
This species is known only from Walton, Okaloosa, and Santa Rosa counties in the western Florida Panhandle. Sites (>60 known) are within the East Bay, Yellow, and Shoal river drainages. Most of its distribution occurs on Eglin Air Force Base. Important distributional references include Bishop (2005).

FOSSIL RECORD
No fossils are known.

SYSTEMATICS AND GEOGRAPHIC VARIATION
Lithobates okaloosae is a member of the American Bullfrog species group (*Aquarana* clade), which consists of seven closely related species (*L. catesbeianus*, *L. clamitans*, *L. grylio* *L. heckscheri*, *L. okaloosae*, *L. septentrionalis*, *L. virgatipes*,) that evolved rapidly from a common ancestor during the late Miocene to early Pliocene (Austin et al., 2003a; Hillis and Wilcox, 2005). The species is actually phylogenetically embedded within *L. clamitans*, which suggests a recent origin for the species (Austin et al., 2003). Polymorphism is low in *L. okaloosae*. Nonpanmictic populations are maintained by continu-

Distribution of *Lithobates okaloosae*

Eggs of *Lithobates okaloosae*. Photo: Ronn Altig

ous genetic structuring along streambeds, with weak isolation-by-distance (Austin et al., 2011). *Lithobates okaloosae* may be the result of hybrid origin, incomplete lineage sorting, or recent hybridization. Hybrids with *L. clamitans* have been reported (Moler, 1992).

ADULT HABITAT

The Florida Bog Frog is found in isolated acid seeps (pH 4.1–5.5), along shallow boggy overflows of larger seepage streams, in headwater streams, stream sections below small impoundments, and rarely along the edge of ponds. Water should be clear, shallow, and not stagnant. For example, Gorman et al. (2009) studied the species biology along a shallow (1–20 cm) headwater stream that varied from 7 to 22 m in width. The species is frequently associated with lush beds of sphagnum moss (*Sphagnum*). Habitats consist of small streams that usually are bordered by black titi (*Cliftonia monophylla*) and sweetbay (*Magnolia virginiana*), and may be associated with Atlantic white cedar (*Chamaecyparis thyoides*). The habitat of clear low-flowing streams in excessively drained sandy soils is limited within the area, thus restricting the distribution of the species.

Florida Bog Frogs are found most often in drainages with nearby mixed-forest wetlands. Surrounding habitats include longleaf pine sandhills and flatwoods. Prescribed burning helps limit encroaching vegetation. In areas where hardwoods have encroached on streams, they may be confined to areas such as power-line rights-of-way.

AQUATIC AND TERRESTRIAL ECOLOGY

This species apparently rarely ventures far from streams and bogs, and adults and juveniles occupy the same habitats. Activity occurs year round. Daily activity occurs within a relatively small area, and Gorman et al. (2009) recorded a mean daily movement of only 1.8 m; the maximum daily movement was 8.9 m. Bishop (2005) estimated the home range to be 37–188 m^2 for *Lithobates okaloosae* depending upon estimator used. Florida Bog Frogs likely bury into sphagnum moss or other available aquatic streamside vegetation to avoid cold temperatures.

CALLING ACTIVITY AND MATE SELECTION

Calls occur throughout the warm months of March to August. The advertisement call is distinctive from that of all other frogs within its range. The call is described as a "series of 3–21 guttural chucks [pulses] repeated at about 5 notes per second" (Moler, 1985) that slows audibly toward the end. Bishop (2005) identified three basic call types (advertisement, single chuck, response) plus a release call that is given when handled. Advertisement calls average 1.4 sec in length, 7.5 pulses,

Comparison of the venters of tadpoles of *Lithobates okaloosae* (*left*) and *Lithobates clamitans* (*right*). Photo: David Bishop

a pulse rate of 0.2, and a dominant frequency of 651–2,466 Hz (mean 1,505 Hz) (Bishop, 2005), but there is a great degree of variation among individual males. Intervals between calls range from 23 to 600 sec (mean 136 sec), and from 16 to 21 notes constitute a call. Advertisement calls do not carry well. Males also have a single quieter call (response) that is issued in response to calls from nearby males. The rates of this call increase in response to playback experiments of conspecifics, but not as strongly to calls of *L. clamitans* (Bishop, 2005). Females also have a soft "chuck" call, the function of which is unknown. Call dominant frequencies are much higher in *L. okaloosae* than in sympatric *L. clamitans*.

Males call from along streamsides and bogs, and their distribution tends to be clumped. The mean number of males calling in Gorman et al.'s (2009) study was 7.3 (range 5 to 10) in a 60 m stretch of headwater stream. Males spaced themselves at a mean distance of 9.2 m, which was actually farther than their nearest neighbor distance with *L. clamitans* at 6.5 m. Satellite males may locate themselves near calling males; these satellites tend to be slightly smaller than calling males. Males usually call from one location for several nights before moving to a different location.

BREEDING SITES

Breeding occurs in shallow sandy-bottomed streams with low but steady surface flow. Calling sites are not randomly positioned, but are closer to the bank (mean 21 cm) and cover (mean 7 cm), adjacent to flowing water, near woody debris, and have higher water temperatures (mean 26°C) when compared with random available sites. Most calling sites have nearby emergent and submerged vegetation (Bishop, 2005; Gorman, 2009). Canopy cover does not influence occupancy, although *L. okaloosae* is generally found in more open habitats.

REPRODUCTION

Eggs are deposited in a single-layer surface mass that is free floating. Occasional masses may be folded or become attached to vegetation. Egg masses average 12.5 cm in length (range 7 to 21) by 9 cm in width (range 6 to 13). Egg masses are deposited in shallow water (2–7 cm) and from 0 to 1,300 cm from the bank (Bishop, 2005). Clutch size is 152–345 eggs/mass (mean 235). Bishop (2005) suggested that females might produce more than one clutch per year.

Males are capable of fertilizing more than one clutch. For example, Bishop (2005) found that 15 males fertilized 35 egg masses, with a mean rate of 1.8 masses per male (range 0 to 9). The number of egg masses fertilized was directly correlated with the number of nights spent calling. The smallest successful male was 37 mm SUL, although males as small as 34.4 mm gave advertisement calls. The smallest gravid female was 37.6 mm SUL. Oviposition sites are located near calling sites, with a mean distance of only 14 cm.

Adult *Lithobates okaloosae*.
Photo: C.K. Dodd Jr.

Habitat of *Lithobates okaloosae*. Photo: David Bishop

LARVAL ECOLOGY
Larvae have been collected throughout the year, and at least some overwinter prior to transformation. Laboratory observations suggest that tadpoles do not swim very much. Recent metamorphs are 18–20 mm SUL (Bishop, 2005).

DIET
Nothing has been reported on the diet of this species. Presumably it eats any available invertebrate it can catch, and prey consumption is likely in proportion to availability.

PREDATION AND DEFENSE
The coloration of this frog makes it difficult to see, thus affording crypsis from visually oriented predators. Upon approach of an intruder, the frog quickly moves to water. The skin secretions of *L. okaloosae* contain antimicrobial peptides, which may assist in defense against microorganisms (Conlon et al., 2007a).

Bishop (2003) reported a water snake (*Nerodia fasciata*) eating two tadpoles and attempting to catch a juvenile. Florida Bog Frogs are likely eaten by a variety of snakes, turtles, birds, and mammals. Larvae are certainly prey of predaceous invertebrates and vertebrates.

POPULATION BIOLOGY
Over a 3 yr period, Bishop (2005) recorded sex ratios of 1.7:1, 1.1:1, and 2.3:1. Only 1 frog was captured in all 3 yrs. He estimated survivorship as 90–96.5% over a 2-month period during a single yr.

COMMUNITY ECOLOGY
This species is sympatric with *Lithobates clamitans*, and the two have similar calling sites and breeding biology. There is no evidence of intraspecific competition, and indeed interactions between the two tend to be positive (Gorman et al., 2009). Call structure and body size are very different between the two species. *L. okaloosae* prefers sites with a greater percentage of submerged vegetation, slow-flowing water, shorter distance to cover, greater amounts of emergent vegetation and woody debris, and shallower water than does *L. clamitans* (Gorman, 2009; Gorman and Haas, 2011), although there is considerable microhabitat overlap. Gorman (2009) could not demonstrate strong evidence for competition between the larvae of these closely related species.

DISEASES, PARASITES, AND MALFORMATIONS
No information is available.

SUSCEPTIBILITY TO POTENTIAL STRESSORS
No information is available.

STATUS AND CONSERVATION
Despite its limited range, most known sites are on Eglin Air Force Base and are thus protected to some extent. A management plan to conserve the species has been implemented. Potential threats include fire suppression, feral pigs, stream siltation, stream impoundment, invasive plants, accidental pollution, and disease. Moler (1992) and Bishop (2005) provided a series of management recommendations to address these threats.

Lithobates onca
(Cope *in* Yarrow, 1875)
Relict Leopard Frog

ETYMOLOGY
onca: a Latin noun meaning 'spotted,' as in a spotted frog.

NOMENCLATURE
Stebbins (2003): *Rana onca*
  Dubois (2006): *Lithobates (Lithobates) onca*
  Synonyms: *Rana draytoni onca, Rana pipiens onca*

IDENTIFICATION
Adults. This is a relatively small brown to olive-green leopard frog with generally smooth skin, although there may be warts or tubercles on the head and neck. Small black spots are bordered by gray. These spots are fewer and the gray border less distinct than in *Lithobates pipiens* or *L. fisheri*. Brown dorsolateral folds are present, but there are no longitudinal folds between the dorsolateral folds. The dorsolateral folds only extend about halfway down the back, making this and *L. fisheri* distinct from other Southwestern leopard frogs. The head is as broad, and it is long, and spots do not occur on the snout. The stripe on the upper lip is faint or absent in front of the eye. The tympanic membrane is large and smooth. Bellies are white, although there may be a yellowish coloration toward the legs. The limbs are short and the hind toes are fully webbed.

Males are smaller than females. At Blue Point Spring, the median size was 53.5 mm SUL for males and 61.5 mm SUL for females (Bradford et al., 2004). Tanner (1931) gave measurements of 2 specimens as 53 and 55 mm SUL. Brennan and Holycross (2006) stated a maximum size of 89 mm SUL.

Larvae. There are no published descriptions of the larvae of this species.

Eggs. The eggs have not been described. Eggs are deposited in a jelly mass, and clutch size is ca. 250 eggs (Brennan and Holycross, 2006).

DISTRIBUTION
Lithobates onca was known historically from at least 24 localities in extreme northwestern Arizona and adjacent Utah and Nevada along the Colorado, Virgin, and Muddy rivers. The historic range likely included about 190 km of river miles between Hurricane, Utah, and Black Canyon below Lake Mead. It now occurs at only five localities in or near Lake Mead National Recreation Area. The species may have occurred in the Pahranagat Valley (Stejneger, 1893), but the identity of these frogs is uncertain. Eaton (1935) reported a single specimen of *L. onca* from Rainbow Bridge Canyon, well to the east of known populations; the specimen was likely *L. pipiens*.

Lithobates onca once was much more widespread within its historic range; populations along the Virgin and Muddy rivers have been extirpated, as have those in Corral and Reber springs along the Colorado River within the last 15 years. Populations likely were extirpated with the building of Boulder Dam and subsequent reservoir filling and the introduction of nonindigenous predators. Important distributional references include Tanner (1931), Cowles and Bogert (1936), Linsdale (1940), Bradford et al. (2004), and Brennan and Holycross (2006).

FOSSIL RECORD
No fossils are known. Holman (2003) noted Miocene (Hemphillian) fossils from Navajo County in northern Arizona that belonged to the *Lithobates pipiens* complex. The relationship between these fossils and *L. onca* is unknown.

Distribution of *Lithobates onca*. Dark gray indicates extant populations; light gray indicates extirpated populations.

Eggs of *Lithobates onca*. Photo: Dana Drake

SYSTEMATICS AND GEOGRAPHIC VARIATION
Lithobates onca is a member of the *Novirana* clade of North American ranid frogs. It is an associate of the mostly lowland and tropical leopard frog group (or *Scurrilirana*) (Hillis and Wilcox, 2005). Its closet relatives in the United States include *L. berlandieri*, *L. blairi*, *L. sphenocephalus*, and especially *L. yavapaiensis*. Although described in 1875, this species has variously been considered a species, a phenotypic variant, or a subspecies of *L. pipiens*. Long thought extinct, the species was rediscovered in 1991, with subsequent molecular and morphologic analysis confirming its phylogenetic uniqueness (Jaeger et al., 2001). The species is probably a Pleistocene–Holocene isolate from ancestral populations to the south. *Lithobates onca* currently exhibits a rather low level of genetic heterozygosity.

Microsatellite markers were characterized by Savage and Jaeger (2009). Oláh-Hemmings et al. (2009) affirmed the close relationship but distinctiveness between *L. onca* and *L. yavapaiensis*. *L. onca* and *L. yavapaiensis* diverged during the early Pleistocene. A population of leopard frogs found in Surprise Canyon in the western Grand Canyon is more closely related to *L. yavapaiensis* than to *L. onca*, despite the close proximity of populations of *L. onca*. *L. onca* and *L. yavapaiensis* have closely related but distinct antimicrobial skin peptides (Conlon et al., 2010).

ADULT HABITAT
The species historically inhabited ponds, springs, and streams within Mohave Desert scrub habitats. The five known localities are all undisturbed spring systems that lack the presence of American Bullfrogs (*L. catesbeianus*), crayfish, and predaceous game fish. Spring pools should be open and free of emergent vegetation such as *Scirpus*. The species prefers open shorelines without dense vegetation. Bradford et al. (2005b) described three areas where the frog occurs or once occurred, and noted differences among the habitats (marshes vs. geothermal spring runs).

AQUATIC AND TERRESTRIAL ECOLOGY
Relict Leopard Frogs likely stay close to water and do not travel great distances between their isolated spring habitats. They are usually observed within a few meters of the shoreline in low riparian vegetation or directly at the water's edge. In a mark-recapture study, Bradford et al. (2004) noted that most frogs were captured within 17.8 m of the previous capture, and that the maximum distance between captures was 120 m. Movements occur along streamcourses. No information is available on overwintering habits, but frogs living near geothermal springs could be active year-round.

CALLING ACTIVITY AND MATE SELECTION
The advertisement call consists of a series of quiet soft clucks.

BREEDING SITES
Breeding occurs in shallow springs, pools, riverside marshes, and spring runs. Water should be lentic or slow moving. Emergent vegetation may be present, but it should not clog the entire wetland.

REPRODUCTION
Calling takes place from January through April, with oviposition occurring usually during February and March. Bradford et al. (2005b) reported breeding in November. As such, this species may have a bimodal breeding season, with most breeding occurring in the spring. Egg masses are attached to vegetation in shallow water just below the water's surface. Bradford et al. (2005b) reported "many hundred eggs" per mass. Embryos hatch in five to seven days (Drake, 2010).

LARVAL ECOLOGY
Nothing is reported in the literature on the larval ecology of this species. Tadpoles likely feed

Tadpoles of *Lithobates onca*. Photo: Dana Drake

Adult *Lithobates onca*. Photo: Dana Drake

on alga, and opportunistic oophagy has been reported (Drake, 2010). This species does not have any of the morphological characteristics of oophagous tadpoles.

DIET
Tanner (1931) reported a damselfly, a beetle, and a wasp from a single specimen collected in June. Like other leopard frogs, it probably consumes a variety of invertebrates in proportion to their availability.

PREDATION AND DEFENSE
Relict Leopard Frogs will sit motionless hidden in riparian vegetation, but take to water at the approach of a predator; there, they seek shelter in vegetation or under rocks. Nothing has been reported concerning predators, but these undoubtedly include snakes (*Thamnophis elegans*), birds, and mammals. Tadpoles have been observed feeding on the eggs of conspecifics (Drake, 2010).

POPULATION BIOLOGY
At Blue Point Spring, Bradford et al. (2004) estimated a 90% monthly survivorship rate during the course of their observations, which extrapolated to a 27% annual survivorship rate. Most populations of this species only contain a small number of adults. Some individuals attain at least 4 yrs of age.

DISEASES, PARASITES, AND MALFORMATIONS
Lithobates onca possesses antimicrobial skin peptides that aid in resistance to some bacteria (*Escherichia coli*, *Candida albicans*) but not others (*Staphylococcus aureus*) (Conlon et al., 2010). Although presented as *Lithobates pipiens*, Parry and Grundman (1965) reported the nematodes *Cosmocercella haberi* and *Physaloptera* sp. from leopard frogs along Santa Clara Creek and the Virgin River in extreme southwestern Utah that were possibly *L. onca*. Other parasites included the cestode *Cysticercus* sp.; the trematodes *Glyp-*

Habitat of *Lithobates onca*. Photo: Dana Drake

STATUS AND CONSERVATION
Small population size combined with limited heterozygosity and a small fragmented range make the status of the species very precarious. Urgent conservation measures were recommended as early as 1980, when the continued presence of leopard frogs along the Virgin River was recognized (Bury et al., 1980). Bradford et al. (2004) estimated the entire population as only 1,100 frogs (range 693 to 1,833) at the 5 known sites.

Populations undoubtedly were lost during dam building along the Colorado River, and populations along the Muddy and Virgin rivers are apparently extinct due to agricultural and water development. Threats to its continued survival include the introduction of nonindigenous predators (game fish, American Bullfrogs, crayfish) and encroachment from emergent vegetation into its spring habitats. In this regard, restricting livestock from some springs actually may hasten the decline of the frog population, as livestock access prevents vegetation encroachment. The aggressive spread of tamarisk is likely detrimental to this species, and this highly invasive tree should be removed whenever encountered. A conservation agreement and strategy has been developed (as of 2005) under the auspices of the Nevada Division of Wildlife (http://www.ndow.org/wild/conservation/frog/Leopard/plan.pdf). Bradford et al. (2005b) reviewed conservation measures needed to ensure the survival of this species.

thelmins quieta, *Haematoloechus coloradensis*, and an unidentified Plagiorchiinae; the protozoans *Chilomastix* sp., *Hexamita intestinalis*, *Karatomorpha swazyi*, *Nyctotherus cordiformis*, *Opalina* sp., *Trichomonas* sp., and *Zelleriella* sp.; and the mite *Hannemania hegeneri* (Parry and Grundman, 1965).

SUSCEPTIBILITY TO POTENTIAL STRESSORS
No information is available.

Lithobates palustris (LeConte, 1825)
Pickerel Frog
Grenouille des marais

ETYMOLOGY
palustris: from the Latin *paluster* meaning 'of the marsh.' The name literally means 'a marsh frog.'

NOMENCLATURE
Conant and Collins (1998): *Rana palustris*
  Dubois (2006): *Lithobates (Lithobates) palustris*

Synonyms: *Rana pardalis*, *Rana palustris mansuetii*

IDENTIFICATION
Frogs of the Leopard Frog complex most closely resemble the Pickerel Frog in shape, size, and coloration. In Pickerel Frogs, the dorsal spots are squared rather than rounded, occur in two well-defined parallel rows down the back between the dorsolateral folds, and are paired rather than scattered about. These frogs also have distinct yellowish to orange color on the underside of the thighs and groin, unlike the leopard frogs. The

American Bullfrog lacks the dorsolateral folds altogether, and the Green Frog has only partial dorsolateral folds, and is unspotted.

Adults. This is a medium to large olive-green to gray, tan, or brownish frog with a light to cream-colored dorsolateral fold that extends down the body from behind the eye to the groin. Two to more than four parallel folds occur between the main dorsolateral folds on the back. There are paired large square dorsal spots on the back and sides; these spots (9–15 normally, but range 7 to 21) normally extend in 2 rows down the back between the dorsolateral folds. The rear limbs have black bars, giving a banded pattern. There is no white spot in the center of the tympanum, but a white line is present on the posterior part of the upper lip. The snout is pointed and the eyes are large. The belly and throat are white, but the underside of the hind limbs and groin are yellow to orange. In some populations, mottling may be present on the ventral surface. Males have paired vocal sacs and enlarged thumbs with thickened pads to clasp the female during amplexus. They also may be slightly smaller and lighter colored than females. Adults are mature by 44 mm TL; maximum size is 87 mm SUL (Gibbs et al., 2007).

Specific size measurements are few in the literature. In Delaware, males were a mean SUL of 58.6 mm (range 47 to 66 mm SUL) and weighed a mean of 16.7 g (range 8.0 to 23.75 g) (Given, 2005). In Ohio, males were 45–58 mm and females were 60–76 mm SUL (Walker, 1946), whereas in Connecticut males were 41–58 mm SUL (mean 50.2 mm) and females were 51–79 mm SUL (mean 60.8 mm) (Klemens, 1993). In Nova Scotia, males were 49.0–63.5 mm SUL (mean 56.3 mm) and females were 50.8–74.6 mm SUL (mean 61.8 mm) (Gilhen, 1984). Males in Louisiana were significantly smaller (mean 52.5 mm, range 42 to 65 mm SUL) than females (mean 64.1 mm, range 48 to 75 mm SUL) (Hardy and Raymond, 1991).

Larvae. The tadpole is large, deep, and full-bodied. The dorsal color is olive green grading to yellow on the sides. The belly is cream colored with white blotches, whereas the dorsum is marked with fine black and yellow spots. The belly is iridescent, and the viscera are visible, but often only slightly. The tail is very dark with black blotches and tail fins that are variously patterned. When viewed from above, the body is very round or oval. Tadpoles of *L. palustris* may be large, to 76 mm TL (Wright, 1929).

The tadpoles of *L. palustris*, *L. pipiens*, *L. clamitans*, and *L. catesbeianus* are difficult to differentiate. The throat of the Northern Leopard Frog tadpole is more extensive and translucent than that of *Rana palustris*. Tadpoles of *R. palustris* usually contain a yellow wash on the sides of the body. Tadpoles are described by Wright (1929) and Altig (1970), and the larval mouthparts are illustrated by Hinckley (1881), Wright (1914, 1929), and Dodd (2004).

Eggs. The eggs are deposited in a firm, spherical cluster 38–100 mm in diameter; each cluster contains <1,000 to 3,000 eggs. Egg masses are placed in deep water at a depth of 7.5–10 cm or greater below the surface, and are usually attached to debris and vegetation. Individual eggs are bicolored, brown above and yellow below, and the vitellus averages 1.7 to 1.85 mm in diameter (range 1.6 to 1.9 mm). The outer envelope averages 4.0 mm (range 3.6 to 5.0 mm) and the inner envelope averages 2.6 mm in diameter (range 2.3 to 3.0 mm). An illustration of the egg in relation to its two membranes is in Tyler (1994), and a description of the eggs is in Livezey and Wright (1947).

DISTRIBUTION

The Pickerel Frog is rather unevenly distributed throughout its range, with many reports referring to its presence as "spotty" or "uncommon." Pickerel frogs are found from the St. Lawrence River

Distribution of *Lithobates palustris*

south to the South Carolina coast at the Savannah River, generally north of the Fall Line in the south Gulf Coast states, and in central North America from eastern Texas north to Wisconsin, the Upper Peninsula of Michigan, and southern Ontario. There are only a few records from the Piedmont of North Carolina (Brown, 1992) and South Carolina (Corrington, 1929; Platt et al., 1999), although Dorcas et al. (2007) give a much wider distribution in the North Carolina Piedmont and Coastal Plain. There are scattered Piedmont localities from Alabama as far as south of the Tallapoosa River, as well as in the deep ravines located in the southern part of the state (Redmond and Mount, 1975; Davis and Jones, 1982). In southeast Oklahoma, their range extends westward up the Red River Valley (Bragg and Taylor, 1968). They are absent from the prairie regions of Illinois, southwest Indiana, and western Kentucky, and from much of central Alabama through northeast Mississippi. In Wisconsin, they are most common in the Driftless Area (Mossman et al., 1998).

Lithobates palustris is only occasionally found on islands, including islands in the Georgian Bay of Lake Huron (Hecnar et al., 2002), Prince Edward Island, Martha's Vineyard, and Nantucket off Cape Cod (Lazell, 1976), Long Island, New York (Overton, 1914), and Wallops Island, Virginia (Conant et al., 1990).

Important distributional references include: Alabama (Mount, 1975; Davis and Jones, 1982), Arkansas (Trauth et al., 2004), Canada (Logier and Toner, 1961; Weller and Green, 1997), Connecticut (Klemens, 1993), Delmarva Peninsula (White and White, 2007), Georgia (Williamson and Moulis, 1994; Jensen et al., 2008), Illinois (Smith, 1961; Redmer, 1998a; Phillips et al., 1999), Iowa (Hemesath, 1998), Indiana (Minton, 2001), Kansas (Smith, 1934; Collins et al., 2010), Louisiana (Dundee and Rossman, 1989), Maine (Hunter et al., 1999), Maryland (Harris, 1975), Massachusetts (Lazell, 1976), Michigan (Thompson, 1912), Mid-Atlantic (Beane et al., 2010), Minnesota (Oldfield and Moriarty, 1994), Missouri (Johnson, 2000; Daniel and Edmond, 2006), New Brunswick (McAlpine, 1997a), New Hampshire (Oliver and Bailey, 1939; Taylor, 1993), New York (Gibbs et al., 2007), North Carolina (Dorcas et al., 2007), Nova Scotia (Bleakney, 1952; Gilhen, 1984), Ohio (Walker, 1946; Pfingsten, 1998; Davis and Menze, 2000), Oklahoma (Sievert and Sievert, 2006), Ontario (Johnson, 1989; Christie, 1997; MacCulloch, 2002), Pennsylvania (Hulse et al., 2001), Québec (Bider and Matte, 1996; Desroches and Rodrigue, 2004), Prince Edward Island (McAlpine et al., 2006), South Carolina (Platt et al., 1999; Montanucci, 2006), Tennessee (Redmond and Scott, 1996), Texas (Smith and Sanders, 1952; Dixon, 2000), Vermont (Andrews, 2001), Virginia (Tobey, 1985; Mitchell and Reay, 1999), West Virginia (Pauley and Green, 1987), Wisconsin (Suzuki, 1951; Vogt, 1981; Mossman et al., 1998).

FOSSIL RECORD

Fossil Pickerel Frogs have been found at a Pliocene site in Texas and in Pleistocene sites in Arkansas, Pennsylvania, Texas, and Virginia. There is some uncertainty as to specific allocation at some of these deposits, as there are only very fine differences between the ilium of this species and that of *L. pipiens* (Holman, 2003).

SYSTEMATICS AND GEOGRAPHIC VARIATION

The Pickerel Frog is most closely related to the Gopher Frog complex (*L. areolata, L. capito, L. sevosa*) and secondarily to the Leopard Frog complex (*L. pipiens* and relatives). A previous suggestion that *L. palustris* is monophyletic with *L. pipiens* and *L. sphenocephalus* (Pytel, 1986) is not supported. As such, they are members of the *Nenirana* clade (*L. palustris* and the gopher frogs) within the *Pantherana* of the North American *Novirana* clade (Hillis and Wilcox, 2005). All members of the *Nenirana* have a snore-like mating call.

Lithobates palustris hybridizes with many other *Lithobates* under laboratory conditions. Larvae are produced in crosses with *L. areolata, L. megapoda, L. sphenocephalus, L. blairi,* and *L. berlandieri* (Mecham, 1969). In a cross with *L. montezumae*, many larvae died, and those that did not were not vigorous. Natural hybridization has been reported with *L. sphenocephalus* in Tennessee (Salthe, 1969) and Maryland (Hardy and Gillespie, 1976). In the latter case, hybridization occurred at a disturbed natural habitat where numerous instances of interspecific amplexus had been observed.

In most populations, the paired dorsal spots are quite distinct, although there may occasionally be some degree of offset between the parallel spot rows. A few frogs may have occasional small spots interspersed among the large square spots. In New Hampshire, Oliver and Bailey (1939) reported a tendency for the spot rows to fuse anteroposteriorally into parallel dorsal stripes of various length; most frogs had at least some fused spots. In New Brunswick, Pickerel Frogs are reported as having a vertebral row of spots in addition to the two main dorsal rows, thus giving them three rows of spots on the back (Cox, 1899). Pickerel Frogs with longitudinal bands instead of paired spots were also reported from Illinois (Garman, 1892). Illustrations of dorsal pattern variation are in Hardy (1964a) and Scaaf and Smith (1970).

Hardy (1964a) described *Lithobates palustris mansuetii* from the Atlantic Coastal Plain of North Carolina. This form was said to possess a small number of dorsal spots that were often fused to form longitudinal stripes, dark mottling on the belly, and melanophore stippling on the vomerine teeth. Subsequently, frogs with this phenotype were found at a number of different locations, including the Delmarva Peninsula in Maryland and Virginia, and the lowlands of the western sections of the Gulf Coastal Plain. However, these characters grade into adjacent populations and the subspecies is not considered valid (Schaaf and Smith, 1970).

ADULT HABITAT

Adults frequent a mosaic of habitat types that includes large ponds and lakes for breeding interspersed within a mostly forested landscape. In the upper Midwest, for example, they are associated with lakes, forested wetlands, and even agricultural landscapes, depending on area (Knutson et al., 2000, 2004), whereas in Ontario, wetland occupancy is correlated with the presence of nearby forest cover (Findlay et al., 2001). They occur in meadows or marshes, along sphagnum bogs, in deep cool ravines (especially in the South), and along cold streams and rivers. However, I have also heard them calling from big-river cypress swamps in the low country of South Carolina. Because of their habitat affinities, Pickerel Frogs may be a biotic indicator of cold-water and spring-fed streams and high-quality wetland habitats (Vogt, 1981; Knutson et al., 2000). Overwintering occurs in springs, in the mud on pond bottoms, in ravines, or near or in caves under stones (Barbour, 1971; DeGraaf and Rudis, 1983; Fenoilio et al., 2005; Sievert and Sievert, 2006). Peak cave use in Oklahoma was from November to December (Fenoilio et al., 2005) when more than 50 were observed within the cave. Adults may occasionally be killed by cold while overwintering (Gorham, 1964).

TERRESTRIAL ECOLOGY

Pickerel Frogs disperse widely in moist cool forested habitats after the breeding season, where they are occasionally observed far from the nearest breeding sites (Huheey and Stupka, 1967; Dodd, 2004). Oldfield and Moriarty (1994) recorded a single Minnesota individual found at least 500 m from the nearest water. There is some discrepancy between such observations and statements in the literature concerning a tendency to remain close to water, but a thorough analysis of movement patterns has not been carried out.

During the nonbreeding season, they occur in damp meadows, in cool ravines, along cold-water streams, in freshwater swamps (Anonymous, 1918), and occasionally in anthropogenic sites such as utility transmission corridors (Fortin et al., 2004b), limestone quarry pools, and golf courses. For example, Gorham (1964) reported they were found in hayfields in New Brunswick during July and August. In much of their range, they are associated with caves, particularly around the entrances (Blanchard, 1925; Bailey, 1933; Burt, 1933; Barr, 1953; Brown and Boschung, 1954; Brode, 1958, 1960; Franz, 1967; Barbour, 1971; Knight, 1972; Black, 1973a; McDaniel and Gardner, 1977; Garton et al., 1993; Buhlmann, 2001; Dodd et al., 2001; Fenoilio et al., 2005), and abandoned mines (Heath et al., 1986; McAllister et al., 1995b; Pauley and Pauley, 2007; Rimer and Briggler, 2010). Pickerel Frogs occur from 30 to 640 m in elevation in New Hampshire (Oliver and Bailey, 1939), but individuals have been found as high as 915 m in the Great Smoky Mountains (Huheey and Stupka, 1967).

Activity normally occurs from midwinter until the autumn, depending on latitude. For example, Pickerel Frogs have been found until October and November in Ohio and other northern locations (Wilcox, 1891; Gilhen, 1984; Klemens, 1993; Gibbs et al., 2007). At the southernmost portions of their range, they also are likely to be active year-round. Indeed, they may be active throughout the winter even in the North to some extent, as Babbitt (1937) found them from December through February in Connecticut, when days were warm and mild. Others have recorded them as being active in springs, caves, and under the ice in winter.

Pickerel Frogs are photopositive in their phototactic response, suggesting they use both sunlight and moonlight when feeding and going about their daily activities (Jaeger and Hailman, 1971, 1973). They are sensitive to light in the blue spectrum ("blue-mode response"), which apparently helps them orient toward areas of increasing illumination, such as the open area above a pond (Hailman and Jaeger, 1974). As such, Pickerel Frogs likely have true color vision.

CALLING ACTIVITY AND MATE SELECTION

The advertisement call is a continuous low-pitched snore. Male *L. palustris* call from both above and below the surface of the water, and the proportion of males calling from underwater varies among sites. If disturbed by a physical stimulus, they will dive below the water's surface and shortly resume calling from underwater. After a few minutes, they return to the water's surface and resume the mating (advertisement) call. They may respond to the calls of nearby *Anaxyrus americanus* by diving underwater and resuming calling, but they do not do so in response to the calls of *Pseudacris crucifer*. However, they will emit aggressive calls in response to the calls of both species (Given, 2008).

Advertisement calls are heard from late winter through early summer, the start of which varies by latitude. In Louisiana and North Carolina, for example, calling begins in February (Murphy, 1963; Dundee and Rossman, 1989; Todd et al., 2003), whereas calls are heard from April to late May in Connecticut (Babbitt, 1937) and New Hampshire (Oliver and Bailey, 1939). Todd et al. (2003) recorded that calling intensity varies among sites. In New Brunswick and Ontario, calling begins in late April, but most choruses do not reach full strength until May (Piersol, 1913; Gorham, 1964). In New York, calling begins in April (Moore, 1939) when water temperatures reach 14–16°C (Wright, 1914). In Delaware, however, Given (2005) reported chorusing at temperatures as low as 8°C, but that calling was curtailed when air temperatures were below 10°C and water temperatures were below 11°C. Calling occurs at night, beginning shortly after dusk, and at least some individuals will call throughout the night until dawn (Todd et al., 2003). Givens (2005) noted that full choruses tend to continue for five or more hours each night.

Advertisement calls have a mean duration of 1.67 sec (range 1.04 to 2.20 sec), a pulse rate of 44.9 pulses/sec (range 29.5 to 61.7 pulses/sec), and a dominant frequency of 1,222 Hz (range 624 to 1,908 Hz) (Given, 2005). These readings are based on a mean body temperature of 15.3°C (range 12.0 to 18.4°C). Larger males tend to produce longer duration advertisement calls than smaller males, suggesting the potential for female discrimination based on male size.

Aggressive calls are of two types, termed a "snicker" and a "growl." These calls, like advertisement calls, can be made from both above and below the water's surface. Male Pickerel Frogs usually produce a combination of snickers and growls toward an intruder. Growls are of longer duration with lower pulse rates than snickers; the dominant frequencies, however, are similar and much lower than those of the advertisement call. These aggressive calls are presumably used to warn other males away from a calling male's territory, and may serve as a prelude to aggressive interactions. Givens (2005) suggested that males might be able to alter their advertisement calls to make them more effective at serving a secondary function of aggression toward males while maintaining their attractiveness toward females.

Amplexus occurs by night or day in shallow water, is pectoral, and may continue for several days following oviposition (Wright, 1914). Occasionally, Pickerel Frogs will attempt to amplex other species, such as *Lithobates clamitans*, *L. pipiens*, *L. sphenocephalus*, or even *Anaxyrus americanus*. Unmated males will also attempt to dislodge an amplexing male. According to Wright (1914), reproductive behavior is quite gregarious within rather small areas, with many unmated individuals and mated pairs in close proximity. Males probably attempt to amplex any incoming female near them.

BREEDING SITES

Lithobates palustris prefers permanent wetlands (Herrmann et al., 2005), usually larger lakes and ponds. In New Hampshire, for example, they are associated with wetlands with long hydroperiods (Babbitt et al., 2003). However, they have been recorded in road-rut ponds in Kentucky, although it is unclear whether they were actually breeding (Adam and Lacki, 1993), and they do breed in ditches and woodland pools (Walker, 1946; Dale et al., 1985; Dodd, 2004) and in small isolated wetlands (Russell et al., 2002a). Pickerel Frogs also are found in anthropogenic sites such as reclaimed mine wetlands (Turner and Fowler, 1981; Lacki et al., 1992), small reservoirs (Gibbons and Bennett, 1974), stormwater-retention ponds (Brand and Snodgrass, 2010), farm ponds, quarries, and sand pits (Klemens, 1993), and in beaver ponds (Metts et al., 2001). Pickerel Frogs may colonize constructed wetlands, but they are rare in such habitats (Briggler, 1998). Breeding sites usually have extensive canopy cover with emergent vegetation around the periphery. However, there is one report of Pickerel Frogs laying eggs in a stream within a cave (Brown, 1984).

Pickerel Frogs are found in ponds that are fish-free, but they also are found in wetlands containing both predaceous and nonpredaceous fishes (Kats et al., 1988; Hecnar, 1997; Hecnar and M'Closkey, 1997b; Baber et al., 2004). Pickerel Frogs are not ubiquitous among breeding sites, but tend to occur only in few wetlands. In Ontario, for example, they occurred in ca. 10% of the wetlands sampled across two large-scale sampling areas, and in only a few or not at all in a third and fourth area (Hecnar and M'Closkey, 1996a, 1998; Hecnar, 1997).

A second type of breeding site is a clear cold-water stream. In Kentucky, Pickerel Frogs deposited eggs in the main channel of a second-order stream, primarily in areas of reduced presence by predaceous fishes. Females also oviposited egg clutches in runs along the edge of the stream, but not in currents or even in adjacent fish-free pools near the stream. They preferred areas farther in the channel and in deeper waters, but not in isolated pools (Holomuzki, 1995). Wright (1914) noted breeding in cold streams, and that Pickerel Frogs often used the plunge pools of waterfalls as breeding sites in New York.

REPRODUCTION

Maturity is reached in the second summer after metamorphosis (i.e., in the third year of life). However, because larvae may overwinter at the northern limit of the species' range, it is possible sexual maturity could be delayed. It is not known whether males and females reach maturity at the same age.

Pickerel Frogs enter the breeding sites 7–30 days prior to when mating begins (Wright, 1914), although Hardy and Raymond (1991) reported immigration as early as October, November, and December for a breeding season in Louisiana that occurred primarily in February and March. In

Eggs of *Lithobates palustris*.
Photo: John Jensen

Tadpole of *Lithobates palustris*. Photo: Stan Trauth

Louisiana, soil and water temperatures of 7–10°C were the best indicators of the initiation of immigration (Hardy and Raymond, 1991). Males and females enter a pond over an extended time, usually during periods of rain, and they enter the wetland synchronously rather than staggered by sex. However, they may not enter in equal numbers.

At a North Carolina breeding pond, the sex ratio was 6.64 males per female over the course of a breeding season (Murphy, 1963), whereas it varied in Louisiana from 1.3 males per female one year to 5.4 males per female the next (Hardy and Raymond, 1991). Adults tend to trickle into a breeding site, with most reproduction occurring after a period of several weeks of immigration, and then tapering off. In Missouri, both sexes initially entered breeding sites in mid-March, and most immigration was finished by late April. A few females, however, continued to enter the pond for about a week after male immigration was complete (Hocking et al., 2008). Adults did not stay at the pond beyond the breeding season, and most emigration was completed by late May. Thus the entire breeding season extended <3 months, with most immigration occurring over a 1.5-month time span. In Delaware, the breeding season extends for about six weeks, with only sporadic calling thereafter (Given, 2005); Murphy (1963) found a similar pattern in North Carolina, as did Hardy and Raymond (1991) in Louisiana.

Males stay at the pond and chorus throughout the breeding season. For example, males called from 1 to 38 nights in Delaware, with a median of 25 nights (Given, 2005). They tend to occupy the same calling sites, or move away only small distances. Intermale aggression includes wrestling and production of the aggressive calls. During the prolonged breeding season, males tend to lose body mass, with larger males losing more body mass than smaller males. These results suggest that prolonged calling and the defense of breeding sites are energetically costly to calling males.

Amplexed pairs move together until the female locates a submerged branch or piece of debris around which she will oviposit her eggs. The female grasps a stick with her hind legs, and flattens her body so that it is horizontal. Her hind limbs are drawn toward her body so that her heels touch in close proximity to the stick. The male's hind limbs are positioned so that they are parallel to the female's, and slightly within them. As the eggs are extruded, the male fertilizes them and thrusts his legs backward, as if pushing the eggs out of the way. During this time, the female holds her arms together. Oviposition is completed in about 3 minutes, with 10–12 fertilizations. Wright (1914) described the mating sequence based on a series of observations.

Eggs are attached to submerged vegetation or rest on the substrate and hatch in about a week to 21 days (Anonymous, 1918), depending on temperature. Eggs develop in water between 8.6 and 30°C (Moore, 1939). However, the eggs must remain well oxygenated, and at temperatures ≥25°C eggs do not survive if they are in the center of the egg mass (Moore, 1940). Ranid frogs living in wetlands with fishes generally have larger clutch sizes than those species living in fish-free ponds. This is true for Pickerel Frogs, with literature reports of clutch sizes of 1,760 (range 960 to 2,943) in Arkansas (Trauth et al., 1990); 2,652 in Kentucky (Westerman et al., 2003b); and 780–1,834 in Nova Scotia (Gilhen, 1984). Reports of egg clutch sizes of 2,000–3,000 appear to originate from Wright (1914). Clutch size is positively correlated with female body size.

LARVAL ECOLOGY

As with many aquatic anurans, larval survivorship and size at metamorphosis are dependent upon which species of predators are present, their abundance, and the density of both conspecific and heterospecific larvae (see Community Ecology). Time to metamorphosis normally is 90–100 or more days (Wright, 1914; Anonymous, 1918). In Missouri, larvae spend from 125 to 157 days in a pond (Hocking et al., 2008). However, larvae may overwinter in the extreme northern portions of the range and metamorphose in July or August of the second summer (Bleakney, 1952). Newly metamorphosed *L. palustris* with no remaining larval tails are 25–28 mm SUL in New York (Munz, 1920), 19–27 mm SUL in the Great Smoky Mountains (Dodd, 2004), 19–26 mm SUL in Ohio (Walker, 1946), and 24.5–29 mm SUL (mean 26.7 mm) in Nova Scotia (Gilhen, 1984).

DISPERSAL

Postmetamorphic Pickerel Frogs generally have an increased aerobic capacity as they begin dispersing from breeding ponds, but they do not develop an endurance capacity (Pough and Kamel, 1984). Both adults and juveniles tend to emigrate during periods of rainfall (Murphy, 1963). Nothing else is reported in the literature concerning the extent of dispersal or habitat use during the nonbreeding season.

DIET

The guts of tadpoles usually contain a mixture of mud and various diatoms, blue-green algae, and both filamentous and nonfilamentous green algae. Some guts may contain pieces of vascular plants or other debris, including particles of epidermis. Juveniles begin feeding at metamorphosis, and eat both aquatic and terrestrial prey, including lepidopteran larvae, fly larvae, flies of various types, small terrestrial and water beetles, weevils, true bugs, snails, and crustacean eggs (Munz, 1920). Recent metamorphs in Nova Scotia ate a great variety of invertebrates, including ants, spiders, true bugs, beetles, sawfly larvae, and moth larvae, whereas adults also fed on sowbugs, centipedes, harvestmen, pseudoscorpions, springtails, mites, damselflies, crickets, aquatic and terrestrial beetles, wasps, earthworms, slugs, ants, and many types of adult and larval flies and moths (Gilhen, 1984). Additional dietary information is in Surface (1913).

PREDATION AND DEFENSE

The larvae of Pickerel Frogs may be distasteful, although they are eaten by certain predaceous fishes. In the presence of fish, larvae tend to reduce their levels of activity, and they prefer shallower waters, presumably avoiding deep waters where larger predaceous fish are more likely to occur (Holomuzki, 1995).

Adult Pickerel Frogs possess a noxious skin secretion that may help to deter predators (Dunn, 1935; Walker, 1946; Wright and Wright, 1949; Formanowicz and Brodie, 1979). The secretion is acrid and distasteful, and is produced by granular glands located along the dorsolateral folds. Garter snakes (*Thamnophis sirtalis*, *T. sauritus*) have been reported to both accept (Manion, 1952;

Adult *Lithobates palustris*.
Photo: C.K. Dodd Jr.

Klemens, 1993) and reject (Dunn, 1935; Babbitt, 1937; Walker, 1946) this species. Garter snakes are known to ingest many poisonous amphibians, such as newts (*Notophthalmus*). Water snakes (*Nerodia sipedon*) also are reported to reject Pickerel Frogs (Babbitt, 1937). Shrews (*Blarina*) also tend to reject this species, with 33% surviving in experimental trials. However, Mulcare (1966) could not demonstrate toxicity in either adults or larvae. Biologists have long been cautioned about putting this species in containers with other frogs, as the confinement often leads to the death of *Lithobates palustris* as well as other amphibians.

When attacked, *L. palustris* inflates the body, faces the predator in a head-down posture, and tilts the body toward the predator. This defensive behavior makes the frog appear bigger than it is, more difficult to attack, and orients the body's granular glands in the predator's direction. Larger frogs are more likely to posture than juveniles. Other defenses include flight, remaining motionless, and crouching and hiding (Marchisin and Anderson, 1978). The Pickerel Frog is an agile jumper and difficult to catch once it takes flight. In addition, the ground coloration is quite cryptic when the frog remains motionless. Pickerel Frogs may give ample warning of their distasteful secretions: the bright ventral coloration of the legs and groin may serve an aposematic function, and an odor described as "like that of alfalfa sprouts or the inside of a latex glove" (Grant, 2001) provides an olfactory cue to alert predators to the noxious quality of the secretion. Vogt (1981) suggested that Northern Leopard Frogs may mimic Pickerel Frogs, but this hypothesis has not been tested.

Specific reports of predators are few, but include the pickerel fish (*Esox*) (Hunter et al., 1999), Bald Eagle (*Haliaeetus leucocephalus*) (Applegate, 1990), Northern Water Snake (*Nerodia sipedon*) (Klemens, 1993), and mink (*Mustela vison*) (Beane, 1990); both American Bullfrogs and Green Frogs eat small Pickerel Frogs (Babbitt, 1937). Other vertebrates are likely to eat adults and juveniles, and both invertebrates and vertebrates likely consume the eggs and larvae. Wright (1914) observed newts (*Notophthalmus viridescens*) eating eggs.

COMMUNITY ECOLOGY

In mixed trials with *Anaxyrus americanus* larvae and predators consisting of newts (*Notophthalmus*) and dragonfly larvae (*Anax*), larval Pickerel Frog survivorship was generally high in the presence of the toad larvae, in the absence of predators, and when the predator was *Notophthalmus*. The presence of *Anax* had much more of an effect on survivorship than the newt, but the mass of the survivors in the *Anax* trial was much greater than in the other trials. By eliminating many conspecifics, the dragonfly larvae ensured that the surviving Pickerel Frog larvae were much larger than in other trials. Presumably, the large *Lithobates palustris* larvae were too large for the newts to handle, although newts ate smaller tadpoles. The presence of the toad larvae offered a degree of competition (reducing survivorship by 5%), since Pickerel Frog survivorship and mass was greater without the toads (Wilbur and Fauth, 1990). These results demonstrated that the presence of the predators served to mediate the effects of competition between *Anaxyrus* and *Lithobates palustris*.

DISEASES, PARASITES, AND MALFORMATIONS

Pickerel Frogs are infected by the nematodes *Abbreviata* sp., *Capillaria tenua*, *Cosmocercoides dukae*, *C. variabilis*, *Oswaldocruzia pipiens*, and *Rhabdias ranae* (Coggins and Sajdak, 1982; Dyer, 1991; reviewed by McAllister et al., 1995c; Bursey and DeWolf, 1998), the trematodes *Alaria mustelae*, *Allassostoma parvus*, *Brachycoelium salamandrae*, *Clinostomum complanatum*, *Euryhelmis monorchis*, *Glypthelmins quieta*, *Gorgodera amplicava*, *G. attenuata*, *Haematoloechus medioplexus*, *Halipegus* sp., and *Haplometrana intestinalis* (Bouchard, 1951; Coggins and Sajdak, 1982; Dyer, 1991; reviewed by McAllister et al., 1995c), and the cestode *Mesocestoides* sp. (McAllister et al., 1995c). Gilhen (1984) noted unidentified nematodes in Pickerel Frogs from Nova Scotia. Adult *Lithobates palustris* can be infected by the mite *Hannemania dunni* (Brandt, 1936; Murphy, 1965; McAllister et al., 1995c). In a study in North Carolina, between 60 and 90% of Pickerel Frogs at a breeding pond were infected over a 3 yr period (Murphy, 1965). Juveniles at the same site had a high incidence rate of spiruroid nematode larvae, and Murphy (1965) noted 12 juveniles with malformed forelimbs and 4 with unflexed hearts. Additional parasites include the protist *Cepedietta virginiensis* (McAllister and Trauth, 1996); other infectious protozoans and protozoan-like organisms include *Myxidium serotinum*, *Nyctotherus cordiformis*, and *Opalina* sp. (McAllister et al., 1995c).

An iridovirus was reported on Pickerel Frogs from Maine, New Hampshire, and Tennessee (Green et al., 2002). In controlled situations, larvae orally inoculated or exposed to water containing Frog Virus 3 or an FV3-like isolate develop disease symptoms, and death occurs in two to six days. Tadpoles exhibited moderate to severe edema, erythema, and had pale livers and kidneys (Hoverman et al., 2010); exposed larvae were 34% more likely to die than uninfected larvae. Pickerel Frogs are susceptible to fungal infections. The virulent amphibian chytrid fungus *Batrachochytrium dendrobatidis* was found on 23% of the Pickerel Frogs sampled in Maine, where it primarily affected the hind toe webbing and skin of the pelvis (Longcore et al., 2007) but at much lower levels in Québec (Ouellet at al., 2005a). Amphibian chytrid also has been found on Pickerel Frogs from Great Smoky Mountains National Park (Todd-Thompson et al., 2009) and in Missouri caves (Rimer and Briggler, 2010). Pickerel Frogs can be infected by an *Ichthyophonus*-like fungal infection (Mikaelian et al., 2000). Under laboratory conditions, Pickerel Frogs can be infected with Frog Virus 3, where it is lethal to embryos (Granoff et al., 1965; Granoff, 1969).

Malformations have been observed in Pickerel Frogs from federal lands in the northeastern United States. Of 212 frogs examined, 5 were malformed. Malformations included missing feet, malformed hind limbs, and an extra hind limb (Converse et al., 2000). Other reports of malformed *Lithobates palustris* include Ryder (1878), Tuckerman (1886), and Murphy (1965).

SUSCEPTIBILITY TO POTENTIAL STRESSORS

Metals. The Pickerel Frog is considered very sensitive to heavy metals, including barium, beryllium, cadmium, copper, iron, magnesium, selenium, silver, and zinc (Westerman et al., 2003a).

pH. pHs between 4 and 4.5 are lethal to Pickerel Frogs, and the critical pH is between 4.3 and 4.5 (Gosner and Black, 1957a). In Nova Scotia, *L. palustris* is found in waters with pHs of 4.5–7.8 (mean 5.2) (Dale et al., 1985), whereas Russell et al. (2002a) reported successful reproduction in isolated wetlands with a pH of 4.3. Hatching success decreases substantially at pHs ≤5.0. Turner and Fowler (1981) also noted the presence of *L. palustris* at reclaimed surface mines with low pH (4.0) in Tennessee.

Chemicals. The Pickerel Frog is considered tolerant of organics (Westerman et al., 2003a). However, the following chemicals cause significant reduction in hatching success of Pickerel Frog eggs (given as mean tested concentration causing significantly decreased hatching): atrazine (20.6 mg/L), carbon tetrachloride (4.98 mg/L), chloroform (40 mg/L), methylene chloride (> 32.1 mg/L), nitrilotriacetic acid (NTA) (222 mg/L), and phenol (1.86 mg/L). $LC_{50\ (4\ day)}$ values were: atrazine (17.96 mg/L), carbon tetrachloride (2.37 mg/L), chloroform (20.55 mg/L), methylene chloride (> 32 mg/L), NTA (134.6 mg/L), and phenol (9.87 mg/L) (Birge et al., 1980).

In a study examining the effects of oil runoff on frogs in Kentucky, there were about the same number of Pickerel Frog egg masses deposited in ponds receiving runoff from pastureland and oil brine pits as in reference ponds. However, sample sizes were small, making conclusions about the effects of these substances on Pickerel Frogs impossible (Westerman et al., 2003b).

Death occurs in about 21 hrs at 30 mg/L of carbaryl, a broad-spectrum insecticide, and significantly reduces larval activity levels at 2.5 mg/L (Bridges and Semlitsch, 2000).

STATUS AND CONSERVATION

Throughout much of its range, the Pickerel Frog is considered uncommon or rare (e.g., Cox, 1899; Brown, 1992; Hecnar and M'Closkey, 1997a; Hemesath, 1998; Mierzwa, 1998; Mossman et al., 1998; Platt et al., 1999; Sargent, 2000; Rice et al., 2001; Brodman, 2003; Baber et al., 2004; Florey and Mullin, 2005; Herrmann et al., 2005). Whether this reflects a naturally spotty distribution or is the result of long-term habitat changes is largely unknown. Redmer (1998a) considered the species not to have declined in Illinois, except in the urbanized Chicago metropolitan area. The species is considered to be declining in Iowa (Christiansen, 1981) and experiencing localized declines in Ontario and Québec, but not in New Brunswick or Nova Scotia (Weller and Green, 1997). The species has not been seen in Kansas since reported by Smith (1934; see Collins, 1993).

In several areas, populations of Pickerel Frogs are isolated in fragmented habitats, such as on the

Breeding habitat of *Lithobates palustris*, Great Smoky Mountains National Park, Tennessee. Photo: C.K. Dodd Jr.

small hills and mountains of southern Québec, otherwise surrounded by agriculture and urban habitats (Ouellet et al., 2005b). The Pickerel Frog is a forest species when not at breeding ponds, and it will move along cool streambeds when dispersing. It is not prone to move through open areas, such as those created by transportation corridors. Thus, the presence of roads near breeding ponds will influence dispersal by this species, and may act to filter movement across a landscape (Gibbs, 1998a). Mortality by highway traffic has been reported by Glista et al. (2008). Unfortunately, the species does not readily colonize restored wetlands (Shulse et al., 2010).

Because Pickerel Frog populations are spottily distributed, it is important to identify and monitor them to ensure their long-term persistence.

Pickerel Frogs have been detected using call surveys in conjunction with regional monitoring programs. In Tennessee, for example, they were heard at 79 of 321 calling stations. Listening for five or ten minutes made no difference in terms of an observer's ability to detect this species (Burton et al., 2006). However, detection of *L. palustris* during call surveys was exceedingly uncommon in Québec and Ontario (Bishop et al., 1997; Lepage et al., 1997) and elsewhere, suggesting other protocols for monitoring might be necessary for this species. Skelly et al. (2003) noted that the probability of determining occupancy for *L. palustris* was low (20%) for surveys conducted annually, and that long-term resurveys of historical sites are necessary to determine the presence and persistence of this species.

Lithobates pipiens (Schreber, 1782)
Northern Leopard Frog
Grenouille léopard

ETYMOLOGY

pipiens: from the Latin *pipiens*, meaning 'peeping.' Apparently the first collector of this species heard Spring Peepers and thought that the leopard frogs that he had just collected were responsible for the loud 'peeps' he heard.

NOMENCLATURE

Conant and Collins (1998) and Stebbins (2003): *Rana pipiens*
  Dubois (2006): *Lithobates (Lithobates) pipiens*
  Hillis (2007): *Rana (Pantherana) pipiens*
  Synonyms: *Rana brachycephala, Rana burnsi, Rana burnsorum, Rana halecina, Rana kandiyohi, Rana noblei, Rana pipiens brachycephala, Rana virescens brachycephala*

There are many scientific publications in which the identification of "leopard frogs" has been confused among the currently recognized eight species within the United States and Canada. This is especially true in physiology and toxicology, where it may be impossible to determine which species was studied. For example, Bachmann (1969) discussed the influence of temperature on development in "*Rana pipiens*," yet included frogs from New Jersey, Vermont, Texas, Florida, Costa Rica, and Mexico in his analysis. Other studies have relied on frogs from commercial suppliers (Bagnara and Frost, 1977), and it cannot be assumed that "leopard frogs" were collected in proximity to the supplier. As such, there is confusion about species identity in the following publications: Moore (1949b), Volpe (1957a), Bresler (1963, 1964), Kaplan and Overpeck (1964), Kaplan and Glaczenski (1965), McLaren (1965), Licht (1967), Levine and Nye (1977), Cole and Casida (1983), Wygoda (1984), Dial and Bauer (1984), and Dial and Bauer Dial (1987). If the original literature does not specify the origin of "*Rana pipiens*," the applicability of the information to every species within the complex is questionable. When consulting much of the scientific literature prior to the early 1970s, readers are cautioned to remember that results and interpretations of numerous past studies involving "leopard frogs," even when localities or sources of individuals are provided, are based on assumptions of clinal or geographic variation rather than presently recognized species-specific differences.

IDENTIFICATION

Adults. This is a medium to large olive-green or brown frog with smooth skin. It possesses a yellow to cream-colored continuous dorsolateral fold that extends the entire length of the body from the snout to the groin. There are scattered large dorsal unpaired oval or roundish spots on the head, back, and sides; these spots may extend in two or three irregular rows down the back between the dorsolateral folds. There are usually 0–1 spots between the posterior extent of the eyes and the snout when viewed dorsally (Bresler, 1963). The rear limbs, especially, have black bars giving a banded pattern. There is a white spot in the center of the tympanum, but the spot may be unclear or inconspicuous, and a white line (supralabial stripe) is present on the upper lip. The snout is pointed and the eyes are large. The belly, throat, and underside of the hind limbs are white.

Sexual dimorphism among Northern Leopard Frogs is not apparent, except during the breeding season when males have paired lateral vocal sacs (evidenced by loose skin between the angle of the jaw and forearm) and swollen thumbs to assist with amplexing females. Males also may be less patterned than females on the head and dorsal regions, but more patterned on the lateral and ventral surfaces (Schueler, 1982a). Unlike some other members of the leopard frog complex, males possess oviducts. Females during the breeding season are usually swollen with eggs, but there is no other distinguishing external characteristic. Females are also usually larger than males, but differences are slight.

Schueler (1979, 1982a) provided an extensive analysis of pattern variation in *Lithobates pipiens* in Canada and the northeastern United States. He concluded that leopard frogs from warmer and moister climates tended to be darker than those from cooler and dryer climates; variation in dermal secretory glands is independent of pattern, but that these glands are more extensive where water is widespread, as well as in the North; linear patterns in coloration and reduced leg spotting are associated with field, lake edge, and marsh habitats; variation in spot number varies considerably within populations, but not between populations; and among-population variation in spot number is inversely correlated with spot area. In addition, the brown morph tended to be associated with marshes and lakes in the North, whereas the green morph was found most often in boreal forest. Whether these trends apply to other populations is unknown.

Adults are 51–111 mm SUL. Specific size records include a mean of 67 mm SUL for females after egg deposition (range 55 to 80 mm SUL) and 63.8 mm SUL for males (range 55 to 75 mm SUL) in Wisconsin, whereas body mass averaged 27.8 g (range 17 to 48 g) for females and 25.8 g for males (17–39 g) (Hine et al., 1981); females from 63 to 90 mm SUL and males 58–77 mm SUL in Ohio (Walker, 1946); a mean of 60.3 mm SUL for females and 58.2 mm SUL for males in West Virginia (Sutton, 2004); males 51–65 (mean 57.2) mm SUL and females 53–65 (mean 57) mm SUL in Connecticut (Klemens, 1993); mean SUL in males 68.3 mm and for females 74.2 mm in New Mexico (*in* Degenhardt et al., 1996); males 57–78 mm SUL and females 66–95.3 mm SUL in Nova Scotia (Gilhen, 1984); males 58–78 mm SUL (mean 69.4 mm) and females 60–91 mm SUL (mean 80.2 mm) on Prince Edward Island (Cook, 1967). Matsuda et al. (2006) give maximum sizes of 111 mm SUL for females and 80 mm SUL for males.

Adult albinos have been reported from New York and Wisconsin (Hensley, 1959) and South Dakota (Browder, 1972). Other unusual color variations include melanistic frogs of both the wild and Burnsi phenotypes (Richards et al., 1969; Richards and Nace, 1983) and blue (axanthic) leopard frogs (Berns, 1966; Berns and Uhler, 1966; Black, 1967).

Larvae. The tadpole is large (normally < 90 mm TL but to 120 mm TL; Claflin, 1962) and deep-bodied, dark brown to dark green dorsally, and covered with small gold spots. The belly is cream colored with a bronze iridescence, the viscera tend to be visible through the skin, and the throat is translucent. The spiracle is sinistral and located below the lateral axis of the body. Tail fins may or may not be heavily marked with speckles or spots (but not large spots), and the dorsal tail fin is rounded. In general, the tail fin is lighter than the body. Tadpoles of frogs that develop the Kandiyohi color pattern as adults are much more heavily pigmented than normal larvae of *L. pipiens* (Volpe, 1955b). Eyes are bronze. The lower jaws are wide. Descriptions of tadpoles and their mouthparts are in Hinckley (1881), Wright (1929), Nichols (1937), and Altig (1970).

There appears to be regional variation in tadpole coloration and pattern. In the East, Northern Leopard Frog larvae are generally darker and more patterned than they are in the West. For example, Scott and Jennings (1985) noted that western *L. pipiens* were relatively pale and not very patterned when compared to larvae of some other members of the *pipiens* complex (*L. chiricahuensis, L. berlandieri, L. yavapaiensis*). Western *L. pipiens* larvae have a deep heart-shaped body, a moderately deep and narrow tail fin that is unpatterned. Larval albinos are reported by Federighi (1938).

Characteristics such as numbers of labial tooth rows, oral papillae condition, mouth shape, visibility of the lateral line system, color of iris, peritoneal coloration, and myomere visibility are not particularly helpful in identifying larvae of species within the Leopard Frog complex. Korky (1978) stated that larvae of *L. pipiens* are "more robust and full bodied" at the posterior of the body and that the snout is more acute than in larvae of *L. blairi* (cylindrical body, more rounded snout); the differences are subtle. He further noted that the oral disk of *L. pipiens* is terminal as opposed to subterminal in *L. blairi,* although it is difficult to discern this in the accompanying photograph. Scott and Jennings (1985) provided a detailed comparison of these features among *L. pipiens, L. berlandieri, L. blairi, L. chiricahuensis*, and *L. yavapaiensis.*

Eggs. Eggs are deposited in a firm jellied mass, with each mass containing 600 to 6,500 eggs (but usually several thousand). The eggs are bicolored, white on the bottom and black on top, average about 2.0 mm (range 1.3 to 2.3 mm) in diameter (Moore, 1939; Livezey and Wright, 1947; Livo, 1981a), and tend to be packed close together. There are two to three jelly envelopes surrounding the egg. Outer jelly envelopes are an average of 5.0 (range 4.2 to 6.0) mm, with inner envelopes averaging 2.25 to 2.4 mm. As expected, egg size depends on time since deposition, as eggs tend to swell after several days. The egg mass is usually attached to grasses or submerged vegetation near the water's surface (within 5 to 20 cm), although occasionally free, and is not as firm as those of *L. sylvaticus* and *L. palustris*. Hatching occurs in from 7–20 days depending on temperature.

DISTRIBUTION

The Northern Leopard Frog is widely distributed in northern sections of eastern and central North America. It ranges from the northern shore of the St. Lawrence River and the Canadian Maritimes west across southern Québec, most of Ontario, southern and central Manitoba, and across Saskatchewan to the northwest corner of the province to Great Slave Lake in the Northwest Territories. An isolated population is found in central Labrador (Lake Melville District). The natural range is ill-defined in the North and West, but the species now occurs from northeast Alberta southwestward to the eastern Rocky Mountain front, across much of eastern and central Montana, Wyoming, southeast Idaho, Utah, and into northern Arizona and New Mexico (but down the Rio Grande Valley as well) (Smith, 2003). A record from Cardenas Marsh in the Grand Canyon (Tomko, 1975; Miller et al., 1982) was probably referable to *L. pipiens*, but the popu-

Distribution of *Lithobates pipiens*. Dark gray indicates extant populations; light gray indicates extirpated populations.

lation has been extirpated. Scattered allopatric populations are found in British Columbia (only one population currently known at Creston Lake), Washington (at Crab Creek), Oregon (at Vale), and (possibly) northern and eastern California. All other historic populations in British Columbia, Washington, and Oregon have been extirpated, as have most native California populations (Jennings and Hayes, 1994b; Jones et al., 2005) and Rocky Mountain (Johnson et al., 2011) populations. The systematic status of some populations remains to be clarified, and some extant populations may have resulted from deliberate introductions, especially in California.

The southern limit of its range encompasses much of Colorado (at elevations of 1,800–2,750 m; highest record at 3,355 m, Hammerson, 1999), central Nebraska (north of the North Platte River) and Iowa, northern Illinois southeast to northern Kentucky, eastward across West Virginia, northwest and central Pennsylvania, and across the New England states. Conant and Collins (1998) showed the species as absent from the Hudson River valley, but populations occur in the lower Hudson and adjacent Housatonic river valleys of New York and Connecticut (Klemens et al., 1987; also see Newman et al., 2012) and in the Albany Pine Bush (Stewart and Rossi, 1981). The species appears absent from most of the northern portions of the Atlantic Coastal Plain; scattered populations, however, are found in New Jersey and Maryland (Klemens et al., 1987). The southernmost extent of the species in the East may have been an isolated population in Cades Cove of the Great Smoky Mountains (King, 1939; Dodd, 2004), but this population is likely extirpated.

Northern Leopard Frogs widely overlap in distribution with other members of the Leopard Frog complex. Early publications tend to lump the ranges of the various species under the name "*Rana pipiens*," although they sometimes acknowledge the morphological variation present. Members of the complex may partition the environment where they occur sympatrically or at least in close proximity. For example, *Lithobates pipiens* is confined to elevations of 1,800–2,750 m in Arizona, well above most of the members of the complex within that state (Frost and Platz, 1983). In New Mexico, it occurs at 1,120–3,050 m (Degenhardt et al., 1996). Distribution maps showing the ranges of various leopard frog species in close proximity to one another are available for

Arizona (Platz, 1976), Colorado (Post and Pettus, 1966), Nebraska (Lynch, 1978), the northern and central Great Plains (Dunlap and Kruse, 1976), and southern Great Plains (Littlejohn and Oldham, 1968).

Northern Leopard Frogs have been introduced to a variety of locations outside their native range (Kraus, 2009). Storer (1925) reported introductions in Tehama County, California, in 1918, and Bury and Luckenbach (1976) summarized the locations of introduced populations west of the Sierra Nevada and in southern California; nearly all of these populations are no longer extant (Jennings and Hayes, 1994b; Bury, personal communication). Other introductions include the Snake Valley, Utah (Hovingh, 1993a), possibly the Cape Cod region of Massachusetts (Lazell, 1976), near Coombs on Vancouver Island, British Columbia (Weller and Green, 1997), a pond on the Washington State University campus (Jones et al., 2005), Baca, Bent, and Yuma counties, Colorado (Hammerson, 1999), and the island of Newfoundland (in 1966; see Maunder, 1983); the Newfoundland population has been extirpated (Maunder, 1997). According to Collins and Wilbur (1979), Northern Leopard Frogs from Vermont were released on the E.S. George Reserve in Michigan. Northern Leopard Frogs also were reintroduced to the Bummer's Flat Conservation Area in British Columbia (Matsuda et al., 2006). Undoubtedly, leopard frogs have been released widely throughout both their native and nonnative ranges after use in schools, experiments, and as pets. It has also been introduced into Germany, Britain, Korea, the Virgin Islands, Hawaii, Nevada, and Oklahoma (Kraus, 2009).

Northern Leopard Frogs occur on islands, including islands in western Lake Erie (Pelee, Kelleys, South Bass, Middle Bass, North Bass, West Sister), but some of these records are old, and populations may have been extirpated (Walker, 1946; Langlois, 1964; King et al., 1997). They occur in the Apostle Archipelago (Lake Superior), on islands in Georgian Bay (Lake Huron), Walpole Island in Lake St. Clair (Woodliffe, 1989), on islands near the outlet to the St. Lawrence River in Lake Ontario (Hecnar et al., 2002), on Île d'Orléans, Île aux Coudres, and Île d'Anticosti in the St. Lawrence estuary (Johansen, 1926; Bleakney, 1954; Fortin et al., 2004a), and on Prince Edward Island (Cook, 1967). Within the Great Lakes, the occupancy rate was ca. 33% of the islands (n =107 visited) surveyed by Hecnar et al. (2002). Gorham (1964) noted them on Grassy Island in the Saint John River, New Brunswick, and Klemens (1993) recorded them on Pasque (Elizabeth Islands, Massachusetts; now probably extirpated), and Aquidneck and Conanicut islands in Narragansett Bay.

Important distributional references include: Alberta (Wagner, 1997; Russell and Bauer, 2000; ANLFRT, 2005), Arizona (Brennan and Holycross, 2006), British Columbia (Matsuda et al., 2006), California (Jennings and Hayes, 1994b), Canada (Bleakney, 1958a; Logier and Toner, 1961; Weller and Green, 1997), Colorado (Hammerson, 1999), Connecticut (Klemens, 1993), Idaho (Slater, 1941b; Tanner, 1941), Illinois (Schmidt and Necker, 1935; Smith, 1961; Redmer, 1996; Phillips et al., 1999), Indiana (Minton, 2001; Brodman, 2003), Labrador (Backus, 1954; Maunder, 1983, 1997), Maine (Hunter et al., 1999), Manitoba (Harper, 1963; Preston, 1982), Maryland (Harris, 1975), Minnesota (Merrell, 1965), Missouri (Daniel and Edmond, 2006), Montana (Black, 1970; Maxwell et al, 2003; Werner et al., 2004), Nebraska (Hudson, 1942; Lynch, 1985; Ballinger et al., 2010; Fogell, 2010), Nevada (Linsdale, 1940), New Brunswick (McAlpine, 1997a), Newfoundland (Maunder, 1983), New Hampshire (Oliver and Bailey, 1939; Taylor, 1993), New Mexico (Van Denburgh, 1924; Degenhardt et al., 1996), North America in general (Pace, 1974), North Dakota (Wheeler and Wheeler, 1966; Hoberg and Gause, 1992), Northwest Territories (Preble, 1908; Harper, 1931; Fournier, 1997, undated), Nova Scotia (Richmond, 1952; Gilhen, 1984), Ohio (Pfingsten, 1998; Davis and Menze, 2000), Ontario (Schueler, 1973; Johnson, 1989), Oregon (Gordon, 1939; Ferguson et al., 1958), Pennsylvania (Hulse et al., 2001), Prince Edward Island (Cook, 1967), Québec (MacCulloch and Bider, 1975; Bider and Matte, 1996; Desroches and Rodrigue, 2004), South Dakota (Peterson, 1974; Fischer, 1998; Ballinger et al., 2000; Smith, 2003; Kiesow, 2006), Utah (Van Denburgh and Slevin, 1915; Tanner, 1931), Vermont (Andrews, 2001), Washington (Metter, 1960; Leonard et al., 1999), West Virginia (Green and Pauley, 1987), Wisconsin (Suzuki, 1951; Vogt, 1981; Mossman et al., 1998), and Wyoming (Baxter and Stone, 1985).

FOSSIL RECORD

Fossils of the Northern Leopard Frog are known from Pleistocene (Rancholabrean) deposits in Ohio and Québec, but Holman (2003) noted that isolated individual fossils of the Leopard Frog complex are extremely difficult to identify. The calcareous fossil from Eardley, Québec, is the best-preserved fossil anuran in North America.

SYSTEMATICS AND GEOGRAPHIC VARIATION

At one time, *L. pipiens* was considered an extremely wide-ranging species that exhibited a great deal of phenotypic variation (Moore, 1944, 1975). Regional variants were sometimes considered subspecies, and a few of these were considered full species, as they are currently (i.e., *L. fisheri*, *L. onca*). Another interpretation suggested variation was due to latitude and elevation-based clines (Ruibal, 1957). Extensive work on call structure, genetics, biochemistry, morphology, coloration, and life history has documented differences among populations of leopard frogs, and has demonstrated that many of the "regional" variants (or morphotypes) merit recognition as species (McAlister, 1962; Post and Pettus, 1966; Littlejohn and Oldham, 1968; Mecham, 1968, 1969; Brown, 1973; Platz and Platz, 1973; Pace, 1974; Dunlap and Kruse, 1976; Frost and Bagnara, 1976; Kruse and Dunlap, 1976; Platz, 1976; Platz and Mecham, 1979; Kruse, 1981b; Platz and Frost, 1984; Hillis, 1988; Platz, 1993). The leopard frog complex now consists of about 20 species occurring from northern Canada through Costa Rica.

The Northern Leopard Frog is a member of the North American *Novirana* clade within the polyphyletic genus *Rana*. Case (1978) recognized all leopard frog species, as well as *Lithobates palustris*, *L. areolata*, *L. sevosa*, *L. capito*, *L. montezumae*, *L. dunni*, and *L. megapoda* within the *pipiens* species group. This arrangement is not currently accepted. Indeed, there has been considerable disagreement as to the phylogenetic placement of this species (Hillis et al., 1983; Pytel, 1986; Hillis, 1988; Hillis and Wilcox, 2005). Mitochondrial DNA analyses have shown that *L. pipiens* is a member of the *Pantherana* clade, which includes a large number of frogs with the "leopard frog" general phenotype and possessing a snore-like mating call functioning solely in mate attraction. The *Pantherina* clade has a long evolutionary history, having diverged from the *Aquarana* (e.g., *Lithobates catesbeianus*) and *Torrentirana* (e.g., *Lithobates tarahumarae*) clades in the Oligocene.

There are two major clades within the Northern Leopard Frog's distribution, an eastern and a western clade separated by the Great Lakes and Mississippi River. These clades likely have been differentiated since the Pliocene, that is, for at least two million years. The western clade has less genetic diversity than the eastern clade, and there is evidence that major range expansions in both clades occurred prior to the last glacial retreat (Hoffman and Blouin, 2004). In addition, there is little protein variation or call differences among western clade populations (Dunlap and Platz, 1981). As with other anurans, isolation by distance and restricted gene flow are the most important factors influencing the genetic structure of this species. In addition, the species underwent a genetic bottleneck as a result of Pleistocene ice advances, whereby remnant allopatric populations were confined to glacial refugia.

There is a considerable amount of genetic differentiation among populations, at least along the northern distribution of the species. Populations > 50 km apart exhibit significant differentiation, primarily due to genetic drift. This diversity is most evident in Ontario and Manitoba and tends to decrease among more western populations. Wilson et al. (2008a) suggested that all populations in the far North share a close evolutionary history and belong to a western haplotype group. These results have conservation implications, especially when managers consider repatriating Northern Leopard Frogs to depleted Alberta localities. Source populations with similar haplotypes may not be readily available.

Lithobates pipiens hybridizes with a number of other members of the *Lithobates pipiens* complex and other *Lithobates* in the laboratory, but most hybrid crosses are not viable (reviewed by Hillis, 1988, and references therein). Development may appear normal in crosses with *L. areolata*, but not after metamorphosis; hybrid inferiority (increased mortality, delayed development, defects, reduced fertility) occurs in crosses with *L. blairi*, *L. forreri*, *L. magnaocularis*, *L. megapoda*, *L. neovolcanica*, *L. palustris*, *L. sphenocephala*, *L. taylori*, and *L. yavapaiensis*; severe hybrid inferiority (high mortality, sterility) occurs with some *L. berland-*

ieri, *L. neovolcanica*, and *L. sphenocephala*; complete hybrid incompatibility (no appreciable development) occurs with other *L. berlandieri* populations. In contrast, Gillis (1975) could find no significant genetic incompatibility between *L. pipiens* and *L. blairi*.

Based on hybridization experiments, Ruibal (1962) considered the *L. pipiens* population at San Felipe Creek, California, to be more closely related to species farther south in Mexico and Central America than to "cold races" further north; this and other California desert populations, no longer extant, are now referred to *L. yavapaiensis* (Jennings and Hayes, 1994b). Natural hybrids may be found where members of the complex co-occur. For example, *L. pipiens* × *L. chiricahuensis* hybrids occur at a rate of about 9% in some Arizona populations (Frost and Platz, 1983). *Lithobates pipiens* × *L. blairi* hybrids occur at a rate of 1–16% in overlapping Nebraska populations (Lynch, 1978), and at 1–4.4% in Iowa and South Dakota populations (Dunlap and Kruse, 1976). Throughout much of the area where contact occurs, there has been asymmetrical genetic swamping of *L. pipiens* nuclear haplotypes by *L. blairi* haplotypes (Di Candia and Routman, 2007). Other hybrid populations have been reported from Illinois (Smith, 1961), Colorado (Post, 1972; Gillis, 1975), and New Mexico (Degenhardt et al., 1996). The percentage of populations in a region that contain hybrids may change through time, perhaps reflecting the extent of interaction between the parental species (Cousineau and Rogers, 1991).

Two polymorphic phenotypes, reflected in both adults and larvae, are often observed in the same breeding ponds in some areas. In West Virginia, about 67% of the leopard frogs are green and 33% brown (Sutton, 2004). Corn (1981) noted that the brown morph was present in about 24–68% of Northern Leopard Frog populations in northern Colorado, and that brown morph larvae had a shorter time to metamorphosis than sympatric green morph larvae (68 vs. 74 days and 83 vs. 86 days in 2 populations). The brown morph is more common than the green morph in early season metamorphs. Corn (1981) suggested that the brown morph was selectively favored over the green morph in ponds experiencing high rates of predation. The color pattern appears genetically stable, with the green morph a simple dominant to the brown morph (Fogelman et al., 1980).

There are two prominent unusual phenotypes within *L. pipiens*, the unpatterned "Burnsi" (Brown and Funk, 1977) and the distinctively mottled "Kandiyohi" (both described as species by Weed, 1922). The Burnsi phenotype is found in western Wisconsin, central and north-central Minnesota, eastern South Dakota, and southeastern North Dakota; it is also reported from Dickinson County, Iowa. It occurs at variable frequencies among different populations. At Block Lake, Minnesota, for example, the frequency of Burnsi frogs was 11.3% from 1967 to 1986, and 18.5% from 1985 to 1999 (McKinnell et al., 2005). At populations in Minnesota, North Dakota, and South Dakota, most frequencies are <10%, but may be as high as 22% (McKinnell et al., 2005). However, Dunlap (1967) found 25% Burnsi phenotypes in Bon Homme County, South Dakota. The Burnsi phenotype appears to be spreading to the West, and it is now 400 km west from its previously identified center of occurrence in Minnesota.

The Kandiyohi pattern is found in west-central and southwest Minnesota and adjacent eastern South Dakota and in southeastern North Dakota (Merrell, 1965). Schueler (1982a) also reported Northern Leopard Frogs with these patterns at Long Point, Ontario. The greatest frequency of the Kandiyohi pattern is also about 10% or less, primarily in northeastern South Dakota; most frequencies are in the 1 to 2% range, however (Breckenridge, 1944; Volpe, 1955b; McKinnell et al., 2005). Kandiyohi frogs in Minnesota are rarely observed (*in* Oldfield and Moriarty, 1994). Distribution maps showing the ranges of the Burnsi and Kandiyohi phenotypes are in McKinnell et al. (2005).

Areas inhabited by frogs displaying these patterns (e.g., the Anoka Sand Plain) were colonized only within the last 10,000 yrs, after the retreat of the Wisconsin glacial ice cover. Both phenotypes are the result of a single dominant gene, although at different loci (Moore, 1942b; Volpe, 1955b, 1956a, 1961), and they do not have any phylogenetic or systematic significance. However, larvae with the Kandiyohi pattern develop more rapidly than normally pigmented *L. pipiens*, suggesting a possible advantage for frogs developing in the prairie habitats where it is found. There is no loss of viability in frogs carrying these genes and associated pigmentation when compared to normally pigmented Northern Leopard Frogs (Merrell, 1972).

In southwestern South Dakota, J.A. Ernst (2001) described a "Conata" morphotype as having the dorsolateral folds broken posteriorly just before the thigh, rugose sides, a pale brown coloration, a pale narrow supralabial stripe, pale brown dorsolateral folds, tympanum with obscure pale dot, dorsal mottling with brown diffuse spots, and obvious yellow coloration in the groin and inner thighs. This phenotype was observed at only 13 locations. Mitochondrial DNA aligned this morphotype with *L. pipiens*, despite a superficial resemblance with *L. blairi*.

In addition to these differences, Northern Leopard Frogs may vary in terms of the number of dorsal spots. In Nova Scotia, there are between 8 and 27 spots (mean 14) (Bleakney, 1952; Gilhen, 1984) and usually 17–21 spots (range 10 to 30) in West Virginia (Sutton, 2004), whereas in Vermont the mean is 5 (Moore, 1944). On Prince Edward Island, spot number ranges between 6 and 20 (mean 13) (Cook, 1967), and in Québec the mean is 12 (Moore, 1944). Further regional specialization may occur; Walker (1946) reported that Northern Leopard Frogs from southern Ohio differed from frogs further north by having narrower heads, smaller overall size, longer legs, and in pattern subtleties.

There are 26 diploid chromosomes in *L. pipiens*, as in all ranids. Di Berardino (1962) and Corcoran and Travis (1980) provided additional information on karyotype.

ADULT HABITAT

Northern Leopard Frogs prefer clear water aquatic habitats to those that are turbid or silt-laden. At a landscape level, their abundance is weakly negatively correlated with the amount of forest cover, and strongly negatively correlated with the volume of traffic, within 2 km of a wetland (Eigenbrod et al., 2008). Occurrence is positively associated with isolated forest remnants in close proximity to wetlands, reflecting the species' preference for open habitats, and connectivity among wetlands (Kolozsvary and Swihart, 1999; Lehtinen et al., 1999). They prefer a mosaic of habitats whereby small- to medium-sized ponds (for breeding) are interspersed among larger water bodies, such as lakes and reservoirs (for overwintering). Open areas for feeding and dispersal should occur between such wetlands. Grassy areas (vegetation less than ca. 30 cm in height), wet meadows, marshy areas, and forest-open habitat ecotones are preferred as feeding sites away from breeding ponds (e.g., Allen, 1868; Gorham, 1964; Sutton, 2004).

In upper New York State and New England, Northern Leopard Frogs are associated with lower elevation sites, sites that have soils with higher pHs, and areas with open water and marshes. They are not as common in swamps (Gibbs et al., 2005). The species is positively associated with grassland or rangeland habitats in the upper Midwest (Lynch, 1978; Knutson et al., 2000), and declines in this habitat type may be responsible, in part, for declining leopard frog populations. Northern Leopard Frogs tend to avoid heavily forested, agricultural, and urban areas, but may breed at such sites if appropriate habitat is nearby (Knutson et al., 2004). Within preferred habitats, Northern Leopard Frogs may not be uniformly distributed, but tend to be clumped within the most favorable areas.

At a landscape scale, *L. pipiens* is relatively ubiquitous throughout much of its range where, for example, it occurred at >40–60% of the sites in Ontario over a period of several years (Hecnar, 1997; Hecnar and M'Closkey, 1996a, 1997a, 1998). There is considerable annual turnover rate in wetland occupancy from one year to the next. As might be expected because of its dispersal capability and generalist habitat requirements, it was one of the species least likely to show nested-assemblage occupancy patterns (i.e., subsets of co-occurring species) when compared with other species found in the same region (Hecnar and M'Closkey, 1997a).

Lithobates pipiens is occasionally found in springs (Allen, 1963), along sandy beaches of large freshwater lakes (Stille, 1952), and at the entrance to caves (Garton et al., 1993).

TERRESTRIAL ECOLOGY

Northern Leopard Frogs are commonly found long distances from water. During the summer, adult Northern Leopard Frogs often are found in distant terrestrial habitats, where they sit concealed in "forms" on the moist forest floor or in grassy moist meadows. Older common names, the grass or meadow frog, refer to its terrestrial habitat preferences. They frequent shallow freshwater marshes, grassy-sedge woods, old fields, wet meadows, unmowed pasture, and even hayfields (Merrell, 1970, 1977; Hine et al., 1981). Beauregard and Leclair (1988) found that preferred habitats in Québec included areas close to a marsh with tall herbaceous vegetation of high species richness and a low percentage of moss cover. The primary factors affecting terrestrial habitat selection are the levels of moisture in leaf litter and soil, a nearness to water, cover, and temperature (Blomquist and Hunter, 2009). Northern Leopard Frogs usually select the warmer habitats when provided a choice.

Frogs may also use crevices or other cavities as retreat sites in dryer situations or when moisture conditions change. Frogs remain in their secretive terrestrial locations for less than 24 hrs to more than 5 days (Dole, 1965a). Most shifts in terrestrial locations are from 5 to 10 m, with movement occurring usually at night or on overcast days and during favorable moisture conditions (rain, heavy dews) than during periods of dry weather.

Northern Leopard Frogs may use the same form or retreat site multiple times, and they usually remain within a defined home range during this period. In Maine, mean home ranges varied between 348 and 1,347 m^2 depending on location. Individual home ranges varied between 13 and 8,425 m^2 (mean 1,096 m^2), and did not vary by sex (Blomquist and Hunter, 2009). The extent and fidelity of frogs to a home range varies, depending on the extent of habitat available and the proximity to water bodies. If little habitat is available, frogs tend to return to terrestrial feeding areas; if much habitat is available, frogs may disperse more widely and not return to the same area as in the previous activity season (Dole, 1965b). Smaller frogs tend to occupy home ranges in wetter areas than larger frogs. Dole (1965b) estimated home ranges at one spatially confined site that were considerably smaller (males 78 m^2, females 113 m^2, subadults 68 m^2) than those found in more extensive habitats (males 362 m^2, females 503 m^2, subadults 283 m^2). There was considerable individual variation, however. Reports in the literature stating that *L. pipiens* does not occupy home ranges (Fitch, 1958) undoubtedly refer to other members of the Leopard Frog Complex.

Occasionally, adult frogs may make extended journeys (100–240 m; Dole, 1965a) to adjacent habitats. Long-distance movements usually occur during nocturnal rains and appear directed, that is, in a straight line without the wandering paths seen during short movements within the home range. Northern Leopard Frogs may move as much as 47 m/hr during long-distance dispersal events (Dole, 1965a). These frogs then remain at the distant locations for several days before returning to the original home range. Long-distance movements usually occur on only one night, although Dole (1965a) recorded one frog that made long-distance movements on two consecutive nights. Such periodic movements by anurans have been documented in a number of species and may facilitate familiarity with adjacent areas. Extended rainfall events and warm temperatures allow for wide dispersal in terrestrial habitats.

Displaced Northern Leopard Frogs are capable of returning to near the place they were captured when displacement distances do not exceed 1 km. At greater distances, they are unable to move back toward the original site, but disperse more randomly. Homing movements are fairly direct, as frogs immediately orient toward the home direction following release; they adjust their movements homeward the farther they travel. Blind frogs are also able to orient toward the home direction when displaced at much lesser distances, although their movements are not as direct. Blind frogs are able to navigate at night and upwind from their release point, as are anosmic animals released in heavy fog, suggesting that vision, brightness, and olfaction may not be required for homing behavior. In additional experiments, displaced anosmic or deaf leopard frogs were able to home (Dole, 1972b). Dole (1968) suggested that homing ability may result from prior familiarity with the habitat surrounding a wetland, or perhaps a combination of general olfactory cues, rather than specific cues from a particular site.

Lithobates pipiens is tolerant of a wide variation of ambient temperatures. The CTmax of adults is 38–39°C, depending upon acclimation temperature. This is not surprising, as Northern Leopard Frogs are heliotropic. Animals acclimated at longer photoperiods also tend to have slightly higher CTmax's than frogs acclimated at shorter or 12:12 photoperiods (Mahoney and Hutchison, 1969). Tolerance for a wide range of temperatures allows for an extended activity season; for example, activity extends from mid-March to late October in New England (Klemens, 1993).

Like many aquatic and semiaquatic amphibians, Northern Leopard Frogs are prone to lose body water while on land. Smaller frogs are more tolerant of water loss per se, but they lose water at a greater rate because of an increase in surface to volume ratio (Thorson, 1955). Rehydration occurs in about 3 hrs for a frog desiccated to a loss of 27–29% body weight, and in 48 hrs for a frog desiccated to 65–75% of its hydrated weight, as long as the soil moisture is at least 20%. Rehydration can occur at lower soil moisture contents, but not to maximum predehydration weight (Dole, 1967a). Adults resorb moisture from soil or from wet vegetation through the highly vascularized skin of the groin (termed the pelvic patch).

Northern Leopard Frogs are photopositive in their phototactic response, suggesting they use both sunlight and moonlight when feeding and going about their daily activities (Jaeger and Hailman, 1971, 1973). They are sensitive to light in the blue spectrum ("blue-mode response"), which apparently helps them orient toward areas of increasing illumination, such as the open area above a pond (Hailman and Jaeger, 1974). Northern Leopard Frogs likely have true color vision.

AQUATIC ECOLOGY

Lithobates pipiens is frequently found in or near water, even during the nonbreeding season. Indeed, Hahn (1968) recorded leopard frogs active in January and February around warm springs in Colorado, and Olson (2010) found a Northern Leopard Frog active in December in northern Wisconsin on ice along the shore. Whereas many frogs disperse, others choose to remain in or around ponds and wetlands. They are heliothermic, basking in the direct sun while sitting on floating vegetation or along the wetland margins. Temperature is controlled by evaporative cooling. Brattstrom (1963) records body temperatures ranging from 18 to 33.5°C, and stated that on cool nights, Northern Leopard Frogs moved to the warmer parts of a pond when water temperatures exceeded air temperatures, or sheltered in small depressions along the shore to reduce the effects of air flow.

Northern Leopard Frogs overwinter in lakes, near the outlet of lakes or other bodies of water, and along stream beds. Overwintering occurs in lakes and streams deep enough not to freeze solid. Dormancy occurs openly on the lake bottom, or in vegetation, detritus, mud, or in other debris on lake or stream substrates. In Ontario, they were found in a stream beneath rock rubble 13–40 cm in diameter and > 85 cm in depth, with a velocity of 22.5 cm/sec (Cunjak, 1986); winter water temperatures were between 0.5 and 2.1°C. There are reports of leopard frogs resting in shallow pits on the muddy bottom of lakes, with a light coating of silt over them (Emery et al., 1972), or wedged into crevices or even under turtles (Ultsch et al., 2000). Emery et al. (1972) recorded them as deep as 3.1 m below the ice. In neither of the latter instances were the frogs concealed, as access to well-oxygenated water took precedence over midwinter predator avoidance. Terrestrial dormancy may not be common, and Northern Leopard Frogs have been shown to choose water over leaf litter on land when laboratory tested at a temperature of 1.5°C (Licht, 1991). However, Waye (2001) reported two radio-tracked leopard frogs in British Columbia that moved to small mammal tunnels and apparently overwintered there, and Wright (1914) reported dormancy under stones in ravines.

When dormant, leopard frogs lie motionless on the substrate with legs and arms extended, and make no effort to move in the cold water; they are capable of sluggish movement if disturbed, however. Leopard frogs prefer dormancy sites that are well-oxygenated. For this reason, they often overwinter near lake outlets or spillways where flowing water supplies fresh oxygen, and there is a limited chance of oxygen depletion. Ultsch et al. (2004) demonstrated that Northern Leopard Frogs are tolerant of a large decrease in oxygen tension (the critical PO_2 at which anaerobic respiration occurs), which permits them to remain in hypoxic icebound ponds for extended periods. Mass winterkills due to oxygen depletion (anoxia) are not uncommon (Manion and Cory, 1952), although for some reason, the Burnsi phenotype of this species is associated with an increased ability to survive low oxygen levels during winter dormancy (Merrell and Rodell, 1968). Frogs may remain dormant for long periods. For example, in Minnesota they remain underwater from October until March.

Unlike other northern frog species, Northern

Leopard Frogs are unable to survive freezing temperatures at the ground surface (Schmid, 1982), and significant mortality may occur if Northern Leopard Frogs are exposed to unusually cold weather. Cold fronts may kill frogs if the fronts occur earlier than normal and catch individuals moving to overwintering locations (Merrell, 1977), or if they occur in the spring as frogs are beginning to leave protected sites and move to breeding ponds. Mortality from oxygen depletion also occurs if snowpack completely blocks sunlight, preventing aquatic photosynthesis. Prolonged exposure to sublethal levels of cold weather reduces the immune response in *L. pipiens* (Maniero and Carey, 1997), which could make them more susceptible to pathogens or toxic chemicals. The immune response is rapidly restored as frogs warm with increasing ambient temperatures.

CALLING ACTIVITY AND MATE SELECTION

Northern Leopard Frogs may emerge from winter habitats as long as a month before calling begins. In West Virginia, for example, they have been observed to emerge as early as the first week in February, even though calling did not begin until mid-March (Sutton, 2004). Ohio specimens also have been found in February (Walker, 1946), and they have been observed in January in Indiana during an exceptional warm spell (Minton, 2001). In Ohio, Zenisek (1963) noted emergence in late March and amplexus by 7 April, but no egg masses until 15 April. Wright (1914) reported emergences from 3 to 18 (mean 7) days prior to calling.

Breeding in Northern Leopard Frogs occurs during the cool spring months of the year, a time that varies by latitude and elevation. In the southern portion of its range, breeding commences in late winter, whereas breeding usually occurs in spring and into the early summer in more northern localities. In the Chihuahuan Desert, eggs or very small tadpoles have been found from April to July, and even in September and October (Scott and Jennings, 1985). Some calling months include: mid-March to mid-April, depending on temperature, in Wisconsin (Hine et al., 1981), Michigan (Cummins, 1920), New England (Klemens, 1993), and West Virginia (Sutton, 2004); March to May in Kentucky (Barbour, 1971); March to early June in Colorado, depending on elevation and year (Post, 1972; Hammerson, 1999); mid-April to mid-May in Nova Scotia (Gilhen, 1984), Québec (Lepage et al., 1997), and South Dakota (Kiesow, 2006); mid-April to late May in South Dakota (Ernst, 2001); late April to early May in Michigan (Dole, 1967b); April to June in New Brunswick (Gorham, 1964) and Prince Edward Island (Cook, 1967); early May to early July in North Dakota (Bowers et al., 1998); and mid-May to June in northern Ontario (Schueler, 1973). Wright (1914) reported that males would occasionally call well after the breeding season, such as during rainfall or on cloudy days. The latest he heard a call was on 14 September; September calling is also reported by Walker (1946).

Increasing air temperature (13–14°C in Minnesota; Merrell, 1977) stimulates adult Northern Leopard Frogs to move from overwintering sites to breeding ponds. Males are first to arrive at breeding locations, with females arriving 3–14 (normally 5–7) days after calling begins. Males begin to call when water and air temperatures reach about 10°C. When air temperatures drop below 10°C, calling activity decreases even if water temperature is above 10°C (Hine et al., 1981), although Ernst (2001) reported calling near 0°C. Because daytime temperatures may remain much more favorable than cool nocturnal temperatures, calling and reproduction are often diurnal in northern populations in the early spring, although nocturnal calling occurs (Wright, 1914; Cummins, 1920). Frogs generally occupy the same calling position, and movements while at the breeding site are minimal. Short movements position adults in warmer locations from which to call, especially on cool or partially overcast days. There is no evidence of territoriality among males.

Leopard frogs are quite vocal and make a variety of sounds under different circumstances (Schmidt, 1968). The mating call of the Northern Leopard Frog, using its paired vocal sacs, has been described as a "staccato snore" (Frost and Platz, 1983), "ir-a-a-a–a-a-h" (Noble and Aronson, 1942), a "long low guttural croak" (Wright and Wright, 1949), a "guttural snore lasting approximately three sec, followed by grunts, squeals, and several clucking notes" (Ernst, 2001), to a chuckle lasting several seconds. Schmidt (1968) reviewed the various calls of the Northern Leopard Frog, and restricted the "chuckle" to a series of sounds following the mating call. The dominant frequency of the mating call is 1.5 kHz, with a duration of 2–4 secs. The pulse rate is temperature-dependent. For example, it is 21 pulses/sec at 17–19°C in Arizona, 14 pulses/sec at 12–16°C in Colorado, 30 pulses/sec at 21°C in South Dakota, and 19 pulses/sec at 18°C in Illinois (Littlejohn and Oldham, 1968; Brown and Brown, 1972a; Dunlap and Kruse, 1976; Frost and Platz, 1983). Frost and Platz (1983) provide a sonogram comparing this species' advertisement call with those of other Southwest leopard frogs.

Males call while floating on the water's surface, with limbs outstretched, and vigorously pursue any approaching frog (Merrell, 1977). Occasionally, frogs may call out of water or from the bottom of a pond. When temperatures are high (20°C), males are unwary during courtship, but if temperatures decrease they become more secretive in calling behavior. Both males and females are approached by a reproductive male; if amplexed, males give a release croak that causes the primary male to release the amplexed male. Females likely choose a male by moving toward or near him, but they are concealed by vegetation so as not to attract other nearby males.

Males readily grasp the rotund females, a body form that indicates a full complement of eggs. Amplexus is pectoral. The male arches his body convexly to fit tightly to the female; the hind limbs are flexed; and the front limbs encircle the female so that the digits are firmly holding the female's venter. Gravid females remain silent when amplexed, as a distended body apparently inhibits the release croak (Diakow, 1977). Females are usually unable to dislodge an amorous male because of the male's tenacity and his ability to grasp her with his swollen thumb pads. Spent females also have a release croak, which appears to be initiated through stimulation of the skin on the body's trunk (Diakow, 1977). There are subtle differences in the release croaks of males and females (e.g., in pulse rate, amplitude, call rate) despite their similar functions, and males tend to call more often than females when employing the release croak (McClelland and Wilczynski, 1989).

Amplexus lasts from several minutes to as long as a day, at least under laboratory conditions (Wright, 1914; Noble and Aronson, 1942). After some time following initial amplexus, the female begins to move backward in position in relation to the position of the male ("backward shuffling"), and extends her thighs backward and laterally in a 45° angle. The female then shifts her hind legs and feet into an oviposition posture forming a diamond-shaped enclosure. The male rotates his hind limbs downward and outward, at which time the pair is in the egg-laying posture. The female begins oviposition through a series of contractions of her abdominal walls, followed by arching her back. The male extends his legs and flexes his body convexly, thus bringing the couple's cloacae into close proximity. As a cluster of eggs is extruded from the female's cloaca, the male fertilizes them (termed ejaculatory pumps). Oviposition usually involves 10–23 distinct cycles of ejaculatory pumps. Total oviposition takes 2–8 minutes. Males usually release females within six minutes after oviposition is completed. Noble and Aronson (1942) provide a detailed description of the entire sequence.

Males sometimes amplex inanimate objects (Merrell, 1977; Livo, 1981a), other frog species (Wright, 1914; Brown and Pierce, 1965), or even fish. Males amplexing spent females release them rapidly, presumably because the slender bodies of the females indicate previous oviposition and because of the female's release croak. Noble and Aronson (1942) describe the contexts of the various types of release croaks issued by this species.

BREEDING SITES

Northern Leopard Frogs choose clear-water breeding sites that range from small (< 0.4 ha; Fischer, 1998) semipermanent ponds (i.e., ponds that rarely dry) to shallow protected sections of lakes (> 2.8 ha; Fischer, 1998) and man-made water bodies such as gravel pits and golf course ponds (Mifsud and Mifsud, 2008); some individuals may breed in the quiet backwater of streams and rivers, or in clear sand-bottomed streams (Lynch, 1978). Frogs appear selective in their choice of available sites, avoiding both large and small wetlands, and they prefer ponds that warm rapidly in spring when compared to deeper and thus cooler ponds. Water depths are usually from 1.5 to 3 m, with bottom substrates of silt, muck, and decaying vegetation. Northern Leopard Frogs prefer breeding sites with >50% cover of the water's surface by submerged and emergent vegetation and with extensive vegetation (cattails, grasses, willows) along the shoreline. Vegetation provides both protection and a place to attach the egg mass. Hine et al. (1981) suggested that open water allowed the pond to warm up faster than sites under canopy cover, and they noted that most reproduction takes place in the warmer sections of a breeding pond. However, Fischer (1998) reported nearly equal percent occupancy in ponds that were open-canopied (<20% cover), hemi-marsh (21–70% cover), and closed-canopied (>70% cover). Fishless ponds are preferred, although leopard frogs are found with both predatory and nonpredatory fishes (Hecnar, 1997; Hecnar and M'Closkey, 1997b). Data on breeding site preference are in Merrell (1968, 1977), Collins and Wilbur (1979), and Fischer (1998).

As with many species of frogs, Northern Leopard Frogs do not inhabit all sites that appear suitable to a human observer. In Wisconsin, for example, occupancy ranged from 16% in 1975 to 32% in 1978 of 83 ponds surveyed, with breeding occurring in only 6% of ponds in 1975 and 23% in 1978 (Hine et al., 1981). Skelly et al. (2003) reported only a 20% probability of detecting this species on the E.S. George Reserve in Michigan over a 5 yr period at 32 sampled ponds. These low numbers may represent a declining regional population, however. *Lithobates pipiens* do not necessarily use the same pond from one year to the next, but instead choose from among a suite of nearby ponds. The pond actually used in any one year reflects the weather conditions or other environmental variables within the region; a pond suitable one year may not be suitable the next, for example, because of drought conditions. In Ontario, Hecnar and M'Closkey (1996b) found Northern Leopard Frogs preferred sites with lower nutrient contents. Brodman (2009) also noted considerable variation in annual occupancy over a 14 yr study in Indiana.

REPRODUCTION

The timing of reproduction varies geographically, as might be expected for such a wide-ranging species occupying varying elevations. In Minnesota, the breeding season extends from mid-March to mid-May, but occurs mostly in April (Merrell, 1977). In Arizona, eggs are observed from mid-April to early June (Frost and Platz, 1983), but these populations are at higher elevations than those of other members of the *L. pipiens* Complex directly to the west in California. Other dates for breeding are mid-March to June (Illinois: P.W. Smith, 1947), mid-March to late April (Ohio: Walker, 1946), late March to April (Colorado: Gillis, 1975; Indiana: Minton, 2001), March to May (Indiana: Brodman and Kilmurry, 1998; New York: Wright, 1914; Ohio: Zenisek, 1963), April to May (Ontario: Piersol, 1913; Québec: Desroches and Rodrigue, 2004; New Hampshire: Oliver and Bailey, 1939; South Dakota: Fischer, 1998; Montana: Werner et al., 2004), late April to late May (Manitoba: Eddy, 1976), April to June, depending on elevation (Idaho: Linder and Fichter, 1977), April to August (Montana: Black, 1970), early to late May, depending on elevation (Colorado: Corn and Livo, 1989), and mid-May to June (New Hampshire: Taylor, 1993).

Females deposit eggs from 2–7 days after arrival, which is about 5–14 days after the males start calling. According to Frost and Platz (1983) most reproduction occurs over a period of 10–14 days, although that can be extended by cold weather. When cold weather interrupts the breeding activity, breeding ceases and resumes once warmer weather appears. Reproduction occurs in shallow water (< 40 cm, Merrell, 1977; mean 13 cm, Corn and Livo, 1989; 25–50 cm, Licht, 2003) near the shoreline, and oviposition occurs rapidly following amplexus. However, Zenisek (1963) reported egg masses as far as 8 m from shore. Eggs are deposited in a gelatinous mass firmly to emergent or submerged vegetation within the pond, and are floated at or near the water's surface. In Wisconsin, egg masses were 40–150 mm in length (mean 87 mm) and 40–110 mm in width (mean 69 mm) (Hine et al., 1981). Masses in Ontario measured 95 × 70 mm (Licht, 2003), whereas masses in West Virginia were a mean of 80 mm in length by 34 mm in width (Sutton, 2004).

There is a great deal of variation in the number of eggs deposited by Northern Leopard Frogs. An egg mass contains from: 300–800 eggs in Ohio (Zenisek, 1963), 2,167–3,767 eggs in Nova Scotia (Gilhen, 1984), 2,000–5,000 eggs in Minnesota (Merrell, 1965), 1,143–5,189 eggs also in Minnesota (Hoppe and McKinnell, 1997), 3,500–5,000 eggs in New York (Wright, 1914), and 645–6,272 (mean 3,045) in Colorado (Corn and Livo, 1989). The number of eggs appears to be proportional to the body length of the female (Livo, 1981a; Gilhen, 1984).

As soon as the eggs are extruded and fertilized, amplexus usually ends rapidly, and females immediately leave the breeding pond. Unfertilized eggs make up between 0.2 and 20% of the clutch (Hoppe and McKinnell, 1997). One clutch is deposited per female annually, with the eggs for the next breeding season maturing in the female's ovaries late in the summer season. In contrast to the females, males may mate more than once in a breeding season (Wilson et al., 2008b). As the season progresses, however, the population of calling males decreases, suggesting that some males have reached their breeding capacity and left the pond for foraging habitats. Other males likely do not breed successfully at all during the season, as calling may continue after all egg clutches have been deposited.

The globular shape of the egg mass may be important in protecting the mass from freezing temperatures in spring. Northern Leopard Frog eggs are killed when temperatures are approximately 2.5°C (Moore, 1939), and development is retarded at 5°C (Atlas, 1935). Temperatures within the mass are usually 2–3°C warmer than those in adjacent water (Merrell, 1970). The dark-colored embryos absorb heat during the day, which may in turn speed up development during the critical early spring breeding season. The maximum temperature that eggs of *L. pipiens* can survive is 28–29°C (Moore, 1939).

Hatching occurs within 4–20 days after deposition (Anonymous, 1918; Livo, 1981a; Licht, 2003; Sutton, 2004) at a size of 5.5–8.4 mm TL (Livo, 1981a).

LARVAL ECOLOGY

Larvae are omnivorous, feeding mostly on plant material but also on dead animal matter. Larvae do not do well on high fat diets, but require water-soluble B and fat-soluble A vitamins and certain proteins for normal development and growth (Emmett and Allen, 1919). Many aspects of the larval ecology of Northern Leopard Frogs, particularly growth rates, time to metamorphosis, size at metamorphosis, and survivorship are reflective of conditions in the breeding pond, such as larval density, food resources, predators, temperature, elevation, and hydroperiod. Some of these factors may be correlated, such as decreased larval growth rates at higher elevations (Baxter, 1952), which are presumably associated with cooler temperatures at breeding ponds. Larvae also tend to grow slower in ponds with a closed canopy because of a general lack of algae for food, competition with Wood Frog larvae, and low dissolved oxygen contents (Werner and Glennemeier, 1999). As a result of these factors, there may be larvae of many different sizes in a breeding pond at any one time (Claflin, 1962).

Crowding reduces growth rates and body size and mass, decreases survivorship, and increases time to metamorphosis (Adolph, 1931a, 1931b; Rugh, 1934; Rose, 1960; Poile, 1982; Distel and Boone, 2011). Growth tends to be retarded under crowded conditions, because tadpoles are less inclined to ingest food, even when food is available (Adolph, 1931b). Crowding induces a physiological stress response resulting in elevated corticosterone levels, which inhibit growth (Glennemeier

Tadpole of *Lithobates pipiens*. Photo: David Dennis

and Denver, 2002). Growth inhibitors also may come from exogenous sources. Many authors (Richards, 1958, 1962; Rose, 1959, 1960; West, 1960; Rose and Rose, 1961) have found that crowding had a subsequent effect on growth rates of tadpoles reared in water previously inhabited by large *L. pipiens* larvae. This suggested the potential for a substance that might inhibit the growth rates of smaller conspecifics, and Richards (1958, 1962) noticed certain cells (perhaps an alga) that appeared to be associated with inhibition that were passed into the water via the feces. The role of these cells in growth inhibition, however, has been disputed (Gromko et al., 1973). Certain types of algae inhibit tadpole and egg development, but it appears that the tadpoles themselves may excrete the primary inhibitory substance (Akin, 1966). In turn, the algae may extend the time the inhibitory substance is effective.

Food, predators, and habitat structure affect activity and behavior. For example, increased levels of food and predator presence independently result in decreased activity and swimming speed in Northern Leopard Frog larvae, and these results are additive (Anholt et al., 2000). As food levels increase, so does body mass regardless of predation threat. As might be expected, smaller larvae are less active when faced by predators than are larger tadpoles (presumably because of predation risk), and larvae tend spatially to avoid predators if possible.

In structurally complex environments, larvae tend to be less conspicuous and less concentrated than they are in open habitats. In these situations, larvae from the complex environment are larger than those from the more open environment, at least in some experimental trials (John and Fenster, 1975). In contrast, experimental trials from a different study incorporating both natural and artificial manipulations have suggested that habitat structure has no effect on Northern Leopard Frog survivorship or mass at metamorphosis. Instead, a more complex habitat structure tended to increase time to metamorphosis (Purrenhage and Boone, 2009). Reasons for the discrepancy in these findings may be because the latter study mimicked natural conditions better than the former; that space and complexity were tested differently; or that different resources were available to the larvae. Because of the very different experimental protocols, the results are hardly comparable.

Different predators also pose different levels of predation risk, which may involve both behavioral and morphological changes. For example, Northern Leopard Frog larvae decrease activity levels in the presence of mudminnows (*Umbra*) and dragonfly larvae (*Anax*), but not Eastern Spotted Newts (*Notophthalmus*) and predaceous beetles (*Dytiscus*) (Relyea, 2001b). In the presence of *Umbra*, larval bodies become shorter and the tails become deeper than normal; this does not occur in the presence of other predators (Relyea, 2001b). Prey handling time, palatability, and capture efficiency by predators of leopard frog larvae also vary by predator (Relyea, 2001a). As such, Northern Leopard Frog larvae do not respond to a generalized predation risk within an aquatic community, but to specific predators. The type and level of response will vary by the predator present.

Northern Leopard Frog larvae do not exhibit any social behavior, such as schooling, and there is no kin recognition. In experimental trials, larvae spend as much time away from siblings as they do in proximity to siblings (Fishwild et al., 1990). However, larvae in mixed-sibship experimental trials generally had greater mass than larvae reared only with siblings, regardless of initial density (Poile, 1982).

A major source of larval mortality is from desiccation, as breeding ponds dry during the summer months. Unusual droughts can cause mass reproductive failure, as occurred in Wisconsin in the mid-1970s (Hine et al., 1981).

There is considerable variation in the size at metamorphosis depending upon the time of metamorphosis, even among adjacent ponds. Early and late froglets tend to be smaller than those metamorphosing during the peak of dispersal. In Wisconsin, the mean length at metamorphosis is 39–40 mm (range 34 to 40 mm), and the body mass ranges from 4.0 to 9.5 g (mean 8.0 g) (Hine et al., 1981). In South Dakota, Hardy (1974) found frogs at various stages of metamorphosis on land from late July to mid-October at 22–34 mm SUL, and in New York, metamorphosis is completed (full tail reabsorbtion) by 20–30 mm SUL (Munz, 1920). Other reports include 21.5 mm in Michigan (Rittschof, 1975), 17–28 mm (mean 24 mm) in Ohio (Walker, 1946), 24–26 mm in Nova Scotia (Gilhen, 1984), 21–36 mm SUL in Colorado (Hammerson, 1999), 35–40 mm in Minnesota (Merrell, 1977), and 25 mm in Québec (Leclair and Castanet, 1987).

Metamorphosis normally occurs from 71 to 111 days after egg deposition (60–80 days after hatching) in nature (Wright, 1914; Sutton, 2004), and developmental growth rates are temperature dependent (Atlas, 1935). At 19°C, for instance, metamorphosis requires 117 days (Adolph, 1931a). The effects of temperature are especially apparent during the early stages of development, when temperatures are low, but less so as development proceeds. Individual body parts are affected differently by temperature. For example, the maximum size of the gills decreases as the developmental temperature decreases (Atlas, 1935). Presumably, larger gills are necessary to extract oxygen from warmer water. Tadpoles may occasionally overwinter prior to metamorphosis if summers are cool or eggs are oviposited late in the season (Oldfield and Moriarty, 1994).

There are many types of stimuli that have been shown experimentally to influence the timing of metamorphosis in anurans, such as pond desiccation and food resources. In Iowa, Bovbjerg (1965) noted that laboratory-reared *L. pipiens* metamorphosed at the same time as did their wild counterparts, even though the two groups experienced different conditions of temperature, exposure to light, day length, precipitation, and rearing density. He concluded that no single environmental variable triggered metamorphosis, but that metamorphosis was initiated by unspecified "internal causes."

Behaviorally, larval Northern Leopard Frogs tend to choose bright, uniformly colored substrates when presented with choices of light, dark, or patterned substrate backgrounds under laboratory conditions (Dunlap and Satterfield, 1985). When disturbed, they tend to flee toward the substrates with lighter backgrounds.

DISPERSAL

Postmetamorphic. Not all postmetamorphic juveniles emerge simultaneously, as emigrants may depart a breeding site gradually or in a series of pulses over a period of several days or weeks. However, the bulk of emigrants may leave about the same time, giving the impression of mass emigration. Bovbjerg and Bovbjerg (1964) observed such a mass emigration over a two-day period in Iowa, where frogs were observed in "the surrounding grassy hills in extremely large numbers." In other years of their study, however, such mass emigration was not observed. Babbitt (1937) also described what appeared to be a mass dispersal of juveniles, having seen "thousands of young frogs about 1.5 to 2 inches on the cement road." In some instances, dispersion does not appear to be triggered by rainfall (Bovbjerg and Bovbjerg, 1964), although it facilitates movements of postmetamorphs in other instances (Dole, 1971). Dole (1971) also recorded large numbers of young in Michigan dispersing over four separate periods in early to mid-August. However, little dispersal occurred on some nights between the main dispersal events.

Emigration normally begins only after the tail is absorbed, although Bovbjerg (1965) reported emigration just prior to complete metamorphosis, suggesting some portion of the tail was still present. Most juveniles disperse from the proximity of where the egg masses were deposited, but this may reflect an opportune location, that is, a shallow-water portion of the pond near the tadpole's preferred habitat with easy egress. Dispersal occurs randomly from some ponds, but in particular directions from others (Dole, 1971).

Orientation is likely influenced by landscape factors, such as physiographic contours and adjacent habitats. For example, they may follow shorelines initially, and they tend to avoid heavily forested habitats. In Alberta, juveniles followed streams for as far as 2.1 km downstream, 1 km upstream, and only 0.4 km overland from the source pond (Seburn et al., 1997). Juveniles may disperse quite far; literature records include movements of 800 m (Rittschof, 1975), 5,200 m (Dole, 1971), and 8,000 m (Seburn et al., 1997). Most dispersal by postmetamorphs occurs over a period of 48–72 hrs, even to distances of 800 m. In Alberta, movements of 1 km were completed by 3 weeks after metamorphosis, and to 2.1 km in <6 weeks (Seburn et al., 1997).

Spring dispersal and summer movement. In northern locations, overwintering adults and juveniles move toward the shoreline as the ice cover melts, presumably in response to increasing water temperature. Under laboratory conditions, frogs kept in water at 1.5°C slowly begin to move and surface when water temperatures were raised to 7–10°C (Licht, 1991). As sunshine elevates the shallow water temperature, frogs become more

Adult *Lithobates pipiens*.
Photo: David Dennis

and more active until they left the water and migrated to the breeding sites. Adults become active earlier than juveniles, although frogs of all ages initially congregated around breeding ponds. Adults quickly occupy a home range around wetlands in preparation for breeding.

Adults may depart breeding sites immediately after to more than a month after the breeding season, and move toward summer home ranges. There is considerable annual variation in timing, which is dependent on weather conditions (Dole, 1967b). In contrast to adults, subadult Northern Leopard Frogs tend to remain at or near overwintering sites for longer periods until they disperse to summer feeding grounds, thus accounting for size-differences in dispersal patterns.

Northern Leopard Frogs disperse toward nearby wet or moist habitats, where they remain for two to three weeks prior to moving to more distant areas. Interpond movement among adjacent small ponds is common (Dole, 1971b; Hardy, 1974). Frogs tend to follow contours that hold moisture, or they may follow the perimeter of ponds or lakes. Most take up feeding sites usually within several hundred meters of the natal pond. Merrell (1977) suggested that the extent of dispersion was inversely proportional to the size of the frog, with smaller frogs (< 40 mm SUL) more likely to remain in the vicinity of water bodies than larger frogs. Since juvenile leopard frogs are prone to desiccation (Thorson, 1955; Schmid, 1965), it is important to remain near water or to disperse rapidly to moist locations.

There is considerable variation in the behavior and movement patterns of individual adult Northern Leopard Frogs after breeding has ceased. Individuals may remain at or in close proximity (within ca. 60 m) to the breeding site, or they may move to slightly more distant terrestrial habitats should moisture and cover permit. Many frogs remain within a particular area along a wetland's perimeter, moving back and forth over short distances as they forage along the shoreline, whereas others move varying distances from one pond to another. However, some frogs move extended distances from breeding sites and do not return, presumably establishing home ranges for summer foraging. Hine et al. (1981) recorded a number of long-distance movements, including one frog that moved 400 m in 5 days from its breeding pond to a saturated area along a river bottom. Movements in excess of 150 m appeared common. These authors also recorded a frog dispersing a maximum of 140 m over a 24 hr period. Other long distance dispersal records for adults include 3,218 m (Merrell, 1970), 160 m (Dole, 1965a), and 1,744 m (Dole, 1968). According to Merrell (1977), an individual Northern Leopard Frog may move from 3 to 6 km over the course of a single year.

Fall movements. In the fall, frogs begin to move to overwintering sites as temperatures start to decrease. The timing of the movement depends on latitude. In Minnesota, for example, fall migration begins in September, whereas in Manitoba, it begins about the first of August (Eddy, 1976). Fall movements take place at night or on heavily overcast rainy days (Merrell, 1970). Movements do not occur en masse, but as a result of similar stimuli affecting the behavior of individual animals. According to Merrell (1977), sexually immature frogs begin the fall migration earlier than adult frogs. Migration continues as long as weather permits and is normally over by autumn. In Minnesota, for example, migration continues until nearly the end of October. Frogs arriving at overwintering sites may choose the warmest bodies of water as temperatures decrease. Instead of immediately entering the water, they may remain around the periphery of a lake, even changing overwintering site location as weather conditions permit.

Another form of fall movement was observed in rivers based on traps set to monitor salmon smolts. These traps picked up relatively large numbers of adult and subadult Northern Leopard Frogs, suggesting that either the frogs were using the river to move downstream toward overwintering sites or that they were captured incidentally as they were searching for overwintering sites among the rubble on the stream bottom (DuBois and Stoll, 1995). The extent to which leopard frogs use lotic waters as dispersal agents is unknown.

DIET

Northern Leopard Frogs eat a wide variety of prey, and very few animals contain empty stomachs during the summer foraging season. They are primarily insectivorous, with a penchant for eating beetles (e.g., tiger, carabid, ladybird). Other insect prey include mayflies, dragonflies, damselflies, roaches, crickets, grasshoppers, water boatmen, water striders, treehoppers, larval and adult lepidoptera, true bugs, flies, mosquitoes (and their eggs), weevils, ants, and bees. Noninsect prey items include worms, spiders, harvestmen, mollusks (gastropod snails, slugs), centipedes, millipedes, isopods (sow bugs), and crayfish. In Drake's (1914) Ohio survey, spiders made up 27% of the prey animals consumed, with insects of various taxa comprising 60%.

Vertebrate prey include shed skins and a juvenile leopard frog (Knowlton, 1944), large numbers of recently transformed leopard frogs (Eddy, 1976), a Wood Frog (Preston, 1982), a Spring Peeper (Sutton, 2004), larval and small *Anaxyrus boreas* (Ruthven and Gaige, 1915), a small garter snake (*Thamnophis*), ruby-throated hummingbird, and a yellow warbler (Breckenridge, 1944). Toad eggs are toxic when injected, however (Licht, 1968). Vegetation may comprise 10–20% of the stomach contents by volume, although most prey consists of invertebrates (Linzey, 1967). Other studies with diet informa-

Adult *Lithobates pipiens*, green Burnsi phenotype. Photo: Robert Newman

tion include Surface (1913), Klugh (1922), Moore and Strickland (1954), Gehlbach and Collette (1959), Whitaker (1961), Rittschof (1975), Eddy (1976), Miller (1978), Hine et al. (1981), Gilhen (1984), Collier et al. (1998), Sutton (2004), and Sutton et al. (2006).

Northern Leopard Frogs in the process of transformation often have mud and algae in their digestive tract, and a majority have pieces of shed epidermis within their guts. Feeding begins when the rapidly reabsorbing tail is 6 mm or less, when very small items (e.g., collembolans) are consumed. As the metamorph grows, the size and variety of the prey increase to become more like that of the adult (Munz, 1920). By the time the tail is < 4 mm, all metamorphs have begun feeding.

In laboratory settings, assemblages of leopard frogs may form a hierarchy during feeding, such that large frogs tend to be more successful than many of their conspecifics. Feeding hierarchies take time to form, however, and whether such hierarchies exist in nature, where frogs are more widely distributed around a pond, is unknown. Boice and Witter (1969) stated that the outcome of feeding interactions were predictable if the frogs had repeated interactions through time.

When feeding, Northern Leopard Frogs move their heads toward prey movement, orient the body toward the prey, and slowly approach. This species also is able to locate familiar prey using olfactory cues, even in the absence of visual cues (Shinn and Dole, 1978). They may walk forward toward the prey for a few steps, and lift or lower their heads as they draw up the hind limbs prior to leaping to capture the prey. They capture prey in a final leap of 15–40 cm. Predation attempts are not always successful. The entire sequence takes a mean of 30 sec, and frogs may travel from 0 to 100 cm (mean 35 cm); the longer a frog takes to approach prey and the farther the distance it travels, the less successful are predation attempts (Wiggins, 1992). Most successful attempts occur when the nearest neighbor is well away from the would-be predator or in solitary situations. An unsuccessful frog was observed to nip a successful frog in laboratory feedings (Boice and Witter, 1969).

PREDATION AND DEFENSE

Northern Leopard Frog eggs are eaten by Eastern Red-spotted Newts (*Notophthalmus viridescens*) (Wright, 1914). Larvae are consumed by a wide variety of vertebrates (fishes, birds, turtles, larval Tiger Salamanders [*Ambystoma tigrinum*]), and invertebrates. Specific examples include larval predaceous diving beetles (Dytiscidae), dragonfly larvae, caddisfly larvae, backswimmers (*Notonecta*), diving beetles (*Dysticus*), and giant water bugs (*Belostoma*). Juvenile frogs are captured by many predators, as are adults, especially northern pike (*Esox lucius*), bass (*Micropterus*), garter snakes (*Thamnophis*), water snakes (*Nerodia*), hognose snakes (*Heterodon*), racers (*Coluber*), bull snake (*Pituophis*), raccoons (*Procyon lotor*), and a wide variety of birds (pied-billed grebe, green heron, great blue heron, sparrow hawk, broad-winged hawk, marsh hawk, great horned owl, burrowing owl). Breckenridge (1944) reported that American merganser ducks eat a considerable number of dormant leopard frogs in winter.

There are both morphological and behavioral means of defense in Northern Leopard Frogs. The dorsal coloration and spot patterning of postmetamorphs allow them to be concealed in grass and other types of vegetation, and to disrupt the outlines of the frog, making them difficult to see. In contrast, the white belly is a form of countershading, making the frogs difficult to see from below by aquatic predators looking up toward the sky. Except while feeding or moving between locations, leopard frogs sit immobile in concealing vegetation and thus do not draw attention to themselves.

In the general vicinity of predators, Northern Leopard Frog juveniles will crouch or maintain a nonmoving posture, making them less likely to be detected and attacked by garter snakes (*Thamnophis*) (Heinen and Hammond, 1997). Davis (1933) noted an instance of a distress cry ("bleating") followed by seizure, then death-feigning, in a leopard frog touched by a garter snake (*Thamnophis*). Northern Leopard Frogs may inflate their body, stand high on their limbs, arch their body, and curl the head between the front legs when attacked by a snake, thus making them difficult to handle and swallow.

When disturbed or if a predator comes too close at a pond or wetland, Northern Leopard Frogs suddenly leap from the shoreline into the water, sometimes emitting a sudden cry at the same time. A single leap can cover 0.5–1.12 m (Black, 1970). Escape is to vegetative cover or by diving into the wetland substrate and remaining quiescent in familiar surroundings. The cry serves to startle and perhaps distract a potential predator, although it is not as obvious in Northern Leopard Frogs as in Southern Leopard Frogs. Frogs captured by a predator also emit a "distress" ("warning" or "fright") call and may empty their bladder; these calls may be trilled or untrilled. Frogs are more wary during daylight hours and on bright moonlit nights than they are at night or when the moon is not visible. In addition, Northern Leopard Frogs in terrestrial situations have longer leaps, greater fight distances, and do not allow predators to approach as closely as do other aquatic frogs (Heinen and Hammond, 1997).

POPULATION BIOLOGY

Estimates of the size of leopard frog breeding populations are not common. Using simple Lincoln-Peterson estimates and adjusting for unequal sex ratios at breeding sites, Merrell (1968) estimated populations from a few hundred to more than 3,700 adults at 6 Minnesota breeding sites of unspecified size. Assuming one male and female per egg mass, effective population sizes (<112 adults per pond, depending on which pond was sampled) were a small fraction of the number of adults actually present. Hoffman et al. (2004) later confirmed that effective population sizes of Northern Leopard Frogs ranged from a few hundred to a few thousand animals by examining changes in allele frequency through time. Population estimates made early in the year were higher than those later in the breeding season, but as the season progressed more recaptures resulted in lower population estimates, as expected. These results led Merrell (1968) to speculate that genetic drift could affect leopard frog populations, despite the high numbers observed during population censuses, a conclusion later confirmed by Hoffman et al.'s (2004) findings of a stable genetic structure over a period of 22–30 yrs. At another site, Merrell (1969) estimated there were 600 adults and as many as 70,980 juveniles. Based on genetic structure and dispersal patterns, it appears unlikely that Northern Leopard Frog populations follow a dynamic (extinction-recolonization) metapopulation model (Smith and Green, 2005), but instead remain in stable equilibrium between drift and gene flow. Pope et al. (2000) noted that the full extent of habitat use (breeding, summer) at a landscape scale is necessary to predict localized population density.

Survivorship is very low from egg deposition to the postmetamorphic juvenile stage. In Wisconsin, for example, Hine et al. (1981) estimated that only 1–6% of frogs survived from egg deposition to 1.5 months postmetamorphosis, assuming 3,500 eggs per egg mass. In Minnesota, most mortality also takes place during the aquatic stage, with resulting ratios of mature to immature frogs at 1:15 to 20 (Merrell, 1977).

At metamorphosis, transforming frogs tend to lose body mass over a four-day period, presumably as energy reserves are used in reconfiguring the body. However, after a month of growth, juveniles in Wisconsin ranged from 40–55 mm SUL and weighed 5–14 g (Hine et al., 1981). Differences resulted from variation in the timing of metamorphosis, as well as the availability of prey from one year to the next. In Minnesota, Merrell (1977) found that Northern Leopard Frogs normally transformed at 35–40 mm SUL, but that considerable variation in body size could result as frogs transformed during drought conditions (recently transformed froglets at 25–30 mm SUL) or after a period of relatively uncrowded larval conditions (at 48–50 mm SUL). There may also be a geographic component to size at transformation, with more northern frogs smaller at transformation than those to the south. In northern Michigan, for example, transformation occurred at 21.5 mm SUL, with frogs growing to 37 mm SUL prior to entering their first winter dormancy (Rittschof, 1975).

In Wisconsin, females reach sexual maturity at approximately 60 mm SUL whereas males reach sexual maturity at 55 mm SUL (Hine et al., 1981). These females likely are 2 yrs old (i.e., in their third season); some females (23%), however, do not mature until 3 yrs old (i.e., in their fourth activity season). Similar times to maturity have been reported for females in other northern populations (Force, 1933; Baxter, 1952; Ryan, 1953; Rittschof, 1975; Eddy, 1976; Merrell, 1977; Leclair and Castanet, 1987), although some males may reach maturity in their second season (Rittschof, 1975), even in southern Michigan. In some regions, elevation plays an important role in the timing of maturity, with higher-elevation females taking an extra year to reach sexual

Adult *Lithobates pipiens*, Kandiyoh phenotype. Photo: Robert Newman

maturity than their lower-elevation counterparts (Baxter, 1952). Based on skeletochronology, Leclair and Castanet (1987) estimated longevity at no more than 4–5 yrs.

Reports of sex ratios generally are male biased, as males remain longer at breeding ponds than females and are more visible while there. Of 104 Northern Leopard Frogs caught at a Wisconsin breeding site, 86% were males and 13% females (Hine et al., 1981). Merrell (1968) reported only 11% of adults were female around a breeding pond in Minnesota. However, sex ratios at other times of the year and away from breeding ponds tended to be 1:1 (Merrell, 1968; Hine et al., 1981). In West Virginia, Sutton (2004) found that males made up 96% of the captures in spring, but only 69% of the captures in fall.

Northern Leopard Frog populations are dynamic and experience considerable annual turnover from one pond or wetland to the next; a pond occupied one year may not be occupied the next. For example, annual occupancy rates varied between 40 to >60% over a 3-year study in Ontario, with considerable variation from one year to the next (Hecnar and M'Closkey, 1996a). Despite high turnover rates, short-term studies have been able to demonstrate apparent regional population declines. However, while populations may be declining in one region, they may be increasing or stable in other nearby regions.

COMMUNITY ECOLOGY

Larval Northern Leopard Frogs often occur in the same wetlands as the larvae of other ranids. Under some circumstances, competition between larval species is apparent (high densities, low food resources), whereas in other situations there appears to be no indication of competition (low densities, high food resources). Indeed, the competitive effects of Northern Leopard Frogs on Wood Frogs may vary annually, so that competition occurs some years but not others (DeBenedictis, 1970, 1974). In the latter situation, these species act as ecological equals.

Under experimental conditions, the effects of the larvae of one anuran species on a heterospecific vary with the rearing conditions in mixed species assemblages. Predators (dragonfly larvae [*Anax*] and mudminnows [*Umbra*]) can affect growth rates, but the effects of predation may be altered by the presence of a second potential prey species. For example, Northern Leopard Frogs grow faster than Wood Frogs when separately raised in the presence of a caged predator. However, when these species are reared together in the presence of the predator, the outcome is reversed. In the presence of competing Wood Frogs, larval *L. pipiens* increase the width of the mouth by 5%, but their tail length remains the same as when reared independently. In the presence of the predator and competitor, however, Northern Leopard

Frog larvae decrease the width of the mouth and the tail length. Thus, predators and prey affect larval phenotypic plasticity both independently or in combination (Relyea, 2000).

Northern Leopard Frogs and other ranids (American Bullfrogs, Green Frogs, Wood Frogs, Mink Frogs, Plains Leopard Frogs) frequently occur within the same pond or wetland or in close proximity to one another. These species are able to coexist via habitat partitioning or by having different reproductive phenology. For example, both Green and Northern Leopard frogs consume the same types and sizes of prey. However, Northern Leopard Frogs generally are found in more dense vegetation than Green Frogs, and Green Frogs are found closer to water when they are on land than Northern Leopard Frogs. In Minnesota, Northern Leopard Frogs (77%) preferred a floating sedge mat over other available habitats in and surrounding a bog, with some individuals (18%) being found in tamarack (Marshall and Buell, 1955); these preferences did not overlap with those of Wood Frogs. In West Virginia, *L. pipiens* was associated with grassy aquatic and associated terrestrial shoreline habitats, whereas few American Bullfrogs were found in such areas (Sutton, 2004). Partitioning the habitat allows the coexistence of these species, and may have resulted from past competition among them (McAlpine and Dilworth, 1989). In yet another example of habitat partitioning, Northern Leopard Frogs prefer clear water streams and wetlands in Nebraska, whereas Plains Leopard Frogs are found in highly turbid and silt-laden habitats (Lynch, 1978). As a result, Plains Leopard Frogs are found in the loess hills regions, whereas Northern Leopard Frogs are found in more sandy habitats. Partitioning of the reproductive season and breeding sites also diminishes the possibility of hybridization or competition with *L. blairi*, although these species can produce fertile offspring and are frequently found syntopically in the summer and autumn in areas of contact on the Great Plains.

DISEASES, PARASITES, AND MALFORMATIONS

Lithobates pipiens is affected by a number of pathogens, including viruses (tadpole edema virus, Frog Virus 3), bacteria, and fungi. Tumors of the kidney have been reported resulting from herpes-like viral infections (Lucké renal adenocarcinoma) (Lucke, 1952; Rafferty, 1964; Granoff et al., 1965; McKinnell and Zambernard, 1968; McKinnell, 1984; McKinnell and Carlson, 2005), and a tumor on the snout, of unknown etiology, is pictured in Sutton (2004). Iridoviruses (*Ranavirus*) have been reported in Northern Leopard Frogs from Minnesota (Green et al., 2002; Uyehara et al., 2010). Echaubard et al. (2010) showed that the severity and effects of infections by *Ranavirus* are density-dependent, affecting growth, survivorship, and larval stress responses.

The bacteria *Aeromonas hydrophila* and *Pseudomonas* (both *P. maculicola* and *P. fulva*) are present in natural populations (Hird et al., 1983; Sutton, 2004). In wild populations, Northern Leopard Frog larvae have a greater tendency to contract the bacterium *A. hydrophila* than do adults (Hird et al., 1981). The bacterium *Flavobacterium indologenes* occurs in captive colonies of leopard frogs (Olson et al., 1992). Peptides in the skin may help in defense against bacterial infections (Goraya et al., 2000).

Although found in many sympatric species, the fungus *Ichthyophonus* was not found on Northern Leopard Frogs in Québec (Mikaelian et al., 2000). However, amphibian chytrid fungus occurs widely in Northern Leopard Frogs. In Maine, Longcore et al. (2007) found it on 25.7% of individuals (n = 74) examined, although there was considerable annual variation in the infection rate. In Québec, it occurred on 10.5% of amphibians sampled between 1960 and 2001 (Ouellet et al., 2005a). These latter authors reported an infection rate of 15.5% (40 of 528) for amphibians examined between 1895 and 2000 in Canada and the United States. Other reports are from Minnesota (Martinez Rodriguez et al., 2009), Colorado (Muths et al., 2008), Iowa (<1%; Loda and Otis, 2009), North Dakota (Green et al., 2002), and the upper Midwest (Sadinski et al., 2010). Amphibian chytrid also has been found in *Lithobates pipiens* shipped from one section of the country to another in the amphibian trade (Gaulke et al., 2010). A *Dermocystidium*-like fungus was reported on these frogs from Minnesota (Green et al., 2002).

The ciliate protozoans *Nyctotherus cordiformis* and *Opalina obtrigonoidea* are known from Ohio Northern Leopard Frogs (Odlaug, 1954). The protozoans *Chilomastix* sp., *Opalina* sp., *Nyctotherus cordiformis*, and *Tritrichomonas batrachorum* are found in Utah *Lithobates pipiens* (Parry and Grundman, 1965). Many enteric protozoans (*Chilomastix* sp., *Endolimax ranarum, Entamoeba ranarum, Euglenamorpha hegneri, Giardia agilis, Hexamita intestinalis, Monocercomonas* sp., *Monocercomonoides* sp., *Nyctotherus cordiformis, Octomastix* sp., *Octomitus* sp., *Opalina* sp., *Phacus pleuronectes, Retortamonas dobelli, Tritrichomonas augusta, T. batrachorum, T. prowazeki, Urophagus* sp.) have been reported from Northern Leopard Frogs in Minnesota (Anderson and Buttrey, 1962). The coccidian protozoan *Isospora lieberkuehni* also occurs in *Lithobates pipiens* (Upton and McAllister, 1988). Blood parasites include *Haemogregarina magna* and *Isospora lieberkuehni* (Levine and Nye, 1977). The following adult and/or larval helminths have been reported from Northern Leopard Frogs: the cestodes *Cylindrotaenia americana, C. quadrijugosa, Mesocestoides* sp., *Nematotaenoides ranae, Ophiotaenia saphena, Proteocephalus saphena, Proteocephalus* sp.; the digeneans *Alaria arisaemoides, Alaria* sp., *Allassostomoides chelydrae, Apharyngostrigea pipientis, Cephalogonimus americanus, Cephalogonimus* sp., *Clinostomum* sp., *C. attenuatum, Diplostomum* sp., *Echinostoma trivolvis, Fibricola* sp., *F. crater, Glypthelmins quieta, Gorgoderina amplicava, G. attenuata, G. simplex, G. translucida, Haematoloechus coloradensis, H. complexus, H. longiplexus, H. medioplexus, H. parviplexus, H. similiplexus, H. varioplexus, Haematoloechus* sp., *Halipegus eccentricus, H. occidualis, Halipegus* sp., *Hichinostometa trivolvis, Megalodiscus americanus, M. temperatus, Ochetostoma* sp., and unidentified Strigeidae; the nematodes *Cosmocercoides dukae, C. variabilis, Cosmocercoides* sp., *Falcaustra ranae, Gyrinicola batrachiensis, Ozwaldocruzia leidyi, O. pipiens, Ozwaldocruzia* sp., *Physaloptera* sp., *P. ranae, Physocephalus* sp., *Rhabdias ranae, Spinitectus gracilis, Spironura ranae, Spiroxys contortus, Spiroxys* sp., and *Strongyloides* sp., unidentified Seuratoidea, and Echinostomatidae; the acanthocephalan *Fessisentis friedi* (Odlaug, 1954; Najarian, 1955; Waitz, 1961; Brooks, 1976; Ulmer and James, 1976a, 1976b; Ashton and Rabalais, 1978; Adamson, 1981c; McAllister and Conn, 1990; Dyer, 1991; McAlpine, 1996, 1997b; McAlpine and Burt, 1998; Goldberg et al., 2001; King et al., 2007, 2008). Most Northern Leopard Frogs are parasitized by nematodes, regardless of life stage (young, juveniles, adults), whereas digeneans infect mostly the adult stage. Leeches (Hirudinea, *Desserobdella picta*) also may parasitize larvae (DeBenedictis, 1970; Bolek and Janovy, 2005) and adults, sometimes to the point of debilitation (Merrell, 1977).

Malformations of Northern Leopard Frogs have been commonly reported in recent years, particularly in the upper Midwest and in northern New England. There are more reports of sites with malformed *Lithobates pipiens* than there are for any other species (Johnson et al., 2000). Malformations may affect larvae, juveniles, and adults, and are not correlated with either the Burnsi or Kandiyohi phenotypes. The most common malformations include missing limbs or digits (amelia), missing segments of limbs (ectromelia), abnormally sized limbs or digits (phocomelia), multiple limbs or digits (polymelia, polydactyly), inappropriately located limbs, webbing between limbs, fusion of digits, many types of gross skeletal abnormalities, or various combinations of these malformations. Other malformations affect eye development, head size, and jaw formation. The vertebral column of larvae and the hind limbs of postmetamorphs are most often affected. Malformations are uncommonly symmetrical. Photographs and discussions of the types and frequencies of different malformations in Northern Leopard Frogs are provided in Meteyer (2000), Meteyer et al. (2000a), and Hoppe (2005).

Malformations involving the hind limbs of the Northern Leopard Frogs likely result, in part, from parasitism by metacercariae (a larval stage) of the trematode *Ribeiroia ondatrae*, which burrow into the tadpole causing abnormal development during crucial periods of skeletal formation. The type of malformation and its effects on larval survivorship is dependent on the timing of infection during development (Schotthoefer et al., 2003). Trematode infection appears mediated by the presence of the herbicide atrazine (Rohr et al., 2008). However, other factors (UVB radiation, disruption of retinoic acid metabolism, chemical exposure) or combinations of factors, may result in malformations, particularly those not affecting the hind limbs. The timing, severity, and sequence of the developmental perturbations influence the type and severity of the malformation in Northern Leopard Frogs.

The history of the discovery of extensive malformations, and the scientific, media, and political response to the malformation crisis in Minnesota and elsewhere, has been told in a popular book (Souder, 2000). Lannoo (2008) summarizes a wealth of information on malformations in this and other species and provides a critique of current hypotheses as to causes. Additional important references include: Rosine (1952), Merrell (1969), Ouellet et al. (1997), Converse et al. (2000), Helgen et al. (1998, 2000), Meteyer (2000a, 2000b), Hoppe (2000, 2005), Rosenberry (2001), and Sutton (2004).

SUSCEPTIBILITY TO POTENTIAL STRESSORS

Metals. Lead at concentrations of 3, 10, and 100 µg/L have deleterious effects on larval *Lithobates pipiens*. Depending on concentration, lead slows tadpole growth, especially in the early stages of development, results in spinal curvature that inhibits swimming ability, and increases time to metamorphosis. It does not have significant effects on percent of larvae that metamorphose, SUL, mortality, or the sex ratio of metamorphs. Most detrimental effects occurred at 100 µg/L. Tissue concentrations of the experimental tadpoles ranged from 0.1 to 224.5 mg/kg dry mass. The authors concluded that EPA regulations requiring lead concentrations at < 2.5 µg/L were sufficient to protect tadpoles (Chen et al., 2006). Tadpoles reared at 10 µg/L cadmium were larger and more advanced after 14 days than controls, and survivorship did not appear to be impaired (Gross et al., 2009). Cadmium chloride, however, has an LC_{50} of 3,700 µg/L for tadpoles (Zettergren et al., 1991).

Other metals known to affect *L. pipiens* include: aluminum (LC_{50} for embryos 400–1,000 µg/L (Freda and McDonald, 1990), arsenic (LC_{50} 0.11 mg/L) (Birge and Just, 1973), mercury ($LC_{50\ (96\ hr)}$ 0.007 mg/L in embryos), silver ($LC_{50\ (96\ hr)}$ 10 µg/L in embryos and larvae), and zinc (LC_{50} 0.05 mg/L) (Birge and Just, 1973; Birge et al., 2000; *in* Sparling, 2003). Copper sulfate does not affect eggs, but concentrations > 0.31 mg/L kill newly hatched tadpoles, and tadpoles reared in concentrations of 0.06–0.16 mg/L weighed less than controls. The LC_{50} of copper sulfate to larvae is 0.15 mg/L (Landé and Guttman, 1973). Birge et al. (2000) provide LC_{50s} for 28 metals that might affect *L. pipiens*.

Glooschenko et al. (1992) reported a tendency of Northern Leopard Frogs in Ontario to avoid wetlands with high metal concentrations as a result of aerial deposition; the presence of Northern Leopard Frogs also was negatively correlated with zinc concentrations.

Other elements. The $LC_{50\ (8\ day)}$ for boron is 130 mg/L for *L. pipiens* embryos (*in* Sparling, 2003).

pH. Low pH adversely affects sperm motility and fertilization rates in *L. pipiens*. At a pH of between 5.5 and 5.8, even fertilized embryos stop development after 48 hrs, and no development occurs at a pH of <4.8. Thus, this species requires waters > pH 6.0 for normal fertilization and development (Schlichter, 1981). A low pH results in a decrease in sodium influx and a significant increase in sodium efflux within *L. pipiens* larvae (Freda and Dunson, 1984). Larvae generally do not survive at a pH <4.0, and a pH of 4.2–4.5 is generally considered the lowest pH tolerated (Karns, 1983; Freda, 1986; Corn et al., 1989; Freda et al., 1991). When given a choice, larvae avoid waters with both lethal and sublethal pH values (Freda and Taylor, 1992). In Nova Scotia, pH at *L. pipiens* breeding ponds ranged between 4.6 and 6.0 (mean 4.91) (Dale et al., 1985). Pope et al. (2000) noted that leopard frog populations tend to increase as pH rises to neutral (7.0), but decrease as ponds become more basic. Presence of Northern Leopard Frogs increases in ponds with a high buffering status; buffering status is inversely correlated with acidity (Glooschenko et al., 1992).

Nitrates and ammonia. Under laboratory conditions, Northern Leopard Frog embryos and larvae exposed to > 1.5 mg/L of un-ionized ammonia (NH_3) had decreased hatching success, decreased survival, increased levels of malformations (especially curling; 60% malformations at ca. 2.25 mg/L), and slower growth and development than untreated embryos and larvae (Jofré and Karasov, 1999; Jofré et al., 2000). These authors speculated that leopard frog larvae and adults might be exposed to detrimental levels of ammonia during winter dormancy or during episodic pulse of ammonia release from sediments. Although PCBs had no effect on hatching success, embryos had reduced hatching success in combination with NH_3. PCBs alone at 50 mg/L had a negative impact on growth and survivorship (Jofré et al., 2000).

Diluted ammonium nitrate (NO_3-N) fertilizer

results in lowered survivorship at a chronic exposure (to 100 days) of 10 mg/L. Acute exposure > 25 mg/L causes mortality in excess of 85%; exposure at 15–20 mg/L causes a much lower mortality rate (22–25%) after 96 hrs (Hecnar, 1995). Changes in behavior and an increase in malformations, particularly bent tails, also were associated with increasing levels of fertilizer. These concentrations are commonly found in natural waters throughout North America (Rouse et al., 1999).

NaCl and *KCl*. Northern Leopard Frog eggs can tolerate an upper salinity limit of 5.0 ppt (semilethal at 3.8–4.5 ppt). Salinities at 13.0 ppt rapidly kill adults.

Chemicals. The fungicide mancozeb (azinphos-methyl) is toxic to Northern Leopard Frog embryos at a $LC_{50\,(96\,hr)}$ of 0.20 mg/L. Mortality occurs at a rate of 60% 1–2 weeks after metamorphosis for leopard frogs exposed as larvae to 8.0 mg/L (Harris et al., 2000).

The herbicide atrazine causes significant mortality at 48.7 mg/L, with decreasing survivorship after 4-days posthatch in concentrations of 13.2 mg/L (Birge et al., 1980). It functions as an immune system disruptor (at 21 ppb for 8-days exposure) in Northern Leopard Frogs by suppressing the recruitment and phagocytic activity of white blood cells (Brodkin et al., 2007). Such immunosuppression results in increasing susceptibility to trematode infection (Rohr et al., 2008). Even at low concentrations, atrazine causes testicular malformations and impedes normal development (Langlois et al., 2010).

In contrast, Allran and Karasov (2001) stated that there were no adverse effects from the exposure of embryos, larvae, and adults to atrazine, although they found a dose-dependent increase in deformed larvae. The larvae were only followed for a period of four days. Atrazine has greater effects on older larvae than younger larvae, however, as larval sensitivity increases with developmental stage (Howe et al., 1998). Chronic effects included edema when larvae are exposed to sublethal doses of atrazine. Tests for lethality after a 30-day exposure suggested values 10- to 20-fold less (5.1 mg/L for early stage larvae, 0.65 mg/L for late-stage larvae) than $LC_{50\,(96\,hr)}$ (47.6 mg/L for early stage larvae, 14.5 mg/L for late-stage larvae). Adding alachlor decreased $LC_{50\,(96\,hr)}$ even further (Howe et al., 1998).

Murphy et al. (2006) found that the incidence of atrazine-induced hermaphrodites was no different in areas affected by agriculture than from those that were not affected. However, atrazine at similar concentrations was found in both types of sites, and there was an increase in testicular oocytes in male juvenile *L. pipiens* that correlated with atrazine concentration, at least in one year. Significant increases in gonadal deformities (testicular oocytes) and in retarded gonadal development also were reported by Hayes et al. (2002, 2003). Atrazine affects the development of gonads by inducing aromatase, an enzyme that converts androgen into estrogen resulting in feminized males. Hayes (2004) has demonstrated that many studies claiming that atrazine does not affect amphibians are seriously flawed in experimental design and sampling procedure.

The herbicide alachlor by itself has detrimental effects on larval Northern Leopard Frogs. $LC_{50\,(96\,hr)}$ exposures were 11.5 mg/L for early stage larvae and 3.5 mg/L for late stage larvae (Howe et al., 1998). Tests for lethality after a 30-day exposure suggested values >5 times less (2.0 mg/L for early stage larvae, 0.47 mg/L for late stage larvae) than $LC_{50\,(96\,hr)}$. These results further suggest that later stage larvae are more susceptible than early developing embryos to the cumulative and synergistic effects of these pesticides. Such chronic exposure results suggest concern for larval amphibians where atrazine is used.

The herbicide Vision™ (glyphosate) reduced larval survivorship at 0.75 and 1.0 mg/L at a pH of 5.5 and 7.5; the less acidic pH actually increased the level of toxicity (Chen et al., 2004). However, it is likely these authors were working with a mixed species assemblage, as some frogs were obtained from the field [New Hampshire] whereas others were obtained from Carolina Biological Supply Company (CBSC) in North Carolina.

Sensitivity (toxicity, expression of an avoidance response) to Vision™ is dose and stage dependent, and may be affected by pH and sediment levels (Wojtaszek et al., 2004). Larval *L. pipiens* are six to eight times more sensitive to Vision™ than embryos, and detrimental effects are largely due to the surfactant polyethoxylated tallow amine. The $LC_{50\,(96\,hr)}$ was 15.1 mg/L at a pH of 6.0 for embryos and 1.8 mg/L for larvae; the $LC_{50\,(96\,hr)}$

was 7.5 mg/L at a pH of 7.5 for embryos and 1.1 mg/L for larvae (Edginton et al., 2004). Thus, the toxicity of Vision™ was increased by an increase in pH. Thompson et al. (2004) and Wojtaszek et al. (2004) concluded that Vision™ did not have toxic effects on *L. pipiens* larvae at environmentally relevant concentrations, although the latter authors noted differences in growth and maximum larval size at different study sites. They did not, however, measure the effects of the herbicide on algae, the tadpole's food source, which could have accounted for the observed variation in tadpole development between sites.

Larval *L. pipiens* show decreased SUL at metamorphosis, increased time to metamorphosis, tail damage, and gonadal abnormalities when exposed to chronic environmentally relevant concentrations of Roundup Original™ (a glyphosate end-use formulation) (Howe et al., 2004). Roundup™ is toxic to tadpoles at 3.8 mg/L, and has serious effects on the tadpole's aquatic community, regardless of the substrate involved (Relyea, 2005a, 2005b). Adding predators may increase the effects of a pesticide. At 1.3 mg/L, Roundup™ reduces larval survivorship and biomass by 29%; in the presence of a predator (*Notophthalmus*) and Roundup™, survivorship decreases by another 21%. The predator affected tadpole growth, but the pesticides did not (Relyea et al., 2005). The herbicide 2,4-D has no appreciable effect on Northern Leopard Frog larvae (Relyea, 2005b), as does the herbicide hexazinone (Berrill et al., 1994, 1997).

Chen et al.'s (2008) report on detrimental effects of the herbicide Release™ (triclopyr) on *Lithobates pipiens* cannot be reliably attributed to *L. pipiens*, as test specimens were obtained from CBSC outside the range of this species. In another series of experiments, however, newly hatched Northern Leopard Frog tadpoles immediately were sensitive to triclorpyr (2.4–4.8 ppm) and the organophosphate insecticide fenitrothion (4–8 ppm), causing either mortality or paralysis (Berrill et al., 1994, 1997). In addition to direct effects, the paralysis would increase mortality by making the tadpole unable to escape predators. The organophosphate insecticide chlorpyrifos has an LC_{50} of 3000 µg/L (Barron and Woodburn, 1995).

The aquatic herbicide diquat had little effect on the eggs of *L. pipiens* at 100 ppm, but larvae showed high mortality within 14 days of administration. In addition, embryos had high rates of exogastrulation when administered diquat at fertilization (Bimber and Mitchell, 1978).

Pyrethroid insecticides known to have adverse effects on Northern Leopard Frogs include permethrin and fenvalerate (Berrill et al., 1997) and esfenvalerate (Cole and Casida, 1983). Adults are less sensitive to pyrethroids than larvae. Esfenvalerate has an $LC_{50\ (96\ hr)}$ of 7.79 µg/L and causes larval convulsions in concentrations as low as 3.6µg/L. Even at 1.3µg/L, activity is decreased (Materna et al., 1995). It should be noted, however, that Materna et al. (1995) did not differentiate between *L. blairi*, *L. pipiens*, and *L. sphenocephalus* in presenting results. Temperature magnifies the effect of the pesticide, and additional effects on frog populations may result from the pesticide's lethal effects on fish and invertebrates. The pyrethroid carbaryl may accelerate the time to metamorphosis at concentrations of 2 mg/L, but does not appear to affect larval growth or survivorship (Bulen and Distel, 2011; Distel and Boone, 2010, 2011). Carbaryl exposed larvae also had a greater body mass at metamorphosis.

Endosulfan, an insecticide, caused direct mortality to a large percentage (84%) of Northern Leopard Frog larvae at 6.4 ppb. As with some other pesticides, metamorphosing *L. pipiens* froglets are more susceptible to this insecticide than larvae. Dosages of 0.47 mg/L were lethal to 33% of frogs exposed during experimental trials, whereas all frogs were killed at > 2.35 mg/L (Harris et al., 2000). Endosulfan also produces skeletal deformities affecting tadpoles, such as tail curvature, which impairs swimming ability.

Another insecticide, diazinon, causes less mortality (24%) at 2.1 ppb than endosulfan, but results in smaller than control-sized metamorphs. Mixtures of 5 insecticides or 5 insecticides and 5 herbicides caused 99% mortality in outdoor mesocosm exposures, even in low individual concentrations (Relyea, 2009). Pesticide exposure can have indirect effects on leopard frog growth and time to metamorphosis by decreasing dietary items, such as periphyton (Relyea, 2009). In areas where hydroperiod is critical to ensure sufficient time for development, the presence of certain pesticides may cause significant indirect mortality by delaying growth, resulting in an inability to transform as ponds desiccate.

The results from other insecticide and herbicide mixtures are variable, depending on the chemicals, dosages, and the presence of food sources and predators. Low doses of the insecticides Sevin™ and malathion actually improve survival of Northern Leopard Frog larvae, but, as above, adding the herbicide Roundup™ completely eliminates larvae (Relyea, 2005b; Relyea et al., 2005). The insecticides eliminated some of the anuran insect predators as well as cladocerans that feed on phytoplankton. As a consequence, phytoplankton increased as a food source for the tadpoles, and predation pressure was reduced. The $LC_{50\ (16\ day)}$ for malathion is 2.4 mg/L (Relyea, 2004b).

In mixtures of pesticides at nonlethal or sublethal levels, *L. pipiens* larvae and juveniles experience decreased proliferation of T lymphocytes, suggesting an adverse effect on immune response (Christin et al., 2003, 2004). This makes larvae and juveniles more susceptible to parasite infection. Parasitic lungworms (*Rhabdias ranae*) are easily able to infect larvae at numbers greater than in nonpesticide treated larvae, and they mature and reproduce faster than they do in untreated larvae. By adversely affecting the larva's immune system, pesticides also facilitate migration and maturation of the worm within the developing larva (Gendron et al., 2003). The combination of pesticide- and parasite-induced immunosuppression increases the susceptibility of juveniles to parasite infestation (Christin et al., 2003).

The lampricide 3-trifluoromethyl-4-nitrophenol (TFM) causes arrested development in embryos at 1.0 mg/L, and has a $LC_{50\ (96\ hr)}$ of 0.95–2.76 mg/L in tadpoles and 3.55 mg/L in adults (Chandler and Marking, 1975; *in* Sparling, 2003). The piscicide rotenone kills both adult and larval *Lithobates pipiens*, although the effects depend on stage of larval metamorphosis, life stage, and concentration (Hamilton, 1941; Farringer, 1972; Burress, 1982). In addition, the following chemicals cause either a significant decrease in hatching success or decreased survival after a period of 4 days posthatching: chloroform (> 11.8 mg/L); nitrilotriacetic acid (NTA) (> 48.8 mg/L); and phenol (> 0.074 mg/L). The LC_{50} for chloroform is 4.16 mg/L, for NTA (39.3 mg/L), for carbon tetrachloride (1.76 mg/L), for methylene chloride (> 85.0 mg/L), and for phenol (0.04 mg/L) (Birge et al., 1980, 2000). Birge et al. (2000) also provide LC_{50s} for aniline, benzene, dichlorobenzene, nitrobenzene, toluene, and m-xylene.

PCBs, polychlorinated dibenzo-*p*-dioxins, and polychlorinated dibenzofurans have been found in the tissues of Northern Leopard Frogs from Wisconsin. Tissue concentrations of PCBs were correlated with sediment concentrations. However, the amount of PCBs in the tissues was considered low (Huang et al., 1999).

DDT causes a variety of larval deformities, respiratory paralysis, and growth inhibition at 1 ppm (Schreiman and Rugh, 1949). Exposure to sublethal doses of pesticides (DDT, malathion, dieldrin) causes an array of effects related to immunosuppression, such as a decreased antibody response. If exposure occurs after the frogs have already been exposed to pathogen antigens, immunosuppression does not occur. In wild populations, significant differences can occur in immune function between populations exposed to pesticides and those that are pesticide free (Gilbertson et al., 2003).

UV light. Under controlled conditions, UV light may have detrimental effects on Northern Leopard Frogs. Ambient levels of UVB light are lethal to *L. pipiens* larvae after several days of direct exposure in shallow water (Higgins and Sheard, 1926; Tietge et al., 2001), and artificially high doses of UVB light can cause significant mortality of newly hatched larvae (Licht, 2003). The effects of UVB also depend on pH, such that a decrease in pH tends to decrease survivorship slightly depending on light intensity (Long et al., 1995). However, it is unlikely that ambient light under natural conditions has detrimental effects on *L. pipiens* hatching and larval survival, because eggs and larvae in natural ponds normally are not exposed to direct UV light for the time and intensity necessary to cause developmental problems (Crump et al., 1999). At 300 nm, the jelly capsule surrounding the egg absorbs about 90% of the UVB light (Licht, 2003).

STATUS AND CONSERVATION

Northern Leopard Frog populations have declined or been extirpated in a number of locations, particularly in the western and northwestern parts of North America (reviewed by Rorabaugh, 2005a). Examples include: (1) they were extirpated from Larimer County, Colorado, between 1973 and 1982, possibly due to the combined effects of drought and small population size (Corn and Fogleman, 1984); (2) they are now absent from the Rocky Mountains at high elevations (Corn et al., 1997; Johnson et al., 2011), and they have declined or disappeared in many lowland areas of Colorado (Hammerson, 1999); (3) most historic populations in eastern Washington have been extirpated (Leonard et al., 1999), and only one remains, and it is in decline (Germaine and Hays, 2009); (4) the species has been extirpated from most locations in western Montana and from 80% of historic locations within the northwestern Plains (Maxwell and Hokit, 1999; Maxwell et al., 2003; Werner et al., 2004; Smith and Keinath, 2007); (5) the species is considered declining in some parts of the upper Midwest (Minton et al., 1982; Brodman et al., 2002), but not in all areas, at least through the early 1980s (Christiansen, 1981); (6) declines have occurred in British Columbia, Alberta, Saskatchewan, Manitoba, and northern Ontario, but not in the eastern Canadian provinces (Wagner, 1997; Weller and Green, 1997; Russell and Bauer, 2000; Adama et al., 2004; ANLFRT, 2005); (7) Minton (2001) considered that numbers had decreased substantially in northern Indiana after about 1970; (8) most populations in California have been extirpated, either from habitat alteration or from the effects of the introduced American Bullfrog (Jennings and Hayes, 1994b; Panik and Barrett, 1994); (9) they have been largely extirpated from the Yellowstone–Grand Teton region (Koch and Peterson, 1995); (10) unexplained declines have occurred in the Laramie Basin and elsewhere in Wyoming (Baxter and Stone, 1985); (11) extant populations were not found at 13 historic locations in Arizona based on surveys from 1983 to 1987 (Clarkson and Rorabaugh, 1989), with only 5 metapopulations remaining by the late 1990s (Sredl, 1998).

In contrast, Hossack et al. (2005) detected no change in status within Theodore Roosevelt National Park in western North Dakota from 1954 to present; Smith and Keinath (2007) considered them common in the Black Hills of South Dakota; Orr et al. (1998) considered them common in northeastern Ohio; they are considered common in northeastern Illinois (Mierzwa, 1998), northwestern Indiana (Brodman and Kilmurry, 1998), and Wisconsin (Casper, 1998; Mossman et al., 1998); and Gibbs et al. (2005) considered them stable or increasing over a 30 yr period in upstate New York, although they had experienced declines in the 1970s and 1980s (Gibbs et al., 2007). In the early part of the 20th Century, they were very common in northeastern Nevada (Ruthven and Gaige, 1915), Nebraska (Hudson, 1942), in the Snake River Valley of Idaho (Linder and Fichter, 1977), and in the Zuni Mountains of northwestern New Mexico (Gehlbach, 1965).

The reason for widespread declines and extirpation are unclear, but may be due to a combination

Breeding habitat of *Lithobates pipiens*, Livingston County, New York. Photo: C.K. Dodd Jr.

Breeding habitat of *Lithobates pipiens*, Housatonic River, Berkshire County, Massachusetts. Photo: Mike Jones

of climate change (particularly drought), stressors (pesticides, extremely cold winters), changes in land-use patterns (Johnson et al., 2011), disease (chytridiomycosis), and the introduction of fish or American Bullfrogs into formerly occupied breeding sites. Although acidification has been suggested as a possible factor influencing population decline in the Rocky Mountains, there is no empirical support for the hypothesis (Corn and Vertucci, 1992; Vertucci and Corn, 1996). Information summarized by Hine et al. (1981) in Wisconsin noted massive mortality events dating back to at least 1958, with widespread mortality by the early to mid-1970s. These events were sometimes associated with winterkill, but the reported symptoms (ventral skin discoloration, dry skin, hard muscles) suggest a chytrid infection. Northern Leopard Frog populations persisted in Wisconsin by the late 1970s, but fewer populations of fewer individuals were noted when compared with previous reports.

In addition to diseases, Northern Leopard Frog populations have been adversely affected by the presence and pervasiveness of severe malformations. Whereas no population is known to have been driven to extinction by malformations, the presence of large percentages of malformed frogs in certain populations is cause for concern. For example, Ouellet et al. (1997) reported 40 of 61 *L. pipiens* at a wetland associated with agriculture (and pesticide application) had severe deformities, including ectromelia and ectrodactyly. It appears that the widely used herbicide atrazine causes immunosuppression within Northern Leopard Frogs, which in turn makes them vulnerable to attacks by malformation-inducing trematodes. This is further enhanced by atrazine's tendency to enhance snail populations by increasing periphytic algae, a food source for the gastropod snails that act as an intermediate host in the life-cycle of the trematode (Rohr et al., 2008).

Like all amphibians, Northern Leopard Frogs are impacted by the destruction or alteration of breeding ponds, overwintering sites, and terrestrial foraging habitats. They will use areas affected by forestry during the spring and summer as long as moisture conditions are favorable, and the clearcuts are near breeding ponds (Blomquist and Hunter, 2009). Fragmentation undoubtedly contributes to loss of individuals and populations, as Northern Leopard Frogs tend to disperse widely among different habitat types. The effects of fragmentation, such as through agriculturally dominated landscapes, may be ameliorated by providing grassy corridors or wooded riparian strips allowing for dispersal (Maisonneuve and Rioux, 2001).

Large numbers are killed on roads and highways (Breckenridge, 1944; Ashley and Robinson, 1996; Hoffman, 2003; Glista et al., 2008), and there are literature reports where roads became slippery with the bodies of killed and injured Northern Leopard Frogs (Breckenridge, 1944). As an example, Ashley and Robinson (1996) recorded 27,846 dead *L. pipiens* based on 4 yrs of road sampling along a 3.6 km section of highway in Ontario; most mortality occurred in August.

According to Merrell (1977), roads may act as barriers to dispersal, as frogs tend to clump together in habitats adjacent to roadways. Frogs move more slowly in the vicinity of roads and tend to take less direct routes when attempting to cross them. Such behavior increases the probability of mortality. For example, mortality was 6% of those frogs trying to cross a road with low traffic volume, but 28% on a road with a much higher traffic volume (Bouchard et al., 2009). Abundance also tends to increase linearly the farther one moves away from a heavily traveled highway, so roads may have an adverse effect (the "road-zone" effect) at considerable distance from the pavement (Eigenbrod et al., 2009). Mitigation measures to reduce or eliminate road-killed Northern Leopard Frogs include physically carrying frogs across roadways (Linck, 2000), the construction of silt fencing to block access (Hoffman, 2003), or providing tunnels or culverts through which they can pass. Woltz et al. (2008) found that fences 0.6–0.9 m in height effectively prevented trespass, and that Northern Leopard Frogs preferred passages to be > 0.5 m in diameter, not too long, and with some light permeability.

Leopard frogs also have been widely used as biological teaching specimens, bait, and for food (Chamberlain, 1897; Anonymous, 1907). Harvest for these uses has been so heavy at times that biologists have voiced concern for the long-term survival of the species over extensive regions (Gibbs et al., 1971). Lannoo et al. (1994) suspected at least a three-fold decline in Northern Leopard Frog populations from the beginning to the end of the 20th Century in Dickinson County, Iowa, because of over-harvest. Early in the century, as many as 20 million frogs were collected annually as food from this one county alone, an extraordinary harvest. Such figures are probably not that uncommon. In 1981, 85,000 Northern Leopard Frogs were collected from Lac St. Pierre and Missisquoi, Québec, for use in school laboratories, as food, and as bait (Bider and Matte, 1996). In 1989, >50,000 frogs were collected in Minnesota for biological supply houses (Oldfield and Moriarty, 1994). Changes in the population size-class structure may signal long-term effects of harvest, such as the general decline of large frogs observed (Hoppe and McKinnell, 1997).

Northern Leopard Frogs usually breed in semi-permanent ponds and disappear if fish are stocked into a formerly fishless site (Bovbjerg, 1965). This may account for much of its disappearance in the Midwest, as state fisheries departments stocked predaceous fishes into prairie potholes and farm ponds for sport fishing. In addition, a number of authors have suggested that the introduction of American Bullfrogs into wetlands inhabited by *L. pipiens* has led to the decline or extirpation

Breeding habitat of *Lithobates pipiens*, Moses Lake, Washington. Photo: Marc Hayes

of the species (Hammerson, 1982; Panik and Barrett, 1994). Livo (1984) noted declines in a leopard frog population after the introduction of bullfrogs. The bullfrogs ate the tadpoles and metamorphic Northern Leopard Frogs and, as a result, there were fewer egg masses deposited, although adults persisted. Although the bullfrogs did not eat the adults to extinction, they effectively eliminated recruitment.

Embryonic and larval *Lithobates pipiens* have poor hatching success and survivorship in apple orchards affected by a variety of pesticides, although identifying the specific cause or causes of effects was not possible in the study by Harris et al. (1998b). These authors concluded that the influx of chemicals from orchards into breeding sites, in addition to factors such as temperature, could act as environmental stressors affecting premetamorphic development.

Northern Leopard Frogs readily colonize restored wetlands (Lehtinen and Galatowitsch, 2001; Brodman et al., 2006). For example, they were found in greater numbers in small wetlands dredged from 30 to 95% of their area than in undredged reference wetlands on Prince Edward Island (Stevens et al., 2002). Captive rearing and reintroduction have been attempted in British Columbia (Adama et al., 2004), but results have not been published. In any case, source populations of similar genetic history may not be available for repatriation (Wilson et al., 2008a). One of the best methods to ensure survival is habitat protection, such as the establishment of reserves within agriculturally dominated areas of the Richelieu Plain in Quèbec (Galois and Ouellet, 2005).

Northern Leopard Frogs are frequently surveyed via road transects, as researchers or volunteers make periodic stops to listen for frog calls (Bishop et al., 1997; Bonin et al., 1997a; Lepage et al., 1997; Bowers et al., 1998; Mossman et al., 1998). Northern Leopard Frogs were detected during 60% of such nocturnal call surveys in Ontario, making them the least detected species. This pattern is repeated in other road-based call surveys. The relatively low detection probability may be due to methodology, rather than mirroring a general population decline or serve as an indication of rarity. Call surveys for this species likely underestimate occupancy, as frogs do not call continuously, and their calls often are not heard across distances (de Solla et al., 2005).

Lithobates septentrionalis (Baird, 1854)
Mink Frog
Grenouille du nord

ETYMOLOGY

septentrionalis: Latin for 'northern,' referring to the northern distribution of this species.

NOMENCLATURE

Conant and Collins (1998): *Rana septentrionalis*
   Dubois (2006): *Lithobates (Aquarana) septentrionalis*
   Synonyms: *Rana sinuata*
   This species has frequently been confused with *Lithobates catesbeianus* and *L. clamitans*, and the literature is replete with identification errors (Hedeen, 1986).

IDENTIFICATION

Adults. Adults are brownish to olive dorsally with slightly rough skin. The blotched or boldly mottled dorsal coloration becomes more purplish toward the rear and lateral sides of the frog. Kramek and Stewart (1980) identified three main patterns: light spotted, dark spotted, and dark reticulated. Large irregular light or dark brownish-black round spots are present that are broken up by coalescing greenish-yellow lines of varying width in no particular pattern. Very dark reticulated patterns are only seen in large females (55–72 mm SUL) (Kramek and Stewart, 1980), but in general frogs become darker with age and size. This occurs as a result of an increase in spot size, but not number. Dorsolateral folds may be present, but they are often faint or absent altogether. The tympanum is large, especially in males. The iris is hazelnut brown. Bellies are grayish white with dusky spots and reticulations on the tibiae, but pale yellow may be present on the lower sides and chins of some animals. The legs are short and without white tuberculations on the back and hind leg skin. The rear toes are fully webbed. Most juveniles have a uniform pattern of dark spots and are similar to light-spotted adults. This species may be very difficult to distinguish from *L. clamitans*, especially among the smaller size classes. In sympatric populations, *L. clamitans* is usually the larger frog.

Males are smaller than females, and frogs from northern populations are larger than those from southern populations (Leclair and Laurin, 1996). The maximum size rarely exceeds 70 mm SUL, but Shirose and Brooks (1997) give the maximum size as 76 mm SUL. Schueler (1975) provides mean SULs for adult males and females and large juvenile females from throughout the range (see Systematics and Geographic Variation). In Maine, males were 30.2–57.2 mm SUL over a 2 yr period (means 51.3 and 51.7 mm) (Bevier et al., 2006). In Minnesota, males were 45–72 mm SUL and females were 54–72 mm SUL (Hedeen, 1972a) in one study, and a mean of 52–56 mm SUL and 54–63 mm SUL, respectively, in another study (Tenneson, 1983). Ontario females averaged 58 to 60 mm SUL over a 6 yr period, with the largest individuals at 66–76 mm SUL (Shirose and Brooks, 1997). In Nova Scotia, males were 49–68 mm SUL (mean 56.3 mm) and females 53.5–65 mm SUL (mean 60.1 mm) (Gilhen, 1984).

Larvae. Most accounts describe the tadpole's dorsum as dark brown to green with small dark markings or spots, although Vogt (1981) reported tadpoles being emerald green or yellowish green with black spots. Wright (1932) indicated that the bright colors tend to become more subdued as the tadpole approaches metamorphosis. Bellies are opaque straw yellow, and the sides are mottled. Tails have pinkish to buff spotting or blotches and the tail musculature is not bicolored; tails terminate to the posterior of the spiracle and are acute. The iris is black and pinkish cinnamon. Larvae can reach 100 mm TL and weigh nearly 16 g (Hedeen, 1971). Tadpoles are described by Altig (1970) and Vogt (1981).

Eggs. The eggs are brown to black dorsally and buff to creamy yellow ventrally. Eggs have two envelopes surrounding the vitellus (Livezey and Wright, 1947). The outer envelope is 5.6–7.0 mm in diameter (mean 6.3 mm) but swells to 8–9 mm; the inner envelope is 2.4–3.0 mm (mean 2.7 mm). The vitellus is 1.3–1.7 mm (mean 1.4 mm) (Wright, 1932; Livezey and Wright, 1947; Moore, 1952). Bishop (*in* Wright, 1932) mentions the possibility of a third envelope. The eggs are deposited in a jelly mass measuring 5 cm × 5 cm × 2.5 cm to 15 cm × 7.5 cm × 5 cm.

DISTRIBUTION

Mink Frogs occur from southern Labrador westward around James Bay, southern Ontario, and into eastern Manitoba (Whiteshell and Mopining Provincial Parks). There are several breeding populations on Newfoundland, but these represent introductions (Warkentin et al., 2003; Campbell et al., 2004). They are found around Lake Superior in northeastern Minnesota, northern Wisconsin, and the Upper Peninsula of Michigan. Populations are found from Nova Scotia and New Brunswick west across the northern New England states to upstate New York (Adirondack and Tug Hill plateaus). The species is absent from extreme southern Ontario. A record from Okak in eastern Labrador (Packard, 1866) is in error (Harper, 1956; Maunder, 1983). Mink Frogs may be found on islands in the Apostle Archipelago of Lake Superior (Hecnar et al., 2002; not observed by Bowen and Beever, 2010), Isle Royale (Ruthven, 1908), Anticosti Island in James Bay (Bleakney,

Distribution of *Lithobates septentrionalis*

1954), and on northern Cape Breton Island (Gilhen, 1984).

Important distributional references include: Range-wide (Hedeen, 1986), eastern Canada (Bleakney, 1958a; Logier and Toner, 1961), Labrador (Maunder, 1983), Maine (Hunter et al., 1999), Manitoba (Cook, 1963; Preston, 1982), Michigan (Ruthven, 1912; Harding and Holman, 1992), Minnesota (Oldfield and Moriarty, 1994), New Brunswick (Cox, 1899; Rowe, 1899), New Hampshire (Oliver and Bailey, 1939; Taylor, 1993), New York (Gibbs et al., 2007), Ontario (Schueler, 1973; MacCulloch, 2002), Québec (Bleakney, 1954; Harper, 1956; MacCulloch and Bider, 1975; Bider and Matte, 1996; Desroches and Rodrigue, 2004), Vermont (Andrews, 2001), and Wisconsin (Suzuki, 1951; Vogt, 1981). Moore (1952) and Hedeen (1986) provided a comprehensive summary of the early distributional literature on this species and noted many misidentifications.

FOSSIL RECORD

There are no fossils reported for this species.

SYSTEMATICS AND GEOGRAPHIC VARIATION

The Mink Frog is a member of the *Lithobates catesbeianus* species group (*Aquarana*) of North American (*Novirana*) ranid frogs (Hillis and Wilcox, 2005). The species appears to have diverged from the ancestral lineage about 11 mya (Austin et al., 2003a). The species undoubtedly assumed its present distribution through a combination of vicariance and dispersal events, but the exact way this occurred is not known. In one scenario, the ancestral lineage may have been present on the Atlantic Coastal Plain, with subsequent isolation and dispersal by *L. septentrionalis* northward after the last Ice Ages. Alternatively, an ancestral species may have been very widespread, with subsequent isolation and dispersal in the Appalachian (*L. catesbeianus–L. clamitans*) and the northern Laurentian (*L. septentrionalis*) regions (Austin et al., 2003a). In this regard, a previous suggestion that the *catesbeianus* species group is paraphyletic (Pytel, 1986) is not currently supported.

Morphological and phenotypic variation is present among Mink Frog populations. For example, Cox (1899) noted that specimens from New Brunswick were larger than more southern forms, and Mink Frogs in northern portions of Québec and Ontario and eastern Manitoba are much larger (female means 60–67 mm SUL; male means 52–64 mm SUL) than those from the southern portions of Ontario and Québec (female means 49–55 mm SUL; male means 44–55 mm SUL) (Schueler, 1975). In northern Québec males average 61.1 mm SUL and females 67.9 mm SUL, whereas in southern Québec males average 52.6 mm SUL and females 58.6 mm SUL (Leclair and Laurin, 1996); in general, the northern frogs were 17% larger than those in the South. New Brunswick *L. septentrionalis* have much blotching on the throat that is not observed in other populations, and Gilhen (1984) noted specimens in Nova Scotia that are strikingly green, bronze, and black.

ADULT HABITAT

Mink Frogs prefer larger wetlands (ponds and lakes > 1.5 ha) with medium to high densities of emergent vegetation (such as *Carex* sp., *Muhlenbergia* sp., *Typha* sp.), a large amount of floating vegetation (such as *Nuphar* sp.), and substantial amounts of mud and silt as substrates (Marshall and Buell, 1955; Courtois et al., 1995; Popescu and Gibbs, 2009). Presence is often associated with beaver ponds. Mink Frogs do not generally occur in areas where the mean temperature in July is >19.5°C (Popescu and Gibbs, 2009), and they are not associated with rocky substrates. This species prefers littoral habitats, although frogs may be found along the shoreline in tamarack (*Larix* sp.) vegetation (Marshall and Buell, 1955).

Some authors have classified the Mink Frog as a boreal species or northern hardwood-white pine associate, but the type of surrounding forest is not important. Instead, *Lithobates septentrionalis* is an inhabitant of what has been termed the humid cold climatic region (Hedeen, 1986). Mink Frogs may be restricted to cold-water lakes because the globular egg mass may not allow sufficient oxygen diffusion for development at warmer temperatures. Although Mink Frog aquatic habitats are surrounded by densely wooded terrestrial habitats (Knutson et al., 2000), this is truly a large wetland species that lives in the forest.

In addition to large wetlands, *L. septentrionalis* occurs in bogs, in quiet streams running through bogs, and even in ditches, pools, and puddles (Hedeen, 1971). For example, Mink Frogs have been reported from a small creek (0.5 m deep ×

1.5 m wide) running through a forest of willows and alders (Schueler, 1973) and from along riverbanks (Cox, 1899). Mink Frogs inhabit a variety of riparian habitat types, including herbaceous, shrubby, and wooded areas (Maisonneuve and Rioux, 2001). Dickerson's (1906) assertion that the Mink Frog is a river frog that is not found in lakes is incorrect and probably obtained from Garnier's (1883) observations.

Mink Frogs occur to 680 m in New Hampshire (Oliver and Bailey, 1939), 872 m in New York (Moore, 1952), and 896 m in Vermont (Hoopes, 1938). Even in northern Maine, the species is not found at elevations < 300 m along the coast.

AQUATIC AND TERRESTRIAL ECOLOGY

Lithobates septentrionalis is a highly aquatic, cold-adapted, species. The species rarely leaves the vicinity of large water bodies and is not found in small potholes or springs. Adults are generally found throughout the deeper water areas, but juveniles are found only in the shallow water and even in temporary pools (Hedeen, 1972a; *in* Kramek and Stewart, 1980; Gibbs et al., 2007). However, Tenneson (1983) noted that both sexes were found in the shallower areas of a Minnesota lake by day, but at night males tended to move into deeper waters while calling. Mink Frogs desiccate easily and cannot tolerate even semimoist conditions. They will die if they lose only 35% of their body water, which is less than 8 other frog species tested by Schmid (1965).

Within a geographical region, Mink Frogs are found at ca. 20–65% of apparently suitable wetlands. For example, they were found at 18 of 28 sites surveyed in Québec (Courtois et al, 1995) and 10 of 46 sites surveyed in upstate New York (Popescu and Gibbs, 2009). At the southern edge of its range in Ontario, however, Mink Frogs were found in only 2 of 180 ponds sampled (Hecnar and M'Closkey, 1996b; Hecnar and M'Closkey, 1997a).

The short activity season of Mink Frogs only about five months at most. They emerge from dormancy in mid-spring, with the earliest record from Nova Scotia on 7 May (Gilhen, 1984). Mink Frogs enter dormancy in late summer to fall. For example, the latest record of activity in Nova Scotia is 30 September (Gilhen, 1984). They overwinter in the muddy bottom of permanent aquatic sites. Although a cold-loving species, they do not possess any physiological adaptations, such as the production of glycerol, which allow supercooling (Schmid, 1982).

Mink Frogs are photopositive in their phototactic response, suggesting they use both sunlight and moonlight when feeding and going about their daily activities (Jaeger and Hailman, 1971, 1973). They are sensitive to light in the blue spectrum ("blue-mode response"), which apparently helps them orient toward areas of increasing illumination, such as the open area above a pond (Hailman and Jaeger, 1974). Mink Frogs likely have true color vision.

CALLING ACTIVITY AND MATE SELECTION

Calling may begin several weeks prior to the commencement of oviposition. Males call from throughout a pond or lake, with most calls occurring from midnight until 09:00 (Bevier et al., 2004). In Nova Scotia, calling occurs both day and night (Gilhen, 1984), and it is likely that diurnal calling occurs more often than farther north in the range. Chorus tenure is not prolonged (a mean of eight nights; Tenneson, 1983); many males may be recorded calling only once during the breeding season. The operational sex ratio during the breeding season is male-biased, although males and females do not, on average, differ in the amount of time they spend at a breeding pond. Only a few reproductively ready females usually are at the pond on any given night (Tenneson, 1983).

Male *L. septentrionalis* do not remain at any particular calling site either nightly or throughout the breeding season (Hedeen, 1972a; Bevier et al., 2006). Tenneson (1983) found that they used on average 9.75 calling sites within a particular area. Males move from 0.6–90 m from previous calling sites, with consecutive night movements a mean of < 15 m (Bevier et al., 2006). Calling males are spaced significantly farther apart (mean 9.3 m) than noncalling males. The mean male activity area was 410 m^2 (range 23 to 1726 m^2) in Bervien et al.'s (2006) study in Maine and 1610 m^2 in Tenneson's (1983) study in Minnesota.

Some authors have asserted that territories are not defended, although males will aggressively fight with one another when they come in close proximity. In contrast, Tenneson (1983) suggested that territories do exist but that only some males (4 of 14 in his study) were territorial; the mean male calling territory was estimated as 272 m². A male will orient toward another male and utter a "cuk" or a "cuk and rumble" call while swimming toward him (Bevier et al., 2004). Fights result in a silent, brief, wrestling match or chase beginning when a male moves within about 1 m from another male. Head-butting, clasping, and "jump attacks" also may be observed (Tenneson, 1983). Intermale distances are inversely proportional to the number of males calling. Females are neither aggressive nor territorial.

Males call while floating on the water among emergent vegetation from near the shoreline to as much as 30 m from shore (Hedeen, 1972a). Taylor (1993) likens the call to "a rapid series of tiny hammer blows." They have two energy peaks, at dominant frequencies of 470–970 Hz and 1,220–1,960 Hz (Tenneson, 1983). Call rates do not vary with temperature (mean air 16°C and water 19°C) or the day of the calling period (Bervier et al., 2004); males appear to put equal amounts of effort into calling whether early or late in the season. Males produce two types of notes, a "cuk" (a single short pulse) and a "rumble" (a variable number of repeated pulses that increase in intensity). The type of call and the number of times it is given changes during the nightly calling period, but the "cuk and rumble" series is given most often during peak calling. It consists of 1–9 "cuks" followed by 1–9 "rumbles." Bervier et al. (2004) provided sonograms of these calls as well as summaries of the note duration, call duration, and dominant frequencies of the calls and call series. The dominant frequencies and note duration are higher than in other sympatric ranids (*L. catesbeianus, L. clamitans, L. virgatipes*).

Preoviposition and oviposition behavior have been described based on laboratory (Aronson, 1943b) and field (Tenneson, 1983) observations. A female moves toward a calling male and, when in close proximity, is amplexed by him. An amplexed pair then swims through the water and alternates swimming with a period of backward shuffling movements. This occurs for up to 3 hrs prior to oviposition. During this time, repeated dives may occur wherein the pair remains under water for up to 5 min.

Oviposition occurs under the surface of the water. When she is ready to oviposit, the female forms her rear legs one or more times into a diamond-shaped flexure into which the eggs will be deposited; the knees rest on the substrate, with the rear toes pointed upward and toward the head. As the female forms her legs into position, the male flexes his hind limbs and thighs, extending almost to a right angle from the anterior part of his body. The female contracts her abdominal muscles and arches her back concavely; some eggs are extruded by this contraction. The male then arches his back convexly causing his cloaca to move forward (termed the upstroke). This causes the male's legs to move forward, with his feet turned outward, as he fertilizes the eggs under water. The female then straightens out her arched back (termed the downstroke). The male follows suit, with his legs posteriorly oriented. Series of upstrokes and downstrokes occur for about 10 min until all eggs are extruded into a mass. As the female moves away from the oviposition site, the male releases his grip and the frogs swim to the surface. In general, this sequence is similar to that observed in *L. pipiens*.

Females do not necessarily stay in the immediate vicinity of her amplexed partner's calling site, but may move up to 7 m to select before choosing an oviposition site (Bervier et al., 2006). Thus, females do not appear to select males based on the resource quality of the calling site. Females also disperse rapidly after oviposition, often to the shallow portions of the pond or lake, and do not remain in the vicinity of the breeding site. If males are inadvertently clasped by other males, they have a "growl" of a release call (Tenneson, 1983).

BREEDING SITES
Breeding occurs in ponds and lakes with slight acidity (pH 5.2–6.4), low conductivity (16–36 µmhos/cm), low alkalinity (0–44 mg $CaCO_3$/L), and low levels of sodium, potassium, calcium, magnesium, chlorine, and sulfate (Dale et al., 1985). However, these values are based on only two sites. Abundant emergent vegetation should be present for the frogs to sit upon and to attach their egg masses. Ponds may contain fish, as might be expected of a species that occupies large bodies of water (Hecnar and M'Closkey, 1997b). Nothing has been reported concerning the characteristics of bog breeding sites, or whether breeding occurs in rivers (unlikely), although tadpoles have been observed in streams. It is more likely the tadpoles entered the stream through a pond inlet or outlet.

Egg mass of *Lithobates septentrionalis*. Photo: David Patrick

REPRODUCTION

Breeding occurs during the warm months of the year and, as such, *L. septentrionalis* is a prolonged breeder. Specific records include May to early August in New York (Gibbs et al., 2007), mid-June in Québec near the Labrador border (Bleakney, 1958a), late June to mid-August in New Hampshire and Minnesota (Oliver and Bailey, 1939; Hedeen, 1972a; Tenneson, 1983), June to August in Canada (Bleakney, 1958a), July in Nova Scotia (Gilhen, 1984). Wright (1932) suggested peak breeding occurred during the second week in July. Note that calling may begin as early as late May in Minnesota, so the calling period is not at the same time as when eggs are deposited. Calling also may extend beyond the actual dates of oviposition (Tenneson, 1983).

Eggs are deposited in a globular, solid jelly mass attached to vegetation (such as water lily stalks or pickerelweed) from just below the water's surface to as much as 1.5 m below the surface. Oviposition sites may be as deep as 2.4 m. This keeps the eggs safe from the occasional freezing temperatures, which may occur into June at northern locations. Placing eggs in thick vegetation also helps secure them from predatory fishes. Clutch sizes vary. Hedeen (1972a) reported a count of 509 eggs from a single mass, and Bider and Matte (1996) give the clutch size as 500–1,500. Clutch size in Nova Scotia ranged from 986 to 1,715 eggs (Gilhen, 1984). Clutch size is positively correlated with female body size (Gilhen, 1984). Hatching occurs in 4–5 days in the laboratory (Moore, 1952) to 11–13 days in the field (Wright, 1932) at 8–10 mm TL.

LARVAL ECOLOGY

Larvae are herbivorous, primarily consuming algae. Like many species, the rasping mouthparts are used to scrape algae from plant stems and submerged debris. Feeding occurs for at least one year. Most larvae overwinter and metamorphose the following summer, although some may spend two winters prior to metamorphosis (Hedeen, 1971). During the winter, growth essentially ceases, and mortality from freezing may be high unless the ponds and lakes are deep enough to prevent freezing solid. Larvae are found in well-oxygenated waters, and Garnier (1883) observed them in fast-flowing riffles, where they were extremely wary.

Under laboratory conditions, embryos will not develop at temperatures >30.4°C (Moore, 1952). No development occurs at 7.8°C, and gill circulation fails to develop at 12.9°C. Moore (1952) suggested that the range of temperature tolerance was 14–30°C, and that *L. septentrionalis* did not possess any specific cold weather adaptations regarding larval development. In nature, however, most waters where this species deposits its eggs remain <21°C. Embryonic growth rates are positively correlated with temperature. Tadpoles from northern populations tend to metamorphose at larger sizes than those from southern populations. Recent metamorphs average 37 mm SUL (range 33 to 42 mm) in Minnesota (Hedeen, 1971) and 35.9 mm SUL (range 28 to 42 mm) in Nova Scotia (Gilhen, 1984).

DIET AND FEEDING BEHAVIOR

Feeding is diurnal. Mink Frogs are sit-and-wait predators that sit in a heads-up posture on lily pads or other floating vegetation, or float at the water's surface with only their eyes and snout above the surface (Hedeen, 1972c). Feeding activity varies, from frogs that essentially let the prey come to them and do not stalk it to those that actively pursue prey. If sitting and waiting does not work, they may change locations or switch to a more active predatory role (Hedeen, 1972a; Kramek, 1976). When in search mode, frogs are able to identify prey as far away as 3 m and readily enter the water, surfacing as they near the prey. As they come within range, a prey item

Tadpole of *Lithobates septentrionalis*. Photo: Mark Roth

is consumed in a rapid lunge. Frogs appear to have favorite feeding stations, and both males and females feed during the breeding season. Capture success may be high (>80%) for slow-moving prey, but low (<20%) for aerial prey (Kramek, 1976). When not looking for prey, the frogs assume a crouched posture on floating vegetation.

Postmetamorphs eat a wide variety of available prey, including spiders, beetles (chrysomelids, predaceous diving beetles, whirligigs, rove, click, blister, ground), dragonflies, damselflies, mayflies, water striders, riffle bugs, katydids, aphids, various other types of flies, ants, slugs, snails, collembolans, lepidoptera (larvae and adults), and ichneumons (Hedeen, 1972c; Kramek, 1972; Stewart and Sandison, 1972: Gilhen, 1984). Feeding is opportunistic, and there is no difference between the sexes in prey choice. Vegetation is frequently found within the gut, which is ingested incidentally while lunging for prey. Dietary differences among sympatric northern ranids (*L. septentrionalis, L. clamitans, L. catesbeianus*) reflect differences in habitat preferences rather than specialization.

PREDATION AND DEFENSE
This species has a strong, odorous skin secretion that presumably is noxious to potential predators. The secretion smells like rotting onions or a mink, hence the species' English common name. Vogt (1981), however, considered it "rather pleasing." Mink Frogs are extremely wary and difficult to approach, and their spotted or reticulated patterns likely aid in crypsis. When disturbed, they skitter across the water's surface and floating vegetation and dive into the substrate to bury in the mud and debris (Garnier, 1883; Hedeen, 1972b); frogs remain hidden from 30 sec to nearly 30 min. As they start their escape, they utter a loud warning cry. When they surface, only the eyes are above water. Like many ranid frogs, this species also utters a distress call when handled or seized (Wright, 1932).

Eggs and hatchlings are eaten by the newt *Notophthalmus viridescens*. Larvae and postmetamorphs are consumed by mammals (raccoons, skunks), birds (Great Blue Herons, Wood Ducks, Mergansers, Herring Gulls), snakes (*Thamnophis sirtalis*), and other frogs (*Lithobates clamitans*) (Wright, 1932; Hedeen, 1967, 1972b; Gibbs et al., 2007). Fish may damage the tails of Mink Frog larvae, even if they do not consume the body. Aquatic insects (giant water bug, *Lethocerus americanus*) and leeches (*Macrobdella decora*) also feed on tadpoles.

POPULATION BIOLOGY
Male *Lithobates septentrionalis* reach sexual maturity in Minnesota at 45–50 mm SUL, a year after metamorphosis. Females, however, reach maturity at 54–59 mm SUL from 1 to 2 years after metamorphosis (Hedeen, 1972a). In northern Québec, maturity in both sexes occurs 2 yrs following metamorphosis, but after only 1 yr in southern Québec (Leclair and Laurin, 1996). Several authors (Hedeen, 1972a; *in* Kramek and Stewart, 1980; Tenneson, 1983; Leclair and Laurin, 1996) have noted that there may be more females than males observed within a wetland. For example, Tenneson (1983) reported there were 1.4 adult females per male and noted some evidence of habitat partitioning between the sexes;

whether the differences were real or an artifact of sampling is unknown. In Québec, the ratio was 1.8 females per male.

Growth is rapid and rates are similar regardless of whether populations are in the northern or southern parts of the range; females grow faster than males. The mean age and longevity are higher in females than males, but only in southern populations. In northern populations, male and female ages are essentially the same, but females are 11% larger than males. Males have higher mortality rates than females in southern populations compared with northern populations. This results in population differences in age, longevity, and sex ratios (Leclair and Laurin, 1996). Taken together, differences in growth rates, delayed sexual maturity, greater mean ages, and size at transformation account for the larger sizes of northern frogs. Maximum longevity is 4 yrs.

Many *L. septentrionalis* may be found in close proximity to one another (Hoopes, 1938). For example, Warkentin et al. (2003) observed 30 Mink Frogs within a 10 m^2 area at one end of a pond. It is not clear whether this represented a breeding chorus, as some authors report relatively small numbers of chorusing males spread out over an entire wetland (Tenneson, 1983; Bervier et al., 2006). In New Brunswick, McAlpine (1997c) used a transect method to estimate there were 0.7–88.8 Mink Frogs/100 m^2 at 5 lakes in June; abundance tended to decrease slightly through the season on 9 lakes. In Minnesota, Tenneson (1983) estimated there were 16, 27, 42, and 178 Mink Frogs at 4 different locations. Clearly some wetlands support only a few adults, whereas others support a rather abundant population.

Widely distributed throughout its range, *L. septentrionalis* may experience fluctuations in abundance from one location or year to the next. For example, estimates ranged from <10 adults at Lake Sasajewun, Ontario, in 1985 and 1992 to >35 individuals in 1987 and 1993 (Shirose and Brooks, 1997). These authors also estimated there were 2.8 transforming frogs for every adult observed.

COMMUNITY ECOLOGY

Mink Frogs are frequently found in lakes occupied by *L. clamitans* and *L. catesbeianus*, and these species breed at the same time of year. When *L. catesbeianus* is absent from lakes, *L. septentrionalis* is more abundant and evenly distributed throughout the water (Courtois et al., 1995); however, the presence of American Bullfrogs is only moderately negatively correlated with the presence of Mink Frogs (Popescu and Gibbs, 2009). These results suggest that American Bullfrogs influence the distribution and abundance of Mink Frogs to some extent, perhaps through competition. Schueler (1975) also speculated that differential size-structuring of Mink Frog populations might be related to competition with *L. catesbeianus*.

Habitat partitioning appears to occur with *L. clamitans*, with this species most often found along the shoreline and adult *L. septentrionalis* found in deeper water littoral habitats. However, habitat use by all three species overlaps considerably, and habitat structure per se is not a good predictor of species composition (Courtois et al., 1995). In addition, the diet of these three species broadly overlaps, with the differences reflecting differences in habitat preferences rather than diet specialization (Stewart and Sandison, 1972).

DISEASES, PARASITES, AND MALFORMATIONS

An intraerythrocytic virus (FEV, frog erythrocytic virus) was reported from 1 of 75 *L. septentrionalis* from Ontario (Desser and Barta, 1984a; Gruia-Gray et al., 1989). A prokaryotic parasite (*Thrombocytozoons ranarum*) has been reported in the blood of Ontario Mink Frogs (Desser and Barta, 1984b). Substantial mortality resulted from an iridovirus outbreak that occurred in Minnesota (Green et al., 2002); a fungus similar to *Dermocystidium* also was reported in connection with the iridovirus mortality event. The virulent pathogen amphibian chytrid fungus (*Batrachochytrium dendrobatidis*) has been found in Mink Frogs from the upper Midwest (Sadinski et al., 2010) and Maine (in 18 of 102 frogs examined; Longcore et al., 2007). Areas infected included toe-webbing, tibiae skin, and pelvic skin. Mortality possibly resulting from the bacterial pathogen *Aeromonas* was reported in Minnesota (Hedeen, 1972b; Tenneson, 1983). Tenneson (1983) suggested that severe winters resulting in low dissolved oxygen content in overwintering sites, coupled with redleg disease, might have been responsible for mortality.

Mink Frogs are parasitized by the trematodes *Cephalogonimus americanus, Glypthelmins quieta, Gorgoderina attenuata, G. simplex, G. translucida, Haematoloechus longiplexus, H. medioplexus, H. similiplexus, Halipegus* sp., *Loxogenes arcanum,* and *Megalodiscus temperatus*. It is also parasitized by the cestode *Cylindrotaenia americana* (Bouchard, 1951).

Malformations in Mink Frogs have been reported from Minnesota (Gardiner and Hoppe, 1999; Helgen et al., 2000; Green et al., 2002). A frequent malformation is skin-webbing, a cutaneous fusion of the skin between the femoral and tibiofibula regions that inhibits swimming (Hoppe, 2000). A Mink Frog also has been reported missing an eye (Converse et al., 2000). Gardiner and Hoppe (1999) did not identify the source of the malformations but speculated that the causative agent disrupted retinoid-sensitive development pathways.

SUSCEPTIBILITY TO POTENTIAL STRESSORS

pH. In Nova Scotia, this species occurs in ponds and lakes with a pH of 5.2–6.4 (Dale et al., 1985).

UV light. Ambient levels of UVB light can be lethal to embryos and larvae of *L. septentrionalis* when shade and refugia are not available (Tietge et al., 2001). Indeed, larvae are more sensitive than embryos, experiencing 100% mortality after 9 days of direct exposure to ambient UVB light. Even exposure at 50–75% ambient UVB light caused substantial mortality.

STATUS AND CONSERVATION

Like all amphibians, habitat destruction and alteration have had substantial effects on Mink Frog populations. Unfortunately, the cumulative loss of wetlands often goes unrecognized, and the results of habitat loss are rarely quantified. There may have been unexplained declines, even in protected areas. For example, Tenneson (1983) noted sharp population declines of Mink Frogs over 15 yr and 25 yr periods at 2 Minnesota localities. In Canada, captures of Mink Frogs at sites mined for peat are substantially less than at unmined sites (Mazerolle, 2003), but the effect on the persistence of highly aquatic Mink Frog populations is unknown. The presence of roads is negatively associated with wetland occupancy by *Lithobates septentrionalis* (Findlay et al., 2001), but the extent to which roads affect this highly aquatic species is unknown.

The species is not considered of conservation concern throughout its range. For example, Weir et al. (2009) found no trends in occupancy for Maine populations followed over a 7 yr period; Weller and Green (1997) reported no evidence of decline in any Canadian province; Moriarty (2000) suggested populations were stable in Minnesota. In general, it is considered a rather common species distributed over a wide area. Because of this species' sensitivity to warm temperatures, an increase in temperature could cause detrimental effects for this species at the southern part of its range. These effects might be offset, however, by an ability to expand its range northward,

Adult *Lithobates septentrionalis.* Photo: Kenny Wray

Habitat of *Lithobates septentrionalis*, Schoolcraft County, Michigan. Photo: Kenny Wray

especially by occupying beaver ponds (Popescu and Gibbs, 2009).

Other than protecting breeding sites, leaving riparian buffer strips through agricultural landscapes allows some Mink Frogs to persist (Maisonneuve and Rioux, 2001). The species slightly prefers shrubby and wooded habitat types to herbaceous riparian vegetation, but even then abundance may not be high along stream or river corridors. Because of its often late-night calling pattern, Mink Frogs are not effectively sampled using call surveys along highways (Bishop et al., 1997; Lepage et al., 1997).

Lithobates sevosus (Goin and Netting, 1940)
Dusky Gopher Frog

ETYMOLOGY
sevosa: from the Latin *sevosa* meaning 'tallow-like,' in reference to the secretions of the skin.

NOMENCLATURE
Conant and Collins (1998): *Rana capito sevosa*
  Dubois (2006): *Lithobates (Lithobates) sevosus*
  Synonyms: *Rana areolata sevosa, Rana capito sevosa*

Information on *L. capito* is sometimes contained in accounts of frogs identified as *R. c. sevosus*.

IDENTIFICATION
Adults. Lithobates sevosus is a very dusky (black to pale gray to light brown) gopher frog with a warty dorsum and spotted venter. Warts are not present on the head, but the entire dorsum and sides are covered with them. These warts may fuse to form short folds. The dark, irregular dorsal spots may blend in with the background coloration, and they are not rimmed with a light-colored halo. The background color of the venter is dirty white to tan, with most spots anteriorly situated; the chin, throat, and pectoral region are heavily spotted. The head and waist are broad, as are the dorsolateral folds that extend from the nostrils to the sacral hump. Forelegs are short and stocky. Dark bars on the hind limbs are wider than the intervening light spaces between them. The rear digits are fully webbed. Males have a gray nuptial pad on the inner side of the first finger, and the vocal sac is dark. Males are more spotted ventrally than females.

Sexual size dimorphism is present, with males smaller than females. In Mississippi, males were 62–84 mm SUL (mean 73.6 mm) whereas females were 73–92.5 mm SUL (mean 82.3 mm) (Goin and Netting, 1940). At Glen's Pond, the mean male size was 63.2–70.2 mm SUL, whereas female means were 78–82.7 mm SUL over a 3 yr period (Richter and Seigel, 2002).

Larvae. The tadpole of *L. sevosus* is large (to 74 mm TL, Volpe, 1957b) and deep-bodied, dark brown to various shades of green (light, dark, olive) dorsally, and covered with small yellow spots. There is usually a vertical white line down the middle of the snout between the nostrils. The belly is a lustrous yellow, the viscera are not visible through the skin, and the throat is translucent. Upper and lower tail fins may or may not be heavily marked with speckles or spots (but not large spots), and the dorsal tail fin is blunt. In general, the tail fin is lighter than the body. Eyes are bronze. The lower jaws are wide. The larvae are similar to those of leopard frogs (*L. sphenocephalus*) and may be impossible to distinguish. A detailed description is in Volpe (1957b).

Eggs. Eggs are relatively large and dark colored. Two gelatinous envelopes are present; the outer envelope averages 4.0 mm in diameter (range 3.4 to 4.5 mm) and the inner averages 2.4 mm (range 2.1 to 2.8 mm). The vitellus is 1.7–2.2 mm in diameter (mean 1.8 mm) (Volpe, 1957b). Eggs are deposited in an ovoid jelly mass and attached to vegetation below the water's surface. Embryonic development was described by Volpe (1957b).

DISTRIBUTION

The Dusky Gopher Frog historically occurred from the Florida Parishes of Louisiana east to Mobile Bay along the northern Gulf Coastal Plain. Netting and Goin (1942) reported on a Dusky Gopher Frog east of Mobile Bay in Baldwin County, Alabama, that was "clearly" *L. sevosus*. In nature, the species is currently known only from two ponds in Harrison County in southern Mississippi. Other important distributional references include: Neill (1957b), Dundee and Rossman (1989), Young and Crother (2001), and Bailey and Means (2004).

FOSSIL RECORD

No fossils are known.

SYSTEMATICS AND GEOGRAPHIC VARIATION

Dusky Gopher Frogs are in the *Nenirana* clade of North American ranid frogs, a group that includes *Lithobates palustris*, *L. capito,* and *L. areolatus* (Hillis and Wilcox, 2005). They are only distantly related to the *Scurrilirana* clade. The species has

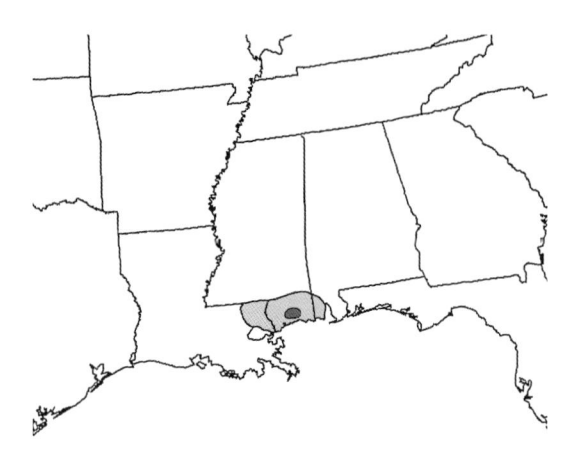

Distribution of *Lithobates sevosus*. Dark gray indicates extant population; light gray indicates extirpated populations.

had a complex nomenclature reflecting differing views based entirely on phenotypic characters that show much variation. At various times it has been considered a full species or subspecies of either *Lithobates capito* or *L. areolatus* (e.g., Neill, 1957b). Young and Crother (2001) recognized *L. sevosus* as a taxon distinct from other gopher frogs, based on fixed differences in several allozyme loci.

ADULT HABITAT

This species historically occurred in the longleaf pine (*Pinus palustris*) savannas of the northern Gulf Coastal Plain. Much of the Gulf Coast longleaf pine forest has been highly degraded through silvicultural and agricultural activities, leading to the loss of many temporary ponds. The only remaining population of *Lithobates sevosus* is found in a longleaf pine-wiregrass (*Aristida beyrichiana*) community that is managed through prescribed burns.

TERRESTRIAL ECOLOGY

Dusky Gopher Frogs are highly fossorial. Historically, they were closely associated with the Gopher Tortoise (*Gopherus polyphemus*), where they would sit at the tortoise's burrow entrance when conditions permitted and wait for prey. Allen (1932) stated that there was only a single frog per Gopher Tortoise burrow. Because of the decline in Gopher Tortoises at the only known population, they currently occupy mammal burrows and any other underground tunnel, such as root cavities and stump holes, which may be available

(Richter et al., 2001). Most frogs remain within 300 m of their breeding pond (mean 173 m), with movements away from breeding ponds taking 1–2 days and occurring during rainfall (Richter et al., 2001). When dispersing, they seek shelter in burrows and under grass clumps and surface litter. During the dry season and cold weather, they move to the deepest portion of the thermally stable burrow to avoid adverse conditions. Foraging may occur for short distances from refugia at night or on cloudy, rainy days. Dusky Gopher Frogs have been captured at drift fences from October to early June, but there is considerable annual variation in capture. Terrestrial activity at burrow entrances likely occurs throughout the year to some extent, weather permitting.

CALLING ACTIVITY AND MATE SELECTION

Males enter breeding sites prior to females and stay longer (females <20 days, males 40 to >80 days depending on conditions) (Richter and Seigel, 2002). The call of *Lithobates sevosus* is a snore that has been compared to a series of hoarse and guttural croaks similar to an outboard motor (Volpe, 1957b). Calling usually occurs in the water, with occasional reports of calling from underwater (*in* Dundee and Rossman, 1989).

BREEDING SITES

Breeding occurs in temporary ponds that are open-canopied with firm substrates dominated by shallow aquatic vegetation (sedges, grasses, rushes). Ponds are usually small and relatively shallow. Glen's Pond, for example, is 1.5 ha with a maximum water depth of 1 m. In Louisiana, breeding ponds were described as clear woodland pools (Volpe, 1957b).

REPRODUCTION

Breeding occurs opportunistically from October to April, with a peak from January to March. Breeding is closely tied to rainfall, which determines when ponds fill and the duration of hydroperiod. Breeding bouts may occur more than once in a season. Eggs are deposited in an ovoid plinth up to 20 cm in length attached to submerged vegetation. The mass is located up to 30 cm below the water's surface. Volpe (1957b) reported a single clutch of 6,000 eggs, but Richter et al. (2003) counted a mean clutch size of only 1,134 eggs.

There are years when no egg masses are deposited due to drought, and most adults breed on an average of only 1.2 yrs. A few, however, breed from 3 to 5 yrs (Richter and Seigel, 2002). The number of egg masses varied at Glen's Pond from 3 to 130 from 1988 to 2001. Egg mortality varies (e.g., 4.7–25.3%; Richter et al., 2003), especially due to predation by caddisfly larvae.

LARVAL ECOLOGY

The larval period is variable, with specific values of 81–179 days at Glen's Pond (Richter et al., 2003). From 1988 to 2001, metamorphosis occurred in only 3 yrs, and the number of metamorphs varied from 221 to 2,488 (Richter et al., 2003). Canopy closure reduces larval growth by as much as 20%, and larvae have lower survivor-

Eggs of *Lithobates sevosus* attacked by caddisfly larvae. Photo: Stephen Richter

Tadpole of *Lithobates sevosus*. Photo: John Tupy

ship and reduced size at metamorphosis in shaded ponds compared with open-canopied ponds. Shaded ponds also produce lower overall larval biomass than open-canopied ponds (Thurgate and Pechmann, 2007). In shaded ponds, the mean larval duration is 176 days, whereas in open-canopied ponds it is 166 days. At metamorphosis, Dusky Gopher Frogs are 19–30 mm SUL (Volpe, 1957b; Thurgate and Pechmann, 2007) or a mean of 29.8–35.7 mm SUL (Richter and Seigel, 2002). Metamorph size differs among years.

DIET
No published information is available. Dusky Gopher frogs likely eat any invertebrate or small vertebrate that comes within range of their mouth.

PREDATION AND DEFENSE
This species is extremely wary and readily dives to the bottom of the pond when breeding or deep into a tortoise burrow at the slightest disturbance. Predators of eggs include caddisfly (Trichoptera) larvae (Richter, 2000). Predators of larvae likely include a variety of predaceous invertebrates, such as predaceous diving beetles, water bugs, dragonfly naiads, and vertebrates such as kinosternid turtles and aquatic snakes. No information is available on predators of postmetamorphs, but they likely include snakes, birds, and mammals, especially raccoons.

POPULATION BIOLOGY
Although breeding may occur every year with winter rains, larvae complete metamorphosis only in years with sufficient rainfall resulting in an extended hydroperiod. For example, metamorphs were produced at Glen's Pond only 3 times between 1996 and 2001 (1997 [221], 1998 [2,248], 2001 [130]), the last time only when water was pumped into the pond to maintain hydroperiod.

The age at maturity is 6–8 months for males (mean 56 mm SUL at maturity) and 2–3 yrs for females (Richter and Seigel, 2002). Annual adult survivorship is from 65–92%, and some frogs return for 4–5 successive yrs to breed. However, the rate at which adult frogs return to a pond to breed is only 16–22% among years. The likelihood of surviving from egg to metamorphosis is very low. For example, Richter et al. (2003) estimated that between 0.35 and 5.36% of eggs survived to metamorphosis in 1997 to 1998 at Glen's Pond. It is clear that no or few metamorphs are produced in most years, with a few good years providing the majority of recruitment. Reproductive failure is quite common, and cohorts likely turn over rapidly. The sex ratio at Glen's Pond is female biased during the breeding season, although it does not differ from 1:1 (Richter and Seigel, 2002). Maximum longevity is < 7 yrs.

DISEASES, PARASITES, AND MALFORMATIONS
Antimicrobial peptides are present in the skin of this species and may provide some protection against bacterial pathogens (Graham et al., 2006). The species has suffered mass larval mortality from a virulent alveolate pathogen (Cook, 2008). No information is available on parasites.

SUSCEPTIBILITY TO POTENTIAL STRESSORS
No information is available.

STATUS AND CONSERVATION
The Dusky Gopher Frog is critically endangered and is known only from two natural breeding ponds in southern Mississippi. Decline has re-

Adult *Lithobates sevosus*. Photo: David Dennis

Breeding habitat of *Lithobates sevosus*, Glen's Pond, Mississippi. Photo: Joe Pechmann

sulted from habitat destruction and alteration, especially from agriculture and silviculture. This species is particularly vulnerable because of its single population's isolation, low recruitment, and frequent years of reproductive failure (Richter and Seigel, 2002). Lack of proper fire regimes contributes to declines, as canopy closure reduces larval survivorship and fitness. As ponds become overgrown, they become unsuitable as breeding sites for *Lithobates sevosus* (Thurgate and Pechmann, 2007). Changes in precipitation patterns and extended drought have caused decreased hydroperiod at breeding ponds causing reproductive failure. Competition does not appear to have contributed to the species' decline.

One breeding pond, Glen's Pond, may not hold water long enough for *L. sevosus* larvae to complete metamorphosis in drought years. As one management tool, well water was pumped into the pond over a seven-week period to extend hydroperiod. This allowed an avoidance of complete larval mortality in 2001, and 130 metamorphs were produced. This was the first recruitment into the population since 1998 (Seigel et al., 2006). Other options have included captive rearing for head-starting, and trials have been conducted using *L. capito* as a surrogate (Braid et al., 1994). Breeding males have been introduced into a second pond, Mike's Pond, to augment the adult population, with subsequent successful reproduction. Captive individuals are held for conservation research at the Audubon Zoo in New Orleans, the Memphis Zoo in Tennessee, and at the Henry Doorly Zoo in Omaha, Nebraska. This species is listed as "Endangered" under the Endangered Species Act of 1973 and is protected by state law.

Lithobates sphenocephalus (Cope, 1886)
Southern Leopard Frog

ETYMOLOGY
sphenocephalus: from the Greek words *sphenos* meaning 'wedge' and *kephalos* meaning 'headed.' The name refers to the frog's pointed head.

NOMENCLATURE
Conant and Collins (1998): *Rana utricularia*
  Dubois (2006): *Lithobates* (*Lithobates*) *sphenocephalus*
  Synonyms: *Rana halecina*, *Rana oxyrhynchus*, *Rana pipiens sphenocephala*, *Rana sphenocephala utricularia*, *Rana utricularia*, *Rana utricularius*, *Rana virescens*

IDENTIFICATION

As noted in the *Lithobates pipiens* account, it is often impossible to determine which species of "leopard frog" is being referred to in many early papers, especially those in physiology and toxicology. Commercial suppliers in the Southeast, for example Carolina Biological Supply Co. (CBSC), are presumed to have supplied Southern Leopard Frogs under the name "*Rana pipiens*." This may not be true in all cases. For example, Bauer Dial and Dial (1995) used *Lithobates berlandieri* obtained from CBSC. In many papers, the origin of the frogs is not specified but, if obtained locally, were probably *L. sphenocephalus* rather than *L. pipiens* as stated. Readers should be cautious in making species attributions in such cases. In the section below, information in papers by Rose (1960), Kaplan and Yoh (1961), Kaplan and Overpeck (1964), Kaplan and Glaczenski (1965), Nash et al. (1970), Dial and Bauer (1984), Dial and Bauer Dial (1987), and Chen et al. (2008) are included, even though they identify the species discussed as "*Rana pipiens*."

Adults. Adult Southern Leopard Frogs are somewhat smaller than their northern counterparts. They are variably colored, generally tan, brown, light or dark to olive green, or bronze with conspicuous white or yellow dorsolateral folds, and they have a noticeable white spot located in the center of the tympanum. Some frogs may be bicolored, having green between the dorsolateral folds and brown along the lateral sides of the body. They have two or three irregular rows of rounded or oval spots between the dorsolateral folds, with fewer rounded spots on the side of the body. These spots generally lack a light border. Several small ridges the same color as the body also may or may not occur between the dorsolateral folds. The legs have dark bars or elongated spots, and there is no orange coloration on the underside of the rear legs and groin, as there is in *L. palustris*. The hind legs are long with webbed toes. As with many other *Lithobates*, there is a light line on the upper jaw. Bellies and the undersides of legs are normally white. Note that during cold weather or after extended stays away from wetlands, Southern Leopard Frogs may appear darker than normal.

Males are slightly smaller than females. In addition, males develop enlarged thumbs during the breeding season, and the male's paired vocal pouch may be visible as skin folds at the corners of the jaw. When calling, these vocal pouches extend on each side of the body, giving the appearance of water wings. Most adults are 50–90 mm SUL, but they can reach 127 mm SUL. Literature records based on actual measurements are few. Maximum SULs are 70 mm for males and 80 mm for females in Virginia (Mitchell, 1986), and Lazell (1989) stated that Southern Leopard Frogs from the lower Florida Keys attain SULs of 120 mm. Duellman and Schwartz (1958) recorded mainland maximum sizes of males 85.2 mm SUL and females 99.6 mm SUL. On Big Pine Key, the largest male was 80.0 mm SUL and the largest female 97.7 mm SUL.

Albinistic adults are reported from Maryland and New York (Hensley, 1959). Bechtel (1995) has a photograph of a hypomelanistic (red) Southern Leopard Frog. Martof (1962c) recorded an axanthic (blue) frog from Sapelo Island, Georgia.

Larvae. Like the tadpole of *L. pipiens*, the tadpole of *L. sphenocephalus* is large, normally to 74 mm TL (Wright, 1929; Lazell, 1989); the largest known example is 83 mm (Siekmann, 1949). *Lithobates sphenocephalus* larvae are deep-bodied, dark brown to various shades of green (light, dark, olive) dorsally, and covered with small gold spots. There is usually a vertical white line down the middle of the snout between the nostrils. The belly is cream colored to pale pink with a bronze iridescence, the viscera tend to be visible through the skin, and the throat is translucent. The spiracle is sinistral and located below the lateral axis of the body. Upper and lower tail fins may or may not be heavily marked with speckles or spots (but not large spots), and the dorsal tail fin is rounded. In general, the tail fin is lighter than the body. Eyes are bronze. The lower jaws are wide.

Tadpoles of Southern Leopard Frogs are easily confused with those of gopher frogs in areas where these overlap in distribution, and there appears to be no characteristic that separates Southern Leopard Frog larvae from Northern Leopard Frog larvae. Albinistic larvae are reported from Georgia, New York, and possibly Oklahoma (Murlett, 1896; Hensley, 1959). Descriptions of the tadpole are in Wright (1929, 1932) and Gregoire (2005).

Eggs. Eggs are deposited in globular jellied masses in shallow water. Egg masses are usually attached to submerged vegetation or to the stalks of emergent vegetation, but they may be left unattached. The number of eggs per egg mass varies between a few hundred and 4,000, but a female may oviposit more than one egg mass per breeding season. As such, total egg counts may run between 2,000 and 5,500 per female. The individual eggs average 1.76 mm in diameter (Trauth, 1989). The diameter of the vitellus ranges between 1.4 and 1.8 mm, the diameter with inner envelope 2.4–3 mm, and the diameter with outer envelope 3.4–4 mm (Wright, 1932). Hatching normally occurs in 7–12 days, but Wright (1932) reported hatching in 3–5 days during the early summer in Georgia.

DISTRIBUTION

Southern Leopard Frogs occur from a few southern counties of New York and Long Island (but see Newman et al., 2012), south through New Jersey, the Delmarva Peninsula, and the Coastal Plain and Piedmont of the southeastern United States. They occupy lowland habitats through east Texas and Oklahoma, southeastern Kansas, and much of Missouri. Their northern range limit extends to central Illinois and Indiana, much of Kentucky, and down the western front of the Appalachians to northern Georgia and South Carolina. The species once occurred in upstate South Carolina near the Blue Ridge Escarpment, but the filling of Lake Jocassee may have eliminated that population (Montanucci, 2006). The species is generally absent from the Alleghenies and Southern Appalachians, north-central and eastern Kentucky, and the foothills of the Appalachians along their eastern front in North Carolina and Virginia. *Lithobates sphenocephalus* barely enters southern Ohio and Iowa.

An introduced population of Southern Leopard Frogs occurs in central New York around the Seneca Army Depot (Gibbs et al., 2007). The species very rarely overlaps naturally with *L. pipiens*, although their ranges are often allopatric. Minton (2001) reported the ranges overlapped in southeastern and central Indiana, and Smith (1961) stated that there was an intergradation zone in central Illinois. It has also been introduced into the Bahamas, California, Connecticut, and Massachusetts (Kraus, 2009).

Southern Leopard Frogs frequently are found on islands, including Long Island, New York

Distribution of *Lithobates sphenocephalus*. The population in western New York is introduced.

(Overton, 1914), Kent Island, Maryland (Grogan and Bystrak, 1973b), Assateague Island in Maryland and Virginia (Lee, 1972; Mitchell and Anderson, 1994), Chincoteague Island, Virginia (Conant et al., 1990), Shackleford Banks, Smith, Bodie, Hatteras, and Roanoke islands in North Carolina (Brimley, 1944; Lewis, 1946; Engels, 1952; Braswell, 1988; Gaul and Mitchell, 2007), Kiawah and Capers islands in South Carolina (Gibbons and Harrison, 1981), Wassaw, Ossabaw, St. Catherines, Blackbeard, Sapelo, Little Cumberland, and Cumberland Island in Georgia (Martof, 1963; Gibbons and Coker, 1978; Laerm et al., 2000; Shoop and Ruckdeschel, 2006), Dauphin Island, Alabama (Jackson and Jackson, 1970), Marco Island and Big Pine, Ramrod, Summerland, Cudjoe, Little Torch, Middle Torch, and Big Torch keys, and St. George and St. Vincent islands, Florida (Duellman and Schwartz, 1958; Lazell, 1989; Irwin et al., 2001). Historically reported populations at Key West may no longer be extant.

Important distributional references include: Alabama (Mount, 1975), Delmarva Peninsula (White and White, 2007), Florida (Bartlett and Bartlett, 1999), Georgia (Williamson and Moulis, 1994; Jensen et al., 2008), Illinois (Smith, 1961; Phillips et al., 1999), Indiana (Minton, 2001; Brodman, 2003), Kansas (Collins, 1993; Collins et al., 2010), Louisiana (Dundee and Rossman, 1989), Maryland (Harris, 1975), Missouri

(Johnson, 2000; Daniel and Edmond, 2006), New Jersey (Conant, 1962; Schwartz and Golden, 2002), North Carolina (Dorcas et al., 2007), Ohio (Davis and Menze, 2000), Oklahoma (Sievert and Sievert, 2006), Tennessee (Redmond and Scott, 1996), Texas (Hardy, 1995; Dixon, 2000) and Virginia (Tobey, 1985; Mitchell and Reay, 1999).

FOSSIL RECORD

Fossils of *L. sphenocephalus* are reported from Pleistocene (Rancholabrean) localities in Florida and Georgia. Since fossils (vertebral column and limbs) of *L. sphenocephalus* and *L. pipiens* are not distinguishable, Holman (2003) made allocation based on zoogeographic criteria.

SYSTEMATICS AND GEOGRAPHIC VARIATION

Lithobates sphenocephalus is a member of the *Novirana* clade of North American ranid frogs, and it is classified within the *Pantherana*, a group that includes the leopard and gopher frogs and comprises the *Rana pipiens* complex. Mating calls within the *Pantherana* are highly complex in structure (used both to advertise presence to females and other males in territorial positioning), and include elements described as chuckles, grunts, and snores (Hillis and Wilcox, 2005). Mating calls are only produced during breeding choruses, whereas other elements of the call are produced at other times of the year in the *Pantherana*. Southern Leopard Frogs appear more closely related phylogenetically to *Lithobates blairi*, *L. berlandieri*, *L. yavapainensis*, *L. onca*, and several Mexican species (the *Scurrilirana*, frogs that produce a chuckle-like call) than they are to *Lithobates pipiens* and *L. chiricahuensis*. A previous suggestion that *L. palustris* is monophyletic with *L. pipiens* and *L. sphenocephalus* (Pytel, 1986) is not supported.

Southern Leopard Frogs have long been considered a part of the *Rana pipiens* complex, and were generally thought to be a southern variant of a gradual morphological cline. Moore (1975) and Hillis (1988) have reviewed the history of the classification of "leopard frogs" (also see the *Lithobates pipiens* account). Genetic and call analyses have clearly established the specific identity of Southern Leopard Frogs. Some authors have identified two subspecies within *L. sphenocephalus*. *Lithobates sphenocephalus sphenocephalus* was said to occur in the more northern portions of the range and not contain vestigial oviducts, whereas *L. sphenocephalus utricularius* occurred mostly in the Florida peninsula and contained vestigial oviducts (Pace, 1974). Hillis and Wilcox (2005) found some weak differentiation between Kansas (*sphenocephalus*) and Florida (*utricularius*) populations, but specific status does not appear warranted.

In the lower Florida Keys, Southern Leopard Frogs are much larger and darker than mainland populations, and frequently have a dark or mottled throat and belly (Duellman and Schwartz, 1958; Lazell, 1989). The dorsolateral fold is bronze. Juveniles may be colored like mainland populations, but as they grow they become increasingly melanistic. Mount (1975) notes that an unspotted "Burnsi" phenotype (see *L. pipiens* account) is occasionally observed in Alabama, making the individual difficult to differentiate from *L. clamitans*. Other unspotted individuals have been reported from New Jersey, Kentucky, Mississippi, Texas, and Illinois (Brown and Funk, 1977; Redmer, 1992).

Using both genomic and mitochondrial DNA, Newman et al. (2012) identified a new species of leopard frog, tentatively here referred to as the Tri-State Leopard Frog, from a small region of southern New York, Staten Island, and northern New Jersey. These authors suggested that this species may extend as far as Connecticut and perhaps northeastern Pennsylvania. This leopard frog, unnamed as of May 2012, appears more closely related to *L. palustris* and *L. sphenocephalus* than to *L. pipiens* and may be sympatric with *L. pipiens*. Hybridization with adjacent species of leopard frogs does not appear to occur, suggesting reproductive isolation.

Stable natural hybrid zones occur between various members of the Leopard Frog complex. For example, Kocher and Sage (1986) showed that while allele frequencies changed annually, the extent of the hybrid zone between *L. sphenocephalus* and *L. berlandieri* in central Texas remained constant. These authors noted that most of the metamorphosing frogs had parental genotypes, and that most hybrids probably did not metamorphose successfully due to developmental instability. Natural hybridization has been reported with *L. palustris* in Tennessee (Salthe, 1969) and Maryland (Hardy and Gillespie, 1976). In the latter case, hybridization occurred at a disturbed natural habitat where numerous instances of interspecific amplexus had been observed. Hybrids also have been reported between *L. sphenocephalus* and *L. berlandieri* (McAlister, 1962; Littlejohn and Oldham, 1968; Sage and Selander, 1979; Kocher and Sage, 1986), *L. blairi* (Axtell, 1976), and *L. pipiens* (Pace, 1974). In general, hybrids make up only a small portion of the frogs observed (2–7%), but there are reports of much higher frequencies (30–40%) in some areas (Sage and Selander, 1979).

Under laboratory conditions, viable hybrids have been produced between ♂ *L. sphenocephalus* and ♀ *L. areolatus*. However, crosses between ♂ *L. areolatus* and ♀ *L. sphenocephalus* are inviable or only produce a small number of larvae (McAlister, 1961; Cuellar, 1971). Note that there may be some confusion about whether *L. sphenocephalus* and *L. berlandieri* were used in the experiments by McAlister (1961). Other laboratory studies have produced a small to large number of larvae in crosses between Southern Leopard Frogs and *L. areolatus, L. blairi, L. megapoda, L. montezumae,* and *L. palustris* (Mecham, 1969).

ADULT HABITAT

Optimal Southern Leopard Frog habitat includes a landscape with a mosaic of different sized wetlands of various hydroperiods within a mesic forested region. Because of its generalist habitat preferences, Southern Leopard Frogs are fairly ubiquitous throughout their range, and often may be the most abundant amphibian at a site (e.g., Dodd et al., 2007). They frequent wet forest in riparian floodplains, where they shelter under woody debris, and they are found commonly in wet meadows, marshes, cypress savannas, cypress-gum ponds, dome swamps, Carolina bays, Atlantic White Cedar swamps, freshwater tidal wetlands, sloughs, sawgrass marsh, wet prairie, swales (seasonally or temporally flooded wetlands), beaver ponds, ditches, canals, and along lakes and impoundments (Wright, 1932; Carr, 1940a; Mount, 1975; Smithberger and Swarth, 1993; Enge and Wood, 1998; Meshaka et al., 2000; Metts et al., 2001; Lichtenberg et al., 2006; White and White, 2007; Liner et al., 2008). Small, isolated ponds may be very important to this species (Russell et al., 2002a). In winter or during cold fronts, dormancy occurs along the shoreline or in the mud on pond bottoms. Activity resumes when the temperature rises to approximately 20°C.

Southern Leopard Frogs are frequently found in riparian habitats along rivers and streams (Rudolph and Dickson, 1990; Muenz et al., 2006), and such habitats may serve as dispersal corridors (Burbrink et al., 1998). These frogs commonly inhabit the steep wooded ravines of the lower Gulf Coastal Plain, habitats that frequently have small cold streams at their bottoms (Enge, 1998b). They have also been found in creeks, ponds, and ditches with saline contents of 0.05 to 21.4 ppt (Pearse, 1936; also see Deckert, 1922; Liner, 1954; Duellman and Schwartz, 1958; Lazell, 1989). Meshaka et al. (2000) reported them in mangrove forest. Christman (1974) noted that coastal populations were more tolerant of saline conditions than conspecifics living in more inland localities. Coastal populations are able to decrease water loss as a result of having more concentrated plasma, which reduces the osmotic gradient between the frog and its environment. He further noted that coastal *L. sphenocephalus* were larger in body mass than frogs from inland populations. The size increase results in changes in the ratio of relative surface area to body water volume and to an increase in skin thickness. As a result, hurricane overwash has little affect on this species (Gunzburger et al., 2010).

Lithobates sphenocephalus prefers habitats with an abundant herbaceous plant understory and a low overhead canopy; these conditions supply cover and retain high levels of humidity. Terrestrial refugia and feeding occur in variety of different types of old fields, deciduous hardwood forests, mixed pine and hardwoods, scrub, and pines adjacent to wetland communities (Harima, 1969; Bennett et al., 1980; Clawson and Baskett, 1982; Buhlmann et al., 1993; Enge, 1998a; Enge and Wood, 1998; Branch and Hokit, 2000; Hanlin et al., 2000; Meshaka et al., 2000). Individuals are occasionally found at the entrance and inside of caves (Brode, 1958; Black, 1973a).

TERRESTRIAL ECOLOGY

Southern Leopard Frogs may be active year-round, depending on temperature. Activity generally ceases below 10–13°C (Kilby, 1945). At the northern extent of the range, they are most likely to be active throughout the late winter to autumn. For example, they have been recorded from late February through mid-October along the Patuxent River in Maryland (Smithberger and Swarth, 1993). However, even in southern Illinois the species is considered to be active year-round (Rossman, 1960). They are not active during periods of drought, as they have high rates of evaporative water loss (Wygoda, 1984). At such times, they retreat to permanent wetlands, stump holes, and burrows of mammals and crayfish. Recent metamorphs of *L. sphenocephalus* actively burrow into the substrate to avoid desiccation or passively resort to existing burrows to minimize evaporative water loss (Parris, 1998); those that use existing burrows tend to lose less water than those that actively burrow. Because they are prone to water loss, adults and juveniles are active during rainfall and at high humidity, usually at night, and are found only in moist conditions and in proximity to a source of water.

After metamorphosis and the breeding season, juvenile and adult Southern Leopard Frogs disperse widely into adjacent terrestrial communities. Rainfall is important in stimulating migrations from wetlands. Terrestrial sites include both upland and bottomland communities, as long as they are moist and humid. In experimental trials, Southern Leopard frogs initially often moved into clearcuts when given a choice at an ecotone between forest and open habitats. However, the farther the frogs moved, the more likely they were to be found within the forest. Of 48 frogs followed in one study, the maximum distance moved in 24 hrs was 351 m (mean 86 m), indicating considerable ability to move rapidly over short periods of time (Graeter, 2005; Graeter et al., 2008). In addition, pathways tended to be direct, rather than wandering, indicating a clear ability at orientation. Dodd (1992) noted that juvenile *L. sphenocephalus* were using a temporary wetland as a stopover during dispersal; it is likely that dispersing adults and juveniles do likewise when such habitats are available.

Southern Leopard Frogs are able to use the position of the sun as a means of orientation between feeding and breeding sites. Displaced frogs use a Y-axis orientation to determine the direction of the shoreline from whence they were taken, which then allows movement back to the point of origin (Jordan et al., 1968a). Presumably, free-ranging frogs can use sun compass orientation as they disperse after the breeding season and to migrate toward breeding sites in the winter and spring.

Southern Leopard Frogs are photopositive in their phototactic response, suggesting they use both sunlight and moonlight when feeding and going about their daily activities (Jaeger and Hailman, 1971, 1973). They are sensitive to light in the blue spectrum ("blue-mode response"), which apparently helps them orient toward areas of increasing illumination, such as the open area above a pond (Hailman and Jaeger, 1974). Southern Leopard Frogs likely have true color vision.

CALLING ACTIVITY AND MATE SELECTION

Males normally call from water while floating half-submerged in shallow vegetation (but see Jensen et al., 2008, noting submerged calling). Calling often begins in the autumn in the North, declines or ceases in the winter, and resumes in the spring; in Virginia, for example, calling occurred from 22 February to 15 April and from 1 to 23 September (Mitchell, 1986). Calling begins in November in Mississippi and Alabama (M.J. Allen, 1932; Mount, 1975; Doody and Young, 1995), January in Arkansas (Trauth, 1989), and in February in North Carolina and Illinois (Phillips et al., 1999; Todd et al., 2003). Occasional calls are heard throughout the summer in North Carolina (e.g., Gaul and Mitchell, 2007), but most calling occurs between fall and spring (November to May, depending on location). On Long Island, one of the most northern populations, calling extends from early March to June (Overton, 1914; Gibbs et al., 2007). In Florida, calling occurs year-round (Carr, 1940b).

Calling may occur at any time of the day, although intensity is much greater from 19:00 to about dawn than it is diurnally (Todd et al., 2003). In North Carolina, calling occurred at rather constant low levels throughout the day, with a slight increase from 18:00 to 24:00 hrs. Calling intensity may vary from one pond to the next, even during the same time period. The optimum calling temperature is ca. 10°C, with a range from <5 to <30°C (Steelman and Dorcas, 2010). Wiest (1982) noted calling at 3–26.1°C over an 87-day period in Texas.

Call characteristics of the Southern Leopard Frog clearly differentiate it from other members of the Leopard Frog complex (Littlejohn and Oldham, 1968), and responses by females are species-specific (Oldham, 1974). In Texas, calls last a mean of 0.41 sec (range 0.31 to 0.52 sec) with a pulse rate of 14.8 pulses/sec (range 14.3 to 15.3 pulses/sec). The pulse duration averages 39.4 msec (range 33 to 50 msec) with a mean pulse rise time of 24.3 msec (range 20 to 30 msec). These values are for frogs calling at temperatures of 20–25°C. The call has been described in many tortured ways, such as various guttural croaks and clucks, chuckles, or even similar to running a thumb over an inflated balloon.

BREEDING SITES

Lithobates sphenocephalus breeds in a wide variety of ponds, both in natural settings (see Adult Habitat) as well as in man-made wetlands adjacent to agricultural, human-disturbed, and formerly mined sites (Myers and Klimstra, 1960; Surdick, 2005; Anderson and Arruda, 2006; Babbitt et al., 2006; Dodd and Barichivich, 2007; Scott et al., 2008). Indeed, larval survival actually may increase in golf course ponds compared to natural ponds, since insect predators are reduced by contaminants (Boone et al., 2008). It uses both temporary and permanent wetlands for breeding, and is often found in large numbers at these sites (Cash, 1994).

Southern Leopard Frogs prefer fishless ponds, although they will breed in ponds containing certain species of fish, such as *Pimephales* and *Notemigonus* (Sexton and Phillips, 1986; Babbitt and Tanner, 2000). It appears that they are sensitive to the presence of predaceous fishes, such as *Lepomis* and other centrarchids, and usually avoid ponds containing these species. In experimental trials, larval *Lithobates sphenocephalus* are readily consumed by predaceous fish, and in the presence of such species larvae reduce their activity. Susceptibility to larval predation likely accounts for their preference of fishless ponds for breeding (Gregoire and Gunzburger, 2008).

L. sphenocephalus breeds in ponds both with open and closed canopies, and experimental studies have demonstrated that the amount of shade does not affect larval survivorship or time to metamorphosis. However, survivorship tends to increase slightly on grass substrates as compared to leaf litter substrates, and the time to metamorphosis decreases on grass substrates compared to leaf litter (Williams et al., 2008).

As with many amphibians, *L. sphenocephalus* is not found in every wetland that might appear to a human observer to be good habitat. For example, Babbitt et al. (2006) found Southern Leopard Frogs at 62% of wetlands surveyed in an agriculturally dominated landscape in south Florida. At these sites, larval abundance was related to low conductivity and pH, where breeding occurred both in seasonally flooded wetlands in close proximity to woodlands and in ponds completely surrounded by pastures. However, they certainly are among the most widespread and ubiquitous anurans throughout much of their range, and large numbers of Southern Leopard Frogs may be found at individual breeding sites; at times, the ground literally may be moving with Southern Leopard Frogs jumping to safety as an observer walks along a preferred wetland's shoreline.

REPRODUCTION

Breeding begins in the autumn and occurs throughout the winter into spring, temperature permitting; in the South, breeding may take place year-round (Kilby, 1945; Bancroft et al., 1983; Semlitsch et al., 1996; Babbitt and Tanner, 2000; Jensen et al., 2008). Although many areas appear

Eggs and pond habitat of *Lithobates sphenocephalus*. Photo: Alan Cressler

to have a biphasic breeding season, most breeding takes place from late winter to spring. According to Mitchell (1986), eggs are oviposited throughout the calling period, but Doody and Young (1995) found different levels of oviposition during discrete breeding events rather than continuous oviposition throughout the breeding season. The mean duration of each breeding event was 2.5 days, with a lag time of 1–3 days between rainfall and oviposition.

Migration to breeding ponds may take place at different times during the same year, even in ponds in close proximity to one another (Hocking et al., 2008). For example, these authors recorded migration between 5 March to 30 April in one area, but between 16 March to 27 May in an adjacent area. This result suggests a prolonged and flexible breeding cycle tailored to local conditions. Migration to breeding sites occurs almost exclusively at night, even though dark rainy days might seem ideal (Pechmann and Semlitsch, 1986; Todd and Winne, 2006).

Eggs are deposited in globular masses approximately 10 cm long by 5 cm in width. If emergent and submergent vegetation is available, egg masses will be attached to the stems; if no vegetation is available, masses will be left free on the substrate. In cooler temperatures, egg masses are deposited in sunny shallow water, but in warmer temperatures, eggs are deposited in deeper waters. Water depth at oviposition sites in Arkansas was 20–50 cm (Trauth, 1989), whereas in Mississippi, eggs were in water 7–13 cm in depth (Doody and Young, 1995).

Egg masses are deposited singly or occur in small clusters of 2–5 placed ca. 20 cm apart, but Trauth (1989) reported a communal oviposition site containing 75 masses in 3 m². Caldwell (1986) suggested that communal egg deposition was more likely to occur during cooler temperatures, perhaps to confer a developmental thermal advantage, as the sun heats up shallow waters quickly. Communal egg deposition declined in warm summer and autumn waters, as females oviposited isolated single masses in cooler, deeper water.

Clutch sizes reported from the literature include a mean of 4,000 eggs per clutch in Virginia (Mitchell, 1986) and 1,054 eggs from the Okefenokee Swamp (Wright, 1932). Trauth (1989) reported a mean clutch size of 2,959 (range 1,700 to 5,537) in Arkansas, but the mean number of eggs per egg mass was only 2,106 (range 1,289 to 3,366). This suggests that females may deposit more than one egg mass during the breeding season. The weight of an egg mass averaged 65.3 g (range 20.3 to 105.3 g). There is a positive correlation between mean egg mass and clutch size and female SUL. Mean egg diameter and mass per egg are inversely correlated with the number of eggs per mass (Trauth, 1989). Clutch size and clutch mass are also positively correlated.

The number of females breeding in a pond is probably a function of wetland size and proximity. In a single South Carolina pond, Semlitsch et al. (1996) recorded that from 40 to 120 females visited a pond annually during 8 of 16 yrs the pond was surveyed. Many fewer females visited the pond during the other years. Reproduction failed completely during 6 yrs, and was negligible during an additional 7 yrs. Recruitment was inhibited by an extended drought, which resulted in reproductive failure throughout most of the study.

LARVAL ECOLOGY

Larvae are pond-adapted, featuring a deep body with less-developed tail musculature that allows them to forage on plankton in lentic pond habitats (Hillis, 1982). The tadpole diet consists mostly of planktonic green algae, and larvae apparently can discriminate and choose foods of higher quality over those of lower quality (Taylor at al., 1995). Larvae hatched in the autumn overwinter and transform the following summer, whereas larvae hatching in spring transform later in the same summer. As such, the larval period varies accordingly, with larvae hatching in spring transforming in 70–90 days, but larvae hatching in the autumn transforming at 270 days (Mitchell, 1986). Gaul and Mitchell (2007) noted the presence of larvae at all times of the year in eastern North Carolina, and a year-round larval presence is probably the norm throughout much of the range of this species because of its extended breeding period. Recent metamorphs are 20–33 mm SUL (Siekmann, 1949; Minton, 2001; Jensen et al., 2008).

A great many juveniles may be produced from breeding ponds, despite high levels of larval predation and the threat of insufficient hydroperiods. The numbers depend on the size of the pond and environmental conditions during the larval period. For example, Seale (1980) estimated there were 10,275 tadpoles in her study pond on

Tadpole of *Lithobates sphenocephalus*. Photo: Stan Trauth

16 May from eggs that had hatched the previous autumn; this estimate was derived immediately prior to metamorphosis. Seale's (1980) estimates, however, may include larvae of both *L. sphenocephalus* and *L. blairi*, as they are sympatric at her study site. Semlitsch et al. (1996) recorded 56,225 metamorphs from a single pond in South Carolina over a 16 yr period; however, nearly all these metamorphs resulted from a single breeding season. Also in South Carolina, Todd and Winne (2006) recorded 210,152 juveniles at a 10 ha wetland over a 1 yr period; these resulted from 638 adults captured at the pond. A series of 8 small temporary ponds produced at least 1,338 metamorphs over 7 yrs in central Florida sandhills (Greenberg and Tanner, 2005b).

Like many frogs, larvae of *L. sphenocephalus* increase their growth rates in relation to food availability and the space available to them to develop. Time to metamorphosis, larval size, and survivorship are all density dependent in the absence of predators (Wilbur et al., 1983; Richter et al., 2009). In laboratory experiments, larvae fed ad libitum grew larger in larger containers than they did in small containers. Density certainly affects growth, as small numbers of tadpoles grow faster than do larger numbers of tadpoles grown in the same sized containers or mesocosms (Nicholson, 1980). However, certain larvae grow fast regardless of the number of smaller larvae within containers of the same size. As such, density affects some individuals, but not all, and the density effects are more pronounced on smaller tadpoles when size classes are mixed. Indeed, density can affect time to metamorphosis even when growth rates are kept constant in experimental trials (Richter et al., 2009).

It appears that larger *L. sphenocephalus* tadpoles are able to produce a species-specific water-based inhibitory substance which further retards the growth of small tadpoles in crowded conditions (Rose, 1960; Steinwascher, 1978b; see *L. pipiens*). Not surprisingly, therefore, small tadpoles tend to avoid larger tadpoles and may occupy different areas within a pond when multiple size classes are present (Alford and Crump, 1982). Growth in crowded conditions is thus subject to a density-dependent feedback loop in which large tadpoles, regardless of food availability, increase their chances of metamorphosis by inhibiting growth of small tadpoles (termed chemically based interference competition). In addition, the largest tadpoles are able to out-compete smaller tadpoles for food, particularly the feces of other larvae (Steinwascher, 1978b).

Within a pond, there may be several size classes of Southern Leopard Frog larvae at the same time, as breeding occurs year-round in some locations. Different sized larvae may partition the habitat differently and, indeed, tend to associate more with larvae of their own size class and species than with heterospecifics (Alford and Crump, 1982; Alford, 1986). Positioning of *L. sphenocephalus* larvae within a habitat may change through the year, as the larvae alter their response to environmental variables, such as the presence of a sand substrate and aquatic vegetation, as they continue to grow and develop (Alford, 1986). Large larvae tend to avoid sandy substrates, but small larvae do not (Alford and Crump, 1982).

The duration of the larval period depends on a suite of characteristics, such as predators, food resources, and the density of conspecifics and heterospecifics. At high densities, tadpoles tend to delay metamorphosis and overwinter prior to transformation (Morin, 1983; Wilbur et al., 1983). Hydroperiod also plays an important role. In experimental trials, longer hydroperiods increased the proportion of larvae that successfully completed metamorphosis, although the test hydroperiods had no effects on the mass of survivors (Ryan and Winne, 2001). Tadpoles are able to accelerate their development in response to drying ponds, which results in early metamorphosis. The resultant metamorphs may, however, be smaller than metamorphs from wetlands with longer hydroperiods.

DIET

Larvae are herbivores, feeding on filamentous algae. The diet of postmetamorphic Southern Leopard Frogs is highly insectivorous, with terrestrial beetles and crickets featuring prominently. Other prey include spiders, snails, roaches, crayfish, true bugs, ants, flies, mayflies, lepidoptera larvae, and grasshoppers (Kilby, 1945; Crawford et al., 2009). A variety of vertebrates may be consumed, including fish (*Aphredoderus, Fundulus, Gambusia, Jordanella, Notemigonus,* small sunfish, and catfish), salamanders (*Desmognathus* sp.), and frogs (*Acris gryllus, Anaxyrus quercicus, Hyla cinerea, H. squirella, Lithobates sphenocephalus, Pseudacris ocularis*) (Kilby, 1945; Duellman and Schwartz, 1958; Jensen et al., 2008). Feeding occurs both nocturnally and diurnally. The frog is an ambush predator that orients toward the prey item and will sometimes leap toward it before extruding the tongue. Small prey are eaten whole using only the tongue, but large prey are manipulated and held using both front feet. Aquatic and terrestrial prey are consumed, but feeding does not occur underwater. Southern Leopard Frogs are capable of feeding inside caves, where they consume flies, beetles, and crickets associated with bat guano (Black, 1973a). Additional information on diet is included in Force (1925), Duellman and Schwartz (1958), and Lewis (1962a).

PREDATION AND DEFENSE

Eggs and embryos are subject to predation by caddisfly larvae (Richter, 2000), other insectivorous invertebrates, crayfish (*Procambarus*) (Saenz et al., 2003), and newts (both *Notophthalmus perstriatus* and *N. viridescens*). Southern Leopard Frogs larvae and adults are eaten by a wide variety of predators, including members of all vertebrate groups, and they likely form a very important seasonal food source to many animals. Larvae are eaten by many types of predaceous invertebrates. Vertebrate records include several species of water snakes (*Nerodia*), crayfish snakes (*Regina*), racers (*Coluber*), garter snakes (*Thamnophis*), cottonmouths (*Agkistrodon piscivorus*), snapping turtles (*Chelydra*), wading birds, grackles, a barn owl, raccoons, and foxes (Wright, 1932; Clark, 1949; Telford, 1952; Mitchell and Anderson, 1994; Briggler, 2000).

The Southern Leopard Frog, like *Lithobates pipiens,* uses both morphological and behavioral means of defense, including remaining motionless, crouching at the approach of a predator, body inflation, and hiding (Marchisin and Anderson, 1978). The dorsal coloration, spot patterning, and bright dorsolateral lines of postmetamorphs allow them to be concealed in grass and other types of vegetation, and to disrupt the outlines of the frog, making them difficult to see. In contrast, the white belly is a form of counter-shading, making the frogs difficult to see from below by aquatic predators looking up toward the sky. Except while feeding or moving between locations, leopard frogs sit immobile in concealing vegetation, and thus do not draw attention to themselves. They sometimes will remain motionless (immobile) when grasped, a reaction that is enhanced by the emission of a loud noise prior to grasping (Nash et al., 1970). Psychologists have interpreted such a reaction as a "fear response," but it undoubtedly has a selective advantage in a predatory encounter.

If approached too quickly, they utter a warning scream and sometimes jump into the water. The scream serves to startle and disorient a would-be predator. They will dive into the concealing substrate, or swim out in a wide arc and circle back under water to the shoreline, where they may sit with just the snout protruding among emergent vegetation (Bragg, 1945a). The immobile posture and cryptic coloration help the frog avoid further detection. More often than not, however, they escape terrestrially rather than venture into the water. Carr (1940a) noted that frightened leopard frogs will return to the shoreline in order to take a terrestrial escape route, jumping six to eight leaps through dense vegetation, all the while pivoting with each landing. He reported, "I have seen them cover some distance in the precipitous retreats without ever touching the ground." Frogs will also scream when grasped by a predator.

COMMUNITY ECOLOGY

Lithobates sphenocephalus sometimes overlaps in geographic range with the leopard frogs *L. blairi, L. pipiens,* and *L. berlandieri*. Interbreeding occurs but is uncommon because of differences in habitat preference, call structure, and reproductive patterns. In Texas, for example, *L. blairi* is most associated with warm, turbid pools, whereas the sympatric *L. sphenocephalus* is associated with clear and cool water (Hillis, 1981). In addition, calling patterns vary when species are in sympatry. Then, *L. berlandieri* switches from a bimodal pattern to breeding from August to December, whereas *L. sphenocephalus* breeds from January to March. In areas of sympatry with *L. berlandieri*, *L. sphenocephalus* in Texas tends to forgo its brief autumn breeding period that occurs in areas of allopatry (Hillis, 1981).

In the complex larval amphibian communities of the Southeast, there is the potential for considerable interaction among many divergent species. One of the most interesting effects is the ability of developing larvae within egg capsules to speed up development in the presence of chemical extracts of crayfish (Saenz et al., 2003). Crayfish are serious predators of amphibian eggs. By speeding up development in the presence of crayfish chemical cues, larvae are able to hatch quicker and presumably increase survivorship once distanced from the mass of developing embryos. A single small larva is much harder for a predator to find in the complex vegetative structure of submerged vegetation than is a mass of larger squirming larvae within a jelly globular mass.

Competition likely plays an important role in pond communities inhabited by Southern Leopard Frogs, depending upon the order of breeding and the resources available. For example, in paired laboratory trials, *L. sphenocephalus* larvae have a negative impact on the number of *Anaxyrus* larvae that survive. As Southern Leopard Frog larvae increase in size, fewer *Anaxyrus* survive, presumably as the leopard frogs outcompete the small toads for food (Wilbur et al., 1983; also see Alford, 1989a). Mean growth rates of *Hyla gratiosa* and *H. chrysoscelis* tadpoles also decrease as the density of *Lithobates sphenocephalus* increases (Morin, 1983; Alford, 1989a). Competition from *L. sphenocephalus* also causes a decrease in survival and body mass of larval *Hyla andersonii* (Pehek, 1995). However, in mixed species trials, the growth rates of *Scaphiopus holbrooki* were correlated positively with densities of *Lithobates sphenocephalus*, even though *Scaphiopus holbrooki* has a negative effect on the growth rates of *Lithobates sphenocephalus* (Wilbur et al., 1983; Alford, 1989a). In other examples, survivorship of Southern Leopard Frogs improves after the removal of American Bullfrog larvae, suggesting release from a competitive effect (Boone et al., 2008), and in experimental ponds, interspecific competition with *L. blairi* and *L. sphenocephalus* resulted in an increased larval period and a decreased body mass of metamorphic *L. areolatus* (Parris and Semlitsch, 1998). The presence of *L. areolatus* seems to facilitate growth of *L. sphenocephalus* larvae.

The precise nature of the larval interspecific interactions depends not only on species involved but also on their breeding phenology. Which species arrives first at a breeding site will define the nature of competitive interactions, and these, in turn, will influence survivorship and traits at metamorphosis. *Anaxyrus americanus* does much better in terms of metamorphic traits when hatching prior to or simultaneous with *Lithobates sphenocephalus*. Likewise, *L. sphenocephalus* does much better when alone or introduced after *Anaxyrus americanus* (Alford and Wilbur, 1985). These results are not consistent with a simple size-based competitive interaction and suggest that arriving females might do well to gauge the presence of heterospecifics already at breeding ponds. The community ecology of potential breeding ponds might thus be responsible for differences in reproductive timing among even nearby breeding populations.

In another example of how breeding phenology affects interspecific interactions and life history, larval *Lithobates sphenocephalus* out-compete other anuran larvae, which delays metamorphosis. As soon as the leopard frogs metamorphose, the other species are released from competition and grow faster and increase in body mass (Alford, 1989a). However, *Scaphiopus* larvae will out-compete *Lithobates* larvae in drying conditions, and thus cause mortality to the leopard frogs unable to grow fast enough to complete metamorphosis (Wilbur et al., 1983). The extent and duration of the competitive effects, thus, will depend on how much the breeding phenology overlaps, and that will vary annually and by region. Breeding phenology sets conditions of density-dependent effects, as well as determines which predators the larvae will be exposed to.

As a result of differences in breeding phenology, different sized Southern Leopard Frog larvae are often in a pond simultaneously with various other species. Larvae tend to associate with same-sized conspecifics rather than be dispersed randomly in relation to the presence of other species. At high densities, especially, larval *Lithobates* do not tend to aggregate with other species. As noted by

Alford (1986), however, aggregation could result from positive associations among *Lithobates* larvae, or avoidance of *Lithobates* larvae by the other species tested.

In aquatic breeding sites, Southern Leopard Frog larvae are presented with a variety of predators of different sizes, in habitats that vary structurally. Large predators usually have an advantage over small predators, since the large size of Southern Leopard Frog larvae may prevent successful attacks upon them by small predators. Larval Southern Leopard Frogs are relatively active in the presence of caged potential predators such as dragonfly naiads (*Tramea*), newts (*Notophthalmus*), and sunfish (*Lepomis*) (Richardson, 2001). This suggests that the larvae either do not detect the presence of the predators, or that they do not modify their behavior perhaps because of their large size.

Under natural conditions, some tadpoles are simply too large to eat, and this is reflected in the results of controlled experiments on body size, survival, and age at metamorphosis; small predators may not have much of a role in larval success. Predation may actually assist *Lithobates sphenocephalus* larvae by removing small competitors or decreasing the effects of conspecific competition. As predators eat smaller animals, the survivors grow larger and faster to a size that reduces the likelihood of predation (Morin, 1983; Wilbur et al., 1983). However, increasing the number of predators, regardless of size, tends to greatly reduce larval survivorship (Morin, 1983).

Large predators (such as dragonfly naiads, *Tramea*), however, readily attack larvae of *Lithobates sphenocephalus*. In experimental situations, attacks by even large naiads were less successful in structurally complex habitats than in more simplified habitats (Babbitt and Tanner, 1998). The structurally complex habitats provided hiding places for larvae and made detection and attack by the naiads less successful. Habitat structure thus plays an important role in larval Southern Leopard Frog survivorship, even in the presence of predators that have no trouble consuming larvae.

Another factor that may influence larval survivorship in the presence of predators is the spatial scale in which the frogs find themselves, or in the scale chosen by an experimenter conducting observations in mesocosms. In experimental situations, Southern Leopard Frog larvae survived equally in mesocosms of varying depth. At initial low densities, predators decreased survival rates, but at high initial densities, the presence of dragonfly naiads (*Tramea*) increased densities of the survivors. The tadpoles from full mesocosms had increased larval periods compared with those from lesser-filled tanks. This result may be due to higher temperatures and more food in the full tanks. Thus, the effects of spatial scale are most likely to be observed on life history attributes of

Adult *Lithobates sphenocephalus*. Photo: C.K. Dodd Jr.

this species that are related to density (Gascon and Travis, 1992).

A number of ranids may breed at a location and metamorphose at approximately the same time and size. Since similarly sized juveniles have a similar gape width, prey may be identical for weeks following transformation. For example, Crawford et al. (2009) found very similar prey between juvenile *Lithobates sphenocephalus* and *L. areolatus* in Oklahoma, and suggested that these species might be competitive until allometric morphological changes in gape width allowed *L. areolatus* to consume larger prey. If one species was rare, the more numerous species could have an initial resource advantage, thus affecting the persistence of the rare species.

DISEASES, PARASITES, AND MALFORMATIONS
Amphibian chytrid fungus (*Batrachochytrium dendrobatidis*) has been found historically (1979–1982) in Southern Leopard Frogs from South Carolina, although a more recent survey (2004) did not detect the disease (Daszak et al., 2005; Peterson et al., 2007). The latter authors suggested that *Lithobates sphenocephalus* might not be susceptible to the fungus, even when inhabiting sites where other species are infected. However, amphibian chytrid is known from North Carolina, Arkansas, and Louisiana (Rothermel et al., 2008). If the fungus causes keratinized tadpole mouthparts to fall out, the result is reduced growth and development rates (Venesky et al., 2010). A *Dermocystidium*-like fungus was reported on *Lithobates sphenocephalus* from Mississippi (Green et al., 2002). A lethal alveolate pathogen is known to infect Southern Leopard Frogs, causing mass larval mortality in Florida and Georgia. It is closely related to organisms in the genus *Perkinsus*, which are known to infect both freshwater and marine organisms (Davis et al., 2007). Larvae have enlarged abdomens, swim erratically, and all organs are infiltrated by the 6-μm organism, particularly the liver and kidneys. Mortality is rapid and affects nearly all larvae at a pond.

Parasitic protozoans include *Cytamoeba bactifera, Entamoeba ranarum, Lankesterella, Leptotheca ohlmacheri, Nyctotherus cordiformis, Octomitus intestinalis, Opalina* sp., *O. carolinensis, O. kennicotti, O. obtrigonoidea, Trichomonas augusta*, and *Trypanosoma rotatorium* (Brandt, 1936). The myxozoan *Myxidium serotinum* also parasitizes *Lithobates sphenocephalus* (McAllister et al., 2008). Trematodes include *Brachycoelium hospitale, Cephalogonimus americanus, Diplodiscus temperatus, Glypthelmins quieta, Gorgoderina attenuata, Gyrodactylus* sp., *Loxogenes bicolor, Megalodiscus temperatus*, and *Ostiolum complexus* (Trowbridge and Hefley, 1934; Brandt, 1936; Manter, 1938; Green and Dodd, 2007; Paetow et al., 2009). Other helminths include proteocephalid cysts, larval nematodes of the order Spirurida within liver tissue (Davis et al., 2007), and the nematodes *Agamascaris odontocephala, Agamonema, Cosmocercoides dukae, Dujardinia, Foleyella* sp., *F. americana, F. ranae, Microfilaria, Ozwaldocruzia collaris, O. pipiens, Physaloptera ranae, Rhabdias, R. ranae*, and *Spironoura catesbeianae* (Trowbridge and Hefley, 1934; Brandt, 1936). Acanathocephalans include *Centrorhynchus*. Mites (*Hannemania penetrans*; species unidentified) also occur on *Lithobates sphenocephalus* (Trowbridge and Hefley, 1934; Brandt, 1936).

Scoliosis (curvature of the tail) has been reported in *L. sphenocephalus* larvae found in an abandoned swimming pool in Maryland (Hardy, 1964b). McCallum (1999a) described malformations of five juvenile *L. sphenocephalus*, including missing limbs, split limbs, complete but abnormal limbs, missing eyes, and partial limbs, at a pond in southern Illinois. He also noted a press report about a malformed Southern Leopard Frog found in Missouri. Other reports of malformed *L. sphenocephalus* come from the Midwest (Converse et al., 2000), and include missing digits, missing limbs, and tail retained and fused.

L. sphenocephalus possesses antimicrobial skin peptides, which may aid in defense against *Escherichia coli* and other microbial pathogens (Conlon et al., 1999).

SUSCEPTIBILITY TO POTENTIAL STRESSORS
Metals. Metals known to have adverse effects on Southern Leopard Frogs include copper ($LC_{50\,(8\,day)}$ 0.05 mg/L in embryos and $LC_{50\,(72\,hr)}$ 6.37 mg/L for adults) (Kaplan and Yoh, 1961). Cadmium causes a decrease in larval survivorship and metamorphosis with increasing concentrations of 5–200 μg Cd/L (James et al., 2005). Very few larvae survived to metamorphosis at high Cd concentrations. However, mass at metamorphosis

was not affected. Amphibians are able to concentrate cadmium, with those individuals exposed to the highest levels concentrating the most cadmium.

Other elements. Fluorine causes eggs to hatch at an earlier stage of development than normal, but at a less advanced stage of development. Concentrations as low as 1–4 ppm retard the developmental rate and the stage of hatching (Cameron, 1940).

pH. The lethal pH for Southern Leopard Frogs is 3.7, with a critical pH of 4.1 (Gosner and Black, 1957a). Experimental acidification had no effect on *Lithobates sphenocephalus* at a mean pH between 5.1–5.2, regardless of loam or clay soil type (Sparling et al., 1995). Low pH (3.9) has no effect on survival, body mass, duration of the larval period, or growth of larval *L. sphenocephalus* (Pehek, 1995).

Chemicals. The herbicide diquat has no obvious effects on either embryos or larvae after exposure for 15 days following exposure at common application levels (2.0 mg/L), but did result in larval mortality at 5–10 mg/L. Paraquat also had no effects on embryos at 1–2.0 ppm, although aquatic exposure posthatching to 15 days at > 2.0 mg/L produced 100% mortality (Dial and Bauer Dial, 1987). Even at lower concentrations, survivorship was decreased, tadpole growth was retarded, and metamorphs were only about 50% of the size of controls (Dial and Bauer, 1984). It is important to note, however, that the test animals were obtained from Carolina Biological Supply Company in the preceding two reports. Thus, the species involved is presumed to be *L. sphenocephalus* rather than *L. pipiens*.

The herbicide triclorpyr (Release®) at concentrations of 0.25–0.50 mg/L^{-1} causes decreased larval survivorship. At higher herbicide concentrations, the decrease in survivorship was enhanced when conducted in water at a pH of 5.5 compared with a pH of 7.5. At low concentrations, pH did not have an effect (Chen et al., 2008). These results were obtained regardless of food concentration. Likewise, the toxicity effects of the herbicide glyphosate (Vision®) were enhanced at a pH of 7.5, but even at a lower pH, the herbicide decreased survivorship, and no larvae survived over the 10-day experimental period (Chen et al., 2004). Since both wild *L. pipiens* and larvae obtained from a southern-based supply house (*L. sphenocephalus*?) were used in the experiment, these results must be used with caution.

The insecticide carbaryl has negligible effect by itself on *L. sphenocephalus* embryo survival and hatching success (Bridges, 2000). Some results also indicate that it has negligible effects on larval survival, at least in experimental trials at 3.5, 5.0, and 7.0 mg/L (Boone and Semlitsch, 2002; Boone et al., 2004b; Mills and Semlitsch, 2004). The age at metamorphosis also is not affected at carbaryl concentrations of as little as 0.16 mg/L. Indeed, body mass and developmental stage may be improved to some effect by the addition of this insecticide. This result is in contrast to the results obtained by Bridges (2000), who found that chronic carbaryl exposure caused significant reductions in mass at metamorphosis, regardless of the stage at which eggs to larvae were exposed. In addition, few larvae survived at 0.4 and 1.0 mg/L when exposed over an extended period, suggesting that there is a sensitive period to carbaryl regardless of when tadpoles are exposed (Bridges, 2000). Bridges (2000) also reported high rates of larval deformities—18% of surviving larvae after 2 weeks of exposure. Whether the pesticide is experienced chronically or in pulsed doses thus becomes important in determining adverse effects.

Death occurs in about 20 hrs at 30 mg/L of carbaryl exposure, and time to death varies significantly among *L. sphenocephalus* populations (Illinois, Missouri, Texas, South Carolina, Virginia) suggesting regional variation in tolerance to the pesticide (Bridges and Semlitsch, 2000). In addition, there appears to be considerable variation in tolerance among both full and half-sib families to this pesticide (Bridges and Semlitsch, 2001). Some families appear more sensitive than others to carbaryl, and small tadpoles are more tolerant than larger tadpoles. The heritability of tolerance to carbaryl was 0.285 in Bridges and Semlitsch's Missouri population.

Simultaneous exposure has little effect on the outcome of predator-prey interactions involving Southern Leopard Frogs and newts (*Notophthalmus*), although Bridges (1999b) noted that asynchronous exposure could disrupt predator-prey dynamics by interfering with normal feeding, activity, and avoidance responses. A concentration of carbaryl at 2.5 mg/L significantly reduces larval activity and interferes with a predator's ability to capture prey. However, carbaryl exposure may actually benefit Southern Leopard Frog larvae by eliminating insect competitors, although elimination of periphyton might cause a reduction in size at metamorphosis (Mills and Semlitsch, 2004), and crowding still has negative effects (Boone et al., 2004b). Clearly, Southern Leopard Frog larvae, their predators and competitors, and pesticides interact in critical ways affecting metamorphosis and survivorship.

The addition of the herbicide atrazine to carbaryl-treated mesocosms did not have an effect on the larvae, despite a decrease in chlorophyll levels. However, as the density of the larvae increased, the addition of atrazine decreased larval survival, indicating a density effect on food resources. As atrazine reduced food, tadpoles in high densities were more likely to starve (Boone and James, 2003).

Organophosphorus insecticides decrease both red and white blood cell counts, an effect that is inversely proportional to dosage; neutrophils decrease whereas lymphocytes increase in proportion to dosage. Some chemicals cause the blood cells to change shape and become distorted. Such insecticides include malathion, TEPP (tetra-ethyl pyrophosphate), parathion, trithion, phosdrin, and OMPA (octamethyl pyrophosphoramide) (Kaplan and Glaczenski, 1965). These changes suggest not only direct toxicity, but perhaps adverse effect on normal immune function. Halogenated hydrocarbons (aldrin, benzene hexachloride, chlordane, dieldrin, endrin, methoxychlor, toxaphene) cause color changes, decreased cardiac and respiratory rates, neuromuscular spasms, tremors and convulsions, salivation, vomiting, cloacal prolapse, changes in heart, liver, and intestinal morphology, and death, depending on compound, in adult leopard frogs (Kaplan and Overpeck, 1964). Tadpoles exposed to the organophosphorus compound phenyl saliginen cyclic phosphate (PSCP) had high levels of mortality at concentrations > 500 ppb; surviving larvae had many abnormalities, including abdominal edema, body blisters, and spinal curvature (Fulton and Chambers, 1985).

The now-banned insecticides endrin and toxaphene are both lethal to larval and subadult *Lithobates sphenocephalus*. However, these pesticides had no effects on eggs, and subadults were less affected by toxicity than larvae (Hall and Swineford, 1980). Toxaphene, in particular, had very severe effects, including delayed mortality, abnormal behavior, and decreased growth rates of survivors, effects that were dose-related. $LC_{50\ (96\ hr)}$ values were 130 ppb (range 95 to 184) for toxaphene and 9 ppb (range 6 to 14) for endrin (Hall and Swineford, 1981).

Residues of the insecticide fenvalerate could not be detected in tissues from Southern Leopard Frogs one week after application (Bennett et al., 1983). Esfenvalerate has an $LC_{50\ (96\ hr)}$ of 7.79 µg/L, and causes larval convulsions in concentrations as low as 3.6 µg/L. Even at 1.3 µg/L, activity is decreased (Materna et al., 1995). It should be noted, however, that Materna et al. (1995) did not differentiate between *L. blairi*, *L. pipiens*, and *L. sphenocephalus* in presenting results. Temperature magnifies the effect of the pesticide, and additional effects on frog populations may result from the pesticide's lethal effects on fish and invertebrates.

Pesticides may have interactive effects with UVB radiation that actually benefits some species of tadpoles. For example, the insecticide carbaryl (5.0 mg/L) had no effect on survival or length of the larval period of Southern Leopard Frogs raised in controlled mesocosms. However, mass at metamorphosis was significantly greater in carbaryl treatments, an effect that was enhanced by increasing levels of UVB light. In carbaryl-treated mesocosms, chlorophyll levels were much higher than in untreated mesocosms, thus providing the larvae with increased food resources. Because larvae were in deeper water than adverse UVB levels would penetrate, and because larvae had a leaf-litter substrate in which to hide, UVB radiation had no effect on larval development. Carbaryl also breaks down rapidly after killing zooplankton, so its effects are immediate but long-term, allowing increases in phytoplankton on which tadpoles feed (Bridges and Boone, 2003).

However, photo-enhanced toxicity has been demonstrated to affect *L. sphenocephalus* larvae when the larvae are exposed to petroleum hydrocarbons. In the absence of enhanced solar radiation, there were no differences between controls and various water-soluble fractions. However, when solar radiation was 17µW/cm^2, UVB, photo-enhanced toxicity was evident with 100% tadpole mortality (Little et al., 2000).

In addition to pesticides, chemicals used to suppress wildfires may have detrimental effects on

Southern Leopard Frogs, particularly when UVB radiation is increased. Calfee and Little (2003) found that mortality increased when tadpoles were raised in separate trials with four different fire retardants (Fire-Trol® GTS-R, Fire-Trol 300-F, Fire-Trol LCA-R, Fire-Trol LCA-F) in the presence of UV radiation and YPS (yellow prussiate of soda, sodium ferrocyanide). YPS is a corrosion inhibitor mixed with the retardants to minimize damage during storage, transport, and delivery systems. However, it is known to be toxic to aquatic organisms in the presence of UV radiation. The chemicals were less toxic in the absence of YPS. In these formulations, the presence of un-ionized ammonia and cyanide could be cause for concern. Photoenhanced toxicity was not observed in trials with Phos-Chek® D75R and Phos-Chek D75F, since they do not contain YPS. Still, they are toxic to Southern Leopard Frog larvae. Toxic effects may persist for several weeks following application, particularly in GTS-R and D75R.

Finally, the piscicide rotenone has an $LC_{50\ (96\ hr)}$ of 0.5 mg/L (range 0.42 to 0.59 mg/L) when administered as Noxfish (rotenone in acetone) (Chandler and Marking, 1982).

STATUS AND CONSERVATION

Southern Leopard Frogs apparently are doing well at most locations. Based on call surveys between 1986 and 1989, for example, Florey and Mullin (2005) suggested populations were increasing in Illinois. At the northern limit of their range in New York State, however, they are considered a species of "Special Concern" (Gibbs et al., 2007).

Because they inhabit a variety of wetlands and environmental conditions, and because they tolerate the presence of fish, they have been considered relatively unimpacted by the effects of urbanization (Delis, 1993; Delis et al., 1996). They even occur in wellfield areas where the hydrology has been significantly altered (Means and Franz, 2005), although calling intensity appears to be less in those areas than in unaffected areas (Guzy et al., 2006). Still, Southern Leopard Frog populations are adversely impacted by both destruction of wetlands and adjacent uplands. In south Florida, for example, decreases in calling intensity between 2000 and 2004 were thought to be linked with increasing urbanization (Pieterson et al., 2006). One possible management option to benefit this species is to construct artificial ponds, which leopard frogs readily colonize (Shulse et al., 2010).

Southern Leopard Frogs can also be affected by natural perturbations, such as flooding, when rising waters may force them away from protected microhabitats and into adverse environmental conditions (Tucker et al., 1995). Drought, which has occurred throughout the Southeast with increasing frequency, may be responsible for declines observed at long-term study sites (Daszak et al., 2005). Disturbances may have significant impacts on populations of Southern Leopard Frogs. Following Hurricanes Ivan and Katrina in Louisiana, Southern Leopard Frog numbers plummeted in marsh and levee habitats, presumably due to saltwater intrusion (Schriever et al., 2009).

Lithobates sphenocephalus is frequently monitored during road-based call surveys along a prescribed survey route (Florey and Mullin, 2005; Burton et al., 2006; Lichtenberg et al., 2006; Pieterson et al., 2006). As such, it is important to note that calling may peak later than when most call surveys are carried out. Hence, estimates of frequency or abundance based on early evening call surveys likely are temporally biased. Stops of 5 min for listening appear to be adequate to detect this species (Burton et al., 2006).

Occupancy of wetlands varies seasonally in the Southern Leopard Frog, with most sites occupied in the autumn and only <20% occupied in spring and summer, at least in south Georgia. This result may be an artifact of sampling, however, as more than three sites per visit are required to verify occupancy by this species. In contrast, monitoring surveys report high detection probabilities in spring and summer, which tend to taper off in autumn (Smith et al., 2006). These results indicate that sampling for monitoring programs should be conducted during the early part of the activity season.

Significant road mortality probably occurs at all wetlands harboring adjacent Southern Leopard Frog populations, although such mortality has been documented only on a limited basis (Smith and Dodd, 2003; Smith et al., 2005). Across Paynes Prairie State Preserve south of Gainesville, Florida, U.S. Highway 441 was reported to become slick with leopard frog carcasses after heavy rains, creating a highway hazard to motorists. On a single night, more than 10,000 *L. sphenocephalus* were reported killed on the 3 km stretch of highway (Smith et al., 2005). Construction of a barrier wall and underpass culvert system significantly reduced anuran mortality on this highway (Dodd et al., 2004).

Southern Leopard Frogs readily colonize restored or created wetlands, and they are among the first species to arrive and breed successfully (Sexton and Phillips, 1986; Briggler, 1998; Merovich and Howard, 2000; Mierzwa, 2000; Pechmann et al., 2001; Palis, 2007; Shulse et al., 2010). If American Bullfrogs are present, however, Southern Leopard Frogs are likely to be much less abundant than in the absence of this predator and competitor (Briggler, 1998). Southern Leopard Frogs are frequently found in open habitats created during forestry operations, such as in canopy gaps and along skidder trails (Cromer et al., 2002). Such areas may contain high insect abundance. They rapidly colonize isolated wetlands surrounded by various silvicultural management treatments, including both clearcuts and clearcuts with mechanical site preparation, as long as source populations remain in close proximity (Enge and Marion, 1986; McLeod, 1995; Clawson et al., 1997; Russell et al., 2002a, 2002b). It should be noted, however, that the wetlands studied by Russell et al. (2002b) were each treated simultaneously, such that each treatment surrounded each pond. In addition, the wetlands themselves were not disturbed. As such, it is not surprising that numbers of Southern Leopard Frogs did not decrease after 2 yrs, since readily suitable habitat surrounded each pond throughout the study. Such a staggered treatment protocol could ensure the persistence of this and other amphibian species. Young of the year are also more frequently found in adjacent intact hardwood forests than in cutover hardwoods, even six to eight years after the cutting (McLeod and Gates, 1998). Vickers et al. (1985) noted they were the most abundant amphibian in both ditched and unditched cypress ponds in pine flatwoods in Florida.

Prescribed burns may impact Southern Leopard Frogs, particularly if small temporary wetlands are burned when frogs do not have access to aquatic refugia. For example, in Maryland few frogs were found on previously burned plots, even a year or more after a prescribed burn (McLeod, 1995; McLeod and Gates, 1998). In contrast, in Mississippi, Southern Leopard Frogs were found commonly on a previously burned site a year after burning (Langford et al., 2007).

Southern Leopard Frogs have been widely used for food, laboratory experiments, and for dissection in countless school biology classrooms. A total of 2,494 Southern Leopard Frogs were collected in Florida between 1990 and 1994 for the pet trade (Enge, 2005a).

COMMENT

Although this is one of the most common and widespread species in the southern United States, much information (age at maturity, longevity, sex ratios, population size-class structure, distance traveled to breeding sites, growth rates, survivorship, clutch size variation, and abundance in different habitats, to name just a few) remains unknown. Literature information is often based on *L. pipiens*, and assumptions that *L. sphenocephalus* have the same life history characteristics as *L. pipiens* are not valid.

Lithobates sylvaticus (LeConte, 1825)
Wood Frog
Grenouille des bois

ETYMOLOGY
sylvaticus: from the Latin *sylvaticus* meaning 'amidst the trees.' The common name and scientific name both refer to its forested habitat.

NOMENCLATURE
Conant and Collins (1998) and Stebbins (2003): *Rana sylvatica*

Dubois (2006): *Lithobates (Aquarana) sylvaticus*

Synonyms: *Rana cantabrigensis, Rana maslini, Rana pennsylvanica, Rana sylvatica cantabrigensis, Rana sylvatica cherokiana, Rana sylvatica latiremis*

IDENTIFICATION

Adults. Wood Frogs are medium-sized tan to reddish-brown to dark brown unspotted frogs with a distinctive dark mask through the eyes. Dorsal coloration is uniform and unblotched. The mask extends from the snout to the top of the forelegs. There is a white stripe on the posterior portion of the upper lip from the eye rearward to the angle of the jaw. Dorsolateral folds are present from just behind the eyes nearly to the posterior of the body; in eastern populations, these match the ground coloration. Two to four dark bars may or may not be present on the upper surfaces of the rear legs. The belly is white to grayish white, sometimes tinged with yellow or a green. Males have a swollen thumb for clasping females during the breeding season (Wright, 1914). Vocal sacs are paired.

In northern populations, Wood Frogs have a lighter-colored indistinct dorsolateral fold, a median light stripe bordered by a dark streak down the center-dorsal body from the tip of the snout to the end of the urostyle, dark blotches or spotting dorsally, and a light line along the posterior faces of the thigh and leg. Wood Frogs from the northern portions of the range generally have smaller limbs than those from the southern portion of the range (Schmidt, 1938; Suzuki, 1957); the legs are not as long and slender as in eastern *Lithobates sylvaticus*. The belly may be faintly mottled. These frogs were previously described as *Rana cantabrigensis* (Baird, 1854) or as *R. c. latiremis* from Alaska (Cope, 1886), but in fact there is a gradation in morphology and coloration between eastern and northwestern Wood Frogs across the continent. Comprehensive descriptions are in Howe (1899), Wright and Wright (1949), and Martof and Humphries (1959).

Maximum mean body size varies geographically, with the largest frogs (> 50 mm body size; largest ca. 82.5 mm) coming from the Southern Appalachians (North Carolina, Georgia) and the eastern United States, and the smallest adults (< 40 mm) from Michigan's upper and lower peninsulas, northeastern Wisconsin, and the northern regions of the Canadian prairie provinces (Alberta, Manitoba, Saskatchewan, as well as southern Northern Territories). There is no clinal gradient in size, however, as adults approaching 50 mm are found in east-central Alaska, the Rocky Mountains, Minnesota, and from Missouri to southern Labrador (Martof and Humphries, 1959). These populations are interspersed by populations whose adults range from 40 to 50 mm. Martof and Humphries (1959) provide a map showing size distribution across North America.

The body coloration may change seasonally, with frogs emerging from overwintering sites much darker and sometimes nearly black (Richards and Nace, 1983) than frogs emerging later in the season. Howard (1981) suggests that males entering ponds become darker in order to match the dark breeding pond water, and that the initially lighter-colored females darken to match the males when amplexed. Dark frogs also lighten when exposed to daylight, often in only a few hours (Allen, 1868). The ability to change color is not sex-dependent.

Wood Frogs in north-central populations have a mid-dorsal white stripe, extending from snout to vent, which is normally lacking in those of the eastern United States and Canada (Browder et al., 1966; Fishbeck and Underhill, 1971). The presence of this heritable (Browder et al., 1966), polymorphic stripe led in part to the description of *R. sylvaticus cantabrigensis*, but Martof and Humphries (1959) found no basis for subspecific recognition. The frequency of the stripe ranges from 0% of populations in the South, to 20% on the edge of the Canadian prairies, 20–30% in northern boreal forests, 50% in the forest-grassland ecotone in southern Manitoba, and >80% along Hudson Bay (Schueler and Cook, 1980). These latter authors associate the lack of the stripe with transformation in marginal habitats (that is, those that are subject to desiccation), whereas the striped morph is more prevalent in ecotonal situations, at least in the far North.

In Minnesota, stripe frequency increased from 0% in the Southeast to 32% in the Northwest (Fishbeck and Underhill, 1971). These authors found no correlation with habitat type, but suggested it might aid in crypsis in coniferous forests and while breeding in emergent vegetation. They also noted that striped frogs were found in areas considered stripeless by Martof and Humphries (1959).

Wood Frogs are sexually size-dimorphic. Adult males (37–55 mm SUL) are smaller than females (46–60 mm) from New Hampshire, Connecticut, and Wisconsin; males to 67 mm and females to

82 mm in North Carolina (Witschi, 1953); males 31–43 mm SUL in northern Ontario with females 31–51 mm SUL (Schueler, 1973); mean SULs in Québec 43.6 mm for males and 48.8 mm for females (Sagor et al., 1998); Nova Scotia males 38–53 mm (mean 46.9 mm) and 48–60 mm for females (mean 53.6 mm) (Gilhen, 1984); Michigan males with mean SULs 38–39 mm and females 43–44 mm SUL (Howard, 1981); Alabama males with mean SULs of 50 mm and females 60 mm (Davis and Folkerts, 1986).

In Maryland and Virginia, lowland females and males are smaller than their mountain counterparts, but males are smaller than females in both groups (lowland ♂♂ mean of 42 mm vs. 55 mm in mountains; lowland ♀♀ mean of 48 mm vs. 64 in mountains; Berven, 1982a). There also are certain color differences between the sexes (see Predation and Defense). Schueler (1973) reported that in northern Ontario females are redder on the dorsum and legs than males.

Larvae. The tadpoles are medium-sized and usually dark gray to brown. The dorsal tail fin terminates at or just anterior to the spiracle, and there are no stripes on either the dorsal fin or tail musculature. They have a faint white, cream, or gold stripe along the upper jaw, giving the impression of a mustache. The venter is light and may be slightly pigmented along the sides, but the gut is not usually visible. There is a considerable amount of slight color and pattern variation, but this is often the only dark tadpole of its size in the spring and early summer. Albino tadpoles have been reported from Wisconsin (Luce and Moriarty, 1999), and a leucistic tadpole from Virginia (Mitchell and White, 2005). Descriptions and illustrations of larval mouthparts are in Hinckley (1882) and Dodd (2004).

Eggs. The bicolored eggs (black above, white below) are deposited in a firm jelly cluster, often bluish to milky white. The eggs are 1.5–3.3 mm in diameter (Moore, 1949a; Wright and Wright, 1949; Herreid and Kinney, 1967; Meeks and Nagel, 1973; Trauth et al., 1989; Camp et al., 1990), with the inner jelly envelope surrounding the egg indistinct and measuring 3.0 to 7.2 mm. The outer jelly envelope is 4.3–17.3 mm and distinct (Livezey and Wright, 1947; Davis, 1980). The largest envelopes are found in Alabama (Davis, 1980). Egg size varies with latitude. Smaller-diameter eggs and associated jelly envelopes are observed in more northern populations (Witschi, 1953; Davis, 1980; Davis and Folkerts, 1986; Camp et al., 1990; Riha and Berven, 1991), for example, 1.6 mm in Alaska (Herreid and Kinney, 1967); 2.25 mm in Tennessee (Meeks and Nagel, 1973); 2.8 mm (range 2.2 to 3.3 mm) in Georgia (Camp et al., 1990); 2.9 mm in Alabama (Davis and Folkerts, 1986). The egg mass measures 38 to 100 mm in diameter and may be as large as 152 mm in length. It assumes its final size and globular shape within 2–3 hrs of deposition (Wright, 1914). Eggs can hatch in as few as five to nine days after oviposition (Wright, 1914; Davis and Folkerts, 1986; Redmer, 2002). Illustrations of eggs and developing embryos (to 11 mm, 164 hrs postfertilization) are in Pollister and Moore (1937).

DISTRIBUTION

Wood Frogs have the most extensive range of any North American anuran (>10.5 million km^2; Martof and Humphries, 1959). Their range is essentially continuous from the Southern Appalachian Mountains of northern Georgia and South Carolina north to northern Québec and Labrador, around Hudson and James Bays, and then northwestward across Ontario, the northern portions of the Canadian prairie provinces, the Yukon, and all the way to the Bering Sea in western Alaska. They occur from the Arctic Ocean north of the Brooks Range to the coastal plain of Virginia along Chesapeake Bay, with isolated populations in Hyde and Tyrrell counties, North Carolina (Dorcas et al., 2007). Isolated populations also occur in Alabama, Georgia (the most southern population in North America is located in Upson County, Georgia, in the lower Piedmont), Arkansas, Missouri, North Dakota, and in the Rocky Mountains of Colorado, Wyoming, and Idaho. Werner et al. (2004) considered records from Montana of questionable validity. The species occurs from near sea level to 3,000 m; even in Denali National Park they are found at 985 m in elevation (Hokit and Brown, 2006).

Wood Frogs occur on some islands, such as those at the mouth of the St. Lawrence River (Fortin et al., 2004a), in James Bay (Hodge, 1976), the St. Lawrence, Georgian Bay, and Apostle archipelagos of the Great Lakes (Hecnar et al., 2002; Bowen and Beever, 2010), Cape Breton Island (Gilhen, 1984), Isle Royale (Ruthven, 1908),

Distribution of *Lithobates sylvaticus*

Prince Edward Island (Cook, 1967), and Long Island, New York (Overton, 1914). Wood Frogs were introduced into Newfoundland in the 1960s and persist in at least 19 sites (Buckle, 1971; Maunder, 1983; Campbell et al., 2004).

Important distributional references include: Alabama (Mount, 1975; Davis, 1980; Davis and Jones, 1982; Davis and Folkers, 1986), Alaska (Loomis and Jones, 1953; Hock, 1957; Hodge, 1976), Alberta (Harper, 1931; Salt, 1979; Russell and Bauer, 2000), Arkansas (Black, 1938; Black and Dellinger, 1938; Trauth et al., 2004), British Columbia (Carl, 1943; Carl and Hardy, 1943; Loomis and Jones, 1953; Mennell, 1997; Matsuda et al., 2006), Canada (Bleakney, 1954, 1958a), Colorado (Maslin, 1947; Hammerson, 1999; Muths et al., 2005), Connecticut (Klemens, 1993), Delaware (White and White, 2007), Georgia (Williamson and Moulis, 1994; Camp and Pyron, 2003; Graham et al., 2007; Jensen et al., 2008); Idaho (Dumas, 1957), Illinois (Schmidt, 1929; Schmidt and Necker, 1935; Smith, 1961; Redmer, 1998a, 2002), Indiana (Minton, 2001; Brodman, 2003), Kentucky (Barbour, 1971), Labrador (Maunder, 1997), Maine (Hunter et al., 1999), Manitoba (Preston, 1982), Maryland (Harris, 1969), Minnesota (Oldfield and Moriarty, 1994), Missouri (Hurter, 1911; Johnson, 2000; Daniel and Edmond, 2006), New Brunswick (McAlpine, 1997a), New Hampshire (Taylor, 1993), New York (Gibbs et al., 2007), North Carolina (Dorcas et al., 2007), North Dakota (Wheeler and Wheeler, 1966; Murphy, 1987; Hoberg and Gause, 1992; Cabarle, 2006), Northwest Territories (Preble, 1908; Loomis and Jones, 1953; Fournier, 1997, undated), Nova Scotia (Gilhen, 1984), Nunavut (Hodge, 1976; Harper, 1963), Ohio (Pfingsten, 1998; Davis and Menze, 2000), Ontario (Schueler, 1973), Pennsylvania (Hulse et al., 2001), Prince Edward Island (Cook, 1967), Québec (Trapido and Clausen, 1938; Harper, 1956; Bider and Matte, 1996), Saskatchewan (Nero and Cook, 1964; Cook, 1965), South Carolina (Quinby, 1954; Platt et al., 1999; Montanucci, 2006), South Dakota (Fischer, 1998; Ballinger et al., 2000), Tennessee (Redmond and Scott, 1996; Samoray and Regester, 2001; Corser, 2008), Vermont (Andrews, 2001), Virginia (Funderburg et al., 1974a; Tobey, 1985; Mitchell and Reay, 1999), West Virginia (Green and Pauley, 1987), Wisconsin (Suzuki, 1951; Vogt, 1981; Mossman et al., 1998), Wyoming (Dunlap, 1977), and the Yukon (Loomis and Jones, 1953; Mennell, 1997).

FOSSIL RECORD
Lithobates sylvaticus is known from the Pliocene of Nebraska and the Pleistocene of Indiana, Kansas, Maryland, Nebraska, Pennsylvania, Tennessee, and West Virginia (Holman, 2003). Most fossils are associated with cave deposits. The ilium of this species is diagnostic and easily recognized by the rounded tubercle on its dorsal prominence (Holman, 2003:183).

SYSTEMATICS AND GEOGRAPHIC VARIATION
Based on 2 kb from the mitochondrial genome, *Lithobates sylvaticus* is a member of the monophyletic *Novirana* clade of North American *Rana* and is most closely phylogenetically related to the *Lithobates catesbeianus* (*Aquarana*) group of eastern North America (including *L. catesbeianus*, *L. clamitans*, *L. grylio*, *L. heckscheri*, *L. okaloosae*, *L. septentrionalis*, *L. virgatipes*) (Hillis and Wilcox, 2005). This relationship is similar to that proposed by Case (1978) and Post and Uzzell (1981), in part, based on immunological comparisons. It differs from affinities proposed by Hillis and Davis (1986), based on nuclear ribosomal RNA restriction sites, and Post and Uzzell (1981), based in part on immunology, who suggested an affinity with the Eurasian *Rana temporaria* group. Experimental crosses between *R. pretiosa* or *R. aurora* and *Lithobates sylvaticus* are not viable (Dumas, 1966).

Several names have been applied to Wood Frogs: *Rana cantabrigensis* (Baird, 1854), *R. maslini* (Porter, 1969a), *R. pennsylvanica* (Harlan, 1826), *R. sylvatica cherokiana* (Witschi, 1953), and *R. sylvatica latiremis* (Schmidt, 1938). The nomenclatural history of Wood Frogs is discussed by Martof and Humphries (1959). Porter (1969a) described *R. maslini* from north-central Colorado and south-central Wyoming. Although clearly allied with *Lithobates sylvaticus*, *Rana maslini* was infertile with Wood Frogs from Manitoba, had minor morphological differences, and had a lower dominant frequency in its mating call (Porter, 1969a, 1969b). In contrast, Bagdonas and Pettus (1976) could find no evidence of genetic incompatibility, and suggested that *R. maslini* was not a valid species. Martof (1970) provides a synonymy.

Based on a suite of characters, Martof and Humphries (1959) recognized five morphotypes (Alaskan, Rocky Mountain, Labrador, Midwest, Appalachian) that do not correspond to historic nomenclature; they did not formally recognize these phenotypes taxonomically. No subspecies are currently recognized despite substantial morphological variation throughout the species' range (Martof and Humphries, 1959). Zeyl (1993) noted genetic differences between northeastern, north-central, and Missouri populations, and attributed them to isolation and recolonization during and after the Wisconsinan glacial period.

At least some Wood Frog populations have a distinct sexual dimorphism in coloration that persists long after capture. King and King (1991) showed that females reflected longer wavelengths of light than males on both the yellow-blue and red-green axes, but the significance of these observations is obscure since both sexes matched at least some types of surface leaves equally well. A variably brown body with a dark eye mask is a common phenotype among litter-dwelling ranid frogs worldwide, being observed in *R. arvalis*, *R. chensinensis*, *R. dalmatina*, *R. iberica*, *R. tsushimensis*, and *R. zhenhaiensis* among many others.

ADULT HABITAT
As their name implies, Wood Frogs are closely associated with closed-canopy deciduous and boreal forests of the eastern and northern portions of North America (Martof, 1970; Guerry and Hunter, 2002; Browne et al., 2009), where their presence and population status have been suggested as measures of forest health (Knutson et al., 2000). When hardwoods and conifers occur together, Wood Frogs usually are found more often in the hardwoods (Roberts and Lewin, 1979; DeGraaf and Rudis, 1990; Constible et al., 2001). They generally avoid open conifer and shrub habitats in Alberta (Browne et al., 2009), but in Alaska, Wood Frogs are found in shrubby tundra often far from the nearest boreal forest (Hokit and Brown, 2006). Occupancy is often quite high in south central Michigan and is related to wetland proximity, but it is difficult to predict which habitat variables are most important to the species (Roloff et al., 2011).

In a 2 yr study in New Hampshire, Wood Frogs were the most common amphibian captured

(1,200 of 2,080) at both upland and streamside locations in red maple (*Acer rubrum*), northern hardwood, and balsam fir (*Aibes balsamea*) forests (DeGraaf and Rudis, 1990). In a bog undergoing succession in Minnesota, Wood Frogs inhabited the tamarack (*Larix laricina*) and black spruce (*Picea mariana*) zones, and especially the balsam fir-ash (*Fraxinus nigra*) zone surrounding the bog. They appeared to occupy habitats exactly opposite of sympatric Northern Leopard Frogs (*Lithobates pipiens*), suggesting habitat partitioning (Marshall and Buell, 1955).

In Alberta, Wood Frogs are found most often along shallow-water lake margins with tall sedges, herbs, and shrubs, or with grass cover. They also prefer areas with willow (*Salix*), alder (*Alnus*), and currant and gooseberry (*Ribes* spp). Similar preferences have been reported elsewhere in Canada and Colorado (Schueler, 1973; Roberts and Lewin, 1979; Haynes and Aird, 1981). These habitats have moist soil conditions and deep deciduous leaf cover. Another favored forest litter type contains a great quantity of aspen (*Populus tremuloides*) leaves, a habitat with abundant carabid beetles, a favorite food source (Moore and Strickland, 1955). Constible et al. (2001) found a negative association between moss and lichen and Wood Frog abundance. Such habitat avoidance may be related to the thermal environment; temperatures are cooler under deciduous leaf litter than under lichens (Heatwole, 1961). The southernmost *Lithobates sylvaticus* populations occur in mesic semideciduous forests along the flood plains of major streams in east-central Alabama and Georgia.

TERRESTRIAL ECOLOGY

Wood Frogs are active both by day and night, depending on temperature, season, and latitude. In Minnesota, they have a biphasic diurnal activity pattern, being more active in mid-morning and late afternoon. In the morning, activity was inversely related to humidity, but not to air and soil temperatures (Bellis, 1962a, also see Shelford, 1913). Although Wood Frogs are forest inhabitants, they forage in more open areas at night or under overcast conditions (Heatwole, 1961). During daylight or when the weather clears, they return to the more concealed forest environment. Rainfall stimulates activity (Bellis, 1962a; Baldwin et al., 2006). In cold weather, snow cover may serve as important insulation for Wood Frogs overwintering in near-surface leaf litter, but an extended period of snow cover may actually delay the initiation of movement to the breeding ponds (Kessel, 1965).

A number of studies have investigated landscape-level habitat associations and the presence of Wood Frogs. For example, Wood Frog populations in northern and western New York are positively correlated with areas of higher elevation, few pastures, and a large amount of nearby deciduous and mixed forest and swamps, but they are negatively associated with evergreen forest (Gibbs et al., 2005). In Maine and Connecticut, the amount of forest cover is positively correlated with the probability of finding Wood Frogs breeding in associated wetlands (Gibbs, 1998b; Guerry and Hunter, 2002). Gibbs (1998b) found that Wood Frogs were present at all sites with >70% forest cover, 60–80% of sites with 40–60% cover, and at no sites with <30% cover. In Ontario, there also was a strong positive association between forest cover and the presence of Wood Frogs (Eigenbrod et al., 2008), especially when forest cover was near breeding ponds. The presence of roads is negatively correlated with Wood Frog wetland occupancy in Ontario (Findlay et al., 2001), and the area of immediate adverse impacts from roads (the "road-zone" effect) extends 600–1,100 m from the pavement (Eigenbrod et al., 2009).

Defining core habitat as a 200 m area around breeding ponds and a "broader landscape context" as extending 200–1,000 m, in Ohio, 58% of the core and 30% of the extended habitats were forested at sites containing Wood Frogs (Porej et al., 2004). These values were significantly higher than corresponding values for wetlands without Wood Frogs. In all, 43% of the 54 wetlands examined had Wood Frogs breeding at them (Porej et al., 2004). Wood Frogs were found even in dry *Pinus resinosa* forest surrounding a Minnesota bog (Marshall and Buell, 1955).

In terrestrial situations, Wood Frogs prefer moist soils (mean 100% moisture content), but are not affected by soil pH (Wyman, 1988). The amount of rainfall is strongly correlated with annual adult survival in terrestrial habitats (Berven, 1990). In forested wetlands in summer, they are found in sphagnum hummocks and riparian margins in moist leaf litter (Baldwin et al., 2006;

Okonkwo, 2011). During cold conditions, Wood Frogs shelter under logs, leaf litter, soil, and other surface debris (Allen, 1868; Wright, 1914; Wright and Wright, 1949; Licht, 1991). In Alaska, they have been reported from shallow depressions in the upper layer of dead vegetation (Hodge, 1976) and from under leaf litter in Michigan (Blanchard, 1933; Heatwole, 1961). They do not overwinter in water, although they are capable of surviving extended periods (ten days) under ice under laboratory conditions (Licht, 1991) and are sometimes seen under ice at breeding ponds. Under laboratory conditions, Lithobates sylvaticus also prefers land to water in which to overwinter, and selects moist soil in which to do so. They curl their body into a form in small depressions, and are able to right themselves even at temperatures of 0–2.5°C (Manis and Claussen, 1986; Licht, 1991).

In upland deciduous woodland in Missouri, overwintering male Wood Frogs were found at a ratio of 8:1 < 65 m from the nearest breeding pond, 4.9 times more likely than females to be close to the ponds (Regosin et al., 2003). Even at distances > 65 m, the sex ratio was male biased (1.6:1). Terrestrial densities varied from 0 to 6.3 Wood Frogs/100 m^2. The frogs began to appear at overwintering sites from early to mid-November. Wood Frogs did not use the area sampled from March and April until November, suggesting that overwintering and summer foraging habitats were different. Overwintering close to breeding ponds ensures that males will arrive first at the ponds and as soon as the weather permits (Zweifel, 1989; Regosin et al., 2003).

The possibility that Wood Frogs in Maine use different terrestrial habitats for foraging immediately after breeding also is suggested by Baldwin at al. (2006). Following 43 frogs using radio-telemetry, these authors found that dispersing Wood Frogs initially selected damp leaf litter retreats near breeding ponds in forested wetlands, which they preferred over forested uplands. As the warm season progressed, 8 frogs moved to closed-canopy habitats at a median of 169 m (range 102 to 340 m) from the breeding pond, occasionally stopping over 1–17 nights along the way. Other frogs moved to adjacent breeding pools, forested wetlands, or sphagnum-bordered forested streams. Summer habitats were shady, moist areas in sphagnum-dominated habitats. The frogs did not prefer the forested but presumably dryer terrestrial uplands. In Missouri, Wood Frogs spend the spring and summer primarily in stream drainages buried in deciduous leaf litter, with only the eyes visible and the pelvic patch firmly pressed against the substrate (Rittenhouse and Semlitsch, 2007a).

Wood Frogs survive extended periods of sub-freezing temperatures by using increased levels of blood glucose as a cryoprotectant (Schmid, 1982; Storey and Storey, 1986, 1987, 1989; Layne, 1995; Layne and Lee, 1987, 1995). Individuals can survive whole body freezing for 8 days at −2.5°C (Storey and Storey, 1987), and the ice crystallization temperature is not dependent on body mass (Layne, 1995). Glucose functions to decrease dehydration and cellular ice content (Storey et al., 1992) and may help to stabilize the protein structure of cellular walls. Although Wood Frogs can survive to −2°C on an unfrozen surface, they quickly freeze if they come into contact with ice crystals (Layne et al., 1990). Plasma glucose is highest in the autumn, and such frogs can survive 2 weeks at −5°C (Layne, 1995). The capacity for producing cryoprotectants decreases in the spring, as glycogen reserves are depleted or used for other functions (Storey and Storey, 1987; Wells and Bevier, 1997). Repeated freeze-thaw episodes deplete blood glucose content (Lee and Costanzo, 1993), suggesting that selection should favor frogs finding protected overwintering sites and becoming dormant as quickly as possible just prior or in response to the first cold weather of the season.

In addition to cryoprotectants, there are other morphological and physiological adaptations allowing Wood Frogs to survive extended periods of cold and freezing weather. Striated muscle function remains intact below freezing (Miller and Dehlinger, 1969), and cardiac activity is unaffected (Lotshaw, 1977). Furthermore, Wood Frogs have reduced rates of oxygen consumption at temperatures approaching freezing (Johansen, 1962). Thawing takes place over a period of several hours, but the heart may resume beating as soon as 1 hr after thawing begins (Layne and First, 1991).

Although Wood Frogs are inhabitants of colder climates, they are able to tolerate temperatures of 35–36°C as adults, 37–38.5°C as juveniles, and nearly 40°C as larvae, depending on acclimation temperature. Frogs acclimated at lower temperatures (10°C, 20°C) have lower tolerances than those acclimated at 30°C, regardless of life history stage (Cupp, 1980).

Wood Frogs are often abundant where they occur. In transect samples, Wood Frogs were the most abundant frogs observed at 17 lakes in northeastern Alberta, representing 75% and 92% of captures, respectively (Constible et al., 2001). No significant differences were found in the abundance of Wood Frogs between logged and naturally burned sites. However, Wood Frogs were more abundant at deciduous sites when compared with coniferous sites, regardless of disturbance. In Alaska, they were found at 45% of sites examined in Denali National Park, Alaska, and Hokit and Brown (2006) described their occupancy as saturating the landscape around Wonder Lake.

In a Minnesota peat bog, Wood Frogs have summer home ranges of about 65 m^2 in their terrestrial habitat, but there is considerable variability (Bellis, 1965). Movements are rather short over the course of a season (mean 11.2 m), and males and females move about the same distances. The extent of movements by smaller frogs (28–42 mm) also did not differ from distances moved by larger frogs (43–55 mm). Occasional adult frogs made long-distance movements, to 71.9 m from the farthest capture location, and juveniles were found as far as 100 m between capture locations. The juveniles appeared to be dispersing through the area. Most frogs captured from one year to the next were within 27 m of the first year's capture location, but some were as far as 88 m from the original capture point (Bellis, 1965). In Maine, frogs also made rather short movements during the summer (median 2.8 m, range 61 m), with longer movements occurring during rainfall (Baldwin et al., 2006).

Wood Frogs are often found at the entrances of caves (Black, 1938; Barr, 1953; Franz, 1967; Garton et al., 1993; Dodd et al., 2001; Prather and Briggler, 2001).

Wood Frogs are photopositive in their phototactic response, suggesting they use both sunlight and moonlight when feeding and going about their daily activities (Pearse, 1910; Jaeger and Hailman, 1973). They are sensitive to light in the blue spectrum ("blue-mode response"), which helps them orient toward areas of increasing illumination, such as the open area above a pond (Hailman and Jaeger, 1974). Wood Frogs likely have true color vision.

CALLING ACTIVITY AND MATE SELECTION

The call of the Wood Frog is an explosive series of clucks. Overton (1914) described it as if hearing a chorus of barnyard ducks clucking, not quacking. Martof (1970) likened it "as a rasping 'craw-aw-auk' or as a hoarse clacking call suggesting the quack of a duck." At breeding choruses, individuals may make 500–600 call notes/hr (Wells and Bevier, 1997). However, Wood Frogs also have a soft one- to three-syllable call that they make from terrestrial overwintering sites (Burroughs, 1913; Schueler, 1973; Zweifel, 1989). Zweifel (1989) heard the calls throughout the autumn and winter from isolated frogs buried beneath the surface litter. Calls were of short duration; on occasion from two to four frogs would call simultaneously. Gorham (1964), Waldman (1982b), and Davis and Folkerts (1986) also describe a "post-breeding" call after oviposition has ceased; Gorham (1964) heard calls as late as 3 December in New Brunswick.

Males call from the water surface as they float or swim around a pond. The front limbs hang down into the water, and the rear limbs are extended, with the digits expanded for maximum swimming capability. Males are very active at breeding ponds, and they interact with other males frequently (Wright, 1914; Noble and Farris, 1929; Wright and Wright, 1949; Davis and Folkerts, 1986). They are extremely wary and quickly dive into the leaf litter when approached or disturbed. In the South, frogs remain buried in the leaf litter by day and generally only call at night; calling occurs throughout the night, with most calling in the first few hours after dark (Davis and Folkerts, 1986; Bowers et al., 1998). However, diurnal calling activity increases with latitude, such that calling at the extreme northern portion of the range is mostly diurnal.

Wood Frogs are among the earliest arrivals at breeding ponds, often moving across snow and ice patches as they emigrate from terrestrial overwintering sites. Males arrive before females and space themselves at somewhat regular intervals on the pond surface (Howard, 1981; Seale, 1982). Calling not only attracts females, but the initial calls of males serve to attract other males to the location of the breeding pond (Bee, 2007b). When male densities are high (0.25 males/m^2), males move less frequently and tend to search for females at the periphery of the breeding aggregation (3.5 males/m^2); at low densities (0.16 males/m^2), males are much more active when searching for females, and they tend to concentrate their activity at the oviposition site (4.3 males/m^2) (Woolbright et al., 1990). Calling occurs from late winter through early spring, when air temperatures reach 5–10°C (Howard, 1981; Bowers et al., 1998), the exact dates depending on latitude, elevation, and weather conditions. In the Mid-Atlantic region, the initiation of breeding spanned six weeks (Berven, 1982b); similar variation in initial breeding date likely occurs at most locations.

Some call dates include early January to early May in the Southern Appalachians (Huheey and Stupka, 1967; Dodd, 2004), late February to early April on Maryland's coastal plain (Smithberger and Swarth, 1993), late March to mid-April at Ithaca, New York (Wright, 1914), early May in Colorado (Haynes and Aird, 1981), mid-April to late May (peak late April to early May) in Québec (Lepage et al., 1997), late May to early June along the south shore of Hudson Bay (Schueler, 1973), and as late as mid-June in Alaska (Herreid and Kinney, 1967).

Wood Frogs do not have lipid reserves at the start of the spring breeding season, apparently using them during overwintering. They do not feed during the breeding season, instead using stored carbohydrates to fuel calling and other reproductive activity. Muscle glycogen reserves are depleted after an evening of calling. The large glycogen reserves in trunk musculature that they have at the beginning of the breeding season (14–22 mg/g) are largely used up (down to 6.4 mg/g) by the end of the breeding period (Wells and Bervier, 1997). These reserves are quickly refilled as frogs begin to feed.

Although early reports suggested a chemoreceptive base for mate selection (Banta, 1914), there is no experimental evidence for it (Cummins, 1920; Noble and Farris, 1929). There appears to be no special recognition prior to amplexus, with males searching for and approaching virtually any other frog in the vicinity as a potential mate. Multiple males may try to mate with a single female (Wells, 1977a; Howard, 1981). Mate recognition only occurs after contact, and movement is necessary to attract a male's attention. Noble and Farris (1929) suggest that the firmness and girth of the amplexed animal provide an ardent male with information on sex and when to release a spent female after egg deposition; males and spent females are smaller in girth and are not as firm as females carrying a full complement of eggs. Voice (the "warning croak" of toads, for example) does not play a part in sex recognition.

Mating occurs both by day and night. Amplexus normally occurs in water (Banta, 1914; Noble and Farris, 1929; Berven, 1981), but it also may occur terrestrially as noncalling males intercept females on their way to breeding ponds (Schueler and Rankin, 1982). Frogs remain in amplexus for short periods (1.6–84.9 hrs, mean 23.1 hrs) (Howard, 1981), although Wright (1914) recorded one pair remaining in amplexus 4 days. Early arrivals tend to remain in amplexus longer than late-arrivals. An amplexed pair may move up to 40 m in order to find a suitable oviposition site (Seale, 1982).

Sexual selection is an important component of Wood Frog reproductive biology, and mating is nonrandom. Since larger females carry more eggs (as many as five times more) than small females, males tend to mate with larger females (Howard and Kluge, 1985). Size is also important for males, and large male size is correlated with mating success, since larger males have a better chance of holding onto large females and completing fertilization than small males (Howard,

Amplexing *Lithobates sylvaticus* with egg mass. Photo: Mike Graziano

1980; Berven, 1981; Howard and Kluge, 1985). Larger males enter the breeding ponds ahead of smaller males, thus enhancing their opportunity for mate choice. Still, approximately 61% of the males at breeding ponds in Michigan did not mate, suggesting strong selection for both size and early arrival.

Females are capable of dislodging amplexing males should they choose (Seale, 1982), indicating some degree of female choice. However, as many as 10–15 males may fight for access to a single female, and Howard (1981) observed 3 females that had been killed by aggressive male suitors prior to oviposition. Males have been observed amplexing dead females that had not oviposited, perhaps as a result of male aggression (Trauth et al., 2000). Males rarely mate with more than one female (Howard, 1981). Amplexus has been observed between *L. sylvaticus* and *L. sphenocephalus*, but development after initial cleavage does not occur (Nelson, 1971; Davis and Folkerts, 1986; Redmer, 2002). Males even have been reported attempting to amplex Spotted Salamanders (*Ambystoma maculatum*) (Kapfer and Muehlfeld, 2006).

BREEDING SITES

Wood Frogs breed in shallow water (< 110 cm), often in only a few cm. Females generally select the warmer portions of a breeding pond in which to oviposit, rather than in adjacent cooler areas (Seale, 1982). Breeding occurs in many types of wetlands, from small shallow temporary ditches, road ruts, woodland pools, and depressions to pools, ponds, bogs, fens, lakes, potholes, or along the margins of beaver (*Castor canadensis*) ponds (Porter, 1969a; Davis and Folkerts, 1986; Camp et al., 1990; Karns, 1983, 1992; Redmer, 2002; Dodd, 2004; Hokit and Brown, 2006; Barry et al., 2008; Karraker and Gibbs, 2009; Okonkwo, 2011). They also use man-made wetlands such as stormwater and other artificial urban ponds (Snodgrass et al., 2008; Simon et al., 2009; Brand and Snodgrass, 2010; Shulse et al., 2010), and can be found in agriculturally disturbed landscapes as long as appropriate habitat is nearby (Knutson et al., 2004). In the Great Smoky Mountains, they use shallow ruts in formerly grazed pasture, roadside borrow pits, and even old wells (Dodd, 2004). Adam and Lacki (1993) reported breeding in road-rut ponds, and Mifsud and Mifsud (2008) found them to use golf course ponds.

In New Hampshire, Wood Frogs rarely are found in breeding ponds with <60% forest cover surrounding them (Herrmann et al., 2005), although the amount of forest cover and road density at 250–2,000 m had no effect on larval density. In a study of 124 breeding ponds in western Rhode Island, hydroperiod was a critical factor in determining the subsequent breeding success of Wood Frogs, and in New Hampshire, Wood Frogs dominated the amphibian communities in wetlands with short- and intermediate-level hydroperiods (Babbitt et al., 2003; Baber et al., 2004). In Alaska, they breed in open-canopied tundra as long as shrubby vegetation is present, and they may be found > 1 km from the nearest boreal forest (Hokit and Brown, 2006). Breeding sites usually had emergent vegetation and woody vegetation (spruce and alder) in riparian zones. Alaskan Wood Frogs tended to avoid breeding in deep water sites, sites with permanent stream connections, sites where peat was predominant, and sites with beaver activity.

Wood Frogs normally breed in wetlands that dry annually or semiannually and that usually lack fish (Collins and Wilbur, 1979; Davis and Folkerts, 1986; Hecnar, 1997; Skelly et al., 1999; Babbitt et al., 2003; Baber et al., 2004; Egan and Paton, 2004; Petranka et al., 2004; Porej et al., 2004; Herrmann et al., 2005; Skidds et al., 2007). For example, egg masses were abundant in Rhode Island breeding pools that held water 6–11 months and that dried between 1 August and 30 November, whereas breeding populations were reduced in ponds that did not dry over a 2 yr period (Egan and Paton, 2004). Breeding only rarely occurs in permanent wetlands (Werner and Glennemeier, 1999), except in Labrador (Murray, 1990).

The size of the pond seems to have mixed influences on its choice as a breeding site. Ponds < 0.2 ha tended to have similar numbers of egg masses in Rhode Island as ponds > 0.2 ha. In Rhode Island, Wood Frogs often used ponds < 0.05 ha (29 of 41 ponds < 0.05 ha were occupied) for breeding. There does not appear to be a correlation between the number of egg masses deposited and the size of the breeding pool for wetlands > 0.15 ha (Egan and Paton, 2004). However, in Massachusetts, Skidds et al. (2007) found a positive correlation between breeding pond size and egg mass counts for ponds < 0.6 ha, although the correlation was not strong among larger-sized ponds. Breeding pond use is positively correlated with invertebrate abundance and richness (Babbitt et al., 2003).

Most of the ponds within an area may be used for breeding, but the number changes annually. Some examples include: 69% of the 124 ponds were used during a 2 yr study in Rhode Island (Egan and Paton, 2004); 88% of 65 ponds examined between 2001 and 2005 in Massachusetts (Skidds et al., 2007); Roberts and Lewin (1979) found breeding at 14 of 24 sites, although Wood Frogs were present at 24 of 25 ponds visited over 2 yrs; 20 of 42 wetlands were inhabited over a 2 yr period in New Hampshire (Babbitt et al., 2003); in Maine, 95 of 116 ponds visited had Wood Frogs, again over a 2 yr survey (Guerry and Hunter, 2002); in Alberta, 20 of 24 ponds had Wood Frog egg masses (Stevens and Paszkowski, 2004); 10 of 16 ponds monitored for 2 yrs in Arkansas had frogs and/or egg masses (Trauth et al., 1989). In Colorado, however, only 34 of 145 surveyed ponds contained Wood Frog adults, larvae, or eggs; all were between 2,400–3,000 m in elevation (Haynes and Aird, 1981).

Egg mass number is positively correlated with pond volume and DOC (Dissolved Oxygen Content), but at Pennsylvania breeding sites, there are no correlations involving a variety of metals, cations, anions, pH, alkalinity, or conductivity (Rowe and Dunson, 1993). Based on a study of 15 ponds in Québec, pH ranged from 3.7 to 6.7, and pH and the amount of organic carbon were negatively associated with the density of egg masses (Gascon and Planas, 1986). In Ontario, Hecnar and M'Closkey (1996b) found a negative association with conductivity at wetlands. However, in Nova Scotia, Dale et al. (1985) found no combination of habitat and water quality variables that could discriminate between presence and absence of Wood Frogs.

Females attach egg masses mostly to live woody vegetation, dead branches, woody edge species, and emergent vegetation in shallow water (mean 45 cm) near the water surface (Egan and Paton, 2004); occasionally, egg masses may be free floating (Morgan, 2010). Greater shallow-water vegetation complexity is usually associated with egg deposition. For example, in Colorado, Wood Frogs deposit egg masses along the shallow northern shores of ponds in areas associated with emergent vegetation, particularly sedges (*Carex* sp.) (Haynes and Aird, 1981). In Massachusetts, egg mass counts were positively correlated with persistent nonwoody plant cover (Skidds et al., 2007), and in Ohio, more egg masses were found in slightly deeper water with emergent vegetation than in very shallow water with less cover (Brindle et al., 2009). Negative influences on egg mass abundance include aquatic bed vegetation, unvegetated areas, fish, and the presence of inlet streams (Egan and Paton, 2004). The presence of fish may deter Wood Frog breeding, but not eliminate it (Haynes and Aird, 1981; Hecnar, 1997; Redmer, 2002). However, a reduction occurs in the number of egg masses deposited where fish are present compared to ponds without fish, and in Rhode Island, Wood Frogs bred in only 3 of 20 ponds that contained fish (Egan and Paton, 2004).

Wood Frogs breed in both open- and closed-canopy wetlands (Redmer, 2002), with optimum conditions at 30–70% canopy cover (Skidds et al., 2007). Canopy closure does not affect larval survivorship (Werner and Glennemeier, 1999), presence (Skelly et al., 1999), or extirpation (Skelly, 2001) per se, but it does affect larval growth rates (Werner and Glennemeier, 1999; Skelly et al., 2002; Skelly, 2004). Wood Frogs in open-canopy (<25% coverage) ponds grew 90% faster than those in closed-canopy (>75% coverage) ponds (Skelly et al., 2002), but in Michigan, Wood Frogs also grew and survived well in closed-canopy situations (Werner and Glennemeir, 1999). These results suggest that frogs breeding in closed-canopy situations, where pond drying is likely, may be at a disadvantage compared to those breeding in open-canopied breeding ponds with greater amounts of phytoplankton. Slow-growing larvae might not survive long enough to complete metamorphosis in drought years in ponds under closed canopies with less abundant food (Skelly, 2004).

Recently metamorphosed Wood Frogs depart nonrandomly toward closed-canopy forest during emigration (deMaynadier and Hunter, 1999). Wood Frog juvenile recruitment appears highest in older (>25 yrs) compared with newly formed

(<10 yrs) beaver ponds in the boreal forests of Alberta (Stevens et al., 2006a). Such ponds are associated with reduced riparian canopy cover and greater abundance of submergent vegetation, a warmer thermal environment, and a higher DOC, which promotes larval growth and development. In contrast, Egan (2001) found that in Rhode Island, Wood Frogs were more likely to occur in habitats with greater canopy cover in deciduous forests and that they continue to breed in closed-canopy forests, especially when no alternatives are available. It seems likely that canopy cover, hydroperiod, and wetland productivity interact in complex ways, depending on environmental conditions, to produce a mosaic of optimal and less than optimal breeding sites.

REPRODUCTION

Wood Frogs arrive at ponds often when ice still covers much of the water surface, and snow persists in the surrounding terrestrial habitat (Allen, 1868). In Alaska, ten or more days with mean temperatures above freezing are necessary before frogs move to ponds (Kessel, 1965).

In Michigan, migration takes place over a period of 33 days and occurs during high humidity (Cummins, 1920), although rainfall is not necessary to stimulate migration. At other locations, heavy late winter rainfall is correlated with the onset of movements to the breeding ponds (Davis and Folkerts, 1986; Trauth et al., 1989). Variation in the timing of breeding migrations led Crouch and Paton (2000) to recommend sampling over at least a three-week period when doing egg mass surveys in Rhode Island.

Wood Frog breeding dates vary with latitude and elevation: January to March in the Southern Appalachians (Huheey and Stupka, 1967; Meeks and Nagel, 1973; Dodd, 2004; Petranka et al., 2004); mid-January to February in Alabama (Davis, 1980; Davis and Folkerts, 1986); February in Georgia (Camp et al., 1990); early February to early March in Arkansas (Trauth et al., 1989); mid-February to mid-March in the mountains of Virginia, the Coastal Plain of Maryland, and in Illinois (Berven, 1982a; Redmer, 2002); mid-February in West Virginia (Wilson, 1945); late March to early April in New Jersey (Seigel, 1983); late March through mid-April at Ithaca, New York (Wright, 1914); late February to late March in Missouri (Guttman et al., 1991); early April in southern Ontario (Logier, 1952), but late March to late April near Toronto (Piersol, 1913); mid-April to early May in New Hampshire (Oliver and Bailey, 1939) and Alberta (Salt, 1979); late April in Nova Scotia (Bleakney, 1952); late April to late May in Alaska (Kessel, 1965; Herreid and Kinney, 1967); early May in Newfoundland (Campbell et al., 2004); May in British Columbia (Cowan, 1939); mid-May to mid-June in Colorado (Corn et al., 2000); late May to July in Québec (Hildebrand, 1949; Harper, 1956); and April to June in the Northwest Territories (Preble, 1908). In Rhode Island, immigration occurred from 9 March to 3 April (most on 10 March), and adults emigrated from 9 March to 3 April (most on 2 April) (Paton et al., 2000). The mean date of breeding initiation increases 5.2 days per degree of latitude (Guttman et al., 1991). Populations within a region may not breed at exactly the same time (Petranka et al., 2004).

Lithobates sylvaticus is an explosive breeder (Banta, 1914; Cowan, 1939; Kessel, 1965; Meeks and Nagel, 1973; Schueler, 1973; Howard, 1980; Seale, 1982; Waldman, 1982b; Seigel, 1983; Davis and Folkerts, 1986; Guttman et al., 1991), with most reproduction occurring over a short period. Petranka et al. (2004) reported that 90% of the populations they studied in the Southern Appalachians bred within 1 week. In Tennessee, all breeding occurred within a three-day period in late February, and all adults left the pond area within eight days (Meeks and Nagel, 1973). Likewise in Rhode Island, most eggs were deposited in a relatively short period (<8 days), with a mean stay of 17.4 days (Crouch and Paton, 2000). Egg deposition took five days in Michigan (Howard, 1981) and two to six nights in Illinois (Redmer, 2002). Explosive (or synchronized) breeding allows all tadpoles within a breeding site to develop at about the same rate and size, thus avoiding cannibalism. When tadpoles are about the same size, Wood Frog tadpoles are not cannibalistic. When large size discrepancies occur, *L. sylvaticus* larvae become cannibalistic on both conspecific eggs and smaller larvae (Petranka and Thomas, 1995).

As expected with an explosive breeder, a great many Wood Frogs congregate at a breeding pond simultaneously. Cowan (1939) recorded several aggregations of from 20 to 60 individuals in areas of not more than 10 ft^2 of water surface. At a

Tadpole of *Lithobates sylvaticus*. Photo: David Dennis

pond in Rhode Island, Paton et al. (2000) monitored the movements of Wood Frogs at a 0.16 ha pond for 7 months, recording a population estimate of 753 (316 ♀, 437 ♂). Also in Rhode Island, Doty (1978) estimated 13–95 frogs at a 0.33 ha pond. In southeastern New Hampshire, Wood Frog density at 20 breeding ponds was 0.48 ± 0.2/m^2 (Babbitt et al., 2003). In Alaska, most egg masses are deposited in the late afternoon and evening (Herreid and Kinney, 1967).

At Alaskan breeding ponds, oviposition took 6–10 days, and frogs stayed 12–22 days (Herreid and Kinney, 1967). If breeding season commenced late in the spring, the duration of egg laying and time at the breeding pond was shorter than when breeding commenced early. Although Wood Frogs first appear from late April to late May in Alaska, depending on weather conditions, oviposition does not begin until four to eight days after breeding commences and four to six days after the frogs first appear (Kessel, 1965; Herreid and Kinney, 1967). Wood Frogs begin to deposit eggs at 5–6°C, with optimal air temperatures of 11–16°C (Wright, 1914; Davis and Folkerts, 1986). In Illinois, maximum soil temperatures are usually 9–10°C when breeding commences (Redmer, 2002).

In Missouri, immigration to breeding ponds occurs from late February to late March in a repatriated population (Guttman et al., 1991). Males outnumbered females (1.57:1) at the breeding site and immigrated earlier than females. Most captures occurred in ravines, suggesting that the frogs use topographical features when moving to ponds, and orientation into and away from the pond was nonrandom. They spent 32 days at the breeding site.

Eggs normally are placed in nonflowing water, ranging from a few cm to a little more than 1 m in depth. In Colorado, eggs are usually placed in 10–15 cm of water about 1–3 m from the shoreline. All sites surveyed by Redmer (2002) in Illinois, whether natural or man-made, were < 95 cm in depth, and in Arkansas, mean maximum water depth was 23.5 cm (15–45 cm) at primary communal clusters (Cartwright et al., 1998). Even in areas of deeper water, the eggs are placed on a platform of submerged vegetation in such a manner that they are still 10–15 cm below the water surface (Haynes and Aird, 1981). Occasionally egg masses are deposited on leaf litter in dry pond basins during drought conditions (Paton et al., 2003); frogs also may relocate to more favorable sites during droughts. In such instances, mortality of both egg masses and adults can be high.

Males amplex females throughout the breeding pond, and amplexed pairs move to a communal oviposition site. Oviposition occurs rapidly, within 30 min under laboratory conditions (Wright, 1914) and 8–15 min under field conditions (Howard, 1981). Wood Frogs tend to oviposit eggs in the same places as other Wood Frogs (Hinckley, 1882; Wright, 1914; Wright and Wright, 1949; Herreid and Kinney, 1967; Porter, 1969a; Meeks and Nagel, 1973; Seale, 1982; Waldman and Ryan, 1983; Cartwright et al., 1998; Stevens and Paszkowski, 2004), with more than 100 females observed depositing egg masses at a single site in Connecticut (Wells, 1977a) and up to 963 masses at a communal site in Pennsylvania (Seale, 1982). In Alaska, 23% of 75 1 m² oviposition sites held 10 or more egg masses, accounting for 87% of all recorded egg masses (Herreid and Kinney, 1967).

Egg mass density in one Rhode Island study was 288 ± 60 egg masses/0.1 ha, with a mean number of 185 masses/pond; most ponds had >50 egg masses (Egan and Paton, 2004). Over a 16 yr period at 18 Rhode Island ponds, there was a mean of 441.5±343.7 egg masses per pond, during which time the population appeared to be increasing slightly (Raithel et al., 2011). In Michigan, Howard (1981) observed 58 of 60 egg masses within 1 m² in a 256 m² pond, and in Arkansas, Cartwright et al. (1998) recorded a mean of 62 egg masses per pond (range 1 to 290). In suboptimal bog and fen sites, however, egg mass densities can be much lower (5.8/100 m², 30.8/100 m², respectively) (Karns, 1992). Petranka et al. (2004) termed the floating egg masses a "raft." Later arrivals deposit eggs at the edge of the existing clumps deposited by earlier arrivals. As many as 16% of the masses may be deposited away from the communal aggregation in ponds containing <100 egg masses (Crouch and Paton, 2000).

The number of egg masses is positively correlated with the number of both females and males, accounting for approximately 92% of the females at a breeding pond in Rhode Island (Crouch and Paton, 2000). In 35 ponds, Crouch and Paton (2000) found a mean of 131 egg masses (± 114.8/pond). In Québec, mean egg mass density was 0.3 egg masses m² (range 0.03 to 2.36) (Gascon and Planas, 1986), which was similar to values provided by Howard (1980; 0.24 egg masses m²) and Waldman (1982b; 0.46 egg masses m²).

The number of eggs per mass is variably reported as between 304 and 874 (mean 575) in Illinois (Redmer, 2002); 386–543 (mean 465) in Tennessee (Meeks and Nagel, 1973); 350–709 (mean 496) in Alabama (Davis, 1980; Davis and Folkerts, 1986); 295–706 (mean 553) in Georgia (Camp et al., 1990); mean of 642–920 in Maryland and Virginia (Berven, 1982a); 700–900 (Moore, 1940); 42–1,570 (mean 778) in Alaska (Herreid and Kinney, 1967); 510–1,433 (mean 867) in Arkansas (Trauth et al., 1989); 746–1,111 (mean 911) in Colorado (Porter, 1969a); 711–1,248 (mean 876) in Wyoming (Corn and Livo, 1989); 1,000–3,000 in Minnesota (Bellis, 1957); 2,000–3,000 in New York (Wright and Wright, 1949); and an annual mean of 431–836 in Michigan, depending upon the age of the female (Berven, 2009).

Differences in literature reports of egg clutch size may be partly related to latitudinal differences in the size of breeding females, with Alaskan Wood Frogs smaller than their more southern counterparts. In addition, Wood Frogs may deposit all of their eggs at one location, at which time the egg mass may appear as two fused egg masses. This is because one mass comes from one ovary, and the other from the second ovary (Davis, 1980). Thus, estimates of the number of eggs per female, as well as the number of breeding females, could be biased by misinterpreting what constitutes a clutch. In general, larger females produce larger clutches of larger eggs (in both size and mass) (Berven, 2009).

Lowland female Wood Frogs deposit smaller clutch sizes (mean 642) in Maryland and Virginia than their Appalachian Mountain counterparts (mean 920) (Berven, 1982a), and they produce significantly smaller eggs (mean diameter 1.82 mm vs. 2.28 mm). However, the total egg volume of mountain females was actually lower than lowland females (Berven, 1982a). Differences in age and size at first reproduction between transplanted individuals (mountain to lowland; lowland to mountain) suggest that they result from both genetic and environmental components (Berven, 1982a). However, egg size, number, and total volume appeared to be genetically controlled.

Even within a population, reproductive traits related to egg size, clutch size, and total egg volume may vary within females of different ages. While these traits are generally positively correlated with female size, egg size and clutch size are inversely correlated with age class after adjusting for body size differences (Berven, 1988, 2009). In Maryland, for example, 1 yr females had the largest clutches but the smallest eggs, whereas 3 yr females had the smallest clutches but the largest eggs. Berven (1988) also reported that these traits varied annually independently of body size and initial size at metamorphosis. These results suggest a trade-off between egg size, clutch size, female body size, and age at first reproduction. Larger and older female Wood Frogs produce larger eggs that will result in larger tadpoles, whereas younger and smaller females compensate by producing a greater number of eggs.

Egg size also plays a role in how well subsequent larvae cope with their environment. Larvae from small eggs had longer larval periods and larger sizes at metamorphosis at low densities and high food levels than larvae from large eggs. At high densities and low food levels, however, larvae from small eggs had longer developmental times and were smaller at metamorphosis. Larvae resulting from small eggs were more sensitive to density than larvae resulting from large eggs. Berven and Chadra (1988) thus concluded that optimal egg size is correlated with environmental factors.

Male Wood Frog mating success is low the first breeding season (23%), but remains steady at 32% during subsequent breeding seasons. Most males and females breed only once, with the proportion breeding multiple times (two to five), declining with age (Berven, 2009). Female mating success is 100%, that is, all females breed. In Michigan, males may successfully fertilize about 600 eggs per year, regardless of age, whereas female egg production increases from 614 during the first reproductive season to ca. 741, as they increase in size with age (Howard, 1988b). There is little correlation of reproductive success with size and age, unlike some of the larger ranid species.

Fertilization rates vary among ponds. In Alaska, the rate was 78% for 53 masses from 8 ponds, with 1 pond having a low rate of 33% (Herreid and Kinney, 1967). Davis and Folkerts (1986) reported high fertilization rates (to 100%), but also recorded 3 masses as infertile. Hatching success is inversely correlated with pH, and hatching success is generally >80%, although it can be much lower (20–58%) (Gascon and Planas, 1986). Seigel (1983) reported a hatching success of 96.6% of 1,361 eggs from 5 egg masses deposited in New Jersey, and >92% at 2 additional breeding sites. Hatching success was 90% in 1 Wyoming population (Corn and Livo, 1989).

Temperature appears important for egg hatching success in Wood Frogs, since hatching success increases in warmer years compared with colder years under field conditions (Karns, 1992). Hatching success also is higher at the center of communal egg masses, and centrally located egg masses generally hold more eggs than those at the periphery. Since the number of eggs is positively correlated with female body size (Berven, 1988; Trauth et al., 1989; Redmer, 2002), this suggests that larger females breed first and select the choice egg deposition sites (Waldman, 1982b), at least in part to ensure an optimal thermal environment (Seale, 1982).

The eggs of *L. sylvaticus* develop more rapidly at colder temperatures than other North American species (Moore, 1939). The temperature of an egg mass at its center averages about 1–3°C higher than the surrounding water, presumably from heat generated by the developing eggs (Herreid and Kinney, 1967; Howard, 1981; Seale, 1982; Waldman and Ryan, 1983). Likewise, egg masses at the center of communal oviposition sites are 4.2–6.8°C warmer than surrounding water, and even egg masses at the periphery are 1–5°C warmer than surrounding water. Singly deposited masses away from communal sites are 0.6–2.2°C warmer than water 10 cm away from the mass (Waldman, 1982b). The temperature of egg masses also varies on sunny (to 1.5°C warmer) vs. cloudy days, and with the overall size of the communal egg mass (large clumps are warmer than small clumps) (Waldman and Ryan, 1983). The egg mass swells allowing for oxygenation and waste elimination during development, and convection of water through the interstitial spaces between eggs further aids in oxygenation (Pinder and Friet, 1994; Seymour, 1995). Optimal developmental temperatures are low (ca. 15°C), and 100% mortality occurs in water >24°C (Moore, 1939, 1940).

Developmental rates depend on water temperature, but appear similar between Alaska and New York populations. Approximately 50% of the eggs survive to hatch at temperatures of 6–24°C, but optimal temperatures are 7.5–20°C. At higher temperatures, developmental abnormalities become prevalent. These optimal temperatures appear similar regardless of breeding location (Moore, 1939; Herreid and Kinney, 1967). Eggs develop at different rates among breeding ponds. In New York, development takes 65–115 (mean 90) days under natural conditions (Wright, 1914) and 70–87 days in Illinois (Redmer, 2002), but this can be accelerated to 49–60 days under laboratory conditions (Moore, 1939; Redmer, 2002).

The major source of reproductive failure of both egg masses and developing larvae is desiccation due to pond drying (Camp et al., 1990; Berven, 1995; DiMauro and Hunter, 2002). Exposed egg masses are resistant to desiccation, however, with some eggs surviving within masses as long as 10–14 days when exposed to prolonged terrestrial conditions (Forester and Lykens, 1988). Communal deposition serves both to stabilize temperature and to provide a cushion against dehydration during the latter stages of development, especially as ponds dry down during the summer. Other sources of egg mortality include fungal infections (common at low temperatures), developmental abnormalities, and heat stress in masses located at the water's surface (Herreid and Kinney, 1967). In Alaska, abnormalities occurred throughout development, and fungal infestations were present on affected embryos prior to hatching. However, some fungal infections may occur only after egg mortality (Davis and Folkerts, 1986).

LARVAL ECOLOGY

In Tennessee, eggs deposited in late February hatched by mid-March (18–20 days) (Meeks and Nagel, 1973), and in Massachusetts, Hinckley (1882) reported a developmental period of 14–24 days. In Georgia, hatching occurs from 18–25 days (Camp et al., 1990), but after only 7–9 days at water temperatures of 5–17°C in Alabama (Davis and Folkerts, 1986). In laboratory situations, hatching can occur within five to eight days (Redmer, 2002). In Michigan, Wood Frog eggs hatched from 9 to 24 days after deposition, with eggs from shaded wetlands taking up to twice as long to hatch as those from much brighter wetlands under natural conditions (Skelly, 2004). One of the reasons eggs may hatch at different times, other than from the effects of temperature, is the presence of predators. In laboratory trials, Smith and Fortune (2009) demonstrated that Wood Frog eggs hatched faster in the presence of a predator (mosquitofish, *Gambusia*) and in the presence of extracts of the predator than controls did in the absence of predators. Thus, embryos may be able to detect predators chemoreceptively, and thus accelerate the time to hatching.

Larval densities and biomass can be extremely high (1,816/m^2 and 290 g/m^2, respectively; Biesterfeldt et al., 1993). This is important because mean body size at metamorphosis and survivorship decreases as tadpole density increases (Lynn and Edelman, 1936; Wilbur, 1976; Berven, 1990). There is a negative correlation between tadpole density and time to metamorphosis (Berven, 1990; but see Lynn and Edelman, 1936), and this relationship is dependent on food availability (Murray, 1990). Tadpoles may even use visual cues to gauge the density of conspecifics (Rot-Nikcevic et al., 2006).

Embryos developing in warm water grow faster and to a larger body mass than those from cool water (Castano et al., 2010), and they can reach hatching stages 18% quicker than those from cool water. Embryos developing in the darkest wetlands developed 12% faster than those from relatively unshaded habitats in experimental conditions of high and low water temperature, however, suggesting the presence of countergradient variation over small spatial scales (Skelly, 2004). Wood Frog larvae from forested wetlands have lower critical thermal maxima than larvae from open beaver-created wetlands, and larvae from open-canopied ponds hatch at lower rates when placed in forest-covered ponds than they normally would (Skelly and Freidenburg, 2000). In addition, Wood Frog larvae from close-canopied forested wetlands tend to have strong temperature preferences that are higher than those of larvae raised in open beaver-created ponds (Freidenburg and Skelly, 2004). These results suggest that Wood Frogs are able to respond rapidly to changing or altered thermal conditions.

In Alaska, only 4% of the eggs initially oviposited survived to produced metamorphs (Herreid and Kinney, 1966), with 50% of the loss occurring prior to hatching and the remainder during the final two-thirds of development. Berven (1990) reported premetamorphic survival at 4.5% in one pond and 0.95% in another; the low rates of survival resulted from pond drying. Seigel (1983) reported 65.7% survivorship from egg deposition to metamorphosis, and Waldman (1982b) gave rates of 41.6–86.3% survivorship.

The mean size of hatchlings is 8–9 mm (Hinckley, 1882; Herreid and Kinney, 1967; Meeks and Nagel, 1973; Camp et al., 1990). Larvae develop rapidly once they are free swimming; tadpoles developed limb buds at 14–17 weeks postoviposition. Maximum larval weights (at Witschi stages 28–29) ranged from 2.15 to 2.85 g at maximum lengths of 5.5–6.0 cm. Maximum larval total length are 50–66 mm (Wright and Wright, 1949; Bellis, 1957; Herreid and Kinney, 1967; Meeks and Nagel, 1973; Camp et al., 1990).

Tadpole growth rates and metamorphic size may be under some degree of genetic control, although growth rates of larval Wood Frogs are not correlated with heterozygosity, at least during the latter stages of premetamorphic development (Wright and Guttman, 1995). Michigan Wood Frog larvae grew faster (0.02 g/day), have shorter larval periods (78 days), and are larger (19.7 mm) than their Maryland and Virginia conspecifics (0.05 g/day in Maryland and 0.005 g/day in Virginia; 92 days in Maryland and 108 days in Virginia; 16.0 mm in Maryland and 18.7 mm in Virginia, respectively) (Riha and Berven, 1991). Developmental rates, however, were similar between these populations. In some Maryland and Virginia populations, developmental time was negatively correlated with density, but not in Michigan populations (Riha and Berven, 1991). In contrast, time to metamorphosis increased at high densities in laboratory experiments, based on other Maryland populations (Lynn and Edelman, 1936). Michigan larvae, however, were more sensitive to the effects of density on metamorphic size. Food level undoubtedly plays some role in the differences observed (Riha and Berven, 1991).

Large tadpoles are more active than small tadpoles, regardless of the amount of food present, because the larger animals are less prone to predation than small tadpoles (Brodie and Formanowicz, 1983; Anholt et al., 2000). For reasons unknown, surfacing tadpoles exhibit a left-handedness, with >59% turning to the left after surfacing to breath air (Oseen et al., 2001).

Alaskan Wood Frog tadpoles are active even at temperatures approaching 0°C (Johansen, 1962), but they are capable of tolerating temperatures near 39°C if they have had sufficient time to acclimate (Cupp, 1980). Tadpoles have not been found to overwinter (Herreid and Kinney, 1967), despite speculation that they might do so (Hildebrand, 1949; Johansen, 1962), such as when food is limited (Murray, 1990). Murray (1990) pointed out that early stage tadpoles have been found in late August and September in permanent ponds in Labrador, suggesting that some larvae could not metamorphose prior to lake freezing.

Under laboratory and experimental pool conditions, larval Wood Frogs clump together with siblings (Waldman, 1984; Cornell et al., 1989; Fishwild et al., 1990) and may even avoid nonsiblings (Waldman, 1986b). In nature, the tadpoles are asocial, suggesting that there is no kin recognition (Waldman, 1991). Wood Frog larvae do not retain the ability to recognize related conspecifics after metamorphosis (Waldman, 1989), and Jasienski (1992) concluded that kinship is not of profound significance in explaining differences in larval life history traits.

In contrast, Wood Frog tadpoles are able to recognize the presence of other tadpole species through chemosensory cues, and can alter their behavior accordingly. For example, Wood Frogs in Illinois decreased their activity levels, increased the use of refugia, and grew slower in response to allotopic over-wintered American Bullfrog (*Lithobates catesbeianus*) tadpoles than to syntopic over-wintered American Bullfrog larvae. Such behavior is associated with antipredator responses (Van Buskirk and Relyea, 1998; Relyea, 2002a, 2004a). There was no such response to syntopic over-wintered American Bullfrogs, suggesting that the syntopic Wood Frog larvae did not perceive them as predators. In the presence of *L. catesbeianus*, however, Wood Frog larval survivorship declined by 10%, likely the result of competition (Walston and Mullin, 2007a).

The free-swimming tadpole stage occupies two-thirds of the developmental period, the length of which is controlled by food, density, prey and predator densities, and temperature. Literature reports of larval period are wide-ranging: 84 days in Tennessee (Meeks and Nagel, 1973), 53–78 days in Alaska (Herreid and Kinney, 1967), 73–113 days in Maryland (Berven, 1982b), 115–130 days in Georgia (Camp et al., 1990), 70–87 days in Illinois (Redmer, 2002), 68–94 days in Michigan (Berven, 2009), and 58 days in Massachusetts (Hinckley, 1882). Metamorphosing Wood Frogs have been found on 22 July in northern Ontario, suggesting a mid-May breeding season at this location (Schueler, 1973). In Colorado, metamorphosis begins as early as late May and is complete by mid-August (Haynes and Aird, 1981). In Nova Scotia, eggs are laid in April and transformation occurs in August (Bleakney, 1952). In Rhode Island, metamorphs left from late June to early July after eggs were deposited in March.

The body length at metamorphosis varies between 12 and 22 m (mean 16 mm) (Wright, 1914; Wright and Wright, 1949; Bellis, 1961). Metamorphosis occurred at a mean of 17.2 mm in Alberta, 17.8 mm (range 15 to 21 mm) in Georgia

(Camp et al., 1990), 14–18 mm in Maryland (Berven, 1990), 17–21 mm in Minnesota (Bellis, 1961), and a body mass of 0.25–0.83 g (mean 0.62 g) in Michigan (Berven, 2009). In Alaska, juveniles collected in August and September ranged from 0.55 to 1.5 g in body mass (Johansen, 1962). Size of juveniles varies significantly from one year to the next.

The size of the metamorphic juvenile is critical to both overwintering after metamorphosis and adult fitness, since larger size is positively correlated with increased survivorship and reproduction. Relyea (2001c) showed that larval body mass and larval period are positively correlated with body mass at metamorphosis. Larval period is also positively correlated with metamorph hind limb length and forelimb length, but negatively correlated with body width. Larval body length also is positively correlated with metamorph forelimb length. Taken together, these attributes suggest that long larval periods produce heavier and larger metamorphosing juveniles with longer limbs (Relyea, 2001c). Since exposure to predators may ameliorate larval morphology, such trait-mediated phenotypic plasticity could alter the future fitness of adults.

Wood Frog larvae can detect differences in food levels in experimental trials and are able to modify their growth patterns accordingly (Jasienski, 1992). Larvae reared under low food conditions increase their growth rates when switched to high levels of food. For example, stunted tadpoles often increase in size dramatically after larger conspecifics metamorphose, with a five-fold increase in size over a two-week period. Larvae reared at high levels of food but switched to low levels likewise reduce their growth rates, but the rate reduction is not as pronounced as the rate increase when switching from low to high food levels (Jasienski, 1992). Temperature also interacts with food availability, affecting both growth and survivorship (Castano et al., 2010).

The interaction of food and larval density is another important life history variable affecting fitness. Tadpoles experimentally raised at lower densities often have larger body mass (to 41%, Goater and Vandenbos, 1997) and subsequent size at metamorphosis than those raised at high densities. In turn, tadpoles at higher body masses were 14.7% greater in body mass at first overwintering than those that metamorphosed at smaller body masses; growth rate was the same between the two groups (Goater and Vandenbos, 1997). Assuming that the amount of food in a wetland is proportional to wetland size, large numbers of Wood Frog larvae developing in very small wetlands, as commonly occurs, could be at a size disadvantage as they begin to overwinter compared to tadpoles that developed from larger wetlands with more food resources.

However, opposite effects of food and larval density were found by Murray (1990) working with a population of *L. sylvaticus* from Labrador. In experimental trials, increasing food levels resulted in tadpoles with greater body masses, regardless of density, although tadpoles raised at the higher densities weighed less than tadpoles at lower densities. The time to metamorphosis was shortest at high food levels and low densities, and longest at high densities and low food levels, as expected. Tadpoles with the highest food levels also had the highest growth rates and body mass. Murray (1990) suggested that methodological differences might explain differences between his results and Wilbur (1977a), but that there also might be differences in certain life history responses between northern and southern *L. sylvaticus* populations.

Other trade-offs affecting life history traits of Wood Frog larvae occur. For example, the presence of predators decreases body mass, but their presence also lengthens the larval period. Thus, body masses of metamorphosing individuals are often nearly identical from both predator-free and predator-present environments, despite differences in developmental rates (Relyea, 2001c). The density of tadpoles also affects growth rates, at least in experimental populations exposed to differing levels of food (Wilbur, 1977a; Murray, 1990; Goater and Vandenbos, 1997), and high densities of conspecifics in natural ponds result in decreased body sizes. This suggests that food may be a limiting factor in larval growth rates and body size, at least under some circumstances (DeBenedictis, 1974; Wilbur, 1977a).

Wood Frog larvae express considerable variation in habitat selection, morphology (termed phenotypic plasticity), growth, and behavior in response to the presence of predators. They may seek shelter, reduce levels of activity and swimming speed, change their behavior, and even their morphology (Anholt et al., 2000;

Richardson, 2001; Awan and Smith, 2007). For example, Wood Frog larvae preferred the least complex habitat in the absence of the predator *Dytiscus verticalis* in experimental laboratory trials. However, they shifted to the most complex habitat, that is, the habitat with the most cover, in the presence of this predaceous diving beetle (Formanowicz and Bobka, 1989). They will also reduce their activity in the presence of abundant food, and this change has the effect of decreasing swimming, which in turns decreases their exposure potential to predation (Anholt et al., 2000). The presence of fish causes larvae to decrease activity and to seek shelter, regardless of the number of larvae present (Awan and Smith, 2007).

The presence of predators can also result in different tadpole morphologies. For example, dragonflies (*Anax*) preferentially prey on tadpoles with shallow and short tail fins and narrow tail musculature. In the presence of these predators, Wood Frog tadpoles have shorter bodies and deeper tail fins (photo in Relyea, 2001a) than those not exposed to predators. When no predators are present, the bodies are larger, but the tail fin is less deep and more streamlined. However, tadpoles with the predator-induced phenotype grow slower than normal tadpoles when raised in the absence of predators. Thus, there is a trade-off between morphological types and growth rates, with one body type more favorable in one set of circumstances, but another body type more favorable in a different set of circumstances. Tadpoles are able to change these morphotypes in response to chemical cues detected from the predator, even without direct physical contact.

Much ecological understanding of the effects of predators on phenotypic plasticity has resulted from research on Wood Frogs. Phenotypic plasticity in tadpole morphology is favored by natural selection, depending on the predators present (Van Buskirk and Relyea, 1998; Relyea, 2002a, 2002b), time in the breeding season (Relyea, 2001b), and geographic location (Relyea, 2002b). Certain predators may cause morphological change (the mudminnow *Umbra*, *Anax*), but others do not (*Dysticus*, the Eastern Red-spotted Newt *Notophthalmus viridescens*) (Relyea, 2001b). This variation in selection pressure has resulted in local populations of Wood Frog larvae, facing different predator-competitor environments, evolving different phenotypically plastic responses to their natal environment (Relyea, 2002b).

A variety of options is important to a growing tadpole because different predators can exert different levels of predation pressure on Wood Frog larvae (Brodie and Formanowicz, 1983). As might be expected, the highest rates of predation occur by predators (e.g., *Umbra*) that do not normally coexist with Wood Frog larvae (Relyea, 2002a). Wood Frogs also may change their behavior and reduce their activity in the presence of some pred-

Adult *Lithobates sylvaticus*, tan phase. Photo: David Dennis

ators (*Umbra, Anax, Notophthalmus*), but not others (*Dysticus*) (Relyea, 2000, 2001b; Fraker, 2010). When exposed to predators in experimental situations, artificially immobilized Wood Frog tadpoles were attacked less often by dragonfly larvae than active tadpoles (Skelly, 1994).

Changes may occur in the oral structures of Wood Frog tadpoles that help them increase feeding efficiency during times of intense competition and predation. For example, the proportion of oral structures that increase feeding efficiency also increase, allowing better utilization of potentially dwindling resources as competition increases (Relyea and Auld, 2005). There is even some indication that tadpoles may use visual cues to assess the potential level of competition (Sutherland et al., 2009). Larval phenotypic plasticity in the arrangement of oral structures also may result from pollution and disease.

Most mortality (92–99%) in the Wood Frog life cycle occurs among premetamorphic individuals. Large juveniles and those metamorphosing early have higher survival rates than small juveniles and those metamorphosing late in the season. As with egg masses, drying ponds are a major source of larval mortality.

DISPERSAL

Adults spend a very short time at the breeding ponds and disperse very soon after egg deposition. In Missouri, adult dispersal away from breeding ponds occurs from early to late March (Guttman et al., 1991; Rittenhouse and Semlitsch, 2007a). In New England, the movement of Wood Frogs does not appear to be influenced by the presence of forest edges, roads, and streambeds (Gibbs, 1998a; deMaynadier and Hunter, 2000). Wood Frogs moved readily across these boundaries, although abundance increases with increasing proximity to interior forested habitats (deMaynadier and Hunter, 1998). Rittenhouse and Semlitsch (2007b) reported that Wood Frogs moved a mean distance of 110 m (range 1 to 394 m) from the nearest breeding site over a 50-day period. Most animals were found within 133 m, and nearly all were within 293 m. Frogs spent a mean 6 to 11 days per location, and most movements were short (< 20 m). Long distance movements (3.8% of total) occurred during rain events (Rittenhouse and Semlitsch, 2007a).

Wood Frogs move away from breeding ponds in a direct fashion and do not require topographic cues with which to orient or serve as corridors. Upon metamorphosis, Wood Frogs at a Maine pond oriented toward a forested wetland rather than to more open habitat (Vasconcelos and Calhoun, 2004), and in Illinois, they preferentially moved toward adjacent forested habitat (Walston and Mullin, 2008). Translocated metamorphs continued to show directionality at the displaced location in a manner appropriate to the natal pond, even though the direction to the forested wetland had been reversed at the release site. Patrick et al. (2007) suggested that dispersing Wood Frog metamorphs used unspecified indirect cues in orientation, a situation they considered maladaptive if the natal breeding was altered. They assumed that Wood Frogs are philopatric, which is likely not the case (Petranka et al., 2004). Indeed, generalized orientation could actually lead returning frogs to adjacent but suitable sites should the original pond be perturbed, as long as ponds are grouped within reasonable traveling distance.

After metamorphosis, juvenile Wood Frogs rapidly increase their endurance and aerobic activity; increases in hematocrit, hemoglobin concentration, and heart mass also help to improve their activity capabilities prior to dispersal (Pough and Kamel, 1984). Juvenile dispersal tends to occur nonrandomly at individual breeding ponds in any one year. However, dispersal over a multiyear time period tends to be more random than a single year of observation might indicate. During an 8 yr study in Massachusetts, recently metamorphosed Wood Frogs dispersed away from breeding ponds nonrandomly, with a positive bias to the west and a negative bias to the southwest. However, there was considerable annual variation among ponds in dispersal orientation, which dampened directional trends (Timm et al., 2007c).

In Massachusetts, dispersal took place over a period of 74 days (early May to mid-July) at 14 ponds monitored over a 5 yr period (Timm et al., 2007b). A total of 51,608 juveniles were captured leaving these ponds. Movements usually had 2.5 pulses, with 10 emigration events required for juvenile frogs to fully depart from breeding ponds. Temperature did not have any importance in the timing of juvenile emigration, although precipitation did (Timm et al., 2007b). Precipitation was especially effective at triggering emigration if a prolonged drought preceded the rainfall event.

Over a 3 yr study in Maine, Wood Frog adult males, females, and juveniles were found 30 m, 150 m, and 300 m respectively from the breeding site in closed-canopy forest (Vasconcelos and Calhoun, 2004). The median body length of the juveniles increased with increasing distance from the wetland, suggesting that the juveniles were feeding as they dispersed. Most Wood Frogs enter their overwintering sites in October and November in the North, but frogs are occasionally active throughout the winter in some areas (Zweifel, 1989).

The cues that Wood Frogs use to colonize new breeding sites are unknown, but Sexton et al. (1998) suggested that a repatriated population in Missouri used valleys and ridge tops as movement corridors between the initial release pond and distant potential breeding ponds. Frogs also dispersed across an interstate highway to reach one new pond.

DIET

Wood Frog tadpoles are facultative suspension feeders that ingest a variety of suspended particles in the water column. They use their keratinized mouthparts to break up algae and other plant material, and they will reingest their own fecal pellets. Food items include diatoms, filamentous algae, blue-green algae, and protozoans (Munz, 1920). When feeding, they maintain a nearly constant ingestion rate at the morphological and physiological limits of their filtering ability (Seale and Wassersug, 1979). Wood Frog tadpoles seem to prefer filtering phytoplankton when food levels are high, but switch to grazing when phytoplankton in the water column falls to low concentration levels.

Wood Frog larvae lack specialized morphological correlates of carnivory (Wassersug et al., 1981), but they will eat eggs and larvae of sympatric salamanders, and occasionally they will eat each other's eggs and larvae. Bleakney (1958a) mentioned that all the tadpoles he observed in a Nova Scotia pond had missing or partially missing tails. He assumed that cannibalism was responsible, although he did not directly observe it. Cannibalism also occurs when size discrepancies exist in conspecific larval size classes (Petranka and Thomas, 1995). Wood Frog tadpoles eat both eggs and developing larvae of the Spotted Salamander (*Ambystoma maculatum*) (Petranka et al., 1998; Dodd, 2004), and appear to select eggs and larvae within clear egg masses in comparison with white egg masses (Petranka et al., 1998). They may also graze the alga *Oophilia amblystomatis* that grows on this salamander's egg mass.

Adults are sit-and-wait predators that rarely move except to snap at invertebrates moving in their immediate vicinity. However, McCallum et al. (2003b) recorded an instance in which sound was used by a Wood Frog to locate and capture a Spring Peeper (*Pseudacris crucifer*). Feeding does not occur during the breeding season (Davis, 1980). In Minnesota, food consists chiefly of ants, beetles, spiders, crane flies, and gastropods (Marshall and Buell, 1955), and in Alberta, arthropods and carabid beetles were most often consumed, although the diet included a great variety of taxa (Moore and Strickland, 1955). In Québec, recently metamorphosed Wood Frogs ate larval lepidoptera, spiders, adult beetles, and adult flies (60% by weight) and ant larvae, adult flies, spiders, adult beetles, and springtails among a wide variety of invertebrates (Leclair and Vallières, 1981). Additional reports of food items are in Gilhen (1984), Davis (1980), and Davis and Folkerts (1986). Food habits are basically similar among frogs living in different habitat types. Martof and Humphries (1959) found the stomachs of Wood Frogs in Alaska stuffed to overflowing, suggesting an intensive amount of feeding during the short activity season.

PREDATION AND DEFENSE

Leeches may attack egg masses (Cory and Manion, 1953; Trauth and Neal, 2004), and Wood Frog eggs and larvae are palatable to a variety of vertebrate and invertebrate predators (Dickerson, 1906; Herreid and Kinney, 1966; Formanowicz and Brodie, 1982; Brodie and Formanowicz, 1983; Stein, 1985; Davis and Folkerts, 1986; Eaton and Paszkowski, 1999; Redmer, 2002; Relyea, 2002a; Jennette, 2010; Szuroczki and Richardson, 2011). Even deer have been observed eating Wood Frog egg masses in the Great Smokies (Tanner Jessel, personal communication). Larvae are palatable (Kats et al., 1988) and are eaten by dragonfly naiads, belostomatid and dyticid beetles, notonectids, waterstriders (Gerridae), crayfish (*Cambarus diogenes*, *Procambarus acutus*), ambystomatid salamanders (*Ambystoma jeffersonianum*, *A. maculatum*, *A. opacum*,

A. *tigrinum*), newts (*Notophthalmus viridescens*), sunfish (*Lepomis*), grackles (*Quiscalus*), Eastern Garter Snakes (*Thamnophis sirtalis*), and fish. The salamanders, in particular, frequently co-occur with Wood Frog larvae during the breeding season.

The palatable Wood Frog larvae are able to minimize adverse encounters with later-arriving vertebrate and invertebrate predators by often being the first amphibian in the pond and growing rapidly to attain a size at which predation pressure is diminished. As tadpoles approach metamorphosis, however, they are avoided by predators as granular glands develop in the skin (Formanowicz and Brodie, 1982).

Larval Wood Frogs have different escape strategies based on their morphological development. At the beginning and very end of larval development, when they are particularly vulnerable to predation due to small size or morphological changes associated with metamorphosis, they tend to turn at sharp angles (>40°) in order to quickly change their swimming trajectory. In the middle portions of larval development, they are more likely to "sprint," that is, swim farther (> 30 cm) and more rapidly (> 9 cm/sec) toward protective cover and not to maneuver sharply (Brown and Taylor, 1995).

Most of the research on the effects of predators on Wood Frog larvae has been conducted in mesocosms or other experimental settings. In outdoor experimental pens, no Wood Frog larvae survived to metamorphosis when confined with Marbled Salamander (*Ambystoma opacum*) larvae (Cortwright and Nelson, 1990). In mesocosms, the survival of Wood Frog tadpoles was negatively correlated with the survival of Jefferson Salamanders (*A. jeffersonianum*) and positively associated with growth of Spotted Salamanders (*A. maculatum*) (Rowe and Dunson, 1995). Under experimental conditions, 66% of Pennsylvania Wood Frog larvae transformed by 56 days, whereas very few survived longer hydroperiods (to 158 days) because of predation by the ambystomatid larvae (Rowe and Dunson, 1995).

The mosquito *Culex territans* feeds on Wood Frogs in the lab but cannot use the frog's call to locate its next blood meal (Bartlett-Healy et al., 2008). Other predators of adults include raccoons (*Procyon lotor*), striped skunks (*Mephitis mephitis*), Eastern and Red-sided Garter Snakes (*Thamnophis sirtalis* ssp.), Northern Water Snakes (*Nerodia sipedon*), kingfishers (*Megaceryle alcyon*), hawks (*Buteo* sp.), unspecified raptors, crows, and ravens (*Corvus* sp.), a mallard duck (*Anas platyrhynchos*), American Bullfrogs (*Lithobates catesbeianus*), fish (*Micropterus salmoides*), crayfish (*Cambarus diogenes*), and the giant water bug (*Lethocerus* sp.) (Huheey and Stupka, 1967; Seale, 1982; Larsen and Gregory, 1988; Thurow, 1994; Cochran, 1999; deMaynadier and Hunter, 1999, 2000; Eaton and Eaton, 2001; Redmer, 2002; Dodd, 2004; Baldwin et al., 2007; Beaudry, 2007). Large mortality events may also result from wading bird (Ciconiformes) predation (Trauth et al., 2000). A peculiar form of attack is recorded by Davis and Folkerts (1980) who noted a beetle (*Eutheola rugiceps*), presumably eaten, that burrowed through the stomach lining of a Wood Frog.

The various shades of adult Wood Frog coloration (Martof and Humphries, 1959) likely serve a crypsis function, thus making the frogs difficult to detect in brown and reddish leaf litter (Heatwole, 1961). The dark eye mask functions as disruptive coloration helping to break up the contours of the frog, again aiding in crypsis. Examining the "Midwest type" coloration of Martof and Humphries (1959), King and King (1991) showed that their test subjects matched various leaf shades well, with females matching the litter substrate more closely than males. The more or less uniformly colored Wood Frogs in the South tend to match forested habitat, whereas the blotchy and striped Wood Frogs in the North are less conspicuous in their more open, grassy environments.

Seib (1977) reported a 35 mm Wood Frog to "death-feign" in response to handling, and adults death-feigned at an Arkansas breeding pond after diving into the leaf litter and being picked up (McCallum et al., 2003a). No defensive postures were observed in response to experimental attacks by shrews (*Blarina brevicauda*) (Formanowicz and Brodie, 1979). However, adult Wood Frogs readily survive attacks by shrews and do not appear to be palatable. When attacked, the experimental frogs frequently (87.5%) emitted a "mercy" cry, which is thought to startle a potential predator (Wright, 1914; Formanowicz and Brodie, 1979). Breeding adults are also sensitive to the presence of potential predators at ponds and will quickly

Adult *Lithobates sylvaticus*, near Fairbanks, Alaska. Photo: C.K. Dodd Jr.

stop calling and submerge if a perceived threat approaches or flies overhead (Banta, 1914). Frogs on the shoreline will dive into the water and submerge in leaf litter or surface at a distance away from the perceived threat (Heatwole, 1961). Bellis (1962b) found that Wood Frogs evaded capture better in tamarack forest, where ground cover was extensive, than in the more open black spruce (*Picea mariana*) stands. Terrestrial frogs seek refuge in mammal burrows and among the roots of trees and shrubs (Schueler, 1973) and appear well aware of the location of emergency escape holes (Bellis, 1962b).

POPULATION BIOLOGY

The best data on the population biology of Wood Frogs are from the 21 yr study reported by Berven (2009). In short, annual survivorship of males and females was negatively correlated with the total number of males and females, and the annual variation in survival rates was best explained by the annual variation in the number of adult males and females. Juvenile survival was negatively correlated with total juvenile population size, but not with juvenile body size, adult population size, or environmental factors. When numbers of juveniles were low, female juveniles matured earlier at a larger body size, and produced more eggs of smaller sizes than when numbers of juveniles were high. Thus, juvenile fitness traits were more attuned to juvenile population sizes than anything related to adult demography. In addition, Berven (2009) found no evidence of larval fitness traits influencing fitness traits of postmetamorphic frogs.

He concluded that density-dependence operated terrestrially to regulate overall Wood Frog population size.

Paton et al. (2000) estimated a recruitment rate of only 0.06 juveniles per breeding female. In Michigan, male juveniles survived at much higher rates than female juveniles over a 21 yr period (males 17.2%, range 4.3 to 33.2%; females 5.5%, range 1.8 to 11.5%) (Berven, 2009). Other estimates include 19.3 per breeding female (Doty, 1978) and 110 per breeding female in Michigan (Berven, 1995). In Maryland, survival to year 1 was 24%, 17% to year 2, and 11% to year 3 in males (Berven, 1990). For females, survivorship was 22% to year 2 and 11% to year 3 (Berven, 1990). These values suggest an overall survival rate of 14% for males and 13% for females. Adult annual survival rates also are low in Michigan, 8.0% for females and 12.5% for males (Howard and Kluge, 1985; Howard, 1988b). In another Michigan study, males and females had similar rates of survival, ranging from ca. 14 to 20% (Berven, 2009). Maximum longevity is 6 yrs (Berven, 2009).

Breeding population estimates are few in the literature. Howard and Kluge (1985) estimated that their Michigan population varied from 5,202 to 8,955 individuals over 4 yrs. Over a 7 yr period in another Michigan study, Wood Frog population size fluctuated by a factor of 10 from one year to the next (Berven, 1990). Annual replacement rates varied from 0.009 to 7.49, and survivorship was constant for juveniles and adults. The longest-running Wood Frog study followed populations

> 20 yrs, during which the breeding population size fluctuated from 594–6,196 (mean 3,030) (Berven, 2009).

In Wood Frogs, females generally mature one year later than males (Collins, 1975; Howard, 1981; Berven, 1990), and this difference results in male-biased (e.g., 3:1 and 5.5:1 in Michigan) sex ratios at breeding ponds (range 1.02 to 12.3:1). Other reported male-biased sex ratios are 3:1 (range 1.02 to 12.3:1) in Maryland (Berven, 1990), 2.64:1 in Rhode Island (Crouch and Paton, 2000), 1.82 in Virginia (Berven and Grudzien, 1990), 2.67:1 in Michigan over a 4 yr period (Howard and Kluge, 1985), and 2.1 to 7.7:1 (mean 4.6:1) over a 21 yr period in Michigan (Berven, 2009). Such highly skewed sex ratios at breeding ponds result in intense male–male competition for access to breeding females.

A considerable amount of variation is present in terms of annual reproductive output from a pond. Berven (2009) found that the mean number of juveniles leaving a Michigan pond over a 21 yr period was 36,162 (range 1,210 to 113,686). In Massachusetts, Timm et al. (2007a) recorded 51,608 newly metamorphosing Wood Frogs emigrating from 14 seasonal ponds over a 4 yr period; the most metamorphs captured at a single pond in a single year was 13,122. The number of metamorphs produced may be a function of wetland type, with semipermanent wetlands producing 1.2 to 23 times the number of metamorphs as temporary wetlands (Karraker and Gibbs, 2009); in addition, metamorph body mass was larger in semipermanent ponds than in temporary ponds.

Transformation occurs rapidly, with most metamorphs leaving within a few days of one another. In Timm et al.'s (2007a) Massachusetts study, approximately 80% of the emigrants emerged over a 5-day period, with emigration taking place from about mid-June to the first week of July, depending on year. In Alaska, frogs disappeared from the vicinity of breeding ponds over a 3-day period (Johansen, 1962). Arkansas metamorphs begin dispersal by late May (Trauth et al., 1989).

Wood Frogs continue to grow throughout their life span, and growth rates vary considerably (Bellis, 1961). Most males reach maturity at one year following metamorphosis (i.e., in their second year of life) whereas females reach maturity in their second year following metamorphosis (i.e., in their third year of life) (Berven, 2009). Males that reach maturity of 2 yrs are smaller than males reaching sexual maturity at 3 yrs, but there is considerable annual variation in size at sexual maturity from one year to the next (Berven, 1982a; Berven, 2009). In Alaska, reproduction begins at 3 yrs (Herreid and Kinney, 1966), but Leclair et al. (2000) suggested > 5 yrs in northern populations. Males from the Atlantic Coast lowlands reach maturity in the first spring after metamorphosis, and 86% of males bred in their first season (Berven, 1982a). A total of 84% of males bred once, 14% twice, and 2% three times; no males lived > 3 breeding seasons. Most lowland females bred in their second year (99%), but a few (1%) may breed even in their first year. 12.6% survive to breed in year 3, and like males, no females survived > 3 yrs.

In Québec, longevity also reached 4 yrs based on skeletochronology (Sagor et al., 1998), and some females can reach 5 yrs (Bastien and Leclair, 1992). However, in Wood Frogs from the Appalachian Mountains, maturity normally occurs 3 yrs after metamorphosis with a few maturing at 2 and 4 yrs (Berven, 1982a). Most mountain-dwelling females (62%) matured at 4 yrs. Latitude also affects the timing of sexual maturity. In Québec, southern populations of *Lithobates sylvaticus* matured later and were larger than southern conspecifics from low elevations studied by Berven (1982a), but they matured earlier and were smaller than Berven's (1982a) southern frogs from higher elevations (Bastien and Leclair, 1992; Sagor et al., 1998).

The results of several studies (Newman and Squire, 2001; Stevens et al., 2006a) suggest that Wood Frogs are not as philopatric as some authors (e.g., Berven and Grudzien, 1990) indicate. Still, Vasconcelos and Calhoun (2004) found that 98% of males and 88% of females returned to their pond of original capture. Most populations at nearby ponds have very similar allele frequencies suggesting significant amounts of gene flow among closely spaced breeding ponds (Newman and Squire, 2001; Squire and Newman, 2002).

In an example of breeding site flexibility, the majority of the population tends to move to adjoining ponds rather than breed in remnant pools when beaver ponds are altered by dam destruction (Petranka et al., 2004). When beavers altered the drainage patterns of formerly isolated ponds that allowed fish to invade, Wood Frog use declined in relation to the extent of fish invasion (more fish, less breeding), until they stopped breeding altogether and shifted to alternate ponds. If fish were present, they bred only in very shallow extensions of the pond, away from fish. When beaver activity was curtailed and the ponds restored to a fish-free presence, Wood Frogs again used the site for breeding (Petranka et al., 2004). Overall, the general population remained stable despite the various habitat shifts over a 10 yr period.

Site-shifting also happens when excessive siltation occurs at formerly favorable breeding sites. If fish are present and fishless ponds are available, Wood Frogs will go to the fishless ponds, even if they formerly bred in the pond now containing fish (Hopey and Petranka, 1994), especially if the ponds were < 6 m apart. Site philopatry is weak when clusters of ponds are close together.

Wood Frog populations fluctuate in size asynchronously in the southern Appalachians, Maryland, Virginia, and Michigan in studies ranging from 7 to 13 yrs, both within a specific geographic area and among populations scattered over a much larger region (Berven, 1995; Petranka et al., 2004). Petranka et al. (2004) suggest this resulted from extensive habitat shifting among locally proximate ponds. Population synchrony was independent of the distances to nearest ponds. Population turnover rates are high, with 12 of 26 pond populations failing to use local ponds in one or more years. No regional population experienced a turnover in either Petranka et al's (2004) or Berven's (1995) studies. Thus, breeding habitats are dynamic and subject to fluctuation even in protected areas. Pond switching occurs, and clusters of local ponds may lack a metapopulation structure, because interpond distances are so small between groups of ponds as to provide a lack of genetic and demographic independence because of high rates of intergroup dispersal (Petranka et al., 2004).

Wood Frog populations routinely seem to go through bottlenecks, followed by localized population recovery. Berven (1995) noted that declines of 50–80% were common in as little as 1 yr. Reproductive failures are not uncommon (22% in Michigan over 13 yrs; 43% in Virginia over 8 yrs; 29% in Maryland over 7 yrs) (Berven, 1995), but recruitment at 5–6 times the average annual replacement rate can occur in some years. For example, females deposited from 13,400 to 819,973 eggs over the 7 yr period at the two ponds studied by Berven (1990). These population fluctuations result from, and contribute to, fluctuations in the number of breeding adults in any particular year, which in turn are dependent on the cyclical and stochastic production of juveniles. The population dynamics of Wood Frogs result from a subtle interplay of both density-dependent (food resources, competition, predation) and density-independent (hydroperiod, drought) sources that vary from one year to the next, and may have different levels of importance from one region to another (Berven, 1995).

Local flexibility in choice of breeding site is confirmed by studies of microsatellite variation of populations in relatively close spatial proximity (< 1 km, and certainly < 400 m) (Newman and Squire, 2001; Squire and Newman, 2002). These authors also found 40% interpopulation movement based on mark-recapture studies. Local ponds are habitat patches. Still, even Wood Frog populations located 200 km distant show substantial similarities in allele frequencies (Squire and Newman, 2002).

COMMUNITY ECOLOGY

Some species of frogs change their behavior or change the timing of developmental processes when exposed to alarm cues excreted by eggs or larvae that have been attacked by predators or reared in water containing predators. Wood frog embryos, however, do not appear to alter time to hatching, size at hatching, or have survivorship affected by chemoreceptive cues, although there is variation among sibships (Anderson and Petranka, 2003; Touchon et al., 2006; Dibble et al., 2009). It may be selectively disadvantageous to delay hatching when developing in ponds that may soon be colonized by a variety of additional predators.

Some of the most important ecological research on predation and competition has involved field and mesocosm studies of larval *Lithobates sylvaticus*. In pond communities, larval Wood Frogs have multiple effects on other amphibian larvae, serving as prey or competitors, and either directly or indirectly affecting the interactions between other amphibians depending on density, species composition, seasonality, predators, and prey base (Morin and Johnson, 1988; Werner, 1992; Petranka et al., 1994; Relyea, 2000). For example, Wood Frog larvae adversely affect some syntopic salamander (*Ambystoma*) larvae by competing with the salamander's invertebrate prey, periphyton, and phytoplankton. When the prey base declines, salamander growth rates and survivorship decrease. In turn, salamander larvae may have no effect on survivorship or length of the larval period of Wood Frogs, but their presence may slow tadpole growth rates as they compete for the same food.

In other situations, the presence of Wood Frog larvae may affect salamanders positively, by serving as a large prey base; Wood Frog larvae are a favorite prey of tiger salamanders, *A. tigrinum*. This interaction may benefit other syntopic salamander larvae living in ponds with Wood Frogs and Tiger Salamanders by reducing tadpole numbers, decreasing competition, and diverting predator focus. With *A. tigrinum* focusing on *Lithobates sylvaticus*, other larval *Ambystoma* have a greater chance of survival in mixed-species assemblages.

Wood Frog larvae are out-competed by Northern Leopard Frog larvae in experimental situations through the suppression of growth rates (Werner, 1992; *in* Relyea, 2000). However, this does not occur in natural situations, in which the reverse happens (Davis and Folkerts, 1986; Werner and Glennemeier, 1999; Relyea, 2000). The competitive effects of Northern Leopard Frogs on Wood Frogs also may vary annually, so that competition occurs some years but not others (DeBenedictis, 1970, 1974). Indeed, predators may also reverse the direction of competition. If Wood Frogs and Northern Leopard Frogs are raised together with the potential predators *Umbra* and *Anax*, it is the Northern Leopard Frog larvae that grow faster than the Wood Frogs (Relyea, 2000).

Petranka et al. (1994) demonstrated that American Toads, which breed four to ten weeks after Wood Frogs in North Carolina, avoid breeding ponds inhabited by Wood Frog larvae. In a series of natural and mesocosm experiments, Wood Frog tadpoles totally consumed eggs and hatchlings of the toad very quickly, and none survived (Petranka et al., 1994). Pond amphibian larval communities inhabited by Wood Frogs, and even ponds in the vicinity not inhabited by them, thus display complex trophic interactions between predators, prey, and competitors, interactions the nature of which change depending on year, season, and species present (Wilbur, 1972).

Fish also exert influences on survivorship of *Lithobates sylvaticus* larvae. In Alberta, for example, larval recruitment is much higher following winters in which severe fish kills occur. Presumably, reduced fish abundance decreases predation pressure on developing Wood frog tadpoles. The effect is most apparent when small-bodied fish are killed (Eaton et al., 2005a). Experiments have demonstrated that small-bodied fish can have significant predatory effects on Wood Frog larvae.

DISEASES, PARASITES, AND MALFORMATIONS

Ranaviruses are present and may cause serious die-offs in Wood Frog larval populations in the eastern United States as well as depress recruitment in subsequent years (Green et al., 2002; Petranka et al., 2003; Dodd, 2004; Gahl, 2007; Uyehara et al., 2010; Brunner et al., 2011). Frog Virus 3-like infections also have been reported in natural populations of Wood Frog larvae from Ontario (Duffus et al., 2008). Mass mortality in larval populations in Rhode Island also has been attributed to the bacterium *Aeromonas hydrophila*. Symptoms include emaciation, petechial hemorrhages, and loss of escape and feeding behavior (Nyman, 1986). No disease was observed at Nyman's pond the following year, but 3 yrs after the mortality event, only one egg mass was observed. Nyman (1986) suggested that subsequent reproductive failure resulted from a lack of recruitment as a result of the previous mortality event. An alveolate pathogen that causes mortality and morbidity in larvae was reported from Maine (Gahl, 2007).

Fungal infections are not uncommon on Wood Frog eggs (Herreid and Kinney, 1967; Davis and Folkerts, 1986; Gascon and Planas, 1986; Dodd, 2004), and the alga *O. amblystomatis* also can infiltrate the egg mass (Dickerson, 1906; Davis and Folkerts, 1986; Trauth et al., 1989). *Saprolegnia* (water mold) colonized Wood Frog eggs at 75% of the breeding ponds surveyed in the Adirondacks of New York (Karraker and Ruthig, 2009). "Mold" (probably the fungus *Saprolegnia* sp.) caused embryo developmental abnormalities and was present on all unhatched eggs in Québec. Mold was observed to penetrate the jelly of viable eggs (Gascon and Planas, 1986). These latter authors speculated that low pH decreases the resistance of the egg membrane to mould penetration. Fungal infections are reported as common on Wood Frog egg masses at cold temperatures (Herreid and Kinney, 1967). An *Ichthyophonus*-like fungal infection has been reported on Wood Frogs examined in Maine, Maryland, and Québec (Mikaelian et al., 2000; Green et al., 2002; Gahl, 2007). The highly virulent fungus *Batrachochytrium dendrobatidis* has been found on Wood Frogs from Rocky Mountain National Park and elsewhere in the Rockies (Rittman et al., 2003; Muths et al., 2008), Québec (Ouellet et al., 2005a), Maine (Gahl, 2007), Michigan (Zellmer et al., 2008), the Great Smokies (Chatfield et al., 2009), the upper Midwest (Martinez Rodriguez et al., 2009; Sadinski et al., 2010), but not in Alaska (Chestnut et al., 2008).

Wood Frogs are parasitized by blowfly larvae (*Lucilla elongata, L. silvarum*) (Roberts, 1998; Bolek and Janovy, 2004; Eaton et al., 2008), a variety of helminths (Bouchard, 1951; Odlaug, 1954; Najarian, 1955; Catalano and White, 1977; Baker, 1978c, 1979b; Muzzall and Preebles, 1991; McAllister et al., 1995a; Yoder and Coggins, 1996; Eaton et al., 2003; Green and Muths, 2005; Yoder and Coggins, 2007), many enteric protozoans (*Chilomastix* sp., *Entamoeba ranarum, Euglenamorpha hegneri, Giardia agilis, Hexamita intestinalis, Monocercomonas* sp., *Monocercomonoides* sp., *Nyctotherus cordiformis, Octomastix* sp., *Octomitus* sp., *Opalina obtrigonoidea, Opalina* sp., *Phacus longicauda, P. monilata, P. torta, P. pleuronectes, Retortamonas dobelli, Rhizomastix* sp., *Tritrichomonas augusta, T. batrachorum, T. prowazeki*, and *Urophagus* sp.) and the myxidium parasite *Myxidium* sp. (Odlaug, 1954; Anderson and Buttrey, 1962; McAllister et al., 1995a). Trematodes include *Alaria* sp., *Apharyngostrigea pipientis, Brachycoelium salamandrae, Echinoparyphium rubrum, Fibricola* sp., *F. texensis, Gorgoderina translucida, Haematoloechus* sp., *H. complexus, H. varioplexus, Lechriorchis tygarti*, and *Telorchis bonnerensis*; nematodes include *Cosmocercoides* sp., *C. dukae, Gyrinicola batrachiensis, Oxysomatium americanum, Ozwaldocruzia pipiens, Rhabdias bakeri, R. ranae*, and Spiruridae sp. (Odlaug, 1954; Adamson, 1981c; Yoder and Coggins, 2007; Pulis et al., 2011).

Although parasitized by *R. ranae*, infestation has no effect on growth and survival of metamorphic juveniles (Goater and Vandenbos, 1997). Cestodes include *Mesocestoides* sp. (Yoder and Coggins, 2007; Pulis et al., 2011).

When Wood Frog tadpoles come into contact with trematode (*Echinostoma* sp.) cercariae, they initiate sudden and forceful angular movements that have the result of dislodging the parasite from the body and tail region (Taylor et al., 2004). Such movements help prevent parasites, such as *Ribeiroia*, from lodging in the inguinal region where the hind limb buds form, and minimize the potential for resulting limb malformations. Limb deformities resulting from infestation by *Ribeiroia* have been reported in breeding populations of Wood Frogs, in which the prevalence of deformities is higher in ponds adjacent to agricultural fields than in natural settings (Kiesecker, 2002). The presence of pesticides results in suppression of immune activity, and thus makes the developing larvae more prone to trematode infestation.

Larvae are parasitized by the leech *Desserobdella picta*. As the numbers of leeches affecting a tadpole increases, the probability of mortality also increases. Although only one leech usually affects a tadpole, its effects are mediated by environmental conditions and tadpole size. For example, Leeches are particularly lethal to small tadpoles under conditions of lower temperatures and food levels under high population densities. Even a single leech can cause mortality. Under conditions of leech infestation, both survivorship and growth rates are decreased, indicating that leeches can have both lethal and sublethal effects on a Wood Frog tadpole community (Berven and Boltz, 2001).

Scoliosis (bent tails) has been reported in laboratory-reared Wood Frog larvae and metamorphosing juveniles; metamorphs sometimes had ankylose rear limbs (Hisaoka and List, 1957). Deformities (missing limbs and feet) were noted in 9 of 36 adult Wood Frogs collected from the Gaspé Peninsula; the deformities could not be attributed to predation (Trapido and Clausen, 1938). In a study of 21,050 specimens from throughout western Canada, only 2% of Wood Frogs had deformities, and included multiple limbs (polymelia), multiple sets of digits on one foot (polyphalangy), partial limbs (ectromelia), and missing limbs (amelia). Most deformities, however, were attributed to trauma (Eaton et al., 2004); Gray and Lethaby (2010) also reported a traumatic malformation. In a survey of 38 National Wildlife Refuges, only 0.4% of Wood Frogs had malformations (Converse et al., 2000). In Minnesota, 4.2% of Wood Frogs examined had malformations, but the sample size was small (Hoppe, 2005). Adult and late-term (105 day) larvae of wood frogs with ovotestes have been reported from Michigan (Cheng, 1929a, 1929b, 1930).

SUSCEPTIBILITY TO POTENTIAL STRESSORS

Metals. Wood Frog tadpoles are susceptible to total aluminum content, especially in conjunction with lower pH levels. Larvae exposed to aluminum may develop fluid-filled blisters over the entire body (Freda and McDonald, 1993), a condition associated with aluminum toxicity in *Lithobates pipiens* (Freda and McDonald, 1990). Swimming ability may be impaired in larvae, even when toxicity per se is low. However, aluminum toxicity in natural pond water may be offset via complexion by dissolved organic compounds (Freda and McDonald, 1993). Copper is extremely toxic to Wood Frog larvae (Horne and Dunson, 1995a). Wood Frog larvae are not susceptible to iron, lead, and zinc (Horne and Dunson, 1995a).

Under laboratory conditions, Wood Frog larvae accumulate heavy metals, transition metals, and nonmetals (arsenic, cadmium, iron, selenium, strontium, and vanadium) when raised in water contaminated with coal combustion wastes. Exposure to polluted water decreased growth and developmental rates, reduced survivorship, and decreased metamorphic success. Furthermore, the duration of the larval period increased by 11%, and the size at metamorphosis decreased by 39% (Snodgrass et al., 2004). These results further suggest severe effects on amphibian larval development in water polluted by coal ash power plants. Selenium and strontium may be the primary toxic agents.

Wood frog larvae exposed to zinc or to tire debris (a source of zinc) had increased time to metamorphosis compared to controls. The body mass of these larvae also was less than that of controls. Camponelli et al. (2009) noted that zinc leached from tire material was available to developing larvae, and that the presence of this metal in ponds near roads could influence larval development.

Other elements. High boron concentrations result in deformed larvae, specifically the curling effect (52% at low concentrations [50 mg L^{-1}] and 78% at high concentrations [100 mg L^{-1}]) (Laposata and Dunson, 1998). There was no effect on hatching success, however.

pH. The Wood Frog is often the most pH tolerant amphibian within its range and frequently breeds at acidic wetlands (Karns, 1983, 1992; Dale et al., 1985; Freda and Dunson, 1986; Ling et al., 1986; Sadinski and Dunson, 1992). Larval tolerance to acidity is not associated with either egg size or the thickness of the egg capsule, but hatching success is negatively correlated with ovum volume at pH 4.0 and capsule thickness at pH 3.75 (Pierce et al., 1987). Results of studies on the effects of acidity on Wood Frog larval development also must be interpreted with caution, since acid tolerance may vary depending on the stage of development (Freda and Dunson, 1985). Under laboratory conditions, larval mortality reaches 100% at pHs of 3.5–4.0; significant mortality occurs at pHs <4.25 (termed the critical pH by Freda, 1986; Freda and Dunson, 1986; Ling et al., 1986; Freda et al., 1991; Grant and Licht, 1993). The LC$_{50}$ is a pH of 3.74 in Wood Frogs, which avoided lethal pHs in laboratory trials although they entered water considered sublethal at a pH of 4.0 (Freda and Taylor, 1992). Sadinski and Dunson (1992) reported an LC$_{50}$ of 4.1, but Karns (1983, 1992), Pierce and Harvey (1987), and Pierce and Wooten (1992) noted regional differences in tolerance to acidic breeding conditions, perhaps explaining the variable results. Genetic factors appear to be important in the ability of Wood Frogs to develop in acidic wetlands (Pierce and Sikand, 1985; Pierce and Wooten, 1992).

In Nova Scotia, Wood Frogs successfully bred in acidic bogs (as low as pH 4.1 in ponds with *Ambystoma maculatum*) (mean pH 5.3, range 4.3 to 7.8) and appeared to be the least sensitive of 11 amphibians to acidic water (Dale et al., 1985). In Minnesota, Wood Frogs successfully hatched at acidic bogs, and were the only amphibian to survive in marginal fens. Abnormalities were common, however, and hatching occurred at higher rates in higher water temperatures (15°C) than in cold (3.5°C) water temperatures (Karns, 1992).

Abnormalities associated with sublethal effects

of low pH result in the "curling defect" in which embryos are tightly curled within the egg. Most such embryos do not hatch, but if they do they are deformed (Freda et al., 1991).

DOC. High levels of DOC, especially coupled with low pH, significantly lengthen the time to metamorphosis in Wood Frog larvae (Horne and Dunson, 1995b). At low DOC (15,000 µg liter^{-1}), pH is a primary toxin, whereas at high DOC (30,000 µg liter^{-1}), metals are more likely to be primary toxins (Horne and Dunson, 1995b).

Hardness. Survivorship increases in water with high levels of hardness (40 µS/cm) compared with low water hardness (27 µS/cm). Aluminum reduces survivorship in both chronic and acute exposure tests, but survivorship increases when larvae are reared in water with high hardness as compared with low water hardness. The presence of Wood Frogs is negatively correlated with conductivity in Ontario (Glooschenko et al., 1992).

Ammonium, Nitrates, Nitrites, and Phosphorus. Laposata and Dunson (1998) could not demonstrate adverse effects of nitrate up to 40 mg L^{-1}. Nitrates do not appear to affect larval survivorship or size at metamorphosis (Smith et al., 2011). Whereas Smith (2007) found no effects of nitrates, nitrites, or phosphorus on Wood Frog larvae in short-term acute exposure experiments, Griffis-Kyle (2005, 2007), and Smith et al. (2011) found that larvae exposed to nitrate throughout the larval period had delayed metamorphosis compared with controls. Developmental stage may be important to this species in terms of susceptibility to these compounds. In addition, Wood Frog tadpoles exposed to high concentrations of ammonium nitrate have lower embryo and early tadpole survivorship, slowed embryonic development, higher rates of deformed hatchlings, and are less active than controls (Burgett et al., 2007; Griffis-Kyle and Ritchie, 2007).

NaCl. Road salts used as deicers have significant affects on Wood Frog larvae. At realistic environmental concentrations, larvae had significantly lower survivorship, decreased time to metamorphosis, reduced body mass (reduced feeding) and activity (less prone to any movement), and increased abnormalities (bent tails) with increasing salt concentrations (Sanzo and Hecnar, 2006). Salt concentrations decreased with increasing distances from roads, and amphibian species richness increased accordingly. The median lethal concentration for Wood Frog larvae was between 2,635.5 and 5,109.2 mg/L NaCl.

Chemicals. Other contaminants shown to have adverse effects on Wood Frog tadpoles include endosulfan (Berrill et al., 1998) and endrin (Hedtke and Puglisi, 1982). Pyrethren insecticides are not lethal per se, but they have serious sublethal effects in that they slow growth and cause tadpoles to twist abnormally rather than to dart away in response to a threat. Larval survivorship actually increased after application of carbaryl (Sevin™) and malathion in mixed herbivore-predator species assemblages, although the basis for the increase was not determined (Relyea, 2005b). The $LC_{50\ (16\ day)}$ for malathion is 1.3 mg/L (Relyea, 2004b). However, death occurs in about 8 hrs at 30 mg/L of carbaryl and significantly reduces larval activity levels at 2.5 mg/L (Bridges and Semlitsch, 2000). This is the most sensitive species to the insecticide carbaryl of all ranids so far tested. Malathion by itself causes delayed metamorphosis but does not affect survivorship (Smith et al., 2011). Indeed, *Lithobates sylvaticus* larvae exposed to malathion are larger at metamorphosis when compared with controls. When combined with nitrate, the negative effects of malathion on larval duration are reduced. Nearly 100% of a Wood Frog population was killed by DDT sprayed at a dose of 0.45 kg per 0.4 ha (Fashingbauer, 1957).

The herbicide glyphosate (marketed as Roundup™ and Rodeo™), which inhibits essential amino acid synthesis, kills both larvae and juveniles at the manufacturer's suggested application rate when combined with a POEA surfactant (Relyea, 2005a, 2005b); 98% of larvae died over a period of 3 weeks, whereas 79% of juveniles died within 1 day. Relyea (2005a) suggested, however, that the POEA surfactant in Roundup™ was actually responsible for the mortality. Wood Frog larvae are sensitive to low levels of the herbicide atrazine (3 ppb), with survivorship dropping sharply at day 25 after exposure and decreasing to 40% after 32 days. High levels of atrazine (30, 100 ppb) had little effect on survivorship (Storrs and Kiesecker, 2004). In contrast, Allran and Karasov (2001) could not demonstrate adverse effects of 2.6–20 mg/L of atrazine on hatchability or mortality, but observations were conducted only for 96 hr posthatching. Wood Frog larvae were not affected by 2,4-D (Relyea, 2005b).

Finally, the chemical reagent thiosemicarbazide (TSC) severely affects the development of Wood Frog larvae, causing curvature of digits, abnormal limb articulations, difficulty in swimming, and mortality. Tadpoles exposed to low doses (10 mg/L) had slower growth than at higher doses (50 mg/L). TSC affects endochondral ossification by speeding up bone formation (Riley and Weil, 1986). However, short exposures (3–6 hrs) had no lasting effects, with increasing duration of exposure leading to increased malformations. Tadpoles at the 24–30 day stage of development are most affected (Riley and Weil, 1987).

UV light. Wood Frog eggs and larvae are not susceptible to adverse of effects of UVA or UVB light under ecologically relevant doses in terms of hatching success, developmental period, duration of metamorphic climax, or body mass at metamorphosis (Grant and Licht, 1995; Starnes et al., 2000). However, artificially high-intensity doses of UVB under laboratory conditions produced 100% embryo (exposed 30 min) and recently metamorphosed juvenile (exposed 2 weeks) mortality (Grant and Licht, 1995). Exposure to UVB at 12°C twice for 15 min produced the curling effect (photograph in Licht and Grant, 1997:140), unlike those exposed at 20°C for 15 min, most of which survived.

STATUS AND CONSERVATION

Wood Frogs are widespread and often abundant in the range occupied. Stable populations are reported in Rocky Mountain National Park, Colorado (Corn et al., 1997), and there are no reports of declines anywhere in Canada (Weller and Green, 1997). In a meta-analysis of 121 time series data sets, Green (2003) reported declines in 74 Wood Frog groups (61.2%) and increases in 47 (38.8%). No population declines, recoveries, or populations with "no change" were recorded. Likewise, call surveys conducted from 1973 to 1980 and again in 2001 to 2002 reported overall stable populations (disappearance from 18 sites, new appearance at 12 sites) in northern and western New York (Gibbs et al., 2005). However, Redmer and Trauth (2005) noted that populations in some states where the frog is peripheral consider it rare or restricted, and that it may be afforded legal protection.

As with all anurans, individual populations are threatened by habitat loss and fragmentation, and habitat modification has undoubtedly led to the disappearance of populations of this wide-ranging species. Rapid urbanization threatens many populations, and existing regulations and protected areas generally are inadequate to conserve this species (Baldwin and deMaynadier, 2009). For example, Windmiller et al. (2008) documented the collapse (by an estimated 94%) of a breeding population of *L. sylvaticus* in Massachusetts after 90% of the surrounding upland habitat was destroyed. Even after 10 yrs, no further occupancy was observed. At another population, abundance declined substantially (by 87%) after 41% of the surrounding upland habitat within 300 m of a breeding pond was destroyed for housing construction. Some animals continue to breed at this site, but abundance has never equaled preconstructions levels.

Populations must be conserved within a landscape context, and Homan et al. (2004) suggested that a critical threshold of 10–20% forest cover at 300 m from a breeding pond would be critical to the survival of this species; 20–30% cover at 100 m was nearly significant, further suggesting the importance of maintaining forested areas near breeding ponds. In addition, Skidds et al. (2007) noted a negative correlation between areas of residential development within 1 km of breeding sites and egg mass counts. The effects of fragmentation, such as in agriculturally dominated landscapes, may be ameliorated by providing shrubby or wooded riparian strips allowing for dispersal (Maisonneuve and Rioux, 2001).

Populations may be particularly susceptible to habitat loss and succession in areas where Wood Frogs were never abundant or are on the periphery of their range. The Wood Frogs of Cedar Bog, Ohio, which may never have been common, likely were adversely affected as the relictual boreal fen's original 2,833 ha were whittled to 2.5 ha over the last century (Lovich and Jaworski, 1988). The last Wood Frog seen in a 100-acre woods in Delaware County, Indiana, was observed in 1959. The reason for the disappearance is unknown (Minton et al., 1982). In addition, the lack of protection afforded very small wetlands could be a threat to this species. Egan and Paton (2004) and Calhoun et al. (2003) demonstrate the importance of even very small vernal pools to the survival of Wood Frogs.

In Rhode Island and elsewhere, the abundance of egg masses in breeding ponds is negatively correlated with nearby road density (Egan and Paton, 2004; Veysey et al., 2011), suggesting an adverse effect on local breeding populations through road-related mortality. Likewise, high levels of traffic are inversely correlated with the presence of Wood Frogs in complex landscapes (Eigenbrod et al., 2008). In contrast, Wood Frog adults and most juveniles crossed road corridors 5 m and 12 m in width (deMaynadier and Hunter, 2000). The tendency to cross roads is likely to expose this species to highway-related mortality, and thus to population declines in the vicinity of roads (Karraker and Gibbs, 2011). Populations living adjacent to roads also might be impacted by road deicing salts (Karraker, 2008) and are subject to decreased hydroperiods and increased isolation (Veysey et al., 2011).

Other transportation corridors may be used by Wood Frogs, however. In Québec, Wood Frogs were not trapped either at the ecotone between transmission line corridors and natural forest or within the corridors themselves. However, they were heard calling from wetlands located within the corridors (Fortin et al, 2004b). In terrestrial situations, Wood Frogs generally avoid open areas, such as clearcuts and power-line rights-of-way (deMaynadier and Hunter, 1999), although they may forage in grassy pastures on overcast days and at night.

As the above examples illustrate, Wood Frog populations are affected by habitat alteration at several spatial scales, from localized effects on breeding ponds to development at a landscape scale at considerable distances from a breeding site. In New England, Egan and Paton (2008) and Clark et al. (2008) demonstrated that Wood Frog populations are positively affected by breeding pond availability, pond size, forested wetlands, and upland forest within 1,000 to 2,000 m of breeding ponds. Breeding is significantly reduced in landscapes with high human population density (Clark et al., 2008) and >7% developed lands and road densities accounting for >4% of the landscape within 1,000 m of breeding sites (Egan and Paton, 2008). These latter authors noted, however, that at least some individuals were detected in areas with as little as 15% forested landscape cover. Low road densities were associated with good Wood Frog populations. Few roads mean decreased levels of mortality during the breeding and dispersal migrations.

Wood Frog calls generally have low detection probabilities during volunteer-based anuran monitoring programs (de Solla et al., 2005). Thus, estimating Wood Frog population status and trends using short-term call-based occupancy models may have limited utility over a wide geographic area. However, chorus ranks (0–3) based on the number of males calling were positively correlated with the number of egg masses in 20

Prairie breeding habitat of *Lithobates sylvaticus* and *Pseudacris maculata*. Photo: Robert Newman

Eastern woodland pool breeding habitat of *Lithobates sylvaticus*, Great Smoky Mountains National Park, Tennessee. Photo: C.K. Dodd Jr.

Alberta ponds (Stevens and Paszkowski, 2004). The timing of the call survey also was important, with only the second of four three- to six-day call-sampling periods being the best significant predictor of egg mass number. Incorporating both diurnal sampling protocols with acoustic surveys and increasing the number of sampling visits may help to offset low detection probabilities associated with Wood Frog call surveys. Egg mass counts also are effective tools to monitor long-term population trends (Raithel et al., 2011).

Wood Frog adults and juveniles prefer closed-canopy mature forest (70–90 yrs) to clearcuts (2–11 yrs) (deMaynadier and Hunter, 1999). Timber harvesting reduced the abundance of Wood Frogs along headwater streams in Maine over a 3 yr period, except where experimental buffers were left in place (Perkins and Hunter, 2006). Frog abundance in buffer zones (11, 23 m) was greater than in adjacent harvested areas, but abundance in the 11 m zone was still significantly less than in preharvest upstream sections. There was no difference in abundance, however, between the 23 m buffer zone and the preharvest upstream section (Perkins and Hunter, 2006). In a retrospective study, Wood Frog captures were significantly lower in clear cuts compared with buffer zones in areas harvested 4–10 yrs prior to sampling. These authors suggest that partial harvesting may have little short-term detrimental effect on Wood Frogs, but that Wood Frogs continued preferentially to occupy unharvested buffers 4–10 yrs after initial clearcuts.

Wood Frogs breed in small wetlands resulting from timber harvesting (termed anthropogenic temporary pools). However, such pools are usually shallower, more exposed, and more prone to desiccation than natural woodland pools, resulting in increased probabilities of mortality, especially in rainfall-deficient years. Wood Frog metamorphs resulting from such pools are smaller than those from natural pools, and they metamorphose faster (DiMauro and Hunter, 2002). DiMauro and Hunter (2002) suggest such silviculturally derived anthropogenic pools serve as ecological traps for Wood Frogs unless they can be enlarged, their depth increased, and an increased amount of canopy cover be left to shade them.

Hannon et al. (2002) found no differences in amphibian abundance in buffers 20–800 m around 12 lakes in Alberta. These authors suggested that Wood Frogs were more of a habitat generalist in the northern portion of their range, and that there, at least, small buffer zones may be effective at ensuring presence. However, even in the "clearcuts" adjacent to their sampling sites, up to 10% of the trees were left uncut, and trees are cut in a checkerboard pattern two-pass system, with the passes 10 yrs apart. Thus, the results of Hannon et al. (2002) may not be directly comparable where trees are clearcut using different protocols. Another strategy would be to have buffer zones that are unharvested, surrounded by secondary buffer zones that use a partial harvest approach (Perkins and Hunter, 2006).

Baldwin et al. (2006) also suggested that traditional buffer zones extending "x" amount of distance around breeding sites may not be necessary if exact movement patterns and habitat use are known in advance. Using GIS to predict conservation corridors, Wood Frog populations might be adequately protected in urbanizing situations for far less money than would be necessary if circular large buffer zones were drawn around each breeding pond. At their study site, Baldwin et al. (2006) estimated that the "habitat element conservation plan" approach could reduce acquisition costs by 66–75% and result in improved conservation planning.

In contrast to some of the previous studies, Harper et al. (2008) noted that extinction risk was high for Wood Frog populations protected by 30–1,000 m buffer zones. The life history characteristics of this species (generally short-lived, high fecundity) are not conducive to survival when faced by habitat loss and fragmentation of remaining populations. Even with 1,000 m buffers, the extinction probability was high (11%) within 20 yrs. In models where 93% of the mortality occurred during the first year, there was a 24% probability that the population would go extinct within 20 yrs. Cumulative extinction probabilities increase rapidly through time for Wood Frogs (Harper et al., 2008).

Peat mining adjacent to boreal bog remnants has a negative impact on Wood Frog abundance, and the greatest abundance of Wood Frogs occurs in unmined bogs (Mazerolle, 2003). This is likely due to the addition of drainage ditches in mining operations, rather than habitat loss per se. Mazerolle (2003) attributes decreases in abundance to loss of connectivity between terrestrial foraging and overwintering sites and breeding ponds. Wood Frogs also prefer natural forested areas to areas converted into black spruce (*P. mariana*) plantations; Wood Frogs in ponds adjacent to both habitats tend to orient toward the natural forest (Waldick et al., 1999).

Wood Frogs will also use restored wetlands, although they avoid wetlands if fish are present (Lehtinen and Galatowitsch, 2001; Petranka et al., 2003; Vasconcelos and Calhoun, 2004; Shulse et al., 2010). On Prince Edward Island, Stevens et al. (2002) found no differences between restored and reference ponds in Wood Frog occupancy and abundance. Petranka et al. (2003) found that Wood Frogs readily used restored ponds shortly after creation, but that there was substantial variation in reproductive output, due to the effects of drought and a ranaviral outbreak. Larval survivorship rather than embryonic mortality accounted for most of the variation in the number of metamorphosing juveniles produced. In Arkansas, Wood Frog occupancy was correlated with the age of newly created ponds, at least for the first 3 yrs after creation (Cartwright et al., 1998). In contrast, Wood Frogs may suffer complete embryo mortality when exposed to toxic chemicals in the sediments of man-made retention ponds (Snodgrass et al., 2008).

There have been two apparently successful reintroductions of Wood Frogs into areas from where they had previously disappeared. Wood Frogs were reintroduced into eastern Missouri in 1980, a region where they had been previously reported but were considered rare (Hurter, 1911). The population at the Tyson Research Center continues to be viable, and Wood Frogs have colonized ponds as far as 2.4 km from the original release site (Sexton et al., 1998). Eggs, larvae, and adults were introduced from Indiana to McDonough County, Illinois, beginning in 1982 and extending to 1991 (Thurow, 1994), near an area where they had been reported historically (Garman, 1892). However, there is some question as to the validity of the Garman (1892) record, and the biological and ethical basis for the introduction is questionable (Redmer, 1998a; Szafoni et al., 1999). The resulting population appears to have increased and expanded, despite drought conditions (Thurow, 1994). In contrast, an effort to reestablish *Lithobates sylvaticus* at the Gateway National Recreation Area in New York has been unsuccessful (Cook, 2008).

Lithobates tarahumarae (Boulenger, 1917)
Tarahumara Frog

ETYMOLOGY
tarahumarae: Latinized name for the Sierra Tarahumare, the type locality of the species.

NOMENCLATURE
Stebbins (2003): *Rana tarahumarae*
   Dubois (2006): *Lithobates (Lithobates) tarahumarae*
   Synonyms: None

IDENTIFICATION
Adults. This large ranid is dark brown dorsally with small black spots. The skin may range from smooth to rough or bumpy. A very obvious fold of skin occurs from the posterior corner of the eye running above and beyond the tympanum and terminates near the front of the foreleg. Dorsolateral folds may be very faint in a few individuals, but most frogs lack dorsolateral folds. The tympanum is small, only one-half the diameter of the eye. Vocal sacs or vocal sac openings are not apparent. The venter is white, although the throat and upper chest may be dark gray. Hind limbs are barred dorsally, and the groin may contain some yellow

coloration. Males are smaller than females, and reach 88 mm SUL; females reach 113 mm SUL.

Larvae. Larvae have a greenish-gray body and tail coloration, and there is very prominent spotting on the body, tail fin, and tail musculature. Tail fins are high and acutely pointed. There are five to six upper tooth rows (Webb and Korky, 1977), although previous authors (Campbell, 1931; Wright and Wright, 1949; Stebbins, 1951; Orton, 1952) report only four. Larvae reach 106 mm TL (Campbell, 1931).

Eggs. Eggs are black dorsally and white ventrally. They are deposited in a round cluster that measures about 5.6 cm × 7.6 cm. There are two envelopes, the outer measuring 3.7–5.0 mm and the inner 2.9–3.4 mm. The vitellus is 2.0–2.2 mm in diameter. Zweifel (1955) estimated there were 2,200 or so eggs per clutch, but this and the other measurements above were based on only a single clutch.

DISTRIBUTION

This species barely entered the United States in southern Arizona in the Pajaritos, Santa Rita, and Tumacacori Mountains, but these populations were extirpated by 1983 (Hale et al., 1995). *Lithobates tarahumarae* has been repatriated into Santa Cruz County, Arizona, and an experimental population was established at Kofa National Wildlife Refuge in western Arizona (Rorabaugh and Elliot, 2006). There are literature records for this species elsewhere in Arizona (Little, 1940), New Mexico (Linsdale, 1933a; Little and Keller, 1937), and Texas (Smith and Taylor, 1948), but these were based on misidentified young *L. cates-*

Distribution of *Lithobates tarahumarae*

beianus or unsupported range extrapolation (Zweifel, 1955). In Mexico, the species occurs in eastern Sonora, western Chihuahua, and northern Sinaloa in western drainages of the Sierra Madre Occidental. The eastern and southern distributional limits in Mexico have not been determined. Distributional references include Campbell (1931), Hale et al. (1977), Rorabaugh and Hale (2005), Brennan and Holycross (2006), and Lemos Espinal and Smith (2007a).

FOSSIL RECORD

No fossils are known.

SYSTEMATICS AND GEOGRAPHIC VARIATION

Lithobates tarahumarae is a member of the *Novirana* clade of North American ranid frogs. It gives its name to the *L. tarahumarae* group (or *Zweifelia*), an assemblage of mostly Mesoamerican ranids that includes *L. zweifeli* and *L. pustulosus* from Mexico (Hillis and Wilcox, 2005). Its closet relatives in the United States are the frogs in the leopard frog complex, especially those of the Southwest.

ADULT HABITAT

This species appears to prefer narrow rocky canyons, arroyos, or barrancas containing permanent or seasonal streams with plunge pools through oak or pine woodlands. In Pena Blanca Springs, *L. tarahumarae* was found sitting around water-filled potholes within the canyon along the course of the then-dry stream channel (Campbell, 1931). Pools should be deep enough to contain water during the driest seasons and contain underwater rocky shelves or boulders that frogs can hide beneath. In Mexico, the Tarahumara Frog has been found in wide (ca. 5.5 m) flowing streams that are relatively shallow (26 cm) through Sinaloan thornscrub and tropical deciduous forest. In Arizona, *L. tarahumarae* is found at an elevation of about 1,220–1,710 m.

AQUATIC AND TERRESTRIAL ECOLOGY

Adult frogs probably do not venture far from water, but Hale et al. (1977) noted that individuals have been observed in "unlikely" canyons during the summer monsoon season. Some individuals seem to prefer to remain at a single location. Adult males have been reported to move up to 1,885 m, and adult females to 651 m (*in*

Egg mass of *Lithobates tarahumarae*. Photo: James Rorabough

Rorabaugh and Hale, 2005). Movements occur from the end of the dry season in June through the rainy season in August. Juveniles have been reported to disperse long distances (to 1,885 m) along watercourses (*in* Rorabaugh and Hale, 2005), usually upstream.

Lithobates tarahumarae may be seen sitting along streams and pools, in rock niches in the canyon walls adjacent to stream channels, or in riffles directly in the stream. Adjacent rock cracks or crevices, as well as the plunge pools, may provide shelter during cold weather. Little information is available on the species' thermal habits, but it has been observed in water 18–27°C (Zweifel, 1955).

CALLING ACTIVITY AND MATE SELECTION

Lithobates tarahumarae has been considered voiceless, but they produce a distress call (a squawk) as they jump into the water (Campbell, 1934), and both sexes have a release call, a low grunt or snore of about 0.5 sec. Reports of voicelessness are in error. Rorabaugh and Elliot (2006) recorded 4 distinct calls, including a snore (326–2,086 Hz), a soft whine (707–1,040 Hz), a vibrant "eep" or "eep-up" (1,113–4,474 Hz), and a throaty croak or squawk (309–2,339 Hz) uttered before a snore. The calls are not loud and easily blend into background noise. Since both juveniles and adults made the snore and whine calls, these likely are not advertisement calls. Rorabaugh and Elliot (2006) noted one amplexing pair making the whine call underwater. All other observations were made out of water. These authors provided spectrograms of the various calls.

Breeding occurs at the end of the dry season in April and May (Rorabaugh and Hale, 2005). However, calling has been reported in the Kofa experimental population from early March into October. Since the calls may not be advertisement calls, they are not indicative of breeding activity. Most calling occurs out of water at dusk and continues into the evening. Calls are heard when water temperatures are >15°C. Males are usually in close proximity to one another when calling, suggesting that pairing occurs haphazardly as frogs approach one another; reproductive males likely attempt amplexus with any frog moving in their proximity.

BREEDING SITES

Breeding occurs in shallow pools within intermittent streambeds in the narrow rocky canyons inhabited by this species. Rorabaugh and Humphrey (2002) noted that the best breeding sites are plunge pools that have low mean flows (< 5.6 l/s) and steep gradients (60 m/km) across a bedrock surface. At Pena Blanca Spring, Tarahumara Frogs inhabited artificial impoundments.

REPRODUCTION

There is a single observation of an egg mass in Pena Blanca Springs on 23 August (Stebbins, 1951). This mass was attached to the rocky bottom of a pool within the dry streambed that received a trickle of water. Stebbins (1951) reported at that time that both gravid and spent females were captured, making it seem likely that egg deposition occurs during the summer. The presence of small tadpoles in late October suggests a fall breeding season as well (Rorabaugh and Hale, 2005). Other masses have been reported attached to the rocky substrate or free-floating (Rorabaugh and Hale, 2005). Clutch size in a partial (?) clutch was 850–900 eggs (Rorabaugh and Humphrey, 2002). In an unpublished report, Hale (*in* Rorabaugh and Hale, 2005) reported a mean clutch size of 1,083 (range 527 to 1,635 eggs/clutch). Eggs hatch in about eight days.

Tadpole *of Lithobates tarahumarae*. Photo: James Rorabough

LARVAL ECOLOGY

Larvae are omnivores, preferring algae. They may drift downstream with stream currents during development, a movement that is counteracted by juvenile dispersal upstream. Metamorphosing tadpoles or large larvae have been found from mid-June (Campbell, 1931) to late October (Zweifel, 1955; Rorabaugh and Hale, 2005). If breeding occurs during the spring to early summer (Stebbins, 1951; Rorabaugh and Hale, 2005), then larvae probably delay metamorphosis until at least the following summer. Zweifel (1955) speculated that larvae might even require 2 yrs before transformation, but Rorabaugh and Humphrey (2002) reported that some larvae in captivity transformed in 86 days; most took >10 months, however. Summer transformation would allow recent metamorphs time to grow. In addition, it might allow juveniles to disperse along stream beds during brief periods of rain during the summer monsoons. Most recent metamorphs are 35–40 mm SUL, but froglets as small as 21 mm SUL have been reported (Rorabaugh and Hale, 2005).

DIET

Lithobates tarahumarae is not a selective predator and will feed on just about any animal it can catch. Prey includes invertebrates (beetles, sphinx moths, water bugs, scorpions, mantids, grasshoppers, wasps, spiders, caddisflies, katydids, moths, centipedes), fish (*Gila ditaenia*), juvenile mud

Adult *Lithobates tarahumarae*. Photo: Carl Gerhardt

Habitat of *Lithobates tarahumarae*, Santa Rita Mountains, Arizona. Photo: James Rorabough

turtles (*Kinosternon sonoriense*), and even snakes (*Tantilla atriceps*) (Zweifel, 1955; *in* Rorabaugh and Hale, 2005). Zweifel (1955) suggested that the prey were indicative of nocturnal feeding habits.

PREDATION AND DEFENSE
This species dives into the water upon the approach of a potential predator and may utter a startle "squawk" (Campbell, 1934). They hide in leaf litter, under rocks, or in woody debris. According to Rorabaugh and Hale (2005), the skin secretions of *Lithobates tarahumarae* are noxious to taste and may cause minor skin irritation. Specific predators have not been recorded, but it seems likely a variety of mammals (*Bassariscus astutus*), birds, and snakes (including the garter snake *Thamnophis cyrtopsis*) prey upon juveniles and adults. Larvae may be eaten by aquatic invertebrates (*Belostoma*, *Lethocerus*), mud turtles, and garter snakes (*in* Rorabaugh and Hale, 2005).

POPULATION BIOLOGY
Virtually nothing is known about the population biology of this species. Campbell (1934) recorded it as "of rather common occurrence" at the permanent waterholes within the region. An estimate of the population size at its peak in the Santa Rita Mountains was 1,020 frogs in 1976, but this fell to 625 by 1977. By 1982, only one to three frogs were observed, with the last frog being sighted in 1983 (Hale et al., 1995). In unpublished data, Hale and Mays (*in* Rorabaugh and Hale, 2005) reported maturity is reached at 2 yrs following metamorphosis, when males are ca. 64 mm SUL and females 67 mm SUL. Longevity may extend to 6 yrs postmetamorphosis.

DISEASES, PARASITES, AND MALFORMATIONS
Amphibian chytrid fungus (*Batrachochytrium dendrobatidis*) has been found in *Lithobates tarahumarae* populations throughout the species' range, and is likely to have played a major role in the disappearance of this frog in Arizona and parts of Mexico (Hale et al., 2005). Some infected populations have not declined, however. Antimicrobial peptides are present in the skin of this species and may provide some protection against the fungal pathogen (Rollins-Smith et al., 2002).

Tarahumara Frogs are parasitized by the trematodes *Glypthelmins quieta*, *Haematoloechus breviplexus*, *Langeronia macrocirra*; the cestode *Ophiotaenia magna*; the nematodes *Falcaustra inglisi*, *F. lowei*, *Foleyellides striatus*, *Oswaldocruzia pipiens*, *Physaloptera* sp., *Rhabdias ranae*, *Subulascaris falcaustriformis*; and unidentified acanthocephalans (Bursey and Goldberg, 2001). Hale and Jarchow (*in* Rorabaugh and Hale, 2005) also reported unspecified nematodes and cestodes in this species.

SUSCEPTIBILITY TO POTENTIAL STRESSORS
Metals. Elevated levels of cadmium and arsenic were found in water where populations of *Lithobates tarahumarae* disappeared in southern Arizona. Hale et al. (1995) speculated that acid-soluble zinc leached from canyon walls, in conjunction with accumulations of insoluble cadmium in sediments, combined to have detrimental effects upon this species. This hypothesis has yet to be empirically examined.

STATUS AND CONSERVATION

This species was extirpated by 1983 from southern Arizona due to undocumented causes, and its precarious status in the United States was recognized as early as 1974, when dead and dying frogs were discovered in the Atascosa-Pajarito Mountains (Bury et al., 1980; Hale et al., 2005). Populations in the Santa Rita Mountains began declining in 1977. Many populations in Mexico also are declining or have been extirpated (Hale et al., 2005), but the species is not protected there. In order to have larvae and frogs to repatriate into formerly occupied habitat, a partial egg mass was imported into the U.S. in 2000 to begin studies on captive breeding (Rorabaugh and Humphrey, 2002). An experimental population later was established at the Kofa National Wildlife Refuge in western Arizona. Tarahumara Frogs now have been repatriated into Big Casa Blanca Canyon in Santa Cruz County. Frogs to be repatriated are treated with itraconazole to counteract the effects of amphibian chytrid. Egg masses, large tadpoles, juveniles, and adults were observed subsequently, indicating reproduction and survival. The repatriated population is protected by state law.

Suggested causes of declines have included habitat loss, the influence of toxic chemicals resulting from copper smelters, acid rain, and disease, especially the amphibian chytrid pathogen *Batrachochytrium dendrobatidis*. This pathogen may be the primary cause of decline, especially since some of the symptoms attributed to metal poisoning (dry skin, lack of righting response, partial paralysis) are also symptomatic of the final stages of chytridiomycosis. However, some populations in Mexico have been infected for at least 17 yrs without declining (Hale et al., 2005), and Hale et al. (1995) noted that no abnormalities were present in histopathological examinations carried out by Elliott Jacobson at the University of Florida. Cold and pollution may increase the severity of the disease or at least make frogs more susceptible to it. The interaction of sublethal effects on the virulence of amphibian chytrid is not well understood.

Much data on the life history and ecology of this species can only be found in unpublished reports. Rorabaugh and Hale (2005) have summarized some of this information.

Lithobates virgatipes (Cope, 1891)
Carpenter Frog

ETYMOLOGY
virgatipes: from the Latin words *virgatus*, meaning 'striped,' and *pes*, meaning 'foot.' The name refers to the striped markings on the rear feet.

NOMENCLATURE
Conant and Collins (1998): *Rana virgatipes*
  Dubois (2006): *Lithobates (Aquarana) virgatipes*
  Synonyms: None

IDENTIFICATION
Adults. Adults are olive, greenish, tan, or brown with blackish mottled spots dorsally. There are two yellowish, reddish-brown, or light brown dorsolateral lines running parallel along the dorsum; these lines begin behind the eye. A second series of lines runs from the edge of the upper jaw, below the tympanum, to the groin; the lower lines may be more distinctive and broader than the upper lines and contain light-colored tubercles. Both sets of lines tend to fade out posteriorly. The tympanum and iris are bronze. Bellies are whitish with dark spots, although some individuals have a dull yellowish coloration on the lower surface of the head and forefeet. Blackish-brown spots or blotches are found along the lateral sides of the body; these are more prominent than the dorsal markings. The rear of the femur has alternating light and dark bars, with the blackish-brown variegated markings of the undersides extending onto the belly. The longest toes of the rear feet extend well beyond the webbing, unlike that of small *Lithobates grylio*, which are similar in appearance to Carpenter Frogs and with which they may be confused. Considerable variation exists among frogs with regard to coloration and extent of dorsal and ventral blotching, and there are no differences between the sexes in color patterns. Fowler (1905) illustrates several examples of ventral pattern variation. Males have a swollen

thumb, vocal sacs, and a tympanum larger than that of females.

Males may be slightly smaller than females. In New Jersey, adult males ranged between 39 and 62 mm SUL (Given, 1988a). Wright and Wright (1949) gave a range of 41 to 63 mm SUL for males and 41–66 mm SUL for females. In Virginia, males averaged 53 mm SUL (range 45 to 63 mm), and females averaged 53 mm SUL (range 42 to 66 mm) (Georgel, 2001). The largest reported male is 68 mm SUL, and the largest female is 73 mm SUL (Standaert, 1967).

Larvae. The dorsal larval coloration is grayish or brown with black spots. There is a distinctive stripe or row of dots formed along the lateral line pores on the dorsal tail fin. The tail musculature is not bicolored, a tail musculature stripe is present, and there are light spots surrounded by dark pigment near the edge of the tail fins. The gut usually is not visible, but it may be indistinct. Lower jaws are relatively narrow. The maximum size of the larva is 100 mm TL (Jensen et al., 2008). The tadpole is described by Altig (1970).

Eggs. The eggs of *L. virgatipes* are black above and creamy white to light tan below; they have only one envelope surrounding the vitellus. The envelope is a mean of 5.4 mm in diameter (range 3.8 to 6.9 mm), and the vitellus is a mean of 1.6 mm (range 1.4 to 1.8 mm) in diameter (Wright, 1932; Livezey and Wright, 1947). Eggs are deposited in globular firm clumps and hatch in about three to seven days.

DISTRIBUTION

Carpenter Frogs occur in scattered populations from the southern half of New Jersey to northern Florida (Osceola National Forest). They are found solely on the Atlantic Coastal Plain, reaching as far inland as the Fall Line in South Carolina and adjacent North Carolina. Important distributional references include: Range-wide (Reed, 1957), Delaware (Conant, 1940; White, 1988; White and White, 2007), Florida (Stevenson, 1969), Georgia (Wright, 1932; Neill, 1952; Williamson and Moulis, 1994; Jensen et al., 2008), Maryland (Conant, 1947; Grogan, 1974; Harris, 1975; Given, 1999a; White and White, 2007), Mid-Atlantic (Beane et al., 2010), New Jersey (McCormick, 1970; Schwartz and Golden, 2002), North Carolina (Brimley, 1944; Dorcas et al., 2007; Gaul and Mitchell, 2007), South Carolina (Neill, 1952), and Virginia (Funderburg et al., 1974b; Grogan, 1974; Tobey, 1985; Mitchell and Reay, 1999).

Distribution of *Lithobates virgatipes*

FOSSIL RECORD

There are no known fossils of this species.

SYSTEMATICS AND GEOGRAPHIC VARIATION

The Carpenter Frog is a member of the *Lithobates catesbeianus* species group (*Aquarana*) of North American (*Novirana*) ranid frogs (Hillis and Wilcox, 2005). The species appears to be the basal member of the *Aquarana* clade and to have evolved from a monophyletic *Aquarana* ancestor on the Atlantic Coastal Plain. All other members of *Aquarana* clade diverged from the ancestral *Lithobates catesbeianus* group during the late Miocene to early Pliocene (Austin et al., 2003a). Changes in sea level during the Tertiary appear to have played a major role in the evolution of the *Aquarana* clade and likely contributed to the species' current distribution. In this regard, a previous suggestion that the *catesbeianus* species group is paraphyletic (Pytel, 1986) is not currently supported.

Laboratory hybridization between ♂ *L. virgatipes* and ♀ *L. pipiens* does not proceed beyond the gastrula stage (Moore, 1949a).

ADULT HABITAT

The Carpenter Frog is a species of highly acidic (pH 4.0–5.0) wetlands on the Atlantic Coastal Plain, although Gosner and Black (1957a) noted that low pH was not required by this species for embryonic development. It is found in a variety of aquatic habitats, including larger rivers, lakes, blackwater creek swamps, small isolated wetlands, vernal pools, cranberry bogs, beaver ponds, seasonally flooded "Delmarva Bays," Carolina Bays, pocosins (shrub bogs), and even man-made impoundments, borrow pits, drainage ditches, and canals, as long as there is an abundance of floating or thick emergent vegetation such as lily pads and green arrow arum (*Peltandra virginica*). These frogs are particularly associated with mats of sphagnum and are often called "sphagnum frogs" (Fowler, 1905).

In Maryland, occupancy is associated with intermediate hydroperiods, acidic wetlands (pH 4.0–5.0), and a surrounding forest cover from 50 to 500 m away from the wetland (Otto et al., 2007b). The amount of forest cover within 50 m of the high-water mark is particularly important. In the New Jersey Pinelands, Carpenter Frogs are found in areas of elevated acidity, conductivity (ca. 50–80 µS cm^{-1}), and nitrate-nitrite levels, where they prefer unaltered habitats where the American Bullfrog is absent (Bunnell and Zampella, 1999; Zampella and Bunnell, 2000).

In New Jersey, *Lithobates virgatipes* is a species of the Pine Barrens, where the dominant pine is pitch pine (*Pinus rigida*) (McCormick, 1970). They are frequently associated with Atlantic white cedar (*Chamaecyparis thyoides*). Dominant trees associated with Atlantic white cedar include bald cypress (*Taxodium distichum*), tupelo gum (*Nyssa aquatica*), and red maple (*Acer rubrum*). In the South Carolina Coastal Plain, the species may be found in isolated small wetlands interspersed within longleaf pine (*Pinus palustris*) forests located on sandhill ridges (Russell et al., 2002a). In all habitats, soils are organic, and standing water usually occurs year-round except during serious droughts.

Carpenter Frogs may inhabit rather small wetlands. For example, Russell et al. (2002a) recorded 307 different Carpenter Frogs over a 2 yr period in 5 small isolated wetlands in South Carolina, and even a 0.38 ha pond contained 30 *Lithobates virgatipes*.

AQUATIC AND TERRESTRIAL ECOLOGY

Carpenter Frogs sit along the shorelines of slow-moving streams and ponds, resting on sphagnum mats partially submerged in the water. There may be some differences in habitat use between the sexes, with males occupying the centers of wetlands and females more likely to be found nearer the shoreline (Standaert, 1967). Males, however, will call from shallow water near the shoreline, moving to open water following breeding. Activity occurs throughout the warm season into the autumn. In Virginia, for example, Carpenter Frogs may be found from March through November (Georgel, 2001).

These frogs can tolerate high temperatures, with the mean CTmax of adults from 35.8 to 38°C, depending upon acclimation temperature (Holzman and McManus, 1973). Overwintering occurs in aquatic habitats. Environmentally based mortality can occur both in summer and winter. For example, Standaert (1967) reported finding dead frogs in shallow pools in spring that probably died during the winter due to anaerobic conditions. He further noted frogs that may have died during high summer water temperatures, which also produced anaerobic conditions. Males seemed to be particularly vulnerable to anaerobic conditions during what he termed the "summer torpor."

Movement patterns of this species are not well understood. Standaert (1967) anecdotally mentioned that small frogs dispersed from a large pond to a nearby smaller pond in fall, remained there during the following spring, then returned to the larger core pond as smaller adjacent ponds dried during the summer. Thus, there may be routine periodic movements among adjacent ponds, depending on hydroperiod or other factors.

Carpenter Frogs are photopositive in their phototactic response, suggesting they use both sunlight and moonlight when feeding and going about their daily activities (Jaeger and Hailman, 1971, 1973). Given (1988a) confirmed night feeding. They likely have true color vision.

CALLING ACTIVITY AND MATE SELECTION

Calling begins in the spring and may extend throughout the summer. In Virginia, for example, calls can be heard from March through September (Georgel, 2001). The advertisement call of this species is a series (1–10) of loud "ca-thunk ca-thunks" that sounds like carpenters hammering nails (dominant frequency of 460–720 Hz; lesser defined peak at 1,500–2,500 Hz; 110 dB) (Given, 1987, 1990). Call intensity increases with male body mass. The calling male may be difficult to locate, despite the loud distinctive call. Most males call every night throughout their breeding season, and Given (1988b) recorded a median residency of 40.5 nights (range 8 to 94).

Calling normally occurs from dusk to dawn throughout the main portion of the breeding season, with peak levels at 23:00–01:00 hrs in New Jersey (Given, 1987) and 21:00 hrs in Virginia (Georgel, 2001). In August and September in Virginia, calling becomes more diurnal and tapers off at night. Not all males call with equal intensity. First year males tend to put less effort into calling than larger, older males. In Given's (1988a) study, 9 frogs called with low intensity, 26 with average intensity, and 7 with high intensity.

Males call from shallow-water habitats located from 0.1 to 0.5 m from the shoreline, while sitting among emergent vegetation. Vegetation must extend at least 1.7 m from the water's edge, but water depth per se is not important to territorial establishment. Males call in a high posture, with their heads up at a slight angle and their lungs inflated. They tend to call nightly from the same locations, and they will defend their calling station in aggressive encounters with other males (Given, 1988b). Such territorial defense lasts about two weeks until a male has completed his breeding season. Males establish from 0 to 4 territories (median 1.5) and spend about 84% of their time in such territories (Given, 1988b). Territories are ca. 1 m^2 in area, and the mean distance a male moves from one territory to establish another is 22.5 m (range 4 to 200 m). Males frequently return to a previously established territory.

When males detect another frog (whether male or female), they switch from the advertisement call to a single-note aggressive call (dominant frequency of 565–720 Hz; secondary frequencies of 1,140–1,380 Hz; 17–32 ms call duration). Intruding males respond by uttering a single-note aggressive call. The aggressive call is higher in primary frequency than the advertisement call, but the secondary frequency of the advertisement call is higher than the secondary frequency of the aggressive call (Given, 1999b). In particular, males have a higher rate of aggressive calls in response to frequencies of 1,500 Hz than to other frequencies. Essentially what the male does is to modify the secondary frequency of his advertisement call toward that of the aggressive call, while maintaining aspects of the advertisement call that are important to female choice (Givens, 1999b). If an intruder is a male, there may be a series of aggressive calls between them, followed by a brief wrestling match that residents usually win. Males have a release call that is uttered by one male toward another if amplexus is attempted.

Females also have a low-intensity, single-note call during courtship that has higher dominant and secondary dominant frequencies than the male aggressive call (dominant frequency of 720–780 Hz; secondary frequencies of 1,480–1,520 Hz; 16–23 msec call duration) (Given, 1993). If the intruder is a female, the male will amplex her, and she will begin searching for an oviposition site. Females initiate courtship by moving toward a calling male and uttering her encounter call. Males respond with a single or multinote aggressive call and move toward her, followed by amplexus. The multinote call is only given in response to a combination of the female's call and vibrations in the water. The mixture of sound and vibrations thus helps the male to home in on the position of the female. In playback experiments, females do not respond to variation in call frequencies or repetition rates, but they may use differences in the harmonic structure to differentiate between large and small males (Given, 1993). *Lithobates virgatipes* thus has an unusual call repertoire in which the same aggressive call is used toward both sexes. The aggressive response of the territorial male is consistent, and whether fighting or mating ensues depends on the response of the frog that approaches him.

Male mating success is correlated with size (greater for larger frogs) and the number of nights a male calls. Losers of aggressive bouts and smaller males may take up positions as satellite males near a large resident. These satellites remain with a resident from one to seven days. They may utter a single-note call simultaneously with the resident, but they assume a more crouched posture (Given, 1988b).

BREEDING SITES

The species breeds in ponds that have intermediate to long hydroperiods (Eason and Fauth, 2001;

Tadpole of *Lithobates virgatipes*. Photo: John Bunnell

Otto et al., 2007b). Most studies suggest optimal breeding sites are permanent wetlands with a long hydroperiod (Given 1988b; Zampella and Bunnell, 2000; Georgel, 2001), as the larvae may overwinter prior to metamorphosis in some populations. Otto et al. (2007b), however, reported Carpenter Frogs in seasonal wetlands. Some breeding ponds can be quite small (i.e., < 1 ha) (Russell et al., 2002a).

REPRODUCTION

Calling occurs from spring throughout the summer, at least in the northern portions of the range. Specific times include mid-March to early August (North Carolina: Gaul and Mitchell, 2007), late April to early August (New Jersey: Fowler, 1905; Conant, 1947; Given, 1987, 1988a, 1990), and a peak from April to August, with calls heard from March to September (Virginia: Georgel, 2001). Wright (1932) heard calls from June to August in the Okefenokee Swamp, but he acknowledged that the breeding season probably began at least in March. I have heard them chorusing in north Florida on 20 April, and there are specimens in the Florida Museum of Natural History that were collected from March to December. After the first several months of the breeding season, calling becomes much more sporadic. It is not clear whether breeding takes place throughout the calling period, or whether calling takes place over a more extended time than actual oviposition.

Eggs masses are oviposited near the surface in shallow water (0.05–1.0 m in depth), attached to vegetation or debris (Wright, 1932; Given, 1990). The sometimes elongate globular mass is ca. 6.5 cm × 6.5 cm to 5 cm × 10 cm and often has a bluish tinge (Livezey and Wright, 1947). There appear to be few actual counts of the number of eggs per mass, with most authorities repeating Wright and Wright's (1949) estimate of 200–600 eggs per mass. Wright (1932) gave counts of 349 and 474 from 2 masses.

LARVAL ECOLOGY

Little information is available on the larval ecology of *L. virgatipes*. In New Jersey, tadpoles first appear in late May within the shore zones of ponds. Density then increases steadily through peak metamorphosis in September, when they can reach nearly 120 per ha (Standaert, 1967). Most tadpoles metamorphose in the autumn (September to October), despite suggestions in the literature (Wright, 1932; Wright and Wright, 1949) that all tadpoles overwinter and metamorphose the following spring or summer. It is likely, however, that some tadpoles do overwinter, especially those resulting from egg deposition late in the season. Wright and Wright (1949) reported that young frogs were 23–31 mm SUL following metamorphosis, but Standaert (1967) noted that the mean metamorphic size in New Jersey was 32 mm SUL (range 28 to 36 mm). These postmetamorphs disperse rapidly with autumn rains and quickly enter overwintering sites.

DIET

Feeding occurs throughout the breeding season, although it appears to be depressed during peak breeding, as calling males concentrate on attracting mates and defending territories. Feeding increases rapidly following the breeding season. In Virginia, insects constituted 85% of the items consumed and 94% of the prey volume (Georgel, 2001). Prey included beetles, collembolans, flies, ants, dragonflies, leafhoppers, true bugs, homopterans, a larvae *L. virgatipes*, and a larval Lesser Siren (*Siren intermedia*) (Georgel, 2001). Males tended to consume more beetles of the species *Donacia rufescens* than females, perhaps consistent with differences in habitat use between the sexes. This beetle frequents lily pads and emergent vegetation in open water habitats. Females also ate prey of a wider variety of taxa (35 vs. 26) than males. A small amount of vegetation may be ingested with the invertebrate prey. Prey volume is correlated with the frog's head width, meaning that large frogs eat more and larger prey than small frogs.

PREDATION AND DEFENSE

These frogs are wary and cryptic as they sit among sphagnum, and quickly sink out of site at the approach of a potential predator. They will swim short distances and seek shelter in submerged vegetation. Predators include water snakes (*Nerodia sipedon*) and black racers (*Coluber constrictor*) (Wright, 1932; Jensen et al., 2008) and undoubtedly many types of birds and mammals. Larvae are likely attacked by a variety of predaceous insects and vertebrates, such as birds, turtles, and snakes. The skin contains antimicrobial peptides that may aid in defense against the bacteria *Escherichia coli* and *Staphylococcus aureus* (Conlon et al., 2005b).

POPULATION BIOLOGY

Tadpoles that metamorphose in summer may reach the minimum size of sexual maturity prior to their first overwintering period and breed the following spring. For those that overwinter because of late-season hatching (August) and metamorphose in spring (April to May), sexual maturity is not reached until mid- to late summer of the first warm season activity period (i.e., about 10–12 months after egg deposition). The size at maturity for males is 40–45 mm SUL in Virginia (Georgel, 2001) and 42–44 mm SUL in New Jersey (Standaert, 1967). Standaert (1967) noted, however, that there was considerable variation in size at sexual maturity. For example, 95% of males were mature by 48 mm SUL, whereas 5% were mature by 40 mm SUL. He considered frogs 38–47 mm SUL as juveniles. Populations may experience considerable turnover from one year to the next. For example, only 3 of the 26 males followed by Given (1988a) were seen the year after first marking.

Growth rates may be extremely rapid for newly metamorphosed frogs, with rates of 7 mm/month recorded (Standaert, 1967). They usually enter their first overwintering period at 34–36 mm SUL, but a few frogs may reach 49 mm SUL by October of their first year. After emergence, growth rates continue at ca. 5 mm/month allowing sexual maturity by mid-June. Male growth rates are lower than those of females, and males reach a smaller asymptotic size than females. Longevity exceeds 3 yrs.

Calling is energetically expensive for males throughout the extended breeding period. As the season progresses, dry body mass and percent lipid both decrease, although both show a sharp increase in August, after the end of the breeding season, presumably as males begin concentrated feeding (Given, 1988a). For example, Given (1988a) mentions a male with a growth rate of 0.13 g/day prior to the breeding season, 0.03 g/day during the breeding season, and 0.10 g/day after the breeding season. Smaller males

Adult *Lithobates virgatipes*.
Photo: David Dennis

(i.e., first year adults) have a lower calling effort than larger males and tend to increase their body mass more than large males. This indicates a trade-off between reproductive effort (calling) and growth among the smaller males.

The sex ratio at breeding ponds is male-biased. After hibernation, Standaert (1967) found that there were as many as 2.5 males per female in April, but by June the ratio had dropped to 1.65:1.00. During the latter part of the activity period, the sex ratio became female-biased. In addition, sex ratios changed by position in the pond and by time of day. These results suggest differences in behavior, habitat choice, and activity between males and females that bias determinations of the sex ratio.

COMMUNITY ECOLOGY

Many authors have noted the association of Carpenter Frogs with acidic wetlands. As noted by Gosner and Black (1957a), normal embryonic and larval development can take place at higher pHs than usually encountered in the boggy or swampy areas frequented by Carpenter Frogs, although a low pH is lethal to many other anuran eggs and larvae. It may be that by occupying acidic habitats, Carpenter Frogs are able to reduce interspecific competition with other anurans, particularly American Bullfrogs and Pickerel Frogs. Other ranids (*Lithobates clamitans*, *L. sphenocephalus*), however, are somewhat acid tolerant and are often found in the same habitats as *L. virgatipes*, whereas the hylids *Acris crepitans* and *Hyla versicolor* are not (Zampella and Bunnell, 2000). Thus, the life-stage (larval, adult, or both) at which competition might occur among species, and whether and how acidity might reduce competition, is not well understood.

When calling, *Lithobates virgatipes* and *L. clamitans* form mixed-species aggregations along a shoreline. There is a slight degree of habitat partitioning when calling, with *L. clamitans* calling from vegetation at the shoreline and *L. virgatipes* calling from locations 0.1–0.5 m into the wetland (Given, 1990). Male *L. clamitans* are capable of displacing male *L. virgatipes* from calling sites. When choruses are mixed, *L. virgatipes* tend to associate with conspecifics rather than the larger and heavier *L. clamitans*; *L. clamitans* shows no such preference.

DISEASES, PARASITES, AND MALFORMATIONS

A single specimen tested for amphibian chytrid fungus at the Savannah River Site in South Carolina was negative (Daszak et al., 2005).

SUSCEPTIBILITY TO POTENTIAL STRESSORS

pH. The lethal and critical pH levels for this species are 3.5 and 3.6–3.8, respectively (Gosner and Black, 1957a). Bunnell and Zampella (1999)

Habitat of *Lithobates virgatipes*, New Jersey Pinelands. Photo: John Bunnell

found Carpenter Frogs in wetlands with a pH of 3.8–4.1 in the Pine Barrens of New Jersey. Carpenter Frogs can tolerate water to a pH of 9.2 with no ill effects.

STATUS AND CONSERVATION

This species occurs over a discontinuous range on the Atlantic Coastal Plain. It may be locally abundant, although its secretive nature means it could be easily overlooked. Some populations undoubtedly have been lost due to habitat destruction or alteration. For example, Brady (1927) recorded this species as widespread in Virginia's Dismal Swamp, but subsequent attempts to locate it there after years of intensive forest cutting have proven unsuccessful (Pague and Mitchell, 1987). However, Russell et al. (2002b) found no effects on resident *L. virgatipes* inhabiting isolated wetlands (to 1.06 ha) up to 1.5 yrs following clearcutting in South Carolina. The wetlands were left intact, and wetland vegetation was undisturbed by forestry treatments, thus allowing this highly aquatic species to persist. In New Jersey, Weir et al. (2009) reported a slight positive trend in occupancy in wetlands monitored from 2001 to 2007.

In Maryland, this species is protected as "in need of conservation." Given (1999a) found the species at only 8 of 22 historically reported populations in that state (a total of 9 of 34 sites surveyed). Otto et al. (2007b) later recorded them at 18 of 40 study sites on the western and southern portions of the Delmarva Peninsula of Maryland. In Virginia, the species is considered of "Special Concern," as there are only six populations known from the state. The largest population is on Ft. A.P. Hill.

COMMENT

Everything that is known concerning the life history of *L. virgatipes* is based on studies in New Jersey and Virginia. Nothing is known about its habits, breeding phenology, activity patterns, or reproductive biology in the more southern part of its range. Life history studies in the South are long overdue.

Lithobates yavapaiensis
(Platz and Frost, 1984)
Lowland Leopard Frog

ETYMOLOGY

yavapaiensis: named for Yavapai County, Arizona, where the holotype and paratype specimens were collected. The county was named for the Yavapai Nation.

NOMENCLATURE

Stebbins (2003): *Rana yavapaiensis*
  Dubois (2006): *Lithobates* (*Lithobates*) *yavapaiensis*
    Synonyms: None
  Leopard frogs have long been recognized for their phenotypic variation. Different phenotypes have been considered to reflect polytypic variation of a wide-ranging species (*L. pipiens*), the result of clinal variation, or as members of a wide-ranging multispecies complex (the Leopard Frog complex) (Ruibal, 1957; Moore, 1975; Hillis, 1988). Some assessments (e.g., Moore, 1944) mixed specimens from a variety of locations, resulting in considerable taxonomic confusion. Much of the scientific literature uses the name *Rana pipiens* for frogs now recognized as *Lithobates yavapaiensis*, or it may be referred to as the "lowland type" or form (e.g., Frost and Platz, 1983). Readers should verify locations when the using older literature.

IDENTIFICATION

Adults. Lithobates yavapaiensis is a smooth-skinned light brown, gray-brown, or olive leopard frog with prominent lighter-colored dorsolateral folds. These folds are poorly defined and interrupted posteriorly and deflected medially. Dark brown spots are present, but they may be pale brown in some specimens; they lack the light halo seen in some other Southwestern leopard frogs. Spots are absent on the head in front of the eyes. An incomplete light stripe is present on the upper lip but does not extend in front of the eyes. The groin region contains yellow pigment that may extend onto the venter and the underside of the legs. There is a dark reticulation pattern on the posterior portion of the thighs, and the dorsal portions of the thighs are barred. Venters are cream colored, and chins are light. Rear toes are well webbed. Males have prominent external vocal sacs and enlarged thumb pads. This species is longer-limbed than the similar *L. onca*.

Males are smaller than females. In the type series, males averaged 54.7 mm SUL (range 50 to 57 mm), and females averaged 63.5 mm SUL (range 59 to 72 mm). Other morphological measurements are provided by Platz and Frost (1984). Degenhardt et al. (1996) give the size range as 46–72 mm SUL for males and 53–87 mm SUL for females.

Larvae. Larval *L. yavapaiensis* have a dusky yellow-olive dorsum. The sides of the tadpoles are paler than the dorsum and have numerous obscure dark spots. The venter is white with a yellow cast to the throat. The tail muscle is olive gray with scattered dark gray spots, and tail fins are shallow and yellowish white, with discrete gray spots. The iris is bronze. The larvae reach a TL of ca. 77 mm, with the tail nearly twice the body length. A detailed description of larvae and mouthparts is in Scott and Jennings (1985).

Eggs. The vitellus is surrounded by three envelopes. Storer (1925) provided the only measurements of the eggs, and noted they had already begun development: vitellus 1.78–2.11 mm (mean 1.97 mm); inner capsule 2–2.5 mm (mean 2.25 mm); middle capsule 2.22–2.72 (mean 2.35 mm); outer capsule 4.23–4.78 (mean 4.48 mm). Eggs are deposited in an irregular jelly mass about 75 mm in diameter.

DISTRIBUTION

Lithobates yavapaiensis historically was found from the Bill Williams River drainage of eastern Mohave County, Arizona (Jones, 1981), southeast in an arc to Hidalgo, Catron, and Grant counties in southwestern New Mexico. Additional populations occur in Santa Cruz and Cochise counties, Arizona, and in Surprise Canyon in the western part of the Grand Canyon. In Mexico, the species occurs in Sonora, but little is known about its status there. It once was widespread in riparian habitats in southern Arizona (Gila and Salt drainages) and possibly along the lower Colorado River into Mexico, but these populations have been extirpated (e.g., in the Santa Catalina Mountains; Lazaroff et al., 2006). Likewise, there have been no reports of Lowland

Distribution of *Lithobates yavapaiensis*. Dark gray indicates extant populations; light gray indicates extirpated populations.

Leopard Frogs in California (San Felipe-Carrizo drainages eastward to the Colorado River in Imperial, Riverside, San Bernardino counties) since 1965, and the species is likely extirpated in the state. There are no reports from New Mexico since 1985. Grismer (2002) stated that Clarkson and DeVos (1986) observed two *L. yavapaiensis* in the border region with Baja California in 1981, but the latter authors reported seeing frogs of the "leopard frog complex." These could have been *L. berlandieri*. Important distributional references include Jones (1981), Platz (1988b), Jennings and Hayes (1994a, 1994b), Degenhardt et al. (1996), Gelczis and Drost (2004), Brennan and Holycross (2006), and Oláh-Hemmings et al. (2009).

FOSSIL RECORD

No fossils are known. Holman (2003) noted Miocene (Hemphillian) fossils from Navajo County in northern Arizona that belonged to the *Lithobates pipiens* complex. The relationship between these fossils and *L. yavapaiensis* is unknown.

SYSTEMATICS AND GEOGRAPHIC VARIATION

Lithobates yavapaiensis is a member of the *Novirana* clade of North American ranid frogs. It is an associate of the mostly lowland and tropical leopard frog group (or *Scurrilirana*) (Hillis and Wilcox, 2005). Its closet relatives in the United States include *Lithobates berlandieri, L. blairi, L. onca,* and *L. sphenocephalus*. Microsatellite markers were characterized by Savage and Jaeger

(2009). Oláh-Hemmings et al. (2009) affirmed the close relationship but distinctiveness between *L. onca* and *L. yavapaiensis* and noted that there are two distinct evolutionary lineages within the species. One group includes the populations in Arizona and northern Mexico, whereas the other group is found only in Surprise Canyon in the western Grand Canyon. *L. onca* and *L. yavapaiensis* diverged during the early Pleistocene. *L. yavapaiensis* and *L. onca* also have closely related but distinct antimicrobial skin peptides (Conlon et al., 2010).

Variation in leopard frog phenotypes from the American Southwest has been recognized for some time. The species can be separated from other leopard frogs by a combination of morphological, biochemical, auditory, and genetic characteristics (Platz and Platz, 1973; Platz and Mecham, 1979; Frost and Platz, 1983). Frost and Bagnara (1976) presented a table comparing various phenotypes among leopard frog populations.

Under laboratory conditions, crosses between ♀ *L. yavapaiensis* and ♂♂ of other members of the leopard frog complex (*L. chiricahuensis*, *L. magnaocularis*, *L. pipiens* from Vermont) can produce relatively high rates of hybrid offspring (Platz and Frost, 1984). These hybrids, however, do not produce offspring when backcrossed with parental species. Crosses between *L. forreri* and *L. yavapaiensis* are unsuccessful, as are crosses between ♂ *L. yavapaiensis* and ♀ *L. pipiens* from Arizona.

ADULT HABITAT

This species occurs in small slow-flowing streams and river, springs, and rock pools in Sonoran Desert scrub and pinyon-juniper communities. Rocky rivers, small intermittent creeks, open marshlands surrounding streams, and even the grassy marshes along large rivers such as the lower Colorado were inhabited. Jennings and Hayes (1994a) noted that man-made stock ponds, ditches, and canals were occupied in the Imperial Valley of California. Preferred microhabitats include small spring runs where they enter large streams, which provide potential refugia under streamside debris piles. Deeper pools with tree root masses, which provide protection against desiccation and predators, also are favored. This species is reported from habitats dominated by bulrushes, cattails, and riparian grasses near or under a partial cottonwood-willow canopy (Jennings and Hayes, 1994a). Jennings and Hayes (1994a) provided historical photographs of former habitat along the lower Colorado River near Needles and at Harper's Well Wash (San Felipe Creek) in Imperial County. In New Mexico, it occurred to an elevation of about 1,700 m.

AQUATIC AND TERRESTRIAL ECOLOGY

There is little information available on habitat use. Individuals likely stay within close proximity to aquatic habitats. Movements are likely short and take place along watercourses, much as in *L. onca*. Seim and Sredl (*in* Sredl, 2005a) found

Eggs of *Lithobates yavapaiensis*. Photo: Cecil Schwalbe

that juveniles were associated with small pools and marshy areas, whereas adults were found more often in larger pools. During drought conditions, frogs retreat into mammal burrows, rock cracks, and fissures, and deep within dry mud cracks (*in* Sredl, 2005a).

CALLING ACTIVITY AND MATE SELECTION

The advertisement call of *L. yavapaiensis* is a series of short chuckles (or trill) with a low number of pulses (12 at 24°C) per note. Brennan and Holycross (2006) characterize it as a "low, grunting noise that resembles the sound of a finger stroking an inflated balloon." The first note is shorter than the following 6–15 notes, and the notes last 3–8 sec depending upon the number of notes and temperature. Pauses between notes and the duration of the notes tend to decrease as the call sequence progresses. Pulse rates average 12/sec (11 pulses in the first note to 3–4 in the last of the series) at a dominant frequency of 1.8 kHz (Frost and Platz, 1983). This produces a somewhat high-pitched call that sounds somewhat like an insect. Frost and Platz (1983) provide a sonogram comparing this species' advertisement call with those of other southwest leopard frogs.

BREEDING SITES

Breeding likely takes place in shallow water in pools and springs. Specific characteristics of breeding sites have not been published.

REPRODUCTION

Breeding in this species is tied to predictable stream flows, when eggs and larvae are not likely to be washed away by floods or adversely affected by drought. This species breeds earlier than the other Southwest ranids. In the California desert, breeding occurred from January to March, when water temperatures were 10–18°C (Ruibal, 1962). In Arizona, the breeding period is from late January through late May (Collins and Lewis, 1979; Frost and Platz, 1983; Degenhardt et al., 1996; Sartorius and Rosen, 2000). In addition, this species is known to deposit eggs in September and October (Platz and Platz, 1973; Collins and Lewis, 1979; Frost and Platz, 1983; Sartorius and Rosen, 2000).

Eggs are deposited in spheroidal masses in shallow water. The masses are attached to vegetation, bedrock, or gravel up to 1 m below the water surface. Most masses, however, are at or just below the surface. Late spring egg masses may be only half the size of those oviposited earlier in the spring. Eggs take 15–18 days to hatch at water 14°C, and most egg masses will produce larvae (Sartorius and Rosen, 2000). Eggs exposed to air suffer the most mortality.

Normal development for most leopard frog populations occurs at water temperatures >8°C. However, the minimum temperature for egg development was 11°C in the California desert (Ruibal, 1962). The maximum temperature that eggs of *L. yavapaiensis* can survive is 28–29°C (Ruibal, 1962).

LARVAL ECOLOGY

There is very little information available on larval ecology. Larvae have been found throughout the year at some localities. Some larvae overwinter,

Tadpole of *Lithobates yavapaiensis*. Photo: Cecil Schwalbe

Adult *Lithobates yavapaiensis*. Photo: Dennis Suhre

perhaps from the autumnal breeding period. Sredl (2005a) gives the larval period as three to four months or as long as nine months. Newly transformed froglets are 25–29 mm SUL (Platz, 1988b).

In experimental situations, larval *L. yavapaiensis* are able to detect the presence of the nonnative sunfish *Lepomis cyanellus* and respond by swimming faster when chemical cues from this predator are present than they do in control treatments (Shah et al., 2010). Although tadpole behavior was altered, it is unknown whether a similar behavior occurs in the wild or whether the altered behavior affects the survivorship of the tadpoles. In any case, the tadpole reaction to the nonindigenous predator suggests that they are capable of rapid learning.

DIET
Larvae are herbivorous. There is no published information on the diet of postmetamorphs of this species. Like other leopard frogs, they probably feed on a variety of invertebrates and even small vertebrates in relation to their availability.

PREDATION AND DEFENSE
Lithobates yavapaiensis seeks refuge among roots and streamside debris along small creeks. Jennings and Hayes (1994a) provided a photograph of a California *L. yavapaiensis* with its rear limbs held into the body and the forelimbs brought up over and shielding the eyes. The frog is crouched against the substrate. Perhaps such an immobile posture did not draw attention to the frog, or made it more difficult for a small predator to attack. Tadpoles may be able to detect chemical cues emanating from predators and alter their behavior accordingly.

There is no information on predation other than Jones's (1990) reports of larvae and adults being consumed by the garter snake *Thamnophis cyrtopsis*.

POPULATION BIOLOGY
Populations of adults breeding in selected pools may be small. For example, Sartorius and Rosen (2000) counted only 3–20 frogs per transect in surveys conducted between February and September at a population in Santa Catalina Mountains. Survivorship is lowest during the autumn and winter months, and highest in spring and summer. Frogs live at least 3 yrs postmetamorphosis (Sredl, 2005a).

DISEASES, PARASITES, AND MALFORMATIONS
Lithobates yavapaiensis possesses antimicrobial skin peptides that may aid in defense against *Escherichia coli* and *Staphylococcus aureus* (Conlon et al., 2009). The virulent pathogen *Batrachochytrium dendrobatidis* has been found in fairly high prevalence (10 of 13 populations examined) in this species in Arizona (Bradley et al., 2002; Schlaepfer et al., 2007). Fully 43% of the Lowland Leopard Frogs examined were positive for amphibian chytrid fungus. Schlaepfer et al. (2007) suggested that the pathogen becomes virulent during the winter in conjunction with other factors, and that thermal springs may offer refuge from the pathogen.

Lowland Leopard Frogs are parasitized by the trematodes *Cephalogonimus brevicirrus*, *Glypthelmins quieta*, *Haematoloechus complexus*, and *Megalodiscus temperatus*. It is parasitized by the nematodes *Falcaustra catesbeiana*, *Physaloptera* sp., and *Rhabdias ranae* (Goldberg et al., 1998a).

SUSCEPTIBILITY TO POTENTIAL STRESSORS

NaCl. Lowland Leopard Frogs were found historically in brackish water along desert streams (Ruibal, 1959). Ruibal (1959) recorded salinities of from 1.75 to 9 ppt at locations with frogs, with reproduction occurring at salinities of 2–2.75 ppt. This population no longer exists.

STATUS AND CONSERVATION

At one time, this frog was quite abundant in southern Arizona. For example, King (1932) noted they were common along the stream in Sabino Canyon in 1930. Populations in the Santa Catalina Mountains, in riparian habitats of southern Arizona (Gila and Salt river drainages), and along the lower Colorado River have been extirpated (Vitt and Ohmart, 1978; Lazaroff et al., 2006), as have populations in the Imperial Valley of California (Jennings and Hayes, 1994a, 1994b). Lowland Leopard Frogs were not found in previously occupied habitats in California during surveys from 1983 to 1987 (Clarkson and Rorabaugh, 1989). Jennings and Hayes (1994a) summarized the historical distribution and decline of this species in that state. The last verified collection of Lowland Leopard Frogs in California was from near Calexico in 1965. *Lithobates yavapaiensis* appears to have disappeared from New Mexico by 1985 (Degenhardt et al., 1996). Clarkson and Rorabaugh (1989) reported that populations in upland areas of Arizona were still largely intact as late as 1987, and Sredl (1998) noted fairly high occupancy rates at historical localities, with many new populations discovered.

The likely causes of decline include a variety of factors, especially habitat alteration and outright destruction as their springs, riparian environments, and rivers were modified and water reallocated for human use. Vegetation changes and the introduction of *L. catesbeianus* into many isolated spring habitats likely hastened population decline. Competition also may have occurred between this species and the introduced Rio Grande Leopard Frog (*L. berlandieri*), which has become widespread in southern Arizona (Clarkson and Rorabaugh, 1989). Bury and Lukenbach (1976) noted that some leopard frogs in the lower California deserts did not appear to be native, and *L. berlandieri* is now known to be widespread in former *L. yavapaiensis* habitat there. Ranid declines in Arizona, including *L. yavapaiensis*, are correlated with the presence of introduced crayfish and the amphibian pathogen *Batrachochytrium dendrobatidis* (Lazaroff et al., 2006; Witte et al., 2008). Declines are thus the product of multiple causes acting independently or cumulatively on a population-by-population basis.

Much information on the ecology and life history of *Lithobates yavapaiensis* is contained in unpublished reports. Sredl (2005a) summarizes some of this information. The species is protected by state law in Arizona.

Habitat of *Lithobates yavapaiensis*. Photo: Dennis Suhre

Rana aurora Baird and Girard, 1852
Northern Red-legged Frog

ETYMOLOGY
aurora: Latin for 'red,' referring to the frog's coloration.

NOMENCLATURE
Dubois (2006): *Rana (Amerana) aurora*
   Synonyms: *Rana agilis aurora, Rana temporaria aurora*

IDENTIFICATION
Adults. This is a large (males to 68 mm SUL, females to 101 mm SUL; Storm, 1960; Jennings and Hayes, 1994b), rather smooth-skinned frog, although some individuals have a granular skin texture. The dorsum is brown, reddish brown, or greenish gray and peppered with small black spots with a few large spots in between. A prominent dorsolateral fold consisting of granular glands is present, and there usually is a light lip-line from the eye to shoulder, bordered above by a dark mask. The venter is heavily infused with gray. The groin has prominent irregular black and gray mottling. The lighter mottling may range from white to red to green to yellow. The venter is yellow with a superimposed tinting of red. The undersides of the arms, legs, and abdomen are red; this red coloration is not apparent in juveniles. Legs are long. Male *Rana aurora* usually lack a vocal sac (Hayes and Krempels, 1986). Dunlap (1955) and Dumas (1966) provide extensive information on morphological comparisons with other Pacific Northwest ranids.

Females are larger than males. Adult females were 72–93 mm SUL (mean 84 mm) and males were 49–65 mm SUL (mean 59 mm) in an Oregon population (Storm, 1960). In British Columbia, males were 45–60 mm SUL and females were 62–80 mm SUL (Licht, 1974).

Larvae. The tadpole is tan to dark brown with numerous scattered clumps of golden flecks. The venter is off-white or brassy and iridescent in coloration, and there are two or three transverse bands of clumped, brassy pigment cells. The dorsal and ventral tail fins are the same depth as the tail musculature, and they may be marked with minute dark spots and have a golden tone. Small tadpoles (< 15 mm TL) have a light gold line along each side of the body that disappears as they grow larger. Total length prior to metamorphosis ranges from 50 mm in ponds with short hydroperiods to 70 mm in permanent ponds.

Eggs. The eggs are black dorsally and creamy white ventrally and are the largest eggs of all North American ranids. They are deposited in a loose and viscid jelly mass 15–25 cm in diameter, with each mass containing 500–1,300 eggs. The bluish-tinged mass is attached to vegetation at the surface of the water. There are three indistinct envelopes surrounding the egg. The outer envelope is 10–14 mm in diameter; the middle envelope is 6.25–7.9 mm (mean 6.8 mm) in diameter; the inner envelope is 4–6.7 mm (mean 5.7 mm) in diameter. The vitellus is 2.3–3.6 mm in diameter. Eggs are described by Livezey and Wright (1947).

DISTRIBUTION
Rana aurora is found from southwestern British Columbia, including Vancouver Island, south through coastal Mendocino County, California. The range extends eastward to the western foothills of the Cascades in Oregon, although

Distribution of *Rana aurora*

Slater (1955) found a single population in the eastern Cascades in western Klickitat County, Washington, in the Columbia River Gorge. Populations of *R. aurora* and *R. draytonii* are found intermingled between Mills Creek in southern Mendocino County and Big River in northern Mendocino County, but all red-legged frogs south of the Mills Creek drainage are *R. draytonii* (Shaffer et al., 2004). Northern Red-legged Frogs occur on islands, including Vancouver Island and the adjacent Gulf Islands, and islands in Puget Sound (Anderson, Bainbridge, Blakely, Camano, Cypress, Fidalgo, Fox, Hairstine, Indian, Marrowstone, Maury, McNeil, Orcas, Raft, Saddlebag, San Juan, Vashon, Whidby) (Brown and Slater, 1939; Slater, 1941a). Important distributional references include Slevin (1928), Dunlap (1955), Slater (1955), Nussbaum et al. (1983), Leonard et al. (1993), Jennings and Hayes (1994b), Adams et al. (1998), Shaffer et al. (2004), Jones et al. (2005), and Matsuda et al. (2006).

FOSSIL RECORD

Fossils referred to as *Rana aurora* have been reported from Pliocene Blancan deposits in Idaho (Chantell, 1970; Sanchiz, 1998).

SYSTEMATICS AND GEOGRAPHIC VARIATION

Rana aurora and *R. draytonii* have long been considered closely related, based on a similar morphology and the red coloration of the legs (Camp, 1917; Slater, 1939a). Although they were described originally as distinct species (Baird and Girard, 1852), most researchers until recently have considered them subspecies within the wide-ranging, polytypic species *R. aurora*. Data from studies on allozymes, mtDNA, karyotypes, vocal sac structure, advertisement calls, skin peptides, and oviposition behavior (Hayes and Miyamoto, 1984; Green, 1985a, 1986a, 1986b; Hayes and Krempels, 1986; Macey et al., 2001; Shaffer et al., 2004; Conlon et al., 2006) have supported specific distinctiveness between them, although various authors have interpreted their results differently, especially within the area of sympatry in Mendocino County (reviewed by Shaffer et al., 2004).

Rana aurora and *R. draytonii* are not sister taxa. Instead, the *R. aurora* lineage is one of five well-supported clades within the monophyletic *R. muscosa* clade (sometimes termed the *Rana boylii* species group; Macey et al., 2001) within the *Amerana* clade of western North American ranid frogs (Post and Uzzell, 1981; Shaffer et al., 2004; Hillis and Wilcox, 2005). *Rana aurora* is most closely related to *R. cascadae* and somewhat more distantly to *R. muscosa*, *R. sierrae*, and *R. draytonii*. Divergence occurred first between the southern red-legged frogs (*draytonii*) and the northern red-legged frogs (*aurora-cascadae*) and later between inland (*cascadae*) and coastal (*aurora*) populations. The contact area between *R. draytonii* and *R. aurora* in Mendocino County is an important contact zone between northern and southern evolutionary lineages for many amphibian species groups.

Despite suggestions of broad to narrow zones of hybridization between *R. aurora* and *R. draytonii*, hybrids between these species in northern California have not been reported. Experimental crosses between *R. aurora* and *Lithobates sylvaticus* or *Rana cascadae* are not successful (Porter, 1961; Dumas, 1966). The diploid number of chromosomes is 26 with 5 pairs of large chromosomes and 8 pairs of small chromosomes (Haertel et al., 1974; Green, 1986a).

ADULT HABITAT

Northern Red-legged Frogs occupy a variety of open-canopied wetlands, including large grassy temporary ponds, permanent ponds, floodplain wetlands, river oxbows, beaver ponds, and slow-moving rivers and streams (Stebbins, 1951; Brown, 1975b; Jennings and Hayes, 1994b; Henning and Schirato, 2006). Frogs along streams may occur under closed, shaded canopies. Hydroperiods vary in duration, but these frogs appear to do well in wetlands with intermediate hydroperiods and much emergent vegetation. They are often found in the presence of fish, particularly nonpredaceous fishes native to the Pacific Northwest.

In terrestrial mature forest, they are associated with coarse woody debris, a greater density of broadleaf trees, and a higher percent cover of midlevel canopy trees (Aubry and Hall, 1991). Such habitats include willow thickets, dense sedge swales, and dense undergrowths of sword ferns and sedges along streamcourses; such habitats retain moisture. *Rana aurora* is generally found in more humid habitats than many of the other ranids in the Pacific Northwest; they both desiccate and rehydrate more rapidly than many of the sympatric pond ranids (Dumas, 1966). This species has been found at elevations from near sea level to 1,427 m (Leonard et al., 1993), but most populations favor low elevations with minimal slopes (Bury et al., 1991a).

AQUATIC AND TERRESTRIAL ECOLOGY

During the nonbreeding season, Northern Redlegged Frogs are largely terrestrial, sitting in the open on dry land or moving among dense shoreline vegetation. Some frogs (<25%), however, will sit in the water. They often are hidden in complex tangles of logs and moist debris, sometimes far from water. Brattstrom (1963) suggested that *R. aurora* uses solar radiation to raise its body temperature, and this undoubtedly would occur in frogs exposed to direct sunlight. When shallow pools dry up as the season progresses, *R. aurora* moves to forest or other terrestrial habitats in the vicinity of slow-moving streams and rivers.

Overwintering occurs in both aquatic and terrestrial environments. Emergence occurs immediately after snow and ice begin melting from rivers, ponds, and terrestrial habitats, and may occur a few weeks prior to the actual start of breeding. Air temperatures >5°C for several days are required for emergence, and frogs will resume dormancy if temperatures decrease after initial emergence. In contrast to the adults, subadult *R. aurora* only emerge several weeks after the breeding season has ceased and air temperatures are consistently >10°C (Licht, 1969c).

Movement to breeding sites begins soon after emergence. Movements occur at night and are facilitated by rain and cloud cover. After breeding, the adults disperse into thick moist vegetation (Twedt, 1993). Like California Red-legged Frogs, Northern Red-legged Frogs are sometimes found at significant distances from water. Dumas (1966), for example, recorded them up to 914 m from the nearest water, whereas Nussbaum et al. (1983) noted terrestrial movements 200–300 m from water. In a mark-recapture study, Hayes et al. (2001) caught both males and females from 1.1 to 2.4 km (straight-line distance) between captures. These results suggest that *R. aurora* can make terrestrial movements of considerable distances either between breeding sites or from breeding sites to terrestrial refugia and foraging habitat.

CALLING ACTIVITY AND MATE SELECTION

Calling normally occurs from midwinter to early spring, although advertisement calls have been heard in September (Leonard et al., 1997b). Males arrive at breeding sites prior to females, and are active diurnally, sitting around the margins of ponds and on stream banks in full sunlight. After about a week, vocalization begins as temperatures increase to >6°C consistently over several days; once initiated, calling can occur at temperatures as low as 4°C (Storm, 1960; Calef, 1973b; Brown, 1975b). Males usually call nocturnally while completely submerged in water usually > 60 cm in depth (range 16 cm to 1.8 m; Licht, 1969c). Occasional calls are made during the day or especially in the early morning. Frogs call from the bottom of the pond or river while concealed in vegetation; they occasionally will surface for air.

Male *R. aurora* position themselves a few meters apart from other males and call from locations more than 1 m from the pond or stream shoreline. They apparently do not defend territories, and overt aggression does not occur. Calling frequency increases in the vicinity of another male, and indeed any movement near a calling male causes him to call continuously for up to 30 sec. Movements by calling males back and forth between various locations within a wetland or pond may occur during the breeding season. Most such movements are for short distances, although Calef (1973b) recorded a few frogs moving > 300 m.

The advertisement call of *R. aurora* is low in volume and has poor carrying capacity in air (normally < 10 m). Licht (1969c) noted that calling males in 60 cm of water cannot be heard by terrestrial observers. Calls given from the surface cannot be heard > 15 m away, although calls given underwater can be detected 10 m distant using underwater hydrophones. Licht (1969c) recorded listening to a "resounding" chorus of *R. aurora* using hydrophones, a chorus that could not be detected out of water! The call consists of two to five notes (usually three to four), with each note consisting of five to six pulses. The dominant frequency ranges between 450 and 1,300 Hz, with some notes extending to 7.5 kHz. Calls last about 1 sec depending on the number of notes. Storm (1960) and Licht (1969c) described the call as a repetitive series of "uh uh uH UH," and Licht (1969c) included a sonogram.

Amplexus is axillary (Storm, 1960). A male may attempt to amplex a female even prior to beginning his vocalizations. When a female is receptive, amplexus leads to egg deposition. Males are tenacious. If unreceptive, the female cannot dislodge a persistent male simply by uttering a release call with its associated abdominal vibrations. Instead, she must roll to her side while extending her back legs. She remains in this position with all legs stiff and extended until the male releases her (Licht, 1969b). Even so, a male may remain clasped to a posturing female for as long as 15 min. Kicking toward the male or assuming an arched body posture is not as effective in discouraging an ardent male. The Northern Red-legged Frog also has a release call that is directed toward conspecific males should amplexus be attempted. Many males may try to amplexus a female simultaneously. Amplexus has been recorded with other ranids, *Ambystoma gracile*, and even an apple (Storm, 1960; Nussbaum et al., 1983).

Males have two different calls that they make during amplexus that are distinct from the advertisement call. The first is directed toward receptive females and is uttered about once per second more or less continuously. This call is a single unpulsed note of 750–1,500 Hz that is directed into the female's ear. A second amplexus call, more like the male's release call, appears directed toward unreceptive females and is uttered in a series of one to eight notes that are structurally different from the amplexus call directed toward receptive females. When a female attempts to displace a male, he increases the speed, intensity, and repetition rate of this amplexus call. Males make this call 2 or 3 times a minute when clasped to an inactive female, but if the female attempts to displace him the repetition rate increases to 10–13 times per minute. Licht (1969b, 1969c) provided sonograms of these clasping calls and photographs of a female's behavior while discouraging a suitor.

BREEDING SITES

Breeding sites are varied, from temporary to permanent wetlands to slow-moving (< 5 cm/sec) streams. Northern Red-legged Frogs tend to be associated with temporary wetlands having a shallow slope and a southerly aspect (Adams, 1999). However, they are found in permanent wetlands. The presence of *R. aurora* is negatively associated with the abundance of exotic fish but not with the amount of emergent vegetation. They particularly like habitats with thin-stemmed rushes, herbs, sedges, and grasses (Richter and Azous, 1995), little (<27%) open water, and an intermediate abundance of insect predators (Pearl et al., 2005b). Open habitats with little emergent

Egg mass of *Rana aurora*.
Photo: Brome McCreary

vegetation are not preferred if American Bullfrogs are present.

REPRODUCTION

Breeding begins several weeks after emergence from winter dormancy, the timing of which is dependent on the melting of snow and ice. As such, the initiation of breeding activities varies annually depending on weather conditions. Storm (1960) recorded that breeding begins in January and extends into February in Oregon, although frogs may move to breeding ponds by mid-December. In Washington and British Columbia, breeding begins in February or March (Licht, 1969c; Brown, 1975b). Calef (1973a) recorded the first egg masses on 21 February, 9 March, and 4 April over a period of three years in a British Columbia population. The breeding season is short, with most females ovipositing within 2 weeks or so of initial reproduction; the entire effective breeding season is <30 days. Egg deposition requires a water temperature >6°C (Licht, 1971; Brown, 1975b) and occurs only at night.

Egg masses are 15–25 cm in diameter, and each mass usually contains >600 eggs (Brown, 1975b). More specifically, Licht (1974) recorded from 194 to 921 (mean 680) eggs per mass, and Storm (1960) counted between 541 and 1,081 (mean 831) eggs per mass. Jones et al. (2005) stated that up to 2,000 eggs could be in a mass, but the basis for this is unclear. Variation in clutch size may reflect variation in female size.

The eggs are tolerant of a range of temperatures from 4 to 21°C (Licht, 1971). Temperatures within egg masses may be 3.6°C warmer than in surrounding water (Licht, 1971), with means about 1°C above ambient temperatures normal (Licht, 1971; Brown, 1975b). Embryonic mortality is low despite the cold temperatures at which eggs are deposited. Licht (1974) found that 90% of the embryos hatched regardless of whether they were oviposited in a pond or river.

Eggs are attached to the stalks of submerged vegetation (cattails, spikerush, willows, branches) in water > 25 cm–5 m in depth and at least a meter (to 4.5 m) from the shoreline of the pond. In British Columbia, Licht (1971) found that egg masses were attached to submerged vegetation in at least 46 cm of water. The egg mass normally is completely submerged, although portions of egg mass could break free of the vegetation and float to the surface. In streams, egg masses are deposited in deep water (60 cm–1.2 m) in the deepest part of the channel, usually on the downstream side of the current. Egg masses may take on a greenish tint due to algal growth as development proceeds. Algae produce oxygen, which is beneficial to the developing embryos.

Egg masses are spaced several meters apart in the immediate vicinity of the calling location of the successful male. Thus, there is an evenness of spacing, although Licht (1969c) recorded 10 masses in a 1.8 m^2 area. Egg masses are positioned so that they will be exposed to direct sunlight in water with virtually no flow. However, some egg masses will be lost as ponds desiccate or due to the sudden onset of freezing temperatures. Eggs also experience increased mortality in habitats containing large quantities of silt, perhaps in conjunction with the physico-chemical properties of the soils involved (Platin and Richter, 1995).

Females likely breed every year and stay at a breeding pond only until they oviposit. Males stay throughout the breeding period.

LARVAL ECOLOGY

The period of embryonic development lasts about 35 days in water that ranged from 4.5 to 7.8°C (Brown, 1975b); Storm (1960) recorded embryonic development over a period of 35 to 49 days under field conditions. The jet black larvae hatch at from 11 to 12 mm TL (Storm, 1960; Brown, 1975b), and the larval period is 110 days in water ranging from 7.3 to 23.5°C (Brown, 1975b). Calef (1973a) estimated the larval period in British Columbia as 11–14 weeks, with tadpole developmental rates directly correlated with the number of degree-days since hatching. Larvae are significant grazers on algae and have an important role in structuring the periphyton community (Dickman, 1968). Transformation occurs at about 29 mm TL, with recently completely metamorphosed juveniles ranging between 20 and 29 mm TL (Storm, 1960; Calef, 1973a; Nussbaum et al., 1983). Brown (1975b) reported that larvae reach a maximum size of 34 mm TL at Gosner stage 39, but Storm (1960) found larvae nearly 60 mm TL.

Because egg masses contain many eggs, and there may be many masses deposited in a pond, tadpole densities immediately after hatching can reach 500 per m^2 (Calef, 1973a). In laboratory tests, tadpoles associated with siblings during

Tadpole of *Rana aurora*. Photo: William Leonard

early development (Blaustein and O'Hara, 1986), but they soon dispersed rather rapidly and evenly across a pond's substrate. Subsequently, no kin-based associations were observed.

Tadpoles are capable of selecting certain substrate features based on early experience, and they can retain their preferences throughout development. They also can develop a substrate preference at any stage of development, so there is no early learning or imprinting. Wiens (1970) suggested that a ready capacity to learn suitable substrate patterns would be advantageous to tadpoles, especially those that hatched initially onto unsuitable substrates. However, tadpoles reared in featureless environments did not show a substrate preference.

Predation can have a significant effect on larval survivorship. In experimental mesocosms in natural lakes, larvae were twice as likely to survive when salamanders (*Taricha granulosa*) were absent as when salamanders were present (Calef, 1973a). Tadpoles are eaten in proportion to their density, although tadpoles reaching a large size have a decreased predation rate and a greater chance of survival than small tadpoles. Food does not appear to be a limiting factor, as tadpole survivorship is independent of density, even at densities of 75 larvae per square meter.

Calef (1973a) estimated that high tadpole mortality occurred in the first four weeks after hatching, with higher survivorship through time as the tadpoles grew. Only 5% of the tadpoles that hatched were estimated to reach metamorphosis. Licht (1974) obtained a similar estimate.

DIET

Northern Red-legged Frogs are mostly insectivorous, with larger frogs eating larger prey than smaller or recently metamorphosed frogs (Licht, 1986b). The diet consists largely of spiders, beetles, aphids, flower flies, leafhoppers, crane flies, spittlebugs, caterpillars, isopods, and other small invertebrates (Fitch, 1936; Licht, 1986b).

PREDATION AND DEFENSE

The most important predators on postmetamorphic *Rana aurora* are garter snakes (*Thamnophis* sp.). Adult *Rana pretiosa* feed upon small *R. aurora* (Licht, 1986b), and juveniles and adults are preyed upon by American Bullfrogs (*Lithobates catesbeianus*) (Twedt, 1993; Kiesecker and Blaustein, 1997a). Birds (herons) and mammals (raccoon, mink) also likely take postmetamorphs. Eggs (Licht, 1969a) and tadpoles are highly palatable. Larvae are eaten by garter snakes (*Thamnophis sirtalis*), larval and paedogenic adult salamanders (*Taricha granulosa*, *Ambystoma gracile*), fish (*Salmo gairdneri*), and predaceous aquatic insects (giant diving bug *Lethocerus americanus*; dragonfly naiads) (Storm, 1960; Calef, 1973a). Eggs are eaten by leeches (*Batrachobdella picta*).

Like many anurans, larval Northern Red-legged Frogs appear to be able to detect predator chemicals in their environment and alter their behavior accordingly. For example, small larvae reduce their levels of activity in the presence of chemicals emanating from newts (*Taricha granulosa*) fed on a diet of tadpoles, but not from newts fed on insects. They also are able to detect chemicals from injured or disturbed conspecifics (Wilson and Lefcort, 1993; Kiesecker et al., 1999). Indeed, disturbed tadpoles show an increased amount of ammonium secretion, and larvae exposed to increased levels of ammonium respond with antipredator behavior similar to that of disturbed tadpoles (Kiesecker et al., 1999). In contrast, Adams and Claeson (1998) could not demonstrate trap avoidance by large *Rana aurora* tadpoles in response to damage release chemicals. Response to these chemicals may be related to size (large tadpoles are not as prone to predation as small tadpoles), or perhaps the chemosensory response does not occur under field conditions. Changing behavior in the presence of an unseen threat would likely enable larvae to decrease their risk of predation, so it seems plausible that a lack of response might be size-related.

Adult Northern Red-legged Frogs are wary, especially during the day. At night, however, males may be approached easily. They remain stationary on land until a certain approach distance has been breached, then quickly jump into the water. This approach distance is farther for large predators (ca. 1 m) than it is for small predators (< 44 cm), and moving objects per se do not necessarily elicit an escape response. The frogs swim rapidly away from shore but then turn around and move toward shallow water without emerging. There, they hide in root tangles, under bank crevices, or in dense submerged vegetation. Some frogs dive under water and disappear completely for several minutes before returning to the surface and floating. They then will remain immobile in the water and reemerge on land in from 8 to 45 min (Gregory, 1979).

Not all populations of *R. aurora* do the same thing when a predator approaches. In contrast to Gregory's (1979) observations, Licht (1986a) stated that *R. aurora* prefers terrestrial escape routes to water, where it flees with 3 to 6 straight long jumps at an initial angle of 45° and hides in vegetation. If a frog jumped toward the water, it tended to swim parallel to the shoreline rather than swim away from it. These differences in escape behavior may be related to habitat, as Gregory (1979) studied escape behavior in a pond, whereas Licht (1986a) studied it in a slow-moving stream. Differences between these observations also might stem from differences in experimental approaches, the presence of other predators (*R. pretiosa*, predaceous fish), or the amount of vegetation along the shoreline. The long limbs and reduced webbing suggest this species is well adapted to terrestrial escape.

Although this frog is quite striking and conspicuous in hand, its reddish coloration blends well into its surroundings of Western Red Cedar (*Thuja plicata*) and Western Hemlock (*Tsuga heterophylla*) needles and woody debris on the forest floor. *Rana aurora* emits a distress call if captured by a predator, but the effectiveness of this call in predation avoidance is unknown. It may serve to startle a hearing predator or to alert conspecifics, especially kin, that might be in the vicinity. Although not predation, mortality has been recorded after a male *Anaxyrus boreas* amplexed a female *Rana aurora* resulting in the female's death by evisceration (Brown, 1977).

COMMUNITY ECOLOGY

Northern Red-legged Frogs were common in the region around Mount St. Helens, Washington, prior to the volcanic eruption of 1980. Essentially all breeding populations were eliminated in the impact area, and it took about 5 yrs before recolonizers were observed at previously occupied sites. By 1995, only 5 sites were occupied of 33 examined in the blowdown and scorch zones; of course, many habitats had changed drastically because of the volcanic activity. By 2000, however, they were recorded at >80 sites in the debris-avalanche zone, with breeding activity at 40 locations (Crisafulli et al., 2005). As time passes and ponds become suitable, *R. aurora*

Adult *Rana aurora*. Photo: David Dennis

should reoccupy virtually all its former territory in the region.

POPULATION BIOLOGY

Many Northern Red-legged Frogs may deposit eggs in a single pond. For example, Calef (1973a, 1973b) counted 618 and 620 egg masses over a 2 yr period at a breeding pond in British Columbia and estimated the male breeding population as 1,770 one year and 3,600 the next. He estimated that there were initially 300,000 tadpoles present, a number which quickly fell to 75,000 due to predation. Only about 15,000 tadpoles (5%) metamorphosed. In another British Columbia study, Licht (1974) found that only 52% of metamorphosing juveniles survived the first year. Once reaching adulthood, survivorship can be high (69%) from one year to the next.

Many authors have speculated that the size at and timing of metamorphosis influence postmetamorphic fitness. Chelgren et al. (2006) tested this hypothesis with *R. aurora* and found that large metamorphs had higher rates of survival than small metamorphs and that survival was inversely proportional to the number of days metamorphosis was delayed after the first metamorphs emigrated from the breeding pond. Thus, ecological factors within the pond that delay growth (e.g., food resources, density, competitors) have important subsequent effects on metamorph survival and fitness.

Juveniles are capable of rapid growth (0.09–0.18 mm/day). Sexual maturity is reached by males at 45–50 mm SUL the first year following metamorphosis; females attain sexual maturity at 60 mm SUL during the second year following metamorphosis (Licht, 1974; Hayes and Hayes, 2003). Several authors have estimated longer times to maturity, such that individuals do not reach maturity until their third or fourth year following metamorphosis (Nussbaum et al., 1983; Leonard et al., 1993; Jennings and Hayes, 1994b); these may be overestimates, although more data are needed from multiple sites. Longevity is > 10 yrs based on records in captivity (Cowan, 1941).

DISEASES, PARASITES, AND MALFORMATIONS

Eggs that do not develop or are placed in very cold water are colonized quickly by fungus. Larvae are infected by the parasitic yeast *Candida humicola* (Lefcort and Blaustein, 1995). Tadpoles infected by this yeast have higher thermal preferences ("behavioral fever") than normal tadpoles, and they exhibit an inability to discriminate chemosensory cues from potential predators. In experimental trials, they suffered increased levels of predation when compared with noninfected tadpoles (Lefcort and Blaustein, 1995). These results suggest that the pathogen has sublethal effects that decrease the ability of the larvae to survive to metamorphosis, regardless of the direct effects of infection.

Iridovirus has been reported from larval *Rana aurora* in northern California (Mao et al., 1999). Amphibian chytrid fungus (*Batrachochytrium dendrobatidis*) has been found in larvae from Redwood National Park in northern California at a rate of 6.4% of those examined, especially those with deformed oral disks (Nieto et al., 2007). Additional reports of suspected or confirmed chytridiomycosis are in Green et al. (2002) and Adams et al. (2010). Unlike tadpoles infected by parasitic yeast, Northern Red-legged Frog larvae experimentally infected with amphibian chytrid fungus show no evidence of behavioral fever or altered thermoregulation (Han et al., 2008).

Parasites of larvae include protozoans (*Apiosoma* sp., *Epistylis* sp., an unidentified trichodinid, *Ichthyobodo* sp.), the trematodes *Brachycoelium lynchi*, *Gyrodactylus aurorae*, and *Ribeiroia* sp., and unidentified leeches in the oral cavity (Ingles, 1936; Mizelle et al., 1969; Nieto et al., 2007). Adults are parasitized by leeches (*Batrachobdella picta*) (Licht, 1974).

The trematode *Ribeiroia ondatrae* has been found in free-living *Rana aurora*. This trematode causes malformations, and 10 different malformations were observed in 7.9% (9 of 114) of the frogs examined from throughout the Pacific Northwest (Johnson et al., 2002). These included missing eyes, multiple limbs and other limb deformities, and cutaneous fusions of the hind limb. From 1 to 78 metacercariae were found per larva or recent metamorph. Other trematodes infecting *R. aurora* include *Gorgoderina multilobata*, *Margeana* (*Glypthelmins*) *californiensis*, and *Megalodiscus microphagas* (Zamparo and Brooks, 2005).

Rana aurora and *R. draytonii* also have closely related but distinct antimicrobial skin peptides (Conlon et al., 2006). These peptides aid in the frog's defense against bacterial pathogens.

SUSCEPTIBILITY TO POTENTIAL STRESSORS

Nitrates and sulfates. Embryos of *R. aurora* suffer 100% mortality at concentrations > 105 mg/L NH_4-N (ammonium nitrate) and 918 mg/L NO_3-N (sodium nitrate) (Schuytema and Nebeker, 1999a). Growth of *R. aurora* tadpoles is inhibited at concentrations > 134 mg/L NH_4-N (Nebeker and Schuytema, 2000), and decreases in length or mass of *R. aurora* embryos are evident at > 13.2 mg/L NH_4-N and < 29.1 mg/L NO_3-N (Schuytema and Nebeker, 1999a). These values are indicative of an intermediate level of sensitivity to ammonium-nitrogen fertilizers and thus suggest that ammonium-nitrate fertilizers by themselves cannot be blamed for the disappearance of Northern Red-legged Frogs in agricultural areas. The ammonium ion contributes the toxic effect of nitrates to amphibians. The median lethal concentration of nitrite for larvae is 1.2 mg N-NO_2/L at 15 days, 4.0 at 7 days, and 5.6 at 4 days (Marco et al., 1999).

Chemicals. Death occurs in about 34 hrs at 30 mg/L of carbaryl, a broad-spectrum insecticide, and significantly reduces larval activity levels at 2.5 mg/L (Bridges and Semlitsch, 2000). *Rana aurora* is the most tolerant ranid species tested toward carbaryl.

Traces of DDE, DDT, and PCBs have been found in *R. aurora* eggs from British Columbia, but the amounts were below levels thought to cause developmental or behavioral problems (de Solla et al., 2002a). This was true for both agricultural and nonagricultural areas.

UV light. Under experimental conditions, the hatching success of *R. aurora* eggs under enhanced levels of UVB light (15–30% above ambient levels) is significantly less (56%) than in ambient UVB levels (90%) or in UVB blocked light (81%) (Ovaska et al., 1997). Enhanced UVB light also significantly reduced larval survival over a period of 2 months (to 2.6%), although there was no effect in the other treatments. It appears that ambient levels of UVB light have no effect on *R. aurora* prior to metamorphosis. Eggs of this species have high concentrations of photolyase, an enzyme that repairs DNA damage from UV light (Blaustein et al., 1996).

STATUS AND CONSERVATION

The historic range of the Northern Red-legged Frog has decreased through the years, with populations declining due to the combined effect of habitat loss and degradation, the introduction of exotic fish and American Bullfrogs, modification of temporary wetlands to a permanent hydroperiod, and the widespread use of toxic chemicals within its habitat (Hayes and Jennings, 1986; Adams, 1999, 2000). In many areas, large shallow temporary wetlands have been replaced by deeper and more permanent ponds, making breeding sites more habitable by exotics. In agricultural areas, hatching success of *R. aurora* eggs is significantly lower than in reference sites (<35% vs. 85%, respectively) and may be due to high concentrations of ammonia and total phosphorus and a high biological oxygen demand (de Solla et al., 2002b). In certain areas, population losses have been substantial. For example, they were found to have disappeared from 69% of 42 historically occupied habitats in the Willamette Valley of Oregon (*in* Kiesecker et al., 2001a). In contrast, however, Jones et al. (2005) considered them still to be common throughout much of their range.

In theory, the introduction of American Bullfrogs (*Lithobates catesbeianus*) may have had adverse effects on populations of *Rana aurora* where competition for resources among sympatric larvae may influence distribution patterns. For example, larval *Lithobates catesbeianus* have negative effects on developing *Rana aurora* larvae in mesocosms, depending upon whether food resources are clumped or more widely scattered. If clumped, the size and mass at metamorphosis of red-legged frog larvae was reduced, whereas scattered resources appeared to have no effect on larval growth and development (Kiesecker et al., 2001a). Interference competition occurs when resources are not evenly distributed, with large *Lithobates catesbeianus* larvae out-competing smaller *Rana aurora* larvae. If human activities alter the distribution of resources, such as by decreasing the extent of shallow water habitat or the amount of emergent vegetation (both preferred by red-legged frogs in the presence of bullfrogs), then competition becomes more severe.

In the presence of adult *Lithobates catesbeianus*, *Rana aurora* larvae tend to decrease their activity and alter their habitat use. In field enclosures, time to metamorphosis increased whereas mass at metamorphosis decreased when either larval or adult bullfrogs were present. Smallmouth bass (*Micropterus* sp.), another nonnative predator, had essentially no effect on larval *Rana aurora* growth. Survivorship, however, was decreased when bullfrog larvae and either bullfrog adults or smallmouth bass were present. These results suggest that both the life stage and composition of predators can affect developing *R. aurora* larvae (Kiesecker and Blaustein, 1998).

In contrast to these results, Adams (2000) could not demonstrate any direct competitive effects of larval American Bullfrogs on larval Northern Red-legged Frogs. Excluding bullfrogs from natural enclosures did not result in increased larval survivorship by the native species in permanent ponds, but the presence of fish completely eliminated *R. aurora* larvae. Since food does not seem limiting, Adams (2000) suggested that direct competition did not play a role in the decline of Northern Red-legged Frogs, but that indirect effects of exotics were important in determining the results of interactions between native and exotic species.

Competition is by no means the only way larval American Bullfrogs could adversely affect the native Northern Red-legged Frog. In syntopic habitats, larval *R. aurora* tend to decrease their activity and increase their use of refugia when in the presence of chemical cues of larval and adult American Bullfrogs (Kiesecker and Blaustein, 1997a). Larvae from allopatric populations do not change their behavior and are subject to increased levels of predation when faced with the novel predator. The difference in behavior suggests that American Bullfrogs have their greatest impact on native frogs shortly after colonizing a new area. If the native red-legged frogs survive the initial onslaught, the survivors adopt behaviors to coexist, that is, they do not move around as much, and they hide.

Interspecific amplexus may be another way adult American Bullfrogs interfere with the mating system of native Pacific Northwest ranids. Instances of interspecific amplexus are not uncommon and involve male *R. aurora* amplexing juvenile, male, or female *Lithobates catesbeianus* (Storm, 1952; Pearl et al., 2005a). Males thus waste their time during the breeding season rather than mate successfully with a conspecific, a form of social interference. Of course, adult American Bullfrogs likely prey on smaller ranid frogs of all species.

The presence of American Bullfrogs clearly does not exclude *Rana aurora*, and the species are sometimes found in the same pond. Co-occurrence does not mean that there is no effect, however, nor does it indicate the nature of past historical interactions. It is likely that American Bullfrogs have different impacts at different localities, depending upon habitat structure, the timing of introduction, life stage, hydroperiod, and the presence of exotic fishes. It is not surprising, therefore, that attempts to correlate the presence

Breeding habitat of *Rana aurora*. Photo: Michael Van Hattem

or absence of American Bullfrogs and Northern Red-legged Frogs are not conclusive (Richter and Azous, 1995; Adams, 1999, 2000; Pearl et al., 2005b).

Habitat fragmentation likely has detrimental effects on populations of Northern Red-legged Frogs as terrestrial foraging and overwintering sites may be located distantly from breeding sites. Silvicultural operations may open up large amounts of habitats that become unsuitable for *R. aurora*, although Cole et al. (1997) found little effect of cutting, burning, and herbicide application on this species. Ashton (2002), however, failed to find *R. aurora* in headwater streams in redwood forests affected by silvicultural practices many years previously, and Gomez and Anthony (1996) noted that *R. aurora* was found more in riparian forested areas than in upslope habitats affected by silviculture. Leaving red alder (*Alnus rubra*) along 75–100 m on each side of riparian zones affected by cutting may help mitigate effects on red-legged frogs (Gomez and Anthony, 1996; Cole et al., 1997).

Northern Red-legged Frogs do not use single trees or small patches of remnant forests for cover, instead preferring large forest patches (> 0.8 ha) located within close proximity to one another and containing a stream. When moving across exposed areas, they do not move from patch to nearest patch, instead preferring a more direct route. Indeed, they will move from one patch to another only if the patches are between 5 and 20 m away; otherwise, forested patches are occupied randomly. Chan-McLeod and Moy (2007) recommended leaving small forested patches (0.8–1.5 ha) clumped together along stream channels as a means to mitigate the impacts of silviculture on the movements of this species, but these must be located on direct migratory corridors.

Northern Red-legged Frogs readily occupy restored wetlands. They appear to do best in such wetlands with intermediate hydroperiods and in the presence of native fishes (Henning and Schirato, 2006).

Rana boylii Baird, 1854
Foothill Yellow-legged Frog

ETYMOLOGY

boylii: Latinized patronym in honor of Dr. Charles E. Boyle (1821–1870). Boyle was a physician who traveled in California and collected specimens of amphibians and reptiles that he sent to the Smithsonian Institution. Included were the first specimens of the Foothill Yellow-legged Frog, which was named in his honor.

NOMENCLATURE

Dubois (2006): *Rana* (*Amerana*) *boylii*
  Synonyms: *Rana boylei*, *Rana pachyderma*

IDENTIFICATION

Adults. The dorsal background coloration is a uniform black to light gray, olive green, reddish, or brown, usually with dark markings. Gray, greenish, and yellowish ground colors are less common than the browns. There is a light patch on top of the head between the nares and eyes, behind which is a dark area crossing the eyelids and merging into the dorsal coloration. Irregular dark spots may be present between dorsolateral folds; the folds may have a reddish appearance. Occasional individuals may be unpatterned. The throat and chest are often spotted with dark markings. The belly and the underside of the legs are bright yellow on a white ground color, and the bright yellow color intensifies with age. In young animals, the yellow coloration may be faint or lacking, or there may be a hint of orange coloration on the yellow thighs. There is no eye mask, and the tympanum is small, roughened, and difficult to see. Dark bars are present on the dorsal surface of the hind legs. The hind feet are fully webbed, and the digits have slightly expanded tips, which aid in clinging to rocks and rock faces. The vocal sacs are subgular and internal. Males have a swollen and darkened thumb during the breeding season. This is the smallest species among the yellow-legged frogs, with a maximum size of 82 mm SUL (Jones et al., 2005). Jones et al. (2005) mention that one albino is known.

Larvae. Tadpoles are olive to beige with diffuse

dark spotting on the tail fin and tail musculature. The upper body surfaces are spotted, but the spots are not evident because of the overall dark coloration. Golden flecks may be present on the tail and dorsal body surface. Tail fins are highest at the midpoint of the tail. The venter is silvery in coloration and fairly opaque. The tadpoles are dorsoventrally flattened, possess dorsally focused eyes, have a muscular tail with narrow tail fin, and have more labial tooth rows on the oral disk (six to seven above the mouth and five to six below) than most other western frogs. The maximum size is 47–70 mm TL. Zweifel (1955) illustrated the tadpole and mouthparts.

Eggs. The eggs are black above and off-white below. There are three distinct envelopes surrounding the vitellus. The outer envelope is 3.9–4.5 mm in diameter; the middle envelope is 2.6–3.4 mm in diameter; the inner envelope is 2.3–3 mm in diameter; the vitellus is 1–2.5 mm. Some variation exists in ovum size, possibly reflecting differences in the date of deposition (Zweifel, 1955). Eggs are deposited in a compact grape-like cluster measuring around 5 cm × 10 cm × 12 cm, and eggs are firmly attached to one another. Information on eggs is contained in Storer (1925), Livezey and Wright (1947), Wright and Wright (1949), and Zweifel (1955). Zweifel (1955) provided a diagram of the egg and jelly capsules.

Distribution of *Rana boylii*. Dark gray indicates extant populations; light gray indicates extirpated populations.

DISTRIBUTION

Rana boylii occurs from Mehana in Marion County, Oregon, south through the San Gabriel River in Los Angeles County, California. Populations are confined to low to medium elevations (sea level to 1,940 m) west of the Cascades crest and in the Coast Ranges and the Sierra Nevada. Isolated populations also were found in Butte and Los Angeles counties, although some of these populations are extinct. Another isolated population once existed in the Sierra San Pedro Martir (2,040 m) in Baja California, but it too may be extinct (Loomis, 1978; Welsh, 1988; Grismer, 2002). Important distributional references include Storer (1925), Nussbaum et al. (1983), Leonard et al. (1993), Jennings and Hayes (1994b), and Jones et al. (2005).

FOSSIL RECORD

No fossils are recognized.

SYSTEMATICS AND GEOGRAPHIC VARIATION

There have been a number of studies that suggested sometimes conflicting phylogenies among Pacific Northwest ranids (Zweifel, 1955; Case, 1978; Farris et al., 1979, 1982; Green, 1986a, 1986b; Macey et al., 2001; Hillis and Wilcox, 2005). At one time, three subspecies were recognized: *boylii*, *muscosa*, and *sierrae*. Zweifel's (1955) separation of *R. boylii* from *R. muscosa* (including *muscosa* and *sierrae*) was supported by Houser and Sutton (1969), based on morphological and karyological data.

This species is considered to be a member of the *R. muscosa* complex (sometimes termed the *Rana boylii* species group; Macey et al., 2001) of the *Amerana* clade of North American ranid frogs. This monophyletic group is approximately 8 my old, and recent levels of genetic divergence largely result from Pleistocene vicariance events over the last few million years. This clade has its affinities with the Palearctic brown frogs rather than the eastern North American ranids, based on immunology (Zweifel, 1955; Case, 1978; Post and Uzzell, 1981). It is most closely related to *Rana luteiventris* and *R. pretiosa* (Macey et al., 2001; Hillis and Wilcox, 2005), despite earlier assertions of a close relationship with *R. muscosa* based on karyotype (Green, 1986a). The diploid chromosome number is 26; Houser and Sutton (1969) and Green (1986a) described the chromosomes.

There is considerable individual and geographic variation in coloration, dorsal spotting, and the extent of mottling in the groin and throat. In general, frogs at the northern end of the range are darker than those in the Sierra Nevada and the southern Coast Ranges, but any single population may have a wide array of morphotypes. Nussbaum et al. (1983) also noted that frogs in extreme southwestern Oregon had brick-red coloration more commonly than those in northern Oregon. Lind et al. (2011) found a moderate amount of genetic variation within the species that correlated, in part, with some of the phenotypic differences noted by earlier workers. In general, genetic variation is low among populations within the largest central clade of *R. boylii*, but there is substantial variation among populations at the edges of its geographic range and from these populations and the more central populations. Zweifel (1955) discussed regional variation in this species. Laboratory hybridization experiments suggest genetic incompatibility with *R. cascadae, R. muscosa,* and *R. sierrae* (Zweifel, 1955).

ADULT HABITAT

The best habitat of *R. boylii* is clear and cool (<25°C) streams and rivers with partial shade, shallow riffles, and substrates containing sand, gravel, or rocks larger than cobbles. They do not like streams with sandy substrates. Optimal streams are typically 2–4 m in width, rarely > 30 cm in depth, except in pools, and do not have a steep gradient. *Rana boylii* inhabits permanent streams, although water flow may be greatly reduced during the dry summer. It may occasionally inhabit intermittent streams, but only if American Bullfrogs are not present. Foothill Yellow-legged Frogs inhabit some large streams and rivers and even some smaller streams, although abundance is not as great as in the intermediate-sized streams. Small streams may provide important protected sites for tadpoles. Adjacent forests are usually of mixed conifers or deciduous trees (alders, sycamores, cottonwoods, willows) that provide intermittent shaded canopy over the streams. Other forest and vegetation types include chaparral, Digger pine-blue oak forest, yellow pine forest, and coastal redwoods. The species prefers habitats that are relatively free from disturbances.

AQUATIC AND TERRESTRIAL ECOLOGY

This is an aquatic species that rarely ventures more than a few meters from streams and rivers. They sit on sand bars and rocks within streams, or they may be found along rocky shorelines. They are active at body temperatures of 11–25°C (Zweifel, 1955; Brattstrom, 1963), and Zweifel (1955) noted that activity does not occur at water temperatures <7°C. Activity begins as early as December or January at low elevations, although most frogs do not become active until February or March. At high elevations, cold temperatures may delay emergence until March or April. In coastal regions, activity continues to October or early November, but at higher elevations activity ceases with the onset of cold weather. Young individuals disperse upstream after metamorphosis (Twitty et al., 1967) and are active later in the season than larger adults.

The overwintering behavior is unknown, but overwintering may occur under rocks in the heavily shaded and cool tributaries to the main stream channels or perhaps even under rocks in nearby forests within a few meters of a stream (Storer, 1925; Zweifel, 1955). Nussbaum et al. (1983) reported unearthing small *R. boylii* at a rock outcrop 50 m from a river in April.

According to Kupferberg (1996), adults are seen in small tributaries prior to the mating season in spring. Storer (1925) suggested that this species does not enter dormancy because of the relatively mild temperatures throughout much of the range. In many of the small creeks, winter rainfall causes considerable scouring flow across rocky substrates, making dormancy within stream channels risky.

Like most frogs, Foothills Yellow-legged Frogs are photopositive in their phototactic response, suggesting they use both sunlight and moonlight when feeding and going about their daily activities (Jaeger and Hailman, 1973). *Rana boylii* likely has true color vision.

CALLING ACTIVITY AND MATE SELECTION

Males have a guttural, grating advertisement call that has one pitch or a slightly rising inflection. Rombough and Hayes (2005) described it as "a series (5 to 7 notes) of distinct, rubbery clucks." According to Stebbins (1951), the frog sometimes utters groups of 4–5 croaks lasting ca. 0.5 sec each with ca. 0.5 sec intervals between croaks. This series may be followed by "a prolonged rattling sound lasting 2.5 seconds." In addition, *R. boylii* produces several different calls while submerged, suggesting an extensive vocabulary that is capable of conveying information to other Foothills Yellow-legged Frogs. MacTague and Northen (1993) reported four different calls produced underwater: a short, unpulsed call; pulsed calls of two intermediate lengths; a long call where groups of pulses form notes. These calls could be used in a territorial context (advertisement and defense) as well as to attract mates. *Rana boylii* also makes a faint short pulsed call when the frogs are handled.

Males arrive at the breeding sites and remain in the same location throughout the breeding season. Females arrive asynchronously throughout the breeding season and do not remain after oviposition has been completed. Wheeler and Welsh (2008) suggested that males defended optimum calling sites that may not be directly associated with an oviposition site; they observed male–male aggression that indicated territorial defense. Not surprisingly, the daily operational sex ratios at breeding sites are male-biased, although the seasonal operational sex ratio actually may be female-biased (Wheeler and Welsh, 2008).

It appears large males have an advantage when mating in comparison to small males, although strict size-assortive mating does not occur. In Wheeler and Welsh's (2008) study, for example, amplexing males were larger than males that never mated within a breeding season. Thus mating is likely not random, but it is not known whether females choose mates or how amplexus is initiated. Occasionally, *R. boylii* will attempt to mate with other species, such as males amplexing female *Lithobates catesbeianus* (Lind et al., 2003).

BREEDING SITES

Breeding occurs only in relatively slow-moving streams on a rock or boulder substrate, which allows for the development of sediment bars and backwater habitats. Preferred sites include river bars offering protection from swift currents and in proximity to the confluence of tributaries where rivers are wide and shallow. Oviposition also occurs in off-channel scour pools or troughs. Frogs may deposit eggs in narrow or deep channels, but eggs oviposited in these areas have poor survival in both wet (from scouring effects) and dry (prone to desiccation) years. Optimal sites may be used repeatedly from one year to the next.

REPRODUCTION

Emergence occurs several weeks prior to the initiation of breeding, and breeding is closely linked with stream hydrology. Reproduction occurs after the high-flow stream discharge resulting from winter rainfall and high-elevation snowmelt; it is correlated with warming air and water temperatures and decreases in stream flow. Reproduction is thus timed to reduce the threat of eggs being washed away via stream scouring.

The breeding season of *Rana boylii* lasts from three to seven weeks, from late March to early June, depending upon elevation, location, and environmental conditions (Storer, 1925; Fitch, 1936; Wright and Wright, 1949; Jennings and Hayes, 1994b; Zweifel, 1955; Jones et al., 2005). For example, the breeding season lasted 19–52 days (mean 49.5) in Del Norte County, California, over a 6 yr period (Wheeler and Welsh, 2008), and Kupferberg (*in* Wheeler and Welsh, 2008) found the breeding season lasted 18–63 days (mean 34) in Mendocino County, California. Plasticity in the timing of reproduction allows this species to take advantage of optimum stream conditions in an environment subject to stochastic changes in rainfall and river flow patterns. In contrast, Rombough and Hayes (2007) suggested that *R. boylii* was an explosive breeder in Oregon, and that the breeding season only lasted about a week. These differences may be related to differences in the availability of habitat and to physical differences in streams between locations.

Breeding occurs over an extended period in spring, but the precise dates of reproduction depend on water flow and temperature. In Oregon, for example, breeding occurs from April to June (Nussbaum et al., 1983). In California, breeding begins from early April to early May, depending on river flows, with river flows of 0.10–0.6 m/s producing the greatest amount of breeding activ-

Eggs of *Rana boylii*. Photo: Koen Breedveld

ity. River flow levels usually are optimal >30 days following the initiation of the breeding season, although reproduction may be halted or delayed by high stream flows that might dislodge eggs or wash them away. The initiation of breeding occurs earlier in low-flow years than in high-flow years and, as such, the amount and timing of precipitation may have direct impact on the duration of the breeding season and indirectly on reproductive output. This is because the number of egg masses produced at a location may be correlated with maximum annual river flow during the breeding season. In addition, water temperatures should be >9.5–12°C (Zweifel, 1955; Kupferberg, 1996).

Eggs are attached to substrates (cobbles and boulders, rarely on bedrock or vegetation) within the stream in shallow water (4–43 cm) of low velocity. Attachment sites are on the downstream sides of cobbles and boulders, where stream velocity is slower than in the main channel. The location of egg deposition depends on water level. In high-flow years, eggs may be deposited along the shoreline; in low-flow years, eggs may be placed as much as 1.25 m from shore (Kupferberg, 1996). The location of the egg mass ensures adequate oxygenation as cool water passes around it. Kupferberg (1996) reported a mean of 19 clutches per site at 15 study sites in northern California. Egg masses are frequently covered in sediment after a few days, at which time they match the brown alga *Nostoc* that is present in the same habitats. The resemblance may help conceal the eggs.

The male amplexes the female and then rides along with her as she selects an oviposition site. Females prepare oviposition sites by scraping a rock surface with their hind or front limbs to remove algae and debris, thus making the rock less slippery. This process normally lasts < 10 min (but up to 50 min) at a rock, but females may scrape several rocks prior to selecting the final oviposition site, thus prolonging site selection to 3 hrs or more. The male then lowers his feet to rest on top of her legs. The female releases her eggs onto the scraped surface while the male rubs up and down her thighs with his feet. This helps to position the eggs into a "basket" formed by the combined hind limbs of the pair. The male lowers his vent and fertilizes the eggs as they are oviposited, a process that takes 1.5–8 min. The male then lifts his legs from the female's thighs, and she crawls away from the egg mass; he releases her a short while afterwards. Egg deposition is described by Wheeler et al. (2003) and Rombough and Hayes (2005). Occasionally other males may attempt to interfere with oviposition by ramming the amplexed pair or by fighting among themselves; in return, the female may kick intruding males (Rombough and Hayes, 2007), or the pair may briefly take refuge among cobbles (Rombough and Hayes, 2005).

A female oviposits 900–1,050 eggs per clutch,

and the grape-like compact egg mass initially measures approximately 30 mm × 40 mm × 15 mm but swells within 20 min to 85 mm × 65 mm × 54 mm (Rombaugh and Hayes, 2005). Storer (1925; repeated by Livezey and Wright, 1947) reported clutch sizes of 919, 952, and 1,037 eggs. Eggs usually hatch in about 2 weeks, depending on water temperature, with Zweifel (1955) recording extremes of 5–30 days. Nussbaum et al. (1983) recorded hatching in 5 days at 20°C, and Jones et al. (2005) gave a figure of 10–28 days. Eggs at the perimeter of a clutch hatch first, with the last eggs to hatch those attached to the substrate. Eggs have been found in water up to 21.5°C (Zweifel, 1955). Larvae are 7.3–7.7 mm TL at hatching.

LARVAL ECOLOGY

Little information is available on larval ecology except for feeding habits. The large number of tooth rows suggests an adaptation to living and feeding in swift-flowing streams, much like *Ascaphus*. Larvae are able to use epiphytic diatoms attached to the filamentous green alga *Cladophora glomerata* that grows in California rivers. The diatoms increase the nutritional value of the alga to tadpoles by increasing the protein content. The diatoms are also able to store

Early stage tadpole of *Rana boylii*. Photo: Koen Breedveld

Late stage tadpole of *Rana boylii*. Photo: Koen Breedveld

Tadpole of *Rana boylii* with copepod parasite. Photo: Alessandro Catenazzi

photosynthate as lipids rather than carbohydrates. When compared to other algae not containing the diatom epiphytes, *Cladophora* with diatoms is a much better larval food source, and larvae grow faster and metamorphose quicker than they would on other algae or on *Cladophora* without diatoms (Kupferberg, 1997b).

Larvae are able to use multiple cues (visual, chemoreceptive, mechanical, and combinations thereof) to detect the presence of native predators, such as the salamander *Taricha granulosa*. When such a native predator is detected, larvae reduce their activity level, but only when multiple cues are involved. They do not respond solely to chemical cues, perhaps because they live in flowing water where chemical cues would be less effective than in lentic water. They tend not to reduce their activity levels in the presence of the nonnative predatory fish *Micropterus dolomieu*, making them particularly sensitive to adverse impacts from this nonindigenous species (Paoletti et al., 2011).

Larvae have been found in streams at water temperatures to 30.2°C (Zweifel, 1955). They tend to avoid swift water, with large larvae seeking shelter at velocities of 10 cm·s^{-1} and smaller larvae tolerating velocities to 20 cm·s^{-1} (Kupferberg et al., 2011). The response also varies by developmental stage, and the proportion of the time spent swimming is influenced by water velocity. Higher water velocities also decrease growth rates and survivorship, even at subcritical velocities of 5–10 cm·s^{-1}. The duration of the larval period is three to five months (Storer, 1925; Jones et al., 2005), and larvae do not overwinter. Recently metamorphosed frogs measure 18 to 30 mm SUL.

DIET

Postmetamorphic individuals likely eat a variety of invertebrates, but there is almost no dietary information available for this species (Storer, 1925; Fitch, 1936). Storer (1925) reported one individual that had fed extensively on grasshoppers.

PREDATION AND DEFENSE

Eggs are eaten by newts (*Taricha granulosa*) (Evenden, 1948), and the salamander *Dicamptodon ensatus* has been observed feeding on tadpoles (Fidenci, 2006). Major predators of postmetamorphs likely include garter snakes (*Thamnophis couchi, T. elegans, T. sirtalis*), birds, and mammals (Zweifel, 1955). Tadpoles are cryptic and easily overlooked. When disturbed, postmetamorphs dive into water and swim to the bottom, where they hide in mud and silt. They also may take refuge under overhanging rocks in clear water (Storer, 1925). The dorsal coloration and pattern aid them as camouflage against the stream substrate and while they are out of water.

POPULATION BIOLOGY

Adults reach sexual maturity at a body size of 40 mm SUL (Zweifel, 1955). Zweifel (1955) suggested that frogs reach maturity the first summer following metamorphosis, with first breeding not occurring until the following spring, at 2 yrs of age. Whether males and females reach maturity at the same time is unknown.

COMMUNITY ECOLOGY

Grazing by frog larvae has important influences on local macroinvertebrate communities through interference and exploitation of diatoms. In field experiments, *Rana boylii* larvae graze on the inedible filamentous alga *Cladophora glomerata* when this species is often covered by nutritious epiphytic diatoms. Detritus is decreased on the alga, and diatom abundance can be reduced by >50%; these effects result in an increase in the extent of *Cladophora* turfs (Kupferberg, 1997c). By feeding on periphyton, *Rana boylii* larvae actually increase area-specific primary productivity by 10%, although biomass-specific productivity decreases. As a result, the abundance of invertebrate consumers and their predators declines in areas grazed by Mountain Yellow-legged Frog larvae.

Rana boylii can experience resource competition from the introduced American Bullfrog (*Lithobates catesbeianus*). In field experiments, bullfrog larvae selectively graze on the most nutritious algae (*Cladophora* with diatoms) and by doing so reduce survivorship and mass at metamorphosis of those sympatric *Rana boylii* larvae that do survive (Kupferberg, 1997a). According to Kupferberg (1997b:155), "a small impact of bullfrogs on algal quality, thus resulted in a large impact on *Rana boylii* because there is such a strong correlation between algal quality and the biomass of *R. boylii* metamorphosing from the enclosures." Thus, negative effects on native ranids can result by the way bullfrog larvae selectively graze food resources. Large overwintering bullfrog larvae can also get a head start on newly hatched *R. boylii* larvae by out-competing them physically for access to high quality food.

DISEASES, PARASITES, AND MALFORMATIONS

Rana boylii is parasitized by the introduced Asian copepod *Lernaea cyprinacea*, outbreaks of which are correlated with unusually warm summers (Kupferberg et al., 2009). In California, outbreaks of this parasite were associated with increased levels of malformations (to 26.5 vs. 1.1% of uninfected larvae), particularly at infestation sites around the hind limbs and cloaca. Malformations included missing hands, limbs, fused digits, abnormal limb development, and extra limbs. Recent metamorphs were generally smaller if infested with copepods as larvae compared with uninfected froglets, and infestation rates were higher in downstream sites compared with upstream sites.

The amphibian chytrid fungus *Batrachochytrium dendrobatidis* has been found in *Rana boylii* from California (Green et al., 2002) and from natural populations in the Pacific Northwest (Adams et al., 2010). Indeed, this fungus was found in museum specimens collected from California as early as 1966 (Padgett-Flohr and Hopkins, 2009). It does not kill postmetamorphic Foothill Yellow-legged Frogs outright, but it does substantially retard growth by as much as 50% (Davidson et al., 2007). *Rana boylii* has peptides in the skin that inhibit the growth of amphibian chytrid and likely prevent substantial mortality in adults. However, the effectiveness of the peptides is reduced when frogs are exposed to certain contaminants such as the pesticide carbaryl.

Foothills Yellow-legged Frogs are parasitized by the trematodes *Clinostomum* sp., *Deropegus* [*Halipegus*] *aspina*, *Glypthelmins quieta*, *Gorgoderina multilobata*, *Haematoloechus* sp., *H. kernensis*, *H. variplexus*, *Megalodiscus microphagus*, and *M. temperatus*; *Distoichometra bufonis* and unidentified dilepinid cestodes; the nematodes *Cosmocercoides variabilis*, *Falcaustra pretiosa*, *F. ranae*, *Hedruris* sp., and *Rhabdias ranae*; and acanthocephalans (centrorhychid and oligocanthorhychid cystacanths) (Ingles and Langston, 1933; Ingles, 1936; Walton, 1941; Lehmann, 1960; Goldberg and Bursey, 2002a; Bursey et al., 2010).

SUSCEPTIBILITY TO POTENTIAL STRESSORS

Metals. Mercury has been found in *Rana boylii* from California at concentrations of 0.030–0.65 µg/g wet weight (Hothem et al., 2010). Values > 0.3 µg/g wet weight exceed EPA standards for mercury concentrations in fish.

Chemicals. Death occurs in about 18 hrs at 30 mg/L of carbaryl, a broad-spectrum insecticide, and significantly reduces larval activity levels at 2.5 mg/L (Bridges and Semlitsch, 2000). In low concentrations (0.48 mg/L), it has no effect on survival. The estimated minimal lethal concentration is 4.8 mg/L (Davidson et al., 2007). However, carbaryl, even at low doses, reduces the effectiveness of skin peptides in repelling amphib-

Adult *Rana boylii*. Photo: Aubrey Heupel

Adult *Rana boylii*, amplexus. Photo: Koen Breedveld

ian chytrid fungus in postmetamorphs and thus has sublethal effects that might be overlooked in dosage experiments alone.

The agricultural pesticides chlorpyrifos, diazinon, malathion, and their oxon derivatives are toxic to larval *R. boylii* at concentrations found in the environment. $LC_{50\ (1-4\ day)}$ values are 3.0 mg/L for chlorpyrifos, 2.14 mg/L for malathion, and 7.49 mg/L for diazinon. Oxon derivatives are 10–100 times more toxic than their parental forms (Sparling and Fellers, 2007). These pesticides depress normal cholinesterase activity.

STATUS AND CONSERVATION

This species has disappeared from much of its historic range (Jennings, 1988; Jennings and Hayes, 1994a; Bury, 2008). Jennings and Hayes (1994a), in particular, provide a review of extinctions in the southern portion of its range. According to Stebbins (2003), populations have disappeared from 55% of its range in Oregon, 45% of its range in California, and 66% of its range in the Sierra Nevada. Moyle (1973) noted that *R. boylii* already had disappeared from much of its historic range along the margins of the San Joaquin Valley and attributed declines to habitat loss and impacts from the introduced American Bullfrog (*Lithobates catesbeianus*). At that time, *Rana boylii* only occurred in slightly higher-elevation streams not inhabited by bullfrogs. Likewise, Bury (2008) attributed the near extirpation of this species in the Willamette Valley in Oregon to rapid urbanization, development, and the effects of nonindigenous species.

Because the Foothills Yellow-legged Frogs lay their eggs in streams, they may be detrimentally affected by stream scouring during periods of excessive or unusual rainfall. They also are vulnerable to anthropogenic changes that alter water velocity, particularly the release of water on dammed rivers (Lind et al., 1996; Kupferberg et al., 2011). In streams with artificial flow regulation, eggs and larvae may be washed away by a sudden increase in water flow that can scour streambeds, or the substrate may be altered, rendering a previously optimum breeding site unusable. In addition, water flow substantially decreases (by 70–90%) throughout much of the year, and riparian vegetation is altered by water regulation. On the Trinity River in California, Lind et al. (1996) documented the detrimental effects of dam construction on this species and recommended that management protocols be instituted that mimic natural stream flow.

Silviculture may have adverse effects on *R. boylii* if streams feeding into its preferred habitats become clogged with silt. However, silvicultural operations also may open up habitats and make them suitable for Foothills Yellow-legged Frogs. For example, far more *R. boylii* are found in second-growth redwood forests than in forest in late-seral stage (Ashton, 2002; Ashton et al.,

Adult *Rana boylii* in stream habitat. Photo: Alessandro Catenazzi

River habitat of *Rana boylii*. Photo: Koen Breedveld

2006). Presumably the canopy removal increased temperature and habitat characteristics in a manner beneficial to the frogs.

The genetic structure of R. boylii populations as described by Lind et al. (2011) indicate that individual river systems as well as larger geographic and hydrologic regions need to be managed separately in order to conserve this species and its genetic variance. It is likely that populations in the south, now extirpated, constituted unique genetic assemblages and represent a significant loss to historic biological diversity of this species.

Rana cascadae Slater, 1939a
Cascades Frog

ETYMOLOGY
cascadae: Latinized name for the Cascade Mountains where the species was first discovered.

NOMENCLATURE
Dubois (2006): *Rana (Amerana) cascadae*
  Synonyms: *Rana aurora cascadae*

This species has sometimes been misidentified as *Rana pretiosa* in the literature (e.g., Grinnell et al., 1930; see Jennings and Hayes, 1994b).

IDENTIFICATION
Adults. The dorsal ground coloration is brown, red brown, or slightly greenish brown. Dorsal spots (a few to > 50) are large and distinct, and the dorsal skin ranges from smooth to rather rough in texture. Dorsolateral folds are prominent, and the jaw has a distinct light stripe. Venters are yellowish or buff colored, and melanophores are absent from the medial abdominal area. A diffuse dark reticulation pattern is evident in the groin, which is yellow green. Most of the ventral yellow coloration is located posteriorly and on the lower limbs. Jennings and Hayes (1994b) noted that the iris is brown with gold iridophores. Males have prominent nuptial pads. Dunlap (1955) provided extensive information on the color pattern and morphology of this species.

Females are slightly larger than males. In Oregon, males reach a maximum SUL of 57 mm whereas females reach a maximum SUL of 66 mm (Briggs and Storm, 1970). Jennings and Hayes (1994) give a maximum size of 75 mm SUL.

Larvae. The tadpole is brown or olive brown dorsally and on the tail musculature. The belly is pale and has a golden iridescence, and the tail often has small blotches. Larvae reach a mean maximum size of 53–66 mm TL. Sype (1975) provides a description of the eggs and larvae from fertilization through metamorphosis. Albino tadpoles were reported by Altig and Brodie (1968) from Oregon, and Brome McCreary observed albino tadpoles and a recent metamorph (photos in these volumes).

Eggs. Eggs are black dorsally and cream colored ventrally. There are three distinct envelopes surrounding the vitellus. The outer envelope is ca. 11.3 mm; the middle envelope is ca. 5.8 mm in diameter; the inner envelope is ca. 4.9 mm in diameter; the vitellus is ca. 2.3 mm (Livezey and Wright, 1947). Eggs are deposited in a solid jelly mass.

DISTRIBUTION
The Cascades Frog is found in three disjunct regions: the Olympic Mountains of Washington, the Cascades Mountains of Washington and Oregon, and the Klamath-Siskiyou Mountains in northern California. The species occurs at elevations of 230–2,740 m. Important distributional references include Slater (1955), Metter (1960), Bury (1973), Hayes and Cliff (1982), Nussbaum et al. (1983), Leonard et al. (1993), Jennings and Hayes (1994b), and McAllister (1995).

FOSSIL RECORD
No fossils are known.

SYSTEMATICS AND GEOGRAPHIC VARIATION
There have been a number of studies that suggested sometimes conflicting phylogenies among Pacific Northwest ranids (Zweifel, 1955; Dumas, 1966; Case, 1978; Farris et al., 1979, 1982; Green, 1986a, 1986b; Macey et al., 2001; Hillis and Wilcox, 2005). This species is a member of the *R. muscosa* complex (sometimes termed the *Rana boylii* species group; Macey et al., 2001) of the *Amerana* clade of North American ranid frogs. It is most closely related to *R. aurora*, and somewhat more distantly to *R. muscosa* and

Distribution of *Rana cascadae*

R. sierrae. Green (1986a) described the chromosomes and hypothesized that *R. cascadae* was related more to *R. aurora*-*R. draytonii* and *R. pretiosa*-*R. luteiventris* than to *R. boylii* and *R. muscosa*. The diploid chromosome number is 26 (Haertel et al., 1974).

Populations of *R. cascadae* show a high degree of genetic differentiation at a local scale, and there is a discordance between the results of molecular analyses depending upon whether mtDNA or nuclear DNA is studied. mtDNA results suggest a 2–3 my separation between the Olympic and Cascades populations, but this is not supported by microsatellite data, which indicate a more recent separation. However, microsatellite results indicate a break between the Oregon and Washington Cascades populations at the Columbia River. Monsen and Blouin (2003) suggested that there were two distinct population segments within *R. cascadae*, Washington/Oregon and California, but that the three disjunct populations should be managed for conservation separately.

Laboratory hybridization experiments suggest genetic incompatibility with *R. draytonii* (Zweifel, 1955). Hybridization occurs with *R. pretiosa* under laboratory conditions, with larvae completing metamorphosis (Haertel and Storm, 1970) or not (Dumas, 1966). Hybridization was more successful when *R. cascadae* was the female parent. Green (1985c) documented hybridization with *R. pretiosa* in nature, but the hybrids were infertile because of a lack of chromosome pairing during meiosis.

ADULT HABITAT

The Cascades Frog is found in boreal habitats (especially associated with firs [*Abies*] and arborvitae [*Thuja*]) in the high Cascade and Olympic Mountains of the Pacific Northwest. At one time, however, they were also known from low elevation sites on the Olympic Peninsula (Leonard et al., 1993). Habitats include ponds, lakes, slow-moving steams through wet meadows, meadow wetlands, and riparian areas along fast-moving, steep mountain streams. Both permanent and ephemeral habitats may be occupied, but ephemeral habitats must have saturated areas for the species to survive. This species prefers old-growth forest habitats, but can be found in less abundance in mature second-growth and young forest stands (Aubry and Hall, 1991). In contrast, Bosakowski (1999) found most Cascades Frogs in open meadows surrounded by conifer saplings. Pearl et al. (2009a) estimated occupancy at 72.4% of historically known sites in Oregon over a 3 yr survey, but they were unable to determine reliable predictor variables (presence of fish, forest vegetation, roads, age of ponds, elevation) associated with occupancy.

AQUATIC AND TERRESTRIAL ECOLOGY

This species spends its entire life in or very close to permanent waters, but it is not often found floating in lentic water as is *R. pretiosa*, for example. *Rana cascadae* are diurnal and are commonly observed along the shoreline of ponds and lakes. Brattstrom (1963) found them sitting in the sun on bare soil or on moss-covered rocks in the middle of cold streams; frogs were found at daytime temperatures of 11.9–29.5°C, although Wollmuth et al. (1987) noted that postmetamorphs selected temperatures <17°C under controlled conditions. Body temperatures at night mirror water temperature, and frogs are more likely to be found in the warmer water than in the cool night air. They may be active terrestrially at temperatures of <8°C, however. Overwintering occurs in mud bottoms within ponds, ponded streams, or in spring-water saturated ground at depths of 0.3–1.0 m (Briggs, 1987).

Movements may occasionally cover very long distances. For example, Crisafulli et al. (2005) reported movements of 0.75 and 1.2 km from the original capture location of 2 Cascades Frogs within areas affected by the 1980 eruption of Mount St. Helens. Clearly this species has the capacity to colonize sites at some distance from known breeding ponds, even though it is thought to rarely move away from permanent water.

CALLING ACTIVITY AND MATE SELECTION

Rana cascadae is considered an explosive breeder. Emergence occurs soon after melting snow and ice create ponded water, and males appear at breeding sites prior to females. Males usually call from matted floating vegetation in water 5–10 cm in depth, but Briggs (1987) noted a single male calling underwater. Although initially dispersed along the shoreline, they tend to form groups of individuals only several cm apart. Calling occurs during daylight hours in full sunlight, and calling decreases or stops completely on cloudy days. Several hundred males may chorus at once.

Briggs (1987) identified 4 types of calls: series cluck (4–39 notes in a continuous series of ca. 7 notes per sec; dominant frequency 400–1,500 Hz; 158.5 pulses/sec; duration of note = 0.6 sec; internote interval of 0.09 sec; usually made by stationary frogs or during short swims); double cluck (series of 2–4 clucks made by stationary frogs; clucks separated by 5–10 sec; short note duration of 0.05 sec; long interval between notes of 0.13 sec; 152.7 pulses/sec); mew call (given at end of series of clucks when male stopped swimming; dominant frequency of 750–1,250 Hz; duration of 0.3 sec; pulse rate of 183/sec); release calls (given by males to avoid amplexus). He provided a sonogram of the series cluck, mew, and double cluck calls.

Males will clasp any frog moving near them. Amplexus is axillary. Males will attempt amplexus with other males, and Briggs (1987) reported from two to seven males attempting to amplex a single female. Such mass amplexus can result in the death of the female. Necrogamy has also been observed (Garwood and Anderson, 2010).

BREEDING SITES

Rana cascadae breeds in permanent or temporary quiet ponds and lakes in high mountain meadows, as well as in slow-moving streams moving through such meadows. Oviposition occurs in shallow areas with maximum solar insolation; oligotrophic ponds are most favorable (Briggs and Storm, 1970). Preferred ponds have deep water, which allows sufficient time for embryonic development prior to a significant decrease in hydroperiod as the summer progresses. However, breeding also may occur in seasonally temporary

Eggs of *Rana cascadae*.
Photo: Brome McCreary

ponds (Walls et al., 1992), and Metter (1960) noted that breeding can take place even in small potholes and ponds fed by snowmelt. Wetlands with fish are generally avoided.

REPRODUCTION

Breeding occurs from early spring throughout the summer months, and the timing depends on elevation and environmental conditions, particularly temperature. Breeding begins sooner (March to April) at low elevations than at high elevations (May and later). For example, breeding begins in mid-March at Mount St. Helens and continues into July (Nussbaum et al., 1983; Karlstrom, 1986). Also in Washington, Slater (1955) reported breeding from mid-May to mid-July, whereas Briggs and Storm (1970) observed breeding in March and April in Oregon.

Frogs select deposition sites in warm waters where ice first disappears. Oviposition is diurnal. Females tend to oviposit in shallow unshaded water (Briggs, 1987) in close proximity to one another, and some clutches may be placed on top of others forming clumps. Jennings and Hayes (1994b) suggested that females might find communal oviposition sites through olfaction. Eggs are deposited at the edge of ponds in loose masses that either float free or become entangled in vegetation; they are not deliberately attached to vegetation. Deposition sites can include grass, rocks, and woody debris on the substrate. Egg masses are positioned in such a way as to receive full sun exposure.

During development, the egg masses can drift away from the original deposition site into deeper and cooler water. Masses can drift relatively long distances (37 m), which occasionally results in mortality (Garwood et al., 2007). The masses are globular and approximately 15 cm. in diameter. Livezey and Wright (1947) gave a single figure of 425 eggs per mass based on data from Slater (1939a); Briggs (1987) reported a mean of 412 (n = 5 clutches); and Sype (1975; repeated by Jennings and Hayes, 1994b) gave a figure of 400–600 eggs per mass.

During early development, the egg masses are exposed to wide temperature fluctuations, especially in those eggs near the water's surface. Eggs tolerate a temperature range of 6 to 27°C during development, and embryos acclimate as temperatures warm, giving them greater thermal tolerance as development continues. The dark-colored ovum absorbs heat during the day, and the jelly surrounding the egg helps to retain heat for several hours after sunset. Indeed, the temperature within an egg mass may be a few degrees warmer than the surrounding water. Eggs initially develop slowly, but the developmental rate increases following the gastrula stage. Embryos in water <6°C fail to develop normally. The optimal temperature

Tadpole aggregation of *Rana cascadae*. Photo: Justin Garwood

Albino tadpoles of *Rana cascadae*. Photo: Brome McCreary

range for development is 8–25°C (Sype, 1975). Mortality of eggs arises from desiccation or freezing temperatures.

LARVAL ECOLOGY

Pond temperatures range between 5–30°C under field conditions. Larvae voluntarily select warm water (mean = 27.3°C; Wollmuth et al., 1987) in order to optimize development, and even in early metamorphosis, they prefer high temperatures (mean = 28.8°C). As they go through metamorphosis, however, temperature preferences drop to 19°C. Larvae tend to occur in deeper water, especially in the morning and afternoon, but move into the shallows, usually within 5 m from shore, during midday (Bancroft et al., 2008). They appear to follow thermal gradients whereby warm water is sought in the daylight shallows, but deeper and presumably warmer waters at night are preferred as the shallows cool at sunset.

Cascades Frog larvae form aggregations (usually <100 individuals/aggregation) that remain in close proximity to the hatching site. These aggregations are especially noticeable in the afternoon and evening. Siblings tend to associate with one another over half- or nonsiblings under field (O'Hara and Blaustein, 1985) and experimental conditions (O'Hara and Blaustein, 1981; Blaustein and O'Hara, 1982). Even early association with nonsiblings does not affect sibling recognition. They prefer maternal cues to paternal cues when aggregating, and both learned and innate components likely play a role in kin recognition. In nature, however, sibling association tends to decrease with increasing aggregation size.

Kinship aggregation does have its costs, however. In laboratory experiments, tadpoles have increased developmental times, are shorter, and weigh less in dense aggregations than they do at lower densities. Individual survivorship also decreases in high-density aggregations. These effects are evident whether or not the conspecifics are related. Indeed, nonkin raised at high densities have a greater mass than kin raised at the same densities, although the variance is also greater. Thus, density-dependent effects are independent of genetic relatedness, and competition may be greater in kin groups than in nonkin aggregations (Hokit and Blaustein, 1994).

The advantage of aggregation may be context dependent under natural conditions. The type of and access to substrates, the distribution of food resources, the thermal environment, the presence of predators, and genetic-relatedness interact to influence tadpole dispersal (Hokit and Blaustein, 1997). Kinship also acts to influence group size, and density and access to the substrate influence growth and survivorship. Tadpole aggregations are not randomly dispersed, and, as might be expected, there is a tradeoff between food, temperature, predation, and competition in where and how aggregations are formed and maintained. Large aggregations in nature facilitate feeding by stirring up food from the substrate, increase temperature for activity due to the large number of compact black bodies, and may serve to alert conspecifics to predators and thus decrease individual risk from predators. Kinship may be important in size-structured aggregations of anuran larvae, although the specific mechanism is not yet understood.

Recent metamorphs of *Rana cascadae*, normal coloration (*left*) and albino (*right*). Photo: Brome McCreary

Larval *R. cascadae* show an alarm reaction to the presence of chemicals emanating from conspecifics that have been injured. Although they do not avoid areas containing damage chemicals, they significantly increase the level of activity in their presence. Larvae do not respond to damage chemicals emanating from *Pseudacris regilla*, however, indicating a species-specific response rather than a generalized response to an injured tadpole (Hews and Blaustein, 1985).

Transformation requires 30–60 days (Nussbaum et al., 1983; Briggs, 1987), although Slater (1955) gave a figure of 85–95 days. Larvae do not overwinter, although tadpoles have been observed in ponds as late as early September.

DIET

Postmetamorphic Cascades Frogs presumably eat a variety of invertebrates, but there appears to be nothing in the literature on its diet.

PREDATION AND DEFENSE

The larvae of *Rana cascadae* are palatable to predaceous insects (*Lethocerus americanus*, *Dysticus*) (Peterson and Blaustein, 1992) and salamander larvae (*Ambystoma gracile*, *Taricha granulosa*) (Peterson and Blaustein, 1991). Tadpoles are cryptic in coloration, which may serve as a primary defense against predation. Suspected predators of postmetamorphs include mammals (black bears, coyotes, raccoons, mink), birds (Sharp-shinned Hawks, Gray Jay, owls, robins), and garter snakes (*Thamnophis*) (Briggs and Storm, 1970).

POPULATION BIOLOGY

Little is known of the population dynamics of this species. Males grow at slower rates than females. In Oregon, males reach sexual maturity at 35 mm SUL and females at 52 mm SUL (Briggs and Storm, 1970). Growth data suggest that males could reach this size at 2 yrs and females at 4 yrs (Jennings and Hayes, 1994b). These authors caution, however, that actual reproduction might not occur for another 1–2 yrs. The smallest male observed in amplexus was 49 mm SUL, whereas the smallest female was 59 mm SUL. Briggs and Storm (1970) further suggested that longevity is > 5 yrs. In Briggs's and Storm's (1970) study, mortality was high over a 2 yr period: 43% for males and 49% for females.

Abundance of this frog can be high at a breeding pond. For example, Briggs and Storm (1970) estimated there were 1,020–1,293 males and 390–722 females at a breeding site in Oregon

Adult *Rana cascadae*. Photo: Justin Garwood

over a 2 yr period. The total pond estimate was 1,975–3,204 frogs. Sex ratios were male biased among year classes, but they varied annually. Much more data are needed on the demography and population structure of this species.

DISEASES, PARASITES, AND MALFORMATIONS

Cascades Frog larvae experimentally infected with amphibian chytrid fungus (*Batrachochytrium dendrobatidis*) showed no evidence of behavioral fever or altered thermoregulation (Han et al., 2008). In addition, uninfected larvae did not avoid infected larvae during laboratory aggregation trials. This virulent pathogen has been reported from natural populations in the Pacific Northwest (Adams et al., 2010). According to Turner (1958a), Ingles's (1936) reports of the parasites *Spironoura pretiosa* and *Gorgoderina multilobata* from "*Rana pretiosa*" were probably from *Rana cascadae*. Malformations in this species have been reported at frequencies of 3.3–6% at certain sites in the Pacific Northwest, with from 1 to 34 *Ribeiroia* metacercariae per frog (Johnson et al., 2002). The most common malformations were missing and malformed digits.

SUSCEPTIBILITY TO POTENTIAL STRESSORS

There are many potential stressors affecting *Rana cascadae*, and the interaction among stressors may be underappreciated. For example, there are no effects on larval activity or survivorship when they are individually tested for effects of low pH, high nitrate concentrations, or ambient levels of UVB light. However, activity and survivorship is reduced when exposed to the three factors simultaneously (Hatch and Blaustein, 2000). In combination, UVB tends to decrease survivorship whereas low pH and high nitrates act to decrease activity. When cold temperatures and amphibian chytrid are combined, survivorship also is reduced (Searle et al., 2010). Thus researchers need to bear in mind that multiple stressors may have adverse effects on Cascades Frogs even when individual stressors do not.

Chemicals. The herbicide Roundup® is toxic to larvae, with an $LC_{50\ (24\ hrs)}$ of 2.11 mg/L (King and Wagner, 2010). The LC_{50} decreases at 15 days to 1.33 mg/L.

Nitrates. Juveniles may be sensitive to urea-based fertilizers. In laboratory experiments, juvenile *R. cascadae* avoided paper towels saturated with urea but not soils dosed with urea. However, juvenile mortality increased significantly over a five-day period on urea-dosed soils when compared with controls. In addition, juveniles exposed to urea had a lowered feeding rate than controls (Hatch et al., 2001). Thus, juveniles may suffer adverse effects because they apparently cannot detect urea-based fertilizers under natural conditions.

UV light. Measurements of ambient UVB light at natural breeding ponds of *R. cascadae* demonstrate significant variation, depending on time of day, location, and water depth (Belden et al., 2003; Bancroft et al., 2008; Romansic et al., 2009). Larvae of *R. cascadae* do not avoid ambient levels of UVB light (Bancroft et al., 2008), nor do levels of the glucocorticoid hormone corticosterone increase with prolonged exposure to UVB. Belden et al. (2003) suggested that larvae cannot detect UVB light, hence the lack of production of stress hormones to prolonged exposure to ambient UVB light levels. In addition, larval *R. cascadae* do not have high levels of photolyase, an enzyme involved in DNA repair resulting from UVB damage (Blaustein et al., 1994b). This might suggest that larvae are susceptible to increasing levels of UVB at high elevations, which in turn could contribute to population declines.

Studies of the potential sublethal effects of UVB have had ambiguous results. Hatching success is lower under ambient UVB than when UVB is screened (Blaustein and Belden, 2003). Growth and development may or may not be affected by UVB under laboratory and field conditions (Romansic et al., 2009; Searle et al., 2010). Likewise, studies on the effects of UVB on susceptibility of larvae to predation have had mixed results (Romansic et al., 2009). Tadpole survival after 42 days of exposure to ambient levels of UVB is lower than in tadpoles shielded from UVB (Belden et al., 2003), in contrast to the findings of Romansic et al. (2009), who determined that survivorship was not affected by UVB; differences in results between studies were attributed to differences in exposure time and nutrition. Malformations of the tail (lateral flexure, curling, fraying) tend to be more prevalent in larvae exposed to UVB than in those shielded from UVB (Romansic et al., 2009).

It is clear that UVB *may* have adverse effects

Habitat of *Rana cascadae*.
Photo: Justin Garwood

on the larvae of *R. cascadae*, but differences in experimental study design hamper definitive conclusions about whether UVB actually impacts populations in nature. In any case, the spectral characteristics of water in amphibian breeding ponds, coupled with oviposition site selection, indicate that adverse effects of UVB radiation would be mediated and not likely reach levels to impact embryos (Palen et al., 2002; Palen and Schindler, 2010). UVB likely plays a minimum role in amphibian declines throughout the West (but see Blaustein et al., 2004).

Adult *R. cascadae* may experience retinal damage from cumulative exposure to solar radiation. *Lithobates pipiens* that have been experimentally light damaged show similar types of retinal damage as observed in natural populations of *Rana cascadae* (Fite et al., 1998). These authors expressed concern that increasing levels of UVB radiation at high elevations may have detrimental effects on postmetamorphic frogs.

STATUS AND CONSERVATION

Like native western ranids throughout the Pacific Northwest, Cascades Frogs have experienced severe population declines and extinctions. Populations in the Klamath-Siskiyou Mountains seem most to have been affected. Although once quite abundant, the species has disappeared from the southern end of its historic range near Lassen Volcanic National Park in California (Fellers and Drost, 1993; Fellers et al., 2007b). Fellers et al. (2007b) observed *R. cascadae* at only 6 sites of 856 surveyed over a 14 yr period, with one additional locality reported to them. Losses have been attributed to the introduction of nonnative predatory fish, prolonged drought, pathogens, toxic substances, loss of habitat, and sometimes mismanagement, even in pristine protected areas. Humans even used to shoot the frogs for "sport" (Briggs and Storm, 1970). It is probable these factors have interacted synergistically to cause the decline of this species.

The introduction of nonnative trout species throughout the Western states has had devastating effects on a number of amphibian species, including the Cascades Frog. In extensive surveys, Welsh et al. (2006) found a negative correlation between the presence of trout and occupancy by Cascades Frogs, even in remote wilderness areas. *Rana cascadae* was three times more likely to be found in fishless areas than in areas containing nonindigenous fishes. However, Cascades Frogs may be able to detect the presence of fish in ponds and possibly move to protected areas that do not contain fish or that minimize exposure to them. Adults are less prone than larvae to fish predation, because they are larger than the mouth gape of most trout.

Rana cascadae was breeding within the 1980

blast zone at Mount St. Helens 5 yrs following the devastating eruption (Karlstrom, 1986); by 1997, it occupied 30 sites and was breeding at 18 sites (Crisafulli et al., 2005). The species fared much more poorly in the debris-avalanche zone, however, and no breeding was recorded through 2000.

Rana draytonii
Baird and Girard, 1852
California Red-legged Frog

ETYMOLOGY
draytonii: a patronym honoring Joseph Drayton (1795–1856), an artist with the United States Exploring Expedition, who collected the type series.

NOMENCLATURE
Dubois (2006): *Rana (Amerana) draytonii*
  Synonyms: *Epirhexis longipes, Rana aurora draytonii, Rana leconteii, Rana longipes, Rana nigricans*

IDENTIFICATION
Adults. This is a relatively large, sexually dimorphic species with prominent dorsolateral folds. Males can reach 116 mm SUL and females 138 mm SUL. The dorsum is variable in coloration, from light yellowish brown to reddish brown to dark brown, with scattered irregular dark brown to black spots throughout; each spot has a lighter center. The tympanum is smaller than the eye, and there is a distinct dorsolateral fold from the posterior of the eye to the rear of the body. The belly and the underside of the legs and feet have a bright salmon-red coloration, but some frogs are more intense than others, and the coloration can extend throughout the body. Some adults lack the red pigment entirely or it may be absent on just the legs and feet. The groin is black with light blotches, ranging from pale yellow to white. The posterior of the thigh is uniform brown with 3 to 12 distinct yellow spots. Males have swollen thumb pads during the breeding season and usually have small paired subgular vocal sacs (Hayes and Krempels, 1986). An albino adult was reported from San Mateo County, California (Hensley (1959).

Females are larger than males. Males range between 78 and 116 mm SUL (mean 101 mm) at San Luis Obispo and 82–108 mm SUL (mean 92 mm) at Santa Barbara. Females range between 91 and 138 mm SUL (mean 120 mm) at San Luis Obispo and 87–129 mm SUL (mean 112 mm) at Santa Barbara (Hayes and Miyamoto, 1984).

Larvae. Tadpoles are large (reaching 83 mm TL; Storer, 1925) and dark brown to dark yellow dorsally. There are diffuse spots on the body that measure more than 1 mm in diameter. The snout is somewhat pointed when viewed from above, and the eyes are close together. The tail is mottled with light spots and is bluntly tapered toward its end. Bellies have a pinkish iridescence with the margins a mixture of black and pink iridescence in a fine pattern. Tadpoles of this species are sometimes confused with the larvae of *Lithobates catesbeianus* (see comparison in Storer, 1925).

Eggs. The eggs are black to dark reddish brown dorsally and creamy white ventrally. They are deposited in a soft viscid jelly mass 15 cm × 10 cm × 10 cm, with each mass containing 800 to 6,000 eggs. The mass is attached to vegetation just below the surface of the water. There are three distinct envelopes surrounding the egg. The outer envelope is 7.6–11.8 mm in diameter (mean 8.5 mm); the middle envelope is 3.9–6.4 mm (mean 4.4 mm) in diameter; the inner envelope is 3.1–5 mm (mean 3.5 mm) in diameter. The vitellus is 2–2.8 mm in diameter. Eggs are described by Storer (1925) and Livezey and Wright (1947).

DISTRIBUTION
Rana draytonii is known from the inland Sierra Nevada of California and from Pacific coastal watersheds of Mendocino County south through Baja California Norte. Populations of *R. aurora* and *R. draytonii* are found intermingled between Mills Creek in southern Mendocino County and Big River in northern Mendocino County, California; all red-legged frogs from Mills Creek south are *R. draytonii* (Shaffer et al., 2004). They historically occurred in four desert drainages on the east side of the Coastal Range (Sheep Creek, Whitewater River, San Felipe Creek, and the Mojave River system), but there have been no

Distribution of *Rana draytonii*

sightings since 1968 (Jennings and Hayes, 1994a). There are reports of this species on Santa Cruz and Santa Rosa in the Channel Islands, but these populations have been extirpated (Schoenherr et al., 1999). Introduced populations of *R. draytonii* have been reported at Duckwater Springs in the Railroad Valley (Green, 1985b) and at Millet (Linsdale, 1938, 1940), both in Nye County, Nevada. The elevation range is from near sea level to 1,500 m (Lemm, 2006). Important distributional references include Storer (1925), Slevin (1928), Jennings and Hayes (1994a, 1994b), Davidson et al. (2001), and Shaffer et al. (2004).

FOSSIL RECORD

Fossils have been reported (as *R. aurora*) from Pleistocene Rancholabrean sites in Orange and Los Angeles counties, California (Sanchiz, 1998; Holman, 2003).

SYSTEMATICS AND GEOGRAPHIC VARIATION

This species and *R. aurora* have long been considered related, based on a similar morphology and the red coloration of the legs (Camp, 1917; Slater, 1939a). Although they were described originally as distinct species (Baird and Girard, 1852), most researchers until recently have considered them subspecies within the wide-ranging, polytypic species *R. aurora*. Data from studies on allozymes, mtDNA, skin peptides, karyotypes, vocal sac structure, advertisement calls, and oviposition behavior (Hayes and Miyamoto, 1984; Green, 1985a, 1986a, 1986b; Hayes and Kremples, 1986; Macey et al., 2001; Shaffer et al., 2004; Conlon et al., 2006) have supported specific distinctiveness between them, although various authors have interpreted their results within the area of sympatry in Mendocino County differently (reviewed by Shaffer et al., 2004).

Rana draytonii and *R. aurora* are not sister taxa. Instead, the *R. draytonii* lineage is one of five well-supported clades within the monophyletic *R. boylii* species group within the *Amerana* clade of western North American ranid frogs (Shaffer et al., 2004; Hillis and Wilcox, 2005). *Rana draytonii* is most closely related to the *R. aurora-R. cascadae* clade and somewhat more distantly to the *R. muscosa-R. sierrae* clade within *Amerana*. Divergence occurred first between the southern red-legged frogs (*draytonii*) and the northern red-legged frogs (*aurora-cascadae*). The contact area between *R. draytonii* and *R. aurora* in Mendocino County is an important contact zone between northern and southern evolutionary lineages for many amphibian species groups.

Despite suggestions of broad to narrow zones of hybridization between *R. aurora* and *R. draytonii*, there appears to be no support for the presence of hybrids between these species in northern California. In laboratory crosses, only 1 of 177 *R. cascadae* eggs fertilized by *R. draytonii* sperm reached metamorphosis (Zweifel, 1955). The diploid number of chromosomes is 26 with 5 pairs of large chromosomes and 8 pairs of small chromosomes (Green, 1986a).

ADULT HABITAT

California Red-legged Frogs inhabit permanent ponds, marshes, slow-moving streams, and reservoirs in the mountainous foothills of the coast and Sierra Nevada Mountains. In marshes, the species is found around the margins in relatively shallow water among emergent vegetation; it generally avoids deep open-water areas lacking emergent vegetation (Cook, 1997b). Streams usually have small deep intermittent pools along their length with emergent plants (*Typha*, *Scirpus*) and shoreline vegetation of willows (*Salix*) (Hayes and Jennings, 1988); frogs tend to occur in low-gradient streams covering a small drainage area. This species appears associated particularly with plunge-pool habitats with willows overhanging the stream channels.

Rana draytonii tolerates native fishes, but tends not to be found in wetlands and streams with non-native predaceous fishes and American Bullfrogs. They forage terrestrially in adjacent ravine, forest, grassland, and range habitats, and will cross varied terrestrial habitats during movements between breeding ponds and terrestrial refugia. Examples of adjacent terrestrial habitats include redwood (*Sequoia*) forests, riparian woodlands (dominated by *Populus, Platanus, Salix,* and *Woodwardia*), and foothill woodlands (dominated by a variety of shrubs, including *Arctostaphylos, Heteromeles,* and *Adenostoma*, with meadows of various grasses and herbs) (Morafka and Banta, 1976).

AQUATIC AND TERRESTRIAL ECOLOGY

California Red-legged Frogs are mostly aquatic and usually remain within 150 m of their aquatic residence sites throughout their lives, even when making occasional terrestrial forays. Many frogs remain at the breeding pond throughout the year, but others leave breeding sites for terrestrial sites or nonbreeding seasonal wetlands. In the vicinity of Point Reyes, California, for example, 66% of females and 25% of males moved to terrestrial areas following the breeding season, and females were more likely than males to disperse even from permanent ponds. There was no difference in the distances moved between males and females, however.

There are several distinct periods of terrestrial activity, including summer, prebreeding season, breeding season, and postbreeding season. During the summer (May to October), frogs normally are found < 30 m (and usually with 6 m) from the shoreline of their resident wetland, except when precipitation is falling. They will move short distances into terrestrial habitats for four to six days following summer rains, but days with precipitation are infrequent events. Long-distance movement may occur if seasonal wetlands dry.

Most terrestrial activity occurs in winter after the onset of winter rains (that is, the first rainfall > 0.5 cm). This normally occurs from September to November in inland populations, and from November to December in coastal populations (Bulger et al., 2003; Tatarian, 2008). The exact timing will depend on weather patterns, which vary annually. Frogs will move onto land from the beginning of the wet season until breeding commences, a median period of about 20–30 (but as many as 50) continuous days over a 1–2 month period. Local movements of 15–25 m away from the wetland occur at this time, with longer forays as the season and conditions permit. They also move farther aquatically at this time (to 107 m) than at other times of the year (Tatarian, 2008). Occasional frogs move up to 100 m or more from water, but a majority of the population is found at distances < 60 m from water (Bulger et al., 2003). The mean prebreeding season movement distance is about 42 m (Tatarian, 2008).

During the mid- to late-winter breeding season (comprising 32% of movements) and during the postbreeding season (comprising only 11% of movements), most frogs only make occasional forays onto land (median 1–4 days), except when migrating, regardless of the often abundant rainfall and favorable ambient temperatures that should not limit terrestrial activity. The mean movement distance during the breeding season is 13.5 m, whereas it is 16 m during the postbreeding season for residents. As the dry season takes hold, the frogs again become closely tied to the resident wetland. Preferred residence wetlands are those with a significant amount of nearby surface cover and under-shelter objects that can be used during terrestrial forays should adverse weather conditions be encountered.

Occasional California Red-legged Frogs will make use of adjacent ravines that have some water flow during both summer and winter. Terrestrial activity in ravines occurs for a median of 14–15 days in summer and 21–25 days in winter. The least activity in ravines occurs after the breeding season through May, when the median of terrestrial activity drops to between three to seven days. Maximum ravine terrestrial activity occurs in winter, when frogs have been reported to remain in these habitats 38–42 days. Most terrestrial activity in ravines occurs < 55 m from the resident wetland, but a maximum distance of 150 m has been recorded (Bulger et al., 2003). Ravines also are used by migrating frogs.

Rana draytonii in some areas are capable of making long-distance terrestrial movements between winter breeding sites and summer resident ponds (Fellers and Kleeman, 2007). These movements in coastal populations begin from late October to late November, with returns from late January to early May. Frogs will travel for several days, and then remain in one location for several days until making the next series of movements. Short-distance movements < 300 m are made over a period of 1–3 days, with longer distances (movements recorded up to 2.8–3.6 km between ponds 1.4–2.8 km apart) requiring as much as 2 months. Movements occur in a straight line, without regard to vegetation or topography, and frogs do not follow riparian corridors.

When *R. draytonii* travel over land, they may be found at distances up to 500 m from the nearest water and move 1.2–1.4 km across terrain without contacting a pond or stream. Bulger et al. (2003) estimated that from 11 to 22% of the adult population migrated terrestrially between breeding sites and summer wetlands, although most frogs remained near one resident wetland. In other populations, there may be no terrestrial overland migration. Instead, frogs move between breeding sites using only aquatic habitats (Tatarian, 2008).

When terrestrial and away from wetlands but not actively moving, California Red-legged Frogs often remain concealed under vegetation, surface debris, or under surface litter. In contrast, Fellers and Kleeman (2007) often found them sitting exposed at night, despite the availability of nearby cover. An abundance of cover objects is particularly important in xeric habitats, where they have been found using ground squirrel burrows (Fellers and Kleeman, 2007; Tatarian, 2008) that are not occupied in more mesic habitats. They occupy terrestrial habitats in adjacent forests, grass, and shrub rangeland and even agricultural land as long as concealing vegetation or other cover objects are present. Tatarian (2008) found that they preferred cooler north-facing slopes compared to south-facing or west-facing slopes.

While at resident wetlands, California Red-legged Frogs often are observed sitting along the shoreline in the direct sunlight, seemingly to bask (Fellers and Kleeman, 2007). Adults and even newly metamorphosed frogs exhibit this behavior either on the shore or in adjacent shallow water. In some instances, frogs on land well away from water also appear to bask.

Metamorphs are 22–42 mm SUL (mean 30.7 mm; Allaback et al., 2010); dispersal occurs during precipitation in November and December. Even if heavy rains occur from January through March, very few metamorphs are observed. The first significant rainfall in the late summer or fall is the trigger for metamorph dispersal (Allaback et al., 2010).

CALLING ACTIVITY AND MATE SELECTION

Males enter breeding ponds prior to females, and call for a period of two to four weeks. Unlike

Egg mass of *Rana draytonii*.
Photo: Chris Dellith

R. aurora, *R. draytonii* advertisement calls are made from the surface of the water (Hayes and Miyamoto, 1984). The call is guttural and composed of three to seven notes. As with *R. aurora*, the call does not have much carrying capacity, so the calls cannot be heard at a distance > 30 m from a breeding pond. Nearly all calling occurs at night, although calls are rarely heard during daylight hours. Females move toward calling males. Amplexus ensues at the approach of a female, as a male will attempt amplexus with virtually any moving object of appropriate size in his vicinity. The amplexed frogs usually stay near the male's calling position, where oviposition occurs.

BREEDING SITES

Breeding occurs mostly in resident wetlands, such as ponds, marshes, and reservoirs with open canopies and shallow-water (ca. 40 cm) areas for egg deposition. Breeding sites typically have dense surrounding emergent and shoreline vegetation, comprised particularly of cattail (*Typha latifolia*) and bulrush (*Scirpus californicus*). Dense growths of spikerush (*Eleocharis*), smartweed (*Polygonum*), aquatic buttercup (*Ranunculus*), water plantain (*Alisma*), water weed (*Elodea*), and pond weed (*Potamogeton*) may occur throughout the wetlands and grow to the surface in summer months; in winter, the dead vegetation offers refugia and sites for egg deposition. Habitats with deep and open water are not preferred breeding sites.

REPRODUCTION

California Red-legged Frogs usually return to the same wetland to breed from one year to the next. Males enter the breeding ponds during the winter rainy season from late November to early December. Females arrive later, from early January to mid-February. Actual oviposition occurs from mid-January to mid-March (Bulger et al., 2003), but it can be delayed until April if freezing or drought conditions occur during the normal breeding season. During the breeding season, the frogs remain in close proximity to the breeding site. Females usually arrive at a breeding site, mate, and depart within a median of 12 days. Males, however, do not depart breeding sites until all oviposition has ceased.

Eggs are deposited in loose gelatinous clusters on the surface of the water, attached to emergent vegetation (Hayes and Miyamoto, 1984). Newly oviposited egg clusters may have a bluish hue, but are later covered by silt and algae. The clutch size ranges between 800 and 6,000 eggs per female (Cook, 1997a; Padgett-Flohr, 2008). Water temperatures must be 16°C or less; eggs do not tolerate high temperatures or salinity (> 9 ppt). Hatching occurs in 1–4 weeks (6–14 days according to Jennings and Hayes, 1994b); the hatchlings are about 8.8–10.3 mm in TL (Storer, 1925).

Tadpole of *Rana draytonii*. Photo: Chris Brown

LARVAL ECOLOGY

Development occurs in cool waters in salinities < 4.5 ppt (Jennings and Hayes, 1994b). Tadpoles are bottom-dwellers and spend much of their time concealed in dense vegetation. Their diet consists of algae, which are grazed from plant surfaces and the substrate. Metamorphosis occurs in three to five months (Storer, 1925), and overwintering does not occur as larvae. Newly metamorphosed *Rana draytonii* are ca. 27 mm SUL. As with most anurans, survivorship from egg to metamorphosis is quite low, with only 1–5% of eggs and larvae surviving through metamorphosis (Calef, 1973a; Licht, 1974).

DIET

The diet is varied, with larger frogs taking larger prey. Indeed, this species is capable of ingesting very large prey, such as adult toads (Hays, 1955). Frogs orient toward movement, but do not discriminate among prey types. A wide variety of invertebrates is consumed, especially beetles, water striders, larval alderflies, spiders, isopods, snails, and sowbugs. Other prey includes Three-spined Stickleback (*Gasterosteus aculeatus*), Pacific Treefrogs (*Pseudacris regilla*), larval *Anaxyrus boreas*, field mice (*Peromyscus californicus*), California voles (*Microtus californicus*), Western Harvest Mouse (*Reithrodontomys megalotis*), and garter snakes (*Thamnophis sirtalis*) (Hayes and Tennent, 1985; Hayes et al., 2006b; Davidson, 2010; Stitt

Adult *Rana draytonii*.
Photo: Dennis Suhre

and Seltenrich, 2010). Feeding occurs primarily at night in adults, but juveniles feed both diurnally and nocturnally. Prey are recognized by movement, and predation is opportunistic.

PREDATION AND DEFENSE
Larvae likely are eaten by a variety of predaceous invertebrates, including diving beetles, odonate larvae, and crayfish. Predators of postmetamorphs include snakes (garter snakes, *Thamnophis sirtalis*, *T. hammondi*), birds (Great Blue Herons, *Ardea herodia*; Bitterns, *Botaurus lentiginosus*; Black-crowned Night Herons, *Nycticorax nycticorax*), and mammals (raccoons, *Procyon lotor*; otters, *Lutra canadensis*; rats, *Rattus* sp.) (Cunningham, 1959; Wharton, 1989; Jennings and Hayes, 1994b; Fellers and Kleeman, 2007). Frogs also may be stepped on and killed by livestock.

California Red-legged Frogs are wary as adults, although juveniles are more conspicuous and much less wary. They appear to detect the movement vibrations of an approaching predator at night and flee in advance of its striking distance. As with Northern Red-legged Frogs, the reddish coloration may aid in concealment during terrestrial activity. When handled, frogs may assume a "hands-up" posture whereby the front limbs are drawn up to the head, and the hands are positioned as to shield the eyes (Wilkinson, 2006).

Rana aurora and *R. draytonii* also have closely related but distinct antimicrobial skin peptides (Conlon et al., 2006). These peptides aid in the frog's defense against bacterial pathogens.

POPULATION BIOLOGY
Adults become sexually mature in 2 (males) or 3 (females) yrs following metamorphosis (Jennings and Hayes, 1985), but both sexes may not actually reproduce for the first time until 3–4 yrs of age (Jennings and Hayes, 1994b).

COMMUNITY ECOLOGY
California Red-legged Frogs and American Bullfrogs frequently occupy the same marsh habitats, where they come into contact. There is some evidence of habitat partitioning, with *R. draytonii* occupying the shoreline and *Lithobates catesbeianus* occupying the deeper water (Twedt, 1993; Cook, 1997b). Preferred breeding sites also differ among these species. Further, they have a different breeding season, which decreases the potential for interspecific interactions of both adults and larvae. Still, there is much overlap in habitat use, and *Rana draytonii* not infrequently winds up in the diet of *Lithobates catesbeianus*. There also have been numerous cases of interspecific amplexus (Twedt, 1993). That American Bullfrogs have a longer larval period than *Rana draytonii* may prevent this species from colonizing temporary wetlands in great numbers.

DISEASES, PARASITES, AND MALFORMATIONS
This species is susceptible to amphibian chytrid fungus (*Batrachochytrium dendrobatidis*), and exposed individuals have been found under natural conditions (e.g., Adams et al., 2010; Tatarian and Tatarian, 2010). Indeed, this pathogen has

been found in a small number of museum specimens collected in California from the 1980s to the 2000s (Padgett-Flohr and Hopkins, 2009). In experimental inoculations, frogs survived the infection and showed no clinical signs of the disease (Padgett-Flohr, 2008). As with American Bullfrogs, California Red-legged Frogs may not be as susceptible to amphibian chytrid infection as are many other species.

California Red-legged Frogs are parasitized by the trematodes *Cephalogonimus americanus*, *Gorgoderina* sp., *G. aurora*, *Haematoloechus* sp., *H. complexus*, *H. kernensis*, and *Margeana californiensis* (*Glypthelmins quieta*) (Cort, 1919; Ingles, 1936; Goldberg and Bursey, 2002a). Cestodes include *Ophiotaenia magna*, and nematodes include *Cosmocercoides variabilis*, *Falcaustra pretiosa*, *Oswaldocruzia pipiens*, and *Rhabdias joaquinensis* (Ingles, 1932a, 1932b, 1936; Walton, 1941; Goldberg and Bursey, 2002a).

SUSCEPTIBILITY TO POTENTIAL STRESSORS
No information is available.

COMMERCIAL USE
California Red-legged Frogs have long been considered an important commercial species (Chamberlain, 1897; Storer, 1933; Jennings and Hayes, 1985). Most frogs were harvested from wild populations between 1888 and 1935, although frogs had been harvested commercially as early as 1849. From 1888 to 1895, a total of 460,704 *Rana draytonii* were taken from wild populations, a harvest which focused on females. In 1895 and 1899, 22,405 kg and 9,383 kg, respectively, were harvested. The numbers of frogs collected after 1895 decreased sharply, suggesting population depletion. The timing of this depletion coincides with the introduction of *Lithobates catesbeianus* in California, ostensibly to meet the continuing demand for frog's legs.

A frog ranching operation was carried out between 1898 and 1907 at El Cerrito that stocked both *Rana draytonii* and *Lithobates catesbeianus*; precise information on the numbers produced is unavailable. Storer (1933) mentions that several other frog farms were established in California between 1900 and 1930, but he does not mention which species were involved. Escaped *Lithobates catesbeianus* likely populated areas previously occupied by a large *Rana aurora* population as the frog farms failed. This makes it difficult to determine the direct effect of *Lithobates catesbeianus* on the past status and current recovery efforts for *Rana aurora*.

STATUS AND CONSERVATION
The California Red-legged Frog is protected under provisions of the federal Endangered Species Act of 1973 (as "Threatened") (USFWS, 1996) and by state law. Former populations in the San Joaquin Valley are now extirpated. According to Jennings and Hayes (1994b) and Davidson et al. (2001), the species has disappeared from >70% of its historic range, including the Sierras (Drost and Fellers, 1996), the Central Valley of California (Fisher and Shaffer, 1996), the desert drainages of southern California (Jennings, 1988; Jennings and Hayes, 1994a) and south of Los Angeles (only one population in Riverside County). The species still is present in Mexico (Grismer, 2002; Fidenci, 2004). Throughout much of its range, only a few small populations remain. Even in 1930, Myers (1930a) thought the species was becoming rare.

The possible reasons for the declines are varied, including habitat destruction and alteration (mining, grazing, urban and agricultural development, dams), climate changes, influences from UVB light, the introduction of exotic predators (predaceous fishes [*Gambusia affinis*, *Lepomis*], American Bullfrog, crayfish [*Procambarus clarkii*]), toxic chemicals and pesticides, pathogens and parasites, and commercial over-harvest (Banta and Morafka, 1966; Jennings and Hayes, 1985; Hayes and Jennings, 1986, 1988; Cook, 1997a; Lawler et al., 1999; Davidson et al., 2001, 2002). The housing boom in California undoubtedly wiped out many populations, although they were still present in the San Francisco Bay area in the early 1960s (Banta and Morafka, 1966). Many factors likely acted synergistically, both locally and regionally. Using an extensive dataset to model population declines, Davidson et al. (2001, 2002) tested a number of hypotheses about the causes for declines. The extensive use of wind-borne agrochemicals coupled with urbanization best fit the historical data, whereas there was weak support for UVB effects and none for climate change.

One of the factors Davidson et al. (2001, 2002) could not test for was the impact of American Bullfrogs on California Red-legged Frogs. Presumably, the presence of adult and larval bullfrogs in a breeding pond could affect *Rana draytonii* in a manner similar to *R. aurora* (see *R. aurora* species account). *Rana aurora* survivorship decreases in the presence of bullfrog larvae at least under experimental conditions (Lawler et al., 1999). As with *R. aurora*, researchers have had mixed results in correlating the presence of bullfrogs with the disappearance of *R. draytonii*. In many areas they co-occur, especially in marshes with high habitat complexity (Twedt, 1993; Cook, 1997b), but in other areas they do not. Doubledee et al. (2003) modeled the presence of these species and noted that serious flash floods occurring at least once every 5 yrs allowed their co-occurrence in ravines. Bullfrogs do not tolerate flooding well in narrow ravines, and presumably the floods prevented bullfrog populations from increasing in abundance to the point that they impacted the native *R. draytonii*. Eliminating livestock ponds or draining them every 2 yrs, in conjunction with shooting bullfrogs, could be successful at reducing the impacts of *Lithobates catesbeianus* on *Rana draytonii* (Doubledee et al., 2003).

The presence of another nonnative predator, *Gambusia affinis*, also may impact *Rana aurora*. Mosquitofish eat small *R. aurora* tadpoles, which in turn decrease their activity levels in the presence of fish. *Gambusia* may considerably injure larger tadpoles without killing them. Lawler et al. (1999) found that metamorphosing larvae weighed 34% less in body mass when in the presence of this species than in experimental ponds without *Gambusia*. This difference may result from decreased activity in the presence of the fish or from injuries received, especially to the tail, which impair tadpole activity. In either case, tadpole feeding behavior likely would be adversely affected.

The behavior and movements of California Red-legged Frogs suggest that some populations can be conserved by maintaining a variety of concealing habitats within 100 m or so of breeding ponds (Bulger et al., 2003), but this may not be a universal solution (Fellers and Kleeman, 2007). Dispersal corridors between breeding sites need to be identified on an individual basis, as movements are generally in straight-line paths and do not of necessity include specific topographical features or riparian areas. In other areas, protection of a contiguous series of wetlands would be appropriate, depending upon wetland hydrology. Buffers should involve resident breeding wetlands, dispersal corridors, and terrestrial habitat. Because of differences in site characteristics and behavior among populations, conservation measures must be undertaken on a site-by-site basis. Human disturbance during the winter terrestrial activity period should be closely monitored in areas near breeding ponds. Since *Rana draytonii* is adept at crossing barriers (Rathbun et al, 1997), special care must be taken when trying to exclude them from roadways, unsuitable ponds, or other areas where they might be at risk.

Habitat of *Rana draytonii*.
Photo: Sean Barry

Rana luteiventris Thompson, 1913
Columbia Spotted Frog

ETYMOLOGY
luteiventris: from the Latin *luteus* meaning 'golden yellow' and *ventris* meaning 'of the belly.' The name refers to the yellow ventral coloration.

NOMENCLATURE
Dubois (2006): *Rana (Amerana) luteiventris*
 Synonyms: *Rana pretiosa luteiventris, Rana temporaria pretiosa*

This species was originally described as a subspecies of *Rana pretiosa* (Thompson, 1913; supported by Dumas, 1966) and much information on *R. luteiventris* is contained in accounts of that species. Readers should consult the location of studies to verify whether the information contained refers to *R. pretiosa* or *R. luteiventris*.

IDENTIFICATION
Adults. Adults are light yellowish brown to dark brown to gray, and the dorsum has small spots with rough skin or tubercles. Occasional individuals even can be green or olive. The dorsum is characterized by irregularly distributed spots with generally light centers. There may be a dark mask on the side of the head bordered by a light stripe on the upper jaw. The eyes appear to be oriented upward, and the tympanum is small. The venter is yellow, orange, yellow orange, orange red, or salmon, and varies geographically (Turner, 1959a); the coloration may extend to the thighs, stomach, and throat. Young frogs have pale, almost white, venters, and there may not be any red coloration on the undersides of the thighs or abdomen. The throat and venter may be mottled. Webbing between the toes is extensive on the hind feet. Males have swollen and darkened thumbs during the breeding season.

Females are larger than males, with sizes ranging to 91 mm SUL. In Tule Valley, Utah, females were 43–66 mm SUL and males were 40–59 mm SUL (Hovingh, 1993b), whereas in northeastern Oregon, females were 59–91 mm SUL and males were 58–74 mm SUL (Bull and Hayes, 2001). At Grady Salt Marsh in Utah, females averaged 60.5 mm SUL (range 51 to 70 mm) and males 51.6 mm SUL (range 45 to 59 mm) (Cuellar, 1994). In the Palouse Region of Washington and Idaho, the median male size was 56.5 mm SUL (range 46 to 71 mm), and the median female size was 67 mm SUL (range 43 to 86 mm) (Davis and Verrell, 2005). Along the Wasatch Front, males averaged 49.9 mm SUL (range 47 to 64 mm) and females 70 mm SUL (range 62 to 79 mm), but sample sizes were small (Morris and Tanner, 1969). In Nevada, males averaged 54 mm SUL (range 33 to 71 mm), and females averaged 57 mm SUL (range 30 to 90 mm) (Reaser, 2000). Both mass and SUL are positively correlated with age.

Larvae. Tadpoles are brownish green with gold flecking dorsally and bronze ventrally toward the edges, and they may be mottled with light and dark brown splotches. As they develop, they become browner to black in color (Jones et al., 2005). Bellies are cream to pale yellow, becoming lighter with age. The pale tail is long (comprising two-thirds of the TL) and contains gold or black flecks or blotches. Tail fins are highest about halfway down the tail. TLs may reach 100 mm at low-elevation sites and 80 mm at high-elevation sites. A figure of the tadpole mouthparts is in Thompson (1913).

Eggs. Eggs are dark brown to black above and pale yellow to light tan ventrally. Eggs are attached to one another by a gelatinous cord. They have two gelatinous envelopes (although Livezey and Wright, 1947, report only one), with the outer envelope normally ranging between 9 and 13 mm and the inner envelope 3.8–6.1 mm (mean 5) (Morris and Tanner, 1969). However, these latter authors reported one egg that reached 21 mm in diameter! The vitellus is 1.8–3 mm in diameter. Eggs are deposited in a firm globular jelly mass that may initially have a bluish cast. Older egg masses may be green from algal growth.

DISTRIBUTION
The Columbia Spotted Frog is a mountain, foothill, and Great Basin species that ranges from northern British Columbia, the Yukon, and the panhandle of southern Alaska (Stikine River) south through southwestern Alberta, western Montana, and northwestern Wyoming. The range includes most of Idaho, eastern Oregon, and eastern Washington. Isolated populations occur in Wyoming (Big Horn Mountains), Utah (Toole County), and Nevada (Elko, Eureka, Nye, and White Pine counties). Many populations have been extirpated, however, such as those in north-

Distribution of *Rana luteiventris*

western Nevada. The species ranges from near sea level in the North to 2,150 m in Alberta, 2,010 m in Glacier National Park, 2,530 m in Yellowstone, and 2,650 m in Nevada.

Important distributional references include: Range-wide (Green et al., 1997; Jones et al., 2005), Alaska (Swarth, 1936; Hock, 1957), Alberta (Salt, 1979; Russell and Bauer, 2000), British Columbia (Logier, 1932; Carl, 1943; Carl and Hardy, 1943; Mennell, 1997; Matsuda et al., 2006), Idaho (Slater, 1941b), Montana (Black, 1970; Franz, 1971; Marnell, 1997; Maxwell et al., 2003; Werner et al., 2004), Nevada (Linsdale, 1940; NVWD, 2003), Utah (Van Denburgh and Slevin, 1915; Tanner, 1931; Hovingh, 1993a, 1993b), Washington (Slater, 1955; Metter, 1960; Leonard et al., 1993; McAllister, 1995), Wyoming (Baxter and Stone, 1985), Yellowstone National Park (Turner, 1955; Koch and Peterson, 1995), and the Yukon (Mennell, 1997).

FOSSIL RECORD
No fossils are recognized.

SYSTEMATICS AND GEOGRAPHIC VARIATION
In contrast to suggestions by previous investigators (e.g., Dumas, 1966), molecular data confirm the distinctiveness of this species (Green et al., 1997). *Rana luteiventris* is a member of the *R. muscosa* complex of the *Amerana* clade of North American ranid frogs. It is most closely related to *R. pretiosa* and somewhat more distantly to *R. boylii*. Within *R. luteiventris*, there are three main evolutionary lineages: Northern, Great Basin, and Utah (Funk et al., 2008). The divergence is enough to suggest significant evolutionary lineages that may be supportive of separation into three species. Even within these lineages, additional variation is evident, such as the highly divergent Deep Creek clade as a subdivision of the Utah clade. Low genetic variation is found in Columbia Spotted Frogs from southeastern Oregon, making them prone to extinction. Experimental crosses between *R. luteiventris* and *R. pretiosa* produce viable larvae through the latter stages of development (Dumas, 1966).

Turner (1959a) described variation in pattern, coloration, and spots of *R. luteiventris* from Yellowstone National Park and compared them with similar variables from frogs in other parts of the species' range. However, some comparisons were with frogs that are now referred to *R. pretiosa* from Oregon and possibly *R. cascadae*. In any case, patterns were quite variable among all characters examined, even within the Yellowstone population. Likewise, Metter (1960) noted much color variation and mottling among populations in eastern Washington, even when populations were in close proximity. There also appears to be regional variation in size, with frogs from Tule Valley, Utah, for example, about 5 mm SUL smaller than those from other locations (Hovingh, 1993b).

ADULT HABITAT
Columbia Spotted Frogs prefer slow-moving or ponded waters that are clear and do not have any canopy cover. Examples include springs, lakes, oxbows, beaver ponds, stock ponds, seeps in wet meadows, and stream or river backwaters in open habitats (Cuellar, 1994; NVDW, 2003). Occupancy is associated with fishless wetlands with high solar radiation. Such areas may be surrounded by either high- or low-density forests, especially Ponderosa Pine (*Pinus ponderosa*), and frogs are not likely to be in close association with anthropogenic development. Sunny wetlands at lower elevations in mostly flat areas offer the best habitats.

Ponds formed from permanent water sources such as springs provide ideal habitat, as they support aquatic and emergent vegetation and contain deep mud bottoms for overwintering. These sites contain a variety of concealing plant species, including sedges (*Carex*), stonewort (*Chara*), mare's tail (*Hippurus*), and cattail (*Typha*). *Spirogyra* provides good hiding places for frogs and developing tadpoles. Columbia Spotted Frogs do not occupy all available habitats within a region. For example, Goldberg and Waits (2009) recorded *Rana luteiventris* in 31% of 105 wetlands surveyed on private lands in Idaho, and predictive models suggested that breeding should occur in only 8–15% of wetlands in their survey area.

Temperatures at Columbia Spotted Frog localities may fluctuate widely (Turner, 1960b; Licht, 1969c; Turner and Dumas, 1972; Hovingh, 1993b), and conditions at some sites can be harsh. At lowland sites in Utah, for example, adults are found in waters at temperatures up to 29°C with conductivities of up to 4,700 μmhos/cm and pH >9.0 in summer. Frogs tend to avoid the warmest waters as the season progresses.

AQUATIC AND TERRESTRIAL ECOLOGY

Adult Columbia Spotted Frogs rarely venture very far from water except when dispersing. Adults are seen most often sitting in shallow water along wetland margins or on the shoreline. Foraging occurs near the shore, either from a terrestrial or aquatic vantage point, although adults forage as far as 10 m from water; juveniles occasionally will venture even farther. There appears to be some spatial segregation between age classes, with first yr juveniles being restricted to flooded marshes and shallow water courses; older juveniles will use the same ponds as adults, but they also are found in adjacent uplands where they presumably forage (Turner, 1960b). After the breeding season, second and third yr juveniles become more numerous around breeding ponds, indicating a later arrival than breeding adults.

Overwintering occurs along streams, headwater springs, and in deep (> 3 m) alpine ponds (Pilliod et al., 2002), depending upon location and severity of winter. Emergence is asynchronous. Frogs become active for several weeks before the start of the breeding season, and they even have been observed under ice. For example, *R. luteiventris* emerges as early as late February to early March in Utah, although eggs are not deposited until late March (Morris and Tanner, 1969; Hovingh, 1993b). Frogs also are more active at warm springs than in cold-water springs. Frogs may be found hiding under objects in water 5°C in Montana, although they are still relatively inactive (Middendorf, 1957), and Turner (1960b) found frogs in water in Yellowstone at temperatures of 1°C. Actual emergence usually coincides with several consecutive days with air temperatures of 13–16°C or after early warm rainstorms. Water temperatures must be >10–11°C (Morris and Tanner, 1969; Salt, 1979). Males emerge several days prior to females, and emergence can be disrupted or delayed by periodic cold weather.

With the onset of breeding season, adult frogs move to the breeding sites. Migration often occurs under very cold conditions, when snow still covers the ground and ice is present on ponds. Nonbreeding animals also make long-distance movements from upland foraging or aquatic overwintering sites, but these occur later than breeding migrations, for which the destination includes wet meadows, temporary pools and streams, and seepage areas within the forest. Most activity occurs during the day between 13:00 and 19:00, as water temperatures reach 10–26°C (Turner, 1960b). After the breeding season, adults may disperse to summer habitats, with most such movements occurring by females (Pilliod et al., 2002).

Movement distances can be considerable, normally covering 183–427 m (at 66–188 m/day over periods of 1–4 days) in Yellowstone National Park, for example. Topography and the location of other pond habitats in proximity to breeding sites undoubtedly influence the propensity for movement and the distances traveled. Reports of long-distance movements are not uncommon: 1,290 m in 92 days and 700 m in 23 days in Yellowstone (Turner, 1960b); 723 m in 24 days in Jackson Hole (Carpenter, 1954a); 560 m in northeastern Oregon (Bull and Hayes, 2001); 424 m for males and 1,033 m for females in Idaho (Pilliod et al., 2002); a straight-line distance of 5 km in the Great Basin of Nevada (Reaser, 1996). In the Idaho study, 2 frogs moved 1,500 and 2,000 m in a straight line across mountain basins. Females tend to move farther distances than males. With the onset of hot summer dry weather (ca. late August), the frogs then return to permanent waters such as creeks and springs, where they remain until the onset of winter. They will remain active until at least late September or October in Utah (Morris and Tanner, 1969; Hovingh, 1993b).

Columbia Spotted Frogs frequently move

among adjacent wetlands (Turner, 1960b). There is a great deal of individual variation in the tendency to move among wetlands, but local environmental conditions and the availability of adjacent wetlands are important factors in determining whether to move. If other water sources are within 100 m, frogs are likely to move among them. However, frogs at isolated ponds tend to remain at the pond. A few frogs will remain in the vicinity of their breeding ponds regardless of the nearby presence of other wetlands.

Juvenile terrestrial dispersal is an important phase in the life cycle of *R. luteiventris*. For example, Funk et al. (2005b) observed a 62% dispersal rate from the vicinity of the native pond, with many frogs traveling > 5 km (maximum 5.75 km) over substantial (to 770 m) gains in elevation. Some frogs even managed to climb 36° slopes for 2 km. Such dispersal distances suggest substantial amounts of gene flow within mountain basins. In contrast, only 4% of marked adults moved > 200 m, although 4 frogs moved 2 km. In another example, 1–32% of marked juveniles moved between breeding ponds and summer habitats in Idaho (Pilliod et al., 2002). When juveniles stop dispersing, they tend to remain in the vicinity of the newly adopted ponds rather than return to their natal pond.

There is a great deal of variation in landscape ecology from one season to the next and among individual frogs. Some frogs remain within the same areas virtually all their lives and repeatedly use the same breeding and overwintering sites. Others remain in an area for 1–2 yrs before moving to a different area, and still other frogs roam at considerable distances across the landscape. Turner (1960b) recaptured frogs as far as 610 m from the point of original capture. Indeed, the activity range of *R. luteiventris* in Yellowstone varied between 230 and 3,312 m^2 and was not related to the sex or age of the frog.

Breeding ponds do tend to be isolated in some areas, with the extent of genetic isolation a function of distance between breeding ponds, regardless of whether distance is measured linearly or along drainages (Goldberg and Waits, 2010). Goldberg and Waits (2010) suggested that urban and rural developed lands were most responsible for genetic isolation and that agricultural lands and grasslands provided the least resistance to isolation in northern Idaho.

CALLING ACTIVITY AND MATE SELECTION

Males arrive at the breeding sites 1–36 days (median 6.5 days; Davis and Verrell, 2005) prior to females, and they establish the oviposition site. Calling occurs by day or at night depending upon population. The call of this species sounds like repeatedly clicking one's tongue against the roof of the mouth (Morris and Tanner, 1969). There are from 4 to 50 notes per call, with the rate of note production directly correlated with temperature; the warmer the temperature, the more notes are produced. Males can produce 300–480 notes per minute, and calls last from 4 to 10 sec. Calls are produced both in air as the frog sits in the shallow water along a pond's edge or while the male is submerged. Calls are not very loud (in contrast to descriptions of deep resonant calls by Svihla, 1935), and a human observer can only hear a call < 25 m from the caller. One male calling often stimulates others to begin, especially after a disturbance.

Amplexus is axillary and occurs in water as soon as a female enters a pond; males are aggressive in pursuing females. Amplexus may last over an extended number of days (Svihla, 1935; Turner, 1958b). Multiple males may attempt amplexus simultaneously and to dislodge an already amplexed male, and Black (1970) described hearing a "chorus of release calls" as large numbers of males amplexed each other at a breeding site. Amplexus even may be attempted with inanimate objects. Males remain calling at a pond throughout the breeding season, but females depart immediately after oviposition.

BREEDING SITES

Breeding sites consist of lentic pools in sunny open meadows and thinly forested areas or in headwater springs. Shallow streams, beaver ponds, or even puddles also may be used. Breeding sites often are surrounded by bulrushes (*Scirpus*) and contain duckweed (*Lemna*), pondweed (*Potamogeton*), and buttercup (*Ranunculus*) that offer the frogs protection and cover. Indeed, the best predictors of breeding sites are shallow waters, abundant emergent vegetation, and locations receiving maximum solar radiation. Pools are only slightly shaded. Breeding sites can include human-modified sites such as dredge tailing ponds, wildlife watering ponds, cattle troughs, and gravel pit ponds (Bull and Hayes, 2000).

Wide temperature fluctuations may occur daily, and occasionally *Rana luteiventris* are exposed to rather harsh conditions. In Utah, for example, some of the breeding pools are slightly saline (Hovingh, 1993b; Cuellar, 1994); at these Great Basin sites, larvae are exposed to temperatures between 1 and 25°C over the course of development, and conductivities of up to 3,200 µmhos/cm and a pH ranging to 9.7.

Oviposition sites are chosen that balance the need for insolation in the cool morning and avoidance of very warm temperatures during the midday sun. Oviposition sites are often located at the western edge of a pond. Ponds should have clear, open waters for egg deposition, as eggs are not oviposited in vegetation. Eggs are deposited over a wide range of temperatures: 4–7°C in western Utah (Hovingh, 1993b); 6°C in British Columbia (Licht, 1969c); and >14°C along the Wasatch Front in Utah (Morris and Tanner, 1969). Tadpoles metamorphose prior to midsummer, when water heats up to lethal temperatures.

As with most frogs, breeding does not occur in every potential breeding site. For example, breeding occurs in 14–26% of potential breeding sites examined in the Greater Yellowstone Ecosystem (Patla, 1997; Corn et al., 2005; Corn, 2007).

REPRODUCTION

Breeding occurs over a wide range of dates, depending upon location, temperature, and elevation. For example, breeding occurs from mid-March to early April in Utah (Hovingh, 1993b; Cuellar, 1994) and the Palouse Region of Idaho and Washington (Davis and Verrell, 2005), and between late March and late May in northeastern Oregon and eastern Washington (Svihla, 1935; Metter, 1960; Bull and Hayes, 2001). Breeding normally occurs in mid- to late May and June in Yellowstone National Park (although adults have been observed in April: Yeager, 1926 [as "northern frogs"]; Turner, 1955, 1958c) and into early July in interior British Columbia (Logier, 1932). Breeding dates also change with elevation. For example, breeding occurs from late March to May at low elevations in Montana, but in June at higher elevations (Black, 1970; Werner et al., 2004). Similar elevation-based trends are reported in Alberta, with breeding occurring from early May to June (Salt, 1979). Even within a localized area, frogs will breed earlier in warm-water spring habitats than in ponds fed by cold-water springs. Breeding also can be accelerated by warm weather or delayed or interrupted by cold weather. During cold weather, an amplexed pair will move to deep water until temperatures rise again.

Males enter breeding ponds prior to females, and females are ready to ovulate as soon as they enter the breeding ponds. Eggs are deposited either day or night in a jelly mass 10–20 cm below the water's surface attached to floating or emergent vegetation. Masses measure 75 to 200 mm in

Egg mass of *Rana luteiventris*. Photo: Brome McCreary

diameter. Within a short time the attachments give way, and the eggs tend to flatten out and float at the surface. Some of the eggs are attached to one another by a small gelatinous cord, and a single egg may be connected to five other eggs through the cord or outer membranes. This species has communal oviposition sites, and the egg masses tend to be in close proximity to one another. For example, Morris and Tanner (1969) recorded 50 egg masses within an area 75 cm in diameter, and Bull and Marx (2002) noted up to 135 egg masses per oviposition site. Koch and Peterson (1995) observed communal sites with 34 and 45 egg masses. Only a few clutches may be oviposited away from the larger mass of eggs, but there may be multiple communal oviposition sites per lake.

Each mass normally contains approximately 325–950 eggs and measures ca. 20 cm × 15 cm. Specific counts include a mean clutch size of 605 eggs (range 430 to 725) one year in Utah and 746 (range 147 to 1,160) the next (Morris and Tanner, 1969); 325–710 (mean 444) eggs per clutch in western Utah (Cuellar, 1994); a mean of 539 eggs per clutch (range 206 to 802) in Wyoming (Turner, 1958b); a range of 150 to 2,000, with most masses containing 500 to 600 eggs (Livo, 1998); 2 masses of 1,000 and 1,500 eggs in eastern Washington (Svihla, 1935); and a maximum clutch size of 2,400 (Livezey and Wright, 1947). Algae may enter the egg capsules, giving them a greenish appearance as development progresses. Egg mortality results chiefly from freezing and desiccation, although embryos within the jelly mass may survive for some time out of water (Hossack, 2006). Presumably, such embryos would have the chance to survive short-term desiccation until reinundated.

Hatching occurs in 12–21 days, with metamorphosis following in about 60 days of further aquatic development (Turner, 1958b), although Black (1970) stated hatching occurs in only 4 days. In interior British Columbia, Logier (1932) suggested that larvae overwinter and do not metamorphose until the following early summer. Hatching and development rates are directly tied to water temperature, with later hatching and slower growth at low temperatures. At hatching, larvae are 7–10 mm TL.

Columbia Spotted Frog females may not produce a clutch of eggs every year, and at most they produce only one egg mass per female. In Yellowstone National Park, for example, successful egg deposition takes place only once every 2–3 yrs, although males breed every year (Turner, 1958b). The number of egg masses at a location also varies annually. For example, Hovingh (1993b) reported from 2 to 33 clutches at a site in Tule Valley, Utah, although at other sites as many as 462 masses were counted at a single location in a single year. In northeastern Oregon, Bull and Hayes (2001) counted from 3 to 39 egg masses at 6 breeding ponds. Whether these values reflect differences in resource availability, environmental conditions, or fluctuations in breeding female population size is unknown.

LARVAL ECOLOGY

Larvae feed on decomposing plant material and green algae. They also may be able to derive nutrition from bacteria in the water (Burke, 1933), and they will feed on carrion such as dead conspecifics during the later stages of development. Tadpoles form aggregations that function to stir up food from the substrate or to conserve heat and thus speed up development (Carpenter, 1953b). These aggregations may produce audible clicking sounds as the tadpoles smack their lips and rise to the surface to gulp air. Larvae can be quite abundant in optimal sites. Salt (1979) reported densities of 150–225 larvae/m^2 at sites in Alberta. At another site, he noted from 10 to 60 larvae in a single square meter. Carpenter (1953b) recorded from 1,000 to 1,500 tadpoles within a 30 cm × 30 cm area.

The normal lethal temperature limit of larval development is 28°C (Licht, 1971), although Brues (1932) reported larvae of this species "swimming about in water ranging from 39.2–41.6°C" in Yellowstone National Park.

There is considerable variance in the maximum size of tadpoles. Morris and Tanner (1969) reported that larvae normally reach 50–55 mm TL just prior to metamorphosis, but that some tadpoles reached 70 mm TL. Turner (1958b) recorded the maximum size as 60 mm TL. Recent metamorphs, however, are 16–25 mm TL (Turner, 1958b; Livo, 1998; Russell and Bauer, 2000). Likewise, the larval period is quite variable, ranging between 70 and 85 days in Wyoming (Turner, 1958b) and 122–209 days in Utah (Morris and

Tanner, 1969). Most authorities state that tadpoles do not overwinter. Still, large tadpoles have been found in January in Montana (Black, 1970).

DIET

This species consumes a variety of invertebrates, particularly snails, *Gammarus* shrimp (or scuds), sow bugs, spiders, beetles (both aquatic and terrestrial), moths, water striders, ants, lepidopteran larvae, and many types of flies and fly larvae (Ruthven and Gaige, 1915; Tanner, 1931; Moore and Strickland, 1955; Turner, 1959b; Miller, 1978). The type of prey may change seasonally, depending on availability. Plant debris and inert detritus may be consumed as prey are captured and forced into the mouth.

No diet specialization is evident, and frogs likely consume whatever small invertebrates come within feeding range. Feeding occurs mostly during the day and at dusk. Although small frogs naturally consume small prey, large frogs eat a wide variety of different-sized prey. Debris may be ingested in conjunction with feeding, and Turner (1960b) recorded an instance of a large stone becoming lodged in the intestine of a *Rana luteiventris*, which undoubtedly would have caused its death.

PREDATION AND DEFENSE

When disturbed in water, Columbia Spotted Frogs dive into the substrate and bury themselves in mud. If on land, Black (1970) reported they might attempt escape by jumping down mouse burrows.

Likely predators of postmetamorphic Columbia Spotted Frogs include mammals (coyote, otters, mink), birds (Ravens, Northern Harriers, Herons, Egrets, Sandhill Cranes, Great Gray Owls, Pygmy Owls, Ring-billed and California Gulls), garter snakes (*Thamnophis*), and trout (Salt, 1979; Koch and Peterson, 1995). Eggs masses are attacked by predaceous diving beetles, crane flies, and leeches (*Haemopis marmorata*) (Hovingh, 1993b). Larvae are probably consumed by predaceous invertebrates (dytiscid beetles), a variety of birds including Gray Jays, and garter snakes (Turner, 1960b; Black, 1970).

POPULATION BIOLOGY

Age at sexual maturity is variable depending upon environmental conditions and elevation. For example, sexual maturity is delayed in this species in Yellowstone National Park, with males breeding first at about at 4 yrs (at > 45 mm SUL) and females at 5–6 yrs (at > 60 mm SUL). In coastal populations, however, maturity of both sexes is reached at 2 yrs. Longevity is approximately 10 yrs for males and 12–13 yrs for females in natural populations at Yellowstone (Turner, 1960b), but only 3–4 yrs in coastal populations and 5 yrs for males and 7 yrs for females in the Nevada Great Basin (Reaser, 2000). The rigors of Yellowstone and other interior regions thus have major consequences on the demography and longevity of Columbia Spotted Frogs.

Young Columbia Spotted Frogs in Yellowstone grow rapidly after metamorphosis and are generally 25–30 mm SUL when they emerge from overwintering the following year. If newly metamorphosed frogs are ca.16 mm SUL, then this indicates an initial growth spurt of 9–14 mm (Turner, 1960b). By the end of the first year, juveniles will have grown 19% of their total body size, and 52% by the end of the second full year after transformation (Turner, 1960c). In Yellowstone, they do not reach 50 mm SUL until their fourth or fifth full season after metamorphosis. Turner (1960b) provided growth curves for his Yellowstone population and speculated that differences in male and female growth rates first become apparent in their fourth year. Some populations of *Rana luteiventris* grow much faster than these rates, perhaps reflecting differences in growing season, resource availability, and environmental conditions among populations.

Survivorship can be relatively high. Funk et al. (2005b) estimated a juvenile survival rate of 33%, indicating that those juveniles that survived dispersal had a relatively good chance of establishing themselves at the ponds to which they immigrated. It is thus not surprising that populations of Columbia Spotted Frogs historically could be substantial. In the early 1950s, for example, Turner (1960b) estimated there were 1,200–1,850 frogs occupying his 28 ha Yellowstone study site. In western Utah, Cuellar (1994) estimated that there were 100 frogs/ha. However, breeding populations are often small.

The sex ratio of frogs in a population can vary depending on age-class (Reaser, 2000). Since males remain at breeding sites longer than females and call from communal oviposition sites, sex ratios at breeding ponds might be expected to be male-biased (Davis and Verrell, 2005). For example, Morris and Tanner (1969) reported five males per every female at their Utah breeding sites. In the Palouse Region, males sometimes exceeded females by 11 to 1 at small breeding ponds (Davis and Verrell, 2005). In contrast, Reaser (2000) reported female-biased sex ratios at most Nevada Great Basin sites she studied and found that that sex ratios varied annually among sites. Likewise in Turner's (1960b) study, the overall sex ratio was female-biased (one male: two females). Sex ratios were 1:1 until the fifth year, when highly skewed female-biased sex ratios were observed. These observations, coupled with Reaser's (2000) analysis of lines of arrested growth, suggest that females live longer than males, perhaps because of differential survivorship. In general, however, overall annual survivorship is high for adults.

In the mountainous West, landscape features play important roles in the genetic structuring of populations of Columbia Spotted Frogs. In an extensive analysis of six microsatellite loci, Funk et al. (2005a) demonstrated that mountain ridges and elevation act as genetic barriers and thus increase genetic differentiation among sites. Most populations consist of multiple ponds, except in the case of isolated wetlands, and genetic relatedness is more evident within mountain basins than between them. Gene flow also occurs more extensively among low elevation populations than among those at high elevations; genetic variation among populations is strongly negatively correlated with elevation. Low elevation populations are important sources of immigration and genetic variation. If these populations are eliminated, their loss could affect the persistence of high elevation populations, even without direct habitat threats.

Columbia Spotted Frogs are easily observed when present during monitoring programs. During surveys of potential habitat, for example, detection probabilities were generally high in Glacier and Yellowstone National Parks (75–95%), and adjusted occupancy rates ranged between 17 and 26% of habitats examined over a 2 yr period (Corn et al, 2005).

DISEASES, PARASITES, AND MALFORMATIONS

Peptides in the skin may help in defense against bacterial infections (Goraya et al., 2000). The amphibian fungal pathogen *Batrachochytrium dendrobatidis* has been detected in this species in the Greater Yellowstone Ecosystem (Corn, 2007), Oregon (Adams et al., 2010), and elsewhere in Idaho, Montana, and Wyoming (Muths et al., 2008); these latter authors found the pathogen in 40% of the individuals sampled. In Oregon,

Adult *Rana luteiventris*.
Photo: Dirk Stevenson

12 of 14 sites had positive indications of chytridiomycosis, and 71 of 198 (35.9%) frogs tested were positive (Adams et al., 2010); one larva also tested positive.

Columbia Spotted Frogs are parasitized by the nematodes *Aplectana gigantica* and *Spironoura pretiosa* and by the trematodes *Glypthelmins* sp., *Gorgoderina tanneri*, *G. translucida*, *Haematoloechus parviplexus*, *H. similiplexus*, *Halipegus* sp., and *Haplometrana intestinalis*, *H. utahensis* (Olsen, 1937, 1938; Turner, 1958a; Waitz, 1961). Males and females are subject to equal parasitic loads. Leeches may attach themselves to the gular region of tadpoles (Carpenter, 1953b).

Malformations of *Rana luteiventris* have been reported at frequencies of 21–31% at certain sites in the Pacific Northwest, with from 82 to 144 *Ribeiroia* metacercariae per frog (Johnson et al., 2002). Malformations included cutaneous skin fusions (extra webbing), missing digits (ectrodactyly), partial limbs (hemimelia), femoral projections, and other miscellaneous deformities.

SUSCEPTIBILITY TO POTENTIAL STRESSORS

Chemicals. The herbicide Roundup® is toxic to larvae, with an $LC_{50\ (24\ hrs)}$ of 1.65 mg/L (King and Wagner, 2010). The LC_{50} decreases at 15 days to 0.98 mg/L.

Metals. The presence of heavy metals, either singly or in combination, may alter the ability of Columbia Spotted Frog larvae to respond to predators. When taken in combination, sublethal effects of metals are evident at lower dosages than they are singly. For example, combinations of cadmium and zinc are much more toxic to larvae than either metal is by itself. Medium levels of zinc and medium to high levels of lead decrease the fright response of tadpoles, and exposure to soils containing metals may delay metamorphosis and decrease larval fright responses (Lefcort et al., 1998). The $LC_{50\ (4\ day)}$ of specific metals is: zinc (28.4 ppm), cadmium (15.8 ppm), zinc + cadmium (4.44–4.52 ppm). These results suggest that sites contaminated by a variety of metals could be both toxic and have sublethal effects on tadpole survival.

UV light. Columbia Spotted Frogs have high levels of photolyase, an enzyme important in DNA repair from UVB radiation. Spotted frogs reared in ambient UVB and in situations where UVB is shielded have similar levels of hatching success and embryo development (Blaustein et al., 1999).

STATUS AND CONSERVATION

Columbia Spotted Frogs are susceptible to habitat loss, as are all frogs. Populations have disappeared from many historic locations, such as along the Wasatch Mountain front in the Bonneville Basin of Utah, the northern Great Basin, and in southwestern Alberta (Weller and Green, 1997; Corn, 2000; Wente et al., 2005). Wente et al. (2005) estimated that *Rana luteiventris* occupied only 53% of its historically documented locations in northern Nevada and eastern Oregon. Suggested causes include the outright destruction of habitats, habitat fragmentation, water modification projects (channelization, diversion, irrigation, floodplain alteration), and the introduction of nonnative species (crayfish, fish, American Bullfrogs, raccoons) (Hovingh, 1993b). Other human-related mortality occurs from roadkill, use as bait, malicious killing, and trampling by human-introduced livestock (Turner, 1960b; Ross et al., 1999).

There is little doubt that the introduction of nonnative salmonid fishes into high-elevation lakes and streams has adversely affected Columbia Spotted Frog populations, even in remote wilderness (Pilliod and Peterson, 2000, 2001). Fish come to occupy nearly all frog habitats (those > 1 ha and > 4 m deep), and many fish-uninhabited sites are too small, shallow (< 1 ha and < 4 m deep), and ephemeral for frog occupancy. Bull and Marx (2002) found fish at 50% of the sites where they found Columbia Spotted Frogs in northeastern Oregon and suggested that fish were not good indicators of habitat occupancy for this species. In fact, the abundance of *R. luteiventris* decreases when fish are present, even if the species is not completely eliminated (Pilliod and Peterson, 2000). *R. luteiventris* will readily colonize artificial ponds, but not those containing nonindigenous fish (Monello and Wright, 1999). These results suggest that fish play an important role in habitat selection by this species.

The pattern of genetic structured populations, high dispersal rates, dependence on lowland populations for immigrants, and complex landscape features makes *R. luteiventris* particularly vulnerable to habitat fragmentation. That small populations of Columbia Spotted Frogs often

Habitat of *Rana luteiventris*, Utah. Photo: Breck Bartholomew

inhabit temporary wetlands in rather dry agricultural areas makes them susceptible to habitat modification and competition from humans and livestock for water (Bull and Hayes, 2000; Davis and Verrell, 2005; Goldberg and Waits, 2009, 2010). The species needs a variety of habitats, from breeding sites to migration and dispersal corridors to overwintering sites that are substantially wetland dependent. Interruption of movement patterns or loss of any required wetlands within a landscape are likely to have repercussions to a resident and possibly peripheral frog population. Coupled with the potential for extreme climatic variation in many areas of the West, dependence on scattered and isolated wetlands makes this species extremely vulnerable to both natural and man-made disturbances. For these and other reason, sensitive wetlands need to be fenced from livestock and protected.

Even in national parks, construction, landscape modification, and water extraction can result in substantial declines. For example, the population studied by F.B. Turner in the 1950s in Yellowstone National Park decreased to an estimated 225–400 frogs after 40 yrs of habitat changes associated with park development (Patla, 1997; Patla and Peterson, 1999). These changes resulted in differences in movement patterns, activity ranges, dispersion, and spatial distribution (Patla, 1997). Reproductive effort (i.e., the number of egg masses) decreased substantially, and reproductive success and recruitment were poor in the mid-1990s compared with the early 1950s. In the late 1990s, however, a wet season produced substantial recruitment, indicating the importance of environmental factors influencing reproduction. Still, major habitat modifications make it unlikely that this important population will ever recover to its 1950s vitality.

The effect of climate changes, particularly drought, has been debated with regard to this species. According to McMenamin et al. (2008, 2009), climate change (increased temperatures, decreased precipitation) was responsible for declines in *R. luteiventris* in Yellowstone National Park. This finding was disputed by Patla et al. (2009) who contended that the previous study's findings resulted from sampling differences and model assumptions rather than an empirical validation of population decline. Indeed, Patla et al. (2009) contended that no decline had taken place and that McMenamin et al.'s (2008) study actually demonstrated persistence in the face of drought. Whereas it seems logical that long-term climate change may affect the persistence of this species at some locations, long-term monitoring at additional sites would help researchers to understand the effects of climate on this and other anurans.

Rana muscosa Camp, 1917
Mountain Yellow-legged Frog

ETYMOLOGY
muscosa: from the Latin word *muscosa* meaning 'mossy' or 'full of moss.' The name refers to the dorsal lichen-like dark dorsal patches.

NOMENCLATURE
Dubois (2006): *Rana (Amerana) muscosa*
  Synonyms: *Rana boylii muscosa*

Little information is available on the biology of *Rana muscosa* since much of the historical literature on "*R. muscosa*" actually refers to *R. sierrae*. This is particularly evident in Zweifel (1955), in which the accounts of morphology and biology include both species under the name *R. muscosa*.

IDENTIFICATION
Adults. The body is dark yellow to brown with a reticulated moss-like dorsal pattern; other color variations are known, however, and may include a mixture of gray, red, or greenish brown. A few individuals may be dark brown and lack the lichen-like dorsal pattern. The undersides are pale to intense yellow with dusky spots on the throat and chest. The upper lip is mottled below the eye and does not have a white stripe. The dorsolateral fold is not distinct and is not pitted anteriorly, and the tympanic area is usually rough. The tips of the toes are expanded. In general, *R. muscosa* have longer limbs in relation to body size and wider heads compared with the closely related species *R. sierrae*. Males are generally smaller than females and lack vocal sacs. Females may have a more obscure dorsal pattern, but this character is not consistent. Maximum size is 80 mm SUL (Jennings and Hayes, 1994b).

Larvae. Larvae are brown with a golden tint resulting from numerous yellow-gold chromatophores. There may or may not be aggregations of dark spots on the tail musculature and tail fin, and the area bordering the tail musculature on the tail fin is clear. The body appears somewhat transparent. The tail musculature is massive and tapers slowly toward the tail tip. The tail fin of the tadpole of *R. muscosa* is about the same height throughout the tail, rather than peaking midway along the tail as it does in *R. boylii*. The venter is faintly yellow and not silvery. The viscera are faintly visible.

Eggs. The eggs are grayish tan to black dorsally and light gray to tan to creamy white ventrally. There are three distinct envelopes surrounding the vitellus. Eggs are deposited in a solid jelly mass with each egg a few mm apart from the next. Algae may grow on the eggs, giving a greenish tint. The diameters of the envelopes and vitellus are probably similar to *R. sierrae*.

DISTRIBUTION
Rana muscosa occurs in the San Gabriel, San Bernardino, and San Jacinto mountains in southern California, and from the region of the Kings River (Fresno County) southward in the Sierra Nevada Mountains to Inyo and Tulare counties. Nine populations are known. No populations occur east of the Sierra crest. There is a contact zone with *R. sierrae* between the Middle and South Forks of the Kings River in a line stretching from Mather Pass to Monarch Divide. Isolated populations occurred on Palomar Mountain in northern San Diego County and on Breckenridge Mountain

Distribution of *Rana muscosa*. Dark gray indicates extant populations; light gray indicates extirpated populations.

in Kern County. A small population remains in the San Bernardino Mountains, despite earlier concerns that the species had been extirpated there. Other populations in the Transverse Ranges may well be extinct. Important distributional references include Camp (1917), Slevin (1928), Jennings and Hayes (1994a, 1994b), Lemm (2006), and Vredenburg et al. (2007).

FOSSIL RECORD
No fossils are known.

SYSTEMATICS AND GEOGRAPHIC VARIATION
There have been several studies that suggested sometimes conflicting phylogenies among Pacific Northwest ranids, whether based on morphological or genetic data (Zweifel, 1955; Case, 1978; Farris et al., 1979, 1982; Green, 1986a, 1986b; Macey et al., 2001; Hillis and Wilcox, 2005). This species is a member of the *R. muscosa* complex (sometimes termed the *Rana boylii* species group; Macey et al., 2001) of the *Amerana* clade of North American ranid frogs. *Rana muscosa* is most closely allied phylogenetically with *R. sierrae* and somewhat more distantly with *R. aurora* and *R. cascadae* (Macey et al., 2001; Hillis and Wilcox, 2005; Vredenburg et al., 2007), despite earlier assertions of a close relationship with *R. boylii* based on karyotype (Green, 1986a). The diploid chromosome number is 26 (Haertel et al., 1974). Haertel et al. (1974) and Green (1986a) described the chromosomes.

Camp (1917) originally described *R. muscosa* as a subspecies of *R. boylii* from the Transverse Ranges of southern California. The name *Rana boylii sierrae* was allocated to the yellow-legged frogs of the Sierra Nevada of California and Nevada. Zweifel (1955) used morphological (including the shape of the pectoral girdle) characters to elevate *R. muscosa* to specific status, and until recently all yellow-legged frogs in the Sierras were considered to be *R. muscosa*. Genetic data have clearly differentiated the species. Analyzing 1,901 base pairs of mitochondrial DNA, acoustic data, and morphological characters, Vredenburg et al. (2007) recognized 2 distinct evolutionary clades within *R. muscosa* that they then differentiated into *R. sierrae* from the northern and central Sierra Nevada and *R. muscosa* from the southern Sierra Nevada and Transverse Ranges. Levels of genetic divergence largely result from Pleistocene vicariance events due to climate factors; the 2 major clades within *R. muscosa* diverged ca. 2.2 mya, whereas subsequent divergence within the 2 clades occurred 1.5 mya (Macey et al., 2001). Four subgroups within *R. muscosa* subsequently diverged ca. 1.4 mya. Each of the mountain ranges where this species still occurs represents a separate evolutionary lineage (Schoville et al., 2011).

In the original description, Camp (1917) noted color pattern differences between *R. muscosa* from the San Gabriel Mountains (generally lighter and thickly marked with lichen-like dark patches) and those from the San Jacinto Mountains (darker in coloration, with some animals approaching a uniform dark brown or having a blotched or spotted pattern similar to *R. sierrae*). Laboratory hybridization experiments suggest genetic incompatibility with *R. boylii* (Zweifel, 1955).

ADULT HABITAT
Rana muscosa inhabits two very different regions and has two very different lifestyles. In the southernmost part of its range in the Transverse Ranges, it inhabits deeply cut canyons. Streams have moderate to steep gradients, and large boulders likely are present. Streams are relatively small (to 4.5 m in width), with maximum depths of 60 cm; gravel banks are numerous, rocks project above the water, and the shorelines are shaded. Adjacent vegetation includes mostly chaparral with coniferous forest at higher elevations. In the Transverse Ranges, populations were found at elevations from 366 to 2,286 m.

In the southern Sierra Nevada, Mountain Yellow-legged Frogs inhabit much the same types of habitats as the closely related *R. sierrae*. They occur in lakes in montane meadows, which may have muddy or sandy shores in contrast to the rocky habitats where the frog is found at lower elevations. They tend to prefer undercut banks with riparian willows and avoid bedrock in summer. However, the use of rocky habitats increases in autumn as the frogs move to overwintering sites. Adjacent vegetation includes meadows, white fir, incense cedar, and forests of lodgepole, sugar, yellow, and Jeffrey pines. *Rana muscosa* occurs from 1,300 to 3,700 m in the Sierra Nevada.

AQUATIC AND TERRESTRIAL ECOLOGY
In the Transverse Ranges, *R. muscosa* will climb canyon walls by day and sit facing the stream, or

they will use rocks projecting from the water to sit in the sun. *Rana muscosa* is capable of dispersing long distances. In a repatriation study, Fellers et al. (2007a) recorded one frog dispersing > 800 m from its release pond, although most frogs dispersed over much shorter distances (50–510 m, mean 80 m). Whereas most interpond movement likely occurs along watercourses, these frogs are capable of making straight-line movements over inhospitable terrain of several hundred meters or more.

Seasonal migrations occur between summer breeding and feeding sites and sites where frogs spend the winter. At high elevations, Mountain Yellow-legged Frogs tended to remain within a lake or adjacent stream, with mean movements of 77 m over the course of August (Matthews and Pope, 1999). In September, movements increased, with mean straight-line movements of 145 m and circuitous movements of 315–466 m. These autumnal movements decreased substantially (to a mean of 43 m) by October, as frogs took up winter residences. Accordingly, home range estimates were 385 m^2 in August, 5,336 m^2 in September, and 53 m^2 in October (Matthews and Pope, 1999). Temperatures during autumn movements can be quite cool (5.5–12.5°C). Movements generally follow watercourses, but frogs occasionally move overland as much as 66 m between wetland sites (Matthews and Pope, 1999).

Overwintering by larvae and postmetamorphs occurs in permanent lakes and streams in the high mountains, where activity may occur only during a brief three to five month summer season. These sites may be several hundred meters from the summer feeding areas, and are usually deep and of large surface area. Matthews and Pope (1999) tracked frogs to near shore habitats under ledges and in deep water crevices in fractured bedrock. The crevices were 0.2–1.2 m below the water surface in Matthews and Pope's (1999) study. Some of the high-elevation lakes and streams may be covered by ice six to nine months of the year, and if winters are severe with oxygen depletion, nearly all Mountain Yellow-legged Frogs will die.

CALLING ACTIVITY AND MATE SELECTION

The calls are comprised of a series of raspy scraping sounds, often ending in a loud, accentuated note (Lang et al., 2009). *Rana muscosa* also makes a call with a stuttered note. Calls appear to require a great deal of effort. The calls of *R. muscosa* and *R. sierrae* have different properties and are easily distinguished using oscillogram and spectrogram analyses (Vredenburg et al., 2007). *R. muscosa*, for example, does not have transitions between pulsed and noted sounds whereas *R. sierrae* does.

BREEDING SITES

Eggs are deposited along streams and in higher lakes, tarns, and meadow ponds. In streams, eggs

Eggs of *Rana muscosa*. Note how silt attaches to the egg membranes. Photo: Adam Backlin

Tadpole of *Rana muscosa*. Photo: Chris Brown

may be attached to rocks on the stream bottom, whereas in lakes they may not be attached and allowed to float free. Eggs also may be oviposited in lentic pools in close proximity to lakes, streams, or meadow wetlands. The species tends to avoid the smallest streams.

REPRODUCTION

Little information is available on reproduction in this species. In the southern part of its range, the breeding season likely extends from early April to mid-May. For example, Storer (1925) reported finding a pair in amplexus in early April; Zweifel (1955) found evidence of breeding from late April to early May; and Stebbins (1951) reported egg deposition in early to mid-May. In the Sierra Nevada, it seems likely that reproduction occurs much later, from mid-May to early July, depending upon elevation. At high elevations, the frogs must wait until ice and snow melts from the breeding sites.

LARVAL ECOLOGY

Larval Mountain Yellow-legged Frogs prefer relatively high water temperatures at breeding sites in the high mountains and actively seek warm (to 27°C) microhabitats in shallow water during the day. At night, they will move to deeper water as the shallows become cool. Larvae may form aggregations similar to those of *R. cascadae* and appear to follow much the same thermal pattern of diel habitat use as that species (Jennings and Hayes, 1994b). Larval *R. muscosa* overwinter in streams and lakes at high elevation sites for one or two winters prior to metamorphosis. During severe winters, nearly all larvae die due to oxygen depletion when waters freeze. However, some larvae can survive nearly anoxic conditions. Larvae have a lower critical oxygen tension than postmetamorphs, allowing them a greater tolerance of low PO_2. In addition, they are able to reduce energy and oxygen consumption at low PO_2 (Bradford, 1983). Despite their diurnal preference for higher water temperatures, larvae thrive in cold water, and Brattstrom (1963) recorded them in water at temperatures of 4–9°C.

In the Transverse Ranges, metamorphosis usually occurs the same season as oviposition, although Zweifel (1955) reported one observation that suggested at least some tadpoles overwinter. Overwintering usually would not be advantageous in the mountain streams because of the potential for scour flooding during the winter wet season. Recently transformed individuals are 20–30 mm SUL.

DIET

Adults are known to consume large quantities of anuran tadpoles, particularly *Anaxyrus canorus* and *Pseudacris regilla* (Pope, 1999; Pope and Matthews, 2002). Other prey include spiders, ticks, harvestmen, beetles, flies, ants, bees, wasps, water striders, and aquatic Hemiptera (true bugs) (Long, 1970). Lemm (2006) stated that cannibalism had been reported, but this observation was likely in reference to *Rana sierrae*. Males and females have similar diets, and, based on gut analyses, some feeding occurs year-round. Prey size is proportional to frog size, and prey items are selected opportunistically.

PREDATION AND DEFENSE

When disturbed, *R. muscosa* dive into the water and submerge their bodies in the muddy substrate. As the frog buries itself, silt is stirred up, making it difficult to locate the well-camouflaged frog. Potential predators include mammals (black bears, coyotes, humans), birds (Brewer's Blackbirds, Ravens, Eared Grebes, American Dippers, American Kestrels), snakes (*Thamnophis elegans*), and lizards (*Elgaria coerulea*) on adults, subadults, and larvae (Fellers et al., 2007a). In experiments, trout rapidly consume both newly hatched and larger tadpoles (Vredenburg, 2004).

POPULATION BIOLOGY

Nothing is known concerning the population biology of this species in the Transverse Ranges. At high elevations, it likely has similar population and demographic characteristics as other high-elevation ranids such as *Rana cascadae* and *R. sierrae*.

COMMUNITY ECOLOGY

Because of the short growing season, adult *R. muscosa* must feed rapidly in order to grow and reproduce during the three- to five-month activity season. Pope and Matthews (2002) noted a correlation between the distribution of larval *Pseudacris regilla* and *Anaxyrus canorus* in Kings Canyon National Park, and suggested that an abundance of anuran larvae was essential to the growth and good body condition of *Rana muscosa*. Adult *R. muscosa* may seek out water bodies with anuran larvae in order to take advantage of this nutritious food source in an otherwise food-limited environment.

DISEASES, PARASITES, AND MALFORMATIONS

The virulent amphibian pathogen *Batrachochytrium dendrobatidis* was reported from *Rana mucosa* tadpoles, metamorphs, and postmetamorphs in the Sierra Nevada as early as 1998 (Fellers et al., 2001; Rachowicz et al., 2006; Woodhams et al., 2007). Green et al. (2002) also reported this pathogen in *R. muscosa* but without locality data, although likely referring to the Fellers et al. (2001) study. As noted below, amphibian chytrid was likely responsible for the mass mortality observed at Kings Canyon National Park instead of the bacterial pathogen *Aeromonas hydrophila*, although this species plus the bacteria *Enterobacter aerogenes* and *E. agglomerans* were isolated from dead specimens (Bradford, 1991). Endoparasites include the trematode *Langeronia brenesi* (Goodman, 1989).

SUSCEPTIBILITY TO POTENTIAL STRESSORS

pH. Tadpoles do not occur in lakes with a pH ≤ 5.2, although adults inhabit lakes with a pH as low as 5.0 (Bradford et al., 1998). Embryos are much more sensitive to low pH than adults.

Chemicals. The pesticides DDE, γ-chlordane, *trans*-nonachlor, chlorpyrifos, and diazinon have been found in frog tissues from the high Sierras in areas that could have received contaminants borne on the wind from the San Joaquin Valley (Fellers et al., 2004, 2007a). Healthy frog populations in a wind-protected site did not have high concentrations of pesticides, although these frogs may have been the closely related *Rana sierrae*. It is uncertain whether organophosphate pesticides, which do not favor bioaccumulation, have played a role in frog declines in the high Sierras.

UVB radiation. UVB radiation does not affect hatching success or developmental rates of embryos (Vredenburg et al., 2010a).

STATUS AND CONSERVATION

The Mountain Yellow-legged Frog has disappeared from most of its historic distribution in the Transverse Ranges and disjunct populations in southern California. In addition, the species has declined in the southern Sierra Nevada, including habitat-protected sites in Kings Canyon and Sequoia National Parks, where more than half of the historic populations have disappeared (Bradford, 1991; Bradford et al., 1994b; Jennings and Hayes, 1994a, 1994b; Vredenburg et al., 2005). Remaining populations are characterized by low

Adult *Rana muscosa*. Photo: Chris Brown

genetic diversity, evidence of genetic bottlenecks, and a high degree of historical isolation with little gene flow (Schoville et al., 2011). The presence of nonnative predaceous fish in alpine lakes is correlated with an absence of *R. muscosa* tadpoles (Bradford, 1989; Bradford et al., 1998) and has fragmented remaining populations (Bradford et al., 1993). When fish are removed, frog populations recover rapidly (Vredenburg, 2004). Schoville et al. (2011) noted that the high degree of isolation will require individual management protocols.

Batrachochytrium dendrobatidis has decimated amphibians throughout the Western states, including *Rana muscosa*, and may be the primary cause of population declines in this species (Bradford et al., 2011). The first documentation of its effects on *R. muscosa* was by Bradford (1991), who noted the extinction of this species at a site in Kings Canyon National Park. He observed that tadpoles and recent metamorphs had symptoms akin to the bacterial pathogen *Aeromonas hydrophila*, but the mortality event was more likely due to amphibian chytrid. By 2005, Mountain Yellow-legged Frogs in Kings Canyon National Park were nearly all infected (96.7%) with *Batrachochytrium dendrobatidis*. Rachowicz et al. (2006) later found that 19% of 144 *Rana muscosa* populations in the southern Sierra Nevada were infected during both years of their survey (2003–2004), with another 16% of the populations uninfected the first year but infected the second year. Following an outbreak, population size could be reduced as much as 88%. Despite the lethality of the fungus in laboratory and field settings, however, it is clear that at least some frogs survive and reproduce to allow persistence of infected frog populations over a considerable period (Briggs et al., 2005).

Infected frogs that survive appear to have low fungal loads, survive between years, and may even lose and regain the infection (Briggs et al., 2010). Survivorship within infected frog populations may result from density-dependent host-pathogen dynamics. This is especially true with regard to the pathogen's life history characteristics, which allow for optimal encysting and the rapid production of zoospores at temperatures of 17–25°C within a host's cell wall. As temperatures drop to 7–10°C, greater numbers of zoospores are produced, which remain infectious for a long time

Habitat of *Rana muscosa*. Photo: Chris Brown

(Woodhams et al., 2008). In effect, the pathogen responds to decreasing temperatures with increased fecundity as maturation rate slows, and infectivity increases as growth decreases (Woodhams et al., 2008). These results help explain why *Batrachochytrium dendrobatidis* is successful at colder temperatures. The fungus also is likely aided in its persistence within frog populations by its infection of the long-lived tadpole and by nonhost fungal reservoirs (Briggs et al., 2005).

Secretions from *Rana muscosa* possess peptide mixtures, which inhibit the growth of the pathogen (Woodhams et al., 2007). The peptides are produced by a variety of symbiotic bacteria and may be important in assisting disease resistance and defense in at least a few frogs. Still, this emerging infectious disease is likely the proximate cause of Mountain Yellow-legged Frog disappearance throughout much of its former range. Whatever its origin, the pathogen seems to have affected frogs initially from 1979 to 1983. Bradford's (1991) population was extinct by 1989.

Because of population losses, Mountain

Yellow-legged Frogs have been repatriated into sites where they occurred formerly, in the hopes of reestablishing the species. In Sequoia National Park, for example, Fellers et al. (2007a) repatriated eggs, tadpoles, subadults, and adults into four previously occupied sites. Although survivorship was good the first year, the repatriation effort soon failed. Since fish were not present at any of the sites, Fellers et al. (2007a) attributed the failure to the effects of the pathogen *Batrachochytrium dendrobatidis* or possibly the continuing effects of contaminants.

The IUCN lists this species as "Endangered," and the species is listed as "Endangered" under the Federal Endangered Species Act of 1973. The state of California also protects this species.

Rana pretiosa Baird and Girard, 1853
Oregon Spotted Frog

ETYMOLOGY

pretiosa: Latin for 'to be prized' or 'worth the effort.' The name may refer to the reward of catching such an attractive frog.

NOMENCLATURE

Dubois (2006): *Rana (Amerana) pretiosa*
 Synonyms: *Rana temporaria pretiosa*

Much of what has been written on *Rana pretiosa* includes information on the Columbia Spotted Frog, *R. luteiventris* (e.g., Nussbaum et al., 1983). The latter species was separated from *R. pretiosa* based on molecular data, and its distinctiveness has been subsequently confirmed (Green et al., 1997; Funk et al., 2008). Readers should consult the location of past studies to verify whether the information contained refers to *R. pretiosa* or *R. luteiventris*.

IDENTIFICATION

Adults. Adults are brown or reddish brown and become redder with age. Very old adults can approach a brick-red coloration covering the dorsum. Juveniles, however, are brown or even olive green. The head and dorsum have black spots with light centers; these spots become darker and larger with age. When viewed from above, the eyes point upward rather than laterally, and they tend to focus dorsally as one might expect from a frog that spends much time in the water. The frog has a tan to orange dorsolateral fold that is lighter than the rest of the dorsum. The fold begins behind the eyes and tends to break up as it extends posteriorly. Juvenile venters are white or cream with red pigments under the legs and abdomen. Pigments become brighter with age. Adults have a bright orange-red pigmentation ventrally that extends from the posterior of the frog toward its chest but not up the sides of the frog. Throats have a tan, brown, or gray mottled appearance on an otherwise white background. Mottling is absent on the groin or comprised of a black or gray coloration on a light background. This mottling pattern does not contain lines and spots of black, green, yellow, or red as it does in *R. aurora*. Webbing is complete on the hind feet. Dunlap (1955) provided extensive information on the color pattern and morphology of this species.

Males are smaller than females. In the Cascades of Washington, males average 57 mm SUL (range 46 to 66 mm) and females average 75 mm SUL (range 59 to 89 mm) (McAllister and Leonard, 1997); in Thurston County, Washington, males averaged 56 mm SUL (range 46 to 65 mm), whereas females averaged 66 mm SUL (range 51 to 76 mm). In British Columbia, males reach 64 mm SUL and females reach 80 mm SUL (Licht, 1974, 1975). The maximum recorded size is 107.5 mm SUL for a female in Washington at a locality where many females exceed 100 mm (Rombaugh et al., 2006).

Larvae. The tadpole is tan to dark brown to greenish, with tails that may have small flecks or blotches. Oregon Spotted Frog larvae are similar to larvae of *Rana aurora* but have a lighter (white- or aluminum-colored) belly. They also may have a pale gold coloration around the margin of the belly. Older tadpoles usually have metallic flecking on the head, body, and anterior tail musculature, which is not accompanied by the deep gold or brassy pigments observed in *R. aurora*. The tail fin and tail musculature are of equal height. This species' larvae may be indistinguishable from those of *R. cascadae*. The tadpole may reach 70 mm TL.

Eggs. Eggs are black dorsally and white ventrally. The outer envelope is large, measuring 10 to 15 mm in diameter. The inner envelope is 5–6 mm in diameter and is indistinct. The vitellus averages 2.3 mm in diameter. Eggs are oviposited in rounded or globular jelly masses, but they are not as firm as in other species. The eggs also appear rather far apart from one another.

DISTRIBUTION

Rana pretiosa historically occurred from northeastern California through southwestern British Columbia. Many populations have disappeared, however, and there are only 3 extant populations in British Columbia (lower Fraser River Valley), 6 in Washington, and 24 in Oregon (Pearl et al., 2009b). All populations in northeastern California (Pit River drainage, five historic localities) and the Willamette Valley in Oregon (Bury, 2008), and most populations from a large area of western Washington (the Puget Trough, Olympic Peninsula, Mount Ranier), have been extirpated.

Important distributional references include: Range-wide (Nussbaum et al., 1983; Green et al., 1997), British Columbia (Weller and Green, 1997; Matsuda et al., 2006), California (Jennings and Hayes, 1994b), and Washington (Meek and Elliot, 1899; Slater, 1955; Leonard et al., 1993; McAllister et al., 1993; McAllister, 1995; McAllister and Leonard, 1997).

FOSSIL RECORD

No fossils are recognized.

SYSTEMATICS AND GEOGRAPHIC VARIATION

Until relatively recently, two subspecies were recognized, *R. p. pretiosa* and *R. p. luteiventris*. Green et al. (1997) demonstrated that these are different taxa. There have been a number of studies that suggested sometimes conflicting phylogenies among Pacific Northwest ranids (Zweifel, 1955; Dumas, 1966; Case, 1978; Farris et al., 1979, 1982; Green, 1986a, 1986b; Macey et al., 2001; Hillis and Wilcox, 2005). For example, Green (1986b) considered the Oregon Spotted Frog to be most closely related to *R. boylii*, but recent evidence suggests it is closely related to *R. luteiventris* and only somewhat more distantly to *R. boylii*. This species is a member of the *R. muscosa* complex (sometimes termed the *Rana boylii* species group; Macey et al., 2001) of the *Amerana* clade of North American ranid frogs. In general, *Rana pretiosa* exhibits a low degree of genetic variation, making it vulnerable to extinction (Funk et al., 2008).

Experimental crosses between *R. luteiventris* and *R. pretiosa* produced viable larvae through the latter stages of development (Dumas, 1966). Hybridization occurs with *R. cascadae* under laboratory conditions, with larvae completing metamorphosis (Haertel and Storm, 1970) or not (Dumas, 1966). Hybridization was less successful when *R. pretiosa* was the female parent. Green (1985c) documented hybridization with *R. cascadae* in nature, but the Oregon hybrids were infertile because of a lack of chromosome pairing during meiosis. Crosses between *R. pretiosa* and *Lithobates sylvaticus* are not viable (Dumas, 1966). The diploid chromosome number is 26 (Haertel et al., 1974).

ADULT HABITAT

Throughout much of its current range, this is a high elevation frog that occurs primarily along the crest and eastern slope of the Oregon Cas-

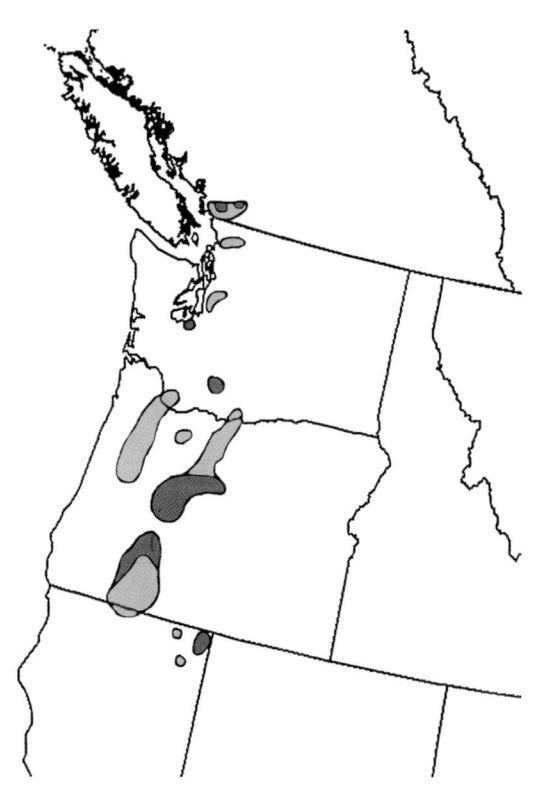

Distribution of *Rana pretiosa*. Dark gray indicates extant populations; light gray indicates extirpated populations.

cades Range. However, historic populations were found at much lower elevations, even near sea level in the Puget Trough. The species requires lakes, ponds, and slow-moving streams with abundant, shallow-water emergent vegetation for concealment. Wetlands with 50–75% open water are preferred. Adults spend most of the summer in these shallow warm-water wetlands, hidden within the emergent vegetation, but they will move back and forth along smaller ephemeral or permanent streams during the wet season. Permanent wetlands must be available that do not dry completely during the dry season.

AQUATIC AND TERRESTRIAL ECOLOGY

Rana pretiosa is an extremely aquatic species that rarely ventures overland away from its home wetland. Adults spend most of their time in the water between the shoreline and the deeper portions of the ponds. Female home ranges are approximately 1.3–5.0 ha, and females move locally at the rate of 6.6–6.8 m/day (Watson et al., 2003); little is known of male home ranges or movements. Long-distance movements (32–111 m/day) are infrequent and occur between distant pools, especially between overwintering and breeding sites. Movements between breeding sites and overwintering sites occur along seasonally saturated or flooded channels. Home range during the breeding season (February to May in Washington) occupies about half the extent of habitat used throughout the rest of the year (Watson et al., 2003).

Movement patterns may differ among locations. Individual frogs move frequently between pools or streams in close proximity to one another, especially during wet weather in Washington (September to January) (Watson et al., 2003). At this time, frogs occupy both ephemeral and permanent streams. During dry weather in Washington (June to August), frogs generally do not move very much and remain at deeper permanent wetlands. Most frogs tend to remain within a particular area, but Watson et al. (2003) found frogs up to 260 m from their point of original capture. In contrast, there was little evidence of among-pool movements in summer at a relocation site in Oregon (Chelgren et al., 2008); it is not known whether relocation curtailed movements at the relatively isolated relocation site. Differences in movement between locations may reflect differences in activity seasons, precipitation patterns, and familiarity with the habitat.

Seasonal movements occur between breeding and overwintering sites. During the course of movements to overwintering sites, frogs may shelter along creeks in willow root complexes or in beaver runs (Shovlain, 2006). Most activity occurs from late winter (February to March) through the autumn (October). Emergence in spring begins as much as a month prior to actual oviposition (Licht, 1969c), and frogs are sometimes seen under ice in shallow water (Leonard et al., 1997a). Frogs spend the winter in shallow water that does not freeze solid, such as springs, flowing water channels, outflow channels, or in deep (1.3 m) open water associated with springs; they bury themselves in the substrate at the base of dense vegetation, and may be active even under 5 cm of ice (Shovlain, 2006). Overwintering frogs often concentrate in spatially limited protected sites such as springs and thus are vulnerable to winter-active predators. By overwintering at the breeding site, they are ready to breed as soon as ice melts from the pond despite the considerable amount of snow that might remain in terrestrial habitats (Leonard et al., 1996).

Feeding occurs either from the water or shoreline. Oregon Spotted Frogs sit with just their eyes exposed and wait for a prey item to come within reach. Frogs also will stalk prey while submerged, or they will crawl semisubmerged through the surface vegetation to put themselves in position to lunge at a prey individual (Pearl et al., 2005c). They also have been observed to stalk toads terrestrially (*Anaxyrus boreas*) (Pearl and Hayes, 2002). Most feeding occurs in water, but they will forage on land through the wet vegetation during or immediately after rainfall. According to Licht (1986b), food is only swallowed in the water, although frogs in captivity readily swallow prey out of water. A number of authors have remarked at how unwary and easy these frogs are to catch.

CALLING ACTIVITY AND MATE SELECTION

Calling normally occurs from late winter into early spring, although advertisement calls have been reported in September and early October (Leonard et al., 1997b). Calls are first heard when air temperatures reach 12°C in direct sunlight (Licht, 1969c). Nocturnal underwater calling may begin several weeks prior to the formation of breeding aggregations, at distances of 80–150 m from where breeding will eventually take place (Bowerman, 2010). These calls are undetectable at the surface and may be heard only 1–4 m away when the calling male is hidden in submerged vegetation. The audible advertisement call of this species is low in volume and has been likened to the "distant tapping of a woodpecker" (McAllister and Leonard, 1997); it consists of 5–50 short, unpulsed, low-pitched hollow notes, with each note lasting about 0.3 sec. Call intervals are 0.11–0.14 sec at 11.4°C with a dominant frequency of 0.5–1.5 kHz. Most calls consist of <12 notes, but as a male approaches another male, the number of notes increases (Licht, 1969c). Call intensity also increases with the number of notes. Calls do not project very far to a human listener, only to about 30 m at most.

Calling occurs both day and night, usually at locations adjacent to oviposition sites, although an amplexed pair may travel 20 m to the communal site. There may be latitudinal variation in temporal aspects of calling, depending on temperature. In British Columbia, for example, most calling occurs during daylight because of the near-freezing temperatures at night. Once breeding has begun, the effective breeding season lasts 11 to 15 days (Licht, 1969c).

Males call in groups, often in close proximity to one another. Calling occurs as the male floats in shallow water or while sitting in mats of vegetation. While calling, they face toward the shore, presumably in anticipation of approaching females. Males are attracted toward any nearby motion and will readily amplex other calling males. These encounters are brief, however. Bouts of amplexus are usually observed early in the breeding season and taper off as the season progresses.

Females move to breeding ponds and remain within the calling distance of males (to ca. 23 m) until ready to oviposit. When she is ready, she then moves into the shallow pool toward the calling males. A male then amplexes the female behind the forelimbs in axillary amplexus. Unreceptive females have a release call and use abdominal vibrations to discourage males. Females choose the oviposition site and frequently deposit their eggs on top of other egg masses (Licht, 1969c). Males also may cluster around the communal oviposition sites waiting for approaching females. Fertilization is external as the egg mass is extruded from her cloaca. Females depart breeding ponds and move immediately toward nonbreeding habitats. Males remain at breeding ponds for several weeks until all oviposition has ceased.

BREEDING SITES

Preferred breeding sites are ponds, lakes, and marshes, with shallow water margins and gradually sloping substrates supporting moderate to dense herbaceous vegetation (Pearl et al., 2009b). Oregon Spotted Frogs also deposit eggs in slow-moving streams that run through flooded marshes. Oviposition may occur in areas of sparse vegetation but at much less frequency. The dominant vegetation usually consists of sedges (*Carex*), rushes (*Juncus*), and grasses. The presence of cover is important in breeding site selection, as the extent of emergent and submergent vegetation increases, the number of egg masses oviposited at a location also increases. Examples of breeding sites include oxbows, beaver ponds, run-off channels, and even anthropogenic wetlands (Pearl et al., 2009b). Breeding sites should be in close proximity to overwintering sites that are free from nonnative predaceous fishes. Most occupied breeding sites are in close proximity to other breeding sites used by the species.

REPRODUCTION

Rana pretiosa is an explosive breeder that begins calling soon after breeding sites thaw in early spring. This usually occurs in mid-February in Washington (McAllister and Leonard, 1997) and in early March in British Columbia (Licht, 1974). At higher elevations (> 560 m), however, breeding does not commence until late March and extends into early June. Oviposition usually occurs within one to two weeks following the initiation of breeding activity. Females breed only once per breeding season, and a single egg mass is deposited annually (Licht, 1974). Clutch size has been reported as 643 (range 249 to 935) in British Columbia (Licht, 1974) and 598 in Washington (McAllister and Leonard, 1997). Dickerson's (1906) report of 1,500 does not appear accurate. Livezey and Wright (1947) reported a clutch size of 1,100–1,500, but they may have been repeating Dickerson's earlier figure.

Eggs are laid communally in shallow (5–30 cm) water, usually near the shoreline, with most ponds containing a single oviposition site; Licht (1971) reported from 5 to 26 masses in communal deposition sites. Some breeding sites may have multiple oviposition sites, however, and breeding sites may be shared with *R. aurora*. Eggs are deposited at night or in daylight, usually in mid-afternoon, at air temperatures of 15–16°C (Licht, 1969c, 1971; McAllister and Leonard, 1997). Pearl et al. (2009b) recorded a mean of 9.7 masses per oviposition site (range 1 to 255 masses), with most oviposition sites containing <20 masses. Breeding ponds usually contain a mean of 10–12 masses, but in the Deschutes River basin, the mean was 78 (range 1 to 576) masses per pond. In Washington, McAllister and Leonard (1997) reported from 10 to 75 masses per site. Eggs deposited in shallow-water pools are prone to desiccation and freezing temperatures, especially since the communal oviposition sometimes leaves egg masses exposed above water. *Rana pretiosa* eggs do not survive at temperatures <7°C (Licht, 1971).

Egg masses are initially small and compact but quickly swell in volume. The species has several adaptations to raise developmental temperatures: the jelly coats retain heat after sundown for about 1.5 hrs, and the temperature within egg masses is greater than in the surrounding water during daylight hours. Hatching occurs in from 14–30 days, depending on water temperature, and tadpoles are 12 mm at hatching (Licht, 1975). Licht (1971)

Egg mass of *Rana pretiosa*. Photo: Betsy Scott

Tadpole of *Rana pretiosa*. Photo: William Leonard

reported a low temperature of 4°C in an egg mass and a high temperature of 27.5°C.

LARVAL ECOLOGY

The larvae of *R. pretiosa* usually develop in the warmer parts of the breeding ponds. They tolerate temperatures of from 6 to 28°C, and they become more thermally tolerant the longer development proceeds (Licht, 1971), much as in *R. cascadae*. Unlike *R. aurora* and *R. cascadae*, *R. pretiosa* larvae do not display any evidence of kin recognition, although they may associate with conspecifics to some degree (O'Hara and Blaustein, 1988). Metamorphosis occurs from 13 to 16 weeks following hatching, and tadpoles can reach a maximum TL of 90 mm (Licht, 1975). Larvae do not overwinter. At metamorphosis, froglets are 30–33 mm SUL (Licht, 1975, 1986b).

DIET

Tadpoles are grazing herbivores (pond and river algae, detritus), although they will feed on carrion. Postmetamorphs dine on invertebrates and consume many different beetles, spiders, syrphid flies, long-legged flies, ants, and water striders. Adult *R. pretiosa* feed mostly as they float in the water or cling to aquatic vegetation and, as a result, most prey are aquatic species. On wet days, however, the frogs will move to land and feed near the shoreline. They do not lunge at prey, but orient and move slowly toward it. If in water, they may look toward prey on land, but they do not leave the water in order to attempt to feed. Prey includes spiders, various hemipterans (aphids, spittlebugs, leafhoppers), beetles, long-legged flies, mosquitoes, caterpillars, other types of flies and fly larvae, wasps, ants, crayfish, snails, and many other small invertebrates (Schonberger, 1945). *Rana pretiosa* also eat tadpoles and newly metamorphosed *Anaxyrus boreas*, *Rana aurora*, and *R. pretiosa* and adult *Pseudacris regilla* (Licht, 1986b; Pearl and Hayes, 2002).

PREDATION AND DEFENSE

Eggs are eaten by various invertebrates (including the leech *Batrachobdella picta*) and salamanders (*Ambystoma*, *Taricha*). Larvae are eaten by both invertebrate and vertebrate predators, including dytiscid beetles (*Dytiscus*), backswimmers (*Notonecta*), dragonfly nymphs (Odonata), giant water bugs (*Lethocerus*), water scorpions (*Ranatra*), leeches, birds (kingfishers, Hooded Mergansers), snakes (*Thamnophis elegans*, *T. sirtalis*), larval and adult salamanders (*Ambystoma gracile*, *Taricha granulosa*), and fish (*Gasterosteus*, *Novumbra*, *Oncorhynchus*, *Salmo*). Postmetamorphs are eaten by American Bullfrogs, garter snakes (*Thamnophis*), birds (herons, kingfishers, Greater Sandhill Cranes), and mammals (raccoons, striped skunks, mink, river otters, feral domestic cats, coyotes) (Licht, 1974, 1986a; Reaser and Dexter, 1996; McAllister and Leonard, 1997; Hayes et al., 2006a; Chelgren et al., 2008).

The Oregon Spotted Frog usually will remain motionless as a predator approaches, at least to about 20 cm. *Rana pretiosa* in the water will dive into the substrate when disturbed, or they will slowly sink beneath the water's surface at the approach of a potential predator. On land, they hop rapidly into the water and seek refuge on the bottom among dense vegetation and soft mud (Licht, 1986a). Frogs may remain submerged up to 17 min, and come to the water's surface with only their eyes above water. Sudden movements cause them to resubmerge. If water is not readily available, they will jump in a circle using short weak jumps.

POPULATION BIOLOGY

Rana pretiosa attains sexual maturity at from 2 yrs (males, a few females) to 3 yrs of age (females); the smallest reported sexually mature male was 32 mm SUL and found in Washington (McAllister and Leonard, 1997), although Licht (1975) estimated sexual maturity was not reached until 45 mm SUL for males and 62 mm SUL for females. In British Columbia, males are 38–58 mm SUL after 1 full yr of growth and 58–64 mm SUL after 2 full yrs of growth; females are 38–62 mm SUL after 1 full yr of growth and 62–77 mm SUL after 2 full yrs of growth (Licht, 1975).

The best population data on this species are those from Licht's (1974) study in British Columbia. Assuming his results are typical of other populations, mortality of hatchlings and tadpoles is high, but much lower in postmetamorphs. Embryonic survival was about 70%. About 5.2% of the oviposited eggs and about 7.3% of the tadpoles survived to metamorphosis, depending upon environmental conditions, and only 0.8% survived from the egg stage through first year following metamorphosis. However, 67% of juveniles survived the first year, and thereafter the annual postmetamorphic survival rate was about 64%. Males had poorer survival rates (45%) than females (68%), but the rates were calculated based on only a 2 yr study.

Sex ratios at breeding ponds are usually male-skewed as they are at many frog breeding sites. Of 760 frogs captured randomly over an extended time, however, McAllister et al. (2004) reported an essentially 1:1 sex ratio. Capture probabilities differ between the sexes seasonally, with male capture probabilities higher than females in spring, but female capture probabilities higher than males during the summer (Chelgren et al., 2008). The differences probably reflect differences in activity patterns, with males moving to breeding ponds en masse in spring and females conspicuously feeding in summer in order to build fat reserves to yolk an egg clutch the following year.

COMMUNITY ECOLOGY

Newly transformed *R. aurora* and *R. pretiosa* overlap substantially in diet, but the different foraging modes of the larger juveniles and adults precludes competition and overlap in the types of prey eaten (Licht, 1986b).

DISEASES, PARASITES, AND MALFORMATIONS

Amphibian chytrid fungus (*Batrachochytrium dendrobatidis*) has been found infecting *Rana pretiosa* in Washington (Hayes et al., 2009; Padgett-Flohr and Hayes, 2011) and Oregon (Adams et al., 2010). However, this species has a diverse array of antimicrobial peptides in its skin, which may assist in developing resistance to the fungus (Conlon et al., 2011; Padgett-Flohr and Hayes, 2011). The leech *Batrachobdella picta* was found on 75% of *Rana pretiosa* during Licht's (1974) study in British Columbia, with frogs having as many as 20 leeches. Lucker (1931) reported the trematode *Haplometrana intestinalis* from *Rana pretiosa* from Bothell, Washington; *R. pretiosa* no longer occurs in this area. This species also is parasitized by the trematode *Gorgoderina multilobata* and the nematode *Falcaustra pretiosa*

Adult *Rana pretiosa*. Photo: David Dennis

Habitat of *Rana pretiosa*, Conboy Lake, Washington. Photo: Marc Hayes

(Ingles, 1936). The trematode *Ribeiroia ondatrae* has been reported from this species in the Pacific Northwest, with 88 metacercariae per frog (Johnson et al., 2002).

SUSCEPTIBILITY TO POTENTIAL STRESSORS

Nitrate and nitrite. *Rana pretiosa* larvae are particularly sensitive to nitrates and nitrites in experimental trials. At high nitrate and nitrite concentrations, larvae reduce feeding, swim less vigorously, exhibit disequilibrium, develop abnormalities (edema, bent tails) and paralysis, and often die (Marco et al., 1999). For nitrate, the $LC_{50\,(15\,day)}$ is 16.45 mg $N-NO_3^-/L$. For nitrite, the effects increase with increasing concentration, and the $LC_{50\,(15\,day)}$ is 0.57 mg $N-NO_2^-/L$. *Rana pretiosa* is 7 times more sensitive to nitrite than *R. aurora* and 20 times more sensitive than *P. regilla*. As a comparison, the EPA standard for nitrite in drinking water is 1.0 mg $N-NO_2^-/L$. The EPA standard of 90 mg $N-NO_3^-/L$ also is highly toxic to larval *R. pretiosa*.

Chemicals. Death occurs in about 30 hrs at 30 mg/L of carbaryl, a broad-spectrum insecticide, and significantly reduces larval activity levels at 2.5 mg/L (Bridges and Semlitsch, 2000).

UV light. Oregon Spotted Frogs have high levels of photolyase, an enzyme important in DNA repair from UVB radiation. Spotted frogs reared in ambient UVB and in situations where UVB is shielded have similar levels of hatching success and embryo development (Blaustein et al., 1999).

STATUS AND CONSERVATION

A number of authors have expressed concern for the status of *R. pretiosa* or recorded population disappearance, and populations appear to have begun their decline >30 years ago (Dumas, 1966; Nussbaum et al., 1983; McAllister et al., 1993; Weller and Green, 1997; Adams et al., 1998; Corn, 2000). The Oregon Spotted Frog has disappeared from >70% of historically occupied sites in the Pacific Northwest. As late as 1997, the species was thought to occur at only 11 historic sites in Washington, 20 extant sites in Oregon, and 1 site in British Columbia (McAllister and Leonard, 1997; Weller and Green, 1997). Fortunately, a few more populations have been discovered in Oregon and British Columbia, but the status of the species remains precarious. Threats include habitat loss and modification (e.g., through grazing), disease, and the introduction of nonnative predators. Dredging and straightening stream channels and the alteration of shallow wetlands have been particularly harmful to this species.

Nonindigenous American Bullfrogs have been implicated in the decline of many Pacific Northwest ranid frogs and occasionally occupy breeding ponds used by *R. pretiosa*. Pearl et al. (2005a)

recorded 18 interspecific cases of amplexus, including both free-ranging examples and animals caught in traps. Male *R. pretiosa* clasped both female and juvenile *Lithobates catesbeianus*, and these pairings frequently left the *L. catesbeianus* dead. Pearl et al. (2005a) hypothesized that prolonged amplexus by the explosive-breeding *Rana pretiosa* could adversely affect spotted frog reproduction by occupying males in futile reproductive attempts.

Declines in two of the three remaining Washington *R. pretiosa* populations are correlated with finding dead or dying individuals infected with amphibian chytrid fungus (Hayes et al., 2009). Chamberlain (1897) stated that *R. pretiosa* (including *R. luteiventris*) was commonly used for food, but there are no statistics on the number of frogs harvested.

Habitat alteration may adversely affect this species, and Shovlain (2006) reported that as grazing pressure increased, frogs preferred ungrazed livestock exclosures. When grazing pressure was low, however, frogs tended to use grazed and ungrazed sections of land in equal proportions. It is likely that the effects of heavy grazing would take a long time to mitigate, as grazing has many cumulative effects (e.g., vegetation structure alteration and the presence of nitrogenous waste and fertilizer).

Rana pretiosa have been successfully established in at least one constructed pond in Oregon (Chelgren et al., 2008). Large frogs had a better rate of survival than small frogs (7–20%), and survivorship was lower for relocated frogs than for frogs that developed from relocated egg masses (50–70%). The greatest mortality occurred within a short time of relocation and during the summer (especially for males); lowest mortality occurred during the winter. Females had higher survival rates than males, and survivorship was lower during the first year following relocation than during subsequent years. Relocated frogs showed no evidence of homing.

Rana sierrae Camp, 1917
Sierra Nevada Yellow-legged Frog

ETYMOLOGY
sierrae: Latin for 'mountain.' The name refers to the Sierra Nevada Mountains where this species is found.

NOMENCLATURE
Synonyms: *Rana boylii sierrae*

Much of the information on the biology of this species has been published using the names *Rana muscosa* or *R. boylii*. This is particularly evident in Zweifel (1955), for example, in which the accounts of morphology and biology include both species under the name *R. muscosa*. Readers should check the locations in the primary literature to verify which species is being discussed.

IDENTIFICATION
Adults. This species is similar in appearance to *R. muscosa*. Dorsal color patterns are varied, but the body usually is dark yellowish brown with a reticulated moss-like dorsal pattern. There is no faint pigmentation patch between the eyes. The undersides are yellow with faint dusky spots on the throat and chest. The upper lip is mottled below the eye and does not have a white stripe. The dorsolateral fold is not distinct and is not pitted anteriorly, and the tympanic area is usually smooth. The hind limbs are not distinctly barred, and the toe tips are dark and not expanded. The rear feet are fully webbed. There are no vocal sacs. In general, *R. sierrae* have shorter limbs in relation to body size and narrower heads compared with the closely related species *R. muscosa*.

Larvae. Larvae are brown with a golden tint, resulting from numerous yellow-gold chromatophores. There may or may not be aggregations of dark spots on the tail musculature and tail fin, and the area bordering the tail musculature on the tail fin is clear. The tail fin of the tadpole of *R. sierrae* is about the same height throughout the tail, rather than peaking midway along the tail as it does in *R. boylii*. The venter is faintly yellow and not silvery. The viscera are faintly visible.

Eggs. Eggs are grayish tan to black dorsally and light gray to tan to creamy white ventrally. There are three distinct envelopes surrounding the vitellus. The outer envelope is 6.4–7.9 mm; the middle envelope is 4.3–5.0 mm in diameter; the inner envelope is 3.8–4.8 mm in diameter; the vitellus is 1.8–2.3 mm (Livezey and Wright, 1947). Eggs are

Distribution of *Rana sierrae*

deposited in a solid jelly mass 2.8 cm × 4.0 cm; the eggs are 7–15 mm apart from one another. Algae may grow on the eggs giving a greenish tint.

DISTRIBUTION
This species is found in the higher elevations in the northern and central Sierra Nevada of California and adjacent Nevada. The historic range extended from Lake Almanor, Plumas County, California, in the north to Matlock Lake, Inyo County, in the south. The species occurred east of the Sierras, northeast of Lake Tahoe in Nevada, and in the Glass Mountains south to Matlock Lake. Populations may not be continuous, and both the northernmost and southernmost populations may be disjunct from the main extent of the species' range. Many historical populations may no longer exist. For example, a population once known from Mt. Rose (Washoe County), Nevada, is extinct. There is a contact zone with *R. muscosa* between the Middle and South Forks of the Kings River in a line stretching from Mather Pass to Monarch Divide. Important distributional references include Storer (1925), Slevin (1928), and Jennings and Hayes (1994b, in part).

FOSSIL RECORD
No fossils are recognized.

SYSTEMATICS AND GEOGRAPHIC VARIATION
Rana sierrae originally was described as a subspecies of *R. boylii* by Camp (1917) from the Sierra Nevada Range of California and Nevada. The name *Rana boylii muscosa* was allocated to the yellow-legged frogs of the Transverse Ranges of southern California. Zweifel (1955) used morphological characters to elevate *R. muscosa* to species status, and until recently, all yellow-legged frogs were considered to be *R. muscosa*. Zweifel's (1955) separation of *R. muscosa* from *R. boylii* was supported by Houser and Sutton (1969) based on morphological and karyological data. The diploid chromosome number is 26 (Haertel et al., 1974).

Analyzing 1,901 base pairs of mitochondrial DNA, acoustic data, and morphological characters, Vredenburg et al. (2007) recognized 2 distinct evolutionary clades within *R. muscosa* that they then differentiated into *R. sierrae* from the northern and central Sierra Nevada and *R. muscosa* from the southern Sierra Nevada and Transverse Ranges. This species is a member of the *R. muscosa* complex of the *Amerana* clade (sometimes termed the *Rana boylii* species group; Macey et al., 2001) of North American ranid frogs. *Rana sierrae* is most closely allied phylogenetically with *R. muscosa* and somewhat more distantly with *R. aurora* and *R. cascadae* (Macey et al., 2001; Hillis and Wilcox, 2005; Vredenburg et al., 2007). Levels of genetic divergence largely result from Pleistocene vicariance events due to climate factors; the 2 major clades leading to *R. muscosa* and *R. sierrae* diverged ca. 2.2 mya, whereas subsequent divergence within the 2 clades occurred 1.5 mya (Macey et al., 2001).

Camp (1917) noted some regional variation in color patterns, extent of blotching, and morphology. For example, *R. sierrae* from Whitney Meadows (Tulare County) possessed well-developed dorsolateral folds. Laboratory hybridization experiments suggest general incompatibility with *R. boylii* (Zweifel, 1955).

ADULT HABITAT
Rana sierrae is found along high-elevation lakes and slow-moving streams in the Sierra Nevada

Mountains of California and Nevada. Streams should have gentle slopes of shallow water rather than slopes that plunge directly into deep pools. The frog does not prefer small shallow streams, those with steep gradients, or lakes with an open shoreline. The presence of Sierra Nevada Yellow-legged Frogs is correlated with a significant amount of vegetation along the shoreline of ponds,lakes, and streams and with the nearby presence of mountain meadows. They may be common in rocky lake outlets that feed alpine streams.

This species occurs most frequently at elevations of approximately 2,500–3,500 m in fishless waters that are > 3 m in depth. The lower limits of elevation are 1,370 m and the upper limits are 3,660 m (Zweifel, 1955; Mullally and Cunningham, 1956b; Vredenburg et al., 2005). At elevations > 3,500 m, occupancy drops off rapidly. The presence of Sierra Nevada Yellow-legged Frogs is correlated with occupancy of nearby ponds; the more isolated a pond is in relation to other ponds, the less likely it will be occupied by this species.

AQUATIC AND TERRESTRIAL ECOLOGY

Rana sierrae is an aquatic species that rarely ventures more than a few meters from water. Most activity is diurnal, especially at high elevations. Some animals may be crepuscular, but activity greatly declines after dark. Activity occurs only during the warmest time of the year, that is, from early June through August, although recently transformed individuals may be found as late as October at the lowest elevations. *Rana sierrae* control their body temperature through behavioral means, and readily bask during cool weather. Adult body temperatures usually range between 15 and 25.6°C, although they have been observed in water 2°C (Mullally and Cunningham, 1956b). The CTmax is between 33 and 37°C.

At high elevations, *R. sierrae* may experience below freezing temperatures at almost any time of the year. When ice forms along the shoreline in summer, the frogs may rest upon it and use it as a basking platform until the sun melts it. Sierra Yellow-legged Frogs may not visually perceive the difference between ice and water. When disturbed, the frogs do not avoid the ice but jump on it as if they were expecting water (Mullally and Cunningham, 1956b). They go sliding off the edge of the ice, but quickly resurface and resume basking. Insolation along the shoreline quickly raises ground temperatures to warm the frogs. When substrate temperatures become too warm, they return to the water or seek shade. Information on movement patterns and dispersal is lacking.

CALLING ACTIVITY AND MATE SELECTION

The calls of *R. muscosa* and *R. sierrae* have different properties and are easily distinguished using oscillogram and spectrogram analyses (Vredenburg et al., 2007). *Rana sierrae*, for example, has transitions between pulsed and noted sounds whereas *R. muscosa* does not.

BREEDING SITES

Eggs are deposited along streams and in higher lakes and meadow ponds. For example, Zweifel (1955) reported instances of eggs being affixed to an undercut bank along a stream in a meadow or attached to sedges at the edge of a pool. In streams, eggs may be attached to rocks on the stream bottom. Eggs also may be oviposited in lentic pools in close proximity to lakes, streams, or meadow wetlands.

REPRODUCTION

Clutch size ranges from 100 to 350 to 1,400 eggs per mass according to Livezey and Wright (1947), although there may be some confusion about species identification in this record. Zweifel (1955) counted a single egg clutch from Calaveras

Egg masses of *Rana sierrae*. Photo: Roland Knapp

Adult *Rana sierrae* with large *Rana sierrae* tadpole. Photo: Ceal Klingler

County, California, of 800 eggs. Water temperatures in cold alpine lakes and meadow ponds vary between 8.5 and 21°C, but mostly at the lower temperatures. Night temperatures may approach freezing, however, so eggs must be able to tolerate a wide range of thermal environments.

LARVAL ECOLOGY

At high elevations, development cannot be completed without overwintering one or more winters prior to metamorphosis, as the larval growing season is only about four months in summer. Larvae of *R. sierrae* overwinter in deep water (> 3 m) lakes and ponds in the high mountains of the Sierra Nevada (Knapp et al., 2003). A great depth allows them to avoid being completely frozen in ice in winter or stranded in drying ponds in summer, as frequently occurs in shallow water bodies. Deep cold water holds oxygen longer and offers the potential to avoid anoxic conditions that occur during the very long winters at high elevations. Larvae prefer cold water, and they are not found in water >27°C (Mullally and Cunningham, 1956b). Still, they may congregate in the warmer reaches of the pond during the day. At lower elevations, transformation occurs during the same season as oviposition, and larvae do not overwinter; at intermediate elevations, it appears that some frogs may overwinter whereas others do not (Zweifel, 1955).

DIET

Very little information is available. Larvae are benthic herbivores. Adults have been reported to feed on the larvae of *Anaxyrus canorus*, unspecified small invertebrates, and dragonfly nymphs (Mullally, 1953). They probably eat any invertebrate they can put into their mouth.

PREDATION AND DEFENSE

This species is extremely wary on land and readily jumps to water at the approach of an observer. They swim to the substrate and bury themselves in mud and vegetative debris or hide under rocks. Mullally and Cunningham (1956b) noted one individual that took refuge in a rodent burrow. In the water, frogs are much less wary, and they easily blend in with vegetation and the substrate. When attacked, this species may scream or puff up its body size to distract or thwart potential predators (Camp, 1917). There is very little information on predators, but they likely include a number of species of birds (crows) and mammals. Coyotes have been reported to feed on its tadpoles (Moore, 1929), and garter snakes (*Thamnophis elegans*) eat postmetamorphs (Mullally and Cunningham, 1956b).

POPULATION BIOLOGY

Nothing is known concerning the population biology of this species. At high elevations, it likely has similar population and demographic characteristics as other high-elevation ranids such as *Rana cascadae*.

COMMUNITY ECOLOGY

The introduction of nonnative trout species throughout the western states has had devastating effects on a number of amphibian species, including the Sierra Nevada Yellow-legged Frog. For example, the distribution of *R. sierrae* was strongly negatively correlated with the presence of trout in Yosemite National Park, based on a survey of 2,655 ponds and marshes; the frog was only found in 282 wetlands (Knapp, 2005). Although adult trout do not eat adult frogs, the trout consume a great amount of insects and insect larvae that would otherwise be available to frogs. Trout also consume *R. sierrae* larvae. Thus, trout adversely affect frogs through trophic interactions as well as predation; in essence, they disrupt natural food webs (Finlay and Vredenburg, 2007).

DISEASES, PARASITES, AND MALFORMATIONS

The virulent pathogen *Batrachochytrium dendrobatidis* was reported from both tadpoles and recent metamorphs based on collections from 1998 to 2000 throughout the Sierra Nevada (Fellers et al., 2001), and it probably acts on *Rana sierrae* populations much as it does on *R. muscosa* populations (see *Rana muscosa* species account). Bradford et al. (2011) suggested that amphibian chytridiomycosis is likely the leading cause of population declines in this species. Larvae typically have abnormalities of the oral disk, including missing keratinized mouthparts, depigmented jaw sheaths, and swollen labial papillae. Tadpoles with abnormalities were found at 70% of the sites and from 4.2 to 100% of the tadpoles examined per site. Sierra Nevada Yellow-legged Frogs in Yosemite National Park were infected at a high percentage (67.5%) with the pathogen during surveys in 2005. Secretions from these frogs possess peptide mixtures that inhibit the growth of the pathogen (Woodhams et al., 2007). The peptides are produced by a variety of symbiotic bacteria and may be important in assisting disease resistance and individual survival.

SUSCEPTIBILITY TO POTENTIAL STRESSORS

pH. Increased levels of acidity have been mentioned as a possible cause of the decline of this species throughout much of its range. The LC_{50} for embryos is a pH of 4.37 and for hatchlings <4.0 (Bradford et al., 1992). However, Bradford et al. (1994a) could find no evidence in support of the acidity hypothesis as a cause of decline in *R. sierrae*.

STATUS AND CONSERVATION

The Sierra Nevada Yellow-legged Frog has experienced severe population declines throughout its

Adult *Rana sierrae*. Photo: Dana Drake

Habitat of *Rana sierrae*.
Photo: Roland Knapp

range. Declines or disappearances have occurred even in protected areas such as Yosemite National Park (Bradford et al., 1994b). Whereas a variety of factors may have contributed to the declines, the introduction of nonnative predatory fishes (various trout), the spread of amphibian chytridiomycosis, and the extensive use of wind-borne agricultural pesticides in the San Joaquin Valley are the primary factors resulting in population extirpation (Knapp and Matthews, 2000; Knapp et al., 2001; Davidson and Knapp, 2007; Bradford et al., 2011).

The use of pesticides is likely the primary reason for the near extinction of this species. Fish are particularly damaging at the local or site level, whereas pesticide use is the primary agent at the landscape level causing population loss. In the extensive survey by Davidson and Knapp (2007), *R. sierrae* was present at only 13% of 6,831 water bodies sampled within its historic range. The distribution of *R. sierrae* generally does not coincide with the presence of predaceous introduced fish, although small numbers of frogs may be found in protected habitats within water bodies where fish occur. *Rana sierrae* recolonizes lakes where fish have been removed, offering hope for population recovery at least in some areas (Knapp et al., 2001).

One important factor that could contribute to declines in this species is changes in climate, particularly rainfall patterns. Recruitment is much greater in ponds that are permanent than it is in ponds that dry even once over a period of 10 yrs (Lacan et al., 2008). Recruitment is also greater in lakes that retain water during any preceding 2 yrs than in lakes that dried over the same period. Thus, changes in precipitation patterns that increased the severity and duration of periodic droughts could cause the species to decline further, especially since the larger lakes may be inhabited by competitive nonnative trout and thus be unfit as breeding sites.

Family Scaphiopodidae

Scaphiopus couchii Baird, 1854
Couch's Spadefoot

ETYMOLOGY
couchii: a patronym honoring Darius N. Couch (1822–1897), a U.S. Army officer who collected vast numbers of natural history specimens in Mexico at his own expense while on leave.

NOMENCLATURE
Synonyms: *Scaphiopus couchii rectifrenis*, *Scaphiopus laticeps*, *Scaphiopus rectifrenis*, *Scaphiopus varius*, *Spea laticeps*

IDENTIFICATION
Adults. Coloration is highly variable from reticulated green and black markings to mottled shades of black, green, and yellow, or brown to nearly uniform green with black spots. There also may be sexual dichromism in color pattern, with males a uniform light green and females a darker yellowish green. Hind limbs are short with rear feet having a black keratinized spade for digging. Venters are cream to dirty white. Eyes are large with a characteristic vertical pupil, and there is no boss between the eyes. Males have a large vocal sac that when distended is nearly three times the size of the head. Additional descriptions are in Strecker (1908). Adults are not sexually size-dimorphic. The mean SUL of adult males in Arizona was 69.2–74.4 mm SUL (Sullivan and Sullivan, 1985); Jennings and Hayes (1994b) gave an adult size range of 45–82 mm SUL in California. The maximum size is 90 mm SUL (Degenhardt et al., 1996).

Larvae. The tadpoles are small and blackish brown, and are lighter ventrally and on the tail than on the body. A profusion of coppery-bronze flecks may provide an iridescent sheen. Larvae have an overall appearance of being short and plump, with the widest part of the body toward the posterior when viewed dorsally. Tails are slender and sometimes blotched or have a dark spotted pattern. Jaws are narrow to medium in width and never cuspate. Larvae reach a maximum size of 25 mm TL. Descriptions are in Wright (1929) and Altig (1970).

Eggs. The eggs are black dorsally and white ventrally with a colorless jelly mass surrounding them. There is one thick gelatinous capsule surrounding the vitellus measuring 2.4–3.5 mm in diameter (Ortenburger, 1924; Grubb, 1972). The vitellus is 1.4–1.6 mm in diameter (Livezey and Wright, 1947).

DISTRIBUTION
Couch's Spadefoot occurs from western Baja California and the Mojave Desert of California eastward across southern Arizona, New Mexico, and Texas to southwest Oklahoma. The range includes much of central Texas west of the Balcones Escarpment (Smith and Buechner, 1947) and extends south well into central Mexico. Isolated populations occur in southeastern Colorado (Petrified Forest National Monument) and east-central Arizona. The species is absent from mountainous regions, such as the Mogollon Rim, southern Rockies in New Mexico, and Sierra Madre in Mexico.

Important distributional references include: Arizona (Brennan and Holycross, 2006), Baja California (Grismer, 2002), California (Tinkham, 1962; Vitt and Ohmart, 1978; Jennings and Hayes, 1994b), Coahuila (Lemos-Espinal and Smith, 2007b), Colorado (Hammerson, 1999), New Mexico (Van Denburgh, 1924; Degenhardt et al., 1996), Oklahoma (Sievert and Sievert, 2006), and Texas (Dixon, 2000).

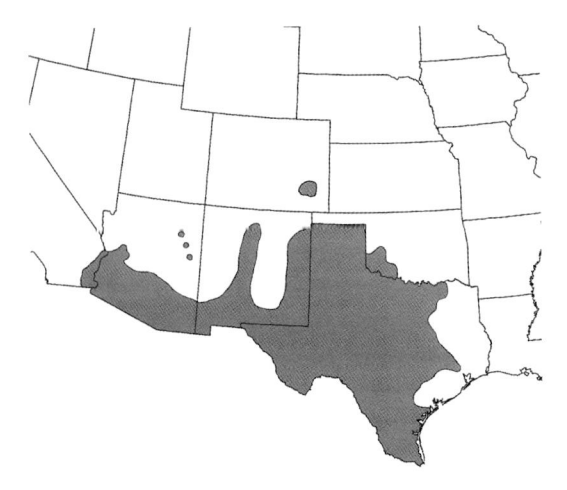

Distribution of *Scaphiopus couchii*

FOSSIL RECORD

Pleistocene fossils of Couch's Spadefoot are known from Arizona, New Mexico, Sonora, and Texas (Holman, 2003). The species is readily separated from other fossil scaphiopodids by a variety osteological characters (see Holman, 2003), and is closely related to the fossil spadefoot *Scaphiopus alexanderi*.

SYSTEMATICS AND GEOGRAPHIC VARIATION

This species is closely related to *S. holbrookii* and *S. hurterii* (Tanner, 1939; Sattler, 1980; García-París et al., 2003). Spadefoots from north-central Mexico were described as *S. rectifrenis* by Cope and said to be larger and more uniformly reticulated in dorsal pattern than *S. couchii*. However, there is no justification for recognizing specimens from west Texas as belonging to this taxon as recommended by Smith and Sanders (1952). García-París et al. (2003) found substantial mtDNA divergence between populations of *S. couchii* from Arizona and Baja California Sur, indicating that these populations may not be conspecific. A range-wide study of the genetics of *S. couchii* is needed.

Natural hybridization between *S. couchii* and *S. hurterii* has been reported in Texas (Wasserman, 1957). In laboratory crosses, ♀ *S. couchii* × ♂ *S. hurterii* pairings produce a high percentage of hybrids. However, ♀ *S. hurterii* × ♂ *S. couchii* crosses are inviable (Wasserman, 1957).

Stroud (1949) described a very light phenotype that inhabits White Sands National Monument in New Mexico.

ADULT HABITAT

Scaphiopus couchii is an arid land species that occupies short-grass prairies, mesquite savannahs, and creosote-dominated communities (e.g., Jameson and Flury, 1949; Dayton et al., 2004). In the Chihuahuan Desert, it is found on alluvial plains covered by creosote bush (*Larrea tridentata*), ocotillo (*Fouquiera splendens*), mesquite (*Prosopis juliflora*), yucca (*Yucca* sp.), grasses, and cacti. In Texas, it is associated with black calcareous soils in the east and with any friable, well-drained sandy soil toward the western part of the range; the species prefers mixed scrub vegetation (Wasserman, 1957; Dayton et al., 2004). In California and New Mexico, it is associated with dry washes and the edges of sand dunes, such as Algodones Dunes and White Sands, presumably where it can burrow down and be in contact with moist soil (Mayhew, 1965; Degenhardt et al., 1996). It is also associated with irrigated agricultural lands.

Dayton and Fitzgerald (2006) developed a habitat suitability model with high predictability for Couch's Spadefoot in Big Bend National Park, Texas. They found that this species was associated with clay loam soils with good water-holding capacity, temporary pools, and alluvial floodplains. Most populations occurred along the Rio Grande River or in the northern regions of the park at low elevations (< 1,200 m). This species has been found at 900–1,800 m in New Mexico (Strecker, 1908; Degenhardt et al., 1996).

TERRESTRIAL ECOLOGY

Scaphiopus couchii is active throughout the warmer months of the year. For example, they have been observed from May to October in Colorado (Hammerson, 1999). Couch's Spadefoot is mostly nocturnal in its habits, although it may be active in the morning or late afternoon as well as on cloudy, rainy days. Most of the time toads remain hidden within or at the mouth of their burrows, and they can remain so for ten months of the year, even during nondrought years. The burrows are located within close proximity to breeding sites and are frequently under vegetation. In New Mexico, Cruesere and Whitford (1976) found nonbreeding adults mostly in sparse grass habitats 100–300 m from water, and Mayhew (1965) recorded young spadefoots 400 m from the nearest water. Burrows of rodents, such as pocket gophers, also may be occupied. Adults are usually active on the surface only after summer thunderstorms, and they do not appear to be active in winter, regardless of rainfall (Mayhew, 1965). Juveniles may be active on warm summer nights, regardless of precipitation (e.g., Minton, 1958).

Couch's Spadefoots can "disappear" for very long periods, until a violent thunderstorm brings them to the surface in large numbers. Like other spadefoots in the desert, this species emerges in response to low frequency sounds (<100 Hz) that accompany rainfall rather than the moisture per se (Dimmitt and Ruibal, 1980b).

Scaphiopus couchii is physiologically adapted to arid conditions (Mayhew, 1965; McClanahan, 1964, 1967). They can store nitrogenous waste as urea, thus enabling them to conserve water over long periods, and their blood and lymph have high osmotic concentrations. Water is conserved within the bladder, which serves as a storage reservoir. Although size per se is not correlated with tolerance to dehydration, large recently metamorphosed spadefoots can resist dehydration better than small recent metamorphs because of their lower mass-specific rate of water loss (Newman and Dunham, 1994). Couch's Spadefoot can loose a maximum of 44–50% of their body water before lethal limits are reached, and can rehydrate 72% of their lost water within 3 hrs. Water uptake occurs directly through the skin by contact with water or moist soil (Walker and Whitford, 1970). Mayhew (1965) noted that spadefoots retrieved from burrows may have multiple layers of unshed skin that could help reduce water loss. They also are able to store energy as fat, which can be mobilized during dormancy.

Couch's Spadefoots are photopositive in their phototactic response, suggesting they use both sunlight and moonlight when feeding and going about their activities (Jaeger and Hailman, 1973). They are sensitive to light in the blue spectrum ("blue-mode response"), which apparently helps them orient toward areas of increasing illumination, such as the open horizon above ponds (Hailman and Jaeger, 1974). Couch's Spadefoots likely have true color vision.

CALLING ACTIVITY AND MATE SELECTION

Scaphiopus couchii is an explosive breeder, with choruses forming only after torrential rainstorms in spring and summer. In the Sonoran Desert of Arizona, the mean chorus duration was 1.0 day and the mean chorus size was 23 males (Sullivan and Sullivan, 1985; Sullivan, 1985a, 1989). Sullivan and Fernandez (1999) recorded males at 2 Arizona breeding sites for only 19 nights over a 6 yr period. No breeding may take place during years of drought. Males call mostly from the banks of a temporary pool (Bragg and Smith, 1942), although they may call while floating in the water. Males outnumber females at breeding pools. There is no correlation between male body size and mating success (Sullivan and Sullivan, 1985).

The call of *S. couchii* is described by Strecker (1908) as a "loud, resonant yē-ŏw repeated at intervals." It has a minimum frequency range of 140 to 235 cps and a maximum frequency range of 160 to 292 cps, depending on temperature. Calls last from 0.5 to 0.9 sec (Blair, 1955c), and the trill rate is 126/sec (McAlister, 1959). As with other spadefoots, Couch's Spadefoot utters a plaintive cry when handled or confined. McAlister (1959) described the morphology of the vocal structures.

Ortenburger (1924) described a scenario whereby females sit around listening to a male before he suddenly approaches one and amplexes her. However, it is likely males will attempt to amplex any moving frog within his vicinity. Amplexus is inguinal. A few Couch's Spadefoots have been observed in amplexus with *S. bombifrons* (Cruesere and Whitford, 1976).

BREEDING SITES

This species breeds in shallow temporary pools and puddles. It has been recorded in flooded fields, roadside ditches, water-filled quarries, and deep muddy pools (Wright and Wright, 1938; Bragg and Smith, 1942). Breeding pools should have dense grass clumps scattered throughout (Cruesere and Whitford, 1976).

REPRODUCTION

Breeding occurs throughout the warm season only after heavy rainfall and is thus opportunistic. The rain does not have to be of great volume, but it has to occur torrentially. For example, as little as 12.7 mm of rainfall can initiate breeding, but it must fall very rapidly (Bragg, 1945b). In Oklahoma and Texas, breeding has been reported between April and August (Wright and Wright, 1938; Bragg and Smith, 1942; Bragg, 1950a; Black, 1973b).

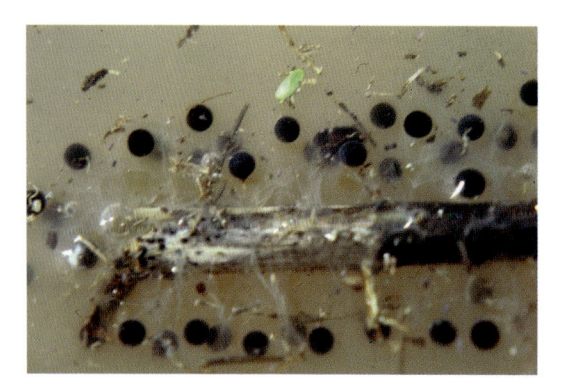

Eggs of *Scaphiopus couchii*. Photo: Cecil Schwalbe

Oviposition requires about 3 hrs (Strecker, 1908), during which eggs are deposited in as many as 8 separate bouts of egg extrusion. Eggs are attached to vegetation or debris in shallow water. Strecker (1908) recorded clutch sizes between 343 and 528 eggs; Woodward (1987a) reported a mean clutch size of 3,310 eggs; Stebbins (1962) gives a clutch size of 300–700 eggs. It appears likely that Strecker was counting eggs during each laying bout. Eggs are deposited in a double or triple row of strings or in a loose irregular cylinder about 6 mm across (Livezey and Wright, 1947). Each string is about 40–65 mm in length and contains from 45 to 125 eggs. Female body weight may or may not be correlated with clutch weight but does appear to be correlated with the number of eggs produced (Woodward, 1987a).

Eggs are attached to grass and other debris in water 75–150 mm in depth (Strecker, 1908), but they readily float on the water's surface. Eggs hatch in as little as 12 hrs to as long as 8–10 days depending upon temperature (Strecker, 1908; Hubbs and Armstrong, 1961; Mayhew, 1965; Ballinger and McKinney, 1966; Zweifel, 1968b; Black, 1973b; Justus et al., 1977). Eggs will not hatch at temperatures <15°C, and eggs at 10°C show no signs of development (Hubbs and Armstrong, 1961; Ballinger and McKinney, 1966). Development will occur at temperatures of 11–15°C, and the percentage of those developing increases with temperature.

LARVAL ECOLOGY

The larval period is extremely short, since spadefoots must complete metamorphosis in temporary pools that may dry rapidly. For example, Morey and Reznick (2004) found that 62% of the breeding pools used by *S. couchii* in California dried before metamorphosis began. The developmental period is not fixed in this species and lasts 7–20 days (Strecker, 1908; Mayhew, 1965; Bragg, 1967; Newman, 1988a, 1989, 1992, 1994; Tinsley and Tocque, 1995; Morey and Reznick, 2004). Tadpole mortality is often high due to desiccation.

Tadpoles form aggregations that are maintained visually (Black, 1973b). Tadpoles developing in ponds with long hydroperiods are able to prolong the larval period in order to take advantage of food resources (Newman, 1989). This is important, since it allows these tadpoles to gain the greater fat reserves that are necessary for post-metamorphic survival. Developmental plasticity in the length of the larval period results from a trade-off between the need to metamorphose rapidly in drying ponds and the need to gain sufficient energy reserves for increased fitness in the terrestrial environment, as indicated by a larger size at metamorphosis. As a result, tadpoles that develop in short hydroperiod ponds tend to have a shorter larval period and to metamorphose at a smaller size than tadpoles that develop in long hydroperiod ponds. Recent metamorphs from quick-drying ponds have no remnants of the larval tail and begin feeding immediately. Those from ponds with longer hydroperiods may have tail remnants and still contain food in the gut garnered while still in the tadpole stage (Morey and Janes, 1994). These latter toadlets also tend to be significantly larger and have more body fat than toadlets from quick-drying pools. Accelerated metamorphosis may allow survival in a drying pond, but there may be a fitness cost immediately post-transformation.

Although the response to hydroperiod is developmentally plastic, there is a genetic component that determines the limits of the plastic response (Newman, 1988b, 1994). The siblings from some spadefoots have a more accelerated developmental response to decreasing hydroperiod than others, but these larvae are less plastic in growth. In contrast, larvae with the slowest developmental rates, hence those best adapted to ponds with long hydroperiods, are also the most plastic in their growth rates. Thus, all larvae have an accelerated response to a drying pond, but some larvae are better adapted than others to conditions of long hydroperiods. The cost of greater plasticity in growth is slower development, and the cost of

Tadpole of *Scaphiopus couchii*. Photo: James Rorabough

rapid development is a decrease in fitness (i.e., a smaller size at metamorphosis) in ponds with long hydroperiods. Thus, there is a trade-off between growth and development in long hydroperiod ponds vs. short hydroperiod ponds (Newman, 1988a, 1989, 1992).

In addition to hydroperiod, density and food resources also affect the developmental period as they do in other anurans. In laboratory experiments, larvae reared in low densities and with high food resources metamorphose at a larger size than larvae in high density or low food resource trials. High food resources, in particular, cause larvae to metamorphose sooner than larvae reared in low density but with low food resources (Newman, 1989).

It appears that there is a threshold size that must be obtained to complete metamorphosis (Morey and Reznick, 2000). Low amounts of food will cause slower growth, which delays the attainment of the threshold size and could make individuals vulnerable to pool desiccation. With sufficient resources, tadpoles gain the threshold size and then have a degree of plasticity in the extent to which they remain in a breeding pool. With long hydroperiods, tadpoles can remain longer and grow to a larger size, even if food resources decrease. As long as the threshold size is reached, metamorphosis is assured. In *Scaphiopus couchii*, the age at metamorphosis is uniformly fast and not associated with growth rate. As a result, the size at metamorphosis may vary considerably, but the age at metamorphosis does not (Morey and Reznick, 2004).

Temperature further plays an important role in the duration of the larval period, with age at metamorphosis being sensitive to temperature in all spadefoot families, whereas size at metamorphosis is sensitive to temperature in some spadefoot families but not others (Newman, 1994). At high temperatures or low food, there is a positive correlation between age and size at metamorphosis. Thus temperature, food, density, and hydroperiod all interact to determine whether tadpoles will successfully metamorphose.

Larval *S. couchii* have cryptic genetic variation when exposed to novel diets. Normally larvae of *S. couchii* are detritivores. If fed on a diet of fairy shrimp, larvae show evidence of diet-dependent heritable variation in growth, developmental rates, and gut length. They also exhibit a wider range of phenotypic variation associated with feeding than they do when fed on a "normal" detritus diet (Ledón-Rettig et al., 2008). Diet-induced hormones, such as corticosterone, appear to be involved in the expression of this cryptic variation. For example, *S. couchii* fed on a diet of fairy shrimp experienced reduced growth and developmental rates and elevated levels of corticosterone when compared with *S. couchii* fed on an "ancestral" detritus diet. This led Ledón-Rettig et al. (2009) to suggest that the evolution of carnivory in spadefoots initially may have had a cost in reduced fitness. *Scaphiopus couchii* has been used as a model for understanding the evolution of carnivory in the larvae of the phylogenetically related genus *Spea* (Ledón-Rettig et al., 2008, 2009, 2010).

The most important causes of mortality in *Scaphiopus couchii* larval communities are desiccation and predation. For example, desiccation accounted for most mortality at 49 of 82 ponds studied by Newman (1987) in Texas. At 16 ponds, predators were the most significant source of mortality. Only 8 ponds produced metamorphs, and even at these ponds significant mortality occurred; the largest numbers of metamorphs produced at a pool were 101 and 118. The predators of larvae included water scavenger beetle larvae (*Hydrophilus* sp.), the turtle *Kinosternon flavescens,* and mammals (spotted skunk, *Spilogale putorius*). Supplemental feeding increased the number of metamorphs slightly, indicating the effect of food on developmental rates. Low-density ponds easily produced metamorphs if sufficient hydroperiod was available and predation was absent or minimal.

Metamorphosis occurs rapidly and simultaneously at a breeding pool. Even before metamorphosis, urea becomes the dominant waste product, which allows larvae to survive outside of water for a short time (Jones, 1980). Ureotelism could enable them to get just enough time to complete metamorphosis in a desiccating pond. Another way to cope with exposed breeding sites in often very hot climates is for larvae to be heat resistant. Such is the case with *Scaphiopus couchii* larvae, with Brown (1969) reporting that larvae could survive 21 hrs of exposure at a temperature of 39.1°C when acclimated at 25°C. The lethal temperature is between 41 and 42°C. Body lengths of recent metamorphs vary with larval period, but in general they are 8–20 mm SUL (Strecker, 1908; Mayhew, 1965; Newman, 1987; Degenhardt et al., 1996).

DIET

Scaphiopus couchii is a sit-and-wait predator. Prey includes earthworms, beetles, moths, termites, lepidopteran larvae, crickets, grasshoppers, spiders, true bugs, hoppers, bees, larval lepidopterans, lacewings, stick insects, mantids, pseudoscorpions, scorpions, centipedes, solpugids, harvestmen, ants, centipedes, mites, various types of flies, and even a whiptail lizard (*Aspidocelis* sp.) (Strecker, 1908; Little and Keller, 1937; Whitaker et al., 1977; Dimmitt and Ruibal, 1980a; Punzo, 1991a; Tocque et al., 1995). Termites in particular made up a large part of the diet of Couch's Spadefoots in California. Assuming termites and beetles were the primary prey and that a spadefoot could eat a large number in a single feeding, Dimmitt and Ruibal (1980a) estimated that an adult would only need 1.7 feedings to obtain the amount of fat metabolized in a single year. A single spadefoot could eat 55% of its body weight in a single feeding. In Arizona, however, *Scaphiopus couchii* ate a much wider variety of species that were available at different times of the season. Spadefoots were very efficient at converting high-energy prey into growth and fat deposition, but less nutritious prey may be necessary for essential nutrients. This suggests that surface feeding occurs over a much more extended time than previously recognized (Tocque et al., 1995). The larvae are filter-feeding detritivores, although some individuals are cannibalistic on weakened recent metamorphs (Mayhew, 1965).

PREDATION AND DEFENSE

When approached in deeper water, Couch's Spadefoots will dive below the surface to seek refuge. This species possesses a noxious skin secretion. To a human observer, the secretion is irritating to the eyes and nose, and produces an allergic reaction similar to hay fever. When handled, they produce a "creaking groan" sound (Hammerson, 1999).

In experimental trials, larvae are eaten by a wide variety of potential predators, indicating no chemically based defensive mechanisms. In addition, they do not change their levels of activity in the presence of predators. Natural predators include the larvae of *S. bombifrons* (Bragg, 1962b) on tadpoles, and ants (*Aphaenogaster cockerelli*) have been observed attacking recent metamorphs (Bonine et al., 2001). There is a report of *Scaphiopus couchii* being eaten by Sonoran Desert Toads (*Ollotis alvaria*) (Gates, 1957) and a badger (*Taxidea taxus*) (Dowler et al., 2010). Eggs are also palatable and are eagerly consumed by mosquitofish (*Gambusia*) in experimental trials (Grubb, 1972). In nature, eggs are eaten by ants (*Forelius mccooki*) (Dayton and Jung, 1999) and by conspecific larvae (Dayton and Wapo, 2002). A lack of antipredator mechanisms may partially explain why this species breeds in temporary rather than permanent pools (Woodward, 1983).

POPULATION BIOLOGY

Postmetamorphic growth occurs rapidly, with juveniles reaching half the adult size within three months (Mayhew, 1965). There is a great deal of individual variation in growth rates, with growth positively correlated with summer rainfall and hence feeding opportunities. Sexual maturity is reached at 2 yrs in males and 3 yrs in females (Tinsley and Tocque, 1995), with most growth occurring during the second season of life. Following maturity, annual survivorship is high. Sullivan and Fernandez (1999) reported up to 6 lines of arrested growth, but Tinsley and Tocque (1995) estimated longevity of 13 yrs for females and 11 yrs for males. Whether these differences represent variation in population age structure due to environmental conditions is unknown. Most of the breeding population is > 5 yrs of age, with about 5% reaching > 10 yrs of age; slower-growing animals tend to live the longest (Tinsley and Tocque, 1995).

COMMUNITY ECOLOGY

Scaphiopus couchii and *S. hurterii* come into close association in Texas, but hybridization is not common. Wasserman (1957) suggested that these species normally did not hybridize because of differences in habitat preferences (grasslands vs. woodlands), breeding phenology, advertisement call location (bank vs. water), advertisement call (Blair, 1955c), hybrid inviability, decreased fertility, and outright sterility. Of these, he considered habitat separation of primary importance. *Scaphiopus couchii* also may be found calling from the same pools as *S. multiplicata* and *S. bombifrons*, with *S. couchii* preferring the deeper parts of the pool, although there appears to be no real habitat segregation (e.g., McAlister, 1958). They may call from different locations (shore vs. water) (Bragg and Smith, 1942), but there is considerable overlap in calling position.

Larval *S. bombifrons* may impact aspects of the life history of *S. couchii*. For example, predaceous *S. bombifrons* larvae may attack and eat many *S. couchii* larvae within a pool. By reducing the number of *S. couchii*, predation actually increases the likelihood that survivors will metamorphose successfully (Bragg, 1962b).

Interspecific effects among larvae may extend across genera. For example, Northern Leopard Frog larvae inhibit the growth of *S. couchii* larvae under crowded experimental conditions, whereas a dense concentration of *Anaxyrus woodhousii* larvae has no effects on *Scaphiopus couchii* larvae (Licht, 1967). However, dense concentrations of *S. couchii* larvae inhibit the growth of *Ollotis nebulifer* larvae but not *Anaxyrus woodhousii* larvae. Decreases in body mass also occurred in larval *A. punctatus, A. speciosus,* and *Gastrophryne olivacea* when reared with *Scaphiopus couchii* larvae (Dayton and Fitzgerald, 2001). In nature, however, *S. couchii* are often the only tadpole within a breeding pool. The larvae are generally more active than those of other species,

Adult *Scaphiopus couchii*.
Photo: Aubrey Heupel

Habitat of *Scaphiopus couchii*, Big Bend National Park, Texas. Photo: Robert Newman

and are also the most susceptible to predation by aquatic predators (Dayton and Fitzgerald, 2001). Resource limitation could be a real problem for species with many tadpoles that breed together in desert pools, and *S. couchii* larvae are able to exclude or out-compete other species.

DISEASES, PARASITES, AND MALFORMATIONS
The trematode *Pseudodiplorchis americanus* has been reported from this species and may drain energy reserves of the spadefoot during dormancy, especially from those animals in poor body condition (Tinsley, 1990; Tocque, 1993; Tocque and Tinsley, 1994). However, there is no correlation between parasite burden and male mating success (Tinsley, 1990). About 50% of Couch's Spadefoots can lose their parasite burden entirely indicating some degree of resistance to the helminths. The species also is parasitized by the cestode *Distoichometra bufonis* and the nematodes *Aplectana incerta* and *Oswaldocruzia pipiens* (Goldberg and Bursey, 1991b).

SUSCEPTIBILITY TO POTENTIAL STRESSORS
pH. Breeding pools in California had a pH of 6.8–7.6 (Morey and Reznick, 2004). No information is available on lethal levels.

STATUS AND CONSERVATION
The species tolerates human activity better than many species, but is still vulnerable to habitat loss, alteration, and fragmentation. Couch's Spadefoots are killed in numbers on roads during their brief movements to breeding ponds and later dispersal (Bragg, 1944a). There does not appear to be widespread indications of population declines, but the secretive nature of this species makes an assessment of its status difficult. Couch's Spadefoots were extremely abundant at one time and may still be so in some areas. Strecker (1908) noted that populations in Waco, Texas, were "simply enormous."

A threat to Couch's Spadefoots is premature emergence in response to sound vibrations that accompany off-road vehicle activity in arid habitats. Vibrations from electric motors have been shown to induce 100% emergence from protected burrows (Dimmitt and Ruibal, 1980b). If spadefoots emerged under inhospitable conditions due to human-induced vibrations, they could expose themselves to direct harm from vehicles or be unable to rebury themselves in dry desert substrates. Breeding sites need to be completely protected, especially in vulnerable areas such as the base of the Algodones Dunes in southern California (Jennings and Hayes, 1994b). The species is considered of "Special Concern" in California.

COMMENT
Popular accounts of spadefoot biology are in Bragg (1955) and, especially, Bragg's (1965) book *Gnomes of the Night*.

Scaphiopus holbrookii (Harlan, 1835)
Eastern Spadefoot

ETYMOLOGY
holbrookii: a patronym in honor of South Carolina physician John Edwards Holbrook (1794–1871), the father of North American herpetology. Between 1836 and 1842, Holbrook published the first major review of the amphibians and reptiles of North America in a series of five separately issued parts (2nd ed., 1842; reprinted 1976).

NOMENCLATURE
Synonyms: *Rana holbrookii*, *Scaphiopus albus*, *Scaphiopus solitarius*. Also see *Scaphiopus hurterii*.

Although commonly called a spadefoot "toad," this species is in the family Scaphiopodidae, not the family Bufonidae, as are the true toads.

IDENTIFICATION
Adults. The Eastern Spadefoot is a medium-sized frog with a short blunt snout and large expressive eyes with vertical pupils. The dorsum is brown and slightly rugose, with light yellow narrow lines running parallel down the back. Parotoid glands are present, but there are no cranial ridges. The belly is white and unmarked, and there are two distinct pectoral glands on the venter. The fingers on the front feet are slightly thickened at the tips. The rear feet of *Scaphiopus holbrookii* are webbed and have obvious black spade-like projections that are used to dig backward into the ground, giving this species its common name. This species has a musty or peppery smell and secretions, to which some people are allergic. Boulenger (1899) provides a detailed description of the skeleton. An albino has been photographed from New Jersey (John Bunnell, these volumes).

Adults are 45–78 mm SUL. Specific records include: males 52–64 mm SUL (mean 59.1 mm) and females 55–67 mm SUL (mean 61.0 mm) in Connecticut (Klemens, 1993); a length of 23–63 mm SUL (mean 34 mm) and body mass 1.0–23 g (mean 6.6 g) in Maryland (Smithberger and Swarth, 1993); males with a mean of 64.0 mm SUL in Charleston, South Carolina (Duellman, 1955); males 51–64 mm SUL (mean 57.7 mm) and females 46–60 mm SUL (mean 53.4 mm) in Gainesville, Florida (Pearson, 1958); males with a mean of 51.7 mm SUL and females 48.1 mm in Key West, Florida (Duellman, 1955); males with a mean of 59.2 mm from Miami, Florida (Duellman, 1955). The largest recorded adult is 78 mm SUL (Lee and Sabette, 2009). Some references indicate that males are slightly smaller than females, but sexual size dimorphism is not pronounced, and adult body lengths are probably not significantly different between the sexes.

Larvae. The tadpole is small and dark colored bronze to brown with small orange spots. The body is round when viewed from above, and the eyes are close set. The head is wide relative to the body width, and the tail is short and rounded. The spiracle is well below the longitudinal body axis. Jaws are narrow in width and are not cuspate. Labial tooth row formulas are so variable as to suggest there is no "typical" count for this species, although 5/5 and 6/5 (upper labial tooth row count/lower labial tooth row count) are most common (Hampton and Volpe, 1963). Additional descriptions of the larvae are in Gosner and Black (1954) and Altig (1970). Albino tadpoles from Virginia were reported by Hensley (1959) and from Florida by Johnston and Johnston (2006).

Eggs. Eastern Spadefoot eggs have a single gelatinous envelope 3.8–5.6 mm in diameter. They are black or brown above and white or cream colored below. The eggs are deposited in a loose irregular cylinder or band about 25–75 mm wide and 25–300 mm long. About 200 eggs are oviposited per cylinder. Livezey and Wright (1947) give a clutch size of 2,330 eggs per female, but larger clutches have been recorded (see Reproduction). The eggs are placed on leaves, stems, or branches in the water, to which they quickly adhere upon oviposition.

DISTRIBUTION
Eastern Spadefoots are found from Massachusetts south to the Florida Keys along the eastern seaboard, and west generally to the Mississippi River in the southern United States. The species crosses the Mississippi River to include the Florida Parishes of Louisiana, northeastern and extreme eastern Arkansas, and southeastern Missouri. Eastern Spadefoots appear to have expanded their range northward via river valleys in many areas. The range extends northward from the mid-South

Distribution of *Scaphiopus holbrookii*

into southern Illinois and Indiana, and southeastern Ohio. An isolated population occurs in west-central Indiana. The species is generally absent from the Appalachian and Allegheny mountains from Tennessee and North Carolina northward, but scattered populations are known from east Tennessee (Irwin et al., 1999), West Virginia (now only two counties; Johnson, 2003), and west-central Virginia (Shenandoah Valley). It seems to prefer the coastal plain in North Carolina and southeastern Virginia, but enters the Piedmont in central Virginia and ranges northward into the mountains of central Pennsylvania via the Susquehanna River valley.

The occurrence of Eastern Spadefoots is not contiguous within its overall range; populations may be widely scattered, and the presence of this species often goes unnoticed, especially in the North, inasmuch as surface activity is infrequent and only occurs during reproductive bouts after very heavy precipitation. In many areas, it is associated with river valley floodplains, areas that might have sandy soils deposited during flood events. Undoubtedly, much remains to be learned about its biogeography.

Eastern Spadefoots have been recorded on the coastal barrier islands of Alabama, North Carolina, South Carolina, and Georgia (Chamberlain, 1939; Engels, 1952; Jackson and Jackson, 1970; Gibbons and Coker, 1978; Gibbons and Harrison, 1981; Kiviat, 1982; Shoop and Ruckdeschel, 2006; Gaul and Mitchell, 2007). There is a specimen from Nantucket (but no recent records), and they occur on Martha's Vineyard near Edgartown (Hargitt, 1888; Lazell, 1976; S. Smyers, personal communication). Reports of their extinction on Martha's Vineyard (e.g., Lazell, 1974) fortunately were premature. There are call records in 2006 from Plum Island in Massachusetts (S. Smyers, personal communication). Klemens (1993) reported that it was still present on eastern Long Island, New York (also see Overton, 1914; Burnley, 1973), but that western island populations were extirpated by urbanization. The furthest south Eastern Spadefoots occurred was Key West (Garman, 1884), and there is a specimen taken from there in 1932 and now in the Florida Museum of Natural History; it is unlikely that they still occur on Key West today (Lazell, 1989). The only other Keys report is from Upper Matecumbe Key (Wright and Wright, 1949), but there are no specimens, and Duellman and Schwartz (1958) questioned the record.

Important distributional references include: Alabama (Mount, 1975), Arkansas (Black and Dellinger, 1938; Bacon and Anderson, 1976; Trauth et al., 2004), Connecticut (Ball, 1936; Klemens, 1993), Delaware (White and White, 2007), Florida (Duellman and Schwartz, 1958; Lazell, 1989), Georgia (Defauw and English, 1994; Williamson and Moulis, 1994; Jensen et al., 2008), Illinois (Smith, 1961; Phillips et al., 1999), Indiana (Minton, 2001; Brodman, 2003), Kentucky (Barbour, 1971), Louisiana (Dundee and Rossman, 1989), Maryland (White and White, 2007; Harris and Crocetti, 2008), Massachusetts (Lazell, 1974, 1976; Klemens, 1993), Missouri (Johnson, 2000; Daniel and Edmond, 2006), New Jersey (Schwartz and Golden, 2002), New York (Stewart and Rossi, 1981; Klemens, 1993; Tierney and Stewart, 2001; Gibbs et al., 2007), North Carolina (Dorcas et al., 2007), Ohio (Green, 1948; Pfingsten, 1998; Davis and Menze, 2000), Pennsylvania (Hulse et al., 2001), Rhode Island (Klemens, 1993), Tennessee (Redmond and Scott, 1996), Virginia (Merkle, 1977; Tobey, 1985; Mitchell and Reay, 1999; White and White, 2007), West Virginia (Green, 1938a; Johnson, 2003).

FOSSIL RECORD

Eastern Spadefoots are found in late Pliocene deposits in Florida (Meylan, 2005) and in

Pleistocene Rancholabrean (10,000–150,000 yrs BP) sites in Arkansas, Florida, Georgia, Pennsylvania, and Virginia. Miocene fossils from Florida attributed to *S. holbrookii* (Auffenberg, 1956) are actually from the genus *Spea*, whereas fossils from Nebraska tentatively identified as *S. holbrookii* (Chantell, 1971) were reassigned to a now extinct species, *S. hardeni*. Holman (2003) summarizes the Pleistocene fossil history and diagnostic osteology.

SYSTEMATICS AND GEOGRAPHIC VARIATION

Scaphiopus holbrookii is most closely related to *S. hurterii* and *S. couchii* and is only distantly related to the western Spadefoot genus *Spea* (Tanner, 1939; Sattler, 1980; García-París et al., 2003). *Scaphiopus hurterii* is often included as a subspecies of *S. holbrookii*. The two are geographically separate, but their calls are essentially the same (Blair, 1958b) and they readily hybridize under laboratory conditions (Wasserman, 1958). Fertility of hybrid backcrosses is high. Zweifel (1956a) noted differences in the skulls between *S. hurterii* and *S. holbrookii*, especially in the fronto-parietal bone. In addition, there are size differences, differences in the presence or absence of a boss (a raised area) between the eyes, and minor tadpole prey differences (see *S. hurterii* account). Finally, García-París et al. (2003) provided genetic evidence supporting recognition of two distinct species. The two spadefoots also are treated as separate species by Lannoo (2005) and Crother (2008).

Scaphiopus albus was described from Key West based on its lighter coloration (broad, irregular white areas on a dorsal pattern of olive brown) and small size (Garman, 1877, 1884). Wright and Wright (1949) reported that it also had a narrower interorbital distance than *S. holbrookii* and retained it as a subspecies of the Eastern Spadefoot. This narrowness is a function of size, however, since the interorbital distance is narrower in small frogs compared with large frogs, as noted by Duellman (1955). Other than its smaller size, there appears to be no reason to recognize *S. albus* or *S. h. albus* as valid taxa. There is some question as to whether it still occurs in the Florida Keys, although Carr (1940a) recorded it as fairly common, especially on the south side of the island.

Eastern Spadefoots do not hybridize naturally with other *Scaphiopus* since they do not co-occur anyplace. In laboratory crosses, *S. holbrookii* females do not produce viable offspring with *S. couchii* males; offspring can be produced in reciprocal crosses, although backcrossed hybrids have reduced fertility. Female hybrids are more fertile than male hybrids (Wasserman, 1963).

ADULT HABITAT

Adults are found associated with a variety of habitats, including pine flatwoods, scrub, sandhill, xeric hammock, mesophytic hammocks, upland hardwoods, steephead ravines, and basin swamps (Carr, 1940a; Dodd, 1992; Smithberger and Swarth, 1993; Enge et al., 1996; Enge, 1998a, 1998b; Enge and Woody, 1998; Baxley and Qualls, 2009). They also may occur in disturbed habitats, such as suburban developments, agricultural fields, pastures, and managed pine plantations (Netting and Wilson, 1940; Neill, 1950a; Pearson, 1955; Bennett et al., 1980; Enge et al., 1996; Leiden et al., 1999; Hanlin et al., 2000; Russell et al., 2002a; Johnson, 2003; Surdick, 2005), as long as the soil remains friable for digging and individuals are not killed outright. Eastern Spadefoots often extend their range along sandy river terraces.

TERRESTRIAL ECOLOGY

Eastern Spadefoots are generally nocturnal in their habits and are most active from 21:00 to 01:00 hrs (Punzo, 1992a). In much of their range, they may be observed only during breeding events and are seldom seen at other times. In Florida (personal observation) and Mississippi (W.R. Allen, 1932), however, they are easily observed on warm nights of high humidity and rainfall as they sit at burrow entrances or forage in nearby leaf litter. During the day or in dry or cold conditions, they are fossorial, hiding in burrows, and may be extremely cryptic and difficult to locate on the forest floor. As might be expected from a fossorial frog, they have no special adaptations to reduce evaporative water loss, and are thus "typical" frogs in that regard (Wygoda, 1984).

In Florida and the Atlantic and Gulf Coastal Plains, Eastern Spadefoots are active all year. For example, they have been observed in March, July, and October in Mississippi (M.J. Allen, 1932), and I have observed them in every month of the year in north-central Florida, especially after heavy rainfall. The activity season becomes restricted to early spring through autumn with increasing north latitude. For example, adults have been observed in Maryland from late March into mid-October, with young-of-the-year observed into early September (Smithberger and Swarth, 1993). Most activity in West Virginia occurred from August to October (Johnson, 2003).

Regardless of latitude, rainfall is the trigger to stimulate nocturnal activity in many areas, whether for foraging or reproduction (but see Johnson, 2003). Eastern spadefoots are also more active when relative humidity is high (>80%) than when it is lower, when air temperatures are around 20.5°C, and when soil temperatures >15°C. They are generally not active at air temperatures <10°C and >29°C (Pearson, 1955; Johnson, 2003). Gosner and Black (1955) reported a minimum daylong temperature of 9.6°C as necessary for emergence in New Jersey, although Ball (1936) recorded activity at 7.5–8.6°C in Connecticut. These differences are slight, but they could reflect latitudinal variation in temperature tolerance.

Eastern Spadefoots disperse extensively into the uplands surrounding breeding sites, and they have been found > 800 m from breeding sites (Pearson, 1955; Dodd, 1996). For example, I found them widely dispersed in sandhill habitats of north-central Florida from 95 to 914 m (mean 539 m) from the nearest breeding site. Adult dispersal occurs exclusively at night (Todd and Winne, 2006). However, dispersal occurs slowly, with individuals usually moving only short distances from one movement to the next. As such, it can take a considerable time to traverse large distances between terrestrial and breeding sites.

Nearly all of their time is spent in burrows about 7 to 30.5 cm in length, and from 2 to 4 cm in diameter (Pearson, 1955). Most animals do not dig in more than 18 cm in depth, at least in the South. The burrow itself does not consist of an open tube, but instead may be filled with loose sand and soil through which the spadefoot makes its way each time it enters and departs from the burrow. Pearson (1955) noted that the burrow is open when the spadefoot is away foraging (photo in Johnson, 2003). They may remain within a burrow for as many as 104 consecutive nights (mean 9.5 nights). Further, Pearson (1955) observed that spadefoots were active on only 8% of the nights sampled, and thus would be foraging or otherwise away from the burrow on only 29 nights during an entire year in Florida. Eastern Spadefoots tend to remain in the same burrow, and only one frog is found per burrow. However, they can also shift burrows during an activity season, the mean distance usually being approximately 3.3 m (range 0.6 to 11.6 m in Florida; Pearson, 1955). Johnson (2003) recorded one spadefoot using 4 burrows within 2.0 m. They may remain in the new burrow, or return to the first one. Pearson (1955) provides examples of how spadefoots used multiple burrows over a period of up to ten months. Spadefoots also overwinter in burrows, according to Pike (1886) on south-facing slopes, at least in the Northeast. In the North, spadefoots have been found to a depth of several meters underground (Pike 1886; Ball, 1936).

Home ranges are used from one year to the next, that is, *S. holbrookii* is rather philopatric in its choice of terrestrial habitat (Pearson, 1957). Home ranges of Eastern Spadefoots in Florida were measured from 0.65 to 82.1 m^2 (mean 9.9 m^2). Males had average home ranges slightly smaller (8.4 m^2) than females (10.3 m^2), but there was such variance as to preclude statistical significance (Pearson, 1955). Home ranges within a sex did not overlap, leading Pearson (1955) to suggest individuals were territorial to members of their own sex. It appears that most time is spent within a particular area within a season, although occasional movements well outside the core area are not uncommon (about 10% of movements are of this type). Movements also vary with habitat type. For example, Pearson (1955) found greater movements in ecotonal habitats than within core hammock (forested) habitats. Shorter-distance movements also occur where spadefoot densities are highest. Patterns of movement and burrow use within home ranges were illustrated by Pearson (1955, 1957). After breeding, Eastern Spadefoots return to the terrestrial home range from which they departed.

Population density appears to be rather high in areas near breeding sites. Pearson (1955) estimated from 493.4 to 976.0 spadefoots/ha on a single 73 m^2 plot over a 1 yr period, depending on month, and from 237 to 511.2 spadefoots/ha on another 5,520 m^2 plot over a 2 yr period. However, he was not confident that the assumptions of mark-recapture studies were met, as the spadefoots were usually inactive and the marked animals did not seem to mix randomly with the total population; he thus considered his estimates as "very inaccurate." Later adjusted counts still gave estimates as high as 631 spadefoots/ha, with considerable monthly variation in estimate (Pearson, 1957). Density decreases during the course of a calendar year within an area, and survivorship is highest as the spadefoots overwinter.

Eastern Spadefoots are photopositive in their phototactic response, suggesting they use both sunlight and moonlight when feeding and going about their daily activities (Jaeger and Hailman, 1973). They are sensitive to light in the blue spectrum ("blue-mode response") (Hailman and Jaeger, 1974). Eastern Spadefoots likely have true color vision.

CALLING ACTIVITY AND MATE SELECTION

Males call both diurnally (reviewed by Neill, 1957a) and nocturnally. Calling begins as soon as individuals arrive at the breeding site and may extend over a period of several days depending on environmental conditions. For example, Church et al. (2002) recorded from 9 to 50 adults at a breeding pond in Virginia over a period of 8 days. The distinctive call is difficult to describe, but has been likened to the second syllable in a long drawn-out meow of a Siamese cat ("owww, owww"), an explosive grunt, languishing moans, a young crow, "waagh" or "waaank" (Palis, 1994). Calling occurs as males float in the water. They arch their heads and bodies convexly backward, allowing the expansion of the vocal pouch. Calls are uttered at 4–6 sec intervals at a fundamental frequency of 170 cps and a trill rate of 177 (McAlister, 1959). Not all males call, as many seem to await the approach of anything that might be a female, which they readily clasp. McAlister (1959) described the vocal anatomy and method of call production.

Amplexus is inguinal. Many if not most females are amplexed as they make their way toward the breeding site rather than at the site. Even as they move toward ephemeral wetlands, females may be clasped by more than one male, although the primary suitor will issue a warning call and vigorously kick rival males that attempt to amplex the pair (Ball, 1936). The female swims during amplexus and determines the oviposition site. Once the eggs are expelled and fertilized, amplexus ends and the adults separate.

Eastern Spadefoot calls have been recorded with dominant frequencies of 1,300–1,550 cps at 18°C for durations from 0.52 to 0.62 sec (Blair, 1958b). The calls of *S. holbrookii* and *S. hurterii* are essentially the same in call characteristics and sound.

BREEDING SITES

Reproduction usually occurs in small shallow (< 14 cm) temporary pools of water. Such pools may be located in both natural habitats, such as

Eggs of *Scaphiopus holbrookii*. Photo: C.K. Dodd Jr.

deciduous forest or in greatly disturbed areas, such as agricultural fields, pastures, retention ponds, or virtually any low-lying area within a landscape that will hold water. However, Duellman and Schwartz (1958) noted breeding in a pond 1.6 km in length by 91 m in width. In pools with deeper water, the eggs are deposited in the shallowest section of the wetland. Both the breeding site hydrological history (past use) and the current hydrology (presence of water in shallow pools) are the best predictors of occupancy and colonization of particular breeding sites (Church, 2008).

REPRODUCTION

For Eastern Spadefoots, successful metamorph recruitment is always a race between the duration of the hydroperiod and sufficient time for development. They only breed after very heavy precipitation, when the ground is saturated and temporary pools form. The trigger for movement may be the precipitation itself, or in combination with maximum changes in barometric pressure (Greenberg and Tanner, 2004). If heavy or torrential rains do not occur, *S. holbrookii* will not breed. Instead, females are capable of carrying ova throughout the year, and males have viable sperm in their testes throughout the year (Hansen, 1958). This allows them to be ready to breed whenever conditions permit. Hansen (1958) suggested that water-uptake caused the female gonads to ovulate by stimulating the anterior lobe of the pituitary to hypersecretion of gonadotropins.

Breeding can occur at almost any time of the year from South Carolina through Florida (Hansen, 1958; Greenberg and Tanner, 2005a), but it occurs during the warmer parts of the year in more northern latitudes. Specific breeding dates include February and March (Arkansas: Trauth et al., 1990), February to July (Louisiana: Dundee and Rossman, 1989), March to July (Massachusetts: Allen, 1868), April to May (Rhode Island: Anonymous, 1918), April to August (Missouri: Johnson, 2000), July to September (South Carolina: Chamberlain, 1939), and from March to August at various areas from Virginia northward (Gosner and Black, 1955; Church et al., 2002). Hansen (1958) provides an extensive compilation of dates of breeding choruses throughout eastern North America. Several authors have recorded sequential breeding at a site within a few months of one another, but different individuals are likely involved rather than females ovipositing multiple clutches (Ball, 1936; Gosner and Black, 1955).

Breeding aggregations can be very large during the short, explosive breeding period. In what must have been a truly spectacular event, Duellman and Schwartz (1958) estimated there were 60,000 spadefoots (4 per 9.2 m^2) breeding in a very large pond after torrential rains in Surfside, Florida, in 1952. Breeding occurred in other nearby wetlands, as well as throughout south Florida that evening, but few were heard in subsequent nights. In South Carolina, the breeding population at one site was composed of 258 and 812 individuals during 2 yrs of a 4 yr study, but only 9 and 20 adults were observed during the other 2 yrs at the same breeding site (Pechmann et al., 2001). In Georgia, 9 Eastern Spadefoots were captured in a drift fence surrounding a potential breeding pond one year, whereas 158 were captured the following year during an 18-month study (Cash, 1994). In Florida, I captured more than 1,400 spadefoots during a single breeding event over a 5 yr study (Dodd, 1992). Greenberg and Tanner (2005a) recorded up to 340 spadefoots at a pond over a 2-day period in central Florida. The low numbers may represent incidental captures of foraging or transient spadefoots, or they may represent only a small number of breeders that year.

Individual females are capable of producing several thousand eggs during a breeding season. Ball (1936) noted that the clutch size was 800–2,000 in Connecticut. Clutch size in Arkansas was 3,522–4,469 (mean 3,838), and the mean ovarian mass was 3.78 g (2.93–4.89 g) (Trauth et al., 1990). Eggs hatch within 24 hr to 15 days depending on water temperature.

LARVAL ECOLOGY

Upon hatching, the tiny larvae remain motionless, attached to the jelly, substrate, or subsurface debris by a ventral sucker for about one to six days. Mouthparts develop, and the external gills are covered by the operculum, the ventral sucker disappears, the tail lengthens, and the tail fins develop. They then become much more characteristically active. As they grow, larval Eastern Spadefoots become very active swimmers and voracious feeders. Initially filter feeders, they quickly become omnivorous, grazing on periphyton, vegetation, amphibian eggs, and dead animals found on subsurface debris and within the substrate. As

Tadpoles of *Scaphiopus holbrookii*. Photo: Steve Bennett

the mouthparts develop, they can even become cannibalistic on injured or dying conspecifics (Ball, 1936), or on other tadpole species.

As the feeding mode changes, somewhat randomly swimming Eastern Spadefoots gradually form dense aggregations, some consisting of thousands of individuals. Richmond (1947) reports observing an aggregation measuring 1.2 m wide, 2.4 m long, and 10 cm thick. Others formed streams 1.2 m wide and 6.1–9.1 m long. Ball (1936) estimated that one aggregation he observed contained 12,000 tadpoles. The aggregations may be quite dense, although the streams are not as thickly populated. Aggregations tend to behave as a single entity moving across the shallow substrate. Single aggregations may divide, and multiple aggregations may come together and unite to form a single large group. Richmond (1947) provides detailed descriptions of tadpole behavior within aggregations, particularly the constant rotation of individuals as they jockey for position within the moving mass. Unlike aggregations of *Anaxyrus americanus*, for example, aggregations of Eastern Spadefoot larvae do not appear to be kin-based. Such dense aggregations facilitate filter feeding by churning up sediments, but they also may have defensive and/or thermal functions.

Temperature plays an important role in embryonic and larval development, and larval growth rates are directly proportional to increasing temperature. For example, the larval period is 22 days at 24°C, whereas at 32°C it is only 15 days (Gomez-Mestre and Buchholz, 2006). Prolonged exposure to very cold water (<4–6°C) will kill developing embryos, but normal development occurs in water >9.6°C (Richmond, 1947; Gosner and Black, 1955). The upper water temperature at which normal development takes place is 29–33°C (Richmond, 1947; Gosner and Black, 1955). Likewise, some larval mortality occurs in water of 8.5°C, but a few larvae may survive at lower temperatures depending on the duration of exposure. The larval CTmax is 37.5°C, but mortality increases substantially at 30°C (Gosner and Black, 1955).

Under laboratory trials, the growth rate of larval *S. holbrookii* is inversely proportional to density, and the mean number of days to metamorphosis decreases with decreasing density (27 days at low density vs. 86 days at high densities) (Semlitsch and Caldwell, 1982). The size of larvae at metamorphosis is curvilinear in relationship to density. Thus, low larval densities produce larger metamorphs, whereas higher tadpole densities produce smaller metamorphs, at least as tadpoles begin transformation. As tadpoles leave, however, the remaining larvae increase their growth and thus their size at transformation once released from the effects of density (Semlitsch and Caldwell, 1982). Survivorship, duration of the larval period, and the size at metamorphosis is

Tadpole school of *Scaphiopus holbrookii*. Photo: Jamie Barichivich

density-dependent in the absence of predators and competitors (Wilbur et al., 1983).

Survivorship under controlled conditions also is inversely proportional to initial conspecific density, such that survivorship where tadpole densities are low approaches 90%. In nature, of course, tadpole densities are often high in the small ephemeral pools favored by the species for development, and hydroperiods are often short. Tadpoles cope with these variables by having rapid growth rates, short larval periods, and generally small body sizes at metamorphosis. Even these traits, however, have a certain degree of phenotypic plasticity depending on both density-dependent and density-independent characteristics of the habitat and larval community.

Metamorphosis occurs in two to three weeks in Rhode Island (Anonymous, 1918), but can take longer depending on temperature and hydroperiod. The duration of the larval period is dependent on the water temperature during development, which in turn may be a function of the date when eggs are oviposited (for example, April vs. August). It can occur in as few as 9–10 days (at 28.5–33°C), or as long as 60–63 days in colder water (Overton, 1914; Driver, 1936). Most figures appear to be in the range of 15–30 days in nature (e.g., Greenberg and Tanner, 2004).

The effect of hydroperiod on fitness may extend beyond the larval period. For example, Gomez-Mestre and Buchholz (2006) found that postmetamorphs resulting from larvae reared in long-hydroperiod regimes had longer limbs and snouts than larvae reared in short hydroperiod regimes. This type of environmentally induced morphological variation appears to be common in all Pelobatoidea and is not a matter of simple allometry.

DISPERSAL

Tremendous numbers of juveniles may be observed following particularly intensive breeding, such as after tropical storms. Literally tens of thousands of young spadefoots may carpet the ground for days as they disperse to terrestrial habitats (Pike, 1886; Abbott, 1904; Allen, 1941; Neill, 1957a). Such events occurred after the hurricanes in Florida in 2004 and 2005, with reports of waves of tiny spadefoots (and Southern Toads) dispersing through suburban and natural habitats throughout the state. However, Pechmann et al. (2001) recorded only 58 juveniles produced by 258 breeding adults in South Carolina, and no recruitment at all following breeding by 812 adults in another year; over a 6 yr period at another Carolina Bay, 3,483 juveniles were produced and 197 females captured, although most metamorphs were produced in only 1 yr (Semlitsch et al., 1996); at 2 other South Carolina ponds, only 11 *S. holbrookii* were observed in 1 yr, but 1,358 the next, nearly all of which (1,321) were in 1 of the 2 ponds (Gibbons and Bennett, 1974).

Metamorphs generally disperse at night, although Todd and Winne (2006) captured about 20% of them during daylight movement in South Carolina. Metamorphs usually completely leave the breeding site within a week of when metamorphosis begins, regardless of precipitation, although a few individuals may remain in the vicinity of the pond for a considerable amount of time if conditions permit (Greenberg and Tanner, 2004). Immigration and emigration patterns are not necessarily the same, although movements appear directed. Pearson (1955) noted that it takes about four to six weeks from the time metamorphs depart from breeding sites until they take up a burrow-based existence. Those dispersing later in the season or in winter likely construct burrows sooner than juveniles emerging during the warmer parts of the year.

FEEDING AND DIET

Eastern Spadefoots forage very near their burrows (for example, 2–204 cm, mean 34.5 cm; Johnson, 2003), usually going a maximum of about 1 m during foraging bouts. Pike (1886) also noted that they may sit at the entrance to their burrows with only the eyes visible, ready to ambush unsuspecting prey. The diet consists of invertebrates, particularly beetles, diptera (flies), ants, orthoptera (grasshoppers, crickets, katydids), myriopoda (millipedes, centipedes), isopterans (termites), isopods, spiders, hemiptera (bugs, cicadas, aphids), oligochaetes (worms), and larval lepidoptera (Carr, 1940a; Pearson, 1955; Whitaker et al., 1977; Punzo, 1992a; Jamieson and Trauth, 1996; Timm and McGarigal, 2010a). The diet is similar between the sexes, but it can change seasonally with changing invertebrate availability. Ants are particularly important in the diet of subadults.

Adult *Scaphiopus holbrookii*, amplexus. Photo: John Bunnell

PREDATION AND DEFENSE

When disturbed, Eastern Spadefoots rapidly burrow into the soil, posterior first. They use the spades on their rear feet to alternately dig and move soil out of the way as they contort and move their body backward. Complete immersion is rapid and occurs within only a few seconds.

When confronted by predators and unable to dig, they curl their body by tucking the chin downward, closing their eyes, and drawing the limbs tightly in to the body with the hands and feet covering the venter. The spadefoot remains immobile in this closed posture (Dodd and Cupp, 1978), thus exposing the predator to the noxious secretion that emanates from the frog's dorsal surface. The secretion has a noxious smell and is likely distasteful. Many people have an allergic reaction to it; they sneeze, their eyes water copiously, and if the secretion enters the eyes or mucous membranes, it produces a burning sensation. Larvae approaching metamorphosis also become immobile when disturbed; immobility duration decreases as metamorphosis is imminent or, as in adults, at increasing temperature (Dodd and Cupp, 1978). Some individuals may attempt to flee, but their hopping is slow, and they may have trouble eluding a predator.

Adult Eastern Spadefoots seem to have few natural predators. These include Southern Toads (*Anaxyrus terrestris*), American Bullfrogs (*Lithobates catesbeianus*), Southern Hognose Snakes (*Heterodon simus*), Northern Water Snakes (*Nerodia sipedon*), Banded Water Snakes (*Nerodia fasciata*), Black Racers (*Coluber constrictor*), Cottonmouths (*Agkistrodon piscivorus*), Cattle Egrets (*Bulbulcus ibis*), owls, gulls (*Larus* sp.), and feral pigs (*Sus scrofa*) (Carr, 1940a; Goin, 1947a, 1955; Hamilton and Pollack, 1956; Holman, 1957; Lynch, 1964b; Jenni, 1969; Gloyd and Conant, 1990; Palis, 1994; Palmer and Braswell, 1995; Beane et al., 1998; Jolley et al., 2010; Timm and McGarigal, 2010b). Metamorphs are consumed by Banded Water Snakes, European Starlings (*Sturnus vulgaris*), and Common Grackles (*Quiscalus quiscula*) (Palis, 2000). Other predators include raccoons (*Procyon lotor*) and opossums (*Didelphis virginianus*) feeding on adults as well as larvae as ponds dry (Lynch, 1964b; Church et al., 2002). Eggs are eaten by leeches, and larvae by aquatic insects.

The newt *Notophthalmus viridescens* is frequently used in experimental trials examining the effects of predation on tadpole life history. Presumably, newts also feed on spadefoot tadpoles in nature when they encounter them. The frequency of tail injuries on *S. holbrookii* larvae can be used to assess the predation pressure exerted by newts on these larvae (Morin, 1985).

Albino *Scaphiopus holbrookii*. Photo: John Bunnell

POPULATION BIOLOGY

Growth occurs rapidly at first, and tends to slow as the spadefoots become larger. Males grow faster than females initially, but as they increase in size, growth rates are essentially the same between the sexes. There is some disagreement in the literature concerning the age when maturity occurs. Duellman and Schwartz (1958) noted that no female < 43 mm SUL was gravid in south Florida. According to Pearson (1955), maturity is reached in about 2 yrs after metamorphosis in northern Florida. However, Greenberg and Tanner (2005a) rarely caught individuals < 50 mm SUL at breeding sites in central Florida, and suggested that breeding began at ages 4–5. If so, they may spend 3–4 yrs in upland habitat prior to first reproduction. Likewise, there are literature reports that they may live as long as 12 yrs in the wild, although Greenburg and Tanner (2005a) conservatively estimated an adult 7 yr life span. Pearson (1955) noted that 15–39% of his population was >6 yrs old and that the population became skewed to older individuals as his study continued without successful reproduction. Longevity allows them to skip years when breeding conditions are unsuitable, as recruitment appears to be unsuccessful during most years.

Not all breeding sites are equally productive in terms of recruitment, with some sites producing more metamorphs than others. Even at "good" sites, there are times when breeding occurs but no successful recruitment occurs. In a central Florida study, breeding events (defined as >25 adults at a pond within a month) occurred 23 times on 9 occasions over a 9 yr period at 7 ponds. Recruitment (defined as >100 metamorphs) occurred on only 5 occasions, however. Greenberg and Tanner (2005a) noted that four ponds acted as source populations, and then only during some years. Some of the source ponds acted as "sinks" in other years, that is, no recruitment occurred despite a breeding event; three ponds always acted as sinks. At a small north-central Florida temporary pond, there was only a single breeding event in 5 yrs, but no metamorphs were produced (Dodd, 1992). Breeding events without any recruitment appears to be common in Eastern Spadefoot populations.

Greenberg and Tanner (2005a) further reported that from 0 to 4,648 metamorphs were produced per breeding event, which apparent population trends reflected breeding effort, that recaptures were rare, and that capture rates fluctuated substantially from one year to the next. Over the 9 yr study, from 0 to 7,370 recruits were captured in any 1 yr, whereas 15,145 metamorphs were captured in total over the study period (Greenberg and Tanner, 2005a). Some interpond movement was recorded, mostly to nearby sites. One spadefoot traveled 416 m between breeding sites, however.

COMMUNITY ECOLOGY

In mixed larval groups, *Scaphiopus holbrookii* larvae have complex and sometimes subtle interactions and effects upon both other anuran larvae and larval predators. For example, *S. holbrookii* is a superior competitor to Cope's Gray Treefrog (*Hyla chrysoscelis*) in laboratory trials, and they actually grow better in the presence of *Anaxyrus americanus* and *Lithobates sphenocephalus* larvae than in their absence. Alford (1989a) suggested that spadefoots were competitively dominant to these species. However, *Anaxyrus* and *Lithobates* larvae scrape periphyton off mesocosm walls and release nutrients that enhance the growth of phytoplankton, a better food source for spadefoot larvae than filamentous algae (Wilbur et al., 1983). This may allow them to grow faster without having a direct negative impact on the other species. In another experiment, the sunfish *Enneacanthus* completely eliminated larval *Scaphiopus holbrookii* (Kurzava and Morin, 1998).

In mesocosms where hydroperiod is controlled

(for 100 days, 50 days), *S. holbrookii* larvae generally survive well at both low and high densities, achieving as high as an 80% survivorship (Wilbur, 1987). However, adding newts (*Notophthalmus viridescens*) to experimental tanks nearly eliminates survivorship, as newts seem to prefer *Scaphiopus* larvae to *Anaxyrus*, *Hyla*, and *Lithobates* tadpoles in mixed-species assemblages (Morin, 1983; Wilbur, 1987). When newts are added to a mesocosm featuring only *Scaphiopus*, the initial density of the tadpoles (low or high) affects the larval period and mass at metamorphosis. At high tadpole densities, the newts tend to become satiated so that survivors are actually larger than they would be at low densities, where food might be more abundant but tadpoles are easy targets for the predators. Likewise, the absence of newts, or their introduction early in development, increases mean larval mass, especially when tadpoles are initially in high densities (Alford, 1989a). Increasing the number of predators further tends to increase the mass of surviving larval spadefoots; although survivorship decreases dramatically, larval period decreases but mean body mass and growth rates increase (Morin, 1983).

DISEASES, PARASITES, AND MALFORMATIONS
The amphibian chytrid fungus (*Batrachochytrium dendrobatidis*) has been found in a single individual from Cape Cod, Massachusetts (Tupper et al., 2011). Parasites include protozoans (*Nyctotherus cordiformis*, *Octomitus intestinalis*, *Opalina oblanceolata*, *O. obtrigonoidea*, *O. carolinensis*, *O. triangulata*, *Trichomonas augusta*), cestodes (*Distoichometra bufonis*, proteocephalid cysts), and nematodes (*Agamonema*, *Cosmocercoides dukae*, *Oswaldocruzia leidyi*, *O. pipiens*, *Physaloptera*, *Rhabdias*, *R. ranae*) (Brandt, 1936).

SUSCEPTIBILITY TO POTENTIAL STRESSORS
No information is available.

STATUS AND CONSERVATION
Determining the status of Eastern Spadefoots can be difficult, since individuals are generally rarely observed, despite intensive sampling, especially in the northern portions of the species' range. For example, the first records of Eastern Spadefoots from Great Smoky Mountains National Park were only made in the late 1990s, despite intensive observations within the park for more than 70 yrs (Irwin et al., 1999). C.C. Abbott (1884) noted that Eastern Spadefoots bred in a nearby sinkhole 10 yrs apart. Nichols (1852) reported breeding at a site only 4 times in 30 yrs. Indeed, the species may be relatively common, despite the rarity of sightings or museum specimens from specific localities (Gentry, 1955). Individual records may

Breeding habitat of *Scaphiopus holbrookii*.
Photo: John Bunnell

be the result of chance observations (Ball, 1936; Brodman, 2003; *in* Montanucci, 2006), whereas intensive sampling at appropriate times of the year may prove them to be more widely distributed than previous observations would suggest (e.g., Rice et al., 2001).

Eastern Spadefoots may have been able to follow sandy river terraces into mountainous regions or into otherwise unsuitable habitat, where they could dig into sandy soils (Netting and Wilson, 1940; Klemens, 1993; Montanucci, 2006). These areas are often affected by river damming and other modifications, as well as development along river terraces. Habitat specificity plus subsequent fragmentation may account for the scarce and disjunct records from uplands habitats and from certain portions of its range. Thus, peripheral populations may be particularly vulnerable to habitat modification or loss, such as on Long Island and in southern Connecticut. This is also true of their temporary shallow breeding sites, which are virtually unprotected throughout the species' range. Some well-documented populations, such as in Ball (1936), are now extirpated.

Although *Scaphiopus holbrookii* may be found in wetlands surrounded by pine plantations, their numbers may be small, and it is unclear as to the extent to which site preparation and planting affects population viability. For example, Russell et al. (2002b) found only 6 spadefoots over a 2 yr period at 5 wetlands embedded within a pine plantation in South Carolina. The low number could reflect habitat disturbance, or perhaps appropriate environmental stimuli such as heavy bouts of rainfall were absent. In Florida, Enge and Marion (1986) found only 26 Eastern Spadefoots in maximum or minimum clearcut and site prepared pine plantations, but 4,696 individuals in an uncut forest wetland over the same 1 yr sampling period. Roller chopping and heavy-duty site preparation would seem particularly detrimental to this fossorial species.

It is clear that large-scale development projects can have very negative effects on Eastern Spadefoots. This is because their unprotected ephemeral breeding sites are particularly prone to destruction, and terrestrial uplands are prime sites for industrial and residential development. Near Tampa, Florida, for example, Eastern Spadefoots were eliminated following the development of a large housing project, even as they remained abundant in adjacent undeveloped areas (Delis, 1993; Delis et al., 1996). This may be because of their inability to burrow into unnatural substrates often used by landscapers, or because urban landscapes are heavily watered (Jansen et al., 2001). They may persist in suburban areas, but only if breeding and nonbreeding habitats are present, and friable soils are available for burrowing. Roads undoubtedly take a serious toll on some populations (Sutherland et al., 2010).

Creating wetlands to replace destroyed habitats may not be a successful conservation strategy for this species. Although adults may visit created ponds, Pechmann et al. (2001) found no successful breeding for 8 yrs following pond creation at a South Carolina restoration site. In Maryland, *S. holbrookii* did not colonize newly created ponds over a 2 yr period following construction (Merovich and Howard, 2000). Also in Maryland, Eastern Spadefoots were found around a number of created wetlands along the Patuxent River (Touré and Middendorf, 2002).

A total of 1,916 *S. holbrookii* were exported from Florida for the pet trade from 1990 to 1994 (Enge, 2005a).

Scaphiopus hurterii Strecker, 1910a
Hurter's Spadefoot

ETYMOLOGY
hurterii: a patronym honoring Julius Hurter (1842–1917), Missouri herpetologist and author of the monograph *Herpetology of Missouri*, published in 1911.

NOMENCLATURE
Conant and Collins (1998): *Scaphiopus holbrookii hurterii*
  Synonyms: *Scaphiopus holbrookii hurterii*

IDENTIFICATION
Adults. The dorsal coloration is greenish brown with pale yellowish lines extending from each eye and converging posteriorly. This species is

similar to *S. holbrookii* but has a narrower head and smaller tympanum. The parotoid glands are round, high, and conspicuous. Venters are yellowish white. As in all spadefoots, there is a distinct digging spade on each rear hind foot.

Larvae. At hatching, tadpoles are black but quickly become dark gray. Chromatophores give the appearance of an hourglass shape when viewed dorsally. Heads are narrow with the small eyes close to one another. Venters are gray but lighter than the dorsum. Tail fins are generally clear and low, and the tail tip is rounded. Tails are longer than the body. Larvae reach a maximum size of 21–27 mm TL. Bragg (1944b) and Bragg et al. (1964) described the tadpole and its mouthparts.

Eggs. Eggs are black dorsally and light gray to white ventrally. There is a single gelatinous capsule surrounding the egg that is ca. 6.7 mm in diameter; the vitellus is ca. 2.3 mm in diameter (Bragg, 1944b). Gelatinous coats are sticky and easily acquire mud or debris, making them difficult to see in muddy water. The gelatinous eggs tend to stick together in various patterns, although they often appear strung together. These egg strings can be short or extend many cm in length.

DISTRIBUTION

Hurter's Spadefoot occurs from eastern Oklahoma and western Arkansas south to northwest Louisiana and the Texas Gulf Coast. It occurs east of the Balcones Escarpment in Texas (Smith and Buechner, 1947). An isolated population occurs in central Arkansas, but the ranges of this species and the Eastern Spadefoot do not overlap. Zweifel (1956a) indicated overlap in Louisiana, but this was based on an erroneous record in the Tulane collection (Wasserman, 1958). This species is limited to the eastern deciduous forest biome, and does not occur in prairies. Important distributional references include: Arkansas (Trauth et al., 2004), northern Louisiana (Dundee and Rossman, 1989), Oklahoma (Sievert and Sievert, 2006), and Texas (Hardy, 1995; Dixon, 2000).

FOSSIL RECORD

Pleistocene fossils of this species were reported by Davis (1973) from Peccary Cave in Arkansas. Although referred to *S. holbrookii*, the cave location is within the range of *S. hurterii*. Holman (1969)

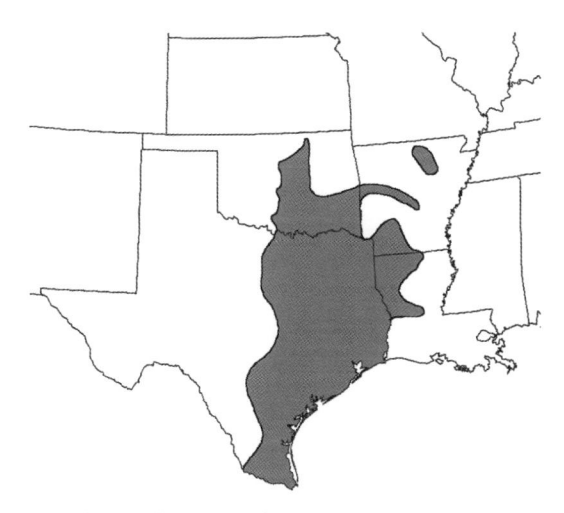

Distribution of *Scaphiopus hurterii*

noted two bones found in a cave in Texas could be referable to either this species or *S. couchii*.

SYSTEMATICS AND GEOGRAPHIC VARIATION

Hurter's Spadefoot was described from a single specimen collected in Texas (Strecker, 1910a). *Scaphiopus hurterii* is closely related to *S. holbrookii* and *S. couchii*, and is only distantly related to the western Spadefoot genus *Spea* (Tanner, 1939; Sattler, 1980; García-París et al., 2003). *Scaphiopus hurterii* is often included as a subspecies of *S. holbrookii*. The two are geographically separate, but their calls are essentially the same (Blair, 1958b), and they readily hybridize under laboratory conditions (Wasserman, 1958). Fertility of hybrid backcrosses is high. However, there are differences between the skulls of *S. hurterii* and *S. holbrookii*, especially in the fronto-parietal bone (Smith, 1937; Tanner, 1939; Zweifel, 1956a). In addition, there are size differences, differences in the presence or absence of a boss (a raised area) between the eyes, and minor tadpole prey differences. Finally, García-París et al. (2003) provided genetic evidence supporting recognition of two distinct species. The two spadefoots also are treated as separate species by Lannoo (2005), Crother (2008), and Elliott et al. (2009).

Natural hybridization between *S. couchii* and *S. hurterii* has been reported in Texas (Wasserman, 1957). In laboratory crosses, ♀ *S. couchii* × ♂ *S. hurterii* pairings produce a high percentage of hybrids. However, ♀ *S. hurterii* × ♂ *S. couchii* crosses are inviable (Wasserman, 1957). Crosses between *S. hurterii* and *S. holbrookii* are fully successful regardless of which species is the parent (Wasserman, 1958).

ADULT HABITAT

Scaphiopus hurterii is a woodland and savanna species that is found in oak-hickory and oak-pine associations. They also may be found in river floodplains that extend into grassland habitats. They prefer sand, sandy-clay, or calcareous soils that afford them the opportunity to burrow. In Oklahoma, they are more likely to be associated with calcareous soils than they are in Texas.

TERRESTRIAL ECOLOGY

Hurter's Spadefoot is a nocturnal species that spends most of its time in well-concealed burrows, emerging only to breed and occasionally forage. The terrestrial ecology and behavior of adult *S. hurterii* likely is similar to *S. holbrookii* but specific information is unavailable. Recent metamorphs emerge en masse and move across the landscape in waves of thousands of tiny toadlets. Dispersal occurs both by night and day until the tiny spadefoots reach their destination. Young *S. hurterii* seek shade and protection under debris and vegetation rather than by burrowing into the mud in order to avoid desiccation (Bragg, 1945b). Mortality in direct sunlight may be high. Dispersing toadlets may emerge from cover at night or after rain showers to feed.

CALLING ACTIVITY AND MATE SELECTION

Males enter breeding pools during or shortly after heavy rainfall and begin calling immediately. Males call mostly while floating in the deeper water of a temporary pool, although they may call from the bank or at the very edge of the pool (Bragg and Smith, 1942). Calling males on the bank tend to remain stationary, whereas those calling from the water move around a great deal. Females move to pools simultaneously with males. Mate selection is by scramble competition, that is, males will move toward and attempt to amplex any spadefoot moving in the pond. However, a female must touch a calling male for him to amplex her (Axtell, 1958). Inguinal amplexus and oviposition occur quickly. Males frequently wrestle with other males in clasping attempts, but the clasped male gives its warning vibration and call, which causes the amplexing male to release its hold.

The male advertisement call is a harsh, guttural loud noise that can be heard at some distance, and directs both males and females to the location of a breeding site. The call of *S. hurterii* is characterized by a minimum frequency range of 164–225 cps and a maximum frequency range of 245–375 cps, depending on temperature. Calls last from 0.3 to 0.6 sec (Blair, 1955c, 1958b) with a trill rate of 185 per sec (McAlister, 1959). The dominant frequency is 1,325–1,710 cps (Blair, 1958b). McAlister (1959) described the morphol-

Adult *Scaphiopus hurterii*, amplexus with eggs. Photo: Stan Trauth

ogy of the vocal structures. Wiest (1982) noted calling at 13–18°C over a 5-day period in Texas.

BREEDING SITES
This species breeds in shallow temporary pools, puddles, gravel pits, muddy pools in woodlands, ponds, and ditches (Bragg and Smith, 1942; Bragg, 1944b). These pools may be devoid of vegetation and exposed to direct sunlight. Algae are usually present and serve as a food source. Bragg (1964a) speculated that chemoreception might be important in breeding pool selection by this species but offered no empirical data.

REPRODUCTION
Spadefoots are the epitome of explosive breeders. Reproduction occurs only after heavy rainfall and for a very short time. Most eggs are oviposited the first night following torrential rains. Breeding may occur at any time of the year there is a favorable combination of warm temperatures and heavy precipitation. In Oklahoma, for example, breeding has been reported between March and September (Bragg and Smith, 1942; Bragg, 1944b, 1950a; Black, 1973b). Females tend to stagger breeding, such that not all females breed at the same time, regardless of favorable conditions. Other records of breeding dates include mid-March to June in Arkansas (Trauth et al., 1990). The mean clutch size of 8 females in Arkansas was 2,494 eggs (range 1,961 to 4,847) (Trauth et al., 1990). Oviposition occurs in shallow water near the shore, with eggs being placed over detritus and vegetation. Hatching occurs from 16 hrs to 2 days (Black, 1973b); Bragg (1944b) described the hatching process in detail.

LARVAL ECOLOGY
The larval period is extremely short, since spadefoots must complete metamorphosis in temporary pools that may dry very rapidly. Developmental periods last from <12–46 days (Bragg, 1945b, 1967; Black, 1973b). Feeding begins by day two, and tadpoles swim continuously in search of food. The small larvae eat almost any organic matter through filter feeding, are predaceous on other tadpoles and invertebrates as they grow larger, and may be cannibalistic, especially at metamorphosis. Bragg (1962b) noted that metamorphs that reenter water after leaving it are attacked by larvae in the final stages of development.

Tadpole of *Scaphiopus hurterii*. Photo: Stan Trauth

Large aggregations of tadpoles may be categorized as feeding, scooping, and metamorphic (Black, 1973b). These aggregations form based on visual cues (Black, 1973b). Feeding usually occurs in small to large aggregations that serve to keep detritus stirred up and maintain an optimal thermal regime (Bragg, 1944b, 1968). These aggregations tend to form a few days after hatching, prior to which larvae feed singly or in small groups. Tadpoles move around a pond within these loose or dense aggregations, churning up the substrate and feeding on organic detritus, presumably as food becomes scarcer. Bragg (1959) noted that large aggregations may not form in exceptionally wet years, perhaps indicating that aggregations are formed in response to decreasing food levels or shortened hydroperiods. Large aggregations also may serve an antipredator function, as predaceous beetle larvae fail to enter such aggregations (Bragg, 1967). Larval aggregations are able to survive declining water levels by burrowing into and deepening cavities in the soft mud bottom of pools, allowing them extra time to complete development (termed scooping aggregations by Bragg, 1959). Finally, larvae tend to aggregate just prior to metamorphosis as they depart en masse from breeding ponds, although in rainy years individuals may depart singly or in small groups rather than en masse (Bragg, 1959). Newly transformed individuals are 8–10 mm SUL (Bragg, 1944b).

DIET
Feeding occurs on land and generally not at the breeding pond. Hurter's Spadefoots are sit-and-wait predators, but once a prey item is observed, the spadefoot moves rapidly toward it and captures it with a final lunge. Postmetamorphs

Adult *Scaphiopus hurterii*.
Photo: David Dennis

eat a wide variety of prey, including a great many beetles, flies, lepidopteran larvae, crickets, grasshoppers, spiders, scorpions, sowbugs, and isopods (Jamieson and Trauth, 1996). Jamieson and Trauth (1996) found that 48% of the stomachs they examined contained food with a mean of 3 prey items per stomach. Bragg (1962c) noted that larvae occasionally feed on mosquito larvae.

PREDATION AND DEFENSE
Like the Eastern Spadefoot, this species has a noxious skin secretion that may deter predators. The secretion is peppery in odor, and many people experience a mild allergic reaction (skin irritation, sneezing) when mucous membranes are exposed to it. Calling Hurter's Spadefoots sink to the bottom of pools at the approach of a predator, but the species is not particularly wary on land. No information is available on predators.

COMMUNITY ECOLOGY
Scaphiopus couchii and *S. hurterii* come into close association in Texas, but hybridization is not common. Wasserman (1957) suggested that these species normally did not hybridize because of differences in habitat preferences (grasslands vs. woodlands), breeding phenology, advertisement call location (bank vs. water), advertisement call (Blair, 1955c), hybrid inviability, decreased fertility, and outright sterility. Of these, he considered habitat separation of primary importance. Larval *Scaphiopus hurterii* also have a growth-inhibiting substance that retards or delays growth in the larvae of *S. bombifrons* in experimental conditions (Black, 1973b). The presence of growth-inhibiting substances may explain why multiple species of spadefoots often do not breed in the same ponds despite similar breeding habitat preferences.

DISEASES, PARASITES, AND MALFORMATIONS
No information is available.

SUSCEPTIBILITY TO POTENTIAL STRESSORS
No information is available.

STATUS AND CONSERVATION
Hurter's Spadefoots face the usual threats from habitat loss, alteration, and fragmentation. Many are killed on the roads during their brief movements to breeding ponds and later dispersal (Bragg, 1944a). In Arkansas, the species is considered rare and susceptible to extirpation (Trauth et al., 2004) because of its limited distribution. Essentially no data are available on its status elsewhere. Popular accounts of spadefoot biology are in Bragg (1955) and, especially, Bragg's (1965) enjoyable book *Gnomes of the Night*.

Spea bombifrons (Cope, 1863)
Plains Spadefoot

ETYMOLOGY
bombifrons: from the Greek *bombos* meaning 'buzzing' and Latin 'frons' meaning 'leaf' or 'frond.' The name refers to the call of the species.

NOMENCLATURE
Synonyms: *Scaphiopus bombifrons*, *Scaphiopus hammondii bombifrons*

As is true for other *Spea*, considerable confusion sometimes exists in the older literature regarding species identification. Information on *Spea multiplicata* and *S. intermontana* is sometimes intermingled with accounts of the biology of this species (e.g., Smith, 1934).

IDENTIFICATION
Adults. The Plains Spadefoot is a gray to brown (rarely greenish) squat frog with irregular dark dorsal markings and numerous orange or red tubercles. Except for the tubercles, the skin usually is smooth and moist, but there is a considerable variation in the extent of rugosity. The dark markings may take the form of spots or blotches. Two to four light cream-colored stripes are present dorsally that curve backward from the eyes. These bands may be poorly defined or indistinct in some animals. The snout is blunt, and the eyes are large, with vertical pupils indicating nocturnal activity. The iris is golden. A bony boss (raised area) is present between the eyes. Venters are white. Like all spadefoots, there are distinct keratinized black digging spades on the hind feet. Males have dusky throats and dark keratinized patches on the first three fingers during the breeding season.

There is no sexual size dimorphism. In Oklahoma, males averaged 55.1 mm SUL (range 47 to 55 mm) and females averaged 49.6 mm SUL (range 36 to 62 mm) (King, 1960). The largest male in Iowa was 55.2 mm SUL, whereas the largest female was 59.4 mm SUL (Mabry and Christiansen, 1991). In Nebraska, males averaged 48 mm SUL (range 41 to 58 mm) whereas females averaged 49 mm SUL (range 42 to 58 mm) (Ballinger et al., 2010). The largest recorded Plains Spadefoot was a Kansas male measuring 64 mm SUL (Collins et al., 2010).

Larvae. The tadpole is tan, brown, or gray and somewhat transparent. Blotching increases with age. Bodies tend to be globular and somewhat depressed, and wider anteriorly more so than posteriorly when viewed from above. Eyes are positioned dorsally. The jaws are wide and frequently cuspate. In carnivorous tadpoles there is a distinctive beak on the upper jaw and a notch into which it fits in the lower jaw; omnivorous tadpoles have a more "typical" oral apparatus. The tail fin arises abruptly from the body and is clear. Larvae can reach a maximum size of 65 mm TL, although Bragg (1941b) observed *S. bombifrons* larvae ready to go through metamorphosis at 42–51 mm TL. Albino tadpoles were reported from Oklahoma (Hensley, 1959). Altig (1970) included them in his tadpole key, and Gilmore (1924) provided detailed illustrations of tadpoles and mouthparts. Bragg and Bragg (1958b) described a considerable amount of variation in the oral morphology of this species.

Eggs. Mature eggs are black dorsally and light colored ventrally; at oviposition, they are dark brown dorsally and pale yellow ventrally. Hoyt (1960) described three envelopes surrounding the egg, with the inner two being much smaller than the outer envelope. Eggs are deposited in small irregular jelly packets. Each packet contains 10–30 pigmented eggs that are attached to submerged vegetation at the edge of the water (Bragg, 1941b). The egg diameter varies among populations and ranges from 0.70 to 1.9 mm (Trowbridge, 1941; Zweifel, 1968b; Pomeroy, 1981; Mabry and Christiansen, 1991). The outer envelope is 2.25–3.5 mm, the middle envelope 1.65–2.3 mm, the innermost envelope 1.55–2.0 mm, and the vitellus 1.4–1.7 mm (Hoyt, 1960). Early development was described by Trowbridge and Trowbridge (1937) and Trowbridge (1941).

DISTRIBUTION
The Plains Spadefoot is a species of the Great Plains. Its range extends from southeastern Alberta across southern Saskatchewan to southwestern Manitoba, thence south through western Texas to southeastern Arizona. Populations occur across New Mexico to northeastern Arizona and southern Utah. The species' range extends eastward along the Missouri River all the way to the Mississippi River. Isolated populations

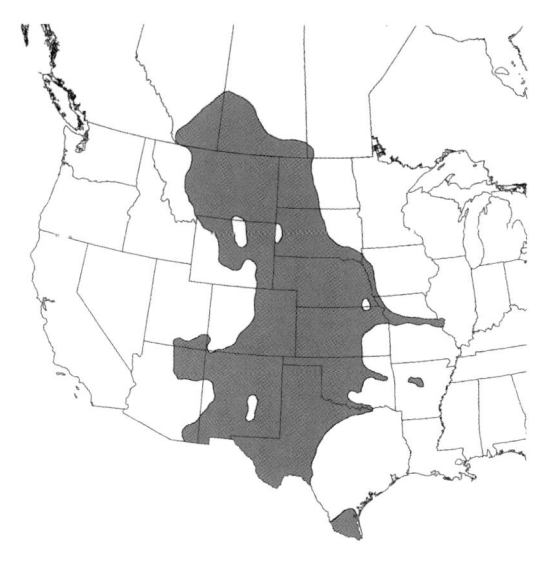

Distribution of *Spea bombifrons*

occur in the Loess Hills of western Iowa (recent range expansion; Christiansen and Mabry, 1985), northwest Missouri, west-central Arkansas (two locations in Arkansas River valley), east Texas, south Texas at the mouth of the Rio Grande, south-central Colorado, and in Mexico.

Important distributional references include: Alberta (Russell and Bauer, 2000), Arkansas (Trauth et al., 2004), Arizona (Brennan and Holycross, 2006), Colorado (Hammerson, 1999), Iowa (Christiansen and Mabry, 1985), Kansas (Collins et al., 2010), Manitoba (Cook and Hatch, 1964; Preston, 1982), Missouri (Johnson, 2000; Daniel and Edmond, 2004), Montana (Black, 1970; Maxell et al., 2003; Werner et al., 2004), Nebraska (Lynch, 1985; Ballinger et al., 2010; Fogell, 2010), New Mexico (Degenhardt et al., 1996), North Dakota (Wheeler and Wheeler, 1966; Hoberg and Gause, 1992), Oklahoma (Sievert and Sievert, 2006), Saskatchewan (Cook, 1965, 1978), South Dakota (Ballinger et al., 2000; Kiesow, 2006), Texas (Dixon, 2000), and Wyoming (Baxter and Stone, 1985).

FOSSIL RECORD

Spea bombifrons is known from Miocene and Pliocene fossil localities in Nebraska and Pleistocene localities in Colorado, Kansas, Nebraska, and New Mexico (Holman, 2003). Fossils are characterized by a lack of a preacetabular fossa on the ilia and by having an obsolete or no dorsal prominence on the ilia. The fossil species *Scaphiopus studeri* is probably synonymous with *Spea bombifrons*.

SYSTEMATICS AND GEOGRAPHIC VARIATION

Phylogenetic relationships within the genus *Spea* have historically been unclear, and there have been differing interpretations of relationships (Tanner, 1939; Bragg, 1945b; Sattler, 1980; Wiens and Titus, 1991; Holman, 2003; García-París et al., 2003). Wiens and Titus (1991) considered *S. bombifrons* as most closely related with clades containing two *S. intermontana* populations, particularly in Colorado. Holman (2003), however, identified two clades within the monophyletic genus *Spea*, an eastern clade consisting of *S. bombifrons* and a western clade consisting of *S. hammondii*, *S. intermontana*, and *S. multiplicata*. In Holman's (2003) scenario, based on fossil evidence, the eastern clade appeared by the middle Miocene, whereas the western clade first appeared in the late Pliocene. This would make *S. hammondii* the oldest member of the western clade, with the other members branching in the Pleistocene. Neither of these phylogenies is supported by mtDNA. *Spea bombifrons* is closely allied with *S. intermontana* and with southern California populations of *S. hammondii*; it is only distantly related to *S. multiplicata* (García-París et al., 2003).

Natural hybridization between *S. multiplicata* and *S. bombifrons* occurs in southeastern Arizona (Simovich, 1985, 1994; Simovich et al., 1991; Pfennig, 2003), the Four Corners region of the Colorado Plateau (Northen, 1970), and from southwestern New Mexico and Texas (Forester, 1973; Sattler, 1985). Experimental hybridization between *S. bombifrons* and *S. multiplicata* indicates a high level of genetic compatibility (Littlejohn, 1959; Brown, 1967; Forester, 1975). Hybrids involving ♀ *S. bombifrons* and ♂ *S. multiplicata* out-perform nonhybrid *S. bombifrons* by reaching metamorphosis faster (Pfennig and Simovich, 2002). These authors suggested that *S. bombifrons* females might actually gain selective advantages through hybridization when breeding pools are very ephemeral.

ADULT HABITAT

The Plains Spadefoot is a species of grassland prairies throughout central North America, especially in the Great Plains. It occurs in other grassland habitats, however, such as in the open savanna ponderosa pine forests north of Flagstaff, Arizona (Aitchison and Tomko, 1974). It is found in short-grass prairies as well as in agricultural pastures and croplands, but not in river bottoms or woodlands. This species prefers sandy, gravelly, or loamy friable soils in which it can dig. The species occurs to 2,134 m in Arizona (Aitchison and Tomko, 1974), from 900 to 2,200 m in New Mexico (Degenhardt et al., 1996), and to 2,440 m in Colorado (Hammerson, 1999).

TERRESTRIAL ECOLOGY

During the nonbreeding season, *S. bombifrons* prefers sparse to dense grassland habitats within 300 m of their breeding site (Cruesere and Whitford, 1976). Nocturnal activity occurs throughout the warmer portions of the year. For example, Plains Spadefoots are observed from mid-January through mid-October in Kansas and Iowa, depending upon the weather, but peak activity is from late April through early July (Heinrich and Kaufman, 1985; Mabry and Christiansen, 1991; Collins et al., 2010). Adults cease activity prior to juveniles; juveniles presumably are active later in the year to maximize foraging and growth prior to dormancy. Spadefoots usually spend their days in underground retreats and emerge only at night to forage. They dig their own burrows backward using their keratinized digging spade, and they leave no trace of the burrow opening on the surface to indicate their presence. They also are frequently found in burrows of the black-tailed prairie dog (*Cynomys ludovicianus*) (Kretzer and Cully, 2001; Lomolino and Smith, 2003), gophers, or spotted ground squirrels. Emergence from burrows occurs at temperatures >11°C (Bragg and Smith, 1942). Overwintering spadefoots dig deep into the soil and have been found 90 cm below the ground surface (Black, 1970). Juvenile (and presumably adult) *Spea bombifrons* are freeze intolerant, but some individuals have been reported to be able to supercool to −4°C (Swanson and Graves, 1995). This could be important for juveniles that might not be able to dig as deep as adults to avoid freezing temperatures.

In the spring, *S. bombifrons* may be active several days prior to actual arrival of a rain event that will initiate breeding. They can move distances of at least 1 km to a breeding site within a single night (Landreth and Christensen, 1971).

In experimental trials, Plains Spadefoots were able to use the sun's position for orientation toward a breeding site and to use celestial cues during postbreeding activities (Landreth and Christensen, 1971). Since they generally move at night to the breeding ponds, prebreeding surface activity may allow them to use celestial cues to orient in the right direction. Once a few individuals reach a breeding site and begin calling, auditory and perhaps celestial cues are used as a guide for later arriving animals.

Upon metamorphosis, young *S. bombifrons* burrow rapidly into the mud in order to avoid desiccation (Bragg, 1945b). They emerge at night or after rain showers to feed but otherwise remain in the burrow. After about a week, the toadlets begin to disperse away from the breeding site, making a new burrow each night. According to Werner et al. (2004), adults may disperse up to 2.3 km from breeding ponds, although most spadefoots remain within 400 m. Hammerson (1999) recorded movements of 60–150 m per night over 1–2 nights of observation.

Plains Spadefoots are photopositive in their phototactic response, suggesting they use both sunlight and moonlight when feeding and going about their activities (Jaeger and Hailman, 1973). They are sensitive to light in the blue spectrum ("blue-mode response") (Hailman and Jaeger, 1974). Plains Spadefoots likely have true color vision.

CALLING ACTIVITY AND MATE SELECTION

Spea bombifrons is an explosive breeder, with choruses forming only after torrential rainstorms in spring and summer, although Gilmore (1924) reported the initiation of breeding after several days of drizzling rain. Females arrive at pools shortly after chorusing has begun. Most breeding occurs at night, although rare breeding choruses form by day, especially in the late afternoon (Gilmore, 1924; Trowbridge and Trowbridge, 1937; Black, 1970). Calling Plains Spadefoots have been observed at temperatures as low as 9–12°C (Bragg, 1945b; Hammerson, 1999).

Males call while floating in the water (rarely from the shore), and they tend to stay in one location throughout the calling period (Bragg and Smith, 1942). Cruesere and Whitford (1976) recorded them most often in sparse vegetation > 12 cm in water depth. Females enter the pool and move directly toward a calling male. Inguinal amplexus is initiated when the female actually touches the male. Even juveniles (28–38 mm SUL) have a well-developed clasping reflex (Bragg, 1958b). Amplexed pairs are wary and readily dive below the water's surface when alarmed. As the male tightly clasps his partner, abdominal contractions force the extrusion of a small number of eggs. At the same time, the male arches his back bringing his cloaca in position to emit semen and fertilize them.

The call of *S. bombifrons* is a trill (likened to a "waah" or "waac" [Black, 1970] or as a "weird plaintive cry" [Gilmore, 1924]) with a frequency of 1,250–1,800 cps, a range of 0.4 to 0.7 sec in duration, 22–47 trills per call, and 41–74 trills per sec (Blair, 1955c; McAlister, 1959; Northen, 1970; Forester, 1973). Specific values will vary with temperature. The call has a long carrying capacity and can be heard for > 1.6 km away. McAlister (1959) described the morphology of the vocal structures. Literature records of different trill rates (fast and slow) within this species (e.g., Pierce, 1976) actually refer to *S. bombifrons* and *S. multiplicata*.

Plains Spadefoots spend only a short period of time at the breeding site. In the Sonoran Desert of Arizona, for example, the mean chorus duration was 1.8 days and the mean chorus size was 27 males (Sullivan and Sullivan, 1985; Sullivan, 1985a, 1989). Most reproductive activity occurs during the first night of breeding. Even if calling males persist after the first night, little if any successful breeding takes place. Other reports (Gilmore, 1924) report chorus sizes in the hundreds. As might be expected, males outnumber females at breeding pools; the latter depart quickly after breeding.

BREEDING SITES

This species breeds in temporary pools and puddles that form after heavy thunderstorms, but it may use slightly deeper water than other spadefoots. Anderson et al. (1999) recorded presence in pools < 30 cm in depth. They breed in roadside ditches, irrigation pits in playa grasslands, flooded shallow fields, agricultural fields, stock tanks, and deep muddy pools (Gilmore, 1924; Bragg and Smith, 1942; Anderson et al., 1999). Plains Spadefoots frequently call from bison wallows, and these sites occasionally may hold water long enough for successful metamorphosis (Bragg and Smith, 1942; Gerlanc, 1999). Breeding ponds often appear barren but may have some vegetation.

REPRODUCTION

Breeding occurs opportunistically only after rainfall and occurs at almost any time during the warm season, depending upon location. Most breeding occurs in late May, June, and early July. Specific records include March to September in Oklahoma (Trowbridge and Trowbridge, 1937; Bragg and Smith, 1942; Bragg, 1950a; Black, 1973b), late March to mid-August in Kansas (Collins et al., 2010), April and June in Arkansas (Trauth et al., 1990, 2004), May to July in Colorado and Wyoming (Gilmore, 1924; Baxter and Stone, 1985), and July to August in New Mexico (Bragg, 1941b). The rain does not have to be of great volume, but it usually occurs torrentially. As little as 12.7 mm of rainfall can initiate breeding, but it must fall very rapidly (Bragg, 1945b); even 50 mm of rain might not stimulate breeding if it occurs in a gentle or nonviolent manner (Bragg and Smith, 1942, but see Gilmore, 1924). Bragg and Smith (1942) reported breeding at 12°C.

Oviposition occurs at night and in the early hours of the morning after sunrise. In some instances, breeding can occur in the late afternoon after heavy rain. Eggs are deposited in small masses attached to vegetation. They may be placed several cm below the water's surface in shallow water. Woodward (1987a) reported a mean clutch size of 1,600 eggs, whereas a single clutch from Arkansas contained 1,697 eggs (Trauth et al., 1990) and one from Kansas 1,909 eggs (Hoyt, 1960). In Iowa, Mabry and Christiansen (1991) recorded a mean clutch size of 2,626 eggs (range 1,572 to 3,844). In contrast, Gilmore (1924) reported clutch sizes of 200–500 eggs per mass, with small masses containing only 10–50 eggs; these figures undoubtedly refer to partial clutches rather than total female fecundity. Body size is positively correlated with the number of eggs (Mabry and Christiansen, 1991). Eggs hatch from 20 hrs (at 30°C) to 12 days (at 10°C) (Justus et al., 1977; Black, 1973b).

LARVAL ECOLOGY

The larval period is extremely short, since spadefoots must complete metamorphosis in temporary pools that may dry very rapidly. It lasts from as few as 13–14 days to ca. 30–40 days, with carnivorous larvae metamorphosing faster than omnivorous larvae (Gilmore, 1924; Bragg, 1945b, 1967; King, 1960; Black, 1973b). Bragg (1966) kept larvae 145–146 days in captivity. Larvae eat almost any organic matter through filter feeding, are predaceous on other tadpoles and invertebrates, and may be aggressively cannibalistic. They tend to perform equally well regardless of diet, which enables them to utilize whatever resources are available in their ephemeral habitats (Ledón-Rettig et al., 2009). Feeding begins immediately after hatching, in large aggregations that serve to keep detritus stirred up and maintain an optimal thermal regime (Bragg and King, 1960; Black, 1973b). Aggregations form in response to food resources, temperature, light intensity, or a combination of these factors (Bragg, 1964b), and they may be maintained visually (Black, 1973b).

Tadpoles respond positively to chemoreceptive signals from different types of food in experimental settings, and avoid attacking larvae of *Anaxyrus speciosus* (Bragg, 1960a). Large aggregations also may serve an antipredator function. Larvae in aggregations are able to survive declining water levels by burrowing into and deepening cavities in the soft mud bottom of pools, allowing them extra time to complete development. They are constantly moving, stirring up silt. Even if most larvae die, a few are able to survive within the mass to metamorphose and leave the pool (Bragg and King, 1960; Black, 1973b). Still, pool desiccation is probably the chief source of larval mortality.

Larvae of *Spea bombifrons* have two distinct larval morphologies, but the diet of diatoms, arthropods, macrophytes, cyanobacteria, and detritus is similar between them (Ghioca, 2005). Larvae of both morphotypes also eat carrion and may attack other tadpoles. That there are two larval morphologies has led to considerable confusion in the literature as to larval identification (Smith, 1934; Bragg, 1941a, 1956; Orton, 1954). Bragg and King (1960) noted that larval aggregations consisted of larvae of different sizes, which was not seen in aggregations of *Scaphiopus*. This observation might have been of larvae with different morphologies rather than larvae at different stages of development. The carnivorous morph is large, with modified mouthparts that are used to attack both conspecific and heterospecific larvae. The omnivorous morph is rounder and smaller, with mouthparts used mostly for grazing and filter feeding. Carnivorous larvae are able to recognize kin and avoid eating them; omnivorous larvae do not recognize kin and occasionally may eat their relatives (Pfennig, 1999). Discrimination by carnivores reduces the potential for eating kin by the larval morphology most likely to do so.

Tadpoles of *Scaphiopus bombifrons*. Photo: Eric Dallalio, courtesy of the Montana Natural Heritage Program

Bragg (1964c) noted the propensity for larvae of various morphotypes to attack conspecifics and heterospecific spadefoots.

Aggregations of *Spea bombifrons* contain larvae with both morphologies, although the majority of larvae are carnivorous. Mixed aggregations occur regardless of whether other *Spea* are present within the same breeding pool (Ghioca, 2005). At metamorphosis, both larval types are the same size, and both have the same developmental period regardless of the adjacent land use. When water is deep, carnivorous larvae may be larger than they are in years when pools are shallower. Carnivorous larvae also have wider and shorter guts than omnivorous larvae. Indeed, there is a certain degree of innate phenotypic plasticity in feeding morphologies, even among sibships, with different diets inducing different phenotypes. For example, tadpoles fed on a diet of fairy shrimp have shorter guts than those fed on detritus, although the variation in gut phenotypes is the same among tadpoles fed on different diets (Ledón-Rettig et al., 2008).

At metamorphosis, toadlets are 12–14 mm in body length but still may contain a considerable amount of tail. Even these toadlets begin feeding immediately. Mabry and Christiansen (1991) reported recent metamorphs at 19–21 mm SUL and Degenhardt et al. (1996) 18–22 mm SUL. Bragg (1962b) noted that metamorphs that reenter water after leaving it are attacked by larvae in the final stages of larval development.

Tadpoles, presumably *S. bombifrons*, were collected in a hot spring in Thermopolis, Wyoming, at a water temperature of 37–38.4°C (Brues, 1932). When disturbed, they scattered into water 45°C before quickly returning to cooler water.

DIET

Larval *S. bombifrons* (both omnivorous and carnivorous morphs) are predaceous on other tadpoles, including *Scaphiopus couchii* (Bragg, 1962b). They consume a variety of organic matter, and a strictly carnivorous diet may not be sufficient for proper development. Bragg (1962c) noted that larvae feed on several species of fairy shrimp, and much experimental work has involved using fairy shrimp as prey. Postmetamorphs eat a variety of invertebrates, including flies, gnats, mosquitoes, ants, moths, beetles, true bugs, thrips, leafhoppers, lacewings, aphids, lepidopteran larvae, collembolans, mole crickets, termites, and spiders (Trowbridge and Trowbridge, 1937; Bragg, 1941b, 1944a; Whitaker et al., 1977; *in* Johnson, 2000). There is a report of a Plains Spadefoot eating a mouse (*Peromyscus*) (*in* Hammerson, 1999). As might be expected, juveniles eat small prey.

PREDATION AND DEFENSE

Spea bombifrons has a sticky, musky smelling secretion that is noxious and probably helps to deter predation. The odor has been reported to smell like roasted nuts, buttered popcorn, corn chips, or garlic (Hammerson, 1999). They can exude a considerable amount of mucous, and some people have a mild allergic reaction to the secretion; they also release fluid from the vent. When handled or confronted by a predator, Plains Spadefoots puff-up and repeatedly vocalize a sound like "oh" or "bat" (Hammerson, 1999). They arch their bodies and close their eyes, extend the front limbs, and tightly tuck the rear limbs into the body.

Predators undoubtedly include mammals (including domestic cats), birds (Swainson's Hawk, Burrowing Owls, Black-crowned Night Heron), rattlesnakes (*Crotalus* sp.), and garter snakes (*Thamnophis* sp.) on postmetamorphs, but few reports are available (Stabler, 1948; Sexton and Marion, 1974; Hammerson, 1999; Werner et al., 2004; Collins et al., 2010). Smith et al. (1965) reported a spotted ground squirrel (*Spermophilus spilosoma*) evicting a Plains Spadefoot Toad from its burrow.

POPULATION BIOLOGY

Postmetamorphic growth is rapid, with Trowbridge and Trowbridge (1937) noting a doubling of body length after three months. Sexual maturity is probably reached by the first summer after metamorphosis, although first breeding likely does not take place until the following year. In Iowa, the smallest sexually mature male was 31.1 mm SUL, although juveniles as large as 38.2 mm SUL were observed; the smallest sexually mature female was 32.5 mm SUL (Mabry and Christiansen, 1991). These authors suggested sexual maturity occurred between 31 and 38 mm SUL for males and 32–40 mm SUL for females.

Adult *Scaphiopus bombifrons*. Photo: David Dennis

Wright and Wright (1949) reported a larger size at sexual maturity, i.e., 38 mm for males and 40 mm for females.

COMMUNITY ECOLOGY

Larval *Spea bombifrons* of different age classes usually are not found in the same breeding pool. This is because the larger larvae out-compete smaller larvae and even have growth-inhibiting factors to retard growth of the smaller larvae (Black, 1973b). In turn, larvae of *S. bombifrons* are growth-inhibited by larval *S. hurterii*. Plains Spadefoot larvae also impact aspects of the life history of other species found within the same breeding pools. For example, predaceous *S. bombifrons* larvae may attack and eat many *Scaphiopus couchii* larvae within a pool. By reducing the number of *S. couchii*, predation actually increases the likelihood that survivors will metamorphose successfully (Bragg, 1962b). In turn, predaceous *Spea bombifrons* have an increased developmental rate when compared to nonpredaceous conspecifics that allows them to transform prior to the desiccation of the breeding pool.

When *S. multiplicata* and *S. bombifrons* breed in the same pool, the frequency of the different tadpole morphotypes is very different than when only one species is found in a pool. When sympatric, *S. multiplicata* larvae nearly all take on the omnivorous morphotype, whereas *S. bombifrons* larvae assume the carnivorous morphotype. This is not surprising, since *S. multiplicata* is a superior competitor on detritus whereas *S. bombifrons* is a superior competitor on fairy shrimp. When allopatric, however, both species' larvae develop both carnivorous and omnivorous morphotypes. The differences in the frequency of the morphotypes appear to have evolved in sympatry in order to reduce competition for scarce food resources. The manner in which these desert species partition resources by changing larval morphology is an example of character displacement (Pfennig and Murphy, 2000, 2002, 2003).

Two or three different species of spadefoot may breed in close proximity to one another (e.g., *S. multiplicata*, *S. bombifrons*, *Scaphiopus couchii*), yet the species often do not breed in the same temporary pool. One species will breed in one pool while a different species breeds in a pool only a short distance away. On the occasions when they do breed together, they often occur in similar numbers, at least in some areas (Bragg, 1941b, 1945b). They may call from different locations (shore vs. water) (Bragg and Smith, 1942), but there is considerable overlap in calling position. Differences in mating calls help keep the species from interbreeding (Forester, 1973). It also seems likely that a priority effect occurs, that is, the species that arrives at a pool first claims that pool and the pool subsequently is avoided by heterospecifics. Conspecifics also may cluster together when calling from the same pool. For example, *S. couchii* may call from the same pools as *Spea multiplicata* and *S. bombifrons* with *Scaphiopus couchii* preferring the deeper parts of the pool, although there appears to be no real habitat segregation (e.g., McAlister, 1958).

Interspecific amplexus between *Spea bombifrons* and *S. multiplicata* is not unusual (Cruesere and Whitford, 1976). Natural hybridization between *S. multiplicata* and *S. bombifrons* occurs in southeastern Arizona. Hybridization occurs more frequently in smaller pools where the different species are likely to come into close contact (Simovich, 1994). Large numbers of F_1 hybrids and the offspring of hybrid backcrosses are observed in the area. Here, *S. bombifrons* females are more likely to hybridize than *S. multiplicata* females or *S. bombifrons* males (Simovich, 1985; Pfennig and Simovich, 2002). Hybrid males are sterile, and hybrid females produce about half the eggs of the parentals. In laboratory experiments, Simovich et al. (1991) found that hybrid tadpoles develop faster and to a larger size than the tadpoles of either parental species; they also have higher survivorship. However, they tend to be intermediate in size at metamorphosis when compared to the parentals. Thus, the life history characteristics of the hybrid offspring may afford them some advantages over the parentals, especially when exposed to conditions of short hydroperiods, despite problems with fertility and fecundity.

Not surprisingly, hybridization also tends to occur most often in the most temporary ponds, and the duration of hydroperiod is inversely correlated with hybridization. Pfennig (2003), however, found that the rates of hybridization between these species had declined over a 27 yr period in this region. After testing a number of hypotheses, she concluded that despite the larval advantages in short hydroperiod ponds, the hybrids are less fit than the parental species, and thus reproductive isolation as a result of the evolution of premating isolating mechanisms (termed reinforcement) best explained decreasing hybridization.

In addition to hybridization, the presence of Plains Spadefoots has specific life history consequences for the less competitive Mexican Spadefoot in areas of sympatry. *Spea multiplicata* females tend to be in poorer body condition when they are in sympatry with *S. bombifrons* than when in allopatry. Since the abundance of *S. bombifrons* is inversely correlated with mean body condition of female *S. multiplicata*, this suggests that severe competition is correlated with poor body condition in the latter species (Pfennig and Martin, 2009). Indeed, competition plays an important role in divergent trait evolution in spadefoots (Rice et al., 2009).

DISEASES, PARASITES, AND MALFORMATIONS
The fungus *Saprolegnia* sp. was reported from tadpoles in Oklahoma (Bragg and Bragg, 1958a).

Breeding habitat of *Scaphiopus bombifrons*.
Photo: Kerry Griffis-Kyle

Plains Spadefoots are parasitized by the trematodes *Neodiplorchis scaphiopodis* and *Polystoma nearcticum* and the nematodes *Aplectana incerta, A. itzocanensis,* and *Physaloptera* sp. (Brooks, 1976; Goldberg and Bursey, 2002b). The mite *Eutrombicula alfreddugesi* is known from Plains Spadefoots in Texas (Mertins et al., 2011).

SUSCEPTIBILITY TO POTENTIAL STRESSORS
No information is available.

STATUS AND CONSERVATION
Because of their secretive behavior, *Spea bombifrons* may be easily overlooked during faunal surveys. For example, historical surveys in the Flint Hills of Kansas did not record this species, although later surveys did (Busby and Parmelee, 1996). In contrast, Hossack et al. (2005) found them in recent surveys in Theodore Roosevelt National Park and noted that they had been found in surveys going back to the 1920s. Detecting this species is not always easy, even when present. Sampling biases occur when using automated frog call devices, as calling spadefoots could easily be missed at their temporary breeding pools. For example, this species has occasionally been misidentified as *Pseudacris maculata* or *Hyla chrysoscelis* during call surveys (Lotz and Allen, 2007). However, they may be detected by recordings when site visits fail to observe them (Corn et al., 2000). Thus the determination of range expansion or population trends is difficult because of the potential for sampling biases.

Undoubtedly, Plains Spadefoots face the usual threats from habitat loss, alteration, and fragmentation. They frequently are killed in large numbers on roads during their brief movements to breeding ponds and later dispersal (Bragg, 1944a). There are no indications, however, that this species has declined throughout much of its range (e.g., Christiansen, 1981; Weller and Green, 1997; Hammerson, 1999). In Arkansas it is considered extremely rare and vulnerable to extinction, and it is considered a "Species of Concern" in Montana because of the relatively few available records (Werner et al., 2004).

The effects of habitat alteration on this species may be unclear or even misleading. For example, larval densities at playa breeding sites are not different among playa habitats with different adjacent land uses, suggesting widespread accommodation to human activity (Anderson et al., 1999; Ghioca, 2005). Abundance may actually appear high in agricultural areas when compared to other land uses (Gray et al., 2004b). On a landscape scale, occupancy of playa wetlands is positively associated with decreasing distance between pools and with increasing interplaya landscape complexity. Such a relationship may be of recent occurrence, as nestedness patterns may result from the inability of this small species to disperse across geometrically complex agricultural landscapes that are found in the Texas Panhandle (Gray, 2002; Gray et al., 2004a, 2004b). In addition, a major predator of spadefoot larvae (*Ambystoma mavortium*) tends to be less common in agricultural settings. Thus, habitat fragmentation historically might have affected this species positively and negatively despite the species' current abundance and ability to persist in agricultural landscapes.

Habitat alteration also may have increased the likelihood of hybridization in areas of contact such as in southeastern Arizona. Simovich (1994) noted that the vegetation and precipitation runoff patterns changed substantially after the introduction of cattle grazing. Runoff previously allowed to form in scattered pools is now diverted to more permanent stock tanks. Spadefoots now breed in these pools, which results in different species breeding in close proximity and the potential for competition. Modification of water-flow patterns coupled with periodic drought could exacerbate hybridization potential.

There are certain trapping biases when sampling for larval *Spea bombifrons* as part of community analyses. Plains Spadefoots are readily captured in funnel traps, but less likely to be captured by seining or dip-netting (Ghioca and Smith, 2007). Thus, trapping tends to overestimate the proportion of Plains Spadefoots within a pool. Traps also tend to catch more of the carnivorous morph larvae compared to omnivorous morphs.

COMMENT
Popular accounts of spadefoot biology are in Bragg (1955) and, especially, Bragg's (1965) book *Gnomes of the Night*.

Spea hammondii (Baird, 1859)
Western Spadefoot

ETYMOLOGY
hammondii: a patronym honoring Dr. John F. Hammond (1820–1886), who captured the type specimen at Fort Redding, California, during the Pacific Railroad Surveys.

NOMENCLATURE
Synonyms: None

Much of the literature on "*Scaphiopus hammondii*" actually refers to *Spea multiplicata* or other *Spea* (e.g., all of A.N. Bragg's work in Oklahoma; Van Denburgh, 1924; Slevin, 1928; Livezey and Wright, 1947; Brown, 1967 [in part]).

IDENTIFICATION
Adults. Adults are greenish, gray, or brown in background coloration with irregular markings and dark orange or reddish-tipped tubercles. There is a faint hourglass marking dorsally, consisting of two to four irregular, light-colored stripes, although this pattern may be absent. Eyes are large, with a distinctive vertical pupil indicative of nocturnal activity; the iris is pale gold. Hind limbs are short. The rear feet have black keratinized spades that are used for digging backward into the soil. Venters are cream colored to off-white. Juveniles may be more spotted than adults. Brown (1976) reported a mean SUL for males of 53.2 mm (range 45 to 58.5 mm). The maximum size is 75 mm SUL (Lemm, 2006).

Larvae. The body is gray and somewhat transparent. It tends to be globular to depressed and wider anteriorly more than posteriorly when viewed from above. The jaws are wide and frequently cuspate. In carnivorous tadpoles, there is a distinctive beak on the upper jaw and a notch into which it fits in the lower jaw; omnivorous tadpoles have a more "typical" oral apparatus. The tail fin does not arise abruptly from the body. Larvae reach a maximum size of 71 mm TL (Storer, 1925). Albino tadpoles were reported from Madera County (Childs, 1953; Hensley, 1959).

Eggs. The dark greenish-olive eggs are small, with two gelatinous envelopes surrounding the vitellus. The diameter of the vitellus is 1.5–1.6 mm, the inner envelope is 1.63–1.88 mm, and the outer envelope is 3.25–3.75 mm (Storer, 1925).

DISTRIBUTION
Western Spadefoots occur from inland north-central California (near Redding in Shasta County) southward along the coastal regions to northwestern Baja California Norte. The species has disappeared from substantial portions of its historic range. Important distributional references include Storer (1925), Jennings and Hayes (1994b), Grismer (2002), and Lemm (2006).

FOSSIL RECORD
Fossils referred to *Spea hammondii* are known from Pliocene deposits in Idaho and Pleistocene deposits in Arizona, Nevada, and New Mexico (Holman, 2003). Additional fossils referred to as either *S. hammondii* or *S. bombifrons* are known from New Mexico and Texas. Clearly, this species had a much wider distribution during the Pliocene–Pleistocene than it does today. Holman (2003) provided identifying characteristics.

SYSTEMATICS AND GEOGRAPHIC VARIATION
Phylogenetic relationships within the genus *Spea* historically have been unclear, and there are differing interpretations of its evolutionary history (Tanner, 1939; Bragg, 1945b; Brown, 1976;

Distribution of *Spea hammondii*

Wiens and Titus, 1991; García-París et al., 2003; Holman, 2003). Brown (1976) noted differences in the mating calls and physiology between California *S. hammondii* and Arizona "*S. hammondii*" (i.e., *S. multiplicata*) and separated these former subspecies into full species. Tanner (1989) considered *S. multiplicata* to be a subspecies of *S. hammondii* and relegated *S. stagnalis* to subspecific status, *Spea hammondii stagnalis*, based on osteological characters. Wiens and Titus (1991) considered *S. h. stagnalis* to be conspecific with *S. multiplicata*.

Holman (2003) identified two clades within the monophyletic genus *Spea*, an eastern clade consisting of *S. bombifrons* and a western clade consisting of *S. hammondii*, *S. intermontana*, and *S. multiplicata*. In Holman's (2003) scenario, based on fossil evidence, the eastern clade appeared by the middle Miocene, whereas the western clade first appeared in the late Pliocene. This would make *S. hammondii* the oldest member of the western clade, with the other members branching in the Pleistocene. Wiens and Titus (1991) considered *S. hammondii* to be most closely related with a clade containing *S. intermontana* and *S. bombifrons* and more distantly related to *S. multiplicata*. This arrangement mirrored Sattler's (1980) results and was confirmed using mtDNA, although García-París et al. (2003) identified three lineages within *S. hammondii*. Two (in San Diego County, California) were closely related, but a third (in Alameda County) was much more distantly separated. This result indicates the potential for cryptic species within *S. hammondii*.

Experimental hybridization between *S. hammondii* and *S. multiplicata*, *S. intermontana*, and *S. bombifrons* indicate a high level of genetic compatibility (Brown, 1967).

ADULT HABITAT

The Western Spadefoot is primarily a grassland species, where it inhabits parklands, chaparral, scrub, and small meadows within the foothill oak woodlands of central and southern California (Morafka and Banta, 1976). Canopy cover is sparse, and climatic extremes are harsh. Soils must be friable to allow digging well below the surface, and sandy soils are preferred. The species occurs from near sea level to 1,410 m (Ervin et al., 2001).

TERRESTRIAL ECOLOGY

Activity can occur throughout the year depending on weather conditions, but most activity occurs from October through May. Emergence occurs at night under favorable weather conditions for foraging, but even a little light can make them dig rapidly into the substrate. For most of the year, however, the species remains in burrows that can be as deep as 1 m below the soil surface (Stebbins, 1972). The Western Spadefoot also uses rodent burrows, such as those of the California Ground Squirrel (*Spermophilus beecheyi*), to avoid drought, heat, and cold. Recently transformed juveniles hide in deep mud cracks in the vicinity of the drying breeding pond.

Western Spadefoots are photopositive in their phototactic response, suggesting they use both sunlight and moonlight when feeding and going about their daily activities (Jaeger and Hailman, 1973). They are sensitive to light in the blue spectrum ("blue-mode response") (Hailman and Jaeger, 1974). Western Spadefoots likely have true color vision.

CALLING ACTIVITY AND MATE SELECTION

Emergence is triggered by a combination of rainfall and favorable temperatures, but much of the breeding biology remains poorly documented. Calls may be heard during much of the year when environmental conditions permit, and breeding may be more opportunistic than literature records suggest. Unlike most spadefoots that breed during the warmest season, *Spea hammondii* frequently breeds during the coolest season of the year in California. Most observations are from the winter and spring rainy season from October to May (Storer, 1925; Ervin et al., 2005), but Ervin and Cass (2007) reported breeding in August, Brown (1976) noted that choruses formed from November to July, and Lemm (2006) reported breeding from January to June. Breeding may be initiated by rainfall, but *S. hammondii* will continue to call for as long as two to three weeks following a previous rain event (Storer, 1925; Brown, 1976; Morey and Reznick, 2004). Breeding aggregations can consist of several to >1,000 males (Jennings and Hayes, 1994b).

Males call while concealed at the edge of a breeding pool, with only part of the body immersed in water (Brown, 1976). They rarely are observed in the open area of the breeding pool.

Eggs of *Spea hammondii*. Photo: Sara Schuster

Calling occurs asynchronously with many males calling at once. The call of *S. hammondii* is a slow trill with a dominant frequency of 1,370 to 1,600 cps, a range of 0.32 to 1.25 sec in duration, 17–45 trills per call, and 28–50 trills per sec (Brown, 1976); specific values vary with temperature. Storer (1925) likened the call to a prolonged low-toned "tirr-r-r-r." The call carries a long distance, and it has a ventriloquist quality that makes it difficult for a human observer to locate a calling male. Females make a somewhat muffled sound, especially when handled.

BREEDING SITES

This species breeds in unshaded shallow temporary pools and puddles, with water temperatures between 9 and 30°C (Brown, 1967). Eggs also may be deposited in temporary pools along or in streamcourses or in road ruts (Ervin et al., 2005). Such pools lack predators such as crayfish, American Bullfrogs, and fish. In laboratory trials, tadpoles grew faster and had much greater survival through metamorphosis in alkaline water than in neutral tap water (Burgess, 1950). Many of the natural breeding pools of spadefoots tend to be alkaline, and Storer (1925) noted that high alkalinity did not seem to be a deterrent to Western Spadefoots.

REPRODUCTION

Eggs are deposited in shallow water (< 500 mm) in irregular clusters of 10–42 eggs per cluster (Storer, 1925), but there appear to be no counts of the total number of eggs deposited. Lemm (2006) reported clutch sizes of as many as 500 eggs. Clusters may or may not be cylindrical. These are attached to detritus and vegetation from 25 to 100 mm below the water's surface, much as in other *Spea*. Hatching occurs in <1–6 days depending on temperature (Brown, 1967).

LARVAL ECOLOGY

The larvae of *S. hammondii* are developmentally plastic in response to ponds of differing hydroperiods. When hydroperiods are short, development is accelerated in order to ensure that some tadpoles reach metamorphosis. In ponds with long hydroperiods, developmental rates are slower, allowing tadpoles to reach a larger size prior to metamorphosis, which presumably confers a higher fitness in terrestrial habitats. Not surprisingly, both larval duration and mass at metamorphosis are positively correlated with hydroperiod (Morey and Reznick, 2004). Specific reports of larval duration include 55–60 (Brown, 1976) and 30–80 (mean 58) days (Morey and Reznick, 2004; Lemm, 2006). Pond drying is likely a major source of larval mortality. Morey and Reznick (2004) found that 15% of the breeding pools used by *Scaphiopus couchii* in California dried before metamorphosis began.

Tadpoles apparently can use the rate of water volume reduction within a pond to gauge the developmental rate necessary in order to metamorphose (Denver et al., 1998). Tadpoles have a continuum of responses rather than a threshold response, with acceleration of development approximating the rate of water reduction. Interestingly, developmental rates can be slowed by adding water to the experimental pond or pool, which results in an increase in larval period and an increase in larval mass. Denver et al. (1998) suggested that tadpoles are able to gauge the swimming volume available to them and to adjust their developmental rates accordingly.

In addition to hydroperiod, food resources play a critical role in the larval development of *Spea hammondii*. It appears that there is a threshold size that must be met to complete metamorphosis (Morey and Reznick, 2000). Low amounts of food cause slower growth, which delays the attainment of the threshold size and could make individuals vulnerable to pool desiccation. With sufficient resources, tadpoles gain the threshold size and then have a degree of plasticity in the extent to which they remain in a breeding pool. With long hydroperiods, tadpoles remain longer and grow to a larger size, even if food resources decrease. As long as the threshold size is reached, metamorphosis is assured. As a result, the age at metamorphosis is more variable at low food levels, whereas the size at metamorphosis is more variable at high food levels (Morey and Reznick, 2004).

Larval development occurs normally at temperatures from 9 to 30°C, with an optimal range of 11 to 28°C (Brown, 1967). Larvae have been found in water temperatures as low as 7°C. However, Brown (1969) reported 50% larval survival even at a temperature of 39.5°C. The rate of development is slower than in the summer-developing *S. multiplicata*, even after adjusting for temperature. As Brown (1976) noted, *S. hammondii* appears to be a more cold-adapted species than *S. multiplicata*.

One of the major sources of larval mortality is desiccation of breeding pools prior to metamorphosis (Ervin et al., 2005). In that regard, *S. hammondii* can begin leaving the water even when a considerable portion of the tail remains. Young spadefoots readily burrow after leaving a pond.

DIET
Postmetamorphs eat a variety of invertebrate prey, including crickets, butterflies, beetles, flies, ants, and earthworms (*in* Jennings and Hayes, 1994b). A single spadefoot can eat 15% of its body weight in a single feeding (Dimmitt and Ruibal, 1980a).

PREDATION AND DEFENSE
Secretive behavior and cryptic coloration make this species difficult to locate throughout much of the year. As with other spadefoots, Western Spadefoots have a distinctive glandular secretion that is noxious to potential predators. The secretion smells like peanuts. Human handlers may

Tadpole of *Spea hammondii*.
Photo: Chris Brown

Adult *Spea hammondii*.
Photo: Chris Brown

Habitat of *Spea hammondii*.
Photo: Sara Schuster

experience allergy-like symptoms of the eyes and mucous membranes after coming in contact with this species.

Tadpoles and postmetamorphs are eaten by dytiscid beetle larvae, raccoons, garter snakes (*Thamnophis* sp.), Great Blue Herons, Burrowing Owls, California Tiger Salamanders (*Ambystoma californiense*), and American Bullfrogs (*Lithobates catesbeianus*) (Childs, 1953; Feaver, 1971; Ervin and Fisher, 2001; Ervin et al., 2007; Stitt and Balfour, 2011).

POPULATION BIOLOGY

Little is known of the population biology of this species. Jennings and Hayes (1994b) speculated that sexual maturity might require 2 yrs to attain.

DISEASES, PARASITES, AND MALFORMATIONS

The nematode *Aplectana incerta* has been reported from *Spea hammondii* (Goldberg et al., 2002b).

SUSCEPTIBILITY TO POTENTIAL STRESSORS

pH. Breeding pools in California had a pH of 7.1–10.1 (Morey and Reznick, 2004). No information is available on lethal levels.

Chemicals. The pesticides parathion, Bayer 37289, guthion, methyl parathion, G-30494, and Ronnel were not toxic to *S. hammondii* larvae in field trials (Mulla, 1962). For some unexplained reason, pesticides were thought desirable to kill anurans because frogs might eat insects and cause noise at night.

STATUS AND CONSERVATION

Spea hammondii has experienced severe habitat loss throughout much of its historic range in California (Fisher and Shaffer, 1996). Populations are extinct in the Sacramento Valley, and populations in the eastern San Joaquin Valley are severely reduced. More than 80% of its habitat in southern California has been altered by development and agriculture as to no longer support populations. In northern and central California, habitat loss was estimated by Jennings and Hayes (1994b) as affecting >30% of the area, but this has undoubtedly increased in the last 16 yrs. Temporary breeding pools are particularly vulnerable, and surviving populations are further threatened by habitat fragmentation. In addition, the introduction of exotic mosquitofish (*Gambusia* sp.) may impact breeding. Breeding sites must be protected for this species to survive. Fire appears to have little effect on the species, at least over a short-term postburn period (Rochester et al., 2010). The Western Spadefoot is considered a "Species of Concern" by the State of California.

Spea intermontana (Cope, 1883)
Great Basin Spadefoot

ETYMOLOGY
intermontana: from the Latin *inter* meaning 'between,' *montis* meaning 'mountain,' and *-anus* meaning 'belonging to.' The name is in reference to the Great Basin, the land between mountains.

NOMENCLATURE
Synonyms: *Scaphiopus intermontanus*, *Scaphiopus hammondii intermontana*

Older literature (e.g., Slevin, 1928; Tanner, 1931; Linsdale, 1938, 1940; Slater 1941b; Thorson, 1955; Metter, 1960) refers to this species as *Spea hammondii*, even through most references to *S. hammondii* refer to *S. multiplicata*. Individuals also have been occasionally misidentified as *S. bombifrons*. It is necessary to check collecting localities in original citations in order to determine which species is being discussed.

IDENTIFICATION
Adults. *Spea intermontana* is tan, gray, or olive with dark markings, and it has a glandular boss (raised area) between the eyes. The skin is smooth in comparison with other spadefoots. There are two light stripes dorsolaterally on each side that create an hourglass pattern on the back. In addition, the dorsum has dark reddish-brown spots, and the upper eyelid has dark brown spots. The venter is white or cream colored with grayish spotting or patches. As in other spadefoots, the pupil of the eye is vertical, and there is a keratinized black spade on each hind foot that is used for digging backward into the ground. Males have a dark throat patch and nuptial pads on the first three fingers of each hand. Vocal sacs are slightly bilobed. Sexual size dimorphism is absent. Logier (1932) reported a size range of 40 to 59 mm SUL (mean 48.25 mm) for a small collection from British Columbia, and Hovingh et al. (1985) recorded an overall adult mean size of 51 mm SUL (n = 254) from throughout its range. Specimens from the Bonneville Basin had a larger mean size (58 mm SUL) than *S. intermontana* from other areas.

Larvae. The body is gray brown, brown, or black and somewhat transparent. It tends to be globular or ovoid to depressed and wider anteriorly more than posteriorly when viewed from above. Gold or brassy flecks occur profusely scattered on the dorsum or in patches on the body. The jaws are wide and frequently cuspate. In carnivorous tadpoles, there is a distinctive beak on the upper jaw and a notch into which it fits in the lower jaw; omnivorous tadpoles have a more "typical" oral apparatus. The eyes are positioned dorsally. Venters have a somewhat golden iridescence, and the gut coils are visible. The tail fin does not arise abruptly from the body and has dark flecking, mostly on the dorsal fin. Tail tips are acute. At hatching, larvae are 5–10 mm TL. Larvae reach a maximum size of 85 mm TL (Hammerson, 1999). Albino tadpoles, likely *S. intermontana*, were observed by Wood (1935) in northeastern Utah. Altig (1970) included *S. intermontana* in his tadpole key. Hall et al. (1997, 2002) described postembryonic larvae and tadpoles in great detail, including illustrations of melanophore pattern and mouthparts.

Eggs. Bicolored eggs are deposited in small irregular (strings, sheets, or clusters) jelly packets 1.5–2.0 cm in diameter. Each packet contains 10–40 pigmented eggs (Hammerson, 1999; Buseck et al., 2005).

DISTRIBUTION
As its name implies, the Great Basin Spadefoot occupies the Great Basin from central British

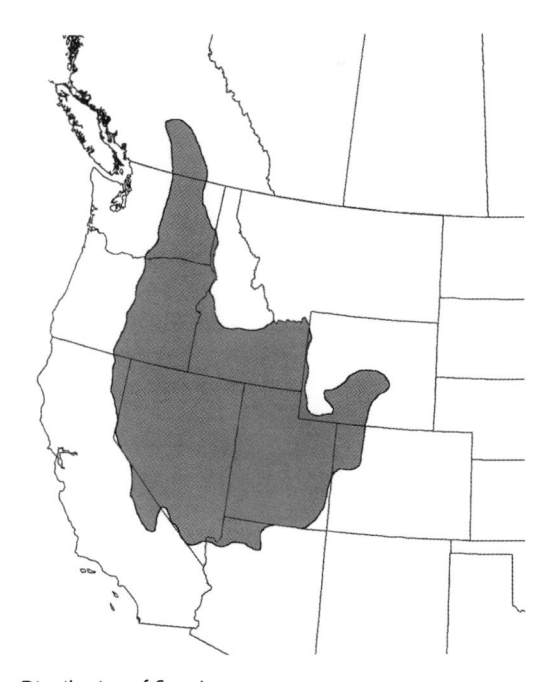

Distribution of *Spea intermontana*

Columbia southward to the Arizona Strip north of the Grand Canyon. The range extends into the Death Valley region of California, throughout much of Nevada, southern Idaho (Snake River drainage), western Utah, northeastern Colorado, and south-central Wyoming west of the Continental Divide. In Washington, the species occurs only east of the Cascades. Gehlbach's (1965) record from New Mexico is a misidentified *S. bombifrons* (Northen, 1970).

Important distributional references include: Range-wide (Northen, 1970), Arizona (Miller et al., 1982; Brennan and Holycross, 2006), British Columbia (Logier, 1932; Matsuda et al., 2006), Colorado (Hammerson, 1999), Idaho (Slater 1941b; Tanner, 1941; Jones et al., 2005), Nevada (Linsdale, 1940; Hovingh, 1997), Oregon (Leonard et al., 1993; Jones et al., 2005), Pacific Northwest (Nussbaum et al., 1983), Utah (Hovingh et al., 1985; Hovingh, 1993a, 1997), Washington (Slater, 1955; Leonard et al., 1993; McAllister, 1995; Jones et al., 2005), and Wyoming (Baxter and Stone, 1985; Buseck et al., 2005).

FOSSIL RECORD

Great Basin Spadefoots are known from Pleistocene fossil deposits in Nevada and Utah (Holman, 2003). The fossils of this species and *S. hammondii* are nearly identical, and identification is based mostly on the modern ranges of these species.

SYSTEMATICS AND GEOGRAPHIC VARIATION

Phylogenetic relationships within the genus *Spea* have been interpreted differently through time (Bragg, 1945b; Wiens and Titus, 1991; García-París et al., 2003; Holman, 2003). Northen (1970) demonstrated that *S. intermontana* was a valid species based on a variety of characters. Wiens and Titus (1991) considered *S. intermontana* most closely related to *S. bombifrons* and more distantly related to *S. hammondii* and *S. multiplicata*. However, there was substantial differentiation between *S. intermontana* samples from Colorado and Oregon, suggesting that recognition of additional species might be warranted. Holman (2003) considered there were two clades within the monophyletic genus *Spea*, an eastern clade consisting of *S. bombifrons* and a western clade consisting of *S. hammondii*, *S. intermontana*, and *S. multiplicata*. In Holman's (2003) scenario, based on fossil evidence, the eastern clade appeared by the middle Miocene, whereas the western clade first appeared in the late Pliocene. This would make *S. hammondii* the oldest member of the western clade, with the other members branching in the Pleistocene. Using mtDNA, García-París et al. (2003) found that *S. intermontana* was most closely related to *S. bombifrons* and two lineages of *S. hammondii*, and more distantly to another *S. hammondii* lineage and *S. multiplicata*. Buseck et al. (2005) reviewed the taxonomic and systematic status of this species. Northen (1970) reported natural hybrids between *S. multiplicata* and *S. intermontana* from southeastern Utah.

Great Basin Spadefoots often match the substrate background coloration closely. For example, individuals in the Diamond Craters area of southeastern Oregon are dark and blend in with the dark-colored basaltic rocks. Specimens from the nearby sagebrush are much lighter (Nussbaum et al., 1983).

ADULT HABITAT

Although *S. intermontana* is found at a range of elevations, most of the time it occurs on valley floors within the Great Basin. Habitats include shrub-steppes, arid sagebrush plains and short-grass prairies, pinyon-juniper woodlands at midelevations, and open savanna-like coniferous forest (spruce-fir, ponderosa pine, Douglas fir) at higher elevations. It also occurs in agricultural areas. *Spea intermontana* requires friable sandy soils for digging. The species occurs from 518 to 2,286 m in Nevada (Linsdale, 1940), to 2,012 m in Utah (Hovingh et al., 1985), to 1,800 m in British Columbia (Matsuda et al., 2006), and to 2,800 m in Wyoming (Buseck et al., 2005). Hovingh et al. (1985) recorded most Great Basin Spadefoots at elevations < 1,600 m in the Bonneville Basin of Utah.

TERRESTRIAL ECOLOGY

The Great Basin Spadefoot is a nocturnal species that spends most of its time in well-concealed burrows, emerging only to breed and occasionally forage. Juveniles emerge at night to feed throughout the summer independently of rain, and may occasionally be observed in some numbers (Kellogg, 1932). Foraging occurs in a variety of modes, from a sit-and-wait position to active pursuit. Spadefoots tend to choose larger prey and may ignore small prey as adults. Movement seems required to attract their attention. Although mostly nocturnal, Linsdale (1938) observed yearling toads foraging openly during daylight, usually in the early morning and late afternoon. Some were active even in the afternoon during periods of high temperatures.

Spea intermontana both digs its own burrows and occupies mammal burrows. It also has been found under surface debris, rocks, and in rock crevices, where multiple individuals take refuge (Svihla, 1953). While temporarily dormant, they are in a horizontal position within the burrow. Ruthven and Gaige (1915) noted they could be induced to stick their heads out of the burrow or even leave it by stamping loudly on the ground. Burrows are dug using the spades on the hind feet, and Great Basin Spadefoots disappear rapidly beneath the surface. The entrance to the burrow may be indicated by slight pretzel-shaped ridges, but these are quickly obliterated by wind, leaving no trace of where the burrow is located.

Great Basin Spadefoots are active at any time during the warm season, including several weeks prior to breeding. For example, Linsdale (1940) noted that most activity occurred from May to July in Nevada, with extremes of late March to early October. At higher latitude and elevation, activity becomes more restricted, from April to September. In Colorado, activity also occurs from April to September (Hammerson, 1999).

Recently metamorphosed Great Basin Spadefoots disperse in large numbers. Logier (1932) reported that movements were directed toward nearby clay bluffs. Dispersal occurs at night. Harestad (1985) recorded recent metamorph dispersal to 110 m along a dry hillside. Movements of adults are little studied, but most probably stay within proximity of breeding ponds. Adults have been found up to 100 m or more from the nearest breeding pool. Hovingh et al. (1985) suggested that they might move up to 5 km to find a suitable breeding site.

The arid-dwelling Great Basin Spadefoot loses water at a rather slow rate, although it also reabsorbs water slowly (Thorson, 1955). Water is absorbed directly through the skin in contact with moist soil, and there is no indication that the spadefoot drinks in order to replenish its body water. This species can tolerate as much as 48.8% of their initial body weight in water loss (Thorson and Svihla, 1943).

Great Basin Spadefoots are photopositive, suggesting they use both sunlight and moonlight when foraging and moving between locations (Jaeger and Hailman, 1973). They are sensitive to light in the blue spectrum ("blue-mode response") (Hailman and Jaeger, 1974). Great Basin Spadefoots likely have true color vision.

CALLING ACTIVITY AND MATE SELECTION

Breeding is frequently associated with heavy rainfall, but rainfall is not a prerequisite for successful breeding, and breeding may not occur

Eggs of *Spea intermontana*.
Photo: Breck Bartholomew

Tadpoles of *Spea intermontana*. Note the algae on one tadpole. Photo: Dana Drake

despite heavy rainfall. *Spea intermontana* can be an explosive breeder, with most breeding occurring on only one to two days in temporary pools. However, breeding may be asynchronous and extend over a period of weeks, as this species is not as tied to rainfall as other spadefoots and often breeds at sites with comparatively long hydroperiods. Males arrive at breeding sites prior to females. Although breeding success may not be correlated with chorus size, females tend to move to large choruses in proportionally greater numbers than they move to small choruses. However, the extent of breeding site fidelity, if any, is unknown. Calling occurs from spring through summer at air temperatures of 3–22°C and water temperatures of 10–24° (Fouquette, 1980). Specific breeding dates include April to June (Nevada: Linsdale, 1938; Oregon: Nussbaum et al., 1983; Utah: Hovingh et al., 1985) and April to July (Wyoming: Buseck et al., 2005).

Calling occurs while males float sprawled out at the water's surface near the shore. They may hold onto emergent vegetation, or, in some cases, simply sit on the bottom of shallow pools. The advertisement call is a low-pitched, guttural, and somewhat rasping series of short rapid calls lasting 0.2 to 1.0 sec each. Buseck et al. (2005) likened it to sounding like "kwaah-kwaah-kwaah" and noted that the call is readily audible 100–200 m distant and can carry 1.5 km. Call characteristics include a mean dominant frequency of 1,380 cps with a mean of 35.4 pulses/sec, but call length and duration is variable among populations (Northen, 1970). Northen (1970) noted calls with intermediate characteristics in natural hybrids between *S. multiplicata* and *S. intermontana*. McAlister (1959) described the morphology of the vocal structures.

Mate selection is by scramble competition, that is, males will move toward and attempt to amplex any female moving near them. Females do not have much opportunity to select mates. Amplexus is inguinal.

BREEDING SITES

Breeding often occurs in shallow temporary pools and puddles after a heavy rainstorm, but breeding also occurs in permanent pools, springs, stream edges, ponds, and even in roadside and irrigation ditches. In Utah, breeding sites ranged from 0.08 m^2 and 10 cm deep to reservoirs with > 1,200 m^3 of water. Both temporary and permanent water sources are used, especially man-made reservoirs (Hovingh et al., 1985). Great Basin Spadefoots readily breed in very muddy and polluted ponds (from livestock manure), stock tanks, and along both intermittent and permanent desert streams (Hovingh et al., 1985; Morey and Reznick, 2004; Buseck et al, 2005). Breeding sites often contain no vegetation. In Nevada, Linsdale (1938) reported breeding in a large field of shallow water formed by excessive snow meltwater along a river. In this case, no rain had fallen for a long period of time, but spadefoots still bred by the hundreds in the flooded pastures. Hovingh et al. (1985) concluded any available standing water might be used as long as the total dissolved solids were < 5,000 mg/L. Not all sites are used; in the Bonneville Basin, Hovingh et al. (1985) found Great Basin Spadefoots at 51 of 169 potential sites over a 3 yr period.

REPRODUCTION

Egg packets, each containing 20–40 eggs, are oviposited in shallow water, either attached to vegetation or the muddy substrate at the bottom of breeding pools. Packets are 15–20 mm in diameter. The total clutch size is usually 300–500 eggs, with a maximum of 800 (Nussbaum et al., 1983; Buseck et al., 2005). Eggs can develop normally at temperatures at least to 25°C (Morey and Reznick, 2000). Hatching occurs in two to three days in warm weather and up to seven days in cool weather. Hatchlings are 5–7 mm TL.

LARVAL ECOLOGY

Larvae forage at the water's surface, on submerged vegetation and along the substrate. Larvae are more active by night than day, although diurnal foraging is not uncommon. Growth is rapid. The duration of the larval period is longer than in other spadefoots but like the other species is dependent on a variety of environmental variables. Larvae take 30–60 days (mean 48 days) to complete development (Morey and Reznick, 2004) under field conditions, with longer durations occurring in cooler habitats. Temperature during development is important. For example, the larval period is 36 days at 23°C and 30 days at 24°C, whereas at 32°C it is only 22 days in laboratory conditions (Brown, 1989; Gomez-Mestre and Buchholz, 2006).

Food resources play a critical role in the larval development of *S. intermontana*, since the duration of the larval period is tied to food availability. In experimental trials, larvae having greater food resources grow more quickly, metamorphose sooner, and have a greater body mass than larvae fed on less food. It appears that there is a threshold size that must be met to complete metamorphosis (Morey and Reznick, 2000). Low amounts of food will cause slower growth, which delays the attainment of the threshold size and could make individuals vulnerable to pool desiccation. With sufficient resources, tadpoles gain the threshold size and then have a degree of plasticity in the extent to which they remain in a breeding pool. With long hydroperiods, tadpoles can remain longer and grow to a larger size, even if food resources decrease. As long as the threshold size is reached, metamorphosis is assured. As a result, the age at metamorphosis is more variable at low food levels, whereas the size at metamorphosis is more variable at high food levels (Morey and Reznick, 2004).

Hydroperiod further affects larval development, with larvae developing in pools with short hydroperiods undergoing metamorphosis more quickly than larvae developing in pools with long hydroperiods. The effect of hydroperiod on

Adult *Spea intermontana*.
Photo: David Dennis

Breeding habitat of *Spea intermontana*. Photo: Dana Drake

fitness may extend beyond the larval period. For example, Gomez-Mestre and Buchholz (2006) found that postmetamorphs from larvae reared in long-hydroperiod regimes had longer limbs and snouts than larvae reared in short hydroperiod regimes. This type of environmentally induced morphological variation appears to be common in all Pelobatoidea and is not a matter of simple allometry.

Newly metamorphosed toadlets vary from 10 to 20 mm TL (Brown, 1989; Jones et al., 2005) to 16–38 mm TL (Hovingh et al., 1985; Brown, 1989). Toadlets may or may not have a remnant portion of the tail at transformation, perhaps accounting for discrepancies in the reported size of metamorphs. The biflagellated alga *Chlorogonium* has been observed as an epzoic symbiont of *S. intermontana* larvae in Utah (Drake and Trauth, 2010).

DIET

Cope (1889) noted that recent metamorphs, some even containing remnants of the larval tail, readily fed on grasshoppers along the shore of a breeding pond. The species also eats ants, beetles, grasshoppers, larval lepidopterans, crickets, and flies (Tanner, 1931; Waye and Shewchuk, 1995). A single spadefoot can eat 18% of its body weight in a single feeding (Dimmitt and Ruibal, 1980a).

As with other spadefoots, *S. intermontana* has both omnivorous and carnivorous tadpoles. Carnivorous morphs eat insects, carrion, and other tadpoles. Omnivorous morphs consume algae, detritus, plankton, organic debris, carrion, and plant material.

PREDATION AND DEFENSE

The secretive habits and cryptic coloration of this species make it difficult to locate. When in chorus, males will cease calling at the approach of an observer. They burrow rapidly into the soil when disturbed, and the openings of the burrows are closed and difficult to locate, even where a toad was present a few seconds before. When handled, *S. intermontana* has a distinctive odor similar to that of peanuts (Waye and Shewchuk, 1995). It exudes a mucous secretion that serves to make the spadefoot unpalatable to predators. This secretion may cause allergic reactions in the mucous membranes of some people, which are similar to those of hay fever.

Predators of larvae include the larvae of the fly *Tabanus punctifer* (Jackman et al., 1983) and possibly other aquatic invertebrates. Adults are eaten by snakes (*Thamnophis* sp.), birds (burrowing owls, crows), and mammals (coyote), but little information is available (Wood, 1935; Harestad, 1985; Hammerson, 1999; *in* Buseck et al., 2005).

POPULATION BIOLOGY

According to Nussbaum et al. (1983), the age at first breeding is 2–3 yrs (i.e., during their third summer). In males, maturity occurs ca. 40 mm SUL, whereas females mature at ca. 45 mm SUL.

DISEASES, PARASITES, AND MALFORMATIONS

The parasitic trematode *Ribeiroia ondatrae* has been reported from a single specimen of *S. intermontana* (Johnson et al., 2002). Other helminths include the trematode *Polystoma nearcticum*, the cestode *Distoichometra bufonis*, and the nematodes *Aplectana incerta*, *Physaloptera* sp., and Acuariidea gen. sp. (Goldberg and Bursey, 2002b). No parasites were observed in five *S. intermontana* from Idaho (Waitz, 1961).

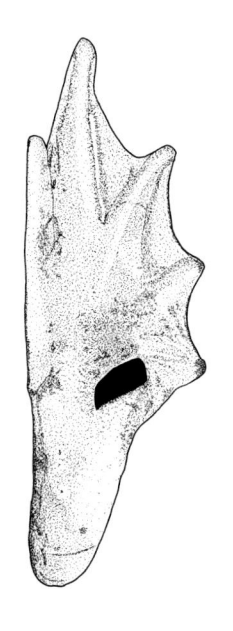

SUSCEPTIBILITY TO POTENTIAL STRESSORS

Other elements. Pesticides pose a serious threat to Great Basin Spadefoots, especially in Canada at the northern limit of the species' range. Cholinesterase activity in tadpoles is depressed in individuals exposed to organophosphorus pesticides (e.g., diazinon), as these pesticides bind with cholinesterase to inhibit neural function. Endosulfan causes slight decreases in larval survival over an eight-day period, larval deformities (kinked tails), and changes in the behavior of survivors at environmentally relevant doses (Westman et al., 2010); it does not affect egg development. At concentrations > 50 ng/L of azinphosmethyl, *S. intermontana* experience a significant decrease in survivorship. Survivorship also significantly decreases as concentrations of diazinon increase. Significant mortality occurs at the highest lethal concentrations of endosulfan and diazinon (Westman et al., 2010).

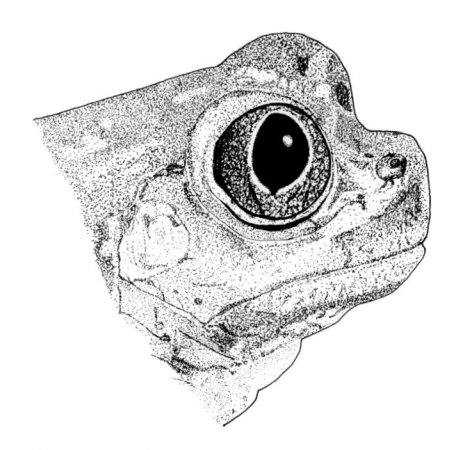

pH. In Utah, breeding occurs in pools with a pH of 7.2–10.4 (Hovingh et al., 1985). Breeding pools in California had a pH of 7.1–9.0 (Morey and Reznick, 2004). There are no reports for lethal pH values.

STATUS AND CONSERVATION

The Great Basin Spadefoot undoubtedly faces threats from habitat loss, alteration, and fragmentation. Other threats include the effects of toxic chemicals and impacts from invasive species, such as introduced fish (Buseck et al., 2005). High levels of road mortality of dispersing Great Basin Spadefoots have been reported (Logier, 1932). This species is protected as "Threatened" in Canada. Buseck et al. (2005) discussed conservation measures necessary to ensure the species' survival in Wyoming. Throughout much of its range, the species is considered relatively common (e.g., Hammerson, 1999), but declines in populations in British Columbia have been noted (Weller and Green, 1997). Because of the stochastic nature of breeding, status surveys frequently fail to detect the presence of Great Basin Spadefoots, even though historical records may be available (e.g., Bradford et al., 2005a).

Hind foot of *Spea intermontana*. Illustration by Breck Bartholomew

Lateral head view of *Spea intermontana*. Illustration by Breck Bartholomew

Spea multiplicata (Cope, 1863)
Mexican Spadefoot

ETYMOLOGY
multiplicata: from the Latin *multi* meaning 'many' and *plicatus* meaning 'folded' or 'braided.'

NOMENCLATURE
Synonyms: *Scaphiopus dugesi, Scaphiopus hammondii multiplicatus, Scaphiopus multiplicatus, Spea hammondii stagnalis, Spea multiplicata stagnalis, Spea stagnalis*

Much of the literature on "*Scaphiopus hammondii*" actually refers to *Spea multiplicata* (e.g., all of A.N. Bragg's work in Oklahoma; Cowles, 1924; Ortenburger, 1924; Slevin, 1928; Wright, 1929; Blair, 1955c; Brown, 1967 [in part]; Ruibal et al., 1969; Aitchison and Tomko, 1974; Whitaker et al., 1977). It is necessary to check collecting localities in original citations in order to determine which species is being discussed.

IDENTIFICATION
Adults. The background coloration is gray, tan, or light brown, with color changes occurring depending on whether the spadefoot is observed by day or night. For example, an individual might appear light brown at night and grayish brown during daylight. Dark pigmentation surrounds the dorsal warts and is on the hind legs. Small dark spots or blotches and red-tipped tubercles may be present dorsally. There is a distinct boss between the eyes, and the pupils are vertical indicating nocturnal activity. The iris is pale copper. No dorsolateral stripes are present. As in other spadefoots, there is a black wedge-shaped keratinized spade on each hind foot. Males have a dark, heavily pigmented vocal sac and keratinized patches on the first three digits of the forelimbs. The vocal sac is slightly bilobed when expanded.

Sexual size-dimorphism is absent. Brown (1976) reported a mean SUL for Arizona males of 46.9 mm (range 41 to 51.9 mm). Mean SULs of adult males in other Arizona populations were 48–51.3 mm SUL (Sullivan and Sullivan, 1985). Degenhardt et al. (1996) reported a maximum size of 65 mm SUL.

Larvae. The body is olive to gray and somewhat transparent. It tends to be globular to depressed and wider anteriorly more than posteriorly when viewed from above. Eyes are close together and dorsally positioned. The jaws are wide and frequently cuspate. In carnivorous tadpoles there is a distinctive beak on the upper jaw and a notch into which it fits in the lower jaw; omnivorous tadpoles have a more "typical" oral apparatus. The tail fin does not arise abruptly from the body. Larvae reach a maximum size of 74 mm TL (Cowles, 1924). Altig (1970) included them in his tadpole key. Detailed descriptions are in Wright (1929) under the name *Scaphiopus hammondii*.

Eggs. Spea multiplicata eggs range in color from light cream to gray black, although most are brown to dark brown. A single clutch may have different colored eggs. The eggs are small with a thin gelatinous capsule; their diameter is 1.16–2 mm with a mean of 1.52 to 1.8 mm (Ortenburger, 1924; Pomeroy, 1981). The smallest eggs are gray whereas the largest eggs are red brown (Pomeroy, 1981). The eggs are attached to one another via a thin stalk 5–10 mm in length. Egg compliments are oviposited in small masses which, according to Ortenburger (1924), are arranged spirally around vegetation.

DISTRIBUTION
The Mexican Spadefoot occurs from eastern Nevada and western Utah south throughout eastern and central Arizona to Guerrero and Oaxaca, Mexico. Populations are found from southeastern Colorado and the Oklahoma and Texas panhandles southward throughout much of central (west of the Balcones Escarpment; Smith and Buechner, 1947) and western Texas. A possibly isolated population occurs in south-central Colorado. Hammerson (1999) noted a number of instances of misidentification, leading to confusion about distribution. Important distributional references include: Arizona (Brennan and Holycross, 2006), Colorado (Hammerson, 1999), Oklahoma (Sievert and Sievert, 2006), New Mexico (Van Denburgh, 1924; Degenhardt et al., 1996), and Texas (Dixon, 2000).

FOSSIL RECORD
Fossils of the Mexican Spadefoot are known from Pleistocene deposits in New Mexico (Holman, 2003). Skeletal features are essentially identical with *S. hammondii* and *S. intermontana*.

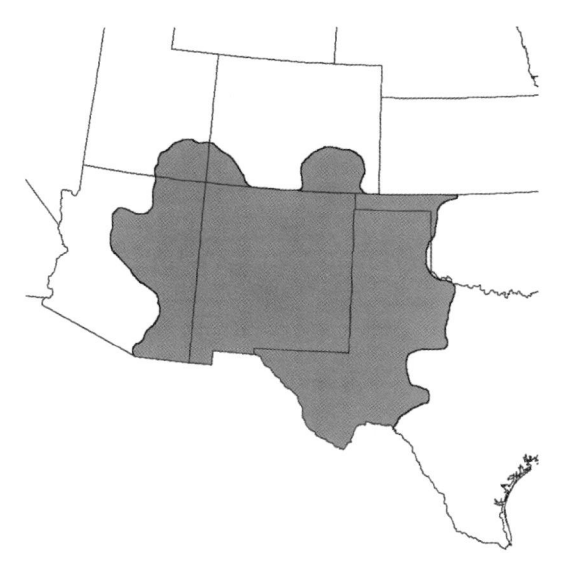

Distribution of *Spea multiplicata*

SYSTEMATICS AND GEOGRAPHIC VARIATION

The phylogenetic relationships within the genus *Spea* historically have been unclear, and there have been differing interpretations of relationships (Tanner, 1939; Bragg, 1945b; Wiens and Titus, 1991; García-París et al., 2003; Holman, 2003). Brown (1976) noted differences in the mating calls and physiology between California *S. hammondii* and Arizona "*S. hammondii*" (i.e., *S. multiplicata*) and separated these former subspecies into full species. Tanner (1989) considered *S. multiplicata* to be a subspecies of *S. hammondii* and relegated *Spea stagnalis* to subspecific status, *Spea hammondii stagnalis*, based on osteological characters. However, Wiens and Titus (1991) considered *S. h. stagnalis* to be conspecific with *S. multiplicata*.

Based on fossils, Holman (2003) identified two clades within the monophyletic genus *Spea*, an eastern clade consisting of *S. bombifrons* and a western clade consisting of *S. hammondii*, *S. intermontana*, and *S. multiplicata*. In Holman's (2003) scenario, the eastern clade appeared by the middle Miocene, whereas the western clade first appeared in the late Pliocene. This would make *S. hammondii* the oldest member of the western clade, with the other members branching in the Pleistocene. Molecular data, however, do not support this scenario. Instead, *S. multiplicata* was likely the first to branch from the ancestral *Spea* (Sattler, 1980; Wiens and Titus, 1991; García-París et al., 2003). In addition, there appear to be two separate lineages (Arizona, Mexico) within *S. multiplicata*, which indicates the presence of additional genetic diversity and could lead to the recognition of cryptic species (García-París et al., 2003).

Natural hybridization between *S. multiplicata* and *S. bombifrons* has been reported in southeastern Arizona (Brown, 1976; Simovich, 1985, 1994; Simovich et al., 1991; Pfennig, 2003), the Four Corners region of the Colorado Plateau (Northen, 1970), and from southwestern New Mexico and Texas (Forester, 1973; Sattler, 1985). Northen (1970) noted intermediate calls of natural hybrids between *S. multiplicata* and *S. intermontana*. Experimental hybridization between *S. multiplicata* and *S. hammondii* indicate a high level of genetic compatibility (Brown, 1967; Forester, 1975), although hybrids involving ♀ *S. multiplicata* and ♂ *S. bombifrons* have lower survival and longer larval periods than nonhybrid *S. multiplicata* (Pfennig and Simovich, 2002). Genetic compatibility is also high for crosses of *S. multiplicata* with *S. intermontana* and *S. bombifrons* (Littlejohn, 1959; Brown, 1967).

ADULT HABITAT

This species occurs in plains grasslands, pinyon-juniper woodlands, sagebrush flats, river valleys, semiarid shrublands, and even in the open ponderosa pine forest savanna of north-central Arizona (Jameson and Flury, 1949; Aitchison and Tomko, 1974). It is found in short-grass prairies as well as in agricultural pastures and croplands. The species occurs to 2,134 m in Arizona (Aitchison and Tomko, 1974) and 900–2,600 m in New Mexico (Degenhardt et al., 1996).

TERRESTRIAL ECOLOGY

Activity may occur at any time during the warm season, when temperatures are 10–24°C (Hammerson, 1999). The Mexican Spadefoot is a nocturnal species that spends most of its time in well-concealed burrows, emerging only to breed and occasionally forage. *Spea multiplicata* enter burrows in the fall (e.g., September in southeastern Arizona; Ruibal et al., 1969) and do not emerge in numbers until heavy summer rains fall the next July. Thus, they spend at least nine months of the year within the burrow. Feeding occurs only rarely once a spadefoot has entered its cool-season burrow. Nonbreeding habitats are in sparse to dense grasslands within 300 m of the breeding site (Cruesere and Whitford, 1976).

Positions within the burrow change seasonally according to temperature and moisture conditions. Mean depths early in the season are shallow (ca. 24 cm below the surface) but by midwinter are a mean of 54 cm below the surface. With the coming of warmer weather in June, spadefoots again move toward the surface at a mean depth of 42 cm. The maximum depth Ruibal et al. (1969) recorded a burrowed Mexican Spadefoot was 91 cm below the surface. During the rainy reason, burrows are very shallow (ca. 4 cm). The spadefoot digs the burrow using its hind foot spade, and the burrow is backfilled so that the vertically positioned spadefoot is enclosed within a small chamber in the tunnel. Ruibal et al. (1969) recorded temperatures of 17–22.4°C within the burrows they examined early in the dormancy season, and 5–15°C in winter. In summer, the shallow burrows can reach 27–39°C.

Emergence occurs in late summer, although a few Mexican Spadefoots may emerge in June or early July after light rain. It appears likely that some individuals are active nocturnally at the mouths of their burrows prior to breeding emergence. Most animals, however, emerge in large numbers to breed only after very heavy (> 5 mm) rainfall, such as occurs during desert thunderstorms. Mexican Spadefoots emerge in response to low frequency sounds (<100 Hz) that accompany rainfall, rather than the moisture per se (Dimmitt and Ruibal, 1980b). After breeding, adults continue to be active at night throughout the summer rainy season. They occupy the shallow burrows during the day and may remain at one location for several days before continuing to forage. While temporarily dormant, they are in a horizontal position.

Recent metamorphs emerge en masse and move across the landscape in waves of thousands of tiny toadlets. Dispersal occurs both by night and day until the tiny spadefoots reach their destination.

Like other desert or arid-adapted spadefoots, the Mexican Spadefoot readily absorbs water from moist soil (Walker and Whitford, 1970). They maintain an osmotic concentration close to that of the surrounding soil, which allows them to remain in a burrow for a long time without losing moisture (Ruibal et al., 1969). Their permeable skin facilitates moisture uptake. Little and Keller (1937) noted that Mexican Spadefoots appeared to have a "gelatinous coat" around them when excavated. Presumably this helps to conserve water during long periods of dormancy.

CALLING ACTIVITY AND MATE SELECTION

Spea multiplicata is an explosive breeder, with choruses forming only after torrential thunderstorms in summer. Calling occurs from the open water of temporary pools or from along pool margins while sitting in mud. Calling also occurs in sparse vegetation at water depths > 12 cm (Cruesere and Whitford, 1976). Most Mexican Spadefoots call while floating in the water with only their heads above the surface, and they tend to move around frequently throughout the calling period. In the Sonoran Desert of Arizona, the mean chorus duration was 1.6 days, and the mean chorus size was 44 males (Sullivan and Sullivan, 1985; Sullivan, 1985a, 1989). Males outnumber females at breeding pools. Sullivan (1989) suggested that skewed sex ratios resulted from male attempts to breed at every opportunity, whereas females breed only once, regardless of how many favorable opportunities arise throughout the breeding season.

Males associate with conspecific males and are attracted to a chorus by the advertisement call of conspecifics (Pfennig et al., 2000). In Arizona, for example, Pfennig et al. (2000) found a mean nearest neighbor distance of 6.6 m, with most spadefoots < 5 m from one another in a 1,886 m^2 pond. Small males use the call rate to judge the quality of other males much the same way as females do; males with fast call rates are preferred. In this way, satellite males can position themselves in proximity to a "high quality" male and thus have an opportunity to intercept a female.

Males move immediately toward any other spadefoot in the vicinity and attempt amplexus. Males frequently wrestle with other males in clasping attempts, but the clasped male gives its warning vibration and call, which causes the amplexing male to release its hold. Females move toward calling males, which readily clasp them at their approach. Eggs are fertilized as they are deposited. There is no correlation between male body size and mating success (Sullivan and Sullivan, 1985).

Ortenburger (1924) colorfully described the advertisement call of *S. multiplicata* as sounding "like the loud purr of a cat but with the metallic sound of grinding gears." The call is a slow trill with a dominant frequency of 1,150 to 1,600 cps, a range of 0.7 to 1.4 sec in duration, 14–25 trills per call, and 18–28 trills/sec (Blair, 1955c; McAlister, 1959; Northen, 1970; Forester, 1973; Brown, 1976). Of course, specific values will vary with temperature. Literature records of different trill rates (fast and slow) within *S. bombifrons* (e.g., Pierce, 1976) actually refer to calls of *S. bombifrons* and *S. multiplicata*. McAlister (1959) described the morphology of the vocal structures.

Mate selection is by scramble competition, that is, males will move toward and attempt to amplex any female moving in the pond. Females, however, can use call characteristics of the males to assess both the identity and the quality of potential mates, using both call rate and pulse rate as identifying features. In areas where this species is not sympatric with *S. bombifrons*, female *S. multiplicata* prefer males with more extreme call values (suggesting different quality males), whereas females from areas of sympatry prefer average values of call characteristics. This difference in preference helps minimize interspecific matings in areas of sympatry where call characteristics may overlap. Thus, females trade-off potential mate quality for correct species recognition when species overlap (Pfennig, 2000). Still, Mexican Spadefoots have been observed in amplexus with *S. bombifrons* (Cruesere and Whitford, 1976). Amplexus is inguinal.

BREEDING SITES

Breeding usually occurs in shallow temporary pools after heavy rainfall. Anderson et al. (1999) recorded presence in pools < 25 cm in depth. They also may breed in irrigation pits in playa grasslands (Anderson et al., 1999) and in stock ponds. Woodward (1987a) reported a single instance of *S. multiplicata* breeding in a permanent pond, and Minton (1958) observed calling Mexican Spadefoots in water 1–1.5 m in depth.

REPRODUCTION

Spadefoots are the epitome of explosive breeders. Reproduction occurs diurnally or nocturnally only after heavy rainfall and for a short period. This usually occurs from May through early July,

Eggs of *Spea multiplicata*. Photo: Breck Bartholomew

but breeding may occur earlier and extend to August in some locations. The rain does not have to be of great volume, but it has to occur torrentially. As little as 12.7 mm of rainfall can initiate breeding, but it must fall rapidly (Bragg, 1945b). Breeding may occur at any time of the year there is a favorable combination of warm temperatures and heavy precipitation. Females tend to stagger breeding, such that not all females breed at the same time, regardless of favorable conditions. In Arizona, for example, Cowles (1924) reported different sizes of tadpoles (24–74 mm) within a pool suggesting multiple breeding events at a single pond should hydroperiod be sufficient.

Eggs are deposited in shallow water in cylindrical masses attached to submerged vegetation and debris. They hatch in from 42 to 48 hrs. Woodward (1987a) reported a mean clutch size of 1,070 eggs. Female body weight does not appear correlated with clutch weight or number of eggs produced (Woodward, 1987a).

LARVAL ECOLOGY

Spea multiplicata larvae form large aggregations that tend to remain in the vicinity of the oviposition site. They associate more often with siblings than nonsiblings, and as such have been said to exhibit kin recognition. However, these aggregations may be based on cues acquired early in development. For example, nonsiblings also tend to aggregate with one another when fed on similar foods early in larval development as opposed to nonsiblings fed on different foods (Pfennig, 1989, 1990b). Pfennig (1990b) suggested that tadpole aggregations may be more attuned to habitat selection than kin per se. Tadpoles that remain

Tadpoles of *Spea multiplicata*. Photo: Scott Trageser

in the vicinity of oviposition sites tend to have greater body mass and more rapid development than larvae that forage in unfamiliar surroundings.

Larvae of *S. multiplicata* have two distinct larval morphologies, which can be identified as early as three days following egg deposition. Indeed, that there are two larval morphologies has sometimes led to confusion in the literature as to larval identification (Bragg, 1941a). One morph is large, grows rapidly, has a very long gut, possesses large jaw muscles, and has modified mouthparts that are used to attack both conspecific and heterospecific larvae. Carnivorous larvae are able to recognize kin and avoid eating them (Pfennig, 1999). The other is slow growing, rounder, and smaller than the carnivorous morph, and has a shorter gut and mouthparts developed for grazing and filter feeding. This phenotype is omnivorous but will take animal prey including conspecific tadpoles and does not have enlarged jaw muscles. Omnivorous larvae do not recognize kin and may eat their relatives. Tadpoles with this latter morphology frequently are seen feeding at the water's surface skimming organic debris as they swim. Thyroxine induces the development of the carnivorous morph by omnivores (Pomeroy, 1981).

Detritus, diatoms, and cyanobacteria form the primary food sources of the tadpoles regardless of tadpole morphology (Ghioca, 2005). MacKay et al. (1990) reported fairy shrimp, tadpole shrimp (*Triops*), and conspecific tadpoles to be important in the diet of larval *S. multiplicata*. They also eat feces, mud, and invertebrates and they will scavenge on carrion. Although omnivorous larvae will attack fairy shrimp, they are highly inefficient at eating them; consumption can take up to 5 min as opposed to 6 sec by carnivorous larvae. Likewise, carnivorous larvae are less efficient eating detritus than are omnivorous larvae (Pomeroy, 1981). Carnivorous larvae spend much more time swimming (presumably searching for prey) and much less time browsing than omnivorous larvae; both types engage in filter feeding (Pomeroy, 1981).

Breeding pools may contain both types of larvae, although the omnivorous morphology was the only morph in the breeding pools studied by Ghioca (2005). Interestingly, it is possible to induce the carnivorous morph by feeding the omnivorous morph fairy shrimp, as well as to change a carnivorous morph into an omnivorous morph (Pomeroy, 1981; Pfennig, 1989). Pfennig (1990a) found that if tadpoles ate a substantial number of shrimp they developed the carnivorous morph. The mere presence of shrimp, however, is not enough to induce the carnivorous morph, and there appears to be a threshold value that once is met results in carnivorous tadpoles.

The type of tadpole morph also is related to the hydrology of the breeding pool in which it develops. In short-lived pools, the carnivorous morph predominates because it can grow fast and presumably metamorphose quickly to escape desiccation. However, in ponds with long hydroperiods, the omnivorous morph is more abundant because its greater lipid reserves enhance post-metamorphic survival (Pfennig, 1989). With both morphs present in a pool, competition among the common morph makes the rare morph more successful (Pfennig, 1989, 1992). Essentially each breeding pool has an independent morph frequency that is stabilized by frequency-dependent morph reversal, which is closely attuned to food resources and the duration of hydroperiod.

In nature, the age and size of both tadpole morphs at metamorphosis is inversely correlated, that is, large tadpoles tend to be younger than small tadpoles and vice versa. This may be why carnivorous tadpoles tend to be larger in breeding pools surrounded by croplands when compared to carnivorous larvae in grassland playas (Ghioca, 2005); agricultural breeding pools tend to hold water for a shorter time than breeding pools in grassland playas. In experimental pools, the age and size at metamorphosis were negatively correlated with high food resources and positively correlated with low food resources under conditions whereby ponds held water over a long duration (Pfennig et al., 1991).

Plasticity in tadpole morphotype appears to have evolved as a response to variable hydroperiod conditions (and indirectly with food resources) in arid environments, but different tadpole feeding morphologies (e.g., in terms of gut length and cellular proliferation) also can be induced by diet (Ledón-Rettig et al., 2008). In *S. multiplicata*, this response is limited when larvae are fed on a pure diet of fairy shrimp, as might be expected for a species historically exposed to shrimp. Although not a novel prey, fairy shrimp may still affect development through their abundance. Tadpoles should grow as fast as possible in temporary ponds to achieve as large a size as possible, depending upon food availability. A negative relationship between age and size at metamorphosis often exists, because *S. multiplicata* larvae may be exposed to a rapidly diminishing food resource (e.g., fairy shrimp), which results in intraspecific competition. Thus, if the food resource is diminished, tadpoles will actually take longer to reach metamorphosis and be at a smaller size than they would if food were not limited.

As expected, there is a considerable range in variation in the larval period and the size of the metamorph, depending upon hydroperiod, environmental conditions, temperature, and food resources. Pfennig et al. (1991) found that larval periods lasted 14–44 days in Arizona. Tadpoles from short hydroperiod ponds (14–30 days) were larger at metamorphosis and grew faster than those from long duration ponds (>30 days). Tadpoles of the carnivorous morph transformed in 13 days in another study in Arizona, whereas omnivore morph tadpoles transformed in 17 days (Pomeroy, 1981). The water temperature also has a direct impact on developmental rates. For example, the larval period is 28 days at 24°C, whereas at 32°C it is only 15 days (Gomez-Mestre and Buchholz, 2006). Further, nutrient input from surrounding habitats can affect the larval period. In Texas, for example, the larval period tends to be shorter in agricultural breeding ponds as opposed to breeding pools in grassland playas (Ghioca, 2005).

The effect of hydroperiod on fitness may extend beyond the larval period. Gomez-Mestre and Buchholz (2006) found that postmetamorphs resulting from larvae reared in long-hydroperiod regimes had longer limbs and snouts than postmetamorphs reared in short hydroperiod regimes. This type of environmentally induced morphological variation appears to be common in all Pelobatoidea and is not a matter of simple allometry. Whether the longer limbs and snout have selective advantages have yet to be determined.

Larval development occurs normally at temperatures from 13 to 32.5°C, with an optimal range of 21 to 31°C (Brown, 1967). One way to cope with exposed breeding sites in often very hot climates is for larvae to be heat resistant. Such is the case with *S. multiplicata* larvae, with Brown (1969) reporting 50% larval survival even at a temperature of 40.5°C. The rate of development is faster in *S. multiplicata* than in *S. hammondii* after adjusting for temperature. Metamorphosis occurs rapidly and simultaneously. Even before metamorphosis, urea becomes the dominant waste product, which allows larvae to survive outside of water for a short period (Jones, 1980). Ureotelism could enable them to get just enough time to complete metamorphosis in a desiccating pond.

Densities in breeding pools may be quite high, ranging between 0.03 and 4,130 tadpoles/m^3 (Pfennig et al., 1991). In later observations, Pfennig

(1990b) observed tadpole densities of >3,000/m³ early in the larval period that gradually decreased in volume to ca. 1,000/m³ as development proceeded. Tadpole aggregations usually remain together throughout the developmental period until about four days prior to metamorphosis.

The SUL of metamorphs was 12–29 mm in Arizona (Pfennig et al., 1991). Pomeroy (1981) found that carnivorous morph tadpoles were 19–31 mm at metamorphosis, whereas omnivore morph tadpoles were 16–26 mm at metamorphosis.

DIET
Spea multiplicata are generalists whose diet includes many types of beetles, lepidopteran larvae, ants, crickets, flies, true bugs, spiders, pillbugs, cicadas, leafhoppers, aphids, and termites (Ruibal et al., 1969; Whitaker et al., 1977; Dimmitt and Ruibal, 1980a; Punzo, 1991a). Termites in particular made up a large part of the diet of Mexican Spadefoots in California. Assuming termites and beetles were the primary prey, and that a spadefoot could eat a large number in a single feeding, Dimmitt and Ruibal (1980a) estimated that an adult would need 6.8 feedings to obtain the amount of fat metabolized in a single year. A single spadefoot could eat 36% of its body weight in a single feeding. Newly metamorphosed juveniles consume beetles (many taxa, especially click and leaf beetles), spiders, hemipterans, ants, and lepidopteran larvae (Smith et al., 2004). Some vegetation also may be ingested while feeding.

PREDATION AND DEFENSE
Calling males readily dive underwater when alarmed. This species possesses a noxious skin secretion. To a human observer, the secretion has an odor of raw peanuts, buttered popcorn, or garlic, is irritating to the eyes and nose, and produces an allergic reaction similar to hay fever. Others have reported the secretion to be actually painful. When handled or confronted by a predator, Mexican Spadefoots puff-up and vocalize "eh,eh,eh,eh,eh" (Hammerson, 1999). Both sexes produce this sound. They also arch their bodies and close their eyes, extend the front limbs, and tightly tuck the rear limbs into the body (Livo et al., 1997).

In experimental trials, larvae are eaten by a wide variety of potential predators, indicating no chemically based defensive mechanisms. In addition, they do not change their levels of activity in the presence of predators. A lack of larval antipredator mechanisms may partially explain why this species breeds in temporary rather than permanent pools (Woodward, 1983). Reported predators include garter snakes (*Thamnophis marcianus*) (Woodward and Mitchell, 1990).

COMMUNITY ECOLOGY
Two or three different species of spadefoot may breed in close proximity to one another (e.g., *Spea multiplicata, S. bombifrons, Scaphiopus couchii*), yet the species often do not breed in the same temporary pool. One species will breed in one pool,

Adult *Spea multiplicata*.
Photo: Rob Schell

while a different species breeds in a pool only a short distance away. On the infrequent occasions when they do breed together, they often occur in rather equal numbers (Bragg, 1945b). Differences in mating calls help keep the species from interbreeding (Forester, 1973). It also seems likely that a priority effect occurs, that is, the species that arrives at a pool first claims that pool, and the pool subsequently is avoided by other spadefoots. Conspecifics may cluster together when calling from the same pool. For example, *S. couchii* may call from the same pools as *Spea multiplicata* and *S. bombifrons*, with *Scaphiopus couchii* preferring the deeper parts of the pool. There appears to be no real habitat segregation among species, however (e.g., McAlister, 1958).

Natural hybridization between *Spea multiplicata* and *S. bombifrons* has been reported in southeastern Arizona, where it tends to be more prevalent at ephemeral pools at lower elevations. Hybridization also occurs more frequently in smaller pools, where the different species are likely to come into close contact (Simovich, 1994). Large numbers of F_1 hybrids and the offspring of hybrid backcrosses occur in the area. *Spea bombifrons* females are more likely to hybridize than *S. multiplicata* females or *S. bombifrons* males (Simovich, 1985; Pfennig and Simovich, 2002). Hybrid males are sterile, and hybrid females produce about half the eggs of the parentals. In laboratory experiments, Simovich et al. (1991) found that hybrid tadpoles develop faster and to a larger size than the tadpoles of either parental species; they also have higher survivorship. However, they tend to be intermediate in size at metamorphosis when compared to the parentals. Thus, the life history characteristics of the hybrid offspring may afford them some advantages over the parentals, especially when exposed to conditions of short hydroperiods, despite problems with fertility and fecundity.

Not surprisingly, hybridization tends to occur most often in the most temporary ponds, and the duration of hydroperiod is inversely correlated with hybridization. Pfennig (2003), however, found that the rates of hybridization between these species had declined over a 27 yr period in this region. After testing a number of hypotheses, Pfennig (2003) concluded that despite the larval advantages in short hydroperiod ponds, the hybrids are less fit than the parental species, and thus reproductive isolation as a result of the evolution of premating isolating mechanisms (termed reinforcement) best explained decreasing hybridization.

In addition to hybridization, the presence of Plains Spadefoots has specific life history consequences for the less competitive Mexican Spadefoot in areas of sympatry. *Spea multiplicata* females tend to be in poorer body condition when they are in sympatry with *S. bombifrons* than they are in allopatry. Since the abundance of *S. bombifrons* is inversely correlated with mean body condition of female *S. multiplicata*, this suggests that severe competition is correlated with poor body condition in the latter species (Pfennig and Martin, 2009). Mean egg size is larger in allopatric *S. multiplicata* females than when in sympatry with *S. bombifrons*, and females with the best body conditions produce the most carnivorous larvae that are most adept at foraging, regardless of whether in sympatry or not. Taken together, these results indicate that that there is a maternal effect that mediates population divergence and character displacement when faced by severe competition (Pfennig and Martin, 2009) and that competition plays an important role in divergent trait evolution in spadefoots (Rice et al., 2009).

When *S. multiplicata* and *S. bombifrons* breed in the same pool, the frequency of the different tadpole morphotypes is very different than when only one species is found in a pool. When sympatric, *S. multiplicata* larvae nearly all take on the omnivorous morphotype whereas *S. bombifrons* larvae assume the carnivorous morphotype. This is not surprising, since *S. multiplicata* is a superior competitor on detritus whereas *S. bombifrons* is a superior competitor on fairy shrimp. When allopatric, however, both species' larvae develop both carnivorous and omnivorous morphotypes. The differences in the frequency of the morphotypes appear to have evolved in sympatry in order to reduce competition for scarce food resources. The manner in which these desert species partition resources by changing larval morphology is an example of character displacement (Pfennig and Murphy, 2000, 2002, 2003).

DISEASES, PARASITES, AND MALFORMATIONS

The Mexican Spadefoot is parasitized by metacercariae of the trematode *Clinostomum attenuatum* (Miller et al., 2004). Adult and juvenile spadefoots living in playa grasslands tend to have much smaller parasite loads of *C. attenuatum* than those living in agricultural areas. This may be because of the differences in hydroperiods (short in agricultural areas, long in playa grasslands) that cause disruptions in the parasite's life cycle (Gray et al., 2007c). Yearly variation in parasite load also is probably related to hydroperiod. Other parasites include the cestode *Distoichometra bufonis*, the nematodes *Aplectana incerta*, *A. itzocanensis*, and *Physaloptera* sp., and the protozoans *Opalina* sp. and *Nyctotherus cordiformis* (Parry and Grundman, 1965; Goldberg et al., 1995). The mite *Eutrombicula alfreddugesi* is known from Mexican Spadefoots in Texas (Mertins et al., 2011).

SUSCEPTIBILITY TO POTENTIAL STRESSORS
Chemicals. *Spea multiplicata* was not affected by the herbicides Roundup WeatherMAX® or Ignite® 280SL in experimental trials. However, juvenile survivorship was reduced when exposed to Roundup Weed and Grass Killer Ready-to-use Plus® (both at 1.33 ml glyphosphate/m^2) for 48 hrs (Dinehart et al., 2009).

STATUS AND CONSERVATION
Undoubtedly, Mexican Spadefoots face the usual threats from habitat loss, alteration, and fragmentation. There are no indications, however, that this species has declined throughout much of its range (e.g., Hammerson, 1999). The effects of habitat alteration on this species may be unclear or even misleading. For example, larval densities at playa breeding sites are not different among playa habitats with different adjacent land uses, suggesting widespread accommodation to human activity (Anderson et al., 1999; Ghioca, 2005). Abundance may actually appear high in agricultural areas when compared to other land uses (Gray et al., 2004b). On a landscape scale, occupancy of playa wetlands is positively associated with decreasing distance between pools and with increasing interplaya landscape complexity. Such a relationship may be of recent occurrence, as such nestedness patterns may result from the inability of this small species to disperse across geometrically complex agricultural landscapes that are found in the Texas Panhandle (Gray, 2002; Gray et al., 2004a, 2004b). In addition, a major predator of spadefoot larvae, *Ambystoma mavortium*, tends to be less common in agricultural settings. Thus, habitat fragmentation historically might have affected this species positively and negatively despite the species' current abundance and ability to persist in agricultural landscapes.

Habitat alteration also may have increased the likelihood of hybridization in areas of contact such as in southeastern Arizona. Simovich (1994) noted that the vegetation and precipitation runoff patterns changed substantially after the introduction of cattle grazing. Runoff previously allowed to form in scattered pools is now diverted to more permanent stock tanks. Spadefoots now breed in these pools, which results in different species breeding in close proximity. Modification of water flow patterns coupled with periodic drought could exacerbate hybridization potential.

A threat to Mexican Spadefoots is premature emergence in response to sound vibrations that accompany off-road vehicle activity in arid habitats. Vibrations from electric motors have been shown to induce 100% emergence from protected burrows (Dimmitt and Ruibal, 1980b). If spadefoots emerged under inhospitable conditions due to human-induced vibrations, they could expose themselves to direct harm from vehicles or be unable to rebury themselves in dry desert substrates.

There are certain trapping biases when sampling for larval *Spea multiplicata* as part of community analyses. Mexican Spadefoots are not readily captured in funnel traps, and are more likely to be captured by seining or dip-netting (Ghioca and Smith, 2007). Thus, trapping tends to underestimate the proportion of Mexican Spadefoots within a pool. Traps also tend to catch more of the carnivorous morph larvae when compared to the proportion of all morphs within a population.

COMMENT
Popular accounts of spadefoot biology are in Bragg (1955) and, especially, Bragg's (1965) book *Gnomes of the Night*.

ESTABLISHED NONNATIVE SPECIES

Family Dendrobatidae

Dendrobates auratus (Girard, 1855)
Green and Black Dart-Poison Frog

ETYMOLOGY

auratus: derived from Latin meaning 'golden.' The species was described from a population in Panama that has metallic gold spots on a brown background.

NOMENCLATURE

Synonyms: *Phyllobates auratus*. Complete synonymy in Amphibian Species of the World 5.1, an online reference.

IDENTIFICATION

Adults. Dendrobates auratus is one of the most beautiful frogs known. It is a small species with a highly variable bright iridescent green and black coloration. Rarely, some individuals will be black and white or purple and white. Others may tend toward yellow or blue. This frog is small, has a pointed snout, and short delicate limbs. The finger disks are large and there is no webbing between the toes. Males are slimmer and have a vocal sac but are otherwise indistinguishable from the females. The normal size range is 25–40 mm SUL for males and 27–42 mm SUL for females (*in* Savage, 2002).

Larvae. Larvae are uniformly colored dark brown to black, including the robust body and tail musculature. No spots or marking are apparent. Snouts are blunt. Tail fins are narrower than the tail musculature. Tails are long and about 1.5 times the length of the body. Maximum size is 30–39 mm TL (but see below).

Eggs. The eggs are small and black, with each egg surrounded by a clear gelatinous capsule. The clutch is oviposited on a cushion of clear jelly, which presumably holds moisture and cushions the eggs.

DISTRIBUTION

This species is native to Central America (Nicaragua to northwestern Colombia). It was introduced to Hawaii in 1932 in the upper Manoa Valley and Waiahole Valley on Oahu, ostensibly for the biocontrol of injurious insects (Hunsaker and Breese, 1967; McKeown, 1996; Kraus and Duvall, 2004). The frogs originally released came from Panama. It occurs only on Oahu, in moist leeward and windward valleys.

LIFE HISTORY

The Green and Black Dart-poison Frog is a terrestrial species of moist and humid leaf litter in shaded forest. They are adept at climbing, with reports from Central America of frogs > 13 m above the ground in the tree canopy (Savage, 2002). They may be especially active in the morning and late afternoon, but less so on bright sunny days. They walk or move in short hops, searching for tiny insects. In the Hawaiian winter (November to April), they are more active than they are during the hottest summer months (May to October),

Distribution of *Dendrobates auratus*

when they will seek refuge under debris and sheltering objects. *Dendrobates auratus* eats small insects, mites, and collembolans, with a preference for ants. Sexual maturity is reached in 15 months in Central America (Savage, 2002).

The bright coloration indicates that the frog is highly toxic. Although not particularly dangerous to humans, care should be taken when handling them. People should wash their hands immediately and never touch their eyes or mucous membranes after touching one of these frogs. It is likely that few predators bother this species. *Dendrobates auratus* is photopositive in their phototactic response, suggesting they use sunlight when feeding and going about their daily activities (Jaeger and Hailman, 1973). They have a flattened spectral curve that is unclassifiable (Hailman and Jaeger, 1974).

BREEDING SITES AND REPRODUCTION

Males call diurnally to attract females, usually in the late afternoon, and they are aggressive in defense of a calling location. Calls are made from holes or hollows at the base of trees. Both males and females are aggressive toward conspecifics. Intruding males are met with a challenge, and likely a wrestling match will ensue. The call is a low, slurred buzz ("cheez-cheez-cheez") lasting 2–4 sec, with 3–5 notes followed by a 5 sec pause before the call is repeated. The dominant frequency is 3.5 kHz (Savage, 2002). Females actually court the males after they reach his calling position and will chase other females from the territory of a perspective mate. Although a male can mate with many females, a territorial female will attack an intruder female.

Courtship lasts several hours, with males and females touching and hopping around one another. Mating occurs in a protected and secluded terrestrial location. Amplexus does not occur. The female deposits her eggs (normally 4–6, but up to 13) in a moist location, and the male then fertilizes them. The female then departs, and the male takes up a position to guard the nest from predators and desiccation. Males will guard multiple clutches and continue to call for females after mating. Females are capable of laying multiple clutches at eight- to ten-day intervals throughout the breeding season. The eggs hatch

Adult *Dendrobates auratus* carrying tadpoles, Costa Rica. Photo: Brian Kubicki

Adult *Dendrobates auratus*, Oahu, Hawaii. Photo: Fred Kraus in 10–16 days, and the male then transports the larvae on his back to standing water (e.g., shallow pools, tree hole cavities) and releases them. The male also may pick up tadpoles and carry them between sites.

The tadpoles are omnivorous, but they readily eat other animals, particularly mosquito larvae, daphnia (water fleas), and drowned insects. For this reason, there is usually only one tadpole in each small water-filled pocket or tree hole, inasmuch as larvae will readily attack and consume conspecifics. They also eat algae and moss. Tadpoles are active by day, but will rest in a vertical position at night with the head pointing upward at the water's surface. In Hawaii, tadpoles normally grow to at least 22 mm TL, but Mc-Keown (1996) reported larvae as large as 45 mm TL. The duration of the larval period is 39–102 days in Central America.

DISEASES, PARASITES, AND MALFORMATIONS
Ticks are reported as ectoparasites in Central America.

IMPACT ON NATIVE SPECIES
This species has a limited distribution, and there is no evidence it has had a detrimental effect on native Hawaiian fauna. There is nothing reported in the literature on the life history of this species in Hawaii. Indeed, some information in the primary literature appears contradictory and may reflect geographic variation in behavior and life history characteristics.

Family Eleutherodactylidae

Eleutherodactylus coqui
Thomas, 1966
Coquí

ETYMOLOGY
coqui: the name refers to the male's advertisement call, a loud CO-QUI (KO-KEE)

IDENTIFICATION
Adults. The ground color of *Eleutherodactylus coqui* is brown to grayish brown, with the dorsum quite variable in pattern, from uniform to mottled or freckled. A broad light dorsolateral band may extend down each side with the dorsum dark and the lateral colors much lighter below the band. Some coquis have a vertebral stripe of varying width extending down the middle of the back. A dark line may be present from the nostril through the eye extending to the front limb. Eyes are large, and the snout is pointed. The thighs are sometimes brown with yellowish-green mottling and darker than the body. Venters are variable in the amount of pigmentation, from a light salt-and-pepper pattern to a nearly even darker granular coloration. Males are smaller than females. In Puerto Rico, males are 29.5–37 mm SUL, whereas females are 35.5–52 mm SUL (Townsend and Stewart, 1994). O'Neill and Beard (2011) recorded a male mean of 27.9–30.7 mm SUL at Hilo and Kona, Hawaii.

Larvae. There is no free-swimming larva. The larval period is passed within the egg and froglets hatch as miniature adults.

Eggs. Eggs are white and unpigmented and tend to stick together in a clump. They are 3.5 mm in diameter and surrounded by a tough egg capsule.

DISTRIBUTION
This species is native to Puerto Rico. Established populations are in Hawaii (Hawaii, Maui, and Oahu; Kraus et al., 1999), but individuals have been found occasionally in Florida and Louisiana. Reports of *Eleutherodactylus marticensis* introduced to Hawaii with nursery stock at Kokomo and Omaopio on Maui (Kraus et al., 1999) were misidentified *E. coqui* (Kraus and Campbell, 2002). The Hawaiian populations originated from northern Puerto Rico, based on an analysis of genomic DNA (Velo-Antón et al., 2007). Dispersal has been facilitated through the nursery trade, with the first reports of coquis occurring in the late 1980s (Kraus and Campbell, 2002). However, frogs were intentionally released in order to establish new populations. It is found associated with nurseries, residential areas, resort hotels, ornamental parks, and natural forest. The species is rapidly expanding its range.

LIFE HISTORY
This species is very common and widespread in Puerto Rico, both in native forest and in human-disturbed areas. It appears to be occupying the same type of habitats in Hawaii, where populations have been reported from 40 to 952 m in elevation. Populations can be very large. For example, Kraus et al. (1999) reported collecting

Distribution of *Eleutherodactylus coqui*

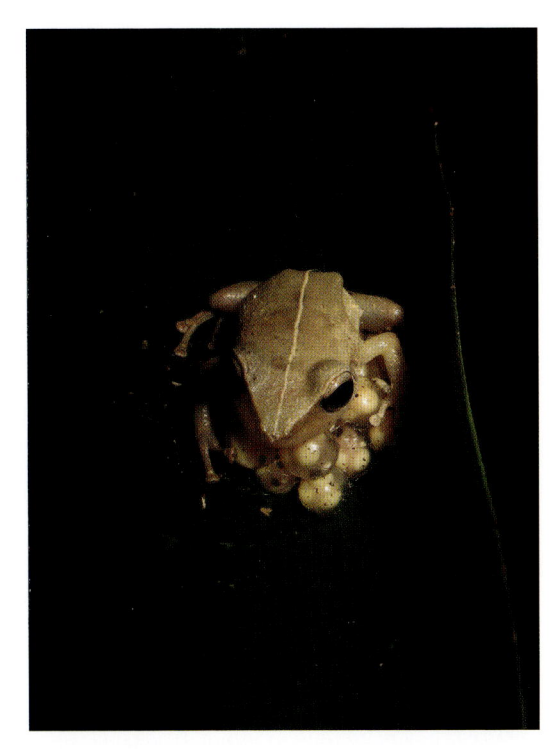

Male *Eleutherodactylus coqui* with eggs, Hawaii. Photo: Jesse Poulos

105 adult coquis in 50 minutes from an area 50 m². Populations are small for about a year following introduction but rapidly increase afterward. Woolbright et al. (2006) reported densities three times greater in Hawaii than in Puerto Rico, with adult estimates ranging from means of 3,413 to 11,800/ha depending upon population and season. There was a mean of 0.7 to 7.5 juveniles per adult, with ratios again changing seasonally. The life history of this species is reviewed by Rivero (1998) and Joglar (1998).

Sex ratios are biased toward males, with mark-recapture estimates ranging from 2.2 to 4.5 males per female (Woolbright et al., 2006). Sexual maturity is reached in about 8 months to 1 yr; populations are capable of rapid turnover, but coquis may live 4–5 yrs (Rivero, 1998). Coquis are largely opportunistic leaf litter feeders that consume prey in proportion to availability. The diet varies among populations, in accordance with invertebrate abundance. Prey consists of spiders, roaches, beetles, collembolans, flies, ants, lepidopteran larvae, amphipods, isopods, with lesser numbers of other invertebrates including native species (Beard and Pitt, 2005; Beard, 2007). Most prey comprised nonnative species, especially ants (30%) and amphipods (22%). Endemic invertebrates (mites, beetles, springtails, flies) and pest species (mosquitoes, termites) did not make up a significant portion of the diet. In turn, coquis made up 6.6% of the prey by weight of mongooses (*Herpestes javanicus*) in one Hawaiian study, but neither rats (*Rattus exulans, R. rattus*) nor Cane Toads (*Rhinella marina*) consumed coquis (Beard and Pitt, 2006). At Hawaii Volcanoes National Park, they have been found to 1,158 m in elevation.

BREEDING SITES AND REPRODUCTION
Reproduction may occur year-round, but peaks during the wet season. The call is a very loud "ko-kee" that is unmistakable. The *ko* part of the call establishes territorial boundaries with other males, whereas the *kee* is the advertisement call to attract females. Calls may be heard at any time, but most calling occurs from dusk to midnight, with an increase in intensity at dawn. Males call from the ground, short bushes and branches, and well into the tree canopy. Females approach the calling male, who then "leads" her to the area where eggs will be deposited. There is a certain degree of variation in call characteristics (fundamental frequency of each call syllable and call duration, and the number of calls per minute), which is correlated positively or negatively with elevation (O'Neill and Beard, 2011). Some of the

Calling male *Eleutherodactylus coqui*, Hawaii. Photo: Steve Johnson

variation can be explained by differences in male body size, which is greater at higher elevations.

Amplexus lasts a considerable period of time (to 12 hrs). Fertilization is internal via cloacal apposition, with the male remaining in amplexus as the eggs are extruded (Rivero, 1998). Eggs are deposited on soil or in leaf axils (see Joglar, 1998) in a moist location in terrestrial litter or under surface debris. In Puerto Rico, females deposit 4–6 clutches per year with a mean clutch size of 28 eggs per clutch (range 16 to 41) (Townsend and Stewart, 2004). Clutch size is positively correlated with female body size, and larger clutches are oviposited in the wet season than in the dry season. A female can oviposit multiple clutches at about eight-week intervals, and males can mate with multiple females. Male mating success is not correlated with SUL. The male guards the eggs from predators and desiccation for the 2–3 weeks (17–26 days) required for development. In doing this, he will position himself directly over the clutch with his arms folded around the eggs (Joglar, 1998). Hatchlings are ca. 5 mm TL.

DISEASES, PARASITES, AND MALFORMATIONS

The amphibian chytrid fungus *Batrachochytrium dendrobatidis* has been reported from *Eleutherodactylus coqui* on Maui and Hawaii (Beard and O'Neill, 2005). The infection rate was 2.4% of the frogs tested.

IMPACT ON NATIVE SPECIES

As noted above, populations of *E. coqui* can be very large in Hawaii, probably due to a lack of predators and an abundance of retreat sites. Much concern has been expressed about the potential effects of *E. coqui* as it expands its range into native rainforest and mesic forest at higher elevations. These areas have a very diverse insect and spider fauna, and it is feared that coquis could impact invertebrate biodiversity (Kraus et al., 1999; Kraus and Campbell, 2002; Beard and Pitt, 2005). Rödder (2009) noted that coquis could spread into higher elevations on Hawaii as a result of climate change as well as to many other habitats around the world, including Florida. It also has been speculated that increased predation on native insects could have detrimental impacts on Hawaii's endangered avifauna, although there is little overlap in habitats used by endemic birds and coquis.

In studies of leaf production, Sin et al. (2008) found that the presence of coquis significantly reduced the invertebrate communities, especially aerial species, herbivores, and leaf litter species, in some habitats. This in turn resulted in lowered herbivory rates, increases in nutrients (NH_4, magnesium, nitrogen, phosphorus, potassium), increased leaf production by some plants, and increased decomposition rates in other plants. As such, coquis could inadvertently make ecosystems

Habitat of *Eleutherodactylus coqui*, Oahu, Hawaii. Photo: Steve Johnson

more vulnerable to invasive plants by making nutrients more available while at the same time decreasing native invertebrate populations.

Hawaiians have often reported the loud advertisement call to be annoying, and the annoyance factor plus concerns about decreased property values due to loud frogs may have spurred more efforts at eradication than concern for biodiversity (e.g., Raloff, 2003). The State of Hawaii bans the sale of this species and has forbidden intentional introduction since 1998. Kraus and Campbell (2002) advocated removal programs, but noted a great amount of indifference to the problem, an inability or unwillingness to control introductions, public misperceptions of beneficial aspects, and outright defiant ignorance (Singer and Grismaijer, 2005).

Hand capture does not appear to be efficient, so research has been undertaken to test the effects of weak detergents, spraying citric acid (not effective), caffeine application (effective), and even the introduction of frog parasites to eradicate the species (not likely to be successful; Marr et al., 2010). The National Park Service has an eradication program at Hawaii Volcanoes National Park that combines spraying citric acid with a frog barrier fence (Tavares, 2008). Despite these efforts, it appears *E. coqui* has become another member of Hawaii's nonindigenous invaders.

Eleutherodactylus planirostris (Cope, 1862)
Greenhouse Frog

ETYMOLOGY
planirostris: from the Latin prefix *plani-* meaning 'flat' and *rostris* meaning 'nose.' The name refers to the flat area anterior to the eyes.

NOMENCLATURE
Frost et al. (2006a): *Euhyas planirostris*
   Synonyms: *Eleutherodactylus ricordii planirostris, Euhyas planirostris, Hylodes planirostris, Lithodytes planirostris*
   Hedges et al. (2008) placed *E. planirostris* within the *Eleutherodactylus (Euhyas) planirostris* Species Series.

IDENTIFICATION
Adults. This is a small frog with granular skin. There are two morphs of Greenhouse Frogs, a mottled and a striped morph. Striped frogs are light to dark brown, reddish brown, or bronze with a light orange to tan stripe running from the nostril dorsolaterally down the body. A triangular area of light orange to tan is present on top of the head, and a dark bar is present between the eyes. The brown coloration continues laterally but tends to fade out ventrally. The venter is light gray to white and speckled with scattered dark brown spots, as is the throat and undersides of the thighs. Eyes are sometimes reddish. Mottled frogs are similar in coloration, but lack the dorsolateral striping. Instead, the dorsum is a uniform light to dark brown, but often mottled with light brown coloration. Dorsal coloration is not correlated with light intensity or with sex. An unusually colored *Eleutherodactylus planirostris* (partial albino or "pinto") was reported from Dade County (Petrovic, 1973).

In north-central Florida, the striped form constituted up to 75% of the patterns observed by Goin (1947b). In south Florida, however, striped individuals comprised only 28.4–44.1% of individuals examined (Duellman and Schwartz, 1958). In Louisiana, the mottled morph is more common than the striped morph (Meshaka et al., 2009c). There is considerable variation in morph frequency among populations as noted by Goin (1947b), based on frequency counts from throughout the species' range. Coloration appears to result from a simple dominant/recessive gene, with the striped morph dominant (Goin, 1947b). Males are smaller than females. For example, in Louisiana, males are 16–25 mm SUL (mean 19.8 mm) and females are 18–29 mm SUL (mean 24.1 mm) (Meshaka et al., 2009c). In the Florida Keys, males are 15–17.5 mm SUL (mean 16.6 mm) and females 19.5–25 mm SUL (mean 22.6 mm) (Duellman and Schwartz, 1958).

Larvae. There is no free-swimming larva. The larval period is passed within the egg and froglets hatch as miniature adults.

Eggs. The eggs are white with a vitellus ca. 2

Distribution of *Eleutherodactylus planirostris*

mm in diameter. There is a single jelly envelope that is 3–4 mm in diameter that encloses a narrow layer of fluid, a wide layer of jelly, and a tough jelly envelope on the outside (Goin, 1947b). Just prior to hatching, the eggs are 5–6 mm in diameter.

DISTRIBUTION

The native range of *E. planirostris* is the Bahamas, the Cayman Islands, and Cuba. It was first reported by Cope (1863) from Key West (often incorrectly attributed to Barbour, 1910) and south Florida, where it may have arrived on ships coming from Cuba. It now occurs throughout urban and natural habitats in peninsular Florida (in north-central Florida by 1933; Van Hyning, 1933), southern Georgia (Jensen et al., 2008), and in scattered locations along the Gulf of Mexico. The species has been found throughout the Florida Keys and on St. George Island, Florida (Duellman and Schwartz, 1958; Lazell, 1989; Irwin et al., 2001). Subsequent range expansion has occurred both naturally in Florida and through the nursery trade elsewhere. The Greenhouse Frog has been reported from Alabama, Hawaii (Hawaii, Oahu), Louisiana, and Mississippi (Plotkin and Atkinson, 1979; Carey, 1982; Kraus et al., 1999; Winn et al., 1999; Dinsmore, 2004; Liner, 2007; Kraus, 2009; Meshaka et al., 2009c).

LIFE HISTORY

The Greenhouse Frog is common in humid and mesic habitats throughout Florida in natural, suburban, silvicultural, and agricultural settings (Goin, 1947b; Duellman and Schwartz, 1958; Dodd, 1992; Delis et al., 1996; Enge, 1998a; Surdick, 2005). This is a species of the leaf litter, although Harper (1935a) recorded it under tree bark on wild tamarind trees in south Florida hardwood hammocks. It hides under surface debris, rocks, woodpiles, and rotting logs and in burrows, leaf mold, and residential gardens within the soil. In cold and dry weather, they occupy burrows in the soil. Similar microhabitats are used in Hawaii, as are subterranean lava tubes (Olson and Beard, 2012).

Activity occurs year-round weather permitting in Florida. In Louisiana, activity has been reported in all months except December, with most activity from May to October (Meshaka et al., 2009c). Most activity occurs at night, especially during warm weather, but Greenhouse Frogs may be active diurnally after rain and on cloudy days. They are rapid jumpers and are readily able to escape into the leaf litter; however, they are more sluggish in winter at temperatures <15.6°C.

Growth is rapid, and sexual maturity is reached within a year of hatching. The diet consists of a wide variety of tiny invertebrates, including ants, springtails, booklice, beetles and beetle larvae, roaches, leafhoppers, true bugs, caterpillars, spiders, mites, centipedes, millipedes, and earthworms (Goin, 1947b; Duellman and Schwartz, 1958; Olson and Beard, 2012). Ants form about 40% of the diet, which consists entirely of terrestrial leaf-litter dwelling species. Predators are undoubtedly many, but there are no specific reports. According to Olson and Beard (2012), one site in Hawaii was estimated to contain 12,500 frogs per ha, in which case they may consume 129,000 invertebrates per ha per night.

BREEDING SITES AND REPRODUCTION

Calling occurs from protected locations in the terrestrial litter and under vegetation. The call is very soft, and has been likened to a tiny bell or a short chirp consisting of four to six notes (Goin, 1947b). Most calling occurs at night, especially after dusk and just before sunrise. When the sky is overcast or during rainfall, calling occurs anytime. Males have been heard calling from April to September in north-central Florida and to December in south Florida. However, egg deposition does not begin until late May and is closely tied to rainfall.

Amplexus is axillary, but courtship has not been described. The eggs are deposited terrestrially under vegetation, leaf mold, rocks, and in rotting wood. They are oviposited individually directly into a slight cavity in the soil in several layers. The female may cover the eggs by kicking soil onto the clutch. The clutch size is 3–26 (mean 16) in Florida, with the largest clutches deposited in midsummer (Goin, 1944, 1947b). In Louisiana, hatching occurs mostly in May to June (Meshaka et al., 2009c), and Liner (2007) recorded a single clutch of 9 eggs. One clutch is deposited per season. Neither parent attends the eggs. Development requires 13–20 days (mean 15.6) in Florida (Goin, 1947b). Young (4.3–5.7 mm TL) are born as miniature adults, using their egg tooth to break free of the egg capsule.

DISEASES, PARASITES, AND MALFORMATIONS
No information is available.

Adult *Eleutherodactylus planirostris*, striped pattern. Photo: Alan Cressler

Adult *Eleutherodactylus planirostris*, nonstriped pattern. Photo: David Dennis

Habitat of *Eleutherodactylus planirostris*, Alachua County, Florida. Photo: C.K. Dodd Jr.

IMPACT ON NATIVE SPECIES

There have been no recorded adverse effects on the native fauna anywhere where this species occurs. It is firmly established in the Floridian fauna, and Kraus et al. (1999) suggested it should have no effect on Hawaiian fauna because it is terrestrial and restricted to low elevations in areas of no great native insect diversity. In the introduced Hawaiian population, Olson and Beard (2012) found these leaf-litter frogs consumed a wider variety of invertebrates compared with conspecifics in their native range. However, they could not determine whether native invertebrates were negatively impacted by this species. Despite this, great concern has sometimes been expressed in the Hawaiian press.

Family Ranidae

Glandirana rugosa
(Temminck and Schlegel, 1838)
Wrinkled Frog

ETYMOLOGY
rugosa: the name *rugosa* is Latin and refers to the frog's wrinkled appearance.

IDENTIFICATION
Adults. The Wrinkled Frog is a relatively small dark green or gray to brown rugose frog. As its name implies, there are numerous parallel folds or ridges on its dorsum, which give the frog a wrinkled appearance. These folds extend onto both the front and rear limbs. The tympanum is about the size of the eye. Venters are mottled gray with a slight suffusion of yellow. Slight parallel dark bars are present on the legs. The rear toes are highly webbed. Males are smaller than females and have nuptial pads during the breeding season. In Japan, males are 37–46 mm SUL (mean 41 mm) and females are 44–53 mm SUL (mean 50 mm) (Maeda and Matsui, 1999). In Hawaii, Oliver and Shaw (1953) reported 7 females at 38.3–48 mm SUL and a single male at 37 mm SUL.

Larvae. Tadpoles are greenish gray to brown with a somewhat lichen-like pattern that blends in well with the substrate. The tail musculature and fins are pigmented like the body, and there are no dark spots as in *Lithobates catesbeianus* larvae. The dorsal and ventral tail fins are equal in depth. The lateral line system is obvious, especially on the head, around the eyes, and snout. According to Goris and Maeda (2004), tadpoles may have silvery spots on the body and tail. Tadpoles reach 38–80 mm TL (McKeown, 1996; Maeda and Matsui, 1999; Goris and Maeda, 2004). Oliver and Shaw (1953) noted that larvae on Oahu at 43 mm TL had rear hind limbs. Svihla (1936) has a figure of the mouthparts, egg, and surrounding envelopes.

Eggs. The eggs are light tan to dark brown dorsally and light tan ventrally. They are 0.8–1.5 mm in diameter. Two envelopes surround the eggs, the first 3 mm in diameter and the second 5.1–6 mm in diameter (Svihla, 1936).

DISTRIBUTION
Glandirana rugosa is indigenous to Honshu, Shikoku, and Kyushu Islands in Japan, Korea, northeastern China, and southern Primorskii, Russia. It was introduced to Hawaii in 1895 or 1896, ostensibly for the biocontrol of injurious insects (Svihla, 1936; Oliver and Shaw, 1953; McKeown, 1996; Kraus, 2009). It is known from the islands of Hawaii, Kauai, Maui, and Oahu.

LIFE HISTORY
The Wrinkled Frog is a species of shaded clear mountain streams at low to middle elevations in Hawaii. It prefers shallow quiet pools in areas free from American Bullfrogs. Adults stay close to water and do not wander extensively. The life history of the introduced populations is unknown,

Distribution of *Glandirana rugosa*

Adult *Glandirana rugosa*, Kauai, Hawaii. Photo: Ronn Altig

Habitat of *Glandirana rugosa* at 1,100 m in the Kohala Mountains, Island of Hawaii. Photo: James Juvik

but in Japan sexual maturity is reached in 1 yr following metamorphosis. Wrinkled Frogs have been observed to bask on stream rocks, but they readily dive into the water upon the approach of an observer. The frog will then remain in the leaf litter for five to ten minutes before resurfacing. The diet is largely insectivorous, but they appear to have a preference for ants, at least in Japan (Oliver and Shaw, 1953; Goris and Maeda, 2004). Wrinkled Frogs have a skin secretion that is distasteful to predators, but there are no reports of predation in Hawaii.

BREEDING SITES AND REPRODUCTION

In Japan, the breeding season is May to September, but breeding in Hawaii is February to August (Oliver and Shaw, 1953) and perhaps longer. Although Svihla (1936) indicated they were voiceless, the advertisement call is a buzz-like croak and consists of notes of 20 pulses that last ca. 0.4 sec. The dominant frequency is 2.7 kHz. Amplexus occurs in water and is axial. Eggs are adhesive and deposited in a jelly clumps of 15–70 eggs at a time, attached to vegetation and debris in slow-moving water, pools, or ponds. Svihla (1936), however, reported clusters containing 130–150 eggs in an area of 30 cm × 33 cm where the eggs were unattached since there was no vegetation within the pool. The total clutch size is 690–2,600 in Japan, with females producing multiple clutches by different males during the breeding season. Eggs hatch in five days. In Japan, larvae overwinter and transform the following year, but data are not available for Hawaii. Tadpoles are common in July and August. Recently metamorphosed froglets are ca. 19.5–26.8 mm SUL (Oliver and Shaw, 1953).

DISEASES, PARASITES, AND MALFORMATIONS
No information is available.

IMPACT ON NATIVE SPECIES
There is no evidence that this species has had any negative impact on the native Hawaiian fauna.

Family Hylidae

Osteopilus septentrionalis
(Duméril and Bibron, 1841)
Cuban Treefrog

ETYMOLOGY

septentrionalis: Latin for 'northern.' The original describers thought that the frog came from Norway.

NOMENCLATURE

Synonyms: *Hyla dominicensis* [in part], *H. septentrionalis*. Complete synonymy in Amphibian Species of the World 5.1, an online reference.

IDENTIFICATION

Adults and juveniles. This is the largest treefrog in Florida, in particular, the adult females. The ground color is gray to light brown; some frogs are uniform in coloration whereas others have brown or green mottling or lichen-like patterns dorsally. The skin is bumpy, and the skin on the head is fused to the skull. This allows Cuban Treefrogs to back into tight places with the head blocking the entrance. Since the skin is fused to the skull, the frog is difficult to handle, and the bony skull prevents water loss. The eyes and toepads are exceptionally large and conspicuous. A yellow wash may be present in the axil of the forelegs. Venters are white to off-white. The male has bilateral vocal sacs. Juveniles have a pale green to light tan ground color with a broad dorsolateral cream to yellow stripe. They are sometimes confused with *Hyla cinerea*, but the expanded toe disks are much larger in *Osteopilus septentrionalis*. In addition, juvenile Cuban Treefrogs often have red eyes, and they have blue bones that are especially evident in the long hind limbs.

Females are substantially larger than males, but the sexual size dimorphism decreases with increasing latitude; females tend to be smaller in the north, but male SULs remain about the same. In central Florida, females are 25–79 mm SUL (mean 53 mm) whereas males are 25–62 mm SUL (mean 43.1 mm). Females have eggs by 53 mm SUL and males have external nuptial pads by 26 mm SUL (McGarrity and Johnson, 2009). In south Florida, males averaged 46.1 to 52.9 mm SUL whereas females averaged 57.3 to 71.3 SUL, depending upon population. Waddle et al. (2010) reported a mean length of 38.6 mm SUL from southwest Florida. In the Everglades, males averaged 46.1 mm SUL (range 28.9 to 59.8 mm) whereas females averaged 64.2 mm SUL (range 44.5–99 mm) (Meshaka, 2001). Meshaka (2001) provides additional information on body size for populations in the lower Keys, Palmdale, Lake Placid, Okeechobee, and Tampa, Florida. In contrast, Duellman and Schwartz (1958) recorded males averaging 53 to 57.5 mm SUL and females averaging 71.7 to 78.6 mm SUL from Miami and Key West, with the largest female measuring 96.5 mm SUL and the largest male 61.6 mm SUL. Cuban Treefrogs within their native range are larger, with males averaging 44 to 56.2 mm and females averaging 68.5 to 77.3 mm SUL. The largest Florida male is 85 mm SUL and the largest Florida female is 122 mm SUL (McGarrity and Johnson, 2009).

Larvae. Larvae retain yolk for several days after hatching. Mature tadpoles are black dorsally and around the eyes with grayish-brown tail musculature. This grayish-brown color extends onto the body laterally. They have a round body and wide transparent tail fin that may contain small brown pigment spots. Venters are light. Tadpoles reach at least 32 mm TL. Duellman and Schwartz (1958) provide a figure of the tadpole and its mouthparts.

Eggs. Eggs average 1.2 mm in diameter (range

1 to 1.5 mm) at oviposition. The outer envelope is 3 mm in diameter, the inner envelope is 2.4 mm in diameter, and the vitellus is 2 mm in diameter within 24 hrs of deposition (Duellman and Schwartz, 1958).

DISTRIBUTION

Cuban Treefrogs are native to Cuba and the island of Hispaniola. The first record of them in Florida was by Barbour (1931), and Duellman and Schwartz (1958) speculated that the species could be considered native to Florida. *Osteopilus septentrionalis* has spread from the Florida Keys as far as north-central Florida (roughly along a line extending from Duval to Alachua to Levy counties) with scattered individuals reported in the Florida Panhandle and coastal Georgia (Savannah, Brunswick). Additional reports occur along transportation corridors (Interstate Highways 75 and 95) as far north as Ontario, Minnesota, and west to Colorado (Livo et al., 1998). There are also unconfirmed reports of an established population in the Houston, Texas, area (M. McGarrity, personal communication). Dispersal is aided by the tendency of Cuban Treefrogs to hide in closed spaces such as on boats, motor homes, camping trailers, and in nursery stock, for example, hiding within palm trees used for landscaping (Meshaka, 1996b). It occurs in both urbanized and natural habitats. Other distribution references include Goin (1944), Trapido (1947), Schwartz (1952), Allen and Neill (1953), Duellman and Schwartz (1958), Austin (1973), Meshaka (2001), and Johnson and McGarrity (2010). The northward spread of *O. septentrionalis* will likely be limited by minimum ambient temperature.

LIFE HISTORY

The life history of *Osteopilus septentrionalis* in Florida is reviewed by Meshaka (2001). The Cuban Treefrog is especially common in human-disturbed habitats such as urban and suburban residential developments and agriculture (Surdick, 2005). It also has moved into mesic hardwood forests, scrubby and mesic pine flatwoods, cypress ponds and domes, wet prairies, hydric hammocks, bottomland forest, swamps, and even into xeric sandhill Florida habitats (McGarrity and Johnson, 2009). Cuban Treefrogs are often present in great abundance. Females grow faster than males and mature later. Meshaka (2001) recorded differences in the age and sex structure among populations.

Osteopilus septentrionalis is active mostly during warm weather (>15°C), as the CTmin is only 6.4°C; the CTmax is 39°C (John-Alder et al., 1988). In the winter during the dry season and in the heat of the day, they take refuge in shelters offering confined spaces and cavities. On Egmont Key, for example, I have found them in arboreal retreats within dead palms and sheltered under oil drums. Meshaka (1996c) reported them using bird boxes, PVC pipe, air conditioning units, burrows in sphagnum, drain pipes, tree holes in a variety of species, palm axils, and in bark and sheet metal folds. As with other arboreal hylids, evaporative water loss is lower than in more terrestrial and aquatic species (Wygoda, 1984), which likely also aids in human-mediated dispersal.

In natural habitats, Cuban Treefrogs favor forested habitats, where they are found most often in tree canopies. If displaced into old-field habitats, they will occupy the shrub and subcanopy layers if trees are not available, but will rapidly move to adjacent forested habitats. Most movements occur at night, when humidity and temperature is high, wind speed is low, and moonlight is subdued. If displaced 200 m to other forested habitats, movements occur more randomly, at least over a short time, with frogs tending to settle into the new area. In old-field to forest movements, frogs averaged 21 m/day (range 11 to 47 m), whereas within

Distribution of *Osteopilus septentrionalis*

forest movements only averaged 8 m/day (range 4 to 19 m) (McGarrity and Johnson, 2010).

Cuban Treefrogs eat just about anything they can stuff into their mouths, including insects, spiders, small frogs, lizards, and even snakes. Heflick (2001), for example, found 11 orders of prey in *O. septentrionalis* guts, including beetles, cockroaches, moths, spiders, and the treefrog *Hyla cinerea*. Meshaka (1996a) found roaches to be the main prey from a small Florida sample, but in a much larger sample (Meshaka, 2001), he found 20 orders of prey in stomachs, including a substantial number of frogs and even lizards, as well as shed skin, stones, and vegetation. Beetles, isopods, lepidopterans, and cockroaches seem particularly favored, with females eating roaches, beetles, earwigs, and spiders, and males and juveniles concentrating on beetles, moths, and spiders. Other prey includes the Florida Striped Scorpion *Centruroides hentzi* (Granatosky et al., 2011) and the snake *Storeria victa* (Maskell et al., 2003). Duellman and Schwartz (1958) also noted beetles and other small invertebrates in the diet. *Osteopilus septentrionalis* does not necessarily prefer Green Treefrogs (*Hyla cinerea*) in prey choice trials, but they readily consume this and other hylids (Wyatt and Forys, 2004).

Osteopilus septentrionalis is eaten by conspecifics, turtles (*Chelydra serpentina*), lizards (*Anolis equestris*), snakes (*Agkistrodon piscivorus, Coluber constrictor, Nerodia fasciata, Pantherophis* sp., *Thamnophis sauritus, T. sirtalis*), and birds (*Egretta caerulea, Corvus brachyrynchos, Quiscalus quiscala, Strix varia, Tyto alba*) (Meshaka and Ferster, 1995; Meshaka, 2001). Tadpoles are consumed by turtles (*C. serpentina*), snakes (*N. fasciata, T. sauritus*), and birds (*E. caerulea*). *Osteopilus septentrionalis* blends in well with its background and wedges itself into hiding places, both of which help avoid detection and predation. If attacked, they kick, scream, inflate their bodies with air, and exude a smelly, sticky secretion that is noxious and probably toxic. Some people that handle Cuban Treefrogs report a mild allergic reaction (sneezing, runny nose) to the secretions or a burning of the mucous membranes if contacted.

Tadpoles can be observed year-round in south Florida. Foraging occurs at night and on cloudy days in open water. Development occurs in water 10–41°C. Growth is rapid, with transformation occurring in ca. 27 days at 25°C. However, metamorphosis can be delayed in cool weather, and Meshaka (2001) reported overwintering from October to March. In cold weather, tadpoles form aggregations and appear to bask. Newly metamorphosed froglets are 10.5–16 mm TL (Duellman and Schwartz, 1958; Meshaka, 2001). Maturity is reached within seven to nine months of transformation, with males growing slightly faster than females. According to Meshaka

Eggs of *Osteopilus septentrionalis*. Photo: Steve Johnson

Adult *Osteopilus septentrionalis*, amplexus. Photo: C.K. Dodd Jr.

(2001), longevity is short with males usually living only about 1 yr.

BREEDING SITES AND REPRODUCTION

The breeding season is primarily during warm weather and long daylight hours from March to October in peninsular Florida but may be year-round in extreme south Florida and the Everglades. Although there is a distinct testicular cycle in males, they are capable of breeding year-round, and amplexed pairs have been observed in winter (Meshaka, 2001). Males with nuptial pads can be found in any month of the year. Gravid females are also found year-round, but most gravid females are observed during the summer wet season.

Cuban Treefrogs use just about any standing water for breeding, including temporary wetlands, cypress swamps, marshes, wet flatwoods, pools in bottomland forests, urban water catchments, garden ponds, swimming pools, and open containers holding water. On Egmont Key, for example, they breed in old sewer system catchments. On Key Vaca, Peterson et al. (1952) reported them breeding in brackish water. However, reproduction is not successful in water with predaceous fish. The species is an explosive breeder, with large numbers breeding simultaneously.

Males have a loud advertisement call described as a "hoarse mraaaak!" (Johnson and McGarrity, 2010) or "waarh-waarh" (Lee, 1969b). Calls consist of 2 elements and are described in detail by Blair (1958a); the frequency ranges from 300 to 5,100 cps over a duration of 0.35 sec. In south Florida, calls occur throughout the year but after dark only from March to October (Meshaka, 2001). Calls also occur diurnally during heavy thunderstorms or during tropical cyclones.

Eggs are deposited in a large surface film of 200–300 (rarely to 1,000) eggs at a time. The film may sink, however, during hard rain. The egg film is sticky, especially after deposition. Total clutch size and ovum diameter are correlated with female SUL, but the relationship between SUL and ovum size varies among populations. In south Florida, clutches average 3,961 eggs (range 1,177 to 16,371) (Meshaka, 2001). Hatching occurs in ca. 27.5–30 hrs. Tadpoles are 5.9 mm TL 48 hrs after hatching.

DISEASES, PARASITES, AND MALFORMATIONS

Parasites of the introduced population in Florida have not yet been examined.

IMPACT ON NATIVE SPECIES

Because of its large size and appetite, there is much concern about the effects Cuban Treefrogs have on native treefrog populations in Florida. In areas where *O. septentrionalis* populations are large, both the abundance and survivorship of native species decreases after habitats are invaded by *O. septentrionalis*. In southwest Florida, Waddle et al. (2010) found that *Hyla cinerea* and *H. squirella* were much less likely to occur

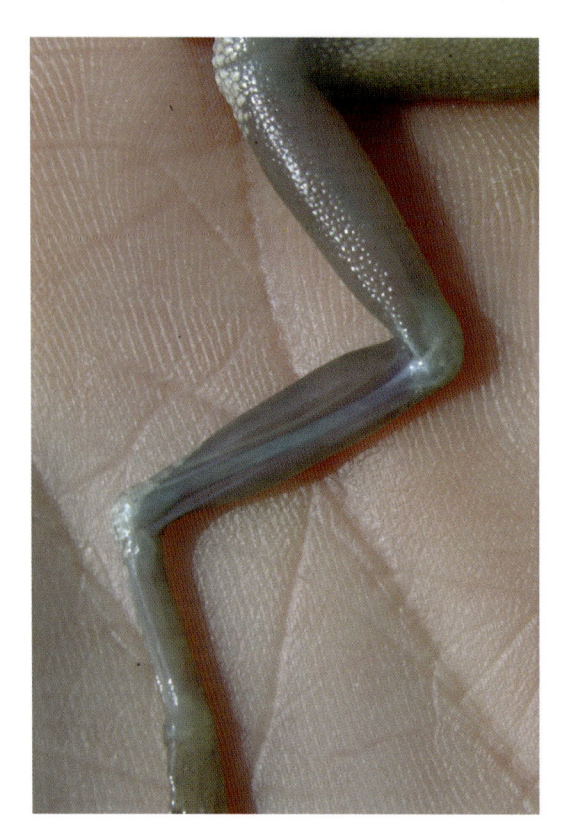

Leg of juvenile *Osteopilus septentrionalis* showing blue bones. Photo: Steve Johnson

at sites where Cuban Treefrogs were present. In central Florida, Hoffman (2007) found a negative correlation between *Osteopilus septentrionalis* abundance and the abundance of *Hyla femoralis*. The good news is that removing Cuban Treefrogs results in an increase in native treefrog (*H. cinerea*, *H. squirella*) abundance (Rice et al., 2011).

Competition may occur for food resources among postmetamorphs, and Cuban Treefrogs readily eat smaller hylids. For example, Meshaka (2001) reported significant dietary overlap between *Osteopilus septentrionalis* and *Hyla cinerea* and *H. squirella*. Competition also may occur for breeding and retreat sites, and they may interfere with the reproduction of native species. However, Hoffman (2007) was unable to demonstrate interference competition for retreat sites in laboratory trials or for inhibition against occupancy by *H. cinerea* and *H. femoralis* in refugia previously occupied by *Osteopilus septentrionalis*.

The tadpoles are superior competitors to some native species, and Cuban Treefrog larvae are known to eat the larvae of other species, such as *Hyla squirella*, at least in captivity (Smith, 2005b). This is of concern, since Cuban Treefrog tadpoles often outnumber the tadpoles of native species because of the high fecundity of *Osteopilus septentrionalis*. In laboratory trials, the presence of *O. septentrionalis* larvae resulted in decreased growth rates and delayed metamorphosis of *Anaxyrus terrestris* and *Hyla cinerea* larvae. In addition, the mass at metamorphosis was less than normal in *Anaxyrus terrestris*. Once *Osteopilus septentrionalis* larvae had metamorphosed, the mass of longer developing *Hyla cinerea* larvae actually increased at metamorphosis (Smith, 2005a). These results suggest that priority at breeding pools can significantly impact larval development of native species.

Temperature will likely play an important role in determining how far *Osteopilus septentrionalis* will disperse in the southern United States.

Osteopilus septentrionalis will breed just about anywhere, including this cistern on Egmont Key, Florida. During cold and dry weather, Cuban Treefrogs sheltered in palm axils and in the interior of dead palms. Photo: C.K. Dodd Jr.

In addition, the small female size in northern dispersers suggests decreased fecundity among colonizing females compared with those farther south (McGarrity and Johnson, 2009). Smaller female size could also mean less probability of large females having a significant impact on native hylids through predation or competition.

It may be impossible to completely eliminate Cuban Treefrogs from Florida's fauna. The University of Florida Extension Service recommends euthanizing Cuban Treefrogs using 20% benzocaine. Johnson et al. (2010) reported that a product called "Stiff 'n' Stop" (IFOAM Specialty Products, Sanford, Florida) is effective at deterring Cuban Treefrogs from taking up residence in buildings and electrical switchgear boxes. Avoidance is based on time-release of the chemical isophorone, which many animals find unpleasant. This is important, since electrical outages caused by Cuban Treefrogs are common in south Florida, causing thousands of dollars in damage.

Family Pipidae

Xenopus laevis (Daudin, 1802)
African Clawed Frog

ETYMOLOGY
laevis: from Latin meaning 'smooth.'

NOMENCLATURE
Synonyms: Complete synonymy in Amphibian Species of the World 5.1, an online reference.

IDENTIFICATION
Adults. This species is streamlined for living in the water, with a narrow dorsoventrally flattened head and smooth globular gray body. The ground color is olive to brown, finely spotted to marbled or with large round or irregular spots in yellow or dark tints. A lateral line system is obvious on the sides of the body, looking like a line of stitches. The eyes face upward. There are no eyelids nor does it possess a tongue or visible tympanum. The front legs are small in relation to the rear legs; front toes are not webbed. The hind legs are long and muscular, and the hind toes are fully webbed with claws on the tips, hence its common name. Venters are immaculate yellowish white to densely spotted. The body is covered with mucous making it difficult to handle.

Males have black roughened areas on the ventral surfaces of the forelimbs during the breeding season; females have large cloacal lips whereas those of males are barely discernible. Females are larger than males, normally averaging 90 to 100 mm SUL whereas males average 60 to 80 mm SUL. In California, McCoid and Fritts (1989, 1995) reported males to 80 mm SUL and females 65–119 mm SUL.

Larvae. The larvae are easily recognized by their tentacles and transparent body. They actually look more like young catfish than tadpoles. Heads are flattened and nearly translucent, and the body is almost completely transparent with only slight indications of pigment (transparent in daylight and darker at night). Viscera are visible through the skin. In high concentrations of algae, the tadpole can take on a green appearance. Eyes are black. Larvae have a very long tail with a whip-like tip and high tail fin. Maximum length is 80 mm.

Eggs. The eggs are pale brown and very small, averaging 1.1 to 1.2 mm in diameter inside a firm capsule 1.6 mm in diameter. Hatching occurs in 24 hrs.

DISTRIBUTION
The African Clawed Frog is indigenous to central and southern Africa, from Uganda and adjacent Sudan to South Africa. A variety of subspecies are recognized, with those in the United States most

Distribution of *Xenopus laevis*

Tadpole of *Xenopus laevis*, lateral view. Photo: Chris Brown

Tadpole of *Xenopus laevis*, dorsal view. Photo: Chris Brown

likely *X. l. laevis*. The species was introduced into the United States in large numbers beginning in the 1930s to 1940s in connection with human pregnancy tests. Others were sold as pets. In 1970–71, for example, more than 200,000 African Clawed Frogs were imported into the United States.

Populations were first reported in California in 1968 (St. Amant and Hoover, 1969; Mahrdt and Knefler, 1972, 1973; St. Amant et al., 1973; Bury and Luckenbach, 1976; Lemm, 2006), with periodic accounts of African Clawed Frogs being captured elsewhere across the continent (Colorado, Florida, Massachusetts, North Carolina, Texas, Virginia, Wisconsin; Zell, 1986; *in* Tinsley and Kobel, 1996; Kraus, 2009). They are now established in most drainages of southern California and in the Tucson and Phoenix areas of Arizona (Brennan and Holycross, 2006). The recent capture of a large female in Titusville, Florida, suggests the possibility of a breeding population. No other breeding populations have been confirmed in the United States. McCoid et al. (1993) reported on the apparent extirpation of a number of populations in California, possibly due to prolonged drought.

LIFE HISTORY

Xenopus laevis is an entirely aquatic species with a highly developed sense of smell and an ability to remove oxygen from water via the skin. As such, it can remain submerged for long periods. The short front limbs help push food into the mouth and circulate water across the lateral line system, which detects both vibrations and chemicals in the water. The rear legs are used for locomotion. In its native range, it occupies ponds and lakes in a wide variety of climatic conditions, from low tropics to high mountains (to 2,300 m). Feral populations in southern California inhabit both lentic and lotic waters, especially man-made impoundments and canals in highly disturbed areas. The biology of *Xenopus* is reviewed by Tinsley and Kobel (1996) and du Preez and Carruthers (2009).

Populations of African Clawed Frogs can be small to large. For example, McCoid and Fritts (1980b) estimated population size at two ponds as 602 and 494 adult frogs. Dispersal may occur downstream within a stream system or during sheet-flooding events, but most dispersal is probably human-mediated. African Clawed Frogs can move overland when ponds dry and move to the next available water, but normally they do not leave the water. Frogs in drying ponds are capable of excavating deeper holes to maintain water longer and lower the surface water temperature (McCoid and Fritts, 1980b). Activity occurs year-round, although aestivation can be induced in the laboratory. When aestivating, frogs maintain a heads-up position within a small vertical chamber that they dig; the chamber retains a connection to the surface of the pond bottom, which allows them to breathe.

Feeding occurs entirely in water, and adults capture a wide variety of prey and are known to scavenge. Juveniles eat crustaceans and mosquito larvae, whereas adults eat about anything they can shove into their mouths. Lenaker (1972) reported mosquito larvae, dragonfly larvae, aquatic beetles and their larvae, earthworms, leeches, sow bugs, snails, a moth, the fish *Gambusia*, toad tadpoles (*Anaxyrus* sp.), and unidentified adult amphibians in stomach contents from southern California frogs. McCoid and Fritts (1980a) reported the diet as consisting of slow-moving invertebrates, including amphipods, cladocerans, copepods, ostracods, various aquatic insect larvae, spiders, and snails. Lafferty and Page (1997) reported predation on the Tidewater Goby (*Eucyclogobius newberryi*) and an Arroyo Chub (*Gila orcutti*). *Xenopus laevis* also eats conspecific eggs and tadpoles as well as crayfish.

Growth rates are unequal, with females growing faster than males and growth occurring even in winter (McCoid and Fritts, 1995). Sexual maturity in California is reported at 6 months following metamorphosis for females and at 34 months for males (McCoid and Fritts, 1989). In its native range, maturity normally is reached in 2 yrs, although in some instances it occurs much faster. Sex ratios are 1:1. *Xenopus* can tolerate salinity up to 14‰ (40% sea water) for a period of several days and have been found in brackish water habitats in California (Munsey, 1972). They occur in water over a wide range of temperatures (7.2–32.2°C).

The tadpoles are obligate midwater suspension feeders with an efficient entrapment process for securing food. They do this by secreting mucous on the gills of the buccal cavity, which traps phyto- and zooplankton. They gulp water for this and for respiratory purposes. Tadpoles are sedentary at first, hovering in water almost perpendicular to the surface. As they grow, they tend to move around more, often forming large midwater schools and coming to the surface periodically to gulp air. Their long tentacles are extremely sensitive to vibrations in the water and alert them to possible predators. The preferred water temperature is ca. 22°C. As with many other amphibians, *X. laevis* tadpoles are sensitive to the herbicide atrazine, which significantly affects gonad development (Tavera-Mendoza et al., 2002).

BREEDING SITES AND REPRODUCTION

Breeding occurs in permanent streams and ponds and continues over an extended period. In California, breeding has been reported in all months except December, but most often from March to June (Lenaker, 1972; McCoid, 1985; McCoid and Fritts, 1989, 1993, 1995). It is possible that reproduction could occur year-round. Mature females have large ovarian masses throughout the

Adult *Xenopus laevis*.
Photo: Ronn Altig

Habitat of *Xenopus laevis* in southern California. Photo: Chris Brown

year, and different sized tadpoles indicate nonsynchronized breeding over an extended period.

Calling and mating occur at dusk or after dark, although occasional calls may be heard by day (McCoid, 1985). The male approaches a nearby female and gives chase. If amplexed, the pair swims to vegetation sites, at which time their intertwined feet pump 2–10 times in 2–3 sec as the female ejects eggs and the male fertilizes them. Amplexed pairs remain together throughout the night, periodically depositing eggs in clumps of vegetation. Nonresponsive females have a "rigor stance," whereby she becomes totally catatonic, and the male soon releases her (McCoid, 1985).

Calling occurs from underwater. This species has a complex vocal repertoire involving advertisement calls, intermale communication to establish dominance, and female response calls (Tobias et al., 2004). The advertisement call is a series of very long trills (also likened to a buzz or undulating snore), the first part of which is higher in pulse rate and in the number of pulses; the second part is more rattling. The fundamental frequency is 2 kHz with a mean of 54 pulses/sec, 20–37 pulses/note, 1–2 notes per sec, and a trill duration lasting 6–28 sec.

Eggs are laid singly or in small clusters of eight to ten eggs, each attached firmly to vegetation or other underwater debris. A female can deposit 15,000 eggs or more in a single breeding season, but these are oviposited on multiple occasions. McCoid and Fritts (1989) gave egg counts of 2,700–17,000, with the number of eggs correlated with female body size. It is likely that reproductive output is much less per bout (1,000–3,500 eggs) than the number of ova counted within a female at any one time (McCoid and Fritts, 1995). Reproductive bouts can occur at about one-month intervals. Males are always active and ready to breed. In captivity, hatching success is 74–76%. In South Africa, metamorphosis requires 49–64 days.

DISEASES, PARASITES, AND MALFORMATIONS

In California, *X. laevis* is parasitized by the protozoans *Balantidium* sp., *Nyctotherus* sp., and *Protoopalina xenopodus*; the trematodes *Clinostomum* sp., *Gyrdicotylus gallieni*, and *Protopolystoma xenopodus*; the cestode *Cephalochlamys namaquensis*; the nematodes *Contracaecum* sp. and *Eustrongylides* sp.; and the acanthocephalan *Acanthocephalus* sp. (Lafferty and Page, 1997; Kuperman et al., 2004). Not surprisingly, several of these have an African origin.

IMPACT ON NATIVE SPECIES

Xenopus laevis could eat substantial numbers of the larvae of native frogs, as they are reported to have a voracious appetite. There has been much speculation about potential negative impacts (e.g., Branson, 1975) but little empirical data. McCoid and Fritts (1980a, 1980b) suggested that *Xenopus* could be a nuisance but not a significant threat to native species. However, Lafferty and Page (1997) reported predation on the endangered Tidewater Goby (*Eucyclogobius newberryi*) and recommended trapping to remove the frog. There has been speculation that the virulent amphibian chytrid fungus (*Batrachocytrium dendrobatidis*) originated in Africa and has been spread worldwide via trade in *Xenopus laevis* for medical research and the pet trade (Weldon et al., 2004; Soto-Azat et al., 2010). This hypothesis is not universally accepted, although it is clear that chytrid has a long evolutionary history in southern Africa.

Larvae and postmetamorphs are eaten by the garter snake *Thamnophis hammondii* (Ervin and Fisher, 2001, 2007) and Black-crowned Night Herons (Crayon and Hothem, 1998) in California. The noxious skin secretions of *Xenopus laevis* induce gaping behavior in some snakes, which presumably allows the frog to escape (Zielinski and Barthalmus, 1989). However, *Thamnophis hammondii* readily eats African Clawed Frogs with no apparent ill effects, nor do the secretions produce any effects on locomotor performance (Foster and Mullin, 2008).

It is illegal to own, transport, or sell *Xenopus laevis* in Arizona, California, Hawaii, Kentucky, Louisiana, Nevada, New Jersey, North Carolina, Oregon, Virginia, and Washington, but it is legal to own *X. laevis* in Canada.

Glossary

Brief definitions of some of the more commonly used words within the text are listed below. For a more comprehensive dictionary of herpetological terms, see H.B. Lillywhite (2008) Dictionary of Herpetology (Krieger Publishing Company, Malabar, FL).

Acuminate	Tapering to a point
Aerobic	With oxygen
Allele	One of a series of possible forms of a gene
Allelochemical	A chemical produced by one organism that has a detrimental effect on another organism; frequently used in reference to plant chemicals that have an effect on other plants and animals
Allozyme	Allelic (different) forms of enzymes that can be identified using a procedure called electrophoresis
Amelanic	Lacking the pigment melanin that results in black or brown coloration
Anaerobic	Without oxygen
Animal pole	The dorsal portion of a frog's egg, usually black or brown in coloration
Anosmic	Inability to smell
Anthrogenic	Resulting from human activities
Aposematic	Characters that convey warning that the animal may be dangerous, poisonous, or distasteful
Assortive mating	Nonrandom pairing of males and females in reproduction, for example, large females preferably mating with large males
Barrancas	Gully or ravine in the desert Southwest
Behavioral fever	Raising the body temperature by basking in the sun in response to disease or infection
Boss	The raised bony area between the eyes in spadefoots
Buccal pumping	Method of moving air via positive pressure into the lungs by rhythmic movements of the throat or floor of the mouth
Celestial cues	Cues based on the sun, moon, and stars
Chromatophore	A skin cell containing a color pigment
Clade	An evolutionary lineage reflecting monophyly of its derived taxa
Conspecific	Consisting of the same species
Cryoprotectant	A substance that can be used to depress the freezing point of intracellular fluids and thus allow survival in very cold weather; most amphibians use glucose as a cryoprotectant, although some use glycerol
Cuspate	Possessing serrations
Detritivore	An animal feeding on a wide variety of organic plant and animal debris (detritis)
Dextral	To the right, on the right side
Ecotone	The abrupt border between two habitat types, such as between a field and forest
Edema	Swelling due to fluid accumulation
Evolution	Organic genetically based change through natural selection. Evolution is the fundamental unifying principle of biology.
Extirpation	Localized extinction or disappearance of a species
Fossorial	Living beneath the ground
Gape limited	Limited by the extent to which the mouth can be opened

Geotaxis	Directed response to gravity. Geotaxes can be positive (moving toward) or negative (moving away from) a source, such as by going up or down a slope.
Gosner stages	Graded stages of embryo and tadpole development. See Gosner (1960)
Heliothermic	Controlling body temperature by basking
Heliotropic	Moving toward (positive) or away from (negative) a heat source
Herpetology	The study of amphibians and reptiles
Heterospecific	Consisting of different species
Hydroperiod	The amount of time a body of water holds water
Hypoxic	Reduced or inadequate oxygen
Intercalary cartilage	A segment of cartilage between the ultimate and penultimate phalanges in the fingers of certain frog species
Introgression	Incorporating the genes of one species into the genes of another species via hybridization
Iridophores	A chromatophore containing guanine. Guanine results in a silvery or white coloration or in reflectance.
Irvingtonian	North American Land Mammal Age corresponding to 1.9 to 0.15 my BP
IUCN	World Conservation Union, known formerly as the International Union for the Conservation of Nature and Natural Resources
Karyotype	The number and form of the chromosomes
Littoral	Area in shallow water adjacent to the shoreline
Loess	Fine-grained soil deposits resulting from wind-blown silt and clay
Melanophores	Chromatophore containing melanin, resulting in dark (brown and black) coloration
Metacecariae	The encysted stage of a fluke larva that can produce developmental abnormalities or infection
Metachrosis	Change in color; the ability to change color
Miocene	Geologic epoch extending roughly from 5.3–23 mya
Monophyly	Evolved from a single ancestor
Myrmecophagy	Ant eating
Nearctic	Biogeographic realm consisting of North America north of the Isthmus of Tehuantepec
Nekton	Animals within the water column capable of independent movement
Nestedness	A measure of interconnectedness among species within a community or region
Neuston	Biotic community at the water's surface
Operational sex ratio	The sex ratio of breeding adults
Operculum	The covering over the gill chamber of tadpoles
Paedomorphosis	Ability to reproduce while retaining larval characteristics
Panmictic	A population where mating is completely random
Paraphyletic	A taxonomic lineage consisting of two or more species and all their common ancestors, but not all the descendants of the common ancestor
Parotoid	Specialized modified granular (mucous) gland that secretes a toxic or noxious substance to deter predators
Periphyton	A complex mixture of algae, cyanobacteria, microbes, and detritus that is attached to submerged surfaces in most aquatic ecosystems
Phagocytosis	The process whereby a cell engulfs foreign material
Phenotype	Physical characters of an organism that are observable
Phenotypic plasticity	Ability to alter the phenotype (morphology, color, life history) in response to environmental or biotic interactions
Philopatric	Remaining in the same location where development was completed; literally 'staying at home'
Photopositive	Orientation toward a light source
Phototaxis	Moving toward (positive) or away from (negative) a light source

GLOSSARY

Physiographic	Pertaining to the surface features of the habitat	Sinistral	To the left, on the left side
Plasma osmolality	Concentration of osmotically active solutes within the plasma	Spiracle	The outside opening to the tadpole's gill chamber
Pleistocene	Geologic epoch from approximately 12,000–2.6 mya	Subacuminate	Tapering to a point on the lower side
Plinth	A firm jelly egg mass that is somewhat rectangular	Sun compass orientation	The ability to use the position of the sun to orient perpendicularly to the shoreline
Pliocene	Geologic epoch extending roughly from 2.6–5.3 mya	Teratogenic	Refers to an agent causing physical defects in a developing embryo
Ploidy	The number of sets of chromosomes; haploid is one, diploid is two, and polyploid is more than two	Thigmotactic	Crevice-dwelling or wedging into tight spaces
		Titi	A shrubby tree of the Southeast; *Cyrilla racemiflora*
Poikilotherm	Refers to organisms whose body temperature is dependent on ambient temperature; such organisms do not generate a constant body temperature through metabolism	Toad	A frog in the Family Bufonidae; all toads are frogs, but not all frogs are toads
		Trophic	Relating to resource (food) use
Polyphenic	Having multiple observable patterns, that is, different phenotypes within a population	Tympanum	The sheath of skin covering the opening to the middle ear in frogs
		USDI	United States Department of Interior
Polyploidy	Condition whereby the number of sets of chromosomes is greater than two. Most polyploids are triploids (3) or tetraploids (4).	Vegetal pole	The ventral portion of a frog's egg, usually white or cream colored; contains the yolk for the developing embryo
Rancholabrean	North American Land Mammal Age corresponding to 150,000 to 10,000 years BP	Vicariance	Speciation due to geographic barriers or isolation
		Vitellus	The egg exclusive of its membranes
Recent	Geologic epoch from 12,000 years ago to present	Xanthophores	Chromatophores containing yellow or orange pigments
Riparian	Habitats along the margins of flowing rivers, streams, and creeks	Y-axis orientation	See 'Sun compass orientation'
Ruderal	Habitats disturbed by human activity		
Silviculture	Management related to the development and use of forests for commercial purposes; associated widely with tree farming		

Bibliography

Abbott, C.C. 1882. Notes on the habits of the "savannah cricket frog." American Naturalist 16:707–711.

Abbott, C.C. 1884. Recent studies of the spade-foot toad. American Naturalist 18:1075–1080.

Abbott, C.C. 1904. One explanation of reported showers of toads. Proceedings of the American Philosophical Society 43:163–164.

AbuBakr, S., and S.S. Crupper. 2010. Prevalence of cadmium resistance in *Staphylococcus sciuri* isolated from the gray treefrog, *Hyla chrysoscelis* (Anura: Hylidae). Phyllomedusa 9:141–146.

Acker, P.M., K.C. Kruse, and E.B. Krehbiel. 1986. Aging *Bufo americanus* by skeletochronology. Journal of Herpetology 20:570–574.

Ackroyd, J.F., and R.L. Hoffman. 1946. An albinistic specimen of *Pseudacris feriarum*. Copeia 1946:257–258.

Adam, M.D., and M.J. Lacki. 1993. Factors affecting amphibian use of road-rut ponds in Daniel Boone National Forest. Transactions of the Kentucky Academy of Science 54:13–16.

Adama, D., K. Lansley, and M.-A. Beaucher. 2004. Northern leopard frog (*Rana pipiens*) recovery: captive rearing and reintroduction in southeast British Columbia, 2003. Columbia Basin Fish and Wildlife Compensation Program, Nelson, British Columbia.

Adams, M.J. 1993. Summer nests of the tailed frog (*Ascaphus truei*) from the Oregon Coast Range. Northwestern Naturalist 74:15–18.

Adams, M.J. 1999. Correlated factors in amphibian decline: exotic species and habitat change in western Washington. Journal of Wildlife Management 63:1162–1171.

Adams, M.J. 2000. Pond permanence and the effects of exotic vertebrates on anurans. Ecological Applications 10:559–568.

Adams, M.J., and S. Claeson. 1998. Field response of tadpoles to conspecific and heterospecifics alarm. Ethology 104:955–961.

Adams, M.J., and R.B. Bury. 2002. The endemic headwater stream amphibians of the American Northwest: associations with environmental gradients in a large forested preserve. Global Ecology & Biogeography 11:169–178.

Adams, M.J., R.B. Bury, and S.A. Swarts. 1998. Amphibians of the Fort Lewis Military Reservation, Washington: sampling techniques and community patterns. Northwestern Naturalist 79:12–18.

Adams, M.J., S.D. West, and L. Kalmbach. 1999. Amphibian and reptile surveys of U.S. Navy lands on the Kitsap and Toandos peninsulas, Washington. Northwestern Naturalist 80:1–7.

Adams, M.J., S. Galvan, R. Scalera, C. Grieco, and R. Sindaco. 2008. *Batrachochytrium dendrobatidis* in amphibian populations in Italy. Herpetological Review 39:324–326.

Adams, M.J., N.D. Chelgren, D. Reinitz, R.A. Cole, L.J. Rachowicz, S. Galvan, B. McCreary, C.A. Pearl, L.L. Bailey, J. Bettaso, E.L. Bull, and M. Leu. 2010. Using occupancy models to understand the distribution of an amphibian pathogen, *Batrachochytrium dendrobatidis*. Ecological Applications 20:289–302.

Adams, S.B., and C.A. Frissell. 2001. Thermal habitat use and evidence of seasonal migration by Rocky Mountain tailed frogs, *Ascaphus montanus*, in Montana. Canadian Field-Naturalist 115:251–256.

Adams, S.B., D.A. Schmetterling, and M.K. Young. 2005. Instream movements by boreal toads (*Bufo boreas boreas*). Herpetological Review 36:27–33.

Adamson, M.L. 1981a. Development and transmission of *Gyrinicola batrachiensis* (Walton, 1929) (Pharyngodonidae: Oxyuroidea). Canadian Journal of Zoology 59:1351–1367.

Adamson, M.L. 1981b. Seasonal changes in populations of *Gyrinicola batrachiensis* (Walton, 1929) in wild tadpoles. Canadian Journal of Zoology 59:1377–1386.

Adamson, M.L. 1981c. *Gyrinicola batrachiensis* (Walton, 1929) n. comb. (Oxyuroidea; Nematoda) from tadpoles in eastern and central Canada. Canadian Journal of Zoology 59:1344–1350.

Ade, C.M., M.D. Boone, and H.J. Puglis. 2010. Effects of an insecticide and potential predators on green frogs and northern cricket frogs. Journal of Herpetology 44:591–600.

Adolph, E.F. 1931a. Body size as a factor in the metamorphosis of tadpoles. Biological Bulletin 61:376–386.

Adolph, E.F. 1931b. The size of the body and the size of the environment in the growth of tadpoles. Biological Bulletin 61:350–375.

AFCC. 1936. A future in frogs. American Frog Canning Company, New Orleans.

Agassiz, L. 1850. Lake Superior: its physical character, vegetation, and animals, compared with those of other and similar regions. With a narrative of the tour by J. Elliot Cabot and contributions by other scientific gentlemen. Part 7. Description of some new species of reptiles from the region of Lake Superior. Gould, Kendall and Lincoln, Boston, 378–382 + plate 6.

Aitchison, S.W., and D.S. Tomko. 1974. Amphibians and reptiles of Flagstaff, Arizona. Plateau 47:18–25.

Akers, E.C. 1997. Effects of predators and water color on growth, shape, and coloration of the tadpoles of *Hyla chrysoscelis* (Anura: Hylidae). M.S. thesis, Mississippi State University, Mississippi State.

Akers, E.C., C.M. Taylor, and R. Altig. 2008. Effects of clay-associated organic material on the growth of *Hyla chrysoscelis* tadpoles. Journal of Herpetology 42:408–410.

Akin, G.C. 1966. Self-inhibition of growth in *Rana pipiens* tadpoles. Physiological Zoology 39:341–356.

Albertini, G., and B. Lanza. 1987. *Rana catesbeiana* Shaw, 1802 in Italy. Alytes 6:117–129.

Alexander, D.G. 1966. An ecological study of the swamp cricket frog, *Pseudacris nigrita feriarum* (Baird), with comparative notes on two other hylids of the Chapel Hill, North Carolina, region. Ph.D. diss., University of North Carolina, Chapel Hill.

Alexander, T.R. 1965. Observations on the feeding behavior of *Bufo marinus* (Linne). Herpetologica 20:255–259.

Alford, R.A. 1985. Effects of parentage on competitive ability and vulnerability to predation in *Hyla chrysoscelis* tadpoles. Oecologia 68:199–204.

Alford, R.A. 1986. Habitat use and positional behavior of anuran larvae in a northern Florida temporary pond. Copeia 1986:408–423.

Alford, R.A. 1989a. Variation in predator phenology affects predator performance and prey community composition. Ecology 70:206–219.

Alford, R.A. 1989b. Competition between larval *Rana palustris* and *Bufo americanus* is not affected by variation in reproductive phenology. Copeia 1989:993–1000.

Alford, R.A., and M.L. Crump. 1982. Habitat partitioning among size classes of larval southern leopard frogs, *Rana utricularia*. Copeia 1982:367–373.

Alford, R.A., and H.W. Wilbur. 1985. Priority effects in experimental pond communities: competition between *Bufo* and *Rana*. Ecology 66:1097–1105.

Ali, M.F., K.R. Lips, F.C. Knoop, B. Fritzsch, C. Miller, and J.M. Conlon. 2002. Antimicrobial peptides and protease inhibitors in the skin secretions of the crawfish frog, *Rana areolata*. Biochimica et Biophysica Acta—Proteins and Proteomics 1601:55–63.

Allaback, M.L., D.M. Laabs, D.S. Keegan, and J.D. Harwayne. 2010. *Rana draytonii* (California Red-legged Frog). Dispersal. Herpetological Review 41:204–206.

Allan, D.M. 1973. Some relationships of vocalization to behavior in the Pacific treefrog, *Hyla regilla*. Herpetologica 29:366–371.

Allard, H.A. 1908. *Bufo fowleri* (Putnam) in northern Georgia. Science 28:655–656.

Allen, A.C. 1963. The amphibia of Wayne County, Ohio. Journal of the Ohio Herpetological Society 4:23–30.

Allen, J.A. 1868. Catalogue of the reptiles and batrachians found in the vicinity of Springfield, Massachusetts, with notices of all other species known to inhabit the state. Proceedings of the Boston Society of Natural History 12:3–38.

Allen, E.R. 1938. Notes on Wright's bullfrog, *Rana heckscheri* (Wright). Copeia 1938:50.

Allen, E.R. 1941. The value of toads. All-Pets Magazine. [reprinted by The Florida Reptile Institute, Silver Springs, FL]

Allen, E.R., and W.T. Neill. 1953. The treefrog, *Hyla septentrionalis*, in Florida. Copeia 1953:127–128.

Allen, M.J. 1932. A survey of the amphibians and reptiles of Harrison County, Mississippi. American Museum Novitates 542:1–20.

Allen, W.R. 1932. Further comment on the activity of the spade-foot toad. Copeia 1932:104.

Allran, J.W., and W.H. Karasov. 2001. Effects of atrazine on embryos, larvae, and adults of anuran amphibians. Environmental Toxicology and Chemistry 20:769–775.

Altig, R. 1970. A key to the tadpoles of the continental United States and Canada. Herpetologica 26:180–207.

Altig, R. 1972a. Notes on the larvae and premetamorphic tadpoles of four *Hyla* and three *Rana* with notes on tadpole color patterns. Journal of the Elisha Mitchell Scientific Society 88:113–119.

Altig, R.A. 1972b. Defensive behavior in *Rana areolata* and *Hyla avivoca*. Quarterly Journal of the Florida Academy of Sciences 35:212–216.

Altig, R., and E.D. Brodie Jr. 1968. Albinistic and cyanistic frogs from Oregon. Wasmann Journal of Biology 26:241–242.

Altig, R., and E.D. Brodie Jr. 1972. Laboratory behavior of *Ascaphus truei* tadpoles. Journal of Herpetology 6:21–24.

Altig, R., and M.T. Christensen. 1981. Behavioral characteristics of the tadpoles of *Rana heckscheri*. Journal of Herpetology 15:151–154.

Altig, R., and C.K. Dodd, Jr. 1987. The status of the Amargosa toad (*Bufo nelsoni*) in the Amargosa

River drainage of Nevada. Southwestern Naturalist 32:276–278.
Altig, R., and R.W. McDiarmid. 1999. Body plan. Development and morphology. Pp. 24–51 *In* R.W. McDiarmid and R.G. Altig (eds.), Tadpoles. The Biology of Anuran Larvae. University of Chicago Press, Chicago.
Anderson, A.L., and W.D. Brown. 2009. Plasticity of hatching in green frogs (*Rana clamitans*) to both egg and tadpole predators. Herpetologica 65: 207–213.
Anderson, A.M., D.A. Haukos, and J.T. Anderson. 1999. Habitat use by anurans emerging and breeding in playa wetlands. Wildlife Society Bulletin 27:759–769.
Anderson, A.R., and J.W. Petranka. 2003. Odonate predator does not affect hatching time or morphology of embryos of two amphibians. Journal of Herpetology 37:65–71.
Anderson, G.A., S.C. Schell, and I. Pratt. 1965. The life cycle of *Bunoderella metteri* (Allocreadidae: Bunoderinae), a trematode parasite of *Ascaphus truei*. Journal of Parasitology 51:579–582.
Anderson, J.D., K.A. Hawthorne, J.M. Galandak, and M.J. Ryan. 1978. A report on the status of the endangered reptiles and amphibians of New Jersey. Bulletin of the New Jersey Academy of Science 23:26–33.
Anderson, J.L., and B.W. Buttrey. 1962. Enteric protozoa of four species of frogs from the Lake Itasca region of Minnesota. Proceedings of the South Dakota Academy of Science 41:73–82.
Anderson, K., and P.E. Moler. 1986. Natural hybrids of the Pine Barrens treefrog, *Hyla andersonii* with *H. cinerea* and *H. femoralis* (Anura: Hylidae): morphological and chromosomal evidence. Copeia 1986:70–76.
Anderson, L.R., and J.A. Arruda. 2006. Land use and anuran biodiversity in southeast Kansas, USA. Amphibian and Reptile Conservation 4:46–59.
Anderson, P.K. 1942. Amphibians and reptiles of Jackson County, Missouri. Bulletin of the Chicago Academy of Science 6:203–220.
Anderson, P.K. 1951. Albinism in tadpoles of *Microhyla carolinensis*. Herpetologica 7:56.
Anderson, P.K. 1954. Studies in the ecology of the narrow-mouthed toad, *Microhyla carolinensis carolinensis*. Tulane Studies in Zoology 2:15–46.
Anderson, P.K., E.A. Liner, and R.E. Etheridge. 1952. Notes on amphibian and reptile populations in a Louisiana pineland area. Ecology 33:274–278.
Anderson, R.C., and G.F. Bennett. 1963. Opthalmic myiasis in amphibians in Algonquin Park, Ontario, Canada. Canadian Journal of Zoology 41:1169–1170.
Andrews, J.S. 2001. The Atlas of the Reptiles and Amphibians of Vermont. Privately Published, Middlebury, VT.
Andrews, K.D., R.L. Lampley, M.A. Gillman, D.T. Corey, S.R. Ballard, M.J. Blasczyk, and W.G. Dyer. 1992. Helminths of *Rana catesbeiana* in southern Illinois with a checklist of helminths in bullfrogs of North America. Transactions of the Illinois State Academy of Sciences 85:147–172.
Angermann, J.E., G.M. Fellers, and F. Matsumura. 2002. Polychlorinated biphenyls and toxaphene in Pacific tree frog tadpoles (*Hyla regilla*) from the California Sierra Nevada, USA. Environmental Toxicology and Chemistry 21:2209–2215.
Anholt, B.R., D.K. Skelly, and E.E. Werner. 1996. Factors modifying antipredator behavior in larval toads. Herpetologica 52:301–313.
Anholt, B.R., E. Werner, and D.K. Skelly. 2000. Effect of food and predators on the activity of four larval ranid frogs. Ecology 81:3509–3521.
ANLFRT (Alberta Northern Leopard Frog Recovery Team). 2005. Alberta Northern Leopard Frog Recovery Plan, 2005–2010. Alberta Sustainable Resource Development, Fish and Wildlife Division, Alberta Species at Risk Recovery Plan No. 7, Edmonton.
Anonymous. 1892. Bullfrogs for market. The Evening Star, Washington, DC [2 January].
Anonymous. 1899. Miss Seldon's frog farm. The Evening Star, Washington, DC [4 February].
Anonymous. 1907. Frogs. Okoboji Protective Association Bulletin 3:5.
Anonymous. 1918. The batrachians of Rhode Island. Part 2. Toads and frogs. Roger Williams Park, Park Museum Bulletin 10(4):93–96.
Anonymous. 2007. Rare blue frog discovered at Corkscrew Swamp Sanctuary. Florida Naturalist (Spring):11.
Anzalone, C.R., L.B. Kats, and M.S. Gordon. 1998. Effects of solar UV-B radiation on embryonic development in *Hyla cadaverina*, *Hyla regilla*, and *Taricha torosa*. Conservation Biology 12:646–653.
Applegate, R.D. 1990. *Rana catesbeiana*, *Rana palustris* (Bullfrog, Pickerel Frog). Predation. Herpetological Review 21:90–91.
Aresco, M.J. 1996. Geographic variation in the morphology and lateral stripe of the green treefrog (*Hyla cinerea*) in the southeastern United States. American Midland Naturalist 135:293–298.
Aresco, M.J. 2004. River frog *Rana heckscheri* (Wright). Pp. 17–18 *In* R.E. Mirarchi, M.A. Bailey, T.M. Haggerty, and T.L. Best (eds.), Alabama Wildlife. Vol. 3. Imperiled Amphibians, Reptiles, Birds, and Mammals. University of Alabama Press, Tuscaloosa.
Arndt, R.G. 1977. A blue variant of the green frog, *Rana clamitans melanota* (Amphibia, Anura, Ranidae) from Delaware. Journal of Herpetology 11:102–103.
Arnold, H.L. 1968. Poisonous Plants of Hawaii. Tong Publishing, Honolulu.

Arnold, S.J., and R.J. Wassersug. 1978. Differential predation on metamorphic anurans by garter snakes (*Thamnophis*): social behavior as a possible defense. Ecology 59:1014–1022.

Aronson, L.R. 1943a. The sexual behavior of Anura. 5. Oviposition in the green frog, *Rana clamitans*, and the bull frog, *Rana catesbeiana*. American Museum Novitates 1224:1–6.

Aronson, L.R. 1943b. The sexual behavior of anura. 4. Oviposition in the mink frog, *Rana septentrionalis* Baird. American Midland Naturalist 29:242–244.

Aronson, L.R. 1944. The sexual behavior of Anura. 6. The mating pattern of *Bufo americanus*, *Bufo fowleri*, and *Bufo terrestris*. American Museum Novitates 1250:1–15.

Asay, M.J., P.G. Harowicz, and L. Su. 2005. Chemically mediated mate recognition in the tailed frog (*Ascaphus truei*). Chemical Signals in Vertebrates 10:24–31.

Ashley, E.P., and J.T. Robinson. 1996. Road mortality of amphibians, reptiles and other wildlife on the Long Point causeway, Lake Erie, Ontario. Canadian Field-Naturalist 110:403–412.

Ashton, A.D., and F.C. Rabalais. 1978. Helminth parasites of some anurans of northwestern Ohio. Proceedings of the Helminthological Society of Washington 45:141–142.

Ashton, D.T. 2002. A comparison of abundance and assemblage of lotic amphibians in late-seral and second-growth redwood forests in Humboldt County, California. M.A. thesis, Humboldt State University, Arcata, CA.

Ashton, D.T., S.B. Marks, and H.H. Welsh Jr. 2006. Evidence of continued effects from timber harvesting on lotic amphibians in redwood forests of northwestern California. Forest Ecology and Management 221:183–193.

Ashton, R.E. Jr., and P.S. Ashton. 1988. Handbook of Reptiles and Amphibians of Florida. Part 3. The Amphibians. Windward Publishing, Inc., Miami.

Ashton, R.E. Jr., S.I. Guttman, and P. Buckley. 1973. Notes on the distribution, coloration, and breeding of the Hudson Bay toad, *Bufo americanus copei* (Yarrow and Henshaw). Journal of Herpetology 7:17–20.

Atlas, M. 1935. The effect of temperature on the development of *Rana pipiens*. Physiological Zoology 8:290–310.

Aubry, K.B., and P.A. Hall. 1991. Terrestrial amphibian communities in the southern Washington Cascade Range. Pp. 327–340 In L.F. Ruggiero, K.B. Aubry, A.B. Carey, and M.H. Huff (coordinators). Wildlife and Vegetation of Unmanaged Douglas-fir Forests. USDA Forest Service General Technical Report PNW-GTR-285, Portland, OR.

Auburn, J.S., and D.H. Taylor. 1979. Polarized light perception and orientation in larval bullfrogs *Rana catesbeiana*. Animal Behaviour 27:658–668.

Auffenberg, W. 1956. Remarks on some Miocene anurans from Florida, with a description of a new species of *Hyla*. Breviora (52):1–11.

Austin, D.F. 1973. Range expansion of the Cuban tree frog in Florida. Florida Naturalist (August):28.

Austin, J.D., and K.R. Zamudio. 2008. Incongruence in the pattern and timing of intra-specific diversification in bronze frogs and bullfrogs (Ranidae). Molecular Phylogenetics and Evolution 48:1041–1053.

Austin, J.D., S.C. Lougheed, L. Niedrauer, A.A. Chek, and P.T. Boag. 2002. Cryptic lineages in a small frog: the post-glacial history of the spring peeper, *Pseudacris crucifer* (Anura: Hylidae). Molecular Phylogenetics and Evolution 25:316–329.

Austin, J.D., S.C. Lougheed, P.E. Moler, and P.T. Boag. 2003a. Phylogenetics, zoogeography, and the role of dispersal and vicariance in the evolution of the *Rana catesbeiana* (Anura: Ranidae) species group. Biological Journal of the Linnean Society 80:601–624.

Austin, J.D., J.A. Dávila, S.C. Lougheed, and P.T. Boag. 2003b. Genetic evidence for female-biased dispersal in the bullfrog, *Rana catesbeiana* (Ranidae). Molecular Ecology 12:3165–3172.

Austin, J.D., S.C. Lougheed, and P.T. Boag. 2004a. Discordant temporal and geographic patterns in maternal lineages of eastern North American frogs, *Rana catesbeiana* (Ranidae) and *Pseudacris crucifer* (Hylidae). Molecular Phylogenetics and Evolution 32:799–816.

Austin, J.D., S.C. Lougheed, and P.T. Boag. 2004b. Controlling for the effects of history and non-equilibrium conditions in gene flow estimates in northern bullfrog (*Rana catesbeiana*) populations. Genetics 168:1491–1506.

Austin, J.D., T.A. Gorman, and D. Bishop. 2011. Assessing fine-scale genetic structure and relatedness in the micro-endemic Florida bog frog. Conservation Genetics 12:833–838.

Awan, A.R., and G.R. Smith. 2007. The effect of group size on the responses of wood frog tadpoles to fish. American Midland Naturalist 158:79–84.

Awbrey, F.T. 1963. Homing and home range in Bufo valliceps. Texas Journal of Science 15:127–141.

Awbrey, F.T. 1965. An experimental investigation of the effectiveness of anuran mating calls as isolating mechanisms. Ph.D. diss., University of Texas, Austin.

Awbrey, F.T. 1972. "Mating call" of a *Bufo boreas* male. Copeia 1972:579–581.

Awbrey, F.T. 1978. Social interaction among chorusing Pacific tree frogs, *Hyla regilla*. Copeia 1978:208–214.

Axtell, C.B. 1976. Comparisons of morphology, lactate dehydrogenase, and distribution of *Rana blairi* and *Rana utricularia* in Illinois and Missouri. Transactions of the Illinois State Academy of Science 69:37–48.

Axtell, R.W. 1958. Female reaction to the male call in two anurans (Amphibia). Southwestern Naturalist 3:70–76.

Axtell, R.W., and N. Haskell. 1977. An interhiatal population of *Pseudacris streckeri* from Illinois, with an assessment of its postglacial dispersion history. Natural History Miscellanea (202):1–8.

Babbitt, K.J., and G.W. Tanner. 1997. Effects of cover and predator identity on predation of *Hyla squirella* tadpoles. Journal of Herpetology 31:128–130.

Babbitt, K.J., and G.W. Tanner. 1998. Effects of cover and predator size on survival and development of *Rana utricularia* tadpoles. Oecologia 114: 258–262.

Babbitt, K.J., and G.W. Tanner. 2000. Use of temporary wetlands by anurans in a hydrologically modified landscape. Wetlands 20:313–322.

Babbitt, K.J., M.J. Baber, and T.L. Tarr. 2003. Patterns of larval amphibian distribution along a wetland hydroperiod gradient. Canadian Journal of Zoology 81:1539–1552.

Babbitt, K.J., M.J. Baber, and G.W. Tanner. 2005. The impact of agriculture on temporary wetland amphibians in Florida. Pp. 48–55 *In* W.E. Meshaka Jr. and K.J. Babbitt (eds.), Amphibians and Reptiles. Status and Conservation in Florida. Krieger Publishing, Malabar, FL.

Babbitt, K.J., M.J. Baber, and L.A. Brandt. 2006. The effect of woodland proximity and wetland characteristics on larval anuran assemblages in an agricultural landscape. Canadian Journal of Zoology 84:510–519.

Babbitt, L.H. 1937. The amphibia of Connecticut. Connecticut State Geological and Natural History Survey Bulletin No. 57.

Baber, M.J. 2001. Understanding anuran community structure in temporary wetlands: the interaction and importance of landscape and biotic processes. Ph.D. diss., Florida International University, Miami.

Baber, M.J., and K.J. Babbitt. 2003. The relative impacts of native and introduced predatory fish on a temporary wetland tadpole assemblage. Oecologia 136:289–295.

Baber, M.J., and K.J. Babbitt. 2004. Influence of habitat complexity on predator-prey interactions between the fish (*Gambusia holbrooki*) and tadpoles of *Hyla squirella* and *Gastrophryne carolinensis*. Copeia 2004:173–177.

Baber, M.J., E. Fleishman, K.J. Babbitt, and T.L. Tarr. 2004. The relationship between wetland hydroperiod and nestedness patterns in assemblages of larval amphibians and predatory macroinvertebrates. Oikos 107:16–27.

Baber, M.J., K.J. Babbitt, F. Jordan, H.L. Jelks, and W.M. Kitchens. 2005. Relationships among habitat type, hydrology, predator composition, and distribution of larval anurans in the Florida Everglades. Pp. 154–160 *In* W.E. Meshaka Jr. and K.J. Babbitt (eds.), Amphibians and Reptiles. Status and Conservation in Florida. Krieger Publishing, Malabar, FL.

Bachmann, K. 1969. Temperature adaptations of amphibian embryos. American Naturalist 103: 115–130.

Backus, R.H. 1954. Notes on the frogs and toads of Labrador. Copeia 1954:226–227.

Bacon, E.J., and Z.M. Anderson. 1976. Distributional records of amphibians and reptiles from Coastal Plain of Arkansas. Proceedings of the Arkansas Academy of Science 30:14–15.

Bagdonas, K.R., and D. Pettus. 1976. Genetic compatibility in wood frogs (Amphibia, Anura, Ranidae). Journal of Herpetology 10:105–112.

Bagnara, J.T., and J.S. Frost. 1977. Leopard frog supply. Science 197:106–107.

Bai, C., T.W.J. Garner, and Y. Li. 2010. First evidence of *Batrachochytrium dendrobatidis* in China: discovery of chytridiomycosis in introduced American bullfrogs and native amphibians in the Yunnan Province, China. EcoHealth 7:127–134.

Bailey, M.A. 1989. Migration of *Rana areolata sevosa* and associated winter-breeding amphibians at a temporary pond in the lower coastal plain of Alabama. M.S. thesis, Auburn University, Auburn, AL.

Bailey, M.A. 1990. Movement of the dusky gopher frog (*Rana areolata sevosa*) at a temporary pond in the lower Coastal Plain of Alabama. Pp. 27–43 *In* C.K. Dodd Jr., R.E. Ashton Jr., R. Franz, and E. Wester (eds.), Burrow Associates of the Gopher Tortoise. Proceedings of the 8th Annual Meeting, Gopher Tortoise Council, Gainesville, FL.

Bailey, M.A. 1991. The dusky gopher frog in Alabama. Journal of the Alabama Academy of Science 62:28–34.

Bailey, M.A., and D.B. Means. 2004. Gopher frog *Rana capito* Leconte and Mississippi gopher frog *Rana sevosa* Goin and Netting. Pp. 15–17 *In* R.E. Mirarchi, M.A. Bailey, T.M. Haggerty, and T.L. Best (eds.), Alabama Wildlife. Vol. 3. Imperiled Amphibians, Reptiles, Birds, and Mammals. University of Alabama Press, Tuscaloosa.

Bailey, M.A., J.N. Holmes, K.A. Buhlmann, and J.C. Mitchell. 2006. Habitat Management Guidelines for Amphibians and Reptiles of the Southeastern United States. Partners in Amphibian and Reptile Conservation, Technical Publication HMG-2, Montgomery, AL.

Bailey, R.M. 1943. Four species new to the Iowa herpetofauna, with notes on their natural histories. Proceedings of the Iowa Academy of Science 50:347–352.

Bailey, V. 1933. Cave life of Kentucky. American Midland Naturalist 14:385–635.

Baird, S.F. 1854. Descriptions of new genera and species of North American frogs. Proceedings of the Academy of Natural Sciences of Philadelphia 7:59–62.

Baird, S.F. 1859. Reptiles of the boundary, with notes by the naturalists of the survey. Pp. 1–35 *In* United States and Mexican Boundary Survey, under the order of Lieut. Col. W.H. Emory. U.S. Government Printing Office, Washington, DC.

Baird, S.F., and C. Girard. 1852. Descriptions on new species of reptiles, collected by the U.S. Exploring Expedition under the command of Capt. Charles Wilkes, U.S.N. Part 1.including the species from the west coast of North America. Proceedings of the Academy of Natural Sciences of Philadelphia 6:174–177.

Baird, S.F., and C.H. Girard. 1853. Descriptions of new species of reptiles collected by the U.S. Exploring Expedition under the command of Capt. Charles Wilkes, U.S.N. 2nd part. Proceedings of the Academy of Natural Sciences of Philadelphia 6:378–379.

Baker, M.R. 1977. Redescription of *Oswaldocruzia pipiens* (Nematoda: Trichostrongylidae) from amphibians of eastern North America. Canadian Journal of Zoology 55:104–109.

Baker, M.R. 1978a. Morphology and taxonomy of *Rhabdias* sp. (Nematoda: Rhabdiasidae) from reptiles and amphibians of southern Ontario. Canadian Journal of Zoology 56:2127–2141.

Baker, M.R. 1978b. Transmission of *Cosmocercoides dukae* (Nematoda: Cosmocercoidea) to amphibians. Journal of Parasitology 64:765–766.

Baker, M.R. 1978c. Development and transmission of *Oswaldocruzia pipiens* Walton, 1929 (Nematoda: Trichostrongylidae) in amphibians. Canadian Journal of Zoology 56:1026–1031.

Baker, M.R. 1979a. The free-living and parasitic development of *Rhabdias* sp. (Nematoda: Rhabdiasidae) in amphibians. Canadian Journal of Zoology 57:161–178.

Baker, M.R. 1979b. Seasonal population changes in *Rhabdias ranae* Walton, 1929 (Nematoda: Rhabdiasidae) in *Rana sylvatica* of Ontario. Canadian Journal of Zoology 57:179–183.

Baker, R.H. 1942. The bullfrog, a Texas wildlife resource. Texas Game, Fish and Oyster Commission, Bulletin No. 23.

Baldwin, R.F., and P.G. deMaynadier. 2009. Assessing threats to pool-breeding amphibian habitat in an urbanizing landscape. Biological Conservation 142:1628–1638.

Baldwin, R.F., A.J.K. Calhoun, and P.G. deMaynadier. 2006. Conservation planning for amphibian species with complex habitat requirements: a case study using movements and habitat selection of the wood frog *Rana sylvatica*. Journal of Herpetology 40:442–453.

Baldwin, R.F., P.G. deMaynadier, and A.J.K. Calhoun. 2007. *Rana sylvatica* (Wood Frog). Predation. Herpetological Review 38:194–195.

Ball, R.W., and D.L. Jameson. 1966. Premating isolating mechanisms in sympatric and allopatric *Hyla regilla* and *Hyla californiae*. Evolution 20:533–551.

Ball, R.W., and D.L. Jameson. 1970. Biosystematics of the canyon tree frog *Hyla cadaverina* Cope (= *Hyla californiae* Gorman). Proceedings of the California Academy of Science 37:363–380.

Ball, S.C. 1936. The distribution and behavior of the spadefoot toad in Connecticut. Transactions of the Connecticut Academy of Arts and Science 32: 351–379.

Ballinger, R.E., and C.O. McKinney. 1966. Developmental temperature tolerance of certain anuran species. Journal of Experimental Zoology 161: 21–28.

Ballinger, R.E., J.W. Meeker, and M. Thies. 2000. A checklist and distribution maps of the amphibians and reptiles of South Dakota. Transactions of the Nebraska Academy of Sciences 26:29–46.

Ballinger, R.E., J.D. Lynch, and G.R. Smith. 2010. Amphibians and Reptiles of Nebraska. Rusty Lizard Press, Oro Valley, AZ.

Bancroft, B.A., N.J. Baker, C.L. Searle, T.S. Garcia, and A.R. Blaustein. 2008. Larval amphibians seek warm temperatures and do not avoid harmful UVB radiation. Behavioral Ecology 19:879–886.

Bancroft, G.T., J.S. Godley, D.T. Gross, N.N. Rojas, D.A. Sutphen, and R.W. McDiarmid. 1983. The herpetofauna of Lake Conway: species accounts. U.S. Army Corps of Engineers Aquatic Plant Control Research Program, Miscellaneous Paper A-83-5.

Bank, M.S., J. Crocker, B. Connery, and A. Amirbahman. 2007. Mercury bioaccumulation in green frog (*Rana clamitans*) and bullfrog (*Rana catesbeiana*) tadpoles from Acadia National Park, Maine, USA. Environmental Toxicology and Chemistry 26: 118–125.

Banta, A.M. 1914. Sex recognition and the mating behavior of the wood frog, *Rana sylvatica*. Biological Bulletin 26:171–183.

Banta, B.H. 1961. On the occurrence of *Hyla regilla* in the lower Colorado River, Clark County, Nevada. Herpetologica 17:106–108.

Banta, B.H. 1974. Death-feigning in *Hyla regilla* on Santa Cruz Island, Santa Barbara County, California. Bulletin of the Maryland Herpetological Society 10:88.

Banta, B.H., and D. Morafka. 1966. An annotated check list of the recent amphibians and reptiles inhabiting the city and county of San Francisco, California. Wasmann Journal of Biology 24:223–236.

Banta, B.H., and G. Carl. 1967. Death-feigning behavior in the eastern gray treefrog *Hyla versicolor versicolor*. Herpetologica 23:317–318.

Barbeau, T.R., and H.B. Lillywhite. 2005. Body wiping behaviors associated with cutaneous lipids in hylid tree frogs of Florida. Journal of Experimental Biology 208:2147–2156.

Barber, P.H. 1999. Phylogeography of the canyon treefrog, *Hyla arenicolor* (Cope) based on mitochondrial DNA sequence data. Molecular Ecology 8:547–562.

Barbour, R.W. 1953. The amphibians of Big Black Mountain, Harlan County, Kentucky. Copeia 1953:84–89.

Barbour, R.W. 1956. *Pseudacris brachyphona* in Tennessee. Copeia 1956:54.

Barbour, R.W. 1957. Observations on the mountain chorus frog, *Pseudacris brachyphona* (Cope) in Kentucky. American Midland Naturalist 57: 125–128.

Barbour, R.W. 1971. Amphibians & Reptiles of Kentucky. University Press of Kentucky, Lexington.

Barbour, R.W., and E.P. Walters. 1941. Notes on the breeding habits of *Pseudacris brachyphona*. Copeia 1941:116.

Barbour, T. 1910. *Eleutherodacylus ricordii* in Florida. Proceedings of the Biological Society of Washington 23:100.

Barbour, T. 1931. Another introduced frog in North America. Copeia 1931:140.

Barnosky, A.D., N. Matzke, S. Tomiya, G.O.U. Wogan, B. Swartz, T.B. Quental, C. Marshall, J.L. McGuire, E.L. Lindsey, K.C. McGuire, B. Mersey, and E.A. Ferrer. 2011. Has the Earth's sixth mass extinction already arrived? Nature 471:51–57.

Barr, T.C. Jr. 1953. Notes on the occurrence of ranid frogs in caves. Copeia 1953:60–61.

Barrasso, D.A., R. Cajade, S.J. Nenda, G. Baloriani, and R. Herrera. 2009. Introduction of the American Bullfrog *Lithobates catesbeianus* (Anura: Ranidae) in natural and modified environments: an increasing conservation problem in Argentina. South American Journal of Herpetology 4:69–75.

Barrett, K., C. Guyer, and D. Watson. 2010. Water from urban streams slows growth and speeds metamorphosis in Fowler's toad (*Bufo fowleri*) larvae. Journal of Herpetology 44:297–300.

Barron, M.G., and K.B. Woodburn. 1995. Ecotoxicology of chlorpyrifos. Reviews of Environmental Contamination and Toxicology 144:1–93.

Barry, D.S., T.K. Pauley, and J.C. Maerz. 2008. Amphibian use of man-made pools on clear-cuts in the Allegheny Mountains of West Virginia, USA. Applied Herpetology 5:121–128.

Barta, J.R., and S.S. Desser. 1984. Blood parasites of amphibians from Algonquin Park, Ontario. Journal of Wildlife Diseases 20:180–189.

Bartelt, P.E. 1998. *Bufo boreas* (Western Toad). Mortality. Herpetological Review 29:96.

Bartelt, P.E. 2000. A biophysical analysis of habitat selection in western toads (*Bufo boreas*) in southwestern Idaho. Ph.D. diss., Idaho State University, Pocatello.

Bartelt, P.E., and C.R. Peterson. 2005. Physically modeling operative temperatures and evaporative rates in amphibians. Journal of Thermal Biology 30:93–102.

Bartelt, P.E., C.R. Peterson, and R.W. Klaver. 2004. Sexual differences in the post-breeding movements and habitats selected by Western toads (*Bufo boreas*) in southeastern Idaho. Herpetologica 60:455–467.

Bartlett, R.D., and P.P. Bartlett. 1999. A Field Guide to Florida Reptiles and Amphibians. Gulf Publishing Company, Houston.

Bartlett-Healy, K., W. Crans, and R. Gaugler. 2008. Phonotaxis to amphibian vocalizations in *Culex territans* (Diptera: Culicidae). Annals of the Entomological Society of America 101:95–103.

Bartram, W. 1791. Travels through North and South Carolina, Georgia, East and West Florida, the Cherokee Country, the Extensive Territories of the Muscoculges, or Creek Confederation, and the Country of the Chactaws. James and Johnson, Philadelphia.

Bastien, H., and R. Leclair, Jr. 1992. Aging wood frogs (*Rana sylvatica*) by skeletochronology. Journal of Herpetology 26:222–225.

Baud, D.R., and M.L. Beck. 2005. Interactive effects of UV-B and copper on spring peeper tadpoles (*Pseudacris crucifer*). Southeastern Naturalist 4: 15–22.

Bauer Dial, C.A., and N.A. Dial. 1995. Lethal effects of the consumption of field levels of paraquat-contaminated plants on frog tadpoles. Bulletin of Environmental Contamination and Toxicology 55:870–877.

Baxley, D., and C. Qualls. 2009. Habitat associations of reptile and amphibian communities in longleaf pine habitats of south Mississippi. Herpetological Conservation and Biology 4:295–305.

Baxter, G.T. 1952. Notes on growth and the reproductive cycle of the leopard frog, *Rana pipiens* Schreber, in southern Wyoming. Journal of the Colorado-Wyoming Academy of Science 4:91.

Baxter, G.T., and M.D. Stone. 1985. Amphibians and Reptiles of Wyoming. 2nd ed. Wyoming Game and Fish Department, Cheyenne.

Baxter, G.T., M.R. Stromberg, and C.K. Dodd Jr. 1982. The status of the Wyoming toad (*Bufo hemiophrys baxteri*). Environmental Conservation 9:348, 338.

Bayless, L.E. 1966. Comparative ecology of two sympatric species of *Acris* (Anura; Hylidae) with emphasis on interspecific competition. Ph.D. diss., Tulane University, New Orleans.

Bayless, L.E. 1969a. Post-metamorphic growth of *Acris crepitans*. American Midland Naturalist 81:590–592.

Bayless, L.E. 1969b. Ecological divergence and distribution of sympatric *Acris* populations (Anura: Hylidae). Herpetologica 25:181–187.

Bayne, K.A. 2004. The natural history and morphology of the eastern cricket frog, *Acris crepitans crepitans*, in West Virginia. M.S. thesis, Marshall University, Huntington.

Beane, J.C. 1990. *Rana palustris* (Pickerel Frog). Predation. Herpetological Review 21:59.

Beane, J.C. 1998. Status of the river frog, *Rana heckscheri* (Anura: Ranidae), in North Carolina. Brimleyana 25:69–79.

Beane, J.C., and L.T. Pusser. 2005. *Bufo terrestris* (Southern Toad). Diet and scavenging. Herpetological Review 36:432.

Beane, J.C., T.J. Thorp, and D.A. Jackan. 1998. *Heterodon simus* (Southern Hognose Snake). Diet. Herpetological Review 29:44–45.

Beane, J.C., A.L. Braswell, J.C. Mitchell, W.M. Palmer, and J.R. Harrison III. 2010. Amphibians & Reptiles of the Carolinas and Virginia. University of North Carolina Press, Chapel Hill.

Beard, K.H. 2007. Diet of the invasive frog, *Eleutherodacylus coqui*, in Hawaii. Copeia 2007:281–291.

Beard, K.H., and E.M. O'Neill. 2005. Infection of an invasive frog *Eleutherodacylus coqui* by the chytrid fungus *Batrachochytrium dendrobatidis* in Hawaii. Biological Conservation 126:591–595.

Beard, K.H., and W.C. Pitt. 2005. Potential consequences of the coqui frog invasion in Hawaii. Diversity and Distributions 11:427–433.

Beard, K.H., and W.C. Pitt. 2006. Potential predators of an invasive frog (*Eleutherodactylus coqui*) in Hawaiian forests. Journal of Tropical Ecology 22:345–347.

Beasley, V.R., S.A. Faeh, B. Wikoff, C. Staehle, J. Eisold, D. Nichols, R. Cole, A.M. Schotthoefer, M. Greenwell, and L.E. Brown. 2005. Risk factors and declines in northern cricket frogs (*Acris crepitans*). Pp. 75–86 *In* M.J. Lannoo (ed.), Amphibian Declines. The Conservation Status of United States Species. University of California Press, Berkeley.

Beauclerc, K.B., B. Johnson, and B.N. White. 2010. Distinctiveness of declining northern populations of Blanchard's cricket frog (*Acris blanchardi*) justifies recovery efforts. Canadian Journal of Zoology 88:553–566.

Beaudry, F. 2007. *Rana sylvatica* (Wood Frog). Predation. Herpetological Review 38:195.

Beauregard, N., and R. Leclair Jr. 1988. Multivariate analysis of the summer habitat structure of *Rana pipiens* Schreber, in Lac Saint Pierre (Québec, Canada). Pp. 129–143 *In* R.C. Szaro, K.E. Severson, and D.R. Patton (technical coordinators), Management of Amphibians, Reptiles, and Small Mammals in North America. Proceedings of a Symposium. USDA Forest Service, General Technical Report RM-166.

Bechtel, H.B. 1995. Reptile and Amphibian Variants. Colors, Patterns, and Scales. Krieger Publishing, Malabar, FL.

Beck, C.W., and J.D. Congdon. 1999. Effects of individual variation in age and size at metamorphosis on growth and survivorship of southern toads (*Bufo terrestris*) metamorphs. Canadian Journal of Zoology 77:944–951.

Beck, C.W., and J.D. Congdon. 2000. Effects of age and size at metamorphosis on performance and metabolic rates of Southern toad, *Bufo terrestris*, metamorphs. Functional Ecology 14:32–38.

Beck, W.M. Jr. 1948. An ecological study of the cold-blooded vertebrates of a north Florida lake. M.S. thesis, University of Florida, Gainesville.

Becker, C.G., C.R. Fonseca, C.F.B. Haddad, R.F. Batista, and P.I. Prado. 2007. Habitat split and the global decline of amphibians. Science 318:1775–1777.

Bee, M.A. 2001. Habituation and sensitization of aggression in bullfrogs (*Rana catesbeiana*): testing the dual-process theory of habituation. Journal of Comparative Psychology 115:307–316.

Bee, M.A. 2002. Territorial male bullfrogs (*Rana catesbeiana*) do not assess fighting ability based on size-related variation in acoustic signals. Behavioral Ecology 13:109–124.

Bee, M.A. 2003. Experience-based plasticity of acoustically evoked aggression in a territorial frog. Journal of Comparative Physiology 189A:485–496.

Bee, M.A. 2007a. Sound source segregation in grey treefrogs: spatial release from masking by the sound of a chorus. Animal Behaviour 74:549–558.

Bee, M.A. 2007b. Selective phonotaxis by male wood frogs (*Rana sylvatica*) to the sound of a chorus. Behavioral Ecology and Sociobiology 61:955–966.

Bee, M.A. 2008a. Parallel female preferences for call duration in a diploid ancestor of an allotetraploid freefrog. Animal Behaviour 76:845–853.

Bee, M.A. 2008b. Finding a mate at a cocktail party: spatial release from masking improves acoustic mate recognition in grey treefrogs. Animal Behaviour 75:1781–1791.

Bee, M.A., and S.A. Perrill. 1996. Responses to conspecific advertisement calls in the green frog (*Rana clamitans*) and their role in male-male communication. Behaviour 133:283–301.

Bee, M.A., and H.C. Gerhardt. 2001a. Neighbour-stranger discrimination by territorial male bullfrogs (*Rana catesbeiana*). 1. Acoustic basis. Animal Behaviour 62:1129–1140.

Bee, M.A., and H.C. Gerhardt. 2001b. Neighbour-stranger discrimination by territorial male bullfrogs (*Rana catesbeiana*). 2. Perceptual basis. Animal Behaviour 62:1141–1150.

Bee, M.A., and H.C. Gerhardt. 2002. Individual voice recognition in a territorial frog (*Rana catesbeiana*). Proceedings of the Royal Society, London 269:1443–1448.

Bee, M.A., and E.M. Swanson. 2007. Auditory masking of anuran advertisement calls by road traffic noise. Animal Behaviour 74:1765–1776.

Bee, M.A., and K.K. Riemersma. 2008. Does common spatial origin promote the auditory grouping of temporally separated signal elements in grey treefrogs? Animal Behaviour 76:831–843.

Bee, M.A., and J.J. Schwartz. 2009. Behavioral measures of signal recognition thresholds in frogs in presence and absence of chorus-shaped noise. Journal of the Acoustical Society of America 126: 2788–2801.

Bee, M.A., S.A. Perrill, and P.C. Owen. 1999. Size assessment in simulated territorial encounters between male green frogs (*Rana clamitans*). Behavioral Ecology and Sociobiology 45:177–184.

Behle, W.H., and R.J. Erwin. 1962. The green frog (*Rana clamitans*) established at West Ogden, Weber County, Utah. Utah Academy of Sciences, Arts, and Letters Proceedings 39:74–76.

Beiswenger, R.E. 1975. Structure and function in aggregations of tadpoles of the American toad, *Bufo americanus*. Herpetologica 31:222–233.

Beiswenger, R.E. 1977. Diel patterns of aggregative behavior in tadpoles of *Bufo americanus*, in relation to light and temperature. Ecology 58:98–108.

Beiswenger, R.E. 1981. Predation by grey jays on aggregating tadpoles of the boreal toad (*Bufo boreas*). Copeia 1981:274–276.

Beiswenger, R.E. 1986. An endangered species, the Wyoming toad *Bufo hemiophrys baxteri*—the importance of an early warning system. Biological Conservation 37:59–71.

Belden, L.K., E.L. Wildy, A.C. Hatch, and A.R. Blaustein. 2000. Juvenile western toads, *Bufo boreas*, avoid chemical cues of snakes fed juvenile, but not larval, conspecifics. Animal Behaviour 59:871–875.

Belden, L.K., I.T. Moore, R.T. Mason, J.C. Wingfield, and A.R. Blaustein. 2003. Survival, the hormonal stress response and UV-B avoidance in Cascades frog tadpoles (*Rana cascadae*) exposed to UV-B radiation. Functional Ecology 17:409–416.

Bellis, E.D. 1957. An ecological study of the wood frog *Rana sylvatica* LeConte. Ph.D. diss., University of Minnesota, Minneapolis.

Bellis, E.D. 1961. Growth of the wood frog, *Rana sylvatica*. Copeia 1961:74–77.

Bellis, E.D. 1962a. The influence of humidity on wood frog activity. American Midland Naturalist 68:139–148.

Bellis, E.D. 1962b. Cover value and escape habits of the wood frog in a Minnesota bog. Herpetologica 17:228–231.

Bellis, E.D. 1965. Home range and movements of the wood frog in a northern bog. Ecology 46:89–98.

Beltz, E. 2007. Scientific and common names of the reptiles and amphibians of North America explained. http://ebeltz.net/herps/etymain.html.

Benard, M.F. 2007. Predators and mates: conflicting selection on the size of male Pacific treefrogs (*Pseudacris regilla*). Journal of Herpetology 41:317–320.

Bennett, L. 2003. The miracle of the toads. Alberta Naturalist 33:72–73.

Bennett, R.S., E.E. Klaas, J.R. Coats, M.A. Mayse, and E.J. Kolbe. 1983. Fenvalerate residues in nontarget organisms from treated cotton fields. Bulletin of Environmental Contamination and Toxicology 31:61–65.

Bennett, S.H., J.W. Gibbons, and J. Glanville. 1980. Terrestrial activity, and diversity of amphibians in differently managed forest types. American Midland Naturalist 103:412–416.

Beringer, J., and T.R. Johnson. 1995. *Rana catesbeiana* (Bullfrog). Diet. Herpetological Review 26:98.

Bernal, X., and S.R. Ron. 2004. *Leptodactylus fragilis* (White-lipped Foamfrog). Courtship. Herpetological Review 35:372–373.

Berns, M.W. 1966. Some genetic and histochemical aspects of the variant blue frog in the genus *Rana*. M.S. thesis, Cornell University, Ithaca, NY.

Berns, M.W., and L.D. Uhler. 1966. Blue frogs of the genus *Rana*. Herpetologica 22:181–183.

Berns, M.W., and K.S. Narayan. 1970. An histochemical and ultrastructural analysis of the dermal chromatophores of the variant ranid blue frog. Journal of Morphology 132:169–180.

Berrill, M., S. Bertram, A. Wilson, S. Louis, and D. Brigham. 1993. Lethal and sublethal impacts of pyrethroid insecticides on amphibian embryos and tadpoles. Environmental Toxicology and Chemistry 12:525–539.

Berrill, M., S. Bertram, L. McGillivray, M. Kolohon, and B. Pauli. 1994. Effects of low concentrations of forest use pesticides on frog embryos and tadpoles. Environmental Toxicology and Chemistry 13:657–664.

Berrill, M., S. Bertram, and B. Pauli. 1997. Effects of pesticides on amphibian embryos and larvae. SSAR Herpetological Conservation 1:233–245.

Berrill, M., D. Coulson, L. McGillivary, and B. Pauli. 1998. Toxicity of endosulfan to aquatic stages of anuran amphibians. Environmental Toxicology and Chemistry 17:1738–1744.

Bertram, S., and M. Berrill. 1997. Fluctuations in a northern population of gray treefrogs, *Hyla versicolor*. SSAR Herpetological Conservation 1:57–63.

Berven, K.A. 1981. Mate choice in the wood-frog, *Rana sylvatica*. Evolution 35:707–722.

Berven, K.A. 1982a. The genetic basis of altitudinal variation in the wood frog, *Rana sylvatica*. 1. An experimental analysis of life history traits. Evolution 36:962–983.

Berven, K.A. 1982b. The genetic basis of altitudinal variation in the wood frog, *Rana sylvatica*. 2. An experimental analysis of larval development. Oecologia 52:360–369.

Berven, K.A. 1988. Factors affecting variation in reproductive traits within a population of wood frogs (*Rana sylvatica*). Copeia 1988:605–615.

Berven, K.A. 1990. Factors affecting population fluctuations in larval and adult stages of the wood frog (*Rana sylvatica*). Ecology 71:1599–1608.

Berven, K.A. 1995. Population regulation in the wood frog, *Rana sylvatica*, from three diverse geographic locations. Australian Journal of Ecology 20:385–392.

Berven, K.A. 2009. Density dependence in the terrestrial stage of wood frogs: evidence from a 21-year population study. Copeia 2009:328–338.

Berven, K.A., and B.G. Chadra. 1988. The relationship among egg size, density and food level on larval development in the wood frog (*Rana sylvatica*). Oecologia 75:67–72.

Berven, K.A., and T.A. Grudzien. 1990. Dispersal in the wood frog (*Rana sylvatica*): implications for genetic population structure. Evolution 44:2047–2056.

Berven, K.A., and R.S. Boltz. 2001. Interactive effects of leech (*Desserobdella picta*) infection on wood frog (*Rana sylvatica*) tadpole fitness traits. Copeia 2001:907–915.

Berven, K.A., D.E. Gill, and S.J. Smith-Gill. 1979. Countergradient selection in the green frog, *Rana clamitans*. Evolution 33:609–623.

Bervier, C.R., K. Larson, K. Reilly, and S. Tat. 2004. Vocal repertoire and calling activity of the mink frog, *Rana septentrionalis*. Amphibia-Reptilia 25:255–264.

Bervier, C.R., D.C. Tierney, L.E. Henderson, and H.E. Reid. 2006. Chorus attendance and site fidelity in the mink frog, *Rana septentrionalis*: are males territorial? Journal of Herpetology 40:160–164.

Bezy, K.B., R.L. Bezy, K. Bolles, and E.F. Enderson. 2004. Breeding behavior of the western chorus frog (*Pseudacris triseriata* complex) in Arizona: do chorus frogs call in the snow on the Colorado Plateau? Sonoran Herpetologist 17:82–85.

Bezy, R.L., W.C. Sherbrooke, and C.H. Lowe. 1966. Rediscovery of *Eleutherodactylus augusti* in Arizona. Herpetologica 22:221–225.

Bider, J.R., and K.A. Morrison. 1981. Changes in toad (*Bufo americanus*) responses to abiotic factors at the northern limit of their distribution. American Midland Naturalist 106:293–304.

Bider, J.R., and S. Matte. 1996. The Atlas of Amphibians and Reptiles of Québec. St. Lawrence Valley Natural History Society and the Ministère de l'Environnment et de la Faune Direction de la faune et des habitats Québec, Sainte-Anne-de-Bellevue, Québec.

Biesterfeldt, J.M., J.W. Petranka, and S. Sherbondy. 1993. Prevalence of chemical interference competition in natural populations of wood frogs, *Rana sylvatica*. Copeia 1993:688–695.

Bimber, D.L., and R.A. Mitchell. 1978. Effects of diquat on amphibian embryo development. Ohio Journal of Science 78:50–51.

Birchfield, G.L. 2002. Green frog (*Rana clamitans*) movement behavior and terrestrial habitat use in fragmented landscapes in central Missouri. Ph.D. diss., University of Missouri, Columbia.

Birchfield, G.L., and J.E. Deters. 2005. Movement paths of displaced northern green frogs (*Rana clamitans melanota*). Southeastern Naturalist 4:63–76.

Birdsall, C.W., C.E. Grue, and A. Anderson. 1986. Lead concentrations in bullfrog *Rana catesbeiana* and green frog *R. clamitans* tadpoles inhabiting highway drainages. Environmental Pollution (Series A) 40:233–247.

Birge, W.J. 1978. Aquatic toxicology of trace elements of coal and fly ash. Pp. 219–240 *In* J.H. Thorp and G.W. Gibbons (eds.), Energy and Environmental Stress in Aquatic Systems, U.S. Department of Energy Symposium Series No. 48.

Birge, W.J., and J.J. Just. 1973. Sensitivity of vertebrate embryos to heavy metals as a criterion of water quality. Water Resources Research Institute, University of Kentucky, Research Report 61.

Birge, W.J., J.A, Black, and A.G. Westerman. 1979. Evaluation of aquatic pollutants using fish and amphibian eggs as bioassay organisms. Pp. 108–118 *In* Animals as Monitors of Environmental Pollutants. National Academy of Sciences, Washington.

Birge, W.J., J.A. Black, and R.A. Kuehne. 1980. Effects of organic compounds on amphibian reproduction. Water Resources Research Institute, University of Kentucky, Research Report 121.

Birge, W.J., J.A. Black, A.G. Westerman, and B.A. Ramey. 1983. Fish and amphibian embryos: a model system for evaluating tetratogenicity. Fundamental and Applied Toxicology 3:237–242.

Birge, W.J., A.G. Westerman, and J.A. Spromberg. 2000. Comparative toxicology and risk assessment of amphibians. Pp. 727–791 *In* D.W. Sparling, G. Linder, and C.A. Bishop (eds.), Ecotoxicology of Amphibians and Reptiles. SETAC Press, Pensacola, FL.

Bishop, C.A., K.E. Pettit, M.E. Gartshore, and D.A. MacLeod. 1997. Extensive monitoring of anuran populations using call counts and road transects in Ontario (1992 to 1993). SSAR Herpetological Conservation 1:149–160.

Bishop, D.C. 2003. *Rana okaloosae* (Florida Bog Frog). Predation. Herpetological Review 34:235.

Bishop, D.C. 2005. Ecology and distribution of the Florida bog frog and flatwoods salamander on Eglin Air Force Base. Ph.D. diss., Virginia Polytechnic Institute and State University, Blacksburg.

Black, I.H., and K.L. Gosner. 1958. The barking tree frog, *Hyla gratiosa* in New Jersey. Herpetologica 13:254–255.

Black, J.D. 1938. Additional records of *Rana sylvatica* in Arkansas. Copeia 1938:48–49.

Black, J.D., and S.C. Dellinger. 1938. Herpetology of Arkansas. Part 2. The Amphibians. Occasional

Papers of the University of Arkansas Museum No. 2:1–30.
Black, J.H. 1967. A blue leopard frog from Montana. Herpetologica 23:314–315.
Black, J.H. 1970. Amphibians of Montana. Montana Fish and Game Department, Animals of Montana Series No. 1.
Black, J.H. 1971. The toad genus *Bufo* in Montana. Northwest Science 45:156–162.
Black, J.H. 1973a. A checklist of the cave fauna of Oklahoma: Amphibia. Proceedings of the Oklahoma Academy of Science 53:33–37.
Black, J.H. 1973b. Ethoecology of *Scaphiopus* (Pelobatidae) larvae in temporary pools in central and southwestern Oklahoma. Ph.D. diss., University of Oklahoma, Norman.
Black, J.H. 1975. The formation of "tadpole nests" by anuran larvae. Herpetologica 31:76–79.
Black, J.H., and J.N. Black. 1968. Frog with a tail. Montana Outdoors 3(3):not paginated (3 pp.).
Black, J.H., and J.N. Black. 1969. Postmetamorphic basking aggregations of the boreal toad, *Bufo boreas boreas*. Canadian Field-Naturalist 83:155–156.
Black, J.H., and R.B. Brunson. 1971. Breeding behavior of the boreal toad, *Bufo boreas boreas* (Baird and Girard), in western Montana. Great Basin Naturalist 31:109–113.
Blair, A.P. 1941a. Variation, isolation mechanisms and hybridization in certain toads. Genetics 26:398–417.
Blair, A.P. 1941b. Isolating mechanisms in tree frogs. Proceedings of the National Academy of Sciences 27:14–17.
Blair, A.P. 1942. Isolating mechanisms in a complex of four species of toads. Biological Symposia 6:235–249.
Blair, A.P. 1943a. Geographical variation of ventral markings in toads. American Midland Naturalist 29:615–620.
Blair, A.P. 1943b. Population structure in toads. American Naturalist 77:563–568.
Blair, A.P. 1946. Description of a six-year-old hybrid toad. American Museum Novitates 1327:1–3.
Blair, A.P. 1947a. Variation in two characteristics in *Bufo fowleri* and *Bufo americanus*. American Museum Novitates 1343:1–5.
Blair, A.P. 1947b. The male warning vibration in *Bufo*. American Museum Novitates 1344:1–7.
Blair, A.P. 1950. Notes on Oklahoma microhylid frogs. Copeia 1950:152.
Blair, A.P. 1951. Note on the herpetology of the Elk Mountains, Colorado. Copeia 1951:239–240.
Blair, A.P. 1955. Distribution, variation, and hybridization in a relict toad (*Bufo microscaphus*) in southwestern Utah. American Museum Novitates 1722:1–38.
Blair, A.P., and H.L. Lindsay Jr. 1961. *Hyla avivoca* (Hylidae) in Oklahoma. Southwestern Naturalist 6:202.
Blair, W.F. 1936. A note on the ecology of *Microhyla olivacea*. Copeia 1936:115.
Blair, W.F. 1953. Growth, dispersal and age at sexual maturity of the Mexican toad (*Bufo valliceps* Wiegmann). Copeia 1953:208–212.
Blair, W.F. 1955a. Size difference as a possible isolation mechanism in *Microhyla*. American Naturalist 89:297–302.
Blair, W.F. 1955b. Mating call and stage of speciation in the *Microhyla olivacea-M. carolinensis* complex. Evolution 9:469–480.
Blair, W.F. 1955c. Differentiation of mating call in spadefoots, genus *Scaphiopus*. Texas Journal of Science 7:183–188.
Blair, W.F. 1956a. The mating calls of hybrid toads. Texas Journal of Science 8:350–355.
Blair, W.F. 1956b. Call difference as an isolation mechanism in Southwestern toads (genus *Bufo*). Texas Journal of Science 8:87–106.
Blair, W.F. 1957. Mating call and relationships of *Bufo hemiophrys* Cope. Texas Journal of Science 9:99–108.
Blair, W.F. 1958a. Call difference as an isolation mechanism in Florida species of hylid frogs. Quarterly Journal of the Florida Academy of Sciences 21:32–48.
Blair, W.F. 1958b. Mating call and stage of speciation of two allopatric populations of spadefoots (*Scaphiopus*). Texas Journal of Science 10:484–488.
Blair, W.F. 1958c. Call structure and species groups in U.S. treefrogs (*Hyla*). Southwestern Naturalist 3:77–89.
Blair, W.F. 1959. Genetic compatibility and species groups in U.S. toads (*Bufo*). Texas Journal of Science 11:427–453.
Blair, W.F. 1960a. A breeding population of the Mexican toad (*Bufo valliceps*) in relation to its environment. Ecology 41:165–174.
Blair, W.F. 1960b. Mating call as evidence of relations in the *Hyla eximia* group. Southwestern Naturalist 5:129–135.
Blair, W.F. 1961a. Further evidence bearing on intergroup and intragroup genetic compatibility in toads (genus *Bufo*). Texas Journal of Science 13:163–175.
Blair, W.F. 1961b. Calling and spawning seasons in a mixed population of anurans. Ecology 42:99–110.
Blair, W.F. 1963a. Intragroup genetic compatibility in the *Bufo americanus* species group of toads. Texas Journal of Science 15:15–34.
Blair, W.F. 1963b. Evolutionary relationships of North American toads of the genus *Bufo*: a progress report. Evolution 17:1–16.
Blair, W.F. 1964. Evidence bearing on the relationships of the *Bufo boreas* group of toads. Texas Journal of Science 16:181–192.
Blair, W.F. 1966. Genetic compatibility in the *Bufo valliceps* and closely related groups of toads. Texas Journal of Science 18:333–351.

Blair, W.F. 1972. Evidence from hybridization. Pp. 196–232 *In* W.F. Blair (ed.), Evolution in the Genus *Bufo*. University of Texas Press, Austin.

Blair, W.F. 1974. Character displacement in frogs. American Zoologist 14:1119–1125.

Blair, W.F., and D. Pettus. 1954. The mating call and its significance in the Colorado River toad (*Bufo alvarius* Girard). Texas Journal of Science 6:72–77.

Blair, W.F., and M.J. Littlejohn. 1960. Stage of speciation of two allopatric populations of chorus frogs (*Pseudacris*). Evolution 14:82–87.

Blair, W.R. 1964. Evidence bearing on the relationships of the *Bufo boreas* group of toads. Texas Journal of Science 16:181–192.

Blanchard, F.N. 1925. A collection of amphibians and reptiles from southern Indiana and adjacent Kentucky. Papers of the Michigan Academy of Science, Arts and Letters 5:367–388.

Blanchard, F.N. 1933. Late autumn collections and hibernating situations of the salamander *Hemidactylium scutatum* (Schlegel) in southern Michigan. Copeia 1933:216.

Blaney, R.M. 1971. An annotated check list and biogeographic analysis of the insular herpetofauna of the Apalachicola region, Florida. Herpetologica 27:406–430.

Blatchley, W.S. 1892. Notes on the batrachians and reptiles of Vigo County, Indiana. Journal of the Indiana Society of Natural History 14:27.

Blatchley, W.S. 1899. Notes on the batrachians and reptiles of Vigo County, Indiana. Annual Report of the Indiana Department of Geology and Natural Resources 24:537–552.

Blaustein, A.R., and R.K. O'Hara. 1982. Kin recognition in *Rana cascadae* tadpoles: maternal and paternal effects. Animal Behaviour 30:1151–1157.

Blaustein, A.R., and R.K. O'Hara. 1986. An investigation of kin recognition in red-legged frogs (*Rana aurora*) tadpoles. Journal of Zoology, London 209:347–353.

Blaustein, A.R., and B. Waldman. 1992. Kin recognition in anuran amphibians. Animal Behaviour 44:207–221.

Blaustein, A.R., and L.K. Belden. 2003. Amphibian defenses against ultraviolet-B radiation. Evolution & Development 5:89–97.

Blaustein, A.R., and P.T.J. Johnson. 2003. The complexity of deformed amphibians. Frontiers in Ecology and the Environment 1:87–94.

Blaustein, A.R., K.S. Chang, H.G. Lefcourt, and R.K. O'Hara. 1990. Toad tadpole kin recognition: recognition of half siblings and the role of maternal cues. Ethology, Ecology and Evolution 2:215–226.

Blaustein, A.R., D.G. Hokit, R.K. O'Hara, and R.A. Holt. 1994a. Pathogenic fungus contributes to amphibian losses in the Pacific Northwest. Biological Conservation 67:251–254.

Blaustein, A.R., P.D. Hoffman, D.G. Hokit, J.M. Kiesecker, S.C. Walls, and J.B. Hays. 1994b. UV repair and resistance to solar UV-B in amphibian eggs: a link to population declines. Proceedings of the National Academy of Science, USA 91:1791–1795.

Blaustein, A.R., P.D. Hoffman, J.M. Kiesecker, and J.B. Hayes. 1996. DNA repair activity and resistance to solar UV-B radiation in eggs of the red-legged frog. Conservation Biology 10:1398–1402.

Blaustein, A.R., J.B. Hays, P.D. Hoffman, D.P. Chivers, J.M. Kiesecker, W.P. Leonard, A. Marco, D.H. Olson, J.K. Reaser, and R.G. Anthony. 1999. DNA repair and resistance to UV-B radiation in western spotted frogs. Ecological Applications 9:1100–1105.

Blaustein, A.R., B. Han, B. Fasy, J. Romansic, E.A. Scheessele, R.G. Anthony, A. Marco, D.P. Chivers, L.K. Belden, J.M. Kiesecker, T. Garcia, M. Lizana, and L.B. Kats. 2004. Variable breeding phenology affects the exposure of amphibian embryos to ultraviolet radiation and optical characteristics of natural waters protect amphibians from UV-B in the U.S. Pacific Northwest: comment. Ecology 85:1747–1754.

Blaustein, A.R., J.M. Romansic, and E.A. Scheessele. 2005. Ambient levels of ultraviolet-B radiation cause mortality in juvenile western toads, *Bufo boreas*. American Midland Naturalist 154:375–382.

Bleakney, S. 1952. The amphibians and reptiles of Nova Scotia. Canadian Field-Naturalist 66:125–129.

Bleakney, S. 1954. Range extensions of amphibians in eastern Canada. Canadian Field-Naturalist 68:165–171.

Bleakney, J.S. 1958a. A zoogeographical study of the amphibians and reptiles of eastern Canada. National Museum of Canada Bulletin 155:1–119.

Bleakney, S. 1958b. Cannibalism in *Rana sylvatica* tadpoles. Herpetologica 14:34.

Bleakney, S. 1959. Postglacial dispersal of the western chorus frog in eastern Canada. Canadian Field-Naturalist 73:197–205.

Bleakney, S. 1963. First North American record of *Bufolucilia silvarum* (Meigen) (Diptera: Calliphoridae) parasitizing *Bufo terrestris americanus* Holbrook. Canadian Entomologist 95:107.

Blem, C.R., J.W. Steiner, and M.A. Miller. 1978. Comparison of jumping abilities of the cricket frogs *Acris gryllus* and *Acris crepitans*. Herpetologica 34:288–291.

Blihovde, W.B. 2006. Terrestrial movements and upland habitat use of gopher frogs in central Florida. Southeastern Naturalist 5:265–276.

Blomquist, S.M. 2005. *Bufo retiformis* Sanders and Smith, 1951. Sonoran Green Toad. Pp. 433–435 *In* M.J. Lannoo (ed.), Amphibian Declines. The Conservation Status of United States Species. University of California Press, Berkeley.

Blomquist, S.M., and M.L. Hunter, Jr. 2009. A multiscale assessment of habitat selection and movement patterns by northern leopard frogs (*Lithobates* [*Rana*] *pipiens*) in a managed forest. Herpetological Conservation and Biology 4:142–160.

Blouin, M.S. 1992a. Comparing bivariate reaction norms among species: time and size at metamorphosis in three species of *Hyla* (Anura: Hylidae). Oecologia 90:288–293.

Blouin, M.S. 1992b. Genetic correlations among morphometric traits and rates of growth and differentiation in the green tree frog, *Hyla cinerea*. Evolution 46:735–744.

Boatwright-Horowitz, S.S., C.A. Cheney, and A.M. Simmons. 1999. Atmospheric and underwater propagation of bullfrog vocalizations. Bioacoustics 9:257–280.

Bogart, J.P., and C.E. Nelson. 1976. Evolutionary implications from karyotypic analysis of frogs of the families Microhylidae and Rhinophrynidae. Herpetologica 32:199–208.

Bogart, J.P., and A.P. Jaslow. 1979. Distribution and call parameters of *Hyla chrysoscelis* and *Hyla versicolor* in Michigan. Royal Ontario Museum Life Sciences Contribution 117, 13 pp.

Bogert, C.M. 1947. Results of the Archbold Expeditions. No. 57. A field study of homing in the Carolina toad. American Museum Novitates 1355:1–24.

Bogert, C.M. 1960. The influence of sound on the behavior of amphibians and reptiles. Pp. 37–320 In W.E. Lanyon and W.N. Tavolga (eds.), Animal Sounds and Communication, American Institute of Biological Sciences Publication No. 7, Washington, DC.

Bogert, C.M. 1962. Isolation mechanisms in toads of the *Bufo debilis* group in Arizona and western Mexico. American Museum Novitates 2100:1–37.

Bohnsack, K.K. 1951. Temperature data on the terrestrial hibernation of the greenfrog, *Rana clamitans*. Copeia 1951:236–239.

Bohnsack, K.K. 1952. Terrestrial hibernation of the bullfrog, *Rana catesbeiana* Shaw. Copeia 1952:114.

Boice, R., and D.W. Witter. 1969. Hierarchical feeding behaviour in the leopard frog (*Rana pipiens*). Animal Behaviour 17:474–479.

Boice, R., and C. Boice. 1970. Interspecific competition in captive *Bufo marinus* and *Bufo americanus* toads. Journal of Biological Psychology 12:32–36.

Bolek, M.G., and J.R. Coggins. 1998. Endoparasites of Cope's gray treefrog, *Hyla chrysoscelis*, and western chorus frog, *Pseudacris t. triseriata*, from southeastern Wisconsin. Journal of the Helminthological Society of Washington 65:212–218.

Bolek, M.G., and J.R. Coggins. 2002. Observations on myiasis by the calliphorid, *Bufolucilia silvarum*, in the eastern American toad (*Bufo americanus americanus*) from southeastern Wisconsin. Journal of Wildlife Diseases 38:598–603.

Bolek, M.G., and J. Janovy Jr. 2004. Observations on myiasis by the calliphorids, *Bufolucilia silvarum* and *Bufolucilia elongata*, in wood frogs, *Rana sylvatica*, from southeastern Wisconsin. Journal of Parasitology 90:1169–1171.

Bolek, M.G., and J. Janovy Jr. 2005. New host and distribution records for the amphibian leech *Desserobdella picta* (Rhynchobdellida: Glossiphoniidae) from Nebraska and Wisconsin. Journal of Freshwater Ecology 20:187–189.

Bolek, M.G., and J. Janovy Jr. 2008. Alternative life cycle strategies of *Megalodiscus temperatus* in tadpoles and metamorphosed anurans. Parasite 15:396–401.

Bonin, J., J.-L. DesGranges, J. Rodrigue, and M. Ouellet. 1997a. Anuran species richness in agricultural landscapes of Québec: foreseeing long-term results of road call surveys. SSAR Herpetological Conservation 1:141–149.

Bonin, J., M. Ouellet, J. Rodrigue, J.-L. DesGranges, F. Gagné, T.F. Sharbel, and L.A. Lowcock. 1997b. Measuring the health of frogs in agricultural habitats subjected to pesticides. SSAR Herpetological Conservation 1:246–257.

Bonine, K.E., G.H. Dayton, and R.E. Jung. 2001. Attempted predation of Couch's spadefoot (*Scaphiopus couchii*) juveniles by ants (*Aphaenogaster cockerelli*). Southwestern Naturalist 46:104–106.

Bonnaterre, M. l'A. 1789. Tableau Encyclopédique et Méthodique des Trois Règnes de la Nature. Erpétologie. Chez Panckoucke, Hôtel de Thou, Paris.

Boone, M.D., and C.M. Bridges. 1999. The effect of temperature on the potency of carbaryl for survival of tadpoles of the green frog (*Rana clamitans*). Environmental Toxicology and Chemistry 18:1482–1484.

Boone, M.D., and R.D. Semlitsch. 2001. Interactions of an insecticide with larval density and predation in experimental amphibian communities. Conservation Biology 15:228–238.

Boone, M.D., and R.D. Semlitsch. 2002. Interactions of an insecticide with competition and pond drying in amphibian communities. Ecological Applications 12:307–316.

Boone, M.D., and S.M. James. 2003. Interactions of an insecticide, herbicide, and natural stressors in amphibian community mesocosms. Ecological Applications 13:829–841.

Boone, M.D., C.M. Bridges, and B.B. Rothermel. 2001. Growth and development of larval green frogs (*Rana clamitans*) exposed to multiple doses of an insecticide. Oecologia 129:518–524.

Boone, M.D., E.E. Little, and R.D. Semlitsch. 2004a. Overwintered bullfrog tadpoles negatively affect salamanders and anurans in native amphibian communities. Copeia 2004:683–690.

Boone, M.D., R.D. Semlitsch, J.F. Fairchild, and B.B. Rothermel. 2004b. Effects of an insecticide on amphibians in large-scale experimental ponds. Ecological Applications 14:685–691.

Boone, M.D., R.D. Semlitsch, and C. Mosby. 2008. Suitability of golf course ponds for amphibian metamorphosis when bullfrogs are removed. Conservation Biology 22:172–179.

Bosakowski, T. 1999. Amphibian macrohabitat associations on a private industrial forest in western Washington. Northwestern Naturalist 80:61–69.

Bosc, L.A.G. 1800. *In* F.M. Daudin, Histoire Naturelle des Quadrupèdes Ovipaires. Livraison 1:10, plate 5, Marchant et Cie, Paris.

Bosc, L.A.G., and F.M. Daudin. 1801. *Laraine ocularie, Hyla ocularis. In* C.S. Sonnini and P.A. Latreille, Histoire Naturelle des Reptiles, avec Figures dissinées d'après Nature. Vol. 2. Deterville, Paris.

Boschulte, D.S. 1993. Toxicity of six commonly used herbicides on larval bullfrogs (*Rana catesbeiana*). M.S. thesis, Illinois State University, Normal.

Bossert, M., M. Draud, and T. Draud. 2003. *Bufo fowleri* (Fowler's toad) and *Malaclemys terrapin terrapin* (northern diamondback terrapin) refugia and nesting. Herpetological Review 34:135.

Botch, P.S., M.C. Beers, and T.M. Judd. 2007. The effects of calcium on the feeding preference of the tadpole of *Bufo americanus*. Bulletin of the Maryland Herpetological Society 43:162–166.

Bouchard, J.L. 1951. The platyhelminthes parasitizing some northern Maine amphibia. Transactions of the American Microscopical Society 70:245–250.

Bouchard, J., A.T. Ford, F.E. Eigenbrod, and L. Fahrig. 2009. Behavioral responses of northern leopard frogs (*Rana pipiens*) to roads and traffic: implications for population persistence. Ecology and Society 14(2):23. http://www.ecologyandsociety.org/Vol14/iss2/art23/.

Boughton, R.G., J. Staiger, and R. Franz. 2000. Use of PVC pipe refugia as a sampling technique for hylid treefrogs. American Midland Naturalist 144:168–177.

Boulenger, G.A. 1882. Descriptions of a new genus and species of frogs of the family Hylidae. Annals and Magazine of Natural History, Series 5, 10:326–328.

Boulenger, G.A. 1899. On the American spade-foot (*Scaphiopus solitarius* Holbrook). Proceedings of the Zoological Society of London 1899(3):790–793.

Boulenger, G.A. 1917. Descriptions of new frogs of the genus *Rana*. Annals of the Magazine of Natural History 20:413–418.

Bovbjerg, R.V. 1965. Experimental studies on the dispersal of the frog, *Rana pipiens*. Proceedings of the Iowa Academy of Science 72:412–418.

Bovbjerg, R.V., and A.M. Bovbjerg. 1964. Summer emigrations of the frog *Rana pipiens* in northwestern Iowa. Proceedings of the Iowa Academy of Science 71:511–518.

Bowen, K.D., and E.A. Beever. 2010. Daytime amphibian surveys in three protected areas in the western Great Lakes. IRCF Reptiles & Amphibians 17:26–31, 34–35.

Bowerman, J. 2010. Submerged calling by Oregon spotted frogs (*Rana pretiosa*) remote from breeding aggregations. IRCF Reptiles & Amphibians 17:84–87.

Bowers, D.G., D.E. Andersen, and N.H. Euliss Jr. 1998. Anurans as indicators of wetland condition in the Prairie Pothole Region of North Dakota: an environmental monitoring and assessment program pilot project. Pp. 369–378 *In* M.J. Lannoo (ed.), Status & Distribution of Midwestern Amphibians. University of Iowa Press, Iowa City.

Bowker, R.W., and B.K. Sullivan. 1991. *Bufo punctatus* × *B. retiformis* (Red-spotted Toad, Sonoran Green Toad). Natural hybridization. Herpetological Review 22:54.

Boyd, C.E. 1964. The distribution of cricket frogs in Mississippi. Herpetologica 20:201–202.

Boyd, C.E., and D.H. Vickers. 1963. Distribution of some Mississippi amphibians and reptiles. Herpetologica 19:202–205.

Boyd, S.H. 1975. Inhibition of fish reproduction by *Rana catesbeiana* larvae. Physiological Zoology 48:225–234.

Boykin, K.G. 2006. Multiscale analysis of habitat, vegetation change, and streamflow as ecological factors affecting population dynamics of *Rana chiricahuensis*. Ph.D. diss., New Mexico State University, Las Cruces.

Boykin, K.G., and K.C. McDaniel. 2008. Simulated potential effects of ecological factors on a hypothetical population of Chiricahua leopard frog (*Rana chiricahuensis*). Ecological Modelling 218:175–181.

Bradford, D.F. 1983. Winterkill, oxygen regulations, and energy metabolism of a submerged dormant amphibian, *Rana muscosa*. Ecology 64:1171–1183.

Bradford, D.F. 1989. Allotopic distribution of native frogs and introduced fishes in high Sierra Nevada lakes of California: implication of the negative effect of fish introductions. Copeia 1989:775–778.

Bradford, D.F. 1991. Mass mortality and extinction in a high-elevation population of *Rana muscosa*. Journal of Herpetology 25:174–177.

Bradford, D.F., C. Swanson, and M.S. Gordon. 1992. Effects of low pH and aluminum on two declining species of amphibians in the Sierra Nevada, California. Journal of Herpetology 26:369–377.

Bradford, D.F., F. Tabatabai, and D.M. Graber. 1993. Isolation of remaining populations of the native frog, *Rana muscosa*, by introduced fishes in Sequoia and Kings Canyon National Parks, California. Conservation Biology 7:882–888.

Bradford, D.F., M.S. Gordon, D.F. Johnson, R.D. Andrews, and W.B. Jennings. 1994a. Acidic deposition as an unlikely cause for amphibian population declines in the Sierra Nevada, California. Biological Conservation 69:155–161.

Bradford, D.F., D.M. Graber, and F. Tabatabai. 1994b. Population declines of the native frog, *Rana muscosa*, in Sequoia and Kings Canyon National Parks, California. Southwestern Naturalist 39: 323–327.

Bradford, D.F., S.D. Cooper, T.M. Jenkins Jr., K. Kratz, O. Sarnelle, and A.D. Brown. 1998. Influences of natural acidity and introduced fish on faunal assemblages in California alpine lakes. Canadian Journal of Fisheries and Aquatic Sciences 55: 2478–2491.

Bradford, D.F., A.C. Neale, M.S. Nash, D.W. Sada, and J.R. Jaeger. 2003. Habitat patch occupancy by toads (*Bufo punctatus*) in a naturally fragmented desert landscape. Ecology 84:1012–1023.

Bradford, D.F., J.R. Jaeger, and R.D. Jennings. 2004. Population status and distribution of a decimated amphibian, the relict leopard frog (*Rana onca*). Southwestern Naturalist 49: 218–228.

Bradford, D.F., J.R. Jaeger, and S.A. Shanahan. 2005a. Distributional changes and population status of amphibians in the eastern Mojave Desert. Western North American Naturalist 65:462–472.

Bradford, D.F., R.D. Jennings, and J.R. Jaeger. 2005b. *Rana onca* Cope, 1875(b). Relict Leopard Frog. Pp. 567–568 *In* M.J. Lannoo (ed.), Amphibian Declines. The Conservation Status of United States Species. University of California Press, Berkeley.

Bradford, D.F., R.A. Knapp, D.W. Sparling, M.S. Nash, K.A. Stanley, N.G. Tallent-Halsell, L.L. McConnell, and S.M. Simonich. 2011. Pesticide distributions and population declines of California, USA, alpine frogs, *Rana muscosa* and *Rana sierrae*. Environmental Toxicology and Chemistry 30:682–691.

Bradley, G.A., P.C. Rosen, M.J. Sredl, T.R. Jones, and J.E. Longcore. 2002. Chytridiomycosis in native Arizona frogs. Journal of Wildlife Diseases 38: 206–212.

Brady, M. 1927. Notes on the amphibians and reptiles of the Dismal Swamp. Copeia (162):26–29.

Brady, M.K., and F. Harper. 1935. A Florida subspecies of *Pseudacris nigrita* (Hylidae). Proceedings of the Biological Society of Washington 48:107–110.

Bragg, A.N. 1936. Notes on the breeding habits, eggs, and embryos of *Bufo cognatus* with a description of the tadpole. Copeia 1936:14–20.

Bragg, A.N. 1937a. A note on the metamorphosis of the tadpoles of *Bufo cognatus*. Copeia 1937: 227–228.

Bragg, A.N. 1937b. Observations on *Bufo cognatus* with special reference to the breeding habits and eggs. American Midland Naturalist 18:273–284.

Bragg, A.N. 1939. Possible hybridization of *Bufo cognatus* and *B.w. woodhousii*. Copeia 1939:173.

Bragg, A.N. 1940a. Observations on the ecology and natural history of Anura. 2. Habits, habitat and breeding of *Bufo woodhousii woodhousii* (Girard) in Oklahoma. American Midland Naturalist 24:306–335.

Bragg, A.N. 1940b. Observations on the ecology and natural history of Anura. 6. The ecological importance of the study of the habits of animals as illustrated by toads. The Wasmann Collector 4:6–16.

Bragg, A.N. 1940c. Observations on the ecology and natural history of Anura. 1. Habits, habitat and breeding of *Bufo cognatus* Say. American Naturalist 74:322–349, 424–438.

Bragg, A.N. 1941a. Tadpoles of *Scaphiopus bombifrons* and *Scaphiopus hammondii*. The Wasmann Collector 4:92–94.

Bragg, A.N. 1941b. Some observations on amphibia at and near Las Vegas, New Mexico. Great Basin Naturalist 2:109–117.

Bragg, A.N. 1941c. Observations on the ecology and natural history of Anura. 9. The invasion of the Canadian River flood plain by two prairie species. Proceedings of the Oklahoma Academy of Science 22:73–74.

Bragg, A.N. 1942. Observations on the ecology and natural history of Anura. 10. The breeding habits of *Pseudacris streckeri* Wright and Wright in Oklahoma including a description of the eggs and tadpoles. Wasmann Collector 5:47–62.

Bragg, A.N. 1943a. Observations on the ecology and natural history of Anura. 15. The hylids and microhylids in Oklahoma. Great Basin Naturalist 4:62–80.

Bragg, A.N. 1943b. Observations on the ecology and natural history of Anura. 16. Life-history of *Pseudacris clarkii* (Baird) in Oklahoma. Wasmann Collector 5:129–140.

Bragg, A.N. 1944a. The spadefoot toads in Oklahoma with a summary of our knowledge of the group. American Naturalist 78:517–533.

Bragg, A.N. 1944b. Breeding habits, eggs, and tadpoles of *Scaphiopus hurterii*. Copeia 1944:230–241.

Bragg, A.N. 1945a. Notes on the psychology of frogs and toads. Journal of General Psychology 32:27–37.

Bragg, A.N. 1945b. The spadefoot toads in Oklahoma with a summary of our knowledge of the group. 2. American Naturalist 79:52–72.

Bragg, A.N. 1948. Observations on the life history of *Pseudacris triseriata* (Wied.) in Oklahoma. The Wasmann Collector 7:149–168.

Bragg, A.N. 1950a. Salientian breeding dates in Oklahoma. Pp. 35–38 *In* Researches on the Amphibia of Oklahoma. University of Oklahoma Press, Norman.

Bragg, A.N. 1950b. Observations on the ecology and natural history of Anura. 17. Adaptations and distribution in accordance with habitats in Oklahoma. Pp. 59–100 *In* Researches on the Amphibia of Oklahoma. University of Oklahoma Press, Norman.

Bragg, A.N. 1950c. Observations on *Microhyla* (Salientia: Microhylidae). Wassman Journal of Biology 8:113–118.

Bragg, A.N. 1950d. The identification of salientia in Oklahoma. Pp. 9–29 *In* Researches on the Amphibia of Oklahoma. University of Oklahoma Press, Norman.

Bragg, A.N. 1950e. Observations on the ecology and natural history of Anura. 14. Growth rates and age at sexual maturity of *Bufo cognatus* under natural conditions in central Oklahoma. Pp. 47–58 In Researches on the Amphibia of Oklahoma. University of Oklahoma Press, Norman.

Bragg, A.N. 1952. Decline in toad populations in central Oklahoma. Proceedings of the Oklahoma Academy of Science 33:70.

Bragg, A.N. 1953. A study of *Rana areolata* in Oklahoma. Wasmann Journal of Biology 11:273–318.

Bragg, A.N. 1955. In quest of spadefoots. New Mexico Quarterly 25(4):345–258.

Bragg, A.N. 1956. Dimorphism and cannibalism in tadpoles of *Scaphiopus bombifrons* (Amphibia, Salientia). Southwestern Naturalist 1:105–108.

Bragg, A.N. 1958a. A melanistic tendency in the Great Plains toad, *Bufo cognatus*. Southwestern Naturalist 3:229–230.

Bragg, A.N. 1958b. The clasping reflex observed in juvenile spadefoots (*Scaphiopus*). Southwestern Naturalist 3:229.

Bragg, A.N. 1959. Behavior of tadpoles of Hurter's spadefoot during an exceptionally rainy season. Wasmann Journal of Biology 17:23–42.

Bragg, A.N. 1960a. Experimental observations on the feeding of spadefoot tadpoles. Southwestern Naturalist 5:201–207.

Bragg, A.N. 1960b. Feeding in the Houston toad. Southwestern Naturalist 5:106.

Bragg, A.N. 1962a. *Saprolegnia* on tadpoles again in Oklahoma. Southwestern Naturalist 7:79–80.

Bragg, A.N. 1962b. Predator-prey relationship in two species of spadefoot tadpoles with notes on some other features of their behavior. Wasmann Journal of Biology 20:81–97.

Bragg, A.N. 1962c. Predation on arthropods by spadefoot tadpoles. Herpetologica 18:144.

Bragg, A.N. 1964a. A hypothesis to explain the almost exclusive use of temporary water by breeding spadefoot toads. Proceedings of the Oklahoma Academy of Science 44:24–25.

Bragg, A.N. 1964b. Mass movements resulting in aggregations of tadpoles of the Plains spadefoot, some of them in response to light and temperature (Amphibia: Salientia). Wasmann Journal of Biology 22:299–305.

Bragg, A.N. 1964c. Further study of predation and cannibalism in spadefoot tadpoles. Herpetologica 20:17–24.

Bragg, A.N. 1965. Gnomes of the Night. The Spadefoot Toads. University of Pennsylvania Press, Philadelphia.

Bragg, A.N. 1966. Longevity of the tadpole stage in the Plains spadefoot (Amphibia: Salientia). Wasmann Journal of Biology 24:71–73.

Bragg, A.N. 1967. Recent studies on the spadefoot toads. Bios 38:75–84.

Bragg, A.N. 1968. The formation of feeding schools in tadpoles of spadefoots. Wasmann Journal of Biology 26:11–16.

Bragg, A.N., and C.C. Smith. 1942. Observations on the ecology and natural history of Anura. 9. Notes on breeding behavior in Oklahoma. Great Basin Naturalist 3:33–50.

Bragg, A.N., and C.C. Smith. 1943. Observations on the ecology and natural history of Anura. 4. The ecological distribution of toads in Oklahoma. Ecology 24:285–309.

Bragg, A.N., and O. Sanders. 1951. A new subspecies of the *Bufo woodhousii* group of toads (Salientia: Bufonidae: *Bufo woodhousii velatus* subsp. nov.). Wasmann Journal of Biology 9:363–378.

Bragg, A.N., and V.E. Dowell. 1954. Leopard frog eggs in October. Proceedings of the Oklahoma Academy of Science 35:41.

Bragg, A.N., and W.N. Bragg. 1958a. Parasitism of spadefoot tadpoles by *Saprolegnia*. Herpetologica 14:34.

Bragg, A.N., and W.N. Bragg. 1958b. Variations in the mouth parts in tadpoles of *Scaphiopus* (*Spea*) *bombifrons* Cope (Amphibia: Salientia). Southwestern Naturalist 3:55–69.

Bragg, A.N., and M. Brooks. 1958. Social behavior in juveniles of *Bufo cognatus* Say. Herpetologica 14:141–147.

Bragg, A.N., and O.M. King. 1960. Aggregational and associated behavior in tadpoles of the Plains spadefoot. Wasmann Journal of Biology 18: 273–289.

Bragg, A.N., and R.J. Taylor. 1968. A range extension of the pickerel frog, *Rana palustris palustris*, in Oklahoma with indications of an extension of the Austroriparian life province. Southwestern Naturalist 13:372–374.

Bragg, A.N., R. Matthews, and R. Kingsinger Jr. 1964. The mouth parts of tadpoles of Hurter's spadefoot. Herpetologica 19:284–285.

Braid, M.R., C.B. Raymond, and W.S. Sanders. 1994. Feeding trials with the dusky gopher frog, *Rana capito sevosa*, in a recirculating water system and other aspects of their culture as part of a "head-starting" effort. Journal of the Alabama Academy of Science 65:249–262.

Branch, L.C., and D.G. Hokit. 2000. A comparison of scrub herpetofauna on two central Florida sand ridges. Florida Scientist 63:108–117.

Brand, A.B, and J.W. Snodgrass. 2010. Value of artificial habitats for amphibian reproduction in altered landscapes. Conservation Biology 24:295–301.

Brand, A.B., J.W. Snodgrass, M.T. Gallagher, R.E. Casey, and R. Van Meter. 2010. Lethal and sublethal effects of embryonic and larval exposure of *Hyla versicolor* to stormwater pond sediments.

Archives of Environmental Contaminants and Toxicology 58:325–331.

Brandon, R.A., and S.R. Ballard. 1998. Status of Illinois chorus frogs in southern Illinois. Pp.102–112 In M.J. Lannoo (ed.), Status & Conservation of Midwestern Amphibians. University of Iowa Press, Iowa City.

Brandt, B.B. 1936. Parasites of certain North Carolina salientia. Ecological Monographs 6:493–532.

Brandt, B.B., and C.F. Walker. 1933. A new species of *Pseudacris* from the southeastern United States. Occasional Papers of the Museum of Zoology, University of Michigan No. 272.

Branson, B.A. 1975. Claude who? Another unwanted exotic species. National Parks and Conservation Magazine (June):17–18.

Brashears, V. Sr., and V. Brashears Jr. 1950. Frog Raising. Brashears Printing Co., Berryville, AR.

Braswell, A. 1988. A survey of amphibians and reptiles of Nags Head Woods Ecological Preserve. Association of Southeastern Biologists Bulletin 35:199–217.

Braswell, A.L. 1993. Status report on *Rana capito capito* Leconte, the Carolina gopher frog in North Carolina. Nongame and Endangered Wildlife Program, North Carolina Wildlife Resources Commission, Raleigh.

Brattstrom, B.H. 1953. The amphibians and reptiles from Rancho La Brea. Transactions of the San Diego Natural History Society 11:365–392.

Brattstrom, B.H. 1958. New records of Cenozoic amphibians and reptiles from California. Bulletin of the Southern California Academy of Sciences 54:1–4.

Brattstrom, B.H. 1962. Thermal control of aggregation behavior in tadpoles. Herpetologica 18:38–46.

Brattstrom, B.H. 1963. A preliminary review of the thermal requirements of amphibians. Ecology 44:238–255.

Brattstrom, B.H., and J.W. Warren. 1955. Observations on the ecology and behavior of the Pacific treefrog, *Hyla regilla*. Copeia 1955:181–191.

Breckenridge, W.J. 1944. Reptiles and Amphibians of Minnesota. University of Minnesota Press, Minneapolis.

Breckenridge, W.J., and J.R. Tester. 1961. Growth, local movements and hibernation of the Manitoba toad, *Bufo hemiophrys*. Ecology 42:637–646.

Breden, F. 1987. The effect of post-metamorphic dispersal on the population genetic structure of Fowler's toad, *Bufo woodhousei fowleri*. Copeia 1987:386–395.

Breden, F. 1988. Natural history and ecology of Fowler's Toad, *Bufo woodhousei fowleri* (Amphibia: Bufonidae), in the Indiana Dunes National Seashore. Fieldiana Zoology, new series (18):1–16.

Breden, F., and C.H. Kelly. 1982. The effect of conspecific interactions on metamorphosis in *Bufo americanus*. Ecology 63:1682–1689.

Breden, F., A. Lum, and R. Wassersug. 1982. Body size and orientation in aggregates of toad tadpoles *Bufo woodhousei*. Copeia 1982:672–680.

Breder, C.M. Jr. 1927. Frog tagging: a method of studying anuran life habits. Zoologica 9:201–229.

Brekke, D.R., S.D. Hillyard, and R.M. Winokur. 1991. Behavior associated with the water absorption response by the toad, *Bufo punctatus*. Copeia 1991:393–401.

Brennan, T.C., and A.T. Holycross. 2006. Amphibians and Reptiles in Arizona. Arizona Game and Fish Department, Phoenix.

Brenner, F.J. 1969. Role of temperature and fat deposition in hibernation and reproduction in two species of frogs. Herpetologica 25:105–113.

Brenowitz, E.A., and G.J. Rose. 1999. Female choice and plasticity of male calling behaviour in the Pacific treefrog. Animal Behaviour 57:1337–1342.

Brenowitz, E.A., W. Wilczynski, and H.H. Zakon. 1984. Acoustic communication in spring peepers: environmental and behavioral aspects. Journal of Comparative Physiology 155A:585–592.

Bresler, J.B. 1963. Pigmentation characteristics of *Rana pipiens*: dorsal region. American Midland Naturalist 70:197–207.

Bresler, J.B. 1964. Pigmentation characteristics of *Rana pipiens*: tympanum spot, line on upper jaw, and spots on upper eyelids. American Midland Naturalist 72:382–389.

Bridges, C.M. 1997. Tadpole swimming performance and activity affected by acute exposure to sublethal levels of carbaryl. Environmental Toxicology and Chemistry 16:1935–1939.

Bridges, C.M. 1999a. Effects of a pesticide on tadpole activity and predator avoidance behavior. Journal of Herpetology 33:303–306.

Bridges, C.M. 1999b. Predator-prey interactions between two amphibian species: effects of insecticide exposure. Aquatic Ecology 33:205–211.

Bridges, C.M. 2000. Long-term effects of pesticide exposure at various life stages of the southern leopard frog (*Rana sphenocephala*). Archives of Environmental Contamination and Toxicology 39:91–96.

Bridges, C.M., and R.D. Semlitsch. 2000. Variation in pesticide tolerance of tadpoles among and within species of Ranidae and patterns of amphibian decline. Conservation Biology 14:1490–1499.

Bridges, C.M., and R.D. Semlitsch. 2001. Genetic variation in insecticide tolerance in a population of southern leopard frogs (*Rana sphenocephala*): implications for amphibian conservation. Copeia 2001:7–13.

Bridges, C.M., and M.D. Boone. 2003. The interactive effects of UV-B and insecticide exposure on tadpole survival, growth and development. Biological Conservation 113:49–54.

Briggler, J.T. 1998. Amphibian use of constructed woodland ponds in the Ouachita National Forest. M.S. thesis, University of Arkansas, Fayetteville.

Briggler, J.T. 2000. *Rana utricularia* (Southern Leopard Frog). Predation. Herpetological Review 31:171.

Briggler, J.T., K.M. Lohraff, and G.L. Adams. 2001. Amphibian parasitism by the leech *Desserobdella picta* at a small pasture pond in northwest Arkansas. Journal of Freshwater Ecology 16:105–111.

Briggs, C.J., V.T. Vredenburg, R.A. Knapp, and L.J. Rachowicz. 2005. Investigating the population-level effects of chytridiomycosis: an emerging infectious disease of amphibians. Ecology 86:3149–3159.

Briggs, C.J., R.A. Knapp, and V.T. Vredenburg. 2010. Enzootic and epizootic dynamics of the chytrid fungal pathogen of amphibians. Proceedings of the National Academy of Sciences, USA 107:9695–9700.

Briggs, J.L. 1975. A case of *Bufolucilla elongata* Shannon 1924 (Diptera: Calliphoridae) myiasis in the American toad, *Bufo americanus* Holbrook 1836. Journal of Parasitology 61:412.

Briggs, J.L. 1987. Breeding biology of the Cascades frog, *Rana cascadae*, with comparisons to *R. aurora* and *R. pretiosa*. Copeia 1987:241–245.

Briggs, J.L., and R.M. Storm. 1970. Growth and population structure of the Cascade frog, *Rana cascadae* Slater. Herpetologica 26:283–300.

Brill, T.M. 1993. Postmetamorphic population structure and the effects of salinity on larval development and size at metamorphosis of *Bufo woodhousei fowleri* Hinckley (Anura: Bufonidae). M.S. thesis, Bloomsburg University, Bloomsburg, PA.

Brimley, C.S. 1926. Revised key and list of the amphibians and reptiles of North Carolina. Journal of the Elisha Mitchell Scientific Society 42:75–93.

Brimley, C.S. 1944. Amphibians and Reptiles of North Carolina. Carolina Biological Supply Company, Elon College, NC.

Brindle, A.A., S.B. Karr, G.R. Smith, and J.A. Rettig. 2009. Within-pond oviposition site selection in the wood frog (*Rana sylvatica*). Bulletin of the Maryland Herpetological Society 45:57–60.

Britson, C.A., and R.E. Kissell Jr. 1996. Effects of food type on developmental characteristics of an ephemeral pond-breeding anuran, *Pseudacris triseriata feriarum*. Herpetologica 52:374–382.

Britson, C.A., and S.T. Threlkeld. 2000. Interactive effects of anthropogenic, environmental, and biotic stressors on multiple endpoints in *Hyla chrysoscelis*. Journal of the Iowa Academy of Science 107:61–66.

Brocchi, P. 1877. Sur quelques batraciens Raniformes et Bufoniformes d l'Amérique Centrale. Bulletin de la Société Philomathique de Parie, Série 7, 1:175–197.

Brocchi, M.P. 1879. Sur divers Batraciens anoures de l'Amérique Centrale. Bulletin de la Société Philomathique de Paris, Série 7, 3(1):19–24.

Brockelman, W.Y. 1968. Natural regulation of density in tadpoles of *Bufo americanus*. Ph.D. diss., University of Michigan, Ann Arbor.

Brockelman, W.Y. 1969. An analysis of density effects and predation in *Bufo americanus* tadpoles. Ecology 50:632–644.

Brode, W.E. 1958. The occurrence of the pickerel frog, three salamanders and two snakes in Mississippi caves. Copeia 1958:47–48.

Brode, W.E. 1959. Territoriality in *Rana clamitans*. Herpetologica 15:140.

Brode, W.E. 1960. Some notes on the occurrence of *Rana palustris* in caves. Journal of the Mississippi Academy of Science 6:233.

Brodie, E.D. Jr. 1968. A case of interbreeding between *Bufo boreas* and *Rana cascadae*. Herpetologica 24:86.

Brodie, E.D. Jr., and D.R. Formanowicz. 1983. Prey size preference of predators: differential vulnerability of larval anurans. Herpetologica 39:67–75.

Brodie, E.D. Jr., and D.R. Formanowicz. 1987. Antipredator mechanisms of larval anurans: protection of palatable individuals. Herpetologica 43:369–373.

Brodie, E.D. Jr., D.R. Formanowicz Jr., and E.D. Brodie III. 1978. The development of noxiousness of *Bufo americanus* tadpoles to aquatic insect predators. Herpetologica 34:302–306.

Brodkin, M.A., H. Madhoun, M. Rameswaran, and I. Vatnick. 2007. Atrazine is an immune disruptor in adult northern leopard frogs (*Rana pipiens*). Environmental Toxicology and Chemistry 26:80–84.

Brodman, R. 2003. Amphibians and reptiles from twenty-three counties of Indiana. Proceedings of the Indiana Academy of Science 112:43–54.

Brodman, R. 2009. A 14-year study of amphibian populations and metacommunities. Herpetological Conservation and Biology 4:106–119.

Brodman, R., and M. Kilmurry. 1998. Status of amphibians in northwestern Indiana. Pp. 125–136 *In* M.J. Lannoo (ed.), Status & Distribution of Midwestern Amphibians. University of Iowa Press, Iowa City.

Brodman, R., S. Cortwright, and A. Resetar. 2002. Historical changes of reptiles and amphibians of northwest Indiana Fish and Wildlife properties. American Midland Naturalist 147:135–144.

Brodman, R., M. Parrish, H. Kraus, and S. Cortwright. 2006. Amphibian biodiversity recovery in a large-scale ecosystem restoration. Herpetological Conservation and Biology 1:101–108.

Broel, A. 1950. Frog Raising for Pleasure and Profit. Marlboro House, New Orleans.

Brooks, D.R. 1975. A review of the genus *Allassostomoides* (Trematoda: Paramphistomidae) with a redescription of *Allassostomoides chelydrae*. Journal of Parasitology 61:882–885.

Brooks, D.R. 1976. Parasites of amphibians of the Great Plains. Part 2. Platyhelminths of amphibians in Nebraska. Bulletin of the University of Nebraska State Museum 10:65–92.

Brooks, D.R., and N.J. Welch. 1976. Parasites of

amphibians of the Great Plains. 1. The cercariae of *Cephalogonimus brevicirrus* Ingles, 1932 (Trematoda: Cephalogonimidae). Proceedings of the Helminthological Society of Washington 43:92–93.

Brooks, G.R. Jr. 1964. An analysis of the food habits of the bullfrog, *Rana catesbeiana*, by body size, sex, month, and habitat. Virginia Journal of Science 15:173–186.

Brothers, D.R. 1965. An annotated list of the amphibians and reptiles of northeastern North Carolina. Journal of the Elisha Mitchell Scientific Society 81:119–124.

Browder, L.W. 1972. Genetic and embryological studies of albinism in *Rana pipiens*. Journal of Experimental Zoology 180:149–156.

Browder, L.W., J.C. Underhill, and D.J. Merrell. 1966. Mid-dorsal stripe in the wood frog. Journal of Heredity 57:65–67.

Brown, B.C. 1950. An annotated check list of the reptiles and amphibians of Texas. Baylor University, Waco, TX.

Brown, C.J., B. Blossey, J.C. Maerz, and S.J. Joule. 2006. Invasive plant and experimental venue affect tadpole performance. Biological Invasions 8:327–338.

Brown, D.R. 1984. *Rana palustris* (Pickerel Frog). Oviposition. Herpetological Review 15:110–111.

Brown, E.E. 1980. Some historical data bearing on the Pine Barrens Treefrog, *Hyla andersoni*, in South Carolina. Brimleyana 3:113–117.

Brown, E.E. 1992. Notes on amphibians and reptiles of the western Piedmont of North Carolina. Journal of the Elisha Mitchell Scientific Society 108:38–54.

Brown, G.P., B.L. Phillips, J.K. Webb, and R. Shine. 2006. Toad on the road: use of roads as dispersal corridors by cane toads (*Bufo marinus*) at an invasion front in tropical Australia. Biological Conservation 133:88–94.

Brown, H.A. 1967. Embryonic temperature adaptations and genetic compatibility in two allopatric populations of the spadefoot toad, *Scaphiopus hammondi*. Evolution 21:742–761.

Brown, H.A. 1969. The heat resistance of some anuran tadpoles (Hylidae and Pelobatidae). Copeia 1969:138–147.

Brown, H.A. 1975a. Embryonic temperature adaptations of the Pacific treefrog, *Hyla regilla*. Comparative Biochemistry and Physiology 51A:863–873.

Brown, H.A. 1975b. Reproduction and development of the red-legged frog, *Rana aurora*, in northwestern Washington. Northwest Science 49:241–252.

Brown, H.A. 1975c. Temperature and development of the tailed frog, *Ascaphus truei*. Comparative Biochemistry and Physiology 50A:397–405.

Brown, H.A. 1976. The status of California and Arizona populations of the western spadefoot toads (genus *Scaphiopus*). Natural History Museum of Los Angeles County Contributions in Science No. 286:1–15.

Brown, H.A. 1977. A case of interbreeding between *Rana aurora* and *Bufo boreas* (Amphibia, Anura). Journal of Herpetology 11:92–94.

Brown, H.A. 1989. Tadpole development and growth of the Great Basin spadefoot toad, *Scaphiopus intermontanus*, from central Washington. Canadian Field-Naturalist 103:531–534.

Brown, H.A. 1990. Morphological variation and age-class determination in overwintering tadpoles of the tailed frog, *Ascaphus truei*. Journal of Zoology 220:171–184.

Brown, J.S. 1956. The frogs and toads of Alabama. Ph.D. diss., University of Alabama, Tuscaloosa.

Brown, J.S., and H.T. Boschung Jr. 1954. *Rana palustris* in Alabama. Copeia 1954:226.

Brown, L.E. 1964. An electrophoretic study of variation in the blood proteins of the toads *Bufo americanus* and *Bufo woodhousei*. Systematic Zoology 13:92–95.

Brown, L.E. 1969. Natural hybrids between two toad species in Alabama. Quarterly Journal of the Florida Academy of Science 32:285–290.

Brown, L.E. 1970. Interspecies interactions as possible causes of racial size differences in the toads *Bufo americanus* and *Bufo woodhousei*. Texas Journal of Science 21:261–267.

Brown, L.E. 1971a. Natural hybridization and reproductive ecology of two toad species in a disturbed environment. American Midland Naturalist 86:78–85.

Brown, L.E. 1971b. Natural hybridization and trend toward extinction in some relict Texas toad populations. Southwestern Naturalist 16:185–199.

Brown, L.E. 1973. Speciation in the *Rana pipiens* complex. American Zoologist 13:73–79.

Brown, L.E. 1974. Behavioral reactions of bullfrogs while attempting to eat toads. Southwestern Naturalist 19:329–340.

Brown, L.E. 1975. The status of the near-extinct Houston toad (*Bufo houstonensis*) with recommendations for its conservation. Herpetological Review 6:37–40.

Brown, L.E. 1978. Subterranean feeding by the chorus frog *Pseudacris streckeri* (Anura: Hylidae). Herpetologica 34:212–216.

Brown, L.E., and J.R. Pierce. 1965. Observations on the breeding behavior of certain anuran amphibians. Texas Journal of Science 26:313–317.

Brown, L.E., and J.R. Pierce. 1967. Male-male interactions and chorusing intensities of the Great Plains toad, *Bufo cognatus*. Copeia 1967:149–154.

Brown, L.E., and M.A. Ewert. 1971. A natural hybrid between the toads *Bufo hemiophrys* and *Bufo cognatus* in Minnesota. Journal of Herpetology 5:78–82.

Brown, L.E., and J.R. Brown. 1972a. Call types of

the *Rana pipiens* complex in Illinois. Science 176: 928–929.

Brown, L.E., and J.R. Brown. 1972b. Mating calls and distributional records of treefrogs of the *Hyla versicolor* complex in Illinois. Journal of Herpetology 6:233–234.

Brown, L.E., and M.J. Littlejohn. 1972. Male release call in the *Bufo americanus* group. Pp. 310–323 *In* W.F. Blair (ed.), Evolution in the Genus *Bufo*. University of Texas Press, Austin.

Brown, L.E., and J.R. Brown. 1973. Notes on breeding choruses of two anurans (*Scaphiopus holbrookii, Pseudacris streckeri*) in southern Illinois. Natural History Miscellanea (192):1–3.

Brown, L.E., and J.R. Brown. 1977. Comparison of environmental and body temperatures as predictors of mating call parameters of spring peepers. American Midland Naturalist 97:209–211.

Brown, L.E., and R.S. Funk. 1977. Absence of dorsal spotting in two species of leopard frogs (Anura: Ranidae). Herpetologica 33:290–293.

Brown, L.E., and D.B. Means. 1984. Fossorial behavior and ecology of the chorus frog *Pseudacris ornata*. Amphibia-Reptilia 5:261–273.

Brown, L.E., and G.B. Rose. 1988. Distribution, habitat, and calling season of the Illinois chorus frog (*Pseudacris streckeri illinoensis*) along the lower Illinois River. Illinois Natural History Survey Biological Notes 132:1–13.

Brown, L.E., and J.E. Cima. 1998. Illinois chorus frogs and the Sand Lake dilemma. Pp. 301–311 *In* M.J. Lannoo (ed.), Status & Conservation of Midwestern Amphibians. University of Iowa Press, Iowa City.

Brown, L.E., and A. Mesrobian. 2005. Houston toads and Texas politics. Pp. 150–167 *In* M.J. Lannoo (ed.), Amphibian Declines. The Conservation Status of United States Species. University of California Press, Berkeley.

Brown, L.E., H.O. Jackson, and J.R. Brown. 1972. Burrowing behavior of the chorus frog, *Pseudacris streckeri*. Herpetologica 28:325–328.

Brown, L.E., M.A. Morris, and T.R. Johnson. 1993. Zoogeography of the plains leopard frog (*Rana blairi*). Bulletin of the Chicago Academy of Sciences 15:1–13.

Brown, M.B., and C.R. Brown. 2009. *Lithobates catesbeianus* (American Bullfrog). Predation on cliff swallows. Herpetological Review 40:206.

Brown, R.E. 1972. Size variation and food habits of larval bullfrogs, *Rana catesbeiana* Shaw, in western Oregon. Ph.D. diss., Oregon State University, Corvallis.

Brown, R.L. 1974. Diets and habitat preferences of selected anurans in southeast Arkansas. American Midland Naturalist 91:468–473.

Brown, R.M., and D.H. Taylor. 1995. Compensatory escape mode trade-offs between swimming performance and maneuvering behavior through larval ontogeny of the wood frog, *Rana sylvatica*. Copeia 1995:1–7.

Brown, W. 1953. Brolo Frog Farm. Privately published, West Frankfort, IL.

Brown, W.C., and J.R. Slater. 1939. The amphibians and reptiles of the islands of the state of Washington. Occasional Papers, Department of Biology, College of Puget Sound No. 4:6–31.

Browne, C.L., and C.A. Paszkowski. 2010. Factors affecting the timing of movements to hibernation sites by western toads (*Anaxyrus boreas*). Herpetologica 66:250–258.

Browne, C.L., C.A. Paszkowski, A.L. Foote, A. Moenting, and S.M. Boss. 2009. The relationship of amphibian abundance to habitat features across spatial scales in the Boreal Plains. Ecoscience 16:209–223.

Brues, C.T. 1932. Further studies on the fauna of North American hot springs. Proceedings of the American Academy of Arts and Science 67:185–303.

Brugger, K.E. 1984. Aspects of reproduction in the squirrel treefrog, *Hyla squirella*. M.S. thesis, University of Florida, Gainesville.

Bruggers, R.L. 1973. Food habits of bullfrogs in northwest Ohio. Ohio Journal of Science 73:185–188.

Bruggers, R.L., and W.B. Jackson. 1974. Eye-lens weight of the bullfrog (*Rana catesbeiana*) related to larval development, transformation, and age of adults. Ohio Journal of Science 74:282–286.

Bruneau, M., and E. Magnin. 1980a. Vie larvaire des ouaouarons *Rana catesbeiana* Shaw (Amphibia Anura) des Laurentides, Québec. Canadian Journal of Zoology 58:169–174.

Bruneau, M., and E. Magnin. 1980b. Croissance, nutrition, et reproduction des ouaouarons *Rana catesbeiana* Shaw (Amphibia Anura) des Laurentides au nord du Montréal. Canadian Journal of Zoology 58:175–183.

Brunner, J.L., K.E. Barnett, C.J. Gosier, S.A. McNulty, M.J. Rubbo, and M.B. Kolozsvary. 2011. *Ranavirus* infection in die-offs of vernal pool amphibians in New York, USA. Herpetological Review 42:76–79.

Buchanan, B.W. 1993. Effects of enhanced lighting on the behaviour of nocturnal frogs. Animal Behaviour 45:893–899.

Buckle, J. 1971. A recent introduction of frogs to Newfoundland. Canadian Field-Naturalist 85:72–74.

Buhlmann, K.A. 2001. A biological inventory of eight caves in northwestern Georgia with conservation implications. Journal of Cave and Karst Studies 63:91–98.

Buhlmann, K.A., and J.C. Mitchell. 1997. Ecological notes on the amphibians and reptiles of the Naval Surface Warfare Center, Dahlgren Laboratory, King George County, Virginia. Banisteria 9:45–51.

Buhlmann, K.A., J.C. Mitchell, and C.A. Pague. 1993.

Amphibian and small mammal abundance and diversity in saturated forested wetlands and adjacent uplands of southeastern Virginia. Pp. 1–7 *In* S.D. Eckles, A. Jennings, A. Spingarn, and C. Wienhold (eds.), Proceedings of a Workshop on Saturated Forested Wetlands in the Mid-Atlantic Region. The State of the Science. U.S. Fish and Wildlife Service, Annapolis, MD.

Bulen, B.J., and C.A. Distel. 2011. Carbaryl concentration gradients in realistic environments and their influence on our understanding of the tadpole food web. Archives of Environmental Contamination and Toxicology 60:343–350.

Bulger, J.B., N.J. Scott Jr., and R.B. Seymour. 2003. Terrestrial activity and conservation of adult California red-legged frogs *Rana aurora draytonii* in coastal forests and grasslands. Biological Conservation 110:85–95.

Bull, E.L. 2006. Sexual differences in the ecology and habitat selection of western toads (*Bufo boreas*) in northeastern Oregon. Herpetological Conservation and Biology 1:27–38.

Bull, E.L. 2009. Dispersal of newly metamorphosed and juvenile Western toads (*Anaxyrus boreas*) in northeastern Oregon, USA. Herpetological Conservation and Biology 4:236–247.

Bull, E.L., and B.E. Carter. 1996. Winter observations of tailed frogs in northeastern Oregon. Northwestern Naturalist 77:45–47.

Bull, E.L., and M.P. Hayes. 2000. Livestock effects on reproduction of the Columbia spotted frog. Journal of Range Management 53:291–294.

Bull, E.L., and M.P. Hayes. 2001. Post-breeding season movements of Columbia spotted frogs (*Rana luteiventris*) in northeastern Oregon. Western North American Naturalist 61:119–123.

Bull, E.L., and D.B. Marx. 2002. Influence of fish and habitat on amphibian communities in high elevation lakes in northwestern Oregon. Northwest Science 76:240–248.

Bull, E.L., and C. Carey. 2008. Breeding frequency of western toads (*Bufo boreas*) in northeastern Oregon. Herpetological Conservation and Biology 3:282–288.

Bull, E.L., and J.L. Hayes. 2009. Selection of diet by metamorphic and juvenile western toads (*Bufo boreas*) in northwestern Oregon. Herpetological Conservation and Biology 4:85–95.

Bullard, A.J. 1965. Additional records of the tree frog *Hyla andersonii* from the Coastal Plain of North Carolina. Herpetologica 21:154–155.

Bundy, D., and C.R. Tracy. 1977. Behavioral response of American toads (*Bufo americanus*) to stressful thermal and hydric environments. Herpetologica 33:455–458.

Bunnell, J.F., and R.A. Zampella. 1999. Acid water anuran pond communities along a regional forest to agro-urban ecotone. Copeia 1999:614–627.

Burbrink, F.T., C.A. Phillips, and E.J. Heske. 1998. A riparian zone in southern Illinois as a potential dispersal corridor for reptiles and amphibians. Biological Conservation 86:107–115.

Burger, W.L. Jr., and A.N. Bragg. 1947. Notes on *Bufo boreas* (B. and G.) from the Gothic region of Colorado. Proceedings of the Oklahoma Academy of Science 27:61–65.

Burger, W.L., P.W. Smith, and H.M. Smith. 1949. Notable records of reptiles and amphibians in Oklahoma, Arkansas, and Texas. Journal of the Tennessee Academy of Science 14:130–134.

Burgess, R.C. Jr. 1950. Development of spade-foot toad larvae under laboratory conditions. Copeia 1950:49–53.

Burgett, A.A., C.D. Wright, G.R. Smith, D.T. Fortune, and S.L Johnson. 2007. Impact of ammonium nitrate on wood frog (*Rana sylvatica*) tadpoles: effects on survivorship and behavior. Herpetological Conservation and Biology 2:29–34.

Burke, V. 1933. Bacteria as food for vertebrates. Science 78:194–195.

Burkett, R.D. 1969. An ecological study of the cricket frog, *Acris crepitans*, in northeastern Kansas. Ph.D. diss., University of Kansas, Lawrence.

Burkett, R.D. 1984. An ecological study of the cricket frog, *Acris crepitans*. Pp. 89–103 *In* R.A. Seigel, L.E. Hunt, J.L. Knight, L. Malaret, and N.L. Zuschlag (eds.), Vertebrate Ecology and Systematics. A Tribute to Henry S. Fitch. Museum of Natural History, University of Kansas, Lawrence.

Burkett, R.D. 1989. Status of diploid/tetraploid gray treefrogs (*Hyla chrysoscelis-Hyla versicolor*) in the mid-South. Pp. 51–57 *In* A.F. Scott (ed.), Proceedings of the Second Annual Symposium on the Natural History of Lower Tennessee and Cumberland Valleys. Center for Field Biology, Austin Peay State University, Clarksville, TN.

Burkett, R.D. 1991. Ecological comparisons of some closely related species of amphibians in Land Between the Lakes and the mid-South. Journal of the Tennessee Academy of Science 66:161–164.

Burkholder, G. 1998. *Hyla chrysoscelis* (Gray Treefrog). Hibernacula. Herpetological Review 29: 231.

Burkholder, L.L., and L.V. Diller. 2007. Life history of postmetamorphic coastal tailed frogs (*Ascaphus truei*) in northwestern California. Journal of Herpetology 41:251–262.

Burmeister, S., W. Wilczynski, and M.J. Ryan. 1999a. Temporal call changes and prior experience affect graded signalling in the cricket frog. Animal Behaviour 57:611–618.

Burmeister, S., J. Konieczka, and W. Wilczynski. 1999b. Agonistic encounters in a cricket frog (*Acris crepitans*) chorus: behavioral outcomes vary with local competition and within breeding season. Ethology 105:335–347.

Burmeister, S., A.G. Ophir, M.J. Ryan, and W. Wilczynski. 2002. Information transfer during cricket frog contests. Animal Behaviour 64:715–725.

Burnett, S. 1997. Colonizing cane toads cause population declines in native predators: reliable anecdotal information and management implications. Pacific Conservation Biology 3.65–72.

Burnley, J.M. 1973. Eastern spadefoots, *Scaphiopus holbrooki*, found on the south fork of Long Island during 1973. Englehardtia 6:10.

Burress, R.M. 1982. Effects of synergized rotenone on nontarget organisms in ponds. U.S. Fish and Wildlife Service, Investigations in Fish Control No 91:1–7.

Burroughs, J. 1913. The Writings of John Burroughs. 17. The Summit of the Years. Houghton Mifflin Co., Boston, MA.

Bursey, C.R., and W.F. DeWolf II. 1998. Helminths of the frogs, *Rana catesbeiana*, *Rana clamitans*, and *Rana palustris*, from Coshocton County, Ohio. Ohio Journal of Science 98:28–29.

Bursey, C.R., and S.R. Goldberg. 2001. *Falcaustra lowei* n. sp. and other helminths from the Tarahumara frog, *Rana tarahumarae* (Anura: Ranidae), from Sonora, Mexico. Journal of Parasitology 87:340–344.

Bursey, C.R., S.R. Goldberg, and J.B. Bettaso. 2010. Persistence and stability of the component helminth community of the foothill yellow-legged frog, *Rana boylii* (Ranidae), from Humboldt County, California, 1964–1965, versus 2004–2007. American Midland Naturalist 163:476–482.

Burt, C.E. 1933. A contribution to the herpetology of Kentucky. American Midland Naturalist 14: 669–679.

Burt, C.E. 1935. Further records of the ecology and distribution of amphibians and reptiles in the Middle West. American Midland Naturalist 16: 311–336.

Burt, C.E. 1936. Contributions to the herpetology of Texas. 1. Frogs of the genus *Pseudacris*. American Midland Naturalist 17:770–775.

Burt, C.E. 1937 (1938). Contributions to Texan herpetology. 6. Narrow-mouthed froglike toads (*Microhyla* and *Hypopachus*). Papers of the Michigan Academy of Science, Arts, and Letters 23:607–610.

Burton, E.C., M.J. Gray, and A.C. Schmutzer. 2006. Comparison of anuran call survey durations in Tennessee wetlands. Proceedings of the Annual Conference of Southeastern Fish and Wildlife Agencies, pp. 15–18.

Burton, E.C., D.L. Miller, E.L. Styer, and M.J. Gray. 2008. Amphibian ocular malformation associated with Frog Virus 3. Veterinary Journal 177:442–444.

Burton, E.C., M.J. Gray, A.C. Schmutzer, and D.L. Miller. 2009. Differential responses of postmetamorphic amphibians to cattle grazing in wetlands. Journal of Wildlife Management 73:269–277.

Bury, R.B. 1968. The distribution of *Ascaphus truei* in California. Herpetologica 24:39–46.

Bury, R.B. 1973. The Cascade frog, *Rana cascadae*, in the North Coast Range of California. Northwest Science 47:228–229.

Bury, R.B. 1983. Differences in amphibian populations in logged and old growth redwood forest. Northwest Science 57:167–178.

Bury, R.B. 1988. Habitat relationships and ecological importance of amphibians and reptiles. Pp. 61–76 *In* K.J. Raedeke (ed.), Streamside Management. Riparian Wildlife and Forestry Interactions. Institute of Forest Resources, University of Washington, Contribution No. 59.

Bury, R.B. 2008. Amphibians in the Willamette Valley, Oregon: survival in a rapidly urbanizing and developing region. Pp. 435–444 *In* J.C. Mitchell, R.E Jung Brown, and B. Bartholomew (eds.), Urban Herpetology. Herpetological Conservation 3. Society for the Study of Amphibians and Reptiles, Salt Lake City, UT.

Bury, R.B., and R.A. Luckenbach. 1976. Introduced amphibians and reptiles in California. Biological Conservation 10:1–14.

Bury, R.B., and J.A. Whelan. 1984. Ecology and management of the bullfrog. U.S. Fish and Wildlife Service Resource Publication 155.

Bury, R.B., and P.S. Corn. 1988a. Responses of aquatic and streamside amphibians to timber harvest: a review. Pp. 165–181 *In* K.J. Raedeke (ed.), Streamside Management. Riparian Wildlife and Forestry Interactions. Institute of Forest Resources, University of Washington, Contribution No. 59.

Bury, R.B., and P.S. Corn. 1988b. Douglas-fir forests in the Oregon and Washington Cascades: relation of the herpetofauna to stand age and moisture. Pp. 11–22 *In* R.C. Szaro, K.E. Severson, and D.R. Patton (eds.), Management of Amphibians, Reptiles, and Small Mammals in North America. USDA Forest Service General and Technical Report RM-166.

Bury, R.B., and M.J. Adams. 1999. Variation in age at metamorphosis across a latitudinal gradient for the tailed frog, *Ascaphus truei*. Herpetologica 55:283–291.

Bury, R.B., C.K. Dodd Jr., and G.M. Fellers. 1980. Conservation of the Amphibia of the United States. A Review. U.S. Fish and Wildlife Service Resource Publication 134.

Bury, R.B., P.S. Corn, and K.B. Aubry. 1991a. Regional patterns of terrestrial amphibian communities in Oregon and Washington. Pp. 341–350 *In* L.F. Ruggiero, K.B. Aubry, A.B. Carey, and M.H. Huff (coordinators), Wildlife and Vegetation of Unmanaged Douglas-fir Forests. USDA Forest Service General Technical Report PNW-GTR-285, Portland, OR.

Bury, R.B., P.S. Corn, K.B. Aubrey, F.F. Gilbert, and L.L.C. Jones. 1991b. Aquatic amphibian communities in Oregon and Washington. Pp. 353–362 *In* L.F. Ruggiero, K.B. Aubry, A.B. Carey, and M.H. Huff (coordinators), Wildlife and Vegetation of Unmanaged Douglas-fir Forests. USDA Forest Service General Technical Report PNW-GTR-285, Portland, OR.

Bury, R.B., P. Loafman, D. Rofkar, and K.I. Mike. 2001. Clutch sizes and nests of tailed frogs from the Olympic Peninsula, Washington. Northwest Science 75:419–422.

Busack, S.D., and R.B. Bury. 1975. Toad in exile. National Parks and Conservation Magazine (March): 15–16.

Busby, W.H., and J.R. Parmelee. 1996. Historical changes in a herpetofaunal assemblage in the Flint Hills of Kansas. American Midland Naturalist 135: 81–91.

Busby, W.H., and W.R. Brecheisen. 1997. Chorusing phenology and habitat associations of the crawfish frog, *Rana areolata* (Anura: Ranidae), in Kansas. Southwestern Naturalist 42:210–217.

Busby, W.H., J.T. Collins, and G. Suleiman. 2005. The Snakes, Lizards, Turtles, and Amphibians of Fort Riley and Vicinity. 2nd rev. ed. Kansas Biological Survey, Lawrence.

Buseck, R.S., D.A. Keinath, and M. Geraud. 2005. Species assessment for Great Basin spadefoot toad (*Spea intermontana*) in Wyoming. Wyoming Natural Diversity Database, University of Wyoming, Laramie.

Bush, F.M., and E.F. Menhinick. 1962. The food of *Bufo woodhousei fowleri* Hinckley. Herpetologica 18:110–114.

Bush, S.L., H.C. Gerhardt, and J. Schul. 2002. Pattern recognition and call preferences in treefrogs (Anura: Hylidae): a quantitative analysis using a no-choice paradigm. Animal Behaviour 63:7–14.

Bushnell, R.J., E.P. Bushnell, and M.V. Parker. 1939. A chromosome study of five members of the Family Hylidae. Journal of the Tennessee Academy of Science 14:209–215.

Butterfield, B.P. 1988. Age structure and reproductive biology of the Illinois chorus frog (*Pseudacris streckeri illinoensis*) from northeastern Arkansas. M.S. thesis, Arkansas State University, Jonesboro.

Butterfield, B.P., W.E. Meshaka, and S.E. Trauth. 1989. Fecundity and egg mass size of the Illinois chorus frog, *Pseudacris streckeri illinoensis* (Hylidae), from northeastern Arkansas. Southwestern Naturalist 34:556–557.

Butterfield, B.P., H.A. Messer, L.D. Bennie, and J.W. Stanley. 2009. Distribution *Pseudacris crucifer* (Spring Peeper). Herpetological Review 40: 234–235.

Buzo, D. 2008. A GIS model for identifying potential breeding habitat for the Houston toad (*Bufo houstonensis*). M.S. thesis, Texas State University-San Marcos, Texas.

Cabarle, K.C. 2006. Geographic distribution: *Rana sylvatica* (Wood Frog). Herpetological Review 37:359.

Cagle, F.R. 1942. Herpetological fauna of Jackson and Union counties, Illinois. American Midland Naturalist 28:164–200.

Cahn, A.R. 1926. A set of albino frog eggs. Copeia (151):107–109.

Caldwell, J.P. 1982. Disruptive selection: a tail color polymorphism in *Acris* tadpoles in response to differential predation. Canadian Journal of Zoology 60:2818–2827.

Caldwell, J.P. 1986. Selection of egg deposition sites: a seasonal shift in the southern leopard frog, *Rana sphenocephala*. Copeia 1986:249–253.

Caldwell, J.P. 1987. Demography and life history of two species of chorus frogs (Anura: Hylidae) in South Carolina. Copeia 1987:114–127.

Caldwell, J.P., J.H. Thorp, and T.O. Jervey. 1980. Predator-prey relationships among larval dragonflies, salamanders, and frogs. Oecologia 46:285–289.

Calef, G.W. 1973a. Natural mortality of tadpoles in a population of *Rana aurora*. Ecology 54:741–758.

Calef, G.W. 1973b. Spatial distribution and "effective" breeding population of red-legged frogs (*Rana aurora*) in Marion Lake, British Columbia. Canadian Field-Naturalist 87:279–284.

Calfee, R.D., and E.E. Little. 2003. The effects of ultraviolet-B radiation on the toxicity of firefighting chemicals. Environmental Toxicology and Chemistry 22:1525–1531.

Calhoun, A.J.K., T.E. Walls, S.S. Stockwell, and M.M. McCollough. 2003. Evaluating several vernal pools as a basis for conservation strategies: a Maine case study. Wetlands 23:70–81.

Cameron, J.A. 1940. Effect of fluorine on hatching time and hatching stage in *Rana pipiens*. Ecology 21:288–292.

Camp, C.D., and R.A. Pyron. 2003. Geographic distribution: *Rana sylvatica* (Wood Frog). Herpetological Review 34:382.

Camp, C.D., C.E. Condee, and D.G. Lovell. 1990. Oviposition, larval development, and metamorphosis in the wood frog, *Rana sylvatica* (Anura: Ranidae), in Georgia. Brimleyana 16:17–21.

Camp, C.L. 1915. *Batrachoseps major* and *Bufo cognatus californicus*, new amphibia from southern California. University of California Publications in Zoology 12:327–334.

Camp, C.L. 1916. Description of *Bufo canorus*, a new toad from the Yosemite National Park. University of California Publications in Zoology 17:59–62.

Camp, C.L. 1917. Notes on the systematic status of the toads and frogs of California. University of California Publications in Zoology 17:115–125.

Campbell, B. 1931. *Rana tarahumarae*, a frog new to the United States. Copeia 1931:164.

Campbell, B. 1934. Report on a collection of reptiles and amphibians made in Arizona during the summer of 1933. Occasional Papers of the Museum of Zoology, University of Michigan No. 289.

Campbell, C.A. 1969. Who cares for the Fowler's toad? Ontario Naturalist 7(4):24–27.

Campbell, C.E., I.G. Warkentin, and K.G. Powell. 2004. Factors influencing the distribution and potential spread on introduced anurans in western Newfoundland. Northeastern Naturalist 11:151–162.

Campbell, J.B. 1970a. Life history of *Bufo boreas boreas* in the Colorado Front Range. Ph.D. diss., University of Colorado, Boulder.

Campbell, J.B. 1970b. Food habits of the boreal toad, *Bufo boreas boreas*, in the Colorado Front Range. Journal of Herpetology 4:83–85.

Campbell, J.B. 1970c. Hibernacula of a population of *Bufo boreas boreas* in the Colorado Front Range. Herpetologica 26:278–282.

Campbell, J.B. 1970d. New elevational records for the boreal toad (*Bufo boreas boreas*). Arctic and Alpine Research 2:157–159.

Campbell, J.B. 1976. Environmental controls on boreal toad populations in the San Juan Mountains. Pp. 289–295 *In* H.W. Steinhoff and J.D. Ives (eds.), Ecological Impacts of Snowpack Augmentation in the San Juan Mountains, Colorado. San Juan Ecology Project, Colorado State University Publishing, Fort Collins.

Campbell, J.B., and W.G. Degenhardt. 1971. *Bufo boreas boreas* in New Mexico. Southwestern Naturalist 16:209–220.

Campbell, K.R., T.S. Campbell, and S.A. Johnson. 2010. The use of PVC pipe refugia to evaluate spatial and temporal distributions of native and introduced treefrogs. Florida Scientist 73:78–88.

Campbell, R.A. 1968. A comparative study of the parasites of certain Salientia from Pocohontas State Park, Virginia. Virginia Journal of Science 19:13–20.

Camponelli, K.M., R.E. Casey, J.W. Snodgrass, S.M. Lev, and E.R. Landa. 2009. Impacts of weathered tire debris on the development of *Rana sylvatica* larvae. Chemosphere 74:717–722.

Capranica, R.R. 1965. The Evoked Vocal Response of the Bullfrog. MIT Press, Cambridge, MA.

Capranica, R.R. 1968. The vocal repertoire of the bullfrog (*Rana catesbeiana*). Behaviour 31:302–325.

Capranica, R.R., L.S. Frishkopf, and E. Nevo. 1973. Encoding of geographic dialects in the auditory system of the cricket frog. Science 182:1272–1275.

Carey, C. 1976. Thermal physiology and energetics of boreal toads, *Bufo boreas boreas*. Ph.D. diss., University of Michigan, Ann Arbor.

Carey, C. 1993. Hypothesis concerning the causes of the disappearance of boreal toads from the mountains of Colorado. Conservation Biology 7:355–362.

Carey, C., G.D. Maniero, and J.F. Stinn. 1996. Effect of cold on immune function and susceptibility to bacterial infection in toads (*Bufo marinus*). Pp. 123–129 *In* F. Geiser, A.J. Hulbert, and S.C. Nicol (eds.), Adaptations to Cold. Tenth International Hibernation Symposium. University of New England Press, Armidale, Australia.

Carey, C., P.S. Corn, M.S. Jones, L.J. Livo, E. Muths, and C.W. Loeffler. 2005. Factors limiting the recovery of boreal toads (*Bufo b. boreas*). Pp. 222–236 *In* M.J. Lannoo (ed.), Amphibian Declines. The Conservation Status of United States Species. University of California Press, Berkeley.

Carey, C., J.E. Bruzgul, L.J. Livo, M.L. Walling, K.A. Kuehl, B.F. Dixon, A.P. Pessier, R.A. Alford, and K.B. Rogers. 2006. Experimental exposures of boreal toads (*Bufo boreas*) to a pathogenic chytrid fungus (*Batrachochytrium dendrobatidis*). EcoHealth 3:5–21.

Carey, S.D. 1982. Geographic distribution. *Eleutherodactylus planirostris*. Herpetological Review 13:130.

Carl, G.C. 1943. The amphibians of British Columbia. British Columbia Provincial Museum, Bulletin No. 2.

Carl, G.C., and G.A. Hardy. 1943. Report on a collecting trip to the Lac La Hache area, British Columbia. Report of the British Columbia Provincial Museum for 1942:25–49.

Carpenter, C.C. 1953a. A study of hibernacula and hibernating associations of snakes and amphibians in Michigan. Ecology 34:74–80.

Carpenter, C.C. 1953b. Aggregation behavior of tadpoles of *Rana p. pretiosa*. Herpetologica 9:77–78.

Carpenter, C.C. 1954a. A study of amphibian movement in the Jackson Hole Wildlife Park. Copeia 1954:197–200.

Carpenter, C.C. 1954b. Feeding aggregation of narrow mouth toads (*Microhyla carolinensis olivacea*). Proceedings of the Oklahoma Academy of Science 35:45.

Carpenter, C.C., and D.E. Delzell. 1951. Road records as indicators of differential spring migrations of amphibians. Herpetologica 7:63–64.

Carpenter, H.L., and E.O. Morrison. 1973. Feeding behavior of the bullfrog, *Rana catesbeiana*, in northcentral Texas. Bios 44:188–193.

Carr, A.F. Jr. 1940a. A contribution to the herpetology of Florida. University of Florida Publication, Biological Science Series 3(1):1–118.

Carr, A.F. Jr. 1940b. Dates of frog choruses in Florida. Copeia 1940:55.

Cartwright, M.E., S.E. Trauth, and J.D. Wilhide. 1998. Wood frog (*Rana sylvatica*) use of wildlife ponds in northcentral Arkansas. Journal of the Arkansas Academy of Science 52:32–34.

Case, S.M. 1978. Biochemical systematics of members of the genus *Rana* native to western North America. Systematic Zoology 27:299–311.

Case, S.M., P.G. Haneline, and M.F. Smith. 1975. Pro-

tein variation in several species of *Hyla*. Systematic Zoology 24:281–295.
Cash, M.N., and J.P. Bogart. 1978. Cytological differentiation of the diploid-tetraploid species pair of North American treefrogs (Amphibia, Anura, Reptilia). Journal of Herpetology 12:555–558.
Cash, W.B. 1994. Herpetofaunal diversity of a temporary wetland in the Southeast Atlantic Coastal Plain. M.S. thesis, Georgia Southern University, Statesboro.
Casper, G.S. 1996. Geographic distributions of the amphibians and reptiles of Wisconsin. An interim report of the Wisconsin Herpetological Atlas Project. Milwaukee Public Museum, Milwaukee, WI.
Casper, G.S. 1998. Review of the status of Wisconsin amphibians. Pp. 199–205 *In* M.J. Lannoo (ed.), Status & Distribution of Midwestern Amphibians. University of Iowa Press, Iowa City.
Castano, B., S. Miely, G.R. Smith, and J.E. Rettig. 2010. Interactive effects of food availability and temperature on wood frog (*Rana sylvatica*) tadpoles. Herpetological Journal 20:209–211.
Catalano, P.A., and A.M. White. 1977. New host records for *Haematoloechus complex* (Seely) Krull, 1933 from *Hyla crucifer* and *Rana sylvatica*. Ohio Journal of Science 77:99.
Cecil, S.G., and J.J. Just. 1979. Survival rate, population density and development of a naturally occurring anuran larvae (*Rana catesbeiana*). Copeia 1979:447–453.
Cely, J.E., and J.A. Sorrow Jr. 1986. Distribution and habitat of *Hyla andersonii* in South Carolina. Journal of Herpetology 20:102–104.
Chamberlain, E.B. 1939. Frogs and toads of South Carolina. Charleston Museum Leaflet No. 12.
Chamberlain, F.M. 1897. Notes on the edible frogs of the United States and their artificial propagation. U.S. Bureau of Fisheries for 1897:249–261.
Chance, K.B. 2002. A telemetric study of winter habitat selection by the American bullfrog, *Rana catesbeiana*, in east-central Kansas. M.S. thesis, Emporia State University, Emporia, KS.
Chandler, J.H., and L.L. Marking. 1975. Toxicity of the lampricide 3-trifluoromethyl-4-nitrophenol (TFM) to selected aquatic invertebrates and frog larvae. U.S. Fish and Wildlife Service, Investigations in Fish Control No. 62:1–7.
Chandler, J.H., and L.L. Marking. 1982. Toxicity of rotenone to selected aquatic invertebrates and frog larvae. Progressive Fish Culturist 44:78–80.
Chan-McLeod, A.C.A., and A. Moy. 2007. Evaluating residual tree patches as stepping stones and short-term refugia for red-legged frogs. Journal of Wildlife Management 71:1836–1844.
Chantell, C.J. 1970. Upper Pliocene frogs from Idaho. Copeia 1970:654–664.
Chantell, C.J. 1971. Fossil amphibians from the Egelhoff Local Fauna of northcentral Nebraska. Contributions from the Museum of Paleontology, University of Michigan 23:239–246.
Chapel, W.L. 1939. Field notes on *Hyla wrightorum* Taylor. Copeia 1939:225–227.
Chatfield, M.W.H., B.B. Rothermel, C.S. Brooks, and J.B. Kay. 2009. Detection of *Batrachochytrium dendrobatidis* in amphibians from the Great Smoky Mountains of North Carolina and Tennessee. Herpetological Review 40:176–179.
Chelgren, N.D., D.K. Rosenberg, S.S. Heppell, and A.I. Gitelman. 2006. Carryover aquatic effects on survival of metamorphic frogs during pond emigration. Ecological Applications 16:250–261.
Chelgren, N.D., C.A. Pearl, M.J. Adams, and J. Bowerman. 2008. Demography and movement in a relocated population of Oregon spotted frogs (*Rana pretiosa*): influence of season and gender. Copeia 2008:742–751.
Chen, C.Y., K.M. Hathaway, and C.L. Folt. 2004. Multiple stress effects of Vision™ herbicide, pH, and food on zooplankton and larval amphibian species from forest wetlands. Environmental Toxicology and Chemistry 23:823–831.
Chen, C.Y., K.M. Hathaway, D.G. Thompson, and C.L. Folt. 2008. Multiple stressor effects of herbicide, pH, and food on wetland zooplankton and a larval amphibian. Ecotoxicology and Environmental Safety 71:209–218.
Chen, G.J., and S.S. Desser. 1989. The Coccidia (Apicomplexa: Eimeriidae) of frogs from Algonquian Park, with descriptions of two new species. Canadian Journal of Zoology 67:1686–1689.
Chen, T.-H., J.A. Gross, and W.H. Karasov. 2006. Sublethal effects of lead on northern leopard frog (*Rana pipiens*) tadpoles. Environmental Toxicology and Chemistry 25:1383–1389.
Cheng, T.C. 1961. Description, life history and developmental pattern of *Glypthelmins pennsylvaniensis* n. sp. (Trematoda: Brachycoeliidae), new parasite of frogs. Journal of Parasitology 47:469–477.
Cheng, T.-H. 1929a. Intersexuality in *Rana cantabrigensis*. Journal of Morphology and Physiology 48:345–369.
Cheng, T.-H. 1929b. A new case of intersexuality in *Rana cantabrigensis*. Biological Bulletin 57:412–421.
Cheng, T.-H. 1930. Intersexuality in tadpoles of *Rana cantabrigensis*. Papers of the Michigan Academy of Science, Arts and Letters 11:353–368.
Chestnut, T., J.E. Johnson, and R.S. Wagner. 2008. Results of amphibian chytrid (*Batrachochytrium dendrobatidis*) sampling in Denali National Park, Alaska, USA. Herpetological Review 39:202–204.
Childs, C. 2000. The Secret Knowledge of Water. Little, Brown and Co., New York.
Childs, H.E. Jr. 1953. Selection by predation on albino and normal spadefoot toads. Evolution 7:228–233.

Chivers, D.P., J.M. Kiesecker, A. Marco, E.L. Wildy, and A.R. Blaustein. 1999. Shifts in life history as a response to predation in Western toads (*Bufo boreas*). Journal of Chemical Ecology 25:2455–2463.

Chrapliwy, P.S., and K.L. Williams. 1957. A species of frog new to the fauna of the United States: *Pternohyla fodiens* Boulenger. Natural History Miscellanea (160):1–2.

Christein, D., and D.H. Taylor. 1978. Population dynamics in breeding aggregations of the American toad, *Bufo americanus* (Amphibia, Anura, Bufonidae). Journal of Herpetology 12:17–24.

Christein, D., S.I. Guttman, and D.H. Taylor. 1979. Heterozygote deficiencies in a breeding population of *Bufo americanus* (Bufonidae: Anura): the test of a hypothesis. Copeia 1979:498–502.

Christian, K.A. 1976. Ontogeny of the food niche of *Pseudacris triseriata*. M.S. thesis, Colorado State University, Fort Collins.

Christiansen, J.L. 1981. Population trends among Iowa's amphibians and reptiles. Proceedings of the Iowa Academy of Science 88:24–27.

Christiansen, J.L. 1998. Perspectives on Iowa's declining amphibians and reptiles. Journal of the Iowa Academy of Science 105:109–114.

Christiansen, J.L. 2001. Non-native amphibians and reptiles in Iowa. Journal of the Iowa Academy of Science 108:210–211.

Christiansen, J.L., and R.R. Burken. 1978. The endangered and uncommon reptiles and amphibians of Iowa. Iowa Science Teachers Journal, Special Issue, 26 pp.

Christiansen, J.L., and C.M. Mabry. 1985. The amphibians and reptiles of Iowa's Loess Hills. Proceedings of the Iowa Academy of Science 92:159–163.

Christiansen, J.L., and R.M. Bailey. 1991. The salamanders and frogs of Iowa. Iowa Department of Natural Resources, Nongame Technical Series 3.

Christiansen, J.L., and H. Feltman. 2000. A relationship between trematode metacercariae and bullfrog limb abnormalities. Journal of the Iowa Academy of Science 107:79–85.

Christie, P. 1997. Reptiles and Amphibians of Prince Edward County Ontario. Natural Heritage Books, Toronto.

Christin, M.-S., A.D. Gendron, P. Brousseau, L. Ménard, D.J. Marcogliese, D. Cyr, S. Ruby, and M. Fournier. 2003. Effects of agricultural pesticides on the immune system of *Rana pipiens* and on its resistance to parasitic infection. Environmental Toxicology and Chemistry 22:1127–1133.

Christin, M.-S., L. Ménard, A.D. Gendron, S. Ruby, D. Cyr, D.J. Marcogliese, L. Rollins-Smith, and M. Fournier. 2004. Effects of agricultural pesticides on the immune system of *Xenopus laevis* and *Rana pipiens*. Aquatic Toxicology 67:33–43.

Christman, S.P. 1970. *Hyla andersoni* in Florida. Quarterly Journal of the Florida Academy of Sciences 33:80.

Christman, S.P. 1974. Geographic variation for salt water tolerance in the frog *Rana sphenocephala*. Copeia 1974:773–778.

Church, D.R. 2008. Role of current versus historical hydrology in amphibian species turnover within local pond communities. Copeia 2008:115–125.

Church, D.R., H.M. Wilbur, S.M. Roble, F.C. Huber, and M.W. Donahue. 2002. Observations on breeding by eastern spadefoots (*Scaphiopus holbrookii*) in Augusta County, Virginia. Banisteria 20:71–72.

Churchill, T.A., and K.B. Storey. 1996. Organ metabolism and cryoprotectant synthesis during freezing in spring peepers *Pseudacris crucifer*. Copeia 1996:517–525.

Claflin, W.J. 1962. Larval growth and development of *Pseudacris nigrita triseriata* and *Rana pipiens* in semi-permanent ponds in south-eastern South Dakota. M.A. thesis, State University of South Dakota, Vermillion.

Clark, C.B., and A.N. Bragg. 1950. A comparison of the ovaries of two species of *Bufo* with different ecological requirements. Pp. 143–152 *In* Researches on the Amphibia of Oklahoma, University of Oklahoma Press, Norman.

Clark, P.J., J.M. Reed, B.G. Tavernia, B.S. Windmiller, and J.V. Regosin. 2008. Urbanization effects on spotted salamander and wood frog presence and abundance. Pp. 67–75 *In* J.C. Mitchell, R.E. Jung Brown, and B. Bartholomew (eds.), Urban Herpetology. Herpetological Conservation 3. Society for the Study of Amphibians and Reptiles, Salt Lake City, UT.

Clark, R.F. 1949. Snakes of the hill parishes of Louisiana. Journal of the Tennessee Academy of Science 24:244–261.

Clark, V.R. 2001. *Rana heckscheri* (River Frog). Ectoparasites. Herpetological Review 32:36.

Clarke, R.D. 1972. The effect of toe clipping on survival in Fowler's toad (*Bufo woodhousei fowleri*). Copeia 1972:182–185.

Clarke, R.D. 1974a. Activity and movement patterns in a population of Fowler's toad, *Bufo woodhousei fowleri*. American Midland Naturalist 92:257–274.

Clarke, R.D. 1974b. Food habits of the toads, genus *Bufo* (Amphibia, Bufonidae). American Midland Naturalist 91:140–147.

Clarke, R.D. 1974c. Postmetamorphic growth rates in a natural population of Fowler's toad, *Bufo woodhousei fowleri*. Canadian Journal of Zoology 52:1489–1498.

Clarke, R.D. 1977. Postmetamorphic survivorship of Fowler's toad, *Bufo woodhousei fowleri*. Copeia 1977:594–597.

Clarke, R.F. 1958. An ecological study of reptiles and

amphibians in Osage County, Kansas. Emporia State Research Studies 7:1–52.
Clarkson, R.W., and J.C. DeVos Jr. 1986. The bullfrog, *Rana catesbeiana* Shaw, in the lower Colorado River, Arizona–California. Journal of Herpetology 20:42–49.
Clarkson, R.W., and J.C. Rorabaugh. 1989. Status of leopard frogs (*Rana pipiens* complex: Ranidae) in Arizona and southeastern California. Southwestern Naturalist 34:531–538.
Claussen, D.L. 1973a. The water relations of the tailed frog, *Ascaphus truei*, and the Pacific treefrog, *Hyla regilla*. Comparative Biochemistry and Physiology 44A:155–171.
Claussen, D.L. 1973b. The thermal relations of the tailed frog, *Ascaphus truei*, and the Pacific treefrog, *Hyla regilla*. Comparative Biochemistry and Physiology 44A:137–153.
Claussen, D.L. 1974. Urinary bladder water reserves in the terrestrial toad, *Bufo fowleri*, and the aquatic frog, *Rana clamitans*. Herpetologica 30:360–367.
Claussen, D.L., and J.R. Layne Jr. 1983. Growth and survival of juvenile toads, *Bufo woodhousei*, maintained on different diets. Journal of Herpetology 17:107–112.
Clawson, M.E., and T.S. Baskett. 1982. Herpetofauna of the Ashland Wildlife Area, Boone County, Missouri. Transactions, Missouri Academy of Science 16:5–16.
Clawson, R.G., B.G. Lockley, and R.H. Jones. 1997. Amphibian responses to helicopter harvesting in forested floodplains of low order, blackwater streams. Forest Ecology and Management 90:225–235.
Cochran, P.A. 1999. *Rana sylvatica* (Wood Frog). Predation. Herpetological Review 30:94.
Cochran, P.A. 2008. An unusual microhabitat for an American toad (*Anaxyrus americanus*). Bulletin of the Chicago Herpetological Society 43:62.
Cochran, P.A., and J.A. Cochran. 2003. *Pseudacris crucifer* (Spring Peeper). Predation. Herpetological Review 34:360.
Cocroft, R.B. 1994. A cladistic analysis of chorus frog phylogeny (Hylidae: *Pseudacris*). Herpetologica 50:420–437.
Coggins, J.R., and R.A. Sajdak. 1982. A survey of helminth parasites in the salamanders and certain anurans from Wisconsin. Proceedings of the Helminthological Society of Washington 49:99–102.
Cohen, N.W., and W.E. Howard. 1958. Bullfrog food and growth at the San Joaquin Experimental Range, California. Copeia 1958:223–225.
Cole, C.J. 1962. Notes on the distribution and food habits of *Bufo alvarius* at the eastern edge of its range. Herpetologica 18:172–175.
Cole, E.C., W.C. McComb, M. Newton, C.L. Chambers, and J.P. Leeming. 1997. Response of amphibians to clearcutting, burning, and glyphosate application in the Oregon Coast Range. Journal of Wildlife Management 61:656–664.
Cole, L.M., and J.E. Casida. 1983. Pyrethroid toxicology in the frog. Pesticide Biochemistry and Physiology 20:217–224.
Collier, A., J.B. Keiper, and L.P. Orr. 1998. The invertebrate prey of the northern leopard frog, *Rana pipiens*, in a northeastern Ohio population. Ohio Journal of Science 98:39–41.
Collins, J.P. 1975. A comparative study of the life history strategies in a community of frogs. Ph.D. diss., University of Michigan, Ann Arbor.
Collins, J.P., and M.A. Lewis. 1979. Overwintering tadpoles and breeding season variation in the *Rana pipiens* complex in Arizona. Southwestern Naturalist 24:371–373.
Collins, J.P., and H.M. Wilbur. 1979. Breeding habits and habitats of the amphibians of the Edwin S. George Reserve, Michigan, with notes on the local distribution of fishes. Occasional Papers of the Museum of Zoology, University of Michigan No. 686.
Collins, J.T. 1993. Amphibians & Reptiles in Kansas. 3rd rev. ed. University of Kansas, Museum of Natural History, Lawrence.
Collins, J.T., and S.L. Collins. 2006. Amphibians, turtles and reptiles of Cheyenne Bottoms. Sternberg Museum of Natural History, Fort Hays State University, Fort Hays, KS.
Collins, J.T., S.L. Collins, and T.W. Taggart. 2010. Amphibians, Reptiles, and Turtles in Kansas. Eagle Mountain Publishing, Eagle Mountain, UT.
Conant, R. 1940. *Rana virgatipes* in Delaware. Herpetologica 1:176–177.
Conant, R. 1947. The carpenter frog in Maryland. Maryland, A Journal of Natural History 17:72–73.
Conant, R. 1962. Notes on the distribution of reptiles and amphibians in the Pine Barrens of southern New Jersey. New Jersey Nature Notes (New Jersey Audubon Society) 17:16–21.
Conant, R. 1977 [1978]. Semiaquatic reptiles and amphibians of the Chihuahuan Desert and their relationships to drainage patterns of the region. Pp. 455–491 *In* R.H. Wauer and D.H. Riskind (eds.), Transactions of the Symposium on the Biological Resources of the Chihuahuan Desert Region, United States and Mexico. U.S. National Park Service Transactions and Proceedings Series No. 3.
Conant, R., and J.T. Collins. 1998. A Field Guide to the Reptiles and Amphibians. Eastern and Central North America. 3rd exp. ed. Houghton Mifflin Co., Boston, MA.
Conant, R., J.C. Mitchell, and C.A. Pague. 1990. Herpetofauna of the Virginia barrier islands. Virginia Journal of Science 41:364–380.
Conaway, C.H., and D.E. Metter. 1967. Skin glands associated with breeding in *Microhyla carolinensis*. Copeia 1967:672–673.

Conlon, J.M., T. Halverson, J. Dulka, J.E. Platz, and F.C. Knoop. 1999. Peptides with antimicrobial activity of the brevinin-1 family isolated from the skin secretions of the southern leopard frog, *Rana sphenocephala*. Journal of Peptide Research 54:522–527.

Conlon, J.M., A. Sonnevend, C. Davidson, D.D. Smith, and P.F. Nielsen. 2004. The ascaphins: a family of antimicrobial peptides from the skin secretions of the most primitive extant frog, *Ascaphus truei*. Biochemical and Biophysical Research Communications 320:170–175.

Conlon, J.M., T. Jouenne, P. Cosette, D. Cosquer, H. Vaudry, C.K. Taylor, and P.W. Abel. 2005a. Bradykinin-related peptides and tryptophyllins in the skin secretions of the most primitive extant frog, *Ascaphus truei*. General and Comparative Endocrinology 143:193–199.

Conlon, J.M., B. Abraham, A. Sonnevend, T. Jouenne, P. Cosette, J. Leprince, H. Vaudry, and C.A. Bevier. 2005b. Purification and characterization of antimicrobial peptides from the skin secretions of the carpenter frog *Rana virgatipes* (Ranidae, Aquarana). Regulatory Peptides 131:38–45.

Conlon, J.M., N. Al-Ghafari, L. Coquet, J. Leprince, T. Jouenne, H. Vaudry, and C. Davidson. 2006. Evidence from peptidomic analysis of skin secretions that the red-legged frogs, *Rana aurora draytonii* and *Rana aurora aurora*, are distinct species. Peptides 27:1305–1312.

Conlon, J.M., L. Coquet, J. Leprince, T. Jouenne, H. Vaudry, J. Kolodziejek, N. Nowotny, C.R. Bevier, and P.E. Moler. 2007a. Peptidomic analysis of skin secretions from *Rana heckscheri* and *Rana okaloosae* provides insight into phylogenetic relationships among frogs of the *Aquarana* species group. Regulatory Peptides 138:87–93.

Conlon, J.M., C.R. Bevier, L. Coquet, J. Leprince, T. Jouenne, H. Vaudry, and B.R. Hossack. 2007b. Peptidomic analysis of skin secretions supports separate species status for the tailed frogs, *Ascaphus truei* and *Ascaphus montanus*. Comparative Biochemistry and Physiology 2D:121–125.

Conlon, J.M., E. Ahmed, L. Coquet, T. Jouenne, J. Leprince, H. Vaudry, and J.D. King. 2009. Peptides with potent cytolytic activity from the skin secretions of the North American leopard frogs, *Lithobates blairi* and *Lithobates yavapaiensis*. Toxicon 53:699–705.

Conlon, J.M., L. Coquet, J. Leprince, T. Jouenne, H. Vaudry, and J.D. King. 2010. Primary structures of skin antimicrobial peptides indicate a close, but not conspecific, phylogenetic relationship between the leopard frogs *Lithobates onca* and *Lithobates yavapaiensis* (Ranidae). Comparative Physiology and Biochemistry 151C:313–317.

Conlon, J.M., M. Mechkarska, E. Ahmed, L. Coquet, T. Jouenne, J. Leprince, H. Vaudry, M.P. Hayes, and G. Padgett-Flohr. 2011. Host defense peptides in skin secretions of the Oregon spotted frog *Rana pretiosa*: implications for species resistance to chytridiomycosis. Developmental and Comparative Immunology 35:644–649.

Constible, J.M., P.T. Gregory, and B.R. Anholt. 2001. Patterns of distribution, relative abundance, and microhabitat use of anurans in a boreal landscape influenced by fire and timber harvest. Ecoscience 8:462–470.

Constible, J.M., P.T. Gregory, and K.W. Larsen. 2010. The pitfalls of extrapolation in conservation: movements and habitat use of a threatened toad are different in the boreal forest. Animal Conservation 13:43–52.

Converse, K.A., J. Mattsson, and L. Eaton-Poole. 2000. Field surveys of Midwestern and Northeastern Fish and Wildlife Service lands for the presence of abnormal frogs and toads. Journal of the Iowa Academy of Science 107:160–167.

Cook, D. 1997a. Biology of the California red-legged frog: a synopsis. Transactions of the Western Section of the Wildlife Society 33:79–82.

Cook, D. 1997b. Microhabitat use and reproductive success of the California red-legged frog (*Rana aurora draytonii*) and bullfrog (*Rana catesbeiana*) in an ephemeral marsh. M.A. thesis, Sonoma State University, Sonoma, CA.

Cook, F.R. 1960. New localities for the Plains spadefoot toad, tiger salamander and the Great Plains toad in the Canadian prairies. Copeia 1960:363–364.

Cook, F.R. 1963. The rediscovery of the mink frog in Manitoba. Canadian Field-Naturalist 77:129–130.

Cook, F.R. 1964a. Additional records and a correction of the type locality for the boreal chorus frog in northwestern Ontario. Canadian Field-Naturalist 78:186–192.

Cook, F.R. 1964b. The rusty colour phase of the Canadian toad, *Bufo hemiophrys*. Canadian Field-Naturalist 78:263–267.

Cook, F.R. 1965. Additions to the known range of some amphibians and reptiles in Saskatchewan. Canadian Field-Naturalist 79:112–120.

Cook, F.R. 1966. A guide to the amphibians and reptiles of Saskatchewan. Saskatchewan Museum of Natural History Popular Series 13:1–40.

Cook, F.R. 1967. An analysis of the herpetofauna of Prince Edward Island. National Museum of Canada Bulletin 212.

Cook, F.R. 1977. Records of the boreal toad from the Yukon and northern British Columbia. Canadian Field-Naturalist 91:185–186.

Cook, F.R. 1978. Amphibians and reptiles of Saskatchewan. Saskatchewan Museum of Natural History, popular series, No. 13.

Cook, F.R. 1983. An analysis of toads of the *Bufo americanus* group in a contact zone in central northern North America. National Museums of Canada, Publications in Natural Sciences 3:i–89.

Cook, F.R., and D.R.M. Hatch. 1964. A spadefoot toad from Manitoba. Canadian Field-Naturalist 78:60–61.

Cook, F.R., and J.C. Cook. 1981. Attempted avian predation by a Canadian toad, *Bufo americanus hemiophrys*. Canadian Field-Naturalist 95:346–347.

Cook, J.C. 2008. Transmission and occurrence of *Dermomycoides* sp. in *Rana sevosa* and other ranids in the north central Gulf of Mexico states. M.S. thesis, University of Southern Mississippi, Hattiesburg.

Cook, R.P. 1989. And the voice of the grey tree frog was heard again in the land. Park Science 9(3):6–7.

Cook, R.P. 2008. Potential and limitations of herpetofaunal restoration in an urban landscape. Pp. 465–478 *In* J.C. Mitchell, R.E. Jung Brown, and B. Bartholomew (eds.), Urban Herpetology. Herpetological Conservation 3, Society for the Study of Amphibians and Reptiles, Salt Lake City, UT.

Cooper, J.E. 1953. Notes on the amphibians and reptiles of southern Maryland. Maryland Naturalist 23:90–100.

Cooper, J.E. 1979. The brothers Brimley: North Carolina naturalists. Brimleyana 1:1–14.

Cope, E.D. 1862. On some new and little known American Anura. Proceedings of the Academy of Natural Sciences of Philadelphia 14:151–159.

Cope, E.D. 1863. On *Trachycephalus, Scaphiopus* and other American Batrachia. Proceedings of the Academy of Natural Sciences of Philadelphia 15:43–54.

Cope, E.D. 1866. On the structures and distribution of the genera of the arciferous anura. Journal of the Academy of Natural Sciences of Philadelphia, series 2, 6:67–112.

Cope, E.D. "1866" (1867). On the reptilia and Batrachia of the Sonoran Province of the Nearctic Region. Proceedings of the Academy of Natural Sciences of Philadelphia 18:300–314. (Dated 1866 but published in 1867. The date attributed to newly described species is thus 1867.)

Cope, E.D. 1875. *Rana onca*, Cope, sp. nov. Pp. 528–529 and Plate 15 *In* H.C. Yarrow, Report upon the collections of batrachians and reptiles made in portions of Nevada, Utah, California, Colorado, New Mexico, and Arizona during 1871, 1872, 1873, and 1874. Ch. 4, pp. 509–584 *In* Report Upon Geographical and Geological Surveys West of the One Hundredth Meridian in Charge of First Lieut. Geo. M. Wheeler. Vol. 5. Zoology. U.S. Government Printing Office, Washington, DC.

Cope, E.D. "1877" (1878). A new genus of Cystignathidae from Texas. American Naturalist 12:253.

Cope, E.D. 1879. Eleventh contribution to the herpetology of tropical America. Proceedings of the American Philosophical Society 18:261–277.

Cope, E.D. 1880. On the zoological position of Texas. Bulletin of the United States National Museum 17:1–51.

Cope, E.D. 1883. Notes on the geographic distribution of Batrachia and Reptilia in western North America. Proceedings of the Academy of Natural Sciences of Philadelphia 35:10–35.

Cope, E.D. 1886. Synonymic list of the North American species of *Bufo* and *Rana*, with descriptions of some Batrachia, from specimens in the National Museum. Proceedings of the American Philosophical Society 23:514–526.

Cope, E.D. 1889. The Batrachia of North America. Bulletin of the United States National Museum 34:1–525.

Cope, E.D. 1891. A new species of frog from New Jersey. American Naturalist 25:1017–1019.

Corcoran, M.F., and J.R. Travis. 1980. A comparison of the karyotypes of the frogs *Rana areolata*, *Rana sphenocephala*, and *Rana pipiens*. Herpetologica 36:296–300.

Corkran, C.C., and C. Thoms. 2006. Amphibians of Oregon, Washington and British Columbia. 2nd ed. Lone Pine Publishing, Edmonton, Alberta.

Corn, P.S. 1980a. Polymorphic reproductive behavior in male chorus frogs (*Pseudacris triseriata*). Journal of the Colorado-Wyoming Academy of Sciences 12:6–7.

Corn, P.S. 1980b. Comment on the occurrence of *Pseudacris clarki* in Montana. Bulletin of the Chicago Herpetological Society 15:77–78.

Corn, P.S. 1981. Field evidence for a relationship between color and developmental rate in the northern leopard frog (*Rana pipiens*). Herpetologica 37:155–160.

Corn, P.S. 1986. Genetic and developmental studies of albino chorus frogs. Journal of Heredity 77:164–168.

Corn, P.S. 1993. *Bufo boreas* (Boreal Toad). Predation. Herpetological Review 24:57.

Corn, P.S. 1994. What we know and don't know about amphibian declines in the West. Pp. 59–67 *In* W.W. Covington and L.F. DeBano (eds.), Sustainable Ecological Systems: Implementing and Ecological Approach to Land Management. USDA Forest Service, General Technical Report RM-247, Ft. Collins, CO.

Corn, P.S. 1998. Effects of ultraviolet radiation on boreal toads in Colorado. Ecological Applications 8:18–26.

Corn, P.S. 2000. Amphibian declines: review of some current hypotheses. Pp. 663–696 *In* D.W. Sparling, G. Linder, and C.A. Bishop (eds.), Ecotoxicology of Amphibians and Reptiles. SETAC Press, Pensacola, FL.

Corn, P.S. 2003. Endangered toads in the Rockies. Pp. 43–51 *In* L. Taylor, K. Martin, D. Hik, and A. Ryall (eds.), Ecological and Earth Sciences in Mountain Areas. The Banff Center, Banff, Alberta.

Corn, P.S. 2005. Climate change and amphibians. Animal Biodiversity and Conservation 28:59–67.

Corn, P.S. 2007. Amphibians and disease: implications for conservation in the Greater Yellowstone Ecosystem. Yellowstone Science 15(2):11–16.

Corn, P.S., and J.C. Fogleman. 1984. Extinction of montane populations of the Northern Leopard Frog (*Rana pipiens*) in Colorado. Journal of Herpetology 18:147–152.

Corn, P.S., and R.B. Bury. 1989. Logging in western Oregon: responses of headwater habitats and stream amphibians. Forest Ecology and Management 29:39–57.

Corn, P.S., and L.J. Livo. 1989. Leopard frog and wood frog reproduction in Colorado and Wyoming. Northwestern Naturalist 70:1–9.

Corn, P.S., and F.A. Vertucci. 1992. Descriptive risk assessment of the effects of acidic deposition on Rocky Mountain amphibians. Journal of Herpetology 26:361–369.

Corn, P.S., and E. Muths. 2002. Variable breeding phenology affects the exposure of amphibian embryos to ultraviolet radiation. Ecology 83:2958–2963.

Corn, P.S., W. Stolzenburg, and R.B. Bury. 1989. Acid precipitation studies in Colorado and Wyoming: interim report of surveys of montane amphibians and water chemistry. U.S. Fish and Wildlife Service Biological Report 80(40.26).

Corn, P.S., M.L. Jennings, and E. Muths. 1997. Survey and assessment of amphibian populations in Rocky Mountain National Park. Northwestern Naturalist 78:34–55.

Corn, P.S., E. Muths, and W.M. Iko. 2000. A comparison in Colorado of three methods to monitor breeding amphibians. Northwestern Naturalist 81:22–30.

Corn, P.S., B.R. Hossack, E. Muths, D.A. Patla, C.R. Peterson, and A.L. Gallant. 2005. Status of amphibians on the Continental Divide: surveys on a transect from Montana to Colorado, USA. Alytes 22:85–94.

Cornell, T.J., K.A. Berven, and G.J. Gamboa. 1989. Kin recognition by tadpoles and froglets of the wood frog *Rana sylvatica*. Oecologia 78:312–316.

Corrington, J.D. 1929. Herpetology of the Columbia, South Carolina, region. Copeia (172):58–83.

Corse, W.A., and D.E. Metter. 1980. Economics, adult feeding and larval growth of *Rana catesbeiana* on a fish hatchery. Journal of Herpetology 14:231–238.

Corser, J.D. 2008. The Cumberland Plateau disjunct paradox and the biogeography and conservation of pond-breeding amphibians. American Midland Naturalist 159:498–503.

Cort, W.W. 1919. A new distome from *Rana aurora*. University of California Publications in Zoology 19:283–298.

Cortwright, S.A. 1998. Ten- to eleven-year population trends of two pond-breeding amphibians species, red-spotted newts and green frogs. Pp. 61–71 *In* M.J. Lannoo (ed.), Status & Distribution of Midwestern Amphibians. University of Iowa Press, Iowa City.

Cortwright, S.A., and C.E. Nelson. 1990. An examination of multiple factors affecting community structure in an aquatic amphibian community. Oecologia 83:123–131.

Cory, L., and J.J. Manion. 1953. Predation on eggs of the woodfrog *Rana sylvatica*, by leeches. Copeia 1953:66.

Cory, L., and J.J. Manion. 1955. Ecology and hybridization in the genus *Bufo* in the Michigan-Indiana region. Evolution 9:42–51.

Costanzo, J.P., M.F. Wright, and R.E. Lee Jr. 1992. Freeze tolerance as an overwintering adaptation in Cope's gray treefrog (*Hyla chrysoscelis*). Copeia 1992:565–569.

Coues, E., and H.C. Yarrow. 1878. Notes on the herpetology of Dakota and Montana. Bulletin of the United States Geological and Geographical Survey 4:259–291.

Counts, C.L. III, and R.W. Taylor. 1977. A xanthoma of indeterminate origin in *Bufo americanus* (Amphibia, Anura, Bufonidae). Journal of Herpetology 11:235–236.

Courtois, D., R. Leclair Jr., S. Lacasse, and P. Magnan. 1995. Habitats préférentiels d'amphibiens ranidés dans les lacs oligotrophes du Bouclier laurentien, Québec. Canadian Journal of Zoology 73:1744–1753.

Cousineau, M., and K. Rogers. 1991. Observations on sympatric *Rana pipiens*, *R. blairi*, and their hybrids in eastern Colorado. Journal of Herpetology 25:114–116.

Cowan, I.M. 1939. The vertebrate fauna of the Peace River District of British Columbia. Occasional Papers of the British Provincial Museum 1:1–102.

Cowan, I.M. 1941. Longevity of the red-legged frog. Copeia 1941:48.

Cowles, R.B. 1924. Notes regarding the breeding habits of *Scaphiopus hammondii* Baird. Pomona College Journal of Entomology and Zoology 16:108–110.

Cowles, R.B., and C.M. Bogert. 1936. The herpetology of the Boulder Dam region (Nev., Ariz., Utah). Herpetologica 1:33–42.

Cowman, D.F. 2005. Pesticides and amphibian declines in the Sierra Nevada Mountains, California. Ph.D. diss., Texas A&M University, College Station.

Cox, P. 1898. Batrachia of New Brunswick. Bulletin of the Natural History Society of New Brunswick 4:64–66.

Cox, P. 1899. The anoura of New Brunswick. Proceedings of the Natural History Association of Miramichi 1:9–19.

Cragin, F.W. 1881. A preliminary catalogue of Kansas

reptiles and batrachians. Transactions of the Kansas Academy of Science 7:112–120.

Crawford, J.A., D.B. Shepard, and C.A. Conner. 2009. Diet composition and overlap between recently metamorphosed *Rana areolata* and *Rana sphenocephala*: implications for a frog of conservation concern. Copeia 2009:642–646.

Crawshaw, G.J. 1997. Diseases in Canadian amphibian populations. SSAR Herpetological Conservation 1:258–270.

Crawshaw, L.I., R.N. Rausch, L.P. Wollmuth, and E.J. Bauer. 1992. Seasonal rhythms of development and temperature selection in larval bullfrogs, *Rana catesbeiana* Shaw. Physiological Zoology 65:346–359.

Crayon, J.J., and R.L. Hothem. 1998. *Xenopus laevis* (African Clawed Frog). Predation. Herpetological Review 29:165–166.

Creel, G.C. 1963. Bat as a food item of *Rana pipiens*. Texas Journal of Science 15:104–106.

Crenshaw, J.W., and W.F. Blair. 1959. Relationships in the *Pseudacris nigrita* complex in southwestern Georgia. Copeia 1959:215–222.

Crisafulli, C.M., L.S. Trippe, C.P. Hawkins, and J.A. MacMahon. 2005. Amphibian responses to the 1980 eruption of Mount St. Helens. Pp. 183–197 *In* V.H. Dale, F.J. Swanson, and C.M. Crisafulli (eds.), Ecological Responses to the 1980 Eruption of Mount St. Helens. Springer, New York.

Croes, S.A., and R.E. Thomas. 2000. Freeze tolerance and cryoprotectant synthesis of the Pacific tree frog *Hyla regilla*. Copeia 2000:863–868.

Cromer, R.B., J.D. Lanham, and H.H. Hanlin. 2002. Herpetofaunal response to gap and skidder-rut wetland creation in a southern bottomland hardwood forest. Forest Science 48:407–413.

Cronin, J.T., and J. Travis. 1986. Size-limited predation on larval *Rana areolata* (Anura: Ranidae) by two species of backswimmer (Insecta: Notonectidae). Herpetologica 42:171–174.

Crosby, C.R., and S.C. Bishop. 1925. A new genus and two new species of spiders collected by *Bufo quercicus*. Florida Entomologist 9:33–36.

Cross, K., and J.M. Hranitz. 1999. *Bufo americanus* (American Toad). Endoparasite. Herpetological Review 31:39.

Crossland, M.R., G.P. Brown, M. Anstis, C. Shilton, and R. Shine. 2008. Mass mortality of native anuran tadpoles in tropical Australia due to the invasive cane toad (*Bufo marinus*). Biological Conservation 141:2387–2394.

Crosswhite, E., and M. Wyman. 1920. [No title: figure and notes on an abnormal toad, presumably *A. boreas*]. Pomona College Journal of Entomology and Zoology 12:78.

Crother, B.I. (Committee Chair). 2008. Scientific and Standard English Names of Amphibians and Reptiles on North America North of Mexico, with Comments Regarding Confidence in Our Understanding. 6th ed. Society for the Study of Amphibians and Reptiles, Herpetological Circular No. 37.

Crouch, W.B., and P.W.C. Paton. 2000. Using egg-mass counts to monitor wood frog populations. Wildlife Society Bulletin 28:895–901.

Cruesere, F.M., and W.G. Whitford. 1976. Ecological relationships in a desert anuran community. Herpetologica 32:7–18.

Crump, D., M. Berrill, D. Coulson, D. Lean, L. McGillivray, and A. Smith. 1999. Sensitivity of amphibian embryos, tadpoles, and larvae to enhanced UV-β radiation in natural pond conditions. Canadian Journal of Zoology 77:1956–1966.

Crump, M.L. 1981a. Energy accumulation and amphibian metamorphosis. Oecologia 49:167–169.

Crump, M.L. 1981b. Intraclutch egg size variability in *Hyla crucifer* (Anura: Hylidae). Copeia 1981:302–308.

Cuellar, H.S. 1971. Levels of genetic compatibility of *Rana areolata* with southwestern members of the *Rana pipiens* complex (Anura: Ranidae). Evolution 25:399–409.

Cuellar, O. 1994. Ecological observations on *Rana pretiosa* in western Utah. Alytes 12:109–121.

Culley, D.D. Jr. 1973. Use of bullfrogs in biological research. American Zoologist 13:85–90.

Cummins, H. 1920. The role of voice and coloration in spring migration and sex recognition in frogs. Journal of Experimental Zoology 30:325–343.

Cunjak, R.A. 1986. Winter habitat of northern leopard frogs, *Rana pipiens*, in a southern Ontario stream. Canadian Journal of Zoology 64:255–257.

Cunningham, J.D. 1955a. A case of cannibalism in the toad *Bufo boreas*. Herpetologica 10:166.

Cunningham, J.D. 1955b. Observations on the ecology of the canyon treefrog, *Hyla californiae*. Herpetologica 20:55–61.

Cunningham, J.D. 1955c. Observations on the natural history of the California toad, *Bufo californicus* Camp. Herpetologica 17:255–260.

Cunningham, J.D. 1959. Reproduction and food of some California snakes. Herpetologica 15:17–19.

Cunningham, J.D., and D.P. Mullally. 1956. Thermal factors in the ecology of the Pacific treefrog. Herpetologica 12:68–79.

Cupp, P.V. Jr. 1980. Thermal tolerance of five salientian amphibians during development and metamorphosis. Herpetologica 36:234–244.

Currie, W., and E.D. Bellis. 1969. Home range and movements of the bullfrog, *Rana catesbeiana* Shaw, in an Ontario pond. Copeia 1969:688–692.

Cushman, S.A. 2006. Effects of habitat loss and fragmentation on amphibians: a review and prospectus. Biological Conservation 128:231–240.

Daigle, C. 1997. Distribution and abundance of the chorus frog *Pseudacris triseriata* in Québec. SSAR Herpetological Conservation 1:73–77.

Dailey, M.D., and S.R. Goldberg. 2000. *Langeronia burseyi* sp. n. (Trematoda: Lecithodendriidae) from the California treefrog, *Hyla cadaverina* (Anura: Hylidae), with revision of the genus *Langeronia* Caballero and Bravo-Hollis, 1949. Comparative Parasitology 67:165–168.

Dale, J.M., B. Freedman, and J. Kerekes. 1985. Acidity and associated water chemistry of amphibian habitats in Nova Scotia. Canadian Journal of Zoology 63:97–105.

Daniel, R., and B. Edmond. 2006. Atlas of Missouri amphibians and reptiles for 2005. http://atlas.moherp.org/pubs/atlas05.pdf.

Dapson, R.W., and L. Kaplan. 1975. Biological half-life and distribution of radiocesium in a contaminated population of green treefrogs *Hyla cinerea*. Oikos 26:39–42.

Da Silva, E.T., E. Pinheiro dos Reis, R.N. Feio, and O.P.R. Filho. 2009. Diet of the invasive frog *Lithobates catesbeianus* in Viçosa, Minas Gerais State, Brazil. South American Journal of Herpetology 4:286–294.

Da Silva, E.T., E. P. dos Reis, P.S. Santos, and R.N. Feio. 2010. *Lithobates catesbeianus* (American Bullfrog). Diet. Herpetological Review 41:475–476.

Da Silva, H.R. 1997. Two character states new for hylines and the taxonomy of the genus *Pseudacris*. Journal of Herpetology 31:609–613.

Daszak, P., A. Strieby, A.A. Cunningham, J.E. Longcore, C. Brown, and D. Porter. 2004. Experimental evidence that the bullfrog (*Rana catesbeiana*) is a potential carrier of chytridiomycosis, an emerging fungal disease of amphibians. Herpetological Journal 14:201–207.

Daszak, P., D.E. Scott, A.M. Kilpatrick, C. Faggioni, J.W. Gibbons, and D. Porter. 2005. Amphibian population declines at Savannah River Site are linked to climate, not chytridiomycosis. Ecology 86:3232–3237.

Datta, S., L. Hansen, L. McConnell, J. Baker, J. LeNoir, and J.N. Selber. 1998. Pesticides and PCB contaminants in fish and tadpoles from the Kaweah River Basin, California. Bulletin of Environmental Contamination and Toxicology 60:829–836.

Daudin, F.M. 1802. Histoire Naturelle des Rainettes, des Grenouilles, et des Crapauds. Bertrandet, Libraire Levrault, Paris.

Daughtery, C.H. 1979. Population ecology and genetics of *Ascaphus truei*: an examination of gene flow and natural selection. Ph.D. diss., University of Montana, Missoula.

Daugherty, C.H., and A.L. Sheldon. 1982a. Age-determination, growth, and life history of a Montana population of the tailed frog (*Ascaphus truei*). Herpetologica 38:461–468.

Daugherty, C.H., and A.L. Sheldon. 1982b. Age-specific movement patterns of the frog *Ascaphus truei*. Herpetologica 38:468–474.

Davidson, C. 2010. *Rana draytonii* (California red-legged frog). Prey. Herpetological Review 41:66.

Davidson, C., and R.A. Knapp. 2007. Multiple stressors and amphibian declines: dual impacts of pesticides and fish on yellow-legged frogs. Ecological Applications 17:587–597.

Davidson, C., H.B. Shaffer, and M.R. Jennings. 2001. Declines of the California red-legged frog: climate, UV-B, habitat, and pesticides hypotheses. Ecological Applications 11:464–479.

Davidson, C., H.B. Shaffer, and M.R. Jennings. 2002. Spatial tests of the pesticide drift, habitat destruction, UV-B, and climate-change hypotheses for California amphibian declines. Conservation Biology 16:1588–1601.

Davidson, C., M.F. Bernard, H.B. Shaffer, J.M. Parker, C. O'Leary, J.M. Conlon, and L.A. Rollins-Smith. 2007. Effects of chytrid and carbaryl exposure on survival, growth and skin peptide defenses in foothill yellow-legged frogs. Environmental Science and Technology 41:1771–1776.

Davis, A.B., and P.A. Verrell. 2005. Demography and reproductive ecology of the Columbia spotted frog (*Rana luteiventris*) across the Palouse. Canadian Journal of Zoology 83:702–711.

Davis, A.K., M.J. Yabsley, M.K. Keel, and J.C. Maerz. 2007. Discovery of a novel alveolate pathogen affecting southern leopard frogs in Georgia: description of the disease and host effects. EcoHealth 4:310–317.

Davis, B.J., and N. Hollenback. 1978. New records of the bird-voiced treefrog, *Hyla avivoca* (Hylidae), from Arkansas and Louisiana. Southwestern Naturalist 23:161–162.

Davis, D.D. 1933. Unusual behavior in a leopard frog. Copeia 1933:223–224.

Davis, J.G., and S.A. Menze. 2000. Ohio Frog and Toad Atlas. Ohio Biological Survey Miscellaneous Contribution No. 6.

Davis, J.G., P.J. Krusling, and J.W. Ferner. 1998. Status of amphibian populations in Hamilton County, Ohio. Pp. 155–165 *In* M.J. Lannoo (ed.), Status & Distribution of Midwestern Amphibians. University of Iowa Press, Iowa City.

Davis, L.C. 1973. The herpetofauna of Peccary Cave, Arkansas. M.S. thesis, University of Arkansas, Fayetteville.

Davis, M.S. 1980. Observations on the life history of the wood frog, *Rana sylvatica* LeConte, in Alabama. M.S. thesis, Auburn University, Auburn, AL.

Davis, M.S., and G.W. Folkerts. 1980. A beetle (*Eutheola rugiceps* LeC.: Scarabaeidae) penetrates the stomach wall of its predator (*Rana sylvatica* LeC.: Amphibia). Coleopterists Bulletin 34:396.

Davis, M.S., and T.R. Jones. 1982. Reassessment of the distribution of three amphibians in Alabama. Journal of the Alabama Academy of Science 53:10–16.

Davis, M.S., and G.W. Folkerts. 1986. Life history of the wood frog, *Rana sylvatica* LeConte (Amphibia: Ranidae), in Alabama. Brimleyana 12:29–50.

Davis, T.M., and P.T. Gregory. 2003. Decline and local extinction of the western toad, *Bufo boreas*, on southern Vancouver Island, British Columbia, Canada. Herpetological Review 34:350–352.

Davis, W.T. 1904. *Hyla andersonii* and *Rana virgatipes* at Lakehurst, New Jersey. American Naturalist 38:893.

Davis, W.T. 1907. Additional observations on *Hyla andersonii* and *Rana virgatipes* in New Jersey. American Naturalist 41:49–51.

Dawson, J.T. 1982. Kin recognition and schooling in the American toad (*Bufo americanus*). Ph.D. diss., State University of New York, Albany.

Dayton, G.H. 2000. *Gastrophryne olivacea* (Narrow-mouthed Toad). Vocalization. Herpetological Review 31:40.

Dayton, G.H., and R.E. Jung. 1999. *Scaphiopus couchii* (Couch's Spadefoot). Predation. Herpetological Review 30:164.

Dayton, G.H., and L.A. Fitzgerald. 2001. Competition, predation, and the distributions of four desert anurans. Oecologia 129:430–435.

Dayton, G.H., and S.D. Wapo. 2002. Cannibalistic behavior in *Scaphiopus couchii*: more evidence for larval anuran oophagy. Journal of Herpetology 36:531–532.

Dayton, G.H., and C.W. Painter. 2005. *Bufo speciosus* Girard, 1854. Texas Toad. Pp. 435–436 *In* M.J. Lannoo (ed.), Amphibian Declines. The Conservation Status of United States Species. University of California Press, Berkeley.

Dayton, G.H., and L.A. Fitzgerald. 2006. Habitat suitability models for desert amphibians. Biological Conservation 132:40–49.

Dayton, G.H., R.E. Jung, and S. Droege. 2004. Large-scale habitat associations of four desert anurans in Big Bend National Park, Texas. Journal of Herpetology 38:619–627.

Dayton, G.H., R. Skiles, and L. Dayton. 2007. Frogs & Toads of Big Bend National Park. Texas A&M University Press, College Station.

DeBenedictis, P.A. 1970. Interspecific competition between tadpoles of *Rana pipiens* and *Rana sylvatica*: an experimental field study. Ph.D. diss., University of Michigan, Ann Arbor.

DeBenedictis, P.A. 1974. Interspecific competition between larvae of *Rana pipiens* and *Rana sylvatica*: an experimental field study. Ecological Monographs 44:129–151.

Deckert, R.F. 1914. Further notes on the salientia of Jacksonville, Fla. Copeia (9):1–3.

Deckert, R.F. 1915. An albino pond frog. Copeia (24):53–54.

Deckert, R.F. 1922. Notes on Dade County salientia. Copeia (112):88.

Defauw, S.L., and P.J. English. 1994. Geographic distribution. *Scaphiopus holbrookii holbrookii* (Eastern Spadefoot). Herpetological Review 25:162.

Degenhardt, W.G., C.W. Painter, and A.H. Price. 1996. Amphibians & Reptiles of New Mexico. University of New Mexico Press, Albuquerque.

Degner, J.F., D.M. Silva, T.D. Hether, J.M. Daza, and E.A. Hoffman. 2010. Fat frogs, mobile genes: unexpected phylogeographic patterns for the ornate chorus frog (*Pseudacris ornata*). Molecular Ecology 19:2501–2515.

DeGraaf, J.D., and D.G. Nein. 2010. Predation of spotted turtle (*Clemmys guttata*) hatchling by green frog (*Rana clamitans*). Northeastern Naturalist 17:667–670.

DeGraaf, R.M., and D.D. Rudis. 1983. Amphibians and Reptiles of New England. Habitats and Natural History. University of Massachusetts Press, Amherst.

DeGraaf, R.M., and D.D. Rudis. 1990. Herpetofaunal species composition and relative abundance among three New England forest types. Forest Ecology and Management 32:155–165.

Deguise, I., and J.S. Richardson. 2009. Prevalence of the chytrid fungus (*Batrachochytrium dendrobatidis*) in western toads in southwestern British Columbia, Canada. Northwestern Naturalist 90:35–38.

Delis, P.R. 1993. Effects of urbanization on the community of anurans of a pine flatwood habitat in west central Florida. M.S. thesis, University of South Florida, Tampa.

Delis, P.R. 2001. *Hyla gratiosa* and *H. femoralis* (Anura: Hylidae) in west central Florida: a comparative study of rarity and commonness. Ph.D. diss., University of South Florida, Tampa.

Delis, P.R., H.R. Mushinsky, and E.D. McCoy. 1996. Decline of some west-central Florida anuran populations in response to habitat degradation. Biodiversity and Conservation 5:1579–1595.

Delvinquier, B.L.J., and S.S. Desser. 1996. Opalinidae (Sarcomastigophora) in North American amphibia: genus *Opalina* Purkinje and Valentin, 1835. Systematic Parasitology 33:33–51.

Delzell, D.E. 1958. Spatial movement and growth of *Hyla crucifer*. Ph.D. diss., University of Michigan, Ann Arbor.

deMaynadier, P.G., and M.L. Hunter Jr. 1998. Effects of silvicultural edges on the distribution and abundance of amphibians in Maine. Conservation Biology 12:340–352.

deMaynadier, P.G., and M.L. Hunter Jr. 1999. Forest canopy closure and juvenile emigration by pool-breeding amphibians in Maine. Journal of Wildlife Management 63:441–450.

deMaynadier, P.G., and M.L. Hunter Jr. 2000. Road effects on amphibian movements in a forested landscape. Natural Areas Journal 20:56–65.

Demlong, M.J. 1997. Head-starting *Rana subaquavocalis* in captivity. Reptiles Magazine (January): 24–28, 30, 32–33.

Denman, N.S., and L. Denman. 1985. Geographic distribution. *Bufo americanus copei* (Hudson Bay Toad). Herpetological Review 16:114.

DeNoyelles, F., W.D. Kettle, C.H. Fromm, M.F. Moffett, and S.L. Dewey. 1989. Use of experimental ponds to assess the effects of a pesticide on the aquatic environment. Pp. 41–56 *In* J.R. Voshell (ed.), Using Mesocosms to Assess the Aquatic Ecological Risk of Pesticides: Theory and Practice. Entomological Society of America, Lanham, MD.

Denver, R.J., N. Mirhadi, and M. Phillips. 1998. Adaptive plasticity in amphibian metamorphosis: response of *Scaphiopus hammondii* tadpoles to habitat desiccation. Ecology 79:1859–1872.

Dernehl, P.H. 1902. Place-modes of *Acris Gryllus* for Madison, Wis. Bulletin of the Wisconsin Natural History Society 2:75–83.

de Solla, S.R., C.A. Bishop, K.E. Pettit, and J.E. Elliott. 2002a. Organochlorine pesticides and polychlorinated biphenyls (PCBs) in eggs of red-legged frogs (*Rana aurora*) and northwestern salamanders (*Ambystoma gracile*) in an agricultural landscape. Chemosphere 46:1027–1032.

de Solla, S.R., K.E. Pettit, C.A. Bishop, K.M. Cheng, and J.E. Elliott. 2002b. Effects of agricultural runoff on native amphibians in the lower Fraser River valley, British Columbia, Canada. Environmental Toxicology and Chemistry 21:353–360.

de Solla, S.R., L.J. Shirose, K.J. Fernie, G.C. Barrett, C.S. Brousseau, and C.A. Bishop. 2005. Effect of sampling effort and species detectability on volunteer based anuran monitoring programs. Biological Conservation 121:585–594.

Desroches, J.-F., and D. Rodrigue. 2004. Amphibiens et Reptiles du Québec et des Maritimes. Éditions Michel Quintin, Waterloo, Québec.

Dessauer, H.C., and E. Nevo. 1969. Geographic variation of blood and liver proteins in cricket frogs. Biochemical Genetics 3:171–188.

Desser, S., and J.R. Barta. 1984a. An intraerythrocytic virus and rickettsia of frogs from Algonquin Park, Ontario. Canadian Journal of Zoology 62:1521–1524.

Desser, S., and J.R. Barta. 1984b. *Thrombocytozoons ranarum* Tchacarof 1863, a prokaryotic parasite in thrombocytes of the mink frog *Rana septentrionalis* in Ontario. Journal of Parasitology 70:454–456.

Desser, S., and S. Jones. 1985. *Hexamita intestinalis* Dujardin in the blood of frogs from southern and central Ontario. Journal of Parasitology 71:841.

Desser, S., J. Lom, and I. Dykova. 1986. Developmental stages of *Sphaerospora ohlmacheri* (Whinery, 1893) n. comb. (Myxozoa: Myxosporea) in the renal tubules of bullfrog tadpoles, *Rana catesbeiana* from Lake of Two Rivers, Algonquin Park, Ontario. Canadian Journal of Zoology 64:2344–2347.

Desser, S., M.E. Siddall, and J.R. Barta. 1989. Ultrastructural observations of the developmental stages of *Lankesterella minima* (Apicomplexa) in experimentally infected *Rana catesbeiana* tadpoles. Journal of Parasitology 76:97–103.

Desser, S., H. Hong, M.E. Siddall, and J.R. Barta. 1993. An ultrastructural study of *Brugerolleia algonquinensis* gen. nov., sp. nov. (Diplomonadina: Diplomonadida), a flagellate parasite in the blood of frogs from Ontario, Canada. European Journal of Protistology 29:72–80.

DeVito, J., D.P. Chivers, J.M. Kiesecker, A. Marco, E.L. Wildy, and A.R. Blaustein. 1998. The effects of snake predation on metamorphosis of western toads, *Bufo boreas* (Amphibia, Bufonidae). Ethology 104:185–193.

DeVito, J., D.P. Chivers, J.M. Kiesecker, L.K. Belden, and A.R. Blaustein. 1999. Effects of snake predation on aggregation and metamorphosis of Pacific treefrog (*Hyla regilla*) larvae. Journal of Herpetology 33:504–507.

de Vlaming, V.L., and R.B. Bury. 1970. Thermal selection in tadpoles of the tailed frog, *Ascaphus truei*. Journal of Herpetology 4:179–189.

Diakow, C. 1977. Initiation and inhibition of the release croak of *Rana pipiens*. Physiology & Behavior 19:607–610.

Dial, N.A., and C.A. Bauer. 1984. Teratogenic and lethal effects of paraquat on developing frog embryos (*Rana pipiens*). Bulletin of Environmental Contamination and Toxicology 33:592–597.

Dial, N.A., and C.A. Bauer Dial. 1987. Lethal effects of diquat and paraquat on developing frog embryos and 15-day-old tadpoles, *Rana pipiens*. Bulletin of Environmental Contamination and Toxicolgy 38:1006–1011.

Díaz de Pascual, A., and C. Guerrero. 2008. Diet composition of bullfrogs, *Rana catesbeiana* (Anura: Ranidae) introduced into the Venezuelan Andes. Herpetological Review 39:425–427.

Dibble, C.J., J.E. Kauffman, E.M. Zuzik, G.R. Smith, and J.E. Rettig. 2009. Effects of potential predator and competitor cues and sibship on wood frog (*Rana sylvatica*) embryos. Amphibia-Reptilia 30: 294–298.

Di Berardino, M.A. 1962. The karyotype of *Rana pipiens* and investigation of its stability during embryonic differentiation. Developmental Biology 5:101–126.

Di Candia, M.R., and E.J. Routman. 2007. Cytonuclear discordance across a leopard frog contact zone. Molecular Phylogenetics and Evolution 45:564–575.

Dickerson, M.C. 1906. The Frog Book. North American Toads and Frogs with a Study of the Habits

and Life Histories of those of the Northeastern States. Doubleday, Page & Co., New York.
Dickey, L.B. 1921. A new amphibian cestode. Journal of Parasitology 7:129–137.
Dickman, M. 1968. The effect of grazing by tadpoles on the structure of a periphyton community. Ecology 49:1188–1190.
Dickson, N.J. 2002. The natural history and possible extirpation of Blanchard's cricket frog, *Acris crepitans blanchardi*, in West Virginia. M.S. thesis, Marshall University, Huntington.
Didiuk, A. 1997. Status of amphibians in Saskatchewan. SSAR Herpetological Conservation 1:110–116.
Diener, R.A. 1965. The occurrence of tadpoles of the green treefrog, *Hyla cinerea cinerea* (Schneider), in Trinity Bay, Texas. British Journal of Herpetology 3:198–199.
Dill, L.M. 1977. "Handedness" in the Pacific tree frog (*Hyla regilla*). Canadian Journal of Zoology 55:1926–1929.
Diller, L.V., and R.L. Wallace. 1999. Distribution and habitat of *Ascaphus truei* in streams on managed, young growth forests in north coastal California. Journal of Herpetology 33:71–79.
DiMauro, D., and M.L. Hunter Jr. 2002. Reproduction of amphibians in natural and anthropogenic temporary pools in managed forests. Forest Science 48:397–406.
Dimmitt, M.A., and R. Ruibal. 1980a. Exploitation of food resources by spadefoot toads (*Scaphiopus*). Copeia 1980:854–862.
Dimmitt, M.A., and R. Ruibal. 1980b. Environmental correlates of emergence in spadefoot toads (*Scaphiopus*). Journal of Herpetology 14:21–29.
Dinehart, S.K., L.M. Smith, S.T. McMurry, T.A. Anderson, P.N. Smith, and D.A. Haukos. 2009. Toxicity of a glufosinate- and several glyphosate-based herbicides to juvenile amphibians from the Southern High Plains, USA. Science of the Total Environment 407:1065–1071.
Dinsmore, A. 2004. Geographic distribution. *Eleutherodactylus planirostris*. Herpetological Review 35:403.
Dirig, R. 1978. A large diurnal breeding assemblage of American toads, *Bufo americanus americanus*, in the Catskill Mountains. Pitch Pine Naturalist 4:1–2.
Distel, C.A., and M.D. Boone. 2009. Effects of aquatic exposure to the insecticide carbaryl and density on aquatic and terrestrial growth and survival in American toads. Environmental Toxicology and Chemistry 28:1963–1969.
Distel, C.A., and M.D. Boone. 2010. Effects of aquatic exposure to the insecticide carbaryl are species-specific across life stages and mediated by heterospecific competitors in anurans. Functional Ecology 24:1342–1352.
Distel, C.A., and M.D. Boone. 2011. Pesticide has asymmetric effects on two tadpole species across density gradient. Environmental Toxicology and Chemistry 30:650–658.
Dixon, J.R. 2000. Amphibians and Reptiles of Texas, with Keys, Taxonomic Synopses, Bibliography, and Distribution Maps. 2nd ed. Texas A & M University Press, College Station.
Dodd, C.K. Jr. 1977. Immobility in juvenile *Bufo woodhousei fowleri*. Journal of the Mississippi Academy of Sciences 22:90–94.
Dodd, C.K. Jr. 1979. A photographic technique to study tadpole populations. Brimleyana 2:131–136.
Dodd, C.K. Jr. 1991. Drift fence-associated sampling bias of amphibians at a Florida sandhills temporary pond. Journal of Herpetology 25:296–301.
Dodd, C.K. Jr. 1992. Biological diversity of a temporary pond herpetofauna in north Florida sandhills. Biodiversity and Conservation 1:125–142.
Dodd, C.K. Jr. 1994. The effects of drought on population structure, activity, and orientation of toads (*Bufo quercicus* and *B. terrestris*) at a temporary pond. Ethology Ecology & Evolution 6:331–349.
Dodd, C.K. Jr. 1995. The ecology of a sandhills population of the eastern narrow-mouthed toad, *Gastrophryne carolinensis*, during a drought. Bulletin of the Florida Museum of Natural History 38:11–41.
Dodd, C.K. Jr. 1996. Use of terrestrial habitats by amphibians in the sandhill uplands of north-central Florida. Alytes 14:42–52.
Dodd, C.K. Jr. 2004. The Amphibians of Great Smoky Mountains National Park. University of Tennessee Press, Knoxville.
Dodd, C.K. Jr. 2009. Amphibian Ecology and Conservation. A Handbook of Techniques. Oxford University Press, Oxford.
Dodd, C.K. Jr., and P.V. Cupp. 1978. The effect of temperature and stage of development on the duration of immobility in selected anurans. British Journal of Herpetology 5:783–788.
Dodd, C.K. Jr., and R.A. Seigel. 1991. Relocation, repatriation and translocation of amphibians and reptiles: are they conservation strategies that work? Herpetologica 47:335–350.
Dodd, C.K. Jr., and B.S. Cade. 1998. Movement patterns and the conservation of amphibians breeding in small, temporary wetlands. Conservation Biology 12:331–339.
Dodd, C.K. Jr., and L.L. Smith. 2003. Habitat destruction and alteration: historical trends and future prospects for amphibians. Pp. 94–112 *In* R.D. Semlitsch (ed.), Amphibian Conservation. Smithsonian Books, Washington, DC.
Dodd, C.K. Jr., and W.J. Barichivich. 2007. Establishing a baseline and faunal history in amphibian monitoring programs: the amphibians of Harris Neck, GA. Southeastern Naturalist 6:125–134.

Dodd, C.K. Jr., M.L. Griffey, and J.D. Corser. 2001. The cave associated amphibians of Great Smoky Mountains National Park: review and monitoring. Journal of the Elisha Mitchell Scientific Society 117:139–149.

Dodd, C.K. Jr., W.J. Barichivich, and L.L. Smith. 2004. Effectiveness of a barrier wall and culverts in reducing wildlife mortality on a heavily traveled highway in Florida. Biological Conservation 118:619–631.

Dodd, C.K. Jr., W.J. Barichivich, S.A. Johnson, and J.S. Staiger. 2007. Changes in a northwestern Florida Gulf Coast herpetofaunal community over a 28-y period. American Midland Naturalist 158:29–48.

Doherty, J.A., and H.C. Gerhardt. 1984. Evolutionary and neurobiological implications of selective phonotaxis in the spring peeper (*Hyla crucifer*). Animal Behaviour 32:875–881.

Dole, J.W. 1965a. Summer movements of adult leopard frogs, *Rana pipiens* Schreber, in northern Michigan. Ecology 46:236–255.

Dole, J.W. 1965b. Spatial relations in natural populations of the leopard frog, *Rana pipiens* Schreber, in northern Michigan. American Midland Naturalist 74:464–478.

Dole, J.W. 1967a. The role of substrate moisture and dew in the water economy of leopard frogs, *Rana pipiens*. Copeia 1967:141–149.

Dole, J.W. 1967b. Spring movements of leopard frogs, *Rana pipiens* Schreber, in northern Michigan. American Midland Naturalist 78:167–181.

Dole, J.W. 1968. Homing in leopard frogs, *Rana pipiens*. Ecology 49:386–399.

Dole, J.W. 1971. Dispersal of recently metamorphosed leopard frogs, *Rana pipiens*. Copeia 1971:221–228.

Dole, J.W. 1972a. Homing and orientation of displaced toads, *Bufo americanus*, to their home sites. Copeia 1972:151–158.

Dole, J.W. 1972b. The role of olfaction and audition in the orientation of leopard frogs, *Rana pipiens*. Herpetologica 28:258–260.

Dole, J.W., B.B. Rose, and K.H. Tachiki. 1981. Western toads (*Bufo boreas*) learn odor of prey insects. Herpetologica 37:63–68.

Donnelly, M.A., C.J. Farrell, M.J. Baber, and J.L. Glenn. 2001. The amphibians and reptiles of the Kissimmee River. 1. Patterns of abundance and occurrence in altered floodplain habitats. Herpetological Natural History 8:161–170.

Doody, J.S., and J.E. Young. 1995. Temporal variation in reproduction and clutch mortality of leopard frogs (*Rana utricularia*) in south Mississippi. Journal of Herpetology 29:614–616.

Dorcas, M.E., S.J. Price, J.C. Beane, and S.C. Owen. 2007. The Frogs and Toads of North Carolina. Field Guide and Recorded Calls. North Carolina Wildlife Resources Commission, Raleigh, NC.

Doty, T.L. 1978. A study of larval amphibian population dynamics in a Rhode Island vernal pond. Ph.D. diss., University of Rhode Island, Kingston.

Doubledee, R.A., E.B. Miller, and R.M. Nisbet. 2003. Bullfrogs, disturbance regimes, and the persistence of California red-legged frogs. Journal of Wildlife Management 67:424–438.

Dougherty, C.K., and G.R. Smith. 2006. Acute effects of road de-icers on the tadpoles of three anurans. Applied Herpetology 3:87–93.

Douglas, C.L. 1966. Amphibians and reptiles of Mesa Verde National Park, Colorado. University of Kansas Publications, Museum of Natural History 15:711–744.

Dowe, B.J. 1979. The effect of time of oviposition and microenvironment on growth of larval bullfrogs (*Rana catesbeiana*) in Arizona. M.S. thesis, Arizona State University, Tempe.

Dowler, R.C., J.K. McCoy, and L.J. Fohn. 2010. *Scaphiopus couchii* (Couch's Spadefoot). Predation. Herpetological Review 41:480.

Drake, C.J. 1914. The food of *Rana pipiens* Shreber. The Ohio Naturalist 14:257–269.

Drake, D.L. 2010. *Lithobates onca* (Relict Leopard Frog). Cannibalistic oophagy. Herpetological Review 41:198–199.

Drake, D.L., and S.E. Trauth. 2010. *Spea intermontana* (Great Basin Spadefoot). Algal symbiosis. Herpetological Review 41:481–482.

Drake, D.L., A. Drayer, and S.E. Trauth. 2007. *Bufo americanus* (American Toad). Algal symbiosis. Herpetological Review 38:435–436.

Dreitz, V.J. 2006. Issues in species recovery: an example based on the Wyoming toad. BioScience 56:765–771.

Driver, E.C. 1936. Observations on *Scaphiopus holbrooki* (Harlan). Copeia 1936:67–69.

Drost, C.A., and G.M. Fellers. 1996. Collapse of a regional frog fauna in the Yosemite area of the California Sierra Nevada. Conservation Biology 10:414–425.

Duarte, A., D.J. Brown, and M.R.J. Forstner. 2011. Estimating abundance of the endangered Houston toad on a primary recovery site. Journal of Fish and Wildlife Management 2:207–215.

Dubois, A. 2006. Naming taxa from cladograms: a cautionary tale. Molecular Phylogenetics and Evolution 42:317–330.

DuBois, R.B., and F.M. Stoll. 1995. Downstream movement of leopard frogs in a Lake Superior tributary exemplifies the concept of a lotic macrodrift. Journal of Freshwater Ecology 10:135–139.

Duellman, W.E. 1951. Notes on the reptiles and amphibians of Greene County, Ohio. Ohio Journal of Science 51:335–341.

Duellman, W.E. 1955. Systematic status of the Key West spadefoot toad, *Scaphiopus holbrookii albus*. Copeia 1955:141–143.

Duellman, W.E. 2001. Hylid Frogs of Middle America. Vol. 1. Society for the Study of Amphibians and Reptiles, Ithaca, NY.

Duellman, W.E., and A. Schwartz. 1958. Amphibians and reptiles of southern Florida. Bulletin of the Florida State Museum, Biological Sciences 3:181–324.

Duffus, A.L.J., B.D. Pauli, K. Wozney, C.R. Brunetti, and M. Berrill. 2008. Frog Virus 3-like infections in aquatic amphibian communities. Journal of Wildlife Diseases 44:109–120.

Dumas, P.C. 1957. *Rana sylvatica* Le Conte in Idaho. Copeia 1957:150–151.

Dumas, P.C. 1966. Studies of the *Rana* species complex in the Pacific Northwest. Copeia 1966:60–74.

Duméril, A.M.C., and G. Bibron. 1841. Erpetologie General, ou Histoire Naturelle Complete des Reptiles, Vol. 8. Roret, Paris.

Dundee, H.A. 1974. Recognition characters for *Rana grylio*. Journal of Herpetology 8:275–276.

Dundee, H.A. 1999. *Gastrophryne olivacea* (Great Plains Narrowmouth Toad). Aggregation with tarantula. Herpetological Review 30:91–92.

Dundee, H.A., and D.A. Rossman. 1989. The Amphibians and Reptiles of Louisiana. Louisiana State University Press, Baton Rouge.

Dunlap, D.G. 1955. Inter- and intraspecific variation in Oregon frogs of the genus *Rana*. American Midland Naturalist 54:314–331.

Dunlap, D.G. 1963. The status if [sic] the gray treefrog, *Hyla versicolor* in South Dakota. Proceedings of the South Dakota Academy of Science 42:136–139.

Dunlap, D.G. 1967. Selected records of amphibians and reptiles from South Dakota. Proceedings of the South Dakota Academy of Science 46:100–106.

Dunlap, D.G. 1977. Wood and western spotted frogs (Amphibia, Anura, Ranidae) in the Big Horn Mountains of Wyoming. Journal of Herpetology 11:85–87.

Dunlap, D.G., and K.C. Kruse. 1976. Frogs of the *Rana pipiens* complex in the northern and central Plains states. Southwestern Naturalist 20:559–571.

Dunlap, D.G., and J.E. Platz. 1981. Geographic variation of proteins and call in *Rana pipiens* from the northcentral United States. Copeia 1981:876–879.

Dunlap, D.G., and C.K. Satterfield. 1985. Habitat selection in larval anurans: early experience and substrate pattern selection in *Rana pipiens*. Developmental Psychobiology 18:37–58.

Dunn, E.R. 1935. The survival value of specific characters. Copeia 1935:85–98.

Dunn, E.R. 1937. The status of *Hyla evittata* Miller. Transactions of the Biological Society of Washington 50:9–10.

du Preez, L., and V. Carruthers. 2009. A Complete Guide to the Frogs of Southern Africa. Struik Nature, Cape Town, South Africa.

Dupuis, L., and D. Steventon. 1999. Riparian management and the tailed frog in northern coastal forests. Forest Ecology and Management 124:35–43.

Dupuis, L., and P. Friele. 2006. The distribution of the Rocky Mountain tailed frog (*Ascaphus montanus*) in relation to the fluvial system: implications for management and conservation. Ecological Research 21:489–502.

Durham, L., and G.W. Bennett. 1963. Age, growth, and homing in the bullfrog. Journal of Wildlife Management 27:107–123.

Dusi, J.L. 1949. The natural occurrence of "red-leg," *Pseudomonas hydrophila*, in a population of American toads, *Bufo americanus*. Ohio Journal of Science 49:70–71.

Dyche, L.L. 1914. Ponds, pond fish, and pond fish culture. Kansas Department of Fish and Game, Bulletin 1.

Dyer, W.G. 1991. Helminth parasites of amphibians from Illinois and adjacent Midwestern states. Transactions of the Illinois State Academy of Science 84:125–143.

Dyrkacz, S. 1981. Recent instances of albinism in North American amphibians and reptiles. SSAR Herpetological Circular No. 11.

Earl, J.E., and H.H. Whiteman. 2009. Effects of pulsed nitrate exposure on amphibian development. Environmental Toxicology and Chemistry 28:1331–1337.

Earl, J.E., and H.H. Whiteman. 2010. Evaluation of phosphate toxicity in Cope's Gray Treefrog (*Hyla chrysoscelis*) tadpoles. Journal of Herpetology 44:201–208.

Eason, G.W. Jr., and J.E. Fauth. 2001. Ecological correlates of anuran species richness in temporary pools: a field study in South Carolina, USA. Israel Journal of Zoology 47:347–365.

Easteal, S. 1981. The history of introductions of *Bufo marinus* (Amphibia: Anura): a natural experiment in evolution. Biological Journal of the Linnaean Society 16:93–113.

Eaton, B.R., and C.A. Paszkowski. 1999. *Rana sylvatica* (Wood Frog). Predation. Herpetological Review 30:164.

Eaton, B.R., and Z.C. Eaton. 2001. An observation of a mallard, *Anas platyrhynchos*, feeding on a wood frog, *Rana sylvatica*. Canadian Field-Naturalist 115:499–500.

Eaton, B.R., C. Grekul, and C.A. Paszkowski. 1999. An observation of interspecific amplexus between boreal, *Bufo boreas*, and Canadian, *Bufo hemiophrys*, toads with a range extension for boreal toad in central Alberta. Canadian Field-Naturalist 113:512–513.

Eaton, B.R., C.A. Paszkowski, and R. Chapman. 2003. *Rana sylvatica* (Wood Frog). Parasite. Herpetological Review 34:55.

Eaton, B.R., S. Eaves, C. Stevens, A. Puchniak, and

C.A. Paszkowski. 2004. Deformity levels in wild populations of the wood frog (*Rana sylvatica*) in three ecoregions of western Canada. Journal of Herpetology 38:283–287.

Eaton, B.R., W.W. Tonn, C.A. Paszkowski, A.J. Danylchuk, and S.M. Boss. 2005a. Indirect effects of fish winterkills on amphibian populations in boreal lakes. Canadian Journal of Zoology 83:1532–1539.

Eaton, B.R., C.A. Paszkowski, K. Kristensen, and M. Hiltz. 2005b. Life-history variation among populations of Canadian toads in Alberta, Canada. Canadian Journal of Zoology 83:1421–1430.

Eaton, B.R., A.E. Moenting, C.A. Paszkowski, and D. Shpeley. 2008. Myiasis by *Lucilla silvarum* (Calliphoridae) in amphibian species in boreal Alberta, Canada. Journal of Parasitology 94:949–952.

Eaton, T.H. Jr. 1935. Report on amphibians and reptiles of the Navajo country. Rainbow Bridge-Monument Valley Expedition Bulletin No. 3, 18 pp.

Eaton, T.H. Jr., and R.M. Imagawa. 1948. Early development of *Pseudacris clarkii*. Copeia 1948: 263–265.

Echaubard, P., K. Little, B. Pauli, and D. Lesbarrères. 2010. Context-dependent effects of ranaviral infection on northern leopard frog life history traits. PLoS One 5(10):e13723.

Echternacht, A.C. 1964. Albino specimens from Arizona. Herpetologica 20:211–212.

Eddy, S.B. 1976. Population ecology of the leopard frog, *Rana pipiens pipiens* Schreber, at Delta Marsh, Manitoba. M.S. thesis, University of Manitoba, Winnipeg.

Edginton, A.N., P.M. Sheridan, G.R. Stephenson, D.G. Thompson, and H.J. Boermans. 2004. Comparative effects of pH and Vision™ herbicide on two life stages of four anuran amphibian species. Environmental Toxicology and Chemistry 23: 815–822.

Edgren, R.A. Jr. 1944. Notes on amphibians and reptiles from Wisconsin. American Midland Naturalist 32:495–498.

Edgren, R.A. 1954. Factors controlling color change in the tree frog, *Hyla versicolor* Wied. Proceedings of the Society for Experimental Biology and Medicine 87:20–23.

Edwards, J.R., K.L. Koster, and D.L. Swanson. 2000. Time course for cryoprotectant synthesis in the freeze-tolerant chorus frog, *Pseudacris triseriata*. Comparative Biochemistry and Physiology 125A:367–375.

Edwards, T.M., K.A. McCoy, T. Barbeau, M.W. McCoy, J.M. Thro, and L.J. Guillette Jr. 2006. Environmental context determines nitrate toxicity in southern toad (*Bufo terrestris*) tadpoles. Aquatic Toxicology 78:50–58.

Egan, R.S. 2001. Within-pond and landscape-level factors influencing the breeding effort of *Rana sylvatica* and *Ambystoma maculatum*. M.S. thesis, University of Rhode Island, Kingston.

Egan, R.S., and P.W.C. Paton. 2004. Within-pond parameters affecting oviposition by wood frogs and spotted salamanders. Wetlands 24:1–13.

Egan, R.S., and P.W.C. Paton. 2008. Multiple scale habitat characteristics of pond-breeding amphibians across a rural-urban gradient. Pp. 53–65 *In* J.C. Mitchell, R.E. Jung Brown, and B. Bartholomew (eds.), Urban Herpetology. Herpetological Conservation 3. Society for the Study of Amphibians and Reptiles, Salt Lake City, UT.

Ehrlich, D. 1979. Predation by bullfrog tadpoles (*Rana catesbeiana*) on eggs and newly hatched larvae of the Plains leopard frog (*Rana blairi*). Bulletin of the Maryland Herpetological Society 15: 25–26.

Eigenbrod, F., S.J. Hecnar, and L. Fahrig. 2008. The relative effects of road traffic and forest cover on anuran populations. Biological Conservation 141: 35–46.

Eigenbrod, F., S.J. Hecnar, and L. Fahrig. 2009. Quantifying the road-effect zone: threshold effects of a motorway on anuran populations in Ontario, Canada. Ecology and Society 14(1):24. http://www.ecologyandsociety.org/Vol14/iss1/art24/.

Einem, G.E., and L.D. Ober. 1956. The seasonal behavior of certain Floridian Salientia. Herpetologica 12:205–212.

Eklöv, P. 2000. Chemical cues from multiple predator-prey interactions induce changes in behavior and growth of anuran larvae. Oecologia 123:192–199.

Elinson, R.P. 1974. A block to cross-fertilization located in the egg jelly of the frog *Rana clamitans*. Journal of Embryology and Experimental Morphology 32:325–335.

Elinson, R.P. 1975a. Viable amphibian hybrids produced by circumventing a block to cross-fertilization (*Rana clamitans* ♀ × *Rana catesbeiana* ♂). Journal of Experimental Zoology 192:323–329.

Elinson, R.P. 1975b. Fertilization of green frog (*Rana clamitans*) eggs in their native jelly by bullfrog (*Rana catesbeiana*) sperm. Journal of Experimental Zoology 193:419–423.

Elliott, L., C. Gerhardt, and C. Davidson. 2009. The Frogs and Toads of North America. Houghton Mifflin Harcourt, Boston.

Ellis, M.M., and J. Henderson. 1913. Amphibia and reptilia of Colorado. Part 1. University of Colorado Studies 10:39–129.

Ellis, M.M., and J. Henderson. 1915. Amphibia and reptilia of Colorado. Part 2. University of Colorado Studies 11:253–263.

Ely, C. 1944. Development of *Bufo marinus* larvae in dilute sea water. Copeia 1944:256.

Emery, A.R., A.H. Berst, and K. Kodaira. 1972. Under-ice observations of wintering sites of leopard frogs. Copeia 1972:123–126.

Emlen, S.T. 1968. Territoriality in the bullfrog, *Rana catesbeiana*. Copeia 1968:240–243.

Emlen, S.T. 1976. Lek organization and mating strategies in the bullfrog. Behavioral Ecology and Sociobiology 1:283–313.

Emlen, S.T. 1977. "Double-clutching" and its possible significance in the bullfrog. Copeia 1977:749–751.

Emmett, A.D., and F.P. Allen. 1919. Nutritional studies on the growth of frog larvae (*Rana pipiens*). Journal of Biological Chemistry 38:325–344.

Enderson, E.F., and R.L. Bezy. 2000. Geographic distribution: *Pternohyla fodiens* (Lowland burrowing treefrog). Herpetological Review 31:251–252.

Engbrecht, N.J., and J.L. Heemeyer. 2010. *Lithobates areolatus circulosus* (Northern Crawfish Frog). *Heterodon platirhinos* (Eastern Hog-nosed Snake). Predation. Herpetological Review 41:168–170.

Engbrecht, N.J., and M.J. Lannoo. 2010. A review of the status and distribution of crawfish frogs (*Lithobates areolatus*) in Indiana. Proceedings of the Indiana Academy of Science 119:64–73.

Enge, K.M. 1998a. Herpetofaunal survey of an upland hardwood forest in Gadsden County, Florida. Florida Scientist 61:141–159.

Enge, K.M. 1998b. Herpetofaunal drift-fence survey of steephead ravines in 2 river drainages. Proceedings of the Annual Conference of Southeastern Association of Fish and Wildlife Agencies 1998:336–348.

Enge, K.M. 2002. Herpetofaunal drift-fence survey of two seepage bogs in Okaloosa County, Florida. Florida Scientist 65:67–82.

Enge, K.M. 2005a. Commercial harvest of amphibians and reptiles in Florida for the pet trade. Pp. 198–211 *In* W.E. Meshaka Jr. and K.J. Babbitt (eds.), Amphibians and Reptiles. Status and Conservation in Florida. Krieger Publishing, Malabar, FL.

Enge, K.M. 2005b. Herpetofaunal drift-fence surveys of steephead ravines in the Florida Panhandle. Southeastern Naturalist 4:657–678.

Enge, K.M., and W.R. Marion. 1986. Effects of clearcutting and site preparation on herpetofauna of a north Florida flatwoods. Forest Ecology and Management 14:177–192.

Enge, K.M., and K.N. Wood. 1998. Herpetofaunal surveys of the Big Bend Wildlife Management Area, Taylor County, Florida. Florida Scientist 61:61–87.

Enge, K.M., and K.N. Wood. 2000. A herpetofaunal survey of Chassahowitzka Wildlife Management Area, Hernando County, Florida. Herpetological Natural History 7:117–144.

Enge, K.M., and K.N. Wood. 2001. Herpetofauna of Chinsegut Nature Center, Hernando County, Florida. Florida Scientist 64:283–305.

Enge, K.M., D.T. Cobb, G.L. Sprandel, and D.L. Francis. 1996. Wildlife captures in a pipeline trench in Gadsden County, Florida. Florida Scientist 59:1–11.

Engels, W.L. 1942. Vertebrate fauna of North Carolina coastal islands. A study in the dynamics of animal distribution. 1. Ocracoke Island. American Midland Naturalist 28:273–304.

Engels, W.L. 1952. Vertebrate fauna of North Carolina coastal islands. 2. Shackleford Banks. American Midland Naturalist 47:702–742.

Engeman, R.M., and E.M. Engeman. 1996. Longevity of Woodhouse's toad in Colorado. Northwestern Naturalist 77:23.

Ernst, C.H. 2001. Some ecological parameters of the wood turtle, *Clemmys insculpta*, in southeastern Pennsylvania. Chelonian Conservation and Biology 4:94–99.

Ernst, J.A. 2001. Assessment of the *Rana pipiens* complex in southwestern South Dakota. M.S. thesis, University of Wisconsin–Stevens Point.

Ervin, E.L. 2005. *Pseudacris cadaverina* (Cope, 1866). California Treefrog. Pp. 467–470 *In* M.J. Lannoo (ed.), Amphibian Declines. The Conservation Status of United States Species. University of California Press, Berkeley.

Ervin, E.L., and R.N. Fisher. 2001. *Thamnophis hammondii* (Two-striped Garter Snake). Prey. Herpetological Review 32:265–266.

Ervin, E.L., and T.L. Cass. 2007. *Spea hammondii* (Western Spadefoot). Reproductive pattern. Herpetological Review 38:196–197.

Ervin, E.L., and R.N. Fisher. 2007. *Thamnophis hammondii* (Two-striped Gartersnake). Foraging behavior. Herpetological Review 38:345–346.

Ervin, E.L., R.N. Fisher, and K. Madden. 2000. *Hyla cadaverina* (California Treefrog). Predation. Herpetological Review 31:234.

Ervin, E.L., A.E. Anderson, T.L. Cas, and R.E. Murcia. 2001. *Spea hammondii* (Western Spadefoot Toad). Elevation record. Herpetological Review 32:36.

Ervin, E.L., S.J. Mullin, M.L. Warburton, and R.N. Fisher. 2003. *Thamnophis hammondii* (Two-striped Garter Snake). Prey. Herpetological Review 34:74–75.

Ervin, E.L., C.D. Smith, and S.V. Christopher. 2005. *Spea hammondii* (Western Spadefoot). Reproduction. Herpetological Review 36:309–310.

Ervin, E.L., D.A. Kisner, and R.N. Fisher. 2006. *Bufo californicus* (Arroyo Toad). Mortality. Herpetological Review 37:199.

Ervin, E.L., A.K. Gonzales, D.S. Johnston, and J.P. Pittman. 2007. *Spea hammondii* (Western Spadefoot). Predation. Herpetological Review 38:197–198.

Estrada, A.R., and R. Ruibal. 1999. A review of Cuban herpetology. Pp. 31–62 *In* B.I. Crother (ed.), Caribbean Amphibians and Reptiles. Academic Press, San Diego, CA.

Evans, H.E., and R.M. Roecker. 1951. Notes on the herpetology of Ontario, Canada. Herpetologica 7:69–71.

Evans, S.E., and M. Borsuk-Bialynicka. 2009. The

Early Triassic stem-frog *Czatkobatrachus* from Poland. Palaeontologia Polonica 65:79–105.

Evenden, F.G. Jr. 1948. Food habits of *Triturus granulosus* in western Oregon. Copeia 1948:219–220.

Evrard, J.O., and S.R. Hoffman. 2000. Amphibians and reptiles captured in drift fences in northwest Wisconsin pine barrens. Journal of the Iowa Academy of Science 107:182–186.

Ewert, M.A. 1969. Seasonal movements of the toads *Bufo americanus* and *B. cognatus* in northwestern Minnesota. Ph.D. diss., University of Minnesota, Minneapolis.

Faeh, S.A., D.K. Nichols, and V.R. Beasley. 1998. Infectious diseases of amphibians. Pp. 259–265 *In* M.J. Lannoo (ed.), Status & Distribution of Midwestern Amphibians. University of Iowa Press, Iowa City.

Fair, J.W. 1970. Comparative rates of rehydration from soil in two species of toads, *Bufo boreas* and *Bufo punctatus*. Comparative Biochemistry and Physiology 34:281–287.

Fairchild, L. 1981. Mate selection and behavioral thermoregulation in Fowler's toads. Science 212:950–951.

Fairchild, L. 1984. Male reproductive tactics in an explosive breeding toad population. American Zoologist 24:407–418.

Faivovich, J., C.F.B. Haddad, P.C.A. Garcia, D.R. Frost, J.A. Campbell, and W.C. Wheeler. 2005. Systematic review of the frog family Hylidae, with special reference to Hylinae: phylogenetic analysis and taxonomic revision. Bulletin of the American Museum of Natural History 294:1–240.

Farrell, M.P. 1971. Effect of temperature and photoperiod acclimations on the water economy of *Hyla crucifer*. Herpetologica 27:41–48.

Farringer, J.E. 1972. The determination of acute toxicity of rotenone and bayer 73 to selected aquatic organisms. M.S. thesis, University of Wisconsin-La Crosse.

Farris, J.S., A. Kluge, and M.F. Margoliash. 1979. Paraphyly of the *Rana boylii* species group. Systematic Zoology 28:627–634.

Farris, J.S., A. Kluge, and M.F. Margoliash. 1982. Immunological distance and the phylogenetic relationships of the *Rana boylii* species group. Systematic Zoology 31:479–491.

Fashingbauer, B.A. 1957. The effect of aerial spraying with DDT on a population of wood frogs. Flicker 29(4):160.

Feaver, P.E. 1971. Breeding pool selection and larval mortality of three California amphibians: *Ambystoma tigrinum californiense* Gray, *Hyla regilla* Baird and Girard, and *Scaphiopus hammondii* Girard. M.A. thesis, Fresno State College, Fresno, CA.

Feder, J.H. 1979. Natural hybridization and genetic divergence between the toads *Bufo boreas* and *Bufo punctatus*. Evolution 33:1089–1097.

Federighi, H. 1938. Albinism in *Rana pipiens* (Schreber). Ohio Journal of Science 38:37–40.

Feliciano, S.L. 2000. A comparison of two northern cricket frog (*Acris crepitans*) populations and their habitats during drought conditions in southeastern New York and the Florida Panhandle. M.S. thesis, University of Florida, Gainesville.

Fellers, G.M. 1975. Behavioral interactions in North American treefrogs (Hylidae). Chesapeake Science 16:218–219.

Fellers, G.M. 1976. Social interactions in North American treefrogs. Ph.D. diss., University of Maryland, College Park.

Fellers, G.M. 1979a. Aggression, territoriality, and mating behaviour in North American treefrogs. Animal Behaviour 27:107–119.

Fellers, G.M. 1979b. Mate selection in the gray treefrog, *Hyla versicolor*. Copeia 1979:286–290.

Fellers, G.M., and C.A. Drost. 1993. Disappearance of the Cascades frog *Rana cascadae* at the southern end of its range, California, USA. Biological Conservation 65:177–181.

Fellers, G.M., and P.M. Kleeman. 2007. California red-legged frog (*Rana draytonii*) movement and habitat use: implications for conservation. Journal of Herpetology 41:276–286.

Fellers, G.M., D.E. Green, and J.E. Longcore. 2001. Oral chytridiomycosis in the mountain yellow-legged frog (*Rana muscosa*). Copeia 2001:945–953.

Fellers, G.M., L.L. McConnell, D. Pratt, and S. Datta. 2004. Pesticides in mountain yellow-legged frogs (*Rana muscosa*) from the Sierra Nevada Mountains of California, USA. Environmental Toxicology and Chemistry 23:2170–2177.

Fellers, G.M., D.F. Bradford, D. Pratt, and L.L. Wood. 2007a. Demise of repatriated populations of mountain yellow-legged frogs (*Rana muscosa*) in the Sierra Nevada of California. Herpetological Conservation and Biology 2:5–21.

Fellers, G.M., K.L. Pope, J.E. Stead, M.S. Koo, and H.H. Welsh Jr. 2007b. Turning population trend monitoring into active conservation: can we save the Cascades frog (*Rana cascadae*) in the Lassen Region of California. Herpetological Conservation and Biology 3:28–39.

Feminella, J.W., and C.P. Hawkins. 1994. Tailed frog tadpoles differentially alter their feeding behavior in response to non-visual cues from four predators. Journal of the North American Benthological Society 13:310–320.

Fenoilio, D.B., G.O. Graening, and J.F. Stout. 2005. Seasonal movement patterns of pickerel frogs (*Rana palustris*) in an Ozark cave and trophic implications supported by stabile isotope evidence. Southwestern Naturalist 50:385–389.

Ferguson, D.E. 1960. Observations on movements and behavior of *Bufo fowleri* in residential areas. Herpetologica 16:112–114.

Ferguson, D.E. 1961. The herpetofauna of Tishomingo County, Mississippi, with comments on its zoogeographic affinities. Copeia 1961:391–396.

Ferguson, D.E. 1963. Orientation in three species of anuran amphibians. Ergebnisse der Biologie 26: 128–134.

Ferguson, D.E. 1966a. Sun-compass orientation in anurans. Pp. 21–33 In R.M. Storm (ed.), Animal Orientation and Navigation. Oregon State University Press, Corvallis.

Ferguson, D.E. 1966b. Evidence of sun-compass orientation in the chorus frog, *Pseudacris triseriata*. Herpetologica 22:106–112.

Ferguson, D.E., and H.F. Landreth. 1966. Celestial orientation of Fowler's toad *Bufo fowleri*. Behaviour 26:107–123.

Ferguson, D.E., and C.C. Gilbert. 1968. Tolerances of three species of anuran amphibians to five chlorinated hydrocarbon insecticides. Journal of the Mississippi Academy of Sciences 13:135–138.

Ferguson, D.E., K.E. Payne, and R.M. Storm. 1958. Notes on the herpetology of Baker County, Oregon. Great Basin Naturalist 18:63–65.

Ferguson, D.E., H.F. Landreth, and M.R. Turnipseed. 1965. Astronomical orientation of the southern cricket frog, *Acris gryllus*. Copeia 1965:58–66.

Ferguson, D.E., H.F. Landreth, and J.P. McKeown. 1967. Sun compass orientation of the Northern Cricket Frog, *Acris crepitans*. Animal Behaviour 15: 45–53.

Ferguson, D.E., J.P. McKeown, O.S. Bosarge, and H.F. Landreth. 1968. Sun-compass orientation of bullfrogs. Copeia 1968:230–235.

Ferguson, J.H., and C.H. Lowe. 1969. Evolutionary relationships in the *Bufo punctatus* group. American Midland Naturalist 81:435–466.

Ferner, J.W., T. Rice, and K.S. Neltner. 1992. Predation on birds, at a birdfeeder, by bullfrogs, *Rana catesbeiana* Shaw. Pp. 61–62 In P.D. Strimple and J.L. Strimple (eds.), Contributions in Herpetology. Greater Cincinnati Herpetological Society, Cincinnati, OH.

Fetkavich, C., and L.J. Livo. 1998. Late-season boreal toad tadpoles. Northwestern Naturalist 79:120–121.

Ficetola, G.F., C. Miaud, F. Pompanon, and P. Taberlet. 2009. Species detection using environmental DNA from water samples. Biology Letters 4:423–425.

Fidenci, P. 2004. The California red-legged frog, *Rana aurora draytonii*, along the Arroyo Santo Domingo, northern Baja California, Mexico. Herpetological Bulletin 88:27–31.

Fidenci, P. 2006. *Rana boylii* (Foothill Yellow-legged Frog). Predation. Herpetological Review 37:208.

Filho, C.B.C., H.C. Costa, E.T Da Silva, O.P. Ribeiro Filho, and R.N. Feio. 2008. *Lithobates catesbeianus* (American Bullfrog). Prey. Herpetological Review 39:338.

Findlay, C.S., L. Lenton, and L. Zheng. 2001. Land-use correlates of anuran community richness and composition in southeastern Ontario wetlands. Ecoscience 8:336–343.

Findley, J.S. 1964. Verification of the occurrence of *Bufo microscaphus* Cope in New Mexico. Southwestern Naturalist 9:102–109.

Finlay, J.C., and V.T. Vredenburg. 2007. Introduced trout sever trophic connections in watersheds: consequences for a declining amphibian. Ecology 88:2187–2198.

Fischer, T.D. 1998. Anura of eastern South Dakota: their distribution and characteristics of their wetland habitats, 1997–1998. M.S. thesis, South Dakota State University, Brookings.

Fishbeck, D.W., and J.C. Underhill. 1971. Distribution of stripe polymorphism in wood frogs, *Rana sylvatica* LeConte, from Minnesota. Copeia 1971: 253–259.

Fisher, H.T., and A. Richards. 1950. The annual ovarian cycle of *Acris crepitans* Baird. Pp. 129–142 In Researches on the Amphibia of Oklahoma, University of Oklahoma Press, Norman.

Fisher, M.C., and T.W.J. Garner. 2007. The relationship between the emergence of *Batrachochytrium dendrobatidis*, the international trade in amphibians and introduced amphibian species. Fungal Biology Reviews 21:2–9.

Fisher, R.N., and H.B. Shaffer. 1996. The decline of amphibians in California's Great Central Valley. Conservation Biology 10:1387–1397.

Fishwild, T.G., R.A. Schemidt, K.M. Jankens, K.A. Berven, G.J. Gamboa, and C.M. Richards. 1990. Sibling recognition by larval frogs (*Rana pipiens*, *R. sylvatica*, and *Pseudacris crucifer*). Journal of Herpetology 24:40–44.

Fitch, H.S. 1936. Amphibians and reptiles of the Rogue River Basin, Oregon. American Midland Naturalist 17:634–652.

Fitch, H.S. 1956a. Temperature responses in free-living amphibians and reptiles of northeastern Kansas. University of Kansas Publications, Museum of Natural History 8:417–476.

Fitch, H.S. 1956b. A field study of the Kansas ant-eating frog, *Gastrophryne olivacea*. University of Kansas Publications, Museum of Natural History 8:275–306.

Fitch, H.S. 1958. Home ranges, territories, and seasonal movements of vertebrates of the Natural History Reservation. University of Kansas Publications, Museum of Natural History 11:63–326.

Fitch, H.S. 1960. Autecology of the copperhead. University of Kansas Publications, Museum of Natural History 13:85–288.

Fite, K.V., A. Blaustein, L. Bengston, and H.E. Hewitt. 1998. Evidence of retinal light damage in *Rana cascadae*: a declining amphibian species. Copeia 1998:906–914.

Fitzgerald, G.J., and J.R. Bider. 1974a. Seasonal activity of the toad *Bufo americanus* in southern Quebec as revealed by a sand transect technique. Canadian Journal of Zoology 52:1–5.

Fitzgerald, G.J., and J.R. Bider. 1974b. Evidence for a relationship between geotaxis and seasonal movements in the toad *Bufo americanus*. Oecologia 17:277–280.

Fitzgerald, G.J., and J.R. Bider. 1974c. Evidence of a relationship between age and activity in the toad *Bufo americanus*. Canadian Field-Naturalist 88:499–501.

Fleming, P.L. 1976. A study of the distribution and ecology of *Rana clamitans* Latreille. Ph.D. diss., University of Minnesota, Minneapolis.

Flores, G. 1978. The reproductive ecology of the northern spring peeper, *Hyla crucifer crucifer*, in southern Connecticut. HERP, Bulletin of the New York Herpetological Society 14:22–25.

Florey, J., and S.J. Mullin. 2005. A survey of anuran breeding activity in Illinois, 1986–1989. Transactions of the Illinois State Academy of Science 98:39–47.

Florida Department of Agriculture. 1952. Bullfrog farming and frogging in Florida. Bulletin (new series) No. 56.

Flowers, M.A. 1994. Feeding ecology and habitat use of juvenile Great Plains toads (*Bufo cognatus*) and Woodhouse's toads (*B. woodhousei*). M.A. thesis, University of South Dakota, Vermillion.

Flowers, M.A., and B.M. Graves. 1995. Prey selectivity and size-specific diet changes in *Bufo cognatus* and *B. woodhousii* during early postmetamorphic ontogeny. Journal of Herpetology 29:608–612.

Flowers, M.A., and B.M. Graves. 1997. Juvenile toads avoid chemical cues from snake predators. Animal Behaviour 53:641–646.

Floyd, T.M., and E.S. Kilpatrick. 2002. *Pseudacris brachyphona* (Mountain Chorus Frog). Verification of historical occurrence. Herpetological Review 33:48.

Floyd, T.M., K.R. Russell, C.E. Moorman, D.H. Van Lear, D.C. Guynn Jr., and J.D. Lanham. 2002. Effects of prescribed fire on herpetofauna within hardwood forests of the upper Piedmont of South Carolina: a preliminary analysis. Pp. 123–127 *In* K.W. Outcalt (ed.), Proceedings of the 11th Biennial Southern Silvicultural Research Conference. USDA Forest Service, Southern Research Station, General, and Technical Report SRS-48, Asheville, NC.

Flury, A.G. 1951. Variations in some local populations of *Hyla versicolor* in central Texas. M.A. thesis, University of Texas, Austin.

Fogell, D.D. 2010. A Field Guide to the Amphibians and Reptiles of Nebraska. Institute of Agriculture and Natural Resources, University of Nebraska-Lincoln.

Fogleman, J.C., P.S. Corn, and D. Pettus. 1980. The genetic basis for a dorsal color polymorphism in *Rana pipiens*. Journal of Heredity 71:439–440.

Fontenot, C.L. Jr. 2011. *Hyla cinerea* (Green Treefrog). Winter aggregation. Herpetological Review 42:84–85.

Force, E.R. 1925. Notes on reptiles and amphibians of Okmulgee County, Oklahoma. Copeia (141).25–27.

Force, E.R. 1930. The amphibians and reptiles of Tulsa County, Oklahoma, and vicinity. Copeia 1930:25–39.

Force, E.R. 1933. The age of attainment of sexual maturity of the leopard frog *Rana pipiens* (Scheber) in northern Michigan. Copeia 1933:128–131.

Ford, L.S., and D.C. Cannatella. 1993. The major clades of frogs. Herpetological Monographs 7:94–117.

Forester, D.C. 1973. Mating call as a reproductive isolating mechanism between *Scaphiopus bombifrons* and *S. hammondii*. Copeia 1973:60–67.

Forester, D.C. 1975. Laboratory evidence for potential gene flow between two species of spadefoot toads, *Scaphiopus bombifrons* and *Scaphiopus hammondii*. Herpetologica 31:282–286.

Forester, D.C., and R. Czarnowsky. 1985. Sexual selection in the spring peeper, *Hyla crucifer* (Amphibia, Anura): role of the advertisement call. Behaviour 92:112–128.

Forester, D.C., and R. Daniel. 1986. Observations on the social behavior of the southern cricket frog, *Acris gryllus* (Anura: Hylidae). Brimleyana 12:5–11.

Forester, D.C., and D.V. Lykens. 1986. Significance of satellite males in a population of spring peepers (*Hyla crucifer*). Copeia 1986:719–724.

Forester, D.C., and D.V. Lykens. 1988. The ability of wood frog eggs to withstand prolonged terrestrial stranding: an empirical study. Canadian Journal of Zoology 66:1733–1735.

Forester, D.C., and K.J. Thompson. 1998. Gauntlet behaviour as a male sexual tactic in the American toad (Amphibia: Bufonidae). Behaviour 135:99–119.

Forester, D.C., S. Knoedler, and R. Sanders. 2003. Life history and status of the mountain chorus frog (*Pseudacris brachyphona*) in Maryland. Maryland Naturalist 46:1–15.

Forester, D.C., J.W. Snodgrass, K. Marsalek, and Z. Lanham. 2006. Post-breeding dispersal and summer home range of female American toads (*Bufo americanus*). Northeastern Naturalist 13:59–72.

Formanowicz, D.R. Jr., and E.D. Brodie Jr. 1979. Palatability and antipredator behavior of selected *Rana* to the shrew *Blarina*. American Midland Naturalist 101:456–458.

Formanowicz, D.R. Jr., and E.D. Brodie Jr. 1982. Relative palatabilities of members of a larval amphibian community. Copeia 1982:91–97.

Formanowicz, D.R. Jr., and M.S. Bobka. 1989. Predation risk and microhabitat preference: an experimental study of the behavioral responses of prey and predator. American Midland Naturalist 121: 379–386.

Forsman, E.D., and J.K. Swingle. 2007. Use of arboreal nests of tree voles (*Arborimus* spp.) by amphibians. Herpetological Conservation and Biology 2:113–118.

Forstner, J.M., M.R.J. Forstner, and J.R. Dixon. 1998. Ontogenetic effects on prey selection and food habits of two sympatric east Texas ranids: the southern leopard frog, *Rana sphenocephala*, and the bronze frog, *Rana clamitans clamitans*. Herpetological Review 29:208–211.

Fortin, C., M. Ouellet, and M.-J. Grimard. 2003. La Rainette faux-grillon boréale (*Pseudacris maculata*): présence officiellement validée au Québec. Le naturaliste canadien 127:71–75.

Fortin, C., M. Ouellet, and P. Galois. 2004a. Les amphibians et les reptiles des îles de l'estuaire du Saint-Laurent: mieux connaître pour mieux conserver. Le naturaliste canadien 128:61–67.

Fortin, C., P. Galois, M. Ouellet, and G.J. Doucet. 2004b. Utilisation des emprises de lignes de transport d'énergie électrique par les amphibians et les reptiles en forêt décidue au Québec. Le naturaliste canadien 128:68–75.

Fortman, J., and R. Altig. 1973. Characters of F_1 hybrid tadpoles between six species of *Hyla*. Copeia 1973:411–416.

Foster, B.J., D.W. Sparks, and J.E. Duchamp. 2004. Urban herpetology. 2. Amphibians and reptiles of the Indianapolis Airport conservation lands. Proceedings of the Indiana Academy of Science 113: 53–59.

Foster, C.D., and S.J. Mullin. 2008. Speed and endurance of *Thamnophis hammondii* are not affected by consuming the toxic frog *Xenopus laevis*. Southwestern Naturalist 53:370–373.

Fouquette, M.J. Jr. 1954. Food competition among four sympatric species of garter snakes, genus *Thamnophis*. Texas Journal of Science 6:172–188.

Fouquette, M.J. Jr. 1960. Call structure in frogs of the Family Leptodactylidae. Texas Journal of Science 12:201–215.

Fouquette, M.J. Jr. 1968. Remarks on the type specimen of *Bufo alvarius* Girard. Great Basin Naturalist 28:70–72.

Fouquette, M.J. Jr. 1975. Speciation in chorus frogs. 1. Reproductive character displacement in the *Pseudacris nigrita* complex. Systematic Zoology 24:16–23.

Fouquette, M.J. Jr. 1980. Effect of environmental temperatures on body temperature of aquatic-calling anurans. Journal of Herpetology 14: 347–352.

Fouquette, M.J. Jr. 2005. *Rhinophrynus dorsalis* Dumeril and Bibron, 1841. Burrowing Toad (Sapo Borracho). Pp. 599–600 *In* M.J. Lannoo (ed.), Amphibian Declines. The Conservation Status of United States Species. University of California Press, Berkeley.

Fournier, M.A. 1997. Amphibians in the Northwest Territories. SSAR Herpetological Conservation 1:100–106.

Fournier, M.A. undated. Amphibians & reptiles in the Northwest Territories. Ecology North, Yellowknife, Northwest Territories. http://www.enr.gov. nt .ca/_live/documents/documentManagerUpload/ amphibians_reptiles_pamphlet.pdf [Accessed 28 July 2010]

Fowler, H.W. 1905. The sphagnum frog of New Jersey—*Rana virgatipes* Cope. Proceedings of the Academy of Natural Sciences of Philadelphia 1905: 662–664.

Fowler, H.W. 1918. An albino spring frog in winter. Copeia (61):84.

Fowler, J.A., and G. Orton. 1947. The occurrence of *Hyla femoralis* in Maryland. Maryland, A Journal of Natural History 17:6–7.

Fowler, J.A., and R.L. Hoffman. 1951. *Gastrophryne carolinensis carolinensis* (Holbrook) in southwestern Virginia. Virginia Journal of Science 2:101.

Fox, S.F., P.A. Shipman, R.E. Thill, J.P. Phelps, and D.M. Leslie Jr. 2004. Amphibian communities under diverse forest management in Ouachita Mountains, Arkansas. Pp. 164–173 *In* J.M. Guldin (compiler), Ouachita and Ozark Mountains Symposium. Ecosystem Management Research. USDA Forest Service General and Technical Report SRS-74, Asheville, North Carolina, USA.

Fraker, M.A. 2010. Risk assessment and anti-predator behavior of wood frog (*Rana sylvatica*) tadpoles: a comparison with green frog (*Rana clamitans*) tadpoles. Journal of Herpetology 44:390–398.

Franz, R. 1967. Annotated checklist of cave-associated reptiles and amphibians of Maryland. Bulletin of the Maryland Herpetological Society 3:51–53.

Franz, R. 1970. Egg development of the tailed frog under natural conditions. Bulletin of the Maryland Herpetological Society 6:27–30.

Franz, R. 1971. Notes on the distribution and ecology of the herpetofauna of northwestern Montana. Bulletin of the Maryland Herpetological Society 7:1–10.

Franz, R. 1972. Feeding in Florida cricket frogs. Bulletin of the Maryland Herpetological Society 8:89–90.

Franz, R. 1986. The Florida gopher frog and the Florida pine snake as burrow associates of the gopher tortoise in northern Florida. Pp. 16–20 *In* D.R. Jackson and R.J. Bryant (eds.), The Gopher Tortoise and its Community. Proceedings of the 5th Annual Meeting of the Gopher Tortoise Council, Florida State Museum, Gainesville.

Franz, R. 2005. *Hyla gratiosa* (Barking Treefrog). Winter retreat. Herpetological Review 36:434–435.

Franz, R., and D.S. Lee. 1970. The ecological and biogeographical distribution of the tailed frog, *Ascaphus truei*, in the Flathead River Drainage of northwestern Montana. Bulletin of the Maryland Herpetological Society 6:62–73.

Franz, R., C.K. Dodd Jr., and C. Jones. 1988. *Rana areolata aesopus* (Florida Gopher Frog). Movement. Herpetological Review 19:82.

Franz, R., C.K. Dodd Jr., and A.M. Bard. 1992. The non-marine herpetofauna of Egmont Key, Hillsborough County, Florida. Florida Scientist 55:179–183.

Freda, J. 1986. The influence of acidic pond water on amphibians: a review. Water, Air and Soil Pollution 30:439–450.

Freda, J., and W.A. Dunson. 1984. Sodium balance of amphibian larvae exposed to low environmental pH. Physiological Zoology 57:435–443.

Freda, J., and W.A. Dunson. 1985. Field and laboratory studies of ion balance and growth rates of ranid tadpoles chronically exposed to low pH. Copeia 1985:415–423.

Freda, J., and W.A. Dunson. 1986. Effects of low pH and other chemical variables on the local distribution of amphibians. Copeia 1986:454–466.

Freda, J., and D.G. McDonald. 1990. Effects of aluminum on the leopard frog, *Rana pipiens*: life stage comparisons and aluminum uptake. Canadian Journal of Fisheries and Aquatic Sciences 47:210–216.

Freda, J., and D.H. Taylor. 1992. Behavioral response of amphibian larvae to acidic water. Journal of Herpetology 26:429–433.

Freda, J., and D.G. McDonald. 1993. Toxicity of amphibian breeding ponds in the Sudbury Region. Canadian Journal of Fisheries and Aquatic Sciences 50:1497–1503.

Freda, J., W.J. Sadinski, and W.A. Dunson. 1991. Long term monitoring of amphibian populations with respect to the effects of acidic precipitation. Water, Air, and Soil Pollution 55:445–462.

Freed, A.N. 1980a. Prey selection and feeding behavior of the green treefrog (*Hyla cinerea*). Ecology 61:461–465.

Freed, A.N. 1980b. An adaptive advantage of basking behavior in an anuran amphibian. Physiological Zoology 53:433–444.

Freed, A.N. 1982. A treefrog's menu: selection for an evening's meal. Oecologia 53:20–26.

Freed, P.S., and K. Neitman. 1988. Notes on predation on the endangered Houston toad, *Bufo houstonensis*. Texas Journal of Science 40:454–456.

Freiburg, R.E. 1951. An ecological study of the narrow-mouthed toad (*Microhyla*) in northeastern Kansas. Transactions of the Kansas Academy of Science 54:374–386.

Freidenburg, L.K., and D.K. Skelly. 2004. Microgeographic variation in thermal preference by an amphibian. Ecology Letters 7:369–373.

Friebele, E., and J. Zambo. 2004. A Guide to the Amphibians and Reptiles of Jug Bay. Jug Bay Wetlands Sanctuary, Lothian, MD.

Friet, S.C. 1993. Aquatic overwintering of the bullfrog (*Rana catesbeiana*) during natural hypoxia in an ice-covered pond. M.S. thesis, Dalhousie University, Halifax, Nova Scotia.

Frost, D. 1983. Past occurrence of *Acris crepitans* (Hylidae) in Arizona. Southwestern Naturalist 28:105.

Frost, D.R., T. Grant, J. Faivovich, R.H. Bain, A. Haas, C.F.B. Haddad, R.O. de Sá, A. Channing, M. Wilkinson, S.C. Donnellan, C.J. Raxworthy, J.A. Campbell, B.L. Blotto, P. Moler, R.C. Drewes, R.A. Nussbaum, J.D. Lynch, D.M. Green, and W.C. Wheeler. 2006a. The amphibian tree of life. Bulletin of the American Museum of Natural History 297:1–370.

Frost, D.R., T. Grant, and J.R. Mendelson III. 2006b. *Ollotis* Cope, 1875 is the oldest name for the genus currently referred to as *Cranopsis* Cope, 1875 (Anura: Hyloides: Bufonidae). Copeia 2006:558.

Frost, D.R., T. Grant, J. Faivovich, R.H. Bain, A. Haas, C.F.B. Haddad, R.O. de Sá, A. Channing, M. Wilkinson, S.C. Donnellan, C.J. Raxworthy, J.A. Campbell, B.L. Blotto, P. Moler, R.C. Drewes, R.A. Nussbaum, J.D. Lynch, D.M. Green, and W.C. Wheeler. 2007. Is *The Amphibian Tree of Life* really fatally flawed? Cladistics 24:385–395.

Frost, J.S., and J.T. Bagnara. 1976. A new species of leopard frog (*Rana pipiens* complex) from northwestern Mexico. Copeia 1976:332–338.

Frost, J.S., and J.T. Bagnara. 1977. Sympatry between *Rana blairi* and the southern form of leopard frog in southeastern Arizona (Anura: Ranidae). Southwestern Naturalist 22: 443–453.

Frost, J.S., and J.E. Platz. 1983. Comparative assessment of modes of reproductive isolation among four species of leopard frogs (*Rana pipiens* complex). Evolution 37:66–78.

Frost, S.W. 1924. Frogs as insect collectors. Bulletin of the New York Entomological Society 32:174–185.

Frost, S.W. 1935. The food of *Rana catesbeiana* Shaw. Copeia 1935:15–18.

Fulk, F.D., and J.O. Whitaker Jr. 1969. The food of *Rana catesbeiana* in three habitats in Owen County, Indiana. Proceedings of the Indiana Academy of Science 78:491–496.

Fuller, T.E., K.L. Pope, D.T. Ashton, and H.H. Welsh Jr. 2011. Linking the distribution of an invasive amphibian (*Rana catesbeiana*) to habitat conditions in a managed river system in northern California. Restoration Ecology 19:204–213.

Fulmer, T., and R. Tumlison. 2004. Important records of the bird-voiced treefrog (*Hyla avivoca*) in the headwaters of the Ouachita River drainage of southwestern Arkansas. Southeastern Naturalist 3:259–266.

Fulton, M.H., and J.E. Chambers. 1985. The toxic

and teratogenic effects of selected organophosphorus compounds on the embryos of three species of amphibians. Toxicology Letters 26:175–180.

Funderburg, J.B., C.H. Hotchkiss, and P. Hertl. 1974a. The wood frog, *Rana sylvatica* LeConte, in the Virginia coastal plain. Bulletin of the Maryland Herpetological Society 10:58.

Funderburg, J.B., P. Hertl, and W.M. Kerfoot. 1974b. A range extension for the carpenter frog, *Rana virgatipes* Cope, in the Chesapeake Bay region. Bulletin of the Maryland Herpetological Society 10:77–79.

Funk, W.C., M.S. Blouin, P.S. Corn, B.A. Maxwell, D.S. Pilliod, S. Amish, and F.W. Allendorf. 2005a. Population structure of Columbia spotted frogs (*Rana luteiventris*) is strongly affected by the landscape. Molecular Ecology 14:483–496.

Funk, W.C., A.E. Greene, P.S. Corn, and F.W. Allendorf. 2005b. High dispersal in a frog species suggests that it is vulnerable to habitat fragmentation. Biology Letters 1:13–16.

Funk, W.C., C.A. Pearl, H.M. Draheim, M.J. Adams, T.D. Mullins, and S.M. Haig. 2008. Range-wide phylogeographic analysis of the spotted frog complex (*Rana luteiventris* and *Rana pretiosa*) in northwestern North America. Molecular Phylogenetics and Evolution 49:198–210.

Funk, W.C., T.S. Garcia, G.A. Cortina, and R.H. Hill. 2011. Population genetics of introduced bullfrogs, *Rana (Lithobates) catesbeianus* [sic], in the Willamette Valley, Oregon, USA. Biological Invasions 13:651–658.

Funkhouser, A. 1976. Observations on pancreas: body weight ratio change during development of the bullfrog, *Rana catesbeiana*. Herpetologica 32:370–371.

Gaertner, J.P., D. McHenry, M.R.J. Forstner, and D. Hahn. 2010. Annual variation of *Batrachochytrium dendrobatidis* in the Houston toad (*Bufo houstonensis*) and a sympatric congener (*Bufo nebulifer*). Herpetological Review 41:456–459.

Gahl, M.K. 2007. Spatial and temporal patterns of amphibian disease in Acadia National Park wetlands. Ph.D. diss., University of Maine, Orono.

Gaige, H.T. 1920. Observations upon the habits of *Ascaphus truei* Stejneger. Occasional Papers of the Museum of Zoology, University of Michigan 84:1–9.

Gaige, H.T. 1931. Notes on *Syrrhophus marnockii* Cope. Copeia 1931:63.

Gaige, H.T. 1932. The status of *Bufo copei*. Copeia 1932:134.

Gallie, J.A., R.L. Mumme, and S.A. Wissinger. 2001. Experience has no effect on the development of chemosensory recognition of predators by tadpoles of the American toad, *Bufo americanus*. Herpetologica 57:376–383.

Galois, P., and M. Ouellet. 2005. Le Grand Bois de Saint-Grégoire, un refuge pour l'herpétofaune dans la plaine montérégienne. Le naturaliste canadien 129:37–43.

Gamble, T., P.B. Berendzen, H.B. Shaffer, D.E. Starkey, and A.M. Simons. 2008. Species limits and phylogeography of North American cricket frogs (*Acris*: Hylidae). Molecular Phylogenetics and Evolution 48:112–125.

Gambs, R.D., and M.J. Littlejohn. 1979. Acoustic behavior of males of the Rio Grande leopard frog (*Rana berlandieri*): an experimental analysis through field playback trials. Copeia 1979:643–650.

García-París, M., D.R. Buchholz, and G. Parra-Olea. 2003. Phylogenetic relationships of Pelobatoidea re-examined using mtDNA. Molecular Phylogenetics and Evolution 28:12–23.

Gardiner, D.M., and D.M. Hoppe. 1999. Environmentally induced limb malformations in mink frogs (*Rana septentrionalis*). Journal of Experimental Zoology 284:207–216.

Garman, H. 1892. A synopsis of the reptiles and amphibians of Illinois. Illinois Laboratory of Natural History Bulletin 3(13):215–388.

Garman, H. 1901. The food of the toad. Kentucky Agricultural Experiment Station, Bulletin No. 91: 60–68.

Garman, S. 1877. On a variation in the colors of animals. Proceedings of the American Association for the Advancement of Science 25:194.

Garman, S. 1884. The North American reptiles and batrachians. Bulletin of the Essex Institute 16:3–64.

Garner, T.W.J., M.W. Perkins, P. Govindarajulu, D. Seglie, S. Walker, A.A. Cunningham, and M.C. Fisher. 2006. The emerging amphibian pathogen *Batrachochytrium dendrobatidis* globally infects introduced populations of the North American bullfrog, *Rana catesbeiana*. Biology Letters 2:455–459.

Garnier, J.H. 1883. The mink or hoosier frog. American Naturalist 17:945–954.

Garrett, J.M., and D.G. Barker. 1987. A Field Guide to the Reptiles and Amphibians of Texas. Texas Monthly Press, Austin.

Garton, E.R., F. Grady, and S.D. Carey. 1993. The vertebrate fauna of West Virginia caves. West Virginia Speleological Survey Bulletin 11.

Garton, J.D., and H.R. Mushinsky. 1979. Integumentary toxicity and unpalatability as an antipredator mechanism in the narrow mouthed toad, *Gastrophryne carolinensis*. Canadian Journal of Zoology 57:1965–1973.

Garton, J.S., and R.A. Brandon. 1975. Reproductive ecology of the green treefrog, *Hyla cinerea*, in southern Illinois (Anura: Hylidae). Herpetologica 31:150–161.

Gartside, D.F. 1980. Analysis of a hybrid zone between chorus frogs of the *Pseudacris nigrita* complex in the southern United States. Copeia 1980: 56–66.

Garwood, J.M., and C.W. Anderson. 2010. *Rana*

cascadae (Cascades Frog). Necrogamy. Herpetological Review 41:204.

Garwood, J.M., C.A. Wheeler, R.M. Bourque, M.D. Larson, and H.H. Welsh Jr. 2007. Egg mass drift increases vulnerability during early development of Cascades frogs (*Rana cascadae*). Northwestern Naturalist 88:95–97.

Gascon, C., and D. Planas. 1986. Spring pond water chemistry and the reproduction of the wood frog, *Rana sylvatica*. Canadian Journal of Zoology 64: 543–550.

Gascon, C., and J. Travis. 1992. Does the spatial scale of experimentation matter? A test with tadpoles and dragonflies. Ecology 73:2237–2243.

Gates, G.O. 1957. A study of the herpetofauna in the vicinity of Wickenburg, Maricopa County, Arizona. Transactions of the Kansas Academy of Science 60: 403–418.

Gatten, R.E. Jr., and C.J. Hill. 1984. Social influence on thermal selection by *Hyla crucifer*. Journal of Herpetology 18:87–88.

Gatz, A.J. Jr. 1981a. Non-random mating by size in American toads, *Bufo americanus*. Animal Behaviour 29:1004–1012.

Gatz, A.J. Jr. 1981b. Selective size mating in *Hyla versicolor* and *Hyla crucifer*. Journal of Herpetology 15:113–114.

Gaudin, A.J. 1964. The tadpole of *Hyla californiae* Gorman. Texas Journal of Science 16:80–84.

Gaudin, A.J. 1965. Larval development of the tree frogs *Hyla regilla* and *Hyla californiae*. Herpetologica 21:117–130.

Gaul, R.W. Jr., and J.C. Mitchell. 2007. The herpetofauna of Dare County, North Carolina: history, natural history, and biogeography. Journal of the North Carolina Academy of Science 123: 65–109.

Gaulke, C.A., R.S. Wagner, J.E. Johnson, and J.T. Irwin. 2010. Northern leopard frogs (*Rana pipiens*) infected with *Batrachochytrium dendrobatidis* in the amphibian trade. Herpetological Review 41:322–323.

Gehlbach, F.R. 1965. Herpetology of the Zuni Mountains region, northwestern New Mexico. Proceedings of the United States National Museum 116:243–332.

Gehlbach, F.R., and B.B. Collette. 1959. Distributional and biological notes on the Nebrakka [sic] herpetofauna. Herpetologica 15:141–143.

Gelczis, L., and C. Drost. 2004. A surprise in Surprise Canyon: a new amphibian species for Grand Canyon National Park. [Grand Canyon National Park] Nature Notes (Fall):1–3.

Geluso, K., and G.D. Wright. 2010. Geographic distribution. *Gastrophryne olivacea* (Western narrow-mouthed toad). Herpetological Review 41:103.

Gendron, A.D., D.J. Marcogliese, S. Barbeau, M.-S. Christin, P. Brousseau, S. Ruby, D. Cyr, and M. Fournier. 2003. Exposure of leopard frogs to a pesticide mixture affects life history characteristics of the lungworm *Rhabdias ranae*. Oecologia 135:469–476.

Gentry, G. 1955. An annonated [sic] check list of the amphibians and reptiles of Tennessee. Journal of the Tennessee Academy of Science 30:168–176.

George, I.D. 1940. A study of the bullfrog, *Rana catesbeiana* Shaw, at Baton Rouge, Louisiana. Ph.D. diss., University of Michigan, Ann Arbor.

Georgel, C.T.W. 2001. Activity, diet, and microhabitat of the carpenter frog (*Rana virgatipes* Cope) in Virginia. M.S. thesis, Christopher Newport University, Newport News, VA.

Gergus, E.W.A., K.B. Malmos, and B.K. Sullivan. 1999. Natural hybridization among distantly related toads (*Bufo alvarius*, *Bufo cognatus*, *Bufo woodhousii*) in central Arizona. Copeia 1999: 281–286.

Gergus, E.W.A., T.W. Reeder, and B.K. Sullivan. 2004. Geographic variation in *Hyla wrightorum*: advertisement calls, allozymes, mtDNA, and morphology. Copeia 2004:758–769.

Gerhardt, H.C. 1973. Reproductive interactions between *Hyla crucifer* and *Pseudacris ornata* (Anura: Hylidae). American Midland Naturalist 89:81–88.

Gerhardt, H.C. 1974a. Mating call differences between eastern and western populations of the treefrog *Hyla chrysoscelis*. Copeia 1974:534–536.

Gerhardt, H.C. 1974b. Behavioral isolation of the tree frogs, *Hyla cinerea* and *Hyla andersonii*. American Midland Naturalist 91:424–433.

Gerhardt, H.C. 1975. Sound pressure levels and radiation patterns of the vocalizations of some North American frogs and toads. Journal of Comparative Physiology 102:1–12.

Gerhardt, H.C. 1978a. Temperature coupling in the vocal communication system of the gray tree frog, *Hyla versicolor*. Science 199:992–994.

Gerhardt, H.C. 1978b. Discrimination of intermediate sounds in a synthetic call continuum by female green tree frogs. Science 199:1089–1091.

Gerhardt, H.C. 1982. Sound pattern recognition in some North American treefrogs (Anura: Hylidae): implications for mate choice. American Zoologist 22:581–595.

Gerhardt, H.C. 1987. Evolutionary and neurobiological implications of selective phonotaxis in the green treefrog, *Hyla cinerea*. Animal Behaviour 35:1479–1489.

Gerhardt, H.C., and J. Rheinlaender. 1982. Localization of an elevated sound source by the green tree frog. Science 217:662–663.

Gerhardt, H.C., and G.M. Klump. 1988. Masking of acoustic signals by the chorus background noise in the green tree frog: a limitation on mate choice. Animal Behaviour 36:1247–1249.

Gerhardt, H.C., S.I. Guttman, and A.A. Karlin. 1980. Natural hybrids between *Hyla cinerea* and *Hyla gratiosa*: morphology, vocalization and electrophoretic analysis. Copeia 1980:577–584.

Gerhardt, H.C., B. Diekamp, and M. Ptacek. 1989. Inter-male spacing in choruses of the spring peeper, *Pseudacris (Hyla) crucifer*. Animal Behaviour 38:1012–1024.

Gerlanc, N.M. 1999. Effects of breeding pool permanence on developmental rate of western chorus frogs, *Pseudacris triseriata*, in tallgrass prairie. M.S. thesis, Kansas State University, Manhattan.

Gerlanc, N.M., and G.A. Kaufman. 2003. Use of bison wallows by anurans on Konza Prairie. American Midland Naturalist 150:158–168.

Germaine, S.S., and D.W. Hays. 2009. Distribution and postbreeding environmental relationships of northern leopard frogs (*Rana* [*Lithobates*] *pipiens*) in Washington. Western North American Naturalist 69:537–547.

Ghioca, D.M. 2005. Effect of landuse on larval amphibian communities in playa wetlands of the Southern High Plains. Ph.D. diss., Texas Tech University, Lubbock.

Ghioca, D.M., and L.M. Smith. 2007. Biases in trapping larval amphibians in Playa wetlands. Journal of Wildlife Management 71:991–995.

Gibbons, J.W., and D.H. Bennett. 1974. Determination of anuran terrestrial activity patterns by a drift fence. Copeia 1974:236–243.

Gibbons, J.W., and J.W. Coker. 1978. Herpetofaunal colonization patterns of Atlantic Coast barrier islands. American Midland Naturalist 99:219–233.

Gibbons, J.W., and J.R. Harrison III. 1981. Reptiles and amphibians of Kiawah and Capers Islands, South Carolina. Brimleyana 5:145–162.

Gibbs, E.L., G.W. Nace, and M.B. Emmons. 1971. The live frog is almost dead. BioScience 21:1027–1034.

Gibbs, J.P. 1998a. Amphibian movements in response to forest edges, roads, and streambeds in southern New England. Journal of Wildlife Management 62:584–589.

Gibbs, J.P. 1998b. Distribution of woodland amphibians along a forest fragmentation gradient. Landscape Ecology 13:263–268.

Gibbs, J.P., K.K. Whiteleather, and F.W. Schueler. 2005. Changes in frog and toad populations over 30 years in New York State. Ecological Applications 15:1148–1157.

Gibbs, J.P., A.R. Breisch, P.K. Ducey, G. Johnson, J.L. Behler, and R.C. Bothner. 2007. The Amphibians and Reptiles of New York State. Oxford University Press, Oxford.

Gibbs, R.H. Jr. 1957. The chorus frog, *Pseudacris nigrita*, at Plattsburgh, New York. Copeia 1957:311–312.

Gilbertson, M.-K., G.D. Haffner, K.G. Drouillard, A. Albert, and B. Dixon. 2003. Immunosuppression in the northern leopard frog (*Rana pipiens*) induced by pesticide exposure. Environmental Toxicology and Chemistry 22:101–110.

Gilhen, J. 1984. Amphibians and Reptiles of Nova Scotia. Nova Scotia Museum, Halifax.

Gillis, J.E. 1975. Characterization of a hybridizing complex of leopard frogs. Ph.D. diss., Colorado State University, Fort Collins.

Gillis, J.E. 1979. Adaptive differences in the water economies of two species of leopard frogs from eastern Colorado (Amphibia, Anura, Ranidae). Journal of Herpetology 13:445–450.

Gilmore, R.J. 1924. Notes on the life history and feeding habits of the spadefoot toad of the western plains. Colorado College Publication, Science Series 13:1–12.

Giovanelli, J.G.R., C.F.B. Haddad, and J. Alexandrino. 2008. Predicting the potential distribution of the alien invasive American bullfrog (*Lithobates catesbeianus*) in Brazil. Biological Invasions 10:585–590.

Girard, C. 1854. A list of the North American bufonids, with diagnoses of new species. Proceedings of the Academy of Natural Sciences of Philadelphia 7:86–88.

Girard, C. 1855. Abstract of a report to Lieutenant James McGillis, U.S.N., upon the reptiles collected during the U.S.N. Astronomical Expedition to Chili. Proceedings of the Academy of Natural Sciences of Philadelphia 7:226–227.

Girard, C. 1859. *In* S.F. Baird, Reptiles of the Boundary. United States and Mexican Boundary Survey, under the order of Lieut. Col. W.H. Emory. United States Government Printing Office, Washington, DC.

Given, M.F. 1987. Vocalizations and acoustic interactions of the carpenter frog, *Rana virgatipes*. Herpetologica 43:467–481.

Given, M.F. 1988a. Growth rate and the cost of calling activity in male carpenter frogs, *Rana virgatipes*. Behavioral Ecology and Sociobiology 22:153–160.

Given, M.F. 1988b. Territoriality and aggressive interactions of male carpenter frogs, *Rana virgatipes*. Copeia 1988:411–421.

Given, M.F. 1990. Spatial distribution and vocal interaction in *Rana clamitans* and *R. virgatipes*. Journal of Herpetology 24:377–382.

Given, M.F. 1993. Male response to female vocalizations in the carpenter frog, *Rana virgatipes*. Animal Behaviour 46:1139–1149.

Given, M.F. 1996. Intensity modulation of advertisement calls in *Bufo woodhousii fowleri*. Copeia 1996:970–977.

Given, M.F. 1999a. Distribution records of *Rana virgatipes* and associated anuran species along Maryland's Eastern Shore. Herpetological Review 30:144–146.

Given, M.F. 1999b. Frequency alteration of the advertisement call in the carpenter frog, *Rana virgatipes*. Herpetologica 55:304–317.

Given, M.F. 2002. Interrelationships among calling effort, growth rate, and chorus tenure in *Bufo fowleri*. Copeia 2002:979–987.

Given, M.F. 2005. Vocalizations and reproductive behavior of male pickerel frogs, *Rana palustris*. Journal of Herpetology 39:223–233.

Given, M.F. 2008. Does physical or acoustical disturbance cause male pickerel frogs, *Rana palustris*, to vocalize underwater? Amphibia-Reptilia 29: 177–184.

Glennemeier, K.A., and L.J. Begnoche. 2002. Organochlorine contamination on amphibian populations in southwestern Michigan. Journal of Herpetology 36:233–244.

Glennemeier, K.A., and R.J. Denver. 2002. Role for corticoids in mediating the response of *Rana pipiens* tadpoles to intraspecific competition. Journal of Experimental Zoology 292:32–40.

Glista, D.J, T.L. DeVault, and J.A. DeWoody. 2008. Vertebrate road mortality predominantly impacts amphibians. Herpetological Conservation and Biology 3:77–87.

Glooschenko, V., W.F. Weller, P.G.R. Smith, R. Alvo, and J.H.G. Archbold. 1992. Amphibian distribution with respect to pond water chemistry near Sudbury, Ontario. Canadian Journal of Fisheries and Aquatic Science 49 (Suppl. 1):114–121.

Gloyd, H.K., and R. Conant. 1990. Snakes of the *Agkistrodon* Complex. A Monographic Review. Society for the Study of Amphibians and Reptiles, Contributions in Herpetology No. 6.

Goater, C.P., and R.E. Vandenbos. 1997. Effects of larval history and lungworm infection on the growth and survival of juvenile wood frogs (*Rana sylvatica*). Herpetologica 53:331–338.

Goater, C.P., M. Mulvey, and G.W. Esch. 1990. Electrophoretic differentiation of two *Halipegus* (Trematoda: Hemiuridae) congeners in an amphibian population. Journal of Parasitology 76:431–434.

Godwin, G.J., and S.M. Roble. 1983. Mating success in male treefrogs, *Hyla chrysoscelis* (Anura: Hylidae). Herpetologica 39:141–146.

Goebel, A.M. 2005. Conservation systematics: the *Bufo boreas species* group. Pp. 210–221 *In* M.J. Lannoo (ed.), Amphibian Declines. The Conservation Status of United States Species. University of California Press, Berkeley.

Goebel, A.M., T.A. Ranker, P.S. Corn, and R.G. Olmstead. 2009. Mitochondrial DNA evolution in the *Anaxyrus boreas* species group. Molecular Phylogenetics and Evolution 50:209–225.

Goin, C.J. 1943. The lower vertebrate fauna of the water hyacinth community in northern Florida. Proceedings of the Florida Academy of Sciences 6:143–154.

Goin, C.J. 1944. *Eleutherodactylus ricordii* at Jacksonville, Florida. Copeia 1944:192.

Goin, C.J. 1947a. A note on the food of *Heterodon simus*. Copeia 1947:275.

Goin, C.J. 1947b. Studies on the life history of *Eleutherodactylus ricordii planirostris* (Cope) in Florida. University of Florida Studies, Biological Science Series 4:1–66.

Goin, C.J. 1948. The peep order in peepers: a swamp water serenade. Quarterly Journal of the Florida Academy of Science 11:59–61.

Goin, C.J., and M.G. Netting. 1940. A new gopher frog from the Gulf Coast, with comments upon the *Rana areolata* group. Annals of the Carnegie Museum 28:137–168.

Goin, C.J., and L.H. Ogren. 1956. Parasitic copepods (Argulidae) on amphibians. Journal of Parasitology 42:154.

Goin, O.B. 1955. The World Outside My Door. Macmillan Press, New York.

Goin, O.B. 1958. A comparison of the nonbreeding habits of two treefrogs, *Hyla squirella* and *Hyla cinerea*. Quarterly Journal of the Florida Academy of Sciences 21:49–60.

Goldberg, C.S. 2002. Habitat, spatial population structure, and methods for monitoring barking frogs (*Eleutherodactylus augusti*) in southern Arizona. M.S. thesis, University of Arizona, Tucson.

Goldberg, C.S., and C.R. Schwalbe. 2004. Considerations for monitoring a rare anuran (*Eleutherodactylus augusti*). Southwestern Naturalist 49:442–448.

Goldberg, C.S., and L.P. Waits. 2009. Using habitat models to determine conservation priorities for pond-breeding amphibians in a privately-owned landscape of northern Idaho, USA. Biological Conservation 142:1096–1104.

Goldberg, C.S., and L.P. Waits. 2010. Comparative landscape genetics of two pond-breeding amphibian species in a highly modified agricultural area. Molecular Ecology 19:3650–3663.

Goldberg, C.S., K.J. Field, and M.J. Sredl. 2004a. Mitochondrial DNA sequences do not support species status of the Ramsey Canyon leopard frog (*Rana subaquavocalis*). Journal of Herpetology 38:313–319.

Goldberg, C.S., B.K. Sullivan, J.H. Malone, and C.R. Schwalbe. 2004b. Divergence among barking frogs (*Eleutherodactylus augusti*) in the southwestern United States. Herpetologica 60:312–320.

Goldberg, C.S., D.S. Pilliod, R.S. Arkle, and L.P. Waits. 2011. Molecular detection of vertebrates in stream waters: a demonstration using Rocky Mountain tailed frogs and Idaho giant salamanders. PLoS One 6(7):e22746.

Goldberg, S.R., and C.R. Bursey. 1991a. Helminths from the red-spotted toad, *Bufo punctatus* (Anura: Bufonidae), from southern Arizona. Journal of the Helminthological Society of Washington 58: 267–269.

Goldberg, S.R., and C.R. Bursey. 1991b. Helminths of three toads, *Bufo alvarius*, *Bufo cognatus* (Bufonidae), and *Scaphiopus couchii* (Pelobatidae), from southern Arizona. Journal of the Helminthological Society of Washington 58:142–146.

Goldberg, S.R., and C.R. Bursey. 1996. Helminths of the oak toad (*Bufo quercicus*, Bufonidae) from Florida (USA). Alytes 14:122–126.

Goldberg, S.R., and C.R. Bursey. 2001a. Helminths of the California treefrog, *Hyla cadaverina* (Hylidae) from southern California. Bulletin of the Southern California Academy of Sciences 100:117–122.

Goldberg, S.R., and C.R. Bursey. 2001b. Persistence of the nematode *Ozwaldocruzia pipiens* (Molineidae), in the Pacific treefrog, *Hyla regilla* (Hylidae), from California. Bulletin of the Southern California Academy of Sciences 100:44–50.

Goldberg, S.R., and C.R. Bursey. 2002a. Helminths of the bullfrog, *Rana catesbeiana* (Ranidae), in California with revisions to the California anuran helminth list. Bulletin of the Southern California Academy of Sciences 101:118–130.

Goldberg, S.R., and C.R. Bursey. 2002b. Helminths of the Plains spadefoot, *Spea bombifrons*, the western spadefoot, *Spea hammondii*, and the Great Basin spadefoot, *Spea intermontana* (Pelobatidae). Western North American Naturalist 62:491–495.

Goldberg, S.R., C.R. Bursey, and I. Ramos. 1995. The component parasite community of three sympatric toad species, *Bufo cognatus*, *Bufo debilis* (Bufonidae), and *Spea multiplicata* (Pelobatidae) from New Mexico. Journal of the Helminological Society of Washington 62:57–61.

Goldberg, S.R., C.R. Bursey, B.K. Sullivan, and Q.A. Truong. 1996a. Helminths of the Sonoran green toad, *Bufo retiformis* (Bufonidae), from southern Arizona. Journal of the Helminthological Society of Washington 63:120–122.

Goldberg, S.R., C.R. Bursey, K.B. Malmos, B.K. Sullivan, and H. Cheam. 1996b. Helminths of the southeastern toad, *Bufo microscaphus*, Woodhouse's toad, *Bufo woodhousii* (Bufonidae), and their hybrids from central Arizona. Great Basin Naturalist 56:369–374.

Goldberg, S.R., C.R. Bursey, E.W.A. Gergus, B.K. Sullivan, and Q.A. Truong. 1996c. Helminths of three treefrogs *Hyla arenicolor*, *Hyla wrightorum*, and *Pseudacris triseriata* (Hylidae) from Arizona. Journal of Parasitology 82:833–835.

Goldberg, S.R., C. R. Bursey and H. Cheam. 1998a. Helminths in two native frog species (*Rana chiricahuensis*, *Rana yavapaiensis*) and one introduced frog species (*Rana catesbeiana*) (Ranidae) from Arizona. Journal of Parasitology 84:175–177.

Goldberg, S.R., C.R. Bursey, and H. Cheam. 1998b. Nematodes of the Great Plains narrow-mouthed toad, *Gastrophryne olivacea* (Microhylidae), from southern Arizona. Journal of the Helminthological Society of Washington 65:102–104.

Goldberg, S.R., C.R. Bursey, and S. Hernandez. 1999a. Helminths of the western toad, *Bufo boreas* (Bufonidae) from southern California. Bulletin of the Southern California Academy of Sciences 98:39–44.

Goldberg, S.R., C.R. Bursey, and G. Galindo. 1999b. Helminths of the lowland burrowing treefrog, *Pternohyla fodiens* (Hylidae), from southern Arizona. Great Basin Naturalist 59:195–197.

Goldberg, S.R., C.R. Bursey, and J.E. Platz. 2000. Helminths of the plains leopard frog, *Rana blairi* (Ranidae). Southwestern Naturalist 45:362–366.

Goldberg, S.R., C.R. Bursey, R.G. McKinnell, and I.S. Tan. 2001. Helminths of northern leopard frogs, *Rana pipiens* (Ranidae), from North Dakota and South Dakota. Western North American Naturalist 61:248–251.

Goldberg, S.R., C.R. Bursey, and C. Wong. 2002. Helminths of the western chorus frog from eastern Alberta, Canada. Northwest Science 76:77–79.

Gomez, D.M., and R.G. Anthony. 1996. Amphibian and reptile abundance in riparian and upslope areas of five forest types in western Oregon. Northwest Science 70:109–119.

Gomez-Mestre, I., and D.R. Buchholz. 2006. Developmental plasticity mirrors differences among taxa in spadefoot toads linking plasticity and diversity. Proceedings of the National Academy of Sciences, USA 103:19021–19026.

Goodman, J.D. 1989. *Langeronia brenesi* n. sp. (Trematoda: Lecithodendriidae) in the mountain yellow-legged frog *Rana muscosa* from southern California. Transactions of the American Microscopical Society 108:387–393.

Goodsell, J.A., and L.B. Kats. 1999. Effect of introduced mosquitofish on Pacific treefrogs and the role of alternate prey. Conservation Biology 13:921–924.

Goraya, J., Y. Wang, Z. Li, M. O'Flaherty, F.C. Knoop, J.E. Platz, and J.M. Conlon. 2000. Peptides with antimicrobial activity from four different families isolated from the skins of the North American frogs *Rana luteiventris*, *Rana berlandieri* and *Rana pipiens*. European Journal of Biochemistry 267:894–900.

Gordon, K. 1939. The amphibia and reptilia of Oregon. Oregon State Monographs, Studies in Zoology No. 1.

Gordon, R.E. 1955. Additional remarks on albinism in *Microhyla carolinensis*. Herpetologica 11:240.

Gorham, S.W. 1964. Notes on the amphibians of Browns Flat Area, New Brunswick. Canadian Field-Naturalist 78:154–160.

Gorham, S.W. 1970. The Amphibians and Reptiles of New Brunswick. New Brunswick Museum, Monographic Series No. 6.

Goris, R.C., and N. Maeda. 2004. Guide to the Amhibians and Reptiles of Japan. Krieger Publishing, Malabar, FL.

Gorman, J. 1960. Treetoad studies. 1. *Hyla californiae*, new species. Herpetologica 16:214–222.

Gorman, T.A. 2009. Ecology of two rare amphibians of the Gulf Coastal Plain. Ph.D. diss., Virginia Polytechnic Institute and State University, Blacksburg.

Gorman, T.A. and C.A. Haas. 2011. Seasonal microhabitat selection and use of syntopic populations of *Lithobates okaloosae* and *Lithobates clamitans clamitans*. Journal of Herpetology 45:313–318.

Gorman, T.A., D.C. Bishop, and C.A. Haas. 2009. Spatial interactions between two species of frogs: *Rana okaloosae* and *R. clamitans clamitans*. Copeia 2009:138–141.

Gorman, W.L. 1986. Patterns of color polymorphism in the cricket frog, *Acris crepitans*, in Kansas. Copeia 1986:995–999.

Gorman, W.L., and M.S. Gaines. 1987. Patterns of genetic variation in the cricket frog, *Acris crepitans*, in Kansas. Copeia 1987:352–360.

Gosner, K.L. 1956. Experimental hybridization between two North American tree frogs. Herpetologica 12:285–289.

Gosner, K.L. 1960. A simplified table for staging anuran embryos and larvae with notes on identification. Herpetologica 16:183–190.

Gosner, K.L., and I.H. Black. 1954. Larval development in *Bufo woodhousei fowleri* and *Scaphiopus holbrooki holbrooki*. Copeia 1954:251–255.

Gosner, K.L., and I.H. Black. 1955. The effects of temperature and moisture on the reproductive cycle of *Scaphiopus h. holbrooki*. American Midland Naturalist 54:192–203.

Gosner, K.L., and I.H. Black. 1956. Notes on amphibians from the upper Coastal Plain of North Carolina. Journal of the Elisha Mitchell Scientific Society 72:40–47.

Gosner, K.L., and I.H. Black. 1957a. The effects of acidity on the development and hatching of New Jersey frogs. Ecology 38:256–262.

Gosner, K.L., and I.H. Black. 1957b. Larval development in New Jersey Hylidae. Copeia 1957:31–36.

Gosner, K.L., and I.H. Black. 1958. Notes on the life history of Brimley's chorus frog. Herpetologica 13:249–254.

Gosner, K.L., and D.A. Rossman. 1959. Observations on the reproductive cycle of the swamp chorus frog, *Pseudacris nigrita*. Copeia 1959:263–266.

Gosner, K.L., and D.A. Rossman. 1960. Eggs and larval development of the treefrogs *Hyla crucifer* and *Hyla ocularis*. Herpetologica 16:225–232.

Gosner, K.L., and I.H. Black. 1967. *Hyla andersonii* Baird. Pine Barrens Treefrog. Catalogue of American Amphibians and Reptiles 54.1–54.2.

Govindarajulu, P., R. Altwegg, and B.R. Anholt. 2005. Matrix model investigation of invasive species control: bullfrogs on Vancouver Island. Ecological Applications 15:2161–2170.

Graeter, G.J. 2005. Habitat selection and movement patterns of amphibians in altered forest habitats. M.S. thesis, University of Georgia, Athens.

Graeter, G.J., B.B. Rothermel, and J.W. Gibbons. 2008. Habitat selection and movement of pond-breeding amphibians in experimentally fragmented pine forests. Journal of Wildlife Management 72:473–482.

Graham, C., S.C. Richter, S. McClean, E. O'Kane, P.R. Flatt, and C. Shaw. 2006. Histamine-releasing and antimicrobial peptides from the skin secretions of the dusky gopher frog, *Rana sevosa*. Peptides 27:1313–1319.

Graham, S.P. 2010. Geographic distribution. *Pseudacris brachyphona* (Mountain Chorus Frog). Herpetological Review 41:241.

Graham, S.P., E.K. Timpe, and L. Giovanetto. 2007. Significant new records for Georgia herpetofauna. Herpetological Review 38:494–495.

Granatosky, M.C., L.M. Wagner, and K.L. Krysko. 2011. *Osteopilus septentrionalis* (Cuban Treefrog). Prey. Herpetological Review 42:90.

Granoff, A. 1969. Viruses of amphibia. Current Topics in Microbiology and Immunology 50:107–137.

Granoff, A., P.E. Came, and K.A. Rafferty. 1965. The isolation and properties of viruses from *Rana pipiens*: their possible relationship to the renal adenocarcinoma of the leopard frog. Annals of the New York Academy of Science 126:237–255.

Grant, J.B. 2001. *Rana palustris* (Pickerel Frog). Production of odor. Herpetological Review 32:183.

Grant, K.P., and L.E. Licht. 1993. Acid tolerance of anuran embryos and larvae from central Ontario. Journal of Herpetology 27:1–6.

Grant, K.P., and L.E. Licht. 1995. Effects of ultraviolet radiation on life-history stages of anurans from Ontario, Canada. Canadian Journal of Zoology 73:2292–2301.

Grasso, R.L., R.M. Coleman, and C. Davidson. 2010. Palatability and antipredator response of Yosemite toads (*Anaxyrus canorus*) to nonnative trout (*Salvelinus fontinalis*) in the Sierra Nevada Mountains of California. Copeia 2010:457–462.

Graves, B.M., and S.H. Anderson. 1987. Habitat Suitability Index Models. Bullfrog. U.S. Fish and Wildlife Service, Biological Report 82(10.138), 22 pp.

Graves, B.M., and J.J. Krupa. 2005. *Bufo cognatus* Say, 1823. Great Plains Toad. Pp. 401–404 *In* M.J. Lannoo (ed.), Amphibian Declines. The Conservation Status of United States Species. University of California Press, Berkeley.

Graves, B.M., C.H. Summers, and K.L. Olmstead. 1993. Sensory mediation of aggregation among postmetamorphic *Bufo cognatus*. Journal of Herpetology 27:315–319.

Gray, B.S., and M. Lethaby. 2010. Observations of

limb abnormalities in amphibians from Erie County, Pennsylvania. Journal of Kansas Herpetology 35:14–16.

Gray, E.P., S. Nunziata, J.W. Snodgrass, D.R. Ownby, and J.E. Havel. 2010. Predation on green frog eggs (*Rana clamitans*) by ostracoda. Copeia 2010:452–456.

Gray, M.J. 2002. Effect of anthropogenic disturbance and landscape structure on body size, demographics, and chaotic dynamics of Southern High Plains amphibians. Ph.D. diss., Texas Tech University, Lubbock.

Gray, M.J., L.M. Smith, and R.I. Leyva. 2004a. Influence of agricultural landscape on a Southern High Plains, USA, amphibian assemblage. Landscape Ecology 19:719–729.

Gray, M.J., L.M. Smith, and R. Brenes. 2004b. Effects of agricultural cultivation on demographics of Southern High Plains amphibians. Conservation Biology 18:1368–1377.

Gray, M.J., D.L. Miller, A.C. Schmutzer, and C.A. Baldwin. 2007a. Frog Virus 3 prevalence in tadpole populations inhabiting cattle-access and non-access wetlands in Tennessee. Diseases of Aquatic Organisms 77:97–103.

Gray, M.J., S. Rajeev, D.L. Miller, A.C. Schmutzer, E.C. Burton, E.D. Rogers, and G.J. Hickling. 2007b. Preliminary evidence that American Bullfrogs (*Rana catesbeiana*) are suitable hosts for *Escherichia coli* O157:H7. Applied and Environmental Microbiology 73:4066–4068.

Gray, M.J., L.M. Smith, D.L. Miller, and C.R. Bursey. 2007c. Influences of agricultural land use on *Clinostomum attenuatum* metacercariae prevalence in Southern Great Plains amphibians, U.S.A. Herpetological Conservation and Biology 2:23–28.

Gray, M.J., D.L. Miller, and J.T. Hoverman. 2009. Ecology and pathology of amphibian ranaviruses. Diseases of Aquatic Organisms 87:243–266.

Gray, P., and E. Stegall. 1986. Distribution and status of Strecker's chorus frog (*Pseudacris streckeri streckeri*) in Kansas. Transactions of the Kansas Academy of Science 89:81–85.

Gray, R.H. 1971. Fall activity and overwintering of the cricket frog, *Acris crepitans*, in central Illinois. Copeia 1971:748–750.

Gray, R.H. 1972. Metachrosis of the vertebral stripe in the cricket frog, *Acris crepitans*. American Midland Naturalist 87:549–551.

Gray, R.H. 1977. Lack of physiological differentiation in three color morphs of the cricket frog (*Acris crepitans*) in Illinois. Transactions of the Illinois State Academy of Science 70:73–79.

Gray, R.H. 1978. Nondifferential predation susceptibility and behavioral selection in three color morphs of Illinois cricket frogs, *Acris crepitans*. Transactions of the Illinois State Academy of Science 71:356–360.

Gray, R.H. 1983. Seasonal, annual and geographic variation in color morph frequencies of the cricket frog, *Acris crepitans*, in Illinois. Copeia 1983:300–311.

Gray, R.H. 1984. Effective breeding size and the adaptive significance of color polymorphism in the cricket frog (*Acris crepitans*) in Illinois, U.S.A. Amphibia-Reptilia 5:101–107.

Gray, R.H. 2000a. Historical occurrence of malformations in the cricket frog, *Acris crepitans*, in Illinois. Transactions of the Illinois State Academy of Science 93:279–284.

Gray, R.H. 2000b. Morphological abnormalities in Illinois cricket frogs, *Acris crepitans*, 1968–1971. Journal of the Iowa Academy of Science 107:92–95.

Gray, R.H., and L.E. Brown. 2005. Decline of northern cricket frogs (*Acris crepitans*). Pp. 47–54 *In* M.J. Lannoo (ed.), Amphibian Declines. The Conservation Status of United States Species. University of California Press, Berkeley.

Graybeal, A. 1993. The phylogenetic utility of cytochrome *b*: lessons from bufonid frogs. Molecular and Phylogenetic Evolution 2:256–269.

Graybeal, A. 1997. Phylogenetic relationships of Bufonid frogs and tests of alternate macroevolutionary hypotheses characterizing their radiation. Zoological Journal of the Linnean Society 119:297–338.

Green, D.E., and E. Muths. 2005. Health evaluation of amphibians in and near Rocky Mountain National Park (Colorado, USA). Alytes 22:109–129.

Green, D.E., and C.K. Dodd Jr. 2007. Presence of amphibian chytrid fungus *Batrachochytrium dendrobatidis* and other amphibian pathogens at warm-water fish hatcheries in southeastern North America. Herpetological Conservation and Biology 2:43–47.

Green, D.E., K.A. Converse, and A.K. Schrader. 2002. Epizootiology of sixty-four amphibian morbidity and mortality events in the USA, 1996–2001. Annals of the New York Academy of Science 969:232–339.

Green, D.M. 1981a (1982). Mating call characteristics of hybrid toads (*Bufo americanus* × *B. fowleri*) at Long Point, Ontario. Canadian Journal of Zoology 60:3293–3297. (This publication is variously cited as 1981 or 1982. Vol. 60 was published in 1981, although issue 12 did not appear until January 1982. The online journal is dated 1981, but the reprints are dated 1982.)

Green, D.M. 1981b. Adhesion and the toe pads of treefrogs. Copeia 1981:790–796.

Green, D.M. 1981c. Theoretical analysis of hybrid zones derived from an examination of two dissimilar zones of hybridization in toads (genus *Bufo*). Ph.D. diss., University of Guelph, Guelph, Ontario.

Green, D.M. 1983. Allozyme variation through a clinal hybrid zone between the toads *Bufo america-*

nus and *B. hemiophrys* in southeastern Manitoba. Herpetologica 39:28–40.

Green, D.M. 1984. Sympatric hybridization and allozyme variation in the toads *Bufo americanus* and *B. fowleri* in southern Ontario. Copeia 1984:18–26.

Green, D.M. 1985a. Differentiation in amount of centromeric heterochromatin between subspecies of the red-legged frog, *Rana aurora*. Copeia 1985:1071–1074.

Green, D.M. 1985b. Biochemical identification of red-legged frogs, *Rana aurora draytoni* (Ranidae) at Duckwater, Nevada. Southwestern Naturalist 30:614–616.

Green, D.M. 1985c. Natural hybrids between the frogs *Rana cascadae* and *Rana pretiosa* (Anura: Ranidae). Herpetologica 41:262–267.

Green, D.M. 1986a. Systematics and evolution of western North American frogs allied to *Rana aurora* and *Rana boylii*: karyological evidence. Systematic Zoology 35:273–282.

Green, D.M. 1986b. Systematics and evolution of western North American frogs allied to *Rana aurora* and *Rana boylii*: electrophoretic evidence. Systematic Zoology 35:283–296.

Green, D.M. 1989. Fowler's toad, *Bufo woodhousii fowleri*, in Canada: biology and population status. Canadian Field-Naturalist 103:486–496.

Green, D.M. 1992. Fowler's toad (*Bufo woodhousei fowleri*) at Long Point, Ontario: changing abundance and implications for conservation. Canadian Wildlife Service Occasional Paper No. 76:37–43.

Green, D.M. 1996. The bounds of species: hybridization in the *Bufo americanus* group of North American toads. Israel Journal of Zoology 42:95–109.

Green, D.M. 1997. Temporal variation in abundance and age structure in Fowler's Toads, *Bufo fowleri*, at Long Point, Ontario. SSAR Herpetological Conservation 1:45–56.

Green, D.M. 2003. The ecology of extinction: population fluctuation and decline in amphibians. Biological Conservation 111:331–343.

Green, D.M. 2005. *Bufo americanus* Holbrook, 1836. American Toad. Pp. 386–390 *In* M.J. Lannoo (ed.), Amphibian Declines. The Conservation Status of United States Species. University of California Press, Berkeley.

Green, D.M., and C. Pustowka. 1997. Correlated morphological and allozyme variation in the hybridizing toads *Bufo americanus* and *Bufo hemiophrys*. Herpetologica 53:218–228.

Green, D.M., and C. Parent. 2003. Variable and asymmetric introgression in a hybrid zone in the toads, *Bufo americanus* and *Bufo fowleri*. Copeia 2003:34–43.

Green, D.M., C.H. Daugherty, and J.P. Bogart. 1980. Karyology and systematic relationships of the tailed frog *Ascaphus truei*. Herpetologica 36:346–352.

Green, D.M., H. Kaiser, T.F. Sharbel, J. Kearsley, and K.R. McAllister. 1997. Cryptic species of spotted frogs, *Rana pretiosa* complex, in western North America. Copeia 1997:1–8.

Green, N.B. 1938a. The eastern spadefoot toad, *Scaphiopus holbrookii* Harlan, in West Virginia. Proceedings of the West Virginia Academy of Science 35:15–19.

Green, N.B. 1938b. The breeding habits of *Pseudacris brachyphona* (Cope) with a description of the eggs and tadpole. Copeia 1938:79–82.

Green, N.B. 1948. The spade-foot toad, *Scaphiopus h. holbrookii*, breeding in southern Ohio. Copeia 1948:65.

Green, N.B. 1952. A study of the life history of *Pseudacris brachyphona* (Cope) in West Virginia with special reference to behavior and growth of marked individuals. Ph.D. diss., The Ohio State University, Columbus.

Green, N.B. 1964. Postmetamorphic growth in the mountain chorus frog, *Pseudacris brachyphona* Cope. Proceedings of the West Virginia Academy of Science 36:34–38.

Green, N.B. 1969. The ratio of crescent and cruciform patterns in populations of the mountain chorus frog, *Pseudacris brachyphona*, in West Virginia. Proceedings of the West Virginia Academy of Science 41:142–144.

Green, N.B., and T.K. Pauley. 1987. Amphibians & Reptiles in West Virginia. University of Pittsburgh Press, Pittsburgh, PA.

Greenberg, C.H. 2001a. Response of reptile and amphibian communities to canopy gaps created by wind disturbance in the southern Appalachians. Forest Ecology and Management 148:135–144.

Greenberg, C.H. 2001b. Spatio-temporal dynamics of pond use and recruitment in Florida gopher frogs (*Rana capito aesopus*). Journal of Herpetology 35:74–85.

Greenberg, C.H., and G.W. Tanner. 2004. Breeding pond selection and movement patterns by eastern spadefoot toads (*Scaphiopus holbrookii*) in relation to weather and edaphic conditions. Journal of Herpetology 38:569–577.

Greenberg, C.H., and G.W. Tanner. 2005a. Spatial and temporal ecology of eastern spadefoot toads on a Florida landscape. Herpetologica 61:20–28.

Greenberg, C.H., and G.W. Tanner. 2005b. Chaos and continuity: the role of isolated ephemeral wetlands on amphibian populations in xeric sand hills. Pp. 79–90 *In* W.E. Meshaka Jr. and K.J. Babbitt (eds.), Amphibians and Reptiles. Status and Conservation in Florida. Krieger Publishing, Malabar, FL.

Greenwell, M., V. Beasley, and L.E. Brown. 1996. The mysterious decline of the cricket frog. Aquaticus 26:48–54.

Gregoire, D.R. 2005. Tadpoles of the southeastern United States coastal plain. U.S. Geological Survey

Report, Southeast Ecological Science Center, Gainesville. http://fl.biology.usgs.gov/armi/Guide_to_Tadpoles/guide_to_tadpoles.html.

Gregoire, D.R., and M.S. Gunzburger. 2008. Effects of predatory fish on survival and behavior of larval gopher frogs (*Rana capito*) and southern leopard frogs (*Rana sphenocephala*). Journal of Herpetology 42:97–103.

Gregory, P.T. 1979. Predator avoidance behavior of the red-legged frog (*Rana aurora*). Herpetologica 35:175–184.

Greuter, K.L. 2004. Early juvenile ecology of the endangered Houston toad, *Bufo houstonensis* (Anura: Bufonidae). M.S. thesis, Texas State University, San Marcos.

Greuter, K.L., and M.R.J. Forstner. 2003. *Bufo houstonensis* (Houston Toad). Growth. Herpetological Review 34:355–356.

Griffin, P.C. 1999. *Bufo californicus*, arroyo toad movement patterns and habitat preferences. M.S. thesis, University of California, San Diego.

Griffis-Kyle, K.L. 2005. Ontogenetic delays in effects of nitrite exposure on tiger salamanders (*Ambystoma tigrinum tigrinum*) and wood frogs (*Rana sylvatica*). Environmental Toxicology and Chemistry 24:1523–1527.

Griffis-Kyle, K.L. 2007. Sublethal effects of nitrite on tiger salamander (*Ambystoma tigrinum tigrinum*) and wood frog (*Rana sylvatica*) embryos and larvae: implications for field populations. Aquatic Ecology 41:119–127.

Griffis-Kyle, K.L., and M.E. Ritchie. 2007. Amphibian survival, growth and development in response to mineral nitrogen exposure and predator cues in the field: an experimental approach. Oecologia 152:633–642.

Griffiths, A.D., and J.L. McKay. 2007. Cane toads reduce the abundance and site occupancy of Merten's water monitor (*Varanus mertensi*). Wildlife Research 34:609–615.

Grinnell, J., J. Dixon, and J.M. Linsdale. 1930. Vertebrate natural history of a section of northern California through the Lassen Park region. University of California Publications in Zoology 35:1–594.

Grismer, L.L. 2002. Amphibians and Reptiles of Baja California, including its Pacific islands and the islands in the Sea of Cortés. University of California Press, Berkeley.

Grogan, C.B., and W.L. Grogan Jr. 2011. *Ollotis alvaria* (Sonoran Desert Toad). Reproduction. Herpetological Review 42:89–90.

Grogan, W.L. Jr. 1974. Notes on *Lampropeltis calligaster rhombomaculata* and *Rana virgatipes*. Bulletin of the Maryland Herpetological Society 10:33–34.

Grogan, W.L. Jr., and P.G. Bystrak. 1973a. Early breeding activity of *Rana sphenocephala* and *Bufo woodhousei fowleri* in Maryland. Bulletin of the Maryland Herpetological Society 9:106.

Grogan, W.L. Jr., and P.G. Bystrak. 1973b. The amphibians and reptiles of Kent Island, Maryland. Bulletin of the Maryland Herpetological Society 9:115–118.

Gromko, M.H., F.S. Mason, and S.J. Smith-Gill. 1973. Analysis of the crowding effect in *Rana pipiens* tadpoles. Journal of Experimental Zoology 186:63–72.

Gross, J.A., P.T.J. Johnson, L.K. Prahl, and W.H. Karasov. 2009. Critical period of sensitivity for effects of cadmium on frog growth and development. Environmental Toxicology and Chemistry 28:1227–1232.

Groves, J.D. 1980. Mass predation on a population of the American toad, *Bufo americanus*. American Midland Naturalist 103:202–203.

Groves, J.D., and F. Groves. 1978. Spider predation on amphibians and reptiles. Bulletin of the Maryland Herpetological Society 14:44–46.

Grubb, J.C. 1970. Orientation in post-reproductive Mexican toads, *Bufo valliceps*. Copeia 1970:674–680.

Grubb, J.C. 1972. Differential predation by *Gambusia affinis* on the eggs of seven species of anuran amphibians. American Midland Naturalist 88:103–108.

Grubb, J.C. 1973. Olfactory orientation in *Bufo woodhousei fowleri*, *Pseudacris clarki* and *Pseudacris streckeri*. Animal Behaviour 21:726–732.

Gruia-Gray, J., and S. Desser. 1992. Cytopathological and epizootiology of frog erythrocytic virus in bullfrogs (*Rana catesbeiana*). Journal of Wildlife Diseases 28:34–41.

Gruia-Gray, J., M. Petric, and S. Desser. 1989. Ultrastructural, biochemical and biophysical properties of an erythrocytic virus of frogs from Ontario, Canada. Journal of Wildlife Diseases 25:497–506.

Guderyahn, L., S.B. Hager, and L. Scott. 2004. Evidence to support the presence of Cope's gray treefrog (*Hyla chrysoscelis*) at Green Wing Environmental Laboratory in northcentral Illinois. Transactions of the Illinois State Academy of Science 97:219–225.

Guerry, A.D., and M.L. Hunter Jr. 2002. Amphibian distributions in a landscape of forests and agriculture: an examination of landscape composition and configuration. Conservation Biology 16:745–754.

Gunzburger, M.S. 2006. Reproductive ecology of the green treefrog (*Hyla cinerea*) in northwestern Florida. American Midland Naturalist 155:321–328.

Gunzburger, M.S., and J. Travis. 2004. Evaluating predation pressure on green treefrog larvae across a habitat gradient. Oecologia 140:422–429.

Gunzburger, M.S., and J. Travis. 2005. Effects of multiple predator species on green treefrog (*Hyla cinerea*) tadpoles. Canadian Journal of Zoology 83:996–1002.

Gunzburger, M.S., W.B. Hughes, W.J. Barichivich, and J.S. Staiger. 2010. Hurricane storm surge and amphibian communities in coastal wetlands of northwestern Florida. Wetlands Ecology and Management 18:651–663.

Guscio, C.G., B.R. Hossack, L.A. Eby, and P.S. Corn. 2007. Post breeding habitat use by adult boreal toads (*Bufo boreas*) after wildfire in Glacier National Park, USA. Herpetological Conservation and Biology 3:55–62.

Guttman, D., J.E. Bramble, and O.J. Sexton. 1991. Observations on the breeding immigration of wood frogs *Rana sylvatica* reintroduced in east-central Missouri. American Midland Naturalist 125:269–274.

Guttman, S.I. 1975. Genetic variation in the genus *Bufo*. Part 2. Isozymes in northern allopatric populations of the American toad *Bufo americanus*. Pp. 679–697 *In* C.L. Markert (ed.), Isozymes. 4. Genetics and Evolution, Academic Press, New York.

Guzy, J.C., T.S. Campbell, and K.R. Campbell. 2006. Effects of hydrological alterations on frog and toad populations at Morris Bridge wellfield, Hillsborough County, Florida. Florida Scientist 69:276–287.

Haber, V.R. 1926. The food of the Carolina treefrog, *Hyla cinerea* Schneider. Journal of Comparative Psychology 6:189–220.

Hadfield, S. 1966. Observations on body temperatures and activity in the toad *Bufo woodhousei fowleri*. Copeia 1966:581–582.

Haertel, J.D., and R.M. Storm. 1970. Experimental hybridization between *Rana pretiosa* and *Rana cascadae*. Herpetologica 26:436–446.

Haertel, J.D., A. Owczarzak, and R.M. Storm. 1974. A comparative study of the chromosomes from five species of the genus *Rana* (Amphibia: Salientia). Copeia 1974:109–114.

Hahn, D. 1968. A biogeographic analysis of the herpetofauna of the San Luis Valley, Colorado. M.S. thesis, Louisiana State University, Baton Rouge.

Hailman, J.P., and R.G. Jaeger. 1974. Phototactic responses to spectrally dominant stimuli and use of colour vision by adult anuran amphibians: a comparative survey. Animal Behaviour 22:757–795.

Hailman, J.P., and R.G. Jaeger. 1978. Phototactic responses of anuran amphibians to monochromatic stimuli of equal quantum intensity. Animal Behaviour 26:274–281.

Hale, S.F., F. Retes, and T.R. Van Devender. 1977. New populations of *Rana tarahumarae* (Tarahumara frog) in Arizona. Journal of the Arizona Academy of Sciences 11:133–134.

Hale, S.F., C.R. Schwalbe, J.L. Jarchow, C.J. May, C.H. Lowe, and T.B. Johnson. 1995. Disappearance of the Tarahumara frog. Pp. 138–140 *In* E.T. LaRoe, G.S. Farris, C.E. Puckett, P.D. Doran, and M.J. Mac (eds.), Our Living Resources. A Report to the Nation on the Distribution, Abundance, and Health of U.S. Plants, Animals, and Ecosystems. National Biological Service, Washington, DC.

Hale, S.F., P.C. Rosen, J.L. Jarchow, and G.A. Bradley. 2005. Effects of the chytrid fungus on the Tarahumara frog (*Rana tarahumarae*) in Arizona and Sonora, Mexico. Pp. 407–411 *In* G.J. Gottfried, B,S, Gebow, L.G. Eskew, C.B. Edminster, and B. Carelton (compilers), Connecting Mountain Islands and Desert Seas. Biodiversity and Management of the Madrean Archipelago. 2. USDA Forest Service Proceedings RMRS-P-36, Fort Collins, CO.

Hall, D.G. 1948. The Blowflies of North America. The Thomas Say Foundation, Baltimore.

Hall, J.A., J.H. Larsen Jr., and R.E. Fitzner. 1997. Postembryonic ontogeny of the spadefoot toad, *Scaphiopus intermontanus* (Anura: Pelobatidae): external morphology. Herpetological Monographs 11:124–178.

Hall, J.A., J.H. Larsen Jr., and R.E. Fitzner. 2002. Morphology of the premetamorphic larva of the spadefoot toad, *Scaphiopus intermontanus* (Anura: Pelobatidae), with an emphasis on the lateral line system and mouthparts. Journal of Morphology 252:114–130.

Hall, R.J. 1990. Accumulation, metabolism and toxicity of parathion in tadpoles. Bulletin of Environmental Contamination and Toxicology 44:629–635.

Hall, R.J., and D. Swineford. 1979. Uptake of methoxychlor from food and water by the American toad (*Bufo americanus*). Bulletin of Environmental Contamination and Toxicology 23:335–337.

Hall, R.J., and E. Kolbe. 1980. Bioconcentration of organophosphorus pesticides to hazardous levels by amphibians. Journal of Toxicological and Environmental Health 6:853–860.

Hall, R.J., and D. Swineford. 1980. Toxic effects of endrin and toxaphene on the southern leopard frog *Rana sphenocephala*. Environmental Pollution (Series A) 23:53–65.

Hall, R.J., and D. Swineford. 1981. Acute toxicities of toxaphene and endrin to larvae of seven species of amphibians. Toxicology Letters 8:331–336.

Hall, R.J., and B.M. Mulhern. 1984. Are anuran amphibians heavy metal accumulators? Pp. 123–133 *In* R.A. Seigel, L.E. Hunt, J.L. Knight, L. Malaret, and N.L. Zuschlag (eds.), Vertebrate Ecology and Systematics. A Tribute to Henry S. Fitch. Museum of Natural History, University of Kansas, Lawrence, KS.

Hallowell, E. 1857 ("1856"). Notice of a collection of reptiles from Kansas and Nebraska, presented to the Academy of Natural Sciences by Dr. Hammond. Proceedings of the Academy of Natural Sciences of Philadelphia 8:238–253.

Hallowell, E. 1859. Explorations and surveys for a railroad route from the Mississippi River to the Pacific Ocean. Report in California for railroad routes to connect with the routes near the 35th and 32d parallels of north latitude, explored by Lieutenant R.S. Williamson, Corps of Topographical Engineers, 1853. Part 4. Zoological Report. No. 1. Report upon reptiles of the route. War Department, Washington, DC. Pp. 1–27 + 10 plates.

Halverson, M.A., D.K. Skelly, J.M. Kiesecker, and L.K. Freidenburg. 2003. Forest mediated light regime linked to amphibian distribution and performance. Oecologia 134:360–364.

Hamer, A.J., and M.J. McDonnell. 2008. Amphibian ecology and conservation in the urbanizing world: a review. Biological Conservation 141:2432–2449.

Hamilton, H.L. 1941. The biological action of rotenone on freshwater animals. Proceedings of the Iowa Academy of Science 48:467–479.

Hamilton, W.J. Jr. 1930. Notes on the food of the American toad. Copeia 1930:45.

Hamilton, W.J. Jr. 1934. The rate of growth of the toad (*Bufo americanus americanus* Holbrook) under natural conditions. Copeia 1934:88–90.

Hamilton, W.J. 1948. The food and feeding behavior of the green frog, *Rana clamitans* Latreille, in New York State. Copeia 1948:203–207.

Hamilton, W.J. Jr. 1955. Notes on the ecology of the oak toad in Florida. Herpetologica 11:205–210.

Hamilton, W.J., and J.A. Pollack. 1956. The food of some colubrid snakes from Fort Benning, Georgia. Ecology 37:519–526.

Hammerson, G.A. 1982. Bullfrog eliminating leopard frogs in Colorado? Herpetological Review 13:115–116.

Hammerson, G.A. 1999. Amphibians and Reptiles in Colorado. A Colorado Field Guide. 2nd ed. University Press of Colorado, Niwot.

Hammerson, G.A., and L.J. Livo. 1999. Conservation status of the northern cricket frog (*Acris crepitans*) in Colorado and adjacent areas at the northwestern extent of the range. Herpetological Review 30:78–80.

Hampton, S.H., and E.P. Volpe. 1963. Development and interpopulation variability of the mouthparts of *Scaphiopus holbrooki*. American Midland Naturalist 70:319–328.

Han, B.A., P.W. Bradley, and A.R. Blaustein. 2008. Ancient behaviors of larval amphibians in response to an emerging fungal pathogen, *Batrachochytrium dendrobatidis*. Behavioral Ecology and Sociobiology 63:241–250.

Hanlin, H.G., F.D. Martin, L.D. Wike, and S.H. Bennett. 2000. Terrestrial activity, abundance and species richness of amphibians in managed forests in South Carolina. American Midland Naturalist 143:70–83.

Hannaca, W.L. 1933. The Frog Industry. Past, Present and Future. Chariton Corporation, Chicago, IL.

Hannon, S.J., C.A. Paszkowski, S. Boutin, J. DeGroot, S.E. Macdonald, M. Wheatley, and B.R. Eaton. 2002. Abundance and species composition of amphibians, small mammals and songbirds in riparian forest buffer strips of varying widths in the boreal mixedwood of Alberta. Canadian Journal of Forest Research 32:1784–1800.

Hansen, K.L. 1957. Movements, area of activity, and growth of *Rana heckscheri*. Copeia 1957:274–277.

Hansen, K.L. 1958. Breeding pattern of the eastern spadefoot toad. Herpetologica 14:57–67.

Hanson, J.A., and J.L. Vial. 1956. Defensive behavior and effects of toxins in *Bufo alvarius*. Herpetologica 12:141–149.

Harding, J.H. 1997. Amphibians and Reptiles of the Great Lakes Region. University of Michigan Press, Ann Arbor, MI.

Harding, J.H., and J.A. Holman. 1992. Michigan Frogs, Toads, and Salamanders. A Field Guide and Pocket Reference. Cooperative Extension Service, Michigan State University, East Lansing, MI.

Hardy, D.G. 1974. Some population dynamics of *Rana pipiens* in an area of southeastern South Dakota. M.S. thesis, University of South Dakota, Vermillion.

Hardy, J.D. 1953. Notes on the distribution of *Mycrohyla* [sic] *carolinensis* in southern Maryland. Herpetologica 8:162–166.

Hardy, J.D. Jr. 1964a. A new frog, *Rana palustris mansuetii*, subsp. nov. from the Atlantic Coastal Plain. Chesapeake Science 5:91–100.

Hardy, J.D. Jr. 1964b. The spontaneous occurrence of scoliosis in tadpoles of the leopard frog, *Rana pipiens*. Chesapeake Science 5:101–102.

Hardy, J.D. Jr., and J.H. Gillespie. 1976. Hybridization between *Rana pipiens* and *Rana palustris* in a modified natural environment. Bulletin of the Maryland Herpetological Society 12:41–55.

Hardy, J.D., and R.J. Burroughs. 1986. Systematic status of the spring peeper, *Hyla crucifer* (Amphibia: Hylidae). Bulletin of the Maryland Herpetological Society 22:68–89.

Hardy, L.M. 1995. Checklist of the amphibians and reptiles of the Caddo Lake watershed in Texas and Louisiana. Bulletin of the Museum of Life Sciences, Louisiana State University in Shreveport, No. 10.

Hardy, L.M. 2004. Genus *Syrrhophus* (Anura: Leptodactylidae) in Louisiana. Southwestern Naturalist 49:263–266.

Hardy, L.M., and L.R. Raymond. 1991. Observations on the activity of the pickerel frog, *Rana palustris* (Anura: Ranidae), in northern Louisiana. Journal of Herpetology 25:220–222.

Harestad, A.S. 1985. *Scaphiopus intermontanus* (Great Basin Spadefoot Toad). Mortality. Herpetological Review 16:24.

Hargis, S.E., M.-K. Harr, C.J. Henderson, W.J. Kim, and G.R. Smith. 2008. Factors influencing the distribution of overwintered bullfrog tadpoles (*Rana catesbeiana*) in two small ponds. Bulletin of the Maryland Herpetological Society 44:39–41.

Hargitt, C.W. 1888. Recent notes on *Scaphiopus holbrookii*. American Naturalist 22:535–537.

Harima, H. 1969. A survey of the herpetofauna of northwestern Florida. M.S. thesis, University of Alabama, Tuscaloosa.

Harkey, G.A., and R.D. Semlitsch. 1988. Effects of temperature on growth, development, and color polymorphism in the ornate chorus frog *Pseudacris ornata*. Copeia 1988:1001–1007.

Harlan, R. 1826. Descriptions of several new species of batrachian reptiles with observations on the larvae of frogs. American Journal of Science and Arts 10:53–66.

Harlan, R. 1835. Medical and Physical Researches, Philadelphia. Lydia R. Bailey, Philadelphia.

Harper, E.B., T.A.G. Rittenhouse, and R.D. Semlitsch. 2008. Demographic consequences of terrestrial habitat loss for pool-breeding amphibians: predicting extinction risks associated with inadequate size of buffer zones. Conservation Biology 22:1205–1215.

Harper, F. 1931. Amphibians and reptiles of the Athabaska and Great Slave Lakes region. Canadian Field-Naturalist 45:68–70.

Harper, F. 1932. A voice from the pines. Natural History 32:280–288.

Harper, F. 1933. A tree-frog new to the Atlantic Coastal Plain. Journal of the Elisha Mitchell Scientific Society 48:228–231.

Harper, F. 1935a. Records of amphibians in the southeastern states. American Midland Naturalist 16:275–310.

Harper, F. 1935b. The name of the gopher frog. Proceedings of the Biological Society of Washington 48:79–82.

Harper, F. 1937. A season with Holbrook's chorus frog (*Pseudacris ornata*). American Midland Naturalist 18:260–272.

Harper, F. 1939a. A southern subspecies of the spring peeper (*Hyla crucifera*). Notulae Naturae 27:1–4.

Harper, F. 1939b. Distribution, taxonomy, nomenclature, and habits of the little tree-frog (*Hyla ocularis*). American Midland Naturalist 22:134–149.

Harper, F. 1947. A new cricket frog (*Acris*) from the middle western states. Proceedings of the Biological Society of Washington 60:39–40.

Harper, F. 1955. A new chorus frog (*Pseudacris*) from the eastern United States. Natural History Miscellanea (150):1–6.

Harper, F. 1956. Amphibians and reptiles of the Ungava Peninsula. Proceedings of the Biological Society of Washington 69:93–104.

Harper, F. 1963. Amphibians and reptiles of Keewatin and northern Manitoba. Proceedings of the Biological Society of Washington 76:159–168.

Harris, H.S. Jr. 1968. A survey of albinism in Maryland amphibians and reptiles. Bulletin of the Maryland Herpetological Society 4:57–60.

Harris, H.S. Jr. 1969. Distributional survey: Maryland and the District of Columbia. Bulletin of the Maryland Herpetological Society 5:97–161.

Harris, H.S. Jr. 1975. Distributional survey (Amphibia/Reptilia): Maryland and the District of Columbia. Bulletin of the Maryland Herpetological Society 11:73–167.

Harris, H.S. Jr., and K. Crocetti. 2008. The eastern spadefoot toad, *Scaphiopus holbrookii*, in Maryland. Bulletin of the Maryland Herpetological Society 44:107–110.

Harris, M.L., C.A. Bishop, J. Struger, M.R. van den Heuvel, G.J. van der Kraak, D.G. Dixon, B. Ripley, and J.P. Bogart. 1998a. The functional integrity of northern leopard frog (*Rana pipiens*) and green frog (*Rana clamitans*) populations in orchard wetlands. 1. Genetics, physiology, and biochemistry of breeding adults and young-of-the-year. Environmental Toxicology and Chemistry 17:1338–1350.

Harris, M.L., C.A. Bishop, J. Struger, B. Ripley, and J.P. Bogart. 1998b. The functional integrity of northern leopard frog (*Rana pipiens*) and green frog (*Rana clamitans*) populations in orchard wetlands. 2. Effects of pesticides and eutrophic conditions on early life stage development. Environmental Toxicology and Chemistry 17:1351–1363.

Harris, M.L., L. Chora, C.A. Bishop, and J.P. Bogart. 2000. Species- and age-related differences in susceptibility to pesticide exposure for two amphibians, *Rana pipiens* and *Bufo americanus*. Bulletin of Environmental Contamination and Toxicology 64:263–270.

Harrison, J.R., J.W. Gibbons, D.H. Nelson, and C.L. Abercrombie III. 1979. Status report: amphibians. Pp. 73–78 *In* D.M. Forsythe, and W.B. Ezell Jr. (eds.), Proceedings of the First South Carolina Endangered Species Symposium. South Carolina Wildlife and Marine Resources Department.

Hartman, F.A. 1906. Food habits of Kansas lizards and batrachians. Transactions of the Kansas Academy of Science 20:225–229.

Hasken, J., J.L. Newby, A.M. Grelle, J. Boling, J. Estes, L.K. Garey, T. Wilmes, R. McKee, D. Gomez, T. Jackson, N. Gibson, E. Davinroy, C.E. Montgomery, and M.I. Kelrick. 2009. Evaluation of chytrid infection level in a newly discovered population of *Anaxyrus boreas* in the Rio Grande National Forest, Colorado, USA. Herpetological Review 40:426–428.

Hatch, A.C., and A.R. Blaustein. 2000. Combined effects of UV-B, nitrate, and low pH reduce the survival and activity level of larval Cascades Frogs (*Rana cascadae*). Archives of Environmental Contamination and Toxicology 39:494–499.

Hatch, A.C., and A.R. Blaustein. 2003. Combined effects of UV-B radiation and nitrate fertilizer on larval amphibians. Ecological Applications 13:1083–1093.

Hatch, A.C., L.K. Belden, E. Scheessele, and A.R.

Blaustein. 2001. Juvenile amphibians do not avoid potentially lethal levels of urea on soil substrate. Environmental Toxicology and Chemistry 20:2328–2335.

Hausfater, G., H.C. Gerhardt, and G.M. Klump. 1990. Parasites and mate choice in gray treefrogs, *Hyla versicolor*. American Zoologist 30:299–311.

Hawkins, C.P., L.J. Gottschalk, and S.S. Brown. 1988. Densities and habitat of tailed frog tadpoles in small streams near Mt. St. Helens following the 1980 eruption. Journal of the North American Benthological Society 7:246–252.

Hay, O.P. 1889. Notes on the life-history of *Chorophilus triseriatus*. American Naturalist 23:770–774.

Hay, O.P. 1892. The batrachians and reptiles of the State of Indiana. Indiana Department of Geology and Natural Resources, Seventeenth Annual Report, Indianapolis. Pp. 409–609.

Hay, R. 1998. Blanchard's cricket frogs in Wisconsin: a status report. Pp. 79–83 *In* M.J. Lannoo (ed.), Status & Conservation of Midwestern Amphibians. University of Iowa Press, Iowa City.

Hayes, F.E. 1989. Antipredator behavior of recently metamorphosed toads (*Bufo a. americanus*) during encounters with garter snakes (*Thamnophis s. sirtalis*). Copeia 1989:1011–1015.

Hayes, M.P., and F.S. Cliff. 1982. A checklist of the herpetofauna of Butte County, the Butte Sink, and Sutter Buttes, California. Herpetological Review 13:85–87.

Hayes, M.P., and M.M. Miyamoto. 1984. Biochemical, behavioral and body size differences between the red-legged frogs, *Rana aurora aurora* and *Rana aurora draytonii*. Copeia 1984:1018–1022.

Hayes, M.P., and M.R. Tennant. 1985. Diet and feeding behavior of the California red-legged frog, *Rana aurora draytonii* (Ranidae). Southwestern Naturalist 30:601–605.

Hayes, M.P., and M.R. Jennings. 1986. Decline of ranid frog species in western North America: are bullfrogs (*Rana catesbeiana*) responsible? Journal of Herpetology 20:490–509.

Hayes, M.P., and D.M. Kremples. 1986. Vocal sac variation among frogs of the genus *Rana* from western North America. Copeia 1986:927–936.

Hayes, M.P., and M.R. Jennings. 1988. Habitat correlates of distribution of the California red-legged frog (*Rana aurora draytonii*) and the foothill yellow-legged frog (*Rana boylii*): implications for management. Pp. 144–158 *In* R.C. Szaro, K.E. Severson, and D.R. Patton (eds.), Management of Amphibians, Reptiles, and Small Mammals in North America. USDA Forest Service General and Technical Report RM-166.

Hayes, M.P., and C.B. Hayes. 2003. *Rana aurora aurora* (Northern Red-legged Frog). Juvenile growth. Male size at maturity. Herpetological Review 34:233–234.

Hayes, M.P., and C.B. Hayes. 2004. *Bufo boreas boreas* (Boreal Toad). Scats and behavior. Herpetological Review 35:369–370.

Hayes, M.P., C.A. Pearl, and C.J. Rombough. 2001. *Rana aurora aurora* (Northern Red-legged Frog). Movement. Herpetological Review 32:35–36.

Hayes, M.P., J.D. Engler, and C.J. Rombaugh. 2006a. *Rana pretiosa* (Oregon Spotted Frog). Predation. Herpetological Review 37:209–210.

Hayes, M.P., M.R. Jennings, and G.B. Rathbun. 2006b. *Rana draytonii* (California Red-legged Frog). Prey. Herpetological Review 37:449.

Hayes, M.P., T. Quinn, D.J. Dugger, T.L. Hicks, M.A. Melchiors, and D.E. Runde. 2006c. Dispersion of coastal tailed frog (*Ascaphus truei*): an hypothesis relating occurrence of frogs in non-fish-bearing headwater basins to their seasonal movements. Journal of Herpetology 40:531–543.

Hayes, M.P., C.J. Rombough, G.E. Padgett-Flohr, L.A. Hallock, J.E. Johnson, R.S. Wagner, and J.D. Engler. 2009. Amphibian chytridiomycosis in the Oregon spotted frog (*Rana pretiosa*) in Washington state, USA. Northwestern Naturalist 90:148–151.

Hayes, T.B. 2004. There is no denying this: defusing the confusion about atrazine. BioScience 54:1138–1149.

Hayes, T.B., K. Haston, M. Tsui, A. Hoang, C. Haeffele, and A. Vonk. 2002. Feminization of male frogs in the wild. Nature 419:895–896.

Hayes, T., K. Haston, M. Tsui, A. Hoang, C. Haeffele, and A. Vonk. 2003. Atrazine-induced hermaphroditism at 0.1 ppb in American leopard frogs (*Rana pipiens*): laboratory and field evidence. Environmental Health Perspectives 111:568–575.

Hayes, T.B., P. Falso, S. Gallipeau, and M. Stice. 2010. The cause of global amphibian declines: a developmental endocrinologist's perspective. Journal of Experimental Biology 213:921–933.

Hayes-Odum, L.A. 1990. Observations on reproduction and embryonic development in *Syrrhophus cystignathoides campi* (Anura: Leptodactylidae). Southwestern Naturalist 35:358–361.

Haynes, C.M., and S.D. Aird. 1981. The distribution and habitat requirements of the wood frog (Ranidae: *Rana sylvatica* Le Conte) in Colorado. Colorado Division of Wildlife, Special Report 50, Fort Collins.

Hays, M.R. 1955. Ultragulosity in the frog *Rana aurora draytoni*. Herpetologica 11:153.

Heard, M. 1904. A California frog ranch. Out West 21:20–27.

Heath, D.R., D.A. Saugey, and G.A. Heidt. 1986. Abandoned mine fauna of the Ouachita Mountains, Arkansas: vertebrate taxa. Proceedings of the Arkansas Academy of Science 40:33–36.

Heatwole, H. 1961. Habitat selection and activity of the wood frog, *Rana sylvatica* Le Conte. American Midland Naturalist 66:301–313.

Heatwole, H., N. Poran, and P. King. 1999. Ontogenetic changes in the resistance of bullfrogs (*Rana catesbeiana*) to the venom of copperheads (*Agkistrodon contortrix contortrix*) and cottonmouths (*Agkistrodon piscivorus piscivorus*). Copeia 1999:808–814.

Hebard, W.B., and R.B. Brunson. 1963. Hind limb anomalies of a western Montana population of the Pacific tree frog, *Hyla regilla* Baird and Girard. Copeia 1963:570–572.

Hecht, M.K., and B.L. Matalas. 1946. A review of middle North American toads of the genus *Microhyla*. American Museum Novitates 1315:1–21.

Hecnar, S.J. 1995. Acute and chronic toxicity of ammonium nitrate fertilizer to amphibians from southern Ontario. Environmental Toxicology and Chemistry 14:2131–2137.

Hecnar, S.J. 1997. Amphibian pond communities in southwestern Ontario. SSAR Herpetological Conservation 1:1–15.

Hecnar, S.J., and R.T. M'Closkey. 1996a. Regional dynamics and the status of amphibians. Ecology 77:2091–2097.

Hecnar, S.J., and R.T. M'Closkey. 1996b. Amphibian species richness and distribution in relation to pond water chemistry in south-western Ontario, Canada. Freshwater Biology 36:7–15.

Hecnar, S.J., and R.T. M'Closkey. 1997a. Patterns of nestedness and species association in a pond-dwelling amphibian fauna. Oikos 80:371–381.

Hecnar, S.J., and R.T. M'Closkey. 1997b. The effects of predatory fish on amphibian species richness and distribution. Biological Conservation 79:123–131.

Hecnar, S.J., and R.T. M'Closkey. 1997c. Changes in the composition of a ranid frog community following bullfrog extinction. American Midland Naturalist 137:145–150.

Hecnar, S.J., and R.T. M'Closkey. 1998. Species richness patterns of amphibians in southwestern Ontario ponds. Journal of Biogeography 25:763–772.

Hecnar, S.J., G.S. Casper, R.W. Russell, D.R. Hecnar, and J.N. Robinson. 2002. Nested species assemblages of amphibians and reptiles on islands in the Laurentian Great Lakes. Journal of Biogeography 29:475–489.

Hedeen, S.E. 1967. Feeding behavior of the great blue heron in Itasca State Park, Minnesota. Loon 39:116–120.

Hedeen, S.E. 1971. Growth of the tadpoles of the mink frog, *Rana septentrionalis*. Herpetologica 27:160–165.

Hedeen, S.E. 1972a. Postmetamorphic growth and reproduction of the mink frog, *Rana septentrionalis* Baird. Copeia 1972:169–175.

Hedeen, S.E. 1972b. Escape behavior and causes of death of the mink frog, *Rana septentrionalis*. Herpetologica 28:261–262.

Hedeen, S.E. 1972c. Food and feeding behavior of the mink frog, *Rana septentrionalis* Baird, in Minnesota. American Midland Naturalist 88:291–300.

Hedeen, S.E. 1976. Scoliosis and hydrops in a *Rana catesbeiana* (Amphibia, Anura, Ranidae) tadpole population. Journal of Herpetology 10:261–262.

Hedeen, S.E. 1986. The southern geographic limit of the mink frog, *Rana septentrionalis*. Copeia 1986:239–244.

Hedges, S.B. 1986. An electrophoretic analysis of Holarctic hylid frog evolution. Systematic Zoology 35:1–21.

Hedges, S.B. 1989. Evolution and biogeography of West Indian frogs of the genus *Eleutherodactylus*: slow-evolving loci and the major groups. Pp. 305–370 *In* C.A. Woods (ed.), Biogeography of the West Indies. Past, Present, and Future. Sandhill Crane Press, Gainesville, FL.

Hedges, S.B., W.E. Duellman, and M.P. Heinicke. 2008. New World direct-developing frogs (Anura: Terrarana): molecular phylogeny, classification, biogeography, and conservation. Zootaxa 1737:1–182.

Hedtke, S.F., and F.A. Puglisi. 1982. Short-term toxicity of five oils to four freshwater species. Archives of Environmental Contamination and Toxicology 11:425–430.

Heemeyer, J.L., J.G. Palis, and M.J. Lannoo. 2010. *Lithobates areolatus circulosus* (Northern Crawfish Frog). Predation. Herpetological Review 41:475.

Hefleck, S.K. 2001. Ecology of the exotic Cuban tree frog, *Osteopilus septentrionalis*, within Brevard County, Florida. M.S. thesis, Florida Institute of Technology, Melbourne.

Heinen, J.T. 1993. Aggregations of newly metamorphosed *Bufo americanus*: tests of two hypotheses. Canadian Journal of Zoology 71:334–338.

Heinen, J.T. 1994. Antipredator behavior of newly metamorphosed American toads (*Bufo a. americanus*), and mechanisms of hunting by eastern garter snakes (*Thamnophis s. sirtalis*). Herpetologica 50:137–145.

Heinen, J.T. 1995. Predator cues and prey responses: a test using eastern garter snakes (*Thamnophis s. sirtalis*) and American toads (*Bufo a. americanus*). Copeia 1995:738–741.

Heinen, J.T., and G. Hammond. 1997. Antipredator behaviors of newly metamorphosed green frogs (*Rana clamitans*) and leopard frogs (*R. pipiens*) in encounters with eastern garter snakes (*Thamnophis s. sirtalis*). American Midland Naturalist 137:136–144.

Heinicke, M.P., W.E. Duellman, and S.B. Hedges. 2007. Major Caribbean and Central American frog faunas originated by ancient oceanic dispersal. Proceedings of the National Academy of Sciences, USA 104:10092–10097.

Heinrich, M.L. 1985. *Pseudacris triseriata triseriata* (Western Chorus Frog). Reproduction. Herpetological Review 16:24.

Heinrich, M.L., and D.W. Kaufman. 1985. Herpetofauna of the Konza Prairie Research Natural Area. Prairie Naturalist 17:101–112.

Hekkala, E.R., R.A. Saumure, J.R. Jaeger, H.-W. Herrman, M.J. Sredl, D.F. Bradford, D. Drabeck, and M.J. Blum. 2011. Resurrecting an extinct species: archival DNA, taxonomy, and conservation of the Vegas Valley leopard frog. Conservation Genetics 12:1379–1385.

Helbing, C.C., K. Ovaska, and L. Ji. 2006. Evaluation of the effect of acetochlor on thyroid hormone receptor gene expression in the brain and behavior of *Rana catesbeiana* tadpoles. Aquatic Toxicology 80:42–51.

Helgen, J.C., R.G. McKinnell, and M.C. Gernes. 1998. Investigation of malformed northern leopard frogs in Minnesota. Pp. 288–297 *In* M.J. Lannoo (ed.), Status & Conservation of Midwestern Amphibians. University of Iowa Press, Iowa City.

Helgen, J.C., M.C. Gernes, S.M. Kersten, J.W. Chirhart, J.T. Canfield, D. Bowers, J. Haferman, R.G. McKinnell, and D.M. Hoppe. 2000. Field investigations of malformed frogs in Minnesota 1993–97. Journal of the Iowa Academy of Science 107:96–112.

Heller, J.A. 1927. Brewer's mole as food of the bullfrog. Copeia (165):116.

Hellman, R.E. 1953. A comparative study of the eggs and tadpoles of *Hyla phaeocrypta* and *Hyla versicolor* in Florida. Publications of the Research Division, Ross Allen's Reptile Institute 1:61–74.

Hemesath, L.M. 1998. Iowa's frog and toad survey, 1991–1994. Pp. 206–216 *In* M.J. Lannoo (ed.), Status & Conservation of Midwestern Amphibians. University of Iowa Press, Iowa City.

Henderson, G.G. Jr. 1961. Reproductive potential of *Microhyla olivacea*. Texas Journal of Science 13:355–356.

Hennig, B.M., and A.J. Remsburg. 2009. Lakeshore vegetation effects on avian and anuran populations. American Midland Naturalist 161:123–133.

Henning, J.A., and G. Schirato. 2006. Amphibian use of Chehalis River floodplain wetlands. Northwestern Naturalist 87:209–214.

Henrich, T.W. 1968. Morphological evidence of secondary intergradation between *Bufo hemiophrys* Cope and *Bufo americanus* Holbrook in eastern South Dakota. Herpetologica 24:1–13.

Hensley, F.R. 1993. Ontogenetic loss of phenotypic plasticity of age at metamorphosis in tadpoles. Ecology 74:2405–2412.

Hensley, M. 1959. Albinism in North American amphibians and reptiles. Publications of the Museum, Michigan State University, Biological Series 1:133–159.

Herreid, C.F. II. 1963. Range extension for *Bufo boreas boreas*. Herpetologica 19:218.

Herreid, C.F., and S. Kinney. 1966. Survival of Alaskan woodfrog (*Rana sylvatica*) larvae. Ecology 47:1039–1041.

Herreid, C.F., and S. Kinney. 1967. Temperature and development of the wood frog, *Rana sylvatica*, in Alaska. Ecology 48:579–590.

Herrmann, H.L., K.J. Babbitt, M.J. Baber, and R.G. Congalton. 2005. Effects of landscape characteristics on amphibian distribution in a forest-dominated landscape. Biological Conservation 123:139–149.

Hewitt, H. 1950. The bullfrog as a predator on ducklings. Journal of Wildlife Management 14:244.

Hews, D.K. 1988. Alarm response in larval western toads, *Bufo boreas*: release of larval chemicals by a natural predator and its effect on predator capture efficiency. Animal Behaviour 36:125–133.

Hews, D.K., and A.R. Blaustein. 1985. An investigation of the alarm responses in *Bufo boreas* and *Rana cascadae* tadpoles. Behavioral and Neural Biology 43:47–57.

Heyer, M.M., W.R. Heyer, and R.O. de Sa. 2006. *Leptodactylus fragilis* (Brocchi). White-lipped Thin-toed Frog. Catalogue of American Amphibians and Reptiles 830.1–830.26.

Heyer, W.R. 1978. Systematics of the fuscus group of the frog genus *Leptodactylus* (Amphibia, Leptodactylidae). Natural History Museum of Los Angeles County, Science Bulletin 29:1–85.

Heyer, W.R. 2002. *Leptodactylus fragilis*, the valid name for the Middle American and northern South American white-lipped frog (Amphibia: Leptodactylidae). Proceedings of the Biological Society of Washington 115:321–322.

Higginbotham, A.C. 1939. Studies on amphibian activity. 1. Preliminary report on the rhythmic activity of *Bufo americanus americanus* Holbrook and *Bufo fowleri* Hinckley. Ecology 20:58–70.

Higgins, C., and C. Sheard. 1926. Effects of ultraviolet radiation on the early development of *Rana pipiens*. Journal of Experimental Zoology 46:333–343.

Hildebrand, H. 1949. Notes on *Rana sylvatica* in the Labrador Peninsula. Copeia 1949:168–172.

Hillis, D.M. 1981. Premating isolating mechanisms among three species of the *Rana pipiens* complex in Texas and southern Oklahoma. Copeia 1981:312–319.

Hillis, D.M. 1982. Morphological differentiation and adaptation of the larvae of *Rana berlandieri* and *Rana sphenocephala* (*Rana pipiens* complex) in sympatry. Copeia 1982:168–174.

Hillis, D.M. 1988. Systematics of the *Rana pipiens* complex: puzzle and paradigm. Annual Review of Ecology and Systematics 19:39–63.

Hillis, D.M. 2007. Constraints in naming parts of the Tree of Life. Molecular Phylogenetics and Evolution 42:331–338.

Hillis, D.M., and S.K. Davis. 1986. Evolution of ribosomal DNA: fifty million years of recorded history in the frog genus *Rana*. Evolution 40:1275–1288.

Hillis, D.M., and T.P. Wilcox. 2005. Phylogeny of the New World true frogs (*Rana*). Molecular Phylogenetics and Evolution 34:299–314.

Hillis, D.M., J.S. Frost, and D.A. Wright. 1983. Phylogeny and biogeography of the *Rana pipiens* complex: a biochemical evaluation. Systematic Zoology 32:132–143.

Hillis, D.M., A.M. Hillis, and R.F. Martin. 1984. Reproductive ecology and hybridization of the endangered Houston toad (*Bufo houstonensis*). Journal of Herpetology 18:56–72.

Hillis, D.M., J.T. Collins, and J.P. Bogart. 1987. Distribution of diploid and tetraploid species of gray tree frogs (*Hyla chrysoscelis* and *Hyla versicolor*) in Kansas. American Midland Naturalist 117:214–217.

Himes, J.G., and T.W. Bryan. 1998. Geographic distribution, *Bufo americanus charlesmithi* (Dwarf American Toad). Herpetological Review 29:246.

Hinckley, M.H. 1880. Notes on eggs and tadpoles of *Hyla versicolor*. Proceedings of the Boston Society of Natural History 21:104–107.

Hinckley, M.H. 1881. On some differences in the mouth structure of tadpoles of the anourous batrachians found in Milton, Mass. Proceedings of the Boston Society of Natural History 21:307–315.

Hinckley, M.H. 1882. Notes on the development of *Rana sylvatica* LeConte. Proceedings of the Boston Society of Natural History 22:85–95.

Hinckley, M.H. 1884. Notes on the peeping frog, *Hyla pickeringii*, Leconte. Memoirs of the Boston Society of Natural History 3:311–318.

Hine, J.S. 1911. Beneficial habits of snakes, toads and related animals. The Agricultural College Extension Bulletin, The Ohio State University 6(7):3–14.

Hine, R.L., B.L. Les, and B.F. Hellmich. 1981. Leopard frog populations and mortality in Wisconsin, 1974–1976. Wisconsin Department of Natural Resources, Technical Bulletin No. 122.

Hinshaw, S.H., and B.K. Sullivan. 1990. Predation on *Hyla versicolor* and *Pseudacris crucifer* during reproduction. Journal of Herpetology 24:196–197.

Hirai, T. 2004. Diet composition of introduced bullfrog, *Rana catesbeiana*, in the Mizorogaike Pond of Kyoto, Japan. Ecological Research 19:375–380.

Hird, D.W., S.L. Diesch, R.G. McKinnell, E. Gorham, F.B. Martin, S.W. Kurtz, and C. Dubrovolny. 1981. *Aeromonas hydrophila* in wild-caught frogs and tadpoles (*Rana pipiens*). Laboratory Animal Science 31:166–169.

Hird, D.W., S.L. Dietsch, R.G. McKinnell, E. Gorham, F.B. Martin, C.A. Meadows, and M. Gasiorowski. 1983. Enterobacteriaceae and *Aeromonas hydrophila* in Minnesota frogs and tadpoles (*Rana pipiens*). Applied and Environmental Microbiology 46:1423–1425.

Hirschfeld, C.J., and J.T. Collins. 1961. Notes on Ohio herpetology. Journal of the Ohio Herpetological Society 3:1–4.

Hisaoka, K.K., and J.C. List. 1957. The spontaneous occurrence of scoliosis in larvae of *Rana sylvatica*. Transactions of the American Microscopical Society 76:381–387.

Höbel, G. 2011. *Hyla cinerea* (Green Treefrog). Cannibalism and defensive posture. Herpetological Review 42:85–86.

Höbel, G., and A. Slocum. 2010. *Hyla cinerea* (Green Treefrog). Morphology. Herpetological Review 41:335–336.

Hoberg, T., and C. Gause. 1992. Reptiles & amphibians of North Dakota. North Dakota Outdoors 55(1):7–19.

Hobson, C.S., and E.C. Moriarty. 2003. Geographic distribution. *Pseudacris nigrita nigrita*. Herpetological Review 34:259–260.

Hobson, J.A. 1966. Electrographic correlates of behavior in the frog with special reference to sleep. Electroencephalography and Clinical Neurophysiology 22:113–121.

Hobson, J.A., C.J. Goin, and O.B. Goin. 1967. Sleep behavior of frogs. Quarterly Journal of the Florida Academy of Sciences 30:184–186.

Hock, R. 1957. Alaskan zoogeography and Alaskan amphibia. Proceedings of the Fourth Alaskan Science Conference 1953, pp. 201–206. Privately Published, Fairbanks.

Hocking, D.J., and R.D. Semlitsch. 2007. Effects of timber harvest on breeding-site selection by gray treefrogs (*Hyla versicolor*). Biological Conservation 138:506–513.

Hocking, D.J., T.A.G. Rittenhouse, B.B. Rothermel, J.R. Johnson, C.A. Conner, E.B. Harper, and R.D. Semlitsch. 2008. Breeding and recruitment phenology of amphibians in Missouri oak-hickory woodlands. American Midland Naturalist 160:41–60.

Hodge, R.P. 1976. Amphibians & Reptiles in Alaska, the Yukon & Northwest Territories. Alaska Northwest Publishing Co., Anchorage.

Hoff, J.G., and S.A. Moss. 1974. A distress call in the bullfrog, *Rana catesbeiana*. Copeia 1974:533–534.

Hoffman, A.S., J.L. Heemeyer, P.J. Williams, J.R. Robb, D.R. Karns, V.C. Kinney, N.J. Engbrecht, and M.J. Lannoo. 2010. Strong site fidelity and a variety of imaging techniques reveal around-the-clock and extended activity patterns in crawfish frogs (*Lithobates areolatus*). BioScience 60:829–834.

Hoffman, E.A., and M.S. Blouin. 2004. Evolutionary history of the Northern Leopard Frog: reconstruction of phylogeny, phylogeography, and historical changes in population demography from mitochondrial DNA. Evolution 58:145–159.

Hoffman, E.A., F.W. Schueler, and M.S. Blouin. 2004. Effective population sizes and temporal stability of genetic structure in *Rana pipiens*, the northern leopard frog. Evolution 58:2536–2545.

Hoffman, K.E. 2007. Testing the influence of Cuban treefrogs (*Osteopilus septentrionalis*) on native treefrog detection and abundance. M.S. thesis, University of Florida, Gainesville.

Hoffman, N. 2003. Frog fence along Vermont Rt. 2 in Sandbar Wildlife Management Area. Collaboration between Vermont Agency of Transportation and Vermont Agency of Natural Resources. International Conference on Ecology and Transportation 2003 Proceedings, pp. 431–432.

Hoffman, R.L. 1946. The voice of *Hyla versicolor* in Virginia. Herpetologica 3:141–142.

Hoffman, R.L. 1955. On the occurrence of two species of hylid frogs in Virginia. Herpetologica 11:30–32.

Hoffman, R.L. 1981. On the occurrence of *Pseudacris brachyphona* (Cope) in Virginia. Catesbeiana 1:9–13.

Hoffman, R.L. 1996. *Hyla chrysoscelis* also crosses the Blue Ridge: sic juvat transcendere montes. Catesbeiana 16:3–8.

Hokit, D.G., and A.R. Blaustein. 1994. The effects of kinship on growth and development in tadpoles of *Rana cascadae*. Evolution 48:1383–1388.

Hokit, D.G., and A.R. Blaustein. 1997. The effects of kinship on interactions between tadpoles of *Rana cascadae*. Ecology 78:1722–1735.

Hokit, D.G., and A. Brown. 2006. Distribution patterns of wood frogs (*Rana sylvatica*) in Denali National Park. Northwestern Naturalist 87:128–137.

Holbrook, J.E. 1836. North American Herpetology. 1st ed., Vol. 1. J. Dobson and Son, Philadelphia, PA.

Holbrook, J.E. 1840. North American Herpetology. 1st ed. Vol. 4. J. Dobson and Son, Philadelphia, PA.

Holbrook, J.E. 1842. North American Herpetology, or a Description of the Reptiles Inhabiting the United States. 2nd ed., Vol. 4. J. Dobson and Son, Philadelphia, PA.

Holland, A.A., K.R. Wilson, and M.S. Jones. 2006. Characteristics of boreal toad (*Bufo boreas*) breeding habitat in Colorado. Herpetological Review 37: 157–159.

Holloway, A.K., D.C. Canatella, H.C. Gerhardt, and D.M. Hillis. 2006. Polyploids with different origins and ancestors form a single sexual polyploid species. American Naturalist 167:E88–E101.

Holman, J.A. 1957. Bullfrog predation on the eastern spadefoot, *Scaphiopus holbrooki*. Copeia 1957:229.

Holman, J.A. 1967. Additional Miocene anurans from Florida. Quarterly Journal of the Florida Academy of Science 30:121–140.

Holman, J.A. 1969. The Pleistocene amphibians and reptiles of Texas. Publications of the Museum, Michigan State University, Biological Series 4:161–192.

Holman, J.A. 2003. Fossil Frogs and Toads of North America. Indiana University Press, Bloomington.

Holman, J.A., H.O. Jackson, and W.H. Hill. 1964. *Pseudacris streckeri illinoensis* Smith from extreme southern Illinois. Herpetologica 20:205.

Holomuzki, J.R. 1995. Oviposition sites and fish-deterrent mechanisms of two stream anurans. Copeia 1995:607–613.

Holomuzki, J.R., and N. Hemphill. 1996. Snail-tadpole interactions in streamside pools. American Midland Naturalist 136:315–327.

Holzman, N., and J.J. McManus. 1973. Effects of acclimation on metaboloic rate and thermal tolerance in the carpenter frog, *Rana vergatipes* [sic]. Comparative Biochemistry and Physiology 45A: 833–842.

Holzwart, J., and K.D. Hall. 1984. The absence of glycerol in the hibernating American toad (*Bufo americanus*). Bios 55:31–36.

Homan, R.N., B.S. Windmiller, and J.M. Reid. 2004. Critical thresholds associated with habitat loss for two vernal pool-breeding amphibians. Ecological Applications 14:1547–1553.

Hoopes, I. 1930. *Bufo* in New England. Bulletin of the Boston Society of Natural History 57:13–20.

Hoopes, I. 1938. Do you know the mink frog? New England Naturalist 1:4–6.

Hopey, M.E., and J.W. Petranka. 1994. Restriction of wood frogs to fish-free habitats: how important is adult choice versus direct predation? Copeia 1994:1023–1025.

Hopkins, W.A., M.T. Mendonca, and J.D. Congdon. 1997. Increased circulating levels of testosterone and corticosterone in Southern toads, *Bufo terrestris*, exposed to coal combustion waste. General and Comparative Endocrinology 108:237–246.

Hopkins, W.A., M.T. Mendonca, C.L. Rowe, and J.D. Congdon. 1998. Elevated trace element concentrations in Southern toads, *Bufo terrestris*, exposed to coal combustion waste. Archives of Environmental Contamination and Toxicology 35:325–329.

Hopkins, W.A., J.K. Ray, and J.D. Congdon. 2000. Incidence and impact of axial malformations in bullfrog larvae (*Rana catesbeiana*) developing in sites polluted by a coal burning power plant. Environmental Toxicology and Contamination 19: 862–868.

Hoppe, D.M. 1978. Thermal tolerance in tadpoles of the chorus frog *Pseudacris triseriata*. Herpetologica 34:318–321.

Hoppe, D.M. 1981. Chorus frogs and their colors. Pp. 7–8 *In* L. Ewell, K. Cram, and C. Johnson (eds.), Ecology of Reptiles and Amphibians in Minnesota. Proceedings of a Symposium. Bald Eagle Outdoor Learning Center, Cass Lake, MN.

Hoppe, D.M. 2000. History of Minnesota frog abnormalities: do recent findings represent a new phenomenon? Journal of the Iowa Academy of Science 107:86–89.

Hoppe, D.M. 2005. Malformed frogs in Minnesota: history and interspecific differences. Pp. 103–108 *In* M.J. Lannoo (ed.), Amphibian Declines. The Conservation Status of United States Species. University of California Press, Berkeley.

Hoppe, D.M., and D. Pettus. 1984. Developmental features influencing color polymorphism in chorus frogs. Journal of Herpetology 18:113–120.

Hoppe, D.M., and R.G. McKinnell. 1997. Observations on the status of Minnesota leopard frog populations. Pp. 38–42 *In* J.J. Moriarty and D. Jones (eds.), Minnesota's Amphibians and Reptiles. Their Conservation and Status. Minnesota Herpetological Society, Serpent's Tale Natural History Book Distributors, Lanesboro, MN.

Horn, S., J.L. Hanula, M.D. Ulyshen, and J.C. Kilgo. 2005. Abundance of green tree frogs and insects in artificial canopy gaps in a bottomland hardwood forest. American Midland Naturalist 153:321–326.

Horne, M.T., and W.A. Dunson. 1995a. Effects of low pH, metals, and water hardness on larval amphibians. Archives of Environmental Contamination and Toxicology 29:500–505.

Horne, M.T., and W.A. Dunson. 1995b. The interactive effects of low pH, toxic metals, and DOC on a simulated temporary pond community. Environmental Pollution 89:155–161.

Hossack, B.R. 2006. *Rana luteiventris* (Columbia Spotted Frog). Reproduction. Herpetological Review 37:208–209.

Hossack, B.R., and P.S. Corn. 2007a. Wildfire effects on water temperature and selection of breeding sites by the boreal toad (*Bufo boreas*) in seasonal wetlands. Herpetological Conservation and Biology 3:46–54.

Hossack, B.R., and P.S. Corn. 2007b. Responses of pond-breeding amphibians to wildfire: short-term patterns in occupancy and colonization. Ecological Applications 17:1403–1410.

Hossack, B.R., P.S. Corn, and D.S. Pilliod. 2005. Lack of significant changes in the herpetofauna of Theodore Roosevelt National Park, North Dakota, since the 1920s. American Midland Naturalist 154:423–432.

Hossack, B.R., S.A. Diamond, and P.S. Corn. 2006a. Distribution of boreal toad populations in relation to estimated UV-B dose in Glacier National Park, Montana, USA. Canadian Journal of Zoology 84:98–107.

Hossack, B.R., P.S. Corn, and D.B. Fagre. 2006b. Divergent patterns of abundance and age-class structure of headwater stream tadpoles in burned and unburned watersheds. Canadian Journal of Zoology 84:1482–1488.

Hothem, R.L., M.R. Jennings, and J.J. Crayon. 2010. Mercury contamination in three species of anuran amphibians from the Cache Creek watershed, California, USA. Environmental Monitoring Assessment 163:433–448.

Houck, W.J., and C. Henderson. 1953. A multiple appendage anomaly in the tadpole of *Rana catesbeiana*. Herpetologica 9:76–77.

Houser, H.L. Jr., and D.A. Sutton. 1969. Morphologic and karyotypic differentiation of the California frogs *Rana muscosa* and *R. boylii*. Copeia 1969:184–188.

Hoverman, J.T., M.J. Gray, and D.L. Miller. 2010. Anuran susceptibilities to ranaviruses: role of species identity, exposure route, and a novel virus isolate. Diseases of Aquatic Organisms 89:97–107.

Hovey, T.E., and E.L. Ervin. 2005. *Pseudacris cadaverina* (California Treefrog). Predation. Herpetological Review 36:304–305.

Hovingh, P. 1993a. Zoogeography and paleozoology of leeches, mollusks, and amphibians in Western Bonneville Basin, Utah, USA. Journal of Paleolimnology 9:41–54.

Hovingh, P. 1993b. Aquatic habitats, life history observations, and zoogeographic considerations of the spotted frog (*Rana pretiosa*) in Tule Valley, Utah. Great Basin Naturalist 53:168–179.

Hovingh, P. 1997. Amphibians in the eastern Great Basin (Nevada and Utah, USA): a geographical study with paleozoological models and conservation implications. Herpetological Natural History 5:97–134.

Hovingh, P., B. Benton, and D. Bornholdt. 1985. Aquatic parameters and life history observations of the Great Basin spadefoot toad in Utah. Great Basin Naturalist 45:22–30.

Howard, R.D. 1978a. The influence of male-defended oviposition sites on early embryo mortality in bullfrogs. Ecology 59:789–798.

Howard, R.D. 1978b. The evolution of mating strategies in bullfrogs, *Rana catesbeiana*. Evolution 32:850–871.

Howard, R.D. 1980. Mating behaviour and mating success in woodfrogs, *Rana sylvatica*. Animal Behaviour 28:705–716.

Howard, R.D. 1981. Sexual dimorphism in bullfrogs. Ecology 62:303–310.

Howard, R.D. 1988a. Sexual selection on male body size and mating behaviour in American toads, *Bufo americanus*. Animal Behaviour 36:1796–1808.

Howard, R.D. 1988b. Reproductive success in two species of anurans. Pp. 99–113 *In* T.H. Clutton-Brock (ed.), Reproductive Success. Studies of Individual Variation in Controlling Breeding Systems. University of Chicago Press, Chicago, IL.

Howard, R.D., and A.G. Kluge. 1985. Proximate mechanisms of sexual selection in wood frogs. Evolution 39:260–277.

Howard, R.D., and J.G. Palmer. 1995. Female choice

in *Bufo americanus*: effects of dominant frequency and call order. Copeia 1995:212–217.

Howard, R.D., and J.R. Young. 1998. Individual variation in male vocal traits and female mating preferences in *Bufo americanus*. Animal Behaviour 55:1165–1179.

Howard, R.D., H.H. Whiteman, and T.I. Schueller. 1994. Sexual selection in American toads: a test of a good-genes hypothesis. Evolution 48:1286–1300.

Howard, W.E. 1950. Birds as bullfrog food. Copeia 1950:152.

Howe, C.M., M. Berrill, B.D. Pauli, C.C. Helbring, K. Werry, and N. Veldhoen. 2004. Toxicity of glyphosate-based pesticides to four North American frog species. Environmental Toxicology and Chemistry 23:1928–1938.

Howe, G.E., R. Gillis, and R.C. Mowbray. 1998. Effect of chemical synergy and larval stage on the toxicity of atrazine and alachlor to amphibian larvae. Environmental Toxicology and Chemistry 17:519–525.

Howe, R.H. Jr. 1899. North American wood frogs. Proceedings of the Boston Society of Natural History 28:369–374.

Hoyt, D.L. 1960. Mating behavior and eggs of the Plains spadefoot. Herpetologica 14:57–67.

Hranitz, J.M. 1989. Ecological and genetic variation between demes of *Bufo woodhousei fowleri* (Hinckley) (Chordata: Amphibia) on Assateague Island and the adjacent Del-mar-va peninsula. M.S. thesis, Bloomsburg University, Bloomsburg, PA.

Hranitz, J.M. 1993. Ontogenetic changes in the expression of isozymes, population genetic structure and heterozygosity-growth relationships in *Bufo woodhousii fowleri* (Anura: Bufonidae). Ph.D. diss., Mississippi State University, Mississippi State.

Hranitz, J.M., T.S. Klinger, F.C. Hill, R.G. Sagar, T. Mencken, and J. Carr. 1993. Morphometric variation between *Bufo woodhousii fowleri* Hinckley (Anura: Bufonidae) on Assateague Island, Virginia, and the adjacent mainland. Brimleyana 19:65–75.

Huang, Y.-W., W.H. Karasov, K.A. Patnode, and C.R. Jefcoate. 1999. Exposure of northern leopard frogs in the Green Bay ecosystem to polychlorinated biphenyls, polychlorinated dibenzo-*p*-dioxins, and polychlorinated bibenzofurans is measured by direct chemistry but not hepatic ethoxyresorufin-*o*-deethylase activity. Environmental Toxicology and Chemistry 18:2123–2130.

Hubbs, C.L. 1918. *Bufo fowleri* in Michigan, Indiana and Illinois. Copeia (55):40–43.

Hubbs, C., and N.E. Armstrong. 1961. Minimum developmental temperature tolerance of two anurans, *Scaphiopus couchi* and *Microhyla olivacea*. Texas Journal of Science 13:358–362.

Hubbs, C., and F.D. Martin. 1967. *Bufo valliceps* breeding in artificial pools. Southwestern Naturalist 12:105–106.

Hubbs, C., T. Wright, and O. Cuellar. 1963. Developmental temperature tolerance of central Texas populations of two anuran amphibians *Bufo valliceps* and *Pseudacris streckeri*. Southwestern Naturalist 8:142–149.

Hudson, G.E. 1942. The amphibians and reptiles of Nebraska. University of Nebraska, Nebraska Conservation Bulletin 24. [reprinted 1972]

Huheey, J.E. 1965a. Unusual behavior in juvenile toads, *Bufo americanus*. Journal of the Ohio Herpetological Society 5:33.

Huheey, J.E. 1965b. A toad in the diet of a frog. Journal of the Ohio Herpetological Society 5:33.

Huheey, J.E., and A. Stupka. 1967. Amphibians and Reptiles of Great Smoky Mountains National Park. University of Tennessee Press, Knoxville.

Hulse, A.C., C.J. McCoy, and E.J. Censky. 2001. Amphibians and Reptiles of Pennsylvania and the Northeast. Cornell University Press, Ithaca, NY.

Hunka, D.L. 1974. Physiological aspects of overwintering in the boreal chorus frog, *Pseudacris triseriata maculata*. M.S. thesis, University of Manitoba, Winnipeg.

Hunsaker, D. II, and P. Breese. 1967. Herpetofauna of the Hawaiian Islands. Pacific Science 21:423–428.

Hunt, R.H. 1980. Toad sanctuary in a tarantula burrow. Natural History 89(3):49–53.

Hunter, M.L. Jr., A.J.K. Calhoun, and M. McCollough. 1999. Maine Amphibians and Reptiles. University of Maine Press, Orono.

Hurter, J. 1911. Herpetology of Missouri. Transactions of the Academy of Science of St. Louis 20(5):59–274.

Hutchison, V.H., and L.G. Hill. 1978. Thermal selection of bullfrog tadpoles (*Rana catesbeiana*) at different stages of development and acclimation temperatures. Journal of Thermal Biology 3:57–60.

Ideker, J. 1968. Secondary intergradation between *Bufo americanus americanus* Holbrook and *Bufo woodhousei woodhousei* Girard in southeastern South Dakota. M.A. thesis, University of South Dakota, Vermillion.

Ideker, J. 1976. Tadpole thermoregulatory behavior facilitates grackle predation. Texas Journal of Science 27:244–245.

Ideker, J. 1979. Adult *Cybister fimbriolatus* are predaceous (Coleoptera: Dytiscidae). Coleopterists Bulletin 33:41–44.

Illingworth, J.F. 1941. Feeding habits of *Bufo marinus*. Proceedings of the Hawaiian Entomological Society 11:51.

Ingles, L.G. 1932a. *Cephalogonimus brevicirrus*, a new species of trematode from the intestine of *Rana aurora* from California. University of California Publications in Zoology 37:203–210.

Ingles, L.G. 1932b. Four new species of *Haematoloechus* (Trematoda) from *Rana aurora draytoni* from

California. University of California Publications in Zoology 37:189–201.

Ingles, L.G. 1936. Worm parasites of California amphibia. Transactions of the American Microscopical Society 55:73–92.

Ingles, L.G., and M.P. Langston. 1933. A new species of bladder fluke from California frogs. Transactions of the American Microscopical Society 52:243–245.

Ingram, W.M., and E.C. Raney. 1943. Additional studies on the movement of tagged bullfrogs, *Rana catesbeiana* Shaw. American Midland Naturalist 29:239–241.

Irwin, J.T. 2005. Overwintering in northern cricket frogs (*Acris crepitans*). Pp. 55–58 *In* M.J. Lannoo (ed.), Amphibian Declines. The Conservation Status of United States Species. University of California Press, Berkeley.

Irwin, J.T., J.P. Costanzo, and R.E. Lee Jr. 1999. Terrestrial hibernation in the northern cricket frog, *Acris crepitans*. Canadian Journal of Zoology 77:1240–1246.

Irwin, K.J., C.K. Dodd Jr., and M.L. Griffey. 1999. Geographic distribution. *Scaphiopus holbrooki* (Eastern Spadefoot Toad). Herpetological Review 30:232.

Irwin, K.J., L. Irwin, T.W. Taggart, J.T. Collins, and S.L. Collins. 2001. An herpetofaunal survey of the Apalachicola barrier islands, Florida: the 2000–2001 season. Kansas Herpetological Society Newsletter (123):12–15.

Irwin, M.S. 1929. A new lung fluke from *Rana clamitans* Latreille. Transactions of the American Microscopical Society 48:74–79.

Isaacs, J.S. 1971. Temporal stability of vertebral stripe color in a cricket frog population. Copeia 1971:551–552.

Iverson, J.B. 1973. *Hyla cinerea* (Greeg treefrog). HISS News Journal 1:153.

Jackman, R., S. Nowicki, D.J. Aneshansley, and T. Eisner. 1983. Predatory capture of toads by fly larvae. Science 222:515–516.

Jackson, A.W. 1952. The effect of temperature, humidity, and barometric pressure on the rate of call in *Acris crepitans* Baird in Brazos County, Texas. Herpetologica 8:18–20.

Jackson, C.G. Jr., and M.M. Jackson. 1970. Herpetofauna of Dauphin Island, Alabama. Quarterly Journal of the Florida Academy of Sciences 33:281–287.

Jackson, J.T., F.W. Weckerly, T.M. Swannack, and M.R.J. Forstner. 2006. Inferring absence of Houston toads given imperfect detection probabilities. Journal of Wildlife Management 70:1461–1463.

Jacobson, N.L. 1989. Breeding dynamics of the Houston toad. Southwestern Naturalist 34:374–380.

Jaeger, J.R., B.R. Riddle, R.D. Jennings, and D.F. Bradford. 2001. Rediscovering *Rana onca*: evidence for phylogenetically distinct leopard frogs from the border region of Nevada, Utah, and Arizona. Copeia 2001:339–354.

Jaeger, R.G., and J.P. Hailman. 1971. Two types of phototactic behaviour in anuran amphibians. Nature 230:189–190.

Jaeger, R.G., and J.P. Hailman. 1973. Effects of intensity on the phototactic responses of adult anuran amphibians: a comparative survey. Zeitschrift für Tierpsychologie 33:352–407.

Jaeger, R.G., and J.P. Hailman. 1976. Ontogenetic shift of spectral phototactic preferences in anuran tadpoles. Journal of Comparative Physiology and Psychology 90:930–945.

James, M.T., and T.P. Maslin. 1947. Notes on myiasis of the toad, *Bufo boreas boreas* Baird and Girard. Journal of the Washington Academy of Sciences 37:366–368.

James, P. 1966. The Mexican burrowing toad, *Rhinophrynus dorsalis*, an addition to the vertebrate fauna of the United States. Texas Journal of Science 18:272–276.

James, S.M., E.E. Little, and R.D. Semlitsch. 2004. Effects of multiple routes of cadmium exposure on the hibernation success of the American Toad (*Bufo americanus*). Archives of Environmental Contamination and Toxicology 46:518–527.

James, S.M., E.E. Little, and R.D. Semlitsch. 2005. Metamorphosis of two amphibian species after chronic cadmium exposure in outdoor aquatic mesocosms. Environmental Toxicology and Chemistry 24:1994–2001.

Jameson, D.L. 1950. The development of *Eleutherodactylus latrans*. Copeia 1950:44–46.

Jameson, D.L. 1954. Social patterns in the leptodactylid frogs *Syrrhophus* and *Eleutherodactylus*. Copeia 1954:36–38.

Jameson, D.L. 1955. The population dynamics of the cliff frog, *Syrrhophus marnocki*. American Midland Naturalist 54:342–381.

Jameson, D.L. 1956. Growth, dispersal and survival of the Pacific tree frog. Copeia 1956:25–29.

Jameson, D.L. 1957. Population structure and homing responses in the Pacific tree frog. Copeia 1957:221–228.

Jameson, D.L., and A.G. Flury. 1949. The reptiles and amphibians of the Sierra Vieja range of southwestern Texas. Texas Journal of Science 1:54–79.

Jameson, D.L., and R.M.E. Myers. 1957. Albino Pacific tree frogs. Herpetologica 13:74.

Jameson, D.L., and S. Pequegnat. 1971. Estimation of relative viability and fecundity of color polymorphism in anurans. Evolution 25:180–194.

Jameson, D.L., and R.C. Richmond. 1971. Parallelism and convergence in the evolution of size and shape in Holarctic *Hyla*. Evolution 25:497–508.

Jameson, D.L., J.P. Mackey, and R.C. Richmond. 1966. The systematics of the Pacific tree frog, *Hyla regilla*. Proceedings of the California Academy of Sciences 33(19):551–620.

Jameson, D.L., W. Taylor, and J. Mountjoy. 1970. Metabolic and morphological adaptation to heterogeneous environments by the Pacific tree toad, *Hyla regilla*. Evolution 24:75–89.

Jameson, D.L., J.P. Mackey, and M. Anderson. 1973. Weather, climate, and the external morphology of Pacific tree toads. Evolution 27:285–302.

Jameson, E.W. Jr. 1947. The food of the western cricket frog. Copeia 1947:212.

Jamieson, D.H., and S.E. Trauth. 1996. Dietary diversity and overlap between two subspecies of spadefoot toads (*Scaphiopus holbrookii holbrookii* and *S. h. hurterii*) in Arkansas. Proceedings of the Arkansas Academy of Science 50:75–78.

Jamieson, D.H., S.E. Trauth, and C.T. McAllister. 1993. Food habits of male bird-voiced treefrogs, *Hyla avivoca* (Anura: Hylidae), in Arkansas. Texas Journal of Science 45:45–49.

Jansen, K.P., A.P. Summers, and P.R. Delis. 2001. Spadefoot toads (*Scaphiopus holbrookii holbrookii*) in an urban landscape: effects of nonnatural substrates on burrowing in adults and juveniles. Journal of Herpetology 35:141–145.

Jasienski, M. 1992. Kinship effects and plasticity of growth in amphibian larvae. Ph.D. diss., Harvard University, Cambridge, MA.

Jaslow, A.P., and R.C. Vogt. 1977. Identification and distribution of *Hyla versicolor* and *Hyla chrysoscelis* in Wisconsin. Herpetologica 33:201–205.

Jay, J.M., and W.J. Pohley. 1981. *Dermosporidium penneri* sp. n. from the skin of the American toad, *Bufo americanus* (Amphibia: Bufonidae). Journal of Parasitology 67:108–110.

Jefferson, D.M., and R.W. Russell. 2008. Ontogenetic and fertilizer effects on stable isotopes in the green frog (*Rana clamitans*). Applied Herpetology 5:189–196.

Jenkins, J.R. 2000. Bioenergetics of freeze-thaw cycles in the chorus frog (*Pseudacris triseriata*). M.S. thesis, University of South Dakota, Vermillion.

Jennette, M.A. 2010. *Lithobates sylvaticus* (Wood Frog). Egg predation. Herpetological Review 41:476–477.

Jenni, D.A. 1969. A study of the ecology of four species of herons during the breeding season at Lake Alice, Alachua County, Florida. Ecological Monographs 39:245–270.

Jennings, M.R. 1988. Natural history and decline of native ranids in California. Pp. 61–72 *In* H.P. De Lisle, P.R. Brown, B. Kaufman, and B.M. McGurty (eds.), Proceedings of the Conference on California Herpetology. Southwestern Herpetologists Society, Special Publication No. 4, Van Nuys, CA.

Jennings, M.R. 2005. *Rana fisheri* Stejneger, 1893. Vegas Valley Leopard Frog. Pp. 554–555 *In* M.J. Lannoo (ed.), Amphibian Declines. The Conservation Status of United States Species. University of California Press, Berkeley.

Jennings, M.R., and M.P. Hayes. 1985. Pre-1900 overharvest of California red-legged frogs (*Rana aurora draytonii*): the inducement for bullfrog (*Rana catesbeiana*) introduction. Herpetologica 41:94–103.

Jennings, M.R., and M.P. Hayes. 1994a. Decline of native ranid frogs in the desert Southwest. Pp. 183–211 *In* P.R. Brown and J.W. Wright (eds.), Herpetology of the North American Deserts. Proceedings of a Symposium. Southwestern Herpetologists Society, Special Publication No. 5.

Jennings, M.R., and M.P. Hayes. 1994b. Amphibian and reptile species of special concern in California. California Department of Fish and Game, Inland Fisheries Division, Rancho Cordova, CA.

Jennings, M.R., D.F. Bradford, and D.F. Johnson. 1992. Dependence of the garter snake *Thamnophis elegans* on amphibians in the Sierra Nevada of California. Journal of Herpetology 26:503–505.

Jennings, R.D., and N.J. Scott Jr. 1993. Ecologically correlated morphological variation in tadpoles of the leopard frog, *Rana chiricahuensis*. Journal of Herpetology 27:285–293.

Jense, G.K., and R.L. Linder. 1970. Food habits of badgers in eastern South Dakota. Proceedings of the South Dakota Academy of Science 49:37–41.

Jensen, J.B. 2004. Little grass frog *Pseudacris ocularis* (Bosc and Daudin). Pp. 27 *In* R.E. Mirarchi, M.A. Bailey, T.M. Haggerty, and T.L. Best (eds.), Alabama Wildlife. Vol. 3. Imperiled Amphibians, Reptiles, Birds, and Mammals. University of Alabama Press, Tuscaloosa.

Jensen, J.B., J.G. Palis, and M.A. Bailey. 1995. *Rana capito sevosa* (Dusky Gopher Frog). Submerged vocalization. Herpetological Review 26:98.

Jensen, J.B., M.A. Bailey, E.L. Blankenship, and C.D. Camp. 2003. The relationship between breeding by the gopher frog, *Rana capito* (Amphibia: Ranidae) and rainfall. American Midland Naturalist 150:185–190.

Jensen, J.B., C.D. Camp, W. Gibbons, and M.J. Elliott (eds.). 2008. Amphibians and Reptiles of Georgia. University of Georgia Press, Athens.

Jenssen, T.A. 1967. Food habits of the green frog, *Rana clamitans*, before and during metamorphosis. Copeia 1967:214–218.

Jenssen, T.A., and W.D. Klimstra. 1966. Food habits of the green frog, *Rana clamitans*, in southern Illinois. American Midland Naturalist 76:169–182.

Jenssen, T.A., and W.B. Preston. 1968. Behavioral responses of the male green frog, *Rana clamitans*, to its recorded call. Herpetologica 24:181–182.

Jilek, R., and R. Wolff. 1978. Occurrence of *Spinitectus gracilis* (Nematoda, Spiuroidea) in the toad

Bufo woodhousei fowleri. Journal of Parasitology 64:619.

Jirků, M., M.G. Bolek, C.M. Whipps, J. Janovy Jr., M.L. Kent, and D. Modrý. 2006. A new species of *Myxidium* (Myxosporea: Myxidiidae), from the western chorus frog, *Pseudacris triseriata triseriata*, and Blanchard's cricket frog, *Acris crepitans blanchardi* (Hylidae), from eastern Nebraska: morphology, phylogeny, and critical comments on amphibian *Myxidium* taxonomy. Journal of Parasitology 92:611–619.

Jofré, M.B., and W.H. Karasov. 1999. Direct effect of ammonia on three species of North American anuran amphibians. Environmental Toxicology and Chemistry 18:1806–1812.

Jofré, M.B., M.L. Rosenshield, and W.H. Karasov. 2000. Effects of PCB 126 and ammonia, alone and in combination, on green frog (*Rana clamitans*) and leopard frog (*R. pipiens*) hatching success, development, and metamorphosis. Journal of the Iowa Academy of Science 107:113–122.

Joglar, R.L. 1998. Los Coquís de Puerto Rico. Su Historia Natural y Conservacion. Editorial de la Universidad de Puerto Rico, San Juan.

Johansen, F. 1926. Occurrences of frogs on Anticosti Island and Newfoundland. Canadian Field-Naturalist 40:16.

Johansen, K. 1962. Observations on the wood frog *Rana sylvatica* in Alaska. Ecology 43:146–147.

John, K.R., and D. Fenster. 1975. The effects of partitions on the growth rates of crowded *Rana pipiens* tadpoles. American Midland Naturalist 93:123–130.

John-Alder, H.B., P.J. Morin, and S.P. Lawler. 1988. Thermal physiology, phenology, and distribution of tree frogs. American Naturalist 132:506–520.

John-Alder, H.B., M.C. Barnhart, and A.F. Bennett. 1989. Thermal sensitivity of swimming performance and muscle contraction in northern and southern populations of tree frogs (*Hyla crucifer*). Journal of Experimental Biology 142:357–372.

Johnson, B. 1989. Familiar Amphibians and Reptiles of Ontario. Natural Heritage/Natural History, Inc., Toronto, Ontario.

Johnson, B.K., and J.L. Christiansen. 1976. The food and food habits of Blanchard's cricket frog, *Acris crepitans blanchardi* (Amphibia, Anura, Hylidae), in Iowa. Journal of Herpetology 10:63–74.

Johnson, C. 1959. Genetic incompatibility in the call races of *Hyla versicolor* Le Conte in Texas. Copeia 1959:327–335.

Johnson, C. 1963. Additional evidence of sterility between the call-types in the *Hyla versicolor* complex. Copeia 1963:139–143.

Johnson, C. 1966. Species recognition in the *Hyla versicolor* complex. Texas Journal of Science 18: 361–364.

Johnson, C.R. 1980. The effects of five organophosphorus insecticides on thermal stress in tadpoles of the Pacific treefrog, *Hyla regilla*. Zoological Journal of the Linnaean Society 69:143–147.

Johnson, C.R., and R.B. Bury. 1965. Food of the Pacific treefrog, *Hyla regilla* Baird and Girard, in northern California. Herpetologica 21:56–58.

Johnson, C.R., and J.E. Prine. 1976. The effects of sublethal concentrations of organophosphorus insecticides and an insect growth regulator on temperature tolerance in hydrated and dehydrated juvenile Western toads, *Bufo boreas*. Comparative Biochemistry and Physiology 53A:147–149.

Johnson, D.H., M.D. Bryant, and A.H. Miller. 1948. Vertebrate animals of the Providence Mountains area of California. University of California Publications in Zoology 48:221–376.

Johnson, D.H., S.C. Fowle, and J.A. Jundt. 2000. The North American reporting center for amphibian malformations. Journal of the Iowa Academy of Science 107:123–127.

Johnson, J.R. 2005. Multi-scale investigations of gray treefrog movements: patterns of migration, dispersal, and gene flow. Ph.D. diss., University of Missouri, Columbia.

Johnson, J.R., and R.D. Semlitsch. 2003. Defining core habitat of local populations of the gray treefrog (*Hyla versicolor*) based on choice of oviposition site. Oecologia 137:205–210.

Johnson, K.A. 2003. Abiotic factors influencing the breeding, movement, and foraging of the eastern spadefoot (*Scaphiopus holbrookii*) in West Virginia. M.S. thesis, Marshall University, Huntington, WV.

Johnson, P.T.R., and D.R. Sutherland. 2003. Amphibian deformities and *Ribeiroia* infection: an emerging helminthiasis. Trends in Parasitology 19:332–335.

Johnson, P.T.R., K.B. Lunde, E.G. Ritchie, and A.E. Launer. 1999. The effect of trematode infection on amphibian limb development and survivorship. Science 284:802–804.

Johnson, P.T.R., K.B. Lunde, R.W. Haight, J. Bowerman, and A.R. Blaustein. 2001a. *Rebeiroia ondatrae* (Trematoda: Digenea) infection induces severe limb malformations in Western toads (*Bufo boreas*). Canadian Journal of Zoology 79:370–379.

Johnson, P.T.R., K.B. Lunde, E.G. Ritchie, J.K. Reaser, and A.E. Launer. 2001b. Morphological abnormality patterns in a California amphibian community. Herpetologica 57:336–352.

Johnson, P.T.R., K.B. Lunde, E.M. Thurman, E.G. Ritchie, S.N. Wray, D.R. Sutherland, J.M. Kapfer, T.J. Frest, J. Bowerman, and A.R. Blaustein. 2002. Parasite (*Riberoia ondatrae*) infection linked to amphibian malformations in the western United States. Ecological Monographs 72:151–168.

Johnson, P.T.R., V.J. McKenzie, A.C. Peterson, J.L. Kerby, J. Brown, A.R. Blaustein, and T. Jackson. 2011. Regional decline of an iconic amphibian associated with elevation, land-use change, and invasive species. Conservation Biology 25:556–566.

Johnson, S.A., and M. McGarrity. 2010. Identification Guide to the Frogs of Florida. University of Florida Extension Service, Gainesville.

Johnson, S.A., M.E. McGarrity, and C.L. Staudhammer. 2010. An effective chemical deterrent for invasive Cuban treefrogs. Human-Wildlife Interactions 4:112–117.

Johnson, T.R. 1977. The amphibians of Missouri. University of Kansas Museum of Natural History, Public Education Series No. 6.

Johnson, T.R. 2000. The Amphibians and Reptiles of Missouri. Missouri Department of Conservation, Jefferson City.

Johnston, G.R., and J.C. Johnston. 2006. *Scaphiopus holbrookii holbrookii* (Eastern Spadefoot). Albinism. Herpetological Review 37:211–212.

Jolley, D.B., S.S. Ditchkoff, B.D. Sparklin, L.B. Hanson, M.S. Mitchell, and J.B. Grand. 2010. Estimate of herpetofauna depredation by a population of wild pigs. Journal of Mammalogy 91:519–524.

Jones, D. 2004. Aquatic and terrestrial use of habitat by the Amargosa Toad (*Bufo nelsoni*). M.S. thesis, University of Nevada, Reno.

Jones, D., E.T. Simandle, C.R. Tracy, and B. Hobbs. 2003. *Bufo nelsoni* (Amargosa Toad). Predation. Herpetological Review 34:229.

Jones, J.M. 1973. Effects of thirty years hybridization on the toads *Bufo americanus* and *Bufo woodhousei fowleri* at Bloomington, Indiana. Evolution 27:435–448.

Jones, K., B. Rincon, and D.A. Steen. 2010. *Bufo terrestris* (Southern Toad). Predation. Herpetological Review 41:334–335.

Jones, K.B. 1981. Distribution, ecology, and habitat management of the reptiles and amphibians of the Hualapai-Aquarius Planning Area, Mohave and Yavapai counties, Arizona. U.S. Department of the Interior, Bureau of Land Management, Technical Note 353.

Jones, K.B. 1990. Habitat use and predatory behavior of *Thamnophis cyrtopsis* (Serpentes: Colubridae) in a seasonally variable aquatic environment. Southwestern Naturalist 35: 115–122.

Jones, L., D.R. Gossett, S.W. Banks, and M.L. McCallum. 2010. Antioxidant defense system in tadpoles of the American bullfrog (*Lithobates catesbeianus*) exposed to paraquat. Journal of Herpetology 44:222–228.

Jones, L.L.C., and M.J. Sredl. 2005. Chiricahua leopard frog status in the Galiuro Mountains, Arizona, with a monitoring framework for the species' entire range. Pp. 88–91 In G.J. Gottfried, B.S. Gebow, L.G. Eskew, C.B. Edminster, and B. Carelton (compilers), Connecting Mountain Islands and Desert Seas: Biodiversity and Management of the Madrean Archipelago. 2. USDA Forest Service Proceedings RMRS-P-36, Fort Collins, CO.

Jones, L.L.C., W.P. Leonard, and D.H. Olson (eds.). 2005. Amphibians of the Pacific Northwest. Seattle Audubon Society, Seattle, WA.

Jones, M.S., J.P. Goettl, and L.J. Livo. 1999. *Bufo boreas* (Boreal Toad). Predation. Herpetological Review 30:91.

Jones, R.M. 1980. Nitrogen excretion by *Scaphiopus* tadpoles in ephemeral ponds. Physiological Zoology 53:26–31.

Jones, T.R., and R.J. Timmons. 2010. *Hyla wrightorum* (Arizona Treefrog). Predation. Herpetological Review 41:473–474.

Jordan, O.R. Jr., W.W. Byrd, and D.E. Ferguson. 1968a. Sun-compass orientation in *Rana pipiens*. Herpetologica 24:335–336.

Jordan, O.R., J.S. Garton, and R.F. Ellis. 1968b. The amphibians and reptiles of a middle Tennessee cedar glade. Journal of the Tennessee Academy of Science 43:72–78.

Joy, J.E., and B.T. Dowell. 1994. *Glypthelmins pennsylvaniensis* (Trematode: Digenea) in the spring peeper, *Pseudacris c. crucifer* (Anura: Hylidae), from southwestern West Virginia. Journal of the Helminthological Society of Washington 61:227–229.

Joy, J.E., and C.A. Bunten. 1997. *Cosmocercoides variabilis* (Nematoda, Cosmocercoidea) populations in the eastern American toad, *Bufo a. americanus* (Salientia, Bufonidae), from western West Virginia. Journal of the Helminthological Society of Washington 64:102–105.

Joy, J.E., E.M. Walker, S.G. Koh, J.M. Bentley, and A.G. Crank. 1996. Intrahepatic larval nematode infection in the Northern Spring Peeper, *Pseudacris crucifer crucifer* (Anura: Hylidae), in West Virginia. Journal of Wildlife Diseases 32:340–343.

Judd, F.W., and K.J. Irwin. 2005. *Hypopachus variolosus* (Cope, 1866). Sheep Frog. Pp. 506–508 In M.J. Lannoo (ed.), Amphibian Declines. The Conservation Status of United States Species. University of California Press, Berkeley.

Judge, K.A., and R.J. Brooks. 2001. Chorus participation by male bullfrogs, *Rana catesbeiana*: a test of the energetic constraint hypothesis. Animal Behaviour 62:849–861.

Jung, R.E. 1993. Blanchard's cricket frogs (*Acris crepitans blanchardi*) in southwest Wisconsin. Transactions of the Wisconsin Academy of Sciences, Arts and Letters 81:79–87.

Jung, R.E., and C.H. Jagoe. 1995. Effects of low pH and aluminum on body size, swimming performance, and susceptibility to predation of green tree frog (*Hyla cinerea*) tadpoles. Canadian Journal of Zoology 73:2171–2183.

Jung, R.E., S. Claeson, J.E. Wallace, and W.C. Welbourn Jr. 2001. *Eleutherodactylus guttilatus* (Spotted Chirping Frog), *Bufo punctatus* (Red-spotted Toad), *Hyla arenicolor* (Canyon Tree Frog), and *Rana berlandieri* (Rio Grande Leopard Frog). Mite infestation. Herpetological Review 32:33–34.

Jung, R.E., K.E. Bonine, M.L. Rosenshield, A. de la Rieza, S. Raimondo, and S. Droege. 2002. Evaluation of canoe surveys for anurans along the Rio Grande in Big Bend National Park, Texas. Journal of Herpetology 36:390–397.

Jungels, J.M., K.L. Griffis-Kyle, and W.J. Boeing. 2010. Low genetic differentiation among populations of the Great Plains toad (*Bufo cognatus*) in southern New Mexico. Copeia 2010:388–396.

Justis, C.S., and D.H. Taylor. 1976. Extraocular photoreception and compass orientation in bullfrogs, *Rana catesbeiana*. Copeia 1976:98–105.

Justus, J.T., M. Sandomir, T. Urquhart, and B.O. Ewan. 1977. Developmental rates of two species of toads from the desert Southwest. Copeia 1977:592–594.

Kaess, W., and F. Kaess. 1960. Perception of apparent motion in the common toad. Science 132:953.

Kagarise Sherman, C. 1980. A comparison of the natural history and mating system of two anurans: Yosemite toads (*Bufo canorus*) and black toads (*Bufo exsul*). Ph.D. diss., University of Michigan, Ann Arbor.

Kagarise Sherman, C., and M.L. Morton. 1984. The toad that stays on its toes. Natural History 93(3):73–78.

Kagarise Sherman, C., and M.L. Morton. 1993. Population declines of Yosemite toads in the eastern Sierra Nevada of California. Journal of Herpetology 27:186–198.

Kalb, H.J., and G.R. Zug. 1990. Age estimates for a population of American toads, *Bufo americanus* (Salientia: Bufonidae), in northern Virginia. Brimleyana 16:79–86.

Kaminsky, S. 1997. *Bufo americanus* (American Toad). Reproduction. Herpetological Review 28:84.

Kapfer, J.M., and T.C. Muehlfeld. 2006. *Rana sylvatica* (Wood Frog) and *Ambystoma maculatum* (Spotted Salamander). Behavior. Herpetological Review 37:210–211.

Kaplan, H.M., and L. Yoh. 1961. Toxicity of copper for frogs. Herpetologica 17:131–135.

Kaplan, H.M., and J.G. Overpeck. 1964. Toxicity of halogenated hydrocarbon insecticides for the frog, *Rana pipiens*. Herpetologica 20:163–169.

Kaplan, H.M., and S.S. Glaczenski. 1965. Hematological effects of organophosphate insecticides in the frog (*Rana pipiens*). Life Sciences 4:1213–1219.

Karlstrom, E.L. 1954. On robins and tadpoles. Yosemite Nature Notes 33:185–188.

Karlstrom, E.L. 1958. Sympatry of the Yosemite and Western toads in California. Copeia 1958:152–153.

Karlstrom, E.L. 1962. The toad genus *Bufo* in the Sierra Nevada of California: ecological and systematic relationships. University of California Publications in Zoology 62:1–104.

Karlstrom, E.L. 1986. Amphibian recovery in the North Fork Toutle River debris avalanche area of Mount St. Helens. Pp. 334–344 *In* S.A.C. Keller (ed.), Mount St. Helens. Five Years Later. Eastern Washington University Press, Cheney.

Karlstrom, E.L., and R.L. Livezey. 1955. The eggs and larvae of the Yosemite toad *Bufo canorus* Camp. Herpetologica 11:221–227.

Karns, D.R. 1983. Toxic bog water in northern Minnesota peatlands: ecological and evolutionary consequences for breeding amphibians. Ph.D. diss., University of Minnesota, Minneapolis.

Karns, D.R. 1992. Effects of acidic bog habitats on amphibian reproduction in a northern Minnesota peatland. Journal of Herpetology 26:401–412.

Karraker, N.E. 2007. Are embryonic and larval green frogs (*Rana clamitans*) insensitive to road deicing salt? Herpetological Conservation and Biology 2:35–41.

Karraker, N.E. 2008. Impacts of road deicing salts on amphibians and their habitats. Pp. 211–223 *In* J.C. Mitchell, R.E. Jung Brown, and B. Bartholomew (eds.), Urban Herpetology. Herpetological Conservation 3. Society for the Study of Amphibians and Reptiles, Salt Lake City, UT.

Karraker, N.E., and G.S. Beyersdorf. 1997. A tailed frog (*Ascaphus truei*) nest site in northwestern California. Northwestern Naturalist 78:110–111.

Karraker, N.E., and J.P. Gibbs. 2009. Amphibian production in forested landscapes in relation to wetland hydroperiod: a case study of vernal pools and beaver ponds. Biological Conservation 142:2293–2302.

Karraker, N.E., and G.R. Ruthig. 2009. Effect of road deicing salt on the susceptibility of amphibian embryos to infection by water molds. Environmental Research 109:40–45.

Karraker, N.E., and J.P. Gibbs. 2011. Contrasting road effect signals in reproduction of long- versus short-lived amphibians. Hydrobiologia 664:213–218.

Karraker, N.E., D.S. Pilliod, M.J. Adams, E.L. Bull, P.S. Corn, L.V. Diller, L.A. Dupuis, M.P. Hayes, B.R. Hossack, G.R. Hodgson, E.J. Hyde, K. Lohman, B.R. Norman, L.M. Ollivier, C.A. Pearl, and C.R. Peterson. 2006. Taxonomic variation in oviposition by tailed frogs (*Ascaphus* spp.). Northwestern Naturalist 87:87–97.

Kats, L.B., and R.G. Van Dragt. 1986. Background color-matching in the spring peeper, *Hyla crucifer*. Copeia 1986:109–115.

Kats, L.B., J.W. Petranka, and A. Sih. 1988. Antipredator defenses and the persistence of amphibian larvae with fishes. Ecology 69:1865–1870.

Kay, F.R. 1969. An albino Pacific tree frog, *Hyla regilla*, from Death Valley, California. Great Basin Naturalist 29:111.

Keeton, D., and C.C. Carpenter. 1955. A study of the variations of dorsal markings of *B. w. woodhousei*.

Proceedings of the Oklahoma Academy of Science 36:81–83.

Kehr, A.I. 1997. Stage-frequency and habitat selection of a cohort of *Pseudacris ocularis* tadpoles (Hylidae: Anura) in a Florida temporary pond. Herpetological Journal 7:103–109.

Keinath, D., and J. Bennett. 2000. Distribution and status of the boreal toad (*Bufo boreas boreas*) in Wyoming. Wyoming Natural Diversity Database, University of Wyoming, Laramie.

Kelleher, K.E., and J.R. Tester. 1969. Homing and survival in the Manitoba toad, *Bufo hemiophrys*, in Minnesota. Ecology 50:1040–1048.

Keller, M.J., and H.C. Gerhardt. 2001. Polyploidy alters advertisement call structure in gray treefrogs. Proceedings of the Royal Society of London B 268:341–345.

Kellogg, R. 1932. Notes on the spadefoot of the western plains (*Scaphiopus hammondii*). Copeia 1932:36.

Kennedy, J.P. 1962. Spawning season and experimental hybridization of the Houston toad, *Bufo houstonensis*. Herpetologica 17:239–245.

Kent, D.M., M.A. Langston, and D.W. Hanf. 1997. Observations of vertebrates associated with gopher tortoise burrows in Orange County, Florida. Florida Scientist 60:197–201.

Kessel, B. 1965. Breeding dates of *Rana sylvatica* at College, Alaska. Ecology 46:206–208.

Kiesecker, J.M. 1996. pH-mediated predator-prey interactions between *Ambystoma tigrinum* and *Pseudacris triseriata*. Ecological Applications 6:1325–1331.

Kiesecker, J.M. 2002. Synergism between trematode infection and pesticide exposure: a link to amphibian limb deformities in nature? Proceedings of the National Academy of Sciences, USA 99:9900–9904.

Kiesecker, J.M., and A.R. Blaustein. 1997a. Population differences in responses of red-legged frogs (*Rana aurora*) to introduced bullfrogs. Ecology 78:1752–1760.

Kiesecker, J.M., and A.R. Blaustein. 1997b. Influences of egg laying behavior on pathogenic infection of amphibian eggs. Conservation Biology 11:214–220.

Kiesecker, J.M., and A.R. Blaustein. 1998. Effects of introduced bullfrogs and smallmouth bass on microhabitat use, growth, and survival of native red-legged frogs (*Rana aurora*). Conservation Biology 12:776–787.

Kiesecker, J.M., D.P. Chivers, and A.R. Blaustein. 1996. The use of chemical cues in predator recognition by western toad tadpoles. Animal Behaviour 32:1237–1245.

Kiesecker, J.M., D.P. Chivers, A. Marco, C. Quilchano, M.T. Anderson, and A.R. Blaustein. 1999. Identification of a disturbance signal in larval red-legged frogs, *Rana aurora*. Animal Behaviour 57:1295–1300.

Kiesecker, J.M., A.R. Blaustein, and C.L. Miller. 2001a. Potential mechanisms underlying the displacement of native red-legged frogs by introduced bullfrogs. Ecology 82:1964–1970.

Kiesecker, J.M., A.R. Blaustein, and L.K. Belden. 2001b. Complex causes of amphibian population declines. Nature 410:681–684.

Kiesow, A.M. 2006. Field Guide to Amphibians and Reptiles of South Dakota. South Dakota Department of Game, Fish and Parks, Pierre, SD.

Kiffney, P.M., and J.S. Richardson. 2001. Interactions among nutrients, periphyton, and invertebrate and vertebrate (*Ascaphus truei*) grazers in experimental channels. Copeia 2001:422–429.

Kilby, J.D. 1936. A biological analysis of the food and feeding habits of *Rana sphenocephala* (Cope) and *Hyla cinerea* (Schneider). M.S. thesis, University of Florida, Gainesville.

Kilby, J.D. 1945. A biological analysis of the food and feeding habits of two frogs. Quarterly Journal of the Florida Academy of Sciences 8:71–104.

Kim, B., T.G. Smith, and S.S. Dresser. 1998. Life history and host specificity of *Hepatozoon clamitae* (Apicomplexia: Adeleorina) and ITS-1 nucleotide sequence variation of *Hepatozoon* species of frogs and mosquitoes from Ontario. Journal of Parasitology 84:789–797.

Kim, J.B., T. Halverson, Y.J. Basir, J. Dulka, F.C. Knoop, P.W. Abel, and J.M. Conlon. 2000. Purification and characterization of antimicrobial and vasorelaxant peptides from skin extracts and skin secretions of the North American pig frog *Rana grylio*. Regulatory Peptides 90:53–60.

Kime, N.M., S.S. Burmeister, and M.J. Ryan. 2004. Female preferences for socially variable call characters in the cricket frog, *Acris crepitans*. Animal Behaviour 68:1391–1399.

King, F.W. 1932. Herpetological records and notes from the vicinity of Tucson, Arizona, July and August, 1930. Copeia 1932:175–177.

King, F.W., and T. Krakauer. 1966. The exotic herpetofauna of Florida. Quarterly Journal of the Florida Academy of Sciences 29:144–154.

King, J.J., and R.S. Wagner. 2010. Toxic effects of the herbicide Roundup® Regular on Pacific Northwest amphibians. Northwestern Naturalist 91:318–324.

King, K.C., J.D. McLaughlin, A.D. Gendron, B.D. Pauli, I. Giroux, B. Rondeau, M. Boily, P. Juneau, and D.J. Marcogliese. 2007. Impacts of agriculture on the parasite communities of northern leopard frogs (*Rana pipiens*) in southern Quebec, Canada. Parasitology 134:2063–2080.

King, K.C., A.D. Gendron, J.D. McLaughlin, I. Giroux, P. Brousseau, D. Cyr, S.M. Ruby, M. Fournier, and D.J. Marcogliese. 2008. Short-term seasonal changes in parasite community structure in northern leopard froglets (*Rana pipiens*) inhabiting agricultural wetlands. Journal of Parasitology 94: 13–22.

King, K.C., J.D. Mclaughlin, M. Boily, and D.J. Marcogliese. 2010. Effects of agricultural landscape and pesticides on parasitism in native bullfrogs. Biological Conservation 143:302–310.

King, O.M. 1960. Observations on Oklahoma toads. Southwestern Naturalist 5:103.

King, R.B., and B. King. 1991. Sexual differences in color and color change in wood frogs. Canadian Journal of Zoology 69:1963–1968.

King, R.B., M.J. Oldham, W.F. Weller, and D. Wynn. 1997. Historic and current amphibian and reptile distributions in the island region of western Lake Erie. American Midland Naturalist 138:153–173.

King, W. 1939. A survey of the herpetology of Great Smoky Mountains National Park. American Midland Naturalist 21:531–582.

Kingsbury, B., and J. Gibson (eds.). 2002. Habitat Management Guidelines for Amphibians and Reptiles of the Midwest. Partners in Amphibian and Reptile Conservation, Technical Publication HMG-1, Montgomery, AL.

Kinney, V.C., J.L. Heemeyer, A.P. Pessier, and M.J. Lannoo. 2011. Seasonal pattern of *Batrachochytrium dendrobatidis* infection and mortality in *Lithobates areolatus*: affirmation of Vredenburg's "10,000 zoospore rule." PLoS One 6:e16708.

Kirkland, A.H. 1897. The habits, food and economic value of the American toad, *Bufo lentiginosus americanus* (LeC.). Hatch Experiment Station of the Massachusetts Agricultural College, Bulletin No. 46.

Kirkland, A.H. 1904. Usefulness of the American toad. U.S. Department of Agriculture Farmer's Bulletin No. 196.

Kirkland, G.L. Jr., H.W. Snoddy, and T.L. Amsler. 1996. Impact of fire on small mammals and amphibians in a central Appalachian deciduous forest. American Midland Naturalist 135:253–260.

Kiviat, E. 1982. Geographic distribution. *Scaphiopus holbrookii holbrooki* (Eastern Spadefoot). Herpetological Review 13:51–52.

Kiviat, E., and J. Stapleton. 1983. *Bufo americanus* (American toad). Estuarine habitat. Herpetological Review 14:46.

Klemens, M.W. 1993. Amphibians and Reptiles of Connecticut and Adjacent Regions. State Geological and Natural History Survey of Connecticut, Bulletin 112.

Klemens, M.W., E. Kiviat, and R.E. Schmidt. 1987. Distribution of the northern leopard frog, *Rana pipiens*, in the lower Hudson and Housatonic River valleys. Northeastern Environmental Science 6:99–101.

Klimstra, W.D., and C.W. Myers. 1965. Foods of the toad, *Bufo woodhousei fowleri* Hinckley. Transactions of the Illinois State Academy of Science 58:11–26.

Kline, J. 1998. Monitoring amphibians in created and restored wetlands. Pp. 360–368 *In* M.J. Lannoo (ed.), Status & Distribution of Midwestern Amphibians. University of Iowa Press, Iowa City.

Klomberg, K.F., and C.A. Marler. 2000. The neuropeptide arginine vasotocin alters male call characteristics involved in social interactions in the grey treefrog, *Hyla versicolor*. Animal Behaviour 59:807–812.

Klugh, A.B. 1922. The economic value of the leopard frog. Copeia 1922:14–15.

Klump, G.M., and H.C. Gerhardt. 1987. Use of non-arbitrary acoustic criteria in mate choice by female gray tree frogs. Nature 326:286–288.

Klymus, K.E., S.C. Humfeld, V.T. Marshall, D. Cannatella, and H.C. Gerhardt. 2010. Molecular patterns of differentiation in canyon treefrogs (*Hyla arenicolor*): evidence for introgressive hybridization with the Arizona treefrog (*H. wrightorum*) and correlations with advertisement call differences. Journal of Evolutionary Biology 23:1425–1435.

Knapp, R.A. 2005. Effects of nonnative fish and habitat characteristics on lentic herpetofauna in Yosemite National Park, USA. Biological Conservation 121:265–279.

Knapp, R.A., and K.R. Matthews. 2000. Non-native fish introductions and the decline of the mountain yellow-legged frog from within protected areas. Conservation Biology 14:428–438.

Knapp, R.A., K.R. Matthews, and O. Sarnelle. 2001. Resistance and resilience of alpine lake fauna to fish introductions. Ecological Monographs 71:401–421.

Knapp, R.A., K.R. Matthews, H.K. Preisler, and R. Jellison. 2003. Developing probabilistic models to predict amphibian site occupancy in a patchy landscape. Ecological Applications 13:1069–1082.

Knight, E.L. 1972. Notes on cave-dwelling *Rana palustris* in Mississippi. Bulletin of the Maryland Herpetological Society 8:88.

Knowlton, G.F. 1944. Some insect food of *Rana pipiens*. Copeia 1944:119.

Knutson, M.G., J.R. Sauer, D.A. Olsen, M.J. Mossman, L.M. Hemesath, and M.J. Lannoo. 2000. Landscape associations of frogs and toad species in Iowa and Wisconsin, U.S.A. Journal of the Iowa Academy of Science 107:134–145.

Knutson, M.G., W.B. Richardson, D.M. Reineke, B.R. Gray, J.R. Parmelee, and S.E. Weick. 2004. Agricultural ponds support amphibian populations. Ecological Applications 14:669–684.

Koch, E.D., and C.R. Peterson. 1995. Amphibians & Reptiles of Yellowstone and Grand Teton National Parks. University of Utah Press, Salt Lake City.

Kocher, T.D., and R.D. Sage. 1986. Further genetic analyses of a hybrid zone between leopard frogs (*Rana pipiens* complex) in central Texas. Evolution 40:21–33.

Koller, R.L., and A.J. Gaudin. 1977. An analysis of helminth infections in *Bufo boreas* (Amphibia: Bufonidae) and *Hyla regilla* (Amphibia: Hylidae) in southern California. Southwestern Naturalist 21:503–509.

Kolozsvary, M.B., and R.K. Swihart. 1999. Habitat fragmentation and the distribution of amphibians: patch and landscape correlates in farmland. Canadian Journal of Zoology 77:1288–1299.

Koprivnikar, J., R.L. Baker, and M.R. Forbes. 2006. Environmental factors influencing trematode prevalence in grey tree frog (*Hyla versicolor*) tadpoles in southern Ontario. Journal of Parasitology 92:997–1001.

Korky, J.D. 1978. Differentiation of the larvae of members of the *Rana pipiens* complex in Nebraska. Copeia 1978:455–459.

Korschgen, L.J. 1970. Soil-food-chain-pesticide wildlife relationships in aldrin-treated fields. Journal of Wildlife Management 34:186–199.

Korschgen, L.J., and D.L. Moyle. 1955. Food habits of the bullfrog in central Missouri farm ponds. American Midland Naturalist 54:332–341.

Korschgen, L.J., and T.S. Baskett. 1963. Foods of impoundment- and stream-dwelling bullfrogs in Missouri. Herpetologica 19:89–99.

Koster, W.J. 1946. The robber frog in New Mexico. Copeia 1946:173.

Krakauer, T. 1968. The ecology of the Neotropical toad, *Bufo marinus*, in south Florida. Herpetologica 24:214–221.

Krakauer, T. 1970a. The invasion of the toads. Florida Naturalist 43(1):12–14.

Krakauer, T. 1970b. Tolerance limits of the toad, *Bufo marinus*, in south Florida. Comparative Biochemistry and Physiology 33:15–26.

Kramek, W.C. 1972. Food of the frog *Rana septentrionalis* in New York. Copeia 1972:390–392.

Kramek, W.C. 1976. Feeding behavior of *Rana septentrionalis* (Amphibia, Anura, Ranidae). Journal of Herpetology 10:251–252.

Kramek, W.C., and M.M. Stewart. 1980. Ontogenetic and sexual differences in the pattern of *Rana septentrionalis*. Journal of Herpetology 14:369–375.

Kramer, D.C. 1973. Movements of western chorus frogs *Pseudacris triseriata triseriata* tagged with Co^{60}. Journal of Herpetology 7:231–235.

Kramer, D.C. 1974. Home range of the western chorus frog *Pseudacris triseriata triseriata*. Journal of Herpetology 8:245–246.

Kraus, F. 2009. Alien Reptiles and Amphibians. A Scientific Compendium and Analysis. Springer, Dordrecht, The Netherlands.

Kraus, F., and E.W. Campbell III. 2002. Human-mediated escalation of a formerly eradicable problem: the invasion of Caribbean frogs in the Hawaiian Islands. Biological Invasions 4:327–332.

Kraus, F., and F. Duvall. 2004. New records of alien reptiles and amphibians in Hawai'i. Bishop Museum Occasional Papers 79:62–64.

Kraus, F., E.W. Campbell, A. Allison, and T. Pratt. 1999. *Eleutherodactylus* frog introductions to Hawaii. Herpetological Review 30:21–25.

Kretzer, J.E., and J.F. Cully Jr. 2001. Effects of black-tailed prairie dogs on reptiles and amphibians in Kansas shortgrass prairie. Southwestern Naturalist 46:171–177.

Krupa, J.J. 1986a. Multiple egg clutch production in the Great Plains toad. Prairie Naturalist 18:151–152.

Krupa, J.J. 1986b. Distribution in Oklahoma of the bird-voiced treefrog (*Hyla avivoca*). Proceedings of the Oklahoma Academy of Science 66:37–38.

Krupa, J.J. 1988. Fertilization efficiency of the Great Plains toad (*Bufo cognatus*). Copeia 1988:800–802.

Krupa, J.J. 1989. Alternative mating tactics in the Great Plains toad. Animal Behaviour 37:1035–1043.

Krupa, J.J. 1990. Advertisement call variation in the Great Plains toad. Copeia 1990:884–886.

Krupa, J.J. 1994. Breeding biology of the Great Plains toad in Oklahoma. Journal of Herpetology 28:217–224.

Krupa, J.J. 1995. *Bufo woodhousii* (Woodhouse's Toad). Fecundity. Herpetological Review 26: 142, 144.

Krupa, J.J. 2002. Temporal shift in diet in a population of American bullfrog (*Rana catesbeiana*) in Carlsbad Caverns National Park. Southwestern Naturalist 47:461–467.

Kruse, K.C. 1981a. Mating success, fertilization potential, and male body size in the American toad *(Bufo americanus)*. Herpetologica 37:228–233.

Kruse, K.C. 1981b. Phonotactic responses of female northern leopard frogs (*Rana pipiens*) to *Rana blairi*, a presumed hybrid, and conspecific mating trills. Journal of Herpetology 15:145–150.

Kruse, K.C., and D.G. Dunlap. 1976. Serum albumins and hybridization in two species of the *Rana pipiens* complex in the north central United States. Copeia 1976:394–396.

Kruse, K.C., and M.G. Francis. 1977. A predation deterrent in larvae of the bullfrog, *Rana catesbeiana*. Transactions of the American Fisheries Society 106:248–252.

Kruse, K.C., and M. Mounce. 1982. The effects of multiple matings on fertilization capability in male American toads (*Bufo americanus*). Journal of Herpetology 16:410–412.

Kruse, K.C., and B.M. Stone. 1984. Largemouth bass (*Micropterus salmoides*) learn to avoid feeding on toad (*Bufo*) tadpoles. Animal Behaviour 32:1035–1039.

Krysko, K.L., and C.M. Sheehy III. 2005. Ecological status of the ocellated gecko, *Sphaerodactylus argus*

argus Gosse 1850, in Florida, with additional herpetological notes from the Florida Keys. Caribbean Journal of Science 41:169–172.

Kuntz, R.E. 1941. The metazoan parasites of some Oklahoma anura. Proceedings of the Oklahoma Academy of Science 21:33–34.

Kuperman, B.I., V.E. Matey, R.N. Fisher, E.L. Ervin, M.L. Warburton, L. Bakhireva, and C.A. Lehman. 2004. Parasites of the African clawed frog, *Xenopus laevis*, in Southern California, U.S.A. Comparative Parasitology 71:229–232.

Kupferberg, S.J. 1996. Hydrologic and geomorphic factors affecting conservation of a river-breeding frog (*Rana boylii*). Ecological Applications 6:1332–1344.

Kupferberg, S.J. 1997a. Bullfrog (*Rana catesbeiana*) invasion of a California river: the role of larval competition. Ecology 78:1736–1751.

Kupferberg, S.J. 1997b. The role of larval diet in anuran metamorphosis. American Zoologist 37:146–159.

Kupferberg, S.J. 1997c. Facilitation of periphyton production by tadpole grazing: functional differences between species. Freshwater Biology 37:427–439.

Kupferberg, S.J., J.C. Marks, and M.E. Power. 1994. Effects of variation in natural algal and detrital diets on larval anuran (*Hyla regilla*) life-history traits. Copeia 1994:446–457.

Kupferberg, S.J., A. Catenazzi, K. Lunde, A.J. Lind, and W.J. Palen. 2009. Parasitic copepod (*Lernaea cyprinacea*) outbreaks in foothill yellow-legged frogs (*Rana boylii*) linked to unusually warm summers and amphibian malformations in northern California. Copeia 2009:529–537.

Kupferberg, S.J., A.J. Lind, V. Thill, and S.M. Yarnell. 2011. Water velocity tolerance in tadpoles of the foothill yellow-legged frog (*Rana boylii*): swimming performance, growth, and survival. Copeia 2011:141–152.

Kurzava, L.M., and P.J. Morin. 1998. Tests of functional equivalence: complementary roles of salamanders and fish in community organization. Ecology 79:477–489.

Kuyt, E. 1991. A communal overwintering site for the Canadian toad, *Bufo americanus hemiophrys*, in the Northwest Territories. Canadian Field-Naturalist 105:119–121.

Labanick, G.M. 1976. Prey availability, consumption and selection in the cricket frog, *Acris crepitans* (Amphibia, Anura, Hylidae). Journal of Herpetology 10:293–298.

Labanick, G.M., and R.A. Schlueter. 1976. Growth rates of recently transformed *Bufo woodhousei fowleri*. Copeia 1976:824–826.

Lacan, I., K. Matthews, and K. Feldman. 2008. Interaction of an introduced predator with future effects of climate change in the recruitment dynamics of the imperiled Sierra Nevada yellow-legged frog (*Rana sierrae*). Herpetological Conservation and Biology 3:211–223.

Lacki, M.J., J.W. Hummer, and H.J. Webster. 1992. Mine-drainage treatment wetland as habitat for herpetofaunal wildlife. Environmental Management 16:513–520.

Laerm, J., N.K. Castleberry, M.A. Menzel, R.A. Moulis, G.K. Williamson, J.B. Jensen, B. Winn, and M.J. Harris. 2000. Biogeography of amphibians and reptiles of the Sea Islands of Georgia. Florida Scientist 63:193–231.

Lafferty, K.D., and C.J. Page. 1997. Predation on the endangered tidewater goby, *Eucyclogobius newberryi*, by the introduced African clawed frog, *Xenopus laevis*, with notes on the frog's parasites. Copeia 1997:589–592.

LaFiandra, E.M., and K.J. Babbitt. 2004. Predator induced phenotypic plasticity in the pinewoods tree frog, *Hyla femoralis*: necessary cues and the cost of development. Oecologia 138:350–359.

Lamb, A.C. III. 1980. Observations on the life history of the pig frog, *Rana grylio* Stejneger, in southwest Georgia. M.S. thesis. Auburn University, Auburn, AL.

Lamb, T. 1984a. Amplexus displacement in the southern toad, *Bufo terrestris*. Copeia 1984:1023–1025.

Lamb, T. 1984b. The influence of sex and breeding condition on microhabitat selection and diet in the pig frog *Rana grylio*. American Midland Naturalist 111:311–318.

Lamb, T. 1987. Call site selection in a hybrid population of treefrogs. Animal Behaviour 35:1140–1144.

Lamb, T., and J.C. Avise. 1986. Directional introgression of mitochondrial DNA in a hybrid population of tree frogs: the influence of mating behavior. Proceedings of the National Academy of Sciences, USA 83:2526–2530.

Lamb, T., and J.C. Avise. 1987. Morphological variability in genetically defined categories of anuran hybrids. Evolution 41:157–163.

Lamb, T., J.M. Novak, and D.L. Mahoney. 1990. Morphological asymmetry and interspecific hybridization: a case study using hylid frogs. Journal of Evolutionary Biology 3:295–309.

Lamb, T., B.K. Sullivan, and K. Malmos. 2000. Mitochondrial gene markers for the hybridizing toads *Bufo microscaphus* and *Bufo woodhousii* in Arizona. Copeia 2000:234–237.

Lamoureux, V.S., and D.M. Madison. 1999. Overwintering habitats of radio-implanted green frogs, *Rana clamitans*. Journal of Herpetology 33:430–435.

Lamoureux, V.S., J.C. Maerz, and D.M. Madison. 2002. Premigratory autumn foraging forays in the green frog, *Rana clamitans*. Journal of Herpetology 36:245–254.

Lampman, B.H. 1946. The Coming of the Pond Fishes. Binfords & Mort Publishers, Portland, OR.

Lance, S.L., and K.D. Wells. 1993. Are spring peeper

satellite males physiologically inferior to calling males? Copeia 1993:1162–1166.
Landé, S.P., and S.I. Guttman. 1973. The effects of copper sulfate on the growth and mortality rate of *Rana pipiens* tadpoles. Herpetologica 29:22–27.
Landreth, H.F., and D.E. Ferguson. 1966a. Behavioral adaptations in the chorus frog, *Pseudacris triseriata*. Journal of the Mississippi Academy of Sciences 12:197–202.
Landreth, H.F., and D.E. Ferguson. 1966b. Celestial orientation of Fowler's toad, *Bufo fowleri*. Behaviour 26:105–123.
Landreth, H.F., and D.E. Ferguson. 1967. Movements and orientation of the tailed frog, *Ascaphus truei*. Herpetologica 23:81–93.
Landreth, H.F., and D.E. Ferguson. 1968. The sun compass of Fowler's toad, *Bufo woodhousei fowleri*. Behaviour 30:27–43.
Landreth, H.F., and M.T. Christensen. 1971. Orientation of the Plains spadefoot toad, *Scaphiopus bombifrons*, to solar cues. Herpetologica 27:454–461.
Lang, E., C. Gerhardt, and C. Davidson. 2009. The Frogs and Toads of North America. A Comprehensive Guide to their Identification, Behavior, and Calls. Houghton Mifflin Harcourt, Boston.
Langford, G.J., J.A. Borden, C.S. Major, and D.H. Nelson. 2007. Effects of prescribed fire on the herpetofauna of a southern Mississippi pine savanna. Herpetological Conservation and Biology 2:135–143.
Langlois, T.H. 1964. Amphibians and reptiles of the Erie islands. Ohio Journal of Science 64:11–25.
Langlois, V.S., A.C. Carew, B.D. Pauli, M.G. Wade, G.M. Cooke, and V.L. Trudeau. 2010. Low levels of the herbicide atrazine alters sex ratios and reduces metamorphic success in *Rana pipiens* tadpoles raised in outdoor mesocosms. Environmental Health Perspectives 118:552–557.
Lannoo, M. 1996. Okoboji Wetlands. A Lesson in Natural History. University of Iowa Press, Iowa City.
Lannoo, M. J. 1998. Amphibian conservation and wetland management in the upper Midwest: a catch-22 for the cricket frog? Pp. 330–339 *In* M.J. Lannoo (ed.), Status & Distribution of Midwestern Amphibians. University of Iowa Press, Iowa City.
Lannoo, M. J. (ed.). 2005. Amphibian Declines. The Conservation Status of United States Species. University of California Press, Berkeley.
Lannoo, M.J. 2008. Malformed Frogs. The Collapse of Aquatic Ecosystems. University of California Press, Berkeley.
Lannoo, M.J., K. Lang, T. Waltz, and G.S. Phillips. 1994. An altered amphibian assemblage: Dickinson County, Iowa, 70 years after Frank Blanchard's survey. American Midland Naturalist 131:311–319.
Lannoo, M.J., J.A. Holman, G.S. Casper, and E. Johnson. 1998. Mummification following winterkill of adult green frogs (Ranidae: *Rana clamitans*). Herpetological Review 29:82–84.
Lannoo, M.J., C. Petersen, R.E. Lovich, P. Nanjappa, C. Phillips, J.C. Mitchell, and I. Macallister. 2011. Do frogs get their kicks from Route 66? Continental U.S. transect reveals spatial and temporal patterns of *Batrachochytrium dendrobatidis* infection. PLoS One 6(7):e22211.
Laposata, M.M., and W.A. Dunson. 1998. Effects of boron and nitrate on hatching success of amphibian eggs. Archives of Environmental Contamination and Toxicology 35:615–619.
La Rivers, I. 1948. Some Hawaiian ecological notes. Wasmann Collector 7:85–110.
Larsen, K.W., and P.T. Gregory. 1988. Amphibians and reptiles in the Northwest Territories. Occasional Papers of the Prince of Wales Northern Heritage Centre No. 3:31–51.
Latham, R. 1968a. Notes on the eating of May beetles by a Fowler's toad. Engelhardtia 1:29.
Latham, R. 1968b. Notes on some hibernating Fowler's toads. Engelhardtia 1:17.
Lawler, S.P. 1989. Behavioural responses to predators and predation risk in four species of larval anurans. Animal Behaviour 38:1039–1047.
Lawler, S.P., and P.J. Morin. 1993. Temporal overlap, competition, and priority effects in larval anurans. Ecology 74:174–182.
Lawler, S.P., D. Dritz, T. Strange, and M. Holyoak. 1999. Effects of introduced mosquitofish and bullfrogs on the threatened California red-legged frog. Conservation Biology 13:613–622.
Layne, J.R. Jr. 1995. Seasonal variation in the cryobiology of *Rana sylvatica* from Pennsylvania. Journal of Thermal Biology 20:349–353.
Layne, J.R. Jr., and M.A. Romano. 1985. Critical thermal minima of *Hyla chrysoscelis*, *H. cinerea*, *H. gratiosa* and natural hybrids (*H. cinerea* × *H. gratiosa*). Journal of Experimental Zoology 41:216–221.
Layne, J.R. Jr., and R.E. Lee Jr. 1987. Freeze tolerance and the dynamics of ice formation in wood frogs (*Rana sylvatica*) from southern Ohio. Canadian Journal of Zoology 65:2062–2065.
Layne, J.R. Jr., and R.E. Lee Jr. 1989. Seasonal variation in freeze tolerance and ice content of the tree frog *Hyla versicolor*. Journal of Experimental Biology 249:133–137.
Layne, J.R. Jr., and M.C. First. 1991. Resumption of physiological functions in the wood frog (*Rana sylvatica*) following freezing. American Journal of Physiology 261:R134-R137.
Layne, J.R. Jr., and R.E. Lee Jr. 1995. Adaptations of frogs to survive freezing. Climate Research 5:53–59.
Layne, J.R., and M.G. Stapleton. 2009. Annual variation in glycerol mobilization and effect of freeze rigor on post-thaw locomotion in the freeze-tolerant frog *Hyla versicolor*. Journal of Comparative Physiology 179B:215–221.

Layne, J.R. Jr., R.E. Lee Jr., and J.L. Huang. 1990. Inoculation triggers freezing at high subzero temperatures in a freeze-tolerant frog (*Rana sylvatica*) and insect (*Eurosta solidaginis*). Canadian Journal of Zoology 68:506–510.

Lazaroff, D.W., P.C. Rosen, and C.H. Lowe Jr. 2006. Amphibians, Reptiles, and their Habitats at Sabino Canyon. University of Arizona Press, Tucson.

Lazell, J.D. Jr. 1968. Mr. Fowler's toad. Massachusetts Audubon 52(2):4 pp.

Lazell, J.D. Jr. 1974. Reptiles & amphibians in Massachusetts. Rev. ed. Massachusetts Audubon Society, Lincoln.

Lazell, J.D. Jr. 1976. This Broken Archipelago. Cape Cod and the Islands, Amphibians and Reptiles. Quadrangle/New York Times Book Company, NY.

Lazell, J.D. Jr. 1989. Wildlife of the Florida Keys. Island Press, Washington, DC.

Lazell, J.D., and T. Mann. 1991. Geographic distribution: *Bufo americanus charlesmithi* (Dwarf American Toad. Herpetological Review 22:62, 64.

Leary, C.J. 1999. Comparison between release vocalizations emitted during artificial and conspecific amplexus in *Bufo americanus*. Copeia 1999:506–508.

Leary, C.J. 2001. Evidence of convergent character displacement in release vocalizations of *Bufo fowleri* and *Bufo terrestris* (Anura: Bufonidae). Animal Behaviour 61:431–438.

Leary, C.J., A.M. Garcia, and R. Knapp. 2006. Stress hormone is implicated in satellite-caller associations and sexual selection in the Great Plains toad. American Naturalist 168:431–440.

Leavitt, D.J., and L.A. Fitzgerald. 2009. Diet of nonnative *Hyla cinerea* in a Chihuahuan Desert wetland. Journal of Herpetology 43:541–545.

Leclair, R. Jr., and L. Vallières. 1981. Régimes alimentaires de *Bufo americanus* (Holbrook) et *Rana sylvatica* LeConte (Amphibia: Anura) nouvellement metamorphoses. Le naturaliste canadien 108:325–329.

Leclair, R. Jr., and J. Castanet. 1987. A skeletochronological assessment of age and growth in the frog *Rana pipiens* Schreber (Amphibia, Anura) from southwestern Quebec. Copeia 1987:361–369.

Leclair, R. Jr., and G. Laurin. 1996. Growth and body size in populations of mink frogs *Rana septentrionalis* from two latitudes. Ecography 19:296–304.

Leclair, R., M.H. Leclair, J. Dubois, and J. Daoust. 2000. Age and size of wood frogs, *Rana sylvatica*, from Kuujjuarapik, northern Quebec. Canadian Field-Naturalist 114:381–387.

LeConte, J. 1825. Remarks on the American species of the genera *Hyla* and *Rana*. Annals of the Lyceum of Natural History of New York 1:278–282.

LeConte, J. 1855. Descriptive catalogue of the Ranina of the United States. Proceedings of the Academy of Natural Sciences of Philadelphia 7:423–431.

LeConte, J. 1857 "1856". Description of a new species of *Hyla* from Georgia. Proceedings of the Academy of Natural Sciences of Philadelphia 8:146.

Ledón-Rettig, C.C., D.W. Pfennig, and N. Nascone-Yoder. 2008. Ancestral variation and the potential for genetic accommodation in larval amphibians: implications for the evolution of novel feeding strategies. Evolution & Development 10:316–325.

Ledón-Rettig, C.C., D.W. Pfennig, and E.J. Crespi. 2009. Stress hormones and the fitness consequences associated with the transition to a novel diet in larval amphibians. Journal of Experimental Biology 212:3743–3750.

Ledón-Rettig, C.C., D.W. Pfennig, and E.J. Crespi. 2010. Diet and hormonal manipulation reveal cryptic genetic variation: implications for the evolution of novel feeding strategies. Proceedings of the Royal Society B 277:3569–3578.

Lee, D.S. 1969a. Notes on the feeding behavior of cave-dwelling bullfrogs. Herpetologica 25:211–212.

Lee, D.S. 1969b. The treefrogs of Florida. Florida Naturalist (July):117–120.

Lee, D.S. 1972. List of amphibians and reptiles of Assateague Island. Bulletin of the Maryland Herpetological Society 8:90–95.

Lee, D.S. 1973a. Seasonal breeding distributions for selected Maryland and Delaware amphibians. Bulletin of the Maryland Herpetological Society 9:101–104.

Lee, D.S. 1973b. Additional reptiles and amphibians from Assateague Island. Bulletin of the Maryland Herpetological Society 9:110–111.

Lee, J.R., and C. Sabette. 2009. *Scaphiopus holbrookii holbrookii* (Eastern Spadefoot). Maximum size. Herpetological Review 40:211.

Lee, R.E. Jr., and J.P. Costanzo. 1993. Integrated physiological responses promoting anuran freeze tolerance. Pp. 501–510 *In* C. Carey, G.L. Florant, B.A. Wunder, and B. Horwitz (eds.), Life in the Cold. 3. Ecological, Physiological, and Molecular Mechanisms. Westview Press, Boulder, CO.

Lefcort, H., and S.M. Eiger. 1993. Antipredatory behaviour of feverish tadpoles: implications for pathogen transmission. Behaviour 126:13–27.

Lefcort, H., and A.R. Blaustein. 1995. Disease, predator avoidance, and vulnerability to predation in tadpoles. Oikos 74:469–474.

Lefcort, H., R.A. Meguire, L.H. Wilson, and W.F. Ettinger. 1998. Heavy metals alter the survival, growth, metamorphosis, and antipredatory behavior of Columbia spotted frog (*Rana luteiventris*) tadpoles. Archives of Environmental Contamination and Toxicology 35:447–456.

Leftwich, K.N., and P.D. Lilly. 1992. The effects of duration of exposure to acidic conditions on survival of *Bufo americanus* embryos. Journal of Herpetology 26:70–71.

Lehmann, D.L. 1960. Some parasites of central California amphibians. Journal of Parasitology 46:10.

Lehtinen, R.M. 2002. A historical study of the distribution of Blanchard's cricket frog (*Acris crepitans blanchardi*) in southeastern Michigan. Herpetological Review 33:194–197.

Lehtinen, R.M., and S.M. Galatowitsch. 2001. Colonization of restored wetlands by amphibians in Minnesota. American Midland Naturalist 145:388–396.

Lehtinen, R.M., and A.A. Skinner. 2006. The enigmatic decline of Blanchard's cricket frog (*Acris crepitans blanchardi*): a test of the habitat acidification hypothesis. Copeia 2006:159–167.

Lehtinen, R.M., S.M. Galatowitsch, and J.R. Tester. 1999. Consequences of habitat loss and fragmentation for wetland amphibian assemblages. Wetlands 19:1–12.

Leiden, Y.A., M.E. Dorcas, and J.W. Gibbons. 1999. Herpetofaunal diversity in coastal plain communities of South Carolina. Journal of the Elisha Mitchell Scientific Society 115:270–280.

Leips, J., M.G. McManus, and J. Travis. 2000. Response of treefrog larvae to drying ponds: comparing temporary and permanent pond breeders. Ecology 81:2997–3008.

Lemm, J.M. 2006. Field Guide to Amphibians and Reptiles of the San Diego Region. University of California Press, Berkeley.

Lemmon, E.M. 2009. Diversification of conspecific signals in sympatry: geographic overlap drives multidimensional reproductive character displacement in frogs. Evolution 63:1155–1170.

Lemmon, E.M., A.R. Lemmon, and D.C. Cannatella. 2007a. Geological and climatic forces driving speciation in the continentally distributed trilling chorus frogs (*Pseudacris*). Evolution 61:2086–2103.

Lemmon, E.M., A.R. Lemmon, J.T. Collins, J.A. Lee-Yaw, and D.C. Cannatella. 2007b. Phylogeny-based delimitation of species boundaries and contact zones in the trilling chorus frogs (*Pseudacris*). Molecular Phylogenetics and Evolution 44:1068–1082.

Lemmon, E.M., A.R. Lemmon, J.T. Collins, and D.C. Cannatella. 2008. A new North American chorus frog species (Amphibia: Hylidae: *Pseudacris*) from the south-central United States. Zootaxa 1675:1–30.

Lemos Espinal, J.A., and H.M. Smith. 2007a. Anfibios y reptiles del estado de Chihuahua, México. Universidad Nacional Autónoma de México and Comisión Nacional para el Conocimiento y Uso de la Biodiversidad, México.

Lemos Espinal, J.A., and H.M. Smith. 2007b. Anfibios y reptiles del estado de Coahuila, México. Universidad Nacional Autónoma de México and Comisión Nacional para el Conocimiento y Uso de la Biodiversidad, México.

Lenaker, R.P. Jr. 1972. *Xenopus* in Orange County. Senior thesis, California State Polytechnic University, Kellogg-Voorhis Campus, Pomona, CA.

Leney, J.L., K.G. Drouillard, and G.D. Haffner. 2006. Does metamorphosis increase the susceptibility of frogs to highly hydrophobic contaminants? Environmental Science and Technology 40:1491–1496.

Leonard, W.P., H.A. Brown, L.L.C. Jones, K.R. McAllister, and R.M. Storm. 1993. Amphibians of Washington and Oregon. Seattle Audubon Society, Seattle.

Leonard, W.P., N.P. Leonard, R.M. Storm, and P.E. Petzel. 1996. *Rana pretiosa* (Spotted Frog). Behavior and reproduction. Herpetological Review 27:195.

Leonard, W.P., L. Hallock, and K.R. McAllister. 1997a. *Rana pretiosa* (Oregon spotted frog). Behavior and reproduction. Herpetological Review 28:28.

Leonard, W.P., K.R. McAllister, and L.A. Hallock. 1997b. Autumn vocalizations by the red-legged frog (*Rana aurora*) and the Oregon spotted frog (*Rana pretiosa*). Northwestern Naturalist 78:73–74.

Leonard, W.P., K.R. McAllister, and R.C. Friesz. 1999. Survey and assessment of northern leopard frog (*Rana pipiens*) populations in Washington state. Northwestern Naturalist 80:51–60.

Lepage, M., R. Courtois, C. Daigle, and S. Matte. 1997. Surveying calling anurans in Québec using volunteers. SSAR Herpetological Conservation 1:128–140.

Letnic, M., J.K. Webb, and R. Shine. 2008. Invasive cane toads (*Bufo marinus*) cause mass mortality of freshwater crocodiles (*Crocodylus johnstoni*) in tropical Australia. Biological Conservation 141:1773–1782.

Lever, C. 2001. The Cane Toad. The History and Ecology of a Successful Colonist. Westbury Academic and Scientific Publishing, Otley, England.

Lever, C. 2003. Naturalized Reptiles and Amphibians of the World. Oxford University Press, Oxford.

Levine, N.D., and R.R. Nye. 1977. A survey of blood and other tissue parasites of leopard frogs *Rana pipiens* in the United States. Journal of Wildlife Diseases 13:17–23.

Lewis, D.L., G.T. Baxter, K.M. Johnson, and M.D. Stone. 1985. Possible extinction of the Wyoming toad, *Bufo hemiophrys baxteri*. Journal of Herpetology 19:166–168.

Lewis, R.J. 1962. Food habits of the leopard frog (*Rana pipiens sphenocephala*) in a minnow hatchery. Transactions of the Illinois State Academy of Science 55:78–79.

Lewis, T.H. 1946. Reptiles and amphibians of Smith Island, N.C. American Midland Naturalist 36:682–684.

Lewis, T.H. 1950. The herpetofauna of the Tularosa Basin and Organ Mountains of New Mexico with notes on some ecological features of the Chihuahuan Desert. Herpetologica 6:1–10.

Lewis, W.M. Jr. 1962. Stomach contents of bullfrogs (*Rana catesbeiana*) taken from a minnow hatchery. Transactions of the Illinois State Academy of Science 55:80–83.

Li, C., and F. Xie. 2004. Invasion of bullfrogs (*Rana catesbeiana* Shaw) in China and its management. Chinese Journal of Applied and Environmental Biology 10:95–98 [in Chinese with English abstract].

Licht, L.E. 1967. Growth inhibition in crowded tadpoles: intraspecific and interspecific effects. Ecology 48:736–745.

Licht, L.E. 1968. Unpalatability and toxicity of toad eggs. Herpetologica 24:93–98.

Licht, L.E. 1969a. Palatability of *Rana* and *Hyla* eggs. American Midland Naturalist 82:296–298.

Licht, L.E. 1969b. Unusual aspects of anuran sexual behavior as seen in the red-legged frog, *Rana aurora aurora*. Canadian Journal of Zoology 47:505–509.

Licht, L.E. 1969c. Comparative breeding behavior of the red-legged frog (*Rana aurora aurora*) and the western spotted frog (*Rana pretiosa pretiosa*) in southwestern British Columbia. Canadian Journal of Zoology 47:1287–1299.

Licht, L.E. 1971. Breeding habits and embryonic thermal requirements of the frogs, *Rana aurora aurora* and *Rana pretiosa pretiosa*, in the Pacific Northwest. Ecology 52:116–124.

Licht, L.E. 1974. Survival of embryos, tadpoles, and adults of the frogs *Rana aurora aurora* and *Rana pretiosa pretiosa* sympatric in southwestern British Columbia. Canadian Journal of Zoology 52: 613–627.

Licht, L.E. 1975. Comparative life history features of the western spotted frog, *Rana pretiosa*, from low- and high-elevation populations. Canadian Journal of Zoology 53:1254–1257.

Licht, L.E. 1976. Sexual selection in toads (*Bufo americanus*). Canadian Journal of Zoology 54: 1277–1284.

Licht, L.E. 1986a. Comparative escape behavior of sympatric *Rana aurora* and *Rana pretiosa*. American Midland Naturalist 115:239–247.

Licht, L.E. 1986b. Food and feeding behavior of sympatric red-legged frogs, *Rana aurora*, and spotted frogs, *Rana pretiosa*, in southwestern British Columbia. Canadian Field-Naturalist 100:22–31.

Licht, L.E. 1991. Habitat selection of *Rana pipiens* and *Rana sylvatica* during exposure to warm and cold temperatures. American Midland Naturalist 125:259–268.

Licht, L.E. 2003. Shedding light on ultraviolet radiation and amphibian embryos. BioScience 53: 551–561.

Licht, L.E., and K.P. Grant. 1997. The effects of ultraviolet radiation on the biology of amphibians. American Zoologist 37:137–145.

Lichtenberg, J.S., S.L. King, J.B. Grace, and S.C. Walls. 2006. Habitat associations of chorusing anurans in the lower Mississippi River Alluvial Valley. Wetlands 26:736–744.

Ligas, F.J. 1960. The Everglades bullfrog life history and management. Proceedings of the 14th Annual Conference, Southeastern Association of Game and Fish Commissioners, pp. 9–14. [slightly modified version entitled "The Everglades Bullfrog" published in Florida Wildlife, 1963, 16(12):14–19]

Ligon, D.B., and P.A. Stone. 2003. *Kinosternon sonoriense* (Sonoran Mud Turtle) and *Bufo punctatus* (Red-spotted Toad). Predator-prey. Herpetological Review 34:141–142.

Lillywhite, H.B. 1970. Behavioral temperature regulation in the bullfrog, *Rana catesbeiana*. Copeia 1970:158–168.

Lillywhite, H.B. 1971a. Temperature selection by the bullfrog, *Rana catesbeiana*. Comparative Biochemistry and Physiology 40A:213–227.

Lillywhite, H.B. 1971b. Thermal modulation of cutaneous mucus discharge as a determinant of evaporative water loss in the frog, *Rana catesbeiana*. Zeitschrift für vergleichende Physiologie 73:84–104.

Lillywhite, H.B. 1975. Physiological correlates of basking in amphibians. Comparative Biochemistry and Physiology 52A:323–330.

Lillywhite, H.B., and R.J. Wassersug. 1974. Comments on a postmetamorphic aggregation of *Bufo boreas*. Copeia 1974:984–986.

Limbaugh, B.A., and E.P. Volpe. 1957. Early development of the Gulf Coast Toad, *Bufo valliceps* Wiegmann. American Museum Novitates 1842:1–32.

Linck, M.H. 2000. Reduction in road mortality in a northern leopard frog population. Journal of the Iowa Academy of Science 107:209–211.

Lind, A.J., H.H. Welsh Jr., and R.A. Wilson. 1996. The effects of a dam on breeding habitat and egg survival of the foothill yellow-legged frog (*Rana boylii*) in northwestern California. Herpetological Review 27:62–67.

Lind, A.J., J.B. Bettaso, and S.M. Yarnell. 2003. *Rana boylii* (Foothill Yellow-legged Frog) and *Rana catesbeiana* (Bullfrog). Reproductive behavior. Herpetological Review 34:234–235.

Lind, A.J., P.Q. Spinks, G.M. Fellers, and H.B. Shaffer. 2011. Rangewide phylogeography and landscape genetics of the western U.S. endemic frog *Rana boylii* (Ranidae): implications for the conservation of frogs and rivers. Conservation Genetics 12: 269–284.

Linder, A.D., and E. Fichter. 1977. The Amphibians and Reptiles of Idaho. Idaho State University Press, Pocatello.

Lindsay, H.L. Jr. 1958. Analysis of variation and factors affecting gene exchange in *Pseudacris clarki* and *Pseudacris nigrita* in Texas. Ph.D. diss., University of Texas, Austin.

Liner, A.E., L.L. Smith, S.W. Golladay, S.B. Castleberry, and J.W. Gibbons. 2008. Amphibian distributions within three types of isolated wetlands in southwest Georgia. American Midland Naturalist 160:69–81.

Liner, E.A. 1954. The herpetofauna of Lafayette, Terrebonne and Vermilion parishes, Louisiana. Proceedings of the Louisiana Academy of Sciences 17:65–85.

Liner, E.A. 2007. Geographic distribution. *Euhyas planirostris*. Herpetological Review 38:214.

Ling, R.W., J.P. Van Amberg, and J.K. Werner. 1986. Pond acidity and its relationship to larval development of *Ambystoma maculatum* and *Rana sylvatica* in upper Michigan. Journal of Herpetology 20:230–236.

Linnaeus, C. 1758. Systema Naturae per Regna Tria Naturae, Secundum Classes, Ordines, Genera, Species cum Characteribus, Differentils, Synonymis, Locis. 10th ed., Vol. 1, L. Salvius, Stockholm, Sweden.

Linsdale, J.M. 1927. Amphibians and reptiles of Doniphan County, Kansas. Copeia (164):75–81.

Linsdale, J.M. 1933a. A specimen of *Rana tarahumarae* from New Mexico. Copeia 1933:222.

Linsdale, J.M. 1933b. Records of *Ascaphus truei* in Idaho. Copeia 1933:223.

Linsdale, J.M. 1938. Environmental responses of vertebrates in the Great Basin. American Midland Naturalist 19:1–206.

Linsdale, J.M. 1940. Amphibians and reptiles in Nevada. Proceedings of the American Academy of Arts and Sciences 73:197–257.

Linzey, D.W. 1967. Food of the leopard frog, *Rana p. pipiens*, in central New York. Herpetologica 23:11–17.

Little, E.E., R. Calfee, L. Cleveland, R. Skinker, A. Zaga-Parkhurst, and M.G. Barron. 2000. Photo-enhanced toxicity in amphibians: synergistic interactions of solar ultraviolet radiation and aquatic contaminants. Journal of the Iowa Academy of Science 107:67–71.

Little, E.L. 1940. Amphibians and reptiles of the Roosevelt Reservoir area, Arizona. Copeia 1940:260–265.

Little, E.L., and J. Keller. 1937. Amphibians and reptiles of the Jornada Experimental Range, New Mexico. Copeia 1937:215–222.

Little, M.L. 1983. The zoogeography of the *Hyla versicolor* complex in the central Appalachians, including physiological and morphological analyses. Ph.D. diss., University of Louisville.

Little, M.L., and T.K. Pauley. 1986. A new record of the diploid species of gray treefrog, *Hyla chrysoscelis*, in West Virginia. Proceedings of the West Virginia Academy of Science 58:57–58.

Little, M.L., B.L. Monroe Jr., and J.E. Wiley. 1989. The distribution of the *Hyla versicolor* complex in the northern Appalachian highlands. Journal of Herpetology 23:299–303.

Littlejohn, M.J. 1958. Mating behavior in the treefrog *Hyla versicolor*. Copeia 1958:222–223.

Littlejohn, M.J. 1959. Artificial hybridization within the Pelobatidae and Microhylidae. Texas Journal of Science 9:57–59.

Littlejohn, M.J. 1961a. Artificial hybridization between some hylid frogs of the United States. Texas Journal of Science 13:176–184.

Littlejohn, M.J. 1961b. Mating call discrimination by females of the spotted chorus frog (*Pseudacris clarki*). Texas Journal of Science 13:49–50.

Littlejohn, M.J. 1971. A reappraisal of mating call differentiation in *Hyla cadaverina* (= *Hyla californiae*) and *Hyla regilla*. Evolution 25:98–102.

Littlejohn, M.J., and T.C. Michaud. 1959. Mating call discrimination by females of Strecker's chorus frog (*Pseudacris streckeri*). Texas Journal of Science 11:86–92.

Littlejohn, M.J., and M.J. Fouquette Jr. 1960. Call discrimination by female frogs of the *Hyla versicolor* complex. Copeia 1960:47–49.

Littlejohn, M.J., and R.S. Oldham. 1968. *Rana pipiens* complex: mating call structure and taxonomy. Science 162:1003–1005.

Liu, X., Y. Li, and M. McGarrity. 2010. Geographical variation in body size and sexual size dimorphism of introduced American bullfrogs in southwestern China. Biological Invasions 12:2037–2047.

Livezey, R.L. 1950. The eggs of *Acris gryllus crepitans* Baird. Herpetologica 6:139–140.

Livezey, R.L. 1952. Some observations on *Pseudacris nigrita triseriata* (Wied) in Texas. American Midland Naturalist 47:372–381.

Livezey, R.L. 1960. Description of the eggs of *Bufo boreas exsul*. Herpetologica 16:48.

Livezey, R.L. 1962. Food of adult and juvenile *Bufo exsul*. Herpetologica 17:267–268.

Livezey, R.L., and A.H. Wright. 1947. A synoptic key to the salientian eggs of the United States. American Midland Naturalist 37:179–222.

Livezey, R.L., and H.M. Johnson. 1948. *Rana grylio* in Texas. Herpetologica 4:164.

Livo, L. 1981a. Leopard frog (*Rana pipiens*) reproduction in Boulder County, Colorado. M.A. thesis, University of Colorado, Denver.

Livo, L. 1984. Dry years along with bullfrogs are creating problems for Colorado's two species of leopard frogs. Colorado Outdoors 33(4):16–18.

Livo, L. 1998. Identification guide to montane amphibians of the southern Rocky Mountains. Colorado Division of Wildlife, Denver.

Livo, L. 1999. The role of predation in the early life history of *Bufo boreas* in Colorado. Ph.D. diss., University of Colorado, Boulder.

Livo, L., and D. Yeakley. 1997. Comparison of current with historical elevational range in the boreal toad, *Bufo boreas*. Herpetological Review 28:143–144.

Livo, L.J., D. Chiszar, and H.M. Smith. 1997. *Spea multiplicata* (New Mexico Spadefoot). Defensive posture. Herpetological Review 28:148.

Livo, L.J., G.A. Hammerson, and H.M. Smith. 1998. Summary of amphibians and reptiles introduced into Colorado. Northwestern Naturalist 79:1–11.

Loda, J.L., and D.L. Otis. 2009. Low prevalence of amphibian chytrid fungus (*Batrachochytrium dendrobatidis*) in northern leopard frog (*Rana pipiens*) populations on north-central Iowa, USA. Herpetological Review 40:428–431.

Loftus-Hills, J.J. 1975. The evidence for reproductive character displacement between the toads *Bufo americanus* and *B. woodhousii fowleri*. Evolution 29:368–369.

Loftus-Hills, J.J., and M.J. Littlejohn. 1992. Reinforcement and reproductive character displacement in *Gastrophryne carolinensis* and *G. olivacea* (Anura: Microhylidae): a reexamination. Evolution 46:896–906.

Logier, E.B.S. 1931. The amphibians and reptiles of Long Point. Transactions of the Royal Canadian Institute 18:229–236.

Logier, E.B.S. 1932. Some account of the amphibians and reptiles of British Columbia. Transactions of the Royal Canadian Institute 18:311–336.

Logier, E.B.S. 1952. The Frogs, Toads and Salamanders of Eastern Canada. Clarke, Irwin & Co., Toronto.

Logier, E.B.S., and G.C. Toner. 1961. Check list of the amphibians & reptiles of Canada & Alaska. Royal Ontario Museum, Toronto, Ontario.

Lomolino, M.A., and G.A. Smith. 2003. Terrestrial vertebrate communities at black-tailed prairie dog (*Cynomys ludovicianus*) towns. Biological Conservation 115:89–100.

Long, C.A. 1964. The badger as a natural enemy of *Ambystoma tigrinum* and *Bufo boreas*. Herpetologica 20:144.

Long, C.A. 1998. Notes on development and mortality in giant toads from Washington Island, Door County, Wisconsin. Bulletin of the Chicago Herpetological Society 33:161–164.

Long, L.E., L.S. Saylor, and M.E. Soulé. 1995. A pH/UV-B synergism in amphibians. Conservation Biology 9:1301–1303.

Long, M.C. 1970. Food habits of *Rana muscosa* (Anura: Ranidae). Herpeton, Journal of the Southwestern Herpetologists Society 5(1):1–8.

Long, Z. 2010. *Anaxyrus boreas* (Western Toad). Advertisement vocalization. Herpetological Review 41:332–333.

Longcore, J.R., J.E. Longcore, A.P. Pessier, and W.A. Halteman. 2007. Chytridiomycosis widespread in anurans of northeastern United States. Journal of Wildlife Management 71:435–444.

Loomis, R.B. 1978. The yellow-legged frog, *Rana boylii*, from the Sierra San Pedro Mártir, Baja California Norte, Mexico. Herpetologica 21:78–80.

Loomis, R.B., and J.K. Jones Jr. 1953. Records of the wood frog, *Rana sylvatica*, from western Canada and Alaska. Herpetologica 9:149–151.

López, L.O., G.A. Woolrich Piña, and J.A. Lemos Espinal. 2009. La familia Bufonidae en México. Universidad Nacional Autónomade de México, México.

Lopez, T.J., and L.R. Maxon. 1990. *Rana catesbeiana* (Bullfrog). Polymely. Herpetological Review 21:90.

Lord, R.D. Jr., and W.B. Davis. 1956. A taxonomic study of the relationship between *Pseudacris nigrita triseriata* Wied and *Pseudacris clarki* Baird. Herpetologica 12:115–120.

Lorraine, R.K. 1984. *Hyla crucifer crucifer* (Northern Spring Peeper). Reproduction. Herpetological Review 15:16–17.

Lotshaw, D.P. 1977. Temperature adaptation and effects of thermal acclimation in *Rana sylvatica* and *Rana catesbeiana*. Comparative Biochemistry and Physiology 56B:287–294.

Lotz, A., and C.R. Allen. 2007. Observer bias in anuran call surveys. Journal of Wildlife Management 71:675–679.

Louisiana Department of Conservation. 1931. Frog industry in Louisiana. Division of Fisheries, Educational Pamphlet No. 2.

Love, E.K., and M.A. Bee. 2010. An experimental test of noise-dependent voice amplitude regulation in Cope's grey treefrog, *Hyla chrysoscelis*. Animal Behaviour 80:509–515.

Lovich, J.E., and T.R. Jaworski. 1988. Annotated list of amphibians and reptiles reported from Cedar Bog, Ohio. Ohio Journal of Science 88:139–143.

Lowcock, L.A., T.F. Sharbel, J. Bonin, M. Ouellet, J. Rodrigue, and J.-L. DesGranges. 1997. Flow cytometric assay for in vivo genotoxic effects of pesticides in green frogs (*Rana clamitans*). Aquatic Toxicology 38:241–255.

Lucas, E.A., and W.A. Reynolds. 1967. Temperature selection by amphibian larvae. Physiological Zoology 40:159–171.

Luce, D., and J.J. Moriarty. 1999. *Rana sylvatica* (Wood Frog). Coloration. Herpetological Review 30:94.

Lucke, B. 1952. Kidney carcinoma in the leopard frog: a virus tumor. Annals of the New York Academy of Science 54:1093–1109.

Lucker, J.T. 1931. A new genus and a new species of trematode worms of the family Plagiorchiidae. Proceedings of the United States National Museum 79:1–8.

Luepschen, L.K. 1981. *Bufo punctatus* (Red-spotted Toad). Larval coloration. Herpetological Review 12:79.

Luhring, T.M. 2008. "Problem species" of the Savannah River Site, such as Brimley's chorus frog (*Pseudacris brimleyi*), demonstrate the hidden biodiversity concept on an intensively studied government reserve. Southeastern Naturalist 7:371–373.

Luja, V.H., and R. Rodríguez-Estrella. 2010. The invasive bullfrog *Lithobates catesbeianus* in oases of Baja California Sur, Mexico: potential effects in a fragile ecosystem. Biological Invasions 12: 2979–2983.

Lutterschmidt, W.I., G.A. Marvin, and V.H. Hutchison. 1996. *Rana catesbeiana* (Bullfrog). Record size. Herpetological Review 27:74–75.

Lykens, D.V., and D.C. Forester. 1987. Age structure in the spring peeper: do males advertise longevity? Herpetologica 43:216–223.

Lynch, J.D. 1964a. The toad *Bufo americanus charlesmithi* in Indiana, with remarks on the range of the subspecies. Journal of the Ohio Herpetological Society 4:103–104.

Lynch, J.D. 1964b. Two additional predators of the spadefoot toad, *Scaphiopus holbrookii* (Harlan). Journal of the Ohio Herpetological Society 4:79.

Lynch, J.D. 1970. A taxonomic revision of the leptodactylid frog genus *Syrrhophus* Cope. University of Kansas Publications, Museum of Natural History 20:1–45.

Lynch, J.D. 1978. The distribution of leopard frogs (*Rana blairi* and *Rana pipiens*) (Amphibia, Anura, Ranidae) in Nebraska. Journal of Herpetology 12:157–162.

Lynch, J.D. 1985. Annotated checklist of the amphibians and reptiles of Nebraska. Transactions of the Nebraska Academy of Sciences 13:33–57.

Lynn, W.G., and A. Edelman. 1936. Crowding and metamorphosis in the tadpole. Ecology 17:104–109.

Mabry, C.M., and J.L. Christiansen. 1991. The activity and breeding cycle of *Scaphiopus bombifrons* in Iowa. Journal of Herpetology 25:116–119.

MacArthur, D.L., and J.W.T. Dandy. 1982. Physiological aspects of overwintering in the boreal chorus frog (*Pseudacris triseriata maculata*). Comparative Biochemistry and Physiology 72A:137–141.

MacCulloch, R.D. 2002. The ROM Field Guide to Amphibians and Reptiles of Ontario. Royal Ontario Museum, Toronto.

MacCulloch, R.D., and J.R. Bider. 1975. New records of amphibians and garter snakes in the James Bay Area of Quebec. Canadian Field-Naturalist 89: 80–82.

Macey, J.R., J.L. Strasburg, J.A. Brisson, V.T. Vredenburg, M. Jennings, and A. Larson. 2001. Molecular phylogenetics of western North American frogs of the *Rana boylii* species group. Molecular Phylogenetics and Evolution 19:131–143.

Maciel, N.M., R.G. Collevatti, G.R. Colli, and E.F. Schwartz. 2010. Late Miocene diversification and phylogenetic relationships of the huge toads in the *Rhinella marina* (Linnaeus, 1758) species group (Anura: Bufonidae). Molecular Phylogenetics and Evolution 57:787–797.

MacKay, W.P., S.J. Loring, T.M. Frost, and W.G. Whitford. 1990. Population dynamics of a playa community in the Chihuahuan Desert. Southwest Naturalist 35:393–402.

MacKenzie, D.I., J.D. Nichols, G.B. Lachman, S. Droege, J.A. Royle, and C.A. Langtimm. 2002. Estimating site occupancy rates when detection probabilities are less than one. Ecology 83:2248–2255.

MacTague, L., and P.T. Northen. 1993. Underwater vocalization by the foothill yellow-legged frog (*Rana boylii*). Transactions of the Western Section of the Wildlife Society 29:1–7.

Maeda, N., and M. Matsui. 1999. Frogs and Toads of Japan. Rev. ed. Bun-Ichi Sogo Shuppan Ltd., Tokyo.

Maerz, J.C., B. Blossey, and V. Nuzzo. 2005a. Green frogs show reduced foraging success in habitats invaded by Japanese knotweed. Biodiversity and Conservation 14:2901–2911.

Maerz, J.C., C.J. Brown, C.T. Chapin, and B. Blossey. 2005b. Can secondary compounds of an invasive plant affect larval amphibians? Functional Ecology 19:970–975.

Mahan, J.T., and C.J. Biggers. 1977. Electrophoretic investigation of blood and parotoid venom proteins in *Bufo americanus* and *Bufo woodhousei fowleri*. Comparative Biochemistry and Physiology 57C:121–126.

Mahaney, P.A. 1994. Effects of freshwater petroleum contamination on amphibian hatching and metamorphosis. Environmental Toxicology and Chemistry 13:259–265.

Mahoney, J.J., and V.H. Hutchison. 1969. Photoperiod acclimation and 24-hour variations in the Critical Thermal Maxima of a tropical and a temperate frog. Oecologia 2:143–161.

Mahrdt, C.R., and F.T. Knefler. 1972. Pet or pest? The African clawed frog. Environment Southwest (446):2–5.

Mahrdt, C.R., and F.T. Knefler. 1973. The clawed frog—again. Environment Southwest (450):1–3.

Maisonneuve, C., and S. Rioux. 2001. Importance of riparian habitats for small mammal and herpetofaunal communities in agricultural landscapes of southern Québec. Agriculture, Ecosystems and Environment 83:165–175.

Majji, S., S. LaPatra, S.M. Long, R. Sample, L. Bryan, A. Sinning, and V.G. Chinchar. 2006. *Rana catesbeiana* virus Z (RCV-Z): a novel pathogenic ranavirus. Diseases of Aquatic Organisms 73:1–11.

Malaret, L. 1978. Herpetofauna of Lacreek National Wildlife Refuge. Transactions of the Kansas Academy of Science 80:145–150.

Malmos, K.B., R. Reed, and B. Starret. 1995. Hybridization between *Bufo woodhousii* and *Bufo punctatus* from the Grand Canyon region of Arizona. Great Basin Naturalist 55:368–371.

Malmos, K.B., B.K. Sullivan, and T. Lamb. 2001. Calling behavior and directional hybridization between two toads (*Bufo microscaphus* × *B. woodhousii*) in Arizona. Evolution 55:626–630.

Maniero, G.D., and C. Carey. 1997. Changes in selected aspects of immune function in the leopard frog, *Rana pipiens*, associated with exposure to cold. Journal of Comparative Physiology 167B: 256–263.

Manion, J.J. 1952. Comparative ecological studies on the amphibians of Cass County, Michigan, and vicinity. Ph.D. diss., University of Notre Dame, South Bend, IN.

Manion, J.J., and L. Cory. 1952. Winter kill of *Rana pipiens* in shallow ponds. Herpetologica 8:32.

Manis, M.L., and D.L. Claussen. 1986. Environmental and genetic influences on the thermal physiology of *Rana sylvatica*. Journal of Thermal Biology 11:31–36.

Manter, H.W. 1938. A collection of trematodes from Florida amphibia. Transactions of the American Microscopical Society 57:26–37.

Mao, J., D.E. Green, G. Fellers, and V.G. Chinchar. 1999. Molecular characterization of iridoviruses isolated from sympatric amphibians and fish. Virus Research 63:45–52.

Marchisin, A., and J.D. Anderson. 1978. Strategies employed by frogs and toads (Amphibia, Anura) to avoid predation by snakes (Reptilia, Serpentes). Journal of Herpetology 12:151–155.

Marco, A., J.M. Kiesecker, D.P. Chivers, and A. Blaustein. 1998. Sex recognition and mate choice by male western toads, *Bufo boreas*. Animal Behaviour 55:1631–1635.

Marco, A., C. Quilchano, and A.R. Blaustein. 1999. Sensitivity of nitate and nitrite in pond-breeding amphibians from the Pacific Northwest. Environmental Toxicology and Chemistry 18:2836–2839.

Marcogliese, D.J., K.C. King, H.M. Salo, M. Fournier, P. Brousseau, P. Spear, L. Champoux, J.D. McLaughlin, and M. Boily. 2009. Combined effects of agricultural activity and parasites on biomarkers in the bullfrog, *Rana catesbeiana*. Aquatic Toxicology 91: 126–134.

Mares, M.A. 1972. Notes on *Bufo marinus* tadpole aggregations. Texas Journal of Science 23:433–435.

Marimon, S. 1923. Notes on the color changes of frogs. Pomona College Journal of Entomology and Zoology 15:27–31.

Markle, T.M., and D.M. Green. 2009. New distributional records of amphibians in Quebec and Labrador. Herpetological Review 40:240–241.

Marnell, L.F. 1997. Herpetofauna of Glacier National Park. Northwestern Naturalist 78:17–33.

Marr, S.K., S.A. Johnson, A.H. Hara, and M.E. McGarrity. 2010. Preliminary evaluation of the potential of the helminth parasite *Rhabdias elegans* as a biological control agent for invasive Puerto Rico coquis (*Eleutherodactylus coqui*) in Hawaii. Biological Control 54:69–74.

Marshall, J.L., and C.D. Camp. 1995. Aspects of the feeding ecology of the little grass frog, *Pseudacris ocularis* (Anura: Hylidae). Brimleyana 22:1–7.

Marshall, V.T., S.C. Humfeld, and M.A. Bee. 2003. Plasticity of aggressive signaling and its evolution in male spring peepers, *Pseudacris crucifer*. Animal Behaviour 65:1223–1234.

Marshall, W.H., and M.F. Buell. 1955. A study of the occurrence of amphibians in relation to a bog succession, Itasca State Park, Minnesota. Ecology 36:381–387.

Martin, C.H. 1940. Life cycle of toads in Yosemite Valley. Yosemite Nature Notes 19:90–92.

Martin, D.L. 1991. Population census of a species of special concern: the Yosemite toad (*Bufo canorus*). Fourth Biennial Conference of Research in California's National Parks. University of California, Davis.

Martin, P.S. 1958. A biogeography of reptiles and amphibians in the Gomez Farias region, Tamaulipas, Mexico. Miscellaneous Publications of the Museum of Zoology, University of Michigan 101:1–102.

Martin, T.R., and D.B. Conn. 1990. The pathogenicity, localization, and cyst structure of echinostomatid metacercariae (Trematoda) infecting the kidneys of the frogs *Rana clamitans* and *Rana pipiens*. Journal of Parasitology 76:414–419.

Martinez, J.L. 1948. Cuban frog leg industry. U.S. Fish and Wildlife Service, Fishery Leaflet 284.

Martinez-Ortiz, M. 2004. Predicting habitat suitability and occurrence for Blanchard's cricket frogs (*Acris crepitans blanchardi*) in northwest Ohio. M.S. thesis, Bowling Green State University, Bowling Green, OH.

Martinez Rodriguez, E., T. Gamble, M.V. Hurt, and S. Cotner. 2009. Presence of *Batrachochytrium dendrobatidis* at the headwaters of the Mississippi River, Itasca State Park, Minnesota, USA. Herpetological Review 40:48–50.

Martof, B.S. 1952. Early transformation of the greenfrog, *Rana clamitans* Latreille. Copeia 1952: 115–116.

Martof, B.S. 1953a. Home range and movements of the green frog, *Rana clamitans*. Ecology 34: 529–543.

Martof, B.S. 1953b. Territoriality in the green frog, *Rana clamitans*. Ecology 34:165–174.

Martof, B.S. 1955. Observations on the life history and ecology of the amphibians of the Athens area, Georgia. Copeia 1955:166–170.

Martof, B.S. 1956a. Growth and development of the green frog, *Rana clamitans*, under natural conditions. American Midland Naturalist 55:101–117.

Martof, B.S. 1956b. Factors influencing the size and composition of populations of *Rana clamitans*. American Midland Naturalist 56:224–245.

Martof, B.S. 1958. Reproductive behavior of the chorus frog, *Pseudacris nigrita*. Behaviour 13:243–257.

Martof, B.S. 1960. Autumnal breeding of *Hyla crucifer*. Copeia 1960:58–59.

Martof, B.S. 1961. Vocalization as an isolating mechanism in frogs. American Midland Naturalist 65:118–126.

Martof, B.S. 1962a. Some observations on the role of olfaction among salientian amphibia. Physiological Zoology 35:270–272.

Martof, B.S. 1962b. The behavior of Fowler's toad under various conditions of light and temperature. Physiological Zoology 35:38–46.

Martof, B.S. 1962c. An unusual color variant of *Rana pipiens*. Herpetologica 17:269–270.

Martof, B.S. 1963. Some observations on the herpetofauna of Sapelo Island, Georgia. Herpetologica 19:70–72.

Martof, B.S. 1970. *Rana sylvatica* Le Conte. Wood frog. Catalogue of American Amphibians and Reptiles 86.1–86.4.

Martof, B.S., and R.L. Humphries. 1955. Observations on some amphibians from Georgia. Copeia 1955:245–248.

Martof, B.S., and R.L. Humphries. 1959. Geographic variation in the wood frog *Rana sylvatica*. American Midland Naturalist 61:350–389.

Martof, B.S., and E.F. Thompson Jr. 1964. A behavioral analysis of the mating call of the chorus frog, *Pseudacris triseriata*. American Midland Naturalist 71:198–209.

Maskell, A.J., J.H. Waddle, and K.G. Rice. 2003. *Osteopilus septentrionalis* (Cuban Treefrog). Diet. Herpetological Review 34:137.

Maslin, T.P. 1947. *Rana sylvatica cantabrigensis* Baird in Colorado. Copeia 1947:158–162.

Maslin, T.P. 1959. An annotated check list of the amphibians and reptiles of Colorado. University of Colorado Studies, Series in Biology No. 6:1–98.

Masta, S.E., B.K. Sullivan, T. Lamb, and E.J. Routman. 2002. Molecular systematics, hybridization, and phylogeography of the *Bufo americanus* complex in eastern North America. Molecular Phylogenetics and Evolution 24:302–314.

Masta, S.E., N.M. Laurent, and E.J. Routman. 2003. Population genetic structure of the toad *Bufo woodhousii*: an empirical assessment of the effects of haplotype extinction on nested cladistic analysis. Molecular Ecology 12:1541–1554.

Materna, E.J., C.F. Rabini, and T.W. LaPoint. 1995. Effects of the synthetic pyrethroid insecticide, esfenvalerate, on larval leopard frogs (*Rana* spp.). Environmental Toxicology and Chemistry 14:613–622.

Mathews, R.C., and A.C. Echternacht. 1984. Herpetofauna of the spruce-fir ecosystem in the southern Appalachian Mountains regions, with emphasis on the Great Smoky Mountains National Park. Pp. 155–167 *In* P.S. White (ed.), The Southern Appalachian Spruce-Fir Ecosystem. National Park Service, Research/Resources Management Report SER-71.

Matson, T.O. 1990. A morphometric comparison of gray treefrogs, *Hyla chrysoscelis* and *H. versicolor*, from Ohio. Ohio Journal of Science 90:98–101.

Matsuda, B.M., and J.S. Richardson. 2005. Movement patterns and relative abundance of coastal tailed frogs in clearcuts and mature forest stands. Canadian Journal of Forest Resources 35:1131–1138.

Matsuda, B.M., D.M. Green, and P.T. Gregory. 2006. Amphibians and Reptiles of British Columbia. Royal BC Museum Handbook, Victoria, British Columbia.

Mattfeldt, S.D., L.L. Bailey, and E.H.C. Grant. 2009. Monitoring multiple species: estimating state variables and exploring the efficacy of a monitoring program. Biological Conservation 142:720–737.

Matthews, K.R., and K.L. Pope. 1999. A telemetric study of the movement patterns and habitat use of *Rana muscosa*, the mountain yellow-legged frog, in a high-elevation basin in Kings Canyon National Park, California. Journal of Herpetology 33:615–624.

Matthews, K.R., K.L. Pope, H.K. Preisler, and R.A. Knapp. 2001. Effects of nonnative trout on Pacific treefrogs (*Hyla regilla*) in the Sierra Nevada. Copeia 2001:1130–1137.

Matthews, T.C. 1971. Genetic changes in a population of boreal chorus frogs (*Pseudacris triseriata*) polymorphic for color. American Midland Naturalist 85:208–221.

Matthews, T., and D. Pettus. 1966. Color inheritance in *Pseudacris triseriata*. Herpetologica 22:269–275.

Maunder, J.E. 1983. Amphibians of the Province of Newfoundland. Canadian Field-Naturalist 97:33–46.

Maunder, J.E. 1997. Amphibians of Newfoundland and Labrador: status changes since 1983. SSAR Herpetological Conservation 1:93–99.

Mautz, W.J., and M.R. Dohm. 2004. Respiratory and behavioral effects of ozone on a lizard and a frog. Comparative Biochemistry and Physiology 139A:371–377.

Maxson, L.R., and A.C. Wilson. 1975. Albumin evolution and organismal evolution in tree frogs (Hylidae). Systematic Zoology 24:1–15.

Maxson, L., E. Pepper, and R.D. Maxon. 1977. Immunological resolution of a diploid-tetraploid species complex of tree frogs. Science 197:1012–1013.

Maxson, R.D., and L.R. Maxson. 1978. Reply to Ralin. Science 202:336.

Maxwell, B.A., and D.G. Hokit. 1999. Amphibians and reptiles. Pp. 2.1–2.29 *In* G. Joslin and H. Youmans (coordinators), Effects of Recreation on Rocky Mountain Wildlife. A Review for Montana. Committee on Effects of Recreation on Wildlife,

Montana Chapter of the Wildlife Society. http://joomla.wildlife.org/Montana/images/Documents/2hp1.pdf.

Maxwell, B.A., J.K. Werner, P. Hendricks, and D.L. Flath. 2003. Herpetology in Montana. Northwest Fauna No. 5.

Mayer, F.L., and M.R. Ellersieck. 1986. Manual of acute toxicity: interpretation and data base for 410 chemicals and 66 species of freshwater animals. U.S. Fish and Wildlife Service Resource Publication 160.

Mayhew, W.W. 1965. Adaptations of the amphibian, *Scaphiopus couchi*, to desert conditions. American Midland Naturalist 74:95–109.

Maynard, E.A. 1934. The aquatic migration of the toad, *Bufo americanus* Le Conte. Copeia 1934: 174–177.

Mazanti, L.E. 1999. The effects of atrazine, metolachlor and chlorpyrifos on the growth and survival of larval frogs under laboratory and field conditions. Ph.D. diss., University of Maryland, College Park.

Mazerolle, M.J. 2003. Detrimental effects of peat mining on amphibian abundance and species richness in bogs. Biological Conservation 113:215–223.

Mazzoni, R., A.A. Cunningham, P. Daszak, A. Apolo, E. Perdomo, and G. Speranza. 2003. Emerging pathogen of wild amphibians in frogs (*Rana catesbeiana*) farmed for international trade. Emerging Infectious Diseases 9:995–998.

Mazzoni, R., A.A.J. de Mesquita, L.F.F. Fleury, W.M.E. Diederichsen de Brito, I.A. Nunes, J. Robert, H. Morales, A.S.G. Coelho, D.L. Barthasson, L. Galli, and M.H.B. Catroxo. 2009. Mass mortality associated with a Frog Virus 3–like *Ranavirus* infection in farmed tadpoles *Rana catesbeiana* from Brazil. Diseases of Aquatic Organisms 86:181–191.

McAlister, W.H. 1958. Species distribution in a mixed *Scaphiopus-Bufo* breeding chorus. Southwestern Naturalist 3:227–229.

McAlister, W.H. 1959. The vocal structures and method of call production in the genus *Scaphiopus* Holbrook. Texas Journal of Science 11:60–77.

McAlister, W.H. 1961. Artificial hybridization between *Rana a. areolata* and *R. p. pipiens* from Texas. Texas Journal of Science 13:423–426.

McAlister, W.H. 1962. Variation in *Rana pipiens* Schreber in Texas. American Midland Naturalist 67:334–363.

McAlister, W.H. 1963. A post-breeding concentration of the spring peeper. Herpetologica 19:293.

McAllister, C.T. 1987. Protozoan and metazoan parasites of Strecker's chorus frog, *Pseudacris streckeri streckeri* (Anura: Hylidae), from north-central Texas. Proceedings of the Helminthological Society of Washington 54:271–274.

McAllister, C.T. 1991. Protozoan, helminth, and arthropod parasites of the spotted chorus frog, *Pseudacris clarkii* (Anura: Hylidae), from north-central Texas. Journal of the Helminthological Society of Washington 58:51–56.

McAllister, C.T., and S.P. Tabor. 1985. *Gastrophryne olivacea* (Great Plains Narrowmouth Toad). Coexistence. Herpetological Review 16:109.

McAllister, C.T., and D.B. Conn. 1990. Occurrence of tetrathyridia of *Mesocestoides* sp. (Cestoidea: Cyclophyllidea) in North American anurans (Amphibia). Journal of Wildlife Diseases 26:540–543.

McAllister, C.T., and S.E. Trauth. 1995. New host records for *Myxidium serotinum* (Protozoa: Myxosporea) from North American amphibians. Journal of Parasitology 81:485–488.

McAllister, C.T., and S.E. Trauth. 1996. Ultrastructure of *Cedpedia virginiensis* (Protista: Haptophryniidae) from the gall bladder of the pickerel frog, *Rana palustris*, in Arkansas. Proceedings of the Arkansas Academy of Science 50:133–136.

McAllister, C.T., S.J. Upton, and D.B. Conn. 1989. A comparative study of endoparasites in three species of sympatric *Bufo* (Anura: Bufonidae), from Texas. Proceedings of the Helminthological Society of Washington 56:162–167.

McAllister, C.T., S.J. Upton, S.E. Trauth, and C.R. Bursey. 1995a. Parasites of wood frogs, *Rana sylvatica* (Ranidae), from Arkansas, with a description of a new species of *Eimera* (Apicomplexa: Eimeriidae). Journal of the Helminthological Society of Washington 62:143–149.

McAllister, C.T., S.E. Trauth, and L.D. Gage. 1995b. Vertebrate fauna of abandoned mines at Gold Mine Springs, Independence County, Arkansas. Proceedings of the Arkansas Academy of Science 49: 184–187.

McAllister, C.T., S.E. Trauth, and C.R. Bursey. 1995c. Parasites of the pickerel frog, *Rana palustris* (Anura: Ranidae), from the southern part of its range. Southwestern Naturalist 40:111–116.

McAllister, C.T., C.R. Bursey, and S.E. Trauth. 2008. New host and geographic records for some endoparasites (Myxosporea, Trematoda, Cestoidea, Nematoda) of amphibians and reptiles from Arkansas and Texas, U.S.A. Comparative Parasitology 75:241–254.

McAllister, K.R. 1995. Distribution of amphibians and reptiles in Washington State. Northwest Fauna 3:81–112.

McAllister, K.R., and W.P. Leonard. 1997. Washington State status report for the Oregon spotted frog. Washington Department of Fish and Wildlife, Olympia.

McAllister, K.R., W.P. Leonard, and R.M. Storm. 1993. Spotted frog (*Rana pretiosa*) surveys in the Puget Trough of Washington, 1989–1991. Northwestern Naturalist 74:10–15.

McAllister, K.R., J.W. Watson, K. Risenhoover, and T. McBride. 2004. Marking and radiotelemetry of

Oregon spotted frogs (*Rana pretiosa*). Northwestern Naturalist 85:20–25.

McAlpine, D.F. 1996. Acanthocephala parasitic in North American amphibians: a review with new records. Alytes 14:115–121.

McAlpine, D.F. 1997a. Historical evidence does not suggest New Brunswick amphibians have declined. SSAR Herpetological Conservation 1:117–127.

McAlpine, D.F. 1997b. Helminth communities in bullfrogs (*Rana catesbeiana*), green frogs (*Rana clamitans*), and leopard frogs (*Rana pipiens*) from New Brunswick, Canada. Canadian Journal of Zoology 75:1883–1890.

McAlpine, D.F. 1997c. A simple transect technique for estimating abundance of aquatic ranid frogs. SSAR Herpetological Conservation 1:180–184.

McAlpine, D.F., and T.G. Dilworth. 1989. Microhabitat and prey size among three species of *Rana* (Anura: Ranidae) sympatric in eastern Canada. Canadian Journal of Zoology 67:2244–2252.

McAlpine, D.F., and M.D.B. Burt. 1998. Helminth communities in bullfrogs (*Rana catesbeiana*), green frogs (*Rana clamitans*), and leopard frogs (*Rana pipiens*) from New Brunswick, Canada. Canadian Field-Naturalist 112:50–68.

McAlpine, D.F., R.W. Harding, and R. Curley. 2006. Occurrence and biogeographic significance of the pickerel frog (*Rana palustris*) on Prince Edward Island, Canada. Herpetological Natural History 10:95–98.

McAlpine, S. 1993. Genetic heterozygosity and reproductive success in the green treefrog, *Hyla cinerea*. Heredity 70:553–558.

McAlpine, S., and M.H. Smith. 1995. Genetic correlates of fitness in the green treefrog, *Hyla cinerea*. Herpetologica 51:393–400.

McAtee, W.L. 1921. Homing and other habits of the bullfrog. Copeia (96):39–40.

McAuliffe, J.R. 1978. Biological survey and management of sport-hunted bullfrog populations in Nebraska. Nebraska Game and Parks Commission, Lincoln.

McCallum, M.L. 1999a. *Rana sphenocephala* (Southern Leopard Frog) malformities found in Illinois with behavioral notes. Transactions of the Illinois State Academy of Science 92:257–264.

McCallum, M.L. 1999b. *Acris crepitans* (Northern Cricket Frog). Death feigning. Herpetological Review 30:90.

McCallum, M.L. 2003. Reproductive ecology and taxonomic status of *Acris crepitans blanchardi* with additional investigations on the Hamilton and Zuk hypotheses. Ph.D. diss., Arkansas State University, Jonesboro.

McCallum, M.L. 2010. Future climate change spells catastrophe for Blanchard's cricket frog, *Acris blanchardi* (Amphibia: Anura: Hylidae). Acta Herpetologica 5:119–130.

McCallum, M.L., and S.E. Trauth. 2001a. Are tadpoles of the Illinois chorus frog (*Pseudacris streckeri illinoensis*) cannibalistic? Transactions of the Illinois Academy of Science 94:171–178.

McCallum, M.L., and S.E. Trauth. 2001b. *Pseudacris streckeri illinoensis* (Illinois Chorus Frog). Terrestrial feeding. Herpetological Review 32:35.

McCallum, M.L., and S.E. Trauth. 2003a. A forty-three year museum study of northern cricket frog (*Acris crepitans*) abnormalities in Arkansas: upward trends and distributions. Journal of Wildlife Diseases 39:522–528.

McCallum, M.L., and S.E. Trauth. 2003b. *Acris crepitans* (Northern Cricket Frog). Communal hibernaculum. Herpetological Review 34:228.

McCallum, M.L., and S.E. Trauth. 2004. Blanchard's cricket frog in Nebraska and South Dakota. Prairie Naturalist 36:129–135.

McCallum, M.L., and S.E. Trauth. 2006. An evaluation of the subspecies *Acris crepitans blanchardi* (Anura, Hylidae). Zootaxa 1104:1–21.

McCallum, M.L., and S.E. Trauth. 2007. Physiological trade-offs between immunity and reproduction in the northern cricket frog (*Acris crepitans*). Herpetologica 63:269–274.

McCallum, M.L., and S.E. Trauth. 2009. *Hyla cinerea* (Green Tree Frog) and *Pseudacris triseriata* (Western Chorus Frog). Calling in amplexus. Herpetological Review 40:204–205.

McCallum, M.L., B.A. Wheeler, and S.E. Trauth. 2001. *Pseudacris streckeri illinoensis* (Illinois Chorus Frog). Dysfunctional vocal sac. Herpetological Review 32:248–249.

McCallum, M.L., T.L. Klotz, and S.E. Trauth. 2003a. *Rana sylvatica* (Wood Frog). Death feigning. Herpetological Review 34:54–55.

McCallum, M.L., T.L. Klotz, and S.E. Trauth. 2003b. *Rana sylvatica* (Wood Frog). Phonotactic stalking. Herpetological Review 34:54.

McCallum, M.L., C. Brooks, R. Mason, and S.E. Trauth. 2011. Growth, reproduction, and life span in Blanchard's cricket frog (*Acris blanchardi*) with notes on the growth of the northern cricket frog (*Acris crepitans*). Herpetology Notes 4:25–35.

McClanahan, L. Jr. 1964. Osmotic tolerance of the muscles of two desert-inhabiting toads, *Bufo cognatus* and *Scaphiopus couchi*. Comparative Biochemistry and Physiology 12:501–508.

McClanahan, L. Jr. 1967. Adaptations of the spadefoot toad, *Scaphiopus couchi*, to desert environments. Comparative Biochemistry and Physiology 20:73–99.

McClanahan, L.L., R. Ruibal, and V.H. Shoemaker. 1994. Frogs and toads in deserts. Scientific American 270(3):82–88.

McClelland, B.E., and W. Wilczynski. 1989. Release call characteristics of male and female *Rana pipiens*. Copeia 1989:1045–1049.

McClelland, B.E., W. Wilczynski, and M.J. Ryan. 1996. Correlations between call characteristics and morphology in male cricket frogs (*Acris crepitans*). Journal of Experimental Biology 199:1907–1919.

McClure, K.A. 1996. Ecology of *Pseudacris brachyphona*: a second look. M.S. thesis, Marshall University, Huntington, WV.

McCoid, M.J. 1985. An observation of reproductive behavior in a wild population of African clawed frogs, *Xenopus laevis*, in California. California Fish and Game 71:245–246.

McCoid, M.J. 2005. *Rana berlandieri* (Rio Grande Leopard Frog). Salinity tolerance. Herpetological Review 36:437–438.

McCoid, M.J., and T.H. Fritts. 1980a. Notes on the diet of a feral population of *Xenopus laevis* (Pipidae) in California. Southwestern Naturalist 25:272–275.

McCoid, M.J., and T.H. Fritts. 1980b. Observations of feral populations of *Xenopus laevis* (Pipidae) in Southern California. Bulletin of the Southern California Academy of Science 79:82–86.

McCoid, M.J., and T.H. Fritts. 1989. Growth and fatbody cycles in feral populations of the African clawed frog, *Xenopus laevis* (Pipidae), in California with comments on reproduction. Southwestern Naturalist 34:499–505.

McCoid, M.J., and T.H. Fritts. 1993. Speculations on colonizing success of the African clawed frog, *Xenopus laevis* (Pipidae), in California. South African Journal of Zoology 28:59–61.

McCoid, M.J., and T.H. Fritts. 1995. Female reproductive potential and winter growth of African clawed frogs (Pipidae: *Xenopus laevis*) in California. California Fish and Game 81:39–42.

McCoid, M.J., G.K. Pregill, and R.M. Sullivan. 1993. Possible decline of *Xenopus* populations in Southern California. Herpetological Review 24:29–30.

McCollum, S.A., and J. Van Buskirk. 1996. Costs and benefits of a predator-induced polyphenism in the gray treefrog *Hyla chrysoscelis*. Evolution 50:583–593.

McCollum, S.A., and J.D. Leimberger. 1997. Predator-induced morphological changes in an amphibian: predation by dragonflies affects tadpole shape and color. Oecologia 109:615–621.

McCormick, J. 1970. The Pine Barrens: a preliminary ecological inventory. Research Report No. 2, New Jersey State Museum, Trenton.

McCoy, C.J. 1969. Diet of bullfrogs (*Rana catesbeiana*) in central Oklahoma farm ponds. Proceedings of the Oklahoma Academy of Science 48:44–45.

McCoy, C.J., and C.J. Durden. 1965. New distribution records of amphibians and reptiles in eastern Canada. Canadian Field-Naturalist 79:156–157.

McCoy, C.J., H.M. Smith, and J.A. Tihen. 1967. Natural hybrid toads, *Bufo punctatus* × *Bufo woodhousei*, from Colorado. Southwestern Naturalist 12:45–54.

McCoy, K.A., L.J. Bortnick, C.M. Campbell, H.J. Hamlin, L.J. Guillette Jr., and C.M. St. Mary. 2008. Agriculture alters gonadal form and function in the toad *Bufo marinus*. Environmental Health Perspectives 116:1526–1532.

McCoy, M.W. 2007. Conspecific density determines the magnitude and character of predator-induced phenotype. Oecologia 153:871–878.

McCoy, M.W., and B.M. Bolker. 2008. Trait-mediated interactions: influence of prey size, density and experience. Journal of Animal Ecology 77:478–486.

McCreary, B., and C.A. Pearl. 2010. *Pseudacris regilla* (Northern Pacific Treefrog). Cavity use. Herpetological Review 41:202–203.

McDaniel, V.R., and J.E. Gardner. 1977. Cave fauna of Arkansas: vertebrate taxa. Proceedings of the Arkansas Academy of Science 31:68–71.

McDonald, D.G., J.L. Ozog, and B.P. Simons. 1984. The influence of low pH environments on ion regulation in the larval stages of the anuran amphibian, *Rana clamitans*. Canadian Journal of Zoology 62:2171–2177.

McGarrity, M.E., and S.A. Johnson. 2009. Geographic trend in sexual size dimorphism and body size of *Osteopilus septentrionalis* (Cuban treefrog): implications for invasion of the southeastern United States. Biological Invasions 11:1411–1420.

McGarrity, M.E., and S.A. Johnson. 2010. A radio telemetry study of invasive Cuban treefrogs. Florida Scientist 73:225–235.

McGrath, E.A., and M.M. Alexander. 1979. Observations on the exposure of larval bullfrogs to fuel oil. Transactions of the Northeast Section of The Wildlife Society 36:45–51.

McHenry, D.J., M.A. Gaston, and M.R.J. Forstner. 2010. *Bufo houstonensis* (Houston Toad). Predation. Herpetological Review 41:193.

McKamie, J.A., and G.A. Heidt. 1974. A comparison of spring food habits of the bullfrog, *Rana catesbeiana*, in three habitats of central Arkansas. Southwestern Naturalist 19:105–119.

McKeever, S. 1977. Observations of *Corethrella* feeding on tree frogs (*Hyla*). Mosquito News 37:522–523.

McKeever, S., and F.E. French. 1991. *Corethrella* (Diptera: Corethrellidae) of eastern North America: laboratory life history and field responses to anuran calls. Annals of the Entomological Society of America 84:493–497.

McKeown, J.P. 1968. The ontogenetic development of Y-axis orientation in four species of anurans. Ph.D. diss., Mississippi State University, Mississippi State.

McKeown, S. 1996. A Field Guide to Reptiles and Amphibians in the Hawaiian Islands. Diamond Head Publishing, Los Osos, CA.

McKinnell, R.G. 1984. Lucké tumor of frogs. Pp.

581–605 *In* G.L. Hoff, F.L. Frye, and E.R. Jacobson (eds.), Diseases of Amphibians and Reptiles. Plenum Press, New York.

McKinnell, R.G., and J. Zambernard. 1968. Virus particles in renal tumors obtained from spring *Rana pipiens* of known geographic origin. Cancer Research 28:684–688.

McKinnell, R.G., and D.L. Carlson. 2005. Lucké renal adenocarcinoma. Pp. 96–102 *In* M.J. Lannoo (ed.), Amphibian Declines. The Conservation Status of United States Species. University of California Press, Berkeley.

McKinnell, R.G., D.M. Hoppe, and B.K. McKinnell. 2005. Monitoring pigment pattern morphs of northern leopard frogs. Pp. 328–337 *In* M.J. Lannoo (ed.), Amphibian Declines. The Conservation Status of United States Species. University of California Press, Berkeley.

McLaren, I.A. 1965. Temperature and frog eggs. Journal of General Physiology 48:1071–1079.

McLeod, D.S. 2005. Nebraska's declining amphibians. Pp. 292–294 *In* M.J. Lannoo (ed.), Amphibian Declines. The Conservation Status of United States Species. University of California Press, Berkeley.

McLeod, R.F. 1995. The effects of timber harvest and prescribed burning on the distribution and abundance of reptiles and amphibians at Remington Farms, Maryland. M.S. thesis, Frostburg State University, Frostburg, MD.

McLeod, R.F., and J.E. Gates. 1998. Response of herpetofaunal communities to forest cutting and burning at Chesapeake Farms, Maryland. American Midland Naturalist 139:164–177.

McMenamin, S.K., E.A. Hadly, and C.K. Wright. 2008. Climatic change and wetland desiccation cause amphibian decline in Yellowstone National Park. Proceedings of the National Academy of Sciences USA 105:16988–16993.

McMenamin, S.K., E.A. Hadly, and C.K. Wright. 2009. Reply to Patla et al.: amphibian habitat and populations in Yellowstone damaged by drought and global warming. Proceedings of the National Academy of Sciences USA 106:E23.

McNicholl, M.K. 1972. An observation of apparent death-feigning by a toad. Blue Jay 30:54–55.

Meacham, W.R. 1958. Factors affecting gene exchange between two allopatric populations of the *Bufo woodhousei* complex. Ph.D. diss., University of Texas, Austin.

Meacham, W.R. 1962. Factors affecting secondary intergradation between two allopatric populations in the *Bufo woodhousei* complex. American Midland Naturalist 67:282–304.

Means, D.B., and C.J. Longden. 1976. Aspects of the biology and zoogeography of the Pine Barrens treefrog (*Hyla andersonii*) in northern Florida. Herpetologica 32:117–130.

Means, D.B., and D. Simberloff. 1987. The peninsula effect: habitat correlated species decline in Florida's herpetofauna. Journal of Biogeography 14:551–568.

Means, D.B., and R.C. Means. 2005. Effects of sand pine silviculture on pond-breeding amphibians in the Woodville Karst Plain of north Florida. Pp. 56–61 *In* W.E. Meshaka Jr. and K.J. Babbitt (eds.), Amphibians and Reptiles. Status and Conservation in Florida. Krieger Publishing, Malabar, FL.

Means, R.C., and R. Franz. 2005. Herpetofauna of impacted wetlands in east Florida: a pre-augmentation assessment. Pp. 23–31 *In* W.E. Meshaka Jr. and K.J. Babbitt (eds.), Amphibians and Reptiles. Status and Conservation in Florida. Krieger Publishing, Malabar, FL.

Mecham, J.S. 1954. Geographic variation in the green frog, *Rana clamitans* Latreille. Texas Journal of Science 6:1–24.

Mecham, J.S. 1957. Some hybrid combinations between Strecker's chorus frog, *Pseudacris streckeri*, and certain related forms. Texas Journal of Science 9:337–345.

Mecham, J.S. 1958. Some Pleistocene amphibians and reptiles from Friesenhahn Cave, Texas. Southwestern Naturalist 3:17–27.

Mecham, J.S. 1959. Experimental evidence of the relationship of two allopatric chorus frogs of the genus *Pseudacris*. Texas Journal of Science 11:343–347.

Mecham, J.S. 1960a. Introgressive hybridization between two southeastern treefrogs. Evolution 14:445–457.

Mecham, J.S. 1960b. Natural hybridization between the tree frogs *Hyla versicolor* and *Hyla avivoca*. Journal of the Elisha Mitchell Scientific Society 76:64–67.

Mecham, J.S. 1961. Isolating mechanisms in anuran amphibians. Pp. 24–61 *In* W.F. Blair (ed.), Vertebrate Speciation. University of Texas Press, Austin.

Mecham, J.S. 1964. Ecological and genetic relationships of the two cricket frogs, genus *Acris*, from Alabama. Herpetologica 20:84–91.

Mecham, J.S. 1965. Genetic relationships and reproductive isolation in southeastern frogs of the genera *Pseudacris* and *Hyla*. American Midland Naturalist 74:269–308.

Mecham, J.S. 1968. Evidence of reproductive isolation between two populations of the frog, *Rana pipiens*, in Arizona. Southwestern Naturalist 13:35–44.

Mecham, J.S. 1969. New information from experimental crosses on genetic relationships within the *Rana pipiens* species group. Journal of Experimental Zoology 170:169–180.

Mecham, J.S., M.J. Littlejohn, R.S. Oldham, L.E. Brown, and J.R. Brown. 1973. A new species of leopard frog (*Rana pipiens* complex) from the plains of the central United States. Occasional Papers of the Museum of Texas Tech University 18:1–11.

Meehan, W.E. 1906. Frog-farming. Pennsylvania Department of Fisheries, Bulletin No. 4.

Meehan, W.E. 1908a. Frog farming an industry. Technical World Magazine 9(3):246–250.

Meehan, W.E. 1908b. Possibilities of frog farming. Country Life in America 13:614–615, 640.

Meehan, W.E., and E.A. Andrews. 1908. Frogs, *Rana* spp., Ranidae. Pp. 394–395 *In* L.H. Bailey (ed.), Cyclopedia of American Agriculture. A Popular Survey of Agricultural Conditions, Practices and Ideals in the United States and Canada. Macmillan, New York.

Meek, S.E., and D.G. Elliot. 1899. Notes on a collection of cold-blooded vertebrates from the Olympic Mountains. Field Columbian Museum Publication 31, Zoological Series 1:225–236.

Meeks, D.E., and J.W. Nagel. 1973. Reproduction and development of the wood frog, *Rana sylvatica*, in eastern Tennessee. Herpetologica 29:188–191.

Meijden, A. van der, M. Vinces, S. Hoegg, R. Boistel, A. Channing, and A. Meyer. 2007. Nuclear gene phylogeny of narrow-mouthed toads (Family: Microhylidae) and a discussion of competing hypotheses concerning their biogeographical origins. Molecular Phylogenetics and Evolution 44:1017–1030.

Mendonça, M.T., P. Licht, M.J. Ryan, and R. Barnes. 1985. Changes in hormone levels in relation to breeding behavior in male bullfrogs (*Rana catesbeiana*) at the individual and population levels. General and Comparative Endocrinology 58: 270–279.

Mennell, L. 1997. Amphibians in southwestern Yukon and northwestern British Columbia. SSAR Herpetological Conservation 1:107–109.

Merkle, D.A. 1977. The occurrence of the eastern spadefoot, *Scaphiopus h. holbrooki*, in the central Piedmont of Virginia. Bulletin of the Maryland Herpetological Society 13:196–197.

Merovich, C.E., and J.H. Howard. 2000. Amphibian use of constructed ponds on Maryland's Eastern Shore. Journal of the Iowa Academy of Science 107:151–159.

Merrell, D.J. 1965. The distribution of the dominant Burnsi gene in the leopard frog, *Rana pipiens*. Evolution 19:69–85.

Merrell, D.J. 1968. A comparison of the estimated size and the "effective size" of breeding populations of the leopard frog, *Rana pipiens*. Evolution 22:274–283.

Merrell, D.J. 1969. Natural selection in a leopard frog population. Journal of the Minnesota Academy of Science 35:86–89.

Merrell, D.J. 1970. Migration and gene dispersal in *Rana pipiens*. American Zoologist 10:47–52.

Merrell, D.J. 1972. Laboratory studies bearing on pigment pattern polymorphisms in wild populations of *Rana pipiens*. Genetics 70:141–161.

Merrell, D.J. 1973. Ecological genetics of anurans as exemplified by *Rana pipiens*. Pp. 329–335 *In* J.L. Vial (ed.), Evolutionary Biology of the Anurans. University of Missouri Press, Columbia.

Merrell, D.J. 1977. Life history of the leopard frog, *Rana pipiens*, in Minnesota. University of Minnesota Bell Museum of Natural History, Occasional Papers No. 15.

Merrell, D.J., and C.F. Rodell. 1968. Seasonal selection in the leopard frog, *Rana pipiens*. Evolution 22:284–288.

Mertins, J.W., S.M. Torrence, and M.C. Sterner. 2011. Chiggers recently infesting *Spea* spp. in Texas, USA, were *Eutrombicula affreddugesi*, not *Hannemania* sp. Journal of Wildlife Diseases 47:612–617.

Meshaka, W.E. Jr. 1993. Hurricane Andrew and the colonization of five invading species in south Florida. Florida Scientist 56:193–201.

Meshaka, W.E. Jr. 1994. Giant toad eaten by red-shouldered hawk. Florida Field Naturalist 22: 54–55.

Meshaka, W.E. Jr. 1996a. Diet and the colonization of buildings by the Cuban treefrog, *Osteopilus septentrionalis* (Anura: Hylidae). Caribbean Journal of Science 32:59–63.

Meshaka, W.E. Jr. 1996b. Vagility and the Florida distribution of the Cuban treefrog (*Osteopilus septentrionalis*). Herpetological Review 27:37–40.

Meshaka, W.E. Jr. 1996c. Retreat use by the Cuban treefrog (*Osteopilus septentrionalis*): implications for successful colonization in Florida. Journal of Herpetology 30:443–445.

Meshaka, W.E. Jr. 2001. The Cuban Treefrog in Florida. University Press of Florida, Gainesville.

Meshaka, W.E. Jr., and B. Ferster. 1995. Two species of snakes prey on Cuban treefrogs in southern Florida. Florida Field Naturalist 23:97–98.

Meshaka, W.E. Jr., and G.E. Woolfenden. 1999. Relation of temperature and rainfall to movements and reproduction of the eastern narrow-mouthed toad (*Gastrophryne carolinensis*) in south-central Florida. Florida Scientist 62:213–221.

Meshaka, W.E. Jr., and A.L. Mayer. 2005. Diet of the southern toad (*Bufo terrestris*) from the southern Everglades. Florida Scientist 68:261–266.

Meshaka, W.E. Jr., and R. Powell. 2010. Diets of the native southern toad (*Anaxyrus terrestris*) and the exotic cane toad (*Rhinella marina*) from a single site in south-central Florida. Florida Scientist 73: 175–179.

Meshaka, W.E. Jr., W.F. Loftus, and T. Steiner. 2000. The herpetofauna of Everglades National Park. Florida Scientist 63:84–103.

Meshaka, W.E. Jr., B.P. Butterfield, and J.B. Hauge. 2004. The Exotic Amphibians and Reptiles of Florida. Krieger Publishing, Malabar, FL.

Meshaka, W.E. Jr., S.D. Marshall, L.R. Raymond, and L.M. Hardy. 2009a. Seasonal activity, reproductive cycles, and growth of the bronze frog (*Lithobates clamitans clamitans*) in northern Louisiana: the long and the short of it. Journal of Kansas Herpetology 29:12–20.

Meshaka, W.E. Jr., J. Boundy, S.D. Marshall, and J. Delahoussaye. 2009b. Seasonal activity, reproductive cycles, and growth of the bronze frog (*Lithobates clamitans clamitans*) in southern Louisiana: an endpoint in its geographic distribution and the variability of its geographic life history traits. Journal of Kansas Herpetology 31:12–17.

Meshaka, W.E. Jr., J. Boundy, and A. Williams. 2009c. The dispersal of the greenhouse frog, *Eleutherodactylus planirostris* (Anura: Eleutherodactylidae), in Louisiana, with preliminary observations on several potential exotic colonizing species. Journal of Kansas Herpetology 32:13–16.

Metcalf, M.M. 1928. The belltoads and their opalinid parasites. American Naturalist 62:5–21.

Metcalf, Z.P. 1921. The food capacity of the toad. Copeia (100):81–82.

Meteyer, C.U. 2000. Field guide to malformations of frogs and toads, with radiographic interpretations. U.S. Geological Survey, Biological Science Report USGS/BRD/BSR-2000–0005.

Meteyer, C.U., I.K. Loeffler, J.F. Fallon, K.A. Converse, E. Green, J.C. Helgen, S. Kersten, R. Levey, L. Eaton-Poole, and J.G. Burkhart. 2000a. Hind limb malformations in free-living northern leopard frogs (*Rana pipiens*) from Maine, Minnesota, and Vermont suggest multiple etiologies. Teratology 62:151–171.

Meteyer, C.U., R.A. Cole, K.A. Converse, D.E. Docherty, M. Wolcott, J.C. Helgren, R. Levey, L. Eaton-Poole, and J.G. Burkhart. 2000b. Defining anuran malformations in the context of a developmental problem. Journal of the Iowa Academy of Science 107:72–78.

Metter, D.E. 1960. The distribution of amphibians in eastern Washington. M.S. thesis, Washington State University, Pullman.

Metter, D.E. 1961. Water levels as an environmental factor in the breeding season of *Bufo boreas boreas* (Baird and Girard). Copeia 1961:488.

Metter, D.E. 1963. Stomach contents of Idaho larval *Dicamptodon*. Copeia 1963:435–436.

Metter, D.E. 1964a. A morphological and ecological comparison of two populations of the tailed frog, *Ascaphus truei* Stejneger. Copeia 1964:181–195.

Metter, D.E. 1964b. On breeding and sperm retention in *Ascaphus*. Copeia 1964:710–711.

Metter, D.E. 1967. Variation in the ribbed frog *Ascaphus truei* Stejneger. Copeia 1967:634–649.

Metter, D.E. 1968. The influence of floods on population structure of *Ascaphus truei* Stejneger. Journal of Herpetology 1:105–106.

Metter, D.E., and R.J. Pauken. 1969. An analysis of the reduction of gene flow in *Ascaphus truei* in the northwest U.S. since the Pleistocene. Copeia 1969:307–310.

Metts, B.S., J.D. Lanham, and K.R. Russell. 2001. Evaluation of herpetofaunal communities on upland streams and beaver-impounded streams in the upper Piedmont of South Carolina. American Midland Naturalist 145:54–65.

Meyer, F.P. 1965. The experimental use of Guthion as a selective fish eradicator. Transactions of the American Fisheries Society 94:203–209.

Meylan, P. 2005. Late Pliocene anurans from Inglis 1A, Citrus County, Florida. Bulletin of the Florida Museum of Natural History 45:171–178.

Micancin, J.P. 2008. Acoustic variation and species discrimination in southeastern sibling species, the cricket frogs *Acris crepitans* and *Acris gryllus*. Ph.D. diss., University of North Carolina, Chapel Hill.

Micancin, J.P. 2010. *Acris gryllus* (Southern Cricket Frog). Dispersal. Herpetological Review 41:192.

Micancin, J.P., and J.T. Mette. 2009. Acoustic and morphological identification of the sympatric cricket frogs *Acris crepitans* and *A. gryllus* and the disappearance of *A. gryllus* near the edge of its range. Zootaxa 2076:1–36.

Micancin, J.P., and J.T. Mette. 2010. *Acris crepitans* (Northern Cricket Frog) and *Acris gryllus* (Southern Cricket Frog). Interspecific agonism. Herpetological Review 41:192.

Michaud, T.C. 1962. Call discrimination by females of the chorus frogs, *Pseudacris clarki* and *Pseudacris nigrita*. Copeia 1962:213–215.

Michaud, T.C. 1964. Vocal variation in two species of chorus frogs, *Pseudacris nigrita* and *Pseudacris clarki*, in Texas. Evolution 18:498–506.

Middendorf, L.J. 1957. Observations on the early spring activities of the western spotted frog (*Rana pretiosa pretiosa*) in Gallatin County, Montana. Proceedings of the Montana Academy of Science 17:55–56.

Mierzwa, K.S. 1998. Status of northeastern Illinois amphibians. Pp. 115–124 *In* M.J. Lannoo (ed.), Status & Distribution of Midwestern Amphibians. University of Iowa Press, Iowa City.

Mierzwa, K.S. 2000. Wetland mitigation and amphibians: preliminary observations at a southwestern Illinois bottomland hardwood forest restoration site. Journal of the Iowa Academy of Science 107:191–194.

Mifsud, D.A., and R. Mifsud. 2008. Golf courses as refugia for herpetofauna in an urban river floodplain. Pp. 303–310 *In* J.C. Mitchell, R.E. Jung Brown, and B. Bartholomew (eds.), Urban Herpetology. Herpetological Conservation 3. Society for the Study of Amphibians and Reptiles, Salt Lake City, UT.

Mikaelian, I., M. Ouellet, B. Pauli, J. Rodrigue, J.C. Harshbarger, and D.M. Green. 2000. *Ichthyophonus*-like infection in wild amphibians from Québec, Canada. Diseases of Aquatic Organisms 40:195–201.

Mikulka, P., J. Highes, and G. Aggerup. 1980. The effect of pretraining procedures and discriminative stimuli on the development of food selection behaviors in the toad (*Bufo terrestris*). Behavioral and Neural Biology 29:52–62.

Milius, S. 1998. Fatal skin fungus found in U.S. frogs. Science News 154:7–8.

Miller, C.E. 1968. Frogs with five legs. Carolina Tips No. 31(1).

Miller, D.G. Jr. 1899. A new treefrog from the District of Columbia. Proceedings of the Biological Society of Washington 13:75–78.

Miller, D.L., C.R. Bursey, M.J. Gray, and L.M. Smith. 2004. Metacercariae of *Clinostomum attenuatum* in *Ambystoma tigrinum marvortium*, *Bufo cognatus* and *Spea multiplicata* from west Texas. Journal of Helminthology 78:373–376.

Miller, D.L., S. Rajeev, M.J. Gray, and C.A. Baldwin. 2007. Frog Virus 3 infection, cultured American Bullfrogs. Emerging Infectious Diseases 13:342–343.

Miller, D.M., R.A. Young, T.W. Gatlin, and J.A. Richardson. 1982. Amphibians and reptiles of the Grand Canyon. Grand Canyon Natural History Association, Monograph 4:i-144.

Miller, J.D. 1978. Observations on the diets of *Rana pretiosa*, *Rana pipiens*, and *Bufo boreas* from western Montana. Northwest Science 52:243–249.

Miller, K., and G.C. Packard. 1977. An altitudinal cline in critical thermal maxima of chorus frogs (*Pseudacris triseriata*). American Naturalist 111:267–277.

Miller, L.K., and P.J. Dehlinger. 1969. Neuromuscular function and low temperatures in frogs from cold and warm climates. Comparative Biochemistry and Physiology B 28:915–921.

Miller, N. 1909a. The American toad (*Bufo lentiginosus americanus*, LeConte). A study in dynamic biology. American Naturalist 43:641–668.

Miller, N. 1909b. The American toad (*Bufo lentiginosus americanus*, LeConte). 2. A study in dynamic biology. American Naturalist 43:730–745.

Miller, W.D., and J. Chapin. 1910. The toads of the northeastern United States. Science 32:315–317.

Mills, N.E., and R.D. Semlitsch. 2004. Competition and predation mediate the indirect effects of an insecticide on southern leopard frogs. Ecological Applications 14:1041–1054.

Millzner, R. 1924. A larval acathocephalid, *Centrorhynchus californicus*, sp. nov., from the mesentery of *Hyla regilla*. University of California Publications in Zoology 26:225–230.

Milstead, W.W., J.S. Mecham, and H. McClintock. 1950. The amphibians and reptiles of the Stockton Plateau in northern Terrell County, Texas. Texas Journal of Science 2:543–562.

Milstead, W.W., A.S. Rand, and M.M. Stewart. 1974. Polymorphism in cricket frogs: an hypothesis. Evolution 28:489–491.

Minowa, S., Y. Senga, and T. Miyashita. 2008. Microhabitat selection of the introduced bullfrogs (*Rana catesbeiana*) in paddy fields in eastern Japan. Current Herpetology 27:55 59.

Minton, J.E. 1949. Coral snake preyed upon by the bullfrog. Copeia 1949:288.

Minton, S.A. Jr. 1958. Observations on amphibians and reptiles of the Big Bend region of Texas. Southwestern Naturalist 3:28–54.

Minton, S.A. Jr. 2001. Amphibians & Reptiles of Indiana. Indiana Academy of Science, Indianapolis.

Minton, S.A., J.C. List, and M.J. Lodato. 1982. Recent records and status of amphibians and reptiles in Indiana. Proceedings of the Indiana Academy of Science 92:489–498.

Mitchell, J.C. 1986. Life history patterns in a central Virginia frog community. Virginia Journal of Science 37:262–271.

Mitchell, J.C. 1990. *Pseudacris feriarum* (Upland Chorus Frog). Predation. Herpetological Review 21:89–90.

Mitchell, J.C. 2002. An overview of amphibian and reptile assemblages on Virginia's Eastern Shore, with comments on conservation. Banisteria 20:31–45.

Mitchell, J.C., and J.M. Anderson. 1994. Amphibians and Reptiles of Assateague and Chincoteague Islands. Virginia Museum of Natural History, Special Publication No. 2.

Mitchell, J.C., and K.A. Buhlmann, 1999. Amphibians and reptiles of the Shenandoah Valley sinkhole pond system in Virginia. Banisteria 13:129–142.

Mitchell, J.C., and K.K. Reay. 1999. Atlas of Amphibians and Reptiles in Virginia. Virginia Department of Game and Inland Fisheries, Wildlife Diversity Division, Special Publication No.1.

Mitchell, J.C., and B. Burwell. 2004. A malformed Fowler's Toad (*Bufo fowleri*) from the Shenandoah Valley of Virginia. Banisteria 24:51–52.

Mitchell, J.C., and C.T. Georgel. 2005. Anopthalmia in an upland chorus frog (*Pseudacris feriarum feriarum*) from southeastern Virginia. Banisteria 25:53–54.

Mitchell, J.C., and L. McGranaghan. 2005. Albinism in American Bullfrog (*Rana catesbeiana*) tadpoles in Virginia. Banisteria 25:51.

Mitchell, J.C., and T.K. Pauley. 2005. *Pseudacris brachyphona* (Cope, 1889). Mountain Chorus Frog. Pp. 465–466 *In* M.J. Lannoo (ed.), Amphibian Declines. The Conservation Status of United States Species. University of California Press, Berkeley.

Mitchell, J.C., and J. White. 2005. Leucistic wood frog (*Rana sylvatica*) tadpole from northern Virginia. Banisteria 25:52–53.

Mitchell, J.C., A.R. Breisch, and K.A. Buhlmann. 2006. Habitat Management Guidelines for Amphibians and Reptiles of the Northeastern United States. Partners in Amphibian and Reptile Conservation, Technical Publication HMG-3, Montgomery, AL.

Mittleman, M.B. 1945. The status of *Hyla phaeocrypta* with notes on its variation. Copeia 1945:31–37.

Mittleman, M.B. 1950. Miscellaneous notes on some amphibians and reptiles from the southeastern United States. Herpetologica 6:20–24.

Mittleman, M.B., and G.S. Myers. 1949. Geographic variation in the ribbed frog, *Ascaphus truei*. Proceedings of the Biological Society of Washington 62:57–68.

Mizelle, J.D., D.C. Kritsky, and H.D. McDougal. 1969. Studies on monogenetic trematodes. 42. New species of *Gyrodactylus* from amphibia. Journal of Parasitology 55:740–741.

Mohr, C.E. 1948. Unique animals inhabit subterranean Texas. National Speleological Society Bulletin 10:15–21, 88.

Mohr, J.R., and M.E. Dorcas. 1999. A comparison of anuran calling patterns at two Carolina Bays in South Carolina. Journal of the Elisha Mitchell Scientific Society 115:63–70.

Moler, P.E. 1981. Notes on *Hyla andersonii* in Florida and Alabama. Journal of Herpetology 15: 441–444.

Moler, P.E. 1985. A new species of frog (Ranidae: *Rana*) from northwestern Florida. Copeia 1985: 379–383.

Moler, P.E. 1992. Florida Bog Frog *Rana okaloosae* Moler. Pp. 30–33 *In* P.E. Moler (ed.), Rare and Endangered Biota of Florida. Vol. 3. Amphibians and Reptiles. University Press of Florida, Gainesville.

Monello, R.J., and R.G. Wright. 1999. Amphibian habitat preferences among artificial ponds in the Palouse Region of northern Idaho. Journal of Herpetology 33:298–303.

Monroe, B.L. Jr., and R.W. Taylor. 1972. Occurrence of the barking treefrog, *Hyla gratiosa*, in Kentucky. Journal of Herpetology 6:78.

Monroe, B.L. Jr., and R.W. Giannini. 1977. Distribution of the barking treefrog in Kentucky. Transactions of the Kentucky Academy of Science 38: 143–144.

Monsen, K.J., and M.S. Blouin. 2003. Genetic structure in a montane ranid frog: restricted gene flow and nuclear-mitochondrial discordance. Molecular Ecology 12:3275–3286.

Monsen-Collar, K., L. Hazard, and R. Dussa. 2010. Comparison of PCR and RT-PCR in the first report of *Batrachochytrium dendrobatidis* in amphibians in New Jersey, USA. Herpetological Review 41:460–462.

Montanucci, R. 2006. A review of the amphibians of the Jim Timmerman Natural Resources Area, Oconee and Pickens counties, South Carolina. Southeastern Naturalist 5 (Monograph 1):1–58.

Moore, H. 1965. A mid-summer trip to Moosonee. The Wood Duck 19:163–165.

Moore, J.A. 1939. Temperature tolerance and rates of development in eggs of Amphibia. Ecology 20:459–478.

Moore, J.A. 1940. Adaptive differences in the egg membranes of frogs. American Naturalist 74:89–93.

Moore, J.A. 1942a. Embryonic temperature tolerance and rates of development in *Rana catesbeiana*. Biological Bulletin 83:375–388.

Moore, J.A. 1942b. An embryological and genetical study of *Rana burnsi* Weed. Genetics 27:408–416.

Moore, J.A. 1944. Geographic variation in *Rana pipiens* Schreber of eastern North America. Bulletin of the American Museum of Natural History 82:345–370.

Moore, J.A. 1949a. Patterns of evolution in the genus *Rana*. Pp. 315–338 *In* G.L. Jepsen, E. Mayr, and G.G. Simpson (eds.), Genetics, Paleontology and Evolution. Princeton University Press, Princeton, NJ.

Moore, J.A. 1949b. Geographic variation of adaptive characters in *Rana pipiens* Schreber. Evolution 3:1–24.

Moore, J.A. 1952. An analytical study of the geographic distribution of *Rana septentrionalis*. American Naturalist 86:5–22.

Moore, J.A. 1955. Abnormal combinations of nuclear and cytoplasmic systems in frogs and toads. Advances in Genetics 7:139–182.

Moore, J.A. 1975. *Rana pipiens*: the changing paradigm. American Zoologist 15:837–849.

Moore, J.E., and E.H. Strickland. 1954. Notes on the food of three species of Alberta amphibians. American Midland Naturalist 52:221–224.

Moore, J.E., and E.H. Strickland. 1955. Further notes on the food of Alberta amphibians. American Midland Naturalist 54:253–256.

Moore, R.D. 1929. *Canis latrans lestes* Merriam feeding on tadpoles and frogs. Journal of Mammalogy 10:255.

Moore, R.G., and B.A. Moore. 1980. Observations on the body temperature and activity in the red spotted toad, *Bufo punctatus*. Copeia 1980:362–363.

Moore, R.H. 1976. Reproductive habits and growth of *Bufo speciosus* on Mustang Island, Texas, with notes on the ecology and reproduction of other anurans. Texas Journal of Science 27:173–178.

Morafka, D.J., and B.H. Banta. 1976. Ecological relationships of the Recent herpetofauna of Pinnacles National Monument, Monterey and San Benito counties, California. Wasmann Journal of Biology 34:304–324.

Morey, S.R., and D.N. Janes. 1994. Variation in larval habitat duration influences metamorphosis in *Scaphiopus couchii*. Pp. 159–165 *In* P.R. Brown and J.W. Wright (eds.), Herpetology of the North American Deserts. Proceedings of a Symposium. Southwestern Herpetologists Society, Special Publication No. 5.

Morey, S., and D. Reznick. 2000. A comparative

analysis of plasticity in larval development in three species of spadefoot toads. Ecology 81:1736–1749.

Morey, S., and D. Reznick. 2004. The relationship between habitat permanence and larval development in California spadefoot toads: field and laboratory comparisons of developmental plasticity. Oikos 104:172–190.

Morgan, D. 2010. *Rana sylvatica* (Wood Frog). Egg mass survey. Herpetological Review 41:206.

Moriarty, E.C., and D.C. Cannatella. 2004. Phylogenetic relationships of the North American chorus frogs (*Pseudacris*: Hylidae). Molecular Phylogenetics and Evolution 30:409–420.

Moriarty, J.J. 2000. Status of amphibians in Minnesota. Pp. 166–168 *In* M.J. Lannoo (ed.), Status & Conservation of Midwestern Amphibians. University of Iowa Press, Iowa City.

Moriarty, J.J., A. Forbes and D. Jones. 1998. Geographic distribution: *Acris crepitans blanchardi* (Blanchard's cricket frog). Herpetological Review 29:172.

Morin, P.J. 1983. Predation, competition, and the composition of larval anuran guilds. Ecological Monographs 53:119–138.

Morin, P.J. 1985. Predation intensity, prey survival and injury frequency in an amphibian predator-prey interaction. Copeia 1985:638–644.

Morin, P.J. 1986. Interactions between intraspecific competition and predation in an amphibian predator-prey system. Ecology 67:713–720.

Morin, P.J., and E.A. Johnson. 1988. Experimental studies of asymmetric competition among anurans. Oikos 53:398–407.

Morin, P.J., S.P. Lawler, and E.A. Johnson. 1988. Competition between aquatic insects and vertebrates: interaction strength and higher order interactions. Ecology 69:1401–1409.

Morin, P.J., S.P. Lawler, and E.A. Johnson. 1990. Ecology and breeding phenology of larval *Hyla andersonii*: the disadvantages of breeding late. Ecology 71:1590–1598.

Morizot, D.C., and N.H. Douglas. 1970. Notes on *Pseudacris streckeri streckeri* in northwest Louisiana. Bulletin of the Maryland Herpetological Society 6:18.

Morris, M.R. 1989. Female choice of large males in the treefrog *Hyla chrysoscelis*: the importance of identifying the scale of choice. Behavioral Ecology and Sociobiology 25:275–281.

Morris, M.R., and S.L. Yoon. 1989. A mechanism for female choice of large males in the treefrog *Hyla chrysoscelis*. Behavioral Ecology and Sociobiology 25:65–71.

Morris, R.L., and W.W. Tanner. 1969. The ecology of the western spotted frog, *Rana pretiosa pretiosa* Baird and Girard: a life history study. Great Basin Naturalist 29:45–81.

Morse, M. 1904. Batrachians and reptiles of Ohio. Proceedings of the Ohio State Academy of Science 4: 93–144.

Morton, M.L., and K.N. Sokolski. 1978. Sympatry in *Bufo boreas* and *Bufo canorus* and additional evidence of natural hybridization. Bulletin of the Southern California Academy of Science 77:52–55.

Morton, M.L., and M.E. Pereyra. 2010. Habitat use by Yosemite toads: life history traits and implications for conservation. Herpetological Conservation and Biology 5:388–394.

Moseley, K.R., S.B. Castleberry, and S.H. Schweitzer. 2003. Effects of prescribed fire on herpetofauna in bottomland hardwood forests. Southeastern Naturalist 2:475–486.

Mossman, M.J., L.M. Hartman, R. Hay, J.R. Sauer, and B.J. Dhuey. 1998. Monitoring long-term trends in Wisconsin frog and toad populations. Pp. 169–198 *In* M.J. Lannoo (ed.), Status & Conservation of Midwestern Amphibians. University of Iowa Press, Iowa City.

Moulton, J.M. 1954. A late August breeding of *Hyla cinerea* in Florida. Herpetologica 10:107.

Mount, R.H. 1975. The Reptiles & Amphibians of Alabama. Agricultural Experiment Station, Auburn University, Auburn, AL.

Moyle, P.B. 1973. Effects of introduced bullfrogs, *Rana catesbeiana*, on the native frogs of the San Joaquin Valley, California. Copeia 1973:18–22.

Muenz, T.K., S.W. Golladay, G. Vellidis, and L.L. Smith. 2006. Stream buffer effectiveness in an agriculturally influenced area, southwestern Georgia: responses of water quality, macroinvertebrates, and amphibians. Journal of Environmental Quality 35:1924–1938.

Muhse, E.F. 1909. The cutaneous glands of the common toads. American Journal of Anatomy 9:322–359.

Mulaik, S. 1937. Notes on *Leptodactylus labialis* (Cope). Copeia 1937:72–73.

Mulaik, S. 1945. New mites in the family Caeculidae. Bulletin of the University of Utah 35(17): 1–23.

Mulaik, S., and D. Sollberger. 1938. Notes on the eggs and habits of *Hypopachus cuneus* Cope. Copeia 1938:90.

Mulcahy, D.G., and J.R. Mendelsohn. 2000. Phylogeography and speciation of the morphologically variable, widespread species *Bufo valliceps*, based on molecular evidence from mtDNA. Molecular Phylogenetics and Evolution 17:173–189.

Mulcare, D.J. 1966. The problem of toxicity in *Rana palustris*. Proceedings of the Indiana Academy of Science 75:319–324.

Mulla, M.S. 1962. Frog and toad control with insecticides! Pest Control 30:20, 64.

Mulla, M.S. 1963. Toxicity of organochlorine insecticides to mosquitofish *Gambusia affinis* and the bullfrog *Rana catesbeiana*. Mosquito News 23: 299–303.

Mullally, D.P. 1952. Habits and minimum temperatures of the toad *Bufo boreas halophilus*. Copeia 1952:274–276.

Mullally, D.P. 1953. Observations on the ecology of the toad *Bufo canorus*. Copeia 1953:182–183.

Mullally, D.P. 1958. Daily period of activity of the Western toad. Herpetologica 14:29–31.

Mullally, D.P., and J.D. Cunningham. 1956a. Aspects of the thermal ecology of the Yosemite toad. Herpetologica 12:57–67.

Mullally, D.P., and J.D. Cunningham. 1956b. Ecological relations of *Rana muscosa* at high elevations in the Sierra Nevada. Herpetologica 12:189–198.

Mullally, D.P., and D.H. Powell. 1958. The Yosemite toad: northern range extension and possible hybridization with the Western toad. Herpetologica 14:31–33.

Munger, J.C., M. Gerber, K. Madrid, M.-A. Carroll, W. Petersen, and L. Heberger. 1998. U.S. National Wetlands Inventory classifications as predictors of the occurrence of Columbia spotted frogs (*Rana luteiventris*) and Pacific treefrogs (*Hyla regilla*). Conservation Biology 12:320–330.

Munsey, L.D. 1972. Salinity tolerance of the African clawed frog, *Xenopus laevis*. Copeia 1972:584–586.

Munz, P.A. 1920. A study of the food habits of the Ithacan species of anura during transformation. Pomona College Journal of Entomology and Zoology 12:33–56.

Murlett, H. 1896. An albino frog. The Oregon Naturalist 3(11):147.

Murphy, C.G. 1994a. Chorus tenure of male barking treefrogs, *Hyla gratiosa*. Animal Behaviour 48:763–777.

Murphy, C.G. 1994b. Determinants of chorus tenure in barking treefrogs (*Hyla gratiosa*). Behavioral Ecology and Sociobiology 34:285–294.

Murphy, J.F., E.T. Simandle, and D.E. Becker. 2003. Population status and conservation of the black toad, *Bufo exsul*. Southwestern Naturalist 48:54–60.

Murphy, M.B., M. Hecker, K.K. Coady, A.R. Tompsett, P.D. Jones, L.H. DuPreez, G.J. Everson, K.R. Solomon, J.A. Carr, E.E. Smith, R.J. Kendall, G. Van Der Kraak, and J.P. Giesy. 2006. Atrazine concentrations, gonadal gross morphology and histology in ranid frogs collected in Michigan agricultural areas. Aquatic Toxicology 76:230–245.

Murphy, P.J., S. St.-Hilaire, S. Bruer, P.S. Corn, and C.R. Peterson. 2009. Distribution and pathogenicity of *Batrachochytrium dendrobatidis* in boreal toads from the Grand Teton area of western Wyoming. EcoHealth 6:109–120.

Murphy, R.K. 1987. Observations of the wood frog in northwestern North Dakota. Prairie Naturalist 19:262.

Murphy, T.D. 1963. Amphibian populations and movements at a small semi-permanent pond in Orange County, North Carolina. Ph.D. diss., Duke University, Durham.

Murphy, T.D. 1965. High incidence of two parasitic infestations and two morphological abnormalities in a population of the frog, *Rana palustris* Le Conte. American Midland Naturalist 74:233–239.

Murray, D.L. 1990. The effects of food and density on growth and metamorphosis in larval wood frogs (*Rana sylvatica*) from central Labrador. Canadian Journal of Zoology 68:1221–1226.

Murray, I., and C.W. Painter. 2003. Geographic distribution: *Eleutherodactylus augusti* (Barking Frog). Herpetological Review 34:161.

Musgrave, M.E. 1930. *Bufo alvarius*, a poisonous toad. Copeia 1930:96–98.

Muths, E. 2003. Home range and movements of boreal toads in undisturbed habitat. Copeia 2003:160–165.

Muths, E., T.L. Johnson, and P.S. Corn. 2001. Experimental repatriation of boreal toads (*Bufo boreas*) eggs, metamorphs, and adults in Rocky Mountain National Park. Southwestern Naturalist 46:106–113.

Muths, E., P.S. Corn, A.P. Pessier, and D.E. Green. 2003. Evidence for disease-related amphibian decline in Colorado. Biological Conservation 110:357–365.

Muths, E., S. Rittmann, J. Irwin, D. Keinath, and R. Scherer. 2005. Wood Frog (*Rana sylvatica*): a technical conservation assessment. USDA Forest Service, Rocky Mountain Region, Species Conservation Project. http//www.fs.fed.us/r2/projects/scp/assessments/woodfrog.pdf (accessed 20 June 2007).

Muths, E., R.D. Scherer, P.S. Corn, and B.A. Lambert. 2006. Estimation of temporary emigration in male toads. Ecology 87:1048–1056.

Muths, E., D.S. Pilliod, and L.J. Livo. 2008. Distribution and environmental limitations of an amphibian pathogen in the Rocky Mountains, USA. Biological Conservation 141:1484–1492.

Muzzall, P.M., and C.R. Peebles. 1991. Helminths of the wood frog, *Rana sylvatica*, and spring peeper, *Pseudacris c. crucifer*, from southern Michigan. Journal of the Helminthological Society of Washington 58:263–265.

Myers, C.W., and W.D. Klimstra. 1963. Amphibians and reptiles of an ecologically disturbed (strip-mined) area in southern Illinois. American Midland Naturalist 70:126–132.

Myers, G.S. 1927. The differential characters of *Bufo americanus* and *Bufo fowleri*. Copeia (163):50–53.

Myers, G.S. 1930a. Notes on some amphibians in western North America. Proceedings of the Biological Society of Washington 43:55–64.

Myers, G.S. 1930b. The status of the Southern California toad, *Bufo californicus* (Camp). Proceedings of the Biological Society of Washington 43:73–78.

Myers, G.S. 1942. The black toad of Deep Springs

Valley, Inyo County, California. Occasional Papers of the Museum of Zoology, University of Michigan 460:1–13.

Najarian, H.H. 1955. Trematode parasites in the Salientia in the vicinity of Ann Arbor, Michigan. American Midland Naturalist 53:195–197.

Nash, R.F., G.G. Gallup Jr., and M.K. McClure. 1970. The immobility reaction in leopard frogs (*Rana pipiens*) as a function of noise-induced fear. Psychonomic Science 21:155–156.

Naugle, D.E., T.D. Fischer, K.F. Higgins, and D.C. Backlund. 2005. Distribution of South Dakota anurans. Pp. 283–291 *In* M.J. Lannoo (ed.), Amphibian Declines. The Conservation Status of United States Species. University of California Press, Berkeley.

Nebeker, A.V., and G.S. Schuytema. 2000. Effects of ammonium sulfate on growth of larval Northwestern salamanders, red-legged and Pacific treefrog tadpoles, and juvenile fathead minnows. Bulletin of Environmental Contamination and Toxicology 64:271–278.

Nebeker, A.V., G.S. Schuytema, W.L. Griffis, and A. Cataldo. 1998. Impact of Guthion on survival and growth of the frog *Pseudacris regilla* and the salamanders *Ambystoma gracile* and *Ambystoma maculatum*. Archives of Environmental Contamination and Toxicology 35:48–51.

Necker, W.L. 1939. Records of amphibians and reptiles of the Chicago region, 1935–1938. Bulletin of the Chicago Academy of Sciences 6:1–10.

Needham, J.G. 1905. The summer food of the bullfrog (*Rana catesbeiana* Shaw) at Saranac Inn. New York State Museum, Bulletin 86:9–15.

Needham, J.G. 1924. Observations on the life of the ponds at the head of Laguna Canyon. Pomona College Journal of Entomology and Zoology 16:1–12.

Neill, W.E., and J.C. Grubb. 1971. Arboreal habits of *Bufo valliceps* in central Texas. Copeia 1971:347–348.

Neill, W.T. 1947. Doubtful type localities in South Carolina. Herpetologica 4:75–76.

Neill, W.T. 1948. A new subspecies of treefrog from Georgia and South Carolina. Herpetologica 4:175–179.

Neill, W.T. 1949a. Hybrid toads from Georgia. Herpetologica 5:30–32.

Neill, W.T. 1949b. The status of Baird's chorus-frog. Copeia 1949:227–228.

Neill, W.T. 1950a. Reptiles and amphibians in urban areas of Georgia. Herpetologica 6:113–116.

Neill, W.T. 1950b. Taxonomy, nomenclature, and distribution of Southeastern cricket frogs, genus *Acris*. American Midland Naturalist 43:152–156.

Neill, W.T. 1951. A bromeliad herpetofauna in Florida. Ecology 32:140–143.

Neill, W.T. 1952. New records of *Rana virgatipes* and *Rana grylio* in Georgia and South Carolina. Copeia 1952:194–195.

Neill, W.T. 1954. Ranges and taxonomic allocations of amphibians and reptiles in the Southeastern United States. Publications of the Research Division, Ross Allen's Reptile Institute 1:75–96.

Neill, W.T. 1957a. Notes on metamorphic and breeding aggregations of the eastern spadefoot, *Scaphiopus holbrooki* (Harlan). Herpetologica 13:185–188.

Neill, W.T. 1957b. The status of *Rana capito stertens* Schwartz and Harrison. Herpetologica 13:47–52.

Neill, W.T. 1957c. Homing by a squirrel treefrog, *Hyla squirella* Latreille. Herpetologica 13:217–218.

Neill, W.T. 1958. The occurrence of amphibians and reptiles in saltwater areas, and a bibliography. Bulletin of Marine Science of the Gulf and Caribbean 8:1–97.

Neill, W.T. 1964. Frogs introduced on islands. Quarterly Journal of the Florida Academy of Sciences 27:127–130.

Nelson, C.E. 1971. Breakdown of ethological isolation between *Rana pipiens* and *Rana sylvatica*. Copeia 1971:344.

Nelson, C.E. 1972. Systematic studies of the North American microhylid genus *Gastrophryne*. Journal of Herpetology 6:111–137.

Nelson, C.E. 1973. Mating calls of the Microhylinae: descriptions and phylogenetic and ecological considerations. Herpetologica 29:163–176.

Nelson, C.E. 1974. Further studies on the systematics of *Hypopachus* (Anura: Microhylidae). Herpetologica 30:250–274.

Nelson, C.E. 1980. What determines the species composition of larval amphibian pond communities in south central Indiana, U.S.A.? Proceedings of the Indiana Academy of Science 89:149.

Nelson, C.E., and H.S. Cuellar. 1968. Anatomical comparison of tadpoles of the genera *Hypopachus* and *Gastrophryne* (Microhylidae). Copeia 1968:423–424.

Nelson, D.H. 1974. Growth and developmental responses of larval toad populations to heated effluent in a South Carolina reservoir. Pp. 264–276 *In* J.W. Gibbons and R.R. Sharitz (eds.), Thermal Ecology. Atomic Energy Commission Symposium Series, CONF-730505.

Nelson, D.W.M., M.R. Cossland, and R. Shine. 2010. Indirect ecological impacts of an invasive toad on predator-prey interactions among native species. Biological Invasions 12:3363–3369.

Nelson, G.L., and B.M. Graves. 2004. Anuran population monitoring: comparison of the North American Amphibian Monitoring Program's calling index with mark-recapture estimates for *Rana clamitans*. Journal of Herpetology 38:355–359.

Nero, R.W. 1967. A possible record of death-feigning in a toad. Blue Jay 25:193–194.

Nero, R.W., and F.R. Cook. 1964. A range extension

for the wood frog in northeastern Saskatchewan. Canadian Field-Naturalist 78:268–269.

Netting, M.G. 1929. Further distinctions between *Bufo americanus* Holbrook and *Bufo fowleri* Garman. Papers of the Academy of Science, Arts and Letters 11:437–443. (Dated 1929, but published 1930.)

Netting, M.G. 1933. The amphibians of Pennsylvania. Proceedings of the Pennsylvania Academy of Science 7:1–11.

Netting, M.G., and L.W. Wilson. 1940. Notes on amphibians from Rockingham County, Virginia. Annals of the Carnegie Museum 28:1–8.

Netting, M.G., and C.J. Goin. 1942. Additional notes on *Rana sevosa*. Copeia 1942:259.

Netting, M.G., and C.J. Goin. 1945. The occurrence of Fowler's toad, *Bufo woodhousii fowleri* Hinckley, in Florida. Proceedings of the Florida Academy of Sciences 7:181–184.

Nevo, E. 1973a. Adaptive color polymorphism in cricket frogs. Evolution 27:353–367.

Nevo, E. 1973b. Adaptive variation in size of cricket frogs. Ecology 54:1271–1281.

Nevo, E., and R.R. Capranica. 1985. Evolutionary origin of ethological reproductive isolation in cricket frogs, *Acris*. Pp. 147–214 *In* M.K. Hecht, B. Wallace, and G.T. Prance (ed.), Evolutionary Biology, Vol. 19. Plenum Press, New York.

Newman, C.E., J.A. Feinberg, L.J. Rissler, J. Burger, and H.B. Shaffer. 2012. A new species of leopard frog (Anura: Ranidae) from the urban northeastern US. Molecular Phylogenetics and Evolution. Online first:doi:10.1016/j.ympev.2012.01.021.

Newman, R.A. 1987. Effects of density and predation on *Scaphiopus couchi* tadpoles in desert ponds. Oecologia 71:301–307.

Newman, R.A. 1988a. Adaptive plasticity in development of *Scaphiopus couchii* tadpoles in desert ponds. Evolution 42:774–783.

Newman, R.A. 1988b. Genetic variation for larval anuran (*Scaphiopus couchii*) development time in an uncertain environment. Evolution 42:763–773.

Newman, R.A. 1989. Developmental plasticity of *Scaphiopus couchii* tadpoles in an unpredictable environment. Ecology 70:1775–1787.

Newman, R.A. 1992. Adaptive plasticity in amphibian metamorphosis. BioScience 42:671–678.

Newman, R.A. 1994. Genetic variation for phenotypic plasticity in the larval life history of spadefoot toads (*Scaphiopus couchii*). Evolution 48:1773–1785.

Newman, R.A., and A.E. Dunham. 1994. Size at metamorphosis and water loss in a desert anuran (*Scaphiopus couchii*). Copeia 1994:372–381.

Newman, R.A., and T. Squire. 2001. Microsatellite variation and fine-scale population structure in the wood frog (*Rana sylvatica*). Molecular Ecology 10:1087–1100.

Nichols, A. 1852. Occurrence of *Scaphiopus solitarius* in Essex County: with some notices of its history, habits, &c. Journal of the Essex County Natural History Society 1:113–117.

Nichols, R.J. 1937. Taxonomic studies on the mouth parts of larval anura. University of Illinois Bulletin, Illinois Biological Monographs 15:1–73.

Nicholson, F.E. 1932. Nature discloses more queer life. The Philadephia Ledger, 30 October. [republished in Literary Digest, 10 December 1932, pp. 31–32 and in the National Speleological Society Bulletin 10 (1948):22–26]

Nicholson, M.C. 1980. The effects of density and predation on the growth rate, larval period and survivorship of *Rana utricularia* and *Bufo terrestris* tadpoles. M.A. thesis, University of South Florida, Tampa.

Nickerson, M.A., and C.E. Mays. 1968. *Bufo retiformis* Sanders and Smith from the Santa Rosa Valley, Pinal County, Arizona. Journal of Herpetology 1:103.

Nickerson, M.A., and J.A. Hutchison. 1971. The distribution of the fungus *Basidiobolus ranarum* Eidam in fish, amphibians and reptiles. American Midland Naturalist 86:500–502.

Nickum, J.G. 1961. Intra-specific variation in the chorus frog *Pseudacris nigrita* Le Conte. M.A. thesis, State University of South Dakota, Vermillion.

Nie, M., J.D. Crim, and G.R. Ultsch. 1999. Dissolved oxygen, temperature, and habitat selection by bullfrog (*Rana catesbeiana*) tadpoles. Copeia 1999:155–162.

Nielsen, H.I., and J. Dyck. 1978. Adaptation of the tree frog, *Hyla cinerea*, to colored backgrounds, and the role of the three chromatophore types. Journal of Experimental Zoology 205:79–94.

Nielson, M., K. Lohman, and J. Sullivan. 2001. Phylogeography of the tailed frog (*Ascaphus truei*): implications for the biogeography of the Pacific Northwest. Evolution 55:147–160.

Nieto, N.C., M.A. Camann, J.E. Foley, and J.O. Reiss. 2007. Disease associated with integumentary and cloacal parasites in tadpoles of northern red-legged frog *Rana aurora aurora*. Diseases of Aquatic Organisms 78:61–71.

Noble, G.K., and R.C. Noble. 1923. The Anderson tree frog (*Hyla andersonii* Baird): observations on its habits and life history. Zoologica 11:413–455.

Noble, G.K., and E.J. Farris. 1929. The method of sex recognition in the wood frog, *Rana sylvatica* Le Conte. American Museum Novitates 363:1–17.

Noble, G.K., and P.G. Putnam. 1931. Observations on the life history of *Ascaphus truei* Stejneger. Copeia 1931:97–101.

Noble, G.K., and W.G. Hassler. 1936. Three salientians of geographic interest from southern Maryland. Copeia 1936:63–64.

Noble, G.K., and L.R. Aronson. 1942. The sexual

behavior of Anura. 1. The normal mating pattern of *Rana pipiens*. Bulletin of the American Museum of Natural History 80:127–142.

Northen, P.T. 1970. The geographic and taxonomic relationships of the Great Basin spadefoot toad, *Scaphiopus intermontanus*, to other members of the subgenus *Spea*. Ph.D. diss., University of Wisconsin, Madison.

Norton, V.M., and M.J. Harvey. 1975. Herpetofauna of Hardeman County, Tennessee. Journal of the Tennessee Academy of Science 50:131–136.

Nussbaum, R.A., E.D. Brodie Jr., and R.M. Storm. 1983. Amphibians & Reptiles of the Pacific Northwest. University Press of Idaho, Moscow.

NVDW (Nevada Division of Wildlife). 2003. Conservation agreement and conservation strategy. Columbia spotted frog (*Rana luteiventris*). Toiyabe Great Basin subpopulation Nevada. Nevada Division of Wildlife, Carson City.

Nyberg, D., and I. Lerner. 2000. Revitalization of ephemeral pools as frog breeding habitat in an Illinois forest preserve. Journal of the Iowa Academy of Science 107:187–190.

Nyman, S. 1986. Mass mortality in larval *Rana sylvatica* attributable to the bacterium *Aeromonas hydrophila*. Journal of Herpetology 20:196–201.

Oberfoell, C.E.C., and J.L. Christiansen. 2001. Identification and distribution of the treefrogs *Hyla versicolor* and *Hyla chrysoscelis* in Iowa. Journal of the Iowa Academy of Science 108:79–83.

Odlaug, T.O. 1954. Parasites of some Ohio batrachians. Ohio Journal of Science 54:126–128.

Odum, R.A., and P.S. Corn. 2005. *Bufo baxteri* Porter, 1988. Wyoming Toad. Pp. 390–392 *In* M.J. Lannoo (ed.), Amphibian Declines. The Conservation Status of United States Species. University of California Press, Berkeley.

O'Hara, R.K., and A.R. Blaustein. 1981. An investigation of sibling recognition in *Rana cascadae* tadpoles. Animal Behaviour 29:1121–1126.

O'Hara, R.K., and A.R. Blaustein. 1982. Kin preference behavior in *Bufo boreas* tadpoles. Behavioral Ecology and Sociobiology 11:43–49.

O'Hara, R.K., and A.R. Blaustein. 1985. *Rana cascadae* tadpoles aggregate with siblings: an experimental field study. Oecologia 67:44–51.

O'Hara, R.K., and A.R. Blaustein. 1988. *Hyla regilla* and *Rana pretiosa* tadpoles fail to display kin recognition behaviour. Animal Behaviour 36:946–948.

Okafor, J.I., D. Testrake, H.R. Mushinsky, and B.G. Yangco. 1984. A *Basidiobolus* sp. and its association with reptiles and amphibians in southern Florida. Sabouraudia, Journal of Medical and Veterinary Mycology 22:47–51.

Okonkwo, G.E. 2011. The use of small ephemeral wetlands and streams by amphibians in the mixedwood forest of boreal Alberta. M.S. thesis, University of Alberta, Edmonton.

Oláh-Hemmings, V., J.R. Jaeger, M.J. Sredl, M.A. Schlaepfer, R.D. Jennings, C.A. Drost, D.F. Bradford, and B.R. Riddle. 2009. Phylogeography of declining relict and lowland leopard frogs in the desert Southwest of North America. Journal of Zoology 280:343–354.

Oldfield, B., and J.J. Moriarty. 1994. Amphibians & Reptiles Native to Minnesota. University of Minnesota Press, Minneapolis.

Oldham, M.J. 1992. Declines in Blanchard's cricket frog in Ontario. Canadian Wildlife Service Occasional Paper No. 76:30–31.

Oldham, R.S. 1966. Spring movements in the American toad, *Bufo americanus*. Canadian Journal of Zoology 44:63–100.

Oldham, R.S. 1967. Orienting mechanisms of the green frog, *Rana clamitans*. Ecology 48:477–491.

Oldham, R.S. 1974. Mate attraction by vocalization in members of the *Rana pipiens* complex. Copeia 1974:982–984.

Oldham, R.S., and H.C. Gerhardt. 1975. Behavioral isolating mechanisms of the treefrogs *Hyla cinerea* and *H. gratiosa*. Copeia 1975:223–231.

Oliver, J.A., and J.R. Bailey. 1939. Amphibians and reptiles of New Hampshire, exclusive of marine forms. New Hampshire Fish and Game Commission, Biological Survey of Connecticut Watershed Report 4:195–217.

Oliver, J.A., and C.E. Shaw. 1953. The amphibians and reptiles of the Hawaiian Islands. Zoologica 38:65–95.

Oliver, J.H. Jr., M.P. Hayes, J.E. Keirans, and D.R. Lavender. 1993. Establishment of the foreign parthenogenetic tick *Amblyomma rotundatum* (Acari: Ixodidae) in Florida. Journal of Parasitology 79:786–790.

Olsen, O.W. 1937. Description and life history of the trematode *Haplometrana utahensis* sp. nov. (Plagiorchiidae) from *Rana pretiosa*. Journal of Parasitology 23:13–28.

Olsen, O.W. 1938. *Aplectana gigantica* (Cosmocercidae), a new species of nematode from *Rana pretiosa*. Transactions of the American Microscopical Society 57:200–203.

Olsen, O.W. 1962. Animal Parasites: Their Biology and Life Cycles. Burgess, Minneapolis, MN.

Olson, C.A., and K.H. Beard. 2012. Diet of the introduced greenhouse frog in Hawaii. Copeia 2012: 121–129.

Olson, D.H. 1988. The ecological and behavioral dynamics of breeding in three sympatric anuran amphibians. Ph.D. diss., Oregon State University, Corvallis.

Olson, D.H. 1989. Predation on breeding Western toads (*Bufo boreas*). Copeia 1989:391–397.

Olson, D.H. 2001. Ecology and management of montane amphibians of the U.S. Pacific Northwest. Biota 2:51–74.

Olson, D.H., A.R. Blaustein, and R.K. O'Hara. 1986. Mating pattern variability among western toad (*Bufo boreas*) populations. Oecologia 70: 351–356.

Olson, E.R. 2010. *Rana pipiens* (Northern Leopard Frog). Winter activity. Herpetological Review 41:206.

Olson, M.E., S. Gard, M. Brown, R. Hampton, and D.W. Morck. 1992. *Flavobacterium indologenes* infection in leopard frogs. Journal of the American Veterinary Medical Association 201:1766–1770.

Olson, R.E. 1959. Notes on some Texas herptiles. Herpetologica 15:48.

O'Neill, E.D. 1995. Amphibian and reptile communities of temporary ponds in a managed pine flatwoods. M.S. thesis, University of Florida, Gainesville.

O'Neill, E.M., and K.H. Beard. 2011. Clinal variation in calls of native and introduced populations of *Eleutherodactylus coqui*. Copeia 2011:18–28.

Oplinger, C.S. 1963. The life history of the northern spring peeper, *Hyla crucifer crucifer* Wied at Ithaca, New York. Ph.D. diss., Cornell University, Ithaca, New York.

Oplinger, C.S. 1966. Sex ratio, reproductive cycles, and time of ovulation in *Hyla crucifer crucifer* Wied. Herpetologica 22:276–283.

Oplinger, C.S. 1967. Food habits and feeding activity of recently transformed and adult *Hyla crucifer crucifer* Wied. Herpetologica 23:209–217.

Orchard, S.A. 1999. The American Bullfrog in British Columbia: the frog who came to dinner. Pp. 289–296 *In* R. Claudi and J.H. Leach (eds.), Nonindigenous Freshwater Organisms. Vectors, Biology, and Impacts. Lewis Publishers, Boca Raton, FL.

Orr, L.P., J. Neumann, E. Vogt, and A. Collier. 1998. Status of northern leopard frogs in northeastern Ohio. Pp. 91–93 *In* M.J. Lannoo (ed.), Status & Conservation of Midwestern Amphibians. University of Iowa Press, Iowa City.

Ortenburger, A.I. 1924. Life history notes—*Scaphiopus*—the spadefoot toad. Proceedings of the Oklahoma Academy of Science 4:19–20.

Orton, G.L. 1943. The tadpole of *Rhinophrynus dorsalis*. Occasional Papers of the Museum of Zoology, University of Michigan 472:1–7 + 1 plate.

Orton, G.L. 1946. Larval development of the eastern narrow-mouthed frog, *Microhyla carolinensis* (Holbrook), in Louisiana. Annals of the Carnegie Museum 30:241–248.

Orton, G.L. 1947. Notes on some hylid tadpoles in Louisiana. Annals of the Carnegie Museum 30: 363–380 + 1 plate.

Orton, G.L. 1951. An example of interspecific mating in toads. Copeia 1951:78.

Orton, G.L. 1952. Key to the genera of tadpoles in the United States and Canada. American Midland Naturalist 47:382–395.

Orton, G.L. 1954. Dimorphism in larval mouth-parts in spadefoot toads of the *Scaphiopus hammondi* group. Copeia 1954:97–100.

Oseen, K.L., and R.J. Wassersug. 2002. Environmental factors influencing calling in sympatric anurans. Oecologia 133:616–625.

Oseen, K.L., L.K.D. Newhook, and R.J. Wassersug. 2001. Turning bias in woodfrog (*Rana sylvatica*) tadpoles. Herpetologica 57:432–437.

Ostergaard, E.C., K.O. Richter, and S.D. West. 2008. Amphibian use of stormwater ponds in the Puget lowlands of Washington, USA. Pp. 259–270 *In* J.C. Mitchell, R.E. Jung Brown, and B. Bartholomew (eds.), Urban Herpetology. Herpetological Conservation 3. Society for the Study of Amphibians and Reptiles, Salt Lake City, UT.

Otani, A., N. Palumbo, and G. Read. 1969. Pharmacodynamics and treatment of mammals poisoned by *Bufo marinus* toxin. American Journal of Veterinary Research 30:1865–1872.

Otto, C.V., J.W. Snodgrass, D.C. Forester, J.C. Mitchell, and R.W. Miller. 2007a. Climatic variation and the distribution of an amphibian polyploid complex. Journal of Animal Ecology 76:1053–1061.

Otto, C.V., D.C. Forester, and J.W. Snodgrass. 2007b. Influences of wetland and landscape characteristics on the distribution of carpenter frogs. Wetlands 27: 261–269.

Ouellet, M., J. Bonin, J. Rodrigue, J.-L. DesGranges, and S. Lair. 1997. Hindlimb deformities (ectromelia, ectrodactyly) in free-living anurans from agricultural habitats. Journal of Wildlife Diseases 33:95–104.

Ouellet, M., I. Mikaelian, B.D. Pauli, J. Rodrigue, and D.M. Green. 2005a. Historical evidence of widespread chytrid infection in North American amphibian populations. Conservation Biology 19:1431–1440.

Ouellet, M., P. Galois, R. Pétel, and C. Fortin. 2005b. Les amphibians et les reptiles des collines montérégiennes: enjeux et conservation. Le naturaliste canadien 129:42–49.

Ouellet, M., C. Fortin, and M.-J. Grimard. 2009. Distribution and habitat use of the boreal chorus frog (*Pseudacris maculata*) at its extreme northeastern range limit. Herpetological Conservation and Biology 4:277–284.

Ovaska, K., T.M. Davis, and I.N. Flamarique. 1997. Hatching success and larval survival of the frogs *Hyla regilla* and *Rana aurora* under ambient and artificially enhanced solar ultraviolet radiation. Canadian Journal of Zoology 75:1081–1088.

Overton, F. 1914. Long Island flora and fauna. 3. The frogs and toads. The Museum of the Brooklyn Institute of Arts and Sciences Science Bulletin 2(3):21–40.

Owen, R.D. 1996. Breeding phenology and microhabitat use among three chorus frog species (*Pseud-*

acris) in east-central Florida. M.S. thesis, University of Central Florida, Orlando.

Pace, A.E. 1974. Systematics and biological studies of the leopard frogs (*Rana pipiens* complex) of the United States. Miscellaneous Publications, University of Michigan Museum of Zoology 148:1–140.

Pack, H.J. 1920. Eggs of the swamp tree frog. Copeia (77):7.

Packard, A.S. 1866. List of vertebrates observed at Okak, Labrador, by Rev. Samuel Weiz, with annotations. Proceedings of the Boston Society of Natural History 10:264–277.

Packard, G.C., J.K. Tucker, and L.D. Lohmiller. 1998. Distribution of Strecker's chorus frogs (*Pseudacris streckeri*) in relation to their tolerance for freezing. Journal of Herpetology 32:437–440.

Padgett-Flohr, G.E. 2008. Pathogenicity of *Batrachochytrium dendrobatidis* in two threatened California amphibians: *Rana draytonii* and *Ambystoma californiense*. Herpetological Conservation and Biology 3:182–191.

Padgett-Flohr, G.E., and R.L. Hopkins III. 2009. *Batrachochytrium dendrobatidis*, a novel pathogen approaching endemism in central California. Diseases of Aquatic Organisms 83:1–9.

Padgett-Flohr, G.E., and M.P. Hayes. 2011. Assessment of the vulnerability of the Oregon spotted frog (*Rana pretiosa*) to the amphibian chytrid fungus (*Batrachochytrium dendrobatidis*). Herpetological Conservation and Biology 6:99–106.

Paetow, L., D.K. Cone, T. Huyse, J.D. McLaughlin, and D.J. Marcogliese. 2009. Morphology and molecular taxonomy of *Gyrodactylus jennyae* n. sp. (Monogenea) from tadpoles of captive *Rana catesbeiana* Shaw (Anura), with a review of the species of *Gyrodactylus* Nordmann, 1832 parasitising amphibians. Systematic Parasitology 73:219–227.

Pague, C.A., and J.C. Mitchell. 1987. The status of amphibians in Virginia. Virginia Journal of Science 38:304–318.

Pais, R.C., S.A. Bonney, and W.C. McComb. 1988. Herpetofaunal species richness and habitat associations in an eastern Kentucky forest. 1988 Proceedings of the Annual Conference of Southeastern Fish and Wildlife Agencies:448–455.

Palen, W.J., and D.E. Schindler. 2010. Water clarity, maternal behavior, and physiology combine to eliminate UV radiation risk to amphibians in a montane landscape. Proceedings of the National Academy of Science, USA 107:9701–9706.

Palen, W.J., D.E. Schindler, M.J. Adams, C.A. Pearl, R.B. Bury, and S.A. Diamond. 2002. Optical characteristics of natural waters protect amphibians from UV-B in the U.S. Pacific Northwest. Ecology 83:2951–2957.

Palis, J. 1994. Night of the spadefoots. Florida Wildlife (Nov/Dec):12–13.

Palis, J.G. 1997. Species profile: Gopher frog (*Rana capito* spp.) on military installations in the Southeastern United States. U.S. Army Corps of Engineers Waterways Experiment Station Technical Report SERDP-97-5.

Palis, J.G. 1998. Breeding biology of the gopher frog, *Rana capito*, in western Florida. Journal of Herpetology 32:217–223.

Palis, J. 2000. *Scaphiopus holbrookii* (Eastern Spadefoot). Predation. Herpetological Review 31:42–43.

Palis, J.G. 2007. If you build it, they will come: herpetofaunal colonization of constructed wetlands and adjacent terrestrial habitat in the Cache River drainage of southern Illinois. Transactions of the Illinois State Academy of Science 100:177–189.

Palis, J.G. 2009. Frog pond, fish pond: temporal co-existence of crawfish frog tadpoles and fishes. Proceedings of the Indiana Academy of Science 118:196–199.

Palis, J.G., and M.J. Aresco. 2007. Immigration orientation and migration distance of four pond-breeding amphibians in northwestern Florida. Florida Scientist 70:251–263.

Palmer, W.M., and A.L. Braswell. 1980. Additional records of albinistic amphibians and reptiles from North Carolina. Brimleyana 3:49–52.

Palmer, W.M., and A.L. Braswell. 1995. Reptiles of North Carolina. University of North Carolina Press, Chapel Hill.

Palmeri-Miles, A.F., K.A. Douville, J.A. Tyson, K.D. Ramsdell, and M.P. Hayes. 2010. Field observations of oviposition and early development of the coastal tailed frog (*Ascaphus truei*). Northwestern Naturalist 91:206–213.

Palorski, R.A. 2008. Relationship between lakeshore development and frog populations on central Wisconsin. Pp. 77–83 *In* J.C. Mitchell, R.E. Jung Brown, and B. Bartholomew (eds.), Urban Herpetology. Herpetological Conservation 3. Society for the Study of Amphibians and Reptiles, Salt Lake City, UT.

Panik, H.R., and S. Barrett. 1994. Distribution of amphibians and reptiles along the Truckee River system. Northwest Science 68:197–204.

Paoletti, D.J., D.H. Olson, and A.R. Blaustein. 2011. Responses of foothill yellow-legged frog (*Rana boylii*) larvae to an introduced predator. Copeia 2011:161–168.

Parker, J.M. 2000. Habitat use and movements of the Wyoming toad, *Bufo baxteri*: a study of wild juvenile, adult, and released captive-raised toads. M.S. thesis, University of Wyoming, Laramie.

Parker, J.M., and S.H. Anderson. 2003. Habitat use and movements of repatriated Wyoming toads. Journal of Wildlife Management 67:439–446.

Parker, J., S.H. Anderson, and F.J. Lindzey. 2000. *Bufo baxteri* (Wyoming Toad). Predation. Herpetological Review 31:167–168.

Parker, M.V. 1937. Some amphibians and reptiles

from Reelfoot Lake. Journal of the Tennessee Academy of Science 12:60–86.

Parker, M.V. 1941. The trematode parasites from a collection of amphibians and reptiles. Journal of the Tennessee Academy of Science 16:27–45.

Parker, M.V. 1951. Notes on the bird-voiced tree frog, *Hyla phaeocrypta*. Journal of the Tennessee Academy of Science 26:208–213.

Parker, W.S. 1973. Observations on a small, isolated population of *Bufo punctatus* in Arizona. Southwestern Naturalist 18:93–114.

Parris, M.J. 1998. Terrestrial burrowing ecology of newly metamorphosed frogs (*Rana pipiens* complex). Canadian Journal of Zoology 76:2124–2129.

Parris, M.J., and R.D. Semlitsch. 1998. Asymmetric competition in larval amphibian communities: conservation implications for the northern crawfish frog, *Rana areolata circulosa*. Oecologia 116: 219–226.

Parris, M.J., and J.G. Beaudoin. 2004. Chytridiomycosis impacts predator-prey interactions in larval amphibian communities. Oecologia 140:626–632.

Parry, J.E., and A.W. Grundman. 1965. Species composition and distribution of the parasites of some common amphibians of Iron and Washington counties, Utah. Proceedings of the Utah Academy of Sciences, Arts, and Letters 42:271–279.

Patch, C.L. 1918. The economic value of batrachians and reptiles. Ottawa Naturalist 32:29–30.

Patch, C.L. 1922. Some amphibians and reptiles from British Columbia. Copeia (111):74–79.

Patla, D.A. 1997. Changes in a population of spotted frogs in Yellowstone National Park between 1953 and 1995: the effects of habitat modification. M.S. thesis, Idaho State University, Pocatello.

Patla, D.A., and C.R. Peterson. 1999. Are amphibians declining in Yellowstone National Park? Yellowstone Science 7:2–11.

Patla, D.A., C.R. Peterson, and P.S. Corn. 2009. Amphibian decline in Yellowstone National Park. Proceedings of the National Academy of Sciences, USA 106:E22.

Paton, P., S. Stevens, and L. Longo. 2000. Seasonal phenology of amphibian breeding and recruitment at a pond in Rhode Island. Northeastern Naturalist 7:255–269.

Paton, P.W.C., R.S. Egan, J.E. Osenkowski, C.J. Raithel, and R.T. Brooks. 2003. *Rana sylvatica* (Wood Frog). Breeding behavior during drought. Herpetological Review 34:236–237.

Patrick, D.A., A.J.K. Calhoun, and M.L. Hunter Jr. 2007. Orientation of juvenile wood frogs, *Rana sylvatica*, leaving experimental ponds. Journal of Herpetology 41:158–163.

Patrick, D.A., C.M. Schalk, J.P. Gibbs, and H.W. Woltz. 2010. Effective culvert placement and design facilitate passage of amphibians across roads. Journal of Herpetology 44:618–626.

Patten, M.A., and S.J. Myers. 1992. Geographic distribution: *Bufo microscaphus californicus*. Herpetological Review 23:124.

Pauken, R.J., and D.E. Metter. 1971. Geographic representation of morphologic variation among populations of *Ascaphus truei* Stejneger. Systematic Zoology 20:434–441.

Paukstis, G.L., and L.E. Brown. 1987. Evolution of the intercalary cartilage in chorus frogs, genus *Pseudacris* (salientia: Hylidae). Brimleyana 13:55–61.

Pauley, B.A., and T.K. Pauley. 2007. Survey of abandoned coal mines for amphibians and reptiles in New River Gorge National River, West Virginia. Proceedings of the West Virginia Academy of Science 79:22–30.

Pauley, T.K., J.C. Mitchell, R.R. Buech, and J.J. Moriarty. 2000. Ecology and management of riparian habitats for amphibians and reptiles. Pp. 169–192 *In* E.S. Verry, J.W. Hornbeck, and C.A. Dolloff (eds.), Riparian Management in Forests of the Continental Eastern United States. Lewis Publishers, Boca Raton, FL.

Paulson, B.K., and V.H. Hutchison. 1987. Origin of the stimulus for muscular spasms at the critical thermal maximum in anurans. Copeia 1987:810–813.

Pauly, G.B., D.M. Hillis, and D.C. Cannatella. 2004. The history of a Nearctic colonization: molecular phylogenetics and biogeography of the Nearctic toads (*Bufo*). Evolution 58:2517–2535.

Peacor, S.D., and E.E. Werner. 2000. Predator effects on an assemblage of consumers through induced changes in consumer foraging behavior. Ecology 81:1998–2010.

Pearl, C.A., and M.P. Hayes. 2002. Predation by Oregon spotted frogs (*Rana pretiosa*) on western toads (*Bufo boreas*) in Oregon. American Midland Naturalist 147:145–152.

Pearl, C.A., and J. Bowerman. 2006. Observations of rapid colonization of constructed ponds by western toads (*Bufo boreas*) in Oregon, USA. Western North American Naturalist 66:397–401.

Pearl, C.A., M.J. Adams, G.S. Schuytema, and A.V. Nebeker. 2003. Behavioral responses of anuran larvae to chemical cues of native and introduced predators in the Pacific Northwestern United States. Journal of Herpetology 37:572–576.

Pearl, C.A., M.J. Adams, R.B. Bury, and B. McCreary. 2004. Asymmetrical effects of introduced bullfrogs (*Rana catesbeiana*) on native ranid frogs in Oregon. Copeia 2004:11–20.

Pearl, C.A., M.P. Hayes, R. Haycock, J.D. Engler, and J. Bowerman. 2005a. Observations of interspecific amplexus between western North American ranid frogs and the introduced American Bullfrog (*Rana catesbeiana*) and an hypothesis concerning breeding interference. American Midland Naturalist 154:126–134.

Pearl, C.A., M.J. Adams, N. Leuthold, and R.B. Bury. 2005b. Amphibian occurrence and aquatic invaders in a changing landscape: implications for wetland mitigation in the Willamette Valley, Oregon, USA. Wetlands 25:76–88.

Pearl, C.A., J. Bowerman, and D. Knight. 2005c. Feeding behavior and aquatic habitat use by Oregon spotted frogs (*Rana pretiosa*) in central Oregon. Northwestern Naturalist 86:36–38.

Pearl, C.A., M.J. Adams, R.B. Bury, W.H. Wente, and B. McCreary. 2009a. Evaluating amphibian declines with site revisits and occupancy models: status of montane anurans in the Pacific Northwest USA. Diversity 1:166–181.

Pearl, C.A., M.J. Adams, and N. Leuthold. 2009b. Breeding habitat and local population size of the Oregon spotted frog (*Rana pretiosa*) in Oregon, USA. Northwestern Naturalist 90:136–147.

Pearman, P.B. 1993. Effects of habitat size on tadpole populations. Ecology 74:1982–1991.

Pearse, A.S. 1910. The reactions of amphibians to light. Proceedings of the American Academy of Arts and Sciences 45:159–208.

Pearse, A.S. 1936. Estuarine animals at Beaufort, North Carolina. Journal of the Elisha Mitchell Scientific Society 52:174–222.

Pearson, K.J. 2009. Status of the Great Plains Toad (*Bufo* [*Anaxyrus*] *cognatus*) in Alberta: update 2009. Alberta Sustainable Resource Development, Wildlife Status Report No. 14 (update 2009), Edmonton.

Pearson, P.G. 1955. Population ecology of the spadefoot toad, *Scaphiopus h. Holbrooki* (Harlan). Ecological Monographs 25:233–267.

Pearson, P.G. 1957. Further notes on the population ecology of the spadefoot toad. Ecology 38:580–586.

Pearson, P.G. 1958. Body measurements of *Scaphiopus holbrooki*. Copeia 1958:215–217.

Pechmann, J.H.K., and R.D. Semlitsch. 1986. Diel activity patterns in the breeding migrations of winter-breeding anurans. Canadian Journal of Zoology 64:1116–1120.

Pechmann, J.H.K., D.E. Scott, J.W. Gibbons, and R.D. Semlitsch. 1989. Influence of wetland hydroperiod on diversity and abundance of metamorphosing juvenile amphibians. Wetlands Ecology and Management 1:3–11.

Pechmann, J.H.K., D.E. Scott, R.D. Semlitsch, J.P. Caldwell, L.J. Vitt, and J.W. Gibbons. 1991. Declining amphibian populations: the problem of separating human impacts from natural fluctuations. Science 253:892–895.

Pechmann, J.H.K., R.A. Estes, D.E. Scott, and J.W. Gibbons. 2001. Amphibian colonization and use of ponds created for trial mitigation of wetland loss. Wetlands 21:93–111.

Pehek, E.L. 1995. Competition, pH, and the ecology of larval *Hyla andersonii*. Ecology 76:1786–1793.

Pemberton, C.E. 1933. Introductions to Hawaii of the tropical American toad *Bufo marinus*. Hawaiian Planters' Record 37:15–16.

Pemberton, C.E. 1934. Local investigations on the introduced tropical American toad *Bufo marinus*. Hawaiian Planters' Record 38:186–192.

Pemberton, C.E. 1949. Longevity of the tropical American toad, *Bufo marinus*. Science 110:512.

Penn, G.H. 1950. Utilization of crawfishes by cold-blooded vertebrates in the eastern United States. American Midland Naturalist 44:643–658.

Perez, M.E. 1951. The food of *Rana catesbeiana* Shaw in Puerto Rico. Herpetologica 7:102–104.

Perkins, D.W., and M.L. Hunter Jr. 2006. Effects of riparian timber management on amphibians in Maine. Journal of Wildlife Management 70: 657–670.

Perrill, S.A. 1984. Male mating behavior in *Hyla regilla*. Copeia 1984:727–732.

Perrill, S.A., and R.E. Daniel. 1983. Multiple egg clutches in *Hyla regilla*, *H. cinerea*, and *H. gratiosa*. Copeia 1983:513–516.

Perrill, S.A., and M. Magier. 1988. Male mating behavior in *Acris crepitans*. Copeia 1988:245–248.

Perrill, S.A., and W.J. Shepherd. 1989. Spatial distribution and male-male communication in the northern cricket frog, *Acris crepitans blanchardi*. Journal of Herpetology 23:237–243.

Perrill, S.A., H.C. Gerhardt, and R. Daniel. 1978. Sexual parasitism in the green tree frog (*Hyla cinerea*). Science 200:1179–1180.

Perrill, S.A., H.C. Gerhardt, and R. Daniel. 1982. Mating strategy shifts in male green treefrogs (*Hyla cinerea*): an experimental study. Animal Behaviour 30:43–48.

Peters, B., and F.S. Ruth. 1962. Albino bullfrog (*Rana catesbeiana*) population found in San Leandro Creek, Alameda County, California. Turtox News 40(1):7.

Peterson, C.R. 1974. A preliminary report on the amphibians and reptiles of the Black Hills of South Dakota and Wyoming. M.S. thesis, University of Illinois, Urbana-Champaign.

Peterson, H.W., R. Garrett, and J.P. Lantz. 1952. The mating period of the giant tree frog *Hyla dominicensis*. Herpetologica 8:63.

Peterson, J.A., and A.R. Blaustein. 1991. Unpalatability in anuran larvae as a defense against natural salamander predators. Ethology Ecology & Evolution 3:63–72.

Peterson, J.A., and A.R. Blaustein. 1992. Relative palatabilities of anuran larvae to natural aquatic insect predators. Copeia 1992:577–584.

Peterson, J.D., M.B. Wood, W.A. Hopkins, J.M. Unrine, and M.T. Mendonça. 2007. Prevalence of *Batrachochytrium dendrobatidis* in American Bullfrog and Southern Leopard Frog larvae from wetlands on the Savannah River Site, South Carolina. Journal of Wildlife Diseases 43:450–460.

Petranka, J.W. 1989. Response of toad tadpoles to conflicting chemical stimuli: predator avoidance versus "optimal" foraging. Herpetologica 45:283–292.

Petranka, J.W., and D.A.G. Thomas. 1995. Explosive breeding reduces egg and tadpole cannibalism in the wood frog, *Rana sylvatica*. Animal Behaviour 50:731–739.

Petranka, J.W., M.E. Hopey, B.T. Jennings, S.D. Baird, and S.J. Boone. 1994. Breeding habitat segregation of wood frogs and American toads: the role of interspecific tadpole predation and adult choice. Copeia 1994:691–697.

Petranka, J.W., A.W. Rushlow, and M.E. Hopey. 1998. Predation by tadpoles of *Rana sylvatica* on embryos of *Ambystoma maculatum*: implications of ecological role reversals by *Rana* (predator) and *Ambystoma* (prey). Herpetologica 54:1–13.

Petranka, J.W., S.S. Murray, and C.A. Kennedy. 2003. Responses of amphibians to restoration of a Southern Appalachian wetland: perturbations confound post-restoration assessment. Wetlands 23:278–290.

Petranka, J.W., C.K. Smith, and A.F. Scott. 2004. Identifying the minimal demographic unit for monitoring pond-breeding amphibians. Ecological Applications 14:1065–1078.

Petrovic, C.A. 1973. An "albino" frog, *Eleutherodactylus planirostris* Cope. Journal of Herpetology 7:49–51.

Pettus, D., and A.W. Spencer. 1964. Size and metabolic differences in *Pseudacris triseriata* (Anura) from different elevations. Southwestern Naturalist 9:20–26.

Pettus, D., and G.M. Angleton. 1967. Comparative reproductive biology of montane and Piedmont chorus frogs. Evolution 21:500–507.

Pfeiffer, E.E. 1949. The toad: the gardener's assistant. Organic Farming Digest 1(12):18–23.

Pfennig, D.W. 1989. Evolution, development, and behavior of alternative amphibian morphologies. Ph.D. diss., University of Texas, Austin.

Pfennig, D. 1990a. The adaptive significance of an environmentally-cued developmental swith in an anuran tadpole. Oecologia 85:101–107.

Pfennig, D.W. 1990b. "Kin recognition" among spadefoot toad tadpoles: a side-effect of habitat selection? Evolution 44:785–798.

Pfennig, D.W. 1992. Polyphenism in spadefoot toad tadpoles as a locally adjusted evolutionarily stable strategy. Evolution 46:1408–1420.

Pfennig, D.W. 1999. Cannibalistic tadpoles that pose the greatest threat to kin are most likely to discriminate kin. Proceedings of the Royal Society, London 266B:57–61.

Pfennig, D.W., and P.J. Murphy. 2000. Character displacement in polyphenic tadpoles. Evolution 54:1738–1749.

Pfennig, D.W., and P.J. Murphy. 2002. How fluctuating competition and phenotypic plasticity mediate species divergence. Evolution 56:1217–1228.

Pfennig, D.W., and P.J. Murphy. 2003. A test of alternative hypotheses for character divergence between coexisting species. Ecology 84:1288–1297.

Pfennig, D.W., and R.A. Martin. 2009. A maternal effect mediates rapid population divergence and character displacement in spadefoot toads. Evolution 63:898–909.

Pfennig, D.W., A. Mabry, and D. Orange. 1991. Environmental causes of correlations between age and size at metamorphosis in *Scaphiopus multiplicatus*. Ecology 72:2240–2248.

Pfennig, K.S. 2000. Female spadefoot toads compromise on mate quality to ensure conspecific matings. Behavioral Ecology 11:220–227.

Pfennig, K.S. 2003. A test of alternative hypotheses for the evolution of reproductive isolation between spadefoot toads: support for the reinforcement hypothesis. Evolution 57:2842–2851.

Pfennig, K.S., and M.A. Simovich. 2002. Differential selection to avoid hybridization in two toad species. Evolution 56:1840–1848.

Pfennig, K.S., K. Rapa, and R. McNutt. 2000. Evolution of male mating behavior: male spadefoot toads preferentially associate with conspecific males. Behavioral Ecology and Sociobiology 48:69–74.

Pfingsten, R.A. 1998. Distribution of Ohio amphibians. Pp. 220–255 *In* M.J. Lannoo (ed.), Status & Distribution of Midwestern Amphibians. University of Iowa Press, Iowa City.

Pham, L., S. Boudreaux, S. Karhbet, B. Price, A.S. Ackleh, J. Carter, and N. Pal. 2007. Population estimates of *Hyla cinerea* (Schneider) (Green Tree Frog) in an urban environment. Southeastern Naturalist 6:203–216.

Phillips, C.A., R.A. Brandon, and E.O. Moll. 1999. Field Guide to the Amphibians and Reptiles of Illinois. Illinois Natural History Survey, Champaign, IL.

Phillips, K.A., and G.C. Boone. 1976. Occurrence of yellow bullfrogs (*Rana catesbeiana* Shaw) in central Pennsylvania. Proceedings of the Pennsylvania Academy of Science 50:163–164.

Phillips, K.M. 1995. *Rana capito capito*, the Carolina gopher frog, in southeast Georgia: reproduction, early growth, adult movement patterns, and tadpole fright response. M.S. thesis, Georgia Southern University, Statesboro.

Phillipsen, I.C., and A.E. Metcalf. 2009. Phylogeography of a stream-dwelling frog (*Pseudacris cadaverina*) in southern California. Molecular Phylogenetics and Evolution 53:152–170.

Pickens, A.L. 1927a. Intermediate between *Bufo fowleri* and *B. americanus*. Copeia (162):25–26.

Pickens, A.L. 1927b. Amphibians of the upper South Carolina. Copeia (165):106–110.

Pierce, B.A., and N. Sikand. 1985. Intrapopulation variation in acid tolerance of Connecticut wood

frogs: genetic and maternal effects. Canadian Journal of Zoology 63:1647–1651.
Pierce, B.A., and J.M. Harvey. 1987. Geographic variation in acid tolerance of wood frogs. Copeia 1987:94–103.
Pierce, B.A., and D.K. Wooten. 1992. Genetic variation in tolerance of amphibians to low pH. Journal of Herpetology 26:422–429.
Pierce, B.A., M.A. Margolis, and L.J. Nirtaut. 1987. The relationship between egg size and acid tolerance in *Rana sylvatica*. Journal of Herpetology 21:178–184.
Pierce, J.R. 1975. Genetic compatibility of *Hyla arenicolor* with other species in the Family, Hylidae. Texas Journal of Science 26:431–441.
Pierce, J.R. 1976. Distribution of two mating call types of the Plains spadefoot, *Scaphiopus bombifrons*, in southwestern United States. Southwestern Naturalist 20:578–582.
Pierce, J.R., and D.B. Ralin. 1972. Vocalizations and behavior of the males of three species in the *Hyla versicolor* complex. Herpetologica 28:329–337.
Piersol, W.H. 1913. Amphibia. Pp. 242–248 In J.H. Faull (ed.), The Natural History of the Toronto Region, Ontario, Canada. The Canadian Institute, Toronto.
Pieterson, C.B., L.M. Addison, J.N. Agobian, B. Brooks-Solveson, J. Cassani, and E.M. Everham III. 2006. Five years of the Southwest Florida Frog Monitoring Network: changes in frog communities as an indicator of landscape change. Florida Scientist 69 (Supplement 2):117–126.
Piha, H., M. Luoto, and J. Merilä. 2007. Amphibian occurrence is influenced by current and historic landscape characteristics. Ecological Applications 17:2298–2309.
Pike, N. 1886. Notes on the hermit spadefoot (*Scaphiopus holbrooki* Harlan; *S. solitarius* Holbr.). Bulletin of the American Museum of Natural History 1(14):213–221.
Pilliod, D.S., and C.R. Peterson. 2000. Evaluating effects of fish stocking on amphibian populations in wilderness lakes. Pp. 328–335 In Wilderness Science in a Time of Change Conference. Vol. 5. Wilderness Ecosystems, Threats, and Management. RMRS-P-15. http://www.fs.fed.us/rm/pubs/rmrs_p015_5/rmrs_p015_5_328_335.pdf.
Pilliod, D.S., and C.R. Peterson. 2001. Local and landscape effects of introduced trout on amphibians in historically fishless watersheds. Ecosystems 4:322–333.
Pilliod, D.S., and E. Wind (eds.). 2008. Habitat Management Guidelines for Amphibians and Reptiles of the Northwestern United States and Western Canada. Partners in Amphibian and Reptile Conservation, Technical Publication HMG-4, Montgomery, AL.
Pilliod, D.S., C.R. Peterson, and P.I. Ritson. 2002. Seasonal migration of Columbia spotted frogs (*Rana luteiventris*) among complementary resources in a high mountain basin. Canadian Journal of Zoology 80:1849–1862.
Pilliod, D.S., R.B. Bury, E.J. Hyde, C.A. Pearl, and P.S. Corn. 2003. Fire and amphibians in North America. Forest Ecology and Management 178:163–181.
Pilliod, D.S., E. Muths, R.D. Scherer, P.E. Bartelt, P.S. Corn, B.R. Hossack, B.A. Lambert, R. McCaffery, and C. Gaughan. 2010. Effects of amphibian chytrid fungus on individual survival probability in wild boreal toads. Conservation Biology 24:1259–1267.
Pimentel, R.A. 1955. Habitat distribution and movements of *Bufo b. boreas*, Baird and Girard. Herpetologica 11:72.
Pinder, A., and S. Friet. 1994. Oxygen transport in egg masses of the amphibians *Rana sylvatica* and *Ambystoma maculatum*: convection, diffusion and oxygen production by algae. Journal of Experimental Biology 197:17–30.
Pinder, R. 2010. *Rana clamitans* (Green Frog). Coloration. Herpetological Review 41:66.
Pine, R.H. 1975. Star-nosed mole eaten by bull frog. Mammalia 39:713–714.
Platin, T.J., and K.O. Richter. 1995. Amphibians as bioindicators of stress associated with watershed urbanization. Pp. 163–170 In Puget Sound Research '95 Proceedings, Vol. 1. Puget Sound Water Quality Authority, Olympia, WA.
Platt, D.R. (Chairman of Conservation Committee). 1973. Rare, endangered and extirpated species in Kansas. 2. Amphibians and reptiles. Transactions of the Kansas Academy of Science 76:185–192.
Platt, S.G., K.R. Russell, W.E. Snyder, L.W. Fontenot, and S. Miller. 1999. Distribution and conservation status of selected amphibians and reptiles in the Piedmont of South Carolina. Journal of the Elisha Mitchell Scientific Society 115:8–19.
Platz, J.E. 1976. Biochemical and morphological variation of leopard frogs in Arizona. Copeia 1976:660–672.
Platz, J.E. 1981. Suture zone dynamics: Texas populations of *Rana berlandieri* and *R. blairi*. Copeia 1981:733–734.
Platz, J.E. 1988a. Geographic variation in mating call among the four subspecies of the chorus frog: *Pseudacris triseriata* (Wied). Copeia 1988:1062–1066.
Platz, J.E. 1988b. *Rana yavapaiensis* Platz and Frost. Lowland leopard frog. Catalogue of American Amphibians and Reptiles 418.1–418.2.
Platz, J.E. 1989. Speciation within the chorus frog *Pseudacris triseriata*: morphometric and mating call analyses of the boreal and western subspecies. Copeia 1989:704–712.
Platz, J.E. 1993. *Rana subaquavocalis*, a remarkable new species of leopard frog (*Rana pipiens* complex)

from southeastern Arizona that calls under water. Journal of Herpetology 27: 154–162.

Platz. J.E., and A.L. Platz. 1973. *Rana pipiens* complex: hemoglobin phenotypes of sympatric and allopatric populations in Arizona. Science 179:1334–1336.

Platz, J.E., and J.S. Mecham. 1979. *Rana chiricahuensis*, a new species of leopard frog (*Rana pipiens* complex) from Arizona. Copeia 1979:383–390.

Platz, J.E., and J.S. Frost. 1984. *Rana yavapaiensis*, a new species of leopard frog (*Rana pipiens* complex). Copeia 1984:940–948.

Platz, J.E., and T.A. Grudzien. 2003. Limited genetic heterozygosity and status of two populations of the Ramsey Canyon leopard frog: *Rana subaquavocalis*. Journal of Herpetology 37:758–761.

Platz, J.E., R.W. Clarkson, J.C. Rorabaugh, and D.M. Hillis. 1990. *Rana berlandieri*: recently introduced populations in Arizona and southeastern California. Copeia 1990:324–333.

Platz, J.E., A. Lathrop, L. Hofbauer, and M. Vradenburg. 1997. Age distribution and longevity in the Ramsey Canyon leopard frog, *Rana subaquavocalis*. Journal of Herpetology 31:552–557.

Plotkin, M., and R. Atkinson. 1979. Geographic distribution. *Eleutherodactylus planirostris*. Herpetological Review 10:59.

Poile, M.E. 1982. The effects of genetic similarity and crowding on intraspecific competition in tadpoles. M.A. thesis, University of South Dakota, Vermillion.

Pollard, G.M., C.J. Biggers, and M.J. Harvey. 1973. Electrophoretic differentiation of the parotoid venoms of *Bufo americanus* and *Bufo woodhousei fowleri*. Herpetologica 29:251–253.

Pollio, C.A., and S.L. Kilpatrick. 2002. Status of *Pseudacris feriarum* in Prince William Forest Park, Prince William County, Virginia. Bulletin of the Maryland Herpetological Society 38:55–61.

Pollister, A.W., and J.A. Moore. 1937. Tables for the normal development of *Rana sylvatica*. Anatomical Record 68:489–496.

Pomeroy, L.V. 1981. Developmental polymorphism in the tadpoles of the spadefoot toad *Scaphiopus multiplicatus*. Ph.D. diss., University of California, Riverside.

Pope, C.H. 1964. Amphibians and Reptiles of the Chicago Area. Chicago Natural History Museum, Chicago.

Pope, K.L. 1999. *Rana muscosa* (Mountain Yellow-legged Frog). Diet. Herpetological Review 30: 163–164.

Pope, K.L., and K.R. Matthews. 2002. Influence of anuran prey on the condition and distribution of *Rana muscosa* in the Sierra Nevada. Herpetologica 58:354–363.

Pope, S.E., L. Fahrig, and H.G. Merriam. 2000. Landscape complementation and metapopulation effects on leopard frog populations. Ecology 81:2498–2508.

Popescu, V.D., and J.P. Gibbs. 2009. Interactions between climate, beaver activity, and pond occupancy by the cold-adapted mink frog in New York State, USA. Biological Conservation 142:2059–2068.

Porej, D., M. Micacchion, and T.E. Hetherington. 2004. Core terrestrial habitat for conservation of local populations of salamanders and wood frogs in agricultural landscapes. Biological Conservation 120:399–409.

Porter, K.R. 1961. Experimental crosses between *Rana aurora aurora* Baird and Girard and *Rana cascadae* Slater. Herpetologica 17:156–165.

Porter, K.R. 1968. Evolutionary status of a relict population of *Bufo hemiophrys* Cope. Evolution 22:583–594.

Porter, K.R. 1969a. Description of *Rana maslini*, a new species of wood frog. Herpetologica 25: 212–215.

Porter, K.R. 1969b. Evolutionary status of the Rocky Mountain population of wood frogs. Evolution 23:163–170.

Porter, K.R., and D.E. Hakanson. 1976. Toxicity of mine drainage to embryonic and larval boreal toads (Bufonidae: *Bufo boreas*). Copeia 1976:327–331.

Post, D.D. 1972. Species differentiation in the *Rana pipiens* complex. Ph.D. diss., Colorado State University, Fort Collins.

Post, D.D., and D. Pettus. 1966. Variation in *Rana pipiens* (Anura: Ranidae) of eastern Colorado. Southwestern Naturalist 11:476–482.

Post, D.D., and D. Pettus. 1967. Sympatry of two members of the *Rana pipiens* complex in Colorado. Herpetologica 23:323.

Post, T.J., and T. Uzzell. 1981. The relationships of *Rana sylvatica* and the monophyly of the *Rana boylii* group. Systematic Zoology 30:170–180.

Pough, F.H., and S. Kamel. 1984. Post-metamorphic change in activity metabolism of anurans in relation to life history. Oecologia 65:138–144.

Pramuk, J.B., T. Robertson, J.W. Sites Jr., and B.P. Noonan. 2007. Around the world in 10 million years: biogeography of the nearly cosmopolitan true toads (Anura: Bufonidae). Global Ecology and Biogeography 17:72–83.

Prather, J.W., and J.T. Briggler. 2001. Use of small caves by anurans during a drought period in the Arkansas Ozarks. Journal of Herpetology 35:675–678.

Preble, E.A. 1908. Reptiles and batrachians of the Athabaska-MacKenzie region. North American Fauna 27:500–502.

Preston, W.B. 1982. The Amphibians and Reptiles of Manitoba. Manitoba Museum of Man and Science, Winnipeg.

Prestwich, K.N., K.E. Brugger, and M. Topping. 1989.

Energy and communication in three species of hylid frogs: power input, power output and efficiency. Journal of Experimental Biology 144:53–80.

Price, A.H. 2003. The Houston toad in Bastrop State Park 1990–2002: a narrative. Texas Parks and Wildlife Department, Open File Report 03–0401.

Price, R.M., and E.R. Meyer. 1979. An amplexus call made by the male American toad, *Bufo americanus americanus* (Amphibia, Anura, Bufonidae). Journal of Herpetology 13:506–509.

Price, S.J., D.R. Marks, R.W. Howe, J.M. Hanowski, and G.J. Niemi. 2004. The importance of spatial scale for conservation and assessment of anuran populations in coastal wetlands of the western Great Lakes, USA. Landscape Ecology 20:441–454.

Priddy, J.M., and D.D. Culley Jr. 1971. The frog culture industry, past and present. Proceedings Annual Conference of the Southeastern Association of Game and Fish Commissioners 25:597–601.

Priestley, A.S., T.A. Gorman, and C.A. Haas. 2010. Comparative morphology and identification of Florida bog frog and bronze frog tadpoles. Florida Scientist 73:20–26.

Provenzano, S.E., and M.D. Boone. 2009. Effects of density on metamorphosis of bullfrogs in a single season. Journal of Herpetology 43:49–54.

Pryor, G.S. 2003. Roles of gastrointestinal symbionts in nutrition, digestion, and development of bullfrog tadpoles (*Rana catesbeiana*). Ph.D. diss., University of Florida, Gainesville.

Pryor, G.S. 2008. Anaerobic bacteria isolated from the gastrointestinal tracts of bullfrog tadpoles (*Rana catesbeiana*). Herpetological Conservation and Biology 3:176–181.

Pulis, E.E., V.V. Tkach, and R.A. Neuman. 2011. Helminth parasites of the wood frog, *Lithobates sylvaticus*, in prairie pothole wetlands of the northern Great Plains. Wetlands 31:675–685.

Punzo, F. 1991a. Feeding ecology of spadefooted toads (*Scaphiopus couchi* and *Spea multiplicata*) in western Texas. Herpetological Review 22:79–80.

Punzo, F. 1991b. Group learning in tadpoles of *Rana heckscheri* (Anura: Ranidae). Journal of Herpetology 25:214–217.

Punzo, F. 1992a. Dietary overlap and activity patterns in sympatric populations of *Scaphiopus holbrooki* (Pelobatidae) and *Bufo terrestris* (Bufonidae). Florida Scientist 55:38–44.

Punzo, F. 1992b. Socially facilitated behavior in tadpoles of *Rana catesbeiana* and *Rana heckscheri*. Journal of Herpetology 26:219–222.

Punzo, F. 1993a. Effect of mercuric chloride on fertilization and larval development in the river frog, *Rana heckscheri* (Wright) (Anura: Ranidae). Bulletin of Environmental Contamination and Toxicology 51:575–581.

Punzo, F. 1993b. Ovarian effects of a sublethal concentration of mercuric chloride in the river frog, *Rana heckscheri* (Anura: Ranidae). Bulletin of Environmental Contamination and Toxicology 50:385–391.

Punzo, F. 1995. An analysis of feeding in the oak toad, *Bufo quercicus* (Holbrook) (Anura: Bufonidae). Florida Scientist 58:16–20.

Punzo, F., and L. Lindstrom. 2001. The toxicity of eggs of the giant toad, *Bufo marinus*, to aquatic predators in a Florida retention pond. Journal of Herpetology 35:693–697.

Punzo, F., J. Laveglia, D. Lohr, and P.A. Dahm. 1979. Organochlorine insecticide residues in amphibians and reptiles from Iowa and lizards from the Southwestern United States. Bulletin of Environmental Contamination and Toxicology 21:842–848.

Purcell, J.W. 1968. Embryonic temperature adaptations in southwestern populations of *Rana pipiens*. M.S. thesis, Texas Tech University, Lubbock.

Purgue, A.P. 1997. Tympanic sound radiation in the bullfrog, *Rana catesbeiana*. Journal of Comparative Physiology 181A:438–445.

Purrenhage, J.L., and M.D. Boone. 2009. Amphibian community response to variation in habitat structure and competitor density. Herpetologica 65:14–30.

Pyburn, W.F. 1958. Size and movements of a local population of cricket frogs (*Acris crepitans*). Texas Journal of Science 10:325–342.

Pyburn, W.F. 1960. Hybridization between *Hyla versicolor* and *H. femoralis*. Copeia 1960:55–56.

Pyburn, W.F. 1961a. The inheritance and distribution of vertebral stripe color in the cricket frog. Pp. 235–261 *In* W.F. Blair (ed.), Vertebrate Speciation, University of Texas Press, Austin.

Pyburn, W.F. 1961b. Inheritance of the green vertebral stripe in *Acris crepitans*. Southwestern Naturalist 6:164–167.

Pyburn, W.F., and J.P. Kennedy. 1960. Artificial hybridization of the Gray Treefrog, *Hyla versicolor* (Hylidae). American Midland Naturalist 64:216–223.

Pytel, B.A. 1986. Biochemical systematics of the eastern North American frogs of the genus *Rana*. Herpetologica 42:273–282.

Quaranta, A., V. Bellantuono, G. Cassano, and C. Lippe. 2009. Why amphibians are more sensitive than mammals to xenobiotics. PLoS ONE 4(11):e7699.

Quinby, J.A. 1954. *Rana sylvatica sylvatica* Le Conte in South Carolina. Herpetologica 10:87–88.

Quinn, H. 1979. The Rio Grande chirping frog, *Syrrhophus cystignathoides campi* (Amphibia, Leptodactylidae), from Houston, Texas. Transactions of the Kansas Academy of Sciences 82:209–210.

Quinn, H.R., and G. Mengden. 1984. Reproduction and growth of *Bufo houstonensis* (Bufonidae). Southwestern Naturalist 29:189–195.

Rachowicz, L.J., R.A. Knapp, J.A.T. Morgan, M.J.

Stice, V.T. Vredenburg, J.M. Parker, and C.J. Briggs. 2006. Emerging infectious disease as a proximate cause of amphibian mass mortality. Ecology 87:1671–1683.

Rafferty, K.A. Jr. 1964. Kidney tumors of the leopard frog: a review. Cancer Research 24:169–185.

Raimondo, S.M., C.L. Rowe, and J.D. Congdon. 1998. Exposure to coal ash impacts swimming performance and predator avoidance in larval bullfrogs (*Rana catesbeiana*). Journal of Herpetology 32:289–292.

Raithel, C.J., P.W.C. Paton, P.S. Pooleand, and F.C. Golet. 2011. Assessing long-term population trends of wood frogs using egg-mass counts. Journal of Herpetology 45:23–27.

Ralin, D.B. 1968. Ecological and reproductive differentiation in the cryptic species of the *Hyla versicolor* complex (Hylidae). Southwestern Naturalist 13:283–300.

Ralin, D.B. 1976 Comparative hybridization of a diploid-tetraploid cryptic species pair of treefrogs. Copeia 1976:191–196.

Ralin, D.B. 1977. Evolutionary aspects of mating call variation in a diploid-tetraploid species complex of treefrogs (Anura). Evolution 31:721–736.

Ralin, D.B. 1978. "Resolution" of the diploid-tetraploid tree frogs. Science 202:335–336.

Ralin, D.B., and J.S. Rogers. 1972. Aspects of tolerance to desiccation in *Acris crepitans* and *Pseudacris streckeri*. Copeia 1972:519–528.

Ralin, D.B., and R.K. Selander. 1979. Evolutionary genetics of diploid-tetraploid species of treefrogs of the genus *Hyla*. Evolution 33:595–608.

Raloff, J. 2003. Hawaii's hated frogs: tiny invaders raise a big ruckus. Science News 163:11–13.

Ramer, J.D., T.A. Jenssen, and C.J. Hurst. 1983. Size-related variation in the advertisement call of *Rana clamitans* (Anura: Ranidae), and its effect on conspecific males. Copeia 1983:141–155.

Raney, E.C. 1940. Summer movements of the bullfrog, *Rana catesbeiana* Shaw, as determined by the jaw-tag method. American Midland Naturalist 23:733–745.

Raney, E.C., and W.M. Ingram. 1941. Growth of tagged frogs (*Rana catesbeiana* Shaw and *Rana clamitans* Daudin) under natural conditions. American Midland Naturalist 26:201–206.

Raney, E.C., and E.A. Lachner. 1947. Studies on the growth of tagged toads (*Bufo terrestris americanus* Holbrook). Copeia 1947:113–116.

Raphael, M.C. 1988. Long-term trends in abundance of amphibians, reptiles, and mammals in Douglas-fir forests of northwestern California. Pp. 23–31 *In* R.C. Szaro, K.E. Severson, and D.R. Patton (eds.), Management of Amphibians, Reptiles, and Small Mammals in North America. USDA Forest Service General and Technical Report RM-166.

Rathbun, G.B., N.J. Scott Jr., and T.G. Murphy. 1997. *Rana aurora draytonii* (California Red-legged Frog). Behavior. Herpetological Review 28:85–86.

Raun, G.G., and F.R. Gehlbach. 1972. Amphibians and reptiles in Texas. Taxonomic synopsis, bibliography, county distribution maps. Dallas Museum of Natural History, Bulletin 2.

Reaser, J.K. 1996. *Rana pretiosa* (Spotted frog). Vagility. Herpetological Review 27:196–197.

Reaser, J.K. 2000. Demographic analysis of the Columbia spotted frog (*Rana luteiventris*): case study in spatiotemporal variation. Canadian Journal of Zoology 78:1158–1167.

Reaser, J.K., and R.E. Dexter. 1996. *Rana pretiosa* (Spotted Frog). Predation. Herpetological Review 27:75.

Recuero, E., I. Martínez-Solano, G. Parra-Olea, and M. García-París. 2006a. Phylogeography of *Pseudacris regilla* (Anura: Hylidae) in western North America, with a proposal for a new taxonomic rearrangement. Molecular Phylogenetics and Evolution 39:293–304.

Recuero, E., I. Martínez-Solano, G. Parra-Olea, and M. García-París. 2006b. Corrigendum to "Phylogeography of *Pseudacris regilla* (Anura: Hylidae) in western North America, with a proposal for a new taxonomic rearrangement." Molecular Phylogenetics and Evolution 41:511.

Redmer, M. 1992. *Rana sphenocephala* (Southern Leopard Frog). Variation. Herpetological Review 23:58–59.

Redmer, M. 1996. Locality records of the northern leopard frog, *Rana pipiens*, in central and southwestern Illinois. Transactions of the Illinois State Academy of Sciences 89:215–219.

Redmer, M. 1998a. Status and distribution of two uncommon frogs, Pickerel Frogs and Wood Frogs, in Illinois. Pp. 83–90 *In* M.J. Lannoo (ed.), Status & Conservation of Midwestern Amphibians. University of Iowa Press, Iowa City.

Redmer, M. 1998b. *Hyla avivoca* (Bird-voiced Treefrog). Amplexus and oviposition. Herpetological Review 29:230–231.

Redmer, M. 2000. Demographic and reproductive characteristics of a southern Illinois population of the crayfish frog, *Rana areolata*. Journal of the Iowa Academy of Science 107:128–133.

Redmer, M. 2002. Natural history of the wood frog (*Rana sylvatica*) in the Shawnee National Forest, southern Illinois. Illinois Natural History Survey Bulletin 36:163–194.

Redmer, M., and S.E. Trauth. 2005. *Rana sylvatica* LeConte, 1825. Wood Frog. Pp. 590–593 *In* M.J. Lannoo (ed.), Amphibian Declines. The Conservation Status of United States Species. University of California Press, Berkeley.

Redmer, M., L.E. Brown, and R.A. Brandon. 1999a. Natural history of the bird-voiced treefrog (*Hyla avivoca*) and green treefrog (*Hyla cinerea*) in southern Illinois. Illinois Natural History Survey Bulletin 36:37–66.

Redmer, M., D.H. Jamieson, and S.E. Trauth. 1999b. Notes on the diet of female bird-voiced treefrogs (*Hyla avivoca*) in southern Illinois. Transactions of the Illinois State Academy of Sciences 92:271–275.

Redmond, W.H., and R.H. Mount. 1975. A biogeographic analysis of the herpetofauna of the Coosa Valley in Alabama. Journal of the Alabama Academy of Science 46:65–81.

Redmond, W.H., and A.F. Scott. 1996. Atlas of Amphibians in Tennessee. Center for Field Biology, Austin Peay State University, Miscellaneous Publication No. 12.

Reed, C.F. 1957. *Rana virgatipes* in southern Maryland, with notes upon its range from New Jersey to Georgia. Herpetologica 13:137–138.

Reeder, A.L., G.L. Foley, D.K. Nichols, L.G. Hansen, B. Wikoff, S. Faeh, J. Eisold, M.B. Wheeler, R. Warner, J.E. Murphy, and V.R. Beasley. 1998. Forms and prevalence of intersexuality and effects of environmental contaminants on sexuality in cricket frogs (*Acris crepitans*). Environmental Health Perspectives 106:261–266.

Reeder, A.L., M.O. Ruiz, A. Pessier, L.E. Brown, J.M. Levengood, C.A. Phillips, M.B. Wheeler, R.E. Warner, and V.R. Beasley. 2005. Intersexuality and the cricket frog decline: historic and geographic trends. Environmental Health Perspectives 113:261–265.

Regosin, J.V., B.S. Windmiller, and J.M. Reed. 2003. Terrestrial habitat use and winter densities of the wood frog (*Rana sylvatica*). Journal of Herpetology 37:390–394.

Reilly, B.O., and P.T.K. Woo. 1982. The biology of *Trympanosoma andersoni* n. sp. and *Trympanosoma grylli* Nigrelli, 1944 (Kinoplastidia) from *Hyla versicolor* LeConte, 1825 (Anura). Canadian Journal of Zoology 60:116–123.

Reimchen, T.E. 1990. Introduction and dispersal of the Pacific treefrog, *Hyla regilla*, on the Queen Charlotte Islands, British Columbia. Canadian Field-Naturalist 105:288–290.

Relyea, R.A. 2000. Trait-mediated indirect effects in larval anurans: reversing competition with the threat of predation. Ecology 81:2278–2289.

Relyea, R.A. 2001a. The relationship between predation risk and antipredator responses in larval anurans. Ecology 82:541–554.

Relyea, R.A. 2001b. Morphological and behavioral plasticity of larval anurans in response to different predators. Ecology 82:523–540.

Relyea, R.A. 2001c. The lasting effects of adaptive plasticity: predator-induced tadpoles become long-legged frogs. Ecology 82:1947–1953.

Relyea, R.A. 2002a. Competitor-induced plasticity in tadpoles: consequences, cues, and connections to predator-induced plasticity. Ecological Monographs 72:523–540.

Relyea, R.A. 2002b. Local population differences in phenotypic plasticity: predator-induced changes in wood frog populations. Ecological Monographs 72:77–93.

Relyea, R.A. 2004a. Fine-tuned phenotypes: tadpole plasticity under 16 combinations of predators and competitors. Ecology 85:172–179.

Relyea, R.A. 2004b. Synergistic impacts of malathion and predatory stress on six species of North American tadpoles. Environmental Toxicology and Chemistry 23:1080–1084.

Relyea, R.A. 2005a. The lethal impact of Roundup on aquatic and terrestrial amphibians. Ecological Applications 15:1118–1124.

Relyea, R.A. 2005b. The impact of insecticides and herbicides on the biodiversity and productivity of aquatic communities. Ecological Applications 15:618–627.

Relyea, R.A. 2009. A cocktail of contaminants: how mixtures of pesticides at low concentrations affect aquatic communities. Oecologia 159:363–376.

Relyea, R.A., and E.E. Werner. 1999. Quantifying the relation between predator-induced behavior and growth performance in larval amphibians. Ecology 80:2117–2124.

Relyea, R.A., and N. Mills. 2001. Predator-induced stress makes the pesticide carbaryl more deadly to gray treefrog tadpoles (*Hyla versicolor*). Proceedings of the National Academy of Sciences, USA 98:2491–2496.

Relyea, R.A., and J.R. Auld. 2005. Predator- and competitor-induced plasticity: how changes in foraging morphology affect phenotypic trade-offs. Ecology 86:1723–1729.

Relyea, R.A., and K. Edwards. 2010. What doesn't kill you makes you sluggish: how sublethal pesticides alter predator-prey interactions. Copeia 2010:558–567.

Relyea, R.A., N.M. Schoeppner, and J.T. Hoverman. 2005. Pesticides and amphibians: the importance of community context. Ecological Applications 15:1125–1134.

Renaud, M. 1977. Polymorphic and polytypic variation in the Arizona treefrog (*Hyla wrightorum*). Ph.D. diss., Arizona State University, Tempe.

Resetarits, W.J. Jr., and H.M. Wilbur. 1989. Choice of oviposition site by *Hyla chrysoscelis*: role of predators and competitors. Ecology 70:220–228.

Resetarits, W.J. Jr., and H.M. Wilbur. 1991a. Calling site choice by *Hyla chrysoscelis*: effects of predators, competitors, and oviposition sites. Ecology 72:778–786.

Resnick, L.E., and D.L. Jameson. 1963. Color polymorphism in Pacific tree frogs. Science 142:1081–1083.

Reynolds, T.D., and T.D. Stephens. 1984. Multiple ectopic limbs in a wild population of *Hyla regilla*. Great Basin Naturalist 44:166–169.

Rhoads, S.N. 1895. Contributions to the zoology of Tennessee. 1. Reptiles and amphibians. Proceedings of the Academy of Natural Sciences, Philadelphia 47:376–407.

Rice, A.M., A.R. Leichty, and D.W. Pfennig. 2009. Parallel evolution and ecological selection: replicated character displacement in spadefoot toads. Proceedings of the Royal Society, London B 276:4189–4196.

Rice, A.N., T.L. Roberts IV, J.G. Pritchard, and M.E. Dorcas. 2001. Historical trends and perceptions of amphibian and reptile diversity in the western Piedmont of North Carolina. Journal of the Elisha Mitchell Scientific Society 117:264–273.

Rice, K.G., J.H. Waddle, M.W. Miller, M.E. Crockett, F.J. Mazzotti, and H.F. Percival. 2011. Recovery of native treefrogs after removal of nonindigenous Cuban treefrogs, *Osteopilus septentrionalis*. Herpetologica 67:105–117.

Richards, C.M. 1958. The inhibition of growth in crowded *Rana pipiens* tadpoles. Physiological Zoology 31:138–151.

Richards, C.M. 1962. The control of tadpole growth by alga-like cells. Physiological Zoology 35:285–296.

Richards, C.M., and G.W. Nace. 1983. Dark pigment variants in anurans: classification, new descriptions, color changes and inheritance. Copeia 1983:979–990.

Richards, C.M., D.T. Tartof, and G.W. Nace. 1969. A melanoid variant in *Rana pipiens*. Copeia 1969:850–852.

Richardson, J.M.L. 2001. A comparative study of activity levels in larval anurans and response to the presence of different predators. Behavioral Ecology 12:51–58.

Richmond, A.M., T. Tyning, and A.P. Summers. 1999. *Bufo americanus* (American Toad). Depth record. Herpetological Review 30:90–91.

Richmond, N.D. 1947. Life history of *Scaphiopus holbrookii holbrookii* (Harlan). 1. Larval development and behavior. Ecology 28:53–67.

Richmond, N.D. 1952. An addition to the herpetofauna of Nova Scotia, and other records of amphibians and reptiles on Cape Breton Island. Annals of the Carnegie Museum 32:331–332.

Richmond, N.D. 1964. The green frog (*Rana clamitans melanota*) developing in one season. Herpetologica 20:132.

Richmond, N.D., and C.J. Goin. 1938. Notes on a collection of amphibians and reptiles from New Kent County, Virginia. Annals of the Carnegie Museum 27:301–310.

Richter, J., L. Martin, and C.K. Beachy. 2009. Increased larval density induces accelerated metamorphosis independently of growth rate in the frog *Rana sphenocephala*. Journal of Herpetology 43:551–554.

Richter, K.O., and A.L. Azous. 1995. Amphibian occurrence and wetland characteristics in the Puget Sound Basin. Wetlands 15:305–312.

Richter, S.C. 2000. Larval caddisfly predation on the eggs and embryos of *Rana capito* and *Rana sphenocephala*. Journal of Herpetology 34:590–593.

Richter, S.C., and R.A. Seigel. 2002. Annual variation in the population ecology of the endangered gopher frog, *Rana sevosa* Goin and Netting. Copeia 2002:962–972.

Richter, S.C., J.E. Young, R.A. Seigel, and G.N. Johnson. 2001. Postbreeding movements of the dark gopher frog, *Rana sevosa* Goin and Netting: implications for conservation and management. Journal of Herpetology 35:316–321.

Richter, S.C., J.E. Young, G.N. Johnson, and R.A. Seigel. 2003. Stochastic variation in reproductive success of a rare frog, *Rana sevosa*: implications for conservation and for monitoring amphibian populations. Biological Conservation 111:171–177.

Riemer, W.J. 1958. Giant toads of Florida. Quarterly Journal of the Florida Academy of Sciences 21:207–211.

Riha, V.F., and K.A. Berven. 1991. An analysis of latitudinal variation in the larval development of the wood frog (*Rana sylvatica*). Copeia 1991:209–221.

Riley, E.E., and M.R. Weil. 1986. The effects of thiosemicarbazide on development in the wood frog, *Rana sylvatica*. 1. Concentration effects. Ecotoxicology and Environmental Safety 12:154–160.

Riley, E.E., and M.R. Weil. 1987. The effects of thiosemicarbazide on development in the wood frog, *Rana sylvatica*. 2. Critical exposure length and age sensitivity. Ecotoxicology and Environmental Safety 13:202–207.

Rimer, R.L., and J.T. Briggler. 2010. Occurrence of the amphibian chytrid fungus (*Batrachochytrium dendrobatidis*) in Ozark caves, Missouri, USA. Herpetological Review 41:175–177.

Ripplinger, J.I., and R.S. Wagner. 2004. Phylogeography of northern populations of the Pacific treefrog, *Pseudacris regilla*. Northwestern Naturalist 85:118–125.

Ritke, M.E., and R.D. Semlitsch. 1991. Mating behavior and determinants of male mating success in the gray treefrog, *Hyla chrysoscelis*. Canadian Journal of Zoology 69:246–250.

Ritke, M.E., J.G. Rabb, and M.K. Ritke. 1990. Life history of the gray treefrog (*Hyla chrysoscelis*) in western Tennessee. Journal of Herpetology 24:135–141.

Ritke, M.E., J.G. Rabb, and M.K. Ritke. 1991. Breeding-site specificity in the Gray Treefrog (*Hyla chrysoscelis*). Journal of Herpetology 25:123–125.

Ritland, K., L.A. Dupuis, F.L. Bunnell, W.L.Y. Hung, and J.E. Carlson. 2000. Phylogeography of the tailed frog (*Ascaphus truei*) in British Columbia. Canadian Journal of Zoology 78:1749–1758.

Rittenhouse, T.A.G., and R.D. Semlitsch. 2007a. Postbreeding habitat use of wood frogs in a Missouri oak-hickory forest. Journal of Herpetology 41:645–653.

Rittenhouse, T.A.G., and R.D. Semlitsch. 2007b. Distribution of amphibians in terrestrial habitat surrounding wetlands. Wetlands 27:153–161.

Rittmann, S.E., E. Muths, and D.E. Green. 2003. *Pseudacris triseriata* (Western Chorus Frog) and *Rana sylvatica* (Wood Frog). Chytridiomycosis. Herpetological Review 34:53.

Rittschof, D. 1975. Some aspects of the natural history and ecology of the leopard frog, *Rana pipiens*. Ph.D. diss., University of Michigan, Ann Arbor.

Rivero, J.A. 1998. Los Anfibios y Reptiles de Puerto Rico. 2nd ed. Editorial de la Universidad de Puerto Rico, San Juan.

Rizkalla, C.E. 2009. First reported detection of *Batrachochytrium dendrobatidis* in Florida, USA. Herpetological Review 40:189–190.

Rizkalla, C.E. 2010. Increasing detections of *Batrachochytrium dendrobatidis* in central Florida, USA. Herpetological Review 41:180–181.

Roberts, W.R. 1998. The calliphorid fly (*Bufolucilia sylvarum*) parasitic on frogs in Alberta. Alberta Naturalist 28:48.

Roberts, W., and V. Lewin. 1979. Habitat utilization and population densities of the amphibians of northeastern Alberta. Canadian Field-Naturalist 93:144–154.

Robins, A., G. Lippolis, A. Bisazza, G. Vallortigara, and L.J. Rogers. 1998. Lateralized agonistic responses and hindlimb use in toads. Animal Behaviour 56:875–881.

Robinson, M., M.P. Donovan and T.D. Schwaner. 1998. Western toad, *Bufo boreas*, in southern Utah: notes on a single population along the east fork of the Sevier River. Great Basin Naturalist 58:87–89.

Roble, S.M. 1979. Dispersal movements and plant associations of juvenile gray treefrogs, *Hyla versicolor*. Transactions of the Kansas Academy of Science 82:235–245.

Roble, S.M. 1985a. A study of the reproductive and population ecology of two hylid frogs in northeastern Kansas. Ph.D. diss., University of Kansas, Lawrence.

Roble, S.M. 1985b. Observations on satellite males in *Hyla chrysoscelis*, *Hyla picta*, and *Pseudacris triseriata*. Journal of Herpetology 19:432–436.

Rochester, C.J., C.S. Brehme, D.R. Clark, D.C. Stokes, S.A. Hathaway, and R.N. Fisher. 2010. Reptile and amphibian responses to large-scale wildfires in southern California. Journal of Herpetology 44:333–351.

Rödder, D. 2009. "Sleepless in Hawaii": does anthropogenic climate change enhance ecological and socioeconomic impacts of the alien invasive *Eleutherodactylus coqui* Thomas 1966 (Anura: Eleutherodactylidae)? North-Western Journal of Zoology 5:16–25.

Rogers, A.M. 1999. Ecology and natural history of *Rana clamitans melanota* in West Virginia. M.S. thesis, Marshall University, Huntington, WV.

Rogers, C.D. 1996. *Rana catesbeiana* (Bullfrog). Predation. Herpetological Review 27:19.

Rogers, J.S. 1972. Discriminant function analysis of morphological relationships within the *Bufo cognatus* species group. Copeia 1972:381–383.

Rogers, J.S. 1973. Protein polymorphism, genic heterozygosity and divergence in the toads *Bufo cognatus* and *B. speciosus*. Copeia 1973:322–330.

Rogers, K.L. 1987. Pleistocene high altitude amphibians and reptiles from Colorado (Alamosa Local Fauna; Pleistocene; Irvingtonian). Journal of Vertebrate Paleontology 7:82–95.

Rogers, K.L., and L. Harvey. 1994. A skeletochronological assessment of fossil and recent *Bufo cognatus* from south-central Colorado. Journal of Herpetology 28:133–140.

Rogers, K.L., C.A. Repenning, R.M. Forester, E.E. Larson, S.A. Hall, G.R. Smith, E. Anderson, and T.J. Brown. 1985. Middle Pleistocene (Late Irvingtonian: Nebraskan) climatic changes in south-central Colorado. National Geographic Research 1:535–563.

Rohr, J.R., A.M. Schotthoefer, T.R. Raffel, H.J. Carrick, N. Halstead, J.T. Hoverman, C.M. Johnson, L.B. Johnson, C. Lieske, M.D. Piwoni, P.K. Schoff, and V.R. Beasley. 2008. Agrochemicals increase trematode infections in a declining amphibian species. Nature 455:1235–1239.

Rollins-Smith, L.A., L.K. Reinert, V. Miera, and J.M. Conlon. 2002. Antimicrobial peptide defenses of the Tarahumara frog, *Rana tarahumarae*. Biochemical and Biophysical Research Communications 297:361–367.

Roloff, G.J., T.E. Grazia, K.F. Millenbah, and A.J. Kroll. 2011. Factors associated with amphibian occupancy in southern Michigan forests. Journal of Herpetology 45:15–22.

Romano, M.A., D.B. Ralin, S.I. Guttman, and J.H. Skillings. 1987. Parallel electromorph variation in the diploid-tetraploid gray treefrog complex. American Naturalist 130:864–878.

Romansic, J.M., A.A. Waggener, B.A. Bancroft, and A.R. Blaustein. 2009. Influence of ultraviolet-B radiation on growth, prevalence of deformities, and susceptibility to predation in Cascades frog (*Rana cascadae*) larvae. Hydrobiologia 624:219–233.

Rombaugh, C.J., M.P. Hayes, and J.D. Engler. 2006. *Rana pretiosa* (Oregon Spotted Frog). Maximum size. Herpetological Review 37:210.

Rombough, C.J. 2010. *Rana catesbeiana* (American Bullfrog). Predation. Herpetological Review 41:204.

Rombough, C.J., and M.P. Hayes. 2005. Novel aspects of oviposition site preparation by foothill yellow-legged frogs (*Rana boylii*). Northwestern Naturalist 86:157–160.

Rombough, C.J., and M.P. Hayes. 2007. *Rana boylii* (Foothill Yellow-legged Frog). Reproduction. Herpetological Review 38:70–71.

Rombough, C., and M. Bradley. 2010. *Pseudacris regilla* (Northern Pacific Treefrog). Predation. Herpetological Review 41:203.

Rorabaugh, J. 2004. Barking frogs (*Eleutherodactylus augusti*) of the Santa Rita Mountains. Sonoran Herpetologist 17:72–73.

Rorabaugh, J.C. 2005a. *Rana pipiens* Screber, 1782. Northern Leopard Frog. Pp. 570–577 In M.J. Lannoo (ed.), Amphibian Declines. The Conservation Status of United States Species. University of California Press, Berkeley.

Rorabaugh, J.C. 2005b. *Rana berlandieri* Baird. Rio Grande Leopard Frog. Pp. 530–532 In M.J. Lannoo (ed.), Amphibian Declines. The Conservation Status of United States Species. University of California Press, Berkeley.

Rorabaugh, J.C., and J. Humphrey. 2002. The Tarahumara frog: return of a native. Endangered Species Technical Bulletin 27(2):24–26.

Rorabaugh, J.C., and S.F. Hale. 2005. *Rana tarahumarae* Boulenger, 1917. Tarahumara Frog. Pp. 593–595 In M.J. Lannoo (ed.), Amphibian Declines. The Conservation Status of United States Species. University of California Press, Berkeley.

Rorabaugh, J.C., and L. Elliot. 2006. Tarahumara frog (*Rana tarahumarae*) call types and characteristics. Sonoran Herpetologist 19:134–136.

Rose, F.L., T.R. Simpson, M.R.J. Forstner, D.J. McHenry, and J. Williams. 2006. Taxonomic status of *Acris gryllus paludicola*: in search of the pink frog. Journal of Herpetology 40:428–434.

Rose, G.J., and E.A. Brenowitz. 1991. Aggressive thresholds of male Pacific treefrogs for advertisement calls vary with amplitude of neighbors' calls. Ethology 89:244–252.

Rose, G.J., and E.A. Brenowitz. 1997. Plasticity of aggressive thresholds in *Hyla regilla*: discrete accommodation to encounter calls. Animal Behaviour 53:353–361.

Rose, G.J., and E.A. Brenowitz. 2002. Pacific treefrogs use temporal integration to differentiate advertisement from encounter calls. Animal Behaviour 63:1183–1190.

Rose, S.M. 1959. Failure of survival of slowly growing members of a population. Science 129:1026.

Rose, S.M. 1960. A feedback mechanism of growth control in tadpoles. Ecology 41:188–199.

Rose, S.M., and F.C. Rose. 1961. Growth-controlling exudates of tadpoles. Symposia of the Society for Experimental Biology 15:207–218.

Rosen, M., and L.E. Lemon. 1974. The vocal behavior of spring peepers, *Hyla crucifer*. Copeia 1974:940–950.

Rosen, P.C., and C.R. Schwalbe. 2002. Widespread effects of introduced species on reptiles and amphibians in the Sonoran Desert region. Pp. 220–240 In B. Tellman (ed.), Invasive Exotic Species in the Sonoran Region. University of Arizona Press and the Arizona-Sonora Desert Museum, Tucson, AZ.

Rosen, P.C., C.R. Schwalbe, D.A. Parizek Jr., P.A. Holm, and C.H. Lowe. 1995. Introduced aquatic vertebrates in the Chiricahua region: effects on declining native ranid frogs. Pp. 251–260 In L.F. DeBano, G.J. Gottfried, R.H. Hamre, C.B. Edminster, P.F. Ffoliot, and A. Ortega-Rubio (eds.), Madrean Archipelago: the Sky Islands of the Southwestern United States and Northwestern Mexico. U.S. Forest Service Rocky Mountain Station General Technical Report RM-GTR-264.

Rosenberry, D.O. 2001. Malformed frogs in Minnesota: an update. USGS Fact Sheet 043–01, Mounds View, MN.

Rosenshield, M.L., M.B. Jofré, and W.H. Karasov. 1999. Effects of polychlorinated biphenyl 126 on green frog (*Rana clamitans*) and leopard frog (*R. pipiens*) hatching success, development, and metamorphosis. Environmental Toxicology and Chemistry 18:2478–2486.

Rosine, W.N. 1952. Notes on the occurrence of polydactylism in a second species of Amphibia in Muskee Lake, Colorado. Journal of the Colorado-Wyoming Academy of Science 4:100.

Ross, D.A., T.C. Esque, R.A. Fridell, and P. Hovingh. 1995. Historical distribution, current status, and a range extension of *Bufo boreas* in Utah. Herpetological Review 26:187–189.

Ross, D.A., J.K. Reaser, P. Kleeman, and D.L. Drake. 1999. *Rana luteiventris* (Columbia Spotted Frog). Mortality and site fidelity. Herpetological Review 30:163.

Rossi, J.V. 1981. *Bufo marinus* in Florida: some natural history and its impact on native vertebrates. M.A. thesis, University of South Florida, Tampa.

Rossi, J.V. 1983. The use of olfactory cues by *Bufo marinus*. Journal of Herpetology 17:72–73.

Rossman, D.A. 1960. Herpetofaunal survey of the Pine Hills area of southern Illinois. Quarterly Journal of the Florida Academy of Science 22:207–225.

Rothermel, B.B. 2004. Migratory success of juveniles: a potential constraint on connectivity for pond-breeding amphibians. Ecological Applications 14:1535–1546.

Rothermel, B.B., and R.D. Semlitsch. 2002. An experimental investigation of landscape resistance of forest versus old-field habitats to emigrating juvenile amphibians. Conservation Biology 16:1324–1332.

Rothermel, B.B., S.C. Walls, J.C. Mitchell, C.K. Dodd Jr., L.K. Irwin, D.E. Green, V.M. Vasquez, J.W. Petranka, and D.J. Stevenson. 2008. Widespread occurrence of the amphibian chytrid fungus (*Batrachochytrium dendrobatidis*) in the Southeastern United States. Diseases of Aquatic Organisms 82:3–18.

Rot-Nikcevic, I., C.N. Taylor, and R.J. Wassersug. 2006. The role of images of conspecifics as visual cues in the development and behavior of larval amphibians. Behavioral Ecology and Sociobiology 60:19–25.

Rouse, J.D., C.A. Bishop, and J. Struger. 1999. Nitrogen pollution: an assessment of its threat to amphibian survival. Environmental Health Perspectives 107:799–803.

Rowe, C.F.B. 1899. Report on Zoology. Batrachians. Bulletin of the New Brunswick Natural History Society 17:169–170.

Rowe, C.L., and W.A. Dunson. 1993. Relationships among abiotic parameters and breeding effort by three amphibians in temporary wetlands of central Pennsylvania. Wetlands 13:237–246.

Rowe, C.L., and W.A. Dunson. 1995. Impacts of hydroperiod on growth and survival of larval amphibians in temporary ponds of central Pennsylvania, USA. Oecologia 102:397–403.

Rowe, C.L., O.M. Kinney, A.P. Fiori, and J.D. Congdon. 1996. Oral deformities in tadpoles (*Rana catesbeiana*) associated with coal ash deposition: effects on grazing ability and growth. Freshwater Biology 36:723–730.

Rowe, C.L., O.M. Kinney, R.D. Nagle, and J.D. Congdon. 1998. Elevated maintenance costs in an anuran (*Rana catesbeiana*) exposed to a mixture of trace elements during the embryonic and early larval periods. Physiological Zoology 71:27–35.

Rowe, C.L., W.A. Hopkins, and V.R. Coffman. 2001. Failed recruitment of southern toads (*Bufo terrestris*) in a trace element-contaminated breeding habitat: direct and indirect effects that may lead to a local population sink. Archives of Environmental Contamination and Toxicology 40:399–405.

Rowe, C.L., A. Heyes, and J. Hilton. 2011. Differential patterns of accumulation and depuration of dietary selenium and vanadium during metamorphosis in the gray treefrog *(Hyla versicolor)*. Archives of Environmental Contamination and Toxicology 60:336–342.

Roy, J.-S. 2009. Structure and dynamics of a natural hybrid zone between the toads, *Anaxyrus americanus* and *Anaxyrus hemiophrys*, in southeastern Manitoba. M.S. thesis, McGill University, Montreal.

Roznik, E.A., and S.A. Johnson. 2007. *Rana capito* (Gopher Frog). Refuge during fire. Herpetological Review 38:442.

Roznik, E.A., and S.A. Johnson. 2009a. Canopy closure and emigration by juvenile gopher frogs. Journal of Wildlife Management 73:260–268.

Roznik, E.A., and S.A. Johnson. 2009b. Burrow use and survival of newly metamorphosed gopher frogs (*Rana capito*). Journal of Herpetology 43:431–437.

Rudolph, D.C., and J.G. Dickson. 1990. Streamside zone width and amphibian and reptile abundance. Southwestern Naturalist 35:472–476.

Rugh, R. 1934. The space factor in the growth rate of tadpoles. Ecology 15:407–411.

Rugh, R. 1941. Experimental studies on the reproductive physiology of the male spring peeper, *Hyla crucifer*. Proceedings of the American Philosophical Society 84:617–633.

Ruibal, R. 1957. An altitudinal and latitudinal cline in *Rana pipiens*. Copeia 1957:212–221.

Ruibal, R. 1959. The ecology of a brackish water population of *Rana pipiens*. Copeia 1959:315–322.

Ruibal, R. 1962. The ecology and genetics of a desert population of *Rana pipiens*. Copeia 1962:189–195.

Ruibal, R., and S. Hillman. 1981. Cocoon structure and function in the burrowing hylid frog, *Pternohyla fodiens*. Journal of Herpetology 15:403–408.

Ruibal, R., L. Tevis Jr., and V. Roig. 1969. The terrestrial ecology of the spadefoot toad *Scaphiopus hammondii*. Copeia 1969:571–584.

Rundquist, E.M. 1978. The spring peeper, *Hyla crucifer* Wied (Anura, Hylidae) in Kansas. Transactions of the Kansas Academy of Science 80:155–158.

Runkle, L.S., K.D. Wells, C.C. Robb, and S.L. Lance. 1994. Individual, nightly, and seasonal variation in calling behavior of the gray tree frog, *Hyla versicolor*: implications for energy expenditure. Behavioral Ecology 5:318–325.

Russell, A.P., and A.M. Bauer. 2000. The Amphibians and Reptiles of Alberta. A Field Guide and Primer of Boreal Herpetology. 2nd ed. University of Calgary Press, Calgary.

Russell, K.R., D.C. Guynn Jr., and H.G. Hanlin. 2002a. Importance of small isolated wetlands for herpetofaunal diversity in managed, young growth forests in the Coastal Plain of South Carolina. Forest Ecology and Management 163:43–59.

Russell, K.R., H.G. Hanlin, T.B. Wigley, and D.C. Guynn Jr. 2002b. Responses of isolated wetland herpetofauna to upland forest management. Journal of Wildlife Management 66:603–617.

Russell, R.W. 2008. Spring peepers and pitcher plants: a case of commensalism? Herpetological Review 39: 154–155.

Russell, R.W., S.J. Hecnar, and G.D. Haffner. 1995. Organochlorine pesticide residues in southern Ontario spring peepers. Environmental Toxicology and Chemistry 14:815–817.

Ruthig, G.R. 2009. Water molds of the genera *Saprolegnia* and *Leptolegnia* are pathogenic to the North American frogs *Rana catesbeiana* and *Pseudacris crucifer*, respectively. Diseases of Aquatic Organisms 84:173–178.

Ruthven, A.G. 1908. The cold-blooded vertebrates of Isle Royale. Ecology of Isle Royale, Michigan Survey 1908:329–333.

Ruthven, A.G. 1912. The herpetology of Michigan.

Michigan Geological and Biological Survey Publication 10:11–166. (The amphibian sections of this monograph were actually written by Crystal Thompson and Helen Thompson.)

Ruthven, A.G., and H.T. Gaige. 1915. The reptiles and amphibians collected in northeastern Nevada by the Walker-Newcomb expedition of the University of Michigan. Occasional Papers of the Museum of Zoology, University of Michigan 8:1–33.

Ryan, M.J. 1978. A thermal property of the *Rana catesbeiana* (Amphibia, Anura, Ranidae) egg mass. Journal of Herpetology 12:247–248.

Ryan, M.J. 1980. The reproductive behavior of the bullfrog (*Rana catesbeiana*). Copeia 1980:108–114.

Ryan, M.J., and W. Wilczynski. 1988. Coevolution of sender and receiver: effect on local mate preference in cricket frogs. Science 240:1786–1788.

Ryan, M.J., and W. Wilczynski. 1991. Evolution of intraspecific variation in the advertisement call of a cricket frog (*Acris crepitans*, Hylidae). Biological Journal of the Linnean Society 44:249–271.

Ryan, M.J., R.B. Cocroft, and W. Wilczynski. 1990. The role of environmental selection in intraspecific divergence of mate recognition signals in the cricket frog, *Acris crepitans*. Evolution 44:1869–1872.

Ryan, M.J., S.A. Perrill, and W. Wilczynski. 1992. Auditory tuning and call frequency predict population-based mating preferences in the cricket frog, *Acris crepitans*. American Naturalist 139: 1370–1383.

Ryan, M.J., K.M. Warkentin, B.E. McClelland, and W. Wilcznyski. 1995. Fluctuating asymmetries and advertisement call variation in the cricket frog, *Acris crepitans*. Behavioral Ecology 6:124–131.

Ryan, R.A. 1953. Growth rates of some ranids under natural conditions. Copeia 1953:73–80.

Ryan, T.J., and C.T. Winne. 2001. Effects of hydroperiod on metamorphosis in *Rana sphenocephala*. American Midland Naturalist 145:46–53.

Ryder, J.A. 1878. A monstrous frog. American Naturalist 12:751–752.

Ryder, J.A. 1891. Notes on the development of *Engystoma*. American Naturalist 25:838–840.

Saber, P.A., and W.A. Dunson. 1978. Toxicity of bog water to embryonic and larval anuran amphibians. Journal of Experimental Zoology 204:33–42.

Sadinski, W.J., and W.A. Dunson. 1992. A multilevel study of effects of low pH on amphibians of temporary ponds. Journal of Herpetology 26:413–422.

Sadinski, W.J., M. Roth, S. Treleven, J. Theyerl, and P. Dummer. 2010. Detection of the chytrid fungus, *Batrachochytrium dendrobatidis*, on recently metamorphosed amphibians in the north-central United States. Herpetological Review 41:170–175.

Saenz, D., J.B. Johnson, C.K. Adams, and G.H. Dayton. 2003. Accelerated hatching of southern leopard frog (*Rana sphenocephala*) eggs in response to the presence of a crayfish (*Procambarus nigrocinctus*) predator. Copeia 2003:646–649.

Saenz, D., B.T. Kavanagh, and M.A. Kwiatkowski. 2010. *Batrachochytrium dendrobatidis* detected in amphibians from National Forests in eastern Texas, USA. Herpetological Review 41:47–49.

Sage, R.D., and R.K. Selander. 1979. Hybridization between species of the *Rana pipiens* complex in central Texas. Evolution 33:1069–1088.

Sagor, E.S., M. Ouellet, E. Barten, and D.M. Green. 1998. Skeletochronology and geographic variation in age structure in the wood frog, *Rana sylvatica*. Journal of Herpetology 32:469–474.

Salinas, C. 2009. The influence of environmental factors on the activity and movement of *Bufo nebulifer* (Coastal Plain Toad) in a disturbed area. M.S. thesis, University of Texas-Pan American, Edinburg, TX.

Salt, J.R. 1979. Some elements of amphibian distribution and biology in the Alberta Rockies. Alberta Naturalist 9:125–136.

Salthe, S.N. 1969. Geographic variation of the lactate dehydrogenase of *Rana pipiens* and *Rana palustris*. Biochemical Genetics 2:271–303.

Salthe, S.N., and E. Nevo. 1969. Geographic variation in lactate dehydrogenase in the cricket frog, *Acris crepitans*. Biochemical Genetics 3:335–341.

Salthe, S.N., and M.L. Crump. 1977. A Darwinian interpretation of hindlimb variability in frog populations. Evolution 31:737–749.

Samollow, P.B. 1980. Selective mortality and reproduction in a natural population of *Bufo boreas*. Evolution 34:18–39.

Samoray, S.T., and K.J. Regester. 2001. Geographic distribution: *Rana sylvatica* (Wood Frog). Herpetological Review 32:190.

Sams, E., and M.D. Boone. 2010. Interactions between recently metamorphosed green frogs and American toads under laboratory conditions. American Midland Naturalist 163:269–279.

Sanabria, E.A., and L.B. Quiroga. 2010. *Lithobates catesbeianus* (American Bullfrog). Diet. Herpetological Review 41:339.

Sanchiz, B. 1998. Handbuch der Paläoherpetologie. 4. Salientia. Verlag Dr. Friedrich Pfeil, Munich, Germany.

Sanders, H.O. 1970. Pesticide toxicities to tadpoles of the western chorus frog *Pseudacris triseriata* and Fowler's toad *Bufo woodhousii fowleri*. Copeia 1970:246–251.

Sanders, O. 1953. A new species of toad, with a discussion of morphology of the bufonid skull. Herpetologica 9:25–47.

Sanders, O. 1986. The heritage of *Bufo woodhousei* Girard in Texas (Salientia: Bufonidae). Occasional Papers of the Strecker Museum (1):1–28.

Sanders, O. 1987. Evolutionary hybridization and speciation in North American indigenous bufonids. Privately published, Dallas, TX.

Sanders, O., and H.M. Smith. 1951. Geographic

variation in toads of the *debilis* group of *Bufo*. Field & Laboratory 19:141–160.
Sanders, O., and J.C. Cross. 1964. Relationships between certain North American toads as shown by cytological study. Herpetologica 19:248–255.
Sanders, O., and H.S. Smith. 1971. Skin tags and ventral melanism in the Rio Grande leopard frog. Journal of Herpetology 5:31–38.
Sanzo, D., and S.J. Hecnar. 2006. Effects of road de-icing salt (NaCl) on larval wood frogs (*Rana sylvatica*). Environmental Pollution 140:247–256.
Sargent, L.G. 2000. Frog and toad population monitoring in Michigan. Journal of the Iowa Academy of Science 107:195–199.
Sartorius, S.S., and P.C. Rosen. 2000. Breeding phenology of the lowland leopard frog (*Rana yavapaiensis*): implications for conservation and ecology. Southwestern Naturalist 45:267–273.
Sattler, P.W. 1980. Genetic relationships among selected species of North American *Scaphiopus*. Copeia 1980:605–610.
Sattler, P.W. 1985. Introgressive hybridization between spadefoot toads *Scaphiopus bombifrons* and *S. multiplicatus* (Salientia: Pelobatidae). Copeia 1985:324–332.
Saumure, R.A. 1993. *Rana clamitans* (Green Frog). Albinism. Herpetological Review 24:31.
Savage, A.E., and J.R. Jaeger. 2009. Isolation and characterization of microsatellite markers in the lowland leopard frog (*Rana yavapaiensis*) and the relict leopard frog (*Rana onca*), two declining frogs of the North American desert southwest. Molecular Ecology Resources 9:199–202.
Savage, J.M. 1954. A revision of the toads of the *Bufo debilis* complex. Texas Journal of Science 6:83–112.
Savage, J.M. 1959. A preliminary biosystematic analysis of the toads of the *Bufo boreas* group in Nevada and California. Yearbook of the American Philosophical Society 1959:251–254.
Savage, J.M. 2002. The Amphibians and Reptiles of Costa Rica. University of Chicago Press, Chicago, IL.
Savage, J.M., and F.W. Schuierer. 1961. The eggs of toads of the *Bufo boreas* group, with descriptions of the eggs of *Bufo exsul* and *Bufo nelsoni*. Bulletin of the Southern California Academy of Sciences 60:93–99.
Say, T. 1823. *In* E. James, Account of an expedition from Pittsburgh to the Rocky Mountains, performed in the years 1819 and '20, by order of the Honorable J.C. Calhoun, Secretary of War, under the command of Major Stephen H. Long. Vol. 2. H.C. Carey and I. Lea, Philadelphia.
Schaaf, R.T., and J.S. Garton. 1970. Raccoon predation on the American toad, *Bufo americanus*. Herpetologica 26:334–335.
Schaaf, R.T. Jr., and P.W. Smith. 1970. Geographic variation in the pickerel frog. Herpetologica 26:240–254.
Schardien, B.J., and J.A. Jackson. 1982. Killdeers feeding on frogs. Wilson Bulletin 94:85–87.
Schaub, D.L., and J.H. Larsen Jr. 1978. The reproductive ecology of the Pacific treefrog (*Hyla regilla*). Herpetologica 34:409–416.
Schechtman, A.M., and J.B. Olson. 1941. Unusual temperature tolerance of an amphibian egg (*Hyla regilla*). Ecology 22:409–410.
Schell, S.C. 1964. *Bunoderella metteri* gen. and sp. n. (Trematoda: Allocreadiidae) and other trematode parasites of *Ascaphus truei* Stejneger. Journal of Parasitology 50:652–655.
Scherer, R.D., E. Muths, B.R. Noon, and P.S. Corn. 2005. An evaluation of weather and disease as causes of decline in two populations of boreal toads. Ecological Applications 15:2150–2160.
Schlaepfer, M.A., M.J. Sredl, P.C. Rosen, and M.J. Ryan. 2007. High prevalence of *Batrachochytrium dendrobatidis* in wild populations of lowland leopard frogs *Rana yavapaiensis* in Arizona. EcoHealth 4:421–427.
Schlefer, E.K., M.A. Romano, S.I. Guttman, and S.B. Ruth. 1986. Effects of twenty years of hybridization in a disturbed habitat on *Hyla cinerea* and *Hyla gratiosa*. Journal of Herpetology 20:210–221.
Schlichter, L.C. 1981. Low pH affects the fertilization and development of *Rana pipiens* eggs. Canadian Journal of Zoology 59:1693–1699.
Schloegel, L.M., A.M. Picco, A.M. Kilpatrick, A.J. Davies, and A.D. Hyatt. 2009. Magnitude of the US trade in amphibians and presence of *Batrachochytrium dendrobatidis* and ranavirus infection in imported North America bullfrogs (*Rana catesbeiana*). Biological Conservation 142:1420–1426.
Schloegel, L.M., C.M. Ferreira, T.Y. James, M. Hipolito, J.E. Longcore, A.D. Hyatt, M. Yabsley, A.M.C.R.P.F. Martins, R. Mazzoni, A.J. Davies, and P. Daszak. 2010. The North American bullfrog as a reservoir for the spread of *Batrachochytrium dendrobatidis* in Brazil. Animal Conservation 13 (Supplement 1):53–61.
Schmetterling, D.A., and M.K. Young. 2008. Summer movements of boreal toads (*Bufo boreas*) in two western Montana basins. Journal of Herpetology 42:111–123.
Schmid, W.D. 1965. Some aspects of the water economies of nine species of amphibians. Ecology 46:261–269.
Schmid, W.D. 1982. Survival of frogs in low temperatures. Science 215:697–698.
Schmid, W.D. 1986. Winter ecology. Ekologiia (Soviet Journal of Ecology) 6:29–35.
Schmidt, K.P. 1924. A list of amphibians and reptiles collected near Charleston, S.C. Copeia (132):67–69.
Schmidt, K.P. 1929. The frogs and toads of the

Chicago area. Field Museum of Natural History, Zoology Leaflet 11.

Schmidt, K.P. 1938. A geographic variation gradient in frogs. Field Museum of Natural History, Zoology Series 20(29):377–382.

Schmidt, K.P. 1946. How to make money from frog-farming. Turtox News 24:169–170.

Schmidt, K.P., and W.L. Necker. 1935. Amphibians and reptiles of the Chicago region. Bulletin of the Chicago Academy of Sciences 5:57–77.

Schmidt, K.P., and T.F. Smith. 1944. Amphibians and reptiles of the Big Bend region of Texas. Zoological Series of Field Museum of Natural History. 29: 75–96.

Schmidt, R.S. 1968. Chuckle calls of the leopard frog (*Rana pipiens*). Copeia 1968:561–569.

Schmutzer, A.C., M.J. Gray, E.C. Burton, and D.L. Miller. 2008. Impacts of cattle on amphibian larvae and the aquatic environment. Freshwater Biology 53:2613–2625.

Schneider, J.G. 1799. Historia Amphibiorum Naturalis et Literarariae. Fasciculus Primus. Continens Ranas, Calamitas, Bufones, Salamandras et Hydros in Genera et Species Descriptos Notisque suis Distinctos. Friederici Frommanni, Jena.

Schoenherr, A.A., C.R. Feldmeth, and M.J. Emerson. 1999. Natural History of the Islands of California. University of California Press, Berkeley.

Schoeppner, N.M., and R.A. Relyea. 2009. When should prey respond to consumed heterospecifics? Testing hypotheses of perceived risk. Copeia 2009: 190–194.

Schonberger, C.F. 1945. Food of some amphibians and reptiles of Oregon and Washington. Copeia 1945: 120–121.

Schooley, J.D., M.R. Schwemm, and J.M. Barkstedt. 2008. *Rana catesbeiana* (American Bullfrog): tadpole gigantism and deformity. Herpetological Review 39:339–340.

Schotthoefer, A.M., A.V. Koehler, C.U. Meteyer, and R.A. Cole. 2003. Influence of *Ribeiroia ondatrae* (Trematoda: Digenea) infection on limb development and survival of northern leopard frogs (*Rana pipiens*): effects of host stage and parasite-exposure level. Canadian Journal of Zoology 81:1144–1153.

Schoville, S.D., T.S. Tustall, V.T. Vredenburg, A.R. Backlin, E. Gallegos, D.A. Wood, and R.N. Fisher. 2011. Conservation genetics of evolutionary lineages of the endangered mountain yellow-legged frog, *Rana muscosa* (Amphibia: Ranidae), in southern California. Biological Conservation 144:2031–2044.

Schreber, H. 1782. Der Naturforscher. Vol. 18. Johann Jacob Gebaur, Halle.

Schreiman, E., and R. Rugh. 1949. Effects of DDT on functional development of larvae of *Rana pipiens* and *Fundulus heteroclitus*. Proceedings of the Society for Experimental Biology and Medicine 70:431–435.

Schreiver, T.A. 2007. Salinity influences on larval *Hyla cinerea* (Green Treefrog) development and alteration to southeastern Louisiana wetlands and its effects on the herpetofauna. M.S. thesis, Southeastern Louisiana University, Hammond.

Schriever, T.A., J. Ramspott, B.I. Crother, and C.L. Fontenot Jr. 2009. Effects of hurricanes Ivan, Katrina and Rita on a southeastern Louisiana herpetofauna. Wetlands 29:112–122.

Schroeder, E.E. 1966. Age determination and population structure in Missouri bullfrogs. Ph.D. diss., University of Missouri, Columbia.

Schroeder, E.E. 1968. Aggressive behavior in *Rana clamitans*. Journal of Herpetology 1:95–96.

Schroeder, E.E. 1976. Dispersal and movement of newly transformed green frogs, *Rana clamitans*. American Midland Naturalist 95:471–474.

Schroeder, E.E., and T.S. Baskett. 1965. Frogs and toads of Missouri. Missouri Conservationist 26: 15–18.

Schroeder, E.E., and T.S. Baskett. 1968. Age estimation, growth rates, and population structure in Missouri bullfrogs. Copeia 1968:583–592.

Schueler, F.W. 1973. Frogs of the Ontario coast of Hudson Bay and James Bay. Canadian Field-Naturalist 87:409–418.

Schueler, F.W. 1975. Geographic variation in the size of *Rana septentrionalis* in Quebec, Ontario, and Manitoba. Journal of Herpetology 9:177–185.

Schueler, F.W. 1979. Geographic variation in skin pigmentation and dermal glands in the northern leopard frog, *Rana pipiens*. Ph.D. diss., University of Toronto.

Schueler, F.W. 1982a. Geographic variation in skin pigmentation and dermal glands in the Northern Leopard Frog, *Rana pipiens*. National Museums of Canada, Publications in Zoology No. 16.

Schueler, F.W. 1982b. Sexual colour differences in Canadian western toads, *Bufo boreas*. Canadian Field-Naturalist 96:329–332.

Schueler, F.W., and F.R. Cook. 1980. Distribution of the middorsal stripe dimorphism in the wood frog, *Rana sylvatica*, in eastern North America. Canadian Journal of Zoology 58:1643–1651.

Schueler, F.W., and R.M. Rankin. 1982. Terrestrial amplexus in the wood frog, *Rana sylvatica*. Canadian Field-Naturalist 96:348–349.

Schuierer, F.W. 1962. Remarks upon the natural history of *Bufo exsul* Myers, the endemic toad of Deep Springs Valley, Inyo County, California. Herpetologica 17:260–266.

Schuierer, F.W. 1963. Notes on two populations of *Bufo exsul* Myers and a commentary on speciation within the *Bufo boreas* group. Herpetologica 18:262–267.

Schuierer, F.W. 1972. The current status of the endangered species *Bufo exsul* Myers, Deep Springs Valley, Inyo County, California. Herpetological Review 4:81–82.

Schuierer, F.W., and S.C. Anderson. 1990. Population status of *Bufo exsul* Myers, 1942. Herpetological Review 21:57.

Schulte-Hostedde, A.I., and C.M.M. Schank. 2009. Secondary sexual traits and individual phenotype in male green frogs (*Rana clamitans*). Journal of Herpetology 43:89–95.

Schurbon, J.M., and J.E. Fauth. 2003. Effects of prescribed burning on amphibian diversity in a southeastern U.S. national forest. Conservation Biology 17:1338–1349.

Schuytema, G.S., and A.V. Nebeker. 1999a. Effects of ammonium nitrate, sodium nitrate, and urea on red-legged frogs, Pacific treefrogs, and African clawed frogs. Bulletin of Environmental Contamination and Toxicology 63:357–364.

Schuytema, G.S., and A.V. Nebeker. 1999b. Comparative effects of ammonium and nitrate compounds on Pacific Treefrog and African clawed frog embryos. Archives of Environmental Contamination and Toxicology 36:200–206.

Schuytema, G.S., A.V. Nebeker, and W.L. Griffis. 1995. Comparative toxicity of Guthion and Guthion 2S to *Xenopus laevis* and *Pseudacris regilla* tadpoles. Bulletin of Environmental Contamination and Toxicology 54:382–388.

Schwalbe, C.R., and P.C. Rosen. 1988. Preliminary report on effect of bullfrogs on wetland herpetofaunas in southeastern Arizona. Pp. 166–173 *In* R.C. Szaro, K.E. Severson, and D.R. Patton (eds.), Management of Amphibians, Reptiles, and Small Mammals in North America. USDA Forest Service General and Technical Report RM-166.

Schwalbe, C.R., and B. Alberti. 1998. Ground-truthing a troll. Studying the barking frog at Coronado National Memorial. Park Science 18:26–27.

Schwaner, T.D., and B.K. Sullivan. 2005. *Bufo microscaphus* Cope 1867 "1866." Arizona Toad. Pp. 422–424 *In* M.J. Lannoo (ed.), Amphibian Declines. The Conservation Status of United States Species. University of California Press, Berkeley.

Schwaner, T.D., and B.K. Sullivan. 2009. Fifty years of hybridization: introgression between the Arizona toad (*Bufo microscaphus*) and Woodhouse's toad (*B. woodhousii*) along Beaver Dam Wash in Utah. Herpetological Conservation and Biology 4:198–206.

Schwaner, T.D., D.R. Hadley, K.R. Jenkins, M.P. Donovan, and B. Al Tait. 1997. Population dynamics and life history studies of toads: short- and long-range needs for understanding amphibian populations in southern Utah. Pp. 197–202 *In* L.M. Hill (ed.), Learning from the Land. Grand Staircase-Escalante National Monument Science Symposium Proceedings, USDI Bureau of Land Management BLM/UT/GI-98/006+1220.

Schwartz, A. 1952. *Hyla septentrionalis* Dumeril and Bibron on the Florida mainland. Copeia 1952:117–118.

Schwartz, A. 1955. The chorus frog *Pseudacris brachyphona* in North Carolina. Copeia 1955:138.

Schwartz, A. 1957. Chorus frogs (*Pseudacris nigrita* LeConte) in South Carolina. American Museum Novitates 1838:1–12.

Schwartz, A., and J.R. Harrison III. 1956. A new subspecies of gopher frog (*Rana capito* Leconte). Proceedings of the Biological Society of Washington 69:135–144.

Schwartz, J.J. 1989. Graded aggressive calls of the spring peeper, *Pseudacris crucifer*. Herpetologica 45:172–181.

Schwartz, J.J., and H.C. Gerhardt. 1998. The neuroethology of frequency preferences in the spring peeper. Animal Behaviour 56:55–69.

Schwartz, J.J., and K.M. Rahmeyer. 2006. Calling behavior and the capacity for sustained locomotory exercise in the gray treefrog (*Hyla versicolor*). Journal of Herpetology 40:164–171.

Schwartz, V., and D.M. Golden. 2002. Field Guide to Reptiles and Amphibians of New Jersey. New Jersey Division of Fish and Wildlife, Woodbine.

Scott, A.F., and D.F. Harker. 1968. First records of the barking treefrog, *Hyla gratiosa* Le Conte, from Tennessee. Herpetologica 24:82–83.

Scott, D.E., B.S. Metts, and J.W. Gibbons. 2008. Enhancing amphibian biodiversity on golf courses with seasonal wetlands. Pp. 285–292 *In* J.C. Mitchell, R.E. Jung Brown, and B. Bartholomew (eds.), Urban Herpetology. Herpetological Conservation 3. Society for the Study of Amphibians and Reptiles, Salt Lake City, UT.

Scott, G.W., and C.C. Carpenter. 1956. Body temperatures of *Bufo w. woodhousei*. Proceedings of the Oklahoma Academy of Science 36:84–85.

Scott, N.J. Jr., and R.D. Jennings. 1985. The tadpoles of five species of New Mexican leopard frogs. Occasional Papers, the Museum of Southwestern Biology 3:1–21.

Seale, D.B. 1980. Influence of amphibian larvae on primary production, nutrient flux, and competition in a pond ecosystem. Ecology 61:1531–1550.

Seale, D.B. 1982. Physical factors influencing oviposition by the wood frog, *Rana sylvatica*, in Pennsylvania. Copeia 1982:627–635.

Seale, D.B., and R.J. Wassersug. 1979. Suspension feeding dynamics of anuran larvae related to their functional morphology. Oecologia 39:259–272.

Seale, D.B., and N. Beckvar. 1980. The comparative ability of anuran larvae (genera: *Hyla*, *Bufo* and *Rana*) to ingest suspended blue-green algae. Copeia 1980:495–503.

Searle, C.L., L.K. Belden, B.A. Bancroft, B.A. Han,

L.M. Biga, and A.R. Blaustein. 2010. Experimental examination of the effects of ultraviolet-B radiation in combination with other stressors on frog larvae. Oecologia 162:237–245.

Seba, A. 1734. Locupletissimi Rerum Naturalium Thesauri Accurata Descriptio, et Iconibus Artificiosissimis Expressio, per Universam Physices Historiam. Janssonio-Waesbergios, & J. Wetstenium, & Gul. Smith, Amsterdam.

Seburn, C.N.L., D.C. Seburn, and C.A. Paszkowski. 1997. Northern leopard frog (*Rana pipiens*) dispersal in relation to habitat. SSAR Herpetological Conservation 1:64–72.

Seburn, D.C., C.N.L. Seburn, and W.F. Weller. 2008. A localized decline in the western chorus frog, *Pseudacris triseriata*, in eastern Ontario. Canadian Field-Naturalist 122:158–161.

Secor, S.M. 1988. Perch sites of calling male bird-voiced treefrogs, *Hyla avivoca*, in Oklahoma. Proceedings of the Oklahoma Academy of Science 68:71–73.

Seib, G.W. 1977. Probable case of death-feigning by wood frog. Blue Jay 35:148.

Seigel, R.A. 1983. Natural survival of eggs and tadpoles of the wood frog, *Rana sylvatica*. Copeia 1983:1096–1098.

Seigel, R.A., A. Dinsmore, and S.C. Richter. 2006. Using well water to increase hydroperiod as a management option for pond-breeding amphibians. Wildlife Society Bulletin 34:1022–1027.

Semlitsch, R.D. 1990. Effects of body size, sibship, and tail injury on the susceptibility of tadpoles to dragonfly predation. Canadian Journal of Zoology 68:1027–1030.

Semlitsch, R.D. 2003. Amphibian Conservation. Smithsonian Books, Washington, DC.

Semlitsch, R.D., and J.P. Caldwell. 1982. Effects of density on growth, metamorphosis, and survivorship in tadpoles of *Scaphiopus holbrooki*. Ecology 63:905–911.

Semlitsch, R.D., and J.W. Gibbons. 1988. Fish predation in size-structured populations of treefrog tadpoles. Oecologia 75:321–326.

Semlitsch, R.D., J.W. Gibbons, and T.D. Tuberville. 1995. Timing of reproduction and metamorphosis in the Carolina gopher frog (*Rana capito capito*) in South Carolina. Journal of Herpetology 29:612–614.

Semlitsch, R.D., D.E. Scott, J.H.K. Pechmann, and J.W. Gibbons. 1996. Structure and dynamics of an amphibian community: evidence from a 16-year study of a natural pond. Pp. 217–248 *In* M.L. Cody and J.A. Smallwood (eds.), Long-term Studies of Vertebrate Communities. Academic Press, San Diego, CA.

Semlitsch, R.D., C.M. Bridges, and A.M. Welch. 2000. Genetic variation and a fitness tradeoff in the tolerance of gray treefrog (*Hyla versicolor*) tadpoles to the insecticide carbaryl. Oecologia 125:179–185.

Semlitsch, R.D., C.A. Conner, D.J. Hocking, T.A.G. Rittenhouse, and E.B. Harper. 2008. Effects of timber harvesting on pond-breeding amphibian persistence: testing the evacuation hypothesis. Ecological Applications 18:283–289.

Semsar, K., K.F. Klomberg, and C. Marler. 1998. Arginine vasotocin increases calling-site acquisition by nonresident male grey treefrogs. Animal Behaviour 56:983–987.

Sessions, S.K., and S.B. Ruth. 1990. Explanation for naturally occurring supernumerary limbs in amphibians. Journal of Experimental Zoology 254:38–47.

Sexton, O.J., and K.R. Marion. 1974. Probable predation by Swainson's Hawk on swimming spadefoot toads. Wilson Bulletin 86:167–168.

Sexton, O.J., and C. Phillips. 1986. A qualitative study of fish-amphibian interactions in 3 Missouri ponds. Transactions, Missouri Academy of Science 20:25–35.

Sexton, O.J., C.A. Phillips, T.J. Bergman, E.B. Wattenberg, and R.E. Preston. 1998. Abandon not hope: status of repatriated populations of spotted salamanders and wood frogs at the Tyson Research Center, St. Louis County, Missouri. Pp. 340–353 *In* M.J. Lannoo (ed.), Status & Distribution of Midwestern Amphibians. University of Iowa Press, Iowa City.

Seyle, C.W. Jr., and S.E. Trauth. 1982. *Pseudacris ornata* (Ornate Chorus Frog). Reproduction. Herpetological Review 13:45.

Seymour, R.S. 1995. Oxygen uptake by embryos in gelatinous egg masses of *Rana sylvatica*: the roles of diffusion and convection. Copeia 1995:626–635.

Shaffer, H.B., G.M. Fellers, A. Magee, and S.R. Voss. 2000. The genetics of amphibian declines: population substructure and molecular differentiation in the Yosemite toad, *Bufo canorus* (Anura, Bufonidae), based on single-stranded conformation polymorphism analysis (SSCP) and mitochondrial DNA sequence data. Molecular Ecology 9:245–257.

Shaffer, H.B., G.M. Fellers, S.R. Voss, J.C. Oliver, and G.B. Pauly. 2004. Species boundaries, phylogeography and conservation genetics of the red-legged frog (*Rana aurora/draytonii*) complex. Molecular Ecology 13:2667–2677.

Shah, A.A., M.J. Ryan, E. Bevilacqua, and M.L. Schlaepfer. 2010. Prior experience alters the behavioral responses of prey to a nonnative predator. Journal of Herpetology 44:185–192.

Shannon, F.A., and C.H. Lowe Jr. 1955. A new subspecies of *Bufo woodhousei* from the inland Southwest. Herpetologica 11:185–190.

Shannon, P.R. 1988. The parasites of breeding male eastern gray treefrogs, *Hyla versicolor* LeConte, 1825, from the Ashland Wildlife Research Area,

Boone County, Missouri. M.S. thesis, University of Missouri, Columbia.

Shaw, G. 1802. General Zoology, or Systematic Natural History. Vol 3, Part I. Amphibia. Thomas Davison, London.

Shelford, V.E. 1913. The reactions of certain animals to gradients of evaporating power of air: a study in experimental ecology. Biological Bulletin 25:79–120.

Shepard, D.B. 2002. Spatial relationships of male green frogs (*Rana clamitans*) throughout the activity season. American Midland Naturalist 148:394–400.

Shepard, D.B. 2004. Seasonal differences in aggression and site tenacity in male green frogs, *Rana clamitans*. Copeia 2004:159–164.

Shepard, D.B., and A.R. Kuhns. 2000. *Pseudacris triseriata* (Western Chorus Frog): calling sites after drought. Herpetological Review 31:235–236.

Shepard, D.B., and L.E. Brown. 2005. *Bufo houstonensis* Sanders, 1953: Houston Toad. Pp. 415–417 In M.J. Lannoo (ed.), Amphibian Declines. The Conservation Status of United States Species. University of California Press, Berkeley.

Shinn, E.A., and J.W. Dole. 1978. Evidence for a role for olfactory cues in the feeding response of leopard frogs, *Rana pipiens*. Herpetologica 34:167–172.

Shinn, E.A., and J.W. Dole. 1979a. Lipid components of prey odors elicit feeding responses in Western toads (*Bufo boreas*). Copeia 1979:275–278.

Shinn, E.A., and J.W. Dole. 1979b. Evidence for a role for olfactory cues in the feeding response of Western toads (*Bufo boreas*). Copeia 1979:163–165.

Shirer, H.W., and H.S. Fitch. 1970. Comparison from radiotracking of movements and denning habits of the raccoon, striped skunk, and opossum in northeastern Kansas. Journal of Mammalogy 51:491–503.

Shirose, L.J., and R.J. Brooks. 1997. Fluctuations in abundance and age structure in three species of frogs (Anura: Ranidae) in Algonquin Park, Canada, from 1985 to 1993. SSAR Herpetological Conservation 1:16–26.

Shirose, L.J., R.J. Brooks, J.R. Barta, and S.S. Desser. 1993. Intersexual differences in growth, mortality, and size at maturity in bullfrogs in central Ontario. Canadian Journal of Zoology 71:2363–2369.

Shively, J.N., J.G. Songer, S. Prchal, M.S. Keasey III, and C.O. Thoen. 1981. *Mycobacterium marinum* infection in Bufonidae. Journal of Wildlife Diseases 17:3–7.

Shoop, C.R., and C. Ruckdeschel. 2003. Herpetological biogeography of the Georgia Barrier islands: an alternative interpretation. Florida Scientist 66:43–51.

Shoop, C.R., and C. Ruckdeschel. 2006. Amphibians and reptiles of Cumberland Island, Georgia. Occasional Publications of the Cumberland Island Museum No. 2, St. Marys, GA.

Shovlain, A.M. 2006. Oregon spotted frog (*Rana pretiosa*) habitat use and herbage (or biomass) removal from grazing at Jack Creek, Klamath County, Oregon. M.S. thesis, Oregon State University, Corvallis.

Shubin, N.H., and F.A. Jenkins Jr. 1995. An Early Jurassic jumping frog. Nature 377:49–52.

Shulse, C.D., R.D. Semlitsch, K.M. Trauth, and A.D. Williams. 2010. Influences of design and landscape placement parameters on amphibian abundance in constructed wetlands. Wetlands 30:915–928.

Sias, J. 2006. Natural history and distribution of the upland chorus frog, *Pseudacris feriarum* Baird, in West Virginia. M.S. thesis, Marshall University, Huntington.

Siekmann, J.M. 1949. A survey of the tadpoles of Louisiana. M.S. thesis, Louisiana State University, Baton Rouge.

Sievert, G., and L. Sievert. 2006. A Field Guide to Oklahoma's Amphibians and Reptiles. Oklahoma Department of Wildlife Conservation, Oklahoma City.

Sievert, L. 1991. Thermoregulatory behaviour in the toads *Bufo marinus* and *Bufo cognatus*. Journal of Thermal Biology 16:309–312.

Simandle, E.T. 2006. Population structure and conservation of two rare toad species (*Bufo exsul* and *Bufo nelsoni*) in the Great Basin, USA. Ph.D. diss., University of Nevada, Reno.

Simmons, R., and J.D. Hardy Jr. 1959. The river-swamp frog, *Rana heckscheri* Wright, in North Carolina. Herpetologica 15:36–37.

Simon, J.A., J.W. Snodgrass, R.E. Casey, and D.W. Sparling. 2009. Spatial correlates of amphibian use of constructed wetlands in an urban landscape. Landscape Ecology 24:361–373.

Simovich, M.A. 1985. Analysis of a hybrid zone between the spadefoot toads *Scaphiopus multiplicatus* and *Scaphiopus bombifrons*. Ph.D. diss., University of California-Riverside.

Simovich, M.A. 1994. The dynamics of a spadefoot toad (*Spea multiplicata* and *S. bombifrons*) hybridization system. Pp. 167–182 In P.R. Brown and J.W. Wright (eds.), Herpetology of the North American Deserts. Proceedings of a Symposium. Southwestern Herpetologists Society, Special Publication No. 5.

Simovich, M.A., C.A. Sassaman, and A. Chovnick. 1991. Post-mating selection of hybrid toads (*Scaphiopus multiplicatus* and *Scaphiopus bombifrons*). Proceedings of the San Diego Society of Natural History No. 5:1–6.

Sin, H., K.H. Beard, and W.C. Pitt. 2008. An invasive frog, *Eleutherodacylus coqui*, increases new leaf production and leaf litter decomposition rates through nutrient cycling in Hawaii. Biological Invasions 10:335–345.

Singer, S.R., and S. Grismaijer. 2005. Panic in Para-

dise. Invasive Species Hysteria and the Hawaiian Coqui Frog War. ISCD Press, Pahoa, HI.
Skelly, D.K. 1994. Activity level and the susceptibility of anuran larvae to predation. Animal Behaviour 47:465–468.
Skelly, D.K. 1995a. A behavioral tradeoff and its consequences for the distribution of *Pseudacris* treefrog larvae. Ecology 76:150–164.
Skelly, D.K. 1995b. Competition and the distribution of spring peeper larvae. Oecologia 103:203–207.
Skelly, D.K. 2001. Distributions of pond-breeding anurans: an overview of mechanisms. Israel Journal of Zoology 47:313–332.
Skelly, D.K. 2004. Microgeographic countergradient variation in the wood frog, *Rana sylvatica*. Evolution 58:160–165.
Skelly, D.K., and E.E. Werner. 1990. Behavioral and life-historical responses of larval American toads to an odonate predator. Ecology 71:2313–2322.
Skelly, D.K., and L.K. Freidenburg. 2000. Effects of beaver on the thermal biology of an amphibian. Ecology Letters 3:483–486.
Skelly, D.K., E.E. Werner, and S.A, Cortwright. 1999. Long-term distributional dynamics of a Michigan amphibian assemblage. Ecology 80:2326–2337.
Skelly, D.K., L.K. Freidenburg, and J.M. Kiesecker. 2002. Forest canopy and the performance of larval amphibians. Ecology 83:983–992.
Skelly, D.K., K.L. Yurewicz, E.E. Werner, and R.A. Relyea. 2003. Estimating decline and distributional change in amphibians. Conservation Biology 17: 744–751.
Skelly, D.K., S.R. Bolden, and K.B. Dion. 2010. Intersex frogs concentrated in suburban and urban landscapes. EcoHealth 7:374–379.
Skidds, D.E., F.C. Golet, P.W.C. Paton, and J.C. Mitchell. 2007. Habitat correlates of reproductive effort in wood frogs and spotted salamanders in an urbanizing watershed. Journal of Herpetology 41:439–450.
Slater, J.R. 1931. The mating of *Ascaphus truei* Stejneger. Copeia 1931:62–63.
Slater, J.R. 1939a. Description and life-history of a new *Rana* from Washington. Herpetologica 1:145–149.
Slater, J.R. 1939b. Some species of amphibians new to the state of Washington. Occasional Papers, Department of Biology, College of Puget Sound No. 2:4–5. (W.C. Brown's name was apparently omitted as a junior author. He appears as such in the table of contents.)
Slater, J.R. 1941a. Island records of amphibians and reptiles for Washington. Occasional Papers, Department of Biology, College of Puget Sound No. 13:74–77.
Slater, J.R. 1941b. The distribution of amphibians and reptiles in Idaho. Occasional Papers, Department of Biology, College of Puget Sound No. 14:78–109.

Slater, J.R. 1955. Distribution of Washington amphibians. Occasional Papers, Department of Biology, College of Puget Sound No. 16:120–154.
Slevin, J.R. 1928. The amphibians of western North America. An account of the species known to inhabit California, Alaska, British Columbia, Washington, Oregon, Idaho, Utah, Nevada, Arizona, Sonora, and Lower California. Occasional Papers of the California Academy of Sciences 16:1–144.
Smith, A.K. 1977. Attraction of bullfrogs (Amphibia, Anura, Ranidae) to distress calls of immature frogs. Journal of Herpetology 11:234–235.
Smith, B.E. 2003. Conservation assessment for the Northern Leopard Frog in the Black Hills National Forest, South Dakota and Wyoming. USDA Forest Service, Black Hills National Forest, Custer, South Dakota. http://www.fs.fed.us/r2/blackhills/projects/planning/assessments/leopard_frog.pdf.
Smith, B.E., and D.A. Keinath. 2007. Northern leopard frog (*Rana pipiens*): a technical conservation assessment. USDA Forest Service, Rocky Mountain Region, Species Conservation Project. http://www.fs.fed.us/r2/projects/scp/assessments/northern-leopardfrog.pdf.
Smith, C.C., and A.N. Bragg. 1949. Observations on the ecology and natural history of anura, 7. Food and feeding habits of the common species of toads in Oklahoma. Ecology 30:333–349.
Smith, D.A. 1953. Northern swamp tree frog, *Pseudacris nigrita septentrionalis* (Boulenger) from Churchill, Manitoba. Canadian Field-Naturalist 67:181–182.
Smith, D.C. 1983. Factors controlling tadpole populations of the chorus frog (*Pseudacris triseriata*) on Isle Royale, Michigan. Ecology 64:501–510.
Smith, D.C. 1987. Adult recruitment in chorus frogs: effects of size and date at metamorphosis. Ecology 68:344–350.
Smith, D.C. 1990. Population structure and competition among kin in the chorus frog (*Pseudacris triseriata*). Evolution 44:1529–1541.
Smith, D.D., and R. Powell. 1983. *Acris crepitans blanchardi* (Blanchard's Cricket Frog). Anomalies. Herpetological Review 14:118–119.
Smith, G.R. 1999. Microhabitat preferences of bullfrog tadpoles (*Rana catesbeiana*) of different ages. Transactions of the Nebraska Academy of Sciences 25:73–76.
Smith, G.R. 2002. *Pseudacris crucifer* (Spring Peeper). Predation. Herpetological Review 33:48.
Smith, G.R. 2007. Lack of effect of nitrate, nitrite, and phosphorus on wood frog (*Rana sylvatica*) tadpoles. Applied Herpetology 4:287–291.
Smith, G.R., and A.K. Jennings. 2004. Spacing of the tadpoles of *Hyla versicolor* and *Rana clamitans*. Journal of Herpetology 38:616–618.
Smith, G.R., and B.L. Doupnik. 2005. Habitat use and activity level of large American bullfrog tad-

poles: choices and repeatability. Amphibia-Reptilia 26:549–552.
Smith, G.R., and J.J. Schulte. 2008. Conservation management implications of substrate choice in recently metamorphosed American toads (*Bufo americanus*). Applied Herpetology 5:87–92.
Smith, G.R., and A.R. Awan. 2009. The roles of predator identity and group size in the antipredator responses of American toad (*Bufo americanus*) and bullfrog (*Rana catesbeiana*) tadpoles. Behaviour 146:225–243.
Smith, G.R., and D.T. Fortune. 2009. Hatching plasticity of wood frog (*Rana sylvatica*) eggs in response to mosquitofish (*Gambusia affinis*) cues. Herpetological Conservation and Biology 4:43–47.
Smith, G.R., J.E. Rettig, G.G. Mittelbach, J.L. Valiulis, and S.R. Schaack. 1999. The effects of fish on assemblages of amphibians in ponds: a field experiment. Freshwater Biology 41:829–837.
Smith, G.R., M.A. Waters, and J.E. Rettig. 2000. Consequences of embryonic UV-B exposure for embryos and tadpoles of the plains leopard frog. Conservation Biology 14:1903–1907.
Smith, G.R., A. Todd, J.E. Rettig, and F. Nelson. 2003. Microhabitat selection by northern cricket frogs (*Acris crepitans*) along a west-central Missouri creek: field and experimental observations. Journal of Herpetology 37:383–385.
Smith, G.R., A. Boyd, C.B. Dayer, and K.E. Winter. 2008. Behavioral responses of American toad and bullfrog tadpoles to the presence of cues from the invasive fish, *Gambusia affinis*. Biological Invasions 10:743–748.
Smith, G.R., A. Boyd, C.B. Dayer, M.E. Ogle, and A.J. Terlecky. 2009. Responses of gray treefrog and American toad tadpoles to the presence of cues from multiple predators. Herpetological Journal 19:79–83.
Smith, G.R., A. Boyd, C.B. Dayer, M.E. Ogle, A.J. Terlecky, and C.J. Dibble. 2010. Effects of sibship and the presence of multiple predators on the behavior of green frog (*Rana clamitans*) tadpoles. Ethology 116:217–225.
Smith, G.R., S.V. Krishnamurthy, A.C. Burger, and L.B. Mills. 2011. Differential effects of malathion and nitrate exposure on American toad and wood frog tadpoles. Archives of Environmental Contamination and Toxicology 60:327–335.
Smith, H.C. 1975. Reproductive cycles of *Bufo woodhousii* and *Acris gryllus* in northern Mississippi, with additional notes on *Pseudacris triseriata* and *Hyla chrysoscelis*. Ph.D. diss., Mississippi State University, Mississippi State.
Smith, H.M. 1932. A report upon amphibians hitherto unknown from Kansas. Transactions of the Kansas Academy of Science 35:93–96.
Smith, H.M. 1933. On the proper name for the brevicipitid frog *Gastrophryne texensis* (Girard). Copeia 1933: 217.
Smith, H.M. 1934. The amphibians of Kansas. American Midland Naturalist 15:377–528.
Smith, H.M. 1937. Notes on *Scaphiopus hurterii* Strecker. Herpetologica 1:104–108.
Smith, H.M. 1947. Subspecies of the Sonoran toad (*Bufo compactilis* Wiegmann). Herpetologica 4:7–13.
Smith, H.M. 1950. Handbook of amphibians and reptiles of Kansas. Miscellaneous Publications of the Museum of Natural History, University of Kansas (2):1–336.
Smith, H.M., and B.C. Brown. 1947. The Texan subspecies of the treefrog, *Hyla versicolor*. Proceedings of the Biological Society of Washington 60:47–50.
Smith, H.M., and H.K. Buechner. 1947. The influence of the Balcones Escarpment on the distribution of amphibians and reptiles in Texas. Bulletin of the Chicago Academy of Sciences 8:1–16.
Smith, H.M., and E.H. Taylor. 1948. An annotated checklist and key to the amphibia of Mexico. Bulletin of the United States National Museum 194:iv–118.
Smith, H.M., and D.A. Langebartel. 1949. The toad *Bufo valliceps* in Arkansas. Copeia 1949:230.
Smith, H.M., and J.C. List. 1951. The occurrence of digital pads as an anomaly in the bullfrog. Turtox News 29:118–119.
Smith, H.M., and O. Sanders. 1952. Distributional data on Texan amphibians and reptiles. Texas Journal of Science 4:204–219.
Smith, H.M., C.W. Nixon, and P.E. Smith. 1948. A partial description of the tadpole of *Rana areolata circulosa* and notes on the natural history of the race. American Midland Naturalist 39:608–614.
Smith, H.M., T.P. Maslin, and R.L. Brown. 1965. Summary of the distribution of the herpetofauna of Colorado. University of Colorado Studies, Series in Biology No. 15:1–52.
Smith, H.M., D. Chiszar, J.T. Collins, and F. van Breukelen. 1998. The taxonomic status of the Wyoming toad, *Bufo baxteri* Porter. Contemporary Herpetology 1. http://www.contemporaryherpetology.org/ch/1998/1/CH_1998_1.pdf.
Smith, K.G. 2005a. Effects of nonindigenous tadpoles on native tadpoles in Florida: evidence of competition. Biological Conservation 123:433–441.
Smith, K.G. 2005b. An exploratory assessment of Cuban treefrog (*Osteopilus septentrionalis*) tadpoles as predators of native and nonindigenous tadpoles in Florida. Amphibia-Reptilia 26:571–575.
Smith, L.L., and C.K. Dodd Jr. 2003. Wildlife mortality on U.S. Highway 441 across Paynes Prairie, Alachua County, Florida. Florida Scientist 66:128–140.
Smith, L.L., K.G. Smith, W.J. Barichivich, C.K. Dodd Jr., and K. Sorensen. 2005. Roads and Florida's her-

petofauna: a review and mitigation case study. Pp. 32–40 *In* W.E. Meshaka Jr. and K.J. Babbitt (eds.), Amphibians and Reptiles. Status and Conservation in Florida. Krieger Publishing, Malabar, FL.

Smith, L.L., W.J. Barichivich, J.S. Staiger, K.G. Smith, and C.K. Dodd Jr. 2006. Detection probabilities and site occupancy estimates for amphibians at Okefenokee National Wildlife Refuge. American Midland Naturalist 155:149–161.

Smith, L.M., M.J. Gray, and A. Quarles. 2004. Diets of newly metamorphosed amphibians in west Texas playas. Southwestern Naturalist 49:257–263.

Smith, M.A., and D.M. Green. 2005. Dispersal and the metapopulation paradigm in amphibian ecology and conservation: are all amphibian populations metapopulations? Ecography 28:110–128.

Smith, P.W. 1947. The reptiles and amphibians of eastern central Illinois. Bulletin of the Chicago Academy of Sciences 8:21–40.

Smith, P.W. 1951. A new frog and a new turtle from the western Illinois sand prairies. Bulletin of the Chicago Academy of Sciences 9:189–199.

Smith, P.W. 1955. *Pseudacris streckeri illinoensis* in Missouri. Transactions of the Kansas Academy of Science 58:411.

Smith, P.W. 1956. The status, correct name, and geographic range of the boreal chorus frog. Proceedings of the Biological Society of Washington 69:169–176.

Smith, P.W. 1957. An analysis of post-Wisconsin biogeography of the Prairie Peninsula region based on distributional phenomena among terrestrial vertebrate populations. Ecology 38:205–218.

Smith, P.W. 1961. The amphibians and reptiles of Illinois. Illinois Natural History Survey Bulletin 28:1–298.

Smith, P.W., and D.M. Smith. 1952. The relationship of the chorus frogs, *Pseudacris nigrita feriarum* and *Pseudacris n. triseriata*. American Midland Naturalist 48:165–180.

Smith, R.E. 1940. Mating and oviposition in the Pacific Coast tree toad. Science 92:379–380.

Smithberger, S.I., and C.W. Swarth. 1993. Reptiles and amphibians of the Jug Bay Wetlands Sanctuary. The Maryland Naturalist 37(3–4):28–46.

Smits, A.W. 1984. Activity patterns and thermal biology of the toad *Bufo boreas halophilus*. Copeia 1984:689–696.

Snider, A.T., and J.K. Bowler. 1992. Longevity of reptiles and amphibians in North American collections. 2nd ed. SSAR Herpetological Circular No. 21.

Snodgrass, J.W., W.A. Hopkins, and J.H. Roe. 2003. Relationships among developmental stage, metamorphic timing, and concentrations of elements in bullfrogs (*Rana catesbeiana*). Environmental Toxicology and Chemistry 22:1597–1604.

Snodgrass, J.W., W.A. Hopkins, J. Broughton, D. Gwinn, J.A. Baionno, and J. Burger. 2004. Species-specific responses of developing anurans to coal combustion waste. Aquatic Toxicology 66:171–182.

Snodgrass, J.W., R.E. Casey, D. Joseph, and J.A. Simon. 2008. Mesocosm investigations of stormwater pond sediment toxicity to embryonic and larval amphibians: variation in sensitivity among species. Environmental Pollution 154:291–297.

Snyder, W.F., and D.L. Jameson. 1965. Multivariate geographic variation of mating call in populations of the Pacific tree frog (*Hyla regilla*). Copeia 1965:129–142.

Sodhi, N.S., D. Bickford, A.C. Diesmos, T.M. Lee, L.P. Koh, B.W. Brook, C.H. Sekercioglu, and C.J.A. Bradshaw. 2008. Measuring the meltdown: drivers of global amphibian extinction and decline. PLoS ONE 2(2):e1636.

Sonnini, C.S., and P.A. Latreille. 1801. Histoire naturelle des reptiles, avec figures designées d'après nature, 2. Chez Deterville, Paris.

Soto-Azat, C., B.T. Clarke, J.C. Poynton, and A.C. Cunningham. 2010. Widespread historical presence of *Batrachochytrium dendrobatidis* in African pipid frogs. Diversity and Distributions 16:126–131.

Souder, W. 2000. A Plague of Frogs. The Horrifying True Story. Hyperion, New York.

Sparling, D.W. 2003. A review of the role of contaminants in amphibian declines. Pp. 1099–1128 *In* D.J. Hoffman, B.A. Rattner, G.A. Burton Jr., and J. Cairns Jr. (eds.), Handbook of Ecotoxicology. Lewis Publishers, Boca Raton, FL.

Sparling, D.W., and T.P. Lowe. 1996. Metal concentrations of tadpoles in experimental ponds. Environmental Pollution 91:149–159.

Sparling, D.W., and G. Fellers. 2007. Comparative toxicity of chlorpyrifos, diazinon, malathion and their oxon derivatives to larval *Rana boylii*. Environmental Pollution 147:535–539.

Sparling, D.W., T.P. Lowe, D. Day, and K. Dolan. 1995. Responses of amphibian populations to water and soil factors in experimentally-treated aquatic mesocosms. Archives of Environmental Contamination and Toxicology 29:455–461.

Sparling, D.W., T.P. Lowe, and A.E. Pinkney. 1997. Toxicity of Abate™ to green frog tadpoles. Bulletin of Environmental Contamination and Toxicology 58:475–481.

Sparling, D.W., G.M. Fellers, and L.L. McConnell. 2001. Pesticides and amphibian population declines in California, USA. Environmental Toxicology and Chemistry 20:1591–1595.

Spear, P.A., M. Boily, I. Giroux, C. DeBlois, M.H. Leclair, M. Levasseur, and R. Leclair. 2009. Study design, water quality, morphometrics and age of the bullfrog, *Rana catesbeiana*, in sub-watersheds of the Yamaska River drainage basin, Québec, Canada. Aquatic Toxicology 91:110–117.

Spear, S.F., and A. Storfer. 2008. Landscape genetic structure of coastal tailed frogs (*Ascaphus truei*) in protected vs. managed forests. Molecular Ecology 17:4642–4656.

Spear, S.F., and A. Storfer. 2010. Anthropogenic and natural disturbance lead to differing patterns of gene flow in the Rocky Mountain tailed frog, *Ascaphus montanus*. Biological Conservation. 143: 778–786.

Spencer, W.A. 1964a. The relationship of dispersal and migration to gene flow in the boreal chorus frog. Ph.D. diss., Colorado State University, Fort Collins.

Spencer, W.A. 1964b. An unusual phoretic host for the clam, *Pisidium*. Journal of the Colorado-Wyoming Academy of Science 5(5):43–44.

Spencer, W.A. 1964c. Movement in a population of *Pseudacris triseriata*. Journal of the Colorado-Wyoming Academy of Science 5(5):44.

Spencer, W.A. 1971. Boreal chorus frogs (*Pseudacris triseriata*) breeding in the alpine in southwestern Colorado. Arctic and Alpine Research 3:353.

Squire, T., and R.A. Newman. 2002. Fine-scale population structure in the wood frog (*Rana sylvatica*) in a northern woodland. Herpetologica 58:119–130.

Sredl, M.J. 1998. Arizona leopard frogs: balanced on the brink? Pp.573–574 *In* M.J. Mac, P.A. Opler, C.E. Puckett, and P.D. Doran (eds.), Status and Trends of the Nation's Biological Resources. Vol. 2. U.S. Geological Survey, Washington, DC.

Sredl, M.J. 2005a. *Rana yavapaiensis* Platz and Frost, 1984: Lowland Leopard Frog. Pp. 596–599 *In* M.J. Lannoo (ed.), Amphibian Declines. The Conservation Status of United States Species. University of California Press, Berkeley.

Sredl, M.J. 2005b. *Pternohyla fodiens* Boulenger, 1882: Lowland Burrowing Treefrog. Pp. 488–489 *In* M.J. Lannoo (ed.), Amphibian Declines. The Conservation Status of United States Species. University of California Press, Berkeley.

Sredl, M.J., and J.P. Collins. 1991. The effect of ontogeny on interspecific interactions in larval amphibians. Ecology 72:2232–2239.

Sredl, M.J., and J.P. Collins. 1992. The interaction of predation, competition, and habitat complexity in structuring an amphibian community. Copeia 1992:607–614.

Sredl, M.J., and J.M. Howland. 1995. Conservation and management of Madrean populations of the Chiricahua leopard frog. Pp. 379–385 *In* L.F. DeBano, G.J. Gottfried, R.H. Hamre, C.B. Edminster, P.F. Ffoliot, and A. Ortega-Rubio (eds.), Madrean Archipelago. The Sky Islands of the Southwestern United States and Northwestern Mexico. U.S. Forest Service Rocky Mountain Station General Technical Report RM-GTR-264.

Sredl, M.J., and R.D. Jennings. 2005. *Rana chiricahuensis* Platz and Mecham, 1979. Chiricahua Leopard Frog. Pp.546–549 *In* M.J. Lannoo (ed.), Amphibian Declines. The Conservation Status of United States Species. University of California Press, Berkeley.

Stabler, R.M. 1948. Prairie rattlesnake eats spadefoot toad. Herpetologica 4:168.

St. Amant, J.A., and F.G. Hoover. 1969. Addition of *Misgurnus anguillicaudatus* (Cantor) to the California fauna. California Fish and Game 55:330–331.

St. Amant, J.A., F.G. Hoover, and G.R. Stewart. 1973. African clawed frog, *Xenopus laevis* (Daudin), established in California. California Fish and Game 59:151–153.

St.-Amour, V., T.W.J. Garner, A.I. Schulte-Hostedde, and D. Lesbarrères. 2010. Effects of two amphibian pathogens on the developmental stability of green frogs. Conservation Biology 24:788–794.

Stanback, M. 2010. *Gambusia holbrooki* predation on *Pseudacris feriarum* tadpoles. Herpetological Conservation and Biology 5:486–489.

Standaert, W.F. 1967. Growth, maturation, and population ecology of the carpenter frog (*Rana virgatipes* Cope). Ph.D. diss., Rutgers University, New Brunswick, NJ.

Stangel, P.W. 1983. Least sandpiper predation on *Bufo americanus* and *Ambystoma maculatum* larvae. Herpetological Review 14:112.

Starnes, S.M., C.A. Kennedy, and J.W. Petranka. 2000. Sensitivity of embryos of Southern Appalachian amphibians to ambient solar UV-β radiation. Conservation Biology 14:277–282.

Stebbins, R.C. 1951. Amphibians of Western North America. 1st ed. University of California Press, Berkeley.

Stebbins, R.C. 1954. Amphibians and Reptiles of Western North America. McGraw-Hill Book Company, New York.

Stebbins, R.C. 1962. Amphibians of Western North America. 2nd printing. University of California Press, Berkeley.

Stebbins, R.C. 1972. Amphibians and Reptiles of California. California Natural History Guides No. 31. University of California Press, Berkeley.

Stebbins, R.C. 2003. A Field Guide to Western Amphibians and Reptiles. 3rd ed. Houghton Mifflin Co., Boston.

Stebbins, R.C., and N.W. Cohen. 1995. A Natural History of Amphibians. Princeton University Press, Princeton.

Steele, C.W., S. Strickler-Shaw, and D.H. Taylor. 1991. Failure of *Bufo americanus* tadpoles to avoid lead-enriched water. Journal of Herpetology 25:241–243.

Steelman, C.K., and M.E. Dorcas. 2010. Anuran calling survey optimization: developing and testing predictive models of anuran calling activity. Journal of Herpetology 44:61–68.

Stein, R.J. 1985. *Rana sylvatica* (Wood Frog). Predation. Herpetological Review 16:109–110.

Steinwascher, K. 1978a. The effect of coprophagy on the growth of *Rana catesbeiana* tadpoles. Copeia 1978:130–134.

Steinwascher, K. 1978b. Interference and exploitation competition among tadpoles of *Rana utricularia*. Ecology 59:1039–1046.

Steinwascher, K. 1981. Competition for two resources. Oecologia 49:415–418.

Stejneger, L. 1893. Annotated list of the reptiles and batrachians collected by the Death Valley Expedition in 1891, with descriptions of new species. North American Fauna 7:159–228.

Stejneger, L. 1899. Description of a new genus and species of discoglossid toad from North America. Proceedings of the United States National Museum 21:899–901 + 1 plate.

Stejneger, L. 1901. A new species of bullfrog from Florida and the Gulf Coast. Proceedings of the United States National Museum 24:211–215.

Stejneger, L. 1915. A new species of tailless batrachian from North America. Proceedings of the Biological Society of Washington 28:131–132.

Stephens, M.R. 2001. Phylogeography of the *Bufo boreas* (Anura, Bufonidae) species complex and the biogeography of California. M.A. thesis, Sonoma State University, Sonoma, CA.

Stevens, C.E., and C.A. Paszkowski. 2004. Using chorus-size ranks from call surveys to estimate reproductive activity of the wood frog (*Rana sylvatica*). Journal of Herpetology 38:404–410.

Stevens, C.E., A.W. Diamond, and T.S. Gabor. 2002. Anuran call surveys on small wetlands in Prince Edward Island, Canada, restored by dredging of sediments. Wetlands 22:90–99.

Stevens, C.E., C.A. Paszkowski, and G.J. Scrimgeour. 2006a. Older is better: beaver ponds on boreal streams as breeding habitat for the wood frog. Journal of Wildlife Management 70:1360–1371.

Stevens, C.E., C.A. Paszkowski, and D. Stringer. 2006b. Occurrence of the western toad and its use of "borrow pits" in west-central Alberta. Northwestern Naturalist 87:107–117.

Stevenson, D.J. 2007. *Rana catesbeiana* (American Bullfrog). Mortality. Herpetological Review 38:71.

Stevenson, D., and D. Crowe. 1992. Geographic distribution: *Pseudacris crucifer bartramiana* (Southern Spring Peeper). Herpetological Review 23:86.

Stevenson, H.M. 1959. Some altitude records of reptiles and amphibians. Herpetologica 15:118.

Stevenson, H.M. 1969. Occurrence of the carpenter frog in Florida. Quarterly Journal of the Florida Academy of Sciences 32:233–235.

Stewart, M.M. 1984. Redescription of the type of *Rana clamitans*. Copeia 1984:210–213.

Stewart, M.M., and P. Sandison. 1972. Comparative food habits of sympatric mink frogs, bull frogs, and green frogs. Journal of Herpetology 6:241–244.

Stewart, M.M., and J. Rossi. 1981. The Albany Pine Bush: a northern outpost for southern species of amphibians and reptiles in New York. American Midland Naturalist 106:282–292.

Stille, W.T. 1952. The nocturnal amphibian fauna of the southern Lake Michigan beach. Ecology 33:149–162.

Stinner, J., N. Zarlinga, and S. Orcutt. 1994. Overwintering behavior of adult bullfrogs, *Rana catesbeiana*, in northeastern Ohio. Ohio Journal of Science 94:8–13.

Stitt, E.W., and C.P. Seltenrich. 2010. *Rana draytonii* (California Red-legged Frog). Prey. Herpetological Review 41:206.

Stitt, E.W., and P.S. Balfour. 2011. *Spea hammondii* (Western Spadefoot). Predation. Herpetological Review 42:91–92.

Stockwell, S., and M.L. Hunter Jr. 1989. Relative abundance of herpetofauna among eight types of Maine peatland vegetation. Journal of Herpetology 23:409–414.

Stone, P.A., and D.B. Ligon. 2011. *Bufo punctatus* (Red-spotted Toad) and *Thamnophis cyrtopsis* (Black-necked Garter Snake). Prey-predator. Herpetological Review 42:82–83.

Stone, W.B., and R.D. Manwell. 1969. Toxoplasmosis in cold blooded hosts. Journal of Protozoology 16:99–102.

Storer, T.I. 1914. The California toad, an economic asset. University of California Journal of Agriculture 2:89–91.

Storer, T.I. 1922. The eastern bullfrog in California. California Fish and Game 8:219–224.

Storer, T.I. 1925. A synopsis of the amphibia of California. University of California Publications in Zoology 27:1–342.

Storer, T.I. 1933. Frogs and their commercial use. California Fish and Game 19:203–213.

Storey, J.M., and K.B. Storey. 1985. Adaptations of metabolism for freeze tolerance in the gray tree frog, *Hyla versicolor*. Canadian Journal of Zoology 63:49–54.

Storey, K.B., and J.M. Storey. 1986. Freeze tolerance and intolerance as strategies of winter survival in terrestrially hibernating amphibians. Comparative Biochemistry and Physiology 83A:613–617.

Storey, K.B., and J.M. Storey. 1987. Persistence of freeze tolerance in terrestrially hibernating frogs after spring emergence. Copeia 1987:720–726.

Storey, K.B., and J.M. Storey. 1989. Freeze tolerance and freeze avoidance in ectotherms. Pp. 51–82 In L.C.H. Wang (ed.), Advances in Comparative and Environmental Physiology. Springer-Verlag, Berlin.

Storey, K.B., J. Bischof, and B. Rubinsky. 1992. Cryomicroscopic analysis of freezing in liver of the freeze-tolerant wood frog. American Journal of Physiology 263:R185-R194.

Storm, R.M. 1952. Interspecific mating behaviour in

Rana aurora and *Rana catesbeiana*. Herpetologica 8:108.

Storm, R.M. 1960. Notes on the breeding biology of the red-legged frog (*Rana aurora aurora*). Herpetologica 16:251–259.

Storrs, S.I., and J.M. Kiesecker. 2004. Survivorship patterns of larval amphibians exposed to low concentrations of atrazine. Environmental Health Perspectives 112:1054–1057.

Storrs Méndez, S.I., D.E. Tillitt, T.A.G. Rittenhouse, and R.D. Semlitsch. 2009. Behavioral response and kinetics of terrestrial atrazine exposure in American toads (*Bufo americanus*). Archives of Environmental Contamination and Toxicology 57:590–597.

Stoutamire, R. 1932. Bullfrog farming and frogging in Florida. State of Florida, Department of Agriculture Bulletin (new series) No. 56.

Straughan, I.R. 1975. An analysis of the mechanisms of mating call discrimination in the frogs *Hyla regilla* and *H. cadaverina*. Copeia 1975:415–424.

Straw, R.M. 1958. Experimental notes on the Deep Springs toad *Bufo exsul*. Ecology 39:552–553.

Strecker, J.K. Jr. 1908. Notes on the life history of *Scaphiopus couchii* Baird. Proceedings of the Biological Society of Washington 21:199–206.

Strecker, J.K. Jr. 1909. Notes on the narrow-mouthed toads (*Engystoma*) and the description of a new species from southeastern Texas. Proceedings of the Biological Society of Washington 22:115–120.

Strecker, J.K. Jr. 1910a. Description of a new solitary spadefoot (*Scaphiopus hurterii*) from Texas, with other herpetological notes. Proceedings of the Biological Society of Washington 23:115–122.

Strecker, J.K. Jr. 1910b. Studies in North American batrachology. Notes on the robber frog (*Lithodytes latrans* Cope). Transactions of the Academy of Science of St. Louis 19:73–82.

Strecker, J.K. Jr. 1926. Chapters from the life-histories of Texas reptiles and amphibians: part one. Contributions from the Baylor University Museum 8:1–12.

Stroud, C.P. 1949. A white spade-foot toad from the New Mexico white sands. Copeia 1949:232.

Stuart, J.N. 1991. Partial albinism in a New Mexico population of *Bufo woodhousei*. Bulletin of the Maryland Herpetological Society 27:33.

Stuart, J.N. 1995. *Rana catesbeiana* (Bullfrog). Diet. Herpetological Review 26:33.

Stuart, J.N., and C.W. Painter. 1993. *Rana catesbeiana* (Bullfrog). Cannibalism. Herpetological Review 24:103.

Stuart, J.N., and C.W. Painter. 1994. A review of the distribution and status of the boreal toad, *Bufo boreas boreas*, in New Mexico. Bulletin of the Chicago Herpetological Society 29:113–116.

Stuart, S., J.S. Chanson, N.A. Cox, B.E. Young, A.S.L. Rodrigues, D.L. Fishman, and R.W. Waller. 2004. Status and trends of amphibian declines and extinctions worldwide. Science 306:1783–1786.

Stumpel, A.H.P. 1992. Successful reproduction of introduced bullfrogs *Rana catesbeiana* in northwestern Europe: a potential threat to indigenous amphibians. Biological Conservation 60:61–62.

Suhre, D.O. 2010. Dispersal and demography of the American bullfrog (*Rana catesbeiana*) in a semi-arid grassland. M.S. thesis, University of Arizona, Tucson.

Sullivan, B.K. 1982a. Sexual selection in Woodhouse's toad (*Bufo woodhousei*). 1. Chorus organization. Animal Behaviour 30:680–686.

Sullivan, B.K. 1982b. Male mating behavior in the Great Plains toad (*Bufo cognatus*). Animal Behaviour 30:939–940.

Sullivan, B.K. 1983. Sexual selection in Woodhouse's toad (*Bufo woodhousei*). 2. Female choice. Animal Behaviour 31:1011–1017.

Sullivan, B.K. 1984. Advertisement call variation and observations on breeding behavior of *Bufo debilis* and *B. punctatus*. Journal of Herpetology 18:406–411.

Sullivan, B.K. 1985a. Sexual selection and mating system variation in anuran amphibians of the Arizona-Sonora Desert. Great Basin Naturalist 45:688–696.

Sullivan, B.K. 1985b. Male calling behavior in response to playback of conspecific advertisement call in two bufonids. Journal of Herpetology 19:78–83.

Sullivan, B.K. 1986a. Hybridization between the toads *Bufo microscaphus* and *Bufo woodhousei* in Arizona: morphological variation. Journal of Herpetology 20:11–21.

Sullivan, B.K. 1986b. Intra-population variation in the intensity of sexual selection in breeding aggregations of Woodhouse's toad (*Bufo woodhousei*). Journal of Herpetology 20:88–90.

Sullivan, B.K. 1986c. Advertisement call variation in the Arizona tree frog, *Hyla wrightorum* Taylor, 1938. Great Basin Naturalist 46:378–381.

Sullivan, B.K. 1987. Sexual selection in Woodhouse's toad (*Bufo woodhousei*). 3. Seasonal variation in male mating success. Animal Behaviour 35:912–919.

Sullivan, B.K. 1989. Desert environments and the structure of anuran mating systems. Journal of Arid Environments 17:175–183.

Sullivan, B.K. 1992a. Sexual selection and calling behavior in the American toad (*Bufo americanus*). Copeia 1992:1–7.

Sullivan, B.K. 1992b. Calling behavior of the southwestern toad (*Bufo microscaphus*). Herpetologica 48:383–389.

Sullivan, B.K. 1993. Distribution of the southwestern toad (*Bufo microscaphus*) in Arizona. Great Basin Naturalist 53:402–406.

Sullivan, B.K. 1995. Temporal stability in hybridization between *Bufo microscaphus* and *Bufo woodhousii* (Anura: Bufonidae): behavior and morphology. Journal of Evolutionary Biology 8:233–247.

Sullivan, B.K., and E.A. Sullivan. 1985. Size-related variation in advertisement calls and breeding behavior of spadefoot toads (*Scaphiopus bombifrons*, *S. couchi* and *S. multiplicatus*). Southwestern Naturalist 30:349–355.

Sullivan, B.K., and G.E. Walsberg. 1985. Call rate and aerobic capacity in Woodhouse's toad (*Bufo woodhousei*). Herpetologica 41:404–407.

Sullivan, B.K., and M.K. Leek. 1986. Acoustic communication in Woodhouse's toad (*Bufo woodhousei*). 1. Response of calling males to variation in spectral and temporal components of advertisement calls. Behaviour 98:305–319.

Sullivan, B.K., and T. Lamb. 1988. Hybridization between the toads *Bufo microscaphus* and *Bufo woodhousii* in Arizona: variation in release calls and allozymes. Herpetologica 44:325–333.

Sullivan, B.K., and S.H. Hinshaw. 1990. Variation in advertisement calls and male calling behavior in the spring peeper (*Pseudacris crucifer*). Copeia 1990:1146–1150.

Sullivan, B.K., and S.H. Hinshaw. 1992. Female choice and selection on male calling behaviour in the grey treefrog *Hyla versicolor*. Animal Behaviour 44:733–744.

Sullivan, B.K., and K.B. Malmos. 1994. Call variation in the Colorado River toad (*Bufo alvarius*): behavioral and phylogenetic implications. Herpetologica 50:146–156.

Sullivan, B.K., and P.J. Fernandez. 1999. Breeding activity, estimated age-structure, and growth in Sonoran Desert anurans. Herpetologica 55:334–343.

Sullivan, B.K., K.B. Malmos, and M.F. Given. 1996a. Systematics of the *Bufo woodhousii* complex (Anura: Bufonidae): advertisement call variation. Copeia 1996:274–280.

Sullivan, B.K., R.W. Bowker, K.B. Malmos, and E.W.A. Gergus. 1996b. Arizona distribution of three Sonoran Desert anurans: *Bufo retiformis*, *Gastrophryne olivacea*, and *Pternohyla fodiens*. Great Basin Naturalist 56:38–47.

Sullivan, B.K., K.B. Malmos, E.W.A. Gergus, and R.W. Bowker. 2000. Evolutionary implications of advertisement call variation in *Bufo debilis*, *B. punctatus*, and *B. retiformis*. Journal of Herpetology 34:368–374.

Sullivan, J.J., and E.E. Byrd. 1970. *Choledocystus pennsylvaniensis*: life history. Transactions of the American Microscopical Society 89:384–396.

Sullivan, S.R., P.E. Bartelt, and C.R. Peterson. 2008. Midsummer ground surface activity patterns of western toads (*Bufo boreas*) in southeastern Idaho. Herpetological Review 39:35–40.

Surdick, J.A. Jr. 2005. Amphibian and avian species composition of forested depressional wetlands and circumjacent habitat: the influence of land use type and intensity. Ph.D. diss., University of Florida, Gainesville.

Surface, H.A. 1913. First report on the economic features of the amphibians of Pennsylvania. Zoological Bulletin, Division of Zoology, Pennsylvania Department of Agriculture 3:66–152.

Sutherland, D. 2005. Parasites of North American frogs. Pp. 109–123 *In* M.J. Lannoo (ed.), Amphibian Declines. The Conservation Status of United States Species. University of California Press, Berkeley.

Sutherland, M.A.B., G.M. Gouchie, and R.J. Wassersug. 2009. Can visual stimulation alone induce phenotypically plastic responses in *Rana sylvatica* tadpole oral structures? Journal of Herpetology 43:165–168.

Sutherland, R.W., P.R. Dunning, and W.M. Baker. 2010. Amphibian encounter rates on roads with different amounts of traffic and urbanization. Conservation Biology 24:1626–1635.

Sutton, W.B. 2004. The ecology and natural history of the Northern Leopard Frog, *Rana pipiens* Schreber, in West Virginia. M.S. thesis, Marshall University, Huntington, WV.

Sutton, W.B., K.E. Rastall, and T.K. Pauley. 2006. Diet analysis and feeding strategies of *Rana pipiens* in a West Virginia wetland. Herpetological Review 37:152–153.

Suzuki, H.K. 1951. Recent additions to the records of the distribution of the amphibians in Wisconsin. Wisconsin Academy of Sciences, Arts and Letters 40:215–234.

Suzuki, H.K. 1957. A study of leg length variations in the wood frog, *Rana sylvatica* Le Conte. Wisconsin Academy of Sciences, Arts and Letters 46:299–303.

Svihla, A. 1935. Notes on the western spotted frog, *Rana pretiosa pretiosa*. Copeia 1935:119–122.

Svihla, A. 1936. *Rana rugosa* Schelegel. Notes on the life history of this interesting frog. Mid-Pacific Magazine (April-June):124–125.

Svihla, A. 1953. Diurnal retreats of the spadefoot toad *Scaphiopus hammondii*. Copeia 1953:186.

Swann, D. 2005. Rock star: canyon treefrog (*Hyla arenicolor* Cope, 1866). Sonoran Herpetologist 18(4):39–42.

Swannack, T.M., and M.R.J. Forstner. 2007. Possible cause for the sex-ratio disparity of the endangered Houston Toad (*Bufo houstonensis*). Southwestern Naturalist 52:386–392.

Swannack, T.M., J.T. Jackson, and M.R.J. Forstner. 2006. *Bufo houstonensis* (Houston Toad). Juvenile dispersal. Herpetological Review 37:199–200.

Swanack [sic], T.M., W.E. Grant, and M.R.J. Forstner. 2009. Projecting population trends of endangered amphibian species in the face of uncertainty: a pattern-oriented approach. Ecological Modelling 220:148–159.

Swanson, D.L., and B.M. Graves. 1995. Supercooling and freeze intolerance in overwintering juvenile spadefoot toads (*Scaphiopus bombifrons*). Journal of Herpetology 29:280–285.

Swanson, D.L., and S.L. Burdick. 2010. Overwintering physiology and hibernacula microclimates of Blanchard's cricket frogs at their northwestern range boundary. Copeia 2010:247–253.

Swanson, D.L., B.M. Graves, and K.L. Koster. 1996. Freezing tolerance/intolerance and cryoprotectant synthesis in terrestrially overwintering anurans in the Great Plains, USA. Journal of Comparative Physiology 166B:110–119.

Swanson, P.L. 1939. Herpetological notes from Indiana. American Midland Naturalist 22:684–695.

Swarth, H.S. 1936. Origins of the fauna of the Sitkan District, Alaska. Proceedings of the California Academy of Sciences 23:59–78.

Sweet, S.S., and B.K. Sullivan. 2005. *Bufo californicus* Camp, 1915. Arroyo Toad. Pp. 396–400 *In* M.J. Lannoo (ed.), Amphibian Declines. The Conservation Status of United States Species. University of California Press, Berkeley.

Sweetman, H.L. 1944. Food habits and molting of the common tree frog. American Midland Naturalist 32:499–501.

Sype, W.E. 1975. Breeding habits, embryonic thermal requirements and embryonic and larval development of the Cascade frog, *Rana cascadae* (Slater). Ph.D. diss., Oregon State University, Corvallis.

Szafoni, R.E., C.A. Phillips, and M. Redmer. 1999. Translocations of amphibian species outside their native range: a comment on Thurow (1994–1997). Transactions of the Illinois State Academy of Science 92:277–283.

Szuroczki, D., and J.M.L. Richardson. 2011. Palatability of the larvae of three species of *Lithobates*. Herpetologica 67:213–221.

Taigen, T.L., and F.H. Pough. 1981. Activity metabolism of the toad (*Bufo americanus*): ecological consequences of ontogenetic change. Journal of Comparative Physiology 144:247–252.

Taigen, T.L., and K.D. Wells. 1985. Energetics of vocalization by an anuran amphibian (*Hyla versicolor*). Journal of Comparative Physiology 155B:163–170.

Taigen, T.L., S.B. Emerson, and F.H. Pough. 1982. Ecological correlates of anuran exercise physiology. Oecologia 52: 49–56.

Tamsitt, J.R. 1962. Notes on a population of the Manitoba toad (*Bufo hemiophrys*) in the Delta Marsh region of Lake Manitoba, Canada. Ecology 43:147–150.

Tanner, V.M. 1931. A synoptical study of Utah amphibia. Utah Academy of Sciences 8:159–198.

Tanner, V.M. 1939. A study of the genus *Scaphiopus*, the spade-foot toads. Great Basin Naturalist 1:3–20.

Tanner, W.W. 1941. The reptiles and amphibians of Idaho No. 1. Great Basin Naturalist 2:87–97.

Tanner, W.W. 1950. Notes on the habits of *Microhyla carolinensis olivacea* (Hallowell). Herpetologica 6:47–48.

Tanner, W.W. 1989. Status of *Spea stagnalis* Cope (1875), *Spea intermontanus* Cope (1889), and a systematic review of *Spea hammondii* Baird (1839) (Amphibia: Anura). Great Basin Naturalist 49:503–510.

Tardell, J.H., R.C. Yates, and D.H. Schiller. 1981. New records and habitat observations of *Hyla andersoni* Baird (Anura: Hylidae) in Chesterfield and Marlboro counties, South Carolina. Brimleyana 6:153–158.

Tatarian, P.J. 2008. Movement patterns of California red-legged frogs (*Rana draytonii*) in an inland California environment. Herpetological Conservation and Biology 3:155–169.

Tatarian, P., and G. Tatarian. 2010. Chytrid infection of *Rana draytonii* in the Sierra Nevada, California, USA. Herpetological Review 41:325–327.

Tavares, K. 2008. Coqui control, monitoring and outreach program. 2008 Annual Report. Hawaii Volcanoes National Park. http://www.hawaiisfishes.com/coqui/2008%20HAVO%20Report/2008_Coqui_Season-lo-res.pdf.

Tavera-Mendoza, L., S. Ruby, P. Brousseau, M. Fournier, D. Cyr, and D. Marcogliese. 2002. Response of the amphibian tadpole (*Xenopus laevis*) to atrazine during sexual differentiation of the testis. Environmental Toxicology and Chemistry 21:527–531.

Taylor, C.L., R. Altig, and C.R. Boyle. 1995. Can anuran tadpoles choose among foods that vary in quality? Alytes 13:81–86.

Taylor, C.N., K.L. Oseen, and R.J. Wassersug. 2004. On the behavioral response of *Rana* and *Bufo* tadpoles to echinostomatoid cercariae: implications to synergistic factors influencing trematode infections in anurans. Canadian Journal of Zoology 82:701–706.

Taylor, D.H., and D.E. Ferguson. 1970. Extraoptic celestial orientation in the southern cricket frog *Acris gryllus*. Science 168:390–392.

Taylor, E.H. 1929. List of reptiles and batrachians of Morton County, Kansas, reporting species new to the state fauna. University of Kansas Science Bulletin 19:63–65.

Taylor, E.H. 1938. Frogs of the *Hyla eximia* group in Mexico, with descriptions of two new species. University of Kansas Science Bulletin 25:421–445.

Taylor, E.H. 1940. Herpetological miscellany. University of Kansas Science Bulletin 26:489–571.

Taylor, E.H., and J.S. Wright. 1932. The toad *Bufo marinus* (Linnaeus) in Texas. University of Kansas Science Bulletin 20:247–249.

Taylor, J. 1993. The Amphibians & Reptiles of New

Hampshire. New Hampshire Fish and Game Department, Concord.

Taylor, P. 2009. An extension of gray treefrog range in Manitoba and into Saskatchewan. Blue Jay 67: 235–241.

Taylor, R.J., and E.D. Michael. 1971. Habitat effects on monthly foods of bullfrogs in eastern Texas. Proceedings of the Annual Conference of the Southeast Association of Game and Fish Commissioners 25:176–186.

Taylor, S.K. 1998. Investigation of mortality of Wyoming toads and the effect of malathion on amphibian disease susceptibility. Ph.D. diss., University of Wyoming, Laramie.

Taylor, S.K., E.S. Williams, and K.W. Mills. 1999a. Effects of malathion on disease susceptibility in Woodhouse's toads. Journal of Wildlife Diseases 35:536–541.

Taylor, S.K., E.S. Williams, and K.W. Mills. 1999b. Mortality of captive Canadian toads from *Basidiobolus ranarum* mycotic dermatitis. Journal of Wildlife Diseases 35:64–69.

Taylor, S.K., E.S. Williams, E.T. Thorne, K.W. Mills, D.I. Withers, and A.C. Pier. 1999c. Causes of mortality of the Wyoming toad. Journal of Wildlife Diseases 35:49–57.

Taylor, S.K., E.S. Williams, A.C. Pier, K.W. Mills, and M.D. Bock. 1999d. Mucormycotic dermatitis in captive adult Wyoming toads. Journal of Wildlife Diseases 35:70–74.

Taylor, S.K., E.S. Williams, and K.W. Mills. 1999e. Experimental exposure of Canadian toads to *Basidiobolus ranarum*. Journal of Wildlife Diseases 35:58–63.

Telford, S.R. Jr. 1952. A herpetological survey in the vicinity of Lake Shipp, Polk County, Florida. Quarterly Journal of the Florida Academy of Science 15:175–185.

Temminck, C.J., and H. Schlegel. 1838. *In* P.F. Von Siebold (ed.), Fauna Japonica sive Descriptio animalium, quae in itinere per Japonianum, jussu et auspiciis superiorum, qui summum in India Batava Imperium tenent, suscepto, annis 1823–1830 colleget, notis observationibus et adumbrationibus illustratis. Vol. 3 (Chelonia, Ophidia, Sauria, Batrachia). J.G. Lalau, Leiden, Germany.

Tenneson, M.G. 1983. Behavioral ecology and population decline of the mink frog, *Rana septentrionalis*. M.S. thesis, University of North Dakota, Grand Forks.

Test, F.C. 1898. A contribution to the knowledge of the variations of the tree frog *Hyla regilla*. Proceedings of the United States National Museum 21:477–492.

Test, F.H. 1958. Butler's garter snake eats amphibian. Copeia 1958:151–152.

Test, F.H., and R.C. McCann. 1976. Foraging behavior of *Bufo americanus* tadpoles in response to high densities of micro-organisms. Copeia 1976: 576–578.

Tester, J.R., and W.J. Breckenridge. 1964a. Population dynamics of the Manitoba toad, *Bufo hemiophrys*, in northwestern Minnesota. Ecology 45:592–601.

Tester, J.R., and W.J. Breckenridge. 1964b. Winter behavior patterns of the Manitoba toad, *Bufo hemiophrys*, in northwestern Minnesota. Annals Academiae Scientiarum Fennicae (Biologica) 71: 421–431.

Tester, J.R., A. Parker, and D.B. Siniff. 1965. Experimental studies on habitat preferences and thermoregulation of *Bufo americanus*, *B. hemiophrys*, and *B. cognatus*. Journal of the Minnesota Academy of Science 33:27–32.

Tevis, L. Jr. 1966. Unsuccessful breeding by desert toads (*Bufo punctatus*) at the limit of their ecological tolerance. Ecology 47:766–775.

Theodorakis, C.W., J. Rinchard, J.A. Carr, J.-W. Park, L. McDaniel, F. Liu, and M. Wages. 2006. Thyroid endocrine disruption in stonerollers and cricket frogs from perchlorate-contaminated streams in east-central Texas. Ecotoxicology 15:31–50.

Thomas, L.A., and J. Allen. 1997. *Bufo houstonensis* (Houston Toad). Behavior. Herpetological Review 28:40–41.

Thomas, R. 1966. New species of Antillean *Eleutherodactylus*. Quarterly Journal of the Florida Academy of Sciences 28:375–391.

Thomas, R.A., S.A. Nadler, and W.L. Jagers. 1984. Helminth parasites of the endangered Houston toad, *Bufo houstonensis* Sanders, 1953 (Amphibia, Bufonidae). Journal of Parasitology 70:1012–1013.

Thompson, C. 1912. The status of *Rana palustris* LeConte in Michigan. Fourteenth Report of the Michigan Academy of Science, p. 190.

Thompson, C. 1915. Notes on the habits of *Rana areolata* Baird and Girard. Occasional Papers of the Museum of Zoology University of Michigan 9:1–7 + 3 plates.

Thompson, D.G., B.F. Wojtaszek, B. Staznik, D.T. Chartrand, and G.R. Stephenson. 2004. Chemical and biomonitoring to assess potential acute effects of Vision™ herbicide on native amphibian larvae in forest wetlands. Environmental Toxicology and Chemistry 23:843–849.

Thompson, E.F. Jr., and B.S. Martof. 1957. A comparison of the physical characteristics of frog calls (*Pseudacris*). Physiological Zoology 30:328–341.

Thompson, H.B. 1913. Description of a new subspecies of *Rana pretiosa* from Nevada. Proceedings of the Biological Society of Washington 26:53–56.

Thompson, P.D. 2004. Observations of boreal toad (*Bufo boreas*) breeding populations in northwestern Utah. Herpetological Review 35:342–344.

Thompson, P.D., R.A. Fridell, K.K. Wheeler, and C.L. Bailey. 2004. Distribution of *Bufo boreas* in Utah. Herpetological Review 35:255–257.

Thompson, Z. 1842. History of Vermont, Natural, Civil and Statistical, in Three Parts with a New Map of the State, and 200 Engravings. Chauncey Goodrich, Burlington.

Thornton, W.A. 1955. Interspecific hybridization in *Bufo woodhousei* and *Bufo valliceps*. Evolution 9:455–468.

Thornton, W.A. 1960. Population dynamics in *Bufo woodhousei* and *Bufo valliceps*. Texas Journal of Science 12:176–200.

Thorson, T.B. 1955. The relationship of water economy to terrestrialism in amphibians. Ecology 36:100–116.

Thorson, T.B., and A. Svihla. 1943. Correlation of habitats of amphibians with their ability to survive loss of body water. Ecology 24:374–381.

Thrall, J. 1971. Excavation of pits by juvenile *Rana catesbeiana*. Copeia 1971:751–752.

Thrall, J. 1972. Food, feeding, and digestive physiology of the larval bullfrog, *Rana catesbeiana* Shaw. Ph.D. diss., Illinois State University, Normal.

Thurgate, N.Y., and J.H.K. Pechmann. 2007. Canopy closure, competition, and the endangered dusky gopher frog. Journal of Wildlife Management 71:1845–1852.

Thurow, G.R. 1994. Experimental return of wood frogs to west-central Illinois. Transactions of the Illinois State Academy of Science 87:83–97.

Thurston, R.V., T.A. Gilfoil, E.L. Meyn, R.K. Zajdel, T.I. Aoki, and G.D. Veith. 1985. Comparative toxicity of 10 organic chemicals to 10 common aquatic species. Water Research 19:1145–1155.

Tierney, D., and M.M. Stewart. 2001. Geographic distribution: *Scaphiopus holbrookii holbrookii* (Eastern Spadefoot). Herpetological Review 32:56.

Tietge, J.E., S.A. Diamond, G.T. Ankley, D.L. DeFoe, G.W. Holcombe, K.M. Jensen, S.J. Degitz, G.E. Elonen, and E. Hammer. 2001. Ambient solar UV radiation causes mortality in larvae of three species of *Rana* under controlled exposure conditions. Photochemistry and Photobiology 74:261–268.

Tihen, J.A. 1937. Additional distributional records of amphibians and reptiles in Kansas counties. Transactions of the Kansas Academy of Science 40:401–409.

Tihen, J.A. 1962a. Osteological observations on New World *Bufo*. American Midland Naturalist 67:157–183.

Tihen, J.A. 1962b. A review of New World fossil bufonids. American Midland Naturalist 68:1–50.

Timken, R.L., and D.G. Dunlap. 1965. Ecological distribution of two species of *Bufo* in southeastern South Dakota. Proceedings of the South Dakota Academy of Science 44:113–117.

Timm, B.C., and K. McGarigal. 2010a. The diets of subadult Fowler's toads (*Bufo fowleri*) and eastern spadefoot toads (*Scaphiopus h. holbrookii*) at Cape Cod National Seashore. Herpetological Review 41:154–156.

Timm, B.C., and K. McGarigal. 2010b. *Scaphiopus holbrookii* (Eastern Spadefoot). Predation. Herpetological Review 41:207.

Timm, B.C., K. McGarigal, and L.R. Gamble. 2007a. Emigration timing of juvenile pond-breeding amphibians in western Massachusetts. Journal of Herpetology 41:243–250.

Timm, B.C., K. McGarigal, and B.W. Compton. 2007b. Timing of large movement events of pond-breeding amphibians in western Massachusetts, USA. Biological Conservation 136:442–454.

Timm, B.C., K. McGarigal, and C.L. Jenkins. 2007c. Emigration orientation of juvenile pond-breeding amphibians in western Massachusetts. Copeia 2007:685–698.

Timoney, K.P. 1996. Canadian toads near their northern limit in Canada: observations and recommendations. Alberta Naturalist 26:49–50.

Timpe, E.K., S.P. Graham, R.W. Gagliardo, R.T. Hill, and M.G. Levy. 2008. Occurrence of the fungal pathogen *Batrachochytrium dendrobatidis* in Georgia's amphibian populations. Herpetological Review 39:447–449.

Ting, H.-P. 1951. Duration of the tadpole stage of the greenfrog, *Rana clamitans*. Copeia 1951:82.

Tinkham, E.R. 1962. Notes on the occurrence of *Scaphiopus couchii* in California. Herpetologica 18:204.

Tinkle, D.W. 1959. Observations of reptiles and amphibians in a Louisiana swamp. American Midland Naturalist 62:189–205.

Tinsley, R.C. 1990. The influence of parasite infection on mating success in spadefoot toads, *Scaphiopus couchii*. American Zoologist 30:313–324.

Tinsley, R.C., and K. Tocque. 1995. The population dynamics of a desert anuran, *Scaphiopus couchii*. Australian Journal of Ecology 20:376–384.

Tinsley, R.C., and H.R. Kobel (eds.). 1996. The Biology of *Xenopus*. Oxford Science Publications, Oxford.

Tobey, F.J. 1985. Virginia's Amphibians and Reptiles. A Distributional Survey. Virginia Herpetological Society, Purcellville.

Tobias, M.L., C. Barnard, R. O'Hagan, S.H. Horng, M. Rand, and D.B. Kelley. 2004. Vocal communication between male *Xenopus laevis*. Animal Behaviour 67:353–365.

Tocque, K. 1993. The relationship between parasite burden and host resources in the desert toad (*Scaphiopus couchii*), under natural environmental conditions. Journal of Animal Ecology 62:683–693.

Tocque, K., and R.C. Tinsley. 1994. The relationship between *Pseudodiplorchis americanus* (Monogenea) density and host resources under controlled environmental conditions. Parasitology 108:175–183.

Tocque, K., R. Tinsley, and T. Lamb. 1995. Ecological constraints on feeding and growth of *Scaphiopus couchii*. Herpetological Journal 5:257–265.

Todd, B.D., and B.B. Rothermel. 2006. Assessing the quality of clearcut habitats for amphibians: effects on abundances versus vital rates in the Southern toad (*Bufo terrestris*). Biological Conservation 133:178–185.

Todd, B.D., and C.T. Winne. 2006. Ontogenetic and interspecific variation in timing of movement and responses to climatic factors during migrations by pond-breeding amphibians. Canadian Journal of Zoology 84:715–722.

Todd, M.J., R.R. Cocklin, and M.E. Dorcas. 2003. Temporal and spatial variation in anuran calling activity in the western Piedmont of North Carolina. Journal of the North Carolina Academy of Science 119:103–110.

Todd-Thompson, M., D.L. Miller, P.E. Super, and M.J. Gray. 2009. Chytridiomycosis-associated mortality in a *Rana palustris* collected in Great Smoky Mountains National Park, Tennessee, USA. Herpetological Review 40:321–323.

Tomko, D.S. 1975. The reptiles and amphibians of the Grand Canyon. Plateau 47:161–166.

Tordoff, W. III. 1980. Selective predation of gray jays, *Perisoreus canadensis*, upon boreal chorus frogs, *Pseudacris triseriata*. Evolution 34:1004–1008.

Tordoff, W. III., and D. Pettus. 1977. Temporal stability of phenotypic frequencies in *Pseudacris triseriata* (Amphibia, Anura, Hylidae). Journal of Herpetology 11:161–168.

Tordoff, W. III, D. Pettus, and T.C. Matthews. 1976. Microgeographic variation in gene frequencies in *Pseudacris triseriata* (Amphibia, Anura, Hylidae). Journal of Herpetology 10:35–40.

Touchon, J.C., I. Gomez-Mestre, and K.M. Warkentin. 2006. Hatching plasticity in two temperate anurans: responses to a pathogen and predation cues. Canadian Journal of Zoology 84:556–563.

Touré, T.A., and G.A. Middendorf. 2002. Colonization of herpetofauna to a created wetland. Bulletin of the Maryland Herpetological Society 38:99–117.

Townsend, D.S., and M.M. Stewart. 1994. Reproductive ecology of the Puerto Rican frog *Eleutherodactylus coqui*. Journal of Herpetology 28:34–40.

Tracy, C.R., K.A. Christian, M.P. O'Connor, and C.R. Tracy. 1993. Behavioral thermoregulation by *Bufo americanus*: the importance of the hydric environment. Herpetologica 49:375–382.

Trapido, H. 1947. Range extension of *Hyla septentrionalis* in Florida. Herpetologica 3:190.

Trapido, H., and R.T. Clausen. 1938. Amphibians and reptiles of eastern Quebec. Copeia 1938:117–125.

Trauth, J.B., S.E. Trauth, and R.L. Johnson. 2006. Best management practices and drought combine to silence the Illinois chorus frog in Arkansas. Wildlife Society Bulletin 34:514–518.

Trauth, J.B., R.L. Johnson, and S.E. Trauth. 2007. Conservation implications of a morphometric comparison between the Illinois chorus frog (*Pseudacris streckeri illinoensis*) and Strecker's chorus frog (*P. s. streckeri*) (Anura: Hylidae) from Arkansas, Illinois, Missouri, Oklahoma, and Texas. Zootaxa 1589:23–32.

Trauth, S.E. 1989. Female reproductive traits of the southern leopard frog, *Rana sphenocephala* (Anura: Ranidae), from Arkansas. Proceedings of the Arkansas Academy of Science 43:105–108.

Trauth, S.E. 1992. Distributional survey of the bird-voiced treefrog, *Hyla avivoca* (Anura: Hylidae), in Arkansas. Proceedings of the Arkansas Academy of Science 46:80–82.

Trauth, S.E., and C.T. McAllister. 1983. Geographic Distribution. *Rana clamitans melanota*. Herpetological Review 14:83.

Trauth, S.E., and J.W. Robinette. 1990. Notes on distribution, mating activity, and reproduction in the bird-voiced treefrog, *Hyla avivoca*, in Arkansas. Bulletin of the Chicago Herpetological Society 25:218–219.

Trauth, S.E., and R.G. Neal. 2004. Geographic range extension and feeding response by the leech *Macrobdella diplotertia* (Annelida: Hirudinea) to wood frog and spotted salamander egg masses. Journal of the Arkansas Academy of Science 58:139–141.

Trauth, S.E., M.E. Cartwright, and W.E. Meshaka. 1989. Reproduction in the wood frog, *Rana sylvatica* (Anura: Ranidae), from Arkansas. Proceedings of the Arkansas Academy of Science 43:114–116.

Trauth, S.E., R.L. Cox, B. Butterfield, D.A. Saugey, and W.E. Meshaka. 1990. Reproductive phenophases and clutch characteristics of selected Arkansas amphibians. Proceedings of the Arkansas Academy of Science 44:107–113.

Trauth, S.E., A. Holt, J. Davis, and P. Daniel. 1992. A new state record for the plains leopard frog, *Rana blairi*, in Arkansas. Bulletin of the Chicago Herpetological Society 27:255.

Trauth, S.E., W.E. Meshaka Jr., and R.L. Cox. 1999. Post-metamorphic growth and reproduction in the eastern narrow-mouthed toad (*Gastrophryne carolinensis*) from northeastern Arkansas. Journal of the Arkansas Academy of Science 53:120–124.

Trauth, S.E., M.L. McCallum, and M.E. Cartwright. 2000. Breeding mortality in the wood frog, *Rana sylvatica* (Anura: Ranidae), from northcentral Arkansas. Journal of the Arkansas Academy of Science 54:154–156.

Trauth, S.E., H.W. Robison, and M.V. Plummer. 2004. The Amphibians and Reptiles of Arkansas. University of Arkansas Press, Fayetteville.

Travis, J. 1980. Phenotypic variation and the outcome of interspecific competition in hylid tadpoles. Evolution 34:40–50.

Travis, J. 1981. Control of larval growth variation in

a population of *Pseudacris triseriata* (Anura: Hylidae). Evolution 35:423–432.

Travis, J. 1983. Variation in development patterns of larval anurans in temporary ponds. 1. Persistent variation within a *Hyla gratiosa* population. Evolution 37:496–512.

Travis, J. 1984. Anuran size at metamorphosis: experimental test of a model based on intraspecific competition. Ecology 65:1155–1160.

Travis, J., and J.C. Trexler. 1986. Interactions among factors affecting growth, development and survival in experimental populations of *Bufo terrestris* (Anura: Bufonidae). Oecologia 69:110–116.

Travis, J., W.H. Keen and J. Juilianna. 1985a. The role of relative body size in a predator-prey relationship between dragonfly naiads and larval anurans. Oikos 45:59–65.

Travis, J., W.H. Keen, and J. Juilianna. 1985b. The effects of multiple factors on viability selection in *Hyla gratiosa* tadpoles. Evolution 39:1087–1099.

Travis, J., S.B. Emerson, and M. Blouin. 1987. A quantitative-genetic analysis of larval life-history traits in *Hyla crucifer*. Evolution 41:145–156.

Treanor, R.R., and S.J. Nicola. 1972. A preliminary study of the commercial and sporting utilization of the bullfrog, *R. catesbeiana* Shaw in California. California Department of Fish and Game, Inland Fisheries Administrative Report 72–4.

Treat, D.A. 1948. Frogs and toads. Audubon Nature Bulletin, Series 18, No. 8.

Trowbridge, A.H., and H.M. Hefley. 1934. Preliminary studies on the parasite fauna of Oklahoma anurans. Proceedings of the Oklahoma Academy of Science 14:16–19.

Trowbridge, A.H., and M.S. Trowbridge. 1937. Notes on the cleavage rate of *Scaphiopus bombifrons* Cope, with additional remarks on certain aspects of its life history. American Naturalist 71:460–480.

Trowbridge, M.S. 1941. Studies on the normal development of *Scaphiopus bombifrons* Cope. 1. The cleavage period. Transactions of the American Microscopical Society 60:508–526.

Trueb, L., and C. Gans. 1983. Feeding specializations of the Mexican burrowing toad, *Rhinophrynus dorsalis* (Anura: Rhinophrynidae). Journal of Zoology 199:189–208.

Tucker, J.K. 1977. Notes on the food habits of Kirtland's water snake, *Clonophis kirtlandi*. Bulletin of the Maryland Herpetological Society 13:193–195.

Tucker, J.K. 1995. Early post-transformational growth in the Illinois chorus frog (*Pseudacris streckeri illinoensis*). Journal of Herpetology 29:314–316.

Tucker, J.K. 1997. Food habits of the fossorial frog *Pseudacris streckeri illinoensis*. Herpetological Natural History 5:83–87.

Tucker, J.K. 1998. Status of Illinois chorus frogs in Madison County, Illinois. Pp. 94–101 In M.J. Lannoo (ed.), Status & Conservation of Midwestern Amphibians. University of Iowa Press, Iowa City.

Tucker, J.K. 2000. Growth and survivorship in the Illinois chorus frog (*Pseudacris streckeri illinoensis*). Transactions of the Illinois State Academy of Science 93:63–68.

Tucker, J.K., and M.E. Sullivan. 1975. Unsuccessful attempts by bullfrogs to eat toads. Transactions of the Illinois State Academy of Science 68:167.

Tucker, J.K., D.W. Soergel, and J.B. Hatcher. 1995. Flood-associated activities of some reptiles and amphibians at Carlyle Lake, Fayette County, Illinois. Transactions of the Illinois State Academy of Science 88:73–81.

Tuckerman, F. 1886. Supernumerary leg in a male frog (*Rana palustris*). Journal of Anatomy and Physiology 20:516–519.

Tumlison, R., and S.E. Trauth. 2006. A novel facultative mutualistic relationship between bufonid tadpoles and flagellated green algae. Herpetological Conservation and Biology 1:51–54.

Tupper, T.A., and R.P. Cook. 2008. Habitat variables influencing breeding effort in northern clade *Bufo fowleri*: implications for conservation. Applied Herpetology 5:101–119.

Tupper, T.A., J.W. Streicher, S.E. Greenspan, B.C. Timm, and R.P. Cook. 2011. Detection of *Batrachochytrium dendrobatidis* in anurans of Cape Cod National Seashore, Barnstable County, Massachusetts. Herpetological Review 42:62–65.

Turner, F.B. 1952. Peculiar aggregations of toadlets on Alum Creek. Yellowstone Nature Notes 26(5):57–58.

Turner, F.B. 1955. Reptiles and amphibians of Yellowstone National Park. Yellowstone Interpretive Series No. 5, Yellowstone National Park, WY.

Turner, F.B. 1958a. Some parasites of the western spotted frog, *Rana p. pretiosa*, in Yellowstone National Park. Journal of Parasitology 44:182.

Turner, F.B. 1958b. Life-history of the western spotted frog in Yellowstone National Park. Herpetologica 14:96–100.

Turner, F.B. 1959a. Pigmentation of the western spotted frog, *Rana pretiosa pretiosa*, in Yellowstone Park, Wyoming. American Midland Naturalist 61:162–176.

Turner, F.B. 1959b. An analysis of the feeding habits of *Rana p. pretiosa* in Yellowstone Park, Wyoming. American Midland Naturalist 61:403–413.

Turner, F.B. 1959c. Some features of the ecology of *Bufo punctatus* in Death Valley, California. Ecology 40:175–181.

Turner, F.B. 1960a. Size and dispersion of a Louisiana population of the cricket frog, *Acris gryllus*. Ecology 41:258–268.

Turner, F.B. 1960b. Population structure and dynamics of the western spotted frog, *Rana p. pretiosa*

Baird and Girard, in Yellowstone Park, Wyoming. Ecological Monographs 30:251–278.
Turner, F.B. 1960c. Postmetamorphic growth in amphibians. American Midland Naturalist 64:327–338.
Turner, F.B., and R.H. Wauer. 1963. A survey of the herpetofauna of the Death Valley area. Great Basin Naturalist 23:119–128.
Turner, F.B., and P.C. Dumas. 1972. *Rana pretiosa*. Catalogue of American Amphibians and Reptiles 119.1–119.4.
Turner, L.J., and D.K. Fowler. 1981. Utilization of surface mine ponds in east Tennessee by breeding amphibians. U.S. Fish and Wildlife Service, Biological Services Program, FWS/OBS-81/08.
Twedt, B. 1993. A comparative ecology of *Rana aurora* Baird and Girard and *Rana catesbeiana* Shaw at Freshwater Lagoon, Humboldt County, California. M.A. thesis, Humboldt State University, Arcata.
Twitty, V., D. Grant, and O. Anderson. 1967. Amphibian orientation: an unexpected observation. Science 155:352–353.
Tyler, M.S. 1994. Stalking amphibians. Maine Naturalist 2:33–44.
Ubelaker, J.E., D.W. Duszynski, and D.L. Beaver. 1967. Occurrence of the trematode, *Glypthelmins pennsylvaniensis* Cheng, 1961, in chorus frogs, *Pseudacris triseriata*, in Colorado. Bulletin of the Wildlife Disease Association 3:177.
Ugarte, C.A. 2004. Human impacts on pig frog (*Rana grylio*) populations in south Florida wetlands: harvest, water management and mercury contamination. Ph.D. diss., Florida International University, Miami.
Ugarte, C.A., K.G. Rice, and M.A. Donnelly. 2005. Variation of total mercury concentrations in pig frogs (*Rana grylio*) across the Florida Everglades, USA. Science of the Total Environment 345:51–59.
Ulmer, M.J. 1970. Studies on the helminth fauna of Iowa. 1. Trematodes of amphibians. American Midland Naturalist 83:38–64.
Ulmer, M.J., and H.A. James. 1976a. Studies on the helminth fauna of Iowa. 2. Cestodes of amphibians. Proceedings of the Helminthological Society of Washington 43:191–200.
Ulmer, M.J., and H.A. James. 1976b. *Nematotaenoides ranae* gen. et sp. n. (Cyclophyllidea: Nematotaeniidae), from the leopard frog (*Rana pipiens*) in Iowa. Proceedings of the Helminthological Society of Washington 43:185–191.
Ultsch, G.R., S.A. Reese, M. Nie, J.D. Crim, W.H. Smith, and C.M. LeBerte. 1999. Influences of temperature and oxygen upon habitat selection by bullfrog tadpoles and three species of freshwater fishes in two Alabama strip mine ponds. Hydrobiologia 416:149–162.
Ultsch, G.R., T.E. Graham, and C.E. Crocker. 2000. An aggregation of overwintering leopard frogs, *Rana pipiens*, and common map turtles, *Graptemys geographica*, in northern Vermont. Canadian Field-Naturalist 114:314–315.
Ultsch, G.R., S.A. Reese, and E.R. Stewart. 2004. Physiology of hibernation in *Rana pipiens*: metabolic rate, critical oxygen tension, and the effects of hypoxia on several plasma variables. Journal of Experimental Zoology 301A:169–176.
Underhill, J.C. 1961. Intraspecific variation in the Dakota toad, *Bufo hemiophrys*, from northeastern South Dakota. Herpetologica 17:220–227.
Une, Y., A. Sakuma, H. Matsueda, K. Nakai, and M. Murakami. 2009. Ranavirus outbreak in North American bullfrogs (*Rana catesbeiana*), Japan, 2008. Emerging Infectious Diseases 15:1146–1147.
Unrine, J.M., W.A. Hopkins, C.S. Romanek, and B.P. Jackson. 2007. Bioaccumulation of trace elements in omnivorous amphibian larvae: implications for amphibian health and contaminant transport. Environmental Pollution 149:182–192.
Upton, S.J., and C.T. McAllister. 1988. The Coccidia (Apicomplexa: Eimeriidae) of Anura, with descriptions of four new species. Canadian Journal of Zoology 66:1822–1830.
USDI (U.S. Department of Interior). 1965. Commercial possibilities and limitations in frog raising. Fishery Leaflet 436.
USFWS (U.S. Fish and Wildlife Service). 1996. Determination of Threatened status for the California Red Legged Frog. Federal Register 61(101):25813–25833.
USFWS (U.S. Fish and Wildlife Service). 1999. Arroyo southwestern toad (*Bufo microscaphus californicus*) recovery plan. U.S. Fish and Wildlife Service, Portland, OR.
Uyehara, I.K., T. Gamble, and S. Cotner. 2010. The presence of *Ranavirus* in anuran populations at Itasca State Park, Minnesota, USA. Herpetological Review 41:177–179.
Vaala, D.A., G.R. Smith, K.G. Temple, and H.A. Dingfelder. 2004. No effect of nitrate on gray treefrog (*Hyla versicolor*) tadpoles. Applied Herpetology 1:265–269.
Vallinoto, M., F. Sequeira, D. Sodré, J.A.R. Bernardi, I. Sampaio, and H. Schneider. 2010. Phylogeny and biogeography of the *Rhinella marina* species complex (Amphibia, Bufonidae) revisited: implications for Neotropical diversification hypotheses. Zoologica Scripta 39:128–140.
Van Allen, B.G., V.S. Briggs, M.W. McCoy, and J.R. Vonesh. 2010. Carry-over effects of the larval environment on post-metamorphic performance in two hylid frogs. Oecologia 164:891–898.
Van Buskirk, J. 2000. The costs of an inducible defense in anuran larvae. Ecology 81:2813–2821.

Van Buskirk, J., and R.A. Relyea. 1998. Selection for phenotypic plasticity in *Rana sylvatica* tadpoles. Biological Journal of the Linnaean Society 65: 301–328.

Van Buskirk, J., S.A. McCollum, and E.E. Werner. 1997. Natural selection for environmentally induced phenotypes in tadpoles. Evolution 51:1983–1992.

Vance, T., and G. Talpin. 1977. Another yellow albino amphibian. Herpetological Review 8:4–5.

Van Denburgh, J. 1898. Herpetological notes. 1. *Bufo boreas* in Alaska. Proceedings of the American Philosophical Society 37:139.

Van Denburgh, J. 1905. The reptiles and amphibians of the islands of the Pacific Coast of North America from the Farallons to Cape San Lucas and the Revilla Gigedos. Proceedings of the California Academy of Sciences, Series 3, Zoology 4:1–40.

Van Denburgh, J. 1912. Notes on *Ascaphus*, the discoglossid toad of North America. Proceedings of the California Academy of Sciences, Series 4, 3:259–264.

Van Denburgh, J. 1924. Notes on the herpetology of New Mexico, with a list of species known from that state. Proceedings of the California Academy of Sciences, Series 4, 13:189–230.

Van Denburgh, J., and J.R. Slevin. 1914. Reptiles and amphibians of the islands of the West Coast of North America. Proceedings of the California Academy of Sciences, Series 4, 4:129–152.

Van Denburgh, J. and J.R. Slevin. 1915. A list of the amphibians and reptiles of Utah, with notes on the species in the collection of the Academy. Proceedings of the California Academy of Sciences, Series 4, 5:99–110.

Vanderburgh, D.J., and R.C. Anderson. 1987a. The relationship between nematodes of the genus *Cosmocercoides* Wilkie, 1930 (Nematoda: Cosmocercoidea) in toads (*Bufo americanus*) and slugs (*Deroceras laeve*). Canadian Journal of Zoology 65:1650–1661.

Vanderburgh, D.J., and R.C. Anderson. 1987b. Preliminary observations on seasonal changes in prevalence and intensity of *Cosmocercoides variabilis* (Nematoda: Cosmocercoidea) in *Bufo americanus* (Amphibia). Canadian Journal of Zoology 65:1666–1667.

Van Devender, T.R. 1969. A record of albinism in the canyon tree frog, *Hyla arenicolor* Cope. Herpetologica 25:69.

Vandewalle, T.J., K.K. Sutton, and J.L. Christiansen. 1996. *Pseudacris crucifer*: an Iowa case history study of an amphibian call survey. Herpetological Review 27:183–185.

Van Hyning, O.C. 1933. Batrachia and reptilia of Alachua County, Florida. Copeia 1933:3–7.

Varhegyi, G., S.M. Mavroidis, B.M. Walton, C.A. Conaway, and A.R. Gibson. 1998. Amphibian surveys in the Cuyahoga Valley National Recreation Area. Pp.137–154 *In* M.J. Lannoo (ed.), Status & Conservation of Midwestern Amphibians. University of Iowa Press, Iowa City.

Vasconcelos, D., and A.J.K. Calhoun. 2004. Movement patterns of adult and juvenile *Rana sylvatica* (LeConte) and *Ambystoma maculatum* (Shaw) in three restored seasonal pools in Maine. Journal of Herpetology 38:551–561.

Vatnick, I., M.A. Brodkin, M.P. Simon, B.W. Grant, C.R. Conte, M. Gleave, R. Myers, and M.M. Sadoff. 1999. The effects of exposure to mild acidic conditions on adult frogs (*Rana pipiens* and *Rana clamitans*): mortality rates and pH preferences. Journal of Herpetology 33:370–374.

Velo-Antón, G., P.A. Burrowes, R.L. Joglar, I. Martínez-Solano, K.H. Beard, and G. Parra-Olea. 2007. Phylogenetic study of *Eleutherodactylus coqui* (Anura: Leptodactylidae) reveals deep genetic fragmentation in Puerto Rico and pinpoints origins of Hawaiian populations. Molecular Phylogenetics and Evolution 45:716–728.

Venesky, M.D., R.J. Wassersug, and M.J. Parris. 2010. The impact of variation in labial tooth number on the feeding kinematics of tadpoles of southern leopard frog (*Lithobates sphenocephalus*). Copeia 2010:481–486.

Vergeer, T. 1948. Frog catches mouse in natural environment. Turtox News 26:91.

Vertucci, F.A., and P.S. Corn. 1996. Evaluation of episodic acidification and amphibian declines in the Rocky Mountains. Ecological Applications 6:449–457.

Vestal, E.H. 1941. Defensive inflation of the body in *Bufo boreas halophilus*. Copeia 1941:183.

Veysey, J.S., S.D. Mattfeldt, and K.J. Babbitt. 2011. Comparative influence of isolation, landscape, and wetland characteristics on egg-mass abundance of two pool-breeding amphibian species. Landscape Ecology 26:661–672.

Vickers, C.R., L.D. Harris, and B.F. Swindel. 1985. Changes in herpetofauna resulting from ditching of cypress ponds in Coastal Plains flatwoods. Forest Ecology and Management 11:17–29.

Viets, B.E. 1993. An annotated list of the herpetofauna of the F.B. and Rena G. Ross Natural History Reservation. Transactions of the Kansas Academy of Science 96:103–113.

Viosca, P. Jr. 1923. An ecological study of the cold blooded vertebrates of southeastern Louisiana. Copeia (115):35–44.

Viosca, P. Jr. 1928. A new species of *Hyla* from Louisiana. Proceedings of the Biological Society of Washington 41:89–92.

Viosca, P. Jr. 1931. Principles of bullfrog (*Rana catesbeiana*) culture. Transactions of the American Fisheries Society 61:262–269.

Viosca, P. Jr. 1934. Principles of bullfrog culture.

Southern Biological Supply Company, New Orleans, LA.

Viosca, P. Jr. 1944. Distribution of certain cold-blooded animals in Louisiana in relationship to the geology and physiography of the state. Proceedings of the Louisiana Academy of Science 8:47–62.

Viparina, S., and J.J. Just. 1975. The life period, growth and differentiation of *Rana catesbeiana* larvae occurring in nature. Copeia 1975:103–109.

Vitt, L.J., and R.D. Ohmart. 1978. Herpetofauna of the lower Colorado River: Davis Dam to the Mexican border. Proceedings of the Western Foundation of Vertebrate Zoology 2:35–71.

Vogel, L.S. 2007. The decline of Fowler's toad (*Bufo fowleri*) in southern Louisiana: molecular genetics, field experiments and landscape studies. Ph.D. diss., University of New Orleans, New Orleans, LA.

Vogel, L.S. and S.G. Johnson. 2008. Estimation of hybridization and introgression frequency in toads (Genus: *Bufo*) using DNA sequence variation at mitochondrial and nuclear loci. Journal of Herpetology 42:61–75.

Vogel, L.S., and J.H.K. Pechmann. 2010. Response of Fowler's toad (*Anaxyrus fowleri*) to competition and hydroperiod in the presence of the invasive Coastal Plain toad (*Incilius nebulifer*). Journal of Herpetology 44:382–389.

Vogt, R.C. 1981. Natural History of Amphibians and Reptiles of Wisconsin. Milwaukee Public Museum, Milwaukee.

Volpe, E.P. 1952. Physiological evidence for natural hybridization of *Bufo americanus* and *Bufo fowleri*. Evolution 6:393–406.

Volpe, E.P. 1955a. Intensity of reproductive isolation between sympatric and allopatric populations of *Bufo americanus* and *Bufo fowleri*. American Naturalist 89:303–318.

Volpe, E.P. 1955b. A taxo-genetic analysis of the status of *Rana kandiyohi* Weed. Systematic Zoology 4:75–82.

Volpe, E.P. 1956a. Mutant color patterns in leopard frogs. Journal of Heredity 47:79–85.

Volpe, E.P. 1956b. Experimental F$_1$ hybrids between *Bufo valliceps* and *Bufo fowleri*. Tulane Studies in Zoology 4:61–75.

Volpe, E.P. 1957a. Embryonic temperature adaptations in highland *Rana pipiens*. American Naturalist 91:303–310.

Volpe, E.P. 1957b. The early development of *Rana capito sevosa*. Tulane Studies in Zoology 5:207–225.

Volpe, E.P. 1957c. Embryonic temperature tolerance and rate of development in *Bufo valliceps*. Physiological Zoology 30:164–176.

Volpe, E.P. 1957d. Genetic aspects of anuran populations. American Naturalist 91:355–372.

Volpe, E.P. 1958. Interspecific gene exchange between Fowler's and the Southern toad. Anatomical Record 131:605–606.

Volpe, E.P. 1959a. Hybridization of *Bufo valliceps* with *Bufo americanus* and *Bufo terrestris*. Texas Journal of Science 11:335–342.

Volpe, E.P. 1959b. Experimental and natural hybridization between *Bufo terrestris* and *Bufo fowleri*. American Midland Naturalist 61:295–312.

Volpe, E.P. 1960. Evolutionary consequences of hybrid sterility and vigor in toads. Evolution 14:181–193.

Volpe, E.P. 1961. Polymorphism in anuran populations. Pp. 221–234 *In* W.F. Blair (ed.), Vertebrate Speciation. University of Texas Press, Austin.

Volpe, E.P., and J.L. Dobie. 1959. The larva of the oak toad, *Bufo quercicus* Holbrook. Tulane Studies in Zoology 7:145–152.

Volpe, E.P., M.A. Wilkens, and J.L. Dobie. 1961. Embryonic and larval development of *Hyla avivoca*. Copeia 1961:340–349.

Voris, H.K., and J.P. Bacon Jr. 1966. Differential predation on tadpoles. Copeia 1966:594–598.

Vredenburg, V.T. 2004. Reversing introduced species effects: experimental removal of introduced fish leads to rapid recovery of a declining frog. Proceedings of the National Academy of Science USA 101:7646–7650.

Vredenburg, V., G.M. Fellers, and C. Davidson. 2005. *Rana muscosa* Camp, 1917. Mountain Yellow-legged Frog. Pp. 563–566 *In* M.J. Lannoo (ed.), Amphibian Declines. The Conservation Status of United States Species. University of California Press, Berkeley.

Vredenburg, V.T., R. Bingham, R. Knapp, J.A.T. Morgan, C. Moritz, and D. Wake. 2007. Concordant molecular and phenotypic data delineate new taxonomy and conservation priorities for the endangered mountain yellow-legged frog. Journal of Zoology 271:361–374.

Vredenburg, V.T., J.M. Romansic, L.M. Chan, and T. Tunstall. 2010a. Does UV-B radiation affect embryos of three high elevation amphibian species in California? Copeia 2010:502–512.

Vredenburg, V.T., R.A. Knapp, T.S. Tunstall, and C.J. Briggs. 2010b. Dynamics of an emerging disease drive large-scale amphibian population declines. Proceedings of the National Academy of Science USA 107:9689–9694.

Waddle, J.H. 2006. Use of amphibians as ecosystem indicator species. Ph.D. diss., University of Florida, Gainesville.

Waddle, J.H., R.M. Dorazio, S.C. Walls, K.G. Rice, J. Beauchamp, M.J. Schulman, and F.J. Mazzotti. 2010. A new parameterization for estimating co-occurrence of interacting species. Ecological Applications 20:1467–1475.

Wagner, G. 1997. Status of the northern leopard frog (*Rana pipiens*) in Alberta. Alberta Wildlife Status Report 9, Alberta Conservation Association.

Wagner, W.E. 1989a. Social correlates of variation in

Wagner, W.E. 1989a. Fighting, assessment, and male calling behavior in Blanchard's cricket frog, *Acris crepitans blanchardi*. Ethology 82:27–45.

Wagner, W.E. 1989b. Fighting, assessment, and frequency alteration in Blanchard's cricket frog. Behavioral Ecology and Sociobiology 25:429–436.

Wagner, W.E. 1989c. Graded aggressive signals in Blanchard's cricket frog: vocal responses to opponent proximity and size. Animal Behaviour 38: 1025–1038.

Wagner, W.E. 1992. Deceptive or honest signalling of fighting ability? A test of alternative hypotheses for the function of changes in call dominant frequency by male cricket frogs. Animal Behaviour 44:449–462.

Wahbe, T.R., and F.L. Bunnell. 2001. Preliminary observations on movements of tailed frog tadpoles (*Ascaphus truei*) in streams through harvested and natural forests. Northwest Science 75:77–83.

Wahbe, T.R., and F.L. Bunnell. 2003. Relations among larval tailed frogs, forest harvesting, stream microhabitat, and site parameters in southwestern British Columbia. Canadian Journal of Forest Resources 33:1256–1266.

Wahbe, T.R, F.L. Bunnell, and R.B. Bury. 2004. Terrestrial movements of juvenile and adult tailed frogs in relation to timber harvest in coastal British Columbia. Canadian Journal of Forest Resources 34:2455–2466.

Wahbe, T.R., C. Ritland, F.L. Bunnell, and K. Ritland. 2005. Population genetic structure of tailed frogs (*Ascaphus truei*) in clearcut and old-growth stream habitats in south coastal British Columbia. Canadian Journal of Zoology 83:1460–1468.

Waitz, J.A. 1961. Parasites of Idaho amphibians. Journal of Parasitology 47:89.

Wake, D.B., and V.T. Vredenburg. 2008. Are we in the midst of the sixth mass extinction? A view from the world of amphibians. Proceedings of the National Academy of Science USA 105:11466–11473.

Waldick, R.C., B. Freedman, and R.J. Wassersug. 1999. The consequences for amphibians of the conversion of natural, mixed-species forests to conifer plantations in southern New Brunswick. Canadian Field-Naturalist 113:408–418.

Waldman, B. 1981. Sibling recognition in toad tadpoles: the role of experience. Zeitschrift für Tierpsychologie 56:341–358.

Waldman, B. 1982a. Sibling association among schooling toad tadpoles: field evidence and implications. Animal Behaviour 30:700–713.

Waldman, B. 1982b. Adaptive significance of communal oviposition in wood frogs (*Rana sylvatica*). Behavioral Ecology and Sociobiology 10:169–174.

Waldman, B. 1984. Kin recognition and sibling association among wood frog (*Rana sylvatica*) tadpoles. Behavioral Ecology and Sociobiology 14:171–180.

Waldman, B. 1985a. Sibling recognition in toad tadpoles: are kinship labels transferred among individuals? Zeitschrift für Tierpsychologie 68:41–57.

Waldman, B. 1985b. Olfactory basis of kin recognition in toad tadpoles. Journal of Comparative Physiology A 156:565–577.

Waldman, B. 1986a. Preference for unfamiliar siblings over familiar non-siblings in American toad (*Bufo americanus*) tadpoles. Animal Behaviour 34:48–53.

Waldman, B. 1986b. Chemical ecology of kin recognition in anuran amphibians. Pp. 225–242 *In* D. Duvall, D. Müller-Schwarze, and D.M. Silverstein (eds.). Chemical Signals in Vertebrates. 4. Ecology, Evolution, and Comparative Biology. Plenum Press, New York.

Waldman, B. 1989. Do anuran larvae retain kin recognition abilities following metamorphosis? Animal Behaviour 37:1055–1058.

Waldman, B. 1991. Kin recognition in amphibians. Pp. 162–219 *In* P.G. Hepper (ed.). Kin Recognition. Cambridge University Press, Cambridge, UK.

Waldman, B., and K. Adler. 1979. Toad tadpoles associate preferentially with siblings. Nature 282:611–613.

Waldman, B., and M.J. Ryan. 1983. Thermal advantages of communal egg mass deposition in wood frogs (*Rana sylvatica*). Journal of Herpetology 17: 70–72.

Waldman, B., J.E. Rice, and R.L. Honeycutt. 1992. Kin recognition and incest avoidance in toads. American Zoologist 32:18–30.

Walker, C.F. 1932. *Pseudacris brachyphona* (Cope), a valid species. Ohio Journal of Science 32: 379–384.

Walker, C.F. 1946. The Amphibians of Ohio. 1. The frogs and toads (Order Salientia). Ohio State Museum Science Bulletin 1(3):1–109.

Walker, J.M. 1963. Amphibians and reptiles of Jackson Parish Louisiana. Proceedings of the Louisiana Academy of Science 26:91–107.

Walker, R.F., and W.G. Whitford. 1970. Soil water absorption capabilities in selected species of anurans. Herpetologica 26:411–418.

Wallace, R.L., and L.V. Diller. 1998. Length of the larval cycle of *Ascaphus truei* in coastal streams of the Redwood region, northern California. Journal of Herpetology 32:404–409.

Walls, S.C., A.R. Blaustein, and J.J. Beatty. 1992. Amphibian biodiversity of the Pacific Northwest with special reference to old-growth stands. Northwest Environmental Journal 8:53–69.

Walston, L.J., and S.J. Mullin. 2007a. Population responses of wood frog (*Rana sylvatica*) tadpoles to overwintered bullfrog (*Rana catesbeiana*) tadpoles. Journal of Herpetology 41:24–31.

Walston, L.J., and S.J. Mullin. 2007b. Responses of a pond-breeding amphibian community to the experimental removal of predatory fish. American Midland Naturalist 157:63–73.

Walston, L.J., and S.J. Mullin. 2008. Variation in amount of surrounding forest habitat influences the initial orientation of juvenile amphibians emigrating from breeding ponds. Canadian Journal of Zoology 86:141–146.

Walters, B. 1975. Studies of interspecific predation within an amphibian community. Journal of Herpetology 9:267–279.

Walton, A.C. 1938. The Nematoda as parasites of Amphibia. 4. Transactions of the American Microscopical Society 57:38–53.

Walton, A.C. 1941. Notes on some helminths from California amphibia. Transactions of the American Microscopical Society 60:53–57.

Walton, A.C. 1947. Parasites of the Ranidae (Amphibia). Anatomical Record 99:684–685.

Wang, Y., and Y. Li. 2009. Habitat selection by the introduced American Bullfrog (*Lithobates catesbeianus*) on Daishan Island, China. Journal of Herpetology 43:205–211.

Ward, R., C.J. Rutledge, and E.G. Zimmerman. 1987. Genetic variation and population subdivision in the cricket frog *Acris crepitans*. Biochemical Systematics and Ecology 15: 377–384.

Warkentin, I.G., C.E. Campbell, K.G. Powell, and T.D. Leonard. 2003. First record of the mink frog (*Rana septentrionalis*) for insular Newfoundland. Canadian Field-Naturalist 117:477–478.

Warkentin, K.M. 1992a. Effects of temperature and illumination on feeding rates of green frog tadpoles (*Rana clamitans*). Copeia 1992:735–730.

Warkentin, K.M. 1992b. Microhabitat use and feeding rate variation in green frog tadpoles (*Rana clamitans*). Copeia 1992:731–740.

Warner, S.C., W.A. Dunson, and J. Travis. 1991. Interaction of pH, density, and priority effects on the survivorship and growth of two species of hylid tadpoles. Oecologia 88:331–339.

Warner, S.C., J. Travis, and W.A. Dunson. 1993. Effect of pH variation on interspecific competition between two species of hylid tadpoles. Ecology 74:183–194.

Washburn, A.M. 1899. A peculiar toad. American Naturalist 33:139–141.

Wasserman, A.O. 1957. Factors affecting interbreeding in sympatric species of spadefoots (genus *Scaphiopus*). Evolution 11:320–338.

Wasserman, A.O. 1958. Relationships of allopatric populations of spadefoots (genus *Scaphiopus*). Evolution 12:311–318.

Wasserman, A.O. 1963. Further studies of hybridization in spadefoot toads (genus *Scaphiopus*). Copeia 1963:115–118.

Wassersug, R.J. 1973. Aspects of social behavior in anuran larvae. Pp. 273–297 *In* J.L. Vial (ed.), Evolutionary Biology of the Anurans. University of Missouri Press, Columbia.

Wassersug, R.J., and D.G. Sperry. 1977. The relationship of locomotion to differential predation on *Pseudacris triseriata* (Anura: Hylidae). Ecology 58:830–839.

Wassersug, R.J., and K. Hoff. 1979. A comparative study of the buccal pumping mechanisms of tadpoles. Biological Journal of the Linnaean Society 12:225–259.

Wassersug, R.J., K.J. Frogner, and R.F. Inger. 1981. Adaptations for life in tree holes by rhycophorid tadpoles from Thailand. Journal of Herpetology 15:41–52.

Waters, D.L., T.J. Hassler, and B.R. Norman. 1998. On the establishment of the Pacific chorus frog, *Pseudacris regilla* (Amphibian, Anura, Hylidae), at Ketchikan, Alaska. Bulletin of the Chicago Herpetological Society 33:124–127.

Watson, J.W., K.R. McAllister, and D.J. Pierce. 2003. Home ranges, movements, and habitat selection of Oregon spotted frogs (*Rana pretiosa*). Journal of Herpetology 37:292–300.

Waye, H.L. 2001. Teflon tubing as radio transmitter belt material for northern leopard frogs (*Rana pipiens*). Herpetological Review 31:88–89.

Waye, H.L., and C.H. Shewchuk. 1995. *Scaphiopus intermontanus* (Great Basin Spadefoot). Production of odor. Herpetological Review 26:98–99.

Weatherby, C.A. 1982. Introgression between the American toad *Bufo americanus* and the southern toad *B. terrestris* in Alabama. Ph.D. diss., Auburn University, Auburn, AL.

Webb, R.G. 1965. Observations on the breeding habits of the squirrel treefrog, *Hyla squirella* Bosc *in* Daudin. American Midland Naturalist 74:500–501.

Webb, R.G., and J.K. Korky. 1977. Variation in tadpoles of frogs of the *Rana tarahumarae* group in western Mexico (Anura: Ranidae). Herpetologica 33:73–82.

Weed, A.C. 1922. New frogs from Minnesota. Proceedings of the Biological Society of Washington 34:107–110.

Weintraub, J.D. 1974. Movement patterns of the red-spotted toad, *Bufo punctatus*. Herpetologica 30:212–215.

Weir, L., I.J. Fiske, and J.A. Royle. 2009. Trends in anuran occupancy from northeastern states of the North American Amphibian Monitoring Program. Herpetological Conservation and Biology 4:389–402.

Weis, J.S. 1975. The effect of DDT on tail regeneration in *Rana pipiens* and *R. catesbeiana* tadpoles. Copeia 1975:765–767.

Weitzel, N.H., and H.R. Panik. 1993. Long-term fluctuations of an isolated population of the Pacific chorus frog (*Pseudacris regilla*) in northwestern Nevada. Great Basin Naturalist 53:379–384.

Welbourne, W.C. Jr., and R.B. Loomis. 1975.

Hannemania (Acarina: Trombiculidae) and their anuran hosts at Fourtynine Palms Oasis, Joshua Tree National Monument, California. Bulletin of the Southern California Academy of Sciences 74:15–19.

Weldon, C., L.H. du Preez, A.D. Hyatt, R. Muller, and R. Speare. 2004. Origin of the amphibian chytrid fungus. Emerging Infectious Diseases 10:2100–2105.

Weller, W.F., and D.M. Green. 1997. Checklist and current status of Canadian amphibians. SSAR Herpetological Conservation 1:309–328.

Wells, K.D. 1976. Multiple egg clutches in the green frog (*Rana clamitans*). Herpetologica 32:85–87.

Wells, K.D. 1977a. The social behaviour of anuran amphibians. Animal Behaviour 25:666–693.

Wells, K.D. 1977b. Territoriality and male mating success in the green frog (*Rana clamitans*). Ecology 58:750–762.

Wells, K.D. 1978. Territoriality in the green frog (*Rana clamitans*): vocalizations and agonistic behaviour. Animal Behaviour 26:1051–1063.

Wells, K.D., and T.L. Taigen. 1984. Reproductive behavior and aerobic capacities of male American toads (*Bufo americanus*): is behavior constrained by physiology? Herpetologica 40:292–298.

Wells, K.D., and T.L. Taigen. 1986. The effect of social interactions on calling energetics in the gray treefrog (*Hyla versicolor*). Behavioral Ecology and Sociobiology 19:9–18.

Wells, K.D., and C.R. Bevier. 1997. Contrasting patterns of energy substrate use in two species of frogs that breed in cold weather. Herpetologica 53:70–80.

Welsh, H.H. 1988. An ecogeographic analysis of the herpetofauna of the Sierra San Pedro Mártir Region, Baja California, with a contribution to the biogeography of the Baja California herpetofauna. Proceedings of the California Academy of Sciences, 4th Series, 46:1–72.

Welsh, H.H. 1990. Relictual amphibians and old-growth forests. Conservation Biology 4:309–319.

Welsh, H.H. Jr., and L.M. Ollivier. 1998. Stream amphibians as indicators of ecosystem stress: a case study from California's redwoods. Ecological Applications 8:1118–1132.

Welsh, H.H. Jr., K.L. Pope, and D. Boiano. 2006. Subalpine amphibian distributions related to species palatability to non-native salmonids in the Klamath mountains of northern California. Diversity and Distributions 12:298–309.

Wendelken, P.W. 1968. Differential predation and behavior as factors in the maintenance of the vertebral stripe polymorphism in *Acris crepitans*. M.S. thesis, University of Texas, Austin.

Wente, W.H., M.J. Adams, and C.A. Pearl. 2005. Evidence of decline for *Bufo boreas* and *Rana luteiventris* in and around the northern Great Basin, western USA. Alytes 22:95–108.

Werler, J.E., and J. McCallion. 1951. Notes on a collection of reptiles and amphibians from Princess Anne County, Virginia. American Midland Naturalist 45:245–252.

Werner, E.E. 1986. Amphibian metamorphosis: growth rate, predation risk, and the optimal size at transformation. American Naturalist 128:319–341.

Werner, E.E. 1991. Nonlethal effects of a predator on competitive interactions between two anuran larvae. Ecology 75:1709–1720.

Werner, E.E. 1992. Competitive interactions between wood frog and northern leopard frog larvae: the influence of size and activity. Copeia 1992:26–35.

Werner, E.E. 1994. Ontogenetic scaling of competitive relations: size-dependent effects and responses in two anuran larvae. Ecology 75:197–213.

Werner, E.E., and M.A. McPeek. 1994. Direct and indirect effects of predators on two anuran species along an environmental gradient. Ecology 75:1368–1382.

Werner, E.E., and K.S. Glennemeier. 1999. Influence of forest canopy cover on the breeding pond distributions of several amphibian species. Copeia 1999:1–12.

Werner, E.E., G.A. Wellborn, and M.A. McPeek. 1995. Diet composition in postmetamorphic bullfrogs and green frogs: implications for interspecific predation and composition. Journal of Herpetology 29:600–607.

Werner, J.K., B.A. Maxwell, P. Hendricks, and D.L. Flath. 2004. Amphibians and Reptiles of Montana. Mountain Press Publishing Co., Missoula.

Wernz, J.G. 1969. Spring mating of *Ascaphus*. Journal of Herpetology 3:167–169.

Wernz, J.G., and R.M. Storm. 1969. Pre-hatching stages of the tailed frog, *Ascaphus truei* Stejneger. Herpetologica 25:86–93.

West, L.B. 1960. The nature of growth inhibitory material from crowded *Rana pipiens* tadpoles. Physiological Zoology 33:232–239.

Westerman, A.G., A.J. Wigginton, D.J. Price, G. Linder, and W.J. Birge. 2003a. Integrating amphibians into ecological risk assessment strategies. Pp. 283–313 *In* G. Linder, S.K. Krest, and D.W. Sparling (eds.), Amphibian Decline: An Integrated Analysis of Multiple Stressor Effects. SETAC Press, Pensacola, FL.

Westerman, A.G., W. van der Schalie, S.L. Levine, B. Palmer, D. Shank, and R.G. Stahl. 2003b. Linking stressors with potential effects on amphibian survival. Pp. 73–109 *In* G. Linder, S.K. Krest, and D.W. Sparling (eds.), Amphibian Decline: An Integrated Analysis of Multiple Stressor Effects. SETAC Press, Pensacola, FL.

Westman, A.D.J., J. Elliott, K. Cheng, G. Van Aggelen, and C.A. Bishop. 2010. Effects of environmentally relevant concentrations of endosulfan, azinphosmethyl, and diazinon on Great Basin spadefoot (*Spea intermontana*) and Pacific treefrog

(*Psudacris regilla*). Environmental Toxicology and Chemistry 29:1604–1612.

Wetzel, E.J., and G.W. Esch. 1996. Seasonal population dynamics of *Halipegus occidualis* and *Halipegus eccentricus* (Diagenea: Hemiuridae) in their amphibian host, *Rana clamitans*. Journal of Parasitology 82:414–422.

Weyrauch, S.L., and J.P. Amon. 2002. Relocation of amphibians to created seasonal ponds in southwestern Ohio. Restoration Ecology 20:31–36.

Wharton, J.C. 1989. Ecological and life history aspects of the San Francisco garter snake (*Thamnophis sirtalis tetrataenia*). M.S. thesis, San Francisco State University.

Wheeler, C.A., H.H. Welsh Jr., and L.L. Heise. 2003. *Rana boylii* (Foothill Yellow-legged Frog). Oviposition behavior. Herpetological Review 34:234.

Wheeler, C.H., and H.H. Welsh Jr. 2008. Mating strategy and breeding patterns of the foothill yellow-legged frog (*Rana boylii*). Herpetological Conservation and Biology 3:128–142.

Wheeler, G.C., and J. Wheeler. 1966. The Amphibians and Reptiles of North Dakota. University of North Dakota, Grand Forks.

Whitaker, J.O. Jr. 1961. Habitat and food of mouse-trapped young *Rana pipiens* and *Rana clamitans*. Herpetologica 17:173–179.

Whitaker, J.O. Jr. 1971. A study of the western chorus frog, *Pseudacris triseriata*, in Vigo County, Indiana. Journal of Herpetology 5:127–150.

Whitaker, J.O., D. Rubin, and J.R. Munsee. 1977. Observations of food habits of four species of spadefoot toads, genus *Scaphiopus*. Herpetologica 33:468–475.

White, J.F. 1988. Amphibians of Delaware. Transactions of the Delaware Academy of Science 16:17–22.

White, J.F. Jr., and A.W. White. 2007. Amphibians and Reptiles of Delmarva. 2nd ed. Tidewater Publishers, Centreville, MD.

Whiting, M.J., and A.H. Price. 1994. Not the last picture show: new collection records from Paris, Texas (and other places). Herpetological Review 25:130.

Whitney, C.L. 1980. The role of the "encounter" call in spacing of Pacific tree frogs, *Hyla regilla*. Canadian Journal of Zoology 58:75–78.

Whitney, C.L. 1981. The monophasic call of *Hyla regilla* (Anura: Hylidae). Copeia 1981:230–233.

Whitney, C.L., and J.R. Krebs. 1975a. Mate selection in Pacific tree frogs. Nature 255:325–326.

Whitney, C.L., and C.J. Krebs. 1975b. Spacing and calling in Pacific tree frogs, *Hyla regilla*. Canadian Journal of Zoology 53:1519–1527.

Wied-Neuwied, M. Prinz zu. 1838. Reise in das innere Nord-Amerika in den Jahren 1832 bis 1834. Vol. 1. J. Hoelscher, Coblenz.

Wiens, J.A. 1970. Effects of early experience on substrate pattern selection in *Rana aurora* tadpoles. Copeia 1970:543–548.

Wiens, J.J., and T.A. Titus. 1991. A phylogenetic analysis of *Spea* (Anura: Pelobatidae). Herpetologica 47:21–28.

Wiese, R.J. 1985. Ecological aspects of the bullfrog in northeastern Colorado. M.S. thesis, Colorado State University, Fort Collins.

Wiese, R.J. 1990. Genetic structure of native and introduced populations of the bullfrog, a successful colonist. Ph.D. diss., Colorado State University, Fort Collins.

Wiest, J.A. Jr. 1982. Anuran succession at temporary ponds in a post oak-savanna region of Texas. Pp. 39–47 *In* N.J. Scott (ed.), Herpetological Communities. U.S. Fish and Wildlife Service, Wildlife Research Report 13.

Wiewandt, T.A. 1969. Vocalization, aggressive behavior, and territoriality in the bullfrog, *Rana catesbeiana*. Copeia 1969:276–285.

Wiggins, D.A. 1992. Foraging success of leopard frogs (*Rana pipiens*). Journal of Herpetology 26:87–88.

Wiggins, I.L. 1943. Additional note on the range of *Bufo canorus* Camp. Copeia 1943:197.

Wilbur, H.M. 1972. Competition, predation, and the structure of the *Ambystoma-Rana sylvatica* community. Ecology 53:3–21.

Wilbur, H.M. 1976. Density-dependent aspects of metamorphosis in *Ambystoma* and *Rana sylvatica*. Ecology 57:1289–1296.

Wilbur, H.M. 1977a. Interactions of food level and population density in *Rana sylvatica*. Ecology 58:206–209.

Wilbur, H.M. 1977b. Density-dependent aspects of growth and metamorphosis in *Bufo americanus*. Ecology 58:196–200.

Wilbur, H.M. 1982. Competition between tadpoles of *Hyla femoralis* and *Hyla gratiosa* in laboratory experiments. Ecology 63:278–282.

Wilbur, H.M. 1987. Regulation of structure in complex systems: experimental temporary pond communities. Ecology 68:1437–1452.

Wilbur, H.M., and R.A. Alford. 1985. Priority effects in experimental pond communities: responses of *Hyla* to *Bufo* and *Rana*. Ecology 66:1106–1114.

Wilbur, H.M., and J.E. Fauth. 1990. Experimental aquatic food webs: interactions between two predators and two prey. American Naturalist 135:176–204.

Wilbur, H.M., D.I. Rubenstein, and L. Fairchild. 1978. Sexual selection in toads: the roles of female choice and male body size. Evolution 32:264–270.

Wilbur, H.M., P.J. Morin, and R.N. Harris. 1983. Salamander predation and the structure of experimental communities: anuran responses. Ecology 64:1423–1429.

Wilcox, E.V. 1891. Notes on Ohio batrachians. Otterbein Aegis 1(9):133–135.

Wilczynski, W., H.H. Zakon, and E.A. Brenowitz. 1984. Acoustic communication in spring peepers:

call characteristics and neurophysiological aspects. Journal of Comparative Physiology A 155:577–584.

Wiley, J.E. 1982. Chromosome banding patterns of treefrogs (Hylidae) of the eastern United States. Herpetologica 38:507–520.

Wilkinson, J.A. 2006. *Rana aurora draytonii* (California Red-legged Frog). Defensive behavior. Herpetological Review 37:207–208.

Wilks, B.J., and H.E. Laughlin. 1962. Artificial hybridization between the microhylid genera *Hypopachus* and *Gastrophryne*. Texas Journal of Science 14:183–187.

Willhite, C., and P.V. Cupp Jr. 1982. Daily rhythms of thermal tolerance in *Rana clamitans* (Anura: Ranidae) tadpoles. Comparative Biochemistry and Physiology 72A:255–257.

Williams, B.K., T.A.G. Rittenhouse, and R.D. Semlitsch. 2008. Leaf litter input mediates tadpole performance across forest canopy treatments. Oecologia 155:377–384.

Williams, D.D., and S.J. Taft. 1980. Helminths of anurans from NW Wisconsin. Proceedings of the Helminthological Society of Washington 47:278.

Williams, K.L., and P.S. Chrapliwy. 1958. Selected records of amphibians and reptiles from Arizona. Transactions of the Kansas Academy of Science 61:299–301.

Williams, P.J., J.R. Robb, R.H. Kappler, T.E. Piening, and D.R. Karns. 2012. Intraspecific density dependence in larval development of the crawfish frog, *Lithobates areolatus*. Herpetological Review 43:36–38.

Williams, P.K. 1969. Ecology of *Bufo hemiophrys* and *B. americanus* tadpoles in northwestern Minnesota. M.S. thesis, University of Minnesota, Minneapolis.

Williamson, G.K., and R.A. Moulis. 1994. Distribution of amphibians and reptiles in Georgia. Vol. 1, maps; Vol. 2, locality data. Special Publication No. 3, Savannah Science Museum, Savannah, GA.

Willis, Y.L., D.L. Moyle, and T.S. Baskett. 1956. Emergence, breeding, hibernation, movements and transformation of the bullfrog, *Rana catesbeiana*, in Missouri. Copeia 1956:30–41.

Wilmhoff, C.D., L. Williams, R. MacDonald, S. Fisher, and J. Slothouber. 1999. Geographic distribution: *Pseudacris brachyphona* (Mountain Chorus Frog). Herpetological Review 30:173.

Wilson, A.K. 2000. Amphibian and reptile surveys in the Kaskaskia River drainage of Illinois during 1997 and 1998. Journal of the Iowa Academy of Science 107:203–205.

Wilson, D.J., and H. Lefcort. 1993. The effect of predator diet on the alarm response of red-legged frog, *Rana aurora*, tadpoles. Animal Behaviour 46:1017–1019.

Wilson, G.A., T.L. Fulton, K. Kendell, G. Scrimgeour, C.A. Paszkowski, and D.W. Coltman. 2008a. Genetic diversity and structure in Canadian northern leopard frog (*Rana pipiens*) populations: implications for reintroduction programs. Canadian Journal of Zoology 86:863–874.

Wilson, G.A., T.L. Fulton, K. Kendell, D.M. Schock, C.A. Paszkowski, and D.W. Coltman. 2008b. Genetic evidence for single season polygyny in the northern leopard frog (*Rana pipiens*). Herpetological Review 39:46–50.

Wilson, L.D., and L. Porras. 1983. The ecological impact of man on the south Florida herpetofauna. University of Kansas, Museum of Natural History Special Publication 9:1–89.

Wilson, L.W. 1945. Amphibians of Droop Mountain State Park. Proceedings of the West Virginia Academy of Science 16:39–41.

Wilson, V.V. 1975. The systematics and paleoecology of two Pleistocene herpetofaunas of the southeastern United States. Ph.D. diss. Michigan State University, East Lansing.

Windmiller, B., R.N. Homan, J.V. Regosin, L.A. Willitts, D.L. Wells, and J.M. Reed. 2008. Breeding amphibian population declines following loss of upland forest habitat around vernal pools in Massachusetts, USA. Pp. 41–51 *In* J.C. Mitchell, R.E. Jung Brown, and B. Bartholomew (eds.), Urban Herpetology. Herpetological Conservation 3. Society for the Study of Amphibians and Reptiles, Salt Lake City, UT.

Winn, B., J.B. Jensen, and S. Johnson. 1999. Geographic distribution. *Eleutherodactylus planirostris*. Herpetological Review 30:49.

Withers, D.I. 1992. The Wyoming toad (*Bufo hemiophrys baxteri*): an analysis of habitat use and life history. M.S. thesis, University of Wyoming, Laramie.

Witschi, E. 1953. The Cherokee frog, *Rana sylvatica cherokiana nom. nov.*, of the Appalachian Mountain region. Proceedings of the Iowa Academy of Science 60:764–769.

Witte, C.L., M.J. Sredl, A.S. Kane, and L.L. Hungerford. 2008. Epidemiologic analysis of factors associated with local disappearances of native ranid frogs in Arizona. Conservation Biology 22:375–383.

Witte, K., K.-C. Chen, W. Wilczynski, and M.J. Ryan. 2000. Influence of amplexus on phonotaxis in the cricket frog *Acris crepitans blanchardi*. Copeia 2000:257–261.

Witte, K., M.J. Ryan, and W. Wilczynski. 2001. Changes in the frequency structure of a mating call decrease its attractiveness to females in the cricket frog *Acris crepitans blanchardi*. Ethology 107:685–699.

Witte, K., H.E. Farris, M.J. Ryan, and W. Wilczynski. 2005. How cricket frog females deal with a noisy world: habitat-related differences in auditory tuning. Behavioral Ecology 16:571–579.

Wojtaszek, B.F., B. Staznik, D.T. Chartrand, G.R.

Stephenson, and D.G. Thompson. 2004. Effects of Vision™ herbicide on mortality, avoidance response, and growth of amphibian larvae in two forest wetlands. Environmental Toxicology and Chemistry 23:832–842.

Wolf, K., G.L. Bullock, C.E. Dunbar, and M.C. Quimby. 1968. Tadpole edema virus: a viscerotropic pathogen for anuran amphibians. Journal of Infectious Disease 118:253–262.

Wollmuth, L.P., L.I. Crawshaw, R.B. Forbes, and D.A. Grahn. 1987. Temperature selection during development in a montane anuran species, *Rana cascadae*. Physiological Zoology 60:472–480.

Woltz, H.W., J.P. Gibbs, and P.K. Ducey. 2008. Road crossing structures for amphibians and reptiles: informing design through behavioral analysis. Biological Conservation 141:2745–2750.

Woo, P.T.K., and J.P. Bogart. 1984. *Trympanosoma* sp. (Protozoa; Kinetoplastidia) in Hylidae (Anura) from eastern North America with notes on their distribution and prevalences. Canadian Journal of Zoology 62:820–824.

Wood, J.T. 1948. *Microhyla c. carolinensis* in an ant nest. Copeia 1948:226.

Wood, K.V., J.D. Nichols, H.F. Percival, and J.H. Hines. 1998. Size-sex variation in survival rates and abundance of pig frogs, *Rana grylio*, in northern Florida wetlands. Journal of Herpetology 32:527–535.

Wood, T.S. 1977. Food habits of *Bufo canorus*. M.S. thesis, Occidental College, Los Angeles, CA.

Wood, W.F. 1935. Encounters with the western spadefoot, *Scaphiopus hammondii*, with a note on a few albino larvae. Copeia 1935:100–102.

Woodford, J.E., and M.W. Meyer. 2003. Impact of lakeshore development on green frog abundance. Biological Conservation 110:277–284.

Woodhams, D.C., V.T. Vredenburg, M.-A. Simon, D. Billheimer, B. Shakhtour, Y. Shyr, C.J. Briggs, L.A. Rollins-Smith, and R.N. Harris. 2007. Symbiotic bacteria contribute to innate immune defenses of the threatened mountain yellow-legged frog, *Rana muscosa*. Biological Conservation 138:390–398.

Woodhams, D.C., R.A. Alford, C.J. Briggs, M. Johnson, and L.A. Rollins-Smith. 2008. Life-history trade-offs influence disease in changing climates: strategies of an amphibian pathogen. Ecology 89:1627–1639.

Woodhouse, S.W. 1854. Report on the natural history of the country passed over by the exploring expedition under the command of Brev. Capt. L. Sitgreaves, United States Topographical Engineers, during the year 1851. Pp. 33–40 *In* Report of an Expedition down the Zuni and Colorado Rivers, by Captain L. Sitgreaves. Beverley Tucker, Senate Printer, Washington, DC.

Woodliffe, P.A. 1989. Inventory, assessment, and ranking of natural areas of Walpole Island. Proceedings of the Eleventh North American Prairie Conference, pp. 47–52.

Woodward, B.D. 1982a. Sexual selection and nonrandom mating patterns in desert anurans (*Bufo woodhousei, Scaphiopus couchi, S. multiplicatus* and *S. bombifrons*). Copeia 1982:351–355.

Woodward, B.D. 1982b. Male persistence and mating success in Woodhouse's toad (*Bufo woodhousei*). Ecology 63:583–585.

Woodward, B.D. 1983. Predator-prey interactions and breeding-pond use of temporary-pond species in a desert anuran community. Ecology 64:1549–1555.

Woodward, B.D. 1984. Arrival to and location of *Bufo woodhousei* in the breeding pond: effect on the operational sex ratio. Oecologia 62:240–244.

Woodward, B.D. 1987a. Clutch parameters and pond use in some Chihuahuan Desert anurans. Southwestern Naturalist 32:13–19.

Woodward, B.D. 1987b. Interactions between Woodhouse's toad tadpoles (*Bufo woodhousei*) of mixed sizes. Copeia 1987:380–386.

Woodward, B.D., and S. Mitchell. 1990. Predation on frogs in breeding choruses. Southwestern Naturalist 35:449–450.

Woodward, B.D., and S. Mitchell. 1992. Temperature variation and species interactions in aquatic systems in Grand Teton National Park. Pp. 140–145 *In* University of Wyoming/National Park Service Research Center, 16th Annual Report.

Woolbright, L.L., E.J. Greene, and G.C. Rapp. 1990. Density-dependent mate searching strategies of male woodfrogs. Animal Behaviour 40:135–142.

Woolbright, L.L., A.H. Hara, C.M. Jacobsen, W.J. Mautz, and F.L. Benevides Jr. 2006. Population densities of the coquí, *Eleutherodactylus coqui* (Anura: Leptodactylidae) in newly invaded Hawaii and in native Puerto Rico. Journal of Herpetology 40:122–126.

Wright, A.H. 1914. Life-histories of the Anura of Ithaca, New York. Carnegie Institution of Washington, Publication 197.

Wright, A.H. 1924. A new bullfrog (*Rana heckscheri*) from Georgia and Florida. Proceedings of the Biological Society of Washington 37:141–152.

Wright, A.H. 1929. Synopsis and description of North American tadpoles. Proceedings of the United States National Museum 74:1–70.

Wright, A.H. 1932. Life-Histories of the Frogs of Okefinokee Swamp, Georgia. Macmillan Co., New York.

Wright, A.H., and A.A. Wright. 1933. Handbook of Frogs and Toads. 1st ed. Comstock Publishing Associates, Ithaca, NY.

Wright, A.H., and A.A. Wright. 1938. Amphibians of Texas. Transactions of the Texas Academy of Science 21:5–44.

Wright, A.H., and A.A. Wright. 1949. Handbook of

Frogs and Toads. 3rd ed. Comstock Publishing Associates, Ithaca, NY.

Wright, H.P., and G.S. Myers. 1927. *Rana areolata* at Bloomington, Indiana. Copeia (159):173–175.

Wright, J.W. 1966. Predation on the Colorado River toad, *Bufo alvarius*. Herpetologica 22:127–128.

Wright, M.F., and S.I. Guttman. 1995. Lack of an association between heterozygosity and growth rate in the wood frog, *Rana sylvatica*. Canadian Journal of Zoology 73:569–575.

Wu, Z., Y.P. Wang, and Y.M. Li. 2004. Natural populations of bullfrog (*Rana catesbeiana*) and their potential threat in the east of Zhejiang province. Biodiversity Science 12:441–446 [in Chinese with English abstract].

Wu, Z., L. Yiming, W. Yanping, and M.J. Adams. 2005. Diet of introduced bullfrogs (*Rana catesbeiana*): predation on and dietary overlap with native frogs on Daishan Island, China. Journal of Herpetology 39:668–674.

Wyatt, J.L., and E.A. Forys. 2004. Conservation implications of predation by Cuban treefrogs (*Osteopilus septentrionalis*) on native hylids in Florida. Florida Scientist 3:695–700.

Wygoda, M.L. 1984. Low cutaneous evaporative water loss in arboreal frogs. Physiological Zoology 57:329–337.

Wylie, K., M.-L. Gramby, and M. Kostalos. 2009. Inhibition of metamorphosis in *Bufo americanus* tadpoles in Nine Mile Run Restoration Area, Pittsburg [sic], Pennsylvania USA. Froglog 91:10–12.

Wylie, S.R. 1981. Effects of basking on the biology of the canyon treefrog, *Hyla arenicolor* Cope. Ph.D. diss. Arizona State University, Tempe.

Wyman, R.L. 1988. Soil acidity and moisture and the distribution of amphibians in five forests of south-central New York. Copeia 1988:394–399.

Yahner, R.H., W.C. Bramble, and W.R. Byrnes. 2001. Response of amphibian and reptile populations to vegetation maintenance of an electric transmission line right-of-way. Journal of Arboriculture 27:215–221.

Yeager, D. 1926. Miscellaneous notes. Yellowstone Nature Notes 3(4):7.

Yeary, C.M. 1979. Support for a mutualistic relationship between the western narrow-mouthed frog, *Gastrophryne olivacea*, and the tarantula, *Dugesiella hentzii*. M.S. thesis, University of Tulsa, Tulsa, OK.

Yoder, H.R., and J.R. Coggins. 1996. Helminth communities in the northern spring peeper, *Pseudacris c. crucifer* Wied, and the wood frog, *Rana sylvatica* LeConte, from southeastern Wisconsin. Journal of the Helminthological Society of Washington 63:211–214.

Yoder, H.R., and J.R. Coggins. 2007. Helminth communities in five species of sympatric amphibians from three adjacent ephemeral ponds in southeastern Wisconsin. Journal of Parasitology 93:755–760.

Young, F.N., and C.C. Goff. 1939. An annotated list of the arthropods found in the burrows of the Florida gopher tortoise, *Gopherus polyphemus* (Daudin). Florida Entomologist 22:53–62.

Young, J.E., and B.I. Crother. 2001. Allozyme evidence for the separation of *Rana areolata* and *Rana capito* and for the resurrection of *Rana sevosa*. Copeia 2001:382–388.

Young, M.K., G.T. Allison, and K. Foster. 2007. Observations of boreal toads (*Bufo boreas boreas*) and *Batrachochytrium dendrobatidis* in south-central Wyoming and north-central Colorado. Herpetological Review 38:146–150.

Youngstrom, K.A., and H.M. Smith. 1936. Description of the larvae of *Pseudacris triseriata* and *Bufo woodhousii woodhousii* (Anura). American Midland Naturalist 17:629–632.

Zacharow, M., W.J. Barichivich, and C.K. Dodd Jr. 2003. Using ground-placed PVC pipes to monitor hylid treefrogs: capture biases. Southeastern Naturalist 2:575–590.

Zaga, A., E.E. Little, C.F. Raben, and M.R. Ellersieck. 1998. Photoenhanced toxicity of a carbamate insecticide to early life stage anuran amphibians. Environmental Toxicology and Chemistry 17:2543–2553.

Zamparo, D., and D.R. Brooks. 2005. Three rarely reported digeneans inhabiting amphibians from Vancouver Island, British Columbia, Canada. Journal of Parasitology 91:1242–1244.

Zampella, R.A., and J.F. Bunnell. 2000. The distribution of anurans in two river systems of a coastal plain watershed. Journal of Herpetology 34:210–221.

Zell, G.A. 1986. The clawed frog: an exotic from South Africa invades Virginia. Virginia Wildlife 47(2):28–29.

Zellmer, A.J., C.L. Richards, and L.M. Martens. 2008. Low prevalence of *Batrachochytrium dendrobatidis* across *Rana sylvatica* populations in southeastern Michigan, USA. Herpetological Review 39:196–199.

Zelmer, D.A., E.J. Wetzel, and G.W. Esch. 1999. The role of habitat in structuring *Halipegus occidualis* metapopulations in the green frog. Journal of Parasitology 85:19–24.

Zenisek, C.J. 1963. A study of the natural history and ecology of the leopard frog, *Rana pipiens* Schreber. Ph.D. diss., The Ohio State University, Columbus.

Zettergren, L.D., B.W. Boldt, D.H. Petering, M.S. Goodrich, D.N. Weber, and J.G. Zettergren. 1991. Effects of prolonged low-level cadmium exposure on the tadpole immune system. Toxicology Letters 55:11–19.

Zeyl, C. 1993. Allozyme variation and divergence

among populations of *Rana sylvatica*. Journal of Herpetology 27:233–236.

Zhang, Q.-Y., F. Xiao, Z.-Q. Li, J.-F. Gui, J. Mao, and V.G. Chinchar. 2001. Characterization of an iridovirus from the cultured pig frog *Rana grylio* with lethal syndrome. Diseases of Aquatic Organisms 48:27–36.

Zielinski, W.J., and G.T. Barthalmus. 1989. African clawed frog skin compounds: antipredatory effects on African and North American water snakes. Animal Behaviour 38:1083–1086.

Zug, G.R., and P.B. Zug. 1979. The marine toad, *Bufo marinus*: a natural history resumé of native populations. Smithsonian Contributions to Zoology (284):1–58.

Zweifel, R.G. 1955. Ecology, distribution, and systematics of frogs of the *Rana boylei* group. University of California Publications in Zoology 54:207–292.

Zweifel, R.G. 1956a. Two pelobatid frogs from the Tertiary of North America and their relationships to fossil and recent forms. American Museum Novitates 1762:1–45.

Zweifel, R.G. 1956b. A survey of the frogs of the *augusti* group, genus *Eleutherodactylus*. American Museum Novitates 1813:1–35.

Zweifel, R.G. 1961. Larval development of the tree frogs *Hyla arenicolor* and *Hyla wrightorum*. American Museum Novitates 2056:1–19.

Zweifel, R.G. 1968a. Effects of temperature, body size and hybridization on mating calls of toads *Bufo a. americanus* and *Bufo woodhousei fowleri*. Copeia 1968:269–285.

Zweifel, R.G. 1968b. Reproductive biology of anurans of the arid Southwest, with adaptation of embryos to temperature. Bulletin of the American Museum of Natural History 140:1–64.

Zweifel, R.G. 1970a. Distribution and mating call of the treefrog, *Hyla chrysoscelis*, at the northeastern edge of its range. Chesapeake Science 11:94–97.

Zweifel, R.G. 1970b. Descriptive notes on larvae of toads of the *debilis* group, genus *Bufo*. American Museum Novitates 2407:1–13.

Zweifel, R.G. 1989. Calling by the frog, *Rana sylvatica*, outside the breeding season. Journal of Herpetology 23:185–186.

Index of Scientific and Common Names

Acris, xi, xii, xvii, xix
   barbouri, 228
   blanchardi, xviii, 18, 205–221, 223, 226–227, 230, 232–233, 329, 396, 417, 501
   crepitans, 205, 207–211, 213–214, 219–228, 232–233, 252, 304, 351, 386, 680
   gryllus, xix, 148, 168, 205–210, 219–221, 224–235, 537, 554, 560, 630
     g. dorsalis, 226, 228
     g. paludicola, 205, 207, 219
Aeromonas, 37, 46, 61, 64, 76, 87, 176, 382
African Clawed Frog, xiii, 828–832
alvaria, *Ollotis*, 50, 79, 85, 142, 168, 177–180, 182, 758
Amargosa Toad, xviii, 132–136
American Bullfrog, xii, xv, xxv, 42, 64–65, 68, 70, 91, 96, 124, 134–136, 173, 179, 183, 215, 258, 303, 312, 317, 410, 412, 479, 486–515, 521, 524, 528, 533, 539, 551–552, 554, 557, 561, 566, 568–569, 576, 599, 605–607, 615, 631, 637, 653, 658, 676, 680, 691–692, 695–697, 699, 704, 706, 717, 720–721, 731, 744, 746, 769, 788, 790, 819
American Toad, ix, 17–42, 97, 106, 110, 113, 155, 318, 506, 536, 662
americanus, *Anaxyrus*, 17–43, 50, 79, 81, 83, 86–87, 96–101, 104–105, 110, 114, 119–121, 129, 138, 157, 159, 167–169, 174, 182–183, 259, 305, 501, 572, 576, 631, 767
Anaxyrus, xi, xv, xvi, xx, xxi
   americanus, 17–43, 50, 79, 81, 83, 86–87, 96–101, 104–105, 110, 114, 119–121, 129, 138, 157, 159, 167–169, 174, 182–183, 259, 305, 501, 572, 576, 631, 767
     a. charlesmithi, 17, 20, 35
     a. copei, 17, 20
   baxteri, xviii, 19, 43–47, 114, 120–121, 168
   boreas, xi, xvii, 19, 47–65, 71, 73, 79, 92–94, 114, 129, 132–135, 138, 169, 595, 693, 719, 741, 744
     b. halophilus, 47–50, 54, 71, 93, 132–133, 169
   californicus, xviii, 19, 65–70, 76, 78, 128–129, 153
   canorus, xviii, 48, 50–51, 70–77, 93, 133, 736–737, 750
   cognatus, 19, 65–66, 78–87, 114, 138, 153–155, 168–169, 171, 175, 179
   compactilis, 153, 252
   debilis, 20, 50, 79, 88–91, 138, 149–150, 175, 182
     d. insidior, 88–89
   defensor, 157
   exsul, 50, 71, 92–96, 132–133
   fowleri, 19–20, 27, 43, 96–113, 116, 121, 157–159, 163, 167–169, 174, 182, 185, 188, 238
   hemiophrys, 19–20, 43–44, 57, 79, 101, 113–121, 136, 157, 168–169, 182
   houstonensis, 120–126, 155
   kelloggi, 89, 138, 150
   microscaphus, 19, 65–66, 99, 114, 127–132, 138, 153–154, 157, 168–169
   nelsoni, xviii, 50, 71, 93, 132–136
   pliocompactilis, 145
   punctatus, 19–20, 50, 79, 89, 114, 130, 136–145, 150, 152–153, 157, 168–169, 759
   quercicus, 79, 144–149, 189, 630
   retiformis, 88–89, 138, 149–152
   speciosus, 20, 50, 66, 79, 125, 127, 129, 138, 152–155, 157, 167, 169, 175, 182, 759, 781
   terrestris, 17, 19–20, 42–43, 79, 89, 97, 99–100, 105, 110, 114, 121, 129, 138, 148, 153, 155–166, 168–169, 182, 188–190, 284, 560, 769, 826
   tiheni, 145
   woodhousii, 18–20, 27, 43, 65–66, 79, 82, 86, 96–101, 103, 113–114, 121, 125, 127–129, 132, 138, 153, 155, 157, 166–176, 178, 182, 185, 188, 259, 344, 759
     w. australis, 166, 168
     w. bexarensis, 168
     w. velatus, 96, 98, 166, 168
andersonii, *Hyla*, 111, 235–239, 246, 264, 271, 275, 304, 631
arenicolor, *Hyla*, 239–246, 253, 264, 275, 289, 298, 310, 312, 322–323, 329, 404
areolatus, *Lithobates*, 461–466, 468, 474, 478, 480, 483, 618, 625, 631, 633
Arizona Toad, 127–132
Arizona Treefrog, xviii, 309–312
Arroyo Toad, xviii, 65–70
Ascaphidae, 1
Ascaphus, xi, xii, xiii, xvi, xvii, xx
   montanus, 1–7, 9–10, 12–14, 16
   truei, 1–3, 5, 7–16
     t. californicus, 2, 7, 9
augusti, *Craugastor*, 192–196
auratus, *Dendrobates*, xiii, xiv, xvi, 809–811
aurora, *Rana*, 506–507, 641, 687–697, 707–708, 715–716, 719–722, 734, 739, 743–746, 748
avivoca, *Hyla*, 236, 245–250, 252–253, 263, 275, 281, 289, 297

Balcones Barking Frog, 192
Barking Frog, 192–196
Barking Treefrog, ix, 266, 273, 280–288
Batrachochytrium dendrobatidis, xvii, 37, 45, 61, 120, 125, 175, 185, 216, 244, 260, 305, 345, 362, 382, 412, 465, 471, 509, 520–521, 540, 577, 615, 633, 663, 673–674, 685–686, 694, 704, 713, 720, 730, 737–739, 745, 751, 771, 814, 832
baxteri, Anaxyrus, xviii, 19, 43–47, 114, 120–121, 168
berlandieri, Lithobates, 462, 466–474, 477–478, 517, 566, 570, 580, 584, 622, 624–625, 630–631, 682, 686
Bird-voiced Treefrog, ix, 245–250
Black Toad, 92–96
blairi, Lithobates, 453, 461–462, 465–468, 471–479, 517, 566, 570, 580, 583–585, 599, 603, 624–625, 629–631, 635, 682
blanchardi, Acris, xviii, 18, 205–221, 223, 226–227, 230, 232–233, 329, 396, 417, 501
Blanchard's Cricket Frog, xviii, 205–219
bombifrons, Spea, 85, 755, 758–759, 776–787, 791–792, 799, 801, 804–805
Boreal Chorus Frog, 371–384, 411
Boreal Toad, 47–65
boreas, Anaxyrus, xi, xvii, 19, 47–65, 71, 73, 79, 92–94, 114, 129, 132–135, 138, 169, 595, 693, 719, 741, 744
boylii, Rana, 508, 510, 688, 697–708, 716, 724, 733–734, 740, 747–748
brachyphona, Pseudacris, 313–318, 320, 334, 350, 358, 368, 374, 386, 396, 417, 422–423
brimleyi, Pseudacris, 314, 319–322, 350, 358, 368, 374, 386, 396, 422
Brimley's Chorus Frog, 319–322
Bronze Frog, 522–524, 526–528, 530, 533, 538, 544
Bufo antecessor, 168
Bufonidae, 17
Bufo planiorum, 168
Burrowing Toad, 458–460

cadaverina, Pseudacris, 239, 298, 322–328, 402–405
California Red-legged Frog, xviii, 689, 715–722
California Toad, 47, 52, 63, 133
California Treefrog, 322–328, 411
californicus, Anaxyrus, xviii, 19, 65–70, 76, 78, 128–129, 153
Canadian Toad, 43, 113–120
Cane Toad, 42, 186–191, 813
canorus, Anaxyrus, xviii, 48, 50–51, 70–77, 93, 133, 736–737, 750
Canyon Treefrog, 239–245
capito, Lithobates, 148, 462, 479–486, 570, 583, 617–618, 621
carolinensis, Gastrophryne, 111, 148, 190, 439–451, 453–454
Carpenter Frog, 674–681
cascadae, Rana, 688, 699, 707–716, 724, 733–734, 736, 739–740, 744, 748, 751
Cascades Frog, 707–715
catesbeianus, Lithobates, xi, 42, 68, 135–136, 173, 215, 304, 317, 410, 453, 462, 477, 486–515, 521, 524, 526, 538–539, 551–553, 555, 557, 560–561, 566, 569, 583, 608, 610, 612, 614–615, 641, 653, 658, 675, 686, 692, 695–696, 700, 704, 706, 715, 720–722, 747, 769, 790, 819
Chiricahua Leopard Frog, xviii, 515–522
chiricahuensis, Lithobates, xviii, 468, 474, 477, 515–522, 547–551, 580, 584, 624, 683
chrysoscelis, Hyla, 35, 155, 175, 185, 241, 246, 250–263, 275, 281, 284, 289, 294–298, 302–305, 310, 631, 770, 785
cinerea, Hyla, 190, 236, 238, 241, 246, 253, 262–273, 275, 281, 285, 287–289, 298, 324, 445, 451, 501, 554, 560, 630, 822, 824–826
clamitans, Lithobates, 36, 259, 345, 462, 480, 489–490, 501, 507, 522–547, 552, 557, 560–564, 569, 572, 608–610, 612, 614–615, 624, 641, 680
clarkii, Pseudacris, 241, 264, 298, 314, 328–331, 350–351, 357–360, 362, 368, 374, 386, 396, 404, 417, 422
Cliff Chirping Frog, 201–204
Coastal Tailed Frog, 7–16
cognatus, Anaxyrus, 19, 65–66, 78–87, 114, 138, 153–155, 168–169, 171, 175, 179
Columbia Spotted Frog, xviii, 723–732, 739
Cope's Gray Treefrog, 250–262, 300, 770
Coqui, 812–815
coqui, Eleutherodactylus, 812–815
couchii, Scaphiopus, 143, 155, 175, 179, 185, 454, 753–760, 763, 773, 776, 782–784, 788, 804–805
Couch's Spadefoot, 753–760
Cranopsis alvaria, 177
Cranopsis nebulifer, 180
Crapaud d'Amerique, 17
Craugastor, xx
 augusti, 192–196
  a. cactorum, 192–195
  a. latrans, 192–195
Craugastoridae, 191
Crawfish Frog, 461–466
crepitans, Acris, 205, 207–211, 213–214, 219–228, 232–233, 252, 304, 351, 386, 680
crucifer, Pseudacris, 163, 253, 313–314, 331–348, 351, 369, 392, 396–397, 400, 417, 423, 426, 501, 537, 572, 657
Cuban Treefrog, 163, 190, 822–827
cystignathoides, Eleutherodactylus, 197–199
Czatkobatrachus, xiii

debilis, Anaxyrus, 20, 50, 79, 88–91, 138, 149–150, 175, 182
Dendrobates, xx
 auratus, xiii, xiv, xvi, 809–811
Dendrobatidae, 809
Dermocystidium, 540, 599, 615, 633
dorsalis, Rhinophrynus, 458–460
draytonii, Rana, xviii, 501, 506, 688, 694, 708, 715–722
Dusky Gopher Frog, xviii, 617–621
Dwarf American Toad, 20

Eastern Narrow-mouthed Frog, xv, xvii, 439–448
Eastern Spadefoot, xiv, 761–772, 776
Eleutherodactylidae, 197, 812
Eleutherodactylus, xii, xiii, xiv, xx
 coqui, 812–815
 cystignathoides, 197–199
 guttilatus, 199–201
 marnockii, 199–204
 planirostris, ix, 815–818

exsul, Anaxyrus, 50, 71, 92–96, 132–133

femoralis, Hyla, 148, 235–236, 241, 247, 250, 252–253, 263, 274–281, 287–289, 826
feriarum, Pseudacris, 313–315, 318, 320, 329, 334, 348–360, 367–369, 371, 374, 376, 385–388, 390, 396, 421–423
fisheri, Lithobates, x, 547–551
Foothill Yellow-legged Frog, 697–707
fouquettei, Pseudacris, 298, 314, 329–331, 349–350, 357–362, 368, 371, 374, 385–387, 417, 420–422
fowleri, Anaxyrus, 19–20, 27, 43, 96–113, 116, 121, 157–159, 163, 167–169, 174, 182, 185, 188, 238
Fowler's Toad, 96–113, 156, 174, 185
fragilis, Leptodactylus, xiii, 436–438

Gastrophryne, xx
  carolinensis, 111, 148, 190, 439–451, 453–454
  olivacea, 155, 175, 252, 441, 447–456, 759
Glandirana, xx
  rugosa, 819–821
Gopher Frog, 147, 479–485, 570
gratiosa, Hyla, 163, 236, 247, 253, 263–264, 266–267, 273, 275, 278–289, 501, 631
Gray Treefrog, 155, 245–246, 294–309, 310, 396
Great Basin Spadefoot, 791–806
Great Plains Toad, 76–87
Green and Black Dart-Poison Frog, 809–811
Green Frog, 36, 345, 522–547, 569, 576, 599
Greenhouse Frog, 815–818
Green Toad, 88–91
Green Treefrog, 238, 262–273, 287, 824
Grenouille des bois, 637
Grenouille des marais, 568
Grenouille du nord, 608
Grenouille leopard, 578
Grenouille verte, 522
grylio, Lithobates, 489, 506, 526, 551–557, 560–561, 641, 674
gryllus, Acris, xix, 148, 168, 205–210, 219–221, 224–235, 537, 554, 560, 630
Gulf Coast Toad, 180–186

guttilatus, Eleutherodactylus, 199–201

hammondii, Spea, 154, 777–778, 786–792, 798–799, 803
heckscheri, Lithobates, xv, 161, 489, 526, 552, 556–561, 641
hemiophrys, Anaxyrus, 19–20, 43–44, 57, 79, 101, 113–121, 136, 157, 168–169, 182
holbrookii, Scaphiopus, 35, 148, 190, 289, 754, 761–774
houstonensis, Anaxyrus, 120–126, 155
Houston Toad, xviii, 120–126
hurterii, Scaphiopus, 754, 759, 761, 763, 765, 772–776, 783
Hurter's Spadefoot, 772–776
Hyla, xii, xx
  affinis, 322
  andersonii, 111, 235–239, 246, 264, 271, 275, 304, 631
  arenicolor, 239–246, 253, 264, 275, 289, 298, 310, 312, 322–323, 329, 404
  avivoca, 236, 245–250, 252–253, 263, 275, 281, 289, 297
    a. ogechiensis, 245–246
  californiae, 322
  chrysoscelis, 35, 155, 175, 185, 241, 246, 250–263, 275, 281, 284, 289, 294–298, 302–305, 310, 631, 770, 785
  cinerea, 190, 236, 238, 241, 246, 253, 262–273, 275, 281, 285, 287–289, 298, 324, 445, 451, 501, 554, 560, 630, 822, 824–826
    c. evittata, 262, 264
  eximia, 309–310
  femoralis, 148, 235–236, 241, 247, 250, 252–253, 263, 274–281, 287–289, 826
  gratiosa, 163, 236, 247, 253, 263–264, 266–267, 273, 275, 278–289, 501, 631
  nebulosa, 322
  squirella, 148, 189, 237, 241, 247, 253, 262–264, 275, 281, 288–295, 297, 310–311, 402, 404, 445, 451, 630, 825–826
  versicolor, 36, 155, 238, 241, 245–246, 250–257, 260, 264, 289, 294, 309–310, 329, 386, 396, 417, 501, 680
    v. sandersi, 250, 253
  wrightorum, xviii, 241, 243–244, 289, 309–312, 402, 501

Hylidae, xii, 205, 822
Hypopachus, xx, 441, 449
  variolosus, 455–457

Ichthyophonus, 345, 540, 577, 599, 663
Ichthyostega, xiii
illinoensis, Pseudacris, xviii, 363–367, 395–396, 417
Illinois Chorus Frog, xviii, 363–367, 417
Incilius coccifer, 157
intermontana, Spea, 130, 777–778, 787, 791–799

kalmi, Pseudacris, 314, 320, 334, 349–350, 358–359, 362, 367–371, 374, 385–386, 421–423

laevis, Xenopus, xiii, 828–832
Leptodactylidae, 436
Leptodactylus, xx
  fragilis, xiii, 436–438
Leptolegnia, 345
Lithobates, xii, xvi, xvii, xx, xxi, 33
  areolatus, 461–466, 468, 474, 478, 480, 483, 618, 625, 631, 633
    a. circulosus, 461–462, 465
  berlandieri, 462, 466–474, 477–478, 517, 566, 570, 580, 584, 622, 624–625, 630–631, 682, 686
  blairi, 453, 461–462, 465–468, 471–479, 517, 566, 570, 580, 583–585, 599, 603, 624–625, 629–631, 635, 682
  capito, 148, 462, 479–486, 570, 583, 617–618, 621
    c. aesopus, 461, 479–480
    c. stertens, 480
  catesbeianus, xi, 42, 68, 135–136, 173, 215, 304, 317, 410, 453, 462, 477, 486–515, 521, 524, 526, 538–539, 551–553, 555, 557, 560–561, 566, 569, 583, 608, 610, 612, 614–615, 641, 653, 658, 675, 686, 692, 695–696, 700, 704, 706, 715, 720–722, 747, 769, 790, 819
  chiricahuensis, xviii, 468, 474, 477, 515–522, 547–551, 580, 584, 624, 683
  clamitans, 36, 259, 345, 462, 480, 489–490, 501, 507, 522–547, 552, 557, 560–

Lithobates (cont.)
564, 569, 572, 608–610, 612, 614–615, 624, 641, 680
  c. melanota, 522, 526
  fisheri, x, 547–551
  grylio, 489, 506, 526, 551–557, 560–561, 641, 674
  heckscheri, xv, 161, 489, 526, 552, 556–561, 641
  maslini, 641
  okaloosae, 489, 526, 539, 552, 557, 561–564, 641
  onca, xviii, 468, 474, 547–548, 565–568, 583, 624, 682–683
  palustris, xvi, 35, 462, 468, 474, 480, 531, 533, 568–578, 580, 583, 618, 622, 624–625
  p. mansuetii, 568, 571
  pipiens, xxii, 36, 39, 185, 259, 304, 379, 419, 461–462, 466–468, 470, 472–478, 480, 501, 510, 513, 515, 517, 524, 526, 538–539, 551–553, 555, 557, 560–561, 566, 569, 583, 608, 610, 612, 614–615, 641, 653, 658, 675, 686, 692, 695–696, 700, 704, 706, 715, 720–722, 747, 769, 790, 819
    "Burnsi," 578, 580, 584, 587, 595, 600, 624
    "Kandiyohi," 578, 580, 584, 600
  septentrionalis, 489, 523, 526, 539, 552, 557, 561, 608–617, 641
  sevosus, xviii, 462, 480, 617–621
  sphenocephalus, 35, 148, 190, 238, 259, 284, 345, 451, 461–462, 465, 468–469, 471–472, 474, 477–478, 480, 560, 566, 570, 572, 603, 618, 621–637, 646, 680, 682
  s. utricularius, 621, 624
  subaquavocalis, 515, 517, 521
  sylvaticus, xiii, 384, 462, 531, 533, 580, 637–669, 688, 740
  tarahumarae, 583, 667–674
  virgatipes, 489, 526, 538–539, 552, 557, 561, 612, 641, 674–681
  yavapaiensis, 566, 580, 583–584, 681–686
Little Grass Frog, 391–395
Lowland Burrowing Treefrog, 428–431
Lowland Leopard Frog, 681–686
luteiventris, Rana, xviii, 379, 699, 708, 723–732, 739–740, 747

maculata, Pseudacris, xiii, 208, 314, 328–329, 349–350, 358–360, 367–368, 371–386, 421–423, 426–427, 429, 667, 785
marina, Rhinella, xvi, 42, 177, 186–191, 272, 813
marnockii, Eleutherodactylus, 199–204
Mexican Spadefoot, 784, 798–806
Mexican Treefrog, 431–435
Mexican White-lipped Frog, 436–438
Microhyla areolata, 439, 441, 448–449
Microhylidae, 439
microscaphus, Anaxyrus, 19, 65–66, 99, 114, 127–132, 138, 153–154, 157, 168–169
Mink Frog, 506, 539, 599, 608–617
montanus, Ascaphus, 1–7, 9–10, 12–14, 16
Mountain Chorus Frog, 313–318
multiplicata, Spea, 759, 777–778, 780, 783–784, 786–787, 789, 791–792, 794, 798–806
muscosa, Rana, xviii, 76, 410, 688, 698–699, 707–708, 716, 724, 733–740, 747–749, 751

nebulifer, Ollotis, 19–20, 79, 100, 110, 121, 123, 125, 153, 157, 168–169, 171, 180–186, 759
nelsoni, Anaxyrus, xviii, 50, 71, 93, 132–136
New Jersey Chorus Frog, 367–370
nigrita, Pseudacris, 253, 314, 320, 328–329, 331, 334, 348, 350–352, 356–360, 367–369, 371, 374, 385–390, 392, 396, 421–422
Northern Crawfish Frog, 461–462, 465
Northern Cricket Frog, 210–211, 219–226, 229
Northern Green Frog, 500, 506
Northern Red-legged Frog, 687–697, 720

Oak Toad, 144–149
ocularis, Pseudacris, 148, 333, 344, 391–395, 630
okaloosae, Lithobates, 489, 526, 539, 552, 557, 561–564, 641
olivacea, Gastrophryne, 155, 175, 252, 441, 447–456, 759
Ollotis, xi, xx
  alvaria, 50, 79, 85, 142, 168, 177–180, 182, 758
  canaliferus, 50, 89, 138

nebulifer, 19–20, 79, 100, 110, 121, 123, 125, 153, 157, 168–169, 171, 180–186, 759
onca, Lithobates, xviii, 468, 474, 547–548, 565–568, 583, 624, 682–683
Oregon Spotted Frog, xviii, 59, 379, 739–747
ornata, Pseudacris, 253, 314, 320, 329, 334, 344, 350, 386, 395–400, 417
Ornate Chorus Frog, 395–400
Osteopilus, xii, xx
  septentrionalis, 163, 190, 822–827
Ouaouaron, 486

palustris, Lithobates, xvi, 35, 462, 468, 474, 480, 531, 533, 568–578, 580, 583, 618, 622, 624–625
Parapseudacris, 334
Pickerel Frog, 568–578, 680
Pig Frog, 486, 506, 551–556
Pine Barrens Treefrog, 235–239
Pine Woods Treefrog, 274–280
Pipidae, 828
pipiens, Lithobates, xxii, 36, 39, 185, 259, 304, 379, 419, 461–462, 466–468, 470, 472–478, 480, 501, 510, 513, 515, 517, 524, 526, 538–539, 551–553, 555, 557, 560–561, 566, 569, 583, 608, 610, 612, 614–615, 641, 653, 658, 675, 686, 692, 695–696, 700, 704, 706, 715, 720–722, 747, 769, 790, 819
Plains Leopard Frog, 472–479, 599
Plains Spadefoot, 777–785, 805
planirostris, Eleutherodactylus, ix, 815–818
Poloka, 186
Poloka lana, 486
pretiosa, Rana, xviii, 59, 507–641, 692–693, 699, 707–708, 713, 723–724, 739–747
Prosalirus bitis, iii
Pseudacris, xi, xii, xiii, xx
  brachyphona, 313–318, 320, 334, 350, 358, 368, 374, 386, 396, 417, 422–423
  brimleyi, 314, 319–322, 350, 358, 368, 374, 386, 396, 422
  cadaverina, 239, 298, 322–328, 402–405
  clarkii, 241, 264, 298, 314, 328–331, 350–351, 357–360,

INDEX OF SCIENTIFIC AND COMMON NAMES 979

362, 368, 374, 386, 396, 404, 417, 422
crucifer, 163, 253, 313–314, 331–348, 351, 369, 392, 396–397, 400, 417, 423, 426, 501, 537, 572, 657
  c. bartramiana, 331, 333
feriarum, 313–315, 318, 320, 329, 334, 348–360, 367–369, 371, 374, 376, 385–388, 390, 396, 421–423
fouquettei, 298, 314, 329–331, 349–350, 357–362, 368, 371, 374, 385–387, 417, 420–422
hypochondriaca, 400, 402–403
illinoensis, xviii, 363–367, 395–396, 417
kalmi, 314, 320, 334, 349–350, 358–359, 362, 367–371, 374, 385–386, 421–423
maculata, xiii, 208, 314, 328–329, 349–350, 358–360, 367–368, 371–386, 421–423, 426–427, 429, 667, 785
nigrita, 253, 314, 320, 328–329, 331, 334, 348, 350–352, 356–360, 367–369, 371, 374, 385–390, 392, 396, 421–422
ocularis, 148, 333, 344, 391–395, 630
ornata, 253, 314, 320, 329, 334, 344, 350, 386, 395–400, 417
regilla, 60, 241, 289, 298, 309, 322–323, 325, 329, 333, 400–416, 501, 507–508, 712, 719, 736–737, 744, 746
sierrae, 400, 402–403
streckeri, 209, 253, 298, 329, 331, 333–334, 363–364, 395–396, 416–421
triseriata, 253, 314–315, 320, 328–329, 331, 334, 344, 348–350, 357–359, 363, 367–368, 371–374, 376, 382, 384–386, 421–428
Pternohyla, 428–429
punctatus, Anaxyrus, 19–20, 50, 79, 89, 114, 130, 136–145, 150, 152–153, 157, 168–169, 759

quercicus, Anaxyrus, 79, 144–149, 189, 630

Rainette crucifere, 331
Rainette faux-grillon boréale, 331
Rainette faux-grillon de l'ouest, 371
Rainette versicolore, 294

Rana, xii, xvii, xx
  aurora, 506–507, 641, 687–697, 707–708, 715–716, 719–722, 734, 739, 743–746, 748
  boylii, 508, 510, 688, 697–708, 716, 724, 733–734, 740, 747–748
  cascadae, 688, 699, 707–716, 724, 733–734, 736, 739–740, 744, 748, 751
  draytonii, xviii, 501, 506, 688, 694, 708, 715–722
  luteiventris, xviii, 379, 699, 708, 723–732, 739–740, 747
  muscosa, xviii, 76, 410, 688, 698–699, 707–708, 716, 724, 733–740, 747–749, 751
  pretiosa, xviii, 59, 507–641, 692–693, 699, 707–708, 713, 723–724, 739–747
  sierrae, xviii, 76, 688, 698–699, 708, 716, 733–737, 747–752
Ranavirus (ranavirus), 345, 508–509, 540–541, 555, 599, 662
Ranidae, 461, 819
Red-spotted Toad, 136–144
regilla, Pseudacris, 60, 241, 289, 298, 309, 322–323, 325, 329, 333, 400–416, 501, 507–508, 712, 719, 736–737, 744, 746
Relict Leopard Frog, xviii, 565–568
retiformis, Anaxyrus, 88–89, 138, 149–152
Rhacophorus, ix
Rhinella, xii, xx
  arenarum, 187
  cerradensis, 187
  jimi, 187
  marina, xvi, 42, 177, 186–191, 272, 813
  paracnemis, 187
  poeppigii, 187
  schneideri, 187
  veredas, 187
Rhinophrynidae, 458
Rhinophrynus, xx
  dorsalis, 458–460
Rio Grande Chirping Frog, 197–199
Rio Grande Leopard Frog, 466–471, 686
River Frog, xv, 486–487, 556–560
Rocky Mountain Tailed Frog, xiv, 1–7
Rocky Mountain Toad, 166
rugosa, Glandirana, 819–821

Scaphiopodidae, 753
Scaphiopus, xii, xvi, xx, 86, 103, 174, 345

  alexanderi, 754
  couchii, 143, 155, 175, 179, 185, 454, 753–760, 763, 773, 776, 782–784, 788, 804–805
  holbrookii, 35, 148, 190, 289, 754, 761–774
  h. albus, 763
  hurterii, 754, 759, 761, 763, 765, 772–776, 783
Scutiger, ix
septentrionalis, Lithobates, 489, 523, 526, 539, 552, 557, 561, 608–617, 641
septentrionalis, Osteopilus, 163, 190, 822–827
sevosus, Lithobates, xviii, 462, 480, 617–621
Sheep Frog, 455–457
sierrae, Rana, xviii, 76, 688, 698–699, 708, 716, 733–737, 747–752
Sierra Nevada Yellow-legged Frog, xviii, 747–752
Smilisca, xii, xx
  baudinii, 241, 253, 431–435
  fodiens, 428–431
Sonoran Desert Toad, 85, 177–180, 758
Sonoran Green Toad, 149–152
Southern Chorus Frog, 356, 359, 367, 385–390
Southern Crawfish Frog, 461–462
Southern Cricket Frog, 226–235
Southern Leopard Frog, 159, 506, 597, 621–637
Southern Toad, 42, 155–166, 768–769
Southwestern Woodhouse's Toad, 166, 168
Spea, xi, xii, xx
  bombifrons, 85, 755, 758–759, 776–787, 791–792, 799, 801, 804–805
  hammondii, 154, 777–778, 786–792, 798–799, 803
  hardeni, 763
  intermontana, 130, 777–778, 787, 791–799
  multiplicata, 759, 777–778, 780, 783–784, 786–787, 789, 791–792, 794, 798–806
  stagnalis, 787, 798–799
  studeri, 778
speciosus, Anaxyrus, 20, 50, 66, 79, 125, 127, 129, 138, 152–155, 157, 167, 169, 175, 182, 759, 781
sphenocephalus, Lithobates, 35, 148, 190, 238, 259, 284, 345, 451, 461–462, 465, 468–469, 471–472, 474,

sphenocephalus, Lithobates (cont.) 477–478, 480, 560, 566, 570, 572, 603, 618, 621–637, 646, 680, 682
Spotted Chirping Frog, 199–201
Spotted Chorus Frog, 328–331, 420
Spring Peeper, 110, 318, 331–348, 369, 399, 426, 537, 578, 595, 657
squirella, Hyla, 148, 189, 237, 241, 247, 253, 262–264, 275, 281, 288–295, 297, 310–311, 402, 404, 445, 451, 630, 825–826
Squirrel Treefrog, 288–294, 295
streckeri, Pseudacris, 209, 253, 298, 329, 331, 333–334, 363–364, 395–396, 416–421
Strecker's Chorus Frog, 416–421
sylvaticus, Lithobates, xiii, 384, 462, 531, 533, 580, 637–669, 688, 740
Syrrhophus campi, 197–198
Syrrhophus gaigae, 200

tarahumarae, Lithobates, 583, 667–674
Tarahumara Frog, 669–674
terrestris, Anaxyrus, 17, 19–20, 42–43, 79, 89, 97, 99–100, 105, 110, 114, 121, 129, 138, 148, 153, 155–166, 168–169, 182, 188–190, 284, 560, 769, 826
Texas Toad, 152–155
Triadobatrachus, xiii
triseriata, Pseudacris, 253, 314–315, 320, 328–329, 331, 334, 344, 348–350, 357–359, 363, 367–368, 371–374, 376, 382, 384–386, 421–428
Tri-State Leopard Frog, 624
truei, Ascaphus, 1–3, 5, 7–16

Upland Chorus Frog, 318, 348–357, 368, 385

variolosus, Hypopachus, 455–457
Vegas Valley Leopard Frog, 547–551
versicolor, Hyla, 36, 155, 238, 241, 245–246, 250–257, 260, 264, 289, 294, 309–310, 329, 386, 396, 417, 501, 680
virgatipes, Lithobates, 489, 526, 538–539, 552, 557, 561, 612, 641, 674–681

Western Barking Frog, 192
Western Chorus Frog, 350, 367–368, 377, 383, 421–428
Western Narrow-mouthed Frog, 448–455
Western Spadefoot, 763, 786–790
Western Toad, xvii, 47–65, 71, 93, 412
Wood Frog, 33, 317–318
Woodhouse's Toad, 86, 168–176
woodhousii, Anaxyrus, 18–20, 27, 43, 65–66, 79, 82, 86, 96–101, 103, 113–114, 121, 125, 127–129, 132, 138, 153, 155, 157, 166–176, 178, 182, 185, 188, 259, 344, 759
wrightorum, Hyla, xviii, 241, 243–244, 289, 309–312, 402, 501
Wrinkled Frog, 819–821
Wyoming Toad, xviii, 43–47

Xenopus, xii, xx
laevis, xiii, 828–832

yavapaiensis, Lithobates, 566, 580, 583–584, 681–686
Yosemite Toad, xviii, 70–77

Index of Potential Stressors

2-chloroethanol, 512
2,2,2-trichloroethanol, 512
2,4-D, 38, 603, 665
3-trifluoromethyl-4-nitrophenol. *See* TFM
6-chloro-2-picolinic acid, 383

Abate®, 62, 544
acetochlor, 512
acridine, 512
alachlor, 602
aldrin, 39, 112, 226, 234, 511, 635
alkalinity, 542
Altosid®, 62
aluminum, 37–38, 77, 225, 272, 306, 330, 346, 447, 452, 542, 601, 664–665, 739
ammonia, xiv, 38, 413, 542, 601, 636, 695
aniline, 604
arsenic, 163–164, 346, 447, 511, 542, 601, 664, 673
atrazine, 39–40, 260, 305–307, 346, 512, 543–544, 577, 600, 602, 606, 635, 665, 830
azinphos-methyl, 414, 511, 602, 717

barium, 346, 577
Basudin®, 543
Bayer37289®, 61, 790
benzene, 604
benzene hexachloride, 112, 635
beryllium, 346, 577
B-naphtal, 512
boron, 38, 112, 601, 664
bromoxynil, 544

cadmium, 38, 143, 260, 346, 447, 511, 542, 577, 633–634, 664, 673, 731
cadmium chloride, 601
carbaryl, 38–39, 176, 307, 346, 466, 478, 512, 543, 577, 603–604, 634–635, 665, 695, 704, 746
carbon tetrachloride, 112, 512, 577, 664

carbophenothion, 383
chloroform, 112, 577, 604
chlorpyrifos, 260, 307, 413–414, 544, 603, 705, 737
chlorpyrifos-ethyl, 62
chlorpyrifos-methyl, 62
chromium, 163, 447, 511
coal ash, 163–164, 511, 541–542, 664
coal oil, 61
cobalt, 423, 447
conductivity, xxiii, 44, 146, 267, 292, 301, 309, 330, 340, 444, 543, 612, 627, 647, 665, 676
copper, 61, 163, 346–347, 447, 511, 542, 577, 633, 664, 674
copper sulfate, 487, 601
crankcase oil, 273
crude oil, 512

DDE, 39, 346, 414, 675, 737
DDT, 61, 112, 216, 218, 226, 346, 383, 414, 511, 513, 544, 604, 665, 695
DEF, 112
diazinon, 413, 543, 603, 705, 737, 797
dichlorobenzene, 604
dieldrin, 39, 112, 226, 234, 346, 383, 511, 604, 635
dioxins, 604
diquat, 603, 634
dissolved oxygen content (DOC), 23, 340, 452, 492, 531, 591, 615, 647, 665
Dithane®, 543

endosulfan, 39, 414, 544, 603, 665, 797
endrin, 112, 225–226, 234, 383, 511, 635, 665
esfenvalerate, 478, 603, 635

fenitrothion, 39, 511, 544, 603
fenvalerate, 112, 273, 544, 603, 635
fire retardants, 636

fluorine, 634
furans, 544

G-30493 (an insecticide), 61
G-30494 (an insecticide), 61, 790
gallium, 346
GC-3582, 511
germanium, 447
Guthion®, 61, 414, 511, 544, 790

hardness. *See* conductivity
heated water (nuclear reservoir), 164
heptachlor, 112, 511
hexachloroethane, 512
hexazinone, 512, 544, 603
hydrothol 112, 191

Ignite 280SL®, 87, 806
imidacloprid, 226, 544
Imidan®, 544
iron, 61, 225, 306, 346, 542, 577, 664

lead, 225, 346, 447, 511, 542, 601, 664, 731
leptophos-com tri-o-totyl phosphate, 260
lindane, 112, 383

magnesium, 225, 306, 346, 542, 577, 612, 814
malathion, 38–39, 112, 176, 307, 346, 383, 413–414, 512, 543, 604, 635, 665, 705
Mancozeb®, 40, 602
manganese, 225, 306, 542
mercuric chloride, 560
mercury, 111, 225, 260, 346, 413, 447, 511, 542, 555, 560, 601, 704
methoxychlor, 39, 112, 383, 635
methylene chloride, 112, 512, 577, 604
methyl mercury, 260, 511, 542
methylparathion, 62
metolachlor, 307, 544

molinate, 112
m-xylene, 604

naled, 383
nickel, 38, 225, 346, 447
nitrates, xxiii, 38, 62, 163, 260, 413, 426, 542, 601, 605, 665, 695, 713, 746
nitrites, xxiii, 413, 665, 746
nitrobenzene, 604
Nova®, 544

octamethyl pyrophosphoramide, 635
ozone, 327

paraoxon, 260
paraquat, 112, 383, 414, 471, 512, 634
parathion, 61, 383, 414, 511, 635, 790
PCB 126, 543–544
pentachlorophenol, 512
perchlorate, 217
permethrin, 512, 544, 603
pH (acidity), xxiii, 7, 14, 25, 37–38, 61, 76–77, 132, 217, 225, 233, 239, 272, 279, 287, 306, 346, 370, 382, 395, 413, 485, 511, 542, 601, 612, 616, 634, 664–665, 676, 680, 713, 737, 751, 760, 790, 797
phenol, 112, 512, 577, 604
phenyl saliginen cyclic phosphate, 260, 447, 635

Phos-Chek D75F, 636
Phos-Chek D75R, 636
phosdrin, 635
phosphate, 260, 438, 447, 511, 603, 635, 737
phosphorus, 62, 413–414, 635, 665, 695, 797, 814
piperonyl butoxide, 383
polybrominated diphenyl ethers, 544
polychlorinated dibenzofuran, 217
polyethoxylated tallowamine (POEA), 39, 543, 665

radiocesium, 272
Release®, 603, 634
Rodeo®, 665
Ronnel®, 61, 790
rotenone, 604, 636
Roundup®, 39, 62, 87, 307, 346, 414, 543, 603–604, 665, 713, 731, 806

salinity, 112, 166, 234, 270, 272–273, 279, 287, 293, 390, 542, 556, 602, 665, 686, 719, 830
salts. See salinity
selenium, 163–164, 306, 447, 511, 542, 577, 664
Sevin®. See carbaryl
silver, 447, 577, 601
silvex, 112, 383
strontium, 225, 511, 542, 664
strychnine, 191
sulfate, 413, 612, 695

TDE, 112, 383
temephos, 414, 544
tetra-ethyl pyrophosphate, 635
TFM (lampricide), 307, 511, 604
thiodan, 61, 511
thiosemicarbazide (TSC), 666
tin, 346
tire debris, 664
titanium, 346
toluene, 604
toxaphene, 61, 112, 226, 383, 414, 511, 635
trans-nonachlor, 737
triallate, 544
triclorpyr, 512, 603, 634
trifluralin, 544
trisodium nitrilotriacetic acid (NTA), 112, 148

urea, 62, 413, 713, 755, 758, 803
UVA radiation, 41, 544, 666
UVB radiation, 40, 62, 64, 77, 217, 260–261, 308, 327, 346–347, 383, 414–415, 479, 512, 544, 600, 604, 616, 635, 666, 695, 713–714, 721, 731, 737, 746

vanadium, 164, 306, 511, 542, 664
Vision®, 39, 543, 602–603, 634

zinc, 61, 447, 542, 577, 601, 664, 673, 731
zirconium, 346